W9-CYY-541

MATHEMATICAL HANDBOOK FOR SCIENTISTS AND ENGINEERS

Definitions, Theorems, and Formulas for Reference and Review

Granino A. Korn

and

Theresa M. Korn

DOVER PUBLICATIONS, INC.
Mineola, New York

Copyright

Bibliographical Note

This Dover edition, first published in 2000, is an unabridged republication of the work originally published in 1968 by McGraw-Hill, Inc., New York. A number of typographical errors have been corrected.

Library of Congress Cataloging-in-Publication Data

Korn, Granino Arthur, 1922–
Mathematical handbook for scientists and engineers : definitions, theorems, and formulas for reference and review / Granino A. Korn and Theresa M. Korn.
p. cm.
Originally published: 2nd, enl. and rev. ed. New York : McGraw-Hill, c1968.
Includes bibliographical references and index.
ISBN 0-486-41147-8 (pbk.)
1. Mathematics—Handbooks, manuals, etc. I. Korn, Theresa M. II. Title.

QA40 .K598 2000
510'.2'1—dc21

00-030318

Manufactured in the United States of America
Dover Publications, Inc., 31 East 2nd Street, Mineola, N.Y. 11501

This book is dedicated to the memory of Arthur Korn (1870–1945), who excelled as a mathematician as well as a theoretical physicist and as a communications engineer.

PREFACE TO THE SECOND EDITION

This new edition of the *Mathematical Handbook* has been substantially enlarged, and much of our original material has been carefully, and we hope usefully, revised and expanded. Completely new sections deal with z transforms, the matrix notation for systems of differential equations (state equations), representation of rotations, mathematical programming, optimal-control theory, random processes, and decision theory. The chapter on numerical computation was almost entirely rewritten, and the revised appendixes include a discussion of Polya's counting theorem, several new tables of formulas and numerical data, and a new, much larger integral table. Numerous illustrations have been added throughout the remaining text.

The handbook is again designed, first, as a comprehensive reference collection of mathematical definitions, theorems, and formulas for scientists, engineers, and students. Subjects of both undergraduate and graduate level are included. The omission of all proofs and the concise tabular presentation of related formulas have made it possible to incorporate a relatively large amount of reference material in one volume.

The handbook is, however, not intended for reference purposes alone; it attempts to present a connected survey of mathematical methods useful beyond specialized applications. Each chapter is arranged so as to permit review of an entire mathematical subject. Such a presentation is made more manageable and readable through the omission of proofs; numerous references provide access to textbook material for more detailed studies. Special care has been taken to point out, by means of suitable introductions, notes, and cross references, the interrelations of various topics and their importance in scientific and engineering applications.

The writers have attempted to meet the individual reader's requirements by arranging the subject matter at three levels:

1. The most important formulas and definitions have been collected in tables and boxed groups permitting rapid reference and review.

2. The main text presents, in large print, a concise, connected review of each subject.
3. More detailed discussions and advanced topics are presented in small print. This arrangement makes it possible to include such material without cluttering the exposition of the main review.

We believe that this arrangement has proved useful in the first edition.

The arrangement of the introductory chapters was left unchanged, although additions and changes were made throughout their text. Chapters 1 to 5 review traditional college material on *algebra, analytic geometry, elementary and advanced calculus,* and *vector analysis;* Chapter 4 also introduces *Lebesgue and Stieltjes integrals,* and *Fourier analysis.* Chapters 6, 7, and 8 deal with *curvilinear coordinate systems, functions of a complex variable, Laplace transforms,* and other functional transforms; new material on finite Fourier transforms and on z transforms was added.

Chapters 9 and 10 deal with *ordinary and partial differential equations* and include *Fourier- and Laplace-transform methods,* the *method of characteristics,* and *potential theory;* eigenvalue problems as such are treated in Chapters 14 and 15. Chapter 11 is essentially new; in addition to *ordinary maxima and minima* and the *classical calculus of variations,* this chapter now contains material on *linear and nonlinear programming* and on *optimal-control theory,* outlining both the maximum-principle and dynamic-programming approaches.

Chapter 12, considerably expanded in this edition, introduces the elements of modern abstract language and outlines the construction of mathematical models such as *groups, fields, vector spaces, Boolean algebras,* and *metric spaces.* The treatment of function spaces continues through Chapter 14 to permit a modest functional-analysis approach to boundary-value problems and eigenvalue problems in Chapter 15, with enough essential definitions to enable the reader to use modern advanced texts and periodical literature.

Chapter 13 treats *matrices;* we have added several new sections reviewing *matrix techniques for systems of ordinary differential equations* (*state equations of dynamical systems*), including an outline of Lyapunov stability theory. Chapter 14 deals with the important topics of *linear vector spaces, linear transformations* (*linear operators*), introduces *eigenvalue problems,* and describes *the use of matrices for the representation of mathematical models.* The material on *representation of rotations* was greatly enlarged because of its importance for space-vehicle design as well as for atomic and molecular physics. Chapter 15 reviews a variety of topics related to *boundary-value problems* and *eigenvalue problems,* including *Sturm-Liouville problems, boundary-value problems in two and three dimensions,* and *linear integral equations,* considering functions as vectors in a normed vector space.

Chapters 16 and 17 respectively outline *tensor analysis* and *differential geometry*, including the description of *plane and space curves, surfaces,* and *curved spaces.*

In view of the ever-growing importance of statistical methods, the completely revised Chapter 18 presents a rather detailed treatment of *probability theory* and includes material on *random processes, correlation functions,* and *spectra.* Chapter 19 outlines the principal methods of *mathematical statistics* and includes extensive tables of formulas involving special sampling distributions. A subchapter on Bayes tests and estimation was added.

The new Chapter 20 introduces *finite-difference methods* and *difference equations* and reviews a number of basic methods of *numerical computation.* Chapter 21 is essentially a collection of formulas outlining the properties of *higher transcendental functions;* various formulas and many illustrations have been added.

The appendixes present *mensuration formulas, plane and spherical trigonometry, combinatorial analysis, Fourier- and Laplace-transform tables,* a new, larger *integral table,* and a new set of *tables of sums and series.* The treatment of combinatorial analysis was enlarged to outline the use of generating functions and a statement of Polya's counting theorem. Several new tables of formulas and functions were added. As before, there is a *glossary of symbols,* and a comprehensive and detailed *index* permits the use of the handbook as a mathematical dictionary.

The writers hope and believe that this handbook will give the reader an opportunity to scan the field of mathematical methods, and thus to widen his background or to correlate his specialized knowledge with more general developments. We are very grateful to the many readers who have helped to improve the handbook by suggesting corrections and additions; once again, we earnestly solicit comments and suggestions for improvements, to be addressed to us in care of the publishers.

Granino A. Korn
Theresa M. Korn

CONTENTS

MATHEMATICAL HANDBOOK FOR SCIENTISTS AND ENGINEERS

CHAPTER 1

REAL AND COMPLEX NUMBERS ELEMENTARY ALGEBRA

1.1. INTRODUCTION THE REAL-NUMBER SYSTEM

1.1-1. This chapter deals with the algebra* of real and complex numbers, i.e., with the study of those relations between real and complex numbers which involve a finite number of additions and multiplications. This is considered to include the solution of equations based on such relations, even though actual exact numerical solutions may require infinite numbers of additions and/or multiplications. The definitions and relations presented in this chapter serve as basic tools in many more general mathematical models (see also Sec. 12.1-1).

1.1-2. Real Numbers. The axiomatic foundations ensuring the self-consistency of the real-number system are treated in Refs. 1.1 and 1.5, and lead to the acceptance of the following rules governing the addition and multiplication of real numbers.

* See also footnote to Sec. 12.1-2.

$$\begin{array}{ll} a+b \text{ and } ab \text{ are real numbers (algebraic numbers, rational numbers, integers, positive integers) if this is true for both } a \text{ and } b & \text{(CLOSURE)} \\ a+b=b+a \qquad ab=ba & \text{(COMMUTATIVE LAWS)} \\ a+(b+c)=(a+b)+c=a+b+c & \\ \qquad a(bc)=(ab)c=abc & \text{(ASSOCIATIVE LAWS)} \\ a\cdot 1=a & \text{(MULTIPLICATIVE IDENTITY)} \\ a(b+c)=ab+ac & \text{(DISTRIBUTIVE LAW)} \\ a+c=b+c \text{ implies } a=b & \\ ca=cb,\quad c\neq 0 \text{ implies } a=b & \text{(CANCELLATION LAWS)} \end{array} \qquad (1.1\text{-}1)$$

The real number 0 (**zero, additive identity**) has the properties

$$a+0=a \qquad a\cdot 0=0 \qquad (1.1\text{-}2)$$

for every real a.

The (unique) **additive inverse** $-a$ and the (unique) **multiplicative inverse (reciprocal)** $a^{-1}=1/a$ of a real number a are respectively defined by

$$a+(-a)=a-a=0 \qquad aa^{-1}=1 \qquad (a\neq 0) \qquad (1.1\text{-}3)$$

Division by 0 *is not admissible.*

In addition to the "algebraic" properties (1), the class of the **positive integers** 1,2, . . . has the properties of being **simply ordered** (Sec. 12.6-2; n is "greater than" m or $n > m$ if and only if $n = m + x$ where x is a positive integer) and **well-ordered** (every nonempty set of positive integers has a smallest element). *A set of positive integers containing* (1) 1 *and the "successor"* $n + 1$ *of each of its elements* n, *or* (2) *all integers less than* n *for any* n, *contains all positive integers* (*Principle of Finite Induction*).

The properties of positive integers may be alternatively defined by *Peano's Five Axioms*, viz., (1) 1 is a positive integer, (2) each positive integer n has a unique successor $S(n)$, (3) $S(n) \neq 1$, (4) $S(n) = S(m)$ implies $n = m$, (5) the principle of finite induction holds. Addition and multiplication satisfying the rules (1) are defined by the "recursive" definitions $n + 1 = S(n)$, $n + S(m) = S(n + m)$; $n \cdot 1 = n$, $n \cdot S(m) = n \cdot m + n$.

Operations on the elements $m - n$ of the **class of all integers** (positive, negative, or zero) are interpreted as operations on corresponding pairs (m, n) of positive integers m, n such that $(m - n) + n = m$, where 0, defined by $n + 0 = n$, corresponds to (n, n), for all n. An integer is **negative** if and only if it is neither positive nor zero. The study of the properties of integers is called **arithmetic.**

Operations on **rational numbers** m/n $(n \neq 0)$ are interpreted as operations on corresponding pairs (m, n) of integers m, n such that $(m/n)n = m$. m/n is positive if and only if mn is positive.

Real **algebraic** (including rational and **irrational**) **numbers,** corresponding to (real) roots of algebraic equations with integral coefficients (Sec. 1.6-3) and real

transcendental numbers, for which no such correspondence exists, may be introduced in terms of limiting processes involving rational numbers (*Dedekind cuts,* Ref. 1.5).

The class of all rational numbers comprises the roots of all linear equations (Sec. 1.8-1) with rational coefficients, and includes the integers. The class of all real algebraic numbers comprises the real roots of all algebraic equations (Sec. 1.6-3) with algebraic coefficients, including the rational numbers. The class of all real numbers contains the real roots of all equations involving a finite or infinite number of additions and multiplications of real numbers and includes real algebraic and transcendental numbers (see also Sec. 4.3-1).

A real number a is greater than the real number b ($a > b$, $b < a$) if and only if $a = b + x$, where x is a positive real number (see also Secs. 1.1-5 and 12.6-2).

1.1-3. Equality Relation (see also Sec. 12.1-3). An equation $a = b$ implies $b = a$ (*symmetry* of the equality relation), and $a + c = b + c$, $ac = bc$ [in general, $f(a) = f(b)$ if $f(a)$ stands for an operation having a unique result]. $a = b$ and $b = c$ together imply $a = c$ (*transitivity* of the equality relation). $ab \neq 0$ implies $a \neq 0$, $b \neq 0$.

1.1-4. Identity Relation. In general, an equation involving operations on a quantity x or on several quantities $x_1, x_2, \ldots$ will hold only for special values of x or special sets of values $x_1, x_2, \ldots$ (see also Sec. 1.6-2). *If it is desired to stress the fact* that an equation holds for *all* values of x or of $x_1, x_2, \ldots$ within certain ranges of interest, the identity symbol $\equiv$ may be used instead of the equality symbol $=$ [EXAMPLE: $(x - 1)(x + 1) \equiv x^2 - 1$], and/or the ranges of the variables in question may be indicated on the right of the equation. $a \equiv b$ (better $a \triangleq b$) is also used with the meaning "a is defined as equal to b."

1.1-5. Inequalities (see also Secs. 12.6-2 and 12.6-3). $a > b$ *implies* $b < a$, $a + c > b + c$, $ac > bc$ $(c > 0)$, $-a < -b$, $1/a < 1/b$ $(a > 0, b > 0)$. *A real number a is positive* $(a > 0)$, *negative* $(a < 0)$, *or zero* $(a = 0)$. *Sums and products of positive numbers are positive.* $a \leq A$, $b \leq B$ *implies* $a + b \leq A + B$. $a \geq b$, $b \geq c$ *implies* $a \geq c$.

1.1-6. Absolute Values (see also Secs. 1.3-2 and 14.2-5). The **absolute value** $|a|$ of a real number a is defined as equal to a if $a \geq 0$ and equal to $-a$ if $a < 0$. Note

$$\left.\begin{array}{ll} |a| \geq 0 \qquad |a| = 0 \text{ implies } a = 0 & \\ ||a| - |b|| \leq |a + b| \leq |a| + |b| & ||a| - |b|| \leq |a - b| \leq |a| + |b| \end{array}\right\} \tag{1.1-4}$$

$$|ab| = |a|\,|b| \qquad \left|\frac{a}{b}\right| = \frac{|a|}{|b|} \qquad (b \neq 0) \tag{1.1-5}$$

$$|a| \leq A, \quad |b| \leq B \text{ implies } |a| + |b| \leq A + B \quad \text{and} \quad |ab| \leq AB \tag{1.1-6}$$

1.2. POWERS, ROOTS, LOGARITHMS, AND FACTORIALS. SUM AND PRODUCT NOTATION

1.2-1. Powers and Roots. The **n^th^ power** of any real number (**base**) a is defined as the product of n factors equal to a, where the **exponent**

n is a positive integer. The resulting relations

$$a^p a^q = a^{p+q} \qquad (a^p)^q = a^{pq} \tag{1.2-1}$$

are postulated to apply for all real values of p and q and thus serve to *define* powers involving exponents other than positive integers:

$$a^{-p} = \frac{1}{a^p} \qquad a^0 = 1 \qquad (a \neq 0) \tag{1.2-2}$$

A **pth root** $a^{1/p} \equiv \sqrt[p]{a}$ of the **radicand** a is a solution of the equation $x^p = a$. $\sqrt[2]{a} \equiv \sqrt{a}$ is the **square root** of a, and $\sqrt[3]{a}$ is its **cube root.** Powers and roots with irrational exponents can be introduced through limiting processes (see also Sec. E-6). In general, $\sqrt[p]{a}$ is not unique, and some or all of the roots of a given real radicand a may not be real numbers (Sec. 1.3-3). If a is real and positive, many authors specifically denote the *real positive* solution values of $x^2 = a,\ x^3 = a,\ x^4 = a, \ldots$ as $\sqrt[2]{a} \equiv \sqrt{a}, \sqrt[3]{a}, \sqrt[4]{a}, \ldots$. The real solutions of $x^2 = a,\ x^4 = a, \ldots$ are then written as $\pm\sqrt{a}, \pm\sqrt[4]{a}, \ldots$. To emphasize a choice of the positive square root, one may write $\underset{+}{\sqrt{a}}$. For $q \neq 0$

$$\begin{gathered}\frac{a^p}{a^q} = a^{p-q} \qquad a^{\frac{p}{q}} = \sqrt[q]{a^p} = (\sqrt[q]{a})^p \qquad \sqrt[p]{\sqrt[q]{a}} = \sqrt[pq]{a} \\ (ab)^p = a^p b^p \qquad \sqrt[p]{ab} = \sqrt[p]{a}\,\sqrt[p]{b} \\ \left(\frac{a}{b}\right)^p = \frac{a^p}{b^p} \qquad \sqrt[p]{\frac{a}{b}} = \frac{\sqrt[p]{a}}{\sqrt[p]{b}} \qquad (b \neq 0)\end{gathered} \tag{1.2-3}$$

1.2-2. Formulas for Rationalizing the Denominators of Fractions.

$$\frac{a}{\sqrt{b}} = \frac{a}{b}\sqrt{b} \qquad \frac{a}{\sqrt[n]{b}} = \frac{a}{b}\sqrt[n]{b^{n-1}} \tag{1.2-4}$$

$$\frac{a}{\sqrt{b} \pm \sqrt{c}} = \frac{a}{b-c}(\sqrt{b} \mp \sqrt{c}) \qquad \frac{a}{\sqrt{b+\sqrt{c}}} = \frac{a}{b^2-c}\sqrt{(b^2-c)(b-\sqrt{c})} \tag{1.2-5}$$

$$\frac{a}{\sqrt[3]{b} \pm \sqrt[3]{c}} = \frac{a}{b \pm c}(\sqrt[3]{b^2} \mp \sqrt[3]{bc} + \sqrt[3]{c^2}) \tag{1.2-6}$$

1.2-3. Logarithms. The **logarithm** $x = \log_c a$ to the **base** $c > 0$ $(c \neq 1)$ of the number (numerus) $a > 0$ may be defined by

$$c^x = a \qquad \text{or} \qquad c^{\log_c a} = a \tag{1.2-7}$$

Refer to Table 7.2-1 and Sec. 21.2-10 for a more general discussion of logarithms. $\log_c a$ may be a transcendental number (Sec. 1.1-2). Note

$$\begin{gathered}\log_c c = 1 \qquad \log_c c^p = p \qquad \log_c 1 = 0\\ \log_c (ab) = \log_c a + \log_c b \qquad \text{(LOGARITHMIC PROPERTY)}\\ \log_c\left(\frac{a}{b}\right) = \log_c a - \log_c b\\ \log_c (a^p) = p \log_c a \qquad \log_c (\sqrt[p]{a}) = \frac{1}{p}\log_c a \quad (p \neq 0)\end{gathered} \tag{1.2-8}$$

$$\log_{c'} a = \log_c a \log_{c'} c = \frac{\log_c a}{\log_c c'} \qquad \log_{c'} c = \frac{1}{\log_c c'} \qquad (c' \neq 1) \qquad \text{(CHANGE OF BASE)} \tag{1.2-9}$$

Of particular interest are the "common" logarithms to the base 10 and the *natural (Napierian) logarithms* to the base

$$e = \lim_{n\to\infty}\left(1 + \frac{1}{n}\right)^n = 2.71828182 \cdots \tag{1.2-10}$$

e is a transcendental number. $\log_e a$ may be written $\ln a$, $\log a$ or $\log$ nat a. $\log_{10} a$ is sometimes written $\log a$. Note

$$\left.\begin{aligned}\log_e a &= \frac{\log_{10} a}{\log_{10} e} = \log_e 10 \log_{10} a = (2.30259 \cdots)\log_{10} a\\ \log_{10} a &= \frac{\log_e a}{\log_e 10} = \log_{10} e \log_e a = (0.43429 \cdots)\log_e a\end{aligned}\right\} \tag{1.2-11}$$

1.2-4. Factorials. The **factorial** $n!$ of any integer $n \geq 0$ is defined by

$$0! = 1 \qquad n! = \prod_{k=1}^{n} k = 1 \cdot 2 \cdot 3 \cdots (n-1)n \qquad (n > 0) \tag{1.2-12}$$

Refer to Sec. 21.4-2 for approximation formulas.

1.2-5. Sum and Product Notation. For any two integers (positive, negative, or zero) n and $m \geq n$

$$\sum_{j=n}^{m} a_j \equiv a_n + a_{n+1} + \cdots + a_{m-1} + a_m \qquad (m - n + 1 \textbf{ terms}) \tag{1.2-13}$$

$$\prod_{j=n}^{m} a_j \equiv a_n a_{n+1} \cdots a_{m-1} a_m \qquad (m - n + 1 \textbf{ factors}) \tag{1.2-14}$$

Note

$$\sum_{i=n}^{m}\sum_{k=n'}^{m'} a_{ik} = \sum_{k=n'}^{m'}\sum_{i=n}^{m} a_{ik} \qquad \prod_{i=n}^{m}\prod_{k=n'}^{m'} a_{ik} = \prod_{k=n'}^{m'}\prod_{i=n}^{m} a_{ik} \tag{1.2-15}$$

$$\sum_{j=1}^{n} j = \frac{n(n+1)}{2} \qquad \sum_{j=1}^{n} j^2 = \frac{n(n+1)(2n+1)}{6} \qquad \sum_{j=1}^{n} j^3 = \frac{n^2(n+1)^2}{4} \tag{1.2-16}$$

Refer to Chap. 4 for infinite series; and see Sec. E-3.

1.2-6. Arithmetic Progression. If a_0 is the first term and d is the *common difference* between successive terms a_j, then

$$a_j = a_0 + jd \quad (j = 0, 1, 2, \ldots)$$

$$s_n = \sum_{j=0}^{n} a_j = \frac{n+1}{2}(2a_0 + nd) = \frac{n+1}{2}(a_0 + a_n) \tag{1.2-17}$$

1.2-7. Geometric Progression. If a_0 is the first term and r is the *common ratio* of successive terms, then (see Sec. 4.10-2 for infinite geometric series)

$$a_j = a_0 r^j \quad (j = 0, 1, 2, \ldots)$$

$$s_n = \sum_{j=0}^{n} a_j = \sum_{j=0}^{n} a_0 r^j = a_0 \frac{1 - r^{n+1}}{1 - r} = \frac{a_0 - a_n r}{1 - r} \tag{1.2-18}$$

1.3. COMPLEX NUMBERS

1.3-1. Introduction (see also Sec. 7.1-1). **Complex numbers** (sometimes called imaginary numbers) are not numbers in the elementary sense used in connection with counting or measuring; they constitute a new class of mathematical objects defined by the properties described below (see also Sec. 12.1-1).

Each complex number c may be made to correspond to a unique pair (a, b) of real numbers a, b, and conversely. The sum and product of two complex numbers $c_1 \leftrightarrow (a_1, b_1)$ and $c_2 \leftrightarrow (a_2, b_2)$ are defined as $c_1 + c_2 \leftrightarrow (a_1 + a_2, b_1 + b_2)$ and $c_1c_2 \leftrightarrow (a_1a_2 - b_1b_2, a_1b_2 + a_2b_1)$, respectively. The real numbers a are "embedded" in this class of complex numbers as the pairs $(a, 0)$. The **unit imaginary number** i defined as $i \leftrightarrow (0, 1)$ satisfies the relations

$$\boxed{i^2 = (-i)^2 = -1 \qquad i = \sqrt{-1} \qquad -i = -\sqrt{-1}} \tag{1.3-1}$$

Each complex number $c \leftrightarrow (a, b)$ may be written as the sum $c = a + ib$ of a real number $a \leftrightarrow (a, 0)$ and a **pure imaginary number** $ib \leftrightarrow (0, b)$. The real numbers $a = \mathrm{Re}\,(c)$ and $b = \mathrm{Im}\,(c)$ are respectively called the **real part** of c and the **imaginary part** of c. Two complex numbers $c = a + ib$ and $c^* = a - ib$ having equal real parts and equal and opposite imaginary parts are called **complex conjugates.**

Two complex numbers $c_1 = a_1 + ib_1$ and $c_2 = a_2 + ib_2$ are **equal** if and only if their respective real and imaginary parts are equal, i.e., $c_1 = c_2$ if and only if $a_1 = a_2, b_1 = b_2$. $c = a + ib = 0$ implies $a = b = 0$. **Addition and multiplication of complex numbers satisfies all rules of Secs. 1.1-2 and 1.2-1,** with

$$i^2 = -1 \qquad i^3 = -i \qquad i^4 = 1 \qquad i^{4n+1} = i \qquad i^{4n+2} = -1$$
$$i^{4n+3} = -i \qquad i^{4n+4} = 1 \qquad (n = 0, 1, 2, \ldots) \tag{1.3-2}$$

$$\begin{array}{l}
c_1 \pm c_2 = (a_1 \pm a_2) + i(b_1 \pm b_2) \\
c_1c_2 = (a_1a_2 - b_1b_2) + i(a_1b_2 + a_2b_1) \\
\dfrac{c_1}{c_2} = \dfrac{a_1 + ib_1}{a_2 + ib_2} = \dfrac{(a_1a_2 + b_1b_2) + i(a_2b_1 - a_1b_2)}{a_2{}^2 + b_2{}^2} \qquad (c_2 \neq 0) \\
(c_1 + c_2)^* = c_1^* + c_2^* \qquad (c_1c_2)^* = c_1^*c_2^* \\
\qquad (c_1/c_2)^* = c_1^*/c_2^* \qquad (c_2 \neq 0) \qquad (c^*)^* = c \\
a = \mathrm{Re}\,(c) = \dfrac{(c + c^*)}{2} \qquad b = \mathrm{Im}\,(c) = \dfrac{(c - c^*)}{2i}
\end{array} \tag{1.3-3}$$

The class of all complex numbers contains the roots of all equations based on additions and multiplications involving complex numbers and includes the real numbers.

1.3-2. Representation of Complex Numbers as Points or Position Vectors. Polar Decomposition (see also Sec. 7.2-2). Complex

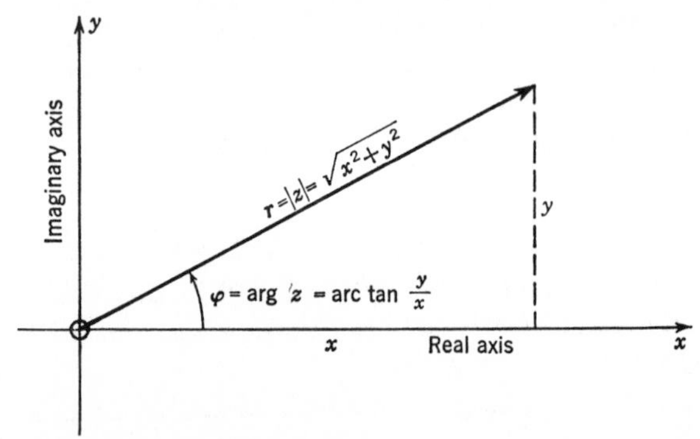

FIG. 1.3-1. Representation of complex numbers as points or position vectors. The x axis and the y axis are called the real axis and the imaginary axis, respectively.

numbers $z = x + iy$ are conveniently represented as points $(z) \equiv (x, y)$ or corresponding position vectors (Secs. 2.1-2 and 3.1-5) in the **Argand or Gauss plane** (Fig. 1.3-1). The (rectangular cartesian, Sec. 2.1-3) x axis and y axis are referred to as the **real axis** and **imaginary axis,** respectively. The abscissa and ordinate of each point (z) respectively represent the real part x and the imaginary part y of z. The corresponding polar coordinates (Sec. 2.1-8)

$$\boxed{\begin{aligned} r &= |z| = \underset{+}{\sqrt{x^2 + y^2}} = \underset{+}{\sqrt{zz^*}} = |z^*| \\ \varphi &= \arg(z) = \arctan(y/x) = -\arg(z^*) \end{aligned}} \tag{1.3-4}$$

are, respectively, the **absolute value (norm, modulus) and the argument (amplitude) of the complex number z.** Note

$$\boxed{x = r\cos\varphi \qquad y = r\sin\varphi \qquad \begin{aligned} z &= x + iy \\ &= r(\cos\varphi + i\sin\varphi) \end{aligned}} \tag{1.3-5}$$

The absolute values of complex numbers satisfy the relations (1.1-4) *to* (1.1-6); if z is a real number, the definition of $|z|$ reduces to that of Sec. 1.1-6.

For any two sets of (real or) complex numbers $\alpha_1, \alpha_2, \ldots, \alpha_n$; $\beta_1, \beta_2, \ldots, \beta_n$ (see also Secs. 14.2-6 and 14.2-6a),

$$\left|\sum_{i=1}^{n} \alpha_i^* \beta_i\right|^2 \leq \sum_{i=1}^{n} |\alpha_i|^2 \sum_{i=1}^{n} |\beta_i|^2 \quad \text{(Cauchy-Schwarz inequality)} \tag{1.3-6}$$

1.3-3. Representation of Addition, Multiplication, and Division. Powers and Roots. Addition of complex numbers corresponds to addition of the corresponding position vectors (see also Secs. 3.1-5 and 5.2-2). Given $z_1 = r_1(\cos\varphi_1 + i\sin\varphi_1)$, $z_2 = r_2(\cos\varphi_2 + i\sin\varphi_2)$,

$$\boxed{\begin{aligned} z_1 z_2 &= r_1 r_2[\cos(\varphi_1 + \varphi_2) + i\sin(\varphi_1 + \varphi_2)] \\ \frac{z_1}{z_2} &= \frac{r_1}{r_2}[\cos(\varphi_1 - \varphi_2) + i\sin(\varphi_1 - \varphi_2)] \qquad (z_2 \neq 0) \\ z^p &= r^p(\cos\varphi + i\sin\varphi)^p = r^p[\cos(p\varphi) + i\sin(p\varphi)] \\ &\qquad \text{(De Moivre's theorem)} \end{aligned}} \tag{1.3-7}$$

(See Sec. 21.2-9 for the case of complex exponents.) **All formulas of Secs. 1.2-1 to 1.2-7 hold for complex numbers** (see also Secs. 1.4-1 to 1.4-3).

Note

$$\left.\begin{aligned}\sqrt[n]{1} &= \cos\frac{2k\pi}{n} + i\sin\frac{2k\pi}{n} \\ \sqrt[n]{-1} &= \cos\frac{(2k+1)\pi}{n} + i\sin\frac{(2k+1)\pi}{n}\end{aligned}\right\}\begin{matrix}(n = 1, 2, \ldots; \\ k = 0, 1, 2, \ldots n-1) \\ (n\ values)\end{matrix} \tag{1.3-8}$$

In particular,

$$\sqrt{1} = \pm 1 \qquad \sqrt{-1} = \pm i \tag{1.3-9}$$

$$\left.\begin{aligned}\sqrt[3]{1} &= \begin{cases}1 \\ \cos 120° + i\sin 120° = \tfrac{1}{2}(-1 + i\sqrt{3}) \\ \cos 120° - i\sin 120° = \tfrac{1}{2}(-1 - i\sqrt{3})\end{cases} \\ \sqrt[3]{-1} &= \begin{cases}\cos 60° + i\sin 60° = \tfrac{1}{2}(1 + i\sqrt{3}) \\ -1 \\ \cos 60° - i\sin 60° = \tfrac{1}{2}(1 - i\sqrt{3})\end{cases}\end{aligned}\right\} \tag{1.3-10}$$

1.4. MISCELLANEOUS FORMULAS

1.4-1. The Binomial Theorem and Related Formulas. If a, b, c are real or complex numbers,

$$\begin{aligned}(a \pm b)^2 &= a^2 \pm 2ab + b^2 \\ (a \pm b)^3 &= a^3 \pm 3a^2b + 3ab^2 \pm b^3 \\ (a \pm b)^4 &= a^4 \pm 4a^3b + 6a^2b^2 \pm 4ab^3 + b^4 \\ &\cdots\cdots\cdots\cdots \\ (a + b)^n &= \sum_{j=0}^{n} \binom{n}{j} a^j b^{n-j} \qquad (n = 1, 2, \ldots) \\ \text{with } \binom{n}{j} &= \frac{n!}{j!(n-j)!} \\ &\qquad (j = 0, 1, 2, \ldots \leq n = 0, 1, 2, \ldots)\end{aligned} \tag{1.4-1}$$

(*Binomial Theorem* for integral exponents n; see also Sec. 21.2-12). The **binomial coefficients** $\binom{n}{j}$ are discussed in detail in Sec. 21.5-1.

$$(a + b + c)^2 = a^2 + b^2 + c^2 + 2ab + 2ac + 2bc \tag{1.4-2}$$

$$\begin{aligned}(a^2 - b^2) &= (a + b)(a - b) \\ (a^2 + b^2) &= (a + ib)(a - ib)\end{aligned} \tag{1.4-3}$$

$$a^n - b^n = (a - b)(a^{n-1} + a^{n-2}b + \cdots + ab^{n-2} + b^{n-1}) \tag{1.4-4a}$$

If n is an *even* positive integer,

$$a^n - b^n = (a + b)(a^{n-1} - a^{n-2}b + \cdots + ab^{n-2} - b^{n-1}) \tag{1.4-4b}$$

If n is an *odd* positive integer,

$$a^n + b^n = (a + b)(a^{n-1} - a^{n-2}b + \cdots - ab^{n-2} + b^{n-1}) \qquad (1.4\text{-}5)$$

Note also

$$a^4 + a^2b^2 + b^4 = (a^2 + ab + b^2)(a^2 - ab + b^2) \qquad (1.4\text{-}6)$$

1.4-2. Proportions. $a:b = c:d$ or $a/b = c/d$ implies

$$\frac{ma + nb}{pa + qb} = \frac{mc + nd}{pc + qd} \qquad \text{(CORRESPONDING ADDITION AND SUBTRACTION)} \qquad (1.4\text{-}7)$$

In particular,

$$\frac{a \pm b}{b} = \frac{c \pm d}{d} \qquad \frac{a - b}{a + b} = \frac{c - d}{c + d} \qquad (1.4\text{-}8)$$

1.4-3. Polynomials. Symmetric Functions. (a) A **polynomial** in (**integral rational function** of) the quantities $x_1, x_2, \ldots, x_n$ is a sum involving a finite number of terms of the form $ax_1^{k1}x_2^{k2} \cdots x_n^{k_n}$, where each k_j is a nonnegative integer. The largest value of $k_1 + k_2 + \cdots + k_n$ occurring in any term is the **degree** of the polynomial. A polynomial is **homogeneous** if and only if all its terms are of the same degree (see also Sec. 4.5-5).

(b) A polynomial in $x_1, x_2, \ldots, x_n$ (and more generally, any function of $x_1, x_2, \ldots, x_n$) is (**completely**) **symmetric** if and only if its value is unchanged by permutations of the $x_1, x_2, \ldots, x_n$ for any set of values $x_1, x_2, \ldots, x_n$. The **elementary symmetric functions** $S_1, S_2, \ldots, S_n$ of $x_1, x_2, \ldots, x_n$ are the polynomials

$$S_1 \equiv x_1 + x_2 + \cdots + x_n \qquad S_2 \equiv x_1x_2 + x_1x_3 + \cdots$$
$$S_3 \equiv x_1x_2x_3 + x_1x_2x_4 + \cdots \qquad \cdots \qquad S_n \equiv x_1x_2 \cdots x_n \qquad (1.4\text{-}9)$$

where S_k is the sum of all $\frac{n!}{(n-k)!k!}$ products combining k factors x_j without repetition of subscripts (see also Table C-2). *Every polynomial symmetric in $x_1, x_2, \ldots, x_n$ can be rewritten as a unique polynomial in $S_1, S_2, \ldots, S_n$*; the coefficients in the new polynomial are algebraic sums of integral multiples of the given coefficients.

Every polynomial symmetric in $x_1, x_2, \ldots, x_n$ can also be expressed as a polynomial in a finite number of the symmetric functions

$$s_0 \equiv n \qquad s_1 \equiv \sum_{i=1}^{n} x_i \qquad s_2 \equiv \sum_{i=1}^{n} x_i^2 \qquad \cdots \qquad s_k \equiv \sum_{i=1}^{n} x_i^k \cdots \qquad (1.4\text{-}10)$$

The symmetric functions (9) and (10) are related by *Newton's formulas*

$$(-1)^k kS_k + (-1)^{k-1}S_{k-1}s_1 + (-1)^{k-2}S_{k-2}s_2 + \cdots = 0 \qquad (k = 1, 2, \cdots) \qquad (1.4\text{-}11)$$

where one defines $S_k = 0$ for $k > n$ and $k < 0$, and $S_0 = 1$ (see also Ref. 1.2 for explicit tabulations of Newton's formulas; and see also Sec. 1.6-4). Note that the relations (11) do not involve n explicitly.

1.5. DETERMINANTS

1.5-1. Definition. The **determinant**

$$D = \det [a_{ik}] = \begin{vmatrix} a_{11} & a_{12} & \cdots & a_{1n} \\ a_{21} & a_{22} & \cdots & a_{2n} \\ \cdots & \cdots & \cdots & \cdots \\ a_{n1} & a_{n2} & \cdots & a_{nn} \end{vmatrix} \tag{1.5-1}$$

of the square array (matrix, Sec. 13.2-1) of n^2 (real or complex) numbers (**elements**) a_{ik} is the sum of the $n!$ terms $(-1)^r a_{1k_1} a_{2k_2} \cdots a_{nk_n}$ each corresponding to one of the $n!$ different ordered sets $k_1, k_2, \ldots, k_n$ obtained by r interchanges of elements from the set $1, 2, \ldots, n$. The number n is the **order** of the determinant (1).

The actual computation of a determinant in terms of its elements is simplified by the use of Secs. 1.5-2 and 1.5-5*a*. Note that

$$|D|^2 \leq \prod_{i=1}^{n} \sum_{k=1}^{n} |a_{ik}|^2 \quad \text{(Hadamard's inequality)} \tag{1.5-2}$$

1.5-2. Minors and Cofactors. Expansion in Terms of Cofactors. The (complementary) **minor** D_{ik} of the element a_{ik} in the nth-order determinant (1) is the $(n-1)^{\text{st}}$-order determinant obtained from (1) on erasing the ith row and the kth column. The **cofactor** A_{ik} of the element a_{ik} is the coefficient of a_{ik} in the expansion of D, or

$$A_{ik} = (-1)^{i+k} D_{ik} = \frac{\partial D}{\partial a_{ik}} \tag{1.5-3}$$

A determinant D may be represented in terms of the elements and cofactors of any one row or column as follows:

$$D = \det[a_{ik}] = \sum_{i=1}^{n} a_{ij} A_{ij} = \sum_{k=1}^{n} a_{jk} A_{jk} \tag{1.5-4}$$
$$(j = 1, 2, \ldots, n) \quad \text{(simple Laplace development)}$$

Note also that

$$\sum_{i=1}^{n} a_{ij} A_{ih} = \sum_{k=1}^{n} a_{jk} A_{hk} = 0 \qquad (j \neq h) \tag{1.5-5}$$

1.5-3. Examples: Second- and Third-order Determinants.

$$\begin{vmatrix} a_{11} & a_{12} \\ a_{21} & a_{22} \end{vmatrix} = a_{11}a_{22} - a_{21}a_{12} \qquad (1.5\text{-}6)$$

$$\begin{vmatrix} a_{11} & a_{12} & a_{13} \\ a_{21} & a_{22} & a_{23} \\ a_{31} & a_{32} & a_{33} \end{vmatrix} = a_{11}a_{22}a_{33} - a_{11}a_{23}a_{32} + a_{12}a_{23}a_{31} - a_{13}a_{22}a_{31} + a_{13}a_{21}a_{32} - a_{12}a_{21}a_{33}$$
$$= a_{11}(a_{22}a_{33} - a_{32}a_{23}) - a_{21}(a_{12}a_{33} - a_{32}a_{13}) + a_{31}(a_{12}a_{23} - a_{22}a_{13})$$
$$= a_{11}(a_{22}a_{33} - a_{32}a_{23}) - a_{12}(a_{21}a_{33} - a_{31}a_{23}) + a_{13}(a_{21}a_{32} - a_{31}a_{22})$$
$$\text{etc.} \qquad (1.5\text{-}7)$$

1.5-4. Complementary Minors. Laplace Development. The mth-order determinant M obtained from the nth-order determinant D by deleting all rows except for the m rows labeled $i_1, i_2, \ldots, i_m$ $(m \leq n)$, and all columns except for the m columns labeled $k_1, k_2, \ldots, k_m$ is an **m-rowed minor of D.** The m-rowed minor M and the $(n - m)$-rowed minor M' of D obtained by deleting the rows and columns conserved in M are **complementary minors;** in the special case $m = n$, $M' = 1$. The **algebraic complement** M'' of M is defined as $(-1)^{i_1+i_2 \cdots +i_m+k_1+k_2 \cdots +k_m}M'$. *Given any m rows (or columns) of D, D is equal to the sum of the products MM'' of all m-rowed minors M using these rows (or columns) and their algebraic complements M'' (Laplace development by rows or columns).*

An nth-order determinant D has $\binom{n}{m}$ m-rowed **principal minors** whose diagonal elements are diagonal elements of D.

1.5-5. Miscellaneous Theorems. **(a)** *The value D of a determinant (1) is not changed by any of the following operations:*

1. *The rows are written as columns, and the columns as rows [interchange of i and k in Eq. (1)].*
2. *An even number of interchanges of any two rows or two columns.*
3. *Addition of the elements of any row (or column), all multiplied, if desired, by the same parameter α, to the respective corresponding elements of another row (or column, respectively).*

EXAMPLES:

$$\begin{vmatrix} a_{11} & a_{12} & \cdots & a_{1n} \\ a_{21} & a_{22} & \cdots & a_{2n} \\ \cdots & \cdots & \cdots & \cdots \\ a_{n1} & a_{n2} & \cdots & a_{nn} \end{vmatrix} = \begin{vmatrix} a_{11} & a_{21} & \cdots & a_{n1} \\ a_{12} & a_{22} & \cdots & a_{n2} \\ \cdots & \cdots & \cdots & \cdots \\ a_{1n} & a_{2n} & \cdots & a_{nn} \end{vmatrix}$$
$$= \begin{vmatrix} a_{11} + \alpha a_{12} & a_{12} & \cdots & a_{1n} \\ a_{21} + \alpha a_{22} & a_{22} & \cdots & a_{2n} \\ \cdots & \cdots & \cdots & \cdots \\ a_{n1} + \alpha a_{n2} & a_{n2} & \cdots & a_{nn} \end{vmatrix} \qquad (1.5\text{-}8)$$

(b) *An odd number of interchanges of any two rows or two columns is equivalent to multiplication of the determinant by -1.*

(c) *Multiplication of all the elements of any one row or column by a factor α is equivalent to multiplication of the determinant by α.*

(d) *If the elements of the jth row (or column) of an nth-order determinant D are represented as sums $\sum_{r=1}^{m} c_{r1}, \sum_{r=1}^{m} c_{r2}, \ldots, \sum_{r=1}^{m} c_{rn}$, D is equal to the sum $\sum_{r=1}^{m} D_r$ of m n^{th}-order determinants D_r. The elements of each D_r are identical with those of D, except for the elements of the jth row (or column, respectively), which are $c_{r1}, c_{r2}, \ldots, c_{rn}$.*

EXAMPLE:

$$\begin{vmatrix} a_{11}+b_{11} & a_{12}+b_{12} & \cdots & a_{1n}+b_{1n} \\ a_{21} & a_{22} & & a_{2n} \\ \cdots & \cdots & \cdots & \cdots \\ a_{n1} & a_{n2} & & a_{nn} \end{vmatrix}$$

$$= \begin{vmatrix} a_{11} & a_{12} & \cdots & a_{1n} \\ a_{21} & a_{22} & \cdots & a_{2n} \\ \cdots & \cdots & \cdots & \cdots \\ a_{n1} & a_{n2} & \cdots & a_{nn} \end{vmatrix} + \begin{vmatrix} b_{11} & b_{12} & \cdots & b_{1n} \\ a_{21} & a_{22} & \cdots & a_{2n} \\ \cdots & \cdots & \cdots & \cdots \\ a_{n1} & a_{n2} & \cdots & a_{nn} \end{vmatrix} \qquad (1.5\text{-}9)$$

(e) *A determinant is equal to zero if*

1. *All elements of any row or column are zero.*
2. *Corresponding elements of any two rows or columns are equal, or proportional with the same proportionality factor.*

1.5-6. Multiplication of Determinants (see also Sec. 13.2-2). The product of two n^{th}-order determinants $\det [a_{ik}]$ and $\det [b_{ik}]$ is

$$\det [a_{ik}] \det [b_{ik}] = \det \left[\sum_{j=1}^{n} a_{ij}b_{jk} \right] = \det \left[\sum_{j=1}^{n} a_{ji}b_{kj} \right]$$
$$= \det \left[\sum_{j=1}^{n} a_{ij}b_{kj} \right] = \det \left[\sum_{j=1}^{n} a_{ji}b_{jk} \right] \qquad (1.5\text{-}10)$$

1.5-7. Changing the Order of Determinants. A given determinant may be expressed in terms of a determinant of higher order as follows:

$$\begin{vmatrix} a_{11} & a_{12} & \cdots & a_{1n} \\ a_{21} & a_{22} & \cdots & a_{2n} \\ \cdots & \cdots & \cdots & \cdots \\ a_{n1} & a_{n2} & \cdots & a_{nn} \end{vmatrix} = \begin{vmatrix} a_{11} & a_{12} & \cdots & a_{1n} & \alpha_1 \\ a_{21} & a_{22} & \cdots & a_{2n} & \alpha_2 \\ \cdots & \cdots & \cdots & \cdots & \cdots \\ a_{n1} & a_{n2} & \cdots & a_{nn} & \alpha_n \\ 0 & 0 & \cdots & 0 & 1 \end{vmatrix} \qquad (1.5\text{-}11)$$

where the α_i are arbitrary. This process can be repeated as desired.

The order of a given determinant may sometimes be reduced through the use of the relation

$$\begin{vmatrix} a_{11} & a_{12} & \cdots & a_{1n} & \alpha_{11} & \alpha_{12} & \cdots & \alpha_{1m} \\ a_{21} & a_{22} & \cdots & a_{2n} & \alpha_{21} & \alpha_{22} & \cdots & \alpha_{2m} \\ \cdots & \cdots & \cdots & \cdots & \cdots & \cdots & \cdots & \cdots \\ a_{n1} & a_{n2} & \cdots & a_{nn} & \alpha_{n1} & \alpha_{n2} & \cdots & \alpha_{nm} \\ 0 & 0 & \cdots & 0 & b_{11} & b_{12} & \cdots & b_{1m} \\ 0 & 0 & \cdots & 0 & b_{21} & b_{22} & \cdots & b_{2m} \\ \cdots & \cdots & \cdots & \cdots & \cdots & \cdots & \cdots & \cdots \\ 0 & 0 & \cdots & 0 & b_{m1} & b_{m2} & \cdots & b_{mm} \end{vmatrix}$$

$$= \begin{vmatrix} a_{11} & a_{12} & \cdots & a_{1n} \\ a_{21} & a_{22} & \cdots & a_{2n} \\ \cdots & \cdots & \cdots & \cdots \\ a_{n1} & a_{n2} & \cdots & a_{nn} \end{vmatrix} \cdot \begin{vmatrix} b_{11} & b_{12} & \cdots & b_{1m} \\ b_{21} & b_{22} & \cdots & b_{2m} \\ \cdots & \cdots & \cdots & \cdots \\ b_{m1} & b_{m2} & \cdots & b_{mm} \end{vmatrix} \qquad (1.5\text{-}12)$$

1.6. ALGEBRAIC EQUATIONS: GENERAL THEOREMS

1.6-1. Introduction. The solution of algebraic equations is of particular importance in connection with the characteristic equations of linear systems in physics (see also Secs. 9.4-1, 9.4-4, and 14.8-5). The general location of the roots needed (e.g., for stability determinations) may be investigated by the methods of Sec. 1.6-6 and/or Sec. 7.6-9. Numerical solutions are discussed in Secs. 20.2-1 to 20.2-3.

1.6-2. Solution of an Equation. Roots. To **solve an equation** (see also Sec. 1.1-3)

$$f(x) = 0 \qquad (1.6\text{-}1)$$

for the unknown x means to find values of x [**roots** of Eq. (1), **zeros of** $f(x)$] which satisfy the given equation. $x = x_1$ is a **root (zero) of order (multiplicity)** m (**multiple** root if $m > 1$; see also Sec. 7.6-1) if and only if, for $x = x_1$, $f(x)/(x - x_1)^{m-1} = 0$ and $f(x)/(x - x_1)^m \neq 0$. A **complete solution** of Eq. (1) specifies all roots together with their orders. Solutions may be verified by substitution.

1.6-3. Algebraic Equations. An equation (1) of the form

$$f(x) = a_0x^n + a_1x^{n-1} + \cdots + a_{n-1}x + a_n = 0 \qquad (a_0 \neq 0) \qquad (1.6\text{-}2)$$

where the **coefficients** a_i are real or complex numbers, is called an **algebraic equation of degree** n **in the unknown** x. $f(x)$ is a **polynomial of degree** n **in** x (**rational integral function;** see also Secs. 4.2-2*d* and 7.6-5). a_n is the **absolute term** of the polynomial (2).

An algebraic equation of degree n has exactly n roots if a root of order m is counted as m roots (Fundamental Theorem of Algebra).

Numbers expressible as roots of algebraic equations with real integral coefficients are algebraic numbers (in general complex, with rational and/or irrational real and imaginary parts); if the coefficients are algebraic, the roots are still algebraic (see also Sec. 1.1-2). General formulas for the roots of algebraic equations in terms of the coefficients and involving only a finite number of additions, subtractions, multiplications, divisions, and root extractions exist *only* for equations of degree one (**linear equations,** Sec. 1.8-1), two (**quadratic equations,** Sec. 1.8-2), three (**cubic equations,** Secs. 1.8-3 and 1.8-4), and four (**quartic equations,** Secs. 1.8-5 and 1.8-6).

1.6-4. Relations between Roots and Coefficients. The symmetric functions S_k and s_k (Sec. 1.4-3) of the roots $x_1, x_2, \ldots, x_n$ of an algebraic equation (2), are related to the coefficients $a_0, a_1, \ldots, a_n$ as follows:

$$\frac{a_k}{a_0} = (-1)^k S_k \qquad (k = 0, 1, 2, \cdots, n) \tag{1.6-3}$$

$$ka_k + a_{k-1}s_1 + a_{k-2}s_2 + \cdots + a_0 s_k = 0 \qquad (k = 1, 2, \cdots) \tag{1.6-4}$$

where one defines $a_k = 0$ for $k > n$ and $k < 0$. The equations (1.6-4) are another version of *Newton's formulas* (1.4-11). Note also

$$\frac{a_k}{a_0} = \frac{(-1)^k}{k!}\begin{vmatrix} s_1 & 1 & 0 & \cdots & \cdots & 0 \\ s_2 & s_1 & 2 & 0 & \cdots & 0 \\ s_3 & s_2 & s_1 & 3 & 0 \cdots & 0 \\ \cdots & \cdots & \cdots & \cdots & \cdots & \cdots \\ s_k & s_{k-1} & \cdots & \cdots & \cdots & s_1 \end{vmatrix}$$

$$s_k = \left(-\frac{1}{a_0}\right)^k \begin{vmatrix} a_1 & a_0 & 0 & \cdots & \cdots & \cdots & 0 \\ 2a_2 & a_1 & a_0 & 0 & \cdots & \cdots & 0 \\ 3a_3 & a_2 & a_1 & a_0 & 0 & \cdots & 0 \\ \cdots & \cdots & \cdots & \cdots & \cdots & \cdots & \cdots \\ ka_k & a_{k-1} & \cdots & \cdots & \cdots & \cdots & a_1 \end{vmatrix} \qquad (k = 1, 2, \cdots, n) \tag{1.6-5}$$

1.6-5. Discriminant of an Algebraic Equation. The **discriminant** Δ of an algebraic equation (2) is the product of a_0^{2n-2} and the squares of all differences $(x_i - x_k)(i > k)$ between the roots x_i of the equation (a multiple root of order m is considered as m equal roots with different subscripts),

$$\Delta = a_0^{2n-2} \prod_{i>k} (x_i - x_k)^2 = a_0^{2n-2} \begin{vmatrix} 1 & x_1 & x_1^2 & \cdots & x_1^{n-1} \\ 1 & x_2 & x_2^2 & \cdots & x_2^{n-1} \\ \cdots & \cdots & \cdots & \cdots & \cdots \\ 1 & x_n & x_n^2 & \cdots & x_n^{n-1} \end{vmatrix}$$

$$= a_0^{2n-2} \begin{vmatrix} s_0 & s_1 & s_2 & \cdots & s_{n-1} \\ s_1 & s_2 & s_3 & \cdots & s_n \\ \cdots & \cdots & \cdots & \cdots & \cdots \\ s_{n-1} & s_n & s_{n+1} & \cdots & s_{2n-2} \end{vmatrix} = (-1)^{\frac{n(n-1)}{2}} \frac{1}{a_0} R(f, f') \tag{1.6-6}$$

where $R(f, f')$ is the resultant (Sec. 1.7-3) of $f(x)$ and its derivative (Sec. 4.5-1) $f'(x)$. Δ is a symmetric function of the roots $x_1, x_2, \ldots, x_n$ and vanishes if and only if $f(x)$ has at least one multiple root [which is necessarily a common root of $f(x)$ and $f'(x)$; see also Sec. 1.6-6*g*]. The second determinant in Eq. (6) is called *Vandermonde's determinant.*

1.6-6. Real Algebraic Equations and Their Roots. An algebraic equation (2) is called **real** if and only if all coefficients a_i are real; the corresponding real polynomial $f(x)$ is real for all real values of x. The following theorems are useful for determining the general location of roots (e.g., prior to numerical solution, Sec. 20.2-1; see also Secs. 9.4-4 and 14.8-5). In theorems (*b*) through (*f*), a root of order m is counted as m roots.

(a) **Complex Roots.** *Complex roots of real algebraic equations occur in pairs of complex conjugates* (Sec. 1.3-1). *A real algebraic equation of odd degree must have at least one real root.*

(b) **Routh-Hurwitz Criterion.** *The number of roots with positive real parts of a real algebraic equation* (2) *is equal to the number of sign changes (disregard vanishing terms) in either one of the sequences*

$$\left.\begin{aligned} &T_0,\ T_1,\ \frac{T_2}{T_1},\ \frac{T_3}{T_2},\ \ldots,\ \frac{T_n}{T_{n-1}} \\ \text{or}\qquad &T_0,\ T_1,\ T_1T_2,\ T_2T_3,\ \ldots,\ T_{n-1}T_{n-2},\ a_n \end{aligned}\right\} \tag{1.6-7}$$

where $T_0 = a_0 > 0 \qquad T_1 = a_1 \qquad T_2 = \begin{vmatrix} a_1 & a_0 \\ a_3 & a_2 \end{vmatrix}$

$$T_3 = \begin{vmatrix} a_1 & a_0 & 0 \\ a_3 & a_2 & a_1 \\ a_5 & a_4 & a_3 \end{vmatrix} \qquad T_4 = \begin{vmatrix} a_1 & a_0 & 0 & 0 \\ a_3 & a_2 & a_1 & a_0 \\ a_5 & a_4 & a_3 & a_2 \\ a_7 & a_6 & a_5 & a_4 \end{vmatrix} \qquad \cdots \tag{1.6-8}$$

Given $a_0 > 0$, all roots have negative real parts if and only if $T_0, T_1, T_2, \ldots, T_n$ are all positive. This is true if and only if all a_i and either all even-numbered T_k or all odd-numbered T_k are positive (Liénard-Chipart Test).

ALTERNATIVE FORMULATION. *All the roots of a real n^{th}-degree equation (2) have negative real parts if and only if this is true for the $(n-1)^{\text{st}}$-degree equation*

$$\begin{aligned} a_0'x^{n-1} + a_1'x^{n-2} + a_2'x^{n-3} + a_3'x^{n-4} + \cdots &\equiv a_1x^{n-1} + a_2x^{n-2} \\ + a_3x^{n-3} + a_4x^{n-4} + \cdots - \frac{a_0}{a_1}a_3x^{n-2} &- \frac{a_0}{a_1}a_5x^{n-4} - \cdots = 0 \end{aligned}$$

This theorem may be applied repeatedly and yields a simple recursion scheme useful, for example, for stability investigations. *The number of roots with negative real parts is precisely equal to the number of negative multipliers $-a_0^{(j)}/a_1^{(j)}$ ($j = 0, 1, 2, \ldots, n-1$; $a_0^{(0)} = a_0 > 0$, $a_1^{(0)} = a_1$) encountered in successive applications of the theorem.* The method becomes more complicated if one of the $a_1^{(j)}$ vanishes (see Ref. 1.6, which also indicates an extension to complex equations).

(c) **Location of Real Roots: Descartes's Rule of Signs.** *The number of positive real roots of a real algebraic equation* (2) *either is equal to the number N_a of sign changes in the sequence $a_0, a_1, \ldots, a_n$ of coefficients, where vanishing terms are disregarded, or it is less than N_a by a positive even integer.* Application of this theorem to $f(-x)$ yields a similar theorem for negative real roots.

(d) Location of Real Roots: An Upper Bound (Sec. 4.3-3a) for the Real Roots. *If the first k coefficients* $a_0, a_1, \ldots, a_{k-1}$ *in a real algebraic equation* (2) *are nonnegative* (a_k *is the first negative coefficient*) *then all real roots of Eq.* (2) *are smaller than* $1 + \sqrt[k]{q/a_0}$, *where q is the absolute value of the negative coefficient greatest in absolute value.* Application of this theorem to $f(-x)$ may similarly yield a lower bound of the real roots.

(e) Location of Real Roots: Rolle's Theorem (see also Sec. 4.7-1a). *The derivative* (*Sec.* 4.5-1) $f'(x)$ *of a real polynomial* $f(x)$ *has an odd number of real zeros between two consecutive real zeros of* $f(x)$.

$f(x) = 0$ has no real root or one real root between two consecutive real roots a, b of $f'(x) = 0$ if $f(a) \neq 0$ and $f(b) \neq 0$ have equal or opposite signs, respectively. At most, one real root of $f(x) = 0$ is greater than the greatest root or smaller than the smallest root of $f'(x) = 0$.

(f) Location of Real Roots: Budan's Theorem. *For any real algebraic equation* (2), *let* $N(x)$ *be the number of sign changes in the sequence of derivatives* (*Sec.* 4.5-1) $f(x), f'(x), f''(x), \ldots, f^{(n)}(x)$, *if vanishing terms are disregarded. Then the number of real roots of Eq.* (2) *located between two real numbers a and* $b > a$ *not themselves roots of Eq.* (2) *is either* $N(a) - N(b)$, *or it is less than* $N(a) - N(b)$ *by a positive even integer.*

The number of real roots of Eq. (2) located between a and b is odd or even if $f(a)$ and $f(b)$ have opposite or equal signs, respectively.

(g) Location of Real Roots: Sturm's Method. *Given a real algebraic equation* (2) *without multiple roots* (*Sec.* 1.6-2), *let* $N(x)$ *be the number of sign changes* (*disregard vanishing terms*) *in the sequence of functions*

$$f_0 = f(x) = g_0(x)f_1(x) - f_2(x) \qquad f_1 = f'(x) = g_1(x)f_2(x) - f_3(x)$$
$$f_2(x) = g_2(x)f_3(x) - f_4(x) \qquad \cdots \qquad (1.6\text{-}9)$$

where for $i > 1$ *each* $f_i(x)$ *is* (-1) *times the remainder* (*Sec.* 1.7-2) *obtained on dividing* $f_{i-2}(x)$ *by* $f_{i-1}(x)$; $f_n(x) \neq 0$ *is a constant. Then the number of real roots of Eq.* (2) *located between two real numbers a and* $b > a$ *not themselves roots of Eq.* (2) *is equal to* $N(a) - N(b)$.

Sturm's method applies even if, for convenience in computation, a function $f_i(x)$ in the above process is replaced by $F_i(x) = f_i(x)/k(x)$, where $k(x)$ is a positive constant or a polynomial in x positive for $a \leq x \leq b$, and the remaining functions are based on $F_i(x)$ instead of on $f_i(x)$. Similar operations may be performed again on any of the $F_j(x)$, etc.

If $f(x)$ has *multiple roots*, $f(x)$ and $f'(x)$ have a common divisor (Sec. 1.7-3); in this case, $f_n(x)$ is not a constant, and $N(a) - N(b)$ is the number of real roots between a and b, where each multiple root is counted *only once*.

1.7. FACTORING OF POLYNOMIALS AND QUOTIENTS OF POLYNOMIALS. PARTIAL FRACTIONS

1.7-1. Factoring of a Polynomial (see also Sec. 7.6-6). If a polynomial $F(x)$ can be represented as a product of polynomials $f_1(x)$, $f_2(x)$, . . . , $f_s(x)$, these polynomials are called **factors (divisors)** of $F(x)$. If $x = x_1$ is a zero of order m of any factor $f_i(x)$, it is also a zero of order $M \geq m$ of $F(x)$. *Every (real or complex) polynomial $f(x)$ of degree n in x can be expressed as a product of a constant and n linear factors $(x - \alpha_k)$ in one and only one way, namely,*

$$f(x) \equiv a_0x^n + a_1x^{n-1} + \cdots + a_{n-1}x + a_n \equiv a_0 \prod_{k=1}^{n} (x - x_k) \tag{1.7-1}$$

where the x_k are the zeros of $f(x)$; a zero x_k of order m_k (Sec. 1.6-2) contributes m_k factors $(x - x_k)$ (Factor Theorem). Pairs of factors $[x - (a_k + i\omega_k)]$, $[x - (a_k - i\omega_k)]$ corresponding to pairs of complex conjugate roots (see also Sec. 1.6-6*a*) $x_k = a_k + i\omega_k$, $x_k = a_k - i\omega_k$ may be combined into real quadratic factors $[(x - a_k)^2 + \omega_k^2]$.

1.7-2. Quotients of Polynomials. Remainder. Long Division. The quotient $F(x)/f(x)$ of a polynomial $F(x)$ of degree N and a polynomial $f(x)$ of degree $n < N$ may be expressed in the form

$$\frac{F(x)}{f(x)} \equiv \frac{A_0x^N + A_1x^{N-1} + \cdots + A_N}{a_0x^n + a_1x^{n-1} + \cdots + a_n} \equiv (b_0x^{N-n} + b_1x^{N-n-1} + \cdots + b_{N-n}) + \frac{r_1(x)}{f(x)} \tag{1.7-2}$$

where the **remainder** $r_1(x)$ is a polynomial of degree smaller than n. The coefficients b_k and the remainder $r_1(x)$ are uniquely determined, e.g., by the process of **long division (division algorithm)** indicated in Fig. 1.7-1.

In Fig. 1.7-1, each product $b_0f(x)$, $b_1f(x)$, . . . is subtracted in turn, with the coefficients b_0, b_1, . . . chosen so as to eliminate the respective coefficients of x^N, x^{N-1}, . . . , in successive differences until the remainder is reached. The remainder $r_1(x)$ vanishes if and only if $f(x)$ is a divisor (Sec. 1.7-1) of $F(x)$.

The remainder obtained on dividing any polynomial $f(x)$ by $(x - c)$ is equal to $f(c)$ (Remainder Theorem).

$$
\begin{array}{l}
(A_0x^N + \quad A_1 x^{N-1} + \qquad\qquad A_2 x^{N-2} + \cdots) \div (a_0x^n + a_1x^{n-1} + a_2x^{n-2} + \cdots) = \frac{A_0}{a_0} x^{N-n} + \left(\frac{A_1}{a_0} - A_0 \frac{a_1}{a_0^2}\right) x^{N-n-1} + \cdots + r_1(x) \\
\underline{-\left(A_0x^N + A_0 \frac{a_1}{a_0} x^{N-1} + \qquad A_0 \frac{a_2}{a_0} x^{N-2} + \cdots\right)} \\
\left(A_1 - A_0 \frac{a_1}{a_0}\right) x^{N-1} + \left(A_2 - A_0 \frac{a_2}{a_0}\right) x^{N-2} + \cdots \\
\cdots\cdots\cdots\cdots\cdots\cdots\cdots\cdots\cdots\cdots \text{etc.}
\end{array}
$$

EXAMPLES

(1) $(2x^3 - 4x^2 - 2x + 3) \div (x - 2) = 2x^2 - 2 - \frac{1}{x - 2}$

$$
\begin{array}{r}
\underline{-(2x^3 - 4x^2)\qquad\qquad} \\
-2x + 3 \\
\underline{-(-2x + 4)} \\
-1
\end{array}
$$

(2) $(x^4 - x^2 + ax - a^4 + a^2 + b) \div (x^2 - a^2) = x^2 + (a^2 - 1) + \frac{ax + b}{x^2 - a^2}$

$$
\begin{array}{l}
\underline{-(x^4 - a^2x^2)} \\
\qquad (a^2 - 1)x^2 + ax - (a^2 - 1)a^2 + b \\
\underline{\quad -[(a^2 - 1)x^2 \qquad\quad - (a^2 - 1)a^2]} \\
\qquad\qquad\qquad ax \qquad\qquad\qquad + b
\end{array}
$$

FIG. 1.7-1. Long division.

1.7-3. Common Divisors and Common Roots of Two Polynomials. If a polynomial $g(x)$ is a **common divisor (factor)** of $F(x)$ and $f(x)$, its zeros are common zeros of $F(x)$ and $f(x)$. In the quotient (2), any common divisor may be factored out and canceled as with numerical fractions.

$F(x)$ and $f(x)$ have at least one common root (and thus a common divisor of degree greater than zero) if and only if the determinant of order $N + n$

$$R(F, f) = \begin{vmatrix} A_0 A_1 A_2 \cdot & \cdot & \cdot & \cdot & \cdot & A_{N-1} A_N & 0 & \cdot & \cdot & 0 \\ 0 \; A_0 A_1 A_2 & \cdot & \cdot & \cdot & \cdot & \cdot & A_{N-1} A_N 0 & \cdot & \cdot & 0 \\ \cdot & \cdot & \cdot & \cdot & \cdot & \cdot & \cdot & \cdot & \cdot & \cdot \\ 0 \cdot & \cdot & \cdot & 0 & A_0 A_1 A_2 & \cdot & \cdot & \cdot & \cdot & A_{N-1} A_N \\ a_0 \, a_1 \, a_2 \cdot & \cdot & a_{n-1} a_n & 0 & \cdot & \cdot & \cdot & \cdot & \cdot & 0 \\ 0 \; a_0 \, a_1 \, a_2 & \cdot & \cdot & a_{n-1} \, a_n 0 & \cdot & \cdot & \cdot & \cdot & \cdot & 0 \\ \cdot & \cdot & \cdot & \cdot & \cdot & \cdot & \cdot & \cdot & \cdot & \cdot \\ 0 \cdot & \cdot & \cdot & \cdot & \cdot & \cdot & 0 & a_0 \, a_1 \, a_2 \cdot & \cdot & a_{n-1} \, a_n \end{vmatrix} \tag{1.7-3}$$

[**resultant** of $F(x)$ and $f(x)$] is equal to zero; otherwise, $F(x)$ and $f(x)$ are *relatively prime.*

The *greatest common divisor* (common factor of greatest degree) of $F(x)$ and $f(x)$ is uniquely defined except for a constant factor and may be obtained as follows: Divide $r_1(x)$ into $f(x)$; divide the resulting remainder $r_2(x)$ into $r_1(x)$, and continue until some remainder, $r_k(x)$, say, vanishes. Then any constant multiple of $r_{k-1}(x)$ is the desired greatest common divisor.

1.7-4. Expansion in Partial Fractions. Any quotient $g(x)/f(x)$ of a polynomial $g(x)$ of degree m and a polynomial $f(x)$ of degree $n > m$, without common roots (Sec. 1.7-3) can be expressed as a sum of n **partial fractions** corresponding to the roots x_k (of respective orders m_k) of $f(x) = 0$ as follows:

$$\begin{aligned} \frac{g(x)}{f(x)} &= \sum_k \sum_{j=1}^{m_k} \frac{b_{kj}}{(x - x_k)^j} \\ &= \sum_k \left[\frac{b_{k1}}{(x - x_k)} + \frac{b_{k2}}{(x - x_k)^2} + \cdots + \frac{b_{km_k}}{(x - x_k)^{m_k}} \right] \end{aligned} \tag{1.7-4}$$

The coefficients b_{kj} are obtained by one of the following methods, or by a combination of these methods:

1. If $m_k = 1$ (x_k is a simple root), then $b_{k1} = g(x_k)/f'(x_k)$.
2. Multiply both sides of Eq. (4) by $f(x)$ and equate coefficients of equal powers of x on both sides.
3. Multiply both sides of Eq. (4) by $f(x)$ and differentiate successively. Let $\varphi_k(x) = f(x)/(x - x_k)^{m_k}$. Then obtain $b_{km_k}, b_{km_k-1}, \ldots$ successively from

$$g(x_k) = b_{km_k}\varphi_k(x_k)$$
$$g'(x_k) = b_{km_k}\varphi_k'(x_k) + b_{km_k-1}\varphi_k(x_k)$$
$$g''(x_k) = b_{km_k}\varphi_k''(x_k) + 2b_{km_k-1}\varphi_k'(x_k) + 2b_{km_k-2}\varphi_k(x_k)$$

. . . .

$$g^{(m_k-1)}(x_k) = b_{km_k}\varphi_k{}^{(m_k-1)}(x_k) + m_k b_{km_k-1}\varphi_k{}^{(m_k-2)}(x_k) + m_k b_{km_k-2}(m_k - 1)\varphi_k{}^{(m_k-3)}(x_k) + \cdots + m_k! b_{k1}\varphi_k(x_k)$$

The partial fractions corresponding to any pair of complex conjugate roots $a_k + i\omega_k$, $a_k - i\omega_k$ of order m_k are usually combined into

$$c_{k1}\frac{x + d_{k1}}{[(x - a_k)^2 + \omega_k{}^2]} + c_{k2}\frac{x + d_{k2}}{[(x - a_k)^2 + \omega_k{}^2]^2} + \cdots + c_{km_k}\frac{x + d_{km_k}}{[(x - a_k)^2 + \omega_k{}^2]^{m_k}} \tag{1.7-5}$$

The coefficients c_{kj} and d_{kj} may be determined directly by method 2 above. If $g(x)$ and $f(x)$ are real polynomials (Sec. 1.6-6), all coefficients b_{kj}, c_{kj}, d_{kj} in the resulting partial-fraction expansion are real.

Every rational function of x (Sec. 4.2-2*c*) *can be represented as a sum of a polynomial and a finite set of partial fractions* (see also Sec. 7.6-8). Partial-fraction expansions are important in connection with integration (Sec. 4.6-6*c*) and integral transforms (Sec. 8.4-5).

1.8. LINEAR, QUADRATIC, CUBIC, AND QUARTIC EQUATIONS

1.8-1. Solution of Linear Equations. The solution of the general equation of the first degree (**linear equation**)

$$ax = b \qquad \text{or} \qquad ax - b = 0 \qquad (a \neq 0) \tag{1.8-1}$$

is

$$x = \frac{b}{a} \tag{1.8-2}$$

1.8-2. Solution of Quadratic Equations. The **quadratic equation**

$$ax^2 + bx + c = 0 \qquad (a \neq 0) \tag{1.8-3}$$

has the roots

$$x_{1,2} = \frac{-b \pm \sqrt{b^2 - 4ac}}{2a} \tag{1.8-4}$$

The roots x_1 and x_2 are real and different, real and equal, or complex conjugates if the discriminant (Sec. 1.6-5) $D = b^2 - 4ac$ is, respectively, positive, zero, or negative. Note $x_1 + x_2 = -b/a$, $x_1x_2 = c/a$.

1.8-3. Cubic Equations: Cardan's Solution. The **cubic equation**

$$x^3 + ax^2 + bx + c = 0 \qquad (1.8\text{-}5)$$

is transformed to the "reduced" form

$$y^3 + py + q = 0 \qquad p = -\frac{a^2}{3} + b \qquad q = 2\left(\frac{a}{3}\right)^3 - \frac{ab}{3} + c \qquad (1.8\text{-}6)$$

through the substitution $x = y - a/3$. The roots y_1, y_2, y_3 of the "reduced" cubic equation (6) are

$$\left.\begin{gathered} y_1 = A + B \qquad y_{2,3} = -\frac{A+B}{2} \pm i\frac{A-B}{2}\sqrt{3} \\ \text{with } A = \sqrt[3]{-\frac{q}{2} + \sqrt{Q}} \qquad B = \sqrt[3]{-\frac{q}{2} - \sqrt{Q}} \\ Q = \left(\frac{p}{3}\right)^3 + \left(\frac{q}{2}\right)^2 \end{gathered}\right\} \qquad (1.8\text{-}7)$$

where the real values of the cube roots are used. *The cubic equation has one real root and two conjugate complex roots, three real roots of which at least two are equal, or three different real roots, if Q is positive, zero, or negative, respectively.* In the latter case ("irreducible" case), the method of Sec. 1.8-4a may be used. Note that the discriminants (Sec. 1.6-5) of Eq. (5) and Eq. (6) are both equal to $-108Q$.

1.8-4. Cubic Equations: Trigonometric Solution. **(a)** If $Q < 0$ ("irreducible" case)

$$\left.\begin{gathered} y_1 = 2\underset{+}{\sqrt{-p/3}} \cos(\alpha/3) \\ y_{2,3} = -2\underset{+}{\sqrt{-p/3}} \cos(\alpha/3 \pm 60°) \\ \text{with} \qquad \cos\alpha = -\frac{q}{2\underset{+}{\sqrt{-(p/3)^3}}} \end{gathered}\right\} \qquad (1.8\text{-}8)$$

(b) If $Q \geq 0$, $p > 0$

$$\left.\begin{gathered} y_1 = -2\underset{+}{\sqrt{p/3}} \cot 2\alpha \qquad y_{2,3} = \underset{+}{\sqrt{p/3}} (\cot 2\alpha \pm i\sqrt{3} \operatorname{cosec} 2\alpha) \\ \text{with} \\ \tan\alpha = \underset{+}{\sqrt[3]{\tan(\beta/2)}} \quad (|\alpha| \leq 45°) \qquad \tan\beta = 2\underset{+}{\sqrt{(p/3)^3}}/q \quad (|\beta| \leq 90°) \end{gathered}\right\} \qquad (1.8\text{-}9a)$$

(c) If $Q \geq 0$, $p < 0$

$$\left.\begin{gathered} y_1 = -2\underset{+}{\sqrt{-p/3}} \operatorname{cosec} 2\alpha \qquad y_{2,3} = \underset{+}{\sqrt{-p/3}} (\operatorname{cosec} 2\alpha \pm i\sqrt{3} \cot 2\alpha) \\ \text{with} \\ \tan\alpha = \underset{+}{\sqrt[3]{\tan(\beta/2)}} \quad (|\alpha| \leq 45°) \qquad \sin\beta = 2\underset{+}{\sqrt{(-p/3)^3}}/q \quad (|\beta| \leq 90°) \end{gathered}\right\} \qquad (1.8\text{-}9b)$$

The real value of the cube root is used.

1.8-5. Quartic Equations: Descartes-Euler Solution. The **quartic equation (biquadratic equation)**

$$x^4 + ax^3 + bx^2 + cx + d = 0 \tag{1.8-10}$$

is transformed to the "reduced" form

$$y^4 + py^2 + qy + r = 0 \tag{1.8-11}$$

through the substitution $x = y - a/4$. The roots y_1, y_2, y_3, y_4 of the "reduced" quartic equation (11) are the four sums

$$\pm \sqrt{z_1} \pm \sqrt{z_2} \pm \sqrt{z_3} \tag{1.8-12}$$

with the signs of the square roots chosen so that

$$\sqrt{z_1}\sqrt{z_2}\sqrt{z_3} = -q/8 \tag{1.8-13}$$

where z_1, z_2, z_3 are the roots of the cubic equation

$$z^3 + \frac{p}{2}z^2 + \frac{p^2 - 4r}{16}z - \frac{q^2}{64} = 0 \tag{1.8-14}$$

1.8-6. Quartic Equations: Ferrari's Solution. Given any root y_1 of the **resolvent cubic equation** corresponding to Eq. (10)

$$y^3 - by^2 + (ac - 4d)y - a^2d + 4bd - c^2 = 0 \tag{1.8-15}$$

the four roots of the quartic equation (10) are given as roots of the two quadratic equations

$$x^2 + \frac{a}{2}x + \frac{y_1}{2} = \pm\sqrt{\left(\frac{a^2}{4} - b + y_1\right)x^2 + \left(\frac{a}{2}y_1 - c\right)x + \frac{y_1^2}{4} - d} \tag{1.8-16}$$

where the radicand on the right is a perfect square. Note that the discriminants (Sec. 1.6-5) of Eq. (10) and Eq. (15) are equal.

1.9. SYSTEMS OF SIMULTANEOUS EQUATIONS

1.9-1. Simultaneous Equations. To solve a suitable set (system) of simultaneous equations

$$f_i(x_1, x_2, \cdots) = 0 \qquad (i = 1, 2, \cdots) \tag{1.9-1}$$

for the unknowns x_1, x_2, . . . means to determine a set of values of x_1, x_2, . . . which satisfy the equations (1) simultaneously. The solution is complete if all such sets are found. One can frequently *eliminate* successive unknowns x_j from a system (1), e.g., by solving one equation for x_j and substituting the resulting expression in the remaining equations. The number of equations and unknowns is thus reduced until a single equation remains to be solved for a single unknown. The procedure is then repeated to yield a second unknown, etc. Solutions may be verified by substitution.

To eliminate x_1, say, from two equations $f_1(x_1, x_2) = 0, f_2(x_1, x_2) = 0$ where $f_1(x_1, x_2)$ and $f_2(x_1, x_2)$ are polynomials in x_1 and x_2 (Sec. 1.4-3), consider both functions as polynomials in x_1 and form their resultant R (Sec. 1.7-3). Then x_2 must satisfy the equation $R = 0$ (*Sylvester's Dialytic Method of Elimination*).

1.9-2. Simultaneous Linear Equations: Cramer's Rule. Consider a set (system) of n linear equations in n unknowns $x_1, x_2, \ldots, x_n$

$$\begin{array}{l} a_{11}x_1 + a_{12}x_2 + \cdots + a_{1n}x_n = b_1 \\ a_{21}x_1 + a_{22}x_2 + \cdots + a_{2n}x_n = b_2 \\ \cdots\cdots\cdots\cdots\cdots\cdots \\ a_{n1}x_1 + a_{n2}x_2 + \cdots + a_{nn}x_n = b_n \end{array} \quad \text{or} \quad \sum_{k=1}^{n} a_{ik}x_k = b_i \quad (i = 1, 2, \cdots, n) \tag{1.9-2}$$

such that at least one of the absolute terms b_i is different from zero. If the **system determinant**

$$D = \det [a_{ik}] = \begin{vmatrix} a_{11} & a_{12} & \cdots & a_{1n} \\ a_{21} & a_{22} & \cdots & a_{2n} \\ \cdots & \cdots & \cdots & \cdots \\ a_{n1} & a_{n2} & \cdots & a_{nn} \end{vmatrix} \tag{1.9-3}$$

differs from zero, the system (2) has the unique solution

$$x_k = \frac{D_k}{D} \qquad (k = 1, 2, \ldots, n) \qquad (\text{Cramer's rule}) \tag{1.9-4}$$

where D_k is the determinant obtained on replacing the respective elements $a_{1k}, a_{2k}, \ldots, a_{nk}$ in the kth column of D by $b_1, b_2, \ldots, b_n$, or

$$D_k = \sum_{i=1}^{n} A_{ik}b_i \qquad (k = 1, 2, \ldots, n) \tag{1.9-5}$$

where A_{ik} is the cofactor (Sec. 1.5-2) of a_{ik} in the determinant D (see also Secs. 13.2-3 and 14.5-3).

1.9-3. Linear Independence (see also Secs. 9.3-2, 14.2-3 and 15.2-1*a*). (**a**) m equations $f_i(x_1, x_2, \ldots, x_n) = 0$ $(i = 1, 2, \ldots, m)$, or m functions $f_i(x_1, x_2, \ldots, x_n)$ are **linearly independent** if and only if

$$\sum_{i=1}^{m} \lambda_i f_i(x_1, x_2, \ldots, x_n) \equiv 0 \text{ implies } \lambda_1 = \lambda_2 = \cdots = \lambda_m = 0 \tag{1.9-6}$$

Otherwise the m equations or functions are **linearly dependent;** i.e., at least one of them can be expressed as a linear combination of the others.

As a trivial special case, this is true whenever one or more of the equations $f_i(x_1, x_2, \ldots, x_n) = 0$ is satisfied identically.

n homogeneous linear functions $\sum_{k=1}^{n} a_{ik}x_k$ *($i = 1, 2, \ldots, n$) are linearly independent if and only if* $\det [a_{ik}] \neq 0$ (see also Sec. 1.9-5).

More generally, *m homogeneous linear functions* $\sum_{k=1}^{n} a_{ik}x_k$ *($i = 1, 2, \ldots, m$) are linearly independent if and only if the $m \times n$ matrix $[a_{ik}]$ is of rank m* (Sec. 13.2-7).

(b) m sets of n numbers $x_1^{(1)}, x_2^{(1)}, \ldots, x_n^{(1)}; x_1^{(2)}, x_2^{(2)}, \ldots, x_n^{(2)}; \ldots; x_1^{(m)}, x_2^{(m)}, \ldots, x_n^{(m)}$ (e.g., solutions of simultaneous equations, or components of m n-dimensional vectors) are **linearly independent** if and only if

$$\sum_{i=1}^{m} \lambda_i x_j^{(i)} = 0 \ (j = 1, 2, \ldots, n) \text{ implies } \lambda_1 = \lambda_2 = \cdots = \lambda_m = 0 \tag{1.9-7}$$

This is true if and only if the $m \times n$ matrix $[x_j^{(i)}]$ is of rank m (Sec. 13.2-7).

1.9-4. Simultaneous Linear Equations: General Theory (see also Sec. 14.8-10). *The system of m linear equations in n unknowns $x_1, x_2, \ldots, x_n$*

$$\sum_{k=1}^{n} a_{ik}x_k = b_i \qquad (i = 1, 2, \ldots, m) \tag{1.9-8}$$

possesses a solution if and only if the matrices

$$\begin{bmatrix} a_{11} & a_{12} & \cdots & a_{1n} \\ a_{21} & a_{22} & \cdots & a_{2n} \\ \cdots & \cdots & \cdots & \cdots \\ a_{m1} & a_{m2} & \cdots & a_{mn} \end{bmatrix}, \quad \begin{bmatrix} a_{11} & a_{12} & \cdots & a_{1n} & b_1 \\ a_{21} & a_{22} & \cdots & a_{2n} & b_2 \\ \cdots & \cdots & \cdots & \cdots & \cdots \\ a_{m1} & a_{m2} & \cdots & a_{mn} & b_m \end{bmatrix} \tag{1.9-9}$$

(**system matrix** *and* **augmented matrix**) *are of equal rank* (Sec. 13.2-7). Otherwise the equations are *inconsistent.*

The unique solution of Sec. 1.9-2 applies if $r = m = n$. If both matrices (9) are of rank $r < m$, the equations (8) are *linearly dependent* (Sec. 1.9-3*a*); $m - r$ equations can be expressed as linear combinations of the remaining r equations and are satisfied by their solution. The r independent equations determine r unknowns as linear functions of the remaining $n - r$ unknowns, which are left arbitrary.

1.9-5. Simultaneous Linear Equations: n Homogeneous Equations in n Unknowns. In particular, *a system of n homogeneous linear equations in n unknowns,*

$$\sum_{k=1}^{n} a_{ik}x_k = 0 \qquad (i = 1, 2, \ldots, n) \tag{1.9-10}$$

has a solution different from the trivial solution $x_1 = x_2 = \cdots = x_n = 0$ if and only if $D = \det [a_{ik}] = 0$ (see also Sec. 1.9-3*a*).

In this case, there exist exactly $n - r$ linearly independent solutions $x_1^{(1)}, x_2^{(1)}, \ldots, x_n^{(1)}; x_1^{(2)}, x_2^{(2)}, \ldots, x_n^{(2)}; \ldots; x_1^{(n-r)}, x_2^{(n-r)}, \ldots, x_n^{(n-r)}$, where r is the rank of the system matrix (Sec. 1.9-4). *The most general solution is, then,*

$$x_i = \sum_{j=1}^{n-r} c_j x_i^{(j)} \qquad (i = 1, 2, \ldots, n) \tag{1.9-11}$$

where the c_j are arbitrary constants (see also Sec. 14.8-10).

In the important special case where $r = n - 1$,

$$x_1 = cA_{k1} \qquad x_2 = cA_{k2} \qquad \ldots \qquad x_n = cA_{kn} \tag{1.9-12}$$

is a solution for any arbitrary constant c, so that all ratios x_i/x_k are uniquely determined; the solutions (12) obtained for different values of k are identical (see also Sec. 14.8-6).

1.10. RELATED TOPICS, REFERENCES, AND BIBLIOGRAPHY

1.10-1. Related Topics. The following topics related to the study of elementary algebra are treated in other chapters of this handbook:

Quadratic and bilinear forms .... Chap. 13
Abstract algebra .... Chap. 12
Matrix algebra .... Chap. 13
Functions of a complex variable .... Chap. 7
Numerical solution of equations, numerical approximations .... Chap. 20

1.10-2. References and Bibliography.

1.1. Aitken, A. C.: *Determinants and Matrices*, 8th ed., Interscience, New York, 1956.
1.2. Birkhoff, G., and S. MacLane: *A Survey of Modern Algebra*, 3d ed., Macmillan, New York, 1965.
1.3. Dickson, L. E.: *New First Course in the Theory of Equations*, Wiley, New York, 1939.
1.4. Kemeny, J. G., et al.: *Introduction to Finite Mathematics*, Prentice-Hall, Englewood Cliffs, N.J., 1957.
1.5. Landau, E.: *The Foundations of Analysis*, Chelsea, New York, 1948.
1.6. Middlemiss, R. R.: *College Algebra*, McGraw-Hill, New York, 1952.
1.7. Uspensky, J. V.: *Theory of Equations*, McGraw-Hill, New York, 1948.

Additional Background Material

1.8. Cohen, L. W., et al.: *The Structure of the Real Number System*, Van Nostrand, Princeton, N.J., 1963.
1.9. Feferman, S.: *The Number Systems: Foundations of Algebra and Analysis*, Addison-Wesley, Reading, Mass., 1964.
1.10. Landin, J., and N. T. Hamilton: *Set Theory: The Structure of Arithmetic*, Allyn and Bacon, Boston, 1961.
1.11. Struik, D. J.: *A Concise History of Mathematics*, 2d ed., Dover, New York, 1948.

(See also Secs. 12.9-2 and 13.7-2.)

CHAPTER **2**

PLANE ANALYTIC GEOMETRY

2.1. INTRODUCTION AND BASIC CONCEPTS

2.1-1. Introduction (see also Sec. 12.1-1). A *geometry* is a mathematical model involving relations between objects referred to as *points.* Each geometry is defined by a self-consistent set of *defining postulates;* the latter may or may not be chosen so as to make the properties of the model correspond to physical space relationships. The study of such models is also called geometry. *Analytic geometry* represents each point by an ordered set of numbers (**coordinates**), so that relations between points are represented by relations between coordinates.

Chapters 2 (Plane Analytic Geometry) and 3 (Solid Analytic Geometry) introduce their subject matter in the manner of most elementary courses: the concepts of Euclidean geometry are assumed to be known and are simply translated into analytical language. A more flexible approach, involving actual construction of various geometries from postulates, is briefly discussed in Chap. 17. The differential geometry of plane curves, including the definition of tangents, normals, and curvature, is outlined in Secs. 17.1-1 to 17.1-6.

2.1-2. Cartesian Coordinate Systems. A **cartesian coordinate system** (cartesian reference system, see also Sec. 17.4-6*b*) associates a unique ordered pair of real numbers (**cartesian coordinates**), the **abscissa** x and the **ordinate** y, with every point $P \equiv (x, y)$ in the finite portion of the Euclidean plane by reference to a pair of directed straight lines (**coordinate axes**) OX, OY intersecting at the **origin** O (Fig. 2.1-1). The parallel to OY through P intersects the **x axis** OX at the point P'. Similarly, the parallel to OX through P intersects the **y axis** OY at P''.

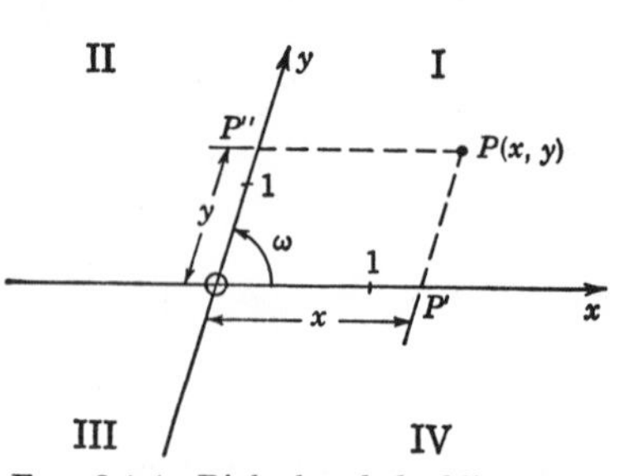

FIG. 2.1-1. Right-handed oblique cartesian coordinate system. The points marked "1" define the coordinate scales used.

The directed distances $OP' = x$ (positive in the positive x axis direction) and $OP'' = y$ (positive in the positive y axis direction) are the cartesian coordinates of the point $P \equiv (x, y)$.

x and y may or may not be measured with equal scales. In a general (oblique) cartesian coordinate system, the angle $XOY = \omega$ between the coordinate axes may be between 0 and 180 deg (**right-handed cartesian coordinate systems**) or between 0 and -180 deg (**left-handed cartesian coordinate systems**).

A system of cartesian reference axes divides the plane into four *quadrants* (Fig. 2.1-1). The abscissa x is positive for points (x, y) in quadrants I and IV, negative for points in quadrants II and III, and zero for points on the y axis. The ordinate y is positive in quadrants I and II, negative in quadrants III and IV, and zero on the x axis. The origin is the point (0, 0).

NOTE: Euclidean analytic geometry *postulates* a reciprocal one-to-one correspondence between the points of a straight line and the real numbers (*coordinate axiom, axiom of continuity*, see also Sec. 4.3-1).

2.1-3. Right-handed Rectangular Cartesian Coordinate Systems.

In a **right-handed rectangular cartesian coordinate system,** the directions of the coordinate axes are chosen so that a rotation of 90 deg in the positive (counterclockwise) sense would make the positive x axis OX coincide with the positive y axis OY (Fig. 2.1-2). The coordinates x and y are thus equal to the respective directed distances between the y axis and the point P, and between the x axis and the point P.

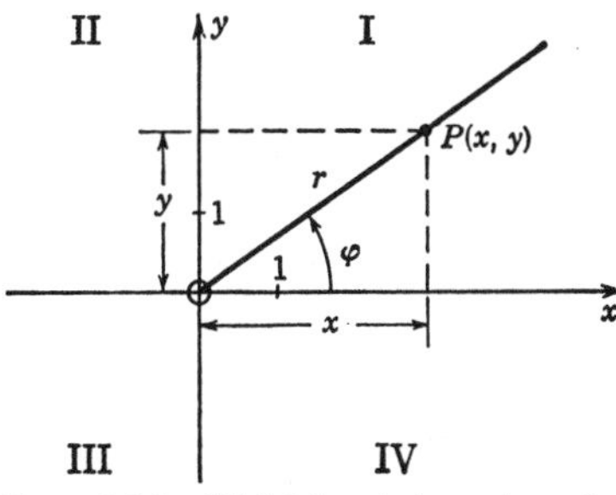

FIG. 2.1-2. Right-handed rectangular cartesian coordinate system and polar-coordinate system.

Throughout the remainder of this chapter, all cartesian coordinates x, y refer to right-handed rectangular cartesian coordinate systems, and equal scale units of unit length are used to measure x and y, unless the contrary is specifically stated.

2.1-4. Basic Relations in Terms of Rectangular Cartesian Coordinates.

In terms of rectangular cartesian coordinates (x, y), the following relations hold:

1. The **distance** d between the points $P_1 \equiv (x_1, y_1)$ and $P_2 \equiv (x_2, y_2)$ is

$$d = \underset{+}{\sqrt{(x_2 - x_1)^2 + (y_2 - y_1)^2}} \tag{2.1-1}$$

2. The oblique **angle** γ between two directed straight-line segments $\overrightarrow{P_1P_2}$ and $\overrightarrow{P_3P_4}$ is given by

$$\cos\gamma = \frac{(x_2 - x_1)(x_4 - x_3) + (y_2 - y_1)(y_4 - y_3)}{\underset{+}{\sqrt{(x_2 - x_1)^2 + (y_2 - y_1)^2}}\;\underset{+}{\sqrt{(x_4 - x_3)^2 + (y_4 - y_3)^2}}} \qquad (2.1\text{-}2)$$

where the coordinates of the points P_1, P_2, P_3, P_4 are denoted by the respective corresponding subscripts.

The **direction cosines** $\cos\alpha_x$ and $\cos\alpha_y$ of a directed line segment $\overrightarrow{P_1P_2}$ are the cosines of the angles α_x and $\alpha_y = 90\text{ deg} - \alpha_x$, respectively.

$$\begin{aligned} \cos\alpha_x &= \frac{x_2 - x_1}{\underset{+}{\sqrt{(x_2 - x_1)^2 + (y_2 - y_1)^2}}} \\ \cos\alpha_y &= \sin\alpha_x = \frac{y_2 - y_1}{\underset{+}{\sqrt{(x_2 - x_1)^2 + (y_2 - y_1)^2}}} \end{aligned} \qquad (2.1\text{-}3)$$

3. The coordinates x, y of **the point** P **dividing the directed line segment between the points** $P_1 \equiv (x_1, y_1)$ **and** $P_2 \equiv (x_2, y_2)$ **in the ratio** $\overrightarrow{P_1P}:\overrightarrow{PP_2} = m:n = \mu:1$ are

$$x = \frac{mx_2 + nx_1}{m + n} = \frac{x_1 + \mu x_2}{1 + \mu} \qquad y = \frac{my_2 + ny_1}{m + n} = \frac{y_1 + \mu y_2}{1 + \mu}$$
$$(-\infty \le \mu \le \infty) \qquad (2.1\text{-}4)$$

Specifically, the coordinates of the **mid-point** of P_1P_2 are

$$x = \frac{x_1 + x_2}{2} \qquad y = \frac{y_1 + y_2}{2} \qquad (2.1\text{-}5)$$

4. The **area** S **of the triangle with the vertices** $P_1 \equiv (x_1, y_1)$, $P_2 \equiv (x_2, y_2)$, $P_3 \equiv (x_3, y_3)$ is

$$S = \tfrac{1}{2}\begin{vmatrix} x_1 & y_1 & 1 \\ x_2 & y_2 & 1 \\ x_3 & y_3 & 1 \end{vmatrix} = \tfrac{1}{2}[x_1(y_2 - y_3) + x_2(y_3 - y_1) + x_3(y_1 - y_2)] \qquad (2.1\text{-}6)$$

This expression is positive if the circumference $P_1P_2P_3$ runs around the inside of the triangle in a positive (counterclockwise) direction. Specifically, if $x_3 = y_3 = 0$,

$$S = \tfrac{1}{2}\begin{vmatrix} x_1 & x_2 \\ y_1 & y_2 \end{vmatrix} = \tfrac{1}{2}(x_1y_2 - x_2y_1) \qquad (2.1\text{-}7)$$

2.1-5. Translation of the Coordinate Axes. Let x, y be the coordinates of any point P with respect to a right-handed rectangular cartesian reference system. Let $\bar{x}$, $\bar{y}$ be the coordinates of the same point P with respect to a second right-handed rectangular cartesian reference system whose axes have the same directions as those of the x, y system, and whose origin has the coordinates $x = x_0$ and $y = y_0$ in the x, y system. If equal scales are used to measure the coordinates in both systems, the coordinates $\bar{x}$, $\bar{y}$ are related to the coordinates x, y by the *transformation equations* (Fig. 2.1-3*a*; see also Chap. 14)

$$\begin{aligned} &\bar{x} = x - x_0 \qquad \bar{y} = y - y_0 \\ \text{or} \qquad &x = \bar{x} + x_0 \qquad y = \bar{y} + y_0 \end{aligned} \tag{2.1-8}$$

The equations (8) permit a second interpretation. If $\bar{x}$, $\bar{y}$ are considered as coordinates referred *to the* x, y *system* of axes, then the point defined by $\bar{x}$, $\bar{y}$ is *translated* by a directed amount $-x_0$ in the x axis direction and by a directed amount $-y_0$ in the y axis direction with respect to the point (x, y). Transformations of this type applied to each point x, y of a plane curve may be used to indicate the translation of the entire curve.

2.1-6. Rotation of the Coordinate Axes. Let x, y be the coordinates of any point P with respect to a right-handed rectangular cartesian reference system. Let $\bar{x}$, $\bar{y}$ be the coordinates of the same point P with respect to a second right-handed rectangular cartesian reference system having the same origin O and rotated with respect to the x, y system so that the angle $XO\bar{X}$ between the x axis OX and $\bar{x}$ axis $O\bar{X}$ is equal to ϑ measured in radians in the positive (counterclockwise) sense (Fig. 2.1-3*b*). If equal scales are used to measure all four coordinates x, y, $\bar{x}$, $\bar{y}$, the coordinates $\bar{x}$, $\bar{y}$ are related to the coordinates x, y by the transformation equations

$$\begin{aligned} &\bar{x} = x \cos \vartheta + y \sin \vartheta \qquad \bar{y} = -x \sin \vartheta + y \cos \vartheta \\ \text{or} \quad &x = \bar{x} \cos \vartheta - \bar{y} \sin \vartheta \qquad y = \bar{x} \sin \vartheta + \bar{y} \cos \vartheta \end{aligned} \tag{2.1-9}$$

A second interpretation of the transformation (9) is the definition of a point $(\bar{x}, \bar{y})$ rotated about the origin by an angle $-\vartheta$ with respect to the point (x, y).

2.1-7. Simultaneous Translation and Rotation of Coordinate Axes. If the origin of the $\bar{x}$, $\bar{y}$ system in Sec. 2.1-6 is not the same as the origin of the x, y system but has the coordinates $x = x_0$ and $y = y_0$ in the x, y system, the transformation equations become

$$\begin{aligned} &\bar{x} = (x - x_0) \cos \vartheta + (y - y_0) \sin \vartheta \\ &\bar{y} = -(x - x_0) \sin \vartheta + (y - y_0) \cos \vartheta \\ \text{or} \qquad &x = x_0 + \bar{x} \cos \vartheta - \bar{y} \sin \vartheta \\ &y = y_0 + \bar{x} \sin \vartheta + \bar{y} \cos \vartheta \end{aligned} \tag{2.1-10}$$

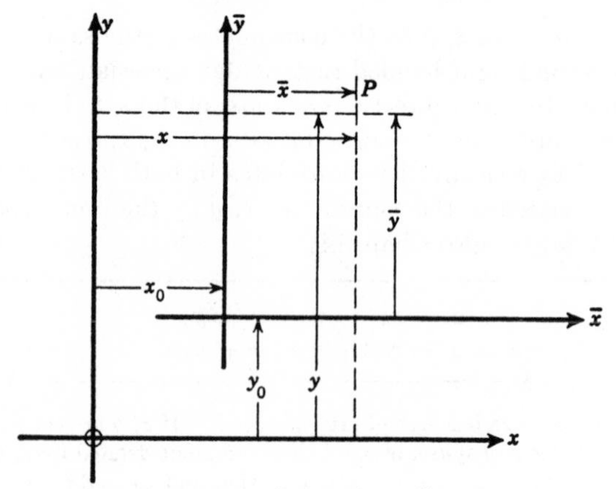

FIG. 2.1-3*a*. Translation of coordinate axes.

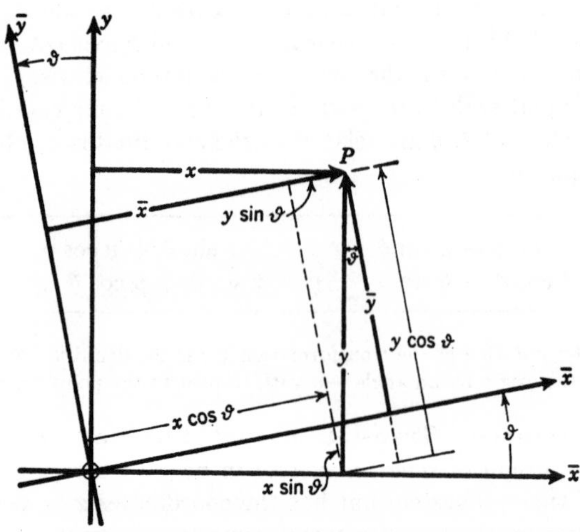

FIG. 2.1-3*b*. Rotation of coordinate axes.

The relations (10) permit one to relate the coordinates of a point in *any* two right-handed rectangular cartesian reference systems if the same scales are used for all coordinate measurements.

The transformation (10) may also be considered as the definition of a point $(\bar{x}, \bar{y})$ translated and rotated with respect to the point (x, y).

NOTE: The transformations (8), (9), and (10) do not affect the value of the distance (1) between two points or the value of the angle given by Eq. (2). *The relations constituting Euclidean geometry are unaffected by (invariant with respect to) translations and rotations of the coordinate system* (see also Secs. 12.1-5, 14.1-4, and 14.4-5).

2.1-8. Polar Coordinates. A **plane polar-coordinate system** associates ordered pairs of numbers r, φ (**polar coordinates**) with each point P of the plane by reference to a directed straight line OX (Fig. 2.1-2), the **polar axis.** Each point P has the polar coordinates r, defined as the directed distance OP, and φ, defined as the angle XOP measured in radians in the counterclockwise sense between OX and OP. The point O is called the **pole** of the polar-coordinate system; r is the **radius vector** of the point P.

Negative values of the angle φ are measured in the clockwise sense from the polar axis OX. Points (r, φ) are by definition identical to the points $(-r, \varphi \pm 180 \text{ deg})$; this convention associates points of the plane with pairs of numbers (r, φ) with negative as well as positive radius vectors r.

NOTE: Unlike a cartesian coordinate system, a polar-coordinate system does not establish a reciprocal one-to-one correspondence between the pairs of numbers (r, φ) and the points of the plane. The ambiguities involved may, however, be properly taken into account in most applications.

If the pole and the polar axis of a polar-coordinate system coincide with the origin and the x axis, respectively, of a right-handed rectangular cartesian coordinate system (Fig. 2.1-2), then the following transformation equations relate the polar coordinates (r, φ) and the rectangular cartesian coordinates (x, y) of corresponding points if equal scales are used for the measurement of r, x, and y:

$$\boxed{\begin{array}{ll} x = r\cos\varphi & y = r\sin\varphi \\ |r| = \underset{+}{\sqrt{x^2 + y^2}} & \varphi = \arctan\dfrac{y}{x} \end{array}} \qquad (2.1\text{-}11)$$

In terms of polar coordinates (r,φ), the following relations hold:

1. The **distance** d between the points (r_1, φ_1) and (r_2, φ_2) is

$$d = \underset{+}{\sqrt{r_1^2 + r_2^2 - 2r_1r_2\cos(\varphi_2 - \varphi_1)}} \qquad (2.1\text{-}12)$$

2. The **area** S **of the triangle** with the vertices $P_1 \equiv (r_1, \varphi_1)$, $P_2 \equiv (r_2, \varphi_2)$, and $P_3 \equiv (r_3, \varphi_3)$ is

$$S = \tfrac{1}{2}[r_1r_2\sin(\varphi_2 - \varphi_1) + r_2r_3\sin(\varphi_3 - \varphi_2) + r_1r_3\sin(\varphi_1 - \varphi_3)] \qquad (2.1\text{-}13)$$

This expression is positive if the circumference $P_1P_2P_3$ runs around the inside of the triangle in the positive (counterclockwise) direction. Specifically, if $r_3 = 0$,

$$S = \tfrac{1}{2}r_1r_2 \sin(\varphi_2 - \varphi_1) \tag{2.1-14}$$

See Chap. 6 for other curvilinear coordinate systems.

2.1-9. Representation of Curves (see also Secs. 3.1-13 and 17.1-1). **(a) Equation of a Curve.** A relation of the form

$$\varphi(x, y) = 0 \qquad \text{or} \qquad y = f(x) \tag{2.1-15}$$

is, in general, satisfied only by the coordinates x, y of points belonging to a special set defined by the given relation. In most cases of interest, the point set will be a *curve* (see also Sec. 3.1-13). Conversely, a given curve will be represented by a suitable equation (15), which must be satisfied by the coordinates of all points (x, y) on the curve. Note that a curve may have more than one branch.

The curves corresponding to the equations

$$\varphi(x, y) = 0 \qquad \lambda\varphi(x, y) = 0$$

where λ is a constant different from zero, are identical.

(b) Parametric Representation of Curves. A plane curve can also be represented by two equations

$$x = x(t) \qquad y = y(t) \tag{2.1-16}$$

where t is a variable parameter.

(c) Intersection of Two Curves. Pairs of coordinates x, y which simultaneously satisfy the equations of two curves

$$\varphi_1(x, y) = 0 \qquad \varphi_2(x, y) = 0 \tag{2.1-17}$$

represent the *points of intersection* of the two curves. In particular, if the equation $\varphi(x, 0) = 0$ has one or more real roots x, the latter are the abscissas of the intersections of the curve $\varphi(x, y) = 0$ with the x axis.

If $\varphi(x, y)$ is a polynomial of degree n (Sec. 1.4-3), the curve $\varphi(x, y) = 0$ intersects the x axis (and any straight line, Sec. 2.2-1) in n points (**nth-order curve**); but some of these points of intersection may coincide and/or be imaginary.

For any real number λ, the equation

$$\varphi_1(x, y) + \lambda\varphi_2(x, y) = 0 \tag{2.1-18}$$

describes a curve passing through all points of intersection (real and imaginary) of the two curves (17).

(d) Given two curves corresponding to the two equations (17), the equation

$$\varphi_1(x, y)\varphi_2(x, y) = 0 \tag{2.1-19}$$

is satisfied by all points of both original curves, and by no other points.

2.2. THE STRAIGHT LINE

2.2-1. The Equation of the Straight Line. Given a right-handed rectangular cartesian coordinate system, every equation linear in x and y, i.e., an equation of the form

$$Ax + By + C = 0 \tag{2.2-1}$$

where A and B must not vanish simultaneously, represents a *straight line* (Fig. 2.2-1). Conversely, every straight line in the finite portion of the plane can be represented by a linear equation (1). The special case $C = 0$ corresponds to a straight line through the origin.

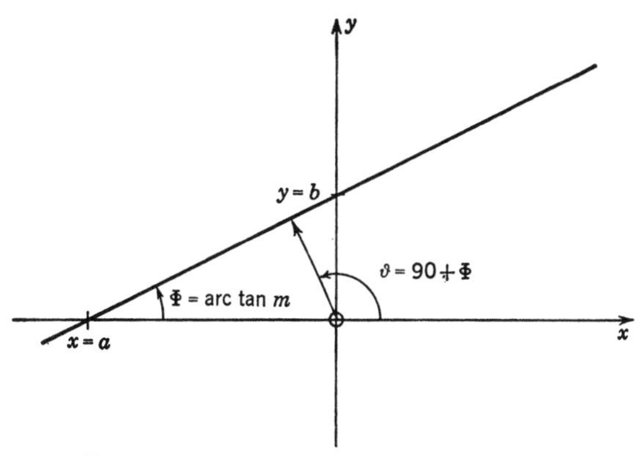

FIG. 2.2-1. The equation of the straight line.

The following special forms of the equation of a straight line are of particular interest:

1. **Slope-intercept form.** Straight line at an angle Φ with the positive x axis and intercepting the y axis at $y = b$:

$$y = mx + b \qquad m = \tan \Phi$$

m is the **slope** of the straight line.

2. **Two-intercept form.** Straight line intercepting the x axis at $x = a$ and the y axis at $y = b$:

$$\frac{x}{a} + \frac{y}{b} - 1 = 0$$

3. **Normal form.** Let p be the length of the directed perpendicular from the origin onto a straight line; let ϑ be the angle between the positive x axis and the directed perpendicular, measured in the positive (counterclockwise) sense. Then the equation of the straight line is

$$x \cos \vartheta + y \sin \vartheta - p = 0$$

4. **Point-slope form.** Straight line through the point (x_1, y_1) and having the slope m:

$$y - y_1 = m(x - x_1)$$

5. **Two-point form.** Straight line through the (noncoincident) points $P_1 \equiv (x_1, y_1)$ and $P_2 \equiv (x_2, y_2)$:

$$\frac{y - y_1}{y_2 - y_1} = \frac{x - x_1}{x_2 - x_1} \qquad \text{or} \qquad \begin{vmatrix} x & y & 1 \\ x_1 & y_1 & 1 \\ x_2 & y_2 & 1 \end{vmatrix} = 0$$

When the equation of a straight line is given in the general form (1), the intercepts a and b, the slope m, the perpendicular distance p from the origin, $\cos \vartheta$, and $\sin \vartheta$ are related to the parameters A, B, and C as follows:

$$a = -\frac{C}{A} \qquad b = -\frac{C}{B} \qquad m = \tan \Phi = -\frac{A}{B} \qquad \Phi = \vartheta - 90 \text{ deg} \tag{2.2-2}$$

$$p = -\frac{C}{\pm\sqrt{A^2 + B^2}} \qquad \cos \vartheta = \frac{A}{\pm\sqrt{A^2 + B^2}} \qquad \sin \vartheta = \frac{B}{\pm\sqrt{A^2 + B^2}} \tag{2.2-3}$$

In order to avoid ambiguity, one chooses the sign of $\pm\sqrt{A^2 + B^2}$ in Eq. (3) so that $p > 0$.

In this case, the directed line segment between origin and straight line defines the direction of the *positive normal* of the straight line; $\cos \vartheta$ and $\sin \vartheta$ are the *direction cosines of the positive normal* (Sec. 17.1-2).

2.2-2. Other Representations of Straight Lines. In terms of a variable parameter t, the rectangular cartesian coordinates x and y of a point on any straight line may be expressed in the form

$$x = c_1 t + c_2 \qquad y = k_1 t + k_2 \tag{2.2-4}$$

with

$$m = \frac{k_1}{c_1} \qquad a = \frac{k_1 c_2 - k_2 c_1}{k_1} \qquad b = \frac{k_2 c_1 - k_1 c_2}{c_1} \tag{2.2-5}$$

and $p = \dfrac{k_1c_2 - k_2c_1}{\pm\sqrt{c_1^2 + k_1^2}} \qquad \cos\vartheta = \dfrac{k_1}{\pm\sqrt{c_1^2 + k_1^2}} \qquad \sin\vartheta = \dfrac{-c_1}{\pm\sqrt{c_1^2 + k_1^2}}$

(2.2-6)

If the sign of $\pm\sqrt{c_1^2 + k_1^2}$ is chosen so that $p > 0$, the origin will always lie on the right of the direction of motion as t increases, i.e., in the direction of the negative normal (Sec. 17.1-2).

In terms of *polar coordinates* r, φ, the equation of any straight line may be expressed in the form

$$r(A\cos\varphi + B\sin\varphi) + C = 0 \tag{2.2-7}$$

or

$$r\cos(\varphi - \vartheta) = p \tag{2.2-8}$$

where A, B, C, p, and ϑ are defined as in Sec. 2.2-1.

2.3. RELATIONS INVOLVING POINTS AND STRAIGHT LINES

2.3-1. Points and Straight Lines. The *directed distance d from the straight line* (2.2-1) *to a point* (x_0, y_0) *is*

$$d = \frac{Ax_0 + By_0 + C}{\pm\sqrt{A^2 + B^2}} \tag{2.3-1}$$

where the sign of $\pm\sqrt{A^2 + B^2}$ is chosen to be opposite to that of C. d is positive if the straight line lies between the origin and the point (x_0, y_0).

Three points (x_1, y_1), (x_2, y_2), *and* (x_3, y_3) *are on a straight line if and only if* (see also Sec. 2.2-1)

$$\begin{vmatrix} x_1 & y_1 & 1 \\ x_2 & y_2 & 1 \\ x_3 & y_3 & 1 \end{vmatrix} = 0 \tag{2.3-2}$$

2.3-2. Two or More Straight Lines. (a) *Two straight lines*

$$A_1x + B_1y + C_1 = 0 \qquad \text{or} \qquad y = m_1x + b_1 \tag{2.3-3a}$$

and

$$A_2x + B_2y + C_2 = 0 \qquad \text{or} \qquad y = m_2x + b_2 \tag{2.3-3b}$$

intersect in the point

$$x = \frac{B_1C_2 - B_2C_1}{A_1B_2 - A_2B_1} = \frac{b_1 - b_2}{m_2 - m_1} \qquad y = \frac{C_1A_2 - C_2A_1}{A_1B_2 - A_2B_1} = \frac{m_2b_1 - m_1b_2}{m_2 - m_1} \tag{2.3-4}$$

(b) *Either angle γ_{12} from the straight line* (3a) *to the line* (3b) is given by

$$\tan\gamma_{12} = \frac{A_1B_2 - A_2B_1}{A_1A_2 + B_1B_2} = \frac{m_2 - m_1}{m_1m_2 + 1} \tag{2.3-5}$$

where γ_{12} is measured counterclockwise from the line (3*a*) to the line (3*b*).

(**c**) The straight lines (3*a*) and (3*b*) are *parallel* if

$$A_1B_2 - A_2B_1 = 0 \qquad \text{or} \qquad m_1 = m_2 \tag{2.3-6}$$

and *perpendicular* to each other if

$$A_1A_2 + B_1B_2 = 0 \qquad \text{or} \qquad m_2 = -\frac{1}{m_1} \tag{2.3-7}$$

(**d**) The equation of a straight line through a point (x_1, y_1) and at an angle γ_{12} (or 180 deg $- \gamma_{12}$) to the straight line (3*a*) is

$$y - y_1 = \frac{B_1 \tan \gamma_{12} - A_1}{A_1 \tan \gamma_{12} - B_1} (x - x_1) \tag{2.3-8}$$

Specifically, the equation of the *normal* to the straight line (3*a*) through the point (x_1, y_1) is

$$y - y_1 = \frac{B_1}{A_1} (x - x_1) = -\frac{1}{m_1} (x - x_1) \tag{2.3-9}$$

(**e**) The equation of any straight line *parallel* to (3*a*) may be expressed in the form

$$A_1x + B_1y + C_2 = 0 \tag{2.3-10}$$

The *distance d between the parallel straight lines* (3*a*) and (10) is

$$d = \frac{C_1 - C_2}{\pm \sqrt{A_1^2 + B_1^2}} \tag{2.3-11}$$

If the sign of $\pm \sqrt{A_1^2 + B_1^2}$ in Eq. (11) is chosen to be the opposite to that of C_1, d will be positive if the straight line (3*a*) is between the origin and the straight line (10).

(**f**) The equation of *every straight line passing through point of intersection of two straight lines* (3*a*) and (3*b*) will be of the form

$$\lambda_1(A_1x + B_1y + C_1) + \lambda_2(A_2x + B_2y + C_2) = 0 \tag{2.3-12}$$

with λ_1, λ_2 not both equal to zero. Conversely, every equation of the form (12) describes a straight line passing through the point of intersection. If the straight lines (3*a*) and (3*b*) are parallel, Eq. (12) represents a straight line parallel to the two. *If the straight lines are given in the normal form* (Sec. 2.2-1), $-\lambda$ is the ratio of the distances (1) between the first and second straight line and any one point on the third straight line, and the straight lines corresponding to $\lambda = 1$ and $\lambda = -1$ *bisect the angles* between the given straight lines.

(g) *Three straight lines*

$$A_1x + B_1y + C_1 = 0 \qquad A_2x + B_2y + C_2 = 0 \qquad A_3x + B_3y + C_3 = 0 \qquad (2.3\text{-}13)$$

intersect in a point or are parallel if and only if

$$\begin{vmatrix} A_1 & B_1 & C_1 \\ A_2 & B_2 & C_2 \\ A_3 & B_3 & C_3 \end{vmatrix} = 0 \qquad (2.3\text{-}14)$$

i.e., if the three equations (13) are *linearly dependent* (Sec. 1.9-3*a*).

2.3-3. Line Coordinates. The equation

$$\xi x + \eta y + 1 = 0 \qquad (2.3\text{-}15)$$

describes a straight line (ξ, η) "labeled" by the **line coordinates** ξ, η. If the point coordinates x, y are considered as constant parameters and the line coordinates ξ, η as variables, Eq. (15) may be interpreted as the *equation of the point* (x, y) [point of intersection of all straight lines (15)]. The symmetry of Eq. (15) in the pairs (x, ξ) and (y, η) results in a correspondence (*duality*) between theorems dealing with the positions of points and straight lines (Secs. 2.3-1 and 2.3-2). An equation $F(\xi, \eta) = 0$ represents a set of straight lines which will, in general, envelop a curve determined by the nature of the function $F(\xi, \eta)$ (see also Table 2.4-2).

2.4. SECOND-ORDER CURVES (CONIC SECTIONS)

2.4-1. General Second-degree Equation. The **second-order curves** or **conic sections (conics)** are represented by the **general second-degree equation**

$$\begin{aligned} &a_{11}x^2 + 2a_{12}xy + a_{22}y^2 + 2a_{13}x + 2a_{23}y + a_{33} = 0 \\ \text{or} \quad &(a_{11}x + a_{12}y + a_{13})x + (a_{21}x + a_{22}y + a_{23})y \\ &\qquad + (a_{31}x + a_{32}y + a_{33}) = 0 \\ \text{with} \quad &a_{ik} = a_{ki} \qquad (i, k = 1, 2, 3) \end{aligned} \qquad (2.4\text{-}1)$$

2.4-2. Invariants. For any equation (1), the three quantities

$$I = a_{11} + a_{22} \qquad D = A_{33} = \begin{vmatrix} a_{11} & a_{12} \\ a_{21} & a_{22} \end{vmatrix} \qquad A = \begin{vmatrix} a_{11} & a_{12} & a_{13} \\ a_{21} & a_{22} & a_{23} \\ a_{31} & a_{32} & a_{33} \end{vmatrix} \qquad (2.4\text{-}2)$$

and the *sign* of the quantity

$$A' = \begin{vmatrix} a_{22} & a_{23} \\ a_{32} & a_{33} \end{vmatrix} + \begin{vmatrix} a_{11} & a_{13} \\ a_{31} & a_{33} \end{vmatrix} \qquad (2.4\text{-}3)$$

are *invariants* with respect to the translation and rotation transformations (2.1-8), (2.1-9), and (2.1-10). Such invariants define properties of the conic which do not depend on its position.

Either A or $\Delta = 8A$ is sometimes called the *discriminant* of Eq. (1).

2.4-3. Classification of Conics. Table 2.4-1 shows the classification of conics in terms of the invariants defined in Sec. 2.4-2.

Table 2.4-1. Classification of Conic Sections (Conics)

			Proper conics $A \neq 0$	**Improper (degenerate) conics** $A = 0$
Central conics $D \neq 0$	$D > 0$	$\frac{A}{I} < 0$	Real **ellipse** (circle if $I^2 = 4D$ or $a_{11} = a_{22}$, $a_{12} = 0$)	
		$\frac{A}{I} > 0$	**No real locus** (imaginary ellipse)	
		$\frac{A}{I} = 0$		**Point** in finite portion of plane (point ellipse; real intersection of two imaginary straight lines)
	$D < 0$		**Hyperbola**	**Two real straight lines** intersecting in finite portion of plane (degenerate hyperbola)
Noncentral conics $D = 0$		$A' > 0$	**Parabola**	**No real locus** (imaginary parallels)
		$A' < 0$		**Two real parallel straight lines**
		$A' = 0$		**One real straight line** (coincident parallels)

2.4-4. Similarity of Proper Conics. Two proper conics ($A \neq 0$) given by equations of the form (1) are *similar* if either $D = 0$ for both equations (i.e., if both conics are parabolas) or if $D \neq 0$ for both equations and the ratios $a_{11}:a_{12}:a_{22}$ are the same for both conics.

2.4-5. Characteristic Quadratic Form and Characteristic Equation. Important properties of conics may be studied in terms of the (symmetric) **characteristic quadratic form**

$$F_0(x, y) = a_{11}x^2 + 2a_{12}xy + a_{22}y^2 \tag{2.4-4}$$

corresponding to Eq. (1). In particular, a proper conic ($A \neq 0$) is a real ellipse, imaginary ellipse, hyperbola, or parabola if $F_0(x, y)$ is, respectively, positive definite, negative definite, indefinite, or semidefinite as determined by the (necessarily real) roots λ_1, λ_2 of the *characteristic equation*

$$\begin{vmatrix} a_{11} - \lambda & a_{12} \\ a_{21} & a_{22} - \lambda \end{vmatrix} = 0 \qquad \text{or} \qquad \lambda^2 - I\lambda + D = 0 \tag{2.4-5}$$

λ_1 and λ_2 are the *eigenvalues* of the matrix a_{ik} (Secs. 13.4-5 and 13.5-2).

2.4-6. Diameters and Centers of Conic Sections. (a) A **diameter** of a conic described by Eq. (1) is the locus of the centers of parallel chords. The diameter **conjugate** to the chords inclined at an angle ϑ with respect to the positive x axis bisects these chords and is a straight line with the equation

$$(a_{11}x + a_{12}y + a_{13}) \cos \vartheta + (a_{21}x + a_{22}y + a_{23}) \sin \vartheta = 0 \tag{2.4-6}$$

(b) All diameters of a conic (1) either intersect at a unique point, the **center** of the conic (see Sec. 2.4-10 for an alternative definition), or they are parallel, according to whether $D \neq 0$ or $D = 0$. In the former case, the conic is a **central conic.**

The coordinates x_0, y_0 of the center are given by

$$a_{11}x_0 + a_{12}y_0 + a_{13} = 0 \qquad a_{21}x_0 + a_{22}y_0 + a_{23} = 0 \tag{2.4-7}$$

so that
$$x_0 = -\frac{1}{D}\begin{vmatrix} a_{13} & a_{12} \\ a_{23} & a_{22} \end{vmatrix} \qquad y_0 = -\frac{1}{D}\begin{vmatrix} a_{11} & a_{13} \\ a_{21} & a_{23} \end{vmatrix} \qquad (D \neq 0) \tag{2.4-8}$$

Given the equation (1) of a central conic, a translation (2.1-8) of the coordinate origin to the center (8) of the conic results in the new equation

$$a_{11}\bar{x}^2 + 2a_{12}\bar{x}\bar{y} + a_{22}\bar{y}^2 + \frac{A}{D} = 0 \tag{2.4-9}$$

in terms of the new coordinates $\bar{x}$, $\bar{y}$.

(c) Two **conjugate diameters** of a central conic each bisect the chords parallel to the other diameter (see also Sec. 2.5-2*e*).

2.4-7. Principal Axes. A diameter perpendicular to its associated (conjugate) chords (principal chords) is a symmetry axis or **principal axis** of the conic. Every (real) noncentral conic has one principal axis; every (real) central conic has either two mutually perpendicular principal axes, or every diameter is a principal axis (circle).

The principal axes are directed along eigenvectors of the matrix $[a_{ik}]$ (Sec. 14.8-6). More specifically, the direction cosines $\cos \vartheta$, $\sin \vartheta$ of the normal to a principal axis (Sec. 2.2-1) satisfy the conditions

$$\left.\begin{aligned} (a_{11} - \lambda) \cos \vartheta + a_{12} \sin \vartheta &= 0 \\ a_{21} \cos \vartheta + (a_{22} - \lambda) \sin \vartheta &= 0 \end{aligned}\right\} \tag{2.4-10}$$

where λ is a nonvanishing root of the characteristic equation (5). The angle Φ between the positive x axis and any principal axis of the conic (1) satisfies the condition

$$\tan 2\Phi = \tan 2\vartheta = \frac{2a_{12}}{a_{11} - a_{22}} \tag{2.4-11}$$

2.4-8. Transformation of the Equation of a Conic to Standard or Type Form. If one introduces a new reference system by combining a *rotation* of the coordinate axes through an angle satisfying Eq. (11) with a suitable *translation* of the origin (Sec. 2.1-7), the equation (1) of any proper conic reduces to one of the *standard or type forms* (sometimes called *canonical forms*) listed below. The parameters a^2, b^2, p appearing in the standard forms are simply related to the invariants A, D, I, and to the roots $\lambda_1 \geq \lambda_2$ of the characteristic equation (5).

$$\frac{x^2}{a^2} + \frac{y^2}{b^2} = 1 \text{ (ELLIPSE)} \qquad \begin{aligned} a^2 &= -\frac{1}{\lambda_2}\frac{A}{D} = -\frac{A}{\lambda_1\lambda_2^2} \\ b^2 &= -\frac{1}{\lambda_1}\frac{A}{D} = -\frac{A}{\lambda_1^2\lambda_2} \end{aligned} \tag{2.4-12a}$$

$$\frac{x^2}{a^2} - \frac{y^2}{b^2} = 1 \text{ (HYPERBOLA)} \qquad \begin{aligned} a^2 &= -\frac{1}{\lambda_1}\frac{A}{D} = -\frac{A}{\lambda_1^2\lambda_2} \\ b^2 &= +\frac{1}{\lambda_2}\frac{A}{D} = +\frac{A}{\lambda_1\lambda_2^2} \end{aligned} \tag{2.4-12b}$$

$$y^2 = 4px \text{ (PARABOLA)} \qquad p = \frac{1}{2I}\sqrt{-\frac{A}{I}} = \frac{1}{2\lambda_1}\sqrt{-\frac{A}{\lambda_1}} > 0 \qquad \lambda_2 = 0 \tag{2.4-12c}$$

The equations of the *improper (degenerate) conics* are similarly transformed to the standard or type forms

$$\left.\begin{aligned} &\frac{x^2}{a^2} + \frac{y^2}{b^2} = 0 \text{ (POINT)} \qquad \frac{x^2}{a^2} - \frac{y^2}{b^2} = 0 \text{ (INTERSECTING STRAIGHT LINES)} \\ &\frac{x^2}{a^2} = 1 \text{ (PARALLEL STRAIGHT LINES)} \qquad x^2 = 0 \text{ (ONE STRAIGHT LINE)} \end{aligned}\right\} \tag{2.4-13}$$

NOTE: Every rotation (2.1-9) through an angle ϑ satisfying Eq. (11) diagonalizes the matrix $[a_{ik}]$ of the characteristic quadratic form (4) (*principal-axis transformation*, see also Sec. 14.8-6). The values of ϑ satisfying Eq. (11) differ by multiples of 90 deg, corresponding to interchanges of x, $-x$, y, and $-y$. The standard forms (12*a*, *b*, *c*) correspond to choices of ϑ such that the foci (Sec. 2.4-9) of the conics lie on the x axis. Equation (11) becomes indeterminate for (real and imaginary) circles and points, which have no definite principal axes.

2.4-9. Definitions of Proper Conics in Terms of Loci. Once the equation of any proper conic is reduced to its standard form (12), a simple translation (Sec. 2.1-5) may be used to introduce a new system of coordinates x, y such that the equation of the conic appears in the form

$$y^2 = 4px - (1 - \epsilon^2)x^2 \tag{2.4-14}$$

The conic passes through the origin of the new x, y system; the x axis is a *symmetry axis* (principal axis) of the conic.

Equation (14) *describes a proper conic as the locus of a point which moves so that the ratio* $\epsilon \geq 0$ (**eccentricity**) *of its distances from a fixed point* (**focus**) *and from a fixed line* (**directrix**) *is a constant. The conic will be an ellipse if* $\epsilon < 1$ *and, specifically, a circle if* $\epsilon = 0$. *The conic will be a hyperbola if* $\epsilon > 1$, *and a parabola if* $\epsilon = 1$.

The equation of the *directrix* of the conic represented by Eq. (14) is

$$x = -\frac{2p}{\epsilon(1+\epsilon)} \tag{2.4-15}$$

The coordinates x and y of the *focus* are

$$x = \frac{2p}{1+\epsilon} \qquad y = 0 \tag{2.4-16}$$

The directrix is perpendicular to the symmetry axis. The latter passes through the focus and also through the **vertex** $x = y = 0$ of the conic. The distance between the focus and the directrix is equal to $2p/\epsilon$.

In the case of a *central conic* (ellipse or hyperbola), the straight line

$$x = \frac{2p}{1-\epsilon^2} = a \tag{2.4-17}$$

is a symmetry axis (principal axis) of the conic, so that two foci and two directrices can be defined.

The **latus rectum** of a proper conic is defined as the length of a chord through the focus and perpendicular to the symmetry axis and is equal to $|4p|$.

NOTE: *All types of improper as well as of proper conics may be obtained as the intersections of a right circular cone with a plane for various inclinations of the plane with respect to the cone.* If the conic is a pair of (distinct, coincident, or imaginary) parallel straight lines (see also Table 2.4-1), then the cone must be regarded as degenerated into a *cylinder* unless the plane is tangent to the cone.

2.4-10. Tangents and Normals of Conic Sections. Polars and Poles. The *equation of the tangent* (Sec. 17.1-1) *to the general conic* (1) *at the point* (x_1, y_1) *of the conic* is

$$\left.\begin{aligned} a_{11}x_1x + a_{12}(y_1x + x_1y) + a_{22}y_1y + a_{13}(x_1 + x) \\ + a_{23}(y_1 + y) + a_{33} = 0 \\ \text{or} \quad (a_{11}x_1 + a_{12}y_1 + a_{13})x + (a_{21}x_1 + a_{22}y_1 + a_{23})y \\ + (a_{31}x_1 + a_{32}y_1 + a_{33}) = 0 \end{aligned}\right\} \tag{2.4-18}$$

The equation of the *normal* (Sec. 17.1-2) to the conic (1) at the point (x_1, y_1) is

$$\frac{x - x_1}{a_{11}x_1 + a_{12}y_1 + a_{13}} = \frac{y - y_1}{a_{21}x_1 + a_{22}y_1 + a_{23}} \tag{2.4-19}$$

Table 2.4-2. Formulas Dealing with Tangents, Normals, Polars, and Poles of Conic Sections

Curve	**Circle** about the origin radius R	**Parabola** directrix $x = -p$ focus $(p, 0)$	**Ellipse,** centered at origin; major axis $2a$ along x axis, minor axis $2b$	**Hyperbola,** centered at origin; transverse axis $2a$ along x axis, conjugate axis $2b$
Equation of **curve**	$x^2 + y^2 = R^2$	$y^2 = 4px$	$\frac{x^2}{a^2} + \frac{y^2}{b^2} = 1 \qquad a \geq b$	$\frac{x^2}{a^2} - \frac{y^2}{b^2} = 1$
Equations of **tangents** from the point (x_1, y_1) to the curve	$\frac{y - y_1}{x - x_1} = \frac{-x_1y_1 \pm R\sqrt{x_1^2 + y_1^2 - R^2}}{R^2 - x_1^2}$	$\frac{y - y_1}{x - x_1} = \frac{y_1 \pm \sqrt{y_1^2 - 4px_1}}{2x_1}$	$\frac{y - y_1}{x - x_1} = \frac{-x_1y_1 \pm \sqrt{b^2x_1^2 + a^2y_1^2 - a^2b^2}}{a^2 - x_1^2}$	$\frac{y - y_1}{x - x_1} = \frac{-x_1y_1 \pm \sqrt{-b^2x_1^2 + a^2y_1^2 + a^2b^2}}{a^2 - x_1^2}$
Equation(s) of **tangent(s)** of slope m to the curve	$y = mx \pm R\sqrt{1 + m^2}$	$y = mx + \frac{p}{m}$	$y = mx \pm \sqrt{m^2a^2 + b^2}$	$y = mx \pm \sqrt{m^2a^2 - b^2}$
Equation of **polar** of a point (x_1, y_1) *or* equation of **tangent** at a point (x_1, y_1) on the curve	$xx_1 + yy_1 = R^2$	$yy_1 = 2p(x + x_1)$	$\frac{xx_1}{a^2} + \frac{yy_1}{b^2} = 1$	$\frac{xx_1}{a^2} - \frac{yy_1}{b^2} = 1$
Coordinates x_1, y_1 of the pole of the straight line $Ax + By + C = 0$ with respect to the curve	$x_1 = -\frac{AR^2}{C}$ $y_1 = -\frac{BR^2}{C}$	$x_1 = \frac{C}{A}$ $y_1 = -\frac{2Bp}{A}$	$x_1 = -\frac{a^2A}{C}$ $y_1 = -\frac{b^2B}{C}$	$x_1 = -\frac{a^2A}{C}$ $y_1 = \frac{b^2B}{C}$
Equation of the **normal** to the curve at a point (x_1, y_1)	$y = \frac{y_1}{x_1}x$	$\frac{y - y_1}{x - x_1} = -\frac{y_1}{2p}$	$\frac{y - y_1}{x - x_1} = \frac{a^2y_1}{b^2x_1}$	$\frac{y - y_1}{x - x_1} = -\frac{a^2y_1}{b^2x_1}$
Condition satisfied by the "line coordinates" ξ, η, of any straight line $\xi x + \eta y + 1 = 0$ tangent to the curve	$\xi^2 + \eta^2 = \frac{1}{R^2}$	$\xi - \eta^2p = 0$	$a^2\xi^2 + b^2\eta^2 = 1$	$a^2\xi^2 - b^2\eta^2 = 1$

Equation (18) defines a straight line called the **polar** of the **pole** (x_1, y_1) with respect to the conic (1), whether the point (x_1, y_1) does or does not lie on the curve. The polar of a point on the conic is the tangent at that point.

The equations of tangents, polars, and normals for proper conic sections described in the standard form (Sec. 2.4-8) are listed in Table 2.4-2.

NOTE the following theorems about polars and poles with respect to conic sections.

1. *If the pole moves along a straight line, the corresponding polar rotates about the pole of the straight line, and conversely.*
2. *If two tangents can be drawn from a point to a conic section, the polar of the point passes through the points of contact.*
3. *For any straight line drawn through the pole P and intersecting the polar at Q and the conic section at R_1 and R_2, the points P and Q* **divide R_1R_2 harmonically,** i.e.

$$\frac{\overrightarrow{PR_1}}{\overrightarrow{PR_2}} = -\frac{\overrightarrow{QR_1}}{\overrightarrow{QR_2}}$$

4. *The tangents at the end points of any chord through a focus intersect on the corresponding directrix.*
5. *Every chord through a focus is perpendicular to the line drawn through the focus and the point of intersection of the tangents at the end points of the chord.*
6. *The tangent at an end of a diameter is parallel to the chords defining the diameter.*
7. *The tangents drawn at the ends of any chord intersect on the diameter conjugate to the chord.*
8. *The pole of any diameter is the point at infinity. The polar of any point on a diameter is parallel to the chords defining the diameter. The center of a conic is the pole of the straight line at infinity (a common definition of the center;* if this definition is used, diameters are defined as straight lines through the center).

2.4-11. Other Representations of Conics. **(a)** *A conic is definitely determined by five of its points if no four of them are collinear* (Sec. 2.3-1). The equation of the conic through five such points (x_1, y_1), (x_2, y_2), (x_3, y_3), (x_4, y_4), (x_5, y_5) is

$$\begin{vmatrix} x^2 & xy & y^2 & x & y & 1 \\ x_1^2 & x_1y_1 & y_1^2 & x_1 & y_1 & 1 \\ x_2^2 & x_2y_2 & y_2^2 & x_2 & y_2 & 1 \\ x_3^2 & x_3y_3 & y_3^2 & x_3 & y_3 & 1 \\ x_4^2 & x_4y_4 & y_4^2 & x_4 & y_4 & 1 \\ x_5^2 & x_5y_5 & y_5^2 & x_5 & y_5 & 1 \end{vmatrix} = 0 \qquad (2.4\text{-}20)$$

The conic will be improper if and only if any three of the given points are collinear. Equation (20) may also be interpreted as a condition that six points lie on a conic.

NOTE: The construction of a conic through five given points is made possible by *Pascal's theorem: For any closed hexagon whose vertices lie on a conic, the intersections of opposite pairs of sides are collinear (or at infinity).*

A conic is also definitely determined by five tangents, if no four of them intersect in a point; *Brianchon's theorem* states that *for any hexagon whose sides are tangent to a conic, the diagonals connecting opposite vertices intersect in a point* (or are parallel).

(b) If the focus of a proper conic is chosen as the pole and the symmetry axis as the polar axis of a polar-coordinate system, the equation of the conic in terms of the polar coordinates r, φ is

$$r = \frac{2p}{1 + \epsilon \cos \varphi} \tag{2.4-21}$$

2.5. PROPERTIES OF CIRCLES, ELLIPSES, HYPERBOLAS, AND PARABOLAS

2.5-1. Special Formulas and Theorems Relating to Circles (see also Table 2.4-2). **(a)** The *general form of the equation of a circle* in rectangular cartesian coordinates is

$$\begin{gathered} x^2 + y^2 + Ax + By + C = 0 \\ \text{or} \qquad (x - x_0)^2 + (y - y_0)^2 = R^2 \\ \text{with} \\ 2x_0 = -A, \quad 2y_0 = -B, \quad R^2 = \tfrac{1}{4}(A^2 + B^2 - 4C) > 0 \end{gathered} \tag{2.5-1}$$

The point (x_0, y_0) is the center of the circle, and R is its radius. The circle (1) touches the x axis or the y axis if $4C = A^2$ or $4C = B^2$, respectively. The *equation of a circle about the origin* is

$$x^2 + y^2 = R^2 \tag{2.5-2}$$

(b) The *equation of the circle through three noncollinear points* (x_1, y_1), (x_2, y_2), (x_3, y_3) is

$$\begin{vmatrix} x^2 + y^2 & x & y & 1 \\ x_1^2 + y_1^2 & x_1 & y_1 & 1 \\ x_2^2 + y_2^2 & x_2 & y_2 & 1 \\ x_3^2 + y_3^2 & x_3 & y_3 & 1 \end{vmatrix} = 0 \tag{2.5-3}$$

Equation (3) is a necessary and sufficient condition that the four points (x, y), (x_1, y_1), (x_2, y_2), (x_3, y_3) lie on a circle.

(c) The length L of each tangent from a point (x_1, y_1) to the circle (1) is

$$L = +\sqrt{(x_1 - x_0)^2 + (y_1 - y_0)^2 - R^2} \tag{2.5-4}$$

(d) Two circles

$$x^2 + y^2 + A_1x + B_1y + C_1 = 0 \qquad x^2 + y^2 + A_2x + B_2y + C_2 = 0 \tag{2.5-5}$$

are *concentric* if and only if $A_1 = A_2$ and $B_1 = B_2$. They are *orthogonal* if and only if $A_1A_2 + B_1B_2 = 2(C_1 + C_2)$.

(e) All circles passing through the (real or imaginary) points of intersection of the two circles (5) have equations of the form (see also Sec. 2.1-9c)

$$(x^2 + y^2 + A_1x + B_1y + C_1 + \,)(x^2 + y^2 + A_2x + B_2y + C_2) = 0 \tag{2.5-6}$$

where λ is a parameter. Note that the curves (6) exist even if the two circles (5) have no real points of intersection.

(f) For $\lambda = -1$, Eq. (6) reduces to the equation of a straight line

$$(A_1 - A_2)x + (B_1 - B_2)y + (C_1 - C_2) = 0 \tag{2.5-7}$$

which is called the **radical axis** of the two circles (5). The radical axis is the locus of points from which tangents of equal length can be drawn to the two circles. If the two circles intersect or touch, the radical axis is the secant or tangent through the common points or point. The radical axis of two concentric circles may be considered as infinitely far away from their common center. The three radical axes associated with three circles intersect in a point (**radical center**). This fact is used for the construction of the radical axis of two nonintersecting circles. If the centers of three circles are collinear, their radical center is at infinity.

(g) In terms of *polar coordinates* r, φ, the equation of a circle of radius R about the point (r_0, φ_0) is

$$r^2 - 2rr_0 \cos(\varphi - \varphi_0) + r_0^2 = R^2 \tag{2.5-8}$$

2.5-2. Special Formulas and Theorems Relating to Ellipses and Hyperbolas (see also Sec. 2.5-3 and Tables 2.4-2 and 2.5-1). (a) For any ellipse (2.4-12a), the respective lengths of the **major and minor axes**

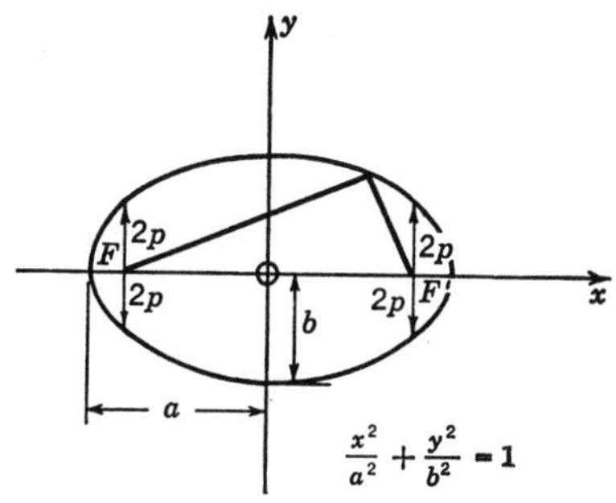

Fig. 2.5-1a

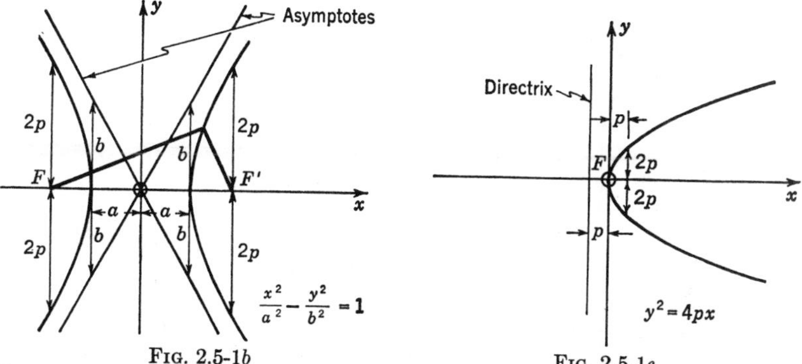

Fig. 2.5-1b Fig. 2.5-1c

Fig. 2.5-1. Graphs of ellipse, hyperbola, and parabola in standard form (Sec. 2.4-8), showing foci, axes, and length of the latus rectum (Secs. 2.4-9 and 2.5-2) for each curve.

Table 2.5-1. Special Formulas Relating to Ellipses, Hyperbolas, and Parabolas Represented in Standard Form (see also Secs. 2.4-9 and 2.5-2)

No.		Ellipse	Hyperbola	Parabola
1	Equation in standard form	$\frac{x^2}{a^2} + \frac{y^2}{b^2} = 1$	$\frac{x^2}{a^2} - \frac{y^2}{b^2} = 1$	$y^2 = 4px$
2	Eccentricity	$\epsilon = \sqrt{1 - \frac{b^2}{a^2}} < 1$	$\epsilon = \sqrt{1 + \frac{b^2}{a^2}} > 1$	$\epsilon = 1$
3	Focus or foci	$(a\epsilon, 0) \quad (-a\epsilon, 0)$	$(a\epsilon, 0) \quad (-a\epsilon, 0)$	$(p, 0)$
4	Equation(s) of directrix or directrices	$x = \frac{a}{\epsilon} \quad x = -\frac{a}{\epsilon}$	$x = \frac{a}{\epsilon} \quad x = -\frac{a}{\epsilon}$	$x = -p$
5	Latus rectum	$\|4p\| = \frac{2b^2}{a}$	$\|4p\| = \frac{2b^2}{a}$	$4p$
6	Focal radius or radii of a point (x_1, y_1) on the curve	$r_1 = a + x_1\epsilon$ $r_2 = a - x_1\epsilon$	$r_1 = a + x_1\epsilon$ $r_2 = -a + x_1\epsilon$	$r = x_1 + p$
7	Equation of diameter conjugate to chords of slope m	$y = -\frac{b^2}{a^2m}x$	$y = \frac{b^2}{a^2m}x$	$y = \frac{2p}{m}$
8	Area of segment between the vertex convex to the left and a chord through (x_1, y_1) and $(x_1, -y_1)$	$\frac{\pi}{2}ab + \frac{b}{a}\left(x_1\sqrt{a^2 - x_1^2} + a^2 \arcsin \frac{x_1}{a}\right)$	$x_1y_1 - ab \log_e\left(\frac{x_1}{a} + \frac{y_1}{b}\right)$	$\frac{4}{3}x_1y_1$
9	Equation in terms of polar coordinates r, φ [Fig. 2.5-1; compare with Eq. (2.4-21)]	$r^2 = \frac{b^2}{1 - \epsilon^2 \cos^2 \varphi}$	$r^2 = \frac{-b^2}{1 - \epsilon^2 \cos^2 \varphi}$	$r = \frac{4p \cos \varphi}{1 - \cos^2 \varphi}$

(Fig. 2.5-1*a*) are $2a$ and $2b$. *The sum of the focal radii* r_1 *and* r_2 (*Table* 2.5-1, 6) *is constant* (*and equal to* $2a$) *for every point on an ellipse.*

(b) A hyperbola (2.4-12*b*) approaches the *asymptotes* (Sec. 17.1-6)

$$\frac{x}{a} + \frac{y}{b} = 0 \qquad \text{and} \qquad \frac{x}{a} - \frac{y}{b} = 0 \tag{2.5-9}$$

The asymptotes are straight lines through the center of the hyperbola; they intersect at an angle equal to $2 \arctan (b/a)$, which is equal to 90 deg for $a = b$ (*rectangular hyperbola*).

(c) For every hyperbola (2.4-12*b*), the length of the **transverse axis** (distance between vertices, Fig. 2.5-1*b*) is equal to $2a$. The **conjugate axis** is the principal axis perpendicular to the transverse axis. The distance between the intersections of the asymptotes with a tangent touching the hyperbola at a vertex equals $2b$ (Fig. 2.5-1*b*). *The difference between the focal radii* r_1 *and* r_2 (*Table* 2.5-1, 6) *is constant and respectively equal to* $2a$ *and* $-2a$ *for points on either branch of a hyperbola.*

(d) If d_1 and d_2 are the lengths of any two conjugate diameters (Sec. 2.4-6*c*) of an ellipse or hyperbola,

$$d_1 d_2 \sin (\arctan m_2 - \arctan m_1) = 4ab \tag{2.5-10}$$

where m_1, m_2 are the respective slopes of the two diameters.

(e) A diameter $y = mx$ of the hyperbola (2.4-12*b*) intersects the hyperbola if and only if $m^2 < b^2/a^2$ and coincides with an asymptote if and only if $m^2 = b^2/a^2$. An asymptote may be considered as a chord intersecting the hyperbola at infinity and identical with its own conjugate diameter.

(f) *For every secant of a hyperbola, the two segments between the hyperbola and an asymptote are equal; for every tangent of a hyperbola the point of contact bisects the segment between the asymptotes. For every point of a hyperbola, the product of the distances to the asymptotes is constant.*

(g) *For every ellipse, the area of the triangle formed by the center and the end points of any two conjugate diameters is constant. For every hyperbola the area of the triangle formed by the asymptotes and any tangent is constant.*

(h) *Two ellipses or hyperbolas are similar if and only if the ratio of the two major axes, or transverse axes, respectively, is equal to the ratio of the two minor axes, or conjugate axes, respectively.*

2.5-3. Construction of Ellipses and Hyperbolas and Their Tangents and Normals. **(a)** *The following procedures may be used for the construction of an ellipse, given the lengths and location of the major and minor axes.*

1. Let O be the center of the ellipse; let $P'OP$ be the major axis and $Q'OQ$ the minor axis. To construct a point of the ellipse, draw any straight line ORS through the center with $OR = b$ and $OS = a$. Then draw a parallel to $P'OP$ through R and a parallel to $Q'OQ$ through S; these two lines will intersect in a point of the ellipse.
2. Obtain the foci F and F' by using the relation $QF = QF' = OP = a$. To construct points of the ellipse, draw any point T on the major axis between O and P. Two circles drawn about F and F' with the radii $P'T$ and PT will intersect in two points of the ellipse.

To construct a hyperbola, given the length and location of the transverse axis $P'OP$ and the location of the foci F and F' (see Table 2.5-1 for the relation between $OF = OF'$, a, and b), draw any point T on the transverse axis such that $OT > OP$. Two circles drawn about F and F' with the radii $P'T$ and PT will intersect in two points of the hyperbola.

(**b**) *The following properties of ellipses and hyperbolas are useful for the construction of tangents and normals to these curves.*

1. For every ellipse (2.4-12*a*), the tangents drawn at a point (x_1, y_1) of the ellipse intersect the x axis at the same point as the tangents of the circle $x^2 + y^2 = a^2$ drawn at the points $(x_1, \pm\sqrt{a^2 - x_1^2})$.
2. The tangent and the normal at any point of an ellipse or hyperbola bisect the angles between straight lines connecting that point to the foci. This theorem implies the "focal property" of the ellipse.
3. The product of the distances between the foci and any tangent of an ellipse or hyperbola is constant and equal to b^2. The perpendiculars drawn from either focus to a tangent meet the latter on the circle drawn with the major axis (transverse axis) as a diameter; this theorem may be used to construct an ellipse or hyperbola by envelopment.

(**c**) *One may draw a portion of an ellipse or hyperbola approximately by drawing the circles of curvature* (Sec. 17.1-4) *for the two vertices.* For any ellipse or hyperbola, the centers of curvature for the vertices lie on the major axis or transverse axis, respectively, and the radius of curvature at either vertex is equal to b^2/a. The centers of curvature for the points at either end of the minor axis of an ellipse lie on the minor axis, and the radius of curvature at either of these points is equal to a^2/b.

2.5-4. Construction of Parabolas and Their Tangents and Normals. (**a**) *The following properties of parabolas may be used for the construction of a parabola* when the (directed) axis, the focus, and the distance $2p$ between the focus and the directrix are given:

1. *The distance between the focus and any point P of a parabola is equal to the distance between the directrix and the point P* (see also Sec. 2.4-9).
2. A straight line perpendicular to the axis of the parabola at a distance p from the focus in the direction of the negative axis is the tangent of the parabola at the vertex. For any point Q on this line, the perpendicular at Q to a line joining Q with the focus will be a tangent of the parabola (construction by envelopment). Conversely, *any tangent and a line perpendicular to it through the focus intersect on the tangent at the vertex.*

(**b**) *The following properties of parabolas are useful in connection with the construction of tangents and normals to a parabola:*

1. *The distance between any point P of a parabola and the focus is equal to the distance between the focus and the intersection of the parabola axis with the tangent at P.*
2. *The tangent and normal at any point P of a parabola bisect the angles formed by the line joining P with the focus and the diameter through P.* Note that the diameter is parallel to the axis; this theorem implies the "focal property" of the parabola.
3. *The normal at any point P of a parabola and the perpendicular dropped from P on the axis intersect the latter a constant distance $2p$ apart.*
4. *For every parabola, the directrix is the locus of the intersections of pairs of tangents which are perpendicular to each other.*

(c) *One may draw a portion of a parabola approximately by drawing the circle of curvature* (Sec. 17.1-4) *for the vertex.* The center of curvature for the vertex lies on the axis, and the radius of curvature at the vertex is equal to $2p$.

2.6. HIGHER PLANE CURVES

2.6-1. Examples of Algebraic Curves (see also Fig. 2.6-1 and Sec. 2.1-9c).

(a) **Neil's parabola:** $y = ax^{3/2}$

(b) **Witch of Agnesi:** $x^2y = 4a^2(2a - y)$

(c) **Conchoid of Nicomedes:**

$$(x^2 + y^2)(x - a)^2 = x^2b^2$$

(d) **Cissoid of Diocles:**

$$y^2(a - x) = x^3 \qquad \text{or} \qquad r = a\left(\frac{1}{\cos\varphi} - \cos\varphi\right)$$

(e) **Lemniscate of Bernoulli:**

$$(x^2 + y^2)^2 - a^2(x^2 - y^2) = 0 \qquad \text{or} \qquad r^2(r^2 - a^2\cos 2\varphi) = 0$$

(f) **Ovals of Cassini:** $(x^2 + y^2 + a^2)^2 - 4a^2x^2 = c^4$

(locus of points for which the product of the distances from $(-a, 0)$ and $(a, 0)$ is equal to c^2)

(g) **Strophoid:** $x^3 + x(a^2 + y^2) = 2a(y^2 + x^2)$

(h) **Cruciform:** $x^2y^2 = a^2(x^2 + y^2)$ or $r = \dfrac{2a}{\sin 2\varphi}$

(i) **Cardioid:**

$$(x^2 + y^2 - ax)^2 = a^2(x^2 + y^2) \qquad \text{or} \qquad r = a(1 + \cos\varphi)$$

(j) **Trisectrix:**

$$y^2 = \frac{x^3(3a - x)}{a + x} \qquad \text{or} \qquad r = a\left(4\cos\varphi - \frac{1}{\cos\varphi}\right)$$

(k) **Astroid:** $x^{2/3} + y^{2/3} = a^{2/3}$

(l) **Leaf of Descartes:** $x^3 + y^3 = 3axy$

(m) **Limaçon of Pascal:** $r = b - a\cos\varphi$

(n) **Lituus:** $r^2\varphi = a^2$

2.6-2. Examples of Transcendental Curves (see also Fig. 2.6-2).

(a) **Catenary:** $y = \dfrac{a}{2}\left(e^{\frac{x}{a}} + e^{-\frac{x}{a}}\right)$

(b) **(Linear) spiral of Archimedes:** $r = a\varphi$

(c) **Parabolic spiral:** $r^2 = 4p\varphi$

(d) **Logarithmic spiral:** $r = ae^{b\varphi}$

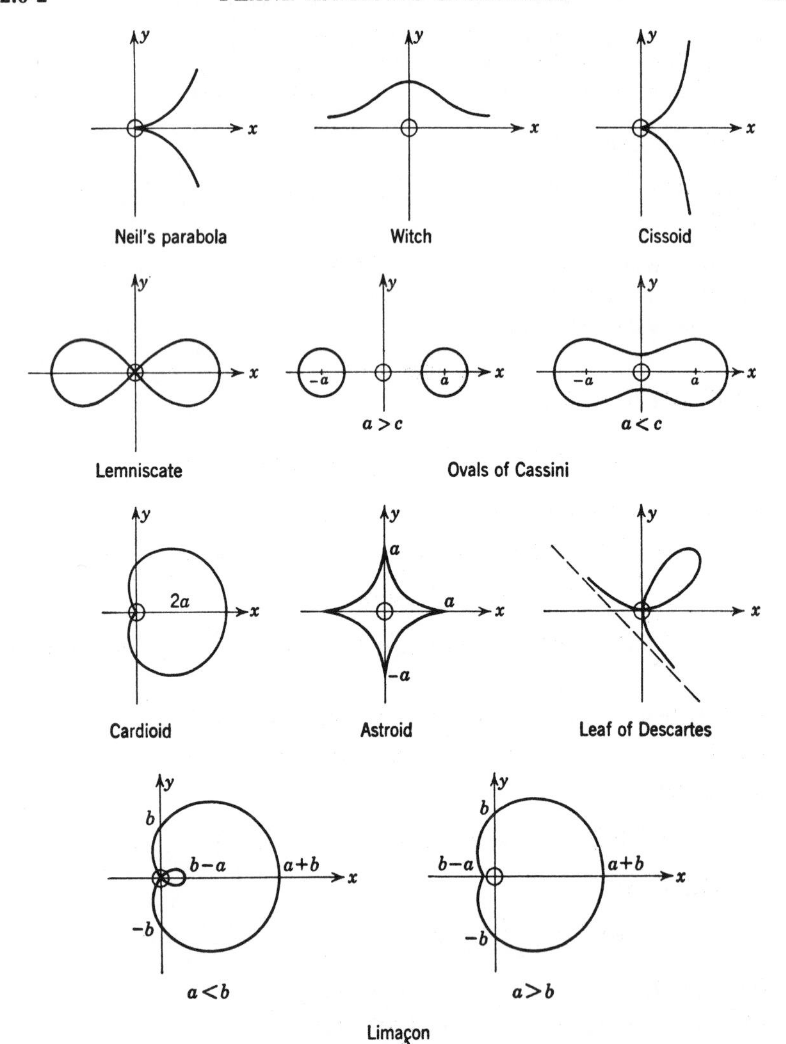

FIG. 2.6-1. Examples of higher algebraic curves.

(e) The locus of a point (x, y) at the distance a_1 from the center of a circle of radius a rolling on the x axis is a **cycloid** and may be represented by

$$x = at - a_1 \sin t, \qquad y = a - a_1 \cos t \tag{2.6-1}$$

if the point (x, y) rolls with the circle.

(f) The locus of a point (x, y) at the distance a_1 from the center of a circle of radius a rolling on the outside of the circle $x^2 + y^2 = b^2$ is an

epicycloid and may be represented by

$$\left.\begin{aligned} x &= (a + b) \sin \frac{at}{b} - a_1 \sin \frac{a + b}{b} t \\ y &= (a + b) \cos \frac{at}{b} - a_1 \cos \frac{a + b}{b} t \end{aligned}\right\} \qquad (2.6\text{-}2)$$

if the point (x, y) rolls with the circle.

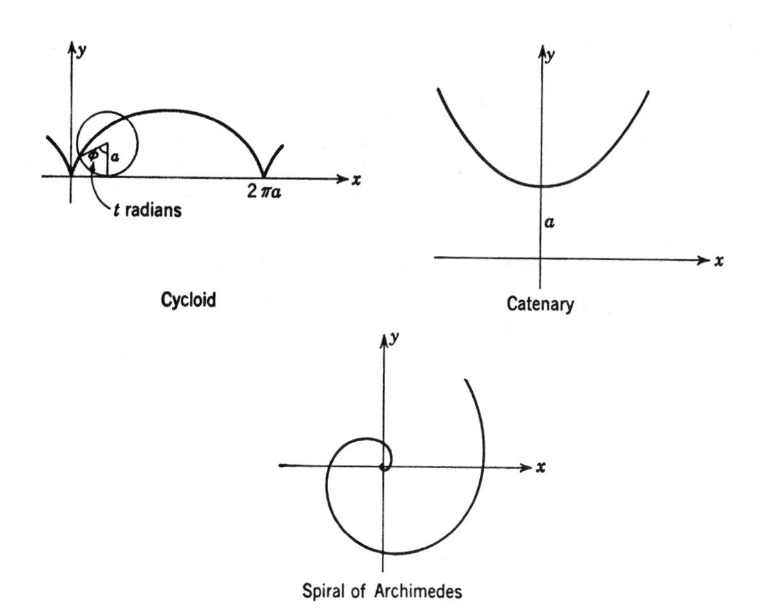

Fig. 2.6-2. Some transcendental curves.

(g) The locus of a point (x, y) at the distance a_1 from the center of a circle of radius a rolling on the inside of the circle $x^2 + y^2 = b^2$ is a **hypocycloid** and may be represented by

$$\left.\begin{aligned} x &= (b - a) \sin \frac{at}{b} - a_1 \sin \frac{b - a}{b} t \\ y &= (b - a) \cos \frac{at}{b} + a_1 \cos \frac{b - a}{b} t \end{aligned}\right\} \qquad (2.6\text{-}3)$$

if the point (x, y) rolls with the circle.

(h) A **tractrix** may be represented by

$$x = a\left(\cos t + \log_e \tan \frac{t}{2}\right) \qquad y = a \sin t \qquad (2.6\text{-}4)$$

2.7. RELATED TOPICS, REFERENCES, AND BIBLIOGRAPHY

2.7-1. Related Topics. The following topics related to the study of plane analytic geometry are treated in other chapters of this handbook:

Formulas from plane geometry.......... Appendix A
Solid analytic geometry.......... Chap. 3
Coordinate transformations.......... Chap. 14
Differential geometry.......... Chap. 17
Algebraic equations.......... Chap. 1
Curvilinear coordinate systems.......... Chap. 6

2.7-2. References and Bibliography.

2.1. Cell, J. W.: *Analytic Geometry,* 3d ed., Wiley, New York, 1960.
2.2. Middlemiss, R. R.: *Analytic Geometry,* 2d ed., McGraw-Hill, New York, 1955.
2.3. Purcell, E. J.: *Analytic Geometry,* Appleton-Century-Crofts, New York, 1958.
2.4. Smith, E. S., et al.: *Analytic Geometry,* Wiley, New York, 1954.

Additional Background Material

2.5. Coxeter, H. S. M.: *Introduction to Geometry,* Wiley, New York, 1961.
2.6. Klein, F.: *Famous Problems of Elementary Geometry,* 2d ed., Dover, New York, 1956.

CHAPTER 3

SOLID ANALYTIC GEOMETRY

3.1. INTRODUCTION AND BASIC CONCEPTS

3.1-1. Introduction (see also Sec. 2.1-1). Chapter 3 deals with the analytic geometry of three-dimensional Euclidean space, corresponding to classical (Euclidean) solid geometry. The elements of the description are **points** labeled by sets of real numbers (**coordinates**) or represented by **position vectors.**

3.1-2. Cartesian Coordinate Systems (see also Sec. 2.1-2). A **cartesian reference system** or **cartesian coordinate system** associates a unique set of three real numbers (**cartesian coordinates**) x, y, z with every point $P \equiv (x, y, z)$ in the finite portion of space by reference to three noncoplanar directed straight lines (**cartesian coordinate axes**) OX, OY, OZ intersecting at the **origin** O (Fig. 3.1-1a).

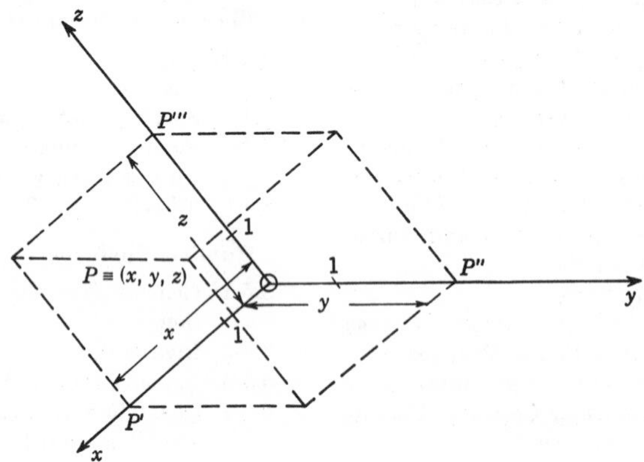

FIG. 3.1-1a. (Oblique) right-handed cartesian coordinate system. The points marked "1" on the x, y, z axes define the coordinate scales used.

Given a (in general *oblique*) system of coordinate axes OX, OY, OZ (Fig. 3.1-1a), the (unique) plane through the point P and parallel to the **yz plane** YOZ intersects the **x axis** OX at P'. Similarly, the planes through P respectively parallel to the **xz plane** XOZ and parallel to the **xy plane** XOY intersect the **y axis** OY at P'' and the **z axis** OZ at P'''. Each directed distance $OP' = x$, $OP'' = y$, and $OP''' = z$ is given the positive sign if its direction coincides with the respective positive direction of the x axis, the y axis, and the z axis. x, y, and z are the *cartesian coordinates* of the point

$P \equiv (x, y, z)$ with respect to the reference system defined by the given coordinate axes and the given (not necessarily equal) scales used to measure x, y, and z.

3.1-3. Right-handed System of Axes (Fig. 3.1-1*a*). The ordered set of three directed straight lines (cartesian axes) OX, OY, OZ is **right-handed** if and only if the rotation needed to turn the x axis OX into the y-axis direction OY through an angle $\sphericalangle XOY < 180$ deg would propel a right-handed screw toward the side (positive side) of the xy plane XOY associated with the positive z axis OZ.

3.1-4. Right-handed Rectangular Cartesian Coordinate System. In a **rectangular** cartesian coordinate system, the coordinate axes are mutually perpendicular (Fig. 3.1-1*b*). The coordinates x, y, z of a point

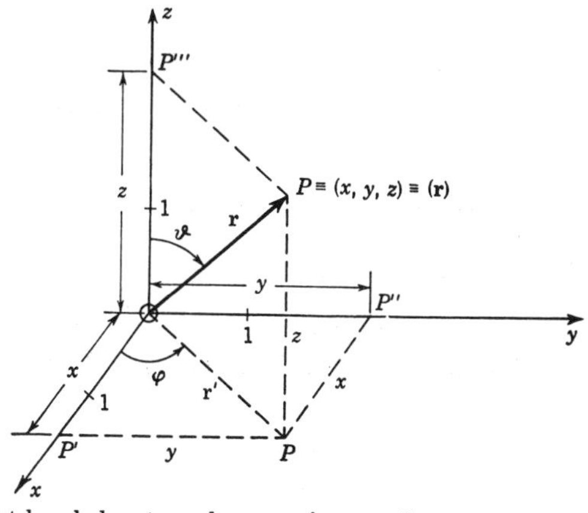

FIG. 3.1-1*b*. Right-handed rectangular cartesian coordinate system, cylindrical coordinate system, and spherical coordinate system.

P are equal to the directed distances, measured on suitable scales, between the origin and the yz plane, the xz plane, and the xy plane, respectively.

Throughout this handbook all coordinate values x, y, z refer to right-handed rectangular cartesian coordinate systems, and equal scale units of unit length are used for measuring x, y, and z, unless otherwise specified.

3.1-5. Position Vectors. Each point $P \equiv (x, y, z) \equiv (\mathbf{r})$ may be represented uniquely by the **position vector**

$$\mathbf{r} = \mathbf{i}x + \mathbf{j}y + \mathbf{k}z \equiv (x, y, z) \tag{3.1-1}$$

which may be represented geometrically by the translation $\overrightarrow{OP}$. The base vectors $\mathbf{i}$, $\mathbf{j}$, $\mathbf{k}$ are unit vectors directed along the x, y, z axes, respectively (see also Secs. 5.1-1 and 5.2-2), of the right-handed rectangular cartesian coordinate system used.

3.1-6. Cylindrical and Spherical Coordinate Systems. Figure 3.1-1*b* also illustrates **cylindrical coordinates** r', φ, z and **spherical coordinates** r, ϑ, φ (**radius vector, colatitude,** and **longitude**) defined so that

$$\left.\begin{aligned} r' &= \underset{+}{\sqrt{x^2 + y^2}} & x &= r' \cos \varphi \\ \varphi &= \arctan \frac{y}{x} & y &= r' \sin \varphi \\ & & z &= z \end{aligned}\right\} \qquad (3.1\text{-}2)$$

$$\left.\begin{aligned} r &= \underset{+}{\sqrt{x^2 + y^2 + z^2}} & x &= r \sin \vartheta \cos \varphi \\ \vartheta &= \arccos \frac{z}{\sqrt{x^2 + y^2 + z^2}} & y &= r \sin \vartheta \sin \varphi \\ \varphi &= \arctan \frac{y}{x} & z &= r \cos \vartheta \end{aligned}\right\} \qquad (3.1\text{-}3)$$

For a further discussion of these and other noncartesian coordinate systems, refer to Chap. 6.

3.1-7. Basic Relations in Terms of Rectangular Cartesian Coordinates and Position Vectors. In terms of rectangular cartesian coordinates x, y, z and position vectors $\mathbf{r} \equiv (x, y, z)$, the following relations hold:

1. The **distance** d between the points P_1 and P_2 specified by

$$P_1 \equiv (x_1, y_1, z_1) \equiv (\mathbf{r}_1), P_2 \equiv (x_2, y_2, z_2) \equiv (\mathbf{r}_2)$$

is

$$\begin{aligned} d &= \underset{+}{\sqrt{(x_2 - x_1)^2 + (y_2 - y_1)^2 + (z_2 - z_1)^2}} \\ &= \underset{+}{\sqrt{(\mathbf{r}_2 - \mathbf{r}_1) \cdot (\mathbf{r}_2 - \mathbf{r}_1)}} = |\mathbf{r}_2 - \mathbf{r}_1| \end{aligned} \qquad (3.1\text{-}4)$$

2. The **angle** γ between the straight lines defined by two directed line segments $\overrightarrow{P_1P_2}$ and $\overrightarrow{P_3P_4}$ is given by

$$\begin{aligned} \cos \gamma &= \\ & \frac{(x_2 - x_1)(x_4 - x_3) + (y_2 - y_1)(y_4 - y_3) + (z_2 - z_1)(z_4 - z_3)}{\underset{+}{\sqrt{(x_2 - x_1)^2 + (y_2 - y_1)^2 + (z_2 - z_1)^2}}\ \underset{+}{\sqrt{(x_4 - x_3)^2 + (y_4 - y_3)^2 + (z_4 - z_3)^2}}} \\ &= \frac{(\mathbf{r}_2 - \mathbf{r}_1) \cdot (\mathbf{r}_4 - \mathbf{r}_3)}{|\mathbf{r}_2 - \mathbf{r}_1|\ |\mathbf{r}_4 - \mathbf{r}_3|} \end{aligned} \qquad (3.1\text{-}5)$$

where the coordinates or position vectors of the points P_1, P_2, P_3, P_4 are denoted by the corresponding subscripts.

If the straight lines P_1P_2 and P_3P_4 do not intersect, γ is defined as the angle between two intersecting straight lines parallel to P_1P_2 and P_3P_4, respectively.

3. The coordinates x, y, z and the position vector $\mathbf{r} \equiv (x, y, z)$ of **the point P dividing the directed line segment $\overrightarrow{P_1P_2}$ in the ratio $\overrightarrow{P_1P}:\overrightarrow{PP_2} = m:n = \mu:1$ are**

$$\left.\begin{aligned} x &= \frac{mx_2 + nx_1}{m+n} = \frac{x_1 + \mu x_2}{1+\mu} \\ y &= \frac{my_2 + ny_1}{m+n} = \frac{y_1 + \mu y_2}{1+\mu} \\ z &= \frac{mz_2 + nz_1}{m+n} = \frac{z_1 + \mu z_2}{1+\mu} \\ \mathbf{r} &= \frac{m\mathbf{r}_2 + n\mathbf{r}_1}{m+n} = \frac{\mathbf{r}_1 + \mu \mathbf{r}_2}{1+\mu} \end{aligned}\right\} \quad (-\infty \leq \mu \leq \infty) \tag{3.1-6}$$

Specifically, the coordinates and the position vector of the **midpoint** of the line segment P_1P_2 are given by

$$x = \frac{x_1 + x_2}{2} \qquad y = \frac{y_1 + y_2}{2} \qquad z = \frac{z_1 + z_2}{2} \qquad \mathbf{r} = \frac{\mathbf{r}_1 + \mathbf{r}_2}{2} \tag{3.1-7}$$

3.1-8. Direction Cosines and Direction Numbers. (a) The **direction cosines** $\cos \alpha_x$, $\cos \alpha_y$, $\cos \alpha_z$ **of a directed line segment $\overrightarrow{P_1P_2}$** are the cosines of the angles α_x, α_y, α_z between $\overrightarrow{P_1P_2}$ and the positive x, y, and z axis, respectively:

$$\left.\begin{aligned} \cos \alpha_x &= \frac{x_2 - x_1}{+\sqrt{(x_2 - x_1)^2 + (y_2 - y_1)^2 + (z_2 - z_1)^2}} \\ \cos \alpha_y &= \frac{y_2 - y_1}{+\sqrt{(x_2 - x_1)^2 + (y_2 - y_1)^2 + (z_2 - z_1)^2}} \\ \cos \alpha_z &= \frac{z_2 - z_1}{+\sqrt{(x_2 - x_1)^2 + (y_2 - y_1)^2 + (z_2 - z_1)^2}} \end{aligned}\right\} \tag{3.1-8}$$

$$\cos^2 \alpha_x + \cos^2 \alpha_y + \cos^2 \alpha_z = 1 \tag{3.1-9}$$

The direction cosines of the directed line segment $\overrightarrow{P_2P_1}$ are $-\cos \alpha_x$, $-\cos \alpha_y$, $-\cos \alpha_z$.

(b) Any set of three numbers a_x, a_y, a_z proportional (with the same proportionality constant) to the direction cosines $\cos \alpha_x$, $\cos \alpha_y$, $\cos \alpha_z$ are called **direction numbers** of the line in question, and

$$\cos \alpha_x = \frac{a_x}{\sqrt{a_x^2 + a_y^2 + a_z^2}} \qquad \cos \alpha_y = \frac{a_y}{\sqrt{a_x^2 + a_y^2 + a_z^2}}$$
$$\cos \alpha_z = \frac{a_z}{\sqrt{a_x^2 + a_y^2 + a_z^2}} \tag{3.1-10}$$

where the sign of the square root is taken to be the same in all three equations; its choice fixes the positive direction of the line. The components of any vector **a** (Sec. 5.2-2) having the same direction as either $\overrightarrow{P_1P_2}$ or $\overrightarrow{P_2P_1}$ are direction numbers of $\overrightarrow{P_1P_2}$. (EXAMPLE: $x_2 - x_1$, $y_2 - y_1$, $z_2 - z_1$.) The direction cosines of $\overrightarrow{P_1P_2}$ are the components of a unit vector directed along $\overrightarrow{P_1P_2}$ (Sec. 5.2-5).

(**c**) The angle γ between two line segments having the respective direction cosines $\cos \alpha_x$, $\cos \alpha_y$, $\cos \alpha_z$ and $\cos \alpha'_x$, $\cos \alpha'_y$, $\cos \alpha'_z$ or the respective direction numbers a_x, a_y, a_z and a'_x, a'_y, a'_z is given by (see also Secs. 3.1-7 and 3.4-1)

$$\begin{aligned}\cos \gamma &= \cos \alpha_x \cos \alpha'_x + \cos \alpha_y \cos \alpha'_y + \cos \alpha_z \cos \alpha'_z \\ &= \frac{a_x a'_x + a_y a'_y + a_z a'_z}{\sqrt{a_x^2 + a_y^2 + a_z^2}\sqrt{a'^2_x + a'^2_y + a'^2_z}}\end{aligned} \qquad (3.1\text{-}11)$$

3.1-9. Projections. The respective *projections* of a line segment $\overrightarrow{P_1P_2}$ having the length (4) and the direction cosines (8) onto the x, y, and z axes are

$$d_x = x_2 - x_1 = d \cos \alpha_x \qquad d_y = y_2 - y_1 = d \cos \alpha_y \qquad d_z = z_2 - z_1 = d \cos \alpha_z$$
$$d_x^2 + d_y^2 + d_z^2 = d^2 \qquad (3.1\text{-}12)$$

The projection of $\overrightarrow{P_1P_2}$ onto a straight line having the direction cosines $\cos \alpha'_x$, $\cos \alpha'_y$, $\cos \alpha'_z$ (Sec. 3.2-1*b*) is $d \cos \alpha$, where $\cos \gamma$ is given by Eq. (11).

The respective projections of $\overrightarrow{P_1P_2}$ onto the yz, xz, and xy planes are

$$d_1 = d \sin \alpha_x \qquad d_2 = d \sin \alpha_y \qquad d_3 = d \sin \alpha_z \qquad d^2 = \tfrac{1}{2}(d_1^2 + d_2^2 + d_3^2) \qquad (3.1\text{-}13)$$

The projection of $\overrightarrow{P_1P_2}$ onto a plane whose normal has the direction cosines $\cos \alpha'_x$, $\cos \alpha'_y$, $\cos \alpha'_z$ (Sec. 3.2-1*b*) is $d \sin \gamma$, where $\cos \gamma$ is given by Eq. (11).

3.1-10. Vector Representation of Areas. A (finite) plane area in three-dimensional Euclidean space may be represented by a vector **A** of magnitude equal to the area A and directed along the positive normal (Sec. 17.3-2) of its plane:

$$\mathbf{A} = A\mathbf{N} \qquad (3.1\text{-}14)$$

where **N** is a unit vector directed along the positive normal. The rectangular cartesian components of **A** are the projections of the area onto the coordinate planes. The area of the parallelogram formed by the vectors **a** and **b** may be represented by $\mathbf{A} = \mathbf{a} \times \mathbf{b}$ (see also Sec. 5.2-7).

The area A of a triangle with the vertices P_1, P_2, P_3 is given by

$$A^2 = \tfrac{1}{4}\left\{\begin{vmatrix} y_1 & z_1 & 1 \\ y_2 & z_2 & 1 \\ y_3 & z_3 & 1 \end{vmatrix}^2 + \begin{vmatrix} z_1 & x_1 & 1 \\ z_2 & x_2 & 1 \\ z_3 & x_3 & 1 \end{vmatrix}^2 + \begin{vmatrix} x_1 & y_1 & 1 \\ x_2 & y_2 & 1 \\ x_3 & y_3 & 1 \end{vmatrix}^2\right\} \qquad (3.1\text{-}15)$$

or

$$A = -\frac{1}{2p}\begin{vmatrix} x_1 & y_1 & z_1 \\ x_2 & y_2 & z_2 \\ x_3 & y_3 & z_3 \end{vmatrix} = -\frac{1}{2p}[\mathbf{r}_1\mathbf{r}_2\mathbf{r}_3] \qquad (3.1\text{-}16)$$

where the coordinates or position vectors of the points P_1, P_2, P_3 are denoted by the respective corresponding subscripts, and where p is the distance between the origin and the plane of the triangle (Sec. 3.2-1*b*). A is positive if a right-handed screw turning in the direction $P_1P_2P_3$ is propelled in the direction of the positive normal of the triangle plane (Sec. 3.2-1*b*).

3.1-11. Computation of Volumes (see also Sec. 5.2-8). **(a)** The volume V of the tetrahedron with the vertices P_1, P_2, P_3, P_4 is

$$V = \frac{1}{6}\begin{vmatrix} x_1 & y_1 & z_1 & 1 \\ x_2 & y_2 & z_2 & 1 \\ x_3 & y_3 & z_3 & 1 \\ x_4 & y_4 & z_4 & 1 \end{vmatrix} = \frac{1}{6}[(\mathbf{r}_2 - \mathbf{r}_1)(\mathbf{r}_3 - \mathbf{r}_1)(\mathbf{r}_4 - \mathbf{r}_1)] \tag{3.1-17}$$

where the coordinates or position vectors of the points P_1, P_2, P_3, P_4 are denoted by the corresponding subscripts. The sign of V depends on the order of the vertices.

(b) The volume V of the parallelepiped formed by three vectors **a**, **b**, **c** is

$$V = [\mathbf{abc}] \tag{3.1-18}$$

3.1-12. Translation and Rotation of Rectangular Cartesian Coordinate Systems. (a) Translation of Coordinate Axes. Let x, y, z be the coordinates of any point P with respect to a right-handed rectangular cartesian reference system. Let $\bar{x}$, $\bar{y}$, $\bar{z}$ be the coordinates of the same point P with respect to a second right-handed rectangular cartesian reference system whose axes have the same directions as the corresponding axes of the x, y, z system, and whose origin has the coordinates $x = x_0$, $y = y_0$, $z = z_0$ in the x, y, z system. If equal scales are used to measure the coordinates in both systems, the coordinates $\bar{x}$, $\bar{y}$, $\bar{z}$ are related to the coordinates x, y, z by the *transformation equations*

$$\begin{array}{ll} \bar{x} = x - x_0 & x = \bar{x} + x_0 \\ \bar{y} = y - y_0 & y = \bar{y} + y_0 \\ \bar{z} = z - z_0 & z = \bar{z} + z_0 \end{array} \tag{3.1-19}$$

(b) Rotation of Coordinate Axes. Given any point $P \equiv (x, y, z)$, let $\bar{x}$, $\bar{y}$, $\bar{z}$ be the coordinates of the same point with respect to a second right-handed rectangular cartesian reference system having the same origin O as the x, y, z system and oriented with respect to the latter so that

The $\bar{x}$ axis has the direction cosines t_{11}, t_{21}, t_{31}.
The $\bar{y}$ axis has the direction cosines t_{12}, t_{22}, t_{32}.
The $\bar{z}$ axis has the direction cosines t_{13}, t_{23}, t_{33}.

Then in the $\bar{x}$, $\bar{y}$, $\bar{z}$ system

The x axis has the direction cosines t_{11}, t_{12}, t_{13}.
The y axis has the direction cosines t_{21}, t_{22}, t_{23}.
The z axis has the direction cosines t_{31}, t_{32}, t_{33}.

If equal scales are used to measure x, y, z, $\bar{x}$, $\bar{y}$, and $\bar{z}$, the transformation equations relating the coordinates $\bar{x}$, $\bar{y}$, $\bar{z}$ to the coordinates x, y, z are

$$\begin{aligned} \bar{x} &= t_{11}x + t_{21}y + t_{31}z \\ \bar{y} &= t_{12}x + t_{22}y + t_{32}z \\ \bar{z} &= t_{13}x + t_{23}y + t_{33}z \end{aligned} \qquad \text{or} \qquad \begin{aligned} x &= t_{11}\bar{x} + t_{12}\bar{y} + t_{13}\bar{z} \\ y &= t_{21}\bar{x} + t_{22}\bar{y} + t_{23}\bar{z} \\ z &= t_{31}\bar{x} + t_{32}\bar{y} + t_{33}\bar{z} \end{aligned} \tag{3.1-20}$$

NOTE: The transformations (20) are *orthogonal* (Secs. 13.3-2 and 14.4-5); *each* t_{ik} *equals the cofactor of* t_{ki} *in the determinant*

$$\det [t_{ik}] = 1 \tag{3.1-21}$$

(Sec. 1.5-2), and

$$\sum_{j=1}^{3} t_{ij}t_{jk} = \sum_{j=1}^{3} t_{ji}t_{kj} = \begin{cases} 1 \text{ if } i = k \\ 0 \text{ if } i \neq k \end{cases} \tag{3.1-22}$$

(c) Simultaneous Translation and Rotation. If the origin of the $\bar{x}$, $\bar{y}$, $\bar{z}$ system is not that of the x, y, z system but has the coordinates $x = x_0$, $y = y_0$, $z = z_0$ in the x, y, z system, the transformation equations become

$$\begin{aligned} \bar{x} &= t_{11}(x - x_0) + t_{21}(y - y_0) + t_{31}(z - z_0) \\ \bar{y} &= t_{12}(x - x_0) + t_{22}(y - y_0) + t_{32}(z - z_0) \\ \bar{z} &= t_{13}(x - x_0) + t_{23}(y - y_0) + t_{33}(z - z_0) \\ \text{or} \qquad x &= t_{11}\bar{x} + t_{12}\bar{y} + t_{13}\bar{z} + x_0 \\ y &= t_{21}\bar{x} + t_{22}\bar{y} + t_{23}\bar{z} + y_0 \\ z &= t_{31}\bar{x} + t_{32}\bar{y} + t_{33}\bar{z} + z_0 \end{aligned} \tag{3.1-23}$$

where the direction cosines t_{ik} satisfy the relations (21) and (22).

The equations (23) relate the coordinates of a point in *any* two right-handed rectangular cartesian reference systems if the same scales are used for all coordinate measurements. More general types of coordinate transformations are discussed in Chap. 6.

(d) Alternative Interpretation of Coordinate Transformations. The transformation equations (23) [of which (19) and (20) are special cases] may also be interpreted as *defining a new point P having the coordinates* $\bar{x}$, $\bar{y}$, $\bar{z}$ *in the* x, y, z *system* and obtained by successive translation and rotation of the original point P (see also Secs. 14.1-3 and 14.5-1).

NOTE: The values of distances between points, angles between line segments, and thus, in fact, all relations between position vectors constituting Euclidean geometry are unaffected by (*invariant* with respect to) translations and rotations of coordinate systems (see also Secs. 12.1-5 and 14.1-4). *Refer to Chap.* 14 *for transformation laws of vector components and base vectors.*

3.1-13. Analytical Representation of Curves (see also Secs. 17.2-1 to 17.2-6). A **continuous curve** in three-dimensional Euclidean space is a set of points $(x, y, z) \equiv (\mathbf{r})$ whose coordinates satisfy a system of parametric equations

$$\begin{aligned} x &= x(t) \\ y &= y(t) \qquad \text{or} \qquad \mathbf{r} = \mathbf{r}(t) \\ z &= z(t) \\ &(-\infty \leq t_1 \leq t \leq t_2 \leq \infty) \end{aligned} \tag{3.1-24}$$

where $x(t)$, $y(t)$, $z(t)$ are continuous functions of the real parameter t throughout the closed interval $[t_1, t_2]$ (**parametric representation of a curve**). Alternatively, a curve may be defined by an equivalent set of two equations

$$\varphi_1(x, y, z) = 0 \qquad \varphi_2(x, y, z) = 0 \tag{3.1-25}$$

A curve can have more than one branch; branches may or may not be connected.

A **simple curve** (simple arc, simple segment) is a (portion of a) continuous curve consisting of a single branch without multiple points, so that the functions (24) are single-valued, and

$$x(\tau_1) = x(\tau_2) \qquad y(\tau_1) = y(\tau_2) \qquad z(\tau_1) = z(\tau_2) \tag{3.1-26}$$

is not satisfied by any pair of values $\tau_1 \neq \tau_2$ of t in the closed interval $[t_1, t_2]$. A **simple closed curve** is a continuous curve without multiple points except for a common initial and terminal point; i.e., the only solutions of Eq. (26) in the closed interval $[t_1, t_2]$ are $\tau_1 = t_1$, $\tau_2 = t_2$. A **regular arc** is a continuous curve which can be represented in terms of some parameter t so that every point of the curve is a **regular point** where $x(t)$, $y(t)$, $z(t)$ have unique continuous derivatives (Sec. 4.5-1) not all equal to zero. A **regular curve** is a simple curve or a simple closed curve composed of a finite number of regular arcs. Such a curve is also called **piecewise smooth.**

NOTE: Analogous definitions apply to curves in two-dimensional spaces (Secs. 2.1-9 and 7.2-1) and also to curves in spaces of dimensionality higher than three (Sec. 17.4-2).

3.1-14. Representation of Surfaces (see also Secs. 17.3-1 to 17.3-13). The coordinates of each point $(x, y, z) \equiv (\mathbf{r})$ on a **continuous surface** in three-dimensional Euclidean space satisfy a set of parametric equations

$$x = x(u, v) \qquad y = y(u, v) \qquad z = z(u, v) \qquad \text{or} \qquad \mathbf{r} = \mathbf{r}(u, v) \tag{3.1-27}$$

for suitable ranges of the real parameters u, v; the functions (27) are continuous (*parametric representation of a surface*). A surface can also be defined by an equation

$$\varphi(x, y, z) = 0 \qquad \text{or} \qquad z = f(x, y) \tag{3.1-28}$$

A surface can have more than one sheet; sheets may or may not be connected. *The surfaces corresponding to the equations*

$$\varphi(x, y, z) = 0 \qquad \text{and} \qquad \lambda\varphi(x, y, z) = 0 \tag{3.1-29}$$

are identical for any constant λ *different from zero.*

A **simple surface (simple surface element)** is a (portion of a) continuous surface consisting of a single sheet without multiple points (Sec. 3.1-13). A **simple closed surface** is a continuous surface which consists of a single sheet and has no multiple points except for the points of one simple closed curve (Sec. 3.1-13); a simple closed surface bounds a connected region of space (Sec. 4.3-6b). Simple surfaces and simple closed surfaces are understood to be *two-sided* (one-sided surfaces such as the Moebius strip are excluded). A **regular point** of a surface (27) is a surface point where, for some choice of the parameters u, v, the functions (27) have continuous partial derivatives such that at least one of the determinants

$$\begin{vmatrix} \frac{\partial x}{\partial u} & \frac{\partial y}{\partial u} \\ \frac{\partial x}{\partial v} & \frac{\partial y}{\partial v} \end{vmatrix} \qquad \begin{vmatrix} \frac{\partial y}{\partial u} & \frac{\partial z}{\partial u} \\ \frac{\partial y}{\partial v} & \frac{\partial z}{\partial v} \end{vmatrix} \qquad \begin{vmatrix} \frac{\partial z}{\partial u} & \frac{\partial x}{\partial u} \\ \frac{\partial z}{\partial v} & \frac{\partial x}{\partial v} \end{vmatrix} \tag{3.1-30}$$

is different from zero. A **regular surface element** is a simple surface element which has only regular points and is bounded by a regular closed curve (Sec. 3.1-13). A **regular surface** is a two-sided simple surface or simple closed surface comprising a finite number of regular surface elements joined along regular arcs (edges) and/or points (vertices). Such a surface is also called **piecewise smooth.**

3.1-15. Special Types of Surfaces. A **ruled surface** is a surface generated by the displacement of a straight line (**generator, generatrix**). A **cylinder** is a surface generated by the displacement of a straight line parallel to itself along a directing curve (**directrix**). In particular, equations of the form $f(x, y) = 0$, $f(x, z) = 0$, $f(y, z) = 0$ describe cylinders whose generators are perpendicular to the xy plane, the xz plane, and the yz plane, respectively. A **cone** is a surface generated by a straight line through a fixed point (**vertex**) and a point of a directing curve (directrix). A **surface of revolution** is the locus of the points of a curve rotated about a straight line (**axis of revolution**). Thus, $f(\sqrt{x^2 + y^2}, z) = 0$ describes a surface of revolution about the z axis.

3.1-16. Surfaces and Curves. (a) The **intersection** of two surfaces

$$\varphi_1(x, y, z) = 0 \qquad \varphi_2(x, y, z) = 0 \tag{3.1-31}$$

is the locus of the points (x, y, z) which satisfy both equations (31) (see also Sec. 3.1-13). A curve of intersection can have more than one branch; or it may degenerate into a set of points where the two surfaces (31) touch.

Specifically, the curves

$$\varphi(0, y, z) = 0, x = 0 \qquad \varphi(x, 0, z) = 0, y = 0 \qquad \varphi(x, y, 0) = 0, z = 0$$

represent the intersections (if they exist) of the surface described by $\varphi(x, y, z) = 0$ with the yz, xz, and xy planes, respectively.

(b) For any real number λ, the equation

$$\varphi_1(x, y, z) + \lambda\varphi_2(x, y, z) = 0 \tag{3.1-32}$$

corresponds to a surface passing through all points of the curve of intersection of the two surfaces (31) if that curve exists.

(c) Given two surfaces corresponding to the two equations (31), the equation

$$\varphi_1(x, y, z)\varphi_2(x, y, z) = 0$$

corresponds to the surface made up of all the points of both original surfaces, and no other points.

(d) The respective equations of the *projections* of a curve described by the equations (31) on the yz, xz, xy planes are obtained by eliminating x, y, z, respectively, from the equations (31).

(e) The coordinates x, y, z of the *point(s) of intersection* of a curve described by Eq. (31) with a surface described by Eq. (28) satisfy all three equations (28) and (31). An *n^{th}-order curve* intersects a plane (Sec. 3.2-1) in n points, some of which may coincide and/or be imaginary. For any real λ_1, λ_2, the equations

$$\varphi(x, y, z) + \lambda_1\varphi_1(x, y, z) = 0 \qquad \varphi(x, y, z) + \lambda_2\varphi_2(x, y, z) = 0 \tag{3.1-33}$$

represent a curve passing through all these points of intersection (if they exist).

(f) A suitable **one-parameter family of curves** described by

$$\varphi_1(x, y, z, \lambda) = 0 \qquad \varphi_2(x, y, z, \lambda) = 0 \tag{3.1-34}$$

generates a surface whose equation may be obtained by eliminating λ from the equations (34).

3.2. THE PLANE

3.2-1. Equation of a Plane. (a) Given a right-handed rectangular cartesian coordinate system, an equation *linear* in x, y, z, i.e., a relation of the general form

$$\boxed{Ax + By + Cz + D = 0 \qquad \text{or} \qquad \mathbf{A}\cdot\mathbf{r} + D = 0} \tag{3.2-1}$$

where A, B, and C must not all vanish simultaneously, represents a **plane;** conversely, every plane situated in the finite portion of space can be represented by a linear equation of the form (1).

A, B, C are direction numbers (Sec. 3.1-8*b*) of the (positive or negative; see below) *normal* to the plane; the vector $\mathbf{A} \equiv (A, B, C)$ is directed along the normal. The special case $D = 0$ corresponds to a plane through the origin.

(b) The following *special forms* of the equation of the plane are of interest:

1. **Intercept form.** Plane intercepting the x axis at $x = a$, the y axis at $y = b$, and the z axis at $z = c$:

$$\frac{x}{a} + \frac{y}{b} + \frac{z}{c} - 1 = 0$$

2. **Normal form.** Let $p > 0$ be the length of the directed perpendicular between the origin and a plane; let $\cos \alpha_x$, $\cos \alpha_y$, $\cos \alpha_z$ be the direction cosines (Sec. 3.1-8a) of this directed perpendicular, which is taken to be the *positive* normal to the plane (see also Sec. 17.3-2). Then the equation of the plane is

$$x \cos \alpha_x + y \cos \alpha_y + z \cos \alpha_z - p = 0$$

3. **Point-direction form.** Plane through the point $(x_1, y_1, z_1) \equiv (\mathbf{r}_1)$; the normal has the direction numbers A, B, C:

$$A(x - x_1) + B(y - y_1) + C(z - z_1) = 0 \quad \text{or} \quad \mathbf{A} \cdot (\mathbf{r} - \mathbf{r}_1) = 0$$

4. **Three-point form.** Plane through the three points $P_1 \equiv (x_1, y_1, z_1) \equiv (\mathbf{r}_1)$, $P_2 \equiv (x_2, y_2, z_2) \equiv (\mathbf{r}_2)$, $P_3 \equiv (x_3, y_3, z_3) \equiv (\mathbf{r}_3)$, which must not lie in a straight line (Sec. 3.4-3a):

$$\begin{vmatrix} x & y & z & 1 \\ x_1 & y_1 & z_1 & 1 \\ x_2 & y_2 & z_2 & 1 \\ x_3 & y_3 & z_3 & 1 \end{vmatrix} = 0 \quad \text{or} \quad [(\mathbf{r} - \mathbf{r}_1)(\mathbf{r} - \mathbf{r}_2)(\mathbf{r} - \mathbf{r}_3)] = 0$$

$$\text{or} \quad \begin{vmatrix} y_1 & z_1 & 1 \\ y_2 & z_2 & 1 \\ y_3 & z_3 & 1 \end{vmatrix} x + \begin{vmatrix} z_1 & x_1 & 1 \\ z_2 & x_2 & 1 \\ z_3 & x_3 & 1 \end{vmatrix} y + \begin{vmatrix} x_1 & y_1 & 1 \\ x_2 & y_2 & 1 \\ x_3 & y_3 & 1 \end{vmatrix} z - \begin{vmatrix} x_1 & y_1 & z_1 \\ x_2 & y_2 & z_2 \\ x_3 & y_3 & z_3 \end{vmatrix} = 0$$

(c) When the equation of a plane is given in the general form (1), the quantities a, b, c, p, $\cos \alpha_x$, $\cos \alpha_y$, $\cos \alpha_z$ defined above are related to the parameters A, B, C, D as follows:

$$a = -\frac{D}{A} \qquad b = -\frac{D}{B} \qquad c = -\frac{D}{C} \tag{3.2-2}$$

$$\left.\begin{aligned} \cos \alpha_x &= \frac{A}{(A^2 + B^2 + C^2)^{1/2}} \qquad & \cos \alpha_y &= \frac{B}{(A^2 + B^2 + C^2)^{1/2}} \\ \cos \alpha_z &= \frac{C}{(A^2 + B^2 + C^2)^{1/2}} \qquad & p &= -\frac{D}{(A^2 + B^2 + C^2)^{1/2}} \end{aligned}\right\} \tag{3.2-3}$$

where the sign of the square root is chosen so that $p > 0$.

3.2-2. Parametric Representation of a Plane. The parametric representation (Sec. 3.1-14) of any plane has the form

$$\begin{array}{l} x = x_1 + a_x u + b_x v \qquad y = y_1 + a_y u + b_y v \qquad z = z_1 + a_z u + b_z v \\ \text{or} \qquad \mathbf{r} = \mathbf{r}_1 + u\mathbf{a} + v\mathbf{b} \end{array} \tag{3.2-4}$$

3.3. THE STRAIGHT LINE

3.3-1. Equations of the Straight Line. **(a)** Two linearly independent linear equations

$$\begin{array}{l} \varphi_1(x, y, z) \equiv A_1 x + B_1 y + C_1 z + D_1 = 0 \\ \varphi_2(x, y, z) \equiv A_2 x + B_2 y + C_2 z + D_2 = 0 \\ \text{or} \qquad \mathbf{A}_1 \cdot \mathbf{r} + D_1 = 0 \qquad \mathbf{A}_2 \cdot \mathbf{r} + D_2 = 0 \end{array} \tag{3.3-1}$$

($\mathbf{A}_1 \times \mathbf{A}_2 \neq 0$; see also Secs. 1.9-3*a*, 5.2-2, and 5.2-7) represent a **straight line** (intersection of two planes, Sec. 3.1-16*a*). Conversely, every straight line situated in the finite portion of space can be represented in the form (1). Equation (1) represents a straight line through the origin if and only if $D_1 = D_2 = 0$.

(b) The following *special forms* of the equations of a straight line are of interest:

1. **Two-point form.** Straight line through the points $(x_1\ y_1, z_1)$ and (x_2, y_2, z_2):

$$\frac{x - x_1}{x_2 - x_1} = \frac{y - y_1}{y_2 - y_1} = \frac{z - z_1}{z_2 - z_1}$$

2. **Point-direction form or symmetric form.** Straight line with direction numbers a_x, a_y, a_z (Sec. 3.1-8*b*) through the point $(x_1, x_2, x_3) \equiv (\mathbf{r}_1)$ (see also Sec. 3.3-2):

$$\frac{x - x_1}{a_x} = \frac{y - y_1}{a_y} = \frac{z - z_1}{a_z} \qquad \text{or} \qquad (\mathbf{r} - \mathbf{r}_1) = \mathbf{a}t$$

(c) The quantities

$$\left.\begin{array}{l} a_x = B_1 C_2 - B_2 C_1 \qquad a_y = C_1 A_2 - C_2 A_1 \qquad a_z = A_1 B_2 - A_2 B_1 \\ \text{(RECTANGULAR CARTESIAN COMPONENTS OF } \mathbf{a} = \mathbf{A}_1 \times \mathbf{A}_2) \end{array}\right\} \tag{3.3-2}$$

are direction numbers (Sec. 3.1-8*b*) of the straight line described by Eq. (1) [*direction numbers of the line of intersection of the two planes* (1)]. The corresponding direction cosines $\cos \alpha_x$, $\cos \alpha_y$, $\cos \alpha_z$, and thus the angles α_x, α_y, α_z between the straight line and the x, y, and z axes, are given by

$$\cos \alpha_x = \frac{1}{M}(B_1C_2 - B_2C_1) \qquad \cos \alpha_y = \frac{1}{M}(C_1A_2 - C_2A_1)$$
$$\cos \alpha_z = \frac{1}{M}(A_1B_2 - A_2B_1)$$
$$M = [(B_1C_2 - B_2C_1)^2 + (C_1A_2 - C_2A_1)^2 + (A_1B_2 - A_2B_1)^2]^{1/2} \qquad (3.3\text{-}3)$$

The (arbitrary) sign of the square root determines the positive direction of the straight line.

Equations (2) and (3) may also be interpreted as yielding the direction numbers and direction cosines of *the straight line normal to two directions described by their respective direction numbers* A_1, B_1, C_1 and A_2, B_2, C_2.

(**d**) The equations of the *planes projecting the straight line* (1) *onto the xy plane, the xz plane, and the yz plane* (i.e., planes through the straight line and perpendicular to the respective coordinate planes) are, respectively,

$$\left.\begin{aligned}(C_1A_2 - C_2A_1)x + (C_1B_2 - C_2B_1)y + (C_1D_2 - C_2D_1) &= 0\\ (B_1A_2 - B_2A_1)x + (B_1C_2 - B_2C_1)z + (B_1D_2 - B_2D_1) &= 0\\ (A_1B_2 - A_2B_1)y + (A_1C_2 - A_2C_1)z + (A_1D_2 - A_2D_1) &= 0\end{aligned}\right\} \qquad (3.3\text{-}4)$$

Any two of the equations (4) describe the straight line (1).

3.3-2. Parametric Representation of a Straight Line. The rectangular cartesian coordinates x, y, z of a point on a straight line satisfy the parametric equations (Sec. 3.1-13)

$$x = x_1 + a_xt \qquad y = y_1 + a_yt \qquad z = z_1 + a_zt \qquad \text{or} \qquad \mathbf{r} = \mathbf{r}_1 + t\mathbf{a} \qquad (3.3\text{-}5)$$

[*straight line through the point* ($\mathbf{r}_1$) *in the direction of the vector* $\mathbf{a}$].

3.4. RELATIONS INVOLVING POINTS, PLANES, AND STRAIGHT LINES

3.4-1. Angles. (**a**) *The angle* γ_1 *between two straight lines* having the direction cosines $\cos \alpha_x$, $\cos \alpha_y$, $\cos \alpha_z$ and $\cos \alpha_x'$, $\cos \alpha_y'$, $\cos \alpha_z'$ is given by (see also Sec. 3.1-8*c*)

$$\cos \gamma_1 = \cos \alpha_x \cos \alpha_x' + \cos \alpha_y \cos \alpha_y' + \cos \alpha_z \cos \alpha_z' \qquad (3.4\text{-}1)$$
$$\begin{aligned}\sin^2 \gamma_1 = (\cos \alpha_y \cos \alpha_z' - \cos \alpha_z \cos \alpha_y')^2 + (\cos \alpha_z \cos \alpha_x'\\ - \cos \alpha_x \cos \alpha_z')^2 + (\cos \alpha_x \cos \alpha_y' - \cos \alpha_y \cos \alpha_x')^2\end{aligned} \qquad (3.4\text{-}2)$$

If the straight lines are given in the parametric form (Sec. 3.3-2) $\mathbf{r} = \mathbf{r}_1 + t\mathbf{a}$, $\mathbf{r} = \mathbf{r}_1' + t\mathbf{a}'$,

$$\cos \gamma_1 = \frac{\mathbf{a} \cdot \mathbf{a}'}{|\mathbf{a}|\,|\mathbf{a}'|} \qquad |\sin \gamma_1| = \frac{|\mathbf{a} \times \mathbf{a}'|}{|\mathbf{a}|\,|\mathbf{a}'|} \qquad (3.4\text{-}3)$$

The two straight lines are *parallel* if $\cos \gamma_1 = 1$ and *mutually perpendicular* if $\cos \gamma_1 = 0$.

(b) *The angle γ_2 between (the normals of) two planes $Ax + By + Cz + D = 0$ and $A'x + B'y + C'z + D' = 0$ or $\mathbf{A} \cdot \mathbf{r} + \mathbf{D} = 0$ and $\mathbf{A}' \cdot \mathbf{r} + \mathbf{D}' = 0$ is given by*

$$\cos \gamma_2 = \frac{AA' + BB' + CC'}{\sqrt{A^2 + B^2 + C^2}\sqrt{A'^2 + B'^2 + C'^2}} = \frac{\mathbf{A} \cdot \mathbf{A}}{|\mathbf{A}|\,|\mathbf{A}'|} \tag{3.4-4}$$

If the planes are given in the parametric form (Sec. 3.2-2) $\mathbf{r} = \mathbf{r}_1 + u^1\mathbf{a} + u^2\mathbf{b}$, $\mathbf{r} = \mathbf{r}_1' + u^1\mathbf{a}' + u^2\mathbf{b}'$, then

$$\cos \gamma_2 = \frac{(\mathbf{a} \times \mathbf{b}) \cdot (\mathbf{a}' \times \mathbf{b}')}{|\mathbf{a} \times \mathbf{b}|\,|\mathbf{a}' \times \mathbf{b}'|} \tag{3.4-5}$$

In particular, the two planes are *parallel* if $\cos \gamma_2 = 1$, and *mutually perpendicular* if $\cos \gamma_2 = 0$.

(c) The angle γ_3 between the straight line

$$(x - x_1)/\cos \alpha_x = (y - y_1)/\cos \alpha_y = (z - z_1)/\cos \alpha_z$$

and (its projection on) the plane $Ax + By + Cz + D = 0$ is given by

$$\sin \gamma_3 = \frac{A \cos \alpha_x + B \cos \alpha_y + C \cos \alpha_z}{(A^2 + B^2 + C^2)^{1/2}} \tag{3.4-6}$$

In particular, the straight line is *parallel to the plane* if $\sin \gamma_3 = 0$ (and lies in the plane if, in addition, $Ax_1 + Bx_1 + Cx_1 + D = 0$) and *perpendicular to the plane* if $\sin \gamma_3 = 1$. The angle between the straight line and the normal to the plane equals $90 \text{ deg} - \gamma_3$.

3.4-2. Distances. **(a)** *Distance d_0 between the point $(x_0, y_0, z_0) = (\mathbf{r}_0)$ and the plane $Ax + By + Cz + D = 0$ or $\mathbf{A} \cdot \mathbf{r} + D = 0$:*

$$d_0 = \frac{Ax_0 + By_0 + Cz_0 + D}{(A^2 + B^2 + C^2)^{1/2}} = \frac{\mathbf{A} \cdot \mathbf{r}_0 + D}{\pm|\mathbf{A}|} \tag{3.4-7}$$

where the sign of the square root is chosen to be opposite to that of D. d_0 is positive if the plane lies between the origin and the point (x_0, y_0, z_0).

If the equation of the plane is given in the parametric form (Sec. 3.2-2) $\mathbf{r} = \mathbf{r}_1 + u\mathbf{a} + v\mathbf{b}$,

$$d_0 = (\mathbf{r}_1 - \mathbf{r}_0) \cdot \frac{\mathbf{a} \times \mathbf{b}}{|\mathbf{a} \times \mathbf{b}|} \tag{3.4-8}$$

(b) *Distance d_0' between the point $(x_0, y_0, z_0) = (\mathbf{r}_0)$ and the straight line $(x - x_1)/\cos \alpha_x = (y - y_1)/\cos \alpha_y = (z - z_1)/\cos \alpha_z$:*

$$d_0' = +\{(x_0 - x_1)^2 + (y_0 - y_1)^2 + (z_0 - z_1)^2 - [(x_0 - x_1)\cos \alpha_x + (y_0 - y_1)\cos \alpha_y + (z_0 - z_1)\cos \alpha_z]^2\}^{1/2} \tag{3.4-9}$$

(c) *(Shortest) distance d_1 between the straight lines $(x - x_1)/\cos \alpha_x = (y - y_1)/\cos \alpha_y = (z - z_1)/\cos \alpha_z$ and $(x - x_1')/\cos \alpha_x' = (y - y_1')/\cos \alpha_y' = (z - z_1')/\cos \alpha_z'$:*

$$d_1 = \left| \frac{1}{\sin \gamma_1} \begin{vmatrix} x_1 - x_1' & \cos \alpha_x & \cos \alpha_x' \\ y_1 - y_1' & \cos \alpha_y & \cos \alpha_y' \\ z_1 - z_1' & \cos \alpha_z & \cos \alpha_z' \end{vmatrix} \right| \tag{3.4-10}$$

where $\sin \gamma_1$ is given by Eq. (2). If the lines are parallel ($\sin \gamma_1 = 0$) d_1 is given by Eq. (9).

NOTE: If x_1, y_1, z_1 are replaced by variable coordinates x, y, z, Eq. (10) describes the plane through the first straight line and parallel to the second. If the two straight lines intersect, this plane contains both of them.

(**d**) *Distance d_2 between two parallel planes $Ax + By + Cz + D = 0$, $Ax + By + Cz + D' = 0$:*

$$d_2 = \left| \frac{D - D'}{(A^2 + B^2 + C^2)^{1/2}} \right| = \frac{|D - D'|}{|\mathbf{A}|} \tag{3.4-11}$$

(**e**) *The distance d_3 between a plane and a parallel straight line*

$$\frac{x - x_0}{\cos \alpha_x} = \frac{y - y_0}{\cos \alpha_y} = \frac{z - z_0}{\cos \alpha_z}$$

is given by Eq. (7).

3.4-3. Special Conditions. (**a**) Note the following special conditions about points:

1. *Three points (x_1, y_1, z_1), (x_2, y_2, z_2), (x_3, y_3, z_3) lie on a straight line* (are **collinear**) if and only if

$$\frac{x_3 - x_1}{x_2 - x_1} = \frac{y_3 - y_1}{y_2 - y_1} = \frac{z_3 - z_1}{z_2 - z_1} \quad \text{or} \quad (\mathbf{r}_3 - \mathbf{r}_1) = \lambda(\mathbf{r}_2 - \mathbf{r}_1) \tag{3.4-12}$$

($\lambda \neq 0$), or [see also Eq. (3.1-15)]

$$\begin{vmatrix} y_1 & z_1 & 1 \\ y_2 & z_2 & 1 \\ y_3 & z_3 & 1 \end{vmatrix} = \begin{vmatrix} z_1 & x_1 & 1 \\ z_2 & x_2 & 1 \\ z_3 & x_3 & 1 \end{vmatrix} = \begin{vmatrix} x_1 & y_1 & 1 \\ x_2 & y_2 & 1 \\ x_3 & y_3 & 1 \end{vmatrix} = 0 \tag{3.4-13}$$

or

$$(\mathbf{r}_2 - \mathbf{r}_1) \times (\mathbf{r}_3 - \mathbf{r}_1) = 0 \tag{3.4-14}$$

2. *Four points (x_1, y_1, z_1), (x_2, y_2, z_2), (x_3, y_3, z_3), (x_4, y_4, z_4) lie in a plane* if and only if [see also Eq. (3.1-17) and Sec. 3.2-1*b*]

$$\begin{vmatrix} x_1 & y_1 & z_1 & 1 \\ x_2 & y_2 & z_2 & 1 \\ x_3 & y_3 & z_3 & 1 \\ x_4 & y_4 & z_4 & 1 \end{vmatrix} = 0 \quad \text{or} \quad [(\mathbf{r}_2 - \mathbf{r}_1)(\mathbf{r}_3 - \mathbf{r}_1)(\mathbf{r}_4 - \mathbf{r}_1)] = 0 \tag{3.4-15}$$

(**b**) Note the following special conditions about planes:

1. *Three planes $Ax + By + Cz + D = 0$, $A'x + B'y + C'z + D' = 0$, $A''x + B''y + C''z + D'' = 0$ intersect in a straight line* if

$$\Delta = \begin{vmatrix} A & B & C \\ A' & B' & C' \\ A'' & B'' & C'' \end{vmatrix} = 0 \tag{3.4-16}$$

unless two of them are parallel. Equation (16) implies that the equations of the three planes are linearly dependent (Sec. 1.9-3*a*).

2. *Four planes* $Ax + By + Cz + D = 0$, $A'x + B'y + C'z + D' = 0$, $A''x + B''y + C''z + D'' = 0$, $A'''x + B'''y + C'''z + D''' = 0$ *intersect in a point* (*or are parallel*) if and only if

$$\begin{vmatrix} A & B & C & D \\ A' & B' & C' & D' \\ A'' & B'' & C'' & D'' \\ A''' & B''' & C''' & D''' \end{vmatrix} = 0 \qquad (3.4\text{-}17)$$

i.e., if the equations of the four planes are linearly dependent (Sec. 1.9-3*a*).

(c) Note the following conditions about straight lines:

1. *Two straight lines lie in a plane* (*i.e., they intersect or are parallel*) if and only if four planes determining them intersect in a point [Eq. (17)].
2. *Three straight lines intersect in a point* (or at least two of them are parallel) if every pair of the three lies in a plane.

3.4-4. Plane Coordinates and the Principle of Duality. The equation

$$\xi x + \eta y + \zeta z + 1 = 0 \qquad (3.4\text{-}18)$$

describes a plane (ξ, η, ζ) "labeled" by the **plane coordinates** ξ, η, ζ. If the point coordinates x, y, z are considered as constant parameters and the plane coordinates ξ, η, ζ as variables, Eq. (18) may be interpreted as the *equation of the point* (x, y, z) [point of intersection of all planes (18)]. Note the *principle of duality* (see also Sec. 2.3-3): *to every theorem involving only the relative positions of points, planes, and straight lines, there corresponds another theorem obtained by interchanging the terms "point" and "plane" in the original theorem.*

3.4-5. Miscellaneous Relations.

(a) If $\varphi_1(x, y, z) = 0$ and $\varphi_2(x, y, z) = 0$ are the equations of two planes, the equation

$$\varphi_1(x, y, z) + \lambda\varphi_2(x, y, z) = 0 \qquad (3.4\text{-}19)$$

describes a plane through their line of intersection (or parallel to both if they are parallel). *If the equations of the first two planes are given in the normal form* (Sec. 3.2-1*b*), then $-\lambda$ is the ratio of the respective distances (7) between the first and second plane and any one point on the third plane; the planes (19) corresponding to $\lambda = 1$ and $\lambda = -1$ *bisect the angles* between the given planes.

(b) The equation of the *normal* through the point (x_0, y_0, z_0) to the plane $Ax + By + Cz + D = 0$ is

$$\frac{x - x_0}{A} = \frac{y - y_0}{B} = \frac{z - z_0}{C} \qquad (3.4\text{-}20)$$

(c) *Direction numbers and direction cosines of the line of intersection of two planes* are given by Eqs. (3.3-2) and (3.3-3).

(d) The *point of intersection of three planes* $Ax + By + Cz + D = 0$,

$A'x + B'y + C'z + D' = 0, \quad A''x + B''y + C''z + D'' = 0$ has the coordinates

$$x = -\frac{1}{\Delta}\begin{vmatrix} D & B & C \\ D' & B' & C' \\ D'' & B'' & C'' \end{vmatrix} \qquad y = -\frac{1}{\Delta}\begin{vmatrix} A & D & C \\ A' & D' & C' \\ A'' & D'' & C'' \end{vmatrix}$$

$$z = -\frac{1}{\Delta}\begin{vmatrix} A & B & D \\ A' & B' & D' \\ A'' & B'' & D'' \end{vmatrix} \qquad \text{with} \qquad \Delta = \begin{vmatrix} A & B & C \\ A' & B' & C' \\ A'' & B'' & C'' \end{vmatrix} \tag{3.4-21}$$

which may also be interpreted as the coordinates of the point of intersection (piercing point) of a plane and a straight line.

3.5. QUADRIC SURFACES

3.5-1. General Second-degree Equation. The following sections deal with the **quadric surfaces** represented by the **general second-degree equation**

$$\begin{gathered} a_{11}x^2 + a_{22}y^2 + a_{33}z^2 + 2a_{12}xy + 2a_{13}xz + 2a_{23}yz \\ + 2a_{14}x + 2a_{24}y + 2a_{34}z + a_{44} = 0 \\ \text{or} \\ (a_{11}x + a_{12}y + a_{13}z + a_{14})x \\ + (a_{21}x + a_{22}y + a_{23}z + a_{24})y \\ + (a_{31}x + a_{32}y + a_{33}z + a_{34})z \\ + (a_{41}x + a_{42}y + a_{43}z + a_{44}) = 0 \\ \text{where} \qquad a_{ik} = a_{ki};\ i, k = 1, 2, 3, 4 \end{gathered} \tag{3.5-1}$$

In vector form, Eq. (1) becomes

$$(\mathbf{Ar}) \cdot \mathbf{r} + 2\mathbf{a} \cdot \mathbf{r} + a_{44} = 0 \tag{3.5-2}$$

where the tensor A has the components $A_k^i = a_{ik}$, and the vector $\mathbf{a}$ has the components $a_i = a_{i4}$ (see also Sec. 16.9-2).

3.5-2. Invariants. For any equation (1), the four quantities

$$I = a_{11} + a_{22} + a_{33} \qquad J = \begin{vmatrix} a_{11} & a_{12} \\ a_{21} & a_{22} \end{vmatrix} + \begin{vmatrix} a_{22} & a_{23} \\ a_{32} & a_{33} \end{vmatrix} + \begin{vmatrix} a_{33} & a_{31} \\ a_{13} & a_{11} \end{vmatrix}$$

$$D = \mathrm{A}_{44} = \begin{vmatrix} a_{11} & a_{12} & a_{13} \\ a_{21} & a_{22} & a_{23} \\ a_{31} & a_{32} & a_{33} \end{vmatrix} \qquad A = \begin{vmatrix} a_{11} & a_{12} & a_{13} & a_{14} \\ a_{21} & a_{22} & a_{23} & a_{24} \\ a_{31} & a_{32} & a_{33} & a_{34} \\ a_{41} & a_{42} & a_{43} & a_{44} \end{vmatrix} \tag{3.5-3}$$

and the *signs* of the quantities*

* A_{ik} denotes the cofactor of a_{ik} in the fourth-order determinant $A = \det |a_{ik}|$ (Sec. 1.5-2).

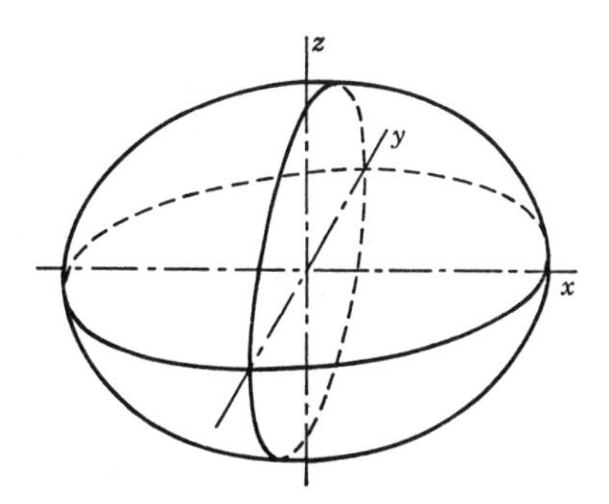

Ellipsoid: $\frac{x^2}{a^2}+\frac{y^2}{b^2}+\frac{z^2}{c^2}=1$

(*a*)

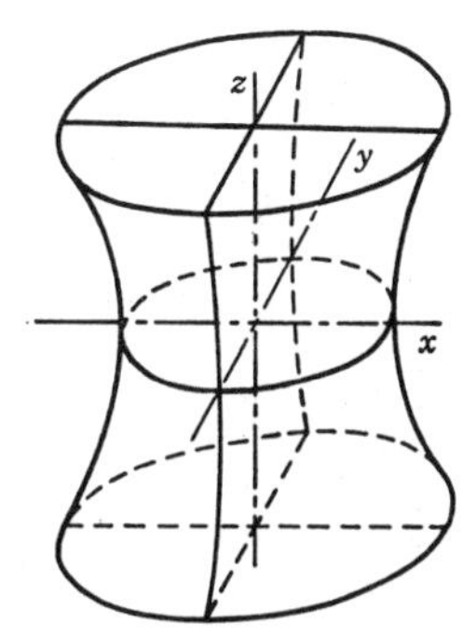

Hyperboloid of one sheet: $\frac{x^2}{a^2}+\frac{y^2}{b^2}-\frac{z^2}{c^2}=1$

(*b*)

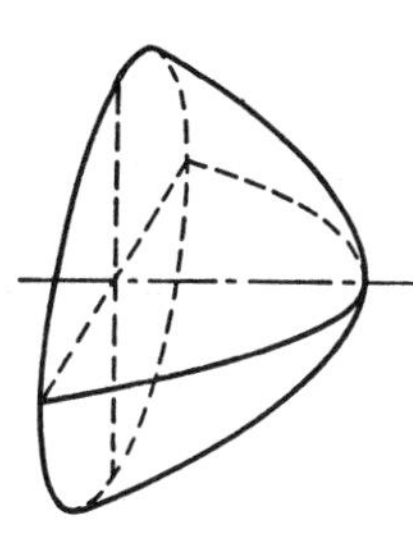

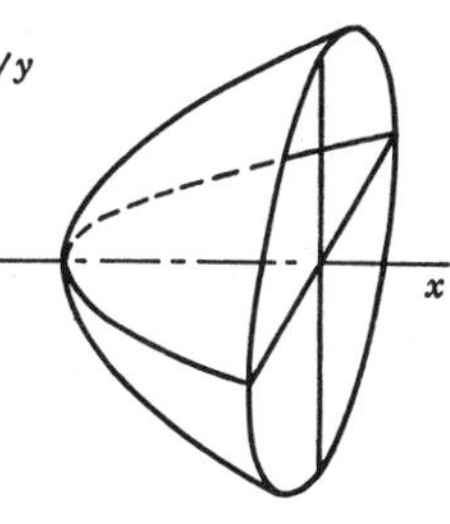

Hyperboloid of two sheets: $\frac{x^2}{a^2}-\frac{y^2}{b^2}-\frac{z^2}{c^2}=1$

(*c*)

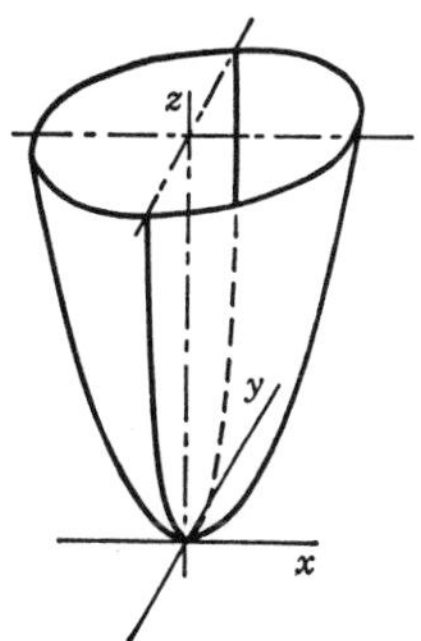

Elliptic paraboloid: $\frac{x^2}{a^2}+\frac{y^2}{b^2}=z$

(*d*)

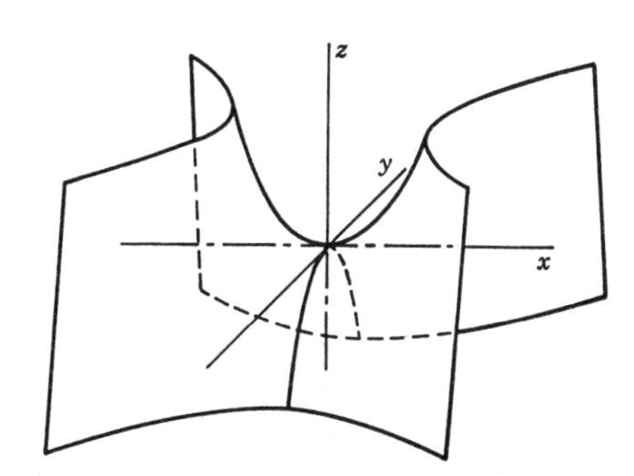

Hyperbolic paraboloid: $\frac{x^2}{a^2}-\frac{y^2}{b^2}=z$

(*e*)

FIG. 3.5-1. Proper quadric surfaces.

Table 3.5-1. Classification of Quadric Surfaces (Quadrics)

		Proper quadrics $A \neq 0$			**Improper (degenerate) quadrics** $A = 0$				
		$A > 0$		$A < 0$	**Cones and cylinders** $A' \neq 0$		**Pairs of planes** (degenerate quadrics) $A' = 0$		
		$A'I$ and J both > 0	$A'I$ and J not both > 0		$A'I$ and J both > 0	$A'I$ and J not both > 0	$A'' > 0$	$A'' < 0$	$A'' = 0$, $A''' \neq 0$
Central quadrics $D \neq 0$	DI and J both > 0	**No real locus** (imaginary ellipsoid)	....	**Real ellipsoid**	**Point** (in finite portion of space; vertex of imaginary elliptic cone; point ellipsoid)	....	....	....	....
	DI and J not both > 0	....	**Hyperboloid of one sheet**	**Hyperboloid of two sheets**	....	**Elliptic cone** (degenerate hyperboloid)	....	....	....
Non-central quadrics $D = 0$	$J > 0$	....	....	**Elliptic paraboloid** (of revolution if $I^2 = 4J$)	**No real locus** (imaginary elliptic cylinder)	**Real elliptic cylinder** (circular if $I^2 = 4J$)	**Straight line** (degenerate elliptic cylinder; intersection of imaginary planes)	....	....
	$J < 0$	....	**Hyperbolic paraboloid**	....	....	**Hyperbolic cylinder**	....	**Two real planes** intersecting in finite portion of space (degenerate hyperbolic cylinder	....
	$J = 0$, $I \neq 0$	....	....	....	....	**Parabolic cylinder**	**No real locus** (imaginary parallel planes)	**Two real parallel planes** (degenerate parabolic cylinder)	**One real plane** (coincident parallel planes)
Rank of the 4th-order square matrix $[a_{ik}]$ (Sec. 13.2-7)		4			3		2		1

$$\left.\begin{aligned} A' &= A_{11} + A_{22} + A_{33} + A_{44} \\ A'' &= \begin{vmatrix} a_{11} & a_{12} \\ a_{21} & a_{22} \end{vmatrix} + \begin{vmatrix} a_{11} & a_{13} \\ a_{31} & a_{33} \end{vmatrix} + \begin{vmatrix} a_{11} & a_{14} \\ a_{41} & a_{44} \end{vmatrix} + \begin{vmatrix} a_{22} & a_{23} \\ a_{32} & a_{33} \end{vmatrix} \\ &\qquad + \begin{vmatrix} a_{22} & a_{24} \\ a_{42} & a_{44} \end{vmatrix} + \begin{vmatrix} a_{33} & a_{34} \\ a_{43} & a_{44} \end{vmatrix} \\ A''' &= a_{11} + a_{22} + a_{33} + a_{44} \end{aligned}\right\} \qquad (3.5\text{-}4)$$

are *invariants* with respect to the translation and rotation transformations (3.1-19), (3.1-20), and (3.1-23) and define properties of the quadric which do not depend on position. The determinant A is called the *discriminant* of Eq. (1).

3.5-3. Classification of Quadrics. Table 3.5-1 shows the classification of quadric surfaces in terms of the invariants defined in Sec. 3.5-2.

3.5-4. Characteristic Quadratic Form and Characteristic Equation. Important properties of quadric surfaces may be studied in terms of the (symmetric) **characteristic quadratic form**

$$F_0(x, y, z) = a_{11}x^2 + a_{22}y^2 + a_{33}z^2 + 2a_{12}xy + 2a_{13}xz + 2a_{23}yz \qquad (3.5\text{-}5)$$

corresponding to Eq. (1). In particular, a proper central quadric ($A \neq 0$, $D \neq 0$) is a real ellipsoid, imaginary ellipsoid, or hyperboloid if $F_0(x, y, z)$ is, respectively, positive definite, negative definite, or indefinite as determined by the (necessarily real) roots $\lambda_1, \lambda_2, \lambda_3$ of the **characteristic equation** (Sec. 13.4-5*a*)

$$\begin{vmatrix} a_{11} - \lambda & a_{12} & a_{13} \\ a_{21} & a_{22} - \lambda & a_{23} \\ a_{31} & a_{32} & a_{33} - \lambda \end{vmatrix} = 0 \qquad \text{or} \qquad \lambda^3 - I\lambda^2 + J\lambda - D = 0 \qquad (3.5\text{-}6)$$

$\lambda_1, \lambda_2, \lambda_3$ are the *eigenvalues* of the matrix corresponding to the quadratic form (5) (Sec. 13.4-2).

3.5-5. Diametral Planes, Diameters, and Centers of Quadric Surfaces. (a) A **diametral plane** of a (proper or improper) quadric surface described by Eq. (1) is the locus of the centers of parallel chords. The diametral plane **conjugate** to (the direction of) the chords having the direction cosines $\cos \alpha_x$, $\cos \alpha_y$, $\cos \alpha_z$ bisects these chords and has the equation

$$\begin{aligned} (a_{11}x + a_{12}y + a_{13}z + a_{14}) \cos \alpha_x + (a_{21}x + a_{22}y + a_{23}z + a_{24}) \cos \alpha_y \\ + (a_{31}x + a_{32}y + a_{33}z + a_{34}) \cos \alpha_z = 0 \end{aligned} \qquad (3.5\text{-}7)$$

(b) The intersection of two diametral planes is called the **diameter** conjugate to the family of planes paralleling the conjugate chords of both diametral planes. The diameter conjugate to the planes whose normals have the direction cosines $\cos \alpha_x$, $\cos \alpha_y$, $\cos \alpha_z$ (Sec. 3.2-1*b*) is represented by

$$\frac{a_{11}x + a_{12}y + a_{13}z + a_{14}}{\cos\alpha_x} = \frac{a_{21}x + a_{22}y + a_{23}z + a_{24}}{\cos\alpha_y} = \frac{a_{31}x + a_{32}y + a_{33}z + a_{34}}{\cos\alpha_z} \tag{3.5-8}$$

(**c**) All diameters of a quadric (1) either intersect at a unique point, the **center** of the quadric (see Sec. 3.5-8*d* for an alternative definition), or they are parallel, according to whether $D \neq 0$ or $D = 0$. In the former case, the quadric is a **central quadric.**

The coordinates x_0, y_0, z_0 of the center are given by

$$\left.\begin{array}{c} a_{11}x_0 + a_{12}y_0 + a_{13}z_0 + a_{14} = 0 \qquad a_{21}x_0 + a_{22}y_0 + a_{23}z_0 + a_{24} = 0 \\ a_{31}x_0 + a_{32}y_0 + a_{33}z_0 + a_{34} = 0 \end{array}\right\} \tag{3.5-9}$$

so that

$$x_0 = -\frac{1}{D}\begin{vmatrix} a_{14} & a_{12} & a_{13} \\ a_{24} & a_{22} & a_{23} \\ a_{34} & a_{32} & a_{33} \end{vmatrix} \qquad y_0 = -\frac{1}{D}\begin{vmatrix} a_{11} & a_{14} & a_{13} \\ a_{21} & a_{24} & a_{23} \\ a_{31} & a_{34} & a_{33} \end{vmatrix} \qquad z_0 = -\frac{1}{D}\begin{vmatrix} a_{11} & a_{12} & a_{14} \\ a_{21} & a_{22} & a_{24} \\ a_{31} & a_{32} & a_{34} \end{vmatrix} \tag{3.5-10}$$

Given the equation (1) of a central quadric, a translation (3.1-19) of the coordinate origin to the center (10) of the quadric results in the new equation

$$a_{11}x^2 + a_{22}y^2 + a_{33}z^2 + 2a_{12}xy + 2a_{13}xz + 2a_{23}yz + \frac{A}{D} = 0 \tag{3.5-11}$$

(**d**) Three diameters of a central quadric are called **conjugate diameters** if and only if each of them is the diameter conjugate to the plane of the two others.

3.5-6. Principal Planes and Principal Axes. (**a**) A diametral plane perpendicular to its conjugate chords is a plane of symmetry or **principal plane (principal diametral plane)** of the quadric; the chords conjugate to a principal plane are called **principal chords.** Every quadric surface has at least two mutually perpendicular principal planes. A central quadric (Sec. 3.5-5*c*) has at least three mutually perpendicular principal planes.

(**b**) The diameter along the line of intersection of two principal planes is a **principal axis** (symmetry axis). Every quadric surface has at least one principal axis; if it has more than one, there exists at least one other principal axis perpendicular to each. A central quadric has at least three mutually perpendicular principal axes, which are necessarily conjugate diameters normal to three corresponding principal planes.

The directions of the normals to the principal planes of a quadric (1), and hence also such principal axes as may exist, are directed along the eigenvectors associated

Table 3.5-2. Equations (in Standard or Type Form) and Principal Properties of Proper Quadric Surfaces
(see also Table 3.5-1; refer to Sec. 3.5-8 for equations of tangent and polar planes)

Surface (see also Fig. 3.5-1)	Equation in standard or type form	Section by plane	Remarks	Parameters a, b, c expressed in terms of the invariants A, D, and λ_1, λ_2, λ_3
(Real) **ellipsoid** [*ellipsoid of revolution* (*spheroid*) if two principal diameters are equal and (*a*) smaller than the third (*prolate spheroid*) or (*b*) greater than the third (*oblate spheroid*); *sphere* if $a^2 = b^2 = c^2$]	$\frac{x^2}{a^2} + \frac{y^2}{b^2} + \frac{z^2}{c^2} = 1$	Real or imaginary **ellipse; point** (tangent plane)	Vertices at $(\pm a, 0, 0)$ if $a^2 > b^2$, $a^2 > c^2$; principal diameters have the lengths $2a$, $2b$, $2c$	$a^2 = -\frac{1}{\lambda_3}\frac{A}{D}$, $b^2 = -\frac{1}{\lambda_2}\frac{A}{D}$, $c^2 = -\frac{1}{\lambda_1}\frac{A}{D}$ $\lambda_1 \geq \lambda_2 \geq \lambda_3 > 0$; $D = \lambda_1\lambda_2\lambda_3$
Hyperboloid of one sheet (*of revolution* if $a^2 = b^2$)	$\frac{x^2}{a^2} + \frac{y^2}{b^2} - \frac{z^2}{c^2} = 1$	**Hyperbola, parabola,** or **ellipse** corresponding to plane parallel to two generators, one generator, or no generator, respectively, of the asymptotic cone	Doubly ruled surface; contains two families of straight lines (generators) $\frac{x}{a} + \frac{z}{c} = \lambda\left(1 + \frac{y}{b}\right), \lambda\left(\frac{x}{a} - \frac{z}{c}\right) = 1 - \frac{y}{b}$ and $\frac{x}{a} + \frac{z}{c} = \mu\left(1 - \frac{y}{b}\right), \mu\left(\frac{x}{a} - \frac{z}{c}\right) = 1 + \frac{y}{b}$ Locus of a straight line intersecting three given straight lines. Asymptotic cone (inside the surface) $x^2/a^2 + y^2/b^2 - z^2/c^2 = 0$	$a^2 = -\frac{1}{\lambda_2}\frac{A}{D}$, $b^2 = -\frac{1}{\lambda_1}\frac{A}{D}$, $c^2 = +\frac{1}{\lambda_3}\frac{A}{D}$ $\lambda_1 \geq \lambda_2 > 0 > \lambda_3$; $D = \lambda_1\lambda_2\lambda_3$
Hyperboloid of two sheets (*of revolution* if $b^2 = c^2$)	$\frac{x^2}{a^2} - \frac{y^2}{b^2} - \frac{z^2}{c^2} = 1$		Vertices at $(\pm a, 0, 0)$; distance between vertices is $2a$. Asymptotic cone (outside the surface) $\frac{x^2}{a^2} - \frac{y^2}{b^2} - \frac{z^2}{c^2} = 0$	$a^2 = -\frac{1}{\lambda_1}\frac{A}{D}$, $b^2 = +\frac{1}{\lambda_3}\frac{A}{D}$, $c^2 = +\frac{1}{\lambda_2}\frac{A}{D}$ $\lambda_1 > 0 > \lambda_2 \geq \lambda_3$; $D = \lambda_1\lambda_2\lambda_3$
Elliptic paraboloid (*of revolution* if $a^2 = b^2$)	$\frac{x^2}{a^2} + \frac{y^2}{b^2} = z$	**Parabola** (diametral plane); real or imaginary **ellipse; point** (tangent plane)	Vertex at origin	$a^2 = \frac{2}{\lambda_2}\sqrt{\mp\frac{A}{J}}$, $b^2 = \frac{2}{\lambda_1}\sqrt{\mp\frac{A}{J}}$ $\lambda_1 \geq \lambda_2 > \lambda_3 = 0$; $J = \lambda_1\lambda_2$
Hyperbolic paraboloid	$\frac{x^2}{a^2} - \frac{y^2}{b^2} = z$	**Parabola** (diametral plane), **hyperbola**	Saddle point at origin. Doubly ruled surface; contains two families of straight lines $\frac{x}{a} + \frac{y}{b} = \lambda, \frac{x}{a} - \frac{y}{b} = \frac{z}{\lambda}$ and $\frac{x}{a} - \frac{y}{b} = \mu, \frac{x}{a} + \frac{y}{b} = \frac{z}{\mu}$	$a^2 = \frac{2}{\lambda_1}\sqrt{\mp\frac{A}{J}}$, $b^2 = -\frac{2}{\lambda_3}\sqrt{\mp\frac{A}{J}}$ $\lambda_1 > \lambda_2 = 0 > \lambda_3$; $J = \lambda_1\lambda_3$

with the matrix $[a_{ik}]$ (Sec. 14.8-6). Their direction cosines $\cos \alpha_x$, $\cos \alpha_y$, $\cos \alpha_z$ satisfy the conditions

$$\left.\begin{aligned} (a_{11} - \lambda)\cos \alpha_x + a_{12}\cos \alpha_y + a_{13}\cos \alpha_z &= 0 \\ a_{21}\cos \alpha_x + (a_{22} - \lambda)\cos \alpha_y + a_{23}\cos \alpha_z &= 0 \\ a_{31}\cos \alpha_x + a_{32}\cos \alpha_y + (a_{33} - \lambda)\cos \alpha_z &= 0 \end{aligned}\right\} \tag{3.5-12}$$

where λ is a root of the characteristic equation (6).

3.5-7. Transformation of the Equation of a Quadric to Standard Form. If a new reference system is introduced by combining

1. A *rotation* of the coordinate axes (Sec. 3.1-12*b*) such that each new coordinate axis is directed either along, or perpendicular to, one of the principal-axis directions specified in Sec. 3.5-6 (*principal-axis transformation*, Sec. 14.8-6)

2. A suitable *translation* (Sec. 3.1-12*a*) of the origin

it is possible to reduce the equation (1) of any proper quadric to one of the *standard or type forms* (sometimes called canonical forms) listed in Table 3.5-2. Table 3.5-2 also shows the principal properties of the individual surfaces as well as the relations between the parameters a^2, b^2, c^2 appearing in the standard forms to the invariants A, D, J, and I of the general equation (1) (Sec. 3.5-2).

The equations of the individual *improper quadrics* (Table 3.5-1) may be similarly reduced to the standard or type forms presented below:

$$\left.\begin{aligned}
&\frac{x^2}{a^2} + \frac{y^2}{b^2} + \frac{z^2}{c^2} = 0 \text{ (POINT)} \qquad \frac{x^2}{a^2} + \frac{y^2}{b^2} - \frac{z^2}{c^2} = 0 \text{ (ELLIPTIC CONE; circular if } a^2 = b^2\text{)} \\
&\frac{x^2}{a^2} + \frac{y^2}{b^2} = 1 \text{ (ELLIPTIC CYLINDER; circular if } a^2 = b^2\text{)} \qquad \frac{x^2}{a^2} - \frac{y^2}{b^2} = 1 \text{ (HYPERBOLIC CYLINDER)} \\
&\frac{x^2}{a^2} + \frac{y^2}{b^2} = 0 \text{ (STRAIGHT LINE)} \qquad \frac{x^2}{a^2} - \frac{y^2}{b^2} = 0 \text{ (TWO INTERSECTING PLANES)} \\
&\frac{x^2}{a^2} = y \text{ (PARABOLIC CYLINDER)} \qquad \frac{x^2}{a^2} = 1 \text{ (TWO PARALLEL PLANES)} \\
&x^2 = 0 \text{ (ONE REAL PLANE)}
\end{aligned}\right\} \tag{3.5-13}$$

In general, the directions of the new x, y, z axes are determined by Eq. (12) except for rotations through integral multiples of 90 deg about any of the new axes, corresponding to interchanges between any of the quantities x, $-x$, y, $-y$, z, $-z$ in the standard forms of Table 3.5-2 and Eq. (13).

3.5-8. Tangent Planes and Normals of Quadric Surfaces; Polar Planes and Poles. **(a)** Refer to Sec. 17.3-2 for the definitions of **tangent planes** and **normals** of suitable three-dimensional surfaces. *The equation of the tangent plane to the general quadric* (1) *at the point* (x_1, y_1, z_1) *of the quadric is*

$$\left.\begin{aligned}&a_{11}x_1x + a_{22}y_1y + a_{33}z_1z + a_{12}(y_1x + x_1y) + a_{13}(z_1x + x_1z)\\&\quad + a_{23}(y_1z + z_1y) + a_{14}(x_1 + x) + a_{24}(y_1 + y)\\&\qquad\qquad + a_{34}(z_1 + z) + a_{44} = 0\\&\text{or}\\&(a_{11}x_1 + a_{12}y_1 + a_{13}z_1 + a_{14})x\\&\qquad + (a_{21}x_1 + a_{22}y_1 + a_{23}z_1 + a_{24})y\\&\qquad + (a_{31}x_1 + a_{32}y_1 + a_{33}z_1 + a_{34})z\\&\qquad\qquad + (a_{41}x_1 + a_{42}y_1 + a_{43}z_1 + a_{44}) = 0\end{aligned}\right\} \tag{3.5-14}$$

(b) The equations of the normal to the quadric (1) at the point (x_1, y_1, z_1) are (see also Sec. 3.3-1*b*)

$$\frac{x - x_1}{a_{11}x_1 + a_{12}y_1 + a_{13}z_1 + a_{14}} = \frac{y - y_1}{a_{21}x_1 + a_{22}y_1 + a_{23}z_1 + a_{24}} = \frac{z - z_1}{a_{31}x_1 + a_{32}y_1 + a_{33}z_1 + a_{34}} \tag{3.5-15}$$

(c) No matter whether the point (x_1, y_1, z_1) in Sec. 3.5-8*b* does or does not lie on the surface (1), Eq. (14) defines a plane called the **polar plane** of the **pole** (x_1, y_1, z_1) with respect to the quadric (1). The polar plane of a point on the quadric surface is the tangent plane at that point.

(d) *The center of a quadric is the pole of the plane at infinity* (*a common definition of the center;* if this definition is used, diametral planes and diameters are defined as planes and straight lines, respectively, through the center).

(e) *The polar planes of all points of any given plane pass through the pole of the given plane.*

(f) *The polar planes of all points of a given straight line pass through a straight line. Conversely, the latter contains the poles of all planes passing through the first straight line.*

(g) *The tangents from any point to a quadric generate an elliptic cone or a pair of planes; the points of tangency lie on the polar plane of the given point.*

3.5-9. Miscellaneous Formulas and Theorems Relating to Quadric Surfaces. **(a)** *Equation of a sphere* of radius r about the point (x_1, y_1, z_1)

$$(x - x_1)^2 + (y - y_1)^2 + (z - z_1)^2 = r^2 \qquad \text{or} \qquad |\mathbf{r} - \mathbf{r}_1| = r \tag{3.5-16}$$

(b) The most general form of the equation of a sphere is

$$A(x^2 + y^2 + z^2) + 2Bx + 2Cy + 2Dz + E = 0 \qquad A \neq 0 \tag{3.5-17}$$

(c) Let P_1 and P_2 be the points of intersection of a sphere and a straight line (secant) through a fixed point $P_0 \equiv (x_0, y_0, z_0)$. Then the products of the directed line segments $\overrightarrow{P_0P_1}$ and $\overrightarrow{P_0P_2}$ are equal for all secants through P_0, and

$$(\overrightarrow{P_0P_1})(\overrightarrow{P_0P_2}) = x_0^2 + y_0^2 + z_0^2 - r^2 \tag{3.5-18}$$

The theorem holds also if P_1 and P_2 coincide (i.e., if the secant becomes a tangent).

(d) Given any three conjugate semidiameters (Sec. 3.5-5*d*) of a given ellipsoid,

1. The sum of their squares is $a^2 + b^2 + c^2$.
2. The parallelepiped having them as adjacent sides has the volume *abc*.

(e) If

$$\frac{x^2}{a^2} \pm \frac{y^2}{b^2} \pm \frac{z^2}{c^2} = 1 \tag{3.5-19}$$

is the equation of an ellipsoid, hyperboloid of one sheet, or hyperboloid of two sheets (see also Table 3.5-2), then

$$\frac{x \cos \alpha_x}{a^2} \pm \frac{y \cos \alpha_y}{b^2} \pm \frac{z \cos \alpha_z}{c^2} = 0 \tag{3.5-20}$$

is the equation of the diameter conjugate to the plane whose normal has the direction cosines $\cos \alpha_x$, $\cos \alpha_y$, $\cos \alpha_z$. The respective direction cosines $\cos \alpha_x$, $\cos \alpha_y$, $\cos \alpha_z$; $\cos \beta_x$, $\cos \beta_y$, $\cos \beta_z$; $\cos \gamma_x$, $\cos \gamma_y$, $\cos \gamma_z$ of three conjugate diameters satisfy the relations

$$\left.\begin{aligned}
\frac{\cos \alpha_x \cos \beta_x}{a^2} \pm \frac{\cos \alpha_y \cos \beta_y}{b^2} \pm \frac{\cos \alpha_z \cos \beta_z}{c^2} = 0 \\
\frac{\cos \alpha_x \cos \gamma_x}{a^2} \pm \frac{\cos \alpha_y \cos \gamma_y}{b^2} \pm \frac{\cos \alpha_z \cos \gamma_z}{c^2} = 0 \\
\frac{\cos \beta_x \cos \gamma_x}{a^2} \pm \frac{\cos \beta_y \cos \gamma_y}{b^2} \pm \frac{\cos \beta_z \cos \gamma_z}{c^2} = 0
\end{aligned}\right\} \tag{3.5-21}$$

The signs in Eqs. (20) and (21) correspond to those in Eq. (19).

3.5-10. Parametric Representation of Quadrics (see also Secs. 3.1-14 and 6.5-1). Note the following parametric representations of quadrics:

$$x = a \sin u \cos v \qquad y = b \sin u \sin v \qquad z = c \cos u \qquad \text{(ELLIPSOID)} \tag{3.5-22}$$

$$x = a \cosh u \cos v \qquad y = b \cosh u \sin v \qquad z = c \sinh u \qquad \text{(HYPERBOLOID OF ONE SHEET)} \tag{3.5-23}$$

$$x = a \cosh u \qquad y = b \sinh u \sin v \qquad z = c \sinh u \cos v \qquad \text{(HYPERBOLOID OF TWO SHEETS)} \tag{3.5-24}$$

$$x = au \cos v \qquad y = bu \sin v \qquad z = u^2 \qquad \text{(ELLIPTIC PARABOLOID)} \tag{3.5-25}$$

$$x = au \cosh v \qquad y = bu \sinh v \qquad z = u^2 \qquad \text{(HYPERBOLIC PARABOLOID)} \tag{3.5-26}$$

$$x = au \cos v \qquad y = bu \sin v \qquad z = cu \qquad \text{(ELLIPTIC CONE)} \tag{3.5-27}$$

Many other parametric representations are possible.

3.6. RELATED TOPICS, REFERENCES, AND BIBLIOGRAPHY

3.6-1. Related Topics. The following topics related to the study of solid analytic geometry are treated in other chapters of this handbook:

Algebraic equations, determinants	Chap. 1
Plane analytic geometry	Chap. 2
Coordinate transformations	Chaps. 2, 6, 14, 16
Curvilinear coordinate systems	Chap. 6
Matrices	Chap. 13
Differential geometry	Chap. 17
Mensuration formulas	Appendix A
Plane and spherical trigonometry	Appendix B

3.6-2. References and Bibliography. Refer to Sec. 2.7-2.

CHAPTER 4

FUNCTIONS AND LIMITS DIFFERENTIAL AND INTEGRAL CALCULUS

4.1. INTRODUCTION

4.1-1. Survey. This chapter is primarily concerned with numerical *functions of real variables.* Such functions furnish "analytical" descriptions of relationships between objects labeled with sets of real numbers (see also Secs. 12.1-1 and 14.1-1). The introduction of *limits of functions* (Secs. 4.4-1 to 4.4-7) permits one to define new mathematical operations (limiting processes), such as addition and multiplications of infinite numbers of terms, differentiation, and integration. Limiting processes are also used to derive numerical approximations.

The *differential calculus* (Secs. 4.5-1 to 4.5-7) describes relations between small changes of suitable variables. The *integral calculus* (Secs. 4.6-1 to 4.6-19, 4.9-3, and 4.9-4) yields measures for over-all or average properties of a set of objects and furnishes techniques for adding many small changes. Sections 4.8-1 to 4.9-2 outline the properties of *infinite series,* and Secs. 4.10-1 to 4.11-8 deal with the representation of functions by *power series, Fourier series,* and *Fourier integrals.*

The definitions and theorems presented in this chapter apply to complex functions and variables as well as to real functions and variables unless a restriction to real quantities is specifically indicated. Analytic functions of a complex variable are discussed in Chap. 7.

4.2. FUNCTIONS

4.2-1. Functions and Variables (see also Sec. 12.1-4). (**a**) Given a *rule of correspondence* which associates a real or complex number

$$y = f(x) \tag{4.2-1}$$

with each given real or complex number x of a set S_x, y is called a (numerical) **function** $y = y(x) = f(x)$ of the **argument** x. Equation (1) specifies a **value** (or values) $y = Y = f(X)$ of the variable y corresponding to each suitable value $x = X$ of the variable x. If the relation (1) is primarily intended to describe the dependence of y on x, x is called the **independent variable,** and y is called the **dependent variable.**

The term "variable x" essentially refers to a *set* of values X, and Eq. (1) symbolizes a set of correspondences relating values X of x and values $Y = f(X)$ of y. In order to conform with the notation employed in most textbooks, the symbol x will be used to denote both the *variable* x and a *value of the variable* x whenever this notation does not result in ambiguities.

If one interprets x and y as plane cartesian coordinates (Sec. 2.1-2), a real function $y = f(x)$ of a real variable x is often represented by a curve (**graph** of y vs. x, see also Sec. 2.1-9).

(**b**) A function

$$y = f(x_1, x_2, \ldots, x_n) \tag{4.2-2}$$

of n variables $x_1, x_2, \ldots, x_n$ similarly associates values of a (dependent) variable y with ordered sets of values of the (independent) variables $x_1, x_2, \ldots, x_n$.

A function may be defined by a *table of function values,* or by rules for computing such a table by means of known operations (*constructive definitions*). A function may also be defined *implicitly* (Sec. 4.5-7), or in terms of *defining properties* described by functional, differential, or integral equations, extreme-value properties (Sec. 11.5-2), behavior for certain values of the argument, etc. Each nonconstructive definition requires an *existence proof* demonstrating, by example or construction, a function with the specified properties.

(**c**) In most applications, the variables x, y or $x_1, x_2, \ldots, x_n$, y label physical objects or quantities, so that *suitable relations* (1) *or* (2) *describe physical relationships* (EXAMPLE: $y = x_1x_2$ if x_1, x_2, and y respectively label the voltage, current, and power in a simple electric circuit).

(**d**) The set S_x of values of x (or of sets of values of $x_1, x_2, \ldots, x_n$) for which the relationship (1) or (2) is defined is the **domain of definition** of the function $f(x)$ or $f(x_1, x_2, \ldots, x_n)$. The corresponding set S_y of values of y is the **range** of the function.

(**e**) A **sequence** of real or complex numbers $s_0, s_1, s_2, \ldots$ represents a function $s_n = s_n(n)$ defined on the set of nonnegative integers n.

4.2-2. Functions with Special Properties (see also Secs. 1.4-3, 7.3-3, 7.6-5, and 7.6-7). **(a)** A function is **single-valued*** wherever a single function value corresponds to the value of the argument. A function is **multiple-valued** wherever two or more function values correspond to the value of the argument. The function $y(x)$ has an **inverse function** $x(y)$ if $y = y(x)$ implies $x = x(y)$ for all x in S_x.

* Many authors categorically define *every* function as single-valued, so that, for example, the two branches $+\sqrt{x}$, $-\sqrt{x}$ of $\pm\sqrt{x}$ are always regarded as *two* functions.

(b) A function $f(x)$ of a real or complex variable x is **even** if and only if $f(-x) \equiv f(x)$; **odd** if and only if $f(-x) \equiv -f(x)$; **periodic** with the **period** T if and only if $f(t + T) \equiv f(t)$.

Every function $f(x)$ defined for the values of x in question can be expressed as the sum of the even function $\frac{1}{2}[f(x) + f(-x)]$ and the odd function $\frac{1}{2}[f(x) - f(-x)]$. A periodic function $f(t)$ with period T is **antiperiodic** if and only if $f(t + T/2) = -f(t)$. Every periodic function $f(t)$ can be expressed as the sum of the antiperiodic function $\frac{1}{2}[f(t) - f(t + T/2)]$ and the function $\frac{1}{2}[f(t) + f(t + T/2)]$, which is periodic with period $T/2$.

(c) $y = f(x)$ is an **algebraic function** of x if and only if x and y satisfy a relation of the form $F(x, y) = 0$, where $F(x, y)$ is a polynomial in x and y (Sec. 1.4-3). In particular, $y = f(x)$ is a **rational (rational algebraic) function** of x if $f(x)$ is a polynomial (**integral rational function**) or a quotient of two polynomials (**fractional rational function**). y is a **linear function** of x if $y = ax + b$.

4.3. POINT SETS, INTERVALS, AND REGIONS

4.3-1. Introduction. When discussing the properties of a function $f(x)$ of a real variable x, one is often required to specify a *set of values* of x such that $f(x)$ is defined and satisfies given conditions. Note that either functions or sets may be described in this manner. It is customary to refer to the values* of a real variable x (or to objects labeled by values of x) as *points* (x) of a line, and to sets of such real numbers as **linear point sets.**

Properties of a function $f(x_1, x_2, \ldots, x_n)$ of n real variables $x_1, x_2, \ldots, x_n$ are similarly related to sets of "points" $(x_1, x_2, \ldots, x_n)$ in an n-dimensional "space" which comprises all points $(x_1, x_2, \ldots, x_n)$ under consideration (Sec. 14.1-2).

The use of geometrical language is prompted by the *Cantor-Dedekind axiom of continuity,* which *postulates* the existence of a one-to-one reciprocal correspondence between the real numbers and the points of a straight line. This "coordinate axiom" (Sec. 2.1-2) is compatible with the properties of real numbers as well as with the postulates defining Euclidean and other geometries.

Sections 4.3-2 to 4.3-6 deal mainly with those properties of point sets which apply directly to the theory of functions of real variables; Secs. 7.2-2 to 7.2-4 deal with regions of the complex number plane. In a more general context, any set (class) of objects (in particular, objects labeled with values of a real variable or variables) may be referred to as a set of points. The properties of such sets are further discussed in Secs. 12.5-1 to 12.5-4.

4.3-2. Properties of Sets. (a) Algebra of Sets (Classes). An object (point) P contained in a set (class) S is an **element** of S ($P \in S$). A set S_1 is a **subset** of another set S_2 (S_1 is **contained in** S_2, $S_1 \subset S_2$) if

* See also Sec. 4.2-1.

and only if each element of S_1 is an element of S_2. S_1 and S_2 are **equal** ($S_1 = S_2$) if and only if both contain the same elements, i.e., if and only if $S_1 \subset S_2$ and $S_2 \subset S_1$. The **empty set** 0 is, by definition, a subset of every set S. A **proper subset (proper part)** of S is a nonempty subset of S not equal to S. The **union (join, logical sum)** $S_1 \cup S_2$ (or $S_1 + S_2$) is the set of all elements contained in *either* S_1 *or* S_2, *or both.* The **intersection (meet, logical product)** $S_1 \cap S_2$ (or S_1S_2) of S_1 and S_2 is the set of all elements contained in *both* S_1 *and* S_2. The **complement** of a set S with respect to a set I containing S is the set of all elements of I *not* contained in S. *The subsets of any set (class) I constitute a Boolean algebra (Sec.* 12.8-1) *under the operations of logical addition and multiplication.*

(b) Cardinal Numbers and Countability. Two sets S_1 and S_2 have the same **cardinal number** if and only if there exists a reciprocal one-to-one correspondence between their respective elements. S is an **infinite set** if it has the same cardinal number as one of its proper subsets; otherwise, S is a **finite set.**

A finite or infinite set S is **countable (enumerable, denumerable)** if and only if it is possible to establish a reciprocal one-to-one correspondence between its elements and those of a set of real integers. *Every finite set is countable;* the cardinal number of a finite set is identical with the number of its elements. *Every subset of a countable set is countable. The union of a countable set of countable sets is a countable set.*

Cardinal numbers corresponding to infinite sets are called **transfinite numbers.** The cardinal number of every countable infinite set is the same as the cardinal number of the set of the positive real integers and is denoted by $\aleph_0$. The set of all real numbers (or the set of points of a continuous line, Sec. 4.3-1) is not countable; the corresponding cardinal number is denoted by $\aleph$.

4.3-3. Bounds. (a) A real number M is an **upper bound** or a **lower bound** of a set S_y of real numbers y if and only if, respectively, $y \leq M$ or $y \geq M$ for all y in S_y. A set of real or complex numbers is **bounded** (has an **absolute bound**) if the set of their absolute values has an upper bound; otherwise the set is **unbounded.**

Every (nonempty) set S_y of real numbers y having an upper bound has a **least upper bound (l.u.b.)** sup y, and every (nonempty) set of real numbers y having a lower bound has a **greatest lower bound (g.l.b.)** inf y. If S_y is finite, sup y is necessarily equal to the **maximum** value max y actually assumed by a number y in S_y, and inf y is equal to the **minimum** min y.

EXAMPLE: The set of all real numbers less than 1 has the least upper bound 1, but no maximum.

(b) A real or complex function $y = f(x)$ or $y = f(x_1, x_2, \ldots, x_n)$ is **bounded** on a set S of "points" (x) or $(x_1, x_2, \ldots, x_n)$ if and only if the corresponding set S_y of function values y is bounded. Similarly, a real function $y = f(x)$ or $y = f(x_1, x_2, \ldots, x_n)$ has an **upper bound, lower bound, least upper bound, greatest lower bound,** (absolute) **maximum,** and/or (absolute) **minimum** on a set S of "points" (x) or $(x_1, x_2, \ldots, x_n)$ if this is true of the corresponding set S_y of function values y.

(c) A real or complex function $f(x, y)$ or $f(x_1, x_2, \ldots, x_n; y)$ is **uniformly bounded** on a set S of "points" (x) or $(x_1, x_2, \ldots, x_n)$ if and only if f as a function of y has an upper bound independent of x or $x_1, x_2, \ldots, x_n$ on S. **Uniform upper bounds** and **uniform lower bounds** are similarly defined.

4.3-4. Intervals (see also Secs. 4.3-3 and 4.3-5). Given a real variable x, the set of all values of x (points) such that

1. $a < x < b$ is the **bounded open interval** (a, b).
2. $a < x$ is the **unbounded open interval** (a, ∞).
3. $x < a$ is the **unbounded open interval** $(-\infty, a)$.
4. $a \leq x \leq b$ is the **bounded closed interval** $[a, b]$.

Sets of points (x) such that $a \leq x < b$, $a < x \leq b$, $a \leq x$, $x \leq a$ may be referred to as **semiclosed intervals.** Every interval I_1 contained in another interval I_2 is a **subinterval** of I_2.

4.3-5. Definition of Neighborhoods. (a) Given any finite real number a, an (open) **δ-neighborhood of the point $(x = a)$ in the space of real numbers** is any open interval $(a - \delta, a + \delta)$ containing $x = a$; or, the set of all points (x) such that $|x - a| < \delta$, for some positive real number δ. A **neighborhood** of $x = a$ is any set containing a δ-neighborhood of $x = a$.

(b) Every set containing all points (x) such that $x > M$ for some real number M is a **neighborhood of plus infinity $(+\infty)$ in the space of real numbers;** every set containing all points (x) such that $x < N$ for some real number N is a **neighborhood of minus infinity $(-\infty)$ in the space of real numbers.**

(c) In a space whose "points" are (described as) ordered sets $(x_1, x_2, \ldots, x_n)$ of real numbers, one may define an (open) δ-neighborhood of the point $(a_1, a_2, \ldots, a_n)$, where $a_1, a_2, \ldots, a_n$ are finite, as the set of all points $(x_1, x_2, \ldots, x_n)$ such that $|x_1 - a_1| < \delta$, $|x_2 - a_2| < \delta$, $\ldots$, and $|x_n - a_n| < \delta$ for some positive real number δ. A neighborhood of the point $(a_1, a_2, \ldots, a_n)$ is any set containing a δ-neighborhood of this point.

NOTE: Definitions of neighborhoods, like those given above, are not necessarily self-evident; they amount to *postulates* defining the "topological" properties of the space in question (see also Sec. 12.5-1). In particular, neighborhoods involving unbounded values of the variables [as in (b) above] may be defined in a variety of ways (thus, $+\infty$ and $-\infty$ may or may not be regarded as the same point, see also Sec. 7.2-2). In applied mathematics, the choice of such definitions will depend on the nature of the objects "represented" by the points (x) or $(x_1, x_2, \ldots, x_n)$. The definition of neighborhoods is closely related to the definition of open sets (Sec. 4.3-6*a*), regions (Sec. 4.3-6*b*), and limits of functions on the space in question (Sec. 4.4-1; see also Sec. 12.5-3).

4.3-6. Open and Closed Sets and Regions. All the following definitions imply a specific definition of neighborhoods in a space containing the sets and regions in question (topology, see also Secs. 12.5-1 to 12.5-4).

(**a**) A point P is a **limit point (cluster point, accumulation point)** of the point set S if and only if every neighborhood of P contains points of S other than P itself. A limit point P is an **interior point** of S if and only if S is a neighborhood of P; otherwise, P is a **boundary point** of S. A point P is an **isolated point** of S if and only if P has a neighborhood in which P is the only point belonging to S.

A point set S is

An **open set** if and only if it contains only interior points.

A **closed set** if and only if it contains all its limit points; a finite set is closed.

A **discrete (isolated) set** if and only if it contains only isolated points; *every discrete set is countable.*

(See also Sec. 12.5-4).

(**b**) In the Euclidean plane (Sec. 2.1-1) or space, a **simply connected open region** D is an open set of points such that every closed curve in D can be continuously contracted into a point without leaving D. If a region is not simply connected, it is said to be **multiply connected.** A point P is on the **boundary** of the region D if every neighborhood of P contains points in D and points not in D; the boundary of a simply connected open region is a simple surface or simple closed surface. Connected regions in D are **subregions** of D. An open region and its boundary or boundaries constitute a **closed region.** A region of a Euclidean plane or space is **bounded** if and only if all its points can be described by bounded cartesian coordinates (see also Sec. 3.1-2).

4.4. LIMITS, CONTINUOUS FUNCTIONS, AND RELATED TOPICS

4.4-1. Limits of Functions and Sequences (see also Secs. 4.8-1 and 12.5-3; see Table 4.7-1 for examples). (**a**) A single-valued function $f(x)$ has (approaches, converges to, tends to)

1. A (*necessarily finite and unique*) **limit** $\lim_{x\to a} f(x) = L$ *as x approaches a finite value $x = a$* [$f(x) \to L$ as $x \to a$] if and only if for each positive real number ϵ there exists a real number $\delta > 0$ such that $0 < |x - a| < \delta$ implies that $f(x)$ is defined and $|f(x) - L| < \epsilon$.
2. A (necessarily finite and unique) limit $\lim_{x\to\infty} f(x) = L$ *as x increases indefinitely* [*increases without bound, tends to infinity;* $f(x) \to L$ as

$x \to \infty$] if and only if for each positive real number ϵ there exists a real number N such that $x > N$ implies that $f(x)$ is defined and $|f(x) - L| < \epsilon$.

(b) A *sequence* of numbers (Sec. 4.2-1*e*) $s_0, s_1, s_2, \ldots$ [$\equiv s(n)$] converges to a (necessarily finite and unique) limit $\lim_{n\to\infty} s_n = S$ if and only if for each positive real number ϵ there exists a real integer N such that $n > N$ implies $|s_n - S| < \epsilon$. Criteria for the convergence of sequences are presented in Sec. 4.9-1 in connection with the convergence of infinite series.

(c) A real function $f(x)$ **increases indefinitely (increases without bound, tends to infinity)**

1. *As x approaches a finite value $x = a$* [$f(x) \to \infty$ as $x \to a$; some authors write $\lim_{x\to a} f(x) = \infty$] if and only if for each positive real number M there exists a real number $\delta > 0$ such that $0 < |x - a| < \delta$ implies that $f(x)$ is defined, and $f(x) > M$.
2. *As x increases indefinitely* [$f(x) \to \infty$ as $x \to \infty$; some authors write $\lim_{x\to\infty} f(x) = \infty$] if and only if for each real number M there exists a real number N such that $x > N$ implies that $f(x)$ is defined and $f(x) > M$.

These definitions apply, in particular, to indefinitely increasing sequences. A real variable x or $f(x)$ **decreases indefinitely** [$x \to -\infty$ or $f(x) \to -\infty$] if, respectively, $-x \to \infty$, or $-f(x) \to \infty$. Section 4.4-1 specifies the mathematical meaning of *infinity* in the context of the *real-number system* (see also Secs. 4.3-5 and 7.2-2).

4.4-2. Operations with Limits (see also Secs. 4.4-6*c* and 4.8-4). If the limits in question exist,

$$\begin{aligned}
&\lim_{x\to a} [f(x) + g(x)] = \lim_{x\to a} f(x) + \lim_{x\to a} g(x)\\
&\lim_{x\to a} [\alpha f(x)] = \alpha \lim_{x\to a} f(x)\\
&\lim_{x\to a} [f(x)g(x)] = \lim_{x\to a} f(x) \lim_{x\to a} g(x)\\
&\lim_{x\to a} \frac{f(x)}{g(x)} = \frac{\lim_{x\to a} f(x)}{\lim_{x\to a} g(x)} \qquad [\lim_{x\to a} g(x) \neq 0]
\end{aligned} \tag{4.4-1}$$

a may be finite or infinite; these rules apply to limits of sequences (Sec. 4.4-1*b*) and also to multiple limits (Sec. 4.4-5).

4.4-3. Asymptotic Relations between Two Functions (see also Sec. 4.8-6). Given two real or complex functions $f(x)$, $g(x)$ of a real or complex variable x, one writes

1. $f(x) = O[g(x)]$ {$f(x)$ is $O[g(x)]$, $f(x)$ is **of the order of** $g(x)$} as $x \to a$ if and only if there exists a neighborhood of $x = a$ such that $|f(x)/g(x)|$ is bounded.

2. $f(x) \sim g(x)$ [$f(x)$ is **asymptotically proportional** to $g(x)$] as $x \to a$ if and only if $\lim_{x \to a} [f(x)/g(x)]$ exists and differs from zero.
3. $f(x) \simeq g(x)$ [$f(x)$ is **asymptotically equal** to $g(x)$] as $x \to a$ if and only if $\lim_{x \to a} [f(x)/g(x)] = 1$; this implies that *the percentage difference between $f(x)$ and $g(x)$ converges to zero* as $x \to a$.
4. $f(x) = o[g(x)]$ as $x \to a$ if and only if $\lim_{x \to a} [f(x)/g(x)] = 0$. This may often be read "$f(x)$ becomes negligible compared with $g(x)$ as $x \to a$."

In each of the above definitions, a may be finite or infinite. Functions **of order 1, 2, . . .** and functions **of exponential order** are functions of the order of x, x^2, . . . , and e^x as $x \to \infty$; functions **of order -1, -2,** . . . are functions of the order of x^{-1}, x^{-2}, . . . as $x \to 0$.

Asymptotic relations often yield estimates or approximations of $f(x)$ in terms of $g(x)$ in a neighborhood of $x = a$; note that $f(x) \simeq g(x)$ implies that the fractional error $\frac{f(x) - g(x)}{f(x)}$ decreases in absolute value as $x \to a$. One writes

$$f(x) = \varphi(x) + O[g(x)] \quad \text{if} \quad f(x) - \varphi(x) = O[g(x)]$$

and

$$f(x) = \varphi(x) + o[g(x)] \quad \text{if} \quad f(x) - \varphi(x) = o[g(x)]$$

4.4-4. Uniform Convergence. (a) A single-valued function $f(x_1, x_2)$ **converges uniformly on a set S of values of x_2**

1. To the (*necessarily finite and unique*) function $\lim_{x_1 \to a} f(x_1, x_2) = L(x_2)$ if and only if for each positive real number ϵ there exists a real number $\delta > 0$ such that $0 < |x_1 - a| < \delta$ implies that $f(x_1, x_2)$ is defined and $|f(x_1, x_2) - L(x_2)| < \epsilon$ *for all x_2 in S* (δ is independent of x_2).
2. To the (necessarily finite and unique) function $\lim_{x_1 \to \infty} f(x_1, x_2) = L(x_2)$ if and only if for each positive real number ϵ there exists a real number N such that $x_1 > N$ implies that $f(x_1, x_2)$ is defined and $|f(x_1, x_2) - L(x_2)| < \epsilon$ *for all x_2 in S.*

(b) A *sequence of functions* $s_0(x)$, $s_1(x)$, $s_2(x)$, . . . converges uniformly on a set S of values of x to the (necessarily finite and unique) function

$$\lim_{n \to \infty} s_n(x) = s(x)$$

if and only if for each positive real number ϵ there exists a real integer N such that $n > N$ implies $|s_n(x) - s(x)| < \epsilon$ *for all x in S* (see also Secs. 4.6-2c and 4.8-2).

4.4-5. Multiple Limits and Iterated Limits. (a) Given the definition of a neighborhood formulated in Sec. 4.3-5c, a single-valued function $f(x_1, x_2)$ has the (*necessarily finite and unique*) limit $\lim_{\substack{x_1 \to a_1 \\ x_2 \to a_2}} f(x_1, x_2) = L$ if and only if for each positive real number ϵ there exists a neighborhood D of the point (a_1, a_2) such that $f(x_1, x_2)$ is defined and $|f(x_1, x_2) - L| < \epsilon$ for all points (x_1, x_2) in D, except possibly at (a_1, a_2). a_1 and/or a_2 may be finite or infinite.

In particular, a *double sequence* $s_{00}, s_{10}, \ldots$ converges to a limit $\lim_{\substack{m \to \infty \\ n \to \infty}} s_{mn} = s$ if and only if for each positive real number ϵ there exists a pair of real numbers M,N such that $\infty > m > M$, $\infty > n > N$ implies $|s_{mn} - s| < \epsilon$. Limits of functions of more than two variables are defined in an analogous manner.

(b) *If there exists a positive real number δ such that*

$$\left.\begin{aligned} &\lim_{x_1 \to a_1} f(x_1, x_2) = f(a_1, x_2) \qquad \textit{for} \qquad 0 < |x_2 - a_2| < \delta \\ &\lim_{x_2 \to a_2} f(x_1, x_2) = f(x_1, a_2) \qquad \textit{for} \qquad 0 < |x_1 - a_1| < \delta \end{aligned}\right\} \tag{4.4-2}$$

and *at least one of these limiting processes converges uniformly in the interval specified, then the three limits* $\lim_{\substack{x_1 \to a_1 \\ x_2 \to a_2}} f(x_1, x_2)$, $\lim_{x_1 \to a_1} [\lim_{x_2 \to a_2} f(x_1, x_2)]$, *and* $\lim_{x_2 \to a_2} [\lim_{x_1 \to a_1} f(x_1, x_2)]$ *exist and are equal.* Analogous theorems apply if a_1 and/or a_2 are infinite.

4.4-6. Continuous Functions. (a) A single-valued function $f(x)$ defined throughout a neighborhood of $x = a$ is **continuous at** $x = a$ [**at the point** $(x = a)$] if and only if $\lim_{x \to a} f(x)$ exists and equals $f(a)$, i.e., if and only if for every positive real number ϵ there exists a real number $\delta > 0$ such that $|x - a| < \delta$ implies $|f(x) - f(a)| < \epsilon$.

Similarly, a single-valued function $f(x_1, x_2, \ldots, x_n)$ defined throughout a neighborhood of the point $(a_1, a_2, \ldots, a_n)$ is continuous at $(a_1, a_2, \ldots, a_n)$ if and only if

$$\lim_{\substack{x_1 \to a_1, x_2 \to a_2, \ldots \\ \ldots, x_n \to a_n}} f(x_1, x_2, \ldots, x_n) = f(a_1, a_2, \ldots, a_n)$$

A single-valued function $f(x_1, x_2, \ldots, x_n)$ is **continuous in** x_1 at the point $(a_1, a_2, \ldots, a_n)$ if and only if $f(x_1, a_2, \ldots, a_n)$ is continuous for $x_1 = a_1$. A function continuous in $x_1, x_2, \ldots,$ and x_n separately at $(a_1, a_2, \ldots, a_n)$ is *not* necessarily continuous at $(a_1, a_2, \ldots, a_n)$.

(b) A function is **continuous on a set of points** (e.g., an interval or region) if and only if it is continuous at each point of the set. *A real function continuous on a bounded closed interval* $[a, b]$ *is bounded on* $[a, b]$ *and assumes every value between and including its g.l.b. and its l.u.b.* (Sec. 4.3-3) *at least once on* $[a, b]$. An analogous theorem holds for a real function of two or more variables continuous on a bounded singly connected closed region.

$f(x)$ is **uniformly continuous on** S if and only if for each positive real number ϵ there exists a real number δ such that $|x - X| < \delta$ implies $|f(x) - f(X)| < \epsilon$ for all X in S. *A function continuous on a bounded closed interval* $[a, b]$ *is uniformly continuous on* $[a, b]$.

(c) *If two functions f and g are continuous at a given point* (x) *or* $(x_1, x_2, \ldots, x_n)$, *the same is true for* $f + g$ *and* fg. *Given* $\lim_{x \to a} y_i(x) = A_i$ $(i = 1, 2, \ldots, n)$ *and a function* $F(y_1, y_2, \ldots, y_n)$ *continuous for* $y_1 = A_1, y_2 = A_2, \ldots, y_n = A_n$,

$$\lim_{x \to a} F[y_1(x), y_2(x), \ldots, y_n(x)] = F[\lim_{x \to a} y_1(x), \lim_{x \to a} y_2(x), \ldots, \lim_{x \to a} y_n(x)] \tag{4.4-3}$$

(*see also* Sec. 4.4-2). *In particular, if each* $y_i(x)$ *is continuous for* $x = a$, *the same is true for* $F[y_1(x), y_2(x), \ldots, y_n(x)]$.

The limit $s(x)$ *of a uniformly convergent sequence of functions* $s_0(x), s_1(x), \ldots$ *all continuous on a set S of values of x is continuous on S.*

4.4-7. One-sided (Unilateral) Limits. Unilateral Continuity. (a) A function $f(x)$ of a real variable x has the (*necessarily finite and unique*) **right-hand limit** $\lim_{x \to a+0} f(x) \equiv \lim_{x \to a+} f(x) \equiv f(a + 0) = L_+$ at $x = a$ if and only if for each positive real number ϵ there exists a real number $\delta > 0$ such that $0 < x - a < \delta$ implies that $f(x)$ is defined, and $|f(x) - L_+| < \epsilon$. $f(x)$ has the **left-hand limit** $\lim_{x \to a-0} f(x) \equiv \lim_{x \to a-} f(x) \equiv f(a - 0) = L_-$ at $x = a$ if and only if for each positive real number ϵ there exists a real number $\delta > 0$ such that $0 < a - x < \delta$ implies that $f(x)$ is defined, and $|f(x) - L_-| < \epsilon$. *If* $\lim_{x \to a} f(x)$ *exists, then*

$$\lim_{x \to a+0} f(x) = \lim_{x \to a-0} f(x) = \lim_{x \to a} f(x)$$

Conversely, $\lim_{x \to a-0} f(x) = \lim_{x \to a+0} f(x)$ *implies the existence of* $\lim_{x \to a} f(x)$.

(b) $f(x)$ is **right continuous** or **left continuous** at $x = a$ if $f(a + 0) = f(a)$ or $f(a - 0) = f(a)$, respectively. A **discontinuity of the first kind** of a real function $f(x)$ is a point $x = a$ such that $f(a + 0)$ and $f(a - 0)$ exist; the greatest difference between two of the numbers $f(a)$, $f(a + 0)$, $f(a - 0)$ is the **saltus** of $f(x)$ at such a discontinuity. *The discontinuities of the first kind of* $f(x)$ *constitute a discrete (and thus countable) set* (Sec. 4.3-6*a*).

(c) $f(x)$ is **piecewise continuous** on an interval I if and only if $f(x)$ is continuous throughout I except for a finite number of discontinuities of the first kind.

$f(x_1, x_2, \ldots, x_n)$ is piecewise continuous on a region V of n-dimensional space if and only if $f(x_1, x_2, \ldots, x_n)$ is continuous throughout V, except possibly on a set of regular hypersurfaces (regular curves for $n = 2$, regular surfaces for $n = 3$, Secs. 3.1-13 and 3.1-14) which divide V into a finite number of subregions such that $f(x_1, x_2, \ldots, x_n)$ has a finite and unique unilateral limit on approaching any boundary point of a subregion from its interior.

4.4-8. Monotonic Functions and Functions of Bounded Variation. (a) A real function $f(x)$ of a real variable x is **strongly monotonic** in (a, b) if $f(x)$ increases as x increases in (a, b) (**increasing** or **positive monotonic function**), or if $f(x)$ decreases as x increases in (a, b)

(**decreasing** or **negative monotonic function**). $f(x)$ is **weakly monotonic** in (a, b) if $f(x)$ does not decrease in (a, b) (**nondecreasing function**), or if $f(x)$ does not increase in (a, b) (**nonincreasing function**). Analogous definitions apply to monotonic *sequences* (Sec. 4.2-1*d*).

(**b**) A real function $f(x)$ of a real variable x is **of bounded variation** in the interval (a, b) if and only if there exists a real number M such that $\sum_{i=1}^{m} |f(x_i) - f(x_{i-1})| < M$ for all partitions $a = x_0 < x_1 < x_2 < \cdots < x_m = b$ of the interval (a, b). $f(x)$ is of bounded variation in (a, b) if and only if it can be expressed in the form $f(x) \equiv f_1(x) - f_2(x)$, where $f_1(x)$ and $f_2(x)$ are bounded and nondecreasing in (a, b) (*alternative definition*). *If $f(x)$ and $g(x)$ are of bounded variation in (a, b), the same is true for $f(x) + g(x)$ and $f(x)g(x)$. $f(x)$ is of bounded variation in every finite open interval where $f(x)$ is bounded and has a finite number of relative maxima and minima* (*Sec.* 11.2-1) *and discontinuities* (*Dirichlet conditions*).

A function of bounded variation in (a, b) is bounded in (a, b), and its only discontinuities are discontinuities of the first kind (Sec. 4.4-7*b*).

In physical applications, the condition that $f(x)$ is of bounded variation in every bounded interval expresses the fact that $f(x)$ is bounded, and that components of very high frequency cannot contribute significantly to its total intensity (Sec. 18.10-1).

4.5. DIFFERENTIAL CALCULUS

4.5-1. Derivatives and Differentiation. (**a**) Let $y = f(x)$ be a real, single-valued function of the real variable x throughout a neighborhood of the point (x). The (**first, first-order**) **derivative** or (**first-order**) **differential coefficient of** $f(x)$ **with respect to** x at the point (x) is the limit

$$\lim_{\Delta x \to 0} \frac{f(x + \Delta x) - f(x)}{\Delta x} \equiv \lim_{\Delta x \to 0} \frac{\Delta y}{\Delta x} \equiv \frac{dy}{dx} \equiv \frac{d}{dx} f(x) \equiv f'(x) \equiv y' \qquad (4.5\text{-}1)$$

The function $dy/dx \equiv f'(x)$ is a measure of the *rate of change of y with respect to x* at each point (x) where the limit (1) exists. On a graph of $y = f(x)$ (Sec. 4.2-1*a*), $f'(x)$ corresponds to the *slope of the tangent* (Sec. 17.1-1).

The corresponding *unilateral* limits (Sec. 4.4-7*a*) are the **left-hand derivative** $f'_-(x)$ and the **right-hand derivative** $f'_+(x)$ of $f(x)$ at the point (x).

(**b**) **The second, third, . . . , n^{th} derivatives** (**second-order, third-order, . . . , n^{th}-order differential coefficients**) **of** $y = f(x)$

with respect to x at the point (x) are respectively defined as

$$\frac{d}{dx}f'(x) \equiv \frac{d^2y}{dx^2} \equiv f''(x) \qquad \frac{d}{dx}f''(x) \equiv \frac{d^3y}{dx^3} \equiv f'''(x), \ldots$$
$$\frac{d}{dx}f^{(n-1)}(x) \equiv \frac{d^ny}{dx^n} \equiv f^{(n)}(x) \qquad (4.5\text{-}2)$$

if the limits in question exist.

Table 4.5-1. Derivatives of Frequently Used Functions (see also Chap. 21 for derivatives of special functions)

(a)

$f(x)$	$f'(x)$	$f^{(r)}(x)$
x^a	ax^{a-1}	$a(a-1)(a-2)\cdots(a-r+1)x^{a-r}$
e^x	e^x	e^x
a^x	$a^x \log_e a$	$a^x(\log_e a)^r$
$\log_e x$	$\frac{1}{x}$	$(-1)^{r-1}(r-1)!\frac{1}{x^r}$
$\log_a x$	$\frac{1}{x}\log_a e$	$(-1)^{r-1}(r-1)!\frac{1}{x^r}\log_a e$
$\sin x$	$\cos x$	$\sin\left(x+\frac{\pi r}{2}\right)$
$\cos x$	$-\sin x$	$\cos\left(x+\frac{\pi r}{2}\right)$

(b)

$f(x)$	$f'(x)$	$f(x)$	$f'(x)$
$\tan x$	$\frac{1}{\cos^2 x}$	$\arcsin x$	$\frac{1}{\sqrt{1-x^2}}$
$\cot x$	$-\frac{1}{\sin^2 x}$	$\arccos x$	$-\frac{1}{\sqrt{1-x^2}}$
$\sec x$	$\frac{\sin x}{\cos^2 x}$	$\arctan x$	$\frac{1}{1+x^2}$
$\operatorname{cosec} x$	$-\frac{\cos x}{\sin^2 x}$	$\operatorname{arccot} x$	$-\frac{1}{1+x^2}$
$\sinh x$	$\cosh x$	$\sinh^{-1} x$	$\frac{1}{\sqrt{x^2+1}}$
$\cosh x$	$\sinh x$	$\cosh^{-1} x$	$\frac{1}{\sqrt{x^2-1}}$
$\tanh x$	$\frac{1}{\cosh^2 x}$	$\tanh^{-1} x$	$\frac{1}{1-x^2}$
$\coth x$	$-\frac{1}{\sinh^2 x}$	$\coth^{-1} x$	$\frac{1}{1-x^2}$
$\operatorname{vers} x$	$\sin x$	x^x	$x^x(1+\log_e x)$

(**c**) The operation of determining $f'(x)$ for a given function $f(x)$ is called **differentiation** of $f(x)$ with respect to x. $f(x)$ is **differentiable** for any value or set of values of x such that $f'(x)$ exists; $f(x)$ is **continuously differentiable (smooth)** wherever $f'(x)$ exists and is continuous. $f(x)$ is **piecewise continuously differentiable** on an interval I if and only if $f(x)$ and $f'(x)$ exist and are piecewise continuous (Sec. 4.4-7*c*) on I. *A function is continuous wherever it is differentiable.*

Derivatives of a number of frequently used functions are tabulated in Table 4.5-1. Additional derivatives may be obtained through use of the differentiation rules of Sec. 4.5-3.

4.5-2. Partial Derivatives. (**a**) Let $y = f(x_1, x_2, \ldots, x_n)$ be a real single-valued function of the real variables $x_1, x_2, \ldots, x_n$ in a neighborhood of the point $(x_1, x_2, \ldots, x_n)$. The (**first-order**) **partial derivative of** $f(x_1, x_2, \ldots, x_n)$ **with respect to** x_1 at the point $(x_1, x_2, \ldots, x_n)$ is the limit

$$\lim_{\Delta x_1 \to 0} \frac{f(x_1 + \Delta x_1, x_2, x_3, \ldots, x_n) - f(x_1, x_2, \ldots, x_n)}{\Delta x_1} \equiv \frac{\partial}{\partial x_1} f \equiv \frac{\partial y}{\partial x_1} \equiv f_{x_1}(x_1, x_2, \ldots, x_n) \qquad (4.5\text{-}3)$$

The function $\dfrac{\partial y}{\partial x_1} \equiv \left(\dfrac{\partial y}{\partial x_1}\right)_{x_1, x_2, \ldots, x_n} \equiv f_{x_1}(x_1, x_2, \ldots, x_n)$ is a measure of the *rate of change of y with respect to x_1 for fixed values of the remaining independent variables* at each point $(x_1, x_2, \ldots, x_n)$ where the limit (3) exists. The partial derivatives $\partial y/\partial x_2, \partial y/\partial x_3, \ldots, \partial y/\partial x_n$ are defined in an analogous manner. *Each partial derivative $\partial y/\partial x_k$ may be found by differentiation of $f(x_1, x_2, \ldots, x_n)$ with respect to x_k while the remaining $n - 1$ independent variables are regarded as constant parameters* [**partial differentiation** of $f(x_1, x_2, \ldots, x_n)$ with respect to x_k].

(**b**) *Higher-order partial derivatives* of $y = f(x_1, x_2, \ldots, x_n)$ are defined by

$$\frac{\partial^2 y}{\partial x_k^2} \equiv f_{x_k x_k} \equiv \frac{\partial}{\partial x_k}\frac{\partial y}{\partial x_k} \qquad \frac{\partial^2 y}{\partial x_i\, \partial x_k} \equiv f_{x_i x_k} \equiv \frac{\partial}{\partial x_i}\frac{\partial y}{\partial x_k}$$
$$\frac{\partial^3 y}{\partial x_i\, \partial x_j\, \partial x_k} \equiv f_{x_i x_j x_k} \equiv \frac{\partial}{\partial x_i}\frac{\partial^2 y}{\partial x_j\, \partial x_k} \qquad \cdots \qquad \text{etc.}$$

if the limits in question exist; in each case, the number of differentiations involved is the **order** of the partial derivative. Note that

$$\frac{\partial y}{\partial x_i\, \partial x_k} = \frac{\partial y}{\partial x_k\, \partial x_i} \qquad (i \neq k) \qquad (4.5\text{-}4)$$

if (1) $\dfrac{\partial y}{\partial x_i\, \partial x_k}$ *exists throughout a neighborhood of the point* $(x_1, x_2, \ldots, x_n)$

and is continuous at $(x_1, x_2, \ldots, x_n)$, *and* (2) $\dfrac{\partial y}{\partial x_k\,\partial x_i}$ *exists at* $(x_1, x_2, \ldots, x_n)$.

(**c**) $y = f(x_1, x_2, \ldots, x_n)$ is **differentiable with respect to** x_k at any point $(x_1, x_2, \ldots, x_n)$ where $\partial y/\partial x_k$ exists. $y = f(x_1, x_2, \ldots, x_n)$ is **differentiable** wherever $\partial y/\partial x_1$, $\partial y/\partial x_2$, . . . , and $\partial y/\partial x_n$ exist. $y = f(x_1, x_2, \ldots, x_n)$ is **continuously differentiable** wherever all these partial derivatives exist and are continuous. $y = f(x_1, x_2, \ldots, x_n)$ is **piecewise continuously differentiable** on a region V if and only if $f(x_1, x_2, \ldots, x_n)$, $\partial y/\partial x_1$, $\partial y/\partial x_2$, . . . , and $\partial y/\partial x_n$ exist and are piecewise continuous on V (Sec. 4.4-7*c*).

4.5-3. Differentials. (**a**) Given an arbitrary small change (increment) dx of the independent variable x (**differential of the independent variable** x), the corresponding term $dy \equiv df \equiv \dfrac{dy}{dx}\,dx \equiv f'(x)\,dx$ in the expansion

$$f(x + dx) - f(x) = f'(x)\,dx + o(dx^2)$$

(Sec. 4.10-4) of a differentiable function $y = f(x)$ is called the (**first-order**) **differential** of the dependent variable y at the point (x). Similarly, the (first-order) differential of a differentiable function

$$y = f(x_1, x_2, \ldots, x_n)$$

of n variables $x_1, x_2, \ldots, x_n$ is

$$dy \equiv df \equiv \frac{\partial f}{\partial x_1}\,dx_1 + \frac{\partial f}{\partial x_2}\,dx_2 + \cdots + \frac{\partial f}{\partial x_n}\,dx_n \tag{4.5-5}$$

The operation of obtaining $dy \equiv df(x_1, x_2, \ldots, x_n)$ is sometimes called **total differentiation,** and df is referred to as a **total differential** made up of the "partial changes" $(\partial f/\partial x_k)\,dx_k$.

(**b**) The differential of each independent variable is regarded as a constant, so that $d^2x \equiv d(dx) \equiv d^3x \equiv d(d^2x) \equiv \cdots \equiv 0$. The differential of a dependent variable is a function of the independent variable or variables. The **second-order, third-order,** . . . **differentials** of suitably differentiable functions are obtained by successive differentiations of the first-order differential, e.g.,

$$d^2f(x) \equiv d(df) \equiv d\left(\frac{dy}{dx}\,dx\right) \equiv \frac{d^2y}{dx^2}\,dx^2 \tag{4.5-6}$$

$$d^2f(x_1, x_2) \equiv d(df) \equiv \frac{\partial^2 f}{\partial x_1^2}\,dx_1^2 + 2\,\frac{\partial^2 f}{\partial x_1\,\partial x_2}\,dx_1\,dx_2 + \frac{\partial^2 f}{\partial x_2^2}\,dx_2^2 \tag{4.5-7}$$

$$d^rf(x_1, x_2) \equiv \sum_{k=0}^{r} \binom{r}{k} \frac{\partial^r f}{\partial x_1^{r-k}\,\partial x_2^{k}}\,dx_1^{r-k}\,dx_2^{k} \tag{4.5-8}$$

Note also

$$d^2f[u_1(x), u_2(x), \ldots, u_n(x)] \equiv d(df) \equiv \frac{d^2f}{dx^2}\,dx^2$$

$$\equiv \sum_{k=1}^{n}\left(d\,\frac{\partial f}{\partial u_k}\,du_k + \frac{\partial f}{\partial u_k}\,d^2u_k\right)$$

$$\equiv \left(\sum_{i=1}^{n}\sum_{k=1}^{n}\frac{\partial^2 f}{\partial u_i\,\partial u_k}\,\frac{du_i}{dx}\,\frac{du_k}{dx} + \sum_{k=1}^{n}\frac{\partial f}{\partial u_k}\,\frac{d^2u_k}{dx^2}\right)dx^2 \qquad (4.5\text{-}9)$$

(**c**) Given a problem involving n independent variables $x_1, x_2, \ldots, x_n$, any function of the order of $dx_1^{r_1}\,dx_2^{r_2}\,\ldots\,dx_n^{r_n}$ as $dx_1 \to 0$, $dx_2 \to 0$, $\ldots$, $dx_n \to 0$ (Sec. 4.4-3) is an **infinitesimal of order** $r_1 + r_2 + \cdots + r_n$. In particular, the rth-order differential d^rf of a suitably differentiable function is an infinitesimal of order r.

4.5-4. Differentiation Rules. (**a**) Table 4.5-2 summarizes the most important differentiation rules. The formulas of Table 4.5-2*a* and *b* apply to *partial differentiation* if $\partial/\partial x_k$ is substituted for d/dx in each case. Thus, if $u_i = u_i(x_1, x_2, \ldots, x_n)$ $(i = 1, 2, \ldots, m)$,

$$\boxed{\frac{\partial}{\partial x_k} f(u_1, u_2, \ldots, u_m) = \sum_{i=1}^{m}\frac{\partial f}{\partial u_i}\,\frac{\partial u_i}{\partial x_k} \qquad (k = 1, 2, \ldots, n)} \qquad (4.5\text{-}10)$$

Multiplication of each formula of Table 4.5-2*a* and *b* by dx or dx^r yields an analogous rule for *total differentiation* (see also Sec. 4.5-3); thus

$$d(u + v) = du + dv \qquad d(uv) = v\,du + u\,dv \qquad (4.5\text{-}11)$$

Refer to Table 4.6-1 for the differentiation of *integrals*, and to Sec. 4.8-4*c* for the differentiation of *infinite series*.

4.5-5. Homogeneous Functions. $f(x_1, x_2, \ldots, x_n)$ is **homogeneous of degree** r in its arguments $x_1, x_2, \ldots, x_n$ if and only if $f(\alpha x_1, \alpha x_2, \ldots, \alpha x_n) \equiv \alpha^r f(x_1, x_2, \ldots, x_n)$ (see also Sec. 1.4-3*a*). *If $f(x_1, x_2, \ldots, x_n)$ is continuously differentiable and homogeneous of degree r, then*

$$x_1\frac{\partial f}{\partial x_1} + x_2\frac{\partial f}{\partial x_2} + \cdots + x_n\frac{\partial f}{\partial x_n} = rf(x_1, x_2, \ldots, x_n)$$

(EULER'S THEOREM ON HOMOGENEOUS FUNCTIONS) (4.5-12)

4.5-6. Jacobians and Functional Dependence (see also Secs. 4.6-12, 6.2-3*b*, and 16.1-2). A set of *transformation equations* (see also Sec. 14.1-3)

$$y_i = y_i(x_1, x_2, \ldots, x_n) \qquad (i = 1, 2, \ldots, n) \qquad (4.5\text{-}13)$$

define a reciprocal one-to-one correspondence between sets $(x_1, x_2, \ldots, x_n)$ and $(y_1, y_2, \ldots, y_n)$ throughout a neighborhood of a "point" $(x_1, x_2,$

Table 4.5-2. Differentiation Rules (Sec. 4.5-4; existence of continuous derivatives is assumed in each case)

(a) Basic Rules

$$\frac{d}{dx} f[u_1(x), u_2(x), \ldots, u_m(x)] = \frac{\partial f}{\partial u_1}\frac{du_1}{dx} + \frac{\partial f}{\partial u_2}\frac{du_2}{dx} + \cdots + \frac{\partial f}{\partial u_m}\frac{du_m}{dx}$$

$$\frac{d}{dx} f[u(x)] = \frac{df}{du}\frac{du}{dx} \qquad \frac{d^2}{dx^2} f[u(x)] = \frac{d^2 f}{du^2}\left(\frac{du}{dx}\right)^2 + \frac{df}{du}\frac{d^2u}{dx^2}$$

(b) Sums, Products, and Quotients. Logarithmic Differentiation

$$\frac{d}{dx}[u(x) + v(x)] = \frac{du}{dx} + \frac{dv}{dx} \qquad \frac{d}{dx}[\alpha u(x)] = \alpha \frac{du}{dx}$$

$$\frac{d}{dx}[u(x)v(x)] = v\frac{du}{dx} + u\frac{dv}{dx} \qquad \frac{d}{dx}\left[\frac{u(x)}{v(x)}\right] = \frac{1}{v^2}\left(v\frac{du}{dx} - u\frac{dv}{dx}\right) \quad [v(x) \neq 0]$$

$$\frac{d}{dx}\log_e y(x) = \frac{y'(x)}{y(x)} \qquad [\text{LOGARITHMIC DERIVATIVE OF } y(x)]$$

NOTE: To differentiate functions of the form $y = \dfrac{u_1(x)u_2(x)\cdots}{v_1(x)v_2(x)\cdots}$, it may be convenient to find the logarithmic derivative first.

$$\frac{d^r}{dx^r}(\alpha u + \beta v) = \alpha\frac{d^r u}{dx^r} + \beta\frac{d^r v}{dx^r} \qquad \frac{d^r}{dx^r}(uv) = \sum_{k=0}^{r}\binom{r}{k}\frac{d^{r-k}u}{dx^{r-k}}\frac{d^k v}{dx^k}$$

(c) Inverse Function Given. If $y = y(x)$ has the unique inverse function $x = x(y)$, and $dx/dy \neq 0$,

$$\frac{dy}{dx} = \left(\frac{dx}{dy}\right)^{-1} \qquad \frac{d^2y}{dx^2} = -\frac{d^2x}{dy^2}\Big/\left(\frac{dx}{dy}\right)^3$$

(d) Implicit Functions (see also Sec. 4.5-7). If $y = y(x)$ is given implicitly in terms of a suitably differentiable relation $F(x, y) = 0$, where $F_y \neq 0$,

$$\frac{dy}{dx} = -\frac{F_x}{F_y} \qquad \frac{d^2y}{dx^2} = -\frac{1}{F_y^{\,3}}(F_{xx}F_y^{\,2} - 2F_{xy}F_xF_y + F_{yy}F_x^{\,2})$$

(e) Function Given in Terms of a Parameter t. Given $x = x(t)$, $y = y(t)$ and $\dot{x}(t) \equiv \dfrac{dx}{dt} \neq 0$, $\dot{y}(t) \equiv \dfrac{dy}{dt}$, $\ddot{x}(t) \equiv \dfrac{d^2x}{dt^2}$, $\ddot{y}(t) \equiv \dfrac{d^2y}{dt^2}$,

$$\frac{dy}{dx} = \frac{\dot{y}(t)}{\dot{x}(t)} \qquad \frac{d^2y}{dx^2} = \frac{\dot{x}(t)\ddot{y}(t) - \ddot{x}(t)\dot{y}(t)}{[\dot{x}(t)]^3}$$

$\ldots, x_n)$ where the functions (13) are single-valued and continuously differentiable, and where the **Jacobian** or **functional determinant**

$$\frac{\partial(y_1, y_2, \ldots, y_n)}{\partial(x_1, x_2, \ldots, x_n)} \equiv \det\left[\frac{\partial y_i}{\partial x_k}\right] \tag{4.5-14}$$

is different from zero.

The Jacobian (14) vanishes whenever the differentials

$$dy_i \equiv \sum_{k=1}^{n} \frac{\partial y_i}{\partial x_k} dx_k \qquad (i = 1, 2, \ldots, n) \tag{4.5-15}$$

are linearly dependent (Sec. 1.9-3). *If the functions* (13) *are continuously differentiable throughout a neighborhood of a point* $(x_1, x_2, \ldots, x_n)$ *where the rank of the matrix* $[\partial y_i/\partial x_k]$ (Sec. 13.2-7) *is everywhere* $m < n$, *then* $(x_1, x_2, \ldots, x_n)$ *has a neighborhood V where the* y_i *are related by a continuously differentiable relation*

$$\Phi(y_1, y_2, \ldots, y_n) = 0$$

such that the $\partial\Phi/\partial y_i$ *do not all vanish simultaneously.* In this case the functions (13) are said to be **functionally dependent** over V (see also Sec. 4.5-7*c*).

4.5-7. Implicit Functions. (**a**) If a function $y = y(x)$ is given **implicitly** in terms of a suitably differentiable relation $F(x, y) = 0$, then

$$\frac{dy}{dx} = -\frac{F_x}{F_y} \qquad \frac{d^2y}{dx^2} = -\frac{1}{F_y^3}(F_{xx}F_y^2 - 2F_{xy}F_xF_y + F_{yy}F_x^2) \qquad (F_y \neq 0) \tag{4.5-16}$$

(**b**) If m functions $y_1 = y_1(x_1, x_2, \ldots, x_n)$, $y_2 = y_2(x_1, x_2, \ldots, x_n)$, $\ldots$, $y_m = y_m(x_1, x_2, \ldots, x_n)$ of n independent variables $x_1, x_2, \ldots, x_n$ are given implicitly in terms of m continuously differentiable relations

$$F_i(y_1, y_2, \ldots, y_m; x_1, x_2, \ldots, x_n) = 0 \qquad (i = 1, 2, \ldots, m) \tag{4.5-17}$$

where the F_i are single-valued functions, then

1. The differentials dy_j and dx_k are related by m linear equations $dF_i = 0$.
2. For each value of $k = 1, 2, \ldots, n$, the m derivatives $\partial y_j/\partial x_k$ may be obtained by Cramer's rule (Sec. 1.9-2) from the m linear equations

$$\sum_{j=1}^{m} \frac{\partial F_i}{\partial y_j}\frac{\partial y_j}{\partial x_k} + \frac{\partial F_i}{\partial x_k} = 0 \qquad (i = 1, 2, \ldots, m) \tag{4.5-18}$$

provided that

$$\frac{\partial(F_1, F_2, \ldots, F_m)}{\partial(y_1, y_2, \ldots, y_m)} \equiv \det\left[\frac{\partial F_i}{\partial y_j}\right] \neq 0 \tag{4.5-19}$$

for the values of $x_1, x_2, \ldots, x_n$ in question. Differentiation of the Eqs. (18) yields relations involving higher derivatives of the y_j.

In particular, two continuously differentiable relations $F(x, y, z) = 0$, $G(x, y, z) = 0$ imply

$$dx:dy:dz = \begin{vmatrix} F_y & F_z \\ G_y & G_z \end{vmatrix} : \begin{vmatrix} F_z & F_x \\ G_z & G_x \end{vmatrix} : \begin{vmatrix} F_x & F_y \\ G_x & G_y \end{vmatrix} \qquad (4.5\text{-}20)$$

whenever the determinant in the denominator is different from zero. The computation of implicit derivatives is of particular importance in thermodynamics (Ref. 4.6).

(c) **An Existence Theorem for Implicit Functions** (see also Sec. 4.2-1b). Given a "point" $P \equiv (x_1, x_2, \ldots, x_n; y_1, y_2, \ldots, y_m)$ such that Eqs. (17) and (19) hold, *m relations* (17) *define the y_j as single-valued continuous functions within a neighborhood of the "point" $(x_1, x_2, \ldots, x_n)$ if all F_i and $\partial F_i/\partial y_j$ exist and are continuous throughout a neighborhood D of P. If, in addition, all $\partial F_i/\partial x_k$ exist and are continuous throughout D, then the derivatives $\partial y_j/\partial x_k$ exist and are continuous within a neighborhood of $(x_1, x_2, \ldots, x_n)$.* If the Jacobian (19) vanishes, then the y_j are not uniquely determined.

4.6. INTEGRALS AND INTEGRATION

4.6-1. Definite Integrals (Riemann Integrals).

(a) A real function $f(x)$ bounded on the bounded closed interval $[a, b]$ is **integrable over** (a, b) **in the sense of Riemann** if and only if the sum $\sum_{i=1}^{m} f(\xi_i)(x_i - x_{i-1})$ tends to a unique finite limit I for every sequence of partitions $a = x_0 < \xi_1 < x_1 < \xi_2 < x_2 \cdots < \xi_m < x_m = b$ as $\max |x_i - x_{i-1}| \to 0$. In this case

$$I = \lim_{\max |x_i - x_{i-1}| \to 0} \sum_{i=1}^{m} f(\xi_i)(x_i - x_{i-1}) = \int_a^b f(x)\,dx \qquad (4.6\text{-}1)$$

is the **definite integral of** $f(x)$ **over** (a, b) **in the sense of Riemann (Riemann integral).** $f(x)$ is called the **integrand;** a and b are the **limits of integration.** *Table* 4.6-1 *summarizes important properties of definite integrals.*

$\int_a^b f(x)\,dx$ represents the *area* bounded by the curve $y = f(x)$ (Sec. 4.2-1a) and the x axis between the lines $x = a$ and $x = b$; areas below the x axis are represented by negative numbers.

(b) *A function $f(x)$ bounded on the bounded closed interval $[a, b]$ is integrable over (a, b) in the sense of Riemann if and only if $f(x)$ is continuous almost everywhere on (a, b)* (Sec. 4.6-14b). This is true, in particular, (1) *if $f(x)$ is continuous on $[a, b]$*; (2) *if $f(x)$ is bounded on (a, b) and continuous on (a, b) except possibly for a discrete set of discontinuities;* (3) *if $f(x)$ is bounded and monotonic on (a, b)*; (4) *if $f(x)$ is of bounded variation in (a, b)* (see also Sec. 4.4-8). *If $f(x)$ is integrable over (on) (a, b), $f(x)$ is necessarily integrable over every subinterval of (a, b).*

4.6-2. Improper Integrals.

(a) Given a function $f(x)$ bounded and integrable on every bounded subinterval of (a, b), the concept of a definite integral $\int_a^b f(x)\,dx$ (Sec. 4.6-1) can be extended to apply even if

Table 4.6-1. Properties of Integrals

(a) Elementary Properties. *If the integrals exist,*

$$\int_a^b f(x)\,dx = -\int_b^a f(x)\,dx \qquad \int_a^b f(x)\,dx = \int_a^c f(x)\,dx + \int_c^b f(x)\,dx$$

$$\int_a^b [u(x) + v(x)]\,dx = \int_a^b u(x)\,dx + \int_a^b v(x)\,dx \qquad \int_a^b \alpha u(x)\,dx = \alpha \int_a^b u(x)\,dx$$

(b) Integration by Parts. *If $u(x)$ and $v(x)$ are differentiable for $a \leq x \leq b$, and if the integrals exist,*

$$\int_a^b u(x)v'(x)\,dx = u(x)v(x)\Big]_a^b - \int_a^b v(x)u'(x)\,dx$$

or

$$\int_a^b u\,dv = uv\Big]_a^b - \int_a^b v\,du$$

(c) Change of Variable (Integration by Substitution). *If $u = u(x)$ and its inverse function $x = x(u)$ are single-valued and continuously differentiable for $a \leq x \leq b$, and if the integral exists,*

$$\int_a^b f(x)\,dx = \int_{u(a)}^{u(b)} f[(x(u)]\,\frac{dx}{du}\,du = \int_{u(a)}^{u(b)} f[x(u)]\left(\frac{du}{dx}\right)^{-1} du$$

(d) Differentiation with Respect to a Parameter. *If $f(x, \lambda)$, $u(\lambda)$, and $v(\lambda)$ are continuously differentiable with respect to λ,*

$$\frac{\partial}{\partial\lambda}\int_a^b f(x, \lambda)\,dx = \int_a^b \frac{\partial}{\partial\lambda} f(x, \lambda)\,dx$$

$$\frac{\partial}{\partial\lambda}\int_{u(\lambda)}^{v(\lambda)} f(x, \lambda)\,dx = \int_{u(\lambda)}^{v(\lambda)} \frac{\partial}{\partial\lambda} f(x, \lambda)\,dx + f(v, \lambda)\,\frac{\partial v}{\partial\lambda} - f(u, \lambda)\,\frac{\partial u}{\partial\lambda}$$

(LEIBNITZ'S RULE)

provided that the integrals exist and, in the case of improper integrals, converge uniformly in a neighborhood of the point (λ).

The second case can often be reduced to the first by a suitable change of variables. Note also

$$\frac{\partial}{\partial\lambda}\int_a^\lambda f(x, \lambda)\,dx = \frac{1}{\lambda - a}\int_a^\lambda \left[f(x, \lambda) + (\lambda - a)\,\frac{\partial f}{\partial\lambda} + (x - a)\,\frac{\partial f}{\partial x}\right] dx$$

(e) Inequalities (see also Sec. 4.6-19). *If the integrals exist,*

$$f(x) \leq g(x) \text{ in } (a, b) \text{ implies } \int_a^b f(x)\,dx \leq \int_a^b g(x)\,dx$$

If $|f(x)| \leq M$ on the bounded interval (a, b), the existence of $\int_a^b f(x)\,dx$ implies the existence of $\int_a^b |f(x)|\,dx$, and

$$\left|\int_a^b f(x)\,dx\right| \leq \int_a^b |f(x)|\,dx \leq M(b - a)$$

1. $f(x)$ is unbounded in a neighborhood of a finite limit of integration $x = a$ or $x = b$ (see also Sec. 4.6-2*b*).
2. The interval (a, b) is unbounded.

Thus, if $f(x)$ is bounded and integrable on every finite interval (a, X) for $a < X < b$, one *defines*

$$\int_a^b f(x)\,dx = \int_a^{b-0} f(x)\,dx = \lim_{X\to b-0} \int_a^X f(x)\,dx \tag{4.6-2a}$$

and, in particular, for $b = \infty$,

$$\int_a^\infty f(x)\,dx = \lim_{X\to\infty} \int_a^X f(x)\,dx \tag{4.6-2b}$$

Similarly

$$\int_a^b f(x)\,dx = \int_{a+0}^b f(x)\,dx = \lim_{X\to a+0} \int_X^b f(x)\,dx$$
$$\int_{-\infty}^b f(x)\,dx = \lim_{X\to\infty} \int_{-X}^b f(x)\,dx \tag{4.6-2c}$$

Each improper integral defined in this manner exists or **converges** if and only if the limit on the right exists. An improper integral over $f(x)$ **converges absolutely** if and only if the corresponding improper integral over $|f(x)|$ converges. *Absolute convergence implies convergence* (see also Secs. 4.6-15*e* and 4.9-3). A convergent improper integral which does not converge absolutely is **conditionally convergent.**

(**b**) *The integration rules of Table* 4.6-1 *apply to suitably convergent improper integrals.* Given a bounded or unbounded interval (a, b) or $[a, b]$ containing a discrete set of points $x = c_1$, $x = c_2$, . . . such that $f(x)$ is unbounded in a neighborhood of $x = c_i (i = 1, 2, \ldots)$, $\int_a^b f(x)\,dx$ may be defined as an improper integral equal to a sum of improper integrals (2); e.g.,

$$\int_a^b f(x)\,dx = \lim_{X_1\to a+0} \int_{X_1}^c f(x)\,dx + \lim_{X_2\to b-0} \int_c^{X_2} f(x)\,dx \qquad (a < c < b) \tag{4.6-3}$$

$$\int_{-\infty}^\infty f(x)\,dx = \lim_{X_1\to\infty} \int_{-X_1}^c f(x)\,dx + \lim_{X_2\to\infty} \int_c^{X_2} f(x)\,dx \tag{4.6-4}$$

$$\int_a^b f(x)\,dx = \lim_{X_1\to c-0} \int_a^{X_1} f(x)\,dx + \lim_{X_2\to c+0} \int_{X_2}^b f(x)\,dx \qquad (a < c < b) \tag{4.6-5}$$

if the limits exist.

Even if the integrals (4) and (5) do not exist, their respective **Cauchy principal values**

$$\lim_{X\to\infty} \int_{-X}^{X} f(x)\,dx \qquad \text{and} \qquad \lim_{\delta\to 0}\left[\int_a^{c-\delta} f(x)\,dx + \int_{c+\delta}^{b} f(x)\,dx\right] \tag{4.6-6}$$

may exist; if either integral exists it is necessarily equal to its principal value.

(c) An improper integral $\int_a^b f(x, y)\,dx$ **converges uniformly on a set S of values of y** if and only if the corresponding limit (Secs. 4.6-2a and b) converges uniformly on S (Sec. 4.4-4). *If $f(x, y)$ is a continuous function, then $\int_a^b f(x, y)\,dx$ is a continuous function of y in every open interval where the integral converges uniformly (Continuity Theorem).*

(d) Criteria for convergence and uniform convergence of improper integrals are listed in Secs. 4.9-3 and 4.9-4.

4.6-3. Arithmetic Means. The **arithmetic means (averages)** of $f(x)$ over the respective intervals (a, b), $(0, \infty)$, and $(-\infty, \infty)$ are defined as

$$\frac{1}{b-a}\int_a^b f(x)\,dx \qquad \lim_{X\to\infty}\frac{1}{X}\int_0^X f(x)\,dx \qquad \text{and} \qquad \lim_{X\to\infty}\frac{1}{2X}\int_{-X}^{X} f(x)\,dx \tag{4.6-7}$$

if these quantities exist.

4.6-4. Indefinite Integrals. A given single-valued function $f(x)$ has an **indefinite integral** $F(x)$ in $[a, b]$ if and only if there exists a function $F(x)$ such that $F'(x) = f(x)$ in $[a, b]$. In this case $F(x)$ is uniquely defined in $[a, b]$ except for an arbitrary additive constant C (**constant of integration;** see also Sec. 9.1-2); one writes

$$F(x) = \int f(x)\,dx + C \qquad (a \leq x \leq b) \tag{4.6-8}$$

Note that $F(x) - F(a) \equiv F(x)\Big]_a^x$ is uniquely defined for $a \leq x \leq b$.

Note also

$$\int [\alpha u(x) + \beta v(x)]\,dx \equiv \alpha \int u(x)\,dx + \beta \int v(x)\,dx \tag{4.6-9}$$

$$\int u(x)v'(x)\,dx \equiv u(x)v(x) - \int u'(x)v(x)\,dx \tag{4.6-10}$$

if the indefinite integrals exist (see also Tables 4.5-2 and 4.6-1).

4.6-5. The Fundamental Theorem of the Integral Calculus. *If $f(x)$ is single-valued, bounded, and integrable on $[a, b]$, and there exists a function $F(x)$ such that $F'(x) = f(x)$ for $a \leq x \leq b$, then*

$$\boxed{\int_a^x f(\xi)\,d\xi = F(x)\Big]_a^x = F(x) - F(a) \qquad (a \leq x \leq b)} \tag{4.6-11}$$

In particular, *if $f(x)$ is continuous in $[a, b]$,*

$$\frac{d}{dx}\int_a^x f(\xi)\,d\xi = f(x) \qquad (a \leq x \leq b) \tag{4.6-12}$$

and Eq. (11) *applies.*

More generally, *whenever $\int_a^b f(\xi)\,d\xi$ exists, the function $\int_a^x f(\xi)\,d\xi$ exists and is con-*

tinuous and of bounded variation (Sec. 4.4-8b) in (a, b), and Eq. (12) holds for almost all x in $[a, b]$ (Sec. 4.6-14b).

NOTE: The fundamental theorem of the integral calculus enables one (1) to evaluate definite integrals by reversing the process of differentiation, and (2) to solve differential equations by numerical evaluation of definite integrals (see also Sec. 4.6-6).

4.6-6. Integration Methods. (a) **Integration** is the operation yielding a (definite or indefinite) integral of a given integrand $f(x)$. Definite integrals may be calculated directly as limits of sums (numerical integration, Secs. 20.6-2 and 20.6-3) or by the calculus of residues (Sec. 7.7-3); more frequently, one attempts to find an indefinite integral which may be inserted into Eq. (11). To obtain an indefinite integral, one must reduce the given integrand $f(x)$ to a sum of known derivatives with the aid of the "integration rules" listed in Table 4.6-1*a*, *b*, *c*.

The remainder of this section deals with integration methods applicable to special types of integrands. **Comprehensive tables of definite and indefinite integrals are presented in Appendix E.**

(b) **Integration of Polynomials.**

$$\int (a_n + a_{n-1}x + a_{n-2}x^2 + \cdots + a_0x^n)\,dx \equiv a_nx + \frac{1}{2}a_{n-1}x^2 + \frac{1}{3}a_{n-2}x^3 + \cdots + \frac{1}{n+1}a_0x^{n+1} + C \tag{4.6-13}$$

(c) **Integration of Rational Functions.** The methods outlined in Secs. 1.7-2 and 1.7-4 will reduce every rational integrand to the sum of a polynomial (Sec. 4.6-6b) and a set of *partial fractions* (1.7-5) and/or (1.7-6). The partial-fraction terms are integrated successively with the aid of the following formulas:

$$\begin{gathered}
\int \frac{dx}{(x - x_1)^m} \equiv \begin{cases} -\dfrac{1}{(m-1)(x-x_1)^{m-1}} + C & (m \neq 1) \\ \log_e (x - x_1) + C & (m = 1) \end{cases} \\
\int \frac{dx}{[(x-a)^2 + \omega^2]} \equiv \frac{1}{\omega}\arctan\frac{x-a}{\omega} + C \\
\int \frac{dx}{[(x-a)^2+\omega^2]^{m+1}} \equiv \frac{x-a}{2m\omega^2[(x-a)^2+\omega^2]^m} + \frac{2m-1}{2m\omega^2}\int\frac{dx}{[(x-a)^2+\omega^2]^m} \\
\int \frac{x\,dx}{[(x-a)^2+\omega^2]^{m+1}} \equiv \frac{a(x-a)-\omega^2}{2m\omega^2[(x-a)^2+\omega^2]^m} + \frac{(2m-1)a}{2m\omega^2}\int\frac{dx}{[(x-a)^2+\omega^2]^m}
\end{gathered} \tag{4.6-14}$$

(d) Integrands Which Can Be Reduced to Rational Functions by a Change of Variables (Table 4.6-1*c*).

1. *If the integrand $f(x)$ is a rational function of sin x and cos x,* introduce $u = \tan (x/2)$, so that

$$\sin x = \frac{2u}{1+u^2} \qquad \cos x = \frac{1-u^2}{1+u^2} \qquad dx = \frac{2du}{1+u^2} \tag{4.6-15}$$

2. *If the integrand $f(x)$ is a rational function of sinh x and cosh x,* introduce $u = \tanh (x/2)$, so that

$$\sinh x = \frac{2u}{1-u^2} \qquad \cosh x = \frac{1+u^2}{1-u^2} \qquad dx = \frac{2du}{1-u^2} \tag{4.6-16}$$

NOTE: If $f(x)$ is a rational function of $\sin^2 x$, $\cos^2 x$, $\sin x \cos x$, and $\tan x$ (or of the corresponding hyperbolic functions), one simplifies the calculation by first introducing $v = x/2$, so that $u = \tan v$ (or $u = \tanh v$).

3. *If the integrand $f(x)$ is a rational function of x and either $\sqrt{1-x^2}$ or $\sqrt{x^2-1}$,* reduce the problem to case 1 or 2 by the respective substitutions $x = \cos v$ or $x = \cosh v$.

4. *If the integrand $f(x)$ is a rational function of x and $\sqrt{x^2+1}$,* introduce $u = x + \sqrt{x^2+1}$, so that

$$x = \frac{1}{2}\left(u - \frac{1}{u}\right) \qquad \sqrt{x^2+1} = \frac{1}{2}\left(u + \frac{1}{u}\right) \qquad dx = \frac{1}{2}\left(1 + \frac{1}{u^2}\right) du \tag{4.6-17}$$

5. *If the integrand $f(x)$ is a rational function of x and $\sqrt{ax^2+bx+c}$,* reduce the problem to case 3 ($b^2 - 4ac < 0$) or to case 4 ($b^2 - 4ac > 0$) through the substitution

$$v = \frac{2ax+b}{\sqrt{|4ac-b^2|}} \qquad x = \frac{v\sqrt{|4ac-b^2|} - b}{2a} \tag{4.6-18}$$

6. *If the integrand $f(x)$ is a rational function of x and $u = \sqrt{\dfrac{ax+b}{cx+d}}$,* introduce u as a new variable.

7. *If the integrand $f(x)$ is a rational function of x, $\sqrt{ax+b}$, and $\sqrt{cx+d}$,* introduce $u = \sqrt{ax+b}$ as a new variable.

Many other substitution methods apply in special cases. Note that the integrals may not be real for all values of x.

(e) Integrands of the form $x^n e^{ax}$, $x^n \log_e x$, $x^n \sin x$, $x^n \cos x$ ($n \neq -1$); $\sin^m x \cos^n x$ ($n + m \neq 0$); $e^{ax} \sin^n x$, $e^{ax} \cos^n x$ yield to repeated *integration by parts* (Table 4.6-1*b*).

(f) Many integrals *cannot* be expressed as finite sums involving only algebraic, exponential, and trigonometric functions and their inverses. One may then expand the integrand as an infinite series (Secs. 4.10-4, 4.11-4, 15.2-6), or one resorts to numerical integration. Refer to Chap. 21 for examples of "new" functions of x defined as integrals of the form $\int_a^x f(\xi)\,d\xi$ or $\int_a^b f(x, \eta)\,d\eta$.

4.6-7. Elliptic Integrals. (a) If $f(x)$ is a rational function of x and $\sqrt{a_0x^4 + a_1x^3 + a_2x^2 + a_3x + a_4}$, $\int_a^b f(x)\,dx$ is called an **elliptic integral**; one may except the trivial case that the equation

$$a_0x^4 + a_1x^3 + a_2x^2 + a_3x + a_4 = 0$$

has multiple roots. *Every elliptic integral can be reduced to a weighted sum of elementary functions and normal elliptic integrals* (*Sec.* 21.6-5); values of the latter are available in tabular form.

4.6-8. Multiple Integrals (see also Secs. 4.6-11 to 4.6-13). (a) *Let $f(x, y)$ be piecewise continuous (Sec. 4.4-7) on a bounded closed region D which is uniquely defined by $a \leq x \leq b$, $g_1(x) \leq y \leq g_2(x)$ as well as by $\alpha \leq y \leq \beta$, $\gamma_1(y) \leq x \leq \gamma_2(y)$, where $g_1(x)$, $g_2(x)$, $\gamma_1(y)$, $\gamma_2(y)$ are piecewise continuous functions.* Then *Fubini's theorem* states

$$\begin{aligned}\int_\alpha^\beta dy \int_{\gamma_1(y)}^{\gamma_2(y)} f(x, y)\,dx &= \int_\alpha^\beta \left[\int_{\gamma_1(y)}^{\gamma_2(y)} f(x, y)\,dx\right] dy \\ &= \int_a^b dx \int_{g_1(x)}^{g_2(x)} f(x, y)\,dy = \iint\limits_D f(x, y)\,dx\,dy \qquad (4.6\text{-}19)\end{aligned}$$

Analogous theorems hold for triple, quadruple, etc., integrals.

The second expression for the integral (19) may be written without the brackets if the meaning of the integration limits is evident.

EXAMPLE: If D is the region bounded by the circle $x^2 + y^2 = 1$, and $f(x, y) = c =$ constant,

$$\iint\limits_D f(x, y)\,dx\,dy = 4\int_0^1 \int_0^{\sqrt{1-y^2}} c\,dx\,dy = 4c \int_0^1 \sqrt{1 - y^2}\,dy = \pi c$$

(b) *If $\int_\alpha^\beta dy \int_a^b f(x, y)\,dx$ exists and equals $\int_a^b dx \int_\alpha^\beta f(x, y)\,dy$ for every $b > a$, then*

$$\int_\alpha^\beta dy \int_a^\infty f(x, y)\,dx = \int_a^\infty dx \int_\alpha^\beta f(x, y)\,dy \qquad (4.6\text{-}20)$$

provided that $\int_a^\infty f(x, y)\,dx$ converges uniformly for $\alpha \leq y \leq \beta$. Similar theorems hold for other improper multiple integrals.

4.6-9. Arc Length of a Rectifiable Curve (see also Secs. 6.2-3, 6.4-3, and 17.2-1). A continuous curve segment is **rectifiable** if and only if every subsegment C_1 in the finite portion of the plane or space has a unique finite **arc length** $s(C_1)$; $s(C_1)$ is the limit of the length of an inscribed polygonal curve as the length of the largest straight-line segment approaches zero.

For a continuous arc C_1 in the Euclidean plane or space, described in terms of *rectangular cartesian* coordinates by

$$\left.\begin{array}{llll} x = x(t) & y = y(t) & \text{in the plane} & \\ \text{or} \quad x = x(t) & y = y(t) & z = z(t) & \text{in space} \end{array}\right\} \qquad (4.6\text{-}21)$$

(Secs. 2.1-9*b* and 3.1-13), the length of an infinitesimal regular arc corresponding to the interval $(t, t + dt)$ is the **element of arc length**

$$ds \equiv \begin{cases} \sqrt{dx^2 + dy^2} \equiv \sqrt{1 + \left(\frac{dy}{dx}\right)^2}\, dx \\ \quad \equiv \sqrt{\left(\frac{dx}{dt}\right)^2 + \left(\frac{dy}{dt}\right)^2}\, dt \equiv \frac{ds}{dt}\, dt \qquad \text{in the plane} \\ \sqrt{dx^2 + dy^2 + dz^2} \\ \quad \equiv \sqrt{\left(\frac{dx}{dt}\right)^2 + \left(\frac{dy}{dt}\right)^2 + \left(\frac{dz}{dt}\right)^2}\, dt \\ \quad \equiv \frac{ds}{dt}\, dt \qquad \text{in space} \end{cases} \qquad (4.6\text{-}22)$$

The sign of ds is assigned arbitrarily, usually so that $ds/dt \geq 0$.

The arc length $s(C_1)$ of a given curve segment C_1 corresponding to a finite interval (t_0, t) is

$$s = s(C_1) = \int_{C_1} ds = \int_{t_0}^{t} \frac{ds}{dt}\, dt = s(t) \qquad (4.6\text{-}23)$$

The arc length $s(C_1)$ is a geometrical object independent of the coordinate system and the particular parameter t used to describe the curve; refer to Secs. 6.2-3 and 6.4-3 for formulas expressing ds (and hence ds/dt) in terms of curvilinear coordinates.

$s(C_1)$ exists if the functions (21) *are of bounded variation on* $[t_0, t]$; in this case ds is defined almost everywhere (Sec. 4.6-14*b*) on C_1. *Every regular curve is rectifiable.*

4.6-10. Line Integrals (see also Secs. 5.4-5, 6.2-3, and 6.4-3 for vector notation and use of curvilinear coordinates). Given a rectifiable arc C described by Eq. (21) for $a \leq t \leq b$, the **line integral** $\int_C f(x, y, z)\, ds$ over a bounded function $f(x, y, z)$ is defined by

$$\int_C f(x, y, z)\, ds = \lim_{\max \Delta s_i \to 0} \sum_{i=1}^{m} f[x(\tau_i), y(\tau_i), z(\tau_i)]\, \Delta s_i$$

with

$$\Delta s_i = \sqrt{[x(t_i) - x(t_{i-1})]^2 + [y(t_i) - y(t_{i-1})]^2 + [z(t_i) - z(t_{i-1})]^2}$$
$$(a = t_0 < \tau_1 < t_1 < \tau_2 < t_2 < \cdots < \tau_m < t_m = b) \quad \text{(4.6-24}a\text{)}$$

if the limit exists (see also Sec. 4.6-1). The line integral (24a) can be computed (or directly defined) as an integral over t:

$$\int_C f(x, y, z)\, ds = \int_a^b f[x(t), y(t), z(t)] \frac{ds}{dt}\, dt \quad \text{(4.6-24}b\text{)}$$

where the element of arc length ds is given by Eq. (22). Omit terms involving z in Eq. (24) for line integrals in the xy plane. Improper line integrals are defined in the manner of Sec. 4.6-2.

4.6-11. Areas and Volumes (see also Secs. 5.4-6, 5.4-7, 6.2-3, 6.4-3, and 17.3-3 for vector notation and use of curvilinear coordinates). **(a)** The area A of the region S bounded by a simple closed curve (Sec. 3.1-13) in the Euclidean plane is the least limit of the area of an inscribed closed polygon as the largest polygon side approaches zero. *This limit exists, in particular, whenever the boundary has a finite arc length.*

The surface area A of a curved surface region S in Euclidean space is the smallest possible limit of the area of an inscribed polyhedral surface as the maximum distance between adjacent vertices decreases. The volume U enclosed by a simple closed surface (Sec. 3.1-14) is the limit of the volume of the inscribed polyhedrons.

(b) Areas and volumes can be computed (or even directly defined) as double and triple integrals over suitable coordinates:

$$A = \int_S dA = \iint_S dx\, dy \quad \text{in the Euclidean plane} \quad \text{(4.6-25)}$$

$$U = \int_V dV = \iiint_V dx\, dy\, dz \quad \text{in Euclidean space} \quad \text{(4.6-26)}$$

Areas and volumes are independent of the particular coordinate system used. In terms of curvilinear coordinates x^1, x^2, x^3 (Sec. 6.2-1)*

$$U = \int_V dV = \iiint_V \frac{\partial(x, y, z)}{\partial(x^1, x^2, x^3)}\, dx^1\, dx^2\, dx^3 \quad \text{(4.6-27)}$$

(see also Secs. 4.6-13, 6.2-3, and 6.4-3).

In terms of curvilinear surface coordinates u, v on a plane or curved surface (Sec. 3.1-14)

$$A = \int_S dA = \iint_S \sqrt{a(u, v)}\, du\, dv \quad \text{(4.6-28)}$$

The function $a(u, v)$ is specified in Sec. 17.3-3c.

* The indices in x^1, x^2, x^3 are superscripts, not exponents.

(c) **Simple Special Cases.** The plane area A bounded by the lines $x = a$ and $x = b$ and two nonintersecting curves $y = g_1(x)$, $y = g_2(x)$ is $\int_a^b [g_2(x) - g_1(x)]\,dx$ [$A > 0$ if $g_1(x) < g_2(x)$; see also Sec. 4.6-1]. The plane area bounded by a suitable closed curve C described by $x = x(t)$, $y = y(t)$ or $r = r(\varphi)$ (Sec. 2.1-8) is

$$A = \frac{1}{2}\oint_C r^2(\varphi)\,d\varphi = \frac{1}{2}\oint_C \left(x\frac{dy}{dt} - y\frac{dx}{dt}\right) dt \qquad (4.6\text{-}29)$$

where the symbol $\oint$ signifies counterclockwise integration around the closed curve.

The surface area A and the internal volume U of a *surface of revolution* (Sec. 3.1-15) generated by rotation of the curve segment $y = f(x) > 0$ $(a \leq x \leq b)$ about the x axis are given by

$$A = 2\pi \int_a^b f(x) \sqrt{1 + \left(\frac{dy}{dx}\right)^2}\, dx \qquad U = \pi \int_a^b [f(x)]^2\, dx$$

(GULDIN'S FORMULAS) (4.6-30)

4.6-12. Surface Integrals and Volume Integrals (see also Secs. 5.4-6, 5.4-7, 6.2-3, and 6.4-3 for vector notation and use of curvilinear coordinates). Surface integrals and volume integrals can be defined as limits of sums in the manner of Secs. 4.6-1 and 4.6-10; or they can be introduced directly as double and triple integrals with the aid of the elements of area and volume defined in Sec. 4.6-11. The **surface integral** of a piecewise continuous function $f(u, v)$ over a suitable bounded surface region S with surface coordinates u, v is

$$\int_S f(u, v)\, dA = \iint_S f(u, v) \sqrt{a(u, v)}\, du\, dv \qquad (4.6\text{-}31)$$

This includes integrals over plane surface regions [and, in particular, also $u = x$, $v = y$, $\sqrt{a(u, v)} \equiv 1$] as special cases.

The **volume integral** of a piecewise continuous function $f(x, y, z) \equiv f[x(x^1, x^2, x^3), y(x^1, x^2, x^3), z(x^1, x^2, x^3)]$ over a bounded region V is

$$\int_V f(x, y, z)\, dV = \iiint_V f(x, y, z)\, dx\, dy\, dz$$
$$= \iiint_V f[x(x^1, x^2, x^3), y(x^1, x^2, x^3), z(x^1, x^2, x^3)]\, \frac{\partial(x, y, z)}{\partial(x^1, x^2, x^3)}\, dx^1\, dx^2\, dx^3$$

(4.6-32)

Improper surface and volume integrals (unbounded functions and regions) are defined in the manner of Sec. 4.6-2. **Refer to Secs. 5.6-1 and 5.6-2 for formulas relating line integrals, surface integrals, and volume integrals.** The following special cases relate line integrals and surface integrals (C is the boundary of S):

$$\left.\begin{aligned}\int_S \left[\frac{\partial Q(x,y)}{\partial x} - \frac{\partial P(x,y)}{\partial y}\right] dA &= \int_C [P(x,y)\,dx + Q(x,y)\,dy] \\ &\text{(GAUSS'S INTEGRAL FORMULA)} \\ \int_S (u\nabla^2 v - v\nabla^2 u)\,dA &= \int_C \left(u\frac{\partial v}{\partial n} - v\frac{\partial u}{\partial n}\right) ds \\ &\text{(TWO-DIMENSIONAL GREEN'S FORMULA)}\end{aligned}\right\} \quad (4.6\text{-}33)$$

where $\nabla^2 v(x,y) \equiv \frac{\partial v^2}{\partial x^2} + \frac{\partial^2 v}{\partial y^2} \qquad \frac{\partial v}{\partial n} ds \equiv \frac{\partial v}{\partial x} dy - \frac{\partial v}{dy} dx$

4.6-13. Change of Variables in Volume Integrals and Surface Integrals. If one introduces new coordinates x^1, x^2, x^3 by a continuously differentiable coordinate transformation

$$x^1 = x^1(\bar{x}^1, \bar{x}^2, \bar{x}^3) \qquad x^2 = x^2(\bar{x}^1, \bar{x}^2, \bar{x}^3) \qquad x^3 = x^3(\bar{x}^1, \bar{x}^2, \bar{x}^3)$$
$$\left[\frac{\partial(x^1, x^2, x^3)}{\partial(\bar{x}^1, \bar{x}^2, \bar{x}^3)} \neq 0\right]$$

(see also Sec. 6.2-1), each volume integral (32) can be rewritten in terms of the new coordinates:

$$\int_V f(x,y,z)\,dV = \iiint\limits_V \varphi(x^1, x^2, x^3)\,dx^1\,dx^2\,dx^3 = \iiint\limits_{\bar{V}} \varphi[x^1(\bar{x}^1, \bar{x}^2, \bar{x}^3), x^2(\bar{x}^1, \bar{x}^2, \bar{x}^3), x^3(\bar{x}^1, \bar{x}^2, \bar{x}^3)] \frac{\partial(x^1, x^2, x^3)}{\partial(\bar{x}^1, \bar{x}^2, \bar{x}^3)} d\bar{x}^1\,d\bar{x}^2\,d\bar{x}^3 \quad (4.6\text{-}34)$$

Surface integrals are transformed in an analogous manner (see also Sec. 17.3-3).

4.6-14. Lebesgue Measure. Measurable Functions. (a) Measure of a Point Set. The **exterior Lebesgue measure** $m_e[S]$ of any bounded linear point set S (Sec. 4.3-3*a*) is the g.l.b. of the combined length of a set of intervals covering S. The **interior Lebesgue measure** $m_i[S]$ of S is the difference between the length $b - a$ of any bounded interval (a, b) containing S and the exterior measure of the complement of S with respect to (a, b) (Sec. 4.3-2*a*). S is a **measurable set** with the **Lebesgue measure** $m[S]$ if and only if $m_e[S] = m_i[S] = m[S]$ (*constructive definition* of the Lebesgue measure). An *unbounded* linear point set S is measurable if and only if $(-X, X) \cap S$ is measurable for all $X > 0$. In this case one defines $m[S] = \lim\limits_{X\to\infty} m[(-X, X) \cap S]$; $m[S]$ may or may not be finite.

More generally, a **measure** $M[S]$ defined on a suitable class (completely additive Boolean algebra, Sec. 12.8-2) of point sets S is a *set function* with the properties

$$M[S] \geq 0 \qquad M[0] = 0 \qquad (4.6\text{-}35)$$

and, for every (finite or infinite) countable set of nonintersecting point sets $S_1, S_2, \ldots$

$$M[S_1 \cup S_2 \cup \cdots] = M[S_1] + M[S_2] + \cdots \tag{4.6-36}$$

The *Lebesgue measure* $m[S]$ of a linear point set has the additional property

$$m[(a, b)] = |b - a| \tag{4.6-37}$$

for every bounded interval (a, b); the Lebesgue measure is thus a generalization of length (*descriptive definition* of the Lebesgue measure; see also Secs. 12.1-1, 12.8-2, and 18.2-2).

The Lebesgue measures of point sets in suitable spaces of two, three, . . . dimensions are defined (either constructively or descriptively) by analogous generalizations of *area* and *volume*.

(b) *Every bounded open set (Sec. 4.3-6a) is measurable.* More generally, a (linear) **Borel set** is obtained by a finite or infinite sequence of unions, intersections, and/or complementations of intervals and of the resulting combinations; *the class of all Borel sets is a completely additive Boolean algebra of measurable sets. Every measurable set is the union of a Borel set and a set of Lebesgue measure zero. Every countable (discrete) set is measurable and has the Lebesgue measure zero.* Analogous theorems apply in the multidimensional case.

A property which holds for every point of a given interval, region, or point set with the possible exception of a set of Lebesgue measure zero is said to hold **almost everywhere** on (at **almost all points** of) the given interval, region, or set.

(c) Measurable Functions. A function $f(x)$ defined on (a, b) is **measurable** on (a, b) if and only if the set of points (x) in (a, b) such that $f(x) \leq c$ is measurable for every real value of c.

In this definition, the condition $f(x) \leq c$ can be replaced by any one of the conditions $f(x) < c$, $f(x) \geq c$, $f(x) > c$.

Every function $f(x)$ continuous on (a, b) is measurable on (a, b). If $f_1(x), f_2(x), \ldots$ are measurable on (a, b), the same is true for $f_1(x) + f_2(x)$, $\alpha f_1(x)$, and $f_1(x) f_2(x)$, and also for $\lim_{n \to \infty} f_n(x)$ if the limit exists on (a, b).

Analogous definitions and theorems apply to measurable functions $f(x_1, x_2, \ldots, x_n)$ defined on a space of "points" $(x_1, x_2, \ldots, x_n)$ permitting the definition of a Lebesgue measure.

4.6-15. Lebesgue Integrals (see also Secs. 4.6-1 and 4.6-2). **(a) The Lebesgue Integral of a Bounded Function.** Given a real function $y = f(x)$ *measurable and bounded* on the bounded interval (a, b), subdivide the range of variation of $f(x)$ on (a, b),

$$y_0 < \eta_1 < y_1 < \eta_2 < \cdots < \eta_n < y_n \tag{4.6-38}$$

and let S_i be the set of points (x) in (a, b) such that $y_{i-1} < f(x) \leq y_i$. *The sum $\sum_{i=1}^{m} \eta_i m[S_i]$ tends to a unique finite limit I for every sequence of par-*

titions (38) *as* max $|y_i - y_{i-1}| \to 0$; the quantity

$$I = \lim_{\max |y_i - y_{i-1}| \to 0} \sum_{i=1}^{m} \eta_i m[S_i] = \int_a^b f(x)\, dx \tag{4.6-39}$$

is the **definite integral of $f(x)$ over (a, b) in the sense of Lebesgue (Lebesgue integral).**

(b) The Lebesgue Integral of an Unbounded Function. If $f(x)$ is measurable and unbounded on the bounded interval (a, b), the Lebesgue integral $\int_a^b f(x)\, dx$ is defined by

$$\left.\begin{aligned} &\int_a^b f(x)\, dx = \lim_{A \to \infty} \int_a^b f_A(x)\, dx + \lim_{B \to \infty} \int_a^b f_B(x)\, dx \\ &\text{with} \\ &f_A(x) = \begin{cases} 0 & \text{if } f(x) \le 0 \\ f(x) & \text{if } 0 < f(x) \le A \\ A & \text{if } f(x) > A \end{cases} \qquad f_B(x) = \begin{cases} 0 & \text{if } f(x) \ge 0 \\ f(x) & \text{if } 0 > f(x) \ge B \\ B & \text{if } f(x) < B \end{cases} \end{aligned}\right\} \tag{4.6-40}$$

$f(x)$ is **integrable in the sense of Lebesgue (summable) over** (a, b) if and only if $\int_a^b f(x)\, dx$ exists.

(c) Lebesgue Integrals over Unbounded Intervals. If $\int_a^X f(x)\, dx$ exists for all $X > a$, one defines the Lebesgue integral $\int_a^\infty f(x)\, dx$ by

$$\left.\begin{aligned} &\int_a^\infty f(x)\, dx = \lim_{X_1 \to \infty} \int_a^{X_1} f_1(x)\, dx + \lim_{X_2 \to \infty} \int_a^{X_2} f_2(x)\, dx \\ &\text{with} \\ &f_1(x) = \begin{cases} 0 & \text{if } f(x) \le 0 \\ f(x) & \text{if } f(x) > 0 \end{cases} \qquad f_2(x) = \begin{cases} 0 & \text{if } f(x) \ge 0 \\ f(x) & \text{if } f(x) < 0 \end{cases} \end{aligned}\right\} \tag{4.6-41}$$

$\int_{-\infty}^b f(x)\, dx$ is similarly defined, and $\int_{-\infty}^\infty f(x)\, dx$ is given by Eq. (4).

(d) Lebesgue Integral over a Point Set. Multiple Lebesgue Integrals. The definitions of Secs. 4.6-15*a* and *b* apply without change to the **Lebesgue integral $\int_S f(x)\, dx$ over any measurable set S of points (x).** Analogous definitions apply to **multiple Lebesgue integrals** over regions or measurable sets of "points" $(x_1, x_2, \ldots, x_n)$.

(e) Existence and Properties of Lebesgue Integrals. Lebesgue Integrals vs. Riemann Integrals. *Every bounded measurable function is summable over any bounded measurable set. A function summable over (on) a measurable set S is summable over every measurable subset of S.*

The definitions of Secs. 4.6-15*a*, *b*, *c*, and *d* imply that *the Lebesgue integral* $\int_S f(x)\,dx$ *exists if and only if the Lebesgue integral* $\int_S |f(x)|\,dx$ *exists. Whenever any proper or improper, single or multiple Riemann integral exists in the sense of absolute convergence (Sec. 4.6-2a),* the corresponding Lebesgue integral exists and is equal to the Riemann integral.*

The theorems of Table 4.6-1 *and Secs.* 4.6-5 *and* 4.7-1*c*, *d apply to Lebesgue integrals.*

For every countable set of nonintersecting measurable point sets $S_1, S_2, \ldots$

$$\int_{S_1 \cup S_2 \cup \cdots} f(x)\,dx = \int_{S_1} f(x)\,dx + \int_{S_2} f(x)\,dx + \cdots \tag{4.6-42}$$

provided that the integrals exist. Every Lebesgue integral over a set of (Lebesgue) measure zero is equal to zero.

NOTE: Lebesgue integration applies to a more general class of functions than Riemann integration and simplifies the statement of many theorems. Many theorems stated in terms of Lebesgue integrals apply directly to absolutely convergent improper Riemann integrals (see also Secs. 4.6-16 and 4.8-4*b*).

4.6-16. Convergence Theorems (Continuity Theorems, see also Sec. 12.5-1*c*). **(a)** *If the sequence of functions* $s_0(x)$, $s_1(x)$, $s_2(x)$, . . . *each bounded and integrable in the sense of Riemann on the bounded interval* (a, b) *converges uniformly to* $s(x)$ *on* $[a, b]$, *then* $\lim_{n\to\infty} \int_a^b s_n(x)\,dx$ *exists and equals* $\int_a^b s(x)\,dx$.

(b) The following more powerful theorem is phrased specifically in terms of Lebesgue integrals: *Let* $s_0(x)$, $s_1(x)$, $s_2(x)$, . . . , *and a positive comparison function* $A(x)$ *be summable over a measurable set* S *such that* $|s_n(x)| \leq A(x)$ *for all* n *and for almost all* x *in* S *(Sec.* 4.6-14*b). Then* $\lim_{n\to\infty} s_n(x) = s(x)$ *for almost all* x *in* S *implies that* $\lim_{n\to\infty} \int_S s_n(x)\,dx$ *exists and equals* $\int_S s(x)\,dx$ *(Lebesgue's Convergence Theorem;* see also Sec. 4.8-4*b*). Note that uniform convergence is not required.

4.6-17. Stieltjes Integrals. (a) Riemann-Stieltjes Integrals. The **Riemann-Stieltjes integral of** $f(x)$ **with respect to** $g(x)$ over the bounded interval $[a, b]$ is defined as

$$\int_a^b f(x)\,dg(x) = \lim_{\max|x_i - x_{i-1}|\to 0} \sum_{i=1}^{m} f(\xi_i)[g(x_i) - g(x_{i-1})] \tag{4.6-43}$$

for an arbitrary sequence of partitions

$$a = x_0 < \xi_1 < x_1 < \xi_2 < x_2 < \cdots < \xi_m < x_m = b$$

* Note that $f(x) \equiv \frac{d}{dx}\left(x^2 \sin \frac{1}{x^2}\right) \equiv 2x \sin \frac{1}{x^2} - \frac{2}{x} \cos \frac{1}{x^2}$ is *not* summable over the interval (0, 1) even though the improper Riemann integral

$$\int_0^1 f(x)\,dx = \lim_{X\to 0} \int_X^1 f(x)\,dx = \sin 1$$

exists.

The limit (43) exists whenever $g(x)$ is of bounded variation (Sec. 4.4-8b), and $f(x)$ is continuous on $[a, b]$.

Improper Riemann-Stieltjes integrals may be defined in the manner of Sec. 4.6-2.

(b) Lebesgue-Stieltjes Integrals. Every function $g(x)$ nondecreasing and right continuous (Sec. 4.4-7b) on the bounded interval (a, b) defines a measure (**Lebesgue-Stieltjes measure**) $M[S]$ for every Borel set (Sec. 4.6-14b) in (a, b) by the relations (35), (36), and

$$M[\alpha < x \leq \beta] = g(\beta) - g(\alpha) \qquad (a < \alpha < \beta < b) \tag{4.6-44}$$

where the expression in the square brackets specifies a set of values of x. Note

$$\left.\begin{array}{ll} M[\alpha \leq x \leq \beta] = g(\beta) - g(\alpha - 0) & M[\alpha < x < \beta] = g(\beta - 0) - g(\alpha) \\ M[\alpha \leq x < \beta] = g(\beta - 0) - g(\alpha - 0) & M[x = \alpha] = g(\alpha) - g(\alpha - 0) \end{array}\right\} \quad (a < \alpha < \beta < b) \tag{4.6-45}$$

Starting with the Lebesgue-Stieltjes measures of bounded intervals, one introduces the Lebesgue-Stieltjes measure $M[S]$ of any measurable set as the common value of an inner and outer measure in the manner of Sec. 4.6-14a (see also Ref. 8.16).

Given a function $y = f(x)$ bounded and measurable on (a, b), the **Lebesgue-Stieltjes integral of $f(x)$ with respect to $g(x)$ over** (a, b) is then defined as

$$\int_a^b f(x)\, dg(x) = \lim_{\max|y_i - y_{i-1}| \to 0} \sum_{i=1}^{m} \eta_i M[S_i] \tag{4.6-46}$$

for an arbitrary subdivision (38) of the range of $f(x)$; S_i is, again, the set of values of x such that $y_{i-1} < f(x) < y_i$.

The Lebesgue-Stieltjes integral of a bounded or unbounded function $f(x)$ over any measurable set may now be defined in the manner of Secs. 4.6-15b, c, and d, with the understanding that $g(x)$ is finite on every bounded set under consideration. In the multidimensional case, $g(x)$ is replaced by a function $g(x_1, x_2, \ldots, x_n)$ nondecreasing with respect to each argument. One may, further, define the Lebesgue-Stieltjes integral $\int_a^b f(x)dg(x)$ with respect to any function $g(x)$ of bounded variation by applying Eq. (48) to the sum of two monotonic functions (Sec. 4.4-8b). ***Each Lebesgue-Stieltjes integral equals the corresponding Riemann-Stieltjes integral if the latter exists in the sense of absolute convergence.***

(c) Properties of Stieltjes Integrals (see also Table 4.6-1). Given a bounded or unbounded interval (a, b) such that the integrals in question exist,

$$\int_a^b f dg = -\int_b^a f dg \qquad \int_a^b f dg = \int_a^c f dg + \int_c^b f dg \tag{4.6-47}$$

$$\int_a^b (f_1 + f_2) dg = \int_a^b f_1 dg + \int_a^b f_2 dg$$

$$\int_a^b f d(g_1 + g_2) = \int_a^b f dg_1 + \int_a^b f dg_2 \tag{4.6-48}$$

$$\int_a^b (\alpha f) dg = \int_a^b f d(\alpha g) = \alpha \int_a^b f dg \tag{4.6-49}$$

$$\int_a^b f dg = fg\Big]_b^a - \int_a^b g df \tag{4.6-50}$$

If $g(x)$ is a nondecreasing function on (a, b), then

$$\left|\int_a^b f dg\right| \leq \int_a^b |f| dg \tag{4.6-51}$$

If $g(x)$ is nondecreasing and $f(x) \leq F(x)$ on (a, b), then

$$\int_a^b f dg \leq \int_a^b F dg \tag{4.6-52}$$

Stieltjes integrals often have an "intuitive" meaning (line integrals, surface integrals, volume integrals; integrals over distributions of mass, charge, and probability, Sec. 18.3-6). *Note that Stieltjes integrals include ordinary integrals and sums as special cases:*

$$\int_a^b f(x)\, dg(x) = \int_a^b f(x)g'(x)\, dx \tag{4.6-53}$$

whenever $g(x)$ is continuously differentiable on (a, b), and

$$\sum_k f(k) = \int_{-\infty}^{\infty} f(x) d \sum_k U_-(x - k) \qquad \left[U_-(x) = \begin{cases} 0 \text{ if } x < 0 \\ 1 \text{ if } x \geq 0 \end{cases}\right] \tag{4.6-54}$$

4.6-18. Convolutions. The **Stieltjes convolution** of two functions $f(x)$ and $g(x)$ over the interval (a, b) is defined as the function

$$\Psi(t) \equiv \int_a^b f(t - x)\, dg(x) \equiv \int_a^b g(t - x)\, df(x) \tag{4.6-55}$$

for all values of t such that the two integrals exist and are equal. The classical convolution of $f(x)$ and $g(x)$ over (a, b) is similarly defined as

$$\Psi(t) \equiv \int_a^b f(t - x)g(x)\, dx \equiv \int_a^b g(t - x)f(x)\, dx \tag{4.6-56}$$

In the literature, either (55) or (56) is often simply referred to as the *convolution* of $f(x)$ and $g(x)$ over (a, b), with either function denoted by $f * g$ or $f \overset{b}{\underset{a}{*}} g$; the precise meaning is usually evident from the context (see also Secs. 4.11-5, 8.3-1, 8.3-3, and 18.5-7). *For $-a = b = \infty$, Eqs. (55) and (56) hold whenever the integrals exist.*

The **convolution of two finite or infinite sequences** $a(0), a(1), a(2), \ldots$ and $b(0), b(1), b(2), \ldots$ is the sequence

$$a * b = \sum_k a(t - k)b(k) \qquad (t = 0, 1, 2, \ldots) \tag{4.6-57}$$

If Eq. (55) or (56) holds,

$$g * f \equiv f * g \qquad f * (g * h) \equiv (f * g) * h \equiv f * g * h \qquad f * (g + h) \equiv f * g + f * h \tag{4.6-58}$$

4.6-19. Minkowski's and Hölder's Inequalities. (a) *Given a bounded or unbounded interval (a, b) such that the integrals on the right exist,*

$$\left[\int_a^b |f_1 + f_2|^p\, dg\right]^{1/p} \le \left[\int_a^b |f_1|^p\, dg\right]^{1/p} + \left[\int_a^b |f_2|^p\, dg\right]^{1/p} \quad (1 \le p < \infty)$$

(MINKOWSKI'S INEQUALITY) (4.6-59)

$$\left|\int_a^b f_1 f_2\, dg\right| \le \left[\int_a^b |f_1|^p\, dg\right]^{1/p} \left[\int_a^b |f_2|^{p/(p-1)}\, dg\right]^{(p-1)/p} \quad (1 \le p < \infty)$$

(HÖLDER'S INEQUALITY) (4.6-60)

These inequalities hold, in particular, for ordinary Riemann and Lebesgue integrals and, more generally, for multidimensional integrals. Equation (60) reduces to the *Cauchy-Schwarz inequality* (15.2-3) for $p = p/(p-1) = 2$.

(**b**) Equations (59) and (60) imply analogous inequalities for sums and for convergent infinite series. *Whenever the sums on the right exist,*

$$\left[\sum_k |a_k + b_k|^p\right]^{1/p} \le \left[\sum_k |a_k|^p\right]^{1/p} + \left[\sum_k |b_k|^p\right]^{1/p} \quad (1 \le p < \infty)$$

(MINKOWSKI'S INEQUALITY) (4.6-61)

$$\left|\sum_k a_k b_k\right| \le \left[\sum_k |a_k|^p\right]^{1/p} \left[\sum_k |b_k|^{p/(p-1)}\right]^{(p-1)/p} \quad (1 \le p < \infty)$$

(HÖLDER'S INEQUALITY) (4.6-62)

Equation (62) reduces to the *Cauchy-Schwarz inequality* for $p = p/(p-1) = 2$ (see also Sec. 1.3-2).

4.7. MEAN-VALUE THEOREMS. VALUES OF INDETERMINATE FORMS. WEIERSTRASS'S APPROXIMATION THEOREMS

4.7-1. Mean-value Theorems (see also Sec. 4.10-4). The following theorems are useful for estimating values and limits of real functions, derivatives, and integrals.

(**a**) *If $f(x)$ is continuous on $[a, b]$ and continuously differentiable on (a, b) there exists a real number X in (a, b) such that*

$$f(b) - f(a) = f'(X)(b - a) \tag{4.7-1}$$

X is often written as $X = a + \vartheta(b - a)$ $(0 < \vartheta < 1)$. For $f(a) = f(b) = 0$ the theorem is known as *Rolle's theorem* (see also Sec. 1.6-6*e*).

If $f(x_1, x_2, \ldots, x_n)$ is continuous for $a_i \le x_i \le b_i$ and continuously differentiable for $a_i < x_i < b_i$ $(i = 1, 2, \ldots, n)$, there exists a set of real numbers $X_1, X_2, \ldots, X_n$ such that $X_i = a_i + \vartheta(b_i - a_i)$ $(0 < \vartheta < 1$, $i = 1, 2, \ldots, n)$ and

$$f(b_1, b_2, \ldots, b_n) - f(a_1, a_2, \ldots, a_n) = \sum_{i=1}^{n} \left.\frac{\partial f}{\partial x_i}\right]_{X_1, X_2, \ldots, X_n} (b_i - a_i) \tag{4.7-2}$$

(**b**) *If* (1) *$u(x)$ and $v(x)$ are continuous on $[a, b]$,* (2) *$v(b) \ne v(a)$, and* (3) *$u'(x)$ and $v'(x)$ exist and do not vanish simultaneously on (a, b), then there exists a real number X in (a, b) such that*

$$\frac{u(b) - u(a)}{v(b) - v(a)} = \frac{u'(X)}{v'(X)} \quad \text{(CAUCHY'S MEAN-VALUE THEOREM)} \tag{4.7-3}$$

(c) *If $f(x)$ is continuous on $[a, b]$ there exists a value X of x in (a, b) such that*

$$\int_a^b f(x)\,dx = f(X)(b-a) \tag{4.7-4}$$

(d) *If $f(x)$ and $g(x)$ are continuous on $[a, b]$, and $g(x) \neq 0$ on $[a, b]$, there exists a value X of x in (a, b) such that*

$$\int_a^b f(x)g(x)\,dx = f(X)\int_a^b g(x)\,dx \tag{4.7-5}$$

If $f(x)$ and $g(x)$ are continuous on $[a, b]$ and $g(x)$ is a nondecreasing or nonincreasing function on (a, b), then there exists a value X of x in (a, b) such that

$$\int_a^b f(x)g(x)\,dx = g(a)\int_a^X f(x)\,dx + g(b)\int_X^b f(x)\,dx \tag{4.7-6}$$

If, in addition, $g(x) > 0$ on (a, b), there exists a value X of x in (a, b) such that

$$\int_a^b f(x)g(x)\,dx = \begin{Bmatrix} g(a)\int_a^X f(x)\,dx \text{ if } g(x) \text{ is nonincreasing} \\ g(b)\int_X^b f(x)\,dx \text{ if } g(x) \text{ is nondecreasing} \end{Bmatrix} \tag{4.7-7}$$

4.7-2. Values of Indeterminate Forms (see Table 4.7-1 for examples). **(a)** Functions $f(x)$ of the form $u(x)/v(x)$, $u(x)v(x)$, $[u(x)]^{v(x)}$, and $u(x) - v(x)$ are not defined for $x = a$ if $f(a)$ takes the form $0/0$, ∞/∞, $0\cdot\infty$, 0^0, ∞^0, 1^∞, or $\infty - \infty$; but $\lim_{x\to a} f(x)$ may exist. In such cases it is often desirable to *define* $f(a) = \lim_{x\to a} f(x)$.

(b) Treatment of 0/0 and ∞/∞. *Let $u(a) = v(a) = 0$. If there exists a neighborhood of $x = a$ such that* (1) *$v(x) \neq 0$, except for $x = a$, and* (2) *$u'(x)$ and $v'(x)$ exist and do not vanish simultaneously, then*

$$\lim_{x\to a}\frac{u(x)}{v(x)} = \lim_{x\to a}\frac{u'(x)}{v'(x)} \tag{4.7-8}$$

whenever the limit on the right exists (L'Hôpital's Rule).

Let $\lim_{x\to a} u(x) = \lim_{x\to a} v(x) = \infty$. If there exists a neighborhood of $x = a$ such that $x \neq a$ implies (1) *$u(x) \neq 0$, $v(x) \neq 0$, and* (2) *$u'(x)$ and $v'(x)$ exist and do not vanish simultaneously, then Eq.* (8) *holds whenever the limit on the right exists.*

If $u'(x)/v'(x)$ is itself an indeterminate form, the above method may be applied to $u'(x)/v'(x)$ in turn, so that

$$\lim_{x\to a}\frac{u(x)}{v(x)} = \lim_{x\to a}\frac{u'(x)}{v'(x)} = \lim_{x\to a}\frac{u''(x)}{v''(x)} \tag{4.7-9}$$

If necessary, this process may be continued.

(c) Treatment of $0\cdot\infty$, 0^0, ∞^0, 1^∞, and $\infty - \infty$. $u(x)v(x)$, $[u(x)]^{v(x)}$, and $u(x) - v(x)$ can often be reduced to the form $\varphi(x)/\psi(x)$ with the aid of one of the following relations:

$$\left.\begin{aligned} u(x)v(x) &\equiv \frac{u(x)}{1/v(x)} \equiv \frac{v(x)}{1/u(x)} \\ [u(x)]^{v(x)} &\equiv e^{g(x)} \quad \left[g(x) \equiv \frac{\log_e u(x)}{1/v(x)} \equiv \frac{v(x)}{1/\log_e u(x)}\right] \\ u(x) - v(x) &\equiv \frac{\frac{1}{v(x)} - \frac{1}{u(x)}}{\frac{1}{u(x)}\,\frac{1}{v(x)}} \equiv \log_e g(x) \quad \left[g(x) \equiv \frac{e^{u(x)}}{e^{v(x)}}\right] \end{aligned}\right\} \tag{4.7-10}$$

so that the methods of Sec. 4.7-2*b* become applicable.

(**d**) It is often helpful to write $\lim_{x\to a} f(x) = \lim_{\Delta x\to 0} f(a + \Delta x)$ and to isolate terms of the order of Δx by algebraic manipulation or by a Taylor-series expansion (Sec. 4.10-4).

(**e**) The methods of Secs. 4.7-2*a*, *b*, *c*, and *d* are readily modified to apply to the *one-sided limits* $\lim_{x\to a-0} f(x)$ and $\lim_{x\to a+0} f(x)$ (Sec. 4.4-7). To find $\lim_{x\to\infty} f(x)$, use $\lim_{x\to\infty} f(x) = \lim_{y\to 0+0} f(1/y)$.

Table 4.7-1. Some Frequently Used Limits (Values of Indeterminate Forms, Sec. 4.7-2)

$$\lim_{n\to\infty}\left(1+\frac{1}{n}\right)^n = e \approx 2.71828 \qquad (n = 1, 2, \ldots) \qquad \lim_{x\to 0}(1+x)^{\frac{1}{x}} = e$$

$$\lim_{x\to 0}\frac{c^x - 1}{x} = \log_e c \qquad \lim_{x\to 0} x^x = 1$$

$$\lim_{x\to 0}\frac{\sin x}{x} = \lim_{x\to 0}\frac{\tan x}{x} = \lim_{x\to 0}\frac{\sinh x}{x} = \lim_{x\to 0}\frac{\tanh x}{x} = 1$$

$$\lim_{x\to 0}\frac{\sin \omega x}{x} = \omega \qquad (-\infty < \omega < \infty)$$

$$\lim_{x\to 0} x^a \log_e x = \lim_{x\to\infty} x^{-a}\log_e x = \lim_{x\to\infty} x^a e^{-x} = 0 \qquad (a > 0)$$

4.7-3. Weierstrass's Approximation Theorems (see also Secs. 4.10-4, 4.11-7, and 12.5-4*b*). *Let $f(x)$ be a real function continuous on the bounded closed interval $[a, b]$. Then for every given positive real number ϵ (maximum error of approximation) there exists*

1. *A real polynomial $P(x) \equiv \sum_{k=0}^{n} a_k x^k$ such that $|f(x) - P(x)| < \epsilon$ for all x in $[a, b]$*
2. *A real* **trigonometric polynomial** $T(x) \equiv \sum_{k=0}^{m} (\alpha_k \cos k\omega x + \beta_k \sin k\omega x)$ *such that $|f(x) - T(x)| < \epsilon$ for all x in $[a, b]$*

Analogous theorems hold for functions of two or more variables. The Weierstrass approximation theorems permit one to derive various properties of continuous functions from corresponding properties of polynomials or trigonometric polynomials.

4.8. INFINITE SERIES, INFINITE PRODUCTS, AND CONTINUED FRACTIONS

4.8-1. Infinite Series. Convergence (see also Sec. 4.4-1). An **infinite series** (infinite sum) $a_0 + a_1 + a_2 + \cdots$ of real or complex numbers (terms) $a_0, a_1, a_2, \ldots$ **converges** if and only if the sequence $s_0, s_1, s_2, \ldots$ of the **partial sums** $s_n = \sum_{k=0}^{n} a_k$ has a (necessarily finite and unique) limit s, i.e., if and only if the sequence of the **remainders** $R_{n+1} = s - s_n$ converges to zero. In this case s is called the **sum** of the infinite series, and it is permissible to write

$$a_0 + a_1 + a_2 + \cdots = \sum_{k=0}^{\infty} a_k = \lim_{n\to\infty} \sum_{k=0}^{n} a_k = \lim_{n\to\infty} s_n = s \qquad (4.8\text{-}1)$$

A (necessarily convergent) infinite series $a_0 + a_1 + a_2 + \cdots$ is **absolutely convergent** if and only if the series $|a_0| + |a_1| + |a_2| + \cdots$ converges. An infinite series which does not converge is **divergent** (**diverges,** see also Sec. 4.8-6). Sections 4.9-1 and 4.9-2 list a number of tests for the convergence of a given infinite series.

4.8-2. Series of Functions. Uniform Convergence (see also Sec. 4.4-4). An infinite series of functions $a_0(x) + a_1(x) + a_2(x) + \cdots$ converges to a function (sum) $s(x)$ for every value of x such that

$$\lim_{n\to\infty} \sum_{k=0}^{n} a_k(x) = s(x)$$

The series **converges uniformly to** $s(x)$ **on a set** S **of values of** x if and only if the sequence of partial sums $s_n(x) \equiv \sum_{k=0}^{n} a_k(x)$ converges uniformly to $s(x)$ on S. Section 4.9-2 lists a number of tests for uniform convergence.

4.8-3. Operations with Convergent Series (see also Table E-1).

(a) Addition and Multiplication by Constants. *If* $\sum_{k=0}^{\infty} a_k$ *and* $\sum_{k=0}^{\infty} b_k$ *are convergent series of real or complex terms, and* α *is a real or complex number, then*

$$\sum_{k=0}^{\infty} a_k + \sum_{k=0}^{\infty} b_k = \sum_{k=0}^{\infty} (a_k + b_k) \qquad \alpha \sum_{k=0}^{\infty} a_k = \sum_{k=0}^{\infty} \alpha a_k \tag{4.8-2}$$

In each case the convergence or absolute convergence of the series on the left implies the same for the series on the right.

(b) Rearrangement of Terms. An **unconditionally convergent** series is a convergent series which converges to the same limit after every rearrangement of its terms; *this is true if and only if the series is absolutely convergent. Every subseries (obtained by omission of terms) of such a series is absolutely convergent.*

The terms of every **conditionally convergent** (*i.e., not absolutely convergent*) *infinite series of real terms can be rearranged* (1) *so that the sequence of partial sums converges to any given limit,* (2) *so that the sequence of partial sums increases indefinitely,* (3) *so that the sequence of partial sums decreases indefinitely, and* (4) *so that the sequence of partial sums oscillates between any two given limits* (*Riemann's Theorem*).

(c) Double Series. A **double series** $\sum_{i=0}^{\infty} \sum_{k=0}^{\infty} a_{ik}$ converges to the limit (sum) s if and only if $\lim_{\substack{m\to\infty \\ n\to\infty}} \sum_{i=0}^{m} \sum_{k=0}^{n} a_{ik} = s$ (Sec. 4.4-5). Note that

$$\sum_{i=0}^{\infty} \sum_{k=0}^{\infty} a_{ik} = \sum_{i=0}^{\infty} \left(\sum_{k=0}^{\infty} a_{ik} \right) = \sum_{k=0}^{\infty} \left(\sum_{i=0}^{\infty} a_{ik} \right) \tag{4.8-3}$$

whenever the three limits exist (*Pringsheim's Theorem on Summation by Columns and Rows*). In particular, $\sum_{i=0}^{\infty} \sum_{k=0}^{\infty} a_{ik}$ converges to a unique limit for *any* order of summation if the double series is **absolutely convergent,** i.e., if $\sum_{i=0}^{\infty} \sum_{k=0}^{\infty} |a_{ik}|$ converges (see also Sec. 4.8-3*b*).

(d) Product of Two Infinite Series. *If* $\sum_{k=0}^{\infty} a_k$ *and* $\sum_{k=0}^{\infty} b_k$ *are absolutely convergent series of real or complex terms, then the double series* $\sum_{i=0}^{\infty} \sum_{k=0}^{\infty} a_i b_k$ *converges absolutely to the limit* $\left(\sum_{k=0}^{\infty} a_k \right) \left(\sum_{k=0}^{\infty} b_k \right)$. More generally,

$$\left(\sum_{k=0}^{\infty} a_k \right) \left(\sum_{k=0}^{\infty} b_k \right) = \sum_{n=0}^{\infty} \left(\sum_{k=0}^{n} a_k b_{n-k} \right) \quad \text{(Cauchy's rule)} \tag{4.8-4}$$

whenever the three infinite series converge (see also Appendix E).

4.8-4. Operations with Infinite Series of Functions. **(a) Addition and Multiplication by Bounded Functions.** *If* $\sum_{k=0}^{\infty} a_k(x)$ *and* $\sum_{k=0}^{\infty} b_k(x)$ *converge uniformly* (*Sec.* 4.8-2) *on a set S of values of x, the same is true for* $\sum_{k=0}^{\infty} [a_k(x) + b_k(x)]$ *and for* $\sum_{k=0}^{\infty} \varphi(x)a_k(x)$, *where* $\varphi(x)$ *is any function bounded for all x in S.*

(b) Limits, Continuity, and Integration (see also Sec. 4.4-5). *Let* $\sum_{k=0}^{\infty} a_k(x)$ *converge uniformly on a bounded open interval* (a, b) *containing* $x = x_0$ *and* $x = x_1$. *Then*

1.
$$\lim_{x\to x_0} \sum_{k=0}^{\infty} a_k(x) = \sum_{k=0}^{\infty} \lim_{x\to x_0} a_k(x)$$
$$\lim_{x\to x_0+0} \sum_{k=0}^{\infty} a_k(x) = \sum_{k=0}^{\infty} \lim_{x\to x_0+0} a_k(x)$$
$$\lim_{x\to x_0-0} \sum_{k=0}^{\infty} a_k(x) = \sum_{k=0}^{\infty} \lim_{x\to x_0-0} a_k(x)$$
provided that the respective limits $\lim_{x\to x_0} a_k(x)$, $\lim_{x\to x_0+0} a_k(x)$, *or* $\lim_{x\to x_0-0} a_k(x)$ *exist for* $k = 0, 1, 2, \ldots$.

2. $\sum_{k=0}^{\infty} a_k(x)$ *is continuous at* $x = x_0$ *if each* $a_k(x)$ *is continuous at* $x = x_0$.

3. $\sum_{k=0}^{\infty} a_k(x)$ *converges uniformly on the closed interval* $[a, b]$ *and*
$$\int_{x_0}^{x_1} \sum_{k=0}^{\infty} a_k(x)\,dx = \sum_{k=0}^{\infty} \int_{x_0}^{x_1} a_k(x)\,dx$$
if each $a_k(x)$ *is continuous throughout* (a, b).

In each of these relations the given conditions imply the convergence of the series on the right.

The following theorem is phrased specifically in terms of *Lebesgue integrals* and may be applied to suitable absolutely convergent improper Riemann integrals (see also

Sec. 4.6-15). *If the infinite series* $\sum_{k=0}^{\infty} a_k(x)$ *of functions summable over a (bounded or unbounded) measurable set (or interval) S converges for almost all x in S, then*

$$\int_S \sum_{k=0}^{\infty} a_k(x)\,dx = \sum_{k=0}^{\infty} \int_S a_k(x)\,dx \tag{4.8-5}$$

provided that there exists a real number A [or a function A(x) summable over S, Sec. 4.6-16] *such that* $\left|\sum_{k=0}^{n} a_k(x)\right| < A$ *for all n and for almost all x in S.* Note that uniform convergence is not required.

(c) Differentiation. *Let* $\sum_{k=0}^{\infty} a_k(x)$ *converge for at least one value of x in (a, b). Then, if* $a_0'(x)$, $a_1'(x)$, $a_2'(x)$, . . . *exist and* $\sum_{k=0}^{\infty} a_k'(x)$ *converges uniformly on (a, b)*

$$\frac{d}{dx}\sum_{k=0}^{\infty} a_k(x) = \sum_{k=0}^{\infty} a_k'(x) \qquad (a < x < b) \tag{4.8-6}$$

whenever the derivative on the left exists. Under the given conditions this is necessarily true if each $a_k'(x)$ *is continuous on (a, b).*

4.8-5. Improvement of Convergence and Summation of Series (see also Secs. 7.7-4 and 20.4-3*a*). **(a) Euler's Transformation.** *If the infinite series on the left converges,*

$$\sum_{k=0}^{\infty} (-1)^k a_k = \sum_{k=0}^{\infty} (-1)^k \frac{\Delta^k a_0}{2^{k+1}} \tag{4.8-7}$$

where the differences $\Delta^k a_0$ *are defined as in Sec.* 20.4-1*a*. The series on the right often converges more rapidly than the series on the left.

(b) Kummer's Transformation. *Given two convergent series* $s = \sum_{k=0}^{\infty} a_k$ *and* $S = \sum_{k=0}^{\infty} b_k$ *such that* $\gamma = \lim_{n\to\infty} (a_n/b_n) \neq 0$,

$$s = \sum_{k=0}^{\infty} a_k = \gamma S + \sum_{k=0}^{\infty} \left(1 - \gamma \frac{b_k}{a_k}\right) a_k \tag{4.8-8}$$

If S is known, Kummer's transformation may be useful in numerical calculations (Ref. 20.11; see Sec. E-5 for useful S series).

(c) Poisson's Summation Formula. *If the infinite series* $\sum_{k=-\infty}^{\infty} f(2\pi k + t)$ *converges uniformly to a function of bounded variation in the interval* $0 \leq t < 2\pi$, *then*

$$\sum_{k=-\infty}^{\infty} f(2\pi k) = \frac{1}{2\pi} \sum_{k=-\infty}^{\infty} \int_{-\infty}^{\infty} f(\tau) e^{-ik\tau}\, d\tau \tag{4.8-9}$$

provided that the integrals on the right exist.

(d) The Euler-MacLaurin Summation Formula. The following relation often yields closed expressions, convergent approximation series, or semiconvergent approximation series (Sec. 4.8-6*a*) for finite sums; the formula is also used for numerical integration (Sec. 20.6-2*b*). *If* $f^{(2m+2)}(x)$ *exists and is continuous for* $0 \leq x \leq n$,

$$\sum_{k=0}^{n} f(k) = \int_0^n f(t)\, dt + \tfrac{1}{2}[f(0) + f(n)] + \sum_{k=1}^{m} \frac{B_{2k}}{(2k)!} [f^{(2k-1)}(n) - f^{(2k-1)}(0)]$$
$$+ \frac{nB_{2m+2}}{(2m+2)!} f^{(2m+2)}(\vartheta n) \qquad (m, n = 1, 2, \ldots;\ 0 < \vartheta < 1) \tag{4.8-10}$$

where the B_j *are the Bernoulli numbers defined in Sec.* 21.5-2. Note also the following special summation formulas:

$$\sum_{k=0}^{n} k = \tfrac{1}{2} n(n+1) \qquad \sum_{k=0}^{n} k^2 = \tfrac{1}{6} n(n+1)(2n+1) \qquad \sum_{k=0}^{n} k^3 = \tfrac{1}{4} n^2(n+1)^2 \tag{4.8-11}$$

$$\sum_{k=0}^{n} k^N = \frac{1}{N+1} \sum_{k=0}^{N} \binom{N+1}{k} n^{N+1-k} B_k \qquad (N = 1, 2, \ldots) \tag{4.8-12}$$

$$\sum_{k=1}^{\infty} \frac{1}{k^{2N}} = \frac{(-1)^{N+1}(2\pi)^{2N}}{2(2N)!} B_{2N} \qquad (N = 1, 2, \ldots) \tag{4.8-13}$$

(see also Secs. 1.2-6 and 1.2-7).

(e) Abel's Lemma. The following lemma is sometimes useful for estimating partial sums. *Given two sequences* $\alpha_0, \alpha_1, \alpha_2, \ldots$ *and* $a_0, a_1, a_2, \ldots$ *of real numbers such that* $\alpha_0 \geq \alpha_1 \geq \alpha_2 \geq \cdots \geq \alpha_n \geq 0$ *and* $M \leq \sum_{k=0}^{n} a_k \leq M'$ *for all n. Then*

$$\alpha_0 M \leq \sum_{k=0}^{n} \alpha_k a_k \leq \alpha_0 M' \text{ for all } n.$$

4.8-6. Divergent Infinite Series.

(a) Semiconvergence. A divergent infinite series $a_0 + a_1 + a_2 + \cdots$ may yield useful approximations to a quantity s if the absolute error $|s - s_n| = \left| s - \sum_{k=0}^{n} a_k \right|$ decreases to a sufficiently small minimum for some value n_0 of n before increasing again (**semiconvergent series**).

(b) Asymptotic Series. A divergent infinite series $\sum_{k=0}^{\infty} a_k(x)$ is an **asymptotic series describing the behavior of** $f(x)$ **as** $x \to x_0$ $\left[f(x) \simeq \sum_{k=0}^{\infty} a_k(x) \text{ as } x \to x_0\right]$ if and only if there exists a positive real number N such that $n > N$ implies

$$\lim_{x \to x_0} \frac{1}{(x - x_0)^n} \left[f(x) - \sum_{k=0}^{n} a_k(x)\right] = 0 \qquad (4.8\text{-}14)$$

A divergent infinite series $\sum_{k=0}^{\infty} a_k(x)$ is an **asymptotic series describing the behavior of** $f(x)$ **as** $x \to \infty$ $\left[f(x) \simeq \sum_{k=0}^{\infty} a_k(x) \text{ as } x \to \infty\right]$ if and only if there exists a positive real number N such that $n > N$ implies

$$\lim_{x \to \infty} x^n \left[f(x) - \sum_{k=0}^{n} a_k(x)\right] = 0 \qquad (4.8\text{-}15)$$

Asymptotic series yield semiconvergent approximations to $f(x)$, with $n_0 \to \infty$ as $x \to x_0$ or $x \to \infty$ (see also Secs. 4.4-3 and 8.4-9).

(c) Summation by Arithmetic Means. A convergent or divergent series $a_0 + a_1 + a_2 + \cdots$ is **summable by arithmetic means [summable by Césaro's means of the first order, summable** C_1, **summable** $(C, 1)$**] to the** C_1**-sum** S_1 if and only if the sequence $\sigma_0, \sigma_1, \sigma_2, \ldots$ of the arithmetic means $\sigma_{n-1} = \frac{1}{n} \sum_{k=0}^{n-1} s_k$ converges to the limit S_1 $\left(s_k = \sum_{j=0}^{k} a_j\right)$. *Every convergent series is summable by arithmetic means, and* $C_1 = s$. *An infinite series of positive real terms is summable by arithmetic means if and only if it converges.* See also Sec. 4.11-7.

4.8-7. Infinite Products. **(a)** An **infinite product**

$$(1 + a_0)(1 + a_1)(1 + a_2) \cdots = \prod_{k=0}^{\infty} (1 + a_k)$$

of real or complex factors $1 + a_k \neq 0$ **converges** to a limit (**value of the infinite product**) $p = \prod_{k=0}^{\infty} (1 + a_k) \neq 0$ if and only if

$$\lim_{n\to\infty} \prod_{k=0}^{n} (1 + a_k) = p$$

This is true if and only if the infinite series $\sum_{k=0}^{\infty} \log_e (1 + a_k)$ *converges to one of the values of* $\log_e p$ (see also Sec. 21.2-10). If $\lim_{n\to\infty} \prod_{k=0}^{n} (1 + a_k) = 0$ the infinite product is said to **diverge to zero.**

(b) A (necessarily convergent) infinite product $\prod_{k=0}^{\infty} (1 + a_k)$ **converges absolutely** if and only if $\prod_{k=0}^{\infty} (1 + |a_k|)$ converges; *this is true if and only if the infinite series* $\sum_{k=0}^{\infty} a_k$ *converges absolutely. An infinite product* **converges unconditionally** (*i.e., its value is independent of the order of factors*) *if and only if it is absolutely convergent* (see also Sec. 4.8-3*b*). See Sec. E-8 for examples of infinite products.

(c) $\prod_{k=0}^{\infty} [1 + a_k(x)]$ **converges uniformly** on a set S of values of x such that $1 + a_k(x) \neq 0$ for all k if and only if the sequence of functions $\prod_{k=0}^{n} [1 + a_k(x)]$ converges uniformly to a function $p(x)$ different from zero on S. *This is, in particular, true if the infinite series* $\sum_{k=0}^{\infty} |a_k(x)|$ *converges uniformly on S.*

4.8-8. Continued Fractions. A sequence of the form

$$s_0 = b_0 \qquad s_1 = b_0 + \frac{a_1}{b_1} \qquad s_2 = b_0 + \cfrac{a_1}{b_1 + \cfrac{a_2}{b_2}} \qquad s_3 = b_0 + \cfrac{a_1}{b_1 + \cfrac{a_2}{b_2 + \cfrac{a_3}{b_3}}} \qquad \cdots \tag{4.8-16}$$

is called a sequence of **continued fractions;** the nth denominator is $b_{n-1} + a_n/b_n$. Continued-fraction expansions of certain functions converge more rapidly than equivalent power-series expansions and are useful in some applications (electrical-network design; see also Sec. 20.5-7 and Ref. 4.7).

4.9. TESTS FOR THE CONVERGENCE AND UNIFORM CONVERGENCE OF INFINITE SERIES AND IMPROPER INTEGRALS

4.9-1. Tests for Convergence of Infinite Series (Sec. 4.8-1). **(a) A Necessary and Sufficient Condition for Convergence.** *A sequence of real or complex numbers* $s_0, s_1, s_2, \ldots$ *(e.g., the partial sums of an* *infinite series, Sec. 4.8-1) converges if and only if for every positive real number* ϵ *there exists a real integer* N *such that* $m > N$, $n > N$ *implies* $|s_n - s_m| < \epsilon$ (*Cauchy's Test* for the convergence of sequences or series).

(b) Tests for Series of Real Positive Terms (useful also as *tests for absolute convergence* of real or complex series, Secs. 4.8-1 and 4.8-3). *An infinite series* $a_0 + a_1 + a_2 + \cdots$ *of real positive terms converges if there exists a real number* N *such that* $n > N$ *implies one or more of the following conditions:*

1. $a_n \leq M_n$ *and/or* $\frac{a_{n+1}}{a_n} \leq \frac{M_{n+1}}{M_n}$ *where* $M_0 + M_1 + M_2 + \cdots$ *is a convergent comparison series of real positive terms* (*comparison tests for convergence*; see Secs. E-5 to E-7 for comparison series).
2. *At least one of the quantities*

$$\frac{a_{n+1}}{a_n}, \qquad \sqrt[n]{a_n}, \qquad n\left(\frac{a_{n+1}}{a_n} - 1\right) + 2, \qquad \left[n\left(\frac{a_{n+1}}{a_n} - 1\right) + 1\right] \log_e n + 2 \qquad (4.9\text{-}1)$$

has an upper bound $A < 1$.

The fourth of these four tests is stronger than the third (*Raabe's Test*), which is, in turn, stronger than the first two (*Cauchy's Ratio and Root Tests*).

3. $a_n \leq f(n)$, *where* $f(x)$ *is a real positive decreasing function whose (improper) integral* $\int_{N+1}^{\infty} f(x)\,dx$ *exists* (*Cauchy's Integral Test for convergence*, see also Sec. 4.9-3*b*).

The infinite series $a_0 + a_1 + a_2 + \cdots$ *diverges if there exists a real number* N *such that* $n > N$ *implies one or more of the following conditions:*

1. $a_n \geq d_n$ *and/or* $\frac{a_{n+1}}{a_n} \geq \frac{d_{n+1}}{d_n}$ *where* $d_0 + d_1 + d_2 + \cdots$ *is a divergent comparison series of real positive terms* (*comparison tests for divergence*).
2. *At least one of the quantities* (1) *has a lower bound* $A \geq 1$.
3. $a_n \geq f(n)$, *where* $f(x)$ *is a real positive decreasing function whose integral* $\int_{N+1}^{\infty} f(x)\,dx$ *diverges* (*Cauchy's Integral Test for divergence*).

NOTE: The series $\sum_{k=1}^{\infty} \frac{1}{k^\lambda}$, which converges for real λ if and only if $\lambda > 1$, is useful as a comparison series.

(c) *An infinite series* $a_0 + a_1 + a_2 + \cdots$ *of real terms converges*

1. *If successive terms are alternatingly positive and negative* (*alternating series*), *decrease in absolute value, and* $\lim_{n\to\infty} a_n = 0$

2. *If the sequence* $s_0, s_1, s_2, \ldots$ *of the partial sums is bounded and monotonic*

(**d**) *Given a decreasing sequence of real positive numbers* $\alpha_0, \alpha_1, \alpha_2, \ldots,$ *the infinite series* $\alpha_0 a_0 + \alpha_1 a_1 + \alpha_2 a_2 + \cdots$ *converges*

1. *If the series* $a_0 + a_1 + a_2 + \cdots$ *converges* (*Abel's Test,* see also Sec. 4.8-5*e*)

2. *If* $\lim_{n\to\infty} \alpha_n = 0$ *and* $\sum_{k=0}^{n} a_k$ *is bounded for all* n (*Dirichlet's Test*)

4.9-2. Tests for Uniform Convergence of Infinite Series (Sec. 4.8-2). **(a) A Necessary and Sufficient Condition for Uniform Convergence.** *A sequence* $s_0(x), s_1(x), s_2(x), \ldots$ *of real or complex functions* (*e.g., partial sums of an infinite series of functions,* Sec. 4.8-2) *converges uniformly on a set* S *of values of* x *if and only if for every positive real number* ϵ *there exists a real number* N *independent of* x *such that* $m > N$, $n > N$ *implies* $|s_n(x) - s_m(x)| < \epsilon$ *for all* x *in* S (*Cauchy's Test* for uniform convergence of sequences or series).

(**b**) *An infinite series* $a_0(x) + a_1(x) + a_2(x) + \cdots$ *of real or complex functions converges uniformly and absolutely on every set* S *of values of* x *such that* $|a_n(x)| \leq M_n$ *for all* n, *where* $M_0 + M_1 + M_2 + \cdots$ *is a convergent comparison series of real positive terms* (*Weierstrass's Test*). The convergence of the comparison series may be tested in the manner of Sec. 4.9-1*b*.

(**c**) *Given a decreasing sequence of real positive terms* $\alpha_0, \alpha_1, \alpha_2, \ldots,$ *the infinite series* $\alpha_0 a_0(x) + \alpha_1 a_1(x) + \alpha_2 a_2(x) + \cdots$ *converges uniformly on a set* S *of values of* x

1. *If the infinite series* $a_0(x) + a_1(x) + a_2(x) + \cdots$ *converges uniformly on* S (*Abel's Test,* see also Sec. 4.9-1*d*)

2. If $\lim_{n\to\infty} \alpha_n = 0$ *and there exists a real number* $A \geq \left|\sum_{k=0}^{n} a_k(x)\right|$ *for all* n *and all* x *in* S (*Dirichlet's Test*)

4.9-3. Tests for Convergence of Improper Integrals (see also Sec. 4.6-2). Sections 4.9-3 and 4.9-4 list convergence criteria for improper integrals of the form $\int_a^\infty f(x)\,dx$ and $\int_a^b f(x)\,dx = \lim_{X\to b-0} \int_a^X f(x)\,dx$. Other improper integrals can be reduced to these forms (Sec. 4.6-2). **It is assumed that** $f(x)$ **is bounded and integrable on every bounded interval** (a, X) **which does not contain the upper limit of integration.**

(**a**) **Necessary and Sufficient Conditions for Convergence (Cauchy's Test).** *The improper integral* $\int_a^\infty f(x)\,dx$ *converges if and* *only if for every positive real number* ϵ *there exists a real number* $M > a$ *such that* $X_2 > X_1 > M$ *implies* $\left| \int_{X_1}^{X_2} f(x)\, dx \right| < \epsilon$.

Similarly, $\int_a^b f(x)\, dx$ *converges if and only if for every positive real number* ϵ *there exists a positive real number* $\delta < b - a$ *such that* $b - X_2 < b - X_1 < \delta$ *implies* $\left| \int_{X_1}^{X_2} f(x)\, dx \right| < \epsilon$.

(b) Tests for Improper Integrals over a Real Nonnegative Function (useful also as *tests for absolute convergence* of real or complex improper integrals; note that *absolute convergence implies convergence*). *Given a real function* $f(x) \geq 0$ *on the interval of integration, the improper integral* $\int_a^\infty f(x)\, dx$ *or* $\int_a^b f(x)\, dx = \lim_{X \to b-0} \int_a^X f(x)\, dx$ *converges if and only if* $\int_a^X f(x)\, dx$ *is bounded for every* X *in the interval of integration.* In particular, *the integral converges if the integration interval contains a real number* M *such that* $x > M$ *implies* $f(x) \leq g(x)$, *where* $g(x)$ *is a real comparison function such that* $\int_a^\infty g(x)\, dx$ $\left[\textit{or } \int_a^b g(x)\, dx, \textit{ respectively} \right]$ *converges (Comparison Test).*

Similarly, if $\int_a^\infty g(x)\, dx$ or $\int_a^b g(x)\, dx$ diverges, $f(x) \geq g(x)$ implies the divergence of the corresponding integral over $f(x)$.

NOTE: $\int_a^\infty \frac{1}{x^\lambda}\, dx$ and $\int_a^b \frac{1}{(x-b)^\lambda}\, dx$ converge for $\lambda > 1$ and diverge for $\lambda \leq 1$. The improper integral $\int_a^\infty f(x)\, dx$ converges absolutely if $f(x) = O(1/x^\lambda) (\lambda > 1)$ as $x \to \infty$; $\int_a^b f(x)\, dx$ converges absolutely if $f(x) = O[1/(x-b)^\lambda] (\lambda > 1)$ as $x \to b - 0$ (see also Sec. 4.4-3).

(c) *The improper integral* $\int_a^\infty \alpha(x) f(x)\, dx$ *or* $\int_a^b \alpha(x) f(x)\, dx$ *converges if the corresponding improper integral* $\int_a^\infty f(x)\, dx$ *or* $\int_a^b f(x)\, dx$ *converges absolutely, and* $\alpha(x)$ *is bounded and integrable on every finite subinterval of the interval of integration.*

(d) *The improper integral* $\int_a^\infty \alpha(x) f(x)\, dx$ *converges if, for every* $X > a$, $\alpha(X)$ *is bounded and monotonic,* $f(x)$ *is bounded and integrable and has a finite number of sign changes on* $[a, X]$, *and*

1. $\int_a^\infty f(x)\, dx$ *converges (analog of Abel's test,* Sec. 4.9-1*d*) *or*
2. $\int_a^X f(x)\, dx$ *is bounded and* $\lim_{x \to \infty} \alpha(x) = 0$ *(analog of Dirichlet's test,* Sec. 4.9-1*d*).

(e) See Sec. 8.2-4 for a number of applications.

4.9-4. Tests for Uniform Convergence of Improper Integrals (see also Sec. 4.6-2c). **(a) Necessary and Sufficient Conditions for Uniform Convergence.** *The Cauchy convergence test of Sec.* 4.9-3a *implies uniform convergence of an improper integral* $\int_a^\infty f(x, y)\,dx$ *or* $\int_a^b f(x, y)\,dx$ *on every set* S *of values of* y *such that the bound* M *or* δ *is independent of* y (see also Sec. 4.9-2a).

(b) *The improper integral* $\int_a^\infty f(x, y)\,dx$ *or* $\int_a^b f(x, y)\,dx$ *converges uniformly and absolutely on every set* S *of values of* y *such that* $|f(x, y)| \leq g(x)$ *on the interval of integration, where* $g(x)$ *is a real comparison function whose integral* $\int_a^\infty g(x)\,dx$ $\left[\text{or } \int_a^b g(x)\,dx, \text{ respectively}\right]$ *converges (analog of Weierstrass's test,* Sec. 4.9-2b).

(c) *The improper integral* $\int_a^\infty \alpha(x, y)f(x, y)\,dx$ *converges uniformly on every set* S *of values of* y *such that, for* $x > a$, $\alpha(x, y)$ *decreases monotonically and uniformly to zero as* $x \to \infty$, $\partial\alpha/\partial x$ *exists and is continuous, and* $\left|\int_a^x f(x, y)\,dx\right|$ *is bounded by a constant* A *independent of* x *and* y *(analog of Dirichlet's test,* Sec. 4.9-2c).

4.10. REPRESENTATION OF FUNCTIONS BY INFINITE SERIES AND INTEGRALS. POWER SERIES AND TAYLOR'S EXPANSION

4.10-1. Representation of Functions by Infinite Series and Integrals. A function $f(x)$ is often represented by a corresponding infinite series $\sum_{k=0}^{\infty} \alpha_k\varphi_k(x)$ because

1. A sequence of partial sums (or arithmetic means, Secs. 4.8-6c and 4.11-7) may yield useful numerical approximations to $f(x)$.
2. It may be possible to describe operations on $f(x)$ in terms of simpler operations on the functions $\varphi_k(x)$ or on the coefficients α_k (transform methods, see also Secs. 4.11-5b and 8.6-5). The functions $\varphi_k(x)$ and the coefficients α_k may have an intuitive (physical) meaning (Sec. 4.11-4b).

Similar advantages apply to representations of functions by (usually improper) integrals $\int_a^b \alpha(\lambda)\varphi(x, \lambda)\,d\lambda$ (see also Secs. 4.11-4c, 4.11-5c, and Chap. 8).

Realization of either or both advantages listed above often requires convergence or uniform convergence of the series or integral (hence the importance of the convergence criteria of Secs. 4.9-1 to 4.9-4), but this is *not always* true (Secs. 4.8-6, 4.11-5, and 4.11-7).

4.10-2. Power Series. **(a)** A **power series** in the (real or complex) variable x is a series of the form

$$a_0 + a_1x + a_2x^2 + \cdots \equiv \sum_{k=0}^{\infty} a_kx^k \tag{4.10-1}$$

where the coefficients $a_0, a_1, a_2, \ldots$ are real or complex numbers. *Given any power series (1), there exists a real number r_c ($0 \leq r_c \leq \infty$) such that the power series converges absolutely and uniformly for $|x| < r_c$ and diverges for $|x| > r_c$.* r_c is called the **radius of convergence** of the power series.

Convergence for $x = x_0$ implies convergence for $|x| < |x_0|$, and divergence for $x = x_0$ implies divergence for $|x| > |x_0|$.

(**b**) *Convergent power series can be added according to Eq.* (4.8-2) *and multiplied according to Cauchy's rule* (4.8-4). *For $|x| < r_c$, the power series* (1) *is a continuous and repeatedly differentiable function of x; the series may be differentiated and integrated term by term, and the resulting power series have the radius of convergence r_c.*

(**c**) *If there exists a positive real number r such that two power series, $\sum_{k=0}^{\infty} a_kx^k$* and *$\sum_{k=0}^{\infty} b_kx^k$, converge to the same sum $f(x)$ for all real x such that $|x| < r$, then $a_0 = b_0$, $a_1 = b_1$, $a_2 = b_2$, . . . (Uniqueness Theorem).*

EXAMPLE: The *infinite geometric series*

$$a_0 + a_0x + a_0x^2 + \cdots \equiv a_0 \sum_{k=0}^{\infty} x^k \equiv a_0 \frac{1}{1-x} \qquad (|x| < 1) \tag{4.10-2}$$

(see also Secs. 1.2-7 and 1.7-2) converges absolutely and uniformly for $|x| < 1$ and diverges for $|x| > 1$.

4.10-3. Abel's and Tauber's Theorems. (**a**) *Given a positive real number r such that $\sum_{k=0}^{\infty} a_kr^k$ converges to the finite limit $f(r)$, $\sum_{k=0}^{\infty} a_kx^k$ converges uniformly for $0 \leq x \leq r$, and $\lim_{x \to r-0} \sum_{k=0}^{\infty} a_kx^k = f(r)$ (Abel's Theorem).*

(**b**) *Given a positive real number r such that $\sum_{k=0}^{\infty} a_kx^k$ converges for $|x| < r$, let $\lim_{x \to r-0} \sum_{k=0}^{\infty} a_kx^k = f(r)$ and $\lim_{k \to \infty} ka_kr^k = 0$. Then $\sum_{k=0}^{\infty} a_kr^k = f(r)$ (Tauber's Theorem); the last condition can be weakened to $k|a_k|r^k \leq c < \infty$ (Karamata's Theorem).*

NOTE: The special cases $r = 1$ and $r = r_c$ are of particular interest.

4.10-4. Taylor's Expansion (see also Sec. 7.5-2). (**a**) *Given a real function $f(x)$ such that $f^{(n)}(x)$ exists for $a \leq x < b$,*

$$\left.\begin{aligned} f(x) &= f(a) + f'(a)(x-a) \\ &\quad + \frac{1}{2!}f''(a)(x-a)^2 + \cdots \\ &\quad + \frac{1}{(n-1)!}f^{(n-1)}(a)(x-a)^{n-1} + R_n(x) \\ \textit{with} \quad |R_n(x)| &\leq \frac{|x-a|^n}{n!} \sup_{a<\xi<x} |f^{(n)}(\xi)| \end{aligned}\right\} (a \leq x < b) \qquad (4.10\text{-}3)$$

More specifically, *there exists a real number* $X = a + \vartheta(x - a)$ *such that* $a < X < x$ (*or* $0 < \vartheta < 1$) *and*

$$R_n(x) = \int_a^x \int_a^\xi \cdots \int_a^\xi f^{(n)}(\xi)\, d\xi^n = \frac{1}{n!} f^{(n)}(X)(x-a)^n \qquad (a \leq x < b)$$

(LAGRANGE'S REMAINDER FORMULA) (4.10-4)

X and ϑ depend on a, x, and n (see also Sec. 4.7-1).

(**b**) *Given a function* $f(x)$ *such that all derivatives* $f^{(k)}(x)$ *exist and* $\lim_{n\to\infty} R_n(x) = 0$ *for* $a \leq x < b$,

$$f(x) = \sum_{k=0}^{\infty} \frac{1}{k!} f^{(k)}(a)(x-a)^k \qquad (a \leq x < b) \qquad (4.10\text{-}5)$$

and the series converges uniformly to $f(x)$ *on every closed subinterval in* $a \leq x < b$ [*Taylor-series expansion of* $f(x)$ *about* $x = a$].

Equation (5) may be written $f(x + \Delta x) = \sum_{k=0}^{\infty} \frac{1}{k!} f^{(k)}(x)\, \Delta x^k$ with $\Delta x = x - a$ (Secs. 4.5-3 and 11.2-1). For $a = 0$, Taylor's series reduces to *MacLaurin's series* $\sum_{k=0}^{\infty} \frac{1}{k!} f^{(k)}(0)x^k$.

(**c**) Every convergent series expansion of $f(x)$ in powers of $(x - a)$ is necessarily identical with Eq. (5) (Sec. 4.10-2*c*). If $f(x)$ is a rational function, series expansions in powers of x or $1/x$ are often obtained by continued long division [Sec. 1.7-2; see also Eq. (2)]. *Refer to Sec.* 21.2-12 *for examples of power-series expansions.*

Power-series expansions permit term-by-term differentiation and integration (Sec. 4.10-2*b*) and are thus useful for the integration of $f(x)$ and for the solution of differential equations (Sec. 9.2-5*b*). Note, however, that the nth partial sum $\sum_{k=0}^{n} \frac{1}{k!} f^{(k)}(a)(x-a)^k$ of a Taylor series is *not* necessarily the most useful n^{th}-degree polynomial approximation to $f(x)$ (see also Sec. 20.5-1).

A function $f(x)$ which can be represented by a convergent power series throughout a neighborhood of $x = a$ is called **analytic** at $x = a$ (see also Sec. 7.3-3).

4.10-5. Multiple Taylor Expansion. (a) Given a real function $f(x_1, x_2, \ldots, x_n)$ such that all partial derivatives of order m exist and are continuous for $a_i \leq x_i < b_i$ $(i = 1, 2, \ldots, n)$, one has

$$f(x_1, x_2, \ldots, x_n) = f(a_1, a_2, \ldots, a_n) + \sum_{i=1}^{n} \frac{\partial f}{\partial x_i}\bigg]_{a_1,a_2,\ldots,a_n} (x_i - a_i)$$
$$+ \frac{1}{2!} \sum_{i=1}^{n} \sum_{j=1}^{n} \frac{\partial^2 f}{\partial x_i \, \partial x_j}\bigg]_{a_1,a_2,\ldots,a_n} (x_i - a_i)(x_j - a_j) + \cdots$$
$$+ R_m(x_1, x_2, \ldots, x_n) \qquad (a_i \leq x_i < b_i;\ i = 1, 2, \ldots, n) \qquad (4.10\text{-}6)$$

The remainder $R_m(x_1, x_2, \ldots, x_n)$ satisfies a relation analogous to Eq. (4); in terms of differentials (Sec. 4.5-3)

$$f(a_1 + dx_1, a_2 + dx_2, \ldots, a_n + dx_n) - f(a_1, a_2, \ldots, a_n)$$
$$= df + \frac{1}{2!} d^2 f + \cdots + \frac{1}{(m-1)!} d^{m-1} f + R_m \qquad (4.10\text{-}7)$$

where all differentials are computed at the point $(a_1, a_2, \ldots, a_n)$, and

$$R_m = \frac{1}{m!} d^m f\bigg]_{X_1,X_2,\ldots,X_n} \qquad (a_i < X_i < a_i + dx_i;\ i = 1, 2, \ldots, n) \qquad (4.10\text{-}8)$$

(b) If all derivatives of $f(x_1, x_2, \ldots, x_n)$ exist and $\lim_{m\to\infty} R_m(x_1, x_2, \ldots, x_n) = 0$ for $a_i \leq x_i < b_i$, then Eq. (6) yields a multiple power-series expansion for $f(x_1, x_2, \ldots, x_n)$ (*multiple Taylor series*). A function $f(x_1, x_2, \ldots, x_n)$ which can be represented by a convergent multiple power series throughout a neighborhood of the point $(x_1, x_2, \ldots, x_n)$ is called **analytic** at $(x_1, x_2, \ldots, x_n)$.

4.11. FOURIER SERIES AND FOURIER INTEGRALS

4.11-1. Introduction. Fourier series and Fourier integrals are used to represent and/or approximate functions (Sec. 4.10-1) in many important applications. Fourier expansions are special instances of *expansions in terms of orthogonal functions* (refer to Secs. 15.2-3 to 15.2-6).

4.11-2. Fourier Series. (a) Given the interval of expansion $-\pi < t < \pi$, the **Fourier series** generated by a real function $f(t)$ such that $\int_{-\pi}^{\pi} |f(\tau)| \, d\tau$ exists* is the infinite **trigonometric series**

$$\tfrac{1}{2}a_0 + \sum_{k=1}^{\infty} (a_k \cos kt + b_k \sin kt) \equiv \sum_{k=-\infty}^{\infty} c_k e^{ikt} \qquad (4.11\text{-}1a)$$

whose coefficients are defined by the *Euler-Fourier formulas*

* In many applications, it suffices to consider the integrals in this section as Riemann integrals (Sec. 4.6-1), but the use of Lebesgue integrals (Secs. 4.6-15) makes the theory more generally applicable. Refer to Secs. 15.2-3 to 15.2-6 for a number of theorems stated specifically in terms of Lebesgue integrals.

$$\left.\begin{aligned} a_k &= \frac{1}{\pi}\int_{-\pi}^{\pi} f(\tau)\cos k\tau\, d\tau \qquad b_k = \frac{1}{\pi}\int_{-\pi}^{\pi} f(\tau)\sin k\tau\, d\tau \\ c_k &= c_{-k}^* = \frac{1}{2}(a_k - ib_k) = \frac{1}{2\pi}\int_{-\pi}^{\pi} f(\tau)e^{-ik\tau}\, d\tau \end{aligned}\right\} (k = 0, 1, 2, \ldots) \tag{4.11-1b}$$

The a_k and b_k are real numbers, and the c_k are, in general, complex numbers (Sec. 4.11-2*d*).

(b) Given the finite interval of expansion $-T/2 < t < T/2$, the Fourier series generated by a real function $f(t)$ such that $\int_{-T/2}^{T/2} |f(\tau)|\, d\tau$ exists is the infinite trigonometric series

$$\frac{1}{2}a_0 + \sum_{k=1}^{\infty}(a_k\cos k\omega_0 t + b_k \sin k\omega_0 t) \equiv \sum_{k=-\infty}^{\infty} c_k e^{ik\omega_0 t} \qquad \left(\omega_0 = \frac{2\pi}{T}\right) \tag{4.11-2a}$$

with

$$\left.\begin{aligned} a_k &= \frac{2}{T}\int_{-T/2}^{T/2} f(\tau)\cos k\omega_0\tau\, d\tau \qquad b_k = \frac{2}{T}\int_{-T/2}^{T/2} f(\tau)\sin k\omega_0\tau\, d\tau \\ c_k &= c_{-k}^* = \frac{1}{2}(a_k - ib_k) = \frac{1}{T}\int_{-T/2}^{T/2} f(\tau)e^{-ik\omega_0 t}\, d\tau \\ &\qquad\qquad \left(\omega_0 = \frac{2\pi}{T};\ k = 0, 1, 2, \ldots\right) \end{aligned}\right\} \tag{4.11-2b}$$

Equation (2) reduces to Eq. (1) in the special case $T = 2\pi$. If $(a, a + T)$ is chosen as the interval of expansion, the integrals in Eq. (2*b*) must be taken between a and $a + T$ instead of between $-T/2$ and $T/2$.

(c) *If* $\int_{-T/2}^{T/2} [f(\tau)]^2\, d\tau$ *exists, the Fourier series (2) is that trigonometric series (2a) whose coefficients* a_k, b_k *minimize each mean-square error*

$$\frac{1}{T}\int_{-T/2}^{T/2} [f(\tau) - s_n(\tau)]^2\, d\tau$$

when successive partial sums (Sec. 4.8-1)

$$s_0(t) = \frac{1}{2}a_0 \qquad s_n(t) \equiv \frac{1}{2}a_0 + \sum_{k=1}^{n}\left(a_k \cos k\frac{2\pi t}{T} + b_k \sin k\frac{2\pi t}{T}\right)$$
$$(n = 1, 2, \ldots)$$

are used as approximations to $f(t)$ (see also Sec. 15.2-6).

If a trigonometric series (2a) converges uniformly to $f(t)$ *in* $(-T/2, T/2)$, *then its coefficients are necessarily the Fourier coefficients (2b) of* $f(t)$ *(Euler's Theorem).*

(d) *If* $|f(t)|$ *is integrable over the interval of expansion, the Fourier coefficients* a_k *and* b_k *exist and tend to zero as* $k \to \infty$ *(Riemann-Lebesgue Theorem).* More specifically, *if, throughout the closed interval of expansion,* $f^{(m-1)}(t)$ *exists and is piecewise continuously differentiable, and* $f^{(m)}(t)$ *is differentiable wherever it is continuous, then* a_k *and* b_k *are of the order of* $k^{-(m+1)}$ *as* $k \to \infty$.

(e) The real coefficients a_k, b_k and the complex coefficients c_k are related by

$$\left.\begin{aligned} a_k &= c_k + c_{-k} & b_k &= i(c_k - c_{-k}) \\ c_k &= \tfrac{1}{2}(a_k - ib_k) & c_{-k} &= \tfrac{1}{2}(a_k + ib_k) \end{aligned}\right\} \quad (k = 0, 1, 2, \ldots) \qquad (4.11\text{-}3)$$

where $b_0 = 0$. The Fourier series (2) of an even or odd function $f(t)$ (Sec. 4.2-2*b*) reduces to a *Fourier cosine series* or a *Fourier sine series*, respectively.

4.11-3. Fourier Integrals and Fourier Transforms. **(a)** The **Fourier Integral** generated by a real function $f(t)$ whose absolute value $|f(t)|$ *is integrable* over the interval* $-\infty < t < \infty$ (interval of expansion) is defined as

$$\begin{aligned} \frac{1}{\pi}\int_0^\infty d\omega \int_{-\infty}^\infty f(\tau)\cos\omega(t-\tau)\,d\tau &\equiv \int_{-\infty}^\infty c(\nu)e^{2\pi i\nu t}\,d\nu \\ &\equiv \int_{-\infty}^\infty F_F(i\omega)e^{i\omega t}\frac{d\omega}{2\pi} \equiv \frac{1}{\sqrt{2\pi}}\int_{-\infty}^\infty C(\omega)e^{i\omega t}\,d\omega \qquad (4.11\text{-}4a) \end{aligned}$$

with

$$\begin{aligned} c(\nu) &\equiv \int_{-\infty}^\infty f(\tau)e^{-2\pi i\nu t}\,d\tau \equiv F_F(i\omega) \\ C(\omega) &\equiv \frac{1}{\sqrt{2\pi}}\int_{-\infty}^\infty f(\tau)e^{-i\omega\tau}\,d\tau \equiv \frac{1}{\sqrt{2\pi}}F_F(i\omega) \equiv \frac{1}{\sqrt{2\pi}}c(\nu) \qquad (\omega = 2\pi\nu) \end{aligned} \qquad (4.11\text{-}4b)$$

The function $c(\nu) \equiv F_F(i\omega)(\omega = 2\pi\nu)$ will be called the **Fourier transform** $\mathfrak{F}[f(t)]$ of $f(t)$. Some authors, especially physicists, refer to $C(\omega) \equiv F_F(i\omega)/\sqrt{2\pi}$ as the Fourier transform of $f(t)$ instead.

(b) The **Fourier cosine and sine integrals** generated by a real function $f(t)$ whose absolute value $|f(t)|$ is integrable over the interval of expansion $0 < t < \infty$ are respectively defined as

$$\left.\begin{aligned} 2\int_0^\infty c_C(\nu)\cos 2\pi\nu t\,d\nu &\equiv 2\int_0^\infty F_C(\omega)\cos\omega t\,\frac{d\omega}{2\pi} \\ &\equiv \sqrt{\frac{2}{\pi}}\int_0^\infty C_C(\omega)\cos\omega t\,d\omega \\ 2\int_0^\infty c_S(\nu)\sin 2\pi\nu t\,d\nu &\equiv 2\int_0^\infty F_S(\omega)\sin\omega t\,\frac{d\omega}{2\pi} \\ &\equiv \sqrt{\frac{2}{\pi}}\int_0^\infty C_S(\omega)\sin\omega t\,d\omega \end{aligned}\right\} \qquad (4.11\text{-}5a)$$

* See footnote to Sec. 4.11-2.

with

$$\left.\begin{aligned}
c_C(\nu) &\equiv 2\int_0^\infty f(\tau)\cos 2\pi\nu\tau\,d\tau \equiv F_C(\omega)\\
C_C(\omega) &\equiv \sqrt{\frac{2}{\pi}}\int_0^\infty f(\tau)\cos\omega\tau\,d\tau \equiv \frac{1}{\sqrt{2\pi}}F_C(\omega) \equiv \frac{1}{\sqrt{2\pi}}c_C(\nu)\\
c_S(\nu) &\equiv 2\int_0^\infty f(\tau)\sin 2\pi\nu\tau\,d\tau \equiv F_S(\omega)\\
C_S(\omega) &\equiv \sqrt{\frac{2}{\pi}}\int_0^\infty f(\tau)\sin\omega\tau\,d\tau \equiv \frac{1}{\sqrt{2\pi}}F_S(\omega) \equiv \frac{1}{\sqrt{2\pi}}c_S(\nu)
\end{aligned}\right\}\quad(\omega = 2\pi\nu) \tag{4.11-5b}$$

The (real) functions $c_C(\nu) \equiv F_C(\omega) \equiv \mathfrak{F}_C[f(t)]$ and $c_S(\nu) \equiv F_S(\omega) \equiv \mathfrak{F}_S[f(t)]$ will be called, respectively, the **Fourier cosine transform** and the **Fourier sine transform** of $f(t)$. Some authors refer to $C_C(\omega)$ and $C_S(\omega)$ as Fourier cosine and sine transforms, instead.

4.11-4. Functions Which Are Actually Equal to Their Fourier Series or Fourier Integrals. Fourier Analysis (see Appendix D for examples). **(a)** *A Fourier series or a Fourier integral generated by a real function $f(t)$ whose absolute value is integrable* over the corresponding expansion interval I*

1. *Converges uniformly to $f(t)$ throughout every open subinterval of I where $f(t)$ is continuous and of bounded variation (Sec. 4.4-8b)*
2. *Converges to $\frac{1}{2}[f(t-0) + f(t+0)]$ throughout every open subinterval of I in which $f(t)$ is of bounded variation (Jordan's Test;* see also Sec. 4.4-8b)

Remember that $f(t)$ is of bounded variation in every finite open interval where $f(t)$ is bounded and has a finite number of relative maxima, minima, and discontinuities (Dirichlet's conditions, Sec. 4.4-8b).

(b) Fourier Analysis (Harmonic Analysis) of Periodic Functions. *Let $f(t)$ be a real periodic function with period T (Sec. 4.2-2b) and such that $\int_{-T/2}^{T/2} |f(\tau)|\,d\tau$ exists. Then*

* See footnote to Sec. 4.11-2.

$$\begin{aligned} f(t) &= \tfrac{1}{2}a_0 + \sum_{k=1}^{\infty} (a_k \cos k\omega_0 t + b_k \sin k\omega_0 t) \\ &= \sum_{k=-\infty}^{\infty} c_k e^{ik\omega_0 t} = c_0 + 2\sum_{k=1}^{\infty} |c_k| \cos (k\omega_0 t + \arg c_k) \\ a_k &= \frac{2}{T}\int_{-T/2}^{T/2} f(\tau) \cos k\omega_0\tau \, d\tau \\ b_k &= \frac{2}{T}\int_{-T/2}^{T/2} f(\tau) \sin k\omega_0\tau \, d\tau \\ c_k &= c_{-k}^* = \frac{1}{T}\int_{-T/2}^{T/2} f(\tau) e^{-ik\omega_0\tau} \, d\tau \\ &\qquad \left(\omega = \frac{2\pi}{T};\, k = 0, 1, 2, \ldots\right) \end{aligned} \tag{4.11-6}$$

throughout every open interval where $f(t)$ is of bounded variation, if one defines $f(t) = \tfrac{1}{2}[f(t-0) + f(t+0)]$ at each discontinuity (see also Sec. 4.4-8*b*). $f(t)$ is thus expressed as the sum of

1. A *constant term* $a_0/2 = c_0$ [*average value* of $f(t)$, see also Secs. 4.6-3 and 18.10-7], and
2. A set of *sinusoidal terms* (*sinusoidal components*) of respective **frequencies** $\nu_0 = 1/T$ (**fundamental** frequency), $2\nu_0 = 2/T$ (**2nd-harmonic** frequency), $3\nu_0 = 3/T$ (**3rd-harmonic** frequency), . . .

The **kth-harmonic component** $2|c_k| \cos\left(k\frac{2\pi t}{T} + \arg c_k\right)$ has the frequency $k\nu_0 = k/T$, the **circular frequency** $k\omega_0 = 2\pi k\nu_0 = 2\pi k/T$, the **amplitude** $2|c_k| = +\sqrt{a_k^2 + b_k^2}$, and the "phase angle" $\arg c_k = -\arctan (b_k/a_k)$.

The odd-harmonic terms in the expansion (6) describe the *antiperiodic part* of $f(t)$ (Sec. 4.2-2*b*). *Note that*

$$\langle f^2 \rangle = \frac{1}{T}\int_{-T/2}^{T/2} [f(\tau)]^2 \, d\tau = \frac{1}{4}a_0^2 + \frac{1}{2}\sum_{k=1}^{\infty} (a_k^2 + b_k^2) = \sum_{k=-\infty}^{\infty} |c_k|^2 \tag{4.11-7}$$

whenever the integral on the left exists (*Parseval's Theorem*, see also Sec. 15.2-4). Table D-1 of Appendix D lists the Fourier coefficients and the *mean-square values* (7) for a number of periodic functions. Refer to Sec. 20.6-6 for numerical harmonic analysis.

(c) Functions Which Can Be Expressed as Fourier Integrals. Let $f(t)$ be a real function such that $\int_{-\infty}^{\infty} |f(\tau)| \, d\tau$ exists. Then

$$\begin{aligned} f(t) &= \int_{-\infty}^{\infty} c(\nu)e^{2\pi i\nu t}\,d\nu = \int_{-\infty}^{\infty} F_F(i\omega)e^{i\omega t}\,\frac{d\omega}{2\pi} \\ &= \frac{1}{\sqrt{2\pi}}\int_{-\infty}^{\infty} C(\omega)e^{i\omega t}\,d\omega \\ \text{with}\quad c(\nu) &\equiv \int_{-\infty}^{\infty} f(\tau)e^{-2\pi i\nu\tau}\,d\tau \equiv F_F(i\omega) \\ C(\omega) &\equiv \frac{1}{\sqrt{2\pi}}\int_{-\infty}^{\infty} f(\tau)e^{-i\omega t}\,d\tau \\ &\equiv \frac{1}{\sqrt{2\pi}}F_F(i\omega) \equiv \frac{1}{\sqrt{2\pi}}c(\nu) \qquad (\omega = 2\pi\nu) \end{aligned} \tag{4.11-8}$$

throughout every open interval where $f(t)$ is of bounded variation, if one defines $f(t) = \frac{1}{2}[f(t-0) - f(t+0)]$ at each discontinuity. A function $f(t)$ having the Fourier transform $c(\nu)$ will be called an **inverse Fourier transform** $\mathfrak{F}^{-1}[c(\nu)]$ of $c(\nu)$; under the given conditions, Eq. (8) defines $\mathfrak{F}^{-1}[c(\nu)]$ uniquely wherever $f(t)$ is continuous.

Equation (8) may be rewritten in different forms, e.g.,

$$\begin{aligned} f(t) &= \frac{1}{\pi}\int_0^{\infty} d\omega \int_{-\infty}^{\infty} f(\tau)\cos\omega(t-\tau)\,d\tau \\ &= \sqrt{\frac{2}{\pi}}\int_0^{\infty} |C(\omega)|\cos[\omega t + \arg C(\omega)]\,d\omega \\ &= 2\int_0^{\infty} [A(\nu)\cos 2\pi\nu t + B(\nu)\sin 2\pi\nu t]\,d\nu \qquad (\omega = 2\pi\nu) \end{aligned} \tag{4.11-9}$$

The real functions $A(\nu)$ and $-B(\nu)$ in Eq. (9) are, respectively, the cosine transform of the even part (Sec. 4.2-2*b*) of $f(t)$, and the sine transform of the odd part of $f(t)$:

$$\left.\begin{aligned} A(\nu) &= \frac{c(\nu) + c(-\nu)}{2} = \mathfrak{F}_C\left[\frac{f(t) + f(-t)}{2}\right] = \int_{-\infty}^{\infty} f(\tau)\cos 2\pi\nu\tau\,d\tau \\ -B(\nu) &= \frac{c(\nu) - c(-\nu)}{2i} = \mathfrak{F}_S\left[\frac{f(t) - f(-t)}{2}\right] = \int_{-\infty}^{\infty} f(\tau)\sin 2\pi\nu\tau\,d\tau \\ c(\nu) &= c^*(-\nu) = A(\nu) - iB(\nu) \end{aligned}\right\} \quad (\nu \geq 0) \tag{4.11-10}$$

$A(\nu) \equiv 0$ *if* $f(t)$ *is odd, and* $B(\nu) \equiv 0$ *if* $f(t)$ *is even* (see also Sec. 4.11-3*b*).

The Fourier-integral expansion describes $f(t)$ as a "sum" of infinitesimal sinusoidal components with frequencies ν or circular frequencies $\omega = 2\pi\nu$ $(\nu \geq 0)$; the functions $2|c(\nu)|$ and $\arg c(\nu)$ respectively define the amplitudes and the phase angles of the sinusoidal components. Note that $c(-\nu) \equiv c^*(\nu)$, $C(-\omega) \equiv C^*(\omega)$, and that

$$\int_{-\infty}^{\infty} [f(\tau)]^2\,d\tau = \int_{-\infty}^{\infty} |c(\nu)|^2\,d\nu = \int_{-\infty}^{\infty} |F_F(i\omega)|^2\,\frac{d\omega}{2\pi} = \int_{-\infty}^{\infty} |C(\omega)|^2\,d\omega \tag{4.11-11}$$

whenever the integral on the left exists (*Parseval's Theorem;* see also Table 4.11-1).

(d) A more general type of function can be expressed as a sum of a function (8) and a set of periodic functions (6), so that both a "band spectrum" and a "line spectrum" exist (see also Sec. 18.10-9). The treatment of Fourier series and Fourier integrals can be formally unified through the introduction of *generalized* (*integrated*) *Fourier transforms* (Sec. 18.10-10).

(e) The Paley-Wiener Theorem. *Given a positive real function $\Phi(\omega)$, there exists a real function $f(t) \not\equiv 0$ such that $\int_{-\infty}^{\infty} |f(t)|^2\,dt$ exists and $f(t) = 0$ for either $t > 0$ or $t < 0$ and*

$$|F_F(i\omega)|^2 \equiv \left| \int_{-\infty}^{\infty} f(t)e^{-i\omega t}\,dt \right|^2 \equiv \Phi(\omega)$$

if $\Phi(-\omega) \equiv \Phi(\omega)$ and both $\int_{-\infty}^{\infty} |\Phi(\omega)|^2\,d\omega$ and $\int_{-\infty}^{\infty} \frac{\log_e \Phi(\omega)}{1+\omega^2}\,d\omega$ exist (Paley-Wiener Theorem).

4.11-5. Representation of Functions and Operations in Terms of Fourier Coefficients or Fourier Transforms (see also Secs. 4.10-1 and 8.3-1). **(a) Uniqueness Theorem.** A suitably integrable function $f(t)$ uniquely defines its Fourier coefficients (2*b*) or its Fourier transform. Conversely, *a complete set of Fourier coefficients or a Fourier transform uniquely defines the corresponding function $f(t)$ almost everywhere (Sec. 4.6-14b) in the interval of expansion;* in particular, *$f(t)$ is uniquely defined at each point of continuity in the interval of expansion.* This uniqueness theorem holds *even if the Fourier series or Fourier integral does not converge* (see also Sec. 4.11-7).

Note that not every trigonometric series (not even every convergent trigonometric series) is a Fourier series, nor is every function $c(\nu)$ a Fourier transform (see also Sec. 4.11-2*c*).

(b) Operations with Fourier Series. *Given $f(t)$ with the Fourier coefficients a_k, b_k, c_k, and $\varphi(t)$ with the Fourier coefficients α_k, β_k, γ_k for the same interval of expansion, let λ and μ be real constants. Then the function $\lambda f(t) + \mu\varphi(t)$ has the Fourier coefficients $\lambda a_k + \mu\alpha_k$, $\lambda b_k + \mu\beta_k$, $\lambda c_k + \mu\gamma_k$ (term-by-term addition and multiplication by constants).*

Term-by-term integration of a Fourier series (2) *over an interval (t_0, t) in the interval of expansion yields a series converging to $\int_{t_0}^{t} f(\tau)\,d\tau$.* The theorem holds for all values of t_0 and t if $f(t)$ is periodic with period T.

Note that these theorems do not require convergence of the given Fourier series. Refer to Sec. 4.8-4*c* for differentiation of infinite series.

(c) Properties of Fourier Transforms. Table 4.11-1 lists the most important properties of Fourier transforms (see also Sec. 8.3-1).

Note also

$$\begin{aligned}
\mathfrak{F}_C[f(t)\cos 2\pi\nu_0 t] &= \tfrac{1}{2}[c_C(\nu+\nu_0) + c_C(\nu-\nu_0)]\\
\mathfrak{F}_C[f(t)\sin 2\pi\nu_0 t] &= \tfrac{1}{2}[c_S(\nu+\nu_0) - c_S(\nu-\nu_0)]\\
\mathfrak{F}_S[f(t)\cos 2\pi\nu_0 t] &= \tfrac{1}{2}[c_S(\nu+\nu_0) + c_S(\nu-\nu_0)]\\
\mathfrak{F}_S[f(t)\sin 2\pi\nu_0 t] &= \tfrac{1}{2}[c_C(\nu-\nu_0) - c_C(\nu+\nu_0)]
\end{aligned}$$

$$\mathfrak{F}_C[f^{(2r)}(t)] = (-1)^r(2\pi\nu)^{2r}\mathfrak{F}_C[f(t)] - 2\sum_{j=0}^{r-1}(-1)^j(2\pi\nu)^{2j}f^{(2r-2j-1)}(0+0)$$

$$\mathfrak{F}_C[f^{(2r+1)}(t)] = (-1)^r(2\pi\nu)^{2r+1}\mathfrak{F}_S[f(t)] - 2\sum_{j=0}^{r}(-1)^j(2\pi\nu)^{2j}f^{(2r-2j)}(0+0)$$

$$\mathfrak{F}_S[f^{(r)}(t)] = -2\pi\nu\mathfrak{F}_C[f^{(r-1)}(t)]$$

Table 4.11-1. Properties of Fourier Transforms
(see also Sec. 4.11-3 and Table 8.3-1)

Let

$$\mathfrak{F}[f(t)] \equiv \int_{-\infty}^{\infty} f(t)e^{-2\pi i\nu t}\,dt$$

$$\equiv c(\nu) \equiv F_F(i\omega) \equiv \sqrt{2\pi}\,C(\omega) \qquad (\omega = 2\pi\nu)$$

$$f(t) \equiv \int_{-\infty}^{\infty} c(\nu)e^{2\pi i\nu t}\,d\nu \equiv \int_{-\infty}^{\infty} F_F(i\omega)e^{i\omega t}\frac{d\omega}{2\pi} \equiv \frac{1}{\sqrt{2\pi}}\int_{-\infty}^{\infty} C(\omega)e^{i\omega t}\,d\omega$$

and assume that the Fourier transforms in question exist.

(*a*) $\mathfrak{F}[\alpha f_1(t) + \beta f_2(t)] \equiv \alpha\mathfrak{F}[f_1(t)] + \beta\mathfrak{F}[f_2(t)]$ (LINEARITY)

$$\mathfrak{F}[f^*(t)] \equiv c^*(-\nu) \equiv F_F^*(-i\omega)$$

$$\mathfrak{F}[f(\alpha t)] \equiv \frac{1}{\alpha}c\left(\frac{\nu}{\alpha}\right) \equiv \frac{1}{\alpha}F_F\left(\frac{i\omega}{\alpha}\right) \qquad \text{(CHANGE OF SCALE, SIMILARITY THEOREM)}$$

$$\mathfrak{F}[f(t+\tau)] \equiv e^{2\pi i\nu\tau}c(\nu) \equiv e^{i\omega\tau}F_F(i\omega) \qquad \text{(SHIFT THEOREM)}$$

(**b**) **Continuity Theorem.** $\mathfrak{F}[f(t,\alpha)] \to \mathfrak{F}[f(t)]$ *as* $\alpha \to a$ *implies* $f(t,\alpha) \to f(t)$ *wherever* $f(t)$ *is continuous.* Analogous theorems apply to Fourier cosine and sine transforms.

(**c**) **Borel's Convolution Theorem.**

$\mathfrak{F}[f_1(t)]\mathfrak{F}[f_2(t)] \equiv \mathfrak{F}[f_1(t) * f_2(t)]$ where

$$f_1(t) * f_2(t) \equiv \int_{-\infty}^{\infty} f_1(\tau)f_2(t-\tau)\,d\tau \equiv \int_{-\infty}^{\infty} f_1(t-\tau)f_2(\tau)\,d\tau$$

$$\mathfrak{F}[f_1(t)f_2(t)] \equiv \int_{-\infty}^{\infty} c_1(\lambda)c_2(\nu-\lambda)\,d\lambda \equiv \int_{-\infty}^{\infty} c_1(\nu-\lambda)c_2(\lambda)\,d\lambda$$

$$\equiv \int_{-\infty}^{\infty} F_{F1}(i\lambda)F_{F2}[i(\omega-\lambda)]\frac{d\lambda}{2\pi} \equiv \int_{-\infty}^{\infty} F_{F1}[i(\omega-\lambda)]F_{F2}(i\lambda)\frac{d\lambda}{2\pi}$$

(**d**) **Parseval's Theorem.** *If* $\int_{-\infty}^{\infty} |f_1(t)|^2\,dt$ and $\int_{-\infty}^{\infty} |f_2(t)|^2\,dt$ *exist, then*

$$\int_{-\infty}^{\infty} \mathfrak{F}^*[f_1(t)]\mathfrak{F}[f_2(t)]\,d\nu = \int_{-\infty}^{\infty} f_1^*(t)f_2(t)\,dt$$

(**e**) **Modulation Theorem.**

$$\mathfrak{F}[f(t)e^{i\omega_0 t}] \equiv F_F[i(\omega-\omega_0)] = c(\nu-\nu_0)$$

$$\mathfrak{F}[f(t)\cos\omega_0 t] \equiv \tfrac{1}{2}\{F_F[i(\omega-\omega_0)] + F_F[i(\omega+\omega_0)]\}$$

$$\equiv \tfrac{1}{2}[c(\nu-\nu_0) + c(\nu+\nu_0)]$$

$$\mathfrak{F}[f(t)\sin\omega_0 t] \equiv \frac{1}{2i}\{F_F[i(\omega-\omega_0)] - F_F[i(\omega+\omega_0)]\}$$

$$\equiv \frac{1}{2i}[c(\nu-\nu_0) - c(\nu+\nu_0)]$$

(**f**) **Differentiation Theorem.** $\mathfrak{F}[f^{(r)}(t)] = (2\pi i\nu)^r\mathfrak{F}[f(t)]$ $(r = 0, 1, 2, \ldots)$ provided that $f^{(r)}(t)$ exists for all t, and that all derivatives of lesser order vanish as $|t| \to \infty$ $(2\pi\nu = \omega)$.

for $r = 0, 1, 2, \ldots$, provided that the derivative on the left exists for $0 < t < \infty$, and that all derivatives of lesser order vanish as $t \to \infty$.

4.11-6. Dirichlet's and Féjér's Integrals (see also Sec. 4.11-7). The partial sums

$$s_0(t) = \frac{1}{2} a_0 \qquad s_n(t) \equiv \frac{1}{2} a_0 + \sum_{k=1}^{n} \left(a_k \cos k \frac{2\pi t}{T} + b_k \sin k \frac{2\pi t}{T} \right)$$
$$(n = 1, 2, \ldots) \qquad (4.11\text{-}12)$$

and the corresponding arithmetic means $\sigma_{n-1}(t) \equiv \frac{1}{n} \sum_{k=0}^{n-1} s_k(t)$ (Sec. 4.8-6c) of a Fourier series (2) may be written

$$s_n(t) = \frac{1}{T} \int_{-T/2}^{T/2} f(t+\tau) \frac{\sin (2n+1) \frac{\pi\tau}{T}}{\sin \frac{\pi\tau}{T}} d\tau$$
$$= \frac{2}{T} \int_0^{T/2} \frac{f(t-\tau) + f(t+\tau)}{2} \frac{\sin (2n+1) \frac{\pi\tau}{T}}{\sin \frac{\pi\tau}{T}} d\tau$$
$$\left(-\frac{T}{2} < t < \frac{T}{2};\ n = 0, 1, 2, \ldots \right) \qquad \text{(DIRICHLET'S INTEGRAL)} \qquad (4.11\text{-}13)$$

$$\sigma_{n-1}(t) = \frac{1}{nT} \int_{-T/2}^{T/2} f(t+\tau) \left(\frac{\sin n \frac{\pi\tau}{T}}{\sin \frac{\pi\tau}{T}} \right)^2 d\tau$$
$$= \frac{2}{nT} \int_0^{T/2} \frac{f(t-\tau) + f(t+\tau)}{2} \left(\frac{\sin n \frac{\pi\tau}{T}}{\sin \frac{\pi\tau}{T}} \right)^2 d\tau$$
$$\left(-\frac{T}{2} < t < \frac{T}{2};\ n = 1, 2, \ldots \right) \qquad \text{(FÉJÉR'S INTEGRAL)} \qquad (4.11\text{-}14)$$

(see also *Dirichlet's integral formula*, Sec. 21.9-4b).

4.11-7. Summation by Arithmetic Means. (a) The partial sums of a Fourier series may not constitute useful approximations to $f(t)$. This may be true, in particular, if the series diverges, or if the partial sums "overshoot" $f(t)$ badly near a discontinuity of $f(t)$ (nonuniform convergence near the discontinuity, *Gibbs phenomenon*). One may then resort to summation by arithmetic means (Sec. 4.8-6c). *Every Fourier series* (2) *is summable by arithmetic means to the sum* $\frac{1}{2}[f(t-0) + f(t+0)]$ *for all* t *in* $(-T/2, T/2)$ *where the latter function exists* (*Féjér's Theorem*). The arithmetic means converge to $f(t)$ almost everywhere in the interval of expansion; they converge uniformly to $f(t)$ on every open subinterval of $(-T/2, T/2)$ where $f(t)$ is continuous.

(b) *Similarly, if* $\int_{-\infty}^{\infty} |f(\tau)|\, d\tau$ *exists, and* $f(t)$ *is continuous in every finite interval, then the arithmetic means*

$$\frac{1}{\lambda\sqrt{2\pi}} \int_0^{\lambda} d\xi \int_{-\xi}^{\xi} C(\omega) e^{i\omega t}\, d\omega = \frac{\lambda}{2\pi} \int_{-\infty}^{\infty} f(t+\tau) \left(\frac{\sin \frac{\lambda\tau}{2}}{\frac{\lambda\tau}{2}} \right)^2 d\tau \qquad (4.11\text{-}15)$$

converge uniformly to $f(t)$ in every finite interval as $\lambda \to \infty$. The uniform convergence extends over $(-\infty, \infty)$ if $f(t)$ is uniformly continuous in $(-\infty, \infty)$.

4.11-8. Multiple Fourier Series and Integrals. (a) Given an n-dimensional region of expansion defined by $a_j < t_j + T_j$, $j = 1, 2, \ldots, n$, the *multiple Fourier series* generated by a function $f(t_1, t_2, \ldots, t_n)$ such that $\int_{a_1}^{a_1+T_1} \int_{a_2}^{a_2+T_2} \cdots \int_{a_n}^{a_n+T_n} |f(\tau_1, \tau_2, \ldots, \tau_n)|\, d\tau_1\, d\tau_2 \ldots d\tau_n$ exists is defined as

$$\left.\begin{aligned} &\sum_{k_1=-\infty}^{\infty} \sum_{k_2=-\infty}^{\infty} \cdots \sum_{k_n=-\infty}^{\infty} c_{k_1k_2\cdots k_n} \exp\left[2\pi i \sum_{j=1}^{n} k_j \frac{t_j}{T_j}\right] \\ &c_{k_1k_2\cdots k_n} = \frac{1}{T_1T_2\cdots T_n} \int_{a_1}^{a_1+T_1} \int_{a_2}^{a_2+T_2} \cdots \int_{a_n}^{a_n+T_n} f(\tau_1, \tau_2, \ldots, \tau_n) \\ &\qquad \exp\left[-2\pi i \sum_{j=1}^{n} k_j \frac{\tau_j}{T_j}\right] d\tau_1\, d\tau_2 \cdots d\tau_n \end{aligned}\right\} \quad (4.11\text{-}16)$$

(b) The *multiple Fourier integral* generated by a function $f(t_1, t_2, \ldots, t_n)$ such that $\int_{-\infty}^{\infty} \int_{-\infty}^{\infty} \cdots \int_{-\infty}^{\infty} |f(\tau_1, \tau_2, \ldots, \tau_n)|\, d\tau_1\, d\tau_2 \ldots d\tau_n$ exists is defined as

$$\left.\begin{aligned} &\frac{1}{\sqrt{2\pi}^n} \int_{-\infty}^{\infty} \int_{-\infty}^{\infty} \cdots \int_{-\infty}^{\infty} C(\omega_1, \omega_2, \ldots, \omega_n) \\ &\qquad \cdot \exp\left[i \sum_{j=1}^{n} \omega_j t_j\right] d\omega_1\, d\omega_2 \ldots d\omega_n \\ &C(\omega_1, \omega_2, \ldots, \omega_n) \equiv \frac{1}{\sqrt{2\pi}^n} \int_{-\infty}^{\infty} \int_{-\infty}^{\infty} \cdots \int_{-\infty}^{\infty} f(\tau_1, \tau_2, \ldots, \tau_n) \\ &\qquad \cdot \exp\left[-i \sum_{j=1}^{n} \omega_j \tau_j\right] d\tau_1\, d\tau_2 \ldots d\tau_n \end{aligned}\right\} \quad (4.11\text{-}17)$$

One may introduce $\nu_j = \omega_j/2\pi$ as in Eq. (4). For regions of expansion defined by $0 < t_j < \infty$ or $-\infty < t_j < 0$ one obtains *multiple Fourier sine or cosine integrals* by analogy to Sec. 4.11-3*b*.

(c) Given the region of expansion $-\infty < t_1 < \infty$, $a_2 < t_2 < a_2 + T_2$, one may write a "mixed" Fourier expansion for $f(t_1, t_2)$:

$$\left.\begin{aligned} &\frac{1}{\sqrt{2\pi}} \int_{-\infty}^{\infty} \sum_{k=-\infty}^{\infty} C_k(\omega_1) e^{i\left(\omega_1 t_1 + k\frac{2\pi t_2}{T_2}\right)} d\omega_1 \\ &C_k(\omega_1) \equiv \frac{1}{T_2\sqrt{2\pi}} \int_{-\infty}^{\infty} \int_{a_2}^{a_2+T_2} f(\tau_1, \tau_2) e^{-i\left(\omega_1\tau_1 + k\frac{2\pi\tau_2}{T_2}\right)} d\tau_1\, d\tau_2 \end{aligned}\right\} \quad (4.11\text{-}18)$$

One may similarly mix Fourier integrals, Fourier series (and also Fourier sine and cosine integrals) in more than two dimensions.

(d) The exponentials in Eqs. (16), (17), and (18) can be expanded into sine and cosine terms through the use of Eq. (21.2-28). *All the theorems of Secs.* 4.11-2 *through* 4.11-7 *can be generalized to apply to multiple Fourier series and integrals.*

4.12. RELATED TOPICS, REFERENCES, AND BIBLIOGRAPHY

4.12-1. Related Topics. The following topics related to the study of functions, limits, and infinite series are treated in other chapters of this handbook:

Functions of a complex variable.. Chap. 7
Divergence theorem, Stokes' theorem.............................. Chap. 5
Metric spaces, convergence.. Chap. 12
Orthogonal expansions... Chap. 15
Numerical approximations.. Chap. 20
Special functions.. Chap. 21

4.12-2. References and Bibliography (see also Secs. 8.7-2 and 12.9-2).

4.1. Apostol, T. M.: *Mathematical Analysis*, Addison-Wesley, Reading, Mass., 1957.
4.2. Bartle, R. G.: *The Elements of Real Analysis*, Wiley, New York, 1964.
4.3. Boas, R. P.: *A Primer of Real Functions*, Wiley, New York, 1960.
4.4. Buck, R. C.: *Advanced Calculus*, 2d ed., McGraw-Hill, New York, 1965.
4.5. Churchill, R. V.: *Fourier Series and Boundary Value Problems*, 2d ed., McGraw-Hill, New York, 1963.
4.6. Dieudonné, J.: *Foundations of Modern Analysis*, Academic, New York, 1960.
4.7. Eggleston, H. G.: *Introduction to Elementary Real Analysis*, Cambridge University Press, New York, 1962.
4.8. Fleming, W.: *Functions of Several Variables*, Addison-Wesley, Reading, Mass., 1966.
4.9. Gelbaum, B. R., and J. M. H. Olmsted: *Counterexamples in Analysis*, Holden-Day, San Francisco, 1964.
4.10. Goffman, C.: *Calculus of Several Variables*, Harper & Row, New York, 1965.
4.11. Goldberg, S.: *Methods of Real Analysis*, Blaisdell, New York, 1964.
4.12. Graves, L. M.: *The Theory of Functions of Real Variables*, 2d ed., McGraw-Hill, New York, 1956.
4.13. Knopp, K.: *Theory and Application of Infinite Series*, Blackie, Glasgow, 1951; also 5th ed. (in German), Springer, Berlin, 1964.
4.14. Natanson, I. P.: *Theory of Functions of a Real Variable* (2 vols.), Unger, New York, 1955/9.
4.15. Papoulis, A.: *The Fourier Integral and Its Application*, McGraw-Hill, New York, 1962.
4.16. Rankin, R. A.: *Introduction to Mathematical Analysis*, Pergamon Press, New York, 1962.
4.17. Rosser, J. B.: Asymptotic Formulas and Series, in E. F. Beckenbach, *Modern Mathematics for the Engineer*, 2d series, McGraw-Hill, New York, 1961.
4.18. Rudin, W.: *Principles of Mathematical Analysis*, 2d ed., McGraw-Hill, New York, 1964.
4.19. Wall, H. S.: *Analytic Theory of Continued Fractions*, Van Nostrand, Princeton, N.J., 1948.
4.20. Widder, D. V.: *Advanced Calculus*, 2d ed., Prentice-Hall, Englewood Cliffs, N.J., 1961.

CHAPTER **5**

VECTOR ANALYSIS

5.1. INTRODUCTION

5.1-1. Euclidean Vectors. Each class of **Euclidean vectors** (e.g., displacements, velocities, forces, magnetic field strengths) permits the definition of operations known as *vector addition* (Sec. 5.2-1), *multiplication of vectors by (real) scalars* (Sec. 5.2-1), and *scalar multiplication of vectors* (Sec. 5.2-6). Each class of (Euclidean) vectors commonly encountered in geometry and physics is, moreover, intimately related to the two- or three-dimensional space of Euclidean geometry:

1. The vectors of each class permit a reciprocal one-to-one representation by *translations* (*displacements, directed line segments*) in the geometrical space. This representation preserves the results of vector addition, multiplication by scalars, and scalar multiplication of vectors (and thus magnitudes and relative directions of vectors; see also Secs. 12.1-6 and 14.2-1 to 14.2-7).
2. In most applications, vectors appear as *functions of position* in geometrical space, so that the vectors are associated with geometrical points (*vector point functions*, Sec. 5.4-1).

Vectors, such as velocities or forces, are usually first introduced in geometrical language as "quantities possessing magnitude and direction" or, somewhat more precisely, as quantities which can be represented by directed line segments subject to a "parallelogram law of addition." Such a geometrical approach, common to most elementary courses, is employed in Secs. 5.2-1 and 5.2-8 to introduce the principal vector operations. Refer to Secs. 12.4-1 and Chap. 14 for a discussion of vectors from a much more general point of view.

Vector analysis is the study of vector (and scalar) functions. Each vector may be specified by a set of numerical functions (*vector components*) in terms of a suitable reference system (Secs. 5.2-2, 5.2-3, 5.4-1, and 6.3-1).

NOTE: The description of a physical situation in terms of vector quantities should not be regarded as merely a kind of shorthand summarizing sets of component equations by single equations, but as an instance of a mathematical model (Sec. 12.1-1) whose essential "building blocks" are not restricted to numbers. Note also that a class of objects admitting a one-to-one reciprocal correspondence with a class of directed line segments is *not* necessarily a vector space unless it has the algebraic properties outlined above (EXAMPLES: finite rotations, directed metal rods).

5.2. VECTOR ALGEBRA

5.2-1. Vector Addition and Multiplication of Vectors by (Real) Scalars. The operation (**vector addition**) of forming the **vector sum** $\mathbf{a} + \mathbf{b}$ of two Euclidean vectors $\mathbf{a}$ and $\mathbf{b}$ of a suitable class is a vector corresponding to the geometrical addition of the corresponding displacements (*parallelogram law*). The **product** of a Euclidean vector $\mathbf{a}$ by a real number (**scalar**) α is a vector corresponding to a displacement α times as long as that corresponding to $\mathbf{a}$, with a reversal in direction if α is negative. The *null vector* $\mathbf{0}$ of each class of vectors corresponds to a displacement of length zero, and $\mathbf{a} + \mathbf{0} = \mathbf{a}$. With these geometrical definitions, vector addition and multiplication by scalars satisfy the relations

$$
\begin{gathered}
\mathbf{a} + \mathbf{b} = \mathbf{b} + \mathbf{a} \qquad \mathbf{a} + (\mathbf{b} + \mathbf{c}) = (\mathbf{a} + \mathbf{b}) + \mathbf{c} = \mathbf{a} + \mathbf{b} + \mathbf{c} \\
\alpha(\beta\mathbf{a}) = (\alpha\beta)\mathbf{a} \qquad (\alpha + \beta)\mathbf{a} = \alpha\mathbf{a} + \beta\mathbf{a} \qquad \alpha(\mathbf{a} + \mathbf{b}) = \alpha\mathbf{a} + \alpha\mathbf{b} \\
(1)\mathbf{a} = \mathbf{a} \qquad (-1)\mathbf{a} = -\mathbf{a} \qquad (0)\mathbf{a} = \mathbf{0} \\
\mathbf{a} - \mathbf{a} = \mathbf{0} \qquad \mathbf{a} + \mathbf{0} = \mathbf{a}
\end{gathered}
\tag{5.2-1}
$$

Refer to Secs. 14.2-1 to 14.2-4 for a more general discussion of vector algebra.

5.2-2. Representation of Vectors in Terms of Base Vectors and Components. m vectors $\mathbf{a}_1, \mathbf{a}_2, \ldots, \mathbf{a}_m$ are **linearly independent** if and only if $\lambda_1\mathbf{a}_1 + \lambda_2\mathbf{a}_2 + \cdots + \lambda_m\mathbf{a}_m = \mathbf{0}$ implies $\lambda_1 = \lambda_2 = \cdots = \lambda_m = 0$; otherwise the set of vectors is **linearly dependent** (see also Secs. 1.9-3 and 14.2-3). Every vector $\mathbf{a}$ of a **three-dimensional vector space** can be represented as a sum

$$
\mathbf{a} = \alpha_1\mathbf{e}_1 + \alpha_2\mathbf{e}_2 + \alpha_3\mathbf{e}_3 \tag{5.2-2}
$$

in terms of three linearly independent vectors $\mathbf{e}_1, \mathbf{e}_2, \mathbf{e}_3$. The coefficients $\alpha_1, \alpha_2, \alpha_3$ are the **components*** of the vector $\mathbf{a}$ with respect to the

* Some authors refer to the **component vectors** $\alpha_1\mathbf{e}_1, \alpha_2\mathbf{e}_2, \alpha_3\mathbf{e}_3$ as components.

reference system defined by the **base vectors** $\mathbf{e}_1$, $\mathbf{e}_2$, $\mathbf{e}_3$ (see also Sec. 14.2-4). Given a suitable reference system, the vectors **a**, **b**, . . . are thus represented by the respective ordered sets $(\alpha_1, \alpha_2, \alpha_3)$, $(\beta_1, \beta_2, \beta_3)$, . . . of their components; note that $\mathbf{a} + \mathbf{b}$ and $\alpha\mathbf{a}$ are respectively represented by $(\alpha_1 + \beta_1, \alpha_2 + \beta_2, \alpha_3 + \beta_3)$ and $(\alpha\alpha_1, \alpha\alpha_2, \alpha\alpha_3)$. *Vector relations can, then, be expressed (represented) in terms of corresponding (sets of) relations between vector components.*

Vectors belonging to **two-dimensional vector spaces** (e.g., plane displacements) are similarly represented by sets of *two* components.

The system of base vectors chosen for the representation of vectors defined at a given point of geometrical space (Sec. 5.4-1) is usually simply related to the coordinate system used for the description of the geometrical space. Chapter 6 deals specifically with the representation of vector relations in terms of "local" base vectors directed along, and perpendicular to, the coordinate lines of curvilinear coordinate systems at each point; the magnitudes and directions of the local base vectors are, in general, different at different points. Transformation equations relating vector components associated with different reference systems are given in Table 6.3-1 and in Sec. 14.6-1.

5.2-3. Rectangular Cartesian Components of a Vector. Given a right-handed rectangular cartesian-coordinate system (Sec. 3.1-4) in the geometrical space, the unit vectors (Sec. 5.2-5) **i**, **j**, **k** respectively directed along the positive x axis, the positive y axis, and the positive z axis form a convenient system of base vectors at each point. The components a_x, a_y, a_z of a vector

$$\boxed{\mathbf{a} = a_x\mathbf{i} + a_y\mathbf{j} + a_z\mathbf{k}} \tag{5.2-3}$$

are the **(right-handed) rectangular cartesian components** of **a**.

Note that

$$a_x = \mathbf{a} \cdot \mathbf{i} \qquad a_y = \mathbf{a} \cdot \mathbf{j} \qquad a_z = \mathbf{a} \cdot \mathbf{k}$$

(Sec. 5.2-6) are direction numbers of the vector **a**; the components $\mathbf{u} \cdot \mathbf{i}$, $\mathbf{u} \cdot \mathbf{j}$, $\mathbf{u} \cdot \mathbf{k}$ of any *unit vector* **u** are its direction cosines (Sec. 3.1-8).

5.2-4. Vectors and Physical Dimensions. Euclidean vectors may also be multiplied by scalars which are not themselves real numbers but are suitably labeled by real numbers (quantities isomorphic with the field of real numbers, Sec. 12.1-6; thus one multiplies a constant velocity vector by a time interval to obtain a displacement vector). *If a vector* (2) *or* (3) *is a physical quantity, one usually associates its physical dimension with the components rather than with the base vectors.* The latter are then regarded as dimensionless and may be used to define a common scheme of reference systems (**scheme of measurements,** see also Sec. 16.1-4) for various classes of vectors having different physical dimensions (e.g., displacements, velocities, forces, etc.; see also Sec. 16.1-4).

5.2-5. Absolute Value (Magnitude, Norm) of a Vector. The **absolute value (magnitude, norm)** $|\mathbf{a}|$ of a Euclidean vector **a** is a scalar proportional to the length of the displacement corresponding to **a** (Sec. 5.1-1; see also Secs. 14.2-5 and 14.2-7 for an abstract definition). Absolute values of vectors satisfy the relations (1.1-3). A vector of magnitude 1 is a **unit vector.** The (mutually perpendicular) base vectors **i, j, k** (Sec. 5.2-3) are defined to be unit vectors, so that $|a_x\mathbf{i}| = |a_x|$, $|a_y\mathbf{j}| = |a_y|$, $|a_z\mathbf{k}| = |a_z|$, and

$$|\mathbf{a}| = \underset{+}{\sqrt{a_x^2 + a_y^2 + a_z^2}} \tag{5.2-4}$$

5.2-6. Scalar Product (Dot Product, Inner Product) of Two Vectors. The **scalar product (dot product, inner product)** $\mathbf{a} \cdot \mathbf{b}$ [alternative notation $(\mathbf{ab})$] of two Euclidean vectors **a** and **b** is the scalar

$$\mathbf{a} \cdot \mathbf{b} = |\mathbf{a}|\,|\mathbf{b}| \cos \gamma \tag{5.2-5}$$

where γ is the angle $\overrightarrow{\sphericalangle \mathbf{a}, \mathbf{b}}$ (see also Secs. 14.2-6 and 14.2-7 for an abstract definition). If **a** and **b** are physical quantities, the physical dimension of the scalar product $\mathbf{a} \cdot \mathbf{b}$ must be observed (see also Sec. 5.2-4). Table 5.2-1 summarizes the principal relations involving scalar products. *Two nonzero vectors* **a** *and* **b** *are perpendicular to each other if and only if* $\mathbf{a} \cdot \mathbf{b} = 0$.

Table 5.2-1. Relations Involving Scalar Products

(a) Basic Relations

$$\mathbf{a} \cdot \mathbf{b} = \mathbf{b} \cdot \mathbf{a} \qquad \mathbf{a} \cdot (\mathbf{b} + \mathbf{c}) = \mathbf{a} \cdot \mathbf{b} + \mathbf{a} \cdot \mathbf{c} \qquad (\alpha\mathbf{a}) \cdot \mathbf{b} = \alpha(\mathbf{a} \cdot \mathbf{b})$$

$$\mathbf{a} \cdot \mathbf{a} = \mathbf{a}^2 = |\mathbf{a}^2| \geq 0 \qquad |\mathbf{a} \cdot \mathbf{b}| \leq |\mathbf{a}|\,|\mathbf{b}| \qquad \cos \gamma = \frac{\mathbf{a} \cdot \mathbf{b}}{\sqrt{\mathbf{a}^2\mathbf{b}^2}}$$

(b) *In terms of rectangular cartesian components (Sec.* 5.2-3)

$$\mathbf{i} \cdot \mathbf{i} = \mathbf{j} \cdot \mathbf{j} = \mathbf{k} \cdot \mathbf{k} = 1 \qquad \mathbf{i} \cdot \mathbf{j} = \mathbf{j} \cdot \mathbf{k} = \mathbf{k} \cdot \mathbf{i} = 0$$

$$\mathbf{a} \cdot \mathbf{b} = (a_x\mathbf{i} + a_y\mathbf{j} + a_z\mathbf{k}) \cdot (b_x\mathbf{i} + b_y\mathbf{j} + b_z\mathbf{k}) = a_xb_x + a_yb_y + a_zb_z$$

$$a_x = \mathbf{a} \cdot \mathbf{i} \qquad a_y = \mathbf{a} \cdot \mathbf{j} \qquad a_z = \mathbf{a} \cdot \mathbf{k}$$

5.2-7. The Vector (Cross) Product. The **vector (cross) product** $\mathbf{a} \times \mathbf{b}$ (alternative notation $[\mathbf{ab}]$) of two vectors **a** and **b** is the vector of magnitude

$$|\mathbf{a} \times \mathbf{b}| = |\mathbf{a}|\,|\mathbf{b}| \sin \gamma \tag{5.2-6}$$

whose direction is perpendicular to both **a** and **b** and such that the axial motion of a right-handed screw turning **a** into **b** is in the direction of $\mathbf{a} \times \mathbf{b}$. *Two vectors are linearly dependent* (*Sec.* 5.2-2) *if and only if* *their vector product is zero.* Table 5.2-2 summarizes the principal relations involving vector products. Refer to Sec. 16.8-4 for a more general definition of the vector product and to Secs. 3.1-10 and 17.3-3*c* for the representation of plane areas as vectors.

Table 5.2-2. Relations Involving Vector (Cross) Products

(a) **Basic Relations**

$$\mathbf{a} \times \mathbf{b} = -(\mathbf{b} \times \mathbf{a})$$

$$\mathbf{a} \times \mathbf{a} = 0 \qquad \mathbf{a} \cdot (\mathbf{a} \times \mathbf{b}) = \mathbf{b} \cdot (\mathbf{a} \times \mathbf{b}) = 0$$

$$(\alpha \mathbf{a}) \times \mathbf{b} = \alpha(\mathbf{a} \times \mathbf{b}) \qquad \mathbf{a} \times (\mathbf{b} + \mathbf{c}) = \mathbf{a} \times \mathbf{b} + \mathbf{a} \times \mathbf{c}$$

$$[(\alpha + \beta)\mathbf{a}] \times \mathbf{b} = (\alpha + \beta)(\mathbf{a} \times \mathbf{b}) = \alpha(\mathbf{a} \times \mathbf{b}) + \beta(\mathbf{a} \times \mathbf{b})$$

(b) *In terms of any basis* $\mathbf{e}_1, \mathbf{e}_2, \mathbf{e}_3$

$$\mathbf{a} = \alpha_1\mathbf{e}_1 + \alpha_2\mathbf{e}_2 + \alpha_3\mathbf{e}_3 \qquad \mathbf{b} = \beta_1\mathbf{e}_1 + \beta_2\mathbf{e}_2 + \beta_3\mathbf{e}_3$$

$$\mathbf{a} \times \mathbf{b} = \begin{vmatrix} \mathbf{e}_2 \times \mathbf{e}_3 & \alpha_1 & \beta_1 \\ \mathbf{e}_3 \times \mathbf{e}_1 & \alpha_2 & \beta_2 \\ \mathbf{e}_1 \times \mathbf{e}_2 & \alpha_3 & \beta_3 \end{vmatrix}$$

(c) *In terms of right-handed rectangular cartesian components*

$$\mathbf{i} \times \mathbf{i} = \mathbf{j} \times \mathbf{j} = \mathbf{k} \times \mathbf{k} = 0 \qquad \mathbf{i} \times \mathbf{j} = \mathbf{k} \qquad \mathbf{j} \times \mathbf{k} = \mathbf{i} \qquad \mathbf{k} \times \mathbf{i} = \mathbf{j}$$

$$\mathbf{a} \times \mathbf{b} = \begin{vmatrix} \mathbf{i} & a_x & b_x \\ \mathbf{j} & a_y & b_y \\ \mathbf{k} & a_z & b_z \end{vmatrix} = \mathbf{i} \begin{vmatrix} a_y & a_z \\ b_y & b_z \end{vmatrix} + \mathbf{j} \begin{vmatrix} a_z & a_x \\ b_z & b_x \end{vmatrix} + \mathbf{k} \begin{vmatrix} a_x & a_y \\ b_x & b_y \end{vmatrix}$$

$$= \mathbf{i}(a_yb_z - a_zb_y) + \mathbf{j}(a_zb_x - a_xb_z) + \mathbf{k}(a_xb_y - a_yb_x)$$

5.2-8. The Scalar Triple Product (Box Product).

$$\begin{aligned} \mathbf{a} \cdot (\mathbf{b} \times \mathbf{c}) \equiv [\mathbf{abc}] &= [\mathbf{bca}] = [\mathbf{cab}] \\ &= -[\mathbf{bac}] = -[\mathbf{cba}] = -[\mathbf{acb}] \end{aligned} \tag{5.2-7}$$

$$\left.\begin{aligned} [\mathbf{abc}]^2 &= [(\mathbf{a} \times \mathbf{b})(\mathbf{b} \times \mathbf{c})(\mathbf{c} \times \mathbf{a})] = \mathbf{a}^2\mathbf{b}^2\mathbf{c}^2 - \mathbf{a}^2(\mathbf{b} \cdot \mathbf{c})^2 \\ &\quad - \mathbf{b}^2(\mathbf{a} \cdot \mathbf{c})^2 - \mathbf{c}^2(\mathbf{a} \cdot \mathbf{b})^2 + 2(\mathbf{a} \cdot \mathbf{b})(\mathbf{b} \cdot \mathbf{c})(\mathbf{a} \cdot \mathbf{c}) \\ &= \begin{vmatrix} \mathbf{a} \cdot \mathbf{a} & \mathbf{a} \cdot \mathbf{b} & \mathbf{a} \cdot \mathbf{c} \\ \mathbf{b} \cdot \mathbf{a} & \mathbf{b} \cdot \mathbf{b} & \mathbf{b} \cdot \mathbf{c} \\ \mathbf{c} \cdot \mathbf{a} & \mathbf{c} \cdot \mathbf{b} & \mathbf{c} \cdot \mathbf{c} \end{vmatrix} \quad \text{(Gram's determinant, see also Sec. 14.2-6)} \end{aligned}\right\} \tag{5.2-8}$$

$$[\mathbf{abc}][\mathbf{def}] = \begin{vmatrix} \mathbf{a} \cdot \mathbf{d} & \mathbf{a} \cdot \mathbf{e} & \mathbf{a} \cdot \mathbf{f} \\ \mathbf{b} \cdot \mathbf{d} & \mathbf{b} \cdot \mathbf{e} & \mathbf{b} \cdot \mathbf{f} \\ \mathbf{c} \cdot \mathbf{d} & \mathbf{c} \cdot \mathbf{e} & \mathbf{c} \cdot \mathbf{f} \end{vmatrix} \tag{5.2-9}$$

In terms of any basis $\mathbf{e}_1, \mathbf{e}_2, \mathbf{e}_3$ (Sec. 5.2-2; see also Secs. 5.2-3 and 6.3-4)

$$[\mathbf{abc}] = \begin{vmatrix} \alpha_1 & \beta_2 & \gamma_1 \\ \alpha_2 & \beta_2 & \gamma_2 \\ \alpha_3 & \beta_3 & \gamma_3 \end{vmatrix} [\mathbf{e}_1\mathbf{e}_2\mathbf{e}_3] \tag{5.2-10}$$

In terms of right-handed rectangular cartesian components (Sec. 5.2-3)

$$[\mathbf{abc}] = \begin{vmatrix} a_x & b_x & c_x \\ a_y & b_y & c_y \\ a_z & b_z & c_z \end{vmatrix} \quad \begin{array}{l}(>0 \text{ if } \mathbf{a}, \mathbf{b}, \mathbf{c} \text{ are directed like right-} \\ \text{handed cartesian axes)}\end{array} \tag{5.2-11}$$

5.2-9. Other Products Involving More Than Two Vectors.

$$\mathbf{a} \times (\mathbf{b} \times \mathbf{c}) = (\mathbf{a} \cdot \mathbf{c})\mathbf{b} - (\mathbf{a} \cdot \mathbf{b})\mathbf{c} = \begin{vmatrix} \mathbf{b} & \mathbf{c} \\ \mathbf{a} \cdot \mathbf{b} & \mathbf{a} \cdot \mathbf{c} \end{vmatrix} \quad \text{(VECTOR TRIPLE PRODUCT)} \tag{5.2-12}$$

$$(\mathbf{a} \times \mathbf{b}) \cdot (\mathbf{c} \times \mathbf{d}) = (\mathbf{a} \cdot \mathbf{c})(\mathbf{b} \cdot \mathbf{d}) - (\mathbf{a} \cdot \mathbf{d})(\mathbf{b} \cdot \mathbf{c}) = \begin{vmatrix} \mathbf{a} \cdot \mathbf{c} & \mathbf{b} \cdot \mathbf{c} \\ \mathbf{a} \cdot \mathbf{d} & \mathbf{b} \cdot \mathbf{d} \end{vmatrix} \tag{5.2-13}$$

$$(\mathbf{a} \times \mathbf{b})^2 = \mathbf{a}^2\mathbf{b}^2 - (\mathbf{a} \cdot \mathbf{b})^2 \tag{5.2-14}$$

$$(\mathbf{a} \times \mathbf{b}) \times (\mathbf{c} \times \mathbf{d}) = [\mathbf{acd}]\mathbf{b} - [\mathbf{bcd}]\mathbf{a} = [\mathbf{abd}]\mathbf{c} - [\mathbf{abc}]\mathbf{d} \tag{5.2-15}$$

5.2-10. Representation of a Vector a as a Sum of Vectors Respectively along and Perpendicular to a Given Unit Vector u.

$$\mathbf{a} = \mathbf{u}(\mathbf{u} \cdot \mathbf{a}) + \mathbf{u} \times (\mathbf{a} \times \mathbf{u}) \tag{5.2-16}$$

5.2-11. Solution of Equations.

(a) $$\left.\begin{array}{r}\mathbf{x} \cdot \mathbf{a} = p \\ \mathbf{x} \times \mathbf{a} = \mathbf{b}\end{array}\right\} \text{ implies } \mathbf{x} = \mathbf{a}\frac{p}{a^2} + (\mathbf{a} \times \mathbf{b})\frac{1}{a^2} \tag{5.2-17}$$

(b) $$\left.\begin{array}{r}\mathbf{x} \cdot \mathbf{a} = p \\ \mathbf{x} \cdot \mathbf{b} = q \\ \mathbf{x} \cdot \mathbf{c} = r\end{array}\right\} \text{ implies } \mathbf{x} = \frac{p(\mathbf{b} \times \mathbf{c}) + q(\mathbf{c} \times \mathbf{a}) + r(\mathbf{a} \times \mathbf{b})}{[\mathbf{abc}]} \tag{5.2-18}$$

(c) $\mathbf{a}x + \mathbf{b}y + \mathbf{c}z + \mathbf{d} = 0$ implies

$$x = -\frac{[\mathbf{dbc}]}{[\mathbf{abc}]} \qquad y = -\frac{[\mathbf{dca}]}{[\mathbf{abc}]} \qquad z = -\frac{[\mathbf{dab}]}{[\mathbf{abc}]} \tag{5.2-19}$$

(d) $(\mathbf{b} \times \mathbf{c})x + (\mathbf{c} \times \mathbf{a})y + (\mathbf{a} \times \mathbf{b})z + \mathbf{d} = 0$ implies

$$x = -\frac{\mathbf{d} \cdot \mathbf{a}}{[\mathbf{abc}]} \qquad y = -\frac{\mathbf{d} \cdot \mathbf{b}}{[\mathbf{abc}]} \qquad z = -\frac{\mathbf{d} \cdot \mathbf{c}}{[\mathbf{abc}]} \tag{5.2-20}$$

5.3. VECTOR CALCULUS: FUNCTIONS OF A SCALAR PARAMETER

5.3-1. Vector Functions and Limits. A vector function $\mathbf{v} = \mathbf{v}(t)$ of a scalar parameter t associates one (single-valued function) or more (multiple-valued function) "values" of the vector $\mathbf{v}$ with every value of the scalar parameter t (independent variable) for which $\mathbf{v}(t)$ is defined (see also Sec. 4.2-1). In terms of rectangular cartesian components

$$\boxed{\mathbf{v} = \mathbf{v}(t) = v_x(t)\mathbf{i} + v_y(t)\mathbf{j} + v_z(t)\mathbf{k}} \tag{5.3-1}$$

A vector function $\mathbf{v}(t)$ is **bounded** if $|\mathbf{v}(t)|$ is bounded. $\mathbf{v}(t)$ has the **limit** (see also Secs. 4.4-1 and 12.5-3) $\mathbf{v}_1 = \lim_{t \to t_1} \mathbf{v}(t)$ if and only if for every positive number ϵ there exists a number $\delta > 0$ such that $|t - t_1| < \delta$ implies $|\mathbf{v}_1 - \mathbf{v}(t)| < \epsilon$. If $\lim_{t \to t_1} \mathbf{v}(t)$ exists,

$$\lim_{t \to t_1} \mathbf{v}(t) = [\mathbf{i} \lim_{t \to t_1} v_x(t) + \mathbf{j} \lim_{t \to t_1} v_y(t) + \mathbf{k} \lim_{t \to t_1} v_z(t)] \tag{5.3-2}$$

Formulas analogous to those of Sec. 4.4-2 (limits of sums, products, etc.) apply to vector sums, scalar products, and vector products. $\mathbf{v}(t)$ is **continuous** for $t = t_1$ if and only if $\lim_{t \to t_1} \mathbf{v}(t) = \mathbf{v}(t_1)$ (see also Secs. 4.4-6 and 12.5-3).

5.3-2. Differentiation. A vector function $\mathbf{v}(t)$ is **differentiable** for $t = t_1$ if and only if the **derivative**

$$\frac{d\mathbf{v}(t)}{dt} = \lim_{\Delta t \to 0} \frac{\mathbf{v}(t + \Delta t) - \mathbf{v}(t)}{\Delta t} \tag{5.3-3}$$

exists and is unique for $t = t_1$ (see also Sec. 4.5-1). If the derivative $d^2\mathbf{v}(t)/dt^2$ of $d\mathbf{v}(t)/dt$ exists, it is called the **second derivative** of $\mathbf{v}(t)$, and so forth. Table 5.3-1 summarizes the principal differentiation rules.

Table 5.3-1. Differentiation of Vector Functions with Respect to a Scalar Parameter

(a) **Basic Rules**

$$\frac{d}{dt}[\mathbf{v}(t) \pm \mathbf{w}(t)] = \frac{d\mathbf{v}}{dt} \pm \frac{d\mathbf{w}}{dt} \qquad \frac{d}{dt}[\alpha\mathbf{v}(t)] = \alpha\frac{d\mathbf{v}}{dt} \ (\alpha \text{ constant})$$

$$\frac{d}{dt}[f(t)\mathbf{v}(t)] = \frac{df}{dt}\mathbf{v} + f\frac{d\mathbf{v}}{dt} \qquad \frac{d}{dt}[\mathbf{v}(t) \cdot \mathbf{w}(t)] = \frac{d\mathbf{v}}{dt} \cdot \mathbf{w} + \frac{d\mathbf{w}}{dt} \cdot \mathbf{v}$$

$$\frac{d}{dt}[\mathbf{v}(t) \times \mathbf{w}(t)] = \frac{d\mathbf{v}}{dt} \times \mathbf{w} + \mathbf{v} \times \frac{d\mathbf{w}}{dt} \qquad \frac{d}{dt}\mathbf{v}[f(t)] = \frac{d\mathbf{v}}{df}\frac{df}{dt}$$

$$\frac{d}{dt}[\mathbf{v}(t)\mathbf{w}(t)\mathbf{u}(t)] = \left[\frac{d\mathbf{v}}{dt}\mathbf{w}\mathbf{u}\right] + \left[\mathbf{v}\frac{d\mathbf{w}}{dt}\mathbf{u}\right] + \left[\mathbf{v}\mathbf{w}\frac{d\mathbf{u}}{dt}\right]$$

(b) *In terms of rectangular cartesian components*

$$\frac{d\mathbf{v}(t)}{dt} = \frac{dv_x(t)}{dt}\mathbf{i} + \frac{dv_y(t)}{dt}\mathbf{j} + \frac{dv_z(t)}{dt}\mathbf{k}$$

(c) If the base vectors $\mathbf{e}_1(t)$, $\mathbf{e}_2(t)$, $\mathbf{e}_3(t)$ are functions of t, and $\mathbf{v}(t) = \alpha_1(t)\mathbf{e}_1(t) + \alpha_2(t)\mathbf{e}_2(t) + \alpha_3(t)\mathbf{e}_3(t)$, then

$$\frac{d\mathbf{v}(t)}{dt} = \left[\frac{d\alpha_1}{dt}\mathbf{e}_1 + \frac{d\alpha_2}{dt}\mathbf{e}_2 + \frac{d\alpha_3}{dt}\mathbf{e}_3\right] + \left[\alpha_1\frac{d\mathbf{e}_1}{dt} + \alpha_2\frac{d\mathbf{e}_2}{dt} + \alpha_3\frac{d\mathbf{e}_3}{dt}\right]$$

Analogous rules apply to the *partial derivatives* $\partial\mathbf{v}/\partial t_1 \equiv \mathbf{v}_{t_1}$, $\partial\mathbf{v}/\partial t_2 \equiv \mathbf{v}_{t_2}$, . . . of a vector function $\mathbf{v} = \mathbf{v}(t_1, t_2, \ldots)$ of two or more scalar parameters $t_1, t_2, \ldots$

NOTE: If $\mathbf{u}(t)$ is a unit vector (of constant magnitude but variable direction) and $\mathbf{v}(t) = v(t)\mathbf{u}(t)$,

$$\frac{d\mathbf{u}}{dt} = \boldsymbol{\omega} \times \mathbf{u} \qquad \text{and} \qquad \frac{d\mathbf{v}}{dt} = \frac{dv}{dt}\mathbf{u} + v\frac{d\mathbf{u}}{dt} = \frac{dv}{dt}\mathbf{u} + \boldsymbol{\omega} \times \mathbf{v} \tag{5.3-4}$$

$\boldsymbol{\omega}$ is directed along the axis about which $\mathbf{u}(t)$ [and thus also $\mathbf{v}(t)$] turns as t varies, so that a right-handed screw turning with $\mathbf{u}(t)$ would be propelled in the direction of $\boldsymbol{\omega}$. Its magnitude is equal to the *angular rate of turn* of $\mathbf{u}(t)$ [and thus also of $\mathbf{v}(t)$] with respect to t (EXAMPLE: *angular velocity vector* in physics; see also Sec. 17.2-3). *Equation* (4) *describes the separate contributions of changes in the magnitude and direction of* $\mathbf{v}(t)$.

5.3-3. Integration and Ordinary Differential Equations. The **indefinite integral** $\mathbf{V}(t) = \int \mathbf{v}(t)\,dt$ of a suitable vector function $\mathbf{v}(t)$ is defined as the solution of the vector differential equation (see also Sec. 9.1-1)

$$\frac{d}{dt}\mathbf{V}(t) = \mathbf{v}(t) \tag{5.3-5}$$

which may be replaced by a set of differential equations for the components of $\mathbf{V}(t)$. Other ordinary differential equations involving differentiation of vectors with respect to a scalar parameter are treated similarly. The **definite integral**

$$\left.\begin{aligned}\int_a^b \mathbf{v}(t)\,dt &= \lim_{\max|t_i - t_{i-1}|\to 0} \sum_{i=1}^{m} \mathbf{v}(\tau_i)(t_i - t_{i-1}) \\ \text{with}\qquad a &= t_0 < \tau_1 < t_1 < \tau_2 < t_2 < \cdots < \tau_m < t_m = b\end{aligned}\right\} \tag{5.3-6}$$

(see also Sec. 4.6-1) may be treated in terms of components:

$$\int_a^b \mathbf{v}(t)\,dt = \mathbf{i}\int_a^b v_x(t)\,dt + \mathbf{j}\int_a^b v_y(t)\,dt + \mathbf{k}\int_a^b v_z(t)\,dt \tag{5.3-7}$$

5.4. SCALAR AND VECTOR FIELDS

5.4-1. Introduction. The remainder of this chapter deals specifically with *scalar and vector functions of position in three-dimensional Euclidean space.* Unless the contrary is stated, the scalar and vector functions of position are assumed to be single-valued, continuous, and suitably differentiable functions of the coordinates, and thus of the position vector $\mathbf{r} \equiv x\mathbf{i} + y\mathbf{j} + k\mathbf{z}$. In Secs. 5.4-2 to 5.7-3 relations involving scalar and vector functions are stated

1. In coordinate-free (invariant) form, and
2. In terms of vector components along *right-handed rectangular* cartesian coordinate axes (Sec. 5.2-3), so that*

$$\mathbf{F}(\mathbf{r}) \equiv \mathbf{F}(x, y, z) \equiv \mathbf{i}F_x(x, y, z) + \mathbf{j}F_y(x, y, z) + \mathbf{k}F_z(x, y, z) \tag{5.4-1}$$

The relations to be described are independent of the coordinate system used to specify position in space. The representation of vector relations in terms of vector components along, or perpendicular to, suitable *curvilinear* coordinate lines (and thus along different directions at different points) is treated in Chap. 6.

* Throughout Chaps. 5 and 6, the subscripts in F_x, F_y, F_z, . . . do *not* indicate differentiation with respect to x, y, z, . . . ; in fact, no scalar function $F(x, y, z)$ is introduced.

5.4-2. Scalar Fields. A **scalar field** is a **scalar function of position (scalar point function)** $\Phi(\mathbf{r}) \equiv \Phi(x, y, z)$ together with its region of definition. The surfaces

$$\boxed{\Phi(\mathbf{r}) \equiv \Phi(x, y, z) = \text{constant}} \tag{5.4-2}$$

(Sec. 3.1-14) are called **level surfaces** of the field and permit its geometrical representation.

5.4-3. Vector Fields. A **vector field** is a **vector function of position (vector point function)** $\mathbf{F}(\mathbf{r}) \equiv F(x, y, z)$ together with its region of definition. The **field lines (streamlines)** of the vector field defined by $\mathbf{F}(\mathbf{r})$ have the direction of the field vector $\mathbf{F}(\mathbf{r})$ at each point $(\mathbf{r})$ and are specified by the differential equations

$$\boxed{d\mathbf{r} \times \mathbf{F}(\mathbf{r}) = 0 \qquad \text{or} \qquad dx{:}dy{:}dz = F_x{:}F_y{:}F_z} \tag{5.4-3}$$

A vector field may be represented geometrically by its field lines, with the relative density of the field lines at each point $(\mathbf{r})$ proportional to the absolute value $|\mathbf{F}(\mathbf{r})|$ of the field vector.

5.4-4. Vector Path Element and Arc Length (see also Sec. 4.6-9). **(a)** The **vector path element (vector element of distance)** $d\mathbf{r}$ along a curve C described by

$$\mathbf{r} = \mathbf{r}(t) \qquad \text{or} \qquad x = x(t) \qquad y = y(t) \qquad z = z(t) \tag{5.4-4}$$

is defined at every regular point $(\mathbf{r}) \equiv [x(t), y(t), z(t)]$ as

$$d\mathbf{r} = \mathbf{i}\,dx + \mathbf{j}\,dy + \mathbf{k}\,dz = \left(\mathbf{i}\frac{dx}{dt} + \mathbf{j}\frac{dy}{dt} + \mathbf{k}\frac{dz}{dt}\right)dt = \frac{d\mathbf{r}(t)}{dt}\,dt \tag{5.4-5}$$

$d\mathbf{r}$ is directed along the tangent to C at each regular point (see also Sec. 17.2-2).

(b) The arc length s on a rectifiable curve (4) (Sec. 4.6-9) is given by

$$\boxed{\begin{aligned} &s = \int_{t_0C}^{t} ds \\ \text{with}\quad &ds = \sqrt{dx^2 + dy^2 + dz^2} \\ & = \sqrt{\left(\frac{dx}{dt}\right)^2 + \left(\frac{dy}{dt}\right)^2 + \left(\frac{dz}{dt}\right)^2}\,dt = \frac{ds}{dt}\,dt \\ & = \sqrt{d\mathbf{r}\cdot d\mathbf{r}} = \sqrt{\frac{d\mathbf{r}}{dt}\cdot\frac{d\mathbf{r}}{dt}}\,dt \\ &\text{at each regular point } (\mathbf{r}) \equiv [x(t), y(t), z(t)] \text{ of the curve}\end{aligned}} \tag{5.4-6}$$

The sign of ds is assigned arbitrarily, e.g., so that $ds/dt > 0$.

5.4-5. Line Integrals (see also Sec. 4.6-10). Given a rectifiable arc C represented by Eq. (4), the scalar line integrals

$$\left.\begin{aligned} \int_C \Phi(\mathbf{r})\,ds &= \int_C \Phi(x, y, z)\,ds = \int_C \Phi[x(t), y(t), z(t)]\frac{ds}{dt}\,dt \\ \int_C d\mathbf{r}\cdot\mathbf{F}(\mathbf{r}) &= \int_C \frac{d\mathbf{r}}{dt}\cdot\mathbf{F}(\mathbf{r})\,dt \\ &= \int_C [F_x(x, y, z)\,dx + F_y(x, y, z)\,dy \\ &\qquad + F_z(x, y, z)\,dz] \\ &= \int_C \left(F_x\frac{dx}{dt} + F_y\frac{dy}{dt} + F_z\frac{dz}{dt}\right)dt \end{aligned}\right\} \tag{5.4-7}$$

can be defined directly as limits of sums in the manner of Sec. 4.6-10; it is, however, more convenient to substitute the functions $x(t)$, $y(t)$, $z(t)$, dx/dt, dy/dt, and dz/dt obtained from Eq. (4) into Eq. (7) and to integrate over t.

One similarly defines the vector line integrals

$$\begin{aligned} \int_C d\mathbf{r}\Phi(\mathbf{r}) &= \int_C \frac{d\mathbf{r}}{dt}\Phi(\mathbf{r})\,dt \\ &= \mathbf{i}\int_C \Phi(x, y, z)\,dx + \mathbf{j}\int_C \Phi(x, y, z)\,dy + \mathbf{k}\int_C \Phi(x, y, z)\,dz \\ &= \int_C \left[\mathbf{i}\Phi(x, y, z)\frac{dx}{dt} + \mathbf{j}\Phi(x, y, z)\frac{dy}{dt} + \mathbf{k}\Phi(x, y, z)\frac{dz}{dt}\right]dt \end{aligned} \tag{5.4-8}$$

and
$$\begin{aligned} \int_C d\mathbf{r}\times\mathbf{F}(\mathbf{r}) &= \int_C \frac{d\mathbf{r}}{dt}\times\mathbf{F}(\mathbf{r})\,dt \\ &= \mathbf{i}\int_C (F_z\,dy - F_y\,dz) + \mathbf{j}\int_C (F_x\,dz - F_z\,dx) \\ &\qquad + \mathbf{k}\int_C (F_y\,dx - F_x\,dy) \\ &= \int_C \left[\mathbf{i}\left(F_z\frac{dy}{dt} - F_y\frac{dz}{dt}\right) + \mathbf{j}\left(F_x\frac{dz}{dt} - F_z\frac{dx}{dt}\right)\right. \\ &\qquad \left. + \mathbf{k}\left(F_y\frac{dx}{dt} - F_x\frac{dy}{dt}\right)\right]dt \end{aligned} \tag{5.4-9}$$

Unless special conditions are satisfied (*Sec.* 5.7-1), *the value of a scalar or vector line integral depends on the path of integration C.*

Refer to Secs. 6.2-3a and 6.4-3a for the use of curvilinear coordinates.

NOTE: It is often useful to introduce the arc length s as a new parameter into the expressions (7) to (9) by means of Eq. (6).

5.4-6. Surface Integrals (see also Secs. 4.6-12 and 17.3-3*c*). **(a)** At each regular point of a two-sided surface represented by $\mathbf{r} = \mathbf{r}(u, v)$ (Sec. 3.1-14), it is possible to define a **vector element of area**

$$d\mathbf{A} = \left(\frac{\partial \mathbf{r}}{\partial u} \times \frac{\partial \mathbf{r}}{\partial v}\right) du\, dv = \left[\mathbf{i}\left(\frac{\partial y}{\partial u}\frac{\partial z}{\partial v} - \frac{\partial z}{\partial u}\frac{\partial y}{\partial v}\right) + \mathbf{j}\left(\frac{\partial z}{\partial u}\frac{\partial x}{\partial v} - \frac{\partial x}{\partial u}\frac{\partial z}{\partial v}\right) + \mathbf{k}\left(\frac{\partial x}{\partial u}\frac{\partial y}{\partial v} - \frac{\partial y}{\partial u}\frac{\partial x}{\partial v}\right)\right] du\, dv \qquad (5.4\text{-}10)$$

at each surface point (u, v). In the case of a closed surface, the sense and order of the surface coordinates u, v are customarily chosen so that the direction of $d\mathbf{A}$ (direction of the positive surface normal, Sec. 17.3-2) is *outward from the bounded volume.*

The **scalar element of area** at the surface point (u, v) is defined as

$$dA = \pm|d\mathbf{A}| = \pm\left|\frac{\partial \mathbf{r}}{\partial u} \times \frac{\partial \mathbf{r}}{\partial v}\right| du\, dv = \sqrt{a(u, v)}\, du\, dv \qquad (5.4\text{-}11)$$

The sign of dA may be arbitrarily assigned (see also Secs. 4.6-11, 4.6-12, 6.4-3*b*, and 17.3-3*c*).

In particular, for $u = x$, $v = y$, $z = z(x, y)$,

$$\left.\begin{aligned} d\mathbf{A} &= \left(-\mathbf{i}\frac{\partial z}{\partial x} - \mathbf{j}\frac{\partial z}{\partial y} + \mathbf{k}\right) dx\, dy \\ dA &= \pm|d\mathbf{A}| = \sqrt{1 + \left(\frac{\partial z}{\partial x}\right)^2 + \left(\frac{\partial z}{\partial y}\right)^2}\, dx\, dy \end{aligned}\right\} \qquad (5.4\text{-}12)$$

(b) In the following it will be assumed that the area $\int_S |d\mathbf{A}|$ (Sec. 4.6-11) of each surface region S under consideration exists; in this case Eq. (10) defines $d\mathbf{A}$ almost everywhere on S (Sec. 4.6-14*b*). The scalar surface integrals

$$\int_S |d\mathbf{A}|\Phi(\mathbf{r}) \qquad \text{and} \qquad \int_S d\mathbf{A} \cdot \mathbf{F}(\mathbf{r}) \qquad (5.4\text{-}13)$$

and the vector surface integrals

$$\int_S d\mathbf{A}\Phi(\mathbf{r}) \qquad \text{and} \qquad \int_S d\mathbf{A} \times \mathbf{F}(\mathbf{r}) \qquad (5.4\text{-}14)$$

of suitable field functions $\Phi(\mathbf{r})$ and $\mathbf{F}(\mathbf{r})$ may then be defined directly as limits of sums in the manner of Secs. 4.6-1, 4.6-10, and 5.3-3. One can, instead, employ Eq. (10) to express each surface integral as a double integral over the surface coordinates u and v (see also Sec. 6.4-3*b*).

Note

$$\int_S d\mathbf{A} \cdot \mathbf{F}(x, y, z) = \iint_S F_x[x(y, z), y, z]\, dy\, dz + \iint_S F_y[x, y(x, z), z]\, dx\, dz + \iint_S F_z[x, y, z(x, y)]\, dx\, dy \qquad (5.4\text{-}15)$$

In the first integral $u = y$, $v = z$ are independent variables; in the second integral $u = z$, $v = x$, etc. [see also Eq. (12)]. Equation (15) must *not* be interpreted to imply

$d\mathbf{A} = \mathbf{i}\,dy\,dz + \mathbf{j}\,dx\,dz + \mathbf{k}\,dx\,dy$ without such qualifications on the meaning of dx, dy, and dz.

5.4-7. Volume Integrals (see also Sec. 4.6-12). Given a simply connected region V of three-dimensional Euclidean space, the scalar volume integral

$$\int_V \Phi(\mathbf{r})\,dV = \iiint\limits_V \Phi(x, y, z)\,dx\,dy\,dz \tag{5.4-16}$$

and the vector volume integral

$$\int_V \mathbf{F}(\mathbf{r})\,dV = \iiint\limits_V [\mathbf{i}F_x(x, y, z) + \mathbf{j}F_y(x, y, z) + \mathbf{k}F_z(x, y, z)]\,dx\,dy\,dz \tag{5.4-17}$$

may be defined as limits of sums in the manner of Sec. 4.6-1, or they may be expressed directly in terms of triple integrals over x, y, and z. Refer to Secs. 6.2-3*b* and 6.4-3*c* for the use of curvilinear coordinates.

5.5. DIFFERENTIAL OPERATORS

5.5-1. Gradient, Divergence, and Curl: Coordinate-free Definitions in Terms of Integrals. The **gradient** grad $\Phi(\mathbf{r}) \equiv \nabla\Phi$ of a scalar point function $\Phi(\mathbf{r}) \equiv \Phi(x, y, z)$ is a vector point function defined at each point $(\mathbf{r}) \equiv (x, y, z)$ where $\Phi(\mathbf{r})$ is suitably differentiable. In coordinate-free form,

$$\operatorname{grad} \Phi(\mathbf{r}) \equiv \nabla\Phi \equiv \lim_{\delta\to 0} \frac{\int_{S_1} d\mathbf{A}\Phi(\boldsymbol{\varrho})}{\int_{V_1} dV} \tag{5.5-1}$$

where V_1 is a region containing the point $(\mathbf{r})$ and bounded by a closed surface S_1 such that the greatest distance between the point $(\mathbf{r})$ and any point of S_1 is less than $\delta > 0$ (see also Sec. 4.3-5*c*).

Given a suitably differentiable vector point function $\mathbf{F}(\mathbf{r}) \equiv \mathbf{F}(x, y, z)$, it is similarly possible to define a scalar point function, the **divergence** of $\mathbf{F}(\mathbf{r})$ at the point $\mathbf{r}$,

$$\operatorname{div} \mathbf{F}(\mathbf{r}) \equiv \nabla \cdot \mathbf{F} = \lim_{\delta\to 0} \frac{\int_{S_1} d\mathbf{A} \cdot \mathbf{F}(\boldsymbol{\varrho})}{\int_{V_1} dV} \tag{5.5-2}$$

and a vector point function, the **curl** (rotational) of $\mathbf{F}(\mathbf{r})$ at the point $\mathbf{r}$,

$$\operatorname{curl} \mathbf{F}(\mathbf{r}) \equiv \nabla \times \mathbf{F} = \lim_{\delta\to 0} \frac{\int_{S_1} d\mathbf{A} \times \mathbf{F}(\boldsymbol{\varrho})}{\int_{V_1} dV} \tag{5.5-3}$$

NOTE: At each point where the vector grad $\Phi \equiv \nabla\Phi$ exists, it has the magnitude

$$|\nabla\Phi| = +\sqrt{\left(\frac{\partial\Phi}{\partial x}\right)^2 + \left(\frac{\partial\Phi}{\partial y}\right)^2 + \left(\frac{\partial\Phi}{\partial z}\right)^2} \tag{5.5-4}$$

of, as well as the direction associated with, the greatest directional derivative $d\Phi/ds$ (Sec. 5.5-3*c*) at that point. $\nabla\Phi$ defines a vector field whose field lines are specified by the differential equations

$$d\mathbf{r} \times (\nabla\Phi) = 0 \qquad \text{or} \qquad dx{:}dy{:}dz = \frac{\partial\Phi}{\partial x}{:}\frac{\partial\Phi}{\partial y}{:}\frac{\partial\Phi}{\partial z} \tag{5.5-5}$$

The **gradient lines** defined by Eq. (5) intersect the level surfaces (5.4-2) perpendicularly.

5.5-2. The Operator ∇.

In terms of rectangular cartesian coordinates, the linear operator ∇ (**del or nabla**) is defined by

$$\nabla \equiv \mathbf{i}\frac{\partial}{\partial x} + \mathbf{j}\frac{\partial}{\partial y} + \mathbf{k}\frac{\partial}{\partial z} \tag{5.5-6}$$

Its application to a scalar point function $\Phi(\mathbf{r})$ or a vector point function $\mathbf{F}(\mathbf{r})$ corresponds formally to a noncommutative multiplication operation with a vector having the rectangular cartesian "components" $\partial/\partial x$, $\partial/\partial y$, $\partial/\partial z$; thus, in terms of right-handed rectangular cartesian coordinates x, y, z,

$$\begin{aligned}
\nabla\Phi(x, y, z) &\equiv \text{grad}\,\Phi(x, y, z) \equiv \mathbf{i}\frac{\partial\Phi}{\partial x} + \mathbf{j}\frac{\partial\Phi}{\partial y} + \mathbf{k}\frac{\partial\Phi}{\partial z} \\
\nabla\cdot\mathbf{F}(x, y, z) &\equiv \text{div}\,\mathbf{F}(x, y, z) \equiv \frac{\partial F_x}{\partial x} + \frac{\partial F_y}{\partial y} + \frac{\partial F_z}{\partial z} \\
\nabla\times\mathbf{F}(x, y, z) &\equiv \text{curl}\,\mathbf{F}(x, y, z) \\
&\equiv \mathbf{i}\left(\frac{\partial F_z}{\partial y} - \frac{\partial F_y}{\partial z}\right) + \mathbf{j}\left(\frac{\partial F_x}{\partial z} - \frac{\partial F_z}{\partial x}\right) + \mathbf{k}\left(\frac{\partial F_y}{\partial x} - \frac{\partial F_x}{\partial y}\right) \\
&\equiv \begin{vmatrix} \mathbf{i} & \frac{\partial}{\partial x} & F_x \\ \mathbf{j} & \frac{\partial}{\partial y} & F_y \\ \mathbf{k} & \frac{\partial}{\partial z} & F_z \end{vmatrix} \\
(\mathbf{G}\cdot\nabla)\mathbf{F} &\equiv G_x\frac{\partial\mathbf{F}}{\partial x} + G_y\frac{\partial\mathbf{F}}{\partial y} + G_z\frac{\partial\mathbf{F}}{\partial z} \\
&\equiv \mathbf{i}(\mathbf{G}\cdot\nabla F_x) + \mathbf{j}(\mathbf{G}\cdot\nabla F_y) + \mathbf{k}(\mathbf{G}\cdot\nabla F_z)
\end{aligned} \tag{5.5-7}$$

Table 5.5-1 summarizes a number of rules for operations with the operator ∇.

Table 5.5-1. Rules for Operations Involving the Operator ∇

(a) Linearity

$$\nabla(\Phi + \Psi) = \nabla\Phi + \nabla\Psi \qquad \nabla(\alpha\Phi) = \alpha\nabla\Phi$$
$$\nabla \cdot (\mathbf{F} + \mathbf{G}) = \nabla \cdot \mathbf{F} + \nabla \cdot \mathbf{G} \qquad \nabla \cdot (\alpha\mathbf{F}) = \alpha\nabla \cdot \mathbf{F}$$
$$\nabla \times (\mathbf{F} + \mathbf{G}) = \nabla \times \mathbf{F} + \nabla \times \mathbf{G} \qquad \nabla \times (\alpha\mathbf{F}) = \alpha\nabla \times \mathbf{F}$$

(b) Operations on Products

$$\nabla(\Phi\Psi) = \Psi\nabla\Phi + \Phi\nabla\Psi$$
$$\nabla(\mathbf{F} \cdot \mathbf{G}) = (\mathbf{F} \cdot \nabla)\mathbf{G} + (\mathbf{G} \cdot \nabla)\mathbf{F} + \mathbf{F} \times (\nabla \times \mathbf{G}) + \mathbf{G} \times (\nabla \times \mathbf{F})$$
$$\nabla \cdot (\Phi\mathbf{F}) = \Phi\nabla \cdot \mathbf{F} + (\nabla\Phi) \cdot \mathbf{F}$$
$$\nabla \cdot (\mathbf{F} \times \mathbf{G}) = \mathbf{G} \cdot \nabla \times \mathbf{F} - \mathbf{F} \cdot \nabla \times \mathbf{G}$$
$$(\mathbf{G} \cdot \nabla)\Phi\mathbf{F} = \mathbf{F}(\mathbf{G} \cdot \nabla\Phi) + \Phi(\mathbf{G} \cdot \nabla)\mathbf{F}$$
$$\nabla \times (\Phi\mathbf{F}) = \Phi\nabla \times \mathbf{F} + (\nabla\Phi) \times \mathbf{F}$$
$$\nabla \times (\mathbf{F} \times \mathbf{G}) = (\mathbf{G} \cdot \nabla)\mathbf{F} - (\mathbf{F} \cdot \nabla)\mathbf{G} + \mathbf{F}(\nabla \cdot \mathbf{G}) - \mathbf{G}(\nabla \cdot \mathbf{F})$$
$$(\mathbf{G} \cdot \nabla)\mathbf{F} = \tfrac{1}{2}[\nabla \times (\mathbf{F} \times \mathbf{G}) + \nabla(\mathbf{F} \cdot \mathbf{G}) - \mathbf{F}(\nabla \cdot \mathbf{G}) + \mathbf{G}(\nabla \cdot \mathbf{F}) - \mathbf{F} \times (\nabla \times \mathbf{G}) - \mathbf{G} \times (\nabla \times \mathbf{F})]$$

Note that vector equations involving $\nabla\Phi$, $\nabla \cdot \mathbf{F}$, and/or $\nabla \times \mathbf{F}$ have a meaning independent of the coordinate system used. Refer to Chap. 6 and Sec. 16.10-7 for transformations expressing $\nabla\Phi$, $\nabla \cdot \mathbf{F}$, and $\nabla \times \mathbf{F}$ in terms of different coordinate systems.

5.5-3. Absolute Differential, Intrinsic Derivative, and Directional Derivative. (a) The change (**absolute differential**) $d\Phi$ of a scalar point function $\Phi(\mathbf{r})$ associated with a change $d\mathbf{r} \equiv \mathbf{i}\,dx + \mathbf{j}\,dy + \mathbf{k}\,dz$ in position is (see also Sec. 4.5-3*a*)

$$d\Phi = \frac{\partial\Phi}{\partial x}dx + \frac{\partial\Phi}{\partial y}dy + \frac{\partial\Phi}{\partial z}dz = d\mathbf{r} \cdot \operatorname{grad}\Phi = (d\mathbf{r} \cdot \nabla)\Phi \tag{5.5-8}$$

(b) The **intrinsic (absolute) derivative** (see also Table 4.5-2*a*) $d\Phi/dt$ of $\Phi(\mathbf{r})$ along the curve $\mathbf{r} = \mathbf{r}(t)$ is, at each point $(\mathbf{r})$ of the curve, the rate of change of $\Phi(\mathbf{r})$ with respect to the parameter t as $\mathbf{r}$ varies as a function of t:

$$\frac{d\Phi}{dt} = \left(\frac{d\mathbf{r}}{dt} \cdot \nabla\right)\Phi = \frac{dx}{dt}\frac{\partial\Phi}{\partial x} + \frac{dy}{dt}\frac{\partial\Phi}{\partial y} + \frac{dz}{dt}\frac{\partial\Phi}{\partial z}$$
$$\text{with} \quad \mathbf{r} = \mathbf{r}(t) \quad \text{or} \quad x = x(t) \quad y = y(t) \quad z = z(t) \tag{5.5-9}$$

NOTE: If Φ depends explicitly on t [$\Phi = \Phi(\mathbf{r}, t)$], then

$$\frac{d\Phi}{dt} = \left(\frac{d\mathbf{r}}{dt} \cdot \nabla\right)\Phi + \frac{\partial\Phi}{\partial t} \tag{5.5-10}$$

(c) The **directional derivative** $d\Phi/ds$ of $\Phi(\mathbf{r})$ at the point $(\mathbf{r})$ is the rate of change of $\Phi(\mathbf{r})$ with the distance s from the point $(\mathbf{r})$ as a function of direction. The directional derivative of $\Phi(\mathbf{r})$ in the direction of the unit vector $\mathbf{u} \equiv \mathbf{i}\cos\alpha_x + \mathbf{j}\cos\alpha_y + \mathbf{k}\cos\alpha_z$ defined by the direction cosines (Sec. 3.1-8*a*) $\cos\alpha_x$, $\cos\alpha_y$, $\cos\alpha_z$ is

$$\frac{d\Phi}{ds} = \cos\alpha_x \frac{\partial\Phi}{\partial x} + \cos\alpha_y \frac{\partial\Phi}{\partial y} + \cos\alpha_z \frac{\partial\Phi}{\partial z} = (\mathbf{u}\cdot\nabla)\Phi \tag{5.5-11}$$

$d\Phi/ds$ is the intrinsic derivative of $\Phi(\mathbf{r})$ with respect to the path length s along a curve directed along $\mathbf{u} = d\mathbf{r}/ds$.

(**d**) The absolute differential, intrinsic derivative, and directional derivative of a vector point function $\mathbf{F}(\mathbf{r})$ are defined in a manner analogous to that for a scalar point function. Thus

$$\left.\begin{aligned}
d\mathbf{F} &= (d\mathbf{r}\cdot\nabla)\mathbf{F} = [\mathbf{i}(d\mathbf{r}\cdot\nabla)F_x + \mathbf{j}(d\mathbf{r}\cdot\nabla)F_y + \mathbf{k}(d\mathbf{r}\cdot\nabla)F_z] \\
&= [\mathbf{i}\,dF_x + \mathbf{j}\,dF_y + \mathbf{k}\,dF_z] \\
\frac{d\mathbf{F}}{dt} &= \left(\frac{d\mathbf{r}}{dt}\cdot\nabla\right)\mathbf{F} = \left[\mathbf{i}\left(\frac{d\mathbf{r}}{dt}\cdot\nabla\right)F_x + \mathbf{j}\left(\frac{d\mathbf{r}}{dt}\cdot\nabla\right)F_y + \mathbf{k}\left(\frac{d\mathbf{r}}{dt}\cdot\nabla\right)F_z\right] \\
&= \left[\mathbf{i}\frac{dF_x}{dt} + \mathbf{j}\frac{dF_y}{dt} + \mathbf{k}\frac{dF_z}{dt}\right] \\
\frac{d\mathbf{F}}{ds} &= (\mathbf{u}\cdot\nabla)\mathbf{F} = [\mathbf{i}(\mathbf{u}\cdot\nabla)F_x + \mathbf{j}(\mathbf{u}\cdot\nabla)F_y + \mathbf{k}(\mathbf{u}\cdot\nabla)F_z] \\
&= \left[\mathbf{i}\frac{dF_x}{ds} + \mathbf{j}\frac{dF_y}{ds} + \mathbf{k}\frac{dF_z}{ds}\right]
\end{aligned}\right\} \tag{5.5-12}$$

5.5-4. Higher-order Directional Derivatives. Taylor Expansion. The nth-order directional derivative of Φ or $\mathbf{F}$ in the $\mathbf{u}$ direction is defined by

$$\frac{d^n}{ds^n}\Phi(\mathbf{r}) = (\mathbf{u}\cdot\nabla)^n\Phi(\mathbf{r}) \qquad \text{and} \qquad \frac{d^n}{ds^n}\mathbf{F}(\mathbf{r}) = (\mathbf{u}\cdot\nabla)^n\mathbf{F}(\mathbf{r}) \tag{5.5-13}$$

respectively. For suitably differentiable functions, one has, if the series in question converges (see also Sec. 4.10-5),

$$\begin{aligned}
\Phi(\mathbf{r}+\Delta\mathbf{r}) &= \Phi(\mathbf{r}) + (\Delta\mathbf{r}\cdot\nabla)\Phi(\mathbf{r})\Big]_{\mathbf{r}} + \frac{1}{2!}(\Delta\mathbf{r}\cdot\nabla)^2\Phi(\mathbf{r})\Big]_{\mathbf{r}} + \cdots = e^{(\Delta\mathbf{r}\cdot\nabla)}\Phi(\mathbf{r})\Big]_{\mathbf{r}} \\
&= \Phi(\mathbf{r}) + \Delta r\frac{d}{ds}\Phi(\mathbf{r})\Big]_{\mathbf{r}} + \frac{1}{2!}(\Delta r)^2\frac{d^2}{ds^2}\Phi(\mathbf{r})\Big]_{\mathbf{r}} + \cdots
\end{aligned} \tag{5.5-14a}$$

and

$$\begin{aligned}
\mathbf{F}(\mathbf{r}+\Delta\mathbf{r}) &= \mathbf{F}(\mathbf{r}) + (\Delta\mathbf{r}\cdot\nabla)\mathbf{F}(\mathbf{r})\Big]_{\mathbf{r}} + \frac{1}{2!}(\Delta\mathbf{r}\cdot\nabla)^2\mathbf{F}(\mathbf{r})\Big]_{\mathbf{r}} + \cdots = e^{(\Delta\mathbf{r}\cdot\nabla)}\mathbf{F}(\mathbf{r})\Big]_{\mathbf{r}} \\
&= \mathbf{F}(\mathbf{r}) + \Delta r\frac{d}{ds}\mathbf{F}(\mathbf{r})\Big]_{\mathbf{r}} + \frac{1}{2!}(\Delta r)^2\frac{d^2}{ds^2}\mathbf{F}(\mathbf{r})\Big]_{\mathbf{r}} + \cdots
\end{aligned} \tag{5.5-14b}$$

where each directional derivative is taken in the direction of $\Delta\mathbf{r}$.

5.5-5. The Laplacian Operator. The **Laplacian operator** $\nabla^2 \equiv (\nabla\cdot\nabla)$ (sometimes denoted by Δ), expressed in terms of rectangular cartesian coordinates by

$$\nabla^2 \equiv (\nabla\cdot\nabla) \equiv \left(\frac{\partial^2}{\partial x^2} + \frac{\partial^2}{\partial y^2} + \frac{\partial^2}{\partial z^2}\right) \tag{5.5-15}$$

(see Chap. 6 and Sec. 16.10-7 for other representations), may be applied to both scalar and vector point functions by noncommutative scalar "multiplication," so that

$$\nabla^2\Phi \equiv \left(\frac{\partial^2}{\partial x^2} + \frac{\partial^2}{\partial y^2} + \frac{\partial^2}{\partial z^2}\right)\Phi \qquad \nabla^2\mathbf{F} \equiv (\mathbf{i}\nabla^2 F_x + \mathbf{j}\nabla^2 F_y + \mathbf{k}\nabla^2 F_z) \tag{5.5-16}$$

Note

$$\nabla^2(\alpha\Phi + \beta\Psi) = \alpha\nabla^2\Phi + \beta\nabla^2\Psi \quad \text{(LINEARITY)} \tag{5.5-17}$$

and

$$\nabla^2(\Phi\Psi) = \Psi\nabla^2\Phi + 2(\nabla\Phi)\cdot(\nabla\Psi) + \Phi\nabla^2\Psi \tag{5.5-18}$$

5.5-6. Repeated Operations. Note the following rules for repeated operations with the operator ∇:

$$\begin{aligned} \text{div grad } \Phi &= \nabla\cdot(\nabla\Phi) = \nabla^2\Phi \\ \text{grad div } \mathbf{F} &= \nabla(\nabla\cdot\mathbf{F}) = \nabla^2\mathbf{F} + \nabla\times(\nabla\times\mathbf{F}) \\ \text{curl curl } \mathbf{F} &= \nabla\times(\nabla\times\mathbf{F}) = \nabla(\nabla\cdot\mathbf{F}) - \nabla^2\mathbf{F} \\ \text{curl grad } \Phi &= \nabla\times(\nabla\Phi) = 0 \\ \text{div curl } \mathbf{F} &= \nabla\cdot(\nabla\times\mathbf{F}) = 0 \end{aligned} \tag{5.5-19}$$

5.5-7. Operations on Special Functions. A number of results of differential operations on scalar and vector functions of the position vector $\mathbf{r} \equiv (x, y, z)$ are tabulated in Tables 5.5-2 and 5.5-3, respectively. Additional formulas may be derived with the aid of Table 5.5-1. Note also

$$[\mathbf{F}(\mathbf{r})\cdot\nabla]\mathbf{r} = \mathbf{F}(\mathbf{r}) \tag{5.5-20}$$

$$\nabla\frac{\mathbf{a}\cdot\mathbf{r}}{r^3} = -\nabla\times\frac{\mathbf{a}\times\mathbf{r}}{r^3} \tag{5.5-21}$$

where **a** is a constant vector.

Table 5.5-2. Operations on Scalar Point Functions
($r \equiv |\mathbf{r}|$; **a** is a constant vector; $n = 0, \pm 1, \pm 2, \ldots$)

Φ	$\nabla\Phi$	$\nabla^2\Phi$
$\mathbf{a}\cdot\mathbf{r}$	$\mathbf{a}$	0
r^n	$nr^{n-2}\mathbf{r}$	$n(n+1)r^{n-2}$
$\log_e r$	$\mathbf{r}/r^2$	$1/r^2$

5.5-8. Functions of Two or More Position Vectors. Many problems involve scalar or vector functions of two or more position vectors (functions depending on

Table 5.5-3. Operations on Vector Point Functions
($r \equiv |\mathbf{r}|$; **a** is a constant vector; $n = 0, \pm 1, \pm 2, \ldots$)

$\mathbf{F}$	$\nabla \cdot \mathbf{F}$	$\nabla \times \mathbf{F}$	$(\mathbf{G} \cdot \nabla)\mathbf{F}$	$\nabla^2\mathbf{F}$	$\nabla\nabla \cdot \mathbf{F}$
$\mathbf{r}$	3	0	$\mathbf{G}$	0	0
$\mathbf{a} \times \mathbf{r}$	0	$2\mathbf{a}$	$\mathbf{a} \times \mathbf{G}$	0	0
$\mathbf{a}r^n$	$nr^{n-2}(\mathbf{r} \cdot \mathbf{a})$	$nr^{n-2}(\mathbf{r} \times \mathbf{a})$	$nr^{n-2}(\mathbf{r} \cdot \mathbf{G})\mathbf{a}$	$n(n+1)r^{n-2}\mathbf{a}$	$nr^{n-2}\mathbf{a} + n(n-2)r^{n-4}(\mathbf{r} \cdot \mathbf{a})\mathbf{r}$
$\mathbf{r}r^n$	$(n+3)r^n$	0	$r^n\mathbf{G} + nr^{n-2}(\mathbf{r} \cdot \mathbf{G})\mathbf{r}$	$n(n+3)r^{n-2}\mathbf{r}$	$n(n+3)r^{n-2}\mathbf{r}$
$\mathbf{a} \log_e r$	$\mathbf{r} \cdot \mathbf{a}/r^2$	$\mathbf{r} \times \mathbf{a}/r^2$	$\dfrac{(\mathbf{G} \cdot \mathbf{r})\mathbf{a}}{r^2}$	$\dfrac{\mathbf{a}}{r^2}$	$\dfrac{\mathbf{a}}{r^2} - \dfrac{2(\mathbf{r} \cdot \mathbf{a})\mathbf{r}}{r^4}$

the positions of two or more points). In the typical case of two position vectors, $\mathbf{r} \equiv (x, y, z)$ and $\boldsymbol{\rho} = (\xi, \eta, \zeta)$, say, functions like $\Phi(\mathbf{r}, \boldsymbol{\rho}) \equiv \Phi(x, y, z; \xi, \eta, \zeta)$ and $\mathbf{F}(\mathbf{r}, \boldsymbol{\rho}) \equiv \mathbf{F}(x, y, z; \xi, \eta, \zeta)$ may be operated on by **two different ∇ operators**, described in terms of right-handed rectangular cartesian "components" by

$$\nabla \equiv \mathbf{i}\frac{\partial}{\partial x} + \mathbf{j}\frac{\partial}{\partial y} + \mathbf{k}\frac{\partial}{\partial z} \qquad \text{and} \qquad \nabla_\rho \equiv \mathbf{i}\frac{\partial}{\partial \xi} + \mathbf{j}\frac{\partial}{\partial \eta} + \mathbf{k}\frac{\partial}{\partial \zeta}$$

Note in particular

$$\left.\begin{aligned} \nabla\Phi(\mathbf{r} - \boldsymbol{\rho}) &= -\nabla_\rho\Phi(\mathbf{r} - \boldsymbol{\rho}) \\ \nabla \cdot \mathbf{F}(\mathbf{r} - \boldsymbol{\rho}) &= -\nabla_\rho \cdot \mathbf{F}(\mathbf{r} - \boldsymbol{\rho}) \\ \nabla \times \mathbf{F}(\mathbf{r} - \boldsymbol{\rho}) &= -\nabla_\rho \times \mathbf{F}(\mathbf{r} - \boldsymbol{\rho}) \end{aligned}\right\} \qquad (5.5\text{-}22)$$

5.6. INTEGRAL THEOREMS

5.6-1. The Divergence Theorem and Related Theorems. (a) Table 5.6-1 summarizes a number of important theorems relating volume integrals over a region V to surface integrals over the boundary surface S of the region V. In the formulas of Table 5.6-1, volume integrals are taken over a bounded, simply connected open region V bounded by a (two-sided) regular closed surface S (Sec. 3.1-14). All functions are assumed to be single-valued throughout V and on S. *The existence of the (proper or improper) volume integrals is assumed. All theorems hold for unbounded regions V as well as for bounded regions if the integrands of the surface integrals are $O(1/r^3)$ in absolute value as $r \to \infty$ (Sec. 4.4-3).* Refer to Chap. 6 and Sec. 17.3-3 for formulas expressing surface and volume elements in terms of curvilinear coordinate systems; see also Secs. 15.6-5 and 15.6-10 for applications.

(b) **Normal-derivative Notation.** The **normal derivative** of a scalar function $\Phi(\mathbf{r})$ at a regular point of the surface S is the directional derivative of $\Phi(\mathbf{r})$ in the direction of the positive normal (usually the outward normal, Sec. 17.3-2), and thus in the direction of the vector $d\mathbf{A}$.

Table 5.6-1. Theorems Relating Volume Integrals and Surface Integrals (see also Sec. 5.6-1)

	Theorem	Vector formulas	Sufficient conditions*	
			(a) Throughout V	(b) On S
1	**Divergence theorem** (Gauss's integral theorem)	$\int_V \nabla \cdot \mathbf{F}(\mathbf{r})\, dV = \int_S d\mathbf{A} \cdot \mathbf{F}(\mathbf{r})$	$\mathbf{F}(\mathbf{r})$, $\Phi(\mathbf{r})$ differentiable with continuous partial derivatives	Existence of integrals is sufficient
2	Theorem of the rotational	$\int_V \nabla \times \mathbf{F}(\mathbf{r})\, dV = \int_S d\mathbf{A} \times \mathbf{F}(\mathbf{r})$		
3	Theorem of the gradient	$\int_V \nabla\Phi(\mathbf{r})\, dV = \int_S d\mathbf{A}\Phi(\mathbf{r})$		
4	**Green's theorems**	$\int_V \nabla\Phi \cdot \nabla\Psi\, dV + \int_V \Psi\nabla^2\Phi\, dV = \int_S d\mathbf{A} \cdot (\Psi\nabla\Phi) = \int_S \Psi \frac{\partial\Phi}{\partial n}\, dA$	$\Phi(\mathbf{r})$, $\Psi(\mathbf{r})$ differentiable with continuous partial derivatives; $\Phi(\mathbf{r})$ twice differentiable with continuous second partial derivatives	$\Psi(\mathbf{r})$ continuous; $\Phi(\mathbf{r})$ differentiable with continuous partial derivatives
5		$\int_V (\Psi\nabla^2\Phi - \Phi\nabla^2\Psi)\, dV = \int_S d\mathbf{A} \cdot (\Psi\nabla\Phi - \Phi\nabla\Psi) = \int_S \left(\Psi \frac{\partial\Phi}{\partial n} - \Phi \frac{\partial\Psi}{\partial n}\right) dA$	$\Phi(\mathbf{r})$, $\Psi(\mathbf{r})$ twice differentiable with continuous second partial derivatives	$\Phi(\mathbf{r})$, $\Psi(\mathbf{r})$ differentiable with continuous partial derivatives
6	Special cases	$\int_V \nabla^2\Phi\, dV = \int_S d\mathbf{A} \cdot \nabla\Phi = \int_S \frac{\partial\Phi}{\partial n}\, dA$ **(Gauss's Theorem)**		
7		$\int_V \lvert\nabla\Phi\rvert^2\, dV + \int_V \Phi\nabla^2\Phi\, dV = \int_S d\mathbf{A} \cdot (\Phi\nabla\Phi) = \int_S \Phi \frac{\partial\Phi}{\partial n}\, dA$		

* Less stringent conditions are discussed in Ref. 5.5.

The normal derivative is customarily denoted by $\partial\Phi/\partial n$, so that

$$\frac{\partial\Phi}{\partial n}\,dA \equiv (dA \cdot \nabla)\Phi$$

5.6-2. Stokes' Theorem and Related Theorems. Given a vector function $\mathbf{F}(\mathbf{r})$ single-valued and differentiable with continuous partial derivatives throughout a finite region V containing a simply connected regular (one-sided) surface segment S bounded by a regular closed curve C,

$$\int_S d\mathbf{A} \cdot [\nabla \times \mathbf{F}(\mathbf{r})] = \oint_C d\mathbf{r} \cdot \mathbf{F}(\mathbf{r}) \qquad \text{(Stokes' theorem)} \tag{5.6-1}$$

i.e., the line integral of $\mathbf{F}(\mathbf{r})$ *around* C *equals the flux of* $\nabla \times \mathbf{F}$ *through the surface bounded by* C.

Under the same conditions as above,

$$\int_S (d\mathbf{A} \times \nabla) \times \mathbf{F}(\mathbf{r}) = \oint_C d\mathbf{r} \times \mathbf{F}(\mathbf{r}) \tag{5.6-2}$$

and for a scalar point function $\Phi(\mathbf{r})$ single-valued and differentiable with continuous partial derivatives throughout V

$$\int_S d\mathbf{A} \times [\nabla\Phi(\mathbf{r})] = \oint_C d\mathbf{r}\,\Phi(\mathbf{r}) \tag{5.6-3}$$

Equations (1), (2), and (3) apply to unbounded regions V if the integrands of the line integrals on the right are $O(1/r^2)$ in absolute value as $r \to \infty$ (Sec. 4.4-3).

5.6-3. Fields with Surface Discontinuities (see also Sec. 15.6-5*b*). Let the scalar field $\Phi(\mathbf{r})$ or the components of $\mathbf{F}(\mathbf{r})$ be continuously differentiable on either side of a regular surface element S but discontinuous on S so that

$$\Phi(\mathbf{r}) = \Phi_+(\mathbf{r}) \qquad \text{or} \qquad \mathbf{F}(\mathbf{r}) = \mathbf{F}_+(\mathbf{r})$$

on the positive side of S (Sec. 17.3-2), while

$$\Phi(\mathbf{r}) = \Phi_-(\mathbf{r}) \qquad \text{or} \qquad \mathbf{F}(\mathbf{r}) = \mathbf{F}_-(\mathbf{r})$$

on the negative side of S. At each point $(\mathbf{r})$ of S one defines the functions

$$\left.\begin{array}{ll} \mathbf{N}(\mathbf{r})[\Phi_+(\mathbf{r}) - \Phi_-(\mathbf{r})] & \text{(surface gradient)} \\ \mathbf{N}(\mathbf{r}) \cdot [\mathbf{F}_+(\mathbf{r}) - \mathbf{F}_-(\mathbf{r})] & \text{(surface divergence)} \\ \mathbf{N}(\mathbf{r}) \times [\mathbf{F}_+(\mathbf{r}) - \mathbf{F}_-(\mathbf{r})] & \text{(surface rotational)} \end{array}\right\} \tag{5.6-4}$$

where $\mathbf{N}(\mathbf{r})$ is the positive unit normal vector of S at the point $(\mathbf{r})$ (Sec. 17.3-2). *The definitions* (4) *permit one to extend the integral theorems of Table* 5.6-1 *to functions with surface discontinuities.*

5.7. SPECIFICATION OF A VECTOR FIELD IN TERMS OF ITS CURL AND DIVERGENCE

5.7-1. Irrotational Vector Fields. A vector point function $\mathbf{F}(\mathbf{r})$ (as well as the field described by it) is called **irrotational (lamellar)**

throughout a region V if and only if, for every point of V,

$$\begin{aligned} &\nabla \times \mathbf{F}(\mathbf{r}) = 0 \\ \text{or} \qquad &\frac{\partial F_y}{\partial z} - \frac{\partial F_z}{\partial y} = \frac{\partial F_z}{\partial x} - \frac{\partial F_x}{\partial z} = \frac{\partial F_x}{\partial y} - \frac{\partial F_y}{\partial x} = 0 \end{aligned} \tag{5.7-1}$$

This is true if and only if $-\mathbf{F}(\mathbf{r})$ *is the gradient* $\nabla\Phi(\mathbf{r})$ *of a scalar point function* $\Phi(\mathbf{r})$ *at every point of* V [*see also Eq.* (5.5-19)]; *in this case*

$$\begin{aligned} d\mathbf{r} \cdot \mathbf{F}(\mathbf{r}) &\equiv F_x(x, y, z)\, dx + F_y(x, y, z)\, dy + F_z(x, y, z)\, dz \\ &\equiv -d\mathbf{r} \cdot \nabla\Phi(\mathbf{r}) \equiv -d\Phi \end{aligned} \tag{5.7-2}$$

is an exact differential (*Sec.* 4.5-3*a*). $\Phi(\mathbf{r})$ is often called the **scalar potential** of the irrotational vector field.

If V *is simply connected* (*Sec.* 4.3-6*b*), $\Phi(\mathbf{r})$ *is a single-valued function uniquely determined by* $\mathbf{F}(\mathbf{r})$ *except for an additive constant, and the line integral*

$$\int_{\substack{\boldsymbol{\varrho}=\mathbf{a} \\ C}}^{\mathbf{r}} d\boldsymbol{\varrho} \cdot \mathbf{F}(\boldsymbol{\varrho}) = -[\Phi(\mathbf{r}) - \Phi(\mathbf{a})] \tag{5.7-3}$$

is independent of the path of integration C *if the latter comprises only points of* V; the line integral $\oint_C d\boldsymbol{\varrho} \cdot \mathbf{F}(\boldsymbol{\varrho})$ around any closed path C ["circulation" of $\mathbf{F}(\mathbf{r})$ around C] in V is zero. If V is multiply connected, $\Phi(\mathbf{r})$ may be a multiple-valued function.

As a special case,

$$\frac{\partial F_x}{\partial y} - \frac{\partial F_y}{\partial x} = 0 \tag{5.7-4}$$

is a necessary and sufficient condition that the line integral

$$\int_C [F_x(x, y)\, dx + F_y(x, y)\, dy]$$

*is independent of the path of integration, i.e., that the integrand is an exact differential.**

5.7-2. Solenoidal Vector Fields. A vector point function $\mathbf{F}(\mathbf{r})$ (as well as the field described by it) is called **solenoidal** throughout a region V if and only if, for every point of V,

$$\nabla \cdot \mathbf{F}(\mathbf{r}) = 0 \qquad \text{or} \qquad \frac{\partial F_x}{\partial x} + \frac{\partial F_y}{\partial y} + \frac{\partial F_z}{\partial z} = 0 \tag{5.7-5}$$

This is true if and only if $\mathbf{F}(\mathbf{r})$ *is the curl* $\nabla \times \mathbf{A}(\mathbf{r})$ *of a vector point function* $\mathbf{A}(\mathbf{r})$ [see also Eq. (5.5-19)], the **vector potential** of the vector field described by $\mathbf{F}(\mathbf{r})$.

* See footnote to Sec. 5.4-1.

5.7-3. Specification of a Vector Point Function in Terms of Its Divergence and Curl. (a) Let V be a finite open region of space, bounded by a regular surface S (Sec. 3.1-14) whose positive normal is uniquely defined and varies continuously at every surface point. *If the divergence and curl of a vector point function* $\mathbf{F}(\mathbf{r})$ *are given at every point* $(\mathbf{r})$ *of* V, *then* $\mathbf{F}(\mathbf{r})$ *may be expressed throughout* V *as the sum of an irrotational vector point function* $\mathbf{F}_1(\mathbf{r})$ *and a solenoidal vector point function* $\mathbf{F}_2(\mathbf{r})$,

$$\left.\begin{aligned} \mathbf{F}(\mathbf{r}) &= \mathbf{F}_1(\mathbf{r}) + \mathbf{F}_2(\mathbf{r}) \\ \text{with} \qquad \nabla \times \mathbf{F}_1(\mathbf{r}) &= \nabla \cdot \mathbf{F}_2(\mathbf{r}) = 0 \end{aligned}\right\} \tag{5.7-6}$$

(Helmholtz's Decomposition Theorem). $\mathbf{F}(\mathbf{r})$ *is uniquely defined throughout* V *if, in addition, the normal component* $\mathbf{F}(\mathbf{r}) \cdot d\mathbf{A}/|d\mathbf{A}|$ *of* $\mathbf{F}(\mathbf{r})$ *is given at every surface point (Uniqueness Theorem).*

The problem of actually computing $\mathbf{F}(\mathbf{r})$ from these data involves the solution of partial differential equations subject to certain boundary conditions. The important special case in which $\nabla \cdot \mathbf{F}(\mathbf{r}) = \nabla \times \mathbf{F}(\mathbf{r}) = 0$ and thus $\mathbf{F}(\mathbf{r}) = -\nabla\Phi(\mathbf{r})$, $\nabla^2\Phi(\mathbf{r}) = 0$ throughout V forms the subject matter of *potential theory* as discussed in Secs. 15.6-1 to 15.6-10.

(b) If suitable functions

$$\nabla \cdot \mathbf{F}(\mathbf{r}) = 4\pi Q(\mathbf{r}) \qquad \text{and} \qquad \nabla \times \mathbf{F}(\mathbf{r}) = 4\pi \mathbf{I}(\mathbf{r}) \tag{5.7-7}$$

are given for every point $(\mathbf{r})$ of space, then Eq. (6) defines $\mathbf{F}_1(\mathbf{r})$ and $\mathbf{F}_2(\mathbf{r})$, and hence $\mathbf{F}(\mathbf{r})$, uniquely except for additive functions $\mathbf{F}_0(\mathbf{r})$ such that $\nabla^2\mathbf{F}_0(\mathbf{r}) = 0$. One has

$$\left.\begin{aligned} \mathbf{F}_1(\mathbf{r}) &= -\nabla\Phi(\mathbf{r}) \qquad & \Phi(\mathbf{r}) &= \int_{\text{all space}} \frac{Q(\boldsymbol{\varrho})}{|\mathbf{r} - \boldsymbol{\varrho}|}\, dV \\ \mathbf{F}_2(\mathbf{r}) &= \nabla \times \mathbf{A}(\mathbf{r}) \qquad & \mathbf{A}(\mathbf{r}) &= \int_{\text{all space}} \frac{\mathbf{I}(\boldsymbol{\varrho})}{|\mathbf{r} - \boldsymbol{\varrho}|}\, dV \end{aligned}\right\} \tag{5.7-8}$$

provided that the integrals on the right (*scalar and vector potentials*) exist (see also Sec. 15.6-5); the integration extends over all points $(\boldsymbol{\varrho})$.

5.8. RELATED TOPICS, REFERENCES, AND BIBLIOGRAPHY

5.8-1. Related Topics. The following topics related to the study of vector analysis are treated in other chapters of this handbook:

Algebra of more general vector spaces .... Chaps. 12, 14
Applications of vector algebra to analytic geometry .... Chap. 3
Applications of vector calculus to the geometry of curves and surfaces (differential geometry) .... Chap. 17
Two-dimensional fields .... Chap. 10
Potential theory .... Chap. 10
General transformation properties of vector point functions .... Chap. 16
Functions of a complex variable .... Chap. 7

5.8-2. References and Bibliography (see also Sec. 6.6-2).

5.1. Brand, L.: *Vector and Tensor Analysis*, Wiley, New York, 1947.
5.2. ———: *Vector Analysis*, Wiley, New York, 1957.
5.3. Dörrie, H.: *Vektoren*, Edwards, Ann Arbor, Mich., 1946.
5.4. Halmos, P. R.: *Finite Dimensional Vector Spaces*, Princeton University Press, Princeton, N.J., 1942.
5.5. Lagally, M.: *Vorlesungen über Vektor-Rechnung*, Edwards, Ann Arbor, Mich., 1947.
5.6. Lass, H.: *Vector and Tensor Analysis*, McGraw-Hill, New York, 1950.
5.7. McQuistan, R. B.: *Scalar and Vector Fields*, Wiley, New York, 1965.
5.8. Sokolnikoff, I. S.: *Tensor Analysis*, 2d ed., Wiley, New York, 1964.
5.9. Weatherburn, C. E.: *Elementary Vector Analysis*, Open Court, LaSalle, Ill., 1948.
5.10. ———: *Advanced Vector Analysis*, Open Court, LaSalle, Ill., 1948.

CHAPTER 6

CURVILINEAR COORDINATE SYSTEMS

6.1. INTRODUCTION

6.1-1. Chapter 6 deals with the description (representation) of scalar and vector functions of position (see also Secs. 5.4-1 to 5.7-3) in terms of *curvilinear coordinates* (Sec. 6.2-1). Vectors will be represented by *components* along, or perpendicular to, the coordinate lines at each point (Secs. 6.3-1 to 6.3-3). The use of curvilinear coordinates simplifies many problems; one may, for instance, choose a curvilinear coordinate system such that a function under consideration is constant on a coordinate surface (Secs. 6.4-3 and 10.4-1*c*).

In accordance with the requirements of many physical applications, Chap. 6 is mainly concerned with *orthogonal coordinate systems* (Secs. 6.4-1 to 6.5-1). The representation of vector relations in terms of nonorthogonal components is treated more elaborately in Chap. 16 in the context of tensor analysis.

6.2. CURVILINEAR COORDINATE SYSTEMS

6.2-1. Curvilinear Coordinates. A **curvilinear coordinate system** defined over a region V of three-dimensional Euclidean space labels each point (x, y, z) with an ordered set of three real numbers x^1, x^2, x^3.* The

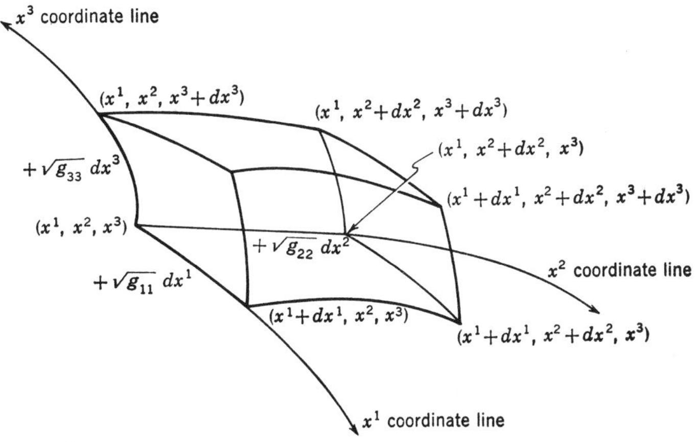

FIG. 6.2-1. Illustrating coordinate line elements, coordinate surface elements, and volume element for a curvilinear coordinate system.

curvilinear coordinates x^1, x^2, x^3 of the point $(x, y, z) \equiv (x^1, x^2, x^3)$ are related to the right-handed rectangular cartesian coordinates x, y, z (Sec. 3.1-4) by three continuously differentiable *transformation equations*

$$x^1 = x^1(x, y, z) \qquad x^2 = x^2(x, y, z) \qquad x^3 = x^3(x, y, z) \quad (6.2\text{-}1)$$

where the functions (1) are single-valued and $\partial(x^1, x^2, x^3)/\partial(x, y, z) \neq 0$ throughout V (*admissible transformation*, see also Secs. 4.5-6 and 16.1-2).

* The indices 1, 2, 3 of x^1, x^2, x^3 are *superscripts*, not exponents; see also Secs. 16.1-2 and 16.1-3.

The x^1, x^2, x^3 coordinate system is *cartesian* (Sec. 3.1-2; not in general rectangular) if and only if the transformation equations (1) are linear equations.

6.2-2. Coordinate Surfaces and Coordinate Lines. The condition $x^i = x^i(x, y, z) = constant$ defines a **coordinate surface.** Coordinate surfaces corresponding to different values of the same coordinate x^i do not intersect in V. Two coordinate surfaces corresponding to different coordinates x^i, x^j intersect on the **coordinate line** corresponding to the third coordinate x^k. Each point $(x, y, z) \equiv (x^1, x^2, x^3)$ of V is represented as the point of intersection of three coordinate surfaces, or of three coordinate lines.

6.2-3. Elements of Distance and Volume (Fig. 6.2-1; see also Secs. 6.4-3 and 17.3-3). **(a)** In terms of curvilinear coordinates x^1, x^2, x^3, the **element of distance** ds between two adjacent points $(x, y, z) \equiv (x^1, x^2, x^3)$ and $(x + dx,\ y + dy,\ z + dz) \equiv (x^1 + dx^1,\ x^2 + dx^2,\ x^3 + dx^3)$ is given by the *quadratic differential form*

$$\begin{aligned} ds^2 &= dx^2 + dy^2 + dz^2 = \sum_{i=1}^{3}\sum_{k=1}^{3} g_{ik}(x^1, x^2, x^3)\, dx^i\, dx^k \\ \text{with } g_{ik}(x^1, x^2, x^3) &= \left[\frac{\partial x}{\partial x^i}\frac{\partial x}{\partial x^k} + \frac{\partial y}{\partial x^i}\frac{\partial y}{\partial x^k} + \frac{\partial z}{\partial x^i}\frac{\partial z}{\partial x^k}\right]_{x^1,x^2,x^3} \\ &= g_{ki}(x^1, x^2, x^3) \qquad (i, k = 1, 2, 3) \end{aligned} \tag{6.2-2}$$

The functions $g_{ik}(x^1, x^2, x^3)$ are the *components of the metric tensor* (Sec. 16.7-1).

(b) The six coordinate surfaces associated with the points (x^1, x^2, x^3) and $(x^1 + dx^1,\ x^2 + dx^2,\ x^3 + dx^3)$ bound the parallelepipedal *volume element*

$$\begin{aligned} dV &= \frac{\partial(x, y, z)}{\partial(x^1, x^2, x^3)}\, dx^1\, dx^2\, dx^3 \\ \text{with} \quad \left[\frac{\partial(x, y, z)}{\partial(x^1, x^2, x^3)}\right]^2 &\equiv \det\,[g_{ik}(x^1, x^2, x^3)] \equiv g \end{aligned} \tag{6.2-3}$$

(see also Secs. 5.4-7, 6.4-3c, and 16.10-10). $\sqrt{g}$ is taken to be positive if the directed increments dx^1, dx^2, dx^3, which define the **positive directions** on the coordinate lines, form a right-handed system (Sec. 3.1-3). The curvilinear coordinate system is then **right-handed** throughout the region V; otherwise the coordinate system is **left-handed** throughout V (see also Sec. 16.7-1).

Refer to Secs. 6.4-3, 17.3-3, and 17.4-2 for the description of *vector path elements* and *surface elements* in terms of curvilinear coordinates.

6.3. REPRESENTATION OF VECTORS IN TERMS OF COMPONENTS

6.3-1. Vector Components and Local Base Vectors. In Chap. 5, a vector function of position was described in terms of *right-handed rectangular cartesian components* by reference to local base vectors **i**, **j**, **k**; the magnitude and direction of each cartesian base vector are the same at every point (x, y, z) (Secs. 5.2-2 and 5.4-1). If the vector function $\mathbf{F}(\mathbf{r})$ is to be described in terms of curvilinear coordinates x^1, x^2, x^3, it is useful to employ local base vectors directed along, or perpendicular to, the coordinate lines at each point (x^1, x^2, x^3). *Such base vectors are themselves vector functions of position.* The following sections describe three different systems of local base vectors associated with each given curvilinear coordinate system.

6.3-2. Representation of Vectors in Terms of Physical Components. Given a curvilinear coordinate system defined as in Sec. 6.2-1, the unit vectors $\mathbf{u}_1(x^1, x^2, x^3)$, $\mathbf{u}_2(x^1, x^2, x^3)$, $\mathbf{u}_3(x^1, x^2, x^3)$ respectively directed along the positive x^1, x^2, x^3 coordinate lines (Sec. 6.2-3*b*) at each point $(x^1, x^2, x^3) \equiv (\mathbf{r})$ may be used as *local base vectors* (see also Sec. 16.8-3). Every vector $\mathbf{F}(\mathbf{r}) \equiv \mathbf{F}(x^1, x^2, x^3)$ can be uniquely represented in the form

$$\boxed{\mathbf{F}(\mathbf{r}) = \mathbf{F}(x^1, x^2, x^3) = \hat{F}_1\mathbf{u}_1 + \hat{F}_2\mathbf{u}_2 + \hat{F}_3\mathbf{u}_3} \tag{6.3-1}$$

at each point $(\mathbf{r}) = (x^1, x^2, x^3)$.

Refer to Table 6.3-1 for the transformation equations relating the functions

$$\begin{aligned} \hat{F}_1 &\equiv \hat{F}_1(x^1, x^2, x^3) \\ \hat{F}_2 &\equiv \hat{F}_2(x^1, x^2, x^3) \\ \hat{F}_3 &\equiv \hat{F}_3(x^1, x^2, x^3) \end{aligned} \tag{6.3-2}$$

(**physical components** of **F** in the coordinate directions, see also Sec. 16.8-3) and the local base vectors $\mathbf{u}_1(x^1, x^2, x^3)$, $\mathbf{u}_2(x^1, x^2, x^3)$, $\mathbf{u}_3(x^1, x^2, x^3)$ to their respective rectangular cartesian counterparts F_x, F_y, F_z and **i**, **j**, **k**. Note that the functions $\frac{1}{\sqrt{g_{ii}}}\frac{\partial x}{\partial x^i}$, $\frac{1}{\sqrt{g_{ii}}}\frac{\partial y}{\partial x^i}$, $\frac{1}{\sqrt{g_{ii}}}\frac{\partial z}{\partial x^i}$ are the direction cosines of $\mathbf{u}_i$ (i.e., of the ith-coordinate line) with respect to the x, y, z axes (see also Fig. 6.3-1).

6.3-3. Representation of Vectors in Terms of Contravariant and Covariant Components. For any curvilinear coordinate system defined as in Sec. 6.2-1, it is possible to introduce the system of local base vectors

$$\mathbf{e}_1(x^1, x^2, x^3) \equiv +\sqrt{g_{11}}\,\mathbf{u}_1 \qquad \mathbf{e}_2(x^1, x^2, x^3) \equiv +\sqrt{g_{22}}\,\mathbf{u}_2 \qquad \mathbf{e}_3(x^1, x^2, x^3) \equiv +\sqrt{g_{33}}\,\mathbf{u}_3 \tag{6.3-3}$$

directed along the coordinate lines, and the system of local base vectors

$$\mathbf{e}^1(x^1, x^2, x^3) \equiv \frac{\mathbf{e}_2 \times \mathbf{e}_3}{[\mathbf{e}_1\mathbf{e}_2\mathbf{e}_3]} \qquad \mathbf{e}^2(x^1, x^2, x^3) \equiv \frac{\mathbf{e}_3 \times \mathbf{e}_1}{[\mathbf{e}_1\mathbf{e}_2\mathbf{e}_3]} \qquad \mathbf{e}^3(x^1, x^2, x^3) \equiv \frac{\mathbf{e}_1 \times \mathbf{e}_2}{[\mathbf{e}_1\mathbf{e}_2\mathbf{e}_3]} \tag{6.3-4}$$

directed perpendicularly to the coordinate surfaces. Each vector $\mathbf{F}(\mathbf{r})$ can then be represented in the respective forms

$$\boxed{\mathbf{F}(\mathbf{r}) = \mathbf{F}(x^1, x^2, x^3) = F^1\mathbf{e}_1 + F^2\mathbf{e}_2 + F^3\mathbf{e}_3 = \mathbf{F}_1\mathbf{e}^1 + \mathbf{F}_2\mathbf{e}^2 + \mathbf{F}_3\mathbf{e}^3} \tag{6.3-5}$$

at each point $(\mathbf{r}) \equiv (x^1, x^2, x^3)$. The magnitudes as well as the directions of the local base vectors $\mathbf{e}_i$ and $\mathbf{e}^i$ change from point to point, unless x^1, x^2, x^3 are cartesian coordinates (Sec. 16.6-1*a*).

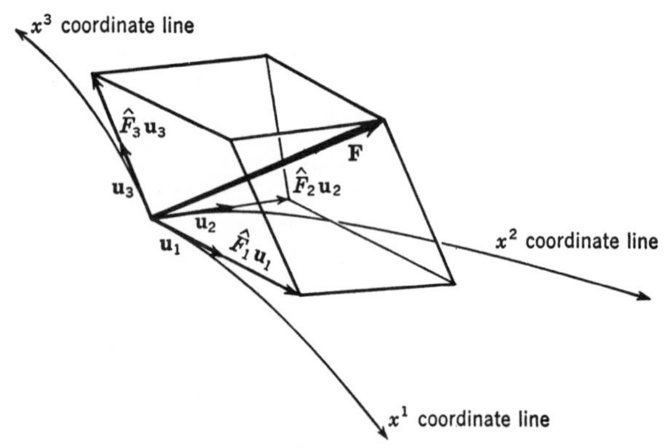

FIG. 6.3-1. Representation of a vector $\mathbf{F}$ in terms of unit vectors $\mathbf{u}_1$, $\mathbf{u}_2$, $\mathbf{u}_3$ and physical components $\hat{F}_1$, $\hat{F}_2$, $\hat{F}_3$.

The **contravariant components** $F^i \equiv F^i(x^1, x^2, x^3) \equiv \mathbf{F} \cdot \mathbf{e}^i$ and the **covariant components** $F_i \equiv F_i(x^1, x^2, x^3) \equiv \mathbf{F} \cdot \mathbf{e}_i$ $(i = 1, 2, 3)$ in the *scheme of measurements* (Sec. 16.1-4) associated with each system of coordinates x^1, x^2, x^3, and the associated base vectors $\mathbf{e}_i$ and $\mathbf{e}^i$ may also be defined directly by the transformations relating them to the rectangular cartesian components F_x, F_y, F_z and base vectors **i, j, k** (Table 6.3-1). Note (see also Secs. 16.6-1 and 16.8-4)

$$\mathbf{e}_1 = \frac{\mathbf{e}^2 \times \mathbf{e}^3}{[\mathbf{e}^1\mathbf{e}^2\mathbf{e}^3]} \qquad \mathbf{e}_2 = \frac{\mathbf{e}^3 \times \mathbf{e}^1}{[\mathbf{e}^1\mathbf{e}^2\mathbf{e}^3]} \qquad \mathbf{e}_3 = \frac{\mathbf{e}^1 \times \mathbf{e}^2}{[\mathbf{e}^1\mathbf{e}^2\mathbf{e}^3]} \tag{6.3-6}$$

$$[\mathbf{e}_1\mathbf{e}_2\mathbf{e}_3] = [\mathbf{e}^1\mathbf{e}^2\mathbf{e}^3]^{-1} = [\mathbf{u}_1\mathbf{u}_2\mathbf{u}_3]\,\underset{+}{\sqrt{g_{11}g_{22}g_{33}}} = \pm\sqrt{g} \tag{6.3-7}$$

$$\mathbf{e}_i \cdot \mathbf{e}_k = g_{ik} \qquad |\mathbf{e}_i| = +\sqrt{g_{ii}} \qquad \mathbf{e}_i \cdot \mathbf{e}^k = \begin{Bmatrix} 1 \text{ if } i = k \\ 0 \text{ if } i \neq k \end{Bmatrix} \tag{6.3-8}$$

Table 6.3-1. Transformations Relating Base Vectors and Vector Components Associated with Different Local Reference Systems

(a) Relations between i, j, k and the Local Base Vectors $\mathbf{u}_i$, $\mathbf{e}_i$, $\mathbf{e}^i$ Associated with a Curvilinear Coordinate System

$$\mathbf{u}_i = \frac{1}{+\sqrt{g_{ii}}}\mathbf{e}_i \qquad \mathbf{e}_i = \frac{\partial x}{\partial x^i}\mathbf{i} + \frac{\partial y}{\partial x^i}\mathbf{j} + \frac{\partial z}{\partial x^i}\mathbf{k}$$

$$\mathbf{e}^i = \frac{\partial x^i}{\partial x}\mathbf{i} + \frac{\partial x^i}{\partial y}\mathbf{j} + \frac{\partial x^i}{\partial z}\mathbf{k} \qquad (i = 1, 2, 3)$$

$$\mathbf{i} = \sum_{i=1}^{3} \frac{\partial x^i}{\partial x}\sqrt{g_{ii}}\,\mathbf{u}_i = \sum_{i=1}^{3} \frac{\partial x^i}{\partial x}\mathbf{e}_i = \sum_{i=1}^{3} \frac{\partial x}{\partial x^i}\mathbf{e}^i$$

$$\mathbf{j} = \sum_{i=1}^{3} \frac{\partial x^i}{\partial y}\sqrt{g_{ii}}\,\mathbf{u}_i = \sum_{i=1}^{3} \frac{\partial x^i}{\partial y}\mathbf{e}_i = \sum_{i=1}^{3} \frac{\partial y}{\partial x^i}\mathbf{e}^i$$

$$\mathbf{k} = \sum_{i=1}^{3} \frac{\partial x^i}{\partial z}\sqrt{g_{ii}}\,\mathbf{u}_i = \sum_{i=1}^{3} \frac{\partial x^i}{\partial z}\mathbf{e}_i = \sum_{i=1}^{3} \frac{\partial z}{\partial x^i}\mathbf{e}^i$$

(b) Relations between F_x, F_y, F_z and the Physical Components $\hat{F}_i$, Contravariant Components F_i, and Covariant Components F^i Associated with a Curvilinear Coordinate System

$$\hat{F}_i = +\sqrt{g_{ii}}\,F^i \qquad F^i = \frac{\partial x^i}{\partial x}F_x + \frac{\partial x^i}{\partial y}F_y + \frac{\partial x^i}{\partial z}F_z$$

$$F_i = \frac{\partial x}{\partial x^i}F_x + \frac{\partial y}{\partial x^i}F_y + \frac{\partial z}{\partial x^i}F_z \qquad (i = 1, 2, 3)$$

$$F_x = \sum_{i=1}^{3} \frac{\partial x}{\partial x^i}\frac{\hat{F}_i}{\sqrt{g_{ii}}} = \sum_{i=1}^{3} \frac{\partial x}{\partial x^i}F^i = \sum_{i=1}^{3} \frac{\partial x^i}{\partial x}F_i$$

$$F_y = \sum_{i=1}^{3} \frac{\partial y}{\partial x^i}\frac{\hat{F}_i}{\sqrt{g_{ii}}} = \sum_{i=1}^{3} \frac{\partial y}{\partial x^i}F^i = \sum_{i=1}^{3} \frac{\partial x^i}{\partial y}F_i$$

$$F_z = \sum_{i=1}^{3} \frac{\partial z}{\partial x^i}\frac{\hat{F}_i}{\sqrt{g_{ii}}} = \sum_{i=1}^{3} \frac{\partial z}{\partial x^i}F^i = \sum_{i=1}^{3} \frac{\partial x^i}{\partial z}F_i$$

(c) Relations between Local Base Vectors and Vector Components Associated with Two Curvilinear Coordinate Systems (barred base vectors and vector components are associated with the $\bar{x}^1$, $\bar{x}^2$, $\bar{x}^3$ system; see also Sec. 16.2-1).

$$\bar{\mathbf{u}}_k = \frac{1}{+\sqrt{\bar{g}_{kk}}}\bar{\mathbf{e}}_k \qquad \bar{\mathbf{e}}_k = \sum_{i=1}^{3} \frac{\partial x^i}{\partial \bar{x}^k}\mathbf{e}_i \qquad \bar{\mathbf{e}}^k = \sum_{i=1}^{3} \frac{\partial \bar{x}^k}{\partial x^i}\mathbf{e}^i \qquad (k = 1, 2, 3)$$

$$\hat{\bar{F}}_k = +\sqrt{\bar{g}_{kk}}\,\bar{F}^k \qquad \bar{F}^k = \sum_{i=1}^{3} \frac{\partial \bar{x}^k}{\partial x^i}F^i \qquad \bar{F}_k = \sum_{i=1}^{3} \frac{\partial x^i}{\partial \bar{x}^k}F_i \qquad (k = 1, 2, 3)$$

$$\bar{g}_{ik} = \sum_{j=1}^{3}\sum_{h=1}^{3} \frac{\partial x^j}{\partial \bar{x}^i}\frac{\partial x^h}{\partial \bar{x}^k}g_{jh} \qquad (i, k = 1, 2, 3)$$

In the special case of a rectangular cartesian coordinate system, $x^1 \equiv x$, $x^2 \equiv y$, $x^3 \equiv z$, and

$$\mathbf{e}_1 = \mathbf{e}^1 = \mathbf{u}_1 = \mathbf{i} \qquad \mathbf{e}_2 = \mathbf{e}^2 = \mathbf{u}_2 = \mathbf{j} \qquad \mathbf{e}_3 = \mathbf{e}^3 = \mathbf{u}_3 = \mathbf{k} \tag{6.3-9}$$

A main advantage of the contravariant and covariant representation of vectors is the relative simplicity of the transformation equations relating contravariant (or covariant) vector components associated with different coordinate systems (Table 6.3-1; see also Sec. 6.3-4).

6.3-4. Derivation of Vector Relations in Terms of Curvilinear Components. In principle, every vector relation given, as in Chap. 5, in terms of rectangular cartesian components can be expressed in terms of curvilinear components with the aid of Table 6.3-1. If the relations in question involve differentiation and/or integration, *note that the base vectors associated with curvilinear-coordinate systems are functions of position.*

Many practically important problems permit the use of *orthogonal coordinate systems* (Sec. 6.4-1). In this case, the formulas of Secs. 6.4-1 to 6.4-3 and Tables 6.4-1 to 6.5-11 yield comparatively simple expressions for many vector relations directly in terms of physical components. When these special methods do not apply, it is usually best to employ contravariant and covariant vector components rather than physical components; **the formulation of vector analysis in terms of contravariant and covariant components is treated in detail in Chap. 16 as part of the more general subject of tensor analysis.**

In particular, one uses the formulas of Secs. 16.8-1 to 16.8-4 for computing scalar and vector products, and the relatively straightforward method of *covariant differentiation* (Secs. 16.10-1 to 16.10-8) yields expressions for differential invariants like $\nabla\Phi$, $\nabla \cdot \mathbf{F}$, and $\nabla \times \mathbf{F}$.

6.4. ORTHOGONAL COORDINATE SYSTEMS. VECTOR RELATIONS IN TERMS OF ORTHOGONAL COMPONENTS

6.4-1. Orthogonal Coordinates. An **orthogonal coordinate system** is a system of curvilinear coordinates x^1, x^2, x^3 (Sec. 6.2-1) chosen so that the functions $g_{ik}(x^1, x^2, x^3)$ satisfy the relations

$$\boxed{g_{ik}(x^1, x^2, x^3) = 0 \qquad \text{if } i \neq k} \tag{6.4-1}$$

at each point (x^1, x^2, x^3). *The coordinate lines, and thus also the local base vectors* $\mathbf{u}_1, \mathbf{u}_2, \mathbf{u}_3$, *of an orthogonal coordinate system are perpendicular to each other at each point; each coordinate line is perpendicular to all coordinate surfaces corresponding to constant values of the coordinate in question.*

6.4-2. Vector Relations. (a) The formulas of Table 6.4-1 express the most important vector relations in terms of orthogonal coordinates and components. The appropriate functions $g_{ii} = g_{ii}(x^1, x^2, x^3)$ for each specific orthogonal coordinate system are obtained from Eq. (6.2-2) or from Tables 6.5-1 to 6.5-11.

Table 6.4-1. Vector Formulas Expressed in Terms of Physical Components for Orthogonal Coordinate Systems (Sec. 6.4-1)

Plus and minus signs refer, respectively, to right-handed and left-handed orthogonal coordinate systems. The appropriate functions $g_{ii}(x^1, x^2, x^3) \equiv |\mathbf{e}_i|^2$ are obtained from Eq. (6.2-2) or from Tables 6.5-1 to 6.5-11

(a) Scalar and Vector Products

$$\mathbf{F}\cdot\mathbf{G} = \hat{F}_1\hat{G}_1 + \hat{F}_2\hat{G}_2 + \hat{F}_3\hat{G}_3 \qquad |\mathbf{F}| = +\sqrt{\hat{F}_1^2 + \hat{F}_2^2 + \hat{F}_3^2}$$

$$\mathbf{F}\times\mathbf{G} = \begin{vmatrix} \mathbf{u}_2\times\mathbf{u}_3 & \hat{F}_1 & \hat{G}_1 \\ \mathbf{u}_3\times\mathbf{u}_1 & \hat{F}_2 & \hat{G}_2 \\ \mathbf{u}_1\times\mathbf{u}_2 & \hat{F}_3 & \hat{G}_3 \end{vmatrix} = \begin{vmatrix} \mathbf{u}_1 & \hat{F}_1 & \hat{G}_1 \\ \mathbf{u}_2 & \hat{F}_2 & \hat{G}_2 \\ \mathbf{u}_3 & \hat{F}_3 & \hat{G}_3 \end{vmatrix} [\mathbf{u}_1\mathbf{u}_2\mathbf{u}_3] = \pm \begin{vmatrix} \mathbf{u}_1 & \hat{F}_1 & \hat{G}_1 \\ \mathbf{u}_2 & \hat{F}_2 & \hat{G}_2 \\ \mathbf{u}_3 & \hat{F}_3 & \hat{G}_3 \end{vmatrix}$$

$$[\mathbf{FGH}] = \begin{vmatrix} \hat{F}_1 & \hat{G}_1 & \hat{H}_1 \\ \hat{F}_2 & \hat{G}_2 & \hat{H}_2 \\ \hat{F}_3 & \hat{G}_3 & \hat{H}_3 \end{vmatrix} [\mathbf{u}_1\mathbf{u}_2\mathbf{u}_3] = \pm \begin{vmatrix} \hat{F}_1 & \hat{G}_1 & \hat{H}_1 \\ \hat{F}_2 & \hat{G}_2 & \hat{H}_2 \\ \hat{F}_3 & \hat{G}_3 & \hat{H}_3 \end{vmatrix}$$

(b) Differential Invariants* ($g \equiv g_{11}g_{22}g_{33}$)

$$\nabla\Phi \equiv \pm\left[\mathbf{u}_1 \frac{1}{\underset{+}{\sqrt{g_{11}}}}\frac{\partial\Phi}{\partial x^1} + \mathbf{u}_2 \frac{1}{\underset{+}{\sqrt{g_{22}}}}\frac{\partial\Phi}{\partial x^2} + \mathbf{u}_3 \frac{1}{\underset{+}{\sqrt{g_{33}}}}\frac{\partial\Phi}{\partial x^3}\right]$$

$$\nabla\cdot\mathbf{F} \equiv \frac{1}{\sqrt{g}}\left[\frac{\partial}{\partial x^1}\left(\hat{F}_1\sqrt{\frac{g}{g_{11}}}\right) + \frac{\partial}{\partial x^2}\left(\hat{F}_2\sqrt{\frac{g}{g_{22}}}\right) + \frac{\partial}{\partial x^3}\left(\hat{F}_3\sqrt{\frac{g}{g_{33}}}\right)\right]$$

$$\nabla\times\mathbf{F} \equiv \pm\frac{1}{\sqrt{g}}\begin{vmatrix} \mathbf{u}_1\sqrt{g_{11}} & \dfrac{\partial}{\partial x^1} & \hat{F}_1\sqrt{g_{11}} \\ \mathbf{u}_2\sqrt{g_{22}} & \dfrac{\partial}{\partial x^2} & \hat{F}_2\sqrt{g_{22}} \\ \mathbf{u}_3\sqrt{g_{33}} & \dfrac{\partial}{\partial x^3} & \hat{F}_3\sqrt{g_{33}} \end{vmatrix}$$

$$(\mathbf{F}\cdot\nabla)\Phi \equiv \pm\left[\frac{\hat{F}_1}{\sqrt{g_{11}}}\frac{\partial\Phi}{\partial x^1} + \frac{\hat{F}_2}{\sqrt{g_{22}}}\frac{\partial\Phi}{\partial x^2} + \frac{\hat{F}_3}{\sqrt{g_{33}}}\frac{\partial\Phi}{\partial x^3}\right]$$

$$\nabla^2\Phi \equiv \frac{1}{\sqrt{g}}\left[\frac{\partial}{\partial x^1}\left(\frac{\sqrt{g}}{g_{11}}\frac{\partial\Phi}{\partial x^1}\right) + \frac{\partial}{\partial x^2}\left(\frac{\sqrt{g}}{g_{22}}\frac{\partial\Phi}{\partial x^2}\right) + \frac{\partial}{\partial x^3}\left(\frac{\sqrt{g}}{g_{33}}\frac{\partial\Phi}{\partial x^3}\right)\right]$$

* To obtain $\nabla^2\mathbf{F}$, use $\nabla^2\mathbf{F} \equiv \nabla(\nabla\cdot\mathbf{F}) - \nabla\times(\nabla\times\mathbf{F})$, or refer to Sec. 16.10-7. To obtain $(\mathbf{F}\cdot\nabla)\mathbf{G}$, refer to Table 5.5-1 or to Sec. 16.10-7.

(b) *For every orthogonal coordinate system,*

$$\left.\begin{array}{c} \mathbf{u}_i\cdot\mathbf{u}_k = \begin{cases} 1 \text{ if } i = k \\ 0 \text{ if } i \neq k \end{cases} \qquad \mathbf{u}_i\times\mathbf{u}_k = \begin{cases} 0 \text{ if } i = k \\ \pm\mathbf{u}_j,\ i \neq k \neq j \neq i \end{cases} \\ [\mathbf{u}_1\mathbf{u}_2\mathbf{u}_3] = \pm 1 \end{array}\right\} \tag{6.4-2}$$

$$\mathbf{u}_i = \frac{1}{+\sqrt{g_{ii}}}\mathbf{e}_i = \pm\sqrt{g_{ii}}\,\mathbf{e}^i \qquad \hat{F}_i = +\sqrt{g_{ii}}\,F^i = \pm\frac{1}{\sqrt{g_{ii}}}F_i \qquad (i = 1, 2, 3) \tag{6.4-3}$$

$$[\mathbf{e}_1\mathbf{e}_2\mathbf{e}_3] = [\mathbf{e}^1\mathbf{e}^2\mathbf{e}^3]^{-1} = [\mathbf{u}_1\mathbf{u}_2\mathbf{u}_3]\underset{+}{\sqrt{g_{11}g_{22}g_{33}}} = \pm\sqrt{g_{11}g_{22}g_{33}} = \pm\sqrt{g} \tag{6.4-4}$$

$$\nabla \equiv \mathbf{e}^1 \frac{\partial}{\partial x^1} + \mathbf{e}^2 \frac{\partial}{\partial x^2} + \mathbf{e}^3 \frac{\partial}{\partial x^3} \equiv \pm \left[\frac{1}{\sqrt{g_{11}}} \mathbf{u}_1 \frac{\partial}{\partial x^1} + \frac{1}{\sqrt{g_{22}}} \mathbf{u}_2 \frac{\partial}{\partial x^2} + \frac{1}{\sqrt{g_{33}}} \mathbf{u}_3 \frac{\partial}{\partial x^3} \right] \tag{6.4-5}$$

where the plus and minus signs refer, respectively, to right-handed and left-handed orthogonal coordinate systems.

NOTE: Expressions for $\nabla\Phi$, $\nabla \cdot \mathbf{F}$, and $\nabla \times \mathbf{F}$ may be derived directly from the definitions of Sec. 5.5-1 with the aid of the volume element shown in Fig. 6.2-1.

6.4-3. Line Integrals, Surface Integrals, and Volume Integrals

(see also Fig. 6.2-1). (**a**) Given a rectifiable curve C (Sec. 5.4-4), the *vector path element* $d\mathbf{r}$ has the physical components $ds_1 = \underset{+}{\sqrt{g_{11}}}\, dx^1$, $ds_2 = \underset{+}{\sqrt{g_{22}}}\, dx^2$, $ds_3 = \underset{+}{\sqrt{g_{33}}}\, dx^3$ in each curvilinear coordinate system (Sec. 6.2-1), i.e.

$$d\mathbf{r} = \frac{d\mathbf{r}}{dt}\, dt = \mathbf{i}\, dx + \mathbf{j}\, dy + \mathbf{k}\, dz = \mathbf{u}_1\, ds_1 + \mathbf{u}_2\, ds_2 + \mathbf{u}_3\, ds_3$$
$$= \mathbf{u}_1 \sqrt{g_{11}}\, dx^1 + \mathbf{u}_2 \sqrt{g_{22}}\, dx^2 + \mathbf{u}_3 \sqrt{g_{33}}\, dx^3 \tag{6.4-6}$$

The appropriate expression (6) for $d\mathbf{r}$ must be substituted in each line integral defined in Sec. 5.4-5. Thus, *for orthogonal coordinates* x^1, x^2, x^3

$$\int_C d\mathbf{r} \cdot \mathbf{F} = \int_C (\hat{F}_1\, ds_1 + \hat{F}_2\, ds_2 + \hat{F}_3\, ds_3)$$
$$= \int_C \left(\hat{F}_1 \frac{ds_1}{dt} + \hat{F}_2 \frac{ds_2}{dt} + \hat{F}_3 \frac{ds_3}{dt} \right) dt \tag{6.4-7}$$

Note that *for every curvilinear coordinate system*

$$d\mathbf{r} = \mathbf{e}_1\, dx^1 + \mathbf{e}_2\, dx^2 + \mathbf{e}_3\, dx^3$$

(see also Sec. 17.2-1); *for orthogonal coordinates* x^1, x^2, x^3,

$$\int_C d\mathbf{r} \cdot \mathbf{F} = \int_C (F_1\, dx^1 + F_2\, dx^2 + F_3\, dx^3) = \int_C (g_{11}F^1\, dx^1 + g_{22}F^2\, dx^2 + g_{33}F^3\, dx^3)$$

(**b**) The description of a surface S and of the vector surface element $d\mathbf{A}$ in terms of surface coordinates is discussed in Secs. 5.4-6 and 17.3-1. In particular, *for orthogonal coordinates* x^1, x^2, x^3 *chosen so that* S *is a portion of an* x^k *coordinate surface* (*Sec.* 6.2-2), *the space coordinates* x^i *and* x^j *are* (*orthogonal*) *surface coordinates on* S, *and*

$$d\mathbf{A} = \mathbf{u}_k \sqrt{g_{ii}g_{jj}}\, dx^i\, dx^j = \mathbf{u}_k\, ds_i\, ds_j = \mathbf{e}_k \frac{\sqrt{g}}{g_{kk}}\, dx^i\, dx^j$$
$$(i \neq j \neq k \neq i;\ k = 1, 2, 3) \tag{6.4-8}$$

The relative simplicity of the expressions (5.4-13) *and* (5.4-14) *for surface integrals resulting from the use of Eq.* (8) *is often the main reason for introducing a curvilinear coordinate system* (*see also Sec.* 10.4-1c). The sign of the square root in Eq. (8) determines the direction of the positive normal (Sec. 17.3-2) and is taken to be positive for right-handed coordinate systems.

(c) The *volume element dV* appearing in the expressions (5.4-16) and (5.4-17) for *volume integrals* is given by Eq. (6.2-3), or

$$dV = \frac{\partial(x, y, z)}{\partial(x^1, x^2, x^3)} dx^1\, dx^2\, dx^3 = \pm \sqrt{g}\, dx^1\, dx^2\, dx^3$$
$$= \pm \sqrt{g_{11}g_{22}g_{33}}\, dx^1\, dx^2\, dx^3 = \pm ds_1\, ds_2\, ds_3 \qquad (6.4\text{-}9)$$

$\sqrt{g}$ is positive wherever the coordinate system is right-handed (Sec. 6.2-3*b*).

6.5. FORMULAS RELATING TO SPECIAL ORTHOGONAL COORDINATE SYSTEMS

6.5-1. Introduction. Tables 6.5-1 to 6.5-11 present formulas relating to a number of special orthogonal coordinate systems. In particular, the functions $g_{ii}(x^1, x^2, x^2) = |\mathbf{e}_i|^2$ are tabulated for each coordinate system and may be substituted in the relations of Table 6.4-1 and of Secs. 6.4-3 and 16.10-1 to 16.10-8 to yield additional formulas.

NOTE: A given family of coordinate surfaces $x^1 = x^1(x, y, z) = constant$, $x^2 = x^2(x, y, z) = constant$, $x^3 = x^3(x, y, z) = constant$ is necessarily common to all systems of coordinates $\bar{x}^1 = \bar{x}^1(x^1)$, $\bar{x}^2 = \bar{x}^2(x^2)$, $\bar{x}^3 = \bar{x}^3(x^3)$. The particular systems described in Tables 6.5-1 to 6.5-11 are representative.

6.6. RELATED TOPICS, REFERENCES, AND BIBLIOGRAPHY

6.6-1. Related Topics. The following topics related to the study of vector relations in terms of curvilinear coordinates are treated in other chapters of this handbook:

Solid analytic geometry .... Chap. 3
Elementary vector analysis .... Chap. 5
Coordinate transformations .... Chaps. 3, 14
Contravariant and covariant vectors .... Chap. 16
Transformation of base vectors .... Chap. 16
Differential invariants .... Chap. 16
Differential geometry .... Chap. 17
Potential theory .... Chap. 15

6.6-2. References and Bibliography (see also Sec. 5.8-2).

6.1. Courant, R., and D. Hilbert: *Methods of Mathematical Physics*, vol. I, Wiley, New York, 1953.

6.2. Kellogg, O. D.: *Foundations of Potential Theory*, Springer, Berlin, 1929.

6.3. MacMillan, W. D.: *Theory of the Potential*, McGraw-Hill, New York, 1930.

6.4. Madelung, E.: *Die mathematischen Hilfsmittel des Physikers*, 7th ed., Springer, Berlin, 1964.

6.5. Magnus, W., and F. Oberhettinger: *Formulas and Theorems for the Functions of Mathematical Physics*, Chelsea, New York, 1954; 3d ed., Springer, Berlin, 1966.

6.6. Margenau, H., and G. M. Murphy: *The Mathematics of Physics and Chemistry*, Van Nostrand, Princeton, N.J., 1943.

6.7. Stratton, J. A.: *Electromagnetic Theory*, Chap. I, McGraw-Hill, New York, 1941.

6.8. Whittaker, E. T., and G. N. Watson: *A Course of Modern Analysis*, Cambridge, New York, 1927.

Table 6.5-1. Vector Formulas in Terms of Spherical and (Circular) Cylindrical Coordinates
(see also Figs. 2.1-2 and 3.1-1*b*)
The formulas for cylindrical coordinates also apply to polar coordinates in the xy plane (Sec. 2.1-8)

	Spherical (polar) coordinates r, ϑ, φ ($r \geq 0, 0 \leq \vartheta \leq \pi$)	(Circular) cylindrical coordinates r', φ, z ($r' \geq 0$)
Coordinate surfaces	$x^2 + y^2 + z^2 = r^2$ (SPHERES) $x^2 + y^2 - z^2 \tan^2 \vartheta = 0$ (CIRCULAR CONES) $y = x \tan \varphi$ (PLANES THROUGH z AXIS)	$x^2 + y^2 = r'^2$ (RIGHT CIRCULAR CYLINDERS) $y = x \tan \vartheta$ (PLANES THROUGH z AXIS) $z = z$ (PLANES PARALLEL TO xy PLANE)
Transformation of coordinates (x, y, z are right-handed rectangular cartesian coordinates)	$r = +\sqrt{x^2 + y^2 + z^2}$ $\quad x = r \sin \vartheta \cos \varphi$ $\vartheta = \arccos \frac{z}{r}$ $\quad y = r \sin \vartheta \sin \varphi$ $\varphi = \arctan \frac{y}{x}$ $\quad z = r \cos \vartheta$	$r' = +\sqrt{x^2 + y^2}$ $\quad x = r' \cos \varphi$ $\varphi = \arctan \frac{y}{x}$ $\quad y = r' \sin \varphi$ $z = z$
Transformation of coordinate differentials	$dx = \sin \vartheta \cos \varphi \, dr + r \cos \vartheta \cos \varphi \, d\vartheta - r \sin \vartheta \sin \varphi \, d\varphi$ $dy = \sin \vartheta \sin \varphi \, dr + r \cos \vartheta \sin \varphi \, d\vartheta + r \sin \vartheta \cos \varphi \, d\varphi$ $dz = \cos \vartheta \, dr - r \sin \vartheta \, d\vartheta$	$dx = \cos \varphi \, dr' - r' \sin \varphi \, d\varphi$ $dy = \sin \varphi \, dr' + r' \cos \varphi \, d\varphi$ $dz = dz$
Square of distance element $ds^2 = (d\mathbf{r})^2 = dx^2 + dy^2 + dz^2$	$ds^2 = dr^2 + r^2 \, d\vartheta^2 + r^2 \sin^2 \vartheta \, d\varphi^2$	$ds^2 = dr'^2 + r'^2 \, d\varphi^2 + dz^2$
$g_{ii} = \|\mathbf{e}_i\|^2$	$g_{rr} = 1 \quad g_{\vartheta\vartheta} = r^2 \quad g_{\varphi\varphi} = r^2 \sin^2 \vartheta$	$g_{r'r'} = 1 \quad g_{\varphi\varphi} = r'^2 \quad g_{zz} = 1$
$\sqrt{g} = \sqrt{g_{11}g_{22}g_{33}}$	$\sqrt{g} = \frac{\partial(x, y, z)}{\partial(r, \vartheta, \varphi)} = r^2 \sin \vartheta$	$\sqrt{g} = \frac{\partial(x, y, z)}{\partial(r', \varphi, z)} = r'$
Christoffel three-index symbols (Sec. 16.10-3; three-index symbols not shown are identically zero)	$[\vartheta\vartheta; r] = -r \quad [\varphi\varphi; \vartheta] = -r^2 \sin \vartheta \cos \vartheta \quad [r\varphi; \varphi] = r \sin^2 \vartheta$ $[\varphi\varphi; r] = -r \sin^2 \vartheta \quad [r\vartheta; \vartheta] = r \quad [\varphi\vartheta; \varphi] = r^2 \sin \vartheta \cos \vartheta$ $\left\{ {r \atop \vartheta\vartheta} \right\} = -r \quad \left\{ {\vartheta \atop r\vartheta} \right\} = \frac{1}{r} \quad \left\{ {\varphi \atop r\varphi} \right\} = \frac{1}{r}$ $\left\{ {r \atop \varphi\varphi} \right\} = -r \sin^2 \vartheta \quad \left\{ {\vartheta \atop \varphi\varphi} \right\} = -\sin \vartheta \cos \vartheta \quad \left\{ {\varphi \atop \varphi\vartheta} \right\} = \cot \vartheta$	$[\varphi\varphi; r'] = -r' \quad [r'\varphi; \varphi] = [\varphi r'; \varphi] = r'$ $\left\{ {r' \atop \varphi\varphi} \right\} = -r' \quad \left\{ {\varphi \atop r'\varphi} \right\} = \left\{ {\varphi \atop \varphi r'} \right\} = \frac{1}{r'}$

Table 6.5-1. Vector Formulas in Terms of Spherical and (Circular) Cylindrical Coordinates
(see also Figs. 2.1-2 and 3.1-1*b*) (*Continued*)

	Spherical (polar) coordinates r, ϑ, φ $(r \geq 0, 0 \leq \vartheta \leq \pi)$	(Circular) cylindrical coordinates r', φ, z $(r' \leq 0)$
Transformation of physical vector components	$\hat{F}_r = F_x \sin\vartheta\cos\varphi + F_y \sin\vartheta\sin\varphi + F_z\cos\vartheta$ $\hat{F}_\vartheta = F_x\cos\vartheta\cos\varphi + F_y\cos\vartheta\sin\varphi - F_z\sin\vartheta$ $\hat{F}_\varphi = -F_x\sin\varphi + F_y\cos\varphi$ $F_x = \hat{F}_r\sin\vartheta\cos\varphi + \hat{F}_\vartheta\cos\vartheta\cos\varphi - \hat{F}_\varphi\sin\varphi$ $F_y = \hat{F}_r\sin\vartheta\sin\varphi + \hat{F}_\vartheta\cos\vartheta\sin\varphi + \hat{F}_\varphi\cos\varphi$ $F_z = \hat{F}_r\cos\vartheta - \hat{F}_\vartheta\sin\vartheta$	$\hat{F}_{r'} = F_x\cos\varphi + F_y\sin\varphi$ $\hat{F}_\varphi = -F_x\sin\varphi + F_y\cos\varphi$ $\hat{F}_z = F_z$ $F_x = \hat{F}_{r'}\cos\varphi - \hat{F}_\varphi\sin\varphi$ $F_y = \hat{F}_{r'}\sin\varphi + \hat{F}_\varphi\cos\varphi$ $F_z = \hat{F}_z$
Gradient in terms of *physical* components $\nabla\Phi = \mathbf{i}\frac{\partial\Phi}{\partial x} + \mathbf{j}\frac{\partial\Phi}{\partial y} + \mathbf{k}\frac{\partial\Phi}{\partial z}$	$\frac{\partial\Phi}{\partial r}\mathbf{u}_r + \frac{1}{r}\frac{\partial\Phi}{\partial\vartheta}\mathbf{u}_\vartheta + \frac{1}{r\sin\vartheta}\frac{\partial\Phi}{\partial\varphi}\mathbf{u}_\varphi$	$\frac{\partial\Phi}{\partial r'}\mathbf{u}_{r'} + \frac{1}{r'}\frac{\partial\Phi}{\partial\varphi}\mathbf{u}_\varphi + \frac{\partial\Phi}{\partial z}\mathbf{u}_z$
Divergence in terms of *physical* components $\nabla\cdot\mathbf{F} = \frac{\partial F_x}{\partial x} + \frac{\partial F_y}{\partial y} + \frac{\partial F_z}{\partial z}$	$\frac{1}{r^2}\frac{\partial}{\partial r}(r^2\hat{F}_r) + \frac{1}{r\sin\vartheta}\frac{\partial}{\partial\vartheta}(\hat{F}_\vartheta\sin\vartheta) + \frac{1}{r\sin\vartheta}\frac{\partial\hat{F}_\varphi}{\partial\varphi}$	$\frac{1}{r'}\frac{\partial}{\partial r'}(r'\hat{F}_{r'}) + \frac{1}{r'}\frac{\partial\hat{F}_\varphi}{\partial\varphi} + \frac{\partial\hat{F}_z}{\partial z}$
Curl in terms of *physical* components $\nabla\times\mathbf{F} = \begin{vmatrix} \mathbf{i} & \frac{\partial}{\partial x} & F_x \\ \mathbf{j} & \frac{\partial}{\partial y} & F_y \\ \mathbf{k} & \frac{\partial}{\partial z} & F_z \end{vmatrix}$	$\frac{1}{r\sin\vartheta}\left[\frac{\partial}{\partial\vartheta}(\hat{F}_\varphi\sin\vartheta) - \frac{\partial\hat{F}_\vartheta}{\partial\varphi}\right]\mathbf{u}_r + \frac{1}{r}\left[\frac{1}{\sin\vartheta}\frac{\partial\hat{F}_r}{\partial\varphi} - \frac{\partial}{\partial r}(r\hat{F}_\varphi)\right]\mathbf{u}_\vartheta$ $+ \frac{1}{r}\left[\frac{\partial}{\partial r}(r\hat{F}_\vartheta) - \frac{\partial\hat{F}_r}{\partial\vartheta}\right]\mathbf{u}_\varphi$	$\left(\frac{1}{r'}\frac{\partial\hat{F}_z}{\partial\varphi} - \frac{\partial\hat{F}_\varphi}{\partial z}\right)\mathbf{u}_{r'} + \left(\frac{\partial\hat{F}_{r'}}{\partial z} - \frac{\partial\hat{F}_z}{\partial r'}\right)\mathbf{u}_\varphi$ $+ \frac{1}{r'}\left[\frac{\partial}{\partial r'}(r'\hat{F}_\varphi) - \frac{\partial\hat{F}_{r'}}{\partial\varphi}\right]\mathbf{u}_z$
Laplacian of a scalar point function* $\nabla^2\Phi = \frac{\partial^2\Phi}{\partial x^2} + \frac{\partial^2\Phi}{\partial y^2} + \frac{\partial^2\Phi}{\partial z^2}$	$\frac{1}{r^2}\frac{\partial}{\partial r}\left(r^2\frac{\partial\Phi}{\partial r}\right) + \frac{1}{r^2\sin^2\vartheta}\frac{\partial}{\partial\vartheta}\left(\sin\vartheta\frac{\partial\Phi}{\partial\vartheta}\right) + \frac{1}{r^2\sin^2\vartheta}\frac{\partial^2\Phi}{\partial\varphi^2}$	$\frac{1}{r'}\frac{\partial}{\partial r'}\left(r'\frac{\partial\Phi}{\partial r'}\right) + \frac{1}{r'^2}\frac{\partial^2\Phi}{\partial\varphi^2} + \frac{\partial^2\Phi}{\partial z^2}$

* To find the Laplacian of a *vector*, use $\nabla^2\mathbf{F} = \nabla(\nabla\cdot\mathbf{F}) - \nabla\times(\nabla\times\mathbf{F})$, or use Eq. (6.4-5) and Table 16.10-1.

Table 6.5-2. General Ellipsoidal Coordinates λ, μ, ν or u, v, w

(a) Coordinate Surfaces (solve each equation to obtain λ, μ, ν in terms of x, y, z)

$$\frac{x^2}{a^2+\lambda}+\frac{y^2}{b^2+\lambda}+\frac{z^2}{c^2+\lambda}=1 \quad \text{(ELLIPSOIDS)}$$

$$\frac{x^2}{a^2+\mu}+\frac{y^2}{b^2+\mu}+\frac{z^2}{c^2+\mu}=1 \quad \text{(HYPERBOLOIDS OF ONE SHEET)}$$

$$\frac{x^2}{a^2+\nu}+\frac{y^2}{b^2+\nu}+\frac{z^2}{c^2+\nu}=1 \quad \text{(HYPERBOLOIDS OF TWO SHEETS)}$$

$$(\lambda > -c^2 > \mu > -b^2 > \nu > -a^2)$$

(b) Transformation to Ellipsoidal Coordinates

$$x^2=\frac{(a^2+\lambda)(a^2+\mu)(a^2+\nu)}{(a^2-b^2)(a^2-c^2)} \qquad y^2=\frac{(b^2+\lambda)(b^2+\mu)(b^2+\nu)}{(b^2-a^2)(b^2-c^2)}$$

$$z^2=\frac{(c^2+\lambda)(c^2+\mu)(c^2+\nu)}{(c^2-a^2)(c^2-b^2)}$$

(c) Alternative System (see Ref. 6.4 for other alternative systems). Introduce u, v, w by

$$\lambda=4\wp(u)-\frac{a^2+b^2+c^2}{3} \qquad \mu=4\wp(v)-\frac{a^2+b^2+c^2}{3}$$

$$\nu=4\wp(w)-\frac{a^2+b^2+c^2}{3}$$

so that (Sec. 21.6-2)

$$du=\frac{d\lambda}{\sqrt{f(\lambda)}} \qquad dv=\frac{d\mu}{\sqrt{f(\mu)}} \qquad dw=\frac{d\nu}{\sqrt{f(\nu)}}$$

with

$$f(t)\equiv(a^2+t)(b^2+t)(c^2+t)$$

(d) $$g_{\lambda\lambda}=\frac{(\lambda-\mu)(\lambda-\nu)}{4f(\lambda)} \qquad g_{\mu\mu}=\frac{(\mu-\lambda)(\mu-\nu)}{4f(\mu)} \qquad g_{\nu\nu}=\frac{(\nu-\lambda)(\nu-\mu)}{4f(\nu)}$$

$$g_{uu}=4[\wp(u)-\wp(v)][\wp(u)-\wp(w)] \qquad g_{vv}=4[\wp(v)-\wp(u)][\wp(v)-\wp(w)]$$

$$g_{ww}=4[\wp(w)-\wp(u)][\wp(w)-\wp(v)]$$

(e) $$\begin{aligned}\nabla^2\Phi &= \frac{4\sqrt{f(\lambda)}}{(\lambda-\mu)(\lambda-\nu)}\frac{\partial}{\partial\lambda}\left[\sqrt{f(\lambda)}\frac{\partial\Phi}{\partial\lambda}\right]+\frac{4\sqrt{f(\mu)}}{(\mu-\lambda)(\mu-\nu)}\frac{\partial}{\partial\mu}\left[\sqrt{f(\mu)}\frac{\partial\Phi}{\partial\mu}\right]\\&\quad+\frac{4\sqrt{f(\nu)}}{(\nu-\lambda)(\nu-\mu)}\frac{\partial}{\partial\nu}\left[\sqrt{f(\nu)}\frac{\partial\Phi}{\partial\nu}\right]\\&=\frac{1}{4}\left\{\frac{1}{[\wp(u)-\wp(v)][\wp(u)-\wp(w)]}\frac{\partial^2\Phi}{\partial u^2}+\frac{1}{[\wp(v)-\wp(u)][\wp(v)-\wp(w)]}\frac{\partial^2\Phi}{\partial v^2}\right.\\&\quad\left.+\frac{1}{[\wp(w)-\wp(u)][\wp(w)-\wp(v)]}\frac{\partial^2\Phi}{\partial w^2}\right\}\end{aligned}$$

Table 6.5-3. Prolate Spheroidal Coordinates σ, τ, φ or u, v, φ
(z axis is axis of revolution; see also Fig. 6.5-1)

(a) Coordinate Surfaces (solve each equation to obtain σ, τ, φ in terms of x, y, z)

$$\frac{x^2 + y^2}{a^2(\sigma^2 - 1)} + \frac{z^2}{a^2\sigma^2} = 1 \quad \text{(PROLATE SPHEROIDS)}$$

$$\frac{x^2 + y^2}{a^2(\tau^2 - 1)} + \frac{z^2}{a^2\tau^2} = 1 \quad \text{(HYPERBOLOIDS OF REVOLUTION, TWO SHEETS)}$$

$$y = x \tan \varphi \quad \text{(PLANES THROUGH } z \text{ AXIS)}$$

$$\sigma \geq 1 \geq \tau \geq -1$$

All spheroids and hyperboloids have common foci $(0, 0, a)$, $(0, 0, -a)$; $2a\sigma$ and $2a\tau$ are, respectively, the sum and difference of the focal radii of the point (x, y).

(b) Transformation to Prolate Spheroidal Coordinates

$$x^2 = a^2(\sigma^2 - 1)(1 - \tau^2) \cos^2 \varphi \qquad y^2 = a^2(\sigma^2 - 1)(1 - \tau^2) \sin^2 \varphi \qquad z = a\sigma\tau$$

(c) Alternative System (removes ambiguities)

$$\sigma = \cosh u \qquad \tau = \cos v$$

(d) $$g_{\sigma\sigma} = a^2 \frac{\sigma^2 - \tau^2}{\sigma^2 - 1} \qquad g_{\tau\tau} = a^2 \frac{\sigma^2 - \tau^2}{1 - \tau^2} \qquad g_{\varphi\varphi} = a^2(\sigma^2 - 1)(1 - \tau^2)$$

$$g_{uu} = g_{vv} = a^2(\sinh^2 u + \sin^2 v) \qquad g_{\varphi\varphi} = a^2 \sinh^2 u \sin^2 v$$

(e) $$\nabla^2\Phi = \frac{1}{a^2(\sigma^2 - \tau^2)} \left\{ \frac{\partial}{\partial \sigma} \left[(\sigma^2 - 1) \frac{\partial \Phi}{\partial \sigma} \right] + \frac{\partial}{\partial \tau} \left[(1 - \tau^2) \frac{\partial \Phi}{\partial \tau} \right] + \frac{\sigma^2 - \tau^2}{(\sigma^2 - 1)(1 - \tau^2)} \frac{\partial^2 \Phi}{\partial \varphi^2} \right\}$$

Table 6.5-4. Oblate Spheroidal Coordinates σ, τ, φ or u, v, φ
(z axis is axis of revolution; see also Fig. 6.5-1)

(a) Coordinate Surfaces (solve each equation to obtain σ, τ, φ in terms of x, y, z)

$$\frac{x^2 + y^2}{a^2(1 + \sigma^2)} + \frac{z^2}{a^2\sigma^2} = 1 \quad \text{(OBLATE SPHEROIDS)}$$

$$\frac{x^2 + y^2}{a^2(1 - \tau^2)} - \frac{z^2}{a^2\tau^2} = 1 \quad \text{(HYPERBOLOIDS OF REVOLUTION, ONE SHEET)}$$

$$y = x \tan \varphi \quad \text{(PLANES THROUGH } z \text{ AXIS)}$$

$$\sigma \geq 0, 1 \geq \tau \geq -1$$

(b) Transformation to Oblate Spheroidal Coordinates

$$x^2 = a^2(1 + \sigma^2)(1 - \tau^2) \cos^2 \varphi \qquad y^2 = a^2(1 + \sigma^2)(1 - \tau^2) \sin^2 \varphi \qquad z = a\sigma\tau$$

(c) Alternative System (removes ambiguities)

$$\sigma = \sinh u \qquad \tau = \cos v$$

(d) $$g_{\sigma\sigma} = a^2 \frac{\sigma^2 + \tau^2}{1 + \sigma^2} \qquad g_{\tau\tau} = a^2 \frac{\sigma^2 + \tau^2}{1 - \tau^2} \qquad g_{\varphi\varphi} = a^2(1 + \sigma^2)(1 - \tau^2)$$

$$g_{uu} = g_{vv} = a^2(\sinh^2 u + \cos^2 v) \qquad g_{\varphi\varphi} = a^2 \cosh^2 u \sin^2 v$$

(e) $$\nabla^2\Phi = \frac{1}{a^2(\sigma^2 + \tau^2)} \left\{ \frac{\partial}{\partial \sigma} \left[(1 + \sigma^2) \frac{\partial \Phi}{\partial \sigma} \right] + \frac{\partial}{\partial \tau} \left[(1 - \tau^2) \frac{\partial \Phi}{\partial \tau} \right] + \frac{\sigma^2 + \tau^2}{(1 + \sigma^2)(1 - \tau^2)} \frac{\partial^2 \Phi}{\partial^2 \varphi} \right\}$$

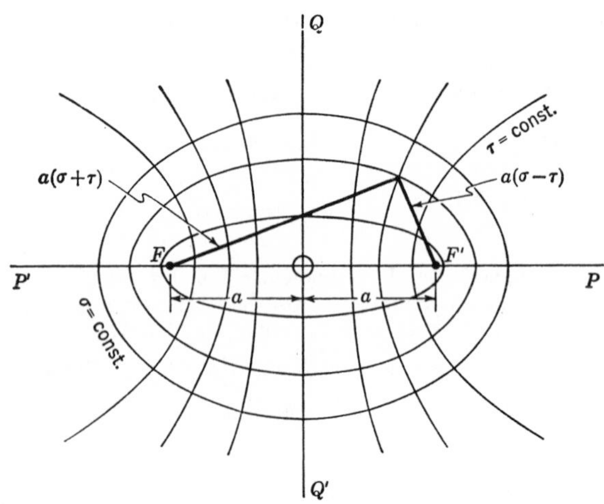

FIG. 6.5-1. An orthogonal system of confocal ellipses and hyperbolas with foci F, F'. Such curves define an elliptic coordinate system in their plane and will generate coordinate surfaces of

1. A *prolate spheroidal coordinate system* if the figure is rotated about the axis $P'OP$ (Table 6.5-3)
2. An *oblate spheroidal coordinate system* if the figure is rotated about the axis $Q'OQ$ (Table 6.5-4)
3. An *elliptic cylindrical coordinate system* if the figure is translated at right angles to the plane of the paper (Table 6.5-5)

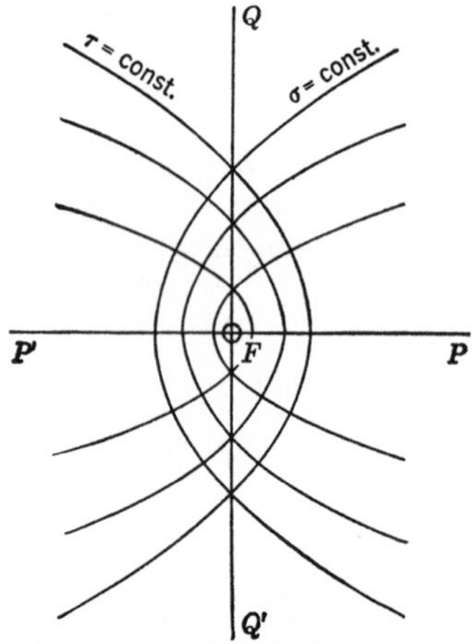

FIG. 6.5-2. An orthogonal system of confocal parabolas with focus F. Such curves define a parabolic coordinate system in their plane and will generate coordinate surfaces of

1. A *parabolic coordinate system* if the figure is rotated about the axis $P'OP$ (Table 6.5-8)
2. A *parabolic cylindrical coordinate system* if the figure is translated at right angles to the plane of the paper (Table 6.5-9)

Table 6.5-5. Elliptic Cylindrical Coordinates σ, τ, z or u, v, z
(also used as *confocal elliptic coordinates* in xy plane; see also Fig. 6.5-1)

(a) Coordinate Surfaces (solve each equation to obtain σ, τ, z in terms of x, y, z)

$$\left.\begin{array}{lr} \dfrac{x^2}{a^2\sigma^2} + \dfrac{y^2}{a^2(\sigma^2 - 1)} = 1 & \text{(RIGHT ELLIPTIC CYLINDERS)} \\ \dfrac{x^2}{a^2\tau^2} + \dfrac{y^2}{a^2(\tau^2 - 1)} = 1 & \text{(RIGHT HYPERBOLIC CYLINDERS)} \\ z = z & \text{(PLANES PARALLEL TO } xy \text{ PLANE)} \end{array}\right\} \sigma \geq 1 \geq \tau \geq -1$$

(b) Transformation to Elliptic Cylindrical Coordinates

$$x = a\sigma\tau \qquad y^2 = a^2(\sigma^2 - 1)(1 - \tau^2) \qquad z = z$$

(c) Alternative System (removes ambiguities)

$$\sigma = \cosh u \qquad \tau = \cos v$$

(d) $$g_{\sigma\sigma} = a^2 \frac{\sigma^2 - \tau^2}{\sigma^2 - 1} \qquad g_{\tau\tau} = a^2 \frac{\sigma^2 - \tau^2}{1 - \tau^2} \qquad g_{zz} = 1$$

$$g_{uu} = g_{vv} = a^2(\sinh^2 u + \sin^2 v) \qquad g_{zz} = 1$$

(e) $$\nabla^2\Phi = \frac{1}{a^2(\sigma^2 - \tau^2)} \left[\sqrt{\sigma^2 - 1} \left(\frac{\partial}{\partial\sigma} \sqrt{\sigma^2 - 1} \frac{\partial\Phi}{\partial\sigma} \right) + \sqrt{1 - \tau^2} \frac{\partial}{\partial\tau} \left(\sqrt{1 - \tau^2} \frac{\partial\Phi}{\partial\tau} \right) \right] + \frac{\partial^2\Phi}{\partial z^2}$$

$$= \frac{1}{a^2(\sinh^2 u + \sin^2 v)} \left(\frac{\partial^2\Phi}{\partial u^2} + \frac{\partial^2\Phi}{\partial v^2} \right) + \frac{\partial^2\Phi}{\partial z^2}$$

Table 6.5-6. Conical Coordinates r, v, w

(a) Coordinate Surfaces (solve each equation to obtain r, v, w in terms of x, y, z)

$$\left.\begin{array}{lr} x^2 + y^2 + z^2 = r^2 & \text{(SPHERES)} \\ \dfrac{x^2}{v^2} + \dfrac{y^2}{v^2 - b^2} + \dfrac{z^2}{v^2 - c^2} = 0 & \text{(CONES)} \\ \dfrac{x^2}{w^2} + \dfrac{y^2}{w^2 - b^2} + \dfrac{z^2}{w^2 - c^2} = 0 & \text{(CONES)} \end{array}\right\} c^2 > v^2 > b^2 > w^2$$

(b) Transformation to Conical Coordinates

$$x = \pm \frac{rvw}{bc} \qquad y^2 = \frac{u^2}{b^2} \frac{(v^2 - b^2)(w^2 - b^2)}{b^2 - c^2} \qquad z^2 = \frac{u^2}{c^2} \frac{(v^2 - c^2)(w^2 - c^2)}{c^2 - b^2}$$

(c) $$g_{rr} = 1 \qquad g_{vv} = -\frac{u^2(v^2 - w^2)}{(v^2 - b^2)(v^2 - c^2)} \qquad g_{ww} = \frac{u^2(v^2 - w^2)}{(w^2 - b^2)(w^2 - c^2)}$$

Table 6.5-7. Paraboloidal Coordinates λ, μ, ν

(a) Coordinate Surfaces (solve each equation to obtain λ, μ, ν in terms of x, y, z)

$$\left.\begin{aligned} \frac{x^2}{\lambda - A} + \frac{y^2}{\lambda - B} &= 2z + \lambda \qquad \text{(ELLIPTIC PARABOLOIDS)} \\ \frac{x^2}{\mu - A} + \frac{y^2}{\mu - B} &= 2z + \mu \qquad \text{(HYPERBOLIC PARABOLOIDS)} \\ \frac{x^2}{\nu - A} + \frac{y^2}{\nu - B} &= 2z + \nu \qquad \text{(ELLIPTIC PARABOLOIDS)} \end{aligned}\right\} \nu > A > \mu > B > \lambda$$

(b) Transformation to Paraboloidal Coordinates

$$x^2 = \frac{(A - \lambda)(A - \mu)(A - \nu)}{(B - A)} \qquad y^2 = \frac{(B - \lambda)(B - \mu)(B - \nu)}{(A - B)}$$

$$z = \frac{1}{2}(A + B - \lambda - \mu - \nu)$$

(c) $$g_{\lambda\lambda} = \frac{1}{4}\frac{(\mu - \lambda)(\nu - \lambda)}{(A - \lambda)(B - \lambda)} \qquad g_{\mu\mu} = \frac{1}{4}\frac{(\nu - \mu)(\lambda - \mu)}{(A - \mu)(B - \mu)}$$

$$g_{\nu\nu} = \frac{1}{4}\frac{(\lambda - \nu)(\mu - \nu)}{(A - \nu)(B - \nu)}$$

Table 6.5-8. Parabolic Coordinates σ, τ, φ
(z axis is axis of revolution; see also Fig. 6.5-2)

(a) Coordinate Surface (solve each equation to obtain σ, τ, φ in terms of x, y, z)

$$\left.\begin{aligned} \frac{x^2 + y^2}{\sigma^2} &= 2z + \sigma^2 \\ \frac{x^2 + y^2}{\tau^2} &= -2z + \tau^2 \end{aligned}\right\} \text{(CONFOCAL PARABOLOIDS OF REVOLUTION; FOCI AT ORIGIN)}$$

$$y = x \tan \varphi \qquad \text{(PLANES THROUGH } z \text{ AXIS)}$$

(b) Transformation to Parabolic Coordinates

$$x = \sigma\tau \cos \varphi \qquad y = \sigma\tau \sin \varphi \qquad z = \tfrac{1}{2}(\tau^2 - \sigma^2)$$

(c) $g_{\sigma\sigma} = g_{\tau\tau} = \sigma^2 + \tau^2 \qquad g_{\varphi\varphi} = \sigma^2\tau^2$

(d) $$\nabla^2\Phi = \frac{1}{\sigma^2 + \tau^2}\left[\frac{1}{\sigma}\frac{\partial}{\partial\sigma}\left(\sigma\frac{\partial\Phi}{\partial\sigma}\right) + \frac{1}{\tau}\frac{\partial}{\partial\tau}\left(\tau\frac{\partial\Phi}{\partial\tau}\right) + \left(\frac{1}{\sigma^2} + \frac{1}{\tau^2}\right)\frac{\partial^2\Phi}{\partial\varphi^2}\right]$$

Table 6.5-9. Parabolic Cylindrical Coordinates σ, τ, z
(also used as *parabolic coordinates* in xy plane; see also Fig. 6.5-2)

(a) Coordinate Surfaces (solve each equation to obtain σ, τ, z in terms of x, y, z)

$$\left.\begin{aligned}\frac{x^2}{\sigma^2} &= 2y + \sigma^2 \\ \frac{x^2}{\tau^2} &= -2y + \tau^2\end{aligned}\right\} \quad \text{(CONFOCAL RIGHT PARABOLIC CYLINDERS)}$$

$$z = z \quad \text{(PLANES PARALLEL TO } xy \text{ PLANE)}$$

(b) Transformation to Parabolic Cylindrical Coordinates

$$x = \sigma\tau \qquad y = \tfrac{1}{2}(\tau^2 - \sigma^2) \qquad z = z$$

(c) $g_{\sigma\sigma} = g_{\tau\tau} = \sigma^2 + \tau^2 \qquad g_{zz} = 1$

(d) $\nabla^2\Phi = \dfrac{1}{\sigma^2 + \tau^2}\left(\dfrac{\partial^2\Phi}{\partial\sigma^2} + \dfrac{\partial^2\Phi}{\partial\tau^2}\right) + \dfrac{\partial^2\Phi}{\partial z^2}$

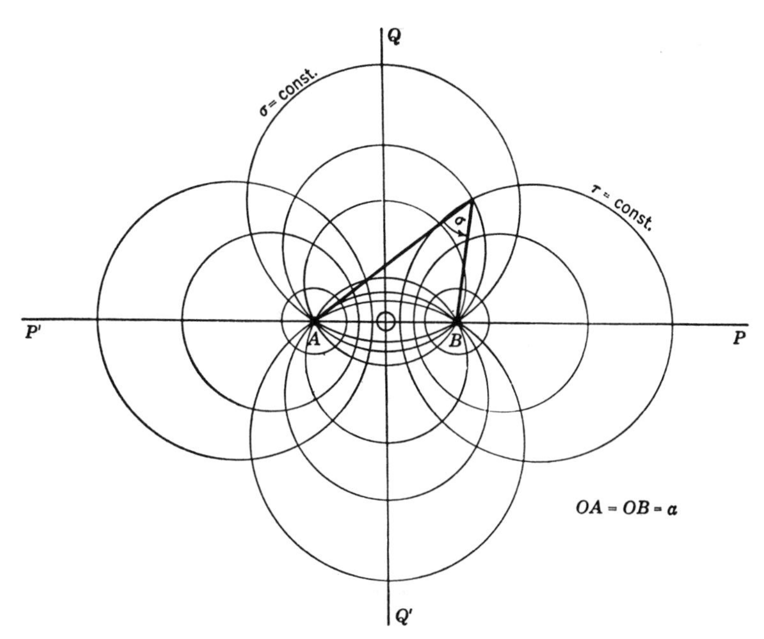

Fig. 6.5-3. A family of circles through two poles A, B, and the family of circles orthogonal to those of the first family. Such curves define a bipolar coordinate system in their plane and will generate coordinate surfaces of

1. A *bipolar coordinate system* if the figure is translated at right angles to the plane of the paper (Table 6.5-10)
2. A *toroidal coordinate system* if the figure is rotated about the axis $Q'OQ$ (Table 6.5-11)

Table 6.5-10. Bipolar Coordinates σ, τ, z
(also used as *bipolar coordinates* in xy plane; see also Fig. 6.5-3)

(a) Coordinate Surfaces

$$x^2 + (y - a \cot \sigma)^2 = \frac{a^2}{\sin^2 \sigma} = a^2(\cot^2 \sigma + 1) \qquad \text{(RIGHT CIRCULAR CYLINDERS)}$$

$$(x - a \coth \tau)^2 + y^2 = \frac{a^2}{\sinh^2 \tau} = a^2(\coth^2 \tau - 1) \qquad \text{(RIGHT CIRCULAR CYLINDERS)}$$

$$z = z \qquad \text{(PLANES PARALLEL TO } xy \text{ PLANE)}$$

For any point (x, y) in the xy plane, σ is the angle subtended at (x, y) by the two *poles* $(-a, 0)$ and $(a, 0)$. e^τ is the ratio of the polar radii to (x, y).

(b) Transformations

$$x = \frac{a \sinh \tau}{\cosh \tau - \cos \sigma} \qquad \sigma = \frac{i}{2} \log_e \frac{x^2 + (y - ia)^2}{x^2 + (y + ia)^2}$$

$$y = \frac{a \sin \sigma}{\cosh \tau - \cos \sigma} \qquad \tau = \frac{1}{2} \log_e \frac{(x + a)^2 + y^2}{(x - a)^2 + y^2} \qquad z = z$$

(c) $$g_{\sigma\sigma} = g_{\tau\tau} = \frac{a^2}{(\cosh \tau - \cos \sigma)^2} \qquad g_{zz} = 1$$

(d) $$\nabla^2 \Phi = \frac{1}{a^2} (\cosh \tau - \cos \sigma)^2 \left(\frac{\partial^2 \Phi}{\partial \sigma^2} + \frac{\partial^2 \Phi}{\partial \tau^2} \right) + \frac{\partial^2 \Phi}{\partial z^2}$$

Table 6.5-11. Toroidal Coordinates σ, τ, φ
(z axis is axis of revolution; see also Fig. 6.5-3)

(a) Coordinate Surfaces

$$(x^2 + y^2) + (z - a \cot \sigma)^2 = \frac{a^2}{\sin^2 \sigma} \qquad \text{(SPHERES)}$$

$$(\sqrt{x^2 + y^2} - a \coth \tau)^2 + z^2 = \frac{a^2}{\sinh^2 \tau} \qquad \text{(TORES OR ANCHOR RINGS)}$$

$$y = x \tan \varphi \qquad \text{(PLANES THROUGH } z \text{ AXIS)}$$

(b) Transformations

$$x = \frac{a \sinh \tau}{\cosh \tau - \cos \sigma} \cos \varphi \qquad \sigma = \frac{i}{2} \log_e \frac{x^2 + y^2 + (z - ia)^2}{x^2 + y^2 + (z + ia)^2}$$

$$y = \frac{a \sinh \tau}{\cosh \tau - \cos \sigma} \sin \varphi \qquad \tau = \frac{1}{2} \log_e \frac{[\sqrt{x^2 + y^2} + a]^2 + z^2}{[\sqrt{x^2 + y^2} - a]^2 + z^2}$$

$$z = \frac{a \sin \sigma}{\cosh \tau - \cos \sigma} \qquad \varphi = \arctan \frac{y}{x}$$

(c) $$g_{\sigma\sigma} = g_{\tau\tau} = \frac{a^2}{(\cosh \tau - \cos \sigma)^2} \qquad g_{\varphi\varphi} = \frac{a^2 \sinh^2 \tau}{(\cosh \tau - \cos \sigma)^2}$$

(d) $$\nabla^2 \Phi = \frac{(\cosh \tau - \cos \sigma)^3}{a^2 \sinh \tau} \left[\frac{\partial}{\partial \tau} \left(\frac{\sinh \tau}{\cosh \tau - \cos \sigma} \frac{\partial \Phi}{\partial \tau} \right) + \frac{\partial}{\partial \sigma} \left(\frac{\sinh \tau}{\cosh \tau - \cos \sigma} \frac{\partial \Phi}{\partial \sigma} \right) + \frac{1}{\sinh \tau (\cosh \tau - \cos \sigma)} \frac{\partial^2 \Phi}{\partial \varphi^2} \right]$$

CHAPTER 7

FUNCTIONS OF A COMPLEX VARIABLE

7.1. INTRODUCTION

7.1-1. The theory of analytic functions of a complex variable furnishes the scientist or engineer with many useful mathematical models. Many mathematical theorems are simplified if the real variables are considered as special values of complex variables. Complex variables are used to describe two-dimensional vectors in physics; analytical functions of a complex variable describe two-dimensional scalar and vector fields (Secs. 15.6-8 and 15.6-9). Finally, analytic functions of a complex variable represent conformal mappings of the points of a plane into another plane (Secs. 7.9-1 to 7.9-5).

7.2. FUNCTIONS OF A COMPLEX VARIABLE. REGIONS OF THE COMPLEX-NUMBER PLANE

7.2-1. Functions of a Complex Variable (see also Secs. 1.3-2 and 4.2-1 and Table 7.2-1; refer to Chap. 21 for additional examples). A complex function

$$w \equiv f(z) \equiv u(x, y) + iv(x, y) \equiv |w|e^{i\vartheta} \qquad (z = x + iy = |z|e^{i\varphi}) \tag{7.2-1}$$

associates one or more values of the complex dependent variable w with each value of the complex independent variable z in a given domain of definition.

Single-valued, multiple-valued, and *bounded* functions of a complex variable are defined as in Secs. 4.2-2*a* and 4.3-3*b*. *Limits of complex functions and sequences* and *continuity of complex functions* as well as *convergence, absolute convergence, and uniform convergence of complex infinite series and improper integrals* are defined as in Chap. 4; **the theorems**

Table 7.2-1. Real and Imaginary Parts, Zeros, and Singularities for a Number of Frequently Used Functions $f(z) = u(x, y) + iv(x, y)$ of a Complex Variable $z = x + iy$ (see also Secs. 1.3-2 and 21.2-9 to 21.2-11)

Note $|f(z)| = \sqrt{u^2 + v^2}$ $\quad \arg f(z) = \arctan \frac{v}{u}$

Function $f(z)$	$u(x, y)$	$v(x, y)$	Zeros (order m)	Isolated singularities
z	x	y	$z = 0 \quad m = 1$	Pole ($m = 1$) at $z = \infty$
z^2	$x^2 - y^2$	$2xy$	$z = 0 \quad m = 2$	Pole ($m = 2$) at $z = \infty$
$\frac{1}{z}$	$\frac{x}{x^2 + y^2}$	$-\frac{y}{x^2 + y^2}$	$z = \infty \quad m = 1$	Pole ($m = 1$) at $z = 0$
$\frac{1}{z^2}$	$\frac{x^2 - y^2}{(x^2 + y^2)^2}$	$\frac{-2xy}{(x^2 + y^2)^2}$	$z = \infty \quad m = 2$	Pole ($m = 2$) at $z = 0$
$\frac{1}{z - (a + ib)}$ (a, b real)	$\frac{(x - a)}{(x - a)^2 + (y - b)^2}$	$-\frac{(y - b)}{(x - a)^2 + (y - b)^2}$	$z = \infty \quad m = 1$	Pole ($m = 1$) at $z = a + ib$
$\sqrt{z}$	$\pm\left(\frac{x + \sqrt{x^2 + y^2}}{2}\right)^{1/2}$	$\pm\left(\frac{-x + \sqrt{x^2 + y^2}}{2}\right)^{1/2}$	Zero of order 1 at $z = 0$ (branch point)	Branch point ($m = 1$) at $z = 0$ Branch point ($m = 1$) at $z = \infty$
e^z	$e^x \cos y$	$e^x \sin y$	....	Essential singularity at $z = \infty$
$\sin z$	$\sin x \cosh y$	$\cos x \sinh y$	$z = k\pi \quad m = 1$ ($k = 0, \pm 1, \pm 2, \pm \ldots$)	Essential singularity at $z = \infty$
$\cos z$	$\cos x \cosh y$	$-\sin x \sinh y$	$z = (k + \frac{1}{2})\pi \quad m = 1$ ($k = 0, \pm 1, \pm 2, \pm \ldots$)	Essential singularity at $z = \infty$
$\sinh z$	$\sinh x \cos y$	$\cosh x \sin y$	$z = k\pi i \quad m = 1$ ($k = 0, \pm 1, \pm 2, \pm \ldots$)	Essential singularity at $z = \infty$
$\cosh z$	$\cosh x \cos y$	$\sinh x \sin y$	$z = (k + \frac{1}{2})\pi i \quad m = 1$ ($k = 0, \pm 1, \pm 2, \pm \ldots$)	Essential singularity at $z = \infty$
$\tan z$	$\frac{\sin 2x}{\cos 2x + \cosh 2y}$	$\frac{\sinh 2y}{\cos 2x + \cosh 2y}$	$z = k\pi \quad m = 1$ ($k = 0, \pm 1, \pm 2, \pm \ldots$)	Essential singularity at $z = \infty$ Poles ($m = 1$) at $z = (k + \frac{1}{2})\pi$ ($k = 0, \pm 1, \pm 2, \pm \ldots$)
$\tanh z$	$\frac{\sinh 2x}{\cosh 2x + \cos 2y}$	$\frac{\sin 2y}{\cosh 2x + \cos 2y}$	$z = k\pi i \quad m = 1$ ($k = 0, \pm 1, \pm 2, \pm \ldots$)	Essential singularity at $z = \infty$ Poles ($m = 1$) at $z = (k + \frac{1}{2})\pi i$ ($k = 0, \pm 1, \pm 2, \pm \ldots$)
$\log_e z$	$\frac{1}{2} \log_e (x^2 + y^2)$	$\arctan\left(\frac{y}{x}\right) + 2k\pi$ ($k = 0, \pm 1, \pm 2, \pm \ldots$)	$z = 1 \quad m = 1$ (branch corresponding to $k = 0$ only)	Branch points of infinite order at $z = 0$, $z = \infty$; both are essential singularities

of Chap. 4 apply to complex functions and variables unless a restriction to real quantities is specifically stated. In particular, *every complex power series* $\sum_{k=0}^{\infty} a_k(z - a)^k$ *has a real radius of convergence* r_c $(0 \leq r_c \leq \infty)$ *such that the series converges uniformly and absolutely for* $|z - a| < r_c$ *and diverges for* $|z - a| > r_c$ (Sec. 4.10-2*a*).

7.2-2. z **Plane and** w **Plane. Neighborhoods. The Point at Infinity** (see also Sec. 4.3-5). Values of the independent variable $z = x + iy$ are associated with unique points (x, y) of an Argand plane (Sec. 1.3-2), the *z plane.* Values of $w = u + iv$ are similarly associated with points (u, v) of a *w plane.*

An (open) **δ-neighborhood** of the point $z = a$ in the finite portion of the plane is defined as the set of points z such that $|z - a| < \delta$ for some $\delta > 0$.

The **point at infinity** $(z = \infty)$ is defined as the point $\bar{z}$ transformed into the origin $(z = 0)$ by the transformation $\bar{z} = 1/z$. A region containing the exterior of any circle is a **neighborhood of the point** $z = \infty$.

7.2-3. Curves and Contours (see also Secs. 2.1-9 and 3.1-13). A **continuous (arc or segment of a) curve** in the z plane is a set of points $z = x + iy$ such that

$$z = z(t) \qquad \text{or} \qquad x = x(t) \qquad y = y(t) \qquad (-\infty \leq t_1 \leq t \leq t_2 \leq \infty) \tag{7.2-2}$$

where $x(y)$ and $y(t)$ are continuous functions of the real parameter t. A (portion of a) continuous curve (2) is a **simple curve (Jordan arc)** if and only if it consists of a single branch without multiple points, so that the functions $x(t)$ and $y(t)$ are single-valued, and the set of equations

$$x(\tau_1) = x(\tau_2) \qquad y(\tau_1) = y(\tau_2)$$

has no distinct solutions τ_1, τ_2 in the closed interval $[t_1, t_2]$. A **simple closed curve (closed Jordan curve)** is a continuous curve consisting of a single branch without multiple points except for a common initial and terminal point.

A simple curve or simple closed curve will be referred to as a (simple) **contour** if and only if it is rectifiable (Sec. 4.6-9).* The element of distance between suitable points z and $z + dz$ on a contour (2) is $ds = |dz| = \sqrt[+]{dx^2 + dy^2}$.

7.2-4. Boundaries and Regions (see also Secs. 4.3-6, 7.9-1*b*, and 12.5-1). The geometry of the complex-number plane (including definitions of distances and angles)

* Some authors restrict the use of the term contour to regular curves (Sec. 3.1-13).

is identical with the geometry of the Euclidean plane of points (x, y) or vectors $\mathbf{r} = \mathbf{i}x + \mathbf{j}y$ for finite values of x and y, but the definition of $z = \infty$ (Sec. 7.2-2) introduces a topology different from that usually associated with plane geometry. The points z of the complex-number plane can be represented homeomorphically (Sec. 12.5-1) by corresponding points of a sphere with longitude arg z and colatitude 2 arccot $(|z|/2)$, (*stereographic projection*), so that $z = 0$ and $z = \infty$ correspond to opposite poles. *The points of every simple closed curve C separate the plane into two singly connected open regions: every continuous curve containing a point of each of the two regions contains a point of their common boundary C (Jordan Separation Theorem).* If C does not contain $z = \infty$ one of the two regions is **bounded** (i.e., situated entirely in the finite portion of the plane, where $|z|$ is bounded) and the other is unbounded; if C contains $z = \infty$ both regions are unbounded.

More generally, the boundary C of a given region or domain D may be a multiplicity of nonintersecting simple closed curves (multiply connected region, see also Sec. 4.3-6). In any case, the **positive direction (positive sense)** on the boundary curve is defined as that leaving the region D (interior of the boundary) on the left (counterclockwise for outside boundaries, see also Fig. 7.5-1). The (open) set of points on one side of a boundary curve C is an open region, and the (closed) set of points on one side and on the boundary is a closed region.

7.2-5. Complex Contour Integrals (see also Secs. 4.6-1 and 4.6-10).

One defines

$$\int_C f(z)\,dz = \lim_{\max|z_i - z_{i-1}| \to 0} \sum_{i=1}^{m} f(\zeta_i)(z_i - z_{i-1}) \qquad (7.2\text{-}3a)$$

where the points $z_0 < \zeta_1 < z_1 < \zeta_2 < z_2 < \cdots < \zeta_m < z_m$ lie on the contour C connecting $z = a = z_0$ and $z = b = z_m$. If the limit (3*a*) exists, then

$$\int_C f(z)\,dz = \int_C [u(x, y)\,dx - v(x, y)\,dy] + i \int_C [v(x, y)\,dx + u(x, y)\,dy] \qquad (7.2\text{-}3b)$$

where the real line integrals are taken over the same path as the complex integral. *The integration rules of Table* 4.6-1 *apply;* in particular, reversal of the sense of integration on the contour C reverses the sign of the integral.

If $f(z)$ is bounded in absolute value by M, and $u(x, y)$ and $v(x, y)$ are of bounded variation (Sec. 4.4-8*b) on a contour C of finite length L, then the integral* (3) *exists, and*

$$\left| \int_C f(z)\,dz \right| \leq ML \qquad (7.2\text{-}4)$$

If C contains $z = \infty$, or if $f(z)$ is not bounded on C, the integral (3) can often be defined as an improper integral in the manner of Sec. 4.6-2.

7.3. ANALYTIC (REGULAR, HOLOMORPHIC) FUNCTIONS

7.3-1. Derivative of a Function (see also Sec. 4.5-1). A function $w = f(z)$ is **differentiable** at the point $z = a$ if and only if the limit

$$\frac{dw}{dz} = f'(z) = \lim_{\Delta z \to 0} \frac{f(z + \Delta z) - f(z)}{\Delta z} \tag{7.3-1}$$

[**derivative of** $f(z)$ **with respect to** z] exists for $z = a$ and is independent of the manner in which Δz approaches zero. A function may be differentiable at a point (e.g., $|z|^2$ at $z = 0$), on a curve, or throughout a region.

7.3-2. The Cauchy-Riemann Equations. $f(z) \equiv u(x, y) + iv(x, y)$ *is differentiable at the point* $z = x + iy$ *if and only if* $u(x, y)$ *and* $v(x, y)$ *are continuously differentiable throughout a neighborhood of* z, *and*

$$\boxed{\frac{\partial u}{\partial x} = \frac{\partial v}{\partial y} \qquad \frac{\partial u}{\partial y} = -\frac{\partial v}{\partial x} \qquad \text{(Cauchy-Riemann equations)}} \tag{7.3-2}$$

at the point z, so that

$$\frac{dw}{dz} = \frac{\partial u}{\partial x} + i\frac{\partial v}{\partial x} = \frac{\partial v}{\partial y} - i\frac{\partial u}{\partial y} \tag{7.3-3}$$

7.3-3. Analytic Functions. (**a**) A single-valued function $f(z)$ shall be called **analytic (regular, holomorphic)*** at the point $z = a$ if and only if $f(z)$ is differentiable *throughout a neighborhood* of $z = a$. $f(z)$ *is analytic at* $z = a$ *if and only if* $f(z)$ *can be represented by a power series* $f(z) = \sum_{k=0}^{\infty} a_k(z - a)^k$ *convergent throughout a neighborhood of* $z = a$ (*alternative definition*). Refer to Secs. 7.4-1 to 7.4-3 for an extension of the definition to multiple-valued functions.

(**b**) $f(z)$ is **analytic at infinity** if and only if $F(\bar{z}) \equiv f(1/\bar{z})$ is analytic at $\bar{z} = 0$. One defines $f'(\infty) = -\bar{z}^2 \left.\frac{dF}{d\bar{z}}\right]_{\bar{z}=0}$.

$f(z)$ is analytic at infinity if and only if $f(z)$ can be expressed as a convergent series of negative powers, $f(z) = \sum_{k=0}^{\infty} b_k(z - a)^{-k}$ *for sufficiently large values of* $|z|$ (see also Sec. 7.5-3).

7.3-4. Properties of Analytic Functions. *Let* $f(z)$ *be analytic throughout an open region* D. *Then, throughout* D,

* The terms *differentiable, analytic, regular,* and *holomorphic* are used interchangeably by some authors.

1. *The Cauchy-Riemann equations* (2) *are satisfied* (*the converse is true*).
2. $u(x, y)$ *and* $v(x, y)$ *are conjugate harmonic functions* (*Sec.* 15.6-8).
3. *All derivatives of* $f(z)$ *with respect to* z *exist and are analytic* (*see also Sec.* 7.5-1).

If the open region D *is simply connected* (*this applies, in particular, to the exterior of a bounded simply connected region*),

4. *The integral* $\int_{a \atop C}^{z} f(\zeta)\, d\zeta$ *is independent of the path of integration, provided that* C *is a contour of finite length situated entirely in* D*; the integral is a single-valued analytic function of* z*, and its derivative is* $f(z)$ (see also Sec. 7.5-1).
5. *The values of* $f(z)$ *on a contour arc or a subregion in* D *define* $f(z)$ *uniquely throughout* D.

All ordinary differentiation and integration rules (*Secs.* 4.5-4 *and* 4.6-1) *apply to analytic functions of a complex variable. If* $f(z)$ *is analytic at* $z = a$ *and* $f'(a) \neq 0$, *then* $f(z)$ *has an analytic inverse function* (*Sec.* 4.2-2*a*) *at* $z = a$. *If* $W = F(w)$ *and* $w = f(z)$ *are analytic, then* W *is an analytic function of* z. *If a sequence* (*or an infinite series, Sec.* 4.8-1) *of functions* $f_i(z)$ *analytic throughout an open region* D *converges uniformly to the limit* $f(z)$ *throughout* D, *then* $f(z)$ *is analytic, and the sequence* (*or series*) *of the derivatives* $f_i'(z)$ *converges uniformly to* $f'(z)$ *throughout* D. *The sequence* (*or series*) *of contour integrals* $\int_C f_i(z)\, dz$ *over any contour* C *of finite length in* D *converges uniformly to* $\int_C f(z)\, dz$.

7.3-5. The Maximum-modulus Theorem. *The absolute value* $|f(z)|$ *of a function* $f(z)$ *analytic throughout a simply connected closed bounded region* D *cannot have a maximum in the interior of* D. If $|f(z)| \leq M$ on the boundary of D, then $|f(z)| < M$ throughout the interior of D unless $f(z)$ is a constant (see also Sec. 15.6-4).

7.4. TREATMENT OF MULTIPLE-VALUED FUNCTIONS

7.4-1. Branches. One extends the theory of analytic functions to suitable multiple-valued functions by considering **branches** $f_1(z)$, $f_2(z)$, . . . of $f(z)$ each defined as a single-valued continuous function throughout its region of definition. Each branch assumes one set of the function values of $f(z)$ (see also Sec. 7.8-1).

7.4-2. Branch Points and Branch Cuts. **(a)** Given a number of branches of $f(z)$ analytic throughout a neighborhood D of $z = a$, except possibly at $z = a$, the point $z = a$ is a **branch point** involving the given branches* of $f(z)$ if and only if $f(z)$ passes from one of these branches to another as a variable point z in D describes a closed circuit about $z = a$. The **order** of the branch point is the number m of branches reached by $f(z)$ before returning to the original or $m + 1^{st}$ branch as z describes successive closed circuits about $z = a$. If $f(z)$ is defined at a branch point $z = a$ the function value $f(a)$ is common to all branches "joining" at $z = a$ (EXAMPLE: $\sqrt[3]{z}$ has a branch point of order 2 at $z = 0$).

The point $z = \infty$ is a branch point of $f(z)$ if and only if the origin is a branch point of $f(1/z)$ (see also Sec. 7.6-3).

Given a function $w = f(z)$ whose inverse function $\Phi(w)$ exists and is single-valued throughout a neighborhood of $w = f(a)$, the point $z = a \neq \infty$ is a branch point of order m of $f(z)$ whenever $\Phi'(w)$ has a zero of order m (Sec. 7.6-1) *or a pole of order $m + 2$ (Sec.* 7.6-2) *at $w = f(a)$.* (EXAMPLES: $w = \sqrt{z}$ and $w = 1/\sqrt{z}$, $a = 0$).

Similarly, *if $\Phi(w)$ exists and is single-valued throughout a neighborhood of $w = f(\infty)$, the point $z = \infty$ is a branch point of order m of $f(z)$ if $\Phi'(w)$ has a zero of order $m + 2$ or a pole of order m at $w = f(\infty)$* (EXAMPLES: $w = \sqrt{z}$ and $w = 1/\sqrt{z}$).

(b) The individual single-valued branches of $f(z)$ are defined in regions bounded by **branch cuts,** which are simple curves chosen so that no closed circuit around a branch point lies within the region of definition of a single branch. The choice of branches and branch cuts for a given function $f(z)$ is not unique, but the branch points and the number of branches are uniquely defined (Fig. 7.4-1).

All branches of a *monogenic analytic function* may be obtained by analytic continuation of successive elements (Sec. 7.8-1).

7.4-3. Riemann Surfaces. It is frequently useful to represent a multiple-valued function $f(z)$ as a single-valued function defined on a Riemann surface which consists of a multiplicity of z planes or "sheets" corresponding to the branches of $f(z)$ and joined along suitably chosen branch cuts. A circuit around a branch point on such a surface transfers the variable point z between two sheets corresponding to two branches of $f(z)$. If both $w = f(z)$ and its inverse are multiple-valued functions, then both the z plane and the w plane may be replaced by suitable Riemann surfaces; $w = f(z)$ will now define a reciprocal one-to-one correspondence (mapping) between the points of the two Riemann surfaces, except at branch points.

NOTE: The Riemann surface for a monogenic analytic function (obtained by analytic continuation, Sec. 7.8-1) must be *connected*; thus the multiple-valued function $f(z) = \pm 1$, although analytic everywhere, is *not* a monogenic analytic function.

Many theorems about single-valued analytic functions apply also to multiple-valued monogenic analytic functions defined on suitable Riemann surfaces without

* Note that $f(z)$ may have other branches which do not join the given branches at $z = a$ and which may or may not be analytic at $z = a$.

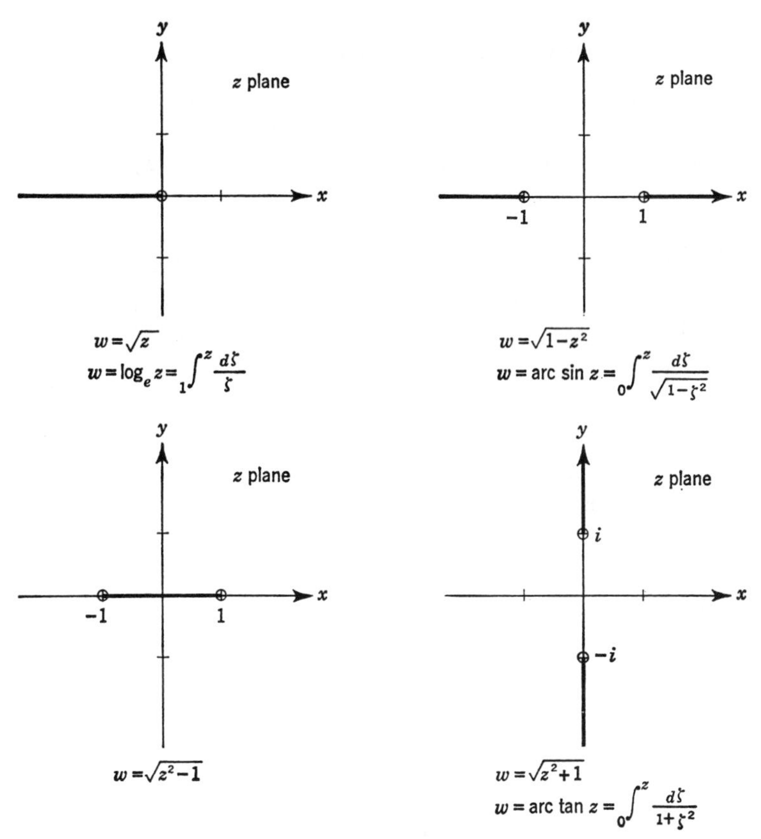

FIG. 7.4-1. Branch points and branch cuts for some elementary functions.

restriction to a single branch. The construction of Riemann surfaces for arbitrary functions may require considerable ingenuity (Refs. 7.8 and 7.15).

Throughout this handbook, statements about analytic functions refer to single-valued analytic functions or to single-valued branches of monogenic analytic functions unless specific reference to multiple-valued functions is made.

7.5. INTEGRAL THEOREMS AND SERIES EXPANSIONS

7.5-1. Integral Theorems. *Let z be a point inside the boundary contour C of a region D throughout which $f(z)$ is analytic, and let $f(z)$ be analytic on C. Then*

$$\oint_C f(\zeta)\,d\zeta = 0 \quad \text{(Cauchy-Goursat integral theorem)} \tag{7.5-1}$$

$$\begin{aligned} f(z) &= \frac{1}{2\pi i}\oint_C \frac{f(\zeta)}{\zeta - z}\,d\zeta \quad \text{(Cauchy's integral formula)} \\ f'(z) &= \frac{1}{2\pi i}\oint_C \frac{f(\zeta)}{(\zeta - z)^2}\,d\zeta \\ f''(z) &= \frac{2!}{2\pi i}\oint_C \frac{f(\zeta)}{(\zeta - z)^3}\,d\zeta \\ &\cdots\cdots\cdots \\ f^{(n)}(z) &= \frac{n!}{2\pi i}\oint_C \frac{f(\zeta)}{(\zeta - z)^{n+1}}\,d\zeta \end{aligned} \tag{7.5-2}$$

Figure 7.5-1 illustrates the application of the Cauchy-Goursat integral theorem (1) to *multiply connected* domains (see also Sec. 7.7-1).

Equation (2) yields $f(z)$ and its derivatives in terms of the boundary values of $f(z)$. Specifically,

$$\oint_C \frac{d\zeta}{\zeta - z} = 2\pi i \tag{7.5-3}$$

Note that $\oint_C \frac{f(\zeta)}{\zeta - z}\,d\zeta$ is an analytic function of z even if $f(z)$ is not analytic throughout D and only continuous on C; Eq. (1) holds if $f(z)$ is also analytic in D.

A continuous single-valued function $f(z)$ is analytic throughout the bounded region D if Eq. (1) *holds for every closed contour C in D and enclosing only points of D (Morera's Theorem).*

7.5-2. Taylor-series Expansion (see also Sec. 4.10-4). **(a)** *If $f(z)$ is analytic inside and on the circle K of radius r about $z = a$, then there exists a unique and uniformly convergent series expansion in powers of $(z - a)$*

$$f(z) = \sum_{k=0}^{\infty} a_k(z-a)^k \qquad (|z-a| \leq r,\ a \neq \infty)$$
$$\text{with} \qquad a_k = \frac{1}{k!} f^{(k)}(a) = \frac{1}{2\pi i} \oint_K \frac{f(\zeta)}{(\zeta - a)^{k+1}}\, d\zeta \qquad (7.5\text{-}4)$$

The largest circle K_c or $|z - a| = r_c$ all of whose interior points are inside the region where $f(z)$ is analytic is the **convergence circle** of the power series (4); r_c is the **radius of convergence** (Sec. 4.10-2*a*). A number of useful power-series expansions are tabulated in Sec. E-7.

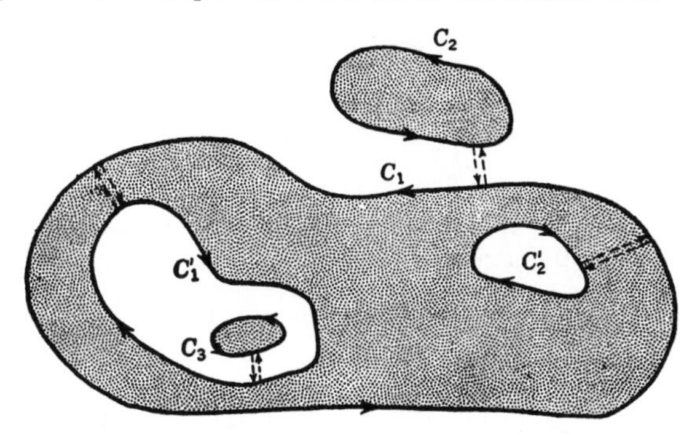

FIG. 7.5-1. Application of Cauchy's integral theorem to a multiply connected region of the z plane (Secs. 7.2-4 and 7.5-1). A region D bounded by exterior contours C_1, C_2, . . . and interior contours C_1', C_2', . . . is made simply connected by cuts (shown in broken lines). The integrals over each cut cancel, and *Cauchy's integral theorem becomes*

$$\sum_j \oint_{C_j} f(\zeta)\, d\zeta - \sum_k \oint_{C_k'} f(\zeta)\, d\zeta = 0$$

where all integrals are taken in the positive (counterclockwise) direction. The same technique applies if D is not bounded.

(b) If $M(r)$ is an upper bound of $|f(z)|$ on K, then

$$|a_n| = \frac{1}{n!} |f^{(n)}(a)| \leq \frac{M(r)}{r^n} \qquad \text{(CAUCHY'S INEQUALITY)} \qquad (7.5\text{-}5)$$

(c) If Taylor's series (4) is terminated with the term $a_{n-1}(z - a)^{n-1}$, the *remainder* $R_n(z)$ is given by

$$R_n(z) = \frac{(z-a)^n}{2\pi i} \oint_K \frac{f(\zeta)\, d\zeta}{(\zeta - a)^n (\zeta - z)} \qquad |R_n(z)| \leq \left(\frac{|z-a|}{r}\right)^n \frac{M(r)r}{r - |z-a|} \qquad (7.5\text{-}6)$$

7.5-3. Laurent-series Expansion. **(a)** *If $f(z)$ is analytic throughout the annular region between and on the concentric circles K_1 and K_2 centered* *at $z = a$ and of radii r_1 and $r_2 < r_1$, respectively, there exists a unique series expansion in terms of positive and negative powers of $(z - a)$,*

$$f(z) = \sum_{k=0}^{\infty} a_k(z-a)^k + \sum_{k=1}^{\infty} b_k(z-a)^{-k} \qquad (r_1 > |r - a| > r_2)$$

$$\text{with} \quad a_k = \frac{1}{2\pi i}\oint_{K_1} \frac{f(\zeta)\,d\zeta}{(\zeta - a)^{k+1}} \qquad b_k = \frac{1}{2\pi i}\oint_{K_2} (\zeta - a)^{k-1} f(\zeta)\,d\zeta \tag{7.5-7}$$

The first term of Eq. (7) is analytic and converges uniformly for $|z - a| \leq r_1$; the second term [**principal part** *of $f(z)$*] *is analytic and converges uniformly for $|z - a| \geq r_2$.*

NOTE: The case $a = \infty$ is treated by using the transformation $\bar{z} = 1/z$, which transforms $z = \infty$ into the origin.

(b) If the first term in Eq. (7) is terminated with $a_{n-1}(z - a)^{n-1}$, the remainder $R_n(z)$ is given by

$$\left.\begin{aligned} R_n(z) &= \frac{(z-a)^n}{2\pi i}\oint_{K_1} \frac{f(\zeta)\,d\zeta}{(\zeta-a)^n(\zeta - z)} \\ \text{with} \qquad |R_n(z)| &\leq \left(\frac{|z-a|}{r_1}\right)^n \frac{M(r_1)r_1}{r_1 - |z-a|} \end{aligned}\right\} \tag{7.5-8}$$

If the second term in Eq. (7) is terminated with $b_{n-1}(z - a)^{-(n-1)}$, the remainder $R'_n(z)$ is given by

$$\left.\begin{aligned} R'_n(z) &= \frac{1}{2\pi i(z-a)^n}\oint_{K_2} \frac{(\zeta-a)^n f(\zeta)}{(z-\zeta)}\,d\zeta \\ \text{with} \qquad |R'_n(z)| &\leq \left(\frac{r_2}{|z-a|}\right)^n \frac{M(r_2)r_2}{|z-a| - r_2} \end{aligned}\right\} \tag{7.5-9}$$

$M(r_1)$ and $M(r_2)$ are upper bounds of $|f(z)|$ on K_1 and K_2, respectively.

7.6. ZEROS AND ISOLATED SINGULARITIES

7.6-1. Zeros (see also Sec. 1.6-2). The points z for which $f(z) = 0$ are called the **zeros of** $f(z)$ [roots of $f(z) = 0$]. A function $f(z)$ analytic at $z = a$ has a **zero of order** m, where m is a positive integer, at $z = a$ if and only if the first m coefficients $a_0, a_1, a_2, \ldots, a_{m-1}$ in the Taylor-series expansion (7.5-4) of $f(z)$ about $z = a$ vanish, so that $f(z)(z - a)^{-m}$ is analytic and different from zero at $z = a$.

The zeros of a function $f(z)$ analytic throughout a region D are either isolated points [i.e., each has a neighborhood (Sec. 7.2-2) throughout which $f(z) \neq 0$ except at the zero itself], or $f(z)$ is equal to zero throughout D.

If $f_1(z)$ and $f_2(z)$ are analytic throughout a simply connected bounded open region D and on its boundary contour C, and if $|f_2(z)| < |f_1(z)| \neq 0$ on C, then $f_1(z)$ and $f_1(z) + f_2(z)$ have the same number of zeros in the region D (Rouché's Theorem). Every polynomial of degree n has n zeros, counting multiplicities (Fundamental Theorem of Algebra, see also Sec. 1.6-3).

7.6-2. Singularities. A **singular point** or **singularity** of the function $f(z)$ is any point where $f(z)$ is not analytic. The point $z = a$ is an **isolated singularity** of $f(z)$ if and only if there exists a real number $\delta > 0$ such that $f(z)$ is analytic for $0 < |z - a| < \delta$ but not for $z = a$. An isolated singularity of $f(z)$ at $z = a \neq \infty$ is

1. A **removable singularity** if and only if $f(z)$ is finite throughout a neighborhood of $z = a$, except possibly at $z = a$ itself; i.e., if and only if all coefficients b_k in the Laurent expansion (7.5-7) of $f(z)$ about $z = a$ vanish.
2. A **pole of order** m ($m = 1, 2, \ldots$) if and only if $(z - a)^m f(z)$ but not $(z - a)^{m-1} f(z)$ is analytic at $z = a$; i.e., if and only if

$$b_m \neq 0 \qquad b_{m+1} = b_{m+2} = \cdots = 0$$

   in the Laurent expansion (7.5-7) of $f(z)$ about $z = a$; or if and only if $1/f(z)$ is analytic and has a zero of order m at $z = a$. In this case, $\lim_{z \to a} |f(z)| = \infty$ no matter how z approaches $z = a$.
3. An **isolated essential singularity** if and only if the Laurent expansion (7.5-7) of $f(z)$ about $z = a$ has an infinite number of terms involving negative powers of $(z - a)$; $|f(z)|$ becomes indefinitely large as z approaches the value $z = a$ for some approach paths but not for others.*
4. A **branch point** if $f(z)$ is a multiple-valued function, and $z = q$ satisfies the conditions of Sec. 7.4-2.

If m branches of a function $f(z)$ join at a branch point $z = a$, that branch point is considered a point of continuity, a zero, a removable singularity, a pole, or an essential singularity of the "complex" $F(z)$ of the m branches in question if and only if the function

$$\Phi(\zeta) = F(\zeta^m + a) \tag{7.6-1}$$

has a point of continuity, a zero, a pole, or an essential singularity, respectively, at $\zeta = 0$; $F(z) = f(z)$ in a neighborhood of $z = a$, except that $F(z)$ can take only values of the m branches joining at $z = a$. Note that $f(z)$ may have other branches at $z = a$ which may or may not join at $z = a$ and whose behavior at $z = a$ may or may

* Sometimes the definition of an isolated essential singularity is extended to cover limit points of poles.

not be different from that of the m branches considered above. If the m branches joining at $z = a$ are the only branches of $f(z)$ at $z = a$, then $F(z) = f(z)$ at $z = a$.

The behavior of different branches of a multiple-valued function $f(z)$ at a point $z = a$ which is not a branch point $f(z)$ must be considered *independently for each branch.*

EXAMPLES: The function $f(z)$ defined by $f(z) = 0$ for $z = 0$, $f(z) = i$ for $z \neq 0$ has a removable singularity at $z = 0$. $\sin 1/z$ has a removable singularity at $z = 0$. $1/(z - 2)^3$ has a pole of order 3 at $z = 2$. $e^{1/z}$ has an essential singularity at $z = 0$. $\sqrt{z}$ has a branch point of order 1 at $z = 0$; this branch point is a zero of order 1 for all branches.

7.6-3. Zeros and Singularities at Infinity. $f(z)$ is analytic at infinity if and only if $f(1/z)$ is analytic at the origin. The point $z = \infty$ is a zero or a singularity of any of the types listed in Sec. 7.6-2 if $f(1/z)$ behaves correspondingly at the origin. The behavior of $f(z)$ at infinity may be investigated with the aid of the Laurent expansion of $f(1/z)$ about the origin.

7.6-4. Weierstrass's and Picard's Theorems. *Let $f(z)$ be a single-valued function having an isolated essential singularity at $z = a$. Then*

1. *For any complex number w, every neighborhood of $z = a$ contains a point z such that $|w - f(z)|$ is arbitrarily small (Weierstrass's Theorem).*
2. *Every neighborhood of $z = a$ contains an infinite set of points z such that $f(z) = w$ for every complex number w with the possible exception of a single value of w (Picard's Theorem).*

7.6-5. Integral Functions. An **integral (entire)** function $f(z)$ is a function whose only singularity is an isolated singularity at $z = \infty$. *If this singularity is a pole of order m, then $f(z)$ must be a polynomial (integral rational function) of degree m. A function $f(z)$ analytic for all values of z is a constant (Liouville's Theorem).*

An integral function $f(z)$ which is not a constant assumes every value w, except possibly one, at at least one point z, and, if $f(z)$ is not a polynomial, at infinitely many points z.

7.6-6. Product Expansion of an Integral Function. For every set of points z_0, z_1, z_2 . . . having no limit point (Sec. 4.3-6*a*) except possibly $z = \infty$, there exists an integral function $f(z)$ whose only zeros are zeros of given orders m_k at the points $z = z_k$. Let $z_0 = 0$, $z_k \neq 0$ $(k > 0)$; if there is no zero for $z = 0$, then $m_0 = 0$. Then $f(z)$ can be represented in the form

$$f(z) = z^{m_0} \prod_{k=1}^{\infty} \left\{ \left(1 - \frac{z}{z_k}\right) \exp\left[\frac{z}{z_k} + \frac{1}{2}\left(\frac{z}{z_k}\right)^2 + \cdots + \frac{1}{r_k}\left(\frac{z}{z_k}\right)^{r_k}\right]\right\}^{m_k} e^{g(z)} \quad (7.6\text{-}2)$$

where $g(z)$ is an arbitrary integral function, and the r_k are (finite) integers chosen so as to make the product converge uniformly throughout every bounded region (*Theorem of Weierstrass;* see Sec. 21.2-13 for examples of product expansions).

7.6-7. Meromorphic Functions. $f(z)$ is **meromorphic** throughout a region D if and only if its only singularities throughout D are poles. The number of such poles in any finite region D is necessarily finite. Many authors alternatively define a function as meromorphic if and only if its only singularities *throughout the finite portion of the plane* are poles.

Every function meromorphic throughout the finite portion of the plane can be expressed as the quotient of two integral functions without common zeros, and thus as the quotient of two products of the type discussed in Sec. 7.6-6. A function meromorphic throughout the entire plane is a *rational algebraic function* expressible as the quotient of two polynomials (see also Sec. 4.2-2c).

7.6-8. Partial-fraction Expansion of Meromorphic Functions (see also Sec. 1.7-4). *Let $f(z)$ be any function meromorphic in the finite portion of the plane and having poles with given principal parts (Sec. 7.5-3) $\sum_{j=1}^{m_k} b_{kj}(z - z_k)^{-j}$ at the points $z = z_k$ of a given finite or infinite set without limit points in the finite portion of the plane. Then it is possible to find polynomials $p_1(z)$, $p_2(z)$, . . . and an integral function $g(z)$ such that*

$$f(z) = \sum_k \left[\sum_{j=1}^{m_k} b_{kj}(z - z_k)^{-j} + p_k(z) \right] + g(z) \tag{7.6-3}$$

and the series converges uniformly in every bounded region where $f(z)$ is analytic (Mittag-Leffler's Theorem).

7.6-9. Zeros and Poles of Meromorphic Functions. *Let $f(z)$ be meromorphic throughout the bounded region inside and continuous on a closed contour C on which $f(z) \neq 0$. Let N be the number of zeros and P the number of poles of $f(z)$ inside C, respectively, where a zero or pole of order m is counted m times. Then*

$$\frac{1}{2\pi i} \oint_C \frac{f'(\zeta)}{f(\zeta)}\, d\zeta = N - P \tag{7.6-4}$$

For $P = 0$, Eq. (4) reduces to the *principle of the argument*

$$N = \frac{\Delta_C \vartheta}{2\pi} \tag{7.6-5}$$

where $\Delta_C \vartheta$ is the variation of the argument ϑ of $f(z)$ around the contour C.

Equation (5) means that $w = f(z)$ maps a moving point z describing the contour C once into a moving point w which encircles the w plane origin $N = 0, 1, 2, \ldots$ times if $f(z)$ has, respectively, 0, 1, 2, . . . zeros inside the contour C in the z plane. Equations (4) and (5) yield important criteria for locating zeros and poles of $f(z)$, such as the famous *Nyquist criterion* (Ref. 7.6).

7.7. RESIDUES AND CONTOUR INTEGRATION

7.7-1. Residues. Given a point $z = a$ where $f(z)$ is either analytic or has an isolated singularity, the **residue** $\text{Res}_f\,(a)$ of $f(z)$ at $z = a$ is the coefficient of $(z - a)^{-1}$ in the Laurent expansion (7.5-7), or

$$\text{Res}_f\,(a) = \frac{1}{2\pi i}\oint_C f(\zeta)\,d\zeta \tag{7.7-1a}$$

where C is any contour enclosing $z = a$ but no singularities of $f(z)$ other than $z = a$.

The residue $\text{Res}_f\,(\infty)$ of $f(z)$ at $z = \infty$ is defined as

$$\text{Res}_f\,(\infty) = \frac{1}{2\pi i}\oint f(\zeta)\,d\zeta \tag{7.7-1b}$$

where the integral is taken in the *negative* sense around any contour enclosing all singularities of $f(z)$ in the finite portion of the plane. Note that

$$\text{Res}_f\,(\infty) = \lim_{z\to\infty}\,[-zf(z)] \tag{7.7-2}$$

if the limit exists.

If $f(z)$ is either analytic or has a removable singularity at $z = a \neq \infty$, then $\text{Res}_f\,(a) = 0$ [see also Eq. (7.5-1)]. *If $z = a \neq \infty$ is a pole of order m, then*

$$\text{Res}_f\,(a) = \frac{1}{(m-1)!}\frac{d^{m-1}}{dz^{m-1}}(z-a)^m f(z)\bigg]_{z=a} \tag{7.7-3}$$

In particular, let $z = a \neq \infty$ be a simple pole of $f(z) \equiv p(z)/q(z)$, where $p(z)$ and $q(z)$ are analytic at $z = a$, and $p(a) \neq 0$. Then

$$\text{Res}_f\,(a) = \frac{p(a)}{q'(a)} \tag{7.7-4}$$

7.7-2. The Residue Theorem (see also Sec. 7.5-1). *For every simple closed contour C enclosing at most a finite number of (necessarily isolated) singularities $z_1, z_2, \ldots, z_n$ of a single-valued function $f(z)$ continuous on C,*

$$\frac{1}{2\pi i}\oint_C f(\zeta)\,d\zeta = \sum_{k=1}^{n}\text{Res}_f\,(z_k) \qquad \text{(RESIDUE THEOREM)} \tag{7.7-5}$$

One of the z_k may be the point at infinity. *Note carefully that the contour C must not pass through any branch cut* (see also Sec. 7.4-2).

7.7-3. Evaluation of Definite Integrals. (a) One can often evaluate a real definite integral $\int_a^b f(x)\,dx$ as a portion of a complex contour integral $\oint_C f(z)\,dz$ such that the contour C includes the interval (a, b) of the real axis. The residue theorem (5) may aid in such computations and may, in particular, relate the unknown integral to one that is already known. Figure 7.7-1 illustrates typical examples.

(b) To evaluate certain integrals of the form $\int_{-\infty}^{\infty} f(x)\,dx$ one applies Eq. (5) to a contour C comprising the interval $(-R, R)$ of the real axis and the arc S of the circle $|z| = R$ in the upper half plane. The following lemmas often yield the integral over S as $R \to \infty$:

1. $\lim_{R\to\infty} \int_S f(\zeta)\,d\zeta = 0$ *whenever the integral exists for all finite values of* R, *and* $zf(z)$ *tends uniformly to zero as* $|z| \to \infty$ with $y \geq 0$
2. *Jordan's Lemma: if* $F(z)$ *is analytic in the upper half plane, except possibly for a finite number of poles, and tends uniformly to zero as* $|z| \to \infty$ *with* $y \geq 0$, *then for every real number* m

$$\lim_{R\to\infty} \int_S F(\zeta)e^{im\zeta}\,d\zeta = 0$$

The contour-integration method may yield the Cauchy principal value (Sec. 4.6-2b) of $\int_{-\infty}^{\infty} f(x)\,dx$ even if the integral itself does not exist. Jordan's lemma is particularly useful for the computation of improper integrals of the form $\int_{-\infty}^{\infty} F(x)e^{imx}\,dx$ and, because of Eq. (21.2-28), integrals of the form $\int_{-\infty}^{\infty} F(x)\cos(mx)\,dx$ and $\int_{-\infty}^{\infty} F(x)\sin(mx)\,dx$ (inverse Laplace and Fourier transforms, Secs. 4.11-3 and 8.2-6).

(c) *If* S' *is any semicircular arc of the circle* $|z - a| = \epsilon$ *about a simple pole* $z = a$ *of* $f(z)$, *then*

$$\lim_{\epsilon\to 0} \int_{S'} f(\zeta)\,d\zeta = \pi i \operatorname{Res}_f(a)$$

This fact is used (1) to evaluate integrals over contours "indented" around simple poles and (2) for computing the Cauchy principal values of certain improper integrals.

(d) One may apply the residue theorem (5) to integrals of the type

$$\int_0^{2\pi} \Phi(\cos\varphi, \sin\varphi)\,d\varphi$$

where Φ is a rational function of $\cos\varphi$ and $\sin\varphi$, with the aid of the transformation

$$\left.\begin{aligned} z &= e^{i\varphi} & d\varphi &= \frac{1}{iz}\,dz \\ \cos\varphi &= \frac{1}{2}\left(z + \frac{1}{z}\right) & \sin\varphi &= \frac{1}{2i}\left(z - \frac{1}{z}\right) \end{aligned}\right\} \tag{7.7-6}$$

(e) **The Method of Steepest Descent (Saddle-point Method).** For a given or suitably deformed contour C such that $|f(z)|$ is small except for a pronounced

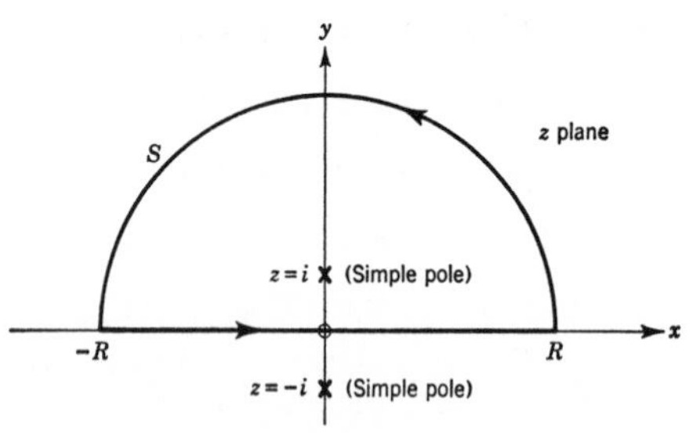

FIG. 7.7-1a. $\int_{-\infty}^{\infty} \frac{dx}{1+x^2} = \lim_{R\to\infty} \int_{-R}^{R} \frac{dx}{1+x^2} = \lim_{R\to\infty} \left[\oint_C - \int_S\right] \frac{dz}{1+z^2}$
$= 2\pi i \operatorname{Res}(z = i) - 0 = \pi$, using Lemma 1, Sec. 7.7-3b.

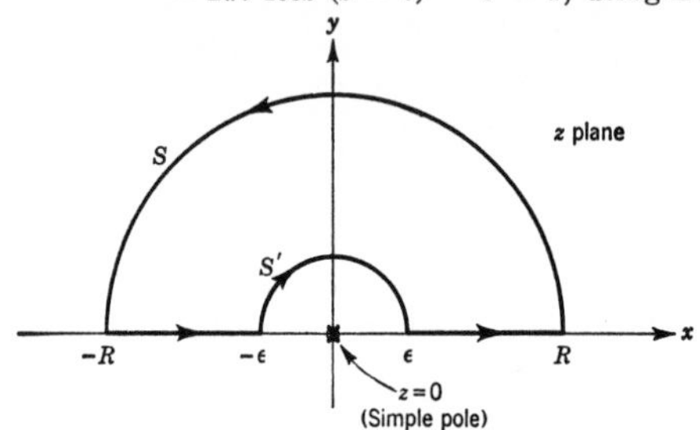

FIG. 7.7-1b. $\int_0^{\infty} \frac{\sin x}{x} dx = \frac{1}{2i} \int_{-\infty}^{\infty} \frac{e^{ix}}{x} dx = \frac{1}{2i} \lim_{\substack{\epsilon\to 0\\ R\to\infty}} \left[\oint_C - \int_S - \int_{S'}\right] \frac{e^{iz}}{z} dz$
$= \frac{1}{2i}[0 - 0 - (-i\pi)] = \frac{\pi}{2}$, using Jordan's Lemma and Sec. 7.7-3c.

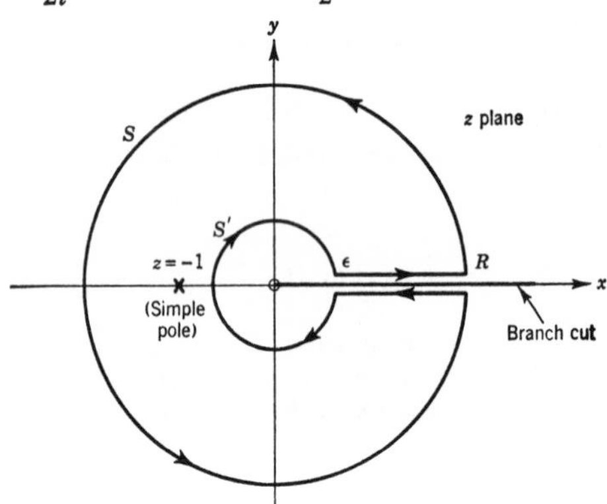

FIG. 7.7-1c. Given $0 < a < 1$,

$$\int_0^{\infty} \frac{x^{a-1}}{1+x} dx = \lim_{\substack{\epsilon\to 0\\ R\to\infty}} \left[\oint_C - \int_S - \int_{S'}\right] \frac{z^{a-1}}{1+z} dz + \int_0^{\infty} \frac{(xe^{2\pi i})^{a-1}}{1+x} dx$$

The integrals over S and S' go to zero, and the integral over C equals $2\pi i \operatorname{Res}(z = -1) = 2\pi i e^{\pi i(a-1)}$. Hence,

$$\int_0^{\infty} \frac{x^{a-1}}{1+x} dx = \frac{1}{1 - e^{2\pi i(a-1)}} \oint_C \frac{z^{a-1}}{1+z} dz = \frac{2\pi i e^{\pi i(a-1)}}{1 - e^{2\pi i(a-1)}} = \frac{\pi}{\sin a\pi} \qquad (0 < a < 1)$$

FIG. 7.7-1. Simple examples of contour-integral evaluation (Sec. 7.7-3; see Refs. 7.11, and 7.16 for more advanced problems).

maximum at $z = z_0$ with $f'(z_0) = 0$ one may attempt an approximation of the form

$$\int_C f(z)\,dz = f(z_0)\int_C \left(1 + \frac{(z - z_0)^2}{2!}\frac{f''(z_0)}{f(z_0)} + \cdots\right) dz$$

$$\approx \mp \sqrt{\frac{2f(z_0)^3}{-f''(z_0)}} \int e^{-s^2}\,ds \approx \mp \sqrt{\frac{2\pi f(z_0)^3}{-f''(z_0)}}$$

The method applies especially to integrals of the form $\int_{z_1}^{z_2} e^{ng(z)}\,dz$ (see also Ref. 6.4).

7.7-4. Use of Residues for the Summation of Series. Given a contour C enclosing the points $z = m, z = m + 1, z = m + 2, \ldots, z = n$, where m is an integer, let $f(z)$ be analytic inside and on C, except possibly for a number of poles $a_1, a_2, \ldots, a_N$ none of which coincide with $z = m, z = m + 1, z = m + 2, \ldots, z = n$. Then

$$\sum_{k=m}^{n} f(k) = \frac{1}{2\pi i}\oint_C \pi f(\zeta) \cot \pi\zeta\,d\zeta - \sum_{j=1}^{N} \text{Res}_g\,(a_j) \qquad (7.7\text{-}7)$$

where $\text{Res}_g\,(a_j)$ is the residue of $\pi f(z) \cot \pi z$ at $z = a_j$; and

$$\sum_{k=m}^{N} (-1)^k f(k) = \frac{1}{2\pi i}\oint_C \pi f(\zeta) \operatorname{cosec} \pi\zeta\,d\zeta - \sum_{j=1}^{n} \text{Res}_g\,(a_j) \qquad (7.7\text{-}8)$$

where $\text{Res}_g\,(a_j)$ is the residue of $\pi f(z) \operatorname{cosec} \pi z$ at $z = a_j$. It is frequently possible to chose the contour C so that the integral on the right of Eq. (7) or (8) vanishes.

7.8. ANALYTIC CONTINUATION

7.8-1. Analytic Continuation and Monogenic Analytic Functions (see also Secs. 7.4-1 to 7.4-3). **(a)** Given a single-valued function $f_1(z)$ defined and analytic throughout a region D_1, the function $f_2(z)$ defined and analytic throughout a region D_2 is an **analytic continuation** of $f_1(z)$ if and only if the intersection of D_1 and D_2 contains a simply connected open region D_C where $f_1(z)$ and $f_2(z)$ are identical.

The analytical continuation $f_2(z)$ is uniquely defined by the values of $f_1(z)$ in D_C (see also Sec. 7.3-3). Moreover, *analytic continuations of $f_1(z)$ satisfy every functional equation and, in particular, every differential equation satisfied by $f_1(z)$ (principle of conservation of functional equations).* One can thus use $f_2(z)$ to extend the region of definition of $f_1(z)$, and conversely: $f_1(z)$ and $f_2(z)$ are regarded as **elements** of a single analytic function $f(z)$ defined throughout D_1 and D_2. $f_1(z)$ and/or $f_2(z)$ may be capable of further analytic continuation leading to additional elements of $f(z)$.

(b) Multiple-valued Functions. Note that $f_1(z)$ and $f_2(z)$ are *not* necessarily identical throughout the entire intersection of their regions of definition. $f_2(z)$ may have an analytic continuation $f_3(z)$ defined on D_1 but *not* identical with $f_1(z)$. *Analytic continuation can yield elements* *belonging to different branches of a multiple-valued analytic function* $f(z)$; *the two values* $f(z_0)$ *obtained by analytic continuation of* $f_1(z)$ *along two different routes* C, C' *are identical if* C *and* C' *do not enclose a branch point of* $f(z)$.

(c) The possible analytic continuations of a given element constitute a **monogenic analytic function** $f(z)$ defined, except at isolated singularities, throughout the plane or on a *connected* region with **natural boundaries.** Whereas the choice of the successive elements defining $f(z)$ is not fixed, *the principle of conservation of functional equations applies to all elements, and any one element uniquely defines all branches, isolated singularities, and natural boundaries of* $f(z)$.

7.8-2. Methods of Analytic Continuation. **(a)** The *standard method of analytic continuation* starts with a function $f(z)$ defined by its power-series expansion (7.5-4) inside some circle $|z - a| = r$. For any point $z = b$ inside the circle, $f(b)$, $f'(b)$, . . . are then known and yield a Taylor-series expansion about $z = b$. The new power series converges inside a circle $|z - b| = r'$ which intersects the first circle but may contain a region not inside the first circle. This process may be continued up to the natural boundaries of the function; each power series is an element of $f(z)$.

NOTE: *A function defined as a power series with a finite radius of convergence has at least one singularity on the circle of convergence.*

(b) *Two functions* $f_1(z)$ *and* $f_2(z)$ *defined and analytic throughout the respective open regions* D_1 *and* D_2 *separated by a contour arc* C *are analytic continuations of each other (elements of the same monogenic analytic function) if they are equal and uniformly continuous on* C.

(c) The Principle of Reflection. *Let* $f(z)$ *be defined and analytic throughout a region* D *intersected by the real axis, and let* $f(z)$ *be real for real* z. *Then, for every value of* a *in* D, $f(z)$ *is defined and analytic at* $z = a^*$, *and*

$$f(a^*) = f^*(a) \tag{7.8-1}$$

More generally, *let* $f_1(z)$ *be defined and analytic throughout a region* D_1 *bounded in part by a straight-line segment* S_z, *where* $f_1(z)$ *is continuous; and let* $w = f_1(z)$ *map* S_z *onto a corresponding straight-line segment* S_w *in the* w *plane. Then the function* $w = f_2(z)$ *mapping the reflection in* S_z *of every point* z *in* D *into the reflection of* $w = f_1(z)$ *in* S_w *is an analytic continuation of* $f_1(z)$.

7.9. CONFORMAL MAPPING

7.9-1. Conformal Mapping. **(a)** A function $w = f(z)$ maps points of the z plane (or Riemann surface, Sec. 7.4-3) into corresponding points of the w plane (or Riemann surface). *At every point z such that $f(z)$ is analytic and $f'(z) \neq 0$ the mapping $w = f(z)$ is* **conformal**; *i.e., the angle between two curves through such a point is reproduced in magnitude and sense by the angle between the corresponding curves in the w plane.*

Infinitely small triangles around such points z are mapped onto similar infinitely small triangles in the w plane; each triangle side is "stretched" in the ratio $|f'(z)|:1$ and rotated through the angle $\arg f'(z)$. The **superficial magnification** (local magnification of small areas) due to the mapping $w = f(z) = u(x, y) + iv(x, y)$ is

$$|f'(z)|^2 = \frac{\partial(u, v)}{\partial(x, y)} = \begin{vmatrix} \frac{\partial u}{\partial x} & \frac{\partial u}{\partial y} \\ \frac{\partial v}{\partial x} & \frac{\partial v}{\partial y} \end{vmatrix} \tag{7.9-1}$$

at every point z where the mapping is conformal. A conformal mapping transforms the lines x = constant, y = constant into families of mutually *orthogonal trajectories* in the w plane. Similarly, the lines $u(x, y)$ = constant, $v(x, y)$ = constant correspond to orthogonal trajectories in the z plane (see also Table 7.2-1).

A region of the z plane mapped onto the entire w plane by $w = f(z)$ is called a **fundamental region** of the function $f(z)$. Points where $f'(z) = 0$ are called **critical points** of the transformation $w = f(z)$.* A mapping which preserves the magnitude but not necessarily the sense of the angle between two curves is called **isogonal** (*example of an isogonal but not conformal mapping: $w = z^*$*).

(b) The mapping $w = f(z)$ is **conformal at infinity** if and only if $w = f(1/\bar{z}) = F(\bar{z})$ maps the origin $\bar{z} = 0$ conformally into the w plane. Two curves are said to intersect at an angle γ at $z = \infty$ if and only if the transformation $\bar{z} = 1/z$ results in two corresponding curves intersecting at an angle γ at $\bar{z} = 0$. Similarly, $w = f(z)$ maps the point $z = a$ conformally into $w = \infty$ if and only if $\bar{w} = 1/f(z)$ maps $z = a$ conformally into the origin $\bar{w} = 0$ (see also Sec. 7.2-3).

7.9-2. Bilinear Transformations. **(a)** A **bilinear transformation (linear fractional transformation, Moebius transformation)**

$$w = \frac{az + b}{cz + d} \qquad (bc - ad \neq 0) \tag{7.9-2}$$

establishes a reciprocal one-to-one correspondence between the points of the z plane and the points of the w plane. In particular, each of the two **invariant points**

$$z = \frac{1}{2c}[(a - d) \pm \sqrt{(a - d)^2 + 4bc}] \tag{7.9-3}$$

(which may or may not be distinct) is mapped onto itself. The mapping is conformal everywhere except at the point $z = -d/c$, which corresponds

* Some authors also refer to the (singular) points where $1/f'(z) = 0$ as critical points of $f(z)$.

to $w = \infty$. *Straight lines and circles in the z plane correspond to straight lines or circles in the w plane, and conversely;* in this connection, every straight line is regarded as a circle of infinite radius through the point at infinity.

For any bilinear transformation mapping the four points z_1, z_2, z_3, z, respectively, into w_1, w_2, w_3, w

$$\frac{z_1 - z}{z_3 - z} \bigg/ \frac{z_1 - z_2}{z_3 - z_2} = \frac{w_1 - w}{w_3 - w} \bigg/ \frac{w_1 - w_2}{w_3 - w_2} \tag{7.9-4}$$

(*invariance of the* **cross ratio** *or* **anharmonic ratio***). Equation (4) defines the *unique bilinear transformation mapping three given points z_1, z_2, z_3, respectively, into three given points w_1, w_2, w_3. There exists a bilinear transformation which will transform a given circle or straight line in the z plane into a given circle or straight line in the w plane* (see also Table 7.9-2 and Sec. 7.10-1).

SPECIAL CASES. The transformation

$$w = Az + B \tag{7.9-5}$$

where A and B are complex numbers, corresponds to a *rotation* through the angle arg A together with a *stretching* or *contraction* by a factor $|A|$, followed by a *translation* through the vector displacement B. *The linear transformation* (5) *is the most general conformal mapping which preserves similarity of geometrical figures.*

The transformation

$$w = 1/z \tag{7.9-6}$$

represents a *geometrical inversion* of the point z with respect to the unit circle about the origin, followed by a *reflection* in the real axis. *The transformation* (6) *maps*

1. *Straight lines through the origin into straight lines through the origin*
2. *Circles through the origin into straight lines which do not contain the origin, and conversely*
3. *Circles which do not pass through the origin into circles which do not pass through the origin*

(**b**) *The bilinear transformations* (2) *constitute a group; inverses and products of bilinear transformations are bilinear transformations* (Sec. 12.2-7). Every bilinear transformation (2) may be expressed as the result (product) of three successive simpler bilinear transformations (see also Sec. 7.9-2*a*):

$$z' = z + d/c \qquad \text{(TRANSLATION)} \tag{7.9-7a}$$

$$z'' = 1/z' \qquad \text{(INVERSION and REFLECTION)} \tag{7.9-7b}$$

$$w = \frac{bc - ad}{c^2} z'' + \frac{a}{c} \qquad \text{(ROTATION and STRETCHING, followed by TRANSLATION)} \tag{7.9-7c}$$

* The cross ratio (4) is real (and admits a geometrical interpretation, Ref. 7.2) if and only if the points z_1, z_2, z_3, z (and hence also the points w_1, w_2, w_3, w) lie on a circle or straight line.

7.9-3. The Transformation $w = \frac{1}{2}(z + 1/z)$. The transformation

$$w = \frac{1}{2}\left(z + \frac{1}{z}\right) \tag{7.9-8a}$$

is equivalent to

$$\frac{w-1}{w+1} = \left(\frac{z-1}{z+1}\right)^2 \qquad \text{or} \qquad z = w + \sqrt{w^2 - 1} \tag{7.9-8b}$$

or

$$u = \frac{1}{2}\left(|z| + \frac{1}{|z|}\right)\cos\varphi \qquad v = \frac{1}{2}\left(|z| - \frac{1}{|z|}\right)\sin\varphi \tag{7.9-8c}$$

The transformation (8) is conformal except at the critical points $z = 1$ and $z = -1$. Both the exterior and the interior of the unit circle $|z| = 1$ are mapped onto the entire plane with the exception of the straight-line segment $(u = -1, u = 1)$, which corresponds to the unit circle $|z| = 1$ itself. Some important properties of this transformation are outlined in Table 7.9-1.

7.9-4. The Schwarz-Christoffel Transformation. The **Schwarz-Christoffel transformation**

$$\left.\begin{aligned} w &= A\int (z - x_1)^{-\frac{\alpha_1}{\pi}}(z - x_2)^{-\frac{\alpha_2}{\pi}} \cdots (z - x_n)^{-\frac{\alpha_n}{\pi}}\,dz + B \\ \text{with} \qquad & \sum_{j=1}^{n} \alpha_j = 2\pi \end{aligned}\right\} \tag{7.9-9}$$

maps the upper half-plane $y > 0$ conformally onto the interior of a polygon in the w plane; the polygon corresponds to the x axis, the vertices $w_1, w_2, \ldots, w_n$ correspond to different points $x_1, x_2, \ldots, x_n$ on the x axis, and the *exterior* angle of the polygon at the vertex w_j equals α_j $(j = 1, 2, \ldots, n)$. For any given polygon in the w plane, three of the x_j's can be chosen arbitrarily; the other x_j's and the parameters A and B are then uniquely determined.

If x_n is chosen to be infinitely large Eq. (9) reduces to

$$\left.\begin{aligned} w &= A'\int (z - x_1')^{-\frac{\alpha_1}{\pi}}(z - x_2')^{-\frac{\alpha_2}{\pi}} \cdots (z - x_{n-1}')^{-\frac{\alpha_{n-1}}{\pi}}\,dz + B' \\ \text{with} \qquad & \sum_{j=1}^{n} \alpha_j = 2\pi \end{aligned}\right\} \tag{7.9-10}$$

where A' and B' are constant parameters, and $x_1', x_2', \ldots, x_{n-1}'$ are new points on the x axis.

Table 7.9-1. Properties of the Transformation $w = \frac{1}{2}\left(z + \frac{1}{z}\right)$

(see also Tables 7.9-2 and 7.10-1)

$z = x + iy = |z|e^{i\varphi}$ $(|z| \geq 0)$; $w = u + iv = re^{i\vartheta}$ $(|w| \geq 0)$

Point, curve(s), or region in z plane	Point, curve(s), or region in w plane	Remarks
Circles about 0 $\lvert z\rvert = e^{\psi} = \text{constant} \neq 1$	Ellipses, foci at ± 1 $\dfrac{u^2}{\cosh^2 \psi} + \dfrac{v^2}{\sinh^2 \psi} = 1$	If φ increases, ϑ increases for $\lvert z\rvert > 1$ and ϑ decreases for $\lvert z\rvert < 1$
Straight-line rays through 0 $\varphi = \text{constant} \neq 0$	Hyperbolas, foci at ± 1 $\dfrac{u^2}{\cos^2 \varphi} - \dfrac{v^2}{\sin^2 \varphi} = 1$	
Unit semicircle $\lvert z\rvert = 1$, $y \geq 0$	Straight-line segment $(+1, -1)$	If φ increases, u decreases
Unit semicircle $\lvert z\rvert = 1$, $y \leq 0$	Straight-line segment $(-1, +1)$	If φ increases, u increases
Straight-line segments of the real axis $y = 0$, specifically the intervals $x = -\infty$ to $x = -1$ $x = -1$ to $x = 0$ $x = 0$ to $x = +1$ $x = +1$ to $x = +\infty$	Straight-line segments of the real axis $v = 0$, specifically $u = -\infty$ to $u = -1$ $u = -1$ to $u = -\infty$ $u = +\infty$ to $u = +1$ $u = +1$ to $u = +\infty$	
"Streamlines" for flow around unit circle with velocity ½ at $z = \infty$ Corresponding lines of constant velocity potential	$v = \text{constant}$ $u = \text{constant}$	

The actual determination of the x_j or x_j' is quite complicated, except in certain "degenerate" cases where one or more of the angles α_j vanish (see also Ref. 7.4). Application of the Schwarz-Christoffel transformation to *parallelograms* and *rectangles* in the w plane yields elliptic functions z of w (Sec. 21.6-1).

The transformations 26 to 30 in Table 7.9-2 are special cases of the Schwarz-Christoffel transformation.

7.9-5. Table of Transformations. Table 7.9-2 illustrates a number of special transformations of interest in various applications.

Table 7.9-2. Table of Transformations of Regions*

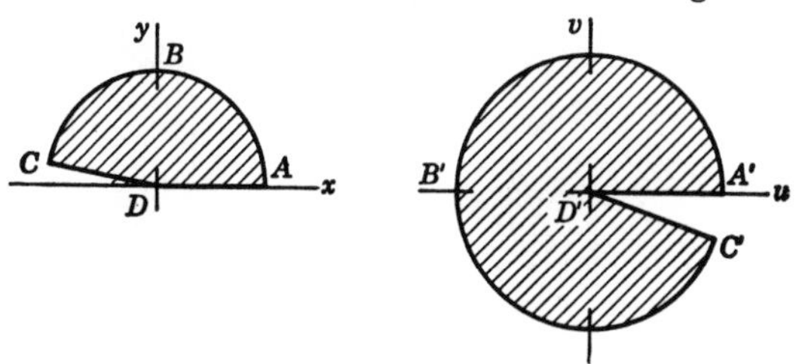

FIG. 1. $w = z^2$.

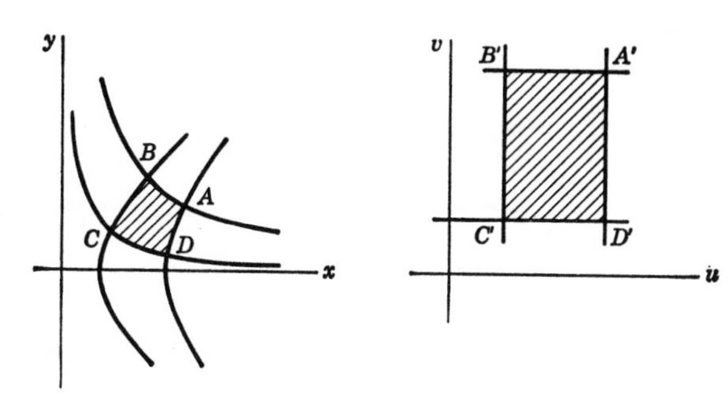

FIG. 2. $w = z^2$.

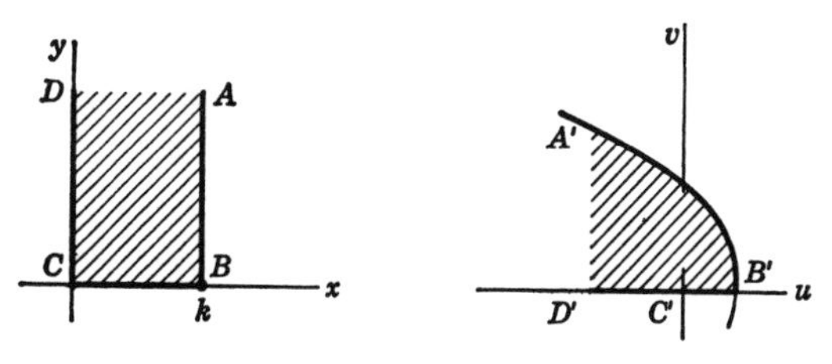

FIG. 3. $w = z^2$;

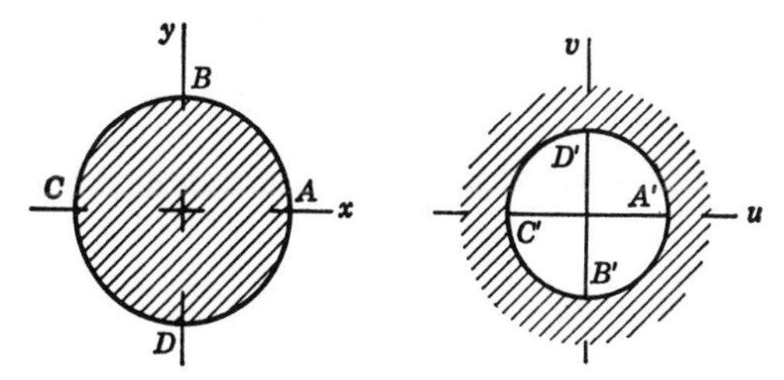

FIG. 4. $w = 1/z$.

* From R. V. Churchill, *Introduction to Complex Variables and Applications*, 2d ed., McGraw-Hill, New York, 1960.

Table 7.9-2. Table of Transformations of Regions (*Continued*)

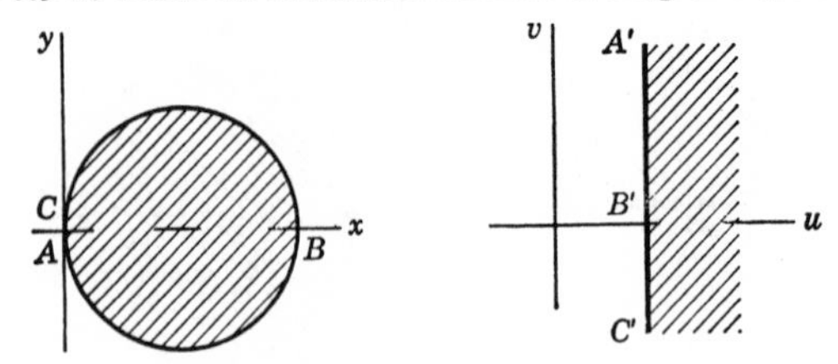

FIG. 5. $w = 1/z$.

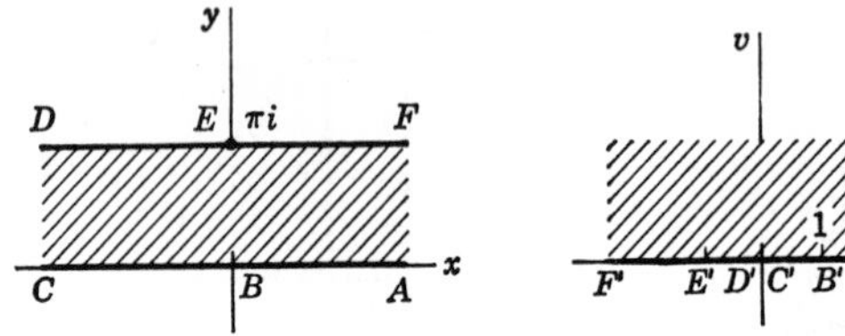

FIG. 6. $w = e^z$.

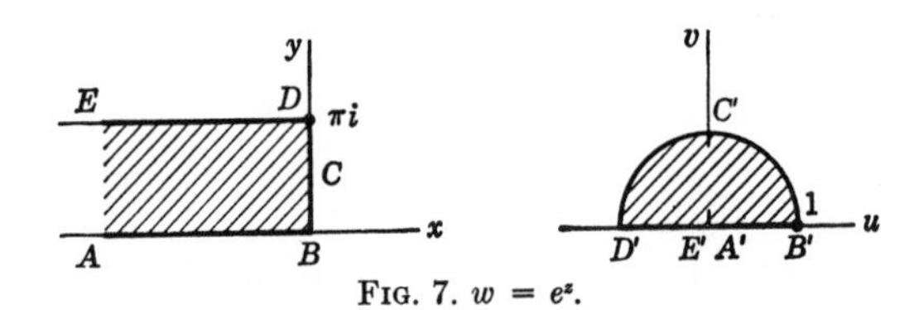

FIG. 7. $w = e^z$.

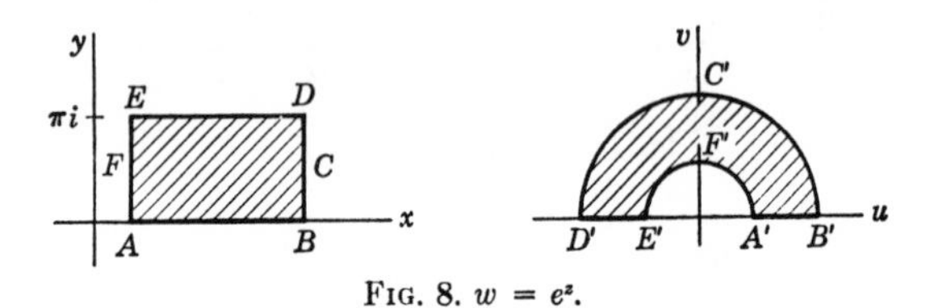

FIG. 8. $w = e^z$.

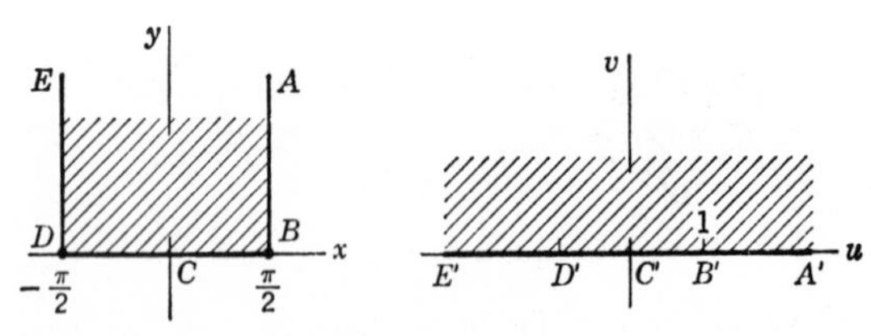

FIG. 9. $w = \sin z$.

Table 7.9-2. Table of Transformations of Regions (*Continued*)

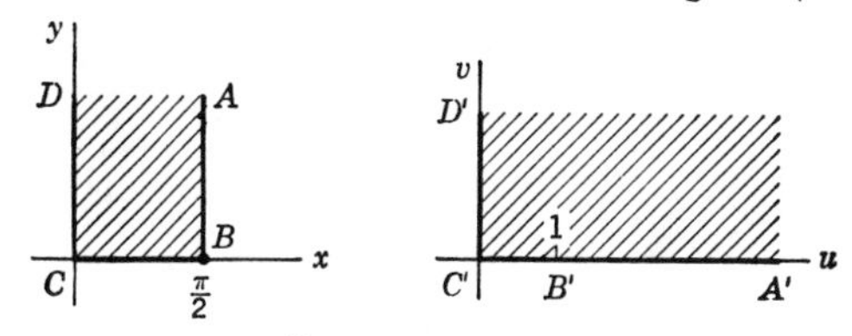

FIG. 10. $w = \sin z$.

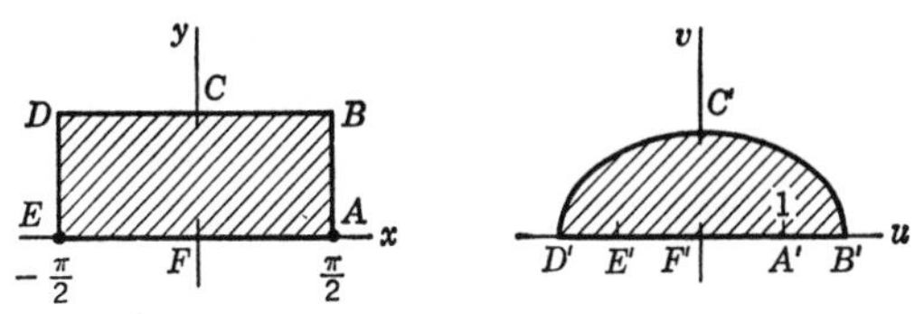

FIG. 11. $w = \sin z$; $BCD: y = k$, $B'C'D'$ is $\left(\frac{u}{\cosh k}\right)^2 + \left(\frac{v}{\sinh k}\right)^2 = 1$.

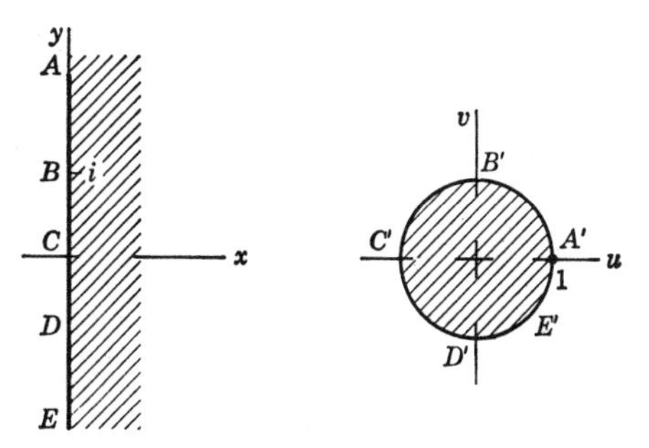

FIG. 12. $w = \dfrac{z - 1}{z + 1}$.

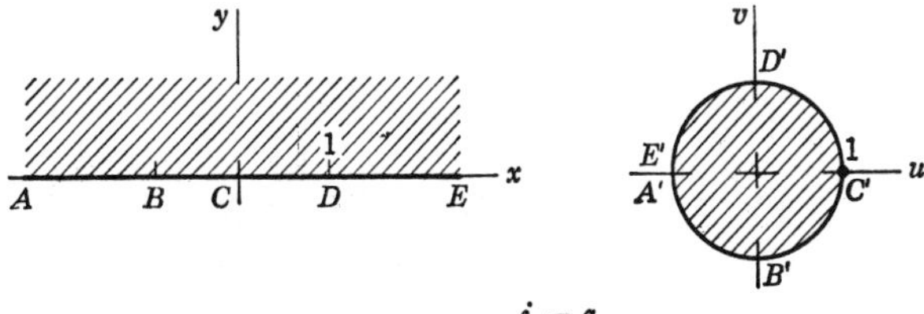

FIG. 13. $w = \dfrac{i - z}{i + z}$.

Table 7.9-2. Table of Transformations of Regions (*Continued*)

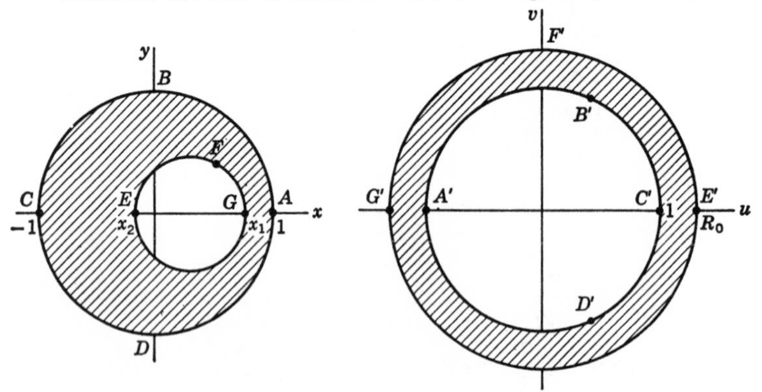

FIG. 14. $w = \frac{z-a}{az-1}$; $a = \frac{1 + x_1x_2 + \sqrt{(1-x_1^2)(1-x_2^2)}}{x_1 + x_2}$;

$R_0 = \frac{1 - x_1x_2 + \sqrt{(1-x_1^2)(1-x_2^2)}}{x_1 - x_2}$ ($a > 1$ and $R_0 > 1$ when $-1 < x_2 < x_1 < 1$).

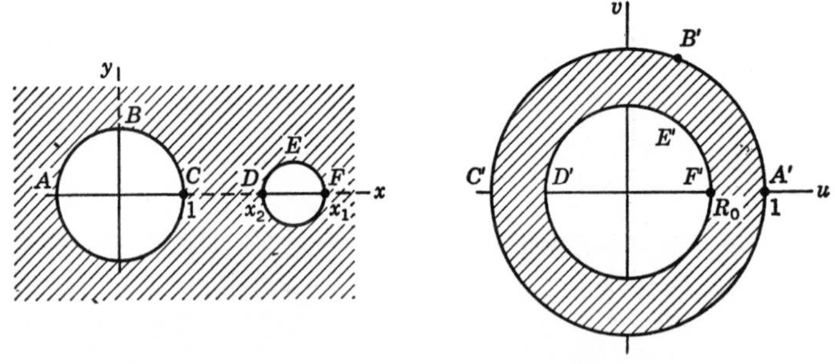

FIG. 15. $w = \frac{z-a}{az-1}$; $a = \frac{1 + x_1x_2 + \sqrt{(x_1^2-1)(x_2^2-1)}}{x_1 + x_2}$;

$R_0 = \frac{x_1x_2 - 1 - \sqrt{(x_1^2-1)(x_2^2-1)}}{x_1 - x_2}$ ($x_2 < a < x_1$ and $0 < R_0 < 1$ when $1 < x_2 < x_1$).

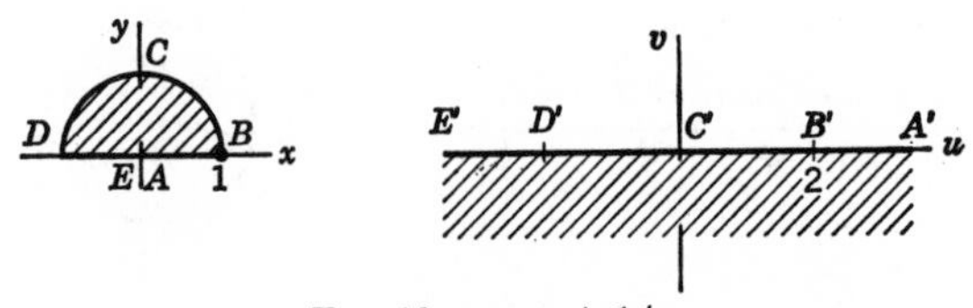

FIG. 16. $w = z + 1/z$.

Table 7.9-2. Table of Transformations of Regions (*Continued*)

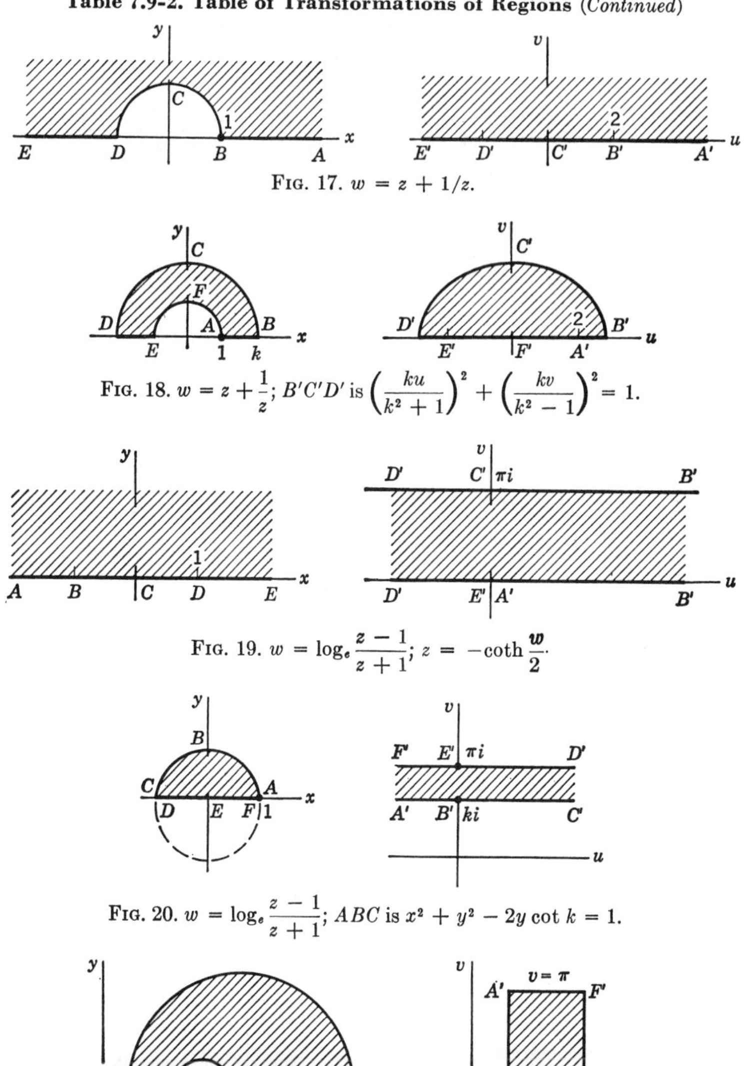

FIG. 17. $w = z + 1/z$.

FIG. 18. $w = z + \frac{1}{z}$; $B'C'D'$ is $\left(\frac{ku}{k^2 + 1}\right)^2 + \left(\frac{kv}{k^2 - 1}\right)^2 = 1$.

FIG. 19. $w = \log_e \frac{z - 1}{z + 1}$; $z = -\coth \frac{w}{2}$.

FIG. 20. $w = \log_e \frac{z - 1}{z + 1}$; ABC is $x^2 + y^2 - 2y \cot k = 1$.

FIG. 21. $w = \log_e \frac{z + 1}{z - 1}$; centers of circles at $z = \coth c_n$, radii: $\operatorname{csch} c_n (n = 1, 2)$.

Table 7.9-2. Table of Transformations of Regions (*Continued*)

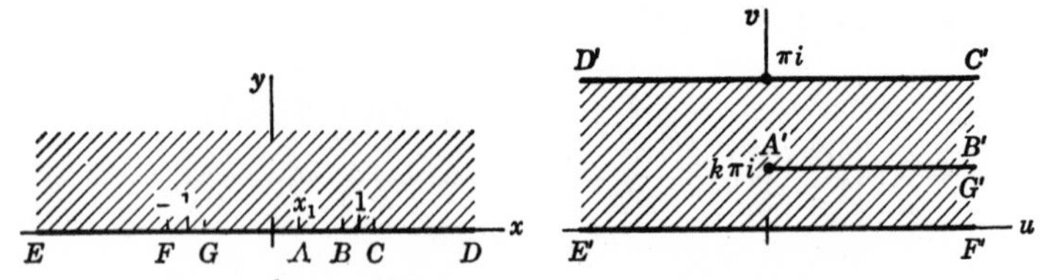

Fig. 22. $w = k \log_e \frac{k}{1-k} + \log_e 2(1-k) + i\pi - k \log_e (z+1)$
$- (1-k) \log_e (z-1)$; $x_1 = 2k - 1$.

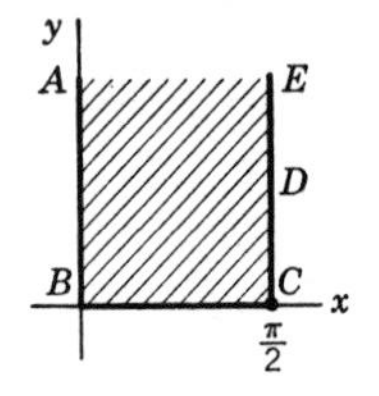

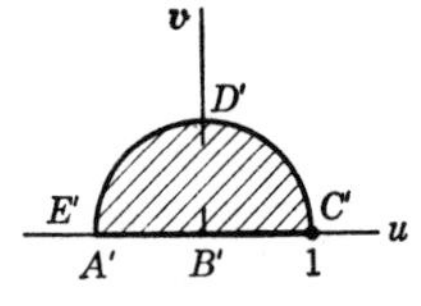

Fig. 23. $w = \tan^2 \frac{z}{2} = \frac{1 - \cos z}{1 + \cos z}$.

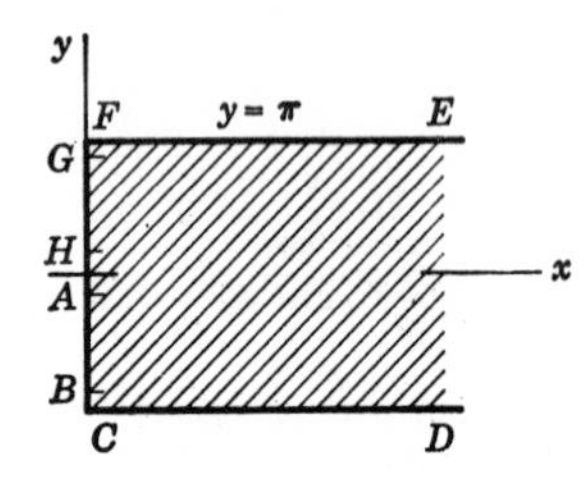

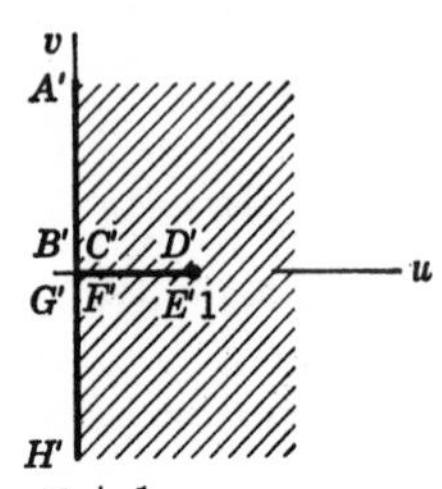

Fig. 24. $w = \coth \frac{z}{2} = \frac{e^z + 1}{e^z - 1}$.

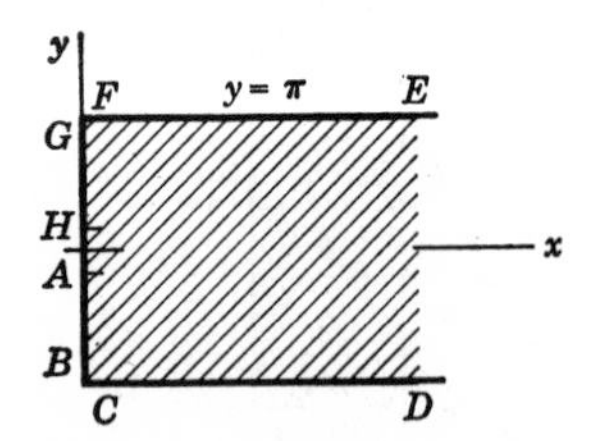

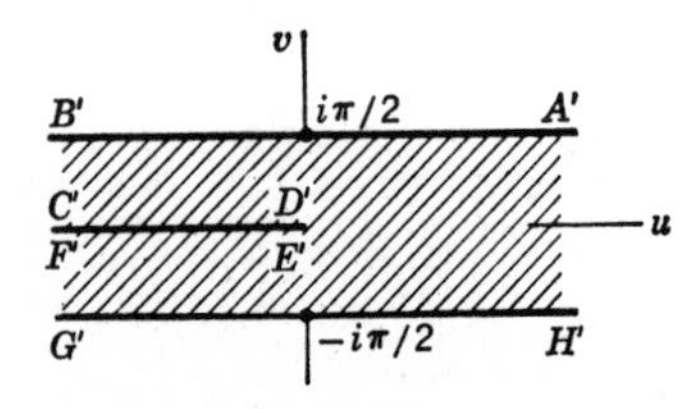

Fig. 25. $w = \log_e \coth \frac{z}{2}$.

Table 7.9-2. Table of Transformations of Regions (*Continued*)

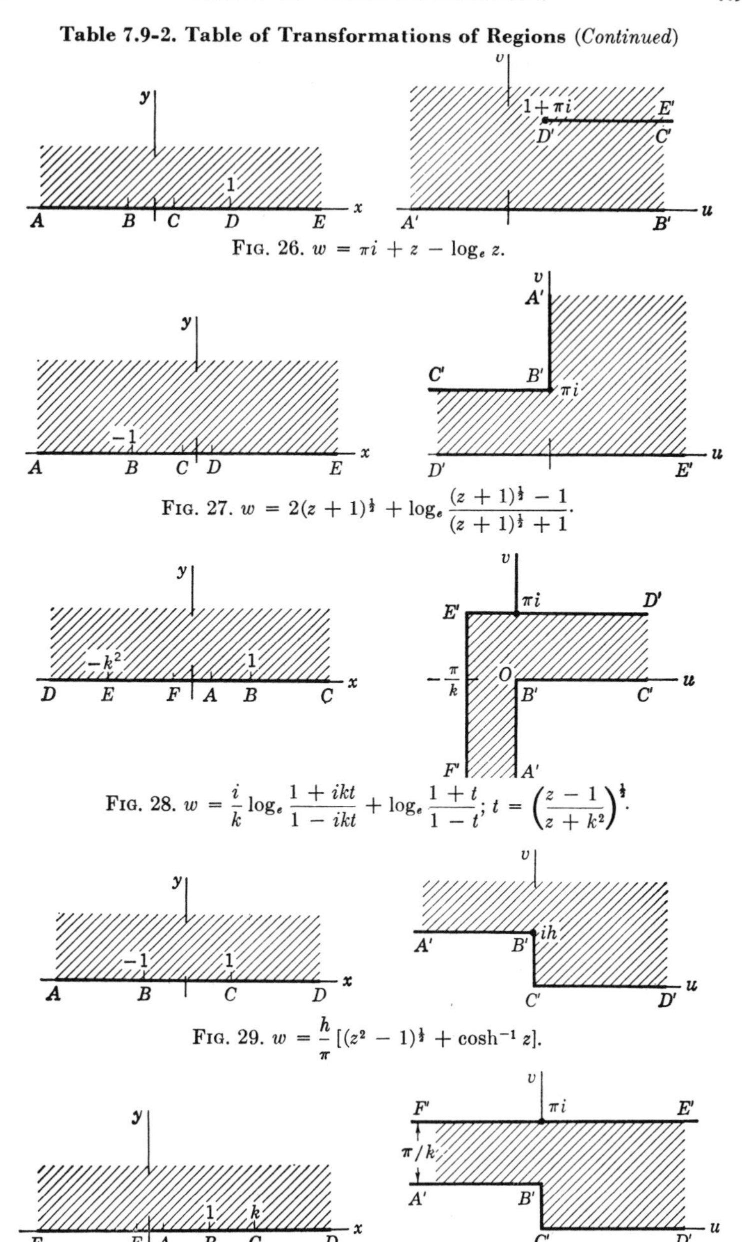

FIG. 26. $w = \pi i + z - \log_e z$.

FIG. 27. $w = 2(z+1)^{\frac{1}{2}} + \log_e \dfrac{(z+1)^{\frac{1}{2}} - 1}{(z+1)^{\frac{1}{2}} + 1}$.

FIG. 28. $w = \dfrac{i}{k} \log_e \dfrac{1+ikt}{1-ikt} + \log_e \dfrac{1+t}{1-t}; \; t = \left(\dfrac{z-1}{z+k^2}\right)^{\frac{1}{2}}$.

FIG. 29. $w = \dfrac{h}{\pi}\left[(z^2-1)^{\frac{1}{2}} + \cosh^{-1} z\right]$.

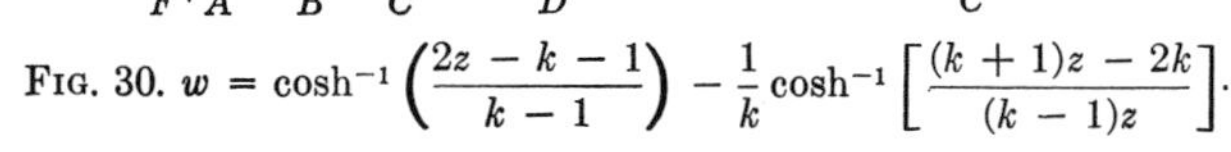

FIG. 30. $w = \cosh^{-1}\left(\dfrac{2z-k-1}{k-1}\right) - \dfrac{1}{k}\cosh^{-1}\left[\dfrac{(k+1)z-2k}{(k-1)z}\right]$.

Table 7.10-1. Table of Transformations Mapping a Specified Region D Conformally onto the Unit Circle ($|w| \leq 1$)

$z = x + iy = |z|e^{i\varphi} \qquad w = u + iv = |w|e^{i\vartheta}$

	Region D in the z plane	Transformation	Remarks
1	Upper half plane $y \geq 0$	$w = e^{i\lambda}\dfrac{z-a}{z-a^*}$ (λ real)	
2	Right half plane $x \geq 0$	$w = e^{b\lambda}\dfrac{z-a}{z+a^*}$ (λ real)	$z = a$ is transformed into the origin $w = 0$
3	Unit circle $\lvert z\rvert \leq 1$	$w = e^{b\lambda}\dfrac{z-a}{a^*z-1}$ (λ real)	
4	Strip of width π $0 \leq x \leq \pi,$ $-\infty \leq y \leq \infty$	$\dfrac{w-1}{w+1} = ie^{iz}$	Special case of the Schwarz-Christoffel transformation (7.9-9)
5	Sector of unit circle $\lvert z\rvert \leq 1,\ 0 \leq \varphi \leq \pi\alpha$	$w = \dfrac{(1+z^{1/\alpha})^2 - i(1-z^{1/\alpha})^2}{(1+z^{1/\alpha})^2 + i(1-z^{1/\alpha})^2}$	
6	z plane cut from $z = 0$ to $z = \infty$ along the positive real axis	$w = \dfrac{\sqrt{z}-i}{\sqrt{z}+i}$	
7	Region outside and on the ellipse $\dfrac{x^2}{a^2} + \dfrac{y^2}{b^2} = 1$	$z = \dfrac{1}{2}\left[(a-b)w + \dfrac{a+b}{w}\right]$	See also Sec. 7.9-3
8	Region outside and on the parabola $\lvert z\rvert \cos^2\dfrac{\varphi}{2} = 1$	$z = \left(\dfrac{2}{w+1}\right)^2$	
9	Region inside and on the parabola $\lvert z\rvert \cos^2\dfrac{\varphi}{2} = 1$	$w = \tan^2\left(\dfrac{\pi}{4}\sqrt{z}\right)$	
10	Semicircle $\lvert z\rvert \leq R,\ x > 0$	$w = i\,\dfrac{z^2 + 2Rz - R^2}{z^2 - 2Rz - R^2}$	
11	Regions bounded by straight lines (polygons)	Combine a Schwarz-Christoffel transformation (Sec. 7.9-4) with transformation 1 above	

7.10. FUNCTIONS MAPPING SPECIFIED REGIONS ONTO THE UNIT CIRCLE

7.10-1. Riemann's Mapping Theorem. *For every simply connected open region D in the z plane, with the exception of the entire z plane and the entire z plane minus one point, there exists a conformal mapping $w = f(z)$ which establishes a reciprocal one-to-one (biunique) correspondence between all points of D and all interior points of the unit circle $|w| = 1$. The analytic function $f(z)$ is uniquely determined if the mapping of a point in D and a direction through that point is specified.* If D is bounded by a regular curve (Sec. 3.1-13) C, then $f(z)$ is continuous on C and establishes a reciprocal one-to-one correspondence between all points of C and all points on the unit circle $|w| = 1$.

NOTE: (1) The transformation thus specified *defines* the analytic function $f(z)$; and (2) with the trivial exceptions noted above, it is possible to map every region D bounded by a simple contour conformally onto any other region D' bounded by a simple contour. The problem of mapping a region D conformally onto the unit circle is closely related to the solution of Dirichlet's boundary-value problem for the region D (Sec. 15.6-9).

Frequently, a desired conformal mapping can be obtained through successive relatively simple transformations.

Table 7.10-1 lists a number of transformations mapping a given region D conformally onto the unit circle.

7.11. RELATED TOPICS, REFERENCES, AND BIBLIOGRAPHY

7.11-1. Related Topics. The following topics related to the study of functions of a complex variable are treated in other chapters of this handbook:

Complex numbers .... Chap. 1
Roots of polynomials .... Chap. 1
Functions, limits, differentiation, integration, infinite series .... Chap. 4
Laplace transformation .... Chap. 8
Two-dimensional potential theory, conjugate harmonic functions .... Chap. 15
Special functions .... Chap. 21

7.11-2. References and Bibliography.

7.1. Ahlfors, L. V.: *Complex Analysis*, 2d ed., McGraw-Hill, New York, 1966.
7.2. Bieberbach, L.: *Einführung in die Konforme Abbildung*, De Gruyter, Berlin, 1927.
7.3. Carrier, G. F., et al.: *Functions of a Complex Variable*, McGraw-Hill, New York, 1966.
7.4. Churchill, R. V.: *Complex Variables and Applications*, 2d ed., McGraw-Hill, New York, 1960.
7.5. Copson, E. T.: *Theory of Functions of a Complex Variable*, Oxford, New York, 1960.

7.6. Cunningham, J.: *Complex-variable Methods in Science and Technology*, Van Nostrand, Princeton, N.J., 1965.
7.7. Hille, E.: *Analytic Function Theory*, 2 vols., Blaisdell, New York, 1959/62.
7.8. Hurwitz, A., and R. Courant: *Allgemeine Funktionentheorie und elliptische Funktionen*, Springer, Berlin, 1964.
7.9. Knopp, K.: *Funktionentheorie*, translated by F. Bagemihl, Dover, New York, 1947.
7.10. Kober, H.: *Dictionary of Conformal Representations*, Dover, New York, 1952.
7.11. McLachlan, N. W.: *Complex Variable and Operational Calculus with Applications*, Macmillan, New York, 1946.
7.12. Nehari, Z.: *Conformal Mapping*, McGraw-Hill, New York, 1952.
7.13. ———: *Introduction to Complex Analysis*, Allyn and Bacon, Boston, 1961.
7.14. Pennisi, L. L.: *Elements of Complex Variables*, Holt, New York, 1963.
7.15. Springer, G.: *Introduction to Riemann Surfaces*, Addison-Wesley, Reading, Mass., 1957.
7.16. Whittaker, E. T., and G. N. Watson: *A Course in Modern Analysis*, Cambridge, New York, 1958.

CHAPTER 8

THE LAPLACE TRANSFORMATION AND OTHER FUNCTIONAL TRANSFORMATIONS

8.1. INTRODUCTION

8.1-1. The Laplace transformation (Sec. 8.2-1) associates a unique function $F(s)$ of a complex variable s with each suitable function $f(t)$ of a real variable t. This correspondence is essentially reciprocal one-to-one for most practical purposes (Sec. 8.2-8); corresponding pairs of functions $f(t)$ and $F(s)$ can often be found by reference to tables. The Laplace transformation is defined so that many relations between, and operations on, the functions $f(t)$ correspond to simpler relations between, and operations on, the functions $F(s)$ (Secs. 8.3-1 to 8.3-4). This applies particularly to the solution of differential and integral equations. It is, thus, often useful to transform a given problem involving functions $f(t)$ into an equivalent problem expressed in terms of the associated Laplace transforms $F(s)$ ("operational calculus" based on Laplace transformations or "transformation calculus," Secs. 9.3-7, 9.4-5, and 10.5-2).

8.2. THE LAPLACE TRANSFORMATION

8.2-1. Definition. The **(one-sided) Laplace transformation**

$$\boxed{\begin{aligned} F(s) \equiv \mathcal{L}[f(t)] &\equiv \int_0^\infty f(t)e^{-st}\,dt \\ &\equiv \lim_{\substack{a\to 0\\ b\to\infty}} \int_a^b f(t)e^{-st}\,dt \qquad (0 < a < b) \end{aligned}} \tag{8.2-1}$$

associates a unique **result or image function** $F(s)$ of the complex variable $s = \sigma + i\omega$ with every single-valued **object or original function** $f(t)$ (t real) such that the improper integral (1) exists. $F(s)$ is called the **(one-sided) Laplace transform** of $f(t)$. The more explicit notation $\mathcal{L}[f(t); s]$ is also used.

8.2-2. Absolute Convergence. *The Laplace transform* (1) *exists for* $\sigma \geq \sigma_0$, *and the improper integral converges absolutely and uniformly* (*Sec.* 4.6-2*a*) *to a function* $F(s)$ *analytic* (*Sec.* 7.3-3) *for* $\sigma > \sigma_0$ *if*

$$\int_0^\infty |f(t)|e^{-\sigma t}\,dt = \lim_{\substack{a\to 0\\ b\to\infty}} \int_a^b |f(t)|e^{-\sigma t}\,dt \qquad (0 < a < b) \qquad (8.2\text{-}2)$$

exists for $\sigma = \sigma_0$. The greatest lower bound σ_a of the real numbers σ_0 for which this is true is called the **abscissa of absolute convergence** of the Laplace transform $\mathfrak{L}[f(t)]$.

Although certain theorems relating to Laplace transforms require only the existence (simple convergence) of the transforms, **the existence of an abscissa of absolute convergence will be implicitly assumed throughout the following sections.** Wherever necessary, it is customary to specify the region of absolute convergence associated with a relation involving Laplace transforms by writing $\sigma > \sigma_a$ to the right of the relation in question, as in Eq. (3).

8.2-3. Extension of the Region of Definition. The region of definition of the analytic function

$$F(s) = \mathfrak{L}[f(t)] \qquad (\sigma > \sigma_a) \qquad (8.2\text{-}3)$$

can usually be extended by analytic continuation (Secs. 7.8-1 and 7.8-2) so as to include the entire s plane with the exception of singular points (Sec. 7.6-2) situated to the left of the abscissa of absolute convergence. Such an extension of the region of definition is implied wherever necessary.

8.2-4. Sufficient Conditions for the Existence of the Laplace Transform. *The Laplace transform* $\mathfrak{L}[f(t)]$ *defined by Eq.* (1) *exists in the sense of absolute convergence* (*Sec.* 8.2-2; *see also Sec.* 4.9-3)

1. *For* $\sigma \geq 0$ *if* $\int_0^\infty |f(t)|\,dt$ *exists*
2. *For* $\sigma > \sigma_a$ *if the* (*possibly improper*) *integral*

$$I = \int_0^{t_1} |f(t)|\,dt$$

exists for every finite $t_1 > 0$, *and* $|f(t)|e^{-\sigma_a t}$ *is uniformly bounded for* $t > t_2 \geq 0$ [$f(t)$ *is of exponential order or* $O(e^{\sigma_a t})$ *as* $t \to \infty$, *Sec.* 4.4-3]

3. *For* $\sigma > 0$ *or* $\sigma > \sigma_a$ (*whichever bound is greater*) *if* $\mathfrak{L}[f'(t)]$ *exists* (*not necessarily in the sense of absolute convergence*) *for* $\sigma > \sigma_a$

8.2-5. Inverse Laplace Transformation. The **inverse Laplace transform** $\mathfrak{L}^{-1}[F(s)]$ of a (suitable) function $F(s)$ of the complex variable $s = \sigma + i\omega$ is a function $f(t)$ whose Laplace transform (1) is $F(s)$. *Not every function* $F(s)$ *has an inverse Laplace transform.*

8.2-6. The Inversion Theorem. *Given* $F(s) = \mathfrak{L}[f(t)]$, $\sigma > \sigma_a$, *then throughout every open interval where* $f(t)$ *is of bounded variation* [*Sec.* 4.4-8*b*; *e.g.*, *an open interval where* $f(t)$ *is bounded and has a finite number of maxima, minima, and discontinuities*]

$$f_I(t) = \frac{1}{2\pi i} \lim_{R\to\infty} \int_{\sigma_1 - iR}^{\sigma_1 + iR} F(s)e^{st}\,ds = \begin{cases} \tfrac{1}{2}[f(t-0) + f(t+0)] & \text{for } t > 0 \\ \tfrac{1}{2}f(0+0) & \text{for } t = 0 \\ 0 & \text{for } t < 0 \end{cases} \quad (\sigma_1 > \sigma_a) \tag{8.2-4a}$$

In particular, for every $t > 0$ where $f(t)$ is continuous

$$f_I(t) = \frac{1}{2\pi i} \lim_{R\to\infty} \int_{\sigma_1 - iR}^{\sigma_1 + iR} F(s)e^{st}\,ds = f(t) \qquad (\sigma_1 > \sigma_a) \tag{8.2-4b}$$

The path of integration in Eq. (4) lies to the right of all singularities of $F(s)$. The **inversion integral** $f_I(t)$ reduces to $\frac{1}{2\pi i}\int_{\sigma_1 - i\infty}^{\sigma_1 + i\infty} F(s)e^{st}\,ds$ if the integral exists; otherwise, $f_I(t)$ is a Cauchy principal value (Sec. 4.6-2b).

8.2-7. Existence of the Inverse Laplace Transform. Note carefully that the existence of the limit (4) for a given function $F(s)$ does *not* in itself imply that $F(s)$ has an inverse Laplace transform [EXAMPLE: $F(s) = e^{s^2}$]. The existence of $\mathcal{L}^{-1}[F(s)]$ should be *checked* [e.g., by Eq. (1)] for every application of the inversion theorem. *The following theorems state sufficient (not necessary) conditions for the existence of $\mathcal{L}^{-1}[F(s)]$.*

1. *If $F(s)$ is analytic for $\sigma \geq \sigma_a$ and of order < -1 (Sec. 4.4-3), then $\mathcal{L}^{-1}[F(s)]$ exists, is continuous for all t and $O(e^{\sigma_a t})$ as $t \to \infty$, and the corresponding abscissa of absolute convergence is σ_a. Note that, under these conditions, Eq. (4b) holds, and $\mathcal{L}^{-1}[F(s)] = 0$ for $t \leq 0$.*
2. *Given $F(s) \equiv \varphi[F_1(s), F_2(s), \ldots, F_n(s)]$ such that $\varphi(z_1, z_2, \ldots, z_n)$ is equal to zero and analytic with respect to each z_k for $z_1 = z_2 = \cdots = z_n = 0$, and*

$$F_k(s) = \mathcal{L}[f_k(t)] \qquad (\sigma > \sigma_{ak};\ k = 1, 2, \ldots, n)$$

then $\mathcal{L}^{-1}[F(s)]$ exists, and the corresponding Laplace transformation possesses an abscissa of absolute convergence.

8.2-8. Uniqueness of the Laplace Transform and Its Inverse. *The Laplace transform (1) is unique for each function $f(t)$ having such a transform. Conversely, two functions $f_1(t)$ and $f_2(t)$ possessing identical Laplace transforms are identical for all $t > 0$, except possibly on a set of measure zero (Sec. 4.6-14); $f_1(t) = f_2(t)$ for all $t > 0$ where both functions are continuous (Lerch's Theorem).* Thus, $f(t)$ is uniquely defined by its Laplace transform for almost all $t > 0$ (Sec. 4.6-14b); a given function $F(s)$ cannot have more than one inverse Laplace transform continuous for all $t > 0$.

Different discontinuous functions may have the same Laplace transform. In particular, the generalized unit step function (see also Sec. 21.9-1) defined by $f(t) = 0$ for $t < 0$, $f(t) = 1$ for $t > 0$ has the Laplace transform $1/s$ regardless of the value assigned to $f(t)$ for $t = 0$.

8.3. CORRESPONDENCE BETWEEN OPERATIONS ON OBJECT AND RESULT FUNCTIONS

8.3-1. Table of Corresponding Operations. Table 8.3-1 lists a number of theorems each establishing a correspondence between an operation on a function $f_1(t)$ and an operation on its Laplace transform $F_1(s)$, and vice versa. These theorems are the basis of Laplace-transform techniques for the simplified representation of operations (operational calculus based on the use of Laplace transforms).

8.3-2. Laplace Transforms of Periodic Object Functions and Amplitude-modulated Sinusoids. (a) *If $f(t)$ is* **periodic with period T** *(Sec. 4.2-2b), and $\int_0^T |f(t)|\,dt$ exists, then*

$$\mathcal{L}[f(t)] = \frac{1}{1 - e^{-Ts}} \int_0^T f(t) e^{-st}\,dt \qquad (\sigma > 0) \tag{8.3-1}$$

NOTE: The integral on the right is an integral function (Sec. 7.6-5), so that $\mathcal{L}[f(t)]$ has no singularities for finite s except possibly for simple poles on the imaginary axis.

(b) *If $f(t)$* is **antiperiodic (Sec. 4.2-2b) with period** T, and $\int_0^{T/2} |f(t)|\,dt$ *exists, then*

$$f\left(t + \frac{T}{2}\right) \equiv -f(t) \qquad \text{and} \qquad \mathcal{L}[f(t)] = \frac{1}{1 + e^{-Ts/2}} \int_0^{T/2} f(t) e^{-st}\,dt \qquad (\sigma > 0) \tag{8.3-2}$$

(c) *If $\varphi_1(t)$ is the result of* **ideal half-wave rectification** *of the antiperiodic function $f(t)$ where both $f(t)$ and $\varphi_1(t)$ are positive for $0 < t < T/2$, then*

$$\varphi_1 = \begin{Bmatrix} f(t) & \text{for } f(t) > 0 \\ 0 & \text{for } f(t) \leq 0 \end{Bmatrix} \qquad \text{and} \qquad \mathcal{L}[\varphi_1(t)] = \frac{1}{1 - e^{-Ts/2}} \mathcal{L}[f(t)] \qquad (\sigma > 0) \tag{8.3-3}$$

(d) *If φ_2 is the result of* **ideal full-wave rectification** *of the antiperiodic function $f(t)$ where both $f(t)$ and $\varphi_2(t)$ are positive for $0 < t < T/2$, then*

$$\varphi_2 = \begin{Bmatrix} f(t) & \text{for } f(t) > 0 \\ -f(t) & \text{for } f(t) \leq 0 \end{Bmatrix} \qquad \text{and} \qquad \mathcal{L}[\varphi_2(t)] = \coth \frac{Ts}{4} \mathcal{L}[f(t)] \qquad (\sigma > 0) \tag{8.3-4}$$

(e) **Transform of an Amplitude-modulated Sinusoid.** *If $F(s)$ is the Laplace transform of $f(t)$, then*

$$\begin{aligned} \mathcal{L}[f(t) \sin \omega_1 t] &\equiv \frac{1}{2i}[F(s - i\omega_1) - F(s + i\omega_1)] \\ \mathcal{L}[f(t) \cos \omega_1 t] &= \tfrac{1}{2}[F(s - i\omega_1) + F(s + i\omega_1)] \end{aligned} \tag{8.3-5}$$

Table 8.3-1. Theorems Relating Corresponding Operations on Object and Result Functions

The following theorems are valid whenever the Laplace transforms $F(s) = \mathcal{L}[f(t)]$ in question exist in the sense of absolute convergence; limits are assumed to exist (see also Secs. 8.3-2 and 8.7-3)

Theorem number	Operation	Object function	Result function
1	**Linearity** (α, β constant)	$\alpha f_1(t) + \beta f_2(t)$	$\alpha F_1(s) + \beta F_2(s)$
	Differentiation of object function*		
2a	. . . if $f'(t)$ exists for all $t > 0$	$f'(t)$	$sF(s) - f(0+0)$
2b	. . . if $f^{(r)}(t)$ exists for all $t > 0$	$f^{(r)}(t)$ $(r = 1, 2, \ldots)$	$s^rF(s) - s^{r-1}f(0+0) - s^{r-2}f'(0+0) - \cdots - f^{(r-1)}(0+0)$
2c	. . . if $f(t)$ is bounded for $t > 0$, and $f'(t)$ exists for $t > 0$ except for $t = t_1, t_2, \ldots$ where $f(t)$ has unilateral limits	$f'(t)$	$sF(s) - f(0+0) - \sum_k e^{-t_ks}[f(t_k+0) - f(t_k-0)]$
3	**Integration of object function** . . . if $f'(t)$ exists for $t > 0$	$\int_0^t f(\tau)\,d\tau + C$	$\frac{F(s)}{s} + \frac{C}{s}$
4	**Change of scale**	$f(at)$ $(a > 0)$	$\frac{1}{a}F\left(\frac{s}{a}\right)$
5	**Translation (shift) of object function** . . . if $f(t) = 0$ for $t \leq 0$	$f(t-b)$ $(b \geq 0)$	$e^{-bs}F(s)$
6	**Convolution of object functions†**	$f_1 * f_2 \equiv \int_0^\infty f_1(\tau)f_2(t-\tau)\,d\tau \equiv f_2 * f_1$	$F_1(s)F_2(s)$

Table 8.3-1. Theorems Relating Corresponding Operations on Object and Result Functions (*Continued*)

Theorem number	Operation	Object function	Result function
7	**Corresponding limits of object and result function (continuity theorem;** α is independent of t and s)	$\lim_{\alpha \to a} f(t, \alpha)$	$\lim_{\alpha \to a} F(s, \alpha)$
8a	**Differentiation and integration with respect to a parameter α independent of t and s**	$\frac{\partial}{\partial \alpha} f(t, \alpha)$	$\frac{\partial}{\partial \alpha} F(s, \alpha)$
8b		$\int_{a_1}^{a_2} f(t, \alpha)\, d\alpha$	$\int_{a_1}^{a_2} F(s, \alpha)\, d\alpha$
9a	**Differentiation of result function**	$-tf(t)$	$F'(s)$
9b		$(-1)^r t^r f(t)$	$F^{(r)}(s)$
10	**Integration of result function** (path of integration situated to the right of the abscissa of absolute convergence)	$\frac{1}{t} f(t)$	$\int_s^{\infty} F(s)\, ds$
11	**Translation of result function**	$e^{at}f(t)$	$F(s - a)$

* The abscissa of absolute convergence for $\mathfrak{L}[f^{(r)}(t)]$ is 0 or σ_a, whichever is greater.

† The existence of $f_1 * f_2$ is assumed; absolute convergence of $\mathfrak{L}[f_1(t)]$ and $\mathfrak{L}[f_2(t)]$ is a sufficient condition for the absolute convergence of $\mathfrak{L}[f_1 * f_2]$. See also Sec. 8.3-3.

8.3-3. Transform of a Product (Convolution Theorem). *Given*

$$F_1(s) \equiv \mathcal{L}[f_1(t)] \quad (\sigma > \sigma_1) \qquad F_2(s) \equiv \mathcal{L}[f_2(t)] \quad (\sigma > \sigma_2)$$

let $\int_0^\infty e^{-2\sigma_1 t}[f_1(t)]^2\,dt$ and $\int_0^\infty e^{-2\sigma_2 t}[f_2(t)]^2\,dt$ *exist. Then, for* $\sigma > \sigma_1$, $\sigma > \sigma_2$, $\sigma > \sigma_1 + \sigma_2$,

$$\mathcal{L}[f_1(t)f_2(t)] \equiv \frac{1}{2\pi i}\int_{\lambda-i\infty}^{\lambda+i\infty} F_1(z)F_2(s-z)\,dz \qquad (\sigma_2 < \lambda < \sigma - \sigma_1) \tag{8.3-6}$$

$$\mathcal{L}[f_1(t)f_2(t)] = \frac{1}{2\pi i}\int_{\lambda-i\infty}^{\lambda+i\infty} F_2(z)F_1(s-z)\,dz \qquad (\sigma_1 < \lambda < \sigma - \sigma_2) \tag{8.3-7}$$

8.3-4. Limit Theorems. *If $F(s)$ is the Laplace transform of $f(t)$ and $\mathcal{L}[f'(t)]$ exists, then*

$$\lim_{s\to\infty} sF(s) = f(0+0) \tag{8.3-8}$$

if the limit on the left exists. If, in addition to the first two conditions, $sF(s)$ is analytic for $\sigma \geq 0$, then

$$\lim_{s\to 0} sF(s) = \lim_{t\to\infty} f(t) \tag{8.3-9}$$

(see also Table 8.3-1,2).

8.4. TABLES OF LAPLACE-TRANSFORM PAIRS AND COMPUTATION OF INVERSE LAPLACE TRANSFORMS

8.4-1. Tables of Laplace-transform Pairs. A number of Laplace-transform pairs are tabulated in Appendix D for reference. In particular, Table D-6 lists a number of Laplace-transform pairs with rational algebraic result functions. Appendix D also shows how tables of Fourier-transform pairs may be used to obtain certain Laplace-transform pairs, and vice versa.

8.4-2. Computation of Inverse Laplace Transforms. Sections 8.4-3 to 8.4-9 describe various procedures for finding the object function $f(t)$ corresponding to a result function $F(s)$ obtained in the course of a problem solution (see also Secs. 9.3-7, 9.4-5, and 10.5-2).

NOTE: Unless the existence of $\mathcal{L}^{-1}[F(s)]$ is definitely known, any results obtained through the use of the inversion integral (8.2-4) should be checked by means of Eq. (8.2-1). Particular caution is recommended in connection with the use of series expansions for $\mathcal{L}^{-1}[F(s)]$; seemingly straightforward asymptotic or even convergent expansions may not be valid unless $F(s)$ and $f(t)$ satisfy certain restrictive conditions. A number of sufficient (but not necessary) conditions for the validity of series expansions of $\mathcal{L}^{-1}[F(s)]$ are presented in Secs. 8.4-6 to 8.4-9. In many problems, a function $f(t)$ "suspected" to be $\mathcal{L}^{-1}[F(s)]$ can be tested, e.g., by resubstitution into a differential equation, so that heuristic methods for finding $\mathcal{L}^{-1}[F(s)]$ may be quite useful.

8.4-3. Use of Contour Integration. Values of the contour integral (8.2-4) may frequently be obtained with the aid of the residue theorem (Secs. 7.7-1 to 7.7-3) and Jordan's lemma (Sec. 7.7-3*b*). If $F(s)$ is a multiple-valued function, the contours used must not cross any branch cuts of $F(s)$ (Sec. 7.4-2; see Refs. 8.5 and 8.10).

8.4-4. Inverse Laplace Transforms of Rational Algebraic Functions: Heaviside Expansion. (a) *If $F(s)$ is a rational algebraic function expressible as the ratio of two polynomials in s,*

$$F(s) = \frac{D_1(s)}{D(s)} \tag{8.4-1}$$

such that the degree of the polynomial $D(s)$ is higher than that of the polynomial $D_1(s)$, then $\mathcal{L}^{-1}[F(s)]$ equals the sum of the residues (Sec. 7.7-1) of $F(s)e^{st}$ at all the singular points (poles) of $F(s)$. To compute the inverse Laplace transform $\mathcal{L}^{-1}[F(s)]$, first find the roots s_k of $D(s) = 0$ [which determine the poles of $F(s)$] by one of the methods described in Secs. 1.8-1 to 1.8-6 or 20.2-2 and 20.2-3. Then

1. If $D(s) \equiv a_0(s - s_1)(s - s_2) \cdots (s - s_n)$ [all roots of $D(s) = 0$ are simple roots], then

$$\mathcal{L}^{-1}[F(s)] \equiv \mathcal{L}^{-1}\left[\frac{D_1(s)}{D(s)}\right] = \sum_{k=1}^{n} \frac{D_1(s)}{D'(s_k)} e^{s_k t} \qquad (t > 0)$$

2. If $D(s) \equiv a_0(s - s_1)^{m_1}(s - s_2)^{m_2} \cdots (s - s_n)^{m_n}$,

$$\mathcal{L}^{-1}[F(s)] \equiv \mathcal{L}^{-1}\left[\frac{D_1(s)}{D(s)}\right] = \sum_{k=1}^{n}\sum_{j=1}^{m_k} H_{kj}t^{m_k-j}e^{s_k t} \qquad (t > 0)$$

with

$$H_{kj} = \frac{1}{(j-1)!(m_k - j)!}\frac{d^{j-1}}{ds^{j-1}}\left[\frac{(s - s_k)^{m_k}D_1(s)}{D(s)}\right]_{s=s_k} \tag{8.4-2}$$

(see also Table D-6).

(b) It is sometimes convenient to obtain an inverse Laplace transform involving a multiple root of $D(s) = 0$ directly by a limiting process leading to the coincidence of distinct simple roots (see also Table 8.3-1,7). EXAMPLE:

$$\mathcal{L}^{-1}\left[\frac{1}{(s-a)^2}\right] = \lim_{h\to 0}\mathcal{L}^{-1}\left[\frac{1}{(s-a-h)(s-a)}\right] = \lim_{h\to 0}\frac{1}{h}[e^{(a+h)t} - e^{at}] = te^{at} \qquad (t > 0)$$

and, more generally,

$$\mathcal{L}^{-1}\left[\frac{1}{(s-a)^m}\right] = \frac{1}{(m-1)!}t^{m-1}e^{at} \qquad (t > 0)$$

(c) **Complex Roots.** In Eq. (2), pairs of terms corresponding to complex-conjugate roots $s = a \pm i\omega_1$ may be combined as follows (see also Sec. 8.4-5):

$$\left.\begin{aligned}(A + iB)t^r e^{(a+i\omega_1)t} &+ (A - iB)t^r e^{(a-i\omega_1)t} \\ &= 2At^r e^{at} \cos \omega_1 t - 2Bt^r e^{at} \sin \omega_1 t \\ &= Rt^r e^{at} \sin (\omega_1 t + \alpha) = Rt^r e^{at} \cos (\omega_1 t + \alpha') \\ \text{with} \quad R = 2\underset{+}{\sqrt{A^2 + B^2}} \quad &\alpha = -\arctan \frac{A}{B} \quad \alpha' = \arctan \frac{B}{A}\end{aligned}\right\} \quad (8.4\text{-}3)$$

A, B, R, α, and α' are real if the coefficients in $D_1(s)$ and $D(s)$ are real; $\mathcal{L}^{-1}[F(s)]$ is then a real function of t.

(d) If one of the roots s_k of $D(s) = 0$ should also be a root of $D_1(s) = 0$, then one or more terms of the expansion (2) will vanish. In general, such common roots of $D(s) = 0$ and $D_1(s) = 0$ can be "factored out" and canceled in Eq. (1).

8.4-5. Inverse Laplace Transforms of Rational Algebraic Functions: Expansion in Partial Fractions. Instead of applying the Heaviside expansion (2) directly, one may expand $F(s) = \dfrac{D_1(s)}{D(s)}$ as a sum of partial fractions by one of the methods described in Sec. 1.7-4. If $D(s)$ and $D_1(s)$ have no common zeros, each real root $s_k = a$ of $D(s) = 0$ will give rise to m_k partial fractions of the form

$$\frac{b_1}{s - a} \qquad \frac{b_2}{(s - a)^2} \qquad \cdots \qquad \frac{b_{m_k}}{(s - a)^{m_k}}$$

where m_k is the order of the root $s_k = a$. Each pair of complex-conjugate roots $s_k = a \pm i\omega_1$ will give rise to m_k partial fractions of the form

$$c_1 \frac{s + d_1}{(s - a)^2 + \omega_1^2} \qquad c_2 \frac{s + d_2}{[(s - a)^2 + \omega_1^2]^2} \qquad \cdots \qquad c_{m_k} \frac{s + d_{m_k}}{[(s - a)^2 + \omega_1^2]^{m_k}}$$

where m_k is the order of the roots $s_k = a \pm i\omega_1$. $\mathcal{L}^{-1}[F(s)]$ is then obtained as the sum of the inverse Laplace transforms of such terms (Table D-6). The method of Sec. 8.4-4*b* may be useful.

8.4-6. Expansions in Series. If the form of $F(s)$ is complicated, or if $F(s)$ is given only implicitly, e.g., as the solution of a differential equation in s (Sec. 9.4-5), it is sometimes possible to obtain $\mathcal{L}^{-1}[F(s)]$ by expanding $F(s)$ into a convergent series and taking the inverse Laplace transform of the latter term by term. Such a procedure may also be useful for *approximating* $\mathcal{L}^{-1}[F(s)]$. Frequently, series-expansion methods are justified by the following theorem. *Let*

$$F_k(s) = \mathcal{L}[f_k(t)] \qquad (\sigma > \sigma_a;\ k = 0, 1, 2, \ldots)$$

and let

$$\sum_{k=0}^{\infty} \int_0^{\infty} |f_k(t)| e^{-\sigma_a t}\, dt$$

converge. Then the series $\sum_{k=0}^{\infty} F_k(s)$ *converges uniformly to a function* $F(s)$

for $\sigma > \sigma_a$; *the series* $\sum_{k=0}^{\infty} f_k(t)$ *converges absolutely to a function* $f(t)$ *for almost all* t *(Sec. 4.6-14b), and*

$$\mathcal{L}[f(t)] = \mathcal{L}\left[\sum_{k=0}^{\infty} f_k(t)\right] = \sum_{k=0}^{\infty} \mathcal{L}[f_k(t)] = \sum_{k=0}^{\infty} F_k(s) = F(s) \qquad (\sigma > \sigma_a) \tag{8.4-4}$$

8.4-7. Expansions in Terms of Powers of *t*. Series expansions of $F(s)$ in descending powers of s,

$$F(s) = \frac{b_1}{s} + \frac{b_2}{s^2} + \cdots \qquad (|s| > r > 0) \tag{8.4-5}$$

may frequently be obtained as Laurent expansions (Sec. 7.5-3) about $s = 0$ or, in the case of rational algebraic functions of the type specified in Sec. 8.4-4, simply by long division (Sec. 1.7-2). If the conditions of Sec. 8.4-6 are satisfied, then, for almost all t in $(0, t_1)$,

$$f(t) = \mathcal{L}^{-1}[F(s)] = b_1 + b_2 t + \frac{b_3}{2} t^2 + \cdots + \frac{b_k}{(k-1)!} t^{k-1} + \cdots \qquad (t_1 > t > 0) \tag{8.4-6}$$

which may furnish (at least) useful approximations to $\mathcal{L}^{-1}[F(s)]$ for $t < t_1$.

If

$$F(s) = \frac{1}{s^\alpha} \sum_{k=0}^{\infty} \frac{c_k}{s^k} \qquad [\text{Re}\,(\alpha) > 0] \tag{8.4-7}$$

converges, then

$$\mathcal{L}^{-1}[F(s)] = t^{\alpha-1} \sum_{k=0}^{\infty} \frac{c_k}{\Gamma(k+\alpha)} t^k \qquad (t \geq 0) \tag{8.4-8}$$

In particular, if the series on the left converges,

$$\mathcal{L}^{-1}\left[\frac{1}{s^{1/2}} \sum_{k=0}^{\infty} \frac{c_k}{z^k}\right] = \frac{1}{\pi t^{1/2}} \left[c_0 + \frac{c_1}{1}(2t) + \frac{c_2}{(1)(3)}(2t)^2 + \cdots\right] \qquad (t \geq 0) \tag{8.4-9}$$

8.4-8. Expansions in Terms of Laguerre Polynomials of *t*. Every function

$$F(s) = \mathcal{L}[f(t)] \qquad (\sigma > \sigma_a)$$

analytic at $s = \infty$ may be expanded into an absolutely convergent series in powers of

$$z = \frac{s - \sigma_a - 1/2}{s - \sigma_a + 1/2} \tag{8.4-10}$$

for $\sigma > \sigma_a$, corresponding to a Taylor series convergent for $|z| < 1$. In particular, for $\sigma_a = 0$, $F(s)$ may be expressed in the form

$$\left.\begin{aligned} F(s) &= (1 - z) \sum_{k=0}^{\infty} c_k z^k = \frac{1}{s + \frac{1}{2}} \sum_{k=0}^{\infty} c_k \left(\frac{s - \frac{1}{2}}{s + \frac{1}{2}}\right)^k \qquad (\sigma > \sigma_a) \\ \text{with} \qquad c_k &= \sum_{j=0}^{k} \binom{k}{j} \frac{1}{j!} F^{(j)}\left(\frac{1}{2}\right) \qquad (k = 0,1,2, \cdot \cdot \cdot) \end{aligned}\right\} \tag{8.4-11}$$

and, *if the conditions of Sec. 8.4-6 are satisfied, then, for almost all* $t \geq 0$,

$$f(t) = \mathfrak{L}^{-1}[F(s)] = e^{-t/2} \sum_{k=0}^{\infty} \frac{c_k L_k(t)}{k!} \qquad (t \geq 0) \tag{8.4-12}$$

where the $L_k(t)$ are the Laguerre polynomials defined in Sec. 21.7-1.

8.4-9. Expansions in Asymptotic Series. Valid approximations to $\mathfrak{L}^{-1}[F(s)]$ may often be obtained in terms of asymptotic series (Sec. 4.8-6*b*). The following type of asymptotic expansion is of importance in connection with the solution of certain partial differential equations (Sec. 10.5-2).

If

$$F(s) = \mathfrak{L}[f(t)] \qquad (\sigma > \sigma_a)$$

can be expanded into a convergent series of the form

$$F(s) = \frac{1}{s}(a_0 + a_1 s^{1/2} + a_2 s + a_3 s^{3/2} + \cdot \cdot \cdot) = \frac{1}{s} \sum_{k=0}^{\infty} a_k s^{k/2} \tag{8.4-13}$$

in a neighborhood of $s = 0$, *then* $f(t)$ *may be represented by the asymptotic series*

$$f(t) \cong a_0 + \frac{1}{\pi \sqrt{t}} \sum_{j=0}^{\infty} (-1)^j a_{2j+1} \frac{\Gamma(j + \frac{1}{2})}{t^j} \qquad \text{as } t \to \infty \tag{8.4-14}$$

provided that either of the following sets of conditions is satisfied (*see also Sec.* 8.5-1 *and Ref.* 8.5):

1. $f(t)$ *is differentiable for* $t > 0$, *continuous for* $t = 0$, *and* $\mathfrak{L}[f'(t)]$ *exists; and there exists a* (*finite*) *real positive number* K *such that*

$$\frac{d}{dt}\{t^j[f(t) - a_0 - S_{j-1}(t)]\} \geq -Kt^{-1/2} \qquad (t > 0, j = 1, 2, \ldots) \tag{8.4-15}$$

   where $S_{j-1}(t)$ *is the* $(j - 1)^{st}$ *partial sum of the series in Eq.* (14).
2. $f(t)$ *is continuous for* $t > t_1 > 0$, *and* $\sigma_a \leq 0$; $F(s)$ *is analytic for* $\sigma \geq 0$ *except for* $s = 0$. *There exists* $\sigma_1 > 0$ *such that for* $\sigma_1 \geq \sigma \geq 0$

$$\lim_{|\omega|\to\infty} F(s) = 0 \textit{ uniformly with respect to } \sigma$$

$$\lim_{|\omega|\to\infty} F^{(k)}(i\omega) = 0 \qquad (k = 1, 2, \ldots)$$

$$\int_0^{\pm\infty} |F^{(j)}(i\omega)|\, d\omega \textit{ exists} \qquad (j = 2, 3, \ldots)$$

$$\textit{and } \int_{-\infty}^{\infty} F(\sigma_1 + i\omega)e^{i\omega t}\, d\omega \textit{ converges uniformly for } t > t_1 > 0.$$

These rather restrictive conditions provide a rigorous basis for a number of asymptotic expansions of $\mathcal{L}^{-1}[\Phi(\sqrt{s})]$ originally derived with the aid of the old Heaviside operational calculus (Refs. 8.3 and 8.10).

8.5. "FORMAL" LAPLACE TRANSFORMATION OF IMPULSE-FUNCTION TERMS

8.5-1. Impulse-function Transforms. (**a**) In Eq. (8.4-13) and in similar series expansions, note that the inverse Laplace transforms of individual terms like a, as, $as^{3/2}$, . . . do not, strictly speaking, exist, since these functions do not tend to zero as $s \to \infty$. In most applications, terms of this nature will appear under summation and integral signs in such a manner that the series or integral does have an inverse Laplace transform.

(**b**) If one applies the Laplace transformation (8.2-1) to the "definitions" of the impulse functions $\delta(t)$ and $\delta_+(t)$ and their "derivatives" (Secs. 9.2 and 21.9-6), one obtains the *formal* results

$$\left.\begin{aligned} &\mathcal{L}[\delta(t)] = \tfrac{1}{2} \qquad \mathcal{L}[\delta_+(t)] = 1 \\ &\mathcal{L}[\delta_+'(t)] \equiv s \qquad \mathcal{L}[\delta_+''(t)] \equiv s^2 \qquad \cdots \\ &\mathcal{L}[\delta(t-a)] \equiv \mathcal{L}[\delta_+(t-a)] \equiv e^{-as} \qquad (a > 0) \\ &\mathcal{L}[\delta^{(k)}(t-a)] \equiv \mathcal{L}[\delta_+{}^{(k)}(t-a)] \equiv e^{-as}s^k \\ &\qquad (a > 0;\ k = 1, 2, \ldots) \end{aligned}\right\} \tag{8.5-1*}$$

and, if $f(t)$ is continuous for $t = a \geq 0$,

$$\mathcal{L}[\delta_+(t-a)f(t)] \equiv e^{-as}f(a) \tag{8.5-2}$$

Eqs. (1) and (2) are useful in many applications, but one must not forget that such relations have no strict mathematical meaning. New results suggested by Eqs. (1) and (2) must always be verified by mathematically legitimate means (see also Secs. 8.2-7 and 21.9-2*a*).

8.6. SOME OTHER INTEGRAL TRANSFORMATIONS

8.6-1. The Laplace transformation (8.2-1) is a *functional transformation* associating "points" $F(s)$ in a *result space* with "points" $f(t)$ in an *object*

* The asymmetrical impulse function $\delta_+(t)$ is more suitable for use in connection with the one-sided Laplace transformation than the symmetrical impulse function $\delta(t)$.

space (see also Secs. 12.1-4 and 15.2-7). Table 8.6-1 and Secs. 8.6-2 to 8.6-4 introduce a number of other functional transformations (see also Sec. 4.11-5, Appendix D, and Refs. 8.8 and 8.10).

8.6-2. The Two-sided (Bilateral) Laplace Transformation. (a) The **two-sided (bilateral) Laplace transformation**

$$\mathcal{L}_B[f(t)] = \int_{-\infty}^{\infty} f(t)e^{-st}\,dt \equiv \mathcal{L}[f(t);\,s] + \mathcal{L}[f(-t);\,-s] \tag{8.6-1}$$

is an attractive generalization of the Laplace transformation applicable, like the Fourier transformation (Secs. 4.11-3 to 4.11-7), to problems where values of $f(t)$ for $t < 0$ are of importance. $\mathcal{L}_B[f(t);\,s]$ *converges absolutely if and only if both* $\mathcal{L}[f(t);\,s]$ *and* $\mathcal{L}[f(-t);\,-s]$ *converge absolutely,* so that the region of absolute convergence, if any, will be a *strip* of the s plane determined by two abscissas of absolute convergence. *Many properties of the two-sided Laplace transformation are simply derived from corresponding properties of the one-sided Laplace transformation by reference to Eq.* (1) (see also Refs. 8.5 and 8.17). In particular, note

$$\left.\begin{aligned} \mathcal{L}_B[f(t)] &\equiv \mathcal{L}[f(t)] && \text{if } f(t) = 0 \text{ for } t \leq 0 \\ \mathcal{L}_B[f(t)] &\equiv \mathcal{L}[f(t)] - \frac{c}{s} && \text{if } f(t) = c \text{ for } t \leq 0 \end{aligned}\right\} \tag{8.6-2}$$

(b) The values of the inverse transform $\mathcal{L}_B^{-1}[F(s)]$ are not restricted to zero for $t < 0$, so that $\mathcal{L}_B^{-1}[F(s)]$ exists for a larger class of functions $F(s)$ than does $\mathcal{L}^{-1}[F(s)]$. *Given*

$$F(s) \equiv \mathcal{L}_B[f(t)] \qquad (\sigma_{a_1} < \sigma < \sigma_{a_2})$$

then, for any value of t having a neighborhood where $f(t)$ is of bounded variation,

$$f_I(t) = \frac{1}{2\pi i}\lim_{R\to\infty}\int_{\sigma_1 - iR}^{\sigma_1 + iR} F(s)e^{st}\,ds = \frac{1}{2}[f(t-0) + f(t+0)]$$

$$(\sigma_{a_1} < \sigma_1 < \sigma_{a_2}) \qquad \text{(INVERSION THEOREM)} \tag{8.6-3}$$

(c) Note also the *convolution theorem*

$$\left.\begin{aligned} \mathcal{L}_B[f_1(t)]\mathcal{L}_B[f_2(t)] &= \mathcal{L}_B[f_1 * f_2] \\ \text{with} \qquad f_1 * f_2 &\equiv \int_{-\infty}^{\infty} f_1(\tau)f_2(t-\tau)\,d\tau \equiv f_2 * f_1 \end{aligned}\right\} \tag{8.6-4}$$

assuming absolute convergence (see also Tables 4.11-1 and 8.3-1).

8.6-3. The Stieltjes-integral Form of the Laplace Transformation. The **Stieltjes-integral form of the Laplace transformation**

$$\mathcal{L}_S[\varphi(t)] \equiv \int_0^{\infty} e^{-st}\,d\varphi(t) \tag{8.6-5}$$

Table 8.6-1. Some Linear Integral Transformations Related to the Laplace Transformation

Each transform is denoted by $F(s)$ (see also Secs. 4.11-3 to 4.11-5, 8.6-1 to 8.6-5, 10.5-1, 15.2-7, 15.3-1, and 18.3-8)

No.	Transform	Definition	Inversion integral	Relation to unilateral Laplace transformation*	Remarks
1	**s-multiplied Laplace transform**	$\int_0^\infty f(t)se^{-st}\,dt$	$\frac{1}{2\pi i}\int_{\sigma_1-i\infty}^{\sigma_1+i\infty}\frac{1}{s}F(s)e^{st}\,ds \quad (t>0)$	$s\mathcal{L}[f(t);\,s]$	
2	**Bilateral Laplace transform**	$\int_{-\infty}^\infty f(t)e^{-st}\,dt$	$\frac{1}{2\pi i}\int_{\sigma_1-i\infty}^{\sigma_1+i\infty}F(s)e^{st}\,ds$	$\mathcal{L}_B[f(t);\,s] \equiv \mathcal{L}[f(t);\,s] + \mathcal{L}[f(-t);\,-s]$	Sec. 8.6-2
3	**Fourier transform**	$\int_{-\infty}^\infty f(t)e^{-i\omega t}\,dt$	$\int_{-\infty}^\infty F(i\omega)e^{i\omega t}\frac{d\omega}{2\pi}$	$\mathcal{L}_B[f(t);\,i\omega]$	Secs. 4.11-3 to 4.11-5
4	**Mellin transform**	$\int_0^\infty f(t)t^{s-1}\,dt$	$\frac{1}{2\pi i}\int_{\sigma_1-i\infty}^{\sigma_1+i\infty}F(s)t^{-s}\,ds \quad (t>0)$	$\mathcal{L}_B[f(e^t);\,-s]$	
5	**Hilbert transform (use principal value)**	$\frac{1}{\pi}\int_{-\infty}^\infty \frac{f(t)}{s-t}\,dt$	$\frac{1}{\pi}\int_{-\infty}^\infty \frac{F(s)}{t-s}\,ds$	$\frac{1}{\pi}\mathcal{L}\{\mathcal{L}[f(-t);\,s'];\,s\} - \frac{1}{\pi}\mathcal{L}\{\mathcal{L}[f(t);\,s'];\,-s\}$	$f(t)$ and $F(t)$ are **quadrature functions,** i.e. $f(t) = \frac{1}{\pi}\int_0^\infty [a(\omega)\cos\omega t + b(\omega)\sin\omega t]\,d\omega$ implies $F(t) = \frac{1}{\pi}\int_0^\infty [b(\omega)\cos\omega t - a(\omega)\sin\omega t]\,d\omega$

* Whenever the transforms in question exist.

permits a more general formulation of many theorems than the ordinary Laplace transformation (see also Sec. 4.6-17 and Refs. 8.5 and 8.18). The other functional transformations listed in Table 8.6-1 may be similarly written in terms of Stieltjes integrals.

Note that $F(s) \equiv se^{-bs}$ can be represented in the form (5) without the use of impulse functions (Sec. 8.5-1). The *s-multiplied Laplace transform* (Table 8.6-1,1) is sometimes employed for similar reasons.

8.6-4. Hankel Transforms and Fourier-Bessel Transforms. (a) Definition and Inversion Theorem. The integral transform

$$\bar{f}(s) \equiv \mathcal{H}_m[f(t)] \equiv \mathcal{H}_m[f(t); s] \equiv \int_0^\infty f(t)tJ_m(st)\,dt \qquad \text{(Hankel transform of order } m) \tag{8.6-6a}$$

where $f(t)$ is a real function, and $J_m(z)$ is the mth-order Bessel function (Sec. 21.8-1), exists in the sense of absolute convergence whenever $\int_0^\infty |f(t)|\,dt$ exists. *If, in addition, $f(t)$ is of bounded variation in a neighborhood of t, one has the inversion formula*

$$f_I(t) = \int_0^\infty \bar{f}(s)sJ_m(st)\,ds = \tfrac{1}{2}[f(t-0) + f(t+0)] \qquad (m \geq -\tfrac{1}{2}) \qquad \text{(Hankel inversion theorem)} \tag{8.6-6b}$$

which determines the inverse transform uniquely wherever it is continuous.

(b) Properties of Hankel Transforms. Note the following relations:

$$\mathcal{H}_m[f(at); s] = \frac{1}{a^2}\mathcal{H}_m\left[f(t); \frac{s}{a}\right] \tag{8.6-7}$$

$$\mathcal{H}_m\left[\frac{1}{t}f(t)\right] = \frac{s}{2m}\{\mathcal{H}_{m-1}[f(t)] + \mathcal{H}_{m+1}[f(t)]\} \tag{8.6-8}$$

$$\mathcal{H}_m[f'(t)] = \frac{s}{2m}\{(m-1)\mathcal{H}_{m+1}[f(t)] - (m+1)\mathcal{H}_{m-1}[f(t)]\} \tag{8.6-9}$$

$$\mathcal{H}_m\left[f''(t) + \frac{1}{t}f'(t) - \frac{m^2}{t^2}f(t)\right] = -s^2\mathcal{H}_m[f(t)] \tag{8.6-10}$$

$$\int_0^\infty s\mathcal{H}_m[f(t)]\mathcal{H}_m[g(t)]\,ds = \int_0^\infty tf(t)g(t)\,dt \qquad (m \geq -\tfrac{1}{2})$$

(Parseval's theorem for Hankel transforms) (8.6-11)

(c) Fourier-Bessel Transforms (see also Secs. 21.8-1 and 21.8-2). The following integral-transform pairs are related to the Hankel-transform pair (6):

$$\left.\begin{aligned} \bar{f}(s) &= \int_0^\infty f(t)tJ_m(st)\,dt \\ f_I(t) &= \int_0^\infty \bar{f}(s)sJ_m(st)\,ds = \tfrac{1}{2}[f(t-0) + f(t+0)] \end{aligned}\right\} \qquad (m = 0, 1, 2, \ldots) \tag{8.6-12}$$

$$\left.\begin{aligned}\bar{f}(s) &= \int_0^\infty f(t)\,\frac{1}{\sqrt{st}}\,J_{m+\frac{1}{2}}(st)t^2\,dt \\ f_I(t) &= \int_0^\infty \bar{f}(s)\,\frac{1}{\sqrt{st}}\,J_{m+\frac{1}{2}}(st)s^2\,ds = \frac{1}{2}\,[f(t-0)+f(t+0)]\end{aligned}\right\} \quad (m = 0, 1, 2, \ldots) \tag{8.6-13}$$

Both integral-transform pairs are referred to as **Fourier-Bessel transform pairs;** for $m = 0$, Eq. (13) reduces to a Fourier sine-transform pair.

8.7. FINITE INTEGRAL TRANSFORMS, GENERATING FUNCTIONS, AND *z* TRANSFORMS

8.7-1. Series as Functional Transforms. Finite Fourier and Hankel Transforms. A finite or convergent series

$$\Phi(x) = \sum_{k=-\infty}^{\infty} f_k \Psi(x,k) \tag{8.7-1}$$

represents a functional transformation of the function (sequence, Sec. 4.2-1) $f_k \equiv f(k)$ defined on the discrete set of integers $k = 0, 1, 2, \ldots$. Note that for suitable f_k and $\Psi(x,k)$ such a series can be written as an integral transform in terms of a Stieltjes integral (Sec. 4.6-17):

$$\Phi(s) = \int_{-\infty}^{\infty} \Psi(x,k)\,d\varphi(k)$$

The series (1) and formulas like (4.10-5), (4.11-6), (7.5-4), and (7.5-7), which relate series coefficients f_k to a function $\Phi(x)$, constitute corresponding inverse functional transformations. Table 8.7-1 lists relations for generalized Fourier-series coefficients regarded as integral transforms with *finite* integration intervals (**finite integral transforms**), and for the analogous finite Hankel transforms. In each case, the transformation of a relevant second-order differential equation is illustrated for use in connection with the solution of boundary-value problems (Refs. 8.15 and 8.16; see also Secs. 10.4-9, 15.2-4, and 21.8-4*c*).

8.7-2. Generating Functions. If the functional transformation (1) takes the form of a finite or convergent *power series*

$$\gamma(s) \equiv \sum_{k=0}^{\infty} f_k s^k \tag{8.7-2}$$

$\gamma(s)$ is called a **generating function** for the sequence of coefficients $f_0, f_1, f_2, \ldots$, while

$$\gamma_E(s) \equiv \sum_{k=0}^{\infty} f_k \frac{s^k}{k!} \tag{8.7-3}$$

Table 8.7-1. Some Finite Integral Transforms (see also Secs. 10.4-2*c* and 15.4-12).
x is a real variable, $\Phi'(x) \equiv d\phi/dx$, $\Phi''(x) \equiv d^2\Phi/dx^2$, and $a > 0$.

(a) Finite Transforms Useful with Boundary Conditions of the Form $\Phi = 0$ or $\Phi' = 0$

	f_k	$\Phi(x)$ or $\frac{1}{2}[\Phi(x-0) + \Phi(x+0)]$	$\lambda_1, \lambda_2, \ldots$ are the positive roots of	Important derivative transformation
Finite cosine transform (Fourier cosine series, Sec. 4.11-2*e*)	$\int_0^a \Phi(x) \cos \lambda_k x \, dx$	$\frac{1}{a}\int_0^a \Phi(x)\, dx + \frac{2}{a}\sum_{k=1}^{\infty} f_k \cos \lambda_k x$	$\tan a\lambda = 0$ $\lambda_k = k\frac{\pi}{a}$ $(k = 1, 2, \ldots)$	$\int_0^a \Phi''(x) \cos \lambda_k x \, dx$ $= -\lambda_k^2 f_k$ $- \Phi'(0) + (-1)^k \Phi'(a)$
Finite sine transform (Fourier sine series, Sec. 4.11-2*e*)	$\int_0^a \Phi(x) \sin \lambda_k x \, dx$	$\frac{2}{a}\sum_{k=1}^{\infty} f_k \sin \lambda_k x$		$\int_0^a \Phi''(x) \sin \lambda_k x \, dx$ $= -\lambda_k^2 f_k$ $+ \lambda_k[\Phi(0) + (-1)^{k+1}\Phi(a)]$
Finite Hankel transform (Hankel series)	$\int_0^a \Phi(x) x J_m(\lambda_k x)\, dx$	$\frac{2}{a^2}\sum_{k=1}^{\infty} f_k \frac{J_m(\lambda_k x)}{[J_m'(\lambda_k a)]^2}$	$J_m(a\lambda) = 0$	$\int_0^a \left[\Phi'' + \frac{\Phi'}{x} - \frac{m^2}{x^2}\right] x J_m(\lambda_k x)\, dx$ $= -\lambda_k^2 f_k - a\lambda_k J_m'(a\lambda_k)\Phi(a)$
Finite annular Hankel transform, with $a' > a > 0$, and $B_m(\lambda x) \equiv J_m(\lambda x)N_m(a\lambda) - N_m(\lambda x)J_m(a\lambda)$	$\int_a^{a'} \Phi(x) x B_m(\lambda_k x)\, dx$	$\frac{\pi^2}{2}\sum_{k=1}^{\infty} f_k \frac{\lambda_k^2 J_m^2(a'\lambda_k)}{J_m^2(a\lambda_k) - J_m^2(a'\lambda_k)} \cdot B_m(\lambda_m x)$	$B_m(a'\lambda) = 0$	$\int_a^{a'} \left[\Phi'' + \frac{\Phi'}{x} - \frac{m^2}{x^2}\right] x B_m(\lambda_k x)\, dx$ $= -\lambda_k^2 f_k + \frac{2}{\pi}\left[\frac{J_m(a\lambda_k)}{J_m(a'\lambda_k)}\Phi(a') - \Phi(a)\right]$

(b) Finite Transforms Useful with Boundary Conditions of the Form $\Phi(a) + b\Phi'(a) = 0 \qquad (b > 0)$

	f_k	$\Phi(x)$ or $\frac{1}{2}[\Phi(x-0) + \Phi(x+0)]$	$\lambda_1, \lambda_2, \ldots$ are the positive roots of	Important derivative transformation
Finite cosine transform	$\int_0^a \Phi(x) \cos \lambda_k x \, dx$	$2 \sum_{k=1}^{\infty} \frac{(\lambda_k^2 + b^2) f_k \cos \lambda_k x}{b + a(\lambda_k^2 + b^2)}$	$\lambda \tan a\lambda = b$	$\int_0^a \Phi''(x) \cos \lambda_k x \, dx = -\lambda_k^2 f_k - \Phi'(0) + [\Phi'(a) + b\Phi(a)] \cos a\lambda_k$
Finite sine transform	$\int_0^a \Phi(x) \sin \lambda_k x \, dx$	$2 \sum_{k=1}^{\infty} \frac{(\lambda_k^2 + b^2) f_k \sin \lambda_k x}{b + a(\lambda_k^2 + b^2)}$	$\lambda \cot a\lambda = -b$	$\int_0^a \Phi''(x) \sin \lambda_k x \, dx = -\lambda_k^2 f_k + \lambda_k \Phi(0) + [\Phi'(a) + b\Phi(a)] \sin a\lambda_k$
Finite Hankel transform	$\int_0^a \Phi(x) x J_m(\lambda_k x) \, dx$	$2 \sum_{k=1}^{\infty} \frac{\lambda_k^2 f_k J_m(\lambda_k x)}{[(\lambda_k^2 + b^2)a^2 - m^2] J_m^2(a\lambda_k)}$	$\lambda J'_m(a\lambda) + b J_m(a\lambda) = 0$	$\int_0^a \left[\Phi'' + \frac{\Phi'}{x} - \frac{m^2}{x^2}\Phi\right] x J_m(\lambda_k x) \, dx = -\lambda_k^2 f_k + [\Phi'(a) + b\Phi(a)] a J_m(a\lambda_k)$

In Tables 8.7-1*a* and *b*, note that $J'_0(\lambda_k a) \equiv -J_1(\lambda_k a)$.

is called an **exponential generating function** for the f_k. Applications and properties of generating functions are further discussed in Sec. 18.3-8 and Appendix C.

EXAMPLE: **Fibonacci numbers** $f_0, f_1, f_2, \ldots$ are defined by

$$f_0 = f_1 = 1 \qquad f_k = f_{k-1} + f_{k-2} \qquad (k = 2, 3, \ldots) \tag{8.7-4}$$

Their generating function is

$$\gamma(s) \equiv \frac{1}{1 - s - s^2} = 1 + s + 2s^2 + 3s^3 + 5s^4 + 8s^5 + \cdots \tag{8.7-5}$$

8.7-3. z Transforms. Definition and Inversion Integral. The **z transform** of a suitable sequence $f_0, f_1, f_2, \ldots$ is defined as

$$\boxed{\mathrm{Z}[f_k;z] = f_0 + \frac{f_1}{z} + \frac{f_2}{z^2} + \cdots = F_Z(z) \qquad (|z| > r_a)} \tag{8.7-6}$$

where z is a complex variable, and the series converges absolutely outside of a **circle of absolute convergence** of radius r_a depending on the given sequence; analytic continuation in the manner of Sec. 8-2.3 can extend the definition. The corresponding inversion integral

$$f_{Ik} = \frac{1}{2\pi i} \oint_C F_Z(z) z^{k-1}\, dz \qquad (k = 0, 1, 2, \ldots) \tag{8.7-7}$$

where C is a closed contour enclosing all singularities of $F_Z(z)$, gives the inverse transform $\mathrm{Z}^{-1}[F_Z(z)] = f_k$ for suitable f_k (Ref. 8.11). Inversion can then utilize the residue calculus in the manner of Sec. 8.4-3, especially if $F_Z(z)$ is a rational function expandable in partial fractions. Inversion is even simpler if $F_Z(z)$ can be expanded directly in terms of powers of $1/z$. Note that the inverse transform must be unique wherever the series (6) converges absolutely (Sec. 4.10-2c).

Table 8.7-2 summarizes the most important properties of z transforms. Their application to the solution of difference equations and sampled-data systems is treated in Sec. 20.4-6, where the relation of z transforms to jump-function Laplace tranforms is also discussed. Table 20.4-1 lists a number of z-transform pairs.

The z transform is related to the Mellin transform of Table 8.6-1. Note the analogy between power series and Mellin transforms and between *Dirichlet series* $\sum_{k=0}^{\infty} y_k e^{-ks}$ and Laplace transforms.

8.8. RELATED TOPICS, REFERENCES, AND BIBLIOGRAPHY

8.8-1. Related Topics. The following topics related to the study of the Laplace transformation and other functional transformations are

Table 8.7-2. Corresponding Operations for z Transforms

The following theorems are valid whenever the z-transform series in question converge absolutely; limits are assumed to exist (see also Table 8.3-1), and $f_{-1} = f_{-2} = \cdots = g_{-1} = g_{-2} = \cdots = 0$

Theorem	Operation	Object sequence	Result function
1	**Linearity** (α, β constant)	$\alpha f_k + \beta g_k$	$\alpha F_Z(z) + \beta G_Z(z)$
2	**Advanced object sequence**	$f_{k+1} \equiv \mathbf{E} f_k$ $f_{k+r} \equiv \mathbf{E}^r f_k \quad (r = 1, 2, \ldots)$	$zF_Z(z) - zf_0$ $z^r F_Z(z) - \sum_{j=0}^{r-1} f_j z^{r-j}$
3	**Delayed object sequence**	$f_{k-1} \equiv \mathbf{E}^{-1} f_k$	$z^{-1} F_Z(z)$
	Finite differences		
4a	forward difference	$f_{k+1} - f_k \equiv \Delta f_k$	$(z-1)F_Z(z) - zf_0$
4b	backward difference	$f_k - f_{k-1} \equiv \nabla f_k$	$\frac{z-1}{z} F_Z(z)$
5	**Finite summation of object sequence**	$\sum_{j=0}^{k} f_j$	$\frac{z}{z-1} F_Z(z)$
6	**Convolution of object sequences**	$\sum_{j=0}^{\infty} f_j g_{k-j}$	$F_Z(z) G_Z(z)$
7	**Corresponding limits (continuity theorem;** α is independent of k and z)	$\lim_{\alpha \to a} f_k(\alpha)$	$\lim_{\alpha \to a} F_Z(z,\alpha)$
8	**Differentiation and integration with respect to a parameter** α **independent of** k **and** z	$\frac{\partial}{\partial \alpha} f_k(\alpha)$ $\int_{\alpha_1}^{\alpha_2} f_k(\alpha)\, d\alpha$	$\frac{\partial}{\partial \alpha} F_Z(z,\alpha)$ $\int_{\alpha_1}^{\alpha_2} F_Z(z,\alpha)\, d\alpha$
9	**Initial and final* values of object sequence**	f_0 $\lim_{k \to \infty} f_k$	$\lim_{z \to \infty} F_Z(z)$ $\lim_{z \to 1} (z-1) F_Z(z)$
10	**Differentiation of result function**	kf_k $k^r f_k \quad (r = 1, 2, \ldots)$	$-z \frac{d}{dz} F_Z(z)$ $-z \frac{d}{dz} \mathcal{Z}[k^{r-1} f_k]$

* $(z-1)F_Z(z)$ is assumed to be analytic for $|z| \geq 1$.

treated in other chapters of this handbook:

Expansion in partial fractions............................ Chap. 1
Limits, integration, and improper integrals................. Chap. 4
Convergent and asymptotic series......................... Chap. 4
Functions of a complex variable, contour integration........ Chap. 7
Transformations... Chaps. 12, 14, 15
Fourier transforms....................................... Chap. 4
Applications of generating functions...................... Chap. 18, Appendix C
z transforms.. Chap. 20
Tables of Fourier, Hankel, and Laplace-transform pairs..... Appendix D

Applications of the Laplace transformation are discussed in other chapters of this handbook as follows:

Ordinary differential equations........................... Chaps. 9, 13
Partial differential equations............................ Chap. 10
Integral equations....................................... Chap. 15

8.8-2. References and Bibliography.

8.1. Campbell, G. A., and R. M. Foster: *Fourier Integrals for Practical Applications*, Van Nostrand, Princeton, N.J., 1958.
8.2. Churchill, R. V.: *Operational Mathematics*, 2d ed., McGraw-Hill, New York, 1958.
8.3. Ditkin, V. A., and A. P. Prudnikov: *Integral Transforms and Operational Calculus*, Pergamon Press, New York, 1965.
8.4. Doetsch, G.: *Anleitung zum Praktischen Gebranch der Laplace-Transformation*, Oldenburg, Munich, 1956.
8.5. ———: *Handbuch der Laplace-Transformation*, 3 vols., Birkhäuser, Basel, 1950.
8.6. ——— and D. Voelker: *Die Zweidimensionale Laplace-Transformation*, Birkhäuser, Basel, 1950.
8.7. ———, H. Kneiss, and D. Voelker: *Tabellen zur Laplace-Transformation und Anleitung zum Gebrauch*, Springer, Berlin, 1947.
8.8. Erdélyi, A., et al.: *Tables of Integral Transforms* (Bateman Project), 2 vols., McGraw-Hill, New York, 1954.
8.9. Jury, E. I.: *Theory and Application of the z-transform Method*, Wiley, New York, 1964.
8.10. McLachlan, N. W.: *Modern Operational Calculus*, Macmillan, New York, 1948.
8.11. Miles, J. W.: Integral Transforms, in E. F. Beckenbach, *Modern Mathematics for the Engineer*, 2d series, McGraw-Hill, New York, 1961.
8.12. Nixon, F. E.: *Handbook of Laplace Transformation*, 2d ed., Prentice-Hall, Englewood Cliffs, N.J., 1965.
8.13. Papoulis, A.: *The Fourier Integral and Its Applications*, McGraw-Hill, New York, 1962.
8.14. Scott, E. J.: *Transform Calculus*, Harper, New York, 1955.
8.15. Sneddon, I. N.: *Fourier Transforms*, McGraw-Hill, New York, 1951.
8.16. Tranter, C. J.: *Integral Transforms in Mathematical Physics*, 2d ed., Wiley, New York, 1956.
8.17. Van Der Pol, B., and H. Bremmer: *Operational Calculus Based on the Two-sided Laplace Integral*, Cambridge University Press, London, 1950.
8.18. Widder, D. V.: *The Laplace Transform*, Princeton University Press, Princeton, N.J., 1941.

CHAPTER 9

ORDINARY DIFFERENTIAL EQUATIONS

9.1. INTRODUCTION

9.1-1. Survey. Differential equations are used to express relations between changes in physical quantities and are thus of great importance in many applications. Sections 9.1-2 to 9.3-10 present a straightforward *classical introduction to ordinary differential equations*, including some *complex-variable theory*. Sections 9.4-1 to 9.4-8 introduce the *linear differential equations with constant coefficients* used in the analysis of vibrations, electric circuits, and control systems, with emphasis on solutions by Laplace-transform methods. Sections 9.5-1 to 9.5-6 deal with *nonlinear second-order equations*. Sections 9.6-1 and 9.6-2 introduce *Pfaffian differential equations*, although these are not ordinary differential equations.

Some naturally related material is treated in other chapters of this handbook, particularly in Chap. 8 and Secs. 13.6-1 to 13.6-7. *Boundary-*

value problems, eigenvalue problems, and *orthogonal-function expansions* of solutions are discussed in Chap. 15, and a number of differential equations defining *special functions* are treated in Chap. 21.

The notation used in the various subdivisions of this chapter has been chosen so as to simplify reference to standard textbooks in different special fields. Thus the usually real variables in Secs. 9.2-1 to 9.2-5 are denoted by x, $y = y(x)$; the frequently complex variables encountered in the general theory of linear ordinary differential equations (Secs. 9.3-1 to 9.3-10) are denoted by z, $w = w(z)$. The variables in Secs. 9.4-1 to 9.5-6 usually represent physical time and various mechanical or electrical variables and are thus introduced as t, $y_k = y_k(t)$.

9.1-2. Ordinary Differential Equations. An **ordinary differential equation of order** r is an equation

$$F[x, y(x), y'(x), y''(x), \ldots, y^{(r)}(x)] = 0 \tag{9.1-1}$$

to be satisfied by the function $y = y(x)$ together with its derivatives $y'(x), y''(x), \ldots, y^{(r)}(x)$ with respect to a single independent variable x. To **solve (integrate)** a given differential equation (1) means to find functions (**solutions, integrals**) $y(x)$ which satisfy Eq. (1) for all values of x in a specified bounded or unbounded interval (a, b). *Note that solutions can be checked by resubstitution.*

The **complete primitive (complete integral, general solution)** of an ordinary differential equation of order r has the form

$$y = y(x, C_1, C_2, \ldots, C_r) \tag{9.1-2}$$

where $C_1, C_2, \ldots, C_r$ are r *arbitrary constants* (**constants of integration,** see also Sec. 4.6-4). Each particular choice of these r constants yields a **particular integral** (2) of the given differential equation. Typical problems require one to find the particular integral (2) subject to r **initial conditions**

$$y(x_0) = y_0 \qquad y'(x_0) = y_0' \qquad y''(x_0) = y_0'' \qquad \cdots \qquad y^{(r-1)}(x_0) = y_0^{(r-1)} \tag{9.1-3}$$

which determine the r constants $C_1, C_2, \ldots, C_r$. Alternatively, one may be given r **boundary conditions** on $y(x)$ and its derivatives for $x = a$ and $x = b$ (see also Sec. 9.3-4).*

Many ordinary differential equations admit additional solutions known as **singular integrals** which are not included in the complete primitive (2) (see also Sec. 9.2-2*b*).

A differential equation is **homogeneous** if and only if $\alpha y(x)$ is a solution for all α whenever $y(x)$ is a solution (see also Secs. 9.1-5 and 9.3-4).

* Strictly speaking, initial and boundary conditions refer to unilateral derivatives (Sec. 4.5-1).

Given an r-parameter family of suitably differentiable functions (2), one can eliminate $C_1, C_2, \ldots, C_r$ from the $r + 1$ equations $y^{(j)} = y^{(j)}(x, C_1, C_2, \ldots, C_r)$ $(j = 0, 1, 2, \ldots, r)$ to obtain an rth-order differential equation describing the family.

NOTE: An ordinary differential equation is a special instance of a **functional equation** imposing conditions on the *functional dependence* $y = y(x)$ for a set of values of x. OTHER EXAMPLES OF FUNCTIONAL EQUATIONS: $y(x_1x_2) = y(x_1) + y(x_2)$ [*logarithmic property*, satisfied by $y(x) \equiv A \log_e x$], partial differential equations (Sec. 10.1-1), integral equations (Sec. 15.3-2), and difference equations (Sec. 20.4-3).

9.1-3. Systems of Differential Equations (see also Secs. 13.6-1 to 13.6-7). A system of ordinary differential equations

$$F_i(x; y_1, y_2, \ldots; y_1', y_2', \ldots) = 0 \qquad (i = 1, 2, \ldots) \tag{9.1-4}$$

involves a set of unknown functions $y_1 = y_1(x)$, $y_2 = y_2(x)$, . . . and their derivatives with respect to a single independent variable x. The **order** r_i of each differential equation (4) is that of the highest derivative occurring. In general, one will require n differential equations (4) to determine n unknown functions $y_k(x)$; and the general solution $y_1 = y_1(x)$, $y_2 = y_2(x)$ will involve a number of arbitrary constants equal to $r = r_1 + r_2 + \cdots + r_n$.

The solution of a system (4) can be reduced to that of a single ordinary differential equation of order r through elimination of $n - 1$ variables y_k and their derivatives. More importantly, *one can reduce every system* (4) *to an equivalent system of r first-order equations* by introducing higher-order derivatives as new variables.

9.1-4. Existence and Desirable Properties of Solutions. A properly posed differential-equation problem requires an **existence proof** indicating the construction of a solution subject to the given type of initial or boundary conditions. The existence of physical phenomena described by a given differential equation may suggest but does *not* prove the existence of a solution; an existence proof checks the *self-consistency of the mathematical model* (see also Secs. 4.2-1*b* and 12.1-1; see Secs. 9.2-1 and 9.3-5 for examples of existence theorems).

It is desirable to design mathematical models involving differential equations so that the solutions are continuous functions of numerical coefficients, initial conditions, etc., so as to avoid excessive errors in solutions due to small errors in numerical data (see also Sec. 9.2-1*a*).

9.1-5. General Hints. (**a**) Substitution of a Taylor series (Sec. 4.10-4) or other series expansion for $y(x)$ in a given differential equation may yield equations for the unknown coefficients (see also Secs. 9.2-5*b* and 9.3-5). Many differential equations can be simplified through transformation of variables (Secs. 9.1-5*b*, 9.1-3, and 9.3-8*c*). Every differential equation or system of differential equations can be reduced to a system of first-order equations; so that the methods of Sec. 9.2-5 apply.

(**b**) *The following special types of differential equations reduce easily to equations of lower order* (see also Secs. 9.2-3 and 9.5-6):

$$F(x, y^{(n)}, y^{(n+1)}, \ldots, y^{(n+m)}) = 0 \qquad \text{(introduce } \bar{y} = y^{(n)}\text{)}$$
$$F(y, y', y'', \ldots, y^{(r)}) = 0 \qquad \text{(introduce } \bar{x} = y, \bar{y} = y'\text{)}$$

If a given differential equation $F(x, y, y', y'', \ldots, y^{(r)}) = 0$ is *homogeneous in the arguments* $y, y', y'', \ldots, y^{(r)}$ (Sec. 4.5-5; this does *not* necessarily imply that the differential equation is homogeneous in the sense of Sec. 9.1-2), introduce $\bar{y} = y'/y$.

9.2. FIRST-ORDER EQUATIONS

9.2-1. Existence and Uniqueness of Solutions. (**a**) *A given first-order differential equation expressible in the form*

$$\frac{dy}{dx} = f(x, y) \tag{9.2-1}$$

has a solution $y = y(x)$ *through every "point"* $(x = x_0, y = y_0)$ *with a neighborhood throughout which* $f(x, y)$ *is continuous.* More specifically, *let* D *be a region of "points"* (x, y), (x, η) *where* $f(x, y)$ *is single-valued, bounded, and continuous and*

$$|f(x, y) - f(x, \eta)| \leq M|y - \eta| \qquad \text{(LIPSCHITZ CONDITION)} \tag{9.2-2}$$

for some real M *independent of* y *and* η. *Then the given differential equation* (1) has a unique solution $y = y(x)$ *through every point* $(x = x_0, y = y_0)$ *of* D, *and* $y(x)$ *is a continuous function of the given value* $y_0 = y(x_0)$. Each solution extends to the boundary of D.

The Lipschitz condition (2) is satisfied, in particular, whenever $f(x, y)$ has a bounded and continuous derivative $\partial f/\partial y$ in D.

(**b**) (See also Sec. 9.1-3). An analogous existence theorem applies to *systems of first-order differential equations*

$$\frac{dy_i}{dx} = f_i(x; y_1, y_2, \ldots, y_n) \qquad (i = 1, 2, \ldots, n) \tag{9.2-3}$$

if the Lipschitz condition (2) is replaced by

$$|f(x; y_1, y_2, \ldots, y_n) - f(x; \eta_1, \eta_2, \ldots, \eta_n)| \leq M \sum_{k=1}^{n} |y_k - \eta_k| \tag{9.2-4}$$

9.2-2. Geometrical Interpretation. Singular Integrals (see also Secs. 17.1-1 to 17.1-7). (**a**) If x, y are regarded as rectangular cartesian coordinates, a first-order differential equation

$$F(x, y, p) = 0 \qquad \left(p \equiv \frac{dy}{dx}\right) \tag{9.2-5}$$

describes a "field" of **line elements** (x, y, p) or elements of straight lines through (x, y) with slope $p = dy/dx = f(x, y)$. Each line element is tangent to a curve of the one-parameter family of solutions

$$y = y(x, \lambda) \qquad \text{or} \qquad \varphi(x, y, \lambda) = 0 \tag{9.2-6}$$

where λ is a constant of integration.

A plot of the field of tangent directions permits at least rough *graphical determination of solutions; the general character of the family of solutions may be further discussed in the manner of Sec.* 9.5-2. It may be helpful to know that the curves $F(x, y, p_1) = 0$ or $f(x, y) = p_1$ are **isoclines** where the solution curves have a specified fixed slope p_1. The curves $\frac{\partial f}{\partial x} + \frac{\partial f}{\partial y} f = 0$ are *loci of points of inflection* (see also Sec. 9.5-2).

(**b**) **Singular Integrals** (see also Sec. 9.1-2). Let $F(x, y, p)$ be twice continuously differentiable with respect to x and y, and let $\partial F/\partial y \neq 0$. Elimination of p from

$$F(x, y, p) = 0 \qquad \frac{\partial F(x, y, p)}{\partial p} = 0 \tag{9.2-7}$$

yields a curve or set of curves called the **p discriminant** of the given differential equation (locus of *singular line elements*). A curve defined by Eq. (7) is a singular integral of the given differential equation if $\frac{\partial F}{\partial x} + \frac{\partial F}{\partial y} p = 0$ on this curve, unless both $\partial F/\partial x$ and $\partial F/\partial y$ vanish at a point of the curve. Geometrically, such singular integrals are frequently *envelopes of the family of solution curves* (6), and may thus be obtained from the complete primitive (6) in the manner of Sec. 17.1-7.

9.2-3. Transformation of Variables (see Sec. 9.2-4 for examples). (**a**) A suitable continuously differentiable transformation

$$\left.\begin{aligned} &x = X(\bar{x}, \bar{y}) \qquad y = Y(\bar{x}, \bar{y}) \qquad \left[\frac{\partial(x, y)}{\partial(\bar{x}, \bar{y})} \neq 0\right] \\ &\text{with} \qquad p \equiv \frac{dy}{dx} = \frac{\dfrac{\partial Y}{\partial \bar{x}} + \dfrac{\partial Y}{\partial \bar{y}} \bar{p}}{\dfrac{\partial X}{\partial \bar{x}} + \dfrac{\partial X}{\partial \bar{y}} \bar{p}} \qquad \left(\bar{p} \equiv \frac{d\bar{y}}{d\bar{x}}\right) \end{aligned}\right\} \tag{9.2-8}$$

will transform the given differential equation (1) or (5) into a new differential equation relating $\bar{x}$ and $\bar{y}$. The new equation may be simpler, or a solution $\bar{y} = \bar{y}(\bar{x})$ may be known. Once $\bar{y} = \bar{y}(\bar{x})$ is found, $y = y(x)$ is given implicitly, or by inverse transformation.

(**b**) **Contact Transformations** (see also Secs. 10.2-5, 10.2-7, and 11.6-8). A set of twice continuously differentiable transformation equations

$$\bar{x} = \bar{x}, (x, y, p) \qquad \bar{y} = \bar{y}(x, y, p) \qquad \bar{p} = \bar{p}(x, y, p) \qquad \left[\frac{\partial(\bar{x}, \bar{y}, \bar{p})}{\partial(x, y, p)} \neq 0\right] \tag{9.2-9a}$$

with the special property

$$d\bar{y} - \bar{p}\, d\bar{x} \equiv g(x, y, p)(dy - p\, dx) \qquad [g(x, y, p) \neq 0] \tag{9.2-9b}$$

or

$$\frac{\partial \bar{x}}{\partial p}\left(\frac{\partial \bar{y}}{\partial x} + p\,\frac{\partial \bar{y}}{\partial y}\right) = \frac{\partial \bar{y}}{\partial p}\left(\frac{\partial \bar{x}}{\partial x} + p\,\frac{\partial \bar{x}}{\partial y}\right) \tag{9.2-9c}$$

defines a **contact transformation** associating line elements (Sec. 9.2-2*a*) (x, y, p) and $(\bar{x}, \bar{y}, \bar{p})$ so that line elements forming regular arcs are mapped onto regular arcs, and contact of regular arcs is preserved. It is then legitimate to write $p \equiv \frac{dy}{dx}$ $\bar{p} = \frac{d\bar{y}}{d\bar{x}}$, and to use suitable contact transformations (9) to simplify differential equation (1) or (5). Once a solution $\bar{y} = \bar{y}(\bar{x})$ of the transformed equation is known, $y = y(x)$ is given implicitly or by inverse transformation.

In particular, $g(x, y, p) \equiv 1$ yields the easily reversible contact transformation

$$\bar{x} = p \qquad \bar{y} = px - y \qquad \bar{p} = x \qquad \text{(Legendre transformation)} \tag{9.2-10}$$

which transforms a given differential equation (5) into

$$F(\bar{p}, \bar{p}\bar{x} - \bar{y}, \bar{x}) = 0 \tag{9.2-11}$$

Equation (11) may be a simpler differential equation or, indeed, an ordinary equation relating $\bar{x}$ and $\bar{y}$.

9.2-4. Solution of Special Types of First-order Equations. (a) The following special types of first-order equations are relatively easy to solve.

1. **The variables are separable:** $y' = f_1(x)/f_2(y)$. Obtain the solution from $\int f_2(y)\, dy = \int f_1(x)\, dx + C$.
2. **"Homogeneous" first-order equations:*** $y' = f(y/x)$. Introduce $\bar{y} = y/x$ to reduce to type 1.
3. **Exact differential equations** can be written in the form

$$P(x, y)\, dx + Q(x, y)\, dy = 0 \tag{9.2-12}$$

where the expression on the left is an exact differential $d\varphi$ $\left(\frac{\partial P}{\partial y} = \frac{\partial Q}{\partial x},\right.$ Sec. 5.7-1$\left.\right)$. Obtain the solution from

$$\varphi(x, y) = \int P(x, y)\, dx + \int \left[Q(x, y) - \frac{\partial}{\partial y}\int P(x, y)\, dx\right] dy = C \tag{9.2-13}$$

* Note that the expression "homogeneous" differential equation is here *not* used in the sense defined in Sec. 9.1-2.

If the expression on the left of Eq. (12) is not an exact differential $\left(\frac{\partial P}{\partial y} \neq \frac{\partial Q}{\partial x}\right)$, one may be able to find an **integrating factor** $\mu = \mu(x, y)$ such that multiplication of Eq. (12) by $\mu(x, y)$ yields an exact differential equation. The integrating factor $\mu(x, y)$ satisfies the partial differential equation

$$\mu\left(\frac{\partial P}{\partial y} - \frac{\partial Q}{\partial x}\right) = Q\frac{\partial \mu}{\partial x} - P\frac{\partial \mu}{\partial y} \tag{9.2-14}$$

4. **The linear first-order equation** $y' + a(x)y = f(x)$ (see also Secs. 9.3-1 and 9.3-3) admits the integrating factor

$$\mu = \mu(x) = \exp \int a(x)\, dx$$

The complete primitive is then

$$y = \frac{1}{\mu(x)}\left[\int f(x)\mu(x)\, dx + C\right] \tag{9.2-15}$$

Many first-order equations can be reduced to one of the above types by transformation of variables (Sec. 9.2-3). In particular

$y' = f(\alpha x + \beta y)$ reduces to type 1 if one introduces $\bar{y} = \alpha x + \beta y$.

$y' = \dfrac{\alpha_1 x + \beta_1 y + \gamma_1}{\alpha_2 x + \beta_2 y + \gamma_2}$ reduces to type 2 by a coordinate translation if $\alpha_1\beta_2 - \alpha_2\beta_1 \neq 0$; otherwise introduce $\bar{y} = \alpha_2 x + \beta_2 y + \gamma_2$ to separate the variables.

$y' = f_1(x)y + f_2(x)y^n$ (BERNOULLI'S DIFFERENTIAL EQUATION) reduces to a linear equation if one introduces $\bar{y} = y^{1-n}$.

(**b**) Given a first-order equation of the form

$$y = h(x, y') \qquad \text{or} \qquad x = h(y, y') \tag{9.2-16}$$

it may be advantageous to differentiate both sides with respect to x. The resulting differential equation

$$y' = \frac{\partial h}{\partial x} + \frac{\partial h}{\partial y'}\frac{dy'}{dx} \qquad \text{or} \qquad 1 = \frac{\partial h}{\partial y}y' + \frac{\partial h}{\partial y'}\frac{dy'}{dx} \tag{9.2-17}$$

might be easy to solve for $y' = y'(x)$ or $y' = y'(y)$, respectively; substitution of this result into the given Eq. (16) yields the desired relation of x and y. If the solution of Eq. (17) takes the form $u(x, y') = 0$ or $u(y, y') = 0$, the desired relation of x and y is given in terms of a parameter $p = y'$.

EXAMPLES: *Clairaut's differential equation* $y = y'x + f(y')$ yields the complete primitive $y = Cx + f(C)$ and the singular integral (in parametric representation) $x = -f'(p)$, $y = -pf'(p) + f(P)$. *Lagrange's differential equation* $y = xf_1(p) + f_2(p)$ is solved in the same manner.

(**c**) **Riccati Equations.** The differential equation

$$y' = a(x)y^2 + b(x)y + c(x) \qquad \text{(GENERAL RICCATI EQUATION)} \tag{9.2-18}$$

is sometimes simplified by the transformation $y = 1/\bar{y}$; alternatively,

$$x = \bar{x} \qquad y = \frac{-\bar{y}'}{a(x)\bar{y}}$$

leads to a homogeneous second-order equation for $\bar{y} = \bar{y}(x)$:

$$\bar{y}'' - \left[\frac{a'(x)}{a(x)} + b(x)\right]\bar{y}' + a(x)c(x)\bar{y} = 0 \tag{9.2-19}$$

If a particular integral $y_1(x)$ of Eq. (18) is known, the transformation

$$y = y_1(x) + \frac{1}{\bar{y}}$$

yields a linear differential equation. If one knows two particular integrals y_1, y_2 or three particular integrals y_1, y_2, y_3, one has, respectively,

$$y = y_1 + \frac{y_2 - y_1}{1 + C\exp\int a(x)(y_2 - y_1)\,dx} \qquad y = \frac{y_1(y_2 - y_3) + Cy_2(y_1 - y_3)}{y_2 - y_3 + C(y_1 - y_3)} \tag{9.2-20}$$

For any four particular integrals y_1, y_2, y_3, y_4, the double ratio $(y_1 - y_2)(y_3 - y_4)/(y_1 - y_3)(y_2 - y_4)$ is constant.

The *special Riccati equation*

$$y' + ay^2 = bx^m \tag{9.2-21}$$

can be reduced to type 1 if $m = 4k/(1 - 2k)$ $(k = 0, \pm 1, \pm 2, \ldots)$. For $k > 0$, the transformation $x = \bar{x}^{1/(m+3)}$, $y = \dfrac{1}{x^2\bar{y}} + \dfrac{1}{ax}$ reduces Eq. (21) to a similar equation

$$\frac{d\bar{y}}{d\bar{x}} + \bar{a}\bar{y}^2 = \bar{b}\bar{x}^{\bar{m}} \tag{9.2-22a}$$

with
$$\bar{a} = \frac{b}{m + 3} \qquad \bar{b} = \frac{a}{m + 3} \qquad \bar{m} = -\frac{m + 4}{m + 3} \tag{9.2-22b}$$

The procedure is repeated until (after k steps) the right side of the differential equation is constant.

Similarly, for $k < 0$, the transformation $x = \bar{x}^{-1/(m+1)}$, $y = \dfrac{b}{\bar{x}(b\bar{x}\bar{y} + m + 1)}$ yields a differential equation of the form (22*a*) with

$$\bar{a} = -\frac{b}{m + 1} \qquad \bar{b} = -\frac{a}{m + 1} \qquad \bar{m} = -\frac{3m + 4}{m + 1} \tag{9.2-22c}$$

9.2-5. General Methods of Solution. (a) Picard's Method of Successive Approximations. To solve the differential equation $y' = f(x, y)$ for a given initial value $y(x_0) = y_0$, start with a trial solution $y^{[0]}(x)$ and compute successive approximations

$$y^{[j+1]}(x) = y_0 + \int_{x_0}^{x} f[x, y^{[j]}(x)]\,dx \qquad (j = 0, 1, 2, \ldots) \tag{9.2-23}$$

to the desired solution $y(x)$. *The process converges subject to the conditions of Sec.* 9.2-1. Picard's method is useful mainly if the integrals in Eq. (23) can be evaluated in closed form, although numerical integration can, in principle, be used.

A completely analogous procedure applies to systems (3) of first-order differential equations.

(b) Taylor-series Expansion (see also Sec. 4.10-4). If the given function $f(x,y)$ is suitably differentiable, obtain the coefficients $y^{(m)}(x_0)/m!$ of the Taylor series

$$y(x) = y(x_0) + y'(x_0)(x - x_0) + \frac{1}{2!} y''(x_0)(x - x_0)^2 + \cdots \qquad (9.2\text{-}24)$$

by successive differentiations of the given differential equation:

$$\begin{aligned} y'(x) &= f(x, y) \\ y''(x) &= \frac{\partial f}{\partial x} + \frac{\partial f}{\partial y} y'(x) = \frac{\partial f}{\partial x} + \frac{\partial f}{\partial y} f(x, y) \\ &\cdots\cdots\cdots\cdots\cdots\cdots \end{aligned}$$

with $x = x_0$, $y = y(x_0) = y_0$.

An analogous procedure applies to systems of first-order equations.

9.3. LINEAR DIFFERENTIAL EQUATIONS

9.3-1. Linear Differential Equations. Superposition Theorems (see also Secs. 10.4-2, 13.6-2, 13.6-3, 14.3-1, and 15.4-2). A **linear** ordinary differential equation of order r relating the real or complex variables z and $w = w(z)$ has the form

$$\boxed{\mathsf{L}w \equiv a_0(z) \frac{d^r w}{dz^r} + a_1(z) \frac{d^{r-1} w}{dz^{r-1}} + \cdots + a_r(z) w = f(z)} \qquad (9.3\text{-}1)$$

where the $a_k(z)$ and $f(z)$ are real or complex functions of z. *The general solution (Sec. 9.1-2) of a linear differential equation* (1) *can be expressed as the sum of any particular integral and the general solution of the homogeneous linear differential equation* (Sec. 9.1-2)

$$\mathsf{L}w \equiv a_0(z) \frac{d^r w}{dz^r} + a_1(z) \frac{d^{r-1} w}{dz^{r-1}} + \cdots + a_r(z) w = 0 \qquad (9.3\text{-}2)$$

For any given nonhomogeneous or "complete" linear differential equation (1), the homogeneous equation (2) is known as the **complementary equation** or **reduced equation,** and its general solution as the **complementary function.**

Let $w_1(z)$ and $w_2(z)$ be particular integrals of the linear differential equation (1) *for the respective "forcing functions" $f(z) \equiv f_1(z)$ and $f(z) \equiv f_2(z)$. Then $\alpha w_1(z) + \beta w_2(z)$ is a particular integral for the forcing function $f(z) \equiv \alpha f_1(z) + \beta f_2(z)$ (Superposition Principle).* In particular, *every linear combination of solutions of a homogeneous linear differential equation* (2) *is also a solution.*

The superposition theorems often represent some physical superposition principle. Mathematically, they permit one to construct solutions of Eq. (1) or (2) subject to given initial or boundary conditions by linear superposition.

Analogous theorems apply to systems of linear differential equations (see also Sec. 9.4-2).

9.3-2. Linear Independence and Fundamental Systems of Solutions (see also Secs. 1.9-3, 14.2-3, and 15.2-1*a*). **(a)** Let $w_1(z)$, $w_2(z)$, . . . , $w_r(z)$ be $r - 1$ times continuously differentiable solutions of a homogeneous linear differential equation (2) with continuous coefficients in a domain D of values of z. The r solutions $w_k(z)$ are *linearly independent* in D if and only if $\sum_{k=1}^{r} \lambda_k w_k(z) \equiv 0$ in D implies $\lambda_1 = \lambda_2 = \cdots = \lambda_r = 0$ (Sec. 1.9-3). *This is true if and only if the* **Wronskian determinant (Wronskian)**

$$W[w_1, w_2, \ldots, w_r] \equiv \begin{vmatrix} w_1(z) & w_2(z) & \cdots & w_r(z) \\ w_1'(z) & w_2'(z) & \cdots & w_r'(z) \\ \cdots & \cdots & \cdots & \cdots \\ w_1^{(r-1)}(z) & w_2^{(r-1)}(z) & \cdots & w_r^{(r-1)}(z) \end{vmatrix} \tag{9.3-3}$$

differs from zero throughout D. $W = 0$ for any z in D implies $W \equiv 0$ for all z in D.*

(b) *A homogeneous linear differential equation* (2) *of order r has at most r linearly independent solutions.* r linearly independent solutions $w_1(z)$, $w_2(z)$, . . . , $w_k(z)$ constitute a **fundamental system** of solutions whose linear combinations $\sum_{k=1}^{r} \alpha_k w_k(z)$ include all particular integrals of Eq. (2).

(c) Use of Known Solutions to Reduce the Order. If $m < r$ linearly independent solutions $w_1(z)$, $w_2(z)$, . . . , $w_m(z)$ of the homogeneous equation (2) are known, then the transformation $\bar{w} = W[w_1, w_2, \ldots, w_m, w]\varphi(x)$ reduces Eq. (2) to a homogeneous linear differential equation of order $r - m$ for any conveniently chosen $\varphi(x)$.

9.3-3. Solution by Variation of Constants. Green's Functions. **(a)** Given r linearly independent solutions $w_1(z)$, $w_2(z)$, . . . , $w_r(z)$ of the homogeneous linear differential equation (2), the general solution of the complete nonhomogeneous equation (1) is

$$w = C_1(z)w_1(z) + C_2(z)w_2(z) + \cdots + C_r(z)w_r(z) \tag{9.3-4}$$

* Note that the theorem in this simple form does *not* apply to every set of $r - 1$ times continuously differentiable functions $w_k(z)$; they must be solutions of a suitable differential equation (2).

$$\left.\begin{aligned} \text{with} \quad \sum_{k=1}^{r} C_k'(z) w_k^{(j)}(z) &= 0 \qquad (j = 0, 1, 2, \ldots, r-2) \\ \sum_{k=1}^{r} C_k'(z) w_k^{(r-1)}(z) &= \frac{f(z)}{a_0(z)} \end{aligned}\right\} \tag{9.3-5}$$

After solving the r simultaneous equations (5) for the r unknown derivatives $C_k'(z)$, one obtains each $C_k(z) = \int C_k'(z)\,dz + K_k$ by a simple integration. In principle, *this procedure reduces the solution of any linear ordinary differential equation to the solution of a homogeneous linear differential equation.*

(**b**) Assuming real variables $z \equiv x$ and $w \equiv w(x)$ for simplicity, particular integrals of the complete differential equation (1) can often be written as

$$w = \int_a^b G(x, \xi) f(\xi)\,d\xi \qquad (a < x < b) \tag{9.3-6}$$

where $G(x, \xi)$ is known as the **Green's function** (sometimes called the **weighting function,** Sec. 9.4-3) yielding the specific particular integral in question. The complete integral of Eq. (1) is then

$$w = \int_a^b G(x, \xi) f(\xi)\,d\xi + \sum_{k=1}^{r} A_k w_k(x) \tag{9.3-7}$$

where the $w_k(z)$ are r linearly independent solutions of Eq. (2), and the A_k are r constants of integration to be determined by suitable initial or boundary conditions.

Any given set of r linearly independent solutions $w_k(x)$ of the complementary equation (2) permits one to construct a particular integral (6) with

$$G(x, \xi) = \frac{1}{f(\xi)} \sum_{k=1}^{r} C_k'(\xi) w_k(x) U(x - \xi) \tag{9.3-8}$$

where the $C_k'(x)$ are obtained from Eq. (5), and $U(x)$ is the unit-step function defined in Sec. 21.9-1.

For linear differential equations of order $r = 2$, *the complete integral is given by Eq.* (4) *or* (7) *with*

$$\left.\begin{aligned} C_1'(x) &= -\frac{f(x)}{a_0(x)} \frac{w_2(x)}{w_1(x)w_2'(x) - w_2(x)w_1'(x)} \\ C_2'(x) &= \frac{f(x)}{a_0(x)} \frac{w_1(x)}{w_1(x)w_2'(x) - w_2(x)w_1'(x)} \\ G(x, \xi) &= -\frac{1}{a_0(\xi)} \frac{w_1(x)w_2(\xi) - w_2(x)w_1(\xi)}{w_1(\xi)w_2'(\xi) - w_2(\xi)w_1'(\xi)} U(x - \xi) \end{aligned}\right\} \tag{9.3-9}$$

(c) While the general solution (7) obtained with the aid of the particular Green's function (8) is only another way of writing Eq. (4), *it is often possible to construct a Green's function $G(x, \xi)$ such that the particular integral* (6) *satisfies the specific initial or boundary conditions of a given problem.* Assuming boundary conditions linear and homogeneous in $w(x)$ and its derivatives, the required Green's function $G(x, \xi)$ must satisfy the given boundary conditions and

$$\mathsf{L}G(x, \xi) = 0 \quad (x \neq \xi) \qquad \int_a^b \mathsf{L}G(x, \xi)\, d\xi = 1 \tag{9.3-10}$$

for x in (a, b), with $\partial^{r-2}G/\partial x^{r-2}$ continuous in (a, b), and

$$\left.\frac{\partial^{r-1}G}{\partial x^{r-1}}\right]_{x=\xi+0} - \left.\frac{\partial^{r-1}G}{\partial x^{r-1}}\right]_{x=\xi-0} = \frac{1}{a_0(\xi)} \qquad (a < \xi < b) \tag{9.3-11}$$

The existence and properties of such Green's functions are discussed from a more general point of view in Sec. 15.5-1; see also Sec. 9.4-3. Table 9.3-1 lists the Green's functions for a number of boundary-value problems.

9.3-4. Reduction of Two-point Boundary-value Problems to Initial-value Problems. The general theory of boundary-value problems and eigenvalue problems involving ordinary differential equations is treated in Secs. 15.4-1 to 15.5-2 (see also Secs. 9.3-3, 20.9-2, and 20.9-3). The following method is often useful in connection with numerical solution methods.

Given an rth-order linear differential equation $\mathsf{L}w = f(z)$ with r suitable boundary conditions to be satisfied by $w(z)$ and its derivatives for $z = a$, $z = b$, write the solution as

$$w = w_0(z) + \sum_{k=1}^{r} \alpha_k w_k(z) \tag{9.3-12a}$$

where the $w_k(z)$ are defined by the $r + 1$ *initial-value problems*

$$\left.\begin{array}{ll} \mathsf{L}w_0(z) = f(z) \quad \text{with} \quad w_0^{(j)}(a) = 0 & (j = 0, 1, 2, \ldots, r-1) \\ \mathsf{L}w_k(z) = 0 \quad \text{with} \quad w_k^{(j)}(a) = \begin{cases} 0 \text{ for } j \neq k-1 \\ 1 \text{ for } j = k-1 \end{cases} & \begin{array}{l}(j = 0, 1, 2, \ldots, r-1; \\ \quad k = 1, 2, \ldots, r)\end{array} \end{array}\right\} \tag{9.3-12b}$$

Apply the r given boundary conditions to the general solution (12a) to obtain r simultaneous equations for the r unknown coefficients α_k.

NOTE: Given a *nonlinear* boundary-value problem like

$$\frac{d^2w}{dz^2} = f(z, w, w') \qquad \text{with} \qquad w(a) = w_a \qquad w(b) = w_b \tag{9.3-13}$$

one can often calculate $w(b)$ for two or three trial values of the unknown initial value $w'(a)$; the correct value of $w'(a)$ is then approximated by interpolation.

9.3-5. Complex-variable Theory of Linear Differential Equations. Taylor-series Solution and Effects of Singularities. **(a)** *A given*

Table 9.3-1. Green's Functions for Linear Boundary-value Problems

Each boundary-value problem listed has the solution $w(x) = \int_a^b G(x, \xi)f(\xi)\,d\xi$. Use $G(x, \xi)$ to obtain solutions for other initial or boundary conditions from Eq. (9.3-7) (see also Secs. 9.3-3, 9.4-3, 10.4-2, 15.4-8, and 15.5-1). The table yields solutions for other intervals (a, b) with the aid of suitable coordinate transformations.

No.	Differential equation	Interval (a, b)	Boundary conditions	$G(x, \xi)$ $(x \le \xi)$ $G(\xi, x)$ $(x \ge \xi)$
1a	$\frac{d^2w}{dx^2} = f(x)$	$a = 0$, $b = 1$	$w(0) = w(1) = 0$	$-(1 - \xi)x$
1b			$w(0) = 0$ $w'(1) = 0$	$-x$
1c			$w(0) = -w(1)$ $w'(0) = -w'(1)$	$-\frac{1}{2}(x - \xi) - \frac{1}{4}$
1d		$a = -1$, $b = 1$	$w(-1) = w(1) = 0$	$-\frac{1}{2}(x - \xi - x\xi + 1)$
1e			$w(-1) = w(1)$ $w'(-1) = w'(1)$	$-\frac{1}{4}(x - \xi)^2 - \frac{1}{2}(x - \xi) - \frac{1}{6}$*
2	$\frac{d^2w}{dx^2} - w = f(x)$	$a = -\infty$ $b = \infty$	w finite in $(-\infty, \infty)$	$-\frac{1}{2}e^{x-\xi}$
3a	$\frac{d^2w}{dx^2} + k^2w = f(x)$	$a = 0$ $b = 1$	$w(0) = w(1) = 0$	$-\frac{\sin kx \sin k(1 - \xi)}{k \sin k}$
3b		$a = -1$ $b = 1$	$w(-1) = w(1)$ $w'(-1) = w'(1)$	$\frac{\cos k(x - \xi + 1)}{2k \sin k}$
4a	$\frac{d^2w}{dx^2} - k^2w = f(x)$	$a = 0$ $b = 1$	$w(0) = w(1) = 0$	$-\frac{\sinh kx \sinh k(1 - \xi)}{k \sinh k}$
4b		$a = -1$ $b = 1$	$w(-1) = w(1)$ $w'(-1) = w'(1)$	$-\frac{\cosh k(x - \xi + 1)}{2k \sinh k}$
5	$\frac{d}{dx}\left(x\frac{dw}{dx}\right) - \frac{m^2}{x}w = f(x)$ (INHOMOGENEOUS BESSEL EQUATION)	$a = 0$ $b = 1$	$w(0)$ finite $w(1) = 0$	$\log_e \xi$ $(m = 0)$ $-\frac{1}{m}\left[\left(\frac{x}{\xi}\right)^m - (x\xi)^m\right]$ $(m = 1, 2, \ldots)$
6	$\frac{d}{dx}\left[(1 - x^2)\frac{dw}{dx}\right] - \frac{m^2}{1 - x^2}w = f(x)$ (INHOMOGENEOUS LEGENDRE EQUATION)	$a = -1$ $b = 1$	$w(-1)$ finite $w(1)$ finite	$\frac{1}{2}\log_e (1 - x)(1 + \xi) - \log_e 2 + \frac{1}{2}$ $(m = 0)$* $-\frac{1}{2m}\left(\frac{1 + x}{1 - x}\cdot\frac{1 - \xi}{1 + \xi}\right)^{m/2}$ $(m = 1, 2, \ldots)$
7	$\frac{d^4w}{dx^4} = f(x)$	$a = 0$ $b = 0$	$w(0) = w'(0) = w(1) = w'(1) = 0$	$-\frac{x^2(\xi - 1)^2}{6}(2x\xi + x - 3\xi)$

* This is a *modified Green's function* in the sense of Sec. 15.5-1b and does not satisfy Eq. (9.3-10)

linear differential equation (1) *has an analytic solution* $w = w(z)$ *at every* **regular point** z *where the functions* $a_k(z)$ *and* $f(z)$ *are analytic* (see also Sec. 7.3-3). *If these functions are single-valued, and if* D *is a singly connected region of regular points, a given set of values* $w(z_0)$, $w'(z_0)$, . . . , $w^{(r-1)}(z_0)$ *for some point* z_0 *in* D *defines a unique solution* $w(z)$ *in* D.

To obtain this solution in Taylor-series form (see also Sec. 4.10-4), substitute

$$w(z) = w(z_0) + w'(z_0)(z - z_0) + \frac{1}{2!} w''(z_0)(z - z_0)^2 + \cdots \qquad (9.3\text{-}14)$$

and

$$f(z) = f(z_0) + f'(z_0)(z - z_0) + \frac{1}{2!} f''(z_0)(z - z_0)^2 + \cdots$$

into the given differential equation (1); comparison of coefficients will yield *recurrence relations* for the unknown coefficients $\frac{1}{k!} w^{(k)}(z_0)$ ($k = r$, $r + 1$, . . .). The series (14) converges absolutely and uniformly within every circle $|z - z_0| < R$ in D.

(**b**) Analytic continuation of any solution $w(z)$ of a linear differential equation around singularities of one or more coefficients $a_k(z)$ will, in general, yield different branches of a multiple-valued solution (see also Secs. 7.4-2, 7.6-2, and 7.8-1).

In particular, one complete circuit around a singularity will transform a fundamental system of solutions $w_1(z)$, $w_2(z)$, . . . , $w_r(z)$ of the homogeneous linear differential equation (2) into a new fundamental system $\bar{w}_1(z)$, $\bar{w}_2(z)$, . . . , $\bar{w}_r(z)$. The two fundamental systems are necessarily related by a nonsingular linear transformation

$$\bar{w}_i(z) = \sum_{k=1}^{r} a_{ik} w_k(z) \qquad (i = 1, 2, \ldots, r) \qquad (9.3\text{-}15)$$

The eigenvalues λ_k of the matrix $[a_{ik}]$ (Sec. 13.4-2) are independent of the particular fundamental system $w_1(z)$, $w_2(z)$, . . . , $w_r(z)$ in question.

9.3-6. Solution of Homogeneous Equations by Series Expansion about a Regular Singular Point. (**a**) A singularity $z = z_1$ of one or more $a_k(z)$ is an **isolated singularity** of the homogeneous linear differential equation (2) if and only if z_1 has a neighborhood containing no other singular point. An isolated singularity $z = z_1$ is a **regular singular point** of the homogeneous linear differential equation if and only if none of its solutions $w(z)$ has an essential singularity at z_1; otherwise $z = z_1$ is an **essential singularity** of the given differential equation. $z = z_1$ *is a regular singular point if and only if* $a_k(z)/a_0(z)$ *has at worst a pole of order* k *at* ($k = 1, 2, \ldots, r$) (*Fuchs's Theorem*). In this case,

Eq. (2) can be rewritten as

$$(z - z_1)^r \frac{d^r w}{dz^r} + (z - z_1)^{r-1} p_1(z) \frac{d^{r-1}w}{dz^{r-1}} + (z - z_1)^{r-2} p_2(z) \frac{d^{r-2}w}{dz^{r-2}} + \cdots + p_r(z)w = 0 \quad (9.3\text{-}16)$$

where all $p_k(z)$ are analytic in a neighborhood D_1 of z_1.

It follows that *a given homogeneous linear differential equation* (2) *admits a solution of the form*

$$w = (z - z_1)^{\mu} \sum_{k=0}^{\infty} a_k (z - z_1)^k \quad (9.3\text{-}17)$$

whenever $z = z_1$ *is a regular point or a regular singular point.* The exponent μ must satisfy the rth-degree algebraic equation

$$\mu(\mu - 1) \cdots (\mu - r + 1) + \mu(\mu - 1) \cdots (\mu - r + 2)p_1(z_1) + \cdots + \mu p_{r-1}(z_1) + p_r(z_1) = 0 \quad \text{(INDICIAL EQUATION)} \quad (9.3\text{-}18)$$

The first coefficient a_0 may be chosen at will, and the other coefficients a_k are found successively from a set of recurrence relations obtained on substitution of the series (17) into Eq. (2) or (16). The series converges absolutely and uniformly within every circle $|z - z_1| < R$ in D_1.

Different roots $\mu = \mu_1, \mu_2, \ldots, \mu_r$ of the indicial equation (18) yield linearly independent solutions (17) of the given differential equation, unless two roots μ_k coincide or differ by an integer. In such cases, one may use the known solutions to reduce the order of the given differential equation in the manner of Sec. 9.3-2*c* or 9.3-7*a*, or use Frobenius's method (Ref. 9.14); see also Sec. 9.3-8*a*.

The exponents μ_k are related to the eigenvalues λ_k obtained from Eq. (15) for one circuit about the regular singular point z_1, with $\lambda_k = e^{2\pi i \mu_k}$.

(b) Regular Singular Points at Infinity (see also Secs. 7.2-1 and 7.6-3). $z = \infty$ is a regular singular point of Eq. (2) if and only if the transformation

$$z = \frac{1}{\bar{z}} \qquad w(z) = w\left(\frac{1}{\bar{z}}\right) = \bar{w}(\bar{z})$$
$$\frac{dw}{dz} = -\bar{z}^2 \frac{d\bar{w}}{d\bar{z}} \quad \frac{d^2w}{dz^2} = \bar{z}^4 \frac{d^2\bar{w}}{d\bar{z}^2} + 2\bar{z}^3 \frac{d\bar{w}}{d\bar{z}} \quad \cdots \quad (9.3\text{-}19)$$

yields a differential equation having a regular singular point at $\bar{z} = 0$. In this case, one may obtain solutions of the transformed equation in the manner of Sec. 9.3-6*a*.

(c) Generalization. If $z = z_1$ is not a regular singular point (e.g., if the functions p_k have poles at $z = z_1$) one may still write solutions similar to Eq. (17) by replacing each power series by a Laurent series (Sec. 7.5-3) admitting negative as well as positive powers of $(z - z_1)$.

9.3-7. Integral-transform Methods. The solution of a linear differential equation (1) with *polynomial coefficients* $a_k(z) \equiv \sum_h a_{hk} z^h$ and given initial conditions $w^{(j)}(0) = w_0^{(j)}$ is often simplified through the use of the unilateral Laplace transformation in the manner of Sec. 9.4-5. Apply the formula

$$\mathcal{L}[z^h w^{(k)}(z); s] = (-1)^h \frac{d^h}{ds^h} \left\{ s^k \mathcal{L}[w(z); s] - \sum_{j=0}^{k-1} s^{k-j-1} w^{(j)}(0+0) \right\} \qquad (9.3\text{-}20)$$

to obtain a new and possibly simpler differential equation for the Laplace transform $\mathcal{L}[w(z); s]$ of the solution. Boundary-value problems can be transformed to initial-value problems by the method of Sec. 9.3-4.

More general integral transformations (Table 8.6-1) may be similarly employed in various special cases (see also Sec. 10.5-1 and Ref. 9.14).

9.3-8. Linear Second-order Equations (see also Secs. 15.4-3c and 15.5-4). **(a)** The theory of Secs. 9.3-1 to 9.3-4 applies to linear second-order equations, so that one is mainly interested in the solution of homogeneous linear second-order equations

$$\boxed{\mathbf{L}w \equiv \frac{d^2w}{dz^2} + a_1(z)\frac{dw}{dz} + a_2(z)w = 0} \qquad (9.3\text{-}21)$$

Equation (21) is equivalent to

$$\left.\begin{aligned} &\frac{d}{dz}\left[p(z)\frac{dw}{dz}\right] + q(z)w = 0 \\ &p(z) = \exp \int a_1(z)\,dz \qquad q(z) = a_2(z)p(z) \end{aligned}\right\} \qquad (9.3\text{-}22)$$

If any solution $w_1(z)$ of Eq. (21) or (22) is known, the complete primitive is

$$w(z) = w_1(z)\left[C_1 + C_2 \int \frac{dz}{w_1^2(z)p(z)}\right] \qquad (9.3\text{-}23)$$

(b) Series Expansion of Solutions (see also Secs. 9.3-5, 9.3-6, 9.3-9, and 9.3-10). In the important special case of a second-order equation expressible in the form

$$(z - z_1)^2 \frac{d^2w}{dz^2} + (z - z_1)p_1(z)\frac{dw}{dz} + p_2(z)w = 0 \qquad (9.3\text{-}24)$$

where $p_1(z)$ and $p_2(z)$ are analytic at $z = z_1$, the indicial equation (18) reduces to

$$\mu^2 + \mu[p_1(z_1) - 1] + p_2(z_1) = 0 \qquad (9.3\text{-}25)$$

The indicial equation has two roots. The root $\mu = \mu_1$ having the larger real part yields a solution of the form

$$w = w_1(z) \equiv (z - z_1)^{\mu_1} \sum_{k=0}^{\infty} a_k(z - z_k)^k \qquad (9.3\text{-}26)$$

The second root μ_2 yields a similar linearly independent solution, which is replaced by

$$w = Aw_1(z) \log_e (z - z_1) + (z - z_1)^{\mu_1} \sum_{k=1}^{\infty} b_k(z - z_1)^k \quad (9.3\text{-}27)$$

if μ_1 and μ_2 are identical or differ by an integer. Substitute each solution (26) or (27) into the given differential equation (24) to obtain recurrence relations for the coefficients.

(c) Transformation of Variables. The following transformations may simplify a given differential equation (21) or reduce it to a differential equation with a known solution.

1. $w = \bar{w} \exp \int \varphi(z)\, dz$ yields

$$\frac{d^2\bar{w}}{dz^2} + [a_1(z) + 2\varphi(z)] \frac{d\bar{w}}{dz} + [a_2(z) + a_1(z)\varphi(z) + \varphi'(z) + \varphi^2(z)]\bar{w} = 0 \quad (9.3\text{-}28)$$

   One attempts to choose $\varphi(z)$ so as to simplify the new differential equation; *in particular,* $\varphi(z) = -a_1(z)/2$ *eliminates the coefficient of* $d\bar{w}/dz$. A suitable substitution $z = z(\bar{z})$ may also yield a simpler differential equation.

2. The substitution

$$w = Ae^{\int \bar{w}(z)\,dz} \qquad \bar{w} = \frac{w'}{w} \quad (9.3\text{-}29)$$

transforms Eq. (21) into a first-order differential equation of the Riccati type (Sec. 9.2-4c).

(d) Existence and Zeros of Solutions for Real Arguments. Let x be a real variable. *The homogeneous linear differential equation*

$$\mathsf{L}w \equiv \frac{d^2w}{dx^2} + a_1(x) \frac{dw}{dx} + a_2(x)w = 0 \quad (9.3\text{-}30)$$

has a solution $w = w(x)$ *in every interval* $[a, b]$ *where* $a_1(x)$ *and* $a_2(x)$ *are real and continuous; the solution is uniquely determined by the values* $w(x_0)$, $w'(x_0)$ *for some* x_0 *in* $[a, b]$. *Unless* $w \equiv 0$ *in* $[a, b]$, $w(x)$ *has at most a finite number of zeros in any finite interval* $[a, b]$; *the zeros of any two linearly independent solutions alternate in* $[a, b]$.

9.3-9. Gauss's Hypergeometric Differential Equation and Riemann's Differential Equation (see also Secs. 9.3-5 and 9.3-6). **(a)** The homogeneous linear differential equation

$$z(1 - z) \frac{d^2w}{dz^2} + [c - (a + b + 1)z] \frac{dw}{dz} - abw = 0$$

(HYPERGEOMETRIC DIFFERENTIAL EQUATION) (9.3-31)

has regular singular points at $z = \infty$ (exponents $\mu = a$, $\mu = b$), $z = 1$ (exponents $\mu = 0$, $\mu = c - a - b$), and $z = 0$ (exponents $\mu = 0$, $\mu = 1$

$- c)$, and no other singularities. The solutions of Eq. (31) include many elementary functions, as well as many of the special transcendental functions of Chap. 21 as special cases.

Series expansion about $z = z_1 = 0$ yields solutions (**hypergeometric functions**) (17) for $\mu = 0$ and $\mu = 1 - c$, with

$$a_{k+1} = \frac{(a + \mu + k)(b + \mu + k)}{(c + \mu + k)(1 + \mu + k)} a_k$$

For $\mu = 0$ one obtains the special hypergeometric function

$$w = F(a, b; c; z) \equiv 1 + \frac{ab}{c} z + \frac{1}{2!} \frac{a(a+1)b(b+1)}{c(c+1)} z^2 + \cdots$$

(HYPERGEOMETRIC SERIES) (9.3-32)

The series converges uniformly and absolutely for $|z| < 1$; the convergence extends to the unit circle if Re $(a + b - c) < 1$, except for the point $z = 1$ if Re $(a + b - c) \geq 0$. The series reduces to a geometric series (Sec. 4.10-2) for $a = 1$, $b = c$, and to a *Jacobi polynomial* (Sec. 21.7-8) if a and/or b equals zero or any negative integer. The function (32) is undefined if one of the denominators c, $c + 1$, . . . equals zero and does not cancel out.

A second (linearly independent) solution of Eq. (31) may be obtained in the manner of Sec. 9.3-8*b*; in particular, the **hypergeometric function of the second kind**

$$\Phi(a, b; c; z) \equiv \frac{\Gamma(a - c + 1)\Gamma(b - c + 1)\Gamma(c - 1)}{\Gamma(a)\Gamma(b)\Gamma(1 - c)} z^{1-c} F(a - c + 1, b - c + 1; 2 - c; z) \quad (9.3\text{-}33)$$

is a solution whenever c is not an integer.

Note the following relations ($|z| < 1$):

$$F(a, b; c; z) = \frac{\Gamma(c)}{\Gamma(a)\Gamma(c - a)} \int_0^1 \zeta^{a-1}(1 - \zeta)^{c-a-1}(1 - z\zeta)^{-b}\, d\zeta$$

$$[\text{Re}\,(c) > \text{Re}\,(a) > 0] \quad (9.3\text{-}34)$$

$$\frac{dF}{dz} = \frac{ab}{c} F(a + 1, b + 1; c + 1; z) \qquad \frac{d\Phi}{dz} = \frac{ab}{c} \Phi(a + 1, b + 1; c + 1; z) \quad (9.3\text{-}35)$$

$$F(a, b; c; 1) = \frac{\Gamma(c)\Gamma(c - a - b)}{\Gamma(c - a)\Gamma(c - b)} \qquad [\text{Re}\,(a + b - c) < 0] \quad (9.3\text{-}36)$$

The following formulas serve for analytic continuation of $F(a, b; c; z)$ outside the unit circle:

$$\begin{aligned} F(a, b; c; z) &= (1 - z)^{-a} F\left(a, c - b; c; \frac{z}{z - 1}\right) \\ &= (1 - z)^{-b} F\left(b, c - a; c; \frac{z}{z - 1}\right) \\ &= (1 - z)^{c-a-b} F(c - a, c - b; c; z) \end{aligned} \quad (9.3\text{-}37)$$

$$\begin{aligned} F(a, b; c; z) &= \frac{\Gamma(c)\Gamma(c - a - b)}{\Gamma(c - a)\Gamma(c - b)} F(a, b; a + b - c + 1; 1 - z) \\ &+ (1 - z)^{c-a-b} \frac{\Gamma(c)\Gamma(a + b - c)}{\Gamma(a)\Gamma(b)} F(c - a, c - b; c - a - b + 1; 1 - z) \end{aligned} \quad (9.3\text{-}38)$$

$$F(a, b; c; z) = \frac{\Gamma(c)\Gamma(b-a)}{\Gamma(b)\Gamma(c-a)}(-z)^{-a}F\left(a, 1-c+a; 1-b+a; \frac{1}{z}\right)$$
$$+ \frac{\Gamma(c)\Gamma(a-b)}{\Gamma(a)\Gamma(c-b)}(-z)^{-b}F\left(b, 1-c+b; 1-a+b; \frac{1}{z}\right) \qquad (9.3\text{-}39)$$

See Table 9.3-1 and Refs. 21.9 and 21.11 for additional formulas.

(b) Riemann's Differential Equation. Paperitz Notation. Equation (31) is a special case of the linear homogeneous differential equation

$$\frac{d^2w}{dz^2} + \left(\frac{1-\alpha-\alpha'}{z-z_1} + \frac{1-\beta-\beta'}{z-z_2} + \frac{1-\gamma-\gamma'}{z-z_3}\right)\frac{dw}{dz}$$
$$+ \left[\frac{\alpha\alpha'(z_1-z_2)(z_1-z_3)}{z-z_1} + \frac{\beta\beta'(z_2-z_3)(z_2-z_1)}{z-z_2} + \frac{\gamma\gamma'(z_3-z_1)(z_3-z_2)}{z-z_3}\right]$$
$$\cdot \frac{w}{(z-z_1)(z-z_2)(z-z_3)} = 0 \quad \text{(RIEMANN'S DIFFERENTIAL EQUATION)} \qquad (9.3\text{-}40)$$

whose only singularities are distinct regular singular points at $z = z_1$ (exponents α, α'), $z = z_2$ (exponents β, β'), and $z = z_3$ (exponents γ, γ'); note

$$\alpha + \alpha' + \beta + \beta' + \gamma + \gamma' = 1$$

A solution of Eq. (40), written in the so-called *Paperitz notation*, is

$$w(z) = P\begin{Bmatrix} z_1 & z_2 & z_3 & \\ \alpha & \beta & \gamma & z \\ \alpha' & \beta' & \gamma' & \end{Bmatrix} = \left(\frac{z-z_1}{z-z_2}\right)^{\alpha}\left(\frac{z-z_3}{z-z_2}\right)^{\gamma} F\left[\alpha+\beta+\gamma, \alpha+\beta'+\gamma;\right.$$
$$\left. 1+\alpha-\alpha'; \frac{(z-z_1)(z_3-z_2)}{(z-z_2)(z_3-z_1)}\right] \qquad (9.3\text{-}41)$$

which reduces to Eq. (32) for $z_1 = 0$, $z_2 = \infty$, $z_3 = 1$ and $\alpha = 0$, $\beta = a$, $\gamma = 0$; $\alpha' = 1 - c$, $\beta' = b$, $\gamma' = c - a - b$. See Table 9.3-2 and Ref. 21.11 for additional formulas.

9.3-10. Confluent Hypergeometric Functions. One can move the singularity $z = 1$ of the hypergeometric differential equation (31) to $z = b$ by substituting z/b for z; the singularity at $z = b$ will then approach the original singularity at $z = \infty$ as $b \to \infty$ (*confluence of singularities*). One thus obtains the new differential equation

$$z\frac{d^2w}{dz^2} + (c-z)\frac{dw}{dz} - aw = 0 \quad \text{(KUMMER'S CONFLUENT HYPERGEOMETRIC DIFFERENTIAL EQUATION)} \qquad (9.3\text{-}42)$$

whose only singularities are a regular singular point at $z = 0$ and an essential singularity at $z = \infty$. Many special transcendental functions are solutions of Eq. (40) for special values of a and c (Chap. 21).

Series expansion about $z = z_1 = 0$ yields solutions (**confluent hypergeometric functions**) (17) for $\mu = 0$ and $\mu = 1 - c$, with

$$a_{k+1} = \frac{a+\mu+k}{(c+\mu+k)(1+\mu+k)}a_k$$

Table 9.3-1. Additional Formulas Relating to Hypergeometric Functions

1. Gauss's Recursion Formulas

$$cF(a, b-1; c; z) - cF(a-1, b; c; z) + (a-b)zF(a, b; c+1; z) = 0$$

$$c(a-b)F(a, b; c; z) - a(c-b)F(a+1, b; c+1; z) + b(c-a)F(a, b+1; c+1; z) = 0$$

$$c(c+1)F(a, b; c; z) - c(c+1)F(a, b; c+1; z) - abzF(a+1, b+1; c+2; z) = 0$$

$$cF(a, b; c; z) - (c-a)F(a, b+1; c+1; z) - a(1-z)F(a+1, b+1; c+1; z) = 0$$

$$cF(a, b; c; z) + (b-c)F(a+1; b; c+1; z) - b(1-z)F(a+1, b+1; c+1; z) = 0$$

$$c(c-bz-a)F(a, b; c; z) - c(c-a)F(a-1, b; c; z) + abz(1-z)F(a+1, b+1; c+1; z) = 0$$

$$c(c-az-b)F(a, b; c; z) - c(c-b)F(a, b-1; c; z) + abz(1-z)F(a+1, b+1; c+1; z) = 0$$

$$cF(a, b; c; z) - cF(a, b+1; c; z) + azF(a+1, b+1; c+1; z) = 0$$

$$cF(a, b; c; z) - cF(a+1, b; c; z) + bzF(a+1, b+1; c+1; z) = 0$$

$$c\{a-(c-b)z\}F(a, b; c; z) - ac(1-z)F(a+1, b; c; z) + (c-a)(c-b)zF(a, b; c+1; z) = 0$$

$$c\{b-(c-a)z\}F(a, b; c; z) - bc(1-z)F(a, b+1; c; z) + (c-a)(c-b)zF(a, b; c+1; z) = 0$$

$$c(c+1)F(a, b; c; z) - c(c+1)F(a, b+1; c+1; z) + a(c-b)zF(a+1, b+1; c+2; z) = 0$$

$$c(c+1)F(a, b; c; z) - c(c+1)F(a+1, b; c+1; z) + b(c-a)zF(a+1, b+1; c+2; z) = 0$$

$$cF(a, b; c; z) - (c-b)F(a, b; c+1; z) - bF(a, b+1; c+1; z) = 0$$

$$cF(a, b; c; z) - (c-a)F(a, b; c+1; z) - aF(a+1, b; c+1; z) = 0$$

2. Miscellaneous Formulas

$$F(a, b; 2b; z) = \left(1 - \frac{z}{2}\right)^{-a} F\left[\frac{a}{2}, \frac{a+1}{2}; b + \frac{1}{2}; \left(\frac{z}{2-z}\right)^2\right]$$

$$(1+z)^{2a}F\left(a, a + \frac{1}{2} - b; b + \frac{1}{2}; z^2\right) = F\left(a, b; 2b; \frac{4z}{(1+z)^2}\right)$$

$$(1+z)^{2a}F(2a, 2a+1-c; c; z) = F\left(a, a+\frac{1}{2}; c; \frac{4z}{(1+z)^2}\right)$$

$$F\left(a, b; a+b+\frac{1}{2}; \sin^2 \vartheta\right) = F\left(2a, 2b; a+b+\frac{1}{2}; \sin^2 \frac{\vartheta}{2}\right)$$

Table 9.3-1. Additional Formulas Relating to Hypergeometric Functions *(Continued)*

3. Some Elementary Functions Expressed in Terms of the Hypergeometric Function (see also Table 21.7-1)

$$(1+x)^n = F(-n, 1; 1; -x)$$

$$\log_e (1+x) = xF(1, 1; 2; -x)$$

$$\arcsin x = xF\left(\frac{1}{2}, \frac{1}{2}; \frac{3}{2}; x^2\right)$$

$$\arctan x = xF\left(\frac{1}{2}, 1; \frac{3}{2}; -x^2\right)$$

$$\sin nx = n \sin x\, F\left(\frac{1+n}{2}, \frac{1-n}{2}; \frac{3}{2}; \sin^2 x\right)$$

$$\cos nx = F\left(\frac{n}{2}, -\frac{n}{2}; \frac{1}{2}; \sin^2 x\right)$$

$$= \cos x\, F\left(\frac{1+n}{2}, \frac{1-n}{2}; \frac{1}{2}; \sin^2 x\right)$$

$$= \cos^n x\, F\left(-\frac{n}{2}, \frac{1-n}{2}; \frac{1}{2}; -\tan^2 x\right)$$

$$\log_e \left(\frac{1+x}{1-x}\right) = 2xF\left(\frac{1}{2}, 1; \frac{3}{2}; x^2\right)$$

$$e^{-nx} = (2\cosh x)^{-n} \frac{\sinh x}{\cosh x} F\left(1+\frac{n}{2}, \frac{1+n}{2}; 1+n; \frac{1}{\cosh^2 x}\right)$$

$$\mathbf{K}(k) = \frac{\pi}{2} F\left(\frac{1}{2}, \frac{1}{2}; 1; k^2\right) \qquad \mathbf{E}(k) = \frac{\pi}{2} F\left(-\frac{1}{2}, \frac{1}{2}; 1; k^2\right)$$

Table 9.3-2. Additional Formulas Relating to Confluent Hypergeometric Functions

$$F(a; 2a; 2z) = 2^{a-\frac{1}{2}} e^{-i\pi(a-\frac{1}{2})/2}\, \Gamma(a+\tfrac{1}{2}) e^z z^{\frac{1}{2}-a} J_{a-\frac{1}{2}}(ze^{i\pi/2})$$

$$F(a; c; z) = e^z F(c-a; c; -z)$$

$$aF(a+1; c+1; z) = (a-c)F(a; c+1; z) + cF(a; c; z)$$

$$aF(a+1; c; z) = (z+2a-c)\, F(a; c; z) + (c-a)F(a-1; c; z)$$

$$\lim_{c\to -n} \frac{1}{\Gamma(c)} F(a; c; z) = \frac{z^{n+1} a(a+1) \cdots (a+n-1)}{(n+1)!} F(a+n+1; n+2; z) \qquad (n = 0, 1, 2, \ldots)$$

$$F(a; c; z) = \frac{\Gamma(c) 2^{1-c}}{\Gamma(a)\Gamma(c-a)} e^{z/2} \int_{-1}^{+1} e^{zt/2}(1-t)^{c-a-1}(1+t)^{a-1}\, dt \qquad (0 < \operatorname{Re} a < \operatorname{Re} c)$$

$$\frac{\Gamma(\alpha+\nu+1)}{\Gamma(\alpha+1)} F(-\nu; \alpha+1; z) = e^z z^{-\alpha/2} \int_0^\infty e^{-t} t^{\nu+\frac{\alpha}{2}} J_\alpha(2\sqrt{zt})\, dt \qquad \left[\operatorname{Re}(\alpha+\nu+1) > 0;\ |\arg z| < \frac{\pi}{2}\right]$$

For $\mu = 0$ one obtains **Kummer's confluent hypergeometric function**

$$w = F(a; c; z) \equiv 1 + \frac{a}{c} z + \frac{1}{2!} \frac{a(a+1)}{c(c+1)} z^2 + \cdots \qquad (|z| < \infty)$$

(CONFLUENT HYPERGEOMETRIC SERIES) (9.3-43)

(see also Sec. 21.7-5).

A second solution may be obtained in the manner of Sec. 9.3-8*b*; in particular, the **confluent hypergeometric function of the second kind**

$$\Phi(a; c; z) \equiv \frac{\Gamma(a-c+1)\Gamma(c-1)}{\Gamma(a)\Gamma(1-c)} z^{1-c} F(a-c+1; 2-c; z) \tag{9.3-44}$$

is a solution whenever c is not an integer.

Note the following relations:

$$F(a; c; z) = \frac{\Gamma(c)}{\Gamma(a)\Gamma(c-a)} \int_0^1 \zeta^{a-1}(1-\zeta)^{c-a-1} e^{z\zeta} \, d\zeta \tag{9.3-45}$$

$$\frac{dF}{dz} = \frac{a}{c} F(a+1; c+1; z) \qquad \frac{d\Phi}{dz} = \frac{a}{c} \Phi(a+1; c+1; z) \tag{9.3-46}$$

$$F(a; c; z) = e^z F(c-a; c; -z) \tag{9.3-47}$$

See Refs. 21.9 and 21.11 for additional formulas.

9.3-11. Pochhammer's Notation. The infinite series (32) and (43) are special cases of

$$\begin{aligned} &{}_mF_n(a_1, a_2, \ldots, a_m; c_1, c_2, \ldots, c_n; z) \\ &\quad \equiv \sum_{k=0}^{\infty} \frac{1}{k!} \frac{(a_1)_k (a_2)_k \cdots (a_m)_k}{(c_1)_k (c_2)_k \cdots (c_n)_k} z^k \qquad (m, n = 1, 2, \ldots) \end{aligned} \tag{9.3-48}$$

where $(x)_k \equiv x(x+1) \cdots (x+k-1)$. *In this notation the hypergeometric function* (32) *becomes* ${}_2F_1\,(a, b, c; z)$, *and the confluent hypergeometric function* (43) *is written as* ${}_1F_1\,(a; c; z)$.

9.4. LINEAR DIFFERENTIAL EQUATIONS WITH CONSTANT COEFFICIENTS

9.4-1. Homogeneous Linear Equations with Constant Coefficients (see also Sec. 9.3-1). **(a)** The *first-order differential equation*

$$a_0 \frac{dy}{dt} + a_1 y = 0 \qquad (a_0 \neq 0) \tag{9.4-1}$$

has the solution

$$y = Ce^{-(a_1/a_0)t} \qquad [C = y(0)] \tag{9.4-2}$$

For $a_0/a_1 > 0$

$$y(a_0/a_1) = \frac{1}{e}\, y(0) \approx 0.37y(0) \qquad y(4a_0/a_1) \approx 0.02y(0)$$

a_0/a_1 is often referred to as the **time constant.**

(**b**) The *second-order equation*

$$a_0 \frac{d^2y}{dt^2} + a_1 \frac{dy}{dt} + a_2 y = 0 \qquad (a_0 \neq 0) \tag{9.4-3}$$

has the solution

$$\left.\begin{aligned} y &= C_1 e^{s_1 t} + C_2 e^{s_2 t} \\ s_{1,2} &= \frac{-a_1 \pm \sqrt{a_1^2 - 4a_0a_2}}{2a_0} \end{aligned}\right\} (a_1^2 - 4a_0a_2 \neq 0) \tag{9.4-4a}$$

$$y = (C_1 + C_2 t)e^{-(a_1/2a_0)t} \qquad (a_1^2 - 4a_0a_2 = 0) \tag{9.4-4b}$$

If a_0, a_1, and a_2 are real, s_1 and s_2 become complex for $a_1^2 - 4a_0a_2 < 0$; in this case, Eq. (4*a*) can be written as

$$y = e^{\sigma_1 t}(A \cos \omega_N t + B \sin \omega_N t) = Re^{\sigma_1 t} \sin (\omega_N t + \alpha) \tag{9.4-4c}$$

where the quantities

$$\sigma_1 = -\frac{a_1}{2a_0} \qquad \omega_N = \frac{\sqrt{4a_0a_2 - a_1^2}}{2a_0} \tag{9.4-4d}$$

are respectively known as the **damping constant** and the **natural (characteristic) circular frequency.** The constants C_1, C_2, A, B, R, and α are chosen so as to match given initial or boundary conditions (see also Sec. 9.4-5*a*).

If $a_0a_2 > 0$, the quantity $\zeta = a_1/2\sqrt{a_0a_2}$ is called the **damping ratio;** for $\zeta > 1$, $\zeta = 1$, $0 < \zeta < 1$ one obtains, respectively, an *overdamped* solution (4*a*), a *critically damped* solution (4*b*), or an *underdamped* (*oscillatory*) solution (4*c*). In the latter case, the **logarithmic decrement** $2\pi\sigma_1/\omega_N$ is the natural logarithm of the ratio of successive maxima of $y(t)$.

Equation (3) is often written in the *nondimensional form*

$$\frac{1}{\omega_1^2}\frac{d^2y}{dt^2} + 2\frac{\zeta}{\omega_1}\frac{dy}{dt} + y = 0$$

with

$$s_{1,2} = -\omega_1\zeta \pm \omega_1\sqrt{\zeta^2 - 1}$$

$\omega_1 = \sqrt{a_2/a_0}$ is called the **undamped natural circular frequency;** for weak damping ($\zeta^2 \ll 1$), $\omega_1 \approx \omega_N$ (see also Fig. 9.4-1).

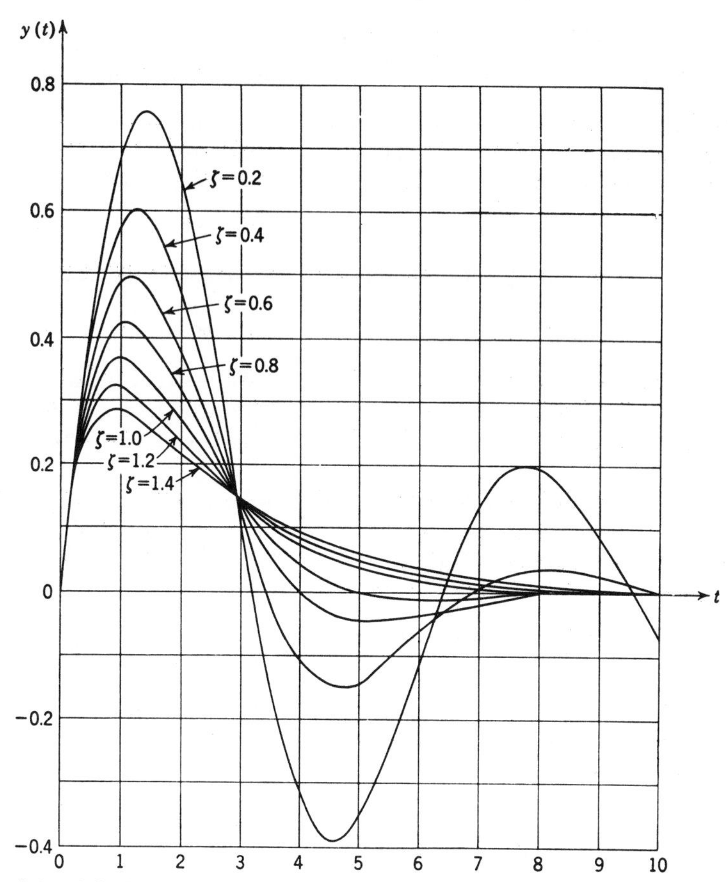

Fig. 9.4-1. Solution of the second-order differential equation

$$\frac{d^2y}{dt^2} + 2\zeta \frac{dy}{dt} + y = 0$$

for $y(0) = 0$, $dy/dt]_0 = 1$. Response is overdamped for $\zeta > 1$, critically damped for $\zeta = 1$, and underdamped for $0 < \zeta < 1$.

(c) To solve the rth-*order differential equation*

$$\mathsf{L}y \equiv a_0 \frac{d^ry}{dt^r} + a_1 \frac{d^{r-1}y}{dt^{r-1}} + \cdots + a_ry = 0 \qquad (a_0 \neq 0) \tag{9.4-5}$$

find the roots of the rth-degree algebraic equation

$$a_0s^r + a_1s^{r-1} + \cdots + a_0 = 0 \quad \text{(CHARACTERISTIC EQUATION)} \qquad (9.4\text{-}6)$$

obtained, for example, on substitution of a trial solution $y = e^{st}$. If the r roots $s_1, s_2, \ldots$ of the characteristic equation (6) are distinct, the given differential equation (5) has the general solution

$$y = C_1e^{s_1t} + C_2e^{s_2t} + \cdots + C_re^{s_rt} \qquad (9.4\text{-}7a)$$

If a root s_k is of multiplicity m_k, replace the corresponding term in Eq. (7a) by

$$(C_k + C_{k1}t + C_{k2}t^2 + \cdots + C_{km_k-1}t^{m_k-1})e^{s_kt} \qquad (9.4\text{-}7b)$$

The various terms of the solution (7) are known as **normal modes** of the given differential equation. The r constants C_k and C_{kj} must be chosen so as to match given initial or boundary conditions (see also Sec. 9.4-5a).

If the given differential equation (5) is real, complex roots of the characteristic equation appear as pairs of complex conjugates $\sigma \pm i\omega$. The corresponding pairs of solution terms will also be complex conjugates and may be combined to form real terms:

$$\begin{aligned} t^mCe^{(\sigma+i\omega)t} + t^mC^*e^{(\sigma-i\omega)t} &= t^me^{\sigma t}(A\cos\omega t + B\sin\omega t) \\ &= Rt^me^{\sigma t}\sin(\omega t + \alpha) \end{aligned} \qquad (9.4\text{-}7c)$$

where A and B, or R and α, are new real constants of integration.

(d) Given a *system of n homogeneous linear differential equations with constant coefficients*

$$\boxed{\varphi_{j1}\left(\frac{d}{dt}\right)y_1 + \varphi_{j2}\left(\frac{d}{dt}\right)y_2 + \cdots + \varphi_{jn}\left(\frac{d}{dt}\right)y_n = 0 \qquad (j = 1, 2, \ldots, n)} \qquad (9.4\text{-}8)$$

where the $\varphi_{jk}\left(\frac{d}{dt}\right)$ are polynomials in d/dt, each of the n solution functions $y_k = y_k(t)$ $(k = 1, 2, \ldots, n)$ has the form (7); the s_k are now the roots of the algebraic equation

$$D(s) \equiv \det[\varphi_{jk}(s)] = 0 \quad \text{[CHARACTERISTIC EQUATION OF THE SYSTEM (8)]} \qquad (9.4\text{-}9)$$

The constants of integration must again be matched to the given initial or boundary conditions (see also Secs. 9.4-5b and 13.6-2).

9.4-2. Nonhomogeneous Equations. Normal Response, Steady-state Solution, and Transients (see also Sec. 9.3-1). **(a)** The superposition theorems and solution methods of Secs. 9.3-1 to 9.3-4 apply to all linear ordinary differential equations. Thus the general solution of the nonhomogeneous differential equation

$$\mathsf{L}y \equiv a_0 \frac{d^r y}{dt^r} + a_1 \frac{d^{r-1}y}{dt^{r-1}} + \cdots + a_r y = f(t) \tag{9.4-10}$$

can be expressed as the sum of the general solution (7) of the reduced equation (5) and any particular integral of Eq. (10).

If, as in many applications, $f(t) = 0$ for $t \leq 0$,* the particular integral $y = y_N(t)$ of Eq. (10) with $y_N = y'_N = y''_N = \cdots = y_N{}^{(r-1)} = 0$ for $t \leq 0$ will be called the **normal response** to the given forcing function $f(t)$. *To solve Eq.* (10) *for* $t > 0$ *with given initial values for* $y, y', y'', \ldots, y^{(r-1)}$, *one adds the solution of the corresponding initial-value problem for Eq.* (5) *to the normal response* $y_N(t)$.

In many applications (stable electric circuits, vibrations), all roots of the characteristic equation (9) have negative real parts, and the complementary function (7) dies out more or less rapidly (stable "transient solution"). In such cases, one is often mainly interested in a suitable nontransient particular integral $y = y_{SS}(t)$, the "steady-state solution" due to the given forcing function $f(t)$. In other cases, $y_{SS}(t)$ is not uniquely defined by the given differential equation but depends on the initial conditions. The normal response $y_N(t)$ may or may not include a transient term.

(**b**) In the same manner, each solution function $y_k = y_k(t)$ of a system of linear differential equations with constant coefficients,

$$\varphi_{j1}\left(\frac{d}{dt}\right) y_1 + \varphi_{j2}\left(\frac{d}{dt}\right) y_2 + \cdots + \varphi_{jn}\left(\frac{d}{dt}\right) y_n = f_j(t) \qquad (j = 1, 2, \ldots, n) \tag{9.4-11}$$

can be expressed as the sum of the corresponding solution function of the complementary homogeneous system (8) and a particular solution function of the given system (11). The **normal response** of the system (11) to a set of forcing functions $f_j(t)$ equal to zero for $t \leq 0$ is the particular solution such that all y_k vanish for $t \leq 0$ together with all derivatives which can be arbitrarily chosen (see also Sec. 13.6-2).

(**c**) If a forcing function contains a periodic term whose frequency equals that of an undamped sinusoidal term in the complementary function (7), then the differential equation or system may not have a finite solution (*resonance,* see also Secs. 9.4-5 and 15.4-12).

9.4-3. Superposition Integrals and Weighting Functions (see also Secs. 9.3-3, 9.4-7c, and 15.5-1). **(a) Physically Realizable Initial-**

* This means one considers only forcing functions of the type $f(t) \equiv f(t)U_+(t)$, where $U_+(t)$ is the *asymmetrical* unit-step function defined in Sec. 21.9-1 (see also the footnote to Sec. 9.4-3).

value Problems. Application of the Green's-function method of Sec. 9.3-3*b* to the differential equation (10) yields the normal-response solution (Sec. 9.4-2) as a weighted mean over past values of $f(t)$ in the form

$$y = y_N(t) = \int_0^t h_+(t - \tau)f(\tau)\, d\tau = \int_0^t h_+(\zeta)f(t - \zeta)\, d\zeta$$

(DUHAMEL'S SUPERPOSITION INTEGRAL, CONVOLUTION INTEGRAL) (9.4-12)

if one assumes (1) $f(t) = 0$ for $t \leq 0$ (*initial-value problem*), and (2) $h_+(t - \tau) = 0$ for $t \leq \tau$, so that "future" values of $f(t)$ cannot affect "earlier" values of $y(t)$, and "instantaneous" effects are also ruled out (*physically realizable systems*). More general problems are considered in Sec. 9.4-3*d*.

If the derivative on the right exists,

$$y_N^{(m)}(t) = \int_0^t \frac{\partial^m}{\partial t^m} h_+(t - \tau)f(\tau)\, d\tau \qquad (m = 0, 1, 2, \ldots, r) \qquad (9.4\text{-}13)$$

The **weighting function** $h_+(t - \tau)$ is the special Green's function defined by

$$\mathsf{L}h_+(t - \tau) = 0 \qquad \int_0^\infty \mathsf{L}h_+(t - \tau)\, d\tau = 1 \qquad (t > \tau) \qquad (9.4\text{-}14a)$$

$$h_+(t - \tau) = h'_+(t - \tau) = \cdots = h_+^{(r-1)}(t - \tau) = 0 \qquad (t \leq \tau) \qquad (9.4\text{-}14b)$$

$h_+(t - \tau)$ is the *normal response to an asymmetrical unit impulse* $\delta_+(t - \tau)$ (Sec. 21.9-6); Eq. (14*a*) may be rewritten as a "symbolic differential equation"

$$\mathsf{L}h_+(t - \tau) = \delta_+(t - \tau) \qquad (t > \tau) \qquad (9.4\text{-}14c)$$

Note that $\int_0^t h_+(t - \tau)\, d\tau$ is the *normal response to the asymmetrical unit-step function* $U_+(t - \tau)$ (Sec. 21.9-1). The "symbolic differential equation" (14*c*) is often easily solved for $h_+(t - \tau)$ by the Laplace-transform method of Sec. 9.4-5 (see also Secs. 8.5-1 and 9.4-7); alternatively, $h_+(t)$ can be found as that solution of the *homogeneous* differential equation

$$\mathsf{L}h_+(t) = 0 \qquad (t > 0) \qquad (9.4\text{-}14d)$$

which satisfies the initial conditions

$$h_+(0 + 0) = h'_+(0 + 0) = \cdots = h_+^{(r-2)}(0 + 0) = 0$$

$$h_+^{(r-1)}(0 + 0) = \frac{1}{a_0} \qquad (9.4\text{-}14e)$$

EXAMPLE: For $\mathsf{L}y \equiv a(dy/dt) + y$, one has $h_+(t) = \frac{1}{a} e^{-t/a} (t > 0)$.

(**b**) Under similar conditions, the normal-response solution of a *system* of linear differential equations (11) can be expressed in the form

$$y_k = \int_0^t (h_+)_{kj}(t - \tau) f_j(\tau) \, d\tau \qquad (k = 1, 2, \ldots, n) \tag{9.4-15}$$

where $(h_+)_{kj}(t - \tau)$ is the kth solution function obtained on substitution of $f_j(t) = \delta_+(t - \tau)$ $(i = 1, 2, \ldots, n)$ in Eq. (11). The weighting-function matrix $[(h_+)_{kj}(t - \tau)]$ is often called a **state-transition matrix** (Sec. 13.6-2).

(**c**) A superposition integral (12) also yields the normal response if the "input function" $f(t)$ and the "output function" $y(t)$ are related by a differential equation of the form

$$a_0 \frac{d^r y}{dt^r} + a_1 \frac{d^{r-1} y}{dt^{r-1}} + \cdots + a_r y = b_0 \frac{d^\rho f}{dt^\rho} + b_1 \frac{d^{\rho-1} f}{dt^{\rho-1}} + \cdots + b_\rho f \tag{9.4-16}$$

Such relations may result, in particular, if a system (11) is reduced to a single differential equation by elimination of all but one of the unknown functions.

Any $(h_+)_{kj}(t)$, and also $h_+(t)$ if relations of the type (16) are considered, may contain delta-function-type singularities (modified Green's function, Sec. 15.5-1*b*). Thus, if $h_+(t) = c_1\delta_+(t - t_1) + c_2\delta_+(t - t_2) + \cdots + h_0(t)$, then Eq. (12) yields the normal response

$$y = y_N(t) = c_1 f(t - t_1) + c_2 f(t - t_2) + \cdots + \int_0^t h_0(t - \tau) f(\tau) \, d\tau \qquad (t > 0)$$

(d) More General Problems. "Symmetrical" vs. "Asymmetrical" Weighting Functions. To deal with forcing functions different from zero for $t \leq 0$, one may introduce the "symmetrical" weighting function $h(t - \tau)$ defined by

$$\mathsf{L}h(t - \tau) = 0 \qquad \int_{-\infty}^{\infty} \mathsf{L}h(t - \tau) \, d\tau = 1 \tag{9.4-17}$$

or $\mathsf{L}h(t - \tau) = \delta(t - \tau)$ (see also Sec. 21.9-2) with suitable initial or boundary conditions. The resulting solution

$$y = \int_{-\infty}^{\infty} h(t - \tau) f(\tau) \, d\tau = \int_{-\infty}^{\infty} h(\zeta) f(t - \zeta) \, d\zeta \tag{9.4-18}$$

will, in particular, satisfy Eq. (10) for $t \geq 0$ with

$$y(0 - 0) = y'(0 - 0) = \cdots = y^{(r-1)}(0 - 0) = 0$$

if $f(t) \equiv f(t)U(t)$ (Sec. 21.9-1), and one adds the condition

$$h(t - \tau) = h'(t - \tau) = \cdots = h^{(r-1)}(t - \tau) = 0 \qquad (t < \tau) \tag{9.4-19a}$$

$h_+(t)$ and $h(t)$, and the solutions (12) and (18), are easily confused. The "asymmetrical" weighting function $h_+(t)$ is particularly convenient for use with the unilateral Laplace transforms employed by most engineers, while $h(t)$ fits the context of Fourier analysis or bilateral Laplace transforms (see also Sec. 18.10-5). *In the usual physical applications, Eq.* (19*a*) *holds since "future" values of $f(t)$ cannot affect the solution;* $h_+(t)$ and $h(t)$ are then identical wherever they are continuous. Frequently, forcing functions cannot even affect the solution instantaneously, so that $h(t - \tau)$ satisfies the stronger condition

$$h(t-\tau) = h'(t-\tau) = \cdots = h^{(r-1)}(t-\tau) = 0 \qquad (t \leq \tau) \quad (9.4\text{-}19b)$$

and $h(t)$ and $h_+(t)$ are identical.

EXAMPLE: In a purely resistive electric circuit, the current $y(t)$ and the voltage $f(t)$ are related by $y(t) = f(t)/R$; so that $h_+(t) = \delta_+(t)/R$ and $h(t) = \delta(t)/R$. But for $\mathbf{L}y \equiv a\dfrac{dy}{dt} + y$, $h_+(t) \equiv h(t) \equiv \dfrac{1}{a}e^{-t/a}$.

9.4-4. Stability. A linear differential equation (10) or a system (11) will be called **completely stable** if and only if all roots of the corresponding characteristic equation (6) or (9) have negative real parts, so that effects of small changes in the initial conditions tend to zero with time (*refer to Sec.* 13.6-5 *for a more general discussion of stability*). The nature of the roots may be investigated with the aid of Secs. 1.6-6 and 7.6-9 (stability criteria for electric circuits and control systems). *A differential equation* (10) *is completely stable if and only if* $\int_0^\infty |h_+(\tau)|\, d\tau$ [*or equivalently* $\int_{-\infty}^\infty |h(\tau)|\, d\tau$] *exists;* a similar condition for every weighting function of a system (11) is necessary and sufficient for complete stability of the system.

9.4-5. The Laplace-transform Method of Solution (see also Secs. 8.1-1, 8.4-1 to 8.4-5, 9.3-7, and 13.6-2). **(a)** To solve a linear differential equation (10) with given initial values $y(0+0)$, $y'(0+0)$, $y''(0+0)$, $\ldots$, $y^{(r-1)}(0+0)$, apply the Laplace transformation (8.2-1) to both sides, and let $\mathcal{L}[y(t)] \equiv Y(s)$, $\mathcal{L}[f(t)] \equiv F(s)$. The resulting linear *algebraic* equation (*subsidiary equation*)

$$\left.\begin{aligned} (a_0s^r + a_1s^{r-1} + \cdots + a_r)Y(s) &= F(s) + G(s) \\ G(s) \equiv y(0+0)(a_0s^{r-1} &+ a_1s^{r-2} + \cdots + a_{r-1}) \\ &+ y'(0+0)(a_0s^{r-2} + a_1s^{r-3} + \cdots + a_{r-2}) \\ &+ \cdots \\ &+ y^{(r-2)}(0+0)(a_0s + a_1) + a_0y^{(r-1)}(0+0) \end{aligned}\right\} \quad (9.4\text{-}20)$$

is easily solved to yield the Laplace transform of the desired solution $y(t)$ in the form

$$Y(s) = \frac{F(s)}{a_0s^r + a_1s^{r-1} + \cdots + a_r} + \frac{G(s)}{a_0s^r + a_1s^{r-1} + \cdots + a_r} \quad (9.4\text{-}21)$$

Here the first term is the *Laplace transform* $Y_N(s)$ *of the normal response* $y_N(t)$ (Sec. 9.4-2*a*), and the second term represents the effects of nonzero initial values of $y(t)$ and its derivatives. The solutions $y(t)$ and $y_N(t)$

are found as inverse Laplace transforms by reference to tables (Appendix D), or by one of the methods of Secs. 8.4-2 to 8.4-9. In particular, each of the r terms in the partial-fraction expansion of $G(s)/(a_0s^r + a_1s^{r-1} + \cdots + a_r)$ (Sec. 8.4-5) yields a corresponding term of the force-free solution (7).

This solution method applies without essential changes to differential equations of the type (16).

(b) In the same manner, one applies the Laplace transformation to a system of linear differential equations (11) to obtain

$$\varphi_{j1}(s)Y_1(s) + \varphi_{j2}(s)Y_2(s) + \cdots + \varphi_{jn}(s)Y_n(s) = F_j(s) + G_j(s) \qquad (j = 1, 2, \ldots, n) \qquad (9.4\text{-}22)$$

where the functions $G_j(s)$ depend on the given initial conditions. The linear *algebraic* equations (22) are solved by Cramer's rule (1.9-4) to yield the unknown solution transforms

$$Y_k(s) = \sum_{j=1}^{n} \frac{A_{jk}(s)}{D(s)} F_j(s) + \sum_{j=1}^{n} \frac{A_{jk}(s)}{D(s)} G_j(s) \qquad (k = 1, 2, \ldots, n) \qquad (9.4\text{-}23)$$

where $A_{jk}(s)$ is the cofactor of $\varphi_{jk}(s)$ in the **system determinant** $D(s) \equiv \det [\varphi_{jk}(s)]$ (see also Sec. 1.9-2). The first sum in Eq. (23) is the Laplace transform of the normal-response solution, while the second sum represents the effect of the initial conditions.

The desired solutions $y_k(t)$ are obtained from Eq. (23) by inverse Laplace transformation.

In problems involving unstable differential equations (Sec. 9.4-4) and/or impulse-type forcing functions, the solutions may contain delta-function-type singularities (see also Secs. 8.5-1 and 21.9-6).

9.4-6. Periodic Forcing Functions and Solutions. The Phasor Method. (a) Sinusoidal Forcing Functions and Solutions. Sinusoidal Steady-state Solutions. *Every system of linear differential equations* (11) *with sinusoidal forcing functions of equal frequency,*

$$f_j(t) \equiv B_j \sin (\omega t + \beta_j) \qquad (j = 1, 2, \ldots, n) \qquad (9.4\text{-}24a)$$

admits a unique particular solution of the form

$$y_k(t) \equiv A_k \sin (\omega t + \alpha_k) \qquad (k = 1, 2, \ldots, n) \qquad (9.4\text{-}24b)$$

In particular, *if all roots of the characteristic equation* (9) *have negative real parts (stable systems, Sec. 9.4-4), the sinusoidal solution* (24*b*) *is the unique steady-state solution obtained after all transients have died out* (Sec. 9.4-2).

(b) The Phasor Method. Given a system of linear differential equations (11) relating sinusoidal forcing functions and solutions (24), *one introduces a reciprocal one-to-one representation of these sinusoids by corresponding complex numbers* (*vectors*, **phasors**)

$$\left.\begin{aligned}\vec{F}_j &= \frac{B_j}{\sqrt{2}} e^{i\beta_j} = \frac{B_j}{\sqrt{2}} \underline{/\beta_j} \qquad (j = 1, 2, \ldots, n) \\ \vec{Y}_k &= \frac{A_k}{\sqrt{2}} e^{i\alpha_k} = \frac{A_k}{\sqrt{2}} \underline{/\alpha_k} \qquad (k = 1, 2, \ldots, n)\end{aligned}\right\} \tag{9.4-25}$$

The absolute value of each phasor equals the root-mean-square value of the corresponding sinusoid, while the phasor argument defines the phase of the sinusoid. *The phasors* (25) *are related by the* (*complex*) *linear algebraic equations* (*phasor equations*)

$$\varphi_{j1}(i\omega)\vec{Y}_1 + \varphi_{j2}(i\omega)\vec{Y}_2 + \cdots + \varphi_{jn}(i\omega)\vec{Y}_n = \vec{F}_j \qquad (j = 1, 2, \ldots, n) \tag{9.4-26}$$

which correspond to Eq. (11) and may be solved for the unknown phasors

$$\vec{Y}_k = \sum_{j=1}^{n} \frac{A_{jk}(i\omega)}{D(i\omega)} \vec{F}_j \qquad (k = 1, 2, \ldots, n) \tag{9.4-27}$$

(see also Sec. 9.4-5*b*). In the case of *resonance* (Sec. 9.4-2*c*), the expression (27) may not exist (may become "infinitely large").

(c) Rotating Phasors. *A set of sinusoidal functions* (24) *satisfies any given system of linear differential equations* (11) *if and only if the same is true for the corresponding set of complex exponential functions* (**rotating phasors**)

$$\left.\begin{aligned}\vec{f}_j(t) &\equiv B_j e^{i(\omega t+\beta_j)} \equiv \vec{F}_j e^{i\omega t}\sqrt{2} \qquad (j = 1, 2, \ldots, n) \\ \vec{y}_k(t) &\equiv A_k e^{i(\omega t+\alpha_k)} \equiv \vec{Y}_k e^{i\omega t}\sqrt{2} \qquad (k = 1, 2, \ldots, n)\end{aligned}\right\} \tag{9.4-28}$$

which are often more convenient to handle than the real sinusoids (24).

(d) More General Periodic Forcing Functions (see also Secs. 4.11-4, 4.11-5, and 9.4-5*c*). Given a stable system (11) with more general periodic forcing functions expressible in the form

$$f(t) = \tfrac{1}{2}a_0 + \sum_{h=1}^{\infty} (a_h \cos h\omega t + b_h \sin h\omega t) \tag{9.4-29}$$

one can apply the phasor method of Sec. 9.4-6*a* separately for each sinusoidal term and superimpose the resulting sinusoidal solutions to obtain the steady-state periodic solution. This procedure may be more convenient than the Laplace-transform method if only a few harmonics of the periodic solution are needed.

9.4-7. Transfer Functions and Frequency-response Functions. **(a) Transfer Functions.** The function

$$H(s) \equiv \frac{1}{a_0 s^r + a_1 s^{r-1} + \cdots + a_r} \equiv \frac{Y_N(s)}{F(s)} \tag{9.4-30}$$

in Eq. (21) is known as a **transfer function.** The transfer function "represents" a linear operator (Sec. 15.2-7) which operates on the forcing function (input) to yield the normal response (output; see also Fig. 9.4-1).

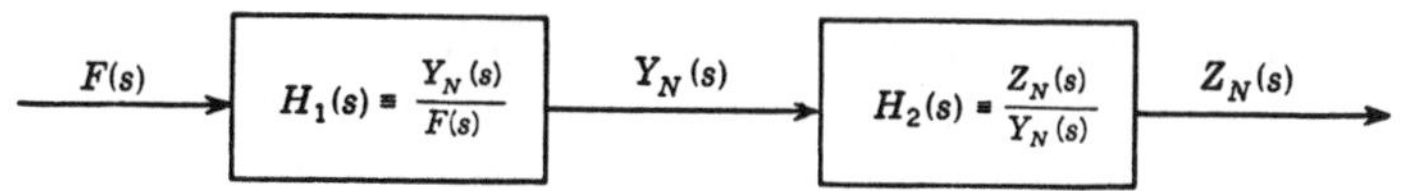

FIG. 9.4-1. Transfer-function representation of linear differential equations with constant coefficients. If $y_N(t)$ in turn serves as the forcing function for a second differential equation to produce the normal response $z_N(t)$, the two transfer functions multiply, i.e., $Z_N(s)/F(s) = H_1(s)H_2(s)$.

More generally, each function $A_{jk}(s)/D(s)$ in Eq. (23) is the transfer function relating the normal-response "output" $y_k(t)$ of the system (11) to the "input" $f_j(t)$ when all other forcing functions vanish identically. The transfer functions $A_{jk}(s)/D(s)$ together constitute the **transfer matrix.**

The transfer function corresponding to Eq. (16) is

$$H(s) \equiv \frac{b_0 s^p + b_1 s^{p-1} + \cdots + b_p}{a_0 s^r + a_1 s^{r-1} + \cdots + a_r} \tag{9.4-31}$$

(b) Frequency-response Functions (see also Sec. 9.4-6*a*). The **frequency-response functions** $H(i\omega)$ and $A_{jk}(i\omega)/D(i\omega)$ similarly relate the phasors representing sinusoidal forcing functions and steady-state solutions of given circular frequency ω. Specifically, *the absolute value and the argument of a frequency-response function respectively relate the amplitudes and the phases of input and output sinusoid;* thus for $f(t) = B \sin(\omega t + \beta)$, $y(t) = A \sin(\omega t + \alpha)$

$$|H(i\omega)| = \frac{A}{B} \qquad \arg H(i\omega) = \alpha - \beta \tag{9.4-32}$$

If frequency-response functions are "cascaded" in the manner of Fig. 9.4-1, the amplitude responses $|H(i\omega)|$ multiply, and the phase responses $\arg H(i\omega)$ add.

(c) Relations between Transfer Functions or Frequency-response Functions and Weighting Functions (see also Secs. 4.11-4*e*, 9.4-3, and the convolution theorem of Table 8.3-1). *The transfer function $H(s)$ is the unilateral Laplace transform of the asymmetrical weighting function $h_+(t)$, and the bilateral Laplace transform* (*Sec.* 8.6-2) *of the symmetrical weighting function $h(t)$:*

$$H(s) = \int_0^\infty h_+(t)e^{-st}\,dt = \int_{-\infty}^\infty h(t)e^{-st}\,dt \qquad (9.4\text{-}33)$$

Hence *the frequency-response function $H(i\omega)$ is related to the symmetrical weighting function $h(t)$ by the Fourier transformation**

$$H(i\omega) = \int_{-\infty}^\infty h(t)e^{-i\omega t}\,dt \qquad (9.4\text{-}34)$$

Equations (33) and (34) indicate the possibility of obtaining weighting functions as inverse Laplace or Fourier transforms of rational functions.

9.4-8. Normal Coordinates and Normal-mode Oscillations. (a) **Free Oscillations.** Small oscillations of undamped mechanical or electrical systems are often described by a set of n linear second-order differential equations of the form

$$\sum_{k=1}^n \left(b_{jk}\frac{d^2}{dt^2} + a_{jk}\right) y_k = 0 \qquad (j = 1, 2, \ldots, n) \qquad (9.4\text{-}35)$$

where the matrices $[a_{jk}]$ and $[b_{jk}]$ are both symmetric, positive-definite (Sec. 13.5-2), and such that the resulting characteristic equation (9) has $2n$ distinct, nonzero, purely imaginary roots $\pm i\omega_1$, $\pm i\omega_2$, . . . , $\pm i\omega_n$. Pairs of these roots correspond to sinusoidal free oscillations at the n **normal-mode frequencies** $\omega_1/2\pi$, $\omega_2/2\pi$, . . . , $\omega_n/2\pi$.

One may introduce **normal coordinates** $\bar{y}_1$, $\bar{y}_2$, . . . , $\bar{y}_n$ for the given system (35) by a linear transformation

$$y_k = \sum_{h=1}^n t_{kh}\bar{y}_h \qquad (k = 1, 2, \ldots, n) \qquad (9.4\text{-}36)$$

with coefficients t_{kh} chosen in the manner of Secs. 13.5-5 and 14.8-7 so as to diagonalize the matrices $[a_{jk}]$ and $[b_{jk}]$ simultaneously; the transformed system takes the simple form

$$\frac{d^2\bar{y}_h}{dt^2} + \omega_h{}^2\bar{y}_h = 0 \qquad (h = 1, 2, \ldots, n) \qquad (9.4\text{-}37)$$

The resulting free sinusoidal normal-mode oscillations

$$\bar{y}_h = \bar{A}_h \sin(\omega_h t + \bar{\alpha}_h) \qquad (h = 1, 2, \ldots, n) \qquad (9.4\text{-}38)$$

do not affect one another (are "uncoupled"). The normal coordinates (38) may have an intuitive physical interpretation.

The problem is a *generalized eigenvalue problem* involving sets of n functions $[\bar{y}_1(t), \bar{y}_2(t), \ldots, \bar{y}_n(t)]$ as eigenvectors (see also Secs. 13.6-2*a*, 14.8-7, and 15.4-5).

* See the footnote to Sec. 4.11-2.

EXAMPLE: For a pair of similar coupled oscillators described by

$$\frac{d^2y_1}{dt^2} = -\omega_0{}^2 y_1 - \alpha^2(y_1 - y_2) \qquad \frac{d^2y_2}{dt^2} = -\omega_0{}^2 y_2 - \alpha^2(y_2 - y_1)$$

the normal coordinates are simply $\bar{y}_1 = y_1 + y_2$, $\bar{y}_2 = y_1 - y_2$. Given $y_1 = 1$, $y_2 = dy_1/dt = dy_2/dt = 0$ for $t = 0$, the normal-mode equations

$$\frac{d^2\bar{y}_1}{dt^2} = -\omega_0{}^2\bar{y}_1 \qquad \frac{d^2\bar{y}_2}{dt^2} = -(\omega_0{}^2 + 2\alpha^2)\bar{y}_2$$

yield

$$\bar{y}_1 = \cos \omega_0 t \qquad \bar{y}_2 = \cos t \sqrt{\omega_0{}^2 + 2\alpha^2}$$

and thus

$$y_1 = \frac{1}{2}(\bar{y}_1 + \bar{y}_2) = \cos \frac{\sqrt{\omega_0{}^2 + 2\alpha^2} - \omega_0}{2} t \cos \frac{\sqrt{\omega_0{}^2 + 2\alpha^2} + \omega_0}{2} t$$

$$y_2 = \frac{1}{2}(\bar{y}_1 - \bar{y}_2) = \sin \frac{\sqrt{\omega_0{}^2 + 2\alpha^2} - \omega_0}{2} t \sin \frac{\sqrt{\omega_0{}^2 + 2\alpha^2} + \omega_0}{2} t$$

For $\alpha^2 \ll \omega_0{}^2$ (weak coupling), this solution describes the so-called resonance phenomenon.

(b) Forced Oscillations. The corresponding forced-oscillation problem

$$\sum_{k=1}^{n} \left(b_{jk} \frac{d^2}{dt^2} + a_{jk}\right) y_k = f_j(t) \qquad (j = 1, 2, \ldots, n) \tag{9.4-39}$$

can, in principle, be solved in the manner of Sec. 14.8-10 through normal-mode expansion of the forcing functions $f_j(t)$. The Laplace-transform method of Sec. 9.4-5 is usually more convenient.

9.5. NONLINEAR SECOND-ORDER EQUATIONS

9.5-1. Introduction. Sections 9.5-2 to 9.5-5 introduce the general terminology and the most easily summarized approximation method of the theory of nonlinear oscillations. References 9.15 to 9.17, 9.22, and 9.23 are recommended for further study; for better or for worse, many solution methods are closely tied to specific applications.

The perturbation method of Sec. 10.2-7*c* is often used to simplify nonlinear problems, especially in celestial mechanics. See Secs. 20.7-4 and 20.7-5 for numerical methods of solution.

9.5-2. The Phase-plane Representation. Graphical Method of Solution (see also Sec. 9.2-2). The second-order differential equation

$$\frac{d^2y}{dt^2} = f\left(t, y, \frac{dy}{dt}\right) \tag{9.5-1}$$

is equivalent to the system of first-order equations

$$\frac{dy}{dt} = \dot{y} \qquad \frac{d\dot{y}}{dt} = f(t, y, \dot{y}) \tag{9.5-2}$$

The general solution $y = y(t)$, $\dot{y} = \dot{y}(t)$ of Eq. (1) or (2) can be represented geometrically by a family of directed **phase-trajectory curves** in the $y\dot{y}$ plane or **phase plane.** The phase-plane representation is most useful if *the given function $f(t, y, dy/dt)$ does not involve the independent variable t explicitly* (e.g., "free" oscillations). In this case, the system (2) is of the general form

$$\frac{dy}{dt} = P(y, \dot{y})(\equiv \dot{y}) \qquad \frac{d\dot{y}}{dt} = Q(y, \dot{y}) \tag{9.5-3}$$

and the phase trajectories satisfy the first-order differential equation

$$\frac{d\dot{y}}{dy} = \frac{Q(y, \dot{y})}{P(y, \dot{y})} \qquad \left[\equiv \frac{Q(y, \dot{y})}{\dot{y}}\right] \tag{9.5-4}$$

which specifies the slope of the solution curve through $(y, \dot{y})$ at that point. The resulting field of tangent directions ("phase-plane portrait" of the given differential equation) permits one to sketch $\dot{y}(y)$ and hence $y(t)$ for given initial values of y and $\dot{y}$; one may begin by drawing loci of constant slope $d\dot{y}/dy = m$ (**isoclines,** Fig. 9.5-1).

9.5-3. Critical Points and Limit Cycles (see also Sec. 9.5-4). **(a) Ordinary and Critical Phase-plane Points.** Given a differential equation (1) reducible to a system (3), a phase-plane point $(y, \dot{y})$ is an **ordinary point** if and only if $P(y, \dot{y})$ and $Q(y, \dot{y})$ are analytic and not both equal to zero; *there exists a unique phase trajectory through each ordinary point.* Phase-plane points $(y_0, \dot{y}_0)$ such that

$$\frac{dy}{dt} = P(y_0, \dot{y}_0) = 0 \qquad \frac{d\dot{y}}{dt} = Q(y_0, \dot{y}_0) = 0 \tag{9.5-5}$$

are **critical points** (or **singular points**) where the trajectory is not uniquely determined. Critical points are classified according to the nature of the phase trajectories in their neighborhood; Fig. 9.5-2 illustrates the most important types. Physically, critical points are *equilibrium points* admitting stable or unstable *equilibrium solutions $y = y_0$* (Sec. 9.5-4).

(b) Periodic Solutions and Limit Cycles. Periodic solutions $y = y(t)$ correspond to closed phase-trajectory curves, and vice versa. A closed phase trajectory C is called a **limit cycle** if each trajectory point has a neighborhood of ordinary points in which all phase trajectories spiral into C (**stable** limit cycle, see also Sec. 9.5-4) or out of C (**unstable** limit cycle), or into C on one side of C and out of C on the other side (**half-stable** limit cycle). For an example, see Secs. 9.5-4*c* and 9.5-5 (see also Fig. 9.5-3).

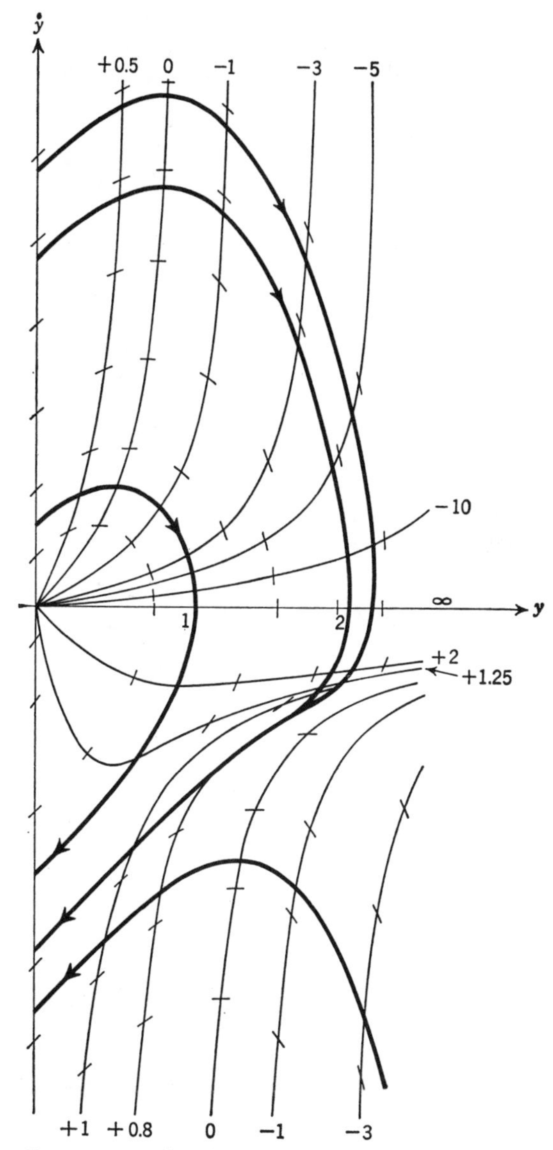

FIG. 9.5-1. Isoclines, tangent directions, and some solutions of the differential equation

$$\frac{d\dot{y}}{dy} = 1 - y^2 - \frac{y}{\dot{y}}$$

corresponding to Van der Pol's differential equation

$$\frac{d^2y}{dt^2} = -y + \mu(1 - y^2)\frac{dy}{dt}$$

with $dy/dt = \dot{y}$, $\mu = 1$. Only the right half-plane is shown.

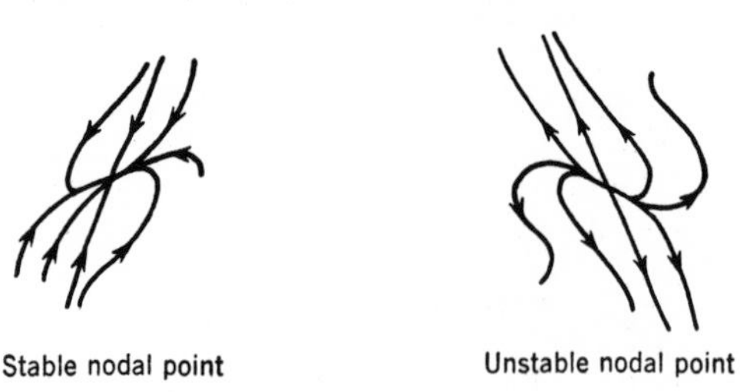

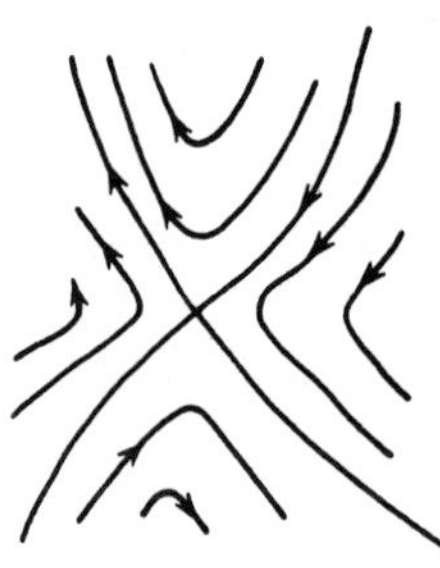

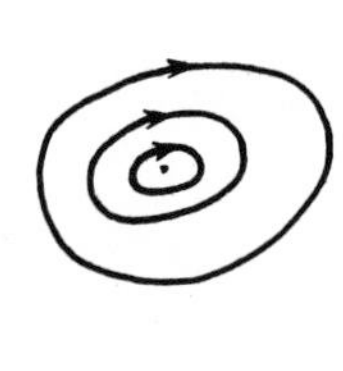

FIG. 9.5-2. Phase trajectories in the neighborhood of six types of critical points (Secs. 9.5-3 and 9.5-4).

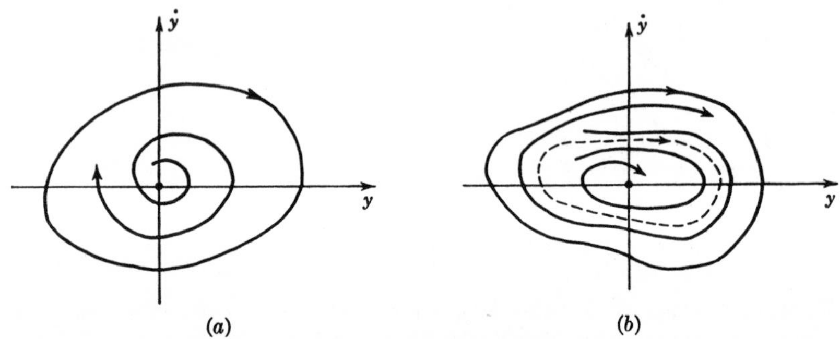

FIG. 9.5-3. (*a*) A stable limit cycle enclosing an unstable critical point at the origin. "Soft" self-excitation of oscillations for arbitrarily small initial values of y and $\dot{y}$. (*b*) A stable limit cycle enclosing a stable critical point at the origin and an unstable limit cycle (shown in broken lines). "Hard" self-excitation of oscillations for initial values outside the unstable limit cycle.

(c) Poincaré's Index and Bendixson's Theorems. The possible existence of limit cycles (stable oscillations) is of interest in many applications. In addition to the analytical criteria of Sec. 9.5-4, the following theory is sometimes helpful.

For any given phase-plane portrait, the **index of a closed curve** C containing only ordinary points is the number of revolutions of the solution tangent as the point $(y, \dot{y})$ makes one complete cycle around C. The **index of an isolated critical point** P is the index of any closed curve enclosing P and no other critical point. Then

1. The index of any closed curve C equals the sum of the indices of all the (isolated) critical points enclosed by C; if C encloses only ordinary points, its index is zero.
2. The index of a nodal point, focal point, or vortex point is 1; the index of a saddle point is -1 (see also Fig. 9.5-2).
3. The index of every closed phase trajectory is 1; hence *a limit cycle must enclose at least one critical point other than a saddle point* (Fig. 9.5-3).

No closed phase trajectories exist within any phase-plane domain where $\partial P/\partial y + \partial Q/\partial \dot{y}$ is of one sign (Bendixson's First Theorem). A trajectory which remains in a bounded region of the phase plane and does not approach any critical point for $0 \leq t < \infty$ is either closed or approaches a closed trajectory asymptotically (Bendixson's Second Theorem).

9.5-4. Poincaré-Lyapunov Theory of Stability (see also Secs. 13.6-5 to 13.6-7). **(a)** The solution $y_i = y_{1i}(t)$ $(i = 1, 2, \ldots, n)$ of a system of ordinary differential equations reduced to the form

$$\frac{dy_i}{dt} = f_i(t; y_1, y_2, \ldots, y_n) \qquad (i = 1, 2, \ldots, n) \tag{9.5-6}$$

is **stable in the sense of Lyapunov** if and only if each $y_i(t) \to y_{1i}(t)$ as $y_i(t_0) \to y_{1i}(t_0)$ $(i = 1, 2, \ldots, n)$, the convergence being uniform for $t > 0$ (Sec. 4.4-4); sufficiently small initial-condition changes then cannot cause large solution changes. A stable solution is **asymptotically stable** if and only if there exists a bound $r(t_0)$ such that $|y_i(t_0) - y_{1i}(t_0)| < r(t_0)$ for all i implies $\lim\limits_{t - t_0 \to \infty} |y_i(t) - y_{1i}(t)| = 0$ for all i. An asymptotically stable solution is **asymptotically stable in the large (completely stable)** if and only if $r(t_0) = \infty$. *This is, in particular, true for every solution of a completely stable system of linear differential equations with constant coefficients* (all roots of characteristic equation have negative real parts, Secs. 9.4-4 to 9.4-7).

(b) Stability of Equilibrium. *For a system of the form*

$$\frac{dy_i}{dt} = f_i(y_1, y_2, \ldots, y_n) \qquad (i = 1, 2, \ldots, n) \tag{9.5-7}$$

with suitably differentiable f_i, an **equilibrium solution**

$$y_i(t) = y_{1i} = \text{const} \qquad \frac{dy_i}{dt} = f_i(y_{11}, y_{12}, \ldots, y_{1n}) = 0 \qquad (i = 1, 2, \ldots, n)$$

is asymptotically stable whenever the linearized system

$$\frac{dy_i}{dt} = \sum_{k=1}^{n} \frac{\partial f_i}{\partial y_k}(y_k - y_{1k}) \qquad (i = 1, 2, \ldots, n) \tag{9.5-8}$$

is completely stable, where the partial derivatives are computed for $y_1 = y_{11}$, $y_2 = y_{12}, \ldots, y_n = y_{1n}$. This is true whenever all roots s of the characteristic equation

$$\det\left[\frac{\partial f_i}{\partial y_k} - s\delta_k^i\right] = 0 \tag{9.5-9}$$

have negative real parts. The equilibrium is unstable if Eq. (9) has a root with positive real part; if the real parts of all roots are negative or zero, one requires a more detailed stability investigation (Sec. 13.6-7).

In particular, for a second-order differential equation (1) reducible to a system (3) the characteristic equation is

$$s^2 + \left(\frac{\partial P}{\partial y} + \frac{\partial Q}{\partial \dot{y}}\right)s + \left(\frac{\partial P}{\partial y}\frac{\partial Q}{\partial \dot{y}} - \frac{\partial P}{\partial \dot{y}}\frac{\partial Q}{\partial y}\right)\left[\equiv s^2 + \left(\frac{\partial P}{\partial y} + 1\right)s + \frac{\partial P}{\partial y}\right] = 0 \tag{9.5-10}$$

where all derivatives are computed at the equilibrium point $(y_0, \dot{y}_0)$. The equilibrium point is

A *stable or unstable nodal point* if both roots s_1 and s_2 of Eq. (10) are real and negative or positive, respectively
A *saddle point* if s_1 and s_2 are real and of opposite sign
A *stable or unstable focal point* if s_1 and s_2 are complex conjugates with negative or positive real parts, respectively
A *vortex point* if s_1 and s_2 are purely imaginary

(see also Sec. 9.5-3*a* and Fig. 9.5-2).

Examples for the six types of equilibrium points listed are most easily obtained from the *linear* differential equation (9.4-3) for different values of the coefficients (Sec. 9.4-1*b*).

(c) Stability of a Periodic Solution. The stability of a periodic solution $y = y_P(t) \not\equiv 0$, $\dot{y} = \dot{y}_P(t)$ of Eq. (3) depends on that of the linearized system

$$\left.\begin{aligned} \frac{d\delta y}{dt} &= \frac{\partial P}{\partial y}\bigg]_{y_P(t),\dot{y}_P(t)} \delta y + \frac{\partial P}{\partial \dot{y}}\bigg]_{y_P(t),\dot{y}_P(t)} \delta\dot{y} \\ \frac{d\delta \dot{y}}{dt} &= \frac{\partial Q}{\partial y}\bigg]_{y_P(t),\dot{y}_P(t)} \delta y + \frac{\partial Q}{\partial \dot{y}}\bigg]_{y_P(t),\dot{y}_P(t)} \delta\dot{y} \end{aligned}\right\} \tag{9.5-11}$$

which is satisfied by small variations (Sec. 11.4-1) δy, $\delta\dot{y}$ of the periodic solution. This is a system of linear differential equations with periodic coefficients having the same period T as the given solution; Eq. (11) admits two linearly independent solutions of the form

$$\delta y = h_{11}(t) \qquad \delta\dot{y} = h_{12}(t) \qquad \delta y = e^{\lambda t}h_{21}(t) \qquad \delta\dot{y} = e^{\lambda t}h_{22}(t) \tag{9.5-12}$$

where the $h_{ik}(t)$ are periodic functions. *The periodic solution is stable if*

$$\lambda = \frac{1}{T}\int_0^T \left[\frac{\partial P}{\partial y} + \frac{\partial Q}{\partial \dot{y}}\right]_{y_P(t),\dot{y}_P(t)} dt \tag{9.5-13}$$

is less than zero, and unstable if $\lambda > 0$; the case $\lambda = 0$ requires additional investigation.

See Sec. 13.6-6 for additional stability criteria.

EXAMPLE: The differential equation

$$\frac{d^2y}{dt^2} + y - \mu(1 - y^2)\frac{dy}{dt} = 0 \quad \text{(Van der Pol's equation)} \tag{9.5-14}$$

yields a stable limit cycle in the neighborhood of the approximate periodic solution $y = y_P(t) \approx 2 \cos t$, $\dot{y} = \dot{y}_P(t) \approx -2 \sin t$, with $\lambda = -\mu[1 + O(\mu)]$ (see also Sec. 9.5-5).

9.5-5. The Approximation Method of Krylov and Bogolyubov.

(a) The First Approximation. Equivalent Linearization. To solve a differential equation of the form

$$\frac{d^2y}{dt^2} + \omega^2 y + \mu f\left(y, \frac{dy}{dt}\right) = 0 \tag{9.5-15}$$

where ω is a given constant, and the last term is a small nonlinear perturbation, write

$$y = r(t) \cos \varphi(t) \tag{9.5-16}$$

Assuming that errors of the order of μ^2 are negligible, the "amplitude" $r(t)$ and the "total phase" $\varphi(t)$ are then obtained from the first-order differential equations

$$\left.\begin{aligned} \frac{dr}{dt} &= \frac{\mu}{2\pi\omega}\int_0^{2\pi} f(r\cos\lambda, -r\omega\sin\lambda)\sin\lambda\, d\lambda = -\frac{r}{2}a_1(r) \\ \frac{d\varphi}{dt} &= \omega + \frac{\mu}{2\pi r\omega}\int_0^{2\pi} f(r\cos\lambda, -r\omega\sin\lambda)\cos\lambda\, d\lambda = \sqrt{a_2(r)} \end{aligned}\right\} \tag{9.5-17}$$

For a given initial value $r(0) = r_0$, the solution (9.4-4) of the *equivalent linear differential equation*

$$\frac{d^2y}{dt^2} + a_1(r_0)\frac{dy}{dt} + a_2(r_0)y = 0 \tag{9.5-18}$$

approximates the solution of the given differential equation (15) with an error of the order of μ^2. For a periodic solution such as a limit cycle (Sec. 9.5-3*b*), the approximate amplitude r_L is obtained from $a_1(r_L) = 0$, and the circular frequency is approximated by $\sqrt{a_2(r_L)}$. The limit cycle is stable if $da_1/dr]_{r=r_L} > 0$ and unstable if $da_1/dr]_{r=r_L} < 0$. For self-excitation from rest, one must have $a_1(0) < 0$.

The first approximation is of considerable interest in connection with periodic nonlinear oscillations. In such cases, the equivalent linear differential equation (18) yields the same energy storage and dissipation per cycle as the given nonlinear equation (15). The equivalent linear equation can therefore be used in many investigations of nonlinear resonance phenomena (Ref. 9.17).

(b) The Improved First Approximation. An improved first-order approximation is given by

$$y = r(t) \cos \varphi(t) + \frac{\mu}{\omega^2}\left\{\tfrac{1}{2}\alpha_0(r) + \sum_{k=2}^{\infty} [\alpha_k(r) \cos k\varphi(t) + \beta_k(r) \sin k\varphi(t)]\right\} \tag{9.5-19}$$

where $r(t)$ and $\varphi(t)$ are given by Eq. (17), and

$$\left.\begin{aligned} \alpha_k(r) &= \frac{1}{\pi(k^2 - 1)} \int_0^{2\pi} f(r \cos \lambda, -r\omega \sin \lambda) \cos k\lambda \, d\lambda \\ &\qquad\qquad (k = 0, 2, 3, \ldots) \\ \beta_k(r) &= \frac{1}{\pi(k^2 - 1)} \int_0^{2\pi} f(r \cos \lambda, -r\omega \sin \lambda) \sin k\lambda \, d\lambda \\ &\qquad\qquad (k = 2, 3, \ldots) \end{aligned}\right\} \tag{9.5-20}$$

EXAMPLE: In the case of Van der Pol's differential equation (14), Eq. (17) yields

$$\frac{dr}{dt} = \frac{\mu r}{2}\left(1 - \frac{r^2}{4}\right) \qquad r(t) = \frac{r_0 e^{\mu t/2}}{\sqrt{1 + \frac{1}{4}r_0^2(e^{\mu t} - 1)}} \tag{9.5-21}$$

There is a stable limit cycle for $r = r_L = 2$. The coefficients (20) all vanish except for β_3; the improved first approximation is

$$y = r(t) \cos (t + \varphi_0) - \mu \frac{r^3(t)}{32} \sin 3(t + \varphi_0) \tag{9.5-22}$$

The Krylov-Bogolyubov approximation (19) is an improvement on an earlier method due to Van der Pol, who derived an approximation of the form

$$y = a(t) \cos \omega t + b(t) \sin \omega t$$

in an analogous manner.

The Krylov-Bogolyubov approximation method can be extended to apply in the case of a periodic *forcing function* on the right side of the differential equation (15) (nonlinear *forced oscillations*, subharmonic resonance, entrainment of frequency). For this and other methods, see Refs. 9.15 to 9.17.

9.5-6. Energy-integral Solution. Differential equations of the form

$$\frac{d^2y}{dt^2} = f(y) \tag{9.5-23}$$

which are of considerable interest in dynamics, can be reduced to first-order equations through multiplication by dy/dt and integration:

$$\left(\frac{dy}{dt}\right)^2 = 2 \int f(y)\, dy + C_1 \tag{9.5-24}$$

$$t = \int \frac{dy}{\sqrt{2\int f(y)\, dy + C_1}} + C_2 \tag{9.5-25}$$

9.6. PFAFFIAN DIFFERENTIAL EQUATIONS

9.6-1. Pfaffian Differential Equations (see also Secs. 3.1-16 and 17.2-2). A **Pfaffian differential equation** (first-order linear **total differential equation**)

$$P(x, y, z)\, dx + Q(x, y, z)\, dy + R(x, y, z)\, dz = 0 \tag{9.6-1}$$

with continuously differentiable coefficients P, Q, R may be interpreted geometrically as a condition $\mathbf{P} \cdot d\mathbf{r} = 0$ on the tangent vector $d\mathbf{r} \equiv (dx, dy, dz)$ of a **solution curve (integral curve)** described by two equations $f(x, y, z) = 0$, $g(x, y, z, C) = 0$, where C is a constant of integration. To find the integral curves lying on an arbitrary regular surface

$$f(x, y, z) = 0 \tag{9.6-2}$$

solve the ordinary differential equation obtained by elimination of z and dz from Eq. (1) and $df(x, y, z) = 0$.

9.6-2. The Integrable Case (see also Sec. 9.2-4). The Pfaffian differential equation (1) is **integrable** if and only if there exists an *integrating factor* $\mu = \mu(x, y, z)$ such that $\mu(P\,dx + Q\,dy + R\,dz)$ is an exact differential $d\varphi(x, y, z)$; this is true if and only if

$$P\left(\frac{\partial Q}{\partial z} - \frac{\partial R}{\partial y}\right) + Q\left(\frac{\partial R}{\partial x} - \frac{\partial P}{\partial z}\right) + R\left(\frac{\partial P}{\partial y} - \frac{\partial Q}{\partial x}\right) = 0 \tag{9.6-3}$$

In this case, every curve on an **integral surface**

$$\varphi(x, y, z) = C \tag{9.6-4}$$

orthogonal to the family of curves described by

$$\frac{dx}{P} = \frac{dy}{Q} = \frac{dz}{R} \tag{9.6-5}$$

is a solution. It follows that the solutions found in the manner of Sec. 9.6-1 from a conveniently chosen *family* of surfaces (usually planes)

$$f(x, y, z) \equiv f(x, y, z; \lambda) = 0 \tag{9.6-6}$$

lie on an integral surface (4) obtainable by elimination of λ from the resulting solution $f(x, y, z; \lambda) = 0$, $g(x, y, z, C; \lambda) = 0$ (*Mayer's Method of Solution*).

To find integral surfaces (4) by another method, *hold z constant* and obtain the solution of the ordinary differential equation $P\,dx + Q\,dy = 0$ in the form $u(x, y, z) - K = 0$. Then

$$\varphi(x, y, z) \equiv u(x, y, z) - \psi(z) = C \tag{9.6-7}$$

describes an integral surface, where $\psi(z)$ is the solution of the ordinary differential equation obtained by elimination of x and y from

$$\frac{1}{P}\frac{\partial u}{\partial x} = \frac{1}{Q}\frac{\partial u}{\partial y} = \frac{1}{R}\left(\frac{\partial u}{\partial z} - \frac{\partial \psi}{dz}\right) \tag{9.6-8}$$

with the aid of Eq. (7). Note that the function (8) is the required integrating factor $\mu(x, y, z)$.

An important application is found in thermodynamics, where the reciprocal of the *absolute temperature* T is an integrating factor for the adiabatic condition $\delta q = 0$ of the form (1); $\delta q/T$ is the (exact) differential of the *entropy*. See Ref. 10.23 for a discussion of total differential equations involving more than three variables.

9.7. RELATED TOPICS, REFERENCES, AND BIBLIOGRAPHY

9.7-1. Related Topics. The following topics related to the study of ordinary differential equations are treated in other chapters of this handbook:

The Laplace transformation .... Chap. 8
Calculus of variations .... Chap. 11
Matrix notation for systems of differential equations .... Chap. 13
Boundary-value problems and eigenvalue problems .... Chap. 15
Numerical solutions .... Chap. 20
Special transcendental functions .... Chap. 21

9.7-2. References and Bibliography.

9.1. Agnew, R. P.: *Differential Equations*, 2d ed., McGraw-Hill, New York, 1960.
9.2. Andronow, A. A., and C. E. Chaikin: *Theory of Oscillations*, Princeton University Press, Princeton, N.J., 1949.
9.3. Bellman, R.: *Stability Theory of Differential Equations*, McGraw-Hill, New York, 1953.
9.4. Bieberbach, L.: *Differentialgleichungen*, 2d ed., Springer, Berlin, 1965.
9.5. Birkhoff, G., and G. Rota: *Ordinary Differential Equations*, Blaisdell, New York, 1962.
9.6. Coddington, E. A.: *An Introduction to Ordinary Differential Equations*, Prentice-Hall, Englewood Cliffs, N.J., 1961.
9.7. ——— and N. Levinson: *Theory of Ordinary Differential Equations*, McGraw-Hill, New York, 1955.
9.8. Ford, L. R.: *Differential Equations*, 2d ed., McGraw-Hill, New York, 1955.
9.9. Golomb, M., and M. E. Shanks: *Elements of Ordinary Differential Equations*, 2d ed., McGraw-Hill, New York, 1965.
9.10. Hale, J. K.: *Oscillations in Nonlinear Systems*, McGraw-Hill, New York, 1963.
9.11. Hartman, P.: *Ordinary Differential Equations*, Wiley, New York, 1964.
9.12. Hochstadt, H.: *Differential Equations*, Holt, New York, 1964.
9.13. Hurewicz, W.: *Lectures on Ordinary Differential Equations*, M.I.T., Cambridge, Mass., 1958.
9.14. Kamke, E.: *Differentialgleichungen, Lösungsmethoden und Lösungen*, vol. I, Chelsea, New York, 1948.
9.15. Krylov, N., and N. Bogolyubov: *Nonlinear Oscillations*, translated by S. Lefschetz, Princeton University Press, Princeton, N.J., 1943.
9.16. Lefschetz, S.: *Differential Equations: Geometric Theory*, 2d ed., Interscience, New York, 1963.
9.17. Minorski, N.: *Nonlinear Oscillations*, Van Nostrand, Princeton, N.J., 1962.
9.18. Petrovsky, I. G.: *Lectures on Partial Differential Equations*, Interscience, New York, 1955.
9.19. Pontryagin, L. S.: *Ordinary Differential Equations*, Addison-Wesley, Reading, Mass., 1962.
9.20. Saaty, T. L., and J. Bram: *Nonlinear Mathematics*, McGraw-Hill, New York, 1964.
9.21. Sansone, G., and R. Conti: *Nonlinear Differential Equations*, Macmillan, New York, 1964.
9.22. Stoker, J. J.: *Nonlinear Vibrations*, Interscience, New York, 1950.
9.23. Struble, R.: *Nonlinear Differential Equations*, McGraw-Hill, New York, 1961.
9.24. Tenenbaum, M., and H. Pollard: *Ordinary Differential Equations*, Harper and Row, New York: 1963.
9.25. Tricomi, F. G.: *Differential Equations*, Hafner, New York, 1961.

CHAPTER 10

PARTIAL DIFFERENTIAL EQUATIONS

10.1. INTRODUCTION AND SURVEY

10.1-1. Sections 10.2-1 to 10.2-7 deal with *partial differential equations of the first order* and their *geometrical interpretation* and include an outline of the *Hamilton-Jacobi theory of canonical equations.* Sections 10.3-1 to 10.3-3 introduce the *characteristics and boundary-value problems of hyperbolic, parabolic, and elliptic second-order equations.* Sections 10.4-1 to 10.5-4 present the solutions of the most important *linear partial differential equations of physics* (heat conduction, wave equation, etc.) from the heuristic point of view of an elementary course and outline the *use of integral-transform methods.* A more sophisticated theory of linear boundary-value problems and eigenvalue problems is described in Chap. 15.

10.1-2. Partial Differential Equations (see also Sec. 9.1-2). **(a)** A **partial differential equation of order** r is a functional equation of the form

$$F\left(x_1, x_2, \ldots, x_n; \Phi; \frac{\partial\Phi}{\partial x_1}, \frac{\partial\Phi}{\partial x_2}, \ldots, \frac{\partial\Phi}{\partial x_n}; \frac{\partial^2\Phi}{\partial x_1^2}, \ldots\right) = 0 \qquad (10.1\text{-}1)$$

which involves at least one r^{th}-order partial derivative of the unknown function $\Phi = \Phi(x_1, x_2, \ldots, x_n)$ of two or more independent variables $x_1, x_2, \ldots, x_n$. A function $\Phi(x_1, x_2, \ldots, x_n)$ which satisfies the given partial differential equation on a specified region of "points" $(x_1, x_2, \ldots, x_n)$ is called a **solution** or **integral** of the partial differential equation.

The **general solution (general integral)** of a given r^{th}-order equation (1) will, in general, involve *arbitrary functions.* Substitution of specific functions yields **particular integrals** corresponding to given accessory conditions, e.g., given conditions on $\Phi(x_1, x_2, \ldots, x_n)$ and/or its derivatives on a curve, surface, etc., in the space of "points" $(x_1, x_2, \ldots, x_n)$ (*boundary conditions, initial conditions*). Many partial differential equations admit additional solutions (**singular integrals**) which are not obtainable through substitution of specific functions for the arbitrary functions in the general integral (Sec. 10.2-1*c*).

(b) A partial differential equation is **homogeneous** if and only if every constant multiple $\alpha\Phi$ of any solution Φ is a solution. A partial differential equation (1) is **linear** if and only if F is a linear function of Φ and its derivatives (see also Secs. 10.4-1 and 10.4-2).

(c) Systems of Partial Differential Equations. Compatibility Conditions. A system of partial differential equations

$$F_i\left(x_1, x_2, \ldots, x_n; \Phi_1, \Phi_2, \ldots; \frac{\partial\Phi_1}{\partial x_1}, \ldots\right) = 0 \qquad (i = 1, 2, \ldots) \qquad (10.1\text{-}2)$$

involves a set of unknown functions $\Phi_1(x_1, x_2, \ldots, x_n)$, $\Phi_2(x_1, x_2, \ldots, x_n)$, . . . and their partial derivatives. *One can reduce every partial differential equation or system of partial differential equations to a system of first-order equations by introducing suitable derivatives as new variables* (see also Sec. 9.1-3).

A system of partial differential equations (2) may admit a solution $\Phi_1, \Phi_2, \ldots$ only if the given functions F_i and their derivatives satisfy a set of **compatibility conditions (integrability conditions)** which ensure that differentiations of two or more equations (2) yield compatible higher-order derivatives. To derive a compatibility condition, eliminate the Φ_k and their derivatives from a set of equations obtained by differentiation of the given partial differential equations (2).

EXAMPLE: Given $\frac{\partial \Phi}{\partial x_1} + f_1(x_1, x_2) = 0$, $\frac{\partial \Phi}{\partial x_2} + f_2(x_1, x_2) = 0$, differentiation yields $\frac{\partial^2 \Phi}{\partial x_1\, \partial x_2} = -\frac{\partial f_1}{\partial x_2}$, $\frac{\partial^2 \Phi}{\partial x_2\, \partial x_1} = -\frac{\partial f_2}{\partial x_1}$, so that the given partial differential equations are compatible only if $\frac{\partial f_1}{\partial x_2} = \frac{\partial f_2}{\partial x_1}$.

(d) **Existence of Solutions.** As with ordinary differential equations (Sec. 9.1-4), the actual existence and uniqueness of solutions for a given partial differential equation or system of partial differential equations require a proof in each case, even if all compatibility conditions are satisfied. See Refs. 10.5 and 10.18 for a number of existence theorems.

10.1-3. Solution of Partial Differential Equations: Separation of Variables (see also Secs. 10.4-2 to 10.4-9). In many important applications, an attempt to write solutions of the form

$$\Phi = \Phi(x_1, x_2, \ldots, x_n) = \varphi_1(x_1)\varphi_0(x_2, x_3, \ldots, x_n) \qquad (10.1\text{-}3)$$

permits one to rewrite a given partial differential equation (1) in the "separated" form

$$F_1\left(x_1, \varphi_1, \frac{d\varphi_1}{dx_1}, \frac{d^2\varphi_1}{dx_1^2}, \ldots\right) = F_0\left(x_2, x_3, \ldots, x_n; \varphi_0; \frac{\partial \varphi_0}{\partial x_2}, \frac{\partial \varphi_0}{\partial x_3}, \ldots\right)$$

Then the unknown functions $\varphi_1(x_1)$ and $\varphi_0(x_2, x_3, \ldots, x_n)$ must satisfy the differential equations

$$F_1\left(x_1, \varphi_1, \frac{d\varphi_1}{dx_1}, \frac{d^2\varphi_1}{dx_1^2}, \ldots\right) = C \qquad (10.1\text{-}4a)$$

$$F_2\left(x_2, x_3, \ldots, x_n; \varphi_0; \frac{\partial \varphi_0}{\partial x_2}, \frac{\partial \varphi_0}{\partial x_3}, \ldots\right) = C \qquad (10.1\text{-}4b)$$

where C is a constant of integration (**separation constant**) to be determined in accordance with suitably given boundary conditions or other accessory conditions. Note that Eq. (4*a*) is an *ordinary* differential equation for the unknown function $\varphi_1(x_1)$; it may be possible to repeat the separation process with Eq. (4*b*).

Separation of variables applies particularly well to many linear homogeneous partial differential equations of physics; sometimes separation becomes possible after an appropriate change of variables (see Secs. 10.4-3 to 10.4-9 for examples).

10.2. PARTIAL DIFFERENTIAL EQUATIONS OF THE FIRST ORDER

10.2-1. First-order Partial Differential Equations with Two Independent Variables. Geometrical Interpretation (see also

Secs. 9.2-2 and 17.3-11). (**a**) Given a first-order partial differential equation

$$F(x, y, z, p, q) = 0 \qquad \left(p \equiv \frac{\partial z}{\partial x}, q \equiv \frac{\partial z}{\partial y}; F_p{}^2 + F_q{}^2 \neq 0\right) \tag{10.2-1}$$

for the unknown function $z = z(x, y)$, let the given function F be single-valued and twice continuously differentiable, and consider x, y, z as rectangular cartesian coordinates. Then every solution $z = z(x, y)$ of the partial differential equation (1) represents a surface whose normal has the direction numbers p, q, -1 at every surface point (x, y, z); the solution surface must touch a **Monge cone** of **characteristic directions** defined by

$$dx:dy:dz = F_p:F_q:(pF_p + qF_q)$$

at every point (x, y, z).

A set of values (x, y, z, p, q) is said to describe a **planar element** associating the direction numbers p, q, -1 (and thus the tangent plane of a hypothetical surface) with a point (x, y, z). A given partial differential equation (1) selects a "field" of planar elements (x, y, z, p, q) tangent to the Monge cones (Fig. 10.2-1).

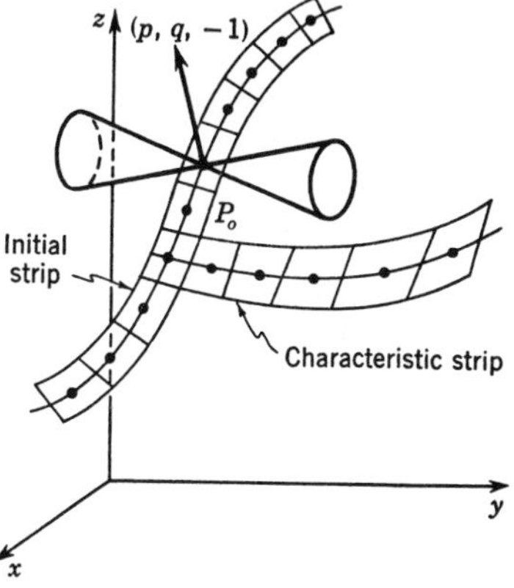

Fig. 10.2-1. The initial strip and one characteristic strip of the solution surface. Note the Monge cone at P_0. (*From Burington, R. S., and C. C. Torrance, Higher Mathematics, McGraw-Hill, New York,* 1939.)

If F_p and F_q do not depend explicitly on p and q (*quasilinear* partial differential equation of the first order), then each Monge cone degenerates into a straight line (**Monge axis**).

(**b**) **Strips and Characteristic Equations.** A set of suitably differentiable functions

$$x = x(t) \qquad y = y(t) \qquad z = z(t) \qquad p = p(t) \qquad q = q(t) \tag{10.2-2}$$

represents the planar elements (points and tangent planes) along a **strip** of a regular surface if the functions (2) satisfy the *strip condition*

$$\frac{dz}{dt} = p\frac{dx}{dt} + q\frac{dy}{dt}$$

Given a first-order partial differential equation (1), every set of functions (2) which satisfies the ordinary differential equations

$$\left.\begin{aligned} \frac{dz}{dt} &= p\frac{dx}{dt} + q\frac{dy}{dt} \qquad \frac{dx}{dt} = F_p \qquad \frac{dy}{dt} = F_q \\ \frac{dp}{dt} &= -(pF_z + F_x) \qquad \frac{dq}{dt} = -(qF_z + F_y) \end{aligned}\right\}$$ [CHARACTERISTIC EQUATIONS associated with the partial differential equation (1)] (10.2-3)

together with Eq. (1) is said to describe a **characteristic strip.** A characteristic strip touches a Monge cone at each point (x, y, z); the associated curve $x = x(t)$, $y = y(t)$, $z = z(t)$ (**characteristic curve, characteristic**) lies on a solution surface and has a characteristic direction at each point. *Solution surfaces can touch one another only along characteristics.*

(c) **Singular Integrals** (see also Secs. 9.2-2b, 10.1-2a, and 10.2-3c). Solutions $z = z(x, y)$ of Eq. (1) derived by elimination of p and q from

$$F(x, y, z, p, q) = 0 \qquad \frac{\partial F(x, y, z, p, q)}{\partial p} = 0 \qquad \frac{\partial F(x, y, z, p, q)}{\partial q} = 0 \qquad (10.2\text{-}4)$$

are *singular integrals;* they do not satisfy the condition $F_p^2 + F_q^2 \neq 0$ and are not contained in a general integral of the partial differential equation (1).

10.2-2. The Initial-value Problem. It is desired to find a solution $z = z(x, y)$ of Eq. (1) subject to the *initial conditions* (*Cauchy-type boundary conditions*)

$$x = x_0(\tau) \qquad y = y_0(\tau) \qquad z = z_0(\tau) \qquad p = p_0(\tau) \qquad q = q_0(\tau) \qquad (10.2\text{-}5a)$$

with

$$F[x_0(\tau), y_0(\tau), z_0(\tau), p_0(\tau), q_0(\tau)] = 0 \qquad \frac{dz_0}{d\tau} = p_0\frac{dx_0}{d\tau} + q_0\frac{dy_0}{d\tau} \qquad (10.2\text{-}5b)$$

which specify an "initial strip" of points and tangent planes of the solution surface along a regular curve C_0; the projection of C_0 onto the xy plane is to be a simple curve (Sec. 3.1-13). To solve this initial-value problem, *find the solution*

$$x = x(t, \tau) \qquad y = y(t, \tau) \qquad z = z(t, \tau) \qquad p = p(t, \tau) \qquad q = q(t, \tau) \qquad (10.2\text{-}6)$$

of the system of characteristic equations (3) *subject to the initial conditions* (5) *for* $t = 0$. *The resulting functions* (6) *satisfy* Eq. (1); find the solution $z = z(x, y)$ or an implicit solution $\varphi(x, y, z) = 0$ by eliminating the parameters t and τ.

The initial-value problem has a unique solution if the given initial conditions (5) *imply*

$$F_p\frac{\partial y}{\partial \tau} - F_q\frac{\partial x}{\partial \tau} \neq 0 \qquad \left[\text{i.e., } \left.\frac{\partial(x, y)}{\partial(t, \tau)}\right]_{t=0} \neq 0\right] \qquad (10.2\text{-}7)$$

Otherwise the problem has a solution only if the given initial conditions (5) *describe a characteristic strip; in this case there are infinitely many solutions.*

NOTE: If the functions (5*a*) are left arbitrary subject to the conditions (5*b*) and (7), then the solution obtained above constitutes a *general integral* of the given partial differential equation (1).

10.2-3. Complete Integrals. Derivation of General Integrals, Particular Integrals, Singular Integrals, and Solutions of the Characteristic Equations. **(a)** A **complete integral** of the first-order partial differential equation (1) is a two-parameter family of solutions

$$z = \Phi(x, y, \lambda, \mu) \qquad \left(\frac{\partial^2\Phi}{\partial x\,\partial\lambda}\frac{\partial^2\Phi}{\partial y\,\partial\mu} - \frac{\partial^2\Phi}{\partial y\,\partial\lambda}\frac{\partial^2\Phi}{\partial x\,\partial\mu} \neq 0\right) \tag{10.2-8}$$

having single-valued derivatives

$$\frac{\partial\Phi}{\partial x} = p(x, y, \lambda, \mu) = p \qquad \frac{\partial\Phi}{\partial y} = q(x, y, \lambda, \mu) = q$$

and continuous second-order derivatives with respect to x, y, λ, μ, except possibly for $\partial^2\Phi/\partial\lambda^2$ and $\partial^2\Phi/\partial\mu^2$; a given set of values (x, y, z, p, q) satisfying Eq. (1) must define a unique set of parameters λ, μ. A complete integral (7) yields a *general integral* if one introduces an arbitrary function $\mu = \mu(\lambda)$ and eliminates λ from

$$z - \Phi[x, y, \lambda, \mu(\lambda)] = 0 \qquad \frac{\partial\Phi}{\partial\lambda} + \frac{\partial\Phi}{\partial\mu}\frac{d\mu}{d\lambda} = 0 \tag{10.2-9}$$

(envelope of a one-parameter family of solutions, Sec. 17.3-11).

(b) Derivation of Particular Integrals. To obtain the *particular integral corresponding to a given set of initial conditions* (5), one must find the correct function $\mu(\lambda)$ to be substituted in the general integral derived in Sec. 10.2-3*a*. Obtain $\mu = \mu(\lambda)$ by eliminating τ from

$$\frac{\partial\Phi(x, y, \lambda, \mu)}{\partial x} = p_0(\tau) \qquad \frac{\partial\Phi(x, y, \lambda, \mu)}{\partial y} = q_0(\tau) \qquad x = x_0(\tau) \qquad y = y_0(\tau) \tag{10.2-10}$$

(c) Derivation of Singular Integrals. Elimination of λ and μ from

$$z - \Phi(x, y, \lambda, \mu) = 0 \qquad \frac{\partial\Phi(x, y, \lambda, \mu)}{\partial\lambda} = 0 \qquad \frac{\partial\Phi(x, y, \lambda, \mu)}{\partial\mu} = 0 \tag{10.2-11}$$

may yield a *singular integral* (envelope of a two-parameter family of solutions).

(d) Solution of the Characteristic Equations. *Every complete integral* (8) *of a first-order partial differential equation* (1) *yields the complete solution of the system of ordinary differential equations* (3). $x = x(t)$, $y = y(t)$ are obtained from

$$\frac{\partial\Phi(x, y, \lambda, \mu)}{\partial\lambda} = t \qquad \frac{\partial\Phi(x, y, \lambda, \mu)}{\partial\mu} = \beta t \tag{10.2-12}$$

where λ, μ, and β are arbitrary constants of integration; $z = z(t)$, $p = p(t)$, and $q = q(t)$ are then obtained by substitution of $x = x(t, \lambda, \mu, \beta)$ and $y = y(t, \lambda, \mu, \beta)$ into

$$z = \Phi(x, y, \lambda, \mu) \qquad p = \frac{\partial\Phi(x, y, \lambda, \mu)}{\partial x} \qquad q = \frac{\partial\Phi(x, y, \lambda, \mu)}{\partial y} \tag{10.2-13}$$

(e) Special Cases. Table 10.2-1 lists complete integrals for a number of frequently encountered types of first-order partial differential equations and permits one to apply the methods of Sec. 10.2-3 to many problems. See also Ref. 10.2 for a general method of deriving complete integrals.

Table 10.2-1. Complete Integrals for Special Types of First-order Partial Differential Equations

No.	Type of differential equation		Complete Integral $z = \Phi(x, y, \lambda, \mu)$
1	x, y, z do not occur explicitly	$F(p, q) = 0$	$z = \lambda x + \lambda' y + \mu$ with $F(\lambda, \lambda') = 0$
2a	Only one of the variables x, y, z occurs explicitly	$p = f(x, q)$	$z = \int f(x, \lambda)\, dx + \lambda y + \mu$
2b		$p = f(z, q)$	$x + \lambda y = \int \frac{dz}{g(z, \lambda)} + \mu$ *
3	The variables are separable	$F_1(x, p) = F_2(y, q)$ ($=\lambda$, Sec. 10.1-3) or $p = f_1(x, \lambda)$ $q = f_2(y, \lambda)$	$z = \int f_1(x, \lambda)\, dx + \int f_2(y, \lambda)\, dy + \mu$
4	Generalized Clairaut equation (see also Sec. 9.2-4)	$z = px + qy + f(p, q)$	$z = \lambda x + \mu y + f(\lambda, \mu)$

10.2-4. Partial Differential Equations of the First Order with n Independent Variables. (a) The Initial-value Problem (see also Secs. 10.2-1 and 10.2-2). It is desired to find the solution

$$z = \Phi(x_1, x_2, \ldots, x_n)$$

of the first-order partial differential equation

$$F(x_1, x_2, \ldots, x_n; z; p_1, p_2, \ldots, p_n) = 0$$
$$\left[p_i \equiv \frac{\partial z}{\partial x_i}, i = 1, 2, \ldots, n; \sum_{k=1}^{n} \left(\frac{\partial F}{\partial p_k}\right)^2 \neq 0\right] \tag{10.2-14}$$

* Where $g(z, \lambda)$ is a solution of $p = f(z, \lambda p)$.

subject to the initial conditions

$$z = z_0(\tau_1, \tau_2, \ldots, \tau_{n-1}) \qquad x_i = x_{i0}(\tau_1, \tau_2, \ldots, \tau_{n-1})$$
$$p_i = p_{i0}(\tau_1, \tau_2, \ldots, \tau_{n-1}) \qquad (i = 1, 2, \ldots, n) \tag{10.2-15a}$$

where $x_i = x_{i0}(\tau_1, \tau_2, \ldots, \tau_{n-1})$ $(i = 1, 2, \ldots, n)$ is to represent a hypersurface free of multiple points, and

$$F(x_{10}, x_{20}, \ldots, x_{n0}; z_0; p_{10}, p_{20}, \ldots, p_{n0}) = 0$$
$$\frac{\partial z_0}{\partial \tau_j} = \sum_{k=1}^{n} p_{k0} \frac{\partial x_{k0}}{\partial \tau_j} \qquad (j = 1, 2, \ldots, n-1) \tag{10.2-15b}$$

To obtain the correct relation between $x_1, x_2, \ldots, x_n$, and z, solve the system of ordinary differential equations

$$\frac{dz}{dt} = \sum_{k=1}^{n} p_k \frac{dx_k}{dt} \qquad \frac{dx_i}{dt} = \frac{\partial F}{\partial p_i} \qquad \frac{dp_i}{dt} = -\frac{\partial F}{\partial x_i} - p_i \frac{\partial F}{\partial z}$$
$$(i = 1, 2, \ldots, n) \quad \text{(CHARACTERISTIC EQUATIONS)} \tag{10.2-16}$$

subject to the initial conditions (15) for $t = 0$, and eliminate the n parameters $\tau_1, \tau_2, \ldots, \tau_{n-1}$, and t.

The initial-value problem has a unique solution if the given initial conditions (15) imply

$$\begin{vmatrix} \frac{\partial F}{\partial p_1} & \frac{\partial F}{\partial p_2} & \cdots & \frac{\partial F}{\partial p_n} \\ \frac{\partial x_1}{\partial \tau_1} & \frac{\partial x_2}{\partial \tau_1} & \cdots & \frac{\partial x_n}{\partial \tau_1} \\ \cdots & \cdots & \cdots & \cdots \\ \frac{\partial x_1}{\partial \tau_{n-1}} & \frac{\partial x_2}{\partial \tau_{n-1}} & \cdots & \frac{\partial x_n}{\partial \tau_{n-1}} \end{vmatrix} \neq 0 \qquad \left[\text{i.e., } \frac{\partial(x_1, x_2, \ldots, x_n)}{\partial(t, \tau_1, \ldots, \tau_{n-1})} \neq 0\right] \tag{10.2-17}$$

(b) Complete Integrals and Solution of the Characteristic Equations (see also Sec. 10.2-3). A complete integral of the first-order partial differential equation (14) is an n-parameter family of solutions

$$z = \Phi(x_1, x_2, \ldots, x_n; \alpha_1, \alpha_2, \ldots, \alpha_n) \qquad \left\{\det\left[\frac{\partial^2 \Phi}{\partial x_i\, \partial \alpha_k}\right] \neq 0\right\} \tag{10.2-18}$$

having single-valued derivatives

$$\partial\Phi/\partial x_i = p_i(x_1, x_2, \ldots, x_n; \alpha_1, \alpha_2, \ldots, \alpha_n)$$

and continuous second-order derivatives $\partial^2\Phi/\partial x_i\, \partial x_k$ and $\partial^2\Phi/\partial x_i\, \partial \alpha_k$; a given set of values $(x_1, x_2, \ldots, x_n; z; p_1, p_2, \ldots, p_n)$ satisfying Eq. (14) must define a unique set of parameters $\alpha_1, \alpha_2, \ldots, \alpha_n$. A complete integral (18) yields a *general integral* if one introduces n arbitrary functions $\alpha_k = \alpha_k(\lambda_1, \lambda_2, \ldots, \lambda_{n-1})$ $(k = 1, 2, \ldots, n)$ and eliminates the $n - 1$ parameters λ_j from the n equations

$$\left.\begin{aligned} z = \Phi[x_1, x_2, \ldots, x_n; \alpha_1(\lambda_1, \lambda_2, \ldots, \lambda_{n-1}),\\ \alpha_2(\lambda_1, \lambda_2, \ldots, \lambda_{n-1}), \ldots, \alpha_n(\lambda_1, \lambda_2, \ldots, \lambda_{n-1})]\\ \sum_{k=1}^{n} \frac{\partial \Phi}{\partial \alpha_k}\frac{\partial \alpha_k}{\partial \lambda_j} = 0 \qquad (j = 1, 2, \ldots, n-1) \end{aligned}\right\} \tag{10.2-19}$$

Every complete integral (18) *yields the complete solution of the system of ordinary differential equations* (16). Obtain $x_i = x_i(t)$ $(i = 1, 2, \ldots, n)$ from

$$\frac{\partial \Phi}{\partial \alpha_k} = \beta_k t \qquad (k = 1, 2, \ldots, n-1) \qquad \frac{\partial \Phi}{\partial \alpha_n} = t \tag{10.2-20}$$

where $\alpha_1, \alpha_2, \ldots, \alpha_n, \beta_1, \beta_2, \ldots, \beta_{n-1}$ are $2n - 1$ arbitrary constants of integration; then find $z = z(t)$ and $p_i = p_i(t)$ $(i = 1, 2, \ldots, n)$ by substituting $x_i = x_i(t; \alpha_1, \alpha_2, \ldots, \alpha_n; \beta_1, \beta_2, \ldots, \beta_{n-1})$ into

$$z = \Phi(x_1, x_2, \ldots, x_n; \alpha_1, \alpha_2, \ldots, \alpha_n)$$

$$p_i = \frac{\partial \Phi}{\partial x_i} \qquad (i = 1, 2, \ldots, n)$$

(c) **Singular Integrals** (see also Secs. 10.2-1*c* and 10.2-3*c*). Singular integrals of the partial differential equation (14) are solutions $z = \Phi(x_1, x_2, \ldots, x_n)$ obtained by elimination of the p_i from the $n + 1$ equations

$$\begin{aligned} F(x_1, x_2, \ldots, x_n; z; p_1, p_2, \ldots, p_n) = 0\\ \frac{\partial F}{\partial p_i} = 0 \qquad (i = 1, 2, \ldots, n) \end{aligned} \tag{10.2-21}$$

subject to $\partial F/\partial x_i + p_i \partial F/\partial z = 0$. A given complete integral (18) may yield a singular integral by elimination of $\alpha_1, \alpha_2, \ldots, \alpha_n$ from the $n + 1$ equations

$$\begin{aligned} z - \Phi(x_1, x_2, \ldots, x_n; \alpha_1 \alpha_2, \ldots, \alpha_n) = 0\\ \frac{\partial \Phi}{\partial \alpha_k} = 0 \qquad (k = 1, 2, \ldots, n) \end{aligned} \tag{10.2-22}$$

10.2-5. Contact Transformations (see also Sec. 9.2-3*b*). Some problems involving first-order partial differential equations can be simplified by a twice continuously differentiable transformation

$$\left.\begin{aligned} \left.\begin{aligned} \bar{x}_i = \bar{x}_i(x_1, x_2, \ldots, x_n; z; p_1, p_2, \ldots, p_n)\\ \bar{p}_i = \bar{p}_i(x_1, x_2, \ldots, x_n; z; p_1, p_2, \ldots, p_n)\\ \bar{z} = \bar{z}(x_1, x_2, \ldots, x_n; z; p_1, p_2, \ldots, p_n) \end{aligned}\right\} (i = 1, 2, \ldots, n)\\ \text{with} \quad \frac{\partial(\bar{x}_1, \bar{x}_2, \ldots, \bar{x}_n; \bar{z}; \bar{p}_1, \bar{p}_2, \ldots, \bar{p}_n)}{\partial(x_1, x_2, \ldots, x_n; z; p_1, p_2, \ldots, p_n)} \neq 0 \end{aligned}\right\} \tag{10.2-23}$$

chosen so that *every complete differential* $dz = \sum_{k=1}^{n} p_k\,dx_k$ *is transformed into a complete differential* $d\bar{z} = \sum_{k=1}^{n} \bar{p}_k\,d\bar{x}_k$ with $\bar{p}_i \equiv \frac{\partial \bar{z}}{\partial \bar{x}_i}$ $(i = 1, 2, \ldots, n)$, *or*

$$d\bar{z} - \sum_{k=1}^{n} \bar{p}_k\,d\bar{x}_k \equiv g(x_1, x_2, \ldots, x_n; z; p_1, p_2, \ldots, p_n)\left(dz - \sum_{k=1}^{n} p_k\,dx_k\right)$$
$$[g(x_1, x_2, \ldots, x_n; z; p_1, p_2, \ldots, p_n) \neq 0] \qquad (10.2\text{-}24)$$

Such a transformation is called a **contact transformation;** a contact transformation necessarily preserves every strip condition and will thus preserve contact (osculation) of regular surface elements for $n = 2$ (see also Sec. 10.2-1).

A contact transformation (23) transforms the given partial differential equation (14) into a new partial differential equation

$$\bar{F}(\bar{x}_1, \bar{x}_2, \ldots, \bar{x}_n; \bar{z}; \bar{p}_1, \bar{p}_2, \ldots, \bar{p}_n) \equiv F(x_1, x_2, \ldots, x_n; z; p_1, p_2, \ldots, p_n) = 0$$
$$\left(\bar{p}_i \equiv \frac{\partial \bar{z}}{\partial \bar{x}_i},\ i = 1, 2, \ldots, n\right) \qquad (10.2\text{-}25)$$

with solutions $\bar{z} = \bar{z}(\bar{x}_1, \bar{x}_2, \ldots, \bar{x}_n)$. It may happen that the new equation (25) does not contain the $\bar{p}_i$ and is thus an ordinary equation.

EXAMPLE: *n-dimensional Legendre transformation* (see also Secs. 9.2-3*b* and 11.5-6).

$$\left.\begin{aligned} &\bar{x}_i = p_i \qquad \bar{p}_i = x_i \qquad (i = 1, 2, \ldots, n) \\ &\bar{z} = \sum_{k=1}^{n} p_k x_k - z + C \end{aligned}\right\} \qquad (10.2\text{-}26)$$

10.2-6. Canonical Equations and Canonical Transformations. **(a) Canonical Equations.** For a first-order partial differential equation

$$G(x_1, x_2, \ldots, x_n; p_1, p_2, \ldots, p_n) = 0 \qquad \left(p_i \equiv \frac{\partial z}{\partial x_i},\ i = 1, 2, \ldots, n\right) \qquad (10.2\text{-}27)$$

which does not contain the dependent variable z explicitly, the characteristic equations (16) take the especially simple form

$$\frac{dz}{dt} = \sum_{k=1}^{n} p_k \frac{dx_k}{dt} \tag{10.2-28}$$

$$\frac{dx_i}{dt} = \frac{\partial G}{\partial p_i} \qquad \frac{dp_i}{dt} = -\frac{\partial G}{\partial x_i} \qquad (i = 1, 2, \ldots, n)$$

(CANONICAL EQUATIONS) (10.2-29)

NOTE: The solution of *every* given first-order partial differential equation (14) can be reduced to the solution of a partial differential equation of the simpler form (27) with $n + 1$ independent variables $x_1, x_2, \ldots, x_n, z$; for every solution

$$u = u(x_1, x_2, \ldots, x_n; z)$$

of the partial differential equation

$$F\left(x_1, x_2, \ldots, x_n; z; -\frac{\frac{\partial u}{\partial x_1}}{\frac{\partial u}{\partial z}}, -\frac{\frac{\partial u}{\partial x_2}}{\frac{\partial u}{\partial z}}, \ldots, -\frac{\frac{\partial u}{\partial x_n}}{\frac{\partial u}{\partial z}}\right) = 0 \tag{10.2-30}$$

yields a corresponding solution $z = z(x_1, x_2, \ldots, x_n)$ of the given partial differential equation (14) such that $u(x_1, x_2, \ldots, x_n; z) = 0$.

(b) Canonical Transformations (see also Sec. 11.6-8). A twice continuously differentiable transformation

$$\left.\begin{array}{l} \left.\begin{array}{l} \bar{x}_i = \bar{x}_i(x_1, x_2, \ldots, x_n; p_1, p_2, \ldots, p_n) \\ \bar{p}_i = \bar{p}_i(x_1, x_2, \ldots, x_n; p_1, p_2, \ldots, p_n) \end{array}\right\} (i = 1, 2, \ldots, n) \\ \text{with} \quad \Delta \equiv \dfrac{\partial(\bar{x}_1, \bar{x}_2, \ldots, \bar{x}_n; \bar{p}_1, \bar{p}_2, \ldots, \bar{p}_n)}{\partial(x_1, x_2, \ldots, x_n; p_1, p_2, \ldots, p_n)} \neq 0 \end{array}\right\} \tag{10.2-31}$$

is a **canonical transformation** if and only if it transforms the canonical equations (29) into a new set of canonical equations

$$\left.\begin{array}{l} \dfrac{d\bar{x}_i}{d\bar{t}} = \dfrac{\partial}{\partial \bar{p}_i} \bar{G}(\bar{x}_1, \bar{x}_2, \ldots, \bar{x}_n; \bar{p}_1, \bar{p}_2, \ldots, \bar{p}_n) \\ \dfrac{d\bar{p}_i}{d\bar{t}} = -\dfrac{\partial}{\partial \bar{x}_i} \bar{G}(\bar{x}_1, \bar{x}_2, \ldots, \bar{x}_n; \bar{p}_1, \bar{p}_2, \ldots, \bar{p}_n) \end{array}\right\} (i = 1, 2, \ldots, n) \tag{10.2-32}$$

for an *arbitrary* twice-differentiable function $G(x_1, x_2, \ldots, x_n; p_1, p_2, \ldots, p_n)$. This is true if and only if

$$\left.\begin{array}{ll} \dfrac{\partial \bar{x}_i}{\partial x_k} = \dfrac{\partial p_k}{\partial \bar{p}_i} & \dfrac{\partial \bar{x}_i}{\partial p_k} = -\dfrac{\partial x_k}{\partial \bar{p}_i} \\ \dfrac{\partial \bar{p}_i}{\partial x_k} = -\dfrac{\partial p_k}{\partial \bar{x}_i} & \dfrac{\partial \bar{p}_i}{\partial p_k} = \dfrac{\partial x_k}{\partial \bar{x}_i} \end{array}\right\} (i, k = 1, 2, \ldots, n) \tag{10.2-33}$$

i.e., if and only if $\sum_{k=1}^{n} (p_k\, dx_k - \bar{p}_k\, d\bar{x}_k)$ is the complete differential $d\Omega$ of a "generating function" $\Omega = \Omega(x_1, x_2, \ldots, x_n; p_1, p_2, \ldots, p_n)$. Note that Eq. (33) implies $\Delta \equiv 1$.

For every canonical transformation (31) the function $\bar{G}$ appearing in the transformed canonical equations (32) is simply given by

$$\bar{G}(\bar{x}_1, \bar{x}_2, \ldots, \bar{x}_n; \bar{p}_1, \bar{p}_2, \ldots, \bar{p}_n) \equiv G(x_1, x_2, \ldots, x_n; p_1, p_2, \ldots, p_n)$$

Given a first-order partial differential equation (27) with the canonical equations (29), the new canonical equations (32) are those associated with the partial differential equation

$$\bar{G}(\bar{x}_1, \bar{x}_2, \ldots, \bar{x}_n; \bar{p}_1, \bar{p}_2, \ldots, \bar{p}_n) = 0 \qquad \left(\bar{p}_i \equiv \frac{\partial \bar{z}}{\partial \bar{x}_i},\ i = 1, 2, \ldots, n\right) \tag{10.2-34}$$

The solution $\bar{z} = \bar{z}(\bar{x}_1, \bar{x}_2, \ldots, \bar{x}_n)$ of the transformed partial differential equation (34) is related to the solution $z = z(x_1, x_2, \ldots, x_n)$ of the original equation (27) by

$$\bar{z} = z - \Omega(x_1, x_2, \ldots, x_n; p_1, p_2, \ldots, p_n) \tag{10.2-35}$$

Equations (31) and (35) together constitute a contact transformation (Sec. 10.2-5).

A canonical transformation can be specified in terms of its generating function $\Omega(x_1, x_2, \ldots, x_n; p_1, p_2, \ldots, p_n)$; the latter is often given indirectly as a function of the x_i and $\bar{x}_i$, or of the p_i and $\bar{p}_i$. In particular, every twice-differentiable function $\Omega = \Psi(x_1, x_2, \ldots, x_n; \bar{x}_1, \bar{x}_2, \ldots, \bar{x}_n)$ defines a canonical transformation (31) such that

$$\sum_{k=1}^{n} (p_k\, dx_k - \bar{p}_k\, d\bar{x}_k) \equiv d\Psi \quad \text{or} \quad p_i = \frac{\partial \Psi}{\partial x_i} \quad \bar{p}_i = -\frac{\partial \Psi}{\partial \bar{x}_i} \qquad (i = 1, 2, \ldots, n) \tag{10.2-36}$$

The canonical transformations (31) constitute a group (see also Sec. 12.2-8).

(c) Poisson Brackets. Given any pair of twice continuously differentiable functions $g(x_1, x_2, \ldots, x_n; p_1, p_2, \ldots, p_n)$, $h(x_1, x_2, \ldots, x_n; p_1, p_2, \ldots, p_n)$ one defines the **Poisson bracket**

$$[g, h] \equiv \sum_{k=1}^{n} \left(\frac{\partial g}{\partial x_k}\frac{\partial h}{\partial p_k} - \frac{\partial h}{\partial x_k}\frac{\partial g}{\partial p_k}\right) \tag{10.2-37}$$

so that

$$[h, g] = -[g, h] \qquad [g, g] = 0 \qquad [g, \text{constant}] = 0 \tag{10.2-38}$$

$$[g_1 + g_2, h] = [g_1, h] + [g_2, h] \qquad [g_1g_2, h] = g_2[g_1, h] + g_1[g_2, h] \tag{10.2-39}$$

$$[f, [g, h]] + [g, [h, f]] + [h, [f, g]] = 0 \quad \text{(POISSON'S IDENTITY)} \tag{10.2-40}$$

Note that $[f, g] = 0$, $[f, h] = 0$ *implies* $[g, h] = 0$.

Given a transformation (31), let

$$\bar{g}(\bar{x}_1, \bar{x}_2, \ldots, \bar{x}_n; \bar{p}_1, \bar{p}_2, \ldots, \bar{p}_n) \equiv g(x_1, x_2, \ldots, x_n; p_1, p_2, \ldots, p_n)$$

$$\bar{h}(\bar{x}_1, \bar{x}_2, \ldots, \bar{x}_n; \bar{p}_1, \bar{p}_2, \ldots, \bar{p}_n) \equiv h(x_1, x_2, \ldots, x_n; p_1, p_2, \ldots, p_n), \text{ and}$$

$$\overline{[\bar{g}, \bar{h}]} \equiv \sum_{k=1}^{n} \left(\frac{\partial \bar{g}}{\partial \bar{x}_k} \frac{\partial \bar{h}}{\partial \bar{p}_k} - \frac{\partial \bar{h}}{\partial \bar{x}_k} \frac{\partial \bar{g}}{\partial \bar{p}_k} \right) \tag{10.2-41}$$

A given transformation (31) *is a canonical transformation if and only if it preserves Poisson brackets, i.e., if and only if* $\overline{[\bar{g}, \bar{h}]} \equiv [g, h]$ *for all twice continuously differentiable functions* g, h.

Now let the variables x_i and p_i be functions of a parameter t such that a set of canonical equations (29) holds. Then

$$\left.\begin{aligned} &[x_i, x_k] = 0 \qquad [p_i, p_k] = 0 \\ &[x_i, p_k] = \begin{cases} 0 \text{ if } i \neq k \\ 1 \text{ if } i = k \end{cases} \end{aligned}\right\} \quad (i, k = 1, 2, \ldots, n) \tag{10.2-42}$$

and for every suitably differentiable function $f(t; x_1, x_2, \ldots, x_n; p_1, p_2, \ldots, p_n)$

$$\frac{df}{dt} = [f, G] + \frac{\partial f}{\partial t} \tag{10.2-43}$$

In particular, $\partial f/\partial t = 0$, $[f, G] = 0$ imply $f =$ constant. Two functions g, h of the $x_i(t)$ and $p_i(t)$ are **canonically conjugate** if and only if $[g, h] = 1$; this is true whenever g and h satisfy a pair of canonical equations (e.g., x_i, p_i, and t, G). *A given transformation* (31) *is a canonical transformation if and only if it preserves the relations* (42), *i.e., if and only if*

$$\left.\begin{aligned} &[\bar{x}_i, \bar{x}_k] = 0 \qquad [\bar{p}_i, \bar{p}_k] = 0 \\ &[\bar{x}_i, \bar{p}_k] = \begin{cases} 0 \text{ if } i \neq k \\ 1 \text{ if } i = k \end{cases} \end{aligned}\right\} \quad (i, k = 1, 2, \ldots, n) \tag{10.2-44}$$

10.2-7. The Hamilton-Jacobi Equation. Solution of the Canonical Equations. (a) An important application of the theory of first-order partial differential equations is the solution of systems of *ordinary* differential equations which can be written as canonical equations associated with a partial differential equation of the special form

$$p + H(x, x_1, x_2, \ldots, x_n; p_1, p_2, \ldots, p_n) = 0$$
$$\left(p \equiv \frac{\partial z}{\partial x}; p_i \equiv \frac{\partial z}{\partial x_i}, i = 1, 2, \ldots, n\right) \tag{10.2-45}$$

(HAMILTON-JACOBI EQUATION)

Note that $n + 1$ independent variables x and x_i are involved. Since Eq. (29) yields $dx/dt = 1$, one can write $x \equiv t$ (assuming $x = 0$ for $t = 0$);

the $2n$ canonical equations (29) for the x_i and p_i become*

$$\left.\begin{aligned} \frac{dx_i}{dt} &= \frac{\partial}{\partial p_i} H(t, x_1, x_2, \ldots, x_n; p_1, p_2, \ldots, p_n) \\ \frac{dp_i}{dt} &= -\frac{\partial}{\partial x_i} H(t, x_1, x_2, \ldots, x_n; p_1, p_2, \ldots, p_n) \end{aligned}\right\} \qquad (i = 1, 2, \ldots, n) \qquad (10.2\text{-}46)$$

Systems of ordinary differential equations having the precise form (46) are of importance in the calculus of variations (Sec. 11.6-8) and in analytical dynamics and optics.

If it is possible to find an n-parameter solution

$$z = \Phi(t, x_1, x_2, \ldots, x_n; \alpha_1, \alpha_2, \ldots, \alpha_n) + \alpha_{n+1} \qquad (10.2\text{-}47)$$

of the Hamilton-Jacobi equation (45) *with* det $[\partial^2\Phi/\partial x_i \partial\alpha_k] \neq 0$, *then the solution* $x_i = x_i(t)$, $p_i = p_i(t)$ $(i = 1, 2, \ldots, n)$ *of the system of* $2n$ *ordinary differential equations* (46) *is given by*

$$\frac{\partial}{\partial \alpha_i} \Phi(t, x_1, x_2, \ldots, x_n; \alpha_1, \alpha_2, \ldots, \alpha_n) = \beta_i \qquad (i = 1, 2, \ldots, n) \qquad (10.2\text{-}48)$$

where the α_k *and* β_i *are* $2n$ *constants of integration.* One first solves the n equations (48) for the $x_i = x_i(t)$; the $p_i = p_i(t)$ are obtained by substitution of $x_i = x_i(t)$ into $p_i = \partial\Phi/\partial x_i$ $(i = 1, 2, \ldots, n)$ (see also Sec. 10.2-4*b*).

(b) Use of Canonical Transformations (see also Sec. 10.2-6*b*). If a complete integral (47) solving the given equations (46) is not known, one may try to introduce a canonical transformation relating the $2n + 2$ variables $x \equiv t, p, x_i, p_i$ to $2n + 2$ new variables $\bar{x} \equiv \bar{t}, \bar{p}, \bar{x}_i, \bar{p}_i$ so that $p + H \equiv \bar{p} + \bar{H}$, and

$$\frac{d\bar{x}_i}{d\bar{t}} = \frac{\partial \bar{H}}{\partial \bar{p}_i} \qquad \frac{d\bar{p}_i}{d\bar{t}} = -\frac{\partial \bar{H}}{\partial \bar{x}_i} \qquad (i = 1, 2, \ldots, n) \qquad (10.2\text{-}49)$$

In this case,

$$\sum_{k=1}^{n} p_k\,dx_k - H\,dt - \left(\sum_{k=1}^{n} \bar{p}_k\,d\bar{x}_k - \bar{H}\,d\bar{t}\right) = d\Omega \qquad (10.2\text{-}50)$$

must be a complete differential of a "generating function" $\Omega = \Omega(t, x_1, x_2, \ldots, x_n; p, p_1, p_2, \ldots, p_n)$. In particular, let $\bar{t} \equiv t$; then every twice continuously differentiable function $\Omega = \Psi(t, x_1, x_2, \cdots, x_n; \bar{x}_1, \bar{x}_2, \cdots, \bar{x}_n)$ with det $[\partial^2\Psi/\partial x_i\,\partial\bar{x}_k] \neq 0$ defines a canonical transformation such that

$$\left.\begin{aligned} p_i &= \frac{\partial \Psi}{\partial x_i} \qquad \bar{p}_i = -\frac{\partial \Psi}{\partial \bar{x}_i} \qquad (i = 1, 2, \ldots, n) \\ \bar{H} &= H + \frac{\partial \Psi}{\partial t} \end{aligned}\right\} \qquad (10.2\text{-}51)$$

* The remaining canonical equation is $\frac{dp}{dt} = -\frac{\partial H}{\partial t}$.

It may be possible to choose this transformation so that $\bar{H}$ does not depend explicitly on the $\bar{x}_i$ (transformation to *cyclic variables* $\bar{x}_i$).

(c) **Perturbation Theory.** Given the solution (47) of the Hamilton-Jacobi equation (45), the generating function

$$\Psi = \Phi(t, x_1, x_2, \ldots, x_n; \bar{x}_1, \bar{x}_2, \ldots, \bar{x}_n)$$

defines a canonical transformation (51) yielding *constant* transformed variables:

$$\bar{x}_i = \alpha_i \qquad \bar{p}_i = -\beta_i \qquad (i = 1, 2, \ldots, n) \tag{10.2-52}$$

As shown in Sec. 10.2-7*a*, the $2n$ equations (52) yield the solution $x_i = x_i(t)$, $p_i = p_i(t)$ of the canonical system (46).

Given such a solution of the "unperturbed" canonical system (46), one often desires to solve the canonical equations

$$\left.\begin{aligned} &\frac{dx_i}{dt} = \frac{\partial K}{\partial p_i} \qquad \frac{dp_i}{dt} = -\frac{\partial K}{\partial x_i} \qquad (i = 1, 2, \ldots, n) \\ &\text{with} \quad K = H(t, x_1, x_2, \ldots, x_n; p_1, p_2, \ldots, p_n) \\ &\qquad\qquad + \epsilon H_1(t, x_1, x_2, \ldots, x_n; p_1, p_2, \ldots, p_n) \end{aligned}\right\} \tag{10.2-53}$$

where ϵH_1 is a small correction term (*perturbation*, e.g., the effect of a small disturbing field in celestial mechanics). Using the known solution (47) of the "unperturbed" Hamilton-Jacobi equation (45), introduce new variables $\bar{x}_i$, $\bar{p}_i$ by a canonical transformation (51) with the generating function

$$\Psi = \Phi(t, x_1, x_2, \ldots, x_n; \bar{x}_1, \bar{x}_2, \ldots, \bar{x}_n)$$

Since Eq. (46) has been replaced by Eq. (53), the $\bar{x}_i$ and $\bar{p}_i$ are no longer constants but satisfy transformed canonical equations

$$\frac{d\bar{x}_i}{dt} = \epsilon \frac{\partial \bar{H}_1}{\partial \bar{p}_i} \qquad \frac{d\bar{p}_i}{dt} = -\epsilon \frac{\partial \bar{H}_1}{\partial \bar{x}_i} \qquad (i = 1, 2, \ldots, n) \tag{10.2-54}$$

which may be easier to solve than the given system (53). If one writes

$$\bar{x}_i = \alpha_i + \epsilon X_i(t) \qquad \bar{p}_i = -\beta_i + \epsilon P_i(t) \qquad (i = 1, 2, \ldots, n)$$

the corrections $\epsilon X_i(t)$, $\epsilon P_i(t)$ to the constants (52) may yield corresponding corrections to the solutions of the unperturbed system (46) by an approximately linear transformation.

10.3. HYPERBOLIC, PARABOLIC, AND ELLIPTIC PARTIAL DIFFERENTIAL EQUATIONS. CHARACTERISTICS

10.3-1. Quasilinear Partial Differential Equations of Order 2 with Two Independent Variables. Characteristics. (a) A partial differential equation of order r is **quasilinear** if and only if it is linear in the rth-order derivatives of the unknown function Φ. Thus a real quasilinear second-order equation with two independent variables x, y has the form

$$a_{11}\frac{\partial^2 \Phi}{\partial x^2} + 2a_{12}\frac{\partial^2 \Phi}{\partial x\, \partial y} + a_{22}\frac{\partial^2 \Phi}{\partial y^2} + B = 0 \tag{10.3-1}$$

where a_{11}, a_{12}, a_{22}, and B are suitably differentiable real functions of x, y, Φ, $\partial\Phi/\partial x$, and $\partial\Phi/\partial y$.

(b) Characteristics. Given a real boundary curve C_0 described by

$$x = x(\tau) \qquad y = y(\tau) \tag{10.3-2a}$$

a set of *Cauchy-type boundary conditions* (called *initial conditions* in Sec. 10.2-2) specifies boundary values*

$$\Phi = z(\tau) \qquad \frac{\partial\Phi}{\partial x} = p(\tau) \qquad \frac{\partial\Phi}{\partial y} = q(\tau) \qquad \left(\frac{dz}{d\tau} = p\frac{dx}{d\tau} + q\frac{dy}{d\tau}\right) \tag{10.3-2b}$$

A given set of suitably differentiable functions (2) uniquely defines the values of $\partial^2\Phi/\partial x^2 = u(\tau)$, $\partial^2\Phi/\partial x\,\partial y = v(\tau)$, $\partial^2\Phi/\partial y^2 = w(\tau)$ (and also the values of higher-order derivatives of Φ) on the curve (2a) at every point P_0 where the functions (2a) do not satisfy the ordinary differential equation

$$a_{11}\left(\frac{dy}{dx}\right)^2 - 2a_{12}\frac{dy}{dx} + a_{22} = 0$$

or

$$\frac{dy}{dx} = \frac{a_{12} \pm \sqrt{a_{12}^2 - a_{11}a_{22}}}{a_{11}} \tag{10.3-4}$$

This is true because the derivatives u, v, w of Φ on C_0 must satisfy Eq. (1) and the "second-order strip conditions"

$$\frac{dp}{d\tau} = u\frac{dx}{d\tau} + v\frac{dy}{d\tau} \qquad \frac{dq}{d\tau} = v\frac{dx}{d\tau} + w\frac{dy}{d\tau} \tag{10.3-5}$$

so that, for instance,

$$v = \frac{a_{11}\,dy\,dp + a_{22}\,dx\,dq + B\,dx\,dy}{a_{11}\,dy^2 - 2a_{12}\,dx\,dy + a_{22}\,dx^2} \tag{10.3-6}$$

Equation (4) holds at P_0 if C_0 is a segment of a **characteristic base curve** (often called a **characteristic**) $y = y(x)$ satisfying Eq. (4), or if C_0 touches such a curve at P_0.

Properly speaking, the **characteristics** associated with the given partial differential equation (1) are curves $x = x(\tau)$, $y = y(\tau)$, $z = z(\tau)$ on the solution surface $z = \Phi(x, y)$ such that $y = y(x)$ satisfies Eq. (4). Since the expression (6) must be

* If one is given the boundary values of the *normal* derivative $\partial\Phi/\partial n$ (Sec. 5.6-1), say

$$\frac{\partial\Phi}{\partial n} \equiv \frac{1}{\sqrt{\left(\frac{dx}{d\tau}\right)^2 + \left(\frac{dy}{d\tau}\right)^2}}\left(q\frac{dx}{d\tau} - p\frac{dy}{d\tau}\right) = P(\tau) \tag{10.3-3}$$

solve Eq. (3) together with $dz/d\tau = p(dx/d\tau) + q(dy/d\tau) = 0$ to obtain $p(\tau)$ and $q(\tau)$.

finite on the solution surface, $p = \partial\Phi/\partial x$ and $q = \partial\Phi/\partial y$ must satisfy the ordinary differential equation

$$a_{11}\,dy\,dp + a_{22}\,dx\,dq + B\,dx\,dy = 0$$

or

$$\frac{dq}{dp} = -\frac{a_{12} \pm \sqrt{a_{12}^2 - a_{11}a_{22}}}{a_{22}} - \frac{B}{a_{22}}\left(\frac{dy}{dp}\right) \tag{10.3-7}$$

on every characteristic defined by Eq. (4), with corresponding plus and minus signs.

NOTE: The second-order derivatives of Φ may be discontinuous (though finite) on a characteristic, so that different solutions can be "patched together" along characteristics.

(c) Hyperbolic, Parabolic, and Elliptic Partial Differential Equations. The given partial differential equation (1) is

Hyperbolic if $a_{11}a_{22} - a_{12}^2 < 0$ in the region of points (x, y) under consideration, so that Eq. (4) describes two distinct families of real characteristic base curves

Parabolic if $a_{11}a_{22} - a_{12}^2 = 0$, so that there exists a single family of real characteristic base curves

Elliptic if $a_{11}a_{22} - a_{12}^2 > 0$, so that no real characteristics exist

10.3-2. Solution of Hyperbolic Partial Differential Equations by the Method of Characteristics (see also Sec. 10.3-4). In the *hyperbolic* case ($a_{11}a_{22} - a_{12}^2 < 0$), simultaneous solution of the four ordinary differential equations (4) and (7) yields $p = \partial\Phi/\partial x$ and $q = \partial\Phi/\partial y$ on the solution surface as functions of x and y, so that $\Phi = \Phi(x, y)$ can be obtained by further integration. In many applications, $\partial\Phi/\partial x$ and $\partial\Phi/\partial y$ rather than $\Phi(x, y)$ are of paramount interest (velocity components); the method forms the basis for many analytical and numerical solution procedures in the theory of compressible flow.

Computations are considerably simplified in special cases. If $B \equiv 0$, one has

$$\left(\frac{dy}{dx}\right)_1\left(\frac{dq}{dp}\right)_2 = -1 \qquad \left(\frac{dy}{dx}\right)_2\left(\frac{dq}{dp}\right)_1 = -1 \tag{10.3-8}$$

where the subscripts refer to the characteristics derived, respectively, with a plus sign and a minus sign in Eqs. (4) and (6). If, in addition, a_{11}, a_{12}, and a_{22} depend only on $\partial\Phi/\partial x$, $\partial\Phi/\partial y$ one need only solve Eq. (7) to obtain the characteristics (e.g., two-dimensional steady supersonic flow). Again, if a_{11}, a_{12}, and a_{22} depend only on x, y one need only solve Eq. (4).

10.3-3. Transformation of Hyperbolic, Parabolic, and Elliptic Differential Equations to Canonical Form. For convenience, let a_{11}, a_{12}, and a_{22} be functions of x and y alone, so that the ordinary differential equation (4) separates into two linear first-order equations

$$\frac{dy}{dx} = \lambda_1(x, y) \qquad \text{with solutions } h_1(x, y) = \alpha_1 \tag{10.3-9a}$$

$$\frac{dy}{dx} = \lambda_2(x, y) \qquad \text{with solutions } h_2(x, y) = \alpha_2 \tag{10.3-9b}$$

where α_1, α_2 are arbitrary constants. Depending on the sign of the function $a_{11}a_{22} - a_{12}^2$ in the region of points (x, y) under consideration,* three cases arise.

1. **Hyperbolic Partial Differential Equation** $(a_{11}a_{22} - a_{12}^2 < 0)$. $\lambda_1(x, y)$ and $\lambda_2(x, y)$ are real and distinct. *There exist two one-parameter families of real characteristics* (9a) *and* (9b); a curve of each family passes through every point (x, y) under consideration. Introduce new coordinates

$$\bar{x} = h_1(x, y) \qquad \bar{y} = h_2(x, y) \tag{10.3-10}$$

to transform the given partial differential equation (1) to the *canonical* form

$$\frac{\partial^2 \Phi}{\partial \bar{x}\, \partial \bar{y}} = f\left(\bar{x}, \bar{y}, \Phi, \frac{\partial \Phi}{\partial \bar{x}}, \frac{\partial \Phi}{\partial \bar{y}}\right) \tag{10.3-11}$$

The alternative coordinate system

$$\xi = \frac{\bar{x} + \bar{y}}{2} \qquad \eta = \frac{\bar{x} - \bar{y}}{2} \tag{10.3-12}$$

yields the second canonical form

$$\frac{\partial^2 \Phi}{\partial \xi^2} - \frac{\partial^2 \Phi}{\partial \eta^2} = g\left(\xi, \eta, \Phi, \frac{\partial \Phi}{\partial \xi}, \frac{\partial \Phi}{\partial \eta}\right) \tag{10.3-13}$$

2. **Parabolic Partial Differential Equation** $(a_{11}a_{22} - a_{12}^2 = 0)$. $\lambda_1(x, y)$ and $\lambda_2(x, y)$ are real and identical. *There exists a single one-parameter family of real characteristics* (9); one characteristic passes through each point (x, y) under consideration. Introduce

$$\bar{x} = h_1(x, y) \qquad \bar{y} = h_0(x, y) \tag{10.3-14}$$

where $h_0(x, y)$ is an arbitrary suitably differentiable function such that $\partial(\bar{x}, \bar{y})/\partial(x, y) \neq 0$. Equation (1) is transformed to the canonical form

$$\frac{\partial^2 \Phi}{\partial \bar{x}^2} = f\left(\bar{x}, \bar{y}, \Phi, \frac{\partial \Phi}{\partial \bar{x}}, \frac{\partial \Phi}{\partial \bar{y}}\right) \tag{10.3-15}$$

3. **Elliptic Partial Differential Equation** $(a_{11}a_{22} - a_{12}^2 > 0)$. $\lambda_1(x, y)$ and $\lambda_2(x, y)$ and hence also $h_1(x, y)$ and $h_2(x, y)$ are complex conjugates; *no real characteristics exist.* Introduce

$$\bar{x} = \frac{h_1(x, y) + h_2(x, y)}{2} \qquad \bar{y} = \frac{h_1(x, y) - h_2(x, y)}{2i} \tag{10.3-16}$$

* In the usual applications, the discriminant $a_{11}a_{22} - a_{12}^2$ does not change sign in the region under consideration. Note also that *the sign of $a_{11}a_{22} - a_{12}^2$ is invariant with respect to any real continuously differentiable coordinate transformation $\bar{x} = \bar{x}(x, y)$, $\bar{y} = \bar{y}(x, y)$ with nonvanishing Jacobian.*

to transform Eq. (1) to the canonical form

$$\frac{\partial^2\Phi}{\partial\bar{x}^2} + \frac{\partial^2\Phi}{\partial\bar{y}^2} = f\left(\bar{x}, \bar{y}, \Phi, \frac{\partial\Phi}{\partial\bar{x}}, \frac{\partial\Phi}{\partial\bar{y}}\right) \tag{10.3-17}$$

These three types of partial differential equations differ significantly with respect to the types of boundary conditions yielding valid and unique integrals (Secs. 10.3-4 and 15.6-2).

10.3-4. Typical Boundary-value Problems for Second-order Equations. (a) Hyperbolic Differential Equations. The *Cauchy initial-value problem* of Sec. 10.3-1*b* requires one to solve the hyperbolic differential equation (1), given Φ, $\partial\Phi/\partial x$, and $\partial\Phi/\partial y$ on an arc C_0 of a regular

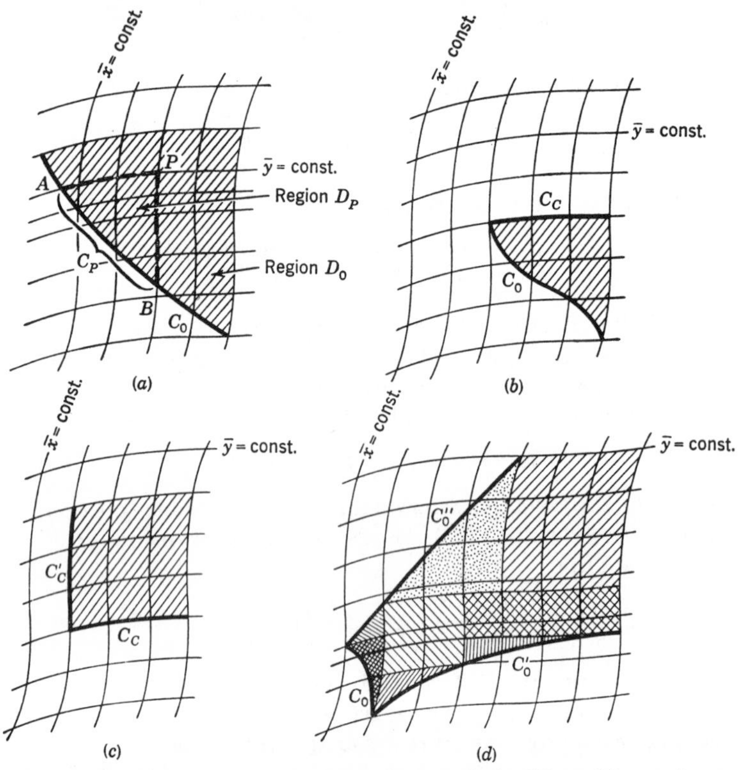

FIG. 10.3-1. Boundary-value problems for hyperbolic differential equations.

curve which is neither a characteristic (6) nor touches any characteristic. Such a curve intersects each characteristic at most once; *the given initial values determine Φ in a triangular region D_0 bounded by C_0 and a characteristic of each family* (Fig. 10.3-1*a*). More specifically, the value of Φ at each point P of D_0 is determined by the values of Φ and its derivatives on the portion C_P of C_0 which is bounded by the characteristics through P.

A *second type* of boundary-value problem prescribes only a linear relation $\alpha(\partial\Phi/\partial n) + \beta\Phi = b(x, y)$ on an arc C_0 specified as above; in addition, Φ is given on a characteristic arc C_C through one end point of C_0 (Fig. 10.3-1*b*).

A *third type* of boundary-value problem prescribes Φ on two intersecting characteristic arcs C_C and C'_C (Fig. 10.3-1*c*).

Combinations of these three types of problems will indicate admissible boundary conditions for more complicated boundaries. Thus in Fig. 10.3-1*d*, Φ, $\partial\Phi/\partial x$, and $\partial\Phi/\partial y$ may be given on C_0, but only one relation of the type $\alpha(\partial\Phi/\partial n) + \beta\Phi = b(x, y)$ can be prescribed on each of C'_0 and C''_0. The solutions in the various regions indicated in Fig. 10.3-1*d* are "patched" together along characteristics ("patching curves") so that Φ is continuous, while $\partial\Phi/\partial x$ and $\partial\Phi/\partial y$ may be discontinuous. Note that *closed boundaries are not admissible.*

EXAMPLE: Initial-value problems for the hyperbolic *one-dimensional wave equation,* Sec. 10.3-5.

(b) Parabolic Differential Equations. There exists one and only one family of characteristics. Although the Cauchy problem can again be solved for a suitable arc C_0, one is usually given Φ on a characteristic $\bar{x}(= t) = 0$ and $\alpha(\partial\Phi/\partial n) + \beta\Phi$ on two curves which do not intersect or touch each other or any characteristic. *Closed boundaries in the xy plane are not admissible.*

EXAMPLE: An admissible boundary-value problem for the parabolic *diffusion equation* $\partial^2\Phi/\partial x^2 + (1/\gamma^2)\,\partial\Phi/\partial t = 0$ specifies $\Phi(x, t) = \Phi_0(x)$ on the characteristic $t = 0$ (*initial conditions*) and $\alpha(x, t)\,\partial\Phi/\partial t + \beta(x, t)\Phi$ on the curves $x = a$ and $x = b$ (*boundary conditions*).

(c) Elliptic Differential Equations (see also Secs. 10.4-1 and 15.6-2). No real characteristics exist; Cauchy-type boundary conditions are not admissible. Typical problems specify $\alpha(x, y)(\partial\Phi/\partial n) + \beta(x, y)\Phi$ on a curve C *enclosing* a bounded or unbounded solution region ("true" boundary-value problems).

10.3-5. The One-dimensional Wave Equation (see also Secs. 10.3-4*a*, 10.4-8*a*, and 10.4-9*b*). The hyperbolic differential equation

$$\frac{\partial^2\Phi(x, t)}{\partial x^2} - \frac{1}{c^2}\frac{\partial^2\Phi(x, t)}{\partial t^2} = 0 \qquad \text{(10.3-18)}$$

(ONE-DIMENSIONAL WAVE EQUATION)

has the general solution

$$\Phi(x, t) = \Phi_1(x - ct) + \Phi_2(x + ct) \tag{10.3-19}$$

which represents a pair of arbitrarily-shaped **waves** respectively propagated in the $+x$ and $-x$ directions with **phase velocity** c. The characteristics $x \pm ct =$ constant are loci of constant phase (Fig. 10.3-2). Sections 10.3-5*a*, *b*, and *c* list solutions for three types of initial-value problems (e.g., waves in a string); in practice, the Fourier-expansion

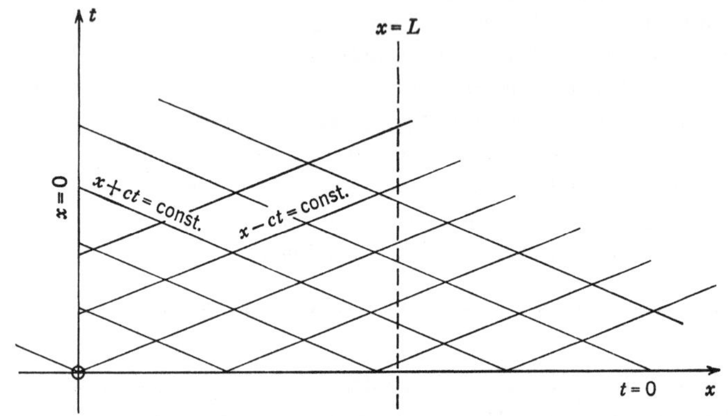

FIG. 10.3-2. Characteristics for the one-dimensional wave equation $\frac{\partial^2 \Phi}{\partial x^2} - \frac{1}{c^2}\frac{\partial^2 \Phi}{\partial t^2} = 0$.

method of Sec. 10.4-9*b* may be preferable, since it applies also to nonhomogeneous differential equations (forced vibrations).

(**a**) The initial conditions

$$\Phi(x, 0) = \Phi_0(x) \qquad \left.\frac{\partial \Phi}{\partial t}\right]_{t=0} = v_0(x) \qquad (-\infty < x < \infty) \tag{10.3-20}$$

specify a true *Cauchy initial-value problem* (Secs. 10.3-1 and 10.3-4*a*; see also Figs. 10.3-1*a* and 10.3-2). The solution is

$$\Phi(x, t) = \frac{1}{2}[\Phi_0(x - ct) + \Phi_0(x + ct)] + \frac{1}{2c}\int_{x-ct}^{x+ct} v_0(\xi)\, d\xi$$

(D'ALEMBERT'S SOLUTION) (10.3-21)

Disturbances initially restricted to any given interval $a < x < b$ affect the solution in the interval $a - ct < x < a + ct$. Discontinuities in $\Phi_0(x)$ are propagated in both directions.

(**b**) The initial conditions

$$\Phi(x, 0) = \Phi_0(x) \qquad \left.\frac{\partial \Phi}{\partial t}\right]_{t=0} = v_0(x) \qquad (x \geq 0) \tag{10.3-22}$$

and the boundary condition

$$\Phi(0, t) = 0 \qquad (t \geq 0) \tag{10.3-23}$$

specify a combination-type boundary-value problem (see also Sec. 10.3-4*a* and Figs. 10.3-1*d* and 10.3-2). The solution is

$$\Phi(x, t) = \frac{1}{2}[P(x - ct) + P(x + ct)] + \frac{1}{2c}\int_{x-ct}^{x+ct} Q(\xi)\,d\xi \tag{10.3-24}$$

$$P(x) = \begin{cases} \Phi_0(x) & (x \geq 0) \\ -\Phi_0(-x) & (x < 0) \end{cases} \qquad Q(x) = \begin{cases} v_0(x) & (x \geq 0) \\ -v_0(-x) & (x < 0) \end{cases} \tag{10.3-25}$$

which corresponds to *superposition of incoming and reflected waves.*

(c) The initial conditions

$$\Phi(x, 0) = \Phi_0(x) \qquad \left.\frac{\partial \Phi}{\partial t}\right]_{t=0} = v_0(x) \qquad (0 \leq x \leq L) \tag{10.3-26}$$

and the boundary conditions

$$\Phi(0, t) = \Phi(L, t) = 0 \qquad (t \geq 0) \tag{10.3-27}$$

define another combination-type problem (see also Figs. 10.3-1*d* and 10.3-2). The solution is given by Eq. (24) if one interprets $P(x)$ and $Q(x)$ as odd periodic functions with period $2L$ and respectively equal to $\Phi_0(x)$ and $v_0(x)$ for $0 \leq x \leq L$.

10.3-6. The Riemann-Volterra Method for Linear Hyperbolic Equations (see also Secs. 10.3-4*a* and 15.5-1). It is desired to solve the Cauchy problem (Sec. 10.3-1) for the real hyperbolic differential equation

$$\mathbf{L}\Phi(x, y) \equiv \frac{\partial^2 \Phi}{\partial x\,\partial y} + a(x, y)\frac{\partial \Phi}{\partial x} + b(x, y)\frac{\partial \Phi}{\partial y} + c(x, y)\Phi = f(x, y) \tag{10.3-28}$$

with Φ and $\partial\Phi/\partial x$, $\partial\Phi/\partial y$ given on a boundary curve C_0 satisfying the conditions of Sec. 10.3-1. Referring to Fig. 10.3-1*a*, the solution at each point P with the coordinates $x = \xi$, $y = \eta$ is given in terms of the "initial values" of $\Phi(x, y)$, $\partial\Phi/\partial x$, and $\partial\Phi/\partial y$ on the boundary-curve segment C_P cut off by the characteristics through P:*

$$\begin{aligned} \Phi(\xi, \eta) &= G_R\Phi\Big]_A - \int_{C_P} G_R\Phi(a\,dy - b\,dx) + \int_{C_P}\left(\Phi\frac{\partial G_R}{\partial y}dy + G_R\frac{\partial \Phi}{\partial x}dx\right) \\ &\qquad + \iint_{D_P} G_R f\,dx\,dy \\ &= G_R\Phi\Big]_B - \int_{C_P} G_R\Phi(a\,dy - b\,dx) \\ &\qquad - \int_{C_P}\left(\Phi\frac{\partial G_R}{\partial x}dx + G_R\frac{\partial \Phi}{\partial y}dy\right) + \iint_{D_P} G_R f\,dx\,dy \end{aligned} \tag{10.3-29}$$

* See also the footnote to Sec. 10.3-1*b*.

where the so-called **Riemann-Green function** $G_R(x, y; \xi, \eta)$ is continuously differentiable on and inside the region D_P bounded by C_P and the characteristics through P and satisfies the conditions of the simpler boundary-value problem

$$\left.\begin{aligned} \mathsf{L}^\dagger G_R &\equiv \frac{\partial^2 G_R}{\partial x\, \partial y} - \frac{\partial}{\partial x}(aG_R) - \frac{\partial}{\partial y}(bG_R) + cG_R = 0 \\ &\qquad\qquad\qquad\qquad (x, y \text{ inside } D_P) \\ G_R &= \exp \int_\eta^y a(\xi, y)\, dy \qquad (x = \xi) \\ G_R &= \exp \int_\xi^x b(x, \eta)\, dx \qquad (y = \eta) \end{aligned}\right\} \qquad (10.3\text{-}30)$$

EXAMPLES: For $a \equiv b \equiv c \equiv 0$, $G_R \equiv 1$; and for $a \equiv b \equiv 0$, $c =$ constant, one has $G_R = J_0[\sqrt{4c(x - \xi)(y - \eta)}]$, where $J_0(z)$ is the zero-order Bessel function of the first kind (Sec. 21.8-1). For many practical applications involving linear hyperbolic differential equations with constant coefficients, the integral-transform methods (Sec. 10.5-1) are preferable.

10.3-7. Equations with Three or More Independent Variables. A real partial differential equation of the form

$$\sum_{i=1}^n \sum_{k=1}^n a_{ik}(x_1, x_2, \ldots, x_n) \frac{\partial^2 \Phi}{\partial x_i\, \partial x_k} + B\left(x_1, \ldots, x_n; \Phi; \frac{\partial \Phi}{\partial x_1}, \ldots, \frac{\partial \Phi}{\partial x_n}\right) = 0 \qquad (10.3\text{-}31)$$

where $\Phi = \Phi(x_1, x_2, \ldots, x_n)$, is an **elliptic** partial differential equation if and only if the matrix $[a_{ik}]$ is positive definite (Sec. 13.5-2) throughout the region of interest.

In many problems involving nonelliptic partial differential equations, the unknown function Φ depends on n "space coordinates" $x_1, x_2, \ldots, x_n$ and a "time coordinate" t; the partial differential equation takes either of the forms

$$\sum_{i=1}^n \sum_{k=1}^n c_{ik} \frac{\partial^2 \Phi}{\partial x_i\, \partial x_k} - \frac{\partial^2 \Phi}{\partial t^2} + B = 0 \qquad \sum_{i=1}^n \sum_{k=1}^n c_{ik} \frac{\partial^2 \Phi}{\partial x_i\, \partial x_k} + B = 0 \qquad (10.3\text{-}32)$$

where the matrix $[c_{ik}] \equiv [c_{ik}(x_1, x_2, \ldots, x_n; t)]$ is real and positive definite and B is a function of the x_i, t, Φ, and the first-order derivatives. The two types of partial differential equations are respectively referred to as **hyperbolic** and **parabolic** differential equations.

Characteristics of the more general partial differential equations (31) and (32) are surfaces or hypersurfaces on which Cauchy-type boundary conditions cannot determine higher-order derivatives of the solution (Ref. 10.5). Elliptic partial differential equations have no real characteristics. The concept of characteristics has also been extended to certain partial differential equations of order higher than two, and to some systems of partial differential equations (Ref. 10.5).

10.4. LINEAR PARTIAL DIFFERENTIAL EQUATIONS OF PHYSICS. PARTICULAR SOLUTIONS

10.4-1. Physical Background and Survey. (a) Many problems of classical physics require one to find a solution $\Phi(x, t)$ or $\Phi(\mathbf{r}, t)$ of a linear partial differential equation on a given space interval or region V (Table 10.4-1). The unknown function Φ and/or its derivatives must, in addition, satisfy given *initial conditions* for $t = 0$ and linear *boundary conditions* on the boundary S of V. Related problems arise in quantum mechanics.

Each partial differential equation listed in Table 10.4-1 is *homogeneous* (Sec. 10.1-2*b*) if $f \equiv 0$. A given boundary condition is, again, homogeneous if and only if it holds for every multiple $\alpha\Phi$ of any function which satisfies the condition. Inhomogeneities represent the action of external influences (forces, heat sources, electric charges or currents) on the physical system under consideration. Typically, an elliptic differential equation describes an *equilibrium situation* (steady-state heat flow, elastic deformation, electrostatic field). Parabolic and hyperbolic differential equations describe *transients* (free vibrations, return to equilibrium after a given initial disturbance) or, if there are time-dependent inhomogeneities ("forcing functions" in the differential equation or boundary conditions), such equations describe the *propagation of disturbances* (forced vibrations, radiation).

One can relate each problem of the type discussed here to an approximating system of ordinary differential equations by replacing each space derivative by a difference coefficient in the manner of Sec. 20.8-3 (*method of difference-differential equations*). This method is not only useful for numerical calculations; the analogy to a discrete-variable problem of the type described in Secs. 9.4-1 to 9.4-8 may give interesting physical insight.

(b) Construction of Solutions by Superposition. The most important methods for the solution of linear differential equations are based on the fundamental *superposition theorems* stated explicitly in Secs. 9.3-1 and 15.4-2. *The most important solution methods superimpose a judiciously chosen set of trial functions to construct solutions which match given forcing functions, given boundary conditions, and/or given initial conditions. Eigenfunction expansions* (Secs. 10.4-2 and 15.4-12) and *integral-transform methods* (Secs. 9.3-7, 9.4-5, and 10.5-1 to 10.5-3) are systematic schemes for constructing such solutions. *Green's-function methods* (Secs. 9.3-3, 9.4-3, 15.5-1 to 15.5-4, 15.6-6, 15.6-9, and 15.6-10)

Table 10.4-1. The Most Important Linear Partial Differential Equations of Classical Physics

Type	Physical background	One-dimensional	Multidimensional	Accessory conditions
Parabolic	**Heat conduction, diffusion**	$\frac{\partial^2\Phi}{\partial x^2} - \frac{1}{\gamma^2}\frac{\partial\Phi}{\partial t} = f(x, t)$	$\nabla^2\Phi - \frac{1}{\gamma^2}\frac{\partial\Phi}{\partial t} = f(\mathbf{r}, t)$	Boundary conditions; initial conditions on Φ
Hyperbolic	**Waves** (strings, membranes, fluids, electromagnetic)	$\frac{\partial^2\Phi}{\partial x^2} - \frac{1}{c^2}\frac{\partial^2\Phi}{\partial t^2} = f(x, t)$	$\nabla^2\Phi - \frac{1}{c^2}\frac{\partial^2\Phi}{\partial t^2} = f(\mathbf{r}, t)$	Boundary conditions; initial conditions on Φ and $\frac{\partial\Phi}{\partial t}$
	Damped waves, transmission lines	$\frac{\partial^2\Phi}{\partial x^2} - a_0\frac{\partial^2\Phi}{\partial t^2} - a_1\frac{\partial\Phi}{\partial t} - a_2\Phi = f(x, t)$	$\nabla^2\Phi - a_0\frac{\partial^2\Phi}{\partial t} - a_1\frac{\partial\Phi}{\partial t} - a_2\Phi = f(\mathbf{r}, t)$	
Elliptic	**Static case**	$\frac{\partial^2\Phi}{\partial x^2} = f(x)$	$\nabla^2\Phi = f(\mathbf{r})$	Boundary conditions only
4th order	**Elastic vibrations**	$\frac{\partial^4\Phi}{\partial x^4} + \frac{1}{c^2}\frac{\partial^2\Phi}{\partial t^2} = f(x, t)$	$\nabla^2\nabla^2\Phi + \frac{1}{c^2}\frac{\partial^2\Phi}{\partial t^2} = f(\mathbf{r}, t)$	Boundary conditions; initial conditions on Φ and $\frac{\partial\Phi}{\partial t}$
	Static case	$\frac{\partial^4\Phi}{\partial x^4} = f(x)$	$\nabla^2\nabla^2\Phi = f(\mathbf{r})$	Boundary conditions only

are superposition schemes which reduce the solution of suitable problems to that of problems with simpler forcing functions or boundary conditions.

The general theory of linear boundary-value problems is treated in Chap. 15; Secs. 10.4-3 to 10.4-8 present useful particular solutions from the heuristic point of view of an elementary course.

(c) Choice of Coordinate System. The system of coordinates x^1, x^2, or x^1, x^2, x^3 used to specify the point $(\mathbf{r})$ is usually chosen so that (1) *separation of variables is possible* (Sec. 10.1-3) and/or (2) *the given boundary S becomes a coordinate line or surface, or a pair of coordinate lines or surfaces.*

10.4-2. Linear Boundary-value Problems (see also Secs. 15.4-1 and 15.4-2). **(a)** Let V be a given three- or two-dimensional region of points $(\mathbf{r})$, and let S be the boundary surface or boundary curve of V. One desires to solve the partial differential equation

$$\mathsf{L}\Phi(\mathbf{r}) = f(\mathbf{r}) \qquad (\mathbf{r} \text{ in } V) \tag{10.4-1a}$$

subject to a set of boundary conditions

$$\mathsf{B}_i\Phi(\mathbf{r}) = b_i(\mathbf{r}) \qquad (i = 1, 2, \ldots, N; \mathbf{r} \text{ in } S) \tag{10.4-1b}$$

where $\mathsf{L}\Phi$ and $\mathsf{B}_i\Phi$ are linear homogeneous functions of the unknown function $\Phi(\mathbf{r})$ and its derivatives. *Every solution of this linear boundary-value problem can be written as the sum $\Phi = \Phi_A + \Phi_B$ of solutions of the simpler boundary-value problems*

$$\mathsf{L}\Phi_A(\mathbf{r}) = 0 \qquad (\mathbf{r} \text{ in } V) \tag{10.4-2a}$$
$$\mathsf{B}_i\Phi_A(\mathbf{r}) = b_i(\mathbf{r}) \qquad (i = 1, 2, \ldots, N; \mathbf{r} \text{ in } S) \tag{10.4-2b}$$

and

$$\begin{aligned} \mathsf{L}\Phi_B(\mathbf{r}) &= f(\mathbf{r}) && (\mathbf{r} \text{ in } V) \\ \mathsf{B}_i\Phi_B(\mathbf{r}) &= 0 && (i = 1, 2, \ldots, N; \mathbf{r} \text{ in } S) \end{aligned} \tag{10.4-3}$$

Note that Eq. (2) involves a homogeneous differential equation, whereas Eq. (3) has homogeneous boundary conditions.

(b) Homogeneous Differential Equation with Nonhomogeneous Boundary Conditions. The particular solutions listed in Secs. 10.4-3 to 10.4-6 often permit one to expand $\Phi_A(\mathbf{r})$ as an infinite series or definite integral

$$\Phi_A(\mathbf{r}) = \sum_\mu \alpha_\mu \Phi_\mu(r) \qquad \text{or} \qquad \Phi_A(\mathbf{r}) = \int_{S_\mu} \alpha(\mu)\Phi_\mu(\mathbf{r})\, d\mu \tag{10.4-4}$$

over suitably chosen solutions $\Phi_\mu(\mathbf{r})$ of Eq. (2*a*); the coefficients α_μ or $\alpha(\mu)$ are chosen so as to satisfy the boundary conditions (2*b*). Frequently the approximation functions $\Phi_\mu(\mathbf{r})$ are *complete orthonormal sets* (Sec. 15.2-4; e.g., Fourier series, Fourier integrals); *one may then expand the* *given functions* $b_i(\mathbf{r})$ *in the form* (4) *and obtain the unknown coefficients* α_μ *or* $\alpha(\mu)$ *by comparison of coefficients* (Sec. 10.4-9).

(c) Nonhomogeneous Differential Equation with Homogeneous Boundary Conditions: Eigenfunction Expansions (see also Secs. 15.4-5 to 15.4-12). For an important class of partial differential equations (1), the solution $\Phi_B(r)$ of Eq. (3) can be constructed by a similar superposition of solutions $\psi(\mathbf{r})$ of the related homogeneous differential equation

$$\mathsf{L}\psi(\mathbf{r}) = \lambda\psi(\mathbf{r}) \qquad (\mathbf{r} \text{ in } V) \tag{10.4-5a}$$

for the different possible values of λ permitting $\Psi(\mathbf{r})$ to satisfy the homogeneous boundary conditions

$$\mathsf{B}_i\psi(\mathbf{r}) = 0 \qquad (i = 1, 2, \ldots, N; \mathbf{r} \text{ in } S) \tag{10.4-5b}$$

In general, such solutions exist only for specific values of the parameter λ (**eigenvalues**); the solutions $\psi = \psi_\lambda(\mathbf{r})$ corresponding to each eigenvalue are called **eigenfunctions** of the boundary-value problem (5).

Sections 10.4-3 to 10.4-8 list particular solutions $\psi(\mathbf{r})$ for a number of partial differential equations of the form (5*a*). These functions may be superimposed to form solutions of the corresponding nonhomogeneous problem (3). Eigenfunctions corresponding to discrete sets of eigenvalues often form convenient orthonormal sets (Sec. 15.2-4) for expansion of forcing functions and solutions in the form

$$f(\mathbf{r}) = \sum_\lambda f_\lambda\psi_\lambda(\mathbf{r}) \qquad \Phi_B = \sum_\lambda \beta_\lambda\psi_\lambda(\mathbf{r}) \tag{10.4-6a}$$

(refer also to Table 8.7-1), whereas continuous sets (continuous **spectra**) S_λ of eigenvalues λ yield *integral-transform expansions.*

$$f(\mathbf{r}) = \int_{S_\lambda} F(\lambda)\psi_\lambda(\mathbf{r})\,d\lambda \qquad \Phi_B = \int_{S_\lambda} \beta(\lambda)\psi_\lambda(\mathbf{r})\,d\lambda \tag{10.4-6b}$$

Substitution of Eq. (6*a*) or (6*b*) into Eq. (3) yields the unknown coefficients β_λ or $\beta(\lambda)$. Refer to Secs. 10.5-1 and 15.4-12 for the general theory of this solution method and its range of applicability. An important alternative method of solving Eq. (3), the so-called *Green's-function method,* is treated in Secs. 15.5-1 to 15.5-4.

(d) Problems Involving the Time Variable. In problems involving the time as well as the space coordinates, one desires to solve a linear differential equation

$$\mathsf{L}\Phi(\mathbf{r}, t) = f(\mathbf{r}, t) \qquad (\mathbf{r} \text{ in } V, t > 0) \tag{10.4-7a}$$

subject to a set of linear *initial conditions*

$$\mathbf{A}_j\Phi(\mathbf{r}, t) = \beta_j(\mathbf{r}) \qquad (j = 1, 2, \ldots, M; \mathbf{r} \text{ in } V; t = 0 + 0) \tag{10.4-7b}$$

as well as the linear *boundary conditions*

$$\mathsf{B}_i\Phi(\mathbf{r}, t) = b_i(\mathbf{r}, t) \qquad (i = 1, 2, \ldots, N; \mathbf{r} \text{ in } S; t > 0) \qquad (10.4\text{-}7c)$$

Since the initial conditions (7*b*) are simply boundary conditions on the "coordinate surface" $t = 0$, the methods of Secs. 10.4-2*a* and *b* apply (Sec. 10.4-9). The following procedures may simplify the treatment of initial conditions:

1. $\Phi_B = \Phi_B(\mathbf{r}, t)$ can be further split into a sum of functions respectively satisfying homogeneous initial conditions and homogeneous boundary conditions.
2. Separation of variables (Secs. 10.1-3, 10.4-7*b*, and 10.4-8).
3. Laplace transformation of the time variable (Secs. 10.5-2 and 10.5-3*a*).
4. Duhamel's method (Sec. 10.5-4).

10.4-3. Particular Solutions of Laplace's Differential Equation: Three-dimensional Case (see also Table 6.5-1, Sec. 10.1-3, and Secs. 15.6-1 to 15.6-9; see Ref. 10.5 for solutions employing other coordinate systems)

(a) Rectangular Cartesian Coordinates x, y, z.

$$\nabla^2\Phi \equiv \frac{\partial^2\Phi}{\partial x^2} + \frac{\partial^2\Phi}{\partial y^2} + \frac{\partial^2\Phi}{\partial z^2} = 0 \qquad (10.4\text{-}8)$$

admits the particular solutions

$$\left.\begin{aligned} \Phi_{k_1k_2k_3}(x, y, z) &= e^{k_1x+k_2y+k_3z} \qquad (k_1^2 + k_2^2 + k_3^2 = 0) \\ \Phi_{000}(x, y, z) &= (a + bx)(\alpha + \beta y)(A + Bz) \end{aligned}\right\} \qquad (10.4\text{-}9)$$

which combine into various products of real linear, exponential, trigonometric, and/or hyperbolic functions.

(b) Cylindrical Coordinates r', φ, z. Let $\Phi = u(\varphi)v(z)w(r')$ (Sec. 10.1-3). Then

$$\nabla^2\Phi \equiv \frac{1}{r'}\frac{\partial}{\partial r'}\left(r'\frac{\partial\Phi}{\partial r'}\right) + \frac{1}{r'^2}\frac{\partial^2\Phi}{\partial\varphi^2} + \frac{\partial^2\Phi}{\partial z^2} = 0 \qquad (10.4\text{-}10)$$

separates into

$$\frac{d^2}{d\varphi^2}u(\varphi) + m^2u(\varphi) = 0 \qquad (10.4\text{-}11)$$

$$\frac{d^2}{dz^2}v(z) + K^2v(z) = 0 \qquad (10.4\text{-}12)$$

$$\frac{d^2}{dr'^2}w(r') + \frac{1}{r'}\frac{d}{dr'}w(r') - \left(K^2 + \frac{m^2}{r'^2}\right)w(r') = 0 \qquad (10.4\text{-}13)$$

where uniqueness requires $u(\varphi + 2\pi) = u(\varphi)$ or $m = 0, \pm 1, \pm 2, \ldots,$ and K is an arbitrary constant (separation constant, Sec. 10.1-3) to be determined by the given boundary conditions. Equation (10) admits particular solutions (**cylindrical harmonics**) of the form

$$\left.\begin{aligned} \Phi_{\pm Km}(r', \varphi, z) &= e^{\pm iKz} Z_m(iKr')(\alpha \cos m\varphi + \beta \sin m\varphi) \\ \Phi_{\pm K0}(r', \varphi, z) &= e^{\pm iKz} Z_0(iKr')(\alpha + \beta\varphi) \\ \Phi_{0m}(r', \varphi, z) &= (a + bz)\left(Ar' + \frac{B}{r'}\right)(\alpha \cos m\varphi + \beta \sin m\varphi) \\ \Phi_{00}(r', \varphi, z) &= (a + bz)(A + B \log_e r')(\alpha + \beta\varphi) \end{aligned}\right\} \qquad (m = 0, 1, 2, \ldots) \tag{10.4-14}$$

where $Z_m(\zeta)$ is a *cylinder function* (Sec. 21.8-1); in particular, if a given problem requires Φ to be analytic for $r = 0$, then $Z_m(\zeta)$ must be a *Bessel function of the first kind* (Sec. 21.8-1). Note that *complex-conjugate solutions* (14) *combine into real particular solutions* like

$$(a \cos Kz + b \sin Kz)[AZ_m(iKr') + A^*Z_m(-iKr')](\alpha \cos m\varphi + \beta \sin m\varphi)$$

for real K. Such solutions can be superimposed to form real Fourier series. Note that $m = 0$ in cases of *axial symmetry*.

(c) Spherical Coordinates r, ϑ, φ. Let $\Phi = u(\varphi)v(\cos\vartheta)w(r)$. Then

$$\boxed{r^2\nabla^2\Phi \equiv \frac{\partial}{\partial r}\left(r^2 \frac{\partial \Phi}{\partial r}\right) + \frac{1}{\sin\vartheta}\frac{\partial}{\partial\vartheta}\left(\sin\vartheta \frac{\partial\Phi}{\partial\vartheta}\right) + \frac{1}{\sin^2\vartheta}\frac{\partial^2\Phi}{\partial\varphi^2} = 0} \tag{10.4-15}$$

separates into

$$\frac{d^2}{d\varphi^2}u(\varphi) + m^2 u(\varphi) = 0 \tag{10.4-16}$$

$$(1 - \zeta^2)\frac{d^2}{d\zeta^2}v(\zeta) - 2\zeta\frac{d}{d\zeta}v(\zeta) + \left[j(j+1) - \frac{m^2}{1-\zeta^2}\right]v(\zeta) = 0 \qquad (\zeta = \cos\vartheta) \tag{10.4-17}$$

$$\frac{d^2}{dr^2}w(r) + \frac{2}{r}\frac{d}{dr}w(r) - \frac{j(j+1)}{r^2}w(r) = 0 \tag{10.4-18}$$

where regularity for $\vartheta = 0$, $\vartheta = \pi$ and uniqueness require that $m = 0, \pm 1, \pm 2, \ldots, \pm j$ and $j = 0, 1, 2, \ldots$. Equation (15) admits particular solutions of the form

$$\Phi_{jm}(r, \vartheta, \varphi) = \left(Ar^j + \frac{B}{r^{j+1}}\right)P_j^m(\cos\vartheta)(\alpha \cos m\varphi + \beta \sin m\varphi) \qquad (j = 0, 1, 2, \ldots;\ m = 0, 1, 2, \ldots, j) \tag{10.4-19}$$

where the $P_j^m(\zeta)$ are *associated Legendre functions of the first kind of degree j*

(Sec. 21.8-10). Combination of such solutions yields more general particular solutions

$$\Phi_j(r, \vartheta\ \varphi,) = \left(Ar^j + \frac{B}{r^{j+1}}\right) Y_j(\vartheta, \varphi) \qquad (j = 0, 1, 2, \ldots) \tag{10.4-20}$$

with

$$\begin{aligned} Y_j(\vartheta\ \varphi,) &= \sum_{m=0}^{j} P_j^m(\cos \vartheta)(\alpha_m \cos m\varphi + \beta_m \sin m\varphi) \\ &= \sum_{m=-j}^{j} \gamma_m P_j^{|m|}(\cos \vartheta)\, e^{im\varphi} \qquad (j = 0, 1, 2, \ldots) \end{aligned} \tag{10.4-21}$$

The functions (21) satisfy Eq. (15) for r = constant and are called **spherical surface harmonics** of degree j (see also Sec. 21.8-12).

There are $2j + 1$ linearly independent spherical surface harmonics of degree j. For orthogonal-series expansion of solutions, note that *the functions*

$$\sqrt{\frac{2j+1}{2\pi}\frac{(j-m)!}{(j+m)!}}\, P_j^m(\cos \vartheta) \cos m\varphi, \quad \sqrt{\frac{2j+1}{2\pi}\frac{(j-m)!}{(j+m)!}}\, P_j^m(\cos \vartheta) \sin m\varphi$$

$(j = 0, 1, 2, \ldots ; m = 0, 1, 2, \ldots , j)$ *or the sometimes more convenient functions*

$$\frac{1}{2}\sqrt{\frac{2j+1}{\pi}\frac{(j-|m|)!}{(j+|m|)!}}\, P_j^{|m|}(\cos \vartheta) e^{im\varphi} \qquad (j = 0, 1, 2, \ldots ; m = 0, \pm 1, \pm 2, \ldots , \pm j)$$

constitute orthonormal sets in the sense of Sec. 21.8-2. These functions are called **tesseral spherical harmonics** (**sectorial spherical harmonics** for $m = j$; see also Secs. 10.4-9*a*, 15.2-6, and 21.8-11). The orthonormal functions

$$\sqrt{\frac{2j+1}{2}}\, P_j(\cos \vartheta)$$

are known as **zonal spherical harmonics.**

If one admits solutions with singularities for $\vartheta = 0$, $\vartheta = \pi$, one must add analogous solutions involving the associated Legendre functions of the second kind (Ref. 21.3).

10.4-4. Particular Solutions for the Space Form of the Three-dimensional Wave Equation

(see also Secs. 10.3-5 and 15.6-10). The differential equation

$$\nabla^2\Phi + k^2\Phi = 0 \qquad \text{(SPACE FORM OF THE WAVE EQUATION, HELMHOLTZ'S EQUATION)} \tag{10.4-22}$$

is obtained, for example, by separation of the time variable t in the three-dimensional wave equation (44). The coefficient k^2 may be negative ($k = i\kappa$, space form of the *Klein-Gordon equation*).

For suitably given *homogeneous* linear boundary conditions (e.g., $\Phi = 0$ on the boundary of S of a bounded region V), Eq. (22) admits solutions only for a corresponding discrete set of values of k^2 (*eigenvalue problem*, Sec. 15.4-5; see Sec. 10.4-9*b* for an example).

(a) Rectangular Cartesian Coordinates x, y, z. Equation (19) has the particular solutions

$$\left.\begin{aligned} \Phi_{k_1k_2k_3}(x, y, z) &= e^{i(k_1x+k_2y+k_3z)} \qquad (k_1^2 + k_2^2 + k_3^2 = k^2) \\ \Phi_{0k_2k_3}(x, y, z) &= (a + bx)e^{i(k_2y+k_3z)} \qquad (k_2^2 + k_3^2 = k^2) \\ \Phi_{00k}(x, y, z) &= (a + bx)(\alpha + \beta y)e^{ikz} \end{aligned}\right\} \quad (10.4\text{-}23)$$

which may be combined into various products of real linear, exponential, trigonometric, and hyperbolic functions.

(b) Cylindrical Coordinates r', φ, z (see also Sec. 10.4-3*b*). Let $\Phi = u(\varphi)v(z)w(r')$. Then Eq. (22) separates into Eq. (11), Eq. (12), and

$$\frac{d^2}{dr'^2}w(r') + \frac{1}{r'}\frac{d}{dr'}w(r') - \left[(K^2 - k^2) + \frac{m^2}{r'^2}\right]w(r') = 0 \quad (10.4\text{-}24)$$

where uniqueness requires $m = 0, \pm 1, \pm 2, \ldots$, and K is an arbitrary separation constant to be determined by the boundary conditions. Equation (22) admits solutions of the form

$$\left.\begin{aligned} \Phi_{\pm Km}(r', \varphi, z) &= e^{\pm iKz}Z_m(r'\sqrt{k^2 - K^2})(\alpha\cos m\varphi + \beta\sin m\varphi) \\ &\qquad (m = 0, 1, 2, \ldots) \\ \Phi_{00}(r', \varphi, z) &= (a + bz)Z_0(kr')(\alpha + \beta\varphi) \end{aligned}\right\} \quad (10.4\text{-}25)$$

Note that $m = 0$ for axial symmetry.

(c) Spherical Coordinates r, ϑ, φ (see also Sec. 10.4-3*c*). Let $\Phi = u(\varphi)v(\cos\vartheta)w(r)$. Then Eq. (22) separates into Eq. (16), Eq. (17), and

$$\frac{d^2}{dr^2}w(r) + \frac{2}{r}\frac{d}{dr}w(r) + \left[k^2 - \frac{j(j+1)}{r^2}\right]w(r) = 0 \quad (10.4\text{-}26)$$

where regularity for $\vartheta = 0$, $\vartheta = \pi$ and uniqueness require that $m = 0, \pm 1, \pm 2, \ldots, \pm j$, and $j = 0, 1, 2, \ldots$. Equation (22) admits particular solutions of the form

$$\left.\begin{aligned} \Phi_{kj}(r, \vartheta, \varphi) &= \frac{1}{\sqrt{r}}Z_{j+\frac{1}{2}}(kr)Y_j(\vartheta, \varphi) \qquad (j = 1, 2, \ldots) \\ \Phi_{k0}(r, \vartheta, \varphi) &= \frac{e^{\pm ikr}}{r} \end{aligned}\right\} \quad (10.4\text{-}27)$$

where $Y_j(\vartheta, \varphi)$ is a spherical surface harmonic (21). In particular, if a given problem requires Φ to be analytic for $r = 0$, the $Z_{j+\frac{1}{2}}(kr)/\sqrt{r}$ are *spherical Bessel functions of the first kind* (Sec. 21.8-8).

10.4-5. Particular Solutions for Two-dimensional Problems (see also Secs. 15.6-7 and 15.6-10*b*). **(a) Laplace's equation**

$$\nabla^2\Phi \equiv \frac{\partial^2\Phi}{\partial x^2} + \frac{\partial^2\Phi}{\partial y^2} \equiv \frac{1}{r}\frac{\partial}{\partial r}\left(r\frac{\partial\Phi}{\partial r}\right) + \frac{1}{r^2}\frac{\partial^2\Phi}{\partial\varphi^2} = 0 \qquad (10.4\text{-}28)$$

admits the particular solutions

$$\Phi_K(x, y) = e^{\pm K(x+iy)} \qquad \Phi_0(x, y) = (a + bx)(\alpha + \beta y) \qquad (10.4\text{-}29)$$

$$\left.\begin{aligned}\Phi_m(r, \varphi) &= \left(Ar^m + \frac{B}{r^m}\right)(\alpha\cos m\varphi + \beta\sin m\varphi) \qquad (m = 0, 1, 2, \ldots)\\ \Phi_0(r, \varphi) &= A + B\log_e r\end{aligned}\right\} \qquad (10.4\text{-}30)$$

where K, like a, b, α, β, A, B, is an arbitrary parameter to be determined by the boundary conditions.

(b) Space Form of the Wave Equation. The two-dimensional space form (22) of the wave equation admits the particular solutions

$$\left.\begin{aligned}\Phi_{k_1k_2}(x, y) &= e^{i(k_1x+k_2y)} \qquad (k_1{}^2 + k_2{}^2 = k^2)\\ \Phi_{0k}(x, y) &= (a + bx)e^{iky}\end{aligned}\right\} \qquad (10.4\text{-}31)$$

$$\left.\begin{aligned}\Phi_{km}(r, \varphi) &= Z_m(kr)(\alpha\cos m\varphi + \beta\sin m\varphi) \qquad (m = 0, 1, 2, \ldots)\\ \Phi_0(r, \varphi) &= Z_0(kr)(a + b\varphi)\end{aligned}\right\} \qquad (10.4\text{-}32)$$

(c) Complex-conjugate solutions (29) and (31) combine into various products of real linear, exponential, trigonometric, and/or hyperbolic functions.

10.4-6. The Static Schrödinger Equation for Hydrogenlike Wave Functions. The three-dimensional partial differential equation

$$\nabla^2\Phi + \left(\frac{2c}{r} - \lambda^2\right)\Phi = 0 \qquad (10.4\text{-}33)$$

admits the particular solutions

$$\left.\begin{aligned}\Phi_{\lambda j}(r, \vartheta, \varphi) &= r^j e^{-\lambda r} L^{2j+1}_{c/\lambda+j}(2\lambda r) Y_j(\vartheta, \varphi)\\ \Phi_{\lambda,-j}(r, \vartheta, \varphi) &= \frac{e^{-\lambda r}}{r^{j+1}} L^{-2j-1}_{c/\lambda-j-1}(2\lambda r) Y_j(\vartheta, \varphi)\end{aligned}\right\} (j = 0, 1, 2, \ldots) \qquad (10.4\text{-}34)$$

where the $L_n{}^k(\zeta)$ are *Laguerre functions* (Sec. 21.7-5). If Φ is to be normalizable (Sec. 15.2-1*b*), one is restricted to the first type of solution (34). Normalizable solutions exist only for *eigenvalues* (Sec. 15.4-5) λ^2 such that $c/\lambda = n = 0, 1, 2, \ldots$. In this case the Laguerre functions reduce to associated Laguerre *polynomials*, and the solutions (34) constitute an orthogonal set in the sense of Secs. 21.7-5 and 21.8-12 (see also Ref. 10.8).

10.4-7. Particular Solutions for the Diffusion Equation (see also Secs. 10.3-4*b*, 10.4-9, 10.5-3, 10.5-4, and 15.5-3). **(a) The one-dimensional diffusion equation**

$$\frac{\partial^2 \Phi}{\partial x^2} - \frac{1}{\gamma^2}\frac{\partial \Phi}{\partial t} = 0 \tag{10.4-35}$$

admits the particular solutions

$$\left.\begin{aligned} \Phi_k(x, t) &= e^{\pm ikx - k^2\gamma^2 t} & \Phi_0(x, t) &= a + bx \\ \Phi(x, t) &= \frac{1}{\sqrt{t}}\, e^{-x^2/4\gamma^2 t} & (t &> 0) \end{aligned}\right\} \tag{10.4-36}$$

where k is an arbitrary separation constant to be determined by the boundary conditions.

(b) The two- or three-dimensional diffusion equation

$$\nabla^2 \Phi - \frac{1}{\gamma^2}\frac{\partial \Phi}{\partial t} = 0 \tag{10.4-37}$$

admits the particular solutions

$$\Phi(\mathbf{r}, t) = \Phi_k(\mathbf{r}) e^{-k^2\gamma^2 t} \tag{10.4-38}$$

where $\Phi_k(\mathbf{r})$ is any particular solution of the corresponding Helmholtz equation (22) (Secs. 10.4-4 and 10.4-5*b*); k is an arbitrary separation constant to be determined by the boundary conditions. Equation (37) also admits the particular solution

$$\Phi(\mathbf{r}, t) = \left\{\begin{aligned} &\frac{1}{t}\, e^{-r^2/4\gamma^2 t} && \text{(TWO-DIMENSIONAL CASE)} \\ &\frac{1}{\sqrt{t^3}}\, e^{-r^2/4\gamma^2 t} && \text{(THREE-DIMENSIONAL CASE)} \end{aligned}\right\} \tag{10.4-39}$$

10.4-8. Particular Solutions for the Wave Equation. Sinusoidal Waves (see also Secs. 4.11-4*b*, 10.3-5, 10.4-9, 10.5-2, and 15.6-10). **(a) The one-dimensional wave equation**

$$\frac{\partial^2 \Phi}{\partial x^2} - \frac{1}{c^2}\frac{\partial^2 \Phi}{\partial t^2} = 0 \tag{10.4-40}$$

admits particular solutions of the form

$$\Phi(x, t) = e^{\pm ikx} e^{\pm i\omega t} = e^{\pm ik(x \pm ct)} \qquad (\omega = kc) \tag{10.4-41}$$

where k is an arbitrary constant to be determined by the boundary conditions. The functions (41) and the corresponding values of k^2 are usually the eigenfunctions and eigenvalues of an eigenvalue problem, Secs. 10.4-2 and 15.4-5.

Solutions of the form (41) combine into real solutions

$$\left.\begin{aligned}\Phi(x, t) &= C\cos(\omega t + \gamma_1)\cos(kx + \gamma_2) \\ &\qquad\qquad \text{(SINUSOIDAL STANDING WAVES)} \\ \Phi(x, t) &= a\cos(\omega t \mp kx) + b\sin(\omega t \mp kx) \\ &= A\cos(\omega t \mp kx + \gamma) \qquad \text{(SINUSOIDAL} \\ &\text{WAVES TRAVELING IN THE } \pm x \text{ DIRECTION)}\end{aligned}\right\} (\omega = kc) \qquad (10.4\text{-}42)$$

The **circular frequency** ω, the **frequency** $\nu = \omega/2\pi$, the **wave number** k, the **wavelength** $\lambda = 2\pi/k$, and the **phase velocity** c of the sinusoidal waves are related by

$$\lambda\nu = \frac{\omega}{k} = c \qquad (10.4\text{-}43)$$

Sinusoidal waves (42) may be superimposed to form Fourier series or Fourier integrals for more general waves.

(b) The two- or three-dimensional wave equation

$$\boxed{\nabla^2\Phi - \frac{1}{c^2}\frac{\partial^2\Phi}{\partial t^2} = 0} \qquad (10.4\text{-}44)$$

admits particular solutions of the form

$$\Phi(\mathbf{r}, t) = \Phi_k(\mathbf{r})e^{\pm i\omega t} \qquad (\omega = kc) \qquad (10.4\text{-}45)$$

where $\Phi_k(\mathbf{r})$ is any particular solution of the corresponding Helmholtz equation (22) (Secs. 10.4-4 and 10.4-5*b*); k is an arbitrary separation constant to be determined by the boundary conditions. Solutions of the form (45) combine into solutions involving real trigonometric functions; in particular, note the following examples:

$$\Phi(x, y, z; t) = A\cos[\omega t \mp (k_1x + k_2y + k_3z) + \gamma]$$

$$(k_1^2 + k_2^2 + k_3^2 = k^2;\ \omega = kc) \quad \text{(SINUSOIDAL PLANE WAVES, WAVE FRONTS NORMAL TO DIRECTION GIVEN BY } k_1, k_2, k_3) \qquad (10.4\text{-}46)$$

$$\Phi(r', \varphi, z; t) = Z_m(r'\sqrt{k^2 - K^2})\cos(\omega t \mp Kz \mp m\varphi + \gamma)$$
$$(m = 0, 1, 2, \ldots;\ \omega = kc)$$
$$\text{(SINUSOIDAL CYLINDRICAL WAVES)} \qquad (10.4\text{-}47)$$

$$\Phi(r, \vartheta, \varphi; t) = \frac{1}{\sqrt{r}} Z_{j+\frac{1}{2}}(kr)Y_j(\vartheta, \varphi)\cos(\omega t + \gamma)$$
$$(j = 0, 1, 2, \ldots;\ \omega = kc)$$
$$\text{(SINUSOIDAL SPHERICAL STANDING WAVES)} \qquad (10.4\text{-}48)$$

$$\Phi(r, \vartheta, \varphi; t) = \frac{A}{r}\cos(\omega t \mp kr + \gamma)$$
$$\text{(POINT-CHARGE RADIATION)} \qquad (10.4\text{-}49)$$

$$\Phi(r, \vartheta, \varphi; t) = A\left[\frac{1}{kr^2}\cos(\omega t \mp kr) \mp \frac{1}{r}\sin(\omega t \mp kr)\right]\cos\vartheta$$
$$\text{(DIPOLE RADIATION)} \qquad (10.4\text{-}50)$$

$$\Phi(r, \varphi; t) = Z_m(kr)(\alpha \cos m\varphi + \beta \sin m\varphi)(a \cos \omega t + b \sin \omega t)$$
$$(m = 0, 1, 2, \ldots ; \omega = kc)$$

(TWO-DIMENSIONAL SINUSOIDAL CIRCULAR STANDING WAVES) (10.4-51)

The cylindrical waves (47) are propagated in the $\pm z$ direction with *phase velocity* $c' = \omega/K = kc/K$, which is seen to depend on ω and K (*dispersion*). One defines the *group velocity* in the z direction as $d\omega/dK = Kc/k$.

(c) The Generalized One-dimensional Damped-wave Equation (Telegrapher's Equation). The transmission-line equation

$$\frac{\partial^2\Phi}{\partial x^2} - a_0 \frac{\partial^2\Phi}{\partial t^2} - a_1 \frac{\partial\Phi}{\partial t} - a_2\Phi = 0 \qquad (10.4\text{-}52)$$

admits particular solutions of the form

$$\Phi(x, t) = e^{\pm ikx}e^{st} \qquad (10.4\text{-}53)$$

where $s = \sigma + i\omega$ is a root of the quadratic equation

$$a_0s^2 + a_1s + (a_2 + k^2) = 0 \qquad (10.4\text{-}54)$$

Complex-conjugate solutions (53) combine into *damped sinusoidal waves* in the manner of Sec. 9.4-1*b*; again, double roots of Eq. (54) are treated as in Sec. 9.4-1*b*. Equation (52) includes Eqs. (35) and (40) as special cases. An analogous generalization applies to the multidimensional case.

10.4-9. Solution of Boundary-value Problems by Orthogonal-series Expansions: Examples (see also Sec. 10.5-3). The following examples illustrate the use of the particular solutions listed in Secs. 10.4-3 to 10.4-8; see also Table 8.7-1.

(a) Dirichlet Problem for a Sphere (see also Secs. 10.4-3*c*, 15.6-2*a*, and 15.6-6*c*). One desires to find the function $\Phi(r)$ which satisfies Laplace's equation $\nabla^2\Phi = 0$ throughout a given sphere $r < R$ and assumes suitably given boundary values $\Phi(R, \vartheta, \varphi) = b(\vartheta, \varphi)$ on the sphere. The use of spherical coordinates r, ϑ, φ is clearly indicated (Sec. 10.4-1*c*). One attempts to write the unknown function as a sum of solutions (19) which remain finite for $r = 0$ ($A = 0$):

$$\Phi = \Phi(r, \vartheta, \varphi) = \sum_{j=0}^{\infty} \sum_{m=0}^{j} \frac{1}{r^{j+1}} P_j^m(\cos \vartheta)(\alpha_{jm} \cos m\varphi + \beta_{jm} \sin m\varphi)$$

The unknown coefficients α_{jm}, β_{jm} are obtained from the given boundary conditions

$$\Phi(R, \vartheta, \varphi) = \frac{1}{R^{j+1}} \sum_{j=0}^{\infty} \sum_{m=0}^{j} P_j^m(\cos \vartheta)(\alpha_{jm} \cos m\varphi + \beta_{jm} \sin m\varphi) = b(\vartheta, \varphi)$$

with the aid of the orthogonality conditions of Sec. 21.8-12:

$$\left.\begin{aligned}
\alpha_{j0} &= \frac{2j+1}{4\pi} R^{j+1} \int_0^{2\pi} d\varphi \int_0^{\pi} b(\vartheta, \varphi)P_j(\cos \vartheta) \sin \vartheta \, d\varphi \\
\alpha_{jm} &= \frac{2j+1}{2\pi} \frac{(j-m)!}{(j+m)!} R^{j+1} \int_0^{2\pi} d\varphi \int_0^{\pi} b(\vartheta, \varphi)P_j^m(\cos \vartheta) \cos m\varphi \sin \vartheta \, d\vartheta \\
\beta_{jm} &= \frac{2j+1}{2\pi} \frac{(j-m)!}{(j+m)!} R^{j+1} \int_0^{2\pi} d\varphi \int_0^{\pi} b(\vartheta, \varphi)P_j^m(\cos \vartheta) \sin m\varphi \sin \vartheta \, d\varphi
\end{aligned}\right\}$$
$$(j = 0, 1, 2, \ldots ; m = 1, 2, \ldots, j) \qquad (10.4\text{-}55)$$

(b) Free Vibrations of an Elastic String (see also Sec. 10.3-5). The lateral displacement $\Phi(x, t)$ of an elastic string satisfies the one-dimensional wave equation

$$\frac{\partial^2 \Phi}{\partial x^2} - \frac{1}{c^2}\frac{\partial \Phi}{\partial t^2} = 0$$

In addition, let

$$\Phi(x, 0) = \Phi_0(x) \qquad \left.\frac{\partial \Phi}{\partial t}\right]_{t=0} = v_0(x) \qquad (0 \leq x \leq L) \quad \text{(INITIAL CONDITIONS)}$$
$$\Phi(0, t) = \Phi_a(t) \qquad \Phi(L, t) = \Phi_b(t) \qquad \text{(BOUNDARY CONDITIONS)}$$

Consider the special case $\Phi_a(t) \equiv \Phi_b(t) \equiv 0$. The standing-wave solutions (42) of the one-dimensional wave equation fit these given boundary conditions whenever $\gamma_2 = -\pi/2$ and the half wavelength $\lambda/2 \equiv \pi/k$ is an integral fraction of the string length L, or

$$k = \frac{n\pi}{L} \qquad \omega = kc = \frac{n\pi c}{L} \qquad (n = 0, 1, 2, \ldots)$$

One next attempts to write the solution as an infinite sum of these particular solutions, i.e.,

$$\Phi(x, t) = \sum_{n=1}^{\infty} \left(a_n \sin \frac{n\pi}{L} x \cos \frac{n\pi c}{L} t + b_n \sin \frac{n\pi}{L} x \sin \frac{n\pi c}{L} t\right)$$

This is seen to be a Fourier series (Sec. 4.11-4*b*) whose coefficients a_n, b_n fit the given initial conditions if

$$a_n = \frac{2}{L}\int_0^L \Phi_0(x) \sin \frac{n\pi}{L} x\, dx \qquad (n = 1, 2, \ldots)$$
$$b_n = \frac{2}{n\pi c}\int_0^L v_0(x) \sin \frac{n\pi}{L} x\, dx \qquad (n = 1, 2, \ldots)$$

The solution is seen to be the sum of harmonic standing waves (modes of vibration) "excited" by the given initial conditions.

(c) Free Oscillations of a Circular Membrane. The transverse displacement $\Phi(\mathbf{r})$ of a membrane clamped along its boundary circle $r = 1$ satisfies the *wave equation* $\nabla^2\Phi - \frac{1}{c^2}\frac{\partial^2\Phi}{\partial t^2} = 0$ $(r < 1)$ together with the *boundary condition* $\Phi = 0$ $(r = 1)$ and a given set of *initial conditions*

$$\Phi = \Phi_0(\mathbf{r}) \qquad \partial\Phi/\partial t = v_0(\mathbf{r}) \qquad \text{for } t = 0 \qquad (r \leq 1)$$

One uses polar coordinates r, φ. Since Φ is to be regular for $r = 0$, one attempts to superimpose solutions (51) involving only Bessel functions of the first kind, i.e., $Z_m(kr) = J_m(kr)$ (Sec. 21.8-1). Such solutions represent *characteristic oscillations* satisfying the given boundary condition $\Phi = 0$ $(r = 1)$ if

$$k = \frac{\omega}{c} = k_{mn} \qquad (n, m = 0, 1, 2, \ldots)$$

where k_{mn} is the nth real positive zero of $J_m(\zeta)$; the problem is an *eigenvalue problem.* The solution appears as a sum of characteristic oscillations

$$\Phi(r, \varphi, t) = \sum_{m=0}^{\infty}\sum_{n=0}^{\infty} J_m(k_{mn}r)(\alpha_{mn} \cos m\varphi + \beta_{mn} \sin m\varphi) \cdot (a_{mn} \cos ck_{mn}t + b_{mn} \sin ck_{mn}t)$$

where the unknown coefficients are determined as ordinary Fourier coefficients upon substitution of the given initial conditions for $t = 0$.

10.5. INTEGRAL-TRANSFORM METHODS

10.5-1. General Theory (see also Secs. 8.1-1, 8.6-1, 9.3-7, 9.4-5, 10.4-2c, and 15.4-12). It is desired to find the solution $\Phi = \Phi(x_1, x_2, \ldots, x_n)$, subject to suitable boundary conditions, of the real linear partial differential equation

$$(\mathbf{L}_1 + \mathbf{L}_2)\Phi \equiv a_0(x_n)\frac{\partial^2\Phi}{\partial x_n^2} + a_1(x_n)\frac{\partial\Phi}{\partial x_n} + a_2(x_n)\Phi + \mathbf{L}_2\Phi = f(x_1, x_2, \ldots, x_n) \qquad (10.5\text{-}1)$$

where $\mathbf{L}_2$ is a linear homogeneous differential operator whose derivatives and coefficients involve $x_1, x_2, \ldots, x_{n-1}$ but not x_n. For simplicity, assume x_n to range over a fixed bounded or unbounded interval (a, b) for all $x_1, x_2, \ldots, x_{n-1}$, so that the boundary S of the solution region consists of two x_n-coordinate surfaces (see also Sec. 10.4-1c). One can often simplify Eq. (1) by applying a linear integral transformation

$$\bar{\varphi}(x_1, x_2, \ldots, x_{n-1}; s) = \int_a^b K(x_n, s)\varphi(x_1, x_2, \ldots, x_n)\, dx_n \qquad (10.5\text{-}2)$$

to both sides. *If the transformation kernel $K(x_n, s)$ is chosen so that*

$$\mathbf{L}_1{}^\dagger K(x_n, s) \equiv \frac{\partial^2}{\partial x_n^2}(a_0K) - \frac{\partial}{\partial x_n}(a_1K) + a_2K = \lambda(s)K \qquad (10.5\text{-}3)$$

then partial integration (generalized Green's formula, Sec. 15.4-3) yields a possibly simpler differential equation

$$\left.\begin{aligned} &[\mathbf{L}_2{}^\dagger + \lambda(s)]\bar{\Phi} = \bar{f}(x_1, x_2, \ldots, x_{n-1}; s) - P(x_1, x_2, \ldots, x_n; s)\Big]_{x_n=a}^{x_n=b} \\ &P(x_1, x_2, \ldots, x_n; s) \equiv a_0\left(K\frac{\partial\Phi}{\partial x_n} - \Phi\frac{\partial K}{\partial x_n}\right) + (a_1 - a_0')\Phi K \end{aligned}\right\} \qquad (10.5\text{-}4)$$

to be satisfied by the unknown integral transform $\bar{\Phi}(x_1, x_2, \ldots, x_{n-1}; s)$. Note that the new differential equation introduces the boundary values of Φ and $\partial\Phi/\partial x_n$ given for $x_n = a$, $x_n = b$ and variable $x_1, x_2, \ldots, x_{n-1}$. The integral-transform method assumes the existence of the relevant integrals (2) and a convergent *inversion formula*

$$\Phi(x_1, x_2, \ldots, x_n) = \int_\alpha^\beta H(x_n, s)\bar{\Phi}(x_1, x_2, \ldots, x_{n-1}; s)\, ds \qquad (10.5\text{-}5)$$

which usually involves complex integration (see also Sec. 15.3-7). Table

8.6-1 lists a number of useful integral transformations and inversion formulas.

Note that constants of integration in solutions of the transformed differential equation (4) must be regarded as arbitrary functions of s; similarly, arbitrary functions of $x_1, x_2, \ldots, x_{n-1}$ must also be arbitrary functions of s.

The integral-transform method may be generalized to apply to differential equations involving higher-order derivatives of x_n; again, the integral transformation may operate on two variables x_n, x_{n-1} simultaneously (Ref. 8.6). If δ-function terms in $H(x_n, s)$ are admitted, Eq. (5) can yield *series expansions* for Φ over finite intervals (a, b) (*finite integral transforms*, Sec. 8.7-1 and Ref. 10.21).

10.5-2. Laplace Transformation of the Time Variable (see also Secs. 8.2-1, 8.3-1, 9.3-7, and 9.4-5). Initial-value problems involving hyperbolic or parabolic differential equations are often simplified by the unilateral Laplace transformation

$$\bar{\varphi}(s) = \int_0^\infty \varphi(t)e^{-st}\,dt$$

which transforms every suitable linear partial differential equation with constant coefficients

$$\nabla^2\Phi - a_0\frac{\partial^2\Phi}{\partial t^2} - a_1\frac{\partial\Phi}{\partial t} - a_2\Phi = f(\mathbf{r}, t) \tag{10.5-6}$$

into a new and possibly simpler differential equation

$$\nabla^2\bar{\Phi} - (a_0s^2 + a_1s + a_2)\bar{\Phi} = \bar{f}(\mathbf{r}, s) - \left[(a_0s + a_1)\Phi + a_0\frac{\partial\Phi}{\partial t}\right]_{t=0+0} \tag{10.5-7}$$

for the unknown integral transform $\bar{\Phi}(\mathbf{r}, s)$ of $\Phi(\mathbf{r}, t)$. The boundary conditions are similarly transformed; note that Eq. (7) includes the effect of given initial conditions.

10.5-3. Solution of Boundary-value Problems by Integral-transform Methods: Examples (see also Sec. 10.4-9). The following examples illustrate the simplest applications of integral transformations to the solution of boundary-value problems (see also Refs. 10.3, 10.18, and 10.19).

(a) **One-dimensional Heat Conduction in a Wall with Fixed Boundary Temperatures: Use of Laplace Transforms.** Duhamel's theorem (Sec. 10.5-4) reduces an important class of one-dimensional heat-conduction problems to the form

$$\frac{\partial^2\Phi(x, t)}{\partial x^2} - \frac{1}{\gamma^2}\frac{\partial\Phi(x, t)}{\partial t} = 0 \qquad (a < x < b;\ t > 0)$$

$$\Phi(x, 0 + 0) = \Phi_0(x) \qquad \text{(INITIAL CONDITIONS)}$$

$$\Phi(a, t) = \Phi_a = \text{constant} \qquad \Phi(b, t) = \Phi_b = \text{constant} \qquad (t > 0) \qquad \text{(BOUNDARY CONDITIONS)}$$

The differential equation transforms into

$$\frac{\partial^2\bar{\Phi}(x, s)}{\partial x^2} - \frac{s}{\gamma^2}\bar{\Phi}(x, s) = -\frac{1}{\gamma^2}\Phi_0(x) \qquad (a < x < b)$$

If, in particular, $\Phi_0(x) = \Phi_0 = \text{constant}$, one has

$$\bar{\Phi}(x, s) = C_1(s)e^{x\sqrt{s}/\gamma} + C_2(s)e^{-x\sqrt{s}/\gamma} + \frac{\Phi_0}{s}$$

where the functions $C_1(s)$ and $C_2(s)$ must be chosen so as to match the transformed boundary conditions

$$\bar{\Phi}(a, s) = \Phi_a/s \qquad \bar{\Phi}(b, s) = \Phi_b/s$$

$\Phi(x, t)$ is obtained as an inverse Laplace transform by one of the methods of Secs. 8.4-2 to 8.4-9. In particular, for $a = 0$, $\Phi_a = 0$, $\Phi_0 = 0$, one has

$$\bar{\Phi}(x, s) = \Phi_b \frac{\sinh \dfrac{x\sqrt{s}}{\gamma}}{s \sinh \dfrac{b\sqrt{s}}{\gamma}}$$

$$\Phi(x, t) = \Phi_b \left[\frac{x}{b} + \frac{2}{\pi} \sum_{j=1}^{\infty} \frac{(-1)^j}{j} e^{-\left(\frac{j\pi\gamma}{b}\right)^2 t} \sin \frac{j\pi x}{b}\right]$$

Note that this problem can also be solved by the method of Sec. 10.4-9.

(b) Heat Conduction into a Wall of Infinite Thickness: Use of Fourier Sine and Cosine Transforms. The method of Sec. 10.5-3*a* still applies if $a = 0$, $b = \infty$, and one is given

$$\frac{\partial^2 \Phi(x, t)}{\partial x^2} - \frac{1}{\gamma^2} \frac{\partial \Phi(x, t)}{\partial t} = 0 \qquad (x > 0;\, t > 0)$$

$$\Phi(x, 0 + 0) = \Phi_0(x) \qquad (x > 0) \qquad \text{(INITIAL CONDITIONS)}$$

$$\left.\begin{aligned} \Phi(0, t) &= \Phi_a(t) \\ \int_0^\infty |\Phi(x, t)|^2\, dx &\text{ exists} \end{aligned}\right\} \quad (t > 0) \qquad \text{(BOUNDARY CONDITIONS)}$$

One may, instead, apply the *Fourier sine transformation*

$$\bar{\varphi}(s, t) = \sqrt{\frac{2}{\pi}} \int_0^\infty \varphi(x, t) \sin sx\, dx$$

(Sec. 4.11-3) with the aid of Sec. 4.11-5*c* to obtain the transformed differential equation

$$\frac{1}{\gamma^2} \frac{\partial \bar{\Phi}(s, t)}{\partial t} + s^2 \bar{\Phi}_0(s, t) = \sqrt{\frac{2}{\pi}}\, s \Phi_a(t)$$

and the transformed initial condition $\bar{\Phi}(s, 0 + 0) = \bar{\Phi}_0(s)$. The transformed problem has the solution

$$\bar{\Phi}(s, t) = \bar{\Phi}_0(s) e^{-\gamma^2 s^2 t} + \sqrt{\frac{2}{\pi}}\, \gamma^2 s \int_0^t \Phi_a(t) e^{-\gamma^2 s^2 (t-\tau)}\, d\tau$$

and $\Phi(x, t) = \sqrt{\frac{2}{\pi}} \int_0^\infty \bar{\Phi}(s, t) \sin sx\, ds$. If, in particular, one is given $\Phi_a(t) = \text{constant} = \Phi_a$ and $\Phi_0(x) \equiv 0$, one has $\bar{\Phi}_0(s) \equiv 0$, and

$$\bar{\Phi}(s, t) = \Phi_a \frac{1 - e^{-\gamma^2 s^2 t}}{s} \sqrt{\frac{2}{\pi}} \qquad \Phi(x, t) = \Phi_a \left(1 - \operatorname{erf} \frac{x}{2\gamma\sqrt{t}}\right)$$

If the given boundary condition specifies, instead,

$$\left.\frac{\partial \Phi(x,\, t)}{\partial x}\right]_{x=0+0} = 0 \qquad (t \geq 0)$$

(zero heat flux through boundary), one uses the *Fourier cosine transformation*

$$\bar{\varphi}(s,\, t) = \sqrt{\frac{2}{\pi}} \int_0^\infty \varphi(x,\, t) \cos sx\, dx$$

and proceeds as before.

10.5-4. Duhamel's Formulas (see also Sec. 9.4-3). (**a**) Let $\mathbf{L}$ be a homogeneous linear operator whose coefficients and derivatives do not involve the time variable t. Let $\Phi(x, t)$ be the solution of the initial-value problem

$$\left.\begin{aligned} &\mathbf{L}\Phi + A_0(x)\frac{\partial^2 \Phi}{\partial t^2} + A_1(x)\frac{\partial \Phi}{\partial t} = 0 && (0 < x < L;\, t > 0) \\ &\Phi(x,\, 0+0) = 0 \qquad \left.\frac{\partial \Phi}{\partial t}\right]_{0+0} = 0 && (0 < x < L) \\ &\alpha \partial\Phi/\partial x + \beta\Phi = b(t) && (x = 0,\, t > 0) \end{aligned}\right\} \tag{10.5-8}$$

with as many *homogeneous* linear boundary conditions for $x = 0$ and/or $x = L$ as needed, and let $\Psi(x, t)$ be the solution of the same problem for $b(t) \equiv 1$ $(t > 0)$. Then

$$\begin{aligned} \Phi(x,\, t) &= \Psi(x,\, t)b(0+0) + \int_0^t \Psi(x,\, t-\tau)b'(\tau)\, d\tau \\ &= \frac{\partial}{\partial t}\int_0^t \Psi(x,\, t-\tau)b(\tau)\, d\tau \qquad (0 < x < L;\, t > 0) \end{aligned} \tag{10.5-9}$$

(**b**) The solution $\Phi(\mathbf{r}, t)$ of the generalized diffusion problem

$$\left.\begin{aligned} \nabla\cdot[k(\mathbf{r})\nabla]\Phi + a(\mathbf{r})\Phi - \frac{1}{\gamma^2}\frac{\partial \Phi}{\partial t} &= f(\mathbf{r},\, t) && (\mathbf{r} \text{ in } V,\, t > 0) \\ \Phi(\mathbf{r},\, 0) &= \Phi_0(\mathbf{r}) && (\mathbf{r} \text{ in } V) \\ \alpha\frac{\partial \Phi}{\partial n} + \beta\Phi &= b(\mathbf{r},\, t) && (\mathbf{r} \text{ in } S,\, t > 0) \end{aligned}\right\} \tag{10.5-10}$$

with time-dependent forcing functions $f(\mathbf{r}, t)$ and $b(\mathbf{r}, t)$ is related to the solution $\Psi(\mathbf{r}, t; \lambda)$ of the simpler problem

$$\left.\begin{aligned} \nabla\cdot[k(\mathbf{r})\nabla]\Psi + a(\mathbf{r})\Psi - \frac{1}{\gamma^2}\frac{\partial \Psi}{\partial t} &= f(\mathbf{r},\, \lambda) && (\mathbf{r} \text{ in } V,\, t > 0) \\ \alpha\frac{\partial \Psi}{\partial n} + \beta\Psi &= b(\mathbf{r},\, \lambda) && (\mathbf{r} \text{ in } S,\, t > 0) \\ \Psi(r,\, 0;\, \lambda) &= \Phi_0(\mathbf{r}) && (\mathbf{r} \text{ in } V) \end{aligned}\right\} \tag{10.5-11}$$

where $f(\mathbf{r}, \lambda)$ and $b(\mathbf{r}, \lambda)$ depend on a fixed parameter λ rather than on the variable t. One obtains $\Phi(\mathbf{r}, t)$ from

$$\Phi(\mathbf{r},\, t) = \frac{\partial}{\partial t}\int_0^t \Psi(\mathbf{r},\, t-\lambda;\, \lambda)\, d\lambda \tag{10.5-12}$$

10.6. RELATED TOPICS, REFERENCES, AND BIBLIOGRAPHY

10.6-1. Related Topics. The following topics related to the study of partial differential equations are treated in other chapters of this handbook:

Fourier series and Fourier integrals . . . . Chap. 4
Curvilinear coordinates . . . . Chap. 6
The Laplace transformation and other integral transformations . . . . Chap. 8
Calculus of variations . . . . Chap. 11
Boundary-value problems and eigenvalue problems . . . . Chap. 15
Numerical solutions . . . . Chap. 20
Special transcendental functions . . . . Chap. 21

10.6-2. References and Bibliography (see also Sec. 15.7-2).

10.1. Bers, L., F. John, and M. Schechter: *Partial Differential Equations*, Wiley, New York, 1963.
10.2. Bieberbach, L.: *Differentialgleichungen*, 2d ed., Springer, Berlin, 1965.
10.3. Churchill, R. V.: *Fourier Series and Boundary-value Problems*, 2d ed., McGraw-Hill, New York, 1963.
10.4. ———: *Operational Mathematics*, 2d ed., McGraw-Hill, New York, 1958.
10.5. Courant, R., and D. Hilbert: *Methoden der mathematischen Physik*, 3 vols., Wiley, New York, 1953/67.
10.6. Duff, G. F., and D. Naylor: *Differential Equations of Applied Mathematics*, Wiley, New York, 1965.
10.7. Epstein, B.: *Partial Differential Equations*, McGraw-Hill, New York, 1962.
10.8. Feshbach, H., and P. Morse: *Methods of Theoretical Physics*, McGraw-Hill, New York, 1953.
10.9. Frank, P., and R. Von Mises: *Die Differentialgleichungen der Mechanik und Physik*, 2d ed., M. S. Rosenberg, New York, 1943.
10.10. Friedman, B.: *Principles and Techniques of Applied Mathematics*, Wiley, New York, 1956.
10.11. Garabedian, P. R.: *Partial Differential Equations*, Wiley, New York, 1964.
10.12. Hopf, L.: *Introduction to the Partial Differential Equations of Physics*, Dover, New York, 1948.
10.13. Kamke, E.: *Differentialgleichungen, Lösungsmethoden und Lösungen*, vol. II, Chelsea, New York, 1948.
10.14. Lebedev, N. N., et al.: *Problems of Mathematical Physics*, Prentice-Hall, Englewood Cliffs, N.J., 1965.
10.15. Petrovsky, E. G.: *Lectures on Partial Differential Equations*, Interscience, New York, 1955.
10.16. Sagan, H.: *Boundary and Eigenvalue Problems in Mathematical Physics*, Wiley, New York, 1961.
10.17. Shapiro, A. H.: *The Dynamics and Thermodynamics of Compressible Fluid Flow*, vol. I, Ronald, New York, 1953.
10.18. Sneddon, I. N.: *Elements of Partial Differential Equations*, McGraw-Hill, New York, 1957.
10.19. ———: *Fourier Transforms*, McGraw-Hill, New York, 1951.

10.20. Sommerfeld, A.: *Partial Differential Equations in Physics*, Academic Press, New York, 1949.

10.21. Tranter, C. J.: *Integral Transforms in Mathematical Physics*, 2d ed., Wiley, New York, 1956.

10.22. Tychonov, A. N., and A. A. Samarski: *Partial Differential Equations in Mathematical Physics*, Holden-Day, San Francisco, 1964.

10.23. Margenau, H., and G. M. Murphy: *The Mathematics of Physics and Chemistry*, 2d ed., Van Nostrand, Princeton, N.J., 1952.

CHAPTER 11

MAXIMA AND MINIMA AND OPTIMIZATION PROBLEMS

11.1. INTRODUCTION

11.1-1. A large class of problems can be stated as *extreme-value problems:* one desires to find parameter values or functions which *maximize* or *minimize* a quantity dependent upon them. In many engineering problems it is, for instance, desirable to maximize a measure of performance or to minimize cost. Again, one can at least approximate the solutions of many problems by choosing unknown parameter values or functions so as to minimize errors in trial solutions; restatement of a problem as an extreme-value problem may then lead to powerful numerical approximation methods.

EXAMPLES: Solution of eigenvalue problems in vibration theory and quantum mechanics (Secs. 15.4-7 and 15.4-8*b*); Hamilton's and Jacobi's principles in dynamics.

11.2. MAXIMA AND MINIMA OF FUNCTIONS OF ONE REAL VARIABLE

11.2-1. Relative Maxima and Minima (see also Sec. 4.3-3). A real function $f(x)$ defined for $x = a$ has a (**relative**) **maximum** or a (**relative**) **minimum** $f(a)$ for $x = a$ if and only if there exists a positive real number δ such that, respectively,

$$\Delta f \equiv f(a + \Delta x) - f(a) < 0 \qquad \text{or} \qquad \Delta f \equiv f(a + \Delta x) - f(a) > 0$$

for all $\Delta x = x - a$ such that $f(a + \Delta x)$ exists and $0 < |\Delta x| < \delta$.* The relative maximum (minimum) is an **interior maximum (interior minimum)** or a **boundary maximum (boundary minimum)** if $x = a$ is, respectively, an interior point or a boundary point of the domain of definition assigned to $f(x)$ (Sec. 4.3-6*a*).†

11.2-2. Conditions for the Existence of Interior Maxima and Minima. (**a**) *If $f'(x)$ exists for $x = a$, then $f(a)$ can be a (necessarily interior) maximum or minimum only if $f(x)$ has a* **stationary value** *for $x = a$, i.e.,*

$$\boxed{f'(a) = 0} \qquad (11.2\text{-}1)$$

(**b**) *If $f(x)$ has a second derivative $f''(x)$ for $x = a$, then $f(a)$ is*

A maximum if $f'(a) = 0$ and $f''(a) < 0$
A minimum if $f'(a) = 0$ and $f''(a) > 0$

(**c**) More generally, *if $f(x)$ has n continuous derivatives $f'(x)$, $f''(x)$, $\ldots$, $f^{(n)}(x)$* for $x = a$, and $f'(a) = f''(a) = \cdots = f^{(n-1)}(a) = 0$, then *$f(a)$ is*

A maximum if n is even and $f^{(n)}(a) < 0$
A minimum if n is even and $f^{(n)}(a) > 0$

* Δf is defined as the change of a given function $f(x)$ resulting from change Δx in the independent variable x. Δf is a function of a and Δx. Δf must not be confused with the *variation* δf introduced in Sec. 11.5-1.

† The problem statement must specify the domain of definition of $f(x)$. Note that $f_1(x) = x(-\infty < x < \infty)$ has no maximum, but $f_2(x) = x$ $(x \leq 1)$ has a boundary maximum for $x = 1$.

If n is odd and $f^{(n)}(a) \neq 0$, then $f(x)$ has neither a maximum nor a minimum for $x = a$, but a *point of inflection* (Sec. 17.1-5).

EXAMPLES: Each of the functions $x^2, x^4, x^6, \ldots$ has a minimum for $x = 0$. Each of the functions $x^3, x^5, x^7, \ldots$ has a point of inflection for $x = 0$.

11.3. MAXIMA AND MINIMA OF FUNCTIONS OF TWO OR MORE REAL VARIABLES

11.3-1. Relative Maxima and Minima.* A real function $f(x_1, x_2, \ldots, x_n)$ defined for $x_1 = a_1, x_2 = a_2, \ldots, x_n = a_n$ has a **(relative) maximum** or a **(relative) minimum** $f(a_1, a_2, \ldots, a_n)$ for $x_1 = a_1, x_2 = a_2, \ldots, x_n = a_n$ if and only if there exists a positive real number δ such that

$$\Delta f \equiv f(a_1 + \Delta x_1, a_2 + \Delta x_2, \ldots, a_n + \Delta x_n) - f(a_1, a_2, \ldots, a_n) \tag{11.3-1}$$

is, respectively, less than zero or greater than zero for all $\Delta x_1, \Delta x_2, \ldots, \Delta x_n$ such that $f(a_1 + \Delta x_1, a_2 + \Delta x_2, \ldots, a_n + \Delta x_n)$ exists and $0 < \Delta x_1^2 + \Delta x_2^2 + \cdots + \Delta x_n^2 < \delta$. The relative maximum (minimum) is an **interior maximum (interior minimum)** or a **boundary maximum (boundary minimum)** if the point $(a_1, a_2, \ldots, a_n)$ is, respectively, an interior point or a boundary point of the region of definition assigned to $f(x_1, x_2, \ldots, x_n)$ (Sec. 4.3-6*a*).

11.3-2. Expansion of Δf. The quantity Δf defined by Eq. (1) is a function of $a_1, a_2, \ldots, a_n$ and $\Delta x_1, \Delta x_2, \ldots, \Delta x_n$. If $f(x_1, x_2, \ldots, x_n)$ is suitably differentiable,

$$\Delta f = \sum_{i=1}^{n} \frac{\partial f}{\partial x_i}\bigg]_{a_1, a_2, \ldots, a_n} \Delta x_i + \frac{1}{2} \sum_{i=1}^{n} \sum_{k=1}^{n} \frac{\partial^2 f}{\partial x_i \, \partial x_k}\bigg]_{a_1, a_2, \ldots, a_n} \Delta x_i \, \Delta x_k + \cdots \tag{11.3-2}$$

(Sec. 4.10-5). The terms of degree 1, 2, . . . in the Δx_i in this expansion respectively constitute the **first-order change (principal part** of Δf) $\Delta^1 f$, the **second-order change** $\Delta^2 f$, . . . of the function $f(x_1, x_2, \ldots, x_n)$, . . . for $x_1 = a_1, x_2 = a_2, \ldots, x_n = a_n$ (see also Sec. 4.5-3*b*).

11.3-3. Conditions for the Existence of Interior Maxima and Minima. (a) *If $f(x_1, x_2, \ldots, x_n)$ is differentiable for $x_1 = a_1, x_2 = a_2, \ldots, x_n = a_n$, $f(a_1, a_2, \ldots, a_n)$ can be a (necessarily interior) maximum or minimum only if the first-order change $\Delta^1 f$ vanishes, i.e., only if*

$$\boxed{\frac{\partial f}{\partial x_1} = 0 \qquad \frac{\partial f}{\partial x_2} = 0 \qquad \cdots \qquad \frac{\partial f}{\partial x_n} = 0} \tag{11.3-3}$$

* See footnotes to Sec. 11.2-1.

for $x_1 = a_1, x_2 = a_2, \ldots, x_n = a_n$. $f(x_1, x_2, \ldots, x_n)$ is then said to have a **stationary value** for $x_1 = a_1, x_2 = a_2, \ldots, x_n = a_n$.

(b) *If $f(x_1, x_2, \ldots, x_n)$ is twice differentiable and satisfies the necessary condition* (3) *for $x_1 = a_1, x_2 = a_2, \ldots, x_n = a_n$, then $f(a_1, a_2, \ldots, a_n)$ is a maximum if the (real symmetric) quadratic form*

$$\Delta^2 f = \frac{1}{2} \sum_{i=1}^{n} \sum_{k=1}^{n} \frac{\partial^2 f}{\partial x_i \, \partial x_k}\bigg]_{a_1, a_2, \ldots, a_n} \Delta x_i \, \Delta x_k \tag{11.3-4}$$

is negative definite (Sec. 13.5-2); and $f(a_1, a_2, \ldots, a_n)$ is a minimum if the quadratic form (4) *is positive definite.*

EXAMPLE: Find the maxima and minima of the function

$$z = 3x^3 - x + y^3 - 3y^2 - 1$$

Here the necessary conditions

$$\frac{\partial z}{\partial x} = 9x^2 - 1 = 0 \qquad \text{and} \qquad \frac{\partial z}{\partial y} = 3y^2 - 6y = 0$$

are satisfied for $x = \frac{1}{3}, y = 0; x = -\frac{1}{3}, y = 0; x = \frac{1}{3}, y = 2; x = -\frac{1}{3}, y = 2$. But $\frac{\partial^2 z}{\partial x^2} = 18x, \frac{\partial^2 z}{\partial y^2} = 6y - 6; \frac{\partial^2 z}{\partial x \, \partial y} = 0$, and inspection of the characteristic equation

$$(18x - \mu)(6y - 6 - \mu) = 0$$

shows that the only extreme values are

a *maximum* $(\mu_1 = -6, \mu_2 = -6)$, $z = -\frac{7}{9}$ for $x = -\frac{1}{3}$ and $y = 0$
and a *minimum* $(\mu_1 = 6, \mu_2 = 6)$, $z = -\frac{47}{9}$ for $x = \frac{1}{3}$ and $y = 2$

(c) If the first-order change $\Delta^1 f$ of a suitably differentiable function $f(x_1, x_2, \ldots, x_n)$ vanishes and $\Delta^2 f$ is *semidefinite* (Sec. 13.5-2), the nature of the stationary value depends on higher-order derivatives. $f(x_1, x_2, \ldots, x_n)$ cannot have a maximum or minimum where $\Delta^2 f$ exists and is *indefinite.*

Given a set of values $x_1 = a_1, x_2 = a_2, \ldots, x_n = a_n$ obtained from Eq. (3), one may investigate the nature of the quadratic form (4) by any of the methods listed in Sec. 13.5-6; or one may test for a maximum or minimum by actual computation of $f(a_1 + \Delta x_1, a_2 + \Delta x_2, \ldots, a_n + \Delta x_n)$ for judiciously chosen combinations of increments $\Delta x_1, \Delta x_2, \ldots, \Delta x_n$.

11.3-4. Extreme-value Problems with Constraints or Accessory Conditions. The Method of Lagrange Multipliers. Maxima and minima of a real function $f(x_1, x_2, \ldots, x_n)$ of n variables $x_1, x_2, \ldots, x_n$ subject to suitably differentiable *constraints* or *accessory conditions* in the form of $m < n$ equations

$$\varphi_1(x_1, x_2, \ldots, x_n) = 0 \qquad \varphi_2(x_1, x_2, \ldots, x_n) = 0 \qquad \cdots \qquad \varphi_m(x_1, x_2, \ldots, x_n) = 0 \tag{11.3-5}$$

may, in principle, be found in the manner of Sec. 11.3-3 after m of the n variables $x_1, x_2, \ldots, x_n$ have been eliminated with the aid of the relations (5). If it is impossible or impractical to eliminate m variables directly, one applies the following *necessary condition for a maximum or minimum of* $f(x_1, x_2, \ldots, x_n)$ *subject to the constraints* (5):

$$\frac{\partial \Phi}{\partial x_1} = \frac{\partial \Phi}{\partial x_2} = \cdots = \frac{\partial \Phi}{\partial x_n} = 0$$
$$\text{with} \quad \Phi(x_1, x_2, \ldots, x_n) \equiv f(x_1, x_2, \ldots, x_n) + \sum_{j=1}^{m} \lambda_j \varphi_j(x_1, x_2, \ldots, x_n) \tag{11.3-6}$$

The m parameters λ_j are called **Lagrange multipliers.** The $n + m$ unknowns $x_i = a_i$ and λ_j are obtained from the $n + m$ equations (5) and (6).

Note that Eq. (6) is a necessary condition for a stationary value of the function $\Phi(x_1, x_2, \ldots, x_n)$ if $x_1, x_2, \ldots, x_n$ are *independent* variables.

EXAMPLE: Find the sides of the rectangle of maximum area inscribed in the circle

$$x^2 + y^2 = r^2$$

The rectangle area A may be expressed in the form

$$A = 4xy$$

Then

$$\Phi(x, y) = 4xy + \lambda(x^2 + y^2 - r^2)$$

The necessary condition for a maximum or minimum yields

$$\frac{\partial \Phi}{\partial x} = 4y + 2\lambda x = 0 \qquad \text{and} \qquad \frac{\partial \Phi}{\partial y} = 4x + 2\lambda y = 0$$

so that $\lambda = -2$, and $x = y$ yields the desired maximum.

11.3-5. Numerical Methods. If, as is often the case, the function to be maximized or minimized depends on many variables, or if explicit differentiation is difficult or impossible, maxima and minima must be found by systematic numerical trial-and-error schemes. The most important numerical methods for problems with and without constraints are outlined in Secs. 20.2-6 and 20.3-2.

11.4. LINEAR PROGRAMMING, GAMES, AND RELATED TOPICS

11.4-1. Linear-programming Problems. (a) Problem Statement. A **linear-programming problem** requires one to determine the values of r variables $X_1, X_2, \ldots, X_r$ which minimize a given linear **objective function (criterion function)**

$$z = F(X_1, X_2, \ldots, X_r) \equiv C_1X_1 + C_2X_2 + \cdots + C_rX_r \tag{11.4-1a}$$

(or maximize $-z$) subject to $n \geq r$ linear **inequality constraints**

$$x_i \equiv A_{i1}X_1 + A_{i2}X_2 + \cdots + A_{ir}X_r - B_i \geq 0 \qquad (i = 1, 2, \ldots, n) \tag{11.4-1b}$$

Figure 11.4-1*a* illustrates a simple example.

In a typical application, the problem is to buy necessarily positive quantities $X_1, X_2, \ldots, X_r$ of r types of raw materials ("inputs") so as to minimize a given total cost (1*a*) while keeping the respective quantities

$$q_i \equiv A_{i1}X_1 + A_{i2}X_2 + \cdots + A_{ir}X_r \qquad (i = 1, 2, \ldots, m)$$

of m "output" products at or above m specified levels $B_1, B_2, \ldots, B_m$; in view of the r conditions $X_k \geq 0$ $(k = 1, 2, \ldots, r)$, one has a total of $n = r + m$ inequality

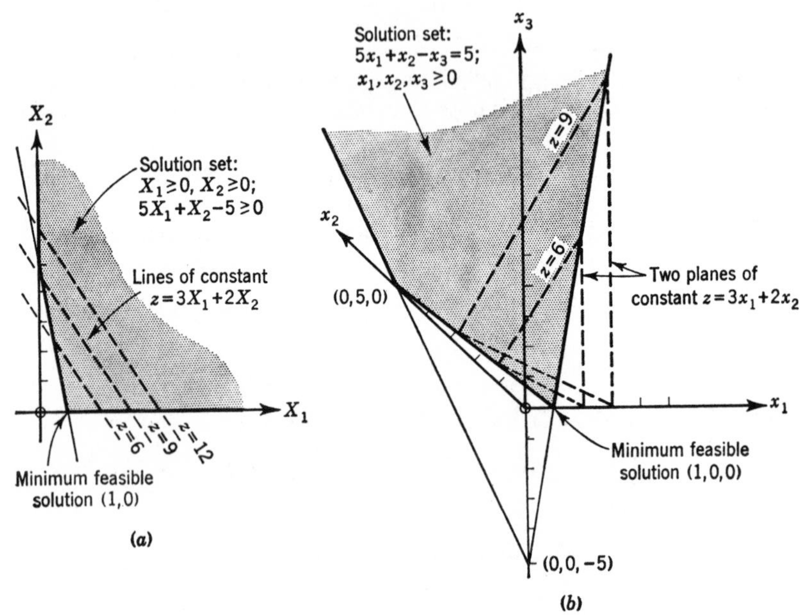

FIG. 11.4-1. Solution sets in (X_1,X_2) space (*a*), and in (x_1,x_2,x_3) space (*b*) for the linear-programming problem

$$z = 3X_1 + 2X_2 = \min$$

with $\qquad x_1 \equiv X_1 \geq 0 \qquad x_2 \equiv X_2 \geq 0 \qquad x_3 \equiv 5X_1 + X_2 - 5 \geq 0$

or $\qquad z = 3x_1 + 2x_2 = \min$

with $\qquad 5x_1 + x_2 - x_3 = 5$

$$x_1 \geq 0 \qquad x_2 \geq 0 \qquad x_3 \geq 0$$

In this case, $r = 2$, $n = 3$, $m = n - r = 1$.

constraints. A large number of applications, especially to management problems, are discussed in Refs. 11.2 to 11.10. Note that a given problem may or may not have a solution.

(b) The Linear-programming Problem in Standard Form. Feasible Solutions. Expressed in terms of the n positive **slack variables** (1*b*), the linear-programming problem (1) requires one to minimize the linear objective function

$$z = f(x_1, x_2, \ldots, x_n) \equiv c_1x_1 + c_2x_2 + \cdots + c_nx_n \qquad (11.4\text{-}2a)$$

subject to $m = n - r < n$ linear constraints

$$a_{i1}x_1 + a_{i2}x_2 + \cdots + a_{in}x_n = b_i \geq 0 \qquad (i = 1, 2, \ldots, m) \qquad (11.4\text{-}2b)$$

and n linear inequality constraints

$$x_k \geq 0 \qquad (k = 1, 2, \ldots, n) \qquad (11.4\text{-}2c)$$

(*standard form of the linear-programming problem*).

A **feasible solution (feasible program)** for a linear-programming problem given in the standard form (2) is an ordered set ("point") $(x_1, x_2, \ldots, x_n)$ which satisfies the constraints (2*b*) and (2*c*); a **minimum feasible solution** actually minimizes the given objective function (2*a*) and thus solves the problem. A **basic feasible solution** is a feasible solution with $N \leq m$ positive x_k; a basic feasible solution is **nondegenerate** if and only if $N = m$.

Whenever a feasible program exists, there exists one with at least $n - m$ of the x_k equal to zero. If, in addition, the objective function z has a lower bound on the solution set, then there exists a minimum feasible solution with at least $n - m$ of the x_k equal to zero (Ref. 11.2).

The set of all feasible solutions to a given linear-programming problem is a **convex set** *in the n-dimensional solution space*; i.e.; every point $(\xi_1, \xi_2, \ldots, \xi_n)$ on a straight-line segment joining two feasible-solution points $(x_1, x_2, \ldots, x_n)$ and $(x_1', x_2', \ldots, x_n')$, or

$$\xi_k = \alpha x_k + (1 - \alpha)x_k' \qquad (0 \leq \alpha \leq 1;\ k = 1, 2, \ldots, n)$$

(see also Sec. 3.1-7) is also a feasible solution. More specifically, *the solution set is a convex polygon or polyhedron* on the plane or hyperplane defined by the constraint equations (2*b*), with boundary lines, planes, or hyperplanes defined by the inequalities (2*c*), as shown in the simple example of Fig. 11.4-1*b*. *If a finite minimum feasible solution exists, the linear-programming problem either has a unique minimum feasible solution at a vertex of the solution polygon or polyhedron, or the same minimal z is obtained on the entire convex set generated by two or more vertices (degenerate solution).*

(c) Duality. The minimization problem

$$\left.\begin{aligned} &z = F(X_1, X_2, \ldots, X_r) \equiv C_1X_1 + C_2X_2 + \cdots + C_rX_r = \min \\ &\text{with} \quad x_i \equiv A_{i1}X_1 + A_{i2}X_2 + \cdots \\ &\qquad\qquad + A_{ir}X_r - B_i \geq 0 \qquad (i = 1, 2, \ldots, m) \\ &\qquad X_k \geq 0 \qquad (k = 1, 2, \ldots, r) \end{aligned}\right\} \qquad (11.4\text{-}3)$$

and the corresponding maximization problem

$$\left.\begin{aligned} w = G(Y_1, Y_2, \ldots, Y_m) \equiv B_1Y_1 + B_2Y_2 + \cdots \\ + B_mY_m = \max \\ \text{with} \quad y_k \equiv A_{1k}Y_k + A_{2k}Y_2 + \cdots \\ + A_{mk}Y_m - C_k \le 0 \qquad (k = 1, 2, \ldots, r) \\ -Y_i \le 0 \qquad (i = 1, 2, \ldots, m) \end{aligned}\right\} \tag{11.4-4}$$

are **dual linear-programming problems.** *If either problem* (3) *or* (4) *has a finite optimum solution, then the same is true for the other problem, and*

$$\min z = \max w \tag{11.4-5}$$

If either problem has an unbounded solution, then the other problem has no feasible solution (Ref. 11.2).

The duality theorem permits replacement of a given linear-programming problem with one having fewer unknowns or fewer given inequalities and is also useful in certain numerical solution methods (Ref. 11.5).

11.4-2. The Simplex Method. (**a**) Linear-programming problems can be solved by the numerical hill-climbing methods outlined in Sec. 20.2-6, but the **simplex method** (Refs. 11.2 to 11.6) takes advantage of the linearity of the expressions (1) and (2). Assuming that a solution of the linear-programming problem (2) exists, any solution-polygon vertex can be found through simultaneous solution of the m linear equations (2b) and $n - m$ linear equations $x_k = 0$. To avoid laborious investigation of an excessive number of nonminimal vertices, the simplex method provides a systematic computing scheme progressing from vertex to vertex in the direction of lower z values.

(**b**) To find a solution vertex, one selects m **basic variables,** say $x_1, x_2, \ldots, x_m$, which are to be greater than zero at the vertex. By systematic elimination of variables (Secs. 1.9-1 and 20.3-1), $x_1, x_2, \ldots, x_m$ can be expressed in terms of the remaining unknowns $x_{m+1}, x_{m+2}, \ldots, x_n$; then Eqs. (2$a$) and (2$b$) can be rewritten in *canonical form relative to the selected basic variables,* viz.,

$$\left.\begin{aligned} x_1 + (\alpha_{1,m+1}x_{m+1} + \cdots + \alpha_{1n}x_n) &= \beta_1 \\ x_2 + (\alpha_{2,m+1}x_{m+1} + \cdots + \alpha_{2n}x_n) &= \beta_2 \\ \cdots\cdots\cdots\cdots\cdots\cdots\cdots\cdots \\ x_m + (\alpha_{m,m+1}x_{m+1} + \cdots + \alpha_{mn}x_n) &= \beta_m \\ z = (\gamma_{m+1}x_{m+1} + \cdots + \gamma_nx_n) + z_0 \end{aligned}\right\} \tag{11.4-6}$$

Now $x_{m+1} = x_{m+2} = \cdots = x_n = 0$ defines a basic feasible solution (Sec. 11.4-1b) if all β_i in Eq. (6) are nonnegative. *If, in addition, all γ_k are nonnegative, then*

$$x_i = \begin{cases} \beta_i & (i = 1, 2, \ldots, m) \\ 0 & (i = m + 1, m + 2, \ldots, n) \end{cases} \tag{11.4-7}$$

is a minimum feasible solution with $z = z_0$; this solution is unique if all γ_k are positive.

(**c**) Consider next the case that Eq. (6) yields a nondegenerate basic feasible solution (i.e., $\beta_1, \beta_2, \ldots, \beta_m > 0$, see also Sec. 11.4-1*b*), which is not optimal, i.e., at least one γ_k is negative. Let γ_K be the smallest (most negative) γ_k. *To obtain a basic feasible solution with a smaller z value, x_K is then introduced as a basic variable in the next iteration cycle; x_K will replace that former basic variable x_I which, referring to Eq.* (6), *reaches zero first as x_K is increased from zero*, i.e., I is associated with the *smallest* possible

$$x_K = \frac{\beta_I}{\alpha_{IK}} \tag{11.4-8a}$$

Then an improved basic feasible solution is given by (8*a*) and

$$\left.\begin{aligned} x_i &= \begin{cases} \beta_i - \alpha_{iK}x_K \geq 0 & (i = 1, 2, \ldots, m;\ \text{note } x_I = 0) \\ 0 & (i = m+1, m+2, \ldots, n \text{ if } i \neq K) \end{cases} \\ z &= z_0 + \gamma_K x_K = z_0 + \gamma_K \frac{\beta_I}{\alpha_{IK}} < z_0 \end{aligned}\right\} \tag{11.4-8b}$$

Now, introduction of

$$x_K = \frac{\beta_I}{\alpha_{IK}} - \frac{1}{\alpha_{IK}}\left(x_I + \sum_{\substack{j=m+1 \\ (j \neq K)}}^{n} \alpha_{Ij}x_j\right) \tag{11.4-9}$$

into Eq. (6) produces the canonical form relative to the new basic variables. The complete simplex algorithm, which permits convenient tabulation (**simplex tableau,** Refs. 11.5 and 11.6) and also adapts readily to machine computation, is repeated and usually progresses to an optimal solution in a finite number of steps. If, in the course of the computation, the basic variables correspond to a *degenerate* basic feasible solution (i.e., if at least one of the basic variables turns out to be zero, see also Sec. 11.4-1*b*), further improvement of the objective function z could, in principle, stop, but this rarely if ever occurs in a practical problem (Ref. 11.5). One can extend the simplex method to degenerate cases, e.g., by introducing small changes in the given coefficients during the computation (perturbation method) (Ref. 11.15). Many practical refinements of the basic simplex method appear in the literature (Refs. 11.2 to 11.7).

NOTE: The simplex algorithm necessarily fails when all α_{iK} are nonpositive; in this case, the objective function z has no lower bound on the solution set.

EXAMPLE (Fig. 11.4-2): The problem

$$z = \tfrac{1}{4}X_1 + \tfrac{1}{3}X_2 = \min$$

with

$$\begin{aligned} x_1 &\equiv X_1 \geq 0 \\ x_2 &\equiv 5X_1 + X_2 - 5 \geq 0 \\ x_3 &\equiv 2X_1 + 5X_2 - 10 \geq 0 \\ x_4 &\equiv X_2 \geq 0 \end{aligned}$$

transforms to the standard form

$$z = \tfrac{1}{4}x_1 + \tfrac{1}{3}x_4 = \min$$

with

$$\begin{aligned} 5x_1 - x_2 + x_4 &= 5 > 0 \\ 2x_1 - x_3 + 5x_4 &= 10 > 0 \\ x_1, x_2, x_3, x_4 &\geq 0 \end{aligned}$$

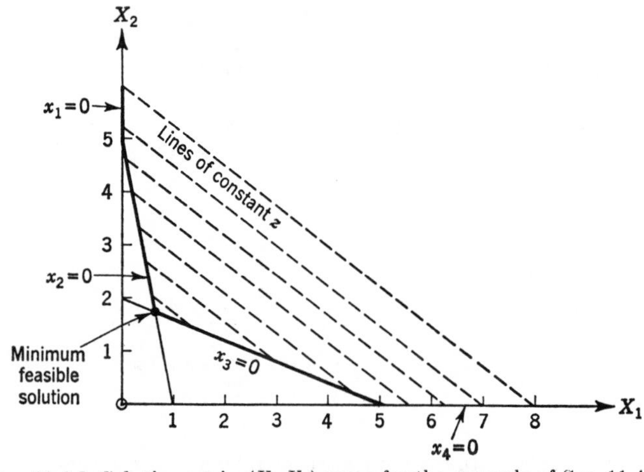

FIG. 11.4-2. Solution set in (X_1,X_2) space for the example of Sec. 11.4-2*c*.

Starting with x_1, x_2 as basic variables (first vertex from the right in Fig. 11.4-2), one obtains the canonical form

$$\begin{aligned} x_1 - \tfrac{1}{2}x_3 + \tfrac{5}{2}x_4 &= 5 > 0 \\ x_2 - \tfrac{5}{2}x_3 + \tfrac{23}{2}x_4 &= 20 > 0 \\ 24z &= 3x_3 - 7x_4 + 30 \end{aligned}$$

The coefficient of x_4 in the last expression is negative; x_4 is, therefore, increased until x_2 reaches zero (this takes place before x_1 reaches zero and corresponds to the second vertex from the right in Fig. 11.4-2). Thus, $K = 4$, $I = 2$, and

$$x_4 = \frac{20}{23/2} = \frac{40}{23} \qquad x_1 = \frac{15}{23} \qquad x_2 = x_3 = 0$$

The new canonical form relative to x_1, x_4 is

$$\begin{aligned} x_1 - \tfrac{5}{23}x_2 + \tfrac{1}{23}x_3 &= \tfrac{15}{23} > 0 \\ x_4 + \tfrac{2}{23}x_2 - \tfrac{5}{23}x_3 &= \tfrac{40}{23} > 0 \\ 12 \times 23z &= 7x_2 + 17x_3 + 205 \end{aligned}$$

which corresponds to the unique minimum feasible solution with

$$z = 205/(12 \times 23) = 205/276.$$

(d) Use of Artificial Variables to Start the Simplex Procedure. The simplex algorithm, as described, presupposes that a suitable choice of the initial basic variables $x_1, x_2, \ldots, x_m$ has produced a basic feasible solution with $\beta_1, \beta_2, \ldots, \beta_m \geq 0$. Such a choice may not be obvious. The following procedure may be used to start the solution, or to decide whether a feasible solution exists.

With the given problem in the standard form (2), introduce m **artificial variables** $x_{n+1}, x_{n+2}, \ldots, x_{n+m}$ and solve the "augmented" linear-programming problem minimizing

$$z + w \equiv z + x_{n+1} + x_{n+2} + \cdots + x_{n+m} \tag{11.4-10a}$$

subject to the constraints

$$\left.\begin{array}{r} a_{11}x_1 + a_{12}x_2 + \cdots + a_{1n}x_n + x_{n+1} = b_1 \geq 0 \\ a_{21}x_1 + a_{22}x_2 + \cdots + a_{2n}x_n + x_{n+2} = b_2 \geq 0 \\ \cdots\cdots\cdots\cdots\cdots\cdots\cdots\cdots \\ a_{m1}x_1 + a_{m2}x_2 + \cdots + a_{mn}x_n + x_{n+m} = b_m \geq 0 \\ x_k \geq 0 \qquad (k = 1, 2, \ldots, n + m) \end{array}\right\} \tag{11.4-10b}$$

The m artificial variables necessarily define a basic feasible solution of the augmented problem; the original problem has no solution unless the minimum of the **feasibility form** $w = x_{n+1} + x_{n+2} + \cdots + x_{n+m}$ is zero.

11.4-3. Nonlinear Programming. The Kuhn-Tucker Theorem. If the objective function and/or one or more inequalities in the linear-programming problem (1) is replaced by an expression nonlinear in the problem variables X_k, one has a *nonlinear-programming problem.* Such a problem results, for example, if the solution-set boundaries and/or the lines of constant z in Fig. 11.4-1a are replaced by nonlinear curves. Nonlinear-programming problems are of considerable practical interest but are, with few exceptions (Ref. 11.11), accessible only to numerical solution methods (see also Sec. 20.2-6).

Assuming suitable differentiability, a necessary (not sufficient) condition for a maximum of

$$z = f(x_1, x_2, \ldots, x_n) \tag{11.4-11}$$

subject to m inequality constraints

$$\phi_i(x_1, x_2, \ldots, x_n) \leq 0 \qquad (i = 1, 2, \ldots, m)$$

is the existence of $m + 1$ *nonnegative Lagrange multipliers* $\lambda_0, \lambda_1, \lambda_2, \ldots, \lambda_m$ *(see also* Sec. 11.3-4) *not all equal to zero and such that*

$$\left.\begin{aligned} &\lambda_i \geq 0 && (i = 0, 1, 2, \ldots, m) \\ &\lambda_i \phi_i = 0 && (i = 1, 2, \ldots, m) \\ &\lambda_0 \frac{\partial f}{\partial x_k} + \sum_{i=1}^{m} \lambda_i \frac{\partial \phi_i}{\partial x_k} = 0 && (k = 1, 2, \ldots, n) \end{aligned}\right\} \tag{11.4-12a}$$

(Fritz John Theorem). λ_0 *is positive and can be set equal to unity in the condition* (12*a*) *whenever there exists a set of n real numbers* $(y_1, y_2, \ldots, y_n)$ *such that, for the values of* $x_1, x_2, \ldots, x_n$ *in question,*

$$\sum_{k=1}^{n} \frac{\partial \phi_i}{\partial x_k} y_k \begin{cases} \leq 0 \text{ if } \phi_i \text{ is a linear function in a neighborhood of } (x_1, x_2, \ldots, x_n) \\ < 0 \text{ otherwise} \end{cases}$$

$$(i = 1, 2, \ldots, m) \tag{11.4-12b}$$

(Abadie's form of the Kuhn-Tucker Theorem, Ref. 11.1).

11.4-4. Introduction to Finite Zero-sum Two-person Games. (a) Games with Pure Strategies. A **finite zero-sum two-person game** is a model of a conflict situation characterized by a finite **payoff matrix,**

$$\begin{bmatrix} a_{11} & a_{12} & \cdots & a_{1m} \\ a_{21} & a_{22} & \cdots & a_{2m} \\ \cdot & \cdot & \cdots & \cdot \\ a_{n1} & a_{n2} & \cdots & a_{nm} \end{bmatrix} \tag{11.4-13}$$

where a_{ik} is the (positive or negative) payment due player A from player B if A chooses the ith of n **pure strategies** $S_1, S_2, \ldots, S_n$ available to him, and B chooses the kth of m pure strategies $S'_1, S'_2, \ldots, S'_m$ possible for him. Neither player knows the other's choice. Note that the sum of the payoffs to both players is zero for each move (hence the name zero-sum game). The game is **symmetrical** if and only if $m = n$, and $a_{ki} = -a_{ik}$ for all i, k.

To win, the **maximizing player** A selects the ith row of the payoff matrix so as to maximize $\min_k a_{ik}$, while the **minimizing player** B attempts to minimize $\max_i a_{ik}$. *For every given payoff matrix* (13)

$$\max_i \min_k a_{ik} \leq \min_k \max_i a_{ik} \tag{11.4-14}$$

If the two quantities in Eq. (14) are *equal* for a (not necessarily unique) pair $i = I, k = K$, the game is said to have the **saddle point** or **solution** I, K and the (necessarily unique) **value** a_{IK}. The optimal strategies for such a game are unaffected if player A knows B's move beforehand, and vice versa.

(b) Games with Mixed Strategies. **Mixed strategies** are defined by *probabilities* (Sec. 18.2-2) $p_1, p_2, \ldots, p_n$ assigned by player A to his n respective strategies $S_1, S_2, \ldots, S_n$, and probabilities $p'_1, p'_2, \ldots, p'_m$ assigned by player B to his strategies $S'_1, S'_2, \ldots, S'_m$. In a mixed-strategy game, A attempts to maximize the minimum of the *expected value* (Sec. 18.3-3)

$$\min_{p_1', p_2', \ldots, p_m'} \sum_{i=1}^{n} \sum_{k=1}^{m} a_{ik} p_i p'_k$$

through his choice of $p_1, p_2, \ldots, p_n$, while B tries to minimize

$$\max_{p_1, p_2, \ldots, p_n} \sum_{i=1}^{n} \sum_{k=1}^{m} a_{ik} p_i p'_k$$

But *for every given payoff matrix* (11),

$$\max_{p_1, p_2, \ldots, p_n} \left[\min_{p_1', p_2', \ldots, p_m'} \sum_{i=1}^{n} \sum_{k=1}^{m} a_{ik} p_i p'_k \right]$$

$$= \min_{p_1', p_2', \ldots, p_m'} \left[\max_{p_1, p_2, \ldots, p_n} \sum_{i=1}^{n} \sum_{k=1}^{m} a_{ik} p_i p'_k \right] \qquad (11.4\text{-}15)$$

(*Minimax Theorem for finite zero-sum two-person games*). The common value of the two quantities (15) is called the **value** v of the game. *Every finite zero-sum two-person game has at least one* **solution** *defined by the optimal strategies* $p_1, p_2, \ldots, p_n; p'_1, p'_2, \ldots, p'_m$. Multiple solutions are possible, but the value of the game is necessarily unique. A **fair** zero-sum two-person game has the value 0.

EXAMPLE: The familiar *penny-matching game* has the payoff matrix

$$\begin{bmatrix} 0 & 1 \\ 1 & 0 \end{bmatrix}$$

where strategy 1 for each player is a bet on heads, and strategy 2 is a bet on tails. The game is symmetrical and fair and has no saddle point. The solution is $p_1 = p'_1 = \frac{1}{2}$, $p_2 = p'_2 = \frac{1}{2}$.

(c) Relation to Linear Programming. While a number of approximation methods yielding solutions to mixed-strategy games have been developed (Ref. 11.10), the most generally applicable approach relates the problem to linear programming.

The solution of a given matrix game is unchanged by addition of a positive constant a to each a_{ik}, so that it is not a restriction to consider only games with positive values $v > 0$. In this case, *the optimal strategies* $p_1, p_2, \ldots, p_n$ and $p'_1, p'_2, \ldots, p'_m$ *and the value* v *for a finite zero-sum two-person game with the payoff matrix* (11) *are given by*

$$\left.\begin{aligned} p_i &= vX_i \qquad (i = 1, 2, \ldots, n) \\ p'_k &= vY_k \qquad (k = 1, 2, \ldots, m) \\ v &= \frac{1}{z_{\max}} = \frac{1}{w_{\min}} \end{aligned}\right\} \tag{11.4-16}$$

where $z_{\max}$, $w_{\min}$, and the X_i and Y_k are determined by the solution of the dual linear-programming problems (Sec. 11.4-1)

$$\left.\begin{aligned} z \equiv X_1 + X_2 + \cdots + X_n &= \min \\ \text{with}\quad a_{1k}X_1 + a_{2k}X_2 + \cdots + a_{nk}X_n &\geq 1 \\ (k = 1, 2, &\ldots, m) \\ X_i \geq 0 \qquad (i = 1, 2, &\ldots, n) \end{aligned}\right\} \tag{11.4-17a}$$

$$\left.\begin{aligned} w = Y_1 + Y_2 + \cdots + Y_m &= \max \\ \text{with}\quad a_{i1}Y_1 + a_{i2}Y_2 + \cdots + a_{im}Y_m &\leq 1 \\ (i = 1, 2, &\ldots, n) \\ -Y_k \leq 0 \qquad (k = 1, 2, &\ldots, m) \end{aligned}\right\} \tag{11.4-17b}$$

11.5. CALCULUS OF VARIATIONS. MAXIMA AND MINIMA OF DEFINITE INTEGRALS

11.5-1. Variations. (a) A **variation** δy of a function $y(x)$ of x is a function of x defined for each value of x as the difference $\delta y \equiv Y(x) - y(x)$ between a new function $Y(x)$ and $y(x)$.

Each variation δy defines *a change in the functional relationship of y and x* and must not be confused with a change Δy in the *value* of a given function $y(x)$ due to a change Δx in the independent variable (Sec. 11.2-1).

(b) Given a function $F[y_1(x), y_2(x), \ldots, y_n(x); x]$, the variation δF corresponding to the respective variations $\delta y_1, \delta y_2, \ldots, \delta y_n$ of the functions $y_1(x), y_2(x), \ldots, y_n(x)$ is

$$\delta F \equiv F(y_1 + \delta y_1, y_2 + \delta y_2, \ldots, y_n + \delta y_n; x) - F(y_1, y_2, \ldots, y_n; x) \tag{11.5-1}$$

If $y(x)$ and δy are differentiable, the variation $\delta y'$ of the derivative $y'(x)$ due to the variation δy is

$$\delta y' \equiv \delta \frac{dy}{dx} \equiv \frac{d}{dx}(\delta y) \equiv Y'(x) - y'(x) \tag{11.5-2}$$

More generally, given $F[y_1(x), y_2(x), \ldots, y_n(x); y'_1(x), y'_2(x), \ldots, y'_n(x); x]$,

$$\delta F \equiv F(y_1 + \delta y_1, y_2 + \delta y_2, \ldots, y_n + \delta y_n; y'_1 + \delta y'_1, y'_2 + \delta y'_2, \ldots, y'_n + \delta y'_n; x) - F(y_1, y_2, \ldots, y_n; y'_1, y'_2, \ldots, y'_n; x) \tag{11.5-3}$$

(c) If F is suitably differentiable, Eq. (3) can be expanded in the form

$$\delta F \equiv \sum_{i=1}^{n} \left\{ \frac{\partial F}{\partial y_i} \delta y_i + \frac{\partial F}{\partial y_i'} \delta y_i' \right\} + \frac{1}{2} \sum_{i=1}^{n} \sum_{k=1}^{n} \left\{ \frac{\partial^2 F}{\partial y_i \, \partial y_k} \delta y_i \, \delta y_k + \frac{\partial^2 F}{\partial y_i \, \partial y_k'} \delta y_i \, \delta y_k' + \frac{\partial^2 F}{\partial y_k' \, \partial y_k'} \delta y_i' \, \delta y_k' \right\} + \cdots \quad (11.5\text{-}4)$$

for each value of x (Sec. 4.10-5). The terms of degree 1, 2, . . . in the δy_i and $\delta y_i'$ in the Taylor expansion (4) respectively constitute the **first-order variation** $\delta^1 F$ of F, the **second-order variation** $\delta^2 F$ of F, ...

(d) The definitions and formulas of Secs. 11.5-1*a*, *b*, and *c* apply without change to functions y, y_i, and F of two or more independent variables $x_1, x_2, \ldots$.

11.5-2. Maxima and Minima of Definite Integrals. While the theory of ordinary maxima and minima (Secs. 11.2-1 to 11.3-5) is concerned with *unknown values of independent variables* x or x_i corresponding to maxima and minima of given functions, it is the objective of the **calculus of variations** to find *unknown functions* $y(x)$ *or* $y_i(x)$ *which will maximize or minimize definite integrals* like

$$I = \int_{x_0}^{x_F} F[y(x), y'(x), x] \, dx \quad (11.5\text{-}5a)$$

or

$$I = \int_{x_0}^{x_F} F[y_1(x), y_2(x), \ldots, y_n(x); y_1'(x), y_2'(x), \ldots, y_n'(x); x] \, dx \quad (11.5\text{-}5b)$$

for a specified function F. I is a functional (Sec. 12.1-4) determined by the function(s) $y(x)$ or $y_i(x)$ together with the integration limits x_1, x_2 and the boundary values $y(x_0)$, $y(x_F)$ or $y_i(x_0)$, $y_i(x_F)$.

A definite integral (5) has a **strong maximum** for given integration limits x_0, x_F and a given function $y(x)$ or a given set of functions $y_i(x)$ if and only if there exists a positive real number ϵ such that the variation

$$\delta I = \delta \int_{x_0}^{x_F} F \, dx = \int_{x_0}^{x_F} \delta F \, dx + \int_{x_0 + \delta x_0}^{x_0} (F + \delta F) \, dx + \int_{x_F}^{x_F + \delta x_F} (F + \delta F) \, dx \quad (11.5\text{-}6)$$

is negative for all increments δx_0, δx_F and all variations δy or δy_1, δy_2, . . . , δy_n whose absolute values are less than ϵ and not identically zero. A **strong minimum** of a definite integral (5) is similarly defined by $\delta I > 0$.

The integral I has a **weak maximum** (or a **weak minimum**) for given x_0, x_F and $y(x)$ or $y_1(x)$, $y_2(x)$, . . . , $y_n(x)$ if and only if there exists a positive real number ϵ such that $\delta I < 0$ (or $\delta I > 0$, respectively) only for all increments δx_0, δx_F and all variations δy or δy_1, δy_2, . . . ,

δy_n *and* $\delta y'$ or $\delta y'_1, \delta y'_2, \ldots, \delta y'_n$ whose absolute values are less than ϵ and not identically zero. A strong maximum (or minimum) is necessarily also a weak maximum (or minimum).

If the region of definition for "admissible" functions $y(x)$ or $y_1(x)$, $y_2(x), \ldots, y_n(x)$ is bounded through inequalities such as

$$y(x) \leq a \qquad y_1^2(x) + y_2^2(x) \leq b \qquad f(y_1, y_2, \ldots, y_n) \geq 0$$

a function $y(x)$ or $y_i(x)$ maximizing or minimizing I can lie wholly in the interior of the region of definition (**interior maximum or minimum**), or wholly or partially on its boundary (**boundary maximum or minimum**, see also Sec. 11.6-6).

In most applications, the maximizing or minimizing functions $y(x)$ or $y_i(x)$ need not be compared with all possible functions of x. **In the following, the existence of the definite integrals in question is assumed wherever necessary, and it is understood that** (1) **maximizing or minimizing functions are to be chosen from the set of all functions having piecewise continuous first derivatives on the interval or region under consideration. In addition, it will be assumed that** (2) **each integrand** F **is twice continuously differentiable on the integration domain** (see also Sec. 11.6-1*c*).

11.5-3. Solution of Variation Problems. Functions $y(x)$ or $y_i(x)$ which maximize or minimize a given definite integral may be found (1) *as solutions of differential equations* which ensure that $\delta^1 F \equiv 0$ (Secs. 11.6-1 to 11.6-7), or (2) *by the "direct" methods* described in Secs. 11.7-1 to 11.7-3.

It is not a trivial observation that *a given problem of the type discussed here may not possess any solution.* Every solution derived with the aid of the necessary (not sufficient) conditions of Secs. 11.6-1 to 11.6-7 *must be tested for actual maximum or minimum properties.* Some *sufficient* conditions for the existence of maxima or minima of definite integrals are discussed in Sec. 11.6-9.

11.6. EXTREMALS AS SOLUTIONS OF DIFFERENTIAL EQUATIONS: CLASSICAL THEORY

11.6-1. Necessary Conditions for the Existence of Maxima and Minima. (a) *A necessary condition for the existence of either a maximum or a minimum of the definite integral*

$$I = \int_{x_0}^{x_F} F[y(x), y'(x), x]\, dx \tag{11.6-1}$$

for fixed x_0, x_F *is*

$$\delta^1 I = \int_{x_0}^{x_F} \delta^1 F\, dx = \frac{\partial F}{\partial y'}\, \delta y \Big]_{x_0}^{x_F} - \int_{x_0}^{x_F} \left[\frac{d}{dx}\left(\frac{\partial F}{\partial y'}\right) - \frac{\partial F}{\partial y} \right] \delta y\, dx = 0$$

for an arbitrary small variation δy. *Hence every maximizing or minimizing function* $y(x)$ *must satisfy the differential equation*

$$\frac{d}{dx}\left(\frac{\partial F}{\partial y'}\right) - \frac{\partial F}{\partial y} = 0 \qquad \text{(EULER-LAGRANGE EQUATION)} \tag{11.6-2}$$

wherever the quantity on the left exists and is continuous (see also Sec. 11.6-1*c*). In addition, $y(x)$ must either assume given boundary values $y(x_0)$ and/or $y(x_F)$, or $y(x)$ must satisfy other conditions determining its boundary values (Sec. 11.6-5).

(**b**) Similarly, *every set of n functions* $y_1(x), y_2(x), \ldots, y_n(x)$ *maximizing or minimizing the definite integral*

$$I = \int_{x_0}^{x_F} F[y_1(x), y_2(x), \ldots, y_n(x); y_1'(x), y_2'(x), \ldots, y_n'(x); x]\, dx \tag{11.6-3}$$

must satisfy the set of n differential equations

$$\frac{d}{dx}\left(\frac{\partial F}{\partial y_i'}\right) - \frac{\partial F}{\partial y_i} = 0 \qquad (i = 1, 2, \ldots, n) \qquad \text{(EULER-LAGRANGE EQUATIONS)} \tag{11.6-4}$$

together with suitably given boundary conditions, wherever all the quantities on the left exist and are continuous (see also Sec. 11.6-1*c*).

(**c**) Functions $y(x)$ or $y_i(x)$ satisfying the Euler equation or equations associated with a given variation problem are called **extremals** for the problem in question.

A further necessary (still not sufficient) condition for a maximum or minimum of I on a given extremal is that the matrix $[\partial^2 F/\partial y_i'\,\partial y_k']$ *be, respectively, negative or positive semidefinite* (Sec. 13.5-2) *on the extremal* (*Legendre's Condition*); this reduces to $\partial F/\partial y'^2 \leq 0$ or $\partial^2 F/\partial y'^2 \geq 0$, respectively, in the one-dimensional case (see also Secs. 11.6-7 and 11.6-10).

In general, the necessary conditions of Secs. 11.6-1*a* and *b* apply only where the $y(x)$ or $y_i(x)$ are twice continuously differentiable, but this is not as restrictive a condition as it might seem. *Let the integrand F be twice continuously differentiable for* $x_0 < x < x_F$. *Then all continuous differentiable functions* $y(x)$ *or* $y_i(x)$ *which actually maximize or minimize I for given* x_0, x_F *and given boundary values necessarily have continuous second-order derivatives for all x in* (x_0, x_F) *where the matrix* $[\partial^2 F/\partial y_i'\,\partial y_k']$ *is negative or positive definite* (Sec. 13.5-2), *or* $\partial^2 F/\partial y'^2 \neq 0$ (*Theorem of Du Bois-Reymond*).

(**d**) The derivation of the conditions (2) and (4) from $\delta^1 I = 0$ is based on the *fundamental lemma of the calculus of variations: if* $f(x)$ *is continuous on the bounded interval* $[x_0, x_F]$ *and* $\int_{x_0}^{x_F} f(x)g(x)\,dx = 0$ *for arbitrary* $g(x)$, *then* $f(x) = 0$ on $[x_0, x_F]$; and on *Du Bois-Reymond's Lemma: a continuous function* $f(x)$ *is necessarily a constant in* (x_0, x_F) *if* $\int_{x_0}^{x_F} f(x)g(x)\,dx = 0$ *for every* $g(x)$ *such that* $\int_{x_0}^{x} g(x)\,dx$ *is a continuous function in* (x_0, x_F), *and* $\int_{x_0}^{x_F} g(x)\,dx = 0$.

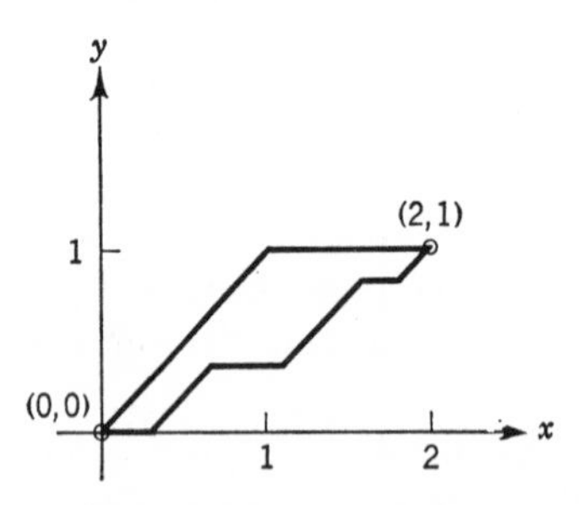

FIG. 11.6-1a. Two of the infinite set of broken-line extremals minimizing $I = \int_0^2 (1 - y')^2 y'^2 \, dx$ with $y(0) = 0$, $y(2) = 1$. The Euler-Lagrange equation yields extremal segments $y = ax + b$, with a and b determined by the boundary conditions and a Weierstrass-Erdmann condition for each corner; it follows that a is either 0 or 1. Note that $\partial^2 F/\partial y'^2 = 12y'^2 - 12y' + 2$ (with undetermined y') can vanish anywhere in (0, 2), but $\partial^2 F/\partial y'^2 > 0$ on each minimizing extremal. Note also that no continuously differentiable extremal yields a smaller value of I (Refs. 11.19 and 11.22).

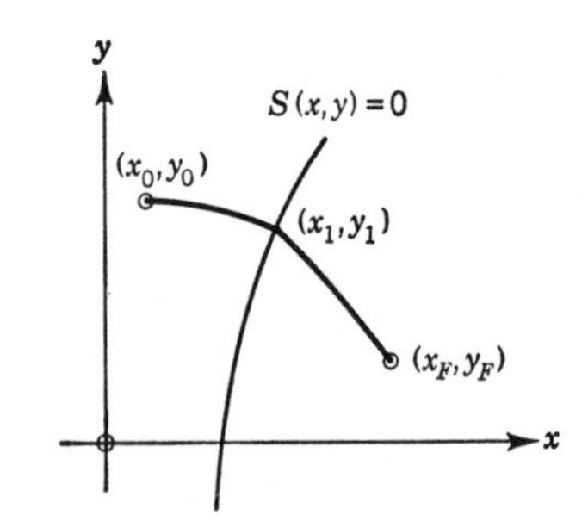

FIG. 11.6-1b. Refraction of extremals.

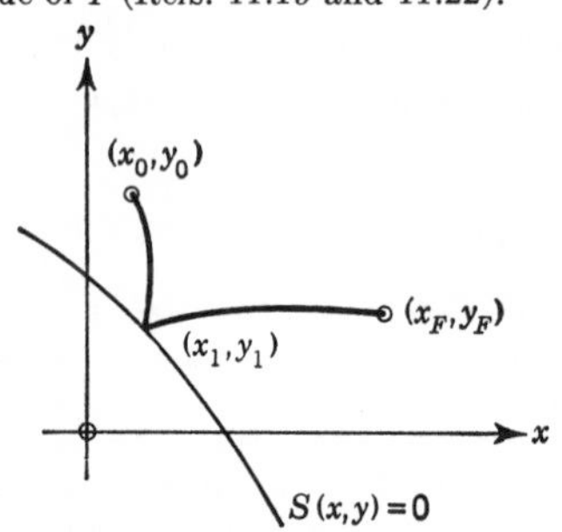

FIG. 11.6-1c. Reflection of extremals from a boundary curve.

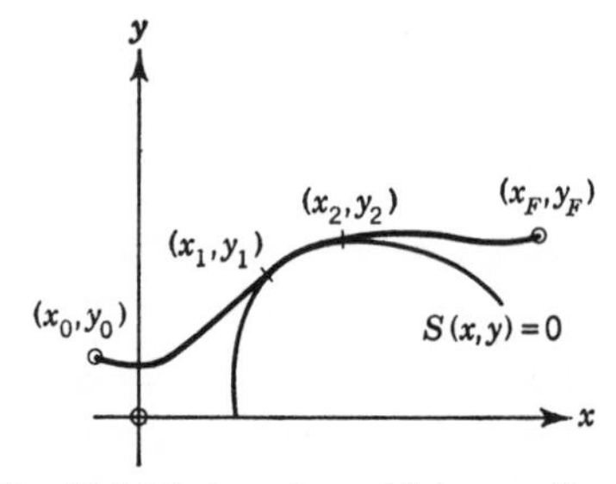

FIG. 11.6-1d. An extremal lying partly on a boundary curve.

FIG. 11.6-1. Extremals with corners.

EXAMPLES OF APPLICATIONS INVOLVING VARIATION PROBLEMS: geodesics in Riemann spaces (Sec. 17.4-3); derivation of Lagrange's equations from Hamilton's principle. See also Sec. 15.4-7.

EXAMPLE: *Brachistochrone in Three Dimensions.* Given a three-dimensional rectangular cartesian coordinate system with the positive x axis vertically downward, this classical problem requires the determination of the space curve $y = y(x)$, $z = z(x)$ which will minimize the time t taken by a particle sliding on the curve without friction under the action of gravity to get from the origin to some specified point $[x_F > 0, y(x_F), z(x_F)]$. Since by conservation of energy

$$gx = \frac{1}{2}v^2 = \frac{1}{2}\left[\left(\frac{dx}{dt}\right)^2 + \left(\frac{dy}{dt}\right)^2 + \left(\frac{dz}{dt}\right)^2\right]$$

where g is the acceleration of gravity, the quantity to be minimized is

$$t\sqrt{2g} = \int_0^{x_F} \frac{1}{x^{1/2}} \sqrt{1 + (y')^2 + (z')^2}\, dx$$

The Euler equations (4) become

$$\frac{d}{dx}\left(\frac{y'}{x^{1/2}\sqrt{1 + (y')^2 + (z')^2}}\right) = \frac{d}{dx}\left(\frac{z'}{x^{1/2}\sqrt{1 + (y')^2 + (z')^2}}\right) = 0$$

or

$$\frac{y'}{x^{1/2}\sqrt{1 + (y')^2 + (z')^2}} = c_1 \qquad \frac{z'}{x^{1/2}\sqrt{1 + (y')^2 + (z')^2}} = c_2$$

where c_1 and c_2 are constants. Since $\frac{dy}{dz} = \frac{c_1}{c_2}$, the curve must lie in a vertical plane, which will be the xy plane if the boundary conditions

$$y(x_F) = y_F \qquad z(x_F) = 0$$

are assumed. In this case, $z' = 0$ and

$$y' = \frac{x}{\sqrt{\frac{x}{c_1^2} - x^2}}$$

so that

$$y = a \arccos\left(1 - \frac{x}{a}\right) - \sqrt{2ax - x^2} + k \qquad \left(a = \frac{1}{2c_1^2}\right)$$

The constant k must vanish, since $y = 0$ for $x = 0$; the constant a will depend on the values of x_F and y_F. The desired curve represents a cycloid (Sec. 2.6-2*e*) with its base along the y axis and its cusp at the origin.

NOTE: The problem of the brachistochrone, like the problem of Sec. 11.6-3, is an example of the minimization of a *line integral* (Sec. 4.6-10). Problems of this type can always be reduced to problems involving integration over a single independent variable.

11.6-2. Variation Problems with Constraints or Accessory Conditions. The Method of Lagrange Multipliers. It is desired to find sets of functions $y_1(x), y_2(x), \ldots, y_n(x)$ which maximize or minimize the definite integral (3) while also subject to $m < n$ suitably differentiable *constraints* or *accessory conditions*

$$\varphi_j(y_1, y_2, \ldots, y_n; x) = 0 \qquad (j = 1, 2, \ldots; m < n) \qquad (11.6\text{-}5)$$

Unless it is possible to eliminate m of the n variables y_i directly with the aid of the relations (5), one obtains the desired sets of functions $y_1(x)$, $y_2(x), \ldots, y_n(x)$ as solutions of the set of differential equations (Euler equations)

$$\frac{d}{dx}\left(\frac{\partial \Phi}{\partial y_i'}\right) - \frac{\partial \Phi}{\partial y_i} = 0 \qquad (i = 1, 2, \ldots, n) \qquad (11.6\text{-}6)$$

subject to the constraints (5), where

$$\Phi \equiv F + \sum_{j=1}^{m} \lambda_j(x)\varphi_j \qquad (11.6\text{-}7)$$

The unknown functions $\lambda_j(x)$ are called **Lagrange multipliers** (see also Sec. 11.3-4). If the problem has a solution, the $n + m$ functions y_i and λ_j can be determined from the $n + m$ relations (6) and (5) together with the given boundary conditions. The differential equations (6) are necessary (but not sufficient) conditions for a maximum or minimum, provided that all the quantities on the left exist and are continuous.

The method of Lagrange multipliers remains applicable in the case of suitably differentiable *nonholonomic constraints* of the form

$$\varphi_j(y_1, y_2, \ldots, y_n; y_1', y_2', \ldots, y_n'; x) = 0 \qquad (j = 1, 2, \ldots, m) \qquad (11.6\text{-}8)$$

11.6-3. Isoperimetric Problems. An *isoperimetric problem* requires one to find sets of functions $y_1(x), y_2(x), \ldots, y_n(x)$ which will maximize or minimize a definite integral (3) subject to the accessory conditions

$$\int_{x_0}^{x_F} \Psi_k(y_1, y_2, \ldots, y_n; y_1', y_2', \ldots, y_n'; x)\, dx = c_k \qquad (k = 1, 2, \ldots, m') \qquad (11.6\text{-}9)$$

where the c_k are given constants. If the given functions Ψ_k are suitably differentiable, the method of Lagrange multipliers applies; the unknown functions $y_i(x)$ must satisfy a set of differential equations (6) subject to the constraints (9), where now

$$\Phi \equiv F + \sum_{k=1}^{m'} \mu_k \Psi_k \qquad (11.6\text{-}10)$$

The m' Lagrange multipliers μ_k are *constants* to be determined together with the n unknown functions $y_i(x)$ from the $m' + n$ relations (9) and (6) and the given boundary conditions.

EXAMPLE: The area o a closed plane curve $x = x(t)$, $y = y(t)$ can be written in the form

$$I = \frac{1}{2}\int_0^{2\pi} \left(x\frac{dy}{dt} - y\frac{dx}{dt}\right) dt$$

if the parameter t is suitably chosen. To maximize I subject to the accessory condition

$$\int_0^{2\pi} \left[\left(\frac{dx}{dt}\right)^2 + \left(\frac{dy}{dt}\right)^2\right] dt = R^2$$

where R is a constant different from zero, write

$$\Phi \equiv \frac{1}{2}\left(x\frac{dy}{dt} - y\frac{dx}{dt}\right) + \mu\left[\left(\frac{dx}{dt}\right)^2 + \left(\frac{dy}{dt}\right)^2\right]$$

The resulting Euler equations (6), viz.,

$$-\frac{dy}{dt} + 2\mu\frac{d^2x}{dt^2} = 0 \qquad \frac{dx}{dt} + 2\mu\frac{d^2y}{dt^2} = 0$$

and the given accessory conditions are satisfied by

$$x = R\cos\frac{t}{2\mu} + x_0 \qquad y = -R\sin\frac{t}{2\mu} + y_0 \qquad \text{with } \mu = \sqrt{\frac{\pi}{2}}$$

i.e., the desired curve is a circle of radius R (see also Sec. 11.7-1).

NOTE: To maximize or minimize the definite integral (3) subject to m accessory conditions (5) *and* r accessory conditions (9), apply the conditions (6) with

$$\Phi \equiv F + \sum_{j=1}^{m} \lambda_j \varphi_j + \sum_{k=1}^{m'} \mu_k \Psi_k$$

11.6-4. Solution of Variation Problems Involving Higher-order Derivatives in the Integrand. To maximize or minimize definite integrals of the form

$$I = \int_{x_1}^{x_2} F(y_1, y_2, \ldots, y_n; y_1', y_2', \ldots, y_n'; y_1'', y_2'', \ldots, y_n''; \ldots; x)\, dx \tag{11.6-11}$$

one may introduce all derivatives of order higher than one as new dependent variables related to one another and to the y_i by constraints $y_i'' = dy_i'/dx$, $y_i''' = dy_i''/dx$, ... The resulting necessary conditions for a maximum or minimum of the definite integral (11) take the form

$$\frac{\partial F}{\partial y_i} - \frac{d}{dx}\left(\frac{\partial F}{\partial y_i'}\right) + \frac{d^2}{dx^2}\left(\frac{\partial F}{\partial y_i''}\right) - \frac{d^3}{dx^3}\left(\frac{\partial F}{\partial y_i'''}\right) \pm \cdots = 0 \qquad (i = 1, 2, \ldots, n) \tag{11.6-12}$$

provided that the integrand F is suitably differentiable, and all the quantities on the left exist and are continuous. In order to solve the problem completely, one must also specify the boundary values of all but the highest derivatives of each function y_i appearing in the integrand.

11.6-5. Variation Problems with Unknown Boundary Values and/ or Unknown Integration Limits. (a) Given Integration Limits, Unknown Boundary Values. To maximize or minimize the definite integral (3) when one or more of the boundary values $y_i(x_0)$ and/or $y_i(x_F)$ are not specified or constrained (Sec. 11.6-5*c*), replace each missing boundary condition by a corresponding *natural boundary condition*

$$\frac{\partial F}{\partial y_i'} = 0 \qquad (x = x_0 \text{ and/or } x = x_F) \tag{11.6-13}$$

(b) Given Boundary Values, Unknown Integration Limit. If one of the integration limits, say x_F, is unknown but all boundary values $y_i(x_F)$ are given, then the extremals $y_i(x)$ must satisfy

$$\sum_{k=1}^{n} \frac{\partial F}{\partial y_k'} y_k' - F = 0 \qquad (x = x_F) \tag{11.6-14}$$

(c) General Transversality Conditions. Frequently, an integration limit, say x_F, and/or one or more of the corresponding boundary values $y_i(x_F)$ will not be given explicitly, but the "point" $[x_F; y_1(x_F),$

$y_2(x_F), \ldots, y_n(x_F)]$ is known to satisfy n_F given continuously differentiable relations

$$G_j[x_F; y_1(x_F), y_2(x_F), \ldots, y_n(x_F)] = 0 \qquad (j = 1, 2, \ldots, n_F \leq n) \tag{11.6-15}$$

The integration limit x_F and the unknown functions $y_i(x)$ which together maximize or minimize the integral (3) must then satisfy the Euler equations (4) and the boundary conditions (15) together with a set of *transversality conditions* expressed by

$$\frac{\partial F}{\partial y_i'} + \sum_{j=1}^{n_F} \Lambda_j \frac{\partial G_j}{\partial y_i} = 0 \qquad (x = x_F) \tag{11.6-16a}$$

for each $y_i(x_F)$ which is not given explicitly, and/or

$$\left[\sum_{k=1}^{n} \frac{\partial F}{\partial y_k'} y_k' - F\right] - \sum_{j=1}^{n_F} \Lambda_j \frac{\partial G_j}{\partial x_F} = 0 \qquad (x = x_F) \tag{11.6-16b}$$

if x_F is not explicitly known. The Λ_j are n_F constant Lagrange multipliers to be determined, together with up to $n + 1$ unknowns $y_i(x_F)$ and/or x_F, from the $n + n_F + 1$ relations (15) and (16). Note that Eqs. (13) and (14) are special cases of Eqs. (16*a*) and (16*b*).

(d) Analogous methods apply if the boundary conditions corresponding to $x = x_0$ are not completely specified.

EXAMPLE: To minimize a line integral

$$I = \int_{t_0}^{t_F} G[x(t), y(t), z(t)]\, ds = \int_{t_0}^{t_F} G[x(t), y(t), z(t)] \sqrt{\left(\frac{dx}{dt}\right)^2 + \left(\frac{dy}{dt}\right)^2 + \left(\frac{dz}{dt}\right)^2}\, dt \qquad [G(x, y, z) > 0]$$

where (x_0, y_0, z_0) and (x_F, y_F, z_F) lie on two suitable (convex) nonintersecting regular surfaces S_0, S_F specified by

$$B_0(x_0, y_0, z_0) = 0 \qquad G_F(x_F, y_F, z_F) = 0$$

the solution extremal must satisfy the transversality conditions

$$\frac{dx}{dt} : \frac{dy}{dt} : \frac{dz}{dt} = \begin{cases} \dfrac{\partial B_0}{\partial x_0} : \dfrac{\partial B_0}{\partial y_0} : \dfrac{\partial B_0}{\partial z_0} & (t = t_0) \\[2ex] \dfrac{\partial G_F}{\partial x_F} : \dfrac{\partial G_F}{\partial y_F} : \dfrac{\partial G_F}{\partial z_F} & (t = t_F) \end{cases}$$

i.e., it must intersect S_0 and S_F orthogonally. If, in particular, $G(x, y, z) \equiv 1$, then I is a distance between the surfaces, and the extremals are straight lines. The actual existence of a minimum depends on the specific nature (convexity) of the given surfaces.

11.6-6. The Problems of Bolza and Mayer. A functional of the form

$$J = \int_{x_0}^{x_F} F[y_1(x), y_2(x), \ldots, y_n(x); y_1'(x), y_2'(x), \ldots, y_n'(x); x]\, dx + h[y_1(x_F)y_2(x_F), \ldots, y_n(x_F); x_F] \tag{11.6-17}$$

to be maximized by a suitable choice of functions $y_i(x)$ (*Problem of Bolza*), subject to suitably given boundary conditions of the form (15), can, for suitably differentiable h, be written as

$$J = \int_{x_0}^{x_F} (F + h')\,dx \tag{11.6-18}$$

$h = h(x)$ is an additional dependent variable satisfying the differential-equation constraint

$$\frac{dh}{dx} = h' \equiv \sum_{k=1}^{n} \frac{\partial h}{\partial y_k} y'_k + \frac{\partial h}{\partial x} \qquad (x_0 < x < x_F) \tag{11.6-19}$$

and the boundary conditions

$$h(0) = 0 \qquad h(x_F) = h[y_1(x_F), y_2(x_F), \ldots, y_n(x_F); x_F] \tag{11.6-20}$$

The $y_i(x)$ must satisfy the usual Euler-Lagrange equations for $x_0 < x < x_F$.

An analogous procedure is used if a function of the boundary values corresponding to $x = x_0$ is added to the given expression (17).

If the $y_i(x)$ are subject to given differential-equation constraints but $F \equiv 0$, so that only the boundary function h remains to be maximized or minimized, one has the *Problem of Mayer*, which will be considered in the context of optimum-control theory in Secs. 11.8-1 to 11.8-6.

11.6-7. Extremals with Corners. Refraction, Reflection, and Boundary Extrema. The functions $y(x)$ or $y_i(x)$ maximizing or minimizing a definite integral (1) or (3) were assumed to be only piecewise continuously differentiable (Sec. 11.5-2). They can have *corners* (discontinuities in first-order derivatives) for values of x such that $\partial^2 F/\partial y'^2 = 0$, or where the matrix $[\partial^2 F/\partial y'_i\,\partial y'_k]$ is only semidefinite for some y, y' or $y_1, y_2, \ldots, y_n; y'_1, y'_2, \ldots, y'_n$, or where F has a discontinuity (Secs. 11.6-1 and 13.5-2). Corners may, in particular, occur

1. For any x over an interval of x values, as illustrated by the example of Fig. 11.6-1*a*.
2. On a curve, surface, or hypersurface

$$S(x, y) = 0 \qquad \text{or} \qquad S(x; y_1, y_2, \ldots, y_n) = 0 \tag{11.6-21}$$

crossed by the extremal curves ("refraction" of extremals, Fig. 11.6-1*b*).

3. On the boundary of a region from which the extremals are excluded by an inequality constraint

$$S(x, y) \le 0 \qquad \text{or} \qquad S(x; y_1, y_2, \ldots, y_n) \le 0 \tag{11.6-22}$$

(Fig. 11.6-1*c* and *d*).

At every "free" corner $[x_1, y(x_1)]$ *or* $[x_1; y_1(x_1), y_2(x_1), \ldots, y_n(x_1)]$ *between extremals* (Fig. 11.6-1*a*, but not *b*, *c*, *d*), *the extremal segments must satisfy*

$$\left.\begin{aligned} &F_{y'}(x_1 - 0) = F_{y'}(x_1 + 0) \qquad [F_{y'}y' - F]_{x=x_1-0} = [F_{y'}y' - F]_{x=x_1+0} \\ \text{or} \quad &F_{y_i'}(x_1 - 0) = F_{y_i'}(x_1 + 0) \qquad (i = 1, 2, \ldots, n) \\ &\left[\sum_{k=1}^{n} F_{y_k'}y'_k - F\right]_{x=x_1-0} = \left[\sum_{k=1}^{n} F_{y_k'}y'_k - F\right]_{x=x_1+0} \end{aligned}\right\}$$

(WEIERSTRASS-ERDMANN CORNER CONDITIONS) (11.6-23)

For refraction (Fig. 11.6-1*b*) and reflection (Fig. 11.6-1*c*) of extremals, the curve, surface, or hypersurface defined by Eq. (21) or (22) acts like an intermediate boundary where each extremal must satisfy Eq. (21) together with the corner conditions

$$\left.\begin{aligned} F_{y_i'}(x_1+0) - F_{y_i'}(x_1-0) &= -\Lambda\left\{\frac{\partial S}{\partial y_i}\bigg]_{x_1+0} - \frac{\partial S}{\partial y_i}\bigg]_{x_1-0}\right\} \\ &\qquad (i = 1, 2, \ldots, n) \\ \left[\sum_{k=1}^{n} \frac{\partial F}{\partial y_k'} y_k' - F\right]_{x=x_1+0} - \left[\sum_{k=1}^{n} \frac{\partial F}{\partial y_k'} y_k' - F\right]_{x=x_1-0} & \\ = \Lambda\left\{\frac{\partial S}{\partial x}\bigg]_{x_1+0} - \frac{\partial S}{\partial x}\bigg]_{x_1-0}\right\} & \end{aligned}\right\} \qquad (11.6\text{-}24)$$

where Λ is a constant Lagrange multiplier. In each case, the corner conditions supplement the given boundary conditions to determine the corner points.

In the boundary-extremum case illustrated by Fig. 11.6-1*d*, *extremals must enter and leave the boundary along a tangent to the boundary* (see also Sec. 11.8-5).

NOTE: In the special case $I = \int_{t_0}^{t_F} G(x, y, z)\, ds$ with $ds^2 = dx^2 + dy^2 + dz^2$, the corner conditions yield the refraction and reflection laws of optics if the function $G(x, y, z)$ is interpreted as the reciprocal of the propagation velocity (Ref. 11.21).

11.6-8. Canonical Equations and Hamilton-Jacobi Equation. The n second-order Euler equations (4) associated with the variation problem of Sec. 11.6-1*b* are equivalent to the $2n$ first-order *canonical equations*

$$y_i' = \frac{\partial H}{\partial p_i} \qquad p_i' = -\frac{\partial H}{\partial y_i} \qquad (i = 1, 2, \ldots, n) \qquad (11.6\text{-}25a)$$

in the $2n$ dependent variables y_i and $p_i = \dfrac{\partial F}{\partial y_i'}$, where

$$\left.\begin{aligned} &H(y_1, y_2, \ldots, y_n; p_1, p_2, \ldots, p_n; x) \\ &\qquad \equiv \sum_{i=1}^{n} p_i y_i' - F(y_1, y_2, \ldots, y_n; y_1', y_2', \ldots, y_n'; x) \\ &\det\left[\frac{\partial^2 F}{\partial y_i'\, \partial y_k'}\right] \neq 0 \end{aligned}\right\} \qquad (11.6\text{-}25b)$$

(contact transformation, see also Secs. 10.2-6 and 10.2-7).

Note that the transversality conditions (13), (14), (16), *and the corner conditions* (23), (24) *are greatly simplified if one introduces the p_i and H.* In classical dynamics, the **canonically conjugate** or **adjoint variables** p_i are interpreted as generalized momenta, and the **Hamiltonian function** H has the dimensions of energy. Sections 11.8-1 to 11.8-6 restate the calculus of variations in terms of canonical equations in the context of optimum-control theory.

One can sometimes simplify the solution of the problem by introducing $2n + 1$ new variables $\bar{y}_i$, $\bar{p}_i$, and $\bar{H}(\bar{y}_1, \bar{y}_2, \ldots, \bar{y}_n; \bar{p}_1, \bar{p}_2, \ldots, \bar{p}_n; x)$ through a suitable *canonical transformation* (Sec. 10.2-6) which yields new canonical equations

$$\bar{y}_i' = \frac{\partial \bar{H}}{\partial \bar{p}_i} \qquad \bar{p}_i' = -\frac{\partial \bar{H}}{\partial \bar{y}_i} \qquad (i = 1, 2, \ldots, n)$$

(b) Assuming that the minimum

$$\begin{aligned}\min_{y_1(x),y_2(x),\ldots,y_n(x)} I &= \min \int_{x_F}^{x_0} F(y_1, y_2, \ldots, y_n; y_1', y_2', \ldots, y_n'; x)\,dx \\ &\equiv S[x_0, y_1(x_0), y_2(x_0), \ldots, y_n(x_0); x_F, y_1(x_F), y_2(x_F), \ldots, y_n(x_F)] \\ &\equiv S[x_0, y_1(x_0), y_2(x_0), \ldots, y_n(x_0)] \equiv S(X, Y_1, Y_2, \ldots, Y_n)\end{aligned}$$

defines unique optimal extremals for fixed x_F, $y_i(x_F)$, and variable $x_0 = X$, $y_i(x_0) = Y_i$ $(i = 1, 2, \ldots, n)$, the function $S(X, Y_1, Y_2, \ldots, Y_n)$ satisfies the first-order partial differential equation

$$\frac{\partial S}{\partial X} + H\left(Y_1, Y_2, \ldots, Y_n; \frac{\partial S}{\partial Y_1}, \frac{\partial S}{\partial Y_2}, \ldots, \frac{\partial S}{\partial Y_n}; X\right) = 0$$

(HAMILTON-JACOBI EQUATION) (11.6-26)

The ordinary differential equations (25*a*) are the corresponding characteristic equations (see also Secs. 10.2-4 and 11.8-7).

11.6-9. Variation Problems Involving Two or More Independent Variables: Maxima and Minima of Multiple Integrals. One often wishes to determine a set of n functions $y_1(x_1, x_2, \ldots, x_m)$, $y_2(x_1, x_2, \ldots, x_m)$, $\ldots$, $y_n(x_1, x_2, \ldots, x_m)$ of m independent variables $x_1, x_2, \ldots, x_m$ so as to maximize or minimize a multiple integral

$$I = \iint_V \cdots \int F\,dx_1\,dx_2 \cdots dx_m \qquad (11.6\text{-}27)$$

where the integrand F is a given function of the $n + m + mn$ variables y_i, x_k, and $y_{i,k} \equiv \partial y_i/\partial x_k$. The boundary S of the region of integration V and the boundary values of the functions y_i may or may not be given. The functions y_i are assumed to be chosen from the set of all functions having continuous second partial derivatives in V (see also Sec. 11.5-2).

If the integrand F is suitably differentiable, *every set of maximizing or minimizing functions* $y_1, y_2, \ldots, y_n$ *must satisfy the set of partial differential equations*

$$\sum_{k=1}^{m} \frac{\partial}{\partial x_k}\left(\frac{\partial F}{\partial y_{i,k}}\right) - \frac{\partial F}{\partial y_i} = 0 \qquad (i = 1, 2, \ldots, n) \qquad (11.6\text{-}28)$$

together with suitable boundary conditions, whenever the quantities on the left exist and are continuous. The Lagrange-multiplier methods of Secs. 11.6-2 and 11.6-3 apply to variation problems involving accessory conditions.

If the boundary values of one or more of the unknown functions y_i are not given, one must use the condition $\delta^1 I = 0$ to derive *natural boundary conditions* analogous to Eq. (13). Thus in the case of two independent variables, x_1, x_2, let the boundary S be a regular closed curve described in terms of its arc length s by $x_1 = x_1(s)$, $x_2 = x_2(s)$;

then the natural boundary conditions are

$$\frac{\partial F}{\partial y_{i,1}}\frac{dx_2}{ds} - \frac{\partial F}{\partial y_{i,2}}\frac{dx_1}{ds} = 0 \qquad (i = 1, 2, \ldots, n) \tag{11.6-29}$$

The case of an unknown boundary is treated in Ref. 11.21.

EXAMPLE: The small displacements of a string of length L may be represented by a function $y(x, t)$ of the coordinate x measured along the string and of the time t. The kinetic and potential energies of the entire string are, respectively,

$$T = \frac{m}{2}\int_0^L y_t^2\,dx \qquad U = \frac{Q}{2}\int_0^L y_x^2\,dx$$

where m is the mass per unit length, and Q is the string tension. For a minimum of the integral

$$\int_{t_0}^{t_F} (T - U)\,dt = \tfrac{1}{2}\int_{t_0}^{t_F}\int_0^L (my_t^2 - Qy_x^2)\,dt\,dx$$

(Hamilton's principle, Ref. 11.22), one must have

$$y_{tt} = \frac{Q}{m}y_{xx}$$

which is the correct wave equation for the vibrating string.

11.6-10. Simple Sufficient Conditions for Maxima and Minima. Consider a one-parameter family of extremals [solutions of the Euler-Lagrange equation (2)] $y = y(x, \mu)$ for the definite integral

$$I = \int_{x_0}^{x_F} F(x, y, y')\,dx \tag{11.6-30}$$

with $y(x_0, \mu) = y_0$ for a parameter range $\mu_1 < \mu < \mu_2$ such that

$$\frac{\partial y(x, \mu)}{\partial \mu} \equiv z(x, \mu)$$

exists. The extremals corresponding to this parameter range all go through the point (x_0, y_0); they will not intersect again in (x_0, x_F) if $z(x, \mu)$ satisfies the differential equation

$$\left(\frac{d}{dx}F_{yy'} - F_{yy}\right)z + \frac{d}{dx}(F_{y'y'}z') = 0 \qquad \text{(JACOBI'S CONDITION)} \tag{11.6-31}$$

in (x_0, x_F) [**extremal field** centered at (x_0, y_0)]. The function

$$P(x, y) \equiv y'(x, \mu)$$

obtained through elimination of μ with the aid of $y = y(x, \mu)$ represents the slope of the extremal field at a point (x, y).

An extremal $y(x, \mu)$ *of the field satisfying Eqs.* (2) *and* (31) *maximizes the given integral* (30) *if the function*

$$E(x, y, y', P) \equiv F(x, y, y') - F(x, y, P) - (y' - P)F_{y'}(x, y, P)$$
$$\text{(WEIERSTRASS } E \text{ FUNCTION)} \tag{11.6-32}$$

is nonpositive for all points (x, y) *sufficiently close to* $y(x, \mu)$*; this is, in particular, true if*

$$F_{y'y'} < 0 \qquad \text{(LEGENDRE'S STRONG CONDITION)} \tag{11.6-33}$$

along the field extremal in question. The maximum is a *strong* maximum (Sec. 11.5-2) if $E \leq 0$ for all (x, y) sufficiently close to the extremal curve and *arbitrary* y'.

$E \leq 0$ along the extremal for every finite y' (*Weierstrass's necessary condition*) is necessary for a strong maximum, but is *not* sufficient by itself. Conditions for a *minimum* of the integral (30) similarly correspond to $E \geq 0$. The multidimensional case is treated in Refs. 11.16 and 11.17.

11.7. SOLUTION OF VARIATION PROBLEMS BY DIRECT METHODS

11.7-1. Direct Methods. Each so-called *direct method* for maximizing or minimizing a given definite integral

$$I = \int_{x_0}^{x_F} F[y(x), y'(x), x]\, dx \qquad (11.7\text{-}1)$$

attempts to approximate the desired function $y(x)$ successively by a sequence of functions $u_1(x)$, $u_2(x)$, . . . selected so as to satisfy the boundary conditions imposed on $y(x)$. Each function $u_r(x)$ is taken to be a differentiable function of x and of r parameters $\alpha_{r1}, \alpha_{r2}, \ldots, \alpha_{rr}$. The latter are then chosen so as to maximize or minimize the *function*

$$I_r(\alpha_{r1}, \alpha_{r2}, \ldots, \alpha_{rr}) \equiv \int_{x_0}^{x_F} F[u_r(x), u_r'(x), x]\, dx \qquad (11.7\text{-}2)$$

with the aid of the relations

$$\frac{\partial I_r}{\partial \alpha_{ri}} = 0 \qquad (i = 1, 2, \ldots, r;\ r = 1, 2, \ldots) \qquad (11.7\text{-}3)$$

(Sec. 11.3-3*a*).

Every tentative solution $y(x)$ obtained in this manner as the limit of a sequence of approximating functions $u_1(x)$, $u_2(x)$, . . . still requires a proof that the definite integral (1) *is actually a maximum or minimum.*

Analogous methods of solution apply to variation problems involving more than one unknown function (Sec. 11.6-1*b*) and/or more than one independent variable (Sec. 11.6-9). If accessory conditions are given (Secs. 11.6-2 and 11.6-3), they are made to apply to each approximating function; the approximating integrals may then be maximized or minimized by the Lagrange-multiplier method of Sec. 11.3-4 (see example).

Direct methods can yield numerical approximations and/or exact solutions. Since, under the conditions listed in Sec. 11.5-2, the solutions of variation problems must satisfy differential equations expressing the condition $\delta^1 I = 0$, *every direct method for the solution of a variation problem is also an approximation method for solving differential equations.*

EXAMPLE: To solve the isoperimetric problem given as an example in Sec. 11.6-3,

$$u_r(t) \equiv \tfrac{1}{2}a_0 + \sum_{k=1}^{r} (a_k \cos kt + b_k \sin kt)$$

$$v_r(t) \equiv \tfrac{1}{2}\alpha_0 + \sum_{k=1}^{r} (\alpha_k \cos kt + \beta_k \sin kt) \qquad (r = 1, 2, \ldots)$$

so that

$$I_r = \frac{1}{2}\int_0^{2\pi}\left(u_r\frac{dv_r}{dt} - v_r\frac{du_r}{dt}\right)dt = \pi\sum_{k=1}^{r} k(a_k\beta_k - b_k\alpha_k) \qquad (r = 1, 2, \ldots)$$

To maximize I_r subject to the accessory condition

$$\int_0^{2\pi}\left[\left(\frac{du_r}{dt}\right)^2 + \left(\frac{dv_r}{dt}\right)^2\right]dt = \pi\sum_{k=1}^{r} k^2(a_k^2 + b_k^2 + \alpha_k^2 + \beta_k^2) = R^2$$

one applies the Lagrange-multiplier method of Sec. 11.3-4 and finds

$$\left.\begin{aligned} \beta_k + 2\lambda_r k a_k &= 0 \qquad & a_k + 2\lambda_r k\beta_k &= 0 \\ -\alpha_k + 2\lambda_r k b_k &= 0 \qquad & -b_k + 2\lambda_r k\alpha_k &= 0 \end{aligned}\right\} \qquad (k = 1, 2, \ldots, r)$$

for $r = 1, 2, \ldots$. Starting with $k = 1$, one obtains $\lambda_r = \frac{1}{2}$, $\beta_1 = -a_1$, $\alpha_1 = b_1$; for $k > 1$, $a_k = b_k = \alpha_k = \beta_k = 0$. In this example, the direct method yields the exact solution

$$x = \tfrac{1}{2}a_0 + a_1\cos t + b_1\sin t \qquad y = \tfrac{1}{2}\alpha_0 - b_1\cos t + a_1\sin t$$

with

$$a_1^2 + b_1^2 = R^2$$

11.7-2. The Rayleigh-Ritz Method. This method attempts to expand the desired solution $y(x)$ as a series in terms of a complete set of functions $\Psi_j(x)$ (Sec. 15.2-4) so that the approximating functions $u_r(x) \equiv \sum_{j=0}^{r} \alpha_{rj}\Psi_j(x)$ satisfy the given boundary conditions. As a rule, the $\Psi_j(x)$ are orthogonal functions, so that the parameters $\alpha_{rj} = \alpha_j$ are independent of r (Sec. 15.2-4), just like in the example of Sec. 11.7-1.

The Rayleigh-Ritz method is useful for numerical solution of certain eigenvalue problems in vibration theory and quantum mechanics (see also Sec. 15.4-7).

11.7-3. Approximation of $y(x)$ by Polygonal Functions. $y(x)$ may be approximated by polygonal functions $u_r(x)$ each defined, say, by its r values $\alpha_{r1} = u_r(x_0 + \Delta x)$, $\alpha_{r2} = u_r(x_0 + 2\Delta x), \ldots, \alpha_{rr} = u_r(x_0 + r\,\Delta x) = u_r(x_F - \Delta x)$. In this case, the conditions (3) yield a difference equation (Sec. 20.4-3) approximating the Euler equation (11.6-2) of the given variation problem.

11.8. CONTROL PROBLEMS AND THE MAXIMUM PRINCIPLE

11.8-1. Problem Statement. (a) State Equations, Controls, and Criterion. In *control problems*, the state of a dynamical (mechanical, electrical, chemical, etc.) system is represented by n **state variables** $x_1(t), x_2(t), \ldots, x_n(t)$ satisfying n first-order differential equations

$$\frac{dx_i}{dt} = f_i(x_1, x_2, \ldots, x_n; u_1, u_2, \ldots, u_r)$$
$$(i = 1, 2, \ldots, n) \qquad \text{(STATE EQUATIONS)} \qquad (11.8\text{-}1)$$

Typical state variables are generalized coordinates and velocities in mechanics, electric currents and voltages, and concentrations of chemicals (Sec. 11.8-3); the independent variable t is usually the time. *The problem is to determine the r* **control variables (controls)** $u_k = u_k(t)$ $(k = 1, 2, \ldots, r)$ *as functions of t for $t_0 \leq t \leq t_F$ so as to minimize a given* **criterion functional**

$$x_0(t_F) = \int_{t_0}^{t_F} f_0(x_1, x_2, \ldots, x_n; u_1, u_2, \ldots, u_r)\, dt \qquad (11.8\text{-}2)$$

(e.g., cost, mean-square error, time to achieve a task, etc.) subject to inequality constraints

$$Q_j(u_1, u_2, \ldots, u_r) \leq 0 \qquad (j = 1, 2, \ldots, N) \qquad (11.8\text{-}3)$$

defining the closed **domain of admissible controls** U. Possible constraints on the admissible states $(x_1, x_2, \ldots, x_n)$ will be discussed in Secs. 11.8-1*d* and 11.8-5.

The optimal control functions $u_k(t)$ will produce an **optimal trajectory** $x_i = x_i(t)$ in the n-dimensional state space. Solution of such a control problem requires suitably given boundary conditions to determine initial and final values $x_i(t_0)$, $x_i(t_F)$; the initial and final times t_0, t_F may themselves be unknowns (Sec. 11.8-1*c*).

(b) Optimum-control Theory and the Calculus of Variations. The methods of Secs. 11.8-2 to 11.8-5 may be regarded as a somewhat generalized calculus of variations applied to the important class of problems defined in Sec. 11.8-1. In the language of earlier sections, both the $x_i(t)$ and the $u_k(t)$ are unknown dependent variables* $y_i(t)$. The state equations (1) are differential-equation constraints, and the variables p_i defined in Sec. 11.8-2 are the corresponding variable Lagrange multipliers. The adjoint equations and maximum principle introduced in Sec. 11.8-2 constitute necessary (not sufficient) conditions for optimal u_k, x_i and are essentially equivalent to the Euler equations (11.6-4) plus the condition $E \geq 0$ (Sec. 11.6-10), where they apply. The maximum principle states the optimizing conditions in an elegant, convenient, and

* As a matter of fact, the entire problem of Secs. 11.6-1 to 11.6-7, i.e., maximization or minimization of integrals

$$I = \int_{x_0}^{x_F} F[y_1(x), y_2(x), \ldots, y_n(x); y_1'(x), y_2'(x), \ldots, y_n'(x); x\; dx$$

with suitable constraints and/or boundary conditions is reformulated as an optimum-control problem if one substitutes

$$x \equiv t \qquad x_0 \equiv t_0 \qquad x_F \equiv t_F$$
$$I = x_0(t_F) \qquad y_i(x) \equiv x_i(t) \text{ with } \frac{dx_i}{dt} \equiv u_i(t) \equiv y_i'(x) \qquad (i = 1, 2, \ldots, n)$$

In this special case, the theory of Sec. 11.8-2 leads to the classical canonical equations of Sec. 11.6-8; this approach may and may not simplify the given problem.

more general form, permitting a relatively straightforward treatment of systems with discontinuous control variables (Sec. 11.8-3).

(c) Initial-state and Terminal-state Manifolds. While the starting time t_0 and the initial values $x_i(t_0)$ of the state variables are given in many control problems, a more general problem statement merely requires the **initial state** $[x_1(t_0), x_2(t_0), \ldots, x_n(t_0)]$ to lie on a given $(n - n_0)$-dimensional *initial-state manifold* (hypersurface, curve, or point in the state space) defined by

$$B_j[x_1(t_0), x_2(t_0), \ldots, x_n(t_0)] = 0 \qquad (j = 1, 2, \ldots, n_0 \leq n) \tag{11.8-4a}$$

The **terminal state** $[x_1(t_F), x_2(t_F), \ldots, x_n(t_F)]$ is similarly constrained to lie on a given $(n - n_F)$-dimensional *terminal-state manifold* defined by

$$G_j[x_1(t_F), x_2(t_F), \ldots, x_n(t_F)] = 0 \qquad (j = 1, 2, \ldots, n_F \leq n) \tag{11.8-4b}$$

In addition, one may have given inequality constraints defining allowable regions on each manifold (4). Note that a single inequality

$$G[x_1(t_F), x_2(t_F), \ldots, x_n(t_F)] \leq 0 \tag{11.8-5}$$

confining, say, the terminal state to the interior of a given n-dimensional state-space region is essentially equivalent to the corresponding equality constraint, since each trajectory entering the terminal-state region must cross its boundary.

(d) Continuity, Differentiability, and Independence Assumptions (see also Sec. 11.5-2 and Ref. 11.25). Unless the contrary is specifically stated, it will be assumed in the following that, for all x_i, u_k, and t under consideration,

1. The given functions $f_0, f_1, f_2, \ldots, f_n$ are continuously differentiable with respect to the state variables x_i, and continuous in the control variables u_k.
2. The functions Q_j are continuously differentiable with nonzero gradients.
3. The n_0 functions B_j and the n_F functions G_j defining the initial- and terminal-state manifolds are continuously differentiable; functional independence is assured by the condition that the ranks (Sec. 13.2-7) of the matrices $[\partial B_j/\partial x_i(t_0)]$ and $[\partial G_j/\partial x_i(t_F)]$ are, respectively, equal to n_0 and n_F.

All admissible controls $u_k(t)$ are to be chosen from the class of functions of bounded variation in (t_0, t_F) (hence, unilateral limits exist everywhere, Sec. 4.4-8), and piecewise continuously differentiable wherever they are continuous. The corresponding $x_i(t)$ will be piecewise continuously differentiable.

Less restrictive assumptions are often possible, but the resulting theorem statements become cumbersome (Ref. 11.25).

(e) Generalizations (see also Secs. 11.8-4 and 11.8-5). *The optimization methods introduced in Secs.* 11.8-1 *to* 11.8-5 *can be made to apply to much more general problems.* In particular,

1. If one or more of the given functions f_i, Q_j, φ_j, Ψ_j depend explicitly on the independent variable t ("nonautonomous" system), one reduces the problem to the simpler case by introducing $t = x_{n+1}$ as an extra state variable, with

$$\frac{dx_{n+1}}{dt} = 1 \equiv f_{n+1} \qquad x_{n+1}(t_0) = t_0 \qquad x_{n+1}(t_F) = t_F \tag{11.8-6}$$

This procedure applies also if a B_j depends explicitly on t_0, or a G_j on t_F.

2. Continuously differentiable *equality constraints* on the state variables,

$$\varphi_j(x_1, x_2, \ldots, x_n) = 0 \qquad (j = 1, 2, \ldots, m) \tag{11.8-7}$$

and *isoperimetric conditions*

$$\int_{t_0}^{t_F} \Psi_j(x_1, x_2, \ldots, x_n)\, dt = c_j \qquad (j = 1, 2, \ldots, m') \tag{11.8-8}$$

with piecewise continuously differentiable Ψ_j can be handled by the Lagrange-multiplier method of Secs. 11.6-2 and 11.6-3. One replaces the function f_0 in the criterion functional (2) with

$$F_0 \equiv f_0 + \sum_{j=1}^{m} \lambda_j(t)\varphi_j + \sum_{k=1}^{m'} \mu_k \Psi_k \tag{11.8-9}$$

where the $\lambda_j(t)$ and u_j are, respectively, variable and constant Lagrange multipliers. Such constraints are not enforced by the control, but modify the definition of the given system. Control-enforced constraints on the x_i will be treated in Sec. 11.8-5.

3. Criterion functionals of the form

$$x_0(t_F) = \int_{t_0}^{t_F} F(x_1, x_2, \ldots, x_n; u_1, u_2, \ldots, u_r)\, dt + h[x_1(t_F), x_2(t_F), \ldots, x_n(t_F)] \tag{11.8-10}$$

reduce to the simpler form (2) if one introduces

$$\left.\begin{aligned} f_0(x_1, x_2, \ldots, x_n; u_1, u_2, \ldots, u_r) &= F + \dot{h}(t) - \dot{h}(t_0) \\ \text{with} \qquad \dot{h}(t) \equiv \frac{d}{dt} h[x_1(t), x_2(t), \ldots, x_n(t)] &\equiv \sum_{i=1}^{n} \frac{\partial h}{\partial x_i} f_i \end{aligned}\right\} \tag{11.8-11}$$

(see also Sec. 11.6-6).

4. If the constraints (4) on the control variables u_k depend explicitly on the state variables x_i (including, possibly, $t = x_{n+1}$), one can usually eliminate this dependence by introducing new control variables.

EXAMPLE: A constraint (4) of the form

$$|u_k| \leq q(x_1, x_2, \ldots, x_n)$$

reduces to the form

$$|v| \leq 1$$

if one introduces the new control variable v defined by $u_k = vq(x_1, x_2, \ldots, x_n)$.

11.8-2. Pontryagin's Maximum Principle. (a) Adjoint Variables and Optimal Hamiltonian. It is convenient to treat the criterion functional (2) as the final value $x_0(t_F)$ of an added state variable $x_0(t)$ satisfying the state equation

$$\frac{dx_0}{dt} = f_0(x_1, x_2, \ldots, x_n; u_1, u_2, \ldots, u_r) \qquad (11.8\text{-}12)$$

and the initial condition

$$x_0(t_0) = 0 \qquad (11.8\text{-}13)$$

Necessary conditions for optimal control are then defined in terms of *Pontryagin's Maximum Principle: Define* $n + 1$ **adjoint variables** $p_0(t)$, $p_1(t), p_2(t), \ldots, p_n(t)$ *as solutions of the* $n + 1$ *first-order differential equations*

$$\frac{dp_i}{dt} = -\sum_{k=0}^{n} p_k \frac{\partial f_k}{\partial x_i} \qquad (i = 0, 1, 2, \ldots, n; t_0 < t < t_F)$$

(ADJOINT EQUATIONS) (11.8-14)

with

$$p_0(t) = \text{constant} \leq 0 \qquad (11.8\text{-}15)$$

Then the optimal control minimizing the criterion functional (2) *is given by that set of admissible control variables* $u_k = u_k(t)$ *which maximize the* **Hamiltonian function**

$$H(x_1, x_2, \ldots, x_n; p_0, p_1, p_2, \ldots, p_n; u_1, u_2, \ldots, u_r) \equiv \sum_{i=0}^{n} p_i f_i \qquad (11.8\text{-}16)$$

for each t *between* t_0 *and* t_F; *moreover,*

$$M(x_1, x_2, \ldots, x_n; p_0, p_1, p_2, \ldots, p_n) \equiv \max_{(u_1, u_2, \ldots, u_r) \text{ in } U} H(x_1, x_2, \ldots, x_n; p_0, p_1, p_2, \ldots, p_n; u_1, u_2, \ldots, u_r) = 0 \qquad (t_0 \leq t \leq t_F) \qquad (11.8\text{-}17)$$

In addition, the optimal $x_i(t)$ *and* $u_k(t)$ *must satisfy the given conditions* (1) *to* (4), *and transversality conditions*

$$p_i + \sum_{j=1}^{n_0} \Lambda'_j \frac{\partial B_j}{\partial x_i} = 0 \qquad (t = t_0; i = 1, 2, \ldots, n) \qquad (11.8\text{-}18a)$$

$$p_i + \sum_{j=1}^{n_F} \Lambda_j \frac{\partial G_j}{\partial x_i} = 0 \qquad (t = t_F; i = 1, 2, \ldots, n) \qquad (11.8\text{-}18b)$$

corresponding to the given boundary conditions (4); the Λ'_j and Λ_j are unknown constants (see also Secs. 11.6-5, 11.6-8, and 11.8-2*b*).

Given the assumptions listed in Sec. 11.8-1*c*, the adjoint variables $p_i(t)$ will be piecewise continuously differentiable functions. They remain continuous even if one admits discontinuities of f_0 and/or $f_1, f_2, \ldots, f_n$ on a hypersurface S given by

$$g(x_1, x_2, \ldots, x_n) = 0$$

provided that g is continuously differentiable, and that the f_i possess unilateral derivatives with respect to $x_1, x_2, \ldots, x_n$ on either side of S ("refraction" of optimal trajectories, see also Sec. 11.6-7). Under these conditions, the optimal Hamiltonian, too, remains continuous on S.

(b) The Boundary-value Problem. Pontryagin's maximum condition yields, in principle, relations expressing each control variable u_k in terms of the x_i and p_i, i.e.,

$$u_k = \hat{u}_k(x_0, x_1, x_2, \ldots, x_n; p_0, p_1, p_2, \ldots, p_n) \qquad (k = 1, 2, \ldots, m) \tag{11.8-19}$$

These relations may be obtained through *solution of an ordinary maximum-of-a-function problem* [possibly with inequality constraints (3)] for each t. Once this has been done, the optimum-control problem reduces to the solution of the $2n + 2$ first-order differential equations (1), (12), and (14), or

$$\frac{dx_i}{dt} = \frac{\partial H}{\partial p_i} \qquad \frac{dp_i}{dt} = -\frac{\partial H}{\partial x_i} \qquad (i = 0, 1, 2, \ldots, n) \tag{11.8-20}$$

subject to Eq. (17) and the boundary conditions. Since the adjoint equations (14) are homogeneous in the p_i, one can arbitrarily choose the constant in Eq. (15), so that

$$p_0(t) = -1 \qquad (t_0 \leq t \leq t_F) \tag{11.8-21}$$

One now has exactly $2n + n_0 + n_F + 2$ boundary conditions (4), (13), (18), and (21) to determine $2n + 2$ unknown constants of integration, $n_0 + n_F$ unknown multipliers Λ'_j, Λ_j, and the unknown time interval $t_F - t_0$. The missing boundary condition is obtained through substitution of $t = t_F$ into Eq. (17). Unless either t_0 or t_F is given explicitly, one introduces the additional state variable x_{n+1} defined by

$$\frac{dx_{n+1}}{dt} = 1 \qquad x_{n+1}(t_0) = t_0$$

(see also Sec. 11.8-1*e*).

NOTE: Whenever a boundary condition (4), say $G_J = 0$, permits explicit determination or elimination of a boundary value, say $x_I(t_F)$, then $p_I(t_F)$ is undetermined by the relations (4) and (18). In this case, the Jth equation (4b) (and hence Λ_J) and the Ith equation (18b) are simply omitted. If, on the other hand, the boundary value of a state variable, say $x_I(t_F)$, is left "free" or undetermined by the terminal-state constraints (4b), then the Ith equation (18b) yields the corresponding "natural boundary condition" $p_I(t_F) = 0$ (see also Sec. 11.6-5). Similar special cases apply to Eqs. (4a) and (18a).

(c) Since the maximum principle, as given, expresses only necessary (not sufficient) conditions for optimum control, the Pontryagin method may yield multiple candidates for the optimal solution, or no solution may exist. Actual solution of the two-point boundary-value problem usually requires numerical iteration methods (see also Secs. 20.9-2 and 20.9-3). In addition, actual derivation of the maximizing functions (19) may also require successive approximations (Refs. 11.23 and 11.24).

11.8-3. Examples. (a) Zermelo's Navigation Problem. A ship with rectangular cartesian coordinates x_1, x_2 runs at a given constant velocity V in a current with given local velocity components $v_1(x_1, x_2)$, $v_2(x_1, x_2)$ $(v_1^2 + v_2^2 = V^2)$. Given $x_1(0) = 0$, $x_2(0) = 0$, one desires to minimize the time

$$x_0(t_F) = t_F = \int_0^{t_F} dt \qquad (f_0 = 1)$$

required to reach a given (attainable) point (x_{1F}, x_{2F}) by choosing the angle $u(t)$ between the instantaneous velocity relative to the water and the x_1 axis.

The state equations are

$$\frac{dx_1}{dt} = v_1(x_1, x_2) + V\cos u \qquad \frac{dx_2}{dt} = v_2(x_1, x_2) + V\sin u$$

$$\frac{dx_0}{dt} = 1 \qquad (x_0 = t)$$

Hence

$$H \equiv p_1(v_1 + V\cos u) + p_2(v_2 + V\sin u) - 1$$

where $p_0 = -1$. For a maximum of H,

$$\cos u = -\frac{p_1}{\sqrt{p_1^2 + p_2^2}} \qquad \sin u = -\frac{p_2}{\sqrt{p_1^2 + p_2^2}}$$

where p_1, p_2 must satisfy the adjoint equations

$$\frac{dp_1}{dt} = -\left(p_1\frac{\partial v_1}{\partial x_1} + p_2\frac{\partial v_2}{\partial x_1}\right) \qquad \frac{dp_2}{dt} = -\left(p_1\frac{\partial v_1}{\partial x_2} + p_2\frac{\partial v_2}{\partial x_2}\right)$$

and $$M \equiv \max_u H = p_1v_1 + p_2v_2 - V\sqrt{p_1^2 + p_2^2} - 1 = 0$$

If, in particular, v_1 and v_2 are constant, then so are p_1, p_2, and u; their values, together with t_F, must satisfy

$$x_1(0) = 0 \qquad x_2(0) = 0 \qquad x_1(t_F) = x_{1F} \qquad x_2(t_F) = x_{2F}$$

(b) Simple Bang-bang Time-optimal Control. Given $x_1(0)$, $x_2(0) = \dot{x}_1(0)$ and the state equations

$$\frac{dx_1}{di} = x_2 \qquad \frac{dx_2}{dt} = u(t)$$

$\left[\text{i.e., } \frac{d^2x_1}{dt^2} = u(t)\right]$, one desires to minimize the time

$$x_0(t_F) = t_F = \int_0^{t_F} dt \qquad (f_0 \equiv 1)$$

required to reach the given terminal state $x_1(t_F) = x_2(t_F) = \dot{x}_1(t_F) = 0$ by choosing an optimal control $u = u(t)$ such that

$$|u(t)| \leq 1$$

Maximization of the Hamiltonian

$$H \equiv p_1x_2 + p_2u - 1$$

subject to $|u| \leq 1$ requires

$$u = \text{sign } p_2 = \begin{cases} 1 & (p_2 > 0) \\ -1 & (p_2 < 0) \end{cases}$$

with

$$\frac{dp_1}{dt} = 0 \qquad \frac{dp_2}{dt} = -p_1 = \text{constant}$$

so that $p_1 = p_1(0)$, $p_2 = p_2(0) - tp_1(0)$. The optimal trajectories in the x_1, x_2 plane (phase plane) are parabolic arcs corresponding to $u = 1$ and $u = -1$. These arcs

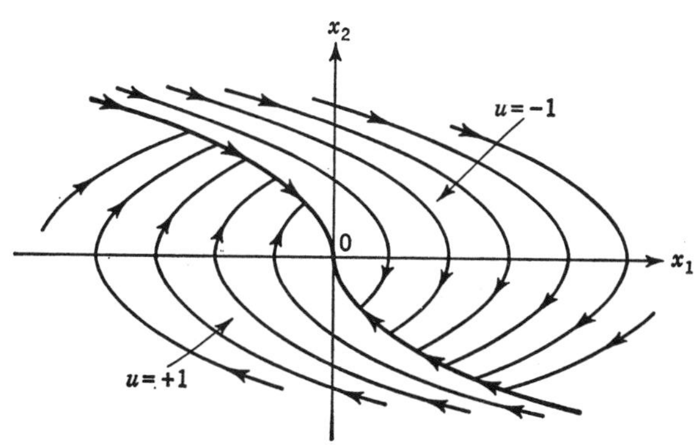

FIG. 11.8-1. Phase-plane trajectories for simple bang-bang control of a simple optimal-control problem (Sec. 11.8-3*b*).

intersect the "switching curve" corresponding to $p_2 = 0$, and each trajectory continues to the origin on the latter (Fig. 11.8-1). Each specific trajectory depends on the unknown parameters $p_1(0)$, $p_2(0)$, which must be determined so that the given boundary conditions $x_1(t_F) = x_2(t_F) = 0$ are satisfied.

(c) A Simple Minimal-time Orbit-transfer and Rendezvous Problem. Referring to Fig. 11.8-2, the motion of a simple rocket-driven vehicle in a vertical plane, assuming a flat earth (constant acceleration of gravity, $-g$) and no air resistance, is given by *the four state equations*

$$\frac{dx}{dt} = \dot{x} \qquad \frac{d\dot{x}}{dt} = \frac{T \cos u}{m_0 + \dot{m}(t - t_0)}$$
$$\frac{dy}{dt} = \dot{y} \qquad \frac{d\dot{y}}{dt} = \frac{T \sin u}{m_0 + \dot{m}(t - t_0)} - g$$

(Fig. 11.8-2). $x \equiv x_1$, $\dot{x} \equiv x_2$, $y \equiv x_3$, and $\dot{y} \equiv x_4$ are state variables; g, T (thrust), m_0 (vehicle mass at start), and $\dot{m} < 0$ (fuel-consumption rate, assumed constant) are

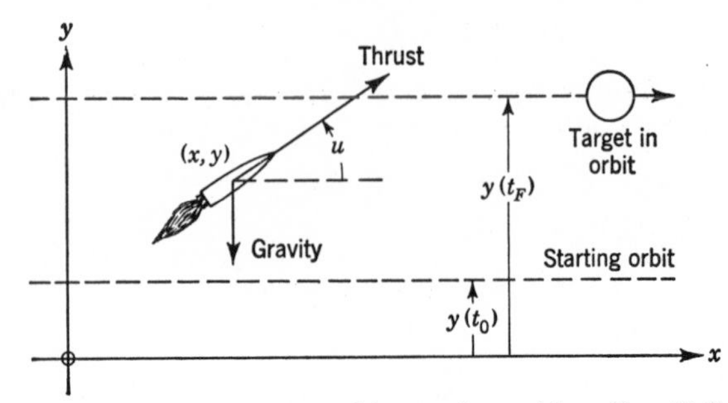

FIG. 11.8-2. Geometry for the orbit-transfer problem (Sec. 11.8-3c).

given constants. It is desired to transfer the vehicle from a given horizontal *starting orbit* defined by

$$x(t_0) = v_0 t_0 \qquad \dot{x}(t_0) = v_0$$
$$y(t_0) = y_0 \qquad \dot{y}(t_0) = 0$$

to the given horizontal orbit of a target, and to match its given position and constant velocity:

$$x(t_F) = v_F t_F + a \qquad \dot{x}(t_F) = v_F$$
$$y(t_F) = y_F \qquad \dot{y}(t_F) = 0$$

This is to be accomplished by programming the control variable (attitude angle) $u = u(t)$ so as to minimize the total fuel consumption, i.e., so as to minimize $\dot{m}(t_F - t_0)$, or simply $\int_{t_0}^{t_F} dt$. Since *the initial- and final-state manifolds depend explicitly on* t_0 *and* t_F, *one introduces* $x_5 = t$ *with* $dx_5/dt = 1$, $x_5(t_0) = t_0$, $x_5(t_F) = t_F$.

Maximization of the Hamiltonian,

$$H = p_1\dot{x} + p_2 \frac{T \cos u}{m_0 + \dot{m}(t - t_0)} + p_3\dot{y} + p_4 \left[\frac{T \sin u}{m_0 + \dot{m}(t - t_0)} - g\right] + p_5 - 1$$

requires $\partial H/\partial u = 0$, or

$$-p_2 \sin u + p_4 \cos u = 0$$

and

$$\cos u = -\frac{p_2}{\sqrt{p_2^2 + p_4^2}} \qquad \sin u = -\frac{p_4}{\sqrt{p_2^2 + p_4^2}}$$

The *adjoint equations*

$$\frac{dp_1}{dt} = \frac{dp_3}{dt} = \frac{dp_5}{dt} = 0 \qquad \frac{dp_2}{dt} = -p_1 \qquad \frac{dp_4}{dt} = -p_3$$

yield

$$p_1(t) = p_1(t_0) \qquad p_2(t) = -p_1(t_0)(t - t_0) + p_2(t_0)$$
$$p_3(t) = p_3(t_0) \qquad p_4(t) = -p_3(t_0)(t - t_0) + p_4(t_0)$$
$$p_5(t) = p_5(t_0)$$

Since the starting and terminal values of $\dot{x}$, y, and $\dot{y}$ are fixed, the only *transversality conditions* (18) of interest are those resulting from

$$x(t_0) - v_0 t_0 = 0 \qquad x(t_F) - v_F t_F - a = 0$$

or

$$p_1(t_0) + \Lambda_1' = 0 \qquad p_1(t_F) + \Lambda_1 = 0$$
$$p_5(t_0) - \Lambda_1' v_0 = 0 \qquad p_5(t_F) - \Lambda_1 v_F = 0$$

With p_1 and p_5 constant, this implies $p_1 = p_5 = 0$. Furthermore, the maximal Hamiltonian is zero at $t = t_F$:

$$-\frac{T}{m_0 + \dot{m}(t - t_0)} [p_2{}^2(t_0) + p_4{}^2(t_0) + p_3{}^2(t_0)(t_F - t_0)^2 - 2p_3(t_0)p_4(t_0)(t_F - t_0)]^{1/2} + gp_3(t_0)(t_F - t_0) - gp_4(t_0) - 1 = 0$$

This, together with the eight given initial and terminal conditions, determines the nine quantities t_0, t_F, $x(t_0)$, $\dot{x}(t_0)$, $y(t_0)$, $\dot{y}(t_0)$, $p_2(t_0)$, $p_3(t_0)$, $p_4(t_0)$, and thus the complete solution.

11.8-4. Matrix Notation for Control Problems. The control problem can be very conveniently expressed in the matrix notation of Sec. 13.6-1 (and also in the corresponding tensor notation, Sec. 14.7-7). One introduces

$x = x(t) \equiv \{x_0, x_1, x_2, \ldots, x_n\}$, $(n + 1) \times 1$ *column matrix* representing the **state vector**

$u = u(t) \equiv \{u_1, u_2, \ldots, u_r\}$, $r \times 1$ *column matrix* representing the **control vector**

$p = p(t) \equiv (p_0, p_1, p_2, \ldots, p_n)$, $1 \times (n + 1)$ *row matrix* representing the **adjoint vector**

(see also Sec. 14.5-2). The relations of Secs. 11.8-1 and 11.8-2 can now be restated compactly; note that $\frac{\partial f}{\partial x} \equiv \left[\frac{\partial f_k}{\partial x_i}\right]$ is an $(n + 1) \times (n + 1)$ matrix. In particular, one has

$$\left.\begin{aligned} &\frac{dx}{dt} = f(x, u) \equiv \frac{\partial H}{\partial p} && \text{(STATE EQUATIONS)} \\ &\frac{dp}{dt} = -p\frac{\partial f}{\partial x} \equiv -\frac{\partial H}{\partial x} && \text{(ADJOINT EQUATIONS)} \\ &H(p, x, u) \equiv pf && \text{(HAMILTONIAN FUNCTION)} \end{aligned}\right\} \tag{11.8-22}$$

The criterion functional (2) may be regarded as the matrix product (inner product, Sec. 14.7-1) $x_0 = cx$ of the column matrix x and the $1 \times (n + 1)$ row matrix $c \equiv (1, 0, 0, \ldots, 0)$. Reference 11.25 treats the case of criterion functionals defined as cx with a more general row matrix c.

11.8-5. Inequality Constraints on State Variables. Corner Conditions (see also Sec. 11.6-6). **(a)** If the domain X of admissible system states $(x_0, x_1, x_2, \ldots, x_n)$ is restricted by continuously differentiable inequality constraints

$$S_j(x_0, x_1, x_2, \ldots, x_n; u_1, u_2, \ldots, u_r) \leq 0 \qquad (j = 1, 2, \ldots, M) \tag{11.8-23}$$

which are to be enforced *by the control* in the course of the optimal-trajectory generation (see also Sec. 11.8-1*e*), the theory of Sec. 11.8-2 applies without change to trajectories and portions of trajectories situated *in the interior of the closed state region X defined by Eq.* (23). For trajectories and portions of trajectories situated *on the boundary* of X, at least one of the constraints (23) becomes an equality. For simplicity, consider a boundary region D_S with only a single equality, say

$$S(x_0, x_1, x_2, \ldots, x_n; u_1, u_2, \ldots, u_r) = 0 \tag{11.8-24}$$

In this boundary region, the Lagrange-multiplier method of Secs. 11.6-2 and 11.8-1*e* applies; i.e., each optimal trajectory satisfies Eq. (24) and the relations of Sec. 11.8-2*a*, with f_0 replaced by

$$F_0 \equiv f_0 + \lambda(t)S \qquad [(x_0, x_1, x_2, \ldots, x_n) \text{ in } D_S] \tag{11.8-25}$$

where λ is a variable Lagrange multiplier.

Alternatively (and equivalently), f_0 can be replaced on D_S by any one of the expressions

$$F_0 \equiv f_0 + \mu(t)S^{(k)} \equiv f_0 + \mu(t) \sum_{i=0}^{n} \frac{\partial S^{(k-1)}}{\partial x_i} f_i \qquad [(x_0, x_1, x_2, \ldots, x_n) \text{ in } D_S; \; k = 1, 2, \ldots] \tag{11.8-26}$$

$$\text{with} \quad S^{(0)} \equiv S = 0, \qquad S^{(1)} \equiv \frac{dS}{dt} \equiv \sum_{i=0}^{n} \frac{\partial S}{\partial x_i} f_i = 0, \qquad \ldots \qquad [(x_0, x_1, x_2, \ldots, x_n) \text{ in } D_S] \tag{11.8-27}$$

where $\mu(t)$ is a variable Lagrange multiplier.

(b) If $S^{(K)}$ is the first of the functions (27) which contains a control variable u explicitly, $S \leq 0$ is termed a **Kth-order inequality constraint.** In this case, an optimal trajectory *entering* the boundary region D_S defined by Eq. (24) from the interior of X at the time $t = t_1$ must satisfy the *corner conditions* (*jump conditions*)

$$\left.\begin{aligned} p_i(t_1 - 0) &= p_i(t_1 + 0) + \sum_{K=0}^{K-2} \nu_k \left.\frac{\partial S^{(k)}}{\partial x_i}\right]_{t=t_1} \\ &\qquad + (\nu_{K-1} + b) \left.\frac{\partial S^{(K-1)}}{\partial x_i}\right]_{t=t_1} \qquad (i = 1, 2, \ldots, n) \\ H\Big]_{t_1-0} &= H\Big]_{t_1+0} \end{aligned}\right\} \tag{11.8-28}$$

where the ν_k are constant Lagrange multipliers, and b is an arbitrary constant.

An optimal trajectory *leaving* the state-space boundary at the time $t = t_2$ for the first time after entering at $t = t_1$ must satisfy the corner conditions

$$\left.\begin{aligned} p_i(t_2 - 0) &= p_i(t_2 + 0) - b \left.\frac{\partial S^{(K-1)}}{\partial x_i}\right]_{t=t_2} \qquad (i = 1, 2, \ldots, n) \\ H\Big]_{t_2-0} &= H\Big]_{t_2+0} \end{aligned}\right\} \tag{11.8-29}$$

The arbitrary constant b occurring in both Eq. (28) and Eq. (29) is usually taken to be zero, so that the $p_i(t)$ are continuous at the exit point. If an optimal trajectory is *reflected* by a boundary region D_S corresponding to a single constraint (24), one has $t_1 = t_2$, and the corner conditions (28) apply with $b = 0$.

If the single constraint (24) is replaced by two or more such constraints, terms corresponding to each new constraint (with additional multipliers) must be added to each sum in Eqs. (25), (26), (28), and (29). Explicit time dependence of constraints is treated in the manner of Sec. 11.8-1*e*.

NOTE: See Ref. 11.25 for exceptions to the corner conditions (28), (29) in special situations.

11.8-6. The Dynamic-programming Approach (see also Sec. 11.9-2). If the optimum-control problem defined in Sec. 11.8-1 defines unique optimal extremals for fixed $x_i(t_F)$ and variable

$$x_i(t_0) = X_i \qquad (i = 1, 2, \ldots, n)$$

then the criterion-function minimum

$$\begin{aligned} &\min_{u_1(t), u_2(t), \ldots, u_r(t)} \int_{t_0}^{t_F} f_0(x_1, x_2, \ldots, x_n; u_1, u_2, \ldots, u_r)\, dt \\ &\quad \equiv S[x_1(t_0), x_2(t_0), \ldots, x_n(t_0); x_1(t_F), x_2(t_F), \ldots, x_n(t_F)] \\ &\quad \equiv S(X_1, X_2, \ldots, X_n) \end{aligned} \tag{11.8-30}$$

satisfies the first-order partial differential equation

$$M\left(X_1, X_2, \ldots, X_n; \frac{\partial S}{\partial X_1}, \frac{\partial S}{\partial X_2}, \ldots, \frac{\partial S}{\partial X_n}\right) = 0$$

(HAMILTON-JACOBI EQUATION) (11.8-31)

where M is the optimal (maximized) Hamiltonian function Eq. (17) (see also Sec. 11.6-8). The ordinary differential equations (20) are the corresponding characteristic equations; their solutions are thus readily determined if a complete integral of the partial differential equation (31) is known (Sec. 10.2-4).

The dynamic-programming approach "imbeds" a given optimum-control problem in an entire class of similar problems with different initial coordinates. The partial differential equation (31), which solves this entire set of problems, expresses the fact that *every portion of an optimal trajectory optimizes the criterion functional for the corresponding initial and terminal points* (*Bellman's Principle of Optimality*). It is possible to derive the entire optimum-control theory under very general assumptions from the principle of optimality. In general, numerical solution of Eq. (31) is difficult for $n > 2$ (see Refs. 11.13 to 11.15, which also discuss dynamic programming in a more general context).

11.9. STEPWISE-CONTROL PROBLEMS AND DYNAMIC PROGRAMMING

11.9-1. Problem Statement. An important class of optimization problems involving stepwise control (this includes control of continuous systems by digital computers) or stepwise optimal resource allocation can be formulated as follows: a system described by a set of state variables $x \equiv (x_1, x_2, \ldots, x_n)$ proceeds through a sequence of states ${}^0x$, ${}^1x$, ${}^2x$, . . . such that each state change is given by the *state equations* (in this case, difference equations, see also Sec. 20.4-3)

$$\left.\begin{aligned} {}^{k+1}x_i &= f_i({}^kx_1, {}^kx_2, \ldots, {}^kx_n; {}^{k+1}u_1, {}^{k+1}u_2, \ldots, {}^{k+1}u_r) \\ & \qquad\qquad\qquad (i = 1, 2, \ldots, n) \\ \text{or} \qquad {}^{k+1}x &= f({}^kx, {}^{k+1}u) \end{aligned}\right\} \tag{11.9-1}$$

where the "control variable" ${}^{k+1}u \equiv ({}^{k+1}u_1, {}^{k+1}u_2, \ldots, {}^{k+1}u_r)$ defines the sequence of decisions (**policy**) changing the kth system state into the $(k + 1)^{\text{st}}$ state. Given the initial state ${}^0x$ (and, possibly, a set of suitable equality or inequality constraints on state and control variables), it is desired to find an **optimal policy** ${}^1u, {}^2u, \ldots, {}^Nu$ which will minimize a given criterion function

$${}^Nx_0 = \sum_{k=0}^{N-1} f_0({}^kx, {}^{k+1}u) + h({}^Nx) = {}^Nx_0({}^0x) \tag{11.9-2}$$

where $N = 1, 2, \ldots$ is the number of steps considered (**dynamic programming**). As in continuous optimal-control problems, the initial and final states may or may not be given by the problem; and it is, again, possible to generalize the problem statement in the manner of Sec. 11.8-1*e*.

11.9-2. Bellman's Principle of Optimality (see also Sec. 11.8-6). *If ${}^1u, {}^2u, \ldots, {}^Nu$ is an optimal policy resulting in the state sequence ${}^0x, {}^1x, {}^2x, \ldots, {}^Nx$ for a given dynamic programming problem with initial state ${}^0x$, then ${}^2u, {}^3u, \ldots, {}^Nu$ is an optimal policy for the same criterion function and final state ${}^Nx$ but initial state ${}^1x$.* If one denotes $\min\limits_u {}^Nx_0(X)$ by ${}^NS(X)$, the optimality principle leads to the fundamental recurrence relation (partial difference equation, Sec. 20.4-3*b*)

$$\left.\begin{aligned} {}^NS(X) &= \min_{{}^1u} \{f_0(X, {}^1u) + {}^{N-1}S[f(X, {}^1u)]\} \qquad (N = 2, 3, \ldots) \\ \text{with} \qquad {}^1S(X) &= \min_{{}^1u} f_0(X, {}^1u) \end{aligned}\right\} \tag{11.9-3}$$

where minima are determined in accordance with any given constraints. Numerical solution of this functional equation for the unknown function ${}^NS(X)$ amounts to stepwise construction of a class of optimal policies for many initial states. The desired optimal policy is "imbedded" in this class. Solution usually requires digital computation; even so, solution of problems with more than two or three state variables x_i is practical only in special cases. References 11.12 to 11.18 describe many examples and approximation methods, and Ref. 11.20 discusses a stepwise-control analogy to the Maximum Principle.

11.10. RELATED TOPICS, REFERENCES, AND BIBLIOGRAPHY

11.10-1. Related Topics. The following topics related to the study of maxima and minima are treated in other chapters of this handbook:

Ordinary differential equations.......... Chap. 9
Partial differential equations, canonical transformations.......... Chap. 10
Eigenvalues of matrices and quadratic forms.......... Chap. 13
Sturm-Liouville problems.......... Chap. 15
Statistical decisions.......... Chap. 19
Numerical methods.......... Chap. 20

11.10-2. References and Bibliography (see also Sec. 20.2-6 for numerical optimization techniques).

Linear and Nonlinear Programming, Theory of Games

11.1. Abadie, J.: On the Kuhn-Tucker Theorem, *ORC 65-18,* Operations Research Center, University of California, Berkeley, 1965.
11.2. Dantzig, G. B.: *Linear Programming and Extensions,* Prentice-Hall, Englewood Cliffs, N.J., 1961.
11.3. Dresher, M.: *Games of Strategy: Theory and Applications,* Prentice-Hall, Englewood Cliffs, N.J., 1961.
11.4. Gale, D.: *The Theory of Linear Economic Models,* McGraw-Hill, New York, 1960.
11.5. Gass, S. I.: *Linear Programming,* 2d ed., McGraw-Hill, New York, 1964.
11.6. Hadley, G.: *Linear Programming,* Addison-Wesley, Reading, Mass., 1962.
11.7. Karlin, S.: *Mathematical Methods and Theory in Games, Programming, and Economics,* 3d ed., Addison-Wesley, Reading, Mass., 1959.
11.8. Kuhn, H. W., and A. W. Tucker, Nonlinear Programming, in *Proc. 2d Berkeley Symp. on Math. Stat. and Prob.,* vol. 5, University of California Press, Berkeley, 1952.
11.9. Luce, R. D., and H. Raiffa: *Games and Decisions,* Wiley, New York, 1957.
11.10. Vajda, S.: *Theory of Games and Linear Programming,* Wiley, New York, 1956.
11.11. Boot, J. C. G.: *Quadratic Programming,* North Holland Publishing Company, Amsterdam, 1964.

See also the articles by L. Rosenberg and E. M. Beale in M. Klerer and G. A. Korn: *Digital Computer User's Handbook,* McGraw-Hill, New York, 1967.

Calculus of Variations, Optimum-control Theory, and Dynamic Programming

11.12. Aris, Rutherford: *Discrete Dynamic Programming,* New York, Wiley, 1963.
11.13. Bellman, R.: *Dynamic Programming,* Princeton University Press, Princeton, N.J., 1957.
11.14. ——— and S. E. Dreyfus: *Applied Dynamic Programming,* Princeton University Press, Princeton, N.J., 1962.
11.15. ——— and R. Kalaba: *Dynamic Programming and Modern Control Theory,* Academic, New York, 1965.
11.16. Bliss, G. A.: *Calculus of Variations,* Open Court, Chicago, 1925.
11.17. Bolza, O.: *Vorlesungen über Variationsrechnung,* Stechert, New York, 1931.

11.18. Dreyfus, S. E.: *Dynamic Programming and the Calculus of Variations*, Academic, New York, 1965.
11.19. Elsgolc, L. E.: *Calculus of Variations*, Pergamon/Addison-Wesley, Reading, Mass., 1962.
11.20. Fan, Liang-Tsen, and Chin-sen Wang: *The Discrete Maximum Principle*, Wiley, New York, 1964.
11.21. Fan, Liang-Tsen: *The Continuous Maximum Principle*, Wiley, New York, 1966.
11.22. Gelfand, I. M., and S. V. Fomin: *Calculus of Variations*, Prentice-Hall, Englewood Cliffs, N.J., 1963.
11.23. Leitmann, G.: *Optimization Techniques*, Academic, New York, 1962.
11.24. Merriam, C. W.: *Optimization Theory and the Design of Feedback Control Systems*, McGraw-Hill, New York, 1964.
11.25. Pontryagin, L. S., et al.: *The Mathematical Theory of Optimal Processes*, Wiley, New York, 1962.
11.26. Weinstock, R.: *Calculus of Variations*, McGraw-Hill, New York, 1952.
11.27. Athans, M., and P. L. Falb: *Optimal Control*, McGraw-Hill, New York, 1966.

CHAPTER **12**

DEFINITION OF MATHEMATICAL MODELS: MODERN (ABSTRACT) ALGEBRA AND ABSTRACT SPACES

12.1. INTRODUCTION

12.1-1. Mathematical Models. Physical processes are, generally speaking, described in terms of *operations* (observations, experiments) relating physical *objects*. The complexity of actual physical situations calls for simplified descriptions in terms of verbal, symbolic, and even physical **models** which "abstract" suitably chosen "essential" properties of physical objects and operations.

Mathematics in the most general sense deals with the definition and manipulation of symbolic models. A **mathematical model** involves a class of undefined (abstract, symbolic) **mathematical objects,** such as numbers or vectors, and **relations** between these objects. A mathematical relation is a hypothetical rule which associates two or more of the undefined objects (see also Secs. 12.1-3 and 12.8-1). Many relations are described in terms of **mathematical operations** associating one or more objects (**operand, operands**) with another object or set of objects (**result**). The abstract model, with its unspecified objects, relations, and operations, is *defined* by a self-consistent set of rules (**defining postulates**) which introduce the operations to be employed and state general relations between their results (**descriptive definition** of a mathematical model in terms of its properties; see Secs. 12.2-1, 12.3-1, 12.4-1, 12.5-2, 12.6-1, 12.8-1, and 1.1-2 for examples). A **constructive definition** introduces a new mathematical model in terms of previously defined mathematical concepts (e.g., definition of matrix addition and multiplication in terms of numerical addition and multiplication, Sec. 13.2-2).

The self-consistency of a descriptive definition must be demonstrated by construction or exhibition of an example satisfying the defining postulates (*existence proof*, see also Secs. 4.2-1*b* and 9.1-4). In addition, it is customary to test the defining postulates for mutual independence.

A mathematical model will reproduce suitably chosen features of a physical situation if it is possible to establish **rules of correspondence** relating specific physical objects and relationships to corresponding mathematical objects and relations. It may also be instructive and/or enjoyable to construct mathematical models which do not match any counterpart in the physical world. The most generally familiar mathematical models are the integral and real-number systems (Sec. 1.1-2) and Euclidean geometry; the defining properties of these models are more or less directly abstracted from physical experience (counting, ordering, comparing, measuring).

Objects and operations of more general mathematical models are often labeled with sets of real numbers, which may be related to the results of physical measurements. The resulting "representations" of mathematical models in terms of numerical operations are discussed specifically in Chaps. 14 and 16.

Questions as to the self-consistency, sufficiency, and applicability of the logical rules used to manipulate mathematical statements form the subject matter of *metamathematics*, which is by no means a finished structure (Ref. 12.16).

12.1-2. Survey. Modern (abstract) **algebra*** deals with mathematical models defined in terms of binary operations ("algebraic" operations, usually referred to as various types of "addition" and "multiplication") which associate *pairs* of mathematical objects (operands, or operator and operand) with corresponding results. Sections 12.2-1 through 12.4-2 introduce some of the most generally useful models of this kind, notably *groups, rings, fields, vector spaces*, and *linear algebras; Boolean algebras* are treated separately in Secs. 12.8-1 through 12.8-6.

The important subject of *linear transformations* (*linear operators*) and their eigenvectors and eigenvalues is introduced in Chap. 14. The *representation of vectors and operators in terms of numerical components and matrices* is discussed in detail in Chaps. 14, 15, and 16.

Sections 12.5-1 through 12.6-3 serve as a brief introduction to mathematical models which permit the definition of *limiting processes* and

* The word *algebra* has three loosely related meanings: (1) a general subject, as used here (*abstract algebra, elementary algebra*); (2) the theory of algebraic operations used in connection with a specific model (*matrix algebra, tensor algebra*); (3) a type of mathematical model (*a linear algebra, a Boolean algebra*).

order. In particular, Secs. 12.5-2 through 12.5-4 deal with *metric spaces.* Sections 12.7-1 through 12.7-5 discuss simple schemes for combining mathematical models (*direct products* and *direct sums*).

12.1-3. "Equality" and Equivalence Relations. (a) **The descriptive definition of each class of mathematical objects discussed in this chapter is understood to imply the existence of a rule stating whether or not two given mathematical objects a, b are "equal" (equivalent or indistinguishable in the context of the model, $a = b$)**; the rule must be such that

1. $a = a$	(REFLEXIVITY OF THE EQUALITY RELATION)
2. $a = b$ implies $b = a$	(SYMMETRY)
3. $a = b$, $b = c$ implies $a = c$	(TRANSITIVITY)

Sections 13.2-2 and 16.3-1 show examples of the definition of equality in *constructively defined models.*

(b) More generally, any relation $a \overset{r}{\sim} b$ between two objects a, b of a class C is called an **equivalence relation** if and only if it is *reflexive* ($a \overset{r}{\sim} a$), *symmetric* ($a \overset{r}{\sim} b$ *implies* $b \overset{r}{\sim} a$), and *transitive* ($a \overset{r}{\sim} b$, $b \overset{r}{\sim} c$ *implies* $a \overset{r}{\sim} c$). *Every equivalence relation defines a* **partition** *of the class* C, i.e., a subdivision of C into subclasses without common elements. Elements of the same subclass are *equivalent* with respect to the properties defining the equivalence relation.

EXAMPLES: Equality, identity of functions (Sec. 1.1-4), congruence and similarity of triangles, metric equality (Sec. 12.5-2), isomorphism (Sec. 12.1-6).

12.1-4. Transformations, Functions, Operations (see also Secs. 4.2-1, 14.1-3, and 14.3-1). A set of rules $x \to x'$ associating an object x' of a class C' with each object x of a class C is called a **transformation (mapping) of C into C'**; x' is a **function**

$$x' = x'(x) = f(x) \tag{12.1-1}$$

of the **argument** x, with **domain (domain of definition)** C and **range** C'. C and C' may or may not be the same class of objects. The correspondence (1) can be regarded as an **operation** on the **operand** x producing the **result (transform)** x'. Given suitable definitions of equality in C and C' (Sec. 12.1-3), an operation (transformation, function) (1) is **well defined** if and only if $x = y$ implies $x' = y'$. **This property will be understood to apply unless the contrary is specifically stated.**

A correspondence (1) is **unique** [$f(x)$ is a **single-valued** function of x] if and only if a unique object x' corresponds to each object x. The correspondence is a **reciprocal one-to-one (biunique)** correspondence of C and C' if and only if the relation (1) maps C uniquely onto all of C' and defines a unique **inverse correspondence (inverse transformation.** Many authors categorically define *every* mapping as unique (see also footnote to Sec. 4.2-2). The set of pairs (x, x') is called the **graph** of the function $f(x)$.

Each object x in Eq. (1) can be a set of objects $x_1, x_2, \ldots$; so that it is possible to define functions $x' = f(x_1, x_2, \ldots)$ of two or more arguments.

A numerical (real or complex) function defined on a set of functions is called a **functional** $\left[\text{e.g., } \int_0^1 \varphi(x)\, dx, \text{ maximum value of } \varphi(x) \text{ in } (a, b)\right]$.

Transformations (functions, operations) can themselves be regarded as new mathematical objects (see, for example, Sec. 14.3-1).

12.1-5. Invariance (see also Secs. 12.2-8, 13.4-1, 14.4-5, 16.1-4, and 16.4-1). Given a transformation (1) of a class (space) C into itself, any function $F(x, y, \ldots)$ such that $F[f(x), f(y), \ldots] = F(x, y, \ldots)$ for all $x, y, \ldots$ in C and any relation $O(x, y, \ldots) = A$ which implies $O[f(x), f(y), \ldots] = A$ for all $x, y, \ldots$ in C is called **invariant with respect to the transformation** (1).

12.1-6. Representation of One Model by Another: Homomorphisms and Isomorphisms. Let M be a mathematical model (Sec. 12.1-1) involving objects $a, b, \ldots$ and operations $O, P, \ldots$ whose results $O(a, b, \ldots), P(a, b, \ldots), \ldots$ are elements of M.* A second model M' is a **homomorphic image** of M with respect to the operation $O(a, b, \ldots)$ if and only if

1. There exists a unique correspondence $a \to a', b \to b', \ldots$ relating one of the elements $a', b', \ldots$ of M' to each element of M.
2. It is possible to define an operation O' on the elements of M' so that $O(a, b, \ldots) \to O'(a', b', \ldots)$.

A correspondence with these properties is called a **homomorphism of** M **to** M' and preserves all relations based on the operation in question; i.e., each such relation between elements $a, b, \ldots$ of M implies a corresponding relation between the elements $a', b', \ldots$ of M'. A homomorphism mapping the set of elements of M into itself (onto a subset of itself) is called an **endomorphism.**

An **isomorphism** is a homomorphism involving reciprocal one-to-one correspondence. Two models M and M' so related are **isomorphic** with respect to the operation in question; in this case both $M \to M'$ and $M' \to M$ are homomorphisms. An isomorphism mapping the set of elements of M onto itself is an **automorphism** of M.

* The elements $a, b, \ldots$ need not belong to a single class of mathematical objects (e.g., there may be vectors and scalars, Sec. 12.4-1).

The concept of a homomorphism, and the related concepts of isomorphism and automorphism, are of the greatest practical importance, since they permit the representation of one model by another. One may, in particular, represent mathematical objects by sets of real numbers (analytical geometry, matrix and tensor representations). Note that isomorphism is an equivalence relation (Sec. 12.1-3*b*) between models: properties of an entire class of isomorphic models may be derived from, or discussed in terms of, the properties of any model of the class.

Some writers require each homomorphism $M \to M'$ to map M onto *all of* M'. In this case the homomorphism defines an isomorphism relating disjoint classes of elements of M to elements of M'.

12.2. ALGEBRA OF MODELS WITH A SINGLE DEFINING OPERATION: GROUPS

12.2-1. Definition and Basic Properties of a Group. (**a**) A class G of objects (elements) $a, b, c, \ldots$ is a **group** if and only if it is possible to define a binary operation (rule of combination) which associates an object (result) $a \odot b$ with every pair of elements a, b of G so that

1. $a \odot b$ is an element of G (CLOSURE under the defining operation)
2. $a \odot (b \odot c) = (a \odot b) \odot c$ (ASSOCIATIVE LAW)
3. G contains a (left) **identity (identity element)** E such that, for each element a of G, $E \odot a = a$
4. For each element a of G, G contains a (left) **inverse** a^{-1} such that $a^{-1} \odot a = E$

Two elements a, b of a group **commute** if and only if $a \odot b = b \odot a$. If all elements a, b of a group G commute, the defining operation of G is **commutative,** and G is a **commutative** or **Abelian group.** A group G containing a finite number g of elements is **of finite order** g; otherwise G is an **infinite group (of infinite order).** In the latter case, G may or may not be a countable set (see Sec. 12.2-11 for a brief discussion of *continuous groups*).

(**b**) *Every group G has a unique left identity and a unique right identity, and the two are identical* ($E \odot a = a \odot E = a$). *Each element a has a unique left inverse and a unique right inverse, and the two are identical* ($a^{-1} \odot a = a \odot a^{-1} = E$).

Hence

$$\left.\begin{aligned} c \odot a = c \odot b \text{ implies } a = b \\ a \odot c = b \odot c \text{ implies } a = b \end{aligned}\right\} \qquad \text{(CANCELLATION LAWS)} \tag{12.2-1}$$

G contains a unique solution x of every equation $c \odot x = b$ *or* $x \odot c = b$; *i.e.*, unique right and left "division" is possible.

(c) The operation defining a group is often referred to as (abstract) *multiplication* (note, however, Sec. 12.2-10); its result is then written as a *product* ab, and the inverse of a is written as a^{-1}. This convention is freely used in the following sections.

Multiple products aa, aaa, . . . are written as *integral powers* a^2, a^3, . . . , with $(a^{-1})^n = a^{-n}$, $a^0 = E$. Note

$$(a^{-1})^{-1} = a \qquad a^m a^n = a^{m+n} \qquad (a^m)^n = a^{mn} \qquad (ab)^{-1} = b^{-1}a^{-1} \tag{12.2-2}$$

12.2-2. Subgroups. A subset G_1 of a group G is a **subgroup** of G if and only if G_1 is a group with respect to the defining operation of G. This is true (1) *if and only if* G_1 *contains all products and inverses of its elements, i.e.*, (2) *if and only if* G_1 *contains the product* ab^{-1} *for each pair of elements* a, b *in* G_1.

The intersection (*Sec.* 4.3-2*a*) *of two subgroups of G is a subgroup of G.* G and E are **improper subgroups** of G; all other subgroups of G are **proper subgroups.** *If G is of finite order g* (*Sec.* 12.1-1*a*), *the order* g_1 *of every subgroup* G_1 *of G is a divisor of g* (*Lagrange's Theorem*); g/g_1 is called the **index** of G_1 (with respect to G).

12.2-3. Cyclic Groups. Order of a Group Element. A **cyclic group** consists of the powers $a^0 = E, a, a^2, \ldots$ of a single element a and is necessarily commutative. *Every group of prime order and every subgroup of a cyclic group is cyclic.*

Each element a of any group G "generates" a cyclic subgroup of G, the **period** of a. The order of this subgroup is called the **order of the element** a and is equal to the least positive integer m such that $a^m = E$.

12.2-4. Products of Complexes. Cosets. **(a)** Every subset C_1 of a group G is called a **complex.** The product* C_1C_2 of two complexes C_1, C_2 of G is the set of all (different) products a_1a_2 of an element a_1 of C_1 and an element a_2 of C_2. $C_1C_1 = C_1$ *if and only if* C_1 *is a subgroup of G. The product* G_1G_2 *of two subgroups of G is a subgroup of G if and only if* $G_1G_2 = G_2G_1$.

(b) A **left coset** xG_1 of a subgroup G_1 of G is the set of all products xa_1 of a given element x of G and any element a_1 of G_1. A **right coset** G_1x is, similarly, the set of all products a_1x. *A coset of* G_1 *is a subgroup of G if and only if x is an element of* G_1; in this case $xG_1 = G_1x = G_1$. *Two left cosets of* G_1 *are either identical or have no common element; the same is true for two right cosets. Every subgroup* G_1 *of G defines a partition* (*Sec.* 12.1-3) *of G into a finite or infinite number* n_1 *of left cosets, and a partition of G into* n_1 *right cosets; if G is of finite order g,* n_1 *equals the index* g/g_1 *of* G_1 (Sec. 12.2-2). *Two elements a, b of G belong to the same left coset of* G_1 *if* G_1 *contains* $a^{-1}b$, *and to the same right coset of* G_1 *if* G_1 *contains* ba^{-1}.

12.2-5. Conjugate Elements and Subgroups. Normal Subgroups. Factor Group. **(a)** Two elements x and x' of a group G are **conjugate** if and only if they are related by a transformation (Sec. 12.1-4).

$$x' = a^{-1}xa \qquad (\text{or } x = ax'a^{-1}) \tag{12.2-3}$$

* The product C_1C_2 of two complexes must not be confused with their *intersection* (*logical product*, Sec. 4.3-2a) $C_1 \cap C_2$, or with a *direct product* (Sec. 12.7-2).

where a is an element of G. x' is then called the **transform of x under conjugation by a.** *Conjugation is an equivalence relation* (*Sec.* 12.1-3*b*) *and defines a partition of G into* **classes** *of conjugate elements.*

(**b**) A transformation (3) transforms each subgroup G_1 of G into a **conjugate subgroup** $G_1' = a^{-1}G_1a$. *G_1 is mapped onto itself ($G_1' = G_1$) for every a in G* (1) *if and only if the elements a_1 of G_1 commute with every element a of G ($aG_1 = G_1a$) or* (2) *if and only if G_1 contains all conjugates of its elements.* A subgroup distinguished by these three properties is called a **normal subgroup (normal divisor, invariant subgroup)** of G.

Every subgroup of index 2 (*Sec.* 12.2-2) *is a normal subgroup of G.* A **simple group,** contains no normal subgroup except itself and E.

(**c**) *The left cosets* (*Sec.* 12.2-4*b*) *of a normal subgroup are identical with the corresponding right cosets and constitute a group with respect to the operation of multiplication defined in Sec.* 12.2-4*a*; this group is the **factor group (quotient group)** G/G_1. *If G is of finite order, the order of G/G_1 equals the index g/g_1 of G_1* (Sec. 12.2-2).

12.2-6. Normal Series. Composition Series. For any group G, a **normal series** is a (finite) sequence of subgroups $G_0 = G, G_1, G_2, \ldots, G_m = E$ such that each G_i is a normal subgroup of G_{i-1}. A normal series is a **composition series** of G if and only if each G_i is a proper normal subgroup of G_{i-1} such that no further normal subgroups can be "interpolated" between G_{i-1} and G_i, i.e., if and only if each **composition factor** G_{i-1}/G_i is a simple group (Sec. 12.2-5). *Given any two composition series of the same group G, there exists a reciprocal one-to-one correspondence between their respective elements such that corresponding elements are isomorphic groups* (*Jordan-Hölder Theorem*). G is a **solvable group** if and only if all its composition factors are cyclic groups (Sec. 12.2-3).

12.2-7. Center. Normalizers. (**a**) *The set of all elements of G which commute with every element of G is a normal subgroup of G, the* **center** *or* **central** *of G.*

(**b**) *The set of all elements of G which commute with a given element a of G is a subgroup of G* (**normalizer of** *a*), *which contains the period of a* (*Sec.* 12.2-3) *as a normal subgroup.*

The set of elements of G which commute with (*every element of*) *a given subgroup G_1 of G is a subgroup of G* (**normalizer of** G_1), *which contains G_1 as a normal subgroup.*

The number of different elements or subgroups of G conjugate (*Sec.* 12.2-5) *to the element or subgroup associated with a normalizer equals the index* (*Sec.* 12.2-2) *of the normalizer.*

12.2-8. Groups of Transformations or Operators (see also Secs. 12.1-4, 14.9-1 to 14.10-7, and 16.1-2). *The set G of all reciprocal one-to-one* (*nonsingular*) *transformations $x' = f(x)$ of any class C onto itself is a group;* the defining operation is the successive application of two transformations (*multiplication of transformations or operators*).

Given any subgroup G_1 of G, two objects x, x' related by a transformation $x' = f(x)$ of G_1 are **equivalent under** G_1. This relationship is an equivalence relation (Sec. 12.1-3), so that each subgroup G_1 defines a partition (classification) of C. *Every* *property invariant (Sec. 12.1-5) with respect to (all transformations of) G_1 is common to all objects x equivalent under G_1.* A set of (usually numerical) functions $F_1(x)$, $F_2(x)$, . . . invariant with respect to G_1 is a **complete set of invariants** with respect to G_1 if and only if the set of function values uniquely defines the equivalence class of any given object x of C.

EXAMPLES OF TRANSFORMATION GROUPS: All $n!$ permutations (see also Appendix C) of n elements (**symmetric group on n elements**); all $n!/2$ **even** permutations corresponding to even numbers of interchanges of n elements (**alternating group on n elements**). Every permutation of a countable set S of objects can be expressed as a product of cyclic permutations of subsets of S so that no two such **cycles** affect the same elements of S.

12.2-9. Homomorphisms and Isomorphisms of Groups. Representation of Groups (see also Secs. 12.2-4, 12.2-5, and 14.9-1 to 14.10-7). (a) A homomorphism or isomorphism (with respect to the defining operation, Sec. 12.1-6) maps the elements a, b, . . . of a given group G onto the elements a', b', . . . of a new group G' so that $(ab)' \to a'b'$. The identity of G is mapped onto the identity of G', and inverses are mapped onto inverses.

For every homomorphism of G onto G', the identity of G' corresponds to a normal subgroup G_E of G, and each element of G' corresponds to a coset of G_E; the factor group G/G_E is isomorphic with G'. G_E is called the **kernel** of the homomorphism. *Every normal subgroup G_1 of G is the kernel of a homomorphism mapping G onto the factor group G/G_1.*

The class of all automorphisms of any group G is a group; the class of all transformations (3) *of G* (**inner automorphisms** *of G*) *is a subgroup of the automorphism group.*

Conjugate subgroups are necessarily isomorphic.

(b) *Every group G can be* **realized (represented)** *by a homomorphism relating the group elements to a group of transformations mapping some class of objects onto itself (Cayley's Theorem;* see also Sec. 12.2-8). In particular, *every group of finite order is isomorphic to a group of permutations (regular representation of a finite group,* Sec. 14.9-1a). **Refer to Secs. 14.9-1 to 14.10-7 for representations of groups in terms of linear transformations and matrices.**

12.2-10. Additive Groups. Residue Classes and Congruence. (a) The defining operation of a *commutative* group (Abelian group, Sec. 12.2-1a) is often referred to as (abstract) addition. Its result may be written as a *sum* $a + b$, and the identity and inverse are written, respectively, as 0 (*zero or null element*) and $-a$, so that

$$a + (-b) = a - b$$

The group is then called an **additive group (addition group, module),** and expressions like $a + a$, $a + a + a$, $-a - a$, . . . may be written as $2a$, $3a$, $-2a$, ...

(b) *Every subgroup of a commutative group is a normal subgroup.* The cosets of a subgroup G_1 of an additive group G are called **residue classes modulo G_1**; two elements a, b of G belonging to the same residue class (i.e., G_1 contains $a - b$; see also Sec. 12.2-4b) are **congruent modulo G_1** $[a \equiv b(G_1)]$. Congruence is an equivalence relation (Sec. 12.1-3; see also Sec. 12.2-4b). The factor group G/G_1 of an additive group G is the group of residue classes modulo G_1 and may be denoted by G modulo G_1.

Two real integers m, n are **congruent modulo** r [$m \equiv n(r)$], where r is a third real integer, if and only if $m - n$ is a multiple of r, so that m/r and n/r have equal remainders.

12.2-11. Continuous and Mixed-continuous Groups. (a) The elements of a **continuous group** G can be labeled with corresponding sets of continuously variable parameters so that the parameters $\gamma_1, \gamma_2, \ldots$ of the product $c = ab$ are continuously differentiable functions of the parameters $\alpha_1, \alpha_2, \ldots$ of a and the parameters $\beta_1, \beta_2, \ldots$ of b. The parameters may, for instance, be the matrix elements in a representation of G (EXAMPLE: the three-dimensional rotation group, Sec. 14.10-7).

(b) A **mixed-continuous group** G is the union of n_1 separated (Sec. 12.5-1*b*) subsets "connected" in the sense of Sec. 12.2-11*a* ($n_1 = 1, 2, \ldots$). The connected subset G_1 containing the identity element is a normal subgroup of G with index n_1; the other connected subsets of G are the cosets associated with G_1 (see also Secs. 12.2-4 and 12.2-5. EXAMPLE: the reflection-rotation group in three-dimensional Euclidean space, see also Sec. 14.10-7).

12.2-12. Mean Values. The **mean value** of a numerical function $F(a)$ defined on the elements a of a group G is a linear functional (Sec. 12.1-4) defined for a group G of finite order g as

$$\text{Mean}\,\{F(a)\} = \frac{1}{g} \sum_{a \text{ in } G} F(a) \tag{12.2-4}$$

More generally, for a suitable numerical function $F(a)$ defined on a continuous or mixed-continuous group G,

$$\text{Mean}\,\{F(a)\} = \frac{\int_G F(a)\, dU(a)}{\int_G dU(a)} \tag{12.2-5a}$$

where the "volume element" $dU(a)$ is defined in terms of the parameters α_i, β_i, and γ_i introduced in Sec. 12.2-11*a* by

$$\left[dU(a) \equiv \frac{\partial(\gamma_1, \gamma_2, \ldots)}{\partial(\alpha_1, \alpha_2, \ldots)}\right]_{b=a^{-1}} d\alpha_1\, d\alpha_2 \cdots \tag{12.2-5b}$$

The definition (5) applies also to countably infinite and finite groups if the integrals are interpreted as Stieltjes integrals (Sec. 4.6-17) and reduces to Eq. (4) for finite groups.

The definition of Mean $\{F(a)\}$ *implies*

$$\text{Mean}\,\{F(a_0 a)\} = \text{Mean}\,\{F(a a_0)\} = \text{Mean}\,\{F(a)\} = \text{Mean}\,\{F(a^{-1})\} \tag{12.2-6}$$

for every fixed element a_0 in G.

12.3. ALGEBRA OF MODELS WITH TWO DEFINING OPERATIONS: RINGS, FIELDS, AND INTEGRAL DOMAINS

12.3-1. Definitions and Basic Theorems. (a) A class R of objects (elements) $a, b, c, \ldots$ is a **ring** if and only if it is possible to define two binary operations, usually denoted as (abstract) *addition* and *multiplication*, such that

1. R is a commutative group with respect to addition (additive group, Sec. 12.2-10); i.e., R is closed under addition, and

$$a + b = b + a \qquad a + (b + c) = (a + b) + c$$
$$a + 0 = a \qquad a + (-a) = a - a = 0$$

2. ab is an element of R (CLOSURE UNDER MULTIPLICATION)
3. $a(bc) = (ab)c$ (ASSOCIATIVE LAW FOR MULTIPLICATION)
4. $a(b + c) = ab + ac \qquad (b + c)a = ba + ca$ (DISTRIBUTIVE LAWS)

Note that $a0 = 0a = 0$ for every element a of R. Two elements p and q of R such that $pq = 0$ are called **left and right divisors of zero.** *For a* **ring without divisors of zero** (*other than zero itself*), $ab = 0$ *implies* $a = 0$, *or* $b = 0$, *or* $a = b = 0$, *and the cancellation laws* (12.2-1) *hold.*

"Multiples" of ring elements like $2a, 3a, \ldots$ (Sec. 12.2-10a) are *not* in general products of ring elements. *Integral powers* of ring elements are defined as in Sec. 12.2-1c.

(**b**) A **ring with identity** (**unity**) is a ring containing a multiplicative (left) identity E such that $Ea = a$ for all a (see also Sec. 12.2-1a). *E is necessarily a unique right identity as well as a unique left identity.* A given element a of a ring with identity may or may not have a multiplicative (left) inverse a^{-1}; *if a^{-1} does exist, it is necessarily a unique right inverse as well as a unique left inverse* (*see also Sec.* 12.2-1).

(**c**) A **field*** is a ring with identity which contains (1) at least one element different from zero and (2) a multiplicative inverse a^{-1} for each element $a \neq 0$. *The nonzero elements of a field F constitute a group with respect to multiplication.*

Given any pair of elements $b, c \neq 0$ of F, the equations $cx = b$ and $xc = b$ have solutions in F. These solutions are unique whenever $b \neq 0$ (*unique left and right division, see also Sec.* 12.2-1b); *they are unique even for $b = 0$ if F is a field without divisors of zero.*

(**d**) A ring or field is **commutative** if and only if $ab = ba$ for all a, b. A commutative field is sometimes simply called a field, as opposed to a **skew** or **noncommutative field.** A **Galois field** is a finite commutative field. An **integral domain** is a commutative ring with identity and without divisors of zero. *Every finite integral domain is a Galois field.*

In every integral domain all nonzero elements of the additive group have the same order (Sec. 12.2-3). This order is referred to as the **characteristic** of the integral domain.

(**e**) Refer to Sec. 12.6-3 for a brief discussion of *ordered fields.*

* Some writers require every field to be an integral domain (Sec. 12.3-1d).

EXAMPLES OF FIELDS: rational numbers, real numbers, complex numbers; continuous functions defined on a finite interval (has zero divisors); polynomials. EXAMPLES OF INTEGRAL DOMAINS: real integers, complex numbers with integral coefficients. EXAMPLE OF A COMMUTATIVE RING WITHOUT IDENTITY: even integers.

12.3-2. Subrings and Subfields. Ideals. (a) A subset R_1 of a ring R is a **subring** of R if and only if it is a ring with respect to the defining operations of R. This is true if and only if R_1 contains $a - b$ and ab for every pair of its elements a, b. Similarly a subset F_1 of a field F is a **subfield** of F if and only if it is a subring of F comprising at least one nonzero element and contains ab^{-1} for every pair of elements a, b of F_1. The nonzero elements of F_1 constitute a subgroup of the multiplicative group of F.

(b) A subset I_1 of a ring R is an **ideal** in R if and only if

1. I_1 is a subgroup of R with respect to addition.
2. I_1 contains all products ab (**left ideal**), or all products ba (**right ideal**), or all products ab and ba (**two-sided ideal**), where a is any element of I_1, and b is any element of R.

12.3-3. Extensions. It is often possible to "imbed" commutative rings or fields as subrings or subfields in "larger" fields (quotient fields, algebraic extension fields, etc.) in a manner analogous to that outlined in Sec. 1.1-2 for the real integers (see also the example in Sec. 12.4-2). The theory of fields, including the so-called Galois theory, deals with the existence of such extensions (APPLICATIONS: investigation of the possibility of constructions with ruler and compass, of solving algebraic equations in terms of radicals, and of constructing latin squares; Refs. 12.1 and 12.9).

12.4. MODELS INVOLVING MORE THAN ONE CLASS OF MATHEMATICAL OBJECTS: LINEAR VECTOR SPACES AND LINEAR ALGEBRAS

12.4-1. Linear Vector Spaces. Let R be a ring with (multiplicative) identity 1 (Sec. 12.3-1b); the elements $\alpha, \beta, \ldots$ of R shall be referred to as **scalars.** A class $\mathcal{V}$ of objects (elements $\mathbf{a}, \mathbf{b}, \mathbf{c}, \ldots$ is a (**linear**) **vector space over the ring** R, and the elements of $\mathcal{V}$ are called **vectors** if and only if it is possible to define two binary operations called **vector addition** and **multiplication of vectors by scalars** such that*

1. $\mathcal{V}$ is a commutative group with respect to vector addition: for every pair of elements $\mathbf{a}$, $\mathbf{b}$ of $\mathcal{V}$, $\mathcal{V}$ contains a **vector sum** $\mathbf{a} + \mathbf{b}$, and

$$\mathbf{a} + \mathbf{b} = \mathbf{b} + \mathbf{a} \qquad \mathbf{a} + (\mathbf{b} + \mathbf{c}) = (\mathbf{a} + \mathbf{b}) + \mathbf{c}$$
$$\mathbf{a} + \mathbf{0} = \mathbf{a} \qquad \mathbf{a} + (-\mathbf{a}) = \mathbf{a} - \mathbf{a} = \mathbf{0}$$

where $\mathbf{0}$ is the additive identity element (**null vector**) of $\mathcal{V}$, and $-\mathbf{a}$ is the additive inverse of $\mathbf{a}$ (Sec. 12.2-10)

* Some authors require that $\mathcal{V}$ contain not only every sum of two vectors, but also every infinite sum $\mathbf{a}_1 + \mathbf{a}_2 + \cdots$ which converges in some specified sense; vector spaces defined in this manner are not in the realm of algebra proper (see also Sec. 14.2-1).

2. Given any vector **a** of $\mathcal{V}$ and any scalar α of R, $\mathcal{V}$ contains a vector $\alpha\mathbf{a}$, the **product of the vector a by the scalar** α (CLOSURE UNDER MULTIPLICATION BY SCALARS)
3. $(\alpha\beta)\mathbf{a} = \alpha(\beta\mathbf{a})$ (ASSOCIATIVE LAW FOR MULTIPLICATION BY SCALARS)
4. $\alpha(\mathbf{a} + \mathbf{b}) = \alpha\mathbf{a} + \alpha\mathbf{b} \qquad (\alpha + \beta)\mathbf{a} = \alpha\mathbf{a} + \beta\mathbf{a}$ (DISTRIBUTIVE LAWS)
5. $1\mathbf{a} = \mathbf{a}$

Note that

$$0\mathbf{a} = \mathbf{0} \qquad (-1)\mathbf{a} = -\mathbf{a} \qquad (-\alpha)\mathbf{a} = -(\alpha\mathbf{a}) \tag{12.4-1}$$

Linear vector spaces are of fundamental importance in applied mathematics; they are treated in detail in Chap. 14 (see also Chaps. 5, 6, 15, 16, and 17).

12.4-2. Linear Algebras. Given a ring R of scalars with identity 1, a class L is a **linear algebra (linear associative algebra, system of hypercomplex numbers) over the ring** R if and only if it is possible to define three binary operations (**addition and multiplication in** L and **multiplication of elements of** L **by scalars**) such that

1. L is a ring
2. L is a linear vector space over the ring R of scalars

The **order** of a linear algebra is its dimension as a vector space (Sec. 14.2-4). A linear algebra is a **division algebra** if and only if it is a field (Sec. 12.3-1*c*). Refer to Sec. 14.9-7 for matrix representations of linear algebras.

An element $\mathsf{A} \neq 0$ of any linear algebra is **idempotent** if and only if $\mathsf{A}^2 = \mathsf{A}$ and **nilpotent** if and only if there exists a real integer $m > 1$ such that $\mathsf{A}^m = 0$. These definitions apply, in particular, to matrices (Sec. 13.2-2) and to linear operators (Sec. 14.3-6).

EXAMPLES (see also Secs. 13.2-2 and 14.4-2): (1) The *field of complex numbers* is a commutative division algebra of order 2 over the field of real numbers. (2) The field of **quaternions** $a, b, \ldots$ is the only extension (Sec. 12.3-3) of the complex-number field which constitutes a noncommutative division algebra over the field of real numbers. Every quaternion a can be represented in the form

$$a = \alpha_0 + i\alpha_1 + j\alpha_2 + k\alpha_3 = (\alpha_0 + i\alpha_1) + (\alpha_2 + i\alpha_3)j \tag{12.4-2}$$

where $\alpha_0, \alpha_1, \alpha_2, \alpha_3$ are real numbers, and i, j, k are special quaternions (**generators**) satisfying the multiplication rules

$$\begin{gathered} i^2 = j^2 = k^2 = -1 \\ jk = -kj = i \qquad ki = -ik = j \qquad ij = -ji = k \end{gathered} \tag{12.4-3}$$

If one defines $a^* = \alpha_0 - i\alpha_1 - j\alpha_2 - k\alpha_3$, then

$$|a|^2 = a^*a = aa^* = \alpha_0^2 + \alpha_1^2 + \alpha_2^2 + \alpha_3^2 \tag{12.4-4}$$
$$a^{-1} = a^*/|a|^2 \tag{12.4-5}$$

(see also Sec. 14.10-6).

12.5. MODELS PERMITTING THE DEFINITION OF LIMITING PROCESSES: TOPOLOGICAL SPACES

12.5-1. Topological Spaces (refer to Sec. 4.3-2 for elementary properties of point sets). (**a**) A class C of objects ("points") x is a **topological space** if and only if it can be expressed as the union of a family $\mathfrak{J}$ of point sets which contains

1. The intersection of every pair of its sets
2. The union of the sets in every subfamily

$\mathfrak{J}$ is a **topology** for the space C, and the elements of $\mathfrak{J}$ are called **open sets** relative to the topology $\mathfrak{J}$. A family $\mathfrak{B}$ of open sets is a **base** for the topology $\mathfrak{J}$ if and only if every set of $\mathfrak{J}$ is the union of sets in $\mathfrak{B}$.

The intersections of any subset C_1 of a topological space C with its open sets constitute a topology for C_1 (**relative topology** of the **subspace** C_1 of C, **relativization** of $\mathfrak{J}$ to C_1).

A given space may admit more than one topology; every space C admits the **indiscrete** (**trivial**) topology comprising only C and the empty set, and the **discrete** topology comprising all subsets of C.

(**b**) For a given topology, a **neighborhood** of a point x in C is any point set in C which contains an open set comprising x. Given the definition of neighborhoods, one defines *limit points, interior points, boundary points*, and *isolated points* of sets in the manner of Sec. 4.3-6a; topological spaces are seen to abstract and generalize certain features of the real-number system.

In any topological space C, *a point set is open if and only if it contains only interior points;* a point set S in C is **closed** (1) if and only if S is the complement (with respect to C) of an open set, or (2) if and only if S contains all its limit points (*alternative definitions*). S is **dense** in (relative to) C if and only if every neighborhood in C contains a point of S. A topological space C is **separable** if and only if it contains a countable dense set; *this is true whenever the topology of C has a countable base.*

A topological space is **compact** if and only if every infinite sequence of points in C has at least one limit point in C; *this is true if and only if every family of open sets which covers (i.e., whose union exhausts) C has a finite subfamily which covers C (alternative definition). Every compact space is separable.*

A point set S in a topological space C is **compact** (**compact in** C) if every infinite sequence of points in S has at least one limit point in C; S is **compact in itself** if every such sequence has a limit point in S.

Separable spaces are of special interest because their points can be labeled with countable sets of coordinates (Sec. 14.5-1). *Compactness* is a generalization of the concept of boundedness (Sec. 4.3-3) in finite-dimensional spaces, where the Bolzano-Weierstrass and Heine-Borel theorems apply (Sec. 12.5-4).

(c) The limit points and the boundary points of a set S respectively constitute its **(first) derived set** S' and its **boundary.** The closed set $S \cup S'$ is the **closure** of S. Two sets are **separated** if and only if neither intersects the closure of the other. A set is **connected** if and only if it cannot be expressed as the union of separated proper subsets (EXAMPLE: connected region in a Euclidean space, Sec. 4.3-6*b*).

(d) Continuity. Homeomorphisms. A correspondence (mapping, transformation, function, operation) $x \to x' = f(x)$ relating points x' of a topological space C' to points x of a topological space C is **continuous at the point** $x = a$ if and only if $f(a)$ exists and every neighborhood of $f(a)$ is the image of a neighborhood of a. $f(x)$ is **continuous** if and only if it is continuous at all points of C, i.e., if and only if every open set in C' is the image of an open set in C.

A continuous reciprocal one-to-one correspondence having a continuous inverse is a **homeomorphism** or **topological transformation;** topological spaces so related are **homeomorphic** or **topologically equivalent.**

(d) **Topology** is the study of relationships invariant with respect to homeomorphisms (**topological invariants;** see also Sec. 12.1-5). Topology has geometrical applications ("rubber-sheet geometry," study of multiply connected surfaces, etc.); more significantly, *suitable topological spaces are models permitting the definition of converging limiting processes* with the aid of the neighborhood concept (see also Secs. 4.3-5, 4.4-1, and 12.5-3). Definitions of specific topologies are often phrased directly as definitions of neighborhoods or convergence.

EXAMPLES: The definition of neighborhoods given in Sec. 4.3-5*a* establishes the "usual" topology of the real-number system and permits the introduction of limits, differentiation, integration, infinite series, etc. Similarly, Sec. 5.3-1 amounts to the definition of a topology for the space of Euclidean vectors (see also Secs. 12.5-3 and 14.2-7).

12.5-2. Metric Spaces. A class C_M of objects ("points") $x, y, z, \ldots$ is a **metric space** if and only if for each ordered pair of points x, y of C it is possible to define a real number $d(x, y)$ (**metric, distance function,** "distance" between x and y) such that for all x, y, z in C_M

1. $d(x, x) = 0$
2. $d(x, y) \leq d(x, z) + d(y, z)$ (TRIANGLE PROPERTY)

This definition implies

$$d(x, y) \geq 0 \qquad d(y, x) = d(x, y) \tag{12.5-1}$$

for all x, y *in* C_M.

x and y are **metrically equal** if and only if $d(x, y) = 0$; this does not necessarily imply $x = y$. *Metric equality is an equivalence relation* (Sec. 12.1-3*b*).

Two metric spaces are **isometric** if and only if they are related by a distance-preserving reciprocal one-to-one correspondence (**isometry**). Properties common to all metric spaces isometric to a given metric space are **metric invariants.**

EXAMPLES: The (finite) real and complex numbers constitute metric spaces with the metric $d(x, y) \equiv |x - y|$. More generally, every normed vector space (Sec. 14.2-5), and thus every unitary vector space, admits the metric $d(\mathbf{x}, \mathbf{y}) \equiv \|\mathbf{x} - \mathbf{y}\|$ (see also Secs. 14.2-7 and 15.2-2).

12.5-3. Topology, Neighborhoods, and Convergence in a Metric Space. (**a**) Given any point a of a metric space C_M, the set (region) of points x in C_M such that $d(a, x) < \delta$ is called an **open ball of radius δ about a.** *The open balls of finite radii constitute a base for a topology in C_M* (Sec. 12.5-1*a*); open sets are unions of open spheres. A **neighborhood** of the point a in C_M is any set containing an open ball of finite radius about a. $d(a, x) \leq r$ defines a **closed ball** in C_M, and $d(a, x) = r$ defines a **sphere.**

(**b**) A sequence of points $x_0, x_1, x_2, \ldots$ of a metric space C_M is said to **converge (metrically) to the limit** a in C_M if and only if $d(x_n, a) \to 0$ as $n \to \infty$ (see also Sec. 14.2-7).

If a variable point $x(\xi)$ of C_M is a function of the real variable ξ, $x(\xi)$ **converges (metrically) to the limit** a in C_M as $\xi \to \alpha$ if and only if $d[x(\xi), a] \to 0$ as $\xi \to \alpha$. $x(\xi)$ is continuous for $\xi = \alpha$ (Sec. 12.5-1*c*) if and only if $x(\alpha)$ exists and $d[x(\xi), x(\alpha)] \to 0$ as $\xi \to \alpha$. A function $x' = f(x)$ relating points x' of a metric space C'_M to points x of a metric space C_M is continuous (Sec. 12.5-1*c*) at the point $x = a$ if and only if $f(a)$ exists and $d[f(x), f(a)] \to 0$ as $d[x, a] \to 0$.

12.5-4. Metric Spaces with Special Properties. Point-set Theory (see also Secs. 4.3-6 and 12.5-1). A metric space C_M is

Complete if and only if every sequence of points $x_1, x_2, \ldots$ of C_M such that $\lim_{\substack{m\to\infty \\ n\to\infty}} d(x_m, x_n) = 0$ (**Cauchy sequence**) converges to a point of C_M (see also Secs. 4.9-1*a*, 4.9-2*a*, 4.9-3*a*, 14.2-7, and 15.2-2).

Precompact if and only if, for every real $\epsilon > 0$, there exists a finite set of balls of radius less than ϵ covering C_M.

Conditionally compact if and only if every infinite set of points in C_M contains a Cauchy sequence.

Boundedly compact or **boundedly conditionally compact** if and only if every closed ball in C_M is, respectively, compact or conditionally compact.

Every closed set in a metric space is a subspace with the same metric. A point set S in a complete metric space C_M is a complete metric subspace if and only if S is closed. Every set compact in itself (*Sec.* 12.5-1) *is bounded. A metric space is compact if and*

only if it is precompact and complete; every precompact metric space is separable. Every finite-dimensional unitary vector space (Euclidean vector space, Sec. 14.2-7) is separable, complete, and boundedly compact; in every Euclidean vector space

1. *Every infinite point set contained in a closed ball K_a of finite radius has at least one limit point in K_a (Bolzano-Weierstrass Theorem).*
2. *Given any rule associating a ball K_a of finite radius with each point x of a closed set S contained in a ball of finite radius, there exists a finite set of these balls which covers (includes) all points of S (Heine-Borel Covering Theorem).*

These theorems apply, in particular, to sets of real or complex numbers, and to point sets in two- or three-dimensional Euclidean spaces.

12.5-5. Examples: Spaces of Numerical Sequences and Functions. The concepts introduced in Secs. 12.5-2 to 12.5-4 offer a concise and suggestive terminology for many problems involving the *approximation* of an arbitrary element of a suitable class C_M by a sequence of elements $x_1, x_2, \ldots$ of C_M. In such instances, the distance $d(x_n, x)$ measures the *error of the approximation*, or the degree to which a system characterized by $x_1, x_2, \ldots$ fails to meet a performance criterion.

Table 12.5-1 lists a number of examples of topological spaces. Especially important applications involve the approximation of a function $f(t)$ by a sequence of functions $s_1(t), s_2(t), \ldots$, such as the partial sums of an infinite series (see also Secs. 4.7-3, 5.3-11, 13.2-1, 13.2-11, 14.2-7, and 15.2-2).

12.5-6. Banach's Contraction-mapping Theorem and Successive Approximations. *Let $x' = f(x)$ be any transformation mapping the*

Table 12.5-1a. Some Spaces of Numerical Sequences
$x \equiv (\xi_1, \xi_2, \ldots), y \equiv (\eta_1, \eta_2, \ldots)$

Common designation	Definition	Metric $d(x, y)$	Remarks
m	Bounded real sequences	$\sup\limits_k \lvert\xi_k - \eta_k\rvert$	Complete; not separable
c	Convergent real sequences	$\sup\limits_k \lvert\xi_k - \eta_k\rvert$	All are complete and separable
l^2	Complex sequences such that $\sum_{k=1}^{\infty} \lvert\xi_k\rvert^2$ exists	$\left[\sum_{k=1}^{\infty} \lvert\xi_k - \eta_k\rvert^2\right]^{1/2}$	
l^p	Complex sequences such that $\sum_{k=1}^{\infty} \lvert\xi_k\rvert^p$ exists $(p = 1, 2, \ldots)$	$\left[\sum_{k=1}^{\infty} \lvert\xi_k - \eta_k\rvert^p\right]^{1/p}$	

Table 12.5-1b. Some Spaces of Functions $x(t)$, $y(t)$ (see also Secs. 14.2-7, 15.2-2, and 18.10-9; the definitions are readily extended to functions of two or more variables)

Common designation	Definition	Metric $d(x, y)$	Remarks
$F[0, 1]$	Real functions defined on $[0,1]$	No metric exists. Topology defined by pointwise convergence	
$C[0, 1]$	Continuous real functions defined on $[0,1]$	$\sup_{0 \leq t \leq 1} \lvert x(t) - y(t) \rvert$	Complete; separable (Sec. 4.7-3); uniform convergence
$L^2(a, b)$	Complex functions such that $\int_a^b \lvert x(t) \rvert^2 \, dt$ exists in the sense of Lebesgue (Sec. 4.6-15)	$\left[\int_a^b \lvert x(t) - y(t) \rvert^2 \, dt \right]^{1/2}$	Complete (Sec. 15.2-2); separable even if (a, b) is not bounded; convergence in mean
$L^p(a, b)$	Complex functions such that $\int_a^b \lvert x(t) \rvert^p \, dt$ exists in the sense of Lebesgue ($p = 1, 2, \ldots$)	$\left[\int_a^b \lvert x(t) - y(t) \rvert^p \, dt \right]^{1/p}$	
$\tilde{L}^2$	Complex functions such that $\lim_{T \to \infty} \frac{1}{T} \int_{-T/2}^{T/2} \lvert x(t) \rvert^2 \, dt$ exists in the sense of Lebesgue	$\left[\lim_{T \to \infty} \frac{1}{T} \int_{-T/2}^{T/2} \lvert x(t) - y(t) \rvert^2 \, dt \right]^{1/2}$	Complete; not separable

closed set S of a complete metric space C_M into itself so that

$$d[f(x), f(y)] \leq \alpha \, d(x, y) \qquad (12.5\text{-}2)$$

for all x, y in S, where α is a positive real number less than unity ("contraction mapping"). Then S contains a unique **fixed point x_f of the mapping** *f such that*

$$f(x_f) = x_f \qquad (12.5\text{-}3)$$

Moreover, the solution x_f of the equation (3) *is the limit of every sequence of successive approximations*

$$x_{n+1} = f(x_n) \qquad (n = 0, 1, 2, \ldots) \qquad (12.5\text{-}4)$$

as $n \to \infty$, for an arbitrary starting point x_0 in S. The rate of convergence of the approximation sequence is given by

$$d(x_n, x_f) \leq \frac{\alpha^n}{1 - \alpha} d(x_1, x_0) \qquad (n = 0, 1, 2, \ldots) \qquad (12.5\text{-}5)$$

The contraction-mapping principle furnishes a powerful method for establishing the convergence of a wide range of approximation techniques (Secs. 20.2-1, 20.2-6, and 20.3-5).

12.6. ORDER

12.6-1. Partially Ordered Sets. (a) **Order.** A class (set) S of objects (elements) $a, b, c, \ldots$ is **partially ordered** if and only if it admits a transitive **ordering relation** (rule of precedence) $a < b$ between some pairs of elements a, b so that

$$a < b, b < c \quad \text{implies} \quad a < c \tag{12.6-1}$$

$a < b$ may or may not preclude $b < a$ (*antisymmetry*). The symbol $\leq$ is used to indicate a *reflexive* ordering relation which satisfies the condition $a \leq a$ for all a.

A partially ordered set permits the definition of **upper bounds, lower bounds, least upper bounds, greatest lower bounds,** maxima, and/or minima of suitable subsets in the manner of Sec. 4.3-3. A partially ordered set S is **order-complete** if and only if every nonempty subset having an upper bound has a least upper bound in S; this is true if and only if every nonempty subset having a lower bound has a greatest lower bound in S (*alternative definitions*).

(b) **Lattices.** A (nonempty) class of objects (elements) $a, b, \ldots$ is a **lattice** if and only if it admits a reflexive ordering relation such that every pair a, b of elements has a unique least upper bound $\sup(a, b)$ and a unique greatest lower bound $\inf(a, b)$. In every lattice one can define "sums" $a + b \equiv \sup(a, b)$ and "products" $ab \equiv \inf(a, b)$. *Sums and products so defined have the properties* 1, 2, 3, 5 *listed in Sec.* 12.8-1 *for Boolean algebras.*

12.6-2. Simply Ordered Sets. A class (set) S of elements $a, b, c, \ldots$ is a **simply (linearly, totally, completely) ordered set (chain)** if and only if it admits an ordering relation having the transitivity property (1) and such that*

$$\begin{aligned} &a < b \text{ or } b < a \text{ for every pair of distinct elements } a, b \\ &a < b, b < a \quad \text{implies} \quad a = b \end{aligned} \tag{12.6-2}$$

The ordering relation may or may not be reflexive, but every simply ordered set admits the reflexive ordering relation defined by $a \leq b$ if $a < b$ or $a = b$.

A simply ordered set S is **well ordered** if and only if every nonempty subset of S has a minimum. *Every countable simply ordered set is well ordered.*

12.6-3. Ordered Fields (see also Secs. 1.1-5 and 12.3-1). (a) A commutative field is an **ordered field** if and only if each element can be uniquely classified as "positive" (>0), "negative" (<0), or zero ($= 0$) in such a way that

$$a > 0, b > 0 \quad \text{implies} \quad a + b > 0, ab > 0 \tag{12.6-3}$$

(b) *Every order-complete* (*Sec.* 12.6-1*a*) *ordered field is isomorphic* (*with respect to addition and multiplication*) *with the field of real numbers. Every ordered integral domain whose positive elements are well ordered* (*Sec.* 12.6-2) *is isomorphic with the field*

* Some authors replace the second condition (2) by "$a < b$ precludes $b < a$."

of real integers. The last theorem expresses the equivalence of binary numbers, decimal numbers, roman numerals, etc.

12.7. COMBINATION OF MODELS: DIRECT PRODUCTS, PRODUCT SPACES, AND DIRECT SUMS

12.7-1. Introduction. Cartesian Products. Sections 12.7-1 to 12.7-5 deal with a class C of mathematical objects described in terms of two or more properties and represented as ordered sets $\{a_1, a_2, \ldots\}$ of objects $a_1, a_2, \ldots$ respectively taken from suitably defined classes $C_1, C_2, \ldots$. The objects $a_1, a_2, \ldots$ may be regarded as properties or attributes of the new object $\{a_1, a_2, \ldots\}$; one defines

$$\{a_1, a_2, \ldots\} = \{b_1, b_2, \ldots\}$$

if and only if $a_1 = b_1, a_2 = b_2, \ldots$. The class C is called the **cartesian product** of the classes $C_1, C_2, \ldots$. *This method of combining mathematical objects into more complicated mathematical objects is associative:*

$$\{a_1, \{a_2, a_3\}\} \equiv \{\{a_1, a_2\}, a_3\} \equiv \{a_1, a_2, a_3\} \qquad (12.7\text{-}1)$$

It remains to relate operations in C to operations in $C_1, C_2, \ldots$.

12.7-2. Direct Products of Groups (see also Sec. 12.2-1). The **direct product $G = G_1 \otimes G_2$ of two groups G_1 and G_2** having the respective elements $a_1, b_1, \ldots$ and $a_2, b_2, \ldots$ is the group comprising all ordered pairs $\{a_1, a_2\}$, with "multiplication" defined by

$$\{a_1, a_2\}\{b_1, b_2\} \equiv \{a_1b_1, a_2b_2\} \qquad (12.7\text{-}2)$$

The order of G equals the product of the orders of G_1 and G_2. G has the identity $E = \{E_1, E_2\}$, where E_1 and E_2 are the respective identities of G_1 and G_2. If G_1 and G_2 have no common elements, one may write $\{a_1, a_2\}$ as an **outer product** a_1a_2, so that $a_1E_2 \equiv a_1$, $E_1a_2 \equiv a_2$, and G_1 and G_2 are subgroups of G.

EXAMPLES: Each scalar quantity in physics is an outer product of a number and a unit of measurement. Expressions like *work* = *mass* × *acceleration* × *displacement* are outer products.

12.7-3. Direct Products of Real Vector Spaces (see also Secs. 12.4-1, 14.2-1, and 14.2-4). The **direct product $\mathcal{V} = \mathcal{V}_1 \otimes \mathcal{V}_2$ of two real vector spaces $\mathcal{V}_1$ and $\mathcal{V}_2$** having the respective elements $\mathbf{a}_1, \mathbf{b}_1, \ldots$ and $\mathbf{a}_2, \mathbf{b}_2, \ldots$ is the real vector space comprising all ordered pairs (outer products) $\{\mathbf{a}_1, \mathbf{a}_2\} \equiv \mathbf{a}_1\mathbf{a}_2$, with vector addition and multiplication of vectors by scalars defined so that

$$\mathbf{a}_1(\mathbf{a}_2 + \mathbf{b}_2) \equiv \mathbf{a}_1\mathbf{a}_2 + \mathbf{a}_1\mathbf{b}_2 \qquad (\mathbf{a}_1 + \mathbf{b}_1)\mathbf{a}_2 \equiv \mathbf{a}_1\mathbf{a}_2 + \mathbf{b}_1\mathbf{a}_2$$
$$\alpha\mathbf{a}_1\mathbf{a}_2 \equiv (\alpha\mathbf{a}_1)\mathbf{a}_2 \equiv \mathbf{a}_1(\alpha\mathbf{a}_2) \qquad (12.7\text{-}3)$$

The linear dimension of the vector space $\mathcal{V}$ *equals the product of the linear dimensions of* $\mathcal{V}_1$ *and* $\mathcal{V}_2$.

EXAMPLE: Construction of tensors as outer products of vectors, Secs. 16.3-6, 16.6-1, and 16.9-1.

12.7-4. Product Space (see also Sec. 12.5-1*a*). The **product space** formed by two topological spaces C_1, C_2 is their cartesian product with the **product topology** (family of open sets in the product space) defined as the family of all cartesian products of S_1 and S_2, where S_1 is an open set in C_1, and S_2 is an open set in C_2.

12.7-5. Direct Sums. (a) The direct sum $\mathcal{V}' = \mathcal{V}_1 \oplus \mathcal{V}_2$ **of two vector spaces** $\mathcal{V}_1$ **and** $\mathcal{V}_2$ having the respective elements $\mathbf{a}_1$, $\mathbf{b}_1$, . . . and $\mathbf{a}_2$, $\mathbf{b}_2$, . . . and admitting the same ring R of scalars is the vector space comprising all pairs $[\mathbf{a}_1, \mathbf{a}_2] \equiv [\mathbf{a}_2, \mathbf{a}_1]$ with vector addition and multiplication of vectors by scalars α of R defined so that

$$[\mathbf{a}_1, \mathbf{a}_2] + [\mathbf{b}_1, \mathbf{b}_2] \equiv [\mathbf{a}_1 + \mathbf{b}_1, \mathbf{a}_2 + \mathbf{b}_2] \qquad \alpha[\mathbf{a}_1, \mathbf{a}_2] \equiv [\alpha\mathbf{a}_1, \alpha\mathbf{a}_2] \qquad (12.7\text{-}4)$$

The dimension of $\mathcal{V}'$ *equals the sum of the dimensions of* $\mathcal{V}_1$ *and* $\mathcal{V}_2$. If $\mathcal{V}_1$ and $\mathcal{V}_2$ have no common elements, one may write $[\mathbf{a}_1, \mathbf{a}_2] \equiv \mathbf{a}_1 + \mathbf{a}_2$, and $\mathcal{V}_1$ and $\mathcal{V}_2$ are subspaces of $\mathcal{V}'$. *Every linear vector space of dimension greater than one can be represented as a direct sum of nonintersecting subspaces.*

(b) The direct sum $R' = R_1 \oplus R_2$ **of two rings** R_1 **and** R_2 having the respective elements $a_1, b_1,$ . . . and $a_2, b_2,$ . . . is the ring comprising all ordered pairs (often referred to as direct sums) $[a_1, a_2]$ with addition and multiplication defined so that

$$[a_1, a_2] + [b_1, b_2] \equiv [a_1 + b_1, a_2 + b_2]$$
$$[a_1, a_2][b_1, b_2] \equiv [a_1b_1, a_2b_2] \qquad (12.7\text{-}5)$$

(c) The **direct sum of two linear algebras** (Sec. 12.4-2) is the linear algebra of ordered pairs, with addition, multiplication, and multiplication by scalars defined as in Eqs. (4) and (5).

12.8. BOOLEAN ALGEBRAS

12.8-1. Boolean Algebras. A **Boolean algebra** is a class $\mathcal{S}$ of objects $A, B, C,$ . . . admitting two binary operations, denoted as (*logical*) *addition and multiplication*, with the following properties:

(**a**) For all A, B, C in $\mathcal{S}$

1. $\mathcal{S}$ contains $A + B$ and AB (CLOSURE)
2. $A + B = B + A$, $AB = BA$ (COMMUTATIVE LAWS)
3. $A + (B + C) = (A + B) + C$, $A(BC) = (AB)C$ (ASSOCIATIVE LAWS)
4. $A(B + C) = AB + AC$, $A + BC = (A + B)(A + C)$ (DISTRIBUTIVE LAWS)
5. $A + A = AA = A$ (IDEMPOTENT PROPERTIES)
6. $A + B = B$ if and only if $AB = A$ (CONSISTENCY PROPERTY)

(**b**) In addition,

7. $\mathcal{S}$ contains elements I and 0 such that, for every A in $\mathcal{S}$,

$$A + 0 = A \qquad AI = A$$

(it follows that $\quad A0 = 0 \qquad A + I = I$)

8. For every element A, $\mathcal{S}$ contains an element $\tilde{A}$ (**complement** of A, also written $\bar{A}$ or $I - A$) such that

$$A + \tilde{A} = I \qquad A\tilde{A} = 0$$

In every Boolean algebra,

$$A(A + B) \equiv A + AB \equiv A \qquad \text{(LAWS OF ABSORPTION)} \qquad (12.8\text{-}1)$$

$$\left.\begin{aligned} \widetilde{(A + B)} &\equiv \tilde{A}\tilde{B} \\ \widetilde{(AB)} &\equiv \tilde{A} + \tilde{B} \end{aligned}\right\} \qquad \text{(DUALITY, OR DE MORGAN'S LAWS)} \qquad (12.8\text{-}2)$$

$$\tilde{\tilde{A}} \equiv A \qquad \tilde{I} = 0 \qquad \tilde{0} = I \qquad (12.8\text{-}3)$$

$$A + \tilde{A}B \equiv A + B \qquad AB + AC + B\tilde{C} \equiv AC + B\tilde{C} \qquad (12.8\text{-}4)$$

If $A + B = B$, one may write $\tilde{A}B$ as $B - A$ (*complement of A with respect to B*). Two or more objects $A, B, C, \ldots$ of a Boolean algebra are **disjoint** if and only if every product involving distinct elements of the set equals 0.

The symbols $\cup$ (cup) and $\cap$ (cap) used in Secs. 4.3-2*a*, 12.8-5*b*, and 18.2-1 to denote union and intersection of sets and events are frequently employed to denote logical addition and multiplication in any Boolean algebra; so that $A \cup B$ stands for $A + B$, and $A \cap B$ stands for AB.

12.8-2. Boolean Functions. Reduction to Canonical Form. Given n **Boolean variables** $X_1, X_2, \ldots, X_n$ each of which can equal any element of a given Boolean algebra, a **Boolean function**

$$Y = F(X_1, X_2, \ldots, X_n)$$

is an expression built up from $X_1, X_2, \ldots, X_n$ through addition, multiplication, and complementation.

In every Boolean algebra, there exist exactly $2^{(2^n)}$ *different Boolean functions of n variables.*

The relations of Sec. 12.8-1 imply

$$\tilde{F}(X, Y, \ldots, +, \cdot) \equiv F(\tilde{X}, \tilde{Y}, \ldots, \cdot, +) \tag{12.8-5}$$

$$\begin{aligned} F(X, \tilde{X}, Y, \ldots) &\equiv XF(I, 0, Y, \ldots) + \tilde{X}F(0, I, Y, \ldots) \\ &\equiv [X + F(0, I, Y, \ldots)][\tilde{X} + F(I, 0, Y, \ldots)] \end{aligned} \tag{12.8-6}$$

$$\left.\begin{aligned} XF(X, \tilde{X}, Y, \ldots) &\equiv XF(I, 0, Y, \ldots) \\ X + F(X, \tilde{X}, Y, \ldots) &\equiv X + F(0, I, Y, \ldots) \end{aligned}\right\} \tag{12.8-7}$$

(**b**) *Every Boolean function is either identical to* 0 *or can be expressed as a unique sum of* **minimal polynomials (canonical minterms)** $Z_1Z_2 \ldots Z_n$, *where* Z_i *is either* X_i *or* $\tilde{X}_i$ *(canonical form of a Boolean*

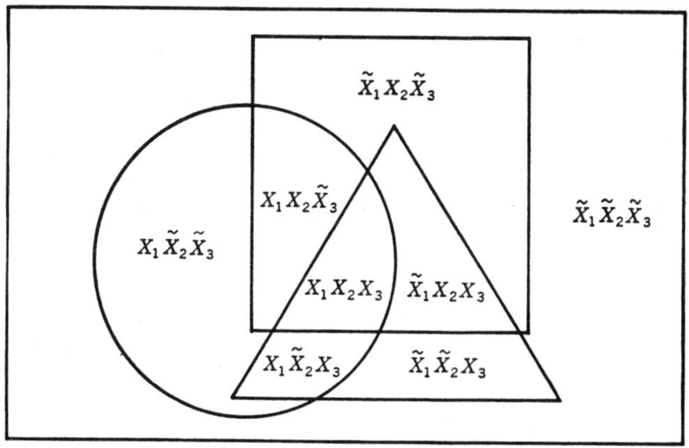

FIG. 12.8-1. A Venn diagram (Euler diagram). Diagrams of this type illustrate relations in an algebra of classes (Sec. 12.8-5). If the rectangle, circle, square, and triangle are respectively labeled by I, X_1, X_2, X_3, the diagram shows how a Boolean function of X_1, X_2, X_3 can be represented as a union of minimal polynomials in X_1, X_2, X_3 (Sec. 12.8-2). Note that there are $2^3 = 8$ different minimal polynomials.

function; see Fig. 12.8-1 for a geometrical illustration). A given Boolean function $Y = F(X_1, X_2, \ldots, X_n)$ may be reduced to canonical form as follows:

1. Use Eq. (2) to expand complements of sums and products.
2. Reduce $F(X_1, X_2, \ldots, X_n)$ to a sum of products with the aid of the first distributive law.
3. Simplify the resulting expression with the aid of the identities $X_iX_i \equiv X_i$, $X_i\tilde{X}_i \equiv 0$, and Eq. (4).
4. If a term f does not contain one of the variables, say X_i, rewrite f as $fX_i + f\tilde{X}_i$.

In many applications, it may be advantageous to omit step 4 and to continue step 3 so as to simplify each term of the expansion as much as possible (see also Sec. 12.8-7).

In view of de Morgan's laws (2), *every Boolean function not identically zero can also be expressed as a unique product of* **canonical maxterms** $Z_1' + Z_2' + \cdots + Z_n'$, *where Z_i is either X_i or $\tilde{X}_i$.* There are, altogether, 2^n minterms and 2^n maxterms.

EXAMPLES:

$$\left.\begin{aligned} \widetilde{(AB + CD)} &= (\tilde{A} + \tilde{B})(\tilde{C} + \tilde{D}) = \tilde{A}\tilde{C} + \tilde{A}\tilde{D} + \tilde{B}\tilde{C} + \tilde{B}\tilde{D} \\ \widetilde{(A + B)(C + D)} &= \tilde{A}\tilde{B} + \tilde{C}\tilde{D} = (\tilde{A} + \tilde{C})(\tilde{A} + \tilde{D})(\tilde{B} + \tilde{C})(\tilde{B} + \tilde{D}) \end{aligned}\right\} \qquad (12.8\text{-}8)$$

12.8-3. The Inclusion Relation (see also Sec. 12.6-1). **(a)** *Either $A + B = B$ or $AB = A$ is equivalent to a reflexive partial ordering relation $A \leq B$ [or $B \geq A$;* **(logical) inclusion relation**].

(b) *In every Boolean algebra, $A \leq B$, $B \leq A$ implies $A = B$, and*

$$A + B \equiv \sup (A, B) \qquad AB \equiv \inf (A, B) \qquad (12.8\text{-}9)$$

where the bounds are defined by the inclusion relation. Every Boolean algebra is a lattice (Sec. 12.6-1*b*).

(c) *Given any element A of a Boolean algebra $\mathcal{S}$, the elements $XA \leq A$ of $\mathcal{S}$ constitute a Boolean algebra in which A takes the place of I.*

12.8-4. Algebra of Classes. The subsets (subclasses) $A, B, \ldots$ of any set (class) I constitute a Boolean algebra (**algebra of classes**) under the operations of logical addition (union), logical multiplication (intersection), and complementation defined in Sec. 4.3-2*a*. The empty set (or any set which contains no element of I) is denoted by 0. The relation $\leq$ (Sec. 12.8-2) becomes the logical inclusion relation $\subset$.

12.8-5. Isomorphisms of Boolean Algebras. Venn Diagrams. *Two Boolean algebras are isomorphic (with respect to addition, multiplication, and complementation,* Sec. 12.1-6) *if and only if they have the same number of elements. Every Boolean algebra is isomorphic to an algebra of classes* (Sec. 12.8-4); Venn diagrams (Euler diagrams) like that shown in Fig. 12.8-1 conveniently illustrate the properties of Boolean algebras in terms of an algebra of classes.

12.8-6. Event Algebras and Symbolic Logic. Event algebras serve as models for the compounding of events (suitably defined outcomes of idealized experiments; see Sec. 18.2-1 for additional discussion). If $E_1, E_2, \ldots$ represent such events, then

$E_1 \cup E_2$ represents the event (proposition) E_1 OR E_2 (or both; **inclusive OR**)

$E_1 \cap E_2$ represents the event (proposition) E_1 AND E_2

$\tilde{E}$ represents the event (proposition) NOT E
I represents a certain event (union of all possible outcomes, Sec. 18.2-1)
0 represents an impossible event

In **two-valued (Aristotelian) logic,** an algebra of hypothetical events (**logical propositions, assertions**) E is related to a simpler Boolean algebra of **truth values** $T[E]$ equal to either 1 (E is **true**) or 0 (E is **false**) by the homomorphism (Sec. 12.1-6)

$$\left.\begin{array}{c} T[I] = 1 \qquad T[0] = 0 \\ T[E_1 \cup E_2] = T[E_1] + T[E_2] \qquad T[E_1 \cap E_2] = T[E_1]T[E_2] \\ T[\tilde{E}] = \widetilde{T[E]} \end{array}\right\} \tag{12.8-10}$$

On the basis of these assumptions, a proposition E is either true or false (*law of the excluded middle*), and the truth value of any proposition E expressible as a Boolean function of ("logically related to") a set of events $E_1, E_2, \ldots$ is given by

$$T[E] = T[F(E_1, E_2, \ldots)] = T[F\{T[E_1], T[E_2], \ldots\}] \tag{12.8-11}$$

with

$$\left.\begin{array}{ccc} 0 + 0 = 0 & 0 + 1 = 1 & 1 + 1 = 1 \\ 0 \cdot 0 = 0 & 0 \cdot 1 = 0 & 1 \cdot 1 = 1 \\ & \tilde{0} = 1 \qquad \tilde{1} = 0 & \end{array}\right\} \tag{12.8-12}$$

12.8-7. Representation of Boolean Functions by Truth Tables. Karnaugh Maps. Given a set of Boolean variables $X_1, X_2, \ldots, X_n$ each capable of taking the values 0 or 1 (Sec. 12.8-6), each of the $2^{(2^n)}$ Boolean functions

$$Y = F(X_1, X_2, \ldots, X_n)$$

is uniquely defined by the corresponding **truth table** (Table 12.8-1) listing the function values for all possible arguments. Table 12.8-1 also lists a common arrange-

Table 12.8-1.* Truth Table for

$$F = \tilde{X}\tilde{Y}Z + \tilde{X}Y\tilde{Z} + \tilde{X}YZ + X\tilde{Y}Z + XY\tilde{Z}$$
$$= (X + Y + Z)(\tilde{X} + Y + Z)(\tilde{X} + \tilde{Y} + \tilde{Z})$$

X	Y	Z	F	Corresponding minterm
0	0	0	0	$\tilde{X}\tilde{Y}\tilde{Z} = m_0$
0	0	1	1	$\tilde{X}\tilde{Y}Z = m_1$
0	1	0	1	$\tilde{X}Y\tilde{Z} = m_2$
0	1	1	1	$\tilde{X}YZ = m_3$
1	0	0	0	$X\tilde{Y}\tilde{Z} = m_4$
1	0	1	1	$X\tilde{Y}Z = m_5$
1	1	0	1	$XY\tilde{Z} = m_6$
1	1	1	0	$XYZ = m_7$

* Based on J. V. Wait: Symbolic Logic and Practical Applications, in M. Klerer and G. A. Korn, *Digital Computer User's Handbook*, McGraw-Hill, New York, 1967.

ment of the corresponding minterms (Sec. 12.8-2); each minterm is assigned the binary number determined by the arrangement of 0's and 1's in X, Y, Z. F is seen to equal the Boolean sum of the minterms corresponding to the function value 1 in the truth table.

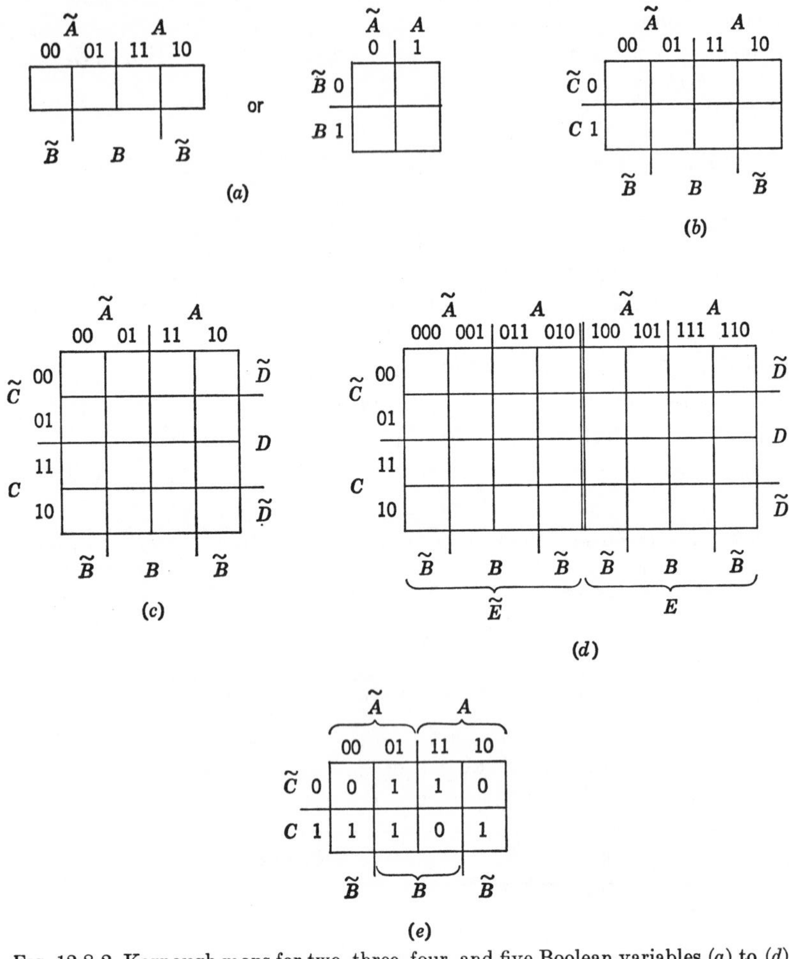

FIG. 12.8-2. Karnaugh maps for two, three, four, and five Boolean variables (*a*) to (*d*), and a two-variable map for the function of Table 12.8-1, (*e*) (*from Ref.* 20.9).

A **Karnaugh map** is a Venn diagram (Sec. 12.8-5) providing for an orderly arrangement of squares corresponding to the 2^n minterms generated by n variables (Fig. 12.8-2). Truth-table values for a given function F are entered into the appropriate squares; the given function F is then the union of all minterms labeled with a 1. For functions of up to perhaps six variables, the Karnaugh map makes it convenient to recombine these minterms into unions and/or intersections so as to minimize, say,

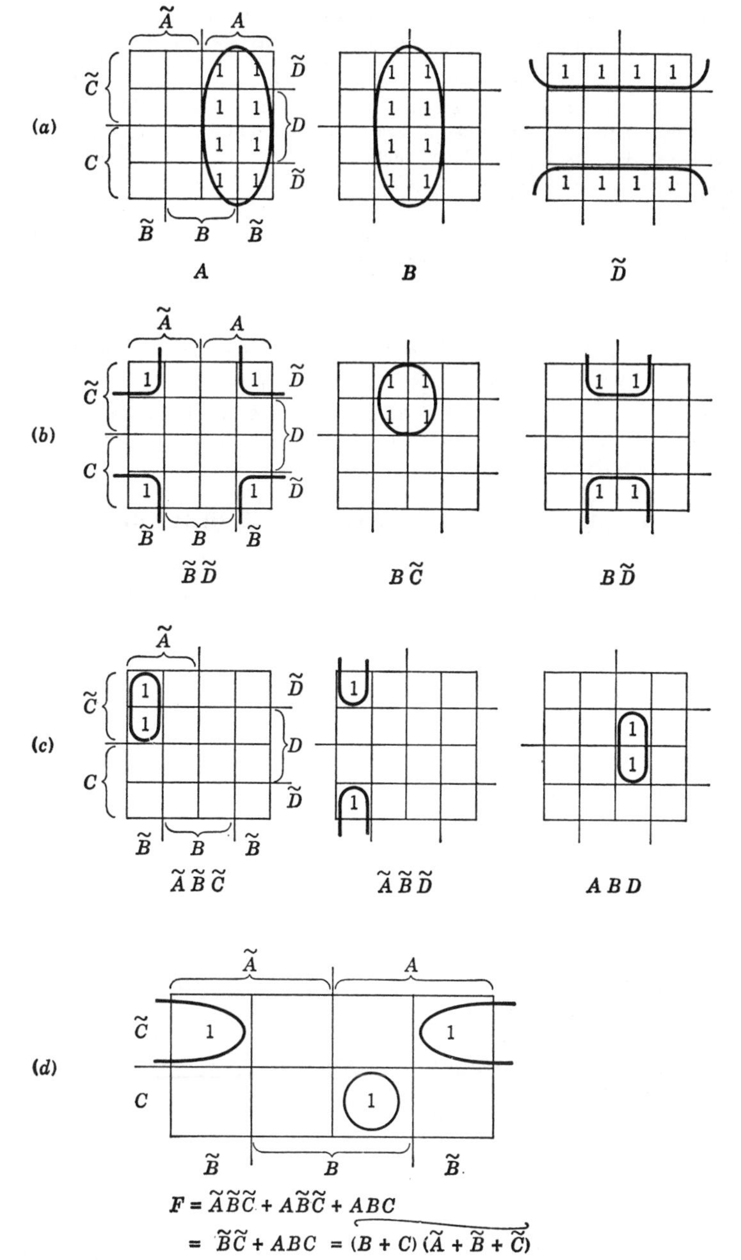

FIG. 12.8-3. Logic simplification with a Karnaugh map. (*Based on Ref.* 20.9.)

the number of logical additions, multiplications, and/or complementations. This is useful for the economical design of digital-computer circuits (Fig. 12.8-3).

12.8-8. Complete Additivity and Measure Algebras (see also Secs. 18.2-1 and 18.2-2). Many algebras of classes and event algebras require one to extend the defining postulates to unions and intersections of infinitely many terms (this is, strictly speaking, outside the realm of algebra proper, Sec. 12.1-2). A Boolean algebra $\mathcal{S}$ is **completely additive** if and only if every infinite sum $A_1 + A_2 + \cdots$ is uniquely defined as an element of $\mathcal{S}$. A completely additive Boolean algebra $\mathcal{S}$ is a **measure algebra** if and only if there exists a real numerical function (**measure**) $M(A)$ defined for all elements A of $\mathcal{S}$ so that

1. $M(A) \geq 0$
2. $M(0) = 0$
3. $M(A_1 + A_2 + \cdots) = M(A_1) + M(A_2) + \cdots$ for every countable set of disjoint elements $A_1, A_2, \ldots$

EXAMPLES OF MEASURES: Cardinal number of a set (Sec. 4.3-2*b*), length, area, volume, Lebesgue and Stieltjes measures (Secs. 4.6-14, 4.6-17), truth value (Sec. 12.8-6), probability (Sec. 18.2-2).

12.9. RELATED TOPICS, REFERENCES, AND BIBLIOGRAPHY

12.9-1. Related Topics. The following topics related to the study of modern algebra and abstract spaces are treated in other chapters of this handbook:

Theory of equations, determinants .... Chap. 1
Point sets .... Chap. 4
Matrices, quadratic and hermitian forms .... Chap. 13
Linear vector spaces, linear operators, eigenvalue problems, matrix representations .... Chap. 14
Function spaces, orthogonal expansions, eigenvalue problems involving differential and integral equations .... Chap. 15
Tensor algebra and analysis .... Chap. 16

12.9-2. References and Bibliography.

Algebra (see also Secs. 1.10-2, 13.7-2, and 14.11-2)

12.1. Barnes, W. E.: *Introduction to Abstract Algebra*, Heath, Boston, 1963.
12.2. Birkhoff, G., and S. MacLane: *A Survey of Modern Algebra*, rev. ed., Macmillan, New York, 1965.
12.3. Herstein, I. N.: *Topics in Algebra*, Blaisdell, New York, 1964.
12.4. Johnson, R. E.: *First Course in Abstract Algebra*, Prentice-Hall, Englewood Cliffs, N.J., 1953.
12.5. McCoy, N. H.: *Introduction to Modern Algebra*, Allyn and Bacon, Boston, 1960.
12.6. Mostow, G. D., et al.: *Fundamental Structures of Algebra*, McGraw-Hill, New York, 1963.
12.7. Schreier, O., and E. Sperner: *Introduction to Modern Algebra and Matrix Theory*, Chelsea, New York, 1955.
12.8. Vander Waerden, B. L.: *Modern Algebra*, rev. ed., 2 vols., Ungar, New York, 1950–1953; 6th German edition, Springer, Berlin, 1964.

Boolean Algebras and Logic

12.9. Church, A.: *Introduction to Mathematical Logic*, Princeton University Press, Princeton, N.J., 1956.
12.10. Copi, I. M.: *Symbolic Logic*, Macmillan, New York, 1954.
12.11. Flegg, H. G.: *Boolean Algebra and Its Applications*, Wiley, New York, 1964.
12.12. Chu, Y.: *Digital Computer Design Fundamentals*, McGraw-Hill, New York, 1962.
12.13. Hohn, F. E.: *Applied Boolean Algebra*, Macmillan, New York, 1960.
12.14. Rosenbloom, P. C.: *The Elements of Mathematical Logic*, Dover, New York, 1951.
12.15. Suppes, P. C.: *Introduction to Logic*, Van Nostrand, Princeton, N.J., 1958.
12.16. Tarski, A.: *Introduction to Logic and to the Methodology of Deductive Sciences*, 2d ed., Oxford, Fair Lawn, N.J., 1946.
12.17. Whitesitt, J. E.: *Boolean Algebra and Its Applications*, Addison-Wesley, Reading, Mass., 1961.

Switching Logic

12.18. Caldwell, S. H.: *Switching Circuits and Logical Design*, Wiley, New York, 1958.
12.19. Marcus, M. P.: *Switching Circuits for Engineers*, Prentice-Hall, Englewood Cliffs, N.J., 1962.
12.20. Miller, R. E.: *Switching Theory*, vol. 1: *Combinatorial Circuits*, Wiley, New York, 1965.

Topology (see also Secs. 4.12-2, 14.11-2, and 15.7-2)

12.21. Aleksandrov, P. S.: *Combinatorial Topology*, 3 vols., Graylock, New York, 1956.
12.22. Bushaw, D.: *Elements of General Topology*, Wiley, New York, 1963.
12.23. Hall, D. W., and G. L. Spencer: *Elementary Topology*, Wiley, New York, 1955.
12.24. Hocking, J., and G. Young: *Topology*, Addison-Wesley, Reading, Mass., 1961.
12.25. Kelley, John L.: *General Topology*, Van Nostrand, Princeton, N.J., 1955.
12.26. Lefschetz, S.: *Introduction to Topology*, Princeton University Press, Princeton, N.J., 1949.
12.27. Liusternik, L. A., and V. J. Sobolev: *Elements of Functional Analysis*, Ungar, New York, 1961.
12.28. Newman, M. H. A.: *Topology of Plane Sets of Points*, Cambridge, New York, 1951.
12.29. Pervin, W. J.: *Foundations of General Topology*, Academic Press, New York, 1964.
12.30. Pontryagin, Lev S.: *Foundations of Combinatorial Topology*, Graylock, Rochester, N.Y., 1952.
12.31. Wallace, A. H.: *Introduction to Algebraic Topology*, Pergamon Press, New York, 1957.

CHAPTER 13

MATRICES. QUADRATIC AND HERMITIAN FORMS

13.1. INTRODUCTION

13.1-1. *Matrices* (Sec. 13.2-1) are the building blocks of an important class of mathematical models. Matrix techniques permit *a simplified representation of various mathematical and physical operations in terms of numerical operations on matrix elements.* Chapter 13 introduces *matrix algebra and calculus* (Secs. 13.2-1 to 13.4-7), *quadratic and hermitian forms* (Secs. 13.5-1 to 13.5-6), and the *matrix/state-variable treatment of ordinary differential equations* (Secs. 13.6-1 to 13.6-7). The general application of matrices to the *representation of vectors, linear transformations (linear operators), and inner products* is reserved for Chap. 14.

Sections 13.5-1 to 13.5-6 similarly introduce *quadratic and hermitian forms* from the viewpoint of simple algebra; their real significance for the representation of *scalar products* is discussed in Secs. 14.7-1 and 14.7-2.

13.2. MATRIX ALGEBRA AND MATRIX CALCULUS

13.2-1. Rectangular Matrices. Norms. **(a)** An array

$$A \equiv \begin{bmatrix} a_{11} & a_{12} & \cdots & a_{1n} \\ a_{21} & a_{22} & \cdots & a_{2n} \\ \cdots & \cdots & \cdots & \cdots \\ a_{m1} & a_{m2} & \cdots & a_{mn} \end{bmatrix} \equiv [a_{ik}] \tag{13.2-1}$$

of "scalars" a_{ik} taken from a commutative field (Sec. 12.3-1) F is called a (**rectangular**) $m \times n$ **matrix** over the field F whenever one of the "matrix operations" defined in Sec. 13.2-2 is to be used. The elements a_{ik} are called **matrix elements;** the matrix element a_{ik} is situated in the

i^{th} **row** and in the kth **column** of the matrix (1). m is the number of rows, and n is the number of columns. A matrix is **finite** if and only if it has a finite number of rows and a finite number of columns; otherwise the matrix is **infinite.**

A finite or infinite matrix (1) over the field of complex numbers is **bounded** if and only if it has a finite **bound (norm)** defined typically as

$$\|A\| \equiv \sup \left| \sum_{i=1}^{m} \sum_{k=1}^{n} a_{ik} \xi_i \eta_k \right| \qquad \left(\sum_{i=1}^{m} |\xi_i|^2 = \sum_{k=1}^{n} |\eta_k|^2 = 1 \right) \qquad (13.2\text{-}2)$$

Table 13.2-1. Some Matrix Norms (n and/or m can be infinite; see also Secs. 12.5-5 and 14.4-1)

(a) Rectangular m × n Matrices $\qquad A \equiv [a_{ik}] \qquad (m, n > 1)$

$$\|A\| \equiv \sup \left| \sum_{i=1}^{m} \sum_{k=1}^{n} a_{ik} \xi_i \eta_k \right| \qquad \left(\sum_{i=1}^{m} |\xi_i|^2 = \sum_{k=1}^{n} |\eta_k|^2 = 1 \right)$$

$$\|A\|_I \equiv \sup_k \sum_{i=1}^{m} |a_{ik}| \qquad \|A\|_{II} \equiv \sup_i \sum_{k=1}^{n} |a_{ik}|$$

$$\|A\|_p \equiv \left(\sum_{i=1}^{m} \sum_{k=1}^{n} |a_{ik}|^p \right)^{1/p} \qquad (p = 1, 2, \ldots)$$

$$\|A\|_1 \equiv \sum_{i=1}^{m} \sum_{k=1}^{n} |a_{ik}| \qquad \text{(TAXICAB NORM)}$$

$$\|A\|_2 \equiv \left(\sum_{i=1}^{m} \sum_{k=1}^{n} |a_{ik}|^2 \right)^{1/2} \qquad \text{(EUCLIDEAN OR FROBENIUS NORM)}$$

(b) Column Matrices $x \equiv \begin{bmatrix} \xi_1 \\ \xi_2 \\ \cdots \\ \xi_n \end{bmatrix}$ **or Row Matrices** $\tilde{x} \equiv (\xi_1, \xi_2, \ldots, \xi_n)$

$$\|x\|_p \equiv \|\tilde{x}\|_p \equiv \left(\sum_{k=1}^{n} |\xi_k|^p \right)^{1/p} \qquad (p = 1, 2, \ldots)$$

$$\|x\|_1 \equiv \|\tilde{x}\|_1 \equiv \sum_{k=1}^{n} |\xi_k| \qquad \text{(TAXICAB NORM)}$$

$$\|x\|_2 \equiv \|\tilde{x}\|_2 \equiv \left(\sum_{k=1}^{n} |\xi_k|^2 \right)^{1/2} \qquad \text{(EUCLIDEAN NORM)}$$

$$\|x\|_\infty \equiv \|\tilde{x}\|_\infty \equiv \sup |\xi_k|$$

(c) Relations. Each norm satisfies the relations

$$\|A + B\| \leq \|A\| + \|B\| \qquad \|\alpha A\| = |\alpha| \|A\| \qquad \|AB\| \leq \|A\| \|B\|$$

In particular,

$$\|Ax\|_2 \leq \|A\|_2 \|x\|_2$$

(see also Table 13.2-1 and Sec. 14.4-1). *A finite matrix over the field of complex numbers is bounded if and only if all its matrix elements are bounded.*

Throughout this handbook all matrices will be understood to be bounded matrices over the field of complex numbers, unless the contrary is specifically stated. A matrix $A \equiv [a_{ik}]$ is **real** if and only if all matrix elements a_{ik} are real numbers.

(b) $n \times 1$ matrices are **column matrices,** and $1 \times n$ matrices are **row matrices.** The following notation will be used:

$$\begin{bmatrix} \xi_1 \\ \xi_2 \\ \cdot \\ \cdot \\ \cdot \\ \xi_n \end{bmatrix} \equiv \{\xi_i\} \equiv x \qquad [\xi_1 \xi_2 \cdot \cdot \cdot \xi_n] \equiv [\xi_k] \equiv \tilde{x} \tag{13.2-3}$$

(see also Secs. 13.3-1, 14.5-1, and 14.5-2).

(c) An $n \times n$ matrix is called a **square matrix** of **order** n. A square matrix $A \equiv [a_{ik}]$ is

Triangular (superdiagonal) if and only if $i > k$ implies $a_{ik} = 0$
Strictly triangular if and only if $i \geq k$ implies $a_{ik} = 0$
Diagonal if and only if $i \neq k$ implies $a_{ik} = 0$
Monomial if and only if each row and column has one and only one element different from zero

13.2-2. Basic Operations. Operations on matrices are defined in terms of operations on the matrix elements.

1. Two $m \times n$ matrices $A \equiv [a_{ik}]$ and $B \equiv [b_{ik}]$ are **equal** ($A = B$) if and only if $a_{ik} = b_{ik}$ for all i, k (see also Sec. 12.1-3).
2. The **sum of two** $m \times n$ **matrices** $A \equiv [a_{ik}]$ and $B \equiv [b_{ik}]$ is the $m \times n$ matrix

$$A + B \equiv [a_{ik}] + [b_{ik}] \equiv [a_{ik} + b_{ik}]$$

3. The **product of the** $m \times n$ **matrix** $A \equiv [a_{ik}]$ **by the scalar** α is the $m \times n$ matrix

$$\alpha A \equiv \alpha[a_{ik}] \equiv [\alpha a_{ik}]$$

4. The **product of the** $m \times n$ **matrix** $A \equiv [a_{ij}]$ **and the** $n \times r$ **matrix** $B \equiv [b_{jk}]$ is the $m \times r$ matrix

$$AB \equiv [a_{ij}][b_{jk}] \equiv \left[\sum_{j=1}^{n} a_{ij}b_{jk}\right]$$

In every matrix product AB the number n of columns of A must match the number of rows of B (A and B must be **conformable**). The existence of AB implies that of BA if and only if A and B are square matrices; in general $BA \neq AB$ (see also Sec. 13.4-4b). Note

$$\begin{array}{c} A + B = B + A \qquad A + (B + C) = (A + B) + C \\ \alpha(\beta A) = (\alpha\beta)A \qquad \alpha(AB) = (\alpha A)B = A(\alpha B) \\ A(BC) = (AB)C \\ \alpha(A + B) = \alpha A + \alpha B \qquad (\alpha + \beta)A = \alpha A + \beta A \\ A(B + C) = AB + AC \qquad (B + C)A = BA + CA \end{array} \tag{13.2-4}$$

$$\|A + B\| \leq \|A\| + \|B\| \qquad \|\alpha A\| = |\alpha|\,\|A\| \qquad \|AB\| \leq \|A\|\,\|B\| \tag{13.2-5}$$

13.2-3. Identities and Inverses. Note the following definitions:

1. The $m \times n$ **null matrix (additive identity)** $[0]$ is the $m \times n$ matrix all of whose elements are equal to zero. Then

$$A + [0] = A \qquad 0A = [0]$$
$$[0]B = C[0] = [0]$$

where A is any $m \times n$ matrix, B is any matrix having n rows, and C is any matrix having m columns.

2. The **additive inverse (negative)** $-A$ of the $m \times n$ matrix $A \equiv [a_{ik}]$ is the $m \times n$ matrix

$$-A \equiv (-1)A \equiv [-a_{ik}]$$

with $A + (-A) = A - A = [0]$.

3. The **identity matrix (unit matrix, multiplicative identity)** I **of order** n is the $n \times n$ diagonal matrix with unit diagonal elements:

$$I \equiv [\delta^i_k] \qquad \text{where} \qquad \delta^i_k = \begin{cases} 0 \text{ if } i \neq k \\ 1 \text{ if } i = k \end{cases}$$

(see also Sec. 16.5-2). Then

$$IB = B \qquad CI = C$$

where B is any matrix having n rows, and C is any matrix having n columns; and for any $n \times n$ matrix A

$$IA = AI = A$$

4. A (necessarily square) matrix A is **nonsingular** (**regular**) if and only if it has a (necessarily unique) bounded **multiplicative inverse** or **reciprocal** A^{-1} defined by

$$AA^{-1} = A^{-1}A = I$$

Otherwise A is a **singular** matrix.

A finite $n \times n$ matrix $A \equiv [a_{ik}]$ is nonsingular if and only if $\det(A) \equiv \det[a_{ik}] \neq 0$; *in this case A^{-1} is the $n \times n$ matrix*

$$A^{-1} \equiv [a_{ik}]^{-1} \equiv \left[\frac{A_{ki}}{\det[a_{ik}]}\right]$$

where A_{ik} is the cofactor of the element a_{ik} in the determinant $\det[a_{ik}]$ (see also Secs. 1.9-2 and 14.5-3).

Products and reciprocals of nonsingular matrices are nonsingular; if A and B are nonsingular, and $\alpha \neq 0$,

$$(AB)^{-1} = B^{-1}A^{-1} \qquad (\alpha A)^{-1} = \alpha^{-1}A^{-1} \qquad (A^{-1})^{-1} = A \tag{13.2-6}$$

(see also Sec. 14.3-5).

A given matrix A is nonsingular if it has a *unique* left or right inverse, or *equal left and right inverses;* the mere existence of one or more left and/or right inverses is not sufficient (see also Sec. 14.3-5). *A given $n \times n$ matrix is nonsingular if and only if it can be partitioned* (Sec. 13.2-8) *into a linearly independent set of row or column matrices.*

13.2-4. Integral Powers of Square Matrices. One defines $A^0 = I$, $A^1 = A$, $A^2 = AA$, $A^3 = AAA$, . . . and, if A is nonsingular,

$$A^{-p} = (A^{-1})^p = (A^p)^{-1} \quad (p = 1, 2, \ldots)$$

The ordinary rules for operations with exponents apply (see also Sec. 14.3-6).

13.2-5. Matrices as Building Blocks of Mathematical Models. The definitions of Secs. 13.2-2 and 13.2-3 (*constructive* definitions, Sec. 12.1-1) imply the following results:

1. Given any pair of (finite) positive integers m, n, *the class of all $m \times n$ matrices over a field F is an mn-dimensional vector space over F* (Secs. 12.4-1 and 14.2-4). In particular, n-element column or row matrices form n-dimensional vector spaces (see also Sec. 14.5-2).
2. *The class of all square matrices of a given (finite) order n over a field F is a linear algebra of order n^2 over F; singular matrices are zero divisors* (Sec. 12.4-2).
3. *The class of all nonsingular square matrices of a given (finite) order n over a field F constitutes a multiplicative group* (*Sec.* 12.2-1) *and, together with the $n \times n$ null matrix, a division algebra of order n^2 over the field F* (Sec. 12.4-2).

Analogous theorems apply to bounded *infinite* matrices over the field of real or complex numbers.

13.2-6. Multiplication by Special Matrices. Permutation Matrices. *Given any $n \times n$ matrix A,*

1. *Premultiplication of A by the matrix obtained on replacing the 1 in the ith row of the $n \times n$ identity matrix by a complex number α multiplies all elements in the ith row of A by α.*
2. *Premultiplication of A by the matrix obtained on replacing the nondiagonal element $\delta_k^i = 0$ of the $n \times n$ identity matrix by 1 adds the kth row of A to the ith row of A.*
3. *Premultiplication of A by the* permutation matrix *formed through a permutation of the rows of the $n \times n$ identity matrix results in an identical permutation of the rows of A.*

The third transposition relation (13.3-1) yields three analogous theorems describing operations on the *columns* of a matrix as results of *postmultiplication* by special matrices.

13.2-7. Rank, Trace, and Determinant of a Matrix (see also Sec. 14.3-2). The **rank** of a given matrix is the largest number r such that at least one r^{th}-order determinant (Sec. 1.5-1) formed from the matrix by deleting rows and/or columns is different from zero. *An $m \times n$ matrix A is nonsingular if and only if $m = n = r$, i.e., if and only if A is square and* $\det(A) \neq 0$ (Sec. 13.2-3).

The **trace (spur)** of an $n \times n$ matrix $A \equiv [a_{ik}]$ is the sum

$$\operatorname{Tr}(A) = \sum_{i=1}^{n} a_{ii}$$

of the diagonal terms. If $n = \infty$, this sum converges whenever A is bounded (Sec. 13.2-1*a*). *For finite matrices A, B*

$$\left.\begin{aligned} \operatorname{Tr}(A + B) &= \operatorname{Tr}(A) + \operatorname{Tr}(B) \qquad & \operatorname{Tr}(\alpha A) &= \alpha \operatorname{Tr}(A) \\ \operatorname{Tr}(BA) &= \operatorname{Tr}(AB) & \operatorname{Tr}(AB - BA) &= 0 \end{aligned}\right\} \tag{13.2-7}$$

$$\det(AB) = \det(BA) = \det(A)\det(B) \tag{13.2-8}$$

The determinant $\det(A)$ of an *infinite* square matrix $A \equiv [a_{ik}]$ is defined as

$$\det(A) = \lim_{n\to\infty} \det[a_{ik}] \qquad (i, k \leq n)$$

if the limit exists. The theorems of Secs. 1.5-1 to 1.5-6 apply to determinants defined in this manner whenever both $\sum_{i\neq k} a_{ik}$ and $\prod_{i=1}^{\infty} a_{ii}$ converge.

13.2-8. Partitioning of Matrices. A matrix having more than one row and column may be **partitioned** into smaller rectangular **submatrices** by lines drawn between rows and/or columns. *One can multiply two similarly partitioned $n \times n$ matrices A and B by entering* *their rectangular submatrices as elements in the ordinary matrix-product formula* (*Sec.* 13.2-2); *the product elements thus obtained are the submatrices of the* $n \times n$ *matrix* AB. This theorem is helpful in numerical computations (Sec. 20.3-4).

13.2-9. Step Matrices. Direct Sums (see also Secs. 13.4-6, 14.8-2, and 14.9-2). A **step matrix** is a square matrix A which can be partitioned into a diagonal matrix (Sec. 13.2-1c) of square submatrices $A_1, A_2, \ldots$, so that

$$A \equiv \left[\begin{array}{c|c|c} A_1 & [0] & \cdots \\ \hline [0] & A_2 & \cdots \\ \hline \cdots & \cdots & \cdots \end{array}\right] \tag{13.2-9}$$

A step matrix is often referred to as the **direct sum**

$$A = A_1 \oplus A_2 \oplus \cdots$$

of the square matrices along its diagonal (see also Sec. 12.7-5). Note that $A^p = A_1{}^p \oplus A_2{}^p \oplus \cdots$ for $p = 0, 1, 2, \ldots$ (and also for $p = -1, -2, \ldots$ if A is nonsingular), and

$$\begin{aligned} \operatorname{Tr}(A) &= \operatorname{Tr}(A_1) + \operatorname{Tr}(A_2) + \cdots \\ \det(A) &= \det(A_1)\det(A_2)\cdots \end{aligned} \tag{13.2-10}$$

13.2-10. Direct Product (Outer Product) of Matrices (see also Sec. 12.7-3). The **direct product (outer product)** $A \otimes B$ of the $m \times n$ matrix $A \equiv [a_{ik}]$ and the $m' \times n'$ matrix $B \equiv [b_{i'k'}]$ is the $mm' \times nn'$ matrix

$$A \otimes B \equiv [c_{jh}] \qquad (c_{jh} = a_{ik}b_{i'k'}) \tag{13.2-11}$$

where j enumerates the pairs (i, i') in the sequence $(1, 1), (1, 2), \ldots, (1, m'), (2, 1), (2, 2), \ldots (m, m')$, and where h enumerates the pairs (k, k') in a similar manner. Note

$$(A \otimes B)(C \otimes D) = AC \otimes BD \tag{13.2-12}$$

$$\operatorname{Tr}(A \otimes B) = \operatorname{Tr}(a)\operatorname{Tr}(b) \tag{13.2-13}$$

13.2-11. Convergence and Differentiation. (**a**) A sequence of matrices $S_0, S_1, S_2, \ldots$ each having the same number of rows and the same number of columns is said to **converge** to a bounded matrix S if and only if every matrix element of S_n converges to the corresponding element of S as $n \to \infty$, i.e., if and only if $\lim\limits_{n\to\infty} \|S - S_n\| = 0$. One similarly defines limits of matrix functions $A = A(t)$ of a scalar parameter t (see also Sec. 12.5-3).

(**b**) If the matrix elements of a matrix $A \equiv [a_{ik}]$ are differentiable functions $a_{ik}(t)$ of a scalar parameter t, one writes

$$\left[\frac{da_{ik}}{dt}\right] \equiv \frac{d}{dt}[a_{ik}(t)] \equiv \frac{d}{dt}A(t) \tag{13.2-14}$$

Partial differentiation and *integration* of matrices are defined in an analogous manner.

13.2-12. Functions of Matrices. *Matrix polynomials* and *algebraic functions of matrices* are defined in terms of the elementary matrix operations. The Cayley-Hamilton theorem (Sec. 13.4-7) reduces every convergent series $\sum_{k=0}^{\infty} \alpha_k A^k$ in powers of an $n \times n$ matrix A (*analytic function* of the matrix A) to an nth-degree polynomial in A.

13.3. MATRICES WITH SPECIAL SYMMETRY PROPERTIES

13.3-1. Transpose and Hermitian Conjugate of a Matrix (see also Secs. 14.4-3 and 14.4-6a). Given any $m \times n$ matrix $A \equiv [a_{ik}]$ over the field of complex numbers,

The **transpose (transposed matrix)** of A is the $n \times m$ matrix $\tilde{A} \equiv [a_{ki}]$.

The **hermitian conjugate** (adjoint, conjugate, associate matrix)* of A is the $n \times m$ matrix $A\dagger \equiv [a_{ki}^*]$.

Note the following relations:

$$\begin{gathered}\widetilde{(A+B)} = \tilde{A} + \tilde{B} \qquad \widetilde{(\alpha A)} = \alpha\tilde{A} \qquad \widetilde{(AB)} = \tilde{B}\tilde{A} \\ \widetilde{(A^{-1})} = (\tilde{A})^{-1} \qquad \tilde{\tilde{A}} = A \qquad \|\tilde{A}\| = \|A\| \\ [\tilde{0}] = [0] \qquad \tilde{I} = I\end{gathered} \tag{13.3-1}$$

$$\begin{gathered}(A+B)\dagger = A\dagger + B\dagger \qquad (\alpha A)\dagger = \alpha^* A\dagger \qquad (AB)\dagger = B\dagger A\dagger \\ (A^{-1})\dagger = (A\dagger)^{-1} \qquad (A\dagger)\dagger = A \qquad \|A\dagger\| = \|A\| \\ [0]\dagger = [0] \qquad I\dagger = I\end{gathered} \tag{13.3-2}$$

A, $\tilde{A}$, and $A\dagger$ are necessarily of equal rank. For every square matrix A,

$$\operatorname{Tr}(\tilde{A}) = \operatorname{Tr}(A) \qquad \det(\tilde{A}) = \det(A) \tag{13.3-3}$$

$$\operatorname{Tr}(A\dagger) = [\operatorname{Tr}(A)]^* \qquad \det(A\dagger) = [\det(A)]^* \tag{13.3-4}$$

13.3-2. Matrices with Special Symmetry Properties (see also Secs. 14.4-4 to 14.4-6). A square matrix $A \equiv [a_{ik}]$ is

* a_{ki}^* is the complex conjugate of a_{ki} (Sec. 1.3-1). The terms *adjoint, conjugate,* and *associate(d)* are used with different meanings (see also Secs. 12.2-5, 14.4-3, 16.7-1, and 16.7-2); some authors refer to the matrix $A^{-1} \det(A)$ (matrix of cofactors with transposed indices, **adjugate** of A) as the adjoint of A. The symbols used for the transpose $\tilde{A}$, the hermitian conjugate $A\dagger$, and the **complex conjugate** $A^* \equiv [a_{ik}^*] \equiv (\tilde{A})\dagger$ of a given matrix A also vary; some authors denote the hermitian conjugate of A by A^*.

Symmetric if and only if $\tilde{A} = A$, i.e., $a_{ik} = a_{ki}$
Skew-symmetric (antisymmetric) if and only if $\tilde{A} = -A$, i.e., $a_{ik} = -a_{ki}$
Hermitian (self-adjoint, self-conjugate) if and only if $A\dagger = A$, i.e., $a_{ik} = a_{ki}^*$
Skew-hermitian (alternating) if and only if $A\dagger = -A$, i.e., $a_{ik} = -a_{ki}^*$
Orthogonal if and only if $\tilde{A}A = A\tilde{A} = I$, i.e., $\tilde{A} = A^{-1}$
Unitary if and only if $A\dagger A = AA\dagger = I$, i.e., $A\dagger = A^{-1}$

A hermitian matrix is symmetric, a skew-hermitian matrix is skew-symmetric, and a unitary matrix is orthogonal if and only if all its matrix elements are real. The diagonal elements of hermitian, skew-hermitian, and skew-symmetric matrices are, respectively, real, pure imaginary, and equal to zero.

The determinant of a hermitian matrix is real. The determinant of a skew-hermitian $n \times n$ matrix is real if n is even, and pure imaginary if n is odd. The determinant of a skew-symmetric matrix of odd order is equal to zero. The determinant of a unitary matrix has the absolute value 1, *and the determinant of an orthogonal matrix equals either* $+1$ *or* -1.

13.3-3. Combination Rules (see also Sec. 14.4-7). **(a)** *If the matrix A is symmetric the same is true for A^p $(p = 0, 1, 2, \ldots)$, A^{-1}, $\tilde{T}AT$, and αA.*

Given any nonsingular matrix T, $\tilde{T}AT$ is symmetric if and only if A is symmetric; hence *for any orthogonal matrix T, $T^{-1}AT$ is symmetric if and only if the same is true for A.*

If A and B are symmetric the same is true for $A + B$. The product AB of two symmetric matrices A and B is symmetric if and only if $BA = AB$.

(b) *If the matrix A is hermitian the same is true for A^p $(p = 0, 1, 2, \ldots)$, A^{-1}, and $T\dagger AT$, and for αA if α is real.*

Given any nonsingular matrix T, $T\dagger AT$ is hermitian if and only if A is hermitian; hence *for any unitary matrix T, $T^{-1}AT$ is hermitian if and only if the same is true for A.*

If A and B are hermitian the same is true for $A + B$. The product AB of two hermitian matrices A and B is hermitian if and only if $BA = AB$.

(c) *If A is an orthogonal matrix the same is true for A^p $(p = 0, 1, 2, \ldots)$, A^{-1}, $\tilde{A}$, and $-A$. If A and B are orthogonal the same is true for AB.*

If A is a unitary matrix the same is true for A^p $(p = 0, 1, 2, \ldots)$, A^{-1}, and $A\dagger$, and for αA if $|\alpha| = 1$. If A and B are unitary the same is true for AB.

13.3-4. Decomposition Theorems. Normal Matrices (see also Secs. 13.4-4*a* and 14.4-8). **(a)** For every square matrix over the field of complex numbers

1. $\frac{1}{2}(A + \tilde{A}) = S_1$ is a *symmetric* matrix, and $\frac{1}{2}(A - \tilde{A}) = S_2$ is *skew-symmetric*. $A = S_1 + S_2$ is the (unique) decomposition of the given matrix A into a *symmetric part* and a *skew-symmetric part.*
2. $\frac{1}{2}(A + A\dagger) = H_1$ and $\frac{1}{2i}(A - A\dagger) = H_2$ are *hermitian* matrices. $A = H_1 + iH_2$ is the (unique) *cartesian decomposition* of the given matrix A into a *hermitian part* and a *skew-hermitian part* (comparable to the cartesian decomposition of complex numbers into real and imaginary parts, Sec. 1.3-1).
3. $A\dagger A$ is *hermitian* (and *nonnegative*, Sec. 13.5-3), and there exists a *polar decomposition* $A = QU$ *of* A into a nonnegative hermitian factor Q and a unitary factor U. Q is uniquely defined by

$$Q^2 = A\dagger A$$

and U is uniquely defined if and only if A is nonsingular (compare this with Sec. 1.3-2).

(b) A square matrix A is a **normal** matrix if and only if $A\dagger A = AA\dagger$, or, equivalently, if and only if $H_2H_1 = H_1H_2$.

13.4. EQUIVALENT MATRICES. EIGENVALUES, DIAGONALIZATION, AND RELATED TOPICS

13.4-1. Equivalent and Similar Matrices (see also Secs. 12.2-5*a*, 13.5-4, 13.5-5, and 14.6-2). **(a)** Two rectangular matrices A and B are **equivalent** if and only if there exist two nonsingular matrices S, T such that A and B are related by the transformation

$$B = SAT \tag{13.4-1}$$

Every matrix B equivalent to a given matrix A has the same number of rows and the same number of columns as A and can be obtained through successive application of the six operations defined in Sec. 13.2-6. *Equivalent matrices are of equal rank; two $m \times n$ matrices of equal rank are necessarily equivalent.*

A and B are equivalent whenever $B = QA$ or $B = AQ$, where Q is a nonsingular matrix.

(b) In particular, two square matrices A and $\bar{A}$ are **similar** (sometimes simply called **equivalent**) if and only if there exists a nonsingular matrix T (*transformation matrix*) such that A and $\bar{A}$ are related by the **similarity transformation (collineatory transformation)**

$$\boxed{\bar{A} = T^{-1}AT \qquad \text{or} \qquad A = T\bar{A}T^{-1}} \tag{13.4-2}$$

A, $\bar{A}$, and T are necessarily square matrices of equal order.

Every similarity transformation (2) *preserves the results of matrix addition, matrix multiplication, and multiplication by scalars* (see also Sec. 12.1-6). *Two similar matrices have the same rank, the same trace, and the same determinant* (see also Sec. 13.4-2*a*).

(c) Two square matrices A and $\bar{A}$ related by a transformation

$$\bar{A} = \tilde{T}AT \tag{13.4-3}$$

where T is nonsingular, are **congruent.** Two square matrices A and $\bar{A}$ related by

$$\bar{A} = T\dagger AT \tag{13.4-4}$$

where T is nonsingular, are **conjunctive.** In either case A, $\bar{A}$, and T are necessarily square matrices of equal order.

(d) *Matrix equivalence, similarity, congruence, and conjunctivity are equivalence relations;* each defines a partition of the class of matrices under consideration (Sec. 12.1-3*b*). In most applications, two or more similar matrices constitute different representations of a linear transformation (linear operator, dyadic) **A** (Sec. 14.6-2). It is, then, of interest (1) to find similarity transformations yielding particularly simple representations of **A** (transformation of a matrix to diagonal or other "canonical" form) and (2) to find properties of matrices which are invariant with respect to similarity transformations and are thus common to each class of similar matrices (e.g., rank, trace, determinant, eigenvalues).

13.4-2. Eigenvalues and Spectra of Square Matrices (see also Sec. 14.8-3). **(a)** The **eigenvalues (proper values, characteristic values, characteristic roots, latent roots)** of a (finite or infinite) square matrix $A \equiv [a_{ik}]$ are those values of the scalar parameter λ for which the **resolvent matrix** $A - \lambda I$ is singular. The **spectrum (eigenvalue spectrum)** of the matrix A is the set of all its eigenvalues.

The eigenvalues of a square matrix A can be defined directly as the eigenvalues of a linear operator represented by A (Sec. 14.8-3); *similar matrices have identical spectra.*

Note that the spectrum of an *infinite* matrix may or may not be a discrete set: e.g., every real number between $-\pi$ and π is an eigenvalue of the infinite matrix $[a_{jk}] \equiv [(-1)^{j-k}\,\delta_k^j/(j-k)]$. Some authors restrict the term *eigenvalue* to values of λ in the discrete spectrum (see also Sec. 14.8-3*d*).

(b) *Given a normal matrix* A ($A\dagger A = AA\dagger$, *Sec.* 13.3-4*b*) *with eigenvalues* λ, $A\dagger$ *has the eigenvalues* λ^*, $H_1 = \frac{1}{2}(A + A\dagger)$ *has the eigenvalues* $\operatorname{Re}(\lambda)$, *and* $H_2 = \frac{1}{2i}(A - A\dagger)$ *has the eigenvalues* $\operatorname{Im}(\lambda)$ (see also Sec. 13.3-4*a*).

All eigenvalues of a given normal matrix are real if and only if the matrix is similar to a hermitian matrix (see also Sec. 14.8-4). In particular, *all eigenvalues of hermitian and real symmetric matrices are real. All eigenvalues of a unitary matrix have absolute values equal to* 1; in particular, *real eigenvalues of real orthogonal matrices equal* $+1$ *or* -1, *and their complex eigenvalues occur in pairs* $e^{\pm i\varphi}$. *A square matrix is nonsingular if and only if all its eigenvalues are different from zero.*

(c) Refer to Secs. 13.4-5*a*, 14.8-5, and 20.3-5 for the *numerical calculation of eigenvalues*, and to Sec. 14.8-9 for the calculation of *bounds for the eigenvalues of a given matrix.*

13.4-3. Transformation of a Square Matrix to Triangular Form. Algebraic Multiplicity of an Eigenvalue (see also Sec. 14.8-3*e*). **(a)** *Given any square matrix A having a purely discrete eigenvalue spectrum, there exists a similarity transformation $\bar{A} = T^{-1}AT$ such that $\bar{A}$ is triangular* (Sec. 13.2-1*c*). *The diagonal elements of every triangular matrix similar to A are eigenvalues of A, and each eigenvalue λ_j of A occurs exactly $m'_j \geq 1$ times as a diagonal element;* m'_j is called the **algebraic multiplicity** of the eigenvalue λ_j.

NOTE: m'_j is *not* necessarily equal to the *degree of degeneracy* m_j defined in Sec. 14.8-3*b*. In the case of infinite matrices, one or more of the m'_j may be infinite.

(b) *If* Tr (A) *exists, and A has a purely discrete eigenvalue spectrum,* Tr (A) *equals the sum of all eigenvalues, each counted a number of times equal to its algebraic multiplicity. If* det (A) *exists (Sec.* 13.2-7), *it equals the similarly computed product of the eigenvalues* (see also Sec. 13.4-5).

13.4-4. Diagonalization of Matrices (see also Sec. 14.8-5). **(a)** *A square matrix A can be* **diagonalized by a similarity transformation** (*i.e., there exists a nonsingular transformation matrix T such that $\bar{A} = T^{-1}AT$ is diagonal, Sec.* 13.2-1*c*) *if and only if A has a purely discrete eigenvalue spectrum and is similar to a normal matrix* (Sec. 13.3-4*b*). More specifically, *a given matrix A having a purely discrete eigenvalue spectrum can be diagonalized by a similarity transformation with unitary transformation matrix T (or with a real orthogonal transformation matrix if A is real) if and only if A is a normal matrix ($A\dagger A = AA\dagger$, Sec.* 13.3-4*b*). In each case the diagonal elements of $\bar{A}$ are eigenvalues of A; every eigenvalue occurs a number of times equal to its algebraic multiplicity. Refer to Sec. 14.8-6 for a procedure yielding transformation matrices T with the desired properties.

SPECIAL CASES OF DIAGONALIZABLE MATRICES. *Hermitian* and *unitary matrices* (and thus *real and symmetric or orthogonal matrices*) are special instances of normal matrices. Every matrix having *only discrete eigenvalues of algebraic multiplicity* 1 is similar to a normal matrix.

(b) *Two hermitian matrices A and B with purely discrete eigenvalue spectra can be diagonalized by the same similarity transformation (and, in particular, by the same similarity transformation with unitary transformation matrix T) if and only if $BA = AB$* (see also Secs. 13.5-5, 14.8-6*e*).

(c) *Given any hermitian matrix A having a purely discrete eigenvalue spectrum, there exists a nonsingular matrix T such that $\bar{A} = T\dagger AT$ is* *diagonal; the diagonal elements of $\bar{A}$ are then real.* In particular, *there exists a nonsingular matrix T such that the diagonal elements of $\bar{A}$ take only the values $+1$, -1, and/or 0.*

Given any real symmetric matrix A having a purely discrete eigenvalue spectrum, there exists a real nonsingular matrix T such that $\bar{A} = \tilde{T}AT$ is diagonal. In particular, *there exists a real nonsingular matrix T such that the diagonal elements of $\bar{A}$ take only the values $+1$, -1, and/or 0.*

Matrices T with the desired properties are obtained from $T = DU$, where U is a unitary (or real orthogonal) matrix such that $U^{-1}AU$ is diagonal, and D is a real diagonal matrix; U is found by the method of Sec. 14.8-6 (see also Sec. 13.5-4*d*).

13.4-5. Eigenvalues and Characteristic Equation of a Finite Matrix. (a) *The eigenvalue spectrum of a finite $n \times n$ matrix $A \equiv [a_{ik}]$ is identical with the set of roots λ of the nth-degree algebraic equation*

$$F_A(\lambda) \equiv \det (A - \lambda I) \equiv \det [a_{ik} - \lambda\delta^i_k] \equiv \begin{vmatrix} a_{11} - \lambda & a_{12} & \cdots & a_{1n} \\ a_{21} & a_{22} - \lambda & \cdots & a_{2n} \\ \cdots & \cdots & \cdots & \cdots \\ a_{n1} & a_{n2} & \cdots & a_{nn} - \lambda \end{vmatrix} = 0 \tag{13.4-5}$$

(CHARACTERISTIC EQUATION OR SECULAR EQUATION OF THE MATRIX A)

The multiplicity (order, Sec. 1.6-2) of each root λ_j equals its algebraic multiplicity m'_j as an eigenvalue, so that $m'_1 + m'_2 + \cdots = n$.

Similar $n \times n$ matrices have identical characteristic equations; the coefficients in Eq. (5) are symmetric functions of the n roots $\lambda_1, \lambda_2, \ldots, \lambda_n$ (Sec. 1.6-4). In particular, the coefficient of λ^{n-1} and the constant term in Eq. (5) are, respectively,

$$(-1)^{n-1}(\lambda_1 + \lambda_2 + \cdots + \lambda_n) = (-1)^{n-1} \operatorname{Tr}(A)$$
$$\lambda_1\lambda_2 \cdots \lambda_n = \det (A) \tag{13.4-6}$$

The coefficient of λ^{n-r} equals $(-1)^r$ times the sum of the $\binom{n}{r}$ r-rowed principal minors (Sec. 1.5-4) of det (A).

(b) (See also Sec. 14.8-3). *Given a finite square matrix A with eigenvalues λ_j, αA has the eigenvalues $\alpha\lambda_j$, and A^p has the eigenvalues λ_j^p ($p = 0, 1, 2, \ldots$; $p = 0, \pm 1, \pm 2, \ldots$ if A is nonsingular). Every polynomial or analytic function $f(A)$ (Sec. 13.2-12) has the eigenvalues $f(\lambda_j)$. A matrix power series $\sum_{k=0}^{\infty} \alpha_k A^k$ converges (Sec. 13.2-11a) if and only if the power series $\sum_{k=0}^{\infty} \alpha_k \lambda_j^k$ converges for every eigenvalue λ_j of A. Given two finite square matrices A and B having the respective eigenvalues λ_j and μ_h, the eigenvalue spectrum of the direct product $A \otimes B$ (Sec. 13.2-10) is the set of products $\lambda_j\mu_h$.*

13.4-6. Eigenvalues of Step Matrices (Direct Sums, Sec. 13.2-9). *The spectrum of a finite or infinite) step matrix (direct sum)* $A = A_1 \oplus A_2 \oplus \cdots$ *is the union of the spectra of* $A_1, A_2, \ldots$; *algebraic multiplicities add.* The contribution of each finite submatrix A_k may be obtained with the aid of its characteristic equation.

13.4-7. The Cayley-Hamilton Theorem and Related Topics.

(a) *Every finite square matrix A satisfies its own characteristic equation* (Sec. 13.4-5*a*), *i.e.*

$$F_A(A) = 0 \quad \text{(Cayley-Hamilton theorem)} \tag{13.4-7}$$

(b) The Cayley-Hamilton theorem permits one to represent every integral power, and hence every analytic function, of a finite $n \times n$ matrix A as a linear function of any n distinct positive integral powers of A (see also Sec. 13.2-12). Specifically,

$$f(A) \equiv \frac{1}{\Delta} \sum_{k=1}^{n} \Delta_{n-k} A^{n-k} \tag{13.4-8}$$

where Δ is the Vandermonde determinant (Sec. 1.6-5) det $[\lambda_i^{k-1}]$, and Δ_j is the determinant obtained on substitution of $f(\lambda_1), f(\lambda_2), \ldots, f(\lambda_n)$ for $\lambda_1^j, \lambda_2^j, \ldots, \lambda_n^j$ in Δ.

If the eigenvalues $\lambda_1, \lambda_2, \ldots, \lambda_n$ of the matrix A are *distinct*, then Eq. (8) can be rewritten as

$$f(A) \equiv \sum_{k=1}^{n} f(\lambda_k) \frac{\prod_{i \neq k} (A - \lambda_i I)}{\prod_{i \neq k} (\lambda_k - \lambda_i)} \quad \text{(Sylvester's theorem)} \tag{13.4-9}$$

13.5. QUADRATIC AND HERMITIAN FORMS

13.5-1. Bilinear Forms. A **bilinear form** in the $2n$ real or complex variables $\xi_1, \xi_2, \ldots, \xi_n, \eta_1, \eta_2, \ldots, \eta_n$ is a homogeneous polynomial of the second degree (Sec. 1.4-3*a*)

$$\tilde{x}Ay \equiv \sum_{i=1}^{n} \sum_{k=1}^{n} a_{ik} \xi_i \eta_k \tag{13.5-1}$$

13.5-2. Quadratic Forms. A **(homogeneous) quadratic form** in n real or complex variables* $\xi_1, \xi_2, \ldots, \xi_n$ is a polynomial

$$\boxed{\tilde{x}Ax \equiv \sum_{i=1}^{n} \sum_{k=1}^{n} a_{ik} \xi_i \xi_k \equiv \tilde{x}A_1 x} \tag{13.5-2}$$

* The theory of Secs. 13.5-2 to 13.5-6 also applies to *quadratic and hermitian forms in a countably infinite set of variables* $\xi_1, \xi_2, \ldots$ $(n = \infty)$, and to the corresponding infinite matrices, provided that the infinite sums $\sum_{i=1}^{\infty} \sum_{k=1}^{\infty} |a_{ik}|^2$, $\sum_{i=1}^{\infty} \sum_{k=1}^{\infty} |b_{ik}|^2$, and $\sum_{i=1}^{\infty} |\xi_i|^2$ converge.

where $A_1 = \frac{1}{2}(A + \tilde{A})$ is the "symmetric part" (Sec. 13.3-4) of the matrix $A \equiv [a_{ik}]$. *The expression* (2) *vanishes identically if and only if A is skew-symmetric* ($a_{ki} = -a_{ik}$, Sec. 13.3-2). A quadratic form (2) is **symmetric** if and only if the matrix $A \equiv [a_{ik}]$ is symmetric ($a_{ki} = a_{ik}$, Sec. 13.3-2) and **real** if and only if A (and thus every a_{ik}, Sec. 13.2-1*a*) is real.

A real symmetric quadratic form (2), and also the corresponding real symmetric matrix A, is called **positive definite, negative definite, nonnegative,** or **nonpositive** if and only if, respectively, $\tilde{x}Ax > 0$, $\tilde{x}Ax < 0$, $\tilde{x}Ax \geq 0$, or $\tilde{x}Ax \leq 0$ for every set of *real* numbers $\xi_1, \xi_2, \ldots, \xi_n$ not all equal to zero. All other real symmetric quadratic forms are **indefinite** (i.e., the sign of $\tilde{x}Ax$ depends on $\xi_1, \xi_2, \ldots, \xi_n$) or identically zero.

A real symmetric quadratic form (2), and also the corresponding real symmetric matrix A, is **positive semidefinite** or **negative semidefinite** if and only if it is, respectively, nonnegative or nonpositive, and $\tilde{x}Ax = 0$ for some set of real numbers $\xi_1, \xi_2, \ldots, \xi_n$ not all equal to zero.

13.5-3. Hermitian Forms. A **hermitian form** in n real or complex variables* $\xi_1, \xi_2, \ldots, \xi_n$ is a polynomial

$$x\dagger Ax \equiv \sum_{i=1}^{n} \sum_{k=1}^{n} a_{ik}\xi_i^*\xi_k \tag{13.5-3}$$

such that the matrix $A \equiv [a_{ik}]$ is hermitian ($a_{ik} = a_{ki}^*$, Sec. 13.3-2). *A form* (3) *is real for every set of complex numbers* $\xi_1, \xi_2, \ldots, \xi_n$ *if and only if A is hermitian* (see also Sec. 14.4-4).

A hermitian form (3), and also the corresponding hermitian matrix $A \equiv [a_{ik}]$, is called **positive definite, negative definite, nonnegative,** or **nonpositive** if and only if, respectively, $x\dagger Ax > 0$, $x\dagger Ax < 0$, $x\dagger Ax \geq 0$, or $x\dagger Ax \leq 0$ for every set of complex numbers $\xi_1, \xi_2, \ldots, \xi_n$ not all equal to zero. All other hermitian forms (or hermitian matrices) are **indefinite** (i.e., the sign of $x\dagger Ax$ depends on $\xi_1, \xi_2, \ldots, \xi_n$) or identically zero.

A hermitian form (3), and also the corresponding hermitian matrix A, is **positive semidefinite** or **negative semidefinite** if and only if it is, respectively, nonnegative or nonpositive, and $x\dagger Ax = 0$ for some set of complex numbers $\xi_1, \xi_2, \ldots, \xi_n$ not all equal to zero.

13.5-4. Transformation of Quadratic and Hermitian Forms. Transformation to Diagonal Form. (**a**) A linear substitution

$$\left.\begin{aligned} \xi_i &= \sum_{k=1}^{n} t_{ik}\bar{\xi}_k \qquad (i = 1, 2, \ldots, n) \\ \text{or} \qquad x &= T\bar{x} \end{aligned}\right\} \quad (\det [t_{ik}] \neq 0) \tag{13.5-4}$$

* See footnote to Sec. 13.5-2.

(nonsingular homogeneous linear transformation of coordinates or vector components, "alias" point of view, Sec. 14.6-1) transforms every quadratic form (2) into a quadratic form in the new variables $\bar{\xi}_1, \bar{\xi}_2, \ldots, \bar{\xi}_n$:

$$\tilde{x}Ax \equiv \sum_{i=1}^{n} \sum_{k=1}^{n} \bar{a}_{ik}\bar{\xi}_i\bar{\xi}_k \equiv \tilde{\bar{x}}\bar{A}\bar{x}$$

with

$$\bar{a}_{ik} = \sum_{j=1}^{n} \sum_{h=1}^{n} a_{jh}t_{ji}t_{hk} \qquad (i, k = 1, 2, \ldots, n) \tag{13.5-5}$$

or

$$\bar{A} = \tilde{T}AT$$

$\bar{A}$ is symmetric if A is symmetric; $\bar{A}$ is real if A and T are real.

The linear substitution (4) transforms every hermitian form (3) into a new hermitian form:

$$x\dagger Ax \equiv \sum_{i=1}^{n} \sum_{k=1}^{n} \bar{a}_{ik}\bar{\xi}_i^*\bar{\xi}_k \equiv \bar{x}\dagger\bar{A}\bar{x}$$

with

$$\bar{a}_{ik} = \sum_{j=1}^{n} \sum_{h=1}^{n} a_{jh}t_{ji}^*t_{hk} \qquad (i, k = 1, 2, \ldots, n) \tag{13.5-6}$$

or

$$\bar{A} = T\dagger AT$$

(**b**) For every given real symmetric quadratic form (2) there exist linear transformations (4) with real coefficients t_{ik} such that the transformed matrix $\bar{A}$ in Eq. (5) is diagonal (see also Sec. 13.4-4*c*), so that

$$\tilde{x}Ax \equiv \tilde{\bar{x}}\bar{A}\bar{x} \equiv \sum_{i=1}^{n} \bar{a}_{ii}\bar{\xi}_i^2 \tag{13.5-7}$$

Similarly, for every given hermitian form (3) there exist linear transformations (4) such that

$$x\dagger Ax \equiv \bar{x}\dagger\bar{A}\bar{x} \equiv \sum_{i=1}^{n} \bar{a}_{ii}|\bar{\xi}_i|^2 \tag{13.5-8}$$

The number r of nonzero coefficients $\bar{a}_{ii}$ in Eq. (7) *or* (8) *is independent of the particular diagonalizing transformation used and equals the rank of the given matrix A; r* is called the **rank** of the given quadratic or hermitian form. *For any given real symmetric quadratic form* (2) *the difference between the respective numbers of positive and negative coefficients $\bar{a}_{ii}$ in Eq.* (7) *is independent of the particular diagonalizing transformation used (Jacobi-Sylvester Law of Inertia);* this number is referred to as the **signature** of the given quadratic form.

(c) In particular, there exists a *real orthogonal* diagonalizing matrix T for every real symmetric quadratic form (2), and a *unitary* diagonalizing matrix T for every hermitian form (3) (see also Sec. 13.4-4). The resulting *principal-axes transformation* (transformation to *normal coordinates* $\bar{\xi}_1, \bar{\xi}_2, \ldots, \bar{\xi}_n$, see also Sec. 9.4-8) yields the *normal form* of the given quadratic or hermitian form, viz.,

$$\tilde{x}Ax \equiv \tilde{\bar{x}}\bar{A}\bar{x} \equiv \sum_{i=1}^{n} \lambda_i \bar{\xi}_i^2 \quad \text{or} \quad x\dagger Ax \equiv \bar{x}\dagger\bar{A}\bar{x} \equiv \sum_{i=1}^{n} \lambda_i |\bar{\xi}_i|^2 \tag{13.5-9}$$

where the set of real numbers λ_i is the eigenvalue spectrum of the given matrix A (Sec. 13.4-2).

(d) The additional transformation $\bar{\xi}_i = \zeta_i/\sqrt{|\lambda_i|}$ $(i = 1, 2, \ldots, n)$ reduces the expressions (9) to their respective *canonical forms*

$$\tilde{x}Ax \equiv \sum_{i=1}^{n} \epsilon_i \zeta_i^2 \quad \text{or} \quad x\dagger Ax \equiv \sum_{i=1}^{n} \epsilon_i |\zeta_i|^2 \tag{13.5-10}$$

where each ϵ_i equals $+1$, -1, or 0 if the corresponding eigenvalue λ_i is positive, negative, or zero.

(e) The calculation of suitable diagonalizing transformation matrices is discussed in Sec. 14.8-6.

13.5-5. Simultaneous Diagonalization of Two Quadratic or Hermitian Forms (see also Secs. 13.4-4*b* and 14.8-7). *Given two real symmetric quadratic forms $\tilde{x}Ax$, $\tilde{x}Bx$, where $\tilde{x}Bx$ is positive definite, it is possible to find a real transformation* (4) *which diagonalizes $\tilde{x}Ax$ and $\tilde{x}Bx$ simultaneously. In particular, there exists a real transformation* (4) *to new coordinates $\bar{\xi}_1, \bar{\xi}_2, \ldots, \bar{\xi}_n$ such that*

$$\tilde{x}Ax \equiv \tilde{\bar{x}}\bar{A}\bar{x} \equiv \sum_{i=1}^{n} \mu_i \bar{\xi}_i^2 \qquad \tilde{x}\bar{B}x \equiv \tilde{\bar{x}}\bar{B}\bar{x} \equiv \sum_{i=1}^{n} \bar{\xi}_i^2 \tag{13.5-11}$$

Similarly, *given two hermitian forms $x\dagger Ax$, $x\dagger Bx$, where $x\dagger Bx$ is positive definite, there exists a transformation* (4) *to new coordinates $\bar{\xi}_1, \bar{\xi}_2, \ldots, \bar{\xi}_n$ such that*

$$x\dagger Ax \equiv \bar{x}\dagger\bar{A}\bar{x} \equiv \sum_{i=1}^{n} \mu_i |\bar{\xi}_i|^2 \qquad x\dagger Bx \equiv \bar{x}\dagger\bar{B}\bar{x} \equiv \sum_{i=1}^{n} |\bar{\xi}_i|^2 \tag{13.5-12}$$

In either case, the set of real numbers $\mu_1, \mu_2, \ldots, \mu_n$ is the eigenvalue spectrum of the matrix $B^{-1}A$, obtainable as the set of roots of the nth-degree algebraic equation

$$\det(A - \mu B) \equiv \det[a_{ik} - \mu b_{ik}] = 0 \tag{13.5-13}$$

The desired transformation matrix T is obtained by the method of Sec. 14.8-7*b*, or from $T = UT_0$, where T_0 is the matrix reducing $\tilde{x}Bx$ or $x\dagger Bx$ to canonical form (Sec. 13.5-4*d*), and U is a unitary matrix which diagonalizes $\tilde{x}Ax$ or $x\dagger Ax$ (Sec. 13.5-4*c*).

NOTE: Two real symmetric quadratic forms $\tilde{x}Ax$, $\tilde{x}Bx$ or two hermitian forms $x\dagger Ax$, $x\dagger Bx$ can be diagonalized simultaneously by the same *unitary* transformation matrix T if and only if $BA = AB$ (see also Secs. 13.4-4*b* and 14.8-6*e*).

13.5-6. Tests for Positive Definiteness, Nonnegativeness, etc. (a) *A real symmetric quadratic form or hermitian form is positive definite, negative definite, nonnegative, nonpositive, indefinite, or identically zero* (*Secs.* 13.5-2 *and* 13.5-3) *if and only if the* (*necessarily real*) *eigenvalues* λ_i *of the matrix* $A \equiv [a_{ik}]$ *are, respectively, all positive, all negative, all nonnegative, all nonpositive, of different signs, or all equal to zero.*

A real symmetric quadratic form or hermitian form is positive semidefinite or negative semidefinite if and only if it is, respectively, nonnegative or nonpositive, and at least one eigenvalue λ_i *of the matrix* $A \equiv [a_{ik}]$ *equals zero.*

Note that the λ_i are the roots of the characteristic equation (13.4-5); the signs of these roots can often be investigated by one of the methods of Sec. 1.6-6.

(b) *A hermitian matrix* $A \equiv [a_{ik}]$ (*and the corresponding hermitian form or real symmetric quadratic form*) *is positive definite if and only if every one of the quantities*

$$a_{11}, \qquad \begin{vmatrix} a_{11} & a_{12} \\ a_{21} & a_{22} \end{vmatrix}, \qquad \begin{vmatrix} a_{11} & a_{12} & a_{13} \\ a_{21} & a_{22} & a_{23} \\ a_{31} & a_{32} & a_{33} \end{vmatrix}, \qquad \ldots, \qquad \det [a_{ik}]$$

is positive (*Sylvester's Criterion*).

(c) *A hermitian matrix A* (*and the corresponding hermitian form or real symmetric quadratic form*) *is negative definite, nonpositive, or negative semidefinite if and only if* $-A$ *is, respectively, positive definite, nonnegative, or positive semidefinite.*

(d) *A matrix A is a nonnegative hermitian matrix if and only if there exists a matrix B such that* $A = B\dagger B$. *A real matrix A is a nonnegative symmetric matrix if and only if there exists a real matrix B such that* $A = \tilde{B}B$. *In either case, A is positive definite if B, and thus A, is nonsingular.*

(e) *If both A and B are positive definite or nonnegative, the same is true for AB. Every positive definite matrix A has a unique pair of square roots* H, $-H$ *defined by* $H^2 = A$; *H is positive definite* (see also Sec. 13.3-4).

13.6. MATRIX NOTATION FOR SYSTEMS OF DIFFERENTIAL EQUATIONS (STATE EQUATIONS). PERTURBATIONS AND LYAPUNOV STABILITY THEORY

13.6-1. Systems of Ordinary Differential Equations. Matrix Notation. As noted in Sec. 9.1-3, a general system of ordinary differential equations (9.1-4) reduces to the first-order form

$$\frac{dy_i}{dt} = f_i(t; y_1, y_2, \ldots, y_n) \qquad (i = 1, 2, \ldots, n) \qquad (13.6\text{-}1a)$$

if appropriate derivatives are introduced as new variables y_i. The system (1a) is written as a single matrix differential equation

$$\frac{dy}{dt} \equiv \frac{d}{dt}\begin{bmatrix} y_1(t) \\ y_2(t) \\ \cdots \\ y_n(t) \end{bmatrix} = \begin{bmatrix} f_1(t; y_1, y_2, \ldots, y_n) \\ f_2(t; y_1, y_2, \ldots, y_n) \\ \cdots\cdots\cdots\cdots \\ f_n(t; y_1, y_2, \ldots, y_n) \end{bmatrix} \equiv f(t, y) \qquad (13.6\text{-}1b)$$

(see also Sec. 13.2-11), where $y(t)$ and $f(t, y)$ are $n \times 1$ column matrices. If the f_i are single-valued and continuous and satisfy a Lipschitz condition (9.2-4) over the domain of interest, then the solution $y(t)$ of Eq. (1) is uniquely determined by the initial condition

$$y(0) \equiv \begin{bmatrix} y_1(0) \\ y_2(0) \\ \cdots \\ y_n(0) \end{bmatrix} = \begin{bmatrix} y_{10} \\ y_{20} \\ \cdots \\ y_{n0} \end{bmatrix} \equiv y_0 \qquad (13.6\text{-}1c)$$

The system (1) is called **autonomous** if and only if f does not depend explicitly on the independent variable t.

More than merely a notational convenience, the matrix notation will be seen to extend intuitive insight gained from studies of simple first-order differential equations to *systems* of first-order equations. Matrix operations needed for solution of linear systems (Sec. 13.6-2) are, moreover, readily implemented with digital computers.

In the most important applications, t represents physical time, and the $y_i(t)$ are *state variables* representing the state of a dynamical system; the system (1) is then called a system of *state equations* (see also Sec. 11.8-4 and Refs. 13.10 to 13.16).*

13.6-2. Linear Differential Equations with Constant Coefficients (Time-invariant Systems). (a) Homogeneous Systems. Normal-mode Solution. The solution of the homogeneous linear system

$$\frac{dy_i}{dt} = \sum_{k=1}^{n} a_{ik}y_k \qquad y_i(0) = y_{i0} \qquad (i = 1, 2, \ldots, n) \qquad (13.6\text{-}2a)$$

or

$$\frac{dy}{dt} = Ay \qquad y(0) = y_0 \qquad (A \equiv [a_{ik}]) \qquad (13.6\text{-}2b)$$

* Many engineering texts refer to the matrix $y(t)$ as a *state vector*. It would be more correct to state that the matrix elements $y_i(t)$ (state variables) *represent* a state vector in a specific scheme of measurements (in the sense of tensor analysis, Chap. 16; see also Ref. 13.15).

with constant coefficients a_{ik} (see also Secs. 9.3-1 and 9.4-1*d*) is *explicitly* given by

$$y(t) = e^{At}y_0 \qquad (t \geq 0) \tag{13.6-3}$$

where the matrix function e^{At} is the $n \times n$ matrix defined in accordance with Secs. 13.2-12 and 13.4-7. Expansion of e^{At} by Eq. (13.4-8) involves cumbersome matrix multiplications but, if the given matrix A has n distinct eigenvalues, expansion of e^{At} by Sylvester's theorem (13.4-9) yields the normal-mode expansion of Sec. 9.4-1.

One can often simplify the solution of a problem (2) by introducing n new dependent variables (state variables) $\bar{y}_h$ by a nonsingular linear transformation

$$y_i = \sum_{h=1}^{n} t_{ih}\bar{y}_h \qquad (i = 1, 2, \ldots, n) \qquad \text{or} \qquad y = T\bar{y} \tag{13.6-4}$$

such that the resulting transformed system

$$\left.\begin{aligned} &\frac{d\bar{y}}{dt} = \bar{A}\bar{y} \qquad \bar{y}(0) = \bar{y}_0 \\ \text{with} \qquad &\bar{A} \equiv T^{-1}AT \qquad \bar{y}_0 \equiv T^{-1}y_0 \end{aligned}\right\} \tag{13.6-5}$$

is simplified (see also Secs. 14.6-1 and 14.6-2). If, in particular, there exists a transformation (4) which *diagonalizes* the given system matrix A (Secs. 13.4-4 and 14.8-6), then the transformed variables $\bar{y}_h$ are **normal coordinates** of the given linear system (see also Sec. 9.4-8): they satisfy "uncoupled" differential equations

$$\frac{d\bar{y}_h}{dt} = \lambda_h\bar{y}_h \qquad (h = 1, 2, \ldots, n) \tag{13.6-6}$$

where $\lambda_1, \lambda_2, \ldots, \lambda_n$ are the eigenvalues of A. If A has n *distinct* eigenvalues, the solution of the original problem (2) is then given by Eq. (4) with

$$\bar{y}_h = \bar{y}_{h0}e^{\lambda_h t} \qquad (h = 1, 2, \ldots, n) \tag{13.6-7}$$

Complex-conjugate terms in a normal-mode solution (4), and also coincident and zero eigenvalues, can be treated in a manner analogous to Sec. 9.4-1. In the general case, one can use a transformation (4) producing a *triangular* matrix $\bar{A}$ (Sec. 13.4-3), so that the $\bar{y}_i(t)$ can be derived one by one (Ref. 13.15).

(b) Nonhomogeneous Equations. The State-transition Matrix. The linear system

$$\frac{dy_i}{dt} = \sum_{k=1}^{n} a_{ik}y_k + f_i(t) \qquad y_i(0) = y_{i0} \qquad (i = 1, 2, \ldots, n) \tag{13.6-8a}$$

$$\text{or} \qquad \frac{dy}{dt} = Ay + f(t) \qquad y(0) = y_0 \qquad (A \equiv [a_{ik}]) \tag{13.6-8b}$$

where $f(t)$ is an $n \times 1$ column matrix, describes the response of a (time-invariant) linear system to the inputs $f_i(t)$. As in Secs. 9.3-1 and 9.4-2, the matrix solution $y(t)$ is obtained by superposition of the homogeneous-system solution (3) and a particular integral (normal response) $y_N(t)$:

$$\left.\begin{aligned} y(t) &= e^{At}y_0 + y_N(t) \\ \text{with}\quad y_N(t) &= \int_0^t h_+(t-\tau)f(\tau)\,d\tau = \int_0^t h_+(\zeta)f(t-\zeta)\,d\zeta \end{aligned}\right\} \quad (t \geq 0) \tag{13.6-9}$$

The $n \times n$ **state-transition matrix** $h_+(t - \tau) \equiv [\{h_+(t-\tau)\}_{ik}]$ for the initial-value problem (8) is a generalization of the one-dimensional weighting function $h_+(t - \tau)$ in Sec. 9.4-3 and satisfies

$$\frac{dh_+(t)}{dt} = Ah_+(t) \quad (t > 0) \qquad h_+(0) = I \tag{13.6-10}$$

so that

$$h_+(t) = e^{At} \qquad (t \geq 0) \tag{13.6-11}$$

$h_+(t)$ is the response to the set of (asymmetrical) unit impulses $f_i(t) = \delta_+(t)$ $(i = 1, 2, \ldots, n$; see also Sec. 9.4-3*d*). Note that the solution (9) is precisely analogous to the solution of the one-dimensional problem $dy/dt = ay + f(t)$, $y(0) = y_0$.

(c) Laplace-transform Solution (see also Sec. 9.4-5). Element-by-element Laplace transformation of the given constant-coefficient matrix equation (8) produces

$$\begin{aligned} sY(s) - y_0 &= AY(s) + F(s) \\ \text{or}\qquad Y(s) &= (sI - A)^{-1}y_0 + (sI - A)^{-1}F(s) \end{aligned} \tag{13.6-12}$$

where $Y(s)$, $F(s)$ denote the respective Laplace transforms of $y(t)$, $f(t)$. The terms in Eq. (12) are the transforms of those in Eq. (9); inverse Laplace transformation of each element $Y_i(s)$ of $Y(s)$ produces $y_i(t)$.

13.6-3. Linear Systems with Variable Coefficients (see also Secs. 9.2-4, 9.3-3, and 18.12-2). **(a)** The most general linear system (1) has the form

$$\frac{dy}{dt} = A(t)y + f(t) \tag{13.6-13}$$

where $A(t) \equiv [a_{ik}(t)]$ is an $n \times n$ matrix, and $f(t)$ is an $n \times 1$ column matrix (linear differential equations with variable coefficients and forcing terms). The solution can again be written as

$$y(t) = w_+(t, 0)y_0 + \int_0^t w_+(t, \lambda)f(\lambda)\,d\lambda \qquad (t \geq 0) \tag{13.6-14}$$

where $w_+(t, \lambda)$ is the $n \times n$ state-transition matrix determined for $t \geq \lambda$ as the solution of

$$\left.\begin{aligned} \frac{dw_+(t, \lambda)}{dt} &= A(t)w_+(t, \lambda) \qquad (t > \lambda) \\ w_+(\lambda, \lambda) &= I \end{aligned}\right\} \tag{13.6-15}$$

or as the response to a set of (asymmetrical) unit impulses $f_i(t) = \delta_+(t - \lambda)$, where $i = 1, 2, \ldots, n$; see also Sec. 9.4-3*d*. For constant-coefficient systems, $w_+(t, \lambda) \equiv h_+(t - \lambda)$.

(b) For any real or complex matrix $A(t)$ with continuous elements, the solution of the homogeneous linear system

$$\frac{dy}{dt} = A(t)y \qquad (13.6\text{-}16)$$

is $Y(t)y(0)$, where $Y = Y(t)$ is an $n \times n$ matrix and the unique solution of the matrix differential equation

$$\frac{dY}{dt} = A(t)Y \qquad Y(0) = I \qquad (13.6\text{-}17)$$

$Y(t)$ is nonsingular; its columns constitute n linearly independent solutions of Eq. (16) (**fundamental-solution matrix,** see also Sec. 9.3-2). $U(t) \equiv [Y^{-1}(t)]\dagger$ is the unique solution of

$$-\frac{dU}{dt} = A\dagger(t)U \qquad \text{with } U(0) = I \qquad (13.6\text{-}18)$$

Equations (17) and (18) are **adjoint linear differential equations.***

The state-transition matrix $w_+(t, \lambda)$ of Sec. 13.6-3*a* is given by

$$w_+(t, \lambda) \equiv Y(t)Y^{-1}(\lambda) \equiv Y(t)U\dagger(\lambda) \qquad (t \geq \lambda) \qquad (13.6\text{-}19)$$

so that Eq. (14) corresponds to a matrix version of the variation-of-constants solution of Sec. 9.3-3.

13.6-4. Perturbation Methods and Sensitivity Equations.

(a) Given a system of differential equations

$$\frac{dy}{dt} = f(t, y; \alpha) \qquad y(0) = y_0 \qquad (13.6\text{-}20)$$

which depends on a set (column matrix) $\{\alpha_1, \alpha_2, \ldots, \alpha_m\}$ of m parameters α_k, let $y_{(1)}(t)$ be the known solution for the parameter values given by $\alpha = \alpha_1 \equiv \{\alpha_{11}, \alpha_{12}, \ldots, \alpha_{1m}\}$. The perturbed solution $y_{(1)}(t) + \delta y(t)$ corresponding to the perturbed parameter matrix $\alpha = \alpha_1 + \delta\alpha$ may be easier to find through solution of

$$\frac{d}{dt}\delta y = f(t, y_{(1)} + \delta y;\ \alpha_1 + \delta\alpha) - f(t, y_{(1)};\ \alpha_1) \qquad \delta y(0) = 0 \qquad (13.6\text{-}21)$$

for the **perturbation** (variation, Sec. 11.5-1) δy than by direct solution

* $I\frac{d}{dt} - A(t)$ and $-I\frac{d}{dt} - A\dagger(t)$ are *adjoint operators* on $n \times 1$ matrix functions $u(t)$ such that $\int_0^\infty u\dagger(t)u(t)\,dt$ exists and $u(0) = 0$ if one defines the inner product of two such functions u, v by $(u,v) = \int_0^\infty u\dagger(t)v(t)\,dt$ (Sec. 14.4-3; see also Sec. 15.4-3).

of Eq. (20). Equation (21) is exact. For suitably differentiable $f(t, y; \alpha)$, one may, however, be able to neglect all but first-order terms in a Taylor-series expansion of Eq. (21) to find an approximation to δy (**first-order perturbation**) by solving the *linear* system

$$\frac{d}{dt}\,\delta y = \frac{\partial f}{\partial y}\,\delta y + \frac{\partial f}{\partial \alpha}\,\delta\alpha \qquad \delta y(0) = 0 \tag{13.6-22}$$

where the elements of the $n \times n$ matrix $\partial f/\partial y \equiv [\partial f_i/\partial y_k]_{y=y_{(1)}}$ and the $n \times m$ matrix $\partial f/\partial \alpha \equiv [\partial f_i/\partial \alpha_k]_{y=y_{(1)}}$ will, in general, depend on the "nominal solution" $y_{(1)}(t)$ and hence on t. If the perturbations δy_i are small compared with the $|y_i|$, one may be able to neglect approximation and numerical errors in the computation of the δy_i.

(**b**) The dependence of the solution $y(t)$ on the parameters α_k is often described by (the $n \times m$ matrix of) the **sensitivity coefficients (parameter-influence coefficients)** $z_{ik} = \partial y_i/\partial \alpha_k$, which form an $n \times m$ matrix $Z \equiv \partial y/\partial \alpha \equiv [\partial y_i/\partial \alpha_k]_{y=y_{(1)}}$. For each given nominal solution $y(t) = y_{(1)}(t)$, the sensitivity coefficients are functions of t and satisfy the mn linear differential equations (**sensitivity equations**) given by

$$\frac{dZ}{dt} = \frac{\partial f}{\partial y}\,Z + \frac{\partial f}{\partial \alpha} \qquad Z(0) = 0 \tag{13.6-23}$$

(**c**) The initial values $y_i(0) = y_{i0}$ may be treated as parameters in perturbation and sensitivity calculations. In this case, the initial conditions $\delta y(0) = 0$ in Eqs. (21) and (22) must be replaced by

$$\delta y(0) = y(0) - y_{(1)}(0) = \delta y_0 \tag{13.6-24}$$

The appropriate initial conditions for the sensitivity equations (23) are

$$z_{ik}(0) = \left.\frac{\partial y_i(0)}{\partial y_{k0}}\right]_{y=y_{(1)}} = \begin{cases} 0 & (i \neq k) \\ 1 & (i = k) \end{cases} \quad (i, k = 1, 2, \ldots, n) \tag{13.6-25}$$

13.6-5. Stability of Solutions: Definitions (see also Sec. 9.5-4). (**a**) Given a system

$$\frac{dy}{dt} = f(t, y) \qquad (t > 0) \tag{13.6-26}$$

different types of stability of a solution $y = y_{(1)}(t)$ can be defined in terms of the effects of various parameter perturbations (Sec. 13.6-4). The following theory is concerned with stability *in the sense of Lyapunov*, which is determined by the effects of small changes

$$\delta y(t_0) \equiv y(t_0) - y_{(1)}(t_0)$$

in initial solution values on the resulting perturbations

$$\delta y(t) \equiv y(t) - y_{(1)}(t)$$

for $t > t_0$.

The solution $y = y_{(1)}(t)$ of the system (26) is

stable in the sense of Lyapunov if and only if for every real $\epsilon > 0$, there exists a real $\Delta(\epsilon, t_0) > 0$ such that $\|\delta y(t_0)\| < \Delta(\epsilon, t_0)$ implies $\|\delta y(t)\| < \epsilon$ for all $t \geq t_0$. Otherwise the solution is **unstable.**

asymptotically stable in a region $D_1(t_0)$ of the "state space" of points $y \equiv \{y_1, y_2, \ldots, y_n\}$ if and only if $y_{(1)}(t)$ is stable, and $y(t_0)$ in $D_1(t_0)$ implies $\lim_{t\to\infty} \delta y(t) = 0$ (i.e., $\|\delta y(t)\| \to 0$ as $t \to \infty$, Sec. 13.2-11).

asymptotically stable in the large (completely stable, globally asymptotically stable) if and only if the entire state space is a region of asymptotic stability.

NOTE: In the above definitions, the norm $\|\delta y\|$ of the $n \times 1$ column matrix $\delta y \equiv \{\delta y_1, \delta y_2, \ldots, \delta y_n\}$, defined in accordance with Eq. (13.2-2) as

$$\|\delta y\| = \sup |\xi_1\, \delta y_1 + \xi_2\, \delta y_2 + \cdots + \xi_n\, \delta y_n| \qquad (\xi_1^2 + \xi_2^2 + \cdots + \xi_n^2 = 1) \tag{13.6-27a}$$

can be conveniently replaced by one of the alternate norms (Table 13.2-1)

$$\|\delta y\|_2 = [(\delta y_1)^2 + (\delta y_2)^2 + \cdots + (\delta y_n)^2]^{1/2} \qquad \text{(EUCLIDEAN NORM)} \tag{13.6-27b}$$

or

$$\|\delta y\|_1 = |\delta y_1| + |\delta y_2| + \cdots + |\delta y_n| \qquad \text{(TAXICAB NORM)} \tag{13.6-27c}$$

Note that these definitions refer to stability of *solutions*, not of *systems* (see also Secs. 9.4-4 and 13.6-7). If a solution is *stable in the sense of Lyapunov, sufficiently small* changes in initial values cannot cause large solution changes at any time. For an *asymptotically stable* solution, the effects of *finite* initial-value changes, up to specified bounds, are nullified after sufficient time has elapsed. If the solution is *asymptotically stable in the large*, even *arbitrarily large* initial-value changes will have negligible long-term effects. Asymptotic stability is a requirement for practical control systems.

(b) An unstable solution $y_{(1)}(t)$ of Eq. (26) has a **finite escape time** T if and only if it becomes unbounded after a finite time $t = T$.

13.6-6. Lyapunov Functions and Stability. (a) Stability of Equilibrium for Autonomous Systems (see also Sec. 9.5-4b). An **equilibrium solution** $y(t) = y_{(1)}(t \geq 0)$ of the autonomous system

$$\frac{dy}{dt} = f(y) \qquad (t > 0) \tag{13.6-28}$$

is defined by

$$f(y_{(1)}) = 0 \tag{13.6-29}$$

It will suffice to consider equilibrium solutions $y(t) = y_{(1)} = 0$, since other equilibrium "points" $y = y_{(1)}$ in state space can be translated to the origin by a simple coordinate transformation.

With reference to the solution $y(t) = 0$ of a given system (28), a **Lyapunov function** is any real function $V(y)$ such that $V(0) = 0$ and, throughout a neighborhood D of the "point" $y = 0$ in the "state space" of "points" $y \equiv \{y_1, y_2, \ldots, y_n\}$, $V(y)$ is continuously differentiable and

$$\left.\begin{aligned} &V[y(t)] > 0 \\ &\frac{d}{dt} V[y(t)] = \sum_{k=1}^{n} \frac{\partial V}{\partial y_k} \frac{dy_k}{dt} = \sum_{k=1}^{n} \frac{\partial V}{\partial y_k} f_k \leq 0 \end{aligned}\right\} \tag{13.6-30}$$

for all $y(t) \neq 0$ satisfying Eq. (28). *The equilibrium solution $y(t) = 0$ is stable in the sense of Lyapunov if* (*and only if*, Ref. 13.11) *there exists a corresponding Lyapunov function. $y(t) = 0$ is asymptotically stable*

1. *If there exists a Lyapunov function $V(y)$ satisfying the stronger condition $dV/dt < 0$ for all solutions $y(t) \neq 0$ of Eq.* (28) *in D* (*Lyapunov's Theorem on Asymptotic Stability*).
2. *If there exists a Lyapunov function $V(y)$ not identically zero on any solution trajectory $y = y(t)$ in D* (*Kalman-Bertram Theorem*).

If the neighborhood D of the origin defining a Lyapunov function $V(y)$ contains a bounded region D_1 such that $V(y) < V_0$, where V_0 is any positive constant, then $y(t) = 0$ is asymptotically stable in D_1. If a Lyapunov function $V(y)$ can be defined for the entire state space, and $V(y) \to \infty$ as $\|y(t)\| \to \infty$, then the solution $y(t) = 0$ is asymptotically stable in the large (*La Salle's Theorem on Asymptotic Stability*).

The equilibrium solution $y(t) = 0$ of Eq. (24) *is unstable if there exists a neighborhood D of $y = 0$, a region D_1 in D, and a real function $U(y)$ such that*

1. *$U(y)$ is continuously differentiable, and*

$$U[y(t)] > 0 \qquad \frac{d}{dt} U[y(t)] > 0$$

   for all solutions $y(t)$ in D_1, except that
2. *$U(y) = 0$ at all boundary points* (Sec. 4.3-6) *of D_1 in D.*
3. *$y = 0$ is a boundary point of D_1* (*Cetaev's Instability Theorem*).

(b) Nonautonomous Systems. Every solution $y = y_{(1)}(t)$ of the system

$$\frac{dy}{dt} = f(t, y) \qquad (t > 0) \tag{13.6-31}$$

can be transformed to the equilibrium solution $\bar{y}(t) = 0$ of a new (generally nonautonomous) system by the transformation $y(t) = \bar{y}(t) + y_{(1)}(t)$.

The equilibrium solution $y(t) = 0$ *of a given system* (26) *is asymptotically stable in the large if there exist a continuously differentiable real function* $V(t, y)$, *two continuous nondecreasing real functions* $V_1(\|y\|)$, $V_2(\|y\|)$, *and a continuous real function* $V_3(\|y\|)$ *such that* $V(t,0) = V_1(0) = V_2(0) = V_3(0) = 0$ *and*

$$V_2(\|y\|) \geq V(t, y) \geq V_1(\|y\|) > 0$$
$$\frac{dV}{dt} \leq -V_3(\|y\|) < 0$$
$$V_1(\|y\|) \to \infty \qquad \text{as} \qquad \|y\| \to \infty$$

for all $y(t) \neq 0$ *satisfying Eq.* (26).

13.6-7. Applications and Examples (see also Sec. 9.5-4). (**a**) Applications such as control-system design motivate the search for Lyapunov functions establishing asymptotic stability in specified state-space regions, or in as large regions as possible ("direct method" of Lyapunov for stability investigations). Lyapunov functions for particular solutions are not unique, and practical search methods are more of an art than a science (Refs. 13.11 to 13.14).

(**b**) As noted in Sec. 9.5-4*a*, *the equilibrium solution* $y(t) = 0$ *of the linear homogeneous constant-coefficient system*

$$\frac{dy}{dt} = Ay \tag{13.6-32}$$

(Sec. 13.6-2*a*) *is asymptotically stable in the large (completely stable) if and only if the system is completely stable in the sense of Sec.* 9.4-4, *i.e., if and only if all eigenvalues of the system matrix* A *have negative real parts. This is true if and only if for an arbitrary positive definite symmetric matrix* Q, *there exists a positive definite symmetric matrix* P *such that*

$$\tilde{A}P + PA = -Q \tag{13.6-33}$$

$V(y) \equiv \tilde{y}Py$ *is then a Lyapunov function for the equilibrium solution* $y(t) = 0$.

(**c**) Duffing's equation

$$\frac{d^2y}{dt^2} + a\frac{dy}{dt} + y + by^3 = 0$$

describes the oscillations of a nonlinear spring. Introducing $y = y_1$, $\dot{y} = y_2$, one has the nonlinear first-order system

$$\frac{dy_1}{dt} = y_2 \qquad \frac{dy_2}{dt} = -ay_2 - y_1 - by_1^3$$

The theory of Sec. 13.6-6*a* indicates that

$$V(y_1, y_2) \equiv \tfrac{1}{4}(by_1^4 + 2y_1^2 + 2y_2^2) \qquad \frac{dV}{dt} = -ay_2^2$$

is a Lyapunov function for the equilibrium solution $y_1(t) = y_2(t) = 0$ when $a > 0$, $b > 0$ ("hard spring"); this solution is asymptotically stable in the large.

For $a > 0$, $b < 0$ ("soft spring"), the equilibrium solution $y_1(t) = y_2(t) = 0$ is asymptotically stable, but not in the large (Fig. 13.6-1).

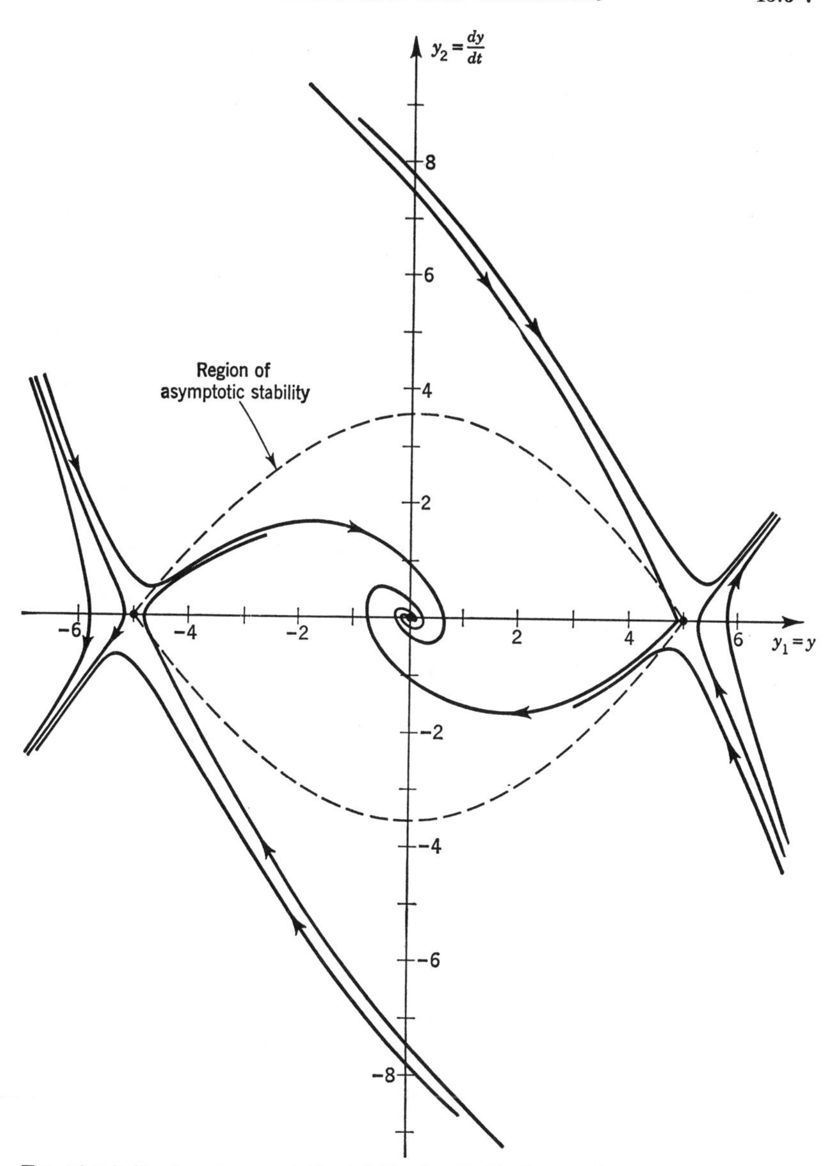

FIG. 13.6-1. Region of asymptotic stability for Duffing's equation

$$\frac{d^2y}{dt^2} + a\frac{dy}{dt} + y + by^3 = 0$$

with $a = 1$, $b = -0{,}04$. (*Based on Ref.* 13.11.)

13.7. RELATED TOPICS, REFERENCES, AND BIBLIOGRAPHY

13.7-1. Related Topics. The following topics related to the study of matrices, quadratic forms, and hermitian forms are treated in other chapters of this handbook:

Linear simultaneous equations.................................. Chap. 1
Systems of ordinary differential equations.......................... Chap. 9
Matrix notation for optimum-control problems........................ Chap. 11
Use of matrices for the representation of vectors, linear transformations (linear operators), scalar products, and group elements................ Chap. 14
Eigenvectors and eigenvalues of linear operators...................... Chap. 14
Matrix techniques for difference equations.......................... Chap. 20
Numerical techniques.. Chap. 20

13.7-2. References and Bibliography (see also Secs. 12.9-2 and 14.11-2).

13.1. Aitken, A. C.: *Determinants and Matrices*, 8th ed., Interscience, New York, 1956.
13.2. Birkhoff, G., and S. MacLane: *A Survey of Modern Algebra*, rev. ed., Macmillan, New York, 1965
13.3. Gantmakher, F. R.: *The Theory of Matrices*, Chelsea, New York, 1959.
13.4. ———: *Applications of the Theory of Matrices*, Inter cience, New York, 1959.
13.5. Hohn, F. E.: *Elementary Matrix Algebra*, 2d ed., Macmillan, New York, 1964.
13.6. Nering, E. D.: *Linear Algebra and Matrix Theory*, Interscience, New York, 1963.
13.7. Shields, P. C.: *Linear Algebra*, Addison-Wesley, Reading, Mass., 1964.
13.8. Thrall, R. M., and L. Tornheim: *Vector Spaces and Matrices*, Wiley, New York, 1957.
13.9. Zurmuehl, R.: *Matrizen*, 2d ed., Springer, Berlin, 1964.

(See also the articles by G. Falk and H. Tietz in vol. II of the *Handbuch der Physik*, Springer, Berlin, 1955. For numerical techniques, see Secs. 20.3-3 to 20.3-5.)

Matrix Techniques for Systems of Differential Equations
(See also Refs. 9.3 and 9.16 in Sec. 9.7-2)

13.10. DeRusso, P., et al.: *State Variables for Engineers*, Wiley, New York, 1965.
13.11. Geiss, G. R.: "The Analysis and Design of Nonlinear Control Systems via Lyapunov's Direct Method," *RTD-TDR*-63-4076, U.S. Air Force Flight Dynamics Laboratory, Wright-Patterson AFB, Ohio, 1964.
13.12. Hahn, W.: *Theory and Application of Lyapunov's Direct Method*, Prentice-Hall, Englewood Cliffs, N.J., 1963.
13.13. Krasovskii, N. N.: *Stability of Motion*, Stanford, Stanford, Calif., 1963.
13.14. Letov, A. M.: *Stability of Nonlinear Control Systems*, Princeton, Princeton, N.J., 1961.
13.15. Schultz, D. G., and J. L. Melsa: *State Functions in Automatic Control*, McGraw-Hill, New York, 1967.
13.16. Tomovič, R.: *Sensitivity Analysis of Dynamic Systems*, McGraw-Hill, New York, 1964.

CHAPTER **14**

LINEAR VECTOR SPACES AND LINEAR TRANSFORMATIONS (LINEAR OPERATORS). REPRESENTATION OF MATHEMATICAL MODELS IN TERMS OF MATRICES

14.1. INTRODUCTION. REFERENCE SYSTEMS AND COORDINATE TRANSFORMATIONS

14.1-1. This chapter reviews the theory of *linear vector spaces* (see also Sec. 12.4-1) and *linear transformations* (*linear operators*). Vectors and linear operators represent physical objects and operations in many important applications.

Most practical problems require a description (representation) of mathematical models (Sec. 12.1-1) in terms of ordered sets of real or complex numbers. In particular, the concepts of homomorphism and isomorphism (Sec. 12.1-6) make it possible to "represent" many mathematical models by corresponding classes of *matrices* (Sec. 13.2-1; see also Sec. 13.2-5), so that *abstract mathematical operations are made to correspond to numerical operations on matrix elements* (EXAMPLES: matrix representations of quantum-mechanical operators and of electrical transducers). Sections 14.5-1 to 14.10-8 describe the use of matrices to represent vectors, linear operators, and group elements.

14.1-2. Numerical Description of Mathematical Models: Reference Systems (see also Secs. 2.1-2, 3.1-2, 5.2-2, 6.2-1, 12.1-1, and 16.1-2). A **reference system (coordinate system)** is a scheme of rules which describes (represents) each object (point) of a class (space, region of a space) C by a corresponding ordered set of (real or complex) numbers **(components, coordinates)** $x_1, x_2, \ldots$. The number of coordinates required to define each point $(x_1, x_2, \ldots)$ is called the **dimension** of the space C (see also Sec. 14.2-4*b*). In many applications, coordinate values are related to physical measurements.

14.1-3. Coordinate Transformations (see also Secs. 2.1-5 to 2.1-8 3.1-12, 6.2-1, and 16.1-2). A **transformation of the coordinates** x_1, x_2, . . . is a set of rules or relations associating each point $(x_1, x_2, \ldots)$ with a new set of coordinates. Coordinate transformations admit two interpretations:

1. **"Alibi" or "active" point of view:** the coordinate transformation

$$x'_1 = x'_1(x_1, x_2, \ldots) \qquad x'_2 = x'_2(x_1, x_2, \ldots) \qquad \cdots \tag{14.1-1}$$

describes an *operation* (*function, mapping,* Sec. 12.1-4) associating a new mathematical object (point) $(x'_1, x'_2, \ldots)$ with each given point $(x_1, x_2, \ldots)$

2. **"Alias" or "passive" point of view:** the coordinate transformation

$$\bar{x}_1 = \bar{x}_1(x_1, x_2, \ldots) \qquad \bar{x}_2 = \bar{x}_2(x_1, x_2, \ldots) \qquad \cdots \tag{14.1-2}$$

introduces *a new description* (*new representation, relabeling*) of each point $(x_1, x_2, \ldots)$ in terms of the new coordinates $\bar{x}_1, \bar{x}_2, \ldots$.

Coordinate transformations permit one to represent abstract mathematical relationships by numerical relations ("alibi" point of view) and to change reference systems ("alias" point of view). A change of reference system often simplifies a given problem (EXAMPLES: principal-axes transformations, Secs. 2.4-8, 3.5-7, 9.4-8, and 17.4-7; contact transformations, Secs. 10.2-5 and 11.5-6; angle and action variables in dynamics).

14.1-4. Invariance (see also Secs. 12.1-5 and 16.2-1, and refer to Secs. 16.1-4 and 16.4-1 for more detailed discussions). A function of the coordinates labeling an object or objects is **invariant with respect to a given coordinate transformation** (1) or (2) if the function value is unchanged on substitution of $x'_i(x_1, x_2, \ldots)$ or $\bar{x}_i(x_1, x_2, \ldots)$ for each x_i. A relation between coordinate values is invariant if it remains valid after similar substitutions.

Invariance with respect to an "alibi"-type coordinate transformation is interpreted in the manner of Sec. 12.1-5. Functions and relations invariant with respect to a suitable class of "alias"-type coordinate transformations may be regarded as functions of, and relations between, the actual objects (**invariants**) represented by different sets of coordinates in different reference systems. A **complete set of invariants** $f_1(x_1, x_2, \ldots), f_2(x_1, x_2, \ldots), \ldots$ uniquely specifies all properties of the object $(x_1, x_2, \ldots)$ which are invariant with respect to a given class (group) of coordinate transformations (see also Sec. 12.2-8).

14.1-5. Schemes of Measurements. The representation of a model involving two or more classes of objects will, in general, require a reference system for each class of objects; the resulting set of reference systems will be called a **scheme of measurements.** A change of the scheme of measurements involves an "alias"-type coordinate transformation for each class of objects; one usually relates these transformations so as to ensure the invariance of important functions and/or relations (see also Secs. 14.6-2, 16.1-4; 16.2-1, and 16.4-1).

14.2. LINEAR VECTOR SPACES

14.2-1. Defining Properties. As already stated in Sec. 12.4-1, a *linear vector space* $\mathcal{V}$ of vectors **a, b, c,** . . . over the ring (with identity, Sec. 12.3-1*b*) R of *scalars* $\alpha, \beta, \ldots$ admits *vector addition* and *multiplication of vectors by scalars* with the following properties:

1. $\mathcal{V}$ is a commutative group with respect to vector addition: for every pair of vectors **a**, **b** of $\mathcal{V}$, $\mathcal{V}$ contains the vector sum $\mathbf{a} + \mathbf{b}$, with

$$\mathbf{a} + \mathbf{b} = \mathbf{b} + \mathbf{a} \qquad \mathbf{a} + (\mathbf{b} + \mathbf{c}) = (\mathbf{a} + \mathbf{b}) + \mathbf{c}$$

and $\mathcal{V}$ contains the additive inverse $-\mathbf{a}$ of each vector **a** and an additive identity (null vector) **0** so that

$$\mathbf{a} + \mathbf{0} = \mathbf{a} \qquad \mathbf{a} + (-\mathbf{a}) = \mathbf{a} - \mathbf{a} = \mathbf{0}$$

2. $\mathcal{V}$ contains the product $\alpha\mathbf{a}$ of every vector **a** of $\mathcal{V}$ by any scalar α of R, with

$$(\alpha\beta)\mathbf{a} = \alpha(\beta\mathbf{a}) \qquad 1\mathbf{a} = \mathbf{a}$$
$$\alpha(\mathbf{a} + \mathbf{b}) = \alpha\mathbf{a} + \alpha\mathbf{b} \qquad (\alpha + \beta)\mathbf{a} = \alpha\mathbf{a} + \beta\mathbf{a}$$

where 1 is the multiplicative identity in R.

Note that

$$0\mathbf{a} = 0 \qquad (-1)\mathbf{a} = -\mathbf{a} \qquad (-\alpha)\mathbf{a} = -(\alpha\mathbf{a}) \tag{14.2-1}$$

Unless the contrary is specifically stated, all linear vector spaces considered in this handbook are understood to be **real vector spaces** or **complex vector spaces** respectively defined as linear vector spaces over the field of real numbers and over the field of complex numbers.

In the case of vector spaces admitting a definition of infinite sums (Sec. 14.2-7*b*), many authors refer to a set of vectors with the above properties as a **linear manifold** and reserve the term **vector space** for a linear manifold which is *closed*, i.e., which contains all its limit points (Sec. 12.5-1*b*; the two terms are equivalent in the case of finite-dimensional manifolds, Sec. 14.2-4).

14.2-2. Linear Manifolds and Subspaces in $\mathcal{V}$. A subset $\mathcal{V}_1$ of a linear vector space $\mathcal{V}$ is a **linear manifold in $\mathcal{V}$** if and only if $\mathcal{V}_1$ is a linear manifold over the same ring of scalars as $\mathcal{V}$; $\mathcal{V}_1$ will be called a **subspace of** $\mathcal{V}$ if it is a closed linear manifold in $\mathcal{V}$ (see also Sec. 14.2-1). A **proper subspace** of $\mathcal{V}$ is a subspace other than **0** or $\mathcal{V}$ itself.

Any given set of vectors $\mathbf{e}_1, \mathbf{e}_2, \ldots$ in $\mathcal{V}$ **generates (spans, determines)** a linear manifold comprising all linear combinations of $\mathbf{e}_1, \mathbf{e}_2, \ldots$.

EXAMPLES: straight lines and planes through the origin in three-dimensional Euclidean space.

14.2-3. Linearly Independent Vectors and Linearly Dependent Vectors (see also Secs. 1.9-3, 5.2-2, and 9.3-2). **(a)** A finite set of vectors $\mathbf{a}_1, \mathbf{a}_2, \ldots$ is **linearly independent** if and only if

$$\lambda_1\mathbf{a}_1 + \lambda_2\mathbf{a}_2 + \cdots = 0 \qquad \text{implies} \qquad \lambda_1 = \lambda_2 = \cdots = 0 \qquad (14.2\text{-}2)$$

Otherwise, the vectors $\mathbf{a}_1, \mathbf{a}_2, \ldots$ are **linearly dependent,** and at least one vector of the set, say $\mathbf{a}_k$, can be expressed as a linear combination $\mathbf{a}_k = \sum_i \mu_i\mathbf{a}_i$ of the other vectors $\mathbf{a}_i$ of the set. As a trivial special case, this is true whenever $\mathbf{a}_k$ is a null vector.

(b) The definitions of Sec. 14.2-3*a* apply to *infinite* sets of vectors $\mathbf{a}_1, \mathbf{a}_2, \ldots$ *if* it is possible to assign a meaning to Eq. (2). In general, this will require the vector space to admit a definition of *convergence* in addition to the algebraic postulates of Sec. 14.2-1 (Secs. 12.5-3 and 14.2-7*b*).

14.2-4. Dimension of a Linear Manifold or Vector Space. Bases and Reference Systems (Coordinate Systems). **(a)** A **(linear) basis** in the linear manifold $\mathcal{V}$ is a set of linearly independent vectors $\mathbf{e}_1, \mathbf{e}_2, \ldots$ of $\mathcal{V}$ such that every vector $\mathbf{a}$ of $\mathcal{V}$ can be expressed as a linear form

$$\boxed{\mathbf{a} = \alpha_1\mathbf{e}_1 + \alpha_2\mathbf{e}_2 + \cdots} \qquad (14.2\text{-}3)$$

in the base vectors $\mathbf{e}_i$. Every set of linearly independent vectors forms a basis for the linear manifold comprising all linear combinations of the given vectors.

(b) *In a* **finite-dimensional** *linear manifold or vector space spanned by n base vectors*

1. *Every set of n linearly independent vectors is a basis.*
2. *No set of $m < n$ vectors is a basis.*
3. *Every set of $m > n$ vectors is necessarily linearly dependent.*

The number n is called the **(linear) dimension** of the vector space. An **infinite-dimensional** vector space does not admit a finite basis.

(c) In every finite-dimensional real or complex vector space, the numbers $\alpha_1, \alpha_2, \ldots, \alpha_n$ are unique **components** or **coordinates** of the vector $\mathbf{a} = \alpha_1\mathbf{e}_1 + \alpha_2\mathbf{e}_2 + \cdots + \alpha_n\mathbf{e}_n$ in a **reference system (coordinate system)** defined by the base vectors $\mathbf{e}_1, \mathbf{e}_2, \ldots, \mathbf{e}_n$. Note that $\mathbf{a} + \mathbf{b}$ has the components $\alpha_i + \beta_i$, and $\alpha\mathbf{a}$ has the components $\alpha\alpha_i$ ($i = 1, 2, \ldots, n$; see also Sec. 5.2-2).

(**d**) Two linear vector spaces $\mathcal{V}$ and $\mathcal{V}'$ over the same ring of scalars $\alpha, \beta, \ldots$ are isomorphic (Sec. 12.1-6) if and only if it is possible to relate their respective vectors $\mathbf{a}, \mathbf{b}, \ldots$ and $\mathbf{a}', \mathbf{b}', \ldots$ by a reciprocal one-to-one correspondence $\mathbf{a} \leftrightarrow \mathbf{a}'$, $\mathbf{b} \leftrightarrow \mathbf{b}'$, . . . such that $\mathbf{a} + \mathbf{b} \leftrightarrow \mathbf{a}' + \mathbf{b}'$, $\alpha\mathbf{a} \leftrightarrow \alpha\mathbf{a}'$. *In the case of finite-dimensional vector spaces, this is true if and only if $\mathcal{V}$ and $\mathcal{V}'$ have the same linear dimension.*

In particular, every n-dimensional real or complex vector space is isomorphic with the space of n-rowed column matrices over the field of real or complex numbers, respectively (*matrix representation*, Sec. 14.5-2).

14.2-5. Normed Vector Spaces. A real or complex vector space $\mathcal{V}$ is a **normed vector space** if and only if for every vector $\mathbf{a}$ of $\mathcal{V}$ there exists a real number $\|\mathbf{a}\|$ (**norm, absolute value, magnitude of a**) such that $\mathbf{a} = \mathbf{b}$ implies $\|\mathbf{a}\| = \|\mathbf{b}\|$, and, for all $\mathbf{a}, \mathbf{b}$ in $\mathcal{V}$,

$$\|\mathbf{a}\| \geq 0 \qquad \|\alpha\mathbf{a}\| = |\alpha|\,\|\mathbf{a}\|$$
$$\|\mathbf{a} + \mathbf{b}\| \leq \|\mathbf{a}\| + \|\mathbf{b}\| \quad \text{(Minkowski's inequality)} \tag{14.2-4}$$

A **unit vector** is a vector of unit magnitude (see also Sec. 5.2-5). Note that $\|-\mathbf{a}\| = \|\mathbf{a}\|$, $\|\mathbf{0}\| = 0$ (see also Sec. 13.2-1).

14.2-6. Unitary Vector Spaces. (**a**) A real or complex vector space $\mathcal{V}_U$ is a **unitary (hermitian) vector space** if and only if it is possible to define a binary operation (**inner** or **scalar multiplication of vectors**) associating a scalar $(\mathbf{a}, \mathbf{b})$, the (**hermitian**) **inner, scalar,** or **dot product** of $\mathbf{a}$ and $\mathbf{b}$, with every pair $\mathbf{a}, \mathbf{b}$ of vectors of $\mathcal{V}_U$, where

1. $(\mathbf{a}, \mathbf{b}) = (\mathbf{b}, \mathbf{a})^*$ (HERMITIAN SYMMETRY)
2. $(\mathbf{a}, \mathbf{b} + \mathbf{c}) = (\mathbf{a}, \mathbf{b}) + (\mathbf{a}, \mathbf{c})$ (DISTRIBUTIVE LAW)
3. $(\mathbf{a}, \alpha\mathbf{b}) = \alpha(\mathbf{a}, \mathbf{b})$ (ASSOCIATIVE LAW)*
4. $(\mathbf{a}, \mathbf{a}) \geq 0$; $(\mathbf{a}, \mathbf{a}) = 0$ implies $\mathbf{a} = 0$ (POSITIVE DEFINITENESS)

It follows that *in every unitary vector space*

$$(\mathbf{b} + \mathbf{c}, \mathbf{a}) = (\mathbf{b}, \mathbf{a}) + (\mathbf{c}, \mathbf{a}) \qquad (\alpha\mathbf{a}, \mathbf{b}) = \alpha^*(\mathbf{a}, \mathbf{b}) \tag{14.2-5}$$

$$|(\mathbf{a}, \mathbf{b})|^2 \leq (\mathbf{a}, \mathbf{a})(\mathbf{b}, \mathbf{b}) \quad \text{(Cauchy-Schwarz inequality)} \tag{14.2-6}$$

* Some authors use the alternative defining postulate $(\alpha\mathbf{a}, \mathbf{b}) = \alpha(\mathbf{a}, \mathbf{b})$ which amounts to an interchange of $\mathbf{a}$ and $\mathbf{b}$ in the definition of $(\mathbf{a}, \mathbf{b})$.

The Cauchy-Schwarz inequality (6) (*see also Sec.* 1.3-2) *reduces to an equation if and only if* **a** *and* **b** *are linearly dependent* (see also Secs. 1.3-2, 4.6-19, and 15.2-1*c*).

m vectors $\mathbf{a}_1, \mathbf{a}_2, \ldots, \mathbf{a}_m$ *of* $\mathcal{V}_U$ *are linearly independent if and only if the* m^{th}*-order determinant* det $[(\mathbf{a}_i, \mathbf{a}_k)]$ *is different from zero* (*Gram's determinant,* see also Secs. 5.2-8 and 15.2-1*a*).

(**b**) *If a unitary vector space is a real vector space, all scalar products* $(\mathbf{a}, \mathbf{b})$ *are real, and scalar multiplication of vectors is commutative, so that*

$$(\mathbf{a}, \mathbf{b}) = (\mathbf{b}, \mathbf{a}) \qquad (\alpha\mathbf{a}, \mathbf{b}) = \alpha(\mathbf{a}, \mathbf{b}) \tag{14.2-7}$$

NOTE: The *inner-product spaces with indefinite metric* used in relativity theory are real or complex vector spaces admitting the definition of an inner product $(\mathbf{a}, \mathbf{b})$ which satisfies conditions (1) to (3), but not condition (4) of Sec. 14.2-6*a*. The vectors of such a space may be classified as vectors of positive, negative, and zero square $(\mathbf{a}, \mathbf{a})$. One defines $\|\mathbf{a}\| = \underset{+}{\sqrt{|(\mathbf{a}, \mathbf{a})|}}$. See also Secs. 14.2-5 and 16.8-1.

14.2-7. Metric and Convergence in Normed Vector Spaces. Banach Spaces and Hilbert Spaces. (**a**) Every normed vector space (Sec. 14.2-5) is a metric space with the metric $d(\mathbf{a}, \mathbf{b}) = \|\mathbf{a} - \mathbf{b}\|$ (Sec. 12.5-2) and permits the definition of *neighborhoods* and *convergence* in the manner of Sec. 12.5-3 (see also Sec. 5.3-1). In this sense, an infinite sum $\mathbf{a}_0 + \mathbf{a}_1 + \mathbf{a}_2 \cdots$ converges to (equals) a vector

$$\mathbf{s} = \lim_{n\to\infty} \sum_{k=0}^{n} \mathbf{a}_k = \sum_{k=0}^{\infty} \mathbf{a}_k \text{ of } \mathcal{V}_U \text{ if and only if } \lim_{n\to\infty} \left\| \mathbf{s} - \sum_{k=0}^{n} \mathbf{a}_k \right\| = 0.$$

A normed vector space $\mathcal{V}$ is complete (Sec. 12.5-4) if and only if every sequence of vectors $\mathbf{s}_0, \mathbf{s}_1, \mathbf{s}_2, \ldots$ of $\mathcal{V}$ such that

$$\lim_{\substack{n\to\infty \\ m\to\infty}} \|\mathbf{s}_n - \mathbf{s}_m\| = 0$$

(*Cauchy sequence,* see also Sec. 4.9-1) converges to a vector **s** of $\mathcal{V}$. Complete normed vector spaces are called **Banach spaces.** *Every finite-dimensional normed vector space is complete.*

(**b**) Every unitary vector space $\mathcal{V}_U$ permits one to introduce the norm (absolute value, magnitude) $\|\mathbf{a}\|$ of each vector **a**, the **distance** $d(\mathbf{a}, \mathbf{b})$ between two "points" **a**, **b** of $\mathcal{V}_U$, and the **angle** γ between any two vectors **a**, **b** by means of the definitions

$$\boxed{\begin{aligned} \|\mathbf{a}\| &= \underset{+}{\sqrt{(\mathbf{a}, \mathbf{a})}} \\ d(\mathbf{a}, \mathbf{b}) &= \|\mathbf{a} - \mathbf{b}\| = \underset{+}{\sqrt{(\mathbf{a} - \mathbf{b}, \mathbf{a} - \mathbf{b})}} \\ \cos\gamma &= \frac{(\mathbf{a}, \mathbf{b})}{\|\mathbf{a}\|\,\|\mathbf{b}\|} \end{aligned}} \tag{14.2-8}$$

The functions $\|\mathbf{a}\|$ *and* $d(\mathbf{a}, \mathbf{b})$ *defined by Eq.* (8) *satisfy all conditions of* Secs. 14.2-5 and 12.5-2. If $\mathcal{V}_U$ is a real unitary vector space, the Cauchy-Schwarz inequality (6) insures that the angle γ is real for all **a**, **b**.

(c) Finite-dimensional real unitary vector spaces are called **Euclidean vector spaces.** *They are separable, complete, and boundedly compact* (Sec. 12.5-4) and serve as models for n-dimensional Euclidean geometries (see also Chaps. 2 and 3 and Secs. 5.2-6 and 17.4-6*d*). Complete infinite-dimensional unitary vector spaces are called **Hilbert spaces.*** The complete sequence and function spaces listed in Table 12.5-1 are all Hilbert spaces (and hence also Banach spaces).

Hilbert spaces preserve many of the properties of Euclidean spaces. In particular, *every separable* (Sec. 12.5-1*b*) *real or complex Hilbert space is isomorphic and isometric to the space* l^2 *of respectively real or complex infinite sequences* $(\xi_1, \xi_2, \ldots)$ *such that* $\|(\xi_1, \xi_2, \ldots)\|^2 = |\xi_1|^2 + |\xi_2|^2 + \cdots$ *converges* (Table 12.5-1*a*). Hence each vector of a separable Hilbert space can be labeled with a countable set of coordinates (or with a column or row matrix, Sec. 14.5-2).

Every linear manifold in a Hilbert space is a complete subspace (see also Sec. 14.2-2) and is, thus, itself a Euclidean vector space or a Hilbert space.

14.2-8. The Projection Theorem. *Given any vector* $\mathbf{x}$ *of a unitary vector space* $\mathcal{V}_U$ *and a complete subspace* $\mathcal{V}_1$, *there exists a unique vector* $\mathbf{y} = \mathbf{x}_p$ *of* $\mathcal{V}_1$ *which minimizes the distance* $\|\mathbf{x} - \mathbf{y}\|$ *for all* $\mathbf{y}$ *in* $\mathcal{V}_1$; *moreover,* $\mathbf{x}_p$ *is the unique vector* $\mathbf{y}$ *of* $\mathcal{V}_1$ *such that* $\mathbf{x} - \mathbf{y}$ *is orthogonal to every vector* $\mathbf{x}_1$ *of* $\mathcal{V}_1$, i.e.,

$$(\mathbf{x} - \mathbf{x}_p, \mathbf{x}_1) = 0 \qquad (\mathbf{x}_1 \text{ in } \mathcal{V}_1) \tag{14.2-9}$$

(see also Sec. 14.7-3*b*). The mapping $\mathbf{x} \to \mathbf{x}_p$ is a bounded linear operation (Sec. 14.4-2) called the **orthogonal projection of** (the points of) $\mathcal{V}_U$ **onto** $\mathcal{V}_1$.

The projection theorem is of the greatest practical importance, for Eq. (9) defines the *optimal approximation* of a vector $\mathbf{x}$ by a vector $\mathbf{y}$ of the "simpler" class $\mathcal{V}_1$ if $\|\mathbf{x} - \mathbf{y}\|^2$ measures the error of the approximation.

EXAMPLES: Projection of points onto planes in Euclidean geometry; orthogonal-function approximations (Secs. 15.2-3, 15.2-6, 20.6-2, and 20.6-3), mean-square regression (Sec. 18.4-6), Wiener filtering and prediction.

14.3. LINEAR TRANSFORMATIONS (LINEAR OPERATORS)

14.3-1. Linear Transformation of a Vector Space. Linear Operators. Given two linear vector spaces $\mathcal{V}$ and $\mathcal{V}'$ over the same field of

* Some authors do not require Hilbert spaces to be infinite-dimensional; others require them to be separable (Sec. 12.5-1*b*) as well as complete. Unitary vector spaces are sometimes referred to as **pre-Hilbert spaces.**

scalars α, β, . . . , a **(homogeneous) linear transformation of $\mathcal{V}$ into $\mathcal{V}'$** is a correspondence

$$\mathbf{x}' = \mathbf{f}(\mathbf{x}) \equiv \mathsf{A}\mathbf{x} \tag{14.3-1}$$

which relates vectors $\mathbf{x}'$ of $\mathcal{V}'$ to vectors $\mathbf{x}$ of $\mathcal{V}$ so as to preserve the "linear" operations of vector addition and multiplication of vectors by scalars:

$$\mathbf{f}(\mathbf{x} + \mathbf{y}) \equiv \mathbf{f}(\mathbf{x}) + \mathbf{f}(\mathbf{y}) \qquad \mathbf{f}(\alpha\mathbf{x}) \equiv \alpha\mathbf{f}(\mathbf{x}) \tag{14.3-2}$$

Each linear transformation can be written as a multiplication by a **linear operator A (linear operation)**, with

$$\mathsf{A}(\mathbf{x} + \mathbf{y}) \equiv \mathsf{A}\mathbf{x} + \mathsf{A}\mathbf{y} \qquad \mathsf{A}(\alpha\mathbf{x}) \equiv \alpha(\mathsf{A}\mathbf{x}) \tag{14.3-3}$$

$\mathbf{f}(\mathbf{x}) \equiv \mathsf{A}\mathbf{x} + \mathbf{a}'$ is called a **linear vector function.** The definition of each linear operator must include that of its domain of definition. In physics, the first relation (3) is often referred to as a *superposition principle* for a class of operations.

14.3-2. Range, Null Space, and Rank of a Linear Transformation (Operator). The *range* (Sec. 12.1-4) of a linear transformation A of $\mathcal{V}$ into $\mathcal{V}'$ is a linear manifold (Sec. 14.2-2) of $\mathcal{V}'$. The **null space** of A is the manifold of $\mathcal{V}$ mapped onto $\mathsf{A}\mathbf{x} = \mathbf{0}$. The **rank** r and the **nullity** r' of a linear transformation A are the respective linear dimensions (Sec. 14.2-4*b*) of its range and null space. *If $\mathcal{V}$ has the finite dimension n, then its range and null space are subspaces, and $r + r' = n$.*

14.3-3. Addition and Multiplication by Scalars. Null Transformation. **(a)** Let A and B be linear transformations (operators) mapping a given domain of definition D in $\mathcal{V}$ into $\mathcal{V}'$. One defines $\mathsf{A} \pm \mathsf{B}$ and $\alpha\mathsf{A}$ as linear transformations of $\mathcal{V}$ into $\mathcal{V}'$ such that

$$(\mathsf{A} \pm \mathsf{B})\mathbf{x} = \mathsf{A}\mathbf{x} \pm \mathsf{B}\mathbf{x} \qquad (\alpha\mathsf{A})\mathbf{x} = \alpha(\mathsf{A}\mathbf{x}) \tag{14.3-4}$$

for all vectors $\mathbf{x}$ in D.

(b) The **null transformation O** of $\mathcal{V}$ into $\mathcal{V}'$ is defined by $\mathsf{O}\mathbf{x} = \mathbf{0}$ for all vectors $\mathbf{x}$ in $\mathcal{V}$.

14.3-4. Product of Two Linear Transformations (Operators). Identity Transformation. **(a)** Let A be a linear transformation (operator) mapping $\mathcal{V}$ into $\mathcal{V}'$, and let B be a linear transformation mapping (the range of A in) $\mathcal{V}'$ into $\mathcal{V}''$. The **product BA** is the linear transformation of $\mathcal{V}$ into $\mathcal{V}''$ obtained by performing the transformations A

and $\mathbf{B}$ successively (see also Sec. 12.2-8):

$$(\mathbf{BA})\mathbf{x} \equiv \mathbf{B}(\mathbf{Ax}) \qquad (14.3\text{-}5)$$

(b) The **identity transformation** $\mathbf{I}$ of any vector space $\mathcal{V}$ transforms every vector $\mathbf{x}$ of $\mathcal{V}$ into itself:

$$\mathbf{Ix} \equiv \mathbf{x} \qquad \mathbf{IA} \equiv \mathbf{AI} = \mathbf{A} \qquad (14.3\text{-}6)$$

14.3-5. Nonsingular Linear Transformations (Operators). Inverse Transformations (Operators). A linear transformation (operator) $\mathbf{A}$ is **nonsingular (regular)** if and only if it defines a reciprocal one-to-one correspondence mapping all of $\mathcal{V}$ onto all of $\mathcal{V}'$ ($\mathcal{V}$ and $\mathcal{V}'$ are then necessarily isomorphic, Sec. 14.2-4*d*). $\mathbf{A}$ is nonsingular if and only if it has a unique **inverse (inverse transformation, inverse operator)** $\mathbf{A}^{-1}$ mapping $\mathcal{V}'$ onto $\mathcal{V}$ so that $\mathbf{x}' = \mathbf{Ax}$ implies $\mathbf{x} = \mathbf{A}^{-1}\mathbf{x}'$ and conversely, or

$$\mathbf{AA}^{-1} = \mathbf{A}^{-1}\mathbf{A} = \mathbf{I} \qquad (14.3\text{-}7)$$

Products and inverses of nonsingular transformations (operators) are nonsingular; if $\mathbf{A}$ and $\mathbf{B}$ are nonsingular, and $\alpha \neq 0$,

$$(\mathbf{AB})^{-1} = \mathbf{B}^{-1}\mathbf{A}^{-1} \qquad (\alpha\mathbf{A})^{-1} = \alpha^{-1}\mathbf{A}^{-1} \qquad (\mathbf{A}^{-1})^{-1} = \mathbf{A} \qquad (14.3\text{-}8)$$

Nonsingular linear transformations (operations) preserve linear independence of vectors and hence also the linear dimensions of transformed manifolds (Secs. 14.2-3 and 14.2-4).

A linear operator $\mathbf{A}$ is nonsingular if it has a *unique* left or right inverse, or if it has *equal left and right inverses;* the mere existence of a left and/or right inverse is not sufficient. *A linear transformation (operator)* $\mathbf{A}$ *defined on a finite-dimensional vector space is nonsingular if and only if* $\mathbf{Ax} = \mathbf{0}$ *implies* $\mathbf{x} = \mathbf{0}$, *i.e., if and only if* $r = n$, $r' = 0$ (*Sec.* 14.3-2).

14.3-6. Integral Powers of Operators. One defines $\mathbf{A}^0 \equiv \mathbf{I}$, $\mathbf{A}^1 = \mathbf{A}$, $\mathbf{A}^2 = \mathbf{AA}$, $\mathbf{A}^3 = \mathbf{AAA}, \ldots$, and, if $\mathbf{A}$ is nonsingular, $\mathbf{A}^{-p} = (\mathbf{A}^{-1})^p = (\mathbf{A}^p)^{-1}$ $(p = 1, 2, \ldots)$. The ordinary rules for operations with exponents apply (see also Sec. 12.4-2).

14.4. LINEAR TRANSFORMATIONS OF A NORMED OR UNITARY VECTOR SPACE INTO ITSELF. HERMITIAN AND UNITARY TRANSFORMATIONS (OPERATORS)

14.4-1. Bound of a Linear Transformation (see also Sec. 13.2-1*a*). A linear transformation $\mathbf{A}$ of a normed vector space (Sec. 14.2-5) $\mathcal{V}$ into a normed vector space $\mathcal{V}'$ is **bounded** if and only if $\mathbf{A}$ has a finite **bound**

(**norm**)

$$\|\mathbf{A}\| = \sup_{\mathbf{x} \text{ in } \mathcal{V}} \frac{\|\mathbf{A}\mathbf{x}\|}{\|\mathbf{x}\|} \tag{14.4-1}$$

Note that

$$\left.\begin{array}{c}\|\mathbf{A}\| \geq 0 \qquad \|\alpha\mathbf{A}\| = |\alpha|\,\|\mathbf{A}\| \qquad \|\mathbf{A} + \mathbf{B}\| \leq \|\mathbf{A}\| + \|\mathbf{B}\| \\ \|\mathbf{A}\mathbf{B}\| \leq \|\mathbf{A}\|\,\|\mathbf{B}\| \qquad \|\mathbf{A}\mathbf{x}\| \leq \|\mathbf{A}\|\,\|\mathbf{x}\|\end{array}\right\} \tag{14.4-2}$$

*Every linear transformation (operator) defined throughout a finite-dimensional normed vector space is bounded.**

If $\mathcal{V}$ and $\mathcal{V}'$ are *unitary* vector spaces (Sec. 14.2-6), then

$$\begin{aligned}\|\mathbf{A}\| &= \sup_{\mathbf{x} \text{ in } \mathcal{V}} \frac{\|\mathbf{A}\mathbf{x}\|}{\|\mathbf{x}\|} = \sup_{\mathbf{x}\neq 0} \frac{\|\mathbf{A}\mathbf{x}\|}{\|\mathbf{x}\|} = \sup_{\|\mathbf{x}\|=1} \|\mathbf{A}\mathbf{x}\| \\ &= \sup_{\substack{\mathbf{x}\neq 0\\ \mathbf{y}\neq 0}} \frac{|(\mathbf{x}, \mathbf{A}\mathbf{y})|}{\|\mathbf{x}\|\,\|\mathbf{y}\|} = \sup_{\|\mathbf{x}\|=\|\mathbf{y}\|=1} |(\mathbf{x}, \mathbf{A}\mathbf{y})|\end{aligned} \tag{14.4-3}$$

14.4-2. The Bounded Linear Transformations of a Normed Vector Space into Itself. (**a**) *The bounded linear transformations (operators)* $\mathbf{A}, \mathbf{B}, \ldots$ *of a normed vector space* $\mathcal{V}$ *into itself constitute a linear algebra (Sec.* 12.4-2). *Within this algebra, the singular transformations are zero divisors (Sec.* 12.3-1*a*); *the nonsingular transformations constitute a multiplicative group and, together with the null transformation (Sec.* 14.3-3*b*), *form a division algebra (Sec.* 12.4-2). If $\mathcal{V}$ has the finite dimension n, the transformation algebra is of order n^2.

The transformation algebra (operator algebra) is *not* in general commutative (see also Sec. 12.4-2). The operator $\mathbf{AB} - \mathbf{BA}$ is called the **commutator** of $\mathbf{A}$ and $\mathbf{B}$.

(**b**) Bounded linear operators defined on complete unitary vector spaces (either finite-dimensional or Hilbert spaces, Sec. 14.2-7*c*) permit the definition of *convergent sequences* and *analytic functions* of operators with the aid of the metric $\|\mathbf{A} - \mathbf{B}\|$ (Sec. 12.5-3) in the manner of Secs. 13.2-11 and 13.2-12.

14.4-3. Hermitian Conjugate of a Linear Transformation (Operator). Every bounded linear transformation† $\mathbf{A}$ of a *complete unitary* vector space $\mathcal{V}_U$ into itself has a unique **hermitian conjugate (adjoint, conjugate, associate operator)** $\mathbf{A}^\dagger$ defined by

$$\boxed{(\mathbf{x}, \mathbf{A}\mathbf{y}) = (\mathbf{A}^\dagger\mathbf{x}, \mathbf{y}) \qquad \text{for all } \mathbf{x}, \mathbf{y} \text{ in } \mathcal{V}_U} \tag{14.4-4}$$

* Many texts define homogeneous linear operators on any normed vector space as operators which satisfy Eq. (14.3-3) *and are bounded* (and hence continuous in the sense of Sec. 12.5-1*c*).

† See footnote to Sec. 14.4-1.

so that

$$\begin{array}{c} (\mathbf{A}+\mathbf{B})\dagger = \mathbf{A}\dagger + \mathbf{B}\dagger \qquad (\alpha\mathbf{A})\dagger = \alpha^*\mathbf{A}\dagger \\ (\mathbf{AB})\dagger = \mathbf{B}\dagger\mathbf{A}\dagger \qquad (\mathbf{A}^{-1})\dagger = (\mathbf{A}\dagger)^{-1} \qquad (\mathbf{A}\dagger)\dagger = \mathbf{A} \\ \|\mathbf{A}\dagger\| = \|\mathbf{A}\| \qquad \|\mathbf{A}\dagger\mathbf{A}\| = \|\mathbf{A}\|^2 \qquad \mathbf{O}\dagger = \mathbf{O} \qquad \mathbf{I}\dagger = \mathbf{I} \end{array} \tag{14.4-5}$$

$$(\mathbf{A}\mathbf{x}, \mathbf{B}\mathbf{y}) \equiv (\mathbf{x}, \mathbf{A}\dagger\mathbf{B}\mathbf{y}) \equiv (\mathbf{B}\dagger\mathbf{A}\mathbf{x}, \mathbf{y}) \tag{14.4-6}$$

(see also Sec. 14.2-6*a*).

14.4-4. Hermitian Operators. A linear operator $\mathbf{A}$ mapping a complete unitary vector space $\mathcal{V}_U$ into itself is a **hermitian** operator (**self-adjoint operator, self-conjugate operator**) if and only if

$$\mathbf{A}\dagger = \mathbf{A} \quad \text{i.e.} \quad (\mathbf{x}, \mathbf{A}\mathbf{y}) = (\mathbf{A}\mathbf{x}, \mathbf{y}) \quad \text{for all } \mathbf{x}, \mathbf{y} \text{ in } \mathcal{V}_U \tag{14.4-7}$$

If $\mathcal{V}_U$ is a complex complete unitary vector space, $\mathbf{A}$ *is hermitian if and only if* $(\mathbf{x}, \mathbf{A}\mathbf{x})$ *is real for all* $\mathbf{x}$, *or*

$$(\mathbf{x}, \mathbf{A}\mathbf{x}) = (\mathbf{A}\mathbf{x}, \mathbf{x})^* = (\mathbf{A}\mathbf{x}, \mathbf{x}) \quad \text{for all } \mathbf{x} \text{ in } \mathcal{V}_U \tag{14.4-8}$$

(*alternative definition*). A transformation (operator) $\mathbf{A}$ such that $\mathbf{A}\dagger = -\mathbf{A}$ is called **skew-hermitian.**

Hermitian operators are of great importance in applications which require $(\mathbf{x}, \mathbf{A}\mathbf{x})$ to be a real quantity (vibration theory, quantum mechanics). A hermitian operator $\mathbf{A}$ is, respectively, **positive definite, negative definite, nonnegative, nonpositive, positive semidefinite, negative semidefinite, indefinite,** or zero if the same is true for the inner product (hermitian form) $(\mathbf{x}, \mathbf{A}\mathbf{x})$ (see also Secs. 13.5-3 and 14.7-1). *If* $\mathbf{A}$ *is nonnegative, there exists a hermitian operator* $\mathbf{Q}$ *such that* $\mathbf{Q}\dagger\mathbf{Q} = \mathbf{Q}\mathbf{Q}\dagger = \mathbf{A}$; $\mathbf{Q}$ *is uniquely defined if* $\mathbf{A}$ *is nonsingular.*

14.4-5. Unitary Transformations (Operators). A linear transformation $\mathbf{A}$ mapping a complete unitary vector space $\mathcal{V}_U$ into itself is **unitary** if and only if

$$\mathbf{A}\dagger\mathbf{A} = \mathbf{A}\mathbf{A}\dagger = \mathbf{I} \quad \text{i.e.} \quad \mathbf{A}\dagger = \mathbf{A}^{-1} \tag{14.4-9}$$

Every unitary transformation $\mathbf{A}$ *is nonsingular and bounded, and* $\|\mathbf{A}\| = 1$. *Every unitary transformation* $\mathbf{x}' = \mathbf{A}\mathbf{x}$ *preserves scalar products:*

$$\begin{array}{ll} (\mathbf{x}', \mathbf{y}') = (\mathbf{A}\mathbf{x}, \mathbf{A}\mathbf{y}) = (\mathbf{x}, \mathbf{y}) & \text{for all } \mathbf{x}, \mathbf{y} \text{ in } \mathcal{V}_U \\ \|\mathbf{x}'\| = \|\mathbf{A}\mathbf{x}\| = \|\mathbf{x}\| & \text{for all } \mathbf{x} \text{ in } \mathcal{V}_U \end{array} \tag{14.4-10}$$

If $\mathcal{V}_U$ is finite-dimensional, each of the relations (10) *implies* that $\mathbf{A}$ is unitary.

Unitary transformations preserve the results of scalar multiplication of vectors as well as those of vector addition and multiplication by scalars; so that absolute values, distances, angles, orthogonality, and orthonormality (Secs. 14.2-7*a* and 14.7-3) are invariant (Sec. 12.1-5).

14.4-6. Symmetric, Skew-symmetric, and Orthogonal Transformations of Real Unitary Vector Spaces. (a) The hermitian conjugate (Sec. 14.4-3) $\mathbf{A}\dagger$ associated with a linear transformation $\mathbf{A}$ of a *real* complete unitary vector space $\mathcal{V}_E$ into itself is often called the **transpose (conjugate, associate operator)** $\tilde{\mathbf{A}}$ of $\mathbf{A}$ and satisfies the relations

$$(\mathbf{x}, \mathbf{A}\mathbf{y}) = (\mathbf{A}\mathbf{y}, \mathbf{x}) = (\tilde{\mathbf{A}}\mathbf{x}, \mathbf{y}) = (\mathbf{y}, \tilde{\mathbf{A}}\mathbf{x}) \qquad \text{for all } \mathbf{x}, \mathbf{y} \text{ in } \mathcal{V}_E \tag{14.4-11}$$

$\tilde{\mathbf{A}}$ *may be substituted for* $\mathbf{A}\dagger$ *in all relations of Sec.* 14.4-3 *whenever the vector space in question is a real unitary vector space.*

(b) A linear transformation $\mathbf{A}$ of a real complete unitary vector space $\mathcal{V}_E$ is **symmetric** if and only if

$$\tilde{\mathbf{A}} = \mathbf{A} \qquad \text{i.e.} \qquad (\mathbf{x}, \mathbf{A}\mathbf{y}) = (\mathbf{y}, \mathbf{A}\mathbf{x}) \qquad \text{for all } \mathbf{x}, \mathbf{y} \text{ in } \mathcal{V}_E \tag{14.4-12}$$

skew-symmetric (antisymmetric) if and only if

$$\tilde{\mathbf{A}} = -\mathbf{A} \quad \text{i.e.} \quad (\mathbf{x}, \mathbf{A}\mathbf{y}) = -(\mathbf{y}, \mathbf{A}\mathbf{x}) \quad \text{for all } \mathbf{x}, \mathbf{y} \text{ in } \mathcal{V}_E \tag{14.4-13}$$

orthogonal if and only if

$$\tilde{\mathbf{A}}\mathbf{A} = \mathbf{A}\tilde{\mathbf{A}} = \mathbf{I} \qquad \text{i.e.} \qquad \tilde{\mathbf{A}} = \mathbf{A}^{-1} \tag{14.4-14}$$

Orthogonal transformations defined on real unitary vector spaces are unitary, so that all theorems of Sec. 14.4-5 *apply.*

14.4-7. Combination Rules (see also Sec. 13.3-3). **(a)** *If the operator* $\mathbf{A}$ *is hermitian, the same is true for* $\mathbf{A}^p$ $(p = 0, 1, 2, \ldots)$, $\mathbf{A}^{-1}$, *and* $\mathbf{T}\dagger\mathbf{A}\mathbf{T}$, *and for* $\alpha\mathbf{A}$ *if* α *is real.*

Given any nonsingular operator $\mathbf{T}$, $\mathbf{T}\dagger\mathbf{A}\mathbf{T}$ *is hermitian if and only if* $\mathbf{A}$ *is hermitian;* hence *for any unitary operator* $\mathbf{T}$, $\mathbf{T}^{-1}\mathbf{A}\mathbf{T}$ *is hermitian if and only if the same is true for* $\mathbf{A}$.

In particular, *if the vector space in question is a real complete unitary vector space, and* $\mathbf{A}$ *is symmetric, the same is true for* $\mathbf{A}^p$ $(p = 0, 1, 2, \ldots)$, $\mathbf{A}^{-1}$, $\tilde{\mathbf{T}}\mathbf{A}\mathbf{T}$, *and* $\alpha\mathbf{A}$. *Given any nonsingular operator* $\mathbf{T}$, $\tilde{\mathbf{T}}\mathbf{A}\mathbf{T}$ *is symmetric if and only if* $\mathbf{A}$ *is symmetric; and for any orthogonal operator* $\mathbf{T}$, $\mathbf{T}^{-1}\mathbf{A}\mathbf{T}$ *is symmetric if and only if the same is true for* $\mathbf{A}$.

(b) *If* $\mathbf{A}$ *and* $\mathbf{B}$ *are hermitian* (*or symmetric*), *the same is true for* $\mathbf{A} + \mathbf{B}$. *The product* $\mathbf{A}\mathbf{B}$ *of two hermitian* (*or symmetric*) *operators* $\mathbf{A}$ *and* $\mathbf{B}$ *is hermitian* (*or symmetric*) *if and only if* $\mathbf{B}\mathbf{A} = \mathbf{A}\mathbf{B}$ (see also Sec. 13.4-4*b*).

(**c**) *If* A *is a unitary transformation (operator), the same is true for* A^p ($p = 0, 1, 2, \ldots$), A^{-1}, *and* $\mathsf{A}\dagger$, *and for* $\alpha\mathsf{A}$ *if* $|\alpha| = 1$. *If* A *and* B *are unitary, the same is true for* AB.

If A *is an orthogonal transformation, the same is true for* A^p ($p = 0, 1, 2, \ldots$), A^{-1}, $\tilde{\mathsf{A}}$, *and* $-\mathsf{A}$. *If* A *and* B *are orthogonal, the same is true for* AB.

14.4-8. Decomposition Theorems. Normal Operators (see also Secs. 13.3-4 and 14.8-4). (**a**) For every linear operator A mapping a complete unitary vector space into itself, $\frac{1}{2}(\mathsf{A} + \mathsf{A}\dagger) = \mathsf{H}_1$ and $\frac{1}{2i}(\mathsf{A} - \mathsf{A}\dagger) = \mathsf{H}_2$ are *hermitian* operators. $\mathsf{A} = \mathsf{H}_1 + i\mathsf{H}_2$ is the (unique) *cartesian decomposition* of the given operator A into a *hermitian part* and a *skew-hermitian part* (comparable to the cartesian decomposition of complex numbers into real and imaginary parts, Sec. 1.3-1).

If A is defined on a *real* complete unitary vector space, the cartesian decomposition reduces to the (unique) decomposition of A into the *symmetric part* $\frac{1}{2}(\mathsf{A} + \tilde{\mathsf{A}})$ and the *skew-symmetric part* $\frac{1}{2}(\mathsf{A} - \tilde{\mathsf{A}})$.

For every linear operator A mapping a complete unitary vector space into itself, $\mathsf{A}\dagger\mathsf{A}$ is *hermitian* and *nonnegative*, and there exists a *polar decomposition* $\mathsf{A} = \mathsf{QU}$ of A into a nonnegative hermitian factor Q and a unitary factor U. Q is uniquely defined by $\mathsf{Q}^2 = \mathsf{A}\dagger\mathsf{A}$, and U is uniquely defined if and only if A is nonsingular (compare this with Sec. 1.3-2).

(**b**) A linear operator A mapping a complete unitary vector space $\mathcal{V}_U$ into itself is a **normal** operator if and only if $\mathsf{A}\dagger\mathsf{A} = \mathsf{A}\mathsf{A}\dagger$ or, equivalently, if and only if $\mathsf{H}_2\mathsf{H}_1 = \mathsf{H}_1\mathsf{H}_2$. *A bounded operator* A *is normal if and only if* $\|\mathsf{A}\mathbf{x}\| = \|\mathsf{A}\|\,\|\mathbf{x}\|$ *for all* $\mathbf{x}$ *in* $\mathcal{V}_U$. Hermitian and unitary operators are normal.

14.4-9. Conjugate (Adjoint, Dual) Vector Spaces. More General Definition of Conjugate (Adjoint) Operators. (**a**) The bounded linear transformations* A of a normed vector space $\mathcal{V}$ into any complete normed vector space (Banach space, Sec. 14.2-7) constitute a complete normed vector space, with addition and multiplication by scalars defined by Eq. (14.3-4), and norm $\|\mathsf{A}\|$. In particular, *the class of bounded, linear, and homogeneous scalar functions* $\varphi(\mathbf{x})$ *defined on a normed vector space* $\mathcal{V}$ *constitute a complete normed vector space,*† *the* **conjugate (adjoint, dual) vector space** $\mathcal{V}\dagger$ *associated with* $\mathcal{V}$.

A bounded linear transformation

$$\mathbf{x}' = \mathsf{A}\mathbf{x}$$

* See footnote to Sec. 14.4-1.

† Note carefully that the *value* of $\varphi(\mathbf{x})$ is a scalar, while the *function* $\varphi(\mathbf{x})$ can be a multidimensional vector. In the context of Chap. 15 (Sec. 15.4-3), the numerical-valued function $\varphi(\mathbf{x})$ is a *functional.*

mapping $\mathcal{V}$ into another normed vector space $\mathcal{V}'$ relates vectors φ, φ' of the corresponding conjugate spaces $\mathcal{V}\dagger$, $\mathcal{V}'\dagger$ by the bounded linear transformation

$$\varphi = \mathbf{A}\dagger\varphi' \tag{14.4-15a}$$

defined by

$$\varphi'(\mathbf{x}') \equiv \varphi'(\mathbf{A}\mathbf{x}) \equiv \varphi(\mathbf{x}) \tag{14.4-15b}$$

$\mathbf{A}\dagger$ is called the **conjugate (adjoint) operator** associated with $\mathbf{A}$; one has

$$\|\mathbf{A}\dagger\| = \|\mathbf{A}\| \tag{14.4-16}$$

Note that $(\mathbf{A}\dagger)\dagger$ is *not* in general identical with $\mathbf{A}$ in this context.

(b) *Every bounded, linear, and homogeneous scalar function* $\varphi(\mathbf{x})$ *defined on a complete unitary vector space (Euclidean space or Hilbert space)* $\mathcal{V}_V$ *can be expressed as an inner product*

$$\varphi(\mathbf{x}) \equiv (\boldsymbol{\phi}, \mathbf{x}) \tag{14.4-17}$$

where $\boldsymbol{\phi}$ is a vector of $\mathcal{V}_V$. The correspondence between the vectors φ of $\mathcal{V}\dagger_V$ and the vectors $\boldsymbol{\phi}$ of $\mathcal{V}_V$ is an isomorphism and, since Sec. 14.4-1 implies

$$\|\varphi(\mathbf{x})\| \equiv \sup_{\|\mathbf{x}\|=1} \varphi(\mathbf{x}) \equiv \|\boldsymbol{\phi}\| \tag{14.4-18}$$

the correspondence is also isometric (Secs. 12.5-2 and 14.2-7b). Hence, complete unitary vector spaces are **self-conjugate,** i.e., identical with their conjugate spaces except for isomorphism and isometry. The definition of operators conjugate to linear transformations of a complete unitary vector space into itself can then be reduced to the simple definition of Hermitian-conjugate operators given in Sec. 14.4-3.

14.4-10. Infinitesimal Linear Transformations (see also Secs. 4.5-3 and 14.10-5). **(a)** An **infinitesimal linear transformation (infinitesimal linear operator, infinitesimal dyadic)** defined on a normed real or complex vector space has the form

$$\mathbf{A} = \mathbf{I} + \epsilon\mathbf{B} \tag{14.4-19}$$

where $\mathbf{B}$ is bounded, and $|\epsilon|^2$ is negligibly small compared to 1 (ϵ is usually a scalar differential).

(b) *For infinitesimal linear transformations,* $\mathbf{A} = \mathbf{I} + \epsilon\mathbf{B}$, $\mathbf{A}_1 = \mathbf{I} + \epsilon_1\mathbf{B}_1$, $\mathbf{A}_2 = \mathbf{I} + \epsilon_2\mathbf{B}_2$

$$\left.\begin{aligned} \mathbf{A}^{-1} = \mathbf{I} - \epsilon\mathbf{B} \qquad \det(\mathbf{A}) = \epsilon \operatorname{Tr}(\mathbf{B}) \\ \mathbf{A}_1\mathbf{A}_2 = \mathbf{I} + (\epsilon_1\mathbf{B}_1 + \epsilon_2\mathbf{B}_2) = \mathbf{A}_2\mathbf{A}_1 \end{aligned}\right\} \tag{14.4-20}$$

Infinitestimal linear transformations (operators) commute.

(c) *An infinitesimal linear transformation* $\mathsf{A} = \mathsf{I} + \epsilon\mathsf{B}$ *defined on a complete unitary vector space is unitary if and only if* $\epsilon\mathsf{B}$ *is skew-hermitian. An infinitesimal linear transformation* $\mathsf{A} = \mathsf{I} + \epsilon\mathsf{B}$ *defined on a real complete unitary vector space is orthogonal if and only if* $\epsilon\mathsf{B}$ *is skew-symmetric.*

14.5. MATRIX REPRESENTATION OF VECTORS AND LINEAR TRANSFORMATIONS (OPERATORS)

14.5-1. Transformation of Base Vectors and Vector Components: "Alibi" Point of View (see also Sec. 14.1-3). Consider a finite-dimensional* real or complex vector space $\mathcal{V}_n$ with a reference system defined by n base vectors $\mathbf{e}_1, \mathbf{e}_2, \ldots, \mathbf{e}_n$ (Sec. 14.2-4). Each vector

$$\mathbf{x} = \xi_1\mathbf{e}_1 + \xi_2\mathbf{e}_2 + \cdots + \xi_n\mathbf{e}_n = \sum_{k=1}^{n} \xi_k\mathbf{e}_k \qquad (14.5\text{-}1)$$

is described by its components $\xi_1, \xi_2, \ldots, \xi_n$. A linear transformation (operator) A mapping $\mathcal{V}_n$ into itself (Sec. 14.3-1) transforms each base vector $\mathbf{e}_k$ into a corresponding vector

$$\mathbf{e}'_k = \mathsf{A}\mathbf{e}_k = a_{1k}\mathbf{e}_1 + a_{2k}\mathbf{e}_2 + \cdots + a_{nk}\mathbf{e}_n = \sum_{i=1}^{n} a_{ik}\mathbf{e}_i \qquad (k = 1, 2, \ldots, n) \qquad (14.5\text{-}2)$$

and each vector $\mathbf{x}$ of $\mathcal{V}_n$ into a corresponding vector $\mathbf{x}'$ of $\mathcal{V}_n$:

$$\mathbf{x}' = \mathsf{A}\mathbf{x} = \mathsf{A}\sum_{k=1}^{n} \xi_k\mathbf{e}_k = \sum_{k=1}^{n} \xi_k\mathbf{e}'_k = \sum_{i=1}^{n} \xi'_i\mathbf{e}_i \qquad (14.5\text{-}3)$$

The components ξ'_i of the vector $\mathbf{x}'$ and the components ξ_k of the vector $\mathbf{x}$, **both referred to the** $\mathbf{e}_1, \mathbf{e}_2, \ldots, \mathbf{e}_n$ **reference system,** are related by the n linear homogeneous transformation equations

$$\left.\begin{aligned} \xi'_1 &= a_{11}\xi_1 + a_{12}\xi_2 + \cdots + a_{1n}\xi_n \\ \xi'_2 &= a_{21}\xi_1 + a_{22}\xi_2 + \cdots + a_{2n}\xi_n \\ &\cdots\cdots\cdots\cdots\cdots\cdots \\ \xi'_n &= a_{n1}\xi_1 + a_{n2}\xi_2 + \cdots + a_{nn}\xi_n \end{aligned}\right\} \quad \begin{matrix}\text{(TRANSFORMATION}\\ \text{OF VECTOR COMPO-}\\ \text{NENTS, "ALIBI"}\\ \text{POINT OF VIEW)}\end{matrix} \qquad (14.5\text{-}4)$$

* The theory of Secs. 14.5-1 to 14.7-7 applies also to certain infinite-dimensional vector spaces (Secs. 14.2-4 and 14.2-7b). Such spaces must permit the definition of countable bases (Sec. 14.2-4), such as orthonormal bases (Sec. 14.7-4), and of convergence (Sec. 14.2-7), so that sums like that in Eq. (1) become convergent infinite series. *This is, in particular, true for every separable Hilbert space* (Sec. 14.2-7c). Vector spaces which do not admit countable bases can be represented by suitable function spaces (Sec. 15.2-1).

14.5-2. Matrix Representation of Vector and Linear Transformations (Operators). For each given reference basis $\mathbf{e}_1, \mathbf{e}_2, \ldots, \mathbf{e}_n$ in $\mathcal{V}_n$

1. The vectors $\mathbf{x} \equiv \xi_1\mathbf{e}_1 + \xi_2\mathbf{e}_2 + \cdots + \xi_n\mathbf{e}_n$ of $\mathcal{V}_n$ are represented on a reciprocal one-to-one basis by the column matrices $\{\xi_k\}$ (Sec. 13.2-1*b*).
2. The linear transformations (operators) mapping $\mathcal{V}_n$ into itself are represented on a reciprocal one-to-one basis by the $n \times n$ matrices $A \equiv [a_{ik}]$ defined by Eq. (2) or (4).

The transition between vectors and operators and the corresponding matrices is an isomorphism (Sec. 12.1-6): sums and products involving scalars, vectors, and transformations correspond to analogous sums and products of matrices. Identities and inverses correspond; nonsingular and unbounded transformations (operators) correspond, respectively, to nonsingular and unbounded matrices, and conversely.

In particular, the coordinate-free vector equation

$$\mathbf{x}' = \mathbf{A}\mathbf{x} \tag{14.5-5}$$

is represented in the $\mathbf{e}_1, \mathbf{e}_2, \ldots, \mathbf{e}_n$ reference system by the matrix equation

$$x' = \begin{bmatrix} \xi'_1 \\ \xi'_2 \\ \cdots \\ \xi'_n \end{bmatrix} = \begin{bmatrix} a_{11} & a_{12} & \cdots & a_{1n} \\ a_{21} & a_{22} & \cdots & a_{2n} \\ \cdots & \cdots & \cdots & \cdots \\ a_{n1} & a_{n2} & \cdots & a_{nn} \end{bmatrix} \begin{bmatrix} \xi_1 \\ \xi_2 \\ \cdots \\ \xi_n \end{bmatrix} = Ax \tag{14.5-6}$$

which is equivalent to the n transformation equations (4); and the product of two linear transformations $\mathbf{A}$ and $\mathbf{B}$ is represented by the product of the corresponding matrices A and B (carefully note Sec. 14.6-3).

NOTE: Transformations of an n-dimensional vector space $\mathcal{V}_n$ into an m-dimensional vector space $\mathcal{V}_m$ may be similarly represented by $m \times n$ matrices. Transformations relating two real vector spaces can always be represented by real matrices.

14.5-3. Matrix Notation for Simultaneous Linear Equations (see also Secs. 1.9-2 to 1.9-5). A set of simultaneous linear equations

$$\sum_{k=1}^{n} a_{ik}x_k = b_i \qquad (i = 1, 2, \ldots, m) \tag{14.5-7}$$

is equivalent to the matrix equation

$$Ax = b \qquad \text{or} \qquad \begin{bmatrix} a_{11} & a_{12} & \cdots & a_{1n} \\ a_{21} & a_{22} & \cdots & a_{2n} \\ \cdots & \cdots & \cdots & \cdots \\ a_{m1} & a_{m2} & \cdots & a_{mn} \end{bmatrix} \begin{bmatrix} x_1 \\ x_2 \\ \cdots \\ x_n \end{bmatrix} = \begin{bmatrix} b_1 \\ b_2 \\ \cdots \\ b_m \end{bmatrix} \tag{14.5-8}$$

The unknowns x_k may be regarded as components of an unknown vector such that the transformation (8) yields the vector represented by the b_i. If, in particular, the matrix $[a_{ik}]$ is nonsingular (Sec. 13.2-3), then the matrix equation (8) can be solved to yield the unique result

$$x = A^{-1}b \tag{14.5-9}$$

which is equivalent to Cramer's rule (1.9-4).

14.5-4. Dyadic Representation of Linear Operators. A linear operator **A** defined on an n-dimensional vector space may also be expressed as a sum of n outer products of pairs of vectors (n **dyads**) in the manner of Sec. 16.9-1. The corresponding $n \times n$ matrix A can be similarly expressed as the sum of n outer products of pairs of row and column matrices (see also Sec. 13.2-10).

14.6. CHANGE OF REFERENCE SYSTEM

14.6-1. Transformation of Base Vectors and Vector Components: "Alias" Point of View

(see also Sec. 14.1-3). (**a**) Given a reference basis $\mathbf{e}_1, \mathbf{e}_2, \ldots, \mathbf{e}_n$ in the finite-dimensional* vector space $\mathcal{V}_n$, *m vectors* $\mathbf{a}_k = \sum_{i=1}^{n} \alpha_{ik}\mathbf{e}_i$ *(k = 1, 2, . . . , m) are linearly independent* (*Sec.* 14.2-3) *if and only if the matrix* $[\alpha_{ik}]$ *is of rank m* (see also Sec. 1.9-3).

In particular, for every reference basis $\bar{\mathbf{e}}_1, \bar{\mathbf{e}}_2, \ldots, \bar{\mathbf{e}}_n$ in $\mathcal{V}_n$

$$\left.\begin{aligned} \bar{\mathbf{e}}_k &= t_{1k}\mathbf{e}_1 + t_{2k}\mathbf{e}_2 + \cdots + t_{nk}\mathbf{e}_n \\ &= \sum_{i=1}^{n} t_{ik}\mathbf{e}_i \qquad (k = 1, 2, \ldots, n) \\ \text{with} & \qquad \det[t_{ik}] \neq 0 \end{aligned}\right\} \quad \begin{pmatrix}\text{TRANSFORMATION}\\ \text{OF BASE VECTORS}\end{pmatrix} \tag{14.6-1}$$

The matrix $T \equiv [t_{ik}]$ represents a (necessarily nonsingular) transformation **T** relating the old base vectors $\mathbf{e}_i$ and the new base vectors $\bar{\mathbf{e}}_k = \mathbf{T}\mathbf{e}_k$ in the manner of Eq. (14.5-2).

(**b**) Now each vector **x** of $\mathcal{V}_n$ can be expressed in terms of vector components ξ_i referred to the $\mathbf{e}_i$ system or in terms of vector components $\bar{\xi}_k$ referred to the $\bar{\mathbf{e}}_k$ system:

$$\mathbf{x} = \sum_{i=1}^{n} \xi_i\mathbf{e}_i = \sum_{k=1}^{n} \bar{\xi}_k\bar{\mathbf{e}}_k \tag{14.6-2}$$

* See the footnote to Sec. 14.5-1.

The vector components ξ_i and $\bar{\xi}_k$ of the same vector **x** are related by the n linear homogeneous transformation equations

$$\left.\begin{aligned}
\xi_1 &= t_{11}\bar{\xi}_1 + t_{12}\bar{\xi}_2 + \cdots + t_{1n}\bar{\xi}_n \\
\xi_2 &= t_{21}\bar{\xi}_1 + t_{22}\bar{\xi}_2 + \cdots + t_{2n}\bar{\xi}_n \\
&\cdots\cdots\cdots\cdots\cdots\cdots \\
\xi_n &= t_{n1}\bar{\xi}_1 + t_{n2}\bar{\xi}_2 + \cdots + t_{nn}\bar{\xi}_n \\
&\text{or in matrix form} \\
x &= \begin{bmatrix} \xi_1 \\ \xi_2 \\ \cdots \\ \xi_n \end{bmatrix} = \begin{bmatrix} t_{11} & t_{12} & \cdots & t_{1n} \\ t_{21} & t_{22} & \cdots & t_{2n} \\ \cdots & \cdots & \cdots & \cdots \\ t_{n1} & t_{n2} & \cdots & t_{nn} \end{bmatrix} \begin{bmatrix} \bar{\xi}_1 \\ \bar{\xi}_2 \\ \cdots \\ \bar{\xi}_n \end{bmatrix} = T\bar{x}
\end{aligned}\right\} \quad \text{(TRANSFORMATION OF VECTOR COMPONENTS, "ALIAS" POINT OF VIEW)} \tag{14.6-3}$$

The meaning of the transformation equations (3) must be carefully distinguished from that of the formally analogous relations (14.5-4) and (14.5-6).

Note also the inverse relations, viz.

$$\mathbf{e}_i = \mathsf{T}^{-1}\bar{\mathbf{e}}_i = \sum_{k=1}^{n} \frac{T_{ik}}{\det [t_{ik}]} \bar{\mathbf{e}}_k \qquad (i = 1, 2, \ldots, n) \tag{14.6-4}$$

$$\left.\begin{aligned}
\bar{\xi}_k &= \sum_{i=1}^{n} \frac{T_{ik}}{\det [t_{ik}]} \xi_i \qquad (k = 1, 2, \ldots, n) \\
\text{or} \qquad \bar{x} &= T^{-1}x
\end{aligned}\right\} \tag{14.6-5}$$

where T_{ik} is the cofactor of t_{ik} in the determinant $\det [t_{ik}]$ (Sec. 1.5-2).

14.6-2. Representation of a Linear Operator in Different Schemes of Measurements. (a) Consider a linear operator A represented by the matrix A in the scheme of measurements (Sec. 14.1-5) associated with the base vectors $\mathbf{e}_i$ (Sec. 14.5-2) and by the matrix $\bar{A}$ in the $\bar{\mathbf{e}}_i$ scheme of measurements, so that for every vector **x** of $\mathcal{V}_n$

$$\mathbf{x}' = \mathsf{A}\mathbf{x} \qquad x' = Ax \qquad \overline{x'} = \bar{A}\bar{x} \tag{14.6-6}$$

Given the transformation matrix T relating the $\mathbf{e}_i$ and $\bar{\mathbf{e}}_k$ reference systems so that

$$x = T\bar{x} \qquad x' = T\overline{x'} \tag{14.6-7}$$

(Sec. 14.6-1), *the matrices A and $\bar{A}$ are related by the similarity transformation* (*Sec.* 13.4-1)

$$\bar{A} = T^{-1}AT \qquad \text{or} \qquad A = T\bar{A}T^{-1} \tag{14.6-8}$$

Conversely, *every matrix* $\bar{A}$ *related to* A *by a similarity transformation* (8) *represents the same linear operator* $\mathbf{A}$ *in a scheme of measurements specified by the base vectors* (1).

(b) *All matrices* (8) *representing the same operator* $\mathbf{A}$ *have the same rank* r; r *equals the rank of* $\mathbf{A}$ (Sec. 14.3-2). *The trace and the determinant of the matrix* A *are also common to all matrices* (8) and are referred to as the **trace** Tr ($\mathbf{A}$) and the **determinant** det ($\mathbf{A}$) **of the operator** $\mathbf{A}$ (see also Secs. 14.1-4 and 13.4-1*b*).

(c) Matrix Transformation of Base Vectors. If one admits row and column matrices of *vectors*, Eqs. (1) and (6) can be respectively written as

$$[\bar{\mathbf{e}}_1 \quad \bar{\mathbf{e}}_2 \quad . \, . \, . \,] = [\mathbf{e}_1 \quad \mathbf{e}_2 \quad . \, . \, . \,]T \tag{14.6-9}$$

$$[\mathbf{e}_1' \quad \mathbf{e}_2' \quad . \, . \, . \,] = [\mathbf{e}_1 \quad \mathbf{e}_2 \quad . \, . \, . \,]A \tag{14.6-10}$$

with

$$\left. \begin{aligned} \mathbf{x} &= [\mathbf{e}_1 \quad \mathbf{e}_2 \quad . \, . \, . \,]x = [\bar{\mathbf{e}}_1 \quad \bar{\mathbf{e}}_2 \quad . \, . \, . \,]\bar{x} \\ \mathbf{x}' &= [\mathbf{e}_1 \quad \mathbf{e}_2 \quad . \, . \, . \,]x' = [\mathbf{e}_1' \quad \mathbf{e}_2' \quad . \, . \, . \,]x \end{aligned} \right\} \tag{14.6-11}$$

(see also Sec. 16.6-2).

14.6-3. Consecutive Operations (see also Secs. 14.5-2 and 14.10-6). **(a)** Consider two consecutive linear operations $\mathbf{A}$, $\mathbf{B}$ defined by the base-vector transformations

$$\left. \begin{aligned} \mathbf{e}_k' &= \mathbf{A}\mathbf{e}_k = \sum_{i=1}^{n} a_{ik}\mathbf{e}_i \\ \mathbf{e}_k'' &= \mathbf{B}\mathbf{e}_k' = \mathbf{B}\mathbf{A}\mathbf{e}_k \end{aligned} \right\} (k = 1, 2, \ldots, n) \tag{14.6-12}$$

where $\mathbf{A}$, $\mathbf{B}$, and $\mathbf{BA}$ are, respectively, represented by the matrices $A \equiv [a_{ik}]$, $B \equiv [b_{ik}]$, and BA in the $\mathbf{e}_i$ scheme of measurements, i.e.,

$$\mathbf{x}' = \mathbf{A}\mathbf{x} \qquad \mathbf{x}'' = \mathbf{B}\mathbf{x}' = \mathbf{B}\mathbf{A}\mathbf{x} \tag{14.6-13}$$

is represented by

$$x' = Ax \qquad x'' = Bx' = BAx \qquad x = (BA)^{-1}x'' \tag{14.6-14}$$

where $x' \equiv \{\xi_1', \xi_2', \ldots\}$ and $x'' \equiv \{\xi_1'', \xi_2'', \ldots\}$ as well as $x \equiv \{\xi_1, \xi_2, \ldots\}$ are columns of vector components measured in the $\mathbf{e}_i$ reference system.

Note carefully that the operation $\mathbf{B}$ *defined by Eqs.* (12) *to* (14) *is, in general, different from the operation defined by*

$$\mathbf{e}_k''' = \sum_{i=1}^{n} b_{ik}\mathbf{e}_i' = \mathbf{A}\mathbf{B}\mathbf{A}^{-1}\mathbf{e}_k' = \mathbf{A}\mathbf{B}\mathbf{e}_k \qquad (k = 1, 2, \ldots, n) \tag{14.6-15}$$

which corresponds to

$$\mathbf{x}''' = \mathbf{A}\mathbf{B}\mathbf{x} \qquad x''' = ABx \tag{14.6-16}$$

since the matrix $B \equiv [b_{ik}]$ represents $\mathbf{ABA}^{-1}$ and *not* $\mathbf{B}$ in the $\mathbf{e}'_k$ scheme of measurements.

(**b**) The column matrices $x \equiv \{\xi_1, \xi_2, \ldots, \xi_n\}$, $\bar{x} \equiv \{\bar{\xi}_1, \bar{\xi}_2, \ldots, \bar{\xi}_n\}$, $\bar{\bar{x}} \equiv \{\bar{\bar{\xi}}_1, \bar{\bar{\xi}}_2, \ldots, \bar{\bar{\xi}}_n\}$ representing *the same* vector

$$\mathbf{x} = \sum_{i=1}^{n} \xi_i \mathbf{e}_i = \sum_{k=1}^{n} \bar{\xi}_k \mathbf{e}'_k = \sum_{k=1}^{n} \bar{\bar{\xi}}_k \mathbf{e}''_k \tag{14.6-17}$$

are related by the alias-type transformations

$$\bar{x} = A^{-1}BA\bar{\bar{x}} \qquad x = A\bar{x} = BA\bar{\bar{x}} \qquad \bar{\bar{x}} = (BA)^{-1}x \tag{14.6-18}$$

Note again that, in general, $\bar{x} \neq B\bar{\bar{x}}$.

14.7. REPRESENTATION OF INNER PRODUCTS. ORTHONORMAL BASES

14.7-1. Representation of Inner Products (see also Secs. 14.2-6, 14.7-6*b*, and 16.8-1). Given a finite-dimensional unitary vector space or separable Hilbert space (Sec. 14.2-7)* $\mathcal{V}_U$, let the vectors $\mathbf{a} = \sum_{i=1}^{n} \alpha_i \mathbf{e}_i$ and $\mathbf{b} = \sum_{i=1}^{n} \beta_i \mathbf{e}_i$ be represented by the respective column matrices $a \equiv \{\alpha_i\}$ and $b \equiv \{\beta_i\}$ in the manner of Sec. 14.5-2 (see also Sec. 13.2-1*b*). Then

$$(\mathbf{a}, \mathbf{b}) = \sum_{i=1}^{n} \sum_{k=1}^{n} g_{ik} \alpha_i^* \beta_k = a\dagger Gb \tag{14.7-1}$$

with

$$G \equiv [g_{ik}] \qquad g_{ik} = (\mathbf{e}_i, \mathbf{e}_k) \qquad (i, k = 1, 2, \ldots, n)$$

The matrix $G \equiv [g_{ik}]$ is necessarily hermitian ($g_{ik} = g_{ki}^*$) and positive definite (Sec. 13.5-3); if $\mathcal{V}_U$ is a real unitary vector space, then G is real and symmetric.

Equation (1) *makes it possible to describe absolute values, angles, distances, convergence, etc., in $\mathcal{V}_U$ in terms of numerical vector components* (see also Sec. 14.2-7). In particular, for every vector $\mathbf{x} = \sum_{i=1}^{n} \xi_i \mathbf{e}_i$ in $\mathcal{V}_U$,

$$\|\mathbf{x}\|^2 = \sum_{i=1}^{n} \sum_{k=1}^{n} g_{ik} \xi_i^* \xi_k = x\dagger Gx \tag{14.7-2}$$

* See the footnote to Sec. 14.5-1.

The hermitian form (Sec. 13.5-3) (2) is called the **fundamental form of $\mathcal{V}_U$ in the scheme of measurements defined by the base vectors $\mathbf{e}_i$.**

14.7-2. Change of Reference System (see also Secs. 14.6-1 and 16.8-1). If one introduces a new set of base vectors $\bar{\mathbf{e}}_k$ such that $a = T\bar{a}$, $b = T\bar{b}$ (Sec. 14.6-1), the invariance (Sec. 14.1-4) of

$$(\mathbf{a}, \mathbf{b}) = a\dagger Gb = \sum_{i=1}^{n} \sum_{k=1}^{n} g_{ik}\alpha_i^* \beta_k = \sum_{i=1}^{n} \sum_{k=1}^{n} \bar{g}_{ik}\bar{\alpha}_i^* \bar{\beta}_k = \bar{a}\dagger \bar{G}\bar{b} \tag{14.7-3}$$

implies

$$\bar{G} \equiv [\bar{g}_{ik}] \equiv [(\bar{\mathbf{e}}_i, \bar{\mathbf{e}}_k)] \equiv T\dagger GT \tag{14.7-4}$$

14.7-3. Orthogonal Vectors and Orthonormal Sets of Vectors. (**a**) Two vectors **a**, **b** of a unitary vector space $\mathcal{V}_U$ are (mutually) **orthogonal** if and only if $(\mathbf{a}, \mathbf{b}) = 0$ ($\gamma = 90$ deg, Sec. 14.2-7*b*). An **orthonormal (normal orthogonal)** set of vectors is a set of mutually orthogonal unit vectors $\mathbf{u}_1, \mathbf{u}_2, \ldots$, so that

$$(\mathbf{u}_i, \mathbf{u}_k) = \delta_k^i = \begin{cases} 0 \text{ if } i \neq k \\ 1 \text{ if } i = k \end{cases} \quad (i, k = 1, 2, \ldots) \tag{14.7-5}$$

Every set of mutually orthogonal nonzero vectors (and, in particular, every orthonormal set) is linearly independent; so that the largest number of vectors in any such set (the **orthogonal dimension** of $\mathcal{V}_U$) cannot exceed the linear dimension of $\mathcal{V}_U$ (see also Secs. 14.2-3 and 14.2-4*b*).

(**b**) **Bessel's Inequality** (see also Sec. 15.2-3*b*). *Given a finite or infinite orthonormal set* $\mathbf{u}_1, \mathbf{u}_2, \ldots$ *and any vector* **a** *in* $\mathcal{V}_U$,

$$\sum_k |(\mathbf{u}_k, \mathbf{a})|^2 \leq \|\mathbf{a}\|^2 \qquad \text{(Bessel's inequality)} \tag{14.7-6}$$

The equal sign applies if and only if the vector **a** *belongs to the linear manifold spanned by the orthonormal set* (see also Sec. 14.7-4). Bessel's inequality is closely related to the projection theorem of Sec. 14.2-8 and is often used to prove the convergence of infinite series.

14.7-4. Orthonormal Bases (Complete Orthonormal Sets). (**a**) In a finite-dimensional unitary vector space of dimension n, every orthonormal set of n vectors is a basis (**orthonormal basis**). More generally, *in every complete unitary vector space* $\mathcal{V}_C$ *(this includes both finite-dimensional unitary vector spaces and Hilbert spaces, Sec.* 14.2-7*c), an orthonormal set of vectors* $\mathbf{u}_1, \mathbf{u}_2, \ldots$ *constitutes an orthonormal basis* (**complete**

orthonormal set, complete orthonormal system) *if and only if it satisfies the following conditions:*

1. Every vector $\mathbf{a}$ of $\mathcal{V}_C$ can be expressed in the form

$$\mathbf{a} = \hat{\alpha}_1\mathbf{u}_1 + \hat{\alpha}_2\mathbf{u}_2 + \cdots \quad \text{with} \quad \hat{\alpha}_k = (\mathbf{u}_k, \mathbf{a}) \quad (k = 1, 2, \ldots)$$

2. Given any vector $\hat{\alpha}_1\mathbf{u}_1 + \hat{\alpha}_2\mathbf{u}_2 + \cdots = \mathbf{a}$,

$$\|\mathbf{a}\|^2 = |\hat{\alpha}_1|^2 + |\hat{\alpha}_2|^2 + \cdots \quad \text{(PARSEVAL'S IDENTITY)}$$

3. Given any pair of vectors $\hat{\alpha}_1\mathbf{u}_1 + \hat{\alpha}_2\mathbf{u}_2 + \cdots = \mathbf{a}$, $\hat{\beta}_1\mathbf{u}_1 + \hat{\beta}_2\mathbf{u}_2 + \cdots = \mathbf{b}$,

$$(\mathbf{a}, \mathbf{b}) = \hat{\alpha}_1^*\hat{\beta}_1 + \hat{\alpha}_2^*\hat{\beta}_2 + \cdots$$

4. The orthonormal set $\mathbf{u}_1, \mathbf{u}_2, \ldots$ is not contained in any other orthonormal set of $\mathcal{V}_C$. If $\mathbf{a}$ is a bounded vector of $\mathcal{V}_C$, then $(\mathbf{u}_k, \mathbf{a}) = 0$ $(k = 1, 2, \ldots)$ implies $\mathbf{a} = 0$.

Each of these four conditions implies the three others. The relative simplicity of the above expressions for $\|\mathbf{a}\|^2$ and $(\mathbf{a}, \mathbf{b})$ makes orthonormal bases especially useful as reference systems. Note that the concept of a complete orthonormal set extends the definition of a basis (Sec. 14.2-4) to suitable infinite-dimensional vector spaces in the following sense: *if* $\mathbf{a}$, $\mathbf{a}'$ *are two vectors with identical components* $\hat{\alpha}_k$, then $\|\mathbf{a} - \mathbf{a}'\| = 0$ (*Uniqueness Theorem*).

(b) Construction of Orthonormal Sets of Vectors. *Given any countable (finite or infinite) set of linearly independent vectors* $\mathbf{e}_1, \mathbf{e}_2, \ldots$ *of a complete unitary vector space, there exists an orthonormal set* $\mathbf{u}_1, \mathbf{u}_2, \ldots$ *spanning the same linear manifold.*

Such an orthonormal set may be constructed with the aid of the following recursion formulas (*Gram-Schmidt orthogonalization process*, see also Sec. 15.2-5):

$$\left.\begin{aligned} \mathbf{u}_i &= \mathbf{v}_i/\|\mathbf{v}_i\| \\ \mathbf{v}_1 &= \mathbf{e}_1 \qquad \mathbf{v}_{i+1} = \mathbf{e}_{i+1} - \sum_{k=1}^{i} (\mathbf{u}_k, \mathbf{e}_{i+1})\mathbf{u}_k \end{aligned}\right\} \quad (i = 1, 2, \ldots) \qquad (14.7\text{-}7)$$

14.7-5. Matrices Corresponding to Hermitian-conjugate Operators (see also Secs. 14.4-3 to 14.4-6, 13.3-1, and 13.3-2). **(a)** Given a linear operator **A** represented by the matrix A, the hermitian conjugate $\mathbf{A}\dagger$ is represented by $G^{-1}A\dagger G$ in the same scheme of measurements.*

(b) For an *orthonormal* reference system $\mathbf{u}_1, \mathbf{u}_2, \ldots$, one has $G = I$ (Sec. 14.7-4), and

$$A \equiv [a_{ik}] \equiv [(\mathbf{u}_i, \mathbf{A}\mathbf{u}_k)] \tag{14.7-8}$$

so that *hermitian-conjugate operators correspond to hermitian-conjugate matrices, and conversely.* Thus hermitian, skew-hermitian, and unitary operators correspond to matrices of the same respective types, and conversely. In particular, symmetric, skew-symmetric, and orthogonal operators defined on real vector spaces correspond, respectively, to symmetric, skew-symmetric, and orthogonal matrices whenever an orthonormal reference system is used.

More generally, unitary or orthogonal operators correspond to matrices of the same respective type whenever the reference base vectors are orthogonal (not necessarily orthonormal).

14.7-6. Reciprocal Bases. **(a)** For every basis $\mathbf{e}_1, \mathbf{e}_2, \ldots, \mathbf{e}_n$ in a finite-dimensional vector space, there exists a uniquely corresponding **reciprocal (dual) basis** $\mathbf{e}^1, \mathbf{e}^2, \ldots, \mathbf{e}^n$ defined by the symmetric relationship

$$(\mathbf{e}^i, \mathbf{e}_k) = \delta^i_k = \begin{Bmatrix} 0 \text{ if } i \neq k \\ 1 \text{ if } i = k \end{Bmatrix} \quad (i, k = 1, 2, \ldots, n) \tag{14.7-9a}$$

so that each $\mathbf{e}_i$ is perpendicular to all $\mathbf{e}_k$ with $k \neq i$, and

$$[(\mathbf{e}^i, \mathbf{e}^k)] = G^{-1} = [(\mathbf{e}_i, \mathbf{e}_k)]^{-1} \tag{14.7-9b}$$

$$\mathbf{e}_k = \sum_{i=1}^{n} g_{ik}\mathbf{e}^i \qquad (k = 1, 2, \ldots, n) \tag{14.7-10}$$

(b) Vectors $\mathbf{a}, \mathbf{b}, \ldots$ represented in the $\mathbf{e}_i$ reference system by column matrices $a, b, \ldots$ are represented in the $\mathbf{e}^i$ system by the column matrices $Ga, Gb, \ldots$,† and

$$(\mathbf{a}, \mathbf{b}) = a\dagger Gb = a\dagger(Gb) = (Ga)\dagger b = (Ga)\dagger G^{-1}(Gb) \tag{14.7-11}$$

In particular, $(\mathbf{e}^i, \mathbf{a}) = \alpha_i$.

A linear operator **A** given by the matrix A in the $\mathbf{e}_i$ scheme of measurements is represented by the matrix GAG^{-1} in the $\mathbf{e}^i$ scheme of measurements.

(c) The reciprocal basis $\bar{\mathbf{e}}^1, \bar{\mathbf{e}}^2, \ldots, \bar{\mathbf{e}}^n$ corresponding to a new set of base vectors

$$\bar{\mathbf{e}}_k = \sum_{i=1}^{n} t_{ik}\mathbf{e}_i \qquad \text{with} \qquad \det [t_{ik}] \neq 0$$

* Note that $\mathbf{A}\dagger$ is represented by the matrix $A\dagger$ in the scheme of measurements corresponding to the reciprocal basis, Sec. 14.7-6.

† One can also represent the vectors $\mathbf{a}, \mathbf{b}, \ldots$ by the *row matrices* $(Ga)\dagger$, $(Gb)\dagger$, $\ldots$ or $(\widetilde{Ga})$, $(\widetilde{Gb})$, $\ldots$, corresponding to a representation by covariant vector components (Secs. 16.2-1 and 16.7-3).

is given by

$$\mathbf{e}^k = \sum_{i=1}^{n} t_{ki}\bar{\mathbf{e}}^i \qquad (14.7\text{-}12)$$

The base vectors $\mathbf{e}^i$ and $\mathbf{e}_k$ are said to transform **contragrediently** (see also Sec. 16.6-2). Similarly

$$x = T\bar{x} \qquad \text{implies} \qquad \bar{G}\bar{x} = T(Gx) \qquad (14.7\text{-}13)$$

(**d**) *Every orthonormal basis (Sec. 14.7-4a) is identical with its reciprocal basis (self-dual), so that* $\mathbf{e}^i = \mathbf{e}_i = \mathbf{u}_i$ $(i = 1, 2, \ldots, n)$, *and*

$$G = I \qquad Gx \equiv x \qquad GAG^{-1} \equiv A \qquad (14.7\text{-}14)$$

14.7-7. Comparison of Notations. In order to permit easy reference to standard textbooks, *subscripts only* have been used throughout Chaps. 12 to 14 to label vector components and matrix elements. The improved *dummy-index notation* employed in tensor analysis and described in Sec. 16.1-3 uses *superscripts* as well as subscripts. Table 14.7-1 reviews the different notations used to describe vectors and linear operators and may be used to translate one notation into the other (see also Sec. 16.2-1).

Table 14.7-1. Comparison of Different Notations Describing Scalars, Vectors, and Linear Operators

Coordinate-free (invariant) notation	Matrix representations	Components (matrix elements)	
		Conventional notation	Dummy-index notation
α (scalar)	α	α	α
$(\mathbf{e}_i, \mathbf{e}_k)$	$[(\mathbf{e}_i, \mathbf{e}_k)] \equiv G$	g_{ik}	g_{ik}
$(\mathbf{e}^i, \mathbf{e}^k)$	$[(\mathbf{e}^i, \mathbf{e}^k)] \equiv G^{-1}$	No special symbol	g^{ik}
$\mathbf{a}$ (vector)	a	α_i	a^i
	Ga	No special symbol	$a_i = g_{ik}a^k$
$\mathbf{y} = \mathbf{A}\mathbf{x}$	$y = Ax$	$\eta_i = \sum_{k=1}^{n} a_{ik}\xi_k$	$y^i = A^i_k x^k$
$\mathbf{A}$ (linear operator, dyadic)	A	a_{ik}	A^i_k
	GA	No special symbols	$A_{ik} = g_{ij}A^j_k$
	AG^{-1}		$A^{ik} = A^i_j g^{jk}$
	GAG^{-1}		$A_i^{\ k} = g_{ij}A^j_h g^{hk}$

14.8. EIGENVECTORS AND EIGENVALUES OF LINEAR OPERATORS

14.8-1. Introduction. The study of eigenvectors and eigenvalues is of outstanding practical interest, because

1. Many relations involving a linear operator are radically simplified if its eigenvectors are introduced as reference base vectors (diagonalization of matrices, quadratic forms, solution of operator equations in spectral form; see also Sec. 15.1-1).
2. The eigenvalues of a linear operator specify important properties of the operator without reference to a particular coordinate system.

In many applications, eigenvectors and eigenvalues of linear operators have a direct geometrical or physical significance; they can usually be interpreted in terms of a maximum-minimum problem (Sec. 14.8-8). The most important applications involve hermitian operators which have *real* eigenvalues (Secs. 14.8-4 and 14.8-10).

14.8-2. Invariant Manifolds. Decomposable Linear Transformations (Linear Operators) and Matrices. The manifold $\mathcal{V}_1$ is a linear vector space $\mathcal{V}$ **is invariant with respect to (reduces)** a given linear transformation $\mathbf{A}$ of $\mathcal{V}$ into itself if and only if $\mathbf{A}$ transforms every vector $\mathbf{x}$ of $\mathcal{V}_1$ into a vector $\mathbf{Ax}$ of $\mathcal{V}_1$. Given a reference basis $\bar{\mathbf{e}}_1, \bar{\mathbf{e}}_2, \ldots, \bar{\mathbf{e}}_m, \bar{\mathbf{e}}_{m+1}, \ldots$ in $\mathcal{V}$ such that $\bar{\mathbf{e}}_1, \bar{\mathbf{e}}_2, \ldots, \bar{\mathbf{e}}_m$ span $\mathcal{V}_1$, $\mathbf{A}$ is represented by a matrix $\bar{A}$ which can be partitioned in the form

$$\bar{A} \equiv \left[\begin{array}{c|c} \bar{A}_1 & \bar{A}' \\ \hline [0] & \bar{A}_2 \end{array} \right] \tag{14.8-1}$$

where $\bar{A}_1$ is the $m \times m$ matrix representing the linear transformation $\mathbf{A}_1$ of $\mathcal{V}_1$ "induced" by $\mathbf{A}$. $\mathbf{A}_1$ may or may not be capable of further reduction.

A linear transformation (linear operator) $\mathbf{A}$ of the vector space $\mathcal{V}$ into itself is **decomposable (reducible, completely reducible*)** if and only if $\mathcal{V}$ is the direct sum $\mathcal{V} = \mathcal{V}_1 \oplus \mathcal{V}_2 \oplus \cdots$ (Sec. 12.7-5*a*) of two or more subspaces $\mathcal{V}_1, \mathcal{V}_2, \ldots$ each invariant with respect to $\mathbf{A}$. In this case, one writes $\mathbf{A}$ as the **direct sum** $\mathbf{A} = \mathbf{A}_1 \oplus \mathbf{A}_2 \oplus \cdots$ **of the linear transformations** $\mathbf{A}_1, \mathbf{A}_2, \ldots$ respectively induced by $\mathbf{A}$ in $\mathcal{V}_1, \mathcal{V}_2, \ldots$.

A square matrix A represents a decomposable operator $\mathbf{A}$ *if and only if A is similar to a step matrix (direct sum of matrices $A_1, A_2, \ldots$ corresponding*

* See the footnote to Sec. 14.9-2*b*.

to $\mathbf{A}_1, \mathbf{A}_2, \ldots$, see also Sec. 13.2-9). A matrix A with this property is also called **decomposable (reducible, completely reducible*).**

14.8-3. Eigenvectors, Eigenvalues, and Spectra (see also Sec. 13.4-2). **(a)** An **eigenvector (proper vector, characteristic vector)** of the linear transformation (linear operator) $\mathbf{A}$ defined on a linear vector space $\mathcal{V}$ is a vector $\mathbf{y}$ of $\mathcal{V}$ such that

$$\boxed{\mathbf{A}\mathbf{y} = \lambda\mathbf{y} \qquad (\mathbf{y} \neq \mathbf{0})} \tag{14.8-2}$$

where λ is a suitably determined scalar called the **eigenvalue (proper value, characteristic value)** of $\mathbf{A}$ associated with the eigenvector $\mathbf{y}$.

(b) *If* $\mathbf{y}$ *is an eigenvector associated with the eigenvalue* λ *of* $\mathbf{A}$, *the same is true for every vector* $\alpha\mathbf{y} \neq \mathbf{0}$. *If* $\mathbf{y}_1, \mathbf{y}_2, \ldots, \mathbf{y}_s$ *are eigenvectors associated with the eigenvalue* λ *of* $\mathbf{A}$, *the same is true for every vector* $\alpha_1\mathbf{y}_1 + \alpha_2\mathbf{y}_2 + \cdots + \alpha_s\mathbf{y}_s \neq 0$; *these vectors span a linear manifold invariant with respect to* $\mathbf{A}$ (Sec. 14.8-2; see also Sec. 14.8-4*c*). This theorem also applies to convergent infinite series of eigenvectors in Hilbert spaces.

An eigenvalue λ associated with exactly $m > 1$ linearly independent eigenvectors is said to be m**-fold degenerate;** m is called the **degree of degeneracy** or **geometrical multiplicity** of the eigenvalue.

Eigenvectors associated with different eigenvalues of a linear operator are linearly independent.

A linear operator defined on an n-dimensional vector space has at most n distinct eigenvalues. Every eigenvalue of a nonsingular operator is different from zero.

(c) *Given a linear operator* $\mathbf{A}$ *with eigenvalues* λ, $\alpha\mathbf{A}$ *has the eigenvalues* $\alpha\lambda$, *and* $\mathbf{A}^p$ *has the eigenvalues* λ^p ($p = 0, 1, 2, \ldots$; $p = 0, \pm 1, \pm 2, \ldots$ *if* $\mathbf{A}$ *is nonsingular*). *Every polynomial* $f(\mathbf{A})$ (*Sec.* 14.4-2*b*) *has the eigenvalues* $f(\lambda)$ (see also Secs. 13.4-5*b*). *All these functions of* $\mathbf{A}$ *have the same eigenvectors as* $\mathbf{A}$.

(d) The Spectrum of a Linear Operator. The **spectrum** of a linear operator $\mathbf{A}$ mapping a complete normed vector space (Banach space, Sec. 14.2-7*a*) $\mathcal{V}$ into itself is the set of all complex numbers (**spectral values,** eigenvalues†) λ such that the vector equation

$$\mathbf{A}\mathbf{x} - \lambda\mathbf{x} = \mathbf{f}$$

does not have a unique solution $\mathbf{x} = (\mathbf{A} - \lambda\mathbf{I})^{-1}\mathbf{f}$ of finite norm for every vector $\mathbf{f}$ of finite norm (see also Sec. 14.8-10). More precisely stated, the operator $\mathbf{A} = \lambda\mathbf{I}$ does not have a unique bounded inverse $(\mathbf{A} - \lambda\mathbf{I})^{-1}$

* See the footnote to Sec. 14.9-2*b*.

† Some authors refer to all spectral values as eigenvalues; others restrict the use of this term to the discrete spectrum.

(resolvent operator). The spectrum may be partitioned into

1. The **discrete spectrum (point spectrum)** defined by Eq. (2) with eigenvectors $\mathbf{y} \neq 0$ of finite norm $\|\mathbf{y}\|$
2. The **continuous spectrum** where $(\mathsf{A} - \lambda \mathsf{I})^{-1}$ is unbounded with domain dense in $\mathcal{V}$
3. The **residual spectrum** where $(\mathsf{A} - \lambda \mathsf{I})^{-1}$ is unbounded with domain not dense in $\mathcal{V}$

The spectrum of a linear operator A contains its **approximate spectrum** defined as the set of all complex numbers λ such that there exists a sequence of unit vectors $\mathbf{u}_1, \mathbf{u}_2, \ldots$ such that $\|(\mathsf{A} - \lambda \mathsf{I})\mathbf{u}_n\| < 1/n$ $(n = 1, 2, \ldots)$. *The approximate spectrum contains both the discrete and the continuous spectrum* (see also Sec. 14.8-4).

The residual spectrum of A *is contained in the discrete spectrum of* $\mathsf{A}\dagger$.

(e) *The spectrum of a linear operator* A *is identical with the spectrum of every matrix A representing* A *in the manner of Sec.* 14.5-2. The **algebraic multiplicity** m_j' of any discrete eigenvalue λ_j of A is the algebraic multiplicity of the corresponding matrix eigenvalue (Sec. 13.4-3*a*). m_j' *is greater than, or equal to, the geometrical multiplicity* m_j *of* λ_j (see also Sec. 14.8-4*c*).

For every linear operator having a purely discrete spectrum and a finite trace, Tr (A) *equals the sum of all eigenvalues, each counted a number of times equal to its algebraic multiplicity. If* det (A) *exists, it equals the similarly computed product of the eigenvalues* (see also Sec. 13.4-3*b*).

The characteristic equation $F_A(\lambda) = 0$ associated with a class of similar finite matrices (Sec. 13.4-5*a*) is the characteristic equation of the corresponding operator A and yields its eigenvalues together with their *algebraic* multiplicities. *The Cayley-Hamilton theorem* $[F_A(\mathsf{A}) = 0$, Sec. 13.4-7*a*] *and the theorems of Sec.* 13.4-7*b apply to linear operators defined on finite-dimensional vector spaces.*

(f) *If* A, $\mathbf{y}$, *and* $\mathbf{z}$ *are bounded, then* $\mathsf{A}\mathbf{y} = \lambda\mathbf{y}$, $\mathsf{A}\dagger\mathbf{z} = \mu\mathbf{z}$ *implies either* $\mu = \lambda^*$ *or* $(\mathbf{y}, \mathbf{z}) = 0$ (see also Sec. 14.8-4).

14.8-4. Eigenvectors and Eigenvalues of Normal and Hermitian Operators (see also Secs. 13.4-2, 13.4-4*a*, 14.4-6, 14.8-8, 15.3-3*b*, 15.3-4, and 15.4-6). (a) *If* A *is a normal operator* $(\mathsf{A}\dagger\mathsf{A} = \mathsf{A}\mathsf{A}\dagger$, Sec. 14.4-8*b*), *then* A *and* $\mathsf{A}\dagger$ *have identical eigenvectors; corresponding eigenvalues of* A *and* $\mathsf{A}\dagger$ *are complex conjugates. The spectrum of every normal operator is identical with its approximate spectrum;* the residual spectrum is empty (see also Sec. 14.8-3*d*).

For every normal operator with eigenvalues λ, *the hermitian operators* $\frac{1}{2}(\mathsf{A} + \mathsf{A}\dagger) = \mathsf{H}_1$, $\frac{1}{2i}(\mathsf{A} - \mathsf{A}\dagger) = \mathsf{H}_2$, *and* $\mathsf{A}\dagger\mathsf{A}$ *have the same eigenvectors as* A *and the respective eigenvalues* $\mathrm{Re}(\lambda)$, $\mathrm{Im}(\lambda)$, *and* $|\lambda|^2$.

(b) *All spectral values of any hermitian operator are real.*

The converse is *not* necessarily true; but every *normal* operator having a real spectrum is hermitian.

The spectrum of every bounded hermitian operator* **A** *is a closed set of real numbers; its largest and smallest values equal* $\sup_{\|\mathbf{x}\|=1} (\mathbf{A}\mathbf{x}, \mathbf{x})$ *and* $\inf_{\|\mathbf{x}\|=1} (\mathbf{A}\mathbf{x}, \mathbf{x})$.

(c) The following *important special properties of normal operators* apply, in particular, to *hermitian* operators:

1. **Orthogonality of Eigenvectors.** *Eigenvectors corresponding to different eigenvalues of a normal operator are mutually orthogonal.*
2. **Completeness Property of Eigenvectors.** *Every bounded normal operator* **A** *defined on a complete unitary vector space* $\mathcal{V}_U$ *is completely reduced (Sec. 14.8-2) by a subspace spanned by a complete orthonormal set of eigenvectors (Sec. 14.7-4) and a subspace orthogonal to every eigenvector of* **A**. *If* $\mathcal{V}_U$ *is separable (in particular, if* $\mathcal{V}_U$ *is finite-dimensional), the orthonormal eigenvectors span* $\mathcal{V}_U$.

Every normal operator **A** *defined on a complete and separable unitary vector space* $\mathcal{V}_U$ *is decomposable into a direct sum (Sec. 14.8-2) of normal operators* $\mathbf{A}_1, \mathbf{A}_2, \ldots$ *defined on corresponding subspaces* $\mathcal{V}_1, \mathcal{V}_2, \ldots$ *of* $\mathcal{V}_U$ *so that each* $\mathbf{A}_j$ *has the single eigenvalue* λ_j. In each subspace $\mathcal{V}_j$, there exists a complete orthonormal set of eigenvectors $\mathbf{y}_i^{(j)}$.

If the eigenvalue λ *of a normal operator has the (finite) algebraic multiplicity* m, *then the degree of degeneracy of* λ *is also equal to* m, *and conversely.*

(d) Spectral Representation. The following properties of normal operators apply, in particular, to hermitian operators. Given an orthonormal set of eigenvectors $\mathbf{y}_1, \mathbf{y}_2, \ldots$ of the bounded normal operator **A** and any vector $\mathbf{x} = \hat{\xi}_1\mathbf{y}_1 + \hat{\xi}_2\mathbf{y}_2 + \cdots$ [$\hat{\xi}_k = (\mathbf{y}_k, \mathbf{x})$, Sec. 14.7-4*a*], note

$$\left.\begin{aligned} \mathbf{A}\mathbf{x} &= \lambda_1\hat{\xi}_1\mathbf{y}_1 + \lambda_2\hat{\xi}_2\mathbf{y}_2 + \cdots \\ &\qquad \text{(SPECTRAL REPRESENTATION OF THE OPERATOR } \mathbf{A}) \\ (\mathbf{x}, \mathbf{A}\mathbf{x}) &= \lambda_1|\hat{\xi}_1|^2 + \lambda_2|\hat{\xi}_2|^2 + \cdots \end{aligned}\right\} \qquad (14.8\text{-}3)$$

where λ_k is the eigenvalue associated with $\mathbf{y}_k$ (see also Sec. 13.5-4*b*).

See Refs. 14.10 and 14.11 for analogous properties of normal operators whose spectra are not discrete. If **A** is an operator of this type, the sums (3) must be replaced by Stieltjes integrals over the spectrum.

14.8-5. Determination of Eigenvalues and Eigenvectors: Finite-dimensional Case (see also Secs. 13.4-5 and 13.5-5; refer to Sec. 20.3-5 for numerical methods). Given the $n \times n$ matrix $A \equiv [a_{ik}]$ representing a linear operator **A** in the scheme of measurements defined by the base vectors $\mathbf{e}_1, \mathbf{e}_2, \ldots, \mathbf{e}_n$, one determines the eigenvalues λ and the eigenvectors $\mathbf{y} = \eta_1\mathbf{e}_1 + \eta_2\mathbf{e}_2 + \cdots + \eta_n\mathbf{e}_n$ of **A** as follows:

* See footnote to Sec. 14.4-1.

1. Find the eigenvalues as roots of the nth-degree algebraic equation (*characteristic equation*)

$$\det(\mathbf{A} - \lambda \mathbf{I}) \equiv \det(A - \lambda I) \equiv \begin{vmatrix} a_{11} - \lambda & a_{12} & \cdots & a_{1n} \\ a_{21} & a_{22} - \lambda & \cdots & a_{2n} \\ \cdots & \cdots & \cdots & \cdots \\ a_{n1} & a_{n2} & \cdots & a_{nn} - \lambda \end{vmatrix} = 0$$

2. Obtain the components $\eta_1^{(j)}, \eta_2^{(j)}, \ldots, \eta_n^{(j)}$ of m_j linearly independent eigenvectors

$$\mathbf{y}^{(j)} = \eta_1^{(j)}\mathbf{e}_1 + \eta_2^{(j)}\mathbf{e}_2 + \cdots + \eta_n^{(j)}\mathbf{e}_n$$

associated with each distinct eigenvalue λ_j by solving the n simultaneous linear equations

$$\begin{aligned} (a_{11} - \lambda_j)\eta_1^{(j)} + a_{12}\eta_2^{(j)} + \cdots + a_{1n}\eta_n^{(j)} &= 0 \\ a_{21}\eta_1^{(j)} + (a_{22} - \lambda_j)\eta_2^{(j)} + \cdots + a_{2n}\eta_n^{(j)} &= 0 \\ \cdots\cdots\cdots\cdots\cdots\cdots\cdots\cdots\cdots \\ a_{n1}\eta_1^{(j)} + a_{n2}\eta_2^{(j)} + \cdots + (a_{nn} - \lambda_j)\eta_n^{(j)} &= 0 \end{aligned}$$

whose system matrix has the rank $n - m_j$ (Sec. 1.9-5).

The m_j column matrices $y^{(j)} \equiv \begin{bmatrix} \eta_1^{(j)} \\ \eta_2^{(j)} \\ \cdots \\ \eta_n^{(j)} \end{bmatrix}$ are called **modal columns** (or simply **eigenvectors**) of the given matrix A, associated with the eigenvalue λ_j. Each modal column may be multiplied by an arbitrary constant different from zero.

14.8-6. Reduction and Diagonalization of Matrices. Principal-axes Transformations (see also Sec. 14.8-2). (**a**) Every finite set of linearly independent eigenvectors $\mathbf{y}_1, \mathbf{y}_2, \ldots, \mathbf{y}_s$ associated with the same eigenvalue of a linear operator $\mathbf{A}$ spans a subspace $\mathcal{V}_1$ invariant with respect to $\mathbf{A}$ (Sec. 14.8-3*b*). If $\mathbf{y}_1, \mathbf{y}_2, \ldots, \mathbf{y}_s$ are introduced as the first s base vectors, $\mathbf{A}$ will be represented by a matrix of the form

$$\bar{A} \equiv \left[\begin{array}{c|c} \bar{A}_1 & \bar{A}' \\ \hline [0] & \bar{A}_2 \end{array} \right] \tag{14.8-4}$$

(**b**) In particular, let $\mathbf{A}$ be a *normal* operator defined on a finite-dimensional vector space of dimension n. Then the procedure of Sec 14.8-5 will yield exactly n linearly independent eigenvectors

$$\mathbf{y}^{(j)} = \eta_1^{(j)}\mathbf{e}_1 + \eta_2^{(j)}\mathbf{e}_2 + \cdots + \eta_n^{(j)}\mathbf{e}_n$$

The n corresponding modal columns form a nonsingular **modal matrix**

$$T \equiv [t_{ik}] \qquad \text{with} \qquad t_{ik} = \eta_i^{(j)} = \eta_{i(k)} \tag{14.8-5}$$

where each distinct eigenvalue λ_j accounts for m_j adjacent columns; the $n = m_1 + m_2 + \cdots$ columns are labeled by successive values of $k = 1, 2, \ldots, n$. *The alias-type coordinate transformation*

$$x = T\bar{x} \tag{14.8-6}$$

(Sec. 14.6-1) introduces the n eigenvectors $\mathbf{y}^{(j)}$ of the normal operator **A** *as a reference basis.* The similarity transformation

$$\bar{A} = T^{-1}AT \tag{14.8-7}$$

yields the matrix $\bar{A}$ representing **A** in the new reference system; $\bar{A}$ is a step matrix

$$\bar{A} \equiv \begin{bmatrix} \bar{A}_1 & [0] & \cdots \\ [0] & \bar{A}_2 & \cdots \\ \cdots & \cdots & \cdots \end{bmatrix} \tag{14.8-8}$$

where each submatrix $\bar{A}_j$ corresponds to a different eigenvalue λ_j of **A** and has exactly m_j rows and m_j columns (see also Sec. 14.8-4*c*).

(c) If the n eigenvectors defining the columns of the modal matrix (5) are *mutually orthogonal*, then the similarity transformation (7) yields a *diagonal* matrix $\bar{A}$ (diagonalization of the given matrix A, Sec. 13.4-4*a*). *To obtain a transformation matrix T which diagonalizes a given matrix A representing a normal operator* **A**, proceed as follows:

1. *If all eigenvalues λ_j are nondegenerate* (this is true whenever the characteristic equation has no multiple roots), every modal matrix (5) diagonalizes A.
2. *If there are degenerate eigenvalues* λ_j, orthogonalize each set of m_j eigenvectors $\mathbf{y}^{(j)} = \eta_1^{(j)}\mathbf{e}_1 + \eta_2^{(j)}\mathbf{e}_2 + \cdots + \eta_n^{(j)}\mathbf{e}_n$ by the Gram-Schmidt process (Sec. 14.7-4*b*). The n modal columns $\bar{y}^{(j)} \equiv \begin{bmatrix} \bar{\eta}_1^{(j)} \\ \bar{\eta}_2^{(j)} \\ \cdots \\ \bar{\eta}_n^{(j)} \end{bmatrix}$ representing the resulting $m_1 + m_2 + \cdots = n$ mutually orthogonal eigenvectors $\bar{\mathbf{y}}^{(j)} = \bar{\eta}_1^{(j)}\mathbf{e}_1 + \bar{\eta}_2^{(j)}\mathbf{e}_2 + \cdots + \bar{\eta}_n^{(j)}\mathbf{e}_n$ form the desired transformation matrix T.

(d) In many applications, the original reference basis $\mathbf{e}_1, \mathbf{e}_2, \ldots, \mathbf{e}_n$ is orthonormal (rectangular cartesian coordinates), so that

$$(\mathbf{x}, \mathbf{A}\mathbf{x}) \equiv x\dagger Ax \equiv \sum_{i=1}^{n}\sum_{k=1}^{n} a_{ik}\xi_i^*\xi_k \tag{14.8-9}$$

(Sec. 14.7-4*a*), and the new reference basis is taken to be an orthonormal set of eigenvectors $\bar{\mathbf{y}}^{(j)}$ (obtained, if necessary, with the aid of the Gram-Schmidt process). Then every modal matrix T formed from the $\bar{\mathbf{y}}^{(j)}$ is a *unitary* matrix. A unitary coordinate transformation (6) introducing n orthonormal eigenvectors as base vectors is called a **principal-axes transformation** for the operator A (see also Secs. 2.4-7 and 3.5-6).

A principal-axes transformation for a hermitian operator A *reduces the corresponding hermitian form* (9) *to its normal form* (13.5-9) (see also Sec. 13.5-4).

(**e**) *Two hermitian operators* A *and* B *can be represented by diagonal matrices in the same scheme of measurements if and only if* $\mathsf{BA} = \mathsf{AB}$ (see also Secs. 13.4-4*b* and 13.5-5).

14.8-7. "Generalized" Eigenvalue Problems (see also Secs. 13.5-5 and 15.4-5). (**a**) Some applications require one to find "eigenvectors" $\mathbf{y}$ and "eigenvalues" μ defined by a relation of the form

$$\mathsf{A}\mathbf{y} = \mu\mathsf{B}\mathbf{y} \qquad (\mathbf{y} \neq 0) \tag{14.8-10}$$

where B is a nonsingular operator. The quantities $\mathbf{y}$ and μ are necessarily the eigenvectors and eigenvalues of the operator $\mathsf{B}^{-1}\mathsf{A}$; the problem reduces to that of Eq. (2) if B is the identity operator. If both A and B are *hermitian*, and B is *positive definite* (Sec. 14.4-4), then

1. *All eigenvalues* μ *are real.*
2. *One can introduce the new inner product*

$$(\mathbf{a}, \mathbf{b})_B \equiv (\mathbf{a}, \mathsf{B}\mathbf{b}) \tag{14.8-11}$$

(see also Sec. 14.2-6*a*). In terms of this new inner product, the operator $\mathsf{B}^{-1}\mathsf{A}$ becomes hermitian, and the orthogonality and completeness theorems of Sec. 14.8-4*c* apply. In particular, eigenvectors $\mathbf{y}$ associated with different eigenvalues μ are *mutually orthogonal relative to the scalar product* (11) (see also Sec. 14.7-3*a*).

(**b**) Consider a finite-dimensional unitary vector space and an orthonormal reference system $\mathbf{u}_1, \mathbf{u}_2, \ldots, \mathbf{u}_n$, so that $(\mathbf{a}, \mathbf{b}) \equiv \sum_{k=1}^{n} \alpha_k^* \beta_k$ (Sec. 14.7-4*a*). Let A and B be represented by hermitian matrices $A \equiv [a_{ik}]$, $B \equiv [b_{ik}]$, where B is positive definite. Then the eigenvalues μ defined by Eq. (10) are identical with the roots of the nth-degree algebraic equation

$$\det(A - \mu B) \equiv \det[a_{ik} - \mu b_{ik}] = 0 \quad \text{(CHARACTERISTIC EQUATION FOR THE "GENERALIZED" EIGENVALUE PROBLEM)} \tag{14.8-12}$$

For each root μ_j of multiplicity m_j, there are exactly m linearly independent eigenvectors $\mathbf{y}^{(j)} = \eta_1^{(j)}\mathbf{u}_1 + \eta_2^{(j)}\mathbf{u}_2 + \cdots + \eta_n^{(j)}\mathbf{u}_n$; the components $\eta_i^{(j)}$ are obtained from the linear simultaneous equations

$$\sum_{k=1}^{n} (a_{ik} - \mu_j b_{ik})\eta_k^{(j)} = 0 \qquad (i = 1, 2, \ldots, n) \qquad (14.8\text{-}13)$$

Application of the Gram-Schmidt process (Sec. 14.7-4*b*) to the $m_1 + m_2 + \cdots = n$ eigenvectors $\mathbf{y}^{(j)}$ yields a complete orthonormal set *relative to the new inner product*

$$(\mathbf{a}, \mathbf{b})_B = a\dagger Bb = \sum_{i=1}^{n} \sum_{k=1}^{n} b_{ik}\alpha_i^*\beta_k \qquad (14.8\text{-}14)$$

If this orthonormal set of eigenvectors is introduced as a reference basis in the manner of Sec. 14.8-6*c*, the hermitian forms $(\mathbf{x}, \mathsf{A}\mathbf{x}) = x\dagger Ax$ and $(\mathbf{x}, \mathsf{B}\mathbf{x}) = x\dagger Bx$ take the form (13.5-12) (*simultaneous diagonalization of two hermitian forms*, Sec. 13.5-5).

(c) Refer to Sec. 15.4-5 for a discussion of analogous "generalized" eigenvalue problems involving an infinite-dimensional vector space.

14.8-8. Eigenvalue Problems as Stationary-value Problems (see also Sec. 15.4-7). **(a)** Consider a hermitian operator A defined on a finite-dimensional unitary vector space $\mathcal{V}_U$; and introduce an orthonormal reference basis* $\mathbf{u}_1, \mathbf{u}_2, \ldots, \mathbf{u}_n$, so that A is represented by a hermitian matrix $A \equiv [a_{ik}]$. *The important problem of finding the eigenvectors* $\mathbf{y}^{(j)} = \eta_1^{(j)}\mathbf{u}_1 + \eta_2^{(j)}\mathbf{u}_2 + \cdots + \eta_n^{(j)}\mathbf{u}_n$ *and the corresponding eigenvalues* λ_j *of* A *is precisely equivalent to each of the following problems:*

1. Find (the components η_i of) each vector $\mathbf{y} \neq \mathbf{0}$ such that

$$\frac{(\mathbf{y}, \mathsf{A}\mathbf{y})}{(\mathbf{y}, \mathbf{y})} \equiv \frac{y\dagger Ay}{y\dagger y} \equiv \frac{\sum_{i=1}^{n} \sum_{k=1}^{n} a_{ik}\eta_i^*\eta_k}{\sum_{i=1}^{n} \eta_i^*\eta_i} \qquad \text{(Rayleigh's quotient)} \qquad (14.8\text{-}15)$$

has a stationary value. $\mathbf{y} = \mathbf{y}^{(j)}$ yields the stationary value λ_j.

2. Find (the components η_i of) each vector $\mathbf{y}$ such that

$$(\mathbf{y}, \mathsf{A}\mathbf{y}) \equiv y\dagger Ay \equiv \sum_{i=1}^{n} \sum_{k=1}^{n} a_{ik}\eta_i^*\eta_k \qquad (14.8\text{-}16)$$

has a stationary value subject to the constraint

$$(\mathbf{y}, \mathbf{y}) \equiv y\dagger y \equiv \sum_{i=1}^{n} \eta_i^*\eta_i = 1 \qquad (14.8\text{-}17)$$

$\mathbf{y} = \mathbf{y}^{(j)}$ yields the stationary value λ_j.

* The orthonormal basis is used for convenience in most applications; in the general case, it is only necessary to substitute $(\mathbf{y}, \mathsf{A}\mathbf{y}) \equiv y\dagger GAy$ and $(\mathbf{y}, \mathbf{y}) \equiv y\dagger Gy$ in Eqs. (15) to (17).

3. Find (the components η_i of) each vector $\mathbf{y} \neq \mathbf{0}$ such that $(\mathbf{y}, \mathbf{y})$ has a stationary value subject to the constraint $(\mathbf{y}, \mathbf{A}\mathbf{y}) = 1$. $\mathbf{y} = \mathbf{y}^{(j)}$ yields the stationary value $1/\lambda_j$.

Let the eigenvalues of $\mathbf{A}$ be arranged in increasing order, with an m-fold degenerate eigenvalue repeated m times, or $\lambda_1 \leq \lambda_2 \leq \cdots \leq \lambda_n$. *The smallest eigenvalue λ_1 is the minimum value of Rayleigh's quotient* (15) *for an arbitrary vector* $\mathbf{y}$ *of* $\mathcal{V}_U$. The rth eigenvalue λ_r in the above sequence similarly is less than or equal to Rayleigh's quotient if $\mathbf{y}$ is an arbitrary nonzero vector orthogonal to all eigenvectors associated with λ_1, $\lambda_2, \ldots, \lambda_{r-1}$; λ_r *is the maximum of* $\min\limits_{\mathbf{y} \text{ in } \mathcal{V}_r} (\mathbf{y}, \mathbf{A}\mathbf{y})/(\mathbf{y}, \mathbf{y})$ *for an arbitrary* $(r-1)$*-dimensional subspace* $\mathcal{V}_r$ of $\mathcal{V}_U$ (*Courant's Minimax Principle*).

The last theorems may be restated for problems 2 and 3, and for maxima instead of minima; note that a minimum in problem 1 or 2 corresponds to a maximum in problem 3, and conversely. The inner product $(\mathbf{y}, \mathbf{A}\mathbf{y})$ usually has direct physical significance. Problem 3 associates the eigenvalues of $\mathbf{A}$ with the principal axes of a second-order hypersurface (see also Secs. 2.4-7 and 3.5-6).

(b) Generalizations. The theory of Sec. 14.8-8*a* may be extended to apply to the "generalized" eigenvalue problem defined by $\mathbf{A}\mathbf{y} = \mu\mathbf{B}\mathbf{y}$, where $\mathbf{A}$ is a hermitian operator, and $\mathbf{B}$ is hermitian and positive definite (Sec. 14.8-7). It is only necessary to replace $(\mathbf{y}, \mathbf{y})$ by $(\mathbf{y}, \mathbf{y})_B \equiv (\mathbf{y}, \mathbf{B}\mathbf{y})$ in each problem statement of Sec. 14.8-8*a*. In particular, Rayleigh's quotient (15) is replaced by

$$\frac{(\mathbf{y}, \mathbf{A}\mathbf{y})}{(\mathbf{y}, \mathbf{B}\mathbf{y})} \equiv \frac{y\dagger Ay}{y\dagger By} \equiv \frac{\sum\limits_{i=1}^{n}\sum\limits_{k=1}^{n} a_{ik}\eta_i^*\eta_k}{\sum\limits_{i=1}^{n}\sum\limits_{k=1}^{n} b_{ik}\eta_i^*\eta_k}$$

(RAYLEIGH'S QUOTIENT FOR THE "GENERALIZED" EIGENVALUE PROBLEM) (14.8-18)

Analogous theorems apply to suitable operators defined on Hilbert spaces; here the inner products $(\mathbf{y}, \mathbf{A}\mathbf{y})$, $(\mathbf{y}, \mathbf{y})$, and $(\mathbf{y}, \mathbf{B}\mathbf{y})$ may be integrals rather than sums, so that the stationary-value problems of Sec. 14.8-8*a* become variation problems (Sec. 15.4-7).

14.8-9. Bounds for the Eigenvalues of Linear Operators (see also Sec. 15.4-10). The following theorems are often helpful for the estimation of eigenvalues.

(a) *Every eigenvalue $z = \lambda$ of a normal linear operator $\mathbf{A}$ represented by a finite $n \times n$ matrix $A \equiv [a_{ik}]$ is contained in the union of the n circles*

$$|z - a_{ii}| \leq \sum_{\substack{k=1 \\ (k \neq i)}}^{n} |a_{ik}| \qquad (i = 1, 2, \ldots, n)$$

(GERSCHGORIN'S CIRCLE THEOREM) (14.8-19)

Re(λ) *lies between the smallest and the largest eigenvalue of* $\frac{1}{2}(A + A\dagger) = H_1$, *and* Im$(\lambda)$ *lies between the smallest and the largest* eigenvalue of $\frac{1}{2i}(A - A\dagger) = H_2$ (see also Sec. 13.3-4a).

(**b**) For *hermitian* matrices and operators

$$|\lambda|^2 \leq \sum_i \sum_k |a_{ik}|^2 \qquad |\lambda| \leq \|\mathbf{A}\| \tag{14.8-20}$$

(**c**) **Comparison Theorems.** Let $\mu_1 \leq \mu_2 \leq \cdots \leq \mu_n$ be the sequence (including multiplicities) of the eigenvalues for a finite-dimensional eigenvalue problem (10), where $\mathbf{A}$ and $\mathbf{B}$ are hermitian, and $\mathbf{B}$ is positive definite. Then

1. *Addition of a positive-definite hermitian operator to* $\mathbf{A}$ *cannot decrease any eigenvalue* μ_r *in the above sequence.*
2. *Addition of a positive-definite hermitian operator to* $\mathbf{B}$ *cannot increase any eigenvalue* μ_r.
3. *If a constraint restricts the vectors* $\mathbf{y}$ *to an* $(n - m)$*-dimensional subspace of* $\mathcal{V}_U$, *then the* $n - m$ *eigenvalues* $\mu'_1 \leq \mu'_2 \cdots \leq \mu'_{n-m}$ *of the constrained problem satisfy the relations*

$$\mu_r \leq \mu'_r \leq \mu_{r+m} \qquad (r = 1, 2, \ldots, n - m) \tag{14.8-21}$$

The constraint usually takes the form of m independent linear equations relating the vector components η_i.

These theorems also apply to operators defined on Hilbert spaces if $\mathbf{A}$ *and* $\mathbf{B}$ *are positive-definite hermitian operators yielding a discrete sequence* $\mu_1 \leq \mu_2 \leq \cdots$ *with finite multiplicities.*

14.8-10. Nonhomogeneous Linear Vector Equations (see also Secs. 1.9-4, 15.3-7, and 15.4-12). (**a**) Given a bounded operator $\mathbf{A}$, the vector equation

$$\mathbf{Ax} - \lambda\mathbf{x} = \mathbf{f} \tag{14.8-22}$$

has a unique solution $\mathbf{x}$ for every given vector $\mathbf{f}$ if and only if the given scalar λ is not contained in the spectrum of $\mathbf{A}$ (Sec. 14.8-3d). *If* λ *equals an eigenvalue* λ_1 *of* $\mathbf{A}$ *in the sense of Eq.* (2), *then Eq.* (22) *has a solution only if the given vector* $\mathbf{f}$ *is orthogonal to every eigenvector of* $\mathbf{A}\dagger$ *associated with the eigenvalue* λ_1^*. In the latter case, there is an infinite number of solutions; every sum of a particular solution and a linear combination of eigenvectors corresponding to the eigenvalue λ_1 is a solution.

(**b**) The important special case

$$\mathbf{Ax} = \mathbf{f} \tag{14.8-23}$$

where **A** is a bounded normal operator, admits a unique solution $\mathbf{x}$ for every given vector $\mathbf{f}$ if and only if $\mathbf{Ax} = 0$ implies $\mathbf{x} = \mathbf{0}$, i.e., if and only if **A** is nonsingular. If **A** is singular, Eq. (23) has a solution only if $\mathbf{f}$ is orthogonal to every eigenvector of $\mathbf{A}\dagger$ associated with the eigenvalue zero.

(c) For a *hermitian* operator $\mathbf{A} = \mathbf{A}\dagger$ having an orthonormal set of eigenvectors $\mathbf{y}_k$ such that $\mathbf{f} = \sum_{k=1}^{\infty} (\mathbf{y}_k, \mathbf{f})\mathbf{y}_k$ the solution of Eq. (22) is given by

$$\mathbf{x} = \sum_{k=1}^{\infty} \frac{(\mathbf{y}_k, \mathbf{f})}{\lambda_k - \lambda} \mathbf{y}_k \tag{14.8-24}$$

where the λ_k are the (not necessarily distinct) eigenvalues corresponding to each $\mathbf{y}_k$.

14.9. GROUP REPRESENTATIONS AND RELATED TOPICS

14.9-1. Group Representations. (a) *Every group (Sec. 12.2-1) can be* **represented** *by a homomorphism (Sec. 12.1-6) relating the group elements to a group of nonsingular linear transformations of a vector space* (**representation space, carrier space**), *and thus to a group of nonsingular matrices* (this is a form of *Cayley's theorem* stated in Sec. 12.2-9*b*). A **representation of degree or dimension** n **of a group** G **in the field** F is a group of $n \times n$ matrices $A, B, \ldots$ over F related to the elements $a, b, \ldots$ of G by a homomorphism $A = A(a), B = B(b), \ldots,$ so that $ab = c$ implies $A(a)B(b) = C(c)$ for all a, b in G (*representation condition*). n equals the linear dimension of the representation space. A representation is **faithful (true)** if and only if it is reciprocal one-to-one (and thus an isomorphism, Sec. 12.1-6).

Every group admits a complex vector space as a representation space; i.e., every group has a representation in the field of complex numbers. Such a representation permits one to describe the defining operation of any group in terms of numerical additions and multiplications (see also Secs. 12.1-1 and 14.1-1). Most applications deal with groups of transformations (Sec. 12.2-8; for examples refer to Sec. 14.10-7).

Every group $a_1, a_2, \ldots, a_g$ of finite order g admits a faithful representation comprising the g linearly independent permutation matrices (Sec. 13.2-6) $A_j(a_j) \equiv [a_{ik}(a_j)]$ defined by

$$a_{ik}(a_j) = \begin{Bmatrix} 1 \text{ if } a_i^{-1}a_ja_k = E \\ 0 \text{ otherwise} \end{Bmatrix} \qquad (i, j, k = 1, 2, \ldots, g) \tag{14.9-1}$$

where E is the identity element of the given group (*regular representation* of the finite group). *Every finite group is thus isomorphic to a group of permutations* (see also Sec. 12.2-8).

(**b**) Two representations $\mathcal{R}$ and $\bar{\mathcal{R}}$ of a group G are **similar** or **equivalent** if and only if all pairs of matrices $A(a)$ of $\mathcal{R}$ and $\bar{A}(a)$ of $\bar{\mathcal{R}}$ are related by the same similarity transformation (Sec. 13.4-1*b*) $\bar{A} = T^{-1}AT$. In this case, the matrices $A(a)$ and $\bar{A}(a)$ are said to describe one and the same linear transformation $\mathbf{A}(a)$ of a representation space common to $\mathcal{R}$ and $\bar{\mathcal{R}}$ (see also Sec. 14.6-2).

A representation $\mathcal{R}$ is **bounded, unitary,** and/or **orthogonal** if and only if all its matrices have the corresponding properties. *Every representation of a finite group and every unitary representation is bounded. For every bounded representation there exists an equivalent unitary representation.*

(**c**) The **rank** of any representation $\mathcal{R}$ is the greatest number of linearly independent matrices* in $\mathcal{R}$.

14.9-2. Reduction of a Representation. (**a**) A representation† $\mathcal{R}$ of a group G is **reducible** if and only if the representation space $\mathcal{V}$ has a proper subspace $\mathcal{V}_1$ invariant with respect to $\mathcal{R}$, i.e., with respect to every linear transformation of $\mathcal{V}$ described by a matrix of $\mathcal{R}$ (Sec. 14.5-2). This is true if and only if there exists a similarity transformation

$$\bar{A} = T^{-1}AT$$

which reduces the matrices $A(a)$, $B(b)$, . . . of $\mathcal{R}$ simultaneously to corresponding matrices of the form

$$\bar{A}(a) \equiv \left[\begin{array}{c|c} A_1(a) & A'(a) \\ \hline [0] & A_2(a) \end{array}\right] \qquad \bar{B}(b) \equiv \left[\begin{array}{c|c} B_1(b) & B'(b) \\ \hline [0] & B_2(b) \end{array}\right] \qquad \cdots \tag{14.9-2}$$

where $A_1(a)$, $B_1(b)$, . . . are square matrices of equal order (*alternative definition*). The matrices $A_1(a)$, $B_1(b)$, . . . constitute a representation $\mathcal{R}_1$ of the given group G, with the representation space $\mathcal{V}_1$. A representation which cannot be reduced in this manner is called **irreducible.**

(**b**) A representation $\mathcal{R}$ will be called **decomposable**‡ if and only if its representation space is a direct sum $\mathcal{V} = \mathcal{V}_1 \oplus \mathcal{V}_2 \oplus \cdots$ (Sec. 12.7-5*a*) of subspaces $\mathcal{V}_1$, $\mathcal{V}_2$, . . . invariant with respect to $\mathcal{R}$. This is true if and only if there exists a similarity transformation $\bar{A} = T^{-1}AT$ which reduces the matrices $A(a)$, $B(b)$, . . . of $\mathcal{R}$ simultaneously to

* Linear independence of matrices is defined in the manner of Sec. 14.2-3, since matrices may be regarded as vectors.

† The definitions of this section also apply to any set of linear transformations of a vector space into itself, or to any corresponding set of matrices (not necessarily a group).

‡ The terms *reducible, decomposable,* and *completely reducible* are variously interchanged by different authors. These terms are, indeed, equivalent in the case of *bounded* matrices, transformations, and representations, and thus for all representations of finite groups (Sec. 14.9-2*d*).

corresponding step matrices

$$\bar{A}(a) \equiv \left[\begin{array}{c|c|c} A_1(a) & [0] & \cdots \\ \hline [0] & A_2(a) & \cdots \\ \hline \cdots & \cdots & \cdots \end{array}\right] \qquad \bar{B}(b) \equiv \left[\begin{array}{c|c|c} B_1(b) & [0] & \cdots \\ \hline [0] & B_2(b) & \cdots \\ \hline \cdots & \cdots & \cdots \end{array}\right] \quad \cdots \tag{14.9-3}$$

(direct sums of matrices, Sec. 13.2-9), where corresponding submatrices are of equal order. Each set of matrices $A_i(a)$, $B_i(b)$, . . . ($i = 1, 2, \ldots$) constitutes a representation $\mathcal{R}_i$ of the given group G with the representation space $\mathcal{V}_i$. $\mathcal{R}$ is written as the direct sum $\mathcal{R} = \mathcal{R}_1 \oplus \mathcal{R}_2 \oplus \cdots$.

(c) A representation $\mathcal{R}$ is **completely reducible** if and only if it is decomposable into irreducible representations (**irreducible components**) $\mathcal{R}^{(1)}, \mathcal{R}^{(2)}, \ldots$.

(d) Conditions for Reducibility (see also Sec. 14.9-5*b*). *Every bounded representation (and, in particular, every representation of a finite group) is either completely reducible or irreducible.*

A group G has a decomposable representation if and only if it is the direct product (Sec. 12.7-2) of simple groups (Sec. 12.2-5b).

A bounded representation $\mathcal{R}$ is completely reducible if and only if there exists a matrix Q, not a multiple of I, which commutes with every matrix of $\mathcal{R}$. Irreducible representations of commutative (Abelian) groups are necessarily one-dimensional.

Whenever all corresponding matrices A, $\bar{A}$ of two irreducible representations $\mathcal{R}$ and $\bar{\mathcal{R}}$ are related by the same transformation $Q\bar{A} = AQ$, then $\mathcal{R}$ and $\bar{\mathcal{R}}$ are either equivalent or $Q = [0]$ (Schur's Lemma).

14.9-3. The Irreducible Representations of a Group. **(a)** *The decomposition $\mathcal{R} = \mathcal{R}^{(1)} \oplus \mathcal{R}^{(2)} \oplus \cdots$ of a given completely reducible representation $\mathcal{R}$ of a group into irreducible components is unique except for equivalence and for the relative order of terms. Every completely reducible representation is uniquely defined by its irreducible components (except for equivalence).* If $\mathcal{R}^{(j)}$ is one of exactly m_j mutually equivalent irreducible components of $\mathcal{R}$ ($j = 1, 2, \ldots$), one may write

$$\mathcal{R} = m_1\mathcal{R}^{(1)} \oplus m_2\mathcal{R}^{(2)} \oplus \cdots$$

(b) *For every group G of finite order g,*

1. *The number m of distinct nonequivalent irreducible representations is finite and equals the number of distinct classes of conjugate elements (Sec. 12.2-5a).*
2. *If n_j is the dimension of the jth irreducible representation, its rank equals n_j^2 (Burnside's Theorem); the rank of every representation $\mathcal{R}$ of G equals the sum of the ranks n_j^2 of the distinct irreducible components in $\mathcal{R}$.*
3. *Each n_j is a divisor of g, and*

$$n_1^2 + n_2^2 + \cdots + n_m^2 = g \tag{14.9-4}$$

4. *The regular representation of G (Sec. 14.8-1a) contains the jth irreducible representation of G exactly n_j times.*

(c) The determination of the complete set of irreducible representations of a group G of operators is of particular interest as a key to the solution of certain eigenvalue problems. *Given any hermitian operator* H *which commutes with every operator of a group* G, *there exists a reciprocal one-to-one correspondence between the distinct eigenvalues* λ_i *of* H *and the nonequivalent irreducible representations* $\mathcal{R}^{(j)}$ *of* G, *and the degree of degeneracy of each* λ_i *equals the dimension of* $\mathcal{R}^{(j)}$ (classification of quantum-mechanical eigenvalues from symmetry considerations, Refs. 14.20 to 14.22).

14.9-4. The Character of a Representation. (a) The **character** of a representation $\mathcal{R}$ is the function

$$\chi(a) \equiv \operatorname{Tr}\,[A(a)] \qquad (14.9\text{-}5)$$

defined on the elements a of the group G represented by $\mathcal{R}$. *Conjugate group elements (Sec. 12.2-5a) have equal character values.*

For every bounded representation,

$$\chi(a^{-1}) \equiv [\chi(a)]^* \qquad (14.9\text{-}6)$$

Two completely reducible representations of the same group G *are equivalent if and only if they have identical characters.*

(b) The characters of irreducible representations are called **simple** or **primitive characters,** and the characters of reducible representations are known as **composite characters.** $\mathcal{R} = \mathcal{R}_1 \oplus \mathcal{R}_2 \oplus \cdots$ *implies* $\chi(a) \equiv \chi_1(a) + \chi_2(a) + \cdots$, *where* $\chi(a)$, $\chi_1(a)$, $\chi_2(a)$, . . . *are the respective characters of* $\mathcal{R}$, $\mathcal{R}_1$, $\mathcal{R}_2$, ...

14.9-5. Orthogonality Relations (see also Sec. 14.9-6). (a) *The primitive characters* $\chi^{(1)}(a)$, $\chi^{(2)}(a)$, . . . *respectively associated with the nonequivalent irreducible representations* $\mathcal{R}^{(1)}$, $\mathcal{R}^{(2)}$, . . . *of a finite group* G *satisfy the relations*

$$\begin{aligned}\text{Mean}\,\{[\chi^{(j)}(a)]^*\chi^{(j')}(a)\} &= \frac{1}{g}\sum_{a \text{ in } G}[\chi^{(j)}(a)]^*\chi^{(j')}(a)\\ &= \delta^j_{j'} = \begin{Bmatrix}0 \text{ if } j \neq j'\\ 1 \text{ if } j = j'\end{Bmatrix} \qquad (j, j' = 1, 2, \ldots, m)\end{aligned} \qquad (14.9\text{-}7)$$

For every completely reducible representation $\mathcal{R} = m_1\mathcal{R}^{(1)} \oplus m_2\mathcal{R}^{(2)} \oplus \cdots$ *of* G,

$$\chi(a) = m_1\chi^{(1)}(a) + m_2\chi^{(2)}(a) + \cdots \qquad (14.9\text{-}8)$$

$$m_j = \text{Mean}\,\{[\chi^{(j)}(a)]^*\chi(a)\} = \frac{1}{g}\sum_{a \text{ in } G}[\chi^{(j)}(a)]^*\chi(a) \qquad (14.9\text{-}9)$$

$$\text{Mean}\,\{|\chi(a)|^2\} = \frac{1}{g}\sum_{a \text{ in } G}|\chi(a)|^2 = m_1^2 + m_2^2 + \cdots \geq 1 \qquad (14.9\text{-}10)$$

Mean $\{|\chi(a)|^2\} = 1$ *whenever* $\mathcal{R}$ *is irreducible.*

(b) Each of the m nonequivalent irreducible representations $\mathcal{R}^{(j)}$ of a finite group G *is equivalent to a corresponding unitary irreducible representation comprising the matrices* $[u_{ik}^{(j)}(a)]$ (see also Sec. 14.9-1b). *The elements of these unitary matrices satisfy the relations*

$$\text{Mean } \{u_{ik}^{(j)}(a)^* u_{i'k'}^{(j')}(a)\} = \frac{1}{g} \sum_{a \text{ in } G} u_{ik}^{(j)}(a)^* u_{i'k'}^{(j')}(a) = \frac{1}{n_j} \delta_{j'}^{j} \delta_{i'}^{i} \delta_{k'}^{k}$$

$$(j, j' = 1, 2, \ldots, m;\ i, k = 1, 2, \ldots, n_j;\ i', k' = 1, 2, \ldots, n_{j'}) \qquad (14.9\text{-}11)$$

(**c**) *The relations* (7) to (11) *apply to countably infinite, continuous, and mixed-continuous groups whenever the mean values* (*Sec.* 12.2-12) *in question exist.* In this case the primitive characters $\chi^{(j)}(a) \equiv \chi^{(j)}[a(\alpha_1, \alpha_2, \ldots)]$ constitute a complete orthogonal set of functions in the sense of Sec. 15.2-4.

14.9-6. Direct Products of Representations. (**a**) Given an n_1-dimensional representation $\mathcal{R}_1$ and an n_2-dimensional representation $\mathcal{R}_2$ of the same group G, the $n_1n_2 \times n_1n_2$ matrices each obtained as a direct product (Sec. 13.2-10) of a matrix of $\mathcal{R}_1$ by a matrix of $\mathcal{R}_2$ constitute a representation of G, the **direct product (Kronecker product)** $\mathcal{R}_1 \otimes \mathcal{R}_2$ of the representations $\mathcal{R}_1$ and $\mathcal{R}_2$ (see also Sec. 12.7-2). Its representation space is the direct product of the representation spaces associated with $\mathcal{R}_1$ and $\mathcal{R}_2$ (Sec. 12.7-3). *The character $\chi(a)$ of $\mathcal{R} = \mathcal{R}_1 \otimes \mathcal{R}_2$ is the product of the respective characters $\chi_1(a)$ of $\mathcal{R}_1$ and $\chi_2(a)$ of $\mathcal{R}_2$:*

$$\chi(a) \equiv \chi_1(a)\chi_2(a) \qquad (14.9\text{-}12)$$

If both $\mathcal{R}_1$ and $\mathcal{R}_2$ are bounded or unitary, the same is true for $\mathcal{R}_1 \otimes \mathcal{R}_2$.

(**b**) *The direct product $\mathcal{R}^{(1)} \otimes \mathcal{R}^{(2)}$ of two bounded irreducible representations $\mathcal{R}^{(1)}$ and $\mathcal{R}^{(2)}$ of G is irreducible if the dimension of $\mathcal{R}^{(1)}$ and/or $\mathcal{R}^{(2)}$ equals* 1; *otherwise, $\mathcal{R}^{(1)} \otimes \mathcal{R}^{(2)}$ is completely reducible.* One may use this last fact to derive new irreducible representations of G from given irreducible representations.

(**c**) *The irreducible representations of the direct product $G_1 \otimes G_2$ of two groups G_1 and G_2* (*Sec.* 12.7-2) *are the direct products $\mathcal{R}_1^{(j)} \otimes \mathcal{R}_2^{(j')}$ of the irreducible representations $\mathcal{R}_1^{(j)}$ of G_1 and $\mathcal{R}_2^{(j')}$ of G_2.*

14.9-7. Representations of Rings, Fields, and Linear Algebras (see also Secs. 12.3-1 and 12.4-2). Rings, fields, and linear algebras may also be represented by suitable classes of matrices or linear transformations. In particular, *a linear algebra of order n^2 over a field F is isomorphic to an algebra of $n \times n$ matrices over F, provided that the given algebra has a multiplicative identity* (*regular representation of a linear algebra,* see also Secs. 14.9-1*a* and 14.10-6).

14.10. MATHEMATICAL DESCRIPTION OF ROTATIONS

14.10-1. Rotations in Three-dimensional Euclidean Vector Space. (**a**) *Every orthogonal linear transformation*

$$\mathbf{x}' = \mathsf{A}\mathbf{x} \qquad (\tilde{\mathsf{A}}\mathsf{A} = \mathsf{A}\tilde{\mathsf{A}} = \mathsf{I}) \qquad (14.10\text{-}1a)$$

of a three-dimensional Euclidean vector space (Sec. 14.2-7*a*) *onto itself preserves absolute values of vectors and angles between vectors* (Secs. 14.4-5 and 14.4-6). Such a transformation is a (proper) **rotation** if and only if det (**A**) = 1, i.e., if and only if the transformation also preserves *the relative orientation of any three base vectors* (and hence *right- and left-handedness of axes, vector products*, and *triple scalar products*). A transformation (1*a*) with det (**A**) = −1 is an **improper rotation,** or a **rotation with reflection.**

(**b**) Given any *orthonormal* basis (Sec. 14.7-4) $\mathbf{u}_1$, $\mathbf{u}_2$, $\mathbf{u}_3$, let

$$\mathbf{x} = \xi_1\mathbf{u}_1 + \xi_2\mathbf{u}_2 + \xi_3\mathbf{u}_3 \qquad \mathbf{x}' = \xi'_1\mathbf{u}_1 + \xi'_2\mathbf{u}_2 + \xi'_3\mathbf{u}_3$$

Every transformation (1*a*) is represented by

$$\begin{aligned} \xi'_1 &= a_{11}\xi_1 + a_{12}\xi_2 + a_{13}\xi_3 \\ \xi'_2 &= a_{21}\xi_1 + a_{22}\xi_2 + a_{23}\xi_3 \\ \xi'_3 &= a_{31}\xi_1 + a_{32}\xi_2 + a_{33}\xi_3 \end{aligned} \qquad (14.10\text{-}1b)$$

or in matrix form

$$x' = Ax \qquad (14.10\text{-}1c)$$

where

$$\det [a_{ik}] = \det (A) = 1 \qquad (14.10\text{-}2)$$

for proper rotations. Since an *orthonormal* reference system is used, the real matrix $A \equiv [a_{ik}]$ describing each rotation is orthogonal ($\tilde{A}A = A\tilde{A} = I$, see also Sec. 14.7-5), i.e.,

$$\sum_{j=1}^{3} a_{ij}a_{kj} = \sum_{j=1}^{3} a_{ji}a_{jk} = \begin{Bmatrix} 0 \text{ if } i \neq k \\ 1 \text{ if } i = k \end{Bmatrix} \qquad (i, k = 1, 2, 3) \qquad (14.10\text{-}3)$$

and each coefficient a_{ik} equals the cofactor of a_{ki} in the determinant det $[a_{ik}]$. Three suitably given coefficients a_{ik} determine all 9.

Geometrically, a_{ik} is the cosine of the angle between the base vector $\mathbf{u}_i$ and the rotated base vector $\mathbf{u}'_k = \mathbf{A}\mathbf{u}_k = \sum_{j=1}^{3} a_{jk}\mathbf{u}_j$ (see also Sec. 14.5-1):

$$a_{ik} = \mathbf{u}_i \cdot \mathbf{u}'_k = \mathbf{u}_i \cdot (\mathbf{A}\mathbf{u}_k) \qquad (i, k = 1, 2, 3) \qquad (14.10\text{-}4)$$

14.10-2. Angle of Rotation. Rotation Axis. (**a**) A rotation (1) rotates the position vector **x** of each point in a three-dimensional Euclidean space through an **angle of rotation** δ about a directed **rotation axis** whose points are invariant. The rotation angle δ and the direction sosines c_1, c_2, c_3 of the positive rotation axis are given by

$$\left.\begin{aligned} \cos\delta &= \tfrac{1}{2}[\operatorname{Tr}(\mathbf{A}) - 1] = \tfrac{1}{2}[\operatorname{Tr}(A) - 1] \\ &= \tfrac{1}{2}(a_{11} + a_{22} + a_{33} - 1) \\ c_1 &= \frac{a_{32} - a_{23}}{2\sin\delta} \qquad c_2 = \frac{a_{13} - a_{31}}{2\sin\delta} \qquad c_3 = \frac{a_{21} - a_{12}}{2\sin\delta} \end{aligned}\right\} \qquad (14.10\text{-}5)$$

so that $\delta > 0$ *corresponds to a rotation in the sense of a right-handed screw propelled in the direction of the positive rotation axis.* Either the sign of δ or the positive direction on the rotation axis may be arbitrarily assigned.

The direction of the positive rotation axis is that of the eigenvector $c_1\mathbf{u}_1 + c_2\mathbf{u}_2 + c_3\mathbf{u}_3$ corresponding to the eigenvalue $+1$ of A and is obtained by a principal-axes transformation of the matrix A (Sec. 14.8-6). The remaining eigenvalues of A are $\cos\delta \pm i\sin\delta = e^{\pm i\delta}$.

(b) The transformation matrix A corresponding to a given rotation described by δ, c_1, c_2, c_3 is

$$A \equiv \begin{bmatrix} a_{11} & a_{12} & a_{13} \\ a_{21} & a_{22} & a_{23} \\ a_{31} & a_{32} & a_{33} \end{bmatrix} = \cos\delta \begin{bmatrix} 1 & 0 & 0 \\ 0 & 1 & 0 \\ 0 & 0 & 1 \end{bmatrix}$$
$$+ (1 - \cos\delta)\begin{bmatrix} c_1{}^2 & c_1c_2 & c_1c_3 \\ c_2c_1 & c_2{}^2 & c_2c_3 \\ c_3c_1 & c_3c_2 & c_3{}^2 \end{bmatrix} + \sin\delta \begin{bmatrix} 0 & -c_3 & c_2 \\ c_3 & 0 & -c_1 \\ -c_2 & c_1 & 0 \end{bmatrix} \qquad (14.10\text{-}6)$$

14.10-3. Euler Parameters and Gibbs Vector. (a) The four **Euler symmetrical parameters**

$$\lambda = c_1 \sin\delta/2 \qquad \mu = c_2\sin\delta/2 \qquad \nu = c_3\sin\delta/2 \qquad \rho = \cos\delta/2$$
$$(\lambda^2 + \mu^2 + \nu^2 + \rho^2 = 1) \qquad (14.10\text{-}7)$$

define the rotation uniquely, since Eq. (6) yields

$$A \equiv \begin{bmatrix} \lambda^2 - \mu^2 - \nu^2 + \rho^2 & 2(\lambda\mu - \nu\rho) & 2(\nu\lambda + \mu\rho) \\ 2(\lambda\mu + \nu\rho) & \mu^2 - \nu^2 - \lambda^2 + \rho^2 & 2(\mu\nu - \lambda\rho) \\ 2(\nu\lambda - \mu\rho) & 2(\mu\nu + \lambda\rho) & \nu^2 - \lambda^2 - \mu^2 + \rho^2 \end{bmatrix} \qquad (14.10\text{-}8)$$

λ, μ, ν, ρ and $-\lambda$, $-\mu$, $-\nu$, $-\rho$ represent the same rotation.

(b) The **Gibbs vector**

$$\left.\begin{aligned} \mathbf{G} &= G_1\mathbf{u}_1 + G_2\mathbf{u}_2 + G_3\mathbf{u}_3 \\ \text{with} \quad G_1 &= c_1\tan\delta/2 = \lambda/\rho \qquad G_2 = c_2\tan\delta/2 = \mu/\rho \\ G_3 &= c_3 \tan\delta/2 = \nu/\rho \end{aligned}\right\} \qquad (14.10\text{-}9)$$

also defines the rotation uniquely. The rotated vector $\mathbf{x}'$ can be written as

$$\mathbf{x}' = \mathsf{A}\mathbf{x} = \cos^2\frac{\delta}{2}\,[(1 - |\mathbf{G}|^2)\mathbf{x} + 2(\mathbf{G}\cdot\mathbf{x})\mathbf{G} + 2\mathbf{G}\times\mathbf{x}] \qquad (14.10\text{-}10)$$

14.10-4. Representation of Vectors and Rotations by Spin Matrices and Quaternions. Cayley-Klein Parameters. (a) Given an orthonormal basis $\mathbf{u}_1$, $\mathbf{u}_2$, $\mathbf{u}_3$, every real vector $\mathbf{x} = \xi_1\mathbf{u}_1 + \xi_2\mathbf{u}_2 + \xi_3\mathbf{u}_3$ may be represented by a (in general complex) hermitian 2×2 matrix

$$H \equiv \begin{bmatrix} \xi_3 & \xi_1 - i\xi_2 \\ \xi_1 + i\xi_2 & -\xi_3 \end{bmatrix} \equiv \xi_1 S_1 + \xi_2 S_2 + \xi_3 S_3 \qquad (14.10\text{-}11)$$

where the hermitian **Pauli spin matrices**

$$S_1 \equiv \begin{bmatrix} 0 & 1 \\ 1 & 0 \end{bmatrix} \qquad S_2 \equiv \begin{bmatrix} 0 & -i \\ i & 0 \end{bmatrix} \qquad S_3 \equiv \begin{bmatrix} 1 & 0 \\ 0 & -1 \end{bmatrix} \qquad (14.10\text{-}12)$$

correspond, respectively, to $\mathbf{u}_1$, $\mathbf{u}_2$, $\mathbf{u}_3$. The correspondence (11) is an isomorphism preserving the results of vector addition and multiplication of vectors by (real) scalars.

For every rotation (1), *the* 2×2 *matrix representing the rotated vector*

$$\mathbf{x}' = \mathbf{A}\mathbf{x} = \xi_1'\mathbf{u}_1 + \xi_2'\mathbf{u}_2 + \xi_3'\mathbf{u}_3$$

is

$$H' = \xi_1'S_1 + \xi_2'S_2 + \xi_3'S_3 = UHU\dagger \qquad (14.10\text{-}13)$$

where U is the (in general complex) unitary 2×2 *matrix with determinant* 1 (**unimodular** 2×2 *matrix) defined by*

$$\left.\begin{aligned} U \equiv \begin{bmatrix} a^* & -b \\ b^* & a \end{bmatrix} \qquad U\dagger \equiv \begin{bmatrix} a & b \\ -b^* & a^* \end{bmatrix} \\ \text{with} \qquad a = \rho + i\nu \qquad b = \mu + i\lambda \\ |a|^2 + |b|^2 = \lambda^2 + \mu^2 + \nu^2 + \rho^2 = 1 \end{aligned}\right\} \qquad (14.10\text{-}14)$$

The complex numbers a, b, determine the corresponding rotation uniquely; but a, b and $-a$, $-b$, and hence U and $-U$, represent the same rotation. Either a, b, $-b^*$, a^* or a^*, ib^*, $-ib$, a are referred to as the **Cayley-Klein parameters** of the rotation.

Geometrically, the complex parameters a, b define the complex-plane transformation

$$u' = \frac{au - b^*}{bu + a^*} \qquad (|a|^2 + |b|^2 = 1) \qquad (14.10\text{-}15)$$

(bilinear transformation, Sec. 7.9-2) relating the stereographic projection u of the point (ξ_1, ξ_2, ξ_3) on a sphere about the origin onto the complex u plane (Sec. 7.2-4) and the stereographic projection u' of the rotated point (ξ_1', ξ_2', ξ_3') (see also Ref. 14.13).

(**b**) The linear combinations of I, iS_1, iS_2, and iS_3 with real coefficients constitute a matrix representation of the quaternion algebra (Sec. 12.4-2), whose scalars correspond to real multiples of I, and whose generators correspond to iS_1, iS_2, iS_3, with

$$\left.\begin{aligned} S_1^2 = S_2^2 = S_3^2 = I \\ S_2S_3 = -S_3S_2 = iS_1 \qquad S_3S_1 = -S_1S_3 = iS_2 \\ S_1S_2 = -S_2S_1 = iS_3 \end{aligned}\right\} \qquad (14.10\text{-}16)$$

Every complex 2×2 matrix can be expressed as such a linear combination; in particular,

$$\left.\begin{aligned} U = \rho I - i(\lambda S_1 + \mu S_2 + \nu S_3) \\ U\dagger = \rho I + i(\lambda S_1 + \mu S_2 + \nu S_3) \end{aligned}\right\} \qquad \text{(QUATERNION REPRESENTATION OF ROTATIONS)} \qquad (14.10\text{-}17)$$

Again, both U and $-U$ define the same rotation uniquely.

14.10-5. Rotations about the Coordinate Axes. The following transformation matrices represent right-handed rotations about the positive coordinate axes:

$$A_1(\psi) \equiv \begin{bmatrix} 1 & 0 & 0 \\ 0 & \cos\psi & -\sin\psi \\ 0 & \sin\psi & \cos\psi \end{bmatrix} \quad \text{(ROTATION THROUGH AN ANGLE } \psi \text{ ABOUT } \mathbf{u}_1) \tag{14.10-18a}$$

$$A_2(\psi) \equiv \begin{bmatrix} \cos\psi & 0 & \sin\psi \\ 0 & 1 & 0 \\ -\sin\psi & 0 & \cos\psi \end{bmatrix} \quad \text{(ROTATION THROUGH AN ANGLE } \psi \text{ ABOUT } \mathbf{u}_2) \tag{14.10-18b}$$

$$A_3(\psi) \equiv \begin{bmatrix} \cos\psi & -\sin\psi & 0 \\ \sin\psi & \cos\psi & 0 \\ 0 & 0 & 1 \end{bmatrix} \quad \text{(ROTATION THROUGH AN ANGLE } \psi \text{ ABOUT } \mathbf{u}_3) \tag{14.10-18c}$$

Note

$$A_i^{-1}(\psi) \equiv \tilde{A}_i(\psi) \equiv A_i(-\psi) \qquad (i = 1, 2, 3) \tag{14.10-19}$$

14.10-6. Euler Angles. **(a)** Every matrix $A \equiv [a_{ik}]$ representing a proper rotation in three-dimensional Euclidean space can be variously expressed as a product of three matrices (18), and in particular as

$$\begin{aligned} A &\equiv A_3(\alpha)A_2(\beta)A_3(\gamma) \\ &\equiv \begin{bmatrix} \cos\alpha & -\sin\alpha & 0 \\ \sin\alpha & \cos\alpha & 0 \\ 0 & 0 & 1 \end{bmatrix} \begin{bmatrix} \cos\beta & 0 & \sin\beta \\ 0 & 1 & 0 \\ -\sin\beta & 0 & \cos\beta \end{bmatrix} \begin{bmatrix} \cos\gamma & -\sin\gamma & 0 \\ \sin\gamma & \cos\gamma & 0 \\ 0 & 0 & 1 \end{bmatrix} \\ &\equiv \begin{bmatrix} \cos\alpha\cos\beta\cos\gamma - \sin\alpha\sin\gamma & -(\cos\alpha\cos\beta\sin\gamma + \sin\alpha\cos\gamma) & \cos\alpha\sin\beta \\ \sin\alpha\cos\beta\cos\gamma + \cos\alpha\sin\gamma & -\sin\alpha\cos\beta\sin\gamma + \cos\alpha\cos\gamma & \sin\alpha\sin\beta \\ -\sin\beta\cos\gamma & \sin\beta\sin\gamma & \cos\beta \end{bmatrix} \equiv A_{32}(\alpha, \beta, \gamma) \end{aligned} \tag{14.10-20}$$

The three **Euler angles** α, β, γ define the rotation uniquely; except for multiples of 2π, they are uniquely determined by a given rotation, unless $\beta = 0$ ("gimbal lock," Sec. 14.10-6*d*).

A set of cartesian x', y', z' axes (moving "body axes" of a rigid body) initially aligned with $\mathbf{u}_1$, $\mathbf{u}_2$, $\mathbf{u}_3$ can be rotated into alignment with $\mathbf{u}_1'$, $\mathbf{u}_2'$, $\mathbf{u}_3'$ by three successive rotations (18) [Fig. 14.10-1; note the discussion of Sec. 14.6-3, which explains the apparently inverted order of the three matrices in Eq. (20)].

1. Rotate about z' axis through the Euler angle α
2. Rotate about y' axis through the Euler angle β
3. Rotate about z' axis through the Euler angle γ

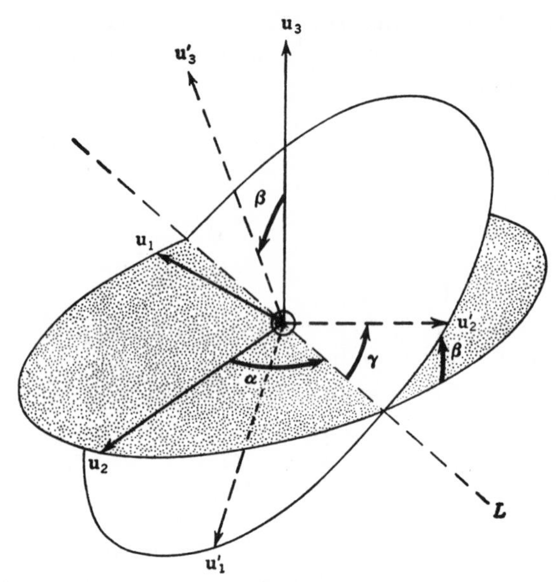

FIG. 14.10-1. The Euler angles α, β, γ. The axis OL of the second rotation (through β) is often called the *line of nodes*. Note that α and β are the spherical polar coordinates for the vector $\mathbf{u}_3'$ in the $\mathbf{u}_1$, $\mathbf{u}_2$, $\mathbf{u}_3$ system.

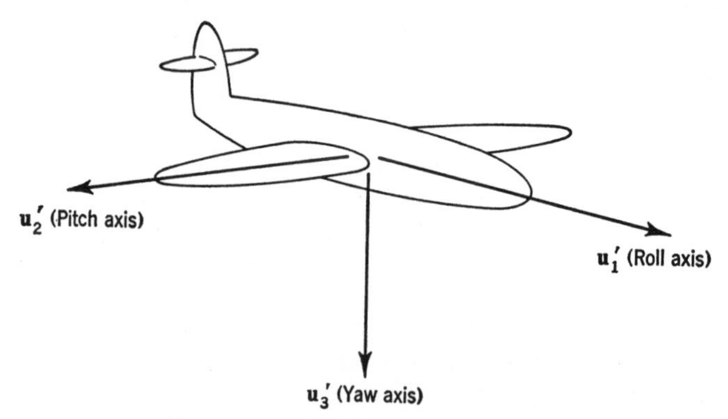

FIG. 14.10-2. Body axes of an aircraft.

The inverse rotation $\mathbf{A}^{-1}$ (which turns $\mathbf{x}'$ back into $\mathbf{x}$) is represented by the matrix

$$A^{-1} \equiv \tilde{A} \equiv A_3(-\gamma)A_2(-\beta)A_3(-\alpha) \equiv A_{32}(-\gamma, -\beta, -\alpha)$$
$$\equiv A_{32}(\pi - \gamma, \beta, \pi - \alpha) \quad (14.10\text{-}21)$$

There are six ways to express a rotation matrix (1) as a product

$$A \equiv A_i(\psi_1)A_k(\psi_2)A_i(\psi_3) \equiv A_{ik}(\psi_1, \psi_2, \psi_3) \qquad (i, k = 1, 2, 3; i \neq k) \quad (14.10\text{-}22)$$

of rotations about two different coordinate axes. Of the different resulting Euler-angle systems, another frequently used one is defined by

$$A \equiv A_3(\alpha')A_1(\beta')A_3(\gamma') \equiv A_{31}(\alpha', \beta', \gamma') \tag{14.10-23}$$

which is related to the system of Eq. (20) by

$$\alpha' = \alpha + \pi/2 \qquad \beta' = \beta \qquad \gamma' = \gamma - \pi/2 \tag{14.10-24}$$

(b) There are, furthermore, six ways to represent a rotation matrix A as a product

$$A \equiv A_i(\vartheta_1)A_j(\vartheta_2)A_k(\vartheta_3) \equiv A_{ijk}(\vartheta_1, \vartheta_2, \vartheta_3) \qquad (i, k = 1, 2, 3;\ i \neq j,\ i \neq k,\ j \neq k) \tag{14.10-25}$$

of rotations about *three* different coordinate axes. In particular,

$$A \equiv A_1(\varphi)A_2(\vartheta)A_3(\psi) \equiv \begin{bmatrix} \cos\vartheta\cos\psi & -\cos\vartheta\sin\psi & \sin\vartheta \\ \sin\varphi\sin\vartheta\cos\psi + \cos\varphi\sin\psi & -\sin\varphi\sin\vartheta\sin\psi + \cos\varphi\cos\psi & -\sin\varphi\cos\vartheta \\ -\cos\varphi\sin\vartheta\cos\psi + \sin\varphi\sin\psi & \cos\varphi\sin\vartheta\sin\psi + \sin\varphi\cos\psi & \cos\varphi\cos\vartheta \end{bmatrix} \equiv A_{123}(\varphi, \vartheta, \psi) \tag{14.10-26}$$

is frequently used to describe the attitude of an aircraft or space vehicle after successive roll φ, pitch ϑ, and yaw ψ about body-centered axes respectively directed forward, to starboard, and toward the bottom of the craft (Fig. 14.10-2).

(c) The profusion of the 12 Euler-angle systems defined above is augmented by the fact that some authors replace one or more of the Euler angles by its negative, and that some of the literature involves left-handed coordinate systems. In addition, the reader is warned to check whether a given Euler-angle transformation is originally defined as an *operation* ("alibi" interpretation, Sec. 14.5-1) or as a *coordinate transformation* ("alias" interpretation, Sec. 14.6-1), since it is possible to confuse A and $A^{-1} = \tilde{A}$. Specifically, the component matrix $x = \{\xi_1, \xi_2, \xi_3\}$ of a vector $\mathbf{x}$ in the fixed $\mathbf{u}_i$ system and the component matrix $\bar{x} \equiv \{\bar{\xi}_1, \bar{\xi}_2, \bar{\xi}_3\}$ in the rotated $\mathbf{u}'_k$ system are related by

$$x = A\bar{x} \qquad \bar{x} = A^{-1}x = \tilde{A}x \tag{14.10-27}$$

and the base-vector matrices (Sec. 14.6-2) transform according to

$$[\mathbf{u}'_1\ \mathbf{u}'_2\ \mathbf{u}'_3] = [\mathbf{u}_1\ \mathbf{u}_2\ \mathbf{u}_3]A \qquad \text{or} \qquad \begin{bmatrix} \mathbf{u}'_1 \\ \mathbf{u}'_2 \\ \mathbf{u}'_3 \end{bmatrix} = \tilde{A} \begin{bmatrix} \mathbf{u}_1 \\ \mathbf{u}_2 \\ \mathbf{u}_3 \end{bmatrix} \tag{14.10-28}$$

Note the discussion of Sec. 14.6-3.

(d) The parameters c_1, c_2, c_3, δ and λ, μ, ν, ρ are readily expressed in terms of Euler angles with the aid of Eq. (5) and the Euler-angle matrix. Thus

$$\left.\begin{aligned} \lambda &= c_1 \sin\frac{\delta}{2} = \sin\frac{\gamma-\alpha}{2}\sin\frac{\beta}{2} \\ \mu &= c_2 \sin\frac{\delta}{2} = \cos\frac{\alpha-\gamma}{2}\sin\frac{\beta}{2} \\ \nu &= c_3 \sin\frac{\delta}{2} = \sin\frac{\alpha+\gamma}{2}\cos\frac{\beta}{2} \\ \rho &= \cos\frac{\delta}{2} = \cos\frac{\alpha+\gamma}{2}\cos\frac{\beta}{2} \end{aligned} \qquad \begin{aligned} a &= e^{i\frac{\alpha+\gamma}{2}}\cos\frac{\beta}{2} \\ b &= e^{-i\frac{\alpha-\gamma}{2}}\sin\frac{\beta}{2} \end{aligned}\right\} \tag{14.10-29}$$
*

Note that addition of 2π to one of the Euler angles changes the sign of all parameters (29) and leaves the rotation matrix A unchanged.

If $\psi_2 = 0$ in Eq. (22) or $\vartheta_2 = \pi/2$ in Eq. (25) (e.g., $\beta = 0$, $\beta' = 0$, or $\vartheta = \pi/2$), then the remaining two Euler angles are no longer uniquely defined by the given rotation ("gimbal lock" in attitude-reference systems). Euler angles are, therefore, employed only for the representation of rotations which are suitably restricted in range.

14.10-7. Infinitesimal Rotations, Continuous Rotation, and Angular Velocity (see also Secs. 5.3-2 and 14.4-9). **(a)** An infinitesimal three-dimensional rotation through a small angle $d\delta$ about a rotation axis with the direction cosines c_1, c_2, c_3 is described by the orthogonal infinitesimal transformation

$$\mathbf{x}' = (\mathsf{I} + \boldsymbol{\Delta})\mathbf{x} \tag{14.10-30}$$

so that $\boldsymbol{\Delta}$ is skew-symmetric (Sec. 14.4-9). In the orthonormal $\mathbf{u}_1, \mathbf{u}_2, \mathbf{u}_3$ reference frame, $\boldsymbol{\Delta}$ is represented by the skew-symmetric matrix†

$$\Delta \equiv \begin{bmatrix} 0 & -c_3 & c_2 \\ c_3 & 0 & -c_1 \\ -c_2 & c_1 & 0 \end{bmatrix} d\delta \tag{14.10-31}$$

obtained from Eq. (6) for $\delta = d\delta \to 0$. It follows that

$$\boldsymbol{\Delta}\mathbf{x} \equiv (\mathbf{c} \times \mathbf{x})\, d\delta \tag{14.10-32}$$

where $\mathbf{c} = c_1\mathbf{u}_1 + c_2\mathbf{u}_2 + c_3\mathbf{u}_3$ is the unit vector in the direction of the

* a and b are defined in Sec. 14.10-4. Note that several different definitions of the Cayley-Klein parameters are in common use.

† In general, if W *is any skew-symmetric linear operator defined on a three-dimensional Euclidean vector space* $\mathcal{V}_E$ *and represented in an orthonormal reference frame* $\mathbf{u}_1, \mathbf{u}_2, \mathbf{u}_3$ *by the skew-symmetric matrix*

$$W \equiv \begin{bmatrix} 0 & -w_3 & w_2 \\ w_3 & 0 & -w_1 \\ -w_2 & w_1 & 0 \end{bmatrix}$$

then, for every vector $\mathbf{x}$ *of* $\mathcal{V}_E$,

$$\mathbf{x}' = \mathsf{W}\mathbf{x} = \mathbf{w} \times \mathbf{x}$$

where $\mathbf{w} = w_1\mathbf{u}_1 + w_2\mathbf{u}_2 + w_3\mathbf{u}_3$ (see also Sec. 16.9-2).

positive rotation axis. This relation is independent of any reference frame (see also Sec. 14.10-3*b*).

(b) For a continuous three-dimensional rotation described by

$$\mathbf{x}'(t) = \mathsf{A}(t)\mathbf{x}$$

where $\mathbf{x}$ is constant, Eq. (32) yields

$$\frac{d\mathbf{x}'(t)}{dt} = \frac{d\mathsf{A}(t)}{dt}\mathbf{x} = \boldsymbol{\omega}(t) \times \mathbf{x}'(t) \equiv \boldsymbol{\omega}(t) \times [\mathsf{A}(t)\mathbf{x}] \qquad (14.10\text{-}33a)$$

The vector $\boldsymbol{\omega}(t)$, given in terms of the fixed and rotating base vectors by

$$\begin{aligned}\boldsymbol{\omega}(t) &\equiv \omega_1(t)\mathbf{u}_1 + \omega_2(t)\mathbf{u}_2 + \omega_3(t)\mathbf{u}_3 \\ &\equiv \bar{\omega}_1(t)\mathbf{u}_1'(t) + \bar{\omega}_2(t)\mathbf{u}_2'(t) + \bar{\omega}_3(t)\mathbf{u}'(t)\end{aligned} \qquad (14.10\text{-}33b)$$

is directed along the **instantaneous axis of rotation** (axis of the rotation $\mathbf{x}' \to \mathbf{x}' + d\mathbf{x}'$), and $|\boldsymbol{\omega}(t)|$ is the **instantaneous absolute rate of rotation** with respect to t. If the parameter t used to describe the rotation is the time, then $\boldsymbol{\omega}(t)$ is the **angular velocity** of the rotation.

Equation (33) yields

$$\frac{d\mathsf{A}(t)}{dt} = \boldsymbol{\Omega}(t)\mathsf{A}(t) \qquad (14.10\text{-}34)$$

where $\boldsymbol{\Omega}$ is the skew-symmetrical operator represented in the $\mathbf{u}_1$, $\mathbf{u}_2$, $\mathbf{u}_3$ and $\mathbf{u}_1'(t)$, $\mathbf{u}_2'(t)$, $\mathbf{u}_3'(t)$ reference systems by the respective matrices

$$\left.\begin{aligned}\Omega(t) &\equiv \begin{bmatrix} 0 & -\omega_3(t) & \omega_2(t) \\ \omega_3(t) & 0 & -\omega_1(t) \\ -\omega_2(t) & \omega_1(t) & 0 \end{bmatrix} \\ \bar{\Omega}(t) &\equiv \begin{bmatrix} 0 & -\bar{\omega}_3(t) & \bar{\omega}_2(t) \\ \bar{\omega}_3(t) & 0 & -\bar{\omega}_1(t) \\ -\bar{\omega}_2(t) & \bar{\omega}_1(t) & 0 \end{bmatrix}\end{aligned}\right\} \qquad (14.10\text{-}35)$$

With the aid of the relations (14.6-8) and $\mathsf{A}^{-1} \equiv \tilde{\mathsf{A}}$, it follows that

$$\left.\begin{aligned} &\frac{dA}{dt} \equiv \Omega A \equiv A\bar{\Omega} \\ &\Omega \equiv \frac{dA}{dt}\tilde{A} \qquad \bar{\Omega} \equiv \tilde{A}\frac{dA}{dt}\end{aligned}\right\} \qquad (14.10\text{-}36)$$

Substitution of Eq. (8), (20), or (26) into Eq. (36) yields relations between angular-velocity components and matrix elements (direction cosines), Euler parameters, or Euler angles. In particular,

$$\frac{dA}{dt} \equiv \frac{d}{dt}[a_{ik}] \equiv \begin{bmatrix} \bar{\omega}_3 a_{12} - \bar{\omega}_2 a_{13} & \bar{\omega}_1 a_{13} - \bar{\omega}_3 a_{11} & \bar{\omega}_2 a_{11} - \bar{\omega}_1 a_{12} \\ \bar{\omega}_3 a_{22} - \bar{\omega}_2 a_{23} & \bar{\omega}_1 a_{23} - \bar{\omega}_3 a_{21} & \bar{\omega}_2 a_{21} - \bar{\omega}_1 a_{22} \\ \bar{\omega}_3 a_{32} - \bar{\omega}_2 a_{33} & \bar{\omega}_1 a_{33} - \bar{\omega}_3 a_{31} & \bar{\omega}_2 a_{31} - \bar{\omega}_1 a_{32} \end{bmatrix} \qquad (14.10\text{-}37)$$

$$\left.\begin{aligned}\omega_1(t) &\equiv 2(\dot\lambda\rho - \lambda\dot\rho - \dot\mu\nu + \mu\dot\nu)\\ &\equiv -\sin\alpha\frac{d\beta}{dt} + \cos\alpha\sin\beta\frac{d\gamma}{dt}\\ \omega_2(t) &\equiv 2(\dot\mu\rho - \mu\dot\rho - \dot\nu\lambda + \nu\dot\lambda)\\ &\equiv \cos\alpha\frac{d\beta}{dt} + \sin\alpha\sin\beta\frac{d\gamma}{dt}\\ \omega_3(t) &\equiv 2(\dot\nu\rho - \nu\dot\rho - \dot\lambda\mu + \lambda\dot\mu)\\ &\equiv \frac{d\alpha}{dt} + \cos\beta\frac{d\gamma}{dt}\end{aligned}\right\}\begin{matrix}\text{[COMPONENTS OF } \boldsymbol{\omega}(t) \text{ IN FIXED}\\ \text{REFERENCE FRAME]}\end{matrix} \qquad (14.10\text{-}38)$$

$$\left.\begin{aligned}\bar\omega_1(t) &\equiv 2(\dot\lambda\rho - \lambda\dot\rho + \dot\mu\nu - \mu\dot\nu)\\ &\equiv -\sin\beta\cos\gamma\frac{d\alpha}{dt} + \sin\gamma\frac{d\beta}{dt}\\ &\equiv \cos\vartheta\cos\psi\frac{d\varphi}{dt} + \sin\psi\frac{d\vartheta}{dt}\\ \bar\omega_2(t) &\equiv 2(\dot\mu\rho - \mu\dot\rho + \dot\nu\lambda - \nu\dot\lambda)\\ &\equiv \sin\beta\sin\gamma\frac{d\alpha}{dt} + \cos\gamma\frac{d\beta}{dt}\\ &\equiv -\cos\vartheta\sin\psi\frac{d\varphi}{dt} + \cos\psi\frac{d\vartheta}{dt}\\ \bar\omega_3(t) &\equiv 2(\dot\nu\rho - \nu\dot\rho + \dot\lambda\mu - \lambda\dot\mu)\\ &\equiv \cos\beta\frac{d\alpha}{dt} + \frac{d\gamma}{dt}\\ &\equiv \sin\vartheta\frac{d\varphi}{dt} + \frac{d\psi}{dt}\end{aligned}\right\}\begin{matrix}\text{[COMPONENTS OF } \boldsymbol{\omega}(t) \text{ IN ROTATING}\\ \text{REFERENCE FRAME]}\end{matrix} \qquad (14.10\text{-}39)$$

$$\left.\begin{aligned}\frac{d\varphi}{dt} &\equiv \frac{1}{\cos\vartheta}(\bar\omega_1\cos\psi - \bar\omega_2\sin\psi)\\ \frac{d\vartheta}{dt} &\equiv \bar\omega_1\sin\psi + \bar\omega_2\cos\psi\\ \frac{d\psi}{dt} &\equiv (\bar\omega_2\sin\psi - \bar\omega_1\cos\psi)\tan\vartheta + \bar\omega_3\end{aligned}\right\} \qquad (14.10\text{-}40)$$

14.10-8. The Three-dimensional Rotation Group and Its Representations (see also Secs. 12.2-1 to 12.2-12 and 14.9-1 to 14.9-6). **(a)** The orthogonal transformations (1) of a three-dimensional Euclidean vector space onto itself are necessarily bounded and nonsingular and constitute a group, the **three-dimensional rotation-reflection group** $R_3^{\pm}$. The proper rotations [det $(\mathbf{A}) = 1$] constitute a normal subgroup of $R_3^{\pm}$, the **three-dimensional rotation group** R_3^+. Neither $R_3^{\pm}$ nor R_3^+ is commutative.

Rotations involving the same absolute rotation angle $|\delta|$ belong to the same class of conjugate elements. The rotations about any fixed axis form a commutative subgroup of R_3^+ (*two-dimensional rotations*).

$R_3^{\pm}$ is a mixed-continuous group, and R_3^+ is a continuous group (normal subgroup of $R_3^{\pm}$ with index 2). $R_3^{\pm}$ is, in turn, a subgroup of the group of all nonsingular linear transformations of the Euclidean vector space onto itself (**full linear group, FLG**). Note that every transformation of FLG is the product of a proper or improper rotation and a nonnegative symmetric transformation (**affine transformation,** stretching or contraction; see also Secs. 14.4-8 and 13.3-4).

(b) The Irreducible Representations of R_3^+. *The* 2×2 *matrices* (14) *constitute an irreducible unitary two-to-one representation of* R_3^+ *in the field of complex numbers;* i.e., the three-dimensional rotation group R_3^+ is represented by the group of unitary transformations with determinant 1 of a two-dimensional unitary vector space onto itself (**two-dimensional unimodular unitary group, special unitary group, SUG**).

More generally, the three-dimensional rotation group R_3^+ has bounded irreducible representations of dimensions $n = 2, 3, 4, \ldots$. The complete set of unitary irreducible representations is conveniently written as $\mathcal{R}^{(1/2)}, \mathcal{R}^{(1)}, \mathcal{R}^{(3/2)}, \ldots$; $\mathcal{R}^{(j)}$ has the dimension $n = 2j + 1$, and the $(2j + 1) \times (2j + 1)$ matrix representing the rotation with Cayley-Klein parameters $a, b, -b^*, a^*$ (Sec. 14.10-4) or Euler angles α, β, γ (Sec. 14.10-6) is

$$[U_{mq}^{(j)}(\alpha, \beta, \gamma)] \equiv \left[\sum_{h=0}^{\infty} \frac{(-1)^h \sqrt{(j+m)!(j-m)!(j+q)!(j-q)!}}{(j-m-h)!(j+q-h)!(h-q+m)!h!} \cdot (a^*)^{j-m-h} a^{j+q-h} (b^*)^{h+m-q} b^h\right]$$

$$\equiv \left[\sum_{h=0}^{\infty} \frac{(-1)^h \sqrt{(j+m)!(j-m)!(j+q)!(j-q)!}}{(j-m-h)!(j+q-h)!(h-q+m)!h!} e^{i(m\alpha+q\gamma)} \cdot \left(\cos\frac{\beta}{2}\right)^{2j-m+q-2h} \left(\sin\frac{\beta}{2}\right)^{m-q+2h}\right]$$

$$\left(j = \frac{1}{2}, 1, \frac{3}{2}, \cdots ; m, q = -j, -j+1, \ldots, j-1, j\right) \qquad (14.10\text{-}41)^*$$

Each sum has only a finite number of terms, since $1/N! = 0$ for $N < 0$. *The representation* $\mathcal{R}^{(j)}$ *is true (one-to-one) for* $j = 1, 2, \ldots$ *and two-to-one for* $j = 1/2, 3/2, \ldots$ (see also Sec. 14.10-4). The character (Sec. 14.9-4) of $\mathcal{R}^{(j)}$ is

$$\chi^{(j)}(\alpha, \beta, \gamma) \equiv \operatorname{Tr}\,[U_{mq}^{(j)}(\alpha, \beta, \gamma)] \equiv \frac{\sin\,(j + \frac{1}{2})\delta}{(\sin\,\delta/2)} \qquad \left(j = 0, \tfrac{1}{2}, 1, \tfrac{3}{2}, \ldots\right) \qquad (14.10\text{-}42)$$

where δ is the angle of rotation defined in Sec. 14.10-2.

The peculiar indices j, m, and q employed to label the $\mathcal{R}^{(j)}$ and $U_{mq}^{(j)}$ are those associated with the *spherical surface harmonics of degree* j (Sec. 21.8-12). For integral values of j, these functions constitute a $(2j + 1)$-dimensional representation space for $\mathcal{R}^{(j)}$, with the functions (21.8-66) as an orthonormal basis.

* Some authors replace the factor $(-1)^h$ by $(-1)^{h-q+m}$, corresponding to multiplication of each matrix by the same diagonal matrix. As written, Eq. (41) reduces to Eq. (14) for $j = 1/2$.

(c) Direct Products of Rotation Groups (see also Secs. 12.7-2 and 14.9-6). Direct products of three-dimensional rotation groups describe, for instance, the composite rotations of dynamical systems comprising two or more rotating bodies (atoms or nuclei in quantum mechanics). $R_3^+ \otimes R_3^+$ is isomorphic with R_3^+; note

$$\begin{aligned}\mathcal{R}^{(j)} \otimes \mathcal{R}^{(j')} &= \mathcal{R}^{(j+j')} \oplus \mathcal{R}^{(j-1)} \otimes \mathcal{R}^{(j'-1)} \\ &= \mathcal{R}^{(j+j')} \oplus \mathcal{R}^{(j+j'-1)} \oplus \cdots \oplus \mathcal{R}^{(|j-j'|)}\end{aligned}$$

(CLEBSCH-GORDAN EQUATION) (14.10-43)

14.11. RELATED TOPICS, REFERENCES, AND BIBLIOGRAPHY

14.11-1. Related Topics. The following topics related to the study of matrix methods are treated in other chapters of this handbook:

Linear simultaneous equations .... Chap. 1
Elementary vector algebra .... Chap. 5
Vector relations in terms of curvilinear coordinates .... Chap. 6
Groups, vector spaces, and linear operators .... Chap. 12
Diagonalization of matrices, eigenvalue problems, quadratic and hermitian forms .... Chap. 14
Tensor algebra .... Chap. 15

14.11-2. References and Bibliography (see also Secs. 12.9-2, 13.7-2, and 15.7-2).

14.1. Birkhoff, G., and S. MacLane: *A Survey of Modern Algebra*, 3d ed., Macmillan, New York, 1965.
14.2. Dunford, N., and J. T. Schwartz: *Linear Operators*, Interscience, New York, 1964.
14.3. Finkbeiner, D. T.: *Introduction to Matrices and Linear Algebra*, Freeman, San Francisco, 1960.
14.4. Halmos, P. R.: *Finite-dimensional Vector Spaces*, 2d ed., Princeton, Princeton, N.J., 1958.
14.5. ———: *Introduction to Hilbert Space and the Theory of Spectral Multiplicity*, Chelsea, New York, 1957.
14.6. Nering, E. D.: *Linear Algebra and Matrix Theory*, Interscience, New York, 1963.
14.7. Shields, P. C.: *Linear Algebra*, Addison-Wesley, Reading, Mass., 1964.
14.8. Stoll, R. R.: *Linear Algebra and Matrix Theory*, McGraw-Hill, New York, 1952.
14.9. Thrall, R. M., and L. Tornheim: *Vector Spaces and Matrices*, Wiley, New York, 1957.
14.10. Von Neumann, J.: *Die Mathematischen Grundlagen der Quantenmechanik*, Springer, Berlin, 1932.
14.11. Zaanen, C.: *Linear Analysis*, North Holland Publishing Company, Amsterdam, 1956.

(See also the articles by G. Falk and H. Tietz in vol. II of the *Handbuch der Physik*, Springer, Berlin, 1955.)

Representation of Rotations

14.12. Goldstein, H.: *Classical Mechanics*, Addison-Wesley, Reading, Mass., 1953.
14.13. Synge, J. L.: Classical Dynamics, in *Handbuch der Physik*, Vol. III, Springer, Berlin, 1960.

Representation of Groups

14.14. Boerner, H.: *Darstellungen von Gruppen mit Beruecksichtigung der Beduerfnisse der modernen Physik*, Springer, Berlin, 1955.
14.15. Hall, M.: *The Theory of Groups*, Macmillan, New York, 1961.
14.16. Kurosh, A. G.: *The Theory of Groups*, 2 vols., 2d ed., Chelsea, New York, 1960.
14.17. Margenau, H., and G. M. Murphy: *The Mathematics of Physics and Chemistry*, 2d ed., Van Nostrand, Princeton, N.J., 1957.
14.18. Murnaghan, F. D.: *The Unitary and Rotation Groups*, Spartan, Washington, D.C., 1962.
14.19. Van der Waerden, G. L.: *Gruppentheoretische Methode in der Quantenmechanik*, Springer, Berlin, 1932.
14.20. Weyl, H.: *The Theory of Groups and Quantum Mechanics*, Dover, New York, 1931.
14.21. ———: *The Classical Groups*, Princeton, Princeton, N.J., 1939.
14.22. Wigner, E.: *Gruppentheorie und ihre Anwendung auf Quantenmechanik der Atomspektren*, Brunswick, 1951.
14.23. Zassenhaus, Hans J.: *The Theory of Groups*, 2d ed., Chelsea, New York, 1958.

CHAPTER **15**

LINEAR INTEGRAL EQUATIONS, BOUNDARY-VALUE PROBLEMS, AND EIGENVALUE PROBLEMS

15.1. INTRODUCTION. FUNCTIONAL ANALYSIS

15.1-1. *Functional analysis* regards suitable classes of functions as "points" in topological spaces (Chap. 12) and, in particular, as *multidimensional vectors* admitting the definition of *inner products* (Sec. 15.2-1) and *expansions in terms of orthogonal functions* (base vectors, Sec. 15.3-4). The resulting elegant and powerful geometrical analogy relates a large class of operations, including linear integral transformations and differentiation, to the theory of linear transformations introduced in Chap. 14. The solution of linear ordinary and partial differential equations and of linear integral equations is thus found to be a more or less simple generalization of the solution of linear simultaneous equations; in particular, the solution may involve an eigenvalue problem.

Sections 15.3-1 to 15.3-10 review *linear integral equations;* Secs. 15.4-1 to 15.4-12 introduce *linear boundary-value problems and eigenvalue problems involving differential equations.* The remainder of the chapter deals with various methods of solving linear boundary-value problems, viz.

1. Eigenfunction expansions (Sec. 15.4-12); this method may be extended to include the various integral-transform methods (Sec. 10.5-1)
2. Green's functions (Secs. 15.5-1, 15.5-3, 15.6-6, and 15.6-9)
3. Reduction to integral equations (Sec. 15.3-2)
4. Variation methods (Sec. 15.4-7; see also Secs. 11.7-1 to 11.7-3)

In particular, Secs. 15.6-1 to 15.6-10 deal with boundary-value problems involving *Laplace's and Poisson's differential equations* (*potential theory*) and the space form of the wave equation.

Although many practical problems will yield only to numerical solution (Sec. 20.9-4), the general and intuitively suggestive viewpoint of functional analysis offers far-reaching insight into the behavior of vibrating systems, atomic phenomena, etc.

15.1-2. Notation (see also Sec. 15.4-1). Throughout Secs. 15.2-1 to 15.5-4, $\Phi(x)$, $f(x)$, $F(x)$, . . . symbolize either functions of a single independent variable x or, for brevity, functions of a set of n independent variables $x^1, x^2, \ldots, x^n$ (see also Secs. 6.2-1 and 16.1-2). In the *one-dimensional case*, dx is a simple differential; in the *multidimensional case*, $dx \equiv dx^1\, dx^2 \ldots dx^n$. An integral

$$I = \int_V f(\xi)\, d\xi \tag{15.1-1}$$

stands either for a one-dimensional definite integral $I = \int_a^b f(\xi)\, d\xi$ over a bounded or unbounded interval $V \equiv (a, b)$, or for an n-dimensional integral

$$I = \int\int \cdots \int_V f(\xi^1, \xi^2, \ldots, \xi^n)\, d\xi^1\, d\xi^2 \cdots d\xi^n \tag{15.1-2}$$

over a region V in n-dimensional space. As a rule it is possible to introduce volume elements $dV(\xi) = \sqrt{|g(\xi)|}\, d\xi$, so that each integral (2) becomes a *volume integral* (Secs. 6.2-3, 15.4-1*b*, and 16.10-10).

15.2. FUNCTIONS AS VECTORS. EXPANSIONS IN TERMS OF ORTHOGONAL FUNCTIONS

15.2-1. Quadratically Integrable Functions as Vectors. Inner Product and Normalization.* (a) A real or complex function $f(x)$ defined on the measurable set E of "points" (x) or $(x^1, x^2, \ldots, x^3)$ is **quadratically integrable** on E if and only if $\int_E |f(\xi)|^2\, d\xi$ exists in the sense of Lebesgue (Sec. 4.6-15). The class L^2 [more accurately $L^2(V)$] of all real or complex functions quadratically integrable on a given interval or region V constitute, respectively, an infinite-dimensional *real or complex unitary vector space* (Sec. 14.2-6) if one regards the functions $f(x)$, $h(x)$, . . . as vectors and defines

The **vector sum of** $f(x)$ **and** $h(x)$ as $f(x) + h(x)$
The **product of** $f(x)$ **by a scalar** α as $\alpha f(x)$
The **inner product of** $f(x)$ **and** $h(x)$ as

* See Sec. 15.1-2 for notation.

$$(f, h) \equiv \int_V \gamma(\xi) f^*(\xi) h(\xi)\, d\xi \tag{15.2-1}$$

where $\gamma(x)$ is a given real nonnegative function (**weighting function**) quadratically integrable on V. A change of weighting function corresponds to a change in the independent variable; in many applications, one has $\gamma(x) \equiv 1$, or $\gamma(\xi)\, d\xi$ is a *volume element* (Sec. 15.4-1*b*).

Linear independence of a set of functions (vectors) in L^2 is defined in the manner of Sec. 1.9-3 (see also Sec. 14.2-3). *A set of quadratically integrable functions* $f_1(x)$, $f_2(x)$, . . . , $f_m(x)$ *are linearly independent if and only if Gram's determinant* det $[(f_i, f_k)]$ *differs from zero* (see also Sec. 14.2-6*a*).

(**b**) As in Sec. 14.2-7, the **norm** of a function (vector) $f(x)$ in L^2 is the quantity

$$\|f\| \equiv \sqrt{(f, f)} \equiv \left[\int_V \gamma(\xi) |f(\xi)|^2\, d\xi\right]^{1/2} \tag{15.2-2}$$

A (necessarily quadratically integrable) function $f(x)$ is **normalizable** if and only if $\|f\|$ exists and is different from zero. Multiplication of a normalizable function $f(x)$ by $1/\|f\|$ yields a function $f(x)/\|f\|$ of unit norm [**normalization** of $f(x)$].

(**c**) The inner product defined by Eq. (1) has all the properties listed in Sec. 14.2-6. In particular, *if* $f(x)$, $h(x)$, *and the real nonnegative weighting function* $\gamma(x)$ *are quadratically integrable on* V

$$|(f, h)|^2 \equiv \left|\int_V \gamma f^* h\, d\xi\right|^2 \leq \int_V \gamma |f|^2\, d\xi \int_V \gamma |h|^2\, d\xi \equiv (f, f)(h, h) \quad \text{(Cauchy-Schwarz inequality)} \tag{15.2-3}$$

and

$$\|f + h\| \equiv \left(\int_V \gamma |f + h|^2\, d\xi\right)^{1/2} \leq \left(\int_V \gamma |f|^2\, d\xi\right)^{1/2} + \left(\int_V \gamma |h|^2\, d\xi\right)^{1/2} \equiv \|f\| + \|h\| \quad \text{(Minkowski's inequality)} \tag{15.2-4}$$

15.2-2. Metric and Convergence in L^2. Convergence in Mean (see also Secs. 12.5-2 to 12.5-5, 14.2-7, and 18.6-3). (**a**) As a unitary vector space, L^2 admits the *distance function* (*metric*, Sec. 12.5-2)

$$d(f, h) \equiv \|f - h\| \equiv \left[\int_V \gamma(\xi) |f(\xi) - h(\xi)|^2\, d\xi\right]^{1/2} \tag{15.2-5}$$

The root-mean-square difference (5) between $f(x)$ and $h(x)$ equals zero (metric equality) if and only if $f(x) = h(x)$ for *almost all* x in V (Sec. 4.6-14*b*).

(**b**) **Convergence in Mean.** The metric (5) leads to the following definition of metric convergence in L^2. For a given interval or region V, a sequence of quadratically integrable functions $s_0(x)$, $s_1(x)$, $s_2(x)$, . . .

converges in mean (with index 2) to the limit $s(x)$ [$s_n(x) \xrightarrow[\text{in mean}]{} s(x)$ as $n \to \infty$] if and only if

$$d^2(s_n, s) \equiv \|s_n - s\|^2 \equiv \int_V \gamma(\xi)|s_n(\xi) - s(\xi)|^2 \, d\xi \to 0 \text{ as } n \to \infty \tag{15.2-6}$$

In this case, the sequence defines its **limit-in-mean** $\underset{n\to\infty}{\text{l.i.m.}}\, s_n(x) = s(x)$ *uniquely almost everywhere in* V. Convergence in mean does *not* necessarily imply actual convergence of the sequence $s_0(x)$, $s_1(x)$, $s_2(x)$, . . . at every point, nor does convergence of a sequence at all points of V imply convergence in mean. Equation (6) does imply convergence of a *subsequence* of $s_n(x)$ for almost every x in V.

In particular, an infinite series $a_0(x) + a_1(x) + a_2(x) + \cdots$ of quadratically integrable functions converges in mean to the limit $s(x)$ if and only if $\sum_{k=0}^{n} a_k(x) \xrightarrow[\text{in mean}]{} s(x)$ as $n \to \infty$. In this case, one writes

$$a_0(x) + a_1(x) + a_2(x) + \cdots \underset{\text{in mean}}{=} s(x)$$

(c) Completeness of L^2: the Generalized Riesz-Fischer Theorem. The space L^2 associated with a given interval or region V is *complete* (Sec. 12.5-4*a*); i.e., *every sequence of quadratically integrable functions* $s_0(x)$, $s_1(x)$, $s_2(x)$, . . . *such that* $\underset{\substack{m\to\infty \\ n\to\infty}}{\text{l.i.m.}}\, |s_m - s_n| = 0$ *(Cauchy sequence) converges in mean to a quadratically integrable function* $s(x)$ *and defines* $s(x)$ *uniquely for almost all* x *in* V *(generalized Riesz-Fischer Theorem).*

NOTE: The completeness property expressed by the generalized Riesz-Fischer theorem establishes L^2 as a *Hilbert space* (Sec. 14.2-7*c*), permitting the introduction of orthonormal bases with all the properties noted in Secs. 14.7-4 and 15.2-4. This is an important reason for the use of Lebesgue integration and convergence in mean.

(d) One defines $f(x, \alpha) \xrightarrow[\text{in mean}]{} F(x)$ as $\alpha \to a$ if and only if $\lim_{\alpha\to a} \|f(x, \alpha) - F(x)\| = 0$.

15.2-3. Orthogonal Functions and Orthonormal Sets of Functions (see also Sec. 14.7-3). **(a)** Two quadratically integrable functions $f(x)$, $h(x)$ are mutually **orthogonal** [orthogonal with respect to the real nonnegative weighting function $\gamma(x)$]* on V if and only if

$$(f, h) \equiv \int_V \gamma(\xi) f^*(\xi) h(\xi) \, d\xi = 0 \tag{15.2-7}$$

A set of functions $u_1(x)$, $u_2(x)$, . . . is an **orthonormal (normal orthogonal, normalized orthogonal) set** if and only if

* Some authors call $f(x)$ and $h(x)$ mutually orthogonal only if also $\gamma(x) \equiv 1$ in V, so that $\int_V f^* h \, d\xi = 0$.

$$(u_i, u_k) \equiv \int_V \gamma u_i^* u_k \, d\xi = \delta_k^i = \begin{Bmatrix} 0 \text{ if } i \neq k \\ 1 \text{ if } i = k \end{Bmatrix} \qquad (i, k = 1, 2, \ldots) \tag{15.2-8}$$

Every set of normalizable mutually orthogonal functions (and, in particular, every orthonormal set) is linearly independent.

(b) Bessel's Inequality. *Given a finite or infinite orthonormal set* $u_1(x), u_2(x), \ldots$ *and any function* $f(x)$ *quadratically integrable over* V,

$$\sum_k |(u_k, f)|^2 \leq (f, f) \qquad \text{(BESSEL'S INEQUALITY)} \tag{15.2-9}$$

The equal sign applies if and only if $f(x)$ *belongs to the linear manifold spanned by* $u_1(x), u_2(x), \ldots$ (see also Secs. 14.2-2, 14.7-3, and 15.2-4).

15.2-4. Complete Orthonormal Sets of Functions (Orthonormal Bases, see also Sec. 14.7-4). An orthonormal set of functions $u_1(x)$, $u_2(x), \ldots$ in $L^2(V)$ is a **complete orthonormal set (orthonormal basis)** if and only if it satisfies the following conditions:

1. Every quadratically integrable function $f(x)$ can be expanded in the form

$$f(x) \underset{\text{in mean}}{=} f_1 u_1(x) + f_2 u_2(x) + \cdots \qquad \text{with } f_k = (u_k, f) \qquad (k = 1, 2, \ldots)$$

2. For every quadratically integrable function $f(x)$ such that $f_1 u_1(x) + f_2 u_2(x) + \cdots \underset{\text{in mean}}{=} f(x)$,

$$(f, f) = |f_1|^2 + |f_2|^2 + \cdots$$ (PARSEVAL'S IDENTITY, COMPLETENESS RELATION; see also Secs. 14.7-3*b* and 15.2-3*b*)

3. For every pair of quadratically integrable functions $f(x)$, $h(x)$ such that $f_1 u_1(x) + f_2 u_1(x) + \cdots \underset{\text{in mean}}{=} f(x)$, $h_1 u_1(x) + h_2 u_2(x) + \cdots \underset{\text{in mean}}{=} h(x)$,

$$(f, h) = f_1^* h_1 + f_2^* h_2 + \cdots$$

4. The orthonormal set $u_1(x), u_2(x), \ldots$ is not contained in any other orthonormal set of $L^2(V)$; i.e., every quadratically integrable function $f(x)$ orthogonal to every $u_k(x)$ equals zero almost everywhere in V.

Each of these four conditions implies the three others.

Given an interval or region V with a complete orthonormal set of functions $u_1(x)$, $u_2(x)$, . . . and any set of complex numbers f_1, f_2, . . . such that $\sum_{k=1}^{\infty} |f_k|^2$ converges, there exists a quadratically integrable function $f(x)$ such that $f_1u_1(x) + f_2u_2(x) + \cdots$ converges in mean to $f(x)$ (Riesz-Fischer Theorem, see also Sec. 15.2-2*c*); the f_k define $f(x)$ uniquely almost everywhere in V, and in particular wherever $f(x)$ is continuous in V (*Uniqueness Theorem,* see also Sec. 4.11-5).

15.2-5. Orthogonalization and Normalization of a Set of Functions (see also Sec. 14.7-4*b*). *Given any countable (finite or infinite) set of linearly independent (Sec.* 1.9-3) *functions $\varphi_1(x)$, $\varphi_2(x)$, . . . normalizable on V, there exists an orthonormal set $u_1(x)$, $u_2(x)$, . . . spanning the same manifold of functions.* Such an orthonormal set may be constructed with the aid of the following recursion formulas (*Gram-Schmidt Orthogonalization Process*):

$$\left.\begin{aligned} u_i(x) &= \frac{v_i(x)}{\|v_i(x)\|} = \frac{v_i(x)}{\sqrt[+]{(v_i, v_i)}} \\ v_1(x) &= \varphi_1(x) \qquad v_{i+1}(x) = \varphi_{i+1}(x) - \sum_{k=1}^{i} (u_k, \varphi_{i+1})u_k(x) \\ &\qquad\qquad (i = 1, 2, \ldots) \end{aligned}\right\} \qquad (15.2\text{-}10)$$

See also Sec. 21.7-1 for examples.

15.2-6. Approximations and Series Expansions in Terms of Orthonormal Functions (see also Secs. 4.11-2*c*, 4.11-4*b*, 15.4-12, 20.6-2, 20.6-3, 20.9-9, and 21.8-12). *Given a quadratically integrable function $f(x)$ and an orthonormal set $u_1(x)$, $u_2(x)$, . . . , an approximation to $f(x)$ of the form*

$$s_n(x) = a_1u_1(x) + a_2u_2(x) + \cdots + a_nu_n(x) \qquad (n = 1, 2, \ldots) \qquad (15.2\text{-}11)$$

yields the least mean-square error $\int_V |s_n(x) - f(x)|^2\,dx$ if $a_k = (u_k, f)$. Note that the choice of the coefficients a_k is independent of n. This property, together with the relative simplicity of the formulas listed in Sec. 15.2-4*a*, establishes the great importance of series expansions

$$f(x) \underset{\text{in mean}}{=} f_1u_1(x) + f_2u_2(x) + \cdots \qquad f_k = (u_k, f) \qquad (k = 1, 2, \ldots)$$

in terms of suitable normalized orthogonal functions $u_1(x)$, $u_2(x)$, ...

15.2-7. Linear Operations on Functions. Sections 8.2-1, 8.6-1 to 8.6-4, 15.3-1, 15.4-1, and 20.4-2 introduce various *linear operations* (Sec. 14.3-1) associating a function

$$\varphi(x) = \mathbf{L}\Phi(\xi) \tag{15.2-12}$$

with a given function $\Phi(\xi)$ so that

$$\left.\begin{aligned} \mathbf{L}[\Phi_1(\xi) + \Phi_2(\xi)] &= \mathbf{L}\Phi_1(\xi) + \mathbf{L}\Phi_2(\xi) \\ \mathbf{L}[\alpha\Phi(\xi)] &= \alpha\mathbf{L}\Phi(\xi) \end{aligned}\right\} \tag{15.2-13}$$

$\Phi(\xi)$ and $\varphi(x)$ may or may not be defined on the same domain. As in Sec. 14.1-3, there exist two distinct interpretations of the *functional transformation* (12):

1. Equation (12) describes an operation on the object function $\Phi(\xi)$ (*"alibi" point of view*).
2. $\Phi(\xi)$ and $\varphi(x)$ *represent* the same abstract vector (in the same sense as two matrices, Sec. 14.6-1*b*), and Eq. (12) describes a *change of representation* (*"alias" point of view,* used especially in quantum mechanics; see also Sec. 8.1-1).

15.3. LINEAR INTEGRAL TRANSFORMATIONS AND LINEAR INTEGRAL EQUATIONS

15.3-1. Linear Integral Transformations.* **(a)** Sections 15.3-1 to 15.3-10 deal with **linear integral transformations**

$$\mathbf{K}f(\xi) \equiv \int_V K(x, \xi)f(\xi)\,d\xi = F(x) \tag{15.3-1}$$

relating a pair of functions $f(\xi)$ and $F(x)$. The function $K(x,\xi)$ is called the **kernel** of the linear integral transformation. All integrals will be assumed to exist in the sense of Lebesgue (Sec. 4.6-15).

The domains of the object function $f(\xi)$ and the result function $F(x)$ in Eq. (1) are not necessarily identical (note, for example, the *Laplace transformation,* Sec. 8.2-1). **Throughout Secs. 15.3-1b to 15.3-10 it is, however, assumed that x and ξ range over the same interval or region V.** The "symbolic" integral transformation

$$\int_V \delta(x, \xi)f(\xi)\,d\xi \equiv f(x)$$

represents the identity transformation (see also Secs. 15.5-1 and 21.9-2).

Linear integral transformations (1) may be interpreted from either the "alibi" or the "alias" point of view (Sec. 15.2-7); in the former case each kernel $K(x, \xi)$ represents a linear operator in the same sense as a matrix (Sec. 14.5-2; see also Sec. **15.3-1***c*).

(b) For a given kernel $K(x, \xi)$, $\tilde{K}(x, \xi) \equiv K(\xi, x)$ is called the **transposed kernel,** and $K\dagger(x, \xi) \equiv K^*(\xi, x)$ is the **adjoint (hermitian conjugate) kernel** (see also Sec. 14.4-3). A given kernel $K(x, \xi)$ is

* Refer again to Sec. 15.1-2 for notation. See also Sec. 15.4-1*b*.

Symmetric if and only if $K(\xi, x) \equiv K(x, \xi)$
Hermitian if and only if $K^*(\xi, x) \equiv K(x, \xi)$
Normalizable if and only if $\int_V \int_V |K(x, \xi)|^2 \, d\xi \, dx$ exists and is different from zero
Continuous in the mean over V if and only if

$$\lim_{\Delta x \to 0} \int_V |K(x + \Delta x, \xi) - K(x, \xi)|^2 \, d\xi = 0$$

(see also Sec. 12.5-1*c*)
Separable (degenerate) if and only if $K(x, \xi)$ can be expressed as a finite sum $K(x, \xi) = \sum_{i=1}^{m} f_i(x)h_i(\xi)$

A normalizable kernel represents a bounded operator (Sec. 14.4-1), so that $F(x)$ is normalizable if $f(\xi)$ is normalizable (Sec. 15.2-1*b*). Separable kernels represent operators of finite rank (Sec. 14.3-2). *If $K(x, \xi)$ is hermitian, normalizable, and continuous in the mean over V, then $F(x)$ is continuous in V whenever $f(\xi)$ is quadratically integrable over V.*

(c) Matrix Representation. Product of Two Integral Transformations. Given a normalizable kernel $K(x, \xi)$ and an orthonormal basis (Sec. 15.2-4) $u_1(x)$, $u_2(x)$, . . . in the space of functions $f(x)$, let

$$\left.\begin{aligned} f(\xi) &\underset{\text{in mean}}{=} \sum_{k=1}^{\infty} f_k u_k(\xi) \qquad F(x) \underset{\text{in mean}}{=} \sum_{k=1}^{\infty} F_k u_k(x) \\ K(x, \xi) &\underset{\text{in mean}}{=} \sum_{i=1}^{\infty} \sum_{k=1}^{\infty} k_{ik} u_i(x) u_k^*(\xi) \\ &\left[k_{ik} = \int_V \int_V u_i^*(x) K(x, \xi) u_k(\xi) \, dx \, d\xi\right] \end{aligned}\right\} \qquad (15.3\text{-}2)$$

Then Eq. (1) is equivalent to the matrix equation $\{F_i\} = [k_{ik}]\{f_k\}$. Note that the matrix product $[m_{ik}][k_{ik}]$ corresponds to the kernel $\int_V M(x, \eta)K(\eta, \xi) \, d\eta$ representing the product of two successive integral transformations (1) whose kernels $K(x, \xi)$, $M(x, \xi)$ correspond to $[k_{ik}]$, $[m_{ik}]$.

If $K(x, \xi)$ is a separable kernel, then it is possible to choose the $u_k(x)$ so that the matrix $[k_{ik}]$ is finite.

15.3-2. Linear Integral Equations: Survey. An **integral equation** is a functional equation (Sec. 9.1-2) involving integral transforms of the unknown function $\Phi(x)$ [if the functional equation also involves a derivative of $\Phi(x)$ one speaks of an *integro-differential equation*]. A given integral equation is **homogeneous** if and only if every multiple $\alpha\Phi(x)$ of any solution $\Phi(x)$ is a solution.

Sections 15.3-2 to 15.3-10 deal with *linear integral equations* of the general form

$$\beta(x)\Phi(x) - \lambda \int_V K(x, \xi)\Phi(\xi) \, d\xi = F(x) \qquad (15.3\text{-}3)$$

where $K(x, \xi)$, $\beta(x)$, and $F(x)$ are given functions; the integration domain V may be given (*Fredholm-type integral equations*) or variable (e.g., *Volterra-type integral equations*, Sec. 15.3-10). Three important types of problems arise:

1. **A linear integral equation of the first kind** [$\beta(x) \equiv 0$, $\lambda = 1$, Sec. 15.3-9] requires one to find an unknown function $\Phi(x)$ with the given integral transform $F(x)$. The corresponding operator equation $\mathbf{K}\Phi = F$ is analogous to a matrix equation $[k_{ik}]\{\Phi_k\} = \{F_i\}$.
2. **A homogeneous linear integral equation of the second kind** [$F(x) \equiv 0$, $\beta(x) \equiv 1$, λ unknown; Secs. 15.3-3 to 15.3-6] represents an *eigenvalue problem*. The corresponding operator equation $\lambda\mathbf{K}\Phi = \Phi$ is analogous to a matrix equation $\lambda[k_{ik}]\{\Phi_k\} = \{\Phi_i\}$.
3. **A nonhomogeneous linear integral equation of the second kind** [$\beta(x) \equiv 1$, λ given; Sec. 15.3-7] may be written as $\Phi - \lambda\mathbf{K}\Phi = F$ and represents a problem of the type discussed in Sec. 14.8-10.

If $\beta(x)$ is a real positive function throughout V, one can reduce the general linear integral equation (3) to a linear integral equation of the second kind with the aid of the transformation

$$\Phi(x) = \frac{\bar{\Phi}(x)}{\sqrt{\beta(x)}} \qquad F(x) = \bar{F}(x)\sqrt{\beta(x)} \qquad K(x, \xi) = \bar{K}(x, \xi)\sqrt{\beta(x)\beta(\xi)} \qquad (15.3\text{-}4)$$

15.3-3. Homogeneous Fredholm Integral Equations of the Second Kind: Eigenfunctions and Eigenvalues (see also Secs. 14.8-3 and 15.4-5). **(a)** A function $\psi = \psi(x)$ which is not identically zero in V and satisfies the *homogeneous Fredholm integral equation of the second kind*

$$\lambda\mathbf{K}\psi(\xi) \equiv \lambda \int_V K(x, \xi)\psi(\xi)\, d\xi = \psi(x) \qquad (15.3\text{-}5)$$

for a suitably determined value of the parameter λ is called an **eigenfunction (characteristic function)** of the linear integral equation (5) or of the kernel $K(x, \xi)$. The corresponding value of λ is an **eigenvalue** of the integral equation.

If $\psi_1(x)$ and $\psi_2(x)$ are eigenfunctions corresponding to the same eigenvalue λ, the same is true for every linear combination $\alpha_1\psi_1(x) + \alpha_2\psi_2(x)$. The number m of linearly independent eigenfunctions corresponding to a given eigenvalue λ is its **degree of degeneracy**; if $K(x, \xi)$ is a normalizable kernel, each m is finite. *Eigenfunctions associated with different eigenvalues of a linear integral equation* (5) *are linearly independent. The total number of linearly independent eigenfunctions is finite if and only if $K(x, \xi)$ is separable.*

If the linear operator $\mathbf{K}$ described by the linear integral transformation (5) has a unique inverse $\mathbf{L} \equiv \mathbf{K}^{-1}$, then the $\psi(x)$ and λ are eigenfunctions and eigenvalues of the nonsingular linear operator **L** (see also Sec. 15.4-5), and all the eigenvalues λ are different from zero. In many applications **L** is a differential operator, and $K(x, \xi)$ is a Green's function (Sec. 15.5-1).

(b) Eigenfunctions and Eigenvalues for Hermitian Kernels (see also Secs. 14.8-4 and 15.4-6). *If $K(x, \xi)$ is a hermitian kernel, then*

1. *All eigenvalues of the integral equation* (5) *are real.*
2. *Eigenfunctions corresponding to different eigenvalues are mutually orthogonal* (*with weighting function* 1, *Sec.* 15.2-3).

If the kernel $K(x, \xi)$ is normalizable as well as hermitian, then

3. *All eigenvalues are different from zero.*
4. *There exists at least one eigenvalue different from zero; the eigenvalues constitute a discrete set comprising at most a finite number of eigenvalues in any finite interval, and every eigenvalue has a finite degree of degeneracy.*

15.3-4. Expansion Theorems. (a) Expansion Theorems for Hermitian Kernels. Given the definitions of inner products (Sec. 15.2-1) and convergence in mean (Sec. 15.2-2) with $\gamma(x) \equiv 1$, *every normalizable hermitian kernel $K(x, \xi)$ admits an orthonormal set of eigenfunctions $\psi_1(x)$, $\psi_2(x)$, . . . such that, for every given function $F(x)$ representable as the integral transform* (1) *of a function $f(\xi)$,*

$$\left.\begin{aligned} F(x) &\underset{\text{in mean}}{=} \sum_{k=1}^{\infty} F_k \psi_k(x) \qquad (x \text{ in } V) \\ \text{with}\qquad F_k &= (\psi_k, F) = \int_V \psi_k^*(\xi) F(\xi)\, d\xi \qquad (k = 1, 2, \ldots) \end{aligned}\right\} \qquad (15.3\text{-}6)$$

The series (6) *converges absolutely and uniformly to $F(x)$ for all x in V* (1) *if $f(\xi)$ is piecewise continuous in V, or* (2) *if $f(\xi)$ is quadratically integrable and $K(x, \xi)$ is continuous in the mean.* A hermitian kernel $K(x, \xi)$ is **complete** if and only if *every* quadratically integrable function $F(x)$ can be represented in the form (1) or (6), so that the eigenfunctions $\psi_k(x)$ constitute a complete orthonormal set (Sec. 15.2-4); this is true, for example, if $K(x, \xi)$ is definite (Sec. 15.3-6).

Every normalizable hermitian kernel $K(x, \xi)$ can be expanded in the form

$$K(x, \xi) \underset{\text{in mean}}{=} \sum_{k=1}^{\infty} \frac{1}{\lambda_k} \psi_k^*(\xi)\psi_k(x) \qquad (x, \xi \text{ in } V) \qquad (15.3\text{-}7)$$

The series converges uniformly to $K(x, \xi)$ for all x, ξ in V (1) *if $K(x, \xi)$ is continuous in the mean, or* (2) *if V is a bounded interval or region such that $K(x, \xi)$ is continuous for all*

x, ξ in V, and either the positive or the negative eigenvalues of Eq. (5) *are finite in number* (*Mercer's Theorem*).

(b) Auxiliary Kernels and Expansion Theorems for Nonhermitian Kernels. For every normalizable kernel $K(x, \xi)$, the *auxiliary hermitian kernels* $\int_V K\dagger(x, \eta)K(\eta, \xi)\, d\eta$ and $\int_V K(x, \eta)K\dagger(\eta, \xi)\, d\eta$ yield identical real eigenvalues μ_k^2 and the respective orthonormal sets of eigenfunctions $v_k(x)$ and $w_k(x)$. Note

$$\mu_k \int_V K(x, \xi)v_k(\xi)\, d\xi = w_k(x) \qquad \mu_k \int_V K\dagger(x, \xi)w_k(\xi)\, d\xi = v_k(x) \qquad (15.3\text{-}8)$$

where $\mu_k > 0$.

Given any normalizable kernel $K(x, \xi)$, every function $F(x)$ expressible as an integral transform (1) *with kernel K or $K\dagger$ can be expanded in the respective form*

$$F(x) \underset{\text{in mean}}{=} \sum_{k=1}^{\infty} b_k w_k(x) \qquad \text{with } b_k = (w_k, F) \qquad (x \text{ in } V) \qquad (15.3\text{-}9a)$$

or

$$F(x) \underset{\text{in mean}}{=} \sum_{k=1}^{\infty} a_k v_k(x) \qquad \text{with } a_k = (v_k, F) \qquad (x \text{ in } V) \qquad (15.3\text{-}9b)$$

Each series converges absolutely and uniformly to $F(x)$ for all x in V (1) *if $f(\xi)$ is piecewise continuous in V, or* (2) *if $f(\xi)$ is quadratically integrable, and $K(x, \xi)$ is continuous in the mean;* the weighting function is 1 (Sec. 15.2-1).

For every normalizable kernel $K(x, \xi)$

$$K(x, \xi) \underset{\text{in mean}}{=} \sum_{k=1}^{\infty} \frac{1}{\mu_k} v_k^*(\xi)w_k(x) \qquad (x, \xi \text{ in } V) \qquad (15.3\text{-}10)$$

15.3-5. Iterated Kernels. The **iterated kernels** $K_p(x, \xi)$ defined by

$$K_1(x, \xi) \equiv K(x, \xi) \qquad K_{p+1}(x, \xi) \equiv \int_V K_p(x, \eta)K(\eta, \xi)\, d\eta \qquad (p = 2, 3, \ldots) \qquad (15.3\text{-}11)$$

represent the powers $\mathbf{K}^p$ of the linear operator $\mathbf{K}$ of Eq. (1). *The linear integral equation*

$$\lambda \mathbf{K}^p \psi(\xi) \equiv \lambda \int_V K_p(x, \xi)\psi(\xi)\, d\xi = \psi(x) \qquad (15.3\text{-}12)$$

has the same eigenfunctions $\psi(x)$ as Eq. (5), *with corresponding eigenvalues* λ^p (see also Sec. 14.8-3*d*). Conversely, *every solution $\psi(x)$ of Eq.* (12) *solves Eq.* (5).

If $K(x, \xi)$ is hermitian and normalizable, then

$$K_p(x, \xi) = \sum_{k=1}^{\infty} \frac{1}{\lambda_k^p} \psi_k^*(\xi)\psi_k(x) \qquad (p = 1, 2, \ldots ; x, \xi \text{ in } V) \qquad (15.3\text{-}13)$$

and this series converges absolutely and uniformly in V.

15.3-6. Hermitian Integral Forms. The Eigenvalue Problem as a Variation Problem (see also Secs. 11.1-1, 13.5-2, 13.5-3, 14.8-8a, and 15.4-7). **(a)** Given a normalizable hermitian kernel $K(x, \xi)$, the (necessarily real) inner product

$$(\Phi, \mathbf{K}\Phi) = \int_V \int_V \Phi^*(x)K(x, \xi)\Phi(\xi)\, dx\, d\xi \tag{15.3-14}$$

is called a **hermitian integral form** or, in particular, a **real symmetric quadratic integral form** if $K(x, \xi)$ is real and symmetric.* The hermitian integral form (14), and also the hermitian kernel $K(x, \xi)$, is **positive definite, negative definite, nonnegative,** or **nonpositive** if and only if the expression (14) is, respectively, positive, negative, nonnegative, or nonpositive for every function $\Phi(x)$ not identically zero in V and such that the integral exists. *The integral form* (14) *is positive definite or negative definite if and only if all eigenvalues of* $K(x, \xi)$ *are, respectively, positive or negative.*

(b) *The problem of finding the eigenfunctions* $\psi(x)$ *and the eigenvalues* λ *for a normalizable hermitian kernel may be expressed as a stationary-value problem in the manner of Sec.* 14.8-8a, *e.g.*,

Find a quadratically integrable function $\Phi(x)$ such that the hermitian integral form (14) has a stationary value subject to the constraint

$$(\Phi, \Phi) = \int_V |\Phi(\xi)|^2\, d\xi = 1$$

$\Phi = \psi_k(x)$ yields the stationary value λ_k.

All the other theorems of Sec. 14.8-8a *apply;* it is only necessary to remember that the operator $\mathbf{K}$ represented by the given kernel $K(x, \xi)$ has the eigenfunctions $\psi_k(x)$ and the eigenvalues $1/\lambda_k$. It is, then, often possible to solve an integral equation of the form (5) exactly or approximately by the methods of the calculus of variations.

15.3-7. Nonhomogeneous Fredholm Equations of the Second Kind (see also Sec. 14.8-10). **(a) Existence and Uniqueness of Solutions.** Fredholm's linear integral equation of the second kind

$$\Phi(x) - \lambda \int_V K(x, \xi)\Phi(\xi)\, d\xi = F(x) \tag{15.3-16}$$

has the following "alternative property":

1. *If the given parameter* λ *is not an eigenvalue of* $K(x, \xi)$, *then Eq.* (16) *has at most one unique solution* $\Phi(x)$.
2. *If* λ *equals an eigenvalue* λ_1 *of Eq.* (5), *then the "adjoint" homogeneous integral equation*

$$\lambda_1^* \int_V K^*(\xi, x)\chi(\xi)\, d\xi = \chi(x) \tag{15.3-17}$$

has a solution $\chi(x)$ *not identically zero in* V, *and the given integral equation* (16) *has a solution* $\Phi(x)$ *only if* $F(x)$ *is orthogonal (with*

* Note that, in terms of the matrix elements defined in Sec. 15.3-1c,

$$(\Phi, \mathbf{K}\Phi) = \sum_{i=1}^{\infty} \sum_{k=1}^{\infty} k_{ik}\Phi_i^*\Phi_k \tag{15.3-15}$$

weighting function 1) *to every solution* $\chi(x)$ *of Eq.* (17). Note that Eqs. (17) and (5) are identical for hermitian kernels.

Solutions subject to the above conditions actually exist, in particular, whenever $K(x, \xi)$ is piecewise continuous and normalizable and $F(x)$ is continuous and quadratically integrable over V. If a solution $\Phi(x)$ exists in case 2, then Eq. (16) has infinitely many solutions, since every sum of a particular solution and a linear combination of eigenfunctions $\psi(x)$ corresponding to λ_1 is a solution; in particular, there exists a unique solution $\Phi(x)$ orthogonal to all these eigenfunctions.

(b) Reduction to an Integral Equation with Hermitian Kernel. If $K(x, \xi)$ is normalizable, every solution $\Phi(x)$ of Eq. (16) is a solution of the integral equation

$$\Phi(x) - |\lambda| \int_V H(x, \xi)\Phi(\xi)\,d\xi = F(x) - \lambda^* \int_V K^*(\xi, x)F(\xi)\,d\xi \qquad (15.3\text{-}18a)$$

where $H(x, \xi)$ is the *hermitian* kernel defined by

$$H(x, \xi) \equiv e^{i \arg \lambda}K + e^{-i \arg \lambda}K^\dagger - |\lambda| \int_V K^*(\eta, x)K(\eta, \xi)\,d\eta \qquad (15.3\text{-}18b)$$

It is thus sufficient to study solution methods for hermitian kernels.

NOTE: $H(x,\xi)$ and $K(x,\xi)$ have identical eigenfunctions, with corresponding eigenvalues $|\lambda_k|$ and λ_k.

(c) The Resolvent Kernel. A solution $\Phi(x)$ of the linear integral equation (16) is conveniently written in the form

$$\Phi(x) = F(x) + \lambda \int_V \Gamma(x, \xi; \lambda)F(\xi)\,d\xi \qquad (15.3\text{-}19)$$

The function $\Gamma(x, \xi; \lambda)$ is called the **resolvent kernel** (sometimes known as the *reciprocal kernel*) for the integral equation (16); Eqs. (16) and (19) represent mutually inverse linear transformations.

Whenever the resolvent kernel $\Gamma(x, \xi; \lambda)$ exists, it satisfies the integral equations

$$\left.\begin{aligned} \Gamma(x, \xi; \lambda) - \lambda \int_V K(x, t)\Gamma(t, \xi; \lambda)\,dt &= K(x, \xi) \\ \Gamma(x, \xi; \lambda) - \lambda \int_V K(t, \xi)\Gamma(x, t; \lambda)\,dt &= K(x, \xi) \\ \Gamma(x, \xi; \lambda) - \Gamma(x, \xi; \lambda') &= (\lambda - \lambda') \int_V \Gamma(x, t; \lambda)\Gamma(t, \xi; \lambda')\,dt \end{aligned}\right\} \qquad (15.3\text{-}20)$$

for arbitrary λ, λ' which are not eigenvalues of $K(x, \xi)$.

15.3-8. Solution of the Linear Integral Equation (16) (see also Sec. 20.8-5 for numerical methods). **(a) Solution by Successive Approximations. Neumann Series.** Starting with a trial solution

$$\Phi^{[0]}(x) = F(x)$$

one may attempt to compute successive approximations

$$\Phi^{[j+1]}(x) = F(x) + \lambda \int_V K(x, \xi)\Phi^{[j]}(\xi)\, d\xi \qquad (j = 0, 1, 2, \ldots) \tag{15.3-21}$$

to the desired solution $\Phi(x)$ of Eq. (16). The functions (21) may be regarded as partial sums of the infinite series

$$F(x) + \lambda \int_V K(x, \xi)F(\xi)\, d\xi + \lambda^2 \int_V K_2(x, \xi)F(\xi)\, d\xi + \cdots$$
$$\equiv (1 + \lambda\mathbf{K} + \lambda^2\mathbf{K}^2 + \cdots)F(x) \quad \text{(NEUMANN SERIES)} \tag{15.3-22}$$

If $K(x, \xi)$ is normalizable, and $F(x)$ is quadratically integrable over V, then there exists a real number $r_C \geq +\left[\int_V \int_V |K(x, \xi)|^2\, d\xi\, dx\right]^{-1/2}$ such that the Neumann series (22) converges in mean to a solution (19) for $|\lambda| < r_C$.

If, in addition, $\int_V |K(x, \xi)|^2\, d\xi$ and $\int_V |K(\xi, x)|^2\, d\xi$ are uniformly bounded in V, then the power series (22) and the corresponding power series for the resolvent kernel

$$\Gamma(x, \xi; \lambda) = K(x, \xi) + \lambda K_2(x, \xi) + \lambda^2 K_3(x, \xi) + \cdots \tag{15.3-23}$$

actually converge uniformly to the indicated limits for x, ξ in V and $|\lambda| < r_C$. The function (23) is then an analytic function of λ for $|\lambda| < r_C$ and can be continued analytically (Sec. 7.8-1) to yield a resolvent kernel for other suitable values of λ. The series (23), as well as the series (22), is known as a Neumann series.

If the normalizable kernel $K(x, \xi)$ is *hermitian*, then the radius of convergence (or mean convergence) r_C is given by $r_C = |\lambda_1|$, where λ_1 is the eigenvalue of $K(x, \xi)$ with the smallest absolute value.

(b) The Schmidt-Hilbert Formula for the Resolvent Kernel (see also Sec. 14.8-10). For every normalizable and hermitian kernel $K(x, \xi)$ the solution of the linear integral equation (16) is given by Eq. (19) with the resolvent kernel

$$\Gamma(x, \xi; \lambda) = K(x, \xi) + \lambda \sum_{k=1}^{\infty} \frac{\psi_k^*(\xi)\psi_k(x)}{\lambda_k(\lambda_k - \lambda)} \qquad (\lambda \neq \lambda_k,\ k = 1, 2, \ldots)$$
$$\text{(SCHMIDT-HILBERT FORMULA)} \tag{15.3-24}$$

where the $\psi_k(x)$ are orthonormal eigenfunctions* of $K(x, \xi)$. The series converges uniformly for x, ξ in V and $\lambda \neq \lambda_k$.

* If λ_i is an m-fold degenerate eigenvalue (Sec. 15.3-3*a*), one writes

$$\lambda_{i+1} = \cdots = \lambda_{i+m-1} = \lambda_i$$

so that the series (24) contains m terms involving the orthonormal eigenfunctions $\psi_i(x), \psi_{i+1}(x), \ldots, \psi_{i+m-1}(x)$ corresponding to λ_i.

$\Gamma(x, \xi; \lambda)$ is a meromorphic function of λ in the finite part of the λ plane; the residues of $\Gamma(x, \xi; \lambda)$ at its poles $\lambda = \lambda_k$ are simply related to the eigenfunctions $\psi_k(x)$ (Secs. 7.6-8 and 7.7-1).

SCHMIDT-HILBERT SOLUTION IF λ EQUALS AN EIGENVALUE λ_i. If the given parameter λ equals an eigenvalue λ_i of the given kernel $K(x, \xi)$, omit the term or terms involving λ_i in the sum (24) (which is now no longer a resolvent kernel) and add an arbitrary eigenfunction $\psi(x)$ associated with λ_i to the right side of Eq. (19). The resulting function $\Phi(x)$ will solve the given integral equation (16), subject to the conditions of case 2 in Sec. 15.3-7*a*.

(c) Fredholm's Formulas for the Resolvent Kernel. If $K(x, \xi)$ is normalizable, the resolvent kernel $\Gamma(x, \xi; \lambda)$ can be expressed as the ratio of two integral functions of λ (see also Sec. 7.6-7), viz.,

$$\Gamma(x, \xi; \lambda) = \frac{D(x, \xi; \lambda)}{D(\lambda)} \tag{15.3-25}$$

where

$$\left.\begin{array}{ll} D(\lambda) = \displaystyle\sum_{k=0}^{\infty} \frac{(-1)^k}{k!} C_k \lambda^k & D(x, \xi; \lambda) = \displaystyle\sum_{k=0}^{\infty} \frac{(-1)^k}{k!} D_k(x, \xi)\lambda^k \\ \text{with} & \\ C_0 = 1 & D_0(x, \xi) = K(x, \xi) \\ \text{and, for } k = 1, 2, \ldots, & \\ C_k = \displaystyle\int_V D_{k-1}(\xi, \xi)\, d\xi & \\ \multicolumn{2}{l}{D_k(x, \xi) = C_k K(x, \xi) - k \displaystyle\int_V K(x, \eta) D_{k-1}(\eta, \xi)\, d\eta} \end{array}\right\} \tag{15.3-26}$$

Both power series converge for all finite λ; the power series for $D(x, \xi; \lambda)$ converges uniformly in V. The poles of $\Gamma(x, \xi; \lambda)$ coincide with the zeros of $D(\lambda)$.

Note the analogy between the functions $D(x, \xi; \lambda)$, $D(\lambda)$ and the determinants used to find the solution x_k of the analogous finite-dimensional problem $\sum_{k=1}^{n} (a_{ik} - \lambda\delta_k^i)x_k = b_i$ $(i = 1, 2, \ldots, n)$ by Cramer's rule (1.9-4).

(d) Singular Kernels. If the given kernel $K(x, \xi)$ becomes unbounded in a neighborhood of $x = \xi$ while the iterated kernel $K_2(x, \xi)$ remains bounded, find the solution $\Phi(x)$ of Eq. (16) by solving the integral equation

$$\Phi(x) - \lambda^2 \int_V K_2(x, \xi)\Phi(\xi)\, d\xi = F(x) + \lambda \int_V K(x, \xi)F(\xi)\, d\xi \tag{15.3-27}$$

Equation (27) is obtained by substitution of

$$\Phi(x) = F(x) + \lambda \int_V K(x, \xi)\Phi(\xi)\, d\xi$$

on the left of Eq. (16). This procedure may be repeated if $K_3(x, \xi)$, $K_4(x, \xi)$, . . . is the first iterated kernel which remains bounded.

15.3-9. Solution of Fredholm's Linear Integral Equation of the First Kind (see also Secs. 14.8-10*b*, 15.4-12, and 15.5-1). If the linear integral equation

$$\int_V K(x, \xi)\Phi(\xi)\, d\xi = F(x) \tag{15.3-28}$$

with a given normalizable kernel $K(x, \xi)$ has a solution $\Phi(x)$, the expansion theorem of Sec. 15.3-4*b* yields

$$\Phi(x) \underset{\text{in mean}}{=} \sum_{k=1}^{\infty} \mu_k\, (w_k, F) v_k(x) + \varphi(x) \tag{15.3-29}$$

where the $v_k(x)$, $w_k(x)$, and μ_k are defined as in Sec. 15.3-4*b*, and $\varphi(x)$ is an arbitrary function orthogonal to all $v_k(x)$.

If the $v_k(x)$ *constitute a complete orthonormal set in* $L_2(V)$ *(Sec.* 15.2-4), *then* $\varphi(x) = 0$ *for almost all* x *in* V, *and the solution* (29) *is unique almost everywhere in* V. If the $w_k(x)$ *as well as the* $v_k(x)$ *form a complete orthonormal set in* $L_2(V)$, *then the integral transformation* (28) *is nonsingular and has the unique inverse*

$$\left.\begin{aligned} \int_V K_{-1}(x, \xi)F(\xi)\, d\xi &= \Phi(x) \\ \text{with} \qquad K_{-1}(x, \xi) &= \sum_{k=1}^{\infty} \mu_k w_k^*(\xi) v_k(x) \end{aligned}\right\} \tag{15.3-30}$$

$K_{-1}(x, \xi)$ is called the **reciprocal kernel** associated with $K(x, \xi)$ and represents the linear operator $\mathbf{K}^{-1}$.

For *hermitian* kernels $K(x, \xi)$ one has $v_k(x) \equiv w_k(x)$, $\mu_k = |\lambda_k|$.

15.3-10. Volterra-type Integral Equations. **(a)** Let x and ξ be *one-dimensional* real variables. Then

1. *Volterra's integral equation of the second kind*

$$\Phi(x) - \lambda \int_0^x H(x, \xi)\Phi(\xi)\, d\xi = F(x) \tag{15.3-31}$$

reduces to a Fredholm-type integral equation (16) over the interval $V \equiv (0, \infty)$ if one introduces the new kernel

$$K(x, \xi) \equiv H(x, \xi) U_+(x - \xi) = \begin{Bmatrix} H(x, \xi) & \text{if } \xi < x \\ 0 & \text{if } \xi \geq x \end{Bmatrix} \tag{15.3-32}$$

2. *Volterra's integral equation of the first kind*

$$\int_0^x \bar{H}(x, \xi)\Phi(\xi)\, d\xi = \bar{F}(x) \tag{15.3-33}$$

reduces by differentiation to the form (31) with

$$H(x, \xi) \equiv \frac{\dfrac{\partial \bar{H}(x, \xi)}{\partial x}}{\bar{H}(x, x)} \qquad F(x) \equiv \frac{\dfrac{d\bar{F}(x)}{dx}}{\bar{H}(x, x)} \tag{15.3-34}$$

(b) The following example illustrates a method of dealing with a class of Volterra-type integral equations having unbounded kernels. To solve

$$\int_a^x \frac{\Phi(\xi)}{(x - \xi)^\alpha}\, d\xi = F(x) \qquad (0 < \alpha < 1) \tag{15.3-35}$$

multiply both sides by $(y - x)^{\alpha-1}$ and integrate with respect to x from a to y; the resulting integral equation has a bounded kernel. It follows that

$$\Phi(x) = \frac{\sin \pi\alpha}{\alpha}\left[\frac{F(a)}{(x - a)^{1-\alpha}} + \int_a^x \frac{F'(\xi)\, d\xi}{(x - \xi)^{1-\alpha}}\right] \tag{15.3-36}$$

In the special case $\alpha = 1/2$, Eq. (35) is known as *Abel's integral equation.*

15.4. LINEAR BOUNDARY-VALUE PROBLEMS AND EIGENVALUE PROBLEMS INVOLVING DIFFERENTIAL EQUATIONS

15.4-1. Linear Boundary-value Problems: Problem Statement and Notation (see also Secs. 9.3-1, 10.3-4, 15.1-2, and 15.2-7). **(a)** A linear homogeneous function of a function $\Phi(x)$ and its derivatives will be written as the product $\mathbf{L}\Phi(x)$ of $\Phi(x)$ and a **linear differential operator $\mathbf{L}$,** such as d/dx or $-\nabla^2 - q(x^1, x^2, x^3)$. One desires to find unknown functions $\Phi(x)$ which satisfy a linear differential equation

$$\boxed{\mathbf{L}\Phi(x) = f(x) \qquad (x \text{ in } V)} \tag{15.4-1a}$$

throughout a given open interval or region V of points (x), subject to a set of N linear boundary conditions

$$\boxed{\mathbf{B}_i\Phi(x) = b_i(x) \qquad (i = 1, 2, \ldots, N;\ x \text{ in } S)} \tag{15.4-1b}$$

to be satisfied on the boundary S of V; each $\mathbf{B}_i\Phi(x)$ is to be a linear homogeneous function of $\Phi(x)$ and its derivatives.

(b) Notation. Volume Integrals and Inner Products (see also Secs. 4.6-12, 15.1-2, 15.2-1, and 16.10-10). In the *one-dimensional* case, x is a real variable, Eq. (1*a*) is an *ordinary differential equation,* and V is a bounded or unbounded open *interval* whose *end points* $x = a$, $x = b$ constitute the boundary S.

In the *n-dimensional* case, x stands for a "point" $(x_1, x_2, \ldots, x_n)$ in the n-dimensional space, and Eq. (1*a*) is a *partial differential equation.* One assumes the possibility of introducing a *volume element*

$$dV(x) \equiv \sqrt{|g(x^1, x^2, \ldots, x^n)|}\, dx^1\, dx^2 \cdots dx^n \equiv \sqrt{|g(x)|}\, dx$$

in the manner of Secs. 6.2-3*b* and 16.10-10; $\sqrt{|g(x)|}$ is to be real and positive throughout V. Then $\int_V \varphi(\xi)\, dV(\xi)$ is an n-dimensional *volume* *integral* over V, and one can define the *inner product* of two functions $u(x)$, $v(x)$ on V (Sec. 15.2-1) as

$$(u, v) \equiv \int_V \gamma_0(\xi)u^*(\xi)v(\xi)\, d\xi \equiv \int_V u^*(\xi)v(\xi)\, dV(\xi) \tag{15.4-2}$$

Note that $\gamma_0(x) \equiv \sqrt{|g(x)|}$ depends on the coordinate system.

One may similarly assume the existence of an appropriate *surface element* $dA(x)$ (x in S) on the boundary hypersurface S; $\int_S \varphi(\xi)dA(\xi)$ is, then, a *surface integral* in the n-dimensional space (see also Secs. 4.6-12 and 6.4-3*b*).

In particular, for $n = 3$, V is a bounded or unbounded open *region of space* with *boundary surface* S; *S is* to be a *regular* surface (Sec. 3.1-14). For $n = 2$, V is usually a *plane region* with (regular) *boundary curve* S.

15.4-2. Complementary Homogeneous Differential Equation and Boundary Conditions for a Linear Boundary-value Problem. Superposition Theorems (see also Secs. 9.3-1, 10.4-2, 14.3-1, and 15.2-7). With each linear boundary-value problem (1), one can associate a unique homogeneous **complementary** or **reduced differential equation**

$$\mathbf{L}\Phi(x) = 0 \qquad (x \text{ in } V) \tag{15.4-3a}$$

and a unique ordered set of homogeneous "complementary boundary conditions"

$$\mathbf{B}_i\Phi(x) = 0 \qquad (i = 1, 2, \ldots, N; x \text{ in } S) \tag{15.4-3b}$$

The following important theorems may be applied in turn to relate solutions $\Phi(x)$ of a given linear boundary-value problem (1) to functions satisfying simpler relations of the form (3).

1. *The solution of every linear boundary-value problem* (1) *can be reduced to that of a linear boundary-value problem involving the same operator* $\mathbf{L}$ *and the homogeneous boundary conditions* (3*b*) (see also Sec. 15.5-4).*
2. *The most general function satisfying a linear differential equation* (1*a*) *can be written as the sum of a particular solution of Eq.* (1*a*) *and the most general solution of the complementary equation* (3*a*).

* Write $\Phi(x) = \bar{\Phi}(x) + v(x)$, where $v(x)$ is a conveniently chosen function satisfying the given boundary conditions (1*b*). Then $\bar{\Phi}(x)$ is the solution of the linear boundary-value problem

$$\left.\begin{aligned} \mathbf{L}\bar{\Phi}(x) &= f(x) - \mathbf{L}v(x) \qquad (x \text{ in } V) \\ \mathbf{B}_i\bar{\Phi}(x) &= 0 \qquad (i = 1, 2, \ldots, N; x \text{ in } S) \end{aligned}\right\} \tag{15.4-4}$$

$\mathbf{L}v(x)$ is determined by the choice of $v(x)$ but may contain δ-function terms; it is often possible to choose $v(x)$ so that $\mathbf{L}v(x) = 0$ (see also Sec. 10.4-2).

3. *A homogeneous linear differential equation is satisfied by every linear combination of solutions.*
4. *Let* $\Phi_1(x)$ *and* $\Phi_2(x)$ *satisfy the respective linear differential equations* $\mathbf{L}\Phi_1(x) = f_1(x)$ *and* $\mathbf{L}\Phi_2(x) = f_2(x)$ *with identical homogeneous linear boundary conditions* $\mathbf{B}_i\Phi(x) = 0$. *Then* $\alpha\Phi_1(x) + \beta\Phi_2(x)$ *satisfies the differential equation* $\mathbf{L}\Phi(x) = \alpha f_1(x) + \beta f_2(x)$ *and the given boundary conditions.*
5. *Let* $\Phi_1(x)$ *and* $\Phi_2(x)$ *satisfy the homogeneous linear differential equation* $\mathbf{L}\Phi(x) = 0$ *subject to the respective linear boundary conditions* $\mathbf{B}_i\Phi_1(x) = b_{1i}(x)$ *and* $\mathbf{B}_i\Phi_2(x) = b_{2i}(x)$. *Then* $\alpha\Phi_1(x) + \beta\Phi_2(x)$ *satisfies the given differential equation and the boundary conditions*

$$\mathbf{B}_i\Phi(x) = \alpha b_{1i}(x) + \beta b_{2i}(x)$$

15.4-3. Hermitian-conjugate and Adjoint Boundary-value Problems. Hermitian Operators (see also Secs. 14.4-3, 14.4-4, and 15.3-1*b*). (a) Given a homogeneous linear boundary-value problem (3) and the definition (2) of inner products, one defines the **hermitian-conjugate boundary-value problem**

$$\mathbf{L}\dagger\chi(x) = 0 \qquad (x \text{ in } V) \tag{15.4-5a}$$
$$\mathbf{B}_i\dagger\chi(x) = 0 \qquad (i = 1, 2, \ldots, N;\, x \text{ in } S) \tag{15.4-5b}$$

by the condition

$$(v, \mathbf{L}u) - (\mathbf{L}\dagger v, u) = \int_V [v^*\mathbf{L}u - u(\mathbf{L}\dagger v)^*]\, dV = 0 \tag{15.4-6}$$

where $u = u(x)$, $v = v(x)$ is any pair of suitably differentiable functions such that $u(x)$ satisfies the given boundary conditions (3*b*), and $v(x)$ satisfies the hermitian-conjugate boundary conditions (5*b*). u and v (or v^*) may then be said to represent vectors in *adjoint vector spaces* (Sec. 14.4-9).

Two linear second-order differential operators $\mathbf{L}$ and $\mathbf{L}\dagger$ will be called **hermitian-conjugate operators** if and only if the function $v^*\mathbf{L}u - u(\mathbf{L}\dagger v)^*$ has the form of an n-dimensional *divergence* $\frac{1}{\sqrt{|g(x)|}} \sum_{k=1}^{n} \frac{\partial}{\partial x^k} [\sqrt{|g(x)|}\, P^k]$ (Table 16.10-1), where each P^k can be a function of u, v, u^*, v^* and their first-order derivatives.* It is then possible to express the volume integral $\int_V [v^*\mathbf{L}u - u(\mathbf{L}\dagger v)^*]\, dV$ as an integral over the boundary S; formulas of this type are known as *generalized Green's formulas* (see also Sec. 15.4-3*c*). Suitable sets of **hermitian-conjugate boundary conditions** $\mathbf{B}_i u = 0$, $\mathbf{B}_i\dagger v = 0$ (x in S) are now defined by the fact that the boundary integral vanishes, so that Eq. (6) holds.

In case of real functions and operators, hermitian-conjugate boundary-value problems, operators, and boundary conditions are usually known as mutually **adjoint.**

* The k's are superscripts, not exponents, as in Chaps. 6 and 16.

(b) Hermitian Operators. A differential operator $\mathbf{L}$ is **hermitian (self-adjoint)** if and only if

$$(v, \mathbf{L}u) - (\mathbf{L}v, u) = \int_V [v^*\mathbf{L}u - u(\mathbf{L}v)^*]\, dV = 0 \qquad (15.4\text{-}7)$$

for every pair of suitably differentiable functions $u = u(x)$, $v = v(x)$ which satisfy identical homogeneous boundary conditions defining a linear manifold of functions. Hermitian-conjugate boundary-value problems with hermitian operators have identical boundary conditions and are thus identical (*self-adjoint boundary-value problems*).

(c) Special Cases. Real Sturm-Liouville Operators and Generalization of Green's Theorem (see also Secs. 5.6-1 and 15.5-4). In the *one-dimensional* case, Eqs. (3*a*) and (5*a*) are ordinary linear differential equations subject to given linear boundary conditions at the end points of an interval $(a, b) \equiv V$. If inner products are defined by $(u, v) = \int_a^b u^*v\, dx$ (Sec. 15.2-1*a*), one has, for *real* second-order differential operators,

$$\left.\begin{aligned} \mathbf{L}u &\equiv a_0(x)\frac{d^2u}{dx^2} + a_1(x)\frac{du}{dx} + a_2(x)u \\ \mathbf{L}^\dagger v &\equiv \frac{d^2}{dx^2}[a_0(x)v] - \frac{d}{dx}[a_1(x)v] + a_2(x)v \\ v\mathbf{L}u - u\mathbf{L}^\dagger v &= \frac{dP(x)}{dx} \\ P(x) &\equiv a_0(x)(vu' - uv') + [a_1(x) - a_0'(x)]uv \end{aligned}\right\} \qquad (15.4\text{-}8)$$

$P(x)$ is sometimes called the **conjunct** of $u(x)$ and $v(x)$ with respect to the operator $\mathbf{L}$.

The condition $a_1(x) \equiv a_0'(x)$ makes $\mathbf{L}$ a self-adjoint **Sturm-Liouville operator**

$$\mathbf{L} \equiv -\frac{d}{dx}\left[p(x)\frac{d}{dx}\right] - q(x) \qquad (15.4\text{-}9)$$

(see also Sec. 15.4-8), and partial integration yields the generalized Green's formula

$$\int_a^b (v\mathbf{L}u - u\mathbf{L}v)\, d\xi = -p(x)(vu' - uv')\Big]_a^b \qquad (15.4\text{-}10)$$

In the *three-dimensional* case, define inner products by Eq. (2), and define the operator ∇^2 in the manner of Tables 6.4-1 or 16.10-1. Then for any differentiable real function $q = q(x^1, x^2, x^3)$, the real differential operator

$$\mathbf{L} \equiv -\nabla^2 - q(x^1, x^2, x^3) \qquad (15.4\text{-}11)$$

is self-adjoint and satisfies the generalized Green's formula

$$\int_V (v\mathbf{L}u - u\mathbf{L}v)\,dV = -\int_S (v\nabla u - u\nabla v)\cdot d\mathbf{A}$$
$$= -\int_S \left(v\frac{\partial u}{\partial n} - u\frac{\partial v}{\partial n}\right) dA \qquad (15.4\text{-}12)$$

(see also Table 5.6-1). An analogous formula may be written for the two-dimensional case.

15.4-4. The Fredholm Alternative Theorem (see also Secs. 14.8-10, 15.3-7*a*, and 15.4-12). *The linear boundary-value problem defined by the differential equation*

$$\mathbf{L}\Phi(x) = f(x) \qquad (x \text{ in } V) \qquad (15.4\text{-}13a)$$

with homogeneous boundary conditions

$$\mathbf{B}_i\Phi(x) = 0 \qquad (i = 1, 2, \ldots, N;\ x \text{ in } S) \qquad (15.4\text{-}13b)$$

cannot have a unique solution $\Phi(x)$ *if the hermitian-conjugate (adjoint) homogeneous boundary-value problem* (5) *has a solution* $\chi(x)$ *other than* $\chi(x) \equiv 0$. *In this case, the given problem* (13) *can be solved only if* $f(x)$ *is orthogonal to every* $\chi(x)$, *i.e., if Eq.* (5) *implies*

$$(\chi, f) = \int_V \chi^* f\,dV = 0 \qquad (15.4\text{-}14)^*$$

If this condition is satisfied, then the given problem (13) *has either no solution or an infinite number of solutions.*

NOTE: In many applications, $\mathbf{L}$ is a hermitian operator, and the hermitian-conjugate problem (5) is identical with the given problem (13).

15.4-5. Eigenvalue Problems Involving Linear Differential Equations (see also Secs. 10.4-2*c*, 14.8-3, and 15.1-1). **(a)** For a given set of linear homogeneous boundary conditions, an **eigenfunction (proper function, characteristic function)** of the linear differential operator $\mathbf{L}$ is a solution $\psi(x)$, not identically zero in V, of the differential equation

$$\mathbf{L}\psi(x) = \lambda\psi(x) \qquad (x \text{ in } V) \qquad (15.4\text{-}15)$$

where λ is a suitably determined constant called the **eigenvalue (proper value, characteristic value)** of $\mathbf{L}$ associated with the eigenfunction $\psi(x)$.

(b) More general eigenvalue problems require one to find eigenfunctions $\psi(x) \not\equiv 0$ and eigenvalues λ satisfying a linear differential equation

$$\mathbf{L}\psi(x) = \lambda B(x)\psi(x) \qquad (x \text{ in } V) \qquad (15.4\text{-}16)$$

* In the *one-dimensional* case, dV is simply identical with dx.

subject to given linear homogeneous boundary conditions; $B(x)$ is to be real and positive throughout V.

(c) *If $\psi(x)$ is an eigenfunction associated with the eigenvalue λ, the same is true for every function $\alpha\psi(x) \not\equiv 0$. If $\psi_1(x), \psi_2(x), \ldots, \psi_s(x)$ are eigenfunctions associated with λ, the same is true for every function*

$$\alpha_1\psi_1 + \alpha_2\psi_2 + \cdots + \alpha_s\psi_s \not\equiv 0$$

This theorem also applies to uniformly convergent infinite series of eigenfunctions.

Eigenfunctions associated with different eigenvalues are linearly independent (Secs. 1.9-3 and 9.3-2). An eigenvalue λ associated with exactly $m > 1$ linearly independent eigenfunctions is said to be m-**fold degenerate;** m is the **degree of degeneracy.**

(d) The Spectrum of a Linear Eigenvalue Problem. Continuous Spectra and Improper Eigenfunctions (see also Secs. 14.8-3*d* and 15.4-12). For a given set of homogeneous linear boundary conditions and $B(x) > 0$, the **spectrum** of a linear eigenvalue problem (16) is the set of all complex numbers λ such that the differential equation

$$\mathbf{L}\Phi(x) - \lambda B(x)\Phi(x) = f(x) \qquad (x \text{ in } V) \tag{15.4-17}$$

with a given normalizable "forcing function" $f(x)$ does not have a unique normalizable solution $\Phi(x)$ subject to the given boundary conditions. The spectrum may comprise *continuous and residual spectra* (Sec. 14.8-3*d*) as well as a *discrete spectrum* defined by Eq. (16) with normalizable eigenfunctions $\psi(x)$. In the special case $B(x) \equiv 1$, one speaks of the *spectrum of the differential operator* $\mathbf{L}$.

If $\mathbf{L}$ is a hermitian operator and $B(x) > 0$, both the discrete and the continuous spectrum of the eigenvalue problem (16) are included in the *approximate spectrum* (Sec. 14.8-3*d*). One can often obtain the approximate spectrum by approximating the given eigenvalue problem with a sequence of eigenvalue problems yielding purely discrete spectra. In the course of such a limit process, the discrete eigenfunctions are replaced by a set of functions labeled by the continuously variable spectral parameter λ; such functions are known as **improper eigenfunctions** and will satisfy Eq. (16).

EXAMPLE: The ordinary differential equation $\dfrac{d^2\Phi(x)}{dx^2} = -\lambda\Phi(x)$ with $V \equiv (0, \infty)$ and the boundary conditions

$$\Phi(0) = 0 \qquad \int_0^\infty |\Phi(\xi)|^2 \, d\xi \text{ exists}$$

yields the continuous spectrum $0 \leq \lambda < \infty$. This spectrum is approximated by the discrete spectrum $\lambda = \left(k\dfrac{\pi}{a}\right)^2$ $(k = 0, 1, 2, \ldots)$ of the eigenvalue problem

$$\frac{d^2\Phi_a(x)}{dx^2} = -\lambda\Phi_a(x) \quad \text{with } V \equiv (0, a),\ \Phi(0) = \Phi(a) = 0$$

as $a \to \infty$. The eigenfunctions $\sin \dfrac{\pi k}{a} x$ $(k = 0, 1, 2, \ldots)$ of the latter problem approximate the improper eigenfunctions $\sin(\sqrt{\lambda}\, x)$ as $a \to \infty$ (transition from Fourier series to Fourier integral).

Similarly, the quantum-mechanical wave functions and energy levels of a particle in free space are approximated by those for particles confined to increasingly larger "boxes."

15.4-6. Eigenvalues and Eigenfunctions of Hermitian Eigenvalue Problems. Complete Orthonormal Sets of Eigenfunctions (see also Secs. 14.8-4, 14.-8-7, 15.2-4, 15.3-3, and 15.4-3*b*). (**a**) *If the operator* L *in Eq.* (15) *is hermitian, then*

1. *All spectral values* λ *are real.*
2. *Normalized eigenfunctions* ψ_i, ψ_k *corresponding to different eigenvalues are mutually orthogonal* in the sense that

$$(\psi_i, \psi_k) = \int_V \psi_i^* \psi_k \, dV = 0 \qquad (i \neq k)$$

If L *is a hermitian operator having a purely discrete spectrum, the following expansion theorem holds:*

3. *There exists an orthonormal set of eigenfunctions* $\psi_1(x)$, $\psi_2(x)$, . . . *permitting a series expansion*

$$\varphi(x) \underset{\text{in mean}}{=} a_1\psi_1(x) + a_2\psi_2(x) + \cdots$$

$$\left(a_k = \int_V \psi_k^* \varphi \, dV, \; k = 1, 2, \ldots\right) \qquad (15.4\text{-}18a)$$

of every quadratically integrable function $\varphi(x)$ *satisfying the boundary conditions of the eigenvalue problem and such that* $\mathsf{L}\varphi(x)$ *exists for almost all* x *in* V.

(**b**) *The same theorems apply to the "generalized" eigenvalue problem* (16), *where* L *is hermitian and* $B(x) > 0$, provided that one redefines orthogonality and normalization in terms of the new inner product*

$$(u, v)_B = \int_V u^* v B \, dV$$

Thus Eq. (18*a*) is replaced by the more general expansion

$$\varphi(x) = a_1\psi_1(x) + a_2\psi_2(x) + \cdots$$

$$\left(a_k = \int_V \psi_k^* \varphi B \, dV, \; k = 1, 2, \ldots\right) \qquad (15.4\text{-}18b)$$

and orthonormality of the eigenfunctions $\psi_k(x)$ is defined by

$$\int_V \psi_i^* \psi_k B \, dV = \delta_k^i = \begin{Bmatrix} 0 \text{ if } i \neq k \\ 1 \text{ if } i = k \end{Bmatrix} \qquad (i, k = 1, 2, \ldots) \qquad (15.4\text{-}19a)$$

These relations apply to the simpler eigenvalue problem (15) as a special case.

* In the *one-dimensional* case, dV is, again, simply identical with dx.

(c) For a hermitian eigenvalue problem with a (necessarily real) continuous spectrum admitting improper eigenfunctions (Sec. 15.4-5*d*), there exists a set of improper eigenfunctions $\psi(x, \lambda)$ such that

$$\int_V \psi^*(\xi, \lambda)\psi(\xi, \lambda')B(\xi)\, dV(\xi) = \delta(\lambda - \lambda') \tag{15.4-19b}$$

15.4-7. Hermitian Eigenvalue Problems as Variation Problems (see also Secs. 11.7-1 to 11.7-3, 14.8-8, 15.3-6*b*, 15.4-8*b*, 15.4-9*b*, and 15.4-10). (a) *The eigenvalue problem* (16) *for a hermitian differential operator* $\mathbf{L}$ *with discrete eigenvalues* $\lambda_1, \lambda_2, \ldots$ *is equivalent to each of the following variation problems:**

1. Find functions $\psi(x) \not\equiv 0$ in V which satisfy the given boundary conditions and reduce the variation (Secs. 11.5-1 and 11.5-2) of

$$\frac{(\psi, \mathbf{L}\psi)}{(\psi, B\psi)} = \frac{\int_V \psi^*\mathbf{L}\psi\, dV}{\int_V |\psi|^2 B\, dV} \qquad \text{(Rayleigh's quotient)} \tag{15.4-20}$$

   to zero.
2. Find functions $\psi(x)$ which satisfy the given boundary conditions and reduce the variation of

$$(\psi, \mathbf{L}\psi) = \int_V \psi^*\mathbf{L}\psi\, dV \tag{15.4-21}$$

   to zero subject to the constraint

$$(\psi, B\psi) = \int_V |\psi|^2 B\, dV = 1 \tag{15.4-22}$$

In each case, $\psi = \psi_k(x)$ yields the stationary value λ_k.

It is, then, possible to utilize the direct methods of the calculus of variations, notably the Rayleigh-Ritz method (Sec. 11.7-2) for the solution of eigenvalue problems involving ordinary or partial differential equations.

(b) Given a hermitian operator $\mathbf{L}$ with a discrete spectrum which includes at most a finite number of negative eigenvalues (Secs. 15.4-8 and 15.4-9), let the eigenvalues be arranged in increasing order, with an m-fold degenerate eigenvalue repeated m times, or $\lambda_1 \leq \lambda_2 \leq \cdots$. *The smallest eigenvalue* λ_1 *is the minimum value of Rayleigh's quotient* (20) *for an arbitrary "admissible"* function $\psi(x)$, i.e., for an arbitrary normalizable $\psi(x)$ which satisfies the given boundary conditions and such that Rayleigh's quotient exists.

Similarly, *the* r^{th} *eigenvalue* λ_r *in the above sequence will not exceed Rayleigh's quotient for any "admissible" function* $\varphi(x)$ *such that*

* In the *one-dimensional* case, $dV = dx$.

$$\int_V \psi_k^* \varphi B \, dV = 0 \tag{15.4-23}$$

for all eigenfunctions ψ_k *associated with* $\lambda_1, \lambda_2, \cdots, \lambda_{r-1}$ *(Courant's Minimax Principle).*

15.4-8. One-dimensional Eigenvalue Problems of the Sturm-Liouville Type. (a) Consider a one-dimensional real variable x, and let V be the bounded interval (a, b). Then the most general real hermitian second-order differential operator $\mathbf{L}$ has the form (9). The real differential equation

$$\mathbf{L}\psi \equiv -\left\{\frac{d}{dx}\left[p(x)\frac{d}{dx}\right] + q(x)\right\}\psi$$
$$\equiv -\left[p(x)\frac{d^2\psi}{dx^2} + p'(x)\frac{d\psi}{dx} + q(x)\psi\right] = \lambda B(x)\psi$$

(HOMOGENEOUS STURM-LIOUVILLE DIFFERENTIAL EQUATION) (15.4-24*a*)

defines a self-adjoint eigenvalue problem if one adds either of the linear homogeneous boundary conditions

$$\mathbf{B}_1\psi \equiv \alpha_1\psi'(a) + \beta_1\psi(a) = 0 \qquad \mathbf{B}_2\psi \equiv \alpha_2\psi'(b) + \beta_2\psi(b) = 0 \tag{15.4-24b}$$

$$\text{or } \mathbf{B}_1(\psi) \equiv \psi(a) - \psi(b) = 0 \qquad \mathbf{B}_2\psi \equiv \psi'(a) - \psi'(b) = 0$$

(PERIODICITY CONDITIONS) (15.4-24*c*)

It will be assumed that $p(x)$, $q(x)$, and $B(x)$ are differentiable in $[a, b]$, that $p(x)$ as well as $B(x)$ is positive in $[a, b]$, and that no eigenfunction corresponding to $\lambda = 0$ satisfies the given boundary conditions. These assumptions ensure that *the eigenvalues* λ *are discrete and positive; they are nondegenerate for the boundary conditions* (24*b*). *If the eigenvalues are arranged in increasing order* $\lambda_1 \leq \lambda_2 \leq \cdots$, *then* λ_n *is asymptotically proportional to* n^2 *as* $n \to \infty$ (Sec. 4.4-3).

Various simplifying transformations and methods for solving homogeneous differential equations of the type (24) are treated in Secs. 9.3-4 to 9.3-10; see also Secs. 21.7-1 to 21.8-12. Eigenvalue problems of the Sturm-Liouville type are met, in particular, upon separation of variables in boundary-value problems involving linear partial differential equations (Secs. 10.4-3 to 10.4-9) and are of great importance in quantum mechanics. *Note that every differential equation of the form*

$$a_0(x)\frac{d^2\psi}{dx^2} + a_1(x)\frac{d\psi}{dx} + c(x)\psi = \lambda h(x)\psi \tag{15.4-25}$$

can be reduced to the form (24*a*) through multiplication by $\exp \int \frac{a_1(x) - a_0'(x)}{a_0(x)} dx$ (see also Sec. 9.3-8*a*). Hence the examples of Secs. 10.4-3 to 10.4-9 may be treated as Sturm-Liouville problems.

(**b**) The generalized Green's formula (10) applies; if one writes $q(x) \equiv q_1'(x) - q_2(x)$,

$$\left.\begin{aligned} \int_a^b u\mathbf{L}u\,d\xi &= \int_a^b (pu'^2 + 2q_1uu' + q_2u^2)\,d\xi - (puu' + q_1u^2)\Big]_a^b \\ \int_a^b u\mathbf{L}v\,d\xi &= \int_a^b [pu'v' + q_1(u'v + v'u) + q_2uv]\,d\xi - (pv' + q_1v)u\Big]_a^b \end{aligned}\right\} \tag{15.4-26}$$

The integrals (26) have physical significance in many applications (see also Sec. 15.4-7).

(**c**) **Generalizations and Related Problems.** Related more general problems involve *unbounded* intervals $V \equiv (0, \infty)$ or $V \equiv (-\infty, \infty)$ with boundary conditions which specify the asymptotic behavior of $\psi(x)$ at infinity or simply demand normalizability. Again, one may admit singularities of $p(x)$, $q(x)$, and/or $B(x)$ for $x = a$ or $x = b$ (Sec. 21.8-10). *If the spectrum is purely discrete,* λ_n *is still asymptotically proportional to* n^2 *even if one admits singularities of* $p(x)$, $q(x)$, *or* $B(x)$ *at the end points of the bounded interval* (a, b).

Nonhomogeneous boundary-value problems involving the Sturm-Liouville operator $\mathbf{L}$ may be solved by the methods of Secs. 9.3-3, 9.3-4, 15.4-12, and 15.5-1.

15.4-9. Sturm-Liouville Problems Involving Second-order Partial Differential Equations (see also Sec. 15.4-3*c*). (**a**) Let $x \equiv (x^1, x^2, \ldots, x^n)$, and define inner products by Eq. (2) with $dV \equiv dx^1\,dx^2 \cdots dx^n$. Then the real partial differential equation

$$\mathbf{L}\psi \equiv -\left[\sum_{k=1}^{n} \frac{\partial}{\partial x^k}\left(p\,\frac{\partial}{\partial x^k}\right) + q\right]\psi = \lambda B\psi \quad \text{(MULTIDIMENSIONAL HOMOGENEOUS STURM-LIOUVILLE EQUATION)} \tag{15.4-27}$$

with

$$\begin{aligned} p &= p(x^1, x^2, \ldots, x^n) > 0 \\ B &= B(x^1, x^2, \ldots, x^n) > 0 \\ q &= q(x^1, x^2, \ldots, x^n) \end{aligned}$$

all differentiable on V and S, defines a self-adjoint eigenvalue problem for given homogeneous boundary conditions of the form $\alpha\,\frac{\partial\psi}{\partial n} + \beta\psi = 0$. *If the given region* V *is bounded, the resulting eigenvalue spectrum is discrete and includes at most a finite number of negative eigenvalues. If the eigenvalues are arranged in a sequence* $\lambda_1 \leq \lambda_2 \leq \cdots$, *then* $\lambda_n \to \infty$ *as* $n \to \infty$.

(**b**) In the *three-dimensional* case, the same theorem applies to the eigenvalues of the differential equation

$$-\nabla^2\psi - q(x^1, x^2, x^3)\psi = \lambda B(x^1, x^2, x^3)\psi$$

with boundary conditions of the form $\alpha\,\frac{\partial\psi}{\partial n} + \beta\psi = 0$ and $B > 0$ and q both differentiable on V and S. The generalized Green's formula (12) applies, and

$$\int_V u\mathbf{L}u\,dV = \int_V [(\nabla u)^2 - qu^2]\,dV - \int_S (u\nabla u)\cdot d\mathbf{A} \tag{15.4-28}$$

The integral (28) has physical significance in many applications (see also Sec. 15.4-7 and Table 5.6-1).

15.4-10. Comparison Theorems (see also Secs. 5.6-1*b*, 14.8-9*c*, and 15.4-7). The following *comparison theorems* apply to ordinary or partial differential equations of the Sturm-Liouville types defined in Secs. 15.4-8 and 15.4-9. *Given a differential equation* (24*a*) *or* (27) *and an interval or region* V *with boundary conditions* (24*b*) *or* (24*c*) ($\beta/\alpha \geq 0$),

1. *An increase in* p, *increase in* q, *and/or decrease in* B *will surely not increase any eigenvalue* λ_k; *similarly, a decrease in* p, *decrease in* q, *and/or increase in* B *will not decrease any eigenvalue* λ_k.
2. *A reduction in the size of the given interval or region* V *will never decrease any eigenvalue* λ_k *generated by the boundary conditions*

$$\psi(a) = \psi(b) = 0 \quad \text{or} \quad \psi(x_1, x_2, \ldots, x_n) = 0 \text{ on } S \tag{15.4-29a}$$

or decrease any eigenvalue λ_k *generated by the boundary conditions*

$$\psi'(a) = \psi'(b) = 0 \qquad \text{or} \qquad \frac{\partial \psi}{\partial n} = 0 \text{ on } S \tag{15.4-29b}$$

3. *Each eigenvalue* λ_k *generated by the boundary conditions*

$$\alpha\psi'(a) + \beta\psi(a) = \alpha\psi'(b) + \beta\psi(b) = 0 \quad \text{or} \quad \alpha\frac{\partial \psi}{\partial n} + \beta\psi = 0 \text{ on } S$$

with $\beta/\alpha \geq 0$ *is a nondecreasing function of* β/α; in particular, the Neumann condition (29*b*) will not yield larger eigenvalues than the Dirichlet condition (29*a*).

4. *A modification of the eigenvalue problem imposing constraints* (*accessory conditions*) *on* ψ *will not decrease any eigenvalue* λ_k.

The comparison theorems are of special interest in vibration theory (effects of mass, stiffness, and geometry on natural frequencies).

EXAMPLES (see also Sec. 10.4-9): $\psi''(x) = -\lambda\psi(x)$ (*vibrating string*) and $\nabla^2\psi(x, y, z) = -\lambda\psi(x, y, z)$ (*vibrating membrane*).

15.4-11. Perturbation Methods for the Solution of Discrete Eigenvalue Problems. Given the eigenvalues λ_k and the *orthonormal* eigenfunctions ψ_k of a hermitian eigenvalue problem defined by

$$\mathbf{L}\psi = \lambda\psi \tag{15.4-30}$$

one desires to approximate the eigenvalues $\bar{\lambda}_k$ and the eigenfunctions $\bar{\psi}_k$ of the "perturbed" hermitian problem

$$\mathbf{L}\bar{\psi} + \epsilon\mathbf{L}'\bar{\psi} = \bar{\lambda}\bar{\psi} \tag{15.4-31}$$

with unchanged boundary conditions; $\epsilon\mathbf{L}'\bar{\psi}$ is to be a small perturbation term ($\epsilon \ll 1$).

(a) Nondegenerate Case. Corresponding to each *nondegenerate* eigenvalue λ_i of the unperturbed problem, one has

$$\left.\begin{aligned}\bar{\lambda}_i &= \lambda_i + \epsilon L'_{ii} + \epsilon^2 \sum_{k \neq i} \frac{|L'_{ik}|^2}{\lambda_i - \lambda_k} + \cdots \\ \bar{\psi}_i &= \psi_i + \epsilon \sum_{k \neq i} \frac{L'_{ki}}{\lambda_i - \lambda_k} \psi_k + \cdots \end{aligned}\right\} \quad (i = 1, 2, \ldots) \tag{15.4-32}$$

where

$$L'_{ik} = \int_V \psi_i^* \mathbf{L}' \psi_k \, dV \qquad (i, k = 1, 2, \ldots) \tag{15.4-33}$$

(b) Degenerate Case. For each *m-fold degenerate* unperturbed eigenvalue λ_j with the m eigenfunctions $\psi_1, \psi_2, \ldots, \psi_m$, one may have up to m distinct eigenvalues of the perturbed operator. The corresponding values of $\Delta\lambda = \dfrac{\bar{\lambda} - \lambda_j}{\epsilon}$ are approximated by the roots of the mth-degree secular equation

$$\begin{vmatrix} L'_{11} - \Delta\lambda & L'_{12} & \cdots & L'_{1m} \\ L'_{21} & L'_{22} - \Delta\lambda & \cdots & L'_{2m} \\ \cdots & \cdots & \cdots & \cdots \\ L'_{m1} & L'_{m2} & \cdots & L'_{mm} - \Delta\lambda \end{vmatrix} = 0 \tag{15.4-34}$$

and may or may not coincide; i.e., a perturbation may remove degeneracies.

The eigenfunction or eigenfunctions corresponding to each value of $\bar{\lambda} = \lambda_j + \Delta\lambda$ are approximated by $\bar{\psi} = \sum_{k=1}^{m} a_k \psi_k$, where the a_k are obtained from

$$\sum_{k=1}^{m} a_k (L'_{ik} - \delta_k^i \, \Delta\lambda) = 0 \qquad (i = 1, 2, \ldots, m) \tag{15.4-35}$$

Note that no eigenfunctions other than the m eigenfunctions of the degenerate eigenvalue λ_j affect the approximations to $\bar{\lambda}$ and $\bar{\psi}$ given for the degenerate case; the approximation given for $\bar{\psi}$ is a "zero-order" approximation not proportional to ϵ. See Ref. 15.18 for higher-order effects.

(c) See Refs. 15.10 and 15.11 for a treatment of continuous spectra.

15.4-12. Solution of Boundary-value Problems by Eigenfunction Expansions (see also Secs. 10.4-2*c*, 14.8-10, 15.4-4, and 15.5-2; refer to Sec. 10.4-9 for examples). **(a)** A very important class of physical boundary-value problems (e.g., elastic vibrations, electromagnetic theory) involves a real linear ordinary or partial differential equation

$$\mathbf{L}\Phi(x) - \lambda B(x)\Phi(x) = f(x) \qquad (x \text{ in } V) \tag{15.4-36}$$

subject to given homogeneous boundary conditions, where $\mathbf{L}$ is a hermitian operator (Sec. 15.4-3*b*), and $B(x) > 0$ in V. If $f(x) \equiv 0$ in V (no applied forces, currents, etc.), then Eq. (36) reduces to the complementary homogeneous differential equation

$$\mathbf{L}\psi(x) = \lambda B(x)\psi(x) \qquad (x \text{ in } V) \tag{15.4-37}$$

which can be satisfied only by proper or improper eigenfunctions $\psi(x)$

with spectral values λ. In the nonhomogeneous case (36) (e.g., forced oscillations), λ will be a given parameter.

Consider first that the eigenvalue problem (37) has a *purely discrete spectrum* of (not necessarily distinct) eigenvalues $\lambda_1, \lambda_2, \ldots$ associated with corresponding orthonormal eigenfunctions $\psi_1(x), \psi_2(x), \ldots$ (Sec. 15.4-6*b*). Assuming that the "forcing function" can be expanded in the form

$$f(x) = B(x) \sum_{k=1}^{\infty} f_k \psi_k(x) \qquad \left(f_k = \int_V \psi_k^* f \, dV, \; k = 1, 2, \ldots \right) \tag{15.4-38}$$

for almost all x in V, the desired solution of Eq. (36) is given by the "normal-mode expansion"

$$\Phi(x) \underset{\text{in mean}}{=} \sum_{k=1}^{\infty} \frac{f_k}{\lambda_k - \lambda} \psi_k(x) \tag{15.4-39}$$

The series (39) defines the solution $\Phi(x)$ uniquely at every point of continuity in V, provided that the given parameter λ does not equal an eigenvalue λ_k (resonance!). In the latter case, a solution exists only if $f(x)$ is orthogonal to all eigenfunctions associated with λ_k, so that $f_k = 0$. One has then an infinite number of solutions given by the series (39) plus any linear combination of eigenfunctions associated with λ_k.

(b) If the eigenvalue problem (37) has a *purely continuous spectrum* D_Λ with improper eigenfunctions $\psi(x, \lambda)$ having the orthonormality property (19*b*) (e.g., Sturm-Liouville operators with singularities in V, or V unbounded), one may attempt to expand the solution $\Phi(x)$ as a "generalized Fourier integral" over the (necessarily real) spectrum D_Λ. One has then, for almost all x in V,

$$\left.\begin{aligned} \Phi(x) &= \int_{D_\Lambda} H(\Lambda)\psi(x, \Lambda) \, d\Lambda \\ H(\Lambda) &= \frac{1}{\Lambda - \lambda} \int_V \psi^*(\xi, \Lambda) f(\xi) \, dV \end{aligned}\right\} \tag{15.4-40}$$

provided that D_Λ does not contain λ (see also Sec. 10.5-1, integral-transform methods).

If the eigenvalue problem (37) has both a discrete and a continuous spectrum, the solution will contain terms of the form (39) as well as an integral (40); both types of terms can be combined into a Stieltjes integral over the spectrum.

(c) The solution methods of Secs. 15.4-12*a* and *b* may apply even if the given operator $\mathbf{L}$ is not hermitian, provided that valid orthonormal-eigenfunction expansions can be shown to exist.

15.5. GREEN'S FUNCTIONS. RELATION OF BOUNDARY-VALUE PROBLEMS AND EIGENVALUE PROBLEMS TO INTEGRAL EQUATIONS

15.5-1. Green's Functions for Boundary-value Problems with Homogeneous Boundary Conditions (see also Secs. 9.3-3, 9.4-3, and 15.4-4; for examples, see Table 9.3-1, Sec. 15.6-6, and Sec. 15.6-9). **(a)** A linear boundary-value problem

$$\mathsf{L}\Phi(x) = f(x) \qquad (x \text{ in } V) \tag{15.5-1a}$$

$$\mathsf{B}_i\Phi(x) = 0 \qquad (i = 1, 2, \ldots, N; x \text{ in } S) \tag{15.5-1b}$$

expresses the given function $f(x)$ as the result of a linear operation on an unknown function $\Phi(x)$ subject to the given boundary conditions. If it is possible to write the corresponding *inverse operation* as a linear integral transformation (Sec. 15.3-1)

$$\Phi(x) = \int_V G(x, \xi)f(\xi)\, dV(\xi) \tag{15.5-2}^*$$

for every suitable $f(x)$, the kernel $G(x, \xi)$ is called the **Green's function for the given boundary-value problem** (1). *$G(x, \xi)$ must satisfy the given homogeneous boundary conditions* (1*b*) *together with*

$$\left.\begin{aligned} \mathsf{L}G(x, \xi) &= 0 && (x, \xi \text{ in } V; x \neq \xi) \\ \int_V \mathsf{L}G(x, \xi)\, dV(\xi) &= 1 && (x \text{ in } V) \end{aligned}\right\} \tag{15.5-3a}$$

or

$$\mathsf{L}G(x, \xi) = \delta(x, \xi) \qquad (x, \xi \text{ in } V) \tag{15.5-3b}$$

where $\delta(x, \xi)$ is the delta function *for the specific coordinate system used* (Sec. 21.9-7).

Equation (2) describes the solution $\Phi(x)$ of the given boundary-value problem (1) as a superposition of elementary solutions $G(x, \xi)f(\xi)$ having singularities at $x = \xi$. These elementary solutions can be interpreted as effects of *impulse forces, point charges*, etc., $f(\xi)\delta(x, \xi)$ at $x = \xi$ (Secs. 9.4-3, 15.5-4, 15.6-6, and 15.6-9).

Green's functions can often be found directly by formal integration of the "symbolic differential equation" (3*b*) subject to the given boundary conditions by the methods of Secs. 9.4-5, 10.5-1, or 15.4-12. See Table 9.3-3, Sec. 15.6-6, and Sec. 15.6-9 for examples of Green's functions.

(b) Modified Green's Functions (see also Sec. 15.4-4). A given boundary-value problem (1) will not admit a Green's function satisfying Eq. (3) if the hermitian-conjugate homogeneous problem

$$\mathsf{L}^\dagger\chi(x) = 0 \qquad (x \text{ in } V) \qquad \mathsf{B}_i{}^\dagger\chi(x) = 0 \qquad (x \text{ in } S) \tag{15.5-4}$$

possesses a solution $\chi(x)$ other than $\chi(x) \equiv 0$ (Sec. 15.4-4). In this case, it may still be possible to find a **modified Green's function** $G(x, \xi)$ permitting the expansion (2) for suitable functions $f(x)$ orthogonal to every $\chi(x)$. $G(x, \xi)$ must satisfy the condition

* In the *one-dimensional case*, $dV(\xi) \equiv d\xi$.

$$\mathbf{L}G(x, \xi) = \delta(x, \xi) - \sum_{k=1}^{\infty} \chi_{0k}^*(\xi)\chi_{0k}(x) \tag{15.5-5}$$

together with the given boundary conditions (1*b*), where the $\chi_{0k}(x)$ are any complete orthonormal set spanning all solutions of the problem (4). Any resulting solution (2) cannot possibly be unique (Sec. 15.4-4), but there exists at most one modified Green's function which satisfies the added orthogonality conditions

$$\int_V \chi_{0k}^*(\xi)G(x, \xi)\, dV(\xi) = 0 \qquad (k = 1, 2, \ldots) \tag{15.5-6}$$

If $\mathbf{L}$ is a hermitian operator, then the χ_{0k} are simply its eigenfunctions for $\lambda = 0$.

(**c**) Green's functions for hermitian-conjugate boundary-value problems (Sec. 15.4-3*a*) are hermitian-conjugate kernels (Sec. 15.3-1*b*). *For every Green's function $G(x, \xi)$ associated with a hermitian operator* $\mathbf{L}$ (Sec. 15.4-3*b*), $G^*(\xi, x) \equiv G(x, \xi)$.

15.5-2. Relation of Boundary-value Problems and Eigenvalue Problems to Integral Equations. Green's Resolvent. (**a**) If there exists a Green's function $G(x, \xi)$ for the boundary-value problem (1), Eq. (2) implies that the more general boundary-value problem

$$\left.\begin{aligned} \mathbf{L}\Phi(x) - \lambda B(x)\Phi(x) = f(x) \qquad (x \text{ in } V) \\ \mathbf{B}_i\Phi(x) = 0 \qquad (i = 1, 2, \ldots, N;\, x \text{ in } S) \end{aligned}\right\} \tag{15.5-7}$$

is equivalent to the linear integral equation

$$\left.\begin{aligned} \Phi(x) - \lambda \int_V K(x, \xi)\Phi(\xi)\, d\xi = F(x) \\ K(x, \xi) = G(x, \xi)B(\xi)\sqrt{|g(\xi)|} \quad F(x) = \int_V G(x, \xi)f(\xi)\, dV(\xi) \end{aligned}\right\} \tag{15.5-8}$$

Note that the integral equation takes the given boundary conditions into account. In the *one-dimensional case*, $dV(\xi) \equiv d\xi$, and $|g(\xi)| \equiv 1$.

The integral-equation representation brings the theory of Secs. 15.3-1 *to* 15.3-9 *and the numerical methods of Sec.* 20.9-10 *to bear on linear boundary-value problems and eigenvalue problems.* In particular, one may employ the Neumann series (15.3-23) and analytic continuation to introduce a resolvent kernel (Sec. 15.3-7*c*) $\Gamma(x, \xi; \lambda)$, so that the solution appears in the form

$$\Phi(x) = F(x) + \lambda \int_V \Gamma(x, \xi; \lambda)F(\xi)\, d\xi \tag{15.5-9}$$

$\Gamma(x, \xi; \lambda)$ is called **Green's resolvent;** it is the kernel of a linear integral transformation representing the resolvent operator $(\mathbf{L} - \lambda)^{-1}$ (Sec. 14.8-3*d*). $\Gamma(x, \xi; \lambda)$ can often be obtained by the methods of Sec. 15.3-8; but note carefully that $K(x, \xi)$ is *not* necessarily a normalizable kernel. *The set of singularities λ of $\Gamma(x, \xi; \lambda)$ is precisely the spectrum of the operator* $\mathbf{L}$ (Sec. 15.4-5*d*), which may include a continuous spectrum.

Specifically, *the poles of* $\Gamma(x, \xi; \lambda)$ *correspond to discrete eigenvalues of* $\mathbf{L}$, *while branch points indicate the presence of a continuous spectrum* (Ref. 15.2).

(b) Problems Involving Hermitian Operators. Eigenfunction Expansion of Green's Function (see also Secs. 15.3-3, 15.3-4, 15.4-6, and 15.4-12). In the important special cases where $\mathbf{L}$ *is a hermitian operator and* $B(x)$ *is real and positive in* V, one may introduce a new unknown function

$$\bar{\Phi}(x) = \Phi(x)\sqrt{B(x)}\sqrt{|g(x)|} \tag{15.5-10}$$

to replace Eq. (8) by a linear integral equation *with hermitian kernel:*

$$\left.\begin{aligned} \bar{\Phi}(x) - \lambda\int_V \bar{K}(x, \xi)\bar{\Phi}(\xi)\,d\xi &= \bar{F}(x) \\ \bar{K}(x, \xi) &= G(x, \xi)\sqrt{B(x)B(\xi)}\sqrt{|g(x)g(\xi)|} \\ \bar{F}(x) &= \sqrt{B(x)}\sqrt{|g(x)|}\int_V G(x, \xi)f(\xi)\,dV(\xi) \end{aligned}\right\} \tag{15.5-11}$$

The hermitian eigenvalue problems

$$\begin{aligned} \mathbf{L}\psi(x) &= \lambda B(x)\psi(x) \qquad (x \text{ in } V) \\ \mathbf{B}_i\psi(x) &= 0 \qquad (i = 1, 2, \ldots, m;\ x \text{ in } S) \end{aligned} \tag{15.5-12a}$$

and

$$\bar{\psi}(x) = \lambda\int_V \bar{K}(x, \xi)\bar{\psi}(\xi)\,d\xi \tag{15.5-12b}$$

yield identical spectra; corresponding eigenfunctions $\psi(x)$ amd $\bar{\psi}(x)$ are related by $\bar{\psi}(x) = \psi(x)\sqrt{B(x)}\sqrt{|g(x)|}$. If the eigenvalue spectrum is purely discrete, there exist complete sets of eigenfunctions $\psi_i(x)$ and $\bar{\psi}_i(x)$ such that

$$\int_V B\psi_i^*\psi_k\,dV = \int_V \bar{\psi}_i^*\bar{\psi}_k\,dV = \delta_k^i \tag{15.5-13}$$

In this case, $\bar{K}(x, \xi)$ is a normalizable kernel, and

$$G(x, \xi) \underset{\text{in mean}}{=} \sum_{k=1}^{\infty} \frac{1}{\lambda_k}\psi_k^*(\xi)\psi_k(x) \qquad (x, \xi \text{ in } V) \tag{15.5-14}$$

which does not depend explicitly on $B(x)$ or $g(x)$. In the case of Sturm-Liouville problems with purely discrete spectra (Secs. 15.4-8*a* and 15.4-9), the series (14) converges absolutely and uniformly in V (see also Mercer's theorem, Sec. 15.3-4).

(c) If the given boundary-value problem (7) is expressed in terms of differential invariants (Sec. 16.10-7), then $G(x, \xi)$, $K(x, \xi)$, and $\bar{K}(x, \xi)$ are point functions invariant with respect to changes in the coordinate system used.

15.5-3. Application of the Green's-function Method to an Initial-value Problem: The Generalized Diffusion Equation (see also Secs. 10.4-7, 10.5-3, and 10.5-4). It is desired to find the solution $\Phi = \Phi(x, t)$ of the initial-value problem

$$\left.\begin{aligned} \nabla^2\Phi - a^2\frac{\partial\Phi}{\partial t} - b\Phi &= f(x, t) \\ \Phi(x, 0) &= \Phi_0(x) \end{aligned}\right\} (x \text{ in } V) \tag{15.5-15}$$

with given constant coefficients a^2, b and homogeneous boundary conditions $\Phi = 0$ or $\partial\Phi/\partial n = 0$ on the boundary of a given n-dimensional region V of points (x), where $n = 1, 2$, or 3. Then

$$\Phi(x, t) = \int_0^t \int_V G(x, t; \xi, \tau) f(\xi, \tau)\, d\tau\, dV(\xi) - a^2 \int_V G(x, t; \xi, 0)\Phi_0(\xi)\, dV(\xi) \qquad (t > 0) \qquad (15.5\text{-}16)$$

where the Green's function $G(x, t; \xi, \tau)$ satisfies the given boundary conditions and

$$\left.\begin{aligned} \nabla^2 G - a^2 \frac{\partial G}{\partial t} - bG &= \delta(x, \xi)\delta(t - \tau) \\ G &= 0 \qquad (t < \tau) \end{aligned}\right\} (x \text{ in } V) \qquad (15.5\text{-}17)$$

Given the eigenfunctions $\psi_k(x)$ and eigenvalues λ_k of the space form of the wave equation

$$\nabla^2\psi(x) + \lambda\psi(x) = 0 \qquad (x \text{ in } V) \qquad (15.5\text{-}18)$$

subject to the given boundary conditions (Secs. 10.4-4 and 10.4-5b), one has

$$G(x, t; \xi, \tau) = -\frac{1}{a^2} \sum_k e^{-\frac{\lambda_k + b}{a^2}(t-\tau)} \psi_k^*(\xi)\psi_k(x) \qquad (t > \tau) \qquad (15.5\text{-}19)$$

If V encompasses the entire space, then

$$G(x, t; \xi, \tau) = -\left[\frac{a^2}{4\pi(t - \tau)}\right]^{n/2} \frac{1}{a^2} \exp\left[-\frac{a^2}{4}\frac{|\mathbf{r} - \boldsymbol{\rho}|^2}{t - \tau} - \frac{b}{a^2}(t - \tau)\right]$$
$$(t > \tau;\ n = 1, 2, 3) \qquad (15.5\text{-}20)$$

where $|\mathbf{r} - \boldsymbol{\rho}|$ is the distance between the points $(x) = (\mathbf{r})$ and $(\xi) = (\boldsymbol{\rho})$. The resulting solution (16) is known as the *Poisson-integral solution* of the diffusion problem.

15.5-4. The Green's-function Method for Nonhomogeneous Boundary Conditions (see also Sec. 15.4-2). **(a)** The solution $\Phi(x)$ of a three-dimensional linear boundary-value problem of the form

$$\mathbf{L}\Phi(x) = 0 \qquad (x \text{ in } V) \qquad \mathbf{B}\Phi(x) = b(x) \qquad (x \text{ in } S) \qquad (15.5\text{-}21)$$

can often be written as a surface integral

$$\Phi(x) = \int_S G_S(x, \xi) b(\xi)\, dA(\xi) \qquad (15.5\text{-}22)$$

for every given function $b(x)$ integrable over S. *$G_S(x, \xi)$ must satisfy the given differential equation for x in V, ξ in S, and*

$$\left.\begin{aligned} \mathbf{B}G_S(x, \xi) &= 0 \qquad (x, \xi \text{ in } S;\ x \neq \xi) \\ \int_S \mathbf{B}G_S(x, \xi)\, dA(\xi) &= 1 \end{aligned}\right\} \qquad (15.5\text{-}23)$$

$G_S(x, \xi)$ is variously defined as a **Green's function of the second kind** or simply as a Green's function (see also Sec. 15.6-6). Analogous relations hold in the *two-dimensional case* (see also Sec. 15.6-9).

Linear boundary-value problems involving nonhomogeneous differential equations as well as nonhomogeneous boundary conditions can often be solved by *superposition of a volume integral* (2) *and a surface integral* (22).

(b) As shown in Sec. 15.4-2, every boundary-value problem (21) can be rewritten as a boundary-value problem of the type (1). Hence, *$G_S(x, \xi)$ can be related to the ordinary Green's function $G(x, \xi)$ defined in the manner of Sec.* 15.5-1 *for the problem with "complementary" homogeneous boundary conditions.* In particular, consider two-dimensional or three-dimensional boundary-value problems involving real self-adjoint differential equations of the form

$$\mathbf{L}\Phi(x) \equiv -[\nabla^2 + q]\Phi(x) = 0 \qquad (x \text{ in } V) \qquad (15.5\text{-}24)$$

where $q = q(x)$ is a real differentiable function. Given a Green's function $G(x, \xi)$ satisfying

$$-[\nabla^2 + q]G(x, \xi) = \delta(x, \xi) \qquad (15.5\text{-}25)$$

for suitably given homogeneous boundary conditions (Sec. 15.5-1), Green's formula (15.4-12) yields

$$\Phi(x) = \int_S \left[G(x, \xi)\frac{\partial \Phi(\xi)}{\partial \nu} - \Phi(\xi)\frac{\partial G(x, \xi)}{\partial \nu} \right] dA(\xi) \qquad (x \text{ in } V) \qquad (15.5\text{-}26)$$

where $\partial/\partial\nu$ denotes normal differentiation with respect to ξ. Hence for boundary conditions of the form

$$\mathbf{B}\Phi(x) \equiv \Phi(x) = b(x) \qquad (x \text{ in } S) \qquad (\text{Dirichlet conditions}) \qquad (15.5\text{-}27)$$

the solution (22) requires

$$G_S(x, \xi) = -\frac{\partial G(x, \xi)}{\partial \nu} \qquad (\xi \text{ in } S) \qquad (15.5\text{-}28)$$

where $G(x, \xi)$ satisfies Eq. (25) in V and vanishes on S.

For boundary conditions of the form

$$\mathbf{B}\Phi(x) \equiv \frac{\partial \Phi}{\partial n} = b(x) \qquad (x \text{ in } S) \qquad (\text{Neumann conditions}) \qquad (15.5\text{-}29)$$

the solution (22) of the differential equation (24) requires one to use the "Neumann function"

$$G_S(x, \xi) = G(x, \xi) \qquad (\xi \text{ in } S) \qquad (15.5\text{-}30)$$

where $G(x, \xi)$ satisfies Eq. (25) in V, and $\partial G(x, \xi)/\partial n = 0$ for x in S.

Sections 15.6-6 and 15.6-9 show the application of these relations to the solution of *true boundary-value problems for elliptic differential equations.* Section 10.3-6 illustrates a similar method for the solution of *initial-value problems for hyperbolic differential equations* (see also Sec. 10.3-5).

15.6. POTENTIAL THEORY

15.6-1. Introduction. Laplace's and Poisson's Differential Equations (see also Secs. 5.7-3, 10.4-3, and 10.4-5). Many important applications involve solutions $\Phi(\mathbf{r})$ of the linear partial differential equations

$$\nabla^2\Phi(\mathbf{r}) = 0 \qquad \text{(Laplace's differential equation)} \tag{15.6-1}$$

$$\nabla^2\Phi(\mathbf{r}) = -4\pi Q(\mathbf{r}) \qquad \text{(Poisson's differential equation)} \tag{15.6-2}$$

where $\Phi(\mathbf{r})$ and $Q(\mathbf{r})$ are functions of position in a three-dimensional Euclidean space of points $(\mathbf{r}) \equiv (x, y, z) \equiv (x^1, x^2, x^3)$, or in a two-dimensional Euclidean space points $(\mathbf{r}) \equiv (x, y) \equiv (x^1, x^2)$. $\Phi(r)$ is most frequently interpreted as the *potential of an irrotational vector field*

$$\mathbf{F}(\mathbf{r}) = -\nabla\Phi(\mathbf{r})$$

due to a distribution of charge or mass such that $\nabla \cdot \mathbf{F}(\mathbf{r}) = 4\pi Q(\mathbf{r})$ (Secs. 5.7-2, 5.7-3, and 15.6-5). The study of such potentials and, in particular, of solutions of Laplace's differential equation (1) is known as **potential theory.**

15.6-2. Three-dimensional Potential Theory: The Classical Boundary-value Problems. (a) The Dirichlet Problem. A bounded region V admitting a solution of the boundary-value problem

$$\nabla^2\Phi(\mathbf{r}) = 0 \quad (\mathbf{r} \text{ in } V) \qquad \Phi(\mathbf{r}) = b(\mathbf{r}) \quad (\mathbf{r} \text{ in } S)$$
$$\text{(Dirichlet problem)} \tag{15.6-3}$$

for any given continuous single-valued function $b(\mathbf{r})$ is called a **Dirichlet region;** *whenever a solution exists, it is necessarily unique.* If V is an unbounded region, one must specify the asymptotic behavior of the solution at infinity, say $\Phi(\mathbf{r}) = O(1/r)$ as $r \to \infty$; *the latter condition ensures uniqueness whenever a solution exists.* Section 15.6-6*d* further discusses the existence of solutions.

The solution $\Phi(\mathbf{r})$ of the Dirichlet problem (3) yields a stationary value of the *Dirichlet integral* $\int_V (\nabla\Phi)^2\, dV$, where $\Phi(\mathbf{r})$ is assumed to be twice continuously differentiable in V and S and to satisfy the given boundary conditions (see also Sec. 15.4-7*a*). Dirichlet problems are of particular importance in electrostatics (Ref. 15.6).

(b) The Neumann Problem. For the second classical boundary-value problem

$$\nabla^2\Phi(\mathbf{r}) = 0 \qquad (\mathbf{r} \text{ in } V) \qquad \frac{\partial\Phi}{\partial n} = b(\mathbf{r}) \qquad (\mathbf{r} \text{ in } S)$$

(NEUMANN PROBLEM) (15.6-4)

where $b(\mathbf{r})$ is a given continuous single-valued function, the existence of solutions requires $\int_S b(\mathbf{r})\,dA = 0$ (see also Gauss's theorem, Table 5.6-1). If V is an unbounded region, one requires $\Phi(\mathbf{r}) = O(1/r)$ and $\partial\Phi/\partial n = O(1/r^2)$ as $r \to \infty$. *The solution of a Neumann problem for a bounded region V is unique except for an additive constant.*

Neumann problems occur, in particular, in connection with the flow of incompressible fluids (Ref. 15.6).

15.6-3. Kelvin's Inversion Theorem. *If $\bar\Phi(\mathbf{r})$ is a solution of Laplace's differential equation in a region $\bar V$ inside the sphere $|\mathbf{r} - \mathbf{a}| = R$,* then

$$\Phi(\mathbf{r}) = \frac{R}{|\mathbf{r} - \mathbf{a}|}\,\bar\Phi\left[\frac{R^2}{|\mathbf{r} - \mathbf{a}|^2}(\mathbf{r} - \mathbf{a}) + \mathbf{a}\right] \tag{15.6-5}$$

is a solution in a corresponding region V outside the sphere, and conversely (Kelvin's Inversion Theorem). In terms of spherical coordinates r, ϑ, φ one may say that, *if $\Phi(r, \vartheta, \varphi)$ is a solution for $r < R$, then $\frac{R}{r}\Phi\left(\frac{R^2}{r}, \vartheta, \varphi\right)$ is a solution for $r > R$, and conversely.*

Hence, the so-called *exterior* boundary-value problem

$$\left.\begin{aligned} \nabla^2\Phi(\mathbf{r}) &= 0 \qquad (\mathbf{r} \text{ outside } V \text{ and } S) \\ \alpha\frac{\partial\Phi}{\partial n} + \beta\Phi &= b(\mathbf{r}) \qquad (\mathbf{r} \text{ in } S) \\ \Phi(\mathbf{r}) &= O(1/r) \qquad \text{as } r \to \infty \end{aligned}\right\} \tag{15.6-6}$$

for a bounded region V can be transformed into a corresponding boundary-value problem for the *interior* of an "inverted" region $\bar V$ through a transformation

$$\bar{\mathbf{r}} - \mathbf{a} = \frac{R^2}{|\mathbf{r} - \mathbf{a}|^2}(\mathbf{r} - \mathbf{a}) \tag{15.6-7}$$

15.6-4. Properties of Harmonic Functions. (a) Mean-value and Maximum-modulus Theorems. Solutions $\Phi(\mathbf{r})$ of Laplace's differential equation (1) are called **harmonic functions.** *Every function $\Phi(\mathbf{r})$ harmonic in the open region V is analytic* (Sec. 4.10-5*b*) *and has harmonic derivatives of every order throughout V. Each value $\Phi(\mathbf{r}_1)$ equals the arithmetic mean* (Sec. 4.6-3) *of $\Phi(\mathbf{r})$ over the surface (and hence over the volume) of any sphere centered at $\mathbf{r} = \mathbf{r}_1$ and contained in V (Mean-value Theorem).* Conversely, a *continuous* function $\Phi(\mathbf{r})$ is harmonic in every open region V where the above mean-value property holds.

A function $\Phi(\mathbf{r})$ harmonic throughout the bounded region V and its boundary S cannot have a maximum or minimum in the interior of V (Maximum-modulus Theorem, see also Sec. 7.3-5). Conversely, a *continuous* function without maxima and minima in the interior of a sphere is harmonic throughout the sphere. If $\Phi(\mathbf{r})$ is harmonic in V, continuous in V and S, and equal to zero on S, then $\Phi(\mathbf{r}) \equiv 0$ in V. If $\Phi(\mathbf{r})$ is harmonic in V, continuously differentiable in V and S, and $\partial\Phi/\partial n = 0$ on S, then $\Phi(\mathbf{r})$ is constant in V. The inversion theorem of Sec. 15.6-3 yields analogous theorems for the unbounded region V' outside S.

(b) Harnack's Convergence Theorems. The following *convergence theorems* are of interest in connection with series approximations for harmonic functions. *If the sequence $s_0(r)$, $s_1(r)$, $s_2(r)$, . . . of functions all harmonic in V and continuous on the boundary S of V converges uniformly on S, then the sequence converges uniformly in V to a function $s(\mathbf{r})$ which is harmonic in V and such that $s(\mathbf{r}) = \lim_{n\to\infty} s_n(\mathbf{r})$ on S. The sequence for any partial derivative of $s_n(\mathbf{r})$ converges uniformly to the corresponding partial derivative of $s(\mathbf{r})$ in every closed subregion of V* (*Harnack's First Convergence Theorem*).

Given a sequence of functions $s_0(\mathbf{r})$, $s_1(\mathbf{r})$, $s_2(\mathbf{r})$, . . . harmonic in V and such that $s_0(\mathbf{r}) \geq s_1(\mathbf{r}) \geq s_2(\mathbf{r}) \geq$ . . . for all $\mathbf{r}$ in V, convergence at any point of V implies convergence throughout V and uniform convergence in every closed subregion of V; the limit is a harmonic function in V (*Harnack's Second Convergence Theorem*).

15.6-5. Solutions of Laplace's and Poisson's Equations as Potentials. (a) Potentials of Point Charges, Dipoles, and Multipoles. The following particular solutions of Laplace's equation have especially simple physical interpretations:

$$\varphi_0(\mathbf{r} - \boldsymbol{\varrho}) \equiv \frac{1}{|\mathbf{r} - \boldsymbol{\varrho}|} \equiv \frac{1}{\sqrt{(x-\xi)^2 + (y-\eta)^2 + (z-\zeta)^2}} \qquad (\mathbf{r} \neq \boldsymbol{\varrho})$$

[POTENTIAL OF A UNIT POINT CHARGE AT THE POINT $(\boldsymbol{\varrho})$] (15.6-8)

$$\varphi_1(\mathbf{r} - \boldsymbol{\varrho}) \equiv -(\mathbf{u}_1 \cdot \nabla)\varphi_0(\mathbf{r} - \boldsymbol{\varrho}) \qquad (\mathbf{r} \neq \boldsymbol{\varrho})$$

[POTENTIAL OF A UNIT DIPOLE DIRECTED ALONG THE UNIT VECTOR $\mathbf{u}_1$ AT THE POINT $(\boldsymbol{\varrho})$] (15.6-9)

More generally, the potential of a **multipole of order** j at the origin $(\boldsymbol{\varrho} = 0)$ is

$$\begin{aligned}\Phi_j(\mathbf{r}) &= (-1)^j(\mathbf{p}_j \cdot \nabla)(\mathbf{p}_{j-1} \cdot \nabla) \cdots (\mathbf{p}_1 \cdot \nabla)\varphi_0(\mathbf{r}) \\ &= \sum\sum_{i+k+l=j}\sum Q_{ikl}^{(j)} \frac{\partial^j \varphi_0(\mathbf{r})}{(\partial x)^i(\partial y)^k(\partial z)^l} \qquad (\mathbf{r} \neq 0) \qquad (15.6\text{-}10)\end{aligned}$$

where the so-called **multipole-moment components** $Q_{ikl}^{(j)}$ are constants determined by the j vectors $\mathbf{p}_1, \mathbf{p}_2, \ldots, \mathbf{p}_j$ defining the multipole. In terms of spherical coordinates r, ϑ, φ,

$$\Phi_j(\mathbf{r}) = \frac{1}{r^{j+1}}\Phi_j\left(\frac{\mathbf{r}}{r}\right) = \frac{1}{r^{j+1}} Y_j(\varphi, \vartheta) \qquad (\mathbf{r} \neq 0) \qquad (15.6\text{-}11)$$

where $Y_j(\varphi, \zeta)$ is a spherical surface harmonic of degree j (Sec. 21.8-12). Multipoles of order 2 and 3 are respectively known as **quadrupoles** and **octupoles.**

(b) Potentials of Charge Distributions. Jump Relations. Other particular solutions of Laplace's equation are obtained by linear superposition or integration of simple and/or dipole potentials. Of particular interest are *volume potentials of charge and dipole distributions,*

$$\int_V Q(\boldsymbol{\varrho})\varphi_0(\mathbf{r}-\boldsymbol{\varrho})\,dV(\boldsymbol{\varrho}) = \int_V \frac{Q(\boldsymbol{\varrho})}{|\mathbf{r}-\boldsymbol{\varrho}|}\,dV(\boldsymbol{\varrho}) \tag{15.6-12}$$

$$-\int_V [\mathbf{p}(\boldsymbol{\varrho})\cdot\nabla]\varphi_0(\mathbf{r}-\boldsymbol{\varrho})\,dV(\boldsymbol{\varrho}) = -\int_V [\mathbf{p}(\boldsymbol{\varrho})\cdot\nabla]\frac{1}{|\mathbf{r}-\boldsymbol{\varrho}|}\,dV(\boldsymbol{\varrho}) \tag{15.6-13}$$

and *surface-distribution potentials for charges and dipoles (potentials of single- and double-layer distributions)*

$$\int_S \sigma(\boldsymbol{\varrho})\varphi_0(\mathbf{r}-\boldsymbol{\varrho})\,dA(\boldsymbol{\varrho}) = \int_S \frac{\sigma(\boldsymbol{\varrho})}{|\mathbf{r}-\boldsymbol{\varrho}|}\,dA(\boldsymbol{\varrho}) \tag{15.6-14}$$

$$\begin{aligned}-\int_S p(\boldsymbol{\varrho})[d\mathbf{A}(\boldsymbol{\varrho})\cdot\nabla]\varphi_0(\mathbf{r}-\boldsymbol{\varrho}) &= -\int_S p(\boldsymbol{\varrho})\frac{\partial\varphi_0}{\partial n}\,dA(\boldsymbol{\varrho})\\ &= -\int_S p(\boldsymbol{\varrho})\frac{\partial}{\partial n}\left(\frac{1}{|\mathbf{r}-\boldsymbol{\varrho}|}\right)dA(\boldsymbol{\varrho})\end{aligned} \tag{15.6-15}$$

Potentials due to *one-dimensional charge distributions* (line integrals) are of interest mainly in two-dimensional potential theory (Sec. 15.6-7).

If the charge density $Q(\mathbf{r})$ is bounded and integrable in V, the simple volume potential (12) and all its derivatives exist and are uniformly continuous for all $\mathbf{r}$; the derivatives may be obtained by differentiation under the integral sign. The potential satisfies Poisson's equation (2) if $Q(\mathbf{r})$ is bounded and continuously differentiable.

If the surface-density functions $\sigma(\mathbf{r})$ and $p(\mathbf{r})$ are twice differentiable on S,

1. The single-layer potential (14) is continuous at every regular point $\mathbf{r}_0$ of the surface S. The same is true for the directional derivative $\partial\Phi/\partial t$ in any given direction tangent to S at $\mathbf{r}_0$; but the directional derivative $\partial\Phi/\partial n$ along the normal of S at $\mathbf{r}_0$ satisfies the *jump relation*

$$\left.\frac{\partial\Phi}{\partial n}\right]_+ - \left.\frac{\partial\Phi}{\partial n}\right]_- = -4\pi\sigma(\mathbf{r}_0) \tag{15.6-16}$$

where the subscripts $+$ and $-$ indicate the respective unilateral limits as $\mathbf{r}\to\mathbf{r}_0$ on the positive-normal side of S and on the negative-normal side of S.

2. The double-layer potential (15) and its tangential derivatives satisfy the jump relations

$$\Phi_+(\mathbf{r}_0) - \Phi(\mathbf{r}_0) = \Phi(\mathbf{r}_0) - \Phi_-(\mathbf{r}_0) = 2\pi p(\mathbf{r}_0) \tag{15.6-17}$$

$$\left.\frac{\partial\Phi}{\partial t}\right]_+ - \left.\frac{\partial\Phi}{\partial t}\right]_{\mathbf{r}=\mathbf{r}_0} = \left.\frac{\partial\Phi}{\partial t}\right]_{\mathbf{r}=\mathbf{r}_0} - \left.\frac{\partial\Phi}{\partial t}\right]_- = 2\pi\frac{\partial p}{\partial t} \tag{15.6-18}$$

at every regular point $\mathbf{r}_0$ of S, while the normal derivative $\partial\Phi/\partial n$ is continuous.

NOTE: In the special case $p(\mathbf{r}) = p =$ constant, the double-layer potential (15) equals p times the *solid angle* subtended by S at the point $(\mathbf{r})$; this angle is taken to be positive if $(\mathbf{r})$ is on the side of the positive normals to S. For a closed surface S such a potential equals $-4\pi p$ if $(\mathbf{r})$ is inside S, and 0 if $(\mathbf{r})$ is outside S.

(c) Multipole Expansion and Gauss's Theorem. Consider the potential $\Phi(\mathbf{r})$ due to any combination of charge distributions confined to a bounded sphere $|\mathbf{r}| \leq R$; let Q_T be the finite total charge. $\Phi(\mathbf{r})$ will be a linear combination of potentials of the types (12) to (15); for $|\mathbf{r}| > R$, one can expand $\Phi(\mathbf{r})$ as a Taylor series (5.5-4) of terms (10), or as a series of spherical harmonics (11). For sufficiently large r, the potential is thus successively approximated by the potential of a point charge Q_T at the origin, by a point-charge potential plus a dipole potential, etc. (*multipole expansion*). For every regular surface enclosing the entire charge distribution, Gauss's theorem (Table 5.6-1) takes the special form

$$\int_S d\mathbf{A} \cdot \nabla\Phi = \int_S \frac{\partial \Phi}{\partial n}\, dA = -4\pi Q_T \tag{15.6-19}$$

(d) General Solutions of Laplace's and Poisson's Equations as Potentials. Let V be a singly connected, bounded or unbounded three-dimensional region with regular boundary surface S, let $\Phi(\mathbf{r})$ be twice continuously differentiable in V and continuously differentiable on S, and let $\Phi(\mathbf{r}) = O(1/r)$ as $r \to \infty$. Then Green's theorem (Table 5.6-1 or Sec. 15.4-3c) permits one to represent $\Phi(\mathbf{r})$ in the form

$$\boxed{\begin{aligned}\Phi(\mathbf{r}) = \frac{1}{4\pi} &\int_S [\varphi_0(\mathbf{r} - \boldsymbol{\varrho})\nabla_\rho\Phi(\boldsymbol{\varrho})] \cdot d\mathbf{A}(\boldsymbol{\varrho}) \\ &- \frac{1}{4\pi}\int_S [\Phi(\boldsymbol{\varrho})\nabla_\rho\varphi_0(\mathbf{r} - \boldsymbol{\varrho}) \cdot d\mathbf{A}(\boldsymbol{\varrho}) \\ &- \frac{1}{4\pi}\int_V \varphi_0(\mathbf{r} - \boldsymbol{\varrho})\nabla^2\Phi(\boldsymbol{\varrho})\, dV(\boldsymbol{\varrho}) \qquad (\mathbf{r} \text{ in } V)\end{aligned}} \tag{15.6-20a}$$

$$\varphi_0(\mathbf{r} - \boldsymbol{\varrho}) = \frac{1}{|\mathbf{r} - \boldsymbol{\varrho}|} \tag{15.6-20b}$$

where the operator ∇_ρ implies differentiations with respect to the components of $\boldsymbol{\varrho}$; note that $\nabla_\rho\varphi_0(\mathbf{r} - \boldsymbol{\varrho}) = \nabla\varphi_0(\mathbf{r} - \boldsymbol{\varrho})$. *Equation* (20) *expresses every solution* $\Phi(\mathbf{r})$ *of Poisson's differential equation* (2) *as a potential due to three distributions:*

1. *A single-layer surface distribution* (14) *of density* $\dfrac{1}{4\pi}\dfrac{\nabla\Phi \cdot d\mathbf{A}}{dA} = \dfrac{1}{4\pi}\dfrac{\partial\Phi}{\partial n}$
2. *A double-layer surface distribution* (15) *of density* $\dfrac{1}{4\pi}\Phi$
3. *A volume distribution* (12) *of density* $-\dfrac{1}{4\pi}\nabla^2\Phi = Q(\mathbf{r})$

The last potential vanishes if $\Phi(\mathbf{r})$ *satisfies Laplace's differential equation* (1) *in* V.

NOTE: The expression (20*a*) vanishes if $\mathbf{r}$ is outside V and S, and equals $\Phi(\mathbf{r})/2$ for $\mathbf{r}$ in S.

15.6-6. Solution of Three-dimensional Boundary-value Problems by Green's Functions. The Green's-function methods of Secs. 15.5-1 and 15.5-4 express the solution $\Phi(\mathbf{r})$ of the Dirichlet problem (3) and the Neumann problem (4) for suitable regions V as surface integrals

$$\Phi(\mathbf{r}) = \int_S G_S(\mathbf{r}, \boldsymbol{\varrho}) b(\boldsymbol{\varrho})\, dA(\boldsymbol{\varrho}) \tag{15.6-21}$$

Section 15.5-4*b* relates each "surface" Green's function $G_S(\mathbf{r}, \boldsymbol{\varrho})$ simply to the ordinary Green's function $G(\mathbf{r}, \boldsymbol{\varrho})$ which yields the solution

$$\Phi(\mathbf{r}) = 4\pi \int_V G(\mathbf{r}, \boldsymbol{\varrho}) Q(\boldsymbol{\varrho})\, dV(\boldsymbol{\varrho}) \tag{15.6-22}$$

of Poisson's differential equation (2) subject to "complementary" homogeneous Dirichlet or Neumann conditions. Superposition of Eqs. (21) and (22) yields solutions of Poisson's equation with given boundary values $b(\boldsymbol{\varrho})$ of Φ or $\partial\Phi/\partial n$ (Sec. 15.4-2). Note $G(\boldsymbol{\varrho}, \mathbf{r}) \equiv G(\mathbf{r}, \boldsymbol{\varrho})$ (Sec. 15.5-1). The Green's functions are easily found in the following special cases; note that the positive surface normals point *outward.*

(a) Green's Function for the Entire Space. If V is the entire three-dimensional space, Eq. (20) yields the unique solution (22) of Poisson's equation subject to the "boundary conditions" $\Phi(\mathbf{r}) = O(1/r)$ as $r \to \infty$. The required Green's function is

$$G(\mathbf{r}, \boldsymbol{\varrho}) = \frac{1}{4\pi} \frac{1}{|\mathbf{r} - \boldsymbol{\varrho}|} = \frac{1}{4\pi} \varphi_0(\mathbf{r} - \boldsymbol{\varrho}) \tag{15.6-23}$$

(see also Sec. 5.7-3).

(b) Infinite Plane Boundary with Dirichlet Conditions. Image Charges. If V is the half-space $z > 0$, the *Green's functions for Dirichlet conditions are*

$$G(\mathbf{r}, \boldsymbol{\varrho}) = \frac{1}{4\pi}\left(\frac{1}{|\mathbf{r} - \boldsymbol{\varrho}|} - \frac{1}{|\mathbf{r} - \bar{\boldsymbol{\varrho}}|}\right) = \frac{1}{4\pi}[\varphi_0(\mathbf{r} - \boldsymbol{\varrho}) - \varphi_0(\mathbf{r} - \bar{\boldsymbol{\varrho}})] \tag{15.6-24}$$

$$G_S(\mathbf{r}, \boldsymbol{\varrho}) = -\frac{\partial G}{\partial \nu} = \left.\frac{\partial G}{\partial \zeta}\right]_{\zeta=0} = \frac{z}{2\pi[(x - \xi)^2 + (y - \eta)^2 + z^2]^{3/2}} \tag{15.6-25}$$

where the point $(\bar{\boldsymbol{\varrho}}) \equiv (\xi, \eta, -\zeta)$ is the *reflected image* of $(\boldsymbol{\varrho}) \equiv (\xi, \eta, \zeta)$ in the boundary plane. The solution (22) can be interpreted as a volume potential due to the given charges and induced *image charges;* equation (21) expresses the effect of nonhomogeneous Dirichlet conditions as the potential of a double layer on the boundary.

(c) Sphere with Dirichlet Conditions. Poisson's Integral Formulas. If V is the *interior* of a sphere $|\mathbf{r}| = R$, the *Green's functions for*

Dirichlet conditions are

$$G(\mathbf{r}, \boldsymbol{\varrho}) = \frac{1}{4\pi}\left(\frac{1}{|\mathbf{r} - \boldsymbol{\varrho}|} - \frac{R}{r}\frac{1}{\left|\frac{R^2}{r^2}\mathbf{r} - \boldsymbol{\varrho}\right|}\right)$$

$$= \frac{1}{4\pi}\left[\varphi_0(\mathbf{r} - \boldsymbol{\varrho}) - \frac{R}{r}\varphi_0\left(\frac{R^2}{r^2}\mathbf{r} - \boldsymbol{\varrho}\right)\right] \quad (15.6\text{-}26)$$

$$G_S(\mathbf{r}, \boldsymbol{\varrho}) = -\frac{\partial G}{\partial \nu} = -\frac{\partial G}{\partial \rho}\bigg]_{\rho=R} = -\frac{1}{4\pi}\frac{r^2 - R^2}{R(r^2 + R^2 - 2Rr\cos\gamma)^{3/2}} \quad (15.6\text{-}27)$$

where γ is the angle between $\mathbf{r}$ and $\boldsymbol{\varrho}$, or

$$\cos\gamma = \cos\vartheta\cos\vartheta' + \sin\vartheta\sin\vartheta'\cos(\varphi - \varphi') \quad (15.6\text{-}28)$$

if the spherical coordinates of the points $(\mathbf{r})$ and $(\boldsymbol{\varrho})$ are respectively denoted by r, ϑ, φ and φ, ϑ', φ'. The second term in Eq. (26) may be regarded as the effect of an induced *image charge* located with the aid of Kelvin's inversion theorem (Sec. 15.6-3). The solution (21) of the Dirichlet problem

$$\nabla^2\Phi(r, \vartheta, \varphi) = 0 \qquad (r < R) \qquad \Phi = b(\vartheta, \varphi) \qquad (r = R)$$

becomes

$$\Phi(r, \vartheta, \varphi) = \frac{R(R^2 - r^2)}{4\pi}\int_0^{2\pi} d\varphi' \int_0^{\pi} \frac{b(\vartheta', \varphi')\sin\vartheta'\,d\vartheta'}{(R^2 + r^2 - 2Rr\cos\gamma)^{3/2}}$$

(POISSON'S INTEGRAL FORMULA) (15.6-29)

which may again be interpreted as a double-layer potential. The expression (27) can be expanded in spherical surface harmonics with the aid of Eq. (21.8-68) (multipole expansion, Sec. 15.6-5*c*; see also Sec. 10.4-9).

Equations (26) to (29) yield solutions in the *interior* of the sphere $(r < R)$. If V is the *exterior* of the sphere $(r > R)$, $G(\mathbf{r}, \boldsymbol{\varrho})$ is still given by Eq. (26), but now $G_S(\mathbf{r}, \boldsymbol{\varrho}) = -\frac{\partial G(\mathbf{r}, \boldsymbol{\varrho})}{\partial \nu} = \frac{\partial G(\mathbf{r}, \boldsymbol{\varrho})}{\partial \rho}\Big]_{\rho=R}$, so that *the signs on the right sides of Eqs.* (27) *and* (29) *must be reversed.*

(d) An Existence Theorem. *A Green's function $G(\mathbf{r}, \boldsymbol{\varrho})$ for Poisson's equation with Dirichlet conditions—and hence a solution of the Dirichlet problem* (3) *with reasonable boundary values—exists for every region V bounded by a finite number of regular surface elements such that every boundary point can be the vertex of a tetrahedron outside V.*

15.6-7. Two-dimensional Potential Theory. Logarithmic Potentials. **(a)** In many three-dimensional potential problems $\Phi(x, y, z)$ is independent of the z coordinate, and V is a right-cylindrical region whose boundary surface S intersects the xy plane in a boundary curve C. V is then represented by a plane region D bounded by C (e.g., potentials and flows about infinite cylinders, plane boundaries). Such problems form the subject matter of *two-dimensional potential theory.* The divergence theorem and Green's theorem take the form (4.6-33); one may use rectangular cartesian coordinates x, y, or plane polar coordinates r, φ to write Laplace's and Poisson's equations as

$$\nabla^2\Phi \equiv \frac{\partial^2\Phi}{\partial x^2} + \frac{\partial^2\Phi}{\partial y^2} \equiv \frac{1}{r}\frac{\partial}{\partial r}\left(r\frac{\partial\Phi}{\partial r}\right) + \frac{1}{r^2}\frac{\partial^2\Phi}{\partial\varphi^2} = 0$$

(LAPLACE'S DIFFERENTIAL EQUATION) (15.6-30)

$$\nabla^2\Phi \equiv \frac{\partial^2\Phi}{\partial x^2} + \frac{\partial^2\Phi}{\partial y^2} \equiv \frac{1}{r}\frac{\partial}{\partial r}\left(r\frac{\partial\Phi}{\partial r}\right) + \frac{1}{r^2}\frac{\partial^2\Phi}{\partial\varphi^2} = -2\pi Q$$

(POISSON'S DIFFERENTIAL EQUATION) (15.6-31)

with $\Phi = \Phi(\mathbf{r})$, $Q = Q(\mathbf{r})$, $\mathbf{r} \equiv (x,y) \equiv (r,\varphi)$. Theorems entirely analogous to those of Sec. 15.6-4 apply to two-dimensional harmonic functions; it is only necessary to substitute *circles* for spheres in each mean-value theorem. Kelvin's inversion formula (5) holds for every *circle* $|\mathbf{r} - \mathbf{a}| = R$; thus, if $\Phi(r, \varphi)$ is a solution for $r < R$, the same is true for $\frac{R}{r}\Phi\left(\frac{R^2}{r}, \varphi\right)$ $(r > R)$, and conversely.

(**b**) The two-dimensional Laplace equation (30) has the elementary particular solution

$$\psi_0(\mathbf{r} - \boldsymbol{\varrho}) = \log_e \frac{1}{|\mathbf{r} - \boldsymbol{\varrho}|} = \log_e \frac{1}{\sqrt{(x - \xi)^2 + (y - \eta)^2}}$$

(LOGARITHMIC POTENTIAL of a point source at $\mathbf{r} = \boldsymbol{\varrho}$) (15.6-32)

which describes the potential of a uniformly charged straight line in the z direction. The function (32) takes the role of *the* function (8) in the two-dimensional theory, so that the *potentials* (12), (14), *and* (15) *of Sec.* 15.6-5*b are replaced by corresponding logarithmic potentials*

$$\int_D Q(\boldsymbol{\varrho})\psi_0(\mathbf{r} - \boldsymbol{\varrho})\, dA(\boldsymbol{\varrho}) = \int_D Q(\boldsymbol{\varrho}) \log_e \frac{1}{|\mathbf{r} - \boldsymbol{\varrho}|}\, dA(\boldsymbol{\varrho}) \tag{15.6-33}$$

$$\int_C \sigma(\boldsymbol{\varrho})\psi_0(\mathbf{r} - \boldsymbol{\varrho})\, ds(\boldsymbol{\varrho}) = \int_C \sigma(\boldsymbol{\varrho}) \log_e \frac{1}{|\mathbf{r} - \boldsymbol{\varrho}|}\, ds(\boldsymbol{\varrho}) \tag{15.6-34}$$

$$-\int_C p(\rho)\frac{\partial\psi_0}{\partial n}\, ds(\boldsymbol{\varrho}) = -\int_C p(\boldsymbol{\varrho})\frac{\partial}{\partial n} \log_e \frac{1}{|\mathbf{r} - \boldsymbol{\varrho}|}\, ds(\boldsymbol{\varrho}) \tag{15.6-35}$$

Equation (20) is replaced by

$$\Phi(\mathbf{r}) = \frac{1}{2\pi}\int_C \psi_0(\mathbf{r} - \boldsymbol{\varrho})\frac{\partial\Phi}{\partial\nu}\, ds(\boldsymbol{\varrho}) - \frac{1}{2\pi}\int_S \Phi(\boldsymbol{\varrho})\frac{\partial\psi_0}{\partial\nu}\, ds(\boldsymbol{\varrho}) - \frac{1}{2\pi}\int_D \psi_0(\mathbf{r} - \boldsymbol{\varrho})\nabla_\rho{}^2\Phi(\boldsymbol{\varrho})\, dA(\boldsymbol{\varrho}) \tag{15.6-36}$$

Note

$$\frac{\partial f(\xi, \eta)}{\partial\nu}\, ds(\boldsymbol{\varrho}) \equiv \frac{\partial f(\xi, \eta)}{\partial\xi}\, d\eta - \frac{\partial f(\xi, \eta)}{\partial\eta}\, d\xi \tag{15.6-37}$$

15.6-8. Two-dimensional Potential Theory: Conjugate Harmonic Functions (see also Secs. 7.3-2 and 7.3-3). (**a**) The xy plane may be regarded as a complex-number plane with points $z = x + iy$. A pair of (necessarily harmonic) functions $\Phi(x, y)$, $\Psi(x, y)$ are **harmonic conjugate functions** in a region D of the plane if and only if

$$\Xi = \Phi(x, y) + i\Psi(x, y)$$

is an analytic function of $z = x + iy$ in D. Harmonic conjugate functions are related by the Cauchy-Riemann equations

$$\frac{\partial \Phi}{\partial x} = \frac{\partial \Psi}{\partial y} \qquad \frac{\partial \Phi}{\partial y} = -\frac{\partial \Psi}{\partial x} \tag{15.6-38}$$

and define each other uniquely throughout D, except for arbitary additive constants. Given a function $\Phi(x, y)$ harmonic in D, one obtains $\psi(x, y)$ as the line integral

$$\Psi(x, y) = \int_{(x_0,y_0)}^{(x,y)} \left(-\frac{\partial \Phi}{\partial y}\, dx + \frac{\partial \Phi}{\partial x}\, dy \right) \tag{15.6-39a}$$

where x_0 and y_0 are arbitrary constants, and the path of integration is located in D. If $\Psi(x, y)$ is given, one has

$$\Phi(x, y) = \int_{(x_0,y_0)}^{(x,y)} \left(\frac{\partial \Psi}{\partial y}\, dx - \frac{\partial \Psi}{\partial x}\, dy \right) \tag{16.6-39b}$$

The curves $\Phi(x,y) = constant$ *and* $\Psi(x,y) = constant$ *are mutually orthogonal families.* These curves have important physical interpretations (*equipotential lines* and *gradient lines* in electrostatics, *lines of constant velocity potential* and *streamlines* for incompressible flow). Ξ is often called the **complex potential.**

(**b**) *Every transformation*

$$\bar{z} = \bar{z}(z) \qquad (z = x + iy, \qquad \bar{z} = \bar{x} + i\bar{y}) \tag{15.6-40}$$

which is analytic and such that $d\bar{z}/dz \neq 0$ *in* V *(conformal mapping, Sec. 7.9-1) transforms the harmonic conjugates* $\Phi(x, y)$, $\Psi(x, y)$ *into a new pair of harmonic conjugates* $\bar{\Phi}(\bar{x}, \bar{y})$, $\bar{\Psi}(\bar{x}, \bar{y})$ *with mutually orthogonal contour lines.* This theorem permits one to simplify boundaries and contour lines by conformal mapping (see also Sec. 15.6-9).

(**c**) Let $\Psi(x, y)$ be the solution of a Neumann problem

$$\nabla^2\Psi = 0 \quad \text{(in } D\text{)} \qquad \frac{\partial \Psi}{\partial n} = b(x, y) \equiv b(s) \quad \text{(on } C\text{)} \tag{15.6-41}$$

such that $\Psi(x, y)$ and its derivatives are continuous on C as well as in D. Then the complex conjugate (39b) of $\Psi(x, y)$ is the solution of the Dirichlet problem

$$\nabla^2\Phi = 0 \qquad (\text{in } D) \qquad \Phi(x, y) = B(x, y) = B(s) \qquad (\text{on } C) \tag{15.6-42}$$

with $$\frac{\partial\Phi}{\partial s} = -\frac{\partial\Psi}{\partial n} \qquad (\text{on } C) \qquad \text{or} \qquad B(s) = -\int b(s)\,ds + \text{constant} \tag{15.6-43}$$

The solution $\Phi(x, y)$ of the Dirichlet problem (42) similarly yields the solution (39*a*) of the Neumann problem (41), provided that

$$-\oint_C b(s)\,ds = \oint_C \frac{\partial B}{\partial s}\,ds = 0 \tag{15.6-44}$$

so that Gauss's theorem is satisfied (see also Table 5.6-1 and Sec. 15.6-5*c*).

15.6-9. Solution of Two-dimensional Boundary-value Problems. Green's Functions and Conformal Mapping (see also Secs. 15.5-1, 15.5-4, and 15.6-6). The Green's-function methods of Secs. 15.5-1 and 15.5-3 express the solution $\Phi(\mathbf{r})$ of Poisson's differential equation (31) with homogeneous linear boundary conditions in the form

$$\Phi(\mathbf{r}) = \int_D G(\mathbf{r}, \boldsymbol{\varrho})Q(\boldsymbol{\varrho})\,dA(\boldsymbol{\varrho}) \tag{15.6-45}$$

and the solution $\Phi(\mathbf{r})$ of Laplace's differential equation (30) with given boundary values $b(\mathbf{r})$ of Φ or $\partial\Phi/\partial n$ as

$$\Phi(\mathbf{r}) = \int_C G_S(\mathbf{r}, \boldsymbol{\varrho})b(\boldsymbol{\varrho})\,ds(\boldsymbol{\varrho}) \tag{15.6-46}$$

Solutions of Poisson's equation subject to nonhomogeneous linear boundary conditions may be obtained by superposition of suitable integrals (45) and (46). Note $G(\mathbf{r}, \boldsymbol{\varrho}) = G(\boldsymbol{\varrho}, \mathbf{r})$. The Green's functions are easily found in the following special cases.

(a) Green's Function for the Entire Plane. If D is the entire plane, Eq. (36) yields the unique solution (45) of Poisson's equation subject to the "boundary conditions" $\Phi(\mathbf{r}) \to 0$ as $r \to \infty$. The required Green's function is

$$G(\mathbf{r}, \boldsymbol{\varrho}) = \frac{1}{2\pi} \log_e \frac{1}{|\mathbf{r} - \boldsymbol{\varrho}|} \tag{15.6-47}$$

(b) Half Plane with Dirichlet Conditions. If D is the half plane $x > 0$, the *Green's functions for Dirichlet conditions* are

$$\begin{aligned} G(\mathbf{r}, \boldsymbol{\varrho}) &= \frac{1}{2\pi}\left[\log_e \frac{1}{|\mathbf{r} - \boldsymbol{\varrho}|} - \log_e \frac{1}{|\mathbf{r} - \bar{\boldsymbol{\varrho}}|}\right] \\ &= \frac{1}{2\pi}\left[\psi_0(\mathbf{r} - \boldsymbol{\varrho}) - \psi_0(\mathbf{r} - \bar{\boldsymbol{\varrho}})\right] \\ &= \frac{1}{4\pi} \log_e \frac{(x + \xi)^2 + (y - \eta)^2}{(x - \xi)^2 + (y - \eta)^2} \end{aligned} \tag{15.6-48}$$

$$G_S(\mathbf{r}, \boldsymbol{\varrho}) = -\frac{\partial G}{\partial \nu} = \left.\frac{\partial G}{\partial \xi}\right]_{\xi=0} = \frac{1}{\pi}\frac{x}{x^2 + (y - \eta)^2} \tag{15.6-49}$$

where the point $(\bar{\varrho}) \equiv (-\xi, \eta)$ is the reflected image of $(\varrho) = (\xi, \eta)$ in the boundary line.

(c) Circle with Dirichlet Conditions. Poisson's Integral Formulas. If D is the *interior* of a circle $r = R$, the *Green's functions for Dirichlet conditions* are

$$G(\mathbf{r}, \varrho) = \frac{1}{4\pi} \log_e \frac{R^2 + \dfrac{r^2\rho^2}{R^2} - 2r\rho \cos(\varphi - \varphi')}{r^2 + \rho^2 - 2r\rho \cos(\varphi - \varphi')} \tag{15.6-50}$$

$$G_S(\mathbf{r}, \varrho) = -\frac{\partial G}{\partial \nu} = -\left.\frac{\partial G}{\partial \rho}\right]_{\rho = R} = \frac{1}{2\pi R} \frac{R^2 - r^2}{R^2 + r^2 - 2Rr \cos(\varphi - \varphi')} \tag{15.6-51}$$

where ρ, φ' are the polar coordinates of the point (ϱ). The solution (45) of the Dirichlet problem

$$\nabla^2\Phi(r, \varphi) = 0 \qquad (r < 0) \qquad \Phi = b(\varphi) \qquad (r = R) \tag{15.6-52}$$

becomes

$$\Phi(r, \varphi) = \frac{1}{2\pi} \int_0^{2\pi} \frac{R^2 - r^2}{R^2 + r^2 - 2Rr \cos(\varphi - \varphi')} b(\varphi')\, d\varphi'$$

(POISSON'S INTEGRAL FORMULA) (15.6-53)

which may be expanded as a Fourier series in φ.

Equations (50) to (53) yield solutions in the *interior* of the circle $(r < R)$. If D is the *exterior* of the circle $(r > R)$, $G(\mathbf{r}, \varrho)$ is still given by Eq. (50), but now $G_S(\mathbf{r}, \varrho) = -\dfrac{\partial G}{\partial \nu} = \left.\dfrac{\partial G}{\partial \rho}\right]_{\rho = R}$, so that the signs on the right sides of Eqs. (51) and (53) must be reversed.

(d) Existence of Green's Functions and Conformal Mapping. Section 15.6-9*c* and Riemann's mapping theorem (Sec. 7.10-1) imply the existence of a Green's function (and hence of a solution of the Dirichlet problem) for any region D which, together with its boundary C, can be mapped conformally onto the unit circle. More specifically, *let $w = w(z)$ be the analytic function mapping the points $z = x + iy$ of D and C onto the unit circle so that the point $z = \zeta = \xi + i\eta$ of D is transformed into the origin, or*

$$w(z) = 1 \quad (z \text{ on } C) \qquad w(\zeta) = 0 \tag{15.6-54}$$

Then the Dirichlet problem for the region D admits the Green's function

$$G(x, y; \xi, \eta) \equiv G(z, \zeta) = \log_e \frac{1}{|w(z)|} \tag{15.6-55}$$

15.6-10. Extension of the Theory to More General Differential Equations. Retarded and Advanced Potentials (see also Sec. 10.4-4).

The theory of Secs. 15.6-5 to 15.6-7 and 15.6-9 attempts to construct solutions of Laplace's and Poisson's differential equations by superposition of simple and dipole potentials. The theory is readily generalized to deal with the more general differential equations

$$\nabla^2\Phi + k^2\Phi = 0 \qquad (15.6\text{-}56)$$

$$\nabla^2\Phi + k^2\Phi = -4\pi q(\mathbf{r}) \qquad (15.6\text{-}57)$$

which include the space form of the wave equation (k real, Sec. 10.4-4) and the space form of the Klein-Gordon equation used in nuclear physics ($k = i\kappa$). The differential equation (57) is of the type discussed in Sec. 15.5-4*b*.

(a) Three-dimensional Case. The three-dimensional equation (56) admits the elementary particular solution

$$\varphi_0(\mathbf{r} - \boldsymbol{\varrho}) = \frac{e^{\pm ik|\mathbf{r}-\boldsymbol{\varrho}|}}{|\mathbf{r} - \boldsymbol{\varrho}|} \qquad (\mathbf{r} \neq \boldsymbol{\varrho}) \qquad (15.6\text{-}58)$$

For real k the positive sign corresponds to outgoing waves, and the negative sign to incoming waves; for imaginary $k = i\kappa$, only negative exponents $-|\kappa|$ are of general interest. *Substitution of the appropriate expression* (58) *for* $\varphi_0(\mathbf{r} - \boldsymbol{\varrho}) = 1/|\mathbf{r} - \boldsymbol{\varrho}|$ *in Eqs.* (8) *to* (15), (23), *and* (24) *yields solutions of the differential equation* (56) *or* (57) *instead of the corresponding solutions of Laplace's or Poisson's equation.*

The resulting solutions of the time-dependent wave equation (Sec. 10.4-8) are special cases of **retarded potentials** [positive sign in Eq. (58)] and **advanced potentials** [negative sign in Eq. (58)].

In particular, if $\Phi(\mathbf{r})$ is a twice continuously differentiable solution of the homogeneous differential equation (56), Eq. (20) is replaced by *Helmholtz's theorem*

$$\Phi(\mathbf{r}) = \frac{1}{4\pi}\int_S \left(\frac{e^{\pm ik|\mathbf{r}-\boldsymbol{\varrho}|}}{|\mathbf{r} - \boldsymbol{\varrho}|}\nabla_\rho\Phi\right)\cdot d\mathbf{A}(\boldsymbol{\varrho}) - \frac{1}{4\pi}\int_S \left(\Phi\nabla_\rho \frac{e^{\pm ik|\mathbf{r}-\boldsymbol{\varrho}|}}{|\mathbf{r} - \boldsymbol{\varrho}|}\right)\cdot d\mathbf{A}(\boldsymbol{\varrho})$$

$$(\mathbf{r} \text{ in } V) \qquad (15.6\text{-}59)$$

(b) Two-dimensional Case. In *the case of two-dimensional differential equations of the form* (56) *or* (57), the elementary particular solution (32) is replaced by

$$\varphi_0(\mathbf{r} - \boldsymbol{\varrho}) = \begin{cases} -\dfrac{\pi i}{2} H_0^{(1)}(k|\mathbf{r} - \boldsymbol{\varrho}|) & \text{(OUTGOING WAVES)} \\ \dfrac{\pi i}{2} H_0^{(2)}(k|\mathbf{r} - \boldsymbol{\varrho}|) & \text{(INCOMING WAVES)} \end{cases} \qquad (15.6\text{-}60)$$

where the $H_0^{(r)}(z)$ are Hankel functions (Sec. 21.8-1).

15.7. RELATED TOPICS, REFERENCES, AND BIBLIOGRAPHY

15.7-1. Related Topics. The following topics related to the study of linear integral equations, boundary-value problems, and eigenvalue problems are treated in other chapters of this handbook:

15.7-2. References and Bibliography (see also Secs. 4.12-2, 10.6-2, 12.9-2, and 14.11-2).

15.1. Akhiezer, N. I.: *Theory of Approximation*, Ungar, New York, 1956.
15.2. ——— and I. M. Glazman: *Theory of Linear Operators in Hilbert Space*, Ungar, New York, 1963.
15.3. Banach, S.: *Théorie des Opérations Linéaires*, Chelsea, New York, 1933.
15.4. Berberian, S. K.: *Introduction to Hilbert Space*, Oxford, Fair Lawn, N.J., 1963.
15.5. Bochner, S.: *Lectures on Fourier Integrals*, Princeton, Princeton, N.J., 1959.
15.6. Courant, R., and D. Hilbert: *Methods of Mathematical Physics*, rev. ed., Wiley, New York, 1953/66.
15.7. Dieudonné, J. A.: *Foundations of Modern Analysis*, Academic, New York, 1960.
15.8. Dunford, N., and J. T. Schwartz: *Linear Operators*, Interscience, New York, 1964.
15.9. Edwards, R. E.: *Functional Analysis*, Holt, New York, 1965.
15.10. Feshbach, H., and P. M. Morse: *Methods of Theoretical Physics* (2 vols.), McGraw-Hill, New York, 1953.
15.11. Friedman, B.: *Principles and Techniques of Applied Mathematics*, Wiley, New York, 1956.
15.12. ———: *Generalized Functions and Partial Differential Equations*, Prentice-Hall, Englewood Cliffs, N.J., 1956.
15.13. Halmos, P. R.: *Introduction to Hilbert Space and the Theory of Spectral Multiplicity*, Chelsea, New York, 1957.
15.14. Kellogg, O. D.: *Foundations of Potential Theory*, Ungar, New York, 1943.
15.15. Kolmogorov, A., and S. V. Fomin: *Elements of the Theory of Functions and Functional Analysis* (2 vols.), Graylock, New York, 1957/61.
15.16. Lanczos, C.: *Linear Differential Operators*, Van Nostrand, Princeton, N.J., 1961.
15.17. Liusternik, L. A., and V. J. Sobolev: *Elements of Functional Analysis*, Ungar, New Yo.k, 1961.
15.18. Lorch, E. R.: *Spectral Theory*, Oxford, Fair Lawn, N.J., 1962.
15.19. Madelung, E.: *Die mathematischen Hilfsmittel des Physikers*, 7th ed., Springer, Berlin, 1964.
15.20. Mikhlin, S. G.: *Integral Equations*, Pergamon, New York, 1957.
15.21. Riesz, I., and B. Nagy: *Functional Analysis*, Ungar, New York, 1955.
15.22. Vulikh, B. Z.: *Functional Analysis for Scientists and Technologists*, Pergamon, New York, 1963.

(See also the articles by F. Schlögl and J. Sneddon in vol. I of the *Handbuch der Physik*, Springer, Berlin, 1956, and the references for Chap. 10.)

CHAPTER **16**

REPRESENTATION OF MATHEMATICAL MODELS: TENSOR ALGEBRA AND ANALYSIS

16.1. INTRODUCTION

16.1-1. Tensors are mathematical objects (see also Sec. 12.1-1) associated as functions of "position" with a class (space) of other objects ("points" labeled with numerical coordinates). Each tensor is described (represented) by an ordered set of numerical functions (tensor components) of the coordinates in such a manner that it is possible to define mathematical relations between tensors independent of (invariant with respect to) the particular scheme of numerical description used.

Tensor algebra was developed by successive generalizations of the theory of vector spaces (Secs. 12.4-1 and 14.2-1 to 14.2-7), linear algebras (Secs. 12.4-2 and 14.3-5), and their representations (Sec. 14.1-2). *Tensor analysis* is particularly concerned with tensors in their aspect as point functions and applies especially to the description of curved spaces (Chap. 17) and of continuous fields in physics. Tensor methods frequently link complicated numerical data measured in different frames of reference to relatively simple abstract models.

16.1-2. Coordinate Systems and Admissible Transformations. Consider a class (*space*, region of a space) of objects (*points*) labeled with corresponding ordered sets of $n < \infty$ continuously variable real numbers (*coordinates*) $x^1, x^2, \ldots, x^n$. A coordinate transformation ("alias" point of view, see also Sec. 14.1-3) "*admissible*" in the sense of the following sections is a *relabeling* of each point with n new coordinates $\bar{x}^1, \bar{x}^2, \ldots, \bar{x}^n$ related to the original coordinates $x^1, x^2, \ldots, x^n$ by n *transformation equations*

$$\bar{x}^k = \bar{x}^k(x^1, x^2, \ldots, x^n) \qquad (k = 1, 2, \ldots, n) \tag{16.1-1}$$

such that, throughout the region of points under consideration, (1) each function $\bar{x}^k(x^1, x^2, \ldots, x^n)$ is single-valued and continuously differentiable and (2) the Jacobian (Sec. 4.5-5) det $[\partial \bar{x}^k/\partial x^i]$ is different from zero.

The class of "admissible" transformations constitutes a group (*Sec.* 12.2-1) *with respect to the operation of forming their "product," i.e., of applying two transformations successively* (*see also Sec.* 12.2-8). *The Jacobian of the product of two transformations is the product of the individual Jacobians. Each admissible transformation* (1) *has a unique inverse whose Jacobian is the reciprocal of the original Jacobian.*

16.1-3. Description (Representation) of Abstract Functions by Components. Dummy (Umbral) Index Notation. Tensor analysis deals with abstract objects associated with the points $(x^1, x^2, \ldots, x^n)$ of an n-dimensional space ($n < \infty$, Sec. 14.1-2) as *point functions* (see also Sec. 5.4-1) defined on a region of the space. Each point function $\mathbf{Q}(x^1, x^2, \ldots, x^n)$, say, will be described or **represented** by an ordered set of $n^R < \infty$ numerical functions (**components** of $\mathbf{Q}$, see also Sec. 14.1-2) $Q(j_1, j_2, \ldots, j_R; x^1, x^2, \ldots, x^n)$ of the coordinates $x^1, x^2, \ldots, x^n$, where each index $j_1, j_2, \ldots, j_R$ runs from 1 to n.

Depending on the type of object (Secs. 16.2-1 and 16.2-2), certain of the indices labeling each component are written as superscripts, and others as subscripts. Thus an object $\mathbf{Q}$ may be described by n^{r+s} components $Q^{i_1 i_2 \cdots i_r}_{i_1' i_2' \cdots i_s'}(x^1, x^2, \ldots, x^n)$, where all indices run from 1 to n. With this notation, it is possible to abbreviate sets of sums like

$$\sum_{k=1}^{n} A^{ik}B_k = c^i \qquad (i = 1, 2, \ldots, n) \qquad \text{by } A^{ik}B_k = c^i$$

$$\sum_{i=1}^{n} \sum_{k=1}^{n} A^i_{kj}B^kC_{ih} = D_{jh} \qquad (j, h = 1, 2, \ldots n) \qquad \text{by } A^i_{kj}B^kC_{ih} = D_{jh}$$

and also sums involving vectors, like

$$\sum_{k=1}^{n} a^k \mathbf{e}_k = \mathbf{a} \qquad \text{by } a^k \mathbf{e}_k = \mathbf{a}$$

through the use of the following conventions.

1. *Summation Convention.* A summation from 1 to n is performed over every **dummy (umbral) index** appearing once as a superscript and once as a subscript.

   Any dummy index may be changed at will, since dummy indices are "canceled" by the summation (EXAMPLE: $A^{ik}B_k = A^{ij}B_j = A^{ih}B_h$). Different summation indices should be used in the case of multiple summations.

2. *Range Convention.* All **free indices** appearing only as superscripts or only as subscripts are understood to run from 1 to n; so that an equation involving R free indices stands for n^R equations.

   The superscripts and subscripts on the two sides of any equation must match.

3. In derivatives like $\partial a^i/\partial x^k$, k is considered as a subscript.

The dummy-index notation defined by the above conventions is used throughout the remainder of this chapter. Thus expressions like $A^{ik}B_k$ are understood to be sums, unless the contrary is explicitly indicated, as in $(A^{ik}B_k)_{\text{no sum}}$.

In many applications, the dummy-index notation is considerably more powerful than the matrix notation employed in Chap. 13 (see Table 14.7-1 for a comparison of notations).

16.1-4. Schemes of Measurements and Induced Transformations. Invariants (see also Secs. 14.1-5 and 16.2-1). A **scheme of measurements** for a class (or a set of classes) of abstract point functions is a reference system (or a set of reference systems, one for each class of functions) for the (biunique) representation of each point function by a corresponding set of numerical components.

Each function will be represented by the same number of components in all schemes of measurements considered. In physics, each component value usually corresponds to the result of a physical measurement.

It is useful to associate a definite scheme of measurements (x scheme of measurements) with each system of space coordinates $x^1, x^2, \ldots, x^n$. The components $Q^{i_1 i_2 \cdots i_r}_{i_1' i_2' \cdots i_s'}(x^1, x^2, \ldots, x^n)$ of a point function $\mathbf{Q}(x^1, x^2, \ldots, x^n)$ in the x scheme of measurements and the components $\bar{Q}^{k_1 k_2 \cdots k_r}_{k_1' k_2' \cdots k_s'}$ $(\bar{x}^1, \bar{x}^2, \ldots, \bar{x}^n)$ of $\mathbf{Q}$ in the $\bar{x}$ scheme of measurements are then related by an **induced transformation**

$$\bar{Q}^{k_1k_2 \cdots k_r}_{k_1'k_2' \cdots k_s'}(\bar{x}^1, \bar{x}^2, \ldots \bar{x}^n)$$
$$= \bar{Q}^{k_1k_2 \cdots k_r}_{k_1'k_2' \cdots k_s'}[Q^{11 \cdots 1}_{11 \cdots 1}(x^1, x^2, \ldots, x^n), \ldots, Q^{nn \cdots n}_{nn \cdots n}(x^1, x^2, \ldots, x^n)] \quad (16.1\text{-}2)$$

associated with (induced by) each admissible coordinate transformation (1).

A class C of abstract point functions $\mathbf{Q}(x^1, x^2, \ldots, x^n)$ described in an x scheme of measurements by numerical components $Q^{i_1i_2 \cdots i_r}_{i_1'i_2' \cdots i_s'}(x^1, x^2, \ldots, x^n)$ constitutes a class of **invariants** or **geometrical objects** (sometimes called **tensors** in the most general sense) *if their mathematical properties can be defined in terms of (abstract) operations independent of the scheme of measurements* (see also Secs. 12.1-1 and 16.4-1). Then *the induced transformations* (2) *are related to the defining operations of* C as follows:

1. The correspondence between any admissible coordinate transformation (1) and the corresponding induced transformation (2) is an isomorphism (Sec. 12.1-6) preserving the operation of forming the product of two transformations.
2. Every induced transformation is an isomorphism preserving the defining operations of C (see also Sec. 16.4-1). Defining operations involving two or more classes $C_1, C_2, \ldots$ of invariants (e.g., scalars and vectors) are also preserved by the induced transformations of $C_1, C_2, \ldots$.

A class C of point functions may also be invariants only with respect to a subgroup of the group of admissible coordinate transformations (Sec. 12.2-8).

16.2. ABSOLUTE AND RELATIVE TENSORS

16.2-1. Definition of Absolute and Relative Tensors in Terms of Their Induced Transformation Laws

(see also Table 16.2-1 and Sec. 6.3-3). *Throughout this handbook (and in practically all applications),* **tensors** *are understood to be* **real absolute and relative tensors** *represented by* **real** *components. Each type of absolute or relative tensor is defined by a linear and homogeneous induced transformation law* (Sec. 16.1-4) *relating the tensor components in different schemes of measurements.* Given a space of points $(x^1, x^2, \ldots, x^n)$ and a group of admissible coordinate transformations (16.1-2),

1. An **(absolute) scalar (scalar invariant; absolute tensor of rank 0)** α is an object represented in an x scheme of measurements by a function $\alpha(x^1, x^2, \ldots, x^n)$ and in an $\bar{x}$ scheme of measurements by a function $\bar{\alpha}(\bar{x}^1, \bar{x}^2, \ldots, \bar{x}^n)$ related to $\alpha(x^1, x^2, \ldots, x^n)$ at each point by the induced transformation

$$\boxed{\bar{\alpha} = \alpha} \quad (16.2\text{-}1)$$

2. An **(absolute) contravariant vector (absolute contravariant tensor of rank** 1) **a** is an object represented in an x scheme of measurements by an ordered set of n functions **(components)**

Table 16.2-1. Definition of the Most Frequently Used Types of Absolute Tensor Quantities in Terms of Their Induced Transformation Laws (Sec. 16.2-1; refer to Sec. 16.1-3 for dummy-index notation)

	Type of tensor quantity	Components in x scheme of measurements	Components in $\bar{x}$ scheme of measurements
1	Absolute **scalar** (scalar invariant), α ($R = r = s = 0$)	$\alpha(x^1, x^2, \ldots, x^n)$	$\bar{\alpha}(\bar{x}^1, \bar{x}^2, \ldots, \bar{x}^n) = \alpha(x^1, x^2, \ldots, x^n)$
2	Absolute **contravariant vector, a** ($R = r = 1$, $s = 0$)	$a^i(x^1, x^2, \ldots, x^n)$	$\bar{a}^k(\bar{x}^1, \bar{x}^2, \ldots, \bar{x}^n) = \dfrac{\partial \bar{x}^k}{\partial x^i} a^i(x^1, x^2, \ldots, x^n)$
3	Absolute **covariant vector, a** ($R = s = 1$, $r = 0$)	$a_i(x^1, x^2, \ldots, x^n)$	$\bar{a}_k(\bar{x}^1, \bar{x}^2, \ldots, \bar{x}^n) = \dfrac{\partial x^i}{\partial \bar{x}^k} a_i(x^1, x^2, \ldots, x^n)$
4	Absolute **contravariant tensor of rank** 2, **A** ($R = r = 2$, $s = 0$)	$A^{ik}(x^1, x^2, \ldots, x^n)$	$\bar{A}^{jh}(\bar{x}^1, \bar{x}^2, \ldots, \bar{x}^n) = \dfrac{\partial \bar{x}^j}{\partial x^i} \dfrac{\partial \bar{x}^h}{\partial x^k} A^{ik}(x^1, x^2, \ldots, x^n)$
5	Absolute **covariant tensor of rank** 2, **A** ($R = s = 2$, $r = 0$)	$A_{ik}(x^1, x^2, \ldots, x^n)$	$\bar{A}_{jh}(\bar{x}^1, \bar{x}^2, \ldots, \bar{x}^n) = \dfrac{\partial x^i}{\partial \bar{x}^j} \dfrac{\partial x^k}{\partial \bar{x}^h} A_{ik}(x^1, x^2, \ldots, x^n)$
6	Absolute **mixed tensor of rank** 2, **A** ($R = 2$, $r = s = 1$)	$A^i_k(x^1, x^2, \ldots, x^n)$	$\bar{A}^j_h(\bar{x}^1, \bar{x}^2, \ldots, \bar{x}^n) = \dfrac{\partial \bar{x}^j}{\partial x^i} \dfrac{\partial x^k}{\partial \bar{x}^h} A^i_k(x^1, x^2, \ldots, x^n)$

$a^i(x^1, x^2, \ldots, x^n)$ and in an $\bar{x}$ scheme of measurements by an ordered set of n components $\bar{a}^k(\bar{x}^1, \bar{x}^2, \ldots, \bar{x}^n)$ related to the $a^i(x^1, x^2, \ldots, x^n)$ at each point by the induced transformation

$$\bar{a}^k = \frac{\partial \bar{x}^k}{\partial x^i} a^i \tag{16.2-2}$$

3. An (**absolute**) **covariant vector** (**absolute covariant tensor of rank** 1) **a** is an object represented in an x scheme of measurements by an ordered set of n functions (components) $a_i(x^1, x^2, \ldots, x^n)$ and in an $\bar{x}$ scheme of measurements by an ordered set of n components $\bar{a}_k(\bar{x}^1, \bar{x}^2, \ldots, \bar{x}^n)$ related to the $a_i(x^1, x^2, \ldots, x^n)$ at each point by the induced transformation

$$\bar{a}_k = \frac{\partial x^i}{\partial \bar{x}^k} a_i \tag{16.2-3}$$

4. An (**absolute**) **tensor A of rank** $r + s$, **contravariant of rank** r **and covariant of rank** s, is an object represented in an x scheme of measurements by an ordered set of n^{r+s} functions (components) $A^{i_1 i_2 \cdots i_r}_{i_1' i_2' \cdots i_s'}(x^1, x^2, \ldots, x^n)$ and in an $\bar{x}$ scheme of measurements by an ordered set of n^{r+s} components $\bar{A}^{k_1 k_2 \cdots k_r}_{k_1' k_2' \cdots k_s'}(\bar{x}^1, \bar{x}^2, \ldots, \bar{x}^n)$ related to the $A^{i_1 i_2 \cdots i_r}_{i_1' i_2' \cdots i_s'}(x^1, x^2, \ldots, x^n)$ at each point by the induced transformation

$$\bar{A}^{k_1 k_2 \cdots k_r}_{k_1' k_2' \cdots k_s'} = \frac{\partial \bar{x}^{k_1}}{\partial x^{i_1}} \frac{\partial \bar{x}^{k_2}}{\partial x^{i_2}} \cdots \frac{\partial \bar{x}^{k_r}}{\partial x^{i_r}} \frac{\partial x^{i_1'}}{\partial \bar{x}^{k_1'}} \frac{\partial x^{i_2'}}{\partial \bar{x}^{k_2'}} \cdots \frac{\partial x^{i_s'}}{\partial \bar{x}^{k_s'}} A^{i_1 i_2 \cdots i_r}_{i_1' i_2' \cdots i_s'} \tag{16.2-4}$$

5. A **relative tensor** (**pseudotensor**) **A of weight** W **and of rank** $r + s$, **contravariant of rank** r **and covariant of rank** s, is an object represented in an x scheme of measurements by n^{r+s} functions (components) $A^{i_1 i_2 \cdots i_r}_{i_1' i_2' \cdots i_s'}(x^1, x^2, \ldots, x^n)$ and in an $\bar{x}$ scheme of measurements by an ordered set of n^{r+s} components $\bar{A}^{k_1 k_2 \cdots k_r}_{k_1' k_2' \cdots k_s'}(\bar{x}^1, \bar{x}^2, \ldots, \bar{x}^n)$ related to the $A^{i_1 i_2 \cdots i_r}_{i_1' i_2' \cdots i_s'}(x^1, x^2, \ldots, x^n)$ at each point by the induced transformation

$$\bar{A}^{k_1 k_2 \cdots k_r}_{k_1' k_2' \cdots k_s'} = \frac{\partial \bar{x}^{k_1}}{\partial x^{i_1}} \frac{\partial \bar{x}^{k_2}}{\partial x^{i_2}} \cdots \frac{\partial \bar{x}^{k_r}}{\partial x^{i_r}} \frac{\partial x^{i_1'}}{\partial \bar{x}^{k_1'}} \frac{\partial x^{i_2'}}{\partial \bar{x}^{k_2'}} \cdots \frac{\partial x^{i_s'}}{\partial \bar{x}^{k_s'}} A^{i_1 i_2 \cdots i_r}_{i_1' i_2' \cdots i_s'} \left[\frac{\partial(x^1, x^2, \ldots, x^n)}{\partial(\bar{x}^1, \bar{x}^2, \ldots, \bar{x}^n)}\right]^W \tag{16.2-5}$$

where W is a real integer. Relative tensors are called **densities** for $W = 1$ and **capacities** for $W = -1$ (EXAMPLES: Volume and surface elements are scalar and vector capacities; see also Secs. 6.2-3b and 17.3-3c).

The defining transformation (5) includes Eqs. (1) to (4) as special cases ($W = 0$ for absolute tensors). A tensor represented by $A^{i_1 i_2 \cdots i_r}_{i_1' i_2' \cdots i_s'}$ is a **mixed tensor** if and only if neither r nor s equals zero. The induced transformation (5) characterizing every absolute or relative tensor quantity is linear and homogeneous in the tensor components. The corresponding inverse transformation is

$$A^{i_1 i_2 \cdots i_r}_{i_1' i_2' \cdots i_s'} = \frac{\partial x^{i_1}}{\partial \bar{x}^{k_1}} \frac{\partial x^{i_2}}{\partial \bar{x}^{k_2}} \cdots \frac{\partial x^{i_r}}{\partial \bar{x}^{k_r}} \frac{\partial \bar{x}^{k_1'}}{\partial x^{i_1'}} \frac{\partial \bar{x}^{k_2'}}{\partial x^{i_2'}} \cdots \frac{\partial \bar{x}^{k_s'}}{\partial x^{i_s'}} \bar{A}^{k_1 k_2 \cdots k_r}_{k_1' k_2' \cdots k_s'} \left[\frac{\partial(\bar{x}^1, \bar{x}^2, \ldots, \bar{x}^n)}{\partial(x^1, x^2, \ldots, x^n)} \right]^W \tag{16.2-6}$$

NOTE: *Relative ordering of superscripts and subscripts is frequently used to conserve symbols.* Thus $A^i{}_k$ and $A_k{}^i$ denote *different sets of components* (see also Secs. 16.7-2 and 16.9-1).

16.2-2. Infinitesimal Displacement. Gradient of an Absolute Scalar (see also Secs. 5.5-2, 5.7-1, 6.2-3, and 16.10-7). The coordinate differentials dx^i represent an absolute contravariant vector, the **infinitesimal displacement** $d\mathbf{r}$.

Given a suitably differentiable absolute scalar α, the components $\partial\alpha/\partial x^i$ represent an absolute covariant vector called the **gradient** $\nabla\alpha$ of α. *A given absolute covariant vector described by* a_i *is the gradient of an absolute scalar if and only if* $\dfrac{\partial a_i}{\partial x^k} - \dfrac{\partial a_k}{\partial x^i} = 0$ *for all* i, k.

16.3. TENSOR ALGEBRA: DEFINITION OF BASIC OPERATIONS

16.3-1. Equality of Tensors. Two tensors $\mathbf{A}$ and $\mathbf{B}$ of the same type, rank, and weight are **equal** ($\mathbf{A} = \mathbf{B}$) *at the point* $(x^1, x^2, \ldots, x^n)$ if and only if their corresponding components in any one scheme of measurements are equal at this point:

$$A^{i_1 i_2 \cdots i_r}_{i_1' i_2' \cdots i_s'}(x^1, x^2, \ldots, x^n) = B^{i_1 i_2 \cdots i_r}_{i_1' i_2' \cdots i_s'}(x^1, x^2, \ldots, x^n) \quad \text{(TENSOR EQUALITY)} \tag{16.3-1}$$

Corresponding components of $\mathbf{A}$ *and* $\mathbf{B}$ *are then equal in every scheme of measurements* (see also Sec. 16.4-1). Tensor equality is symmetric, reflexive, and transitive (Sec. 12.1-3).

NOTE: Relations between the "values" of tensor point functions *at different points* are not defined in ordinary tensor algebra. Some such relations are discussed, for the special case of tensors defined on Riemann spaces, in Sec. 16.10-9.

16.3-2. Null Tensor. The **null tensor 0** of any given type, rank, and weight is the tensor whose components in any one scheme of measurements are all equal to zero. Thus $\mathbf{A} = \mathbf{0}$ at the point $(x^1, x^2, \ldots, x^n)$ if and only if

$$A^{i_1 i_2 \cdots i_r}_{i_1' i_2' \cdots i_s'}(x^1, x^2, \ldots, x^n) = 0 \quad \text{(NULL TENSOR)} \tag{16.3-2}$$

All components of $\mathbf{A}$ *are then equal to zero in every scheme of measurements.*

16.3-3. Tensor Addition. Given a suitable class of tensors all of the same type, rank, and weight, the **sum** $\mathbf{C} = \mathbf{A} + \mathbf{B}$ **of two tensors** $\mathbf{A}$

and B is the tensor described in any one scheme of measurements (*and hence in all schemes of measurements*) by the sums of corresponding components of **A** and **B**:

$$C^{i_1 i_2 \cdots i_r}_{i_1' i_2' \cdots i_s'} = A^{i_1 i_2 \cdots i_r}_{i_1' i_2' \cdots i_s'} + B^{i_1 i_2 \cdots i_r}_{i_1' i_2' \cdots i_s'} \quad \text{(TENSOR ADDITION)} \tag{16.3-3}$$

A + **B** is of the same rank, type, and weight as **A** and **B**. Tensor addition is commutative and associative.

16.3-4. Multiplication of a Tensor by an Absolute Scalar. The *product* **B** = α**A** of a tensor **A** and an absolute scalar α is the tensor represented in every scheme of measurements by the products of the components of **A** and the scalar α:

$$B^{i_1 i_2 \cdots i_r}_{i_1' i_2' \cdots i_s'} = \alpha A^{i_1 i_2 \cdots i_r}_{i_1' i_2' \cdots i_s'} \quad \text{(MULTIPLICATION BY SCALARS)} \tag{16.3-4}$$

α**A** is of the same rank, type, and weight as **A**. Multiplication by scalars is commutative, associative, and distributive with respect to both tensor and scalar addition. In particular, (-1)**A** $\equiv$ $-$**A** is the negative (additive inverse) of **A**, with **A** $-$ **A** $= 0$.

16.3-5. Contraction of a Mixed Tensor. One may **contract** a mixed tensor **A** described by $A^{i_1 i_2 \cdots i_r}_{i_1' i_2' \cdots i_s'}$ by equating a superscript to a subscript and summing over the pair. The resulting n^{r+s-2} sums describe a tensor of the same weight as **A**, contravariant of rank $r - 1$ and covariant of rank $s - 1$. In general, a mixed tensor can be contracted in more than one way and/or more than once.

EXAMPLE: An absolute or relative mixed tensor **A** of rank 2 described by A^i_k may be contracted to form the absolute or relative scalar (**trace** of **A**)

$$A^i_i = A^1_1 + A^2_2 + \cdots + A^n_n$$

16.3-6. (Outer) Product of Two Tensors (see also Sec. 12.7-3). The **(outer) product C = AB of two tensors A and B**, respectively, of weight W and W' and represented by $A^{i_1 i_2 \cdots i_r}_{i_1' i_2' \cdots i_s'}$ and $B^{k_1 k_2 \cdots k_p}_{k_1' k_2' \cdots k_q'}$ is the tensor described by

$$C^{i_1 i_2 \cdots i_r k_1 k_2 \cdots k_p}_{i_1' i_2' \cdots i_s' k_1' k_2' \cdots k_q'} = A^{i_1 i_2 \cdots i_r}_{i_1' i_2' \cdots i_s'} B^{k_1 k_2 \cdots k_p}_{k_1' k_2' \cdots k_q'} \quad \text{(OUTER MULTIPLICATION)} \tag{16.3-5}$$

AB is contravariant of rank $r + p$ and covariant of rank $s + q$ and is of weight $W + W'$. Outer multiplication is associative; it is distributive with respect to tensor addition. It is not in general commutative, since the relative order of the indices in Eq. (5) must be observed.

EXAMPLES: $a^i b^k = A^{ik}$; $a^i b_k = A^i_k$; $a_i b_k = A_{ik}$. The product (4) is a special case of an outer product.

16.3-7. Inner Products. If the outer product of two tensors **A** and **B**, described by Eq. (5), can be contracted (Sec. 16.3-5) so that one or more *superscripts* of $A^{i_1 i_2 \cdots i_r}_{i_1' i_2' \cdots i_s'}$ are summed against one or more *subscripts* of $B^{k_1 k_2 \cdots k_p}_{k_1' k_2' \cdots k_q'}$ and/or conversely, the resulting sums represent an **inner product** of the tensors **A** and **B**. In general, several such inner products can be formed.

Every inner product of two tensors **A** and **B** is a tensor of the same weight as **AB**. The rank of the inner product is equal to that of **AB** diminished by twice the number of index pairs summed. Inner multiplication is distributive with respect to tensor addition.

EXAMPLES: $a^i b_i = \gamma$; $A^i_k a^k = c^i$; $A^i_k b_i = d_k$; $B^{ik} b_i = f^k$; $C_{ik} a^k = h_i$.

16.3-8. Indirect Tests for Tensor Character. Let **Q** be an object described in an x scheme of measurements by n^R components $Q(j_1, j_2, \ldots, j_R; x^1, x^2, \ldots, x^n)$, and let **X** be a tensor described by $X^{i_1 i_2 \cdots i_r}_{i_1' i_2' \cdots i_s'}(x^1, x^2, \ldots, x^n)$. The *outer product* **QX** is defined, as in Sec. 16.3-6, by the n^{R+r+s} components $Q(j_1, j_2, \ldots, j_R) X^{i_1 i_2 \cdots i_r}_{i_1' i_2' \cdots i_s'}$. *Inner products* of **Q** and **X** are represented by sums formed through contraction (Sec. 16.3-5) of **QX**, so that one or more indices of $Q(j_1, j_2, \ldots, j_R)$ are summed against one or more superscripts and/or subscripts of $X^{i_1 i_2 \cdots i_r}_{i_1' i_2' \cdots i_s'}$.

The object **Q** *is a tensor if and only if the outer product* **QX**, *or a given type of inner product of* **Q** *and* **X**, *is a tensor* **Y** *of fixed rank, type, and weight for any arbitrary tensor* **X** *of fixed rank, type, and weight.*

An analogous theorem results if **X** is, instead, *the outer product of R distinct arbitrary vectors of fixed types and weights.* In either case one infers the rank and type of **Q** by matching superscripts and/or subscripts. The weight of **Q** must be the difference of the respective weights of **Y** and **X**.

EXAMPLES: (1) **Q** is an absolute tensor contravariant of rank r and covariant of rank s if, for every absolute vector **a** represented by a^i,

$$\sum_{j_{r+s}=1}^{n} Q(j_1, j_2, \ldots, j_{r+s}) a^{j_{r+s}} = B^{j_1 j_2 \cdots j_r}_{j_{r+1} j_{r+2} \cdots j_{r+s-1}}$$

where the components on the right describe an absolute tensor depending on **a**.

(2) **Q** is an absolute tensor contravariant of rank r and covariant of rank s if, for every absolute tensor **A** represented by $A^{i_1 i_2 \cdots i_s}_{i_1' i_2' \cdots i_r'}$,

$$\sum_{j_1=1}^{n} \sum_{j_2=1}^{n} \cdots \sum_{j_{r+s}=1}^{n} Q(j_1, j_2, \ldots, j_{r+s}) A^{j_1 j_2 \cdots j_s}_{j_{s+1} j_{s+2} \cdots j_{r+s}} = \alpha$$

where α represents an absolute scalar depending on **A**.

NOTE: An object **Q** described by n^2 components Q_{ik} is an absolute covariant tensor of rank 2 if and only if

$$\sum_{i=1}^{n} \sum_{k=1}^{n} Q_{ik} a^i a^k = \alpha$$

represents a scalar invariant for every absolute contravariant vector **a** described by a^i, *and* $Q_{ik} = Q_{ki}$.

16.4. TENSOR ALGEBRA: INVARIANCE OF TENSOR EQUATIONS

16.4-1. Invariance of Tensor Equations. *For each admissible coordinate transformation* (16.1-1), *the induced transformation laws used to define tensor components preserve the results of tensor addition, contraction, and outer (and hence also inner) multiplication, as well as tensor equality.* Every relation between tensors expressible in terms of combinations of such operations (including convergent limiting processes) is *invariant with respect to the group of admissible coordinate transformations.* If the relation applies to tensor components in any one scheme of measurements, it holds in all schemes of measurements (see also Secs. 12.1-5 and 16.1-4).

EXAMPLE: $A^{i_1 i_2 \cdots i_r}_{i_1' i_2' \cdots i_s'} + B^{i_1 i_2 \cdots i_r}_{i_1' i_2' \cdots i_s'} = C^{i_1 i_2 \cdots i_r}_{i_1' i_2' \cdots i_s'}$ implies $\bar{A}^{k_1 k_2 \cdots k_r}_{k_1' k_2' \cdots k_s'} + \bar{B}^{k_1 k_2 \cdots k_r}_{k_1' k_2' \cdots k_s'} = \bar{C}^{k_1 k_2 \cdots k_r}_{k_1' k_2' \cdots k_s'}$, and conversely; this relation may be symbolized by the abstract equation $\mathbf{A} + \mathbf{B} = \mathbf{C}$.

One may, then, speak of tensors and tensor operations without reference to a specific scheme of measurements. Each suitable class of tensor point functions is a class of *invariants* and constitutes an abstract model definable (to within an isomorphism, Sec. 12.1-6) in terms of mathematical operations without reference to components (see also Secs. 12.1-1 and 16.1-4).

Thus suitable classes of absolute tensors of rank 0, 1, and 2, respectively, constitute scalar fields (Sec. 12.3-1*c*), vector spaces (Sec. 12.4-1), and linear algebras (Sec. 12.4-2; see also Sec. 16.9-2). Classes of tensors of rank 2, 3, . . . may be built up as direct products of vector spaces (Sec. 12.7-3; see also Sec. 16.6-1*c*).

16.5. SYMMETRIC AND SKEW-SYMMETRIC TENSORS

16.5-1. Symmetry and Skew-symmetry. An object **Q** described by n^R components $Q(j_1, j_2, \ldots, j_R)$ each labeled with an ordered set of R indices $j_1, j_2, \ldots, j_R$ is

1. **Symmetric with respect to any pair of indices, say j_1 and j_2,** if and only if

$$Q(j_1 = i, j_2 = k, j_3, \ldots, j_R) = Q\,(j_1 = k, j_2 = i, j_3, \ldots, j_R)$$

2. **Skew-symmetric (antisymmetric) with respect to any pair of indices, say i_1 and i_2,** if and only if

$$Q(j_1 = i, j_2 = k, j_3, \ldots, j_R) = -Q(j_1 = k, j_2 = i, j_3, \ldots, j_R)$$

for all sets of values of $i, k, j_3, \ldots, j_R$, each running from 1 to n. **Q** is (**completely**) **symmetric** or (**completely**) **skew-symmetric with**

respect to a set of indices if and only if **Q** is, respectively, symmetric or skew-symmetric with respect to every pair of indices of the set.

The symmetry or skew-symmetry of an absolute or relative tensor with respect to any pair of superscripts or any pair of subscripts is invariant with respect to the group of admissible coordinate transformations.

16.5-2. Kronecker Deltas. The (generalized) **Kronecker delta** of rank $2r$ is the absolute tensor represented by n^{2r} components $\delta^{i_1 i_2 \cdots i_r}_{k_1 k_2 \cdots k_r}$ defined as follows:

1. $\delta^{i_1 i_2 \cdots i_r}_{k_1 k_2 \cdots k_r} = +1$ or -1 if all superscripts $i_1, i_2, \ldots, i_r$ are different, and the ordered set of subscripts $k_1, k_2, \ldots, k_r$ is obtained, respectively, by an even or odd permutation (even or odd number of transpositions) of the ordered set $i_1, i_2, \ldots, i_r$
2. $\delta^{i_1 i_2 \cdots i_r}_{k_1 k_2 \cdots k_r} = 0$ for all other combinations of superscripts and subscripts

Of particular interest is the Kronecker delta of rank 2 described by

$$\delta^i_k = \begin{cases} 0 \text{ if } i \neq k \\ 1 \text{ if } i = k \end{cases} \tag{16.5-1}$$

Contraction of any mixed tensor **A** by summation over a superscript i and a subscript i' (Sec. 16.3-5) is equivalent to inner multiplication (Sec. 16.3-7) of **A** by $\delta^{i'}_i$.

Each Kronecker delta is completely skew-symmetric (Sec. 16.5-1) with respect to both the set of superscripts and the set of subscripts. Kronecker deltas of rank $2r > 2n$ are zero.

If $A^{i_1 i_2 \cdots i_r}_{i_1' i_2' \cdots i_s'}$ is symmetric with respect to any pair of superscripts, then

$$\delta^{k_1 k_2 \cdots k_r}_{i_1 i_2 \cdots i_r} A^{i_1 i_2 \cdots i_r}_{i_1' i_2' \cdots i_s'} = 0 \tag{16.5-2}$$

If $A^{i_1 i_2 \cdots i_r}_{i_1' i_2' \cdots i_s'}$ is symmetric with respect to any pair of subscripts, then

$$\delta^{i_1' i_2' \cdots i_s'}_{k_1 k_2 \cdots k_s} A^{i_1 i_2 \cdots i_r}_{i_1' i_2' \cdots i_s'} = 0 \tag{16.5-3}$$

Note also

$$\delta^{i_1 i_2 \cdots i_r}_{k_1 k_2 \cdots k_r} A^{k_1 k_2 \cdots k_r} = r! A^{i_1 i_2 \cdots i_r} \tag{16.5-4}$$

$$\delta^{i_1 i_2 \cdots i_r}_{k_1 k_2 \cdots k_r} = \frac{(n-s)!}{(n-r)!} \delta^{i_1 i_2 \cdots i_r i_{r+1} \cdots i_s}_{k_1 k_2 \cdots k_r i_{r+1} \cdots i_s} \qquad \delta^{i_1 i_2 \cdots i_r}_{i_1 i_2 \cdots i_r} = \frac{n!}{(n-r)!} \tag{16.5-5}$$

$$\frac{\partial x^i}{\partial x^k} = \delta^i_k \tag{16.5-6}$$

16.5-3. Permutation Symbols (see also Sec. 16.7-2*c*). The **permutation symbols**

$$\epsilon^{i_1 i_2 \cdots i_n} = \delta^{i_1 i_2 \cdots i_n}_{12 \cdots n} \quad \text{and} \quad \epsilon_{i_1 i_2 \cdots i_n} = \delta^{12 \cdots n}_{i_1 i_2 \cdots i_n} \tag{16.5-7}$$

represent completely skew-symmetric relative tensors of rank n and of weight $+1$ and -1, respectively. Note that

1. $\epsilon^{i_1 i_2 \cdots i_n} = \epsilon_{i_1 i_2 \cdots i_n} = 0$ if two or more of the indices $i_1, i_2, \ldots, i_n$ are equal.

2. $\epsilon^{i_1 i_2 \cdots i_n} = \epsilon_{i_1 i_2 \cdots i_n} = 1$ if the ordered set $i_1, i_2, \ldots, i_n$ is obtained by an even permutation of the set $1, 2, \ldots, n$.
3. $\epsilon^{i_1 i_2 \cdots i_n} = \epsilon_{i_1 i_2 \cdots i_n} = -1$ if the ordered set $i_1, i_2, \ldots, i_n$ is obtained by an odd permutation of the set $1, 2, \ldots, n$.

And note

$$\epsilon^{i_1 i_2 \cdots i_r i_{r+1} \cdots i_n} \epsilon_{k_1 k_2 \cdots k_r i_{r+1} \cdots i_n} = (n - r)! \delta^{i_1 i_2 \cdots i_r}_{k_1 k_2 \cdots k_r} \qquad (16.5\text{-}8)$$

$$\epsilon^{i_1 i_2 \cdots i_n} \epsilon_{i_1 i_2 \cdots i_n} = n! \qquad (16.5\text{-}9)$$

$$\epsilon^{i_1 i_2 \cdots i_n} A^1_{i_1} A^2_{i_2} \cdots A^n_{i_n} = \epsilon_{i_1 i_2 \cdots i_n} A^{i_1}_1 A^{i_2}_2 \cdots A^{i_n}_n = \det [A^i_k] \qquad (16.5\text{-}10)$$

16.5-4. Alternating Product of Two Vectors (see also Secs. 16.8-4 and 16.10-6). The **alternating product** (sometimes called bivector) of two contravariant vectors and the alternating product of two covariant vectors are skew-symmetric tensors of rank two respectively represented by

$$V^{ij} = a^i b^j - a^j b^i \qquad \text{and} \qquad V_{ij} = a_i b_j - a_j b_i \qquad (16.5\text{-}11)$$

The weight of the alternating product is the sum of the weights of the two factors.

16.6. LOCAL SYSTEMS OF BASE VECTORS

16.6-1. Representation of Vectors and Tensors in Terms of Local Base Vectors.

(a) Given an x scheme of measurements, each (absolute or relative) contravariant vector described by $a^i(x^1, x^2, \ldots, x^n)$ may be represented as an invariant linear form

$$\mathbf{a} = a^i(x^1, x^2, \ldots, x^n)\mathbf{e}_i(x^1, x^2, \ldots, x^n) \qquad (16.6\text{-}1)$$

(see Sec. 16.1-3 for umbral-index notation) in the n absolute **contravariant local base vectors** (Sec. 14.2-3) $\mathbf{e}_1(x^1, x^2, \ldots, x^n)$, $\mathbf{e}_2(x^1, x^2, \ldots, x^n)$, $\ldots$, $\mathbf{e}_n(x^1, x^2, \ldots, x^n)$ associated with the x scheme of measurements. *The ith base vector $\mathbf{e}_i$ has the components $\delta^1_i, \delta^2_i, \ldots, \delta^n_i$.*

(b) Similarly, each (absolute or relative) covariant vector $\mathbf{b}$ described by $b_i(x^1, x^2, \ldots, x^n)$ may be represented in the form

$$\mathbf{b} = b_i(x^1, x^2, \ldots, x^n)\mathbf{e}^i(x^1, x^2, \ldots, x^n) \qquad (16.6\text{-}2)$$

in terms of the n absolute **covariant local base vectors** $\mathbf{e}^1(x^1, x^2, \ldots, x^n)$, $\mathbf{e}^2(x^1, x^2, \ldots, x^n)$, $\ldots$, $\mathbf{e}^n(x^1, x^2, \ldots, x^n)$ associated with the x scheme of measurements. *The ith base vector $\mathbf{e}^i$ has the components $\delta^i_1, \delta^i_2, \ldots, \delta^i_n$.* Note that the vectors (1) and the vectors (2) will, in general, belong to different vector spaces.

(c) Every absolute or relative tensor A described by $A^{i_1 i_2 \cdots i_r}_{i_1' i_2' \cdots i_s'}$ may be represented as an invariant form

$$\mathbf{A} = A^{i_1 i_2 \cdots i_r}_{i_1' i_2' \cdots i_s'} \mathbf{e}_{i_1} \mathbf{e}_{i_2} \cdots \mathbf{e}_{i_r} \mathbf{e}^{i_1'} \mathbf{e}^{i_2'} \cdots \mathbf{e}^{i_s'} \tag{16.6-3}$$

in the local base vectors $\mathbf{e}_i$ and $\mathbf{e}^i$.

16.6-2. Relations between Local Base Vectors Associated with Different Schemes of Measurements. The $2n$ local base vectors $\mathbf{e}_i(x^1, x^2, \ldots, x^n)$ and $\mathbf{e}^i(x^1, x^2, \ldots, x^n)$ may be thought of as *defining* the x scheme of measurements (Sec. 16.1-4) in invariant language. New local base vectors $\bar{\mathbf{e}}_k(\bar{x}^1, \bar{x}^2, \ldots, \bar{x}^n)$ and $\bar{\mathbf{e}}^k(\bar{x}^1, \bar{x}^2, \ldots, \bar{x}^n)$ associated with an $\bar{x}$ scheme of measurements have the components δ^i_k and δ^k_i, respectively, in the $\bar{x}$ scheme of measurements and are related to their respective counterparts associated with the x scheme of measurements as follows:

$$\begin{aligned} \bar{\mathbf{e}}_k(\bar{x}^1, \bar{x}^2, \ldots, \bar{x}^n) &= \frac{\partial x^i}{\partial \bar{x}^k} \mathbf{e}_i(x^1, x^2, \ldots, x^n) \\ \bar{\mathbf{e}}^k(\bar{x}^1, \bar{x}^2, \ldots, \bar{x}^n) &= \frac{\partial \bar{x}^k}{\partial x^i} \mathbf{e}^i(x^1, x^2, \ldots, x^n) \end{aligned} \tag{16.6-4}$$

Note that $\mathbf{e}_i(x^1, x^2, \ldots, x^n)$ and $\bar{\mathbf{e}}_i(\bar{x}^1, \bar{x}^2, \ldots, \bar{x}^n)$ are, in general, *different vectors* (of the same vector space), *not* different descriptions of the same vector. The base vectors $\mathbf{e}_i$ transform formally like (*cogrediently* with) absolute covariant vector components (Sec. 16.2-1), whereas the $\mathbf{e}^i$ transform like absolute contravariant vector components. Absolute contravariant and covariant vector components a^i and b_i (and hence contravariant and covariant base vectors $\mathbf{e}_i$ and $\mathbf{e}^i$) transform *contragrediently*, so that inner products like a^ib_i are *invariant* (see also Sec. 16.4-1).

16.7. TENSORS DEFINED ON RIEMANN SPACES. ASSOCIATED TENSORS

16.7-1. Riemann Space and Fundamental Tensors. Riemann spaces permit the definition of scalar products of vectors in such a manner that the resulting definitions of distances and angles (Sec. 16.8-1; see also Sec. 14.2-7) lead to useful generalizations of Euclidean geometry (see also Secs. 17.4-1 to 17.4-7). A finite-dimensional space of points labeled by ordered sets of real* coordinates $x^1, x^2, \ldots, x^n$ is a **Riemann space** if it is possible to define an absolute covariant tensor of rank 2 (Sec. 16.2-1) described (in an x scheme of measurements) by components $g_{ik}(x^1, x^2, \ldots, x^n)$ having the following (invariant) properties throughout the region under consideration:

* The theory presented in Secs. 16.7-1 to 16.10-11 applies to vectors and tensors with *real* components defined on Riemann spaces described by *real* coordinates. The theory also applies to the Riemann spaces considered in relativity theory, where the introduction of an imaginary coordinate (Sec. 17.4-6) is essentially a notational convenience.

1. Each $g_{ik}(x^1, x^2, \ldots, x^n)$ is a real single-valued function of the coordinates and possesses continuous partial derivatives.
2. $g_{ik}(x^1, x^2, \ldots, x^n) = g_{ki}(x^1, x^2, \ldots, x^n)$
3. $g = g(x^1, x^2, \ldots, x^n)$
$= \det [g_{ik}(x^1, x^2, \ldots, x^n)] \neq 0$

The matrix $[g_{ik}(x^1, x^2, \ldots, x^n)]$ is frequently, but not necessarily, positive definite (Sec. 13.5-2); the indefinite case is of interest in relativity theory (see also Secs. 16.8-1 and 17.4-6).

The **metric tensor** (see also Sec. 17.4-2) described by the $g_{ik}(x^1, x^2, \ldots, x^n)$ and the absolute symmetric tensor of rank 2 (**conjugate or associated metric tensor**) whose components $g^{ik}(x^1, x^2, \ldots, x^n)$ are defined by

$$g^{ik}g_{kj} = \delta^i_j \qquad \text{or} \qquad g^{ik} = \frac{G^{ik}}{g} \tag{16.7-1}$$

where G^{ik} $(= G^{ki})$ is the cofactor (Sec. 1.5-2) of g_{ik} in the determinant $\det [g_{ik}]$, are the **fundamental tensors** of the Riemann space.

The components of either fundamental tensor define the *element of distance ds* and the entire *intrinsic differential geometry* of the Riemann space (Secs. 17.4-1 to 17.4-7). A system of coordinates $x^1, x^2, \ldots, x^n$ is, respectively, **right-handed** or **left-handed** if the scalar density $\sqrt{|g(x^1, x^2, \ldots, x^n)|}$ is positive or negative; an arbitrary choice of sign for *any one* coordinate system defines every admissible coordinate system as either right-handed or left-handed, since

$$\sqrt{|\bar{g}(\bar{x}^1, \bar{x}^2, \ldots, \bar{x}^n)|} = \sqrt{|g(x^1, x^2, \ldots, x^n)|}\, \frac{\partial(x^1, x^2, \ldots, x^n)}{\partial(\bar{x}^1, \bar{x}^2, \ldots, \bar{x}^n)}$$

(see also Secs. 6.2-3*b* and 6.4-3*c*).

16.7-2. Associated Tensors.* **Raising and Lowering of Indices.** An (absolute or relative) contravariant vector represented by a^i and a covariant vector represented by a_k, defined on a Riemann space and related at every point $(x^1, x^2, \ldots, x^n)$ by

$$a^k = a_i g^{ik} \qquad \text{and hence} \qquad a_i = g_{ik}a^k \tag{16.7-2}$$

are called **associated vectors.** More generally, an **associated tensor** of a given tensor described by $A^{i_1i_2\cdots i_r}_{k_1k_2\cdots k_s}$ is obtained by **raising a subscript** k through inner multiplication by g^{ki}, or by **lowering a superscript** i through inner multiplication by g_{ji}; or by any combination of such operations. A tensor of rank greater than one has more than one associated tensor. Since it is desirable to denote the components of all

* See also footnote to Sec. 13.3-1.

tensors associated with a given tensor **A** by the same symbol A, *it is necessary to order superscripts and subscripts with respect to each other* (see also Sec. 16.2-1). Thus the result of raising the subscript k_2 in $A^{i_1 i_2 \cdots i_r}_{k_1 k_2 \cdots k_s}$ is denoted by

$$A^{i_1 i_2 \cdots i_r \ \ j}_{\qquad \ \ k_1 \ k_3 k_4 \cdots k_s} = g^{j k_2} A^{i_1 i_2 \cdots i_r}_{k_1 k_2 \cdots k_s} \tag{16.7-3}$$

Raising of previously lowered indices and/or lowering of previously raised indices restores the original tensor components.

NOTE: The contravariant and covariant permutation symbols (Sec. 16.5-3) are *not* associated relative tensors but are related by

$$\epsilon_{i_1 i_2 \cdots i_n} = \frac{1}{g} g_{i_1 k_1} g_{i_2 k_2} \cdots g_{i_n k_n} \epsilon^{k_1 k_2 \cdots k_n} \qquad \epsilon^{k_1 k_2 \cdots k_n} = g g^{k_1 i_1} g^{k_2 i_2} \cdots g^{k_n i_n} \epsilon_{i_1 i_2 \cdots i_n} \tag{16.7-4}$$

16.7-3. Equivalence of Associated Tensors. The correspondence between associated tensors defined on a Riemann space is an equivalence relation partitioning the class of all tensors (Sec. 12.1-3*b*). The components of any associated tensor of a tensor **A** defined on a Riemann space are, then, considered as **a different description (representation) of the same tensor A** (see also Sec. 16.9-1).

In particular, *vector components a^k and a_i related by Eq.* (2) *are interpreted as the contravariant and covariant representations of the same vector* **a** in the x scheme of measurements used. In the notation of Sec. 16.6-1,

$$\mathbf{a} = a^k \mathbf{e}_k = a_i \mathbf{e}^i \tag{16.7-5}$$

so that the base vectors $\mathbf{e}_1, \mathbf{e}_2, \ldots, \mathbf{e}_n$ and $\mathbf{e}^1, \mathbf{e}^2, \ldots, \mathbf{e}^n$ defined on a Riemann space may be regarded as (reciprocal) bases of *the same vector space* (see also Sec. 16.8-2). They are related formally like associated vector components:

$$\mathbf{e}^k = g^{ik} \mathbf{e}_i \qquad \mathbf{e}_i = g_{ik} \mathbf{e}^k \tag{16.7-6}$$

Substitution of the appropriate expression (6) for some $\mathbf{e}^k$ or $\mathbf{e}_i$ in the expansion (16.6-3) of any tensor **A** corresponds, respectively, to raising or lowering the index in question (see also Sec. 16.9-1).

16.7-4. Operations with Tensors Defined on Riemann Spaces. If the tensors in question are defined on a Riemann space

1. Any two tensors having the same rank and weight may be *added* in the manner of Sec. 16.3-3, after their components have been reduced to the same configuration of superscripts and subscripts through raising and/or lowering of indices.
2. A tensor may be *contracted over any pair of indices* in the manner of Sec. 16.3-5, after one of the indices has been appropriately raised or lowered. Contraction over two superscripts i, k corresponds, then, to inner multiplication by g_{ik}; contraction over two subscripts i, k corresponds to inner multiplication by g^{ik}.

3. *Inner products* of two tensors **A** and **B** are defined as contractions of their outer product **AB** over an index (or indices) of **A** and a corresponding index (or indices) of **B** as in (2) above.

16.8. SCALAR PRODUCTS AND RELATED TOPICS

16.8-1. Scalar Product (Inner Product) of Two Vectors Defined on a Riemann Space. In accordance with Sec. 16.7-4, it is possible to define the **scalar product (inner product,** see also Secs. 5.2-6, 6.4-2a, and 14.2-6) $\mathbf{a} \cdot \mathbf{b}$ of any two absolute or relative vectors $\mathbf{a}$ and $\mathbf{b}$ represented by (real) components a^i or a_k and b^i or b_k:

$$\begin{aligned}\mathbf{a} \cdot \mathbf{b} &= g_{ik}(x^1, x^2, \ldots, x^n)a^i(x^1, x^2, \ldots, x^n) b^k(x^1, x^2, \ldots, x^n) \\ &= a_k b^k = a^i b_i = g^{ik} a_i b_k = \mathbf{b} \cdot \mathbf{a}\end{aligned} \tag{16.8-1}$$

The **magnitude (norm, absolute value;** Sec. 14.2-5) $|\mathbf{a}|$ of an absolute or relative vector $\mathbf{a}$ described by (real) components a^i or a_k is the absolute or relative scalar invariant

$$|\mathbf{a}| = \sqrt[+]{|\mathbf{a}^2|} \qquad \mathbf{a}^2 = \mathbf{a} \cdot \mathbf{a} = g_{ik} a^i a^k = a^i a_i = g^{ik} a_i a_k \tag{16.8-2}$$

A **unit vector** is an absolute vector of magnitude one. The cosine of the **angle** γ between two absolute or relative vectors $\mathbf{a}$ and $\mathbf{b}$ is the absolute scalar invariant

$$\cos \gamma = \frac{\mathbf{a} \cdot \mathbf{b}}{|\mathbf{a}|\,|\mathbf{b}|} \tag{16.8-3}$$

Equations (2) and (3) imply the elementary definition (5.2-5) of the scalar product.

NOTE: If the quadratic form $g_{ik}a^i a^k$ is indefinite (*indefinite metric,* see also Sec. 17.4-4) at a point $(x^1, x^2, \ldots, x^n)$, the square $\mathbf{a} \cdot \mathbf{a}$ of an absolute or relative vector $\mathbf{a}$ represented by components a^i at that point is positive, negative, or zero depending on the sign of $g_{ik}a^i a^k$, and $|\mathbf{a}| = 0$ does not necessarily imply $\mathbf{a} = 0$.

16.8-2. Scalar Products of Local Base Vectors. Orthogonal Coordinate Systems (see also Secs. 6.3-3, 6.4-1, and 17.4-7a). The magnitudes of, and angles between, the local base vectors $\mathbf{e}_1, \mathbf{e}_2, \ldots, \mathbf{e}_n$ and $\mathbf{e}^1, \mathbf{e}^2, \ldots, \mathbf{e}^n$ at each point $(x^1, x^2, \ldots, x^n)$ (Sec. 16.6-1) are given by

$$\mathbf{e}_i \cdot \mathbf{e}_k = g_{ik}(x^1, x^2, \ldots, x^n) \qquad \mathbf{e}^i \cdot \mathbf{e}^k = g^{ik}(x^1, x^2, \ldots, x^n) \qquad \mathbf{e}^i \cdot \mathbf{e}_k = \delta^i_k \tag{16.8-4}$$

The vectors $\mathbf{e}_i$ are directed along the corresponding coordinate lines, and the $\mathbf{e}^k$ are directed along the normals to the coordinate hypersurfaces. Each $\mathbf{e}^k$ is perpendicular to all $\mathbf{e}_i$ except $\mathbf{e}_k$. A system of coordinates $x^1, x^2, \ldots, x^n$ is an **orthogonal coordinate system** if and only if $g_{ik}(x^1, x^2, \ldots, x^n) \equiv 0$ for $i \neq k$; in this case the $\mathbf{e}_i$ (and thus also the $\mathbf{e}^k$) are mutually orthogonal. Not every Riemann space admits orthogonal coordinates.

NOTE: Two sets of base vectors $\mathbf{e}^i$ and $\mathbf{e}_k$ satisfying the relation $\mathbf{e}^i \cdot \mathbf{e}_k = \delta^i_k$ constitute **reciprocal bases** of the vector space in question (see also Sec. 14.7-6).

16.8-3. Physical Components of a Tensor (see also Sec. 6.3-4). The **local unit vectors $\mathbf{u}_i$ in the coordinate directions corresponding to the subscript** are related to the $\mathbf{e}_i$ and $\mathbf{e}^k$ by

$$\mathbf{u}_i = \frac{1}{+\sqrt{|g_{ii}|}}\,\mathbf{e}_i = \frac{1}{+\sqrt{|g_{ii}|}}\,g_{ik}\mathbf{e}^k \qquad \mathbf{e}_i = +\sqrt{|g_{ii}|}\,\mathbf{u}_i \qquad \mathbf{e}^k = g^{ik}\mathbf{e}_i \tag{16.8-5}$$

The **physical components** $\hat{a}_j$ and $\hat{A}_{j_1 j_2 \cdots j_R}$ of vectors and tensors are defined by

$$\mathbf{a} = \sum_{j=1} \hat{a}_j \mathbf{u}_j \qquad \hat{a}_j = \underset{+}{\sqrt{|g_{jj}|}}\,a^j \tag{16.8-6}$$

$$\mathsf{A} = \sum_{j_1=1}^{n} \sum_{j_2=1}^{n} \cdots \sum_{j_R=1}^{n} \hat{A}_{j_1 j_2 \cdots j_R} \mathbf{u}_{j_1} \mathbf{u}_{j_2} \cdots \mathbf{u}_{j_R} \tag{16.8-7}$$

where the $\hat{A}_{j_1 j_2 \cdots j_R}$ are obtained by comparison of Eqs. (7) and (16.6-3). The physical component of a vector $\mathbf{a}$ in the direction of another vector $\mathbf{b}$ is defined as $\mathbf{a} \cdot \mathbf{b}/|\mathbf{b}|$.

16.8-4. Vector Product and Scalar Triple Product (see also Secs. 5.2-7, 5.2-8, 6.3-3, and 6.4-2). The **vector product $\mathbf{a} \times \mathbf{b}$** of two absolute or relative vectors $\mathbf{a}$ and $\mathbf{b}$ defined on a **three-dimensional** Riemann space ($n = 3$) is the vector represented by the components (see also Sec. 16.5-3)

$$\left.\begin{aligned} &\frac{1}{\sqrt{|g|}}\,\epsilon^{ijk} a_i b_j = \frac{1}{2\sqrt{|g|}}\,\epsilon^{ijk}(a_i b_j - a_j b_i) \\ \text{or} \qquad &\sqrt{|g|}\,\epsilon_{ijk} a^i b^j = \tfrac{1}{2}\sqrt{|g|}\,\epsilon_{ijk}(a^i b^j - a^j b^i) \end{aligned}\right\} \tag{16.8-8}$$

so that

$$\mathbf{a} \times \mathbf{b} = \frac{1}{\sqrt{|g|}} \begin{vmatrix} \mathbf{e}_1 & a_1 & b_1 \\ \mathbf{e}_2 & a_2 & b_2 \\ \mathbf{e}_3 & a_3 & b_3 \end{vmatrix} = \sqrt{|g|} \begin{vmatrix} \mathbf{e}^1 & a^1 & b^1 \\ \mathbf{e}^2 & a^2 & b^2 \\ \mathbf{e}^3 & a^3 & b^3 \end{vmatrix} = -\mathbf{b} \times \mathbf{a} \tag{16.8-9}$$

Note

$$\left.\begin{aligned} \mathbf{e}^1 &= \frac{\mathbf{e}_2 \times \mathbf{e}_3}{[\mathbf{e}_1\mathbf{e}_2\mathbf{e}_3]} & \mathbf{e}^2 &= \frac{\mathbf{e}_3 \times \mathbf{e}_1}{[\mathbf{e}_1\mathbf{e}_2\mathbf{e}_3]} & \mathbf{e}^3 &= \frac{\mathbf{e}_1 \times \mathbf{e}_2}{[\mathbf{e}_1\mathbf{e}_2\mathbf{e}_3]} \\ \mathbf{e}_1 &= \frac{\mathbf{e}^2 \times \mathbf{e}^3}{[\mathbf{e}^1\mathbf{e}^2\mathbf{e}^3]} & \mathbf{e}_2 &= \frac{\mathbf{e}^3 \times \mathbf{e}^1}{[\mathbf{e}^1\mathbf{e}^2\mathbf{e}^3]} & \mathbf{e}_3 &= \frac{\mathbf{e}^1 \times \mathbf{e}^2}{[\mathbf{e}^1\mathbf{e}^2\mathbf{e}^3]} \end{aligned}\right\} \tag{16.8-10}$$

where the **scalar triple product** $[\mathbf{abc}]$ is defined, as in Sec. 5.2-8, by

$$\left.\begin{aligned} [\mathbf{abc}] = \mathbf{a} \cdot (\mathbf{b} \times \mathbf{c}) &= \frac{1}{\sqrt{|g|}} \begin{vmatrix} a_1 & b_1 & c_1 \\ a_2 & b_2 & c_2 \\ a_3 & b_3 & c_3 \end{vmatrix} = \sqrt{|g|} \begin{vmatrix} a^1 & b^1 & c^1 \\ a^2 & b^2 & c^2 \\ a^3 & b^3 & c^3 \end{vmatrix} \\ &= \frac{1}{\sqrt{|g|}}\,\epsilon^{ijk} a_i b_j c_k = \sqrt{|g|}\,\epsilon_{ijk} a^i b^j c^k \end{aligned}\right\} \tag{16.8-11}$$

so that

$$[\mathbf{e}_1\mathbf{e}_2\mathbf{e}_3] = \sqrt{|g|} \qquad [\mathbf{e}^1\mathbf{e}^2\mathbf{e}^3] = \frac{1}{\sqrt{|g|}} \tag{16.8-12}$$

The formulas of Table 5.2-2 and Sec. 5.2-9 hold.

NOTE: The definition (8) of the vector product implies the elementary relation (5.2-6) and defines the vector product of two absolute vectors as an absolute vector. Some authors replace $\sqrt{|g|}$ by 1 in the definition (8), so that the vector product of two absolute vectors becomes an "axial" vector (as contrasted to "polar" or absolute vectors) described either as a relative contravariant vector of weight $+1$, or as a relative covariant vector of weight -1.

16.9. TENSORS OF RANK TWO (DYADICS) DEFINED ON RIEMANN SPACES

16.9-1. Dyadics. Absolute or relative tensors of rank 2 (**dyadics**) A, B, . . . defined on Riemann spaces and described, for instance, in terms of their respective mixed components A^i_k, B^i_k, . . . (see also Sec. 16.7-2 for ordering of indices) are of interest in many applications. They are sometimes thought to warrant the special notation outlined in the following sections.

Every dyadic A may be represented as a sum of n **dyads** (outer products of two vectors):

$$\mathsf{A} = \mathbf{p}_j\mathbf{q}^j = (p^i_j\mathbf{e}_i)(q^j_k\mathbf{e}^k) \qquad A^i_k = p^i_j q^j_k \tag{16.9-1}$$

Either the **antecedents** $\mathbf{p}_j$ or the **consequents** $\mathbf{q}^j$ may be arbitrarily assigned as long as they are linearly independent (Sec. 14.2-3). In particular

$$\mathsf{A} = A^i_k\mathbf{e}_i\mathbf{e}^k = A_{ik}\mathbf{e}^i\mathbf{e}^k = A^{ik}\mathbf{e}_i\mathbf{e}_k = A_i{}^k\mathbf{e}^i\mathbf{e}_k \tag{16.9-2}$$

$$\mathsf{A} = \mathbf{e}_i\mathbf{A}^i = \mathbf{A}_k\mathbf{e}^k \qquad \mathbf{A}^i = A^i_k\mathbf{e}^k \qquad \mathbf{A}_k = A^i_k\mathbf{e}_i \tag{16.9-3}$$

In terms of *physical components* $\hat{A}_{ik}$ (Sec. 16.8-3)

$$\mathsf{A} = \sum_{i=1}^{n}\sum_{k=1}^{n} \hat{A}_{ik}\mathbf{u}_i\mathbf{u}_k \qquad \hat{A}_{ik} = (A^{ik}\underset{+}{\sqrt{|g_{ii}g_{kk}|}})_{\text{no sum}} \qquad A^{ik} = g^{kj}A^i_j \tag{16.9-4}$$

In the case of orthogonal coordinates (Sec. 16.8-2)

$$\hat{A}_{ik} = \left(A^i_k \underset{+}{\sqrt{\frac{g_{ii}}{g_{kk}}}}\right)_{\text{no sum}} \tag{16.9-5}$$

16.9-2. Inner-product Notation (see also Sec. 16.8-1). The following notation is sometimes useful for the description of inner products involving (real) dyadics and vectors defined on Riemann spaces:

$$\mathsf{A}\cdot\mathbf{a} = \mathbf{p}_j(\mathbf{q}^j\cdot\mathbf{a}) \qquad \textit{vector described by } A^i_k a^k \tag{16.9-6}$$

$$\mathbf{a}\cdot\mathsf{A} = (\mathbf{a}\cdot\mathbf{p}_j)\mathbf{q}^j \qquad \textit{vector described by } A^i_k a_i \tag{16.9-7}$$

$$\mathsf{A}\cdot\mathsf{B} = \mathbf{p}_j(\mathbf{q}^j\cdot\mathbf{p}'_h)\mathbf{q}'^h \qquad \textit{dyadic described by } A^i_j B^j_k \tag{16.9-8}$$

where $\qquad \mathsf{B} = \mathbf{p}'_j\mathbf{q}'^j = B^i_k\mathbf{e}_i\mathbf{e}^k$

With these definitions, the algebra of dyadics is precisely the algebra of linear operators described in Secs. 14.3-1 to 14.3-6; a dyadic associates a linear transformation with each point $(x^1, x^2, \ldots, x^n)$. *Table* 14.7-1 *relates the tensor notation to the "classical" notation of Chap.* 14 *and to the matrix notation for dyadics and vectors.*

Symmetry and skew-symmetry of a dyadic are defined in the manner of Sec. 16.5-1. Thus the dyadic A is symmetric if and only if $A^{ki} = A^{ik}$, or if and only if $A_{ki} = A_{ik}$; but this does *not* necessarily imply that $A^k_i = A^i_k$, nor does the last relation imply the symmetry of A (see also Sec. 14.7-5).

$\mathrm{Tr}(\mathsf{A}) = A_j^j = \mathbf{p}_j \cdot \mathbf{q}^j$ is called the **scalar of the dyadic** (1). The **double-dot product** of A and B is the scalar $\mathsf{A} \cdot\cdot \mathsf{B} = \sum_{i=1}^{3} \sum_{k=1}^{3} (\mathbf{p}_i \cdot \mathbf{p}'_k)(\mathbf{q}^i \cdot \mathbf{q}'^k)$.

For $n = 3$, it is possible to define the **cross products**

$$\left.\begin{aligned} \mathbf{a} \times \mathsf{A} = (\mathbf{a} \times \mathbf{p}_j)\mathbf{q}^j \qquad \mathsf{A} \times \mathbf{a} = \mathbf{p}_j(\mathbf{q}^j \times \mathbf{a}) \\ \mathsf{A} \times \mathsf{B} = \mathbf{p}_j(\mathbf{q}^j \times \mathbf{p}'_h)\mathbf{q}'_h \end{aligned}\right\} \tag{16.9-9}$$

The vector $\mathbf{v}_A = \mathbf{p}_j \times \mathbf{q}^j$ is called the **vector of the dyadic** (1). $\mathbf{v}_A = \mathbf{0}$ if and only if A is symmetric. If A is skew-symmetric, then for every vector $\mathbf{a}$

$$\mathbf{a} \cdot \mathsf{A} = \tfrac{1}{2}\mathbf{v}_A \times \mathbf{a} = -\mathsf{A} \cdot \mathbf{a} \tag{16.9-10}$$

so that *vector multiplication is equivalent to inner multiplication by a skew-symmetric dyadic.*

16.9-3. Eigenvalue Problems (see also Sec. 14.8-3). Eigenvalues and eigenvectors of dyadics are defined *at each point* $(x^1, x^2, \ldots, x^n)$ in the manner of Sec. 14.8-3. The coefficients appearing in the characteristic equation (Sec. 14.8-5) corresponding to a dyadic are absolute or relative scalars. *Given any symmetric dyadic* A *in Euclidean space with continuously differentiable components and a region* V *where* $\det [A_k^i] \neq 0$, *there exists an orthogonal coordinate system* (*Sec.* 16.8-2) *such that the matrix* $[A_k^i]$ *is diagonal throughout* V (*normal coordinates,* see also Sec. 17.4-7).

A symmetric dyadic A defined on a three-dimensional Euclidean space may be represented geometrically by a quadric surface (3.5-1), with $a_{ik} = A_k^i$, for $i, k = 1, 2, 3$ (see also Sec. 3.5-1).

16.10. THE ABSOLUTE DIFFERENTIAL CALCULUS. COVARIANT DIFFERENTIATION

16.10-1. Absolute Differentials. (a) A small change or differential of a tensor quantity cannot be defined directly as the difference between "values" of the tensor function resulting from changes $dx^1, dx^2, \ldots, dx^n$ in the coordinates $x^1, x^2, \ldots, x^n$, since the tensor algebra of Secs. 16.3-1 to 16.3-7 does not define relations between tensor "values" at different points. The **absolute differentials** $d\alpha$, $d\mathbf{a}$, $d\mathbf{b}$, $d\mathsf{A}$, $d\mathsf{B}$, . . . of absolute scalars, vectors, and tensors α, $\mathbf{a}$, $\mathbf{b}$, A, B, . . . defined on a Riemann space in terms of suitably differentiable components are, instead, defined by the following postulates:

1. The absolute differentials $d\alpha$, $d\mathbf{a}$, and $d\mathsf{A}$ are absolute tensor quantities of the same respective ranks and types as α, $\mathbf{a}$, and A.
2. The absolute differential $d\alpha$ of an absolute scalar α is represented in the x scheme of measurements by

$$D\alpha \equiv \frac{\partial \alpha}{\partial x^j}\, dx^j \equiv \alpha_{,j}\, dx^j \tag{16.10-1}$$

3. The following differentiation rules hold:

$$\begin{gathered} d(\mathbf{a}\cdot\mathbf{b}) = \mathbf{a}\cdot d\mathbf{b} + \mathbf{b}\cdot d\mathbf{a} \\ d(\mathsf{A}+\mathsf{B}) = d\mathsf{A} + d\mathsf{B} \qquad d(\alpha\mathsf{A}) = \mathsf{A}\,d\alpha + \alpha\,d\mathsf{A} \\ d(\mathsf{AB}) = \mathsf{A}\,d\mathsf{B} + \mathsf{B}\,d\mathsf{A} \end{gathered} \tag{16.10-2}$$

In particular, Eq. (2) implies

$$d(\mathbf{a}+\mathbf{b}) = d\mathbf{a} + d\mathbf{b} \qquad d(\alpha\mathbf{a}) = \mathbf{a}\,d\alpha + \alpha\,d\mathbf{a}$$

so that

$$\begin{aligned} d\mathbf{a} = d(a^i\mathbf{e}_i) &= \mathbf{e}_i\,da^i + a^i\,d\mathbf{e}_i \\ &= a^i_{,j}\,(x^1, x^2, \ldots, x^n)\,dx^j\,\mathbf{e}_i \\ = d(a_i\mathbf{e}^i) &= \mathbf{e}^i\,da_i + a_i\,d\mathbf{e}^i \\ &= a_{i,j}(x^1, x^2, \ldots, x^n)\,dx^j\,\mathbf{e}^i \end{aligned} \tag{16.10-3}$$

i.e., $d\mathbf{a}$ is described by the components

$$Da^i \equiv a^i_{,j}\,dx^j \qquad \text{or} \qquad Da_i \equiv a_{i,j}\,dx^j$$

The postulates listed above result in a self-consistent and invariant (Secs. 16.4-1 and 16.10-7) generalization of the vector calculus described in Chap. 5; the postulates are satisfied if one chooses

$$\begin{aligned} a^i_{,j} &\equiv \frac{\partial a^i}{\partial x^j} + a^k \begin{Bmatrix} i \\ k\ j \end{Bmatrix} \\ a_{i,j} &\equiv \frac{\partial a_i}{\partial x^j} - a_k \begin{Bmatrix} k \\ i\ j \end{Bmatrix} \equiv g_{ih}a^h_{,j} \end{aligned} \tag{16.10-4}$$

where the functions $\begin{Bmatrix} k \\ i\ j \end{Bmatrix} \equiv \Gamma_i{}^k{}_j(x^1, x^2, \ldots, x^n)$ are the **Christoffel three-index symbols of the second kind** defined in Sec. 16.10-3.

NOTE: Equations (3) and (4) express each component of the absolute differential $d\mathbf{a}$ as the sum of a "relative differential" da^i or da_i and a term *due to the point-to-point changes in the base vectors (i.e., in the metric).* One may define *vector derivatives* $\partial\mathbf{a}/\partial x^j$ by

$$d\mathbf{a} = \frac{\partial\mathbf{a}}{\partial x^j}\,dx^j \tag{16.10-5}$$

with
$$\begin{aligned} \frac{\partial\mathbf{a}}{\partial x^j} &= \frac{\partial}{\partial x^j}(a^i\mathbf{e}_i) = \frac{\partial a^i}{\partial x^j}\mathbf{e}_i + a^i\frac{\partial\mathbf{e}_i}{\partial x^j} = a^i_{,j}\mathbf{e}_i \\ &= \frac{\partial}{\partial x^j}(a_i\mathbf{e}^i) = \frac{\partial a_i}{\partial x^j}\mathbf{e}^i + a_i\frac{\partial\mathbf{e}^i}{\partial x^j} = a_{i,j}\mathbf{e}^i \end{aligned} \tag{16.10-6}$$

$$\frac{\partial\mathbf{e}_i}{\partial x^j} = \begin{Bmatrix} k \\ i\ j \end{Bmatrix}\mathbf{e}_k = [ij;k]\mathbf{e}^k = \frac{\partial\mathbf{e}_j}{\partial x^i} \qquad \frac{\partial\mathbf{e}^i}{\partial x^j} = -\begin{Bmatrix} i \\ k\ j \end{Bmatrix}\mathbf{e}^k \tag{16.10-7}$$

(b) The postulates of Sec. 16.10-1a imply that the absolute differential of any absolute tensor A defined on a Riemann space in terms of suitably differentiable components $A^{i_1 i_2 \cdots i_r}_{i_1' i_2' \cdots i_s'}$ is an absolute tensor of the same type and rank, with components $DA^{i_1 i_2 \cdots i_r}_{i_1' i_2' \cdots i_s'}$ given by

$$\begin{aligned} d\mathsf{A} &= d(A^{i_1 i_2 \cdots i_r}_{i_1' i_2' \cdots i_s'} \mathbf{e}_{i_1} \mathbf{e}_{i_2} \cdots \mathbf{e}_{i_r} \mathbf{e}^{i_1'} \mathbf{e}^{i_2'} \cdots \mathbf{e}^{i_s'}) \\ &= DA^{i_1 i_2 \cdots i_r}_{i_1' i_2' \cdots i_s'} \mathbf{e}_{i_1} \mathbf{e}_{i_2} \cdots \mathbf{e}_{i_r} \mathbf{e}^{i_1'} \mathbf{e}^{i_2'} \cdots \mathbf{e}^{i_s'} \\ DA^{i_1 i_2 \cdots i_r}_{i_1' i_2' \cdots i_s'} &\equiv A^{i_1 i_2 \cdots i_r}_{i_1' i_2' \cdots i_s',j}\, dx^j \end{aligned} \tag{16.10-8}$$

with*

$$\begin{aligned} A^{i_1 i_2 \cdots i_r}_{i_1' i_2' \cdots i_s',j} &\equiv \frac{D}{\partial x^j} A^{i_1 i_2 \cdots i_r}_{i_1' i_2' \cdots i_s'} \equiv \frac{\partial}{\partial x^j} A^{i_1 i_2 \cdots i_r}_{i_1' i_2' \cdots i_s'} \\ &- \begin{Bmatrix} k \\ i_1' \; j \end{Bmatrix} A^{i_1 i_2 \cdots i_r}_{k i_2' i_3' \cdots i_s'} - \cdots - \begin{Bmatrix} k \\ i_s' \; j \end{Bmatrix} A^{i_1 i_2 \cdots i_r}_{i_1' i_2' \cdots i_{s-1}' k} \\ &+ \begin{Bmatrix} i_1 \\ k \; j \end{Bmatrix} A^{k i_2 i_3 \cdots i_r}_{i_1' i_2' \cdots i_s'} + \cdots + \begin{Bmatrix} i_r \\ k \; j \end{Bmatrix} A^{i_1 i_2 \cdots i_{r-1} k}_{i_1' i_2' \cdots i_s'} \end{aligned} \tag{16.10-9}$$

16.10-2. Absolute Differential of a Relative Tensor. Absolute differentials of relative tensor quantities (Sec. 16.2-1) are defined in the manner of Sec. 16.10-1. In particular, the absolute differential $d\alpha$ of a relative scalar α of weight W is represented in the x scheme of measurements by

$$D\alpha = \alpha_{,j}\, dx^j \qquad \text{with} \qquad \alpha_{,j} = \frac{\partial \alpha}{\partial x^j} - W \begin{Bmatrix} k \\ j \; k \end{Bmatrix} \alpha \tag{16.10-10}$$

Equation (10) reduces to Eq. (1) for $W = 0$. The absolute differential $d\mathsf{A}$ of any (suitably differentiable) relative tensor of weight W takes the form (8) with a term

$$-W \begin{Bmatrix} k \\ j \; k \end{Bmatrix} A^{i_1 i_2 \cdots i_r}_{i_1' i_2' \cdots i_s'} \tag{16.10-11}$$

added in the expression (9) for $A^{i_1 i_2 \cdots i_r}_{i_1' i_2' \cdots i_s',j}$, so that

$$\frac{D}{\partial x^j} A^{i_1 i_2 \cdots i_r}_{i_1' i_2' \cdots i_s'} = \sqrt{|g|}^W \frac{D}{\partial x^j} \left[\frac{1}{\sqrt{|g|}^W} A^{i_1 i_2 \cdots i_r}_{i_1' i_2' \cdots i_s'} \right] \tag{16.10-12}$$

$d\mathsf{A}$ is a relative tensor of the same rank, type, and weight as A.

16.10-3. Christoffel Three-index Symbols. (a) The **Christoffel three-index symbols of the first kind** $[ij;\, k] \equiv \Gamma_{ij;k}(x^1, x^2, \cdots, x^n)$ and the **Christoffel three-index symbols of the second kind** $\begin{Bmatrix} k \\ i \; j \end{Bmatrix} \equiv \Gamma_i{}^k{}_j(x^1, x^2, \ldots, x^n)$ associated with an x scheme of measurements in a Riemann space are functions of the coordinates $x^1, x^2, \ldots, x^n$, viz.,

* Some authors use the notation

$$\frac{\partial A^{i_1 i_2 \cdots i_r}_{i_1' i_2' \cdots i_s'}}{\partial x^j} \equiv A^{i_1 i_2 \cdots i_r}_{i_1' i_2' \cdots i_s',j} \qquad \frac{DA^{i_1 i_2 \cdots i_r}_{i_1' i_2' \cdots i_s'}}{\partial x^j} \equiv A^{i_1 i_2 \cdots i_r}_{i_1' i_2' \cdots i_s';j} \text{ (or } A^{i_1 i_2 \cdots i_r}_{i_1' i_2' \cdots i_s'|j}\text{)}$$

$$[ij;k] \equiv \frac{1}{2}\left(\frac{\partial g_{ik}}{\partial x^j} + \frac{\partial g_{jk}}{\partial x^i} - \frac{\partial g_{ij}}{\partial x^k}\right) \qquad \left\{ {k \atop i\ j} \right\} \equiv g^{kh}[ij;h] \qquad (16.10\text{-}13)$$

where g_{ik} and g^{ik} are the fundamental-tensor components of the Riemann space in the x scheme of measurements. *The Christoffel three-index symbols are not in general tensor components* but transform according to the *Christoffel transformation equations*

$$\left.\begin{aligned} \overline{[hk;r]} &= \frac{\partial x^i}{\partial \bar{x}^h}\frac{\partial x^j}{\partial \bar{x}^k}\frac{\partial x^s}{\partial \bar{x}^r}[ij;s] + \frac{\partial^2 x^i}{\partial \bar{x}^h \partial \bar{x}^k}\frac{\partial x^j}{\partial \bar{x}^r} g_{ij} \\ \overline{\left\{ {r \atop h\ k} \right\}} &= \frac{\partial x^i}{\partial \bar{x}^h}\frac{\partial x^j}{\partial \bar{x}^k}\frac{\partial \bar{x}^r}{\partial x^s}\left\{ {s \atop i\ j} \right\} + \frac{\partial^2 x^s}{\partial \bar{x}^h \partial \bar{s}^k}\frac{\partial \bar{x}^r}{\partial x^s} \end{aligned}\right\} \qquad (16.10\text{-}14)$$

where the functions $\overline{[hk;r]} \equiv \Gamma_{hk;r}(\bar{x}^1, \bar{x}^2, \ldots, \bar{x}^n)$ and $\overline{\left\{ {r \atop h\ k} \right\}} \equiv \bar{\Gamma}_h{}^r{}_k(\bar{x}^1, \bar{x}^2, \ldots, \bar{x}^n)$ are the Christoffel three-index symbols associated with an $\bar{x}$ scheme of measurements related to the x scheme of measurements by a suitably differentiable coordinate transformation.

(**b**) The Christoffel three-index symbols satisfy the following relations:

$$[ij;k] = [ji;k] \qquad \left\{ {k \atop i\ j} \right\} = \left\{ {k \atop j\ i} \right\} \qquad (16.10\text{-}15)$$

$$[ij;k] = g_{hk}\left\{ {h \atop i\ j} \right\} \qquad \left\{ {k \atop i\ j} \right\} = g^{kh}[ij;h] \qquad (16.10\text{-}16)$$

$$\frac{\partial g_{ij}}{\partial x^k} = [ik;j] + [jk;i] = g_{hj}\left\{ {h \atop i\ k} \right\} + g_{ih}\left\{ {h \atop j\ k} \right\} \qquad (16.10\text{-}17)$$

$$\frac{\partial g^{ij}}{\partial x^k} = -g^{ih}\left\{ {j \atop h\ k} \right\} - g^{jh}\left\{ {i \atop h\ k} \right\} \qquad (16.10\text{-}18)$$

$$\frac{\partial}{\partial x^i}\log_e \sqrt{|g|} = \left\{ {h \atop i\ h} \right\} \qquad g \equiv \det[g_{ik}] \qquad (16.10\text{-}19)$$

$$\left\{ {k \atop i\ j} \right\} = \frac{\partial \mathbf{e}_i}{\partial x^j}\cdot \mathbf{e}^k = \frac{\partial \mathbf{e}_j}{\partial x^i}\cdot \mathbf{e}^k = -\frac{\partial \mathbf{e}^k}{\partial x^j}\cdot \mathbf{e}_i \qquad [ij;k] = \frac{\partial \mathbf{e}_i}{\partial x^j}\cdot \mathbf{e}_k \qquad (16.10\text{-}20)$$

(**c**) In the important special case of *orthogonal coordinates* $x^1, x^2, \ldots, x^n$ (Secs. 6.4-1 and 16.8-2), one has $g_{ik} = g_{ii}\delta^i_k$, so that

$$\left.\begin{aligned} &[ij;k] \equiv \left\{ {k \atop i\ j} \right\} \equiv 0 \qquad (i \neq j \neq k \neq i) \\ &[ii;k] \equiv -\frac{1}{2}\frac{\partial g_{ii}}{\partial x^k} \qquad (i \neq k) \\ &[ij;i] \equiv [ji;i] \equiv \frac{1}{2}\frac{\partial g_{ii}}{\partial x^j} \\ &\left\{ {k \atop i\ i} \right\} \equiv -\frac{1}{2g_{kk}}\frac{\partial g_{ii}}{\partial x^k} \qquad (i \neq k) \\ &\left\{ {k \atop i\ k} \right\} \equiv \left\{ {k \atop k\ i} \right\} \equiv \frac{1}{2g_{kk}}\frac{\partial g_{kk}}{\partial x^i} \equiv \frac{1}{2}\frac{\partial \log_e g_{kk}}{\partial x^i} \end{aligned}\right\} \qquad (16.10\text{-}21)$$

16.10-4. Covariant Differentiation. Because of the analogy between Eq. (4) or (9) and ordinary partial differentiation, the operation of obtaining the functions $A^{i_1i_2\cdots i_r}_{i_1'i_2'\cdots i_s',j} \equiv \frac{D}{\partial x^j} A^{i_1i_2\cdots i_r}_{i_1'i_2'\cdots i_s'}$ from the tensor components $A^{i_1i_2\cdots i_r}_{i_1'i_2'\cdots i_s'}$ is called **covariant differentiation** (with respect to the metric described by the g_{ik}). If the components $A^{i_1i_2\cdots i_r}_{i_1'i_2'\cdots i_s'}$ represent an absolute tensor $\mathbf{A}$, then the components $A^{i_1i_2\cdots i_r}_{i_1'i_2'\cdots i_s',j}$ describe an absolute tensor

$$\nabla \mathbf{A} = A^{i_1i_2\cdots i_r}_{i_1'i_2'\cdots i_s',j}\mathbf{e}_{i_1}\mathbf{e}_{i_2}\cdots\mathbf{e}_{i_r}\mathbf{e}^{i_1'}\mathbf{e}^{i_2'}\cdots\mathbf{e}^{i_s'}\mathbf{e}^{j} \qquad (16.10\text{-}22)$$

contravariant of rank r and covariant of rank $s + 1$. $\nabla\mathbf{A}$ is commonly called the **covariant derivative** of $\mathbf{A}$.

Note that, in general, neither the functions $\frac{\partial}{\partial x^j} A^{i_1i_2\cdots i_r}_{i_1'i_2'\cdots i_s'}$ *nor the "relative" differentials* $dA^{i_1i_2\cdots i_r}_{i_1'i_2'\cdots i_s'}$ *are tensor components.*

16.10-5. Rules for Covariant Differentiation (see also Sec. 16.10-7). Computations are frequently simplified by the fact that the ordinary rules for differentiation of sums and products (Table 4.5-2) apply formally to covariant differentiation. Note

$$\begin{aligned}
&\frac{D}{\partial x^j}\left(A^{i_1i_2\cdots i_r}_{i_1'i_2'\cdots i_s'} + B^{i_1i_2\cdots i_r}_{i_1'i_2'\cdots i_s'}\right) = A^{i_1i_2\cdots i_r}_{i_1'i_2'\cdots i_s',j} + B^{i_1i_2\cdots i_r}_{i_1'i_2'\cdots i_s',j}\\
&\frac{D}{\partial x^j}\left(A^{i_1i_2\cdots i_r}_{i_1'i_2'\cdots i_s'}B^{k_1k_2\cdots k_r}_{k_1'k_2'\cdots k_s'}\right) = A^{i_1i_2\cdots i_r}_{i_1'i_2'\cdots i_s'}B^{k_1k_2\cdots k_r}_{k_1'k_2'\cdots k_s',j}\\
&\qquad\qquad + B^{k_1k_2\cdots k_r}_{k_1'k_2'\cdots k_s'}A^{i_1i_2\cdots i_r}_{i_1'i_2'\cdots i_s',j}\\
&\frac{D}{\partial x^j} A^{i_1i_2\cdots k\cdots i_r}_{i_1'i_2'\cdots k\cdots i_s'} = A^{i_1i_2\cdots k\cdots i_r}_{i_1'i_2'\cdots k\cdots i_s',j} \qquad \text{(CONTRACTION RULE)}
\end{aligned} \qquad (16.10\text{-}23)$$

The last two rules apply to the covariant differentiation of inner products (Secs. 16.3-7 and 16.8-1). Note also

$$\begin{aligned}
g_{ik,j} &= g^{ik}_{,j} = 0 \qquad \text{(RICCI'S THEOREM)}\\
g_{,j} &= 0\\
\nabla\mathbf{e}_i &= \nabla\mathbf{e}^i = \mathbf{0}
\end{aligned} \qquad (16.10\text{-}24)$$

Equation (24) shows that *the fundamental tensors behave like constants with respect to covariant differentiation. The covariant derivative of every associated tensor of* $\mathbf{A}$ *is an associated tensor of* $\nabla\mathbf{A}$ (see also Secs. 16.7-2 and 16.7-3).

The covariant derivative of every Kronecker delta and permutation symbol (*Secs.* 16.5-2 *and* 16.5-3) *is zero.*

16.10-6. Higher-order Covariant Derivatives. Successive covariant differentiations* of tensor components yield *higher-order covariant derivatives* described by components like $A^{i_1 i_2 \cdots i_r}_{i_1' i_2' \cdots i_s', j_1 j_2 \cdots j_m}$. These tensors are *not* in general symmetric with respect to pairs of subscripts j; *the order of covariant differentiation may be inverted if and only if the Riemann space in question is flat* (Sec. 17.4-6c). For any vector described by a_i,

$$a_{i,jk} - a_{i,kj} = R^r_{ijk} a_r \tag{16.10-25}$$

where R^r_{ijk} are the components of the *mixed curvature tensor* (Sec. 17.4-5) of the Riemann space.

Table 16.10-1. Differential Invariants Defined on Riemann Spaces

($\nabla \equiv \mathbf{e}^j \frac{D}{x^j}$; see also Secs. 16.2-2, 16.10-1 to 16.10-7, and Tables 6.4-1 to 6.5-11)

(a) The **gradient of an absolute scalar** α is the absolute vector

$$\nabla \alpha = \mathbf{e}^j \frac{\partial \alpha}{\partial x^j} = \mathbf{e}_i g^{ji} \frac{\partial \alpha}{\partial x^j}$$

(b) The covariant derivative $\nabla \mathbf{a}$ of an absolute vector $\mathbf{a}$ is called the **gradient** or **local dyadic** of $\mathbf{a}$. The **divergence of an absolute vector a** is the absolute scalar $\nabla \cdot \mathbf{a}$ represented by

$$\frac{Da^i}{\partial x^i} = \frac{\partial a^i}{\partial x^i} + a^k \begin{Bmatrix} i \\ k\ i \end{Bmatrix} = \frac{1}{\sqrt{|g|}} \frac{\partial}{\partial x^i} (\sqrt{|g|}\, a^i)$$

NOTE: The appropriate formulas of Table 5.5-1 apply to gradients and divergences defined on Riemann spaces (see also Sec. 16.10-5).

(c) The **Laplacian operator** $\nabla^2 \equiv \nabla \cdot \nabla$ is the invariant scalar operator represented by $g^{ik} \frac{D}{x^i} \frac{D}{x^k}$. In particular, the Laplacian $\nabla^2 \alpha$ of an absolute scalar α is represented by

$$g^{ik} \frac{D}{\partial x^i} \frac{\partial \alpha}{\partial x^k} \equiv g^{ik} \left(\frac{\partial^2 \alpha}{\partial x^i\, \partial x^k} - \begin{Bmatrix} j \\ i\ k \end{Bmatrix} \frac{\partial \alpha}{\partial x^j} \right) = \frac{1}{\sqrt{|g|}} \frac{\partial}{\partial x^i} \left(g^{ik} \sqrt{|g|} \frac{\partial \alpha}{\partial x^k} \right)$$

(d) *Given an absolute vector* $\mathbf{a}$, *the skew-symmetric absolute tensor represented by*

$$C_{ij} \equiv \frac{Da_i}{\partial x^j} - \frac{Da_j}{\partial x^i} \equiv \frac{\partial a_i}{\partial x^j} - \frac{\partial a_j}{\partial x^i}$$

is identically zero if and only if $\mathbf{a}$ *is the gradient of an absolute scalar.*

NOTE: For $n = 3$, it is possible to define the **curl of an absolute vector a** as the absolute vector $\nabla \times \mathbf{a}$ represented by the components

$$\frac{1}{\sqrt{|g|}} \frac{Da_j}{\partial x^i} \epsilon^{ijk} \equiv \frac{1}{2\sqrt{|g|}} \left(\frac{Da_j}{\partial x^i} - \frac{Da_i}{\partial x^j} \right) \epsilon^{ijk} \equiv \frac{1}{2\sqrt{|g|}} \left(\frac{\partial a_j}{\partial x^i} - \frac{\partial a_i}{\partial x^j} \right) \epsilon^{ijk}$$

(see also Sec. 16.8-4). The formulas of Table 5.5-1 and Eq. (5.5-19) apply to absolute vectors defined on three-dimensional Riemann spaces.

* It is understood that the components g_{ik} of the metric tensor as well as the tensor components to be differentiated are repeatedly differentiable.

16.10-7. Differential Operators and Differential Invariants (see also Secs. 5.5-2 and 5.5-5). (**a**) If one defines

$$\frac{D}{\partial x^j} A^{i_1 i_2 \cdots i_r}_{i_1' i_2' \cdots i_s'} \mathbf{e}_{i_1} \mathbf{e}_{i_2} \cdots \mathbf{e}_{i_r} \mathbf{e}^{i_1'} \mathbf{e}^{i_2'} \cdots \mathbf{e}^{i_s'} \equiv A^{i_1 i_2 \cdots i_r}_{i_1' i_2' \cdots i_s' ,j} \mathbf{e}_{i_1} \mathbf{e}_{i_2} \cdots \mathbf{e}_{i_r} \mathbf{e}^{i_1'} \mathbf{e}^{i_2'} \cdots \mathbf{e}^{i_s'}$$

the covariant derivative (22) of a tensor **A** may be written as an "outer product" (Sec. 16.3-6) of **A** and the (invariant) *vector differential operator*

$$\nabla \equiv \mathbf{e}^j \frac{D}{\partial x^j} \tag{16.10-26}$$

(**del** or **nabla**) whose "components" $D/\partial x^j$ transform like covariant vector components (Sec. 16.2-1). For every admissible transformation (16.1-1) of coordinates describing the same Riemann space

$$\frac{D}{\partial \bar{x}^k} = \frac{\partial x^j}{\partial \bar{x}^k} \frac{D}{\partial x^j} \tag{16.10-27}$$

(**b**) *Tensor relations involving covariant differentiation as well as tensor addition and multiplication are invariant with respect to the group of admissible coordinate transformations* (see also Secs. 16.4-1 and 16.10-1). Tensor quantities obtained through outer and/or inner multiplication of other tensor quantities by the invariant operator (26) are called **differential invariants.** Table 16.10-1 lists the most useful differential invariants.

16.10-8. Absolute (Intrinsic) and Directional Derivatives (see also Sec. 5.5-3). Given a regular arc in Riemann space,

$$x^i = x^i(t) \qquad (t_1 \leq t \leq t_2) \tag{16.10-28}$$

the components dx^i/dt represent a contravariant vector $d\mathbf{r}/dt$ "directed" along the tangent to the given curve (Sec. 17.4-2). The **absolute (intrinsic) derivative** $d\mathbf{A}/dt$ of a suitably differentiable absolute or relative tensor **A** with respect to the parameter t along the given curve is a tensor of the same rank, type, and weight as **A**:

$$\frac{d\mathbf{A}}{dt} = \left(\frac{d\mathbf{r}}{dt} \cdot \nabla\right) \mathbf{A}, \text{ with components} \quad \frac{D}{dt} A^{i_1 i_2 \cdots i_r}_{i_1' i_2' \cdots i_s'} \equiv A^{i_1 i_2 \cdots i_r}_{i_1' i_2' \cdots i_s' ,j} \frac{dx^j}{dt} \tag{16.10-29}$$

If the components of **A** depend explicitly as well as implicitly on t,

$$\frac{d\mathbf{A}}{dt} = \frac{\partial \mathbf{A}}{\partial t} + \left(\frac{d\mathbf{r}}{dt} \cdot \nabla\right) \mathbf{A} \qquad \frac{DA^{i_1 i_2 \cdots i_r}_{i_1' i_2' \cdots i_s'}}{dt} = \frac{\partial A^{i_1 i_2 \cdots i_r}_{i_1' i_2' \cdots i_s'}}{\partial t} + A^{i_1 i_2 \cdots i_r}_{i_1' i_2' \cdots i_s' ,j} \frac{dx^j}{dt} \tag{16.10-30}$$

The **directional derivative** dA/ds of A in the direction of the given curve ($ds \neq 0$, Sec. 17.4-2) is the absolute derivative of A with respect to the arc length s along the curve.

16.10-9. Tensors Constant along a Curve. Equations of Parallelism. A tensor A is *defined* as **constant** along a regular curve arc (28) (i.e., its "values" at neighboring points on the curve are "equal") if and only if its absolute derivative (30) (and thus also its absolute differential $dA = \frac{dA}{ds}\,ds$) along the curve is zero. *Sums and products of such tensors are also constant along the curve in question*, so that, for example, the absolute values of, and the angles between, constant vectors are constant. Every vector **a** whose components $a^i(x^1, x^2, \ldots, x^n)$ or $a_i(x^1, x^2, \ldots, x^n)$ satisfy the differential equations

$$\boxed{Da^i = 0 \quad \text{or} \quad Da_i = 0 \qquad \text{(EQUATIONS OF PARALLELISM)}} \qquad (16.10\text{-}31)$$

as the coordinates $x^1, x^2, \ldots, x^n$ vary along a curve (28) undergoes a "parallel displacement" along the curve.

NOTE: A vector obtained by "parallel displacement" of a given vector along a closed curve is *not* in general equal to the original vector when the starting point is reached (see also Sec. 17.4-6).

16.10-10. Integration of Tensor Quantities. Volume Element. Integrals of tensor quantities over curves in Riemann space may be defined in terms of scalar integrals over a suitable parameter as in Secs. 5.4-5 and 6.2-3*a*. The **volume element** dV is the scalar capacity (Sec. 16.2-1) defined by (see also Secs. 6.2-3*b* and 6.4-3*c*)

$$dV = \sqrt{|g|}\, dx^1\, dx^2 \cdots dx^n \qquad (16.10\text{-}32)$$

Volume integrals over scalar invariants are scalar invariants, but volume integrals over tensors of rank $R > 0$ are not, in general, tensors.

Volume elements in suitable subspaces take the place of surface elements in three-dimensional space. Generalization of the integral theorems of Secs. 5.6-1 and 5.6-2 exist (Sec. 16.10-11 and Ref. 16.6, Chap. 7).

16.10-11. Differential Invariants and Integral Theorems for Dyadics (see also Sec. 16.9-1). The **divergence** of a suitably differentiable dyadic A defined on a Riemann space is the vector $\nabla \cdot A$, where the operator ∇ (Sec. 16.10-7) acts like a covariant vector. Note

$$\nabla \cdot (\alpha A) = \alpha \nabla \cdot A + A \cdot \nabla \alpha \qquad (16.10\text{-}33)$$

$$\nabla \cdot (\mathbf{a} \cdot A) = (\nabla \mathbf{a}) \cdot\cdot A + (\nabla \cdot \tilde{A}) \cdot \mathbf{a} \qquad \nabla \cdot (A \cdot \mathbf{a}) = (\nabla \mathbf{a}) \cdot\cdot \tilde{A} + (\nabla \cdot A) \cdot \mathbf{a} \qquad (16.10\text{-}34)$$

where $\tilde{A}$ is the *transposed dyadic* of A ($\tilde{A}^i_k = A^k_i$). The dyadic $\nabla \mathbf{a}$ is the gradient of **a** (Table 16.10-1).

The following *integral theorems* analogous to those of Secs. 5.6-1 and 5.6-2 hold for suitable functions, surfaces, and curves:

$$\int_V \nabla \cdot \mathbf{A}\, dV = \int_S d\mathbf{A} \cdot \mathbf{A} \tag{16.10-35}$$

$$\int_V \nabla \mathbf{a}\, dV = \int_S d\mathbf{A}\, \mathbf{a} \qquad \int_V \nabla \times \mathbf{A}\, dV = \int_S d\mathbf{A} \times \mathbf{A} \tag{16.10-36}$$

$$\int_V \nabla \cdot (\mathbf{A} \cdot \mathbf{a})\, dV = \int_S d\mathbf{A} \cdot (\mathbf{A} \cdot \mathbf{a}) \tag{16.10-37}$$

$$\int_S d\mathbf{A} \cdot (\nabla \times \mathbf{A}) = \int_C d\mathbf{r} \cdot \mathbf{A} \tag{16.10-38}$$

16.11 RELATED TOPICS, REFERENCES, AND BIBLIOGRAPHY

16.11-1. Related Topics. The following topics related to the study of tensor analysis are treated in other chapters of this handbook:

Vector analysis . . . . Chap. 5
Abstract algebra . . . . Chap. 12
Linear transformations . . . . Chap. 14
Rotation of a vector . . . . Chap. 14
Differential geometry . . . . Chap. 17
Partial differentiation . . . . Chap. 4
Special coordinate systems . . . . Chap. 6

16.11-2. References and Bibliography.

16.1. Brillouin, L.: *Les Tenseurs en mécanique et en elasticité*, Masson, Paris, 1949.
16.2. Eisenhart, L. P.: *An Introduction to Differential Geometry*, Princeton University Press, Princeton, N.J., 1947.
16.3. Lagally, M.: *Vorlesungen über Vektor-Rechnung*, Edwards, Ann Arbor, Mich., 1947.
16.4. Lichnerowicz, A.: *Elements of Tensor Calculus*, Wiley, New York, 1962.
16.5. Phillips, H. B.: *Vector Analysis*, Wiley, New York, 1933.
16.6. Rainich, G. Y.: *Mathematics of Relativity*, Wiley, New York, 1950.
16.7. Sokolnikov, I. S.: *Tensor Analysis*, 2nd Ed., Wiley, New York, 1964.
16.8. Synge, J. L., and A. Schild: *Tensor Calculus*, University of Toronto Press, 1949.

CHAPTER **17**

DIFFERENTIAL GEOMETRY

17.1. CURVES IN THE EUCLIDEAN PLANE

17.1-1. Tangent to a Plane Curve. Given a plane curve C represented by

$$y = f(x) \tag{17.1-1a}$$

or

$$\varphi(x, y) = 0 \tag{17.1-1b}$$

or

$$x = x(t) \qquad y = y(t) \tag{17.1-1c}$$

(Sec. 2.1-9) in terms of suitably differentiable functions, the **tangent** to C at the point $P_1 \equiv (x_1, y_1)$ is defined as the limit of a straight line (secant) through P_1 and a neighboring point P_2 as P_2 approaches P_1. *The curve* (1) *has a unique tangent described by*

$$y - y_1 = \frac{dy}{dx}(x - x_1) \tag{17.1-2a}$$

or

$$\frac{\partial \varphi}{\partial x}(x - x_1) + \frac{\partial \varphi}{\partial y}(y - y_1) = 0 \tag{17.1-2b}$$

or

$$x = \frac{dx}{dt}(t - t_1) + x_1 \qquad y = \frac{dy}{dt}(t - t_1) + y_1 \tag{17.1-2c}$$

at every **regular point** (x_1, y_1) *where it is possible to choose a parameter t so that $x(t)$ and $y(t)$ have unique continuous derivatives not both equal to zero; or, equivalently, where $\varphi(x, y)$ has unique continuous first partial derivatives not both equal to zero.* The *slope* (Sec. 2.2-1) of the tangent (2) is

$$\boxed{\tan \vartheta = \frac{dy}{dx} = -\frac{\partial \varphi}{\partial x}\Big/\frac{\partial \varphi}{\partial y} = \frac{dy}{dt}\Big/\frac{dx}{dt}} \tag{17.1-3}$$

17.1-2. Normal to a Plane Curve. The **normal** to the curve (1) at a regular point $P_1 \equiv (x_1, y_1)$ is the straight line through P_1 and perpendicular to the tangent at P_1:

$$y - y_1 = -\frac{1}{dy/dx}(x - x_1) \qquad (17.1\text{-}4)$$

The direction of the *positive* normal is arbitrarily fixed with respect to the common positive direction of curve and tangent. The positive direction on a curve (1) is arbitrarily fixed by some convention (e.g., direction of increasing t, increasing x, etc.; see also Sec. 2.2-1).

17.1-3. Singular Points. Given a curve (1*b*) such that the nth-order derivatives of $\varphi(x, y)$ exist and do not vanish simultaneously at P_1, let all $(n - 1)^{\text{st}}$-order derivatives of $\varphi(x, y)$ be equal to zero at P_1. Then the curve has n tangents at P_1; two or more of these tangents may coincide, and an even number of them may be imaginary. Thus if all first-order derivatives but not all second-order derivatives of $\varphi(x, y)$ vanish at $P_1 \equiv (x_1, y_1)$, the slopes dy/dx of the two tangents are obtained as roots of the quadratic equation

$$\left(\frac{dy}{dx}\right)^2 \frac{\partial^2 \varphi}{\partial y^2} + 2\frac{dy}{dx}\frac{\partial^2 \varphi}{\partial x\, \partial y} + \frac{\partial^2 \varphi}{\partial x^2} = 0 \qquad (x = x_1,\ y = y_1) \qquad (17.1\text{-}5)$$

The roots of Eq. (5), and hence the two tangents, may be *real and different* (*double point*), *coincident* (*cusp, or self-osculation point*, see also Sec. 17.1-5), or *imaginary* (*isolated point*). The properties of a curve at a singular point can be similarly described in terms of discontinuous or multiple-valued derivatives of $f(x)$ or of $x(t)$ and $y(t)$.

17.1-4. Curvature of a Plane Curve. The **circle of curvature** (**osculating circle**) of a plane curve C at the point P_1 is the limit of a circle through P_1 and two other distinct points P_2 and P_3 of C as P_2 and P_3 approach P_1. The center of this circle (**center of curvature** of C corresponding to the curve point P_1) is located on the normal to C at P_1. The coordinates of the center of curvature are

$$\left.\begin{aligned} x_\kappa &= x_1 - \frac{dy}{dx}\left[1 + \left(\frac{dy}{dx}\right)^2\right] \Big/ \frac{d^2y}{dx^2} = x_1 - \frac{\dot{y}(\dot{x}^2 + \dot{y}^2)}{\dot{x}\ddot{y} - \dot{y}\ddot{x}} \\ y_\kappa &= y_1 + \left[1 + \left(\frac{dy}{dx}\right)^2\right] \Big/ \frac{d^2y}{dx^2} = y_1 + \frac{\dot{x}(\dot{x}^2 + \dot{y}^2)}{\dot{x}\ddot{y} - \dot{y}\ddot{x}} \end{aligned}\right\} \qquad (17.1\text{-}6)$$

where all derivatives are computed for $x = x_1$ $(t = t_1)$; dots indicate differentiation with respect to t.* The radius ρ_κ of the circle of curvature (**radius of curvature** of C at P_1) equals the reciprocal of the **curvature** κ of C at P_1 defined as the rate of turn of the tangent with respect to the arc length s along C (Sec. 4.6-9):

* Equations (6) and (7) may be rewritten in terms of the partial derivatives of $\varphi(x, y)$ with the aid of Eq. (4.5-16).

$$\kappa = \frac{1}{\rho_\kappa} = \frac{d\vartheta}{ds} = \frac{d^2y}{dx^2} \Big/ \underset{+}{\sqrt{1 + \left(\frac{dy}{dx}\right)^2}}^{3} = \frac{\dot{x}\ddot{y} - \dot{y}\ddot{x}}{\underset{+}{\sqrt{\dot{x}^2 + \dot{y}^2}}^{3}} \tag{17.1-7}$$

where all derivatives are computed for $x = x_1$ $(t = t_1)$.* A given curve C is, respectively, *concave* or *convex* in the direction of the positive y axis wherever d^2y/dx^2 and thus κ is positive or negative. Many authors introduce $|\kappa|$ rather than κ as the curvature, as in Sec. 17.2-3.

In terms of *polar coordinates* r, φ (Sec. 2.1-8), the distance element ds and the angle μ between the tangent to a curve $r = r(\varphi)$ and the radius vector at each regular point $P_1 \equiv (r_1, \varphi_1)$ are given by

$$ds^2 = dr^2 + r^2\, d\varphi^2 \qquad \tan \mu = r \Big/ \frac{dr}{d\varphi} \qquad (\varphi = \varphi_1) \tag{17.1-8}$$

and

$$\kappa = \frac{1}{\rho_\kappa} = \frac{d\vartheta}{ds} = \frac{d\varphi}{ds} + \frac{d\mu}{ds} = \frac{r^2 + 2\left(\frac{dr}{d\varphi}\right)^2 - r\frac{d^2r}{d\varphi^2}}{\underset{+}{\sqrt{r^2 + \left(\frac{dr}{d\varphi}\right)^2}}^{3}} \qquad (\varphi = \varphi_1) \tag{17.1-9}$$

17.1-5. Osculation. A point $P_1 \equiv (x_1, y_1)$ is an **osculation point (point of contact) of order** n of two curves described by suitably differentiable equations $y = f(x)$ and $y = g(x)$ if and only if

$$f(x_1) = g(x_1) \qquad f'(x_1) = g'(x_1) \qquad \cdots \qquad f^{(n)}(x_1) = g^{(n)}(x_1)$$
$$f^{(n+1)}(x_1) \neq g^{(n+1)}(x_1) \tag{17.1-10}$$

Such a point may be regarded as the limit of $n + 1$ real or imaginary points of intersection approaching one another. At every osculation point the tangents to the two curves coincide; the curves intersect (cross) if and only if n is even. A point where a curve intersects its own tangent is a **point of inflection.** At every point of inflection $\kappa = 0$.

17.1-6. Asymptotes. A straight line is an **asymptote** of the given curve C (C **approaches the straight line asymptotically**) if and only if the distance between the straight line and a point $P \equiv (x, y)$ on the curve tends to zero as $x^2 + y^2 \to \infty$. If C is a regular arc the asymptote is the limiting case of the tangent at P.

17.1-7. Envelope of a Family of Plane Curves. The **envelope** of a suitable one-parameter family of plane curves described by

$$\varphi(x, y, \lambda) = 0 \tag{17.1-11}$$

osculates, or contains a singular point of, every curve (11). One obtains the equation of the envelope by elimination of the parameter λ from Eq. (11) and

$$\frac{\partial\varphi(x, y, \lambda)}{\partial\lambda} = 0 \tag{17.1-12}$$

* Equations (6) and (7) may be rewritten in terms of the partial derivaties of $\varphi(x, y)$ with the aid of Eq. (4.5-16).

The envelope exists in a region of values of x, y, and λ such that

$$\frac{\partial(\varphi, \varphi_\lambda)}{\partial(x, y)} \neq 0 \qquad \varphi_{\lambda\lambda} \neq 0 \tag{17.1-13}$$

Equations (11) and (12) define the locus of *limiting points of intersection* of $\varphi(x, y, \lambda_1) = 0$ and $\varphi(x, y, \lambda_2) = 0$ as $\lambda_2 \to \lambda_1$. This locus (**λ discriminant**) includes loci of cusps and nodes as well as envelopes.

17.1-8. Isogonal Trajectories. The family of curves intersecting a given family $\varphi(x, y, \lambda) = 0$ at a given angle γ is described by the differential equation

$$\left(\frac{\partial\varphi}{\partial x}\cos\gamma - \frac{\partial\varphi}{\partial y}\sin\gamma\right)dx + \left(\frac{\partial\varphi}{\partial x}\sin\gamma + \frac{\partial\varphi}{\partial y}\cos\gamma\right)dy = 0 \tag{17.1-14}$$

For $\gamma = \pi/2$, Eq. (14) yields **orthogonal trajectories.**

17.2. CURVES IN THREE-DIMENSIONAL EUCLIDEAN SPACE

17.2-1. Introduction (see also Sec. 3.1-13). Sections 17.2-1 to 17.2-6 deal with the geometry of a *regular arc* C described by

$$\left.\begin{array}{l} \mathbf{r} = \mathbf{r}(t) \\ \text{or} \quad x = x(t) \qquad y = y(t) \qquad z = z(t) \end{array}\right\} \quad (t_1 \leq t \leq t_2) \tag{17.2-1}$$

where the functions (1) have unique continuous first derivatives and $d\mathbf{r}/dt \neq 0$ for $t_1 \leq t \leq t_2$. Higher-order derivatives will be assumed to exist as needed. It is convenient to introduce the *arc length* $s \equiv \int_{t_{1C}}^{t} ds \equiv \int_{t_{1C}}^{t} \sqrt{d\mathbf{r}\cdot d\mathbf{r}} \equiv \int_{t_{1C}}^{t} \sqrt{dx^2 + dy^2 + dz^2}$ (Sec. 5.4-4) as a new parameter; the sign of ds is arbitrarily fixed to determine the *positive direction* of curve and tangents (see also Secs. 17.2-2 and 17.2-3). Differentiation with respect to s will be indicated by primes, so that, for example,

$$x' \equiv \frac{dx}{ds} \equiv \frac{dx}{dt}\Big/\frac{ds}{dt}$$

The representation of curves in terms of *curvilinear coordinates* (Chap. 6) is briefly discussed in Sec. 17.4-1.

17.2-2. The Moving Trihedron (see also Secs. 17.2-3 and 17.2-4). **(a) Tangent to a Curve.** The **tangent** to a regular arc C at the point $P_1 \equiv (\mathbf{r}_1) \equiv (x_1, y_1, z_1)$ is the limit of a straight line (secant) through P_1 and another point P_2 of C as P_2 approaches P_1. A unique tangent exists at every point of a regular arc. The positive tangent direction coincides with the positive direction of C at P_1.

(b) Osculating Circle and Plane. Principal Normal. The **osculating circle** or **circle of curvature** of C at the curve point P_1 is the limit of a circle through P_1 and two other distinct points P_2 and P_3

of C as P_2 and P_3 approach P_1. The plane of this circle (**osculating plane** of C at P_1) contains the tangent to C at P_1. The directed straight line from the curve point P_1 to the center of the osculating circle (**center of curvature**) is perpendicular to the tangent and is called the **principal normal** of C at P_1.

(c) **Binormal. Normal and Rectifying Planes.** The **binormal** of C at P_1 is the directed straight line through P_1 such that *the positive tangent, principal normal, and binormal form a system of right-handed rectangular cartesian axes* (Sec. 3.1-3). These axes determine the "moving trihedron" comprising the osculating plane and the **normal and rectifying planes** respectively normal to the tangent and to the principal normal of C at P_1 (Fig. 17.2-1).

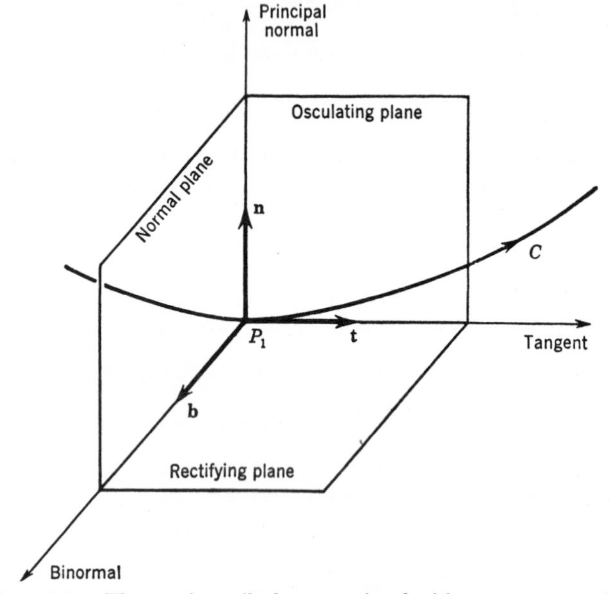

FIG. 17.2-1. The moving trihedron associated with a space curve C.

(d) The unit vectors **t**, **n**, and **b** respectively directed along the positive tangent, the principal normal, and the binormal are given by

$$\left.\begin{aligned} \mathbf{t} &= \mathbf{r}' && \text{(UNIT TANGENT VECTOR)} \\ \mathbf{n} &= \frac{\mathbf{r}''}{|\mathbf{r}''|} = \frac{1}{\kappa}\mathbf{r}'' && \text{(UNIT PRINCIPAL-NORMAL VECTOR)} \\ \mathbf{b} &= \mathbf{t} \times \mathbf{n} && \text{(UNIT BINORMAL VECTOR)} \end{aligned}\right\} \qquad (17.2\text{-}2)$$

at each suitable point of the curve. The vector $\kappa\mathbf{n} = \mathbf{r}''$ is called the **curvature vector**; κ is the *curvature* further discussed in Sec. 17.2-3.

17.2-3. Serret-Frenet Formulas. Curvature and Torsion of a Space Curve (see also Secs. 17.2-4 and 17.2-5). (**a**) The unit vectors (2) satisfy the relations

$$\mathbf{t}' = \kappa\mathbf{n} \qquad \mathbf{n}' = -\kappa\mathbf{t} + \tau\mathbf{b} \qquad \mathbf{b}' = -\tau\mathbf{n} \quad \text{(Serret-Frenet formulas)}$$
$$\text{with} \quad \kappa = \frac{1}{\rho_\kappa} = |\mathbf{t}'| = |\mathbf{r}''| \qquad \tau = \frac{1}{\rho_\tau} = \frac{1}{\kappa^2}[\mathbf{r}'\mathbf{r}''\mathbf{r}'''] \tag{17.2-3}$$

at each curve point P_1. As s increases, the point P_1 moves along the curve C and

1. The tangent rotates about the instantaneous binormal direction at the (positive) angular rate κ (**curvature** of C at P_1).
2. The binormal rotates about the instantaneous tangent direction at the angular rate τ (**torsion** of C at P_1; τ is positive wherever the curve turns in the manner of a right-handed screw).
3. The entire moving trihedron rotates about the instantaneous direction of the **Darboux vector** $\boldsymbol{\Omega} = \tau\mathbf{t} + \kappa\mathbf{b}$ at the (positive) angular rate $|\boldsymbol{\Omega}| = \sqrt{\tau^2 + \kappa^2}$ (**total curvature** of C at P_1).

The instantaneous rotation of $\mathbf{t}$, $\mathbf{n}$, and $\mathbf{b}$ becomes more evident on rewriting the Serret-Frenet formulas as follows (see also Secs. 5.3-2 and 14.10-5):

$$\left.\begin{aligned} \mathbf{t}' = \boldsymbol{\Omega} \times \mathbf{t} = (\kappa\mathbf{b}) \times \mathbf{t} \qquad \mathbf{n}' = \boldsymbol{\Omega} \times \mathbf{n} \\ \mathbf{b}' = \boldsymbol{\Omega} \times \mathbf{b} = (\tau\mathbf{t}) \times \mathbf{b} \end{aligned}\right\} \tag{17.2-4}$$

$\rho_\kappa = 1/\kappa$ is the radius of the circle of curvature (**radius of curvature**) of C at P_1; $\rho_\tau = 1/\tau$ is called the **radius of torsion.**

(**b**) The scalar functions $\kappa = \kappa(s)$ and $\tau = \tau(s)$ together define the curve C uniquely, except for its position and orientation in space (*intrinsic equations of a space curve*). C is a *plane curve* if and only if its torsion τ vanishes identically, and a *straight line* if and only if its curvature κ vanishes identically.

(**c**) In terms of the more general parameter t,

$$\kappa = \frac{1}{\rho_\kappa} = \frac{|\dot{\mathbf{r}} \times \ddot{\mathbf{r}}|}{|\dot{\mathbf{r}}|^3} \qquad \tau = \frac{1}{\rho_\tau} = \frac{[\dot{\mathbf{r}}\ddot{\mathbf{r}}\dddot{\mathbf{r}}]}{|\dot{\mathbf{r}} \times \ddot{\mathbf{r}}|^2} \tag{17.2-5}$$

where the dots indicate differentiations with respect to t. Note also

$$\ddot{\mathbf{r}} = \ddot{s}\mathbf{t} + \frac{1}{\rho_\kappa}\dot{s}^2\mathbf{n} = \ddot{s}\mathbf{t} + \left(\frac{\dot{s}}{\rho_\kappa}\mathbf{b}\right) \times \dot{\mathbf{r}} \tag{17.2-6}$$

(*decomposition of the acceleration of a moving point into tangential and normal components,* see also Sec. 5.3-2).

17.2-4. Equations of the Tangent, Principal Normal, and Binormal, and of the Osculating, Normal, and Rectifying Planes. (**a**) The tangent, principal normal, and binormal of C at the point $P_1 \equiv (\mathbf{r}_1) \equiv (x_1, y_1, z_1)$ are respectively described by

$$\mathbf{r} = \mathbf{r}_1 + u\mathbf{t} \qquad \mathbf{r} = \mathbf{r}_1 + u\mathbf{n} \qquad \mathbf{r} = \mathbf{r}_1 + u\mathbf{b} \tag{17.2-7}$$

where u is a variable parameter. The equations of the osculating, normal, and rectifying planes are, respectively,

$$(\mathbf{r} - \mathbf{r}_1) \cdot \mathbf{b} = 0 \qquad (\mathbf{r} - \mathbf{r}_1) \cdot \mathbf{t} = 0 \qquad (\mathbf{r} - \mathbf{r}_1) \cdot \mathbf{n} = 0 \tag{17.2-8}$$

(**b**) The right-handed rectangular cartesian components of the unit vectors (2) are

$$t_x = x' \qquad t_y = y' \qquad t_z = z' \qquad \text{(DIRECTION COSINES OF THE TANGENT)} \tag{17.2-9a}$$

$$n_x = \frac{1}{\kappa} x'' \qquad n_y = \frac{1}{\kappa} y'' \qquad n_z = \frac{1}{\kappa} z'' \qquad \text{(DIRECTION COSINES OF THE PRINCIPAL NORMAL)} \tag{17.2-9b}$$

$$b_x = \frac{1}{\kappa}(y'z'' - y''z') \qquad b_y = \frac{1}{\kappa}(z'x'' - z''x') \qquad b_z = \frac{1}{\kappa}(x'y'' - x''y')$$

$$\text{(DIRECTION COSINES OF THE BINORMAL)} \tag{17.2-9c}$$

and

$$\kappa = \frac{1}{\rho_\kappa} = \underset{+}{\sqrt{x''^2 + y''^2 + z''^2}} \qquad \tau = \frac{1}{\rho_\tau} = \frac{1}{\kappa^2}\begin{vmatrix} x' & x'' & x''' \\ y' & y'' & y''' \\ z' & z'' & z''' \end{vmatrix} \tag{17.2-10}$$

Substitution of the correct direction cosines (9) in

$$\frac{x - x_1}{\cos \alpha_x} = \frac{y - y_1}{\cos \alpha_y} = \frac{z - z_1}{\cos \alpha_z} \tag{17.2-11}$$

yields the equations of the tangent, principal normal, and binormal in terms of rectangular cartesian coordinates, and

$$(x - x_1)\cos \alpha_x + (y - y_1)\cos \alpha_y + (z - z_1)\cos \alpha_z = 0 \tag{17.2-12}$$

represents the osculating, normal, and rectifying planes.

17.2-5. Additional Topics. (**a**) The center of curvature associated with the curve point P_1 has the position vector

$$\mathbf{r}_\kappa = \mathbf{r}_1 + \rho_\kappa \mathbf{n} \tag{17.2-13}$$

(**b**) The limit of a sphere through four distinct points P_1, P_2, P_3, P_4 of C as P_2, P_3, and P_4 approach P_1 is the **osculating sphere** of C at P_1. Its center lies on the directed straight line in the positive binormal direction through the center of curvature (**axis of curvature** or **polar line** of C at P_1). The radius ρ_s of the osculating sphere and the position vector $\mathbf{r}_s$ of its center are

$$\rho_s^2 = \rho_\kappa^2 + (\rho_\tau \rho_\kappa')^2 \qquad \mathbf{r}_s = \mathbf{r}_\kappa + \rho_\tau \rho_\kappa' \mathbf{b} = \mathbf{r}_1 + \rho_\kappa \mathbf{n} + \rho_\tau \rho_\kappa' \mathbf{b} \tag{17.2-14}$$

C lies on a sphere of radius R if and only if $\rho_\tau \equiv R$.

The polar lines of C are tangent to the **polar curve** defined as the locus of the centers of the osculating spheres of C. The **polar surface** (**polar developable**) of C is the ruled surface (Sec. 3.1-15) generated by the polar lines.

(**c**) **Involutes and Evolutes.** The tangents of C generate a ruled surface (**tangent surface, tangential developable**) of two sheets, which are tangent to one another at the given curve. The **involutes** of the given curve C are those curves on the tangent surface which are orthogonal to the generating tangents. Given the position vector $\mathbf{r} = \mathbf{r}(s)$ of a moving point P_s on C, the points $P_I \equiv (\boldsymbol{\varrho})$ of every involute of C are given by

$$\boldsymbol{\varrho} = \boldsymbol{\varrho}(s) = \mathbf{r}(s) + (d - s)\mathbf{t}(s) \tag{17.2-15}$$

Each involute corresponds to a specific value of d. Note that d is the constant sum of s and the tangent P_sP_I (*string property of involutes*).

A curve C' is an **evolute** of C if the tangents of C' are normal to C, i.e., if C is an involute of C'. The evolutes of C lie on its polar surface (see also Ref. 17.7).

17.2-6. Osculation (see also Sec. 17.1-5). A point $P_1 \equiv (\mathbf{r}_1) \equiv (x_1, y_1, z_1)$ is an **osculation point (point of contact)** of order n of two regular arcs represented by $\mathbf{r} = \mathbf{f}(s)$ and $\mathbf{r} = \mathbf{g}(s)$ if and only if at P_1 one has $s = s_1$ such that

$$\mathbf{f}(s_1) = \mathbf{g}(s_1) = \mathbf{r}_1 \qquad \mathbf{f}'(s_1) = \mathbf{g}'(s_1) \qquad \cdots \qquad \mathbf{f}^{(n)}(s_1) = \mathbf{g}^{(n)}(s_1)$$
$$\mathbf{f}^{(n+1)}(s_1) \neq \mathbf{g}^{(n+1)}(s_1) \qquad (17.2\text{-}16)$$

17.3. SURFACES IN THREE-DIMENSIONAL EUCLIDEAN SPACE

17.3-1. Introduction (see also Sec. 3.1-14; see Sec. 3.5-10 for examples). Sections 17.3-1 to 17.3-14 deal with the geometry of a *regular surface element* S described by

$$\mathbf{r} = \mathbf{r}(u, v) \qquad (17.3\text{-}1a)$$

or

$$x = x(u, v) \qquad y = y(u, v) \qquad z = z(u, v) \qquad (17.3\text{-}1b)$$

or

$$\varphi(x, y, z) = 0 \qquad (17.3\text{-}1c)$$

in some region of values of the parameters (**surface coordinates**) u, v (Sec. 3.1-14). The functions (1) are to have continuous first partial derivatives such that the rank of the matrix

$$\begin{bmatrix} \frac{\partial x}{\partial u} & \frac{\partial y}{\partial u} & \frac{\partial z}{\partial u} \\ \frac{\partial x}{\partial v} & \frac{\partial y}{\partial v} & \frac{\partial z}{\partial v} \end{bmatrix} \qquad (17.3\text{-}2)$$

equals 2 (Sec. 13.2-7), i.e., such that the three functional determinants $\partial(x, y)/\partial(u, v)$, $\partial(y, z)/\partial(u, v)$, $\partial(z, x)/\partial(u, v)$ do not vanish simultaneously, or

$$\mathbf{r}_u \times \mathbf{r}_v \neq 0 \qquad \left(\mathbf{r}_u \equiv \frac{\partial \mathbf{r}}{\partial u}, \mathbf{r}_v \equiv \frac{\partial \mathbf{r}}{\partial v}\right) \qquad (17.3\text{-}3)$$

Higher partial derivatives will be assumed to exist as needed.

The conditions listed above ensure the existence and linear independence of the vectors $\mathbf{r}_u$ and $\mathbf{r}_v$ directed, respectively, along the tangents to the **surface-coordinate lines** u = constant and v = constant through the surface point (u, v). Surface points where the three determinants exist but vanish for every choice of the parameters u, v are *singular points* corresponding to edges, vertices, etc.

17.3-2. Tangent Plane and Surface Normal. (a) At every surface point $P_1 \equiv (\mathbf{r}_1) \equiv (x_1, y_1, z_1) \equiv (u_1, v_1)$ satisfying the conditions of Sec. 17.3-1 (*regular point* of the surface) there exists a unique **tangent plane** defined as the limit of a plane through three distinct surface points P_1, P_2, P_3 as P_2 and P_3 approach P_1 along curves which do not have a common tangent at P_1. This tangent plane contains the tangents of all regular surface arcs through P_1; the equation of the tangent plane is

$$[(\mathbf{r} - \mathbf{r}_1)\mathbf{r}_u\mathbf{r}_v] = 0 \qquad \text{or} \qquad \begin{vmatrix} x - x_1 & \dfrac{\partial x}{\partial u} & \dfrac{\partial x}{\partial v} \\ y - y_1 & \dfrac{\partial y}{\partial u} & \dfrac{\partial y}{\partial v} \\ z - z_1 & \dfrac{\partial z}{\partial u} & \dfrac{\partial z}{\partial v} \end{vmatrix} = 0 \tag{17.3-4}$$

where all derivatives are taken for $u = u_1$, $v = v_1$.

(b) The **surface normal** of S at the regular surface point $P_1 \equiv (\mathbf{r}_1) \equiv (x_1, y_1, z_1) \equiv (u_1, v_1)$ is the straight line through P_1 normal to the tangent plane at that point. This surface normal is described by the parametric equation

$$\mathbf{r} - \mathbf{r}_1 = t\mathbf{N} \tag{17.3-5}$$

where

$$\mathbf{N} = \frac{\mathbf{r}_u \times \mathbf{r}_v}{|\mathbf{r}_u \times \mathbf{r}_v|} = \frac{\mathbf{i}\,\dfrac{\partial(y, z)}{\partial(u, v)} + \mathbf{j}\,\dfrac{\partial(z, x)}{\partial(u, v)} + \mathbf{k}\,\dfrac{\partial(x, y)}{\partial(u, v)}}{\sqrt[+]{\left[\dfrac{\partial(y, z)}{\partial(u, v)}\right]^2 + \left[\dfrac{\partial(z, x)}{\partial(u, v)}\right]^2 + \left[\dfrac{\partial(x, y)}{\partial(u, v)}\right]^2}} \tag{17.3-6}$$

is the **unit normal vector** of S at the point P_1; all derivatives are taken for $u = u_1$, $v = v_1$. The direction of $\mathbf{N}$ is the **direction of the positive normal** at P_1; note that the positive u direction, the positive v direction, and the positive normal form a right-handed system of axes (Sec. 3.1-3) directed along $\mathbf{r}_u$, $\mathbf{r}_v$, and $\mathbf{N}$.

17.3-3. The First Fundamental Form of a Surface. Elements of Distance and Area. (a) The vector path element (Sec. 5.4-4) along a *surface curve*

$$\mathbf{r} = \mathbf{r}[u(t), v(t)] \qquad \text{or} \qquad u = u(t) \qquad v = v(t) \tag{17.3-7}$$

is

$$d\mathbf{r} = \mathbf{r}_u\, du + \mathbf{r}_v\, dv \tag{17.3-8}$$

and the square of the **element of distance** $ds = |d\mathbf{r}|$ on the surface (1) is given at each surface point (u, v) by

$$\left.\begin{aligned} ds^2 &= |d\mathbf{r}|^2 \\ &= E(u, v)\, du^2 + 2F(u, v)\, du\, dv + G(u, v)\, dv^2 \\ \text{with}& \\ E(u, v) &= \mathbf{r}_u \cdot \mathbf{r}_u = \left(\frac{\partial x}{\partial u}\right)^2 + \left(\frac{\partial y}{\partial u}\right)^2 + \left(\frac{\partial z}{\partial u}\right)^2 \\ F(u, v) &= \mathbf{r}_u \cdot \mathbf{r}_v = \frac{\partial x}{\partial u}\frac{\partial x}{\partial v} + \frac{\partial y}{\partial u}\frac{\partial y}{\partial v} + \frac{\partial z}{\partial u}\frac{\partial z}{\partial v} \\ G(u, v) &= \mathbf{r}_v \cdot \mathbf{r}_v = \left(\frac{\partial x}{\partial v}\right)^2 + \left(\frac{\partial y}{\partial v}\right)^2 + \left(\frac{\partial z}{\partial v}\right)^2 \end{aligned}\right\} \tag{17.3-9}$$

(FIRST FUNDAMENTAL FORM OF THE SURFACE)

At every regular point of a real surface (1) described in terms of real surface coordinates u, v the quadratic form (9) is positive definite (Sec. 13.5-2), i.e.,

$$E > 0 \qquad G > 0 \qquad EG - F^2 > 0 \tag{17.3-10}$$

(**b**) The angle γ between two regular surface arcs

$$\mathbf{r} = \mathbf{R}_1(t) \quad \text{or} \quad u = U_1(t) \quad v = V_1(t)$$

and

$$\mathbf{r} = \mathbf{R}_2(t) \quad \text{or} \quad u = U_2(t) \quad v = V_2(t)$$

through the surface point (u, v) is given by

$$\begin{aligned}\cos\gamma &= \frac{d\mathbf{R}_1 \cdot d\mathbf{R}_2}{|d\mathbf{R}_1|\,|d\mathbf{R}_2|}\\ &= \frac{E\,dU_1\,dU_2 + F(dU_1\,dV_2 + dV_1\,dU_2) + G\,dV_1\,dV_2}{\sqrt{E\,dU_1^2 + 2F\,dU_1\,dV_1 + G\,dV_1^2}\,\sqrt{E\,dU_2^2 + 2F\,dU_2\,dV_2 + G\,dV_2^2}}\\ &= E\frac{dU_1}{ds}\frac{dU_2}{ds} + F\left(\frac{dU_1}{ds}\frac{dV_2}{ds} + \frac{dV_1}{ds}\frac{dU_2}{ds}\right) + G\frac{dV_1}{ds}\frac{dV_2}{ds}\end{aligned} \tag{17.3-11}$$

In particular, the angle γ_1 between the surface-coordinate lines u = constant and v = constant through the surface point (u, v) is given by

$$\cos\gamma_1 = \frac{F}{\sqrt{EG}} \qquad \sin\gamma_1 = \frac{\sqrt{EG - F^2}}{\sqrt{EG}} \tag{17.3-12}$$

The surface coordinates u, v are *orthogonal* if and only if $F \equiv 0$ (see also Secs. 6.4-1 and 16.8-2).

(**c**) The **vector element of area** $d\mathbf{A}$ and the **scalar element of area** dA at the regular surface point (u, v) are defined by

$$\begin{aligned} d\mathbf{A} &= (\mathbf{r}_u \times \mathbf{r}_v)\,du\,dv = \mathbf{N}|d\mathbf{A}|\\ dA &= \pm|d\mathbf{A}| = \sqrt{a(u, v)}\,du\,dv\\ \text{with}\quad a(u, v) &= |\mathbf{r}_u \times \mathbf{r}_v|^2 = EG - F^2\\ &= \left[\frac{\partial(y, z)}{\partial(u, v)}\right]^2 + \left[\frac{\partial(z, x)}{\partial(u, v)}\right]^2 + \left[\frac{\partial(x, y)}{\partial(u, v)}\right]^2 \end{aligned} \tag{17.3-13}$$

The sign of dA may be fixed arbitrarily (see also Secs. 4.6-11, 5.4-6*a*, and 6.4-3*b*).

17.3-4. Geodesic and Normal Curvature of a Surface Curve. Meusnier's Theorem. (**a**) At each point (u, v) of a regular surface arc C described by

$$\mathbf{r} = \mathbf{r}[u(s), v(s)] \qquad \text{or} \qquad u = u(s) \qquad v = v(s) \tag{17.3-14}$$

the curvature vector $\mathbf{r}'' = \kappa\mathbf{n}$ (Sec. 17.2-2*d*) has a unique component in the tangent plane (**geodesic** or **tangential curvature vector**) and a unique component along the surface normal (**normal curvature vector**), i.e.,

$$\mathbf{r}'' = \kappa\mathbf{n} = \kappa_G(\mathbf{N} \times \mathbf{r}') + \kappa_N\mathbf{N} \tag{17.3-15}$$

where the primes indicate differentiation with respect to the arc length

s along C. At each point $[u(s), v(s)]$,

$$\kappa_G = \kappa[\mathbf{r}'\mathbf{n}\mathbf{N}] = [\mathbf{r}'\mathbf{r}''\mathbf{N}] \quad \text{[GEODESIC OR TANGENTIAL CURVATURE OF } C \text{ AT } (u, v)] \qquad (17.3\text{-}16)$$

is the curvature of the projection of C onto the tangent plane (see also Sec. 17.4-2*d*), and

$$\kappa_N = \kappa(\mathbf{n} \cdot \mathbf{N}) = \mathbf{r}'' \cdot \mathbf{N} = -\mathbf{r}' \cdot \mathbf{N}' \quad \text{[NORMAL CURVATURE OF } C \text{ AT } (u, v)] \qquad (17.3\text{-}17)$$

is the curvature of the **normal section** (intersection of the surface and a plane containing the surface normal) through the tangent of C (see also Sec. 17.3-5).

(b) *At a given surface point (u, v), the curvature κ of every surface curve whose osculating plane (Sec. 17.2-2b) makes an angle α with the normal-section plane through the same tangent is*

$$\kappa = \left| \frac{\kappa_N}{\cos \alpha} \right| \qquad (17.3\text{-}18)$$

where κ_N is the normal-section curvature given by Eq. (17) *or* (20) (*Meusnier's Theorem*). Equation (18) yields, in particular, the curvature κ of any *oblique surface section* in terms of the curvature κ_N of the normal section through the same tangent.

17.3-5. The Second Fundamental Form. Principal Curvatures, Gaussian Curvature, and Mean Curvature. **(a)** To write Eq. (17) in terms of the surface coordinates u, v, let $d\mathbf{N} = \mathbf{N}_u du + \mathbf{N}_v dv$. Then

$$\left.\begin{aligned} &-d\mathbf{r} \cdot d\mathbf{N} \\ &= L(u, v)\, du^2 + 2M(u, v)\, du\, dv + N(u, v)\, dv^2 \\ &\text{with} \\ &L(u, v) = -\mathbf{r}_u \cdot \mathbf{N}_u = \frac{[\mathbf{r}_{uu}\mathbf{r}_u\mathbf{r}_v]}{\sqrt{EG - F^2}} \\ &M(u, v) = -\mathbf{r}_u \cdot \mathbf{N}_v = -\mathbf{r}_v \cdot \mathbf{N}_u = \frac{[\mathbf{r}_{uv}\mathbf{r}_u\mathbf{r}_v]}{\sqrt{EG - F^2}} \\ &N(u, v) = -\mathbf{r}_v \cdot \mathbf{N}_v = \frac{[\mathbf{r}_{vv}\mathbf{r}_u\mathbf{r}_v]}{\sqrt{EG - F^2}} \end{aligned}\right\} \qquad (17.3\text{-}19)$$

(SECOND FUNDAMENTAL FORM OF THE SURFACE)

where all derivatives are taken at the surface point (u, v); the box products can be expanded by Eq. (5.2-11). *At a given surface point* $(\mathbf{r}) \equiv (u, v)$ *the curvature of the normal section containing the adjacent surface point* $(\mathbf{r} + d\mathbf{r}) \equiv (u + du, v + dv)$ *is*

$$\kappa_N = -\frac{d\mathbf{r}\cdot d\mathbf{N}}{ds^2} = L\left(\frac{du}{ds}\right)^2 + 2M\frac{du}{ds}\frac{dv}{ds} + N\left(\frac{dv}{ds}\right)^2$$
$$= \frac{L\,du^2 + 2M\,du\,dv + N\,dv^2}{E\,du^2 + 2F\,du\,dv + G\,dv^2} \qquad (17.3\text{-}20)$$

(**b**) Unless κ_N has the same value for every normal section through (u, v) ($L:M:N = E:F:G$, **umbilic point**), there exist two **principal normal sections** respectively associated with the largest value κ_1 and the smallest value κ_2 of κ [**principal curvatures** of S at the surface point (u, v)]. *The planes of the principal normal sections are mutually perpendicular; for any normal section through (u, v) whose plane forms the oblique angle ϑ with the plane of the first principal normal section,*

$$\kappa_N = \kappa_1 \cos^2 \vartheta + \kappa_2 \sin^2 \vartheta \qquad \text{(EULER'S THEOREM)} \qquad (17.3\text{-}21)$$

κ_1 and κ_2 are eigenvalues of the type discussed in Sec. 14.8-7 (see also Sec. 14.8-8); they are obtainable as roots of the characteristic equation

$$\begin{vmatrix} L - \kappa E & M - \kappa F \\ M - \kappa F & N - \kappa G \end{vmatrix} = 0 \qquad (17.3\text{-}22)$$

The symmetric functions $H(u, v) \equiv \frac{1}{2}(\kappa_1 + \kappa_2)$ and $K(u, v) \equiv \kappa_1\kappa_2$ are respectively known as the **mean curvature** and the **Gaussian curvature** of the surface S at the point (u, v); note

$$H \equiv \frac{1}{2}(\kappa_1 + \kappa_2) \equiv \frac{1}{2}\frac{EN - 2FM + GL}{EG - F^2} \qquad \text{(MEAN CURVATURE)} \qquad (17.3\text{-}23)$$

$$K \equiv \kappa_1\kappa_2 \equiv \frac{LN - M^2}{EG - F^2} \qquad \text{(GAUSSIAN CURVATURE)} \qquad (17.3\text{-}24)$$

(see also Secs. 17.3-8 and 17.3-13). κ_1, κ_2, H, and K are surface point functions independent of the particular surface coordinates u, v used.

Depending on whether the quadratic form (19) is definite, semidefinite, or indefinite (Sec. 13.5-2) at the surface point (u, v), the latter is

An **elliptic point** with $K = \kappa_1\kappa_2 > 0$ (normal sections all convex or all concave; surface does not intersect its tangent plane. EXAMPLE: any point of an ellipsoid)
A **parabolic point** with $K = \kappa_1\kappa_2 = 0$ (e.g., any point of a cylinder)
A **hyperbolic point (saddle point)** with $K = \kappa_1\kappa_2 < 0$ (both convex and concave normal sections; surface intersects its tangent plane. EXAMPLE: any point of a hyperboloid of one sheet)

An umbilic point ($\kappa_1 = \kappa_2$, Sec. 17.3-5*b*) is necessarily either elliptic or parabolic.

17.3-6. Special Directions and Curves on a Surface. Minimal Surfaces. (**a**) A **line of curvature** is a surface curve whose tangent at each point belongs to a principal normal section; the differential equation

$$\begin{vmatrix} dv^2 & -du\,dv & du^2 \\ E & F & G \\ L & M & N \end{vmatrix} = 0 \qquad (17.3\text{-}25)$$

defines the two mutually perpendicular lines of curvature $v = v(u)$ through each point (u, v).

(b) An **asymptotic line** is a surface curve of zero normal curvature (20), so that

$$L\,du^2 + 2M\,du\,dv + N\,dv^2 = 0 \qquad (17.3\text{-}26)$$

(EXAMPLE: any straight line on the surface). The tangent to an asymptotic line defines an **asymptotic direction** on the surface. Note that the asymptotic lines at elliptic surface points are imaginary curves $v = v(u)$. *The lines of curvature bisect the asymptotic directions.*

(c) Two regular surface arcs $u = U_1(t)$, $v = V_1(t)$ and $u = U_2(t)$, $v = V_2(t)$ have **mutually conjugate directions** at the surface point (u, v) if and only if the tangent of one arc is the limiting intersection of the tangent planes at two points approaching (u, v) on the other arc, or

$$L\,dU_1\,dU_2 + M(dU_1\,dV_2 + dU_2\,dV_1) + N\,dV_1\,dV_2 = 0 \qquad (17.3\text{-}27)$$

This relationship is necessarily reciprocal [EXAMPLE: directions of the lines of curvature at (u, v)].

(d) The parametric lines u = constant, v = constant are

Orthogonal if and only if $F \equiv 0$
Conjugate if and only if $M \equiv 0$
Lines of curvature if and only if $F \equiv M \equiv 0$
Asymptotic lines if and only if $L \equiv N \equiv 0$

(e) A **minimal surface** is a surface such that $H(u, v) \equiv 0$; *this is true if and only if the asymptotic lines form an orthogonal net.* Surfaces having minimum area for a given reasonable boundary curve are minimal curves (Plateau's problem, soap-film models).

17.3-7. Surfaces as Riemann Spaces. Three-index Symbols and Beltrami Parameters. A regular surface element S with fundamental form (9) is a two-dimensional *Riemann space* of points (u, v) with *metric-tensor components* E, F, G (Secs. 16.7-1 and 17.4-1; see also Sec. 17.3-12); $a(u, v) \equiv EG - F^2$ is the *metric-tensor determinant* (Sec. 16.7-1). One can define *surface vectors and tensors, surface scalar products,* and *surface covariant differentiation* on the Riemann space S in the manner of Secs. 16.2-1, 16.8-1, and 16.10-1. The *Christoffel three-index symbols of the second kind* (Sec. 16.10-3) for the surface S take the form

$$\left.\begin{aligned}
&\left\{ {1 \atop 1\;1} \right\}_S \equiv \frac{GE_u - 2FF_u + FE_v}{2(EG - F^2)} &\quad &\left\{ {2 \atop 1\;1} \right\}_S \equiv \frac{-FE_u + 2EF_u - EE_v}{2(EG - F^2)} \\
&\left\{ {1 \atop 1\;2} \right\}_S \equiv \left\{ {1 \atop 2\;1} \right\}_S \equiv \frac{GE_v - FG_u}{2(EG - F^2)} &\quad &\left\{ {2 \atop 1\;2} \right\}_S \equiv \left\{ {2 \atop 2\;1} \right\}_S \equiv \frac{EG_u - FE_v}{2(EG - F^2)} \\
&\left\{ {1 \atop 2\;2} \right\}_S \equiv \frac{-FG_v + 2GF_v - GG_u}{2(EG - F^2)} &\quad &\left\{ {2 \atop 2\;2} \right\}_S \equiv \frac{EG_v - 2FF_v + FG_u}{2(EG - F^2)}
\end{aligned}\right\} \qquad (17.3\text{-}28)$$

where the subscript S has been introduced to distinguish the *surface* Christoffel symbols (28) clearly from the Christoffel symbols $\{{}_i{}^k{}_j\}$ associated with the surrounding space.

For suitably differentiable functions $\Phi(u, v)$, $\Psi(u, v)$, the functions

$$\left.\begin{aligned}\nabla_S(\Phi, \Psi) &\equiv \frac{E\Phi_v\Psi_v - F(\Phi_u\Psi_v + \Phi_v\Psi_u) + G\Phi_u\Psi_u}{EG - F^2} && \text{(Beltrami's first differential parameter)}\\ \nabla_S^2(\Phi) &\equiv \frac{1}{\sqrt{EG - F^2}}\left[\frac{\partial}{\partial u}\frac{G\Phi_u - F\Phi_v}{\sqrt{EG - F^2}} + \frac{\partial}{\partial v}\frac{E\Phi_v - F\Phi_u}{\sqrt{EG - F^2}}\right] && \text{(Beltrami's second differential parameter)}\end{aligned}\right\} \quad (17.3\text{-}29)$$

are surface differential invariants respectively analogous to $\nabla\Phi \cdot \nabla\Psi$ and $\nabla^2\Phi$ as defined in Table 16.10-1. Since the two-dimensional Riemann space S is "embedded" in three-dimensional Euclidean space, one can interpret covariant differentiation on a curved surface by actual comparison of surface vectors at adjacent surface points.

17.3-8. Partial Differential Equations Satisfied by the Coefficients of the Fundamental Forms. Gauss's Theorema Egregium. (a) The following relations describe the change of the linearly independent vectors $\mathbf{r}_u$, $\mathbf{r}_v$, $\mathbf{N}$ ("local trihedron") along the surface-coordinate directions:

$$\left.\begin{aligned}\mathbf{r}_{uu} &= \left\{{1 \atop 1\ 1}\right\}_S \mathbf{r}_u + \left\{{2 \atop 1\ 1}\right\}_S \mathbf{r}_v + L\mathbf{N}\\ \mathbf{r}_{uv} &= \left\{{1 \atop 1\ 2}\right\}_S \mathbf{r}_u + \left\{{2 \atop 1\ 2}\right\}_S \mathbf{r}_v + M\mathbf{N}\\ \mathbf{r}_{vv} &= \left\{{1 \atop 2\ 2}\right\}_S \mathbf{r}_u + \left\{{2 \atop 2\ 2}\right\}_S \mathbf{r}_v + N\mathbf{N}\end{aligned}\right\} \quad \text{(Gauss's equations)} \quad (17.3\text{-}30)$$

$$\left.\begin{aligned}\mathbf{N}_u &= \frac{1}{EG - F^2}[(FM - GL)\mathbf{r}_u + (FL - EM)\mathbf{r}_v]\\ \mathbf{N}_v &= \frac{1}{EG - F^2}[(FN - GM)\mathbf{r}_u + (FM - EN)\mathbf{r}_v]\end{aligned}\right\} \quad \text{(Weingarten equations)} \quad (17.3\text{-}31)$$

(b) The following relations are compatibility relations (integrability conditions, Sec. 10.1-2c) ensuring that $\mathbf{r}_{uuv} = \mathbf{r}_{vuu}$, $\mathbf{r}_{vuv} = \mathbf{r}_{uvv}$:

$$\left.\begin{aligned}L_v + \left\{{1 \atop 1\ 1}\right\}_S M + \left\{{2 \atop 1\ 1}\right\}_S N &= M_u + \left\{{1 \atop 1\ 2}\right\}_S L + \left\{{2 \atop 1\ 2}\right\}_S M\\ M_v + \left\{{1 \atop 1\ 2}\right\}_S M + \left\{{2 \atop 1\ 2}\right\}_S N &= N_u + \left\{{1 \atop 2\ 2}\right\}_S L + \left\{{2 \atop 2\ 2}\right\}_S M\end{aligned}\right\} \quad \text{(Mainardi-Codazzi equations)} \quad (17.3\text{-}32)$$

$$\frac{1}{F}\left[\frac{\partial}{\partial u}\left\{{1 \atop 1\ 2}\right\}_S - \frac{\partial}{\partial v}\left\{{1 \atop 1\ 1}\right\}_S + \left\{{2 \atop 1\ 2}\right\}_S\left\{{1 \atop 1\ 2}\right\}_S - \left\{{2 \atop 1\ 1}\right\}_S\left\{{1 \atop 2\ 2}\right\}_S\right] = \frac{LN - M^2}{EG - F^2} = K \quad (17.3\text{-}33)$$

(c) Equation (33) expresses K solely in terms of E, F, and G and their derivatives: *the Gaussian curvature $K(u, v)$ of a surface is a* **bending invariant** *unchanged by deformations which preserve the first fundamental form (Gauss's Theorema Egregium).*

17.3-9. Definition of a Surface by E, F, G, L, M, and N. Three given functions $E(u, v) > 0$, $G(u, v) > 0$, and $F(u, v)$ such that $EG - F^2 > 0$ define the metric properties (intrinsic geometry) of the surface. *Six functions E, F, G, L, M, N satisfying the above inequalities and the* *compatibility conditions* (32) *and* (33) *uniquely define a corresponding real surface* $\mathbf{r} = \mathbf{r}(u, v)$ *except for its position and orientation in space (Fundamental Theorem of Surface Theory).*

17.3-10. Mappings. (a) A suitably differentiable one-to-one transformation

$$\bar{u} = \bar{u}(u, v) \qquad \bar{v} = \bar{v}(u, v) \qquad \left[\frac{\partial(\bar{u}, \bar{v})}{\partial(u, v)} \neq 0\right] \tag{17.3-34}$$

maps the points (u, v) of a given regular surface element $\mathbf{r} = \mathbf{r}(u, v)$ onto corresponding points $(\bar{u}, \bar{v})$ of another regular surface element $\mathbf{r} = \bar{\mathbf{r}}(\bar{u}, \bar{v})$. In the following, barred symbols refer to the second surface. *The mapping is*

Isometric *(preserves all metric properties) if and only if* $\bar{E}(\bar{u}, \bar{v}) \equiv E(u, v)$, $\bar{F}(\bar{u}, \bar{v}) \equiv F(u, v)$, $\bar{G}(\bar{u}, \bar{v}) \equiv G(u, v)$

Conformal *(preserves angles) if and only if* $\bar{E}(\bar{u}, \bar{v}):\bar{F}(\bar{u}, \bar{v}):\bar{G}(\bar{u}, \bar{v}) \equiv E(u, v):F(u, v):G(u, v)$

Equiareal *(preserves areas) if and only if* $\dfrac{\partial(\bar{u}, \bar{v})}{\partial(u, v)} \equiv \sqrt{\dfrac{EG - F^2}{\bar{E}\bar{G} - \bar{F}^2}}$

(b) To obtain a conformal mapping (34), one maps each surface conformally onto a plane and relates the two planes by an analytic complex-variable transformation (Sec. 7.9-1).

To map the surface $\mathbf{r} = \mathbf{r}(u, v)$ conformally onto a plane with rectangular cartesian coordinates $\xi(u, v)$, $\eta(u, v)$, solve the differential equation

$$E(u, v)\,du^2 + 2F(u, v)\,du\,dv + G(u, v)\,dv^2 = 0 \tag{17.3-35a}$$

or, equivalently,

$$\frac{dv}{du} = \frac{1}{G}(-F + i\sqrt{EG - F^2}) \qquad \text{and} \qquad \frac{dv}{du} = \frac{1}{G}(-F - i\sqrt{EG - F^2}) \tag{17.3-35b}$$

to obtain the *complex* surface curves (**isotropic** or **minimal** surface "curves" defined by $ds^2 = 0$)

$$U(u, v) = \text{constant} \qquad V(u, v) = \text{constant}$$

Then the *real* orthogonal surface coordinates

$$\xi(u, v) = \frac{1}{2}(U + V) \qquad \eta(u, v) = \frac{1}{2i}(U - V) \tag{17.3-36}$$

(**isometric** or **isothermic** surface coordinates) reduce the first fundamental form of the surface to

$$ds^2 = \psi(\xi, \eta)(d\xi^2 + d\eta^2) \tag{17.3-37}$$

which is proportional to the first fundamental form $d\xi^2 + d\eta^2$ of a plane with rectangular cartesian coordinates ξ, η.

17.3-11. Envelopes (see also Secs. 10.2-3 and 17.1-7). Let the equations

$$\varphi(x, y, z, \lambda) = 0 \tag{17.3-38}$$

represent a one-parameter family of surfaces such that

$$\nabla\varphi \times \nabla\varphi_\lambda \neq 0 \qquad \varphi_{\lambda\lambda} \neq 0 \tag{17.3-39}$$

in some region V of space. Then V contains a surface (**envelope** of the given surfaces) which touches (has a common tangent plane with) each surface (38) along a curve (**characteristic**) described by

$$\varphi(x, y, z, \lambda) = 0 \qquad \varphi_\lambda(x, y, z, \lambda) = 0 \tag{17.3-40}$$

Elimination of λ from Eq. (40) yields the equation of the envelope.

If, in addition to the condition (39), $[\nabla\varphi\nabla\varphi_\lambda\nabla\varphi_{\lambda\lambda}] \neq 0$ in V, then the envelope has an **edge of regression**

$$\varphi(x, y, \lambda) = 0 \qquad \varphi_\lambda(x, y, z, \lambda) = 0 \qquad \varphi_{\lambda\lambda}(x, y, z, \lambda) = 0 \tag{17.3-41}$$

which touches each characteristic (40) at a **focal point** obtainable by elimination of λ from Eq. (41).

Equation (40) defines the locus of the *limiting curves of intersection* of $\varphi(x, y, z, \lambda_1) = 0$ and $\varphi(x, y, z, \lambda_2) = 0$ as $\lambda_2 \to \lambda_1$. This locus (λ **discriminant**) may include loci of edges and nodes, as well as envelopes. An edge of regression is, similarly, a locus of *limiting points of intersection* of three surfaces (38).

17.3-12. Geodesics (see also Sec. 17.4-3). A **geodesic** on the regular surface element S is a regular arc whose geodesic curvature (Sec. 17.3-4*a*) is identically zero; a geodesic is either a straight line, or its principal normal coincides with the surface normal at each point. Every geodesic

$$u = u(s) \qquad v = v(s) \qquad \text{or} \qquad v = v(u)$$

satisfies the differential equations

$$\left.\begin{aligned} &\frac{d^2u}{ds^2} + \begin{Bmatrix} 1 \\ 1\ 1 \end{Bmatrix}_S \left(\frac{du}{ds}\right)^2 + 2\begin{Bmatrix} 1 \\ 1\ 2 \end{Bmatrix}_S \frac{du}{ds}\frac{dv}{ds} + \begin{Bmatrix} 1 \\ 2\ 2 \end{Bmatrix}_S \left(\frac{dv}{ds}\right)^2 = 0 \\ &\frac{d^2v}{ds^2} + \begin{Bmatrix} 2 \\ 1\ 1 \end{Bmatrix}_S \left(\frac{du}{ds}\right)^2 + 2\begin{Bmatrix} 2 \\ 1\ 2 \end{Bmatrix}_S \frac{du}{ds}\frac{dv}{ds} + \begin{Bmatrix} 2 \\ 2\ 2 \end{Bmatrix}_S \left(\frac{dv}{ds}\right)^2 = 0 \end{aligned}\right\} \tag{17.3-42a}$$

$$\text{or} \quad \frac{d^2v}{du^2} = \begin{Bmatrix} 1 \\ 2\ 2 \end{Bmatrix}_S \left(\frac{dv}{du}\right)^3 + \left[2\begin{Bmatrix} 1 \\ 1\ 2 \end{Bmatrix}_S - \begin{Bmatrix} 2 \\ 2\ 2 \end{Bmatrix}_S\right]\left(\frac{dv}{du}\right)^2 + \left[\begin{Bmatrix} 1 \\ 1\ 1 \end{Bmatrix}_S - 2\begin{Bmatrix} 2 \\ 1\ 2 \end{Bmatrix}_S\right]\frac{dv}{du} - \begin{Bmatrix} 2 \\ 1\ 1 \end{Bmatrix}_S \tag{17.3-42b}$$

These relations define a unique geodesic through any given point in any given direction.

Geodesics on a curved surface have many properties analogous to those of straight lines on a plane (see also Sec. 17.4-3). *If there exists a surface curve of smallest or greatest arc length joining two given surface points, then that curve is a geodesic.* The actual existence of a geodesic through two given surface points requires a separate investigation in each case.

17.3-13. Geodesic Normal Coordinates. Geometry on a Surface (see also Sec. 17.4-7). **(a)** For a system of **geodesic normal coordinates** u, v the u coordinate lines are orthogonal trajectories of a "field" of geodesics v = constant. The u coordinate lines are then **geodesic parallels** cutting off equal increments of $u = s$ on each geodesic v = constant, and

$$ds^2 = du^2 + G(u, v)\, dv^2 \qquad (17.3\text{-}43)$$

In terms of geodesic normal coordinates u, v

$$K = -\frac{1}{\sqrt{G}}\frac{\partial^2\sqrt{G}}{\partial u^2} = \frac{G_u{}^2 - 2GG_{uu}}{4G^2} \qquad (17.3\text{-}44)$$

(**b**) In the special case of **geodesic polar coordinates,** the geodesics v = constant intersect at a point (*origin, pole*), and v is the angle (Sec. 17.3-3*b*) between the geodesic labeled by v and the geodesic $v = 0$. Each u coordinate line is a **geodesic circle of radius** u intersecting all v geodesics at right angles.

The "circular arc" of "radius" u corresponding to an angular increment dv is

$$\sqrt{G(u, v)}\, dv = [u - \tfrac{1}{6}K_0u^3 + o(u^3)]\, dv \qquad (17.3\text{-}45)$$

where K_0 is the Gaussian curvature at the origin. The quantity (445) is less than, equal to, or greater than $u\,dv$ if, respectively, $K_0 > 0$, $K_0 = 0$, or $K_0 < 0$. The circumference $C_G(u)$ and the area $A_G(u)$ of a small geodesic circle of radius u about the origin are related to the circumference $2\pi u$ and the area πu^2 of a plane circle of equal radius by

$$\frac{3}{\pi}\lim_{u\to 0}\frac{2\pi u - C_G(u)}{u^3} = \frac{12}{\pi}\lim_{u\to 0}\frac{\pi u^2 - A_G(u)}{u^4} = K_0 \qquad (17.3\text{-}46)$$

(**c**) For any geodesic triangle on a surface of constant Gaussian curvature K, the excess of the sum of the angles A, B, C over π and the triangle area S_T are related by

$$A + B + C - \pi = KS_T \qquad (17.3\text{-}47)$$

The resulting surface geometry is Euclidean for $K = 0$, **elliptic** for $K > 0$, and **hyperbolic** for $K < 0$. *Surfaces of equal constant Gaussian curvature are isometric (Minding's Theorem).*

EXAMPLES: On a sphere of radius R, $K = 1/R^2$ (see also Sec. B-6). A surface with constant $K < 0$ is obtained by rotation of the *tractrix*

$$x = \frac{1}{\sqrt{-K}}\sin t \qquad y = \frac{1}{\sqrt{-K}}\left(\log_e \tan\frac{t}{2} + \cos t\right)$$

(*pseudosphere*, Ref. 17.7).

17.3-14. The Gauss-Bonnet Theorem. *Let $K(u, v)$ be continuous on a simply connected surface region S whose boundary C consists of n regular arcs of geodesic curvature $K_G(u, v)$. Then the sum Θ of the n exterior angles of the boundary is related to the* **integral curvature** $\int_S K\,dA$ *of the surface region S by*

$$\int_C K_G\,ds + \int_S K\,dA = 2\pi - \Theta \quad \text{(Gauss-Bonnet theorem)} \qquad (17.3\text{-}48)$$

The first integral vanishes if all the boundary segments are geodesics; Eq. (47) is a special case of Eq. (48).

17.4. CURVED SPACES

17.4-1. Introduction. The theory of Secs. 17.4-2 to 17.4-7 treats geometrical concepts like distance, angle, and curvature in terms of general curvilinear coordinates and extends these concepts to a class of multidimensional spaces, viz., the *Riemann spaces* introduced in Chap. 16. The tensor notation of Chap. 16 is employed.

17.4-2. Curves, Distances, and Directions in Riemann Space (see also Secs. 4.6-9, 5.4-4, and 6.2-3).* (**a**) The metric properties (see also Sec. 12.5-2) associated with an n-dimensional Riemann space of points $(x^1, x^2, \ldots, x^n)$ are specified by its *metric-tensor components* $g_{ik}(x^1, x^2, \ldots, x^n)$ (Sec. 16.7-1) which define *scalar products*, and thus magnitudes and directions of vectors at each point (Sec. 16.8-1).

(**b**) A **regular arc** C is described by n parametric equations

$$x^i = x^i(t) \qquad (t_1 \leq t \leq t_2) \tag{17.4-1}$$

with unique continuous derivatives dx^i/dt which do not vanish simultaneously. The components $dx^i \equiv (dx^i/dt)\,dt$ represent a vector $d\mathbf{r}$ along (tangent to) the curve C at each of its points $(x^1, x^2, \ldots, x^n)$, and the **element of distance** between two neighboring points $(x^1, x^2, \ldots, x^n)$ and $(x^1 + dx^1, x^2 + dx^2, \ldots, x^n + dx^n)$ on C is defined as

$$\boxed{\begin{aligned} ds &\equiv \sqrt{|g_{ik}(x^1, x^2, \ldots, x^n)\,dx^i\,dx^k|} \\ &\equiv \sqrt{\left|g_{ik}(x^1, x^2, \ldots, x^n)\frac{dx^i}{dt}\frac{dx^k}{dt}\right|}\,dt \end{aligned}} \tag{17.4-2}$$

The sign of ds is chosen so that $ds \geq 0$ for $dt > 0$ (*positive direction* on the curve C). The **arc length** s along the curve C, measured from the curve point corresponding to $t = t_0$, is

$$s \equiv \int_{t_0}^{t} ds \equiv \int_{t_0}^{t} \sqrt{\left|g_{ik}\frac{dx^i}{dt}\frac{dx^k}{dt}\right|}\,dt \tag{17.4-3}$$

The value of the integral (3) is independent of the particular parameter t used.

(**c**) The direction of the curve (1) at each of its points $(x^1, x^2, \ldots, x^n)$ is that of the vector $d\mathbf{r}$; i.e., the angle γ between any vector $\mathbf{a}$ defined at $(x^1, x^2, \ldots, x^n)$ and the curve is given by $\cos\gamma = \mathbf{a} \cdot d\mathbf{r}/|\mathbf{a}|\,|ds|$ (Sec. 16.8-1). In particular, the angle γ between two regular arcs $x^i = X_1^i(t)$ and $x^i = X_2^i(t)$ is given by

* Equations (4) to (6) apply directly whenever $ds \neq 0$. The case of null directions ($d\mathbf{r} \neq 0$, $ds = |d\mathbf{r}| = 0$) in Riemann spaces with indefinite metric is briefly discussed in Sec. 17.4-4.

$$\cos \gamma = g_{ik} \frac{dX_1^i}{ds} \frac{dX_2^k}{ds} \tag{17.4-4}$$

The points common to $n - 1$ of the n **coordinate hypersurfaces** x^i = constant ($i = 1, 2, \ldots, n$) through a given point $(x^1, x^2, \ldots, x^n)$ lie on a **coordinate line** associated with the nth coordinate (see also Sec. 6.2-2). The cosine of the angle between the ith and the kth coordinate line through the point $(x^1, x^2, \ldots, x^n)$ equals $g_{ik}/\sqrt{g_{ii}g_{kk}}$.

(**d**) The unit vector $d\mathbf{r}/ds$ (represented by dx^i/ds) is the **unit tangent** vector of C at each curve point $(x^1, x^2, \ldots, x^n)$. The **first curvature vector** $d^2\mathbf{r}/ds^2$ $\left[\text{represented by } \frac{D}{ds}\left(\frac{dx^i}{ds}\right)\right]$ is perpendicular to the curve (**principal-normal direction,** see also Sec. 17.2-2*b*); the absolute value

$$|\kappa| = \left|\frac{d^2\mathbf{r}}{ds^2}\right| = \sqrt[+]{\left| g_{ik} \frac{D}{ds}\left(\frac{dx^i}{ds}\right) \frac{D}{ds}\left(\frac{dx^k}{ds}\right)\right|} \tag{17.4-5}$$

is the **absolute geodesic curvature (absolute first curvature)** of C at $(x^1, x^2, \ldots, x^n)$ (see also Sec. 17.3-7).

17.4-3. Geodesics (see also Sec. 17.3-12).* (**a**) A **geodesic** in a Riemann space is a regular arc whose geodesic curvature is identically zero, so that the unit tangent vector $d\mathbf{r}/ds$ is constant along the curve (parallel to itself, Sec. 16.10-9), i.e.,

$$\frac{d^2\mathbf{r}}{ds^2} = 0 \quad \text{or} \quad \frac{D}{ds}\left(\frac{dx^i}{ds}\right) \equiv \frac{d^2x^i}{ds^2} + \begin{Bmatrix} i \\ j\,k \end{Bmatrix} \frac{dx^j}{ds}\frac{dx^k}{ds} = 0 \tag{17.4-6}$$

The n second-order differential equations (6) *define a unique geodesic $x^i = x^i(s)$ through any given point $[x^i = x^i(s_1)]$ in any given direction (given values of the dx^i/ds for $s = s_1$).*

More generally, the differential equations

$$\frac{d^2x^i}{dt^2} + \begin{Bmatrix} i \\ j\,k \end{Bmatrix} \frac{dx^j}{dt}\frac{dx^k}{dt} = \lambda(t)\frac{dx^i}{dt} \tag{17.4-7}$$

(with suitable initial conditions on the x^i and dx^i/dt) define a geodesic $x^i = x^i(t)$; the choice of the function

$$\lambda(t) \equiv -\frac{d^2t}{ds^2}\Big/\left(\frac{dt}{ds}\right)^2 \tag{17.4-8}$$

amounts to a choice of the parameter t used to describe the geodesic but does not affect the curve as such.

(**b**) Geodesics have many of the properties of the straight lines in Euclidean geometry (see also Sec. 17.4-6). *If there exists a curve of*

* See footnote to Sec. 17.4-2.

smallest or greatest arc length (3) *joining two given points of a Riemann space, then that curve is a geodesic.* More generally, the differential equations (6) or (7) of a geodesic may be regarded as Euler equations (Secs. 11.6-1 and 11.6-2) which ensure that the first variation of the arc length (3) along a curve joining the given points is equal to zero. The actual existence of a geodesic through two given points of a Riemann space requires a separate investigation in each case.

17.4-4. Riemann Spaces with Indefinite Metric. Null Directions and Null Geodesics. If the fundamental quadratic form $g_{ik}(x^1, x^2, \ldots, x^n)\,dx^i\,dx^k$ of a Riemann space is indefinite (Sec. 13.5-2) at a point $(x^1, x^2, \ldots, x^n)$, the square $|\mathbf{a}|^2 = g_{ik}a^ia^k$ of a vector $\mathbf{a}$ may be positive, negative, or zero, and $|\mathbf{a}| = 0$ does not necessarily imply $\mathbf{a} = 0$. At any point $(x^1, x^2, \ldots, x^n)$ the direction of a vector $\mathbf{a} \neq 0$ such that $|\mathbf{a}|^2 = g_{ik}a^ia^k = 0$ is a **null direction.** For a vector displacement $d\mathbf{r} \neq 0$ in a null direction, $ds = |d\mathbf{r}| = 0$; note that the points $(x^1, x^2, \ldots, x^n)$ and $(x^1 + dx^1, x^2 + dx^2, \ldots, x^n + dx^n)$ separated by such a **null displacement** $d\mathbf{r}$ are *not* identical. A curve $x^i = x^i(t)$ such that

$$g_{ik}\frac{dx^i}{dt}\frac{dx^k}{dt} = 0 \qquad (17.4\text{-}9)$$

has a null direction at each of its points (curve of zero arc length, **null curve, minimal curve,** see also Sec. 17.3-10*b*). A curve $x^i = x^i(t)$ satisfying Eq. (7) as well as Eq. (9) is a **null geodesic (geodesic null line).** *Every null direction defines a unique null geodesic through a given point* $(x^1, x^2, \ldots, x^n)$ (*application:* light paths in relativity theory).

17.4-5. Specification of Space Curvature. (a) The **Riemann-Christoffel curvature tensor** of a given Riemann space is the absolute tensor of rank 4 defined by the mixed components

$$R^i_{jkh} \equiv \frac{\partial}{\partial x^k}\begin{Bmatrix} i \\ j\;h \end{Bmatrix} - \frac{\partial}{\partial x^h}\begin{Bmatrix} i \\ j\;k \end{Bmatrix} + \begin{Bmatrix} i \\ r\;k \end{Bmatrix}\begin{Bmatrix} r \\ j\;h \end{Bmatrix} - \begin{Bmatrix} i \\ r\;h \end{Bmatrix}\begin{Bmatrix} r \\ j\;k \end{Bmatrix} \qquad (17.4\text{-}10)$$

or by the covariant components

$$\begin{aligned} R_{ijkh} &\equiv g_{ir}R^r_{jkh} \equiv \frac{\partial}{\partial x^k}[jh; i] - \frac{\partial}{\partial x^h}[jk; i] + \begin{Bmatrix} r \\ j\;k \end{Bmatrix}[ih; r] - \begin{Bmatrix} r \\ j\;h \end{Bmatrix}[ik; r] \\ &\equiv \frac{1}{2}\left[\frac{\partial^2 g_{ih}}{\partial x^j\,\partial x^k} + \frac{\partial^2 g_{jk}}{\partial x^i\,\partial x^h} - \frac{\partial^2 g_{ik}}{\partial x^j\,\partial x^h} - \frac{\partial^2 g_{jh}}{\partial x^i\,\partial x^k}\right] \\ &\qquad + g^{rs}\{[jk; s][ih; r] - [jh; s][ik; r]\} \end{aligned} \qquad (17.4\text{-}11)$$

The components of the curvature tensor satisfy the following relations:

$$R^i_{jkh} = -R^i_{jhk} = g^{ri}R_{rjkh} \qquad R^i_{jkh} + R^i_{khj} + R^i_{hjk} = 0 \qquad (17.4\text{-}12)$$

$$R_{ijkh} = R_{khij} = -R_{jikh} = -R_{ijhk} \qquad R_{ijkh} + R_{ikhj} + R_{ihjk} = 0 \qquad (17.4\text{-}13)$$

In an n-dimensional Riemann space there exist at most $n^2(n^2 - 1)/12$ distinct non-vanishing covariant components R_{ijkh}. Note also

$$\left.\begin{aligned} R^i_{jkh,r} + R^i_{jhr,k} + R^i_{jrk,h} &= 0 \\ R_{ijkh,r} + R_{ijhr,k} + R_{ijrk,h} &= 0 \end{aligned}\right\} \quad \text{(BIANCHI IDENTITIES)} \qquad (17.4\text{-}14)$$

(**b**) The **Ricci tensor** of a Riemann space is the absolute tensor of rank 2 defined by the covariant components

$$R_{ij} \equiv R_{ji} \equiv R^k_{ijk}$$
$$\equiv \frac{\partial^2 \log_e \sqrt{|g|}}{\partial x^i \, \partial x^j} - \frac{\partial}{\partial x^k} \begin{Bmatrix} k \\ i \; j \end{Bmatrix} - \begin{Bmatrix} h \\ i \; j \end{Bmatrix} \frac{\partial}{\partial x^h} \log_e \sqrt{|g|} + \begin{Bmatrix} k \\ h \; j \end{Bmatrix} \begin{Bmatrix} h \\ i \; k \end{Bmatrix} \tag{17.4-15}$$

or by the mixed components $R^i_j \equiv g^{ik}R_{kj}$. *In an n-dimensional Riemann space there exist at most $n(n + 1)/2$ distinct nonvanishing covariant components R_{ij}.*

The eigenvector directions (Sec. 14.8-3) of the Ricci tensor are the **Ricci principal directions** at each point $(x^1, x^2, \ldots, x^n)$ of the Riemann space.

(**c**) The **curvature invariant** or **scalar curvature** of a Riemann space is the absolute scalar invariant

$$R \equiv R^i_i \equiv g^{ik}R_{ik} \equiv g^{ik}R^j_{ikj} \tag{17.4-16}$$

(**d**) The **Einstein tensor** of a Riemann space is the absolute tensor of rank 2 defined by the components

$$G^i_j \equiv R^i_j - \tfrac{1}{2}R\delta^i_j \qquad \text{or} \qquad G_{ij} \equiv g_{ik}G^k_j \equiv R_{ij} - \tfrac{1}{2}Rg_{ij} \tag{17.4-17}$$

The divergence (Sec. 16.10-7) $G^i_{j,i}$ *of the Einstein tensor vanishes identically.*

(**e**) Refer to Sec. 17.3-7 for the special case of a two-dimensional Riemann space (curved surface in three-dimensional Euclidean space).

17.4-6. Manifestations of Space Curvature. Flat Spaces and Euclidean Spaces. (**a**) **Parallel Propagation of a Vector.** Parallel propagation of a vector **a** ($Da^i = 0$, Sec. 16.10-9) around the infinitesimal closed circuit $(x^1, x^2, \ldots, x^n) \rightarrow (x^1 + dx^1, x^2 + dx^2, \ldots, x^n + dx^n) \rightarrow (x^1 + dx^1 + d\xi^1, x^2 + dx^2 + d\xi^2, \ldots, x^n + dx^n + d\xi^n) \rightarrow (x^1 + d\xi^1, x^2 + d\xi^2, \ldots, x^n + d\xi^n) \rightarrow (x^1, x^2, \ldots, x^n)$ changes each component a^i of a by

$$\delta a^i = -R^i_{jkh}a^j \, dx^k \, d\xi^h \tag{17.4-18}$$

[see also Eq. (16.10-25)].

(**b**) **Geodesic Parallels. Geodesic Deviation.** If a one-parameter family of geodesics $x^i = x^i(s, \lambda)$ has orthogonal trajectories,* the latter are **geodesic parallels** cutting off equal increments of the arc length s on each geodesic of the given family (see also Secs. 17.3-12 and 17.4-7).

Adjacent geodesics $x^i = x^i(s, \lambda)$ and $x^i = x^i(s, \lambda + d\lambda)$ are then related by

$$x^i(s, \lambda + d\lambda) = x^i(s, \lambda) + \eta^i(s, \lambda)$$

The **geodesic-deviation vector** represented by the $\eta^i(s, \lambda)$ is normal to the geodesic $x^i(s, \lambda)$, or $g_{ik}\eta^i p^k = 0$, where the p^k are the unit-tangent-

* Note that the notation $x^i = x^i(s, \lambda)$ excludes null geodesics.

vector components of the geodesic $x^i = x^i(s, \lambda)$. *Along any one of the given geodesics (λ = constant), the η^i change with s so that*

$$\frac{D^2\eta^i}{ds^2} = -R^i_{jkh}\eta^k p^j p^h \quad \text{(EQUATION OF GEODESIC DEVIATION)} \qquad (17.4\text{-}19)$$

(c) **Flat Spaces.** A Riemann space is **flat**

1. If and only if all components R^i_{jkh} or R_{ijkh} of the curvature tensor vanish identically, so that successive covariant differentiations commute [Eq. (16.10-25)], and parallel propagation around infinitesimal circuits leaves all tensor components constant (Secs. 16.10-9 and 17.4-6*a*)
2. If and only if the Riemann space admits a system of *rectangular cartesian coordinates* $\xi^1, \xi^2, \ldots, \xi^n$ such that, at every point of the space,

$$ds^2 = \epsilon_1(\xi^1)^2 + \epsilon_2(\xi^2)^2 + \cdots + \epsilon_n(\xi^n)^2 \qquad (17.4\text{-}20)$$

where each ϵ_i is constant and equals either $+1$ or -1.

In every scheme of measurements defined by a cartesian coordinate system all Christoffel symbols (Sec. 16.10-3) vanish identically; covariant differentiation reduces to ordinary differentiation, and every geodesic or geodesic null line can be described by *linear* parametric equations $\xi^i = a^i t + b^i$.

Equation (20) may be formally simplified by the introduction of *homogeneous coordinates* $\zeta^i = \underset{+}{\sqrt{\epsilon_i}}\,\xi^i$; ζ^i is imaginary if $\epsilon_i = -1$ (this convention is used in relativity theory, Ref. 17.4).

(d) A **Euclidean space** is a flat Riemann space having a positive-definite metric, so that all ϵ_i in Eq. (20) are equal to $+1$ (*Euclidean geometry*, see also Chaps. 2 and 3). Note that the topology (Sec. 12.5-1) of a Euclidean space may differ from the "usual" topology employed in elementary geometry (EXAMPLES: surfaces of cylinders and cones).

17.4-7. Special Coordinate Systems. Because of the invariance of tensor equations (Secs. 16.1-4, 16.4-1, and 16.10-7*b*), it is often permissible to simplify mathematical arguments by using one of the following *special coordinate systems.*

(a) Not every Riemann space admits *orthogonal coordinates* ($g_{ik} \equiv 0$ for $i \neq k$, Sec. 16.8-2), but it is possible to choose *one* of the coordinates, say x^n, so that its coordinate lines are normal to all others, or

$$ds^2 = \sum_{i=1}^{n-1}\sum_{k=1}^{n-1} g_{ik}\,dx^i\,dx^k + g_{nn}(dx^n)^2 \qquad (17.4\text{-}21)$$

at every point $(x^1, x^2, \ldots, x^n)$. It is always possible to choose such a coordinate system so that either $g_{nn} \equiv 1$ or $g_{nn} \equiv -1$; x^n is, then, measured by the arc length s along its coordinate lines, and the latter are geodesics normal to every hypersurface x^n = constant (**geodesic normal coordinates,** see also Sec. 17.3-13).

(b) Every Riemann space admits **local rectangular cartesian coordinates** $\xi^1, \xi^2, \ldots, \xi^n$ such that the metric is given by Eq. (20) at any *one* given point $(\xi^1, \xi^2, \ldots, \xi^n)$. Hence every Riemann space is "locally flat"; i.e., every sufficiently small portion of the space is flat (Euclidean if the metric is positive definite).

(c) **Riemann coordinates with origin** O are defined as $x^i = sp^i$, where the p^i are unit-tangent-vector components at O of the geodesic joining O and the point $(x^1, x^2, \ldots, x^n)$, and s is the geodesic distance between these points. *Every Riemann space admits Riemannian coordinates for any given origin* O; note that all Christoffel symbols vanish at O.

17.5. RELATED TOPICS, REFERENCES, AND BIBLIOGRAPHY

17.5-1. Related Topics. The following topics related to the study of differential geometry are treated in other chapters of this handbook:

Plane analytic geometry........ Chap. 2
Solid analytic geometry........ Chap. 3
Vector analysis........ Chap. 5
Curvilinear coordinates........ Chap. 6
Tensor analysis, Riemann spaces........ Chap. 16

17.5-2. References and Bibliography.

17.1. Blaschke, W., and H. Reichardt: *Vorlesungen über Differentialgeometrie,* 2d ed., Springer, Berlin, 1960.
17.2. Eisenhart, L. P.: *An Introduction to Differential Geometry,* Princeton University Press, Princeton, N.J., 1947.
17.3. ———: *Riemannian Geometry,* Princeton, Princeton, N.J., 1949.
17.4. Guggenheimer, H. W.: *Differential Geometry,* McGraw-Hill, New York, 1963.
17.5. Kreyszig, E.: *Differential Geometry,* 2d ed., University of Toronto Press, Toronto, Canada, 1963.
17.6. Rainich, G. Y.: *Mathematics of Relativity,* Wiley, New York, 1950.
17.7. Sokolnikov, I. S.: *Tensor Analysis,* 2d ed., Wiley, New York, 1964.
17.8. Spain, B.: *Tensor Calculus,* Oliver & Boyd, London, 1953.
17.9. Struik, D. J.: *Lectures on Classical Differential Geometry,* 2d ed., Addison-Wesley, Reading, Mass., 1961.
17.10. Synge, J. L., and A. Schild: *Tensor Calculus,* University of Toronto Press, Toronto, Canada, 1949.
17.11. Willmore, T. J.: *Introduction to Differential Geometry,* Oxford, Fair Lawn, N.J., 1959.

(See also the article by H. Tietz in vol. II of the *Handbuch der Physik,* Springer, Berlin, 1955.)

CHAPTER **18**

PROBABILITY THEORY AND RANDOM PROCESSES

18.1. INTRODUCTION

18.1-1. Mathematical probabilities are values of a real numerical function defined on a class of idealized events, which represent results of an experiment or observation. Mathematical probabilities are *not* defined directly in terms of "likelihood" or relative frequency of occurrence; they are introduced by a set of defining postulates (Sec. 18.2-2; see also Sec. 12.1-1) which abstract essential properties of statistical relative frequencies (Sec. 19.2-1). The concept of probability can, then, often be related to reality by the assumption that, in practically every sequence of independently repeated experiments, the relative frequency of each event tends to a limit represented by the corresponding probability (Sec. 19.2-1).* Theories based on the probability concept may, however, be useful even if they are not subject to direct statistical interpretation.

Probability theory deals with the definition and description of models involving the probability concept. The theory is especially concerned with methods for calculating the probability of an event from the known or postulated probabilities of other events which are logically related to the first event. Most applications of probability theory may be interpreted as special cases of *random processes* (Secs. 18.8-1 to 18.11-5).

* Whenever this proposition is justified, it must be regarded as a *law of nature;* it should not be confused with *mathematical theorems* like Bernoulli's theorem or the mathematical law of large numbers (Sec. 18.6-5).

18.2. DEFINITION AND REPRESENTATION OF PROBABILITY MODELS

18.2-1. Algebra of Events Associated with a Given Experiment. Each probability model describes a specific idealized experiment or observation having a class $\mathcal{S}\dagger$ of theoretically possible results (events, states) E permitting the following definitions.

1. The **union (logical sum)** $E_1 \cup E_2 \cup \cdots$ (or $E_1 + E_2 + \cdots$) of a countable (finite or infinite) set of events $E_1, E_2, \ldots$ is the event of realizing *at least one* of the events $E_1, E_2, \ldots$.
2. The **intersection (logical product)** $E_1 \cap E_2$ (or E_1E_2) of two events E_1 and E_2 is the *joint event* of realizing *both* E_1 *and* E_2.
3. The **(logical) complement** $\tilde{E}$ of an event E is the event of *not* realizing E ("opposite" or complementary event of E).
4. I is the **certain event** of realizing at least one of the events of $\mathcal{S}\dagger$.
5. 0 is the **impossible event** of realizing no one of the events of $\mathcal{S}\dagger$.

In each case, the class $\mathcal{S}$ *of events comprising* $\mathcal{S}\dagger$ *and* 0 *is to constitute a completely additive Boolean algebra* (**algebra of events associated with the given experiment or observation**) *having all the properties outlined in Secs.* 12.8-1 *and* 12.8-4. Either $E_1 \cup E_2 = E_1$ or $E_1 \cap E_2 = E_2$ implies the **logical inclusion relation** $E_2 \subset E_1$ (E_2 *implies* E_1); note $0 \subset E \subset I$. E_1 and E_2 are **mutually exclusive (disjoint)** if and only if $E_1 \cap E_2 = 0$. The set $\mathcal{S}_1$ of joint events $E \cap E_1$ is the algebra of events associated with the given experiment *under the hypothesis that* E_1 *occurs;* $E_1 \cap E_1 = E_1$ is the certain event in $\mathcal{S}_1$ (see also Sec. 12.8-3).

18.2-2. Mathematical Definition of Probabilities. Conditional Probabilities. It is possible to assign a (mathematical) **probability** $P[E]$ (**probability of** E, **probability of realizing the event** E) to each event E of the class $\mathcal{S}$ (event algebra, Sec. 18.2-1) associated with a given experiment if and only if one can define a single-valued real function $P[E]$ on $\mathcal{S}$ so that

1. $P[E] \geq 0$ for every event E of $\mathcal{S}$
2. $P[E] = 1$ if the event E is *certain*
3. $P[E_1 \cup E_2 \cup \cdots] = P[E_1] + P[E_2] + \cdots$ for every countable (finite or infinite) set of *mutually exclusive* events $E_1, E_2, \ldots$

Postulates 1 *to* 3 *imply* $0 \leq P[E] \leq 1$; *in particular,* $P[E] = 0$ *if* E *is an impossible event.* Note carefully that $P[E] = 1$ or $P[E] = 0$ do *not* necessarily imply that E is, respectively, certain or impossible.

A fourth defining postulate relates the "absolute" probability $P[E]$ associated with the given experiment to the "conditional" probabilities $P[E|E_1]$ referring to a "simpler" experiment restricted by the hypothesis that E_1 occurs. The **conditional probability** $P[E|E_1]$ **of** E **on (relative to) the hypothesis that the event** E_1 **occurs** is defined by the postulate

> 4. The probability of the joint event $E \cap E_1$ is
> $$P[E \cap E_1] = P[E_1]P[E|E_1] \quad \text{(MULTIPLICATION LAW, LAW OF COMPOUND PROBABILITIES)}$$

$P[E|E_1]$ *is not defined if* $P[E_1] = 0$.

In the context of the restricted experiment, the quantities $P[E|E_1]$ are ordinary probabilities associated with the joint events $E \cap E_1$ constituting the event algebra $\mathcal{S}_1$ of the restricted experiment (Sec. 18.2-1). In practice, every probability can be interpreted as a conditional probability relative to some hypothesis implied by the experiment under consideration.

18.2-3. Statistical Independence. Two events E_1 and E_2 are **statistically independent (stochastically independent)** if and only if

$$P[E_1 \cap E_2] = P[E_1]P[E_2] \tag{18.2-1}$$

so that $P[E_1|E_2] = P[E_1]$ if $P[E_2] \neq 0$, and $P[E_2|E_1] = P[E_2]$ if $P[E_1] \neq 0$.

N events $E_1, E_2, \ldots, E_N$ are statistically independent if and only if not only each pair of events E_i, E_k but also each pair of possible joint events is statistically independent:

$$\left.\begin{array}{ll} P[E_i \cap E_j] = P[E_i]P[E_j] & (1 \leq i < j \leq N) \\ P[E_i \cap E_j \cap E_k] = P[E_i]P[E_j]P[E_k] & (1 \leq i < j < k \leq N) \\ \cdots\cdots\cdots\cdots\cdots\cdots\cdots & \\ P[E_1 \cap E_2 \cap \cdots \cap E_N] = P[E_1]P[E_2] \cdots P[E_N] & \end{array}\right\} \tag{18.2-2}$$

18.2-4. Compound Experiments, Independent Experiments, and Independent Repeated Trials. Frequently an experiment appears as a combination of component experiments (see also Secs. 18.7-3 and 18.8-1). Let $E', E'', E''', \ldots$ denote any result associated, respectively, with the first, second, third, . . . component experiment. The results of the compound experiment can be described as joint events $E = E' \cap E'' \cap E''' \cap \ldots$; their probabilities will, in general, depend on the nature and interaction of all component experiments. The probability $P[E']$ of realizing the component result E' *in the course of a given compound experiment* is, in general, different from the probability associated with E' in an independently performed component experiment.

Two or more component experiments of a given compound experiment are **independent** if and only if their respective results $E', E'', E''', \ldots$ obtained *in the course of the compound experiment* are statistically independent, i.e.,

$$P[E' \cap E''] = P[E']P[E''] \qquad P[E' \cap E'' \cap E'''] = P[E']P[E'']P[E'''] \qquad \ldots$$

for all $E', E'', E''', \ldots$ (Sec. 18.2-3). If a component experiment is independent of all others, the probability of realizing each of its results in the course of the given compound experiment is equal to the corresponding probability for the independently performed component experiment.

Repeated independent trials are independent experiments each having the same set of possible results E and the same set of associated probabilities $P[E]$. *The probability of obtaining the sequence of results $E_1, E_2, \ldots E_n$ in the compound experiment corresponding to a sequence of n repeated independent trials is*

$$P[E_1, E_2, \ldots, E_n] = P[E_1]P[E_2] \cdots P[E_n] \qquad (18.2\text{-}3)$$

18.2-5. Combination Rules (see also Secs. 18.7-1 to 18.7-3).

Each of the theorems in Table 18.2-1 expresses the probability of an event in terms of the (possibly already known) probabilities of other events logically related to the first event.

More generally, *the probability of realizing* **at least** m *and* **exactly** m *of* N (*not necessarily statistically independent*) *events* $E_1, E_2, \ldots, E_N$ is, respectively

$$P_m = \sum_{j=m}^{N} (-1)^{j-m} \binom{j-1}{m-1} S_j \qquad P_{[m]} = \sum_{j=m}^{N} (-1)^{j-m} \binom{j}{m} S_j \qquad (m = 1, 2, \ldots, N) \qquad (18.2\text{-}4)$$

where

$$S_1 = \sum_i P[E_i] \qquad S_2 = \sum_{i<k}\sum P[E_i \cap E_k] \qquad \cdots$$

$$S_N = P[E_1 \cap E_2 \cap \cdots \cap E_N] \qquad (18.2\text{-}5)$$

Note that

$$S_j = \sum_{k=j}^{N} \binom{k-1}{j-1} P_k = \sum_{k=j}^{N} \binom{k}{j} P_{[k]} \qquad (j = 1, 2, \ldots, N) \qquad (18.2\text{-}6)$$

If $E_1, E_2, \ldots, E_N$ are statistically independent, the quantities (5) reduce to the symmetric functions (1.4-9) of the $P[E_i]$ (Table 18.2-1*b*).

EXAMPLES: If the probability of each throw with a die is $1/6$, then

The probability of throwing *either* 1 *or* 6 is $1/6 + 1/6 = 1/3$.

The probability of *not* throwing 6 is $1 - 1/6 = 5/6$.

The probability of throwing 6 *at least once* in two throws is $1/6 + 1/6 - 1/36 = 11/36$.

The probability of throwing 6 *exactly once* in two throws is $1/3 - 2/36 = 5/18$.

The probability of throwing 6 twice in two throws is $1/36$; etc.

18.2-6. Bayes's Theorem (see also Sec. 18.4-5*b*).

Let $H_1, H_2, \ldots$ be a set of mutually exclusive events such that $H_1 \cup H_2 \cup \cdots = I$. Then, for each pair of events H_i, E,

$$P[H_i|E] = \frac{P[H_i \cap E]}{P[E]} = \frac{P[H_i]P[E|H_i]}{P[E]} = \frac{P[H_i]P[E|H_i]}{\sum_i P[H_i]P[E|H_i]} \qquad \text{(BAYES'S THEOREM)} \qquad (18.2\text{-}7)$$

Equation (7) can be used to relate the "a priori" probability $P[H_i]$ of a hypothetical cause H_i of the event E to the "a posteriori" probability $P[H_i|E]$ if (and only if) the H_i are "random" events permitting the definition of probabilities $P[H_i]$.

Table 18.2-1. Probabilities of Logically Related Events

(a)	Probability of **not** realizing the event E	$P[\tilde{E}] = 1 - P[E]$
	Probability of realizing **at least one of two events** E_1 and E_2 (E_1 or E_2 or both)	$P[E_1 \cup E_2] = P[E_1] + P[E_2] - P[E_1 \cap E_2]$
	Probability of realizing **all of** N **events** $E_1, E_2, \ldots, E_N$	$P[E_1 \cap E_2 \cap \cdots \cap E_N] = P[E_1]P[E_2 \vert E_1]P[E_3 \vert E_1 \cap E_2] \cdots P[E_N \vert E_1 \cap E_2 \cap \cdots \cap E_{N-1}]$
(b)	Probability of realizing **at least one of** N **statistically independent events** $E_1, E_2, \ldots, E_N$	$P[E_1 \cup E_2 \cup \cdots \cup E_N] = 1 - \{1 - P[E_1]\}\{1 - P[E_2]\} \cdots \{1 - P[E_N]\}$
	Probability of realizing **all of** N **statistically independent events** $E_1, E_2, \ldots, E_N$	$P[E_1 \cap E_2 \cap \cdots \cap E_N] = P[E_1]P[E_2] \cdots P[E_N]$

18.2-7. Representation of Events as Sets in a Sample Space. Every class S of events E permitting the definition of probabilities $P[E]$ can be described in terms of a set $\mathcal{I}$ of mutually exclusive events $\hat{E} \neq 0$ such that each event E is the union of a corresponding subset of $\mathcal{I}$. $\mathcal{I}$ is called a **sample space** or **fundamental probability set** associated with the given experiment; each set of **sample points (simple events, elementary events, phases)** $\hat{E}$ of $\mathcal{I}$ corresponds to an event E. In particular, $\mathcal{I}$ itself corresponds to a certain event, and an empty subset of $\mathcal{I}$ corresponds to an impossible event.

The probabilities $P[E]$ can then be regarded as values of a *set function*, the **probability function** defining the **probability distribution** of the sample space. *Each probability $P[E]$ is the sum of the probabilities attached to the simple events included in the event E.*

The event algebra S is thus represented isomorphically by an algebra of measurable sets (see also Secs. 4.6-17*b* and 12.8-4). The fundamental probability set associated with the conditional probabilities $P[E|E_1]$ is the subset of I representing E_1. Conversely, a sample space associated with any given experiment may be regarded as a subset "embedded" in a space of events associated with a more general experiment (see also Secs. 18.2-1 and 18.2-2).

18.2-8. Random Variables. A random variable (stochastic variable, chance variable, variate) is any (not necessarily numerical)* variable $\mathbf{x}$ whose "values" $\mathbf{x} = \mathbf{X}$ constitute a fundamental probability set (sample space, Sec. 18.2-7) of simple events $[\mathbf{x} = \mathbf{X}]$, or whose values label the points of a sample space on a reciprocal one-to-one basis. The associated probability distribution is the **distribution of the random variable** x. The definition of any random variable must specify its distribution.

Every single-valued measurable function (*Sec.* 4.6-14*c*) $\mathbf{x}$ *defined on any fundamental probability set* $\mathcal{I}$ *is a random variable; its distribution is defined by the probabilities of the events* (*measurable subsets of* $\mathcal{I}$, *Sec.* 18.2-7) *corresponding to each set of values of* $\mathbf{x}$.

18.2-9. Representation of Probability Models in Terms of Numerical Random Variables and Distribution Functions. The simple events (sample points) $\hat{E}$ of the fundamental probability set associated with a given problem are frequently *labeled* with corresponding values (*sample values*) X of a real numerical random variable x. Each sample value of x may, for instance, correspond to the result of a measurement defining a simple event. Compound events, like $[x \leq a]$, $[\sin x > 0.5]$, or $[x = \arctan 2]$, correspond to measurable sets of values of x (see also Sec. 18.2-8).

More generally, each simple event may be labeled by a corresponding (ordered) set $\mathbf{X} \equiv (X_1, X_2, \ldots)$ of real numbers $X_1, X_2, \ldots$ which

* The boldface type used to denote a multidimensional random variable $\mathbf{x}$ does *not* necessarily imply that $\mathbf{x}$ is a vector.

constitutes a "value" of a *multidimensional* random variable $\mathbf{x} \equiv (x_1, x_2, \ldots)$. Each of the real variables $x_1, x_2, \ldots$ is itself a random variable (see also Sec. 18.4-1).

Given a random variable x or $\mathbf{x}$ labeling the simple events of the given fundamental probability set on a one-to-one basis, *the probabilities associated with the corresponding experiment are uniquely described by the probability distribution of the random variable.*

Throughout this handbook, all real numerical random variables are understood to range from $-\infty$ to $+\infty$; values of a numerical random variable which do not label a possible simple event $\hat{E}$ are treated as impossible events and are assigned the probability zero.

The distribution (or the probability function, Sec. 18.2-7) of any real numerical random variable x is uniquely described by its (**cumulative**) **distribution function**

$$\Phi_x(X) \equiv \Phi(X) \equiv P[x \leq X] \tag{18.2-8}$$

Similarly, the distribution of a multidimensional random variable $\mathbf{x} \equiv (x_1, x_2, \ldots)$ is uniquely described by its (cumulative) distribution function

$$\Phi_{\mathbf{x}}(X_1, X_2, \ldots) \equiv \Phi(X_1, X_2, \ldots) \equiv P[x_1 \leq X_1, x_2 \leq X_2, \ldots] \tag{18.2-9}$$

Conversely, the distribution function corresponding to a given probability distribution is uniquely defined for all values of the random variable in question. Every distribution function is a nondecreasing function of each of its arguments, and

$$\Phi_x(-\infty) = 0 \qquad \Phi_x(\infty) = 1 \tag{18.2-10}$$

$$\Phi_{\mathbf{x}}(-\infty, X_2, X_3, \ldots) = \Phi_{\mathbf{x}}(X_1, -\infty, X_3, \ldots) = \cdots = 0$$
$$\Phi_{\mathbf{x}}(\infty, \infty, \ldots) = 1 \tag{18.2-11}$$

18.3. ONE-DIMENSIONAL PROBABILITY DISTRIBUTIONS

18.3-1. Discrete One-dimensional Probability Distributions (see Tables 18.8-1 to 18.8-7 for examples). The real numerical random variable x is a **discrete random variable** (has a **discrete probability distribution**) if and only if the probability

$$p_x(X) \equiv p(X) \equiv P[x = X] \tag{18.3-1}$$

is different from zero only on a countable set of **spectral values** $X = X_{(1)}, X_{(2)}, \ldots$ (**spectrum** of the discrete random variable x). Each discrete probability distribution is defined by the function (1), or by the corresponding *(cumulative) distribution function* (Sec. 18.2-9)

$$\Phi_x(X) \equiv \Phi(X) \equiv P[x \leq X] \equiv \sum_{X_{(i)} \leq X} p(X_{(i)}) \qquad (18.3\text{-}2a)^*$$

Throughout this handbook, the notation $\sum_x y(x)$ **will be used to signify summation of a function** $y(x)$ **over all spectral values** $X_{(i)}$ **of a discrete random variable** x (see also Sec. 18.3-6). Note

$$\sum_x p(x) = \Phi(\infty) = 1 \qquad (18.3\text{-}3)$$

18.3-2. Continuous One-dimensional Probability Distributions (see Table 18.8-8 for examples). The real numerical random variable x is a **continuous random variable** (has a **continuous probability distribution**) if and only if its (cumulative) distribution function $\Phi_x(X) \equiv \Phi(X)$ is continuous and has a piecewise continuous derivative, the **frequency function (probability density, differential distribution function)** of x

$$\varphi_x(X) \equiv \varphi(X) \equiv \lim_{\Delta x \to 0} \frac{P[X < x \leq X + \Delta x]}{\Delta x} \equiv \frac{d\Phi}{dX} \qquad (18.3\text{-}4)$$

for all X.† $P[X < x \leq X + dx] = d\Phi = \varphi(X)\,dx$ is called a **probability element (probability differential).** Note

$$\begin{aligned} P[x \leq X] &\equiv \Phi(X) \equiv \int_{-\infty}^{X} \varphi(x)\,dx \\ P[a < x \leq b] &= \Phi(b) - \Phi(a) = \int_a^b \varphi(x)\,dx \end{aligned} \qquad (18.3\text{-}5)$$

$$\int_{-\infty}^{\infty} \varphi(x)\,dx = \Phi(\infty) = 1 \qquad (18.3\text{-}6)$$

If x is a continuous random variable, each event $[x = X]$ has the probability zero but *is not necessarily impossible.* The **spectrum** of a continuous random variable x is the set of values $x = X$ where $\varphi(X) \neq 0$.

* In terms of the step function $U_-(t)$ $[U_-(t) = 0$ if $t < 0$, $U_-(t) = 1$ if $t \geq 0$, Sec. 21.9-1],

$$\Phi(X) \equiv p(X_{(1)})U_-(X - X_{(1)}) + p(X_{(2)})U_-(X - X_{(2)}) + \cdots \qquad (18.3\text{-}2b)$$

† Some authors call a probability distribution continuous whenever its distribution function is continuous.

NOTE: A random variable can be continuous (i.e., have a piecewise continuous frequency function) over part of its range, while it is discrete elsewhere (see also Sec. 18.3-6).

18.3-3. Expected Values and Variance. Characteristic Parameters of One-dimensional Probability Distributions (see also Sec. 18.3-6). (a) The **expected value (mean, mean value, mathematical expectation)** of a function $y(x)$ of a discrete or continuous random variable x is

$$E\{y(x)\} = \begin{cases} \sum_x y(x)p(x) & (x \text{ discrete}) \\ \int_{-\infty}^{\infty} y(x)\varphi(x)\,dx & (x \text{ continuous}) \end{cases} \tag{18.3-7}$$

if this expression exists in the sense of absolute convergence (see also Secs. 4.6-2 and 4.8-1).

(b) In particular, the **expected value (mean, mean value, mathematical expectation)** $E\{x\} = \xi$ and the **variance** $\text{Var}\,\{x\} = \sigma^2$ of a discrete or continuous one-dimensional random variable x are defined by

$$\begin{gathered} E\{x\} = \xi = \begin{cases} \sum_x xp(x) & (x \text{ discrete}) \\ \int_{-\infty}^{\infty} x\varphi(x)\,dx & (x \text{ continuous}) \end{cases} \\ \text{Var}\,\{x\} = \sigma^2 = E\{(x-\xi)^2\} \\ = \begin{cases} \sum_x (x-\xi)^2 p(x) & (x \text{ discrete}) \\ \int_{-\infty}^{\infty} (x-\xi)^2\varphi(x)\,dx & (x \text{ continuous}) \end{cases} \end{gathered} \tag{18.3-8}$$

For computation purposes note (see also Sec. 18.3-10)

$$\text{Var}\,\{x\} = \sigma^2 = E\{x^2\} - \xi^2 = E\{x(x-1)\} - \xi(\xi - 1) \tag{18.3-9}$$

Whenever $E\{x\}$ *and* Var $\{x\}$ *exist, the mean square deviation*

$$E\{(x - X)^2\} = \sigma^2 + (\xi - X)^2$$

of the random variable x *from one of its values* X *is least (and equal to* σ^2*) for* $X = \xi$.

(c) $E\{x\}$ and Var $\{x\}$ are *not* functions of x; they are functionals (Sec. 12.1-4) describing properties of the *distribution* of x. $E\{x\}$ is a *measure of location*, and Var $\{x\}$ is a *measure of dispersion* (*or concentration*) of the probability distribution of x. A number of other numerical "characteristic parameters" describing specific properties of one-dimensional probability distributions are defined in Table 18.3-1 and in Secs. 18.3-7 and 18.3-9. Note that one or more parameters like $E\{x\}$, Var $\{x\}$, $E\{|x - \xi|\}$, . . . may not exist for a given probability distribution.

(d) Tables 18.8-1 to 18.8-8 list mean values and variances for a number of frequently used probability distributions.

Table 18.3-1. Numerical Parameters Describing Properties of One-dimensional Probability Distributions
(see also Secs. 18.3-3, 18.3-7, and 18.3-9)

(a) The Fractiles. The P **fractiles (quantiles of order** P) of a one-dimensional probability distribution are values x_P of x such that

$$P[x \leq x_P] = \Phi(x_P) = P \qquad (0 < P < 1)$$

$x_{1/2}$ is the **median** of the distribution. The **quartiles** $x_{1/4}$, $x_{1/2}$, $x_{3/4}$, the **deciles** $x_{0.1}$, $x_{0.2}$, . . . , $x_{0.9}$, and the **percentiles** $x_{0.01}$, $x_{0.02}$, . . . , $x_{0.99}$ respectively divide the range of x into 4, 10, and 100 intervals corresponding to (compound) events having equal probabilities.

The various fractiles exist for every probability distribution but are not necessarily unique. Tables of fractiles are widely used in statistics (Secs. 19.5-3 and 19.5-4).

(b) Measures of Location

1. The *expected value* $E\{x\}$ (Sec. 18.3-3)
2. The *median* $x_{1/2}$ (see above)
3. A **mode of a** *continuous* probability distribution is a value ξ_{mode} of x such that $\varphi(\xi_{\text{mode}})$ is a relative maximum (Sec. 11.2-1). A mode of a *discrete* probability distribution is a spectral value ξ_{mode} preceded and followed by spectral values associated with probabilities smaller than $p(\xi_{\text{mode}})$.

A distribution having one, two, or more modes is respectively called **unimodal, bimodal,** or **multimodal.**

(c) Measures of Dispersion

1. The *variance* $\sigma^2 = \text{Var}\{x\}$ (Sec. 18.3-3)
2. The **standard deviation (root-mean-square deviation, dispersion)** $\sigma = \sqrt{\text{Var}\{x\}}$
3. The **coefficient of variation (relative dispersion)** σ/ξ
4. The **mean (absolute) deviation** $E\{|x - \xi|\}$
5. The **interquartile range** $x_{3/4}-x_{1/4}$ and the **10–90-percentile range** $x_{0.9}-x_{0.1}$
6. The **range** (difference between the largest and the smallest spectral value)
7. The **half width** of a unimodal continuous distribution is half the difference between the two values of x where $\varphi(x)$ reaches one-half its maximum value

(d) Measures of Skewness and Excess (or Curtosis). The first two of the following measures are defined in terms of the moments discussed in Sec. 18.3-7 (see also Sec. 18.3-9).

1. The **coefficient of skewness** $\gamma_1 = \mu_3/\mu_2^{3/2} = \kappa_3/\kappa_2^{3/2}$
2. The **coefficient of excess** $\gamma_2 = \mu_4/\mu_2^2 - 3 = \kappa_4/\kappa_2^2$
3. **Pearson's measure of skewness** for unimodal distributions $(\xi - \xi_{\text{mode}})/\sigma$ (see also Sec. 19.2-4)

The quantities γ_1^2 and $\gamma_2 + 3$ or $\dfrac{\gamma_2 + 3}{2}$ are also used instead of γ_1 and γ_2

18.3-4. Normalization. Given a function $\psi(x) \geq 0$ known to be proportional to the function $p(x)$ associated with a discrete random variable x (Sec. 18.3-1),*

$$p(x) = \frac{1}{\kappa}\psi(x) = \frac{\psi(x)}{\sum_x \psi(x)} \qquad E\{y(x)\} = \frac{\sum_x y(x)\psi(x)}{\sum_x \psi(x)} \tag{18.3-10}$$

Given a function $\psi(x) \geq 0$ known to be proportional to the frequency function $\varphi(x)$ of a continuous random variable x (Sec. 18.3-2),

$$\varphi(x) = \frac{1}{\kappa}\psi(x) = \frac{\psi(x)}{\int_{-\infty}^{\infty}\psi(x)\,dx} \qquad E\{y(x)\} = \frac{\int_{-\infty}^{\infty} y(x)\psi(x)\,dx}{\int_{-\infty}^{\infty}\psi(x)\,dx} \tag{18.3-11}$$

In either case, κ is called the **normalization factor.** Analogous procedures apply to multidimensional distributions (Sec. 18.4-1).

18.3-5. Chebyshev's Inequality and Related Formulas. The following formulas specify upper bounds for the probability that a random variable x, or its absolute deviation $|x - \xi|$ from the mean value $\xi = E\{x\}$, exceeds a given value $a > 0$.

$$P[x \geq a] \leq \frac{\xi}{a} \qquad (a > 0) \tag{18.3-12}$$

$$P[|x - \xi| \geq a] \leq \frac{\sigma^2}{a^2} \qquad (a > 0) \qquad \text{(CHEBYSHEV'S INEQUALITY)} \tag{18.3-13}$$

If x has a continuous distribution with a single mode (Table 18.3-1) ξ_{mode}, one has the stronger inequality

$$P[|x - \xi| \geq a] \leq \frac{4}{9}\,\frac{1 + \Sigma^2}{(a/\sigma - |\Sigma|)^2} \tag{18.3-14}$$

where Σ is Pearson's measure of skewness (Table 18.3-1); note that $\Sigma = 0$ if the distribution is symmetrical about the mode.

18.3-6. Improved Description of Probability Distributions: Use of Stieltjes Integrals. The treatment of discrete and continuous probability distributions is unified if one expresses the probability of each event $[X - \Delta X < x \leq X + \Delta X]$ as a Lebesgue-Stieltjes integral (Sec. 4.6-17)

$$P[X - \Delta X < x \leq X + \Delta X] \equiv \int_{X-\Delta X}^{X+\Delta X} d\Phi(x) \tag{18.3-15}$$

* In order to conform with the notation used in many textbooks, the *values* $x = X$ of a random variable x will be denoted simply by x whenever this notation does not lead to ambiguities.

where $\Phi(X) \equiv P[x \leq X]$ is the *cumulative distribution function* (Secs. 18.2-9, 18.3-1, and 18.3-2) defining the distribution of the random variable x. For continuous distributions the Stieltjes integral (15) reduces to a Riemann integral. For a discrete distribution, $\Phi(X)$ is given by Eq. (2), and $P[X - \Delta X < x \leq X + \Delta X]$ reduces to the function $p(X)$ defined in Sec. 18.3-1.

In terms of the Stieltjes-integral notation,

$$E\{x\} = \int_{-\infty}^{\infty} x \, d\Phi(x) \qquad E\{y(x)\} = \int_{-\infty}^{\infty} y(x) \, d\Phi(x)$$
$$\text{Var}\,\{x\} = \int_{-\infty}^{\infty} (x - \xi)^2 \, d\Phi(x) \tag{18.3-16}$$

for both discrete and continuous distributions. The Stieltjes-integral notation applies also to probability distributions which are partly discrete and partly continuous. An analogous notation is used for multidimensional distributions (Secs. 18.4-4 and 18.4-8).

Discrete distributions may be *formally* represented in terms of a "probability density" involving impulse functions $\delta_-(X - X_{(i)})$ (see also Secs. 18.3-1 and 21.9-6).

18.3-7. Moments of a One-dimensional Probability Distribution

(see also Secs. 18.3-6 and 18.3-10). **(a)** The **moment of order** $r \geq 0$ (r^{th} **moment**) **about** $x = X$ of a given random variable x is the mean value $E\{(x - X)^r\}$, if this quantity exists in the sense of absolute convergence (Sec. 18.3-3).

(b) In particular, the rth moment of x about $X = 0$ is

$$\alpha_r = E\{x^r\} = \int_{-\infty}^{\infty} x^r \, d\Phi(x) = \begin{cases} \sum_x x^r p(x) & (x \text{ discrete}) \\ \int_{-\infty}^{\infty} x^r \varphi(x) \, dx & (x \text{ continuous}) \end{cases} \tag{18.3-17}$$

and the rth moment of x about its mean value ξ (**central moment of order** r) is

$$\mu_r = E\{(x - \xi)^r\} = \int_{-\infty}^{\infty} (x - \xi)^r \, d\Phi(x) = \begin{cases} \sum_x (x - \xi)^r p(x) & (x \text{ discrete}) \\ \int_{-\infty}^{\infty} (x - \xi)^r \varphi(x) \, dx & (x \text{ continuous}) \end{cases} \tag{18.3-18}$$

The existence of α_r or μ_r implies the existence of all moments α_k and μ_k of order $k \leq r$; the divergence of α_r or μ_r implies the divergence of all moments α_k and μ_k of order $k \geq r$.

If the probability distribution is symmetric about its mean, all (existing) central moments μ_r of odd order r are equal to zero.

(c) The rth **factorial moment of x about $X = 0$** is

$$\alpha_{[r]} = E\{x^{[r]}\} = E\{x(x-1)\cdots(x-r+1)\} \tag{18.3-19}$$

The rth **central factorial moment of x** is $E\{(x-\xi)^{[r]}\}$. The rth **absolute moment of x about $X = 0$** is $\beta_r = E\{|x|^r\}$. Note

$$\left.\begin{array}{ll} E\{(x-X)^2\} = \mu_2 + (\xi - X)^2 \geq \mu_2 & \sqrt[r]{\beta_r} \leq \sqrt[r+1]{\beta_{r+1}} \\ \alpha_0 = \mu_0 = \alpha_{[0]} = \beta_0 = 1 & \mu_1 = 0 \end{array}\right\} \tag{18.3-20}$$

(d) *A one-dimensional probability distribution is uniquely defined by its moments* $\alpha_0, \alpha_1, \alpha_2, \ldots$ *if they all exist and are such that the series* $\sum_{k=0}^{\infty} \alpha_k s^k/k!$ *converges absolutely for some* $|s| > 0$ [see also Eq. (28) and the footnote to Sec. 18.3-8b].

(e) Refer to Tables 18.8-1 to 18.8-7 for examples, and to Sec. 18.3-10 for relations connecting the α_r, μ_r, and $\alpha_{[r]}$.

18.3-8. Characteristic Functions and Generating Functions (see also Sec. 18.3-6; refer to Tables 18.8-1 to 18.8-8 for examples).*

(a) The probability distribution of any one-dimensional random variable x uniquely defines its (generally complex-valued) **characteristic function**

$$\chi_x(q) \equiv E\{e^{iqx}\} \equiv \int_{-\infty}^{\infty} e^{iqx}\, d\Phi(x) \equiv \begin{cases} \sum_x e^{iqx} p(x) & (x \text{ discrete}) \\ \int_{-\infty}^{\infty} e^{iqx} \varphi(x)\, dx & (x \text{ continuous}) \end{cases} \tag{18.3-21}$$

where q is a real variable ranging between $-\infty$ and ∞.

(b) The probability distribution of a random variable x uniquely defines its **moment-generating function**

$$\mathrm{M}_x(s) \equiv E\{e^{sx}\} \equiv \int_{-\infty}^{\infty} e^{sx}\, d\Phi(x) \equiv \begin{cases} \sum_x e^{sx} p(x) & (x \text{ discrete}) \\ \int_{-\infty}^{\infty} e^{sx} \varphi(x)\, dx & (x \text{ continuous}) \end{cases} \tag{18.3-22}$$

and its **generating function** (see also Sec. 8.6-5)

* See footnote to Sec. 18.3-4.

$$\gamma_x(s) \equiv E\{s^x\} \equiv \int_{-\infty}^{\infty} s^x \, d\Phi(x) \equiv \begin{cases} \sum_x s^x p(x) & (x \text{ discrete}) \\ \int_{-\infty}^{\infty} s^x \varphi(x)\, dx & (x \text{ continuous}) \end{cases} \tag{18.3-23}$$

for each value of the complex variable s such that the function in question exists in the sense of absolute convergence.

(**c**) *The characteristic function* $\chi_x(q)$ *defines the probability distribution of* x *uniquely.** *The same is true for each of the functions* $M_x(s)$ *and* $\gamma_x(s)$ *if it exists, in the sense of absolute convergence, throughout an interval of the real axis including* $s = 0$ *in the case of* $M_x(s)$, *and* $s = 1$ *in the case of* $\gamma_x(s)$. Specifically, if x is a discrete or continuous random variable,

$$\begin{aligned} p(x) &= \lim_{Q\to\infty} \frac{1}{2Q} \int_{-Q}^{Q} e^{-iqx}\chi_x(q)\,dq & (x \text{ discrete}) \\ \varphi(x) &= \frac{1}{2\pi} \int_{-\infty}^{\infty} e^{-iqx}\chi_x(q)\,dq & (x \text{ continuous}) \end{aligned} \tag{18.3-24}$$

Eq. (24) also yields $p(x)$ or $\varphi(x)$ in terms of $M_x(s)$, since

$$M_x(iq) = \chi_x(q) \tag{18.3-25}$$

(**d**) In many problems it is much easier to obtain a description of a probability distribution in terms of $\chi_x(q)$, $M_x(s)$, or $\gamma_x(s)$ than to compute $\Phi(x)$, $p(x)$, or $\varphi(x)$ directly (Secs. 18.5-3*b*, 18.5-7, and 18.5-8). Again, the methods of Sec. 18.3-10 permit one to compute mean values, variances, and moments by simple differentiations if $\chi_x(q)$, $M_x(s)$, or $\gamma_x(s)$ are known. The linear integral transformations (21) to (24) can often be made with the aid of tables of Fourier or Laplace transform pairs (Appendix D).

(**e**) The generating function $\gamma_x(s)$ is particularly useful in problems involving discrete distributions with spectral values 0, 1, 2, . . . , for then

$$\gamma_x(s) = \sum_{x=0}^{\infty} s^x p(x) \qquad p(x) = \frac{1}{x!}\gamma_x^{(x)}(0) \qquad (x = 0, 1, 2, \ldots)$$

whenever the series converges (see also Sec. 18.8-1; see Ref. 18.4 for a number of interesting applications).

18.3-9. Semi-invariants (see also Sec. 18.3-10). Given a one-dimensional probability distribution such that the rth moment α_r exists, the first r **semi-invariants (cumulants)** $\kappa_0, \kappa_1, \kappa_2, \ldots, \kappa_r$ of the distribu-

* $\Phi(x)$ is, then, uniquely defined except possibly on a set of measure zero; $\Phi(x)$ is unique wherever it is continuous (see also Sec. 18.2-9).

tion exist and are defined by

$$\log_e \chi_x(q) = \sum_{k=1}^{r} \kappa_k \frac{(iq)^k}{k!} + o(q^r) \tag{18.3-26}$$

Under the conditions of Sec. 18.3-7d all semi-invariants $\kappa_1, \kappa_2, \ldots$ exist and define the distribution uniquely.

18.3-10. Computation of Moments and Semi-invariants from $\chi_x(q)$, $M_x(s)$, and $\gamma_x(s)$. Relations between Moments and Semi-invariants. Many properties of a distribution can be computed directly from $\chi_x(q)$, $M_x(s)$, or $\gamma_x(s)$ without previous computation of $\Phi(x)$, $\varphi(x)$, or $p(x)$. If the quantities in question exist,

$$\boxed{\begin{aligned} \alpha_r &= i^{-r}\chi_x^{(r)}(0) = M_x^{(r)}(0) \qquad \alpha_{[r]} = \gamma_x^{(r)}(1) \\ \kappa_r &= i^{-r}\frac{d^r}{dq^r}\log_e \chi_x(q)\bigg]_{q=0} = \frac{d^r}{ds^r}\log_e M_x(s)\bigg]_{s=0} \end{aligned}} \tag{18.3-27}$$

Note

$$M_x(s) = \sum_{k=0}^{\infty} \alpha_k \frac{s^k}{k!} \qquad \log_e M_x(s) = \sum_{k=0}^{\infty} \kappa_k \frac{s^k}{k!} \qquad \gamma_x(s+1) = \sum_{k=0}^{\infty} \alpha_{[k]} \frac{s^k}{k!} \tag{18.3-28}$$

provided that the function on the left (respectively the moment generating function, the **semi-invariant-generating function,** and the **factorial-moment-generating function** of x) is analytic throughout a neighborhood of $s = 0$.

Equations (28) yield $E\{x\}$ and Var $\{x\}$ with the aid of the relations

$$\boxed{\begin{aligned} E\{x\} &= \xi = \alpha_1 = \alpha_{[1]} = \kappa_1 \\ \text{Var}\,\{x\} &= \sigma^2 = \mu_2 = \alpha_2 - \xi^2 = \alpha_{[2]} - \xi(\xi - 1) = \kappa_2 \end{aligned}} \tag{18.3-29}$$

Table 18.3-1 lists other parameters which can be expressed in terms of moments.

The following additional formulas relate moments and semi-invariants:

$$\mu_r = \sum_{k=0}^{r} (-1)^k \binom{r}{k} \alpha_{r-k}\xi^k \qquad \alpha_r = \sum_{k=0}^{r} \binom{r}{k} \mu_{r-k}\xi^k \qquad (r = 0, 1, 2, \ldots) \tag{18.3-30}$$

$$\alpha_{[r]} = \sum_{k=0}^{r-1} S_k^r \alpha_{r-k} \qquad (r = 0, 1, 2, \ldots ; \text{ see also Sec. 21.5-3}) \tag{18.3-31}$$

$$\begin{aligned} \alpha_2 &= \alpha_{[2]} + \alpha_{[1]} &&= \kappa_2 + \kappa_1^2 \\ \alpha_3 &= \alpha_{[3]} + 3\alpha_{[2]} + \alpha_{[1]} &&= \kappa_3 + 3\kappa_1\kappa_2 + \kappa_1^3 \\ \alpha_4 &= \alpha_{[4]} + 6\alpha_{[3]} + 7\alpha_{[2]} + \alpha_{[1]} &&= \kappa_4 + 6\kappa_1^2\kappa_2 + 4\kappa_1\kappa_3 + 3\kappa_2^2 + \kappa_1^4 \end{aligned} \tag{18.3-32}$$

$$\mu_3 = \kappa_3 \qquad \mu_4 = \kappa_4 + 3\kappa_2^2 \tag{18.3-33}$$

$$\begin{aligned}\kappa_4 &= \mu_4 - 3\mu_2^2 \\ \kappa_5 &= \mu_5 - 10\mu_2\mu_3 \\ \kappa_6 &= \mu_6 - 15\mu_2\mu_4 - 10\mu_3^2 + 30\mu_2^3\end{aligned} \qquad (18.3\text{-}34)$$

18.4. MULTIDIMENSIONAL PROBABILITY DISTRIBUTIONS

18.4-1. Joint Distributions (see also Sec. 18.2-9). The probability distribution of a multidimensional random variable $\mathbf{x} \equiv (x_1, x_2, \ldots)$ is described as a **joint distribution** of real numerical random variables $x_1, x_2, \ldots$. Each simple event (point of the multidimensional sample space) $[\mathbf{x} = \mathbf{X}] \equiv [x_1 = X_1, x_2 = X_2, \ldots]$ may be regarded as a result of a compound experiment in which each of the variables $x_1, x_2, \ldots$ is measured. Each joint distribution is completely defined by its (cumulative) joint distribution function.

18.4-2. Two-dimensional Probability Distributions. Marginal Distributions. The joint distribution of two random variables x_1, x_2 is defined by its (cumulative) distribution function

$$\Phi_{\mathbf{x}}(X_1, X_2) \equiv \Phi(X_1, X_2) \equiv P[x_1 \leq X_1, x_2 \leq X_2] \qquad (18.4\text{-}1)$$

The distribution of x_1 and x_2 (**marginal distributions** derived from the joint distribution of x_1 and x_2) are described by the corresponding **marginal distribution functions**

$$\begin{aligned}\Phi_1(X_1) &\equiv P[x_1 \leq X_1] \equiv P[x_1 \leq X_1, x_2 \leq \infty] \equiv \Phi(X_1, \infty) \\ \Phi_2(X_2) &\equiv P[x_2 \leq X_2] \equiv P[x_1 \leq \infty, x_2 \leq X_2] \equiv \Phi(\infty, X_2)\end{aligned} \qquad (18.4\text{-}2)$$

18.4-3. Discrete and Continuous Two-dimensional Probability Distributions. (a) A two-dimensional random variable $\mathbf{x} \equiv (x_1, x_2)$ is a **discrete random variable** (has a **discrete probability distribution**) if and only if the **joint probability**

$$p_{\mathbf{x}}(X_1, X_2) \equiv p(X_1, X_2) \equiv P[x_1 = X_1, x_2 = X_2] \qquad (18.4\text{-}3)$$

is different from zero only for a countable set (**spectrum**) of "points" (X_1, X_2), i.e., if and only if both x_1 and x_2 are discrete random variables (Sec. 18.3-1). The **marginal probabilities** respectively associated with the marginal distributions of x_1 and x_2 (Sec. 18.4-2) are

$$\begin{aligned}p_1(X_1) &\equiv P[x_1 = X_1] \equiv \sum_{x_2} p(X_1, X_2) \\ p_2(X_2) &\equiv P[x_2 = X_2] \equiv \sum_{x_1} p(X_1, X_2)\end{aligned} \qquad (18.4\text{-}4)$$

(b) A two-dimensional random variable $\mathbf{x} \equiv (x_1, x_2)$ is a **continuous random variable** (has a **continuous probability distribution**) if and only if (1) $\Phi(X_1, X_2)$ is continuous for all X_1, X_2, and (2) the **joint frequency function (probability density)**

$$\varphi_{\mathbf{x}}(X_1, X_2) \equiv \varphi(X_1, X_2) \equiv \frac{\partial^2 \Phi(X_1, X_2)}{\partial X_1 \, \partial X_2} \tag{18.4-5}$$

exists and is piecewise continuous everywhere.* $\varphi(X_1, X_2)\, dx_1\, dx_2$ is called a **probability element.** The **spectrum** of a continuous two-dimensional probability distribution is the set of "points" (X_1, X_2) where the frequency function (5) is different from zero. The **marginal frequency functions** respectively associated with the (necessarily continuous) marginal distributions of x_1 and x_2 (Sec. 18.4-2) are

$$\begin{aligned} \varphi_1(X_1) &\equiv \frac{\partial \Phi_1(X_1)}{\partial X_1} \equiv \int_{-\infty}^{\infty} \varphi(x_1, x_2)\, dx_2 \\ \varphi_2(X_2) &\equiv \frac{\partial \Phi_2(X_2)}{\partial X_2} \equiv \int_{-\infty}^{\infty} \varphi(x_1, x_2)\, dx_1 \end{aligned} \tag{18.4-6}$$

(c) Note

$$\begin{aligned} &\sum_{x_1} \sum_{x_2} p(x_1, x_2) = \sum_{x_1} p_1(x_1) = \sum_{x_2} p_2(x_2) = 1 \\ &\int_{-\infty}^{-\infty} \int_{-\infty}^{-\infty} \varphi(x_1, x_2)\, dx_1\, dx_2 = \int_{-\infty}^{\infty} \varphi_1(x_1)\, dx_1 \\ &\qquad = \int_{-\infty}^{\infty} \varphi_2(x_2)\, dx_2 = 1 \end{aligned} \tag{18.4-7}$$

18.4-4. Expected Values, Moments, Covariance, and Correlation Coefficient. **(a)** The **expected value (mean value, mathematical expectation)** of a function $y = y(x_1, x_2)$ of two random variables x_1, x_2 with respect to their joint distribution is

$$\begin{aligned} E\{y(x_1, x_2)\} &= \int_{-\infty}^{\infty} \int_{-\infty}^{\infty} y(x_1, x_2)\, d\Phi(x_1, x_2) \\ &= \begin{cases} \displaystyle\sum_{x_1} \sum_{x_2} y(x_1, x_2) p(x_1, x_2) & \text{for discrete distributions} \\ \displaystyle\int_{-\infty}^{\infty} \int_{-\infty}^{\infty} y(x_1, x_2) \varphi(x_1, x_2)\, dx_1\, dx_2 & \text{for continuous distributions} \end{cases} \end{aligned} \tag{18.4-8}$$

* See footnote to Sec. 18.3-2.

if this expression exists in the sense of absolute convergence (see also Sec. 18.3-3).

NOTE: If y is a function of x_1 alone, the mean value (8) is identical with the mean value (*marginal expected value*) with respect to the marginal distribution of x_1.

(b) The mean values $E\{x_1\} = \xi_1$, $E\{x_2\} = \xi_2$ define a "point" (ξ_1, ξ_2) called the **center of gravity** of the joint distribution. The quantities $E\{(x_1 - X_1)^{r_1}(x_2 - X_2)^{r_2}\}$ are called **moments of order** $r_1 + r_2$ **about the "point"** (X_1, X_2). In particular, the quantities

$$\alpha_{r_1r_2} = E\{x_1^{r_1}x_2^{r_2}\} \qquad \mu_{r_1r_2} = E\{(x_1 - \xi_1)^{r_1}(x_2 - \xi_2)^{r_2}\} \tag{18.4-9}$$

are, respectively, the moments about the origin and the moments about the center of gravity (**central moments**) of order $r_1 + r_2$ (see also Sec. 18.3-7*b*).

(c) The *second-order central moments* are of special interest and warrant a special notation. Note the following definitions:

$$\begin{aligned}
\lambda_{11} &= E\{(x_1 - \xi_1)^2\} = \text{Var}\,\{x_1\} = \sigma_1^2 \\
\lambda_{22} &= E\{(x_2 - \xi_2)^2\} = \text{Var}\,\{x_2\} = \sigma_2^2 \\
&\qquad\qquad (\text{VARIANCES OF } x_1 \text{ AND } x_2) \\
\lambda_{12} &= \lambda_{21} = E\{(x_1 - \xi_1)(x_2 - \xi_2)\} = \text{Cov}\,\{x_1, x_2\} \\
&\qquad\qquad (\text{COVARIANCE OF } x_1 \text{ AND } x_2) \\
\rho_{12} &= \rho_{21} = \rho\{x_1, x_2\} = \frac{\lambda_{12}}{+\sqrt{\lambda_{11}\lambda_{22}}} \\
&= \frac{\text{Cov}\,\{x_1, x_2\}}{+\sqrt{\text{Var}\,\{x_1\}\,\text{Var}\,\{x_2\}}} = E\left\{\frac{x_1 - \xi_1}{\sigma_1}\,\frac{x_2 - \xi_2}{\sigma_2}\right\} \\
&\qquad\qquad (\text{CORRELATION COEFFICIENT OF } x_1 \text{ AND } x_2)
\end{aligned} \tag{18.4-10}$$

(see also Sec. 18.4-8). Note $-1 \leq \rho_{12} \leq 1$, and

$$\text{Cov}\,\{x_1, x_2\} = E\{x_1x_2\} - \xi_1\xi_2 = \rho_{12}\sigma_1\sigma_2 \tag{18.4-11}$$

18.4-5. Conditional Probability Distributions Involving Two Random Variables. **(a)** The joint distribution of two random variables x_1, x_2 defines a **conditional distribution of** x_1 **relative to the hypothesis that** $x_2 = X_2$ for each value X_2 of x_2 and a conditional distribution of x_2 relative to each hypothesis $x_1 = X_1$. The conditional distributions of x_1 and x_2 derived from a *discrete* joint distribution (Sec. 18.4-3*a*) are discrete and may be described by the respective *conditional probabilities* (Sec. 18.2-2)

$$\begin{aligned}
p_{1|2}(X_1|X_2) &\equiv P[x_1 = X_1|x_2 = X_2] \equiv \frac{p(X_1, X_2)}{p_2(X_2)} \\
p_{2|1}(X_2|X_1) &\equiv P[x_2 = X_2|x_1 = X_1] \equiv \frac{p(X_1, X_2)}{p_1(X_1)}
\end{aligned} \tag{18.4-12}$$

The conditional distributions of x_1 and x_2 derived from a *continuous* joint distribution (Sec. 18.4-3*b*) are continuous and may be described by the respective **conditional frequency functions**

$$\varphi_{1|2}(X_1|X_2) \equiv \frac{\varphi(X_1, X_2)}{\varphi_2(X_2)} \qquad \varphi_{2|1}(X_2|X_1) \equiv \frac{\varphi(X_1, X_2)}{\varphi_1(X_1)} \tag{18.4-13}$$

(**b**) Note

$$\sum_{x_1} p_{1|2}(x_1|x_2) = \sum_{x_2} p_{2|1}(x_2|x_1) = 1$$

$$\int_{-\infty}^{\infty} \varphi_{1|2}(x_1|x_2)\,dx_1 = \int_{-\infty}^{\infty} \varphi_{2|1}(x_2|x_1)\,dx_2 = 1 \tag{18.4-14}$$

$$\left.\begin{aligned}
p_{1|2}(X_1|X_2) &\equiv \frac{p_1(X_1)p_{2|1}(X_2|X_1)}{\sum\limits_{x_1} p_1(x_1)p_{2|1}(X_2|x_1)} \\
p_{2|1}(X_2|X_1) &\equiv \frac{p_2(X_2)p_{1|2}(X_1|X_2)}{\sum\limits_{x_2} p_2(x_2)p_{1|2}(X_1|x_2)} \\
\varphi_{1|2}(X_1|X_2) &\equiv \frac{\varphi_1(X_1)\varphi_{2|1}(X_2|X_1)}{\int_{-\infty}^{\infty} \varphi_1(x_1)\varphi_{2|1}(X_2|x_1)\,dx_1} \\
\varphi_{2|1}(X_2|X_1) &\equiv \frac{\varphi_2(X_2)\varphi_{1|2}(X_1|X_2)}{\int_{-\infty}^{\infty} \varphi_2(x_2)\varphi_{1|2}(X_1|x_2)\,dx_2}
\end{aligned}\right\} \begin{array}{l}\text{(Bayes's theorem,}\\ \text{see also Sec. 18.2-6)}\end{array} \tag{18.4-15}$$

(**c**) Given a discrete or continuous joint distribution of two random variables x_1 and x_2, the **conditional expected value of a function** $y(x_1, x_2)$ **relative to the hypothesis that** $x_1 = X_1$ is

$$E\{y(x_1, x_2)|X_1\} = \left\{\begin{array}{ll} \sum\limits_{x_2} y(X_1, x_2)p_{2|1}(x_2|X_1) & \\ & \text{for discrete distributions} \\ \int_{-\infty}^{\infty} y(X_1, x_2)\varphi_{2|1}(x_2|X_1)\,dx_2 & \\ & \text{for continuous distributions} \end{array}\right\} \tag{18.4-16}$$

if this expression exists in the sense of absolute convergence. Note that $E\{y(x_1, x_2)|X_1\}$ is a function of X_1.

EXAMPLE: The **conditional variances** of x_1 and x_2 are the respective functions

$$\text{Var}\,\{x_1|X_2\} = E\{(x_1 - E\{x_1|X_2\})^2|X_2\} \qquad \text{Var}\,\{x_2|X_1\} = E\{(x_2 - E\{x_2|X_1\})^2|X_1\} \tag{18.4-17}$$

18.4-6. Regression (see also Secs. 18.4-9 and 19.7-2). (**a**) Given the joint distribution of two random variables x_1 and x_2, a **regression of** x_2 **on** x_1 is any function $g_2(x_1)$ used to approximate the *statistical* dependence of x_2 on x_1 by a *deterministic* relation $x_2 \approx g_2(x_1)$. More specifically, x_2 is written as a sum of two random variables,

$$x_2 \equiv g_2(x_1) + h_2(x_1, x_2) \qquad (18.4\text{-}18)$$

where $h_2(x_1, x_2)$ is regarded as a correction term. In particular, the function

$$g_2(x_1) \equiv E\{x_2|x_1\} \quad \text{(MEAN-SQUARE REGRESSION OF } x_2 \text{ ON } x_1\text{, REGRESSION OF THE MEAN OF } x_2 \text{ ON } x_1) \qquad (18.4\text{-}19)$$

often simply called the *regression of* x_2 *on* x_1, minimizes the mean-square deviation

$$E\{[h_2(x_1, x_2)]^2\} = E\{[x_2 - g_2(x_1)]^2\} \qquad (18.4\text{-}20)$$

The corresponding curve $x_2 = E\{x_2|x_1\}$ is the (theoretical) **mean-square regression curve** of x_2.

(**b**) It is often sufficient to approximate the regression (19) by the linear function

$$g_2(x_1) \equiv g_2^{(1)}(x_1) \equiv \xi_2 + \beta_{21}(x_1 - \xi_1) \qquad \beta_{21} = \rho_{12}\frac{\sigma_2}{\sigma_1} \quad \text{(LINEAR MEAN-SQUARE REGRESSION OF } x_2 \text{ ON } x_1) \qquad (18.4\text{-}21)$$

Equation (21) describes a straight line, the **mean-square regression line** of x_2; β_{21} is the **regression coefficient** of x_2 on x_1. Equation (21) represents the linear function $ax_1 + b$ whose coefficients a, b minimize the mean-square deviation

$$E\{[x_2 - (ax_1 + b)]^2\} = \sigma_2^2 + a^2\sigma_1^2 - 2a\rho_{12}\sigma_1\sigma_2 + [\xi_2 - (a\xi_1 + b)]^2$$

The resulting minimum mean-square deviation is $\sigma_2^2(1 - \rho_{12}^2)$; *the correlation coefficient* ρ_{12} *is seen to measure the quality of the "best" linear approximation.*

(**c**) The mean-square regression (19) may be approximated more closely by a polynomial of degree m (**parabolic mean-square regression of order** m) or by other approximating functions, with coefficients or parameters chosen so as to minimize (20).

(**d**) If x_2 is regarded as the independent variable, one has similarly

$$g_1(x_2) \equiv E\{x_1|x_2\} \quad \text{(MEAN-SQUARE REGRESSION OF } x_1 \text{ ON } x_2) \qquad (18.4\text{-}22)$$

$$g_1^{(1)}(x_2) \equiv \xi_1 + \beta_{12}(x_2 - \xi_2) \qquad \beta_{12} = \rho_{12}\frac{\sigma_1}{\sigma_2} \quad \text{(LINEAR MEAN-SQUARE REGRESSION OF } x_1 \text{ ON } x_2) \qquad (18.4\text{-}23)$$

Note that in general neither (19) and (22) nor (21) and (23) are inverse functions. All mean-square regression curves and mean-square regression lines pass through the center of gravity (ξ_1, ξ_2) of the joint distribution.

The above definitions apply, in particular, if either of the two random variables, say $x_1 = t$, becomes a given independent variable, and $x_2(t)$ describes a random process (Sec. 18.9-1).

18.4-7. n-dimensional Probability Distributions. (**a**) The joint distribution of n random variables $x_1, x_2, \ldots, x_n$ is uniquely described by its (cumulative) joint distribution function

$$\begin{aligned}\Phi_{\mathbf{x}}(X_1, X_2, \ldots, X_n) &\equiv \Phi(X_1, X_2, \ldots, X_n) \\ &\equiv P[x_1 \leq X_1, x_2 \leq X_2, \ldots, x_n \leq X_n]\end{aligned} \tag{18.4-24}$$

(Sec. 18.2-9). The joint distribution of $m < n$ of the variables $x_1, x_2, \ldots, x_n$ is an m-dimensional **marginal distribution** derived from the original joint distribution. One obtains the corresponding **marginal distribution function** from the joint distribution function (24) by substituting $X_j = \infty$ for each of the $n - m$ arguments X_j which do not occur in the marginal distribution, e.g.,

$$\begin{aligned}\Phi_{12}(X_1, X_2) &\equiv \Phi(X_1, X_2, \infty, \ldots, \infty) \\ \Phi_2(X_2) &\equiv \Phi(\infty, X_2, \infty, \ldots, \infty) \equiv \Phi_{12}(\infty, X_2) \qquad \text{etc.}\end{aligned}$$

(**b**) An n-dimensional random variable $\mathbf{x} \equiv (x_1, x_2, \ldots, x_n)$ is a **discrete random variable** (has a **discrete probability distribution**) if and only if the **joint probability**

$$\begin{aligned}p_{\mathbf{x}}(X_1, X_2, \ldots, X_n) &\equiv p(X_1, X_2, \ldots, X_n) \\ &\equiv P[x_1 = X_1, x_2 = X_2, \ldots, x_n = X_n]\end{aligned} \tag{18.4-25}$$

differs from zero only for a countable set (**spectrum**) of "points" $(X_1, X_2, \ldots, X_n)$, i.e., if and only if each of the n random variables $x_1, x_2, \ldots, x_n$ is discrete (see also Secs. 18.3-1 and 18.4-3a).

Marginal probabilities and **conditional probabilities** are defined in the manner of Secs. 18.4-3a and 18.4-5a, e.g.,

$$\begin{aligned}p_{12}(X_1, X_2) &\equiv \sum_{x_3}\sum_{x_4}\cdots\sum_{x_n} p(X_1, X_2, x_3, \ldots, x_n) \\ p_2(X_2) &\equiv \sum_{x_1}\sum_{x_3}\cdots\sum_{x_n} p(x_1, X_2, x_3, \ldots, x_n) \equiv \sum_{x_1} p_{12}(x_1, X_2) \qquad \text{etc.}\end{aligned}$$

and

$$p_{1|2}(X_1|X_2) \equiv \frac{P_{12}(X_1, X_2)}{P_2(X_2)} \qquad p_{1|23}(X_1|X_2, X_3) \equiv \frac{P_{123}(X_1, X_2, X_3)}{P_{23}(X_2, X_3)}$$

$$p_{23|1}(X_2, X_3|X_1) \equiv \frac{p_{123}(X_1, X_2, X_3)}{P_1(X_1)} \qquad \text{etc.}$$

(**c**) An n-dimensional random variable $\mathbf{x} \equiv (x_1, x_2, \ldots, x_n)$ is a **continuous random variable** (has a **continuous probability dis-**

tribution if and only if (1) $\Phi(X_1, X_2, \ldots, X_n)$ is continuous for all $X_1, X_2, \ldots, X_n$ and (2) the **joint frequency function (probability density)**

$$\varphi_{\mathbf{x}}(X_1, X_2, \ldots, X_n) \equiv \varphi(X_1, X_2, \ldots, X_n) \equiv \frac{\partial^n \Phi(X_1, X_2, \ldots, X_n)}{\partial X_1\, \partial X_2 \cdots \partial X_n} \tag{18.4-26}$$

exists and is piecewise continuous everywhere.* $\varphi(X_1, X_2, \ldots, X_n)\, dx_1\, dx_2 \cdots dx_n$ is called a **probability element** (see also Secs. 18.3-2 and 18.4-3*b*). The **spectrum** of a continuous probability distribution is the set of "points" $(X_1, X_2, \ldots, X_n)$ where the frequency function (26) is different from zero.

(**d**) Note

$$\begin{gathered}\sum_{x_1}\sum_{x_2} \cdots \sum_{x_n} p(x_1, x_2, \ldots, x_n) = 1 \\ \int_{-\infty}^{\infty}\int_{-\infty}^{\infty} \cdots \int_{-\infty}^{\infty} \varphi(x_1, x_2, \ldots, x_n)\, dx_1\, dx_2 \cdots dx_n = 1\end{gathered} \tag{18.4-27}$$

(**e**) The frequency functions associated with the (necessarily continuous) *marginal and conditional distributions* derived from a continuous n-dimensional probability distribution are defined in the manner of Secs. 18.4-3*b* and 18.4-5*a*, e.g.,

$$\begin{aligned}\varphi_{12}(X_1, X_2) &\equiv \frac{\partial^2 \Phi_{12}(X_1, X_2)}{\partial X_1\, \partial X_2} \\ &\equiv \int_{-\infty}^{\infty}\int_{-\infty}^{\infty} \cdots \int_{-\infty}^{\infty} \varphi(X_1, X_2, x_3, \ldots, x_n)\, dx_3\, dx_4 \cdots dx_n \\ \varphi_{1|2}(X_1|X_2) &\equiv \frac{\varphi_{12}(X_1, X_2)}{\varphi_2(X_2)}\end{aligned}$$

$$\varphi_{1|23}(X_1|X_2, X_3) \equiv \frac{\varphi_{123}(X_1, X_2, X_3)}{\varphi_{23}(X_2, X_3)} \qquad \varphi_{12|3}(X_1, X_2|X_3) \equiv \frac{\varphi_{123}(X_1, X_2, X_3)}{\varphi_3(X_3)}$$

(**f**) The *joint distribution of two or more multidimensional random variables* $\mathbf{x} \equiv (x_1, x_2, \ldots)$, $\mathbf{y} \equiv (y_1, y_2, \ldots)$, . . . is the joint distribution of the random variables $x_1, x_2, \ldots ; y_1, y_2, \ldots ; \ldots$.

NOTE: A joint distribution may be discrete with respect to one or more of the random variables involved, and continuous with respect to one or more of the others; and each random variable may be partly discrete and partly continuous.

18.4-8. Expected Values and Moments (see also Sec. 18.4-4). (**a**) The **expected value (mean value, mathematical expectation)** of a function $y = y(x_1, x_2, \ldots, x_n)$ of n random variables $x_1, x_2, \ldots, x_n$ with respect to their joint distribution is

* See footnote to Sec. 18.3-2.

$$E\{y(x_1, x_2, \ldots, x_n)\} = \int_{-\infty}^{\infty}\int_{-\infty}^{\infty}\cdots\int_{-\infty}^{\infty} y(x_1, x_2, \ldots, x_n)\, d\Phi(x_1, x_2, \ldots, x_n)$$

$$= \begin{cases} \sum\limits_{x_1}\sum\limits_{x_2}\cdots\sum\limits_{x_n} y(x_1, x_2, \ldots, x_n)p(x_1, x_2, \ldots, x_n) & \text{(for discrete distributions)} \\ \int_{-\infty}^{\infty}\int_{-\infty}^{\infty}\cdots\int_{-\infty}^{\infty} y(x_1, x_2, \ldots, x_n)\,\varphi(x_1, x_2, \ldots, x_n)\, dx_1\, dx_2 \cdots dx_n & \text{(for continuous distributions)} \end{cases} \tag{18.4-28}$$

if this expression exists in the sense of absolute convergence.

NOTE: If y is a function of only $m < n$ of the n random variables $x_1, x_2, \ldots, x_n$, then the mean value (28) is identical with the mean value of y with respect to the joint distribution (marginal distribution, Sec. 18.4-7) of the m variables in question.

(**b**) The n mean values $E\{x_1\} = \xi_1$, $E\{x_2\} = \xi_2, \ldots E\{x_n\} = \xi_n$ define a "point" $(\xi_1, \xi_2, \ldots, \xi_n)$ called the **center of gravity** of the joint distribution. The quantities $E\{(x_1 - X_1)^{r_1}(x_2 - X_2)^{r_2} \cdots (x_n - X_n)^{r_n}\}$ are the **moments of order** $r_1 + r_2 + \cdots + r_n$ **about the "point"** $(X_1, X_2, \ldots, X_n)$. In particular, the quantities

$$\left.\begin{aligned} \alpha_{r_1r_2\cdots r_n} &= E\{x_1^{r_1}x_2^{r_2} \cdots x_n^{r_n}\} \\ \mu_{r_1r_2\cdots r_n} &= E\{(x_1 - \xi_1)^{r_1}(x_2 - \xi_2)^{r_2} \cdots (x_n - \xi_n)^{r_n}\} \end{aligned}\right\} \tag{18.4-29}$$

are, respectively, the moments about the origin and the moments about the center of gravity (**central moments**).

(**c**) The *second-order central moments* are again of special interest and warrant a special notation; the quantities

$$\lambda_{ik} = \lambda_{ki} = E\{(x_i - \xi_i)(x_k - \xi_k)\} = \begin{cases} \text{Var}\,\{x_i\} = \sigma_i^2 & \text{if } i = k \\ \text{Cov}\,\{x_i, x_k\} & \text{if } i \neq k \end{cases} \quad (i, k = 1, 2, \ldots, n) \tag{18.4-30}$$

define the **moment matrix** $[\lambda_{ik}] \equiv \Lambda$ and its reciprocal (Sec. 13.2-3)*

$$[\Lambda_{ik}] \equiv [\lambda_{ik}]^{-1} \equiv \Lambda^{-1}$$

det $[\lambda_{ik}]$ is the **generalized variance** of the joint distribution. The (total) **correlation coefficients**

* Note that some authors denote the cofactor matrix $[\lambda_{ik}]^{-1}$ det $[\lambda_{ik}]$ by $[\Lambda_{ik}]$. The notation chosen here simplifies some expressions.

$$\rho_{ik} = \rho\{x_i, x_k\} = \frac{\lambda_{ik}}{+\sqrt{\lambda_{ii}\lambda_{kk}}} = E\left\{\frac{x_i - \xi_i}{\sigma_i}\frac{x_k - \xi_k}{\sigma_k}\right\} \qquad (i, k = 1, 2, \ldots, n) \tag{18.4-31}$$

(see also Sec. 18.4-4*c*) define the **correlation matrix** $[\rho_{ik}]$ of the joint distribution. $+\sqrt{\det[\rho_{ik}]} = +\sqrt{\det[\lambda_{ik}]}/\sigma_1\sigma_2 \cdots \sigma_n$ is sometimes called the **scatter coefficient.**

The matrices $[\lambda_{ik}]$ and $[\rho_{ik}]$ are real, symmetric, and nonnegative (Secs. 13.3-2 and 13.5-2). Their common rank (Sec. 13.2-7) r is the **rank of the joint distribution.** The **ellipsoid of concentration** corresponding to a given n-dimensional probability distribution is the n-dimensional "ellipsoid"

$$\sum_{i=1}^{n}\sum_{k=1}^{n} \Lambda_{ik}x_ix_k = n + 2 \tag{18.4-32}$$

defined so that a uniform distribution of a unit probability "mass" inside the hypersurface has the moment matrix $[\lambda_{ik}]$. The ellipsoid of concentration illustrates the "concentration" of the distribution in different "directions"; the "volume" of the ellipsoid is proportional to the square root of the generalized variance. For $r < n$, the probability distribution is **singular:** its spectrum (Sec. 18.4-7) is restricted to an r-dimensional linear manifold (straight line, plane, hyperplane) in the n-dimensional space of "points" $(x_1, x_2, \ldots, x_n)$, and the same is true for its ellipsoid of concentration. Thus the spectrum of a two-dimensional probability distribution is restricted to a straight line if $r = 1$, and to a point if $r = 0$.

18.4-9. Regression. Multiple and Partial Correlation Coefficients (see also Secs. 18.4-6 and 19.7-2). **(a)** Given the joint distribution of n random variables $x_1, x_2, \ldots, x_n$, one may study the dependence of one of the variables, say x_1, on the remaining $n - 1$ variables by writing

$$x_1 = g_1(x_2, x_3, \ldots, x_n) + h_1(x_1, x_2, \ldots, x_n) \tag{18.4-33}$$

where $h_1(x_1, x_2, \ldots, x_n)$ is regarded as a correction term. The function

$$g_1(x_2, x_3, \ldots, x_n) \equiv E\{x_1|x_2, x_3, \ldots, x_n\}$$
$$= \begin{cases} \sum_{x_1} x_1 p_{1|23\cdots n}(x_1|x_2, x_3, \ldots, x_n) & \text{if } x_1 \text{ is discrete} \\ \int_{-\infty}^{\infty} x_1\varphi_{1|23\cdots n}(x_1|x_2, x_3, \ldots, x_n)\,dx_1 & \text{if } x_1 \text{ is continuous} \end{cases} \tag{18.4-34}$$

(mean-square regression of x_1 on $x_2, x_3, \ldots, x_n$) minimizes the mean-square deviation $E[x_1 - g_1(x_2, x_3, \ldots, x_n)]^2$; $E\{x_1|X_2, X_3, \ldots, X_n\}$ is the **conditional mean of x_1 relative to the hypothesis that** $x_2 = X_2, x_3 = X_3, \ldots, x_n = X_n$ (see also Sec. 18.4-5*c*).

(b) The mean-square regression of any variable x_i on the remaining $n - 1$ variables is often approximated by the linear function

$$g_i^{(1)} \equiv \xi_i + \sum_{k \neq i} \beta_{ik}(x_k - \xi_k) \qquad \beta_{ik} = -\frac{\Lambda_{ik}}{\Lambda_{ii}} \tag{18.4-35}$$

(LINEAR MEAN-SQUARE REGRESSION OF x_i ON THE REMAINING $n - 1$ VARIABLES)

(see also Sec. 18.4-6).* The **regression coefficients** β_{ik} are uniquely determined if the distribution is nonsingular (Sec. 18.4-8). The **multiple correlation coefficient**

$$\rho\{x_i, g_i^{(1)}\} = +\sqrt{1 - \frac{1}{\lambda_{ii}\Lambda_{ii}}} \tag{18.4-36}$$

is a measure of the correlation between x_i and the remaining $n - 1$ variables.

(c) The random variable $h_i^{(1)} \equiv x_i - g_i^{(1)}$ (difference between x_i and its "linear estimate" $g_i^{(1)}$ for $\Lambda_{ii} \neq 0$) is the **residual** of x_i with respect to the remaining $n - 1$ variables. Note

$$\text{Cov}\,\{h_i^{(1)}, x_k\} = 0 \quad (i \neq k) \qquad \text{Var}\,\{h_i^{(1)}\} = 1/\Lambda_{ii}$$

(RESIDUAL VARIANCE) (18.4-37)

(d) Regressions and residuals may be similarly defined in connection with a suitable *marginal distribution* (Sec. 18.4-7*a*) of $m < n$ variables, say $x_1, x_2, \ldots, x_m$. The quantities analogous to $\beta_{12}, \beta_{13}, \ldots ; h_1^{(1)}, h_2^{(1)}, \ldots$ are then respectively denoted by $\beta_{12\cdot34\cdots m}, \beta_{13\cdot24\cdots m}, \ldots ; h_{1\cdot23\cdots m}^{(1)}, h_{2\cdot13\cdots m}^{(1)}, \ldots$; in each case, there is a subscript corresponding to each variable of the marginal distribution.

(e) The **partial correlation coefficient of x_1 and x_2 with respect to $x_3, x_4, \ldots, x_n$,**

$$\rho_{12\cdot34\cdots n} = \rho\{h_{1\cdot34\cdots n}^{(1)}, h_{2\cdot34\cdots n}^{(1)}\} = -\frac{\Lambda_{12}}{\sqrt{\Lambda_{11}\Lambda_{12}}} \tag{18.4-38}$$

measures the correlation of x_1 and x_2 after removal of the linearly approximated effects of $x_3, x_4, \ldots, x_n$. In particular, for $n = 3$,

$$\rho_{12\cdot3} = \frac{\rho_{12} - \rho_{13}\rho_{23}}{\sqrt{(1 - \rho_{13}^2)(1 - \rho_{23}^2)}}$$

18.4-10. Characteristic Functions (see also Sec. 18.3-8). The probability distribution of an n-dimensional random variable $\mathbf{x} \equiv (x_1, x_2, \ldots, x_n)$ uniquely defines the corresponding **characteristic function (joint characteristic function** of $x_1, x_2, \ldots, x_n$)

* See footnote to Sec. 18.4-8*b*.

$$\chi_{\mathbf{x}}(\mathbf{q}) \equiv \chi_{\mathbf{x}}(q_1, q_2, \ldots, q_n) \equiv E\left\{\exp\left(i \sum_{k=1}^{n} q_k x_k\right)\right\}$$
$$\equiv \int_{-\infty}^{\infty} \int_{-\infty}^{\infty} \cdots \int_{-\infty}^{\infty} \exp\left[i \sum_{k=1}^{n} q_k x_k\right] d\Phi_{\mathbf{x}}(x_1, x_2, \ldots, x_n) \tag{18.4-39}$$

and conversely. For continuous distributions,

$$\varphi_{\mathbf{x}}(x_1, x_2, \ldots, x_n) \equiv \frac{1}{(2\pi)^n} \int_{-\infty}^{\infty} \int_{-\infty}^{\infty} \cdots$$
$$\int_{-\infty}^{\infty} \exp\left[-i \sum_{k=1}^{n} q_k x_k\right] \chi_{\mathbf{x}}(q_1, q_2, \ldots, q_n)\, dq_1\, dq_2 \cdots dq_n \tag{18.4-40}$$

The joint characteristic function corresponding to the marginal distribution of $m < n$ of the n variables $x_1, x_2, \ldots, x_n$ is obtained by substitution of $q_k = 0$ in Eq. (39) whenever x_k does not occur in the marginal distribution; thus $\chi_{12}(q_1, q_2) \equiv \chi_{\mathbf{x}}(q_1, q_2, 0, \ldots, 0)$.

Moments and *semi-invariants* of suitable multidimensional probability distributions can be obtained as coefficients in multiple series expansions of $\chi_{\mathbf{x}}$ and $\log_e \chi_{\mathbf{x}}$ in a manner analogous to that of Sec. 18.3-10.

18.4-11. Statistically Independent Random Variables (see also Secs. 18.2-3 and 18.5-7).* **(a)** A set of random variables $x_1, x_2, \ldots, x_n$ are **statistically independent** if and only if the events $[x_1 \in S_1]$, $[x_2 \in S_2], \ldots, [x_n \in S_n]$ are statistically independent for every collection of real-number sets $S_1, S_2, \ldots, S_n$. This is true if and only if

$$\Phi(x_1, x_2, \ldots, x_n) \equiv \Phi_1(x_1)\Phi_2(x_2) \cdots \Phi_n(x_n) \tag{18.4-41}$$

or, in the respective cases of discrete and continuous random variables, if and only if

$$\left.\begin{aligned} p(x_1, x_2, \ldots, x_n) &\equiv p_1(x_1)p_2(x_2) \cdots p_n(x_n) \\ \varphi(x_1, x_2, \ldots, x_n) &\equiv \varphi_1(x_1)\varphi_2(x_2) \cdots \varphi_n(x_n) \end{aligned}\right\} \tag{18.4-42}$$

The joint distribution of statistically independent random variables is completely defined by their individual marginal distributions. Statistically independent random variables $x_1, x_2, \ldots$ are **uncorrelated,** i.e., $\rho_{ik} = 0$ for all $i \neq k$ (Sec. 18.4-8*c*), but the converse is *not* necessarily true (see also Sec. 18.8-8).

(b) Statistical independence of *multidimensional* random variables $\mathbf{x}_1, \mathbf{x}_2, \ldots$ is defined by Eqs. (41) or (42) on substitution of $\mathbf{x}_1, \mathbf{x}_2, \ldots$ for $x_1, x_2, \ldots$.

* See footnote to Sec. 18.3-4.

EXAMPLE: The multidimensional random variables (x_1, x_2) and (x_3, x_4, x_5) are statistically independent if and only if

$$\Phi_{12345}(x_1, x_2, x_3, x_4, x_5) \equiv \Phi_{12}(x_1, x_2)\Phi_{345}(x_3, x_4, x_5) \tag{18.4-43}$$

Note that Eq. (43) implies the statistical independence of x_2 and x_5, x_1 and (x_3, x_4), (x_1, x_2) and (x_3, x_5), etc.

(c) Given a joint distribution of n discrete or continuous random variables $x_1, x_2, \ldots, x_n$ such that $(x_1, x_2, \ldots, x_m)$ is statistically independent of $(x_{m+1}, x_{m+2}, \ldots, x_n)$, note

$$\left.\begin{array}{l} p_{12\cdots m|m+1\cdots n}(x_1, x_2, \ldots, x_m|x_{m+1}, \ldots, x_n) \equiv p_{12\cdots m}(x_1, x_2, \ldots, x_m) \\ \text{or} \\ \varphi_{12\cdots m|m+1\cdots n}(x_1, x_2, \ldots, x_m|x_{m+1}, \ldots x_n) \equiv \varphi_{12\cdots m}(x_1, x_2, \ldots, x_m) \end{array}\right\} \tag{18.4-44}$$

(d) *Two random variables x_1 and x_2 are statistically independent if and only if their joint characteristic function is the product of their individual (marginal) characteristic functions* (*Sec.* 18.4-10), *i.e.*,

$$\chi_{12}(q_1, q_2) \equiv \chi_1(q_1)\chi_2(q_2) \tag{18.4-45}$$

An analogous theorem applies for multidimensional random variables (see also Sec. 18.5-7).

(e) *If the random variables $x_1, x_2, \ldots$ are statistically independent, the same is true for the random variables $y_1(x_1), y_2(x_2), \ldots$.* An analogous theorem holds for multidimensional random variables.

18.4-12. Entropy of a Probability Distribution, and Related Topics. **(a)** The **entropy** associated with the probability distribution of a one-dimensional random variable x is defined as

$$H\{x\} = \left\{\begin{array}{lll} E\left\{\log_2 \dfrac{1}{p(x)}\right\} = -\displaystyle\sum_x p(x) \log_2 p(x) & & (x \text{ discrete}) \\ E\left\{\log_2 \dfrac{1}{\varphi(x)}\right\} = -\displaystyle\int_{-\infty}^{\infty} \varphi(x) \log_2 \varphi(x)\, dx & & (x \text{ continuous}) \end{array}\right\} \tag{18.4-46}$$

$H\{x\}$ (entropy of x) is a measure of the *expected uncertainty* involved in a measurement of x. In the case of discrete probability distributions, $H\{x\} \geq 0$, with $H\{x\} = 0$ if and only if x has a causal distribution (Table 18.8-1). *The continuous distribution having the largest entropy for a given variance σ^2 is the normal distribution* (Sec. 18.8-3), with $H\{x\} = \log_2 \sqrt{2\pi e\sigma^2}$.

(b) In connection with the discrete or continuous joint distribution of two random variables x_1, x_2, one defines the **joint entropy**

$$H\{x_1, x_2\} = \left\{\begin{array}{ll} -E\{\log_2 p_{12}(x_1, x_2)\} & (\text{discrete distribution}) \\ -E\{\log_2 \varphi_{12}(x_1, x_2)\} & (\text{continuous distribution}) \end{array}\right\} \tag{18.4-47}$$

and the **conditional entropies**

$$H_{x_2}\{x_1\} = \left\{\begin{array}{ll} -E\{\log_2 p_{1|2}(x_1|x_2)\} & (\text{discrete distribution}) \\ -E\{\log_2 \varphi_{1|2}(x_1|x_2)\} & (\text{continuous distribution}) \end{array}\right\} \tag{18.4-48}$$

and $H_{x_1}\{x_2\}$ (these are *not* conditional expected values, Sec. 18.4-5c), so that

$$H\{x_1, x_2\} = H\{x_1\} + H_{x_1}\{x_2\} = H\{x_2\} + H_{x_2}\{x_1\} \leq H\{x_1\} + H\{x_2\} \tag{18.4-49}$$

The equality on the right applies if and only if x_1 and x_2 are statistically independent (Sec. 18.4-11). The nonnegative quantity

$$\begin{aligned} I\{x_1, x_2\} &= H\{x_1\} + H\{x_2\} - H\{x_1, x_2\} \\ &= H\{x_1\} - H_{x_2}\{x_1\} = H\{x_2\} - H_{x_1}\{x_2\} \end{aligned} \tag{18.4-50}$$

is a measure of the "statistical dependence" of x_1 and x_2. The functionals (46), (47), (48), and (50) have intuitive significance in statistical mechanics and in the theory of communications.

18.5. FUNCTIONS OF RANDOM VARIABLES. CHANGE OF VARIABLES

18.5-1. Introduction. The following relations permit one to calculate probability distributions of suitable functions of random variables and, in particular, to change the random variables employed to describe a given set of events.

18.5-2. Functions (or Transformations) of a One-dimensional Random Variable. (a) *Given a transformation $y = y(x)$ associating a unique value of a random variable y with each value of the random variable x, the probability distribution of y is uniquely determined by that of x* [see also Sec. 18.2-8; $y(x)$ must be a measurable function].

(b) *Let the random variables x and y be related by a reciprocal one-to-one transformation $y = y(x)$, with $x = x(y)$. Then*

1. *If $y(x)$ is an increasing function,*

$$\begin{matrix} \Phi_y(Y) \equiv \Phi_x[x(Y)] & \Phi_x(X) \equiv \Phi_y[y(X)] & \\ y_P = y(x_P) & x_P = x(y_P) & (0 < P < 1) \end{matrix} \tag{18.5-1}$$

Note that either $y(x)$ or $-y(x)$ is necessarily an increasing function. In either case, the *medians* $x_{1/2}$ and $y_{1/2}$ are related by $y_{1/2} = y(x_{1/2})$.

2. *If x and y are continuous random variables,*

$$\varphi_y(Y)|dy| = \varphi_x[x(Y)]|dx| \quad \text{or} \quad \varphi_y(Y) = \varphi_x[x(Y)]\left|\frac{dx}{dy}\right|_{y=Y} \tag{18.5-2}$$

for all values Y of y such that dx/dy exists and is continuous.

NOTE: If $x(y)$ is multiple-valued, one writes $\varphi_y(Y) = \varphi_1(Y) + \varphi_2(Y) + \cdots$, where $\varphi_1(Y), \varphi_2(Y), \ldots$ are the frequency functions obtained from Eq. (2) for the respective single-valued "branches" $x_1(y), x_2(y), \ldots$ of $x(y)$. EXAMPLE: If $y = x^2$, let $x_1(y) = +\sqrt{y}$, $x_2(y) = -\sqrt{y}$. Then

$$\left.\varphi_y(Y) = 0 \text{ if } Y < 0 \qquad \varphi_y(Y) = \frac{1}{2\sqrt{Y}}\varphi_x(\sqrt{Y}) + \frac{1}{2\sqrt{Y}}\varphi_x(-\sqrt{Y}) \qquad (Y \geq 0)\right\} \tag{18.5-3}$$

(c) For single-valued, measurable $y(x)$, $f(y)$,

$$E\{f(y)\} = E\{f[y(x)]\} = \int_{-\infty}^{\infty} f[y(x)]\, d\Phi_x(x) \tag{18.5-4}$$

whenever this expected value exists; note that neither reciprocal one-to-one correspondence nor differentiability has been assumed for $y(x)$. In particular, substitution of $f(y) \equiv e^{sy}$ in Eq. (4) yields the moment-generating function $M_y(s) \equiv E\{e^{sy}\}$, and substitution of $f(y) \equiv e^{iqy}$ produces the characteristic function $\chi_y(q) \equiv E\{e^{iqy}\}$ (Sec. 18.3-8). If the integrals can be calculated, one may then use Eq. (18.3-25) to find $\varphi_y(y)$ or $p_y(y)$.

EXAMPLE: Let

$$y = \sin(x + a)$$

where a is a constant, and x is uniformly distributed between 0 and 2π. Then

$$\begin{aligned}\chi_y(q) &\equiv \frac{1}{2\pi}\int_0^{2\pi} e^{iqy(x)}\, dx \equiv \frac{1}{\pi}\int_{-\pi/2}^{\pi/2} e^{iqy(x)}\, dx \\ &\equiv \frac{1}{\pi}\int_{-1}^{1} e^{iqy}\frac{dy}{\sqrt{1-y^2}} \equiv \int_{-\infty}^{\infty} e^{iqy}\varphi_y(y)\, dy\end{aligned}$$

where we have used the symmetry properties of $\sin x$ and the fact that $dy = \cos(x + a)\, dx = \sqrt{1 - y^2}\, dx$. It follows that

$$\varphi_y(y) = \begin{cases}\dfrac{1}{\pi}\dfrac{1}{\sqrt{1-y^2}} & (|y| < 1) \\ 0 & (|y| > 1)\end{cases}$$

(see also Sec. 18.11-1*b*).

(d) By an extension of the convolution theorem of Sec. 8.3-3 to bilateral Laplace transforms (Sec. 8.6-2), Eq. (4) can be rewritten as

$$E\{f(y)\} = \frac{1}{2\pi i}\int_{\sigma_1 - i\infty}^{\sigma_1 + i\infty} M_x(s)\left[\int_{-\infty}^{\infty} f[y(x)]e^{-sx}\, dx\right] ds \tag{18.5-5}$$

where the integration contour parallels the imaginary axis in a suitable absolute-convergence strip; the quantity in square brackets is seen to be the bilateral Laplace transform of $f[y(x)]$ (see also Secs. 8.6-2 and Table 8.6-1). The complex contour integral (5) may be easier to compute than the integral (4).

(e) Note that, in general, $E\{y(x)\} \neq y(E\{x\})$ (see also Sec. 18.5-3).

18.5-3. Linear Functions (or Linear Transformations) of a One-dimensional Random Variable.

(a) *If x is a continuous random variable, and $y = ax + b$, then*

$$\varphi_y(Y) \equiv \frac{1}{|a|}\varphi_x\left(\frac{Y - b}{a}\right) \tag{18.5-6}$$

(b) *If the mean values in question exist,*

$$E\{ax + b\} = aE\{x\} + b \qquad \text{Var}\,\{ax + b\} = a^2\,\text{Var}\,\{x\} \tag{18.5-7}$$

$$\left.\begin{aligned}E\{(ax + b)^r\} &= a^r\alpha_r + \binom{r}{1}a^{r-1}b\alpha_{r-1} + \cdots + b^r \\ \chi_{ax+b}(q) \equiv e^{ibq}\chi_x(aq) \qquad M_{ax+b}(s) &\equiv e^b M_x(as) \qquad \gamma_{ax}(s) \equiv \gamma_x(s^a)\end{aligned}\right\} \tag{18.5-8}$$

The semi-invariants (Sec. 18.3-9) $\kappa_i\dagger$ of $y = ax + b$ are related to the semi-invariants κ_i of x by $\kappa_1\dagger = a\kappa_1 + b$, $\kappa_r\dagger = a^r\kappa_r$ $(r > 1)$.

(c) Of particular interest is the linear *transformation to standard units*

$$x' = \frac{x - \xi}{\sigma} \qquad \text{with } E\{x'\} = 0, \text{ Var }\{x'\} = 1 \tag{18.5-9}$$

x' is called a **standardized random variable** (see also Sec. 18.8-3).

(d) If $y = y(x)$ is *approximately linear* throughout most of the spectrum of x, it is sometimes permissible to use the approximations

$$y \approx y(\xi) + (x - \xi)y'(\xi) \qquad E\{y(x)\} \approx y(\xi) \qquad \text{Var }\{y(x)\} \approx [y'(\xi)]^2 \text{ Var }\{x\} \tag{18.5-10}$$

where $y'(x) = dy/dx$.

18.5-4. Functions and Transformations of Multidimensional Random Variables.

(a) *If the random variables*

$$y_1 = y_1(x_1, x_2, \ldots, x_n) \qquad y_2 = y_2(x_1, x_2, \ldots, x_n) \qquad \cdots \tag{18.5-11}$$

are single-valued measurable functions of the n random variables $x_1, x_2, \ldots, x_n$ *for all* $x_1, x_2, \ldots, x_n$, *then the probability distribution of each random variable* y_i *is uniquely determined by the joint distribution of* $x_1, x_2, \ldots, x_n$, *and the same is true for each joint or conditional distribution involving a finite set of random variables* y_i.

Thus the distribution function of y_i and the joint distribution function of y_i and y_k are, respectively,

$$\Phi_{y_i}(Y_i) \equiv \underset{y_i(x_1,x_2,\ldots,x_n)\le Y_i}{\int\int\cdots\int} d\Phi_{\mathbf{x}}(x_1, x_2, \ldots, x_p) \tag{18.5-12}$$

$$\Phi_{y_iy_k}(Y_i, Y_k) \equiv \underset{\substack{y_i(x_1,x_2,\ldots,x_n)\le Y_i \\ y_k(x_1,x_2,\ldots,x_n)\le Y_k}}{\int\int\cdots\int} d\Phi_{\mathbf{x}}(x_1, x_2, \ldots, x_p) \tag{18.5-13}$$

(b) *If* $\mathbf{x} \equiv (x_1, x_2, \ldots, x_n)$ *and* $\mathbf{y} \equiv (y_1, y_2, \ldots, y_n)$ *are continuous random variables related by a reciprocal one-to-one (nonsingular) transformation* (11), *their respective frequency functions* $\varphi_{\mathbf{x}}(X_1, X_2, \ldots, X_n)$ *and* $\varphi_{\mathbf{y}}(Y_1, Y_2, \ldots, Y_n)$ *are related by*

$$\begin{aligned} \varphi_{\mathbf{y}}(Y_1, Y_2, \ldots, Y_n)&|dy_1\,dy_2 \cdots dy_n| \\ &= \varphi_{\mathbf{x}}(X_1, X_2, \ldots, X_n)|dx_1\,dx_2 \cdots dx_n| \end{aligned}$$

or

$$\begin{aligned} \varphi_{\mathbf{y}}(Y_1, Y_2, &\ldots, Y_n) \\ &= \varphi_{\mathbf{x}}(X_1, X_2, \ldots, X_n)\left|\frac{\partial(x_1, x_2, \ldots, x_n)}{\partial(y_1, y_2, \ldots, y_n)}\right| \end{aligned} \tag{18.5-14}$$

where

$$X_i = x_i(Y_1, Y_2, \ldots, Y_n) \qquad (i = 1, 2, \ldots, n)$$

for all $Y_1, Y_2, \ldots, Y_n$ *such that the Jacobian exists and is continuous.*

If $\mathbf{x}(\mathbf{y})$ is multiple-valued, $\varphi_{\mathbf{y}}(Y_1, Y_2, \ldots, Y_n)$ may be computed in a manner analogous to that outlined in Sec. 18.5-2*b*.

(c) For single-valued, measurable $y_i = y_i(x_1, x_2, \ldots, x_n)$ $(i = 1, 2, \ldots, m)$ and $f(y_1, y_2, \ldots, y_m)$,

$$E\{f(y_1, y_2, \ldots, y_m)\} = \int_{-\infty}^{\infty}\int_{-\infty}^{\infty}\cdots\int_{-\infty}^{\infty} f\, d\Phi_{\mathbf{x}}(x_1, x_2, \ldots, x_n) \tag{18.5-15}$$

whenever this expected value exists. As in Sec. 18.5-2*c*, neither reciprocal one-to-one correspondence nor differentiability has been assumed.

Table 18.5-1. Distribution of the Sum $x = x_1 + x_2 + \cdots + x_n$ of n Independent Random Variables (see also Secs. 18.5-7, 18.6-5, and 19.3-3)

The distribution of a sum $x = x_1 + x_2 + \cdots + x_n$ of n statistically independent random variables $x_1, x_2, \ldots, x_n$ with mean values $\xi_1, \xi_2, \ldots, \xi_n$ and variances $\sigma_1^2, \sigma_2^2, \ldots, \sigma_n^2$ is often conveniently approximated by a normal distribution (Sec. 18.8-3) plus correction terms even for small n (see also Sec. 18.6-5, Central Limit Theorem).

Let $\xi = E\{x\} = \xi_1 + \xi_2 + \cdots + \xi_n$, and let $\sigma^2 = \text{Var}\,\{x\} = \sigma_1^2 + \sigma_2^2 + \cdots + \sigma_n^2 = \mu_2, \mu_3, \mu_4, \ldots$ be the central moments of the sum x.

$\varphi_x(x)$ can be approximated by the Gram-Charlier series of Sec. 19.3-3 if the series converges (true, e.g., whenever $|x| > a$ implies $\varphi_x = 0$ for some finite real $a > 0$, as for all physical variables). In this case, Eq. (18.8-6) implies

$$\varphi_x(x) = \frac{1}{\sigma\sqrt{2\pi}} e^{-\frac{x'^2}{2}}\left[1 + \frac{1}{3!}\gamma_1(x'^3 - 3x') + \frac{1}{4!}\gamma_2(x'^4 - 6x'^2 + 3) + \cdots\right]\left(x' = \frac{x - \xi}{\sigma}\right)$$

where $\gamma_1 = \mu_3/\sigma^3$, $\gamma_2 = \mu_4/\sigma^4 - 3$ are the coefficients of skewness and excess of x (Table 18.3-1). The normal approximation is especially good if each x_i is symmetrically distributed about ξ_i, so that $\gamma_1 = 0$.

Substitution of $f = \exp\,(s_1y_1 + s_2y_2 + \cdots + s_my_m)$ yields the joint moment-generating function of $y_1, y_2, \ldots, y_m$, and substitution of $f = \exp\,(iq_1y_1 + iq_2y_2 + \cdots + iq_my_m)$ yields the joint characteristic function. Transform methods analogous to Eq. (5) may be useful. Such methods have been successfully applied to special random-process problems (Sec. 18.12-5).

(d) For any two random variables x_1, x_2,

$$E\{x_1x_2\} = E\{x_1\}E\{x_2\} - \text{Cov}\{x_1, x_2\} \qquad (18.5\text{-}16)$$

if this quantity exists. If $x_1, x_2, \ldots, x_n$ are *statistically independent,* then

$$E\{x_1x_2 \ldots x_n\} = E\{x_1\}E\{x_2\} \ldots E\{x_n\}$$

if this quantity exists.

(e) If $y = x_1x_2$, and $\varphi_{x_1}(x_1) = 0$ for $x_1 < 0$, then

$$\varphi_{x_1,y}(x_1, y) \equiv \varphi_{x_1,x_2}\left(x_1, \frac{y}{x_1}\right)\left|\frac{1}{x_1}\right|$$

(Sec. 18.5-4*b*), and

$$\varphi_y(y) \equiv \int_0^\infty \varphi_{x_1,y}(x_1, y)\, dx_1 \equiv \int_0^\infty \varphi_{x_1,x_2}\left(x_1, \frac{y}{x_1}\right)\left|\frac{1}{x_1}\right| dx_1 \qquad (18.5\text{-}17)$$

Other suitable functions $y = y(x_1, x_2)$ can be treated in a similar manner.

18.5-5. Linear Transformations (see also Secs. 14.5-1 and 14.6-1). For every nonsingular linear transformation

$$y_i = \eta_i + \sum_{k=1}^{n} a_{ik}(x_k - \xi_k) \qquad (i = 1, 2, \ldots, n)$$

$$[\text{in matrix form} \qquad y = \eta + A(x - \xi)] \qquad (18.5\text{-}18)$$

the respective joint distributions of $x_1, x_2, \ldots, x_n$ and $y_1, y_2, \ldots, y_n$ are of equal rank (Sec. 18.4-8*c*), and

$$E\{y_i\} = \eta_i \qquad (i = 1, 2, \ldots, n) \qquad (\text{in matrix form} \qquad E\{y\} = \eta) \qquad (18.5\text{-}19)$$

$$\lambda'_{ik} = E\{(y_i - \eta_i)(y_k - \eta_k)\} = \sum_{j=1}^{n}\sum_{h=1}^{n} \lambda_{jh}a_{ij}a_{kh} \qquad (i, k = 1, 2, \ldots, n)$$

$$(\text{in matrix form } \Lambda' = A\Lambda\tilde{A}) \qquad (18.5\text{-}20)$$

if the quantities in question exist. $\Lambda' \equiv [\lambda'_{ik}]$ is the moment matrix (Sec. 18.4-8*c*) of $(y_1, y_2, \ldots, y_n)$. The methods of Sec. 13.5-5 make it possible to find

1. An orthogonal transformation (18) such that the new moment matrix $[\lambda'_{ik}]$ (and hence also the correlation matrix $\left[\frac{\lambda'_{ik}}{\sqrt{\lambda'_{ii}\lambda'_{kk}}}\right]$) is *diagonal* (*transformation to uncorrelated variables* y_i).
2. A transformation (18) such that $\eta_1 = \eta_2 = \cdots = \eta_n = 0$ and $\lambda'_{ik} = \delta^i_k$ (*transformation to uncorrelated standardized variables* y_i (see also Secs. 18.8-6*b* and 18.8-8). The matrix $[E\{x_i^* x_k\}]$ must be nonsingular.

18.5-6. Mean and Variance of a Sum of Random Variables. (a) For any two (not necessarily statistically independent) random variables x_1, x_2

$$\begin{aligned} E\{x_1 \pm x_2\} &= E\{x_1\} \pm E\{x_2\} = \xi_1 \pm \xi_2 \\ \text{Var}\,\{x_1 \pm x_2\} &= \text{Var}\,\{x_1\} + \text{Var}\,\{x_2\} \pm 2\,\text{Cov}\,\{x_1, x_2\} \\ &= \sigma_1^2 + \sigma_2^2 \pm 2\rho_{12}\sigma_1\sigma_2 \quad \text{(VARIANCE LAW)} \end{aligned} \tag{18.5-21}$$

if the quantities in question exist.

(b) More generally,

$$\left.\begin{aligned} E\left\{a_0 + \sum_{i=1}^{n} a_i x_i\right\} &= a_0 + \sum_{i=1}^{n} a_i \xi_i \\ \text{Var}\left\{a_0 + \sum_{i=1}^{n} a_i x_i\right\} &= \sum_{i=1}^{n}\sum_{k=1}^{n} a_i a_k \rho_{ik}\sigma_i\sigma_k \quad \text{(GENERAL VARIANCE LAW)} \end{aligned}\right\} \tag{18.5-22}$$

(c) If $y = y(x_1, x_2, \ldots, x_n)$ is *approximately linear* throughout most of the joint spectrum of $(x_1, x_2, \ldots, x_n)$, it may be permissible to use the approximation

$$y = y(\xi_1, \xi_2, \ldots, \xi_n) + \sum_{k=1}^{n} \left.\frac{\partial y}{\partial x_k}\right]_{x_1=\xi_1, x_2=\xi_2, \ldots, x_n=\xi_n} (x_k - \xi_k)$$

and to compute approximate values of $E\{y\}$ and $\text{Var}\,\{y\}$ by means of Eqs. (19) and (20) (see also Sec. 18.5-7).

18.5-7. Sums of Statistically Independent Random Variables (refer to Sec. 18.8-9 for examples). (a) *If x_1 and x_2 are statistically independent random variables, then*

$$\begin{aligned} \Phi_{x_1+x_2}(X) &\equiv \Phi_1(X) * \Phi_2(X) \equiv \int_{-\infty}^{\infty} \Phi_2(X - x_1)\, d\Phi_1(x_1) \\ &\equiv \int_{-\infty}^{\infty} \Phi_1(X - x_2)\, d\Phi_2(x_2) \\ \chi_{x_1+x_2}(q) &\equiv \chi_1(q)\chi_2(q) \\ p_{x_1+x_2}(X) &\equiv p_1(X) * p_2(X) \equiv \sum_{x_1} p_2(X - x_1)p_1(x_1) \\ &\equiv \sum_{x_2} p_1(X - x_2)p_2(x_2) \quad (x_1, x_2 \text{ discrete}) \\ \varphi_{x_1+x_2}(X) &\equiv \varphi_1(X) * \varphi_2(X) \equiv \int_{-\infty}^{\infty} \varphi_2(X - x_1)\varphi_1(x_1)\, dx_1 \\ &\equiv \int_{-\infty}^{\infty} \varphi_1(X - x_2)\varphi_2(x_2)\, dx_2 \quad (x_1, x_2 \text{ continuous}) \end{aligned} \tag{18.5-23}$$

where the subscripts 1 and 2 refer to the respective distributions of x_1 and x_2 as in Secs. 18.4-2, 18.4-3, and 18.4-7 (see also Table 18.5-1).

(**b**) More generally, if $x = x_1 + x_2 + \cdots + x_n$ is the sum of $n < \infty$ statistically independent random variables $x_1, x_2, \ldots, x_n$,

$$\Phi_x(X) \equiv \Phi_1(X) * \Phi_2(X) * \cdots * \Phi_n(X) \qquad \chi_x(q) \equiv \chi_1(q)\chi_2(q) \cdots \chi_n(q) \tag{18.5-24}$$

and, if the quantities in question exist,

$$p_x(X) \equiv p_1(X) * p_2(X) * \cdots * p_n(X) \qquad \varphi_x(X) \equiv \varphi_1(X) * \varphi_2(X) * \cdots * \varphi_n(X) \tag{18.5-25}$$

$$\mathrm{M}_x(s) \equiv \mathrm{M}_1(s)\mathrm{M}_2(s) \cdots \mathrm{M}_n(s) \qquad \gamma_x(s) \equiv \gamma_1(s)\gamma_2(s) \cdots \gamma_n(s) \tag{18.5-26}$$

$$E\{x\} = \xi = \sum_{i=1}^{n} E\{x_i\} = \sum_{i=1}^{n} \xi_i \qquad \operatorname{Var}\{x\} = \sigma^2 = \sum_{i=1}^{n} \operatorname{Var}\{x_i\} = \sum_{i=1}^{n} \sigma_i^2 \tag{18.5-27}$$

$$E\{(x - \xi)^3\} = \sum_{i=1}^{n} E\{(x_i - \xi_i)^3\} \tag{18.5-28}$$

$$\kappa_r = \sum_{i=1}^{n} \kappa_r^{(i)} \tag{18.5-29}$$

where $\kappa_r^{(i)}$ is the rth-order semi-invariant of x_i. Equations (24) and (26) permit the computation of higher-order moments with the aid of the relations given in Sec. 18.3-10.

(**c**) The distribution of the sum $\mathbf{z} \equiv (z_1, z_2, \ldots) = \mathbf{x} + \mathbf{y}$ of two suitable statistically independent *multidimensional* random variables $\mathbf{x} \equiv (x_1, x_2, \ldots)$ and $\mathbf{y} \equiv (y_1, y_2, \ldots)$ is described by

$$\Phi_{\mathbf{z}}(Z_1, Z_2, \cdots) \equiv \int_{-\infty}^{\infty} \int_{-\infty}^{\infty} \cdots \int_{-\infty}^{\infty} \Phi_{\mathbf{y}}(Z_1 - x_1, Z_2 - x_2, \ldots)\, d\Phi_{\mathbf{x}}(x_1, x_2, \cdots) \tag{18.5-30}$$

$$\chi_{\mathbf{z}}(q_1, q_2, \ldots) \equiv \chi_{\mathbf{x}}(q_1, q_2, \ldots)\chi_{\mathbf{y}}(q_1, q_2, \ldots) \tag{18.5-31}$$

18.5-8. Compound Distributions. Let $x_1, x_2, \ldots$ be independent random variables each having the same probability distribution, and let k be a discrete random variable with spectral values 0, 1, 2, . . . ; let k be statistically independent of $x_1, x_2, \ldots$. If the generating functions $\gamma_{x_1}(s)$ and $\gamma_k(s)$ exist, the distribution of the sum $x = x_1 + x_2 + \cdots + x_k$ is given by its generating function

$$\gamma_x(s) \equiv \gamma_k[\gamma_{x_1}(s)] \tag{18.5-32}$$

18.6. CONVERGENCE IN PROBABILITY AND LIMIT THEOREMS

18.6-1. Sequences of Probability Distributions. Convergence in Probability (see also Sec. 18.6-2). A sequence of random variables $y_1, y_2, \ldots$ **converges in probability** to the random variable y (y_n converges in probability to y as $n \to \infty$) if and only if the probability that y_n differs from y by any finite amount converges to zero as $n \to \infty$, or

$$y_n \xrightarrow[\text{in } p]{} y \text{ as } n \to \infty \text{ if and only if} \quad \lim_{n \to \infty} P[|y - y_n| > \epsilon] = 0 \quad \text{for all } \epsilon > 0 \tag{18.6-1}$$

An m-dimensional random variable $\mathbf{y}_n$ converges in probability to the m-dimensional random variable $\mathbf{y}$ as $n \to \infty$ if and only if each component variable of $\mathbf{y}_n$ converges in probability to the corresponding component variable of $\mathbf{y}$.

If the m random variables $y_{n1}, y_{n2}, \ldots, y_{nm}$ converge in probability to the respective constants $a_1, a_2, \ldots, a_m$ as $n \to \infty$, then any function $g(y_{n1}, y_{n2}, \ldots, y_{nm})$ expressible as a positive power of a rational function of $y_{n1}, y_{n2}, \ldots, y_{nm}$ converges in probability to $g(a_1, a_2, \ldots, a_m)$, provided that this quantity is finite.

18.6-2. Limits of Distribution Functions, Characteristic Functions, and Generating Functions. Continuity Theorems. **(a)** *y_n converges in probability to y as $n \to \infty$ if and only if the sequence of distribution functions $\Phi_{y_n}(Y)$ converges to the limit $\Phi_y(Y)$ for all Y such that $\Phi_y(Y)$ is continuous.*

(b) *y_n converges in probability to y as $n \to \infty$ if and only if the sequence of characteristic functions $\chi_{y_n}(q)$ converges to a limit continuous for $q = 0$; in this case $\chi_y(q) \equiv \lim_{n\to\infty} \chi_{y_n}(q)$ (Continuity Theorem for Characteristic Functions).*

(c) *A sequence of discrete random variables $y_1, y_2, \ldots$ converges in probability to the discrete random variable y as $n \to \infty$ if and only if*

$$\lim_{n\to\infty} p_{y_n}(Y) \equiv p_y(Y) \tag{18.6-2}$$

If the random variables $y_1, y_2, \ldots$ all have nonnegative integral spectral values $0, 1, 2, \ldots$ and possess generating functions $\gamma_{y_1}(s), \gamma_{y_2}(s), \ldots$, then Eq. (2) holds if and only if $\lim \gamma_{y_n}(s) = \gamma_y(s)$ for all real s such that $0 \leq s < 1$ (Continuity Theorem for Generating Functions). Note that a sequence of discrete random variables may converge in probability to a random variable which is not discrete (see, for example, Table 18.8-3).

(d) Analogous definitions apply if $y(n)$ converges in probability as a function of a *continuous* parameter n.

(e) Analogous theorems apply to multidimensional probability distributions.

18.6-3. Convergence in Mean (see also Sec. 12.5-3). Given a random variable y having a finite mean and variance and a sequence of random variables $y_1, y_2, \ldots$ all having finite mean values and variances, y_n **converges in mean (in mean square)** to y as $n \to \infty$ if and only if

$$\lim_{n\to\infty} E\{y - y_n\} = 0 \qquad \lim_{n\to\infty} \text{Var}\,\{y - y_n\} = 0 \tag{18.6-3}$$

Convergence in mean implies convergence in probability, but the converse is not true; $y_n \xrightarrow[\text{in } p]{} y$ *as* $n \to \infty$ does not even imply that $E\{y\}$ or Var $\{y\}$ exists.

18.6-4. Asymptotically Normal Probability Distributions (refer to Table 18.8-3 and Sec. 19.5-3 for examples). The (probability distribution of a) random variable y_n with the distribution function $\Phi_{y_n}(Y, n)$ is **asymptotically normal with mean η_n and variance σ_n^2** if and only if there exists a sequence of pairs of real numbers η_n, σ_n^2 such that the random variable $(y_n - \eta_n)/\sigma_n$ converges in probability to a standardized normal variable (Sec. 18.8-3). This is true if and only if for all $a, b > a$

$$\lim_{n\to\infty} P[\eta_n + a\sigma_n < y_n \leq \eta_n + b\sigma_n] = \Phi_u(b) - \Phi_u(a) \tag{18.6-4}$$

Equation (4) permits one to approximate the probability distribution of y_n by a normal distribution with mean η_n and variance σ_n^2 for sufficiently large n. Note that Eq. (4) does *not* imply that η_n and σ_n^2 are the mean and variance of y_n, that the sequence $y_1, y_2, \ldots$ converges in probability, or that $E\{y_n\}$ and η_n, or Var $\{y_n\}$ and σ_n^2 converge to the same limits; indeed, these limits may not exist.

18.6-5. Limit Theorems. **(a)** *For every class of events E permitting the definition of probabilities P[E]* (Sec. 18.2-2)

1. *The relative frequency* $h[E] = n_E/n$ (Sec. 19.2-1) *of realizing the event E in n independent repeated trials* (Sec. 18.2-4) *is a random variable which converges to P[E] in mean, and thus also in probability, as* $n \to \infty$ *(Bernoulli's Theorem).*
2. $h[E]$ *is asymptotically normal with mean P[E] and variance* $\frac{1}{n}P[E]\{1 - P[E]\}$ (see also Table 18.8-3).

Note that (see also Table 18.8-3)*

$$\left.\begin{aligned} p(n_E) &= \binom{n}{n_E}\{P[E]\}^{n_E}\{1 - P[E]\}^{n-n_E} \qquad (n_E = 0, 1, 2, \ldots, n) \\ E\left\{\frac{n_E}{n}\right\} &= P[E] \qquad \text{Var}\left\{\frac{n_E}{n}\right\} = \frac{1}{n}P[E]\{1 - P[E]\} \end{aligned}\right\} \tag{18.6-5}$$

(b) *Let* $x_1, x_2, \ldots$ *be a sequence of statistically independent random variables all having the same probability distribution with (finite) mean value* ξ. *Then, as* $n \to \infty$,

1. *The random variable* $\bar{x} = \frac{1}{n}(x_1 + x_2 + \cdots + x_n)$ *converges in probability to* ξ *(Khinchine's Theorem, Law of Large Numbers).*
2. $\bar{x}$ *is asymptotically normal with mean* ξ *and variance* σ^2/n, *provided that the common variance* σ^2 *of* $x_1, x_2, \ldots$ exists *(Lindeberg-Lévy Theorem, Central Limit Theorem;* see also Secs. 19.2-3 and 19.5-2).

(c) *Let* $x_1, x_2, \ldots$ *be any sequence of statistically independent random variables having (finite) mean values* $\xi_1, \xi_2, \ldots$ *and variances* $\sigma_1^2, \sigma_2^2, \ldots$. *Then, as* $n \to \infty$,

1. $\sigma_n^2 \to 0$ *implies* $x_n - \xi_n \xrightarrow[\text{in } p]{} 0$ *(Chebyshev's Theorem).*

* See footnote to Sec. 18.3-4.

2. $\frac{1}{n}\sum_{i=1}^{n} x_i - \frac{1}{n}\sum_{i=1}^{n} \xi_i = \frac{1}{n}\sum_{i=1}^{n} (x_i - \xi_i) \underset{\text{in } p}{\longrightarrow} 0$ *provided that*

$$\lim_{n\to\infty} \frac{1}{n^2}\sum_{i=1}^{n} \sigma_i^2 = 0 \qquad \textit{(Law of Large Numbers).}$$

3. *The random variable* $\sum_{i=1}^{n} x_i$ *is asymptotically normal with mean* $\sum_{i=1}^{n} \xi_i$ *and variance* $\sum_{i=1}^{n} \sigma_i^2$, *provided that, for every positive real number* ϵ,

$$\lim_{n\to\infty} \frac{\operatorname{Var}\left\{\sum_{i=1}^{n} z_i\right\}}{\sum_{i=1}^{n} \sigma_i^2} = 1 \qquad \text{where } z_i = \begin{cases} x_i \text{ if } x_i^2 \le \epsilon \sum_{i=1}^{n} \sigma_i^2 \\ 0 \text{ if } x_i^2 > \epsilon \sum_{i=1}^{n} \sigma_i^2 \end{cases} \tag{18.6-6}$$

(Central Limit Theorem, Lindeberg conditions).

The Lindeberg conditions are satisfied, in particular, if there exist two positive real numbers a and b such that $E\{|x_i|^{2+a}\}$ exists and is less than $b\sigma_i^2$ for $i = 1, 2, \ldots$ *(Lyapunov's Condition).* See also Table 18.5-1.

NOTE: The limit theorems are of special importance in statistics (Secs. 19.2-1 and 19.2-3).

18.7. SPECIAL TECHNIQUES FOR SOLVING PROBABILITY PROBLEMS

18.7-1. Introduction. Most probability problems require one to compute the distribution of a random variable x (or the distributions of several random variables) from given conditions specifying the distributions of other random variables $x_1, x_2, \ldots$. As a rule, the simple events labeled by values of x are compound events corresponding to various logical combinations of values of $x_1, x_2, \ldots$. *The first step in the solution of any such problem must be the unequivocal definition of the fundamental probability set labeled by each random variable.* The probabilities of compound events may then be computed by the methods of Secs. 18.2-2 to 18.2-6 and 18.5-1 to 18.7-3. Equation (18.3-3), (18.3-6), (18.4-7), or (18.4-27) may be used to check computations.

18.7-2. Problems Involving Discrete Probability Distributions: Counting of Simple Events and Combinatorial Analysis. Each fundamental probability set labeled by the spectral values of a discrete random variable (Sec. 18.3-1) is a countable set of simple events. The following relations (either alone or in combination with the relations of Secs. 18.2-2 to 18.2-6) aid in computing probabilities of compound events:

(**a**) If, as in many games of chance, *equal* probabilities are assigned to each of the N simple events of a given *finite* fundamental probability set, then the probability of realizing a compound event ("success") defined as the union (Sec. 18.2-1) of N_1 specified simple events ("favorable" simple events) can be computed as

$$\text{Probability of success} = \frac{\text{number of favorable simple events}}{\text{total number of simple events}} = \frac{N_1}{N} \tag{18.7-1}$$

(**b**) Given a countable (finite or infinite) fundamental probability set, let an event E be defined as the union of N_1 simple events each having the probability p_1, N_2 simple events each having the probability p_2, . . . ; then

$$P[E] = N_1p_1 + N_2p_2 + \cdots \tag{18.7-2}$$

$N_1 + N_2 + \cdots$ need not be finite.

(**c**) Given N_1 simple events E', N_2 simple events E'', . . . , and N_n simple events $E^{(n)}$ respectively associated with n independent component experiments (Sec. 18.2-4), there exist exactly $N_1N_2 \cdots N_n$ simple experiments $[E' \cap E'' \cap \cdots \cap E^{(n)}] \equiv [E', E'', \cdots, E^{(n)}]$.

(**d**) In many problems, the simple events under consideration are various possible arrangements of a given set or sets of elements, so that the numbers N_1, N_2, . . . in (*a*), (*b*), and (*c*) above are numbers of *permutations, combinations, etc. The most important relevant definitions and formulas are given in Appendix C.*

18.7-3. Problems Involving Discrete Probability Distributions: Successes and Failures in Component Experiments. Compound events are often described in terms of the results obtained in component experiments each admitting only two possible outcomes ("success" and "failure"). The probabilities of various compound events can be computed by the methods of Secs. 18.2-2 to 18.2-6 from the respective probabilities ϑ_1, ϑ_2, . . . of success in the first, second, . . . component experiment.

The methods of Secs. 18.5-6 to 18.5-8 may become applicable if one labels the events "success" and "failure" in the kth-component experiment with the respective spectral values 1 and 0 of a discrete random variable x_k whose distribution is described by

$$p_{x_k}(1) = \vartheta_k \qquad p_{x_k}(0) = 1 - \vartheta_k \tag{18.7-3}$$

$$E\{x_k\} = \vartheta_k \qquad \text{Var}\,\{x_k\} = \vartheta_k(1 - \vartheta_k) \tag{18.7-4}$$

$$\chi_{x_k}(q) \equiv (1 - \vartheta_k) + \vartheta_k e^{iq} \qquad M_{x_k}(s) \equiv (1 - \vartheta_k) + \vartheta_k e^{s} \qquad \gamma_{x_k}(s) \equiv (1 - \vartheta_k) + \vartheta_k s \tag{18.7-5}$$

Successes in two or more *independent* experiments are, by definition, statistically independent events (Sec. 18.2-4). Repeated independent trials (Sec. 18.2-4) each having only two possible outcomes are called **Bernoulli trials** ($\vartheta_1 = \vartheta_2 = \cdots = \vartheta$). The probability of realizing exactly $x = x_1 + x_2 + \cdots + x_n$ successes in n Bernoulli trials is given by the **binomial distribution** (Table 18.8-3). If the trials are independent, but the ϑ_k are not all equal, one obtains the **generalized binomial distribution of Poisson.**

A subsequence of r successes or failures in any sequence of n trials is called a **run of length** r of successes or failures (see also Ref. 18.4, Chap. 13).

18.8. SPECIAL PROBABILITY DISTRIBUTIONS

18.8-1. Discrete One-dimensional Probability Distributions.* Tables 18.8-1 to 18.8-7 describe a number of discrete one-dimensional distributions of interest, for instance, in connection with sampling problems and games of chance. The generating function rather than the characteristic function or the moment-generating function is tabulated: the latter two functions are easily obtained from

$$\chi_x(q) \equiv \gamma_x(e^{iq}) \qquad \mathrm{M}_x(s) \equiv \gamma_x(e^s)$$

(see also Sec. 18.3-8). Moments not tabulated are also easily derived by the methods of Sec. 18.3-10.

Table 18.8-1. The Causal Distribution (see also Table 18.8-8)

(a) $p(x) = \delta_\xi^x = \begin{Bmatrix} 1 \text{ if } x = \xi \\ 0 \text{ if } x \neq \xi \end{Bmatrix}$ $\quad (x = 0, \pm 1, \pm 2, \ldots ; \xi \text{ is a real integer})$

(b) $E\{x\} = \xi \qquad \mathrm{Var}\,\{x\} = 0 \qquad \gamma_x(s) \equiv s^\xi$

Table 18.8-2. The Hypergeometric Distribution

(a) $$p(x) = \frac{\dbinom{N_1}{x}\dbinom{N - N_1}{n - x}}{\dbinom{N}{n}} \qquad (x = 0, 1, 2, \ldots , n;\ N \geq n \geq 0,\ N \geq N_1 = \vartheta N \geq 0)$$

(b) $$E\{x\} = \frac{nN_1}{N} = n\vartheta \qquad \mathrm{Var}\,\{x\} = \frac{nN_1(N - N_1)}{N^2}\left(1 - \frac{n-1}{N-1}\right) = n\vartheta(1 - \vartheta)\left(1 - \frac{n-1}{N-1}\right)$$

(c) Typical Interpretation. $p(x)$ is the probability that a random sample of size n *without replacement* (Sec. 19.5-5) contains exactly x objects of type 1, if the sample is taken from a population of N objects of which $N_1 = \vartheta N$ are of type 1.

(d) Approximations. As $N \to \infty$ while n and $\vartheta = N_1/N$ remain fixed, the hypergeometric distribution approaches a *binomial distribution* (Table 18.8-3; sampling with and without replacement becomes approximately equivalent if n/N is small). The binomial approximation is usually permissible if $n/N < 0.1$. The binomial distribution may, in turn, be suitable for approximation by a *normal distribution* (see Table 18.8-3) or by a *Poisson distribution* (Table 18.8-4).

* See footnote to Sec. 18.3-4.

Table 18.8-3. The Binomial Distribution (Fig. 18.8-1; see also Sec. 18.7-3)

(a) $p(x) = \binom{n}{x} \vartheta^x (1 - \vartheta)^{n-x} \qquad (x = 0, 1, 2, \ldots; 0 \leq \vartheta \leq 1)$

$p(x)$ is largest when x equals the largest integer $\leq (n + 1)\vartheta$, so that the sequence $p(0), p(1), p(2), \ldots$ *increases monotonically for* $\vartheta > n/(n + 1)$ and *decreases monotonically for* $\vartheta < 1/(n + 1)$; otherwise the binomial distribution is unimodal (Fig. 18.8-1). Note also

$$\Phi_x(x) = \sum_{i=0}^{x} \binom{n}{i} \vartheta^i (1 - \vartheta)^{n-i} = 1 - I_\vartheta\left(\frac{m_1}{2}, \frac{m_2}{2}\right) = 1 - \Phi_{v^2(m_1,m_2)}\left(\frac{m_2}{m_1}\frac{\vartheta}{1 - \vartheta}\right)$$

$$(x = 0, 1, 2, \ldots)$$

with $m_1 = 2(x + 1)$, $m_2 = 2(n - x)$ (refer to Secs. 19.5-3 and 21.4-5).

(b) $E\{x\} = n\vartheta \qquad \text{Var}\{x\} = n\vartheta(1 - \vartheta) \qquad \gamma_x(s) = (\vartheta s + 1 - \vartheta)^n$

$$\alpha_2 = n\vartheta + n(n - 1)\vartheta^2 \qquad \alpha_3 = n(n - 1)(n - 2)\vartheta^3 + 3n(n - 1)\vartheta^2 + n\vartheta$$

$$\mu_3 = n\vartheta(1 - \vartheta)(1 - 2\vartheta) \qquad \mu_4 = n\vartheta(1 - \vartheta)[1 + 3(n - 2)\vartheta(1 - \vartheta)]$$

$$\gamma_1 = \frac{1 - 2\vartheta}{\sqrt{n\vartheta(1 - \vartheta)}} \qquad \gamma_2 = \frac{1 - 6\vartheta(1 - \vartheta)}{n\vartheta(1 - \vartheta)}$$

(c) Typical Interpretation. $p(x)$ is

1. The probability that a random sample of size n *with replacement* (Sec. 19.5-5) contains exactly x objects of type 1 if the sample is taken from a population of N objects of which ϑN are of type 1.
2. The probability of realizing an event ("success") exactly x times in n independent repeated trials (Bernoulli trials, Sec. 18.7-3) such that the probability of success in each trial is ϑ.

(d) Approximations. As $n \to \infty$, the binomial variable x is asymptotically normal with mean $n\vartheta = \xi$, and variance $n\vartheta(1 - \vartheta) = \sigma^2$ (*De Moivre-Laplace Limit Theorem;* a special case of the Central Limit Theorem, Sec. 18.6-5).

For $0 < \vartheta < 1$,

$$p(x) = \binom{n}{x} \vartheta^x (1 - \vartheta)^{n-x} \simeq \varphi_u\left(\frac{x - \xi}{\sigma}\right)$$

$$\simeq \Phi_u\left(\frac{x + \frac{1}{2} - \xi}{\sigma}\right) - \Phi_u\left(\frac{x - \frac{1}{2} - \xi}{\sigma}\right) \qquad \text{as } \frac{(x - \xi)^3}{\sigma^4} \to 0$$

$$P[X_1 \leq x \leq X_2] \simeq \Phi_u\left(\frac{X_2 + \frac{1}{2} - \xi}{\sigma}\right) - \Phi_u\left(\frac{X_1 - \frac{1}{2} - \xi}{\sigma}\right)$$

$$\text{as } \frac{(X_1 - \xi)^3}{\sigma^4} \to 0, \ \frac{(X_2 - \xi)^3}{\sigma^4} \to 0$$

$$P\left[a \leq \frac{x - \xi}{\sigma} \leq b\right] \to \Phi_u(b) - \Phi_u(a) \qquad \text{as } n \to \infty \text{ for fixed } a, b$$

Approximations based on these relations are usually permissible for

$$\sigma^2 = n\vartheta(1 - \vartheta) \geq 9$$

See Refs. 18.4 and 19.8 for discussions of the resulting errors. See Table 18.8-4 for the *Poisson-distribution approximation* to the binomial distribution.

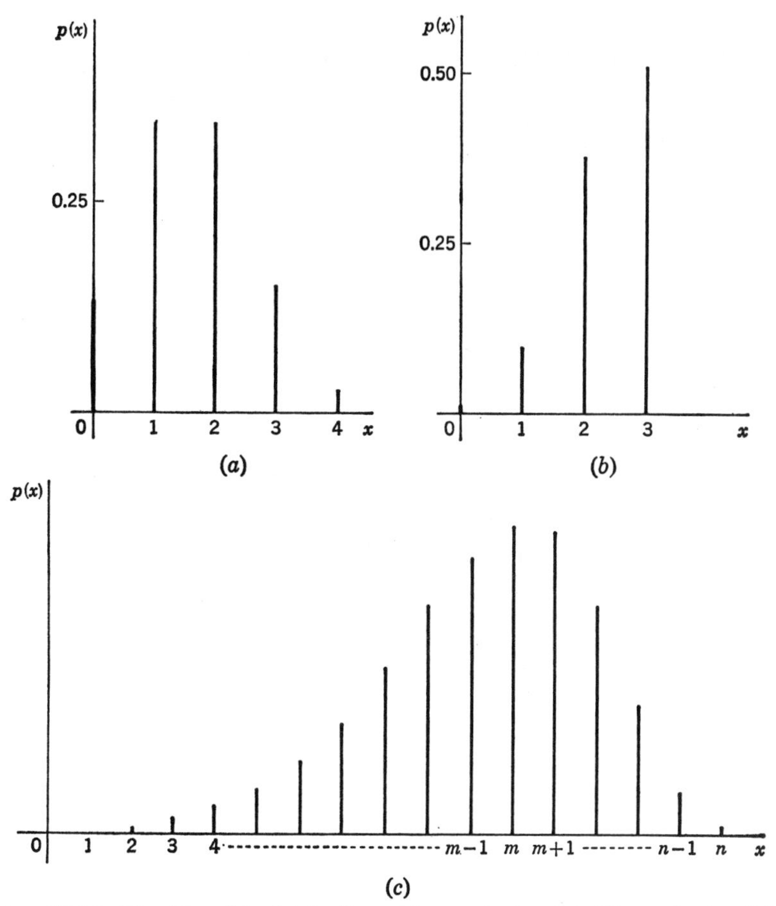

FIG. 18.8-1. The binomial distribution: (*a*) $n = 4$, $\vartheta = 0.4$; (*b*) $n = 3$, $\vartheta = 0.8$; (*c*) $n = 16$, $\vartheta = 0.7$. m is the mode. (*From Mood, A.M., Introduction to the Theory of Statistics, McGraw-Hill, New York,* 1950.)

18.8-2. Discrete Multidimensional Probability Distributions (see also Sec. 18.4-2). **(a) A multinomial distribution** is described by

$$p(x_1, x_2, \ldots, x_n) = \frac{N!}{x_1!x_2! \cdots x_n!} \vartheta_1^{x_1}\vartheta_2^{x_2} \cdots \vartheta_n^{x_n} \quad (x_1, x_2, \ldots, x_n = 0, 1, 2, \ldots; \; x_1 + x_2 + \cdots + x_n = N) \tag{18.8-1}$$

where $\vartheta_1, \vartheta_2, \ldots, \vartheta_n$ are positive real numbers such that

$$\vartheta_1 + \vartheta_2 + \cdots + \vartheta_n = 1$$

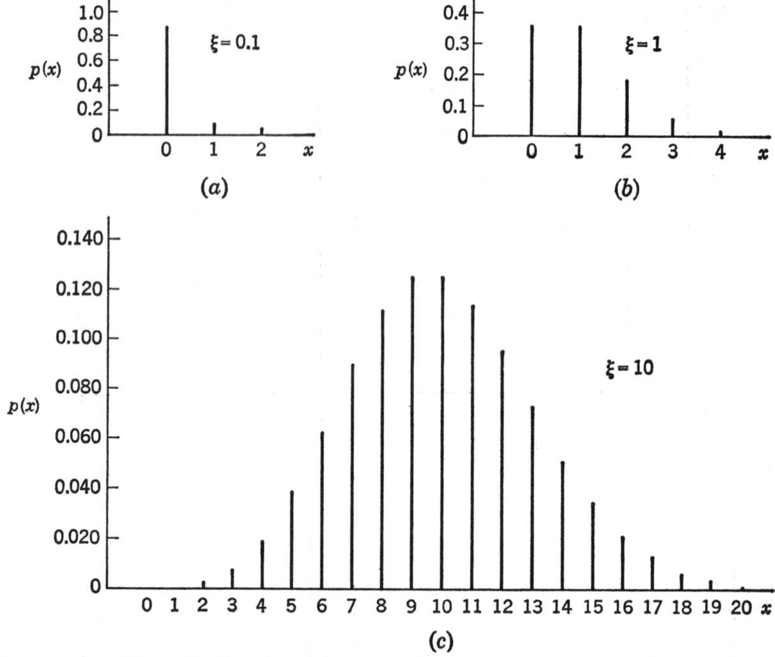

FIG. 18.8-2. The Poisson distribution. (*From Goode, H. H., and R. E. Machol, System Engineering, McGraw-Hill, New York,* 1957.)

Table 18.8-4. The Poisson Distribution (Fig. 18.8-2; see also Sec. 18.9-3)

(a) $p(x) = e^{-\xi}\dfrac{\xi^x}{x!} \qquad (x = 0, 1, 2, \ldots ; \xi > 0)$

(b) $E\{x\} = \text{Var}\{x\} = \xi \qquad \gamma_x(s) \equiv e^{\xi s - \xi}$

$\alpha_2 = \xi(\xi + 1) \qquad \alpha_3 = \xi(\xi^2 + 3\xi + 1) \qquad \alpha_4 = \xi(\xi^3 + 6\xi^2 + 7\xi + 1)$
$\mu_3 = \xi \qquad \mu_4 = 3\xi^2 + \xi$
$\gamma_1 = \xi^{-\frac{1}{2}} \qquad \gamma_2 = \xi^{-1}$

(c) The Poisson distribution approximates a hypergeometric distribution (Table 18.8-2) or a binomial distribution (Table 18.8-3) as $\vartheta N \to \infty$, $n \to \infty$, $\vartheta/n \to 0$ in such a manner that ϑn has a finite limit ξ (*Law of Small Numbers*). The approximation is often useful for $\vartheta \leq 0.1$, $n\vartheta \geq 1$. *The most important applications of the Poisson distribution appear in connection with random processes of the type discussed in Sec.* 17.9-3.

Given an experiment having n mutually exclusive results $E_1, E_2, \ldots, E_n$ with respective probabilities $\vartheta_1, \vartheta_2, \ldots, \vartheta_n$ such that $\vartheta_1 + \vartheta_2 + \cdots + \vartheta_n = 1$, the expression (1) is the probability that the respective events $E_1, E_2, \ldots, E_n$ occur exactly $x_1, x_2, \ldots, x_n$ times in N independent repeated trials (see also Sec. 18.7-3). In classical statistical mechanics, $x_1, x_2, \ldots, x_n$ are the occupation numbers of n independent states with respective a priori probabilities $\vartheta_1, \vartheta_2, \ldots, \vartheta_n$.

Table 18.8-5. The Geometric Distribution

(a) $p(x) = \vartheta(1 - \vartheta)^x \qquad (x = 0, 1, 2, \ldots ; 0 \leq \vartheta \leq 1)$

(b) $E\{x\} = \dfrac{1 - \vartheta}{\vartheta} \qquad \text{Var}\,\{x\} = \dfrac{1 - \vartheta}{\vartheta^2} \qquad \gamma_x(s) \equiv \dfrac{\vartheta}{1 - (1 - \vartheta)s}$

(c) **Typical Interpretation:** $p(x)$ is the probability of realizing an event ("success") for the first time *after exactly* x Bernoulli trials with probability of success ϑ. $\Phi(x) = 1 - (1 - \vartheta)^{x+1}$ $(x = 0, 1, 2, \ldots)$ is the probability that the first success occurs *after at most* x trials (see also Table 18.8-6).

Table 18.8-6. Pascal's Distribution

(a) $p(x) = \dbinom{m + x - 1}{x} \vartheta^m (1 - \vartheta)^x$

$$(x = 0, 1, 2, \ldots ; m = 0, 1, 2, \ldots ; 0 \leq \vartheta \leq 1)$$

(b) $E\{x\} = m\dfrac{1 - \vartheta}{\vartheta} \qquad \text{Var}\,\{x\} = m\dfrac{1 - \vartheta}{\vartheta^2} \qquad \gamma_x(s) \equiv \left(\dfrac{\vartheta}{1 - (1 - \vartheta)s}\right)^m$

(c) **Typical Interpretation:** $p(x)$ is the probability of realizing an event ("success") for the mth time *after exactly* $m + x - 1$ Bernoulli trials with probability of success ϑ. $\Phi(x)$ is the probability that the mth success occurs *after at most* $m + x - 1$ trials. For $m = 1$ Pascal's distribution reduces to the geometric distribution (Table 18.8-5).

Table 18.8-7. Polya's Distribution (Negative Binomial Distribution)

(a) $p(x) = \left(\dfrac{\xi}{1 + \beta\xi}\right)^x \dfrac{1(1 + \beta) \cdots [1 + (x - 1)\beta]}{x!} p(0) \quad (x = 1, 2, \ldots)$

$p(0) = (1 + \beta\xi)^{-\frac{1}{\beta}} \qquad (\xi > 0, \beta \geq 0)$

(b) $E\{x\} = \xi \qquad \text{Var}\,\{x\} = \xi(1 + \beta\xi)$

$$\gamma_x(s) \equiv (1 + \beta\xi - \beta\xi s)^{-\frac{1}{\beta}}$$

(c) Polya's distribution reduces to the Poisson distribution (Table 18.8-4) for $\beta = 0$, and to the geometric distribution (Table 18.8-5) for $\beta = 1$. See Ref. 18.4 for an interpretation in terms of a random process ("contagion").

(b) A **multiple Poisson distribution** is described by

$$p(x_1, x_2, \ldots, x_n) = e^{-(\xi_1+\xi_2+\cdots+\xi_n)} \frac{\xi_1^{x_1}\xi_2^{x_2} \cdots \xi_n^{x_n}}{x_1!x_2! \cdots x_n!}$$
$$(x_1, x_2, \ldots, x_n = 0, 1, 2, \ldots ; \xi_k > 0, k = 1, 2, \ldots, n) \qquad (18.8\text{-}2)$$

18.8-3. Continuous Probability Distributions: The Normal (Gaussian) Distribution. A continuous random variable x is **normally distributed (normal) with mean ξ and variance σ^2** [or **normal with parameters** ξ, σ^2; normal with parameters ξ, σ; normal (ξ, σ^2); normal (ξ, σ)] if

$$\varphi(X) \equiv \frac{1}{\sqrt{2\pi}\,\sigma} e^{-\frac{1}{2}\left(\frac{X-\xi}{\sigma}\right)^2} \equiv \frac{1}{\sigma}\varphi_u\left(\frac{X-\xi}{\sigma}\right)$$

$$\Phi(X) \equiv \frac{1}{\sqrt{2\pi}\,\sigma}\int_{-\infty}^{X} e^{-\frac{1}{2}\left(\frac{x-\xi}{\sigma}\right)^2} dx \equiv \Phi_u\left(\frac{X-\xi}{\sigma}\right)$$

$$\equiv \frac{1}{2}\left[1 + \operatorname{erf}\left(\frac{1}{\sqrt{2}}\frac{X-\xi}{\sigma}\right)\right] \qquad (18.8\text{-}3)$$

The distribution of the **standardized normal variable (normal deviate)** $u = \frac{x - \xi}{\sigma}$ (see also Sec. 18.5-3*c*) is given by

$$\varphi_u(U) \equiv \frac{1}{\sqrt{2\pi}} e^{-\frac{U^2}{2}} \qquad \text{(NORMAL FREQUENCY FUNCTION)}$$

$$\Phi_u(U) \equiv \frac{1}{\sqrt{2\pi}}\int_{-\infty}^{U} e^{-\frac{U^2}{2}} du \equiv \frac{1}{2}\left[1 + \operatorname{erf}\left(\frac{U}{\sqrt{2}}\right)\right] \qquad (18.8\text{-}4)$$

$$\text{(NORMAL DISTRIBUTION FUNCTION)}$$

$$E\{u\} = 0 \qquad \operatorname{Var}\{u\} = 1$$

(see also Fig. 18.8-3 and Sec. 18.8-4). erf z is the frequently tabulated **error function (normal error integral, probability integral;** see also Sec. 21.3-2)

$$\operatorname{erf} z \equiv -\operatorname{erf}(-z) \equiv \frac{2}{\sqrt{\pi}}\int_0^z e^{-\zeta^2} d\zeta \equiv 2\Phi_u(z\sqrt{2}) - 1 \qquad (18.8\text{-}5)$$

$\varphi(X)$ has points of inflection for $X = \xi \pm \sigma$. Note

$$\varphi_u^{(k)}(U) \equiv (-1/\sqrt{2})^k H_k(U/\sqrt{2})\varphi_u(U) \qquad (18.8\text{-}6)$$

where $H_k(z)$ is the kth Hermite polynomial (Sec. 21.7-1).

Every normal distribution is symmetric about its mean value ξ; ξ is the median and the (single) mode. The coefficients of skewness and excess are zero, and

$$\mu_{2k} = 1 \cdot 3 \cdots (2k-1)\sigma^{2k} \qquad \mu_{2k-1} = 0 \qquad (k = 1, 2, \ldots) \qquad (18.8\text{-}7)$$

$$\kappa_1 = \xi \qquad \kappa_2 = \sigma^2 \qquad \kappa_3 = \kappa_4 = \cdots = 0 \qquad (18.8\text{-}8)$$

$$\chi_x(q) \equiv \exp\left(-\frac{\sigma^2}{2}q^2 + i\xi q\right) \qquad (18.8\text{-}9)$$

The moments α_r about the origin may be computed by the methods of Sec. 18.3-10.

The normal distribution is of particular importance in many applications, especially in statistics (Secs. 19.3-1 and 19.5-3).

18.8-4. Normal Random Variables: Distribution of Deviations from the Mean. (**a**) For any normal random variable x with mean ξ and variance σ^2,

$$P[a < x < b] = \Phi_u\left(\frac{b-\xi}{\sigma}\right) - \Phi_u\left(\frac{a-\xi}{\sigma}\right) \qquad (18.8\text{-}10)$$

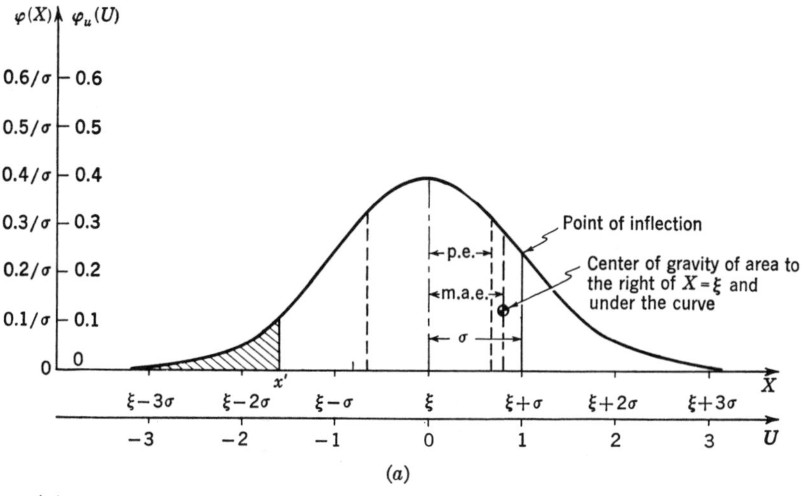

(a)

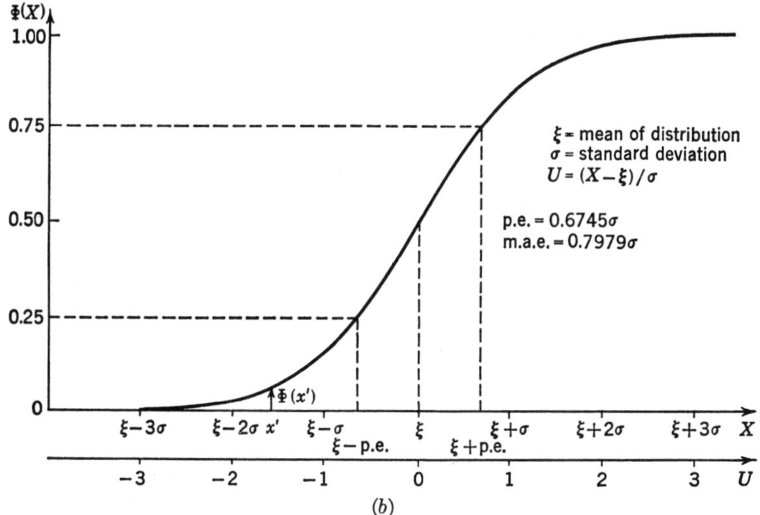

(b)

FIG. 18.8-3. (a) The normal frequency function

$$\varphi(X) = \frac{1}{\sqrt{2\pi}\,\sigma} e^{-\frac{1}{2}\left(\frac{x-\xi}{\sigma}\right)^2} = \frac{1}{\sigma}\varphi_u(U) \qquad \left(U = \frac{X-\xi}{\sigma}\right)$$

and (b) the normal distribution function

$$\Phi(X) = \frac{1}{\sqrt{2\pi}\,\sigma} \int_{-\infty}^{X} e^{-\frac{1}{2}\left(\frac{x-\xi}{\sigma}\right)^2} dx = \Phi_u(U) \qquad \left(U = \frac{X-\xi}{\sigma}\right)$$

(From Burington, R. S., and D. C. May, Handbook of Probability and Statistics, McGraw-Hill, New York, 1953.)

and, for $Y \geq 0$,

$$P[|x - \xi| < Y\sigma] = P[\xi - Y\sigma < x < \xi + Y\sigma]$$
$$= \frac{1}{\sqrt{2\pi}} \int_{-Y}^{Y} e^{-\frac{u^2}{2}} du = 2\Phi_u(Y) - 1 = \operatorname{erf}\left(\frac{Y}{\sqrt{2}}\right) = \Phi_{|u|}(Y) \quad (18.8\text{-}11)$$

$$P[|x - \xi| > Y\sigma] = \frac{1}{\sqrt{2\pi}} \int_{Y}^{\infty} e^{-\frac{u^2}{2}} du = 2[1 - \Phi_u(Y)]$$
$$= 1 - \operatorname{erf}\left(\frac{Y}{\sqrt{2}}\right) = 1 - \Phi_{|u|}(Y) \quad (18.8\text{-}12)$$

(**b**) The fractiles

$$|u|_P = u_{\frac{1+P}{2}} = |u|_{1-\alpha} = u_{1-\frac{\alpha}{2}} \quad (18.8\text{-}13)$$

defined by

$$P[|x - \xi| \leq |u|_P\sigma] = P = 1 - \alpha \quad (18.8\text{-}14)$$

Table 18.8-8. Continuous Probability Distributions

No.	Distribution	Frequency function $\varphi(x)$	Distribution function $\Phi(x)$
1	**Causal distribution**	$\delta(x - \xi) = \begin{cases} 0 & (x \neq \xi) \\ \infty & (x = \xi) \end{cases}$	$U(x - \xi)$
2	**Rectangular or uniform distribution**	$\frac{1}{2\alpha} \quad (\|x - \xi\| < \alpha)$ $0 \quad (\|x - \xi\| > \alpha)$	$0 \quad (x \leq \xi - \alpha)$ $\frac{1}{2\alpha}(x - \xi + \alpha) \quad (\xi - \alpha \leq x \leq \xi + \alpha)$ $1 \quad (x \geq \xi + \alpha)$
3	**Cauchy's distribution**	$\frac{1}{\pi\alpha} \frac{1}{1 + \left(\frac{x - \xi}{\alpha}\right)^2}$	$\frac{1}{2} + \frac{1}{\pi} \arctan \frac{x - \xi}{\alpha}$
4	**Laplace's distribution**	$\frac{1}{2\beta} e^{-\frac{\|x-\xi\|}{\beta}}$	$\frac{1}{2} e^{-\frac{\|x-\xi\|}{\beta}} \quad (x \leq \xi)$ $1 - \frac{1}{2} e^{-\frac{\|x-\xi\|}{\beta}} \quad (x \geq \xi)$
5	**Beta distribution**	$0 \quad (x \leq 0,\ x \geq 1)$ $\frac{\Gamma(\alpha + \beta)}{\Gamma(\alpha)\Gamma(\beta)} x^{\alpha-1}(1 - x)^{\beta-1} (0 < x < 1)$ $(\alpha > 0,\ \beta > 0)$	$0 \quad (x \leq 0)$ $I_x(\alpha, \beta) \quad (0 \leq x \leq 1)$ $1 \quad (x \geq 1)$
6	**Gamma distribution**	$0 \quad (x \leq 0)$ $\frac{1}{\beta^\alpha \Gamma(\alpha)} x^{\alpha-1} e^{-x/\beta} \quad (x > 0)$ $(\alpha > 0,\ \beta > 0)$	$0 \quad (x \leq 0)$ $\frac{1}{\Gamma(\alpha)} \Gamma_{x/\beta}(\alpha) \quad (x \geq 0)$

* See footnote to Sec. 18.3-4.

are often referred to as *tolerance limits of the normal deviate* u or as α *values of the normal deviate* (see also Sec. 19.6-4). Note

$$|u|_{0.95} = u_{0.975} \approx 1.96 \qquad |u|_{0.99} = u_{0.995} \approx 2.58 \qquad |u|_{0.999} = u_{0.9995} \approx 3.29 \qquad (18.8\text{-}15)$$

(c) Note the following measures of dispersion for normal distributions (see also Table 18.3-1):

The *mean deviation* (m.a.e) $E\{|x - \xi|\} = \sigma E\{|u|\} = \sqrt{\frac{2}{\pi}}\,\sigma \approx 0.798\sigma$

The *probable deviation* (p.e., median of $|x - \xi|$) $\quad |u|_{1/2}\sigma = -u_{1/4}\sigma = u_{3/4}\sigma \approx 0.674\sigma$
One-half the half width $\sqrt{2 \log_e 2}\,\sigma \approx 1.177\sigma$

The lower and upper *quartiles* $x_{1/4} = \xi - u_{1/4}\sigma = \xi - |u|_{1/2}\sigma$

$$x_{3/4} = \xi + u_{3/4}\sigma = \xi + |u|_{1/2}\sigma$$

The *precision measure* $h = \dfrac{1}{\sqrt{2}\,\sigma}$

(see also Secs. 18.8-3, 19.3-4, 19.3-5, and 19.5-3)

Mean value $E\{x\}$	Variance Var $\{x\}$	Characteristic function $\chi_x(q)$	Remarks
ξ	0	$e^{i\xi q}$	x is almost always (Sec. 4.6-14*b*) equal to ξ. Note that the rectangular, Cauchy, and Laplace distributions approximate a causal distribution as $\alpha \to 0$ or $\beta \to 0$ (see also Table 18.8-1 and Sec. 21.9-2)
ξ	$\dfrac{\alpha^2}{3}$	$\dfrac{\sin(\alpha q)}{\alpha q} e^{i\xi q}$	x is uniformly distributed over the interval $(\xi - \alpha, \xi + \alpha)$
$E\{x\}$ and Var $\{x\}$ do not exist; the Cauchy principal value (Sec. 4.6-2*c*) of $E\{x\}$ is ξ		$e^{i\xi q - \alpha\|q\|}$	Distribution of $x = \xi + \alpha \tan y$ if y is uniformly distributed between $y = -\pi/2$ and $y = \pi/2$ (rectangular distribution). Cauchy's distribution is symmetric about $x = \xi$. Half width and interquartile range are both equal to 2α
ξ	$2\beta^2$	$\dfrac{e^{i\xi q}}{1 + \beta^2 q^2}$	For $\xi = 0$, the characteristic function is proportional to the frequency function of a Cauchy distribution with $\alpha = 1/\beta$
$\dfrac{\alpha}{\alpha + \beta}$	$\dfrac{\alpha\beta}{(\alpha + \beta)^2(\alpha + \beta + 1)}$	$F(\alpha; \alpha + \beta; iq)$	$I_x(\alpha, \beta)$ is the incomplete beta-function ratio (Sec. 21.4-5; see also Sec. 19.5-3 and Table 18.8-3). Unique mode $(\alpha - 1)/(\alpha + \beta - 2)$ for $\alpha > 1$, $\beta > 1$. $\alpha_r = \dfrac{\Gamma(\alpha + r)\Gamma(\alpha + \beta)}{\Gamma(\alpha)\Gamma(\alpha + \beta + r)}$
$\alpha\beta$	$\alpha\beta^2$	$(1 - \beta iq)^{-\alpha}$	$\Gamma_x(\alpha)$ is the incomplete gamma function (Sec. 21.4-5)

18.8-5. Miscellaneous Continuous One-dimensional Probability Distributions. Table 18.8-8 describes a number of continuous one-dimensional probability distributions (see also Secs. 19.3-4, 19.3-5, and 19.5-3).

18.8-6. Two-dimensional Normal Distributions. (a) A **two-dimensional normal distribution** is a continuous probability distribution described by a frequency function of the form

$$\varphi(x_1, x_2) \equiv \frac{1}{2\pi\sigma_1\sigma_2\sqrt{1-\rho_{12}^2}} \exp\left\{-\frac{1}{2(1-\rho_{12}^2)}\left[\left(\frac{x_1-\xi_1}{\sigma_1}\right)^2 - 2\rho_{12}\frac{x_1-\xi_1}{\sigma_1}\frac{x_2-\xi_2}{\sigma_2} + \left(\frac{x_2-\xi_2}{\sigma_2}\right)^2\right]\right\}$$
$$(\sigma_1 > 0, \sigma_2 > 0, |\rho_{12}| \leq 1) \tag{18.8-16}$$

The marginal distributions of x_1 and x_2 are both normal with respective mean values ξ_1, ξ_2 and variances σ_1^2, σ_2^2; ρ_{12} is the correlation coefficient of x_1 and x_2. The five parameters ξ_1, ξ_2, σ_1, σ_2, ρ_{12} define the distribution completely.

The *conditional distributions* of x_1 and x_2 are both normal, with

$$E\{x_1|x_2\} = \xi_1 + \rho_{12}\frac{\sigma_1}{\sigma_2}(x_2 - \xi_2) \qquad \operatorname{Var}\{x_1|x_2\} = \sigma_1^2(1-\rho_{12}^2) \tag{18.8-17}$$

$$E\{x_2|x_1\} = \xi_2 + \rho_{12}\frac{\sigma_2}{\sigma_1}(x_1 - \xi_1) \qquad \operatorname{Var}\{x_2|x_1\} = \sigma_2^2(1-\rho_{12}^2) \tag{18.8-18}$$

so that the regression curves are identical with the mean-square regression lines (Sec. 18.4-6). *x_1 and x_2 are statistically independent if and only if they are uncorrelated* ($\rho_{12} = 0$, see also Sec. 18.4-11). Note

$$\left.\begin{aligned} P[x_1 \geq \xi_1, x_2 \geq \xi_2] = P[x_1 \leq \xi_1, x_2 \leq \xi_2] &= \frac{1}{4} + \frac{1}{2\pi}\arcsin\rho_{12} \\ P[x_1 \geq \xi_1, x_2 \leq \xi_2] = P[x_1 \leq \xi_1, x_2 \geq \xi_2] &= \frac{1}{4} - \frac{1}{2\pi}\arcsin\rho_{12} \end{aligned}\right\} \tag{18.8-19}$$

(b) Every two-dimensional normal distribution (16) can be described in terms of standardized normal variables u_1, u_2 with the correlation coefficient ρ_{12}, or in terms of statistically independent standardized normal variables (Sec. 18.5-5). Thus

$$\begin{aligned}\varphi(x_1, x_2)\,dx_1\,dx_2 &\equiv \frac{1}{2\pi\sqrt{1-\rho_{12}^2}}\exp\left[-\frac{1}{2(1-\rho_{12}^2)}(u_1^2 - 2\rho_{12}u_1u_2 + u_2^2)\right]du_1du_2 \\ &\equiv \varphi_u(u_1)\varphi_u(u_2')\,du_1\,du_2' \equiv \varphi_u(u_1')\varphi_u(u_2)\,du_1'\,du_2\end{aligned} \tag{18.8-20}$$

with

$$u_1 = \frac{x_1-\xi_1}{\sigma_1} \qquad u_2 = \frac{x_2-\xi_2}{\sigma_2} \qquad u_1' = \frac{u_1-\rho_{12}u_2}{\sqrt{1-\rho_{12}^2}} \qquad u_2' = \frac{u_2-\rho_{12}u_1}{\sqrt{1-\rho_{12}^2}} \tag{18.8-21}$$

(c) The distribution (16) is represented graphically by the **contour ellipses** $\varphi(x_1, x_2) = constant$, or

$$\frac{1}{1-\rho_{12}^2}\left[\left(\frac{x_1-\xi_1}{\sigma_1}\right)^2 - 2\rho_{12}\frac{x_1-\xi_1}{\sigma_1}\frac{x_2-\xi_{12}}{\sigma_2} + \left(\frac{x_2-\xi_2}{\sigma_2}\right)^2\right] = \lambda^2 = constant \tag{18.8-22}$$

The probability that the "point" (x_1, x_2) is inside the contour ellipse (22) is

$$P = \Phi_{\chi^2(2)}(\lambda^2)$$

i.e., $\lambda^2 = \chi_P{}^2(2)$ (Table 19.5-1). The two mean-square regression lines respectively defined by Eqs. (17) and (18) bisect all contour-ellipse chords in the x_1 and x_2 directions, respectively (see also Sec. 2.4-6).

18.8-7. Circular Normal Distributions.

Equation (16) represents a **circular normal distribution** with **dispersion** σ about the center of gravity (ξ_1, ξ_2) if and only if $\rho_{12} = 0$, $\sigma_1 = \sigma_2 = \sigma$. The contour ellipses (22) become circles corresponding to fractiles of the **radial deviation (radial error)** $r = \sqrt{(x_1-\xi_1)^2 + (x_2-\xi_2)^2}$. The distribution of r is given by

$$\begin{aligned}\varphi_r(r) &= \frac{2r}{\sigma^2}\varphi_{\chi^2(2)}\left(\frac{r^2}{\sigma^2}\right) = \frac{r}{\sigma^2}e^{-\frac{r^2}{2\sigma^2}} \qquad (r \geq 0)\\ \Phi_r(R) &\equiv P[(x_1-\xi_1)^2 + (x_2-\xi_2)^2 \leq R^2] \equiv \int_0^R \varphi_r(r)\,dr\\ &\equiv \Phi_{\chi^2(2)}\left(\frac{r^2}{\sigma^2}\right)\end{aligned} \tag{18.8-23}$$

(see also Sec. 18.11-1*b* and Table 19.5-1).

Circular normal distributions are of particular interest in problems related to gunnery; *circular probability paper* shows contour circles for equal increments of $\Phi_r(R)$. Note

$$r_{1/2} = \sqrt{\chi_{1/2}{}^2(2)}\,\sigma \approx 1.1774\sigma \quad \text{(CIRCULAR PROBABLE ERROR, CPE, CEP)} \tag{18.8-24}$$

$$E\{r\} = \sqrt{\frac{\pi}{2}}\,\sigma \approx 1.2533\sigma \quad \text{(MEAN RADIAL DEVIATION, MEAN RADIAL ERROR)} \tag{18.8-25}$$

18.8-8. *n*-Dimensional Normal Distributions.*

The joint distribution of n random variables $x_1, x_2, \ldots, x_n$ is an ***n*-dimensional normal distribution** if and only if it is a continuous probability distribution having a frequency function of the form

$$\varphi(x_1, x_2, \ldots, x_n) \equiv \frac{1}{\sqrt{(2\pi)^n \det[\lambda_{jk}]}} \exp\left[-\frac{1}{2}\sum_{j=1}^{n}\sum_{k=1}^{n}\Lambda_{jk}(x_j-\xi_j)(x_k-\xi_k)\right] \tag{18.8-26}$$

* See footnote to Sec. 18.4-8.

Each normal distribution is completely defined by its center of gravity (ξ_1, $\xi_2, \ldots, \xi_n$) *and its moment matrix* $[\lambda_{jk}] \equiv [\Lambda_{jk}]^{-1}$, *or by the corresponding variances and correlation coefficients* (Sec. 18.4-8). The characteristic function is

$$\chi_x(q_1, q_2, \ldots, q_n) \equiv \exp\left[i \sum_{j=1}^{n} \xi_j q_j - \tfrac{1}{2} \sum_{j=1}^{n} \sum_{k=1}^{n} \lambda_{jk} q_j q_k\right] \qquad (18.8\text{-}27)$$

Each marginal and conditional distribution derived from a normal distribution is normal. All mean-square regression hypersurfaces are identical with the corresponding mean-square regression hyperplanes (Sec. 18.4-9). *n random variables* $x_1, x_2, \ldots, x_n$ *having a normal joint distribution are statistically independent if and only if they are uncorrelated* (see also Sec. 18.4-11).

Each n-dimensional normal distribution can be described as the joint distribution of n statistically independent standardized normal variables related to the original variables by a linear transformation (18.5-15).

18.8-9. Addition Theorems for Special Probability Distributions* (see also Sec. 18.5-7 and Table 19.5-1). (**a**) The *binomial distribution* (Table 18.8-3), the *Poisson distribution* (Table 18.8-4), and the *Cauchy distribution* (Table 18.8-8) "reproduce themselves" on addition of independent variables. *If the random variable x is defined as the sum*

$$x = x_1 + x_2 + \cdots + x_n$$

of n statistically independent random variables $x_1, x_2, \ldots, x_n$, then

$$p_i(x_i) = \binom{n_i}{x_i} \vartheta^{x_i}(1 - \vartheta)^{n_i - x_i} \text{ implies } p_x(x) = \binom{n_0}{x} \vartheta^{x}(1 - \vartheta)^{n_0 - x}$$
$$(n_0 = n_1 + n_2 + \cdots + n_n) \qquad (18.8\text{-}28)$$

$$p_i(x_i) = e^{-\xi_i} \frac{\xi_i^{x_i}}{x_i!} \qquad \text{implies } p_x(x) = e^{-\xi} \frac{\xi^x}{x!}$$
$$(\xi = \xi_1 + \xi_2 + \cdots + \xi_n) \qquad (18.8\text{-}29)$$

$$\varphi_i(x_i) = \frac{1}{\pi\alpha} \frac{1}{1 + \left(\dfrac{x_i - \xi_i}{\alpha}\right)^2} \text{ implies } \varphi_x(x) = \frac{1}{\pi\alpha} \frac{1}{1 + \left(\dfrac{x - \xi}{\alpha}\right)^2}$$
$$(\xi = \xi_1 + \xi_2 + \cdots + \xi_n) \qquad (18.8\text{-}30)$$

(**b**) *The sum $x = x_1 + x_2 + \cdots + x_n$ of n statistically independent random variables $x_1, x_2, \ldots, x_n$ is a normal variable if and only if x_1, $x_2, \ldots, x_n$ are normal variables. In this case,*

$$\xi = \xi_1 + \xi_2 + \cdots + \xi_n \qquad \sigma^2 = \sigma_1^2 + \sigma_2^2 + \cdots + \sigma_n^2 \qquad (18.8\text{-}31)$$

If $x_1, x_2, \ldots, x_n$ are (not necessarily statistically independent) normal variables, then $x = a_1x_1 + a_2x_2 + \cdots + a_nx_n$ is a normal variable whose mean and variance are given by Eq. (18.5-19).

* See footnote to Sec. 18.3-4.

18.9. MATHEMATICAL DESCRIPTION OF RANDOM PROCESSES

18.9-1. Random Processes. Consider a variable x capable of assuming different values $x(t)$ for different values of an independent variable t. A **random process (stochastic process)** selects a specific **sample function** $x(t)$ from a given theoretical population (Sec. 19.1-2) or **ensemble** of possible sample functions. More specifically, the functions $x(t)$ are said to describe a random process if and only if the **sample values** $x_1 = x(t_1)$, $x_2 = x(t_2)$, . . . are random variables admitting definition of a joint probability distribution for every finite set of values (sampling times) $t_1, t_2, \ldots$ (Fig. 19.8-1). The random process is **discrete** or **continuous** if the joint distribution of $x(t_1)$, $x(t_2)$, . . . is, respectively, discrete or continuous for every finite set $t_1, t_2, \ldots$. The process is a **random series** if the independent variable t assumes only a countable set of values. More generally, a random process may be described by a multidimensional variable $\mathbf{x}(t) \equiv [x(t), y(t), \ldots]$.

The definition of a random process implies the existence of a probability distribution on the (in general, infinite-dimensional) sample space (Sec. 18.2-7) of possible *functions* $x(t)$. Each particular function $x(t) \equiv X(t)$ constitutes a simple event [sample point, "value" of the multidimensional random variable $x(t)$].

In most applications the independent variable t is the *time*, and the variable $x(t)$ or $\mathbf{x}(t)$ labels the *state* of a physical system. EXAMPLES: Results of successive observations, states of dynamical systems in Gibbsian statistical mechanics or quantum mechanics, messages and noise in communications systems, economic time series.

18.9-2. Mathematical Description of Random Processes. (a) To describe a random process, one must specify the distribution of $x(t_1)$ and the respective joint distributions of $[x(t_1), x(t_2)]$, $[x(t_1), x(t_2), x(t_3)]$, . . . for every finite set of values $t_1, t_2, t_3, \ldots$ (**first, second, third, . . . probability distributions** associated with the random process). These distributions are described by the corresponding **first, second, . . .** (or **first-order, second-order, . . .**) **distribution functions** (see also Sec. 18.4-7)

$$\begin{aligned} &\Phi_{(1)}(X_1, t_1) \equiv P[x(t_1) \leq X_1] \\ &\Phi_{(2)}(X_1, t_1; X_2, t_2) \equiv P[x(t_1) \leq X_1, x(t_2) \leq X_2] \quad \ldots \end{aligned} \tag{18.9-1a}$$

or, respectively for discrete and continuous random processes, by the corresponding probabilities and frequency functions

$$\begin{aligned} &p_{(1)}(X_1, t_1) \equiv P[x(t_1) = X_1] \\ &p_{(2)}(X_1, t_1; X_2, t_2) \equiv P[x(t_1) = X_1, x(t_2) = X_2] \quad \ldots \\ &\varphi_{(1)}(X_1, t_1) \equiv \frac{\partial \Phi_{(1)}}{\partial X_1}, \ \varphi_{(2)}(X_1, t_1; X_2, t_2) \equiv \frac{\partial^2 \Phi_{(2)}}{\partial X_1 \, \partial X_2}, \quad \ldots \end{aligned} \tag{18.9-1b}$$

NOTE: The sequence of distribution functions (1*a*) describes the random process in increasing detail, since each distribution function $\Phi_{(n)}$ completely defines all preceding ones as marginal distribution functions (Sec. 18.4-7). The same is true for each sequence (1*b*). *Each of the functions* (1) *is symmetric with respect to (unaffected by) interchanges of pairs* X_i, t_i *and* X_k, t_k.

(**b**) *Conditional probability distributions* descriptive of the random process are related to the functions (1*b*) in the manner of Sec. 18.4-7; thus

$$p(X_1, t_1; \ldots ; X_m, t_m | X_{m+1}, t_{m+1}; \ldots ; X_n, t_n) \equiv \frac{p_{(n)}(X_1, t_1; \ldots ; X_n, t_n)}{p_{(n-m)}(X_{m+1}, t_{m+1}; \ldots ; X_n, t_n)} \qquad (18.9\text{-}2a)$$

$$\varphi(X_1, t_1; \ldots ; X_m, t_m | X_{m+1}, t_{m+1}; \ldots ; X_n, t_n) \equiv \frac{\varphi_{(n)}(X_1, t_1; \ldots ; X_n, t_n)}{\varphi_{(n-m)}(X_{m+1}, t_{m+1}; \ldots ; X_n, t_n)} \qquad (18.9\text{-}2b)$$

NOTE: The functions (2) are *not* in general symmetric with respect to interchanges of pairs X_i, t_i and X_k, t_k separated by the bar.

(**c**) A *multidimensional random process*, say one generating a pair of sample functions $x(t)$, $y(t)$, is similarly defined in terms of joint distributions of sample values $x(t_i)$, $y(t_k)$. In particular,

$$\Phi_{(2)}(X_1, t_1; Y_2, t_2) \equiv P[x(t_1) \leq X_1, y(t_2) \leq Y_2] \qquad (18.9\text{-}3)$$

18.9-3. Ensemble Averages. (a) General Definitions. The **ensemble average (statistical average, mathematical expectation)** of a suitable function $f[x(t_1), x(t_2), \ldots, x(t_n)]$ of n sample values $x(t_1)$, $x(t_2), \ldots, x(t_n)$ (*statistic*, see also Sec. 19.1-1) is the expected value (Sec. 18.4-8*a*)

$$E\{f\} \equiv E\{f[x(t_1), x(t_2), \ldots, x(t_n)]\} \equiv \int_{-\infty}^{\infty} \int_{-\infty}^{\infty} \cdots \int_{-\infty}^{\infty} f(X_1, X_2, \ldots, X_n) \, d\Phi_{(n)}(X_1, t_1; X_2, t_2; \ldots ; X_n, t_n) \qquad (18.9\text{-}4)$$

if this limit exists in the sense of absolute convergence. Integration in Eq. (4) is over $X_1, X_2, \ldots, X_n$; $E\{f\}$ *is a function of* $t_1, t_2, \ldots, t_n$.

Similarly, for a multidimensional random process described by $x(t)$, $y(t)$,

$$E\{f[x(t_1), y(t_2); x(t_3), y(t_4); \ldots]\} = \int_{-\infty}^{\infty} \int_{-\infty}^{\infty} \cdots \int_{-\infty}^{\infty} f(X_1, Y_2; X_3, Y_4; \ldots) \, d\Phi(X_1, t_1; Y_2, t_2; X_3, t_3; Y_4, t_4; \ldots) \qquad (18.9\text{-}5)$$

if the limit exists in the sense of absolute convergence.

(b) Ensemble Correlation Functions and Mean Squares. The ensemble averages $E\{x(t_1)\} = \xi(t_1)$, $E\{x^2(t_1)\}$, and

$$R_{xx}(t_1, t_2) \equiv E\{x(t_1)x(t_2)\}$$

(ENSEMBLE AUTOCORRELATION FUNCTION) (18.9-6*a*)

$$R_{xy}(t_1, t_2) \equiv E\{x(t_1)y(t_2)\}$$

(ENSEMBLE CROSSCORRELATION FUNCTION) (18.9-6*b*)

are of special interest. They abstract important properties of the random process and are frequently all that is known about the process: note that

$$\left.\begin{aligned} E\{x^2(t_1)\} &\equiv R_{xx}(t_1, t_1) \\ \text{Var}\,\{x(t_1)\} &\equiv R_{xx}(t_1, t_1) - [E\{x(t_1)\}]^2 \\ \text{Cov}\,\{x(t_1), y(t_2)\} &\equiv R_{xy}(t_1, t_2) - E\{x(t_1)\}E\{y(t_2)\} \end{aligned}\right\} \qquad (18.9\text{-}7)$$

The definitions (6) and Eq. (7) apply to *real* $x(t)$, $y(t)$. If $x(t)$ and/or $y(t)$ is a *complex* variable (really a two-dimensional random variable), then one defines

$$\boxed{\begin{aligned} R_{xx}(t_1, t_2) &\equiv E\{x^*(t_1)x(t_2)\} \equiv R^*_{xx}(t_2, t_1) \\ R_{xy}(t_1, t_2) &\equiv E\{x^*(t_1)y(t_2)\} \equiv R^*_{yx}(t_2, t_1) \end{aligned}} \qquad (18.9\text{-}8)$$

which includes (6) as a special case; R_{xy} is necessarily real for real x and y.

Note that, for real or complex x, y,

$$R_{xx}(t_1, t_1) \equiv E\{|x(t_1)|^2\} \qquad (18.9\text{-}9)$$

$$|R_{xy}(t_1, t_2)|^2 \leq E\{|x(t_1)|^2\}E\{|y(t_2)|^2\} \qquad (18.9\text{-}10)$$

Existence of the quantities on the right implies that of the correlation functions on the left.

(c) Characteristic Functions. The nth **characteristic function** corresponding to the nth distribution function (1*a*) of the random process (see also Sec. 18.4-10) is

$$\chi_{(n)}(q_1, t_1; q_2, t_2; \ldots ; q_n, t_n) \equiv E\left\{\exp\left[i \sum_{k=1}^{n} q_k x(t_k)\right]\right\} \qquad (18.9\text{-}11)$$

Joint characteristic functions for $x(t)$, $y(t)$, . . . are similarly defined. Characteristic functions can yield moments like $E\{x(t_1)\}$, $E\{x^2(t_1)\}$, $R_{xx}(t_1, t_2)$, . . . by differentiation in the manner of Secs. 18.3-10 and 18.4-10.

(d) Ensemble Averages of Integrals and Derivatives (see also Sec. 18.6-3). Random integrals of the form

$$y = \int_a^b f(t)x(t)\,dt \qquad (18.9\text{-}12)$$

are defined in the sense of convergence in probability (Sec. 18.6-1) or, if possible, in the mean-square sense of Sec. 18.6-3. *The integral converges in mean* (in the sense of Sec. 18.6-3) *if and only if*

$$E\{|y|^2\} = \int_a^b dt_1\, f^*(t_1) \int_a^b f(t_2)\, \mathrm{R}_{xx}(t_1, t_2)\, dt_2 \tag{18.9-13}$$

exists. If $\int_a^b |f(t)|E\{|x(t)|\}\, dt$ *exists, then the integral* (12) *exists in the sense of absolute convergence for each sample function* $x(t)$, *except possibly for a set of probability* 0, *and*

$$E\left\{\int_a^b f(t)x(t)\, dt\right\} = \int_a^b f(t)E\{x(t)\}\, dt \tag{18.9-14}$$

The important relation (14) is needed, in particular, to derive the input-output relations of Sec. 18.12-2 (see also Refs. 18.13 to 18.17).

The random process generating $x(t)$ is **continuous in the mean (mean-square continuous)** at $t = t_0$ in the sense of Sec. 18.6-3 if and only if

$$\lim_{t \to t_0} E\{|x(t) - x(t_0)|^2\} = 0$$

this is true if and only if $\mathrm{R}_{xx}(t_1,t_2)$ *exists and is continuous for* $t_1 = t_2 = t_0$. The random process generating $\dot{x}(t)$ will be called the **mean-square derivative** of a random process generating $x(t)$ if and only if

$$\lim_{\Delta t \to 0} E\left\{\left|\frac{x(t + \Delta t) - x(t)}{\Delta t} - \dot{x}(t)\right|^2\right\} = 0 \tag{18.9-15}$$

This is true if and only if $\partial^2\mathrm{R}_{xx}(t_1, t_2)/\partial t_1\, \partial t_2$ exists and equals $\partial^2\mathrm{R}_{xx}(t_1, t_2)/\partial t_2 \partial t_1$ for all $t_1 = t_2$. It follows that

$$\left.\begin{aligned} &E\{\dot{x}(t)\} \equiv \frac{d}{dt}E\{x(t)\} \equiv \xi(t) \\ &\mathrm{R}_{\dot{x}x}(t_1, t_2) \equiv \frac{\partial}{\partial t_1}\mathrm{R}_{xx}(t_1, t_2) \qquad \mathrm{R}_{\dot{x}\dot{x}}(t_1, t_2) = \frac{\partial^2}{\partial t_1 \partial t_2}\mathrm{R}_{xx}(t_1, t_2) \end{aligned}\right\} \tag{18.9-16}$$

(see also Sec. 18.12-2).

18.9-4. Processes Defined by Random Parameters. It is often possible to represent each sample function of a random process as a deterministic function $x = x(t; \eta_1, \eta_2, \ldots)$ of t and a set of random parameters $\eta_1, \eta_2, \ldots$. The process is then defined by the joint distribution of $\eta_1, \eta_2, \ldots$; in this case,

$$\begin{aligned} E\{f[x(t_1), \ldots, x(t_n)]\} \equiv \int_{-\infty}^{\infty}\int_{-\infty}^{\infty} \cdots \int_{-\infty}^{\infty} f[x(t_1; \eta_1, \eta_2, \ldots), \\ \ldots, x(t_n; \eta_1, \eta_2, \ldots)]\, d\Phi_{\eta_1 \eta_2 \cdots}(\eta_1, \eta_2, \ldots) \end{aligned} \tag{18.9-17}$$

In particular, each probability distribution of such a random process is uniquely defined by its characteristic function (Sec. 18.4-10)

$$\chi_{(n)}(q_1, t_1; \ldots ; q_n, t_n) \equiv \int_{-\infty}^{\infty} \int_{-\infty}^{\infty} \cdots \int_{-\infty}^{\infty} \exp\left[i \sum_{k=1}^{n} q_k x(t_k; \eta_1, \eta_2, \ldots)\right] d\Phi_{\eta_1\eta_2\ldots}(\eta_1, \eta_2, \ldots) \qquad (18.9\text{-}18)$$

18.9-5. Orthonormal-function Expansions. Given a real or complex random process $x(t)$ with $E\{x(t)\}$ finite and $R_{xx}(t_1, t_2)$ bounded and continuous on the closed observation interval $[a, b]$, there exist complete orthonormal sets of functions $u_1(t)$, $u_2(t)$, . . . (Sec. 15.2-4) such that

$$\left.\begin{gathered} x(t) = \sum_{k=1}^{\infty} u_k(t) \\ \text{with} \quad \int_a^b u_i^*(t) u_k(t)\, dt = \delta_k^i \qquad c_k = \int_a^b u_k^*(t) x(t)\, dt \qquad (i, k = 1, 2, \ldots) \end{gathered}\right\} \qquad (18.9\text{-}19)$$

where the series and the integral for each c_k converges in mean in the sense of Sec. 18.6-3 (see also Sec. 18.9-3*d*). The random process is, then, represented by the set of random coefficients $c_1, c_2, \ldots$; the first n coefficients may give a useful approximate representation. In particular, *there exists a complete orthonormal set* $u_k(t) \equiv \psi_k(t)$ *such that all the* c_k *are uncorrelated standardized random variables, i.e.,*

$$E\{c_k\} = 0 \qquad E\{c_i^* c_k\} = \delta_k^i \qquad (i, k = 1, 2, \ldots) \qquad (18.9\text{-}20)$$

(*Karhunen-Loéve Theorem*). Specifically, the required $\psi_k(t)$ are the eigenfunctions of the integral equation

$$\lambda \int_a^b R_{xx}(t_1, t_2) \psi(t_2)\, dt_2 = \psi(t_1) \qquad (18.9\text{-}21)$$

(see also Sec. 15.3-3). The corresponding eigenvalues λ_k are nonnegative and have at most a finite degree of degeneracy (by Mercer's theorem, Sec. 15.3-4), and

$$E\left\{\int_a^b |x(t)|^2\, dt\right\} = \int_a^b R_{xx}(t, t)\, dt = \sum_{k=1}^{\infty} \frac{1}{\lambda_k} \qquad (18.9\text{-}22)$$

The Karhunen-Loéve theorem constitutes a generalization of the theorem of Sec. 18.5-5.

EXAMPLES: Periodic random processes (Sec. 18.11-1), band-limited flat-spectrum noise (Sec. 18.11-2*b*). Although explicit analytical solution of the integral equation (14) is rarely possible, the theorem is useful in detection theory (Ref. 19.24).

18.10. STATIONARY RANDOM PROCESSES. CORRELATION FUNCTIONS AND SPECIAL DENSITIES

18.10-1. Stationary Random Processes. A random process, or the corresponding ensemble of functions $x(t)$, is **stationary** if and only if each of its probability distributions is unchanged when t is replaced by

$t + t_0$, so that

$$\Phi_{(n)}(X_1, t_1 + t_0; X_2, t_2 + t_0; \ldots ; X_n, t_n + t_0) \equiv \Phi_{(n)}(X_1, t_1; X_2, t_2; \ldots ; X_n, t_n) \quad (-\infty < t_0 < \infty; n = 1, 2, \ldots) \tag{18.10-1}$$

i.e., the nth probability distribution depends only on a set of $n - 1$ *differences*

$$\tau_1 = t_2 - t_1 \qquad \tau_2 = t_3 - t_1 \qquad \cdots \qquad \tau_{n-1} = t_n - t_1 \tag{18.10-2}$$

of sampling times t_k. Similarly, two or more random processes generating $x(t)$, $y(t)$, . . . are **jointly stationary** if and only if their joint probability distributions are unchanged when t is replaced by $t + t_0$.

For stationary and jointly stationary random processes, each ensemble average (18.9-4) or (18.9-5) depends only on $n - 1$ differences (2):

$$E\{f[x(t_1)x(t_2), \ldots , x(t_n)]\} = E\{f[x(0), x(\tau_1), \ldots , x(\tau_{n-1})]\} \tag{18.10-3}$$

for every t_1 (see also Sec. 18.10-2).

18.10-2. Ensemble Correlation Functions (see also Sec. 18.9-3*b*). **(a)** For stationary $x(t)$ [and jointly stationary $x(t)$, $y(t)$], the expected values

$$E\{x(t)\} \equiv E\{x\} = \xi \qquad E\{|x(t)|^2\} \equiv E\{|x|^2\} \qquad E\{y(t)\} \equiv E\{y\} = \eta \ \ldots$$

are constant, and the ensemble correlation functions (18.9-8) reduce to functions of the delay $t_2 - t_1 = \tau$ separating t_1 and t_2. In this case,

$$\begin{aligned} \mathrm{R}_{xx}(\tau) &\equiv E\{x^*(t)x(t + \tau)\} \equiv \mathrm{R}^*_{xx}(-\tau) \\ \mathrm{R}_{xy}(\tau) &\equiv E\{x^*(t)y(t + \tau)\} \equiv \mathrm{R}^*_{yx}(-\tau) \end{aligned} \tag{18.10-4}$$

$$|\mathrm{R}_{xx}(\tau)| \leq \mathrm{R}_{xx}(0) = E\{|x|^2\} \tag{18.10-5}$$

$$|\mathrm{R}_{xy}(\tau)|^2 \leq \mathrm{R}_{xx}(0)\mathrm{R}_{yy}(0) = E\{|x|^2\}E\{|y|^2\} \tag{18.10-6}$$

Again, *existence of the quantities on the right implies existence of the correlation functions on the left. If* $\mathrm{R}_{xx}(\tau)$ *is continuous for* $\tau = 0$, *it is continuous for all* τ (Ref. 18.17).

$[\mathrm{R}_{xx}(t_i - t_k)]$ is a positive-semidefinite hermitian matrix (Sec. 13.5-3) for every finite set $t_1, t_2, \ldots , t_n$.

(b) Normalized ensemble correlation functions are defined by

$$\rho_{xx}(\tau) \equiv \frac{\mathrm{R}_{xx}(\tau) - |\xi|^2}{E\{|x|^2\} - |\xi|^2} \qquad \rho_{xy}(\tau) = \frac{\mathrm{R}_{xy}(\tau) - \xi^*\eta}{(E\{|x|^2\} - |\xi|^2)^{1/2}(E\{|y|^2\} - |\eta|^2)^{1/2}} \tag{18.10-7}$$

Note $|\rho_{xx}| \leq 1$, $|\rho_{xy}| \leq 1$. For *real* stationary x, y, ρ_{xx} and ρ_{xy} are real correlation coefficients (Sec. 18.4-4), and Eq. (4) implies

$$\mathrm{R}_{xx}(\tau) \equiv \mathrm{R}_{xx}(-\tau) \qquad \mathrm{R}_{xy}(\tau) \equiv \mathrm{R}_{yx}(-\tau) \tag{18.10-8}$$

Random processes which are not stationary or jointly stationary but have constant $E\{x(t)\}$, $E\{y(t)\}$ and "stationary correlation functions" satisfying Eq. (4) are often called **stationary, or jointly stationary, in the wide sense.**

18.10-3. Ensemble Spectral Densities. If $x(t)$ is generated by a stationary random process, and $x(t)$, $y(t)$ by jointly stationary random processes, the **ensemble power spectral density** $\Phi_{xx}(\omega)$ and the **ensemble cross-spectral density** $\Phi_{xy}(\omega)$ are defined by

$$\begin{aligned} \Phi_{xx}(\omega) &\equiv \int_{-\infty}^{\infty} \mathrm{R}_{xx}(\tau)e^{-j\omega\tau}\,d\tau \equiv \Phi_{xx}^*(\omega) \\ \Phi_{xy}(\omega) &\equiv \int_{-\infty}^{\infty} \mathrm{R}_{xy}(\tau)e^{-j\omega\tau}\,d\tau \equiv \Phi_{yx}^*(\omega) \end{aligned} \tag{18.10-9}$$

(WIENER-KHINCHINE RELATIONS)

Assuming suitable convergence, this implies

$$\begin{aligned} \mathrm{R}_{xx}(\tau) &\equiv \int_{-\infty}^{\infty} \Phi_{xx}(\omega)e^{j\omega\tau}\,\frac{d\omega}{2\pi} \equiv \mathrm{R}_{xx}^*(-\tau) \\ \mathrm{R}_{xy}(\tau) &\equiv \int_{-\infty}^{\infty} \Phi_{xy}(\omega)e^{j\omega\tau}\,\frac{d\omega}{2\pi} \equiv \mathrm{R}_{yx}^*(-\tau) \end{aligned} \tag{18.10-10}$$

The Fourier transforms (9) are introduced, essentially, to simplify the relations between input and output correlation functions in linear time-invariant systems (Sec. 18.12-3). Existence of the transforms (9) requires, besides the existence of $E\{|x|^2\}$ and $E\{|y|^2\}$ (Sec. 18.9-3*b*), that $\mathrm{R}_{xx}(\tau)$ or $\mathrm{R}_{xy}(\tau)$ decays sufficiently quickly as $\tau \to \infty$. In the case of periodic and d-c processes, one extends the definitions of spectral densities to include delta-function terms chosen so that Eq. (10) is satisfied (Sec. 18.10-9).

18.10-4. Correlation Functions and Spectra of Real Processes. The relations (9) and (10) apply to both real and complex random processes $x(t)$, $y(t)$. Note that the power spectral density $\Phi_{xx}(\omega)$ is *always* real, even if x is complex; but the cross-spectral density $\Phi_{xy}(\omega)$ may be a complex function even for real x, y. If x and y are real, the same is true for the correlation functions $R_{xx}(\tau)$, $R_{xy}(\tau)$. In this case,

$$\Phi_{xx}(\omega) \equiv \int_{-\infty}^{\infty} R_{xx}(\tau)e^{-j\omega\tau}\,d\tau \equiv 2\int_0^{\infty} R_{xx}(\tau)\cos\omega\tau\,d\tau \equiv \Phi_{xx}(-\omega) \tag{18.10-11}$$

$$R_{xx}(\tau) \equiv \int_{-\infty}^{\infty} \Phi_{xx}(\omega)e^{j\omega\tau}\,\frac{d\omega}{2\pi} \equiv 2\int_0^{\infty} \Phi_{xx}(\omega)\cos\omega\tau\,\frac{d\omega}{2\pi} \equiv R_{xx}(-\tau) \tag{18.10-12}$$

$$\Phi_{xy}(\omega) \equiv \Phi_{yx}(-\omega) = \Phi_{xy}^*(-\omega) \qquad R_{xy}(\tau) \equiv R_{yx}(-\tau) \tag{18.10-13}$$

Note again that Eqs. (11) to (13) apply to *real* x, y.

18.10-5. Spectral Decomposition of Mean "Power" for Real Processes. For *real* $x(t)$, substitution of $\tau = 0$ in Eqs. (11) and (12) yields

$$E\{x^2\} = R_{xx}(0) = \int_{-\infty}^{\infty} \Phi_{xx}(\omega)\,\frac{d\omega}{2\pi} = 2\int_0^{\infty} \Phi_{xx}(\omega)\,\frac{d\omega}{2\pi} \tag{18.10-14}$$

This is interpreted as a *spectral decomposition* of $E\{x^2\}$ (mean "power"). In the first integral, contributions to $E\{x^2\}$ are "distributed" over both positive and negative frequencies with density $\Phi_{xx}(\omega)$ ("two-sided" power spectral density), measured in $(x \text{ units})^2/\text{cps}$, since $\omega/2\pi$ is frequency in cps. Alternatively, we can consider $E\{x^2\}$ as distributed only over nonnegative ("real") frequencies with the "one-sided" power spectral density $2\Phi_{xx}(\omega)$ $(x \text{ units})^2/\text{cps}$.

Intuitive interpretation of the—in general complex—cross-spectral density $\Phi_{xy}(\omega)$ is not quite so simple. For *real* $x(t)$, $y(t)$, substitution of $\tau = 0$ in Eq. (10) yields

$$E\{xy\} = R_{xy}(0) = \int_{-\infty}^{\infty} \Phi_{xy}(\omega)\,\frac{d\omega}{2\pi} = 2\int_0^{\infty} \text{Re}\,\Phi_{xy}(\omega)\,\frac{d\omega}{2\pi} \tag{18.10-15}$$

$\text{Re}\,\Phi_{xy}(\omega)$ is often called a **cross-power spectral density.** $\text{Im}\,\Phi_{xy}(\omega)$ (**cross-quadrature spectral density**) does not contribute to the mean "power" (15).

18.10-6. Some Alternative Ensemble Spectral Densities. Other spectral-density functions found in the literature are

$$S_{xx}(\nu) \equiv \Phi_{xx}(2\pi\nu) \qquad \text{with} \qquad E\{|x|^2\} = \int_{-\infty}^{\infty} S_{xx}(\nu)\,d\nu \tag{18.10-16}$$

($\nu = \omega/2\pi$; *two-sided spectral density in x units*^{2}*/cps*)

$$g_{xx}(\omega) \equiv \frac{1}{2\pi}\Phi_{xx}(\omega) \qquad \text{with} \qquad E\{x^2\} = \int_{-\infty}^{\infty} g_{xx}(\omega)\,d\omega \tag{18.10-17}$$

(*two-sided spectral density in x-units*^{2}*/radian/sec*)

and the one-sided spectral densities

$$\Gamma_{xx}(\nu) \equiv 2\mathrm{S}_{xx}(\nu) = 2\Phi_{xx}(2\pi\nu) \qquad (\nu \geq 0) \qquad (18.10\text{-}18)$$

$$\mathrm{G}_{xx}(\omega) \equiv 2\mathrm{g}_{xx}(\omega) = \frac{1}{\pi}\Phi_{xx}(\omega) \qquad (\omega \geq 0) \qquad (18.10\text{-}19)$$

Note that $\Gamma_{xx}(\nu)$ and $\mathrm{G}_{xx}(\omega)$ are defined *only* for nonnegative frequencies. Similar definitions also apply to cross-spectral densities. *Note that symbols and definitions vary greatly in the literature*; the correct definition should be restated and referred to in each case.

18.10-7. *t* Averages and Ergodic Processes. (a) ***t* Averages.** Given any function $x(t)$, the *t* **average (average over** *t*, frequently a time average) of a measurable function $f[x(t_1), x(t_2), \ldots, x(t_n)]$ is defined as

$$<f[x(t_1), x(t_2), \ldots, x(t_n)]> \equiv \lim_{T\to\infty} \frac{1}{2T}\int_{-T}^{T} f[x(t_1 + t), x(t_2 + t), \ldots, x(t_n + t)]\, dt \qquad (18.10\text{-}20)$$

if the limit exists.* *If $x(t)$ describes a random process, then $<f>$ is (like f, but unlike $E\{f\}$) a random variable (statistic) for each given set of values $t_1, t_2, \ldots, t_n$.* Note that

$$E\{<f>\} = E\{f\} \qquad (18.10\text{-}21)$$

whenever the integrals exist.

(b) **Ergodic Processes.** A (necessarily stationary) random process generating $x(t)$ is **ergodic** if and only if the probability associated with every stationary subensemble is either 0 or 1. Every ergodic process has the **ergodic property:** *the t average* (20) *of every measurable function $f[x(t_1), x(t_2), \ldots, x(t_n)]$ equals its ensemble average* (18.9-4) *with probability one*, i.e.,

$$\boxed{P[<f> = E\{f\}] = 1} \qquad (18.10\text{-}22)$$

whenever these averages exist. Any one of the functions $x(t)$ will then define the random process uniquely with probability one, e.g., in terms of the characteristic functions (18.9-11) computed from $x(t)$ by means of Eq. (21). *Each t average, such as $<x>$, $<x^2>$, or $R_{xx}(\tau)$, will then*

* The notation $\bar{f}$ is sometimes used instead of $<f>$, as well as instead of $E\{f\}$; but the symbol $\bar{f}$ is preferably reserved for the *sample average*

$$\bar{f} = \frac{1}{n}({}^1f + {}^2f + \cdots + {}^nf)$$

where ${}^kf$ is the value of f obtained from one of an empirical random sample of n sample functions $x(t) = {}^kx(t)$ $(k = 1, 2, \ldots, n$; see also Sec. 19.8-4).

represent, with probability one, a property common to the entire ensemble of functions $x(t)$.

Two or more jointly stationary random processes are **jointly ergodic** if and only if the probability associated with every stationary joint subensemble is either 0 or 1. The ergodic theorem applies to averages computed from sample values of jointly ergodic processes.

18.10-8. Non-ensemble Correlation Functions and Spectral Densities. Given the real or complex functions $x(t)$, $y(t)$ (which may or may not be sample functions of a random process) such that

$$\left.\begin{aligned} <|x(0)|^2> &= \lim_{T\to\infty} \frac{1}{2T} \int_{-T}^{T} |x(t)|^2\, dt \\ <|y(0)|^2> &= \lim_{T\to\infty} \frac{1}{2T} \int_{-T}^{T} |y(t)|^2\, dt \end{aligned}\right\} \tag{18.10-23}$$

exist, the t averages

$$<x^*(0)x(\tau)> \equiv \lim_{T\to\infty} \frac{1}{2T} \int_{-T}^{T} x^*(t)x(t+\tau)\, dt \equiv R_{xx}(\tau)$$

[AUTOCORRELATION FUNCTION OF $x(t)$] (18.10-24)

$$<x^*(0)y(\tau)> \equiv \lim_{T\to\infty} \frac{1}{2T} \int_{-T}^{T} x^*(t)y(t+\tau)\, dt \equiv R_{xy}(\tau)$$

[CROSSCORRELATION FUNCTION OF $x(t)$ AND $y(t)$] (18.10-25)

exist. *These correlation functions satisfy all the relations listed in Sec.* 18.10-2, *if each ensemble average (expected value) is replaced by the corresponding t average.* Again, the (non-ensemble) **power spectral density** $\Psi_{xx}(\omega)$ and the cross-spectral density $\Psi_{xy}(\omega)$ are introduced through the Wiener-Khinchine relations

$$\left.\begin{aligned} \Psi_{xx}(\omega) &\equiv \int_{-\infty}^{\infty} R_{xx}(\tau)e^{-i\omega\tau}\, d\tau \equiv \Psi_{xx}^*(\omega) \\ \Psi_{xy}(\omega) &\equiv \int_{-\infty}^{\infty} R_{xy}(\tau)e^{-i\omega\tau}\, d\tau \equiv \Psi_{yx}^*(\omega) \end{aligned}\right\} \tag{18.10-26}$$

If these "individual" spectral densities exist (one formally admits delta-function terms, Sec. 18.10-9), *they satisfy relations analogous to those listed in Secs.* 18.10-3 *to* 18.10-5. Alternative non-ensemble spectral densities can be defined in the manner of Sec. 18.10-6.

If $x(t)$, $y(t)$ are sample functions of jointly stationary random processes, then the correlation functions (24), (25) and the spectral densities (26) are random variables whose expected values equal the corresponding ensemble functions whenever they exist. If $x(t)$, $y(t)$ are jointly ergodic, then the correlation functions (24), (25) and the spectral densities (26) are identical to the corresponding ensemble quantities with probability one.

As an alternative definition, spectral densities are sometimes introduced by the formal relation

$$\Psi_{xy}(\omega) = \lim_{T\to\infty} \frac{1}{2T} a_T^*(\omega) b_T(\omega) \tag{18.10-27a}$$

where $a_T(\omega)$ and $b_T(\omega)$ are Fourier transforms of the "truncated" functions $x_T(t)$, $y_T(t)$ respectively equal to $x(t)$, $y(t)$ for $|t| < T$ and zero for $|t| > T$:

$$a_T(\omega) = \int_{-T}^{T} x(t)e^{-i\omega t}\,dt \qquad b_T(\omega) = \int_{-T}^{T} y(t)e^{-i\omega t}\,dt \tag{18.10-27b}$$

The corresponding ensemble spectral density $\Phi_{xy}(\omega)$ may then be defined by $\Phi_{xy}(\omega) = E\{\Psi_{xy}(\omega)\}$, and the Wiener-Khinchine relations (26) follow from Borel's convolution theorem (Table 4.11-1). In general, however, Eq. (27) is valid only if both sides appear in an integral over ω (in particular, spectral densities often contain delta-function terms, Secs. 18.10-9 and 18.11-5; see also Sec. 18.10-10).

18.10-9. Functions with Periodic Components (see also Sec. 18.11-1). Like other t averages, non-ensemble correlation functions and spectra are of interest mainly if they happen to equal the corresponding ensemble quantities with probability one (this is true for *all* t averages in the case of ergodic processes, Sec. 18.10-7*b*). When this is true, the single integrals (24), (25) may be easier to compute than the double integrals (4). The ergodic property also permits interpretation of, say, $\Phi_{xx}(\omega)$ in terms of the "frequency content" of a single "typical" sample function $x(t)$, since $\Phi_{xx}(\omega) \equiv \Psi_{xx}(\omega)$ with probability one.

Without recourse to probability theory, non-ensemble correlation functions and spectra can be computed only for functions $x(t)$, $y(t)$ representable as sums of periodic components (except for the trivial case that the correlation function or spectral density is identically zero). In particular, for

$$x(t) = a\cos(\omega_1 t + \varphi) \qquad y(t) = \alpha\cos(\omega_2 t + \psi) \tag{18.10-28a}$$

one has

$$R_{xx}(\tau) = \frac{a^2}{2}\cos\omega_1\tau \qquad R_{xy}(\tau) = \begin{cases} \tfrac{1}{2}a\alpha\cos(\omega_1\tau + \psi - \varphi) & (\omega_1 = \omega_2) \\ 0 \text{ otherwise} \end{cases} \tag{18.10-28b}$$

More generally, let $x(t)$ be a real function and of bounded variation in every finite interval and such that $\langle |x(0)|^2 \rangle$ exists. Then $x(t)$ can be represented almost everywhere (Sec. 4.6-14*b*) as the sum of its average value $\langle x(0) \rangle = c_0$, a countable set of periodic components, and an aperiodic component* $p(t)$:

$$\begin{aligned} x(t) &= \sum_{k=-\infty}^{\infty} c_k e^{i\omega_k t} + p(t) = c_0 + \sum_{k=1}^{\infty} (a_k\cos\omega_k t + b_k\sin\omega_k t) + p(t) \\ &= c_0 + \sum_{k=1}^{\infty} A_k\cos(\omega_k t + \varphi_k) + p(t) \end{aligned} \tag{18.10-29a}$$

* The aperiodic component $p(t)$ may be expressible as a Fourier integral $[\langle |x(0)|^2 \rangle = 0]$, or $\langle |x(0)|^2 \rangle$ may be different from zero ("random" component); or $p(t)$ may be a sum of both types of terms.

with $\omega_0 = 0,\ \omega_k = -\omega_k > 0\ (k = 1, 2, \ldots)$, so that

$$\left.\begin{aligned}
&\lim_{T\to\infty} \frac{1}{2T}\int_{-T}^{T} x(t)e^{-i\omega t}\,dt = \begin{cases} c_0 \text{ if } \omega = 0 \\ c_k = c_{-k}^* \text{ if } \omega = \omega_k \qquad (k = \pm 1, \pm 2, \ldots) \\ 0 \text{ otherwise}\end{cases} \\
&\lim_{T\to\infty} \frac{1}{2T}\int_{-T}^{T} x(t)\cos\omega t\,dt = \begin{cases} c_0 \text{ if } \omega = 0 \\ \frac{1}{2}a_k \text{ if } \omega = \omega_k \qquad (k = 1, 2, \ldots) \\ 0 \text{ otherwise}\end{cases} \\
&\lim_{T\to\infty} \frac{1}{2T}\int_{-T}^{T} x(t)\sin\omega t\,dt = \begin{cases} \frac{1}{2}b_k \text{ if } \omega = \omega_k \qquad (k = 1, 2, \ldots) \\ 0 \text{ otherwise}\end{cases} \\
&c_k = \tfrac{1}{2}(a_k - ib_k) = \tfrac{1}{2}A_k e^{i\varphi_k} \qquad (k = 1, 2, \ldots)
\end{aligned}\right\} \tag{18.10-29b}$$

$$\lim_{T\to\infty} \frac{1}{2T}\int_{-T}^{T} p(t)e^{-i\omega t}\,dt = 0 \tag{18.10-29c}$$

Let $y(t)$ be another real function $y(t)$ satisfying the same conditions as $x(t)$, so that

$$\begin{aligned}
y(t) &= \sum_{k=-\infty}^{\infty} \gamma e_k^{i\omega_k t} + q(t) = \gamma_0 + \sum_{k=1}^{\infty} (\alpha_k\cos\omega_k t + \beta_k\sin\omega_k t) + q(t) \\
&= \gamma_0 + \sum_{x=1}^{\infty} B_k\cos(\omega_k t + \psi_k) + q(t)
\end{aligned} \tag{18.10-30}$$

The set of circular frequencies $\omega_1, \omega_2, \ldots$ is understood to include the periodic-component frequencies of both $x(t)$ and $y(t)$. Then

$$\left.\begin{aligned}
R_{xx}(\tau) &= \sum_{k=-\infty}^{\infty} |c_k|^2 e^{i\omega_k\tau} + R_{pp}(\tau) \\
&= c_0^2 + \tfrac{1}{2}\sum_{k=1}^{\infty} A_k^2\cos\omega_k\tau + R_{pp}(\tau) \\
\Psi_{xx}(\omega) &= 2\pi\sum_{k=-\infty}^{\infty} |c_k|^2\delta(\omega - \omega_k) + \Psi_{pp}(\omega) \\
&= 2\pi c_0^2\delta(\omega) + \frac{\pi}{2}\sum_{k=1}^{\infty} A_k^2[\delta(\omega - \omega_k) + \delta(\omega + \omega_k)] + \Psi_{pp}(\omega)
\end{aligned}\right\} \tag{18.10-31}$$

$$(A_k^2 = a_k^2 + b_k^2,\ k = 1, 2, \ldots)$$

and
$$\left.\begin{aligned}
R_{xy}(\tau) &= \sum_{k=-\infty}^{\infty} c_k^*\gamma_k e^{i\omega_k\tau} + R_{pq}(\tau) \\
&= c_0\gamma_0 + \tfrac{1}{2}\sum_{k=1}^{\infty} [(a_k\alpha_k + b_k\beta_k)\cos\omega_k\tau \\
&\qquad + (a_k\beta_k - b_k\alpha_k)\sin\omega_k\tau] + R_{pq}(\tau) \\
&= c_0\gamma_0 + \tfrac{1}{2}\sum_{k=1}^{\infty} A_k B_k\cos(\omega_k\tau + \psi_k - \varphi_k) + R_{pq}(\tau) \\
\Psi_{xy}(\omega) &= 2\pi\sum_{k=-\infty}^{\infty} c_k^*\gamma_k\delta(\omega - \omega_k) + \Psi_{pq}(\omega)
\end{aligned}\right\} \tag{18.10-32}$$

The crosscorrelation function $R_{xy}(\tau)$ measures the "coherence" of $x(t)$ and $y(t)$ or the "serial correlation" between the function values $x(t)$ and $y(t + \tau)$ separated by a *delay* τ. $x(t)$ and $y(t)$ are **uncorrelated** if and only if $R_{xy}(\tau) \equiv 0$.

NOTE: The (real) functions $x(t)$, $y(t)$ belong to a complex unitary vector space with inner product $(u, v) = \langle u^*(0)v(0)\rangle$ (Sec. 14.2-6). Note the useful orthogonality relations

$$\lim_{T\to\infty} \frac{1}{2T}\int_{-T}^{T} e^{-i\omega t} e^{i\Omega t}\, dt = \begin{cases} 1 & (\omega = \Omega) \\ 0 & (\omega \neq \Omega) \end{cases} \tag{18.10-33}$$

$$\lim_{T\to\infty} \frac{1}{2T}\int_{-T}^{T} \cos(\omega t + \alpha)\cos(\Omega t + \beta)\, dt = \begin{cases} \tfrac{1}{2}\cos(\alpha - \beta) & (\omega = \Omega) \\ 0 & (\omega \neq \Omega) \end{cases} \tag{18.10-34}$$

18.10-10. Generalized Fourier Transforms and Integrated Spectra. (**a**) To avoid the difficulties associated with delta-function terms in the Fourier transforms and spectral densities of periodic functions, one may introduce the **generalized or integrated Fourier transform** $X_{INT}(i\omega)$ of $x(t)$, defined (to within an additive constant) by

$$X_{INT}(i\omega) - X_{INT}(i\omega_0) = \int_{-\infty}^{\infty} x(t)\frac{e^{-i\omega t} - e^{-i\omega_0 t}}{-2\pi i t}\, dt \tag{18.10-35}$$

The corresponding inversion integral is the Stieltjes integral (Sec. 4.6-17)

$$x(t) = \int_{-\infty}^{\infty} e^{i\omega t}\, dX_{INT}(i\omega) \tag{18.10-36}$$

If the Fourier transform $X_F(i\omega)$ of $x(t)$ exists, then

$$X_{INT}(i\omega) - X_{INT}(i\omega_0) = \int_{\omega_0}^{\omega} X_F(i\omega)\frac{d\omega}{2\pi} \qquad X_F(i\omega) = \frac{dX_{INT}(i\omega)}{d(\omega/2\pi)} \tag{18.10-37}$$

If $x(t)$ can be represented as $\sum_{k=-\infty}^{\infty} c_k e^{i\omega_k t}$ (this is, in particular, true for periodic functions; see also Sec. 18.11-1), then $X_{INT}(i\omega)$ is a step function (Sec. 21.9-1).

(**b**) The **integrated power spectrum** $\Phi_{INT}(\omega)$ of a stationary or wide-sense stationary random process generating $x(t)$ is the generalized Fourier transform of its autocorrelation function:

$$\left.\begin{aligned} \Phi(i\omega)_{INT} - \Phi_{INT}(i\omega_0) &= \int_{-\infty}^{\infty} \mathrm{R}_{xx}(\tau)\frac{e^{-i\omega\tau} - e^{-i\omega_0\tau}}{-2\pi i\tau}\, d\tau = \int_{\omega_0}^{\omega} \Phi_{xx}(\omega)\frac{d\omega}{2\pi} \\ \text{with} \qquad \mathrm{R}_{xx}(\tau) &= \int_{-\infty}^{\infty} e^{i\omega\tau}\, d\Phi_{INT}(\omega) \end{aligned}\right\} \tag{18.10-38}$$

Analogous relations can be written for non-ensemble correlation functions and spectra.

(**c**) Note the following generalizations of the Wiener-Khinchine relations (9) and (26) for real stationary (or wide-sense stationary) $x(t)$.

$$\begin{aligned} \frac{1}{2\epsilon} E\{|X_{INT}[i(\omega + \epsilon)] - X_{INT}[i(\omega - \epsilon)]|^2\} &= \frac{1}{2\epsilon}[\Phi_{INT}(\omega + \epsilon) - \Phi_{INT}(\omega - \epsilon)] \\ &= \frac{1}{2\epsilon}\int_{\omega-\epsilon}^{\omega+\epsilon} \Phi_{xx}(\omega)\frac{d\omega}{2\pi} = \frac{1}{2\pi}\int_{-\infty}^{\infty} \mathrm{R}_{xx}(\tau)\frac{\sin \epsilon\tau}{\epsilon\tau} e^{-i\omega t}\, d\tau \end{aligned} \tag{18.10-39}$$

$$\begin{aligned} 2\pi \lim_{\epsilon\to 0} \frac{1}{2\epsilon}\int_{-\infty}^{\infty} &|X_{INT}[i(\omega + \epsilon)] - X_{INT}[i(\omega - \epsilon)]|^2 \cos \omega\tau\, d\omega \\ &= \lim_{T\to\infty}\frac{1}{2T}\int_{-T}^{T} x(t)x(t + \tau)\, dt = R_{xx}(\tau) \end{aligned} \tag{18.10-40}$$

For $\tau = 0$, Eq. (40) yields *Wiener's Quadratic-variation Theorem*

$$\lim_{\epsilon \to 0} \frac{2\pi}{2\epsilon} \int_{-\infty}^{\infty} |X_{INT}[i(\omega + \epsilon)] - X_{INT}[i(\omega - \epsilon)]|^2 \, d\omega = \lim_{T \to \infty} \frac{1}{2T} \int_{-T}^{T} |x(t)|^2 \, dt \quad (18.10\text{-}41)$$

If the non-ensemble power spectral density $\Psi_{xx}(\omega)$ exists, Eq. (40) reduces to the Wiener-Khinchine relation (26), with

$$\Psi_{xx}(\omega) = 2\pi \lim_{\epsilon \to 0} \frac{1}{2\epsilon} |X_{INT}[i(\omega + \epsilon)] - X_{INT}[i(\omega - \epsilon)]|^2 \quad (18.10\text{-}42)$$

18.11 SPECIAL CLASSES OF RANDOM PROCESSES. EXAMPLES

18.11-1. Processes with Constant and Periodic Sample Functions.

(a) Constant Sample Functions (Fig. 18.11-1*a*). If each sample function $x(t)$ is identically equal to a constant random parameter a with given probability distribution, the latter determines the resulting random process uniquely. The process is stationary; but it is not ergodic. If $E\{a^2\}$ exists,

$$E\{x(t)\} = E\{a\} \qquad \mathrm{R}_{xx}(\tau) = E\{a^2\} \quad (18.11\text{-}1a)$$

while

$$\langle x(t) \rangle = a \qquad R_{xx}(\tau) = a^2 \quad (18.11\text{-}1b)$$

(b) Random-phase Sine Waves. Let

$$x(t) = a \sin(\omega t + \alpha) \quad (18.11\text{-}2)$$

(Fig. 18.11-1*b*) where a is a given constant, and the phase angle α is a random variable uniformly distributed between 0 and 2π. The process is stationary and ergodic, with

$$\left.\begin{aligned} \varphi_{(1)}[x(t)] &= \frac{1}{\pi\sqrt{a^2 - x^2}} \qquad (|x| \le a) \\ E\{x(t)\} &= \langle x(t) \rangle = 0 \qquad \mathrm{R}_{xx}(\tau) \equiv R_{xx}(\tau) = \frac{a^2}{2} \cos \omega\tau \end{aligned}\right\} \quad (18.11\text{-}3)$$

If the amplitude a of the random-phase sine wave is not a constant, but is itself a (positive) random variable independent of α (as in amplitude modulation), the process is stationary but not in general ergodic. Now

$$\left.\begin{aligned} \varphi_{(1)}[x(t)] &= \frac{1}{\pi} \int_0^{\infty} \frac{\varphi_a(a)\, da}{\sqrt{a^2 - x^2}} \\ E\{x(t)\} &= 0 \qquad \mathrm{R}_{xx}(\tau) = \tfrac{1}{2} E\{a^2\} \cos \omega\tau \end{aligned}\right\} \quad (18.11\text{-}4)$$

If, in particular, the amplitude a has a *Rayleigh distribution* defined by

$$\varphi_a(a) = \begin{cases} ae^{-a^2/2} & (a \ge 0) \\ 0 & (a < 0) \end{cases} \quad (18.11\text{-}5)$$

(circular normal distribution with $\sigma^2 = 1$, Sec. 18.8-7), then the random process is Gaussian (Sec. 18.11-3).

If the phase angle α is not uniformly distributed between 0 and 2π, then the process is nonstationary even if the amplitude a is fixed.

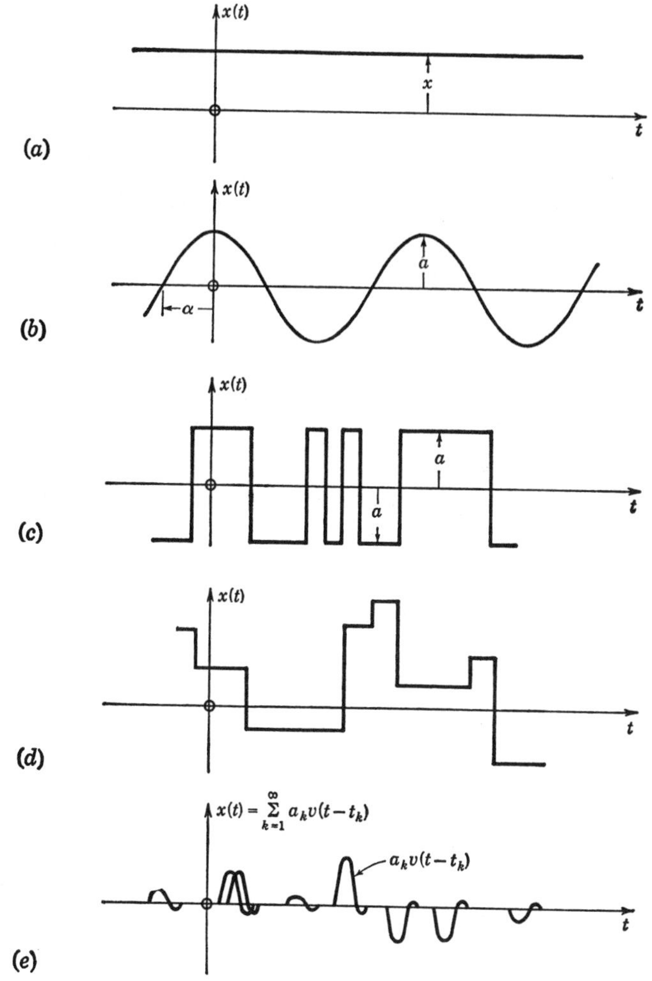

FIG. 18.11-1. Sample functions $x(t)$ for five examples of random processes. In Fig. 18.11-1e, $x(t)$ is the *sum* of the individual pulses $a_k v(t - t_k)$ shown.

(c) More General Periodic Processes (see also Sec. 18.10-9). The random-phase sine wave is a special case of the general **random-phase periodic process** represented by

$$x(t) = c_0 + \sum_{k=1}^{\infty} [a_k \cos k(\omega_0 t + \alpha) + b_k \sin k(\omega_0 t + \alpha)] \quad (18.11\text{-}6)$$

where α is uniformly distributed between 0 and 2π; it is assumed that the series converges in mean square in the sense of Sec. 18.6-3. The

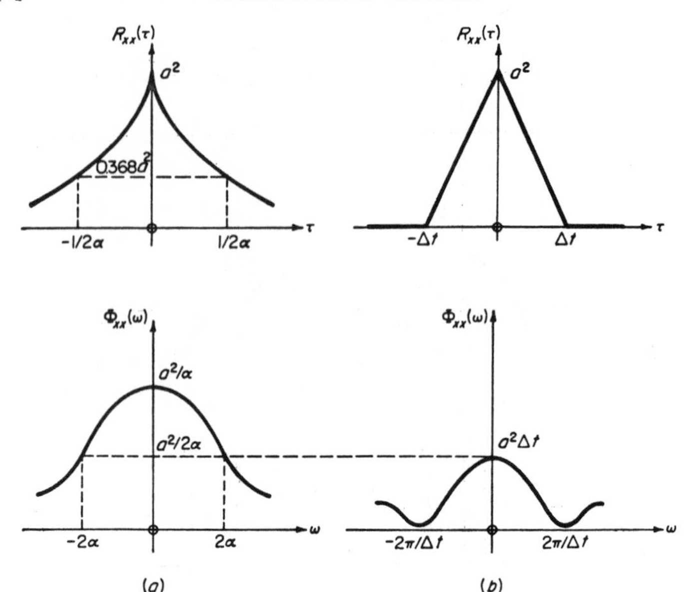

Fig. 18.11-2. Autocorrelation function and power spectrum for a random telegraph wave (*a*) and a coin-tossing sample-hold process (*b*) having equal mean count rates $\alpha = 1/2\Delta t$, both with zero mean and mean square a^2. Note that different ω scales are used in (*a*) and (*b*). (*From G. A. Korn, Random-process Simulation and Measurements, McGraw-Hill, New York,* 1966.)

process is stationary and ergodic, with

$$\left.\begin{aligned} E\{x(t)\} &= \langle x(t)\rangle = c_0 \\ \mathrm{R}_{xx}(\tau) \equiv R_{xx}(\tau) &= c_0^2 + \frac{1}{2}\sum_{k=1}^{\infty}(a_k^2 + b_k^2)\cos k\omega_0\tau \\ \Phi_{xx}(\omega) \equiv \Psi_{xx}(\omega) &\equiv 2\pi c_0^2\delta(\omega) \\ &\quad + \frac{\pi}{2}\sum_{k=1}^{\infty}(a_k^2 + b_k^2)[\delta(\omega - k\omega_0) + \delta(\omega + k\omega_0)] \end{aligned}\right\} \tag{18.11-7}$$

A still more general periodic process is defined by the Fourier series

$$x(t) = c_0 + \sum_{k=1}^{\infty}(a_k \cos k\omega_0 t + b_k \sin k\omega_0 t) \tag{18.11-8}$$

with real random coefficients c_0, a_k, b_k, assuming that the series converges in mean square. Such a process is wide-sense stationary if and only if

$$\left.\begin{aligned} E\{a_k\} &= E\{b_k\} = 0 \qquad E\{a_k^2\} = E\{b_k^2\} \\ E\{c_0a_i\} &= E\{c_0b_k\} = E\{a_ib_k\} = 0 \\ E\{a_ia_k\} &= E\{b_ib_k\} = 0 \qquad (i \neq k) \end{aligned}\right\} \tag{18.11-9}$$

In this case, Eq. (8) is an orthogonal-series expansion in the sense of Sec. 18.9-5, and

$$E\{x(t)\} = E\{c_0\} \qquad R_{xx}(\tau) = E\{c_0^2\} + \tfrac{1}{2}\sum_{k=1}^{\infty} E\{a_k^2 + b_k^2\} \cos k\omega_0\tau \qquad (18.11\text{-}10)$$

18.11-2. Band-limited Functions and Processes. Sampling Theorems. (a) A function $x(t)$ is **band-limited** between $\omega = 0$ and $\omega = 2\pi B$ if and only if its Fourier transform $X_F(i\omega)$ (Sec. 4.11-3) exists and equals zero for $|\omega| > 2\pi B$; B (measured in cycles per second if t is measured in seconds) is the **bandwidth** associated with $x(t)$. *For every band-limited* $x(t)$

$$x(t) = \sum_{k=-\infty}^{\infty} x(t_k) \frac{\sin 2\pi B(t - t_k)}{2\pi B(t - t_k)} \qquad \left(t_k = \frac{k}{2B},\ k = 0, \pm 1, \pm 2, \ldots\right) \qquad (18.11\text{-}11)$$

i.e., $x(t)$ is uniquely determined for all t by samples $x(t_k)$ spaced $1/2B$ t-units apart (Nyquist-Kotelnikov-Shannon Sampling Theorem).

The functions (Fig. 18.11-3)

$$u_k(t) \equiv \sqrt{2B}\,\text{sinc}\,2B(t - t_k) \equiv \sqrt{2B}\,\frac{\sin 2\pi B(t - t_k)}{2\pi B(t - t_k)} \qquad (k = 0, \pm 1, \pm 2, \ldots) \qquad (18.11\text{-}12)$$

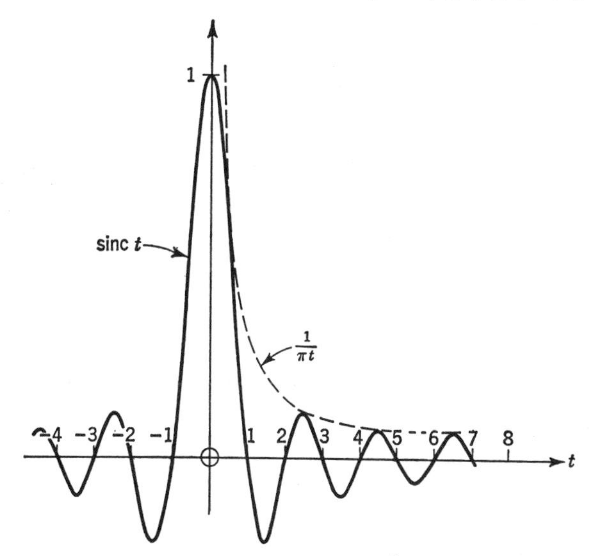

FIG. 18.11-3. The sampling function $\text{sinc}\, t \equiv \dfrac{\sin \pi t}{\pi t}$ (see also Table F-21).

constitute a complete orthonormal set for the space of functions $x(t)$ band-limited between $\omega = 0$ and $\omega = 2\pi B$ (Sec. 15.2-4); note

$$x_k = x(t_k) = 2B \int_{-\infty}^{\infty} x(t) \operatorname{sinc} 2B(t - t_k)\, dt \qquad (k = 0, \pm 1, \pm 2, \ldots)$$

("SAMPLING PROPERTY" OF THE sinc FUNCTION) (18.11-13)

$$\int_{-\infty}^{\infty} \operatorname{sinc}(\lambda - i) \operatorname{sinc}(\lambda - k)\, d\lambda = 2B \int_{-\infty}^{\infty} \operatorname{sinc} 2B(t - t_i) \operatorname{sinc} 2B(t - t_k)\, dt$$

$$= \delta_k^i = \begin{Bmatrix} 1 & (i = k) \\ 0 & (i \neq k) \end{Bmatrix} \qquad (i, k = 0, \pm 1, \pm 2, \ldots) \tag{18.11-14}$$

(b) A stationary or wide-sense stationary random process with sample functions $x(t)$ is **band-limited** between $\omega = 0$ and $\omega = 2\pi B$ if and only if its ensemble power spectral density $\Phi_{xx}(\omega)$ exists and equals zero for $|\omega| > 2\pi B$. In this case, *the expansion* (11) *applies in the sense of mean-square convergence* (Sec. 18.6-3), *i.e.*,

$$E\left\{\left[x(t) - \sum_{k=-\infty}^{\infty} x_k \operatorname{sinc} 2B(t - t_k)\right]^2\right\} = 0 \tag{18.11-15}$$

and Eq. (11) *represents each sample function* $x(t)$ *in terms of its sample values* $x_k = x(k/2B)$ *with probability one.*

NOTE: In the special case of a stationary band-limited "flat-spectrum" process with

$$\Phi_{xx}(\omega) = \begin{cases} \Phi_0 & (|\omega| \leq 2\pi B) \\ 0 & (|\omega| > 2\pi B) \end{cases} \qquad R_{xx}(\tau) = 2\Phi_0 B \frac{\sin 2\pi B\tau}{2\pi B\tau} \tag{18.11-16}$$

the sample values $x_k = x(k/2B)$ have zero mean and are uncorrelated.

18.11-3. Gaussian Random Processes (see also Secs. 18.8-3 to 18.8-8, and 18.12-6). A real random process is **Gaussian** if and only if all its probability distributions are normal distributions for all $t_1, t_2, \ldots$. *Every Gaussian process is uniquely defined by its (necessarily normal) second-order probability distribution, and hence by the ensemble autocorrelation function* $R_{xx}(t_1, t_2) \equiv E\{x(t_1)x(t_2)\}$ *together with* $\xi(t) \equiv E\{x(t)\}$. Specifically, the joint distribution of every set of sample values $x_1 = x(t_1)$, $x_2 = x(t_2), \ldots, x_n = x(t_n)$ is a normal distribution with probability density

$$\varphi(x_1, x_2, \ldots, x_n)$$

$$\equiv \frac{1}{\sqrt{(2\pi)^n \det[\lambda_{jk}]}} \exp\left[-\frac{1}{2} \sum_{j=1}^{n} \sum_{k=1}^{n} \Lambda_{jk}(x_j - \xi_j)(x_k - \xi_k)\right] \tag{18.11-17}$$

with

$$\xi_j = E\{x(t_j)\} \qquad \lambda_{jk} = R_{xx}(t_j, t_k) - \xi_j \xi_k \qquad (j, k = 1, 2, \ldots, n) \tag{18.11-18}$$

$$[\Lambda_{ik}] \equiv [\lambda_{ik}]^{-1} \tag{18.11-19}$$

Processes obtained through addition of Gaussian processes and/or linear operations on their sample functions are Gaussian (Sec. 18.12-2). Coefficients in orthogonal-function expansions of a Gaussian process (Sec. 18.9-5) are jointly Gaussian random variables.

18.11-4. Markov Processes and the Poisson Process. **(a) Random Process of Order n.** **A random process of order n** is a random process completely specified by its nth (n^{th}-order) distribution function $\Phi_{(n)}$ (Sec. 18.9-2), but not by $\Phi_{(n-1)}$.

(b) Purely Random Processes. A random process described by $x(t)$ is a **purely random process** if and only if the random variables $x(t_1)$, $x(t_2)$, . . . are statistically independent for every finite set t_1, t_2, ... *A purely random process is completely specified by* $\Phi_{(1)}(X_1, t_1)$, $p_{(1)}(X_1, t_1)$, *or* $\varphi_{(1)}(X_1, t_1)$.

EXAMPLES: Successive independent observations, Bernoulli trials, and random samples in statistics (Sec. 19.1-2) represent *purely random series.* Purely random *continuous-parameter* processes imply sample functions of unlimited bandwidth and cannot, strictly speaking, describe real physical phenomena.

(c) Markov Processes. A discrete or continuous random process described by $x(t)$ is a (simple) **Markov process** if and only if, for every finite set $t_1 < t_2 < \cdots < t_{n-1} < t_n$,

$$p(X_n, t_n|X_1, t_1; \ldots ; X_{n-1}, t_{n-1}) = p(X_n, t_n|X_{n-1}, t_{n-1}) \qquad (18.11\text{-}20a)$$

$$\text{or} \quad \varphi(X_n, t_n|X_1, t_1; \ldots ; X_{n-1}, t_{n-1}) = \varphi(X_n, t_n|X_{n-1}, t_{n-1}) \qquad (18.11\text{-}20b)$$

respectively. If $x(t_{n-1}) = X_{n-1}$ is given, knowledge of $x(t_{n-2})$, $x(t_{n-3})$, . . . contributes nothing to one's knowledge of the distribution of $x(t_n)$. *A Markov process is completely specified by its second-order probability distribution* and hence by its first-order probability distribution together with the "transition probabilities" given by

$$p(X_2, t_2|x, t) \qquad \text{or} \qquad \varphi(X_2, t_2|x, t) \qquad (t < t_2) \qquad (18.11\text{-}21)$$

A Markovian random series is often called a **Markov chain.** Every purely random process is a Markov process.

Many physical processes can be described as Markov processes. *An important class of problems involves the determination of the functions* (21) *from their given "initial values" specified for* $t = t_1$. The defining property (20) of a Markov process implies the *Chapman-Kolmogorov-Smoluchovski equation*

$$p(X_2, t_2|X_1, t_1) = \sum_{x(t)} p(X_2, t_2|x, t)p(x, t|X_1, t_1) \qquad (t_1 \leq t \leq t_2) \qquad (18.11\text{-}22a)$$

$$\text{or} \quad \varphi(X_2, t_2|X_1, t_1) = \int_{-\infty}^{\infty} \varphi(X_2, t_2|x, t)\varphi(x, t|X_1, t_1)\, dx \qquad (t_1 \leq t \leq t_2) \qquad (18.11\text{-}22b)$$

Equation (22) is a first-order difference equation (Sec. 20.4-3) which may be solved for the unknown function (21) of the independent variable t whenever $p(x, t|X_1, t_1)$ or $\varphi(x, t|X_1, t_1)$ is suitably given. If $p_{(1)}(X_1, t_1)$ or $\varphi_{(1)}(X_1, t_1)$ is known, the Markov process is now completely determined for all $t > t_1$.

(d) The Poisson Process. In many problems involving random searches, waiting lines, radioactive decay, etc., $x(t)$ is a discrete random variable capable of assuming the spectral values 0, 1, 2, . . . ("counting process"; number of "successes," telephone calls, disintegrations, etc.). A frequently useful model assumes the Markov property (20a) and

$$p(X_2, t_2|x, t) = \begin{Bmatrix} 0 \text{ if } X_2 < x \\ 1 - \alpha\,\Delta t - o(\Delta t) \text{ if } X_2 = x \\ \alpha\,\Delta t + o(\Delta t) \text{ if } X_2 = x + 1 \\ o(\Delta t) \text{ if } X_2 > x + 1 \end{Bmatrix}$$
$$(\Delta t = t_2 - t > 0;\ x, X = 0, 1, 2, \ldots) \quad (18.11\text{-}23)$$

where $o(\Delta t)$ denotes a term such that $o(\Delta t)/\Delta t$ becomes negligible as $\Delta t \to 0$ (Sec. 4.4-3). To find

$$p(X, t|X_1, t_1) \equiv P(K, T)\ (K = X - X_1 = 0, 1, 2, \ldots;\ T = t - t_1)$$

substitute the given transition probabilities (23) into the Smoluchovski equation (22a) for $t_2 = t + \Delta t$ to obtain the difference equation

$$\frac{P(K, T + \Delta t) - P(K, T)}{\Delta t} = -\alpha P(K, T) + \alpha P(K - 1, T) + \frac{o(\Delta t)}{\Delta t}$$
$$(K = 0, 1, 2, \ldots) \quad (18.11\text{-}24)$$

with $P(-1, T) \equiv 0$. As $\Delta t \to 0$, this reduces to an ordinary differential equation

$$\frac{\partial}{\partial T} P(K, T) = -\alpha P(K, T) + \alpha P(K - 1, T) \qquad (K = 0, 1, 2, \ldots) \quad (18.11\text{-}25)$$

for each K. These differential equations are solved successively for $P(0, T), P(1, T), P(2, T), \ldots$, with initial conditions given by

$$P(0, 0) = p(X_1, t_1|X_1, t_1) = 1 \qquad P(K, 0) = p(X_1 + K, t_1|X_1, t_1) = 0$$
$$(K > 0) \quad (18.11\text{-}26)$$

It follows that

$$P(K, T) = e^{-\alpha T}\frac{(\alpha T)^K}{K!} \qquad (T \geq 0;\ K = 0, 1, 2, \ldots) \quad (18.11\text{-}27)$$

Thus, once the process is started, *the number K of state changes in every time interval of length T has the Poisson distribution* (Table 18.8-4). α is called the **mean count rate** of the Poisson process.

The probability that *no* state changes take place is

$$P(0, T) = e^{-\alpha T} \qquad (T \geq 0) \tag{18.11-28}$$

so that the probability that *at least one* state change takes place is

$$1 - P(0, T) = 1 - e^{-\alpha T} \quad (T \geq 0) \tag{18.11-29}$$

The time interval T_1 between successive state changes is a random variable with probability density

$$\varphi_{T_1}(T_1) = \alpha e^{-\alpha T_1} \tag{18.11-30}$$

and expected value $1/\alpha$.

Within any finite time interval of length T, a Poisson process is also uniquely defined by the joint distribution of the $K + 1$ statistically independent random variables K, $t_1, t_2, \ldots, t_K$, where K is the number of state changes during the time T, and $t_1, t_2, \ldots, t_K$ are now the respective times of the 1st, 2nd, . . . , Kth state change during this time interval. One has

$$\left.\begin{aligned} p_K(K) &= P(K, T) = \frac{(\alpha T)^K}{K!} e^{-\alpha T} \qquad (K = 0, 1, 2, \ldots) \\ \varphi_{t_k|K}(t_k|K) &= \frac{1}{T} \qquad (K = 1, 2, \ldots ; \ k = 1, 2, \ldots, K) \\ \varphi_{t_1 t_2 \cdots t_k|K}(t_1, t_2, \ldots, t_K|K) &= \frac{1}{T^K} \qquad (K = 1, 2, \ldots) \end{aligned}\right\} \tag{18.11-31}$$

(e) See Refs. 18.15, 18.16, and 18.17 for treatments of more general Markov processes.

18.11-5. Some Random Processes Generated by a Poisson Process.

(a) Random Telegraph Wave (Fig. 18.11-1*c*). $x(t)$ equals either a or $-a$, with sign changes generated by the state changes of a Poisson process of mean count rate α (Sec. 18.11-4*d*). The process is stationary and ergodic if started at $t = -\infty$, and

$$\left.\begin{aligned} E\{x(t_1)\} &= 0 \qquad R_{xx}(\tau) = a^2 e^{-2\alpha|\tau|} \\ \Phi_{xx}(\omega) &= \frac{4a^2\alpha}{\omega^2 + (2\alpha)^2} \end{aligned}\right\} \tag{18.11-32}$$

(b) Process Generated by Poisson Sampling (Fig. 18.11-1*d*). $x(t)$ changes value at each state change of a Poisson process with mean count rate α; between state changes, $x(t)$ is constant and takes continuously distributed random values x with given mean ξ and variance σ^2. The process is stationary and ergodic if started at $t = -\infty$, and

$$\left.\begin{aligned} E\{x(t_1)\} &= \xi \qquad R_{xx}(\tau) = \xi^2 + \sigma^2 e^{-\alpha|\tau|} \\ \Phi_{xx}(\omega) &= 2\pi\xi^2\delta(\omega) + \frac{2\sigma^2\alpha}{\omega^2 + \alpha^2} \end{aligned}\right\} \tag{18.11-33}$$

(c) Impulse Noise and Campbell's Theorem (Fig. 18.11-1*e*). $x(t)$ is the sum of many similarly shaped transient pulses,

$$x(t) = \sum_{k=1}^{\infty} a_k v(t - t_k) \tag{18.11-34}$$

whose shape is given by $v = v(t)$, with

$$\int_{-\infty}^{\infty} v(t)e^{-i\omega t}\,dt = V_F(i\omega) \qquad (18.11\text{-}35)$$

while the pulse amplitude a_k is a random variable with finite variance, and the times t_k are random incidence times determined by the state changes of a Poisson process with mean count rate α. The process is stationary and ergodic if started at $t = -\infty$; it approximates a Gaussian random process if many pulses overlap. One has

$$\left.\begin{aligned} E\{x(t_1)\} &= \xi = \alpha E\{a_k\}\int_{-\infty}^{\infty} v(t)\,dt \\ E\{x^2(t_1)\} &= \xi^2 + \alpha E\{a_k^2\}\int_{-\infty}^{\infty} v^2(t)\,dt \\ R_{xx}(\tau) &= \xi^2 + \alpha E\{a_k^2\}\int_{-\infty}^{\infty} v(t)v(t+\tau)\,dt \\ \Phi_{xx}(\omega) &= 2\pi\xi^2\delta(\omega) + \alpha E\{a_k^2\}|V_F(i\omega)|^2 \end{aligned}\right\} \qquad (18.11\text{-}36)$$

In the special case where a_k is a fixed constant, the formulas (36) are known as *Campbell's theorem.*

18.11-6. Random Processes Generated by Periodic Sampling. Certain measuring devices sample a stationary and ergodic random variable $q(t)$ periodically and then hold their output $x(t)$ for a constant sampling interval Δt. The resulting random process is stationary and ergodic if the timing of the periodic sampling commands is random and uniformly distributed between 0 and Δt. A sample function $x(t)$ will be similar to Fig. 18.11-1*d except that state changes must be separated by integral multiples of* Δt. If q is a binary random variable capable of assuming only the values a and $-a$ with probabilities ½, ½, then $x(t)$ will resemble the random telegraph wave of Fig. 18.11-1*c*, except that state changes are, again, separated by integral multiples of Δt ("coin-tossing" sample-hold process).

If different samples of q are statistically independent, then

$$\left.\begin{aligned} R_{xx}(\tau) &= [E\{q\}]^2 = [E\{x\}]^2 = \xi^2 \qquad (|\tau| > \Delta t) \\ R_{xx}(\tau) &= E\{q^2\}\ \text{Prob}\,[t, t+\tau \text{ are in same sampling interval}] \\ &\quad + [E\{q\}]^2\ \text{Prob}\,[t, t+\tau \text{ are not in same sampling interval}] \\ &= \text{Var}\,\{x\}\left(1 - \frac{|\tau|}{\Delta t}\right) + \xi^2 \qquad (|\tau| \le \Delta t) \end{aligned}\right\} \qquad (18.11\text{-}37)$$

and hence

$$\Phi_{xx}(\omega) = \text{Var}\,\{x\}\,\Delta t\left[\frac{\sin(\omega\,\Delta t/2)}{\omega\,\Delta t/2}\right]^2 + 2\pi\xi^2\delta(\omega) \qquad (18.11\text{-}38)$$

Figure 18.11-2 compares $R_{xx}(\tau)$ and $\Phi_{xx}(\omega)$ for a random telegraph wave and a coin-tossing sample-hold process with equal mean count rates $\alpha = 1/2\Delta t$, zero mean, and $E\{x^2\} = a^2$.

18.12. OPERATIONS ON RANDOM PROCESSES

18.12-1. Correlation Functions and Spectra of Sums. Let $x(t)$, $y(t)$ be generated by real or complex random processes. For

$$z(t) = \alpha x(t) + \beta y(t) \tag{18.12-1}$$

with real or complex α, β, the correlation functions $R_{xz}(t_1, t_2)$, $R_{zx}(t_1, t_2)$, $R_{zz}(t_1, t_2)$ are given by

$$\begin{gathered} R_{xz} = \alpha R_{xx} + \beta R_{xy} \qquad R_{zx} = \alpha^* R_{xx} + \beta^* R_{yx} \\ R_{zz} = |\alpha|^2 R_{xx} + |\beta|^2 R_{yy} + \alpha^* \beta R_{xy} + \alpha\beta^* R_{yx} \end{gathered} \tag{18.12-2}$$

These relations also apply to the correlation functions $R_{xz}(\tau)$, $R_{zx}(\tau)$, $R_{zz}(\tau)$ of stationary random processes; the corresponding spectral densities are

$$\begin{gathered} \Phi_{xz} = \alpha \Phi_{xx} + \beta \Phi_{xy} \qquad \Phi_{zx} = \alpha^* \Phi_{xx} + \beta^* \Phi_{yx} \\ \Phi_{zz} = |\alpha|^2 \Phi_{xx} + |\beta|^2 \Phi_{yy} + \alpha^* \beta \Phi_{xy} + \alpha\beta^* \Phi_{yx} \end{gathered} \tag{18.12-3}$$

18.12-2. Input-Output Relations for Linear Systems. (a) Consider a real linear system with real input $x(t)$ and output

$$y(t) = \int_{-\infty}^{\infty} w(t, \lambda) x(\lambda)\, d\lambda = \int_{-\infty}^{\infty} h(t, \zeta) x(t - \zeta)\, d\zeta \tag{18.12-4}$$

where the weighting function (Green's function, Secs. 9.3-3 and 9.4-3) is the system response to a unit-impulse input $\delta(t - \lambda)$ (impulse applied at $t = \lambda$), and $h(t, \zeta) \equiv w(t, t - \zeta)$.

In the most important applications, t represents time, and $w(t, \lambda) = 0$ for $t < \lambda$, since physically realizable systems cannot respond to future inputs (see also Sec. 9.4-3).

(b) If $x(t)$ is generated by a real random process, and if $E\{x^2(t)\}$ and $E\{y^2(t)\}$ exist, then

$$E\{y(t)\} = \int_{-\infty}^{\infty} w(t, \lambda) E\{x(\lambda)\}\, d\lambda \tag{18.12-5}$$

$$\begin{aligned} R_{xy}(t_1, t_2) &= R_{yx}(t_2, t_1) = \int_{-\infty}^{\infty} w(t_2, \lambda) R_{xx}(t_1, \lambda)\, d\lambda \\ R_{yy}(t_1, t_2) &= \int_{-\infty}^{\infty} w(t_2, \mu) R_{yx}(t_1, \mu)\, d\mu \end{aligned} \tag{18.12-6}$$

(GENERALIZED WIENER-LEE RELATIONS)

$$E\{y^2(t)\} = \int_{-\infty}^{\infty} w(t, \mu) R_{yx}(t, \mu)\, d\mu \tag{18.12-7}$$

If $x(t)$ is Gaussian, $y(t)$ is also Gaussian and completely determined by Eqs. (5) *to* (7).

18.12-3. The Stationary Case. (a) *If the input $x(t)$ is stationary, and*

$$\left.\begin{aligned} w(t, \lambda) \equiv h(t - \lambda) \qquad h(t, \zeta) \equiv h(\zeta) &\equiv \int_{-\infty}^{\infty} H(i\omega)e^{i\omega\zeta} \frac{d\omega}{2\pi} \\ \text{with} \qquad H(i\omega) &= \int_{-\infty}^{\infty} h(\zeta)e^{-i\omega\zeta}\, d\zeta \end{aligned}\right\} \tag{18.12-8}$$

(**time-invariant** linear system, see also Sec. 9.4-3), *then the system output $y(t)$ is also stationary;* $y(t)$ will be ergodic (Sec. 18.10-7*b*) if this is true for $x(t)$. The input-output relations (4) to (7) for real $x(t)$, $y(t)$ reduce to

$$y(t) = \int_{-\infty}^{\infty} h(t - \lambda)x(\lambda)\, d\lambda = \int_{-\infty}^{\infty} h(\zeta)x(t - \zeta)\, d\zeta \tag{18.12-9}$$

$$E\{y\} = E\{x\} \int_{-\infty}^{\infty} h(\zeta)\, d\zeta \tag{18.12-10}$$

$$\begin{aligned} R_{xy}(\tau) &= R_{yx}(-\tau) \\ &= \int_{-\infty}^{\infty} h(\tau - \lambda)R_{xx}(\lambda)\, d\lambda = \int_{-\infty}^{\infty} h(\zeta)R_{xx}(\tau - \zeta)\, d\zeta \\ R_{yy}(\tau) & \\ &= \int_{-\infty}^{\infty} h(\tau - \mu)R_{yx}(\mu)\, d\mu = \int_{-\infty}^{\infty} h(\zeta)R_{yx}(\tau - \zeta)\, d\zeta \end{aligned} \tag{18.12-11}$$

(WIENER-LEE RELATIONS)

$$E\{y^2\} = \int_{-\infty}^{\infty} h(\zeta)R_{xy}(\zeta)\, d\zeta \tag{18.12-12}$$

In most applications, physical realizability requires $h(\zeta) = 0$ for $\zeta < 0$ (see also Sec. 9.4-3).

(**b**) The important input-output relations (11) are greatly simplified if they are expressed in terms of spectral densities (Sec. 18.10-3):

$$\begin{aligned} \Phi_{xy}(\omega) = H(i\omega)\Phi_{xx}(\omega) \qquad \Phi_{yx}(\omega) = H^*(i\omega)\Phi_{xx}(\omega) \\ \Phi_{yy}(\omega) = H^*(i\omega)\Phi_{xy}(\omega) = H(i\omega)\Phi_{yx}(\omega) = |H(i\omega)|^2\Phi_{xx}(\omega) \end{aligned} \tag{18.12-13}$$

(**c**) Note also

$$\left.\begin{aligned} R_{yy}(\tau) &= \int_{-\infty}^{\infty} \varphi_{hh}(\lambda)R_{xx}(\tau - \lambda)\, d\lambda \\ \text{with} \qquad \varphi_{hh}(\lambda) &\equiv \int_{-\infty}^{\infty} h(\mu)h(\mu + \lambda)\, d\mu \equiv \int_{-\infty}^{\infty} |H(i\omega)|^2 e^{i\omega\lambda} \frac{d\omega}{2\pi} \end{aligned}\right\} \tag{18.12-14}$$

In the special case of stationary white-noise input with $R_{xx}(\tau) \equiv \Phi_0\delta(\tau)$ (Sec. 18.11-4b), note

$$R_{xy}(\tau) \equiv \Phi_0 h(\tau) \tag{18.12-15}$$

$$R_{yy}(\tau) \equiv \Phi_0 \varphi_{hh}(\tau) \tag{18.12-16}$$

$$E\{y^2\} \equiv \Phi_0\varphi_{hh}(0) = \Phi_0 \int_{-\infty}^{\infty} h^2(\zeta)\, d\zeta \tag{18.12-17}$$

18.12-4. Relations for t Correlation Functions and Non-ensemble Spectra. *The relations* (2), (4), *and* (10) *to* (17) *all hold if each ensemble average, correlation function, and spectral density is replaced by the corresponding t average, t correlation function, and non-ensemble spectral density* (Secs. 18.10-7 to 18.10-9), *whenever these quantities exist.*

18.12-5. Nonlinear Operations. Given a random process generating $x(t)$ and a single-valued, measurable function $y = y(x)$, the functions

$$y = y(x) = y[x(t)] \tag{18.12-18}$$

represent a new random process produced by a (generally nonlinear) zero-memory operation on the $x(t)$; $y(x)$ does not depend explicitly on t. Distributions and ensemble averages of the y process are obtained by the methods of Secs. 18.5-2 and 18.5-4. In particular, the autocorrelation function of the "output" y is, for real variables,

$$R_{yy}(t_1, t_2) = \int_{-\infty}^{\infty} \int_{-\infty}^{\infty} y(x_1)y(x_2)\varphi_{x_1,x_2}(x_1, x_2)\, dx_1\, dx_2 \tag{18.12-19}$$

where $x_1 = x(t_1)$, $x_2 = x(t_2)$; $y_1 = y(x_1)$, $y_2 = y(x_2)$.

If this turns out to be more convenient, $R_{yy}(t_1, t_2)$ can be obtained in the form

$$\left.\begin{aligned} R_{yy}(t_1, t_2) &= -\frac{1}{(2\pi)^2} \int_{C_1} \int_{C_2} M_{x_1x_2}(s_1, s_2) Y(s_1) Y(s_2)\, ds_1\, ds_2 \\ \text{with} \qquad Y(s) &= \int_{-\infty}^{\infty} y(x) e^{-xs}\, dx \qquad (\sigma_1 < \operatorname{Re} s < \sigma_2) \end{aligned}\right\} \tag{18.12-20}$$

where the integration contours C_1, C_2 parallel the imaginary axis in suitable absolute-convergence strips (Ref. 18.15). The "transform method" is especially useful in connection with certain practically important transfer characteristics $y(x)$, e.g., limiters, half-wave detectors, quantizers, etc. (Refs. 18.13 and 18.15).

18.12-6. Nonlinear Operations on Gaussian Processes. **(a) Price's Theorem** (Ref. 18.17). *Given two jointly normal random variables x_1, x_2 with covariance λ_{12} and a function $f(x_1, x_2)$ such that*

$$|f(x_1, x_2)| < ae^{(x_1{}^b + x_2{}^b)}$$

for some real $a > 0$, $b < 2$, then

$$\frac{\partial^n}{\partial \lambda_{12}{}^n} E\{f(x_1, x_2)\} = E\left\{\frac{\partial^{2n} f(x_1, x_2)}{\partial x_1{}^n\, \partial x_2{}^n}\right\} \qquad (n = 1, 2, \ldots) \tag{18.12-21}$$

Price's theorem yields ensemble averages (and, in particular, correlation functions) in the form

$$E\{f(x_1, x_2)\} = \int_0^{\lambda_{12}} E\left\{\frac{\partial^2 f(x_1, x_2)}{\partial x_1\, \partial x_2}\right\} d\lambda_{12} + C \qquad (18.12\text{-}22)$$

where C is the value of $E\{f(x_1, x_2)\}$ for $\lambda_{12} = 0$, i.e., for uncorrelated x_1, x_2.

Price's theorem also leads to the useful recursion formula

$$E\{x_1^n x_2^m\} = mn \int_0^{\lambda_{12}} E\{x_1^{m-1} x_2^{n-1}\}\, d\lambda_{12} + E\{x_1^m\}E\{x_2^n\} \qquad (n, m = 1, 2, \ldots) \qquad (18.21\text{-}23)$$

In particular,

$$E\{x_1^2 x_2^2\} = 2\lambda_{12}^2 + 4\lambda_{12}E\{x_1\}E\{x_2\} + E\{x_1^2\}E\{x_2^2\} \qquad (18.12\text{-}24)$$

(b) Series Expansion. Given a stationary Gaussian process $x(t)$ with $E\{x\} = 0$, $R_{xx}(\tau) = \sigma^2 \rho_{xx}(\tau)$ and a function $y = y(x)$ such that $R_{yy}(\tau)$ exists, then

$$\left.\begin{aligned} &E\{y\} = a_0 \qquad R_{xy}(\tau) = \frac{a_1}{\sigma} R_{xx}(\tau) \qquad R_{yy}(\tau) = \sum_{k=0}^{\infty} a_k^2 \rho_{xx}^k(\tau) \\ &\text{with} \quad a_k = \frac{1}{\sqrt{2^k \pi k!}} \int_{-\infty}^{\infty} y(\sigma v \sqrt{2}) e^{-v^2} H_k(v)\, dv \qquad (k = 0, 1, 2, \ldots) \end{aligned}\right\} \qquad (18.12\text{-}25)$$

where the $H_k(v)$ are the Hermite polynomials defined in Table 21.7-1.

18.13. RELATED TOPICS, REFERENCES, AND BIBLIOGRAPHY

18.13-1. Related Topics. The following topics related to the study of probability theory and random processes are treated in other chapters of this handbook:

Measure, Lebesgue integrals, Stieltjes integrals, Fourier analysis....... Chap. 4
Construction of mathematical models, abstract spaces, Boolean algebras. Chap. 12
Orthogonal-function expansions.. Chap. 15
Mathematical statistics, random-process measurements and tests...... Chap. 19
Permutations and combinations.. Appendix C

18.13-2. References and Bibliography (see also Sec. 19.9-2).

18.1. Arley, N., and K. R. Buch: *Introduction to the Theory of Probability and Statistics*, Wiley, New York, 1950.
18.2. Burington, R. S., and D. C. May: *Handbook of Probability and Statistics*, 2d ed., McGraw-Hill, New York, 1967.
18.3. Cramér, H.: *Mathematical Methods of Statistics*, Princeton, Princeton, N.J., 1951.
18.4. ———: *The Elements of Probability Theory and Some of Its Applications*, Wiley, New York, 1955.
18.5. Feller, W.: *An Introduction to Probability Theory and Its Applications*, vol. I, 2d ed., Wiley, New York, 1958; vol. II, 1966.
18.6. Gnedenko, B. V.: *Theory of Probability*, Chelsea, New York, 1962.
18.7. ——— and A. I. Khinchine: *An Elementary Introduction to the Theory of Probability*, Dover, New York, 1961.

18.8. Loéve, M. M.: *Probability Theory,* 3d ed., Van Nostrand, Princeton, N.J., 1963.
18.9. Parzen, E.: *Modern Probability Theory and Its Applications,* Wiley, New York, 1960.
18.10. Richter, H.: *Wahrscheinlichkeitstheorie,* 2d ed., Springer, Berlin, 1967.

Random Processes

18.11. Bailey, N. T. J.: *The Elements of Stochastic Processes with Applications to the Natural Sciences,* Wiley, New York, 1964.
18.12. Bharucha-Reid, A. J.: *Elements of the Theory of Markov Processes and Their Applications,* McGraw-Hill, New York, 1960.
18.13. Davenport, W. B., Jr., and W. L. Root: *Introduction to Random Signals and Noise,* McGraw-Hill, New York, 1958.
18.14. Doob, J. L.: *Stochastic Processes,* Wiley, New York, 1953.
18.15. Middleton, D.: *An Introduction to Statistical Communication Theory,* McGraw-Hill, New York, 1960.
18.16. Parzen, E.: *Stochastic Processes,* Holden-Day, San Francisco, 1962.
18.17. Papoulis, A.: *Probability, Random Variables, and Stochastic Processes,* McGraw-Hill, New York, 1965.
18.18. Rosenblatt, M.: *Random Processes,* Oxford, New York, 1962.
18.19. Saaty, T. L.: *Elements of Queueing Theory with Applications,* McGraw-Hill, New York, 1961.

CHAPTER 19

MATHEMATICAL STATISTICS

19.1. INTRODUCTION TO STATISTICAL METHODS

19.1-1. Statistics. In the most general sense of the word, **statistics** is the art of using quantitative empirical data to describe experience and to infer and test propositions (numerical estimates, hypothetical correspondences, predicted results, decisions). More specifically, statistics deals (1) with the *statistical description* of processes or experiments, and (2) with the induction and testing of corresponding mathematical models involving the *probability concept.* The relevant portions of probability theory constitute the field of *mathematical statistics.* These techniques extend the possibility of scientific prediction and rational decisions to many situations where deterministic prediction fails because essential parameters cannot be known or controlled with sufficient accuracy.

Statistical description and probability models apply to physical processes exhibiting the following empirical phenomenon. *Even though individual measurements of a physical quantity x cannot be predicted with sufficient accuracy, a suitably determined function $y = y(x_1, x_2, \ldots)$ of a set* (**sample**) *of repeated measurements $x_1, x_2, \ldots$ of x can often be predicted with substantially better accuracy,* and the prediction of y may still yield useful decisions. Such a function y of a set of sample values is called a **statistic,** and the incidence of increased predictability is known as *statistical regularity.* Statistical regularity, in each individual situation, is an empirical physical law which, like the law of gravity or the induction law, is ultimately derived from experience and *not* from mathematics. Frequently a statistic can be predicted with increasing accuracy as the **size** n of the sample $(x_1, x_2, \ldots, x_n)$ increases (*physical laws of large numbers*). The best-known statistics are *statistical relative frequencies* (Sec. 19.2-1) and *sample averages* (Sec. 19.2-3).

19.1-2. The Classical Probability Model: Random-sample Statistics. Concept of a Population (Universe). (**a**) In an important class of applications, a continuously variable physical quantity (observable) x is regarded as a one-dimensional random variable with the inferred or estimated probability density $\varphi(x)$. Each sample $(x_1, x_2, \ldots, x_n)$ of measurements of x is postulated to be the result of n *repeated independent measurements* (Sec. 18.2-4). Hence $x_1, x_2, \ldots, x_n$ are *statistically independent random variables with identical probability density.* A sample $(x_1, x_2, \ldots, x_n)$ defined in this manner is called a **random sample of size** n and constitutes an n-dimensional random variable. The probability density in the n-dimensional sample space of "sample points" $(x_1, x_2, \ldots, x_n)$ is the **likelihood function**

$$L(x_1, x_2, \ldots, x_n) = \varphi(x_1)\varphi(x_2) \cdots \varphi(x_n) \tag{19.1-1}$$

Every random-sample statistic defined as a measurable function

$$y = y(x_1, x_2, \ldots, x_n)$$

of the sample values is a random variable whose probability distribution (**sampling distribution** of y) is uniquely determined by the likelihood function, and hence by the distribution of x. Each sampling distribution will, in general, depend on the sample size n.

While the assumptions made in this section do not apply to every physical situation, the model is capable of considerable generalization. The distribution of x need not be continuous (games, quality control). The sample may be of infinite size and may not even be a countable set; the x_k may not have the same probability distribution and may not be statistically independent (random-process theory, Sec. 18.9-2). Finally, x, and thus each x_k, can be replaced by a multidimensional variable (Secs. 19.7-1 to 19.7-6).

(b) As the size n of a random sample increases, many sample statistics converge in probability to corresponding parameters of the theoretical distribution of x; in particular, statistical relative frequencies converge in mean to the corresponding probabilities (Sec. 19.2-1). Thus one considers each sample drawn from an infinite (theoretical) **population (universe, ensemble)** whose sample distribution (Sec. 19.2-2) is identical with the theoretical probability distribution of x. The probability distribution is then referred to as the **population distribution,** and its parameters are **population parameters.** In many applications, the theoretical population is an idealization of an actual population from which samples are drawn.

19.1-3. Relation between Probability Model and Reality: Estimation and Testing. **(a) Estimation of Parameters.** Statistical methods use empirical data (sample values) to infer specifications of a *probability model*, e.g., to estimate the probability density $\varphi(x)$ of a random variable x. An important application of such inferred models is to make *decisions* based on inferred probabilities of future events. In most applications, statistical relative frequencies (Sec. 19.2-1) are used directly only for rough *qualitative* (graphical) estimates of the population distribution. Instead, one infers (postulates) the general form of the theoretical distribution, say

$$\varphi = \varphi(x; \eta_1, \eta_2, \ldots) \tag{19.1-2}$$

where $\eta_1, \eta_2, \ldots$ are unknown *population parameters* to be estimated on the basis of the given random sample $(x_1, x_2, \ldots, x_n)$. Sections 19.3-1 to 19.3-5 list a number of "general-purpose" frequency functions (2) to be chosen in accordance with the physical background, the form of the sample distribution, and convenience in computations.

The parameters $\eta_1, \eta_2, \ldots$ usually measure specific properties of the theoretical distribution of x (e.g., population mean, population variance, skewness; see also Table 18.3-1). In general, one attempts to estimate values of the parameters $\eta_1, \eta_2, \ldots$ "fitting" a given sample $(x_1, x_2, \ldots, x_n)$ by the empirical values of corresponding sample statistics $y_1(x_1, x_2, \ldots, x_n), y_2(x_1, x_2, \ldots, x_n), \ldots$ which measure analogous properties of the sample (e.g., sample average, sample variance, Secs. 19.2-3 to 19.2-6). "Fitting" is interpreted subjectively and not necessarily uniquely; in particular, one prefers estimates $y(x_1, x_2, \ldots, x_n)$ which converge in probability to η as $n \to \infty$ (**consistent estimates**), whose expected value equals η (**unbiased estimates**), whose sampling distribution has a small variance, and/or which are easy to compute (Secs. 19.4-1 to 19.4-5).

(b) Testing Statistical Hypotheses. Tests of a *statistical hypothesis* specifying some property of a theoretical distribution (say, an inferred set of parameter values $\eta_1, \eta_2, \ldots$) are based on the likelihood (1) of a *test sample* $(x_1, x_2, \ldots, x_n)$ when the hypothetical probability density (2) is used to compute $L(x_1, x_2, \ldots, x_n)$. Generally speaking, the test will *reject* the hypothesis if the test sample $(x_1, x_2, \ldots, x_n)$ falls into a region of small likelihood; or equivalently, if the corresponding value of a *test statistic* $y(x_1, x_2, \ldots, x_n)$ is improbable on the basis of the hypothetical likelihood function. The choice of specific conditions of rejection is again subjective and is ultimately based on the penalties of false rejection and/or acceptance and, to some extent, on the cost of obtaining test samples of various sizes (Secs. 19.6-1 to 19.6-9).

Nonparametric tests test hypothetical distribution properties other than parameter values (e.g., identity of two distributions, statistical independence of two random variables, Secs. 19.6-8 and 19.7-6) and are particularly convenient in suitable applications, since no specific form (2) of the population distribution need be inferred.

NOTE: Incorrect use of statistical methods can lead to grave errors and seriously wrong conclusions. All (possibly tacit) assumptions regarding a theoretical distribution must be checked. *Never use the same sample for estimation and testing.* Finally, remember that statistical tests cannot *prove* any hypothesis; they can only demonstrate a "lack of disproof."

19.2. STATISTICAL DESCRIPTION. DEFINITION AND COMPUTATION OF RANDOM-SAMPLE STATISTICS

19.2-1. Statistical Relative Frequencies. (a) Definition and Basic Properties. Consider an event E which occurs if and only if a measurement of the random variable x yields a value in some measurable set S_E (usually a *class interval*, Sec. 19.2-2*b*). Given a random sample $(x_1, x_2, \ldots, x_n)$ of x, let n_E denote the number of times a sample value x_k implies the occurrence of the event E. The **statistical relative frequency** of the event E obtained from the given random sample is

$$h[E] = \frac{n_E}{n} \tag{19.2-1}$$

where n is the size of the sample.

The definition of statistical relative frequencies implies the existence of an event algebra (Sec. 18.2-1) for the experiment or observation in question. The defining properties of mathematical probabilities (Sec. 18.2-2) are abstracted from corresponding properties of statistical relative frequencies. Thus the relative frequencies of mutually exclusive events add, the relative frequency of a certain event is 1, etc.

(b) Mean and Variance. Since the random sample may be regarded as a set of n Bernoulli trials (Sec. 18.7-3) which yield or do not yield E, the random variable n_E has a binomial distribution (Table 18.8-3) where $\vartheta = P[E] = \int_{S_E} d\Phi(x)$ is the *probability* associated with the event E, and

$$E\{h[E]\} = P[E] \qquad \text{Var}\,\{h[E]\} = \frac{P[E]\{1 - P[E]\}}{n} \tag{19.2-2}$$

The statistical relative frequency $h[E]$ *is an unbiased, consistent estimate of the corresponding probability* $P[E]$; *as* $n \to \infty$, $h[E]$ *is asymptotically normal with the parameters* (2) (Secs. 18.6-4 and 18.6-5*a*).

19.2-2. The Distribution of the Sample. Grouped Data. (a) The Empirical Cumulative Distribution Function. For a given random sample $(x_1, x_2, \ldots, x_n)$, the **empirical cumulative distribution function**

$$F(X) = h[x \leq X] \tag{19.2-3}$$

is a nondecreasing step function, with $F(-\infty) = 0, F(\infty) = 1$. $F(X)$ *is an unbiased, consistent estimate of the cumulative distribution function* $\Phi(X) = P[x \leq X]$ (Sec. 18.2-9) and defines the **distribution (frequency distribution) of the sample (empirical distribution** based on the given sample).

(b) Class Intervals and Grouped Data. Let the range of the random variable x be partitioned into a finite or infinite number of conveniently chosen **class intervals (cells)** $X_j - \frac{\Delta X_j}{2} < x \leq X_j + \frac{\Delta X_j}{2}$ $(j = 1, 2, \ldots)$ respectively of length $\Delta X_1, \Delta X_2, \ldots$ and centered at $x = X_1 < X_2 < \cdots$. For a given random sample, the **class frequency (occupation number)** n_j is the number of times an x_k falls into the jth class interval (description of the sample in terms of **grouped data**). The statistical relative frequencies $h_j = n_j/n$ **(relative frequencies of observations in the** j^{th} **class interval)** must add up to unity and are consistent, unbiased estimates of the corresponding probabilities

$$p_j = \int_{X_j - \Delta X_j/2}^{X_j + \Delta X_j/2} d\Phi(x)$$

The **cumulative frequencies** N_j and the **cumulative relative frequencies** F_j are defined by

$$N_j = \sum_{i=1}^{j} n_i \qquad F_j = \sum_{i=1}^{j} h_i = \frac{1}{n}\sum_{i=1}^{j} n_i = \frac{N_j}{n} = F\left(X_j + \frac{\Delta X_j}{2}\right) \tag{19.2-4}$$

Sample statistics can be calculated from the statistical relative frequencies $h_j = n_j/n$ just as corresponding population parameters are calculated from probabilities. The roundoff implicit in numerical computation of statistics groups data with a class-interval width equal to one least-significant digit. *Grouping into much larger class intervals may be economically advantageous, since "quantization errors" due to grouping very often average out or are easily corrected* (Sec. 19.2-5 and Ref. 19.25). The statistics $F(X)$, n_j, h_j, N_j, and F_j also yield various graphical representations of sample distributions and hence of estimated population distributions (bar charts, histograms, frequency polygons, probability graph paper, etc.; see Refs. 19.1 and 19.8).

(c) Sample Fractiles (see also Table 18.3-1 and Sec. 19.5-2*b*). The **sample** P **fractiles (sample quantiles)** X_P are defined by

$$h[x \leq X_P] = F(X_P) = P \qquad (0 < P < 1) \tag{19.2-5}$$

Equation (5) does not define X_P uniquely but brackets it by two adjacent sample values x_k. $X_{1/2}$ is the **sample median,** and $X_{1/4}$, $X_{1/2}$, $X_{3/4}$ are **sample quartiles,** with analogous definitions for **sample deciles** and **sample percentiles.**

19.2-3. Sample Averages (see also Secs. 18.3-3, 19.2-5, and 19.5-3). **(a) The Sample Average of** x. Given a random sample $(x_1, x_2, \ldots, x_n)$, the **sample average** of x is

$$\bar{x} = \frac{1}{n}(x_1 + x_2 + \cdots + x_n) = \frac{1}{n}\sum_{k=1}^{n} x_k \tag{19.2-6}$$

In terms of the sample distribution over a set of class intervals centered at $x = X_1, X_2, \ldots, X_m$ (Sec. 19.2-2), $\bar{x}$ is approximated by

$$\bar{x}_G = \frac{1}{n}(n_1X_1 + n_2X_2 + \cdots + n_mX_m) = \frac{1}{n}\sum_{j=1}^{m} n_jX_j$$

(SAMPLE AVERAGE FROM GROUPED DATA) (19.2-7)

$\bar{x}$ is a *measure of location* of the sample distribution. Note

$$E\{\bar{x}\} = \xi \qquad \text{Var}\,\{\bar{x}\} = \frac{\sigma^2}{n} \tag{19.2-8}$$

$$E\{(\bar{x} - \xi)^3\} = \frac{\mu_3}{n^2} \qquad E\{(\bar{x} - \xi)^4\} = \frac{3(n-1)\sigma^4 + \mu_4}{n^3} \tag{19.2-9}$$

whenever the quantity on the right exists. *$\bar{x}$ is an unbiased, consistent estimate of the population mean $\xi = E\{x\}$; if σ^2 exists, $\bar{x}$ is asymptotically normal with the parameters* (8) *as $n \to \infty$* (Secs. 18.6-4 and 19.5-2).

(b) The Sample Average of $y(x)$. The **sample average of a function $y(x)$ of the random variable x** is

$$\bar{y} = \frac{1}{n}[y(x_1) + y(x_2) + \cdots + y(x_n)] = \frac{1}{n}\sum_{k=1}^{n} y(x_k) \tag{19.2-10}$$

or, from *grouped data*

$$\bar{y}_G = \frac{1}{n}\sum_{j=1}^{m} n_j y(X_j) \tag{19.2-11}$$

Estimates based on Eq. (11) are sometimes improved by correction terms (Sec. 19.2-5).

19.2-4. Sample Variances and Moments (see also Secs. 18.3-3, 18.3-7, 19.2-5, 19.4-2, and 19.5-2). **(a) The Sample Variances.** The **sample variances**

$$\begin{aligned} s^2 &= \overline{(x - \bar{x})^2} = \frac{1}{n}\sum_{k=1}^{n}(x_k - \bar{x})^2 \\ S^2 &= \frac{n}{n-1}s^2 = \frac{1}{n-1}\sum_{k=1}^{n}(x_k - \bar{x})^2 \end{aligned} \tag{19.2-12}$$

are *measures of dispersion* of the sample distribution; s is called **sample standard deviation** or **sample dispersion.** Note

$$E\{s^2\} = \frac{n-1}{n}\sigma^2 \qquad E\{S^2\} = \sigma^2 \tag{19.2-13}$$

$$\text{Var}\,\{S^2\} = \frac{1}{n}\left(\mu_4 - \frac{n-3}{n-1}\sigma^4\right) \tag{19.2-14}$$

whenever the quantity on the right exists. *S^2 is an unbiased, consistent estimate of the population variance $\sigma^2 = \text{Var}\,\{x\}$* and is thus often more useful than s^2.

(b) Sample Moments. The **sample moments a_r** and **sample central moments m_r of order r** are defined by

$$a_r = \overline{x^r} = \frac{1}{n}\sum_{k=1}^{n} x_k{}^r \qquad m_r = \overline{(x - \bar{x})^r} = \frac{1}{n}\sum_{k=1}^{n}(x_k - \bar{x})^r \tag{19.2-15}$$

Note

$$E\{a_r\} = \alpha_r \qquad \text{Var}\,\{a_r\} = \frac{\alpha_{2r} - \alpha_r^2}{n} \tag{19.2-16}$$

$$E\{m_r\} = \mu_r + O\left(\frac{1}{n}\right)$$

$$\text{Var}\,\{m_r\} = \frac{1}{n}\,(\mu_{2r} - 2r\mu_{r-1}\mu_{r+1} - \mu_r^2 + r^2\mu_2\mu_{r-1}) + O\left(\frac{1}{n^2}\right) \tag{19.2-17}$$

whenever the quantity on the right exists. *a_r is an unbiased, consistent estimate of the corresponding population moment $\alpha_r = E\{x^r\}$. If α_{2r} exists, a_r is asymptotically normal with the parameters* (16) *as $n \to \infty$.* m_r is a consistent (but not unbiased) estimate of μ_r.

The random variables

$$\frac{n^2}{(n-1)(n-2)}\,m_3 \qquad \text{and} \qquad \frac{n(n^2 - 2n + 3)m_4 - 3n(2n - 3)m_2^2}{(n-1)(n-2)(n-3)}$$

are, respectively, consistent, unbiased estimates of μ_3 and μ_4.

(c) **Measures of Skewness and Excess** (see also Table 18.3-1 and Sec. 19.3-5). The statistics

$$g_1 = \frac{m_3}{m_2^{3/2}} \qquad g_2 = \frac{m_4}{m_2^2} - 3 \tag{19.2-18}$$

respectively measure the **skewness** and the **excess** (**curtosis,** flatness) of the sample and are consistent estimates of the corresponding population parameters (see also Sec. 19.4-2). Roughly speaking, $g_1 > 0$ indicates a longer "tail" to the right. Some authors introduce g_1^2 and $g_2 + 3$ or $(g_2 + 3)/2$ instead of g_1 and g_2. Several other measures of skewness have been used (Ref. 19.1).

19.2-5. Simplified Numerical Computation of Sample Averages and Variances. Corrections for Grouping (see Refs. 19.3 and 19.8 for calculation-sheet layouts). (a) For numerical computations, it is convenient to choose a computing origin X_0 near the center of the sample distribution ("guessed mean") and to compute

$$\bar{x} = X_0 + \frac{1}{n}\sum_{k=1}^{n}(x_k - X_0) \tag{19.2-19}$$

or, for grouped data,

$$\bar{x}_G = X_0 + \frac{1}{n}\sum_{j=1}^{m} n_j(X_j - X_0) \tag{19.2-20}$$

(b) The sample variances s^2 and S^2 may be computed from

$$s^2 = \frac{n-1}{n}\,S^2 = \frac{1}{n}\sum_{k=1}^{n} x_k^2 - \bar{x}^2 \tag{19.2-21}$$

which is approximated for grouped data by

$$s_G^2 = \frac{n-1}{n} S_G^2 = \frac{1}{n} \sum_{j=1}^{m} n_j X_j^2 - \bar{x}_G^2 \tag{19.2-22}$$

(c) Computations with grouped data are simplified if all class intervals are of equal length ΔX, and if one of the class-interval mid-points $X_j = X_0$ is taken to be the computing origin, so that

$$X_j = X_0 + Y_j \, \Delta X \tag{19.2-23}$$

where the Y_j are "coded" class-interval centers which take integral values 0, ± 1, ± 2, ... One has then

$$\bar{x}_G = X_0 + \bar{y}_G \, \Delta X \qquad \bar{y}_G = \frac{1}{n} \sum_{j=1}^{m} n_j Y_j \tag{19.2-24}$$

$$s_G^2 = \frac{n-1}{n} S_G^2 = \frac{1}{n} \left(\sum_{j=1}^{m} n_j Y_j^2 - n\bar{y}_G^2 \right) (\Delta X)^2 \tag{19.2-25}$$

One may check computations by introducing a different computing origin X_0, say $X_0 = 0$.

(d) **Sheppard's Corrections for Grouping.** Let all class intervals be of equal length ΔX. Then, if the characteristic function (Sec. 18.3-8) $\chi_x(q)$ and its derivatives are small for $|q| \geq 2\pi/\Delta x$, one may improve the grouped-data approximation s_G^2 to the true sample variance s^2 by adding *Sheppard's correction* $-(\Delta X)^2/12$. Analogous corrections for the grouped-data sample moments

$$a_{rG} = \frac{1}{n} \sum_{j=1}^{m} n_j X_j^r \qquad m_{rG} = \frac{1}{n} \sum_{j=1}^{m} n_j (X_j - \bar{x}_G)^r \tag{19.2-26}$$

yield the improved estimates

$$\left. \begin{aligned} a_1' &= a_{1G} & & \\ a_2' &= a_{2G} - \tfrac{1}{12}(\Delta X)^2 & m_2' &= m_{2G} - \tfrac{1}{12}(\Delta X)^2 \\ a_3' &= a_{3G} - \tfrac{1}{4}a_{1G}(\Delta X)^2 & m_3' &= m_{3G} \\ a_4' &= a_{4G} - \tfrac{1}{2}a_{2G}(\Delta X)^2 & & \\ &\quad + \tfrac{7}{240}(\Delta X)^4 & m_4' &= m_{4G} - \tfrac{1}{2}m_{2G}(\Delta X)^2 + \tfrac{7}{240}(\Delta X)^4 \end{aligned} \right\} \tag{19.2-27}$$

More generally,

$$a_r' = \sum_{k=0}^{r} \binom{r}{k} (2^{1-k} - 1) B_k a_{(r-k)G} (\Delta X)^k \qquad (r = 1, 2, \ldots) \tag{19.2-28}$$

where the B_k are the Bernoulli numbers (Sec. 21.5-2). These corrections become *exact* if $\chi_x(q) = 0$ for $|q| \geq 2\pi/\Delta x - \epsilon$ ($\epsilon > 0$). *For normal variables with* $\Delta X \leq 2\sigma$, $a_1' = a_1$ *within* $2.3 \times 10^{-3} \Delta X$ ($\xi \neq 0$), *and* $-(\Delta X)^2/12$ *is within* $3.1 \times 10^{-2}\sigma^2$ *of the exact correction if* $\xi = 0$.

NOTE: Sheppard's corrections often yield useful estimates of errors due to the use of *rounded-off sample values* in the exact formulas (12) and (15). Thus, if Sheppard's correction applies, a mean round-off error $\Delta X/2$ in the x_k affects s^2 only as $(\Delta X)^2/12$.

19.2-6. The Sample Range (see also Sec. 19.5-4). The **sample range** w for a random sample $(x_1, x_2, \ldots, x_n)$ is the difference of the largest and the smallest sample value x_k. The sample range has physical significance (quality control) and serves as a rough but conveniently calculated estimate of population parameters for specific theoretical distributions (Sec. 19.5-4). The sample range and the smallest and largest sample values are examples of **rank (order) statistics.**

19.3. GENERAL-PURPOSE PROBABILITY DISTRIBUTIONS

19.3-1. Introduction. The probability distributions described in Secs. 18.8-1 to 18.8-9 and, in particular, the normal, binomial, hypergeometric, and Poisson distributions can often serve as theoretical population distributions in statistical models. The applicability of a particular type of probability distribution (with suitably fitted parameters) may be inferred either theoretically or from the graph of an empirical distribution.

Normal distributions are particularly convenient. Each normal distribution is completely defined by its first- and second-order moments; again, the use of normal populations favors the computation of exact sampling distributions for use in statistical tests (Secs. 19.6-4 and 19.6-6). The use of a normal distribution is frequently justified by the *central limit theorem* (Sec. 18.6-5); in particular, *errors in measurements* are often regarded as normally distributed sums of many independent "elementary errors."

19.3-2. Edgeworth-Kapteyn Representation of Theoretical Distributions.* It is often desirable to fit the empirical distribution of a random variable x with a probability distribution described by

$$\Phi_x(x) = \Phi_u\left[\frac{g(x) - \mu}{\sigma_g}\right] \qquad \varphi_x(x) = \frac{1}{\sigma_g\sqrt{2\pi}}e^{-\frac{1}{2}\left[\frac{g(x)-\mu}{\sigma_g}\right]^2}\cdot\left|\frac{dg}{dx}\right| \tag{19.3-1}$$

where $g(x)$ is a function of x selected so as to be normally distributed with parameters (μ, σ_g^2). Once $g(x)$ has been chosen (e.g., from theoretical considerations), only two parameters, μ and σ_g, remain to be estimated, and much of the theory of normal populations is applicable.

Any random variable x described by Eq. (1) may be regarded as the limit $x = \lim\limits_{r\to\infty} y_r$ of a sequence of random variables $y_1 = y_0 + z_1h(y_0)$, $y_2 = y_1 + z_2h(y_1)$, . . . , where $z_1, z_2, \ldots$ are random variables satisfying the conditions of Sec. 18.6-5*c*, and $y_1 + y_2 + \cdots = \int_{x_0}^{x} \frac{dy}{h(y)} = g(x)$. The $z_1, z_2, \ldots$ may be considered as "reaction intensities" in a physical process which successively generates $y_1, y_2, \ldots$.

EXAMPLES: (1) $h(y)$ = constant yields a normal distribution for x. (2) $h(y) = y - \alpha$ results in a **logarithmic normal distribution** defined by

$$\left.\begin{aligned}\varphi(x) &= 0 \qquad (x \le \alpha)\\ \varphi(x) &= \frac{1}{(x-\alpha)\sigma\sqrt{2\pi}}e^{-\frac{1}{2}\left[\frac{\log_e(x-\alpha)-\mu}{\sigma}\right]^2} \qquad (x > \alpha)\end{aligned}\right\} \tag{19.3-2}$$

* See footnote to Sec. 18.3-4.

19.3-3. Gram-Charlier and Edgeworth Series Approximations. It is frequently convenient to approximate the frequency function of a *standardized* random variable $x = \frac{z - E\{z\}}{\sqrt{\text{Var}\{z\}}}$ (Sec. 18.5-5) in the form

$$\begin{aligned}\varphi_x(x) &\approx \varphi_u(x) - \frac{1}{3!}\frac{\mu_3}{\sigma^3}\varphi_u^{(3)}(x) \\ &\quad + \left[\frac{1}{4!}\left(\frac{\mu_4}{\sigma^4} - 3\right)\varphi_u^{(4)}(x) + \frac{10}{6!}\left(\frac{\mu_3}{\sigma^3}\right)^2 \varphi_u^{(6)}(x)\right] \\ &= \varphi_u(x) - \frac{\gamma_1}{3!}\varphi_u^{(3)}(x) + \left[\frac{\gamma_2}{4!}\varphi_u^{(4)}(x) + \frac{10\gamma_1^2}{6!}\varphi_u^{(6)}(x)\right] \qquad (19.3\text{-}3)\end{aligned}$$

where the parameters μ_3, μ_4, γ_1, γ_2 refer to the theoretical distribution of z (Sec. 18.3-7*b* and Table 18.3-1). An analogous expression for the distribution function $\Phi_x(x)$ is obtained by substitution of $\Phi_u^{(k)}(x)$ for $\varphi_u^{(k)}(x)$ in Eq. (3). Note that Eq. (3) expresses $\varphi_x(x)$ in terms of the widely tabulated derivatives of the normal frequency function $\varphi_u(x)$ (Sec. 18.8-3), and that the parameters (coefficients) can be estimated as functions of moments (Sec. 19.4-3); but the computation of sampling distributions is not easy. See also Table 18.5-1.

For a rather restricted class of distributions, the approximation (3) comprises the first terms of the orthogonal expansion (Sec. 15.2-6)

$$\Phi_x(x) = \Phi_u(x) + \frac{\eta_3}{3!}\Phi_u^{(3)}(x) + \frac{\eta_4}{4!}\Phi_u^{(4)}(x) + \cdots \qquad (19.3\text{-}4a)$$

$$\varphi_x(x) = \varphi_u(x) + \frac{\eta_3}{3!}\varphi_u^{(3)}(x) + \frac{\eta_4}{4!}\varphi_u^{(4)}(x) + \cdots \qquad (19.3\text{-}4b)$$

with

$$\eta_k = (-1/\sqrt{2})^k \int_{-\infty}^{\infty} H_k(x/\sqrt{2})\varphi_x(x)\,dx \qquad (19.3\text{-}4c)$$

(*Gram-Charlier Type A series*), where $H_k(x)$ is the kth Hermite polynomial (Sec. 21.7-1). The series (4*a*) converges to $\Phi_x(x)$ if $\mu_1, \mu_2, \ldots$ exist and $\int_{-\infty}^{\infty} e^{\frac{x^2}{4}}\,d\Phi_x(x)$ converges; the series (4*b*) will then converge to $\varphi_x(x)$ at all points of continuity if $\varphi_x(x)$ is of bounded variation (Sec. 4.4-8*b*) in $(-\infty, \infty)$. For a much larger class of distributions, the approximation (3) can be based on *Edgeworth's asymptotic series* (Ref. 19.4).

19.3-4. Truncated Normal Distributions and Pareto's Distribution. (a) Truncated Normal Distributions. If all events $[x \leq \xi_a]$ are removed from a normal population with mean μ and variance σ^2 (Sec. 19.1-2), the remaining population has a **one-sided truncated normal distribution** such that $\varphi(x) = \Phi(x) = 0$ for $x \leq \xi_a$, and

$$\varphi(x) = \frac{1}{\sigma(1 - \tau_a)}\varphi_u\left(\frac{x - \mu}{\sigma}\right) \qquad \Phi(x) = \frac{\Phi_u\left(\frac{x-\mu}{\sigma}\right) - \tau_a}{1 - \tau_a} \qquad (x > \xi_a) \quad (19.3\text{-}5)$$

where $\tau_a = \Phi_u\left(\frac{\xi_a - \mu}{\sigma}\right)$ (**degree of truncation**) is the fraction of the original population removed by the truncation. If ξ_a is known, one may use

$$\left.\begin{aligned}\alpha_1 &= \xi = E\{x\} = \mu + \sigma^2\varphi(\xi_a) \\ \alpha_2 &= \mu^2 + \sigma^2\varphi(\xi_a)(\xi_a + \mu) + \sigma^2\end{aligned}\right\} \tag{19.3-6}$$

to estimate μ and σ by the moment method (Sec. 19.4-3) with the aid of published tables (Ref. 19.15) expressing μ and σ in terms of α_1 and α_2.

(b) Pareto's distribution is defined by $\varphi(x) = \Phi(x) = 0$ for $x \leq \xi_a$ and

$$\varphi(x) = \frac{\alpha}{\xi_a}\left(\frac{\xi_a}{x}\right)^{\alpha+1} \qquad \Phi(x) = 1 - \left(\frac{\xi_a}{x}\right)^{\alpha} \qquad (x > \xi_a;\ \alpha > 0) \tag{19.3-7}$$

with
$$\xi = E\{x\} = \frac{\alpha}{\alpha - 1}\,\xi_a \qquad (\alpha > 1) \qquad x_{1/2} = \xi_a\sqrt[\alpha]{2} \tag{19.3-8}$$

19.3-5. Pearson's General-purpose Distributions. The frequency functions $\varphi(x)$ of many continuous probability distributions can be described as solutions of the differential equation

$$\frac{d\varphi}{dx} = \frac{x + \eta_1}{\eta_2 + \eta_3 x + \eta_4 x^2}\,\varphi(x) \tag{19.3-9}$$

whose four parameters define each distribution completely. Each parameter η_k can be estimated as a function of the first four moments α_r or μ_r (Sec. 19.4-3), but the computation of sampling distributions is, at best, difficult. The distributions defined by Eq. (9) can be classified according to the nature of the roots of $\eta_2 + \eta_3 x + \eta_4 x^2 = 0$ and include most of the continuous distributions described in Table 18.8-8, Sec. 18.8-3, and Sec. 19.5-3 as special cases (see also Ref. 19.7).

For Pearson's distributions, Pearson's measure of skewness (Table 18.3-1) is

$$\frac{\xi - \xi_{\text{mode}}}{\sigma} = \frac{\gamma_1(\gamma_2 + 6)}{2(5\gamma_2 - 6\gamma_1^2 + 6)} \tag{19.3-10}$$

so that for small γ_1, γ_2 one has $\xi_{\text{mode}} \approx \xi - \gamma_1\sigma/2$.

19.4. CLASSICAL PARAMETER ESTIMATION

19.4-1. Properties of Estimates (see also Sec. 19.1-3). **(a)** A sample statistic $y(x_1, x_2, \ldots, x_n)$ is a **consistent estimate** of the theoretical population parameter η if and only if y converges to η in probability as the sample size n increases, i.e., if and only if the probability of realizing any finite deviation $|y - \eta|$ converges to zero as $n \to \infty$ (Sec. 18.6-1).

(b) The **bias** of an estimate y for the parameter η is the difference $b(\eta) \equiv E\{y\} - \eta$. y is an **unbiased estimate** of η if and only if $E\{y\} = \eta$ for all values of η.

(c) Asymptotically Efficient Estimates and Efficient Estimates. It is desirable to employ estimates whose sampling distributions cluster as densely as possible about the desired parameter value, i.e., estimates with small variance Var $\{y\}$, or small standard deviation (**standard error of the estimate**) $\sqrt{\text{Var}\,\{y\}}$. For the important special class of estimates $y(x_1, x_2, \ldots, x_n)$ whose sampling distributions are *asymptotically normal* with mean η and variance constant$/n = \lambda/n$ (see also

Sec. 19.5-2),

$$\lambda = \lim_{n \to \infty} n \operatorname{Var}\{y\} \geq \lambda_{\min} = \frac{1}{E\left\{\left(\frac{\partial \log \varphi}{\partial \eta}\right)^2\right\}} \qquad (19.4\text{-}1)$$

The **asymptotic efficiency** $e_\infty\{y\} = \lambda_{\min}/\lambda$ of such an estimate y measures the concentration of the asymptotic sampling distribution about the parameter η; y is an **asymptotically efficient estimate** of the parameter η if and only if $\lambda = \lambda_{\min}$. *Every asymptotically efficient estimate is consistent.*

More generally, the "relative efficiency" of the estimate $y(x_1, x_2, \ldots, x_n)$ of η from a given sample size n is measured by the reciprocal of the mean-square deviation $E\{(y - \eta)^2\}$. Under quite general conditions (Ref. 19.4, Chap. 32), the mean-square deviations $E\{(y - \eta)^2\}$ of the various possible estimates y of a given parameter η have a lower bound given by

$$E\{(y - \eta)^2\} \geq \frac{\lambda_{\min}}{n} = \frac{\left(1 + \frac{db}{d\eta}\right)^2}{nE\left\{\left(\frac{\partial \log \varphi}{\partial \eta}\right)^2\right\}} \qquad (19.4\text{-}2)$$

(*Cramér-Rao Inequality*). For *unbiased estimates y*, Eq. (2) reduces to

$$\operatorname{Var}\{y\} \geq \frac{\lambda_{\min}}{n} = \frac{1}{nE\left\{\left(\frac{\partial \log \varphi}{\partial \eta}\right)^2\right\}} \qquad (19.4\text{-}3)$$

The **efficiency** $e\{y\} = \frac{\lambda_{\min}}{n \operatorname{Var}\{y\}}$ of an unbiased estimate satisfying Eq. (3) measures the concentration of the sampling distribution about $\eta = E\{y\}$. $\lim_{n \to \infty} e\{y\}$ is again called the asymptotic efficiency, if this quantity exists. A (necessarily unbiased and consistent) estimate y is an **efficient estimate** of η if and only if $\operatorname{Var}\{y\}$ exists and equals the lower bound $\lambda_{\min}/n$.

(d) Sufficient Estimates. An estimate $y(x_1, x_2, \ldots, x_n)$ of the parameter η is a **sufficient estimate** if and only if the likelihood function for one sample (Sec. 19.1-2) can be written in the form

$$L(X_1, X_2, \ldots, X_n; \eta) \equiv L_1(Y, \eta) L_2(X_1, X_2, \ldots, X_n)$$
$$Y \equiv y(X_1, X_2, \ldots, X_n) \qquad (19.4\text{-}4)$$

where L_2 is functionally independent (Sec. 4.5-6) of Y. In this case, the conditional probability distribution of $(x_1, x_2, \ldots, x_n)$ on the hypothesis $[y = Y]$ is independent of η, so that *a sufficient estimate y of η embodies all the information about η which the given sample can supply. Efficient estimates are necessarily sufficient.*

(e) Generalizations. Equations (1) to (3) apply to *discrete probability distributions* if the probability $p(x_1, x_2, \ldots, x_n)$ is substituted for the probability density $\varphi(x_1, x_2, \ldots, x_n)$. The theory of Secs. 19.4-1*a* to d applies without change to *populations described by multidimensional random variables.*

A set of m suitable unbiased estimates $y_1, y_2, \ldots, y_m$ are **joint efficient estimates** of m corresponding population parameters $\eta_1, \eta_2, \ldots, \eta_m$ if the concentration ellipsoid (Sec. 18.4-8*c*) of the joint sampling distribution coincides with a "maximum concentration ellipsoid" analogous to the minimum variance defined by Eq. (3). **Joint asymptotically efficient estimates** are similarly defined in terms of the asymptotic sampling distribution. The reciprocal of the generalized variance (Sec. 18.4-8*c*) associated with the joint sampling distribution of $y_1, y_2, \ldots, y_m$ is a measure of the *relative joint efficiency* of the set of estimates.

To define a set of m **joint sufficient estimates** $y_1, y_2, \ldots, y_m$ it is only necessary to replace the random variable y in Eq. (4) by the m-dimensional variable $(y_1, y_2, \ldots, y_m)$.

19.4-2. Some Properties of Statistics Used as Estimates (see also Sec. 19.2-4). **(a) Functions of Moments.** *Every statistic expressible as a power of a rational function of the sample moments a_r is a consistent estimate of the same function of the corresponding population moments α_r, provided that the α_r's in question exist and yield a finite function value* (see also Sec. 18.3-7). Multiplication of a biased consistent estimate by a suitable function of n will often yield an unbiased consistent estimate.

An entirely analogous theorem applies to functions of the sample moments $a_{r_1r_2}\ldots$ of multivariate samples. In particular, g_1, g_2, l_{ik}, r_{ik}, and $r_{12\cdot34\ldots m}$ (Sec. 19.7-2) are consistent estimates of the corresponding population parameters $\gamma_1, \gamma_2, \lambda_{ik}, \rho_{ik}$, and $\rho_{12\cdot34\ldots m}$.

(b) For samples taken from a *normal population*

1. $\bar{x}$ is an efficient estimate of ξ.
2. $\bar{x}$ and s^2 are joint asymptotically efficient estimates of ξ and σ^2, but s^2 is biased; $\bar{x}$ and $S^2 = \dfrac{n}{n-1} s^2$ are joint sufficient and asymptotically efficient estimates of ξ and σ^2. S^2 has the efficiency $\dfrac{n-1}{n}$.
3. If ξ is known, $\hat{s}^2 = \overline{(x-\xi)^2}$ is an efficient estimate of σ^2.
4. The sample median $x_{1/2}$ has the asymptotic efficiency $\dfrac{2}{\pi}$.

For samples taken from a *binomial distribution* (Table 18.8-3), $\bar{x}$ is an efficient estimate of ξ.

For samples taken from a *two-dimensional normal distribution* (Sec. 18.8-6) with known center of gravity, the sample central moments l_{11}, l_{12}, and l_{22} (Sec. 19.7-2) are joint asymptotically efficient estimates of λ_{11}, λ_{12}, and λ_{22}.

19.4-3. Derivation of Estimates: The Method of Moments. If the population distribution is described by a given function $\Phi(x; \eta_1, \eta_2, \ldots)$, $\varphi(x; \eta_1, \eta_2, \ldots)$, or $p(x; \eta_1, \eta_2, \ldots)$, where $\eta_1, \eta_2, \ldots$ are parameters to be determined, each population characteristic like $E\{x\}$, Var $\{x\}$, α_r, etc., is a function of the parameters $\eta_1, \eta_2, \ldots$. In particular

$$\alpha_1 = \alpha_1(\eta_1, \eta_2, \ldots) \qquad \alpha_2 = \alpha_2(\eta_1, \eta_2, \ldots) \qquad \cdots$$

if these quantities exist. The *method of moments* defines (joint) estimates $y_1(x_1, x_2, \ldots, x_n)$, $y_2(x_1, x_2, \ldots, x_n)$, $\ldots$, $y_m(x_1, x_2, \ldots, x_n)$ of m corresponding population parameters $\eta_1, \eta_2, \ldots, \eta_m$ by the m equations

$$\alpha_r(y_1, y_2, \ldots, y_m) = a_r(x_1, x_2, \ldots, x_n) \qquad (r = 1, 2, \ldots, m) \qquad (19.4\text{-}5)$$

obtained on *equating the first m sample moments a_r to the corresponding population moments α_r.* The resulting estimates y_k are necessarily functions of the sample moments (see also Sec. 19.4-2*a*).

19.4-4. The Method of Maximum Likelihood. For any given sample $(x_1, x_2, \ldots, x_n)$, the value of the likelihood function $L(x_1, x_2, \ldots, x_n)$ (Sec. 19.1-2) is a function of the unknown parameters $\eta_1, \eta_2, \ldots$. The *method of maximum likelihood* estimates each parameter η_k by a corresponding trial function $y_k(x_1, x_2, \ldots, x_n)$ chosen so that $L(x_1, x_2, \ldots, x_n; y_1, y_2, \ldots)$ is as large as possible for each sample $(x_1, x_2, \ldots, x_n)$. One attempts to obtain a set of m (joint) **maximum-likelihood estimates** $y_1(x_1, x_2, \ldots, x_n)$, $y_2(x_1, x_2, \ldots, x_n)$, $\ldots$, $y_m(x_1, x_2, \ldots, x_n)$ as nontrivial solutions of m equations

$$\frac{\partial}{\partial y_k} \log_e L(x_1, x_2, \ldots, x_n; y_1, y_2, \ldots, y_m) = 0$$
$$(k = 1, 2, \ldots, m) \quad \text{(MAXIMUM-LIKELIHOOD EQUATIONS)} \qquad (19.4\text{-}6)$$

which constitute necessary conditions for a maximum of the likelihood function if the latter is suitably differentiable (Sec. 11.3-3).

Although the maximum-likelihood method often involves more complicated computations than the moment method (Sec. 19.4-3), maximum-likelihood estimates may be preferable, particularly in the case of small samples, because

1. *If an efficient estimate (or a set of joint efficient estimates) exists, it will appear as a unique solution of the likelihood equation or equations* (6).
2. *If a sufficient estimate (or a set of sufficient estimates) exists, every solution of the likelihood equation or equations* (6) *will be a function of this estimate or estimates.*

In addition, under quite general conditions (Ref. 19.4, Chap. 33), the likelihood equations (6) have a solution yielding consistent, asymptotically normal, and asymptotically efficient estimates.

EXAMPLE: If x is normally distributed, the maximum-likelihood estimate $X = \bar{x}$ for ξ is an efficient estimate and minimizes $\sum_{i=1}^{n} (x_i - X)^2$ (*method of least squares* in the theory of errors).

Note that the maximum-likelihood method applies also to multidimensional populations and applies even in the case of nonrandom samples.

19.4-5. Other Methods of Estimation. A number of methods usually employed to test the goodness of fit of an estimate or estimates can be modified to infer parameter values (Sec. 19.6-7; see also Secs. 19.9-1 to 19.9-5).

19.5. SAMPLING DISTRIBUTIONS

19.5-1. Introduction (see also Secs. 19.1-2 and 19.1-3). Section 19.5-1 lists properties of a class of *statistics frequently used as consistent estimates of corresponding population parameters.* Section 19.5-2 deals with the *approximate computation of sampling distributions for large samples,* while Secs. 19.5-3 and 19.5-4*b* are concerned with the distributions of statistics derived from *normal populations.*

19.5-2. Asymptotically Normal Sampling Distributions (see also Sec. 18.6-4). The following theorems are derived from the limit theorems of Sec. 18.6-5 and permit one to approximate the sampling distributions of many statistics by normal distributions if the sample size is sufficiently large.

(a) *Let $y(x_1, x_2, \ldots, x_n)$ be any statistic expressible as a function of the sample moments m_k such that $y = f(m_1, m_2, \ldots)$ exists and is twice continuously differentiable throughout a neighborhood of $m_1 = \mu_1$, $m_2 = \mu_2$, ... Then the sampling distribution of y is asymptotically normal as $n \to \infty$, with mean $f(\mu_1, \mu_2, \ldots)$ and variance* $\dfrac{\text{constant}}{n} + O\left(\dfrac{1}{n^2}\right)$. The theorem applies, in particular, to the sample mean $\bar{x}$, to the sample variances s and S, and to the sample moments a_r and m_r (Sec. 19.2-4). *An analogous theorem applies to multidimensional populations.*

(b) *The distribution of each sample fractile X_P is asymptotically normal with mean x_P and variance $\dfrac{P(1-P)}{n\varphi^2(x_P)}$, provided that* (1) *the population fractile x_P is unique, and* (2) *$\varphi'(x)$ exists and is continuous throughout a neighborhood of $x = x_P$.* This theorem applies, in particular, to the sample median $X_{1/2}$. Under analogous conditions, any joint distribution of the sample fractiles (and hence, for example, the sample interquartile range) is also asymptotically normal.

19.5-3. Samples Drawn from Normal Populations. The χ^2, t, and v^2 Distributions. **(a)** In the case of samples drawn from a *normal* population (normal samples), all sample values are normal variables, and many sampling distributions can be calculated explicitly with the aid of Secs. 18.5-7 and 18.8-9. The assumption of a normal population can often be justified by the central-limit theorem (Sec. 18.6-5, e.g., errors in measurements), or the method of Sec. 19.3-2 can be used.

For any sample of size n drawn from a normal population with mean ξ and variance σ^2

1. $\dfrac{\bar{x} - \xi}{\sigma/\sqrt{n}}$ has a **standardized normal distribution** (**u distribution,** Sec. 18.8-4).
2. $\dfrac{\bar{x} - \xi}{S/\sqrt{n}} = \dfrac{\bar{x} - \xi}{s/\sqrt{n-1}}$ (**Student's ratio**) has a **t distribution with $n - 1$ degrees of freedom** (Table 19.5-2).
3. $\dfrac{(n-1)S^2}{\sigma^2} = \dfrac{ns^2}{\sigma^2} = \dfrac{1}{\sigma^2}\sum_{k=1}^{n}(x_k - \bar{x})^2$ **has a χ^2 distribution with $n - 1$ degrees of freedom** (Table 19.5-1).

and

4. $\dfrac{x_i - \xi}{\sigma}$ has a **standardized normal distribution.**
5. $\dfrac{x_i - \xi}{S} = \dfrac{x_i - \xi}{s}\sqrt{\dfrac{n-1}{n}}$ has a **t distribution with $n - 1$ degrees of freedom.**

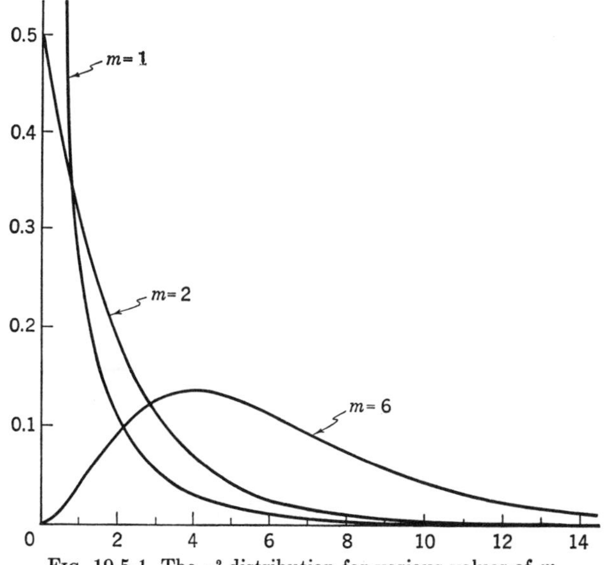

FIG. 19.5-1. The χ^2 distribution for various values of m.

Table 19.5-1. The χ^2 Distribution with m Degrees of Freedom
(Fig. 19.5-1; see also Secs. 18.8-7, 19.5-3, and 19.6-7)

(a) $$\varphi_y(Y) \equiv \varphi_{\chi^2(m)}(Y) = \begin{cases} 0 & \text{for } Y < 0 \\ \dfrac{1}{\Gamma\left(\dfrac{m}{2}\right)\sqrt{2^m}} Y^{(m-2)/2} e^{-Y/2} & \text{for } Y > 0 \end{cases}$$

(b) $E\{y\} = m$ Var $\{y\} = 2m$

Mode	$m - 2 \quad (m \geq 2)$	coefficient of skewness	$2\sqrt{\dfrac{2}{m}}$
r^{th} moment about $y = 0$	$m(m+2) \cdots (m+2r-2)$	coefficient of excess	$\dfrac{12}{m}$
r^{th} semi-invariant	$2^{r-1}(r-1)!m$	characteristic function	$(1 - 2iq)^{-m/2}$

(c) Typical Interpretation. Given any m statistically independent standardized normal variables $u_k = (x_k - \xi_k)/\sigma_k$, the sum $\chi^2 = \sum_{k=1}^{m} u_k^2$ has a χ^2 distribution with m degrees of freedom.

If $\sum_{k=1}^{m} u_k^2$ *is expressed as a sum of r quadratic forms $y_j(u_1, u_2, \ldots, u_m)$ of respective ranks* (Sec. 13.5-4*b*) *m_j, then $y_1, y_2, \ldots, y_r$ are statistically independent and distributed like χ^2 with $m_1, m_2, \ldots, m_r$ degrees of freedom if and only if $m_1 + m_2 + \cdots + m_r = m$ (Partition Theorem).*

The sum of r independent variables $y_1, y_2, \ldots, y_r$ respectively distributed like χ^2 with $m_1, m_2, \ldots, m_r$ degrees of freedom has a χ^2 distribution with $m = m_1 + m_2 + \cdots + m_r$ degrees of freedom (Addition Theorem or *reproductive property,* see also Sec. 18.8-9).

(d) The **fractiles** of y will be denoted by χ^2_P or $\chi^2_P(m)$; published tables frequently show $\chi^2_{1-\alpha}(m)$ vs. α.

(e) Approximations. As $m \to \infty$,
y is asymptotically normal with mean m and variance $2m$
y/m is asymptotically normal with mean 1 and variance $2/m$
$\sqrt{2y}$ is asymptotically normal with mean $\sqrt{2m-1}$ and variance 1
A useful approximation based on the last item listed above is

$$\chi^2_P(m) \approx \tfrac{1}{2}(\sqrt{2m-1} + u_P)^2 \qquad (m > 30)$$

This approximation is worst if P is either small or large. A better approximation is given by

$$\chi^2_P(m) \approx m\left(1 - \frac{2}{9m} + u_P\sqrt{\frac{2}{9m}}\right)^3$$

6. $\dfrac{x_i - \bar{x}}{S}\sqrt{\dfrac{n}{n-1}} = \dfrac{x_i - \bar{x}}{s}$ has an r **distribution** with $m = n - 2$ (Sec. 19.7-4).

7. $\dfrac{n\hat{s}^2}{\sigma^2} = \dfrac{1}{\sigma^2}\sum_{i=1}^{n}(x_i - \xi)^2$ **has a χ^2 distribution with n degrees of freedom.**

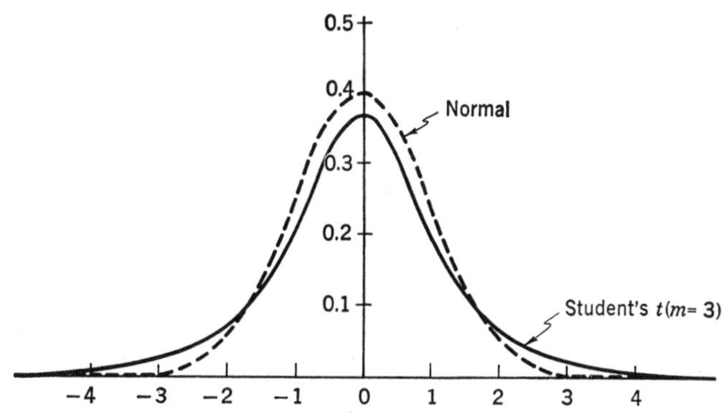

FIG. 19.5-2. Student's t distribution compared to the standardized normal distribution.

Table 19.5-2. Student's t Distribution with m Degrees of Freedom
(Fig. 19.5-2; see also Secs. 19.5-3 and 19.6-6)

(a) $$\varphi_y(Y) \equiv \varphi_{t(m)}(Y) \equiv \frac{\Gamma\left(\frac{m+1}{2}\right)}{\Gamma\left(\frac{m}{2}\right)\sqrt{m\pi}}\left(1+\frac{Y^2}{m}\right)^{-(m+1)/2}$$

(b) $E\{y\} = 0 \qquad (m > 1)$ $\qquad$ $\operatorname{Var}\{y\} = \dfrac{m}{m-2} \qquad (m > 2)$

Mode $\quad 0$ $\qquad$ coefficient of skewness 0

$2r^{\text{th}}$ moment about $y = 0$ $\quad \dfrac{1 \cdot 3 \cdots (2r-1)}{(m-2)(m-4)\cdots(m-2r)} m^r \quad (2r < m)$ $\qquad$ coefficient of excess $\dfrac{3(m-2)}{m-4} \quad (m > 4)$

(c) **Typical Interpretation.** y is distributed like the ratio

$$y = t = \frac{x_0}{\sqrt{\frac{1}{m}(x_1^2 + x_2^2 + \cdots + x_m^2)}}$$

where $x_0, x_1, x_2, \ldots, x_m$ are $m + 1$ statistically independent normal variables, each having the mean 0 and the variance σ^2. Note that t is independent of σ^2.

(d) The **fractiles** y_P of y will be denoted by t_P; note $t_P = -t_{1-P}$. The distribution of $|y| = |t|$ is related to the t distribution by

$$P[|y| < Y] = P[-Y < y < Y] = \int_{-Y}^{Y} \varphi_{t(m)}(y)\,dy = 2\Phi_{t(m)}(Y) - 1 = \Phi_{|t(m)|}(Y)$$

$$P[|y| > Y] = 2\int_{Y}^{\infty} \varphi_{t(m)}(y)\,dy = 2[1 - \Phi_{t(m)}(Y)] = 1 - \Phi_{|t(m)|}(Y)$$

The fractiles

$$|t|_{1-\alpha} = t_{1-\alpha/2}$$

defined by $P[|y| > |t|_{1-\alpha}] = \alpha$ (**α values of t**) are often tabulated for use in statistical tests; note that some published tables denote $|t|_{1-\alpha}$ by t_α.

(e) **Approximations.** As $m \to \infty$, y is asymptotically normal with mean 0 and variance 1, so that $t_P \approx u_P$ and $|t|_{1-\alpha} = t_{1-\alpha/2} \approx |u|_{1-\alpha} = u_{1-\alpha/2}$ (Sec. 18.8-4b) for $m > 30$.

Table 19.5-3. The Variance-ratio Distribution (v^2 Distribution, Snedecor F Distribution, w^2 Distribution) and Related Distributions

(See also Sec. 19.6-6)

(a) $$\varphi_y(Y) \equiv \varphi_{v^2(m,m')}(Y) = \begin{cases} 0 & \text{for } Y < 0 \\ \left(\frac{m}{m'}\right)^{\frac{m}{2}} \dfrac{\Gamma\left(\frac{m+m'}{2}\right)}{\Gamma\left(\frac{m}{2}\right)\Gamma\left(\frac{m'}{2}\right)} Y^{\frac{m-2}{2}} \cdot \left(1 + \frac{m}{m'} Y\right)^{-\frac{m+m'}{2}} & \text{for } Y > 0 \end{cases}$$

(b) $E\{y\} = \dfrac{m'}{m'-2}$ $(m' > 2)$ $\qquad$ $\operatorname{Var}\{y\} = \dfrac{2m'^2(m+m'-2)}{m(m'-2)^2(m'-4)}$ $(m' > 4)$

Mode $\dfrac{m'(m-2)}{m(m'+2)}$ $\qquad$ r^{th} moment about $y = 0$ $\quad \dfrac{\Gamma\left(\frac{m}{2}+r\right)\Gamma\left(\frac{m'}{2}-r\right)}{\Gamma\left(\frac{m}{2}\right)\Gamma\left(\frac{m'}{2}\right)}\left(\dfrac{m'}{m}\right)^r$

(c) **Typical Interpretation.** y is distributed like the ratio v^2 of two random variables having χ^2 distributions with m and m' degrees of freedom (Table 19.5-1), or

$$y = v^2 = \frac{\frac{1}{m}(x_1^2 + x_2^2 + \cdots + x_m^2)}{\frac{1}{m'}(x_1'^2 + x_2'^2 + \cdots + x_{m'}'^2)} = F = e^{2z} = \frac{m'}{m}\frac{g^2}{1-g^2}$$

where the x_i and x_k' are $m + m'$ statistically independent normal variables each having the mean 0 and the variance σ^2. Note that v^2 is independent of σ^2.

(d) **Change of Variables.** The distribution of the variance ratio v^2 is also described by other random variables, viz.

(1) v, with $$\varphi_{v(m,m')}(V) = \begin{cases} 0 & \text{for } V < 0 \\ 2\left(\frac{m}{m'}\right)^{\frac{m}{2}} \dfrac{\Gamma\left(\frac{m+m'}{2}\right)}{\Gamma\left(\frac{m}{2}\right)\Gamma\left(\frac{m'}{2}\right)} V^{m-1}\left(1 + \frac{m}{m'}V^2\right)^{-\frac{m+m'}{2}} & \text{for } V \geq 0 \end{cases}$$

(2) $z = \log_e v = \frac{1}{2}\log v^2$, with

$$\varphi_{z(m,m')}(Z) \equiv 2\left(\frac{m}{m'}\right)^{\frac{m}{2}} \frac{\Gamma\left(\frac{m+m'}{2}\right)}{\Gamma\left(\frac{m}{2}\right)\Gamma\left(\frac{m'}{2}\right)} e^{mZ}\left(1 + \frac{m}{m'}e^{2Z}\right)^{-\frac{m+m'}{2}}$$

(FISHER'S z DISTRIBUTION)

(3) $g^2 = \dfrac{m}{m'}v^2\left(1 + \dfrac{m}{m'}v^2\right)^{-1}$ has the *beta distribution* (Table 18.8-8).

(e) *Published tables* usually present the fractiles $v^2_{1-\alpha}(m, m') = \dfrac{1}{v^2_\alpha(m, m')}$, $v_{1-\alpha}(m, m')$, or $z_{1-\alpha}(m, m')$ as functions of α for various values of m, m'.

Table 19.5-3. The Variance-ratio Distribution (v^2 Distribution, Snedecor F Distribution, w^2 Distribution) and Related Distributions (*Continued*)

Special cases:

$$v^2{}_P(1, m) = t_{\frac{1+P}{2}}(m) \qquad v^2{}_P(m, 1) = \frac{1}{t^2{}_{P/2}(m)}$$

$$v^2{}_P(m, \infty) = \frac{1}{m}\chi^2{}_P(m) \qquad v^2{}_P(\infty, m) = \frac{1}{m\chi^2{}_{1-P}(m)}$$

$$v^2{}_P(1, \infty) = u^2{}_{\frac{1+P}{2}} \qquad v^2{}_P(\infty, 1) = \frac{1}{u^2{}_{P/2}}$$

(f) Approximations. As $m \to \infty$, $m' \to \infty$, z is asymptotically normal with mean $\frac{m - m'}{2mm'}$ and variance $\frac{m + m'}{2mm'}$. This approximation is often useful for $m > 30$, $m' > 30$.

Table 19.6-1 lists additional formulas. Tables 19.5-1 to 19.5-3 detail the properties of the χ^2, t, and v^2 distributions; fractiles of these functions are available in tabular form.

(**b**) *The sample mean $\bar{x}$ and the sample variance s^2 are statistically independent if and only if the sample in question is drawn from a normal population.*

For every normal sample, $\bar{x}$, s^2, and $m_r m_2^{-r/2}$ are statistically independent for all r, and $g_1 = m_3 m_2^{-3/2}$, $g_2 = m_4 m_2^{-2} - 3$ yield

$$E\{g_1\} = 0 \qquad \text{Var}\,\{g_1\} = \frac{6(n-2)}{(n+1)(n+3)} \tag{19.5-1}$$

$$E\{g_2\} = -\frac{6}{n+1} \qquad \text{Var}\,\{g_2\} = \frac{24n(n-2)(n-3)}{(n+1)^2(n+3)(n+5)} \tag{19.5-2}$$

19.5-4. The Distribution of the Sample Range (see also Secs. 19.2-6 and 19.7-6). (**a**) For every continuous distribution, the frequency function of the range w for a random sample of size n is

$$\varphi(w) = n(n-1)\int_{-\infty}^{\infty} [\Phi(x + w) - \Phi(x)]^{n-2}\varphi(x)\varphi(x + w)\,dx \tag{19.5-3}$$

This function has been tabulated for a number of population distributions (Ref. 19.8).

(**b**) *For normal populations, both the mean and the dispersion of w are multiples of the population dispersion σ:*

$$E\{w\} = k_n\sigma \qquad \text{Var}\,\{w\} = c_n{}^2\sigma^2 \tag{19.5-4}$$

k_n, c_n, and c_n/k_n have been tabulated as functions of n (Ref. 19.3); w/k_n is an unbiased estimate of σ. The average range

$$\bar{w} = \frac{(w_1 + w_2 + \cdots + w_m)}{m}$$

obtained from a sample of m random samples of size n is asymptotically normal with mean $k_n\sigma$ and variance $c_n{}^2\sigma^2/m$ as $m \to \infty$; w/k_n is an unbiased, consistent estimate of σ.

(**c**) *For a uniform (rectangular) population distribution in the interval* (a, b)

$$E\{w\} = (b - a)\frac{n-1}{n+1} \qquad \text{Var}\,\{w\} = (b-a)^2\frac{2(n-1)}{(n+1)^2(n+2)} \tag{19.5-5}$$

and $w\dfrac{n+1}{n-1}$ is a consistent, unbiased estimate of $b - a$. Note also that the arithmetic mean of the smallest and the largest sample value is a consistent, unbiased estimate of $E\{x\}$ (Ref. 19.4).

(**d**) For any continuous population distribution, the probability that at least a fraction q of the population lies between the extreme values $x_{\min}$, $x_{\max}$ of a given random sample of size n is

$$1 - nq^{n-1} + (n-1)q^n \tag{19.5-6}$$

19.5-5. Sampling from Finite Populations. Given a finite population of N elements (events, results of observations) $E_1, E_2, \ldots, E_N$ each labeled with one of the $M \leq N$ spectral values $X_{(1)}, X_{(2)}, \ldots, X_{(M)}$ of a (necessarily discrete) random variable x, let $p(X_{(i)}) = N_i/N$, where N_i is the number of elements E_k labeled by $X_{(i)}$. For a random sample of size $n \leq N$ (without replacement), n/N is called the **sampling ratio,** and one has

$$\left.\begin{aligned} E\{x\} &= \xi = \frac{1}{N}\sum_{i=1}^{M} N_i X_{(i)} \\ \text{Var}\,\{x\} &= \sigma^2 = \frac{1}{N}\sum_{i=1}^{M} N_i (X_{(i)} - \xi)^2 \end{aligned}\right\} \tag{19.5-7}$$

$$E\{\bar{x}\} = \xi \qquad \text{Var}\,\{\bar{x}\} = \frac{\sigma^2}{n}\,\frac{N-n}{N-1} \tag{19.5-8}$$

$$E\{\mathring{x}^3\} = \frac{\alpha_3}{\sqrt{n}}\,\frac{N-2n}{N-2}\sqrt{\frac{N-1}{N-n}} \tag{19.5-9}$$

$$\begin{aligned} E\{\mathring{x}^4\} = \frac{N-1}{n(N-2)(N-3)(N-n)}\,[(N^2 - 6Nn + N + 6n^2)\alpha_4 \\ + 3N(N-n-1)(n-1)\alpha_2{}^2] \end{aligned} \tag{19.5-10}$$

These formulas reduce to those given in Secs. 19.2-3 and 19.2-4 for $N = \infty$.

19.6. CLASSICAL STATISTICAL TESTS

19.6-1. Statistical Hypotheses. Consider a "space" of samples (sample points) $(x_1, x_2, \ldots, x_n)$, where $x_1, x_2, \ldots, x_n$ are numerical random variables. Every self-consistent set of assumptions involving the joint distribution of $x_1, x_2, \ldots, x_n$ is a **statistical hypothesis.** A hypothesis H is a **simple** statistical hypothesis if it defines the probability distribution uniquely; otherwise it is a **composite** statistical hypothesis.

More specifically, let the joint distribution of $x_1, x_2, \ldots, x_n$ be defined by $\Phi(x_1, x_2, \ldots, x_n; \eta_1, \eta_2, \ldots)$, $\varphi(x_1, x_2, \ldots, x_n; \eta_1, \eta_2, \ldots)$, or $p(x_1, x_2, \ldots, x_n; \eta_1, \eta_2, \ldots)$, where $\eta_1, \eta_2, \ldots$ are parameters (see also Sec. 19.1-3). Then a *simple statistical hypothesis* assigns definite values $\eta_{10}, \eta_{20}, \ldots$ to the respective parameters $\eta_1, \eta_2, \ldots$ ("point" in *parameter space*), whereas a *composite statistical hypothesis* confines the "points" $(\eta_1, \eta_2, \ldots)$ to a set or region in parameter space. The class of admissible statistical hypotheses (admissible parameter combinations) is restricted by the context of the problem in question.

If $(x_1, x_2, \ldots, x_n)$ is a random sample drawn from a single theoretical population (i.e., if the x_k are statistically independent with identical marginal distributions), *statistical hypotheses refer to values of, or relations between, population parameters.* Note, however, that the theory of Secs. 19.6-1 to 19.6-2 is *not* limited to random samples $(x_1, x_2, \ldots, x_n)$ but applies to samples from any random process.

19.6-2. Fixed-sample Tests: Definitions. Given a fixed sample size n, a **test of the statistical hypothesis** H is a rule rejecting or accepting the hypothesis H on the basis of a test sample $(X_1, X_2, \ldots, X_n)$. Each test specifies a **critical set (critical region, rejection region)** S_C of "points" $(x_1, x_2, \ldots, x_n)$ such that H will be *rejected* if the test sample $(X_1, X_2, \ldots, X_n)$ belongs to the critical set; otherwise H is *accepted.*

Such rejection or acceptance does not constitute a *logical* disproof or proof, even if the sample is infinitely large. Four possible events arise:

1. H is *true* and is *accepted* by the test.
2. H is *false* and is *rejected* by the test.
3. H is *true* but is *rejected* by the test (**error of the first kind**).
4. H is *false* but is *accepted* by the test (**error of the second kind**).

For any set of true (actual) parameter values $\eta_1, \eta_2, \ldots$, the probability that a critical region S_C will *reject* the hypothesis tested is

$$\begin{aligned}\pi_{S_C}(\eta_1, \eta_2, \ldots) &\equiv P[(x_1, x_2, \ldots, x_n) \in S_C; \eta_1, \eta_2, \ldots] \\ &\equiv \int_{S_C} d\Phi(x_1, x_2, \ldots, x_n; \eta_1, \eta_2, \ldots)\end{aligned}$$

(POWER FUNCTION OF THE CRITICAL REGION S_C) (19.6-1)

Whenever the hypothesis $H_1 \equiv [\eta_1 = \eta_{11}, \eta_2 = \eta_{21}, \ldots]$ contradicts the hypothesis tested, the rejection probability $\pi_{S_C}(\eta_{11}, \eta_{21}, \ldots)$ is called the **power (power function) of the test defined by S_C with respect to the alternative (simple) hypothesis** H_1 (see also Fig. 19.6-1*a*). A graph of the *correct-acceptance* probability $1 - \beta$ against the *false-rejection* probability α is called the **operating characteristic** of the test (Fig. 19.6-1*b*; see also Sec. 19.6-3).

19.6-3. Level of Significance. Neyman-Pearson Criteria for Choosing Tests of Simple Hypotheses. (a) It is, generally speaking, desirable to use a critical region S_C such that $\pi_{S_C}(\eta_1, \eta_2, \ldots)$ is small for parameter combinations $\eta_1, \eta_2, \ldots$ admitted by the hypothesis tested, and as large as possible for other parameter combinations. Given a critical region S_C used to test the *simple* hypothesis $H_0 \equiv [\eta_1 = \eta_{10}, \eta_2 = \eta_{20}, \ldots]$ ("null hypothesis"), let H_0 be *true*. Then the probability of falsely rejecting H_0 (error of the first kind) is $\pi_{S_C}(\eta_{10}, \eta_{20}, \ldots) = \alpha$. α is called the **level of significance** of the test: the critical region S_C tests the simple hypothesis H_0 *at the level of significance* α.

In the case of *discrete* random variables $x_1, x_2, \ldots, x_n$, one cannot, in general, specify α at will, but an upper bound for α may be given. A critical region used to test a *composite* hypothesis $H \equiv [(\eta_1, \eta_2, \ldots) \in D]$ will, in general, yield different levels of significance for different simple hypotheses (parameter combinations $\eta_1, \eta_2, \ldots$) admitted by H; one may specify the least upper bound of these levels of significance.

(b) For each given sample size n and level of significance α

1. A **most powerful test** of the simple hypothesis $H_0 \equiv [\eta_1 = \eta_{10}, \eta_2 = \eta_{20}, \ldots]$ relative to the simple alternative $H_1 \equiv [\eta_1 = \eta_{11}, \eta_2 = \eta_{21}, \ldots]$ is defined by the critical region S_C which yields the largest value of $1 - \beta = \pi_{S_C}(\eta_{11}, \eta_{12}, \ldots)$.
2. A **uniformly most powerful test** is most powerful relative to *every* admissible alternative hypothesis H_1; such a test does not always exist.
3. A test is **unbiased** if $\pi_{S_C}(\eta_{11}, \eta_{21}, \ldots) \geq \alpha$ for every alternative simple hypothesis H_1; otherwise the test is **biased.** A **most powerful unbiased test** relative to a given alternative H_1 and a **uniformly most powerful unbiased test** may be defined as above.

To construct the critical region S_C for a *most powerful test,* use all sample points $(x_1, x_2, \ldots, x_n)$ such that the **likelihood ratio** $\varphi(x_1, x_2, \ldots, x_n; \eta_{10}, \eta_{20}, \ldots)/\varphi(x_1, x_2, \ldots, x_n; \eta_{11}, \eta_{21}, \ldots)$ or $p(x_1, x_2, \ldots, x_n; \eta_{10}, \eta_{20}, \ldots)/p(x_1, x_2, \ldots, x_n; \eta_{11}, \eta_{21}, \ldots)$ is less than some fixed constant c; different values of c will yield "best" critical regions at different levels of significance α. *Uniformly most powerful tests* are of particular interest if one desires to test H_0 against a *composite* alternative hypothesis. A *uniformly most powerful unbiased test* may exist even though no uniformly most powerful test exists (see also Ref. 19.4). In practice, ease of computation may be the deciding factor in a choice among several possible tests; one can usually increase the power of each test by increasing the sample size n (see also Sec. 19.6-9).

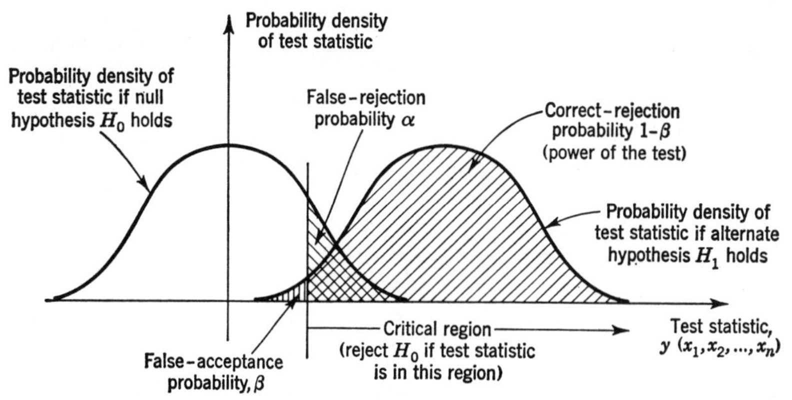

FIG. 19.6-1a. Test of the null hypothesis H_0 against a simple alternate hypothesis H_1 in terms of a test statistic $y = y(x_1, x_2, \ldots, x_n)$.

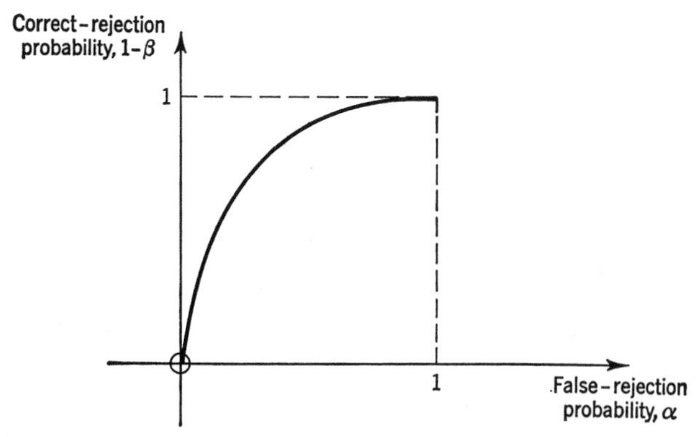

FIG. 19.6-1b. Operating characteristic of a test (see also Fig. 19.6-1a).

Table 19.6-1. Some Tests of Significance Relating to the Parameters ξ, σ^2 of a samples. Obtain fractiles from published tables; the approximations given in Tables

No.	Hypothesis to be tested	Test statistic $y = y(x_1, x_2, \ldots, x_n)$	Critical region S_C rejecting the hypothesis at the level of significance α ($\leq\alpha$ for composite hypotheses)	
1	$\xi = \xi_0$ (σ known)	$\dfrac{\bar{x} - \xi_0}{\sigma/\sqrt{n}}$	$\|y\| > \|u\|_{1-\alpha} = u_{1-\alpha/2}$	
2	$\xi \leq \xi_0$ (σ known)		$y > u_{1-\alpha}$	
3	$\xi \geq \xi_0$ (σ known)		$y < u_\alpha = -u_{1-\alpha}$	
4	$\xi = \xi_0$	$\dfrac{\bar{x} - \xi_0}{S/\sqrt{n}}$	$\|y\| > \|t\|_{1-\alpha} = t_{1-\alpha/2}$	$(m = n - 1)$
5	$\xi \leq \xi_0$		$y > t_{1-\alpha}$	$(m = n - 1)$
6	$\xi \geq \xi_0$		$y < t_\alpha = -t_{1-\alpha}$	$(m = n - 1)$
7	$\sigma^2 = \sigma_0^2$	$(n - 1)\dfrac{S^2}{\sigma_0^2}$	$y < \chi^2_{\alpha/2}$ $y > \chi^2_{1-\alpha/2}$	$(m = n - 1)$
8	$\sigma^2 \leq \sigma_0^2$		$y > \chi^2_{1-\alpha}$	$(m = n - 1)$
9	$\sigma^2 \geq \sigma_0^2$		$y < \chi^2_\alpha$	$(m = n - 1)$

* Note that published tables often tabulate $\chi^2_{1-\alpha}$ rather than χ^2_α; check carefully

19.6-4. Tests of Significance. Many applications require one to test a hypothetical *population property* specified in terms of a set of parameter values $\eta_1 = \eta_{10}$, $\eta_2 = \eta_{20}$, . . . against a corresponding *sample property* described by the respective estimates $y_1(x_1, x_2, \ldots, x_n)$, $y_2(x_1, x_2, \ldots, x_n)$, . . . of $\eta_1, \eta_2, \ldots$. One attempts to construct a "test statistic"

$$y = y(x_1, x_2, \ldots, x_n; \eta_{10}, \eta_{20}, \ldots) \equiv g(y_1, y_2, \ldots; \eta_{10}, \eta_{20}, \ldots) \tag{19.6-2}$$

whose values measure a deviation or ratio comparing the sample property to the hypothetical population property for each test sample $(x_1, x_2, \ldots, x_n)$. The simple hypothesis $H_0 \equiv [\eta_1 = \eta_{10}, \eta_2 = \eta_{20}, \ldots]$ is then *rejected* at a given level of significance α (the "deviation" is *significant*) whenever the sample value of y falls *outside* an acceptance interval $[y_{P_1} \leq y \leq y_{P_2}]$ such that

$$P[y_{P_1} \leq y \leq y_{P_2}] = P_2 - P_1 = 1 - \alpha \tag{19.6-3}$$

Normal Population (see also Secs. 19.5-3a and 19.6-4). Tests are based on random 19.5-1 and 19.5-2 apply if the sample size n is large*.

Power function $\pi_{S_C}(\xi, \sigma)$	Remarks
$\Phi_u\left(u_{\frac{\alpha}{2}} - \frac{\xi - \xi_0}{\sigma/\sqrt{n}}\right) + \Phi_u\left(u_{\frac{\alpha}{2}} + \frac{\xi - \xi_0}{\sigma/\sqrt{n}}\right)$	Simple hypothesis. Uniformly most powerful *unbiased* test; no uniformly most powerful test exists
$\Phi_u\left(u_\alpha + \frac{\lvert\xi - \xi_0\rvert}{\sigma/\sqrt{n}}\right)$	Composite hypotheses. If the admissible hypotheses are restricted to $\xi \geq \xi_0$ for test 1, and to $\xi \leq \xi_0$ for test 2, either test is a uniformly most powerful test for the simple hypothesis $\xi = \xi_0$
Use Eq. (19.6-1)	Simple hypothesis; "two-tailed t test." CAUTION: many published tables denote $\lvert t\rvert_{1-\alpha}$ by t_α
	Composite hypotheses; "one-tailed t tests"
$P\left[\chi^2 < \frac{\sigma^2_0}{\sigma^2}\chi^2_{\frac{\alpha}{2}}\right] + P\left[\chi^2 > \frac{\sigma_0^2}{\sigma^2}\chi^2_{1-\alpha/2}\right]$ $(m = n - 1)$	Simple hypothesis. No uniformly most powerful test exists
$P\left[\chi^2 > \frac{\sigma_0^2}{\sigma^2}\chi^2_{1-\alpha}\right]$ $(m = n - 1)$	Composite hypotheses. If the admissible hypotheses are restricted to $\sigma^2 \geq \sigma_0^2$ for test 8, and to $\sigma^2 \leq \sigma_0^2$ for test 9, either test is a uniformly most powerful test for the simple hypothesis $\sigma^2 = \sigma^2_0$
$P\left[\chi^2 < \frac{\sigma_0^2}{\sigma^2}\chi^2_\alpha\right]$ $(m = n - 1)$	

on the notation used in each case.

Tests defined in this manner are often called **tests of significance.** *Equation* (3) *specifies* $y_{P_1} = y_{P_1}(\eta_{10}, \eta_{20}, \ldots)$ *and* $y_{P_2} = y_{P_2}(\eta_{10}, \eta_{20}, \ldots)$ *as fractiles of the sampling distribution of* $y(x_1, x_2, \ldots, x_n; \eta_{10}, \eta_{20}, \ldots)$. It is frequently possible to choose a test statistic y such that its fractiles y_P are independent of $\eta_{10}, \eta_{20}, \ldots$. Table 19.6-1 and Secs. 19.6-6 and 19.6-7 list a number of important examples (see also Fig. 19.6-1).

In quality-control applications, the acceptance limits y_{P_1} and y_{P_2} defined by Eq. (3) are called **tolerance limits** at the tolerance level α, and the interval $[y_{P_1}, y_{P_2}]$ is called a **tolerance interval** (see also Fig. 19.6-1).

19.6-5. Confidence Regions. (a) Assume that one has constructed a family of critical regions $S_\alpha(\eta_1, \eta_2, \ldots)$ capable of testing a corresponding set of simple hypotheses (admissible parameter combinations) $(\eta_1, \eta_2, \ldots)$ at some given level of significace α.† Then for any fixed test sample $(x_1 = X_1, x_2 = X_2, \ldots, x_n = X_n)$, the set $D_\alpha(X_1, X_2, \ldots,$

† In the case of discrete distributions, the critical regions $S_\alpha(\eta_1, \eta_2, \ldots)$ will, in general, be defined for a level of significance *less than or equal to* α.

X_n) of parameter combinations ($\eta_1, \eta_2, \ldots$) ("points" in parameter space) accepted on the basis of the given sample is a **confidence region at the confidence level** α; $1 - \alpha$ is called the **confidence coefficient.** The confidence region comprises all admissible parameter combinations whose acceptance on the basis of the given sample is associated with a probability $P[(X_1, X_2, \ldots, X_n)$ not in $S_\alpha(\eta_1, \eta_2, \ldots)]$ at least equal to the confidence coefficient $1 - \alpha$.

The method of confidence regions relates a given sample to a corresponding set of "probable" parameter combinations without the necessity of assigning "inverse probabilities" to essentially nonrandom parameter combinations (see also Sec. 19.6-9).

(b) Confidence Intervals Based on Tests of Significance (see also Sec. 19.6-4). To find confidence regions relating values of one of the unknown parameters, say η_1, to the given sample value $Y = y(X_1, X_2, \ldots, X_n)$ of a suitable test statistic y, refer to Fig. 19.6-1. Plot lower

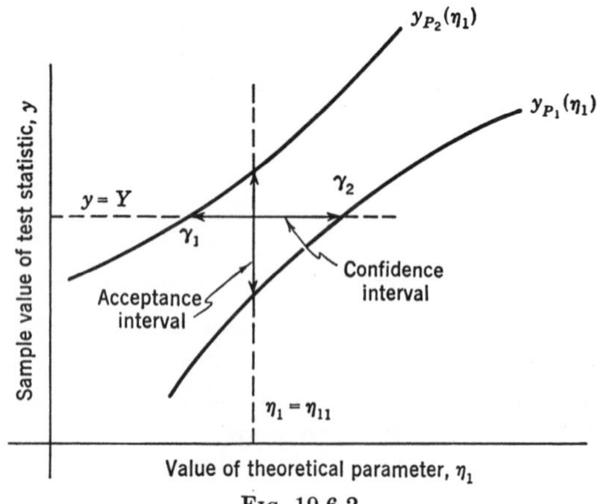

FIG. 19.6-2.

and upper acceptance limits (tolerance limits) $y_{P_1}(\eta_1)$ and $y_{P_2}(\eta_1)$ against η_1 for a given level of significance α.* The intersections of these acceptance-limit curves with each line $y = Y$ define upper and lower **confidence limits (fiducial limits)** $\eta_1 = \gamma_2(Y)$, $\eta_1 = \gamma_1(Y)$ bounding a **confidence interval** $D_\alpha \equiv [\gamma_1, \gamma_2]$ at the **confidence level** α. The confidence interval comprises those values of η_1 whose acceptance on the basis of the sample value $y = Y$ is associated with a probability $P[y_{P_1}(\eta_1) \leq Y \leq y_{P_2}(\eta_1)]$ at least equal to the confidence coefficient $1 - \alpha$.

* See footnote to Sec. 19.6-5*a*.

Table 19.6-2. Confidence Intervals for Normal Populations
Use the approximations given in Tables 19.5-1 to 19.5-3 for large n

No.	Parameter, η_1	Confidence interval $[\gamma_1 \le \eta_1 \le \gamma_2]$, confidence coefficient $1 - \alpha = P_2 - P_1$
1	ξ (σ^2 known)	$\bar{x} - u_{P_2}\dfrac{\sigma}{\sqrt{n}} \le \xi \le \bar{x} - u_{P_1}\dfrac{\sigma}{\sqrt{n}}$
2	ξ (σ^2 unknown)	$\bar{x} - t_{P_2}\dfrac{S}{\sqrt{n}} \le \xi \le \bar{x} - t_{P_1}\dfrac{S}{\sqrt{n}}$
3	σ^2	$S^2\dfrac{n-1}{\chi^2_{P_2}} \le \sigma^2 \le S^2\dfrac{n-1}{\chi^2_{P_1}}$

19.6-6. Tests for Comparing Normal Populations. Analysis of Variance. **(a) Pooled-sample Statistics.** Consider r statistically independent random samples $(x_{i1}, x_{i2}, \ldots, x_{in_i})$ drawn, respectively, from r theoretical populations with means ξ_i and variances σ_i^2 ($i = 1, 2, \ldots, r$). The ith sample is of size n_i and has the mean and variance

$$\bar{x}_i = \frac{1}{n_i}\sum_{k=1}^{n_i} x_{ik} \qquad S_i^2 = \frac{n_i}{n_i - 1}s_i^2 = \frac{1}{n_i - 1}\sum_{k=1}^{n_i}(x_{ik} - \bar{x}_i)^2$$
$$(i = 1, 2, \ldots, r) \qquad (19.6\text{-}4)$$

The r samples may be regarded as a "pooled sample" whose size, mean, and variance are given by

$$n = \sum_{i=1}^{r} n_i \qquad \bar{x} = \frac{1}{n}\sum_{i=1}^{r}\sum_{k=1}^{n_i} x_{ik} = \frac{1}{n}\sum_{i=1}^{r} n_i\bar{x}_i \qquad (19.6\text{-}5)$$

$$\left.\begin{aligned} S^2 &= \frac{n}{n-1}s^2 = \frac{1}{n-1}\sum_{i=1}^{r}\sum_{k=1}^{n_i}(x_{ik} - \bar{x})^2 \\ &= \frac{1}{n-1}[(n - r)S_0^2 + (r - 1)S_A^2] \\ S_0^2 &= \frac{1}{n-r}\sum_{i=1}^{r}(n_i - 1)S_i^2 \qquad S_A^2 = \frac{1}{r-1}\sum_{i=1}^{r} n_i(\bar{x}_i - \bar{x})^2 \end{aligned}\right\} \qquad (19.6\text{-}6)$$

The statistics S_0^2 (**pooled variance**) and S_A^2 are measures of dispersion *within samples* and *between samples*, respectively.

(b) Comparison of Normal Populations (see also Tables 19.5-1 to 19.5-3). If the r independent samples are drawn from *normal* populations with means ξ_i and identical variances $\sigma_i^2 = \sigma^2$, then S^2, S_0^2, and S_A^2 are all consistent and unbiased estimates (Sec. 19.4-1) of the (usually unknown) population variance σ^2. $(n - r)S_0^2/\sigma^2$ and $(r - 1)S_A^2/\sigma^2$ are statistically independent random variables respectively distributed like χ^2 with $n - r$ and $r - 1$ degrees of freedom. Note the sampling distributions of the following test statistics:

1. $\dfrac{(\bar{x}_i - \bar{x}_k) - (\xi_i - \xi_k)}{\sigma}\sqrt{\dfrac{n_in_k}{n_i + n_k}}$ has a *standardized normal distribution.*

2. $\dfrac{(\bar{x}_i - \bar{x}_k) - (\xi_i - \xi_k)}{S_0}\sqrt{\dfrac{n_in_k}{n_i + n_k}}$ has a *t distribution with* $n_i + n_k - 2$ *degrees of freedom.*

3. $\left.\begin{cases} S_i^2/S_k^2 & \text{has a } v^2 \textit{ distribution} \\ \log_e (S_i/S_k) & \text{has a } z \textit{ distribution} \end{cases}\right\} (m = n_i - 1,\ m' = n_k - 1)$

4. $\left.\begin{cases} S_A^2/S_0^2 & \text{has a } v^2 \textit{ distribution} \\ \log_e (S_A/S_0) & \text{has a } z \textit{ distribution} \end{cases}\right\} (m = r - 1,\ m' = n - r)$

Table 19.6-3 shows the use of these test statistics in tests of significance (Sec. 19.6-4) comparing normal populations. It is also possible to calculate confidence limits (Sec. 19.6-5) for the difference $\xi_i - \xi_k$ from the t distribution. The case of normal populations with *different* variances is discussed in Ref. 19.8.

(c) **Analysis of Variance.** The third test of Table 19.6-3 compares the mean values ξ_i by partitioning the over-all variance S^2 into components S_0^2 and S_A^2 respectively due to *statistical fluctuation* within samples and to *differences between samples.* This technique is known as **analysis of variance;** the particular case in question involves a *one-way classification* of samples corresponding to values of the index i. Many similar tests are used to analyze the effects of different medications, soil treatments, etc. (Refs. 19.3, 19.4, and 19.8).

Analysis of Variance for Two-way Classification (Randomized Blocks). Consider rq sample values x_{ik} arranged in an array

$$\begin{array}{cccc} x_{11} & x_{12} & \cdots & x_{1q} \\ x_{21} & x_{22} & \cdots & x_{2q} \\ \cdot & \cdot & \cdots & \cdot \\ x_{r1} & x_{r2} & \cdots & x_{rq} \end{array}$$

and introduce row averages $\overline{x_{i\cdot}}$, column averages $\overline{x_{\cdot k}}$, and the over-all average $\bar{x}$ by

$$\overline{x_{i\cdot}} = \frac{1}{q}\sum_{k=1}^{q} x_{ik} \qquad \overline{x_{\cdot k}} = \frac{1}{r}\sum_{i=1}^{r} x_{ik} \qquad \bar{x} = \frac{1}{rq}\sum_{i=1}^{r}\sum_{k=1}^{q} x_{ik} \tag{19.6-7}$$

The over-all variance is partitioned as follows:

$$\left.\begin{aligned}
S^2 &= \frac{1}{rq-1}\sum_{i=1}^{r}\sum_{k=1}^{q}(x_{ik}-\bar{x})^2 \\
&= \frac{1}{rq-1}[(r-1)(q-1)S_0^2 + (r-1)qS_{\text{Row}}^2 + r(q-1)S_{\text{Col}}^2] \\
S_0^2 &= \frac{1}{(r-1)(q-1)}\sum_{i=1}^{r}\sum_{k=1}^{q}(x_{ik}-\overline{x_{i\cdot}}-\overline{x_{\cdot k}}+\bar{x})^2 \\
&\qquad \text{(VARIANCE DUE TO FLUCTUATIONS WITHIN ROWS AND COLUMNS)} \\
S_{\text{Row}}^2 &= \frac{1}{r-1}\sum_{i=1}^{r}(\overline{x_{i\cdot}}-\bar{x})^2 \qquad \text{(VARIANCE BETWEEN ROWS)} \\
S_{\text{Col}}^2 &= \frac{1}{q-1}\sum_{k=1}^{q}(\overline{x_{\cdot k}}-\bar{x})^2 \qquad \text{(VARIANCE BETWEEN COLUMNS)}
\end{aligned}\right\} \tag{19.6-8}$$

If the random variables x_{ik} are normal with identical variances, S_0^2, S_{Row}^2, and S_{Col}^2 are statistically independent random variables respectively distributed like χ^2 with $(r-1)(q-1)$, $r-1$, and $q-1$ degrees of freedom. The test statistic S_{Row}^2/S_0^2 is, then, distributed like v^2 with $m = r - 1$, $m' = (r-1)(q-1)$ and serves to test the equality of the row means $E\{\overline{x_{i\cdot}}\} = \xi_{i\cdot}$ in the manner of Table 19.6-3. Similarly, S_{Col}^2/S_0^2 is distributed like v^2 with $m = q - 1$, $m' = (r-1)(q-1)$ and serves to test the equality of the column means $E\{\overline{x_{\cdot k}}\} = \xi_{\cdot k}$.

Table 19.6-3. Significance Tests for Comparing Normal Populations
(refer to Sec. 19.6-6; see also Tables 19.5-2 and 19.5-3)

	Hypothesis to be tested	**Critical region** rejecting the hypothesis at a level of significance $\leq \alpha$	
1a 1b	$\sigma_i^2 = \sigma_k^2$ $\sigma_i^2 \leq \sigma_k^2$	$\lvert\log_e (S_i/S_k)\rvert > z_{1-\alpha/2}$ $\log_e (S_i/S_k) > z_{1-\alpha}$ or $S_i^2/S_k^2 > v^2_{1-\alpha}$	$(m = n_i - 1,\ m' = n_k - 1)$ (FISHER'S z TEST AND v^2 TEST)
2a 2b	$\xi_i = \xi_k$ (given $\sigma_i^2 = \sigma_k^2$) $\xi_i \leq \xi_k$ (given $\sigma_i^2 = \sigma_k^2$)	$\left\lvert \frac{\bar{x}_i - \bar{x}_k}{S_0}\sqrt{\frac{n_i n_k}{n_i + n_k}} \right\rvert > t_{1-\alpha/2}$ $\frac{\bar{x}_i - \bar{x}_k}{S_0}\sqrt{\frac{n_i n_k}{n_i + n_k}} > t_{1-\alpha}$	$(m = n_i + n_k - 2)$ (t **test;** note the special case $n_i = 1$)
3	$\xi_1 = \xi_2 = \cdots = \xi_r$ (given $\sigma_1^2 = \sigma_2^2 = \cdots = \sigma_r^2$)	$\lvert\log_e (S_A/S_0)\rvert > z_{1-\alpha/2}$	$(m = r - 1,\ m' = n - r)$ (FISHER'S z TEST)

19.6-7. The χ^2 Test for Goodness of Fit (see also Table 19.5-1). **(a)** The χ^2 test checks the "fit" of the hypothetical probabilities

$p_k = p[E_k]$ associated with r simple events $E_1, E_2, \ldots, E_r$ to their relative frequencies $h_k = h[E_k] = n_k/n$ in a sample of n independent observations. In many applications, each E_k is the event that some random variable x falls within a class interval (Sec. 19.2-2), so that the test compares the hypothetical theoretical distribution of x with its empirical distribution.

The goodness of fit is measured by the test statistic

$$y = n \sum_{k=1}^{r} \frac{(h_k - p_k)^2}{p_k} = \sum_{k=1}^{r} \frac{(n_k - np_k)^2}{np_k} \qquad (19.6\text{-}9)$$

y converges in probability to χ^2 with $m = r - 1$ degrees of freedom as $n \to \infty$. If all $np_k > 10$ (pool some class intervals if necessary), the resulting test *rejects* the hypothetical probabilities $p_1, p_2, \ldots, p_r$ at the level of significance α whenever the test sample yields $y > \chi_{1-\alpha}(m)$; for $m > 30$ one may replace the χ^2 distribution by a normal distribution with the indicated mean and variance.

(b) The χ^2 Test with Estimated Parameters. If the hypothetical probabilities p_k depend on a set of q unknown population parameters $\eta_1, \eta_2, \ldots, \eta_q$, obtain their joint maximum-likelihood estimates from the given sample (Sec. 19.4-4) and insert the resulting values of $p_k = p_k(\eta_1, \eta_2, \ldots, \eta_q)$ into Eq. (9). *The test statistic y will then converge in probability to χ^2 with $m = r - q - 1$ degrees of freedom under very general conditions (see below), and the χ^2 test applies with $m = r - q - 1$.* Tests of this type check the applicability of a normal distribution, Poisson distribution, etc., with unspecified parameters.

The theorem applies whenever the given functions $p_k(\eta_1, \eta_2, \ldots, \eta_q)$ satisfy the following conditions throughout a neighborhood of the joint maximum-likelihood "point" $(\eta_1, \eta_2, \ldots, \eta_q)$:

1. $\sum_{k=1}^{r} p_k(\eta_1, \eta_2, \ldots, \eta_q) = 1.$
2. The $p_k(\eta_1, \eta_2, \ldots, \eta_q)$ have a common positive lower bound, are twice continuously differentiable, and the matrix $[\partial p_k/\partial \eta_j]$ is of rank q.

19.6-8. Parameter-free Comparison of Two Populations: The Sign Test. One desires to test the hypothesis that *two random variables x and y have identical probability distributions,* or

$$P[x - y > 0] = P[x - y < 0] = \tfrac{1}{2}$$

assuming that it is known that $P[x = y] = 0$. Consider a random sample of n pairs $x_1, y_1; x_2, y_2; \ldots; x_n, y_n$; *neglect* any pairs such that $x_i = y_i$ (*ties*) in computing the sample size n. The probability that more than

m differences $x_i - y_i$ are positive is

$$p(m) = \frac{1}{2^n}\left[\binom{n}{m+1} + \binom{n}{m+1} + \cdots + \binom{n}{n}\right]$$

Now let m_α be the smallest value of m such that $p(m) \leq \alpha$ and

1. Reject the hypothesis (1) at the level of significance $\leq\alpha$ whenever the number of positive differences $x_i - y_i$ exceeds m_α (*One-tailed Sign Test*), or
2. Reject the hypothesis at the level of significance 2α if the number of positive or negative differences exceeds m_α (*Two-tailed Sign Test*).

m_α has been tabulated against α and n (Ref. 19.15). The sign test can also be used to test (1) the symmetry of a probability distribution; (2) the hypothesis that $x = X$ is the median of a distribution (Ref. 19.15).

19.6-9. Generalizations (Ref. 19.17). The fixed-sample tests of Secs. 19.6-1 to 19.6-7 admit only two alternatives, viz., acceptance or rejection of a given hypothesis on the basis of a test sample. *Sequential tests* permit an increase in the sample size (additional observations) as a third possible decision; it is then possible to specify *both* the level of significance *and* the power of the test (relative to some alternative hypothesis) when the test hypothesis will finally be accepted or rejected.

Schemes of fixed-sample and sequential tests are special examples of **statistical decision functions** or rules of behavior associating a *decision* (set of parameters $\eta_1, \eta_2, \ldots,$ "point" in parameter space) with each given sample $(x_1, x_2, \ldots)$ of some observed qualities. In practice (operations research, detection theory), the decision function is designed so as to maximize the expected value of some measure of effectiveness (payoff function) involving the gains due to correct decisions, the losses due to incorrect decisions, and the cost of testing samples of various sizes.

19.7. SOME STATISTICS, SAMPLING DISTRIBUTIONS, AND TESTS FOR MULTIVARIATE DISTRIBUTIONS

19.7-1. Introduction. Sections 19.7-2 to 19.7-7 are in no sense an exhaustive survey of multivariate statistics but present a number of frequently used definitions, formulas, and tests for convenient reference. Note that multivariate statistics often serve to estimate and test *stochastic relationships* between two or more random variables.

19.7-2. Statistics Derived from Multivariate Samples. (a) Given a multidimensional random variable $\mathbf{x} \equiv (x_1, x_2, \ldots, x_\nu)$ (Sec. 18.2-9), one proceeds by analogy with Secs. 19.1-2 and 19.2-2 to 19.2-4 to introduce a **random sample of size** n, $(\mathbf{x}_1, \mathbf{x}_2, \ldots, \mathbf{x}_n) \equiv (x_{11}, x_{21}, \ldots, x_{\nu 1}; x_{12}, x_{22}, \ldots, x_{\nu 2}; \ldots; x_{1n}, x_{2n}, \ldots, x_{\nu n})$ and the statistics

$$\bar{x}_i = \frac{1}{n}\sum_{k=1}^{n} x_{ik} \qquad (i = 1, 2, \ldots, \nu) \text{ (SAMPLE AVERAGE OF } x_i) \qquad (19.7\text{-}1)$$

$$\overline{f(x_1, x_2, \ldots, x_\nu)} = \frac{1}{n} \sum_{k=1}^{n} f(x_{1k}, x_{2k}, \ldots, x_{\nu k})$$

[SAMPLE AVERAGE OF $f(x_1, x_2, \ldots, x_\nu)$] (19.7-2)

$$l_{ij} = \overline{(x_i - \bar{x}_i)(x_j - \bar{x}_j)} = l_{ji} \qquad (i, j = 1, 2, \ldots, \nu)$$

(SAMPLE VARIANCES s_i^2 for $i = j$, SAMPLE COVARIANCES for $i \neq j$) (19.7-3)

$$r_{ij} = \frac{l_{ij}}{\sqrt{l_{ii}l_{jj}}} = r_{ji} \qquad (i, j = 1, 2, \ldots, \nu)$$

(SAMPLE CORRELATION COEFFICIENTS) (19.7-4)

The "point" corresponding to $\bar{\mathbf{x}} \equiv (\bar{x}_1, \bar{x}_2, \ldots, \bar{x}_\nu)$ is the **sample center of gravity,** and the matrix $L \equiv [l_{ij}]$ is the **sample moment matrix;** det $[l_{ij}]$ is the **generalized variance** of the sample.

(**b**) Given a two-dimensional random sample ($x_{11}, x_{21}; x_{12}, x_{22}; \ldots ; x_{1n}, x_{2n}$) of a bivariate random variable (x_1, x_2), one defines the **sample regression coefficients**

$$b_{21} = \frac{l_{12}}{l_{11}} = r_{12}\frac{s_2}{s_1} \qquad b_{12} = \frac{l_{12}}{l_{22}} = r_{12}\frac{s_1}{s_2} \tag{19.7-5}$$

The **empirical linear mean-square regression of x_2 on x_1**

$$g_2(x_1) = \bar{x}_2 + b_{21}(x_1 - \bar{x}_1) \tag{19.7-6}$$

is that linear function $ax_1 + b$ which minimizes the sample mean-square deviation

$$\overline{[x_2 - (ax_1 + b)]^2} = \frac{1}{n} \sum_{k=1}^{n} [x_{2k} - (ax_{1k} + b)]^2 \tag{19.7-7}$$

(see also Secs. 18.4-6*b* and 19.7-4*c*).

For ν-dimensional populations, *empirical multiple and partial correlation coefficients* and *regression coefficients* are derived from the sample moments l_{ij} by analogy with Eqs. (18.4-35) to (18.4-38) and serve as estimates of the corresponding population parameters. See Refs. 19.4 and 19.8 for the complete theory.

(**c**) The statistics (1) to (4) can be approximated by *grouped-data estimates* in the manner of Secs. 19.2-3 to 19.2-5. In particular, Sheppard's corrections for grouped-data estimates l_{ikG} of the second-order central moments λ_{ik} are given by

$$l'_{ii} = l_{iiG} - \tfrac{1}{12}(\Delta X_i)^2 \qquad l'_{ik} = l_{ikG} \quad (i \neq k)$$

where the ΔX_i are the constant class-interval lengths. These formulas often render errors due to grouping negligible if ΔX_i is less than one-eighth the range of x_i. Practical computation schemes are given in Refs. 19.3 and 19.8.

19.7-3. Estimation of Parameters (see also Sec. 19.4-1). Since the theorem of Sec. 19.4-2a applies to multivariate distributions, *the statistics* (3) *to* (5), *as well as the sample averages* $\bar{x}_i$, *are consistent estimates of the corresponding population parameters.* The mean value and variance of each sample average $\bar{x}_i$ and sample variance $l_{ii} = s_i^2$ is given by Eqs. (19.2-8), (19.2-13), and (19.2-14); in addition, note

$$E\{l_{ij}\} = \frac{n-1}{n}\lambda_{ij} = \lambda_{ij} + O\left(\frac{1}{n}\right) \qquad (19.7\text{-}8)$$

$$\text{Var}\,\{l_{ij}\} = \frac{1}{n}[E\{(x_i - \xi_i)^2(x_j - \xi_j)^2\} - \lambda_{ij}^2] + O\left(\frac{1}{n^2}\right) \qquad (19.7\text{-}9)$$

$$\text{Cov}\,\{l_{ii}, l_{jj}\} = \frac{1}{n}[E\{(x_i - \xi_i)^2(x_j - \xi_j)^2\} - \lambda_{ii}\lambda_{jj}] + O\left(\frac{1}{n^2}\right) \qquad (19.7\text{-}10)$$

$$\text{Cov}\,\{l_{ii}, l_{ij}\} = \frac{1}{n}[E(x_i - \xi_i)^3(x_j - \xi_j)\} - \lambda_{ii}\lambda_{ij}] + O\left(\frac{1}{n^2}\right) \qquad (19.7\text{-}11)$$

$$E\{r_{ij}\} = \rho_{ij} + O\left(\frac{1}{n}\right) \qquad (19.7\text{-}12)$$

Var $\{r_{ij}\}$ is of the order of $1/n$ as n increases (Ref. 19.4).

For multidimensional *normal* distributions (see also Sec. 19.7-4) only,

$$\left.\begin{aligned} E\{\det[l_{ij}]\} &= \frac{(n-1)(n-2)\cdots(n-\nu)}{n^\nu}\det[\lambda_{ij}] \\ \text{Var}\,\{\det[l_{ij}]\} &= \frac{\nu(2n+1-\nu)}{(n-\nu)(n-\nu+1)} \\ &\quad \frac{(n-1)^2(n-2)^2\cdots(n-\nu)^2}{n^{2\nu}}\det{}^2[\lambda_{ij}] \end{aligned}\right\} (n > \nu - 2) \qquad (19.7\text{-}13)$$

19.7-4. Sampling Distributions for Normal Populations (see also Secs. 18.8-6 and 18.8-8). For random samples drawn from multivariate *normal* populations, sampling distributions and tests involving only the sample averages $\bar{x}_i$ and the sample variances $l_{ii} = s_i^2$ are obtained simply from Sec. 19.5-3. It remains to investigate statistics which describe *stochastic relationships* of different random variables x_i, in particular sample correlation and regression coefficients.

(a) Distribution of the Sample Correlation Coefficient. For simplicity, consider a random sample $(x_{11}, x_{21}; x_{12}, x_{22}; \ldots ; x_{1n}, x_{2n})$ drawn from a two-dimensional normal population described by

$$\varphi(x_1, x_2) = \frac{1}{2\pi\sigma_1\sigma_2\sqrt{1-\rho^2}}\exp\left\{-\frac{1}{2(1-\rho^2)}\left[\left(\frac{x_1-\xi_1}{\sigma_1}\right)^2 - 2\rho\frac{x_1-\xi_1}{\sigma_1}\frac{x_2-\xi_2}{\sigma_2} + \left(\frac{x_2-\xi_2}{\sigma_2}\right)^2\right]\right\}$$
$$(\sigma_1 > 0, \sigma_2 > 0; -1 < \rho = \rho_{12} < 1) \qquad (19.7\text{-}14)$$

(Sec. 18.8-6). The probability density of the sample correlation coefficient $r_{12} = r$ (Sec. 19.7-2) is

$$\varphi_{r(n)}(r) = \frac{2^{n-3}}{\pi(n-3)!}(1-\rho^2)^{\frac{n-1}{2}}(1-r^2)^{\frac{n-4}{2}}\sum_{k=0}^{\infty}\Gamma^2\left(\frac{n+k-1}{2}\right)\frac{(2\rho r)^k}{k!}$$

$$= \frac{n-2}{\pi}(1-\rho^2)^{\frac{n-1}{2}}(1-r^2)^{\frac{n-4}{2}}\int_0^1 \frac{\vartheta^{n-2}}{(1-\rho r\vartheta)^{n-1}}\frac{d\vartheta}{\sqrt{1-\vartheta^2}}$$

$$(-1 < r < 1) \qquad (19.7\text{-}15)$$

and equals zero for $|r| > 1$; note

$$E\{r\} = \rho + O\left(\frac{1}{n}\right) \qquad \operatorname{Var}\{r\} = \frac{(1-\rho^2)^2}{n} + O\left(\frac{1}{n^{3/2}}\right) \qquad (19.7\text{-}16)$$

Note that $\varphi_{r(n)}(r)$ is independent of $\xi_1, \xi_2, \sigma_1, \sigma_2$. For $n = 2$, one has $\varphi_{r(2)}(r) = 0$ $(-1 < r < 1)$, since r is either 1 or -1.

It is useful to introduce the new random variable

$$y = \frac{1}{2}\log_e\frac{1+r}{1-r} \qquad (19.7\text{-}17)$$

which for $n \geq 10$ may be regarded as approximately normal with the approximate mean and variance

$$E\{y\} \approx \frac{1}{2}\log_e\frac{1+\rho}{1-\rho} + \frac{\rho}{2(n-1)} \qquad \operatorname{Var}\{y\} \approx \frac{1}{n-3} \qquad (19.7\text{-}18)$$

Figure 19.7-1 illustrates the behavior of the statistics r and y for different values of ρ and n.

If y and y' are values of the statistic (17) calculated from two independent random samples of respective sizes n, n' from *the same* normal population, then $y - y'$ is approximately normal with mean 0 and variance $1/(n-3) + 1/(n'-3)$.

(b) The r Distribution. Test for Uncorrelated Variables. In the important special case $\rho = 0$ (test for uncorrelated variables!), the frequency function (15) reduces to

$$\varphi_{r(n)}(r) = \frac{1}{\sqrt{\pi}}\frac{\Gamma\left(\frac{n-1}{2}\right)}{\Gamma\left(\frac{n-2}{2}\right)}(1-r^2)^{\frac{n-4}{2}} \qquad (-1 < r < 1) \qquad (19.7\text{-}19)$$

In this case, $t = r\sqrt{n-2}/\sqrt{1-r^2}$ *has a t distribution with* $n - 2$ *degrees of freedom* (Table 19.5-2). The statistic $r\sqrt{n-1}$ is said to have an **r distribution with $n - 2$ degrees of freedom.** The r distribution has been tabulated (Ref. 19.3) and is asymptotically normal with mean 0 and variance 1 as $n \to \infty$. *Either the t distribution or the r distribution yields tests for the hypothesis* $\rho = 0$.

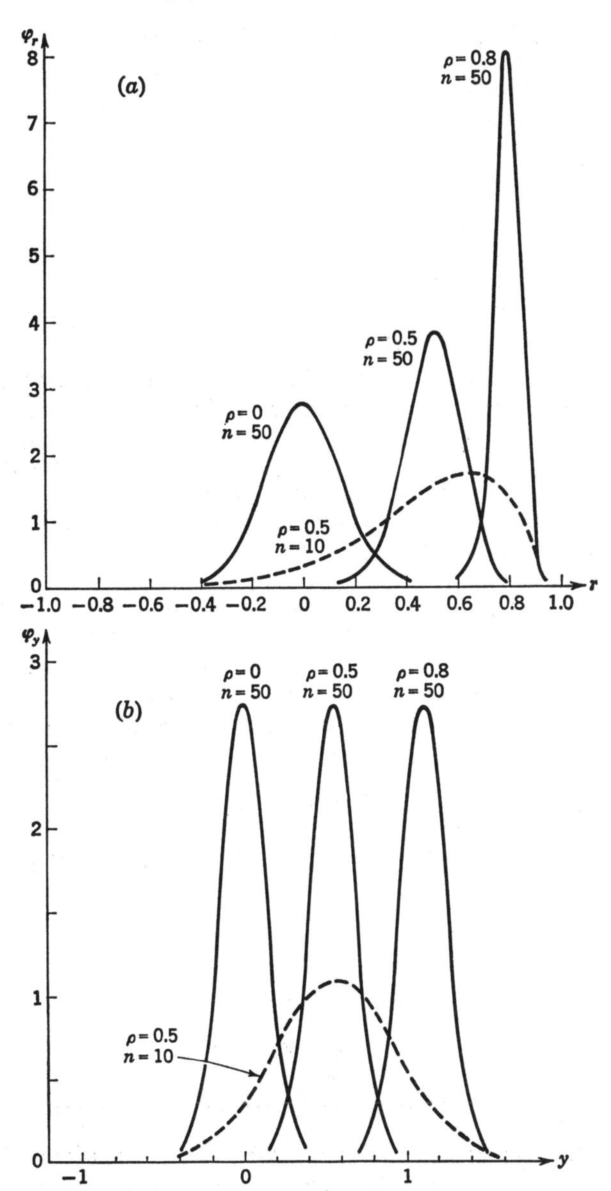

FIG. 19.7-1. Probability densities of the statistics r and y used to estimate the correlation coefficient ρ of a multidimensional normal distribution. (*From Burington and May, Handbook of Probability and Statistics, McGraw-Hill, New York,* 1953.)

(c) Test for Hypothetical Values of the Regression Coefficient (see also Sec. 19.7-2). Given a random sample drawn from a bivariate normal population described by Eq. (14), *one tests hypothetical values of the population regression coefficient* $\beta_{21} = \lambda_{12}/\lambda_{11} = \rho_{12}\sigma_2/\sigma_1$ (Sec. 18.4-6*b*) *by means of the test statistic*

$$t' = \frac{s_1\sqrt{n-2}}{s_2\sqrt{1-r_{12}^2}}(b_{21}-\beta_{21}) \tag{19.7-20}$$

which is distributed like t with $n - 2$ *degrees of freedom.* This test is far more convenient than one using the sample distribution of the sample regression coefficient b_{21} (Ref. 19.8).

(d) ν-dimensional Samples. For random samples drawn from a ν-dimensional normal population described by the probability density (18.8-26), the joint distribution of the sample averages $\bar{x}_1, \bar{x}_2, \ldots, \bar{x}_\nu$ is normal with mean values $\xi_1, \xi_2, \ldots, \xi_\nu$ and moment matrix Λ/n. *The joint distribution of the* $\bar{x}_i$ *is statistically independent of the joint distribution of the* $\nu(\nu+1)/2$ *sample moments* l_{ij} (*Generalized Fisher's Theorem*, see also Sec. 19.5-3*b*).

19.7-5. The Sample Mean-square Contingency. Contingency-table Test for Statistical Independence of Two Random Variables (see also Sec. 19.6-7). **(a)** Given a two-dimensional random sample $(x_1, y_1; x_2, y_2; \ldots; x_n, y_n)$ of a pair of random variables x, y, a **contingency table** arranges the n sample-value pairs (x_k, y_k) in a matrix of s x class intervals and r y class intervals. Let there be

$n_{i\cdot}$ pairs (x_k, y_k) in the ith x class interval
$n_{\cdot j}$ pairs (x_k, y_k) in the kth y class interval
n_{ij} pairs (x_k, y_k) both in the ith x class interval and in the jth y class interval

The statistic

$$f^2 = \sum_{i=1}^{r}\sum_{j=1}^{s}\frac{\left(\frac{n_{ij}}{n} - \frac{n_{i\cdot}}{n}\frac{n_{\cdot j}}{n}\right)^2}{\frac{n_{i\cdot}}{n}\frac{n_{\cdot j}}{n}} = \sum_{i=1}^{r}\sum_{j=1}^{s}\frac{n_{ij}^2}{n_{i\cdot}n_{\cdot j}} - 1$$

(SAMPLE MEAN-SQUARE CONTINGENCY) (19.7-21)

measures the "degree of association" (statistical dependence) between x and y. f^2 ranges between 0 and min $(r, s) - 1$ and reaches the latter value if and only if each row $(r \geq s)$ or each column $(r \leq s)$ contains only one element different from zero.

(b) *If* x *and* y *are statistically independent, then the test statistic* nf^2 *converges in probability to* χ^2 *with* $m = (r-1)(s-1)$ *degrees of freedom* *as* $n \to \infty$ (Table 19.5-1). If all $n_{ij} > 10$ (pool some class intervals if necessary), the hypothesis of statistical independence is rejected at the level of significance α by the critical region $nf^2 > \chi^2_{1-\alpha}(m)$ (Sec. 19.6-3).

(c) The special case $r = s = 2$ (success or failure, two-by-two contingency table) is of special interest; in this case

$$f^2 = \frac{(n_{11}n_{22} - n_{12}n_{21})^2}{n_{1\cdot}n_{2\cdot}n_{\cdot 1}n_{\cdot 2}} \qquad (19.7\text{-}22)$$

(see also Ref. 19.4).

19.7-6. Spearman's Rank Correlation. A Nonparametric Test of Statistical Dependence. Suppose that a random sample of n observation pairs $(x_1, y_1; x_2, y_2; \ldots ; x_n, y_n)$ yields only the information that x_k is the A_kth largest value of x_k in the sample, and y_k is the B_kth largest value of y_k $(k = 1, 2, \ldots , n)$. *If* x *and* y *are statistically independent, the test statistic*

$$R = 1 - \frac{6}{n(n^2 - 1)} \sum_{k=1}^{n} (A_k - B_k)^2$$

(SPEARMAN'S RANK CORRELATION) (19.7-23)

is asymptotically normal with mean 0 *and variance* $1/(n - 1)$ *as* $n \to \infty$. For any value of $n > 1$, the hypothesis of statistical independence is rejected at the level of significance $\leq \alpha$ (Sec. 19.6-3) if

$$R > \frac{u_{1-\alpha}}{\sqrt{n - 1}} \qquad (19.7\text{-}24)$$

(one-tailed test) or

$$|R| > \frac{u_{1-\alpha/2}}{\sqrt{n - 1}} \qquad (19.7\text{-}25)$$

(two-tailed test).

NOTE: If x and y have a *normal* joint distribution, then $2 \sin \pi R/6$ is a consistent estimate of their correlation coefficient ρ_{xy}.

19.8. RANDOM-PROCESS STATISTICS AND MEASUREMENTS

19.8-1. Simple Finite-time Averages. Let

$$f(t_1, t_2, \ldots , t_n) \equiv f[x(t_1), x(t_2), \ldots , x(t_n); y(t_1), y(t_2), \ldots , y(t_n); \ldots] \qquad (19.8\text{-}1)$$

be a given measurable function of the sample values $x(t_i)$, $y(t_i)$, . . . generated by the one-dimensional or multidimensional random process

$x(t), y(t), \ldots$ (Sec. 18.9-1). The **finite-time averages**

$$[f]_n = \frac{1}{n}\sum_{k=1}^{n} f(t_1 + k\,\Delta t, t_2 + k\,\Delta t, \ldots, t_n + k\,\Delta t) = \frac{1}{n}\sum_{k=1}^{n} f(k\,\Delta t) \tag{19.8-2}$$

$$\langle f\rangle_T = \frac{1}{T}\int_0^T f(t_1 + \lambda, t_2 + \lambda, \ldots, t_n + \lambda)\,d\lambda = \frac{1}{T}\int_0^T f(\lambda)\,d\lambda \tag{19.8-3}$$

can be obtained, respectively, through sampled-data and continuous averaging of finite real data. $[f]_n$ and $\langle f\rangle_T$ are random variables whose distributions are determined by the given random process. If $x(t)$ represents a *stationary* (but not necessarily ergodic) random process, then

$$E\{[f]_n\} = E\{\langle f\rangle_T\} = E\{f\} \tag{19.8-4}$$

The finite-time averages $[f]_n$ and $\langle f\rangle_T$ are, then, *unbiased estimates* of the unknown expected value $E\{f\}$ and will be useful for estimating $E\{f\}$ if their random fluctuations about their expected value (4) are reasonably small. More specifically, the mean-square error associated with estimation of $E\{f\}$ by a measured value of $[f]_n$ is

$$\begin{aligned}\text{Var}\,\{[f]_n\} &= \frac{1}{n^2}\sum_{i=1}^{n}\sum_{k=1}^{n} \text{Cov}\,\{f(i\,\Delta t), f(k\,\Delta t)\}\\ &= \frac{1}{n}\,\text{Var}\,\{f\} + \frac{2}{n}\sum_{k=1}^{n-1}\left(1 - \frac{k}{n}\right)\text{Cov}\,\{f(0), f(k\,\Delta t)\}\end{aligned} \tag{19.8-5}$$

If one introduces $k\,\Delta t = \lambda$ and lets $\Delta t \to 0$, $n = T/\Delta t \to \infty$, then the sampled-data average (2) converges to the continuous average $\langle y\rangle_T$ if the latter exists. A similar limiting process applied to Eq. (3) yields

$$\text{Var}\,\{\langle y\rangle_T\} = \frac{2}{T}\int_0^T \left(1 - \frac{\lambda}{T}\right)\text{Cov}\,\{f(0), f(\lambda)\}\,d\lambda \tag{19.8-6}$$

Depending on the nature of the **autocovariance function**

$$\begin{aligned}\text{Cov}\,\{f(0), f(\lambda)\} &= E\{[f(0) - E\{f\}][f(\lambda) - E\{f\}]\}\\ &= E\{f(0)f(\lambda)\} - [E\{f\}]^2 = R_{ff}(\lambda) - R_{ff}(0) = K_{ff}(\lambda)\end{aligned} \tag{19.8-7}$$

the mean-square error (4) or (5) may or may not decrease to an acceptably small value with increasing sample size n or integration time T.

19.8-2. Averaging Filters. Time averaging is often accomplished by low-pass *filters* implemented as electrical networks, by electromechanical measuring devices with inertia and damping, or by digital sampled-data operations. Consider, in particular, a general time-invariant averaging filter with stationary input $f(t) \equiv f(t_1 + t, t_2 + t, \ldots, t_n + t)$, bounded weighting function $h(\zeta)$, and frequency-response function $H(i\omega)$ (Sec. 9.4-7). If the filter input is applied between $t = 0$ and $t = T$ (*averaging time*), the filter output is

$$z(T) = \int_{-\infty}^{T} h_T(T - \lambda)f(\lambda)\,d\lambda = \int_0^\infty h_T(\zeta)f(T - \zeta)\,d\zeta \tag{19.8-8}$$

where

$$h_T(\zeta) = \begin{cases} h(\zeta) & (0 \le t < T) \\ 0 & \text{otherwise} \end{cases} \tag{19.8-9}$$

Hence

$$E\{z(T)\} = E\{f\}\int_0^T h_T(\zeta)\,d\zeta = a(T)E\{f\} \tag{19.8-10}$$

so that $z(T)/a(T)$ is an unbiased estimate of $E\{f\}$.

The estimate variance is given by

$$\text{Var}\,\{z(T)\} = E\{[z(t) - E\{z(t)\}]^2\} = \int_{-\infty}^{\infty} \varphi_{h_Th_T}(\lambda)\{R_{ff}(\lambda) - [E\{f\}]^2\}\,d\lambda \tag{19.8-11}$$

where (see also Sec. 18.12-3)

$$\varphi_{h_Th_T}(\lambda) = \int_{-\infty}^{\infty} h_T(\zeta)h_T(\zeta + \lambda)\,d\zeta \tag{19.8-12}$$

As $T \to \infty$, Var $\{z(T)\}$ will not in general go to zero but, rather, approaches the stationary output variance

$$\begin{aligned} \text{Var}\,\{z(\infty)\} &= \int_{-\infty}^{\infty} \varphi_{hh}(\lambda)\{R_{ff}(\lambda) - [E\{f\}]^2\}\,d\lambda \\ &= \int_{-\infty}^{\infty} |H(i\omega)|^2\Phi(\omega)\,\frac{d\omega}{2\pi} \approx 2|H(0)|^2\Phi(0)B_{EQ} \end{aligned} \tag{19.8-13}$$

where $\Phi(\omega)$ is the power spectral density (Sec. 18.10-3) of $f(t) - E\{f\}$, and B_{EQ} is the bandwidth of an equivalent "rectangular" low-pass filter having the frequency response

$$H_{\text{EQ}}(j\omega) = \begin{cases} H(0) & (|\omega| \le 2\pi B_{\text{EQ}}) \\ 0 & (|\omega| > 2\pi B_{\text{EQ}}) \end{cases}$$

B_{EQ} is a useful measure of the variance-reducing properties of a given averaging filter. Table 19.8-1 lists $H(i\omega)$ and B_{EQ} for some practical filters (flat-spectrum input).

Table 19.8-1. Averaging Filters (Ref. 19.25)

	$H(i\omega)$	B_{EQ}
First-order filter	$\frac{1}{i\omega T_0 + 1}$	$\frac{1}{4T_0}$
Second-order filters:		
Overdamped system	$\frac{1}{(i\omega T_1 + 1)(i\omega T_2 + 1)}$	$\frac{1}{4(T_1 + T_2)}$
Critically damped system	$\frac{1}{(i\omega T_0 + 1)^2}$	$\frac{1}{8T_0}$
Underdamped system	$\frac{\alpha^2 + \omega_1^2}{(i\omega + \alpha)^2 + \omega_1^2}$	$\frac{\alpha^2 + \omega_1^2}{8\alpha}$

19.8-3. Examples (Ref. 19.25). **(a) Measurement of Mean Value.** It is desired to measure the mean value $E\{x\} = \xi$ of a stationary random voltage $x(t)$ with

$$R_{xx}(\tau) = a^2 e^{-\alpha|\tau|} + \xi^2 \qquad \Phi_{xx}(\omega) = \frac{2\alpha a^2}{\omega^2 + \alpha^2} + 2\pi\xi^2\delta(\omega) \tag{19.8-14}$$

(white noise passed through a simple filter with -3-db bandwidth $\alpha/2\pi$ cps, or random telegraph wave with counting rate $\alpha/2$, Sec. 18.11-5). In this case,

$$\text{Var}\,\{\langle x\rangle_T\} = \frac{2a^2}{T}\int_0^T \left(1 - \frac{\lambda}{T}\right) e^{-\alpha\lambda}\, d\lambda = \frac{2a^2}{(\alpha T)^2}(\alpha T - 1 + e^{-\alpha T}) \leq \frac{2a^2}{\alpha T} \tag{19.8-15}$$

and for the first-order averaging filter of Table 19.8-1 with $T \gg T_0$ ($T > 4T_0$ for most practical purposes),

$$\text{Var}\,\{z(T)\} \approx \text{Var}\,\{z(\infty)\} = \frac{a^2\alpha}{\pi}\int_{-\infty}^{\infty} \frac{d\omega}{[(\omega T_0)^2 + 1](\omega^2 + \alpha^2)} = \frac{a^2}{\alpha T_0 + 1} \leq \frac{a^2}{\alpha T_0} \tag{19.8-16}$$

(b) Measurement of Mean Square. It is desired to measure $E\{f\} = E\{x^2\}$ for *Gaussian* noise $x(t)$ satisfying Eq. (14). In this case,

$$\left.\begin{aligned} R_{ff}(\tau) &= 2a^4 e^{-2\alpha|\tau|} + (a^2 + \xi^2)^2 \\ \Phi_{ff}(\tau) &= \frac{8\alpha a^2}{\omega^2 + (2\alpha)^2} + 2\pi(a^2 + \xi^2)^2\delta(\omega) \end{aligned}\right\} \tag{19.8-17}$$

$$\text{Var}\,\{\langle x^2\rangle_T\} = \frac{a^4}{(\alpha T)^2}(2\alpha T - 1 + e^{-2\alpha T}) \leq \frac{2a^4}{\alpha T} \tag{19.8-18}$$

and for the first-order averaging filter with $T \gg T_0$ and $f \equiv x^2$,

$$\text{Var}\,\{z(T)\} \approx \text{Var}\,\{z(\infty)\} = \frac{2a^4}{2\alpha T_0 + 1} \leq \frac{a^4}{\alpha T_0} \tag{19.8-19}$$

(c) Measurement of Correlation Functions (see also Sec. 18.9-3*b*). The variances of *correlation-function estimates* for jointly stationary $x(t)$, $y(t)$ are given by Eqs. (5), (6), and (11) with $f(t) \equiv x(t)y(t + \tau)$. Unfortunately, each variance depends on

$$E\{f(0)f(\lambda)\} = E\{x(t)x(t + \lambda)y(t + \tau)y(t + \tau + \lambda)\} \tag{19.8-20}$$

and this fourth-order moment of the joint distribution of $x(t)$ and $y(t)$ is hardly ever known. In the special case of jointly Gaussian and stationary signals $x(t)$, $y(t)$ with zero mean values,

$$E\{f(0)f(\lambda)\} = R_{xx}(\lambda)R_{yy}(\lambda) + R_{xy}^2(\tau) + R_{xy}(\tau + \lambda)R_{xy}(\tau - \lambda) \tag{19.8-21}$$

but even this involves the unknown correlation function $R_{xy}(\tau)$ itself and hence yields useful information only in simple special cases.

For stationary Gaussian $x(t)$,

$$\text{Var}\,\{\langle x(t)x(t + \tau)\rangle_T\} \leq \frac{4}{T}\int_0^\infty R_{xx}^2(\lambda)\, d\lambda \qquad (|\tau| < T) \tag{19.8-22}$$

To illustrate the dependence of the autocorrelation-function estimate

$$\langle x(t)x(t + \tau)\rangle_T = \frac{1}{T}\int_0^T x(\lambda)x(\lambda + \tau)\, d\lambda \tag{19.8-23}$$

on signal bandwidth and delay, consider again *Gaussian* noise $x(t)$ satisfying Eq. (14). In this case,

$$\text{Var}\,\{\langle x(t)x(t+\tau)\rangle_T\} = \frac{a^4}{2(\alpha T)^2}\,\{2\alpha T - 1 + 2e^{-2\alpha T} + [(2\alpha\tau + 1)(2\alpha T - 1) - 2(\alpha\tau)^2]e^{-2\alpha\tau}\} \qquad (T > \tau \geq 0) \qquad (19.8\text{-}24)$$

For $\tau = 0$, this agrees with Eq. (18). For observation times T large compared to the reciprocal signal bandwidth,

$$\frac{a^4}{\alpha T} \leq \text{Var}\,\{\langle x(t)x(t+\tau)\rangle_T\} \leq 2\,\frac{a^4}{\alpha T} \qquad (|\tau| \leq T) \qquad (19.8\text{-}25)$$

(within 1 per cent for $\alpha T \geq 10^4$).

For the more general case of a stationary Gaussian signal $x(t)$ with

$$R_{xx}(\tau) = a^2 e^{-\alpha|\tau|}\cos\omega_1\tau + [E\{x\}]^2 \qquad (19.8\text{-}26)$$

one similarly obtains

$$\text{Var}\,\{\langle x(t)x(t+\tau)\rangle_T\} \leq 2\,\frac{a^4}{\alpha T} \qquad (\alpha T \geq 10^4,\ |\tau| \leq T) \qquad (19.8\text{-}27)$$

19.8-4. Sample Averages. Different sample functions generated by the same random process will be denoted by ${}^1x(t)$, ${}^2x(t)$, . . . (Fig. 18.9-1) and are regarded as statistically independent; i.e., every finite set of samples ${}^ix(t_1)$, ${}^ix(t_2)$, . . . is statistically independent of every set of samples from a different sample function ${}^kx(t)$. If one can realize a set of sample functions ${}^1x(t)$, ${}^2x(t)$, . . . , ${}^nx(t)$ in independent repeated experiments, then the sample values ${}^1x(t_1)$, ${}^2x(t_1)$, . . . , ${}^nx(t_1)$ constitute a classical *random sample of size n*; i.e., the ${}^kx(t_1)$ are statistically independent random variables with identical probability distributions. Similarly, ${}^1x(t_1)$, ${}^2x(t_2)$; ${}^2x(t_1)$, ${}^2x(t_2)$; . . . ; ${}^nx(t_1)$, ${}^nx(t_2)$ or ${}^1x(t_1)$, ${}^1y(t_2)$; ${}^2x(t_1)$, ${}^2y(t_2)$; . . . ; ${}^nx(t_1)$, ${}^ny(t_2)$ constitute bivariate random samples.

Sample averages, like

$$\left.\begin{aligned} \overline{x(t_1)} &= \frac{1}{n}\,[{}^1x(t_1) + {}^2x(t_1) + \cdots + {}^nx(t_1)] \\ \overline{f[x(t_1), x(t_2), \ldots]} &= \frac{1}{n}\sum_{k=1}^{n} {}^kf \\ &= \frac{1}{n}\sum_{k=1}^{n} f[{}^kx(t_1), {}^kx(t_2), \ldots] \\ \overline{x(t_1)y(t_2)} &= \frac{1}{n}\sum_{k=1}^{n} {}^kx(t_1)\,{}^ky(t_2) \end{aligned}\right\} \qquad (19.8\text{-}28)$$

are, then, random-sample statistics in the sense of classical statistical theory. Sample averages must be obtained from repeated or multiple experiments, but it is usually much simpler to derive variances and probability distributions for sample averages than it is for time averages. In particular,

$$\text{Var}\,\{\overline{x(t_1)}\} = \frac{1}{n}\,\text{Var}\,\{x(t_1)\} \qquad \text{Var}\,\{\bar{f}\} = \frac{1}{n}\,\text{Var}\,\{f\}$$

$$\text{Var}\,\{\overline{x(t_1)y(t_2)}\} = \frac{1}{n}\,\text{Var}\,\{x(t_1)y(t_2)\} \qquad (19.8\text{-}29)$$

just as in Sec. 19.2-3 (see also Fig. 19.8-1).

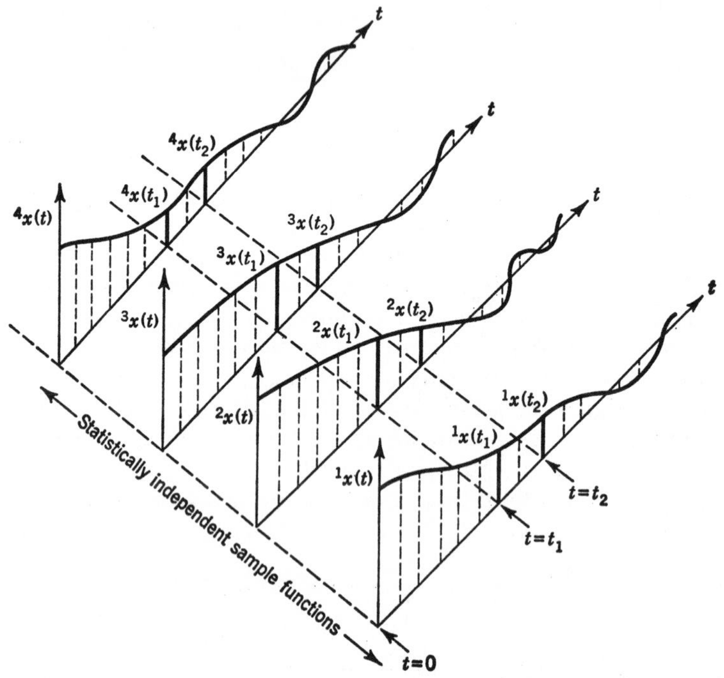

Fig. 19.8-1. Four sample functions $x(t) = {}^kx(t)$ from a continuous random process represented by $x(t)$.

19.9. TESTING AND ESTIMATION WITH RANDOM PARAMETERS

19.9-1. Problem Statement. A practically important class of decision situations can be represented by the model of Fig. 19.9-1. The *cost* C (*risk*) associated with one operation of the system shown is a function of the *state of the environment* represented by the m-dimensional random variable $(s_1, s_2, \ldots, s_m) \equiv \mathbf{s}$ and by a *decision* (response) represented by the r-dimensional variable $(y_1, y_2, \ldots, y_r) \equiv \mathbf{y}$, i.e.,

$$C = C(\mathbf{s}, \mathbf{y}) \tag{19.9-1}$$

The decision maker (man, machine, or system) arrives at the decision y on the basis of *received, observed, or measured data* represented by the

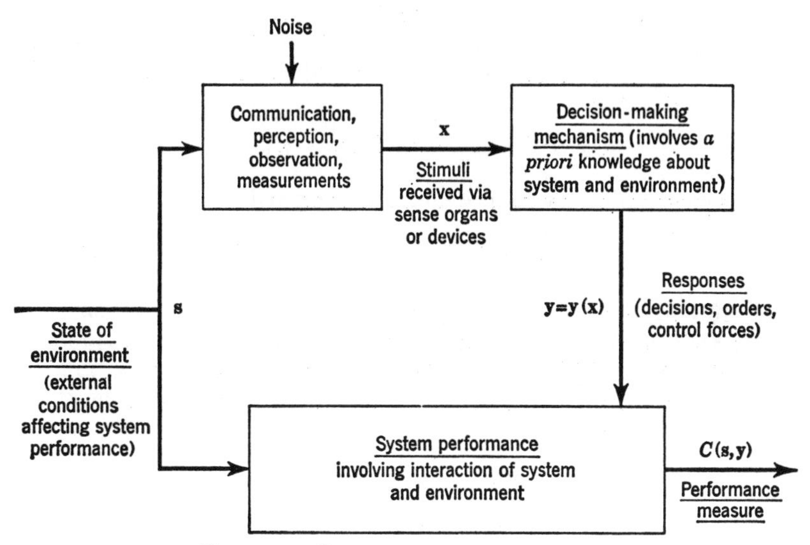

FIG. 19.9-1. Context for statistical decisions.

n-dimensional random variable $(x_1, x_2, \ldots, x_n) \equiv \mathbf{x}$, which is related to the state of the environment through the given joint distribution of $\mathbf{s}$ and $\mathbf{x}$. The decision maker forms $\mathbf{y}$ as a deterministic function (**decision function,** see also Sec. 19.6-9) of the received data:*

$$\mathbf{y} = \mathbf{y}(\mathbf{x}) \tag{19.9-2}$$

Given the joint distribution of $\mathbf{s}$ and $\mathbf{x}$ together with the cost function (1) representing the system performance for each combination of environment state and decision, one desires to minimize the **expected risk**

$$E\{C(\mathbf{s}, \mathbf{y})\} = \int_{\mathbf{s}} \int_{\mathbf{x}} C[\mathbf{s}, \mathbf{y}(\mathbf{x})]\, d\Phi(\mathbf{s}, \mathbf{x}) \tag{19.9-3}$$

through an optimal choice of the **decision function $\mathbf{y}(\mathbf{x})$.** $\mathbf{s}$, $\mathbf{x}$, and $\mathbf{y}$ may be discrete or continuous variables.

* Random or partially random selection of decisions (as in games with mixed strategies, Sec. 11.4-4*b*) will not be considered here.

19.9-2. Bayes Estimation and Tests. If the environment-state parameters $s_1, s_2, \ldots, s_m$ are regarded as parameters of the unknown probability distribution of the observed sample $(x_1, x_2, \ldots, x_n)$, the problem is somewhat similar to the classical problems of testing and estimation; the essential difference is that the parameters $s_1, s_2, \ldots, s_m$ are now *random variables.*

For continuous variables **s**, **x**, the decision maker's knowledge of the environment state on the basis of a received sample **x** is, then, expressed by the conditional probability density $\varphi(\mathbf{s}|\mathbf{x})$. Minimization of the expected risk (3), or

$$E\{C(\mathbf{s}, \mathbf{y})\} = \int_{\mathbf{x}} d\Phi(\mathbf{x}) \int_{\mathbf{s}} C[\mathbf{s}, \mathbf{y}(\mathbf{x})]\, d\Phi(\mathbf{s}|\mathbf{x}) \tag{19.9-4}$$

then requires minimization of the **conditional risk**

$$E\{C(\mathbf{s}, \mathbf{y}|\mathbf{x})\} = \int_{\mathbf{s}} C[\mathbf{s}, \mathbf{y}(\mathbf{x})]\, d\Phi(\mathbf{s}|\mathbf{x}) \tag{19.9-5}$$

for each sample **x** through a proper choice of the decision function $\mathbf{y}(\mathbf{x})$. If one is given the "a priori" distribution of **s** and the conditional probability density $\varphi(\mathbf{x}|\mathbf{s})$, then the "a posteriori" probability distribution required for Eq. (5) is obtained with the aid of Bayes's theorem (Secs. 18.2-6 and 18.4-5) as

$$d\Phi(\mathbf{s}|\mathbf{x}) = \frac{\varphi(\mathbf{x}|\mathbf{s})\, d\Phi(\mathbf{s})}{\int_{\mathbf{s}} \varphi(\mathbf{x}|\mathbf{s})\, d\Phi(\mathbf{s})} \tag{19.9-6}$$

Decision processes based on such minimization of the expected risk are known as *Bayes estimation,* and as **Bayes tests** if **y** is a discrete variable. Note that not all the unknown parameters s_i need affect $C[\mathbf{s}, \mathbf{y}]$ explicitly (e.g., signal-carrier phase in amplitude-modulated radio transmission, Ref. 19.23).

If, as is often the case, one has no reliable knowledge of the cost function $C(\mathbf{s}, \mathbf{y})$ and/or of the "a priori" distribution of **s**, then Bayes estimation and testing becomes impossible. One may assume "worst-case" $C(\mathbf{s}, \mathbf{y})$ and $\varphi(\mathbf{s})$ (*minimax tests,* Refs. 19.24 and 19.26), or one returns to the "classical" procedures of maximum-likelihood estimation (Sec. 19.4-4) and Neyman-Pearson tests (Sec. 19.6-3).

19.9-3. Binary State and Decision Variables: Statistical Tests or Detection (see also Secs. 19.6-1 to 19.6-4). Assume that there exist only *two environment states,* $s = 0$ (*null hypothesis*) and $s = 1$ (alternate hypotheses corresponding, say, to the absence and presence of a target to be detected). Assume *two possible decisions* $y = 0$, $y = 1$, which correspond to acceptance or rejection of the null hypothesis on the basis of the observed data sample $(x_1, x_2, \ldots, x_n)$. The problem amounts to the choice of a critical region (rejection region) S_C of sample points $(x_1, x_2, \ldots, x_n)$ which will minimize the expected risk

$$\begin{aligned} E\{C(s, y)\} &= C(s = 0, y = 0)p_0 \int_{\tilde{S}_C} \varphi(x_1, x_2, \ldots, x_n|s = 0)\, dx_1\, dx_2 \ldots dx_n \\ &+ C(s = 0, y = 1)p_0 \int_{S_C} \varphi(x_1, x_2, \ldots, x_n|s = 0)\, dx_1\, dx_2 \ldots dx_n \\ &+ C(s = 1, y = 0)(1 - p_0) \int_{\tilde{S}_C} \varphi(x_1, x_2, \ldots, x_n|s = 1)\, dx_1\, dx_2 \ldots dx_n \\ &+ C(s = 1, y = 1)(1 - p_0) \int_{S_C} \varphi(x_1, x_2, \ldots, x_n|s = 1)\, dx_1\, dx_2 \ldots dx_n \end{aligned} \tag{19.9-7}$$

where $p_0 = P[s = 0]$, and $\tilde{S}_C$ is the complement of S_C (acceptance region). *$E\{C(s, y)\}$ will be minimized if one rejects the null hypothesis whenever the likelihood ratio*

$$\Lambda(x_1, x_2, \ldots, x_n) \equiv \frac{\varphi(x_1, x_2, \ldots, x_n|s = 1)}{\varphi(x_1, x_2, \ldots, x_n|s = 0)} \tag{19.9-8}$$

(see also Sec. 19.6-3) *exceeds the critical value*

$$\begin{aligned} \Lambda_C &= \frac{p_0}{1 - p_0} \frac{C(s = 0, y = 1) - C(s = 0, y = 0)}{C(s = 1, y = 0) - C(s = 1, y = 1)} \\ &= \frac{p_0}{1 - p_0} \frac{\text{added cost of false rejection (false alarm)}}{\text{added cost of false acceptance (miss)}} \end{aligned} \tag{19.9-9}$$

Note that (1) any increasing or decreasing function of the likelihood ratio (8) can replace the latter as a **test statistic,** and (2) the likelihood ratio (8) is itself a monotonic function of the "a posteriori" conditional probability $p(s|x_1, x_2, \ldots, x_n)$, which may be regarded as a basic test statistic.

EXAMPLE: *Signal Detection with Additive Flat-spectrum Gaussian Noise.* The objective is to decide whether a received signal $x(t)$ of bandwidth B is due to noise alone [$s = 0$, or $x(t) = n(t)$] or to signal and additive noise [$s = 1$, or $x(t) = s(t) + n(t)$]; in view of the finite bandwidth, one can describe signals and noise in terms of sample values

$$x_k = x(k\, \Delta t) \qquad s_k = s(k\, \Delta t) \qquad n = n(k\, \Delta t)$$

with

$$\Delta t = \frac{1}{2B} \qquad (k = 1, 2, \ldots, 2BT)$$

where T is the observation time (Sec. 18.11-2). One is given

$$\varphi_{\mathbf{x}|s}(x_1, x_2, \ldots, x_{2BT}|s = 0) = (2\pi P_N)^{-BT} \exp\left[-\frac{1}{2P_N}\sum_{k=1}^{2BT} x_k^2\right]$$

$$\varphi_{\mathbf{x}|s}(x_1, x_2, \ldots, x_{2BT}|s = 1) = (2\pi P_N)^{-BT} \exp\left[-\frac{1}{2P_N}\sum_{k=1}^{2BT} (x_k - s_k)^2\right]$$

so that

$$\Lambda(x_1, x_2, \ldots, x_n) = \exp\left[-\frac{1}{2P_N}\sum_{k=1}^{2BT} s_k^2 + \frac{1}{P_N}\sum_{k=1}^{2BT} s_k x_k\right]$$

Since

$$\frac{1}{2B}\sum_{k=1}^{2BT} s_k x_k = \sum_{k=1}^{2BT} s(k\,\Delta t)x(k\,\Delta t)\,\Delta t \approx \int_0^T s(t)x(t)\,dt = z(x_1, x_2, \ldots, x_{2BT})$$

is an increasing function of the likelihood ratio $\Lambda(x_1, x_2, \ldots, x_{2BT})$, z rather than Λ may be used as a test statistic. The receiver forms $z(x_1, x_2, \ldots, x_{2BT})$, either by discrete summation or by continuous integration, and compares it with a threshold value z_C determined by p_0 and $C(s, y)$; $z > z_C$ corresponds to the decision $y = 1$ (*crosscorrelation detector or matched-filter detector*). See Secs. 19.22, 19.23, and 19.26 to 19.28 for additional examples and applications.

19.9-4. Estimation (Signal Extraction, Regression). In a typical measurement situation, the environment has a continuously distributed set of states represented by values of $\mathbf{s} \equiv (s_1, s_2, \ldots, s_n)$, and the m components y_k of the m-dimensional decision function $\mathbf{y} \equiv (y_1, y_2, \ldots, y_m)$ must be chosen so as to approximate the corresponding s_k as closely as possible in some sense specified by the cost function $C(\mathbf{s}, \mathbf{y})$. In the practically important case of *least-square estimation,* one assumes a cost function of the form

$$C(\mathbf{s}, \mathbf{y}) \equiv C(s_1, s_2, \ldots, s_m; y_1, y_2, \ldots, y_m) \equiv \sum_{k=1}^{m} (s_k - y_k)^2 \tag{19.9-10}$$

In this case, $E\{C(\mathbf{s}, \mathbf{y})\}$ is minimized if each y_k equals the conditional expected value of s_k for the measured sample $(x_1, x_2, \ldots, x_n)$:

$$y_k = E\{s_k|x_1, x_2, \ldots, x_n\} = \int_{-\infty}^{\infty} s_k\varphi(s_k|x_1, x_2, \ldots, x_n)\,ds_k \qquad (k = 1, 2, \ldots, m) \tag{19.9-11}$$

(see also Secs. 18.4-5, 18.4-6, and 18.4-9), where $\varphi(s_k|x_1, x_2, \ldots, x_n)$ must be obtained from $\varphi(x_1, x_2, \ldots, x_n|s_1, s_2, \ldots, s_m)$ with the aid of Bayes's formula

$$\varphi_{\mathbf{s}|\mathbf{x}}(s_1, s_2, \ldots, s_m|x_1, x_2, \ldots, x_n) = \frac{\varphi_{\mathbf{x}|\mathbf{s}}(x_1, x_2, \ldots, x_n|s_1, s_2, \ldots, s_m)\varphi_{\mathbf{s}}(s_1, s_2, \ldots, s_m)}{\int_{\mathbf{s}} \varphi_{\mathbf{x}|\mathbf{s}}(x_1, x_2, \ldots, x_n|s_1, s_2, \ldots, s_m) \cdot \varphi_{\mathbf{s}}(s_1, s_2, \ldots, s_m)\,ds_1\,ds_2 \ldots ds_m} \tag{19.9-12}$$

EXAMPLE: *D-c Measurement with Additive Gaussian Noise.* The objective is to measure a single random quantity s from a sample $(x_1, x_2, \ldots, x_n)$ of measured values $x_k = s + n_k$, given

$$\varphi_s(s) = \frac{1}{\sigma_s\sqrt{2\pi}} e^{-\frac{1}{2\sigma_s^2}(s-\xi_s)^2} \qquad \text{("A PRIORI" PROBABILITY DENSITY OF } s\text{)}$$

$$\varphi_{x|s}(x_1, x_2, \ldots, x_n|s) = \left(\frac{1}{2\pi P_N}\right)^{n/2} \exp\left[-\frac{1}{2P_N}\sum_{k=1}^{n}(x_k - s)^2\right]$$

which corresponds to additive-noise samples n_k statistically independent of each other and of s. Bayes's formula yields

$$\varphi_{s|x}(s|x_1, x_2, \ldots, x_n) = \frac{1}{\sigma\sqrt{2\pi}} e^{-\frac{1}{2\sigma^2}(s-\xi)^2}$$

with

$$\xi = E\{s|x_1, x_2, \ldots, x_n\} = \sigma^2\left(\frac{n\bar{x}}{P_N} + \frac{\xi_s}{\sigma_s^2}\right)$$

$$\sigma^2 = \text{Var}\,\{s|x_1, x_2, \ldots, x_n\} = \left(\frac{n}{P_N} + \frac{1}{\sigma_s^2}\right)^{-1}$$

$$\bar{x} = \frac{1}{n}\sum_{k=1}^{n} x_k$$

$E\{(y - s)^2\}$ is then minimized by the least-squares estimate

$$y = \xi = \frac{\sigma_s^2\bar{x} + \xi_s\dfrac{P_N}{n}}{\sigma_s^2 + \dfrac{P_N}{n}}$$

which depends on the sample values x_k only by way of the statistic $\bar{x}$ (sample average) and is *biased* by *a priori* knowledge of $\varphi_s(s)$; the bias increases with increasing P_N and decreasing sample size n, both of which reduce the "information" in the measured sample $(x_1, x_2, \ldots, x_n)$. The resulting expected risk is

$$E\{(y - s)^2\} = \frac{P_N}{n}\left(\frac{\sigma_s^2}{\sigma_s^2 + \dfrac{P_N}{n}}\right) \le \frac{P_N}{n}$$

References 19.22, 19.23, and 19.26 to 19.28 describe additional applications.

19.10. RELATED TOPICS, REFERENCES, AND BIBLIOGRAPHY

19.10-1. Related Topics. The following topics related to the study of mathematical statistics are treated in other chapters of this handbook:

19.10-2. References and Bibliography (see also Sec. 18.13-2).

19.1. Brownlee, K. A.: *Statistical Theory and Methodology in Science and Engineering*, Wiley, New York, 1961.

19.2. Brunk, H. D.: *An Introduction to Mathematical Statistics*, 2d ed., Blaisdell, New York, 1964.

19.3. Burington, R. S., and D. C. May: *Handbook of Probability and Statistics with Tables*, 2d ed., McGraw-Hill, New York, 1967.

19.4. Cramér, H.: *Mathematical Methods of Statistics*, Princeton, Princeton, N.J., 1951.

19.5. Dixon, W. J., and F. J. Massey, Jr.: *An Introduction to Statistical Analysis*, 2d ed., McGraw-Hill, New York, 1957.

19.6. Eisenhart, G. C., M. W. Hastay, and W. A. Wallis: *Techniques of Statistical Analysis*, McGraw-Hill, New York, 1947.

19.7. Elderton, W. P.: *Frequency Curves and Correlation*, 3d ed., Cambridge, New York, 1938.

19.8. Hald, A.: *Statistical Theory with Engineering Applications*, Wiley, New York, 1952.

19.9. Hoel, P. G.: *Elementary Statistics*, 2d ed., Wiley, New York, 1966.

19.10. Hogg, R. V., and A. T. Craig: *Introduction to Mathematical Statistics*, Macmillan, New York, 1959.

19.11. Lehmann, E. L.: *Testing Statistical Hypotheses*, Wiley, New York, 1959.

19.12. Mood, A. M., and F. A. Graybill: *Introduction to the Theory of Statistics*, 2d ed., McGraw-Hill, New York, 1963.

19.13. Neyman, J.: *First Course in Probability and Statistics*, Holt, New York, 1950.

19.14. Scheffé, H.: *The Analysis of Variance*, Wiley, New York, 1959.

19.15. Van der Waerden, B. L.: *Mathematische Statistik*, 2d ed., Springer, Berlin, 1965.

19.16. Walsh, J. E.: *Handbook of Nonparametric Statistics* (2 vols.), Van Nostrand, Princeton, N.J., 1960/65.

19.17. Weiss, L.: *Statistical Decision Theory*, McGraw-Hill, New York, 1961.

19.18. Wilks, S. S.: *Mathematical Statistics*, 2d ed., Wiley, New York, 1962.

Random-process Statistics and Decision Theory

19.19. Bendat, J. S.: *Principles and Applications of Random Noise Theory*, Wiley, New York, 1958.

19.20. ——— and A. G. Piersol: *Measurement and Analysis of Random Data*, Wiley, New York, 1966.

19.21. Blackman, R. B., and J. W. Tukey: *The Measurement of Power Spectra*, Dover, New York, 1958.

19.22. Davenport, W. B., and W. L. Root: *Introduction to Random Signals and Noise*, McGraw-Hill, New York, 1958.

19.23. Hancock, J. C.: *Signal Detection*, McGraw-Hill, New York, 1966.

19.24. Helstrom, C. W.: *Statistical Theory of Signal Detection*, Pergamon Press, New York, 1960.

19.25. Korn, G. A.: *Random-process Simulation and Measurements*, McGraw-Hill, New York, 1966.

19.26. Middleton, D.: *An Introduction to Statistical Communication Theory*, McGraw-Hill, New York, 1960.

19.27. Wainstein, L. A., and V. D. Zubakov: *Extraction of Signals from Noise*, Prentice-Hall, Englewood Cliffs, N.J., 1962.

19.28. Wozencraft, J. M., and I. M. Jacobs: *Principles of Communication Engineering*, Wiley, New York, 1965.

CHAPTER **20**

NUMERICAL CALCULATIONS AND FINITE DIFFERENCES

20.1. INTRODUCTION

20.1-1. Chapter 20 outlines a number of the best-known numerical computation schemes, with emphasis on mathematical methods rather than on detailed layout or programming. Sections 20.4-1 to 20.5-7 introduce the calculus of finite differences and the solution of difference equations. Finite-difference methods are of great interest not only for the numerical solution of differential equations but also in connection with many mathematical models whose variables change in discrete steps by nature rather than by necessity.

20.1-2. Errors. Aside from possible **mistakes** (blunders) in the course of numerical computations, there may be **errors in the initial data, round-off errors** due to the use of a finite number of digits, and **truncation errors** due to finite approximations of limiting processes. Approximate effects of small errors Δx_i or fractional errors $\Delta x_i/x_i$ on a result $f(x_1, x_2, \ldots)$ can often be studied through differentiation (Sec. 4.5-3); thus

$$\left.\begin{array}{c}\Delta(x_1 + x_2) = \Delta x_1 + \Delta x_2 \qquad \Delta(x_1 - x_2) = \Delta x_1 - \Delta x_2 \\ \Delta(x_1x_2) \approx x_1\,\Delta x_2 + x_2\,\Delta x_1\end{array}\right\} \tag{20.1-1}$$

$$\left|\frac{\Delta(x_1 - x_2)}{x_1 - x_2}\right| \leq \frac{|\Delta x_1| + |\Delta x_2|}{|x_1 - x_2|} \qquad \frac{\Delta(x_1x_2)}{x_1x_2} = \frac{\Delta x_1}{x_1} + \frac{\Delta x_2}{x_2} \tag{20.1-2}$$

The generation and propagation of errors in more complicated computations are subject to continuing studies; most analytical results of such studies are restricted to very specialized classes of computations (Refs. 20.4 to 20.7, 20.10, 20.44, 20.49, 20.58, and 20.59). It is desirable to check on mistakes and errors by various *checking routines* (e.g., resubstitution of approximate solutions into given equations) carried along between appropriate steps of the main computation. As a very rough rule of thumb, one may carry two significant figures in excess of those justified by the precision of the initial data or by the required accuracy. Convergent iteration schemes (Secs. 20.2-2, 20.2-4, 20.3-2, 20.8-2*c*, and 20.9-3) will tend to check and reduce the effects of errors, unless the errors affect the convergence.

A computing scheme (program of numerical operations, algorithm) is said to be **numerically stable** if it does not increase the (absolute or relative) effects of round-off errors in the given data and in the calculations. More precise definitions of numerical stability can often be related to the asymptotic stability of solutions of difference equations (recurrence relations) in the manner of Sec. 20.8-5.

20.2. NUMERICAL SOLUTION OF EQUATIONS

20.2-1. Introduction. The numerical solution of any equation

$$f(z) = 0 \tag{20.2-1}$$

should be preceded by a rough (often graphical) survey yielding information regarding the existence and position of real roots, estimates for trial solutions, etc. (see also Secs. 1.6-6 and 7.6-9). Solutions may be checked by resubstitution.

20.2-2. Iteration Methods. Newton-Raphson Method and Regula Falsi. The following procedures apply, in particular, to the solution of transcendental equations.

(a) Rewrite the given equation (1) in the form

$$z = \varphi(z) \tag{20.2-2}$$

(see also Sec. 20.2-2*b*). Starting with a *trial solution* $z^{[0]}$, compute *successive approximations*

$$z^{[j+1]} = \varphi(z^{[j]}) \qquad (j = 0, 1, 2, \ldots) \tag{20.2-3}$$

The convergence of such approximations to a desired solution z requires separate investigation. *The Contraction-mapping Theorem of Sec.* 12.5-6 *can often be applied to establish convergence and rate of convergence.* The iteration is terminated when $|z^{[j]} - z^{[j-1]}|/|z^{[j]}|$ is sufficiently small.

Collatz's Convergence Criterion and Error Estimate (Ref. 20.3). *If it is possible to choose a complex-plane region D such that* (1) *there exists a positive real number $M < 1$ with the property*

$$|\varphi(Z_1) - \varphi(Z_2)| \le M|Z_1 - Z_2| \qquad \text{for all } Z_1, Z_2 \text{ in } D$$

and (2) *D contains $z^{[0]}$, $z^{[1]}$, and every point z having the property*

$$|z - z^{[j]}| \le \frac{M}{1 - M}|z^{[j]} - z^{[j-1]}| \tag{20.2-4}$$

for some value of $j > 0$, then the approximations (3) *converge to a solution z of Eq.* (1) *which satisfies Eq.* (4) *and is unique in D.* Note that the right side of Eq. (4) is an upper bound for the error of the jth approximation $z^{[j]}$ (see also Sec. 12.5-6).

(b) In general, *there are many possible ways of rewriting the given equation* (1) *in the form* (2). Particular choices of $\varphi(z)$ yield the following special forms of the iteration formula (3):

$$z^{[j+1]} = z^{[j]} - kf(z^{[j]}) \tag{20.2-5}$$

$$z^{[j+1]} = z^{[j]} - \frac{f(z^{[j]})}{f'(z^{[j]})} \qquad \text{(Newton-Raphson method)} \tag{20.2-6}$$

$$z^{[j+1]} = z^{[j]} - \frac{f(z^{[j]})}{f'(z^{[j]})} - \frac{1}{2}\frac{[f(z^{[j]})]^2 f''(z^{[j]})}{[f'(z^{[j]})]^3} \tag{20.2-7}$$

These iteration schemes are especially useful for the computation of real roots. To find complex roots of real equations, one must start with a complex trial solution $z^{[0]}$.

The convergence considerations and error criterion of Sec. 20.2-2*a* apply in each case. Equation (6) is a special case of the general Newton-Raphson-Kantorovich scheme of Sec. 20.2-7. *If $f'(z)$ exists and is continuously differentiable in the region under consideration, and one can find positive real bounds A, B, C such that*

$$\left|\frac{1}{f'(z^{[0]})}\right| \leq A \qquad \left|\frac{f(z^{[0]})}{f'(z^{[0]})}\right| \leq B \tag{20.2-8a}$$

and

$$|f''(z)| \leq C \leq \frac{1}{2AB} \tag{20.2-8b}$$

whenever

$$|z - z^{[0]}| \leq \frac{1}{AC}(1 - \sqrt{1 - 2ABC}) \tag{20.2-8c}$$

then the desired root z satisfies the last inequality, and the convergence rate of the Newton-Raphson iteration is given by

$$|z - z^{[j]}| \leq \frac{B}{2^{j-1}}(2ABC)^{2^{j-1}} \qquad (j = 1, 2, \ldots) \tag{20.2-9}$$

Note the relatively rapid convergence under these circumstances (see also Sec. 20.2-8 and Refs. 20.43 and 20.47).

USEFUL EXAMPLES: Application of the Newton-Raphson scheme (6) to $1/z - a = 0$, $z^2 - a = 0$, $1/z^2 - 1/a^2 = 0$, and $z^n - a = 0$ yields iteration routines for $1/a$, $\sqrt{a}$, and $\sqrt[n]{a}$:

$$z^{[j+1]} = z^{[j]}(2 - az^{[j]}) \to 1/a \text{ as } j \to \infty \qquad (a > 0) \tag{20.2-10}$$

$$z^{[j+1]} = \tfrac{1}{2}\left(z^{[j]} + \frac{a}{z^{[j]}}\right) \to \sqrt{a} \text{ as } j \to \infty \qquad (a > 0)$$

(HERON'S ALGORITHM) (20.2-11)

$$z^{[j+1]} = z^{[j]}\left[1 + \frac{a - (z^{[j]})^2}{2a}\right] \to \sqrt{a} \text{ as } j \to \infty \qquad (a > 0) \tag{20.2-12}$$

$$z^{[j+1]} = \left(1 - \frac{1}{n}\right)z^{[j]} + \frac{a}{n(z^{[j]})^{n-1}} \to \sqrt[n]{a} \text{ as } j \to \infty \qquad (a > 0) \tag{20.2-13}$$

The iteration (10) converges for all $z^{[0]}$ between 0 and $2/a$; (11), (12,) and (13) require only $z^{[0]} > 0$.

(c) The Regula Falsi. The following iteration scheme applies particularly well to real equations and real roots and is used when $f'(z)$ is not easily computed. Given Eq. (1), start with two trial values $z = z^{[0]}$, $z = z^{[1]}$ and obtain successive approximations

$$z^{[j+1]} = z^{[j]} - \frac{z^{[j]} - z^{[k]}}{f(z^{[j]}) - f(z^{[k]})} f(z^{[j]})$$

$$(j = 0, 1, 2, \ldots; k < j) \tag{20.2-14}$$

For continuous real functions $f(z)$, one attempts to bracket each real root between approximations $z^{[j]}$ and $z^{[k]}$ such that $f(z^{[j]})$ and $f(z^{[k]})$ have opposite signs. Each $z^{[j]}$ can then be obtained by graphical straight-line interpolation; the scale of the graph should be appropriately increased at each step.

(d) Aitken-Steffens Convergence Acceleration. If $f(z)$ is real and three times continuously differentiable for real z in a neighborhood of a real root where $f'(z) \neq 1$, then one can improve on the convergence of the iteration sequence (3) by the following iteration scheme

$$\left.\begin{array}{lll} z^{[1]} = f(z^{[0]}) & z^{[2]} = f(z^{[1]}) & \\ z^{[3]} = z^{[0]} - \dfrac{(z^{[1]} - z^{[0]})^2}{z^{[2]} - 2z^{[1]} + z^{[0]}} & & z^{[4]} = f(z^{[3]}) \\ z^{[5]} = f(z^{[4]}) & z^{[6]} = z^{[3]} - \dfrac{(z^{[4]} - z^{[3]})^2}{z^{[5]} - 2z^{[4]} + z^{[3]}} & \\ z^{[7]} = f(z^{[6]}) & z^{[8]} = f(z^{[7]}) & \\ \cdots & \cdots & \cdots \end{array}\right\} \tag{20.2-15}$$

The iteration is terminated when one of the denominators turns out to be zero (in general, the desired accuracy is obtained before this if the sequence converges: see also Ref. 20.5). This method, like the *regula falsi,* may substitute for the fast-converging Newton-Raphson scheme if computation of $f'(z)$ is too cumbersome.

(e) Multiple Roots. Iteration schemes based on Eq. (6) (Newton-Raphson method) and Eq. (7) will not converge in the neighborhood of a multiple root of the given equation. Note that multiple zeros of $f(z)$ are zeros of $f'(z)$; for algebraic equations, the greatest common divisor of $f(z)$ and $f'(z)$ can be obtained by the method of Sec. 1.7-3.

(f) Interval Halving. If a single simple real root of a real equation (1) is known to lie between $z = a$ and $z = b$, compute $f\left(\frac{a+b}{2}\right)$. If this has the same sign as, say, $f(b)$, compute $f\left[\frac{1}{2}\left(\frac{a+b}{2} + a\right)\right]$, etc.

20.2-3. Numerical Solution of Algebraic Equations: Computation of Polynomial Values. (a) Successive Multiplications. To compute values of a polynomial

$$f(z) \equiv a_0 z^n + a_1 z^{n-1} + \cdots + a_{n-1} z + a_n \tag{20.2-16}$$

for use in the iteration methods of Sec. 20.2-3*b*, compute successively $a_0 z + a_1$, $z(a_0 z + a_1) + a_2$, . . . ; or obtain the desired quantities $f(c)$, $f'(c)$, $f''(c)/2!$, . . . by *Horner's scheme.*

(b) Horner's Scheme. Long division of $f(z)$ by $(z - c)$ yields a new polynomial $f_1(z)$ and the remainder $f(c)$ (Sec. 1.7-2); long division of $f_1(z)$ by $(z - c)$ in turn yields a new polynomial $f_2(z)$ and the remainder $f'(c)$. Continuing in this manner, one obtains successive remainders $f''(c)/2!$, $f'''(c)/3!$, ... Note that these remainders are the coefficients of the polynomial

$$F(u) \equiv f(u + c) \equiv f(c) + f'(c)u + \frac{1}{2!} f''(c)u^2 + \cdots \equiv f(z) \tag{20.2-17}$$

Horner's Scheme for Complex Arguments. Given a real polynomial $f(z)$, let $c = a + ib$. Then $f(c) = Ac + B = (Aa + B) + iAb$, where $Az + B$ is the (real) remainder of the quotient $f(z)/(z^2 - 2az + a^2 + b^2)$ (Ref. 20.3).

20.2-4. Solution of Algebraic Equations: Iteration Methods. (a) General Methods. To compute single simple roots of an algebraic equation

$$f(z) \equiv a_0 z^n + a_1 z^{n-1} + \cdots + a_{n-1} z + a_n = 0 \qquad (20.2\text{-}18)$$

one may

1. Use the *Newton-Raphson method* of Eq. (6).
2. Employ Eq. (5) with $k = 1/a_{n-1}$, calculating successive values of the polynomial
$$z^{[j+1]} = z^{[j]} - f(z^{[j]})/a_{n-1} \qquad (20.2\text{-}19)$$
by Horner's scheme. If $a_{n-1} = 0$, introduce $u = z - c$ and use Eq. (17) to rewrite Eq. (18) as $F(u) = 0$.
3. Attempt to bracket the root between argument values yielding function values of opposite signs, or use the *regula falsi* (Sec. 20.2-2*c*).

Convergence of these and the following iteration methods depends on the initial trial value. This may be based on a priori information, on approximate solutions obtained by one of the methods of Sec. 20.2-5, or on a root of a second-order polynomial approximating $f(z)$. In the absence of such information, the trial value may simply be zero or a small real or complex number. In the latter case, Horner's scheme for complex arguments may be employed if $f(z)$ is a real polynomial.

If $f(z)$ is a real polynomial, one may simply attempt to find trial values $a + ib$ so as to minimize the quantities A and B in Horner's scheme for complex arguments. See also Ref. 20.3.

As successive roots z_i are found, the corresponding factors $z - z_i$ are divided out (Sec. 1.7-2). z_i or $z_i(1 + i)$ may be tried as a starting value for iteration on the next root.

The Muller and Bairstow methods simplify calculations for complex roots and tend to converge better than the Newton-Raphson method when roots are close together.

(b) Muller's Method (Ref. 20.5). Substitution of parabolic for linear interpolation in the derivation of the *regula falsi* (Sec. 20.2-2*c*) produces the iteration algorithm

$$z^{[j+1]} = z^{[j]} - (z^{[j]} - z^{[j-1]}) \frac{2C_j}{B_j \pm \sqrt{B_j^2 - 4A_jC_j}} \qquad (20.2\text{-}20a)$$

where the sign in the denominator is chosen so as to make its absolute value as large as possible, and

$$\left.\begin{aligned} A_j &= q_j f_j - q_j(1 + q_j) f_{j-1} + q_j^2 f_{j-2} \\ B_j &= (2q_j + 1) f_j - (1 + q_j)^2 f_{j-1} + q_j^2 f_{j-2} \\ C_j &= (1 + q_j) f_j \\ q_j &= \frac{z^{[j]} - z^{[j-1]}}{z^{[j-1]} - z^{[j-2]}} \qquad f_j = f(z^{[j]}) \end{aligned}\right\} \quad (j = 0, 1, 2, \ldots) \qquad (20.2\text{-}20b)$$

The solution may be started with $z^{[0]} = -1$, $z^{[1]} = 1$, and $z^{[2]} = 0$ (hence $q_2 = -\frac{1}{2}$), and f_0, f_1, f_2 respectively replaced with $a_n - a_{n-1} + a_{n-2}$, $a_n + a_{n-1} + a_{n-2}$, a_n.

Muller's method applies also to nonalgebraic equations.

(c) **The Bairstow Method** (Refs. 20.5, 20.9, and 20.54). Assume that the given equation (16) admits a quadratic factor $x^2 - ux - v$ defining two distinct simple roots, and that one has trial values $u^{[0]}$, $v^{[0]}$ which approximate u and v sufficiently well. Then a sequence of improved approximations converging to u, v are obtained from

$$\left.\begin{aligned} u^{[j+1]} &= u^{[j]} + \frac{b_n{}^{[j]}c_{n-3}^{[j]} - b_{n-1}^{[j]}c_{n-2}^{[j]}}{(c_{n-2}^{[j]})^2 - c_{n-1}^{[j]}c_{n-3}^{[j]}} \\ v^{[j+1]} &= v^{[j]} + \frac{b_n{}^{[j]}c_{n-1}^{[j]} - b_{n-1}^{[j]}c_{n-2}^{[j]}}{(c_{n-2}^{[j]})^2 - c_{n-1}^{[j]}c_{n-3}^{[j]}} \end{aligned}\right\} \qquad (20.2\text{-}21a)$$

where the $b_k{}^{[j]}$, $c_k{}^{[j]}$ are consecutively obtained for each j for

$$\left.\begin{aligned} &\left.\begin{aligned} b_k{}^{[j]} &= a_k{}^{[j]} + u^{[j]}b_{k-1}^{[j]} + v^{[j]}b_{k-2}^{[j]} \\ c_k{}^{[j]} &= b_k{}^{[j]} + u^{[j]}c_{k-1}^{[j]} + v^{[j]}c_{k-2}^{[j]} \end{aligned}\right\} (k = 0, 1, 2, \ldots, n) \\ &b_{-1}{}^{[j]} = b_{-2}{}^{[j]} = c_{-1}{}^{[j]} = c_{-2}{}^{[j]} = 0 \end{aligned}\right\} \qquad (20.2\text{-}21b)$$

This method has been found to work best for polynomials of even degree.

20.2-5. Additional Methods for Solving Algebraic Equations. (a) **The Quotient-difference Algorithm** (Ref. 20.5). The following scheme (a generalization of the classical method of Bernoulli, Refs. 20.5 and 20.9) can produce approximations to all roots of suitable algebraic equations (16) in one sequence of calculations. The method may be useful for finding trial values to be used in iteration methods.

For a (suitable) given equation (16), write the tabular arrangement of Fig. 20.2-1, where successive entries are obtained from

$$\left.\begin{aligned} Q_{i+1}^{[k]} &= Q_i{}^{[k]} + (E_i{}^{[k]} - E_{i+1}^{[k-1]}) \\ E_{i+1}^{[k]} &= E_i{}^{[k]} \frac{Q_i{}^{[k+1]}}{Q_{i+1}^{[k]}} \end{aligned}\right\} \quad (\text{"RHOMBUS RULES"}) \qquad (20.2\text{-}22a)$$

with $$E_i{}^{[0]} = E_i{}^{[n]} = 0 \qquad (i = 1, 2, \ldots) \qquad (20.2\text{-}22b)$$

This procedure breaks down if a $Q_{i+1}^{[k]}$ or $E_i{}^{[k]}$ equals zero. The quotient-difference scheme exists, however, in many useful special cases, in particular, if all n roots $z_1, z_2, \ldots, z_n$ of the given equation (16) are positive; or if they are simple roots with distinct nonzero absolute values. In the latter case,

$$\lim_{i \to \infty} Q_i{}^{[k]} = z_k \qquad (20.2\text{-}23)$$

$$\lim_{i \to \infty} E_i{}^{[k]} = 0 \qquad (20.2\text{-}24)$$

$$\begin{array}{ccccccc}
Q_0^{[1]} = -\dfrac{a_1}{a_0} & & Q_{-1}^{[2]} = 0 & & Q_{-2}^{[3]} = 0 \;\cdots\cdot & & Q_{1-n}^{[n]} = 0 \\
& E_0^{[1]} = \dfrac{a_2}{a_1} & & E_{-1}^{[2]} = \dfrac{a_3}{a_2} & \cdots\cdot & E_{2-n}^{[n-1]} = \dfrac{a_n}{a_{n-1}} & \\
Q_1^{[1]} & & Q_0^{[2]} & & Q_{-1}^{[3]} \;\cdots\cdot & & Q_{2-n}^{[n]} \\
& E_1^{[1]} & & E_0^{[2]} & \cdots\cdot & E_{3-n}^{[n-1]} & \\
Q_2^{[1]} & & Q_1^{[2]} & & Q_0^{[3]} \;\cdots\cdot & & Q_{3-n}^{[n]} \\
& E_2^{[1]} & & E_1^{[2]} & & E_{4-n}^{[n-1]} & \\
\vdots & \vdots & \vdots & \vdots & \vdots & \vdots & \vdots
\end{array}$$

FIG. 20.2-1. A quotient-difference table.

for each k, so that *each Q column in Fig 20.2-1 yields an approximation to a root.*

More generally, *let $|z_1| \geq |z_2| \geq \cdots \geq |z_n| > 0$. Assuming that the quotient-difference scheme does not break down, Eq.* (23) *still yields every root z_k which differs in absolute value from its neighbors in the above sequence; and Eq.* (24) *still holds whenever $|z_k| > |z_{k+1}|$.* This helps to identify roots with equal absolute values, such as complex-conjugate roots; see Ref. 20.5 for a refined procedure which actually approximates such roots.

(b) Graeffe's Root-squaring Process. Given the real algebraic equation

$$f(z) \equiv \prod_{k=1}^{n} (z - z_k) \equiv z^n + a_1 z^{n-1} + \cdots + a_n = 0$$

obtain the coefficients $a_i^{(1)}$ of

$$\begin{aligned} f(z)f(-z) &\equiv (-1)^n \prod_{k=1}^{n} (z^2 - z_k^2) \\ &\equiv (-1)^n (z^2)^n + a_1^{(1)} (z^2)^{n-1} + \cdots + a_n^{(1)} \end{aligned}$$

by writing the array

$k = n$	$k = n - 1$	$k = n - 2$	$k = n - 3$	(Column number)
a_n a_n	a_{n-1} $-a_{n-1}$	a_{n-2} a_{n-2}	a_{n-3} $-a_{n-3}$	$\cdots$
a_n^2	$-a_{n-1}^2$ $+2a_n a_{n-2}$	a_{n-2}^2 $+2a_n a_{n-4}$ $-2a_{n-1}a_{n-3}$	$-a_{n-3}^2$ $+2a_n a_{n-6}$ $-2a_{n-1}a_{n-5}$ $+2a_{n-2}a_{n-4}$	$\cdots$
$a_n^{(1)}$	$a_{n-1}^{(1)}$	$a_{n-2}^{(1)}$	$a_{n-3}^{(1)}$	$\cdots$

Repeat this process, obtaining successively the coefficients $a_i^{(j)}$ of

$$(-1)^n \prod_{k=1}^{n} (z^{2^j} - z_k^{2^j}) \equiv (-1)^n (z^{2^j})^n + a_1^{(j)} (z^{2^j})^{n-1} + \cdots + a_n^{(j)}$$

As j increases, the array usually assumes a definite pattern: (1) the double products in a column may become negligible, so that successive column entries become squares with equal signs (*regular column**); all entries of a column may have equal signs and absolute values equal to a definite fraction of the squared entry above (*fractionally regular column*); (2) column entries may have regularly alternating signs (*fluctuating column*); and (3) a column may be entirely *irregular*.

Each pair of regular columns (say, k and $k - r$) separated by $r - 1$ nonregular columns corresponds to a set of r roots z of equal magnitude such that

$$\left| \frac{a_k^{(j)}}{a_{k-r}^{(j)}} \right| \to |z|^{2^j r} \qquad \text{as } j \to \infty \tag{20.2-25}$$

These r roots are all either real or pure imaginary if the $r - 1$ separating columns are fractionally regular. Specifically,

1. Two adjacent regular columns (say k and $k - 1$) yield a *simple real root* z such that

$$\left| \frac{a_k^{(j)}}{a_{k-1}^{(j)}} \right| \to |z|^{2^j} \qquad \text{as } j \to \infty$$

2. Two regular columns (k and $k - 2$) separated by a fluctuating column indicate a *pair of simple complex-conjugate roots* z such that

$$\left| \frac{a_k^{(j)}}{a_{k-2}^{(j)}} \right| \to |z|^{2^{j+1}} \qquad \text{as } j \to \infty$$

In practice, one first finds the real and purely imaginary roots; determine signs by substitution or with the aid of Sec. 1.6-6. *Lehmer's method*, a refined version of Graeffe's scheme, permits direct determination of signs.

(c) A Matrix Method. It is possible to compute the roots of an nth-degree algebraic equation (11) with $a_0 = 1$ as eigenvalues of the $(n + 1) \times (n + 1)$ "companion matrix"

$$A \equiv \begin{bmatrix} 0 & 1 & 0 & \cdots & 0 & 0 \\ 0 & 0 & 1 & \cdots & 0 & 0 \\ \cdot & \cdot & \cdot & \cdots & \cdot & \cdot \\ 0 & 0 & 0 & \cdots & 0 & 1 \\ -a_n & -a_{n-1} & -a_{n-2} & \cdots & -a_2 & -a_1 \end{bmatrix} \tag{20.2-26}$$

by one of the methods outlined in Sec. 20.3-5.

* M. B. Reed and G. B. Reed, *Mathematical Methods in Electrical Engineering*, Harper, New York, 1951.

(d) Horner's Method. *Horner's method* (for real roots) evaluates the coefficients of successive polynomials $F_1(u) \equiv f(u + c_1)$, $F_2(u) \equiv F_1(u + c_2)$, . . . by Horner's scheme, where $c_1, c_2, \ldots$ are chosen so as to reduce the absolute values of the remainders. If one succeeds in obtaining $F_j(c_j) \approx 0$, then the desired root is approximated by $c_1 + c_2 + \cdots + c_j$.

20.2-6. Systems of Equations and the Problem of Finding Maxima and Minima (see also Secs. 11.1-1, 12.5-6, 20.2-2, and 20.2-3). **(a) Problem Statement. Iteration Methods.** The problem of solving n simultaneous equations

$$f_i(x_1, x_2, \ldots, x_n) = 0 \qquad (i = 1, 2, \ldots, n) \qquad (20.2\text{-}27)$$

for n unknowns $x_1, x_2, \ldots, x_n$ is equivalent to the problem of *minimizing the function*

$$F(x_1, x_2, \ldots, x_n) \equiv \sum_{i=1}^{n} |f_i(x_1, x_2, \ldots, x_n)|^2 \qquad (20.2\text{-}28)$$

or some other increasing real function of the absolute values $|f_i|$ of the n **residuals (errors)** $f_i = f_i(x_1, x_2, \ldots, x_n)$. The problem of minimizing (or maximizing) a given function of n variables is of great practical importance in its own right.

A useful class of *iteration methods* starts with a trial solution $x_i^{[0]}$ $(i = 1, 2, \ldots, n)$ and attempts to construct successive approximations

$$x_i^{[j+1]} = x_i^{[j]} + \lambda^{[j]} v_i^{[j]} \qquad (i = 1, 2, \ldots, n;\, j = 0, 1, 2, \ldots) \qquad (20.2\text{-}29)$$

which converge to a solution x_i as $j \to \infty$.

Once the ratios $v_1^{[j]} : v_2^{[j]} : \cdots : v_n^{[j]}$ ("direction" of the jth step) are chosen, one may minimize $F(x_1^{[j+1]}, x_2^{[j+1]}, \ldots, x_n^{[j+1]})$ as a function $F(\lambda^{[j]})$ of the parameter $\lambda^{[j]}$ which determines the *step size.* For this purpose, $F(\lambda^{[j]})$ may be approximated by a Taylor series or by an interpolation polynomial (Sec. 20.5-1) based on three to five trial values of $\lambda^{[j]}$. *The latter method also applies to the computation of maxima and minima of a tabulated function* $F(x_1, x_2, \ldots, x_n)$.

The convergence of such iteration methods can, again, be investigated with the aid of the Contraction-mapping Theorem of Sec. 12.5-6; the trial solutions $(x_1, x_2, \ldots, x_n)$ are regarded as vectors with the norm $(x_1^2 + x_2^2 + \cdots + x_n^2)^{1/2}$ or $|x_1| + |x_2| + \cdots + |x_n|$ (see also Ref. 20.5).

(b) Varying One Unknown at a Time. Attempt to minimize F by varying only one of the x_i at each step, either cyclically or so as to reduce the largest absolute residual (see also Sec. 20.3-2*c*, relaxation). Special *one-dimensional search methods* are discussed in Refs. 20.9 and 20.57.

(c) Search Methods for Finding Maxima and Minima. *Pure random searches,* which merely select the largest or smallest value $F(x_1, x_2, \ldots, x_n)$ for a number of randomly chosen points $(x_1, x_2, \ldots, x_n)$ are useful mainly for finding starting approximations for iteration methods (Ref. 20.9). *Random-perturbation methods* start with a parameter point $(x_1, x_2, \ldots, x_n)$ and proceed to apply sets of random perturbations $\Delta x_1, \Delta x_2, \ldots, \Delta x_n$ to all unknowns at once until an improvement in the criterion function F is obtained. The resulting perturbed parameter point constitutes the next approximation, and the search continues. This technique which, unlike a pure random search, can take advantage of the continuity of a criterion function, is a highly useful hill-climbing method when gradient methods are frustrated by adverse features of the multidimensional terrain, such as "ridges," "canyons," "flat spots," and multiple maxima and minima. Random-perturbation methods have been considerably refined by strategies involving step-size changes and preferential treatment of certain directions in parameter space depending on past successes or failures (Refs. 20.9 and 20.57).

(d) Treatment of Constraints (see also Secs. 11.3-4, 11.4-1, and 11.4-3). Maxima and minima *with constraints* $\varphi_i(x_1, x_2, \ldots, x_n) = 0$ $(i = 1, 2, \ldots, r)$ can be treated either by a *penalty-function technique* which adds a term of the form $\sum_{k=1}^{r} K_k|\varphi_k(x_1, x_2, \ldots, x_n)|$ or $\sum_{k=1}^{r} K_k\varphi_k^m(x_1, x_2, \ldots, x_n)$ to the given criterion function $F(x_1, x_2, \ldots, x_n)$, where each K_k is a large positive constant for minimization (negative for maximization), and m is an even positive integer. Another technique projects the computed gradient vector on the constraint surface, so that the gradient search proceeds along the latter (Refs. 20.9 and 20.57). The most useful digital-computer routines for optimization include options for changing the search strategy when convergence is slow.

20.2-7. Steepest-descent Methods (Gradient Methods). (a) Method of Steepest Descent. Choose $v_i^{[j]} = -\partial F/\partial x_i$, where all derivatives are computed for $x_i = x_i^{[j]}$, and reduce the step size $\lambda^{[j]}$ as the minimum of F is approached (Fig. 20.2-2).

For analytical functions F and small f_i, a Taylor-series approximation for $F(\lambda^{[j]})$ yields the optimum step size

$$\lambda^{[j]} = \frac{\sum_{k=1}^{n} \left(\frac{\partial F}{\partial x_k}\right)^2}{\sum_{k=1}^{n} \sum_{h=1}^{n} \frac{\partial^2 F}{\partial x_k\, \partial x_h} \frac{\partial F}{\partial x_k} \frac{\partial F}{\partial x_h}} \qquad (j = 0, 1, 2, \ldots) \qquad (20.2\text{-}30)$$

where all derivatives are computed for $x_i = x_i^{[j]}$. Interpolation-polynomial approximations for $F(\lambda^{[j]})$ may be more convenient.

(b) Descent with Computed Gradient Components. Frequently, derivation of the gradient components $\partial F/\partial x_k$ needed for a gradient descent is impossible or impractical. The required derivatives can then be approximated by difference coefficients $\Delta F/\Delta x_k$ obtained through perturbations of one unknown x_k at a time. Since this takes n gradient-measuring steps for each "working step" in the gradient direction, one may prefer to continue in this gradient direction until the criterion function F is no longer improved and remeasure the gradient only then. A number of refined gradient-descent techniques are discussed in Refs. 20.9 and 20.57.

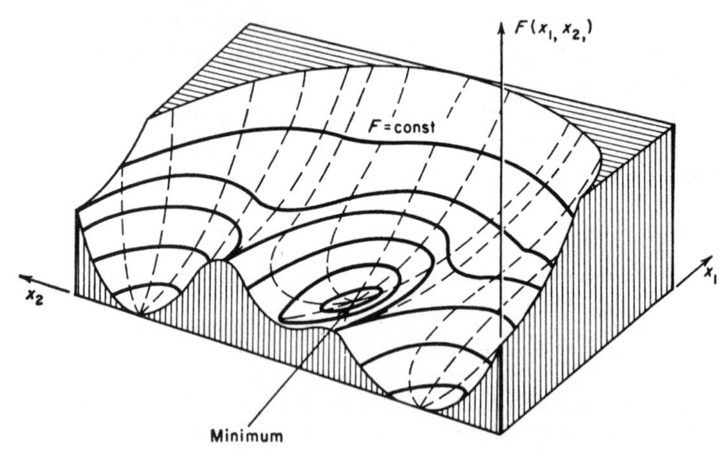

FIG. 20.2-2. Criterion-function surface for a two-variable minimum problem, showing three minima, level lines, and gradient lines. In this example, all three minima of $F(x_1,x_2)$ have the same value, zero. (*Based on L. D. Kovach and H. Meissinger: Solution of Algebraic Equations, Linear Programming, and Parameter Optimization, in H. D. Huskey and G. A. Korn, Computer Handbook, McGraw-Hill, New York, 1962.*)

20.2-8. The Newton-Raphson Method and the Kantorovich Theorem (see also Secs. 20.2-2 and 20.9-3). **(a) The Newton-Raphson Method.** Start with a trial solution $x_i^{[0]}$, and obtain successive approximations $x_i^{[j+1]}$ by solving the simultaneous linear equations

$$f_i + \sum_{k=1}^{n} \frac{\partial f_i}{\partial x_k} (x_k^{[j+1]} - x_k) = 0 \qquad (i = 1, 2, \ldots, n) \qquad (20.2\text{-}31)$$

with $x_k = x_k^{[j]}$ $(j = 0, 1, 2, \ldots)$.

(b) Kantorovich's Convergence Theorem (see also Ref. 20.52). Solution of the linear equations (31) implies that the matrix $[\partial f_i/\partial x_k]$ has an inverse (Sec. 13.2-3) $[\partial f_i/\partial x_k]^{-1} \equiv [\Gamma_{ik}(x_1, x_2, \ldots, x_n)]$ for the x_k

values in question. *Let* A, B, C *be positive real bounds* (matrix norms, Sec. 13.2-1) *such that*

$$\max_i \sum_{k=1}^{n} |\Gamma_{ik}| \leq A \qquad \max_i \sum_{k=1}^{n} |\Gamma_{ik} f_k| \leq B \qquad C \leq \frac{1}{2AB} \tag{20.2-32a}$$

if the initial trial values $x_i = x_i^{[0]}$ *are substituted in the* f_i, $\partial f_i/\partial x_k$, *and* Γ_{ik}. *Also, let the* $f_i(x_1, x_2, \ldots, x_n)$ *be twice continuously differentiable and satisfy*

$$\sum_{j=1}^{n} \sum_{k=1}^{n} \left| \frac{\partial^2 f_i}{\partial x_j \, \partial x_k} \right| \leq C \qquad (i = 1, 2, \ldots, n) \tag{20.2-32b}$$

for all $(x_1, x_2, \ldots, x_n)$ *in the cubical region defined by*

$$\max_i |x_i - x_i^{[0]}| \leq \frac{1}{AC} (1 - \sqrt{1 - 2ABC}) \tag{20.2-33}$$

Then the given system (27) *has a solution* $(x_1, x_2, \ldots, x_n)$ *in the region defined by Eq.* (33), *and the rate of convergence is given by*

$$\max_i |x_i - x_i^{[j]}| \leq \frac{B}{2^{j-1}} (2ABC)^{2^j - 1} \tag{20.2-34}$$

which again indicates relatively rapid convergence.

20.3. LINEAR SIMULTANEOUS EQUATIONS, MATRIX INVERSION, AND MATRIX EIGENVALUE PROBLEMS

20.3-1. "Direct" or Elimination Methods for Linear Simultaneous Equations. (a) Introduction. The solution of a suitable system of linear equations

$$\begin{aligned} a_{11}x_1 + a_{12}x_2 + \cdots + a_{1n}x_n &= b_1 \\ a_{21}x_1 + a_{22}x_2 + \cdots + a_{2n}x_n &= b_2 \\ \cdots\cdots\cdots\cdots\cdots\cdots\cdots\cdots\cdots \\ a_{n1}x_1 + a_{n2}x_2 + \cdots + a_{nn}x_n &= b_n \end{aligned} \tag{20.3-1}$$

by Cramer's rule (Sec. 1.9-2) requires too many multiplications for practical use if $n > 4$. The following procedures solve a given system (1) by successive elimination of unknowns.

(b) Basic Elimination Procedure (Gauss's Pivotal-condensation Method). Let a_{IJ} be the coefficient having the largest absolute value. To eliminate x_J from the ith equation ($i \neq I$), multiply the Ith equation by a_{iJ}/a_{IJ} and subtract from the ith equation. Repeat the process to eliminate a second unknown from the remaining $n - 1$ equations, etc.

(c) An Improved Elimination Scheme (Banachiewicz-Cholesky-Crout Method). Obtain $x_n, x_{n-1}, \ldots, x_1$, in that order, by "back substitution" from the equations

$$\begin{aligned} x_1 + \alpha_{12}x_2 + \alpha_{13}x_3 + \cdots + \alpha_{1n}x_n &= \beta_1 \\ x_2 + \alpha_{23}x_3 + \cdots + \alpha_{2n}x_n &= \beta_2 \\ \cdots\cdots\cdots\cdots\cdots\cdots & \\ x_n &= \beta_n \end{aligned} \qquad (20.3\text{-}2)$$

after calculating successive elements of the $(n + 1) \times n$ array

$$\begin{array}{ccccc|c} \gamma_{11} & \alpha_{12} & \alpha_{13} & \cdots & \alpha_{1n} & \beta_1 \\ \gamma_{21} & \gamma_{22} & \alpha_{23} & \cdots & \alpha_{2n} & \beta_2 \\ \cdots & \cdots & \cdots & \cdots & \cdots & \cdots \\ \gamma_{n1} & \gamma_{n2} & \gamma_{n3} & \cdots & \alpha_{nn} & \beta_n \end{array}$$

with the aid of the recursion formulas

$$\begin{aligned} &\gamma_{i1} = a_{i1} \quad (i = 1, 2, \ldots, n) \qquad \alpha_{1k} = \frac{a_{1k}}{a_{11}} \quad (k = 2, 3, \ldots, n) \\ &\gamma_{ik} = a_{ik} - \sum_{j=1}^{k-1} \gamma_{ij}\alpha_{jk} \quad (i, k = 2, 3, \ldots, n; i \geq k) \\ &\alpha_{ik} = \frac{1}{\gamma_{ii}}\left(a_{ik} - \sum_{j=1}^{i-1} \gamma_{ij}\alpha_{jk}\right) \quad (i, k = 2, 3, \ldots, n; i < k) \\ &\beta_1 = \frac{b_1}{a_{11}} \qquad \beta_i = \frac{1}{\gamma_{ii}}\left(b_i - \sum_{j=1}^{i-1} \gamma_{ij}\beta_j\right) \quad (i = 2, 3, \ldots, n) \end{aligned} \qquad (20.3\text{-}3)$$

For the *computation of determinants*, note

$$\det [a_{ik}] = \gamma_{11}\gamma_{22} \cdots \gamma_{nn} \qquad (20.3\text{-}4)$$

The Banachiewicz-Cholesky-Crout method requires less intermediate recording than the Gauss method and becomes particularly simple if the matrix $[a_{ik}]$ is symmetric ($a_{ki} = a_{ik}$). In this important special case,

$$\alpha_{ik} = \frac{\gamma_{ki}}{\gamma_{ii}} \quad (i < k) \qquad (20.3\text{-}5)$$

(d) Direct Methods in Matrix Form. Gaussian elimination (Sec. 20.3-1*b*) can be interpreted as a transformation of the given system (1), i.e.,

$$Ax = b$$

to the form

$$T_nT_{n-1} \cdot \cdot \cdot T_1Ax = T_nT_{n-1} \cdot \cdot \cdot T_1b$$

by successive nonsingular transformation matrices $T_1, T_2, \ldots, T_n$ chosen so that the resulting system matrix $T_nT_{n-1} \cdot \cdot \cdot T_1A$ has a triangular form similar to that of Eq. (2).

Again, the solution scheme of Sec. 20.3-1*c* amounts to the definition of a single transformation matrix T such that

$$TAx = Tb$$

has the triangular form (2). Various alternative methods employ postmultiplication or similarity transformations instead of the premultiplication used here.

Several modified triangularization schemes associated with the names of Crout, Doolittle, Givens, Householder, and Schmidt are described in Refs. 20.9, 20.11, 20.15, 20.22, and 20.25. Some of the modifications help to reduce round-off-error effects (Ref. 20.58). Wilkinson (Ref. 20.9) gives a comparison of methods on the basis of numerical stability and number of multiplications.

20.3-2. Iteration Methods for Linear Equations. (a) Introduction. Iteration methods (see also Secs. 20.2-6 and 20.2-7) are usually preferred to direct methods for solving linear equations only if the system matrix $[a_{ik}]$ is "sparse," i.e., if most of the a_{ik}, especially those not near the main diagonal, are equal to zero. Precisely this is true for the (possibly very large) systems of linear equations generated by partial-differential-equation problems (Sec. 20.9-4). Given a system of linear equations (1) with real coefficients, each of the following iteration methods approximates the solution x_i $(i = 1, 2, \ldots, n)$ by successive approximations of the form (20.2-20). The residuals obtained at each step of the iteration will be denoted by

$$f_i^{[j]} \equiv \sum_{k=1}^{n} a_{ik}x_k^{[j]} - b_i \qquad (i = 1, 2, \ldots, n; j = 0, 1, 2, \ldots) \tag{20.3-6}$$

NOTE: Many iteration schemes require the given coefficient matrix $A \equiv [a_{ik}]$ to be normal with a positive-definite symmetric part (Sec. 13.3-4), or to be symmetric and positive-definite (Sec. 13.5-2). If this is not true for the given coefficient matrix, one may attempt to rearrange or recombine the given equations so as to obtain relatively large positive diagonal coefficients, or one may solve the equivalent system

$$\sum_{i=1}^{n} \sum_{k=1}^{n} a_{hi}a_{ik}x_k = \sum_{i=1}^{n} a_{hi}b_i \qquad (h = 1, 2, \ldots, n) \tag{20.3-7}$$

whose coefficient matrix is necessarily symmetric and positive-definite whenever the system (1) is nonsingular.

(b) Gauss-Seidel-type Iteration. Rearrange the given system (1) (possibly by recombining equations and/or multiplication by constants) to obtain as large positive diagonal coefficients a_{ii} as practicable. Starting with a trial solution $x_i^{[0]}$ $(i = 1, 2, \ldots, n)$, compute successive approximations

$$x_i^{[j+1]} = x_i^{[j]} - \frac{1}{a_{ii}} f_i^{[j]} = x_i^{[j]} - \frac{1}{a_{ii}} \left(\sum_{k=1}^{n} a_{ik} x_k^{[j]} - b_i \right)$$

$$(i = 1, 2, \ldots, n; j = 0, 1, 2, \ldots) \qquad (20.3\text{-}8)$$

or use

$$x_i^{[j+1]} = x_i^{[j]} - \frac{1}{a_{ii}} \left(\sum_{k=1}^{i-1} a_{ik} x_k^{[j+1]} + \sum_{k=i}^{n} a_{ik} x_k^{[j]} - b_i \right)$$

$$(i = 1, 2, \ldots, n; j = 0, 1, 2, \ldots) \qquad (20.3\text{-}9)$$

Both schemes are simple, but convergence is assured only if the matrix $[a_{ik}]$ is positive definite (Sec. 13.5-2); and convergence may be slow.

(c) **Relaxation methods** depend on the (manual) computer's judgment to obtain rapid convergence of an iteration process. Starting with a trial solution $x_i^{[0]}$ (frequently simply $x_1^{[0]} = 1$, $x_2^{[0]} = x_3^{[0]} = \cdots = x_n^{[0]} = 0$), one attempts to introduce successive approximations $x_i^{[j]}$ in a loosely systematic fashion so as to reduce the n residuals (6) to zero. One tabulates the residuals $f_i^{[j]}$ at each step and combines the following procedures:

1. *Basic Relaxation Procedure.* At each step, "liquidate" the residual $f_I^{[j]}$ having the greatest absolute value by adjusting x_I alone:

$$x_I^{[j+1]} \approx x_I^{[j]} - \frac{f_I^{[j]}}{a_{II}} \qquad x_i^{[j+1]} = x_i^{[j]} \qquad (i \neq I) \qquad (20.3\text{-}10)$$

   Only rough values of $x_I^{[j+1]}$ are required for the initial steps.
2. *Block Relaxation and Group Relaxation.* Apply equal increments $x_i^{[j+1]} - x_i^{[j]}$ to a set ("block") of $x_i^{[j]}$'s so as to liquidate one of the residuals $f_i^{[j+1]}$, or so as to reduce the sum of all residuals to zero. The latter procedure is useful particularly if all the initial residuals $f_i^{[0]}$ are of equal sign.

   Group relaxation applies *different* increments to a chosen set of $x_i^{[j]}$'s for similar purposes.
3. *Overrelaxation.* One can often improve the convergence of a relaxation process by *changing the sign* of the residual operated on while

reducing its absolute value, without liquidating it entirely at this stage.

Relaxation methods are particularly useful for the (manual) solution of the large sets of simple linear difference equations frequently used to approximate a partial differential equation (Sec. 20.9-3). Relaxation methods at their best exploit the manual computer's skill, experience, and knowledge of the underlying physical situation. Many special tricks apply to particular applications (Refs. 20.21 to 20.30).

(d) Systematic overrelaxation is employed in an important class of iteration methods for solving the "sparse" equation systems generated by partial-differential-equation problems. Consider a system of equations (1) rearranged so that all diagonal coefficients equal unity ($a_{ii} = 1$), and modify the iteration (9) to read

$$x_i^{[j+1]} = x_i^{[j]} - \omega_j \left(\sum_{k=1}^{i-1} a_{ik} x_k^{[j+1]} + \sum_{k=i}^{n} a_{ik} x_k^{[j]} - b_i \right)$$

$$(i = 1, 2, \ldots, n; j = 0, 1, 2, \ldots)$$

where ω_j is a real **overrelaxation factor** > 1 chosen (either once and for all or at each step) so as to speed up the convergence. The choice of ω_j is discussed in Refs. 20.15, 20.22, 20.23, and 20.30; Ref. 20.9 contains a large bibliography.

(e) Steepest-descent methods (gradient methods) minimize a positive function like $F \equiv \sum_{i=1}^{n} |f_i|$, $F \equiv \sum_{i=1}^{n} |f_i|^2, \ldots$ in the manner of Sec. 20.2-4*d*. If the given coefficient matrix $[a_{ik}]$ is symmetric and positive definite, one may use

$$x_i^{[j+1]} = x_i^{[j]} - \frac{\sum_{k=1}^{n} f_k^{[j]2}}{\sum_{k=1}^{n} \sum_{h=1}^{n} a_{kh} f_k^{[j]} f_h^{[j]}} f_i^{[j]}$$

$$(i = 1, 2, \ldots, n; j = 0, 1, 2, \ldots) \qquad (20.3\text{-}11)$$

If the convergence is of an oscillatory nature, it may be possible to accelerate the convergence by multiplying the last term in Eq. (11) by 0.9 for some values of j.

(f) A Conjugate-gradient Method. Given a system of linear equations (1) with a real, symmetric, and positive definite coefficient matrix $[a_{ik}]$, start with a trial solution $x_i^{[0]}$. Let $v_i^{[0]} = f_i^{[0]}$, and compute successively

$$\left.\begin{aligned}
x_i^{[j+1]} &= x_i^{[j]} - \frac{\sum_{k=1}^{n} f_k^{[j]} v_k^{[j]}}{\sum_{k=1}^{n}\sum_{h=1}^{n} a_{kh} v_k^{[j]} v_h^{[j]}} v_i^{[j]} \\
f_i^{[j+1]} &= f_i^{[j]} - \frac{\sum_{k=1}^{n} f_k^{[j]} v_k^{[j]}}{\sum_{k=1}^{n}\sum_{h=1}^{n} a_{kh} v_k^{[j]} v_h^{[j]}} \sum_{k=1}^{n} a_{ik} v_k^{[j]} \\
v_i^{[j+1]} &= f_i^{[j+1]} - \frac{\sum_{k=1}^{n}\sum_{h=1}^{n} a_{kh} f_k^{[j+1]} v_h^{[j]}}{\sum_{h=1}^{n}\sum_{h=1}^{n} a_{kh} v_k^{[j]} v_h^{[j]}} v_i^{[j]}
\end{aligned}\right\}$$

$$(i = 1, 2, \ldots, n; j = 0, 1, 2, \ldots) \qquad (20.3\text{-}12)$$

The resulting $f_i^{[j]}$ satisfy Eq. (6). Without round-off errors, this procedure yields the *exact* solution $x_i = x_i^{[n]}$ after n steps; in addition, the method has all the advantages of an iteration scheme (Sec. 20.1-2), but it requires many multiplications.

It is possible to write conjugate-gradient algorithms directly applicable to *any* given coefficient matrix $[a_{ik}]$; such schemes amount essentially to a solution of Eq. (7) by the method of Eq. (12).

20.3-3. Matrix Inversion (see also Secs. 13.2-3 and 14.5-3). **(a)** The methods of Secs. 20.3-1 and 20.3-2 apply directly to the numerical inversion of a given nonsingular $n \times n$ matrix $A \equiv [a_{ik}]$, and also to the calculation of det $[a_{ik}]$.

(b) An Iteration Scheme for Matrix Inversion. To find the inverse A^{-1} of a given $n \times n$ matrix A, start with an $n \times n$ trial matrix $X^{[0]}$ (say $X^{[0]} = I$) and compute successive approximations

$$X^{[j+1]} = X^{[j]}(2I - AX^{[j]}) \qquad (j = 0, 1, 2, \ldots) \qquad (20.3\text{-}13)$$

If the sequence converges, the limit is A^{-1} (see also Sec. 20.2-2*b*).

(c) (See also Sec. 13.4-7.) For every nonsingular $n \times n$ matrix A,

$$\left.\begin{aligned}
A^{-1} &= -\frac{1}{a_n}(A^{n-1} + a_1 A^{n-2} + \cdots + a_{n-1} I) \\
&\text{with} \\
a_1 &= -\operatorname{Tr}(A) \\
a_j &= -\frac{1}{j}[a_{j-1}\operatorname{Tr}(A) + a_{j-2}\operatorname{Tr}(A^2) + \cdots + a_1 \operatorname{Tr}(A^{j-1}) \\
&\qquad + \operatorname{Tr}(A^j)] \qquad (j = 2, 3, \ldots, n)
\end{aligned}\right\} \qquad (20.3\text{-}14)$$

20.3-4. A Partitioning Scheme for the Solution of Simultaneous Linear Equations or for Matrix Inversion (see also Sec. 13.2-8). **(a)** A system of linear equations (1) can be written in matrix form as

$$Ax \equiv \begin{bmatrix} a_{11} & a_{12} & \cdots & a_{1n} \\ a_{21} & a_{22} & \cdots & a_{2n} \\ \cdots & \cdots & \cdots & \cdots \\ a_{n1} & a_{n2} & \cdots & a_{nn} \end{bmatrix} \begin{bmatrix} x_1 \\ x_2 \\ \cdots \\ x_n \end{bmatrix} = \begin{bmatrix} b_1 \\ b_2 \\ \cdots \\ b_n \end{bmatrix} \equiv b \qquad (20.3\text{-}15)$$

(Sec. 14.5-3). Equation (15) can be rewritten in the partitioned form

$$Ax \equiv \left[\begin{array}{c|c} A_{11} & A_{12} \\ \hline A_{21} & A_{22} \end{array}\right] \left[\begin{array}{c} X_1 \\ \hline X_2 \end{array}\right] = \left[\begin{array}{c} B_1 \\ \hline B_2 \end{array}\right] \equiv b$$

with

$$A_{11} \equiv \begin{bmatrix} a_{11} & a_{12} & \cdots & a_{1m} \\ a_{21} & a_{22} & \cdots & a_{2m} \\ \cdots & \cdots & \cdots & \cdots \\ a_{m1} & a_{m2} & \cdots & a_{mm} \end{bmatrix}$$

$$X_1 \equiv \begin{bmatrix} x_1 \\ x_2 \\ \cdots \\ x_m \end{bmatrix} \quad B_1 \equiv \begin{bmatrix} b_1 \\ b_2 \\ \cdots \\ b_m \end{bmatrix} \quad (m < n) \qquad (20.3\text{-}16)$$

The resulting *matrix* equations

$$\begin{aligned} A_{11}X_1 + A_{12}X_2 &= B_1 \\ A_{21}X_1 + A_{22}X_2 &= B_2 \end{aligned} \qquad (20.3\text{-}17)$$

yield

$$(A_{11} - A_{12}A_{22}^{-1}A_{21})X_1 = B_1 - A_{12}A_{22}^{-1}B_2 \qquad (20.3\text{-}18)$$

which, once the $(n - m) \times (n - m)$ matrix A_{22} has been inverted, furnishes m linear equations for the first m unknowns $x_1, x_2, \ldots, x_m$. This procedure may be useful, in particular, if only the first m unknowns are of interest.

(b) The inverse matrix A^{-1} is obtained in the partitioned form

$$A^{-1} \equiv \left[\begin{array}{c|c} C_{11} & C_{12} \\ \hline C_{21} & C_{22} \end{array}\right]$$

with

$$\begin{aligned} C_{11} &= (A_{11} - A_{12}A_{22}^{-1}A_{21})^{-1} \qquad & C_{21} &= -A_{22}^{-1}A_{21}C_{11} \\ C_{22} &= (A_{22} - A_{21}A_{11}^{-1}A_{12})^{-1} \qquad & C_{12} &= -A_{11}^{-1}A_{12}C_{22} \end{aligned} \qquad (20.3\text{-}19)$$

so that the inversion of the $n \times n$ matrix A is reduced to the inversion of smaller $m \times m$ and $(n - m) \times (n - m)$ matrices.

20.3-5. Eigenvalues and Eigenvectors of Matrices (see also Secs. 13.4-2 to 13.4-6, 14.8-5, and 14.8-9). **(a) The Characteristic Equation.** The eigenvalues λ_k of a given $n \times n$ matrix $A \equiv [a_{ik}]$ can be obtained as roots of the characteristic equation

$$F_A(\lambda) \equiv \det [a_{ik} - \lambda\delta_k^i] = 0 \qquad (20.3\text{-}20)$$

by one of the methods outlined in Secs. 20.2-1 to 20.2-3. To avoid direct expansion of the determinant, one may evaluate $F_A(\lambda)$ for $n + 1$ selected values of λ (say $\lambda = 0, 1, 2, \ldots, n$), so that the nth-degree polynomial $F_A(\lambda)$ is given (exactly) by one of the interpolation formulas of Sec. 20.5-3.

(b) An Iteration Method for Hermitian Matrices. Let A be a hermitian matrix, so that all eigenvalues are real; in many applications A is real and symmetric. Assuming that the eigenvalue λ_M with the largest absolute value (**dominant eigenvalue**) is nondegenerate, start with a trial column matrix $x^{[0]}$ such as $\{1, 0, 0, \ldots, 0\}$ and compute successive matrix products

$$x^{[j+1]} = \alpha_j A x^{[j]} \qquad (j = 0, 1, 2, \ldots) \qquad (20.3\text{-}21)$$

where α_j is a convenient numerical factor chosen, for instance, so that the component of $x^{[j+1]}$ largest in absolute value equals 1. *As j increases, $x^{[j]}$ will approximate an eigenvector associated with the dominant eigenvalue* λ_M; the latter may then be obtained from

$$\lambda_M = \frac{x\dagger Ax}{x\dagger x} = \frac{\sum_{i=1}^{n}\sum_{k=1}^{n} a_{ik}x_i^* x_k}{\sum_{i=1}^{n} |x_i|^2} \qquad (20.3\text{-}22)$$

This process converges more rapidly if $|\lambda_M|$ is substantially larger than the absolute values of all other eigenvalues of A, and if the direction of $x^{[0]}$ is already close to that of the desired eigenvector; if $x^{[0]} = \{1, 0, 0, \ldots, 0\}$ does not work well, try $\{0, 1, 0, 0, \ldots, 0\}$, etc. One can accelerate the convergence by using A^2 or A^3 instead of A in Eq. (21).

Additional eigenvectors and eigenvalues are found after reduction of the given matrix (Sec. 14.8-6); see also Refs. 20.7 and 20.48 for other procedures.

If λ_M is m-fold *degenerate,* the vectors described by Eq. (21) will not in general converge, but for large j they will tend to stay in a subspace spanned by the eigenvectors associated with λ_M. Hence m linearly independent matrices $x^{[j]}$ will yield m mutually orthogonal eigenvectors by orthogonalization (Sec. 14.7-4*b*).

(c) A Relaxation Method. Starting with a trial vector represented by the column matrix x, compute $y = Ax$ and adjust the matrix elements ξ_i of x so that the greatest absolute difference between two of the quotients $\eta_1/\xi_1, \eta_2/\xi_2, \ldots, \eta_n/\xi_n$ becomes as small as possible: the η_i are the matrix elements of y.

(d) Jacobi-Von Neumann-Goldstine Rotation Process for Real Symmetric Matrices. Given a real symmetric matrix $A \equiv [a_{ik}] \equiv A^{[0]}$, begin by eliminating the nondiagonal element a_{IK} having the largest absolute value through the orthogonal transformation

$$\left.\begin{aligned} A^{[1]} &= T_1^{-1}A^{[0]}T_1 \qquad (T_1 \equiv [t_{ik}]) \\ \text{with} \\ t_{ik} &= \delta_k^i[1 + (\cos\vartheta_1 - 1)(\delta_I^i + \delta_K^i)] + \sin\vartheta_1(\delta_K^i\delta_k^I - \delta_I^i\delta_k^K) \\ &\qquad (i, k = 1, 2, \ldots, n) \\ \tan 2\vartheta_1 &= \frac{2a_{IK}}{a_{II} - a_{KK}} \end{aligned}\right\} \qquad (20.3\text{-}23)$$

(rotation through an angle ϑ_1 in the "plane" spanned by $\mathbf{e}_I$ and $\mathbf{e}_K$, see also Secs. 2.1-6 and 14.10-2*b*). If $a_{II} = a_{KK}$, then take $\vartheta_1 = \pm\pi/4$, the sign being that of a_{IK}. Apply an analogous procedure to $A^{[1]}$ to obtain $A^{[2]}$, and repeat the process. The product $T_1T_2T_3 \ldots$ of the transformation matrices converges to an orthogonal matrix which diagonalizes the given matrix A even if there are multiple eigenvalues. Variations of the Jacobi method simply reduce all off-diagonal elements a_{ik} with absolute values greater than a predetermined threshold in succession (Ref. 20.9).

(e) Other Methods. A number of other numerical methods for solving matrix eigenvalue problems, including those involving nonhermitian matrices, will be found in Refs. 20.9, 20.22 to 20.26, and 20.31.

20.4. FINITE DIFFERENCES AND DIFFERENCE EQUATIONS

20.4-1. Finite Differences and Central Means. (a) Let $y = y(x)$ be a function of the real variable x. Given a set of equally spaced argument values $x_k = x_0 + k\,\Delta x$ $(k = 0, \pm1, \pm2, \ldots; \Delta x = h > 0)$ and a corresponding set or table of function values $y_k = y(x_k) = y(x_0 + k\,\Delta x)$, one defines the **forward differences**

$$\left.\begin{aligned} \Delta y_k &= y_{k+1} - y_k \qquad \text{(FIRST-ORDER FORWARD DIFFERENCES)} \\ \Delta^2 y_k &= \Delta y_{k+1} - \Delta y_k = y_{k+2} - 2y_{k+1} + y_k \qquad \text{(SECOND-ORDER FORWARD DIFFERENCES)} \\ &\cdots\cdots\cdots\cdots \\ \Delta^r y_k &= \Delta^{r-1}y_{k+1} - \Delta^{r-1}y_k = \sum_{j=0}^{r} (-1)^j \binom{r}{j} y_{k+r-j} \\ &(r = 2, 3, \ldots)(r^{\text{th}}\text{-ORDER FORWARD DIFFERENCES}) \\ &(k = 0, \pm1, \pm2, \ldots) \end{aligned}\right\} \qquad (20.4\text{-}1)$$

and the **backward differences**

$$\left.\begin{aligned}\nabla y_k &= y_k - y_{k-1} = \Delta y_{k-1}\\ \nabla^r y_k &= \nabla^{r-1}y_k - \nabla^{r-1}y_{k-1} = \Delta^r y_{k-r} \qquad (r = 2, 3, \ldots)\end{aligned}\right\}$$
$$(k = 0, \pm 1, \pm 2, \ldots) \tag{20.4-2}$$

(**b**) Even though the function values $y_k = y(x_0 + k\,\Delta x)$ may not be known for half-integral values of k, one can calculate the **central differences**

$$\begin{aligned}\delta y_k &= y_{k+\frac{1}{2}} - y_{k-\frac{1}{2}} = \Delta y_{k-\frac{1}{2}}\\ \delta^r y_k &= \delta^{r-1}y_{k+\frac{1}{2}} - \delta^{r-1}y_{k-\frac{1}{2}} = \Delta^r y_{k-r/2} \qquad (r = 2, 3, \ldots)\end{aligned} \tag{20.4-3}$$

and the **central means**

$$\left.\begin{aligned}\mu y_k &= \tfrac{1}{2}(y_{k-\frac{1}{2}} + y_{k+\frac{1}{2}})\\ \mu^r y_k &= \tfrac{1}{2}(\mu^{r-1}y_{k-\frac{1}{2}} + \mu^{r-1}y_{k+\frac{1}{2}}) \qquad (r = 2, 3, \ldots)\end{aligned}\right\} \tag{20.4-4}$$

for $k = \pm\frac{1}{2}, \pm\frac{3}{2}, \ldots$ if the order r is odd, and for $k = 0, \pm 1, \pm 2, \ldots$ if r is even.

(**c**) Finite differences are conveniently tabulated in arrays like

$$\begin{array}{llllll}
x_{-1} & y_{-1} & & & & \\
 & & \Delta y_{-1} & & & \\
x_0 & y_0 & & \Delta^2 y_{-1} & & \\
 & & \Delta y_0 & & \Delta^3 y_{-1} & \\
x_1 & y_1 & & \Delta^2 y_0 & & \Delta^4 y_{-1}\\
 & & \Delta y_1 & & \Delta^3 y_0 & \\
x_2 & y_2 & & \Delta^2 y_1 & & \\
 & & \Delta y_2 & & & \\
x_3 & y_3 & & & &
\end{array} \tag{20.4-5}$$

$$\begin{array}{llllll}
x_{-3} & y_{-3} & & & & \\
 & & \nabla y_{-2} & & & \\
x_{-2} & y_{-2} & & \nabla^2 y_{-1} & & \\
 & & \nabla y_{-1} & & \nabla^3 y_0 & \\
x_{-1} & y_{-1} & & \nabla^2 y_0 & & \nabla^4 y_1\\
 & & \nabla y_0 & & \nabla^3 y_1 & \\
x_0 & y_0 & & \nabla^2 y_1 & & \\
 & & \nabla y_1 & & & \\
x_1 & y_1 & & & &
\end{array} \tag{20.4-6}$$

$$\begin{array}{llllll}
x_{-2} & y_{-2} & & & & \\
 & & \delta_{-\frac{3}{2}} & & & \\
x_{-1} & y_{-1} & & \delta^2 y_{-1} & & \\
 & & \delta y_{-\frac{1}{2}} & & \delta^3 y_{-\frac{1}{2}} & \\
x_0 & y_0 & & \delta^2 y_0 & & \delta^4 y_0\\
 & & \delta y_{\frac{1}{2}} & & \delta^3 y_{\frac{1}{2}} & \\
x_1 & y_1 & & \delta^2 y_1 & & \\
 & & \delta y_{\frac{3}{2}} & & & \\
x_2 & y_2 & & & &
\end{array} \tag{20.4-7}$$

Note that (1) *the computation of an rth-order difference requires $r + 1$ function values,* and (2) *the nth-order differences of an nth-degree polynomial $y(x)$ are constant.*

20.4-2. Operator Notation. **(a) Definitions.** Given a suitably defined function $y = y(x)$ of the real variable x and a fixed increment $\Delta x = h$ of x, one defines the **displacement operator (shift operator)** E by

$$\mathsf{E}y(x) \equiv y(x + \Delta x) \qquad \mathsf{E}^r y(x) \equiv y(x + r\,\Delta x) \tag{20.4-8}$$

where r is any real number.

The **difference operators** Δ, ∇, δ and the **central-mean operator** μ are defined by

$$\begin{aligned} \Delta y(x) &\equiv y(x + \Delta x) - y(x) \\ &\qquad\qquad \text{(FORWARD-DIFFERENCE OPERATOR)} \\ \nabla y(x) &\equiv y(x) - y(x - \Delta x) \\ &\qquad\qquad \text{(BACKWARD-DIFFERENCE OPERATOR)} \\ \delta y(x) &\equiv y\left(x + \frac{\Delta x}{2}\right) - y\left(x - \frac{\Delta x}{2}\right) \\ &\qquad\qquad \text{(CENTRAL-DIFFERENCE OPERATOR)} \end{aligned} \tag{20.4-9}$$

$$\mu y(x) \equiv \frac{1}{2}\left[y\left(x - \frac{\Delta x}{2}\right) + y\left(x + \frac{\Delta x}{2}\right)\right] \qquad \text{(CENTRAL-MEAN OPERATOR)} \tag{20.4-10}$$

Note that, for $u_k = u(x_0 + k\,\Delta x)$, $v_k = v(x_0 + k\,\Delta x)$

$$\begin{aligned} \Delta(u_k + v_k) &= \Delta u_k + \Delta v_k \qquad \Delta(\alpha u_k) = \alpha\,\Delta u_k \\ \Delta(u_k v_k) &= u_k\,\Delta v_k + v_k\,\Delta u_k + \Delta u_k\,\Delta v_k \end{aligned} \tag{20.4-11}$$

Note also that, for $\Delta x = 1$,

$$\left.\begin{aligned} \Delta x^{[r]} = \Delta[x(x-1)\cdots(x-r+1)] &= rx(x-1)\cdots(x-r+2) = rx^{[r-1]} \\ \Delta\binom{x}{r} &= \binom{x}{r-1} \end{aligned}\right\} \quad (r = 1, 2, \ldots) \tag{20.4-12}$$

(b) Operator Relations (see also Sec. 14.3-1 and Ref. 20.36).

$$\left.\begin{aligned} &\Delta = \mathsf{E} - 1 = \mathsf{E}\nabla = \mathsf{E}^{1/2}\delta \qquad \nabla = 1 - \mathsf{E}^{-1} = \mathsf{E}^{-1}\Delta = \mathsf{E}^{-1/2}\delta \\ &\delta = \mathsf{E}^{1/2} - \mathsf{E}^{-1/2} = \mathsf{E}^{-1/2}\Delta = \mathsf{E}^{1/2}\nabla \qquad \mu = \tfrac{1}{2}(\mathsf{E}^{1/2} + \mathsf{E}^{-1/2}) \\ &\mathsf{E} = 1 + \Delta \qquad \nabla\Delta = \Delta\nabla = \delta^2 \end{aligned}\right\} \tag{20.4-13}$$

As an aid to memory, note

$$\Delta^r = (\mathsf{E} - 1)^r = \sum_{j=0}^{r} (-1)^j \binom{r}{j} \mathsf{E}^{r-j} \qquad (r = 1, 2, \ldots) \tag{20.4-14}$$

together with $\nabla = \mathsf{E}^{-1}\Delta$ and $\delta = \mathsf{E}^{-1/2}\Delta$. Note also

$$\mathsf{E}^r = (1 + \Delta)^r = 1 + \binom{1}{r}\Delta + \binom{2}{r}\Delta^2 + \cdots \tag{20.4-15}$$

If r is a positive integer, the series (15) terminates; otherwise, its convergence requires investigation.

(c) For suitably differentiable operands,

$$\mathsf{E}^r = e^{r\,\Delta x\,\mathsf{D}} = 1 + r\,\Delta x\,\mathsf{D} + \frac{1}{2!}(r\,\Delta x\,\mathsf{D})^2 + \cdots \qquad \left(\mathsf{D} \equiv \frac{d}{dx}\right) \tag{20.4-16}$$

(operator notation for Taylor's series, Sec. 4.10-4; see also Table 8.3-1).

20.4-3. Difference Equations. (a) An **ordinary difference equation of order** r is an equation

$$G(x_k, y_k, y_{k+1}, y_{k+2}, \ldots, y_{k+r}) \equiv G(x_k, y_k, \mathsf{E}y_k, \mathsf{E}^2y_k, \ldots, \mathsf{E}^ry_k) = 0$$
$$(k = 0, \pm1, \pm2, \ldots;\ r = 1, 2, \ldots) \tag{20.4-17}$$

relating values $y_k = y(x_k) = y(x_0 + k\,\Delta x)$ of a function $y = y(x)$ defined on a discrete set of values $x = x_k = x_0 + k\,\Delta x$, where Δx is a fixed increment. It is often convenient to introduce $k = (x - x_0)/\Delta x = 0$, $\pm1, \pm2, \ldots$ as a new independent variable. *An ordinary difference equation of order r relates y_k and a set of finite differences* $\Delta^i y_k$, $\nabla^i y_k$, or $\delta^i y_k$ *of order up to and including r;* or the difference equation may relate y_k and **difference coefficients** $\dfrac{\Delta^i y_k}{(\Delta x)^i}$, $\dfrac{\nabla^i y_k}{(\Delta x)^i}$, or $\dfrac{\delta^i y_k}{(\Delta x)^i}$ of order i up to and including r.

To **solve** a difference equation (17) means to find **solutions** $y = y(x)$ such that the sequence of the y_k satisfies the given equation for some range of values of k. The complete solution of an ordinary difference equation of order r will, in general, involve r arbitrary constants; the latter must be determined by accessory conditions on the y_k, such as initial conditions or boundary conditions. The solution of a difference equation over any finite range amounts, in principle, to that of a set of simultaneous equations.

Difference equations are used (1) to approximate differential equations (Secs. 20.9-2 and 20.9-4) and (2) to deal with situations represented by discrete-variable models.

A SIMPLE EXAMPLE: SUMMATION OF SERIES. The problem of solving a first-order difference equation (*recurrence relation*) of the form

$$\Delta y_{k-1} = \nabla y_k = \alpha_k \qquad \text{or} \qquad y_k = y_{k-1} + \alpha_k \tag{20.4-18}$$

with a given initial value $y_0 = \alpha_0$ is identical with the problem of *summation of series* (Sec. 4.8-5):

$$y_k = y_{k-1} + \alpha_k = \sum_{j=0}^{k} \alpha_k \qquad (k = 0, 1, 2, \ldots) \tag{20.4-19}$$

The problem is analogous to the integration of a differential equation $y' = f(x)$. Σ and ∇ are inverse operators; note Eq. (12) and

$$\sum_{k=m}^{n} u_k \, \Delta v_k = (u_{n+1}v_{n+1} - u_m v_m) - \sum_{k=m}^{n} v_{k+1} \, \Delta u_k \quad \text{(SUMMATION BY PARTS)} \qquad (20.4\text{-}20)$$

(b) A **partial difference equation** relates values $\Phi_{ij\ldots} = \Phi(x_0 + i\,\Delta x, y_0 + j\,\Delta y, \ldots)$ $(i, j, \ldots = 0, \pm 1, \pm 2, \ldots)$ of a function $\Phi = \Phi(x, y, \ldots)$; the order of the partial difference equation is the largest difference between i values, j values, . . . occurring in the equation. Refer to Sec. 20.9-5 for formulas expressing various partial-difference operators in terms of function values $\Phi_{ij\ldots}$, and for the use of partial difference equations to approximate solutions of partial differential equations.

20.4-4. Linear Ordinary Difference Equations. (a) Superposition Theorems

(see also Secs. 9.3-1 and 15.4-2). A **linear difference equation** is linear in the values and differences of the unknown function. Thus a *linear ordinary difference equation of order r* has the form

$$\begin{aligned} a_0(k)y_{k+r} &+ a_1(k)y_{k+r-1} + \cdots + a_r(k)y_k \\ &\equiv [a_0(k)\mathsf{E}^r + a_1(k)\mathsf{E}^{r-1} + \cdots + a_r(k)]y_k = f(k) \end{aligned} \qquad (20.4\text{-}21)$$

where the $a_i(k)$ and $f(k)$ are given functions of $k = 0, \pm 1, \pm 2, \ldots$. As in Sec. 9.3-1, *solutions y_k corresponding to $f(k) = \alpha f_1(k) + \beta f_2(k)$ are the corresponding linear combinations of solutions corresponding to the individual forcing functions $f_1(k)$ and $f_2(k)$ (Superposition Principle). The complete solution of Eq. (21) can be expressed as the sum of any particular solution and the complete solution of the linear and homogeneous "complementary equation"*

$$[a_0(k)\mathsf{E}^r + a_1(k)\mathsf{E}^{r-1} + \cdots + a_r(k)]y_k = 0 \qquad (20.4\text{-}22)$$

Again, *every linear combination of solutions of a linear homogeneous difference equation* (22) *is itself a solution of Eq.* (22).

The theory of ordinary difference equations parallels that of ordinary differential equations in many details (see also Ref. 20.67). In particular, *a homogeneous linear difference equation* (22) *admits at most r solutions $y_{(1)k}, y_{(2)k}, \ldots$ linearly independent on the set $k = 0, 1, 2, \ldots$. r such solutions are linearly independent if and only if the* **Casoratian determinant**

$$K[y_{(1)k}, y_{(2)k}, \ldots, y_{(r)k}] \equiv \begin{vmatrix} y_{(1)k} & y_{(2)k} & \cdots & y_{(r)k} \\ y_{(1)k+1} & y_{(2)k+1} & \cdots & y_{(r)k+1} \\ \cdots & \cdots & \cdots & \cdots \\ y_{(1)k+r-1} & y_{(2)k+r-1} & \cdots & y_{(r)k+r-1} \end{vmatrix} \qquad (20.4\text{-}23)$$

is not identically zero for $k = 0, 1, 2, \ldots$. The Casoratian is analogous to the Wronskian in Sec. 9.3-2.

(b) The Method of Variation of Constants (see also Sec. 9.3-3). Assuming that r linearly independent solutions $y_{(1)k}, y_{(2)k}, \ldots, y_{(r)k}$ of the complementary equation (22) are known, the complete solution of the nonhomogeneous linear difference equation (21) is given by

$$y_k = C_1(k)y_{(1)k} + C_2(k)y_{(2)k} + \cdots + C_r(k)y_{(r)k} \tag{20.4-24a}$$

with

$$\left.\begin{aligned} \sum_{h=1}^{r} y_{(h)k+j}\, \Delta C_h(k) &= 0 \qquad (j = 1, 2, \ldots, r-1) \\ \sum_{h=1}^{r} y_{(h)k+r}\, \Delta C_h(k) &= f(k) \end{aligned}\right\} \tag{20.4-24b}$$

After solving the r simultaneous linear equations (24*b*) for the $\Delta C_h(k)$, obtain each $C_h(k)$ by summation as in Eq. (19).

20.4-5. Linear Ordinary Difference Equations with Constant Coefficients: Method of Undetermined Coefficients (see also Secs. 9.4-1 to 9.4-8). **(a)** The complete solution of the linear *homogeneous* difference equation

$$a_0y_{k+r} + a_1y_{k+r-1} + \cdots + a_ry_k \equiv (a_0\mathbf{E}^r + a_1\mathbf{E}^{r-1} + \cdots + a_r)y_k = 0 \tag{20.4-25}$$

with given constant coefficients $a_0, a_1, \ldots, a_r$ can be expressed in terms of normal-mode terms completely analogous to those of Sec. 9.4-1*c*, viz.,

$$y_k = C_1\lambda_1^k + C_2\lambda_2^k + \cdots + C_r\lambda_r^k \tag{20.4-26}$$

where $\lambda_1, \lambda_2, \ldots, \lambda_r$ are the roots of the algebraic equation

$$a_0\lambda^r + a_1\lambda^{r-1} + \cdots + a_r = 0 \tag{20.4-27}$$

provided that all r roots are different. If a root, say λ_1, is an m-fold root, then the corresponding term in the solution (27) becomes $(C_1 + kC_2 + k^2C_3 + \cdots + k^{m-1}C_m)\lambda_1^k$. Two terms corresponding to complex conjugate roots $\lambda = \alpha \pm i\beta$ may be combined into $|\lambda|^k(A \cos k\varphi + B \sin k\varphi)$ ($\varphi = \arctan \beta/\alpha$). The coefficients $C_j, A, B, \ldots$ must be chosen so that the solution matches suitably given initial or boundary conditions on the y_k.

The solution of the *nonhomogeneous* difference equation

$$a_0y_{k+r} + a_1y_{k+r-1} + \cdots + a_ry_k \equiv (a_0\mathbf{E}^r + a_1\mathbf{E}^{r-1} + \cdots + a_r)y_k = f(k) \tag{20.4-28}$$

can often be derived in the manner of Sec. 20.4-4, but the special methods of Secs. 20.4-6 and 20.4-7*a* may be more convenient.

20.4-6. Transform Methods for Linear Difference Equations with Constant Coefficients. (a) The *z*-transform Method (see also Secs. 8.7-3 and 9.4-5). Given a difference equation (28), z transformation of both sides with the aid of the shift theorem 2 of Table 8.7-2 yields the z transform

$$\mathcal{Z}[y_k; z] \equiv Y_Z(z) \equiv y_0 + \frac{y_1}{z} + \frac{y_2}{z^2} + \cdots \tag{20.4-29}$$

of the unknown solution sequence $y_0, y_1, y_2, \ldots$ in the form

$$Y_Z(z) = \frac{F_Z(z)}{a_0z^r + a_1z^{r-1} + \cdots + a_r} + \frac{G_Z(z)}{a_0z^r + a_1z^{r-1} + \cdots + a_r} \tag{20.4-30}$$

where the first term on the right, as in Sec. 9.4-5, represents a "normal response" to the given forcing sequence $f(0), f(1), f(2), \ldots$, and the second term represents the effect of r given initial values $y_0, y_1, y_2, \ldots, y_{r-1}$. Specifically,

$$\begin{aligned} G_Z(z) \equiv\ & y_0(a_0z^r + a_1z^{r-1} + \cdots + a_{r-1}z) \\ & + y_1(a_0z^{r-1} + a_1z^{r-2} + \cdots + a_{r-2}z) + \cdots + y_{r-1}a_0z \end{aligned} \tag{20.4-31}$$

The unknown y_k may be obtained as coefficients of $1/z^k$ in the power-series expansion of $Y_Z(z)$, or one may utilize a table of z-transform pairs (Table 20.4-1). As in the case of Laplace transforms, $Y_Z(z)$ can be reduced to a sum of simpler forms by a partial-fraction expansion, which yields normal modes corresponding to the roots of the characteristic equation (27).

(b) Sampled-data Representation in Terms of Impulse Trains and Jump Functions. Laplace-transform Method. If one formally admits the asymmetric impulse function $\delta_+(t)$ and step-function differentiation in the sense of Sec. 21.9-6, a sampled-data sequence $u_0, u_1, u_2, \ldots$ can be represented, on a reciprocal one-to-one basis, by a corresponding **impulse train***

$$u\dagger(t) \equiv u_0\delta_+(t) + u_1\delta_+(t - T) + u_2\delta_+(t - 2T) + \cdots \qquad (t > 0) \tag{20.4-32}$$

where t is a real variable, and T a real positive constant (sampling interval).

If the sampled-data sequences $y_0, y_1, y_2, \ldots$ and $f(0), f(1), f(2), \ldots$ satisfy any difference equation (28), the corresponding *functions* $y\dagger(t), f\dagger(t)$ satisfy the functional equation (difference equation for functions)

$$a_0y\dagger(t + r\Delta t) + a_1y\dagger[t + (r-1)\Delta t] + \cdots + a_ry\dagger(t) = f\dagger(t) \qquad (t > 0) \tag{20.4-33}$$

with appropriate initial conditions. Unlike Eq. (28), this relation admits Laplace transformation, with

$$\mathcal{L}[u\dagger(t); s] = u_0 + u_1e^{-Ts} + u_2e^{-2Ts} + \cdots \qquad (\sigma > \sigma_a) \tag{20.4-34}$$

(see also Sec. 8.2-2). The resulting transform method is analogous to the z-transform method with $z = e^{sT} = e^{s\Delta t}$ where $\Delta t = T$.

To avoid the use of symbolic functions, one can, instead, represent the sequence $u_0, u_1, u_2, \ldots$ by the corresponding **jump function**

$$\int_+ u(t) = \int_0^t [u\dagger(t) - u\dagger(t - \Delta t)]\,dt = u_k \qquad \text{for } kT < t \le (k+1)T$$
$$(t > 0;\ k = 0, 1, 2, \ldots) \tag{20.4-35}$$

* $u\dagger(t)$ is usually denoted by $u^*(t)$ in the literature on sampled-data systems.

Table 20.4-1. A Short Table of z Transforms

	Sequence of sample values $y_k (k = 0, 1, 2, \ldots)$	z transform $\mathcal{Z}[y_k; z] = \sum_{j=0}^{\infty} \frac{y_j}{z^j}$
1	1	$\frac{z}{z-1}$
2	k	$\frac{z}{(z-1)^2}$
3	k^2	$\frac{z(z+1)}{(z-1)^3}$
4	$\binom{k}{n} \quad (n = 0, 1, 2, \ldots)$	$\frac{z}{(z-1)^{n+1}}$
5	a^k	$\frac{z}{z-a}$
6	ka^k	$\frac{az}{(z-a)^2}$
7	$\binom{k}{n} a^{k-n} \quad (n = 0, 1, 2, \ldots)$	$\frac{z}{(z-a)^{n+1}}$
8	$\frac{a^k - b^k}{a - b} \quad (a \neq b)$	$\frac{z}{(z-a)(z-b)}$
9	$a^k \sin bk$	$\frac{az \sin b}{z^2 - 2az \cos b + a^2}$
10	$a^k \cos bk$	$\frac{z(z - a \cos b)}{z^2 - 2az \cos b + a^2}$

which can be physically interpreted as the output of a "zero-order data-hold" device. $\int_+ y(t)$ and $\int_+ f(t)$ satisfy the same functional equation as $y\dagger(t)$ and $f\dagger(t)$, so that Laplace transformation is again possible.

20.4-7. Systems of Ordinary Difference Equations (State Equations). Matrix Notation. Just as in the case of differential equations, one may be given a **system of ordinary difference equations** involving two or more unknown sequences $y(x_k) = y_k$, $z(x_k) = z_k$, ... One can reduce any r^{th}-order difference equation (17) to a set of r first-order equations by introducing the $\mathsf{E}^i y_k$ or $\Delta^i y_k$ ($i = 1, 2, \ldots, r - 1$) as new variables (state variables, see also Sec. 13.6-1). Again, as in Sec. 13.6-1, a given system of linear first-order difference equations (recurrence relations, state equations)

$$\left.\begin{array}{l} y_{k+1} = a_{11}y_k + a_{12}z_k + \cdots + f_1(k) \\ z_{k+1} = a_{21}y_k + a_{22}z_k + \cdots + f_2(k) \\ \cdots\cdots\cdots\cdots\cdots\cdots\cdots \end{array}\right\} \qquad (20.4\text{-}36a)$$

with constant coefficients a_{ij} may be rewritten as a matrix equation

$$Y_{k+1} = AY_k + F(k) \qquad (20.4\text{-}36b)$$

(see also Sec. 14.5-3), where Y_k is the column matrix $\{y_k, z_k, \ldots\}$, $F(k)$ is the column matrix $\{f_1(k), f_2(k), \ldots\}$, and $A \equiv [a_{ij}]$. Given $Y_0 \equiv \{y_0, z_0, \ldots\}$, the solution is

$$Y_k = A^k Y_0 + \sum_{h=0}^{k-1} A^{k-h-1}F(h) \qquad (k = 0, 1, 2, \ldots) \qquad (20.4\text{-}37)$$

where A^k, in analogy with Sec. 13.6-2*b*, is called the **state-transition matrix** for the system (36). The powers A^k required for the solution can be computed with the aid of Sylvester's theorem (Sec. 13.4-7*b*), so that each eigenvalue of A, i.e., each root λ of the characteristic equation,

$$\det (A - \lambda I) = 0 \qquad (20.4\text{-}38)$$

again corresponds to a normal mode (see also Secs. 13.6-2 and 20.4-5).

This method applies, in particular, to the solution of the rth-order linear difference equation (28) if it is first reduced to a system of the form (36) through the introduction of $y_{k+1}, y_{k+2}, \ldots, y_{k+r-1}$ as new variables, so that

$$\begin{gathered} Y_k \equiv \{y_{k+r-1}, y_{k+r-2}, \ldots, y_k\} \\ F(k) \equiv \left\{\frac{1}{a_0} f(k), 0, 0, \ldots, 0\right\} \\ A \equiv \begin{bmatrix} -\dfrac{a_1}{a_0} & -\dfrac{a_2}{a_0} & \cdots & -\dfrac{a_{r-1}}{a_0} & -\dfrac{a_r}{a_0} \\ 1 & 0 & \cdots & 0 & 0 \\ 0 & 1 & \cdots & 0 & 0 \\ \cdots & \cdots & \cdots & \cdots & \cdots \\ 0 & 0 & \cdots & 1 & 0 \end{bmatrix} \end{gathered} \qquad (20.4\text{-}39)$$

20.4-8. Stability. (a) By analogy with Sec. 9.4-4, a linear difference equation (28) or a linear system (36) with constant coefficients will be called **completely stable** if and only if all roots of the corresponding characteristic equation (27) or (38) have absolute values less than unity; this ensures that the effects of small changes in the initial conditions tend to zero as k increases.

Reference 20.63 presents *Jury's version of the Schur-Cohn test* for stability, which is analogous to the Routh-Hurwitz criterion (Sec. 1.6-6*b*) for roots with negative real parts.

(b) Lyapunov's definitions (Sec. 13.6-5) and the related theorems (Sec. 13.6-6) on the stability and asymptotic stability of solutions are readily extended to the solution sequences $Y_0, Y_1, Y_2, \ldots$ of linear and nonlinear autonomous difference-equation systems

$$Y_{k+1} = F(Y_k) \qquad (k = 0, 1, 2, \ldots) \tag{20.4-40}$$

Generally speaking, conditions like $dV/dt \leq 0$ for continuous-system Lyapunov functions $V(y)$ simply translate into analogous conditions $\Delta V(Y_k) \leq 0$ for discrete-parameter Lyapunov functions $V(Y_k)$ (Ref. 20.63).

20.5. APPROXIMATION OF FUNCTIONS BY INTERPOLATION

20.5-1. Introduction (see also Secs. 12.5-4*b* and 15.2-5). Given $n + 1$ function values $y(x_k) = y_k$, an **interpolation formula** approximates the function $y(x)$ by a suitable known function $Y(x) \equiv Y(x; \alpha_0, \alpha_1, \alpha_2, \ldots, \alpha_n)$ depending on $n + 1$ parameters α_j chosen so that $Y(x_k) = y(x_k) = y_k$ for the given set of $n + 1$ argument values x_k. In particular, Secs. 20.5-2 to 20.5-6 deal with *polynomial interpolation*. Other interpolation methods are discussed in Secs. 20.5-7 and 20.6-6.

The use of interpolation-type approximations (and, in particular, of polynomial interpolation) is not always justified. Thus, in the case of empirical functions, one may wish to *smooth* fluctuations in the $y(x_k)$ due to random errors (Sec. 20.6-3).

20.5-2. General Formulas for Polynomial Interpolation (Argument Values Not Necessarily Equally Spaced). An **n^{th}-order polynomial interpolation formula** approximates the function $y(x)$ by an nth-degree polynomial $Y(x)$ such that $Y(x_k) = y(x_k) = y_k$ for a given set of $n + 1$ argument values x_k.

(a) Lagrange's Interpolation Formula. Given $y_0 = y(x_0)$, $y_1 = y(x_1), y_2 = y(x_2), \ldots, y_n = y(x_n)$

$$\begin{aligned} Y(x) = {} & \frac{(x - x_1)(x - x_2) \cdots (x - x_n)}{(x_0 - x_1)(x_0 - x_2) \cdots (x_0 - x_n)} y_0 \\ & + \frac{(x - x_0)(x - x_2) \cdots (x - x_n)}{(x_1 - x_0)(x_1 - x_2) \cdots (x_1 - x_n)} y_1 + \cdots \\ & + \frac{(x - x_0)(x - x_1) \cdots (x - x_{n-1})}{(x_n - x_0)(x_n - x_1) \cdots (x_n - x_{n-1})} y_n \end{aligned} \tag{20.5-1}$$

(LAGRANGE'S INTERPOLATION FORMULA)

(b) Divided Differences and Newton's Interpolation Formula. One defines the **divided differences**

$$\left.\begin{aligned} \Delta_1(x_0, x_1) &\equiv \frac{y_1 - y_0}{x_1 - x_0} \\ \Delta_r(x_0, x_1, x_2, \ldots, x_r) & \\ \equiv \frac{\Delta_{r-1}(x_1, x_2, \ldots, x_r) - \Delta_{r-1}(x_0, x_1, \ldots, x_{r-1})}{x_r - x_0} & \\ (r = 2, 3, \ldots) & \end{aligned}\right\} \qquad (20.5\text{-}2)$$

Then

$$\begin{aligned} Y(x) = y_0 &+ (x - x_0)\Delta_1(x_0, x_1) \\ &+ (x - x_0)(x - x_1)\Delta_2(x_0, x_1, x_2) + \cdots \\ &+ \left[\prod_{k=0}^{n-1} (x - x_k)\right] \Delta_n(x_0, x_1, x_2, \ldots, x_n) \end{aligned} \qquad (20.5\text{-}3)$$

(NEWTON'S INTERPOLATION FORMULA)

Unlike in Eq. (1), the addition of a new pair of values x_{n+1}, y_{n+1} requires merely the addition of an extra term. It is convenient to tabulate the divided differences (2) for use in Eq. (3) in the manner of Eq. (20.4-5).

Taking a divided difference is a linear operation (Sec. 15.2-7) on the function $y(x)$. Each function (2) is completely symmetric in its arguments.

(c) Aitken's Iterated-interpolation Method. The following scheme may be useful if one desires values of the interpolation polynomial $Y(x)$ rather than a simple expression for $Y(x)$. Let $Y_{ijk\ldots}$ be the interpolation polynomial through (x_i, y_i), (x_j, y_j), (x_k, y_k), . . . , so that $Y_{012\ldots n} = Y(x)$. Interpolation polynomials of increasing order are then obtained successively as follows:

$$Y_{01} = \frac{1}{x_1 - x_0}\begin{vmatrix} x - x_0 & y_0 \\ x - x_1 & y_1 \end{vmatrix} \qquad Y_{12} = \frac{1}{x_2 - x_1}\begin{vmatrix} x - x_1 & y_1 \\ x - x_2 & y_2 \end{vmatrix} \quad \cdots$$

$$Y_{012} = \frac{1}{x_2 - x_0}\begin{vmatrix} x - x_0 & Y_{01} \\ x - x_2 & Y_{12} \end{vmatrix} \quad \cdots$$

$$Y_{0123} = \frac{1}{x_3 - x_0}\begin{vmatrix} x - x_0 & Y_{012} \\ x - x_3 & Y_{123} \end{vmatrix} \quad \cdots$$

$$\cdots\cdots\cdots\cdots\cdots\cdots\cdots\cdots$$

The process may be terminated whenever two interpolation polynomials of successive orders agree to the desired number of significant figures.

(d) The Remainder. If $y(x)$ is suitably differentiable, the **remainder** (error) $R_{n+1}(x)$ involved in the use of any polynomial-interpolation formula based on the $n + 1$ function values $y_0, y_1, y_2, \ldots, y_n$ may be

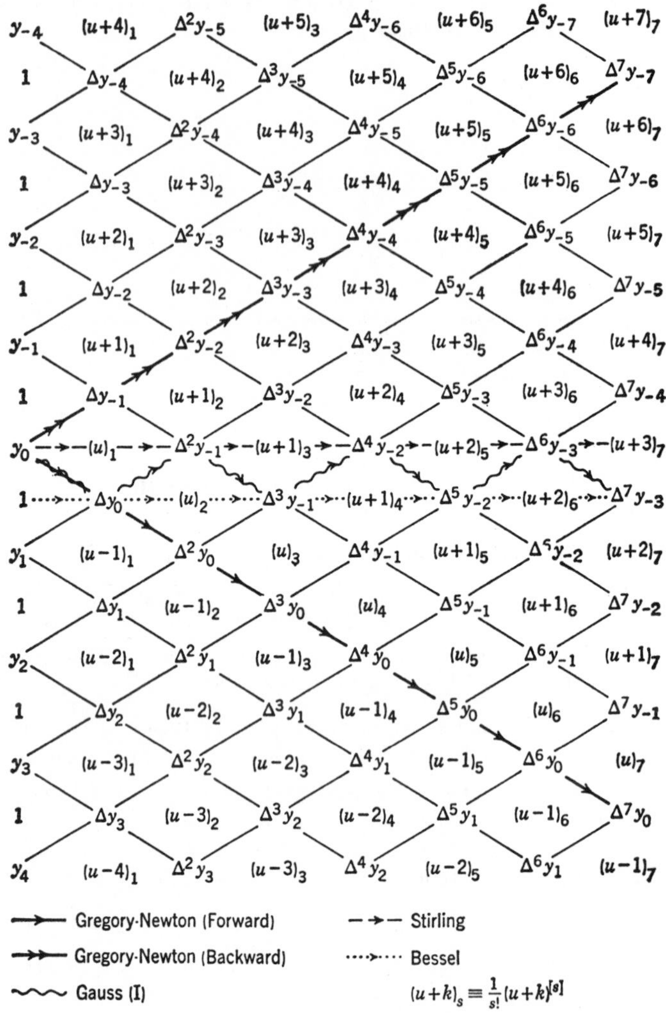

FIG. 20.5-1. Lozenge diagram for interpolation formulas. The abbreviated notation $(u + k)_s = \frac{1}{s!}(u + k)^{[s]} = \binom{u + k}{s}$ is used. To convert a path through the lozenge to an interpolation formula, the following rules are formulated:

1. Each time a difference column is crossed from left to right a term is added.
2. If a path enters a difference column (from the left) at a positive slope, the term added is the product of the difference, say $\Delta^k y_{-p}$, at which the column is crossed and the factorial $(u + p - 1)_k$ lying just below that difference.
3. If a path enters a difference column (from the left) at a negative slope, the term added is the product of the difference, say $\Delta^k y_{-p}$, at which the column is crossed and the factorial $(u + p)_k$ lying just above that difference.
4. If a path enters a difference column horizontally (from the left), the term added

estimated from

$$R_{n+1}(x) \equiv y(x) - Y(x) = \frac{1}{(n+1)!} f^{(n+1)}(\xi) \prod_{k=0}^{n} (x - x_k) \qquad (20.5\text{-}4)$$

where ξ lies in the smallest interval I containing $x_0, x_1, x_2, \ldots, x_n$, and x (see also Sec. 4.10-4; ξ will, in general, depend on x), and

$$|R_{n+1}(x)| \leq \frac{1}{(n+1)!} \max_{X \text{ in } I} |f^{(n+1)}(X)| \prod_{k=0}^{n} |x - x_k| \qquad (20.5\text{-}5)$$

20.5-3. Interpolation Formulas for Equally Spaced Argument Values. Lozenge Diagrams. Let $y_k = y(x_k)$, $x_k = x_0 + k\,\Delta x$ $(k = 0, \pm 1, \pm 2, \ldots)$, where Δx is a fixed increment as in Sec. 20.4-1, and introduce the abbreviation $(x - x_0)/\Delta x = u$.

(a) Newton-Gregory Interpolation Formulas. Given $y_0, y_1, y_2, \ldots$ or $y_0, y_{-1}, y_{-2}, \ldots$, Eq. (3) becomes, respectively,

$$\left.\begin{aligned} Y(x) &= y_0 + \binom{u}{1} \Delta y_0 + \binom{u}{2} \Delta^2 y_0 + \cdots \\ Y(x) &= y_0 + \frac{u}{1!} \nabla y_0 + \frac{u(u+1)}{2!} \nabla^2 y_0 + \cdots \end{aligned}\right\} \begin{matrix}\text{(NEWTON-} \\ \text{GREGORY} \\ \text{INTERPOLATION} \\ \text{FORMULAS)}\end{matrix} \qquad (20.5\text{-}6)$$

(b) Symmetric Interpolation Formulas. More frequently, one is given $y_0, y_1, y_2, \ldots$ and $y_{-1}, y_{-2}, \ldots$; Table 20.5-1 lists the most useful interpolation formulas for this case (see also Fig. 20.5-1). Note that Everett's and Steffensen's formulas are of particular interest for use with printed tables, since only even-order or only odd-order differences need be tabulated.

---

is the product of the difference, say $\Delta^k y_{-p}$, at which the column is crossed and the average of the two factorials $(u + p)_k$ and $(u + p - 1)_k$ lying, respectively, just above and just below that difference.

5. If a path crosses a difference column (from left to right) between two differences, say $\Delta^k y_{-(p+1)}$ and $\Delta^k y_{-p}$, then the added term is the product of the average of these two differences and the factorial $(u + p)_k$ at which the column is crossed.

6. Any portion of a path traversed from right to left gives rise to the same terms as would arise from going along this portion from left to right except that the sign of each term is changed.

7. The zero-difference column corresponding to the tabulated values may be treated by the same rules as any other difference columns provided one thinks of the lozenge as being entered just to the left of this column. Thus this column can be crossed by a path making a positive, negative, or zero slope, just as is true of the other columns. (*Reprinted by permission from K. S. Kunz, Numerical Analysis, McGraw-Hill, New York,* 1957.)

Table 20.5-1. Symmetric Interpolation Formulas

One is given an odd number $n + 1 = 2m + 1$ of function values $y_k = y(x_0 + k\,\Delta x)$ $(k = 0, \pm 1, \pm 2, \ldots, \pm m)$, where Δx is a fixed increment; $u = (x - x_0)/\Delta x$

No.		Interpolation polynomial, $Y(x)$	Remainder, $R_{n+1}(x) \equiv R_{2m+1}(x)$ (ξ lies in the smallest interval containing x and every $x_0 + k\,\Delta x$ used)
1	**Stirling's interpolation formula***	$y_0 + \sum_{k=0}^{m-1} \binom{u+k}{2k+1} \frac{\delta^{2k+1}y_{-1/2} + \delta^{2k+1}y_{1/2}}{2} + \sum_{k=1}^{m} \frac{u}{2k} \binom{u+k-1}{2k-1} \delta^{2k}y_0$	$\binom{u+m}{2m+1} y^{(2m+1)}(\xi)\,\Delta x^{2m+1}$
2	**Bessel's interpolation formula†**	$\frac{y_0 + y_1}{2} + \sum_{k=0}^{m-1} \frac{u - 1/2}{2k+1} \binom{u+k-1}{2k} \delta^{2k+1}y_{1/2} + \sum_{k=1}^{m-1} \binom{u+k-1}{2k} \frac{\delta^{2k}y_0 + \delta^{2k}y_1}{2}$	$\binom{u+m-1}{2m} y^{(2m)}(\xi)\,\Delta x^{2m}$
3	**Everett's interpolation formula***	$(1-u)y_0 + uy_1 + \sum_{k=1}^{m-1} \left\{ \binom{u+k}{2k+1} \delta^{2k}y_1 - \binom{u+k-1}{2k+1} \delta^{2k}y_0 \right\}$	$\binom{u+m-1}{2m} y^{(2m)}(\xi)\,\Delta x^{2m}$
4	**Steffensen's interpolation formula**	$y_0 + \sum_{k=1}^{m} \left\{ \binom{u+k}{2k} \delta^{2k-1}y_{1/2} - \binom{k-u}{2k} \delta^{2k-1}y_{-1/2} \right\}$	$\binom{u+m}{2m+1} y^{(2m+1)}(\xi)\,\Delta x^{2m+1}$

* Note that $\binom{u+k}{2k+1} = \frac{u(u^2 - 1^2)(u^2 - 2^2) \cdots (u^2 - k^2)}{(2k+1)!}$

† *Bessel's modified formula*

$$Y(x) = y_0 + u\,\delta y_{1/2} + \frac{u(u-1)}{2} \frac{\delta^2 y_0 + \delta^2 y_1}{2} + \frac{u(u - 1/2)(u-1)}{6} \left(\delta^3 y_{1/2} - \frac{13}{120}\delta^5 y_{1/2}\right) + \frac{(u+2)(u+1)u(u-1)}{24} \left(\frac{\delta^4 y_0 + \delta^4 y_1}{2} - \frac{191}{924} \frac{\delta^6 y_0 + \delta^6 y_1}{2}\right)$$

gives a simplified polynomial including the effect of sixth-order differences (see also Refs. 20.13 and 20.16).

(c) Use of Lozenge Diagrams. Many interpolation formulas can be obtained with the aid of a *lozenge diagram* (*Fraser diagram*) like that shown in Fig. 20.5-1. One can derive more sophisticated interpolation formulas (e.g., Everett's formula) by averaging two or more equivalent interpolation polynomials.

20.5-4. Inverse Interpolation. (a) Given $x_k = x_0 + k\,\Delta x$, $y_k = y(x_k)$ for $k = 0, 1, 2, \ldots, n$, it is desired to find a value ξ of x between x_0 and x_1 so that $y(x)$ assumes a given value η; Δx is assumed to be so small that ξ is unique. One may apply the general interpolation formulas (1) or (3) by reversing the roles of x and y. Alternatively, one may apply one of the iteration schemes of Sec. 20.2-2 to solve the equation

$$Y(\xi) - \eta = 0$$

where $Y(x)$ is a suitable interpolation polynomial approximating $y(x)$; one uses linear interpolation for the first step of the iteration, second-order interpolation for the second step, etc.

(b) Inverse Interpolation by Reversion of Series. Use any desired interpolation formula and write it as a power series $Y(x) = a_0 + a_1x + a_2x^2 + \cdots \approx y$. Then

$$\left.\begin{aligned}
&x \approx \frac{y - a_0}{a_1} + C_2\left(\frac{y - a_0}{a_1}\right)^2 + \cdots \\
&\text{with} \\
&C_2 = -\frac{a_2}{a_1} \qquad C_3 = -\frac{a_3}{a_1} + 2\left(\frac{a_2}{a_1}\right)^2 \\
&C_4 = -\frac{a_4}{a_1} + 5\left(\frac{a_2a_3}{a_1^2}\right) - 5\left(\frac{a_2}{a_1}\right)^3 \\
&C_5 = -\frac{a_5}{a_1} + 6\,\frac{a_2a_4}{a_1^2} + 3\left(\frac{a_3}{a_1}\right)^2 - 21\,\frac{a_2^2a_3}{a_1^3} + 14\left(\frac{a_2}{a_1}\right)^4 \\
&\cdots\cdots\cdots\cdots\cdots\cdots\cdots\cdots
\end{aligned}\right\} \tag{20.5-7}$$

20.5-5. "Optimal-interval" Interpolation (see also Sec. 20.6-3). The nth-degree polynomial $Y(x)$ which equals $y(x)$ at $n + 1$ points $x = x_j$ $(j = 0, 1, 2, \ldots, n)$ of $[a, b]$ and minimizes $\max\limits_{a\le x\le b}\left|\prod\limits_{k=0}^{n}(x - x_k)\right|$ will approximately minimize the maximum absolute value of the interpolation error (4) in $[a, b]$. The required polynomial $Y(x)$ is given by

$$\left.\begin{aligned}
&Y(x) = \frac{1}{2}A_0 + \sum_{k=1}^{n} A_kT_k\left(\frac{2x - b - a}{b - a}\right) \qquad (n = 1, 2, \ldots) \\
&\text{with} \\
&A_k = \frac{2}{n+1}\sum_{j=0}^{n} y(x_j)\cos\frac{(2j+1)k\pi}{2n+2} \qquad (k = 0, 1, 2, \ldots, n)
\end{aligned}\right\} \tag{20.5-8a}$$

where $T_k(\xi)$ is the kth-degree Chebyshev polynomial (Sec. 21.7-4), and

$$x_j = \frac{a+b}{2} + \frac{b-a}{2}\cos\frac{(2j+1)\pi}{2n+2} \qquad (j = 0, 1, 2, \ldots, n) \tag{20.5-8b}$$

20.5-6. Multiple Polynomial Interpolation. To approximate $z = z(x, y)$ by a polynomial $Z(x, y)$ such that $Z(x, y) = z(x, y)$ for a given set of "points" (x_i, y_k), one may first interpolate with respect to x to approximate $z(x, y_k)$ for a set of values of k; interpolation with respect to y will then yield $Z(x, y)$.

Alternatively, one can substitute an interpolation formula for y into a formula for interpolation with respect to x. Thus, if $\Delta x = \Delta y = h$ is a given fixed increment, and

$$\left.\begin{aligned} z(x_0 + j\,\Delta x, y_0 + k\,\Delta y) &= z_{jk} \qquad (j, k = 0, \pm 1, \pm 2, \ldots) \\ u = \frac{x - x_0}{\Delta x} &\qquad v = \frac{y - y_0}{\Delta y} \end{aligned}\right\} \tag{20.5-9}$$

Table 20.5-2. Interpolation Coefficients*

LAGRANGE FIVE-POINT INTERPOLATION

$$f_s \approx L_{-2}(s)f_{-2} + L_{-1}(s)f_{-1} + L_0(s)f_0 + L_1(s)f_1 + L_2(s)f_2$$

For negative s, use lower column labels

s	$L_{-2}(s)$	$L_{-1}(s)$	$L_0(s)$	$L_1(s)$	$L_2(s)$	
.0	.000000	.000000	1.000000	.000000	.000000	.0
.1	.007838	−.059850	.987525	.073150	−.008663	−.1
.2	.014400	−.105600	.950400	.158400	−.017600	−.2
.3	.019338	−.136850	.889525	.254150	−.026163	−.3
.4	.022400	−.153600	.806400	.358400	−.033600	−.4
.5	.023438	−.156250	.703125	.468750	−.039063	−.5
.6	.022400	−.145600	.582400	.582400	−.041600	−.6
.7	.019338	−.122850	.447525	.696150	−.040163	−.7
.8	.014400	−.089600	.302400	.806400	−.033600	−.8
.9	.007838	−.047850	.151525	.909150	−.020663	−.9
1.0	.000000	.000000	.000000	1.000000	.000000	−1.0
1.1	−.008663	.051150	−.146475	1.074150	.029838	−1.1
1.2	−.017600	.102400	−.281600	1.126400	.070400	−1.2
1.3	−.026163	.150150	−.398475	1.151150	.123338	−1.3
1.4	−.033600	.190400	−.489600	1.142400	.190400	−1.4
1.5	−.039063	.218750	−.546875	1.093750	.273438	−1.5
1.6	−.041600	.230400	−.561600	.998400	.374400	−1.6
1.7	−.040163	.220150	−.524475	.849150	.495338	−1.7
1.8	−.033600	.182400	−.425600	.638400	.638400	−1.8
1.9	−.020663	.111150	−.254475	.358150	.805838	−1.9
2.0	.000000	.000000	.000000	.000000	1.000000	−2.0
	$L_2(s)$	$L_1(s)$	$L_0(s)$	$L_{-1}(s)$	$L_{-2}(s)$	s

NOTE: All coefficients become exact if each terminal 8 is replaced by 75, and each terminal 3 by 25.

* From F. B. Hildebrand, *Introduction to Numerical Analysis*, McGraw-Hill. New York, 1956.

NEWTON INTERPOLATION

$$f_s \approx f_0 + s\,\Delta f_0 + C_2(s)\,\Delta^2 f_0 + C_3(s)\,\Delta^3 f_0 + C_4(s)\,\Delta^4 f_0 + C_5(s)\,\Delta^5 f_0$$
$$f_{n-s} \approx f_n - s\,\nabla f_n + C_2(s)\,\nabla^2 f_n - C_3(s)\,\nabla^3 f_n + C_4(s)\,\nabla^4 f_n - C_5(s)\,\nabla^5 f_n$$

(s positive for interpolation)

s	$C_2(s)$	$C_3(s)$	$C_4(s)$	$C_5(s)$
−1.0	1.00000	−1.00000	1.00000	−1.00000
−.9	.85500	−.82650	.80584	−.78972
−.8	.72000	−.67200	.63840	−.61286
−.7	.59500	−.53550	.49534	−.46562
−.6	.48000	−.41600	.37440	−.34445
−.5	.37500	−.31250	.27344	−.24609
−.4	.28000	−.22400	.19040	−.16755
−.3	.19500	−.14950	.12334	−.10607
−.2	.12000	−.08800	.07040	−.05914
−.1	.05500	−.03850	.02984	−.02447
0	.00000	.00000	.00000	.00000
.1	−.04500	.02850	−.02066	.01612
.2	−.08000	.04800	−.03360	.02554
.3	−.10500	.05950	−.04016	.02972
.4	−.12000	.06400	−.04160	.02995
.5	−.12500	.06250	−.03906	.02734
.6	−.12000	.05600	−.03360	.02285
.7	−.10500	.04550	−.02616	.01727
.8	−.08000	.03200	−.01760	.01126
.9	−.04500	.01650	−.00866	.00537
1.0	.00000	.00000	.00000	.00000

STIRLING INTERPOLATION

$$f_s \approx f_0 + s\,\mu\delta f_0 + C_2(s)\,\delta^2 f_0 + C_3(s)\,\mu\delta^3 f_0 + C_4(s)\,\delta^4 f_0$$

s	$C_2(s)$	$C_3(s)$	$C_4(s)$	s
0	.00000	.00000	.00000	0
.1	.00500	−.01650†	−.00041	−.1
.2	.02000	−.03200†	−.00160	−.2
.3	.04500	−.04550†	−.00341	−.3
.4	.08000	−.05600†	−.00560	−.4
.5	.12500	−.06250†	−.00781	−.5
.6	.18000	−.06400†	−.00960	−.6
.7	.24500	−.05950†	−.01041	−.7
.8	.32000	−.04800†	−.00960	−.8
.9	.40500	−.02850†	−.00641	−.9
1.0	.50000	.00000	.00000	−1.0

† Change sign when reading s from right-hand column.

Bessel Interpolation

$$f_s \approx \mu f_{\frac{1}{2}} + (s - \tfrac{1}{2})\, \delta f_{\frac{1}{2}} + C_2(s)\, \mu\delta^2 f_{\frac{1}{2}} + C_3(s)\, \delta^3 f_{\frac{1}{2}} + C_4(s)\, \mu\delta^4 f_{\frac{1}{2}} + C_5(s)\, \delta^5 f_{\frac{1}{2}}$$

s	$C_2(s)$	$C_3(s)$	$C_4(s)$	$C_5(s)$	s
0	.00000	.00000	.00000	.00000	1.0
.1	−.04500	.00600†	.00784	−.00063†	.9
.2	−.08000	.00800†	.01440	−.00086†	.8
.3	−.10500	.00700†	.01934	−.00077†	.7
.4	−.12000	.00400†	.02240	−.00045†	.6
.5	−.12500	.00000	.02344	.00000	.5

† Change sign when reading s from right-hand column.

Everett Interpolation

$$f_s \approx (1 - s)f_0 + C_2(s)\, \delta^2 f_0 + C_4(s)\, \delta^4 f_0 + s f_1 + C_2(1 - s)\, \delta^2 f_1 + C_4(1 - s)\, \delta^4 f_1$$

s	$C_2(s)$	$C_4(s)$
0	.00000	.00000
.1	−.02850	.00455
.2	−.04800	.00806
.3	−.05950	.01044
.4	−.06400	.01165
.5	−.06250	.01172
.6	−.05600	.01075
.7	−.04550	.00890
.8	−.03200	.00634
.9	−.01650	.00329
1.0	.00000	.00000

Steffensen Interpolation

$$f_s \approx f_0 + C_1(s)\, \delta f_{\frac{1}{2}} + C_3(s)\, \delta^3 f_{\frac{1}{2}} - C_1(-s)\, \delta f_{-\frac{1}{2}} - C_3(-s)\, \delta^3 f_{-\frac{1}{2}}$$

s	$C_1(s)$	$C_3(s)$
−.5	−.12500	.02344
−.4	−.12000	.02240
−.3	−.10500	.01934
−.2	−.08000	.01440
−.1	−.04500	.00784
0	.00000	.00000
.1	.05500	−.00866
.2	.12000	−.01760
.3	.19500	−.02616
.4	.28000	−.03360
.5	.37500	−.03906

one may use Bessel's interpolation formula (Table 20.5-1) twice to obtain

$$\begin{aligned} Z(x, y) &= \tfrac{1}{4}(z_{00} + z_{10} + z_{01} + z_{11}) + \tfrac{1}{2}(u - \tfrac{1}{2})(z_{10} - z_{00} + z_{11} - z_{01}) \\ &\quad + \tfrac{1}{2}(v - \tfrac{1}{2})(z_{01} - z_{00} + z_{11} - z_{10}) + (u - \tfrac{1}{2})(v - \tfrac{1}{2})(z_{11} - z_{10} - z_{01} + z_{00}) \\ &\quad + \cdots \end{aligned}$$

(BESSEL'S FORMULA FOR TWO-WAY INTERPOLATION) (20.5-10)

Analogous methods apply to functions of three or more variables.

20.5-7. Reciprocal Differences and Rational-fraction Interpolation. Given $y(x_k) = y_k$ ($k = 0, 1, 2, \ldots$), where $x_0, x_1, x_2, \ldots$ are not necessarily equally spaced, one defines the **reciprocal differences**

$$\left.\begin{aligned} &\rho_1(x_0, x_1) \equiv \frac{x_0 - x_1}{y_0 - y_1} \qquad \rho_2(x_0, x_1, x_2) \equiv \frac{x_0 - x_2}{\rho_1(x_0, x_1) - \rho_1(x_1, x_2)} + y_1 \\ &\rho_r(x_0, x_1, x_2, \ldots, x_r) \equiv \frac{x_0 - x_r}{\rho_{r-1}(x_0, x_1, \ldots, x_{r-1}) - \rho_{r-1}(x_1, x_2, \ldots, x_r)} \\ &\qquad + \rho_{r-2}(x_1, x_2, \ldots, x_{r-1}) \qquad (r = 3, 4, \ldots) \end{aligned}\right\} \tag{20.5-11}$$

Then

$$Y(x) = y_1 + \frac{x - x_1}{\rho_1(x, x_1)} \tag{20.5-12a}$$

where one successively substitutes

$$\left.\begin{aligned} \rho_1(x, x_1) &= \rho_1(x_1, x_2) + \frac{x - x_2}{\rho_2(x, x_1, x_2) - y_1} \\ \rho_2(x, x_1, x_2) &= \rho_2(x_1, x_2, x_3) + \frac{x - x_3}{\rho_3(x, x_1, x_2, x_3) - \rho_1(x_1, x_2)} \\ &\cdots\cdots\cdots\cdots\cdots\cdots \end{aligned}\right\} \tag{20.5-12b}$$

yields a continued-fraction expansion approximating $y(x)$ by a rational function $Y(x)$ which assumes the given function values $y_0, y_1, y_2, \ldots$ for $x = x_0, x_1, x_2, \ldots$ (*Thiele's Interpolation Formula*, see also Ref. 20.36). The continued fraction terminates whenever $y(x)$ is a rational function.

20.6. APPROXIMATION BY ORTHOGONAL POLYNOMIALS, TRUNCATED FOURIER SERIES, AND OTHER METHODS

20.6-1. Introduction. In practice, polynomial interpolation is best suited for analytic functions uncorrupted by noise or random errors, since high-order interpolation polynomials tend to follow the latter too closely, and low-order interpolation polynomials may waste essential information. In such cases, it is preferable to employ "smoothed" polynomial or rational-fraction approximations designed to minimize either a weighted mean-square approximation error or the maximum absolute approximation error over a specified interval $a < x < b$. By contrast, Taylor-series approximations, which approximate an analytic function closely in the immediate neighborhood of one given point are, in general, useful for numerical approximation only if convergence is extraordinarily rapid.

20.6-2. Least-squares Polynomial Approximation over an Interval (see also Sec. 15.2-6). It is desired to approximate the given function $f(x)$ by

$$F(x) = a_0\varphi_0(x) + a_1\varphi_1(x) + a_2\varphi_2(x) + \cdots + a_n\varphi_n(x) \qquad (20.6\text{-}1)$$

so as to minimize the weighted mean-square error

$$e^2 = \int_a^b \gamma(x)[F(x) - f(x)]^2\,dx = 0 \qquad (20.6\text{-}2)$$

over the expansion interval (a, b), where $\gamma(x)$ is a given nonnegative weighting function. If the $\varphi_k(x)$ are mutually orthogonal real functions such that

$$\int_a^b \gamma(x)\varphi_i(x)\varphi_j(x)\,dx = 0 \qquad (i \neq j) \qquad (20.6\text{-}3)$$

then the desired coefficients a_i are given by

$$a_i = \frac{\int_a^b \gamma(x)f(x)\varphi_i(x)\,dx}{\int_a^b \gamma(x)\varphi_i^2(x)\,dx} \qquad (i = 0, 1, 2, \ldots) \qquad (20.6\text{-}4)$$

Orthogonal-function approximations, exemplified by orthogonal-polynomial expansions (Secs. 20.6-2 to 20.6-4) and truncated Fourier series (Sec. 20.6-6), have the striking advantage that *an improvement of the approximation through addition of an extra term* $a_{n+1}\varphi_{n+1}(x)$ *does not affect the previously computed coefficients* $a_0, a_1, a_2, \ldots, a_n$. Substitution of $x = \alpha z + \beta$, $dx = \alpha\,dz$ in Eqs. (1) to (4) yields rescaled and/or shifted expansion intervals. Note that computation of the coefficients (4) requires one to know $f(x)$ throughout the entire expansion interval (a, b).

20.6-3. Least-squares Polynomial Approximation over a Set of Points. A different type of least-squares approximation of the form (1) requires $f(x)$ to be given only at a discrete set of $m + 1$ points, $x_0, x_1, x_2, \ldots, x_m$. One minimizes the weighted mean-square error

$$e'^2 = \sum_{k=0}^{m} \gamma_k[F(x_k) - f(x_k)]^2 \qquad (20.6\text{-}5)$$

where the γ_k are given positive weights. This is again relatively simple if the $\varphi_k(x)$ are chosen to be polynomials of degree k and are mutually orthogonal in the sense that

$$\sum_{k=0}^{m} \gamma_k\varphi_i(x_k)\varphi_j(x_k) = 0 \qquad (i \neq j) \qquad (20.6\text{-}6)$$

Such polynomials can be obtained from $1, x, x^2, \ldots$ through the Gram-Schmidt orthogonalization procedure of Sec. 14.7-4. The coefficients a_i are given by

$$a_i = \frac{\sum_{k=0}^{m} \gamma_k f(x_k)\varphi_i(x_k)}{\sum_{k=0}^{m} \gamma_k \varphi_i^2(x_k)} \qquad (i = 0, 1, 2, \ldots, n \leq m) \qquad (20.6\text{-}7)$$

The resulting polynomial will be an interpolation polynomial if $n = m$; if $n < m$, addition of extra terms $a_i\varphi_i(x)$ again leaves the earlier terms unchanged. Two special cases are of particular interest.

(a) **Optimum-interval Tabulation.** If one is free to choose the $m + 1$ argument values $x_0, x_1, x_2, \ldots, x_m$, say in the interval $(-1, 1)$, one can pick the $m + 1$ roots of the m + 1st Chebyshev polynomial $T_{m+1}(x)$ (Sec. 21.7-4), i.e.,

$$x_k = \cos \frac{2k + 1}{2m + 2}\pi \qquad (k = 0, 1, 2, \ldots, m) \qquad (20.6\text{-}8)$$

The resulting approximation polynomials defined by Eq. (6) *for unit weights* $\gamma_k = 1$ *will be the Chebyshev polynomials* $T_i(x)$ (see also Sec. 20.5-5).

(b) **Evenly Spaced Data** (see also Ref. 20.6). If one has $m + 1 = 2M + 1$ arguments x_k evenly spaced in an expansion interval (a, b), so that

$$x_k = \frac{a + b}{2} + k\,\Delta x \qquad \left(\Delta x = \frac{b - a}{2M};\ k = 0, \pm 1, \pm 2, \ldots, \pm M\right) \qquad (20.6\text{-}9)$$

then the polynomials $\varphi_i(x)$ satisfying Eq. (7) for unit weights $\gamma_k = 1$ are given by

$$\varphi_i(x) \equiv p_i\left(\frac{2x - a - b}{b - a}M, 2M\right) \qquad (i = 0, 1, 2, \ldots, 2M;\ M = 1, 2, \ldots) \qquad (20.6\text{-}10)$$

where the $p_i(t, 2M)$ are the **Gram polynomials***

$$p_i(t, 2M) \equiv \sum_{k=0}^{\infty} (-1)^{i+k} \frac{(i + k)^{[2k]}}{(k!)^2} \frac{(M + t)^{[k]}}{(2M)^{[k]}} \qquad (i = 0, 1, 2, \ldots, 2M;\ M = 1, 2, \ldots) \qquad (20.6\text{-}11a)$$

* The normalization (12) chosen is that of Ref. 20.6. Other normalization factors are used by different authors.

with

$$\left.\begin{array}{lll} z^{[k]} \equiv z(z-1)(z-2)\cdots(z-k+1) & & (k=1, 2, \ldots) \\ z^{[0]} \equiv 1 \quad (z \geq 0) & 0^{[k]} \equiv 0 & (k=1, 2, \ldots) \end{array}\right\} \tag{20.6-11b}$$

and $$\sum_{k=-M}^{M} p_i^2(k, 2M) = \frac{(2M+i+1)!(2M-i)!}{(2i+1)[(2M)!]^2}$$

$$(i = 0, 1, 2, \ldots, 2M; M = 1, 2, \ldots) \tag{20.6-12}$$

The first five Gram polynomials are

$$\left.\begin{aligned} p_0(t, 2M) &= 1 \\ p_1(t, 2M) &= \frac{t}{M} \\ p_2(t, 2M) &= \frac{3t^2 - M(M+1)}{M(2M-1)} \\ p_3(t, 2M) &= \frac{5t^3 - (3M^2 + 3M - 1)t}{M(M-1)(2M-1)} \\ p_4(t, 2M) &= \frac{35t^4 - 5(6M^2 + 6M - 5)t^2 + 3M(M^2-1)(M+2)}{2M(M-1)(2M-1)(2M-3)} \\ p_5(t, 2M) &= \\ &\frac{63t^5 - 35(2M^2 + 2M - 3)t^3 + (15M^4 + 30M^3 - 35M^2 - 50M + 12)t}{2M(M-1)(M-2)(2M-1)(2M-3)} \end{aligned}\right\} \tag{20.6-13}$$

In particular, for $M = 2$ (five data points),

$$\left.\begin{array}{lll} p_0(t) = 1 \qquad p_1(t) = \frac{1}{2}t & & p_2(t) = \frac{1}{2}(t^2 - 2) \\ p_3(t) = \frac{1}{6}(5t^3 - 17t) & & p_4(t) = \frac{1}{12}(35t^4 - 155t^2 + 72) \end{array}\right\} \tag{20.6-14}$$

If we denote the given function values $f(x_k)$ by f_k, Gram-polynomial approximation with $M = 2$ yields the following smoothed approximation-polynomial values, where $F_i \equiv F(x_i) \equiv G_i(f_0, f_1, f_{-1}, \ldots) \equiv G_i(f_0, f_{-1}, f_1, \ldots)$

$$\left.\begin{aligned} F_{-2} &= \tfrac{1}{70}(69f_{-2} + 4f_{-1} - 6f_0 + 4f_1 - f_2) \equiv f_{-2} - \tfrac{1}{70}\delta^4 f_0 \\ F_{-1} &= \tfrac{1}{35}(2f_{-2} + 27f_{-1} + 12f_0 - 8f_1 + 2f_2) \equiv f_{-1} + \tfrac{2}{35}\delta^4 f_0 \\ F_0 &= \tfrac{1}{35}(-3f_{-2} + 12f_{-1} + 17f_0 + 12f_1 - 3f_2) \equiv f_0 - \tfrac{3}{35}\delta^4 f_0 \\ &\cdots\cdots\cdots\cdots\cdots\cdots\cdots\cdots \end{aligned}\right\} \tag{20.6-15}$$

("smoothing by fourth differences"; $\delta^k f_0$ is defined in Sec. 20.4-1).

20.6-4. Least-maximum-absolute-error Approximations. (a) The computation of the coefficients a_i for an approximation polynomial (1) which minimizes the *maximum absolute error* $|F(x) - f(x)|$ on (a, b) for

$f(x)$ given either on the entire interval or on a discrete set of points requires a fairly laborious iterative procedure. Chebyshev's classical method is described in Ref. 20.61, while Hastings (Ref. 20.41) has evolved a heuristic method relying heavily on the investigator's intuition in relocating the zero-error points in successive plots of the error $F(x) - f(x)$ by iterative parameter changes. Hasting's method is also applicable to more general nonpolynomial approximations of the form $F(x) = F(x; \alpha_1, \alpha_2, \ldots, \alpha_n)$, where $\alpha_1, \alpha_2, \ldots, \alpha_n$ are suitably adjusted parameters.

In practice, approximations of the form (1) in terms of Chebyshev polynomials (shifted and rescaled as needed, see also Sec. 20.6-4*b*), although really derived from least-square considerations, are so close to the ideal least-absolute-error polynomial approximation that most texts on numerical analysis omit the laborious derivation of the latter. This is easily seen whenever it is reasonable to assume that the error of a Chebyshev-polynomial approximation

$$F(x) \equiv \sum_{k=0}^{n} a_k T_k(\alpha x) \qquad [T_k(x) = \cos k\vartheta, \cos \vartheta = x]$$

is essentially equal to the omitted term $a_{n+1}T_{n+1}(\alpha x)$ for x in (a, b). $T_{n+1}(\alpha x)$ oscillates with amplitude 1 in the expansion interval. It follows that the maximum excursions of the absolute error in the expansion interval are all approximately equal to $|a_{n+1}|$.

(**b**) It is often convenient to employ the **shifted Chebyshev polynomials** $T_n^*(x)$ defined by

$$T_n^*(x) \equiv T_n(2x - 1) \equiv \cos n\vartheta \qquad (\cos \vartheta \equiv 2x - 1, n = 0, 1, 2, \ldots) \tag{20.6-16}$$

or

$$T_0^* = 1 \qquad T_{n+1}^*(x) = 2(2x - 1)T_n^*(x) - T_{n-1}^*(x) \qquad (n = 0, 1, 2, \ldots) \tag{20.6-17}$$

(Refs. 20.4 and 20.12).

(**c**) Table 20.6-1 lists the first few polynomials $T_n(x)$ and $T_n^*(x)$ together with expressions for $1, x, x^2, \ldots$ in terms of these polynomials (see also Sec. 21.7-4). Tables 20.6-2 to 20.6-4 list some useful polynomial approximations.

20.6-5. Economization of Power Series (Refs. 20.4 and 20.12). If computation of orthogonal-function-expansion coefficients is laborious, but a truncated-power-series approximation to $f(x)$ is readily available, we can reduce its degree with little loss of accuracy by expressing higher powers of x in terms of lower powers and a Chebyshev polynomial. For

Table 20.6-1. Chebyshev Polynomials $T_n(x)$ and $T_n^*(x)$, and Powers of x

$T_0 = 1$	$1 = T_0$
$T_1 = x$	$x = T_1$
$T_2 = 2x^2 - 1$	$x^2 = \frac{1}{2}(T_0 + T_2)$
$T_3 = 4x^3 - 3x$	$x^3 = \frac{1}{4}(3T_1 + T_3)$
$T_4 = 8x^4 - 8x^2 + 1$	$x^4 = \frac{1}{8}(3T_0 + 4T_2 + T_4)$
$T_5 = 16x^5 - 20x^3 + 5x$	$x^5 = \frac{1}{16}(10T_1 + 5T_3 + T_5)$
$T_6 = 32x^6 - 48x^4 + 18x^2 - 1$	$x^6 = \frac{1}{32}(10T_0 + 15T_2 + 6T_4 + T_6)$
$T_7 = 64x^7 - 112x^5 + 56x^3 - 7x$	$x^7 = \frac{1}{64}(35T_1 + 21T_3 + 7T_5 + T_7)$
$T_8 = 128x^8 - 256x^6 + 160x^4 - 32x^2 + 1$	$x^8 = \frac{1}{128}(35T_0 + 56T_2 + 28T_4 + 8T_6 + T_8)$
$T_9 = 256x^9 - 576x^7 + 432x^5 - 120x^3 + 9x$	$x^9 = \frac{1}{256}(126T_1 + 84T_3 + 36T_5 + 9T_7 + T_9)$
$T_0^* = 1$	$1 = T_0^*$
$T_1^* = 2x - 1$	$x = \frac{1}{2}(T_0^* + T_1^*)$
$T_2^* = 8x^2 - 8x + 1$	$x^2 = \frac{1}{8}(3T_0^* + 4T_1^* + T_2^*)$
$T_3^* = 32x^3 - 48x^2 + 18x - 1$	$x^3 = \frac{1}{32}(10T_0^* + 15T_1^* + 6T_2^* + T_3^*)$
$T_4^* = 128x^4 - 256x^3 + 160x^2 - 32x + 1$	$x^4 = \frac{1}{128}(35T_0^* + 56T_1^* + 28T_2^* + 8T_3^* + T_4^*)$

example, the left-hand side of Table 20.6-1 yields

$$x^9 = \frac{1}{256}(-9x + 120x^3 - 432x^5 + 576x^7) + \frac{1}{256}T_9(x)$$

The last term fluctuates with the relatively small amplitude $\frac{1}{256}$ in $(-1, 1)$. We omit the last term and substitute the rest for x^9, leaving a seventh-degree polynomial; the procedure may then be repeated for x^7, etc.

Instead, we could express all powers in terms of Chebyshev polynomials and neglect the higher-degree polynomials, which will tend to have small coefficients. The resulting expansions on (0, 1) or $(-1, 1)$ are, in general, *not* identical with the true least-squares polynomial expansion of $f(x)$, but the method is convenient.

20.6-6. Numerical Harmonic Analysis and Trigonometric Interpolation (see also Secs. 4.11-4*b* and 20.6-1). (**a**) Given m function values $y(x_k) = y_k$ for $x_k = kT/m$ $(k = 0, 1, 2, \ldots, m-1)$, it is desired to approximate $y(x)$ within the interval $(0, T)$ by a trigonometric polynomial

$$Y(x) = \frac{1}{2}A_0 + \sum_{j=1}^{n}\left(A_j \cos j\frac{2\pi x}{T} + B_j \sin j\frac{2\pi x}{T}\right) \qquad \left(n < \frac{m}{2}\right) \tag{20.6-18}$$

so as to minimize the mean-square error $\sum_{k=0}^{m-1} [Y(x_k) - y_k]^2$. The required

Table 20.6-2. Some Polynomial Approximations

$f(x)$	Approximation, $F(x)$	Parameters	Maximum absolute error	Reference
e^{-x}	$1 + a_1x + a_2x^2$	$a_1 = -.9664 \quad a_2 = .3536$	3×10^{-3}	Carlson and Goldstein (Ref. 20.38)
$(0 \leq x \leq \log_e 2 = .693 \ldots)$	$1 + a_1x + a_2x^2 + a_3x^3 + a_4x^4$	$a_1 = -.99986\ 84$, $a_2 = .49829\ 26$, $a_3 = -.15953\ 32$, $a_4 = .02936\ 41$	3×10^{-5}	
	$1 + a_1x + a_2x^2 + a_3x^3 + a_4x^4 + a_5x^5 + a_6x^6 + a_7x^7$	$a_1 = -.99999\ 99995$, $a_2 = .49999\ 99206$, $a_3 = -.16666\ 53019$, $a_4 = .04165\ 73475$, $a_5 = -.00830\ 13598$, $a_6 = .00132\ 98820$, $a_7 = -.00014\ 13161$	2×10^{-10}	
$\log_e(1 + x)$ $(0 \leq x \leq 1)$	$a_1x + a_2x^2 + a_3x^3 + a_4x^4 + a_5x^5$	$a_1 = .99949\ 556$, $a_2 = -.49190\ 896$, $a_3 = .28947\ 478$, $a_4 = -.13606\ 275$, $a_5 = .03215\ 845$	10^{-5}	Hastings (Ref. 20.41)
	$a_1x + a_2x^2 + a_3x^3 + a_4x^4 + a_5x^5 + a_6x^6 + a_7x^7 + a_8x^8$	$a_1 = .99999\ 64239$, $a_2 = -.49987\ 41238$, $a_3 = .33179\ 90258$, $a_4 = -.24073\ 38084$, $a_5 = .16765\ 40711$, $a_6 = -.09532\ 93897$, $a_7 = .03608\ 84937$, $a_8 = -.00645\ 35442$	3×10^{-8}	
10^x $(0 \leq x \leq 1)$	$(1 + a_1x + a_2x^2 + a_3x^3 + a_4x^4)^2$	$a_1 = 1.14991\ 96$, $a_2 = .67743\ 23$, $a_3 = .20800\ 30$, $a_4 = .12680\ 89$	7×10^{-4}	
	$(1 + a_1x + a_2x^2 + a_3x^3 + a_4x^4 + a_5x^5 + a_6x^6 + a_7x^7)^2$	$a_1 = 1.15129\ 277603$, $a_2 = .66273\ 088429$, $a_3 = .25439\ 357484$, $a_4 = .07295\ 173666$, $a_5 = .01742\ 111988$, $a_6 = .00255\ 491796$, $a_7 = .00093\ 264267$	5×10^{-8}	

Table 20.6-2. Some Polynomial Approximations (*Continued*)

$f(x)$	Approximation, $F(x)$	Parameters	Maximum absolute error	Reference
$\frac{\sin x}{x}$ $\left(0 \le x \le \frac{\pi}{2}\right)$	$1 + a_2x^2 + a_4x^4$	$a_2 = -.16605 \quad a_4 = .00761$	2×10^{-4}	Carlson and Goldstein (Ref. 20.38)
	$1 + a_2x^2 + a_4x^4 + a_6x^6 + a_8x^8 + a_{10}x^{10}$	$a_2 = -.16666\ 66664$, $a_4 = .00833\ 33315$, $a_6 = -.00019\ 84090$, $a_8 = .00000\ 27526$, $a_{10} = -.00000\ 00239$	2×10^{-9}	
$\cos x$ $\left(0 \le x \le \frac{\pi}{2}\right)$	$1 + a_2x^2 + a_4x^4$	$a_2 = -.49670 \quad a_4 = .03705$	9×10^{-4}	
	$1 + a_2x^2 + a_4x^4 + a_6x^6 + a_8x^8 + a_{10}x^{10}$	$a_2 = -.49999\ 99963$, $a_4 = .04166\ 66418$, $a_6 = -.00138\ 88397$, $a_8 = .00002\ 47609$, $a_{10} = -.00000\ 02605$	2×10^{-9}	
$\frac{\tan x}{x}$ $\left(0 \le x \le \frac{\pi}{4}\right)$	$1 + a_2x^2 + a_4x^4$	$a_2 = .31755 \quad a_4 = .20330$	10^{-3}	
	$1 + a_2x^2 + a_4x^4 + a_6x^6 + a_8x^8 + a_{10}x^{10} + a_{12}x^{12}$	$a_2 = .33333\ 14036$, $a_4 = .13339\ 23995$, $a_6 = .05337\ 40603$, $a_8 = .02456\ 50893$, $a_{10} = .00290\ 05250$, $a_{12} = .00951\ 68091$	2×10^{-8}	
$x \cot x$ $\left(0 \le x \le \frac{\pi}{4}\right)$	$1 + a_2x^2 + a_4x^4$	$a_2 = -.332867 \quad a_4 = -.024369$	3×10^{-5}	Carlson and Goldstein (Ref. 20.38)
	$1 + a_2x^2 + a_4x^4 + a_6x^6 + a_8x^8 + a_{10}x^{10}$	$a_2 = -.33333\ 33410$, $a_4 = -.02222\ 20287$, $a_6 = -.00211\ 77168$, $a_8 = -.00020\ 78504$, $a_{10} = -.00002\ 62619$	4×10^{-10}	

Function	Approximation	Coefficients	Error	Source
$\arcsin x$ $(0 \leq x \leq 1)$	$\frac{\pi}{2} - (1-x)^{1/2}(a_0 + a_1x + a_2x^2 + a_3x^3)$	$a_0 = 1.57072\ 88$, $a_1 = -.21211\ 44$, $a_2 = .07426\ 10$, $a_3 = -.01872\ 93$	5×10^{-5}	Hastings (Ref. 20.41)
	$\frac{\pi}{2} - (1-x)^{1/2}(a_0 + a_1x + a_2x^2 + a_3x^3 + a_4x^4 + a_5x^5 + a_6x^6 + a_7x^7)$	$a_0 = 1.57079\ 63050$, $a_1 = -.21459\ 88016$, $a_2 = .08897\ 89874$, $a_3 = -.05017\ 43046$, $a_4 = .03089\ 18810$, $a_5 = -.01708\ 81256$, $a_6 = .00667\ 00901$, $a_7 = -.00126\ 24911$	2×10^{-8}	
$\arctan x$ $(-1 \leq x \leq 1)$	$a_1x + a_3x^3 + a_5x^5 + a_7x^7 + a_9x^9$	$a_1 = .99986\ 60$, $a_3 = -.33029\ 95$, $a_5 = .18014\ 10$, $a_7 = -.08513\ 30$, $a_9 = .02083\ 51$	10^{-5}	Hastings (Ref. 20.41)
$\frac{\arctan x}{x}$ $(-1 \leq x \leq 1)$	$1 + a_2x^2 + \cdots + a_8x^8$	$a_2 = -.33333\ 14528$, $a_4 = .19993\ 55085$, $a_6 = -.14208\ 89944$, $a_8 = .10656\ 26393$, $a_{10} = -.07528\ 96400$, $a_{12} = .04290\ 96138$, $a_{14} = -.01616\ 57367$, $a_{16} = .00286\ 62257$	2×10^{-8}	Carlson and Goldstein (Ref. 20.38)
$\Gamma(x+1) = x!$ $(0 \leq x \leq 1)$	$1 + a_1x + a_2x^2 + a_3x^3 + a_4x^4 + a_5x^5$	$a_1 = -.57486\ 46$, $a_2 = .95123\ 63$, $a_3 = -.69985\ 88$, $a_4 = .42455\ 49$, $a_5 = -.10106\ 78$	5×10^{-5}	Hastings (Ref. 20.41)
	$1 + a_1x + a_2x^2 + \cdots + a_8x^8$	$a_1 = -.57719\ 1652$, $a_2 = .98820\ 5891$, $a_3 = -.89705\ 6937$, $a_4 = .91820\ 6857$, $a_5 = -.75670\ 4078$, $a_6 = .48219\ 9394$, $a_7 = -.19352\ 7818$, $a_8 = .03586\ 8343$	3×10^{-7}	

Table 20.6-3. Some Approximations for Cylinder Functions*

$$J_0(x) = 1 - 2.24999\ 97(x/3)^2 + 1.26562\ 08(x/3)^4 - .31638\ 66(x/3)^6 + .04444\ 79(x/3)^8 - .00394\ 44(x/3)^{10} + .00021\ 00(x/3)^{12} + \epsilon$$

$$|\epsilon| < 5 \times 10^{-8} \qquad (-3 \le x \le 3)$$

$$N_0(x) = (2/\pi)\log_e(\tfrac{1}{2}x)J_0(x) + .36746\ 691 + .60559\ 366(x/3)^2 - .74350\ 384(x/3)^4 + .25300\ 117(x/3)^6 - .04261\ 214(x/3)^8 + .00427\ 916(x/3)^{10} - .00024\ 846(x/3)^{12} + \epsilon$$

$$|\epsilon| < 1.4 \times 10^{-8} \qquad (0 < x \le 3)$$

$$J_0(x) = x^{-1/2}f_0 \cos\theta_0 \qquad Y_0(x) = x^{-1/2}f_0 \sin\theta_0$$

$$f_0 = .79788\ 456 - .00000\ 077(3/x) - .00552\ 740(3/x)^2 - .00009\ 512(3/x)^3 + .00137\ 237(3/x)^4 - .00072\ 805(3/x)^5 + .00014\ 476(3/x)^6 + \epsilon$$

$$|\epsilon| < 1.6 \times 10^{-8}$$

$$\theta_0 = x - .78539\ 816 - .04166\ 397(3/x) - .00003\ 954(3/x)^2 + .00262\ 573(3/x)^3 - .00054\ 125(3/x)^4 - .00029\ 333(3/x)^5 + .00013\ 558(3/x)^6 + \epsilon'$$

$$|\epsilon'| < 7 \times 10^{-8} \qquad (3 \le x < \infty)$$

$$J_1(x) = x^{-1/2}f_1 \cos\theta_1, \qquad N_1(x) = x^{-1/2}f_1 \sin\theta_1$$

$$f_1 = .79788\ 456 + .00000\ 156(3/x) + .01659\ 667(3/x)^2 + .00017\ 105(3/x)^3 - .00249\ 511(3/x)^4 + .00113\ 653(3/x)^5 - .00020\ 033(3/x)^6 + \epsilon$$

$$|\epsilon| < 4 \times 10^{-8}$$

$$\theta_1 = x - 2.35619\ 449 + .12499\ 612(3/x) + .00005\ 650(3/x)^2 - .00637\ 879(3/x)^3 + .00074\ 348(3/x)^4 + .00079\ 824(3/x)^5 - .00029\ 166(3/x)^6 + \epsilon'$$

$$|\epsilon'| < 9 \times 10^{-8} \qquad (3 \le x < \infty)$$

$$x^{-1}J_1(x) = \tfrac{1}{2} - .56249\ 985(x/3)^2 + .21093\ 573(x/3)^4 - .03954\ 289(x/3)^6 + .00443\ 319(x/3)^8 - .00031\ 761(x/3)^{10} + .00001\ 109(x/3)^{12} + \epsilon$$

$$|\epsilon| < 1.3 \times 10^{-8} \qquad (-3 \le x \le 3)$$

$$xN_1(x) = (2/\pi)x\log_e(\tfrac{1}{2}x)J_1(x) - .63661\ 98 + .22120\ 91(x/3)^2 + 2.16827\ 09(x/3)^4 - 1.31648\ 27(x/3)^6 + .31239\ 51(x/3)^8 - .04009\ 76(x/3)^{10} + .00278\ 73(x/3)^{12} + \epsilon$$

$$|\epsilon| < 1.1 \times 10^{-7} \qquad (0 < x \le 3)$$

* From E. E. Allen, Analytical Approximations, *Math. Tables and Other Aids to Computation*, **8:** 240–241 (1954); Polynomial Approximations to Some Modified Bessel Functions, *Math. Tables and Other Aids to Computation*, **10:** 162–164 (1956) (with permission).
Formula arrangement and error bounds from Ref. 20.1.

Table 20.6-4. Chebyshev-polynomial Approximations*

$[T_n^*(x) \equiv \cos n\vartheta \text{ with } \cos \vartheta \equiv 2x - 1]$

$f(x)$	Approximation series	n	A_n	n	A_n
e^{-x} $(0 \le x \le 1)$	$\sum_{n=0}^{\infty} A_n T_n^*(x)$	0	.64503 5270	4	.00019 9919
		1	−.31284 1606	5	−.00000 9975
		2	.03870 4116	6	.00000 0415
		3	−.00320 8683	7	−.00000 0015
e^{x} $(0 \le x \le 1)$	$\sum_{n=0}^{\infty} A_n T_n^*(x)$	0	1.75338 7654	4	.00054 3437
		1	.85039 1654	5	.00002 7115
		2	.10520 8694	6	.00000 1128
		3	.00872 2105	7	.00000 0040
				8	.00000 0001
$\log_e (1 + x)$ $(0 \le x \le 1)$	$\sum_{n=0}^{\infty} A_n T_n^*(x)$	0	.37645 2813	6	−.00000 8503
		1	.34314 5750	7	.00000 1250
		2	−.02943 7252	8	−.00000 0188
		3	.00336 7089	9	.00000 0029
		4	−.00043 3276	10	−.00000 0004
		5	.00005 9471	11	.00000 0001
$\cos \frac{1}{2}\pi x$ $(-1 \le x \le 1)$	$\sum_{n=0}^{\infty} A_n T_n^*(x^2)$	0	.47200 1216	3	−.00059 6695
		1	−.49940 3258	4	.00000 6704
		2	.02799 2080	5	−.00000 0047
$\sin \frac{1}{2}\pi x$ $(1 \le x \le 1)$	$x \sum_{n=0}^{\infty} A_n T_n^*(x^2)$	0	1.27627 8962	3	−.00013 6587
		1	−.28526 1569	4	.00000 1185
		2	.00911 8016	5	−.00000 0007
$\arctan x$ $(-1 \le x \le 1)$	$x \sum_{n=0}^{\infty} A_n T_n^*(x^2)$	0	.88137 3587	6	.00000 3821
		1	−.10589 2925	7	−.00000 0570
		2	.01113 5843	8	.00000 0086
		3	−.00138 1195	9	− 00000 0013
		4	.00018 5743	10	.00000 0002
		5	−.00002 6215		
		For $\lvert x \rvert > 1$, use $\arctan x = \frac{1}{2}\pi - \arctan (1/x)$			
$\arcsin x$ $-\frac{1}{2}\sqrt{2} \le x \le \frac{1}{2}\sqrt{2}$	$x \sum_{n=0}^{\infty} A_n T_n^*(2x^2)$	0	1.05123 1959	5	.00000 5881
		1	.05494 6487	6	.00000 0777
		2	.00408 0631	7	.00000 0107
		3	.00040 7890	8	.00000 0015
		4	.00004 6985	9	.00000 0002
		Note: $\arccos x = \pi/2 - \arcsin x$ For $\frac{1}{2}\sqrt{2} \le x \le 1$, use $\arcsin x = \arccos (1 - x^2)^{1/2}$, $\arccos x = \arcsin (1 - x^2)^{1/2}$			

* Numerical data from C. W. Clenshaw, Polynomial Approximations to Elementary Functions, *Math. Tables and Other Aids to Computation,* **8**:143(1954) (with permission). For more extensive tables, see C. W. Clenshaw, Chebyshev Series for Mathematical Functions, in *Natl. Phys. Lab. Math. Tables,* vol. 5, London, 1962.

coefficients A_j, B_j are

$$A_j = \frac{2}{m} \sum_{k=0}^{m-1} y_k \cos j \frac{2\pi k}{m} \qquad B_j = \frac{2}{m} \sum_{k=0}^{m-1} y_k \sin j \frac{2\pi k}{m} \qquad \left(0 \leq j < \frac{m}{2}\right) \tag{20.6-19}$$

In the special case $n = m/2$, Eqs. (18) and (19) together with

$$A_n = A_{m/2} = \frac{1}{m} \sum_{k=0}^{m-1} (-1)^k y_k \tag{20.6-20}$$

yield $Y(x_k) = y(x_k)$ (*trigonometric interpolation*) for arbitrary B_n.

See Ref. 20.62 for similar formulas applicable when the x_k are not equidistant. Numerical methods for *multidimensional* Fourier analyses and syntheses are discussed in Ref. 20.2. Reference 20.70 discusses "fast" Fourier-analysis routines.

(b) The 12-ordinate Scheme. The calculation of the sums (20*b*) is simplified whenever m is divisible by 4. Table 20.6-5 shows a convenient computation scheme for $m = 12$.

If no harmonics higher than the third are required, note the simpler formulas

$$\begin{aligned} 6A_0 &= y_0 + y_1 + y_2 + \cdots + y_{11} & 6A_3 &= y_0 - y_2 + y_4 - y_6 + y_8 - y_{10} \\ 4A_2 &= y_0 - y_3 + y_6 - y_9 & 6B_3 &= y_1 - y_3 + y_5 - y_7 + y_9 - y_{11} \\ A_1 &= \tfrac{1}{2}(y_0 - y_6) + A_3 & B_1 &= \tfrac{1}{2}(y_3 - y_9) + B_3 \end{aligned} \tag{20.6-21}$$

and

$$4B_2 = y(T/8) - y(3T/8) + y(5T/8) - y(7T/8) \tag{20.6-22}$$

The four additional function values required for Eq. (22) can often be read directly from a graph of $y(x)$ vs. x.

See Ref. 20.62 for a 24-*ordinate scheme.*

(c) Determination of Unknown Periodic Components (Prony's Method). Given N values $f_0 = f(0), f_1 = f(1), f_2 = f(2), \ldots, f_{N-1} = f(N-1)$ of an empirical function $f(u)$ assumed to have the form

$$f = A_1 \cos \omega_1 u + B_1 \sin \omega_1 u + A_2 \cos \omega_2 u + B_2 \sin \omega_2 u + \cdots + B_m \sin \omega_m u \qquad (3m + 1 \leq N)$$

then $\cos \omega_1, \cos \omega_2, \ldots, \cos \omega_m$ are the m roots of the (algebraic) equation

$$\cos m\omega - \alpha_1 \cos (m-1)\omega - \cdots - \alpha_{m-1} \cos \omega - \tfrac{1}{2}\alpha_m = 0$$

whose coefficients α_k must satisfy the linear equations

$$\epsilon_i \equiv \sum_{k=1}^{m-1} (f_{i+k-1} + f_{2m+i-k-1})\alpha_k + f_{m+i-1}\alpha_m - f_{i-1} - f_{2m+i-1} = 0 \qquad (i = 1, 2, \ldots, N-2m)$$

For the best least-squares fit, solve the m linear equations

$$\frac{\partial}{\partial \alpha_k} \sum_{i=1}^{N-2m} \epsilon_i^2 = 0 \qquad (k = 1, 2, \ldots, m)$$

Table 20.6-5. 12-ordinate Scheme for Harmonic Analysis

$y_k = y\left(\frac{kT}{12}\right)$; refer to Sec. 20.6-6

Line	How obtained														
1	Given function values	y_0	y_1	y_2	y_3	y_4	y_5	y_6							
2		...	y_{11}	y_{10}	y_9	y_8	y_7	...							
3	Sum of 1 and 2	s_0	s_1	s_2	s_3	s_4	s_5	s_6							
4	Difference of 1 and 2	...	d_1	d_2	d_3	d_4	d_5	...							
5	Copy	s_0	s_1	s_2	s_3	d_1	d_2	d_3							
6		s_6	s_5	s_4	...	d_5	d_4	...							
7	Sum of 5 and 6	s_0'	s_1'	s_2'	s_3'	s_1''	s_2''	s_3''							
8	Difference of 5 and 6	d_0'	d_1'	d_2'	...	d_1''	d_2''	...							
9	Copy and multiply	s_0'	s_1'	d_0'	...	s_0'	$-s_3'$	d_0'	d_2'	s_3''	...	...	...	s_1''	s_3''
10		s_2'	s_3'	...	...	...	...	...	...	...	...	...	...	...	...
11		...	...	$\frac{1}{2}d_2'$	...	$-\frac{1}{2}s_2'$	$\frac{1}{2}s_1'$	...	...	$\frac{1}{2}s_1''$	...	...	...	...	...
12		...	...	...	$\frac{\sqrt{3}}{2}d_1'$	...	...	...	...	...	$\frac{\sqrt{3}}{2}s_2''$	$\frac{\sqrt{3}}{2}d_1''$	$\frac{\sqrt{3}}{2}d_2''$	...	...
13	Add each column (9 to 12)	S_1	T_1	S_2	T_2	S_3	T_3	S_4	T_4	S_5	T_5	S_6	T_6	S_7	T_7
14	$S_i + T_i$	$6A_0$		$6A_1$		$6A_2$		...		$6B_1$		$6B_2$		...	
15	$S_i - T_i$	$12A_6$		$6A_5$		$6A_4$		$6A_3$		$6B_5$		$6B_4$		$6B_3$	

for the m coefficients α_k. Once the ω_k are known, it is relatively easy to find the A_k, B_k for the best least-squares fit in the manner indicated in Sec. 20.6-6*a*.

The frequencies of sinusoidal components in statistical time series can often be identified by inspection of the empirically determined *autocorrelation function* (Secs. 18.10-9 and 19.8-3*c*).

20.6-7. Miscellaneous Approximations. (**a**) More general approximation methods are not restricted to sums of the form (1) but employ a rational or otherwise readily computable approximation function $F(x) \equiv F(x;\ \alpha_1, \alpha_2, \ldots, \alpha_n)$ with parameters $\alpha_1, \alpha_2, \ldots, \alpha_n$ determined so as to minimize a weighted mean-square error or the maximum absolute approximation error.

(**b**) The *Padé method* (Refs. 20.15 and 20.16) produces rational-fraction approximations (**Padé-table entries**)

$$R_{mn}(x) \equiv \frac{a_0 + a_1x + \cdots + a_mx^m}{1 + b_1x + \cdots + b_nx^n} \qquad (m, n = 0, 1, 2, \ldots) \qquad (20.6\text{-}23)$$

to a given suitably differentiable function $f(x)$ by matching the $m + n + 1$ polynomial coefficients of the identity

$$\left[f(0) + f'(0)x + \frac{1}{2!}f''(0)x^2 + \cdots + \frac{1}{m!}f^{(m)}(0)x^m\right](1 + b_1x + \cdots + b_nx^n)$$
$$\equiv a_0 + a_1x + \cdots + a_mx^m \qquad (20.6\text{-}24)$$

The resulting equations determine the $m + n + 1$ coefficients a_i and b_k, if a solution exists. It is understood that the numerator and denominator in Eq. (23) have no common factor. The $R_{m0}(x)$ are simply truncated MacLaurin series (Sec. 4.10-4).

EXAMPLE: For $f(x) \equiv e^x$, $R_{nm}(x) \equiv 1/R_{mn}(-x)$, and

$$R_{40}(x) \equiv 1 + x + \tfrac{1}{2}x^2 + \tfrac{1}{6}x^3 + \tfrac{1}{24}x^4$$
$$R_{31}(x) \equiv \frac{24 + 18x + 6x^2 + x^3}{24 - 6x}$$
$$R_{22}(x) \equiv \frac{12 + 6x + x^2}{12 - 6x + x^2}$$

(**c**) Padé approximations, like truncated Taylor or MacLaurin series, suffer from errors increasing with distance away from a specific expansion point. By contrast, *Maehly's method* (Refs. 20.15, 20.42, and 20.43) derives rational-fraction approximations as ratios of two truncated Chebyshev series. As another alternative, Ref. 20.41 describes a quite generally applicable heuristic iteration method for finding minimum-absolute-error approximations of the general form $F(x;\ \alpha_1, \alpha_2, \ldots, \alpha_n)$.

Further discussions of the use of approximation methods in connection with digital computation will be found in Refs. 20.38 to 20.46. The optimal form of the approximation chosen depends not only on the function to be approximated but also on the hardware and software configuration of the computer used. A number of examples of various types are presented in Table 20.6-6. In general, approximations computed for successive expansion intervals are "pieced together," and it is again possible to trade expansion-interval size for arithmetic complication. For frequently used, well-known functions, an approximation routine is, in general, substantially faster as well as more economical of computer storage than interpolation from a stored table. For digital computers incorporating fast division hardware, rational-function approximations are, in general, superior to polynomial approximations.

Table 20.6-6. Miscellaneous Approximations*

$f(x)$	Approximation, $F(x)$	Parameters	Maximum absolute error
$\log_{10} x$ $\left(\frac{1}{\sqrt{10}} \le x \le \sqrt{10}\right)$	$a_1 t + a_3 t^3 \qquad t = (x-1)/(x+1)$	$a_1 = .86304 \quad a_3 = .36415$	6×10^{-4}
	$a_1 t + a_3 t^3 + a_5 t^5 + a_7 t^7 + a_9 t^9 \qquad t = (x-1)/(x+1)$	$a_1 = .86859\ 1718 \quad a_7 = .09437\ 6476$ $a_3 = .28933\ 5524 \quad a_9 = .19133\ 7714$ $a_5 = .17752\ 2071$	10^{-7}
$\arctan x$ $(-1 \le x \le 1)$	$\frac{x}{1 + .28x^2}$	—	5×10^{-3}
$\operatorname{erf} x \equiv \frac{2}{\sqrt{\pi}} \int_0^x e^{-\lambda^2}\, d\lambda$ $(0 \le x < \infty)$	$1 - (a_1 t + a_2 t^2 + a_3 t^3) e^{-x^2} \qquad t = \frac{1}{1 + px}$	$p = .47047 \quad a_1 = .34802\ 42$ $a_2 = -.09587\ 98 \quad a_3 = .74785\ 56$	2.5×10^{-5}
	$1 - (a_1 t + a_2 t^2 + a_3 t^3 + a_4 t^4 + a_5 t^5) e^{-x^2}$ $t = \frac{1}{1 + px}$	$p = .32759\ 11 \quad a_1 = .25482\ 9592$ $a_2 = -.28449\ 6736 \quad a_3 = 1.42141\ 3741$ $a_4 = -1.45315\ 2027 \quad a_5 = 1.06140\ 5429$	1.5×10^{-7}
	$1 - \frac{1}{[1 + a_1 x + a_2 x^2 + a_3 x^3 + a_4 x^4]^4}$	$a_1 = .278393 \quad a_2 = .230389$ $a_3 = .000972 \quad a_4 = .078108$	5×10^{-4}
	$1 - \frac{1}{[1 + a_1 x + a_2 x^2 + \cdots + a_6 x^6]^{16}}$	$a_1 = .07052\ 30784 \quad a_2 = .04228\ 20123$ $a_3 = .00927\ 05272 \quad a_4 = .00015\ 20143$ $a_5 = .00027\ 65672 \quad a_6 = .00004\ 30638$	3×10^{-7}
$\frac{1}{\sqrt{2\pi}} e^{-x^2/2}$ $(0 \le x \le \infty)$	$(a_0 + a_2 x^2 + a_4 x^4 + a_6 x^6)^{-1}$	$a_0 = 2.490895 \quad a_4 = -.024393$ $a_2 = 1.466003 \quad a_6 = .178257$	2.7×10^{-3}
	$(b_0 + b_2 x^2 + b_4 x^4 + b_6 x^6 + b_8 x^8 + b_{10} x^{10})^{-1}$	$b_0 = 2.50523\ 67 \quad b_6 = .13064\ 69$ $b_2 = 1.28312\ 04 \quad b_8 = -.02024\ 90$ $b_4 = .22647\ 18 \quad b_{10} = .00391\ 32$	2.3×10^{-4}
$\frac{1}{\sqrt{2\pi}} \int_{-\infty}^x e^{-\lambda^2/2}\, d\lambda$ $(0 \le x \le \infty)$	$1 - \frac{1}{2}(1 + c_1 x + c_2 x^2 + c_3 x^3 + c_4 x^4)^{-4}$	$c_1 = .196854 \quad c_3 = .000344$ $c_2 = .115194 \quad c_4 = .019527$	2.5×10^{-4}
	$1 - \frac{1}{2}(1 + d_1 x + d_2 x^2 + d_3 x^3 + d_4 x^4 + d_5 x^5 + d_6 x^6)^{-16}$	$d_1 = .04986\ 73470 \quad d_4 = .00003\ 80036$ $d_2 = .02114\ 10061 \quad d_5 = .00004\ 88906$ $d_3 = .00327\ 76263 \quad d_6 = .00000\ 53830$	1.5×10^{-7}

* Formulas from C. Hastings, *Approximations for Digital Computers*, Princeton, Princeton, N.J.; copyright The Rand Corporation, Santa Monica, Calif., 1955.

20.7. NUMERICAL DIFFERENTIATION AND INTEGRATION

20.7-1. Numerical Differentiation. Numerical differentiation is subject to errors due to insufficient data, truncation, etc., and should be used with caution.

(a) Use of Difference Tables (equidistant argument values, see also Sec. 20.4-1). For suitably differentiable operands $y(x)$,

$$\frac{d^r}{dx^r} = D^r = \left[\frac{1}{\Delta x}\log_e(1+\Delta)\right]^r = \frac{1}{(\Delta x)^r}\left(\Delta - \frac{1}{2}\Delta^2 + \frac{1}{3}\Delta^3 - \cdots\right) \tag{20.7-1}$$

so that, if $x_k = x_0 + k\,\Delta x$ $(k = 0, \pm 1, \pm 2, \ldots)$,

$$\left.\begin{aligned} y'_k &= y'(x_k) = Dy_k = \frac{1}{\Delta x}\left(\Delta y_k - \frac{1}{2}\Delta^2 y_k + \frac{1}{3}\Delta^3 y_k - \cdots\right) \\ y''_k &= y''(x_k) = D^2 y_k = \frac{1}{(\Delta x)^2}\left(\Delta^2 y_k - \Delta^3 y_k + \frac{11}{12}\Delta^4 y_k \right. \\ &\qquad\qquad \left. - \frac{5}{6}\Delta^5 y_k + \cdots\right) \end{aligned}\right\} \tag{20.7-2}$$

Differentiation of Stirling's and Bessel's interpolation formulas (Table 20.5-1) yields, respectively,

$$\left.\begin{aligned} y'_k &= \frac{\mu}{\Delta x}\left(\delta y_k - \frac{1}{6}\delta^3 y_k + \frac{1}{30}\delta^5 y_k - \cdots\right) \\ y''_k &= \frac{1}{(\Delta x)^2}\left(\delta^2 y_k - \frac{1}{12}\delta^4 y_k + \frac{1}{90}\delta^6 y_k - \cdots\right) \end{aligned}\right\} \tag{20.7-3}$$

$$\left.\begin{aligned} y'_k &= \frac{1}{\Delta x}\left(\delta y_k - \frac{1}{24}\delta^3 y_k + \cdots\right) \\ y''_k &= \frac{\mu}{(\Delta x)^2}\left(\delta^2 y_k - \frac{5}{24}\delta^4 y_k + \cdots\right) \end{aligned}\right\} \tag{20.7-4}$$

Many similar formulas, and also formulas for approximating higher-order derivatives, may be derived by differentiation of suitable interpolation formulas (see also Fig. 20.9-1).

Note also the following explicit *three-point differentiation formulas with error estimates*

$$\left.\begin{aligned} y'_{-1} &= \frac{1}{2\Delta x}(-3y_{-1} + 4y_0 - y_1) + \frac{\Delta x^2}{3}y'''(\xi) \\ y'_0 &= \frac{1}{2\Delta x}(-y_{-1} + y_1) - \frac{\Delta x^2}{6}y'''(\xi) \\ y'_1 &= \frac{1}{2\Delta x}(y_{-1} - 4y_0 + 3y_1) + \frac{\Delta x^2}{3}y'''(\xi) \end{aligned}\right\} \tag{20.7-5}$$

where $y_{-1} < \xi < y_1$. See Ref. 20.17 for analogous five-point formulas.

(b) Use of Divided Differences (see also Sec. 20.5-2*b*). Given the function values $y(x_k)$ for a set of not necessarily equally spaced argument values $x_0, x_1, x_2, \ldots,$ differentiation of Newton's interpolation formula (20.5-3) yields

$$\left.\begin{aligned} y^{(r)}(x) &\approx F_r^{(r)}(x)\Delta_r(x_0, x_1, \ldots, x_r) + F_{r+1}^{(r)}(x)\Delta_{r+1}(x_0, x_1, \ldots, x_{r+1}) \\ &\qquad + \cdots \qquad (r = 0, 1, 2, \ldots) \\ \text{with} \qquad F_j(x) &\equiv \prod_{k=0}^{j-1} (x - x_k) \qquad (j = 0, 1, 2, \ldots) \end{aligned}\right\} \tag{20.7-6}$$

See Ref. 20.6 for a discussion of the errors due to the interpolation-polynomial approximation.

(c) Numerical Differentiation after Smoothing. The following formulas are based on differentiation of smoothed polynomial approximations rather than on interpolation polynomials and may thus be less affected by random errors in empirical data (Ref. 20.12).

$$y_i' \approx \frac{\sum_{k=-n}^{n} k y_{i+k}}{2 \sum_{k=1}^{n} k^2 \, \Delta x} \tag{20.7-7}$$

For $n = 2$, this yields

$$y_i' \approx \frac{1}{10\Delta x}(-2y_{i-2} - y_{i-1} + y_{i+1} + 2y_{i+2}) \tag{20.7-8}$$

Some other formulas are

$$\left.\begin{aligned} y_k' &\approx \frac{1}{12\Delta x}(3y_{k+1} + 10y_k - 18y_{k-1} + 6y_{k-2} - y_{k-3}) \\ y_k' &\approx \frac{1}{12\Delta x}[(y_{k-2} - y_{k+2}) - 8(y_{k-1} - y_{k+1})] \\ y_k' &\approx \frac{1}{12\Delta x}(y_{k+3} - 6y_{k+2} + 18y_{k+1} - 10y_k - 3y_{k-1}) \end{aligned}\right\} \tag{20.7-9}$$

(d) Approximate Differentiation of Truncated Fourier Series. If $y(x)$ is approximated by an $(n + 1)$-term truncated Fourier series (20.6-18), Lanczos (Ref. 20.12) suggests estimation of $y'(x)$ from

$$y'(x) \approx \frac{Y\left(x + \frac{\pi}{n}\right) - Y\left(x - \frac{\pi}{n}\right)}{\frac{2\pi}{n}} \tag{20.7-10}$$

20.7-2. Numerical Integration Using Equally Spaced Argument Values. (a) Newton-Cotes Quadrature Formulas. *Quadrature for-*

mulas of the closed Newton-Cotes type (Table 20.7-1) use the approximation

$$\left.\begin{aligned}&\int_{x_0}^{x_0+n\,\Delta x} y(x)\,dx \approx a_0y_0 + a_1y_1 + a_2y_2 + \cdots + a_ny_n\\ &\text{with}\\ &\qquad a_k = \frac{(-1)^{n-k}\,\Delta x}{k!(n-k)!}\int_0^n \frac{\lambda(\lambda-1)(\lambda-2)\cdots(\lambda-n)}{(\lambda-k)}\,d\lambda\end{aligned}\right\} \tag{20.7-11}$$

where the $y_k = y(x_k)$ are given function values for $n + 1$ equally-spaced argument values $x_k = x_0 + k\,\Delta x$ ($k = 0, 1, 2, \ldots, n$); the resulting error vanishes if $y(x)$ is a polynomial of degree not greater than n. Instead of using values of $n > 6$, one adds m sums (6) of $n \leq 6$ terms for successive subintervals:

$$\int_{x_0}^{x_0+mn\,\Delta x} y(x)\,dx = \int_{x_0}^{x_0+n\,\Delta x} y(x)\,dx + \int_{x_0+n\,\Delta x}^{x_0+2n\,\Delta x} y(x)\,dx + \cdots \tag{20.7-12}$$

(b) Gregory's Formula. The "symmetrical" formula

$$\begin{aligned}\int_{x_0}^{x_0+n\Delta x} y(x)\,dx \approx \Delta x[\tfrac{1}{2}y_0 + y_1 + y_2 + \cdots + y_{n-1} + \tfrac{1}{2}y_n\\ + \tfrac{1}{12}(\Delta f_0 - \Delta f_{n-1}) - \tfrac{1}{24}(\Delta^2 f_0 + \Delta^2 f_{n-2})\\ + \tfrac{19}{720}(\Delta^3 f_0 + \Delta^3 f_{n-3}) - \tfrac{3}{160}(\Delta^4 f_0 + \Delta^4 f_{n-4}) \pm \cdots]\end{aligned}$$

(GREGORY'S QUADRATURE FORMULA) (20.7-13)

adds correction terms to the trapezoidal rule of Table 20.7-1. The

Table 20.7-1. Quadrature Formulas of the Closed Newton-Cotes Type
$[y_k = y(x_k) = y(x + k\,\Delta x),\ k = 0, 1, 2, \ldots, n]$

No.		$I' \approx \int_{x_0}^{x_0+n\,\Delta x} y(x)\,dx = I$ (add analogous expressions for m successive subintervals)	Error, $I - I'$ ($x_0 < \xi < x_0 + n\,\Delta x$)
1	**Trapezoidal rule** ($n = 1$)	$\frac{\Delta x}{2}(y_0 + y_1)$	$-\frac{1}{12}(n\,\Delta x)^3 y^{(2)}(\xi)$
2	**Simpson's rule** ($n = 2$)	$\frac{\Delta x}{3}(y_0 + 4y_1 + y_2)$	$-\frac{1}{90}\left(\frac{n\,\Delta x}{2}\right)^5 y^{(4)}(\xi)$
3	**Weddle's rule** ($n = 6$)*	$\frac{3}{10}\Delta x(y_0 + 5y_1 + y_2 + 6y_3 + y_4 + 5y_5 + y_6)$	$\frac{\vartheta}{212310}\left(\frac{n\,\Delta x}{2}\right)^7 y^{(6)}(\xi)$ ($0 < \vartheta < 1$)

* In Weddle's rule, the correct Newton-Cotes coefficient $\frac{41}{140}$ of $\Delta^6 y_0$ has been replaced by $\frac{3}{10}$.

formula is derived in the manner of Sec. 20.7-4 and is suitably truncated in each case. The formula is exact for polynomials $y(x)$ of up to the $(2m + 1)^{\text{st}}$ degree if up to $2m^{\text{th}}$-order differences are included ($m = 0, 1, 2, \ldots$).

In particular, omission of first-order and higher differences yields the trapezoidal-rule formula

$$\int_{x_0}^{x_0+n\,\Delta x} y(x)\,dx \approx \Delta x[\tfrac{1}{2}y_0 + y_1 + \cdots + y_{n-1} + \tfrac{1}{2}y_n] \qquad (20.7\text{-}14)$$

which is exact for first-degree polynomials $y(x)$. Omission of third-order and higher differences produces

$$\int_{x_0}^{x_0+n\,\Delta x} y(x)\,dx \approx \Delta x[\tfrac{9}{24}y_0 + \tfrac{28}{24}y_1 + \tfrac{23}{24}y_2 + y_3 + \cdots + \cdots + y_{n-3} + \tfrac{23}{24}y_{n-2} + \tfrac{28}{24}y_{n-1} + \tfrac{9}{24}y_n] \qquad (20.7\text{-}15)$$

which is exact for third-degree polynomials $y(x)$.

(c) Use of Euler-MacLaurin Formula. The *Euler-MacLaurin summation formula* (4.8-10) yields a number of quadrature formulas involving values $y'_k = y'(x_k)$, $y''_k = y''(x_k)$, . . . of the *derivatives* of $y(x)$ as well as of $y(x)$. Thus,

$$\int_{x_0}^{x_0+\Delta x} y(x)\,dx = \frac{\Delta x}{2}(y_0 + y_1) - \frac{\Delta x^2}{12}(y'_1 - y'_0) + \frac{\Delta x^4}{720}(y'''_1 - y'''_0) \mp \cdots \qquad (20.7\text{-}16)$$

again adds correction terms to the trapezoidal rule.

20.7-3. Gauss and Chebyshev Quadrature Formulas. (a) Rewrite the given definite integral $\int_a^b y(x)\,dx$ as $\int_{-1}^{1} \eta(\xi)\,d\xi$ with the aid of the transformation

$$x = \frac{b-a}{2}\xi + \frac{a+b}{2} \qquad \eta(\xi) \equiv \frac{b-a}{2}y(x) \qquad (20.7\text{-}17)$$

and approximate the latter integral by

$$\left.\begin{aligned} &\int_{-1}^{1} \eta(\xi)\,d\xi \approx \sum_{k=1}^{n} a_k\eta(\xi_k) \\ &\text{with} \\ &a_k = \frac{2}{(1-\xi_k^2)[P'_n(\xi_k)]^2} \qquad (n = 1, 2, \ldots) \end{aligned}\right\} \qquad \text{(Gauss quadrature formula)} \qquad (20.7\text{-}18)$$

where the n argument values ξ_k are the n zeros of the nth-degree Legendre polynomial $P_n(\xi)$ (Sec. 21.7-1). Table 20.7-2 lists the ξ_k and a_k for a number of values of n.

The error due to the use of the Gauss quadrature formula (18) is

$$E = \frac{(n!)^4(b-a)^{2n+1}}{(2n+1)[(2n!)]^3} y^{(2n)}(X) \qquad (a < X < b) \tag{20.7-19}$$

(**b**) A simpler class of quadrature formulas is obtained with the aid of the transformation (9) and an approximation of the form

$$\int_{-1}^{1} \eta(\xi)\, d\xi \approx \frac{2}{n} [\eta(\xi_1') + \eta(\xi_2') + \cdots + \eta(\xi_n')]$$

$(n = 2, 3, 4, 5, 6, 7, 9)$ (CHEBYSHEV QUADRATURE FORMULA) (20.7-20)

Table 20.7-2 lists the ξ_k' for a number of values of n. The use of equal weights minimizes the probable error if $y(x)$ is affected by normally distributed random errors. The derivation of the ξ_k' is discussed in Refs. 20.6 and 20.10.

Table 20.7-2. Abscissas and Weights for Gauss and Chebyshev Quadrature Formulas (Sec. 20.7-3; adapted from Ref. 20.11)

(a) Abscissas ξ_k and Weights a_k for the Gauss Quadrature Formula (18)

n	Abscissas	Weights
2	±0.577350	1
3	0	8/9
	±0.774597	5/9
4	±0.339981	0.652145
	±0.861136	0.347855
5	0	0.568889
	±0.538469	0.478629
	±0.906180	0.236927

(b) Abscissas ξ_k' for the Chebyshev Quadrature Formula (20)

n	Abscissas	n	Abscissas
2	±0.577350	7	0
			±0.323912
3	0		±0.529657
	±0.707107		±0.883862
4	±0.187592		
	±0.794654	9	0
			±0.167906
5	0		±0.528762
	±0.374541		±0.601019
	±0.832497		±0.911589
6	±0.266635		
	±0.422519		
	±0.866247		

Table 20.7-2. Abscissas and Weights for Gauss and Chebyshev Quadrature Formulas (*Continued*)

(c) Abscissas ξ_k and Weights a_k for the Gauss-Laguerre Quadrature Formula (21)

n	Abscissas	Weights
2	0.585786	0.853553
	3.414214	0.146447
3	0.415775	0.711093
	2.294280	0.278518
	6.289945	0.0103893
4	0.322548	0.603154
	1.745761	0.357419
	4.536620	0.0388879
	9.395071	0.000539295
5	0.263560	0.521756
	1.413403	0.398667
	3.596426	0.0759424
	7.085810	0.00361176
	12.640801	0.0000233700

(d) Abscissas ξ_k and Weights a_k for the Gauss-Hermite Quadrature Formula (22)

n	Abscissas	Weights
2	±0.707107	0.886227
3	0	1.181636
	±1.224745	0.295409
4	±0.524648	0.804914
	±1.650680	0.0813128
5	0	0.945309
	±0.958572	0.393619
	±2.020183	0.0199532

For $n = 3$, the error due to the use of the Chebyshev quadrature formula (20) is $\frac{1}{360}\left(\frac{b-a}{2}\right)^5 y^{(4)}(X)$ $(a < X < b)$.

(c) Note also

$$\left.\begin{aligned} &\int_0^\infty e^{-\xi}\eta(\xi)\,d\xi \approx \sum_{k=1}^{n} a_k\eta(\xi_k) \\ &\text{with} \\ &a_k = \frac{(n!)^2}{\xi_k[L_n'(\xi_k)]^2} \qquad (n = 1, 2, \ldots) \end{aligned}\right\} \quad \text{(Gauss-Laguerre quadrature formula)} \tag{20.7-21}$$

where the ξ_k are the zeros of the nth-degree Laguerre polynomial $L_n(\xi)$ (Sec. 21.7-1); and

$$\left.\begin{aligned}&\int_{-\infty}^{\infty} e^{-\xi^2}\eta(\xi)\,d\xi \approx \sum_{k=1}^{n} a_k\eta(\xi_k)\\ &\text{with}\\ &a_k = \frac{2^{n+1}n!\sqrt{\pi}}{[H_n'(\xi_k)]^2} \qquad (n = 1, 2, \ldots)\end{aligned}\right\} \qquad \text{(Gauss-Hermite quadrature formula)} \tag{20.7-22}$$

where the ξ_k are the zeros of the nth-degree Hermite polynomial $H_n(\xi)$ (Sec. 21.7-1).

Again,

$$\left.\begin{aligned}&\int_{-1}^{1} \frac{\eta(\xi)}{\sqrt{1-\xi^2}}\,d\xi \approx \sum_{k=1}^{n} a_k\eta(\xi_k)\\ &\text{with}\\ &a_k = \frac{\pi}{n} \qquad \xi_k = \cos\frac{2k-1}{2n}\pi \qquad (k = 1, 2, \ldots, n)\end{aligned}\right\}$$

(Gauss-Chebyshev quadrature formula) (20.7-23)

The ξ_k are seen to be zeros of the nth Chebyshev polynomial (Sec. 21.7-4).

Many similar formulas exist (Refs. 20.6 and 20.56). Table 20.7-2 lists some of the ξ_k and a_k for Eqs. (21) and (22).

20.7-4. Derivation and Comparison of Integration Formulas. (a) In general, one can derive unknown coefficients a_k and/or abscissas x_k in an integration formula

$$\int_a^b y(x)\gamma(x)\,dx \approx a_0y(x_0) + a_1y(x_1) + a_2y(x_2) + \cdots + a_ny(x_n) \tag{20.7-24}$$

as follows (Ref. 20.4):

1. The formula is required to hold *exactly* for $y(x) = 1, x, x^2, \ldots, x^m (m < n)$. This yields m equations

$$\int_a^b x^r\gamma(x)\,dx = a_0x_0^r + a_1x_1^r + a_2x_2^r + \cdots + a_nx_n^r \qquad (r = 0, 1, 2, \ldots, m < n) \tag{20.7-25}$$

   for the unknown a_k and x_k.
2. One may *prescribe* some or all of the x_k (Newton-Cotes formulas, Gregory's formula) and/or the a_k (Chebyshev Quadrature).
3. One may *impose constraints* on (relations between) the a_k either for symmetry (Gregory's formula) or to minimize round-off-error effects. *For the latter purpose, all a_k should be positive* (subtraction increases round-off-error effects, Sec. 20.1-2).

The relative desirability of these measures depends on the application. The x_k may or may not be in the integration interval. The same method also applies to formulas containing added derivative terms like $b_k y'(x_k)$, as in Sec. 20.7-2c.

(b) The Gauss-type integration formulas, like Eq. (18), are exact if $\eta(\xi)$ is a polynomial of degree $\leq 2n - 1$, while Newton-Cotes formulas are exact only for polynomials of degree $\leq n$. In this sense, Gaussian formulas are best for functions $\eta(\xi)$ possessing derivatives of high order. If $\eta(\xi)$ has only a piecewise-continuous first-order derivative, then repeated use of the simple trapezoidal rule is about as good as any quadrature formula. See Ref. 20.56 for modified Gaussian quadrature formulas exact for trigonometric polynomials and other special functions.

20.7-5. Numerical Evaluation of Multiple Integrals. Multiple integrals may be evaluated by repeated application of the procedures outlined in Secs. 20.7-2 and 20.7-3; or divide the domain of integration into subregions separated by coordinate lines and use the approximations

$$\int_{-h}^{h}\int_{-h}^{h} f(x, y)\, dx\, dy \approx \frac{2h^2}{3}(2f_{00} + f_{10} + f_{01} + f_{-1,0} + f_{0,-1})$$

(WOOLLEY'S APPROXIMATION) (20.7-26)

$$\int_{-h}^{h}\int_{-h}^{h} f(x, y)\, dx\, dy \approx \frac{h^2}{9}[16f_{00} + 4(f_{10} + f_{01} + f_{-1,0} + f_{0,-1}) + f_{11} + f_{1,-1} + f_{-1,1} + f_{-1,-1}]$$

(SIMPSON'S APPROXIMATION) (20.7-27)

where $f_{jk} = f(j\,\Delta x, k\,\Delta y)$, $\Delta x = \Delta y = h \quad (j, k = 0, \pm 1)$, and

$$\int_{-h}^{h}\int_{-h}^{h}\int_{-h}^{h} f(x, y, z)\, dx\, dy\, dz \approx \tfrac{4}{3}h^3(f_{100} + f_{010} + f_{001} + f_{-1,00} + f_{0,-1,0} + f_{00,-1}) \tag{20.7-28}$$

where $f_{ijk} = f(i\,\Delta x, j\,\Delta y, k\,\Delta z)$, $\Delta x = \Delta y = \Delta z = h$ $(i, j, k = 0, \pm 1)$.

A simple *two-dimensional Gauss-type integration formula* (Ref. 20.56) is

$$\left.\begin{aligned} &\int_{-1}^{1}\int_{-1}^{1} f(\xi, \eta)\, d\xi\, d\eta \approx \sum_{i=1}^{3}\sum_{k=1}^{3} a_i a_k f(\lambda_i, \lambda_k) \\ &\text{with} \quad \lambda_1 = -\sqrt{3/5} \qquad \lambda_2 = 0 \qquad \lambda_3 = \sqrt{3/5} \\ &\qquad\quad\; a_1 = 5/9 \qquad a_2 = 8/9 \qquad a_3 = 5/9 \end{aligned}\right\} \tag{20.7-29}$$

Reference 20.56 also gives integration formulas for s-dimensional cubes, spheres, and other regions. Monte-Carlo methods (Sec. 20.10-1) are also of interest for multidimensional integration.

20.8. NUMERICAL SOLUTION OF ORDINARY DIFFERENTIAL EQUATIONS

20.8-1. Introduction and Notation. Refer to Chap. 9 for ordinary methods of solution. A rough graphical solution (Sec. 9.5-2) may precede the numerical solution for orientation purposes. Sections 20.8-2 to 20.8-8 deal with initial-value problems; the solution of boundary-value problems is discussed in Secs. 20.9-1 and 20.9-3.

To solve the first-order differential equation

$$y' = f(x, y) \tag{20.8-1}$$

for a given initial value $y(x_0) = y_0$, consider fixed increments $\Delta x = h$ of the independent variable x. Let $x_k = x_0 + k\,\Delta x$ $(k = 0, 1, 2, \ldots)$, and denote successive samples of the *computed* (in general, approximate) solution $y(x)$ and its derivative $y'(x)$ by

$$\left.\begin{aligned} y_k &\approx y(x_k) = y(x_0 + k\,\Delta x) \\ y'_k &= f_k = f(x_k, y_k) \approx y'(x_k) \end{aligned}\right\} \quad (k = 0, 1, 2, \ldots) \tag{20.8-2}$$

In the absence of round-off errors, the difference $y_{k+1} - y_{\text{TRUE}}(x_{k+1})$ is the **truncation error** due to a stepwise approximation of continuous integration. If exact solution values $y_{\text{TRUE}}(x_k), y_{\text{TRUE}}(x_{k-1}), \ldots$ are substituted for $y_k, y_{k-1}, \ldots$, then $y_{k+1} - y_{\text{TRUE}}(x_{k+1})$ is the **local truncation error** for the given integration formula. The true truncation error, however, is affected by **propagated errors** from earlier solution steps as well as by the local truncation error (Sec. 20.8-5).

20.8-2. One-step Methods for Initial-value Problems: Euler and Runge-Kutta-type Methods. (a) For sufficiently small increments $\Delta x = h$ and sufficiently small $k = 0, 1, 2, \ldots$, the following simple recursion formulas produce stepwise approximations y_k to the solution $y = y(x)$ of Eq. (1):

$$y_{k+1} = y_k + f_k\,\Delta x \quad (\text{EULER-CAUCHY POLYGONAL APPROXIMATION}) \tag{20.8-3}$$

$$y_{k+1} = y_k + f\left(x_k + \frac{\Delta x}{2}, y_k + f_k\frac{\Delta x}{2}\right)\Delta x \tag{20.8-4}$$

$$y_{k+1} = y_k + \tfrac{1}{2}[f_k + f(x_{k+1}, y_k + f_k\,\Delta x)]\,\Delta x \tag{20.8-5}$$

(b) Table 20.8-1 lists *Runge-Kutta-type routines* for the numerical solution of Eq. (1). Methods (*a*) and (*b*) in Table 20.8-1 are usually called "third-order" methods,* because the formulas for y_{k+1} are exact for $f(x, y) = 1, x, x^2, x^3$; for suitably differentiable $f(x, y)$, the local truncation error is $O[(\Delta x)^4]$ as $x \to 0$ (Sec. 4.4-3). Methods (*c*), (*d*), and (*e*) are, by an analogous definition, "fourth-order" methods. See Refs. 20.11 and 20.64 for higher-order Runge-Kutta formulas (see also Sec. 20.8-5).

20.8-3. Multistep Methods for Initial-value Problems. (a) Starting the Solution. Given the initial value y_0, each of the following solution schemes requires one to compute the first three to five function values $y_1, y_2, \ldots$ by one of the methods of Secs. 9.2-5, 20.8-2, or 20.8-4.

* Some authors call (*a*) and (*b*) "fourth-order" methods, and (*c*) and (*d*) "fifth-order" methods instead.

This "starting solution" should be computed more accurately than the required solution by at least a factor of 10. If the Runge-Kutta method is used to start the solution, employ a step size $\Delta x = h$ smaller than that required for the subsequent difference scheme.

Table 20.8-1. Some Runge-Kutta-type Methods for Ordinary Differential Equations (Sec. 20.8-2) or Systems of Differential Equations (Sec. 20.8-5)

In each formula, $k_1 = f(x_k, y_k)\,\Delta x = f_k\,\Delta x$

(a) $y_{k+1} = y_k + \frac{1}{6}(k_1 + 4k_2 + k_3)$

$$k_2 = f\left(x_k + \frac{\Delta x}{2}, y_k + \frac{k_1}{2}\right)\Delta x$$

$$k_3 = f(x_k + \Delta x, y_k + 2k_2 - k_1)\,\Delta x$$

(b) $y_{k+1} = y_k + \frac{1}{4}(k_1 + 3k_3)$

$$k_2 = f\left(x_k + \frac{\Delta x}{3}, y_k + \frac{k_1}{3}\right)\Delta x$$

$$k_3 = f\left(x_k + \frac{2}{3}\Delta x, y_k + \frac{2}{3}k_2\right)\Delta x$$

(Runge-Kutta-Heun method)

(c) $y_{k+1} = y_k + \frac{1}{6}(k_1 + 2k_2 + 2k_3 + k_4)$

$$k_2 = f\left(x_k + \frac{\Delta x}{2}, y_k + \frac{k_1}{2}\right)\Delta x$$

$$k_3 = f\left(x_k + \frac{\Delta x}{2}, y_k + \frac{k_2}{2}\right)\Delta x$$

$$k_4 = f(x_{k+1}, y_k + k_3)\,\Delta x$$

(d) $y_{k+1} = y_k + \frac{1}{8}(k_1 + 3k_2 + 3k_3 + k_4)$

$$k_2 = f\left(x_k + \frac{\Delta x}{3}, y_k + \frac{k_1}{3}\right)\Delta x$$

$$k_3 = f\left(x_k + \frac{2}{3}\Delta x, y_k - \frac{k_1}{3} + k_2\right)\Delta x$$

$$k_4 = f(x_k + \Delta x, y_k + k_1 - k_2 + k_3)\,\Delta x$$

(e) $y_{k+1} = \frac{1}{6}[k_1 + 2(1 - \sqrt{\frac{1}{2}})k_2 + 2(1 + \sqrt{\frac{1}{2}})k_3 + k_4]$

$$k_2 = f\left(x_k + \frac{\Delta x}{2}, y_k + \frac{k_1}{2}\right)\Delta x$$

$$k_3 = f\left[x_k + \frac{\Delta x}{2}, y_k - \left(\tfrac{1}{2} - \sqrt{\tfrac{1}{2}}\right)k_1 + \left(1 - \sqrt{\tfrac{1}{2}}\right)k_2\right]\Delta x$$

$$k_4 = f\left[x_k + \Delta x, y_k - \sqrt{\tfrac{1}{2}}\,k_2 + \left(1 + \sqrt{\tfrac{1}{2}}\right)k_3\right]\Delta x$$

(Runge-Kutta-Gill method)

(b) Simple Extrapolation Schemes. Given $y_k, y_{k-1}, y_{k-2}, \ldots,$ one approximates successive solution values

$$y(x_{k+1}) \approx y_k + \int_{x_k}^{x_{k+1}} f(x, y)\,dx \tag{20.8-6}$$

by integrating an extrapolation polynomial through $f_k, f_{k-1}, f_{k-2}, \ldots$ instead of $f(x, y)$. Using the second interpolation formula (20.5-6), one obtains

$$y_{k+1} = y_k + (f_k + \tfrac{1}{2}\nabla f_k + \tfrac{5}{12}\nabla^2 f_k + \tfrac{3}{8}\nabla^3 f_k + \tfrac{251}{720}\nabla^4 f_k + \tfrac{95}{288}\nabla^5 f_k + \cdots)\,\Delta x \qquad (20.8\text{-}7)$$

Truncation of the general "open" integration formula (7) after successively higher-order difference terms yields first the Euler formula (3) and the *open trapezoidal rule*

$$y_{k+1} = y_k + \tfrac{1}{2}(3f_k - f_{k-1})\,\Delta x \qquad (20.8\text{-}8)$$

both used in digital differential analyzers, and then the third-order formula

$$y_{k+1} = y_k + \tfrac{1}{12}(23f_k - 16f_{k-1} + 5f_{k-2})\,\Delta x \qquad (20.8\text{-}9)$$

and the fourth-order Adams-Bashforth predictor of Table 20.8-2.

(c) **Predictor-corrector Methods and Step-size Changes.** Denoting the "predicted" value (7) of y_{k+1} as y_{k+1}^{PRED}, one can improve the approximation of Sec. 20.8-3*b* by requiring the interpolation polynomial to assume the value $f_{k+1}^{\text{PRED}} = f(x_{k+1}, y_{k+1}^{\text{PRED}})$. The resulting "corrected" y_{k+1} is obtained by using the predicted f_{k+1} in the "closed" integration formula

$$y_{k+1} = y_{k+1}^{\text{CORRECTED}} = y_k + (f_{k+1} - \tfrac{1}{2}\nabla f_{k+1} - \tfrac{1}{12}\nabla^2 f_{k+1} - \tfrac{1}{24}\nabla^3 f_{k+1} - \tfrac{19}{720}\nabla^4 f_{k+1} - \tfrac{3}{160}\nabla^5 f_{k+1} - \cdots) \qquad (20.8\text{-}10)$$

The **corrector** (10) is truncated like the predictor (7). The resulting difference $y_{k+1}^{\text{CORRECTED}} - y_{k+1}^{\text{PREDICTED}}$ measures the local truncation error of the corrected approximation and is reduced to some preassigned value by suitable selection of the increment Δx. *To halve the step size* for fourth-order formulas, use the interpolation formulas

$$\left.\begin{aligned} y_{k-1/2} &= \tfrac{1}{128}[45y_k + 72y_{k-1} + 11y_{k-2} + (-9f_k + 36f_{k-1} + 3f_{k-2})\,\Delta x] \\ y_{k-3/2} &= \tfrac{1}{128}[11y_k + 72y_{k-1} + 45y_{k-2} - (3f_k + 36f_{k-1} - 9f_{k-2})\,\Delta x] \end{aligned}\right\} \qquad (20.8\text{-}11)$$

To double the step size, one may restart, or use past solution values stored for this purpose (see also Ref. 20.4).

A corrected solution value y_{k+1} can be successively improved further if it is resubstituted into the corrector formula as a new predicted value. In practice, step-size reduction and/or the use of modifiers (Sec. 20.8-4*b*) is often preferred to such iteration.

20.8-4. Improved Multistep Methods. (a) More general "open" integration formulas (useful as predictors) and "closed" integration formulas (useful as correctors) can be written in the respective forms

$$y_{k+1} = A_0 y_k + A_1 y_{k-1} + A_2 y_{k-2} + A_3 y_{k-3} + (B_0 f_k + B_1 f_{k-1} + B_2 f_{k-2} + B_3 f_{k-3})\,\Delta x \qquad \text{(PREDICTOR)} \qquad (20.8\text{-}12)$$

$$y_{k+1} = a_0 y_k + a_1 y_{k-1} + a_2 y_{k-2} + (b_{-1} f_{k+1} + b_0 f_k + b_1 f_{k-1} + b_2 f_{k-2})\,\Delta x \qquad \text{(CORRECTOR)} \qquad (20.8\text{-}13)$$

Instead of determining all the coefficients by making each formula exact for $f(x, y) = 1, x, x^2, \ldots$, one usually prefers to require only "fourth-order" matching (up to and including x^4); the coefficients thus left undetermined are chosen so as to improve error propagation and/or simplify computation (Sec. 20.8-5). Table 20.8-2 lists a number of useful fourth-order predictor and corrector formulas.

Table 20.8-2. Some Fourth-order Predictor-corrector Methods

Each predictor-corrector scheme may be used with or without modifiers, and $f_{k+1}^{\mathrm{MOD}} = f(x_{k+1}, y_{k+1}^{\mathrm{MOD}})$. In each case, the magnitude of the correction on the last line is an upper bound for the local truncation error.

(a) Adams-Bashforth Predictor with Adams-Moulton Corrector and (Optional) Modifier

$$\begin{aligned}
y_{k+1}^{\mathrm{PRED}} &= y_k + \tfrac{1}{24}(55f_k - 59f_{k-1} + 37f_{k-2} - 9f_{k-3})\,\Delta x \\
y_{k+1}^{\mathrm{MOD}} &= y_{k+1}^{\mathrm{PRED}} + \tfrac{251}{270}(y_k^{\mathrm{CORR}} - y_k^{\mathrm{PRED}}) \\
y_{k+1}^{\mathrm{CORR}} &= y_k + \tfrac{1}{24}(9f_{k+1}^{\mathrm{MOD}} + 19f_k - 5f_{k-1} + f_{k-2})\,\Delta x \\
y_{k+1} &= y_{k+1}^{\mathrm{CORR}} - \tfrac{19}{270}(y_{k+1}^{\mathrm{CORR}} - y_{k+1}^{\mathrm{PRED}})
\end{aligned}$$

(b) Hamming's Method

$$\begin{aligned}
y_{k+1}^{\mathrm{PRED}} &= y_{k-3} + \tfrac{4}{3}(2f_k - f_{k-1} + 2f_{k-2})\,\Delta x \\
y_{k+1}^{\mathrm{MOD}} &= y_{k+1}^{\mathrm{PRED}} + \tfrac{112}{121}(y_k^{\mathrm{CORR}} - y_k^{\mathrm{P\ \ ED}}) \\
y_{k+1}^{\mathrm{CORR}} &= \tfrac{1}{8}(9y_k - y_{k-2}) + \tfrac{3}{8}(f_{k+1}^{\mathrm{MOD}} + 2f_k - f_{k-1})\,\Delta x \\
y_{k+1} &= y_{k+1}^{\mathrm{CORR}} - \tfrac{9}{121}(y_{k+1}^{\mathrm{CORR}} - y_{k+1}^{\mathrm{PRED}})
\end{aligned}$$

(c) Milne's Method, here shown with Hamming's modifier, has relatively low local truncation error but is unstable if $\partial f/\partial y$ is a negative real number, or a real matrix having a negative eigenvalue.

$$\begin{aligned}
y_{k+1}^{\mathrm{PRED}} &= y_{k-3} + \tfrac{4}{3}(2f_k - f_{k-1} + 2f_{k-2})\,\Delta x \\
y_{k+1}^{\mathrm{MOD}} &= y_{k+1}^{\mathrm{PRED}} + \tfrac{28}{29}(y_k^{\mathrm{CORR}} - y_k^{\mathrm{PRED}}) \\
y_{k+1}^{\mathrm{CORR}} &= y_{k-1} + \tfrac{1}{3}(f_{k+1}^{\mathrm{MOD}} + 4f_k + f_{k-1})\,\Delta x \\
y_{k+1} &= y_{k+1}^{\mathrm{CORR}} - \tfrac{1}{29}(y_{k+1}^{\mathrm{CORR}} - y_{k+1}^{\mathrm{PRED}})
\end{aligned}$$

See Ref. 20.4 for additional formulas.

(b) Use of Modifiers (Ref. 20.4). In each predictor-corrector method, the difference $y_{k+1}^{\mathrm{CORRECTED}} - y_{k+1}^{\mathrm{PREDICTED}}$ is roughly proportional to the local truncation error. Hence, one may improve the solution by adding a fraction $\alpha(y_k^{\mathrm{CORRECTED}} - y_k^{\mathrm{PREDICTED}})$ of the preceding difference to $y_{k+1}^{\mathrm{PREDICTED}}$ before substitution in the corrector, and subtracting $(1 - \alpha)(y_{k+1}^{\mathrm{CORRECTED}} - y_{k+1}^{\mathrm{PREDICTED}})$ from $y_{k+1}^{\mathrm{CORRECTED}}$ to obtain y_{k+1} (Table 20.8-2).

(c) The Fourth-order *method of Clippinger and Dimsdale* (Ref. 20.11) requires iteration of

$$\left.\begin{aligned} y_{k+1} &= \tfrac{1}{2}(y_k + y_{k+2}) + \tfrac{1}{4}(f_k - f_{k+2})\,\Delta x \\ y_{k+2} &= y_k + \tfrac{1}{3}(f_k + 4f_{k+1} + f_{k+2})\,\Delta x \end{aligned}\right\} \qquad (20.8\text{-}14)$$

starting with a trial value for y_{k+2}, say $y_{k+2} \approx y_k + 2f_k\,\Delta x$, and permits step-size changes by simple substitution of new values of Δx. The method is self-starting, but once the solution is started, one can save iterations by employing extrapolation to predict y_{k+2}.

20.8-5. Discussion of Different Solution Methods, Step-size Control, and Stability. (a) Integration formulas specifically selected for low *local* truncation error can emphasize error propagation over successive solution steps. *The design of differential-equation-solving routines requires a compromise between local truncation error, stability, and computation time. In addition, formula coefficients which produce summation terms of equal sign and not too different absolute values are preferred, so as to reduce fractional errors due to round-off.* The final choice depends on the application and on the computer used. Double-precision accumulation of dependent variables is frequently indicated.

If the given $f(x, y)$ is at all complicated, the principal cost (time) of the computation is associated with calculations of derivative values $f(x, y)$. For differential-equation problems with reasonably smooth (repeatedly differentiable) integrator inputs $f(x, y)$, *multistep integration schemes* require relatively few derivative calculations and conveniently permit timesaving automatic step-size control in terms of $|y_{k+1}^{\text{CORRECTED}} - y_{k+1}^{\text{PREDICTED}}|$. *Runge-Kutta-type methods* tend to be very stable (Sec. 20.8-5*b*) and do not require a separate starting routine; they are, therefore, often preferred for problems involving frequent step inputs. Runge-Kutta schemes require relatively more derivative computations per step, and efficient step-size control is more difficult (see also Refs. 20.4, 20.15, and 20.52).

To estimate the local truncation error for step-size control in Runge-Kutta routines, one can compare results obtained for different step sizes (using stored derivative values whenever possible), or investigate a suitable function of the k_i. For the frequently used Runge-Kutta method (*c*) in Table 20.8-1, the quantity

$$e = 3f_{k+1}\,\Delta x + k_1 - 2k_3 - 2k_4 \qquad (20.8\text{-}15)$$

is a rough measure of the local truncation error preferable to $|k_2 - k_3|/|k_1 - k_2|$ and $k_1 + k_4 = 2k_3$, which are also used (Ref. 20.67).

(b) The stability of the approximate solution $y_0, y_1, y_2, \ldots$ of Eq. (1) computed, say, with the aid of a general multipoint formula

$$\begin{aligned} y_{k+1} = A_0 y_k + A_1 y_{k-1} + \cdots + A_r y_{k-r} \\ + (B_{-1}f_{k+1} + B_0 f_k + B_1 f_{k-1} + \cdots + B_r f_{k-r})\,\Delta x \end{aligned} \qquad (20.8\text{-}16)$$

depends on the stability of the corresponding linearized difference equation for the approximate error sequence $e_0, e_1, e_2, \ldots$, viz.,

$$e_{k+1} \approx A_0 e_k + A_1 e_{k-1} + \cdots + A_r e_{k-r} + \frac{\partial f}{\partial y}\bigg]_{x=x_k} (B_{-1}e_{k+1} + B_0 e_k + B_1 e_{k-1} + \cdots + B_r e_{k-r})\,\Delta x \qquad (20.8\text{-}17)$$

(Sec. 20.4-8). The associated characteristic equation

$$\left(-1 + B_{-1}\frac{\partial f}{\partial y}\bigg]_{x_k}\Delta x\right) z^{r+1} + \left(A_0 + B_0\frac{\partial f}{\partial y}\bigg]_{x_k}\Delta x\right) z^r + \left(A_1 + B_1\frac{\partial f}{\partial y}\bigg]_{x_k}\Delta x\right) z^{r-1} + \cdots + \left(A_r + B_r\frac{\partial f}{\partial y}\bigg]_{x_k}\Delta x\right) = 0 \qquad (20.8\text{-}18)$$

will have a root $z_1 = \exp \frac{\partial f}{\partial y}\Big]_{x_k}\Delta x + O\left(\frac{\partial f}{\partial y}\Big]_{x_k}\Delta x\right)$. If $\frac{\partial f}{\partial y}\Big]_{x_k} > 0$, then the corresponding normal mode of the error sequence is unstable, but so is the true solution $y(x)$ near $x = x_k$, so that the fractional error may be acceptable for small Δx. For $r = 0$ (simple one-step method), z_1 is the only root. For $r > 0$, Eq. (18) will have additional roots corresponding to *spurious modes in the computed solution* because of the higher-order difference approximation. **Relative stability** of the computed solution requires that all these extra roots lie within the unit circle $|z| = 1$ (Sec. 20.4-8) for the values of Δx of interest. The worst hazards exist for step inputs (which, like the round-off noise, may excite spurious modes) and near stability boundaries of the original differential equation. Such situations, if suspected, may be tested by artificially introduced perturbations.

The stability of a predictor-corrector scheme will depend on both the predictor and the corrector formula but is affected more strongly by the latter if the correction is small. Integration routines of order higher than the fourth require careful stability investigation, but their use may be economical in trajectory computations with large step sizes (Ref. 20.64).

20.8-6. Ordinary Differential Equations of Order Higher Than the First and Systems of Ordinary Differential Equations.

(a) Each ordinary differential equation of the second or higher order is equivalent to a system of first-order equations (Sec. 9.1-3). If the latter is written in the matrix form of Sec. 13.6-1, *each solution method of Secs.* 20.8-2 *to* 20.8-4 *yields an analogous method for the numerical solution of systems of differential equations.*

(b) Specifically, consider a system of first-order differential equations

$$y' = f(x, y, z, \ldots) \qquad z' = g(x, y, z, \ldots) \qquad (20.8\text{-}19)$$

with solution $y = y(x)$, $z = z(x)$, ... Note that solution by the *Taylor-series* and *Picard* methods is essentially analogous to the procedures outlined in Secs. 9.2-5*a* and 9.2-5*b*.

The *Runge-Kutta method* is analogous to the scheme of Sec. 20.8-2*b*:

$$y_{k+1} = y_k + \tfrac{1}{6}(k_1 + 2k_2 + 2k_3 + k_4) \qquad z_{k+1} = z_k + \tfrac{1}{6}(m_1 + 2m_2 + 2m_3 + m_4) \qquad \ldots$$

with

$$\left.\begin{aligned} k_1 &= f(x_k, y_k, z_k, \ldots)\,\Delta x \qquad m_1 = g(x_k, y_k, z_k, \ldots)\,\Delta x \qquad \ldots \\ k_2 &= f\left(x_k + \frac{\Delta x}{2}, y_k + \frac{k_1}{2}, z_k + \frac{m_1}{2}, \ldots\right)\Delta x \\ m_2 &= g\left(x_k + \frac{\Delta x}{2}, y_k + \frac{k_1}{2}, z_k + \frac{m_1}{2}, \ldots\right)\Delta x \qquad \ldots \\ k_3 &= f\left(x_k + \frac{\Delta x}{2}, y_k + \frac{k_2}{2}, z_k + \frac{m_2}{2}, \ldots\right)\Delta x \\ m_3 &= g\left(x_k + \frac{\Delta x}{2}, y_k + \frac{k_2}{2}, z_k + \frac{m_2}{2}, \ldots\right)\Delta x \qquad \ldots \\ k_4 &= f(x_k + \Delta x, y_k + k_3, z_k + m_3, \ldots)\,\Delta x \\ m_4 &= g(x_k + \Delta x, y_k + k_3, z_k + m_3, \ldots)\,\Delta x \qquad \ldots \end{aligned}\right\} \qquad (20.8\text{-}20)$$

Any one of the *multistep schemes* of Sec. 20.8-3 may be applied to each equation (19); one writes

$$\left.\begin{aligned} y(x_k) &= y_k & z(x_k) &= z_k & \ldots \\ f(x_k, y_k, z_k, \ldots) &= f_k & g(x_k, y_k, z_k, \ldots) &= g_k & \ldots \end{aligned}\right\} \qquad (20.8\text{-}21)$$

(c) The stability of the correct solution $y = y(x)$ of a *system* (matrix equation)

$$\frac{dy}{dx} = f(x, y)$$

will depend on the *matrix* $[\partial f/\partial y]$ (Sec. 13.6-5). Equation (17) becomes a matrix difference equation, whose characteristic roots must be compared to the eigenvalues of the matrix $[\partial f/\partial y]_{x_k}$ to determine relative stability (Refs. 20.4, 20.6, and 20.52).

20.8-7. Special Formulas for Second-order Equations. Because of the practical importance of second-order differential equations, the following schemes for the numerical solution of the differential equation

$$y'' = f(x, y, y') \qquad (20.8\text{-}22)$$

with given initial values $y(x_0) = y_0$, $y'(x_0) = y'_0$ are of interest; in each case,

$$\begin{aligned} x_0 + k\,\Delta x = x_k \qquad y(x_k) &= y_k \qquad y'(x_k) = y'_k \\ f(x_k, y_k, y'_k) &= f_k \qquad (k = 0, 1, 2, \ldots) \end{aligned} \qquad (20.8\text{-}23)$$

These methods may be extended to apply to the solution of *systems* of two or more second-order equations in the manner of Sec. 20.8-6.

(a) A Runge-Kutta Method for Eq. (22).

$$\left.\begin{aligned} y_{k+1} &= y_k + y'_k\,\Delta x + \tfrac{1}{6}(k_1 + k_2 + k_3)\,\Delta x \\ y'_{k+1} &= y'_k + \tfrac{1}{6}(k_1 + 2k_2 + 2k_3 + k_4) \\ &\text{with} \\ k_1 &= f(x_k, y_k, y'_k)\,\Delta x \\ k_2 &= f\left(x_k + \frac{\Delta x}{2}, y_k + y'_k\frac{\Delta x}{2}, y'_k + \tfrac{1}{2}k_1\right)\Delta x \\ k_3 &= f\left(x_k + \frac{\Delta x}{2}, y_k + y'_k\frac{\Delta x}{2} + \frac{k_1}{4}\,\Delta x, y'_k + \tfrac{1}{2}k_2\right)\Delta x \\ k_4 &= f(x_k + \Delta x, y_k + y'_k\,\Delta x + \tfrac{1}{2}k_2\,\Delta x, y'_k + k_3)\,\Delta x \end{aligned}\right\} \qquad (20.8\text{-}24)$$

(b) An Interpolation-iteration Scheme. Starting with a trial value for f_{k+1}, iterate

$$\left.\begin{aligned} y_{k+1} &= 2y_k - y_{k-1} + (f_k + \tfrac{1}{12}\nabla^2 f_{k+1})\,\Delta x^2 \\ y'_{k+1} &= y'_{k-1} + (2f_k + \tfrac{1}{3}\nabla^2 f_{k+1})\,\Delta x \end{aligned}\right\} \tag{20.8-25}$$

(c) Prediction-correction Schemes.

$$\left.\begin{aligned} y'_{k+1} &= y'_{k-3} + \tfrac{4}{3}(2f_k - f_{k-1} + 2f_{k-2})\,\Delta x && \text{(PREDICTOR)} \\ y_{k+1} &= y_{k-1} + \tfrac{1}{3}(y'_{k+1} + 4y'_k + y'_{k-1})\,\Delta x \\ y'_{k+1} &= y'_{k-1} + \tfrac{1}{3}(f_{k+1} + 4f_k + f_{k-1})\,\Delta x && \text{(CORRECTOR)} \end{aligned}\right\} \tag{20.8-26}$$

If $f(x, y, y')$ does not contain y' explicitly, transform the given differential equation in the manner of Sec. 9.1-5*b*, or use the following prediction-correction scheme:

$$\left.\begin{aligned} y_{k+1}^{\text{PRED}} &= 2y_{k-1} - y_{k-3} + \tfrac{4}{3}(f_k + f_{k-1} + f_{k-2})(\Delta x)^2 \\ &\qquad\qquad \text{(PREDICTOR)} \\ y_{k+1}^{\text{CORR}} &= 2y_k - y_{k-1} + \tfrac{1}{12}(f_{k+1}^{\text{PRED}} + 10f_k + f_{k-1})(\Delta x)^2 \\ &\qquad\qquad \text{(CORRECTOR)} \end{aligned}\right\} \tag{20.8-27}$$

Reference 20.4 gives modifier formulas for this scheme.

(d) Numerov's Method for Linear Equations (Refs. 20.4 and 20.11). Linear differential equations of the form

$$y'' = f(x)y + g(x) \tag{20.8-28}$$

can be solved with the corrector of Eq. (27) alone; substitution of Eq. (28) yields

$$\left.\begin{aligned} u_{k+1} &= 2u_k - u_{k-1} + [f_k y_k + g_k + \tfrac{1}{12}(g_{k+1} - 2g_k + g_{k-1})](\Delta x)^2 \\ \text{with}\qquad y_k &= \frac{u_k}{1 - \tfrac{1}{12}f_k(\Delta x)^2} \end{aligned}\right\} \tag{20.8-29}$$

20.8-8. Frequency-response Analysis. Given a stable integration formula (12), a complex-sinusoid input $f_k \equiv e^{i\omega x_k}$ will eventually produce a steady-state sinusoidal solution $y_k = H(i\omega)e^{i\omega x_k}$ (just as in Sec. 9.4-6). Substitution yields the **sampled-data frequency-response function**

$$H(i\omega) \equiv G(z) \equiv \frac{B_0 z^{-1} + B_1 z^{-2} + \cdots}{1 - A_0 z^{-1} - A_1 z^{-2} - \cdots}\,\Delta x \qquad (z \equiv e^{i\omega\Delta x}) \tag{20.8-30}$$

Integration formulas can now be designed so as to approximate the ideal integrator response $1/i\omega$ as closely as possible in amplitude and phase. To reduce round-off-noise propagation, it is desirable that the error $|H(i\omega) - 1/i\omega|$ decrease with frequency (Ref. 20.4). The same type of analysis applies also to double-integration formulas (Sec. 20.8-7).

20.9. NUMERICAL SOLUTION OF BOUNDARY-VALUE PROBLEMS, PARTIAL DIFFERENTIAL EQUATIONS, AND INTEGRAL EQUATIONS

20.9-1. Introduction. This chapter describes numerical methods applicable to the solution of boundary-value problems involving ordinary or partial differential equations and to the solution of hyperbolic and parabolic partial differential equations. Alternative numerical-solution methods outlined in this handbook include

Reduction of partial differential equations to ordinary differential equations by separation of variables (Secs. 10.1-3 and 10.4-9), solution of characteristic equations (Secs. 10.2-2 and 10.2-4), and the method of characteristics (Sec. 10.3-2)

Reduction to a variation problem (Secs. 11.1-1, 11.7-1, and 15.4-7) and solution by direct methods such as the Rayleigh-Ritz method (Sec. 11.7-2)

Reduction to an integral equation (Sec. 15.5-2) which may be solved directly or by means of one of the approximation methods of Sec. 20.9-5

Perturbation methods for the solution of eigenvalue problems are described in Sec. 15.4-11. Many combinations of the various solution methods are possible, and vast numbers of specialized procedures have been developed for special applications (Refs. 20.9, 20.21 to 20.31, and 20.50).

The *quasilinearization method* for solving two-point boundary-value problems is outlined in Sec. 20.9-3.

20.9-2. Two-point Boundary-value Problems Involving Ordinary Differential Equations: Difference Technique (see also Secs. 9.3-3, 15.4-1, and 15.5-2). Two-point boundary-value problems requiring the solution of a given ordinary differential equation subject to boundary conditions at the end points of an interval (a, b) can often be reduced to initial-value problems by the method of Sec. 9.3-4. The following *finite-difference method* may be more convenient.

Divide the given interval (a, b) into subintervals of equal length by a net

$$x_0 = a \qquad x_1 = x_0 + \Delta x \qquad x_2 = x_0 + 2\Delta x \qquad \ldots \qquad x_n = x_0 + n\,\Delta x = b$$

and replace each derivative in the given differential equation and in the boundary conditions by a corresponding difference approximation (Sec. 20.7-1*a*, Fig. 20.9-1, and Fig. 20.9-6) of equal (or higher) order. The given ordinary differential equation is thus approximated by a difference equation. The numerical solution of the difference equation amounts to the solution of a set of simultaneous equations for the unknown function values $y_k = y(x + k\,\Delta x)$ (Secs. 20.2-5 to 20.3-2). In typical linear problems, the system matrix is usually "sparse" (Sec. 20.3-2*a*), and the problem is thus suitable for the iteration methods of Sec. 20.3-2. In particular, relaxation methods (Sec. 20.3-2*c*) are useful for "manual" computation.

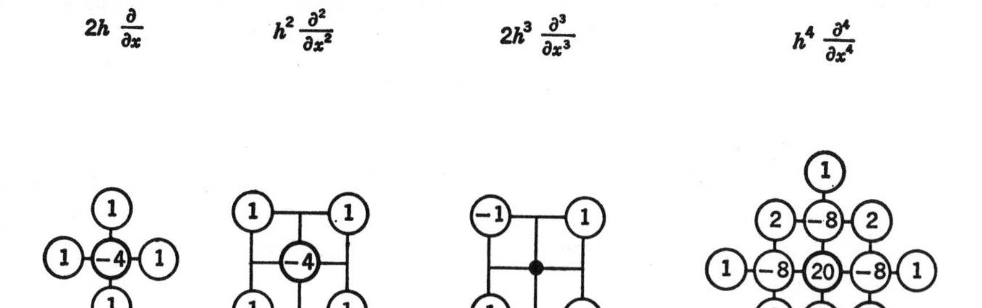

FIG. 20.9-1. Operators for central-difference approximations (rectangular cartesian coordinates: $\Delta x = \Delta y = h$). Percentage errors are of the order of h^2.

EXAMPLE: To solve $\dfrac{d^2y}{dx^2} + \dfrac{1}{5}y = 0$ subject to the boundary conditions $y(0) = 1$, $y(1) = 0$, divide the interval (0, 1) into n subintervals of length $\Delta x = 1/n$ and use $d^2y/dx^2 \approx \Delta^2 y/\Delta x^2$ to obtain the difference equation

$$y_{k+1} - 2y_k + y_{k-1} + (\tfrac{1}{5}\Delta x^2)y_k = 0 \qquad (k = 1, 2, \ldots, n-1)$$

with $y_0 = 1$, $y_n = 0$. For $n = 4$ ($k = 1, 2, 3$), the simultaneous equations to be solved for y_1, y_2, y_3 are

$$\begin{aligned} -1.99y_1 + \quad y_2 \qquad\qquad &= -1 \\ y_1 - 1.99\,y_2 + \quad y_3 &= 0 \\ y_2 - 1.99y_3 &= 0 \end{aligned}$$

20.9-3. The Generalized Newton-Raphson Method (**Quasilinearization**). While both nonlinear and linear two-point boundary-value problems can be treated numerically by the methods of Secs. 9.3-4 and 20.9-2, both these methods are most easily applied to linear differential equations. One can often find solutions of the important class of *nonlinear* two-point boundary-value problems of the form

$$\left.\begin{aligned} y'' &= f(x, y', y'') \qquad (a < x < b) \\ \text{with} \qquad \varphi[y(a), y'(a)] &= 0 \qquad \psi[y(b), y'(b)] = 0 \end{aligned}\right\} \tag{20.9-1}$$

by the following so-called **quasilinearization method.** Start with a trial olution $y^{[0]}(x)$ which satisfies the given boundary conditions, and obtain successive approximations $y^{[1]}(x)$, $y^{[2]}(x)$, . . . by solving linear problems

$$\left.\begin{aligned} y''^{[j+1]} = f(x, y^{[j]}, y'^{[j]}) &+ f_y(x, y^{[j]}, y'^{[j]})(y^{[j+1]} - y^{[j]}) \\ &+ f_{y'}(x, y^{[j]}, y'^{[j]})(y'^{[j+1]} - y'^{[j]}) \\ \text{with} \quad \varphi_y[y^{[j]}(a), y'^{[j]}(a)][y^{[j+1]}(a) &- y^{[j]}(a)] \\ &+ \varphi_{y'}[y^{[j]}(a), y'^{[j]}(a)][y'^{[j+1]}(a) - y'^{[j]}(a)] = 0 \\ \psi_y[y^{[j]}(b), y'^{[j]}(b)][y^{[j+1]}(b) &- y^{[j]}(b)] \\ &+ \psi_{y'}[y^{[j]}(b), y'^{[j]}(b)][y'^{[j+1]}(b) - y'^{[j]}(b)] = 0 \end{aligned}\right\} \tag{20.9-2}$$

where the subscripts denote partial differentiation. This technique is readily generalized to apply to systems of differential equations if one introduces the matrix notation of Sec. 13.6-1 and constitutes a generalization of the Newton-Raphson method of Sec. 20.2-8: like the latter, it may converge quite rapidly. While fairly general convergence conditions can be formulated, convergence is commonly tested by trial and error (Refs. 20.47 and 20.60).

20.9-4. Finite-difference Methods for Numerical Solution of Partial Differential Equations with Two Independent Variables. Methods analogous to that of Sec. 20.9-2 apply to the solution of problems involving partial differential equations. Introduce an appropriate net of coordinate values $x_i = x_0 + i\,\Delta x$, $y_k = y_0 + k\,\Delta y$ $(i, k = 0, \pm 1, \pm 2, \ldots)$; the unknown function $\Phi(x, y)$ will be represented by a discrete set of function values $\Phi(x_i, y_k) = \Phi_{ik}$. Approximate each differential operator by a corresponding difference operator, so that every derivative is approximated by a corresponding partial difference coefficient. The resulting difference equation will yield a system of simultaneous equations to be satisfied by the unknown function values Φ_{ik}. The most important problems lead to the following situations.

1. *A boundary-value problem for an elliptic partial differential equation* (e.g., a Dirichlet problem for $\nabla^2\Phi = 0$, Sec. 15.6-2) produces a set of N simultaneous equations for N unknowns Φ_{ik}.

The large number of equations and unknowns associated with such difference techniques makes many practical partial-differential-equation problems a challenge for even the largest digital computers. As the only redeeming feature, the nature of the difference operators resulting from the usual second-order and fourth-order linear partial differential equations (Table 10.4-1) again leads to systems of linear equations whose system matrices are "sparse," i.e., which have few nonvanishing terms except near the main diagonal. *Such problems are, then, well suited for solution by the iteration methods of Secs.* 20.3-2.

2. A *linear eigenvalue problem* (e.g., $\nabla^2\Phi = \lambda\Phi$ with suitable boundary conditions, Sec. 15.4-1) yields a matrix eigenvalue problem.
3. *Initial-value problems involving a parabolic or hyperbolic partial differential equation* (Sec. 10.3-4) also lead to sets of simultaneous equations if the differencing scheme chosen relates each unknown Φ_{ik} to other unknown as well as to known function values (*implicit methods* for solving initial-value problems).
4. If forward-difference approximations (Sec. 20.7-1) are used for the time derivatives in an initial-value problem, one obtains a set of recursion formulas for successive Φ_{ik}'s, starting with the given initial values (*explicit methods* for solving initial-value problems).

In general, explicit methods will require less computation than implicit methods for solving initial-value problems; but the recursion schemes tend to involve error-propagation or stability problems similar to those associated with the numerical solution of ordinary differential equations. As in the latter case, approximations may be improved through *predictor-corrector methods* analogous to those of Sec. 20.8-3 (Refs. 20.9, 20.16, and 20.20; see also Sec. 20.9-8).

A vast number of special techniques, including coordinate transformations reducing suitable boundaries and differencing nets to more convenient shapes, will be found in Refs. 20.9, 20.16, 20.17, and 20.20 to 20.31. Reference 20.9, in particular, contains an extensive bibliography.

20.9-5. Two-dimensional Difference Operators. Figures 20.9-1 to 20.9-6 list the most frequently needed linear difference operators. In

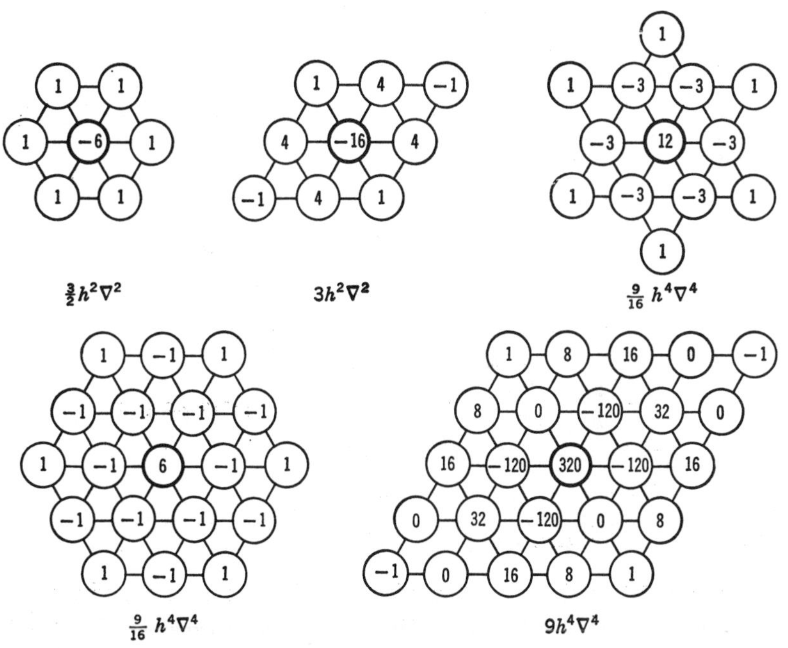

FIG. 20.9-2. Central-difference approximations for equilateral-triangle nets (side h). Percentage errors are of the order of h^2.

particular, *each diagram of Figs.* 20.9-1 *to* 20.9-5 *yields a specified central-difference expression for the center of the "star" or "molecule" as the weighted sum of the function values* Φ_{ik} *"above," "below," "to the right," etc., of the center, each weighting coefficient being indicated in its proper position.* Thus, Fig. 20.9-1 yields

$$h^2\nabla^2\Phi]_{x=x_i,y=y_k} \approx \Phi(x_i + h, y_k) + \Phi(x_i - h, y_k) + \Phi(x_i, y_k + h) + \Phi(x_i, y_k - h) - 4\Phi(x_i, y_k) = \Phi_{i+1,k} + \Phi_{i-1,k} + \Phi_{i,k+1} + \Phi_{i,k-1} - 4\Phi_{ik}$$

where $h = \Delta x = \Delta y$ is the mesh width used for both x and y.

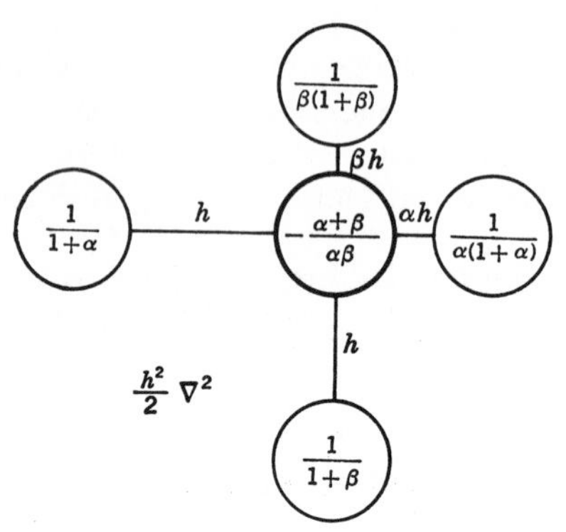

Fig. 20.9-3. A central-difference operator suitable for use with graded nets, or for transition to odd mesh widths near a boundary. The percentage error is of the order of h.

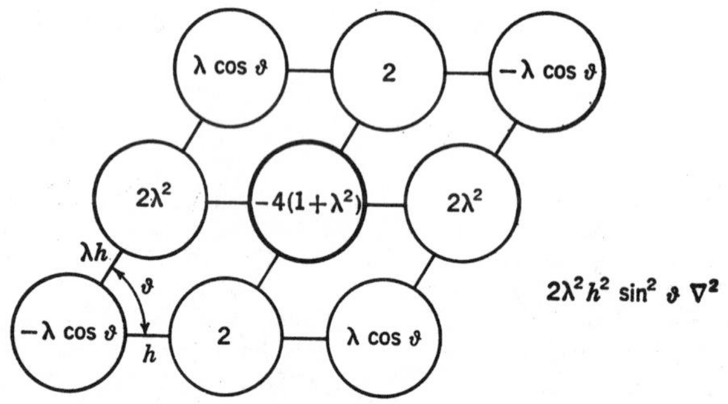

Fig. 20.9-4. A central-difference operator for use with oblique cartesian-coordinate nets. The percentage error is of the order of h^2.

Figure 20.9-1 applies to rectangular cartesian coordinates and equally spaced points. Figure 20.9-3 is useful if one wishes to change mesh width either (1) to accommodate irregular boundaries or (2) to increase the accuracy of the computation in some region of special interest (use of *graded nets*). In any case, the net used can be refined after an initial rough computation.

Figures 20.9-2, 20.9-4, and 20.9-5 show difference operators for nets other than rectangular cartesian. Figure 20.9-6 shows a few forward- and backward-difference operators.

For computation purposes, the net points (x_i, y_k) are often labeled in some simple sequence 1, 2, . . . ; the corresponding function values Φ_{ik} are then denoted by Φ_1, Φ_2, . . . (see, for example, Fig. 20.9-7). In "manual" relaxation-type computations (Sec. 20.3-2) involving problems of type 2, it is customary to use a large plan of the region and to enter

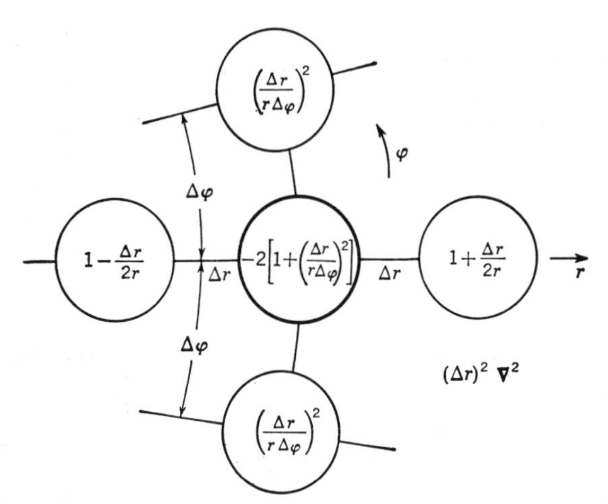

FIG. 20.9-5. A central-difference operator for use with polar coordinates r, φ. Percentage errors are of the order of $r\Delta r\Delta\varphi$; a useful graded net is given by $\Delta r = r\,\Delta\varphi$.

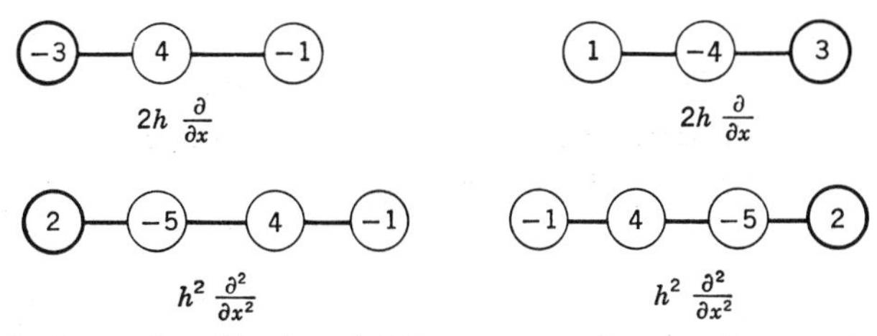

FIG. 20.9-6. Forward- and backward-difference approximations (rectangular cartesian coordinates, $\Delta x = \Delta y = h$). Percentage errors are of the order of h^2.

function values and residuals directly at each net point. Function values and residuals from earlier steps of the relaxation process are simply crossed out or erased.

NOTE: *In problems of types* 2 *and* 3, *one cannot always replace a given differential operator by a difference operator of higher order,* since the resulting difference-equation solution might contain spurious oscillation modes. On the other hand, finite-difference solutions of problems of type 1 (initial-value problems associated with hyperbolic or paraubolic differential eqations) often employ difference operators of higher orders than those of the corresponding derivatives, just as this was one in Secs. 20.9-1 to 20.9-5 (see also Refs. 20.17, 20.23, and 20.48 for a number of special methods).

20.9-6. Representation of Boundary Conditions (Fig. 20.9-7). Approximate the given boundary by mesh lines of the net used; introduce a graded net (Fig. 20.9-3 and Sec. 20.9-3*b*) if necessary. Then

1. If boundary values of the unknown function Φ are given, enter them at the appropriate boundary points.
2. If boundary values of derivatives (such as $\partial\Phi/\partial n$, Sec. 5.6-1*b*) are given, extend the net beyond the given boundary, and approximate the boundary conditions by corresponding difference relations. The resulting equations, as well as the difference relations approximating the given differential equation at points near the boundary, will involve function values at points outside the boundary; but all such function values can be eliminated by algebraic manipulation.

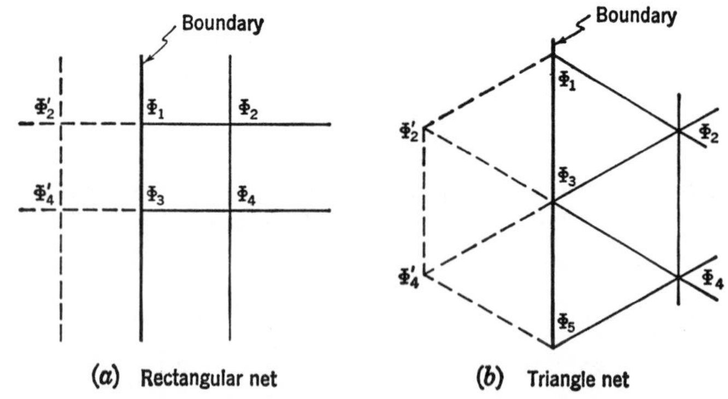

FIG. 20.9-7. Representation of boundary conditions for $\nabla^2\Phi = 0$ or $\nabla^4\Phi = 0$. The net is continued to the left of the given boundary, and "image values" $\Phi_2' = \Phi_2$, $\Phi_4' = \Phi_4$ are introduced to yield difference equations representing $\partial\Phi/\partial n = 0$ at the boundary points shown. The boundary condition $\Phi = 0$, $\nabla^2\Phi = 0$ (e.g., free edge of an elastic plate) is similarly represented by $\Phi_1 = \Phi_3 = 0$, $\Phi_2' = -\Phi_2$, $\Phi_4' = -\Phi_4$.

EXAMPLE: Figure 20.9-7*b* shows how the boundary condition $\partial\Phi/\partial n = 0$ is approximated through "reflection" of function values in the boundary, so that $\Phi_2' - \Phi_2 = 0$, $\Phi_4' - \Phi_4 = 0$. For a given differential equation, say $\nabla^2\Phi = 0$, apply the first difference operator of Fig. 20.9-2 to obtain

$$\Phi_2' + \Phi_4' + \Phi_5 + \Phi_4 + \Phi_2 + \Phi_1 - 6\Phi_3 = 0$$

or, since the boundary condition implies $\Phi_2' = \Phi_2$, $\Phi_4' = \Phi_4$,

$$2\Phi_2 + 2\Phi_4 + \Phi_5 + \Phi_1 - 6\Phi_3 = 0$$

The last equation involves only points inside or on the boundary.

20.9-7. Problems Involving Three or More Independent Variables. Analogous methods apply to problems involving three or more variables. In particular,

$$\left.\begin{aligned} h^2\nabla^2\Phi(x, y, z) \approx \Phi(x+h, y, z) + \Phi(x-h, y, z) \\ + \Phi(x, y+h, z) + \Phi(x, y-h, z) \\ + \Phi(x, y, z+h) + \Phi(x, y, z-h) - 6\Phi(x, y, z) \end{aligned}\right\} \qquad (20.9\text{-}3)$$

The number of independent variables can often be reduced through separation of variables (Sec. 10.1-3).

20.9-8. Validity of Finite-difference Approximations. Some Stability Conditions. (a) Every difference-approximation solution requires an investigation of its validity. The error propagation resulting from two different difference-equation approximations to the same differential equation may differ radically even though the same mesh widths are used. It is, moreover, not always possible to improve the approximation by decreases in mesh widths, even if exact solutions of the difference equations (with zero relaxation residuals) are available. For a proper finite-difference approximation, the solution of the difference equation should converge to that of the given differential equation in a nonoscillatory manner. Implicit-solution schemes for linear partial differential equations (Sec. 20.9-4) rarely lead to difficulties of this type, but in explicit methods for parabolic and hyperbolic differential equations, the various coordinate increments may have to satisfy special *stability conditions.*

(b) In particular, *the solution of the difference equation*

$$\frac{\Phi(x+\Delta x, t) - 2\Phi(x, t) + \Phi(x-\Delta x, t)}{\Delta x^2} = \frac{1}{a^2}\,\frac{\Phi(x, t+\Delta t) - \Phi(x, t)}{\Delta t}$$

or (in recursion-relation form)

$$\Phi(x, t+\Delta t) = \frac{a^2\,\Delta t}{\Delta x^2}\{\Phi(x-\Delta x, t) + \Phi(x+\Delta x, t)\} + \left(1 - \frac{2a^2\,\Delta t}{\Delta x^2}\right)\Phi(x, t)$$

converges to the solution of the partial differential equation (*heat-conduction equation,* Sec. 10.4-1)

$$\frac{\partial^2\Phi}{\partial x^2} = \frac{1}{a^2}\frac{\partial\Phi}{\partial t}$$

if $\Delta t \to 0$, $\Delta x \to 0$ *so that* $\Delta t < \frac{1}{2a^2}\Delta x^2$. $\Delta t = \frac{1}{6a^2}\Delta x^2$ is a useful combination, which will also improve the approximation for finite increments (Ref. 20.9).

By contrast, the *implicit* method based on the *Crank-Nicholson approximation*

$$-\frac{a^2\,\Delta t}{2\Delta x^2}\left[\Phi(x-\Delta x,\,t+\Delta t)+\Phi(x+\Delta x,\,t+\Delta t)\right]+\left(1+\frac{a^2\,\Delta t}{\Delta x^2}\right)\Phi(x,\,t+\Delta t)$$
$$=\left(1-\frac{a^2\,\Delta t}{\Delta x^2}\right)\Phi(x,\,t)+\frac{a^2\,\Delta t}{\Delta x^2}\left[\Phi(x-\Delta x,\,t)+\Phi(x+\Delta x,\,t)\right]$$

will yield approximations which converge to the correct solution of the heat-conduction equations whenever $\Delta x \to 0$, $\Delta t \to 0$.

(c) Again, *the solution of the difference equation*

$$\frac{\Phi(x+\Delta x,\,t)-2\Phi(x,\,t)+\Phi(x-\Delta x,\,t)}{\Delta x^2}=\frac{1}{c^2}\,\frac{\Phi(x,\,t+\Delta t)-2\Phi(x,\,t)+\Phi(x,\,t-\Delta t)}{\Delta t^2}$$

converges to the solution of the wave equation

$$\frac{\partial^2\Phi}{\partial x^2}=\frac{1}{c^2}\frac{\partial^2\Phi}{\partial t^2}$$

if $\Delta t \to 0$, $\Delta x \to 0$ *so that* $\Delta t < \frac{1}{c}\Delta x$.

(d) Reference 20.9 lists numerous other approximation formulas and stability conditions for the heat-conduction and wave equations.

20.9-9. Approximation-function Methods for Numerical Solution of Boundary-value Problems. (a) Approximation by Functions Which Satisfy the Boundary Conditions Exactly. *Let x be either a one-dimensional variable or a multidimensional variable $x = (x_1, x_2, \ldots, x_n)$* (Sec. 15.4-1). It is desired to solve an ordinary or partial differential equation

$$\mathbf{L}\Phi(x) = f(x) \qquad (x \text{ in } V) \tag{20.9-4a}$$

subject to a boundary condition

$$\mathbf{B}\Phi(x) = b(x) \qquad (x \text{ in } S) \tag{20.9-4b}$$

on the boundary S of a given region V (Sec. 15.4-1). Approximate the desired solution $\Phi(x)$ by an *approximation function*

$$\varphi = \varphi(x;\, \alpha_1, \alpha_2, \ldots, \alpha_m) \tag{20.9-5a}$$

which satisfies the given boundary conditions and depends on m parameters $\alpha_1, \alpha_2, \ldots, \alpha_m$. In many applications, the equations (2) are linear, and one approximates $\Phi(x)$ by a linear combination of known functions

$$\varphi = \sum_{k=1}^{m} \alpha_k \varphi_k(x) \tag{20.9-5b}$$

where $\varphi_1(x)$ satisfies the given boundary condition (4b), and $\varphi_2(x)$, $\varphi_3(x)$, . . . , $\varphi_m(x)$ satisfy the associated homogeneous boundary condition $\mathbf{B}\Phi(x) = 0$ (x in S) (see also Sec. 15.4-2). The error due to the use of the approximation (5a) or (5b) is a function of the α_k, given by

$$E(x; \alpha_1, \alpha_2, \ldots, \alpha_m) \equiv \mathbf{L}\varphi(x; \alpha_1, \alpha_2, \ldots, \alpha_m) - f(x) \tag{20.9-6}$$

One determines the unknown parameters $\alpha_1, \alpha_2, \ldots, \alpha_m$ in Eq. (5a) or (5b) by one of the following schemes:

1. **Collocation.** $\varphi(x; \alpha_1, \alpha_2, \ldots, \alpha_m)$ is made to satisfy the given differential equation (4a) exactly at m chosen points $x = X_1$, $x = X_2, \ldots, x = X_m$; the resulting m equations

$$E(X_i; \alpha_1, \alpha_2, \ldots, \alpha_m) = 0 \qquad (i = 1, 2, \ldots, m) \tag{20.9-7}$$

are solved for $\alpha_1, \alpha_2, \ldots, \alpha_m$.

2. **Least-squares Approximations** (see also Sec. 20.6-2). Choose the α_k so as to minimize the mean-square error

$$J(\alpha_1, \alpha_2, \ldots, \alpha_m) \equiv \int_V |E(\xi; \alpha_1, \alpha_2, \ldots, \alpha_m)|^2 \, d\xi \tag{20.9-8}$$

or the cruder mean-square error

$$J'(\alpha_1, \alpha_2, \ldots, \alpha_m) \equiv \sum_{h=1}^{N} b_h |E(X_h; \alpha_1, \alpha_2, \ldots, \alpha_m)|^2 \tag{20.9-9}$$

where the b_h are suitably determined weighting coefficients associated with N chosen points $X_1, X_2, \ldots, X_N$ in V. One may, for instance, space the points X_h evenly and let

$$b_1 = b_2 = \cdots = b_N = 1$$

The α_k are determined by the m conditions

$$\frac{\partial J}{\partial \alpha_k} = 0 \qquad \text{or} \qquad \frac{\partial J'}{\partial \alpha_k} = 0 \qquad (k = 1, 2, \ldots, m) \tag{20.9-10}$$

3. **Galerkin's Weighting-function Method.** Choose m linearly independent "weighting functions" $\Psi_1(x), \Psi_2(x), \ldots, \Psi_m(x)$ (one frequently uses $\Psi_k = \varphi_k$), and determine the α_k so that

$$\int_V \Psi_i^*(\xi) E(\xi; \alpha_1, \alpha_2, \ldots, \alpha_m) \, d\xi = 0 \qquad (i = 1, 2, \ldots, m) \tag{20.9-11}$$

If the given equations (4) are linear (linear boundary-value problem), then an approximation of the form (5*b*) will yield *linear* equations (7), (10), or (11).

(b) Approximation by Functions Which Satisfy the Differential Equation Exactly. It is frequently preferable to use an approximation function (5) which satisfies the given differential equation (3*a*) exactly for all x in V and to determine the parameters α_k so as to match the given boundary conditions in the sense of the collocation method (*boundary collocation*), the least-squares method, or the Galerkin method.

20.9-10. Numerical Solution of Integral Equations (see also Sec. 15.3-2). **(a)** The solution $\Phi(x)$ of a linear integral equation

$$\Phi(x) - \lambda \int_V K(x, \xi)\Phi(\xi)\, d\xi = F(x) \qquad (20.9\text{-}12)$$

can often be approximated by the *iteration method* of Sec. 15.3-8*a*; or one may *approximate the given kernel* $K(x, \xi)$ by a polynomial or other degenerate kernel (Sec. 15.3-1*b*) to simplify the solution. Eigenvalue problems ($F \equiv 0$) can often be treated as variation problems (Sec. 15.3-6).

(b) The *approximation-function methods* of Sec. 20.9-9 also apply directly to the numerical solution of a linear integral equation (12). Use an approximation of the form (5*b*) and compute the m functions

$$f_k(x) = \varphi_k(x) - \lambda \int_V K(x, \xi)\varphi_k(\xi)\, d\xi \qquad (k = 1, 2, \ldots, m) \qquad (20.9\text{-}13)$$

Then

1. The *collocation method* yields the m linear equations

$$\sum_{k=1}^{m} \alpha_k f_k(X_i) = F(X_i) \qquad (i = 1, 2, \ldots, m) \qquad (20.9\text{-}14)$$

2. The *least-squares method* yields the m linear equations

$$\left.\begin{aligned} \sum_{k=1}^{m} A_{ik}\alpha_k &= \beta_i \qquad (i = 1, 2, \ldots, m) \\ \text{with} \qquad A_{ik} &= \sum_{h=1}^{N} b_h f_i^*(X_h) f_k(X_h) \\ \beta_i &= \sum_{h=1}^{N} b_h f_i^*(X_h) F(X_h) \end{aligned}\right\} \qquad (20.9\text{-}15)$$

3. *Galerkin's method* yields the m linear equations

$$\left.\begin{aligned} \sum_{k=1}^{m} A_{ik}\alpha_k &= \beta_i \qquad (i = 1, 2, \ldots, m) \\ \text{with} \qquad A_{ik} &= \int_V \Psi_i^*(\xi) f_k(\xi)\, d\xi \\ \beta_i &= \int_V \Psi_i^*(\xi) F(\xi)\, d\xi \end{aligned}\right\} \tag{20.9-16}$$

20.10. MONTE-CARLO TECHNIQUES

20.10-1. Monte-Carlo Methods (Refs. 20.42 and 20.46). In principle, every **Monte-Carlo computation** may be considered as the estimation of a suitable integral

$$I = \int_{-\infty}^{\infty} \int_{-\infty}^{\infty} \cdots \int_{-\infty}^{\infty} f(\lambda_1, \lambda_2, \ldots, \lambda_N)\, d\Phi(\lambda_1, \lambda_2, \ldots, \lambda_N) \tag{20.10-1}$$

by a random-sample average

$$\overline{f(x_1, x_2, \ldots, x_N)} = \frac{1}{n} \sum_{k=1}^{n} f({}^kx_1, {}^kx_2, \ldots, {}^kx_N) \tag{20.10-2}$$

where $(x_1, x_2, \ldots, x_N)$ is a (generally multidimensional, $N > 1$) random variable with known distribution function $\Phi(x_1, x_2, \ldots, x_N)$.

Monte-Carlo techniques are widely useful in investigations of random-process phenomena too complicated for explicit solution by probability theory, viz., neutron-diffusion problems, detection and communication problems, and a vast variety of operations-research studies. In addition, it often pays to recast other types of problems, especially those involving complicated multidimensional integrals, in a form suitable for Monte-Carlo solution.

For simplicity, the following discussion is based on Monte-Carlo estimation of the *one-dimensional* integral

$$I = \int_{-\infty}^{\infty} f(\lambda)\, d\Phi(\lambda) \tag{20.10-3}$$

by

$$\overline{f(x)} = \frac{1}{n} [f({}^1x) + f({}^2x) + \cdots + f({}^nx)] \tag{20.10-4}$$

The variance of the estimate (4) of (3) on the basis of a random sample $({}^1x, {}^2x, \ldots, {}^nx)$, is

$$\text{Var}\,\{\overline{f(x)}\} = \frac{1}{n} \text{Var}\,\{f(x)\} \tag{20.10-5}$$

so that the rms fluctuation decreases only as $1/\sqrt{n}$ with increasing n (Sec. 19.2-3). The estimate variance is due to the random fluctuation in the distribution of different samples $({}^1x, {}^2x, \ldots, {}^nx)$.

20.10-2. Two Variance-reduction Techniques (Ref. 20.46). The following techniques attempt to "doctor" the sample $({}^1x, {}^2x, \ldots, {}^nx)$ so as to reduce the variance of the sample mean while still preserving the relation

$$E\{\overline{f(x)}\} = E\{f(x)\} = I \tag{20.10-6}$$

i.e., without biasing the estimate.

(a) Stratified Sampling. One divides the range of the random variable x into a number of suitably chosen class intervals $\xi_{j-1} < x \leq \xi_j$ and agrees to fix the number n_j of otherwise independent sample values ${}^kx = {}^ix_j (i = 1, 2, \ldots, n_j)$ falling into the jth class interval. Assuming a priori knowledge of the probabilities

$$P_j = P[\xi_{j-1} < x \leq \xi_j] = \Phi(\xi_j) - \Phi(\xi_{j-1}) \tag{20.10-7}$$

associated with the class intervals (e.g., on the basis of symmetry, uniform distribution, etc.), one can employ the *stratified-sample average*

$$\overline{f(x)}_{\text{STRAT}} = \sum_j P_j \frac{1}{n_j} \left[\sum_{i=1}^{n_j} f({}^ix_j) \right] \tag{20.10-8}$$

as an unbiased estimate of I, with

$$\text{Var}\,\{\overline{f(x)}_{\text{STRAT}}\} = \sum_j \frac{P_j{}^2}{n_j} \text{Var}\,\{f({}^ix_j)\} \tag{20.10-9}$$

Note that repeated stratified samples will differ only within class intervals. The variance (9) can be smaller than the random-sample variance Var $\{\overline{f(x)}\}/n$ with $n = \sum_j n_j$ if a priori information permits a favorable choice of the ξ_j and n_j. In principle, it would be best to choose class intervals for equal variances

$$\text{Var}\,\{f({}^ix_j)\} = \frac{1}{P_j} \int_{\xi_{j-1}}^{\xi_j} f^2(\lambda)\, d\Phi(\lambda) - \left[\frac{1}{P_j} \int_{\xi_{j-1}}^{\xi_j} f(\lambda)\, d\Phi(\lambda) \right]^2 \tag{20.10-10}$$

and then to assign the theoretically correct number of samples to each class interval, i.e.,

$$n_j = nP_j \tag{20.10-11}$$

In this ideal case, one should have the relatively small estimate variance

$$\text{Var}\,\{\overline{f(x)}_{\text{STRAT}}\} = \frac{1}{n} \text{Var}\,\{f({}^ix_j)\} \tag{20.10-12}$$

As the class intervals are decreased, the stratified-sampling techniques will produce results analogous to that of an integration formula, but ordinarily the class intervals are larger; practical applications are usually multidimensional, so that simple symmetry relations may yield favorable class intervals.

(b) Use of Correlated Samples. If individual sample values ${}^kx$ are not statistically independent (as they would be in a true random sample), the expression (9) for the estimate variance is replaced by

$$\text{Var}\,\{\overline{f(x)}_{\text{CORREL}}\} = \frac{1}{n} \text{Var}\,\{x\} + \frac{2}{n^2} \sum_{i<k} \sum \text{Cov}\,\{{}^ix, {}^kx\} \tag{20.10-13}$$

(see also Sec. 19.8-1). Judiciously introduced *negative correlation* between selected sample-value pairs ${}^ix$, ${}^kx$ will produce negative covariance terms in Eq. (13) and may reduce the variance well below the random-sample variance Var $\{f(x)\}/n$ without biasing the estimate.

As a simple example (Ref. 20.42), let x be uniformly distributed between $x = 0$ and $x = 1$, and let $f(x)$ be the monotonic function $(e^x - 1)/(e - 1)$. One designs the sample so that n is even, and ${}^2x = 1 - {}^1x$, ${}^4x = 1 - {}^3x$, . . . , ${}^nx = 1 - {}^{n-1}x$, with sample values otherwise independent. Since $f(x)$ and $f(1 - x)$ are negatively correlated,

$$\mathrm{Var}\,\{\overline{f(x)}_{\mathrm{CORREL}}\} \approx \tfrac{1}{31}\,\mathrm{Var}\,\{\overline{f(x)}\}$$

so that the rms fluctuation is reduced by a factor of about 5.6. In addition, the correlated sample requires one to generate fewer random numbers. More interesting applications are, again, to multidimensional problems. Note that stratified sampling, in effect, also introduces negative correlation between sample values: ${}^{k+1}x$ can no longer fall into a given class interval if ${}^kx$ has filled the latter.

20.10-3. Use of A Priori Information: Importance Sampling. As a matter of principle, Monte-Carlo computations often can and should be simplified through judicious application of partial a priori knowledge of results. As a case in point, *importance-sampling* techniques attempt to estimate an integral (1) by a sample average $[\overline{f(y)/g(y)}]$, where y is a random variable with probability density

$$\varphi_y(y) = g(y)\,\frac{d\Phi(y)}{dy} \tag{20.10-14}$$

The estimate is easily seen to be unbiased. The function $g(y)$ is chosen so that

$$\mathrm{Var}\left\{\frac{f(y)}{g(y)}\right\} = E\left\{\left[\frac{f(y)}{g(y)} - I\right]^2\right\} \tag{20.10-15}$$

is small, subject to the constraint $\int_{-\infty}^{\infty} \varphi_y(y)\,dy = 1$. In particular, $g(y) = f(y)/I$ would reduce the variance (15) to zero, but this would require knowledge of the unknown quantity I. Importance sampling permits one to "concentrate" sampling near values of y of special interest, e.g., where $f(y)$ varies rapidly.

20.10-4. Some Random-number Generators. Tests for Randomness (Refs. 20.9, 20.51, and 20.55). **Congruential methods** for generating pseudo-random numbers x_i less than a given nonnegative **modulus** m start with any nonnegative $x_0 < m$ and compute successive values

$$x_i = [ax_{i-1} + c]_{\mathrm{mod}\ m} \qquad (i = 1, 2, \ldots; 0 < a < m, 0 \le c < m) \tag{20.10-16}$$

where modulo-m addition is defined in the manner of Sec. 12.2-10. On a binary computing machine, the modulus m is conveniently chosen equal to $2^{\text{computer word length}}$. For $c = 0$, the generator is called a **multiplicative congruential generator;** otherwise, it is a **mixed congruential generator.**

Sequences obtained in this manner are not truly random but may have "pseudo-random" properties, such as a uniform distribution between 0 and m, zero corrrelation between different x_i, random-appearing runs of odd and even numbers, etc. The uniform distribution may be tested with a χ^2 test (Sec. 19.6-7), and serial corelation may be tested in the manner of Sec. 19.7-4. Even with zero correlation, samples $(x_1, x_2, \ldots, x_n)$ taken from a pseudo-random-number sequence will *not* be statistically independent, a fact which, depending on the specific application, can result in disagreeable surprises. It may well be wise to obtain true random samples for

Monte-Carlo computations by analog-to-digital conversion of true analog noise (Ref. 20.55).

Pseudo-random numbers having other than uniform distributions are readily obtained as functions $F(x_i)$ of uniformly distributed pseudo-random variables. A number of other methods are discussed in Ref. 20.9.

20.11. RELATED TOPICS, REFERENCES, AND BIBLIOGRAPHY

20.11-1. Related Topics. The following topics related to the study of numerical computations are treated in other chapters of this handbook.

Theory of equations.................................... Chap. 1
Ordinary differential equations.................................... Chaps. 9, 13
Partial differential equations.................................... Chap. 10
Linear and nonlinear programming, dynamic programming,
calculus of variations.................................... Chap. 11
Matrices, diagonalization.................................... Chaps. 13, 14
Eigenvalue problems.................................... Chaps. 14, 15
Boundary-value problems.................................... Chap. 15
Regression.................................... Chaps. 18, 19

20.11-2. References and Bibliography.

General

20.1. Abramowitz, M., and I. A. Stegun (eds.): *Handbook of Mathematical Functions* National Bureau of Standards, Washington, D.C., 1964.
20.2. Booth, A. D.: *Numerical Methods,* Butterworth, London, 1955.
20.3. Collatz, L.: Numerical Methods, in *Handbuch der Physik,* vol. 2, Springer, Berlin, 1955.
20.4. Hamming, R. W.: *Numerical Methods for Engineers and Scientists,* McGraw-Hill, New York, 1962.
20.5. Henrici, P.: *Elements of Numerical Analysis,* Wiley, New York, 1964.
20.6. Hildebrand, F. B.: *Introduction to Numerical Analysis,* McGraw-Hill, New York, 1956.
20.7. Householder, A. S.: *Principles of Numerical Analysis,* McGraw-Hill, New York, 1953.
20.8. Jennings, W.: *First Course in Numerical Methods,* Macmillan, New York, 1964.
20.9. Klerer, M., and G. A. Korn (eds.): *Digital Computer User's Handbook,* McGraw-Hill, New York, 1967.
20.10. Kopal, Z.: *Numerical Analysis,* Wiley, New York, 1955.
20.11. Kunz, K. S.: *Numerical Analysis,* McGraw-Hill, New York, 1957.
20.12. Lanczos, C.: *Applied Analysis,* Prentice-Hall, Englewood Cliffs, N.J., 1956.
20.13. Milne, W. E.: *Numerical Solution of Differential Equations,* Wiley, New York, 1953.
20.14. Noble, B.: *Numerical Methods,* Oliver & Boyd, London, 1964.
20.15. Ralston, A.: *A First Course in Numerical Analysis,* McGraw-Hill, New York, 1964.
20.16. ——— and H. S. Wilf: *Mathematical Methods for Digital Computers,* 2 vols., Wiley, New York, 1960 and 1967.

20.17. Salvadori, M. G., and M. L. Baron: *Numerical Methods in Engineering*, 2d ed., Prentice-Hall, Englewood Cliffs, N.J., 1961.
20.18. Scarborough, J. B.: *Numerical Mathematical Analysis*, 5th ed., Johns Hopkins, Baltimore, 1962.
20.19. Stiefel, E. L.: *An Introduction to Numerical Mathematics*, Academic, New York, 1963.
20.20. Todd, J.: *Survey of Numerical Analysis*, McGraw-Hill, New York, 1962.

Linear Equations, Matrix Problems, and Partial Differential Equations

20.21. Allen, D. N.: *Relaxation Methods*, McGraw-Hill, New York, 1954.
20.22. Fadeev, D. K., and V. N. Fadeeva: *Computational Methods in Linear Algebra*, Freeman, San Francisco, 1963.
20.23. Forsythe, G. E., and W. R. Wasow: *Finite-difference Methods for Partial Differential Equations*, Wiley, New York, 1960.
20.24. Fox, L.: *An Introduction to Numerical Linear Algebra*, Oxford, Fair Lawn, N.J., 1964.
20.25. Householder, A. S.: *The Theory of Matrices in Numerical Analysis*, Blaisdell, New York, 1964.
20.26. Paige, L. J., and O. Taussky: *Simultaneous Linear Equations and the Determination of Eigenvalues*, National Bureau of Standards Applied Mathematics Series 29, 1953.
20.27. Shaw, F. S.: *An Introduction to Relaxation Methods*, Dover, New York, 1953.
20.28. Southwell, R. V.: *Relaxation Methods in Engineering Science*, Oxford, Fair Lawn, N.J., 1940.
20.29. ———: *Relaxation Methods in Theoretical Physics*, Oxford, Fair Lawn, N.J., 1946.
20.30. Varga, R. S.: *Matrix Iterative Analysis*, Prentice-Hall, Englewood Cliffs, N.J., 1962.
20.31. Wilkinson, J. H.: *The Algebraic Eigenvalue Problem*, Oxford, Fair Lawn, N.J., 1965.

(See also the articles by J. H. Wilkinson and by W. K. Karplus and V. Vemuri in Ref. 20.9.)

Finite Differences and Difference Equations

20.32. Goldberg, S.: *Introduction to Difference Equations*, Wiley, New York, 1958.
20.33. Jolley, L. B.: *Summation of Series*, Chapman & Hall, London, 1925; reprinted, Dover, New York, 1960.
20.34. Jordan, C.: *Calculus of Finite Differences*, Chelsea, New York, 1947.
20.35. Jury, E. I.: *Theory and Application of the Z-transform Method*, Wiley, New York, 1964.
20.36. Milne-Thomson, L. N.: *The Calculus of Finite Differences*, Macmillan, London, 1951.
20.37. Ragazzini, J. R., and G. F. Franklin: *Sampled-data Control Systems*, McGraw-Hill, New York, 1958.

Approximation Methods

20.38. Carlson, B., and M. Goldstein: Rational Approximation of Functions, Los Alamos Scientific Laboratory, *Rept. LA-1943*, 1955.
20.39. Davis, P. J.: *Interpolation and Approximation*, Blaisdell, New York, 1963.

20.40. Dunham, C. B.: Convergence Problems in Maehly's Second Method, *J. ACM*, April, 1965.
20.41. Hastings, C.: *Approximations for Digital Computers*, Princeton, Princeton, N.J., 1955.
20.42. Maehly, H.: *First Interim Progress Report on Rational Approximations*, Project NR-044-196, Princeton University, 1958.
20.43. ———: Methods for Fitting Rational Approximations, *J. ACM*, **10**: 257 (1963).
20.44. Meinardus, G.: *Approximation von Funktionen und ihre numerische Behandlung*, Springer, Berlin, 1964.
20.45. Rice, J. R.: *The Approximation of Functions*, Addison-Wesley, Reading, Mass., 1964.
20.46. Snyder, M. A.: *Chebyshev Methods in Numerical Approximation*, Prentice-Hall, Englewood Cliffs, N.J., 1966.

Miscellaneous

20.47. Bellman, R. E., and R. E. Kalaba: *Quasilinearization and Nonlinear Boundary-value Problems*, Elsevier, New York, 1965.
20.48. Collatz, L.: *Eigenwertaufgaben mit technischen Anwendungen*, Akademische Verlagsgesellschaft m.b.H., Leipzig, 1949.
20.49. Fox, L.: *Numerical Solution of Two-point Boundary-value Problems*, Oxford, Fair Lawn, N.J., 1957.
20.50. ———: *Numerical Solution of Ordinary and Partial Differential Equations*, Pergamon Press, New York, 1962.
20.51. Hammersley, J. M., and D. C. Handscomb: *Monte Carlo Methods*, Wiley, New York, 1964.
20.52. Henrici, P.: *Discrete-variable Methods in Ordinary Differential Equations*, Wiley, New York, 1963.
20.53. ———: *Error Propagation for Difference Methods*, Wiley, New York, 1963.
20.54. Ostrowski, A. M.: *Solution of Equations and Systems of Equations*, Academic, New York, 1960.
20.55. Schreider, Y. A.: *Method of Statistical Testing (Monte Carlo Method)*, Elsevier, New York, 1964.
20.56. Stroud, A. H., and D. Secrest: *Gaussian Quadrature Formulas*, Prentice-Hall, Englewood Cliffs, N.J., 1966.
20.57. Wilde, D. J.: *Optimum-seeking Methods*, Prentice-Hall, Englewood Cliffs, N.J., 1964.
20.58. Wilkinson, J. H.: *Rounding Errors in Algebraic Processes*, Prentice-Hall, Englewood Cliffs, N.J., 1964.
20.59. Rall, L. B. (ed.): *Error in Digital Computation*, Wiley, New York, 1965.
20.60. Roberts, S. M., and J. S. Shipman: The Kantorovich Theorem and Two-point Boundary-value Problems, *IBM J. Res.*, September, 1966.
20.61. Remez, E.: General Computation Methods for Chebychev Approximation, *Izv. Akad. Nauk Ukrainsk. SSR*, Kiev, 1957.
20.62. Whittaker, E., and G. Robinson: *The Calculus of Observations*, 4th ed., Blackie, Glasgow, 1944.
20.63. Lindorff, D. P.: *Theory of Sampled-data Control Systems*, Wiley, New York, 1965.
20.64. Fehlberg, E.: New One-step Integration Methods of High-order Accuracy, *NASA TR R-248*, George C. Marshall Space Flight Center, Huntsville, Ala., 1966.

20.65. Huskey, H. D., and G. A. Korn (eds.): *Computer Handbook*, McGraw-Hill, New York, 1962.
20.66. Korn, G. A., and T. M. Korn: *Electronic Analog and Hybrid Computers*, McGraw-Hill, New York, 1964.
20.67. Brand, L.: *Differential and Difference Equations*, Wiley, New York, 1966.
20.68. Warten, R. M.: Automatic Step-size Control for Runge-Kutta Integration, *IBM J. Res.*, October, 1963.
20.69. Beckett, R., and J. Hurt: *Numerical Calculations and Algorithms*, McGraw-Hill, New York, 1967.
20.70. Cooley, J. W., and J. W. Tukey: An Algorithm for the Machine Calculation of Complex Fourier Series, *Math. Comp.*, **19**:297, April, 1965.

CHAPTER **21**

SPECIAL FUNCTIONS

21.1. INTRODUCTION

21.1-1. Chapter 21 is essentially a collection of formulas relating to special functions. Refer to Chap. 7 for the relevant complex-variable theory, and to Chaps. 9, 10, and 15 for a treatment of differential equations. References 21.3 and 21.9 deal with the less frequently encountered special transcendental functions.

21.2. THE ELEMENTARY TRANSCENDENTAL FUNCTIONS

21.2-1. The Trigonometric Functions (see also Secs. 21.1-1 and 21.2-12 and Table 7.2-1). **(a)** The trigonometric functions $w = \sin z$, $w = \cos z$ are defined by their power series (Sec. 21.2-12), as solutions of the differential equation $\frac{d^2w}{dz^2} + w = 0$, by $z = \arcsin w$, $z = \arccos w$ (integral representation, Sec. 21.2-4), or, for real z, in terms of right-triangle geometry (goniometry, Fig. 21.2-1). The remaining trigonometric functions are defined by

$$\tan z = \frac{\sin z}{\cos z} \qquad \cot z = \frac{1}{\tan z} = \frac{\cos z}{\sin z} \tag{21.2-1}$$

$$\sec z = \frac{1}{\cos z} \qquad \operatorname{cosec} z = \frac{1}{\sin z} \tag{21.2-2}$$

(b) $\sin z$ and $\cos z$ are periodic with period 2π; $\tan z$ and $\cot z$ are periodic with period π. $\sin z$, $\tan z$, and $\cot z$ are odd functions, whereas $\cos z$ is an even function. Figure 21.2-2 shows graphs of $\sin z$, $\cos z$, $\tan z$, and $\cot z$ for real arguments. Figure 21.2-3 shows triangles which

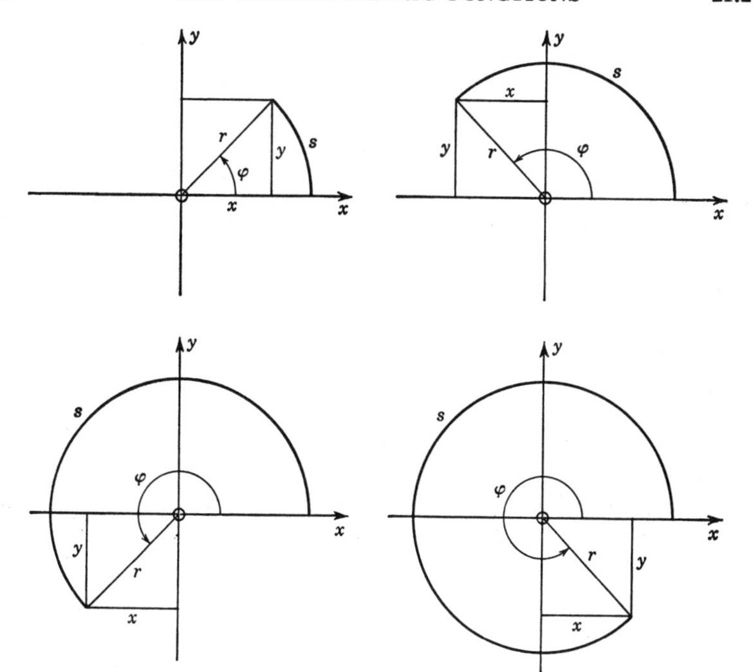

FIG. 21.2-1. Definitions of circular measure and trigonometric functions for a given angle φ:

$$\varphi = \frac{s}{r} \text{ (in radians)}$$

$$\sin \varphi = \frac{y}{r} \qquad \cos \varphi = \frac{x}{r}$$

$$\tan \varphi = \frac{y}{x} = \frac{\sin \varphi}{\cos \varphi} \qquad \cot \varphi = \frac{x}{y} = \frac{\cos \varphi}{\sin \varphi}$$

$$\sec \varphi = \frac{r}{x} = \frac{1}{\cos \varphi} \qquad \operatorname{cosec} \varphi = \frac{r}{y} = \frac{1}{\sin \varphi}$$

serve as memory aids for the derivation of function values for $z = \pi/6 = 30$ deg, $\pi/4 = 45$ deg, and $\pi/3 = 60$ deg (see also Table 21.2-1).

(c) The relations

$$\begin{aligned} \sin z &= \cos\left(\frac{\pi}{2} - z\right) & \cos z &= \sin\left(\frac{\pi}{2} - z\right) \\ \tan z &= \cot\left(\frac{\pi}{2} - z\right) & \cot z &= \tan\left(\frac{\pi}{2} - z\right) \end{aligned} \tag{21.2-3}$$

permit one to express trigonometric functions of any real argument in

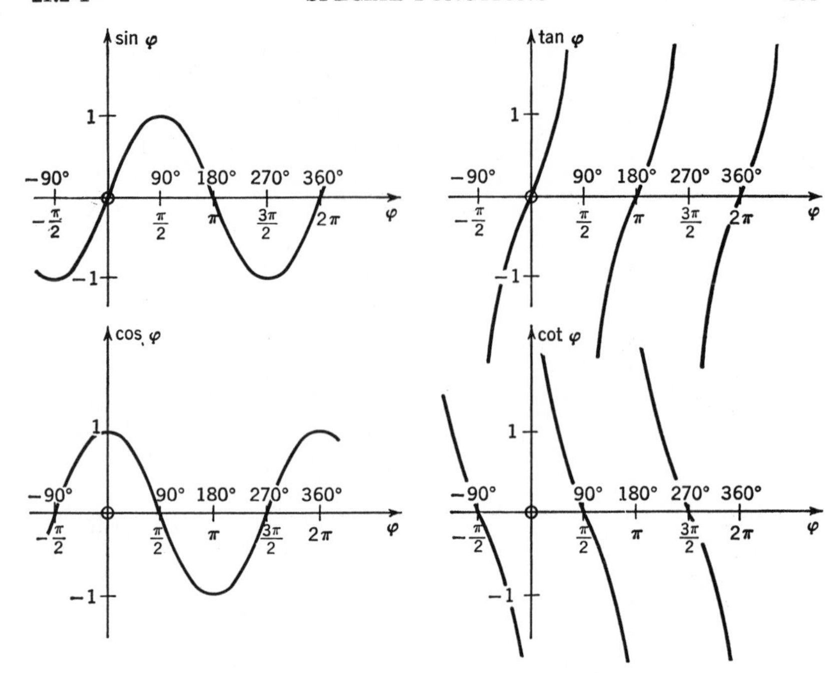

FIG. 21.2-2. Plots of the trigonometric functions for real arguments $z = \varphi$.

Table 21.2-1. Special Values of Trigonometric Functions

A (degrees)	0° 360°	30°	45°	60°	90°	180°	270°
A (radians)	0	$\frac{\pi}{6}$	$\frac{\pi}{4}$	$\frac{\pi}{3}$	$\frac{\pi}{2}$	π	$\frac{3\pi}{2}$
$\sin A$	0	$\frac{1}{2}$	$\frac{1}{\sqrt{2}}$	$\frac{1}{2}\sqrt{3}$	1	0	-1
$\cos A$	1	$\frac{1}{2}\sqrt{3}$	$\frac{1}{\sqrt{2}}$	$\frac{1}{2}$	0	-1	0
$\tan A$	0	$\frac{1}{\sqrt{3}}$	1	$\sqrt{3}$	$\pm\infty$	0	$\pm\infty$
$\cot A$	$\pm\infty$	$\sqrt{3}$	1	$\frac{1}{\sqrt{3}}$	0	$\pm\infty$	0

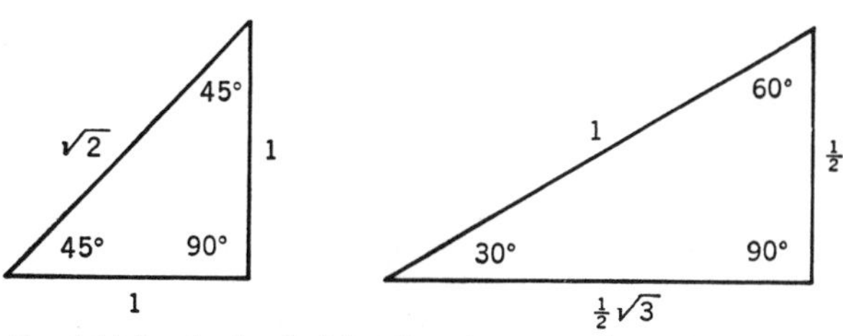

FIG. 21.2-3. Special triangles for deriving the trigonometric functions of 30 deg, 45 deg, and 60 deg.

Table 21.2-2. Relations between Trigonometric Functions of Different Arguments

	$-A$	$90° \pm A$	$180° \pm A$	$270° \pm A$	$n360° \pm A$
sin	$-\sin A$	$\cos A$	$\mp \sin A$	$-\cos A$	$\pm \sin A$
cos	$\cos A$	$\mp \sin A$	$-\cos A$	$\pm \sin A$	$\cos A$
tan	$-\tan A$	$\mp \cot A$	$\pm \tan A$	$\mp \cot A$	$\pm \tan A$
cot	$-\cot A$	$\mp \tan A$	$\pm \cot A$	$\mp \tan A$	$\pm \cot A$

terms of function values for arguments between 0 and $\pi/2 = 90$ deg (Table 21.2-2 and Fig. 21.2-1).

21.2-2. Relations between the Trigonometric Functions (see also Sec. 21.2-6). The basic relations

$$\begin{gathered}\sin^2 z + \cos^2 z = 1 \\ \frac{\sin z}{\cos z} = \tan z = \frac{1}{\cot z}\end{gathered} \tag{21.2-4}$$

yield

$$\begin{aligned}
\sin z &= \pm\sqrt{1-\cos^2 z} = \frac{\tan z}{\pm\sqrt{1+\tan^2 z}} = \frac{1}{\pm\sqrt{1+\cot^2 z}} \\
\cos z &= \pm\sqrt{1-\sin^2 z} = \frac{1}{\pm\sqrt{1+\tan^2 z}} = \frac{\cot z}{\pm\sqrt{1+\cot^2 z}} \\
\tan z &= \frac{\sin z}{\pm\sqrt{1-\sin^2 z}} = \frac{\pm\sqrt{1-\cos^2 z}}{\cos z} = \frac{1}{\cot z} \\
\cot z &= \frac{\sqrt{1-\sin^2 z}}{\sin z} = \frac{\cos z}{\pm\sqrt{1-\cos^2 z}} = \frac{1}{\tan z}
\end{aligned} \tag{21.2-5}$$

21.2-3. Addition Formulas and Multiple-angle Formulas. The basic relation

$$\boxed{\sin (A + B) = \sin A \cos B + \sin B \cos A} \tag{21.2-6}$$

yields

$$\left.\begin{aligned} \sin (A \pm B) &= \sin A \cos B \pm \cos A \sin B \\ \cos (A \pm B) &= \cos A \cos B \mp \sin A \sin B \\ \tan (A \pm B) &= \frac{\tan A \pm \tan B}{1 \mp \tan A \tan B} \\ \cot (A \pm B) &= \frac{\cot A \cot B \mp 1}{\cot A \pm \cot B} \end{aligned}\right\} \tag{21.2-7}$$

$$\left.\begin{aligned} \sin 2A &= 2 \sin A \cos A \\ \cos 2A &= \cos^2 A - \sin^2 A = 2 \cos^2 A - 1 = 1 - 2 \sin^2 A \\ \tan 2A &= \frac{2 \tan A}{1 - \tan^2 A} \\ \cot 2A &= \frac{\cot^2 A - 1}{2 \cot A} = \frac{1}{2} (\cot A - \tan A) \end{aligned}\right\} \tag{21.2-8}$$

$$\left.\begin{aligned} \sin \frac{A}{2} &= \pm \sqrt{\frac{1 - \cos A}{2}} \\ \cos \frac{A}{2} &= \pm \sqrt{\frac{1 + \cos A}{2}} \\ \tan \frac{A}{2} &= \frac{\sin A}{1 + \cos A} = \frac{1 - \cos A}{\sin A} \\ \cot \frac{A}{2} &= \frac{\sin A}{1 - \cos A} = \frac{1 + \cos A}{\sin A} \end{aligned}\right\} \tag{21.2-9}$$

$$\left.\begin{aligned} a \sin A + b \cos A = r \sin (A + B) = r \cos (90^\circ - A - B) \\ r = +\sqrt{a^2 + b^2} \qquad \tan B = \frac{b}{a} \end{aligned}\right\} \tag{21.2-10}$$

$$\left.\begin{aligned} \sin A \pm \sin B &= 2 \sin \frac{A \pm B}{2} \cos \frac{A \mp B}{2} \\ \cos A + \cos B &= 2 \cos \frac{A + B}{2} \cos \frac{A - B}{2} \\ \cos A - \cos B &= -2 \sin \frac{A + B}{2} \sin \frac{A - B}{2} \\ \tan A \pm \tan B &= \frac{\sin (A \pm B)}{\cos A \cos B} \\ \cot A \pm \cot B &= \frac{\sin (B \pm A)}{\sin A \sin B} \end{aligned}\right\} \tag{21.2-11}$$

$$\left.\begin{aligned} 2 \cos A \cos B &= \cos (A - B) + \cos (A + B) \\ 2 \sin A \sin B &= \cos (A - B) - \cos (A + B) \\ 2 \sin A \cos B &= \sin (A - B) + \sin (A + B) \\ 2 \cos^2 A &= 1 + \cos 2A \\ 2 \sin^2 A &= 1 - \cos 2A \end{aligned}\right\} \tag{21.2-12}$$

$$\left.\begin{aligned}\sin nA &= \binom{n}{1}\cos^{n-1}A\sin A - \binom{n}{3}\cos^{n-3}A\sin^3 A\\ &\qquad + \binom{n}{5}\cos^{n-5}A\sin^5 A \mp \cdots\\ \cos nA &= \cos^n A - \binom{n}{2}\cos^{n-2}A\sin^2 A\\ &\qquad + \binom{n}{4}\cos^{n-4}A\sin^4 A \mp \cdots\end{aligned}\right\}\qquad(21.2\text{-}13)$$

If n is an odd integer,

$$\left.\begin{aligned}\sin^n z &= \left(\frac{1}{2i}\right)^{n-1}\left[\sin nz - \binom{n}{1}\sin(n-2)z + \binom{n}{2}\sin(n-4)z\right.\\ &\qquad \left. - \binom{n}{3}\sin(n-6)z + \cdots (-1)^{\frac{n-1}{2}}\binom{n}{\frac{n-1}{2}}\sin z\right]\\ \cos^n z &= \left(\frac{1}{2}\right)^{n-1}\left[\cos nz + \binom{n}{1}\cos(n-2)z + \binom{n}{2}\cos(n-4)z\right.\\ &\qquad \left. + \cdots + \binom{n}{\frac{n-1}{2}}\cos z\right]\end{aligned}\right\}\qquad(21.2\text{-}14)$$

If n is an even integer,

$$\left.\begin{aligned}\sin^n z &= \frac{(-1)^{\frac{n}{2}}}{2^{n-1}}\left[\cos nz - \binom{n}{1}\cos(n-2)z + \binom{n}{2}\cos(n-4)z\right.\\ &\qquad \left. - \cdots + (-1)^{\frac{n-2}{2}}\binom{n}{\frac{n-2}{2}}\cos 2z\right] + \binom{n}{\frac{n}{2}}\frac{1}{2^n}\\ \cos^n z &= \left(\frac{1}{2}\right)^{n-1}\left[\cos nz + \binom{n}{1}\cos(n-2)z + \binom{n}{2}\cos(n-4)z\right.\\ &\qquad \left. + \cdots + \binom{n}{\frac{n-2}{2}}\cos 2z\right] + \binom{n}{\frac{n}{2}}\frac{1}{2^n}\end{aligned}\right\}\qquad(21.2\text{-}15)$$

21.2-4. The Inverse Trigonometric Functions (see also Table 7.2-1).* **(a)** The **inverse trigonometric functions** $w = \arcsin z$, $w = \arccos z$, $w = \arctan z$, $w = \text{arccot}\, z$ are respectively defined by

$$z = \sin w \qquad z = \cos w \qquad z = \tan w \qquad z = \cot w$$

* The functions $\arcsin z$, $\arccos z$, $\arctan z$, and $\text{arccot}\, z$ are often denoted by $\sin^{-1} z$, $\cos^{-1} z$, $\tan^{-1} z$, and $\cot^{-1} z$, respectively. This notation tends to be misleading and is not recommended.

or by

$$\left.\begin{aligned} \arcsin z &= \int_0^z \frac{dz}{\sqrt{1-z^2}} \qquad & \arccos z &= -\int_1^z \frac{dz}{\sqrt{1-z^2}} \\ \arctan z &= \int_0^z \frac{dz}{1+z^2} \qquad & \operatorname{arccot} z &= -\int_{-\infty}^z \frac{dz}{1+z^2} \end{aligned}\right\} \tag{21.2-16}$$

Figure 21.2-4 shows plots of the inverse trigonometric functions for real arguments; note that $\arcsin z$ and $\arccos z$ are real if and only if z is real and $|z| \leq 1$. All four functions are infinitely-many-valued because

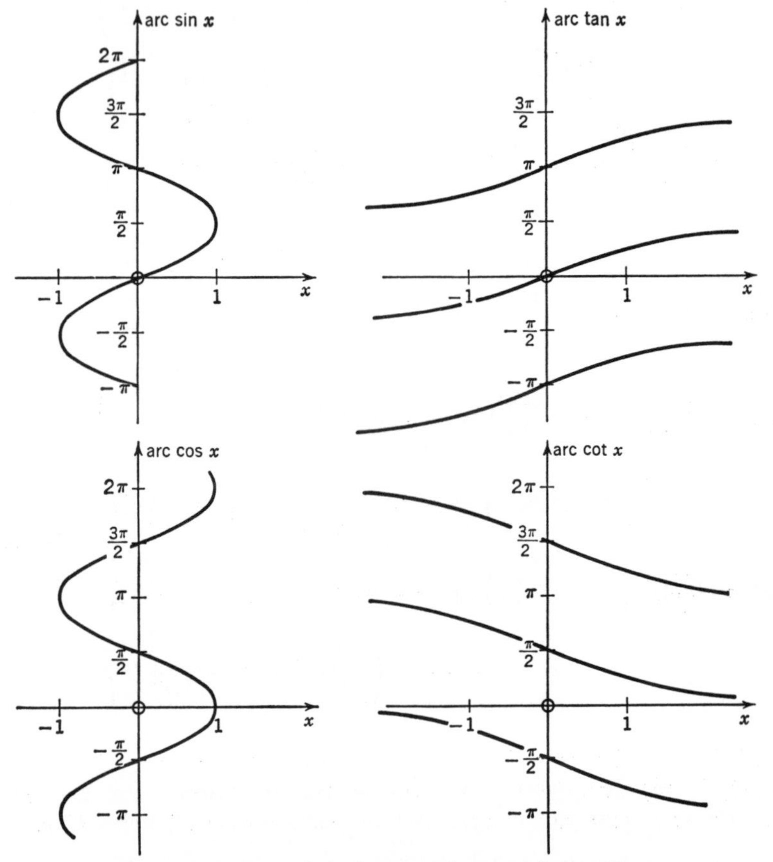

Fig. 21.2-4. Plots of the inverse trigonometric functions.

of the periodicity of the trigonometric functions. For real arguments, the **principal value** of $\arcsin z$ and $\arctan z$ is that between $-\pi/2$ and $\pi/2$ (see also Fig. 21.2-4); the principal value of $\arccos z$ and $\operatorname{arccot} z$ is that between 0 and π (see also Fig. 7.4-1).

(b) Note

$$\left.\begin{aligned}\arcsin a \pm \arcsin b &= \arcsin\,(a\sqrt{1-b^2} \pm b\sqrt{1-a^2})\\ &= \arccos\,(\sqrt{1-a^2}\sqrt{1-b^2} \mp ab)\\ \arccos a \pm \arccos b &= \arccos\,(ab \mp \sqrt{1-a^2}\sqrt{1-b^2})\\ &= \arcsin\,(b\sqrt{1-a^2} \pm a\sqrt{1-b^2})\\ \arctan a \pm \arctan b &= \arctan \frac{a \pm b}{1 \mp ab}\end{aligned}\right\} \qquad (21.2\text{-}17)$$

21.2-5. Hyperbolic Functions (see also Fig. 21.2-5 and Table 7.2-1). The **hyperbolic functions*** $w = \sinh z$, $w = \cosh z$ are defined by the power series (21.2-42), as solutions of the differential equation

$$\frac{d^2w}{dz^2} - w = 0$$

or simply by

$$\boxed{\sinh z = \frac{e^z - e^{-z}}{2} \qquad \cosh z = \frac{e^z + e^{-z}}{2}} \qquad (21.2\text{-}18)$$

Four additional hyperbolic functions are defined as

$$\left.\begin{aligned}\tanh z &= \frac{\sinh z}{\cosh z} \qquad & \coth z &= \frac{\cosh z}{\sinh z}\\ \operatorname{sech} z &= \frac{1}{\cosh z} \qquad & \operatorname{cosech} z &= \frac{1}{\sinh z}\end{aligned}\right\} \qquad (21.2\text{-}19)$$

Geometrical Interpretation of sinh t and cosh t for Real t. If $t/2$ is the area bounded by the rectangular hyperbola (Sec. 2.5-2b) $x^2 - y^2 = 1$, the x axis, and the radius vector of the point (x, y) on the hyperbola, then $y = \sinh t$, $x = \cosh t$. Note that, if the hyperbola is replaced by the circle $x^2 + y^2 = 1$, then $y = \sin t$, $x = \cos t$.

21.2-6. Relations between the Hyperbolic Functions (see also Sec. 21.2-8). The basic relations

$$\boxed{\begin{aligned}\cosh^2 z - \sinh^2 z &= 1\\ \frac{\sinh z}{\cosh z} = \tanh z &= \frac{1}{\coth z}\end{aligned}} \qquad (21.2\text{-}20)$$

yield

$$\left.\begin{aligned}\sinh z &= \pm\sqrt{\cosh^2 z - 1} = \frac{\tanh z}{\pm\sqrt{1-\tanh^2 z}} = \frac{1}{\pm\sqrt{\coth^2 z - 1}}\\ \cosh z &= \pm\sqrt{1+\sinh^2 z} = \frac{1}{\pm\sqrt{1-\tanh^2 z}} = \frac{\coth z}{\pm\sqrt{\coth^2 z - 1}}\\ \tanh z &= \frac{\sinh z}{\pm\sqrt{1+\sinh^2 z}} = \frac{\pm\sqrt{\cosh^2 z - 1}}{\cosh z} = \frac{1}{\coth z}\\ \coth z &= \frac{\pm\sqrt{1+\sinh^2 z}}{\sinh z} = \frac{\cosh z}{\pm\sqrt{\cosh^2 z - 1}} = \frac{1}{\tanh z}\end{aligned}\right\} \qquad (21.2\text{-}21)$$

* The symbols Sin z, Cos z, Tan z, Cot z are also used.

21.2-7. Formulas Relating Hyperbolic Functions of Compound Arguments (these formulas may also be derived from the corresponding formulas for trigonometric functions by using the relations of Sec. 21.2-9).

$$\left.\begin{aligned} \sinh(A \pm B) &= \sinh A \cosh B \pm \cosh A \sinh B \\ \cosh(A \pm B) &= \cosh A \cosh B \pm \sinh A \sinh B \\ \tanh(A \pm B) &= \frac{\tanh A \pm \tanh B}{1 \pm \tanh A \tanh B} \\ \coth(A \pm B) &= \frac{\coth A \coth B \pm 1}{\coth B \pm \coth A} \end{aligned}\right\} \qquad (21.2\text{-}22)$$

$$\left.\begin{aligned} \sinh 2A &= 2 \cosh A \sinh A \\ \cosh 2A &= \cosh^2 A + \sinh^2 A \\ \tanh 2A &= \frac{2 \tanh A}{1 + \tanh^2 A} \\ \coth 2A &= \frac{\coth^2 A + 1}{2 \coth A} \end{aligned}\right\} \qquad (21.2\text{-}23)$$

$$\left.\begin{aligned} \sinh \frac{A}{2} &= \pm \sqrt{\frac{\cosh A - 1}{2}} \\ \cosh \frac{A}{2} &= \pm \sqrt{\frac{\cosh A + 1}{2}} \\ \tanh \frac{A}{2} &= \frac{\sinh A}{\cosh A + 1} = \frac{\cosh A - 1}{\sinh A} \\ \coth \frac{A}{2} &= \frac{\sinh A}{\cosh A - 1} = \frac{\cosh A + 1}{\sinh A} \end{aligned}\right\} \qquad (21.2\text{-}24)$$

$$\left.\begin{aligned} \sinh A \pm \sinh B &= 2 \sinh \frac{A \pm B}{2} \cosh \frac{A \mp B}{2} \\ \cosh A + \cosh B &= 2 \cosh \frac{A + B}{2} \cosh \frac{A - B}{2} \\ \cosh A - \cosh B &= 2 \sinh \frac{A + B}{2} \sinh \frac{A - B}{2} \\ \tanh A \pm \tanh B &= \frac{\sinh(A \pm B)}{\cosh A \cosh B} \\ \coth A \pm \coth B &= \frac{\sinh(B \pm A)}{\sinh A \sinh B} \end{aligned}\right\} \qquad (21.2\text{-}25)$$

$$\left.\begin{aligned} 2 \cosh A \cosh B &= \cosh(A + B) + \cosh(A - B) \\ 2 \sinh A \sinh B &= \cosh(A + B) - \cosh(A - B) \\ 2 \sinh A \cosh B &= \sinh(A + B) + \sinh(A - B) \\ 2 \cosh^2 A &= 1 + \cosh 2A \\ 2 \sinh^2 A &= \cosh 2A - 1 \end{aligned}\right\} \qquad (21.2\text{-}26)$$

21.2-8. Inverse Hyperbolic Functions (see also Sec. 21.2-4). The inverse hyperbolic functions $w = \sinh^{-1} z$, $w = \cosh^{-1} z$, $w = \tanh^{-1} z$ are respectively defined by $z = \sinh w$, $z = \cosh w$, $z = \tanh w$,* or by integrals in the manner of Sec. 21.2-4. Note

$$\left.\begin{aligned} \sinh^{-1} a \pm \sinh^{-1} b &= \sinh^{-1} (a \sqrt{b^2+1} \pm b \sqrt{a^2+1}) \\ &= \cosh^{-1} (\sqrt{a^2+1} \sqrt{b^2+1} \pm ab) \\ \cosh^{-1} a \pm \cosh^{-1} b &= \cosh^{-1} (ab \pm \sqrt{a^2-1} \sqrt{b^2-1}) \\ &= \sinh^{-1} (b \sqrt{a^2-1} \pm a \sqrt{b^2-1}) \\ \tanh^{-1} a \pm \tanh^{-1} b &= \tanh^{-1} \frac{a \pm b}{1 \pm ab} \end{aligned}\right\} \qquad (21.2\text{-}27)$$

21.2-9. Relations between Exponential, Trigonometric, and Hyperbolic Functions (see also Secs. 1.2-3 and 21.2-12 and Table 7.2-1).

$$\boxed{\begin{gathered} e^{iz} = \cos z + i \sin z \\ \cos z = \frac{e^{iz} + e^{-iz}}{2} \qquad \sin z = \frac{e^{iz} - e^{-iz}}{2i} \end{gathered}} \qquad (21.2\text{-}28)$$

$$e^{-iz} = \cos z - i \sin z \qquad (21.2\text{-}29)$$

$$e^{z} = \cosh z + \sinh z \qquad e^{-z} = \cosh z - \sinh z \qquad (21.2\text{-}30)$$

$$\cosh z = \frac{e^z + e^{-z}}{2} \qquad \sinh z = \frac{e^z - e^{-z}}{2} \qquad (21.2\text{-}31)$$

$$\left.\begin{aligned} \cos z &= \cosh iz & \cosh z &= \cos iz \\ \sin z &= -i \sinh iz & \sinh z &= -i \sin iz \\ \tan z &= -i \tanh iz & \tanh z &= -i \tan iz \\ \cot z &= i \coth iz & \coth z &= i \cot iz \end{aligned}\right\} \qquad (21.2\text{-}32)$$

$$\left.\begin{aligned} a^{ix} &= e^{ix \log_e a} = \cos (x \log_e a) + i \sin (x \log_e a) \\ x^{i} &= e^{i \log_e x} = \cos (\log_e x) + i \sin (\log_e x) \\ i^{x} &= e^{x \log_e i} = \cos \frac{\pi x}{2} + i \sin \frac{\pi x}{2} \\ e^{2n\pi i} &= 1 \qquad e^{(2n+1)\pi i} = -1 \qquad (n = 0, \pm 1, \pm 2, \ldots) \\ i^{i} &= e^{-\pi/2} \end{aligned}\right\} \qquad (21.2\text{-}33)$$

21.2-10. Decomposition of the Logarithm (see also Secs. 1.2-3 and 12.2-12 and Table 7.2-1).

$$\boxed{\log_e z = \log_e |z| + i \arg (z)} \qquad (21.2\text{-}34)$$

$$\left.\begin{aligned} \log_e (ix) &= \log_e x + (2n + \tfrac{1}{2})\pi i \\ \log_e (-x) &= \log_e x + (2n + 1)\pi i \end{aligned}\right\} (n = 0, \pm 1, \pm 2, \ldots) \qquad (21.2\text{-}35)$$

* This notation is the usual one in English-speaking countries, although it tends to be misleading (see also Sec. 21.2-4). An alternative notation is ar sinh z, ar cosh z, ar tanh z, or Ar Sin z, Ar Cos z, Ar Tan z.

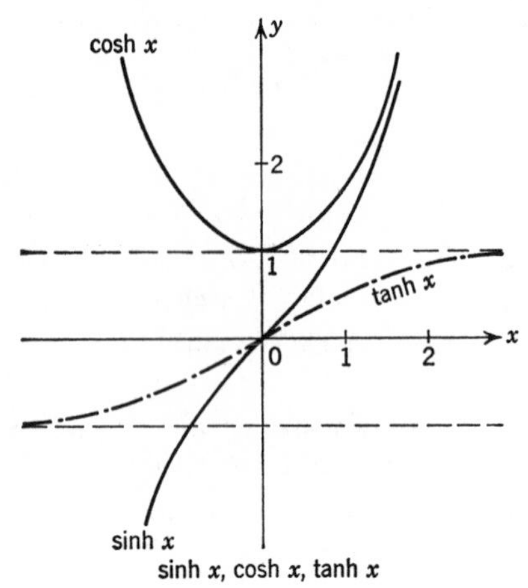

FIG. 21.2-5. Hyperbolic functions. (*From Baumeister and Marks, Mechanical Engineers' Handbook, 6th ed., McGraw-Hill, New York,* 1958.)

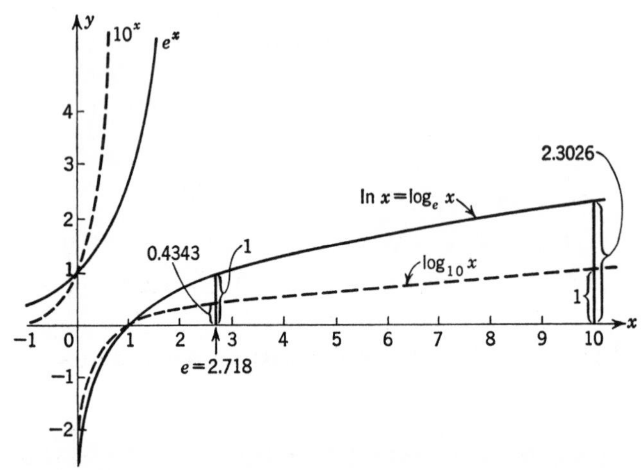

FIG. 21.2-6. Exponential functions and logarithms. (*From Baumeister and Marks, Mechanical Engineers' Handbook, 6th ed., McGraw-Hill, New York,* 1958.)

21.2-11. Relations between Inverse Trigonometric, Inverse Hyperbolic, and Logarithmic Functions.

$$\left.\begin{array}{ll} \arccos z = i \cosh^{-1} z & \cosh^{-1} z = i \arccos z \\ \arcsin z = -i \sinh^{-1} iz & \sinh^{-1} z = -i \arcsin iz \\ \arctan z = -i \tanh^{-1} iz & \tanh^{-1} z = -i \arctan iz \\ \operatorname{arccot} z = i \coth^{-1} iz & \coth^{-1} z = i \operatorname{arccot} iz \end{array}\right\} \tag{21.2-36}$$

$$\left.\begin{array}{ll} \arccos z = -i \log_e (z + i\sqrt{1 - z^2}) & \cosh^{-1} z = \log_e (z + \sqrt{z^2 - 1}) \\ \arcsin z = -i \log_e (iz + \sqrt{1 - z^2}) & \sinh^{-1} z = \log_e (z + \sqrt{z^2 + 1}) \\ \arctan z = -\frac{i}{2} \log_e \frac{1 + iz}{1 - iz} & \tanh^{-1} z = \frac{1}{2} \log_e \frac{1 + z}{1 - z} \\ \operatorname{arccot} z = -\frac{i}{2} \log_e \frac{iz - 1}{iz + 1} & \coth^{-1} z = \frac{1}{2} \log_e \frac{z + 1}{z - 1} \end{array}\right\} \tag{21.2-37}$$

21.2-12. Power Series and Other Expansions. Power-series expansions, as well as some product and continued-fraction expansions for the elementary transcendental functions, are tabulated in Secs. E-7 to E-9 of Appendix E. See also Secs. 20.6-1 to 20.6-5 and Tables 20.6-2 to 20.6-4 for other numerical approximations.

21.2-13. Some Useful Inequalities (see also Figs. 21.2-2 and 21.2-6). For real x,

$$\left.\begin{array}{ll} \sin x \le x \le \tan x & \left(0 \le x \le \frac{\pi}{2}\right) \\ \sin x > \frac{2x}{\pi} & \left(-\frac{\pi}{2} < x < \frac{\pi}{2}\right) \\ \cos x \le \frac{\sin x}{x} \le 1 & (0 \le x \le \pi) \end{array}\right\} \tag{21.2-38}$$

$$\left.\begin{array}{ll} e^x > 1 + x & \\ e^x < \frac{1}{1 - x} & (x < 1) \\ e^{-\frac{x}{1-x}} < 1 - x < e^{-x} & (x < 1) \\ x < (e^x - 1) < \frac{x}{1 - x} & (x < 1) \\ \frac{x}{1 + x} < 1 - e^{-x} < x & (x > -1) \end{array}\right\} \tag{21.2-39}$$

$$\left.\begin{array}{ll} \frac{x}{1 + x} < \log_e (1 + x) < x & (-1 < x \ne 0) \\ x < -\log_e (1 - x) < \frac{x}{1 - x} & (0 \ne x < 1) \\ |\log_e (1 - x)| < \frac{3x}{2} & (0 < x \le 0.5828) \end{array}\right\} \tag{21.2-40}$$

21.3. SOME FUNCTIONS DEFINED BY TRANSCENDENTAL INTEGRALS

21.3-1. Sine, Cosine, Exponential, and Logarithmic Integrals (see also Fig. 21.3-1). One defines

$$\begin{aligned} \mathrm{Si}\,(x) &\equiv \int_0^x \frac{\sin x}{x}\,dx \equiv \frac{\pi}{2} - \int_x^\infty \frac{\sin x}{x}\,dx \equiv \frac{\pi}{2} + \mathrm{si}\,(x) \\ &= x - \frac{1}{3!}\frac{x^3}{3} + \frac{1}{5!}\frac{x^5}{5} \mp \cdots \qquad \text{(SINE INTEGRAL)} \end{aligned} \tag{21.3-1}$$

$$\begin{aligned} \mathrm{Ci}\,(x) &\equiv -\int_x^\infty \frac{\cos x}{x}\,dx = C + \log_e x - \int_0^x \frac{1-\cos x}{x}\,dx \\ &= C + \log_e x - \frac{1}{2!}\frac{x^2}{2} + \frac{1}{4!}\frac{x^4}{4} \mp \cdots \qquad (x > 0) \\ &\qquad \text{(COSINE INTEGRAL)} \end{aligned} \tag{21.3-2}$$

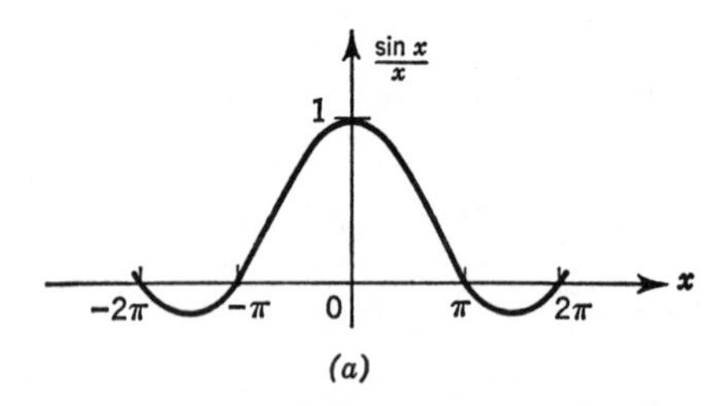

(a)

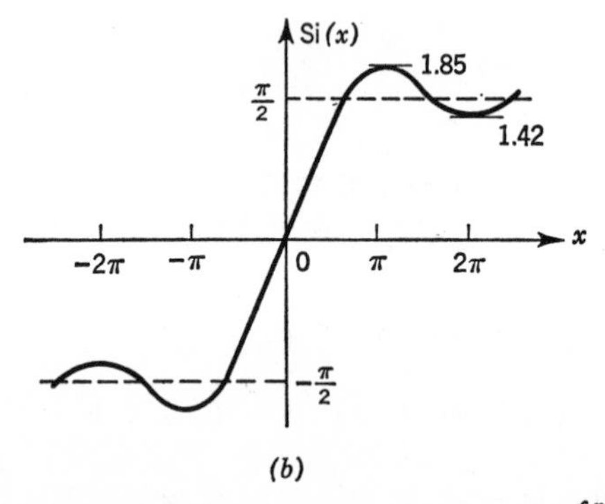

(b)

Fig. 21.3-1. $\sin x/x \equiv \mathrm{sinc}\,(x/\pi)$ (a), and the sine integral $\int_0^x \frac{\sin x}{x}\,dx \equiv \mathrm{Si}\,x$ (b). *(From M. Schwartz, Information Transmission, Modulation, and Noise, McGraw-Hill, New York, 1959.)*

$$\text{Ei}(-x) \equiv -\int_x^\infty \frac{e^{-x}}{x}\,dx \quad (x > 0) \qquad \text{(EXPONENTIAL INTEGRAL)} \qquad (21.3\text{-}3)$$

$$\text{li}(x) \equiv \int_0^x \frac{dx}{\log_e x} \qquad (x > 0) \qquad \text{(LOGARITHMIC INTEGRAL)} \qquad (21.3\text{-}4)$$

where $C \approx 0.577216$ is the Euler-Mascheroni constant defined in Sec. 21.4-1*b*.

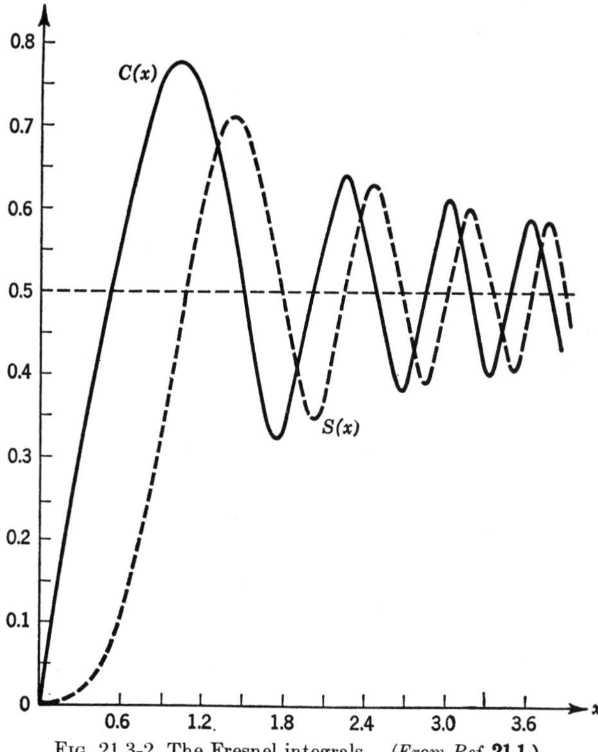

FIG. 21.3-2. The Fresnel integrals. (*From Ref.* **21.1.**)

It is customary to introduce an alternative exponential integral $\overline{\mathrm{Ei}}(z)$ so that, for real x, y,

$$\left.\begin{aligned}\overline{\mathrm{Ei}}(x) &= \lim_{y\to 0+0} \mathrm{Ei}(x \pm iy) \pm i\pi = \mathrm{li}(e^x)\\ \overline{\mathrm{Ei}}\,(xe^{\pm i\pi}) &= \mathrm{Ei}\,(-x) \pm i\pi\\ \overline{\mathrm{Ei}}\,(\pm\, iy) &= \mathrm{Ci}\,(y) \pm i\,\mathrm{Si}\,(y) \pm i\frac{\pi}{2}\end{aligned}\right\} \tag{21.3-5}$$

21.3-2. Fresnel Integrals and Error Function (see also Sec. 18.8-3 and Fig. 21.3-2). One defines

$$\left.\begin{aligned}C(x) &\equiv \int_0^x \cos\frac{\pi}{2}x^2\,dx = J_{1/2}\left(\frac{\pi}{2}x^2\right)\\ &\quad + J_{5/2}\left(\frac{\pi}{2}x^2\right) + J_{9/2}\left(\frac{\pi}{2}x^2\right) + \cdots\\ S(x) &\equiv \int_0^x \sin\frac{\pi}{2}x^2\,dx = J_{3/2}\left(\frac{\pi}{2}x^2\right)\\ &\quad + J_{7/2}\left(\frac{\pi}{2}x^2\right) + J_{11/2}\left(\frac{\pi}{2}x^2\right) + \cdots\end{aligned}\right\} \text{(FRESNEL INTEGRALS)} \tag{21.3-6}$$

$$\operatorname{erf} x \equiv \frac{2}{\sqrt{\pi}}\int_0^x e^{-x^2}\,dx = \frac{2}{\sqrt{\pi}}\left(x - \frac{x^3}{3} + \frac{1}{2!}\frac{x^5}{5} - \frac{1}{3!}\frac{x^7}{7} \pm \cdots\right)$$

(ERROR FUNCTION) (21.3-7)

where the $J_{n/2}(z)$ are the half-integral-order Bessel functions discussed in Sec. 21.8-1*e*. Note

$$C(x) - iS(x) \equiv \frac{1}{1+i}\operatorname{erf}\left(\frac{1+i}{2}x\sqrt{\pi}\right) \tag{21.3-8}$$

$$\left.\begin{aligned}C(x) &= \frac{1}{2} + \frac{1}{\pi x}\sin\frac{\pi}{2}x^2 + O\left(\frac{1}{x^2}\right)\\ S(x) &= \frac{1}{2} - \frac{1}{\pi x}\cos\frac{\pi}{2}x^2 + O\left(\frac{1}{x^2}\right)\end{aligned}\right\} \text{as } x \to \infty \tag{21.3-9}$$

Note also

$$\left.\begin{aligned}\int_0^x C(x)\,dx &= xC(x) - \frac{1}{\pi}\sin\frac{\pi}{2}x^2\\ \int_0^x S(x)\,dx &= xS(x) + \frac{1}{\pi}\cos\frac{\pi}{2}x^2 - \frac{1}{\pi}\\ \int_0^x \operatorname{erf} x\,dx &= x\operatorname{erf} x + \frac{1}{\sqrt{\pi}}(e^{-x^2} - 1)\end{aligned}\right\} \tag{21.3-10}$$

The related integrals

$$\left.\begin{aligned}\frac{1}{2\sqrt{x}}\int_0^x \frac{\sin x}{\sqrt{x}}\,dx &= \frac{x}{3} - \frac{1}{7}\frac{x^3}{3!} + \frac{1}{11}\frac{x^5}{5!} \mp \cdots\\ \frac{1}{2\sqrt{x}}\int_0^x \frac{\cos x}{\sqrt{x}}\,dx &= 1 - \frac{1}{5}\frac{x^2}{2!} + \frac{1}{9}\frac{x^4}{4!} - \frac{1}{13}\frac{x^6}{6!} \pm \cdots\end{aligned}\right\} \tag{21.3-11}$$

are also sometimes known as Fresnel integrals.

The function

$$\operatorname{erfc} z = 1 - \operatorname{erf} z = \frac{2}{\sqrt{\pi}} \int_z^\infty e^{-\zeta^2}\, d\zeta \qquad (21.3\text{-}12)$$

is known as the **complementary error function.**

21.4. THE GAMMA FUNCTION AND RELATED FUNCTIONS

21.4-1. The Gamma Function. (a) Integral Representations. The **gamma function** $\Gamma(z)$ is most frequently defined by

$$\Gamma(z) = \int_0^\infty e^{-t} t^{z-1}\, dt \qquad [\mathrm{Re}\,(z) > 0] \qquad (21.4\text{-}1)$$

(EULER'S INTEGRAL OF THE SECOND KIND)

or for Re $(z) < 0$ by

$$\frac{1}{\Gamma(z)} = \frac{1}{2\pi i} \oint_{C_0} e^t t^{-z}\, dt \quad \text{(HANKEL'S INTEGRAL REPRESENTATION)} \qquad (21.4\text{-}2)$$

where the contour C_0 starts at $-\infty$, skims below the negative x axis, surrounds the origin, and returns just above the negative x axis. The

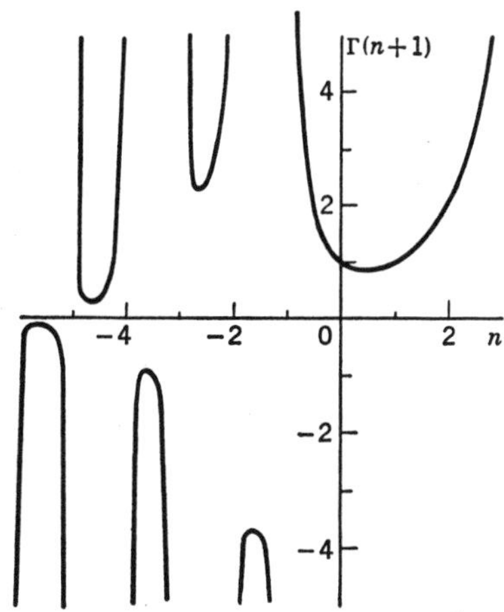

FIG. 21.4-1. $\Gamma(n+1)$ vs. n for real n. Note $\Gamma(n+1) = n!$ for $n = 0, 1, 2, \ldots$, and the alternating maxima and minima given approximately by $\Gamma(1.462) = 0.886$, $\Gamma(-0.5040) = -3.545$, $\Gamma(-1.573) = 2.302$, $\Gamma(-2.611) = -0.888, \ldots$.

definition can be extended by analytic continuation (Sec. 7.8-1). The only singularities of $\Gamma(z)$ in the finite portion of the z plane are simple poles with residues $(-1)^n/n!$ for $z = -n$ $(n = 0, 1, 2, \ldots)$; $1/\Gamma(z)$ is an integral function.

Figure 21.4-1 shows a graph of $\Gamma(x)$ vs. x for real x. Note

$$\Gamma(\tfrac{1}{2}) = \sqrt{\pi} \qquad \Gamma(1) = 1$$
$$\Gamma(n+1) = n! \qquad (n = 0, 1, 2, \ldots) \tag{21.4-3}$$

(b) Other Representations of $\Gamma(z)$.

$$\Gamma(z) = \lim_{n\to\infty} \frac{n!}{z(z+1)(z+2)\cdots(z+n-1)} n^{z-1}$$
(EULER'S DEFINITION) (21.4-4)

$$\frac{1}{\Gamma(z)} = ze^{Cz} \prod_{k=1}^{\infty} \left(1 + \frac{z}{k}\right) e^{-z/k}$$
(WEIERSTRASS'S PRODUCT REPRESENTATION) (21.4-5)

C is the **Euler-Mascheroni constant** defined by

$$\begin{aligned} C &= \lim_{n\to\infty} \left(\sum_{k=1}^{n} \frac{1}{k} - \log_e n\right) \\ &= -\int_0^\infty e^{-t} \log_e t\, dt = -\int_0^1 \log_e \left(\log_e \frac{1}{\tau}\right) d\tau \\ &\approx 0.5772157 \end{aligned} \tag{21.4-6}$$

(c) Functional Equations.

$$\Gamma(z+1) = z\Gamma(z) \tag{21.4-7}$$

$$\Gamma(z)\Gamma(1-z) = \frac{\pi}{\sin \pi z} \qquad \Gamma\left(\frac{1}{2} + z\right)\Gamma\left(\frac{1}{2} - z\right) = \frac{\pi}{\cos \pi z} \tag{21.4-8}$$

$$\Gamma(nz) = \sqrt{\frac{n^{2nz-1}}{(2\pi)^{n-1}}}\, \Gamma(z)\Gamma\left(z + \frac{1}{n}\right)\Gamma\left(z + \frac{2}{n}\right) \cdots \Gamma\left(z + \frac{n-1}{n}\right)$$
$(n = 2, 3, \ldots)$ (GAUSS'S MULTIPLICATION THEOREM) (21.4-9)

21.4-2. Stirling's Expansions for $\Gamma(z)$ and $n!$ (see also Secs. 4.4-3, 4.8-6b, and 21.5-4).

$$\Gamma(z) = e^{-z}z^{z-1/2}\sqrt{2\pi}\left[1 + \frac{1}{12z} + \frac{1}{288z^2} - \frac{139}{51840z^3} - \frac{571}{2488320z^4} + O(z^{-5})\right] \quad (|\arg z| < \pi)$$
(STIRLING'S SERIES) (21.4-10)

Stirling's series is especially useful for large $|z|$; *for real positive z, the absolute value of the error is less than that of the last term used.* Note, in particular,

$$\lim_{n\to\infty} \frac{n!}{n^n e^{-n}\sqrt{2\pi n}} = 1 \qquad \text{or} \qquad n! \simeq n^n e^{-n}\sqrt{2\pi n}$$
as $n \to \infty$ (STIRLING'S FORMULA) (21.4-11)

The fractional error in Stirling's formula is less than 10 per cent for $n = 1$ and decreases as n increases; this asymptotic formula applies particularly to computations of the *ratio* of two factorials or gamma functions, since in such cases the *fractional* error is of paramount interest.

More specifically

$$n^n e^{-n} \sqrt{2\pi n} < n! < n^n \sqrt{2\pi n}\, e^{-n+1/12n} \tag{21.4-12}$$

$$n! \simeq n^n \sqrt{2\pi n} \exp\left(-n + \frac{1}{12n} - \frac{1}{360n^3} + \cdots\right) \qquad \text{as } n \to \infty \tag{21.4-13}$$

21.4-3. The Psi Function (Digamma Function).

$$\psi(z) = \frac{d}{dz} \log_e \Gamma(z) = \sum_{k=0}^{\infty} \left(\frac{1}{k+1} - \frac{1}{z+k}\right) - C \tag{21.4-14}$$

$$\psi(z) = \int_0^\infty \left(\frac{e^{-t}}{t} - \frac{e^{-zt}}{1-e^{-t}}\right) dt = -\int_0^1 \left(\frac{1}{\log_e t} + \frac{t^{z-1}}{1-t}\right) dt \qquad [\text{Re}\,(z) > 0] \tag{21.4-15}$$

Note

$$\psi(1) = -C \qquad \psi(z+1) = \psi(z) + \frac{1}{z} \tag{21.4-16}$$

21.4-4. Beta Functions. The **(complete) beta function** is defined as

$$\text{B}(p, q) \equiv \frac{\Gamma(p)\Gamma(q)}{\Gamma(p+q)} \equiv \text{B}(q, p) \tag{21.4-17}$$

or by analytic continuation of

$$\text{B}(p, q) = \int_0^1 t^{p-1}(1-t)^{q-1}\, dt \qquad [\text{Re}\,(p) > 0,\ \text{Re}\,(q) > 0]$$

(EULER'S INTEGRAL OF THE FIRST KIND) (21.4-18)

$$\text{B}(p, q) = \int_0^\infty \frac{t^{p-1}}{(1+t)^{p+q}}\, dt = 2\int_0^{\pi/2} \sin^{2p-1}\vartheta \cos^{2q-1}\vartheta\, d\vartheta \tag{21.4-19}$$

Note

$$\text{B}(p, q)\text{B}(p+q, r) = \text{B}(q, r)\text{B}(q+r, p) \tag{21.4-20}$$

$$\frac{1}{\text{B}(n, m)} = m\binom{n+m-1}{n-1} = n\binom{n+m-1}{m-1} \qquad (n, m = 1, 2, \ldots) \tag{21.4-21}$$

21.4-5. Incomplete Gamma and Beta Functions. The **incomplete gamma function** $\Gamma_z(p)$ and the **incomplete beta function** $\text{B}_z(p, q)$ are respectively defined by analytic continuation of

$$\Gamma_z(p) = \int_0^z t^{p-1}e^{-t}\, dt \qquad [\text{Re}\,(p) > 0] \tag{21.4-22}$$

$$\text{B}_z(p, q) = \int_0^z t^{p-1}(1-t)^{q-1}\, dt \qquad [\text{Re}\,(p) > 0,\ \text{Re}\,(q) > 0;\ 0 \le z \le 1] \tag{21.4-23}$$

$I_z(p, q) \equiv \text{B}_z(p, q)/\text{B}(p, q)$ is called the **incomplete-beta-function ratio.**

See Appendix E and Ref. 20.6 for additional definite and indefinite integrals related to the gamma function.

21.5. BINOMIAL COEFFICIENTS AND FACTORIAL POLYNOMIALS. BERNOULLI POLYNOMIALS AND BERNOULLI NUMBERS

21.5-1. Binomial Coefficients and Factorial Polynomials (see also Sec. 1.4-1). Table 21.5-1 summarizes the definition and principal properties of the *binomial coefficients* $\binom{x}{n}$. The expression

$$\begin{aligned} x^{[n]} &\equiv \binom{x}{n} n! \equiv x(x-1)\cdots(x-n+1) \\ &\equiv S_0^{(n)}x^n + S_1^{(n)}x^{n-1} + \cdots + S_{n-1}^{(n)}(x) \\ 0^{[n]} &= 1 \qquad (n = 1, 2, \ldots) \end{aligned} \tag{21.5-1}$$

is called a **factorial polynomial of degree** n. The coefficients $S_k^{(n)}$ are known as **Stirling numbers of the first kind** and can be obtained with the aid of the recursion formulas

$$S_k^{(n+1)} = S_k^{(n)} - nS_{i-1}^{(n)} \tag{21.5-2}$$

or from Eq. (17). Note

$$(x+y)^{[n]} = \sum_{k=0}^{n} \binom{n}{k} x^{[k]}y^{[n-k]} \qquad (n = 0, 1, 2, \ldots)$$

(VANDERMONDE'S BINOMIAL THEOREM) (21.5-3)

If one defines $x^{[0]} = 1$, $x^{[r]} = 1/(x+1)(x+2)\cdots(x-r)$ $(r = -1, -2, \ldots)$, then the relation

$$x^{[r]}(x-r)^{[m]} = x^{[r+m]}$$

and Eq. (20.4-12) hold for all integral **pseudo exponents** r, m. Note $0^{[0]} = 1$.

21.5-2. Bernoulli Polynomials and Bernoulli Numbers (see also Sec. 21.2-12). **(a) Definitions.** The **Bernoulli polynomials** $B_k^{(n)}(x)$ **of order** $n = 0, 1, 2, \ldots$ and degree k are defined by their generating function (Sec. 8.6-5)

$$\frac{t^n e^{xt}}{(e^t-1)^n} = \sum_{k=0}^{\infty} B_k^{(n)}(x)\frac{t^k}{k!} \qquad (n = 0, 1, 2, \ldots) \tag{21.5-4}$$

Note

$$B_k^{(0)}(x) = x^k \qquad B_k^{(n+1)}(x) = \frac{k!}{n!}\frac{d^{n-k}}{dx^{n-k}}[(x-1)(x-2)\cdots(x-n)] \qquad (n > k) \tag{21.5-5}$$

The $B_k^{(n)}(0) = B_k^{(n)}$ are called **Bernoulli numbers of order** n, with

$$\frac{t^n}{(e^t-1)^n} = \sum_{k=0}^{\infty} B_k^{(n)}\frac{t^k}{k!} \qquad (n = 0, 1, 2, \ldots) \tag{21.5-6}$$

Table 21.5-1. Definition and Properties of the Binomial Coefficients
(see also Secs. 1.4-1 and 21.5-4)

(a) Definitions and General Properties. If x, y, z are real numbers, and n is an integer,

$$\binom{x}{n} = \begin{cases} \dfrac{x(x-1)\cdots(x-n+1)}{n!} & \text{for } n > 0 \\ 1 & \text{for } n = 0 \\ 0 & \text{for } n < 0 \end{cases} \quad \text{(DEFINITION)}$$

$$\binom{x+y}{n} = \sum_{j=0}^{n} \binom{x}{n-j}\binom{y}{j} \quad (n > 0) \qquad \binom{x}{n} = \sum_{j=0}^{n} \binom{z}{j}\binom{x-z}{n-j} \quad (n > 0)$$

(ADDITION THEOREM)

$$\binom{x+1}{n} = \binom{x}{n} + \binom{x}{n-1} \qquad \binom{-x}{n} = (-1)^n \binom{x+n-1}{n} \quad (x > 0)$$

(b) In particular, **if N and n are positive integers,**

$$\binom{N}{n} = \binom{N}{N-n} = \begin{cases} \dfrac{N(N-1)\cdots(N-n+1)}{n!} = \dfrac{N!}{(N-n)!n!} & \text{for } N \geq n \\ 0 & \text{for } N < n \end{cases}$$

$$\binom{N}{N} = 1 \qquad \binom{N}{N-1} = \binom{N}{1} = N$$

$$\binom{2N}{N} = (-1)^N 2^{2N} \binom{-\frac{1}{2}}{N} = \sum_{j=0}^{N} \binom{N}{j}^2$$

$$\sum_{j=0}^{N} \binom{N}{j} = 2^N \qquad \sum_{j=0}^{N} (-1)^j \binom{N}{j} = 0$$

(c) If M, N, n are positive integers such that $M \geq n$, $N \geq n$,

$$\binom{N+1}{n+1} = \sum_{j=n}^{N} \binom{j}{n} \qquad \binom{N-1}{n-1} = \sum_{j=n}^{N} (-1)^{j-n} \binom{N}{j}$$

$$\binom{N+n}{n} = \sum_{j=0}^{n} \binom{N}{j}\binom{n}{j} \qquad \binom{M+N}{N-n} = \sum_{j=0}^{N-n} \binom{N}{n+j}\binom{M}{j}$$

$$\left.\begin{aligned} &B_0^{(n)} = 1 \qquad B_1^{(n)} = -\tfrac{1}{2}n \qquad B_2^{(n)} = \tfrac{1}{12}n(3n-1) \\ &B_3^{(n)} = -\tfrac{1}{8}n^2(n-1) \\ &B_4^{(n)} = \tfrac{1}{240}n(15n^3 - 30n^2 + 5n + 2) \\ &B_5^{(n)} = -\tfrac{1}{96}n^2(n-1)(3n^2 - 7n - 2) \\ &B_6^{(n)} = \tfrac{1}{4032}n(63n^5 - 315n^4 + 315n^3 + 91n^2 - 42n - 16) \end{aligned}\right\} \quad (21.5\text{-}7)$$

Bernoulli numbers of order 1 are often simply called **Bernoulli numbers** $B_k^{(1)} = B_k$; $B_k = 0$ for all odd $k > 1$, and

$$B_0 = 1 \qquad B_1 = -\tfrac{1}{2} \qquad B_2 = \tfrac{1}{6} \qquad B_4 = -\tfrac{1}{30} \qquad B_6 = \tfrac{1}{42} \qquad \ldots \tag{21.5-8}$$

with alternating signs. The Bernoulli numbers may also be obtained from the recursion formulas

$$B_0 = 1 \qquad 1 + \binom{k}{1} B_1 + \binom{k}{2} B_2 + \cdots + \binom{k}{k-1} B_{k-1} = 0 \qquad (k = 1, 2, \ldots) \tag{21.5-9}$$

or in *determinant form* by solution of the linear equations (6).

(b) Miscellaneous Properties. Note

$$\frac{d}{dx} B_k^{(n)}(x) = kB_{k-1}^{(n)}(x) \qquad \int_a^x B_k^{(n)}(\xi)\, d\xi = \frac{1}{k+1} [B_{k+1}^{(n)}(x) - B_{k+1}^{(n)}(a)] \tag{21.5-10}$$

$$\Delta B_m^{(n)}(k) = mB_{m-1}^{(n-1)}(k) \qquad \Delta^n B_m^{(n)}(k) = m(m-1) \cdots (m-n+1)x^{m-n}$$
$$(k = 0, \pm 1, \pm 2, \cdots ; m \geq n) \tag{21.5-11}$$

$$B_k^{(n)}(x+1) = B_k^{(n)}(x) + kB_{k-1}^{(n-1)}(x) \qquad B_k^{(n)}(1) = B_k^{(n)} + kB_{k-1}^{(n-1)} \tag{21.5-12}$$

$$\int_x^{x+1} B_k^{(n)}(\xi)\, d\xi = \frac{1}{k+1} \Delta B_{k+1}^{(n)}(x) = B_k^{(n-1)}(x) \qquad \int_0^1 B_k^{(n)}(\xi)\, d\xi = B_k^{(n-1)} \tag{21.5-13}$$

$$B_k^{(n)}(n - x) = (-1)^k B_k^{(n)}(x) \qquad \text{(COMPLEMENTARY-ARGUMENT THEOREM)} \tag{21.5-14}$$

$$\left.\begin{aligned} B_k^{(n+1)}(x) &= \left(1 - \frac{k}{n}\right) B_k^{(n)}(x) + k\left(\frac{x}{n} - 1\right) B_{k-1}^{(n)}(x) \\ B_k^{(n+1)}(1) &= \left(1 - \frac{k}{n}\right) B_k^{(n)} \end{aligned}\right\} \tag{21.5-15}$$

$$\left.\begin{aligned} &B_k^{(1)}(mx) = m^{k-1} \sum_{j=0}^{m-1} B_k^{(1)}\left(x + \frac{j}{m}\right) \qquad \text{(MULTIPLICATION THEOREM)} \\ &B_{2k}^{(1)}(x) = (-1)^{k-1} \sum_{j=1}^{\infty} \frac{2 \cos 2j\pi x}{(2j\pi)^{2k}} \qquad B_{2k+1}^{(1)}(x) = (-1)^{k-1} \sum_{j=1}^{\infty} \frac{2 \sin 2j\pi x}{(2j\pi)^{2k+1}} \\ &(k = 0, 1, 2, \ldots) \end{aligned}\right\} \tag{21.5-16}$$

21.5-3. Formulas Relating Polynomials and Factorial Polynomials. *Bernoulli polynomials and Bernoulli numbers relate factorial polynomials* (Sec. 21.5-1) *to powers of x;* such relations aid in the solution of difference equations and, in particular, in the summation of series (Sec. 4.8-5*d*). Note

$$x^{[n]} \equiv \binom{x}{n} n! \equiv x(x-1) \cdots (x-n+1)$$
$$\equiv B_n^{(n+1)}(x+1) \equiv \sum_{k=0}^{n} \binom{n-1}{k-1} B_{n-k}^{(n)}\, x^k \qquad (n = 0, 1, 2, \ldots) \tag{21.5-17}$$

$$\int_0^x (\xi - 1)(\xi - 2) \cdots (\xi - 2r + 1)\, d\xi = \int_0^x (\xi - 1)^{[2r-1]}\, d\xi$$

$$= \frac{1}{2r} [B_{2r}^{(2r)}(x) - B_{2r}^{(2r)}] \qquad (r = 1, 2, \ldots) \qquad (21.5\text{-}18)$$

21.5-4. Approximation Formulas for $\binom{N}{n}$ (see also Sec. 21.4-2). If N is a positive integer and $z = \frac{2}{N}\left|\frac{N}{2} - n\right| \ll \sqrt[3]{\frac{1}{N}}$, then

$$\binom{N}{n} \approx \frac{2^{N+1}}{\sqrt{2\pi N}} e^{-\frac{N-1}{2}z^2} e^{\alpha\frac{Nz^3}{1-z}}$$

$$= \frac{2^{N+1}}{\sqrt{2\pi N}} e^{-\frac{N-1}{2}z^2} \left(1 + \alpha\frac{Nz^3}{1-z} + \cdots\right) \qquad \left(-\frac{1}{2} < \alpha < \frac{1}{2}\right) \qquad (21.5\text{-}19)$$

If $n \ll N$, use

$$\binom{N}{n} \approx \frac{N^n}{n!} e^{-\alpha\frac{n^2}{N-n}} \qquad \left(0 < \alpha < \frac{1}{2}\right) \qquad (21.5\text{-}20)$$

For large values of N, n, and $N - n$, use Stirling's formula (21.4-11).

21.6. ELLIPTIC FUNCTIONS, ELLIPTIC INTEGRALS, AND RELATED FUNCTIONS

21.6-1. Elliptic Functions: General Properties. A function $w = f(z)$ of the complex variable z is an **elliptic function** if and only if

1. $f(z)$ is **doubly periodic** with two finite **primitive periods** (smallest periods) ω_1, ω_2 whose ratio is not a real number, i.e.,

$$f(z + m\omega_1 + n\omega_2) \equiv f(z)$$
$$\left[m, n = 0, \pm 1, \pm 2, \ldots;\ \mathrm{Im}\left(\frac{\omega_1}{\omega_2}\right) \neq 0\right] \qquad (21.6\text{-}1)$$

and

2. The only singularities of $f(z)$ in the finite portion of the z plane are poles (see also Secs. 7.6-7 to 7.6-9).

A doubly periodic function $f(z)$ repeats the values it takes in any **period parallelogram,** say that defined by the points 0, ω_1, ω_2, $\omega_1 + \omega_2$, where the two sides joining the latter three points are excluded as belonging to adjacent parallelograms. The **order** of an elliptic function is the number of poles (counting multiplicities) in any period parallelogram.

A doubly periodic integral function is necessarily a constant. The residues in any period parallelogram add up to zero; hence the simplest nontrivial elliptic function is of order 2. *An elliptic function $f(z)$ of order r assumes* *every desired value w exactly r times in each period parallelogram, if one counts multiple roots of the equation $f(z) - w = 0$ (Liouville's Theorems,* see also Sec. 7.6-5).

Elliptic functions are usually encountered in connection with integrals or differential equations which involve square roots of third- or fourth-degree polynomials (e.g., arc length of an ellipse, equation of motion of a pendulum; see also Secs. 4.6-7 and 21.6-4). *Weierstrass's elliptic functions and normal elliptic integrals* are constructed in terms of simpler functions with known singularities for theoretical simplicity (Secs. 21.6-2, 21.6-3, and 21.6-5*b*). For numerical calculations, one prefers *Jacobi's elliptic functions* (Sec. 21.6-7), which may be regarded as generalizations of trigonometric functions; *Legendre's normal elliptic integrals,* which are closely related to the inverses of Jacobi's functions, are widely tabulated (Secs. 21.6-5 and 21.6-6).

21.6-2. Weierstrass's $\wp$ Function. (a) $\wp(z) \equiv \wp(z|\omega_1, \omega_2)$ is an even elliptic function of order 2 with periods ω_1, ω_2 and 2nd-order poles at $z = m\omega_1 + n\omega_2 (n, m = 0, \pm 1, \pm 2, \ldots)$; in terms of partial fractions (Sec. 7.6-8),

$$\wp(z) = \wp(z|\omega_1, \omega_2) = \frac{1}{z^2} + \sum_{\substack{m \quad n \\ (m^2+n^2 \neq 0)}} \sum \left[\frac{1}{(z - m\omega_1 - n\omega_2)^2} - \frac{1}{(m\omega_1 + n\omega_2)^2} \right]$$

$$= \wp(-z) \qquad \left[\operatorname{Im}\left(\frac{\omega_2}{\omega_1}\right) > 0 \right] \tag{21.6-2}$$

where the summation ranges over all integer pairs m, n except for 0, 0.

$w = \wp(\pm z - C|\omega_1, \omega_2)$ satisfies the differential equation

$$\left(\frac{dw}{dz}\right)^2 = 4w^3 - g_2 w - g_3 \equiv 4(w - e_1)(w - e_2)(w - e_3) \tag{21.6-3}$$

with

$$e_1 = \wp\left(\frac{\omega_1}{2}\right) \qquad e_2 = \wp\left(\frac{\omega_1 + \omega_2}{2}\right) \qquad e_3 = \wp\left(\frac{\omega_2}{2}\right) \tag{21.6-4}$$

$$e_1 + e_2 + e_3 = 0 \qquad e_1e_2 + e_1e_3 + e_2e_3 = -\tfrac{1}{4}g_2 \qquad e_1e_2e_3 = \tfrac{1}{4}g_3 \tag{21.6-5}$$

The parameters g_2, g_3 determine the constants ω_1, ω_2 associated with each $\wp$ function and are known as **invariants** of $\wp(z) \equiv \wp(z|\omega_1, \omega_2) \equiv \wp(z; g_2, g_3)$; note

$$\wp(z; g_2, g_3) \equiv m^2 \wp\left(mz; \frac{g_2}{m^4}, \frac{g_3}{m^6}\right) \qquad \left(m^2 = \frac{g_3}{g_2}\right) \tag{21.6-6}$$

The points $w = e_1, e_2, e_3$ and $w = \infty$ are the branch points of the inverse function

$$z = \int_\infty^w \frac{dw}{\sqrt{4w^3 - g_2 w - g_3}} \qquad \text{(Weierstrass's normal elliptic integral of the first kind)} \tag{21.6-7}$$

of $\wp(z|\omega_1, \omega_2)$ (see also Sec. 21.6-5*b*).

Note the series expansions

$$\left.\begin{aligned} \wp(z) &= \frac{1}{z^2} + \frac{g_2}{20} z^2 + \frac{g_3}{28} z^4 + \frac{g_2^{\,2}}{1200} z^6 + \frac{3g_2g_3}{6160} z^8 + \cdots \\ &= \frac{1}{z^2} + \sum_{k=2}^{\infty} a_k z^{2k-2} \qquad [0 < |z| < \min(|\omega_1|, |\omega_2|)] \\ a_k &= \frac{3}{(k-3)(2k+1)} (a_2 a_{k-2} + a_3 a_{k-3} + \cdots + a_{k-2} a_2) \end{aligned}\right\} \tag{21.6-8}$$

$$\left.\begin{aligned} g_2 &= 60 \sum_{\substack{m \; n \\ (m^2+n^2 \neq 0)}}\sum \frac{1}{(m\omega_1 + n\omega_2)^4} = \left(\frac{2\pi}{\omega_2}\right)^4 \left(\frac{1}{12} + 20 \sum_{k=1}^{\infty} \frac{k^3 q^{2k}}{1 - q^{2k}}\right) \\ g_3 &= 140 \sum_{\substack{m \; n \\ (m^2+n^2 \neq 0)}}\sum \frac{1}{(m\omega_1 + n\omega_2)^6} = \left(\frac{2\pi}{\omega_2}\right)^6 \left(\frac{1}{216} - \frac{7}{3} \sum_{k=1}^{\infty} \frac{k^5 q^{2k}}{1 - q^{2k}}\right) \end{aligned}\right\}$$

$$(q = e^{i\pi\omega_1/\omega_2}) \tag{21.6-9}$$

and the *addition formula*

$$\wp(A + B) = -\wp(A) - \wp(B) + \frac{1}{4}\left[\frac{\wp'(A) - \wp'(B)}{\wp(A) - \wp(B)}\right]^2 \tag{21.6-10}$$

(**b**) *Every elliptic function $f(z)$ with periods ω_1, ω_2 can be represented as a rational function of* $\wp(z; \omega_1, \omega_2)$ and $\wp'(z; \omega_1, \omega_2)$. More specifically, $f(z)$ can be written in the form

$$f(z) = R_1[\wp(z)] + \wp'(z) R_2[\wp(z)] \tag{21.6-11}$$

where R_1 and R_2 are rational functions; $\wp'(z)$ is an odd elliptic function of order 3.

21.6-3. Weierstrass's ζ and σ Functions. (**a**) Weierstrass's ζ and σ functions are *not* elliptic functions but may be used to construct elliptic functions with easily recognizable singularities. They are defined by

$$\left.\begin{aligned} \zeta(z) &= \zeta(z|\omega_1, \omega_2) = \frac{1}{z} + \sum_{\substack{m \; n \\ (m^2+n^2 \neq 0)}}\sum \left[\frac{1}{z - m\omega_1 - n\omega_2} + \frac{1}{m\omega_1 + n\omega_2} + \frac{z}{(m\omega_1 + n\omega_2)^2}\right] \\ &= -\zeta(-z) \\ \sigma(z) &= \sigma(z|\omega_1, \omega_2) \\ &= z \prod_{\substack{m \; n \\ (m^2+n^2 \neq 0)}}\prod \left(1 - \frac{z}{m\omega_1 + n\omega_2}\right) \exp\left[\frac{z}{m\omega_1 + n\omega_2} + \frac{z^2}{2(m\omega_1 + n\omega_2)^2}\right] \\ &= -\sigma(-z) \end{aligned}\right\}$$

$$\left[\operatorname{Im}\left(\frac{\omega_2}{\omega_1}\right) > 0\right] \tag{21.6-12}$$

where the sums and products range over all integer pairs m, n except for 0, 0. $\zeta(z)$ has simple poles, and $\sigma(z)$ has simple zeros at the points $z = m\omega_1 + n\omega_2$, and

$$\zeta'(z) = -\wp(z) \qquad \frac{\sigma'(z)}{\sigma(z)} = \zeta(z) \tag{21.6-13}$$

Equations (8) and (13) yield Laurent expansions of $\zeta(z)$ and $\sigma(z)$ about $z = 0$. Note also

$$\zeta(z) = \int_{\text{const.}}^{\wp(z)} \frac{w\,dw}{\sqrt{4w^3 - g_2 w - g_3}} \qquad \begin{pmatrix}\text{WEIERSTRASS'S NORMAL ELLIPTIC} \\ \text{INTEGRAL OF THE SECOND KIND}\end{pmatrix} \tag{21.6-14}$$

and the *addition theorem*

$$\zeta(A + B) = \zeta(A) + \zeta(B) + \frac{1}{2}\frac{\wp'(A) - \wp'(B)}{\wp(A) - \wp(B)} \tag{21.6-15}$$

Given $2\zeta(\omega_1/2) = \eta_1$, $2\zeta(\omega_2/2) = \eta_2$, one has $\eta_1\omega_2 - \eta_2\omega_1 = 2\pi i$, and

$$\zeta(z + m\omega_1 + n\omega_2) = \zeta(z) + m\eta_1 + n\eta_2 \qquad (m, n = 0, \pm 1, \pm 2, \ldots) \tag{21.6-16}$$

and

$$\sigma(z + m\omega_1 + n\omega_2) = \pm\sigma(z) \exp\left[(m\eta_1 + n\eta_2)\left(z + \frac{m\omega_1 + n\omega_2}{2}\right)\right] \tag{21.6-17}$$

where the positive sign applies only if both m and n are even integers.

(**b**) *Every elliptic function $f(z)$ of order r* (Sec. 21.6-1) *can be represented in the form*

$$f(z) = \frac{\prod_{j=1}^{r} \sigma(z - a_j)}{\prod_{j=1}^{r} \sigma(z - b_j)} \qquad \left(\sum_{j=1}^{r} a_j = \sum_{j=1}^{r} b_j\right) \tag{21.6-18}$$

21.6-4. Elliptic Integrals (see also Sec. 4.6-7).

The function

$$F(z) = \int_a^z f(z)\,dz$$

is an **elliptic integral** whenever $f(z)$ is a rational function of z and the square root $\sqrt{G(z)}$ of a polynomial

$$\begin{aligned} G(z) &\equiv a_0 z^4 + a_1 z^3 + a_2 z^2 + a_3 z + a_4 \\ &\equiv a_0(z - \alpha_1)(z - \alpha_2)(z - \alpha_3)(z - \alpha_4) \end{aligned} \tag{21.6-19}$$

without multiple zeros; one includes the case of third-degree polynomials $G(z) \equiv G_3(z)$ as well as fourth-degree polynomials $G(z) \equiv G_4(z)$ by introducing $\alpha_4 = \infty$ whenever $a_0 = 0$, so that formally $a_0(z - \alpha_4) \equiv a_1$. Every elliptic integral is a multiple-valued function of z; different integration paths yield an infinite number of function values. The points $z = \alpha_1$, $z = \alpha_2$, $z = \alpha_3$, $z = \alpha_4$ are *branch points*. One joins α_1, α_2 and α_3, α_4 by two suitably defined branch cuts to obtain a Riemann surface (Sec. 7.4-3) connected like the surface of a torus.

An **elliptic integral of the first kind** is finite for all z; its only singularities are branch points at $z = \alpha_1, \alpha_2, \alpha_3, \alpha_4$. An **elliptic integral of the second kind** is analytic throughout the z plane except for branch points at $\alpha_1, \alpha_2, \alpha_3, \alpha_4$ and a finite number of poles. An **elliptic integral of the third kind** has a logarithmic singularity (see also Secs. 21.6-5 and 21.6-6).

21.6-5. Reduction of Elliptic Integrals. The following procedures reduce every elliptic integral $\int f(z)\,dz$ to a weighted sum of elementary functions and three so-called *normal elliptic integrals* (see also Refs. 21.2, 21.11, and 21.12 for alternative procedures; Ref. 21.2 contains a very comprehensive collection of explicit formulas expressing elliptic integrals in terms of normal elliptic integrals).

(a) Formal Reduction Procedure. Note that even powers of $\sqrt{G(z)}$ are polynomials in z, and rewrite

$$f(z) \equiv \frac{P_1(z) + P_2(z)\sqrt{G(z)}}{P_3(z) + P_4(z)\sqrt{G(z)}} \equiv \frac{(P_1 + P_2\sqrt{G})(P_3 - P_4\sqrt{G})}{(P_3)^2 - (P_4)^2G} \equiv R_1(z) + \frac{R_2(z)}{\sqrt{G(z)}} \tag{21.6-20}$$

where the $P_i(z)$ are polynomials, and $R_1(z)$, $R_2(z)$ are rational functions. $R_1(z)$ can be integrated to yield elementary functions (Sec. 4.6-6*c*).

Partial-fraction expansion of the rational function $R_2(z)$ (Sec. 1.7-4) reduces the evaluation of $\int \frac{R_2(z)}{\sqrt{G(z)}}\,dz$ to that of integrals of the form

$$I_n = \int \frac{(z-c)^n}{\sqrt{G(z)}}\,dz \qquad (n = 0, \pm 1, \pm 2, \ldots) \tag{21.6-21}$$

Every one of these integrals can be expressed in terms of I_0, I_1, I_2, *and* I_{-1} *alone* with the aid of the recursion formula

$$\begin{aligned}(2n+6)b_0I_{n+4} &+ (2n+5)b_1I_{n+3} + (2n+4)b_2I_{n+2} \\ &+ (2n+3)b_3I_{n+1} + (2n+2)b_4I_n = 2(z-c)^{n+1}\sqrt{G(z)} \\ &\qquad (n = 0, \pm 1, \pm 2, \ldots)\end{aligned} \tag{21.6-22}$$

where the coefficients b_k are defined by the identity

$$\begin{aligned}G(z) &\equiv a_0z^4 + a_1z^3 + a_2z^2 + a_3z + a_4 \\ &\equiv b_0(z-c)^4 + b_1(z-c)^3 + b_2(z-c)^2 + b_3(z-c) + b_4\end{aligned} \tag{21.6-23}$$

In addition, (22) yields I_2 explicitly in terms of I_0, I_1, and I_{-1} if $a_0 = 0$, or if c is a root of $G(z) = 0$ (i.e., $b_4 = 0$). Even if c is not such a root, one can rewrite I_2 as

$$\int \frac{(z-c)^2}{\sqrt{G(z)}}\,dz = \int \frac{(z-c')^2}{\sqrt{G(z)}}\,dz + \alpha \int \frac{z\,dz}{\sqrt{G(z)}} + \beta \int \frac{dz}{\sqrt{G(z)}} \tag{21.6-24}$$

where $z = c'$ is a root of $G(z) = 0$. Hence *every elliptic integral can be expressed as a weighted sum of elementary functions and three relatively simple types of elliptic integrals of the first, second, and third kinds* (Sec.

21.6-4), *viz.*,

$$\int \frac{dz}{\sqrt{G(z)}} \qquad \int \frac{z\,dz}{\sqrt{G(z)}} \qquad \int \frac{dz}{(z-c)\sqrt{G(z)}} \tag{21.6-25}$$

The first of these integrals is usually referred to as a **normal elliptic integral of the first kind;** it is often convenient not to use the other two integrals (25) directly but to introduce suitable linear combinations as **normal elliptic integrals of the second and third kind** (Secs. 21.6-2, 21.6-3, and 21.6-6).

(b) Change of Variables. Weierstrass's and Riemann's Normal Forms. At any convenient stage of the reduction procedure, one may introduce a new integration variable $\bar{z} = \bar{z}(z)$ to transform the elliptic integrals (21) or (25) into new elliptic integrals involving a more convenient polynomial $\bar{G}(\bar{z})$ and, possibly, a simpler recursion formula (22). In particular, a bilinear transformation

$$z = \frac{A\bar{z} + B}{C\bar{z} + D} \qquad (AD - BC \neq 0) \tag{21.6-26}$$

(Sec. 7.9-2) chosen by substitution of corresponding values of z and $\bar{z}$ so as to map the branch points $z = \alpha_1, \alpha_2, \alpha_3, \alpha_4$ into $\bar{z} = e_1, e_2, e_3, \infty$ yields elliptic integrals in **Weierstrass's normal form** with $\bar{G}(\bar{z}) \equiv 4\bar{z}^3 - g_2\bar{z} - g_3$. These integrals are related to Weierstrass's $\wp$ function (Sec. 21.6-2). Again, a transformation (26) mapping $z = \alpha_1, \alpha_2, \alpha_3, \alpha_4$ into $\bar{z} = 0, 1, 1/k, -1/k$, where k is a real number between 0 and 1, yields elliptic integrals in **Riemann's normal form,** with $\bar{G}(\bar{z}) \equiv \bar{z}(1 - \bar{z})(1 - k^2\bar{z})$.

(c) Reduction to Legendre's Normal Form. More frequently, one desires to transform a real elliptic integral $\int_a^x f(x)\,dx$ to **Legendre's normal form** with $\bar{G}(\bar{z}) \equiv (1 - \bar{z}^2)(1 - k^2\bar{z}^2)$ where k^2 is a real number between 0 and 1. The reduction procedure will then yield (real) Legendre's normal integrals (Sec. 21.6-6) whose values are available in published tables.

Let $G(x)$ be a real polynomial greater than zero in (a, x); since $\int_a^x f(x)\,dx$ is to be real, the integration interval cannot include a real root of $G(x) = 0$. Table 21.6-1 (pages 712–713) lists transformations $x = x(\varphi)$ mapping the real integration interval (a, x) into a corresponding range of real angles φ between 0 and $\pi/2$ so that

$$\frac{dx}{\sqrt{G(x)}} \equiv \mu \frac{d\varphi}{\sqrt{1 - k^2 \sin^2 \varphi}} \qquad (0 < k^2 < 1) \tag{21.6-27}$$

for the various possible types of real fourth-degree polynomials $G(x) \equiv$

$G_4(x)$ and third-degree polynomials $G(x) \equiv G_3(x)$. The correct values of the constant parameters k^2 and μ are also tabulated.

In each case, the leading coefficient (a_0 or a_1) of $G(x)$ is taken to be either 1 or -1. In the case of real roots, it is assumed that $\alpha_1 > \alpha_2$, $\alpha_3 > \alpha_4$; complex roots are denoted by $b_1 \pm ic_1$ and $b_2 \pm ic_2$, with $b_1 \geq b_2$, $c_1 > 0$, $c_2 > 0$. The following auxiliary quantities have been introduced:

$$\left.\begin{aligned} \alpha_{ik} &= \alpha_k - \alpha_i \qquad (i, k = 1, 2, 3, 4) \\ (\alpha, \beta, \gamma, \delta) &= \frac{\alpha - \gamma}{\alpha - \delta}\frac{\beta - \delta}{\beta - \gamma} \\ \tan \vartheta_1 &= \frac{\alpha_1 - b_1}{c_1} \qquad \tan \vartheta_2 = \frac{\alpha_2 - b_1}{c_1} \\ \tan \vartheta_3 &= \frac{c_1 + c_2}{b_1 - b_2} \qquad \tan \vartheta_4 = \frac{c_1 - c_2}{b_1 - b_2} \\ (\tan \vartheta_5/2)^2 &= \frac{\cos \vartheta_3}{\cos \vartheta_4} \\ \nu &= \tan\,[(\vartheta_2 - \vartheta_1)/2] \tan\,[(\vartheta_1 + \vartheta_2)/2] \end{aligned}\right\} \qquad (21.6\text{-}28)$$

21.6-6. Legendre's Normal Elliptic Integrals (see also Secs. 26.6-4 and 26.6-5). **(a) Definitions. Legendre's (incomplete) normal elliptic integrals** are defined as

$$\begin{aligned} F(k, \varphi) &\equiv \int_0^\varphi \frac{d\varphi}{\sqrt{1 - k^2 \sin^2 \varphi}} \\ &\equiv \int_0^z \frac{dz}{\sqrt{(1-z^2)(1-k^2z^2)}} \equiv \bar{F}(k, z) \quad \text{(Legendre's normal elliptic integral of the first kind)} \\ E(k, \varphi) &\equiv \int_0^\varphi \sqrt{1 - k^2 \sin^2 \varphi}\, d\varphi \equiv \int_0^z \sqrt{\frac{1 - k^2z^2}{1 - z^2}}\, dz \\ &\equiv \bar{E}(k, z) \quad \text{(Legendre's normal elliptic integral of the second kind)} \end{aligned} \qquad (21.6\text{-}29a)$$

$$\begin{aligned} \pi(c, k, \varphi) &\equiv \int_0^\varphi \frac{d\varphi}{(\sin^2 \varphi - c)\sqrt{1 - k^2 \sin^2 \varphi}} \\ &\equiv \int_0^z \frac{dz}{(z^2 - c)\sqrt{(1 - z^2)(1 - k^2z^2)}} \\ &\equiv \bar{\pi}(c, k, z) \quad \text{(Legendre's normal elliptic integral of the third kind)} \end{aligned} \qquad (21.6\text{-}29b)$$

where $z = \sin \varphi$. k is a complex number called the **modulus (module)** of the elliptic integral. The elliptic integrals (29) are real for real values of the **amplitude** φ between $-\pi/2$ and $\pi/2$ if k^2 is a real number between 0 and 1; $F(k, \varphi)$ and $E(k, \varphi)$ have been tabulated for $0 \leq \varphi \leq \pi/2$ and real values of k^2 between 0 and 1 (see also Fig. 21.6-1). c is called the **characteristic** of the elliptic integral (29).

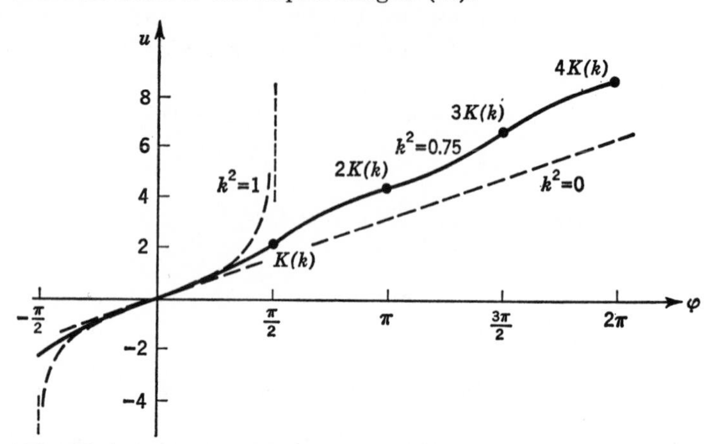

FIG. 21.6-1. Variation of the elliptic integral u of the first kind with φ, the amplitude of u for three values of the modulus k. (*From J. Cunningham, Introduction to Nonlinear Analysis, McGraw-Hill, New York,* 1958.)

For real values of the modulus k, the **modular angle**

$$\alpha = \arcsin k \tag{21.6-29c}$$

is often tabulated as an argument instead of k (Fig. 21.6-2); one writes

$$F(k, \varphi) \equiv F(\varphi\diagdown\alpha) \qquad E(k, \varphi) \equiv E(\varphi\diagdown\alpha) \qquad \pi(c, k, \varphi) \equiv \pi(c; \varphi\diagdown\alpha) \tag{21.6-29d}$$

$90° - \alpha = \arccos k$ is the **complementary modular angle.** $k^2 = m$ is often called the **parameter** of a normal elliptical integral.

(b) Legendre's Complete Normal Elliptic Integrals (see also Fig. 21.6-3). The functions

$$\begin{aligned} \mathbf{K} = \mathbf{K}(k) &\equiv \int_0^{\pi/2} \frac{d\varphi}{\sqrt{1 - k^2 \sin^2 \varphi}} \equiv F(k, \pi/2) \\ &\equiv F(90°\diagdown\alpha) \\ \mathbf{E} = \mathbf{E}(k) &\equiv \int_0^{\pi/2} \sqrt{1 - k^2 \sin^2 \varphi}\, d\varphi \equiv E(k, \pi/2) \\ &\equiv E(90°\diagdown\alpha) \\ \text{with} \qquad & \alpha = \arcsin k \end{aligned} \tag{21.6-30}$$

are respectively known as **Legendre's complete elliptic integrals of the first and second kind.** $k = \sin \alpha$ and $k' = \sqrt{1 - k^2} = \cos \alpha$

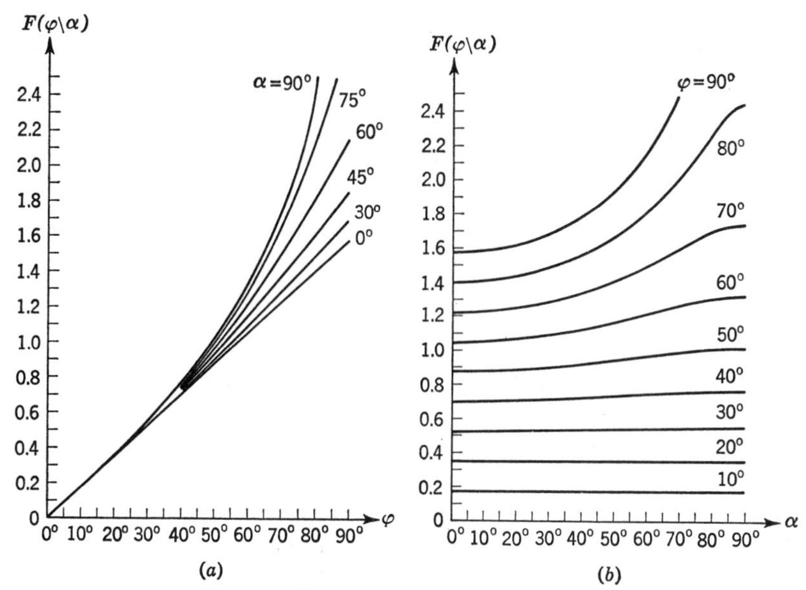

FIG. 21.6-2. The incomplete elliptic integral of the first kind, $F(k,\varphi) \equiv F(\varphi/\alpha)$, plotted against φ (*a*), and against the modular angle $\alpha = \arcsin k$ (*b*). (*From Ref.* 21.1.)

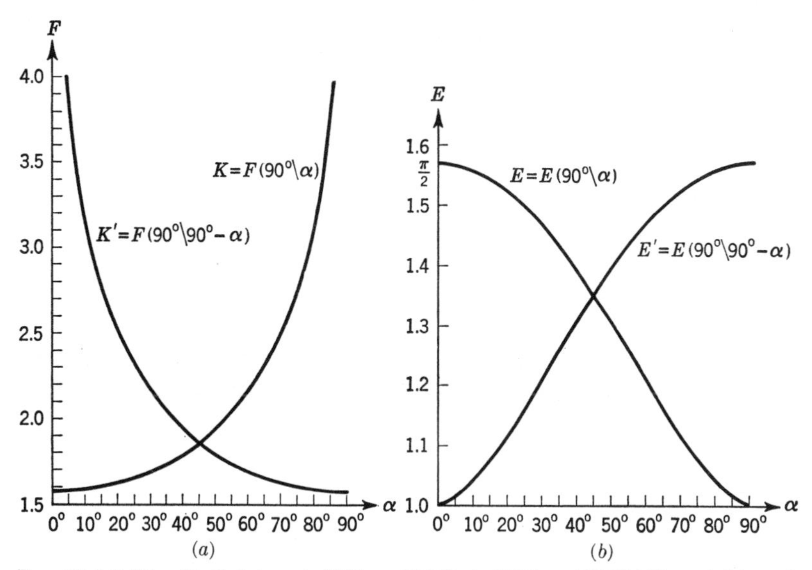

FIG. 21.6-3. The elliptic integrals $\mathbf{K}(k) \equiv F(90°\backslash\alpha)$, $\mathbf{K}'(k) \equiv F(90°\backslash 90° - \alpha)$ (*a*), and $\mathbf{E}(k) \equiv E(90°\backslash\alpha)$, $\mathbf{E}'(k) \equiv E(90°\backslash 90° - \alpha)$ (*b*) ($k = \sin\alpha$; *from Ref.* 21.1).

are called **complementary moduli.** $\mathbf{K}(k)$ and $\mathbf{K}'(k) \equiv \mathbf{K}(k')$ are **associated elliptic integrals of the first kind; $\mathbf{E}(k)$ and $\mathbf{E}'(k) \equiv \mathbf{E}(k')$** are **associated elliptic integrals of the second kind.** Note

$$\mathbf{EK}' + \mathbf{E}'\mathbf{K} - \mathbf{KK}' = \frac{\pi}{2} \qquad \text{(Legendre's relation)} \qquad (21.6\text{-}31)$$

and $\mathbf{K}(0) = \mathbf{K}'(1) = \pi/2$, $\mathbf{K}(1) = \mathbf{K}'(0) = \infty$.

Different values of the multiple-valued elliptic integral $F(k, \varphi)$ differ by $4m\mathbf{K} + 2ni\mathbf{K}'$; different values of $E(k, \varphi)$ differ by $4m\mathbf{E} + 2ni(\mathbf{K}' - \mathbf{E}')$ $(m, n = 0, \pm 1, \pm 2, \ldots$; see also Sec. 21.6-7*b*).

$\mathbf{K}(k)$ and $\mathbf{E}(k)$ satisfy the differential equations

$$\left.\begin{aligned} k(1 - k^2)\frac{d^2\mathbf{K}}{dk^2} + (1 - 3k^2)\frac{d\mathbf{K}}{dk} - k\mathbf{K} &= 0 \\ k(1 - k^2)\frac{d^2\mathbf{E}}{dk^2} + (1 - k^2)\frac{d\mathbf{E}}{dk} + k\mathbf{E} &= 0 \end{aligned}\right\} \qquad (21.6\text{-}32)$$

so that, for real $k^2 < 1$,

$$\left.\begin{aligned} \mathbf{K}(k) &= \frac{\pi}{2}F\left(\frac{1}{2}, \frac{1}{2}; 1; k^2\right) = \frac{\pi}{2}\left[1 + \left(\frac{1}{2}\right)^2 k^2 + \left(\frac{1\cdot 3}{2\cdot 4}\right)^2 k^4 + \cdots\right] \\ \mathbf{E}(k) &= \frac{\pi}{2}F\left(-\frac{1}{2}, \frac{1}{2}; 1; k^2\right) = \frac{\pi}{2}\left[1 - \left(\frac{1}{2}\right)^2 k^2 - \left(\frac{1\cdot 3}{2\cdot 4}\right)^2 \frac{k^4}{3} - \cdots\right] \end{aligned}\right\} \qquad (21.6\text{-}33)$$

where $F(a, b; c; z)$ is the hypergeometric function defined in Sec. 9.3-9.

(c) Transformations. Legendre's normal elliptic integrals (29) with moduli k and $\dot{k} = 1/k,\ k',\ 1/k',\ ik/k',\ k'/ik,\ (1 - k')/(1 + k')$, $2\sqrt{k}/(1 + k)$ are connected by the relations listed in Table 21.6-2, where

$$\sqrt{1 - k^2 \sin^2 \varphi} \equiv \Delta(\varphi, k) \qquad (21.6\text{-}34)$$

(see also Sec. 21.6-7*a*). Table 21.6-3 lists similar relations for the complete normal elliptic integrals (30).

In particular,

$$\mathbf{K}(k) = \frac{2}{1 + k'}\mathbf{K}\left(\frac{1 - k'}{1 + k'}\right) \qquad (21.6\text{-}35)$$

Successive substitution of

$$k_0' = k' \qquad k_{n+1}' = \frac{2\sqrt{k_n'}}{1 + k_n'} \qquad (n = 0, 1, 2, \ldots) \qquad (21.6\text{-}36)$$

for k' in Eq. (35) yields $(1 - k_n')/(1 + k_n') \to 0$; since $\mathbf{K}(0) = \pi/2$, one obtains

$$\mathbf{K}(k) = \frac{\pi}{2}\prod_{n=0}^{\infty}\frac{2}{1 + k_n'} \qquad (21.6\text{-}37)$$

which may be useful for numerical computation of $\mathbf{K}(k)$. $K(k, \varphi)$ can be obtained in analogous fashion.

Table 21.6-1. Transformation to Legendre's Normal Form*
All zeros of $G(x)$ real

$G(x)$, zeros	Leading coefficient	Interval	Transformation $x =$	$\sin^2 \varphi =$	Corresponding values		k^2	μ
					x	φ		
$G_4(x)$, four real zeros	+1	$\alpha_1 \leq x$ or $x \leq \alpha_4$	$\dfrac{\alpha_1\alpha_{42} - \alpha_2\alpha_{41}\sin^2\varphi}{\alpha_{42} - \alpha_{41}\sin^2\varphi}$	$\dfrac{\alpha_{42}}{\alpha_{41}}\dfrac{x - \alpha_1}{x - \alpha_2}$	α_1 α_4	0 $\pi/2$	$(\alpha_1, \alpha_2, \alpha_4, \alpha_3)$	$\dfrac{2}{(\alpha_{31}\alpha_{42})^{1/2}}$
		$\alpha_3 \leq x \leq \alpha_2$	$\dfrac{\alpha_3\alpha_{42} - \alpha_4\alpha_{32}\sin^2\varphi}{\alpha_{42} - \alpha_{32}\sin^2\varphi}$	$\dfrac{\alpha_{42}}{\alpha_{32}}\dfrac{x - \alpha_3}{x - \alpha_4}$	α_3 α_2	0 $\pi/2$		
	−1	$\alpha_4 \leq x \leq \alpha_3$	$\dfrac{\alpha_4\alpha_{31} + \alpha_1\alpha_{43}\sin^2\varphi}{\alpha_{31} + \alpha_{43}\sin^2\varphi}$	$\dfrac{\alpha_{31}}{\alpha_{43}}\dfrac{x - \alpha_4}{\alpha_1 - x}$	α_4 α_3	0 $\pi/2$	$(\alpha_3, \alpha_2, \alpha_4, \alpha_1)$	
		$\alpha_2 \leq x \leq \alpha_1$	$\dfrac{\alpha_2\alpha_{31} - \alpha_3\alpha_{21}\sin^2\varphi}{\alpha_{31} - \alpha_{21}\sin^2\varphi}$	$\dfrac{\alpha_{31}}{\alpha_{21}}\dfrac{x - \alpha_2}{x - \alpha_3}$	α_2 α_1	0 $\pi/2$		
$G_3(x)$, three real zeros	+1	$\alpha_3 \leq x \leq \alpha_2$	$\alpha_3 + \alpha_{32}\sin^2\varphi$	$\dfrac{x - \alpha_3}{\alpha_{32}}$	α_3 α_2	0 $\pi/2$	$\dfrac{\alpha_{32}}{\alpha_{31}}$	$\dfrac{2}{(\alpha_{31})^{1/2}}$
		$\alpha_1 \leq x$	$\dfrac{\alpha_1 - \alpha_2\sin^2\varphi}{1 - \sin^2\varphi}$	$\dfrac{x - \alpha_1}{x - \alpha_2}$	α_1 ∞	0 $\pi/2$		
	−1	$x \leq \alpha_3$	$\alpha_1 - \dfrac{\alpha_{31}}{\sin^2\varphi}$	$\dfrac{\alpha_{31}}{\alpha_1 - x}$	$-\infty$ α_3	0 $\pi/2$	$\dfrac{\alpha_{21}}{\alpha_{31}}$	
		$\alpha_2 \leq x \leq \alpha_1$	$\dfrac{\alpha_2\alpha_{31} - \alpha_3\alpha_{21}\sin^2\varphi}{\alpha_{31} - \alpha_{21}\sin^2\varphi}$	$\dfrac{\alpha_{31}}{\alpha_{21}}\dfrac{x - \alpha_2}{x - \alpha_3}$	α_2 α_1	0 $\pi/2$		

Table 21.6-1. Transformation to Legendre's Normal Form* *(Continued)*
$G(x)$ has complex zeros

$G(x)$, zeros	Leading coefficient	Interval	Transformation	Auxiliary quantities	Corresponding values		k^2	μ
					x	φ		
$G_4(x)$, two real and two complex zeros	1	$\alpha_1 \leq x$ or $x \leq \alpha_2$	$x = \dfrac{\alpha_1 + \alpha_2}{2} - \dfrac{\alpha_1 - \alpha_2}{2} \dfrac{\nu - \cos\varphi}{1 - \nu\cos\varphi}$	θ_1 acute θ_2 obtuse	α_1	0	$[\sin \frac{1}{2}(\theta_1 - \theta_2)]^2$	$\dfrac{(-\cos\theta_1 \cos\theta_2)^{\frac{1}{2}}}{c_1}$
	-1	$\alpha_2 \leq x \leq \alpha_1$	$(\tan \frac{1}{2}\varphi)^2 = \dfrac{\cos\theta_1}{\cos\theta_2} \dfrac{\alpha_1 - x}{x - \alpha_2}$	θ_1, θ_2 acute	α_2	π		$-\dfrac{(\cos\theta_1 \cos\theta_2)^{\frac{1}{2}}}{c_1}$
$G_3(x)$, two complex zeros	1	$\alpha_1 \leq x$	$x = \alpha_1 - \dfrac{c_1}{\cos\theta_1} \dfrac{1 - \cos\varphi}{1 + \cos\varphi}$	θ_1 obtuse	α_1	0	$[\sin(\frac{1}{2}\theta_1 + \frac{1}{4}\pi)]^2$	$\left(\dfrac{-\cos\theta_1}{c_1}\right)^{\frac{1}{2}}$
	-1	$x \leq \alpha_1$	$(\tan \frac{1}{2}\varphi)^2 = \dfrac{\cos\theta_1}{c_1}(\alpha_1 - x)$	θ_1 acute	∞	π		$-\left(\dfrac{\cos\theta_1}{c_1}\right)^{\frac{1}{2}}$
$G_4(x)$, four complex zeros $b_1 > b_2$	1	$-\infty < x < \infty$	$x = b_1 + c_1 \tan(\varphi + \frac{1}{2}\theta_3 + \frac{1}{2}\theta_4)$ $\tan(\varphi + \frac{1}{2}\theta_3 + \frac{1}{2}\theta_4) = (x - b_1)/c_1$	θ_3, θ_4, $\frac{1}{2}\theta_5$ acute	$-\infty$	$-\pi/2 - \frac{1}{2}\theta_3 - \frac{1}{2}\theta_4$	$\sin^2\theta_5$	$\left(\dfrac{\cos\theta_5}{c_1 c_2}\right)^{\frac{1}{2}}$
					b_1	$-\frac{1}{2}\theta_3 - \frac{1}{2}\theta_4$		
$G_4(x)$, four complex zeros $b_1 = b_2$ $c_1 > c_2$			$x = b_1 - c_1 \cot\varphi$ $\tan\varphi = \dfrac{c_1}{b_1 - c_1}$	$\theta_3 = \theta_4 = \pi/2$	∞	$\pi/2 - \frac{1}{2}\theta_3 - \frac{1}{2}\theta_4$	$1 - \left(\dfrac{c_2}{c_1}\right)^2$	$\dfrac{1}{c_1}$

* From A. Erdélyi *et al.*, *Higher Transcendental Functions*, vol. 2, McGraw-Hill, New York, 1953.

Table 21.6-2. Transformations of Elliptic Integrals*

$\dot{k}$	$\sin \dot{\varphi}$	$\cos \dot{\varphi}$	$F(\dot{\varphi}, \dot{k})$	$E(\dot{\varphi}, \dot{k})$
$\frac{1}{k}$	$k \sin \varphi$	$\Delta(\varphi, k)$	$kF(\varphi, k)$	$\frac{1}{k}[E(\varphi, k) - k'^2F(\varphi, k)]$
k'	$-i \tan \varphi$	$\sec \varphi$	$-iF(\varphi, k)$	$i[E(\varphi, k) - F(\varphi, k) - \tan \varphi\Delta(\varphi, k)]$
$\frac{1}{k'}$	$-ik' \tan \varphi$	$\frac{\Delta(\varphi, k)}{\cos \varphi}$	$-ik'F(\varphi, k)$	$\frac{i}{k'}[E(\varphi, k) - k'^2F(\varphi, k) - \tan \varphi\Delta(\varphi, k)]$
$\frac{ik}{k'}$	$\frac{k' \sin \varphi}{\Delta(\varphi, k)}$	$\frac{\cos \varphi}{\Delta(\varphi, k)}$	$k'F(\varphi, k)$	$\frac{1}{k'}\left[E(\varphi, k) - k^2\frac{\sin \varphi \cos \varphi}{\Delta(\varphi, k)}\right]$
$\frac{k'}{ik}$	$-\frac{ik \sin \varphi}{\Delta(\varphi, k)}$	$\frac{1}{\Delta(\varphi, k)}$	$-ikF(\varphi, k)$	$\frac{i}{k}\left[E(\varphi, k) - F(\varphi, k) - k^2\frac{\sin \varphi \cos \varphi}{\Delta(\varphi, k)}\right]$
$\frac{1 - k'}{1 + k'}$	$\frac{(1 + k') \sin \varphi \cos \varphi}{\Delta(\varphi, k)}$	$\frac{\cos^2 \varphi - k' \sin^2 \varphi}{\Delta(\varphi, k)}$	$(1 + k')F(\varphi, k)$	$\frac{2}{1 + k'}[E(\varphi, k) + k'F(\varphi, k)] - (1 - k')\frac{\sin \varphi \cos \varphi}{\Delta(\varphi, k)}$

* From A. Erdélyi *et al.*, *Higher Transcendental Functions*, vol. 2, McGraw-Hill, New York, 1953.

Table 21.6-3. Transformations of Complete Elliptic Integrals*

$\dot{k}$	$\mathbf{K}(\dot{k})$	$\mathbf{K}'(\dot{k})$	$\mathbf{E}(\dot{k})$	$\mathbf{E}'(\dot{k})$
$\frac{1}{k}$	$k(\mathbf{K} + i\mathbf{K}')$	$k\mathbf{K}'$	$\frac{1}{k}(\mathbf{E} + i\mathbf{E}' - k'^2\mathbf{K} - ik^2\mathbf{K}')$	$\frac{1}{k}\mathbf{E}'$
k'	$\mathbf{K}'$	$\mathbf{K}$	$\mathbf{E}'$	$\mathbf{E}$
$\frac{1}{k'}$	$k'(\mathbf{K}' + i\mathbf{K})$	$k'\mathbf{K}$	$\frac{1}{k'}(\mathbf{E}' + i\mathbf{E} - k^2\mathbf{K}' - ik'^2\mathbf{K})$	$\frac{1}{k'}\mathbf{E}$
$\frac{ik}{k'}$	$k'\mathbf{K}$	$k'(\mathbf{K}' - i\mathbf{K})$	$\frac{1}{k'}\mathbf{E}$	$\frac{1}{k'}(\mathbf{E}' + i\mathbf{E} - k^2\mathbf{K}' - ik'^2\mathbf{K})$
$\frac{k'}{ik}$	$k\mathbf{K}'$	$k(\mathbf{K} + i\mathbf{K}')$	$\frac{1}{k}\mathbf{E}'$	$\frac{1}{k}(\mathbf{E} - i\mathbf{E}' - k'^2\mathbf{K} + ik^2\mathbf{K}')$
$\frac{1 - k'}{1 + k'}$	$\frac{1 + k'}{2}\mathbf{K}$	$(1 + k')\mathbf{K}'$	$\frac{\mathbf{E} + k'\mathbf{K}}{1 + k'}$	$\frac{2\mathbf{E}' - k^2\mathbf{K}'}{1 + k'}$
$\frac{2k^{1/2}}{1 + k}$	$(1 + k)\mathbf{K}$	$\frac{1 + k}{2}\mathbf{K}'$	$\frac{2\mathbf{E} - k'^2\mathbf{K}}{1 + k}$	$\frac{\mathbf{E}' + k\mathbf{K}'}{1 + k}$

* From A. Erdélyi *et al.*, *Higher Transcendental Functions*, vol. 2, McGraw-Hill, New York, 1953.

21.6-7. Jacobi's Elliptic Functions. **(a) Definitions.** Inversion of the elliptic integrals $z = F(k, \varphi)$ and $z = \bar{F}(k, w)$ (Sec. 21.6-6*a*) yields the functions am z (**amplitude** of z) and sn z (**sinus amplitudinis** of z), i.e.,

$$\begin{aligned} &\varphi = \operatorname{am} z \qquad z = \int_0^\varphi \frac{d\varphi}{\sqrt{1 - k^2 \sin^2 \varphi}} = F(k, \varphi) \\ &w = \operatorname{sn} z = \sin(\operatorname{am} z) \\ &\qquad\qquad z = \int_0^w \frac{dw}{\sqrt{(1 - w^2)(1 - k^2 w^2)}} = \bar{F}(k, w) \end{aligned} \tag{21.6-38}$$

In addition, one defines the functions cn z (**cosinus amplitudinis of** z) and dn z (**delta amplitudinis** of z) by

$$\begin{aligned} \operatorname{cn} z &= \cos(\operatorname{am} z) = \sqrt{1 - \operatorname{sn}^2 z} \\ \operatorname{dn} z &= \Delta(\operatorname{am} z, k) = \sqrt{1 - k^2 \operatorname{sn}^2 z} \end{aligned} \tag{21.6-39}$$

sn z, cn z, and dn z are **Jacobi's elliptic functions.** A given value of the parameter k is implied in each definition; if required, one writes sn (z, k), cn (z, k), dn (z, k). k', **K**, **K'**, **E**, and **E'** are defined as in Sec. 21.6-6*b*. Jacobi's elliptic functions are real for real z and real k^2 between

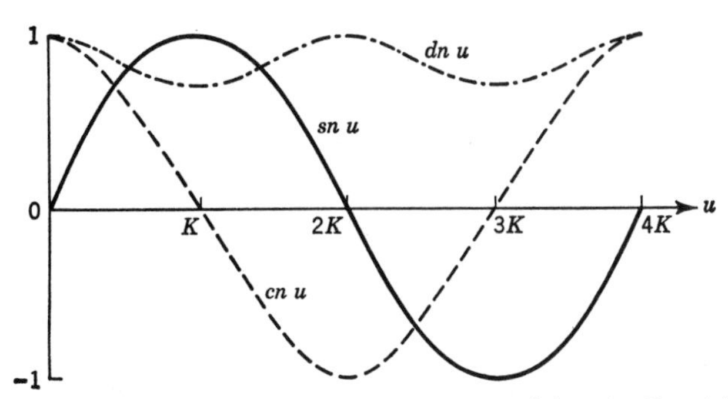

FIG. 21.6-4. The Jacobian elliptic functions sn u, cn u, and dn u for $k^2 = \frac{1}{2}$. (*From Ref.* 21.1.)

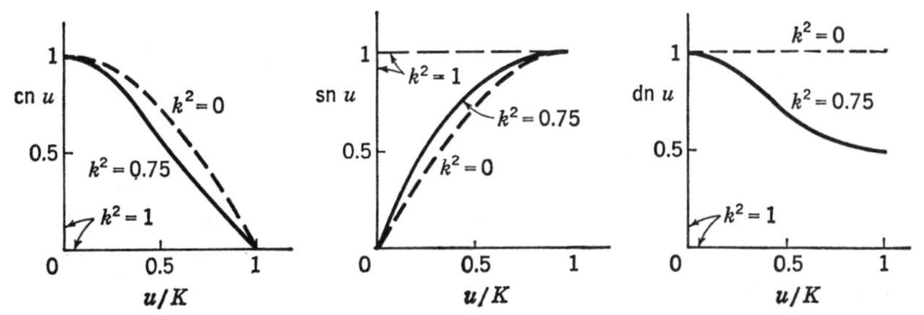

FIG. 21.6-5. One-quarter of a complete cycle of the elliptic functions cn u, sn u, and dn u, plotted against the normalized abscissa $u/\mathbf{K}$, for three values of the modulus k. (*From J. Cunningham, Introduction to Nonlinear Analysis, McGraw-Hill, New York,* 1958.)

Table 21.6-4. Periods, Zeros, Poles, and Residues of Jacobi's Elliptic Functions*
m and n are integers

Function	Primitive Periods	Zeros	Poles	Residues
sn (u, k)	$4\mathbf{K}$ $2i\mathbf{K}'$	$2m\mathbf{K} + 2ni\mathbf{K}'$	$2m\mathbf{K} + (2n+1)i\mathbf{K}'$	$\frac{(-1)^m}{k}$
cn (u, k)	$4\mathbf{K}$ $2\mathbf{K} + 2i\mathbf{K}'$	$(2m+1)\mathbf{K} + 2ni\mathbf{K}'$	$2m\mathbf{K} + (2n+1)i\mathbf{K}'$	$\frac{(-1)^{m+n}}{ik}$
dn (u, k)	$2\mathbf{K}$ $4i\mathbf{K}'$	$(2m+1)\mathbf{K} + (2n+1)i\mathbf{K}'$	$2m\mathbf{K} + (2n+1)i\mathbf{K}'$	$(-1)^{n+1}i$

* From A. Erdélyi *et al.*, *Higher Transcendental Functions*, vol. 2, McGraw-Hill, New York, 1953.

0 and 1 and reduce to elementary functions for $k^2 = 0$ and $k^2 = 1$; in particular,

$$\text{sn}\,(z, 0) = \sin z \qquad \text{cn}\,(z, 0) = \cos z \tag{21.6-40}$$

Jacobi's elliptic functions can also be defined in terms of $\wp$, ζ, σ, or ϑ functions by the relations of Sec. 21.6-9.

(b) Miscellaneous Properties and Special Values. Jacobi's elliptic functions are of order 2 (Sec. 26.1-1); their periods, their (simple) zeros, and their (simple) poles are listed in Table 21.6-4. sn z is an odd function, while cn z and dn z are even functions of z. Table 21.6-5 lists special function values. Table 21.6-6 shows the effects of argument changes by quarter and half periods, using the convenient notation

$$\text{sn}\,(z, k) = s \qquad \text{cn}\,(z, k) = c \qquad \text{dn}\,(z, k) = d \tag{21.6-41}$$

Note also

$$\text{sn}^2 z + \text{cn}^2 z = k^2\text{sn}^2 z + \text{dn}^2 z = 1 \qquad \text{dn}^2 z - k^2\,\text{cn}^2 z = k'^2 \tag{21.6-42}$$

$$\text{sn}\,(-z) = -\text{sn}\,z \qquad \text{cn}\,(-z) = \text{cn}\,z \qquad \text{dn}\,(-z) = \text{dn}\,(z) \tag{21.6-43}$$

$$\left.\begin{aligned} \text{sn}\,(2\mathbf{K} - z) &= \text{sn}\,z & \text{sn}\,(2i\mathbf{K}' - z) &= -\text{sn}\,(z) \\ \text{cn}\,(2\mathbf{K} - z) &= -\text{cn}\,z & \text{cn}\,(2i\mathbf{K}' - z) &= -\text{cn}\,(z) \\ \text{dn}\,(2\mathbf{K} - z) &= \text{dn}\,z & \text{dn}\,(2i\mathbf{K}' - z) &= -\text{dn}\,(z) \end{aligned}\right\} \tag{21.6-44}$$

$$\bar{E}(k, \text{sn}\,z) = \int_0^z \text{dn}^2 z\,dz \tag{21.6-45}$$

References 21.2 and 21.11 contain additional formulas.

Table 21.6-5. Special Values of Jacobian Elliptic Functions*

$\frac{1}{2}ni\mathbf{K}'$ \ $\frac{1}{2}m\mathbf{K}$	0	$\frac{1}{2}\mathbf{K}$	$\mathbf{K}$	$\frac{3}{2}\mathbf{K}$
		sn $(\frac{1}{2}m\mathbf{K} + \frac{1}{2}ni\mathbf{K}')$		
0	0	$(1 + k')^{-\frac{1}{2}}$	1	$(1 + k')^{-\frac{1}{2}}$
$\frac{1}{2}i\mathbf{K}'$	$ik^{-\frac{1}{2}}$	$(2k)^{-\frac{1}{2}}[(1 + k)^{\frac{1}{2}} + i(1 - k)^{\frac{1}{2}}]$	$k^{-\frac{1}{2}}$	$(2k)^{-\frac{1}{2}}[(1 + k)^{\frac{1}{2}} - i(1 - k)^{\frac{1}{2}}]$
$i\mathbf{K}'$	∞	$(1 - k')^{\frac{1}{2}}$	k^{-1}	$(1 - k')^{-\frac{1}{2}}$
$\frac{3}{2}i\mathbf{K}'$	$-ik^{-\frac{1}{2}}$	$(2k)^{-\frac{1}{2}}[(1 + k)^{\frac{1}{2}} - i(1 - k)^{\frac{1}{2}}]$	$k^{-\frac{1}{2}}$	$(2k)^{-\frac{1}{2}}[(1 + k)^{\frac{1}{2}} + i(1 - k)^{\frac{1}{2}}]$
		cn $(\frac{1}{2}m\mathbf{K} + \frac{1}{2}ni\mathbf{K}')$		
0	1	$k'^{\frac{1}{2}}(1 + k')^{-\frac{1}{2}}$	0	$-k'^{\frac{1}{2}}(1 + k')^{-\frac{1}{2}}$
$\frac{1}{2}i\mathbf{K}'$	$k^{-\frac{1}{2}}(1 + k)^{\frac{1}{2}}$	$k'^{\frac{1}{2}}(2k)^{-\frac{1}{2}}(1 - i)$	$-ik^{-\frac{1}{2}}(1 - k)^{\frac{1}{2}}$	$-k'^{\frac{1}{2}}(2k)^{-\frac{1}{2}}(1 + i)$
$i\mathbf{K}'$	∞	$-ik'^{\frac{1}{2}}(1 - k')^{-\frac{1}{2}}$	$-ik^{-1}k'$	$-ik'^{\frac{1}{2}}(1 - k')^{-\frac{1}{2}}$
$\frac{3}{2}i\mathbf{K}'$	$-k^{-\frac{1}{2}}(1 + k)^{\frac{1}{2}}$	$-k'^{\frac{1}{2}}(2k)^{-\frac{1}{2}}(1 + i)$	$-ik^{-\frac{1}{2}}(1 - k)^{\frac{1}{2}}$	$k'^{\frac{1}{2}}(2k)^{-\frac{1}{2}}(1 - i)$
		dn $(\frac{1}{2}m\mathbf{K} + \frac{1}{2}ni\mathbf{K}')$		
0	1	$k'^{\frac{1}{2}}$	k'	$k'^{\frac{1}{2}}$
$\frac{1}{2}i\mathbf{K}'$	$(1 + k)^{\frac{1}{2}}$	$(\frac{1}{2}k')^{\frac{1}{2}}[(1 + k')^{\frac{1}{2}} - i(1 - k')^{\frac{1}{2}}]$	$(1 - k)^{\frac{1}{2}}$	$(\frac{1}{2}k')^{\frac{1}{2}}[(1 + k')^{\frac{1}{2}} + i(1 - k')^{\frac{1}{2}}]$
$i\mathbf{K}'$	∞	$-ik'^{\frac{1}{2}}$	0	$ik'^{\frac{1}{2}}$
$\frac{3}{2}i\mathbf{K}'$	$-(1 + k)^{\frac{1}{2}}$	$-(\frac{1}{2}k')^{\frac{1}{2}}[(1 + k')^{\frac{1}{2}} + i(1 - k')^{\frac{1}{2}}]$	$-(1 - k)^{\frac{1}{2}}$	$-(\frac{1}{2}k')^{\frac{1}{2}}[(1 + k')^{\frac{1}{2}} - i(1 - k')^{\frac{1}{2}}]$

* From A. Erdélyi *et al.*, *Higher Transcendental Functions*, vol. 2, McGraw-Hill, New York, 1953.

Table 21.6-6. Change of the Variable by Quarter and Half Periods. Symmetry*

sn $(m\mathbf{K} + ni\mathbf{K}' \pm u)$

$ni\mathbf{K}'$ \ $m\mathbf{K}$	$-\mathbf{K}$	0	$\mathbf{K}$	$2\mathbf{K}$	$3\mathbf{K}$
$-i\mathbf{K}'$	$-d/(kc)$	$\pm 1/(ks)$	$d/(kc)$	$\mp 1/(ks)$	$-d/(kc)$
0	$-c/d$	$\pm s$	c/d	$\mp s$	$-c/d$
$i\mathbf{K}'$	$-d/(kc)$	$\pm 1/(ks)$	$d/(kc)$	$\mp 1/(ks)$	$-d/(kc)$
$2i\mathbf{K}'$	$-c/d$	$\pm s$	c/d	$\mp s$	$-c/d$

cn $(m\mathbf{K} + ni\mathbf{K}' \pm u)$

$ni\mathbf{K}'$ \ $m\mathbf{K}$	$-\mathbf{K}$	0	$\mathbf{K}$	$2\mathbf{K}$	$3\mathbf{K}$
$-i\mathbf{K}'$	$-ik'/(kc)$	$\pm id/(ks)$	$ik'/(kc)$	$\mp id/(ks)$	$-ik'/(kc)$
0	$\pm k's/d$	c	$\mp k's/d$	$-c$	$\pm k's/d$
$i\mathbf{K}'$	ik'/kc	$\mp id/(ks)$	$-ik'/(kc)$	$\pm id/(ks)$	$ik'/(kc)$
$2i\mathbf{K}'$	$\mp k's/d$	$-c$	$\pm k's/d$	c	$\mp k's/d$

dn $(m\mathbf{K} + ni\mathbf{K}' \pm u)$

$ni\mathbf{K}'$ \ $m\mathbf{K}$	$-\mathbf{K}$	0	$\mathbf{K}$	$2\mathbf{K}$
$-i\mathbf{K}'$	$\mp ik's/c$	$\pm ic/s$	$\mp ik's/c$	$\mp ic/s$
0	k'/d	d	k'/d	d
$i\mathbf{K}'$	$\pm ik's/c$	$\mp ic/s$	$\pm ik's/c$	$\mp ic/s$
$2i\mathbf{K}'$	$-k'/d$	$-d$	$-k'/d$	$-d$
$3i\mathbf{K}'$	$\mp ik's/c$	$\pm ic/s$	$\mp ik's/c$	$\pm ic/s$

* From A. Erdélyi *et al.*, *Higher Transcendental Functions*, vol. 2, McGraw-Hill, New York, 1953.

Table 21.6-7. Transformations of the First Order of Jacobi's Elliptic Functions*

Transformation	$\dot{u}$	$\dot{k}$	$\dot{k}'$	$\dot{\mathbf{K}}$	$\dot{\mathbf{K}}'$	sn $(\dot{u}, \dot{k})$	cn $(\dot{u}, \dot{k})$	dn $(\dot{u}, \dot{k})$
A	$k'u$	$\frac{ik}{k'}$	$\frac{1}{k'}$	$k'\mathbf{K}$	$k'(\mathbf{K}' - i\mathbf{K})$	$k'\frac{\text{sn}(u, k)}{\text{dn}(u, k)}$	$\frac{\text{cn}(u, k)}{\text{dn}(u, k)}$	$\frac{1}{\text{dn}(u, k)}$
B	$-iu$	k'	k	$\mathbf{K}'$	$\mathbf{K}$	$-i\frac{\text{sn}(u, k)}{\text{cn}(u, k)}$	$\frac{1}{\text{cn}(u, k)}$	$\frac{\text{dn}(u, k)}{\text{cn}(u, k)}$
C	ku	$\frac{1}{k}$	$\frac{k'}{ik}$	$k(\mathbf{K} + i\mathbf{K}')$	$k\mathbf{K}'$	$k\,\text{sn}(u, k)$	$\text{dn}(u, k)$	$\text{cn}(u, k)$
D	$-ik'u$	$\frac{1}{k'}$	$\frac{k}{ik'}$	$k'(\mathbf{K}' + i\mathbf{K})$	$k'\mathbf{K}$	$-ik'\frac{\text{sn}(u, k)}{\text{cn}(u, k)}$	$\frac{\text{dn}(u, k)}{\text{cn}(u, k)}$	$\frac{1}{\text{cn}(u, k)}$
E	$-iku$	$\frac{k'}{ik}$	$\frac{1}{k}$	$k\mathbf{K}'$	$k(\mathbf{K} + i\mathbf{K}')$	$-ik\frac{\text{sn}(u, k)}{\text{dn}(u, k)}$	$\frac{1}{\text{dn}(u, k)}$	$\frac{\text{cn}(u, k)}{\text{dn}(u, k)}$

* From A. Erdélyi *et al.*, *Higher Transcendental Functions*, vol. 2, McGraw-Hill, New York, 1953.

(c) Addition Formulas.

$$\left.\begin{aligned} \text{sn}(A+B) &= \frac{\text{sn}\,A\,\text{cn}\,B\,\text{dn}\,B + \text{sn}\,B\,\text{cn}\,A\,\text{dn}\,A}{1 - k^2\,\text{sn}^2 A\,\text{sn}^2 B} \\ \text{cn}(A+B) &= \frac{\text{cn}\,A\,\text{cn}\,B - \text{sn}\,A\,\text{dn}\,A\,\text{sn}\,B\,\text{dn}\,B}{1 - k^2\,\text{sn}^2 A\,\text{sn}^2 B} \\ \text{dn}(A+B) &= \frac{\text{dn}\,A\,\text{dn}\,B - k^2\,\text{sn}\,A\,\text{cn}\,A\,\text{sn}\,B\,\text{cn}\,B}{1 - k^2\,\text{sn}^2 A\,\text{sn}^2 B} \end{aligned}\right\} \quad (21.6\text{-}46)$$

(d) Differentiation.

$$\left.\begin{aligned} \frac{d(\text{sn}\,z)}{dz} &= \text{cn}\,z\,\text{dn}\,z = \sqrt{(1 - \text{sn}^2 z)(1 - k^2\,\text{sn}^2 z)} \\ \frac{d(\text{cn}\,z)}{dz} &= -\text{sn}\,z\,\text{dn}\,z = \sqrt{(1 - \text{cn}^2 z)(k'^2 + k^2\,\text{cn}^2 z)} \\ \frac{d(\text{dn}\,z)}{dz} &= -k^2\,\text{sn}\,z\,\text{cn}\,z = \sqrt{(\text{dn}^2 z - 1)(k'^2 - \text{dn}^2 z)} \end{aligned}\right\} \quad (21.6\text{-}47)$$

(e) Transformations. Table 21.6-7 shows relations between Jacobi's elliptic functions with moduli k and ik/k', k', $1/k$, $1/k'$, k'/ik (see also Sec. 21.6-6*c*).

(f) Series Expansions.

$$\left.\begin{aligned} \text{sn}\,z &= z - (1 + k^2)\frac{z^3}{3!} + (1 + 14k^2 + k^4)\frac{z^5}{5!} - \cdots \\ \text{cn}\,z &= 1 - \frac{z^2}{2!} + (1 + 4k^2)\frac{z^4}{4!} - (1 + 44k^2 + 16k^4)\frac{z^6}{6!} + \cdots \\ \text{dn}\,z &= 1 - k^2\frac{z^2}{2!} + k^2(4 + k^2)\frac{z^4}{4!} - k^2(16 + 44k^2 + k^4)\frac{z^6}{6!} + \cdots \end{aligned}\right\}$$

$$|z| < \min(|\mathbf{K}'|, |2\mathbf{K} + i\mathbf{K}'|, |2\mathbf{K} - i\mathbf{K}'|) \quad (21.6\text{-}48)$$

21.6-8. Jacobi's Theta Functions. (**a**) Given a complex variable v and a complex parameter $q = e^{i\pi\tau}$ such that τ has a positive imaginary part, the four **theta functions***

$$\left.\begin{aligned}
\vartheta_1(v) = \vartheta_1(v|\tau) &= 2\sum_{n=0}^{\infty} (-1)^n q^{(n+\frac{1}{2})^2} \sin\,(2n+1)\pi v \\
&= i\sum_{n=-\infty}^{\infty} (-1)^n q^{(n-\frac{1}{2})^2} (e^{i\pi v})^{2n-1} \\
\vartheta_2(v) = \vartheta_2(v|\tau) &= 2\sum_{n=0}^{\infty} q^{(n+\frac{1}{2})^2} \cos\,(2n+1)\pi v \\
&= \sum_{n=-\infty}^{\infty} q^{(n-\frac{1}{2})^2} (e^{i\pi v})^{2n-1} \\
\vartheta_3(v) = \vartheta_3(v|\tau) &= 1 + 2\sum_{n=1}^{\infty} q^{n^2} \cos 2n\pi v = \sum_{n=-\infty}^{\infty} q^{n^2}(e^{i\pi v})^{2n} \\
\vartheta_4(v) = \vartheta_4(v|\tau) &= 1 + 2\sum_{n=1}^{\infty} (-1)^n q^{n^2} \cos 2n\pi v \\
&= \sum_{n=-\infty}^{\infty} (-1)^n q^{n^2} (e^{i\pi v})^{2n}
\end{aligned}\right\} \qquad (21.6\text{-}49)$$

are (simply) periodic integral functions of v with the respective periods 2, 2, 1, 1. The four theta functions (49) have zeros at $v = m + n\tau$, $m + n\tau + \frac{1}{2}$, $m + (n + \frac{1}{2})\tau + \frac{1}{2}$, and $m + (n + \frac{1}{2})\tau$, respectively, where $m, n = 0, \pm 1, \pm 2, \ldots$; these zeros yield infinite-product representations (7.6-2) (Refs. 21.3 and 21.11).

The theta functions are *not* elliptic functions. *The very rapidly converging series* (49) *permit one to compute various elliptic functions and elliptic integrals with the aid of the relations of Sec.* 21.6-9; in their own right, the theta functions are solutions of the partial differential equation

$$\frac{\partial^2\Phi}{\partial v^2} - 4\pi i \frac{\partial\Phi}{\partial\tau} = 0 \qquad (21.6\text{-}50)$$

which is related to the diffusion equation (Sec. 10.3-4*b*).

(**b**) Note

$$\left.\begin{aligned}
\vartheta_1(v + \tfrac{1}{2}) &= \vartheta_2(v) \qquad & \vartheta_2(v + \tfrac{1}{2}) &= -\vartheta_1(v) \\
\vartheta_3(v + \tfrac{1}{2}) &= \vartheta_4(v) \qquad & \vartheta_4(v + \tfrac{1}{2}) &= \vartheta_3(v)
\end{aligned}\right\} \qquad (21.6\text{-}51)$$

$$\left.\begin{aligned}
\vartheta_1\left(v + \frac{\tau}{2}\right) &= ie^{-i\pi\left(\frac{\tau}{4}+v\right)}\vartheta_4(v) \qquad & \vartheta_2\left(v + \frac{\tau}{2}\right) &= e^{-i\pi\left(\frac{\tau}{4}+v\right)}\vartheta_3(v) \\
\vartheta_3\left(v + \frac{\tau}{2}\right) &= e^{-i\pi\left(\frac{\tau}{4}+v\right)}\vartheta_2(v) \qquad & \vartheta_4\left(v + \frac{\tau}{2}\right) &= ie^{-i\pi\left(\frac{\tau}{4}+v\right)}\vartheta_1(v)
\end{aligned}\right\} \qquad (21.6\text{-}52)$$

* Some authors denote $\vartheta_4(v)$ by $\vartheta_0(v)$ or $\vartheta(v)$.

(c) To find $\vartheta_i\left(\frac{v}{C\tau + D}\middle|\frac{A\tau + B}{C\tau + D}\right)$ (and thus similarly transformed elliptic functions, Sec. 21.6-9), use

$$\left.\begin{aligned} \vartheta_1(v|\tau + 1) &= e^{i\pi/4}\vartheta_1(v|\tau) \qquad & \vartheta_2(v|\tau + 1) &= e^{i\pi/4}\vartheta_2(v|\tau) \\ \vartheta_3(v|\tau + 1) &= \vartheta_4(v|\tau) & \vartheta_4(v|\tau + 1) &= \vartheta_3(v|\tau) \end{aligned}\right\} \tag{21.6-53}$$

$$\left.\begin{aligned} \vartheta_1\left(\frac{v}{\tau}\middle| -\frac{1}{\tau}\right) &= \frac{1}{i}\sqrt{\frac{\tau}{i}}\, e^{i\pi v^2/\tau}\, \vartheta_1(v|\tau) \\ \vartheta_2\left(\frac{v}{\tau}\middle| -\frac{1}{\tau}\right) &= \sqrt{\frac{\tau}{i}}\, e^{i\pi v^2/\tau}\, \vartheta_4(v|\tau) \\ \vartheta_3\left(\frac{v}{\tau}\middle| -\frac{1}{\tau}\right) &= \sqrt{\frac{\tau}{i}}\, e^{i\pi v^2/\tau}\, \vartheta_3(v|\tau) \\ \vartheta_4\left(\frac{v}{\tau}\middle| -\frac{1}{\tau}\right) &= \sqrt{\frac{\tau}{i}}\, e^{i\pi v^2/\tau}\, \vartheta_2(v|\tau) \end{aligned}\right\} \tag{21.6-54}$$

(d) The values of the four theta functions and their derivatives for $v = 0$ (**zero-argument values**) are simply denoted by $\vartheta_i = \vartheta_i(0)$, $\vartheta_i' = \vartheta_i'(0)$, . . . $(i = 1, 2, 3, 4)$ and satisfy the relations

$$\vartheta_1' = \pi\vartheta_2\vartheta_3\vartheta_4 \qquad \frac{\vartheta_1'''}{\vartheta_1'} = \frac{\vartheta_2''}{\vartheta_2} + \frac{\vartheta_3''}{\vartheta_3} + \frac{\vartheta_4''}{\vartheta_4} \tag{21.6-55}$$

21.6-9. Relations between Jacobi's Elliptic Functions, Weierstrass's Elliptic Functions, and Theta Functions. If the various parameters implicit in the definitions of sn z, cn z, dn z, $\wp(z)$, $\zeta(z)$, $\sigma(z)$, and $\vartheta_j(z)$ (Secs. 21.6-2, 21.6-3, 21.6-6, 21.6-7, and 21.6-8) are related by

$$k = \sqrt{\frac{e_2 - e_3}{e_1 - e_3}} = \left(\frac{\vartheta_2}{\vartheta_3}\right)^2 \qquad k' = \sqrt{\frac{e_1 - e_2}{e_1 - e_3}} = \left(\frac{\vartheta_4}{\vartheta_3}\right)^2 \tag{21.6-56}$$

$$\left.\begin{aligned} \mathbf{K} = \frac{\omega_1}{2}\sqrt{e_1 - e_3} = \frac{\pi}{2}\vartheta_3{}^2 \qquad i\mathbf{K}' = \frac{\omega_2}{2}\sqrt{e_1 - e_3} = \tau\mathbf{K} \\ \tau = \frac{i\mathbf{K}'}{\mathbf{K}} = \frac{\omega_2}{\omega_1} \qquad (\text{Im}\,\tau > 0) \end{aligned}\right\} \tag{21.6-57}$$

and

$$w = z\sqrt{e_1 - e_3} = 2\mathbf{K}v \qquad v = \frac{z}{\omega_1} = \frac{w}{2\mathbf{K}} \tag{21.6-58}$$

then

$$\left.\begin{aligned} \text{sn}\, w &= \sqrt{\frac{e_1 - e_3}{\wp(z) - e_3}} = \frac{\vartheta_3}{\vartheta_2}\frac{\vartheta_1(v)}{\vartheta_4(v)} \\ \text{cn}\, w &= \sqrt{\frac{\wp(z) - e_1}{\wp(z) - e_3}} = \frac{\vartheta_4}{\vartheta_2}\frac{\vartheta_2(v)}{\vartheta_4(v)} \\ \text{dn}\, w &= \sqrt{\frac{\wp(z) - e_2}{\wp(z) - e_3}} = \frac{\vartheta_4}{\vartheta_3}\frac{\vartheta_3(v)}{\vartheta_4(v)} \end{aligned}\right\} \tag{21.6-59}$$

$$\wp(z) = e_1 + (e_1 - e_3)\frac{\text{cn}^2\, w}{\text{sn}^2\, w} = e_2 + (e_1 - e_3)\frac{\text{dn}^2\, w}{\text{sn}^2\, w} = e_3 + \frac{e_1 - e_3}{\text{sn}^2\, w} \tag{21.6-60}$$

21.7. ORTHOGONAL POLYNOMIALS

21.7-1. Survey. The orthogonal polynomials discussed in Secs. 21.7-1 to 21.7-8 are special solutions of linear homogeneous second-order differential equations related to the hypergeometric differential equation (9.3-31) (Legendre, Chebychev, and Jacobi polynomials) or to the confluent hypergeometric differential equation (9.3-42) (Laguerre and Hermite polynomials). These special solutions are generated by special homogeneous boundary conditions: each class of orthogonal polynomials is a set of eigenfunctions for an *eigenvalue problem* reducible to the generalized Sturm-Liouville type (Secs. 15.4-8*a* and *c*). Only real $z = x$ are of interest in most applications.

The polynomials $\psi_0(x)$, $\psi_1(x)$, $\psi_2(x)$, . . . of each type are, then, defined except for multiplicative constants, which are usually (but not always) chosen so that *the coefficient of* x^n *in the* n^{th}*-degree polynomial* $\psi_n(x)$ *is unity.* Successive polynomials $\psi_0(x)$, $\psi_1(x)$, $\psi_2(x)$, . . . of each type can be derived

1. In terms of the appropriate *hypergeometric series* (Sec. 9.3-9*a*) or *confluent hypergeometric series* (Sec. 9.3-10)
2. By means of *recursion formulas* derived from the differential equation
3. By successive differentiations of a *generating function* $\gamma(x, s)$ (see also Sec. 8.6-5)
4. Through Gram-Schmidt *orthogonalization* of the powers 1, x, x^2, . . . with the appropriate inner product (Sec. 15.2-5)
5. From an *integral representation* (Sec. 21.7-7), which is usually related to an integral-transform solution of the differential equation or to the Taylor- or Laurent-series coefficients of the generating function

Table 21.7-1 lists the principal formulas.

Series expansions in terms of orthogonal polynomials are derived in the manner of Sec. 15.2-4*a* and yield useful approximations which minimize appropriately defined mean-square errors (Sec. 15.2-6; see also Secs. 20.5-1 and 20.6-3).

21.7-2. Real Zeros of Orthogonal Polynomials. *All zeros of each orthogonal polynomial discussed in Secs.* 21.7-1 *to* 21.7-8 *are simple and located in the interior of the expansion domain. Two consecutive zeros of* $\psi_n(x)$ *bracket exactly one zero of* $\psi_{n+1}(x)$, *and at least one zero of* $\psi_m(x)$ *for each* $m > n$ (Ref. 21.9).

21.7-3. Legendre Functions (see also Secs. 21.8-10, 21.8-11, and 21.8-13; and Refs. 21.3, 21.9, and 21.11). The differential equation (*Legendre's differential equation*) and the recursion formulas for the Legendre polynomials (Table 21.7-1) are satisfied not only by the **Legendre polynomials of the first kind** $P_n(z)$ of Table 21.7-1 but also by the **Legendre functions of the second kind**

$$Q_0(z) = \frac{1}{2}\log_e \frac{1+z}{1-z} \qquad Q_1(z) = \frac{z}{2}\log_e \frac{1+z}{1-z} - 1$$

$$Q_2(z) = \frac{1}{4}(3z^2 - 1)\log_e \frac{1+z}{1-z} - \frac{3}{2}z \qquad \cdots \qquad (21.7\text{-}1)$$

More generally, the method of Sec. 9.3-8*b* permits one to derive linearly independent solutions $P_\alpha(z)$, $Q_\alpha(z)$ of Legendre's differential equation (**Legendre functions of the first and second kind**) for nonintegral positive or negative or complex values of $n = \alpha$; solutions for $n = \alpha$ and $n = -\alpha - 1$ are necessarily identical.

21.7-4. Chebyshev Polynomials of the First and Second Kind. The differential equation and the recursion formulas for the Chebyshev polynomials (Table 21.7-1) are satisfied not only by the **Chebyshev polynomials of the first kind**

$$T_n(x) \equiv \cos(n \arccos x) \qquad (n = 0, 1, 2, \ldots) \qquad (21.7\text{-}2)$$

of Table 21.7-1, but also by the Chebyshev "polynomials" of the second kind

$$U_0(x) \equiv \arcsin x$$

$$U_n(x) \equiv \sin(n \arccos x) \equiv \frac{\sqrt{1-x^2}}{n}\frac{d}{dx}T_n(x) \qquad (n = 1, 2, \ldots) \qquad (21.7\text{-}3)$$

While the functions $U_n(x)$ are not polynomials in x, the functions $\frac{1}{n+1}\frac{dT_{n+1}}{dx}$ and $\frac{dT_n}{dx}$ are polynomials; both are also sometimes referred to as Chebyshev polynomials of the second kind. Note

$$T_n(x) = -\frac{\sqrt{1-x^2}}{n}\frac{d}{dx}U_n(x) \qquad (n = 0, 1, 2, \ldots) \qquad (21.7\text{-}4)$$

$$\int_{-1}^{1} \frac{U_n(x)U_{n'}(x)\,dx}{\sqrt{1-x^2}} = \begin{cases} 0 \text{ for } n' \neq n \text{ or } n' = n = 0 \\ \pi/2 \text{ for } n' = n \neq 0 \end{cases} \qquad (21.7\text{-}5)$$

21.7-5. Associated Laguerre Polynomials and Functions (see also Secs. 9.3-10 and 10.4-6). **(a)** The **associated** (or **generalized**) **Laguerre polynomials* of degree $n - k$ and order k,**

$$L_n^k(x) = \frac{d^k}{dx^k}L_n(x) = (-1)^k n!\binom{n}{k}F(k-n; k+1; x)$$

$$(n = 1, 2, \ldots; k = 0, 1, 2, \ldots, n) \qquad (21.7\text{-}6)$$

satisfy the differential equation

$$z\frac{d^2w}{dz^2} + (k + 1 - z)\frac{dw}{dz} + (n - k)w = 0 \qquad (21.7\text{-}7)$$

* This notation is used in most physics books. Some authors use alternative associated Laguerre polynomials $L_n^{(k)}(x) \equiv (-1)^k L_{n+k}^k(x)/(n+k)!$ of degree n, which satisfy the differential equation

$$z\frac{d^2w}{dz^2} + (k + 1 - z)\frac{dw}{dz} + nw = 0 \qquad (n, k = 0, 1, 2, \ldots)$$

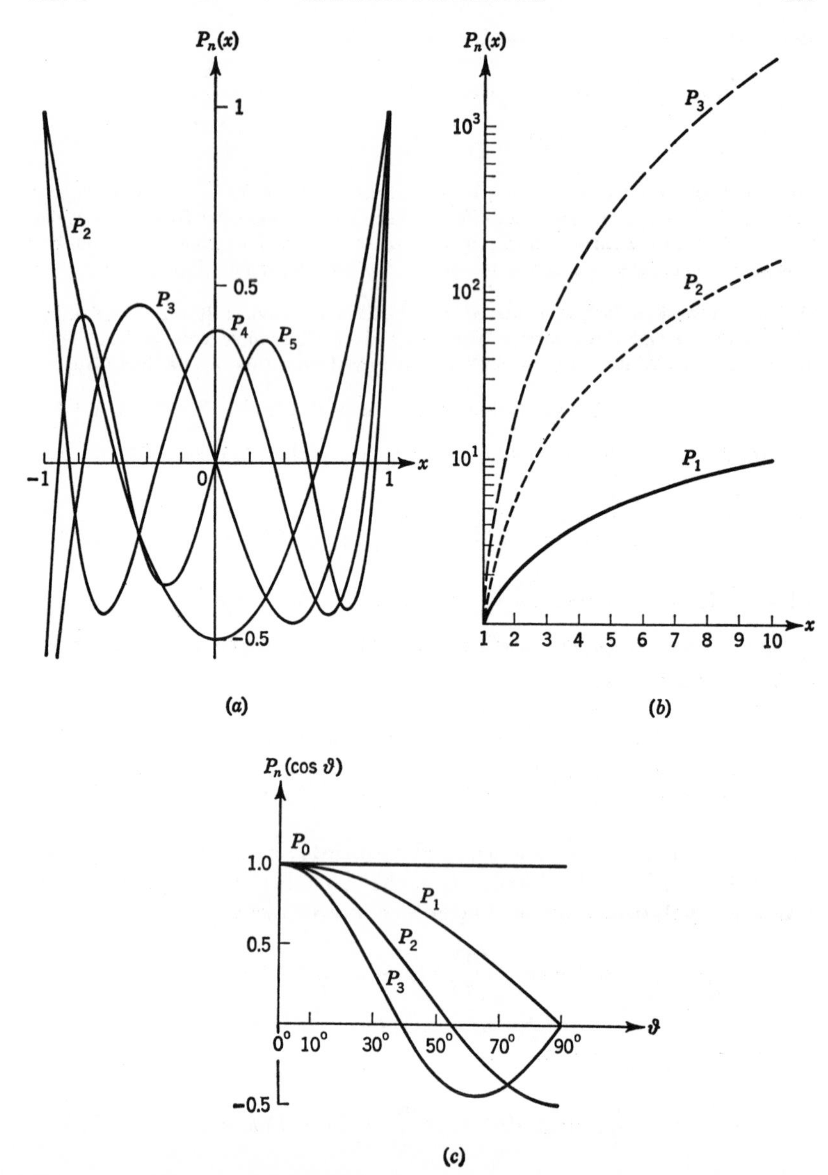

FIG. 21.7-1. Legendre polynomials $P_n(x)$ as functions of x (*a*), (*b*), and as functions of $\theta = \arccos x$ (*c*). (*From Ref.* 21.1.)

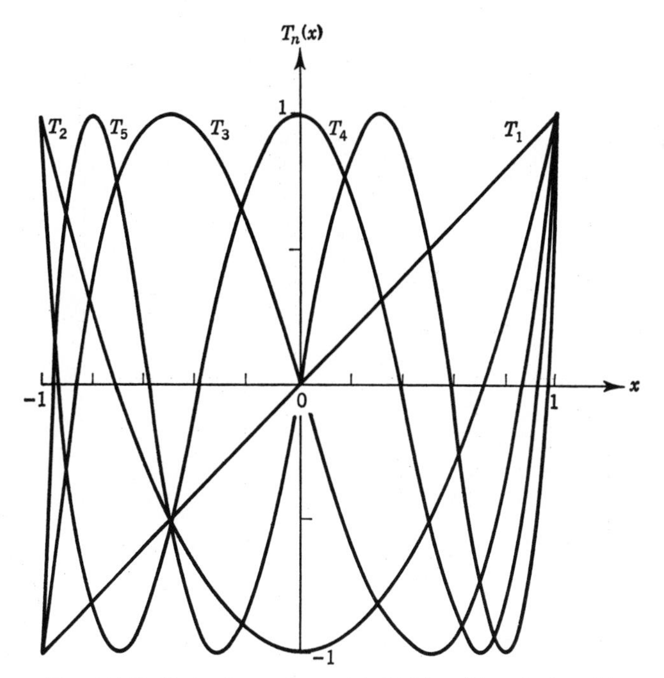

FIG. 21.7-2. Chebyshev polynomials $T_n(x)$. (*From Ref.* 21.1.)

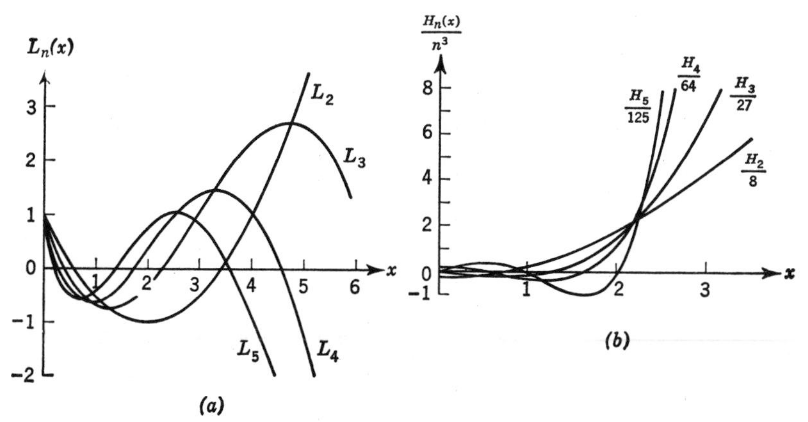

FIG. 21.7-3. Laguerre polynomials (*a*), and Hermite polynomials (*b*). (*From Ref.* 21.1.)

Table 21.7-1. Orthogonal Polynomials of Legendre, Chebyshev, Laguerre, and Hermite (see also Secs. 21.7-1 to 21.7-7)

No.		Legendre polynomials $P_n(x)$ (Fig. 21.7-1)	Chebyshev polynomials $T_n(x)$ (Fig. 21.7-2)
1	Differential equation	$(1 - z^2)\frac{d^2w}{dz^2} - 2z\frac{dw}{dz} + n(n+1)w = 0$	$(1 - z^2)\frac{d^2w}{dz^2} - z\frac{dw}{dz} + n^2w = 0$
2	(Real) domain and boundary conditions	$w(z)$ is unique and analytic for real $z = x$ in $(-1, 1)$; $\int_{-1}^{1} w^2(x)\,dx$ exists	$w(z)$ is unique and analytic for real $z = x$ in $(-1, 1)$; $\int_{-1}^{1} \frac{w^2(x)\,dx}{\sqrt{1 - x^2}}$ exists
3	Orthogonality and normalization	$\int_{-1}^{1} P_n(x)P_{n'}(x)\,dx = \begin{cases} 0 & (n' \neq n) \\ \frac{2}{2n+1} & (n' = n) \end{cases}$	$\int_{-1}^{1} \frac{T_n(x)T_{n'}(x)\,dx}{\sqrt{1 - x^2}} = \begin{cases} 0 & (n' \neq n) \\ \pi/2 & (n' = n \neq 0) \\ \pi & (n' = n = 0) \end{cases}$
4	Series (9.3-32) or (9.3-43)	$P_n(x) = F\left(-n, n+1; 1; \frac{1-x}{2}\right)$ $= \frac{2n!}{2^n(n!)^2} x^n F\left(-\frac{n}{2}, \frac{1-n}{2}; \frac{1}{2} - n; \frac{1}{x^2}\right)$	$T_n(x) = F\left(n, -n; \frac{1}{2}; \frac{1-x}{2}\right)$
5	Recursion formulas	$P_{n+1}(x) = \frac{2n+1}{n+1} xP_n(x) - \frac{n}{n+1} P_{n-1}(x)$ $= xP_n(x) + \frac{x^2 - 1}{n+1}\frac{dP_n}{dx}$	$T_{n+1}(x) = 2xT_n(x) - T_{n-1}(x)$
6	Generalized Rodrigues's formula	$P_n(x) = \frac{1}{2^n n!}\frac{d^n}{dx^n}(x^2 - 1)^n$	$T_n(x) = \frac{(-2)^n n!}{(2n)!}\sqrt{1 - x^2}\frac{d^n}{dx^n}(1 - x^2)^{n-\frac{1}{2}}$
7	Generating function	$\frac{1}{\sqrt{1 - 2sx + s^2}} = \begin{cases} \sum_{n=0}^{\infty} P_n(x)s^n & (\lvert s\rvert < 1) \\ \sum_{n=0}^{\infty} P_n(x)s^{-n-1} & (\lvert s\rvert > 1) \end{cases}$	$\frac{1 - sx}{1 - 2sx + s^2} = \sum_{n=0}^{\infty} T_n(x)s^n \quad (\lvert s\rvert < 1)$
8	First five polynomials	1 $x = \cos\vartheta$ $\frac{1}{2}(3x^2 - 1) = \frac{1}{4}(3\cos 2\vartheta + 1)$ $\frac{1}{2}(5x^3 - 3x) = \frac{1}{8}(5\cos 3\vartheta + 3\cos\vartheta)$ $\frac{1}{8}(35x^4 - 30x^2 + 3) = \frac{1}{64}(35\cos 4\vartheta + 20\cos 2\vartheta + 9)$ $\frac{1}{8}(63x^5 - 70x^3 + 15x)$	1 $x = \cos\vartheta$ $2x^2 - 1 = \cos 2\vartheta$ $4x^3 - 3x = \cos 3\vartheta$ $8x^4 - 8x^2 + 1 = \cos 4\vartheta$ $16x^5 - 20x^3 + 5x = \cos 5\vartheta$ . . . . $T_n(\cos\vartheta) = \cos n\vartheta$

Table 21.7-1. Orthogonal Polynomials of Legendre, Chebyshev, Laguerre, and Hermite (see also Secs. 21.7-1 to 21.7-7) (*Continued*)

No.		Laguerre polynomials* $L_n(x)$ (Fig. 21.7-3*a*)	Hermite polynomials† $H_n(x)$ (Fig. 21.7-3*b*)
1	Differential equation	$z\frac{d^2w}{dz^2} + (1 - z)\frac{dw}{dz} + nw = 0$	$\frac{d^2w}{dz^2} - 2z\frac{dw}{dz} + 2nw = 0$
2	(Real) domain and boundary conditions	$w(z)$ is unique and analytic for real $z = x$ in $(0, \infty)$; $\int_0^\infty e^{-x}w^2(x)\,dx$ exists	$w(z)$ is unique and analytic for real $z = x$ in $(-\infty, \infty)$; $\int_{-\infty}^\infty e^{-x^2}w^2(x)\,dx$ exists
3	Orthogonality and normalization	$\int_0^\infty e^{-x}L_n(x)L_{n'}(x)\,dx = \begin{cases} 0 & (n' \neq n) \\ (n!)^2 & (n' = n) \end{cases}$	$\int_{-\infty}^\infty e^{-x^2}H_n(x)H_{n'}(x)\,dx = \begin{cases} 0 & (n' \neq n) \\ 2^n n!\sqrt{\pi} & (n' = n) \end{cases}$
4	Series (9.3-32) or (9.3-43)	$L_n(x) = n!F(-n; 1; x)$ $= (-1)^n\left[x^n - n^2x^{n-1} + \frac{n^2(n-1)^2}{2!}x^{n-2} \mp \cdots\right]$	$H_{2n}(x) = (-1)^n 2^n(2n-1)(2n-3)\cdots 3\cdot 1F(-n; \tfrac{1}{2}; x^2)$ $H_{2n+1}(x) = (-1)^n 2^{n+1}(2n+1)(2n-1)\cdots 3\cdot 1xF(-n; \tfrac{3}{2}; x^2)$
5	Recursion formulas	$L_{n+1}(x) = (2n + 1 - x)L_n(x) - n^2L_{n-1}(x)$ $\frac{dL_{n+1}}{dx} = (n+1)\left[\frac{dL_n}{dx} - L_n(x)\right]$	$H_{n+1}(x) = 2xH_n(x) - 2nH_{n-1}(x)$ $\frac{dH_n}{dx} = 2nH_{n-1}(x)$
6	Generalized Rodrigues's formula	$L_n(x) = e^x\frac{d^n}{dx^n}(x^ne^{-x})$	$H_n(x) = (-1)^n e^{x^2}\frac{d^n}{dx^n}(e^{-x^2})$
7	Generating function	$\frac{e^{-x\frac{s}{1-s}}}{1-s} = \sum_{n=0}^{\infty} L_n(x)\frac{s^n}{n!} \quad (0 \leq x \leq \infty)$	$e^{-s^2+2sx} = \sum_{n=0}^{\infty} H_n(x)\frac{s^n}{n!}$
8	First five polynomials	1 $-x + 1$ $x^2 - 4x + 2$ $-x^3 + 9x^2 - 18x + 6$ $x^4 - 16x^3 + 72x^2 - 96x + 24$ $-x^5 + 25x^4 - 200x^3 + 600x^2 - 600x + 120$	1 $2x$ $4x^2 - 2$ $8x^3 - 12x$ $16x^4 - 48x^2 + 12$ $32x^5 - 160x^3 + 120x$

* Some authors denote the polynomial $\frac{1}{n!}L_n(x)$ by $L_n(x)$.

† Some authors use alternative Hermite polynomials $\mathrm{He}_n(x) \equiv 2^{-n/2}H_n(x/\sqrt{2})$, which satisfy the differential equation $\frac{d^2w}{dz^2} - z\frac{dw}{dz} + nw = 0$.

Table 21.7-2. Coefficients for Orthogonal Polynomials, and for x^n in Terms of Orthogonal Polynomials*

(a) Legendre Polynomials: $P_n(x) = a_n^{-1} \sum_{m=0}^{n} c_m x^m \qquad x^n = b_n^{-1} \sum_{m=0}^{n} d_m P_m(x)$

		x^0	x^1	x^2	x^3	x^4	x^5	x^6	x^7
	a_n \ b_n	1	1	3	5	35	63	231	429
P_0	1	1 1		1		7		33	
P_1	1		1 1		3		27		143
P_2	2	−1		3 2		20		110	
P_3	2		−3		5 2		28		182
P_4	8	3		−30		35 8		72	
P_5	8		15		−70		63 8		88
P_6	16	−5		105		−315		231 16	
P_7	16		−35		315		−693		429 16

$P_6(x) = \frac{1}{16}[231x^6 - 315x^4 + 105x^2 - 5] \qquad x^6 = \frac{1}{231}[33P_0 + 110P_2 + 72P_4 + 16P_6]$

(b) Chebyshev Polynomials: $T_n(x) = \sum_{m=0}^{n} c_m x^m \qquad x^n = b_n^{-1} \sum_{m=0}^{n} d_m T_m(x)$

	x^0	x^1	x^2	x^3	x^4	x^5	x^6	x^7
b_n	1	1	2	4	8	16	32	64
T_0	1 1		1		3		10	
T_1		1 1		3		10		35
T_2	−1		2 1		4		15	
T_3		−3		4 1		5		21
T_4	1		−8		8 1		6	
T_5		5		−20		16 1		7
T_6	−1		18		−48		32 1	
T_7		−7		56		−112		64 1

$T_6(x) = 32x^6 - 48x^4 + 18x^2 - 1 \qquad x^6 = \frac{1}{32}[10T_0 + 15T_2 + 6T_4 + T_6]$

* Abridged from M. Abramowitz and I. A. Stegun (eds.), *Handbook of Mathematical Functions*, National Bureau of Standards, Washington, D.C., 1964.

Table 21.7-2. Coefficients for Orthogonal Polynomials, and for x^n in Terms of Orthogonal Polynomials (*Continued*)

(c) **Laguerre Polynomials:** $L_n(x) = \sum_{m=0}^{n} c_m x^m \qquad x^n = b_n^{-1} \sum_{m=0}^{n} d_m L_m(x)$

	x^0	x^1	x^2	x^3	x^4	x^5	x^6	x^7
b_n	1	1	2	6	24	120	720	5040
L_0	1 1	1	2	6	24	120	720	5040
L_1	1	−1 −1	−4	−18	−96	−600	−4320	−35280
L_2	2	−4	1 2	18	144	1200	10800	105840
L_3	6	−18	9	−1 −6	−96	−1200	−14400	−176400
L_4	24	−96	72	−16	1 24	600	10800	176400
L_5	120	−600	600	−200	25	−1 −120	−4320	−105840
L_6	720	−4320	5400	−2400	450	−36	1 720	35280
L_7	5040	−35280	52920	−29400	7350	−882	49	−1 −5040

$L_6(x) = x^6 - 36x^5 + 450x^4 - 2400x^3 + 5400x^2 - 4320x + 720$

$x^6 = \frac{1}{720}[720L_0 - 4320L_1 + 10800L_2 - 14400L_3 + 10800L_4 - 4320L_5 + 720L_6]$

(d) **Hermite Polynomials:** $H_n(x) = \sum_{m=0}^{n} c_m x^m \qquad x^n = b_n^{-1} \sum_{m=0}^{n} d_m H_m(x$

	x^0	x^1	x^2	x^3	x^4	x^5	x^6	x^7
b_n	1	2	4	8	16	32	64	128
H_0	1 1		2		12		120	
H_1		2 1		6		60		840
H_2	−2		4 1		12		180	
H_3		−12		8 1		20		420
H_4	12		−48		16 1		30	
H_5		120		−160		32 1		42
H_6	−120		720		−480		64 1	
H_7		−1680		3360		−1344		128 1

$H_6(x) = 64x^6 - 480x^4 + 720x^2 - 120 \qquad x^6 = \frac{1}{64}[120H_0 + 180H_2 + 30H_4 + H_6]$

for integral values $n = 1, 2, \ldots ; k = 0, 1, 2, \ldots, n$. Equation (7) reduces to the differential equation of the Laguerre polynomials (Table 21.7-1) for $k = 0$.

Note the generating function

$$\left(\frac{s}{s-1}\right)^k \frac{e^{\frac{sx}{s-1}}}{1-s} = \sum_{n=k}^{\infty} L_n^k(x) \frac{s^n}{n!} \qquad (k = 0, 1, 2, \ldots) \qquad (21.7\text{-}8)$$

and the orthogonality and normalization given by

$$\int_0^\infty x^k e^{-x} L_n^k(x) L_{n'}^k(x)\, dx = \frac{(n!)^3}{(n-k)!} \delta_{n'}^n \qquad (21.7\text{-}9)$$

If nonintegral real or complex values of n and k are admitted, the differential equation (7) defines the **generalized Laguerre functions.** These functions, of which the polynomials (6) are special cases, are confluent hypergeometric functions (Sec. 9.3-10) with $a = k - n$, $c = k + 1$.

(b) The functions $\psi_{nj}(x) = x^j e^{-x/2} L_{n+j}^{2j+1}(x)$ $(n = 1, 2, \ldots ; j = 0, 1, 2, \ldots, n-1)$, which are often referred to as **associated Laguerre functions,** satisfy the differential equation

$$z\frac{d^2w}{dz^2} + 2\frac{dw}{dz} + \left[n - \frac{z}{4} - \frac{j(j+1)}{z}\right] w = 0$$
$$(n = 1, 2, \ldots ; j = 0, 1, 2, \ldots, n-1) \qquad (21.7\text{-}10)$$

and

$$\int_0^\infty \psi_{nj}^2(x) x^2\, dx = \frac{2n[(n+j)!]^3}{(n-j-1)!} \qquad (21.7\text{-}11)$$

(see also Sec. 10.4-6).

21.7-6. Hermite Functions. The functions $\psi_n(x) = e^{-x^2/2} H_n(x)$ $(n = 0, 1, 2, \ldots)$, which are often referred to as **Hermite functions,** satisfy the differential equation

$$\frac{d^2w}{dz^2} + (2n + 1 - z^2)w = 0 \qquad (n = 0, 1, 2, \ldots) \qquad (21.7\text{-}12)$$

and

$$\int_{-\infty}^{\infty} \psi_n(x)\psi_{n'}(x)\, dx = 2^n n! \sqrt{\pi}\, \delta_{n'}^n \qquad (21.7\text{-}13)$$

21.7-7. Some Integral Formulas (see Refs. 21.3 and 21.11 for additional formulas).

$$P_n(\cos \vartheta) = \frac{1}{\pi} \int_0^\pi (\cos \vartheta + i \sin \vartheta \cos t)^n\, dt \qquad (n = 0, 1, 2, \ldots)$$
$$\text{(Laplace's integral)} \qquad (21.7\text{-}14)$$

$$P_n(x) = \frac{1}{2\pi i} \oint \frac{(t^2 - 1)^n}{2^n (t - x)^{n+1}}\, dt \qquad (n = 0, 1, 2, \ldots)$$
$$\text{(Schlaefli's integral)} \qquad (21.7\text{-}15)$$

The integration contour in Eq. (15) surrounds the point $z = x$.

$$H_n(x) = \sqrt{\frac{2^{n-1}}{\pi}} \int_{-\infty}^{\infty} (x\sqrt{2} + it)^n e^{-t^2/2}\, dt \qquad (n = 0, 1, 2, \ldots) \qquad (21.7\text{-}16)$$

$$\left.\begin{aligned}\int_0^\infty x^k e^{-x}[L_n^k(x)]^2\,dx &= \frac{(n!)^3}{(n-k)!}\\ \int_0^\infty x^{k+1} e^{-x}[L_n^k(x)]^2\,dx &= \frac{(n!)^3}{(n-k)!}(2n-k+1)\\ \int_0^\infty x^{k+2} e^{-x}[L_n^k(x)]^2\,dx &= \frac{(n!)^3}{(n-k)!}(6n^2-6nk+k^2+6n-3k+2)\\ &(n = 1, 2, \ldots; k = 0, 1, 2, \ldots, n-1)\end{aligned}\right\} \quad (21.7\text{-}17)$$

$$\int_{-\infty}^{\infty} xe^{-x^2}H_n(x)H_{n'}(x)\,dx = 2^{n-1}n!\sqrt{\pi}\,\delta_{n'}^{n-1} + 2^n(n+1)!\sqrt{\pi}\,\delta_{n'}^{n+1} \quad (21.7\text{-}18)$$

21.7-8. Jacobi Polynomials and Gegenbauer Polynomials (see Ref. 21.11 for a detailed discussion). **(a) Jacobi's (hypergeometric) polynomials** are special instances of hypergeometric functions

$$\begin{aligned}\psi_n(x) &= F(-n, \alpha+n; \gamma; x)\\ &= \frac{x^{1-\gamma}(1-x)^{\gamma-\alpha}}{\gamma(\gamma+1)\cdots(\gamma+n-1)}\frac{d^n}{dx^n}[x^{\gamma+n-1}(1-x)^{\alpha-\gamma+n}]\end{aligned} \quad (21.7\text{-}19)$$

(Sec. 9.3-9) and satisfy the orthogonality conditions

$$\begin{aligned}\int_0^1 x^{\gamma-1}(1-x)^{\alpha-\gamma}\psi_n(x)\psi_{n'}(x)\,dx = \delta_{n'}^n\,\frac{\Gamma(\gamma)\Gamma(\alpha-\gamma+1)}{\Gamma(\alpha)}\\ \frac{(\alpha-\gamma+1)(\alpha-\gamma+2)\cdots(\alpha-\gamma+n)}{\alpha(\alpha+1)\cdots(\alpha+n-1)\gamma(\gamma+1)\cdots(\gamma+n-1)}\,\frac{n!}{\alpha+2n}\\ [\operatorname{Re}\gamma > 0; \operatorname{Re}(\alpha-\gamma) > -1]\end{aligned} \quad (21.7\text{-}20)$$

(b) The functions

$$C_n^\alpha(x) = \frac{\Gamma(n+2\alpha)}{\Gamma(n+1)\Gamma(2\alpha)}\,F\left(n+2\alpha, -n; \alpha+\frac{1}{2}; \frac{1-x}{2}\right) \quad (21.7\text{-}21)$$

are known as **Gegenbauer (ultraspherical) polynomials.** They constitute a generalization of the Legendre polynomials (Table 21.7-1), to which they reduce for $\alpha = \frac{1}{2}$. The Gegenbauer polynomials satisfy the differential equation

$$(z^2-1)\frac{d^2w}{dz^2} + (2\alpha+1)\,z\,\frac{dw}{dz} - n(n+2\alpha)w = 0 \quad (21.7\text{-}22)$$

and the orthogonality condition

$$\int_{-1}^{1}(1-x^2)^{\alpha-\frac{1}{2}}C_n^\alpha(x)C_{n'}^\alpha(x)\,dx = \delta_{n'}^n\,\frac{\pi\Gamma(2\alpha+n)}{2^{2\alpha-1}(\alpha+n)n![\Gamma(\alpha)]^2} \quad (21.7\text{-}23)$$

The Gegenbauer polynomials can be generated as coefficients of the power series

$$(1-2sx+s^2)^{-\alpha} = \sum_{n=0}^{\infty} C_n^\alpha(x)s^n \quad (21.7\text{-}24)$$

21.8. CYLINDER FUNCTIONS, ASSOCIATED LEGENDRE FUNCTIONS, AND SPHERICAL HARMONICS

21.8-1. Bessel Functions and Other Cylinder Functions. (a) **A cylinder function (circular-cylinder function) of order** m **is a** solution $w = Z_m(z)$ of the linear differential equation

$$\frac{d^2w}{dz^2} + \frac{1}{z}\frac{dw}{dz} + \left(1 - \frac{m^2}{z^2}\right) w = 0 \quad \text{(Bessel's differential equation)} \tag{21.8-1}$$

where m is any real number; one usually imposes the recurrence relations

$$\begin{aligned} Z_{m+1}(z) &= \frac{2m}{z} Z_m(z) - Z_{m-1}(z) = \frac{m}{z} Z_m(z) - \frac{d}{dz} Z_m(z) \\ &= -z^m \frac{d}{dz} [z^{-m} Z_m(z)] \end{aligned} \tag{21.8-2}$$

as additional defining conditions. The functions $e^{\pm i(Kz \pm m\varphi)} Z_m(iKr')$ are solutions of Laplace's partial differential equation in cylindrical coordinates r', φ, z (*cylindrical harmonics*, Sec. 10.4-3*b*). Cylinder functions of nonintegral order are multiple-valued (Sec. 9.3-5*b*); one defines the principal branch by $|\arg z| < \pi$ (branch cut from $z = 0$ to $z = -\infty$, Sec. 7.4-2).

(b) The most generally useful cylinder functions of order m satisfying the recurrence relations (2) are (see also Figs. 21.8-2 to 21.8-4)

$$J_m(z) = \left(\frac{z}{2}\right)^m \sum_{k=0}^{\infty} \frac{(-1)^k}{k!\Gamma(m + k + 1)} \left(\frac{z}{2}\right)^{2k} \quad (|\arg z| < \pi) \tag{21.8-3}$$

(Bessel functions of the first kind)

$$\left.\begin{aligned} N_m(z) &\equiv \frac{1}{\sin m\pi} [J_m(z) \cos m\pi - J_{-m}(z)] \\ &\qquad (m \neq 0, \pm 1, \pm 2, \ldots) \\ N_m(z) &= (-1)^{-m} N_{-m}(z) \\ &= \frac{2}{\pi} J_m(z) \left(\log_e \frac{z}{2} + C\right) \\ &\quad - \frac{1}{\pi}\left(\frac{z}{2}\right)^m \sum_{k=0}^{\infty} \frac{(-1)^k}{k!(m + k)!} \left(\frac{z}{2}\right)^{2k} \left(\sum_{j=1}^{k} \frac{1}{j} + \sum_{j=1}^{m+k} \frac{1}{j}\right) \\ &\quad - \frac{1}{\pi}\left(\frac{z}{2}\right)^{-m} \sum_{k=0}^{m-1} \frac{(m - k - 1)!}{k!} \left(\frac{z}{2}\right)^{2k} \\ &\qquad (m = 0, 1, 2, \ldots ; |\arg z| < \pi) \end{aligned}\right\} \tag{21.8-4}$$

(Neumann's Bessel functions of the second kind)

$$H_m^{(1)}(z) \equiv J_m(z) + iN_m(z) \qquad H_m^{(2)}(z) \equiv J_m(z) - iN_m(z) \tag{21.8-5}$$

(Hankel functions of the first and second kind)

The last three sums in Eq. (4) are given the value zero whenever the lower limit exceeds the upper limit; and $C \approx 0.577216$ is Euler's constant (21.4-6). Note that every function $N_m(z)$ has a singularity at the origin.*

(c) Analytic Continuation. To obtain values of the cylinder functions for $|\arg z| > \pi$, use

$$\left.\begin{aligned} J_m(e^{in\pi}z) &= e^{imn\pi}J_m(z) \\ N_m(e^{in\pi}z) &= e^{-imn\pi}N_m(z) + 2iJ_m(z) \sin mn\pi \cot m\pi \end{aligned}\right\} \quad (n = 0, 1, 2, \ldots) \qquad (21.8\text{-}6)$$

where one uses $\sin mn\pi \cot m\pi = (-1)^{mn}n$ for $m = \pm n$; and

$$\left.\begin{aligned} H_m^{(1)}(e^{i\pi}z) &= -e^{-im\pi}H_m^{(2)}(z) = -H_{-m}^{(2)}(z) \\ H_m^{(2)}(e^{-i\pi}z) &= -e^{im\pi}H_m^{(1)}(z) = -H_{-m}^{(1)}(z) \end{aligned}\right\} \qquad (21.8\text{-}7)$$

Note that *cylinder functions of integral order are single-valued integral functions* (Sec. 7.6-5).

(d) *Every cylinder function of order m can be expressed as a linear combination of $J_m(z)$ and $N_m(z)$ and as a linear combination of $H_m^{(1)}(z)$ and $H_m^{(2)}(z)$:*

$$Z_m(z) = aJ_m(z) + bN_m(z) = \alpha H_m^{(1)}(z) + \beta H_m^{(2)}(z) \qquad (21.8\text{-}8)$$

(*fundamental systems,* Sec. 9.3-2). $J_m(z)$ and $J_{-m}(z)$ constitute a fundamental system unless $m = 0, \pm 1, \pm 2, \ldots$, since then $J_{-m}(z) \equiv (-1)^m J_m(z)$. The three fundamental systems have the respective Wronskians (Sec. 9.3-2) $2/\pi z$, $-4i/\pi z$, and $-2 \sin (m\pi)/\pi z$; the first two Wronskians are independent of m. Note

$$J_m(z) = \frac{1}{2}[H_m^{(1)}(z) + H_m^{(2)}(z)] \qquad N_m(z) = \frac{1}{2i}[H_m^{(1)}(z) - H_m^{(2)}(z)] \qquad (21.8\text{-}9)$$

$$J_{-m}(z) = \tfrac{1}{2}[e^{im\pi}H_m^{(1)}(z) + e^{-im\pi}H_m^{(2)}(z)] \qquad (21.8\text{-}10)$$

(e) Cylinder functions with $m = \pm\frac{1}{2}, \pm\frac{3}{2}, \ldots$ can be written as elementary transcendental functions (see also Sec. 21.8-8):

$$J_{1/2}(z) \equiv \sqrt{\frac{2}{\pi}}\frac{\sin z}{\sqrt{z}} \qquad J_{-1/2}(z) \equiv \sqrt{\frac{2}{\pi}}\frac{\cos z}{\sqrt{z}} \qquad (21.8\text{-}11)$$

$$J_{3/2}(z) \equiv \sqrt{\frac{2}{\pi}}\left(-\frac{\cos z}{\sqrt{z}} + \frac{\sin z}{z\sqrt{z}}\right) \qquad J_{-3/2}(z) \equiv \sqrt{\frac{2}{\pi}}\left(-\frac{\sin z}{\sqrt{z}} - \frac{\cos z}{z\sqrt{z}}\right) \qquad (21.8\text{-}12)$$

$$J_{k+1/2}(z) \equiv \sqrt{\frac{2}{\pi}}\, z^{k+1/2}\left(-\frac{1}{z}\frac{d}{dz}\right)^k \frac{\sin z}{z} \qquad (k = 1, 2, \ldots) \qquad (21.8\text{-}13)$$

$$H_{1/2}^{(1)}(z) \equiv \sqrt{\frac{2}{\pi}}\frac{1}{i}\frac{e^{iz}}{\sqrt{z}} \qquad H_{-1/2}^{(1)}(z) \equiv \sqrt{\frac{2}{\pi}}\frac{e^{iz}}{\sqrt{z}} \qquad (21.8\text{-}14)$$

$$H_{1/2}^{(2)}(z) \equiv -\sqrt{\frac{2}{\pi}}\frac{1}{i}\frac{e^{-iz}}{\sqrt{z}} \qquad H_{-1/2}^{(2)}z \equiv \sqrt{\frac{2}{\pi}}\frac{e^{-iz}}{\sqrt{z}} \qquad (21.8\text{-}15)$$

* The Neumann functions $N_m(z)$ are sometimes denoted by $Y_m(z)$; some authors refer to them as *Weber's Bessel functions of the second kind.*

21.8-2. Integral Formulas (see Sec. 8.6-4, Table D-7, and Refs. 21.3 and 21.11 for additional relations). **(a) Integral Representations of** $J_0(z), J_1(z), J_2(z), \ldots$.

$$J_m(z) = \frac{1}{\pi}\int_0^\pi \cos(mt - z\sin t)\,dt \qquad (m = 0, 1, 2, \ldots)$$

(BESSEL'S INTEGRAL FORMULA) (21.8-16)

$$\left.\begin{aligned} J_{2m}(z) &= \frac{2}{\pi}\int_0^{\pi/2} \cos(z\sin t)\cos 2mt\,dt \\ J_{2m+1}(z) &= \frac{2}{\pi}\int_0^{\pi/2} \sin(z\sin t)\sin(2m+1)t\,dt \end{aligned}\right\} \begin{matrix}(m = 0, \\ 1, 2, \ldots)\end{matrix} \qquad (21.8\text{-}17)$$

$$J_m(z) = \frac{1}{2\pi}\int_{-\pi}^{\pi} e^{iz\cos t}e^{im(t-\pi/2)}\,dt = \frac{(-i)^m}{\pi}\int_0^\pi e^{iz\cos t}\cos mt\,dt$$

$(m = 0, 1, 2, \ldots)$ (HANSEN'S INTEGRAL FORMULA) (21.8-18)

$$J_m(z) = \frac{1}{2\pi i}\left(\frac{z}{2}\right)^m \oint t^{-m-1}e^{t-(z^2/4t)}\,dt \qquad (m = 0, 1, 2, \ldots) \qquad (21.8\text{-}19)$$

where the integration contour encloses the origin, and $|\arg z| < \pi$.

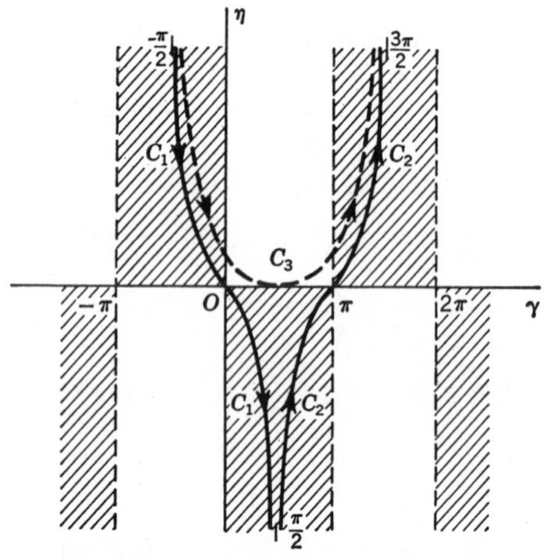

FIG. 21.8-1. Integration contours for Sommerfeld's integrals; $t = \gamma + i\eta$. (*J. A. Stratton, Electromagnetic Theory, McGraw-Hill, New York, 1941.*)

(b) Sommerfeld's and Poisson's Formulas. Referring to Fig. 21.8-1, the complex contour integral

$$Z_m(z) = \frac{e^{-im(\pi/2)}}{\pi}\int_C e^{i(z\cos t + mt)}\,dt \quad \text{(SOMMERFELD'S INTEGRAL)} \qquad (21.8\text{-}20)$$

equals $H_m^{(1)}(z)$ for the contour $C \equiv C_1$, $H_m^{(2)}(z)$ for the contour $C \equiv C_2$, and $2J_m(z)$ for the contour $C \equiv C_3$. The contours may be deformed at will, provided that they start and terminate by approaching $t = \infty$ in the shaded areas indicated. $t = 0$ and

$t = \pi$ may be used as saddle points (Sec. 7.7-3e) for C_1 and C_2 in computations of $Z_m(z)$ for large values of z (see also Sec. 21.8-9).

Note also

$$J_m(z) = \frac{2(z/2)^m}{\sqrt{\pi}\,\Gamma(m + \frac{1}{2})} \int_0^{\pi/2} \cos(z \cos t) \sin^{2m} t \, dt \qquad (m > -\tfrac{1}{2})$$

(POISSON'S INTEGRAL FORMULA) (21.8-21)

(c) Miscellaneous Integral Formulas Involving Cylinder Functions (see also Sec. 21.8-4c).

$$\left.\begin{aligned} \int_0^x J_m(\alpha x) J_m(\beta x) x \, dx &= \frac{x}{\alpha^2 - \beta^2} [\alpha J_m(\beta x) J_{m+1}(\alpha x) - \beta J_m(\alpha x) J_{m+1}(\beta x)] \\ &= \frac{x}{\alpha^2 - \beta^2} [\beta J_{m-1}(\beta x) J_m(\alpha x) - \alpha J_{m-1}(\alpha x) J_m(\beta x)] \qquad (\alpha^2 - \beta^2 \neq 0) \\ \int_0^x [J_m(\alpha x)]^2 x \, dx &= \frac{x^2}{2} [J_m'(\alpha x)]^2 + \frac{1}{2}\left(x^2 - \frac{m^2}{\alpha^2}\right) [J_m(\alpha x)]^2 \end{aligned}\right\}$$

$(m > -1)$ (LOMMEL'S INTEGRALS) (21.8-22)

$$\int_0^\infty x^{-m+n} J_m(\alpha x) \, dx = 2^{n-m} \alpha^{m-n-1} \frac{\Gamma\left(\frac{n+1}{2}\right)}{\Gamma\left(m - \frac{n-1}{2}\right)} \qquad (-1 < n < 2m + 1)$$

(21.8-23)

21.8-3. Zeros of Cylinder Functions. **(a)** *All zeros of cylinder functions are simple zeros, except possibly $z = 0$. Consecutive positive or nega-*

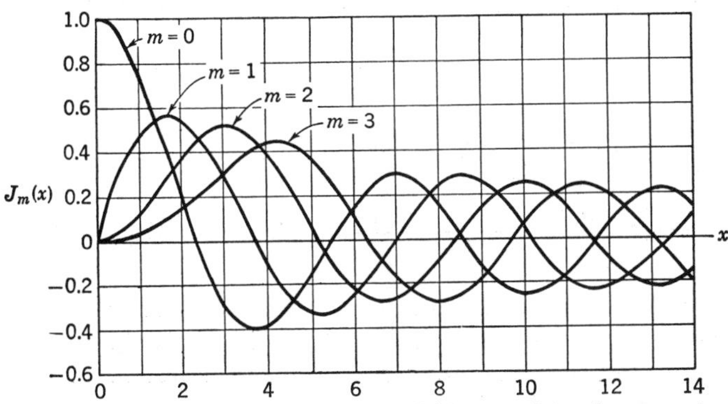

FIG. 21.8-2. The Bessel functions $J_0(x)$, $J_1(x)$, $J_2(x)$, . . . for real arguments. Note that $J_m(-x) = (-1)^m J_m(x)$.

tive real zeros of two linearly independent real cylinder functions of order m alternate; $z = 0$ is the only possible common zero.

(b) (See also Fig. 21.8-2). *The function $J_m(z)$ has an infinite number of real zeros; for $m > -1$, all its zeros are real. For $m = 0, \frac{1}{2}, 1, \frac{3}{2}, 2,$ . . . and $n = 1, 2, \ldots, J_m(z)$ and $J_{m+n}(z)$ have no common zeros.*

For $m = 1, 2, \ldots,$ consecutive positive or negative real zeros of $J_m(z)$

are separated by unique real zeros of $J_{m+1}(z)$ *and by unique real zeros of* $J'_m(z)$*; and consecutive positive or negative real zeros of* $z^{-m}J_m(z)$ *are separated by unique real zeros of* $z^{-m}J_{m+1}(z)$.

21.8-4. The Bessel Functions $J_0(z)$, $J_1(z)$, $J_2(z)$, . . . (a) Generation by Series Expansions. Bessel functions of nonnegative integral order $m = 0, 1, 2, \ldots$ are single-valued integral functions of z. They may be "generated" (see also Sec. 8.6-5) as coefficients of the Laurent series (Sec. 7.5-3)

$$e^{\frac{z}{2}\left(s-\frac{1}{s}\right)} = J_0(z) + \sum_{m=1}^{\infty} [s^m + (-s)^{-m}]J_m(z) \qquad (21.8\text{-}24)$$

or as coefficients of the Fourier series

$$\left.\begin{aligned} \cos(z \sin t) &= J_0(z) + 2\sum_{k=1}^{\infty} J_{2k}(z) \cos 2kt \\ \sin(z \sin t) &= 2\sum_{k=1}^{\infty} J_{2k-1}(z) \sin(2k-1)t \end{aligned}\right\} \qquad (21.8\text{-}25a)$$

$$\begin{aligned} e^{\pm iz \sin t} &= \sum_{m=-\infty}^{\infty} J_m(z)e^{\pm imt} \\ &= J_0(z) + 2\sum_{k=1}^{\infty} [J_{2k}(z) \cos 2kt \pm iJ_{2k-1}(z) \sin(2k-1)t] \end{aligned}$$

(JACOBI-ANGER FORMULA) (21.8-25b)

Note also

$$1 = J_0(z) + 2\sum_{k=1}^{\infty} J_{2k}(z) = J_0^2(z) + 2\sum_{k=1}^{\infty} J_k^2(z) \qquad (21.8\text{-}26)$$

$$z^n = 2^n \sum_{k=0}^{\infty} \frac{(n+2k)(n+k-1)!}{k!} J_{2k+n}(z) \qquad (n = 1, 2, \ldots) \qquad (21.8\text{-}27)$$

(b) Behavior for Real Arguments. For real z, $J_0(z)$, $J_1(z)$, $J_2(z)$, . . . are real functions of $z = x$; Fig. 21.8-2 illustrates their zeros, their maxima and minima, and their asymptotic behavior for $x \to \infty$ (see also Secs. 21.8-3 and 21.8-9).

(c) Orthogonality Relations (see also Secs. 15.4-6 and 21.8-2c). Given any two real zeros x_i, x_k of $J_m(z)$, one has the orthogonality relations

$$\int_0^1 J_m(x_j\xi)J_m(x_k\xi)\xi\, d\xi = \begin{cases} 0 \text{ if } j \neq k \\ \frac{1}{2}[J'_m(x_j)]^2 \text{ if } j = k \end{cases} \qquad (21.8\text{-}28)$$

which yield an orthogonal-function expansion. Note also

$$\alpha \int_0^\infty J_m(\alpha\xi)J_m(\beta\xi)\xi\, d\xi = \delta(\alpha - \beta) \qquad (21.8\text{-}29)$$

where $\delta(x)$ is the impulse function introduced in Sec. 21.9-2.

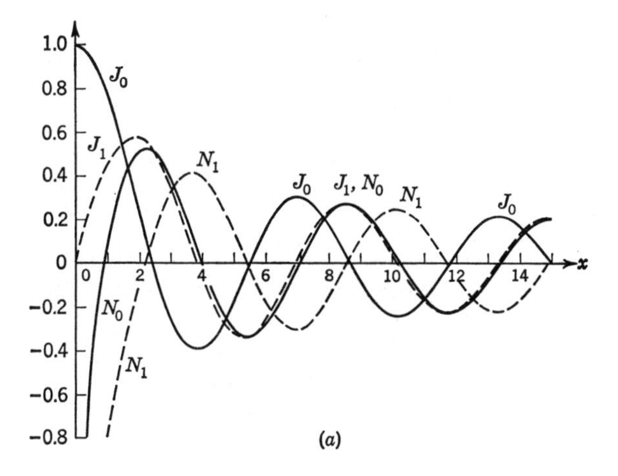

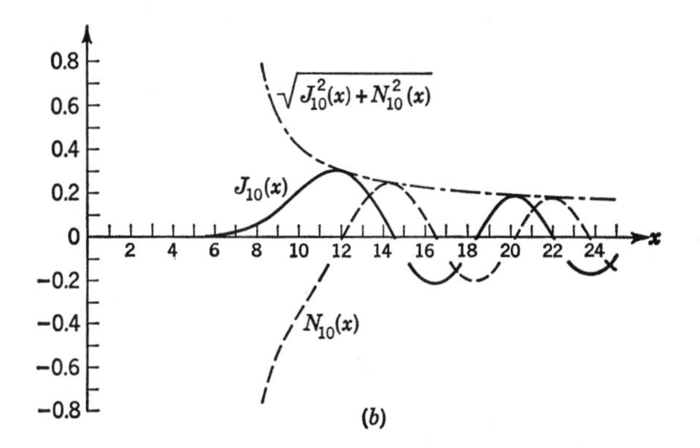

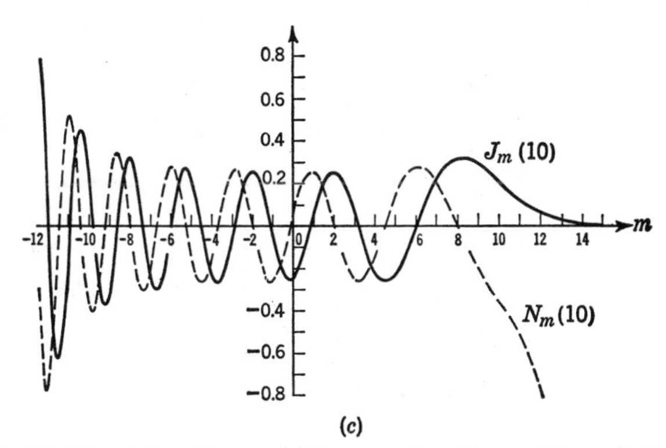

FIG. 21.8-3. Bessel functions and Neumann functions. (*From Ref.* 21.1.)

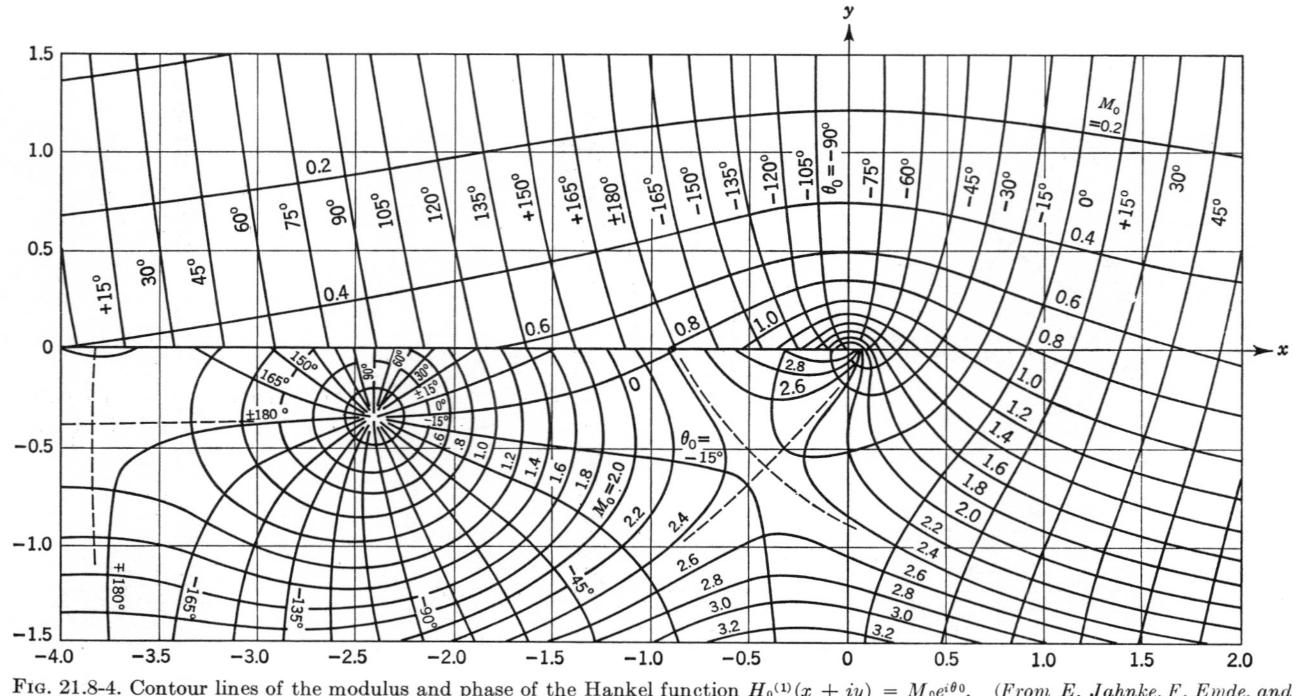

FIG. 21.8-4. Contour lines of the modulus and phase of the Hankel function $H_0^{(1)}(x + iy) = M_0e^{i\theta_0}$. (*From E. Jahnke, F. Emde, and F. Lösch, Tables of Higher Functions, McGraw-Hill, New York,* 1960.)

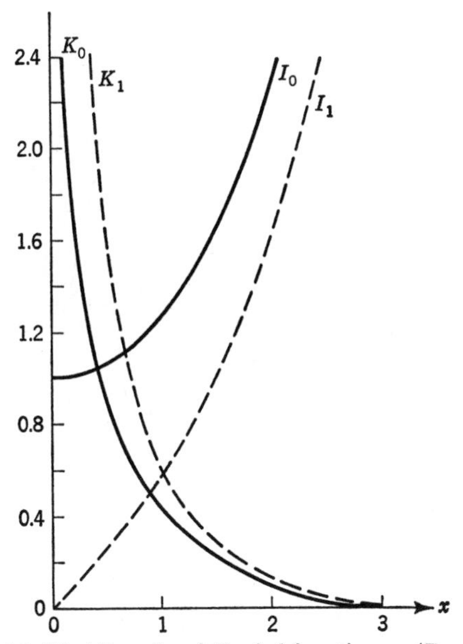

FIG. 21.8-5. Modified Bessel and Hankel functions. (*From Ref.* 21.1.)

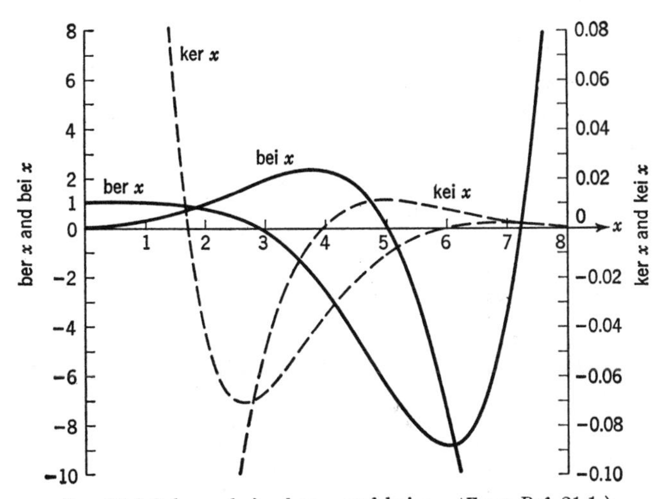

FIG. 21.8-6. ber x, bei x, ker x, and kei x. (*From Ref.* 21.1.)

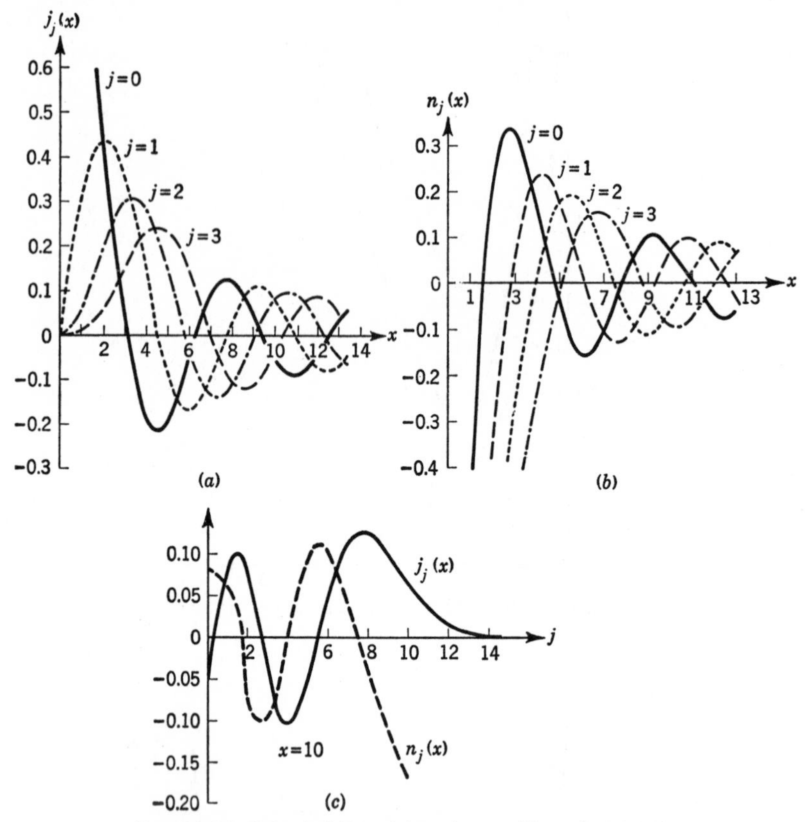

FIG. 21.8-7. Spherical Bessel functions. (*From Ref.* 21.1.)

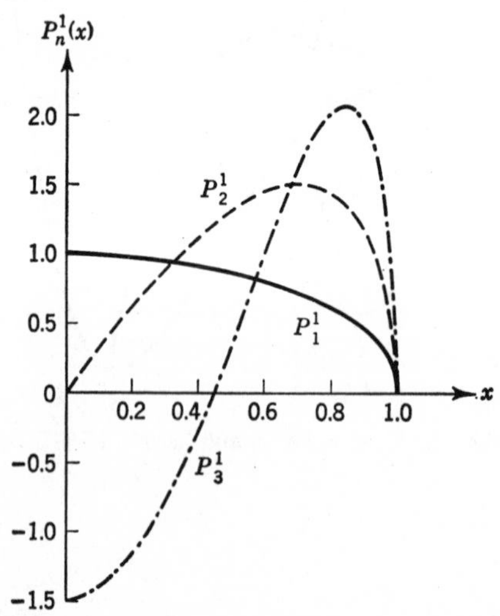

FIG. 21.8-8. The associated Legendre functions $P_1^1(x)$, $P_2^1(x)$, $P_3^1(x)$. (*From Ref.* 21.1.)

21.8-5. Solution of Differential Equations in Terms of Cylinder Functions and Related Functions. The linear differential equation

$$\frac{d^2w}{dz^2} + \frac{1-2a}{z}\frac{dw}{dz} + \left[(bcz^{c-1})^2 + \frac{a^2 - m^2c^2}{z^2}\right]w = 0 \quad (21.8\text{-}30)$$

has solutions of the form

$$w = z^a Z_m(bz^c) \quad (21.8\text{-}31)$$

Many special cases of Eq. (30) are of interest (Secs. 21.8-6 to 21.8-8, Refs. 21.6 and 21.9).

21.8-6. Modified Bessel and Hankel Functions. The **modified cylinder functions of order** m,

$$\left.\begin{aligned} I_m(z) &= i^{-m}J_m(iz) && \text{(MODIFIED BESSEL FUNCTIONS)} \\ K_m(z) &= \frac{\pi}{2}i^{m+1}H_m^{(1)}(iz) && \text{(MODIFIED HANKEL FUNCTIONS)} \end{aligned}\right\} \quad (21.8\text{-}32)$$

are defined with the aid of Eqs. (3) to (5); the definition is extended by means of Eqs. (6) and (7). The functions (32) are linearly independent solutions of the differential equation

$$\frac{d^2w}{dz^2} + \frac{1}{z}\frac{dw}{dz} - \left(1 + \frac{m^2}{z^2}\right)w = 0 \qquad \text{(MODIFIED BESSEL'S DIFFERENTIAL EQUATION)} \quad (21.8\text{-}33)$$

(see also Sec. 10.4-3*b*) and satisfy the recursion formulas

$$\left.\begin{aligned} I_{m+1}(z) &= I_{m-1}(z) - \frac{2m}{z}I_m(z) = 2\frac{d}{dz}I_m(z) - I_{m-1}(z) \\ K_{m+1}(z) &= K_{m-1}(z) + \frac{2m}{z}K_m(z) = -2\frac{d}{dz}K_m(z) - K_{m-1}(z) \end{aligned}\right\} \quad (21.8\text{-}34)$$

$I_m(z)$ and $K_m(z)$ are real monotonic functions for $m = 0, \pm 1, \pm 2, \ldots$ and real z.

21.8-7. The Functions $\operatorname{ber}_m z$, $\operatorname{bei}_m z$, $\operatorname{her}_m z$, $\operatorname{hei}_m z$, $\operatorname{ker}_m z$, $\operatorname{kei}_m z$. The functions $\operatorname{ber}_m z$, $\operatorname{bei}_m z$, $\operatorname{her}_m z$, $\operatorname{hei}_m z$, $\operatorname{ker}_m z$, $\operatorname{kei}_m z$, defined by

$$\left.\begin{aligned} J_m(i^{\pm 3/2}z) &\equiv \operatorname{ber}_m z \pm i \operatorname{bei}_m z \\ H_m^{(1)}(i^{+3/2}z) &\equiv \operatorname{her}_m z + i \operatorname{hei}_m z \end{aligned}\right\} \quad (21.8\text{-}35)$$

$$i^{\mp m}K_m(i^{\pm 1/2}z) \equiv \operatorname{ker}_m z \pm i \operatorname{kei}_m z \quad (21.8\text{-}36)$$

are real for real values of z. The subscript m is omitted for $m = 0$, e.g., $\operatorname{ber}_0 z \equiv \operatorname{ber} z$. Note that

$$\operatorname{ker}_m z \equiv -\frac{\pi}{2}\operatorname{hei}_m z \qquad \operatorname{kei}_m z \equiv \frac{\pi}{2}\operatorname{her}_m z$$

$\operatorname{ber} z$, $\operatorname{bei} z$, $\operatorname{her} z$, and $\operatorname{hei} z$ as well as $J_0(i^{\pm 3/2}z)$ and $H_0^{(1)}(i^{3/2}z)$ are solutions of the differential equation

$$\frac{d^2w}{dz^2} + \frac{1}{z}\frac{dw}{dz} - iw = 0 \quad (21.8\text{-}37)$$

Note

$$\left.\begin{aligned}\operatorname{ber} z &= 1 - \frac{(\frac{1}{2}z)^4}{(2!)^2} + \frac{(\frac{1}{2}z)^8}{(4!)^2} - \cdots \\ &= 1 - \frac{z^4}{2^2\cdot 4^2} + \frac{z^8}{2^2\cdot 4^2\cdot 6^2\cdot 8^2} \mp \cdots \\ \operatorname{bei} z &= \left(\frac{1}{2}z\right)^2 - \frac{(\frac{1}{2}z)^6}{(3!)^2} + \cdots = \frac{z^2}{2^2} - \frac{z^6}{2^2\cdot 4^2\cdot 6^2} \pm \cdots\end{aligned}\right\} \qquad (21.8\text{-}38)$$

In some applications, it is convenient to introduce $|J_m(i^{3/2}z)|$, $|K_m(i^{3/2}z)|$, $\arg J_m(i^{3/2}z)$, and $\arg\,[i^{-m}K_m(i^{3/2}z)]$ as special functions instead of or together with $\operatorname{ber}_m z$, $\operatorname{bei}_m z$, $\ker_m z$, and $\operatorname{kei}_m z$ (Ref. 21.6). All these functions are real for real values of z.

21.8-8. Spherical Bessel Functions. The **spherical Bessel functions of the first, second, third, and fourth kind**

$$\left.\begin{aligned} j_j(z) &\equiv \sqrt{\frac{\pi}{2z}}\,J_{j+\frac{1}{2}}(z) & n_j(z) &\equiv \sqrt{\frac{\pi}{2z}}\,N_{j+\frac{1}{2}}(z) \\ h_j^{(1)}(z) &\equiv \sqrt{\frac{\pi}{2z}}\,H^{(1)}_{j+\frac{1}{2}}(z) & h_j^{(2)}(z) &\equiv \sqrt{\frac{\pi}{2z}}\,H^{(2)}_{j+\frac{1}{2}}(z)\end{aligned}\right\} \qquad (21.8\text{-}39)$$

satisfy the differential equation

$$\frac{d^2w}{dz^2} + \frac{2}{z}\frac{dw}{dz} + \left[1 - \frac{j(j+1)}{z^2}\right]w = 0 \qquad (21.8\text{-}40)$$

(see also Sec. 10.4-4*c*) and the recursion formulas

$$w_{j+1}(z) = \frac{2j+1}{z}\,w_j(z) - w_{j-1}(z) = -z^j\frac{d}{dz}[z^{-j}w_j(z)] \qquad (21.8\text{-}41)$$

For integral values of j, the spherical Bessel functions are elementary transcendental functions (see also Sec. 21.8-1*e*):

$$\left.\begin{aligned} j_0(z) &\equiv \frac{\sin z}{z} & h_0^{(1)}(z) &\equiv -\frac{ie^{iz}}{z} & h_0^{(2)}(z) &\equiv \frac{ie^{-iz}}{z} \\ j_{-1}(z) &\equiv \frac{\cos z}{z} & h_{-1}^{(1)}(z) &\equiv \frac{e^{iz}}{z} & h_{-1}^{(2)}(z) &\equiv \frac{e^{-iz}}{z}\end{aligned}\right\} \qquad (21.8\text{-}42)$$

$$j_j(z) \equiv z^j\left(-\frac{1}{z}\frac{d}{dz}\right)^j\frac{\sin z}{z} \qquad n_j(z) \equiv (-1)^{j+1}j_{-j-1}(z) \qquad (j = 1, 2, \ldots) \qquad (21.8\text{-}43)$$

21.8-9. Asymptotic Expansion of Cylinder Functions and Spherical Bessel Functions for Large Absolute Values of z (see also Secs. 4.4-3 and 4.8-6). As $z \to \infty$,

$$\left.\begin{aligned} J_m(z) &\simeq \sqrt{\frac{2}{\pi z}}\left[A_m(z)\cos\left(z - \frac{m\pi}{2} - \frac{\pi}{4}\right)\right. \\ &\qquad \left. - B_m(z)\sin\left(z - \frac{m\pi}{2} - \frac{\pi}{4}\right)\right] \\ N_m(z) &\simeq \sqrt{\frac{2}{\pi z}}\left[A_m(z)\sin\left(z - \frac{m\pi}{2} - \frac{\pi}{4}\right)\right. \\ &\qquad \left. + B_m(z)\cos\left(z - \frac{m\pi}{2} - \frac{\pi}{4}\right)\right]\end{aligned}\right\} \quad (|\arg|\, z < \pi) \qquad (21.8\text{-}44)$$

where $A_m(z)$ and $B_m(z)$ stand for the asymptotic series

$$\left.\begin{aligned} A_m(z) &= 1 - \frac{(4m^2-1)(4m^2-9)}{2!(8z)^2} \\ &\quad + \frac{(4m^2-1)(4m^2-9)(4m^2-25)(4m^2-49)}{4!(8z)^4} \mp \cdots \\ B_m(z) &= \frac{4m^2-1}{8z} - \frac{(4m^2-1)(4m^2-9)(4m^2-25)}{3!(8z)^3} \\ &\quad \pm \cdots \end{aligned}\right\} \quad (21.8\text{-}45)$$

Substitution of Eqs. (44) and (45) into Eq. (5) yields corresponding asymptotic expansions of $H_m^{(1)}(z)$ and $H_m^{(2)}(z)$.

For $|z| \gg |m|$ as $z \to \infty$,

$$\left.\begin{aligned} J_m(z) &\simeq \sqrt{\frac{2}{\pi z}} \cos\left(z - \frac{m\pi}{2} - \frac{\pi}{4}\right) \\ N_m(z) &\simeq \sqrt{\frac{2}{\pi z}} \sin\left(z - \frac{m\pi}{2} - \frac{\pi}{4}\right) \\ H_m^{(1)}(z) &\simeq \sqrt{\frac{2}{\pi z}}\, e^{i\left(z - \frac{m\pi}{2} - \frac{\pi}{4}\right)} \qquad H_m^{(2)}(z) \simeq \sqrt{\frac{2}{\pi z}}\, e^{-i\left(z - \frac{m\pi}{2} - \frac{\pi}{4}\right)} \end{aligned}\right\} \quad (21.8\text{-}46)$$

$$J_{m+\frac{1}{2}}(z) \simeq \sqrt{\frac{2}{\pi z}} \sin\left(z - \frac{m\pi}{2}\right) \qquad (m = 0, 1, 2, \ldots) \quad (21.8\text{-}47)$$

$$\left.\begin{aligned} j_j(z) &\simeq \frac{1}{z} \cos\left(z - \frac{j+1}{2}\pi\right) \\ n_j(z) &\simeq \frac{1}{z} \sin\left(z - \frac{j+1}{2}\pi\right) \\ h_j^{(1)}(z) &\simeq \frac{1}{z}(-i)^{j+1}e^{iz} \qquad h_j^{(2)}(z) \simeq \frac{1}{z}\, i^{j+1}e^{-iz} \end{aligned}\right\} \quad (21.8\text{-}48)$$

Note that the asymptotic relationship of $J_m(z)$, $N_m(z)$, $H_m^{(1)}(z)$, and $H_m^{(2)}(z)$ is analogous to the relationship of the more familiar trigonometric and exponential functions.

21.8-10. Associated Legendre Functions and Polynomials (see also Secs. 21.7-1 and 21.7-3). **(a)** The **associated Legendre functions of degree** j **and order** m are solutions of the differential equation

$$(1 - z^2)\frac{d^2w}{dz^2} - 2z\frac{dw}{dz} + \left[j(j+1) - \frac{m^2}{1-z^2}\right]w = 0 \quad (21.8\text{-}49)$$

where j and m are complex numbers; Eq. (49) reduces to Legendre's differential equation (Table 21.7-1 and Sec. 21.7-3) for $m = 0$. The general theory of the associated Legendre functions is found in Refs. 21.3 and 21.9. In the most important applications (Secs. 10.4-3c and 21.8-12), *j and m are effectively restricted to the real integral values* 0, 1, 2, . . . , *while $z = x$ is the cosine of a real angle ϑ and hence a real number* *between* -1 *and* 1. Under these conditions, Eq. (49) is satisfied by the **associated Legendre "polynomials" of the first kind***

$$\begin{aligned} P_j^m(x) &= \frac{1}{2^m}\frac{(j+m)!}{(j-m)!m!}(1-x^2)^{m/2} \\ &\qquad \cdot F\left(m-j,\, m+j+1;\, m+1;\, \frac{1-x}{2}\right) \\ &= (1-x^2)^{m/2}\frac{d^m}{dx^m}P_j(x) \\ &= \frac{(1-x^2)^{m/2}}{2^j j!}\frac{d^{j+m}}{dx^{j+m}}(x^2-1)^j \\ &= (-1)^{j+m}P_j^m(-x) \qquad (\text{real } x \text{ between } -1 \text{ and } 1; \\ &\qquad j = 0, 1, 2, \ldots;\ m = 0, 1, 2, \ldots, j) \end{aligned} \tag{21.8-50}$$

(see also Sec. 9.3-9), with $P_j^0(x) \equiv P_j(x)$, and $P_j^m(x) \equiv 0$ for $m > j$. In particular,

$$P_1^1(x) = \sqrt{1-x^2} = \sin\vartheta \tag{21.8-51}$$

$$\left.\begin{aligned} P_2^1(x) &= 3x\sqrt{1-x^2} = \tfrac{3}{2}\sin 2\vartheta \\ P_2^2(x) &= 3(1-x^2) = \tfrac{3}{2}(1-\cos 2\vartheta) \end{aligned}\right\} \tag{21.8-52}$$

$$\left.\begin{aligned} P_3^1(x) &= \tfrac{3}{2}(5x^2-1)\sqrt{1-x^2} = \tfrac{3}{8}(\sin\vartheta + 5\sin 3\vartheta) \\ P_3^2(x) &= 15x(1-x^2) = \tfrac{15}{4}(\cos\vartheta - \cos 3\vartheta) \\ P_3^3(x) &= 15(1-x^2)\sqrt{1-x^2} = \tfrac{15}{4}(3\sin\vartheta - \sin 3\vartheta) \end{aligned}\right\} \tag{21.8-53}$$

$$\begin{aligned} P_j^j(x) &= 1\cdot 3\cdot 5\cdots(2j-1)(1-x^2)^{j/2} \\ &= 1\cdot 3\cdot 5\cdots(2j-1)\sin^j\vartheta \qquad (j = 0, 1, 2, \ldots) \end{aligned} \tag{21.8-54}$$

where $\cos\vartheta = x$ (see also Sec. 21.8-12).

(b) The associated Legendre "polynomials" defined by Eq. (50) satisfy the following recurrence relations ($-1 < x < 1$):

$$(2j+1)xP_j^m(x) - (j-m+1)P_{j+1}^m(x) - (j+m)P_{j-1}^m(x) = 0 \qquad (0 \le m \le j-1) \tag{21.8-55}$$

$$(x^2-1)\frac{d}{dx}P_j^m(x) - (j-m+1)P_{j+1}^m(x) + (j+1)xP_j^m(x) = 0 \qquad (0 \le m \le j) \tag{21.8-56}$$

$$\begin{aligned} P_j^{m+2}(x) &- 2(m+1)\frac{x}{\sqrt{1-x^2}}P_j^{m+1}(x) \\ &+ [j(j+1) - m(m+1)]P_j^m(x) = 0 \qquad (0 \le m \le j-2) \end{aligned} \tag{21.8-57}$$

$$P_{j+1}^m(x) - P_{j-1}^m(x) = (2j+1)\sqrt{1-x^2}\,P_j^{m-1}(x) \qquad (0 \le m \le j-1) \tag{21.8-58}$$

$$\begin{aligned} (j+m)(j+m+1)P_{j-1}^m(x) &- (j-m)(j-m+1)P_{j+1}^m(x) \\ &= (2j+1)\sqrt{1-x^2}\,P_j^{m+1}(x) \qquad (0 \le m \le j-1) \end{aligned} \tag{21.8-59}$$

* Some authors reserve the symbol $P_j^m(x)$ for the functions here denoted by $(-1)^m P_j^m(x)$ or by $(-1)^{j+m}P_j^m(x)$. Note that not all $P_j^m(x)$ are actually polynomials in x.

(c) Asymptotic Behavior. As $j \to \infty$,

$$P_j^m(\cos\vartheta) = (-j)^m \sqrt{\frac{2}{\pi j \sin\vartheta}} \cos\left[\left(j + \frac{1}{2}\right)\vartheta - \frac{\pi}{4} + \frac{m\pi}{2}\right] + O(j^{-3/2})$$

$$(0 < \vartheta < \pi) \qquad (21.8\text{-}60)$$

21.8-11. Integral Formulas Involving Associated Legendre Polynomials (see also Sec. 21.7-7).

$$P_j^m(x) = (-1)^{m/2} \frac{(j+m)!}{j!\pi} \int_0^\pi (x + \sqrt{x^2 - 1}\cos t)^j \cos mt\, dt$$

$(j = 0, 1, 2, \ldots ; m = 0, 1, 2, \ldots , j)$ (HEINE'S INTEGRAL FORMULA) (21.8-61)

$$\int_{-1}^1 P_j^m(x) P_{j'}^m(x)\, dx = \frac{2}{2j+1} \frac{(j+m)!}{(j-m)!} \delta_{j'}^j$$

$$(j, j' = 0, 1, 2, \ldots ; m = 0, 1, 2, \ldots , j < j') \qquad (21.8\text{-}62)$$

$$\int_0^1 [P_j^m(x)]^2\, dx = \frac{1}{2j+1} \frac{(j+m)!}{(j-m)!} \qquad \int_0^1 \frac{[P_j^m(x)]^2\, dx}{1 - x^2} = \frac{1}{2m} \frac{(j+m)!}{(j-m)!}$$

$$(j = 0, 1, 2, \ldots ; m = 1, 2, \ldots , j) \qquad (21.8\text{-}63)$$

21.8-12. Spherical Harmonics. Orthogonality (see also Secs. 10.4-3*c*, 14.10-7*b*, and 15.2-6). **(a)** Solutions $\Phi(r, \vartheta, \varphi)$ of Laplace's partial differential equation in spherical coordinates (10.4-15) are known as **spherical harmonics. Spherical surface harmonics of degree** j are solutions $Y_j(\vartheta, \varphi)$ of the partial differential equation

$$\frac{\partial^2 Y}{\partial \vartheta^2} + \cot\vartheta \frac{\partial Y}{\partial \vartheta} + \frac{1}{\sin^2\vartheta} \frac{\partial^2 Y}{\partial \varphi^2} + j(j+1)Y = 0 \qquad (21.8\text{-}64)$$

obtained on separation of variables in Eq. (10.4-15). If one imposes the "boundary" conditions of regularity and uniqueness for $0 \leq \vartheta \leq \pi$, $0 \leq \varphi \leq 2\pi$ together with $Y_j(\vartheta, \varphi + 2\pi) \equiv Y_j(\vartheta, \varphi)$, the problem becomes an eigenvalue problem (Sec. 15.4-5) admitting only integral values of j. One may disregard negative values of j, since $-j - 1$ and j yield identical values of $j(j + 1)$. The resulting real eigenfunctions

$$\sqrt{\frac{2j+1}{2\pi} \frac{(j-m)!}{(j+m)!}} P_j^m(\cos\vartheta) \cos m\varphi$$

$$\sqrt{\frac{2j+1}{2\pi} \frac{(j-m)!}{(j+m)!}} P_j^m(\cos\vartheta) \sin m\varphi$$

$$(j = 0, 1, 2, \ldots ; m = 0, 1, 2, \ldots , j) \qquad (21.8\text{-}65)$$

are known as **tesseral spherical harmonics** of degree j and order m; they are periodic on the surface of a sphere and change sign along "node lines" ϑ = constant and φ = constant (Fig. 21.8-9). For $m = j$, one has **sectorial** spherical harmonics, and for $m = 0$, **zonal** spherical harmonics.

Both the functions (65) and the frequently more convenient complex functions

$$\frac{1}{2}\sqrt{\frac{2j+1}{\pi}\frac{(j-|m|)!}{(j+|m|)!}}P_j^{|m|}(\cos\vartheta)e^{im\varphi}$$
$$(j = 0, 1, 2, \ldots; m = 0, \pm 1, \pm 2, \ldots, \pm j) \qquad (21.8\text{-}66)$$

are orthonormal sets of eigenfunctions in terms of the inner product

$$(f, h) = \int_0^{2\pi} d\varphi \int_0^{\pi} f^*(\vartheta, \varphi)h(\vartheta, \varphi)\sin\vartheta\, d\vartheta \qquad (21.8\text{-}67)$$

(Sec. 15.4-6*b*). $(f, h) = 0$ for every pair of functions (65) or of functions (66) unless $f \equiv h$, in which case the inner product equals one. There are exactly $2j + 1$ linearly independent spherical surface harmonics of degree j.

(**b**) Every twice continuously differentiable, suitably periodic real function $\Phi(\vartheta, \varphi)$ defined on the surface of a sphere admits the absolutely convergent expansion

$$\begin{aligned}\Phi(\vartheta, \varphi) &= \sum_{j=0}^{\infty}\left[\tfrac{1}{2}\alpha_{j0}P_j(\cos\vartheta) + \sum_{m=1}^{j}P_j^m(\cos\vartheta)(\alpha_{jm}\cos m\varphi + \beta_{jm}\sin m\varphi)\right] \\ &= \sum_{j=0}^{\infty}\sum_{m=-j}^{j}\gamma_{jm}P_j^{|m|}(\cos\vartheta)e^{im\varphi} \qquad (21.8\text{-}68a)\end{aligned}$$

with

$$\left.\begin{aligned}\alpha_{jm} &= \frac{2j+1}{2\pi}\frac{(j-m)!}{(j+m)!}\int_0^{2\pi} d\varphi\cos m\varphi \int_0^{\pi}\Phi(\vartheta, \varphi)P_j^m(\cos\vartheta)\sin\vartheta\, d\vartheta \\ \beta_{jm} &= \frac{2j+1}{2\pi}\frac{(j-m)!}{(j+m)!}\int_0^{2\pi} d\varphi\sin m\varphi \int_0^{\pi}\Phi(\vartheta, \varphi)P_j^m(\cos\vartheta)\sin\vartheta\, d\vartheta \\ \gamma_{jm} &= \gamma_{j,-m}^* = \tfrac{1}{2}(\alpha_{jm} - i\beta_{jm}) \\ &= \frac{2j+1}{4\pi}\frac{(j-m)!}{(j+m)!}\int_0^{2\pi} d\varphi e^{im\varphi}\int_0^{2\pi}\Phi(\vartheta, \varphi)P_j^m(\cos\vartheta)\sin\vartheta\, d\vartheta\end{aligned}\right\}$$
$$(j = 0, 1, 2, \ldots; m = 0, 1, 2, \ldots, j) \qquad (21.8\text{-}68b)$$

Expansions of the form (68) can be physically interpreted as *multipole expansions* of potentials (Secs. 15.6-5*a* and *c*).

21.8-13. Addition Theorems. (a) Addition Theorem for Cylinder Functions. Let P_1 and P_2 be two points of a plane, with polar coordinates (r_1, φ_1), (r_2, φ_2). Referring to Fig. 21.8-10*a*, let $r_1 > r_2$, so that $0 \leq |\psi| < \pi/2$, and

$$\left.\begin{aligned}d^2 &= r_1^2 + r_2^2 - 2r_1r_2\cos(\varphi_1 - \varphi_2) \\ e^{2i\psi} &= \frac{r_1 - r_2e^{-i(\varphi_1-\varphi_2)}}{r_1 - r_2e^{i(\varphi_1-\varphi_2)}}\end{aligned}\right\} \qquad (21.8\text{-}69)$$

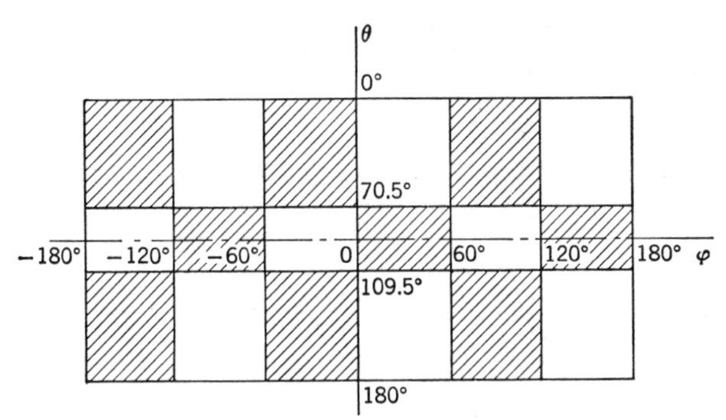

FIG. 21.8-9. Nodes of the function P_5^3 (cos ϑ) sin 3φ on the developed surface of a sphere. The function is negative on the shaded areas. (*J. A. Stratton, Electromagnetic Theory, McGraw-Hill, New York, 1941.*)

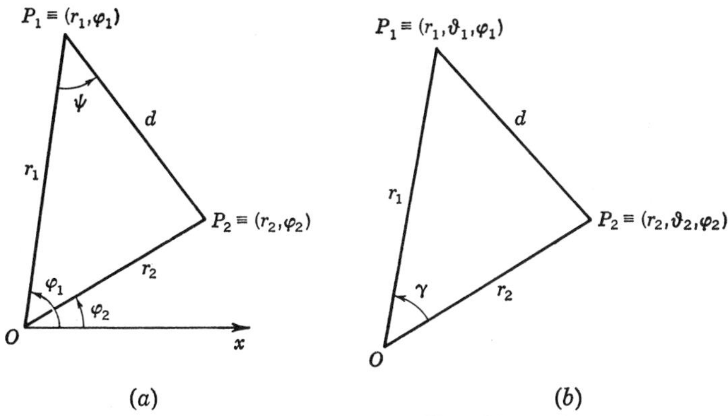

FIG. 21.8-10. Geometry for addition theorems. The addition theorems are useful for expressing effects at P_2 of a source of potential, radiation, etc., at P_1, or vice versa (see also Secs. 15.6-5 and 15.6-6).

Then, for every cylinder function $Z_m(z)$ satisfying Eqs. (1) and (2),

$$Z_m(\alpha d)e^{im\psi} = \sum_{k=-\infty}^{\infty} Z_{m+k}(\alpha r_1)J_k(\alpha r_2)e^{ik(\varphi_1-\varphi_2)}$$

(ADDITION THEOREM FOR CYLINDER FUNCTIONS) (21.8-70)

where α is any complex number. In particular, $\varphi_1 - \varphi_2 = \pi$ yields

$$Z_m[\alpha(r_1 + r_2)] = \sum_{k=-\infty}^{\infty} Z_k(\alpha r_1)J_{m-k}(\alpha r_2) \qquad (21.8\text{-}71)$$

(b) Addition Theorems for Spherical Bessel Functions and Legendre Polynomials. Let P_1 and P_2 be two points in space, with spherical coordinates $(r_1, \vartheta_1, \varphi_1)$, $(r_2, \vartheta_2, \varphi_2)$. Referring to Fig. 21.8-10*b*, let $\vartheta_1 + \vartheta_2 < \pi$. One has

$$\begin{aligned} d^2 &= r_1{}^2 + r_2{}^2 - 2r_1r_2 \cos \gamma \\ \cos \gamma &= \cos \vartheta_1 \cos \vartheta_2 + \sin \vartheta_1 \sin \vartheta_2 \cos (\varphi_1 - \varphi_2) \end{aligned} \qquad (21.8\text{-}72)$$

and

$$\left.\begin{aligned} j_0(\alpha d) &= \frac{\sin \alpha d}{\alpha d} = \sum_{k=0}^{\infty} (2k + 1) j_k(\alpha r_1) j_k(\alpha r_2) P_k(\cos \gamma) \\ h_0{}^{(1)}(\alpha d) &= \frac{e^{i\alpha d}}{i\alpha d} \\ &= \sum_{k=0}^{\infty} (2k + 1) h_k{}^{(1)}(\alpha r_1) j_k(\alpha r_2) P_k(\cos \gamma) \qquad (r_1 > r_2) \end{aligned}\right\}$$

(ADDITION THEOREM FOR ZERO-ORDER SPHERICAL BESSEL FUNCTIONS) (21.8-73)

$$\begin{aligned} P_j(\cos \gamma) &= P_j(\cos \vartheta_1) P_j(\cos \vartheta_2) \\ &\quad + 2 \sum_{m=1}^{j} \frac{(j - m)!}{(j + m)!} P_j^m(\cos \vartheta_1) P_j^m(\cos \vartheta_2) \cos m(\varphi_1 - \varphi_2) \end{aligned}$$

(ADDITION THEOREM FOR LEGENDRE POLYNOMIALS) (21.8-74)

21.9. STEP FUNCTIONS AND SYMBOLIC IMPULSE FUNCTIONS

21.9-1. Step Functions (see also Secs. 4.6-17*c*, 18.3-1, and 20.4-5*c*). (a) A **step function** of the real variable x is a function which changes its value only on a discrete set of discontinuities (necessarily of the first kind, Sec. 4.4-7*b*). The function values at the discontinuities may or may not be defined. The most frequently useful step functions are*

$$U(x) = \begin{cases} 0 \text{ for } x < 0 \\ \tfrac{1}{2} \text{ for } x = 0 \\ 1 \text{ for } x > 0 \end{cases} \qquad \text{(SYMMETRICAL UNIT-STEP FUNCTION)}$$

$$\left.\begin{aligned} U_-(x) &= \begin{cases} 0 \text{ for } x < 0 \\ 1 \text{ for } x \geq 0 \end{cases} \\ U_+(x) &= \begin{cases} 0 \text{ for } x \leq 0 \\ 1 \text{ for } x > 0 \end{cases} \end{aligned}\right\} \qquad \text{(ASYMMETRICAL UNIT-STEP FUNCTIONS)} \qquad (21.9\text{-}1)$$

(see also Fig. 21.9-1).

* The notations employed to denote the various unit-step functions vary; use caution when referring to different texts.

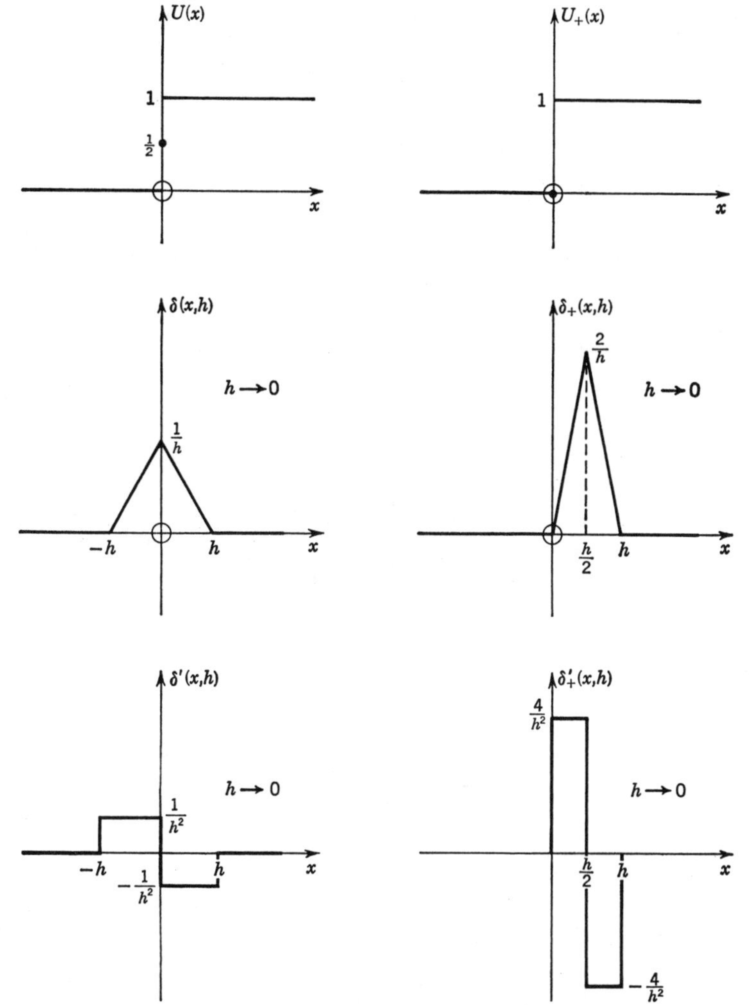

FIG. 21.9-1. The unit-step functions $U(x)$ and $U_+(x)$ and approximations to the impulse functions $\delta(x)$, $\delta_+(x)$, $\delta'(x)$, and $\delta'_+(x)$.

Note

$$U(0-0) = 0 \qquad U(0+0) = 1 \qquad U(e^x) \equiv 1 \tag{21.9-2}$$

$$U[(x-a)(x-b)] = U[x - \max(a, b)] + U[\min(a, b) - x] \tag{21.9-3}$$

Every step function can be expressed (except possibly at its discontinuities $x = x_k$) as a sum of the form $\sum_k a_k U(x - x_k)$, $\sum_k a_k U_-(x - x_k)$, or $\sum_k a_k U_+(x - x_k)$ (EXAMPLES: $\operatorname{Sgn} x \equiv 2U(x) - 1$; and the "jump" functions of Sec. 20.4-5c).

(b) Approximation of Step Functions by Continuous Functions.

$$U(x) = \lim_{\alpha \to \infty} \left[\frac{1}{2} + \frac{1}{\pi} \arctan (\alpha x)\right] \tag{21.9-4}$$

$$U(x) = \lim_{\alpha \to \infty} \tfrac{1}{2}[\text{erf } (\alpha x) + 1] \tag{21.9-5}$$

$$U(x) = \lim_{\alpha \to \infty} (2^{-e^{-\alpha x}}) \tag{21.9-6}$$

$$U(x) = \lim_{\alpha \to \infty} \frac{1}{\pi} \int_{-\infty}^{\alpha x} \frac{\sin \tau}{\tau}\, d\tau \tag{21.9-7}$$

(c) Fourier-integral Representations (see also Sec. 4.11-6). The complex contour integral $\frac{1}{2\pi i} \int_{-\infty}^{\infty} \frac{e^{i\omega t}}{\omega}\, d\omega$ is respectively equal to $U(t)$ or $-U(-t)$ if the integration contour passes *below* or *above* the origin. The Cauchy principal value of the integral (Sec. 4.6-2*b*) equals $U(x) - \frac{1}{2}$. Note also

$$U(t) = \frac{1}{2\pi} \int_{-\infty}^{\infty} \frac{\sin \omega t}{\omega}\, d\omega + \frac{1}{2} \tag{21.9-8}$$

$$U(1 - t) = \frac{1}{\pi} \int_{-\infty}^{\infty} \frac{\sin \omega \cos \omega t}{\omega}\, d\omega \qquad (t \geq 0) \tag{21.9-9}$$

21.9-2. The Symbolic Dirac Delta Function. (a) The **symmetrical unit-impulse function** or **Dirac delta function** $\delta(x)$ of a real variable x is "defined" by

$$\int_a^b f(\xi)\delta(\xi - X)\, d\xi = \begin{cases} 0 \text{ if } X < a \text{ or } X > b \\ \tfrac{1}{2}f(X) \text{ if } X = a \text{ or } X = b \\ f(X) \text{ if } a < X < b \end{cases} \qquad (a < b) \tag{21.9-10a}$$

where $f(x)$ is an arbitrary function continuous for $x = X$. More generally, one "defines" $\delta(x)$ by

$$\int_a^b f(\xi)\delta(\xi - X)\, d\xi = \begin{cases} 0 \text{ if } X < a \text{ or } X > b \\ \tfrac{1}{2}f(X + 0) \text{ if } X = a \\ \tfrac{1}{2}f(X - 0) \text{ if } X = b \\ \tfrac{1}{2}[f(X - 0) + f(X + 0)] \text{ if } a < X < b \end{cases} \qquad (a < b) \tag{21.9-10b}$$

where $f(x)$ is an arbitrary function of bounded variation in a neighborhood of $x = X$. $\delta(x)$ is *not* a true function, since the "definition" (10) implies the inconsistent relations

$$\delta(x) = 0 \qquad (x \neq 0) \qquad \int_{-\infty}^{\infty} \delta(\xi)\, d\xi = 1 \tag{21.9-10c}$$

$\delta(x)$ is a "symbolic function" permitting the formal representation of the functional identity transformation $f(\xi) \to f(x)$ as an integral transformation (Sec. 15.3-1*a*). The "formal" use of $\delta(x)$ furnishes a convenient notation permitting suggestive generalizations of many mathematical relations (see also Secs. 8.5-1, 15.3-1a, 15.5-1, and 18.3-1). Although no functions having the exact properties (10) exist, it is possible to "approximate" $\delta(x)$ by functions exhibiting the desired properties to any degree of approximation (Sec. 21.9-4).

One can usually avoid the use of impulse functions by introducing Stieltjes integrals (Sec. 4.6-17); thus

$$\int_a^b f(\xi)\delta(\xi - X)\,d\xi = \int_a^b f(\xi)\,dU(\xi - X)$$

It is possible to introduce a generalizing redefinition of the concepts of "function" and "differentiation" (*Schwarz's theory of distributions*, Refs. 21.13 to 21.18). Otherwise, *mathematical arguments involving the use of impulse functions should be regarded as heuristic and require rigorous verification* (see also Sec. 8.5-1).

(b) "Formal" Relations Involving $\delta(x)$.

$$\delta(ax) = \frac{1}{a}\,\delta(x) \quad (a > 0) \qquad \delta(-x) = \delta(x) \tag{21.9-11}$$

$$f(x)\delta(x - a) = \tfrac{1}{2}[f(a - 0) + f(a + 0)]\delta(x - a) \qquad x\delta(x) = 0 \tag{21.9-12}$$

$$\delta(x^2 - a^2) = \frac{1}{2a}\,[\delta(x - a) + \delta(x + a)] \qquad (a > 0) \tag{21.9-13}$$

$$\int_{-\infty}^{\infty} \delta(a - x)\delta(x - b)\,dx = \delta(a - b) \tag{21.9-14}$$

21.9-3. "Derivatives" of Step Functions and Impulse Functions (see also Sec. 8.5-1). Equations (10) and (19) and also the relation $\mathcal{L}[\delta(t - a)] = e^{-as} = s\mathcal{L}[U(t - a)]$ $(a > 0)$ suggest the symbolic relationship

$$\delta(x) = \frac{d}{dx}\,U(x) \tag{21.9-15}$$

The impulse functions $\delta'(x), \delta''(x), \ldots, \delta^{(r)}(x)$ are "defined" by

$$\int_a^b f(\xi)\delta^{(r)}(\xi - X)\,d\xi = \begin{cases} 0 \text{ if } X < a \text{ or } X > b \\ \tfrac{1}{2}(-1)^r f^{(r)}(X + 0) \text{ if } x = a \\ \tfrac{1}{2}(-1)^r f^{(r)}(X - 0) \text{ if } x = b \\ \tfrac{1}{2}(-1)^r[f^{(r)}(X - 0) + f^{(r)}(X + 0)] \text{ if } a < X < b \end{cases} \qquad (a < b) \tag{21.9-16}$$

for an arbitrary function $f(x)$ such that the unilateral limits $f^{(r)}(X - 0)$ and $f^{(r)}(X + 0)$ exist. The functions $\delta^{(r)}(\xi - X)$ are kernels of linear integral transformations (Sec. 15.2-1) representing repeated differentiations. Note also the symbolic relation

$$x^r\delta^{(r)}(x) = (-1)^r r!\delta(x) \qquad (r = 0, 1, 2, \ldots) \tag{21.9-17}$$

21.9-4. Approximation of Impulse Functions (see also Fig. 21.9-1). **(a) Continuously Differentiable Functions Approximating** $\delta(x)$. One can approximate $\delta(x)$ by the continuously differentiable functions

$$\delta(x, \alpha) = \frac{\alpha}{\pi(\alpha^2x^2 + 1)} \qquad \text{as } \alpha \to \infty \qquad (21.9\text{-}18a)$$

$$\delta(x, \alpha) = \frac{\alpha}{\sqrt{\pi}} e^{-\alpha^2x^2} \qquad \text{as } \alpha \to \infty \qquad (21.9\text{-}18b)$$

$$\delta(x, \alpha) = \frac{\alpha}{\pi}\frac{\sin \alpha x}{\alpha x} \qquad \text{as } \alpha \to \infty \qquad (21.9\text{-}18c)$$

in the sense that $\lim_{\alpha \to \infty} \delta(x, \alpha) = 0 \; (x \neq 0)$, and

$$\lim_{\alpha \to \infty} \int_{-\infty}^{\infty} f(\xi)\delta(x - \xi, \alpha)\, d\xi = \tfrac{1}{2}[f(x - 0) + f(x + 0)]$$

wherever $f(x - 0)$ and $f(x + 0)$ exist; note also $\lim_{\alpha \to \infty} \int_{-\infty}^{\infty} \delta(\xi, \alpha)\, d\xi = 1$.

Integration of the approximating functions (18) yields the corresponding step-function approximations (4) and (5). $\mathcal{L}[\delta(x - a, \alpha)]$ $(a > 0)$ converges to $e^{-as} = \mathcal{L}[\delta(x - a)]$ as $\alpha \to \infty$ for each function (18).

(b) Discontinuous Functions Approximating $\delta(x)$. $\delta(x)$ is often approximated by the central-difference coefficient (Sec. 20.4-3)

$$\delta(x, h) = \frac{U(x + h) - U(x - h)}{2h} \qquad \text{as } h \to 0 \qquad (21.9\text{-}19)$$

Note also

$$\lim_{\omega \to \infty} \frac{1}{\pi} \int_{-\infty}^{\infty} f(\xi) \frac{\sin \omega(X - \xi)}{X - \xi}\, d\xi = \frac{1}{2}[f(X - 0) + f(X + 0)]$$

$$(-\infty < X < \infty) \qquad \text{(Dirichlet's integral formula)} \qquad (21.9\text{-}20)$$

and

$$\lim_{\alpha \to 1-0} \frac{1}{2\pi} \int_{-\pi}^{\pi} f(\xi) \frac{1 - \alpha^2}{1 - 2\alpha \cos (X - \xi) + \alpha^2}\, d\xi = \frac{1}{2}[f(X - 0) + f(X + 0)]$$

$$(-\pi < X < \pi) \qquad (21.9\text{-}21)$$

if $f(x)$ is of bounded variation in a neighborhood of $x = X$ (see also Sec. 4.11-6).

(c) Functions Approximating $\delta'(x), \delta''(x), \ldots, \delta^{(r)}(x)$. Successive differentiations of Eq. (18a) yield the approximation functions

$$\delta'(x, \alpha) = -\frac{2}{\pi} \frac{\alpha^3 x}{(\alpha^2x^2 + 1)^2} \qquad \text{as } \alpha \to \infty \qquad (21.9\text{-}22)$$

$$\delta^{(r)}(x, \alpha) = \frac{(-1)^r r!}{\pi} \frac{\sin \left[(r + 1) \arctan \frac{\alpha}{x}\right]}{(x^2 + \alpha^2)^{(r+1)/2}} \qquad \text{as } \alpha \to 0$$

$$(r = 0, 1, 2, \ldots) \qquad (21.9\text{-}23)$$

Note also

$$\delta'(x, h) = \frac{U(x + h) - 2U(x) + U(x - h)}{h^2} \qquad \text{as } h \to 0 \qquad (21.9\text{-}24)$$

21.9-5. Fourier-integral Representations. Note the formal relations

$$\delta(x - X) = \frac{1}{2\pi} \int_{-\infty}^{\infty} e^{-i\omega X} e^{i\omega x} \, d\omega \tag{21.9-25}$$

$$\delta^{(r)}(x - X) = \frac{1}{2\pi} \int_{-\infty}^{\infty} (i\omega)^r e^{-i\omega X} e^{i\omega x} \, d\omega \tag{21.9-26}$$

$$\frac{1}{2} [\delta(x - X) + \delta(x + X)] = \frac{1}{\pi} \int_0^{\infty} \cos \omega X \cos \omega x \, d\omega \tag{21.9-27}$$

21.9-6. Asymmetrical Impulse Functions (see also Secs. 8.5-1 and 9.4-3). **(a)** The **asymmetrical impulse functions** $\delta_+(x)$, $\delta'_+(x)$, . . . , $\delta_+^{(r)}(x)$ are "defined" by

$$\int_{a+0}^{b} f(\xi)\delta_+(\xi - X) \, d\xi = \begin{Bmatrix} 0 & \text{if } X < a \text{ or } X \geq b \\ f(X + 0) & \text{if } a \leq X < b \end{Bmatrix} \quad (a < b) \tag{21.9-28}$$

$$\int_{a+0}^{b} f(\xi)\delta_+^{(r)}(\xi - X) \, d\xi = \begin{Bmatrix} 0 \text{ if } X < a \text{ or } X \geq b \\ (-1)^r f^{(r)}(X + 0) \text{ if } a \leq X < b \end{Bmatrix}$$
$$(a < b; r = 1, 2, \ldots) \tag{21.9-29}$$

One may write

$$\delta_+(x) \equiv 2\delta(x)U(x) \equiv \frac{d}{dx} U_+(x) \tag{21.9-30}$$

One way to obtain approximation functions for $\delta_+(x)$ is to substitute the approximation functions of Sec. 21.9-4 into one of the relations (30), e.g.,

$$\delta_+(x, h) = \frac{U(x) - U(x - h)}{h} \qquad \text{as } h \to 0 \tag{21.9-31}$$

Note also

$$\delta'_+(x, h) = 4 \frac{U(x) - 2U(x - h/2) + U(x - h)}{h^2} \qquad \text{as } h \to 0 \tag{21.9-32}$$

(b) One may introduce $\delta_+(-x) \equiv \delta_-(x)$ as a second asymmetrical impulse function corresponding to the "derivative" of the asymmetrical step function $U_-(x)$ (Sec. 21.9-1).

21.9-7. Multidimensional Delta Functions (see also Sec. 15.5-1). For an n-dimensional space of "points" $(x^1, x^2, \ldots, x^n)$ with a volume element defined as

$$dV = dV(x^1, x^2, \ldots, x^n) = \sqrt{|g(x^1, x^2, \ldots, x^n)|} \, dx^1 \, dx^2 \cdots dx^n$$

(Sec. 16.10-10), the **n-dimensional delta function** $\delta(x^1, \xi^1; x^2, \xi^2; \ldots ; x^n, \xi^n)$ must satisfy

$$\int_V f(\xi^1, \xi^2, \ldots, \xi^n)\delta(X^1, \xi^1; X^2, \xi^2; \ldots ; X^n, \xi^n) \, dV(\xi^1, \xi^2, \ldots, \xi^n)$$
$$= f(X^1, X^2, \ldots, X^n) \tag{21.9-33}$$

for every "point" $(X^1, X^2, \ldots, X^n)$ in V where $f(x^1, x^2, \ldots, x^n)$ is continuous. *Note that the definition of $\delta(x^1, \xi^1; x^2, \xi^2; \ldots ; x^n, \xi^n)$ depends on the coordinate system used* and is meaningless wherever $dV = 0$. In particular, for rectangular cartesian coordinates x, y, z, one has $dV = dx \, dy \, dz$, and

$$\delta(x, \xi; y, \eta; z, \zeta) \equiv \delta(x - \xi)\delta(y - \eta)\delta(z - \zeta) \tag{21.9-34}$$

21.10. REFERENCES AND BIBLIOGRAPHY

21.1. Abramowitz, M., and I. A. Stegun (eds.): *Handbook of Mathematical Functions*, National Bureau of Standards Applied Mathematics Series 55, Washington, D.C., 1964.

21.2. Byrd, P. F., and M. D. Friedman: *Handbook of Elliptic Integrals*, Springer, Berlin, 1954.

21.3. Erdélyi, A.: *Higher Transcendental Functions*, vols. 1 and 2 (Bateman Project), McGraw-Hill, New York, 1953.

21.4. Jahnke, E., and F. Emde: *Tables of Functions with Formulae and Curves*, Dover, New York, 1945.

21.5. Hurwitz, A., and R. Courant: *Vorlesungen über allgemeine Funktionentheorie und elliptische Funktionen*, 4th ed., Springer, Berlin, 1964.

21.6. McLachlan, N. W.: *Bessel Functions for Engineers*, Oxford, Fair Lawn, N.J., 1946.

21.7. Schäfke, F. W.: *Einführung in die Theorie der speziellen Funktionen der mathematischen Physik*, Springer, Berlin, 1963.

21.8. Sneddon, I. N.: *The Special Functions of Physics and Chemistry*, Oliver & Boyd, Edinburgh, 1956.

21.9. Whittaker, E. T., and G. N. Watson: *Modern Analysis*, Macmillan, New York, 1943.

21.10. ———: *A Course in Modern Analysis*, Cambridge, New York, 1946.

21.11. Oberhettinger, F., and W. Magnus: *Formulas and Theorems for the Functions of Mathematical Physics*, Chelsea, New York, 1954; 3d ed., Springer, Berlin, 1966.

21.12. Tricomi, F. G.: *Elliptische Funktionen*, Akademische Verlagsgesellschaft, Leipzig, 1948.

(See also the articles by J. Lense and J. Meixner in vol. I of *Handbuch der Physik*, Springer, Berlin, 1956; and see also Sec. 10.6-2).

Generalized Functions and the Theory of Distributions

21.13. Arsac, J.: *Fourier Transforms and the Theory of Distributions*, Prentice-Hall, Englewood Cliffs, N.J., 1966.

21.14. Friedman, A.: *Generalized Functions and Partial Differential Equations*, Prentice-Hall, Englewood Cliffs, N.J., 1956.

21.15. Gelfand, I. M., et al.: *Generalized Functions*, Academic, New York, 1964.

21.16. Lighthill, M. J.: *Introduction to Fourier Analysis and Generalized Functions*, Cambridge, New York, 1958.

21.17. Schwartz, L.: *Théorie des Distributions*, 2d ed., Hermann & Cie, Paris, 1957.

21.18. Zemanian, A. H.: *Distribution Theory and Transform Analysis*, McGraw-Hill, New York, 1965.

APPENDIX **A**

FORMULAS DESCRIBING PLANE FIGURES AND SOLIDS

A-1. The Trapezoid (sides a, b, c, d; a and b are parallel; the **altitude** h is the distance between a and b). The area S is given by

$$S = \frac{1}{2}(a + b)h \qquad h = \frac{2}{a - b}\sqrt{s(s - a + b)(s - c)(s - d)}$$

$$s = \frac{a - b + c + d}{2} \qquad \text{(A-1)}$$

The trapezoid is a **parallelogram** if $a = b$ and a **rhombus** if $a = b = c = d$.

A-2. Regular Polygons (length of side equal to a)

Number of sides, n	Regular polygon	Radius of circumscribed circle, $r = a \Big/ 2 \sin \frac{\pi}{n}$	Radius of inscribed circle, $\rho = a \Big/ 2 \tan \frac{\pi}{n}$	Area, $S = \frac{1}{2} n a \rho$
3	Equilateral triangle	$\frac{a}{3}\sqrt{3}$	$\frac{a}{6}\sqrt{3}$	$\frac{a^2}{4}\sqrt{3}$
4	Square	$\frac{a}{2}\sqrt{2}$	$\frac{a}{2}$	a^2
5	Regular pentagon	$a\sqrt{\frac{1}{2} + \frac{1}{10}\sqrt{5}}$	$a\sqrt{\frac{1}{4} + \frac{1}{10}\sqrt{5}}$	$\frac{a^2}{4}\sqrt{25 + 10\sqrt{5}}$
6	Regular hexagon	a	$\frac{a}{2}\sqrt{3}$	$\frac{3}{2}a^2\sqrt{3}$
8	Regular octagon	$a\sqrt{1 + \frac{1}{2}\sqrt{2}}$	$\frac{a}{2}\left(1 + \sqrt{2}\right)$	$2a^2\left(1 + \sqrt{2}\right)$
10	Regular decagon	$\frac{a}{2}\left(1 + \sqrt{5}\right)$	$\frac{a}{2}\sqrt{5 + 2\sqrt{5}}$	$\frac{5}{2}a^2\sqrt{5 + 2\sqrt{5}}$

A-3. The Circle (radius r, see also Sec. 2.5-1). **(a)**

$$\text{Circumference} = 2\pi r \qquad \text{area} = \pi r^2$$

(b) A central angle of φ radians subtends

an *arc* of length $r\varphi$ a *sector* of area $\frac{1}{2}r^2\varphi$

a *chord* of length $2r \sin \frac{\varphi}{2}$ a *segment* of area $\frac{1}{2}r^2(\varphi - \sin \varphi)$

(c) The area between a circle of radius r_1 and an enclosed (not necessarily concentric) circle of radius r_2 is $\pi(r_1 + r_2)(r_1 - r_2)$.

A-4. Prisms, Pyramids, Cylinders, and Cones. **(a)** *Volume of a prism or cylinder* (bounded by a plane parallel to the base of area S_1, altitude h) hS_1

(b) *Volume of a pyramid or cone* (base area S_1, altitude h) $\frac{1}{3}hS_1$

(c) *Volume of the frustrum of a pyramid or cone* (bounded by parallel planes; base areas S_1, S_2, altitude h) $\frac{1}{3}h(S_1 + \sqrt{S_1S_2} + S_2)$

(d) *Curved surface area of a right circular cone* (base radius r, altitude h) $\pi r\sqrt{r^2 + h^2}$

A-5. Solids of Revolution

No.	Solid	Surface area	Volume
1	**Sphere of radius** r	$4\pi r^2$	$\frac{4}{3}\pi r^3$
2	**Oblate spheroid** $\left(\text{axes } 2a \geq 2b,\ \epsilon = \sqrt{1 - \frac{b^2}{a^2}}\right)$	$2\pi a^2 + \pi \frac{b^2}{\epsilon} \log_e \frac{1+\epsilon}{1-\epsilon}$	$\frac{4}{3}\pi a^2 b$
3	**Prolate spheroid** $\left(\text{axes } 2a \geq 2b,\ \epsilon = \sqrt{1 - \frac{b^2}{a^2}}\right)$	$2\pi b^2 + 2\pi \frac{ab}{\epsilon} \arcsin \epsilon$	$\frac{4}{3}\pi a b^2$
4	**Torus (anchor ring)** circle of radius r rotated about an axis at a distance R from the center	$4\pi^2 Rr$	$2\pi^2 Rr^2$
5	**Zone or segment of a sphere of radius** r between parallel planes at a distance h; base radii r_1, r_2	$2\pi rh + \pi(r_1^2 + r_2^2)$	$\frac{\pi}{6} h(3r_1^2 + 3r_2^2 + h^2)$

A-6. The Five Regular Polyhedra (length of side equal to a; the respective numbers F of surfaces, E of vertices, and K of edges are related by $E + F - K = 2$).

Regular polyhedron	Number and type of surfaces	Radius of circumscribed sphere	Radius of inscribed sphere	Surface area	Volume
Tetrahedron	4 equilateral triangles	$\frac{a}{4}\sqrt{6}$	$\frac{a}{12}\sqrt{6}$	$a^2\sqrt{3}$	$\frac{a^3}{12}\sqrt{2}$
Cube	6 squares	$\frac{a}{2}\sqrt{3}$	$\frac{a}{2}$	$6a^2$	a^3
Octahedron	8 equilateral triangles	$\frac{a}{2}\sqrt{2}$	$\frac{a}{6}\sqrt{6}$	$2a^2\sqrt{3}$	$\frac{a^3}{3}\sqrt{2}$
Dodecahedron	12 regular pentagons	$\frac{a}{4}(1+\sqrt{5})\sqrt{3}$	$\frac{a}{4}\sqrt{10+\frac{22}{\sqrt{5}}}$	$3a^2\sqrt{5(5+2\sqrt{5})}$	$\frac{a^3}{4}(15+7\sqrt{5})$
Icosahedron	20 equilateral triangles	$\frac{a}{4}\sqrt{2(5+\sqrt{5})}$	$\frac{a}{4\sqrt{3}}(3+\sqrt{5})$	$5a^2\sqrt{3}$	$\frac{5}{12}a^3(3+\sqrt{5})$

APPENDIX **B**

PLANE AND SPHERICAL TRIGONOMETRY

PLANE TRIGONOMETRY

B-1. Introduction. Plane trigonometry describes relations between the sides and angles of plane triangles in terms of trigonometric functions (Secs. 21.2-1 to 21.2-4); note that all plane figures bounded by straight lines may be regarded as combinations of triangles. Since all plane triangles may be resolved into right triangles, *the most important trigonometric relations are those relating the sides and angles of right triangles.*

B-2. Right Triangles. In every right triangle (Fig. B-1) with sides a, b and hypotenuse c,

$$
\boxed{\begin{array}{ll}
A + B = 90^\circ \quad a^2 + b^2 = c^2 & (\text{THEOREM OF PYTHAGORAS}) \\
\sin A = \cos B = \dfrac{a}{c} & \sin B = \cos A = \dfrac{b}{c} \\
\tan A = \cot B = \dfrac{a}{b} & \tan B = \cot A = \dfrac{b}{a}
\end{array}} \tag{B-1}
$$

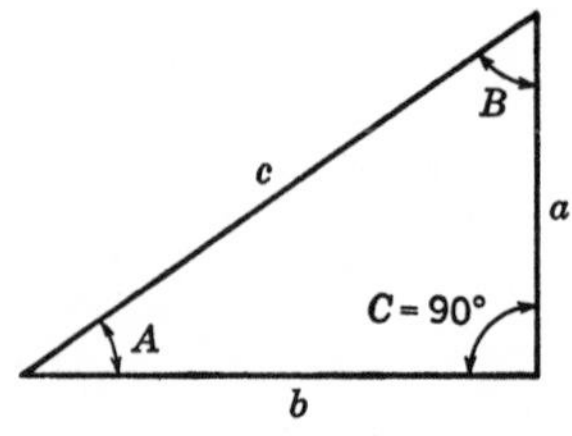

FIG. B-1. Right triangle.

B-3. Properties of Plane Triangles. In every plane triangle (Fig. B-2), the sum of the angles equals 180 deg. The sum of any two sides is greater than the third, and the greater of two sides opposes the greater

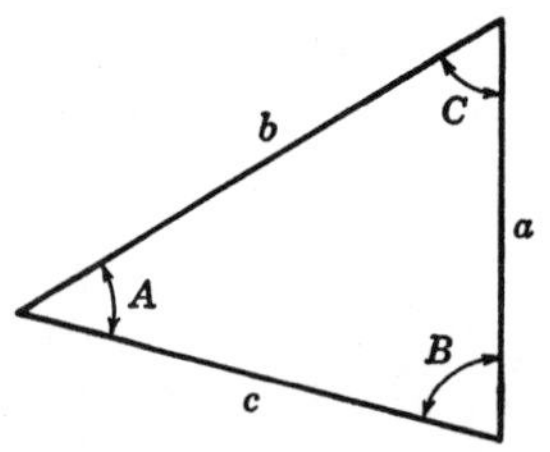

FIG. B-2. Oblique triangle.

of two angles. A plane triangle is *uniquely determined* (except for symmetric images) by

1. Three sides
2. Two sides and the included angle
3. One side and two angles
4. Two sides and the angle opposite the greater side

In every plane triangle, the three bisectors of angles intersect in the center M of the inscribed circle. The three perpendicular bisectors of the sides intersect in the center F of the circumscribed circle. The three medians intersect in the center of gravity G of the triangle. The three altitudes intersect in a point H collinear with the last two points, so that $\overrightarrow{HG}:\overrightarrow{GF} = 2$. The mid-points of the sides, the footpoints of the perpendiculars from the vertices to the sides, and the mid-points of the straight-line segments joining H to each vertex lie on a circle whose radius is half that of the circumscribed circle (**nine-point circle,** or **Feuerbach circle**). The center of the nine-point circle is the mid-point of the straight-line segment HF.

B-4. Formulas for Triangle Computations. In the following relations, A, B, C are the angles opposite the respective sides a, b, c of a plane triangle (Fig. B-2). The triangle area is denoted by S; r and ρ are the respective radii of the circumscribed and inscribed circles, and $s = \frac{1}{2}(a + b + c)$. *Additional formulas are obtained by simultaneous cyclic permutation of A, B, C and a, b, c.* Table B-1 permits the computation of the sides and angles of any plane triangle from three suitable sides and/or angles.

$$a^2 = b^2 + c^2 - 2bc \cos A = (b + c)^2 - 4bc \cos^2 \frac{A}{2} \quad \text{(LAW OF COSINES)} \tag{B-2}$$

$$\frac{a}{\sin A} = \frac{b}{\sin B} = \frac{c}{\sin C} = 2r \quad \text{(LAW OF SINES)} \tag{B-3}$$

$$c = a \cos B + b \cos A \quad \text{(PROJECTION THEOREM)} \tag{B-4}$$

$$\left.\begin{aligned} \sin \frac{A}{2} &= +\sqrt{\frac{(s-b)(s-c)}{bc}} \qquad \cos \frac{A}{2} = +\sqrt{\frac{s(s-a)}{bc}} \\ \tan \frac{A}{2} &= +\sqrt{\frac{(s-b)(s-c)}{s(s-a)}} \\ \sin A &= +\frac{2}{bc}\sqrt{s(s-a)(s-b)(s-c)} \end{aligned}\right\} \tag{B-5}$$

$$\frac{b+c}{b-c} = \frac{\tan \dfrac{B+C}{2}}{\tan \dfrac{B-C}{2}} \tag{B-6}$$

$$(b + c) \sin \frac{A}{2} = a \cos \frac{B-C}{2} \qquad (b - c) \cos \frac{A}{2} = a \sin \frac{B-C}{2} \tag{B-7}$$

$$\begin{aligned} \rho &= (s-a) \tan \frac{A}{2} = (s-b) \tan \frac{B}{2} = (s-c) \tan \frac{C}{2} \\ &= 4r \sin \frac{A}{2} \sin \frac{B}{2} \sin \frac{C}{2} = s \tan \frac{A}{2} \tan \frac{B}{2} \tan \frac{C}{2} \end{aligned} \tag{B-8}$$

$$s = \frac{a+b+c}{2} = 4r \cos \frac{A}{2} \cos \frac{B}{2} \cos \frac{C}{2} \tag{B-9}$$

$$\begin{aligned} S &= \tfrac{1}{2}ab \sin C = \tfrac{1}{2}bc \sin A = \tfrac{1}{2}ac \sin B \\ &= [s(s-a)(s-b)(s-c)]^{1/2} \\ &= 2r^2 \sin A \sin B \sin C = \frac{abc}{4r} = \rho s \end{aligned} \tag{B-10}$$

$$\text{Length of } \textit{altitude } h_a = \frac{bc}{2r} \qquad \left(S = \frac{1}{2} a h_a\right)$$

$$\text{Length of } \textit{angular bisector } w_a = \frac{1}{b+c} \sqrt{bc(a+b+c)(b+c-a)}$$

$$\text{Length of } \textit{median } m_a = \tfrac{1}{2} \sqrt{2b^2 + 2c^2 - a^2}$$

Table B-1. Solution of Plane Triangles. *Obtain all other cases by cyclic permutation* (refer to formulas of Sec. B-4 and to Fig. B-2)

Case	Given	Example	**Formulas used** Note $A + B + C = 180°$	**Conditions for the existence of a solution** (see also Sec. B-3)
1	Three sides	a, b, c	A, B, C from (2) or (3)	Sum of two sides must be greater than the third
2	Two sides and the included angle	b, c, A	$\frac{B+C}{2} = 90° - \frac{A}{2}; \frac{B-C}{2}$ from (6) or (7), hence B and C; or B, C from (3) and (4): $\tan B = \frac{b \sin A}{c - b \cos A}$ a from (3) or (4)	
3	One side and two angles	a, B, C	b, c from (3); $A = 180° - B - C$	
4	Two sides and an opposite angle	b, c, B	From (3), $a = \frac{b \sin A}{\sin B}$, $\sin C = \frac{c \sin B}{b}$; $A = 180° - B - C$	Problem has one solution if $b \geq c$; two solutions if $b < c$, $c \sin B < b$

SPHERICAL TRIGONOMETRY

B-5. Spherical Triangles: Introduction. On the surface of a sphere, the shortest distance between two points is measured along a **great circle,** i.e., a circle whose plane passes through the center of the sphere (geodesic, Sec. 17.3-12). The vertices of a **spherical triangle** are the intersections of three directed straight lines passing through the center of the sphere and the spherical surface. The **sides** a, b, c of the spherical triangle are those three angles between the three directed straight lines which are less than 180 deg. Corresponding to each triangle side, there is a great-circle segment on the surface of the sphere (Fig. B-3). The **angles** A, B, C of the spherical triangle opposite the sides a, b, c, respectively, are the angles less than 180 deg between the great-circle segments corresponding to the triangle sides, or the corresponding angles between the three planes defined by the three given straight lines.

Spherical trigonometry is the study of relations between the sides and angles of spherical triangles (e.g., on the surface of the earth and on the celestial sphere).

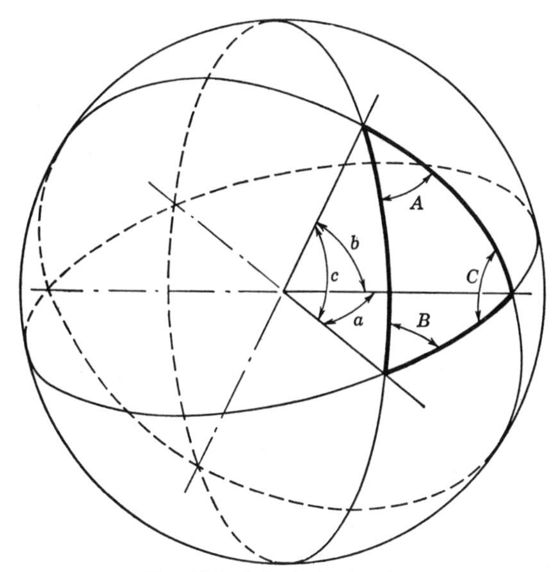

FIG. B-3. Spherical triangle.

In many problems, physicists and engineers will prefer the use of the rotation transformations (Sec. 14.10-1) to the use of spherical trigonometry.

B-6. Properties of Spherical Triangles. Each side or angle of a spherical triangle is, by definition, smaller than 180 deg. The geometry on the surface of a sphere is non-Euclidean (see also Sec. 17.3-13); in every spherical triangle, the sum of the sides will be between 0 and 360 deg, and the sum of the angles will be between 180 and 540 deg. In every spherical triangle, the greater of two sides opposes the greater of two angles. The sum of any two sides is greater than the third, and the sum of any two angles is less than 180 deg plus the third angle. A spherical triangle is *uniquely determined* (except for symmetric images) by

1. Three sides
2. Three angles
3. Two sides and the included angle
4. Two angles and the included side
5. Two sides and an opposite angle, given that the other opposite angle is less than, equal to, or greater than 90 deg
6. Two angles and an opposite side, given that the other opposite side is less than, equal to, or greater than 90 deg

NOTE: In every spherical triangle, it is possible to define great circles as perpendicular bisectors of sides, bisectors of angles, medians, and altitudes. The planes of the three great circles of each type intersect in a straight line.

In analogy to the circumscribed circle of a plane triangle, there exists a *circumscribed right circular cone* containing the three straight lines defining the triangle; the axis of this cone is the straight line formed by the intersection of the planes of the perpendicular bisectors. There is also an *inscribed right circular cone* touching the three planes corresponding to the spherical triangle; the axis of this cone is the straight line formed by the intersection of the planes of the bisectors of the angles. The "radius" r of the circumscribed circle and the "radius" ρ of the inscribed circle are *angles* defined as half the vertex angles of the respective cones.

Given the radius R of the sphere, the *area* S_R of a spherical triangle is given by

$$S_R = R^2\epsilon \tag{B-11}$$

where ϵ is the **spherical excess**

$$\epsilon = A + B + C - \pi \tag{B-12}$$

measured in *radians*. The quantity $d = 2\pi - (a + b + c)$ is called **spherical defect.**

The **polar triangle** corresponding to a given spherical triangle is defined by three directed straight lines perpendicular to the planes associated with the sides of the original triangle. The sides of the polar triangle are equal to the supplements of the corresponding angles of the original triangle, and conversely. Thus every theorem or formula dealing with the sides and angles of the original triangle may be transformed into one dealing with the angles and sides of the polar triangle.

B-7. The Right Spherical Triangle.

In a right spherical triangle, at least one angle, C, say, is equal to 90 deg; the opposite side, c, is called the **hypotenuse.** All important relations between the sides and angles of the right spherical triangle may be derived from *Napier's rules*, two convenient aids to memory:

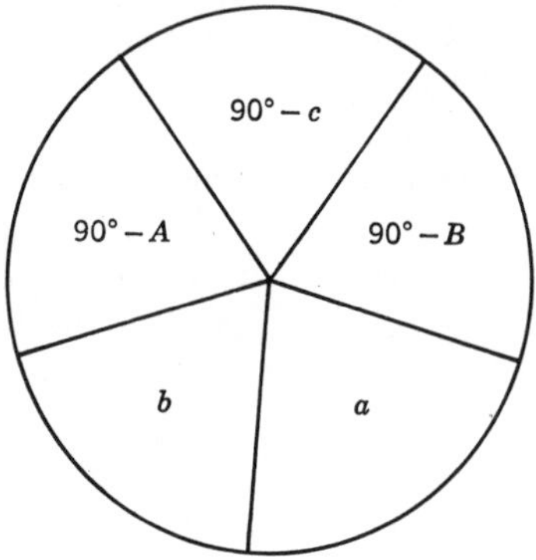

FIG. B-4. Napier's rules.

Napier's Rules: In the diagram of Fig. B-4, *the sine of any of the angles shown is equal*

1. *To the product of the tangents of the two angles adjoining it in the diagram*
2. *To the product of the cosines of the two angles opposing it in the diagram*

EXAMPLE: To compute the sides and angles of a right spherical triangle with the hypotenuse c, given c and a.

This problem has a solution only if $\sin a \leq \sin c$; then

$$\cos b = \frac{\cos c}{\cos a} \qquad \cos B = \frac{\tan a}{\tan c} \qquad \sin A = \frac{\sin a}{\sin c}$$

NOTE: If a is less than, equal to, or greater than 90 deg, so is A, and conversely. If b is less than, equal to, or greater than 90 deg, so is B.

If a and A are given, the problem has a solution only if the above condition is satisfied and $\sin a \leq \sin A$; unless $a = A$, there are two solutions. The situation is analogous if b and B are given.

If A and B are given, the problem has a solution only if $90 < A + B < 270$ deg and $-90 \text{ deg} < A - B < 90$ deg (Sec. B-6).

A spherical triangle having a side equal to 90 deg is called a **quadrantal triangle** and may be treated as the polar triangle (Sec. B-6) of a right spherical triangle.

For all problems involving the spherical-triangle computations (right or oblique triangles), it is strongly recommended that a sketch be drawn which roughly indicates whether the various angles and sides will be less than, equal to, or greater than 90 *deg.*

B-8. Formulas for Triangle Computations (see also Fig. B-3). In the following relations, A, B, C are the angles opposite the respective sides a, b, c of a spherical triangle. The respective "radii" of the circumscribed and inscribed cones are denoted by r and ρ. *Additional formulas are obtained by simultaneous cyclic permutation of A, B, C and a, b, c.* Table B-2 permits the computation of the sides or angles of any spherical triangle from three suitable sides and/or angles. The inequalities noted in Sec. B-6 must be observed in order to avoid ambiguous results in triangle computations.

$$\frac{\sin a}{\sin A} = \frac{\sin b}{\sin B} = \frac{\sin c}{\sin C} \qquad \text{(LAW OF SINES)} \qquad \text{(B-13)}$$

$$\cos a = \cos b \cos c + \sin b \sin c \cos A \qquad \text{(LAW OF COSINES FOR THE SIDES)} \qquad \text{(B-14)}$$

$$\cos A = -\cos B \cos C + \sin B \sin C \cos a \qquad \text{(LAW OF COSINES FOR THE ANGLES)} \qquad \text{(B-15)}$$

$$\left.\begin{aligned}
\tan \frac{b+c}{2} \cos \frac{B+C}{2} &= \tan \frac{a}{2} \cos \frac{B-C}{2} \\
\tan \frac{b-c}{2} \sin \frac{B+C}{2} &= \tan \frac{a}{2} \sin \frac{B-C}{2} \\
\tan \frac{B+C}{2} \cos \frac{b+c}{2} &= \cot \frac{A}{2} \cos \frac{b-c}{2} \\
\tan \frac{B-C}{2} \sin \frac{b+c}{2} &= \cot \frac{A}{2} \sin \frac{b-c}{2}
\end{aligned}\right\} \qquad \text{(NAPIER'S ANALOGIES)} \qquad \text{(B-16)}$$

$$\left.\begin{aligned}\sin\frac{A}{2}\sin\frac{b+c}{2} &= \sin\frac{a}{2}\cos\frac{B-C}{2}\\ \sin\frac{A}{2}\cos\frac{b+c}{2} &= \cos\frac{a}{2}\cos\frac{B+C}{2}\\ \cos\frac{A}{2}\sin\frac{b-c}{2} &= \sin\frac{a}{2}\sin\frac{B-C}{2}\\ \cos\frac{A}{2}\cos\frac{b-c}{2} &= \cos\frac{a}{2}\sin\frac{B+C}{2}\end{aligned}\right\}\quad\text{(Delambre's or Gauss's analogies)}\qquad\text{(B-17)}$$

$$\left.\begin{aligned}&s = \frac{a+b+c}{2} \qquad S = \frac{A+B+C}{2}\\ &\sin\frac{A}{2} = \sqrt{\frac{\sin(s-b)\sin(s-c)}{\sin b\sin c}}\\ &\cos\frac{A}{2} = \sqrt{\frac{\sin s\sin(s-a)}{\sin b\sin c}}\\ &\sin\frac{a}{2} = \sqrt{\frac{-\cos S\cos(S-A)}{\sin B\sin C}}\\ &\cos\frac{a}{2} = \sqrt{\frac{\cos(S-B)\cos(S-C)}{\sin B\sin C}}\end{aligned}\right\}\quad\text{(half-angle formulas)}\qquad\text{(B-18)}$$

$$\left.\begin{aligned}\sin A &= \frac{+2\sqrt{\sin s\sin(s-a)\sin(s-b)\sin(s-c)}}{\sin b\sin c}\\ \sin a &= \frac{+2\sqrt{-\cos S\cos(S-A)\cos(S-B)\cos(S-C)}}{\sin B\sin C}\end{aligned}\right\}\qquad\text{(B-19)}$$

$$\left.\begin{aligned}&\cot r = \sqrt{\frac{\cos(S-A)\cos(S-B)\cos(S-C)}{-\cos S}}\\ &\tan\rho = \sqrt{\frac{\sin(s-a)\sin(s-b)\sin(s-c)}{\sin s}}\\ &\cot\frac{A}{2} = \frac{\sin(s-a)}{\tan\rho} \qquad \tan\frac{a}{2} = \frac{\cos(S-A)}{\cot r}\end{aligned}\right\}\qquad\text{(B-20)}$$

$$\tan\frac{\epsilon}{4} = \sqrt{\tan\frac{s}{2}\tan\frac{s-a}{2}\tan\frac{s-b}{2}\tan\frac{s-c}{2}}\qquad\text{(B-21)}$$

$$\tan\left(\frac{A}{2}-\frac{\epsilon}{4}\right) = +\sqrt{\frac{\tan\frac{s-b}{2}\tan\frac{s-c}{2}}{\tan\frac{s}{2}\tan\frac{s-a}{2}}}\quad\text{(L'Huilier's equation)}\qquad\text{(B-22)}$$

B-9. Formulas Expressed in Terms of the Haversine Function. Certain trigonometrical relations become particularly suitable for logarithmic computations if they are expressed in terms of the new trigonometric functions **versed sine, versed cosine,** and **haversine,** defined by

$$\text{vers } A = 1 - \cos A \qquad \text{covers } A = 1 - \sin A \qquad \text{hav } A = \tfrac{1}{2}(1 - \cos A)\qquad\text{(B-23)}$$

Thus, if tables of the haversine function are available, one may use the following formulas for spherical-triangle computations:

$$\text{hav } A = \frac{\sin(s-b)\sin(s-c)}{\sin b \sin c} \qquad \text{hav } a = \text{hav }(b-c) + \sin b \sin c \text{ hav } A \tag{B-24}$$

Other similar relations may be obtained by cyclic permutation.

Table B-2. Solution of Spherical Triangles (refer to formulas of Sec. B-8 and to Fig. B-3)

Case	Given	Example (obtain other cases by cyclic permutation)	Formulas used	Conditions for the existence of a solution (see also Sec. B-6)
1	Three sides	a, b, c	A, B, C from (18) and cyclic permutation	Sum of two sides must be greater than the third
2	Three angles	A, B, C	a, b, c from (18) and cyclic permutation	$540° > A + B + C > 180°$; sum of two angles must be less than 180° plus the third angle
3	Two sides and the included angle	b, c, A	$\frac{B+C}{2}$ and $\frac{B-C}{2}$ from (16), hence B and C; a from (17), (18), or (14)	
4	Two angles and the included side	B, C, a	$\frac{b+c}{2}$ and $\frac{b-c}{2}$ from (16), hence b and c; A from (17), (18), or (15)	
5	Two sides and an opposite angle	b, c, B	C from (13); A and a from (16)	Problem has either one or two solutions if $\sin c \sin B \leq \sin b$. Retain the values of C which make $A - B$ and $a - b$ of like sign; and $A + B - 180°$ and $a + b - 180°$ of like sign
6	Two angles and an opposite side	B, C, b	c from (13); A and a from (16)	Problem has either one or two solutions if $\sin b \sin C \leq \sin B$. Retain the values of c which make $A - B$ and $a - b$ of like sign, and $A + B - 180°$ and $a + b - 180°$ of like sign

Bibliography

Kells, L. M., et al.: *Plane and Spherical Trigonometry*, 3d ed., McGraw-Hill, New York, 1951.

Palmer, C. I., et al.: *Plane and Spherical Trigonometry*, 5th ed., McGraw-Hill, New York, 1950.

APPENDIX **C**

PERMUTATIONS, COMBINATIONS, AND RELATED TOPICS

Refer to Sec. 1.2-4 and Table 21.5-1 for definitions and properties of factorials and binomial coefficients. Stirling's formula (Sec. 21.4-2) is useful in numerical computations.

Table C-1. Permutations and Partitions

1	Number of different orderings (**permutations**) of a set of n distinct objects	$n!$
2	i. Number of distinguishable sequences of N objects comprising $n \leq N$ indistinguishable objects of type 1 and $N - n$ indistinguishable objects of type 2, **or** ii. Number of distinguishable *partitions* of a set of N distinct objects into 2 classes of $n \leq N$ and $N - n$ objects, respectively	$\binom{N}{n} = \frac{N!}{(N-n)!n!}$ (**binomial coefficient,** Sec. 21.5-1)
3	i. Number of distinguishable sequences of $N = N_1 + N_2 + \cdots + N_r$ objects comprising N_1 indistinguishable objects of type 1, N_2 indistinguishable objects of type 2, . . . , and N_r indistinguishable objects of type r, **or** ii. Number of distinguishable partitions of a set of $N = N_1 + N_2 + \cdots + N_r$ distinct objects into r classes of $N_1, N_2, \ldots, N_r$ objects, respectively	$\frac{N!}{N_1!N_2! \cdots N_r!}$ (**multinomial coefficient**)

Table C-2. Combinations and Samples (see also Table C-3 and Sec. 18.7-2). Each formula holds for $N < n$, $N = n$, and $N > n$

1	Number of distinguishable unordered **combinations** of N distinct types of objects taken n at a time: i. Each type of object may occur *at most once* in any combination (*combinations without repetition;* see also Table C-1, 2) ii. Each type of object may occur 0, 1, 2, . . . , or n times in any combination (*combinations with repetition*) iii. Each type of object must occur *at least once* in each combination	$\binom{N}{n}$ $\binom{N+n-1}{n} = \binom{N+n-1}{N-1}$ $\binom{n-1}{N-1}$
2	Number of distinguishable **samples** (sequences, ordered sets, variations) of size n taken from a population of N distinct types of objects: i. Each type of object may occur *at most once* in any sample (*samples without replacement,* sequences without repetition) ii. Each type of object may occur 0, 1, 2, . . . , or n times in any sample (*samples with replacement,* sequences with repetition)	$N(N-1) \cdots (N-n+1) = \binom{N}{n} n!$ N^n

EXAMPLES: Given a set of $N = 3$ distinct types of elements a, b, c. For $n = 2$, there exist 3 *combinations without repetition* (ab, ac, bc); 6 *combinations with repetition* (aa, ab, ac, bb, bc, cc); 6 distinguishable *samples without replacement* (ab, ac, ba, bc, ca, cb); and 9 distinguishable *samples with replacement* (aa, ab, ac, ba, bb, bc, ca, cb, cc).

Table C-3. Occupancy of Cells or States (see also Table C-2 and Sec. 18.7-2). Each formula holds for $N < n$, $N = n$, and $N > n$

1	Number of distinguishable arrangements of n *indistinguishable* objects in N distinct cells (states):	
	i. *No cell may contain more than one* object	$\binom{N}{n}$
	ii. Any cell may contain 0, 1, 2, . . . , or n objects	$\binom{N+n-1}{n} = \binom{N+n-1}{N-1}$
	iii. Each cell must contain *at least one* object	$\binom{n-1}{N-1}$
2	Number of distinguishable arrangements of n *distinct* objects in N distinct cells:	
	i. *No cell may contain more than one* object	$N(N-1) \cdots (N-n+1) = \binom{N}{n} n!$
	ii. Any cell may contain 0, 1, 2, . . . , or n objects	N^n

C-1. Use of Generating Functions (Refs. C-1, C-3). **(a) Combinations.** *The combinations of n distinct objects $A_1, A_2, \ldots, A_n$, taken k at a time without repetition*, are all "exhibited" as the coefficients α_k generated by the generating function (Sec. 8.)

$$F(s) = \sum_{k=0}^{n} \alpha_k s^k = (1 + A_1 s)(1 + A_2 s) \cdots (1 + A_n s) \qquad \text{(C-1)}$$

The *numbers* of such combinations of n objects are the coefficients $a_k = \binom{n}{k}$ generated by the **enumerating generating function (enumerator)**

$$F_{\text{ENUM}}(s) = \sum_{k=0}^{n} a_k s^k = (1 + s)^n \qquad \text{(C-2)}$$

This "product model" for combinations can be generalized. Thus, if *the object A_1 may be repeated* 0, r_1, *or* r_2 *times while* $A_2, A_3, \ldots, A_n$ *may each occur once or not at all*, the corresponding generating functions take the form

$$F(s) = (1 + A_1^{r_1} s^{r_1} + A_1^{r_2} s^{r_2})(1 + A_2 s) \cdots (1 + A_n s) \qquad \text{(C-3)}$$
$$F_{\text{ENUM}}(s) = (1 + s^{r_1} + s^{r_2})(1 + s)^{n-1} \qquad \text{(C-4)}$$

If there is *no restriction on the number of repetitions*,

$$F_{\text{ENUM}}(s) = (1 + s + s^2 + \cdots)^n = \left(\frac{1}{1-s}\right)^n = \sum_{k=0}^{\infty} \binom{n+k-1}{k} s^k \qquad \text{(C-5)}$$

If *each of the n objects must occur at least once,*

$$F_{\text{ENUM}}(s) = (s + s^2 + \cdots)^n = \left(\frac{s}{1-s}\right)^n = \sum_{k=n}^{\infty} \binom{k-1}{k-n} s^k \qquad \text{(C-6)}$$

(b) Permutations. To enumerate the *permutations of n distinct objects taken k at a time without repetition,* one may use the *exponential* generating function (Sec. 8.7-2.)

$$G_{\text{ENUM}}(s) = \sum_{k=0}^{n} \frac{b_k}{k!} s_k = (1 + s)^n \qquad b_k = k! \binom{n}{k} \qquad \text{(C-7)}$$

If *one of the n objects can be repeated* 0, r_1, or r_2, *times,* then

$$G_{\text{ENUM}}(s) = \left(1 + \frac{s^{r_1}}{r_1!} + \frac{s^{r_2}}{r_2!}\right)(1 + s)^{n-1} \qquad \text{(C-8)}$$

If *any number of repetitions are allowed,*

$$G_{\text{ENUM}}(s) = \left(1 + s + \frac{s^2}{2!} + \cdots\right)^n = e^{ns} \qquad \text{(C-9)}$$

If *each object must occur at least once,*

$$G_{\text{ENUM}}(s) = \left(s + \frac{s^2}{2!} + \cdots\right)^n = (e^s - 1)^n \qquad \text{(C-10)}$$

C-2. Polya's Counting Theorem. Consider a finite set D of n "points" p each to be associated with one of the elements (**figures**) f of a second finite set R (**figure store, figure collection**); more than one p may be associated with the same f. One desires to investigate the class of such arrangements or **configurations (patterns,** mappings from D into R, Fig. C-1) subject to a given group G of permutations of the

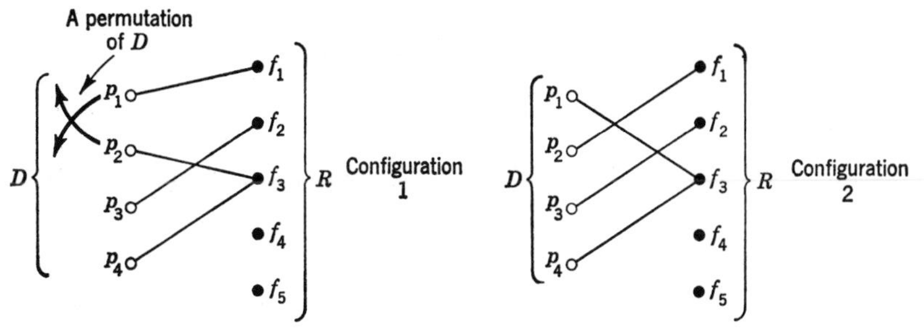

FIG. C-1. Two *configurations* are shown. The right-hand one is obtained from the left-hand one through a permutation of the points p_i. Two such configurations can be *equivalent* by virtue of some previously defined symmetry in the point set D and/or because two (or more) of the figures (in this case, f_1 and f_3) are indistinguishable.

points p (Sec. 12.2-8). Two configurations C_1, C_2 are **equivalent** with respect to G if and only if a permutation in G transforms C_1 into C_2; equivalent configurations necessarily contain the same figures.

The **cycle index** $Z_G(s_1, s_2, \ldots, s_n)$ of the permutation group G is a generating function defined as follows. Every permutation P of G classifies the points p of D into uniquely determined subsets (cycles) such that P produces only cyclic permutations of each subset (Sec. 12.2-8). Let b_k be the number of such cycles of length k for a given permutation $(b_1 + 2b_2 + \cdots + nb_n = n)$. Then the cycle index is defined as the polynomial

$$Z_G(s_1, s_2, \ldots, s_n) = \frac{1}{g} \sum_{\text{all } P \text{ in } G} g_{b_1 b_2 \ldots} s_1^{b_1} s_2^{b_2} \ldots s_n^{b_n} \qquad \text{(C-11)}$$

where g is the total number of permutations in (i.e., the order of) G, and $g_{b_1 b_2 \ldots}$ is the number of permutations, with b_1 cycles of length 1, b_2 cycles of length 2, etc.

Assuming that each type of figure f in R is associated with a nonnegative integer w (**weight** or **content** of f), let a_w be the number of distinguishable f's of weight w. The **figure-counting series (store enumerator)** is the generating function

$$a(s) = \sum_{w=0}^{\infty} a_w s^w \qquad \text{(C-12)}$$

The **content (weight) of a configuration** is the sum of its figure weights; equivalent configurations have equal content. Let A_w be the number of nonequivalent configurations of content w; then the **configuration-counting series (pattern enumerator)** is the generating function

$$A(s) = \sum_{w=0}^{\infty} A_w s^w \qquad \text{(C-13)}$$

The generating functions (12) *and* (13) *are related by*

$$A(s) = Z_G[a(s), a(s^2), \ldots, a(s^n)] \quad \text{(POLYA'S COUNTING THEOREM)} \qquad \text{(C-14)}$$

In particular, *the* **configuration inventory** $A(1) = \sum_{w=0}^{\infty} A_w$ *is related to the* **figure inventory** $a(1) = \sum_{w=0}^{\infty} a_w$ *by*

$$A(1) = Z_G[a(1), a(1), \ldots, a(1)] \qquad \text{(C-15)}$$

The theorem can be generalized to apply to situations where figures and configurations are labeled with two or more weights (Ref. C-1).

References and Bibliography

C-1. Beckenbach, E. F. (ed.): *Applied Combinatorial Mathematics*, Wiley, New York, 1964.
C-2. MacMahon, P. A.: *Combinatory Analysis* (2 vols.), Cambridge, London, 1915–1916.
C-3. Riordan, J.: *An Introduction to Combinatorial Analysis*, Wiley, New York, 1958.
C-4. Ryser, H. J.: *Combinational Mathematics*, Wiley, New York, 1963.

APPENDIX **D**

TABLES OF FOURIER EXPANSIONS AND LAPLACE-TRANSFORM PAIRS

D-1. Tables D-1 to D-7 present a number of Fourier expansions (Sec. 4.11-4), Hankel transforms (Sec. 8.6-4), and Laplace transform pairs (Secs. 8.2-1, 8.2-6, and 8.4-1) for reference.

D-2. Fourier-transform Pairs and Laplace-transform Pairs. (a) For suitable functions $f(t)$ (Secs. 4.11-4 and 8.2-6),

$$\left.\begin{aligned} f(t) &= \frac{1}{\sqrt{2\pi}}\int_{-\infty}^{\infty} C(\omega)e^{i\omega t}\,d\omega = \int_{-\infty}^{\infty} c(\nu)e^{2\pi i\nu t}\,d\nu \quad (-\infty < t < \infty) \\ C(\omega) &\equiv \frac{1}{\sqrt{2\pi}}\int_{-\infty}^{\infty} f(t)e^{-i\omega t}\,dt \qquad c(\nu) \equiv F_F(i\omega) \equiv \int_{-\infty}^{\infty} f(t)e^{-2\pi i\nu t}\,dt \end{aligned}\right\} \quad (\omega = 2\pi\nu) \qquad \text{(D-1)}$$

$$\left.\begin{aligned} f(t) &= \frac{1}{2\pi i}\int_{\sigma_1 - i\infty}^{\sigma_1 + i\infty} F(s)e^{st}\,ds \qquad (t > 0,\ \sigma_1 \geq 0) \\ F(s) &= \int_0^{\infty} f(t)e^{-st}\,dt \qquad (\sigma \geq 0) \end{aligned}\right\} \qquad \text{(D-2)}$$

so that tables of Laplace-transform pairs (2) (*Tables D-6 and D-7*) *may be used to obtain many Fourier-transform pairs* (1):

1. *Given* $f(t)$, obtain, if possible,
   $F(s) \equiv \mathfrak{L}[f(t)]$ and $F_1(s) \equiv \mathfrak{L}[f(-t)]$. *If* $F(s)$ *and* $F_1(s)$ *are analytic for* $\sigma > 0$, *then*

$$\left.\begin{aligned} C(\omega) &\equiv \frac{1}{\sqrt{2\pi}}[F(i\omega) + F_1(-i\omega)] \\ c(\nu) &\equiv F_F(i\omega) \equiv F(i\omega) + F_1(-i\omega) \quad (\omega = 2\pi\nu) \end{aligned}\right\} \qquad \text{(D-3)}$$

2. *Given* $C(\omega)$ *or* $c(\nu) \equiv F_F(i\omega)$,

$$f(t) = \sqrt{2\pi}\,\mathfrak{L}^{-1}\left[C\left(\frac{s}{i}\right)\right] = \mathfrak{L}^{-1}\left[c\left(\frac{s}{2\pi i}\right)\right] = \mathfrak{L}^{-1}[F_F(s)] \qquad (t > 0) \qquad \text{(D-4)}$$

provided that this expression exists, so that $C(s/i)$ *or* $F_F(s)$ *is analytic for* $\sigma \geq 0$, *and* $f(t) = 0$ *for* $t < 0$.

(**b**) *The following procedures permit one to obtain many Laplace-transform pairs* (2) *from tables of Fourier-transform pairs* (1) (*Table D-2 and Ref.* 8.1):

1. *Given* $f(t)$ *such that* $f(t)$ *is real and* $f(t) = 0$ *for* $t < 0$, use the table of Fourier-transform pairs (or any other method) to obtain $C(\omega)$ or $c(\nu) \equiv F_F(i\omega)$. Then

$$\mathfrak{L}[f(t)] = F(s) = \sqrt{2\pi}\,C\left(\frac{s}{i}\right) = c\left(\frac{s}{2\pi i}\right) = F_F(s) \qquad (\sigma > 0) \qquad \text{(D-5)}$$

2. *Given* $f(t)$ *such that* $f(t)$ *is real and even,* $f(-t) = f(t)$,*

$$\mathfrak{L}[f(t)] = F(s) = \sqrt{\frac{\pi}{2}}\,C\left(\frac{s}{i}\right) = \frac{1}{2}c\left(\frac{s}{2\pi i}\right) = \frac{1}{2}F_F(s) \qquad (\sigma > 0) \qquad \text{(D-6)}$$

3. *Given* $F(s)$ *analytic for* $\sigma \geq 0$, obtain $\mathfrak{L}^{-1}[F(s)]$ for $t > 0$ as the function $f(t)$ corresponding to

$$C(\omega) \equiv \frac{1}{\sqrt{2\pi}}F(i\omega) \qquad \text{or} \qquad c(\nu) \equiv F_F(i\omega) \equiv F(2\pi i\nu) \qquad (\omega = 2\pi\nu) \qquad \text{(D-7)}$$

* Note that *every* function can be rewritten as the sum of even and odd parts (Sec. 4.3-2), and that $f(-t)$ is even whenever $f(t)$ is odd.

Table D-1. Fourier Coefficients and Mean-square Values of Periodic Functions $\left[\text{Sec. 4.11-4}a;\ \operatorname{sinc}(x) \equiv \frac{\sin \pi x}{\pi x}\right]$

	Periodic function, $f(t) = f(t+T)$	Fourier coefficients (for phasing as shown in diagram)	Average value $\langle f \rangle = \frac{a_0}{2}$	Mean-square value $\langle f^2 \rangle$
1	Rectangular pulses	$a_n = 2A \frac{T_0}{T} \operatorname{sinc}\left(\frac{nT_0}{T}\right)$ $b = 0$	$A \frac{T_0}{T}$	$A^2 \frac{T_0}{T}$
2	Symmetrical triangular pulses	$a_n = A \frac{T_0}{T} \operatorname{sinc}^2\left(\frac{nT_0}{2T}\right)$ $b_n = 0$	$A \frac{T_0}{2T}$	$A^2 \frac{T_0}{3T}$
3	Symmetrical trapezoidal pulses	$a_n = 2A \frac{T_0 + T_1}{T} \operatorname{sinc}\left(\frac{nT_1}{T}\right) \operatorname{sinc}\left[\frac{n(T_0+T_1)}{T}\right]$ $b_n = 0$	$A \frac{T_0 + T_1}{T}$	$A^2 \frac{3T_0 + 2T_1}{3T}$
4	Half-sine pulses*†	$a_n = A \frac{T_0}{T} \left\{ \operatorname{sinc}\left[\frac{1}{2}\left(\frac{2nT_0}{T} - 1\right)\right] + \operatorname{sinc}\left[\frac{1}{2}\left(\frac{2nT_0}{T} + 1\right)\right] \right\}$ $b_n = 0$	$\frac{2}{\pi} A \frac{T_0}{T}$	$A^2 \frac{T_0}{2T}$
5	Clipped sinusoid $A = A_0\left(1 - \cos \frac{\pi T_0}{T}\right)$	$a_n = \frac{A_0 T_0}{T} \left\{ \operatorname{sinc}\left[(n-1)\frac{T_0}{T}\right] + \operatorname{sinc}\left[(n+1)\frac{T_0}{T}\right] - 2 \cos \frac{\pi T_0}{T} \operatorname{sinc}\left(\frac{nT_0}{T}\right) \right\}$	$\frac{1}{\pi} A_0 \left(\sin \frac{\pi T_0}{T} - \frac{\pi T_0}{T} \cos \frac{\pi T_0}{T} \right)$	$\frac{1}{2\pi} A_0^2 \left(\frac{\pi T_0}{T} - \frac{3}{2} \sin \frac{2\pi T_0}{T} + \frac{2\pi T_0}{T} \cos^2 \frac{\pi T_0}{T} \right)$
6	Triangular waveform	$\left.\begin{matrix} a_n = 0 \\ b_n = -\frac{A}{n\pi} \end{matrix}\right\} n = 1, 2, \ldots$	$\frac{A}{2}$	$\frac{A^2}{3}$

* For $T_0 = \frac{T}{2} = \frac{\pi}{\omega}$, $f(t) = \frac{2}{\pi} A \left(\frac{1}{2} + \frac{\pi}{4} \cos \omega t + \frac{1}{3} \cos 2\omega t - \frac{1}{15} \cos 4\omega t + \frac{1}{35} \cos 6\omega t \pm \cdots \right)$ (HALF-WAVE RECTIFIED SINUSOID).

† For $T_0 = T = \frac{2\pi}{\omega}$, $f(t) = \frac{4}{\pi} A \left(\frac{1}{2} + \frac{1}{3} \cos 2\omega t - \frac{1}{15} \cos 4\omega t + \frac{1}{35} \cos 6\omega t \pm \cdots \right)$ (FULL-WAVE RECTIFIED SINUSOID).

Table D-2a. Fourier-transform Pairs*

	$f(t) = \int_{-\infty}^{\infty} F_F(i\omega) e^{i\omega t} \frac{d\omega}{2\pi}$	$F_F(i\omega) = \int_{-\infty}^{\infty} f(t) e^{-i\omega t}\, dt$	
	$\operatorname{rect} \frac{t}{T} = \begin{cases} 1 & (\lvert t\rvert < T/2) \\ 0 & (\lvert t\rvert > T/2) \end{cases}$	$T \operatorname{sinc} \frac{\omega T}{2\pi} \equiv T \frac{\sin \frac{\omega T}{2}}{\frac{\omega T}{2}}$	
	$\operatorname{sinc} \frac{t}{T} \equiv \frac{\sin \frac{\pi t}{T}}{\frac{\pi t}{T}}$	$T \operatorname{rect} \frac{\omega T}{2\pi} = \begin{cases} 0 & \left(\lvert\omega\rvert < \frac{\pi}{T}\right) \\ T & \left(\lvert\omega\rvert > \frac{\pi}{T}\right) \end{cases}$	
	$\begin{cases} 1 - \frac{\lvert t\rvert}{T} & (\lvert t\rvert < T) \\ 0 & (\lvert t\rvert \geq T) \end{cases}$	$T \operatorname{sinc}^2 \frac{\omega T}{2\pi} \equiv T \left(\frac{\sin \frac{\omega T}{2}}{\frac{\omega T}{2}} \right)^2$	
	$e^{-\frac{\lvert t\rvert}{T}}$	$\frac{2T}{(\omega T)^2 + 1}$	
	$e^{-\frac{1}{2}\left(\frac{t}{T}\right)^2}$	$\sqrt{2\pi}\, T e^{-\frac{1}{2}(\omega T)^2}$	
	$\delta(t - T)$	$e^{-i\omega T}$	(Complex)
	$\cos \omega_0 t$	$\pi[\delta(\omega - \omega_0) + \delta(\omega + \omega_0)]$	
	$\sin \omega_0 t$	$\frac{\pi}{i} [\delta(\omega - \omega_0) - \delta(\omega + \omega_0)]$	(Imaginary)
	$\sum_{k=-\infty}^{\infty} \delta(t - kT) \equiv \frac{1}{T} \sum_{j=-\infty}^{\infty} e^{2\pi i j \frac{t}{T}}$	$\frac{2\pi}{T} \sum_{j=-\infty}^{\infty} \delta\left(\omega - \frac{2\pi j}{T}\right) \equiv \sum_{k=-\infty}^{\infty} e^{ik\omega T}$	

* Reprinted from G. A. Korn, *Basic Tables in Electrical Engineering*, McGraw-Hill, New York, 1965.

Table D-2b. Fourier Transforms*

$$f(x) = \frac{1}{\sqrt{2\pi}} \int_{-\infty}^{\infty} C(\xi) e^{-i\xi x}\, d\xi \quad C(\xi) = \frac{1}{\sqrt{2\pi}} \int_{-\infty}^{\infty} f(x) e^{i\xi x}\, dx$$

$f(x)$	$C(\xi)$
$\dfrac{\sin(ax)}{x}$	$\left(\frac{\pi}{2}\right)^{\frac{1}{2}} \quad \lvert\xi\rvert < a$; $0 \quad \lvert\xi\rvert > a$
$e^{i\omega x} \quad p < x < q$; $0 \quad x < p, x > q$	$\dfrac{i}{(2\pi)^{\frac{1}{2}}} \dfrac{e^{ip(\omega+\xi)} - e^{iq(\omega+\xi)}}{\xi}$
$e^{-cx+i\omega x} \quad x > 0$; $0 \quad x < 0$	$\dfrac{i}{(2\pi)^{\frac{1}{2}}(\omega + \xi + ic)}$
$e^{-px^2} \quad \mathrm{Re}(p) > 0$	$(2p)^{-\frac{1}{2}} e^{-\xi^2/4p}$
$\cos(px^2)$	$(2p)^{-\frac{1}{2}} \cos\left(\dfrac{\xi^2}{4p} - \dfrac{1}{4}\pi\right)$
$\sin(px^2)$	$(2p)^{-\frac{1}{2}} \sin\left(\dfrac{\xi^2}{4p} + \dfrac{1}{4}\pi\right)$
$\lvert x\rvert^{-s} \quad 0 < \mathrm{Re}(s) < 1$	$\dfrac{2^{\frac{1}{2}}\Gamma(1-s)\sin(\frac{1}{2}s\pi)}{\pi^{\frac{1}{2}}\lvert\xi\rvert^{1-s}}$
$\dfrac{1}{\lvert x\rvert}$	$\dfrac{1}{\lvert\xi\rvert}$
$\dfrac{e^{-a\lvert x\rvert}}{\lvert x\rvert^{\frac{1}{2}}}$	$\dfrac{[(a^2+\xi^2)^{\frac{1}{2}} + a]^{\frac{1}{2}}}{(a^2+\xi^2)^{\frac{1}{2}}}$
$\dfrac{\cosh(ax)}{\cosh(\pi x)} \quad -\pi < a < \pi$	$\left(\dfrac{2}{\pi}\right)^{\frac{1}{2}} \dfrac{\cos(\frac{1}{2}a)\cosh(\frac{1}{2}\xi)}{\cosh(\xi) + \cos(a)}$
$\dfrac{\sinh(ax)}{\sinh(\pi x)} \quad -\pi < a < \pi$	$\left(\dfrac{1}{2\pi}\right)^{\frac{1}{2}} \dfrac{\sin(a)}{\cosh(\xi) + \cos(a)}$
$(a^2 - x^2)^{-\frac{1}{2}} \quad \lvert x\rvert < a$; $0 \quad \lvert x\rvert > a$	$(\frac{1}{2}\pi)^{\frac{1}{2}} J_0(a\xi)$
$\dfrac{\sin[b(a^2+x^2)^{\frac{1}{2}}]}{(a^2+x^2)^{\frac{1}{2}}}$	$0 \quad \lvert\xi\rvert > b$; $(\frac{1}{2}\pi)^{\frac{1}{2}} J_0(a\sqrt{b^2-\xi^2}) \quad \lvert\xi\rvert < b$
$P_n(x) \quad \lvert x\rvert < 1$; $0 \quad \lvert x\rvert > 1$	$i^n \pi^{-\frac{1}{2}} J_{n+\frac{1}{2}}(\xi)$
$\dfrac{\cos(b\sqrt{a^2-x^2})}{(a^2-x^2)^{\frac{1}{2}}} \quad \lvert x\rvert < a$; $0 \quad \lvert x\rvert > a$	$(\frac{1}{2}\pi)^{\frac{1}{2}} J_0(a\sqrt{\xi^2+b^2})$
$\dfrac{\cosh(b\sqrt{a^2-x^2})}{(a^2-x^2)^{\frac{1}{2}}} \quad \lvert x\rvert < a$; $0 \quad \lvert x\rvert > a$	$(\frac{1}{2}\pi)^{\frac{1}{2}} J_0(a\sqrt{\xi^2-b^2})$

* From I. A. Sneddon, *Fourier Transforms*, McGraw-Hill, New York, 1951.

Table D-3. Fourier Cosine Transforms*†

$$f(x) = \sqrt{\frac{2}{\pi}} \int_0^\infty C_C(\xi) \cos(\xi x) d\xi \qquad C_C(\xi) = \sqrt{\frac{2}{\pi}} \int_0^\infty f(x) \cos(\xi x) dx$$

$f(x)$	$C_C(\xi)$
$1 \quad 0 < x < a$ $0 \quad x > a$	$\left(\frac{2}{\pi}\right)^{\frac{1}{2}} \frac{\sin(\xi a)}{\xi}$
$x^{p-1} \quad 0 < p < 1$	$\left(\frac{2}{\pi}\right)^{\frac{1}{2}} \Gamma(p) \xi^{-p} \sin\left(\frac{1}{2} p\pi\right)$
$\cos(x) \quad 0 < x < a$ $0 \quad x > a$	$\left(\frac{1}{2\pi}\right)^{\frac{1}{2}} \left\{\frac{\sin[a(1-\xi)]}{1-\xi} + \frac{\sin[a(1+\xi)]}{1+\xi}\right\}$
e^{-x}	$\left(\frac{2}{\pi}\right)^{\frac{1}{2}} \frac{1}{1+\xi^2}$
$\operatorname{sech}(\pi x)$	$\frac{1}{1+\xi^4}$
e^{-x^2}	$e^{-\xi^2}$
$\cos(\frac{1}{2}x^2)$	$\frac{1}{\sqrt{2}}\left[\cos\left(\frac{1}{2}\xi^2\right) + \sin\left(\frac{1}{2}\xi^2\right)\right]$
$\sin(\frac{1}{2}x^2)$	$\frac{1}{\sqrt{2}}\left[\cos\left(\frac{1}{2}\xi^2\right) - \sin\left(\frac{1}{2}\xi^2\right)\right]$
$(1-x^2)^\nu \quad 0 < x < 1$ $0 \quad x > 1,$ $\nu > -\frac{3}{2}$	$2^\nu \Gamma(\nu+1) \xi^{-\nu-\frac{1}{2}} J_{\nu+\frac{1}{2}}(\xi)$

* Three general rules are worthy of notice:

1. If $C_C(\xi)$ is the Fourier cosine transform of $f(x)$, then $f(\xi)$ is the Fourier cosine transform of $C_C(x)$.

2. If $f(x)$ is an even function of x in $(-\infty, \infty)$, then the Fourier cosine transform of $f(x)$ $(0 \le x < \infty)$ is $C(\xi)$.

3. The Fourier cosine transform of $f(x/a)$ is $aC_C(\xi a)$.

† From I. A. Sneddon, *Fourier Transforms*, McGraw-Hill, New York, 1951.

Table D-4. Fourier Sine Transforms*†

$$f(x) = \sqrt{\frac{2}{\pi}} \int_0^\infty C_S(\xi) \sin(\xi x) d\xi \qquad C_S(\xi) = \sqrt{\frac{2}{\pi}} \int_0^\infty f(x) \sin(\xi x) dx$$

$f(x)$	$C_S(\xi)$
e^{-x}	$\left(\frac{2}{\pi}\right)^{\frac{1}{2}} \frac{1}{1+\xi^2}$
$xe^{-\frac{1}{2}x^2}$	$e^{-\frac{1}{2}\xi^2}$
$\frac{\sin(x)}{x}$	$\frac{1}{(2\pi)^{\frac{1}{2}}} \log \left\lvert\frac{1+\xi}{1-\xi}\right\rvert$
$x(1-x^2)^\nu \quad 0 < x < 1,\ \nu > -1$ $0 \quad x > 1$	$2^\nu \Gamma(\nu+1) \xi^{-\frac{1}{2}-\nu} J_{\nu+\frac{3}{2}}(\xi)$
$x^{p-1} \quad 0 < p < 1$	$\left(\frac{2}{\pi}\right)^{\frac{1}{2}} \Gamma(p) \sin\left(\frac{1}{2}p\pi\right) \xi^{-p}$
$x^n e^{-px}$	$\frac{2^{n+\frac{1}{2}} p^n n! \xi}{\pi^{\frac{1}{2}}(p^2+\xi^2)^{n+1}}$
$\cos(ax^2)$	$-a^{-\frac{1}{2}}\left[\cos\left(\frac{\xi^2}{4a}\right) S\left(\frac{\xi}{\sqrt{2\pi a}}\right) - \sin\left(\frac{\xi^2}{4a}\right) C\left(\frac{\xi}{\sqrt{2\pi a}}\right)\right]$
$x^{-\frac{1}{2}} e^{-ax^{-\frac{1}{2}}}$	$\xi^{-\frac{1}{2}}[\cos(2a\xi)^{\frac{1}{2}} - \sin(2a\xi)^{\frac{1}{2}}]$
$0 \quad 0 < x < a$ $(x^2-a^2)^{-\frac{1}{2}} \quad x > a$	$\left(\frac{\pi}{2}\right)^{\frac{1}{2}} J_0(a\xi)$

* In the calculation of Fourier sine transforms we may make use of the rules:

1. If $C_S(\xi)$ is the Fourier sine transform of $f(x)$ then $f(\xi)$ is the Fourier sine transform of $C_S(x)$.
2. If $f(x)$ is an odd function of x in $(-\infty, \infty)$, then the Fourier sine transform of $f(x)$ $(0 < x < \infty)$ is $-iC(\xi)$.
3. The Fourier sine transform of $f(x/a)$ is $aC_S(a\xi)$.

† From I. A. Sneddon, *Fourier Transforms*, McGraw-Hill, New York, 1951.

Table D-5. Hankel Transforms*

$$f(x) = \int_0^\infty \xi \bar{f}(\xi) J_\nu(\xi x)\, d\xi \qquad \bar{f}(\xi) = \int_0^\infty x f(x) J_\nu(\xi x)\, dx$$

$f(x)$	ν	$\bar{f}(\xi)$
$x^\nu \quad 0 < x < a$ $0 \quad x > a$	> -1	$\frac{a^{\nu+1}}{\xi} J_{\nu+1}(\xi a)$
$1 \quad 0 < x < a$ $0 \quad x > a$	0	$\frac{a}{\xi} J_1(a\xi)$
$(a^2 - x^2) \quad 0 < x < a,$ $0, \quad x > a$	0	$\frac{4a}{\xi^3} J_1(\xi a) - \frac{2a^2}{\xi^2} J_0(\xi a)$
$x^{\mu-2} e^{-px^2}$	> -1	$\frac{\xi^\nu \Gamma(\frac{1}{2}\nu + \frac{1}{2}\mu)}{2^{\nu+1} p^{\frac{1}{2}\mu + \frac{1}{2}\nu} \Gamma(1 + \nu)} {}_1F_1 \left(\frac{1}{2}\nu + \frac{1}{2}\mu; \nu + 1; -\frac{\xi^2}{4p}\right)$
$x^\nu e^{-px^2}$	> -1	$\frac{\xi}{(2p)^{\nu+1}} e^{-\xi^2/4p}$
$x^{\mu-1} e^{-px}$	> -1	$\frac{2^\mu \xi^\nu \Gamma(\frac{1}{2}\mu + \frac{1}{2}\nu + \frac{1}{2}) \Gamma(1 + \frac{1}{2}\mu + \frac{1}{2}\nu)}{(\xi^2 + p^2)^{\frac{1}{2}\mu + \frac{1}{2}\nu + \frac{1}{2}} \Gamma(\nu + 1) \Gamma(\frac{1}{2})}$ $\times {}_2F_1 \left(\frac{1}{2}\mu + \frac{1}{2}\nu + \frac{1}{2}; \frac{1}{2}\nu - \frac{1}{2}\mu; 1 + \nu; \frac{\xi^2}{\xi^2 + p^2}\right)$
$x^{\mu-1}$	> -1	$\frac{2^\mu \Gamma(\frac{1}{2} + \frac{1}{2}\mu + \frac{1}{2}\nu)}{\xi^{\mu+1} \Gamma(\frac{1}{2} - \frac{1}{2}\mu + \frac{1}{2}\nu)}$
$\frac{e^{-px}}{x}$	0	$(\xi^2 + p^2)^{-\frac{1}{2}}$
e^{-px}	0	$p(\xi^2 + p^2)^{-\frac{3}{2}}$
$x^{-2} e^{-px}$	1	$\frac{(\xi^2 + p^2)^{\frac{1}{2}} - p}{\xi}$
$\frac{e^{-px}}{x}$	1	$\frac{1}{\xi} - \frac{p}{\xi(\xi^2 + p^2)^{\frac{1}{2}}}$
e^{-px}	1	$\xi(\xi^2 + p^2)^{-\frac{3}{2}}$
$\frac{a}{(a^2 + x^2)^{\frac{3}{2}}}$	0	$e^{-a\xi}$
$\frac{\sin(ax)}{x}$	0	$0 \quad \xi > a$ $(a^2 - \xi^2)^{-\frac{1}{2}} \quad 0 < \xi < a$
$\frac{\sin(ax)}{x}$	1	$\frac{a}{\xi(\xi^2 - a^2)^{\frac{1}{2}}} \quad \xi > a$ $0 \quad \xi < a$
$\frac{\sin(x)}{x^2}$	0	$\sin^{-1}\left(\frac{1}{\xi}\right) \quad \xi > 1$ $\frac{1}{2}\pi \quad \xi < 1$

* From I. A. Sneddon, *Fourier Transforms*, McGraw-Hill, New York, 1951.

Table D-6. Laplace-transform Pairs Involving Rational Algebraic Functions $F(s) = D_1(s)/D(s)$

Each formula holds for complex as well as for real polynomials $D_1(s)$ and $D(s)$ (Sec. 8.4-4); but the latter case is of greater practical interest. In this case the roots of $D(s) = 0$ are either real or they occur as pairs of complex conjugates, and the functions $f(t)$ are real. Note

$$(s-a)^2 + \omega_1^2 = [s - (a + i\omega_1)][s - (a - i\omega_1)]$$

and

$$K_1 \sin \omega t + K_2 \cos \omega t = \sqrt{K_1^2 + K_2^2} \sin (\omega t + \alpha), \text{ with } \alpha = \arctan K_2/K_1$$

No.	$F(s)$	$f(t)$ $(t > 0)$	
1.1	$\frac{1}{s}$	1	
1.2	$\frac{1}{s-a}$	e^{at}	
1.3	$\frac{1}{s(s-a)}$	$Ae^{at} + K$	$A = \frac{1}{a} \qquad K = -\frac{1}{a}$
1.4	$\frac{s+d}{s(s-a)}$		$A = \left(1 + \frac{d}{a}\right) \qquad K = -\frac{d}{a}$
1.5	$\frac{1}{(s-a)(s-b)}$	$Ae^{at} + Be^{bt}$	$A = \frac{1}{a-b} \qquad B = \frac{1}{b-a}$
1.6	$\frac{s+d}{(s-a)(s-b)}$		$A = \frac{a+d}{a-b} \qquad B = \frac{b+d}{b-a}$
1.7	$\frac{1}{s(s-a)(s-b)}$	$Ae^{at} + Be^{bt} + K$	$A = \frac{1}{a(a-b)} \qquad B = \frac{1}{b(b-a)} \qquad K = \frac{1}{ab}$
1.8	$\frac{s+d}{s(s-a)(s-b)}$		$A = \frac{a+d}{a(a-b)} \qquad B = \frac{b+d}{b(b-a)} \qquad K = \frac{d}{ab}$
1.9	$\frac{s^2+gs+d}{s(s-a)(s-b)}$		$A = \frac{a^2+ga+d}{a(a-b)} \qquad B = \frac{b^2+gb+d}{b(b-a)} \qquad K = \frac{d}{ab}$

1.10	$\dfrac{1}{(s-a)(s-b)(s-c)}$	$Ae^{at} + Be^{bt} + Ce^{ct}$	$A = \dfrac{1}{(a-b)(a-c)}$ $B = \dfrac{1}{(b-a)(b-c)}$ $C = \dfrac{1}{(c-a)(c-b)}$
1.11	$\dfrac{s+d}{(s-a)(s-b)(s-c)}$		$A = \dfrac{a+d}{(a-b)(a-c)}$ $B = \dfrac{b+d}{(b-a)(b-c)}$ $C = \dfrac{c+d}{(c-a)(c-b)}$
1.12	$\dfrac{s^2+gs+d}{(s-a)(s-b)(s-c)}$		$A = \dfrac{a^2+ag+d}{(a-b)(a-c)}$ $B = \dfrac{b^2+bg+d}{(b-a)(b-c)}$ $C = \dfrac{c^2+cg+d}{(c-a)(c-b)}$
2.1	$\dfrac{1}{(s-a)^2+\omega_1^2}$	$Ae^{at} \sin (\omega_1 t + \alpha)$	$A = \dfrac{1}{\omega_1}$ $\alpha = 0$
2.2	$\dfrac{s+d}{(s-a)^2+\omega_1^2}$		$A = \dfrac{1}{\omega_1}[(a+d)^2+\omega_1^2]^{1/2}$ $\alpha = \arctan \dfrac{\omega_1}{a+d}$
2.3	$\dfrac{1}{s[(s-a)^2+\omega_1^2]}$	$Ae^{at} \sin (\omega_1 t + \alpha) + K$	$A = \dfrac{1}{\omega_1} \dfrac{1}{(a^2+\omega_1^2)^{1/2}}$ $K = \dfrac{1}{a^2+\omega_1^2}$ $\alpha = -\arctan \dfrac{\omega_1}{a}$
2.4	$\dfrac{s+d}{s[(s-a)^2+\omega_1^2]}$		$A = \dfrac{1}{\omega_1}\left[\dfrac{(a+d)^2+\omega_1^2}{a^2+\omega_1^2}\right]^{1/2}$ $K = \dfrac{d}{a^2+\omega_1^2}$ $\alpha = \arctan \dfrac{\omega_1}{a+d} - \arctan \dfrac{\omega_1}{a}$
2.5	$\dfrac{s^2+gs+d}{s[(s-a)^2+\omega_1^2]}$		$A = \dfrac{1}{\omega_1}\left[\dfrac{(a^2-\omega_1^2+ag+d)^2+\omega_1^2(2a+g)^2}{a^2+\omega_1^2}\right]^{1/2}$ $K = \dfrac{d}{a^2+\omega_1^2}$ $\alpha = \arctan \dfrac{\omega_1(2a+g)}{a^2-\omega_1^2+ag+d} - \arctan \dfrac{\omega_1}{a}$

Table D-6. Laplace-transform Pairs Involving Rational Algebraic Functions $F(s) = D_1(s)/D(s)$ (*Continued*)

No.	$F(s)$	$f(t)\ (t > 0)$	
2.6	$\dfrac{1}{(s-b)[(s-a)^2+\omega_1^2]}$	$Ae^{at}\sin(\omega_1 t+\alpha)+Be^{bt}$	$A = \dfrac{1}{\omega_1}\dfrac{1}{[(a-b)^2+\omega_1^2]^{1/2}} \qquad B = \dfrac{1}{(a-b)^2+\omega_1^2}$ $\alpha = -\arctan\dfrac{\omega_1}{a-b}$
2.7	$\dfrac{s+d}{(s-b)[(s-a)^2+\omega_1^2]}$		$A = \dfrac{1}{\omega_1}\left[\dfrac{(a+d)^2+\omega_1^2}{(a-b)^2+\omega_1^2}\right]^{1/2} \qquad B = \dfrac{b+d}{(a-b)^2+\omega_1^2}$ $\alpha = \arctan\dfrac{\omega_1}{a+d} - \arctan\dfrac{\omega_1}{a-b}$
2.8	$\dfrac{s^2+gs+d}{(s-b)[(s-a)^2+\omega_1^2]}$		$A = \dfrac{1}{\omega_1}\left[\dfrac{(a^2-\omega_1^2+ag+d)^2+\omega_1^2(2a+g)^2}{(a-b)^2+\omega_1^2}\right]^{1/2} \qquad B = \dfrac{b^2+bg+d}{(a-b)^2+\omega_1^2}$ $\alpha = \arctan\dfrac{\omega_1(2a+g)}{a^2-\omega_1^2+ag+d} - \arctan\dfrac{\omega_1}{a-b}$
2.9	$\dfrac{1}{s(s-b)[(s-a)^2+\omega_1^2]}$	$Ae^{at}\sin(\omega_1 t+\alpha)+Be^{bt}+K$	$A = \dfrac{1}{\omega_1}\dfrac{1}{(a^2+\omega_1^2)^{1/2}[(a-b)^2+\omega_1^2]^{1/2}}$ $B = \dfrac{1}{b[(b-a)^2+\omega_1^2]} \qquad K = -\dfrac{1}{b(a^2+\omega_1^2)}$ $\alpha = -\arctan\dfrac{\omega_1}{a-b} - \arctan\dfrac{\omega_1}{a}$
2.10	$\dfrac{s+d}{s(s-b)[(s-a)^2+\omega_1^2]}$		$A = \dfrac{1}{\omega_1(a^2+\omega_1^2)^{1/2}}\left[\dfrac{(d+a)^2+\omega_1^2}{(a-b)^2+\omega_1^2}\right]^{1/2}$ $B = \dfrac{b+d}{b[(b-a)^2+\omega_1^2]} \qquad K = -\dfrac{d}{b(a^2+\omega_1^2)}$ $\alpha = \arctan\dfrac{\omega_1}{a+d} - \arctan\dfrac{\omega_1}{a-b} - \arctan\dfrac{\omega_1}{a}$

2.11	$\dfrac{s^2+gs+d}{s(s-b)[(s-a)^2+\omega_1^2]}$		$A=\dfrac{1}{\omega_1}\left\{\dfrac{(a^2-\omega_1^2+ag+d)^2+\omega_1^2(2a+g)^2}{(a^2+\omega_1^2)[(a-b)^2+\omega_1^2]}\right\}^{1/2}$ $B=\dfrac{b^2+bg+d}{b[(b-a)^2+\omega_1^2]}\qquad K=-\dfrac{d}{b(a^2+\omega_1^2)}$ $\alpha=\arctan\dfrac{\omega_1(2a+g)}{a^2-\omega_1^2+ag+d}-\arctan\dfrac{\omega_1}{a-b}-\arctan\dfrac{\omega_1}{a}$
2.12	$\dfrac{1}{[(s-a)^2+\omega_1^2](s^2+\omega_2^2)}$		$A=\dfrac{1}{\omega_1}\dfrac{1}{[(a^2+\omega_1^2-\omega_2^2)^2+4a^2\omega_2^2]^{1/2}}$ $B=\dfrac{1}{\omega_2}\dfrac{1}{[(a^2+\omega_1^2-\omega_2^2)^2+4a^2\omega_2^2]^{1/2}}$ $\alpha=-\arctan\dfrac{2a\omega_1}{a^2-\omega_1^2+\omega_2^2}$ $\beta=\arctan\dfrac{2a\omega_2}{a^2+\omega_1^2-\omega_2^2}$
2.13	$\dfrac{s+d}{[(s-a)^2+\omega_1^2](s^2+\omega_2^2)}$	$Ae^{at}\sin(\omega_1 t+\alpha)+B\sin(\omega_2 t+\beta)$	$A=\dfrac{1}{\omega_1}\left[\dfrac{(a+d)^2+\omega_1^2}{(a^2+\omega_1^2-\omega_2^2)^2+4a^2\omega_2^2}\right]^{1/2}$ $B=\dfrac{1}{\omega_2}\left[\dfrac{d^2+\omega_2^2}{(a^2+\omega_1^2-\omega_2^2)^2+4a^2\omega_2^2}\right]^{1/2}$ $\alpha=\arctan\dfrac{\omega_1}{a+d}-\arctan\dfrac{2a\omega_1}{a^2-\omega_1^2+\omega_2^2}$ $\beta=\arctan\dfrac{\omega_2}{d}+\arctan\dfrac{2a\omega_2}{a^2+\omega_1^2-\omega_2^2}$
2.14	$\dfrac{s^2+gs+d}{[(s-a)^2+\omega_1^2](s^2+\omega_2^2)}$		$A=\dfrac{1}{\omega_1}\left[\dfrac{(a^2-\omega_1^2+ag+d)^2+\omega_1^2(2a+g)^2}{(a^2+\omega_1^2-\omega_2^2)^2+4a^2\omega_2^2}\right]^{1/2}$ $B=\dfrac{1}{\omega_2}\left[\dfrac{(d-\omega_2^2)^2+g^2\omega_2^2}{(a^2+\omega_1^2-\omega_2^2)^2+4a^2\omega_2^2}\right]^{1/2}$ $\alpha=\arctan\dfrac{\omega_1(2a+g)}{a^2-\omega_1^2+ag+d}-\arctan\dfrac{2a\omega_1}{a^2-\omega_1^2+\omega_2^2}$ $\beta=\arctan\dfrac{g\omega_2}{d-\omega_2^2}+\arctan\dfrac{2a\omega_2}{a^2+\omega_1^2-\omega_2^2}$
3.1	$\dfrac{1}{s^2}$		

Table D-6. Laplace-transform Pairs Involving Rational Algebraic Functions $F(s) = D_1(s)/D(s)$ (*Continued*)

No.	$F(s)$	$f(t)\ (t > 0)$			
3.2	$\dfrac{1}{(s-a)^2}$	$(A + A_1t)e^{at}$	$A = 0$	$A_1 = 1$	
3.3	$\dfrac{s+d}{(s-a)^2}$		$A = 1$	$A_1 = a + d$	
3.4	$\dfrac{1}{s^2(s-a)}$	$Ae^{at} + K + K_1t$	$A = \dfrac{1}{a^2}$	$K = -A$	$K_1 = -\dfrac{1}{a}$
3.5	$\dfrac{s+d}{s^2(s-a)}$		$A = \dfrac{a+d}{a^2}$	$K = -A$	$K_1 = -\dfrac{d}{a}$
3.6	$\dfrac{s^2+gs+d}{s^2(s-a)}$		$A = \dfrac{a^2+ag+d}{a^2}$	$K = 1 - A$	$K_1 = -\dfrac{d}{a}$
3.7	$\dfrac{1}{s(s-a)^2}$	$(A + A_1t)e^{at} + K$	$A = -\dfrac{1}{a^2}$	$A_1 = \dfrac{1}{a}$	$K = -A$
3.8	$\dfrac{s+d}{s(s-a)^2}$		$A = -\dfrac{d}{a^2}$	$A = \dfrac{a+d}{a}$	$K = -A$
3.9	$\dfrac{s^2+gs+d}{s(s-a)^2}$		$A = \dfrac{a^2-d}{a^2}$	$A_1 = \dfrac{a^2+ag+d}{a}$	$K = 1 - A$
3.10	$\dfrac{1}{(s-a)^2(s-b)}$	$(A + A_1t)e^{at} + Be^{bt}$	$A = -\dfrac{1}{(a-b)^2}$	$A_1 = \dfrac{1}{a-b}$	$B = -A$
3.11	$\dfrac{s+d}{(s-a)^2(s-b)}$		$A = -\dfrac{b+d}{(a-b)^2}$	$A_1 = \dfrac{a+d}{a-b}$	$B = -A$

3.12	$\dfrac{s^2 + gs + d}{(s - a)^2(s - b)}$		$A = \dfrac{a^2 - 2ab - bg - d}{(a - b)^2}$ $\quad A_1 = \dfrac{a^2 + ag + d}{a - b}$ $\quad B = \dfrac{b^2 + bg + d}{(a - b)^2}$
3.13	$\dfrac{1}{s^2(s - a)(s - b)}$		$A = \dfrac{1}{a^2(a - b)}$ $\quad B = \dfrac{1}{b^2(b - a)}$ $K = \dfrac{a + b}{a^2b^2}$ $\quad K_1 = \dfrac{1}{ab}$
3.14	$\dfrac{s + d}{s^2(s - a)(s - b)}$	$Ae^{at} + Be^{bt} + K + K_1t$	$A = \dfrac{a + d}{a^2(a - b)}$ $\quad B = \dfrac{b + d}{b^2(b - a)}$ $K = \dfrac{ab + d(a + b)}{a^2b^2}$ $\quad K_1 = \dfrac{d}{ab}$
3.15	$\dfrac{s^2 + gs + d}{s^2(s - a)(s - b)}$		$A = \dfrac{a^2 + ag + d}{a^2(a - b)}$ $\quad B = \dfrac{b^2 + bg + d}{b^2(b - a)}$ $K = \dfrac{abg + d(a + b)}{a^2b^2}$ $\quad K_1 = \dfrac{d}{ab}$
3.16	$\dfrac{1}{s(s - a)^2(s - b)}$		$A = \dfrac{b - 2a}{a^2(a - b)^2}$ $\quad A_1 = \dfrac{1}{a(a - b)}$ $B = \dfrac{1}{b(a - b)^2}$ $\quad K = -\dfrac{1}{a^2b}$
3.17	$\dfrac{s + d}{s(s - a)^2(s - b)}$	$(A + A_1t)e^{at} + Be^{bt} + K$	$A = -\dfrac{a^2 + 2ad - bd}{a^2(a - b)^2}$ $\quad A_1 = \dfrac{a + d}{a(a - b)}$ $B = \dfrac{b + d}{b(a - b)^2}$ $\quad K = -\dfrac{d}{a^2b}$
3.18	$\dfrac{s^2 + gs + d}{s(s - a)^2(s - b)}$		$A = -\dfrac{a^2(b + g) + d(2a - b)}{a^2(a - b)^2}$ $\quad A_1 = \dfrac{a^2 + ag + d}{a(a - b)}$ $B = \dfrac{b^2 + bg + d}{b(a - b)^2}$ $\quad K = -\dfrac{d}{a^2b}$

Table D-6. Laplace-transform Pairs Involving Rational Algebraic Functions $F(s) = D_1(s)/D(s)$ (*Continued*)

No.	$F(s)$	$f(t)\ (t > 0)$		
3.19	$\dfrac{1}{(s-a)^2(s-b)(s-c)}$	$(A + A_1t)e^{at} + Be^{bt} + Ce^{ct}$	$A = \dfrac{(b+c) - 2a}{(a-b)^2(a-c)^2}$ $B = \dfrac{1}{(b-a)^2(b-c)}$	$A_1 = \dfrac{1}{(a-b)(a-c)}$ $C = \dfrac{1}{(c-a)^2(c-b)}$
3.20	$\dfrac{s+d}{(s-a)^2(s-b)(s-c)}$		$A = -\dfrac{a^2 + 2ad - d(b+c) - bc}{(a-b)^2(a-c)^2}$ $B = \dfrac{b+d}{(b-a)^2(b-c)}$	$A_1 = \dfrac{a+d}{(a-b)(a-c)}$ $C = \dfrac{c+d}{(c-a)^2(c-b)}$
3.21	$\dfrac{1}{s^2(s-a)^2}$	$(A + A_1t)e^{at} + K + K_1t$	$A = -\dfrac{2}{a^3}$ $K = -A$	$A_1 = \dfrac{1}{a^2}$ $K_1 = \dfrac{1}{a^2}$
3.22	$\dfrac{s+d}{s^2(s-a)^2}$		$A = -\dfrac{a+2d}{a^3}$ $K = -A$	$A_1 = \dfrac{a+d}{a^2}$ $K_1 = \dfrac{d}{a^2}$
3.23	$\dfrac{s^2 + gs + d}{s^2(s-a)^2}$		$A = -\dfrac{ag+2d}{a^3}$ $K = -A$	$A_1 = \dfrac{a^2 + ag + d}{a^2}$ $K_1 = \dfrac{d}{a^2}$
3.24	$\dfrac{1}{(s-a)^2(s-b)^2}$		$A = -\dfrac{2}{(a-b)^3}$ $B = -A$	$A_1 = \dfrac{1}{(a-b)^2}$ $B_1 = \dfrac{1}{(a-b)^2}$

3.25	$\frac{s+d}{(s-a)^2(s-b)^2}$	$(A + A_1t)e^{at} + (B + B_1t)e^{bt}$	$A = -\frac{a+b+2d}{(a-b)^3}$ $\quad A_1 = \frac{a+d}{(a-b)^2}$ $B = -A$ $\quad B_1 = \frac{b+d}{(a-b)^2}$
3.26	$\frac{s^2+gs+d}{(s-a)^2(s-b)^2}$		$A = -\frac{(a+b)g + 2(ab+d)}{(a-b)^3}$ $\quad A_1 = \frac{a^2+ag+d}{(a-b)^2}$ $B = -A$ $\quad B_1 = \frac{b^2+bg+d}{(a-b)^2}$
3.27	$\frac{1}{(s-a)^3(s-b)}$	$(A + A_1t + A_2t^2)e^{at} + Be^{bt}$	$A = \frac{1}{(a-b)^3}$ $\quad A_1 = -\frac{1}{(a-b)^2}$ $\quad A_2 = \frac{1}{2(a-b)}$ $\quad B = -A$
3.28	$\frac{s+d}{(s-a)^3(s-b)}$		$A = \frac{b+d}{(a-b)^3}$ $\quad A_1 = -\frac{b+d}{(a-b)^2}$ $\quad A_2 = \frac{a+d}{2(a-b)}$ $\quad B = -A$
3.29	$\frac{s^2+gs+d}{s^2(s-a)^3}$	$(A + A_1t + A_2t^2)e^{at} + K + K_1t$	$A = \frac{ag+3d}{a^4}$ $\quad A_1 = -\frac{ag+2d}{a^3}$ $\quad A_2 = \frac{a^2+ag+d}{2a^2}$ $K = -A$ $\quad K_1 = -\frac{d}{a^3}$
4.1	$\frac{1}{s^2[(s-a)^2+\omega_1^2]}$	$Ae^{at}\sin(\omega_1t + \alpha) + K + K_1t$	$A = \frac{1}{\omega_1(a^2+\omega_1^2)}$ $\quad \alpha = -2\arctan\frac{\omega_1}{a}$ $K = \frac{2a}{(a^2+\omega_1^2)^2}$ $\quad K_1 = \frac{1}{a^2+\omega_1^2}$
4.2	$\frac{s+d}{s^2[(s-a)^2+\omega_1^2]}$		$A = \frac{[(a+d)^2+\omega_1^2]^{1/2}}{\omega_1(a^2+\omega_1^2)}$ $\quad \alpha = \arctan\frac{\omega_1}{a+d} - 2\arctan\frac{\omega_1}{a}$ $K = \frac{a^2+\omega_1^2+2ad}{(a^2+\omega_1^2)^2}$ $\quad K_1 = \frac{d}{a^2+\omega_1^2}$
4.3	$\frac{s^2+gs+d}{s^2[(s-a)^2+\omega_1^2]}$		$A = \frac{[(a^2-\omega_1^2+ag+d)^2+\omega_1^2(2a+g)^2]^{1/2}}{\omega_1(a^2+\omega_1^2)}$ $\quad \alpha = \arctan\frac{\omega_1(2a+g)}{a^2-\omega_1^2+ag+d} - 2\arctan\frac{\omega_1}{a}$ $K = \frac{(a^2+\omega_1^2)g+2ad}{(a^2+\omega_1^2)^2}$ $\quad K_1 = \frac{d}{a^2+\omega_1^2}$

Table D-6. Laplace-transform Pairs Involving Rational Algebraic Functions $F(s) = D_1(s)/D(s)$ (Continued)

No.	$F(s)$	$f(t)\ (t > 0)$		
4.4	$\dfrac{1}{(s-b)^2[(s-a)^2+\omega_1^2]}$	$Ae^{at}\sin(\omega_1 t+\alpha)+(B+B_1 t)e^{bt}$	$A = \dfrac{1}{\omega_1[(a-b)^2+\omega_1^2]}$ $B = \dfrac{2(a-b)}{[(a-b)^2+\omega_1^2]^2}$	$\alpha = -2\arctan\dfrac{\omega_1}{a-b}$ $B_1 = \dfrac{1}{(a-b)^2+\omega_1^2}$
4.5	$\dfrac{s+d}{(s-b)^2[(s-a)^2+\omega_1^2]}$	$Ae^{at}\sin(\omega_1 t+\alpha)+(B+B_1 t)e^{bt}$	$A = \dfrac{[(a+d)^2+\omega_1^2]^{1/2}}{\omega_1[(a-b)^2+\omega_1^2]}$ $B = \dfrac{(a-b)^2+\omega_1^2+2(a-b)(b+d)}{[(a-b)^2+\omega_1^2]^2}$	$\alpha = \arctan\dfrac{\omega_1}{a+d} - 2\arctan\dfrac{\omega_1}{a-b}$ $B_1 = \dfrac{b+d}{(a-b)^2+\omega_1^2}$
4.6	$\dfrac{s^2+gs+d}{(s-b)^2[(s-a)^2+\omega_1^2]}$	$Ae^{at}\sin(\omega_1 t+\alpha)+(B+B_1 t)e^{bt}$	$A = \dfrac{[(a^2-\omega_1^2+ag+d)^2+\omega_1^2(2a+g)^2]^{1/2}}{\omega_1[(a-b)^2+\omega_1^2]}$ $B = \dfrac{[(a-b)^2+\omega_1^2](2b+g)+2(a-b)(b^2+bg+d)}{[(a-b)^2+\omega_1^2]^2}$	$\alpha = \arctan\dfrac{\omega_1(2a+g)}{a^2-\omega_1^2+ag+d} - 2\arctan\dfrac{\omega_1}{a-b}$ $B_1 = \dfrac{b^2+bg+d}{(a-b)^2+\omega_1^2}$
4.7	$\dfrac{1}{[(s-a)^2+\omega_1^2]^2}$	$\dfrac{1}{2\omega_1^3}e^{at}(\sin\omega_1 t - \omega_1 t\cos\omega_1 t)$		
4.8	$\dfrac{s-a}{[(s-a)^2+\omega_1^2]^2}$	$\dfrac{1}{2\omega_1}te^{at}\sin\omega_1 t$		
4.9	$\dfrac{(s-a)^2-\omega_1^2}{[(s-a)^2+\omega_1^2]^2}$	$te^{at}\cos\omega_1 t$		

Table D-7. Table of Laplace Transforms†

	$F(s)$	$f(t) \quad (t > 0)$
1	$\frac{1}{s}$	1
2	$\frac{1}{s^2}$	t
3	$\frac{1}{s^n} \; (n = 1, 2, \ldots)$	$\frac{t^{n-1}}{(n-1)!}$
4	$\frac{1}{\sqrt{s}}$	$\frac{1}{\sqrt{\pi t}}$
5	$s^{-\frac{3}{2}}$	$2\sqrt{\frac{t}{\pi}}$
6	$s^{-(n+\frac{1}{2})} \; (n = 1, 2, \ldots)$	$\frac{2^n t^{n-\frac{1}{2}}}{1 \times 3 \times 5 \cdots (2n-1)\sqrt{\pi}}$
7	$\frac{\Gamma(k)}{s^k} \; (k > 0)$	t^{k-1}
8	$\frac{1}{s-a}$	e^{at}
9	$\frac{1}{(s-a)^2}$	te^{at}
10	$\frac{1}{(s-a)^n} \; (n = 1, 2, \ldots)$	$\frac{1}{(n-1)!} t^{n-1} e^{at}$
11	$\frac{\Gamma(k)}{(s-a)^k} \; (k > 0)$	$t^{k-1} e^{at}$
12*	$\frac{1}{(s-a)(s-b)}$	$\frac{1}{a-b}(e^{at} - e^{bt})$
13*	$\frac{s}{(s-a)(s-b)}$	$\frac{1}{a-b}(ae^{at} - be^{bt})$
14*	$\frac{1}{(s-a)(s-b)(s-c)}$	$-\frac{(b-c)e^{at} + (c-a)e^{bt} + (a-b)e^{ct}}{(a-b)(b-c)(c-a)}$
15	$\frac{1}{s^2 + a^2}$	$\frac{1}{a} \sin at$
16	$\frac{s}{s^2 + a^2}$	$\cos at$

* Here a, b, and (in 14) c represent distinct constants.

† From Ruel V. Churchill, *Operational Mathematics*, 2d ed., McGraw-Hill, New York, 1958.

Table D-7. Table of Laplace Transforms (*Continued*)

	$F(s)$	$f(t) \quad (t > 0)$
17	$\dfrac{1}{s^2 - a^2}$	$\dfrac{1}{a} \sinh at$
18	$\dfrac{s}{s^2 - a^2}$	$\cosh at$
19	$\dfrac{1}{s(s^2 + a^2)}$	$\dfrac{1}{a^2} (1 - \cos at)$
20	$\dfrac{1}{s^2(s^2 + a^2)}$	$\dfrac{1}{a^3} (at - \sin at)$
21	$\dfrac{1}{(s^2 + a^2)^2}$	$\dfrac{1}{2a^3} (\sin at - at \cos at)$
22	$\dfrac{s}{(s^2 + a^2)^2}$	$\dfrac{t}{2a} \sin at$
23	$\dfrac{s^2}{(s^2 + a^2)^2}$	$\dfrac{1}{2a} (\sin at + at \cos at)$
24	$\dfrac{s^2 - a^2}{(s^2 + a^2)^2}$	$t \cos at$
25	$\dfrac{s}{(s^2 + a^2)(s^2 + b^2)} \quad (a^2 \neq b^2)$	$\dfrac{\cos at - \cos bt}{b^2 - a^2}$
26	$\dfrac{1}{(s - a)^2 + b^2}$	$\dfrac{1}{b} e^{at} \sin bt$
27	$\dfrac{s - a}{(s - a)^2 + b^2}$	$e^{at} \cos bt$
28	$\dfrac{3a^2}{s^3 + a^3}$	$e^{-at} - e^{at/2} \left(\cos \dfrac{at\sqrt{3}}{2} - \sqrt{3} \sin \dfrac{at\sqrt{3}}{2}\right)$
29	$\dfrac{4a^3}{s^4 + 4a^4}$	$\sin at \cosh at - \cos at \sinh at$
30	$\dfrac{s}{s^4 + 4a^4}$	$\dfrac{1}{2a^2} \sin at \sinh at$
31	$\dfrac{1}{s^4 - a^4}$	$\dfrac{1}{2a^3} (\sinh at - \sin at)$
32	$\dfrac{s}{s^4 - a^4}$	$\dfrac{1}{2a^2} (\cosh at - \cos at)$
33	$\dfrac{8a^3s^2}{(s^2 + a^2)^3}$	$(1 + a^2t^2) \sin at - at \cos at$
34	$\dfrac{1}{s} \left(\dfrac{s - 1}{s}\right)^n$	$\dfrac{e^t}{n!} \dfrac{d^n}{dt^n} (t^n e^{-t})$
35	$\dfrac{s}{(s - a)^{\frac{3}{2}}}$	$\dfrac{1}{\sqrt{\pi t}} e^{at}(1 + 2at)$
36	$\sqrt{s - a} - \sqrt{s - b}$	$\dfrac{1}{2\sqrt{\pi t^3}} (e^{bt} - e^{at})$

Table D-7. Table of Laplace Transforms (*Continued*)

	$F(s)$	$f(t) \qquad (t > 0)$
37	$\dfrac{1}{\sqrt{s} + a}$	$\dfrac{1}{\sqrt{\pi t}} - ae^{a^2 t} \operatorname{erfc}(a\sqrt{t})$
38	$\dfrac{\sqrt{s}}{s - a^2}$	$\dfrac{1}{\sqrt{\pi t}} + ae^{a^2 t} \operatorname{erf}(a\sqrt{t})$
39	$\dfrac{\sqrt{s}}{s + a^2}$	$\dfrac{1}{\sqrt{\pi t}} - \dfrac{2a}{\sqrt{\pi}} e^{-a^2 t} \int_0^{a\sqrt{t}} e^{\lambda^2}\, d\lambda$
40	$\dfrac{1}{\sqrt{s}\,(s - a^2)}$	$\dfrac{1}{a} e^{a^2 t} \operatorname{erf}(a\sqrt{t})$
41	$\dfrac{1}{\sqrt{s}\,(s + a^2)}$	$\dfrac{2}{a\sqrt{\pi}} e^{-a^2 t} \int_0^{a\sqrt{t}} e^{\lambda^2}\, d\lambda$
42	$\dfrac{b^2 - a^2}{(s - a^2)(b + \sqrt{s})}$	$e^{a^2 t}[b - a \operatorname{erf}(a\sqrt{t})] - be^{b^2 t} \operatorname{erfc}(b\sqrt{t})$
43	$\dfrac{1}{\sqrt{s}\,(\sqrt{s} + a)}$	$e^{a^2 t} \operatorname{erfc}(a\sqrt{t})$
44	$\dfrac{1}{(s + a)\sqrt{s + b}}$	$\dfrac{1}{\sqrt{b - a}} e^{-at} \operatorname{erf}(\sqrt{b - a}\,\sqrt{t})$
45	$\dfrac{b^2 - a^2}{\sqrt{s}\,(s - a^2)(\sqrt{s} + b)}$	$e^{a^2 t}\left[\dfrac{b}{a} \operatorname{erf}(a\sqrt{t}) - 1\right] + e^{b^2 t} \operatorname{erfc}(b\sqrt{t})$
46*	$\dfrac{(1 - s)^n}{s^{n+\frac{1}{2}}}$	$\dfrac{n!}{(2n)!\sqrt{\pi t}} H_{2n}(\sqrt{t})$
47	$\dfrac{(1 - s)^n}{s^{n+\frac{3}{2}}}$	$-\dfrac{n!}{\sqrt{\pi}\,(2n + 1)!} H_{2n+1}(\sqrt{t})$
48†	$\dfrac{\sqrt{s + 2a}}{\sqrt{s}} - 1$	$ae^{-at}[I_1(at) + I_0(at)]$
49	$\dfrac{1}{\sqrt{s + a}\,\sqrt{s + b}}$	$e^{-\frac{1}{2}(a+b)t} I_0\left(\dfrac{a - b}{2} t\right)$
50	$\dfrac{\Gamma(k)}{(s + a)^k (s + b)^k} \quad (k > 0)$	$\sqrt{\pi}\left(\dfrac{t}{a - b}\right)^{k-\frac{1}{2}} e^{-\frac{1}{2}(a+b)t} \times I_{k-\frac{1}{2}}\left(\dfrac{a - b}{2} t\right)$
51	$\dfrac{1}{(s + a)^{\frac{1}{2}}(s + b)^{\frac{3}{2}}}$	$te^{-\frac{1}{2}(a+b)t}\left[I_0\left(\dfrac{a - b}{2} t\right) + I_1\left(\dfrac{a - b}{2} t\right)\right]$
52	$\dfrac{\sqrt{s + 2a} - \sqrt{s}}{\sqrt{s + 2a} + \sqrt{s}}$	$\dfrac{1}{t} e^{-at} I_1(at)$

* $H_n(x)$ is the Hermite polynomial.

† $I_n(x) = i^{-n} J_n(ix)$, where J_n is Bessel's function of the first kind.

Table D-7. Table of Laplace Transforms (*Continued*)

	$F(s)$	$f(t) \quad (t > 0)$
53	$\dfrac{(a-b)^k}{(\sqrt{s+a}+\sqrt{s+b})^{2k}} \; (k > 0)$	$\dfrac{k}{t} e^{-\frac{1}{2}(a+b)t} I_k\left(\dfrac{a-b}{2} t\right)$
54	$\dfrac{(\sqrt{s+a}+\sqrt{s})^{-2\nu}}{\sqrt{s}\sqrt{s+a}} \; (\nu > -1)$	$\dfrac{1}{a^\nu} e^{-\frac{1}{2}at} I_\nu\left(\dfrac{1}{2} at\right)$
55	$\dfrac{1}{\sqrt{s^2+a^2}}$	$J_0(at)$
56	$\dfrac{(\sqrt{s^2+a^2}-s)^\nu}{\sqrt{s^2+a^2}} \; (\nu > -1)$	$a^\nu J_\nu(at)$
57	$\dfrac{1}{(s^2+a^2)^k} \; (k > 0)$	$\dfrac{\sqrt{\pi}}{\Gamma(k)} \left(\dfrac{t}{2a}\right)^{k-\frac{1}{2}} J_{k-\frac{1}{2}}(at)$
58	$(\sqrt{s^2+a^2}-s)^k \; (k > 0)$	$\dfrac{ka^k}{t} J_k(at)$
59	$\dfrac{(s-\sqrt{s^2-a^2})^\nu}{\sqrt{s^2-a^2}} \; (\nu > -1)$	$a^\nu I_\nu(at)$
60	$\dfrac{1}{(s^2-a^2)^k} \; (k > 0)$	$\dfrac{\sqrt{\pi}}{\Gamma(k)} \left(\dfrac{t}{2a}\right)^{k-\frac{1}{2}} I_{k-\frac{1}{2}}(at)$
61	$\dfrac{e^{-ks}}{s}$	$\begin{cases} 0 \text{ when } 0 < t < k \\ 1 \text{ when } t > k \end{cases}$
62	$\dfrac{e^{-ks}}{s^2}$	$\begin{cases} 0 & \text{when } 0 < t < k \\ t-k & \text{when } t > k \end{cases}$
63	$\dfrac{e^{-ks}}{s^\mu} \; (\mu > 0)$	$\begin{cases} 0 & \text{when } 0 < t < k \\ \dfrac{(t-k)^{\mu-1}}{\Gamma(\mu)} & \text{when } t > k \end{cases}$
64	$\dfrac{1-e^{-ks}}{s}$	$\begin{cases} 1 \text{ when } 0 < t < k \\ 0 \text{ when } t > k \end{cases}$
65	$\dfrac{1}{s(1-e^{-ks})} = \dfrac{1+\coth \frac{1}{2}ks}{2s}$	$1 + [t/k] = n$ when $(n-1)k < t < nk$ $(n = 1, 2, \ldots)$ (Fig. D-1)
66	$\dfrac{1}{s(e^{ks}-a)}$	$\begin{cases} 0 \quad \text{when } 0 < t < k \\ 1 + a + a^2 + \cdots + a^{n-1} \\ \quad \text{when } nk < t < (n+1)k \\ \quad (n = 1, 2, \ldots) \end{cases}$
67	$\dfrac{1}{s} \tanh ks$	$M(2k,t) = (-1)^{n-1}$ when $2k(n-1) < t < 2kn$ $(n = 1, 2, \ldots)$ (Fig. D-2)
68	$\dfrac{1}{s(1+e^{-ks})}$	$\dfrac{1}{2} M(k,t) + \dfrac{1}{2} = \dfrac{1-(-1)^n}{2}$ when $(n-1)k < t < nk$
69	$\dfrac{1}{s^2} \tanh ks$	$H(2k,t)$ (Fig. D-3)

Table D-7. Table of Laplace Transforms *(Continued)*

	$F(s)$	$f(t) \qquad (t > 0)$
70	$\dfrac{1}{s \sinh ks}$	$F(t) = 2(n-1)$ when $(2n-3)k < t < (2n-1)k$ $(t > 0)$
71	$\dfrac{1}{s \cosh ks}$	$M(2k, t+3k) + 1 = 1 + (-1)^n$ when $(2n-3)k < t < (2n-1)k$ $(t > 0)$
72	$\dfrac{1}{s} \coth ks$	$F(t) = 2n - 1$ when $2k(n-1) < t < 2kn$
73	$\dfrac{k}{s^2+k^2} \coth \dfrac{\pi s}{2k}$	$\lvert\sin kt\rvert$
74	$\dfrac{1}{(s^2+1)(1-e^{-\pi s})}$	$\begin{cases} \sin t \text{ when} \\ \quad (2n-2)\pi < t < (2n-1)\pi \\ 0 \quad \text{when} \\ \quad (2n-1)\pi < t < 2n\pi \end{cases}$
75	$\dfrac{1}{s} e^{-(k/s)}$	$J_0(2\sqrt{kt})$
76	$\dfrac{1}{\sqrt{s}} e^{-(k/s)}$	$\dfrac{1}{\sqrt{\pi t}} \cos 2\sqrt{kt}$
77	$\dfrac{1}{\sqrt{s}} e^{k/s}$	$\dfrac{1}{\sqrt{\pi t}} \cosh 2\sqrt{kt}$
78	$\dfrac{1}{s^{\frac{3}{2}}} e^{-(k/s)}$	$\dfrac{1}{\sqrt{\pi k}} \sin 2\sqrt{kt}$
79	$\dfrac{1}{s^{\frac{3}{2}}} e^{k/s}$	$\dfrac{1}{\sqrt{\pi k}} \sinh 2\sqrt{kt}$
80	$\dfrac{1}{s^{\mu}} e^{-(k/s)} \ (\mu > 0)$	$\left(\dfrac{t}{k}\right)^{(\mu-1)/2} J_{\mu-1}(2\sqrt{kt})$
81	$\dfrac{1}{s^{\mu}} e^{k/s} \ (\mu > 0)$	$\left(\dfrac{t}{k}\right)^{(\mu-1)/2} I_{\mu-1}(2\sqrt{kt})$
82	$e^{-k\sqrt{s}} \ (k > 0)$	$\dfrac{k}{2\sqrt{\pi t^3}} \exp\left(-\dfrac{k^2}{4t}\right)$
83	$\dfrac{1}{s} e^{-k\sqrt{s}} \ (k \geqq 0)$	$\operatorname{erfc}\left(\dfrac{k}{2\sqrt{t}}\right)$
84	$\dfrac{1}{\sqrt{s}} e^{-k\sqrt{s}} \ (k \geqq 0)$	$\dfrac{1}{\sqrt{\pi t}} \exp\left(-\dfrac{k^2}{4t}\right)$
85	$s^{-\frac{3}{2}} e^{-k\sqrt{s}} \ (k \geqq 0)$	$2\sqrt{\dfrac{t}{\pi}} \exp\left(-\dfrac{k^2}{4t}\right)$ $- k \operatorname{erfc}\left(\dfrac{k}{2\sqrt{t}}\right)$
86	$\dfrac{ae^{-k\sqrt{s}}}{s(a+\sqrt{s})} \ (k \geqq 0)$	$-e^{ak}e^{a^2t} \operatorname{erfc}\left(a\sqrt{t} + \dfrac{k}{2\sqrt{t}}\right)$ $+ \operatorname{erfc}\left(\dfrac{k}{2\sqrt{t}}\right)$

Table D-7. Table of Laplace Transforms (*Continued*)

	$F(s)$	$f(t) \qquad (t > 0)$
87	$\dfrac{e^{-k\sqrt{s}}}{\sqrt{s}\,(a + \sqrt{s})} \quad (k \geqq 0)$	$e^{ak}e^{a^2t} \operatorname{erfc}\left(a\sqrt{t} + \dfrac{k}{2\sqrt{t}}\right)$
88	$\dfrac{e^{-k\sqrt{s(s+a)}}}{\sqrt{s(s+a)}}$	$\begin{cases} 0 & \text{when } 0 < t < k \\ e^{-\frac{1}{2}at}I_0(\frac{1}{2}a\sqrt{t^2 - k^2}) & \text{when } t > k \end{cases}$
89	$\dfrac{e^{-k\sqrt{s^2+a^2}}}{\sqrt{s^2 + a^2}}$	$\begin{cases} 0 & \text{when } 0 < t < k \\ J_0(a\sqrt{t^2 - k^2}) & \text{when } t > k \end{cases}$
90	$\dfrac{e^{-k\sqrt{s^2-a^2}}}{\sqrt{s^2 - a^2}}$	$\begin{cases} 0 & \text{when } 0 < t < k \\ I_0(a\sqrt{t^2 - k^2}) & \text{when } t > k \end{cases}$
91	$\dfrac{e^{-k(\sqrt{s^2+a^2}-s)}}{\sqrt{s^2 + a^2}} \quad (k \geqq 0)$	$J_0(a\sqrt{t^2 + 2kt})$
92	$e^{-ks} - e^{-k\sqrt{s^2+a^2}}$	$\begin{cases} 0 & \text{when } 0 < t < k \\ \dfrac{ak}{\sqrt{t^2 - k^2}} J_1(a\sqrt{t^2 - k^2}) & \text{when } t > k \end{cases}$
93	$e^{-k\sqrt{s^2-a^2}} - e^{-ks}$	$\begin{cases} 0 & \text{when } 0 < t < k \\ \dfrac{ak}{\sqrt{t^2 - k^2}} I_1(a\sqrt{t^2 - k^2}) & \text{when } t > k \end{cases}$
94	$\dfrac{a^\nu e^{-k\sqrt{s^2+a^2}}}{\sqrt{s^2 + a^2}\,(\sqrt{s^2 + a^2} + s)^\nu} \quad (\nu > -1)$	$\begin{cases} 0 & \text{when } 0 < t < k \\ \left(\dfrac{t-k}{t+k}\right)^{\frac{1}{2}\nu} J_\nu(a\sqrt{t^2 - k^2}) & \text{when } t > k \end{cases}$
95	$\dfrac{1}{s}\log s$	$\Gamma'(1) - \log t \quad [\Gamma'(1) = -0.5772]$
96	$\dfrac{1}{s^k}\log s \quad (k > 0)$	$t^{k-1}\left\{\dfrac{\Gamma'(k)}{[\Gamma(k)]^2} - \dfrac{\log t}{\Gamma(k)}\right\}$
97	$\dfrac{\log s}{s - a} \quad (a > 0)$	$e^{at}[\log a - \operatorname{Ei}(-at)]$
98	$\dfrac{\log s}{s^2 + 1}$	$\cos t \operatorname{Si} t - \sin t \operatorname{Ci} t$
99	$\dfrac{s \log s}{s^2 + 1}$	$-\sin t \operatorname{Si} t - \cos t \operatorname{Ci} t$
100	$\dfrac{1}{s}\log(1 + ks) \quad (k > 0)$	$-\operatorname{Ei}\left(-\dfrac{t}{k}\right)$

Table D-7. Table of Laplace Transforms (*Continued*)

	$F(s)$	$f(t) \qquad (t > 0)$
101	$\log \frac{s-a}{s-b}$	$\frac{1}{t}(e^{bt} - e^{at})$
102	$\frac{1}{s} \log (1 + k^2 s^2)$	$-2 \operatorname{Ci}\left(\frac{t}{k}\right)$
103	$\frac{1}{s} \log (s^2 + a^2) \ (a > 0)$	$2 \log a - 2 \operatorname{Ci}(at)$
104	$\frac{1}{s^2} \log (s^2 + a^2) \ (a > 0)$	$\frac{2}{a}[at \log a + \sin at - at \operatorname{Ci}(at)]$
105	$\log \frac{s^2 + a^2}{s^2}$	$\frac{2}{t}(1 - \cos at)$
106	$\log \frac{s^2 - a^2}{s^2}$	$\frac{2}{t}(1 - \cosh at)$
107	$\arctan \frac{k}{s}$	$\frac{1}{t} \sin kt$
108	$\frac{1}{s} \arctan \frac{k}{s}$	$\operatorname{Si}(kt)$
109	$e^{k^2 s^2} \operatorname{erfc}(ks) \ (k > 0)$	$\frac{1}{k\sqrt{\pi}} \exp\left(-\frac{t^2}{4k^2}\right)$
110	$\frac{1}{s} e^{k^2 s^2} \operatorname{erfc}(ks) \ (k > 0)$	$\operatorname{erf}\left(\frac{t}{2k}\right)$
111	$e^{ks} \operatorname{erfc} \sqrt{ks} \ \ (k > 0)$	$\frac{\sqrt{k}}{\pi \sqrt{t}\,(t + k)}$
112	$\frac{1}{\sqrt{s}} \operatorname{erfc}(\sqrt{ks})$	$\begin{cases} 0 & \text{when } 0 < t < k \\ (\pi t)^{-\frac{1}{2}} & \text{when } t > k \end{cases}$
113	$\frac{1}{\sqrt{s}} e^{ks} \operatorname{erfc}(\sqrt{ks}) \ (k > 0)$	$\frac{1}{\sqrt{\pi(t + k)}}$
114	$\operatorname{erf}\left(\frac{k}{\sqrt{s}}\right)$	$\frac{1}{\pi t} \sin (2k\sqrt{t})$
115	$\frac{1}{\sqrt{s}} e^{k^2/s} \operatorname{erfc}\left(\frac{k}{\sqrt{s}}\right)$	$\frac{1}{\sqrt{\pi t}} e^{-2k\sqrt{t}}$
116	$K_0(ks)$	$\begin{cases} 0 & \text{when } 0 < t < k \\ (t^2 - k^2)^{-\frac{1}{2}} & \text{when } t > k \end{cases}$
117	$K_0(k\sqrt{s})$	$\frac{1}{2t} \exp\left(-\frac{k^2}{4t}\right)$
118	$\frac{1}{s} e^{ks} K_1(ks)$	$\frac{1}{k}\sqrt{t(t + 2k)}$
119	$\frac{1}{\sqrt{s}} K_1(k\sqrt{s})$	$\frac{1}{k} \exp\left(-\frac{k^2}{4t}\right)$
120	$\frac{1}{\sqrt{s}} e^{k/s} K_0\left(\frac{k}{s}\right)$	$\frac{2}{\sqrt{\pi t}} K_0(2\sqrt{2kt})$

Table D-7. Table of Laplace Transforms (*Continued*)

	$F(s)$	$f(t) \quad (t > 0)$
121	$\pi e^{-ks} I_0(ks)$	$\begin{cases} [t(2k-t)]^{-\frac{1}{2}} & \text{when } 0 < t < 2k \\ 0 & \text{when } t > 2k \end{cases}$
122	$e^{-ks} I_1(ks)$	$\begin{cases} \dfrac{k-t}{\pi k \sqrt{t(2k-t)}} & \text{when } 0 < t < 2k \\ 0 & \text{when } t > 2k \end{cases}$
123	$-e^{as} \operatorname{Ei}(-as)$	$\dfrac{1}{t+a} \quad (a > 0)$
124	$\dfrac{1}{a} + se^{as} \operatorname{Ei}(-as)$	$\dfrac{1}{(t+a)^2} \quad (a > 0)$
125	$\left(\dfrac{\pi}{2} - \operatorname{Si} s\right) \cos s + \operatorname{Ci} s \sin s$	$\dfrac{1}{t^2+1}$

FIG. D-1

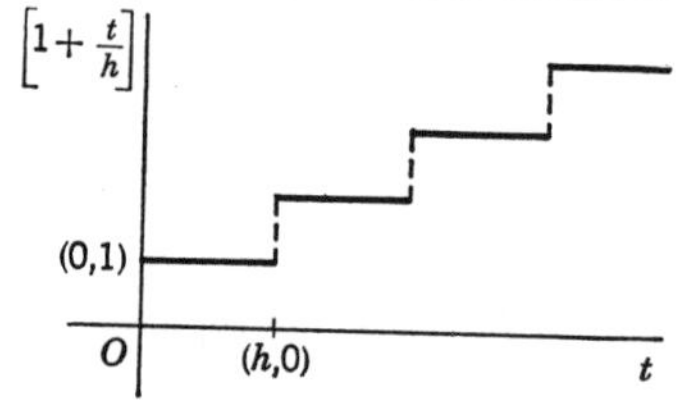

FIG. D-2

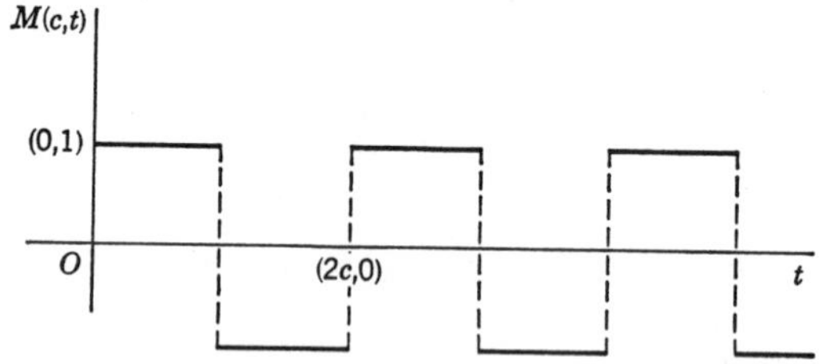

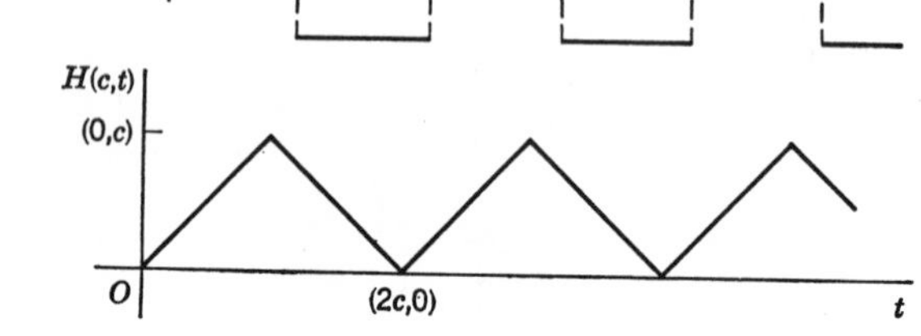

FIG. D-3

BIBLIOGRAPHY

Campbell, C. A., and Foster: *Fourier Integrals for Practical Applications*, Van Nostrand, Princeton, N.J., 1948.

Erdélyi, A.: *Integral Transforms*, vols. 1 and 2 (Bateman Project), McGraw-Hill, New York, 1954.

Oberhettinger, F.: *Tabellen zur Fourier Transformation*, Springer, Berlin, 1957.

Smith, J. J.: Tables of Green's Functions, Fourier Series, and Impulse Functions for Rectangular Coordinates, *Trans. AIEE*, **70**, 22, 1951.

Ditkin, V. A., and A. P. Prudnikov: *Integral Transforms and Operational Calculus*, Pergamon Press, New York, 1965.

APPENDIX **E**

INTEGRALS, SUMS, INFINITE SERIES AND PRODUCTS, AND CONTINUED FRACTIONS

Integral Tables

E-1. Elementary Indefinite Integrals. Add constant of integration in each case.

1. $\int x^n\,dx = \frac{x^{n+1}}{n+1} \qquad (n \neq -1)$

2. $\int \frac{dx}{x} = \log_e |x| \qquad (x \neq 0)$

3. $\int \sin x\,dx = -\cos x$

4. $\int \cos x\,dx = \sin x$

5. $\int \tan x\,dx = -\log_e |\cos x|$

6. $\int \cot x\,dx = \log_e |\sin x|$

7. $\int \frac{dx}{\cos^2 x} = \tan x$

8. $\int \frac{dx}{\sin^2 x} = -\cot x$

9. $\int \frac{dx}{a^2 + x^2} = \frac{1}{a} \arctan \frac{x}{a} \qquad (a \neq 0)$

10. $\int \frac{dx}{a^2 - x^2} = \frac{1}{a} \tanh^{-1} \frac{x}{a} = \frac{1}{2a} \log_e \frac{a + x}{a - x} \qquad (|x| < a)$

11. $\int \frac{dx}{x^2 - a^2} = -\frac{1}{a} \coth^{-1} \frac{x}{a} = \frac{1}{2a} \log_e \frac{x - a}{x + a} \qquad (|x| > a)$

12. $\int e^x \, dx = e^x$

13. $\int a^x \, dx = \frac{a^x}{\log_e a} \qquad (a > 0, a \neq 1)$

14. $\int \sinh x \, dx = \cosh x$

15. $\int \cosh x \, dx = \sinh x$

16. $\int \tanh x \, dx = \log_e \cosh x$

17. $\int \coth x \, dx = \log_e |\sinh x|$

18. $\int \frac{dx}{\cosh^2 x} = \tanh x$

19. $\int \frac{dx}{\sinh^2 x} = -\coth x$

20. $\int \frac{dx}{\sqrt{a^2 - x^2}} = \arcsin \frac{x}{a} \qquad (a \neq 0)$

21. $\int \frac{dx}{\sqrt{a^2 + x^2}} = \sinh^{-1} \frac{x}{a} = \log_e |x + \sqrt{a^2 + x^2}| + C_1$

22. $\int \frac{dx}{\sqrt{x^2 - a^2}} = \cosh^{-1} \frac{x}{a} = \log_e |x + \sqrt{x^2 - a^2}| + C_1$

E-2. Indefinite Integrals.* Add constants of integration as needed. Note: As customary in integral tables, interpret $\log_e f(x)$ as $\log_e |f(x)|$ whenever it occurs on the right-hand side of an integral formula and $f(x)$ is negative. m, n are integers.

(a) *Integrals containing* $ax + b \qquad (a \neq 0)$

1. $$\int (ax+b)^n\,dx = \frac{1}{a(n+1)}(ax+b)^{n+1} \qquad (n \neq -1)$$

2. $$\int \frac{dx}{ax+b} = \frac{1}{a}\log_e(ax+b)$$

3. $$\int x(ax+b)^n\,dx = \frac{1}{a^2(n+2)}(ax+b)^{n+2} - \frac{b}{a^2(n+1)}(ax+b)^{n+1} \qquad (n \neq -1, -2)$$

4. $$\int x^m(ax+b)^n\,dx$$
$$= \frac{1}{a(m+n+1)}\left[x^m(ax+b)^{n+1} - mb\int x^{m-1}(ax+b)^n\,dx\right]$$
$$= \frac{1}{m+n+1}\left[x^{m+1}(ax+b)^n + nb\int x^m(ax+b)^{n-1}\,dx\right]$$
$$(m > 0,\ m+n+1 \neq 0)$$

5. $$\int \frac{x\,dx}{ax+b} = \frac{x}{a} - \frac{b}{a^2}\log_e(ax+b)$$

6. $$\int \frac{x\,dx}{(ax+b)^2} = \frac{b}{a^2(ax+b)} + \frac{1}{a^2}\log_e(ax+b)$$

7. $$\int \frac{x\,dx}{(ax+b)^3} = \frac{b}{2a^2(ax+b)^2} - \frac{1}{a^2(ax+b)}$$

8. $$\int \frac{x\,dx}{(ax+b)^n} = \frac{1}{a^2}\left[\frac{b}{(n-1)(ax+b)^{n-1}} - \frac{1}{(n-2)(ax+b)^{n-2}}\right] \qquad (n \neq 1, 2)$$

* Adapted, in part, from I. Bronstein and K. Semendjajev, *Pocketbook of Mathematics*, 6th ed., published by the Soviet Government, Moscow, 1956.

9. $\displaystyle\int \frac{x^2\,dx}{ax+b} = \frac{1}{a^3}\left[\frac{1}{2}(ax+b)^2 - 2b(ax+b) + b^2 \log_e (ax+b)\right]$

10. $\displaystyle\int \frac{x^2\,dx}{(ax+b)^2} = \frac{1}{a^3}\left[(ax+b) - 2b \log_e (ax+b) - \frac{b^2}{ax+b}\right]$

11. $\displaystyle\int \frac{x^2\,dx}{(ax+b)^3} = \frac{1}{a^3}\left[\log_e (ax+b) + \frac{2b}{ax+b} - \frac{b^2}{2(ax+b)^2}\right]$

12. $\displaystyle\int x^2(ax+b)^n\,dx$

$$= \frac{1}{a^3}\left[\frac{(ax+b)^{n+3}}{n+3} - 2b\,\frac{(ax+b)^{n+2}}{n+2} + b^2\,\frac{(ax+b)^{n+1}}{n+1}\right]$$

$(n \neq -1, -2, -3)$

13. $\displaystyle\int \frac{dx}{x(ax+b)} = \frac{1}{b}\log_e \frac{x}{ax+b}$

14. $\displaystyle\int \frac{dx}{x(ax+b)^2} = \frac{1}{b(ax+b)} - \frac{1}{b^2}\log_e \frac{ax+b}{x}$

15. $\displaystyle\int \frac{dx}{x(ax+b)^3} = \frac{1}{b^3}\left[\frac{1}{2}\left(\frac{ax+2b}{ax+b}\right)^2 + \log_e \frac{x}{ax+b}\right]$

16. $\displaystyle\int \frac{dx}{x^2(ax+b)} = -\frac{1}{bx} + \frac{a}{b^2}\log_e \frac{ax+b}{x}$

17. $\displaystyle\int \frac{dx}{x^2(ax+b)^2} = -\frac{b+2ax}{b^2x(ax+b)} + \frac{2a}{b^3}\log_e \frac{ax+b}{x}$

18. $\displaystyle\int \frac{dx}{x^3(ax+b)} = \frac{2ax-b}{2b^2x^2} + \frac{a^2}{b^3}\log_e \frac{x}{ax+b}$

Let $ax + b \equiv X$ $(a \neq 0)$. Then

19. $\displaystyle\int \frac{x^3\,dx}{X} = \frac{1}{a^4}\left(\frac{X^3}{3} - \frac{3bX^2}{2} + 3b^2X - b^3 \log_e X\right)$

20. $\displaystyle\int \frac{x^3\,dx}{X^2} = \frac{1}{a^4}\left(\frac{X^2}{2} - 3bX + 3b^2 \log_e X + \frac{b^3}{X}\right)$

21. $\displaystyle\int \frac{x^3\,dx}{X^3} = \frac{1}{a^4}\left(X - 3b \log_e X - \frac{3b^2}{X} + \frac{b^3}{2X^2}\right)$

22. $$\int \frac{x^3\,dx}{X^4} = \frac{1}{a^4}\left(\log_e X + \frac{3b}{X} - \frac{3b^2}{2X^2} + \frac{b^3}{3X^3}\right)$$

23. $$\int \frac{x^3\,dx}{X^n}$$

$$= \frac{1}{a^4}\left[\frac{-1}{(n-4)X^{n-4}} + \frac{3b}{(n-3)X^{n-3}} - \frac{3b^2}{(n-2)X^{n-2}} + \frac{b^3}{(n-1)X^{n-1}}\right]$$

$(n \neq 1, 2, 3, 4)$

24. $$\int \frac{dx}{xX^n} = -\frac{1}{b^n}\left[\log_e \frac{X}{x} - \sum_{k=1}^{n-1} \binom{n-1}{k} \frac{(-a)^k x^k}{kX^k}\right] \qquad (n \geq 1)$$

25. $$\int \frac{dx}{x^2X^3} = -a\left[\frac{1}{2b^2X^2} + \frac{2}{b^3X} + \frac{1}{ab^3x} - \frac{3}{b^4}\log_e \frac{X}{x}\right]$$

26. $$\int \frac{dx}{x^2X^n} = -\frac{1}{b^{n+1}}\left[-\sum_{k=2}^{n} \binom{n}{k} \frac{(-a)^k x^{k-1}}{(k-1)X^{k-1}} + \frac{X}{x} - na\log_e \frac{X}{x}\right]$$

$(n \geq 2)$

27. $$\int \frac{dx}{x^3X^2} = -\frac{1}{b^4}\left[3a^2\log_e \frac{X}{x} + \frac{a^3x}{X} + \frac{X^2}{2x^2} - \frac{3aX}{x}\right]$$

28. $$\int \frac{dx}{x^3X^3} = -\frac{1}{b^5}\left[6a^2\log_e \frac{X}{x} + \frac{4a^3x}{X} - \frac{a^4x^2}{2X^2} + \frac{X^2}{2x^2} - \frac{4aX}{x}\right]$$

29. $$\int \frac{dx}{x^3X^n} = -\frac{1}{b^{n+2}}\left[-\sum_{k=3}^{n+1} \binom{n+1}{k} \frac{(-a)^k x^{k-2}}{(k-2)X^{k-2}} + \frac{a^2X^2}{2x^2} - \frac{(n+1)aX}{x} + \frac{n(n+1)a^2}{2}\log_e \frac{X}{x}\right] \qquad (n \geq 3)$$

30. $$\int \frac{dx}{x^mX^n} = -\frac{1}{b^{m+n-1}}\sum_{k=0}^{m+n-2} \binom{m+n-2}{k} \frac{X^{m-k-1}(-a)^k}{(m-k-1)x^{m-k-1}}$$

[terms with $(m-k-1)=0$ are replaced by $\binom{m+n-2}{m-1}(-a)^{m-1}\log_e \frac{X}{x}$]

(b) *Integrals containing* $ax+b$ *and* $cx+d$ $\qquad (a \neq 0, c \neq 0)$

31. $$\int \frac{ax+b}{cx+d}\,dx = \frac{a}{c}x + \frac{bc-ad}{c^2}\log_e (cx+d)$$

32. $\displaystyle\int \frac{dx}{(ax+b)(cx+d)} = \frac{1}{bc-ad}\log_e \frac{cx+d}{ax+b} \qquad (bc-ad \neq 0)$

33. $\displaystyle\int \frac{x\,dx}{(ax+b)(cx+d)}$

$$= \frac{1}{bc-ad}\left[\frac{b}{a}\log(ax+b) - \frac{d}{c}\log_e(cx+d)\right] \qquad (bc-ad \neq 0)$$

34. $\displaystyle\int \frac{dx}{(ax+b)^2(cx+d)}$

$$= \frac{1}{bc-ad}\left[\frac{1}{ax+b} + \frac{c}{bc-ad}\log_e \frac{cx+d}{ax+b}\right] \qquad (bc-ad \neq 0)$$

(c) *Integrals containing* $a + x$ *and* $b + x$ $\qquad (a \neq b)$

35. $\displaystyle\int \frac{x\,dx}{(a+x)(b+x)^2} = \frac{b}{(a-b)(b+x)} - \frac{a}{(a-b)^2}\log_e \frac{a+x}{b+x}$

36. $\displaystyle\int \frac{x^2\,dx}{(a+x)(b+x)^2} = \frac{b^2}{(b-a)(b+x)} + \frac{a^2}{(b-a)^2}\log_e(a+x)$

$$+ \frac{b^2-2ab}{(b-a)^2}\log_e(b+x)$$

37. $\displaystyle\int \frac{dx}{(a+x)^2(b+x)^2}$

$$= \frac{-1}{(a-b)^2}\left(\frac{1}{a+x} + \frac{1}{b+x}\right) + \frac{2}{(a-b)^3}\log_e \frac{a+x}{b+x}$$

38. $\displaystyle\int \frac{x\,dx}{(a+x)^2(b+x)^2}$

$$= \frac{1}{(a-b)^2}\left(\frac{a}{a+x} + \frac{b}{b+x}\right) + \frac{a+b}{(a-b)^3}\log_e \frac{a+x}{b+x}$$

39. $\displaystyle\int \frac{x^2\,dx}{(a+x)^2(b+x)^2}$

$$= \frac{-1}{(a-b)^2}\left(\frac{a^2}{a+x} + \frac{b^2}{b+x}\right) + \frac{2ab}{(a-b)^3}\log_e \frac{a+x}{b+x}$$

(d) *Integrals containing* $ax^2 + bx + c \qquad (a \neq 0)$

40. $$\int \frac{dx}{ax^2 + bx + c} = \begin{cases} \dfrac{1}{\sqrt{b^2 - 4ac}} \log_e \dfrac{2ax + b - \sqrt{b^2 - 4ac}}{2ax + b + \sqrt{b^2 - 4ac}} & (b^2 > 4ac) \\ \dfrac{2}{\sqrt{4ac - b^2}} \arctan \dfrac{2ax + b}{\sqrt{4ac - b^2}} & (b^2 < 4ac) \\ -\dfrac{2}{2ax + b} & (b^2 = 4ac) \end{cases}$$

In numbers 41 through 47, let $b^2 - 4ac \neq 0$.

41. $$\int \frac{dx}{(ax^2 + bx + c)^2} = \frac{2ax + b}{(4ac - b^2)(ax^2 + bx + c)} + \frac{2a}{(4ac - b^2)} \int \frac{dx}{ax^2 + bx + c}$$

42. $$\int \frac{dx}{(ax^2 + bx + c)^{n+1}} = \frac{2ax + b}{n(4ac - b^2)(ax^2 + bx + c)^n} + \frac{2(2n - 1)a}{n(4ac - b^2)} \int \frac{dx}{(ax^2 + bx + c)^n}$$

43. $$\int \frac{x\,dx}{ax^2 + bx + c} = \frac{1}{2a} \log_e (ax^2 + bx + c) - \frac{b}{2a} \int \frac{dx}{ax^2 + bx + c}$$

44. $$\int \frac{x^2\,dx}{ax^2 + bx + c} = \frac{x}{a} - \frac{b}{2a^2} \log_e (ax^2 + bx + c) + \frac{b^2 - 2ac}{2a^2} \int \frac{dx}{ax^2 + bx + c}$$

45. $$\int \frac{x^n\,dx}{ax^2 + bx + c} = \frac{x^{n-1}}{(n - 1)a} - \frac{c}{a} \int \frac{x^{n-2}\,dx}{ax^2 + bx + c} - \frac{b}{a} \int \frac{x^{n-1}\,dx}{ax^2 + bx + c} \qquad (n \neq 1)$$

46. $$\int \frac{x\,dx}{(ax^2 + bx + c)^{n+1}} = \frac{-(2c + bx)}{n(4ac - b^2)(ax^2 + bx + c)^n} - \frac{b(2n - 1)}{n(4ac - b^2)} \int \frac{dx}{(ax^2 + bx + c)^n}$$

47. $$\int \frac{x^m\,dx}{(ax^2+bx+c)^{n+1}} = -\frac{x^{m-1}}{a(2n-m+1)(ax^2+bx+c)^n} - \frac{n-m+1}{2n-m+1}\cdot\frac{b}{a}\int\frac{x^{m-1}\,dx}{(ax^2+bx+c)^{n+1}} + \frac{m-1}{2n-m+1}\cdot\frac{c}{a}\int\frac{x^{m-2}\,dx}{(ax^2+bx+c)^{n+1}} \qquad (m \neq 2n+1)$$

Let $ax^2 + bx + c \equiv X$ $(a \neq 0)$. Then

48. $$\int \frac{x^{2n-1}\,dx}{X^n} = \frac{1}{a}\int\frac{x^{2n-3}\,dx}{X^{n-1}} - \frac{c}{a}\int\frac{x^{2n-3}\,dx}{X^n} - \frac{b}{a}\int\frac{x^{2n-2}\,dx}{X^n}$$

49. $$\int \frac{dx}{xX} = \frac{1}{2c}\log_e\frac{x^2}{X} - \frac{b}{2c}\int\frac{dx}{X}$$

50. $$\int \frac{dx}{xX^n} = \frac{1}{2c(n-1)X^{n-1}} - \frac{b}{2c}\int\frac{dx}{X^n} + \frac{1}{c}\int\frac{dx}{xX^{n-1}}$$

51. $$\int \frac{dx}{x^2X} = \frac{b}{2c^2}\log_e\frac{X}{x^2} - \frac{1}{cx} + \left(\frac{b^2}{2c^2} - \frac{a}{c}\right)\int\frac{dx}{X}$$

52. $$\int \frac{dx}{x^mX^n} = -\frac{1}{(m-1)cx^{m-1}X^{n-1}} - \frac{(2n+m-3)a}{(m-1)c}\int\frac{dx}{x^{m-2}X^n} - \frac{(n+m-2)b}{(m-1)c}\int\frac{dx}{x^{m-1}X^n} \qquad (m > 1)$$

53. $$\int \frac{dx}{(fx+g)X} = \frac{1}{2(cf^2-gbf+g^2a)}\left[f\log_e\frac{(fx+g)^2}{X}\right] + \frac{2ga-bf}{2(cf^2-gbf+g^2a)}\int\frac{dx}{X}$$

(e) Integrals containing $a^2 \pm x^2$, with

$$X \equiv a^2 + x^2 \qquad Y \equiv \arctan\frac{x}{a}$$

or

$$X \equiv a^2 - x^2 \qquad Y \equiv \tanh^{-1}\frac{x}{a} \equiv \begin{cases}\dfrac{1}{2}\log_e\dfrac{a+x}{a-x} & (|x| < a)\\[2ex] \dfrac{1}{2}\log_e\dfrac{x+a}{x-a} & (|x| > a)\end{cases}$$

Where $\pm$ or $\mp$ appears in a formula, the upper sign refers to $X \equiv a^2 + x^2$ and the lower sign to $X \equiv a^2 - x^2$ $(a \neq 0)$.

54. $$\int \frac{dx}{X} = \frac{1}{a} Y$$

55. $$\int \frac{dx}{X^2} = \frac{x}{2a^2X} + \frac{1}{2a^3} Y$$

56. $$\int \frac{dx}{X^3} = \frac{x}{4a^2X^2} + \frac{3x}{8a^4X} + \frac{3}{8a^5} Y$$

57. $$\int \frac{dx}{X^{n+1}} = \frac{x}{2na^2X^n} + \frac{2n-1}{2na^2} \int \frac{dx}{X^n}$$

58. $$\int \frac{x\,dx}{X} = \pm \frac{1}{2} \log_e X$$

59. $$\int \frac{x\,dx}{X^2} = \mp \frac{1}{2X}$$

60. $$\int \frac{x\,dx}{X^3} = \mp \frac{1}{4X^2}$$

61. $$\int \frac{x\,dx}{X^{n+1}} = \mp \frac{1}{2nX^n} \qquad (n \neq 0)$$

62. $$\int \frac{x^2\,dx}{X} = \pm x \mp aY$$

63. $$\int \frac{x^2\,dx}{X^2} = \mp \frac{x}{2X} \pm \frac{1}{2a} Y$$

64. $$\int \frac{x^2\,dx}{X^3} = \mp \frac{x}{4X^2} \pm \frac{x}{8a^2X} \pm \frac{1}{8a^3} Y$$

65. $$\int \frac{x^2\,dx}{X^{n+1}} = \mp \frac{x}{2nX^n} \pm \frac{1}{2n} \int \frac{dx}{X^n} \qquad (n \neq 0)$$

66. $$\int \frac{x^3\,dx}{X} = \pm \frac{x^2}{2} - \frac{a^2}{2} \log_e X$$

67. $$\int \frac{x^3\,dx}{X^2} = \frac{a^2}{2X} + \frac{1}{2} \log_e X$$

68. $$\int \frac{x^3\,dx}{X^3} = -\frac{1}{2X} + \frac{a^2}{4X^2}$$

69. $\int \frac{x^3\,dx}{X^{n+1}} = -\frac{1}{2(n-1)X^{n-1}} + \frac{a^2}{2nX^n} \qquad (n > 1)$

70. $\int \frac{dx}{xX} = \frac{1}{2a^2}\log_e \frac{x^2}{X}$

71. $\int \frac{dx}{xX^2} = \frac{1}{2a^2X} + \frac{1}{2a^4}\log_e \frac{x^2}{X}$

72. $\int \frac{dx}{xX^3} = \frac{1}{4a^2X^2} + \frac{1}{2a^4X} + \frac{1}{2a^6}\log_e \frac{x^2}{X}$

73. $\int \frac{dx}{x^2X} = -\frac{1}{a^2x} \mp \frac{1}{a^3}Y$

74. $\int \frac{dx}{x^2X^2} = -\frac{1}{a^4x} \mp \frac{x}{2a^4X} \mp \frac{3}{2a^5}Y$

75. $\int \frac{dx}{x^2X^3} = -\frac{1}{a^6x} \mp \frac{x}{4a^4X^2} \mp \frac{7x}{8a^6X} \mp \frac{15}{8a^7}Y$

76. $\int \frac{dx}{x^3X} = -\frac{1}{2a^2x^2} \mp \frac{1}{2a^4}\log_e \frac{x^2}{X}$

77. $\int \frac{dx}{x^3X^2} = -\frac{1}{2a^4x^2} \mp \frac{1}{2a^4X} \mp \frac{1}{a^6}\log_e \frac{x^2}{X}$

78. $\int \frac{dx}{x^3X^3} = -\frac{1}{2a^6x^2} \mp \frac{1}{a^6X} \mp \frac{1}{4a^4X^2} \mp \frac{3}{2a^8}\log_e \frac{x^2}{X}$

79. $\int \frac{dx}{(b+cx)X} = \frac{1}{a^2c^2 \pm b^2}\left[c\log_e(b+cx) - \frac{c}{2}\log_e X \pm \frac{b}{a}Y\right]$

(f) Integrals containing $a^3 \pm x^3$, with

$$X \equiv a^3 \pm x^3 \qquad (a \neq 0)$$

Where $\pm$ or $\mp$ appears in a formula, the upper sign refers to $X \equiv a^3 + x^3$ and the lower sign to $X \equiv a^3 - x^3$.

80. $\int \frac{dx}{X} = \pm\frac{1}{6a^2}\log_e \frac{(a \pm x)^2}{a^2 \mp ax + x^2} + \frac{1}{a^2\sqrt{3}}\arctan \frac{2x \mp a}{a\sqrt{3}}$

81. $\int \frac{dx}{X^2} = \frac{x}{3a^3X} + \frac{2}{3a^3}\int \frac{dx}{X}$

82. $\int \frac{x\,dx}{X} = \frac{1}{6a} \log_e \frac{a^2 \mp ax + x^2}{(a \pm x)^2} \pm \frac{1}{a\sqrt{3}} \arctan \frac{2x \mp a}{a\sqrt{3}}$

83. $\int \frac{x\,dx}{X^2} = \frac{x^2}{3a^3X} + \frac{1}{3a^3} \int \frac{x\,dx}{X}$

84. $\int \frac{x^2\,dx}{X} = \pm \frac{1}{3} \log_e X$

85. $\int \frac{x^2\,dx}{X^2} = \mp \frac{1}{3X}$

86. $\int \frac{x^3\,dx}{X} = \pm x \mp a^3 \int \frac{dx}{X}$

87. $\int \frac{x^3\,dx}{X^2} = \mp \frac{x}{3X} \pm \frac{1}{3} \int \frac{dx}{X}$

88. $\int \frac{dx}{xX} = \frac{1}{3a^3} \log_e \frac{x^3}{X}$

89. $\int \frac{dx}{xX^2} = \frac{1}{3a^3X} + \frac{1}{3a^6} \log_e \frac{x^3}{X}$

90. $\int \frac{dx}{x^2X} = -\frac{1}{a^3x} \mp \frac{1}{a^3} \int \frac{x\,dx}{X}$

91. $\int \frac{dx}{x^2X^2} = -\frac{1}{a^6x} \mp \frac{x^2}{3a^6X} \mp \frac{4}{3a^6} \int \frac{x\,dx}{X}$

92. $\int \frac{dx}{x^3X} = -\frac{1}{2a^3x^2} \mp \frac{1}{a^3} \int \frac{dx}{X}$

93. $\int \frac{dx}{x^3X^2} = -\frac{1}{2a^6x^2} \mp \frac{x}{3a^6X} \mp \frac{5}{3a^6} \int \frac{dx}{X}$

(g) *Integrals containing* $a^4 \pm x^4$ $\quad (a \neq 0)$

94. $\int \frac{dx}{a^4 + x^4} = \frac{1}{4a^3\sqrt{2}} \log_e \frac{x^2 + ax\sqrt{2} + a^2}{x^2 - ax\sqrt{2} + a^2} + \frac{1}{2a^3\sqrt{2}} \arctan \frac{ax\sqrt{2}}{a^2 - x^2}$

95. $\int \frac{x\,dx}{a^4 + x^4} = \frac{1}{2a^2} \arctan \frac{x^2}{a^2}$

96. $\int \frac{x^2\,dx}{a^4 + x^4} = -\frac{1}{4a\sqrt{2}} \log_e \frac{x^2 + ax\sqrt{2} + a^2}{x^2 - ax\sqrt{2} + a^2} + \frac{1}{2a\sqrt{2}} \arctan \frac{ax\sqrt{2}}{a^2 - x^2}$

97. $$\int \frac{x^3\,dx}{a^4 + x^4} = \frac{1}{4} \log_e (a^4 + x^4)$$

98. $$\int \frac{dx}{a^4 - x^4} = \frac{1}{4a^3} \log_e \frac{a + x}{a - x} + \frac{1}{2a^3} \arctan \frac{x}{a}$$

99. $$\int \frac{x\,dx}{a^4 - x^4} = \frac{1}{4a^3} \log_e \frac{a^2 + x^2}{a^2 - x^2}$$

100. $$\int \frac{x^2\,dx}{a^4 - x^4} = \frac{1}{4a} \log_e \frac{a + x}{a - x} - \frac{1}{2a} \arctan \frac{x}{a}$$

101. $$\int \frac{x^3\,dx}{a^4 - x^4} = -\frac{1}{4} \log_e (a^4 - x^4)$$

(h) *Integrals containing* $\sqrt{x}$ *and* $a^2 + b^2x$ $(a, b \neq 0)$

102. $$\int \frac{\sqrt{x}\,dx}{a^2 + b^2x} = \frac{2\sqrt{x}}{b^2} - \frac{2a}{b^3} \arctan \frac{b\sqrt{x}}{a}$$

103. $$\int \frac{x\sqrt{x}\,dx}{a^2 + b^2x} = \frac{2x\sqrt{x}}{3b^2} - \frac{2a^2\sqrt{x}}{b^4} + \frac{2a^3}{b^5} \arctan \frac{b\sqrt{x}}{a}$$

104. $$\int \frac{\sqrt{x}\,dx}{(a^2 + b^2x)^2} = -\frac{\sqrt{x}}{b^2(a^2 + b^2x)} + \frac{1}{ab^3} \arctan \frac{b\sqrt{x}}{a}$$

105. $$\int \frac{x\sqrt{x}\,dx}{(a^2 + b^2x)^2} = \frac{2b^2x\sqrt{x} + 3a^2\sqrt{x}}{b^4(a^2 + b^2x)} - \frac{3a}{b^5} \arctan \frac{b\sqrt{x}}{a}$$

106. $$\int \frac{dx}{(a^2 + b^2x)\sqrt{x}} = \frac{2}{ab} \arctan \frac{b\sqrt{x}}{a}$$

107. $$\int \frac{dx}{(a^2 + b^2x)x\sqrt{x}} = -\frac{2}{a^2\sqrt{x}} - \frac{2b}{a^3} \arctan \frac{b\sqrt{x}}{a}$$

108. $$\int \frac{dx}{(a^2 + b^2x)^2\sqrt{x}} = \frac{\sqrt{x}}{a^2(a^2 + b^2x)} + \frac{1}{a^3b} \arctan \frac{b\sqrt{x}}{a}$$

(i) *Integrals containing* $\sqrt{x}$ *and* $a^2 - b^2x > 0$ $(a, b \neq 0)$

109. $$\int \frac{\sqrt{x}\,dx}{a^2 - b^2x} = -\frac{2\sqrt{x}}{b^2} + \frac{a}{b^3} \log_e \frac{a + b\sqrt{x}}{a - b\sqrt{x}}$$

110. $$\int \frac{x\sqrt{x}\,dx}{a^2 - b^2x} = -\frac{2x\sqrt{x}}{3b^2} - \frac{2a^2\sqrt{x}}{b^4} + \frac{a^3}{b^5} \log_e \frac{a + b\sqrt{x}}{a - b\sqrt{x}}$$

111. $$\int \frac{\sqrt{x}\,dx}{(a^2-b^2x)^2} = \frac{\sqrt{x}}{b^2(a^2-b^2x)} - \frac{1}{2ab^3}\log_e\frac{a+b\sqrt{x}}{a-b\sqrt{x}}$$

112. $$\int \frac{x\sqrt{x}\,dx}{(a^2-b^2x)^2} = \frac{-2b^2x\sqrt{x}+3a^2\sqrt{x}}{b^4(a^2-b^2x)} - \frac{3a}{2b^5}\log_e\frac{a+b\sqrt{x}}{a-b\sqrt{x}}$$

113. $$\int \frac{dx}{(a^2-b^2x)\sqrt{x}} = \frac{1}{ab}\log_e\frac{a+b\sqrt{x}}{a-b\sqrt{x}}$$

114. $$\int \frac{dx}{(a^2-b^2x)x\sqrt{x}} = -\frac{2}{a^2\sqrt{x}} + \frac{b}{a^3}\log_e\frac{a+b\sqrt{x}}{a-b\sqrt{x}}$$

115. $$\int \frac{dx}{(a^2-b^2x)^2\sqrt{x}} = \frac{\sqrt{x}}{a^2(a^2-b^2x)} + \frac{1}{2a^3b}\log_e\frac{a+b\sqrt{x}}{a-b\sqrt{x}}$$

116. $$\int \frac{dx}{(a^2-b^2x)^2x\sqrt{x}} = -\frac{2}{a^2(a^2-b^2x)\sqrt{x}} + \frac{3b^2\sqrt{x}}{a^4(a^2-b^2x)} + \frac{3b}{2a^5}\log_e\frac{a+b\sqrt{x}}{a-b\sqrt{x}}$$

(j) *Other integrals containing* $\sqrt{x}$ $\quad (a > \sqrt{x} > 0)$

117. $$\int \frac{\sqrt{x}\,dx}{a^4+x^2} = -\frac{1}{2a\sqrt{2}}\log_e\frac{x+a\sqrt{2x}+a^2}{x-a\sqrt{2x}+a^2} + \frac{1}{a\sqrt{2}}\arctan\frac{a\sqrt{2x}}{a^2-x}$$

118. $$\int \frac{dx}{(a^4+x^2)\sqrt{x}} = \frac{1}{2a^3\sqrt{2}}\log_e\frac{x+a\sqrt{2x}+a^2}{x-a\sqrt{2x}+a^2} + \frac{1}{a^3\sqrt{2}}\arctan\frac{a\sqrt{2x}}{a^2-x}$$

119. $$\int \frac{\sqrt{x}\,dx}{a^4-x^2} = \frac{1}{2a}\log_e\frac{a+\sqrt{x}}{a-\sqrt{x}} - \frac{1}{a}\arctan\frac{\sqrt{x}}{a}$$

120. $$\int \frac{dx}{(a^4-x^2)\sqrt{x}} = \frac{1}{2a^3}\log_e\frac{a+\sqrt{x}}{a-\sqrt{x}} + \frac{1}{a^3}\arctan\frac{\sqrt{x}}{a}$$

(k) *Integrals containing* $\sqrt{ax+b}$***, with***

$$X \equiv ax+b \qquad Y \equiv fx+g \qquad \Delta \equiv bf-ag \qquad (a \neq 0)$$

121. $\int \sqrt{X}\,dx = \frac{2}{3a}\sqrt{X^3}$

122. $\int x\sqrt{X}\,dx = \frac{2(3ax - 2b)\sqrt{X^3}}{15a^2}$

123. $\int x^2\sqrt{X}\,dx = \frac{2(15a^2x^2 - 12abx + 8b^2)\sqrt{X^3}}{105a^3}$

124. $\int \frac{dx}{\sqrt{X}} = \frac{2\sqrt{X}}{a}$

125. $\int \frac{x\,dx}{\sqrt{X}} = \frac{2(ax - 2b)}{3a^2}\sqrt{X}$

126. $\int \frac{x^2\,dx}{\sqrt{X}} = \frac{2(3a^2x^2 - 4abx + 8b^2)\sqrt{X}}{15a^3}$

127. $\int \frac{dx}{x\sqrt{X}} = \begin{cases} -\frac{2}{\sqrt{b}}\tanh^{-1}\sqrt{\frac{X}{b}} = \frac{1}{\sqrt{b}}\log_e\frac{\sqrt{X} - \sqrt{b}}{\sqrt{X} + \sqrt{b}} & (b > 0) \\ \frac{2}{\sqrt{-b}}\arctan\sqrt{\frac{X}{-b}} & (b < 0) \end{cases}$

128. $\int \frac{\sqrt{X}}{x}\,dx = 2\sqrt{X} + b\int \frac{dx}{x\sqrt{X}}$

129. $\int \frac{dx}{x^2\sqrt{X}} = -\frac{\sqrt{X}}{bx} - \frac{a}{2b}\int \frac{dx}{x\sqrt{X}}$

130. $\int \frac{\sqrt{X}}{x^2}\,dx = -\frac{\sqrt{X}}{x} + \frac{a}{2}\int \frac{dx}{x\sqrt{X}}$

131. $\int \frac{dx}{x^n\sqrt{X}} = -\frac{\sqrt{X}}{(n-1)bx^{n-1}} - \frac{(2n-3)a}{(2n-2)b}\int \frac{dx}{x^{n-1}\sqrt{X}} \qquad (n \neq 1)$

132. $\int \sqrt{X^3}\,dx = \frac{2\sqrt{X^5}}{5a}$

133. $\int x\sqrt{X^3}\,dx = \frac{2}{35a^2}(5\sqrt{X^7} - 7b\sqrt{X^5})$

134. $\int x^2\sqrt{X^3}\,dx = \frac{2}{a^3}\left(\frac{\sqrt{X^9}}{9} - \frac{2b\sqrt{X^7}}{7} + \frac{b^2\sqrt{X^5}}{5}\right)$

135. $\displaystyle\int \frac{\sqrt{X^3}}{x}\,dx = \frac{2\sqrt{X^3}}{3} + 2b\sqrt{X} + b^2 \int \frac{dx}{x\sqrt{X}}$

136. $\displaystyle\int \frac{x\,dx}{\sqrt{X^3}} = \frac{2}{a^2}\left(\sqrt{X} + \frac{b}{\sqrt{X}}\right)$

137. $\displaystyle\int \frac{x^2\,dx}{\sqrt{X^3}} = \frac{2}{a^3}\left(\frac{\sqrt{X^3}}{3} - 2b\sqrt{X} - \frac{b^2}{\sqrt{X}}\right)$

138. $\displaystyle\int \frac{dx}{x\sqrt{X^3}} = \frac{2}{b\sqrt{X}} + \frac{1}{b}\int \frac{dx}{x\sqrt{X}} \qquad (b \neq 0)$

139. $\displaystyle\int \frac{dx}{x^2\sqrt{X^3}} = -\frac{1}{bx\sqrt{X}} - \frac{3a}{b^2\sqrt{X}} - \frac{3a}{2b^2}\int \frac{dx}{x\sqrt{X}} \qquad (b \neq 0)$

140. $\displaystyle\int X^{\pm n/2}\,dx = \frac{2X^{(2\pm n)/2}}{a(2 \pm n)} \qquad (n \neq \mp 2)$

141. $\displaystyle\int xX^{\pm n/2}\,dx = \frac{2}{a^2}\left(\frac{X^{(4\pm n)/2}}{4 \pm n} - \frac{bX^{(2\pm n)/2}}{2 \pm n}\right) \qquad (n \neq \mp 2, \mp 4)$

142. $\displaystyle\int x^2X^{\pm n/2}\,dx = \frac{2}{a^3}\left(\frac{X^{(6\pm n)/2}}{6 \pm n} - \frac{2bX^{(4\pm n)/2}}{4 \pm n} + \frac{b^2X^{(2\pm n)/2}}{2 \pm n}\right)$

$(n \neq \mp 2, \mp 4, \mp 6)$

143. $\displaystyle\int \frac{X^{n/2}\,dx}{x} = \frac{2X^{n/2}}{n} + b\int \frac{X^{(n-2)/2}}{x}\,dx \qquad (n \neq 0)$

144. $\displaystyle\int \frac{dx}{xX^{n/2}} = \frac{2}{(n-2)bX^{(n-2)/2}} + \frac{1}{b}\int \frac{dx}{xX^{(n-2)/2}} \qquad (n \neq 2, b \neq 0)$

145. $\displaystyle\int \frac{dx}{x^2X^{n/2}} = -\frac{1}{bxX^{(n-2)/2}} - \frac{na}{2b}\int \frac{dx}{xX^{n/2}} \qquad (b \neq 0)$

146. $$\int \frac{dx}{\sqrt{XY}} = \begin{cases} \dfrac{2}{\sqrt{-af}} \arctan \sqrt{-\dfrac{fX}{aY}} & (af < 0) \\[2ex] \dfrac{2}{\sqrt{af}} \tanh^{-1} \sqrt{\dfrac{fX}{aY}} = \dfrac{2}{\sqrt{af}} \log_e (\sqrt{aY} + \sqrt{fX}) + C_1 & (af > 0) \end{cases}$$

147. $\displaystyle\int \frac{x\,dx}{\sqrt{XY}} = \frac{\sqrt{XY}}{af} - \frac{ag + bf}{2af}\int \frac{dx}{\sqrt{XY}}$

148. $$\int \frac{dx}{\sqrt{X}\sqrt{Y^3}} = -\frac{2\sqrt{X}}{\Delta\sqrt{Y}}$$

149. $$\int \frac{dx}{Y\sqrt{X}} = \begin{cases} \dfrac{2}{\sqrt{-f\Delta}} \arctan \dfrac{f\sqrt{X}}{\sqrt{-f\Delta}} & (f\Delta < 0) \\ \dfrac{1}{\sqrt{f\Delta}} \log_e \dfrac{f\sqrt{X} - \sqrt{f\Delta}}{f\sqrt{X} + \sqrt{f\Delta}} & (f\Delta > 0) \end{cases}$$

150. $$\int \sqrt{XY}\, dx = \frac{\Delta + 2aY}{4af}\sqrt{XY} - \frac{\Delta^2}{8af}\int \frac{dx}{\sqrt{XY}}$$

151. $$\int \sqrt{\frac{Y}{X}}\, dx = \frac{1}{a}\sqrt{XY} - \frac{\Delta}{2a}\int \frac{dx}{\sqrt{XY}}$$

152. $$\int \frac{\sqrt{X}\, dx}{Y} = \frac{2\sqrt{X}}{f} + \frac{\Delta}{f}\int \frac{dx}{Y\sqrt{X}} \qquad (f \neq 0)$$

153. $$\int \frac{Y^n\, dx}{\sqrt{X}} = \frac{2}{(2n+1)a}\left(\sqrt{X}Y^n - n\Delta \int \frac{Y^{n-1}\, dx}{\sqrt{X}}\right)$$

154. $$\int \frac{dx}{\sqrt{X}Y^n} = -\frac{1}{(n-1)\Delta}\left\{\frac{\sqrt{X}}{Y^{n-1}} + \left(n - \frac{3}{2}\right) a \int \frac{dx}{\sqrt{X}Y^{n-1}}\right\}$$

$$(\Delta \neq 0, n \neq 1)$$

155. $$\int \sqrt{X}Y^n\, dx = \frac{1}{(2n+3)f}\left(2\sqrt{X}Y^{n+1} + \Delta \int \frac{Y^n\, dx}{\sqrt{X}}\right) \qquad (f \neq 0)$$

156. $$\int \frac{\sqrt{X}\, dx}{Y^n} = \frac{1}{(n-1)f}\left(-\frac{\sqrt{X}}{Y^{n-1}} + \frac{a}{2}\int \frac{dx}{\sqrt{X}Y^{n-1}}\right) \qquad (f \neq 0, n \neq 1)$$

(l) *Integrals containing $\sqrt{a^2 - x^2}$, with*

$X \equiv a^2 - x^2 \qquad (a > 0)$

157. $$\int \sqrt{X}\, dx = \frac{1}{2}\left(x\sqrt{X} + a^2 \arcsin \frac{x}{a}\right)$$

158. $$\int x\sqrt{X}\, dx = -\frac{1}{3}\sqrt{X^3}$$

159. $\int x^2\sqrt{X}\,dx = -\frac{x}{4}\sqrt{X^3} + \frac{a^2}{8}\left(x\sqrt{X} + a^2 \arcsin\frac{x}{a}\right)$

160. $\int x^3\sqrt{X}\,dx = \frac{\sqrt{X^5}}{5} - a^2\frac{\sqrt{X^3}}{3}$

161. $\int \frac{\sqrt{X}}{x}\,dx = \sqrt{X} - a\log_e\frac{a+\sqrt{X}}{x}$

162. $\int \frac{\sqrt{X}}{x^2}\,dx = -\frac{\sqrt{X}}{x} - \arcsin\frac{x}{a}$

163. $\int \frac{\sqrt{X}}{x^3}\,dx = -\frac{\sqrt{X}}{2x^2} + \frac{1}{2a}\log_e\frac{a+\sqrt{X}}{x}$

164. $\int \frac{dx}{\sqrt{X}} = \arcsin\frac{x}{a}$

165. $\int \frac{x\,dx}{\sqrt{X}} = -\sqrt{X}$

166. $\int \frac{x^2\,dx}{\sqrt{X}} = -\frac{x}{2}\sqrt{X} + \frac{a^2}{2}\arcsin\frac{x}{a}$

167. $\int \frac{x^3\,dx}{\sqrt{X}} = \frac{\sqrt{X^3}}{3} - a^2\sqrt{X}$

168. $\int \frac{dx}{x\sqrt{X}} = -\frac{1}{a}\log_e\frac{a+\sqrt{X}}{x}$

169. $\int \frac{dx}{x^2\sqrt{X}} = -\frac{\sqrt{X}}{a^2x}$

170. $\int \frac{dx}{x^3\sqrt{X}} = -\frac{\sqrt{X}}{2a^2x^2} - \frac{1}{2a^3}\log_e\frac{a+\sqrt{X}}{x}$

171. $\int \sqrt{X^3}\,dx = \frac{1}{4}\left(x\sqrt{X^3} + \frac{3a^2x}{2}\sqrt{X} + \frac{3a^4}{2}\arcsin\frac{x}{a}\right)$

172. $\int x\sqrt{X^3}\,dx = -\frac{1}{5}\sqrt{X^5}$

173. $\int x^2\sqrt{X^3}\,dx = -\frac{x\sqrt{X^5}}{6} + \frac{a^2x\sqrt{X^3}}{24} + \frac{a^4x\sqrt{X}}{16} + \frac{a^6}{16}\arcsin\frac{x}{a}$

174. $\int x^3\sqrt{X^3}\,dx = \frac{\sqrt{X^7}}{7} - \frac{a^2\sqrt{X^5}}{5}$

175. $\int \frac{\sqrt{X^3}}{x}\,dx = \frac{\sqrt{X^3}}{3} + a^2\sqrt{X} - a^3\log_e\frac{a+\sqrt{X}}{x}$

176. $\int \frac{\sqrt{X^3}}{x^2}\,dx = -\frac{\sqrt{X^3}}{x} - \frac{3}{2}x\sqrt{X} - \frac{3}{2}a^2\arcsin\frac{x}{a}$

177. $\int \frac{\sqrt{X^3}}{x^3}\,dx = -\frac{\sqrt{X^3}}{2x^2} - \frac{3\sqrt{X}}{2} + \frac{3a}{2}\log_e\frac{a+\sqrt{X}}{x}$

178. $\int \frac{dx}{\sqrt{X^3}} = \frac{x}{a^2\sqrt{X}}$

179. $\int \frac{x\,dx}{\sqrt{X^3}} = \frac{1}{\sqrt{X}}$

180. $\int \frac{x^2\,dx}{\sqrt{X^3}} = \frac{x}{\sqrt{X}} - \arcsin\frac{x}{a}$

181. $\int \frac{x^3\,dx}{\sqrt{X^3}} = \sqrt{X} + \frac{a^2}{\sqrt{X}}$

182. $\int \frac{dx}{x\sqrt{X^3}} = \frac{1}{a^2\sqrt{X}} - \frac{1}{a^3}\log_e\frac{a+\sqrt{X}}{x}$

183. $\int \frac{dx}{x^2\sqrt{X^3}} = \frac{1}{a^4}\left(-\frac{\sqrt{X}}{x} + \frac{x}{\sqrt{X}}\right)$

184. $\int \frac{dx}{x^3\sqrt{X^3}} = -\frac{1}{2a^2x^2\sqrt{X}} + \frac{3}{2a^4\sqrt{X}} - \frac{3}{2a^5}\log_e\frac{a+\sqrt{X}}{x}$

(m) Integrals containing $\sqrt{x^2+a^2}$, with

$X \equiv x^2 + a^2 \qquad (a > 0)$

185. $\int \sqrt{X}\,dx = \frac{1}{2}\left(x\sqrt{X} + a^2\sinh^{-1}\frac{x}{a}\right)$

$$= \frac{1}{2}\left[x\sqrt{X} + a^2\log_e(x+\sqrt{X})\right] + C_1$$

186. $\int x\sqrt{X}\,dx = \frac{1}{3}\sqrt{X^3}$

187. $\int x^2\sqrt{X}\,dx = \frac{x}{4}\sqrt{X^3} - \frac{a^2}{8}\left(x\sqrt{X} + a^2 \sinh^{-1}\frac{x}{a}\right)$

$$= \frac{x}{4}\sqrt{X^3} - \frac{a^2}{8}[x\sqrt{X} + a^2 \log_e (x + \sqrt{X})] + C_1$$

188. $\int x^3\sqrt{X}\,dx = \frac{\sqrt{X^5}}{5} - \frac{a^2\sqrt{X^3}}{3}$

189. $\int \frac{\sqrt{X}}{x}\,dx = \sqrt{X} - a \log_e \frac{a + \sqrt{X}}{x}$

190. $\int \frac{\sqrt{X}}{x^2}\,dx = -\frac{\sqrt{X}}{x} + \sinh^{-1}\frac{x}{a}$

$$= -\frac{\sqrt{X}}{x} + \log_e (x + \sqrt{X}) + C_1$$

191. $\int \frac{\sqrt{X}}{x^3}\,dx = -\frac{\sqrt{X}}{2x^2} - \frac{1}{2a}\log_e \frac{a + \sqrt{X}}{x}$

192. $\int \frac{dx}{\sqrt{X}} = \sinh^{-1}\frac{x}{a} = \log_e (x + \sqrt{X}) + C_1$

193. $\int \frac{x\,dx}{\sqrt{X}} = \sqrt{X}$

194. $\int \frac{x^2\,dx}{\sqrt{X}} = \frac{x}{2}\sqrt{X} - \frac{a^2}{2}\sinh^{-1}\frac{x}{a}$

$$= \frac{x}{2}\sqrt{X} - \frac{a^2}{2}\log_e (x + \sqrt{X}) + C_1$$

195. $\int \frac{x^3\,dx}{\sqrt{X}} = \frac{\sqrt{X^3}}{3} - a^2\sqrt{X}$

196. $\int \frac{dx}{x\sqrt{X}} = -\frac{1}{a}\log_e \frac{a + \sqrt{X}}{x}$

197. $\int \frac{dx}{x^2\sqrt{X}} = -\frac{\sqrt{X}}{a^2x}$

198. $$\int \frac{dx}{x^3\sqrt{X}} = -\frac{\sqrt{X}}{2a^2x^2} + \frac{1}{2a^3}\log_e \frac{a+\sqrt{X}}{x}$$

199. $$\int \sqrt{X^3}\,dx = \frac{1}{4}\left(x\sqrt{X^3} + \frac{3a^2x}{2}\sqrt{X} + \frac{3a^4}{2}\sinh^{-1}\frac{x}{a}\right)$$
$$= \frac{1}{4}\left[x\sqrt{X^3} + \frac{3a^2x}{2}\sqrt{X} + \frac{3a^4}{2}\log_e(x+\sqrt{X})\right] + C_1$$

200. $$\int x\sqrt{X^3}\,dx = \frac{1}{5}\sqrt{X^5}$$

201. $$\int x^2\sqrt{X^3}\,dx = \frac{x\sqrt{X^5}}{6} - \frac{a^2x\sqrt{X^3}}{24} - \frac{a^4x\sqrt{X}}{16} - \frac{a^6}{16}\sinh^{-1}\frac{x}{a}$$
$$= \frac{x\sqrt{X^5}}{6} - \frac{a^2x\sqrt{X^3}}{24} - \frac{a^4x\sqrt{X}}{16} - \frac{a^6}{16}\log_e(x+\sqrt{X}) + C_1$$

202. $$\int x^3\sqrt{X^3}\,dx = \frac{\sqrt{X^7}}{7} - \frac{a^2\sqrt{X^5}}{5}$$

203. $$\int \frac{\sqrt{X^3}}{x}\,dx = \frac{\sqrt{X^3}}{3} + a^2\sqrt{X} - a^3\log_e\frac{a+\sqrt{X}}{x}$$

204. $$\int \frac{\sqrt{X^3}}{x^2}\,dx = -\frac{\sqrt{X^3}}{x} + \frac{3}{2}x\sqrt{X} + \frac{3}{2}a^2\sinh^{-1}\frac{x}{a}$$
$$= -\frac{\sqrt{X^3}}{x} + \frac{3}{2}x\sqrt{X} + \frac{3}{2}a^2\log_e(x+\sqrt{X}) + C_1$$

205. $$\int \frac{\sqrt{X^3}}{x^3}\,dx = -\frac{\sqrt{X^3}}{2x^2} + \frac{3}{2}\sqrt{X} - \frac{3}{2}a\log_e\left(\frac{a+\sqrt{X}}{x}\right)$$

206. $$\int \frac{dx}{\sqrt{X^3}} = \frac{x}{a^2\sqrt{X}}$$

207. $$\int \frac{x\,dx}{\sqrt{X^3}} = -\frac{1}{\sqrt{X}}$$

208. $$\int \frac{x^2\,dx}{\sqrt{X^3}} = -\frac{x}{\sqrt{X}} + \sinh^{-1}\frac{x}{a}$$
$$= -\frac{x}{\sqrt{X}} + \log_e(x+\sqrt{X}) + C_1$$

209. $$\int \frac{x^3\,dx}{\sqrt{X^3}} = \sqrt{X} + \frac{a^2}{\sqrt{X}}$$

210. $$\int \frac{dx}{x\sqrt{X^3}} = \frac{1}{a^2\sqrt{X}} - \frac{1}{a^3}\log_e \frac{a+\sqrt{X}}{x}$$

211. $$\int \frac{dx}{x^2\sqrt{X^3}} = -\frac{1}{a^4}\left(\frac{\sqrt{X}}{x} + \frac{x}{\sqrt{X}}\right)$$

212. $$\int \frac{dx}{x^3\sqrt{X^3}} = -\frac{1}{2a^2x^2\sqrt{X}} - \frac{3}{2a^4\sqrt{X}} + \frac{3}{2a^5}\log_e \frac{a+\sqrt{X}}{x}$$

(n) Integrals containing $\sqrt{x^2 - a^2}$, with

$X \equiv x^2 - a^2 \qquad (a > 0)$

213. $$\int \sqrt{X}\,dx = \frac{1}{2}\left(x\sqrt{X} - a^2\cosh^{-1}\frac{x}{a}\right)$$
$$= \frac{1}{2}[x\sqrt{X} - a^2\log_e(x+\sqrt{X})] + C_1$$

214. $$\int x\sqrt{X}\,dx = \frac{1}{3}\sqrt{X^3}$$

215. $$\int x^2\sqrt{X}\,dx = \frac{x}{4}\sqrt{X^3} + \frac{a^2}{8}\left(x\sqrt{X} - a^2\cosh^{-1}\frac{x}{a}\right)$$
$$= \frac{x}{4}\sqrt{X^3} + \frac{a^2}{8}[x\sqrt{X} - a^2\log_e(x+\sqrt{X})] + C_1$$

216. $$\int x^3\sqrt{X}\,dx = \frac{\sqrt{X^5}}{5} + \frac{a^2\sqrt{X^3}}{3}$$

217. $$\int \frac{\sqrt{X}}{x}\,dx = \sqrt{X} - a\arccos\frac{a}{x}$$

218. $$\int \frac{\sqrt{X}}{x^2}\,dx = -\frac{\sqrt{X}}{x} + \cosh^{-1}\frac{x}{a}$$
$$= -\frac{\sqrt{X}}{x} + \log_e(x+\sqrt{X}) + C_1$$

219. $$\int \frac{\sqrt{X}}{x^3}\,dx = -\frac{\sqrt{X}}{2x^2} + \frac{1}{2a}\arccos\frac{a}{x}$$

220. $$\int \frac{dx}{\sqrt{X}} = \cosh^{-1}\frac{x}{a} = \log_e(x + \sqrt{X}) + C_1$$

221. $$\int \frac{x\,dx}{\sqrt{X}} = \sqrt{X}$$

222. $$\int \frac{x^2\,dx}{\sqrt{X}} = \frac{x}{2}\sqrt{X} + \frac{a^2}{2}\cosh^{-1}\frac{x}{a}$$
$$= \frac{x}{2}\sqrt{X} + \frac{a^2}{2}\log_e(x + \sqrt{X}) + C_1$$

223. $$\int \frac{x^3\,dx}{\sqrt{X}} = \frac{\sqrt{X^3}}{3} + a^2\sqrt{X}$$

224. $$\int \frac{dx}{x\sqrt{X}} = \frac{1}{a}\arccos\frac{a}{x}$$

225. $$\int \frac{dx}{x^2\sqrt{X}} = \frac{\sqrt{X}}{a^2x}$$

226. $$\int \frac{dx}{x^3\sqrt{X}} = \frac{\sqrt{X}}{2a^2x^2} + \frac{1}{2a^3}\arccos\frac{a}{x}$$

227. $$\int \sqrt{X^3}\,dx = \frac{1}{4}\left(x\sqrt{X^3} - \frac{3a^2x}{2}\sqrt{X} + \frac{3a^4}{2}\cosh^{-1}\frac{x}{a}\right)$$
$$= \frac{1}{4}\left[x\sqrt{X^3} - \frac{3a^2x}{2}\sqrt{X} + \frac{3a^4}{2}\log_e(x + \sqrt{X})\right] + C_1$$

228. $$\int x\sqrt{X^3}\,dx = \frac{1}{5}\sqrt{X^5}$$

229. $$\int x^2\sqrt{X^3}\,dx = \frac{x\sqrt{X^5}}{6} + \frac{a^2x\sqrt{X^3}}{24} - \frac{a^4x\sqrt{X}}{16} + \frac{a^6}{16}\cosh^{-1}\frac{x}{a}$$
$$= \frac{x\sqrt{X^5}}{6} + \frac{a^2x\sqrt{X^3}}{24} - \frac{a^4x\sqrt{X}}{16}$$
$$+ \frac{a^6}{16}\log_e(x + \sqrt{X}) + C_1$$

230. $\displaystyle\int x^3\sqrt{X^3}\,dx\,\frac{\sqrt{X^7}}{7}+\frac{a^2\sqrt{X^5}}{5}$

231. $\displaystyle\int \frac{\sqrt{X^3}}{x}\,dx = \frac{\sqrt{X^3}}{3} - a^2\sqrt{X} + a^3 \operatorname{arccos}\frac{a}{x}$

232. $\displaystyle\int \frac{\sqrt{X^3}}{x^2}\,dx = -\frac{\sqrt{X^3}}{2} + \frac{3}{2}x\sqrt{X} - \frac{3}{2}a^2\cosh^{-1}\frac{x}{a}$

$$= -\frac{\sqrt{X^3}}{2} + \frac{3}{2}x\sqrt{X} - \frac{3}{2}a^2\log_e(x+\sqrt{X}) + C_1$$

233. $\displaystyle\int \frac{\sqrt{X^3}}{x^3}\,dx = -\frac{\sqrt{X^3}}{2x^2} + \frac{3\sqrt{X}}{2} - \frac{3}{2}a\operatorname{arccos}\frac{a}{x}$

234. $\displaystyle\int \frac{dx}{\sqrt{X^3}} = -\frac{x}{a^2\sqrt{X}}$

235. $\displaystyle\int \frac{x\,dx}{\sqrt{X^3}} = -\frac{1}{\sqrt{X}}$

236. $\displaystyle\int \frac{x^2\,dx}{\sqrt{X^3}} = -\frac{x}{\sqrt{X}} + \cosh^{-1}\frac{x}{a}$

$$= -\frac{x}{\sqrt{X}} + \log_e(x+\sqrt{X}) + C_1$$

237. $\displaystyle\int \frac{x^3\,dx}{\sqrt{X^3}} = \sqrt{X} - \frac{a^2}{\sqrt{X}}$

238. $\displaystyle\int \frac{dx}{x\sqrt{X^3}} = -\frac{1}{a^2\sqrt{X}} - \frac{1}{a^3}\operatorname{arccos}\frac{a}{x}$

239. $\displaystyle\int \frac{dx}{x^2\sqrt{X^3}} = -\frac{1}{a^4}\left(\frac{\sqrt{X}}{x} + \frac{x}{\sqrt{X}}\right)$

240. $\displaystyle\int \frac{dx}{x^3\sqrt{X^3}} = \frac{1}{2a^2x^2\sqrt{X}} - \frac{3}{2a^4\sqrt{X}} - \frac{3}{2a^5}\operatorname{arccos}\frac{a}{x}$

(o) *Integrals containing $\sqrt{ax^2+bx+c}$, with*

$$X \equiv ax^2 + bx + c$$

$$\Delta = 4ac - b^2 \qquad k = \frac{4a}{4ac-b^2} \qquad (a \neq 0)$$

$$241.\quad \int \frac{dx}{\sqrt{X}} = \begin{cases} \dfrac{1}{\sqrt{a}} \log_e (2\sqrt{aX} + 2ax + b) + C & (a > 0) \\ \dfrac{1}{\sqrt{a}} \sinh^{-1} \dfrac{2ax + b}{\sqrt{\Delta}} + C_1 & (a > 0, \Delta > 0) \\ \dfrac{1}{\sqrt{a}} \log_e (2ax + b) & (a > 0, \Delta = 0) \\ -\dfrac{1}{\sqrt{-a}} \arcsin \dfrac{2ax + b}{\sqrt{-\Delta}} & (a < 0, \Delta < 0) \end{cases}$$

$$242.\quad \int \frac{dx}{X\sqrt{X}} = \frac{2(2ax + b)}{\Delta\sqrt{X}} \qquad (\Delta \neq 0)$$

$$243.\quad \int \frac{dx}{X^2\sqrt{X}} = \frac{2(2ax + b)}{3\Delta\sqrt{X}}\left(\frac{1}{X} + 2k\right) \qquad (\Delta \neq 0)$$

$$244.\quad \int \frac{dx}{X^{(2n+1)/2}} = \frac{2(2ax + b)}{(2n - 1)\Delta X^{(2n-1)/2}} + \frac{2k(n - 1)}{2n - 1} \int \frac{dx}{X^{(2n-1)/2}} \qquad (\Delta \neq 0)$$

$$245.\quad \int \sqrt{X}\, dx = \frac{(2ax + b)\sqrt{X}}{4a} + \frac{1}{2k} \int \frac{dx}{\sqrt{X}}$$

$$246.\quad \int X\sqrt{X}\, dx = \frac{(2ax + b)\sqrt{X}}{8a}\left(X + \frac{3}{2k}\right) + \frac{3}{8k^2} \int \frac{dx}{\sqrt{X}}$$

$$247.\quad \int X^2\sqrt{X}\, dx = \frac{(2ax + b)\sqrt{X}}{12a}\left(X^2 + \frac{5X}{4k} + \frac{15}{8k^2}\right) + \frac{5}{16k^3} \int \frac{dx}{\sqrt{X}}$$

$$248.\quad \int X^{(2n+1)/2}\, dx = \frac{(2ax + b)X^{(2n+1)/2}}{4a(n + 1)} + \frac{2n + 1}{2k(n + 1)} \int X^{(2n-1)/2}\, dx$$

$$249.\quad \int \frac{x\, dx}{\sqrt{X}} = \frac{\sqrt{X}}{a} - \frac{b}{2a} \int \frac{dx}{\sqrt{X}}$$

$$250.\quad \int \frac{x\, dx}{X\sqrt{X}} = -\frac{2(bx + 2c)}{\Delta\sqrt{X}} \qquad (\Delta \neq 0)$$

$$251.\quad \int \frac{x\, dx}{X^{(2n+1)/2}} = -\frac{1}{(2n - 1)aX^{(2n-1)/2}} - \frac{b}{2a} \int \frac{dx}{X^{(2n+1)/2}}$$

252. $$\int \frac{x^2\,dx}{\sqrt{X}} = \left(\frac{x}{2a} - \frac{3b}{4a^2}\right)\sqrt{X} + \frac{3b^2 - 4ac}{8a^2}\int \frac{dx}{\sqrt{X}}$$

253. $$\int \frac{x^2\,dx}{X\sqrt{X}} = \frac{(2b^2 - 4ac)x + 2bc}{a\Delta\sqrt{X}} + \frac{1}{a}\int \frac{dx}{\sqrt{X}} \qquad (\Delta \neq 0)$$

254. $$\int x\sqrt{X}\,dx = \frac{X\sqrt{X}}{3a} - \frac{b(2ax + b)}{8a^2}\sqrt{X} - \frac{b}{4ak}\int \frac{dx}{\sqrt{X}}$$

255. $$\int xX\sqrt{X}\,dx = \frac{X^2\sqrt{X}}{5a} - \frac{b}{2a}\int X\sqrt{X}\,dx$$

256. $$\int xX^{(2n+1)/2}\,dx = \frac{X^{(2n+3)/2}}{(2n + 3)a} - \frac{b}{2a}\int X^{(2n+1)/2}\,dx$$

257. $$\int x^2\sqrt{X}\,dx = \left(x - \frac{5b}{6a}\right)\frac{X\sqrt{X}}{4a} + \frac{5b^2 - 4ac}{16a^2}\int \sqrt{X}\,dx$$

258. $$\int \frac{dx}{x\sqrt{X}} = \begin{cases} -\dfrac{1}{\sqrt{c}}\log_e\left(\dfrac{2\sqrt{cX}}{x} + \dfrac{2c}{x} + b\right) + C & (c > 0) \\ -\dfrac{1}{\sqrt{c}}\sinh^{-1}\dfrac{bx + 2c}{x\sqrt{\Delta}} + C_1 & (c > 0, \Delta > 0) \\ -\dfrac{1}{\sqrt{c}}\log_e\dfrac{bx + 2c}{x} & (c > 0, \Delta = 0) \\ \dfrac{1}{\sqrt{-c}}\arcsin\dfrac{bx + 2c}{x\sqrt{-\Delta}} & (c < 0, \Delta < 0) \end{cases}$$

259. $$\int \frac{dx}{x^2\sqrt{X}} = -\frac{\sqrt{X}}{cx} - \frac{b}{2c}\int \frac{dx}{x\sqrt{X}}$$

260. $$\int \frac{\sqrt{X}\,dx}{x} = \sqrt{X} + \frac{b}{2}\int \frac{dx}{\sqrt{X}} + c\int \frac{dx}{x\sqrt{X}}$$

261. $$\int \frac{\sqrt{X}\,dx}{x^2} = -\frac{\sqrt{X}}{x} + a\int \frac{dx}{\sqrt{X}} + \frac{b}{2}\int \frac{dx}{x\sqrt{X}}$$

262. $$\int \frac{X^{(2n+1)/2}}{x}\,dx = \frac{X^{(2n+1)/2}}{2n + 1} + \frac{b}{2}\int X^{(2n-1)/2}\,dx + c\int \frac{X^{(2n-1)/2}}{x}\,dx$$

(p) *Other irrational forms* $(a > 0, b \neq 0)$

263. $$\int \frac{dx}{x\sqrt{ax^2 + bx}} = -\frac{2}{bx}\sqrt{ax^2 + bx}$$

264. $$\int \frac{dx}{\sqrt{2ax - x^2}} = \arcsin \frac{x - a}{a}$$

265. $$\int \frac{x\,dx}{\sqrt{2ax - x^2}} = -\sqrt{2ax - x^2} + a \arcsin \frac{x - a}{a}$$

266. $$\int \sqrt{2ax - x^2}\,dx = \frac{x - a}{2}\sqrt{2ax - x^2} + \frac{a^2}{2}\arcsin \frac{x - a}{a}$$

267. $$\int \frac{dx}{(ax^2 + b)\sqrt{fx^2 + g}} = \frac{1}{\sqrt{b}\sqrt{ag - bf}} \arctan \frac{x\sqrt{ag - bf}}{\sqrt{b}\sqrt{fx^2 + g}} \qquad (ag - bf > 0)$$

$$= \frac{1}{2\sqrt{b}\sqrt{bf - ag}} \log_e \frac{\sqrt{b}\sqrt{fx^2 + g} + x\sqrt{bf - ag}}{\sqrt{b}\sqrt{fx^2 + g} - x\sqrt{bf - ag}} \qquad (ag - bf < 0)$$

268. $$\int \sqrt[n]{ax + b}\,dx = \frac{n(ax + b)}{(n + 1)a}\sqrt[n]{ax + b}$$

269. $$\int \frac{dx}{\sqrt[n]{ax + b}}\,dx = \frac{n(ax + b)}{(n - 1)a}\frac{1}{\sqrt[n]{ax + b}}$$

270. $$\int \frac{dx}{x\sqrt{x^n + a^2}} = -\frac{2}{na}\log_e \frac{a + \sqrt{x^n + a^2}}{\sqrt{x^n}}$$

271. $$\int \frac{dx}{x\sqrt{x^n - a^2}} = \frac{2}{na}\arccos \frac{a}{\sqrt{x^n}}$$

272. $$\int \frac{\sqrt{x}\,dx}{\sqrt{a^3 - x^3}} = \frac{2}{3}\arcsin \sqrt{\left(\frac{x}{a}\right)^3}$$

(q) *Recursion formulas* **(m, n, p *are integers*)**

273. $$\int x^m(ax^n + b)^p\,dx$$

$$= \frac{1}{m + np + 1}\left[x^{m+1}(ax^n + b)^p + npb\int x^m(ax^n + b)^{p-1}\,dx\right]$$

$$= \frac{1}{bn(p+1)}\left[-x^{m+1}(ax^n+b)^{p+1} + (m+n+np+1)\int x^m(ax^n+b)^{p+1}\,dx\right]$$

$$= \frac{1}{(m+1)b}\left[x^{m+1}(ax^n+b)^{p+1} - a(m+n+np+1)\int x^{m+n}(ax^n+b)^p\,dx\right]$$

$$= \frac{1}{a(m+np+1)}\left[x^{m-n+1}(ax^n+b)^{p+1} - (m-n+1)b\int x^{m-n}(ax^n+b)^p\,dx\right]$$

(r) *Integrals containing the sine function* $(a \neq 0)$

274. $\int \sin ax\,dx = -\frac{1}{a}\cos ax$

275. $\int \sin^2 ax\,dx = \frac{1}{2}x - \frac{1}{4a}\sin 2ax$

276. $\int \sin^3 ax\,dx = -\frac{1}{a}\cos ax + \frac{1}{3a}\cos^3 ax$

277. $\int \sin^4 ax\,dx = \frac{3}{8}x - \frac{1}{4a}\sin 2ax + \frac{1}{32a}\sin 4ax$

278. $\int \sin^n ax\,dx = -\frac{\sin^{n-1} ax\cos ax}{na} + \frac{n-1}{n}\int \sin^{n-2} ax\,dx \quad (n > 0)$

279. $\int x\sin ax\,dx = \frac{\sin ax}{a^2} - \frac{x\cos ax}{a}$

280. $\int x^2\sin ax\,dx = \frac{2x}{a^2}\sin ax - \left(\frac{x^2}{a} - \frac{2}{a^3}\right)\cos ax$

281. $\int x^3\sin ax\,dx = \left(\frac{3x^2}{a^2} - \frac{6}{a^4}\right)\sin ax - \left(\frac{x^3}{a} - \frac{6x}{a^3}\right)\cos ax$

282. $\int x^n\sin ax\,dx = -\frac{x^n}{a}\cos ax + \frac{n}{a}\int x^{n-1}\cos ax\,dx$

283. $\int \frac{\sin ax}{x}\,dx = ax - \frac{(ax)^3}{3\cdot 3!} + \frac{(ax)^5}{5\cdot 5!} - \frac{(ax)^7}{7\cdot 7!} + \cdots$

284. $$\int \frac{\sin ax}{x^2}\,dx = -\frac{\sin ax}{x} + a\int \frac{\cos ax\,dx}{x}$$

285. $$\int \frac{\sin ax}{x^n}\,dx = -\frac{1}{n-1}\frac{\sin ax}{x^{n-1}} + \frac{a}{n-1}\int \frac{\cos ax}{x^{n-1}}\,dx$$

286. $$\int \frac{dx}{\sin ax} = \int \operatorname{cosec} ax\,dx = \frac{1}{a}\log_e \tan\frac{ax}{2}$$
$$= \frac{1}{a}\log_e (\operatorname{cosec} ax - \cot ax)$$

287. $$\int \frac{dx}{\sin^2 ax} = -\frac{1}{a}\cot ax$$

288. $$\int \frac{dx}{\sin^3 ax} = -\frac{\cos ax}{2a\sin^2 ax} + \frac{1}{2a}\log_e \tan\frac{ax}{2}$$

289. $$\int \frac{dx}{\sin^n ax} = -\frac{1}{a(n-1)}\frac{\cos ax}{\sin^{n-1} ax} + \frac{n-2}{n-1}\int \frac{dx}{\sin^{n-2} ax} \qquad (n > 1)$$

290. $$\int \frac{x\,dx}{\sin ax} = \frac{1}{a^2}\left[ax + \frac{(ax)^3}{3\cdot 3!} + \frac{7(ax)^5}{3\cdot 5\cdot 5!} + \frac{31(ax)^7}{3\cdot 7\cdot 7!} + \frac{127(ax)^9}{3\cdot 5\cdot 9!} + \cdots\right]$$

291. $$\int \frac{x\,dx}{\sin^2 ax} = -\frac{x}{a}\cot ax + \frac{1}{a^2}\log_e \sin ax$$

292. $$\int \frac{x\,dx}{\sin^n ax} = -\frac{x\cos ax}{(n-1)a\sin^{n-1} ax} - \frac{1}{(n-1)(n-2)a^2\sin^{n-2} ax} + \frac{n-2}{n-1}\int \frac{x\,dx}{\sin^{n-2} ax} \qquad (n > 2)$$

293. $$\int \frac{dx}{1+\sin ax} = -\frac{1}{a}\tan\left(\frac{\pi}{4} - \frac{ax}{2}\right)$$

294. $$\int \frac{dx}{1-\sin ax} = \frac{1}{a}\tan\left(\frac{\pi}{4} + \frac{ax}{2}\right)$$

295. $$\int \frac{x\,dx}{1+\sin ax} = -\frac{x}{a}\tan\left(\frac{\pi}{4} - \frac{ax}{2}\right) + \frac{2}{a^2}\log_e \cos\left(\frac{\pi}{4} - \frac{ax}{2}\right)$$

296. $$\int \frac{x\,dx}{1-\sin ax} = \frac{x}{a}\cot\left(\frac{\pi}{4} - \frac{ax}{2}\right) + \frac{2}{a^2}\log_e \sin\left(\frac{\pi}{4} - \frac{ax}{2}\right)$$

297. $$\int \frac{\sin ax\, dx}{1 \pm \sin ax} = \pm x + \frac{1}{a} \tan\left(\frac{\pi}{4} \mp \frac{ax}{2}\right)$$

298. $$\int \frac{dx}{\sin ax\,(1 \pm \sin ax)} = \frac{1}{a} \tan\left(\frac{\pi}{4} \mp \frac{ax}{2}\right) + \frac{1}{a} \log_e \tan \frac{ax}{2}$$

299. $$\int \frac{dx}{(1 + \sin ax)^2} = -\frac{1}{2a} \tan\left(\frac{\pi}{4} - \frac{ax}{2}\right) - \frac{1}{6a} \tan^3\left(\frac{\pi}{4} - \frac{ax}{2}\right)$$

300. $$\int \frac{dx}{(1 - \sin ax)^2} = \frac{1}{2a} \cot\left(\frac{\pi}{4} - \frac{ax}{2}\right) + \frac{1}{6a} \cot^3\left(\frac{\pi}{4} - \frac{ax}{2}\right)$$

301. $$\int \frac{\sin ax\, dx}{(1 + \sin ax)^2} = -\frac{1}{2a} \tan\left(\frac{\pi}{4} - \frac{ax}{2}\right) + \frac{1}{6a} \tan^3\left(\frac{\pi}{4} - \frac{ax}{2}\right)$$

302. $$\int \frac{\sin ax\, dx}{(1 - \sin ax)^2} = -\frac{1}{2a} \cot\left(\frac{\pi}{4} - \frac{ax}{2}\right) + \frac{1}{6a} \cot^3\left(\frac{\pi}{4} - \frac{ax}{2}\right)$$

303. $$\int \frac{dx}{1 + \sin^2 ax} = \frac{1}{2\sqrt{2}a} \arcsin\left(\frac{3 \sin^2 ax - 1}{\sin^2 ax + 1}\right)$$

304. $$\int \frac{dx}{1 - \sin^2 ax} = \int \frac{dx}{\cos^2 ax} = \frac{1}{a} \tan ax$$

305. $$\int \sin ax \sin bx\, dx = \frac{\sin (a - b)x}{2(a - b)} - \frac{\sin (a + b)x}{2(a + b)}$$

$(|a| \neq |b|;$ for $|a| = |b|$ see 275)

306. $$\int \frac{dx}{b + c \sin ax} = \frac{2}{a\sqrt{b^2 - c^2}} \arctan \frac{b \tan (ax/2) + c}{\sqrt{b^2 - c^2}} \qquad (b^2 > c^2)$$

$$= \frac{1}{a\sqrt{c^2 - b^2}} \log_e \frac{b \tan (ax/2) + c - \sqrt{c^2 - b^2}}{b \tan (ax/2) + c + \sqrt{c^2 - b^2}}$$

$(b^2 < c^2)$

307. $$\int \frac{\sin ax\, dx}{b + c \sin ax} = \frac{x}{c} - \frac{b}{c} \int \frac{dx}{b + c \sin ax}$$

308. $$\int \frac{dx}{\sin ax\,(b + c \sin ax)} = \frac{1}{ab} \log_e \tan \frac{ax}{2} - \frac{c}{b} \int \frac{dx}{b + c \sin ax}$$

309. $$\int \frac{dx}{(b + c \sin ax)^2} = \frac{c \cos ax}{a(b^2 - c^2)(b + c \sin ax)} + \frac{b}{b^2 - c^2} \int \frac{dx}{b + c \sin ax}$$

310. $$\int \frac{\sin ax\,dx}{(b + c\sin ax)^2} = \frac{b\cos ax}{a(c^2 - b^2)(b + c\sin ax)} + \frac{c}{c^2 - b^2}\int \frac{dx}{b + c\sin ax}$$

311. $$\int \frac{dx}{b^2 + c^2\sin^2 ax} = \frac{1}{ab\sqrt{b^2 + c^2}}\arctan\frac{\sqrt{b^2 + c^2}\tan ax}{b} \qquad (b > 0)$$

312. $$\int \frac{dx}{b^2 - c^2\sin^2 ax} = \frac{1}{ab\sqrt{b^2 - c^2}}\arctan\frac{\sqrt{b^2 - c^2}\tan ax}{b} \qquad (b^2 > c^2,\ b > 0)$$

$$= \frac{1}{2ab\sqrt{c^2 - b^2}}\log_e\frac{\sqrt{c^2 - b^2}\tan ax + b}{\sqrt{c^2 - b^2}\tan ax - b} \qquad (c^2 > b^2,\ b > 0)$$

(s) *Integrals containing the cosine function* $(a \neq 0)$

313. $$\int \cos ax\,dx = \frac{1}{a}\sin ax$$

314. $$\int \cos^2 ax\,dx = \frac{1}{2}x + \frac{1}{4a}\sin 2ax$$

315. $$\int \cos^3 ax\,dx = \frac{1}{a}\sin ax - \frac{1}{3a}\sin^3 ax$$

316. $$\int \cos^4 ax\,dx = \frac{3}{8}x + \frac{1}{4a}\sin 2ax + \frac{1}{32a}\sin 4ax$$

317. $$\int \cos^n ax\,dx = \frac{\cos^{n-1} ax\sin ax}{na} + \frac{n-1}{n}\int \cos^{n-2} ax\,dx$$

318. $$\int x\cos ax\,dx = \frac{\cos ax}{a^2} + \frac{x\sin ax}{a}$$

319. $$\int x^2\cos ax\,dx = \frac{2x}{a^2}\cos ax + \left(\frac{x^2}{a} - \frac{2}{a^3}\right)\sin ax$$

320. $$\int x^3\cos ax\,dx = \left(\frac{3x^2}{a^2} - \frac{6}{a^4}\right)\cos ax + \left(\frac{x^3}{a} - \frac{6x}{a^3}\right)\sin ax$$

321. $$\int x^n\cos ax\,dx = \frac{x^n\sin ax}{a} - \frac{n}{a}\int x^{n-1}\sin ax\,dx \qquad (n > 0)$$

322. $$\int \frac{\cos ax}{x}\,dx = \log_e (ax) - \frac{(ax)^2}{2 \cdot 2!} + \frac{(ax)^4}{4 \cdot 4!} - \frac{(ax)^6}{6 \cdot 6!} + \cdots$$

323. $$\int \frac{\cos ax}{x^2}\,dx = -\frac{\cos ax}{x} - a \int \frac{\sin ax\,dx}{x}$$

324. $$\int \frac{\cos ax}{x^n}\,dx = -\frac{\cos ax}{(n-1)x^{n-1}} - \frac{a}{n-1} \int \frac{\sin ax\,dx}{x^{n-1}} \qquad (n \neq 1)$$

325. $$\int \frac{dx}{\cos ax} = \frac{1}{a} \log_e \tan\left(\frac{ax}{2} + \frac{\pi}{4}\right) = \frac{1}{a} \log_e (\sec ax + \tan ax)$$

326. $$\int \frac{dx}{\cos^2 ax} = \frac{1}{a} \tan ax$$

327. $$\int \frac{dx}{\cos^3 ax} = \frac{\sin ax}{2a \cos^2 ax} + \frac{1}{2a} \log_e \tan\left(\frac{\pi}{4} + \frac{ax}{2}\right)$$

328. $$\int \frac{dx}{\cos^n ax} = \frac{1}{a(n-1)} \frac{\sin ax}{\cos^{n-1} ax} + \frac{n-2}{n-1} \int \frac{dx}{\cos^{n-2} ax} \qquad (n > 1)$$

329. $$\int \frac{x\,dx}{\cos ax} = \frac{1}{a^2}\left[\frac{(ax)^2}{2} + \frac{(ax)^4}{4 \cdot 2!} + \frac{5(ax)^6}{6 \cdot 4!} + \frac{61(ax)^8}{8 \cdot 6!} + \frac{1{,}385(ax)^{10}}{10 \cdot 8!} + \cdots\right]$$

330. $$\int \frac{x\,dx}{\cos^2 ax} = \frac{x}{a} \tan ax + \frac{1}{a^2} \log_e \cos ax$$

331. $$\int \frac{x\,dx}{\cos^n ax} = \frac{x \sin ax}{(n-1)a \cos^{n-1} ax} - \frac{1}{(n-1)(n-2)a^2 \cos^{n-2} ax} + \frac{n-2}{n-1} \int \frac{x\,dx}{\cos^{n-2} ax} \qquad (n > 2)$$

332. $$\int \frac{dx}{1 + \cos ax} = \frac{1}{a} \tan \frac{ax}{2}$$

333. $$\int \frac{dx}{1 - \cos ax} = -\frac{1}{a} \cot \frac{ax}{2}$$

334. $$\int \frac{x\,dx}{1 + \cos ax} = \frac{x}{a} \tan \frac{ax}{2} + \frac{2}{a^2} \log_e \cos \frac{ax}{2}$$

335. $$\int \frac{x\,dx}{1 - \cos ax} = -\frac{x}{a} \cot \frac{ax}{2} + \frac{2}{a^2} \log_e \sin \frac{ax}{2}$$

336. $$\int \frac{\cos ax\, dx}{1 + \cos ax} = x - \frac{1}{a} \tan \frac{ax}{2}$$

337. $$\int \frac{\cos ax\, dx}{1 - \cos ax} = -x - \frac{1}{a} \cot \frac{ax}{2}$$

338. $$\int \frac{dx}{\cos ax\,(1 + \cos ax)} = \frac{1}{a} \log_e \tan\left(\frac{\pi}{4} + \frac{ax}{2}\right) - \frac{1}{a} \tan \frac{ax}{2}$$

339. $$\int \frac{dx}{\cos ax\,(1 - \cos ax)} = \frac{1}{a} \log_e \tan\left(\frac{\pi}{4} + \frac{ax}{2}\right) - \frac{1}{a} \cot \frac{ax}{2}$$

340. $$\int \frac{dx}{(1 + \cos ax)^2} = \frac{1}{2a} \tan \frac{ax}{2} + \frac{1}{6a} \tan^3 \frac{ax}{2}$$

341. $$\int \frac{dx}{(1 - \cos ax)^2} = -\frac{1}{2a} \cot \frac{ax}{2} - \frac{1}{6a} \cot^3 \frac{ax}{2}$$

342. $$\int \frac{\cos ax\, dx}{(1 + \cos ax)^2} = \frac{1}{2a} \tan \frac{ax}{2} - \frac{1}{6a} \tan^3 \frac{ax}{2}$$

343. $$\int \frac{\cos ax\, dx}{(1 - \cos ax)^2} = \frac{1}{2a} \cot \frac{ax}{2} - \frac{1}{6a} \cot^3 \frac{ax}{2}$$

344. $$\int \frac{dx}{1 + \cos^2 ax} = \frac{1}{2\sqrt{2}a} \arcsin\left(\frac{1 - 3\cos^2 ax}{1 + \cos^2 ax}\right)$$

345. $$\int \frac{dx}{1 - \cos^2 ax} = \int \frac{dx}{\sin^2 ax} = -\frac{1}{a} \cot ax$$

346. $$\int \cos ax \cos bx\, dx = \frac{\sin (a - b)x}{2(a - b)} + \frac{\sin (a + b)x}{2(a + b)}$$

($|a| \neq |b|$; for $|a| = |b|$, see 314)

347. $$\int \frac{dx}{b + c \cos ax} = \frac{2}{a\sqrt{b^2 - c^2}} \arctan \frac{(b - c)\tan (ax/2)}{\sqrt{b^2 - c^2}} \qquad (b^2 > c^2)$$

$$= \frac{1}{a\sqrt{c^2 - b^2}} \log_e \frac{(c - b)\tan (ax/2) + \sqrt{c^2 - b^2}}{(c - b)\tan (ax/2) - \sqrt{c^2 - b^2}}$$

$(b^2 < c^2)$

348. $$\int \frac{\cos ax\, dx}{b + c \cos ax} = \frac{x}{c} - \frac{b}{c} \int \frac{dx}{b + c \cos ax}$$

349. $$\int \frac{dx}{\cos ax\,(b + c\cos ax)} = \frac{1}{ab}\log_e \tan\left(\frac{ax}{2} + \frac{\pi}{4}\right) - \frac{c}{b}\int \frac{dx}{b + c\cos ax}$$

350. $$\int \frac{dx}{(b + c\cos ax)^2} = \frac{c\sin ax}{a(c^2 - b^2)(b + c\cos ax)} - \frac{b}{c^2 - b^2}\int \frac{dx}{b + c\cos ax}$$

351. $$\int \frac{\cos ax\,dx}{(b + c\cos ax)^2} = \frac{b\sin ax}{a(b^2 - c^2)(b + c\cos ax)} - \frac{c}{b^2 - c^2}\int \frac{dx}{b + c\cos ax}$$

352. $$\int \frac{dx}{b^2 + c^2\cos^2 ax} = \frac{1}{ab\sqrt{b^2 + c^2}}\arctan\frac{b\tan ax}{\sqrt{b^2 + c^2}} \qquad (b > 0)$$

353. $$\int \frac{dx}{b^2 - c^2\cos^2 ax} = \frac{1}{ab\sqrt{b^2 - c^2}}\arctan\frac{b\tan ax}{\sqrt{b^2 - c^2}} \qquad (b^2 > c^2, b > 0)$$

$$= \frac{1}{2ab\sqrt{c^2 - b^2}}\log_e\frac{b\tan ax - \sqrt{c^2 - b^2}}{b\tan ax + \sqrt{c^2 - b^2}} \qquad (c^2 > b^2, b > 0)$$

(t) *Integrals containing both sine and cosine* $(a \neq 0)$

354. $$\int \sin ax\cos ax\,dx = \frac{1}{2a}\sin^2 ax$$

355. $$\int \sin^2 ax\cos^2 ax\,dx = \frac{x}{8} - \frac{\sin 4ax}{32a}$$

356. $$\int \sin^n ax\cos ax\,dx = \frac{1}{a(n+1)}\sin^{n+1} ax \qquad (n \neq -1)$$

357. $$\int \sin ax\cos^n ax\,dx = -\frac{1}{a(n+1)}\cos^{n+1} ax \qquad (n \neq -1)$$

358. $$\int \sin^n ax\cos^m ax\,dx$$

$$= -\frac{\sin^{n-1} ax\cos^{m+1} ax}{a(n+m)} + \frac{n-1}{n+m}\int \sin^{n-2} ax\cos^m ax\,dx \qquad (m, n > 0)$$

$$= \frac{\sin^{n+1} ax \cos^{m-1} ax}{a(n+m)} + \frac{m-1}{n+m} \int \sin^n ax \cos^{m-2} ax \, dx$$

$(m, n > 0)$

359. $$\int \frac{dx}{\sin ax \cos ax} = \frac{1}{a} \log_e \tan ax$$

360. $$\int \frac{dx}{\sin^2 ax \cos ax} = \frac{1}{a} \left[\log_e \tan \left(\frac{\pi}{4} + \frac{ax}{2} \right) - \frac{1}{\sin ax} \right]$$

361. $$\int \frac{dx}{\sin ax \cos^2 ax} = \frac{1}{a} \left(\log_e \tan \frac{ax}{2} + \frac{1}{\cos ax} \right)$$

362. $$\int \frac{dx}{\sin^3 ax \cos ax} = \frac{1}{a} \left(\log_e \tan ax - \frac{1}{2 \sin^2 ax} \right)$$

363. $$\int \frac{dx}{\sin ax \cos^3 ax} = \frac{1}{a} \left(\log_e \tan ax + \frac{1}{2 \cos^2 ax} \right)$$

364. $$\int \frac{dx}{\sin^2 ax \cos^2 ax} = -\frac{2}{a} \cot 2ax$$

365. $$\int \frac{dx}{\sin^2 ax \cos^3 ax} = \frac{1}{a} \left[\frac{\sin ax}{2 \cos^2 ax} - \frac{1}{\sin ax} + \frac{3}{2} \log_e \tan \left(\frac{\pi}{4} + \frac{ax}{2} \right) \right]$$

366. $$\int \frac{dx}{\sin^3 ax \cos^2 ax} = \frac{1}{a} \left(\frac{1}{\cos ax} - \frac{\cos ax}{2 \sin^2 ax} + \frac{3}{2} \log_e \tan \frac{ax}{2} \right)$$

367. $$\int \frac{dx}{\sin ax \cos^n ax} = \frac{1}{a(n-1) \cos^{n-1} ax} + \int \frac{dx}{\sin ax \cos^{n-2} ax}$$

$(n \neq 1)$; (see 361, 363)

368. $$\int \frac{dx}{\sin^n ax \cos ax} = -\frac{1}{a(n-1) \sin^{n-1} ax} + \int \frac{dx}{\sin^{n-2} ax \cos ax}$$

$(n \neq 1)$; (see 360, 362)

369. $$\int \frac{dx}{\sin^n ax \cos^m ax}$$

$$= -\frac{1}{a(n-1)} \cdot \frac{1}{\sin^{n-1} ax \cos^{m-1} ax} + \frac{n+m-2}{n-1} \int \frac{dx}{\sin^{n-2} ax \cos^m ax}$$

$(m > 0, n > 1)$

$$= \frac{1}{a(m-1)} \cdot \frac{1}{\sin^{n-1} ax \cos^{m-1} ax} + \frac{n+m-2}{m-1} \int \frac{dx}{\sin^n ax \cos^{m-2} ax}$$

$(n > 0, m > 1)$

370. $$\int \frac{\sin ax\,dx}{\cos^2 ax} = \frac{1}{a\cos ax}$$

371. $$\int \frac{\sin ax\,dx}{\cos^3 ax} = \frac{1}{2a\cos^2 ax} = \frac{1}{2a}\tan^2 ax + C_1$$

372. $$\int \frac{\sin ax\,dx}{\cos^n ax} = \frac{1}{a(n-1)\cos^{n-1} ax}$$

373. $$\int \frac{\sin^2 ax\,dx}{\cos ax} = -\frac{1}{a}\sin ax + \frac{1}{a}\log_e \tan\left(\frac{\pi}{4} + \frac{ax}{2}\right)$$

374. $$\int \frac{\sin^2 ax\,dx}{\cos^3 ax} = \frac{1}{a}\left[\frac{\sin ax}{2\cos^2 ax} - \frac{1}{2}\log_e \tan\left(\frac{\pi}{4} + \frac{ax}{2}\right)\right]$$

375. $$\int \frac{\sin^2 ax\,dx}{\cos^n ax} = \frac{\sin ax}{a(n-1)\cos^{n-1} ax} - \frac{1}{n-1}\int \frac{dx}{\cos^{n-2} ax}$$

$(n \neq 1)$ (see 325, 326, 328)

376. $$\int \frac{\sin^3 ax\,dx}{\cos ax} = -\frac{1}{a}\left(\frac{\sin^2 ax}{2} + \log_e \cos ax\right)$$

377. $$\int \frac{\sin^3 ax\,dx}{\cos^2 ax} = \frac{1}{a}\left(\cos ax + \frac{1}{\cos ax}\right)$$

378. $$\int \frac{\sin^3 ax\,dx}{\cos^n ax} = \frac{1}{a}\left[\frac{1}{(n-1)\cos^{n-1} ax} - \frac{1}{(n-3)\cos^{n-3} ax}\right]$$

$(n \neq 1, n \neq 3)$

379. $$\int \frac{\sin^n ax}{\cos ax}\,dx = -\frac{\sin^{n-1} ax}{a(n-1)} + \int \frac{\sin^{n-2} ax\,dx}{\cos ax} \qquad (n \neq 1)$$

380. $$\int \frac{\sin^n ax}{\cos^m ax}\,dx = \frac{\sin^{n+1} ax}{a(m-1)\cos^{m-1} ax} - \frac{n-m+2}{m-1}\int \frac{\sin^n ax}{\cos^{m-2} ax}\,dx$$

$(m \neq 1)$

$$= -\frac{\sin^{n-1} ax}{a(n-m)\cos^{m-1} ax} + \frac{n-1}{n-m}\int \frac{\sin^{n-2} ax\,dx}{\cos^m ax}$$

$(m \neq n)$

$$= \frac{\sin^{n-1} ax}{a(m-1)\cos^{m-1} ax} - \frac{n-1}{m-1}\int \frac{\sin^{n-1} ax\,dx}{\cos^{m-2} ax}$$

$(m \neq 1)$

381. $\displaystyle\int \frac{\cos ax\,dx}{\sin^2 ax} = -\frac{1}{a \sin ax}$

382. $\displaystyle\int \frac{\cos ax\,dx}{\sin^3 ax} = -\frac{1}{2a \sin^2 ax}$

383. $\displaystyle\int \frac{\cos ax\,dx}{\sin^n ax} = -\frac{1}{a(n-1)\sin^{n-1} ax} \qquad (n \neq 1)$

384. $\displaystyle\int \frac{\cos^2 ax\,dx}{\sin ax} = \frac{1}{a}\left(\cos ax + \log_e \tan \frac{ax}{2}\right)$

385. $\displaystyle\int \frac{\cos^2 ax\,dx}{\sin^3 ax} = -\frac{1}{2a}\left(\frac{\cos ax}{\sin^2 ax} - \log_e \tan \frac{ax}{2}\right)$

386. $\displaystyle\int \frac{\cos^2 ax\,dx}{\sin^n ax} = -\frac{1}{n-1}\left(\frac{\cos ax}{a \sin^{n-1} ax} + \int \frac{dx}{\sin^{n-2} ax}\right)$

$(n \neq 1)$; (see 289)

387. $\displaystyle\int \frac{\cos^3 ax\,dx}{\sin ax} = \frac{1}{a}\left(\frac{\cos^2 ax}{2} + \log_e \sin ax\right)$

388. $\displaystyle\int \frac{\cos^3 ax\,dx}{\sin^2 ax} = -\frac{1}{a}\left(\sin ax + \frac{1}{\sin ax}\right)$

389. $\displaystyle\int \frac{\cos^3 ax\,dx}{\sin^n ax} = \frac{1}{a}\left[\frac{1}{(n-3)\sin^{n-3} ax} - \frac{1}{(n-1)\sin^{n-1} ax}\right]$

$(n \neq 1,\ n \neq 3)$

390. $\displaystyle\int \frac{\cos^n ax}{\sin ax}\,dx = \frac{\cos^{n-1} ax}{a(n-1)} + \int \frac{\cos^{n-2} ax\,dx}{\sin ax} \qquad (n \neq 1)$

391. $\displaystyle\int \frac{\cos^n ax\,dx}{\sin^m ax} = -\frac{\cos^{n+1} ax}{a(m-1)\sin^{m-1} ax} - \frac{n-m+2}{m-1}\int \frac{\cos^n ax\,dx}{\sin^{m-2} ax}$

$(m \neq 1)$

$$= \frac{\cos^{n-1} ax}{a(n-m)\sin^{m-1} ax} + \frac{n-1}{n-m}\int \frac{\cos^{n-2} ax\,dx}{\sin^m ax}$$

$(m \neq n)$

$$= -\frac{\cos^{n-1} ax}{a(m-1)\sin^{m-1} ax} - \frac{n-1}{m-1}\int \frac{\cos^{n-2} ax\,dx}{\sin^{m-2} ax}$$

$(m \neq 1)$

392. $$\int \frac{dx}{\sin ax\,(1 \pm \cos ax)} = \pm\frac{1}{2a(1 \pm \cos ax)} + \frac{1}{2a}\log_e \tan\frac{ax}{2}$$

393. $$\int \frac{dx}{\cos ax\,(1 \pm \sin ax)} = \mp\frac{1}{2a(1 \pm \sin ax)} + \frac{1}{2a}\log_e \tan\left(\frac{\pi}{4} + \frac{ax}{2}\right)$$

394. $$\int \frac{\sin ax\,dx}{\cos ax\,(1 \pm \cos ax)} = \frac{1}{a}\log_e \frac{1 \pm \cos ax}{\cos ax}$$

395. $$\int \frac{\cos ax\,dx}{\sin ax\,(1 \pm \sin ax)} = -\frac{1}{a}\log_e \frac{1 \pm \sin ax}{\sin ax}$$

396. $$\int \frac{\sin ax\,dx}{\cos ax\,(1 \pm \sin ax)} = \frac{1}{2a(1 \pm \sin ax)} \pm \frac{1}{2a}\log_e \tan\left(\frac{\pi}{4} + \frac{ax}{2}\right)$$

397. $$\int \frac{\cos ax\,dx}{\sin ax\,(1 \pm \cos ax)} = -\frac{1}{2a(1 \pm \cos ax)} \pm \frac{1}{2a}\log_e \tan\frac{ax}{2}$$

398. $$\int \frac{\sin ax\,dx}{\sin ax \pm \cos ax} = \frac{x}{2} \mp \frac{1}{2a}\log_e (\sin ax \pm \cos ax)$$

399. $$\int \frac{\cos ax\,dx}{\sin ax \pm \cos ax} = \pm\frac{x}{2} + \frac{1}{2a}\log_e (\sin ax \pm \cos ax)$$

400. $$\int \frac{dx}{\sin ax \pm \cos ax} = \frac{1}{a\sqrt{2}}\log_e \tan\left(\frac{ax}{2} \pm \frac{\pi}{8}\right)$$

401. $$\int \frac{dx}{1 + \cos ax \pm \sin ax} = \pm\frac{1}{a}\log_e \left(1 \pm \tan\frac{ax}{2}\right)$$

402. $$\int \frac{dx}{b\sin ax + c\cos ax} = \frac{1}{a\sqrt{b^2 + c^2}}\log_e \tan\frac{ax + \theta}{2}$$

$$\sin\theta = \frac{c}{\sqrt{b^2 + c^2}} \qquad \tan\theta = \frac{c}{b}$$

403. $$\int \frac{\sin ax\,dx}{b + c\cos ax} = -\frac{1}{ac}\log_e (b + c\cos ax)$$

404. $$\int \frac{\cos ax\,dx}{b + c\sin ax} = \frac{1}{ac}\log_e (b + c\sin ax)$$

405. $$\int \frac{\sin^2 ax\,dx}{b + c\cos^2 ax} = \frac{1}{ac}\sqrt{\frac{b + c}{b}}\arctan\left(\sqrt{\frac{b}{b + c}}\tan ax\right) - \frac{x}{c}$$

406. $\displaystyle\int \frac{\sin ax \cos ax\, dx}{b\cos^2 ax + c\sin^2 ax} = \frac{1}{2a(c-b)}\log_e (b\cos^2 ax + c\sin^2 ax)$ $\quad (c \neq b)$

407. $\displaystyle\int \frac{dx}{b^2\cos^2 ax + c^2\sin^2 ax} = \frac{1}{abc}\arctan\left(\frac{c}{b}\tan ax\right)$

408. $\displaystyle\int \frac{dx}{b^2\cos^2 ax - c^2\sin^2 ax} = \frac{1}{2abc}\log_e \frac{c\tan ax + b}{c\tan ax - b}$

409. $\displaystyle\int \sin ax \cos bx\, dx = -\frac{\cos(a+b)x}{2(a+b)} - \frac{\cos(a-b)x}{2(a-b)}$ $\quad (a^2 \neq b^2$; for $a = b$ see 354)

410. $\displaystyle\int \frac{dx}{b + c\cos ax + d\sin ax}$

$$= \begin{cases} \dfrac{-1}{a\sqrt{b^2 - c^2 - d^2}}\arcsin\dfrac{c^2 + d^2 + b(c\cos ax + d\sin ax)}{\sqrt{c^2 + d^2}(b + c\cos ax + d\sin ax)} & (b^2 > c^2 + d^2,\ |ax| < \pi) \\ \dfrac{1}{a\sqrt{c^2 + d^2 - b^2}} \log_e \dfrac{c^2 + d^2 + b(c\cos ax + d\sin ax) + \sqrt{c^2 + d^2 - b^2}(c\sin ax - d\cos ax)}{\sqrt{c^2 + d^2}(b + c\cos ax + d\sin ax)} & (b^2 < c^2 + d^2,\ |ax| < \pi) \\ \dfrac{1}{ab}\left[\dfrac{b - (c+d)\cos ax + (c-d)\sin ax}{b + (c-d)\cos ax + (c+d)\sin ax}\right] & (b^2 = c^2 + d^2) \end{cases}$$

(u) Integrals containing tangent and cotangent functions $(a \neq 0)$

411. $\displaystyle\int \tan ax\, dx = -\frac{1}{a}\log_e \cos ax$

412. $\displaystyle\int \tan^2 ax\, dx = \frac{1}{a}\tan ax - x$

413. $\displaystyle\int \tan^3 ax\, dx = \frac{1}{2a}\tan^2 ax + \frac{1}{a}\log_e \cos ax$

414. $\displaystyle\int \tan^n ax\, dx = \frac{1}{a(n-1)}\tan^{n-1} ax - \int \tan^{n-2} ax\, dx$ $\quad (n > 1)$

415. $$\int \frac{dx}{b + c \tan ax} = \int \frac{\cot ax\, dx}{b \cot ax + c}$$
$$= \frac{1}{b^2 + c^2}\left[bx + \frac{c}{a}\log_e (b \cos ax + c \sin ax)\right]$$

416. $$\int \frac{dx}{\sqrt{b + c \tan^2 ax}} = \frac{1}{a\sqrt{b - c}} \arcsin\left(\sqrt{\frac{b - c}{b}} \sin ax\right)$$
$(b > 0,\ b^2 > c^2)$

417. $$\int \frac{\tan^n ax}{\cos^2 ax}\, dx = \frac{\tan^{n+1} ax}{a(n + 1)} \qquad (n \neq -1)$$

418. $$\int \cot ax\, dx = \frac{1}{a} \log_e \sin ax$$

419. $$\int \cot^2 ax\, dx = \int \frac{dx}{\tan^2 ax} = -\frac{1}{a}\cot ax - x$$

420. $$\int \cot^3 ax\, dx = -\frac{1}{2a}\cot^2 ax - \frac{1}{a}\log_e \sin ax$$

421. $$\int \cot^n ax\, dx = \int \frac{dx}{\tan^n ax} = -\frac{1}{a(n - 1)}\cot^{n-1} ax$$
$$- \int \cot^{n-2} ax\, dx \qquad (n > 1)$$

422. $$\int \frac{dx}{b + c \cot ax} = \int \frac{\tan ax\, dx}{b \tan ax + c}$$
$$= \frac{1}{b^2 + c^2}\left[bx - \frac{c}{a}\log_e (c \cos ax + b \sin ax)\right]$$

423. $$\int \frac{\cot^n ax}{\sin^2 ax}\, dx = \frac{-\cot^{n+1} ax}{a(n + 1)} \qquad (n \neq -1)$$

(v) *Integrals containing hyperbolic functions* $(a \neq 0)$

424. $$\int \sinh x\, dx = \cosh x$$

425. $$\int \sinh^2 x\, dx = \frac{\sinh 2x}{4} - \frac{x}{2}$$

426. $$\int \frac{dx}{\sinh x} = \log_e \tanh\left(\frac{x}{2}\right)$$

427. $\int \frac{dx}{\sinh^2 x} = -\coth x$

428. $\int \cosh x\, dx = \sinh x$

429. $\int \cosh^2 x\, dx = \frac{\sinh 2x}{4} + \frac{x}{2}$

430. $\int \frac{dx}{\cosh x} = 2 \arctan e^x = \arctan (\sinh x)$

431. $\int \frac{dx}{\cosh^2 x} = \tanh x$

432. $\int \frac{\sinh x}{\cosh^2 x}\, dx = -\frac{1}{\cosh x}$

433. $\int \frac{\cosh x}{\sinh^2 x}\, dx = -\frac{1}{\sinh x}$

434. $\int x \sinh x\, dx = x \cosh x - \sinh x$

435. $\int x \cosh x\, dx = x \sinh x - \cosh x$

436. $\int \tanh x\, dx = \log_e \cosh x$

437. $\int \tanh^2 x\, dx = x - \tanh x$

438. $\int \coth x\, dx = \log_e \sinh x$

439. $\int \coth^2 x\, dx = x - \coth x$

440. $\int \sinh^n ax\, dx$

$$= \begin{cases} \frac{1}{an} \sinh^{n-1} ax \cosh ax - \frac{n-1}{n} \int \sinh^{n-2} ax\, dx & (n > 0) \\ \frac{1}{a(n+1)} \sinh^{n+1} ax \cosh ax - \frac{n+2}{n+1} \int \sinh^{n+2} ax\, dx & (n < -1) \end{cases}$$

441. $\int \cosh^n ax\, dx$

$$= \begin{cases} \dfrac{1}{an} \sinh ax \cosh^{n-1} ax + \dfrac{n-1}{n} \displaystyle\int \cosh^{n-2} ax\, dx & (n > 0) \\ -\dfrac{1}{a(n+1)} \sinh ax \cosh^{n+1} ax + \dfrac{n+2}{n+1} \displaystyle\int \cosh^{n+2} ax\, dx & \\ & (n < -1) \end{cases}$$

442. $\int \sinh ax \sinh bx\, dx = \dfrac{\sinh (a+b)x}{2(a+b)} - \dfrac{\sinh (a-b)x}{2(a-b)}$

443. $\int \cosh ax \cosh bx\, dx = \dfrac{\sinh (a+b)x}{2(a+b)} + \dfrac{\sinh (a-b)x}{2(a-b)}$ $\quad (a^2 \neq b^2)$ (applies to 442–444)

444. $\int \sinh ax \cosh bx\, dx = \dfrac{\cosh (a+b)x}{2(a+b)} + \dfrac{\cosh (a-b)x}{2(a-b)}$

445. $\int \sinh ax \sin ax\, dx = \dfrac{1}{2a} (\cosh ax \sin ax - \sinh ax \cos ax)$

446. $\int \cosh ax \cos ax\, dx = \dfrac{1}{2a} (\sinh ax \cos ax + \cosh ax \sin ax)$

447. $\int \sinh ax \cos ax\, dx = \dfrac{1}{2a} (\cosh ax \cos ax + \sinh ax \sin ax)$

448. $\int \cosh ax \sin ax\, dx = \dfrac{1}{2a} (\sinh ax \sin ax - \cosh ax \cos ax)$

(w) *Integrals containing exponential functions*

449. $\int e^{ax}\, dx = \dfrac{1}{a} e^{ax}$

450. $\int xe^{ax}\, dx = \dfrac{e^{ax}}{a^2} (ax - 1)$

451. $\int x^2 e^{ax}\, dx = e^{ax} \left(\dfrac{x^2}{a} - \dfrac{2x}{a^2} + \dfrac{2}{a^3} \right)$

452. $\int x^n e^{ax}\, dx = \dfrac{1}{a} x^n e^{ax} - \dfrac{n}{a} \int x^{n-1} e^{ax}\, dx \qquad (n > 0)$

453. $\int \dfrac{e^{ax}}{x}\, dx = \log_e x + \dfrac{ax}{1 \cdot 1!} + \dfrac{(ax)^2}{2 \cdot 2!} + \dfrac{(ax)^3}{3 \cdot 3!} + \cdots$

454. $\int \frac{e^{ax}}{x^n}\,dx = \frac{1}{n-1}\left(-\frac{e^{ax}}{x^{n-1}} + a\int \frac{e^{ax}}{x^{n-1}}\,dx\right) \qquad (n > 1)$

455. $\int \frac{dx}{1+e^{ax}} = \frac{1}{a}\log_e \frac{e^{ax}}{1+e^{ax}}$

456. $\int \frac{dx}{b+ce^{ax}} = \frac{x}{b} - \frac{1}{ab}\log_e (b + ce^{ax})$

457. $\int \frac{e^{ax}\,dx}{b+ce^{ax}} = \frac{1}{ac}\log_e (b + ce^{ax})$

458. $\int \frac{dx}{be^{ax}+ce^{-ax}} = \frac{1}{a\sqrt{bc}}\arctan\left(e^{ax}\sqrt{\frac{b}{c}}\right) \qquad (bc > 0)$

$= \frac{1}{2a\sqrt{-bc}}\log_e \frac{c + e^{ax}\sqrt{-bc}}{c - e^{ax}\sqrt{-bc}} \qquad (bc < 0)$

459. $\int \frac{xe^{ax}\,dx}{(1+ax)^2} = \frac{e^{ax}}{a^2(1+ax)}$

460. $\int e^{ax}\log_e x\,dx = \frac{1}{a}e^{ax}\log_e x - \frac{1}{a}\int \frac{e^{ax}}{x}\,dx$

461. $\int e^{ax}\sin bx\,dx = \frac{e^{ax}}{a^2+b^2}(a\sin bx - b\cos bx)$

462. $\int e^{ax}\cos bx\,dx = \frac{e^{ax}}{a^2+b^2}(a\cos bx + b\sin bx)$

463. $\int xe^{ax}\sin bx\,dx = \frac{xe^{ax}}{a^2+b^2}(a\sin bx - b\cos bx)$

$- \frac{e^{ax}}{(a^2+b^2)^2}[(a^2-b^2)\sin bx - 2\,ab\cos bx]$

464. $\int xe^{ax}\cos bx\,dx = \frac{xe^{ax}}{a^2+b^2}(a\cos bx + b\sin bx)$

$- \frac{e^{ax}}{(a^2+b^2)^2}[(a^2-b^2)\cos bx + 2\,ab\sin bx]$

465. $\int e^{ax}\sin bx\sin cx\,dx = \frac{e^{ax}[(b-c)\sin(b-c)x + a\cos(b-c)x]}{2[a^2+(b-c)^2]}$

$- \frac{e^{ax}[(b+c)\sin(b+c)x + a\cos(b+c)x]}{2[a^2+(b+c)^2]}$

466. $$\int e^{ax} \cos bx \cos cx \, dx = \frac{e^{ax}[(b-c)\sin(b-c)x + a\cos(b-c)x]}{2[a^2+(b-c)^2]} + \frac{e^{ax}[(b+c)\sin(b+c)x + a\cos(b+c)x]}{2[a^2+(b+c)^2]}$$

467. $$\int e^{ax} \sin bx \cos cx \, dx = \frac{e^{ax}[a\sin(b-c)x - (b-c)\cos(b-c)x]}{2[a^2+(b-c)^2]} + \frac{e^{ax}[a\sin(b+c)x - (b+c)\cos(b+c)x]}{2[a^2+(b+c)^2]}$$

468. $$\int e^{ax} \sin bx \sin(bx+c) \, dx = \frac{e^{ax}\cos c}{2a} - \frac{e^{ax}[a\cos(2bx+c) + 2b\sin(2bx+c)]}{2(a^2+4b^2)}$$

469. $$\int e^{ax} \cos bx \cos(bx+c) \, dx = \frac{e^{ax}\cos c}{2a} + \frac{e^{ax}[a\cos(2bx+c) + 2b\sin(2bx+c)]}{2(a^2+4b^2)}$$

470. $$\int e^{ax} \sin bx \cos(bx+c) \, dx = -\frac{e^{ax}\sin c}{2a} + \frac{e^{ax}[a\sin(2bx+c) - 2b\cos(2bx+c)]}{2(a^2+4b^2)}$$

471. $$\int e^{ax} \cos bx \sin(bx+c) \, dx = \frac{e^{ax}\sin c}{2a} + \frac{e^{ax}[a\sin(2bx+c) - 2b\cos(2bx+c)]}{2(a^2+4b^2)}$$

472. $$\int e^{ax} \sin^n bx \, dx = \frac{e^{ax}\sin^{n-1} bx\,(a\sin bx - nb\cos bx)}{a^2+n^2b^2} + \frac{n(n-1)b^2}{a^2+n^2b^2}\int e^{ax}\sin^{n-2} bx \, dx$$

473. $$\int e^{ax} \cos^n bx \, dx = \frac{e^{ax}\cos^{n-1} bx\,(a\cos bx + nb\sin bx)}{a^2+n^2b^2} + \frac{n(n-1)b^2}{a^2+n^2b^2}\int e^{ax}\cos^{n-2} bx \, dx$$

(x) *Integrals containing logarithmic functions* $(a \neq 0)$

474. $$\int \log_e ax\,dx = x\log_e ax - x$$

475. $$\int (\log_e ax)^2\,dx = x(\log_e ax)^2 - 2x\log_e ax + 2x$$

476. $$\int (\log_e ax)^n\,dx = x(\log_e ax)^n - n\int (\log_e ax)^{n-1}\,dx \qquad (n \neq -1)$$

477. $$\int \frac{dx}{\log_e ax} = \frac{1}{a}\left[\log_e(\log_e ax) + \log_e ax + \frac{(\log_e ax)^2}{2\cdot 2!} + \frac{(\log_e ax)^3}{3\cdot 3!} + \cdots\right]$$

478. $$\int x\log_e ax\,dx = \frac{x^2}{2}\log_e ax - \frac{x^2}{4}$$

479. $$\int x^2\log_e ax\,dx = \frac{x^3}{3}\log_e ax - \frac{x^3}{9}$$

480. $$\int x^n\log_e ax\,dx = x^{n+1}\left[\frac{\log_e ax}{n+1} - \frac{1}{(n+1)^2}\right] \qquad (n \neq -1)$$

481. $$\int x^n(\log_e ax)^m\,dx = \frac{x^{n+1}}{n+1}(\log_e ax)^m - \frac{m}{n+1}\int x^n(\log_e ax)^{m-1}\,dx \qquad (m, n \neq -1)$$

482. $$\int \frac{(\log_e ax)^n}{x}\,dx = \frac{(\log_e ax)^{n+1}}{n+1} \qquad (n \neq -1)$$

483. $$\int \frac{\log_e x}{x^n}\,dx = -\frac{\log_e x}{(n-1)x^{n-1}} - \frac{1}{(n-1)^2x^{n-1}} \qquad (n \neq 1)$$

484. $$\int \frac{(\log_e x)^m}{x^n}\,dx = -\frac{(\log_e x)^m}{(n-1)x^{n-1}} + \frac{m}{n-1}\int \frac{(\log_e x)^{m-1}}{x^n}\,dx \qquad (n \neq 1)$$

485. $$\int \frac{x^n\,dx}{\log_e ax} = \frac{1}{a^{n+1}}\left[\log_e(\log_e ax) + (n+1)\log_e ax + \frac{(n+1)^2(\log_e ax)^2}{2\cdot 2!} + \frac{(n+1)^3(\log_e ax)^3}{3\cdot 3!} + \cdots\right]$$
$$= \frac{1}{a^{n+1}}\int \frac{e^y\,dy}{y} \qquad [y = (n+1)\log_e ax]$$

486. $$\int \frac{x^n\,dx}{(\log_e ax)^m} = \frac{-x^{n+1}}{(m-1)(\log_e ax)^{m-1}} + \frac{n+1}{m-1}\int \frac{x^n\,dx}{(\log_e ax)^{m-1}}$$

$(m \neq 1)$

487. $$\int \frac{dx}{x \log_e ax} = \log_e (\log_e ax)$$

488. $$\int \frac{dx}{x(\log_e ax)^n} = -\frac{1}{(n-1)(\log_e ax)^{n-1}}$$

489. $$\int \sin (\log_e ax)\,dx = \frac{x}{2}[\sin (\log_e ax) - \cos (\log_e ax)]$$

490. $$\int \cos (\log_e ax)\,dx = \frac{x}{2}[\sin (\log_e ax) + \cos (\log_e ax)]$$

491. $$\int e^{ax} \log_e bx\,dx = \frac{1}{a} e^{ax} \log_e bx - \frac{1}{a}\int \frac{e^{ax}}{x}\,dx$$

(y) *Integrals containing inverse trigonometric and hyperbolic functions* $(a > 0)$

492. $$\int \arcsin \frac{x}{a}\,dx = x \arcsin \frac{x}{a} + \sqrt{a^2 - x^2}$$

493. $$\int x \arcsin \frac{x}{a}\,dx = \left(\frac{x^2}{2} - \frac{a^2}{4}\right) \arcsin \frac{x}{a} + \frac{x}{4}\sqrt{a^2 - x^2}$$

494. $$\int x^2 \arcsin \frac{x}{a}\,dx = \frac{x^3}{3} \arcsin \frac{x}{a} + \frac{1}{9}(x^2 + 2a^2)\sqrt{a^2 - x^2}$$

495. $$\int x^n \arcsin \frac{x}{a}\,dx = \frac{x^{n+1}}{n+1} \arcsin \frac{x}{a} - \frac{1}{n+1}\int \frac{x^{n+1}}{\sqrt{a^2 - x^2}}\,dx$$

$(n \neq -1)$

496. $$\int \frac{\arcsin \frac{x}{a}\,dx}{x} = \frac{x}{a} + \frac{1}{2 \cdot 3 \cdot 3}\frac{x^3}{a^3} + \frac{1 \cdot 3}{2 \cdot 4 \cdot 5 \cdot 5}\frac{x^5}{a^5} + \frac{1 \cdot 3 \cdot 5}{2 \cdot 4 \cdot 6 \cdot 7 \cdot 7}\frac{x^7}{a^7} + \cdots$$

$(x^2 < a^2)$

497. $$\int \frac{\arcsin \frac{x}{a}\,dx}{x^2} = -\frac{1}{x} \arcsin \frac{x}{a} - \frac{1}{a} \log_e \frac{a + \sqrt{a^2 - x^2}}{x}$$

498. $$\int \left(\arcsin \frac{x}{a}\right)^2 dx = x\left(\arcsin \frac{x}{a}\right)^2 + 2\left(\sqrt{a^2 - x^2} \arcsin \frac{x}{a} - x\right)$$

499. $$\int \arccos \frac{x}{a}\, dx = x \arccos \frac{x}{a} - \sqrt{a^2 - x^2}$$

500. $$\int x \arccos \frac{x}{a}\, dx = \left(\frac{x^2}{2} - \frac{a^2}{4}\right) \arccos \frac{x}{a} - \frac{x}{4}\sqrt{a^2 - x^2}$$

501. $$\int x^2 \arccos \frac{x}{a}\, dx = \frac{x^3}{3} \arccos \frac{x}{a} - \frac{1}{9}(x^2 + 2a^2)\sqrt{a^2 - x^2}$$

502. $$\int x^n \arccos \frac{x}{a}\, dx = \frac{x^{n+1}}{n+1} \arccos \frac{x}{a} + \frac{1}{n+1}\int \frac{x^{n+1}}{\sqrt{a^2 - x^2}}\, dx \qquad (n \neq -1)$$

503. $$\int \frac{\arccos \frac{x}{a}\, dx}{x} = \frac{\pi}{2}\log_e x - \frac{x}{a} - \frac{1}{2 \cdot 3 \cdot 3}\frac{x^3}{a^3} - \frac{1 \cdot 3}{2 \cdot 4 \cdot 5 \cdot 5}\frac{x^5}{a^5} - \frac{1 \cdot 3 \cdot 5}{2 \cdot 4 \cdot 6 \cdot 7 \cdot 7}\frac{x^7}{a^7} - \cdots \qquad (x^2 < a^2)$$

504. $$\int \frac{\arccos \frac{x}{a}\, dx}{x^2} = -\frac{1}{x} \arccos \frac{x}{a} + \frac{1}{a}\log_e \frac{a + \sqrt{a^2 - x^2}}{x}$$

505. $$\int \left(\arccos \frac{x}{a}\right)^2 dx = x\left(\arccos \frac{x}{a}\right)^2 - 2\left(\sqrt{a^2 - x^2} \arccos \frac{x}{a} + x\right)$$

506. $$\int \arctan \frac{x}{a}\, dx = x \arctan \frac{x}{a} - \frac{a}{2}\log_e (a^2 + x^2)$$

507. $$\int x \arctan \frac{x}{a}\, dx = \frac{1}{2}(x^2 + a^2) \arctan \frac{x}{a} - \frac{ax}{2}$$

508. $$\int x^2 \arctan \frac{x}{a}\, dx = \frac{x^3}{3} \arctan \frac{x}{a} - \frac{ax^2}{6} + \frac{a^3}{6}\log_e (a^2 + x^2)$$

509. $$\int x^n \arctan \frac{x}{a}\, dx = \frac{x^{n+1}}{n+1} \arctan \frac{x}{a} - \frac{a}{n+1}\int \frac{x^{n+1}\, dx}{a^2 + x^2} \qquad (n \neq -1)$$

510. $$\int \frac{\arctan \frac{x}{a}\, dx}{x} = \frac{x}{a} - \frac{x^3}{3^2a^3} + \frac{x^5}{5^2a^5} - \frac{x^7}{7^2a^7} + \cdots \quad (|x| < |a|)$$

511. $$\int \frac{\arctan \frac{x}{a}\, dx}{x^2} = -\frac{1}{x} \arctan \frac{x}{a} - \frac{1}{2a} \log_e \frac{a^2 + x^2}{x^2}$$

512. $$\int \frac{\arctan \frac{x}{a}\, dx}{x^n} = -\frac{1}{(n-1)x^{n-1}} \arctan \frac{x}{a} + \frac{a}{n-1} \int \frac{dx}{x^{n-1}(a^2 + x^2)} \quad (n \neq 1)$$

513. $$\int \operatorname{arccot} \frac{x}{a}\, dx = x \operatorname{arccot} \frac{x}{a} + \frac{a}{2} \log_e (a^2 + x^2)$$

514. $$\int x \operatorname{arccot} \frac{x}{a}\, dx = \frac{1}{2} (x^2 + a^2) \operatorname{arccot} \frac{x}{a} + \frac{ax}{2}$$

515. $$\int x^2 \operatorname{arccot} \frac{x}{a}\, dx = \frac{x^3}{3} \operatorname{arccot} \frac{x}{a} + \frac{ax^2}{6} - \frac{a^3}{6} \log_e (a^2 + x^2)$$

516. $$\int x^n \operatorname{arccot} \frac{x}{a}\, dx = \frac{x^{n+1}}{n+1} \operatorname{arccot} \frac{x}{a} + \frac{a}{n+1} \int \frac{x^{n+1}\, dx}{a^2 + x^2} \quad (n \neq -1)$$

517. $$\int \frac{\operatorname{arccot} \frac{x}{a}\, dx}{x} = \frac{\pi}{2} \log_e x - \frac{x}{a} + \frac{x^3}{3^2a^3} - \frac{x^5}{5^2a^5} + \frac{x^7}{7^2a^7} - \cdots$$

518. $$\int \frac{\operatorname{arccot} \frac{x}{a}\, dx}{x^2} = -\frac{1}{x} \operatorname{arccot} \frac{x}{a} + \frac{1}{2a} \log_e \frac{a^2 + x^2}{x^2}$$

519. $$\int \frac{\operatorname{arccot} \frac{x}{a}\, dx}{x^n} = -\frac{1}{(n-1)x^{n-1}} \operatorname{arccot} \frac{x}{a} - \frac{a}{n-1} \int \frac{dx}{x^{n-1}(a^2 + x^2)} \quad (n \neq 1)$$

520. $$\int \sinh^{-1}\frac{x}{a}\,dx = x\sinh^{-1}\frac{x}{a} - \sqrt{x^2+a^2}$$

521. $$\int x\sinh^{-1}\frac{x}{a}\,dx = \frac{1}{2}\left(x^2+\frac{a^2}{2}\right)\sinh^{-1}\frac{x}{a} - \frac{x}{4}\sqrt{x^2+a^2}$$

522. $$\int \cosh^{-1}\frac{x}{a}\,dx = x\cosh^{-1}\frac{x}{a} \mp \sqrt{x^2-a^2}$$

$$\left(\text{upper sign for } \cosh^{-1}\frac{x}{a} > 0\right)$$

523. $$\int \tanh^{-1}\frac{x}{a}\,dx = x\tanh^{-1}\frac{x}{a} + \frac{a}{2}\log_e(a^2-x^2)$$

524. $$\int x\tanh^{-1}\frac{x}{a}\,dx = \frac{x^2-a^2}{2}\tanh^{-1}\frac{x}{a} + \frac{ax}{2}$$

525. $$\int \coth^{-1}\frac{x}{a}\,dx = x\coth^{-1}\frac{x}{a} + \frac{a}{2}\log_e(x^2-a^2)$$

E-3. Definite Integrals (*see also Secs. 21.4-1 and 21.6-6 and Appendix D*). m, n are integers.

(a) *Integrals containing algebraic functions*

1. $$\int_1^\infty \frac{dx}{x^n} = \frac{1}{n-1} \qquad (n > 1)$$

2. $$\int_0^\infty \frac{a\,dx}{a^2+x^2} = \begin{cases} \dfrac{\pi}{2} & (a > 0) \\ 0 & (a = 0) \\ -\dfrac{\pi}{2} & (a < 0) \end{cases}$$

3. $$\int_0^1 x^\alpha(1-x)^\beta\,dx = 2\int_0^1 x^{2\alpha+1}(1-x^2)^\beta\,dx$$

$$= \int_0^\infty \frac{x^\alpha}{(1+x)^{\alpha+\beta+2}}\,dx$$

$$= \frac{\Gamma(\alpha+1)\Gamma(\beta+1)}{\Gamma(\alpha+\beta+2)} = B(\alpha+1, \beta+1)$$

4. $\int_0^\infty \frac{dx}{(1+x)x^a} = \frac{\pi}{\sin a\pi} \qquad (a < 1)$

5. $\int_0^\infty \frac{x^{a-1}\,dx}{1+x} = \frac{\pi}{\sin a\pi} \qquad (0 < a < 1)$

6. $\int_0^\infty \frac{dx}{(1-x)x^a} = -\pi \cot a\pi \qquad (a < 1)$

7. $\int_0^\infty \frac{x^{a-1}}{1+x^b}\,dx = \frac{\pi}{b \sin(a\pi/b)} \qquad (0 < a < b)$

8. $\int_0^1 \frac{dx}{\sqrt{1-x^a}} = \frac{\sqrt{\pi}\,\Gamma(1/a)}{a\Gamma\left(\frac{2+a}{2a}\right)}$

(b) Integrals containing trigonometric functions *(see also Secs. E-3c and d)*

9. $\int_0^\pi \sin mx \cos nx\,dx = \begin{cases} 0 & (m - n \text{ even}) \\ \frac{2m}{m^2 - n^2} & (m - n \text{ odd}) \end{cases}$

10. $\int_0^\pi \sin mx \sin(nx + \vartheta)\,dx = \int_0^\pi \cos mx \cos(nx + \vartheta)\,dx$

$$= \begin{cases} 0 & (m \neq n) \\ \frac{\pi}{2}\cos\vartheta & (m = n) \end{cases}$$

11. $\int_0^{\pi/2} \sin^a x\,dx = \int_0^{\pi/2} \cos^a x\,dx = \frac{1}{2}\sqrt{\pi}\,\frac{\Gamma\left(\frac{a+1}{2}\right)}{\Gamma\left(\frac{a}{2}+1\right)} \qquad (a > -1)$

$$= \begin{cases} \frac{1 \cdot 3 \cdot 5 \cdots (a-1)}{2 \cdot 4 \cdot 6 \cdots a} \cdot \frac{\pi}{2} & (a = 2, 4, \ldots) \\ \frac{2 \cdot 4 \cdot 6 \cdots (a-1)}{1 \cdot 3 \cdot 5 \cdot 7 \cdots a} & (a = 3, 5, \ldots) \end{cases}$$

12. $\int_0^{\pi/2} \sin^{2\alpha+1} x \cos^{2\beta+1} x\,dx = \frac{\Gamma(\alpha+1)\Gamma(\beta+1)}{2\Gamma(\alpha+\beta+2)}$

$$= \frac{1}{2}B(\alpha+1, \beta+1) \qquad (\alpha, \beta \neq -1)$$

$$= \frac{\alpha!\beta!}{2(\alpha+\beta+1)!} \qquad (\alpha, \beta \text{ integers} > 0)$$

Note: Use formula (12) to obtain

$$\int_0^{\pi/2} \sqrt{\sin x}\, dx \qquad \int_0^{\pi/2} \sqrt[3]{\sin x}\, dx \qquad \int_0^{\pi/2} \frac{dx}{\sqrt[3]{\cos x}} \qquad \text{etc.}$$

13. $$\int_0^{\infty} \frac{\sin ax}{x}\, dx = \begin{cases} \dfrac{\pi}{2} & (a > 0) \\ -\dfrac{\pi}{2} & (a < 0) \end{cases}$$

14. $$\int_0^{\alpha} \frac{\cos ax\, dx}{x} = \infty \qquad (\alpha \neq 0)$$

15. $$\int_0^{\infty} \frac{\tan ax\, dx}{x} = \begin{cases} \dfrac{\pi}{2} & (a > 0) \\ -\dfrac{\pi}{2} & (a < 0) \end{cases}$$

16. $$\int_0^{\infty} \frac{\cos ax - \cos bx}{x}\, dx = \log_e \frac{b}{a} \qquad (a, b \neq 0)$$

17. $$\int_0^{\infty} \frac{\sin x \cos ax}{x}\, dx = \begin{cases} \dfrac{\pi}{2} & (|a| < 1) \\ \dfrac{\pi}{4} & (|a| = 1) \\ 0 & (|a| > 1) \end{cases}$$

18. $$\int_0^{\infty} \frac{\sin x\, dx}{\sqrt{x}} = \int_0^{\infty} \frac{\cos x\, dx}{\sqrt{x}} = \sqrt{\frac{\pi}{2}}$$

19. $$\int_0^{\infty} \frac{x \sin bx}{a^2 + x^2}\, dx = \frac{\pi}{2} e^{-|ab|} \qquad (b > 0)$$

20. $$\int_0^{\infty} \frac{\cos ax}{1 + x^2}\, dx = \frac{\pi}{2} e^{-|a|}$$

21. $$\int_0^{\infty} \frac{\sin^2 ax}{x^2}\, dx = \frac{\pi}{2} |a|$$

22. $$\int_0^{\infty} \frac{\sin ax \sin bx}{x^2}\, dx = \frac{\pi a}{2} \qquad (a < b)$$

23. $$\int_0^{\infty} \cos (x^2)\, dx = \int_0^{\infty} \sin (x^2)\, dx = \frac{1}{2}\sqrt{\frac{\pi}{2}}$$

24. $$\int_0^\pi \sin^2 mx\,dx = \int_0^\pi \cos^2 mx\,dx = \frac{\pi}{2}$$

25. $$\int_0^\pi \frac{dx}{a + b\cos x} = \frac{\pi}{\sqrt{a^2 - b^2}} \qquad (a > b > 0)$$

26. $$\int_0^{\pi/2} \frac{dx}{a + b\cos x} = \frac{\arccos(b/a)}{\sqrt{a^2 - b^2}} \qquad (a > b)$$

27. $$\int_0^{\pi/2} \frac{dx}{a^2\cos^2 x + b^2\sin^2 x} = \frac{\pi}{2ab} \qquad (ab \neq 0)$$

28. $$\int_0^{\pi/2} \frac{dx}{(a^2\cos^2 x + b^2\sin^2 x)^2} = \frac{\pi(a^2 + b^2)}{4a^3b^3}$$

29. $$\int_0^\pi \frac{(a - b\cos x)\,dx}{a^2 - 2ab\cos x + b^2} = \begin{cases} 0 & (a^2 < b^2) \\ \dfrac{\pi}{2a} & (a = b) \\ \dfrac{\pi}{a} & (a^2 > b^2) \end{cases}$$

30. $$\int_0^\pi \frac{\cos nx\,dx}{1 - 2b\cos x + b^2} = \frac{\pi b^n}{1 - b^2} \qquad (n \geq 0,\ |b| < 1)$$

31. $$\int_0^1 \frac{dx}{1 + 2x\cos a + x^2} = \frac{a}{2\sin a} \qquad \left(0 < a < \frac{\pi}{2}\right)$$

32. $$\int_0^\infty \frac{dx}{1 + 2x\cos a + x^2} = \frac{a}{\sin x} \qquad \left(0 < a < \frac{\pi}{2}\right)$$

33. $$\int_0^{\pi/2} \frac{\sin x\,dx}{\sqrt{1 - k^2\sin^2 x}} = \frac{1}{2k}\log_e \frac{1 + k}{1 - k}$$

34. $$\int_0^{\pi/2} \frac{\cos x\,dx}{\sqrt{1 - k^2\sin^2 x}} = \frac{1}{k}\arcsin k$$

35. $$\int_0^{\pi/2} \frac{\sin^2 x\,dx}{\sqrt{1 - k^2\sin^2 x}} = \frac{1}{k^2}(\mathbf{K} - \mathbf{E})$$

36. $$\int_0^{\pi/2} \frac{\cos^2 x\,dx}{\sqrt{1 - k^2\sin^2 x}} = \frac{1}{k^2}[\mathbf{E} - (1 - k^2)\mathbf{K}]$$

(33–36: $(|k| < 1)$)

(c) Integrals containing exponential and hyperbolic functions $(a > 0)$

37. $$\int_0^\infty e^{-ax}\,dx = \frac{1}{a}$$

38. $$\int_0^\infty x^b e^{-ax}\,dx = \frac{\Gamma(b+1)}{a^{b+1}} \qquad (a > 0,\ b > -1)$$
$$= \frac{b!}{a^{b+1}} \qquad (a > 0,\ b = 0, 1, 2, \ldots)$$

39. $$\int_0^\infty x^b e^{-ax^2}\,dx = \frac{\Gamma\left(\frac{b+1}{2}\right)}{2a^{(b+1)/2}} \qquad (a > 0,\ b > -1)$$
$$= \begin{cases} \dfrac{1\cdot 3 \cdots (b-1)\sqrt{\pi}}{2^{(b/2)+1}a^{(b+1)/2}} & (a > 0,\ b = 0, 2, 4, \ldots) \\[2ex] \dfrac{\left(\frac{b-1}{2}\right)!}{2a^{(b+1)/2}} & (a > 0,\ b = 1, 3, 5, \ldots) \end{cases}$$

40. $$\int_0^\infty e^{-a^2x^2}\,dx = \frac{\sqrt{\pi}}{2a}$$

41. $$\int_0^\infty xe^{-x^2}\,dx = \frac{1}{2}$$

42. $$\int_0^\infty x^2e^{-x^2}\,dx = \frac{\sqrt{\pi}}{4}$$

43. $$\int_0^\infty \sqrt{x}e^{-ax}\,dx = \frac{1}{2a}\sqrt{\frac{\pi}{a}}$$

44. $$\int_0^\infty \frac{e^{-ax}}{\sqrt{x}}\,dx = \sqrt{\frac{\pi}{a}}$$

45. $$\int_0^\infty e^{(-x^2-a^2/x^2)}\,dx = \frac{1}{2}e^{-2a}\sqrt{\pi}$$

46. $$\int_0^\infty \frac{e^{-ax} - e^{-bx}}{x}\,dx = \log_e \frac{b}{a} \qquad (a, b > 0)$$

47. $\displaystyle\int_0^\infty \frac{x\,dx}{e^x - 1} = \frac{\pi^2}{6}$

48. $\displaystyle\int_0^\infty \frac{x\,dx}{e^x + 1} = \frac{\pi^2}{12}$

49. $\displaystyle\int_0^\infty e^{-ax} \cos bx\,dx = \frac{a}{a^2 + b^2}$

50. $\displaystyle\int_0^\infty e^{-ax} \sin bx\,dx = \frac{b}{a^2 + b^2}$

51. $\displaystyle\int_0^\infty e^{-ax} \cosh bx\,dx = \frac{a}{a^2 - b^2} \qquad (a > b \geq 0)$

52. $\displaystyle\int_0^\infty e^{-ax} \sinh bx\,dx = \frac{b}{a^2 - b^2} \qquad (a > b \geq 0)$

53. $\displaystyle\int_0^\infty xe^{-ax} \sin bx\,dx = \frac{2ab}{(a^2 + b^2)^2}$

54. $\displaystyle\int_0^\infty xe^{-ax} \cos bx\,dx = \frac{a^2 - b^2}{(a^2 + b^2)^2}$

55. $\displaystyle\int_0^\infty e^{-a^2x^2} \cos bx\,dx = \frac{\sqrt{\pi} \cdot e^{-b^2/4a^2}}{2a}$

56. $\displaystyle\int_0^\infty x^2e^{-ax} \sin bx\,dx = \frac{2b(3a^2 - b^2)}{(a^2 + b^2)^3}$

57. $\displaystyle\int_0^\infty x^2e^{-ax} \cos bx\,dx = \frac{2a(a^2 - 3b^2)}{(a^2 + b^2)^3}$

58. $\displaystyle\int_0^\infty \frac{e^{-ax} \sin x}{x}\,dx = \arctan \frac{1}{a}$

59. $\displaystyle\int_0^\infty \frac{dx}{\cosh ax} = \frac{\pi}{2a}$

60. $\displaystyle\int_0^\infty \frac{x\,dx}{\sinh ax} = \frac{\pi^2}{4a^2}$

(d) Integrals containing logarithmic functions

61. $\displaystyle\int_0^1 \log_e |\log_e x|\,dx = \int_0^\infty e^{-x} \log_e x\,dx = -C = -0.577\,2157 \cdots$

62. $\int_0^1 \frac{\log_e x}{1-x}\,dx = -\frac{\pi^2}{6}$

63. $\int_0^1 \frac{\log_e x}{1+x}\,dx = -\frac{\pi^2}{12}$

64. $\int_0^1 \frac{\log_e x}{1-x^2}\,dx = -\frac{\pi^2}{8}$

65. $\int_0^1 \log_e\left(\frac{1+x}{1-x}\right)\cdot\frac{dx}{x} = \int_0^\infty \log_e\left(\frac{e^x+1}{e^x-1}\right)dx = \frac{\pi^2}{4}$

66. $\int_0^1 \frac{\log_e x}{\sqrt{1-x^2}}\,dx = -\frac{\pi}{2}\log_e 2$

67. $\int_0^1 x\log_e(1-x)\,dx = -\frac{3}{4}$

68. $\int_0^1 x\log_e(1+x)\,dx = \frac{1}{4}$

69. $\int_0^1 \frac{\log_e(1+x)}{x}\,dx = \frac{\pi^2}{12}$

70. $\int_0^1 \frac{\log_e(1+x)}{x^2+1}\,dx = \frac{\pi}{8}\log_e 2$

71. $\int_0^1 \frac{x^b-x^a}{\log_e x}\,dx = \log_e\frac{1+b}{1+a} \qquad (a, b > -1)$

72. $\int_0^1 \left(\log_e\frac{1}{x}\right)^{1/2} dx = \frac{\sqrt{\pi}}{2}$

73. $\int_0^1 \left(\log_e\frac{1}{x}\right)^{-1/2} dx = \sqrt{\pi}$

74. $\int_0^1 \left(\log_e\frac{1}{x}\right)^{a} dx = \Gamma(a+1) \qquad (a > -1)$

75. $\int_0^1 (\log_e x)^n\,dx = (-1)^n\cdot n! \qquad (n = 1, 2, \ldots)$

76. $\int_0^1 x^n\left(\log_e\frac{1}{x}\right)^{a} dx = \frac{\Gamma(a+1)}{(n+1)^{a+1}} \qquad (a, n > -1)$

77. $\int_0^{\pi/2} \log_e \sin x\,dx = \int_0^{\pi/2} \log_e \cos x\,dx = -\frac{\pi}{2}\log_e 2$

78. $\displaystyle\int_0^\pi x \log_e \sin x \, dx = -\frac{\pi^2}{2} \log_e 2$

79. $\displaystyle\int_0^\pi \log_e (a \pm b \cos x) \, dx = \pi \log_e \left(\frac{a + \sqrt{a^2 - b^2}}{2} \right) \qquad (a \geq b)$

80. $\displaystyle\int_0^\pi \frac{\log_e (1 + \sin a \cos x)}{\cos x} dx = \pi a$

81. $\displaystyle\int_0^{\pi/2} \sin x \log_e \sin x \, dx = \log_e 2 - 1$

82. $\displaystyle\int_0^{\pi/2} \log_e \tan x \, dx = 0$

83. $\displaystyle\int_0^\pi \log_e (a^2 - 2ab \cos x + b^2) \, dx = \begin{cases} 2\pi \log_e a & (a \geq b > 0) \\ 2\pi \log_e b & (b \geq a > 0) \end{cases}$

84. $\displaystyle\int_0^{\pi/4} \log_e (1 + \tan x) \, dx = \frac{\pi}{8} \log_e 2$

SUMS AND INFINITE SERIES

E-4. Some Finite Sums

1. $1 + 2 + 3 + \cdots + (n - 1) + n = \dfrac{n(n + 1)}{2}$
2. $p + (p + 1) + (p + 2) + \cdots + (q - 1) + q = \dfrac{(q + p)(q - p + 1)}{2}$
3. $1 + 3 + 5 + \cdots + (2n - 3) + (2n - 1) = n^2$
4. $2 + 4 + 6 + \cdots + (2n - 2) + 2n = n(n + 1)$
5. $1^2 + 2^2 + 3^2 + \cdots + (n - 1)^2 + n^2 = \dfrac{n(n + 1)(2n + 1)}{6}$
6. $1^3 + 2^3 + 3^3 + \cdots + (n - 1)^3 + n^3 = \dfrac{n^2(n + 1)^2}{4}$
7. $1^2 + 3^2 + 5^2 + \cdots + (2n - 1)^2 = \dfrac{n(4n^2 - 1)}{3}$
8. $1^3 + 3^3 + 5^3 + \cdots + (2n - 1)^3 = n^2(2n^2 - 1)$
9. $1^4 + 2^4 + 3^4 + \cdots + n^4 = \dfrac{n(n + 1)(2n + 1)(3n^2 + 3n - 1)}{30}$
10. $a_0 + (a_0 + d) + (a_0 + 2d) + \cdots + (a_0 + nd) = \dfrac{n + 1}{2} (2a_0 + nd)$ (FINITE ARITHMETIC SERIES)

11. $1 + a + a^2 + \cdots + a^n = \dfrac{1 - a^{n+1}}{1 - a}$ (FINITE GEOMETRIC SERIES)

12. $\displaystyle\sum_{k=1}^{n} \frac{1}{k(k+1)} = 1 - \frac{1}{n+1} = \frac{n}{n+1}$

13. $\displaystyle\sum_{k=1}^{n} \frac{1}{k(k+1)(k+2)} = \frac{1}{2}\left[\frac{1}{1 \cdot 2} - \frac{1}{(n+1)(n+2)}\right]$

14. $\displaystyle\sum_{k=0}^{n} \sin(ka + b) = \frac{\sin\left(\frac{n+1}{2}a\right)\sin\left(\frac{n}{2}a + b\right)}{\sin\frac{a}{2}}$

15. $\displaystyle\sum_{k=0}^{n} \cos(ka + b) = \frac{\sin\left(\frac{n+1}{2}a\right)\cos\left(\frac{n}{2}a + b\right)}{\sin\frac{a}{2}}$

16. $\displaystyle\sum_{k=0}^{2n-1} \cos\frac{k\pi}{n} = \sum_{k=0}^{2n-1} \sin\frac{k\pi}{n} = 0$

(12–16: $(n = 1, 2, \ldots)$)

E-5. Miscellaneous Infinite Series

1. $1 + \dfrac{1}{1!} + \dfrac{1}{2!} + \dfrac{1}{3!} + \cdots + \dfrac{1}{n!} + \cdots = e$
2. $1 - \dfrac{1}{1!} + \dfrac{1}{2!} - \dfrac{1}{3!} + \cdots \pm \dfrac{1}{n!} \mp \cdots = \dfrac{1}{e}$
3. $1 - \dfrac{1}{2} + \dfrac{1}{3} - \dfrac{1}{4} + \cdots \pm \dfrac{1}{n} \mp \cdots = \log_e 2$
4. $1 + \dfrac{1}{2} + \dfrac{1}{4} + \dfrac{1}{8} + \cdots + \dfrac{1}{2^n} + \cdots = 2$
5. $1 - \dfrac{1}{2} + \dfrac{1}{4} - \dfrac{1}{8} + \cdots \pm \dfrac{1}{2^n} \mp \cdots = \dfrac{2}{3}$
6. $1 - \dfrac{1}{3} + \dfrac{1}{5} - \dfrac{1}{7} + \dfrac{1}{9} - \cdots \pm \dfrac{1}{2n-1} \mp \cdots = \dfrac{\pi}{4}$
7. $\dfrac{1}{1 \cdot 2} + \dfrac{1}{2 \cdot 3} + \dfrac{1}{3 \cdot 4} + \cdots + \dfrac{1}{n(n+1)} + \cdots = 1$
8. $\dfrac{1}{1 \cdot 3} + \dfrac{1}{3 \cdot 5} + \dfrac{1}{5 \cdot 7} + \cdots + \dfrac{1}{(2n-1)(2n+1)} + \cdots = \dfrac{1}{2}$

9. $\frac{1}{1 \cdot 3} + \frac{1}{2 \cdot 4} + \frac{1}{3 \cdot 5} + \cdots + \frac{1}{(n-1)(n+1)} + \cdots = \frac{3}{4}$

10. $\frac{1}{3 \cdot 5} + \frac{1}{7 \cdot 9} + \frac{1}{11 \cdot 13} + \cdots + \frac{1}{(4n-1)(4n+1)} + \cdots = \frac{1}{2} - \frac{\pi}{8}$

11. $\frac{1}{1 \cdot 2 \cdot 3} + \frac{1}{2 \cdot 3 \cdot 4} + \cdots + \frac{1}{n(n+1)(n+2)} + \cdots = \frac{1}{4}$

12. $\frac{1}{1 \cdot 2 \ldots l} + \frac{1}{2 \cdot 3 \ldots (l+1)} + \cdots + \frac{1}{n \ldots (n+l-1)} + \cdots = \frac{1}{(l-1)(l-1)!}$

13. $1 + \frac{1}{2^2} + \frac{1}{3^2} + \cdots = \frac{\pi^2}{6}$

14. $1 + \frac{1}{2^4} + \frac{1}{3^4} + \cdots = \frac{\pi^4}{90}$

15. $1 + \frac{1}{2^6} + \frac{1}{3^6} + \cdots = \frac{\pi^6}{945}$

16. $1 + \frac{1}{2^8} + \frac{1}{3^8} + \cdots = \frac{\pi^8}{9450}$

NOTE: The series $\sum_{k=1}^{\infty} k^{-z}$ is referred to as **Riemann's zeta function** of the variable z.

17. $1 + a + a^2 + \cdots = \frac{1}{1-a}$ ($|a| < 1$, INFINITE GEOMETRIC SERIES)

18. $\sum_{k=1}^{\infty} \frac{1}{k(k+1)} = 1$

19. $\sum_{k=1}^{\infty} \frac{1}{k(k+1) \cdots (k+m-1)} = \frac{1}{m-1} \frac{1}{(m-1)!}$ $(m = 2, 3, \ldots)$

E-6. Power Series: Binomial Series

All series 1–6 hold for $(|x| \leq 1)$.

1. $(1 \pm x)^m$

$$= 1 \pm mx + \frac{m(m-1)}{2!}x^2 \pm \frac{m(m-1)(m-2)}{3!}x^3 + \cdots + (\pm 1)^n \frac{m(m-1)\ \ldots\ (m-n+1)}{n!}x^n + \cdots \qquad (m > 0)$$

2. $(1 \pm x)^{1/4}$

$$= 1 \pm \frac{1}{4}x - \frac{1 \cdot 3}{4 \cdot 8}x^2 \pm \frac{1 \cdot 3 \cdot 7}{4 \cdot 8 \cdot 12}x^3 - \frac{1 \cdot 3 \cdot 7 \cdot 11}{4 \cdot 8 \cdot 12 \cdot 16}x^4 \pm \cdots$$

3. $(1 \pm x)^{1/3}$

$$= 1 \pm \frac{1}{3}x - \frac{1 \cdot 2}{3 \cdot 6}x^2 \pm \frac{1 \cdot 2 \cdot 5}{3 \cdot 6 \cdot 9}x^3 - \frac{1 \cdot 2 \cdot 5 \cdot 8}{3 \cdot 6 \cdot 9 \cdot 12}x^4 \pm \cdots$$

4. $(1 \pm x)^{1/2}$

$$= 1 \pm \frac{1}{2}x - \frac{1 \cdot 1}{2 \cdot 4}x^2 \pm \frac{1 \cdot 1 \cdot 3}{2 \cdot 4 \cdot 6}x^3 - \frac{1 \cdot 1 \cdot 3 \cdot 5}{2 \cdot 4 \cdot 6 \cdot 8}x^4 \pm \cdots$$

5. $(1 \pm x)^{3/2}$

$$= 1 \pm \frac{3}{2}x + \frac{3 \cdot 1}{2 \cdot 4}x^2 \mp \frac{3 \cdot 1 \cdot 1}{2 \cdot 4 \cdot 6}x^3 + \frac{3 \cdot 1 \cdot 1 \cdot 3}{2 \cdot 4 \cdot 6 \cdot 8}x^4 \mp \cdots$$

6. $(1 \pm x)^{5/2}$

$$= 1 \pm \frac{5}{2}x + \frac{5 \cdot 3}{2 \cdot 4}x^2 \pm \frac{5 \cdot 3 \cdot 1}{2 \cdot 4 \cdot 6}x^3 - \frac{5 \cdot 3 \cdot 1 \cdot 1}{2 \cdot 4 \cdot 6 \cdot 8}x^4 \mp \cdots$$

7. $(1 \pm x)^{-m}$
$$= 1 \mp mx + \frac{m(m+1)}{2!}x^2 \mp \frac{m(m+1)(m+2)}{3!}x^3 + \cdots + (\pm 1)^n \frac{m(m+1)\cdots(m+n-1)}{n!}x^n \pm \cdots \quad (m > 0)$$

8. $(1 \pm x)^{-1/4}$
$$= 1 \mp \frac{1}{4}x + \frac{1 \cdot 5}{4 \cdot 8}x^2 \mp \frac{1 \cdot 5 \cdot 9}{4 \cdot 8 \cdot 12}x^3 + \frac{1 \cdot 5 \cdot 9 \cdot 13}{4 \cdot 8 \cdot 12 \cdot 16}x^4 \mp \cdots$$

9. $(1 \pm x)^{-1/3}$
$$= 1 \mp \frac{1}{3}x + \frac{1 \cdot 4}{3 \cdot 6}x^2 \mp \frac{1 \cdot 4 \cdot 7}{3 \cdot 6 \cdot 9}x^3 + \frac{1 \cdot 4 \cdot 7 \cdot 10}{3 \cdot 6 \cdot 9 \cdot 12}x^4 \mp \cdots$$

10. $(1 \pm x)^{-1/2}$
$$= 1 \mp \frac{1}{2}x + \frac{1 \cdot 3}{2 \cdot 4}x^2 \mp \frac{1 \cdot 3 \cdot 5}{2 \cdot 4 \cdot 6}x^3 + \frac{1 \cdot 3 \cdot 5 \cdot 7}{2 \cdot 4 \cdot 6 \cdot 8}x^4 \mp \cdots$$

11. $(1 \pm x)^{-1}$
$$= 1 \mp x + x^2 \mp x^3 + x^4 \mp \cdots$$

12. $(1 \pm x)^{-3/2}$
$$= 1 \mp \frac{3}{2}x + \frac{3 \cdot 5}{2 \cdot 4}x^2 \mp \frac{3 \cdot 5 \cdot 7}{2 \cdot 4 \cdot 6}x^3 + \frac{3 \cdot 5 \cdot 7 \cdot 9}{2 \cdot 4 \cdot 6 \cdot 8}x^4 \mp \cdots$$

13. $(1 \pm x)^{-2}$
$$= 1 \mp 2x + 3x^2 \mp 4x^3 + 5x^4 \mp \cdots$$

14. $(1 \pm x)^{-5/2}$
$$= 1 \mp \frac{5}{2}x + \frac{5 \cdot 7}{2 \cdot 4}x^2 \mp \frac{5 \cdot 7 \cdot 9}{2 \cdot 4 \cdot 6}x^3 + \frac{5 \cdot 7 \cdot 9 \cdot 11}{2 \cdot 4 \cdot 6 \cdot 8}x^4 \mp \cdots$$

15. $(1 \pm x)^{-3}$
$$= 1 \mp \frac{1}{1 \cdot 2}(2 \cdot 3x \mp 3 \cdot 4x^2 + 4 \cdot 5x^3 \mp 5 \cdot 6x^4 + \cdots)$$

16. $(1 \pm x)^{-4}$
$$= 1 \mp \frac{1}{1 \cdot 2 \cdot 3}(2 \cdot 3 \cdot 4x \mp 3 \cdot 4 \cdot 5x^2 + 4 \cdot 5 \cdot 6x^3 \mp 5 \cdot 6 \cdot 7x^4 + \cdots)$$

17. $(1 \pm x)^{-5}$
$$= 1 \mp \frac{1}{1 \cdot 2 \cdot 3 \cdot 4}(2 \cdot 3 \cdot 4 \cdot 5x \mp 3 \cdot 4 \cdot 5 \cdot 6x^2 + 4 \cdot 5 \cdot 6 \cdot 7x^3 \mp 5 \cdot 6 \cdot 7 \cdot 8x^4 + \cdots)$$

(Formulas 7–17 hold for $|x| < 1$.)

NOTE: The series for $(1 \pm x)^m$ reduces to finite sums (Binomial Theorem, Sec. 1.4-1) if $m = 1, 2, 3, \ldots$.

E-7. Power Series for Elementary Transcendental Functions

Valid for $(|z| < \infty)$:

1. $e^z = 1 + z + \frac{z^2}{2!} + \frac{z^3}{3!} + \cdots$
2. $\sin z = z - \frac{z^3}{3!} + \frac{z^5}{5!} \mp \cdots$
3. $\cos z = 1 - \frac{z^2}{2!} + \frac{z^4}{4!} \mp \cdots$
4. $\sinh z = z + \frac{z^3}{3!} + \frac{z^5}{5!} + \cdots$
5. $\cosh z = 1 + \frac{z^2}{2!} + \frac{z^4}{4!} + \cdots$
6. $\sin (z + a) = \sin a + z \cos a - \frac{z^2 \sin a}{2!} - \frac{z^3 \cos a}{3!} + \frac{z^4 \sin a}{4!} + \cdots + \frac{z^n \sin\left(a + \frac{n\pi}{2}\right)}{n!} \cdots$
7. $\cos (z + a) = \cos a - z \sin a - \frac{z^2 \cos a}{2!} + \frac{z^3 \sin a}{3!} + \frac{z^4 \cos a}{4!} - \cdots + \frac{z^n \cos\left(a + \frac{n\pi}{2}\right)}{n!} \pm \cdots$

Valid for $(|z| < 1)$:

8. $\log_e (1 + z) = z - \frac{z^2}{2} + \frac{z^3}{3} - \frac{z^4}{4} \pm \cdots$
9. $\arcsin z = z + \frac{1}{2} \cdot \frac{z^3}{3} + \frac{1}{2} \cdot \frac{3}{4} \cdot \frac{z^5}{5} + \frac{1}{2} \cdot \frac{3}{4} \cdot \frac{5}{6} \cdot \frac{z^7}{7} + \cdots$
10. $\sinh^{-1} z = z - \frac{1}{2} \cdot \frac{z^3}{3} + \frac{1}{2} \cdot \frac{3}{4} \cdot \frac{z^5}{5} - \frac{1}{2} \cdot \frac{3}{4} \cdot \frac{5}{6} \cdot \frac{z^7}{7} \pm \cdots$
11. $\arctan z = z - \frac{z^3}{3} + \frac{z^5}{5} \mp \cdots$
12. $\tanh^{-1} z = \frac{1}{2} \log_e \frac{1 + z}{1 - z} = z + \frac{z^3}{3} + \frac{z^5}{5} + \cdots$

NOTE: Series 11 converges also for $|z| = 1$, unless $z = \pm i$.

13. $$\tan z = \sum_{k=1}^{\infty} \frac{2^{2k}(2^{2k} - 1)(-1)^{k-1}B_{2k}}{(2k)!} z^{2k-1} = z + \frac{1}{3} z^3 + \frac{2}{15} z^5 + \frac{17}{315} z^7 + \cdots \quad \left(|z| < \frac{\pi}{2}\right)$$

14. $\cot z = \frac{1}{z}\sum_{k=0}^{\infty}(-1)^k \frac{2^{2k}B_{2k}}{(2k)!}z^{2k} = \frac{1}{z} - \frac{1}{3}z - \frac{1}{45}z^3 - \frac{2}{945}z^5 - \cdots$

$(0 < |z| < \pi)$

where the B_k are the Bernoulli numbers defined in Sec. 21.5-2. 13 and 14 yield similar series for $\tanh z$ and $\coth z$ with the aid of Eq. (21.2-32).

15. $\arctan z = \frac{\pi}{2} - \frac{1}{z} + \frac{1}{3z^3} - \frac{1}{5z^5} + \cdots \qquad (|z| > 1 \text{ and } z^2 \neq -1)$

16. $\arctan z = \frac{z}{1+z^2}\left[1 + \frac{2}{3}\frac{z^2}{1+z^2} + \frac{2\cdot 4}{3\cdot 5}\left(\frac{z^2}{1+z^2}\right)^2 + \cdots\right]$

$(z^2 \neq -1)$

17. $\coth^{-1} z = \frac{1}{z} + \frac{1}{3z^3} + \frac{1}{5z^5} + \cdots \qquad (|z| > 1)$

INFINITE PRODUCTS AND CONTINUED FRACTIONS

E-8. Some Infinite Products (see also Sec. 7-6)

$$\left.\begin{aligned}
&1.\ \sin z = z\prod_{k=1}^{\infty}\left[1 - \left(\frac{z}{k\pi}\right)^2\right]\\
&2.\ \cos z = \prod_{k=1}^{\infty}\left\{1 - \left[\frac{2z}{\pi(2k-1)}\right]^2\right\}\\
&3.\ \sinh z = z\prod_{k=1}^{\infty}\left[1 + \left(\frac{z}{k\pi}\right)^2\right]\\
&4.\ \cosh z = \prod_{k=1}^{\infty}\left\{1 + \left[\frac{2z}{\pi(2k-1)}\right]^2\right\}
\end{aligned}\right\}(|z| < \infty)$$

E-9. Some Continued Fractions (see also Secs. 4.8-8 and 20.5-7)

$$\tan z = \frac{z}{1-}\,\frac{z^2}{3-}\,\frac{z^2}{5-}\,\frac{z^2}{7-}\cdots \qquad \left(z \neq \frac{\pi}{2} \pm n\pi\right)$$

$$\left.\begin{aligned}
e^z &= \frac{1}{1-}\,\frac{z}{1+}\,\frac{z}{2-}\,\frac{z}{3+}\,\frac{z}{2-}\,\frac{z}{5+}\,\frac{z}{2-}\cdots\\
&= 1 + \frac{z}{1-}\,\frac{z}{2+}\,\frac{z}{3-}\,\frac{z}{2+}\,\frac{z}{5-}\,\frac{z}{2+}\,\frac{z}{7-}\cdots
\end{aligned}\right\}(|z| < \infty)$$

$$\log_e(1+z) = \frac{z}{1+}\,\frac{z}{2+}\,\frac{z}{3+}\,\frac{4z}{4+}\,\frac{4z}{5+}\,\frac{9z}{6+}\cdots$$ (z in the plane cut from -1 to $-\infty$)

Table E-1. Operations with Series*

Let $s_1 = 1 + a_1x + a_2x^2 + a_3x^3 + a_4x^4 + \cdots$

$s_2 = 1 + b_1x + b_2x^2 + b_3x^3 + b_4x^4 + \cdots$

$s_3 = 1 + c_1x + c_2x^2 + c_3x^3 + c_4x^4 + \cdots$

Operation	c_1	c_2	c_3	c_4
$s_3 = s_1^{-1}$	$-a_1$	$a_1^2 - a_2$	$2a_1a_2 - a_3 - a_1^3$	$2a_1a_3 - 3a_1^2a_2 - a_4 + a_2^2 + a_1^4$
$s_3 = s_1^{-2}$	$-2a_1$	$3a_1^2 - 2a_2$	$6a_1a_2 - 2a_3 - 4a_1^3$	$6a_1a_3 + 3a_2^2 - 2a_4 - 12a_1^2a_2 + 5a_1^4$
$s_3 = s_1^{1/2}$	$\frac{1}{2}a_1$	$\frac{1}{2}a_2 - \frac{1}{8}a_1^2$	$\frac{1}{2}a_3 - \frac{1}{4}a_1a_2 + \frac{1}{16}a_1^3$	$\frac{1}{2}a_4 - \frac{1}{4}a_1a_3 - \frac{1}{8}a_2^2 + \frac{3}{16}a_1^2a_2 - \frac{5}{128}a_1^4$
$s_3 = s_1^{-1/2}$	$-\frac{1}{2}a_1$	$\frac{3}{8}a_1^2 - \frac{1}{2}a_2$	$\frac{3}{4}a_1a_2 - \frac{1}{2}a_3 - \frac{5}{16}a_1^3$	$\frac{3}{4}a_1a_3 + \frac{3}{8}a_2^2 - \frac{1}{2}a_4 - \frac{15}{16}a_1^2a_2 + \frac{35}{128}a_1^4$
$s_3 = s_1^n$	na_1	$\frac{1}{2}(n-1)c_1a_1 + na_2$	$c_1a_2(n-1)$ $+ \frac{1}{6}c_1a_1^2(n-1)(n-2) + na_3$	$na_4 + c_1a_3(n-1) + \frac{1}{2}n(n-1)a_2^2$ $+ \frac{1}{2}(n-1)(n-2)c_1a_1a_2$ $+ \frac{1}{24}(n-1)(n-2)(n-3)c_1a_1^3$
$s_3 = s_1s_2$	$a_1 + b_1$	$b_2 + a_1b_1 + a_2$	$b_3 + a_1b_2 + a_2b_1 + a_3$	$b_4 + a_1b_3 + a_2b_2 + a_3b_1 + a_4$
$s_3 = s_1/s_2$	$a_1 - b_1$	$a_2 - (b_1c_1 + b_2)$	$a_3 - (b_1c_2 + b_2c_1 + b_3)$	$a_4 - (b_1c_3 + b_2c_2 + b_3c_1 + b_4)$
$s_3 = \exp(s_1 - 1)$	a_1	$a_2 + \frac{1}{2}a_1^2$	$a_3 + a_1a_2 + \frac{1}{6}a_1^3$	$a_4 + a_1a_3 + \frac{1}{2}a_2^2 + \frac{1}{2}a_2a_1^2 + \frac{1}{24}a_1^4$
$s_3 = 1 + \log_e s_1$	a_1	$a_2 - \frac{1}{2}a_1c_1$	$a_3 - \frac{1}{3}(a_2c_1 + 2a_1c_2)$	$a_4 - \frac{1}{4}(a_3c_1 + 2a_2c_2 + 3a_1c_3)$

* From M. Abramowitz and I. A. Stegun (eds.), *Handbook of Mathematical Functions*, National Bureau of Standards, Washington, D.C., 1964.

BIBLIOGRAPHY

Abramowitz, M., and I. A. Stegun: *Handbook of Mathematical Functions*, National Bureau of Standards, Washington, D.C., 1964.

Bierens de Haan, D.: *Nouvelles tables d'intégrales définies*, Stechert, New York, 1939.

Byrd, P. F., and M. D. Friedman: *Handbook of Elliptic Integrals for Engineers and Physicists*, Springer, Berlin, 1954.

Gröbner and Hofreiter: *Integral Tafel*, 2d ed., Springer, Vienna, 1958.

Lindman, C. F.: *Examen des novelles tables d'intégrales définies de M. Bierens de Haan*, 1944.

Luke, Y. I.: *Integrals of Bessel Functions*, McGraw-Hill, New York, 1962.

Petit Bois, G.: *Tables of Indefinite Integrals*, Dover, New York, 1961.

Ryshik, I. M., and I. S. Gradstein: *Tables of Series, Products, and Integrals*, Academic Press, New York, 1964.

APPENDIX **F**

NUMERICAL TABLES

The following numerical tables are intended less for extensive numerical computations than as quantitative background material indicating the behavior of the most important transcendental functions.

The following *numerical constants* are frequently useful:

$$\pi = 3.1415927 \qquad \log_{10} \pi = 0.4971499$$

$$2\pi = 6.2831853 \qquad \log_{10} 2\pi = 0.7981799$$

$$\frac{1}{2\pi} = 0.1591549 \qquad \log_{10} \frac{1}{2\pi} = 9.2018201 - 10$$

$$\pi^2 = 9.8696044 \qquad \log_{10} \pi^2 = 0.9942997$$

$$e = 2.71828183 \qquad \log_{10} e = \frac{1}{\log_e 10} = 0.43429448$$

$$\log_e 10 = \frac{1}{\log_{10} e} = 2.302585093$$

Table F-1. Table of Squares*

	0	1	2	3	4	5	6	7	8	9
0	0	1	4	9	16	25	36	49	64	81
1	100	121	144	169	196	225	256	289	324	361
2	400	441	484	529	576	625	676	729	784	841
3	900	961	1,024	1,089	1,156	1,225	1,296	1,369	1,444	1,521
4	1,600	1,681	1,764	1,849	1,936	2,025	2,116	2,209	2,304	2,401
5	2,500	2,601	2,704	2,809	2,916	3,025	3,136	3,249	3,364	3,481
6	3,600	3,721	3,844	3,969	4,096	4,225	4,356	4,489	4,624	4,761
7	4,900	5,041	5,184	5,329	5,476	5,625	5,776	5,929	6,084	6,241
8	6,400	6,561	6,724	6,889	7,056	7,225	7,396	7,569	7,744	7,921
9	8,100	8,281	8,464	8,649	8,836	9,025	9,216	9,409	9,604	9,801
10	10,000	10,201	10,404	10,609	10,816	11,025	11,236	11,449	11,664	11,881
11	12,100	12,321	12,544	12,769	12,996	13,225	13,456	13,689	13,924	14,161
12	14,400	14,641	14,884	15,129	15,376	15,625	15,876	16,129	16,384	16,641
13	16,900	17,161	17,424	17,689	17,956	18,225	18,496	18,769	19,044	19,321
14	19,600	19,881	20,164	20,449	20,736	21,025	21,316	21,609	21,904	22,201
15	22,500	22,801	23,104	23,409	23,716	24,025	24,336	24,649	24,964	25,281
16	25,600	25,921	26,244	26,569	26,896	27,225	27,556	27,889	28,224	28,561
17	28,900	29,241	29,584	29,929	30,276	30,625	30,976	31,329	31,684	32,041
18	32,400	32,761	33,124	33,489	33,856	34,225	34,596	34,969	35,344	35,721
19	36,100	36,481	36,864	37,249	37,636	38,025	38,416	38,809	39,204	39,601
20	40,000	40,401	40,804	41,209	41,616	42,025	42,436	42,849	43,264	43,681
21	44,100	44,521	44,944	45,369	45,796	46,225	46,656	47,089	47,524	47,961
22	48,400	48,841	49,284	49,729	50,176	50,625	51,076	51,529	51,984	52,441
23	52,900	53,361	53,824	54,289	54,756	55,225	55,696	56,169	56,644	57,121
24	57,600	58,081	58,564	59,049	59,536	60,025	60,516	61,009	61,504	62,001
25	62,500	63,001	63,504	64,009	64,516	65,025	65,536	66,049	66,564	67,081
26	67,600	68,121	68,644	69,169	69,696	70,225	70,756	71,289	71,824	72,361
27	72,900	73,441	73,984	74,529	75,076	75,625	76,176	76,729	77,284	77,841
28	78,400	78,961	79,524	80,089	80,656	81,225	81,796	82,369	82,944	83,521
29	84,100	84,681	85,264	85,849	86,436	87,025	87,616	88,209	88,804	89,401
30	90,000	90,601	91,204	91,809	92,416	93,025	93,636	94,249	94,864	95,481
31	96,100	96,721	97,344	97,969	98,596	99,225	99,856	100,489	101,124	101,761
32	102,400	103,041	103,684	104,329	104,976	105,625	106,276	106,929	107,584	108,241
33	108,900	109,561	110,224	110,889	111,556	112,225	112,896	113,569	114,244	114,921
34	115,600	116,281	116,964	117,649	118,336	119,025	119,716	120,409	121,104	121,801

* This table is reprinted from Table XXVII of Fisher and Yates, *Statistical Tables for Biological, Agricultural, and Medical Research,* published by Oliver & Boyd, Ltd., Edinburgh, by permission of the authors and publishers.

Table F-1. Table of Squares *(Continued)*

	0	1	2	3	4	5	6	7	8	9
35	122,500	123,201	123,904	124,609	125,316	126,025	126,736	127,449	128,164	128,881
36	129,600	130,321	131,044	131,769	132,496	133,225	133,956	134,689	135,424	136,161
37	136,900	137,641	138,384	139,129	139,876	140,625	141,376	142,129	142,884	143,641
38	144,400	145,161	145,924	146,689	147,456	148,225	148,996	149,769	150,544	151,321
39	152,100	152,881	153,664	154,449	155,236	156,025	156,816	157,609	158,404	159,201
40	160,000	160,801	161,604	162,409	163,216	164,025	164,836	165,649	166,464	167,281
41	168,100	168,921	169,744	170,569	171,396	172,225	173,056	173,889	174,724	175,561
42	176,400	177,241	178,084	178,929	179,776	180,625	181,476	182,329	183,184	184,041
43	184,900	185,761	186,624	187,489	188,356	189,225	190,096	190,969	191,844	192,721
44	193,600	194,481	195,364	196,249	197,136	198,025	198,916	199,809	200,704	201,601
45	202,500	203,401	204,304	205,209	206,116	207,025	207,936	208,849	209,764	210,681
46	211,600	212,521	213,444	214,369	215,296	216,225	217,156	218,089	219,024	219,961
47	220,900	221,841	222,784	223,729	224,676	225,625	226,576	227,529	228,484	229,441
48	230,400	231,361	232,324	233,289	234,256	235,225	236,196	237,169	238,144	239,121
49	240,100	241,081	242,064	243,049	244,036	245,025	246,016	247,009	248,004	249,001
50	250,000	251,001	252,004	253,009	254,016	255,025	256,036	257,049	258,064	259,081
51	260,100	261,121	262,144	263,169	264,196	265,225	266,256	267,289	268,324	269,361
52	270,400	271,441	272,484	273,529	274,576	275,625	276,676	277,729	278,784	279,841
53	280,900	281,961	283,024	284,089	285,156	286,225	287,296	288,369	289,444	290,521
54	291,600	292,681	293,764	294,849	295,936	297,025	298,116	299,209	300,304	301,401
55	302,500	303,601	304,704	305,809	306,916	308,025	309,136	310,249	311,364	312,481
56	313,600	314,721	315,844	316,969	318,096	319,225	320,356	321,489	322,624	323,761
57	324,900	326,041	327,184	328,329	329,476	330,625	331,776	332,929	334,084	335,241
58	336,400	337,561	338,724	339,889	341,056	342,225	343,396	344,569	345,744	346,921
59	348,100	349,281	350,464	351,649	352,836	354,025	355,216	356,409	357,604	358,801
60	360,000	361,201	362,404	363,609	364,816	366,025	367,236	368,449	369,664	370,881
61	372,100	373,321	374,544	375,769	376,996	378,225	379,456	380,689	381,924	383,161
62	384,400	385,641	386,884	388,129	389,376	390,625	391,876	393,129	394,384	395,641
63	396,900	398,161	399,424	400,689	401,956	403,225	404,496	405,769	407,044	408,321
64	409,600	410,881	412,164	413,449	414,736	416,025	417,316	418,609	419,904	421,201
65	422,500	423,801	425,104	426,409	427,716	429,025	430,336	431,649	432,964	434,281
66	435,600	436,921	438,244	439,569	440,896	442,225	443,556	444,889	446,224	447,561
67	448,900	450,241	451,584	452,929	454,276	455,625	456,976	458,329	459,684	461,041
68	462,400	463,761	465,124	466,489	467,856	469,225	470,596	471,969	473,344	474,721
69	476,100	477,481	478,864	480,249	481,636	483,025	484,416	485,809	487,204	488,601
70	490,000	491,401	492,804	494,209	495,616	497,025	498,436	499,849	501,264	502,681
71	504,100	505,521	506,944	508,369	509,796	511,225	512,656	514,089	515,524	516,961
72	518,400	519,841	521,284	522,729	524,176	525,625	527,076	528,529	529,984	531,441
73	532,900	534,361	535,824	537,289	538,756	540,225	541,696	543,169	544,644	546,121
74	547,600	549,081	550,564	552,049	553,536	555,025	556,516	558,009	559,504	561,001

Table F-1. Table of Squares (*Continued*)

	0	1	2	3	4	5	6	7	8	9
75	562,500	564,001	565,504	567,009	568,516	570,025	571,536	573,049	574,564	576,081
76	577,600	579,121	580,644	582,169	583,696	585,225	586,756	588,289	589,824	591,361
77	592,900	594,441	595,984	597,529	599,076	600,625	602,176	603,729	605,284	606,841
78	608,400	609,961	611,524	613,089	614,656	616,225	617,796	619,369	620,944	622,521
79	624,100	625,681	627,264	628,849	630,436	632,025	633,616	635,209	636,804	638,401
80	640,000	641,601	643,204	644,809	646,416	648,025	649,636	651,249	652,864	654,481
81	656,100	657,721	659,344	660,969	662,596	664,225	665,856	667,489	669,124	670,761
82	672,400	674,041	675,684	677,329	678,976	680,625	682,276	683,929	685,584	687,241
83	688,900	690,561	692,224	693,889	695,556	697,225	698,896	700,569	702,244	703,921
84	705,600	707,281	708,964	710,649	712,336	714,025	715,716	717,409	719,104	720,801
85	722,500	724,201	725,904	727,609	729,316	731,025	732,736	734,449	736,164	737,881
86	739,600	741,321	743,044	744,769	746,496	748,225	749,956	751,689	753,424	755,161
87	756,900	758,641	760,384	762,129	763,876	765,625	767,376	769,129	770,884	772,641
88	774,400	776,161	777,924	779,689	781,456	783,225	784,996	786,769	788,544	790,321
89	792,100	793,881	795,664	797,449	799,236	801,025	802,816	804,609	806,404	808,201
90	810,000	811,801	813,604	815,409	817,216	819,025	820,836	822,649	824,464	826,281
91	828,100	829,921	831,744	833,569	835,396	837,225	839,056	840,889	842,724	844,561
92	846,400	848,241	850,084	851,929	853,776	855,625	857,476	859,329	861,184	863,041
93	864,900	866,761	868,624	870,489	872,356	874,225	876,096	877,969	879,844	881,721
94	883,600	885,481	887,364	889,249	891,136	893,025	894,916	896,809	898,704	900,601
95	902,500	904,401	906,304	908,209	910,116	912,025	913,936	915,849	917,764	919,681
96	921,600	923,521	925,444	927,369	929,296	931,225	933,156	935,089	937,024	938,961
97	940,900	942,841	944,784	946,729	948,676	950,625	952,576	954,529	956,484	958,441
98	960,400	962,361	964,324	966,289	968,256	970,225	972,196	974,169	976,144	978,121
99	980,100	982,081	984,064	986,049	988,036	990,025	992,016	994,009	996,004	998,001

Exact squares of four-figure numbers can be quickly calculated from the identity $(a \pm b)^2 = a^2 \pm 2ab + b^2$. Thus $693.3^2 = 480249 + 415.8 + 0.09 = 480664.89$.

Table F-2. Five-place Common Logarithms of Numbers†

100–155

No.	L	0	1	2	3	4	5	6	7	8	9
100	00	000	043	087	130	173	217	260	303	346	389
101		432	475	518	561	604	647	689	732	775	817
102		860	903	945	988	*030	*072	*115	*157	*199	*242
103	01	284	326	368	410	452	494	536	578	620	662
104		703	745	787	828	870	912	953	995	*036	*078
105	02	119	160	202	243	284	325	366	408	449	490
106		531	572	612	653	694	735	776	816	857	898
107		938	979	*019	*060	*100	*141	*181	*222	*262	*302
108	03	342	383	423	463	503	543	583	623	663	703
109		743	782	822	862	902	941	981	*021	*060	*100
110	04	139	179	218	258	297	336	376	415	454	493
111		532	571	610	650	689	727	766	805	844	883
112		922	961	999	*038	*077	*115	*154	*192	*231	*269
113	05	308	346	385	423	461	500	538	576	614	652
114		690	729	767	805	843	881	918	956	994	*032
115	06	070	108	145	183	221	258	296	333	371	408
116		446	483	521	558	595	633	670	707	744	781
117		819	856	893	930	967	*004	*041	*078	*115	*151
118	07	188	225	262	298	335	372	408	445	482	518
119		555	591	628	664	700	737	773	809	846	882
120		918	954	990	*027	*063	*099	*135	*171	*207	*243
121	08	279	314	350	386	422	458	493	529	565	600
122		636	672	707	743	778	814	849	884	920	955
123		991	*026	*061	*096	*132	*167	*202	*237	*272	*307
124	09	342	377	412	447	482	517	552	587	621	656
125		691	726	760	795	830	864	899	934	968	*003
126	10	037	072	106	140	175	209	243	278	312	346
127		380	415	449	483	517	551	585	619	653	687
128		721	755	789	823	857	890	924	958	992	*025
129	11	059	093	126	160	193	227	261	294	327	361
130		394	428	461	494	528	561	594	628	661	694
131		727	760	793	826	860	893	926	959	992	*024
132	12	057	090	123	156	189	222	254	287	320	353
133		385	418	450	483	516	548	581	613	646	678
134		710	743	775	808	840	872	905	937	969	*001
135	13	033	066	098	130	162	194	226	258	290	322
136		354	386	418	450	481	513	545	577	609	640
137		672	704	735	767	799	830	862	893	925	956
138		988	*019	*051	*082	*114	*145	*176	*208	*239	*270
139	14	301	333	364	395	426	457	489	520	551	582
140		613	644	675	706	737	768	799	829	860	891
141		922	953	983	*014	*045	*076	*106	*137	*168	*198
142	15	229	259	290	320	351	381	412	442	473	503
143		534	564	594	625	655	685	715	746	776	806
144		836	866	897	927	957	987	*017	*047	*077	*107
145	16	137	167	197	227	256	286	316	346	376	406
146		435	465	495	524	554	584	613	643	673	702
147		732	761	791	820	850	879	909	938	967	997
148	17	026	056	085	114	143	173	202	231	260	289
149		319	348	377	406	435	464	493	522	551	580
150		609	638	667	696	725	754	782	811	840	869
151		898	926	955	984	*013	*041	*070	*099	*127	*156
152	18	184	213	241	270	299	327	355	384	412	441
153		469	498	526	554	583	611	639	667	696	724
154		752	780	808	837	865	893	921	949	977	*005
155	19	033	061	089	117	145	173	201	229	257	285
No.	L	0	1	2	3	4	5	6	7	8	9

Proportional parts

	44	43	42
1	4.4	4.3	4.2
2	8.8	8.6	8.4
3	13.2	12.9	12.6
4	17.6	17.2	16.8
5	22.0	21.5	21.0
6	26.4	25.8	25.2
7	30.8	30.1	29.4
8	35.2	34.4	33.6
9	39.6	38.7	37.8

	41	40	39
1	4.1	4.0	3.9
2	8.2	8.0	7.8
3	12.3	12.0	11.7
4	16.4	16.0	15.6
5	20.5	20.0	19.5
6	24.6	24.0	23.4
7	28.7	28.0	27.3
8	32.8	32.0	31.2
9	36.9	36.0	35.1

	38	37	36
1	3.8	3.7	3.6
2	7.6	7.4	7.2
3	11.4	11.1	10.8
4	15.2	14.8	14.4
5	19.0	18.5	18.0
6	22.8	22.2	21.6
7	26.6	25.9	25.2
8	30.4	29.6	28.8
9	34.2	33.3	32.4

	35	34	33
1	3.5	3.4	3.3
2	7.0	6.8	6.6
3	10.5	10.2	9.9
4	14.0	13.6	13.2
5	17.5	17.0	16.5
6	21.0	20.4	19.8
7	24.5	23.8	23.1
8	28.0	27.2	26.4
9	31.5	30.6	29.7

	32	31	30
1	3.2	3.1	3.0
2	6.4	6.2	6.0
3	9.6	9.3	9.0
4	12.8	12.4	12.0
5	16.0	15.5	15.0
6	19.2	18.6	18.0
7	22.4	21.7	21.0
8	25.6	24.8	24.0
9	28.8	27.9	27.0

Proportional parts

* Indicates change in the first two decimal places.

† From R. H. Perry, *Engineering Manual*, McGraw-Hill, New York, 1959.

Table F-2. Five-place Common Logarithms of Numbers (*Continued*)

155–210

No.	L	0	1	2	3	4	5	6	7	8	9
155	19	033	061	089	117	145	173	201	229	257	285
156		312	340	368	396	424	451	479	507	535	562
157		590	618	645	673	700	728	756	783	811	838
158		866	893	921	948	976	*003	*030	*058	*085	*112
159	20	140	167	194	222	249	276	303	330	358	385
160		412	439	466	493	520	548	575	602	629	656
161		683	710	737	763	790	817	844	871	898	925
162		952	978	*005	*032	*059	*085	*112	*139	*165	*192
163	21	219	245	272	299	325	352	378	405	431	458
164		484	511	537	564	590	617	643	669	696	722
165		748	775	801	827	854	880	906	932	958	985
166	22	011	037	063	089	115	141	168	194	220	246
167		272	298	324	350	376	401	427	453	479	505
168		531	557	583	608	634	660	686	712	737	763
169		789	814	840	866	891	917	943	968	994	*019
170	23	045	070	096	121	147	172	198	223	249	274
171		300	325	350	376	401	426	452	477	502	528
172		553	578	603	629	654	679	704	729	754	776
173		805	830	855	880	905	930	955	980	*005	*030
174	24	055	080	105	130	155	180	204	229	254	279
175		304	329	353	378	403	428	452	477	502	527
176		551	576	601	625	650	674	699	724	748	773
177		797	822	846	871	895	920	944	969	993	*018
178	25	042	066	091	115	139	164	188	212	237	261
179		285	310	334	358	382	406	431	455	479	503
180		527	551	575	600	624	648	672	696	720	744
181		768	792	816	840	864	888	912	935	959	983
182	26	007	031	055	079	102	126	150	174	198	221
183		245	269	293	316	340	364	387	411	435	458
184		482	505	529	553	576	600	623	647	670	694
185		717	741	764	788	811	834	858	881	905	928
186		951	975	998	*021	*045	*068	*091	*114	*138	*161
187	27	184	207	231	254	277	300	323	346	370	393
188		416	439	462	485	508	531	554	577	600	623
189		646	669	692	715	738	761	784	807	830	853
190		875	898	921	944	967	990	*012	*035	*058	*081
191	28	103	126	149	172	194	217	240	262	285	308
192		330	353	375	398	421	443	466	488	511	533
193		556	578	601	623	646	668	691	713	735	758
194		780	803	825	847	870	892	914	937	959	981
195	29	003	026	048	070	092	115	137	159	181	203
196		226	248	270	292	314	336	358	380	403	425
197		447	469	491	513	535	557	579	601	623	645
198		667	688	710	732	754	776	798	820	842	863
199		885	907	929	951	973	994	*016	*038	*060	*081
200	30	103	125	146	168	190	211	233	255	276	298
201		320	341	363	384	406	428	449	471	492	514
202		535	557	578	600	621	643	664	685	707	728
203		750	771	792	814	835	856	878	899	920	942
204		963	984	*006	*027	*048	*069	*091	*112	*133	*154
205	31	175	197	218	239	260	281	302	323	345	366
206		387	408	429	450	471	492	513	534	555	576
207		597	618	639	660	681	702	723	744	765	785
208		806	827	848	869	890	911	931	952	973	994
209	32	015	035	056	077	098	118	139	160	181	201
210		222	243	263	284	305	325	346	366	387	408
No.	L	0	1	2	3	4	5	6	7	8	9

Proportional parts

	29	28	27	26	25	24	23	22
1	2.9	2.8	2.7	2.6	2.5	2.4	2.3	2.2
2	5.8	5.6	5.4	5.2	5.0	4.8	4.6	4.4
3	8.7	8.4	8.1	7.8	7.5	7.2	6.9	6.6
4	11.6	11.2	10.8	10.4	10.0	9.6	9.2	8.8
5	14.5	14.0	13.5	13.0	12.5	12.0	11.5	11.0
6	17.4	16.8	16.2	15.6	15.0	14.4	13.8	13.2
7	20.3	19.6	18.9	18.2	17.5	16.8	16.1	15.4
8	23.2	22.4	21.6	20.8	20.0	19.2	18.4	17.6
9	26.1	25.2	24.3	23.4	22.5	21.6	20.7	19.8

Proportional parts

* Indicates change in the first two decimal places.

Table F-2. Five-place Common Logarithms of Numbers (*Continued*)

210–265

No.	L	0	1	2	3	4	5	6	7	8	9
210	32	222	243	263	284	305	325	346	366	387	408
211		428	449	469	490	511	531	552	572	593	613
212		634	654	675	695	715	736	756	777	797	818
213		838	858	879	899	919	940	960	980	*001	*021
214	33	041	062	082	102	122	143	163	183	203	224
215		244	264	284	304	325	345	365	385	405	425
216		445	465	486	506	526	546	566	586	606	626
217		646	666	686	706	726	746	766	786	806	826
218		846	866	885	905	925	945	965	985	*005	*025
219	34	044	064	084	104	124	143	163	183	203	223
220		242	262	282	301	321	341	361	380	400	420
221		439	459	479	498	518	537	557	577	596	616
222		635	655	674	694	713	733	753	772	792	811
223		830	850	869	889	908	928	947	967	986	*005
224	35	025	044	064	083	102	122	141	160	180	199
225		218	238	257	276	295	315	334	353	372	392
226		411	430	449	468	488	507	526	545	564	583
227		603	622	641	660	679	698	717	736	755	774
228		793	813	832	851	870	889	908	927	946	965
229		984	*003	*021	*040	*059	*078	*097	*116	*135	*154
230	36	173	192	211	229	248	267	286	305	324	342
231		361	380	399	418	436	455	474	493	511	530
232		549	568	586	605	624	642	661	680	698	717
233		736	754	773	791	810	829	847	866	884	903
234		922	940	959	977	996	*014	*033	*051	*070	*088
235	37	107	125	144	162	181	199	218	236	254	273
236		291	310	328	346	365	383	401	420	438	457
237		475	493	511	530	548	566	585	603	621	639
238		658	676	694	712	731	749	767	785	803	822
239		840	858	876	894	912	931	949	967	985	*003
240	38	021	039	057	075	093	112	130	148	166	184
241		202	220	238	256	274	292	310	328	346	364
242		382	399	417	435	453	471	489	507	525	543
243		561	579	596	614	632	650	668	686	703	721
244		739	757	775	792	810	828	846	863	881	899
245		917	934	952	970	987	*005	*023	*041	*058	*076
246	39	094	111	129	146	164	182	199	217	235	252
247		270	287	305	322	340	358	375	393	410	428
248		445	463	480	498	515	533	550	568	585	602
249		620	637	655	672	690	707	724	742	759	777
250		794	811	829	846	863	881	898	915	933	950
251		967	985	*002	*019	*037	*054	*071	*088	*106	*123
252	40	140	157	175	192	209	226	243	261	278	295
253		312	329	346	364	381	398	415	432	449	466
254		483	500	518	535	552	569	586	603	620	637
255		654	671	688	705	722	739	756	773	790	807
256		824	841	858	875	892	909	926	943	960	976
257		993	*010	*027	*044	*061	*078	*095	*111	*128	*145
258	41	162	179	196	212	229	246	263	280	296	313
259		330	347	364	380	397	414	430	447	464	481
260		497	514	531	547	564	581	597	614	631	647
261		664	681	697	714	731	747	764	780	797	814
262		830	847	863	880	896	913	929	946	963	979
263		996	*012	*029	*045	*062	*078	*095	*111	*127	*144
264	42	160	177	193	210	226	243	259	275	292	308
265		325	341	357	374	390	406	423	439	456	472
No.	L	0	1	2	3	4	5	6	7	8	9

Proportional parts

	21	20	19	18
1	2.1	2.0	1.9	1.8
2	4.2	4.0	3.8	3.6
3	6.3	6.0	5.7	5.4
4	8.4	8.0	7.6	7.2
5	10.5	10.0	9.5	9.0
6	12.6	12.0	11.4	10.8
7	14.7	14.0	13.3	12.6
8	16.8	16.0	15.2	14.4
9	18.9	18.0	17.1	16.2

Proportional parts

* Indicates change in the first two decimal places.

Table F-2. Five-place Common Logarithms of Numbers (*Continued*)

265–320

No.	L	0	1	2	3	4	5	6	7	8	9
265	42	325	341	357	374	390	406	423	439	456	472
266		488	504	521	537	553	570	586	602	619	635
267		651	667	684	700	716	732	749	765	781	797
268		813	830	846	862	878	894	911	927	943	959
269		975	991	*008	*024	*040	*056	*072	*088	*104	*120
270	43	136	152	169	185	201	217	233	249	265	281
271		297	313	329	345	361	377	393	409	425	441
272		457	473	489	505	521	537	553	569	584	600
273		616	632	648	664	680	696	712	727	743	759
274		775	791	807	823	838	854	870	886	902	917
275		933	949	965	981	996	*012	*028	*044	*059	*075
276	44	091	107	122	138	154	170	185	201	217	232
277		248	264	279	295	311	326	342	358	373	389
278		404	420	436	451	467	483	498	514	529	545
279		560	576	592	607	623	638	654	669	685	700
280		716	731	747	762	778	793	809	824	840	855
281		871	886	902	917	932	948	963	979	994	*010
282	45	025	040	056	071	086	102	117	133	148	163
283		179	194	209	225	240	255	271	286	301	317
284		332	347	362	378	393	408	423	439	454	469
285		484	500	515	530	545	561	576	591	606	621
286		637	652	667	682	697	712	728	743	758	773
287		788	803	818	834	849	864	879	894	909	924
288		939	954	969	984	*000	*015	*030	*045	*060	*075
289	46	090	105	120	135	150	165	180	195	210	225
290		240	255	270	285	300	315	330	345	359	374
291		389	404	419	434	449	464	479	494	509	523
292		538	553	568	583	598	613	627	642	657	672
293		687	702	716	731	746	761	776	790	805	820
294		835	850	864	879	894	909	923	938	953	967
295		982	997	*012	*026	*041	*056	*070	*085	*100	*115
296	47	129	144	159	173	188	202	217	232	246	261
297		276	290	305	319	334	349	363	378	392	407
298		422	436	451	465	480	494	509	524	538	553
299		567	582	596	611	625	640	654	669	683	698
300		712	727	741	756	770	784	799	813	828	842
301		857	871	886	900	914	929	943	958	972	986
302	48	001	015	029	044	058	073	087	101	116	130
303		144	159	173	187	202	216	230	245	259	273
304		287	302	316	330	344	359	373	387	402	416
305		430	444	458	473	487	501	515	530	544	558
306		572	586	601	615	629	643	657	671	686	700
307		714	728	742	756	770	785	799	813	827	841
308		855	869	883	897	911	926	940	954	968	982
309		996	*010	*024	*038	*052	*066	*080	*094	*108	*122
310	49	136	150	164	178	192	206	220	234	248	262
311		276	290	304	318	332	346	360	374	388	402
312		415	429	443	457	471	485	499	513	527	541
313		554	568	582	596	610	624	638	651	665	679
314		693	707	721	734	748	762	776	790	803	817
315		831	845	859	872	886	900	914	927	941	955
316		969	982	996	*010	*024	*037	*051	*065	*079	*092
317	50	106	120	133	147	161	174	188	202	215	229
318		243	256	270	284	297	311	325	338	352	365
319		379	393	406	420	433	447	461	474	488	501
320		515	529	542	556	569	583	596	610	623	637
No.	L	0	1	2	3	4	5	6	7	8	9

Proportional parts

	17	16	15	14
1	1.7	1.6	1.5	1.4
2	3.4	3.2	3.0	2.8
3	5.1	4.8	4.5	4.2
4	6.8	6.4	6.0	5.6
5	8.5	8.0	7.5	7.0
6	10.2	9.6	9.0	8.4
7	11.9	11.2	10.5	9.8
8	13.6	12.8	12.0	11.2
9	15.3	14.4	13.5	12.6

* Indicates change in the first two decimal places.

Table F-2. Five-place Common Logarithms of Numbers (*Continued*)

320–375

No.	L	0	1	2	3	4	5	6	7	8	9
320	50	515	529	542	556	569	583	596	610	623	637
321		651	664	678	691	705	718	732	745	759	772
322		786	799	813	826	840	853	866	880	893	907
323		920	934	947	961	974	987	*001	*014	*028	*041
324	51	055	068	081	095	108	121	135	148	162	175
325		188	202	215	228	242	255	268	282	295	308
326		322	335	348	362	375	388	402	415	428	441
327		455	468	481	495	508	521	534	548	561	574
328		587	601	614	627	640	654	667	680	693	706
329		720	733	746	759	772	786	799	812	825	838
330		851	865	878	891	904	917	930	943	957	970
331		983	996	*009	*022	*035	*048	*061	*075	*088	*101
332	52	114	127	140	153	166	179	192	205	218	231
333		244	257	271	284	297	310	323	336	349	362
334		375	388	401	414	427	440	453	466	479	492
335		504	517	530	543	556	569	582	595	608	621
336		634	647	660	673	686	699	711	724	737	750
337		763	776	789	802	815	827	840	853	866	879
338		892	905	917	930	943	956	969	982	994	*007
339	53	020	033	046	058	071	084	097	110	122	135
340		148	161	173	186	199	212	224	237	250	263
341		275	288	301	314	326	339	352	365	377	390
342		403	415	428	441	453	466	479	491	504	517
343		529	542	555	567	580	593	605	618	631	643
344		656	668	681	694	706	719	732	744	757	769
345		782	795	807	820	832	845	857	870	883	895
346		908	920	933	945	958	970	983	995	*008	*020
347	54	033	045	058	070	083	095	108	120	133	145
348		158	170	183	195	208	220	233	245	258	270
349		283	295	307	320	332	345	357	370	382	394
350		407	419	432	444	456	469	481	494	506	518
351		531	543	555	568	580	593	605	617	630	642
352		654	667	679	691	704	716	728	741	753	765
353		777	790	802	814	827	839	851	864	876	888
354		900	913	925	937	949	962	974	986	998	*011
355	55	023	035	047	060	072	084	096	108	121	133
356		145	157	169	182	194	206	218	230	242	255
357		267	279	291	303	315	328	340	352	364	376
358		388	400	413	425	437	449	461	473	485	497
359		509	522	534	546	558	570	582	594	606	618
360		630	642	654	666	678	691	703	715	727	739
361		751	763	775	787	799	811	823	835	847	859
362		871	883	895	907	919	931	943	955	967	979
363		991	*003	*015	*027	*038	*050	*062	*074	*086	*098
364	56	110	122	134	146	158	170	182	194	205	217
365		229	241	253	265	277	289	301	313	324	336
366		348	360	372	384	396	407	419	431	443	455
367		467	478	490	502	514	526	538	549	561	573
368		585	597	608	620	632	644	656	667	679	691
369		703	714	726	738	750	761	773	785	797	808
370		820	832	844	855	867	879	891	902	914	926
371		937	949	961	972	984	996	*008	*019	*031	*043
372	57	054	066	078	089	101	113	124	136	148	159
373		171	183	194	206	217	229	241	252	264	276
374		287	299	310	322	334	345	357	368	380	392
375		403	415	426	438	449	461	473	484	496	507
No.	L	0	1	2	3	4	5	6	7	8	9

Proportional parts

	14
1	1.4
2	2.8
3	4.2
4	5.6
5	7.0
6	8.4
7	9.8
8	11.2
9	12.6

	13
1	1.3
2	2.6
3	3.9
4	5.2
5	6.5
6	7.8
7	9.1
8	10.4
9	11.7

	12
1	1.2
2	2.4
3	3.6
4	4.8
5	6.0
6	7.2
7	8.4
8	9.6
9	10.8

* Indicates change in the first two decimal places.

Table F-2. Five-place Common Logarithms of Numbers (*Continued*)

375–430

No.	L	0	1	2	3	4	5	6	7	8	9
375	57	403	415	426	438	449	461	473	484	496	507
376		519	530	542	553	565	577	588	600	611	623
377		634	646	657	669	680	692	703	715	726	738
378		749	761	772	784	795	807	818	830	841	852
379		864	875	887	898	910	921	933	944	956	967
380		978	990	*001	*013	*024	*035	*047	*058	*070	*081
381	58	093	104	115	127	138	149	161	172	184	195
382		206	218	229	240	252	263	275	286	297	309
383		320	331	343	354	365	377	388	399	411	422
384		433	444	456	467	478	490	501	512	524	535
385		546	557	569	580	591	602	614	625	636	647
386		659	670	681	692	704	715	726	737	749	760
387		771	782	794	805	816	827	838	850	861	872
388		883	894	906	917	928	939	950	961	973	984
389		995	*006	*017	*028	*040	*051	*062	*073	*084	*095
390	59	106	118	129	140	151	162	173	184	195	207
391		218	229	240	251	262	273	284	295	306	318
392		329	340	351	362	373	384	395	406	417	428
393		439	450	461	472	483	494	506	517	528	539
394		550	561	572	583	594	605	616	627	638	649
395		660	671	682	693	704	715	726	737	748	759
396		770	780	791	802	813	824	835	846	857	868
397		879	890	901	912	923	934	945	956	966	977
398		988	999	*010	*021	*032	*043	*054	*065	*076	*086
399	60	097	108	119	130	141	152	163	173	184	195
400		206	217	228	239	249	260	271	282	293	304
401		314	325	336	347	358	369	379	390	401	412
402		423	433	444	455	466	477	487	498	509	520
403		531	541	552	563	574	584	595	606	617	627
404		638	649	660	670	681	692	703	713	724	735
405		746	756	767	778	788	799	810	821	831	842
406		853	863	874	885	895	906	917	927	938	949
407		959	970	981	991	*002	*013	*023	*034	*045	*055
408	61	066	077	087	098	109	119	130	140	151	162
409		172	183	194	204	215	225	236	247	257	268
410		278	289	300	310	321	331	342	352	363	374
411		384	395	405	416	426	437	448	458	469	479
412		490	500	511	521	532	542	553	563	574	584
413		595	606	616	627	637	648	658	669	679	690
414		700	711	721	731	742	752	763	773	784	794
415		805	815	826	836	847	857	868	878	888	899
416		909	920	930	941	951	962	972	982	993	*003
417	62	014	024	034	045	055	066	076	086	097	107
418		118	128	138	149	159	170	180	190	201	211
419		221	232	242	252	263	273	284	294	304	315
420		325	335	346	356	366	377	387	397	408	418
421		428	439	449	459	469	480	490	500	511	521
422		531	542	552	562	572	583	593	603	614	624
423		634	644	655	665	675	685	696	706	716	726
424		737	747	757	767	778	788	798	808	818	829
425		839	849	859	870	880	890	900	910	921	931
426		941	951	961	972	982	992	*002	*012	*022	*033
427	63	043	053	063	073	083	094	104	114	124	134
428		144	155	165	175	185	195	205	215	225	236
429		246	256	266	276	286	296	306	317	327	337
430		347	357	367	377	387	397	407	417	428	438
No.	L	0	1	2	3	4	5	6	7	8	9

Proportional parts

	11
1	1.1
2	2.2
3	3.3
4	4.4
5	5.5
6	6.6
7	7.7
8	8.8
9	9.9

	10
1	1.0
2	2.0
3	3.0
4	4.0
5	5.0
6	6.0
7	7.0
8	8.0
9	9.0

Proportional parts

* Indicates change in the first two decimal places.

Table F-2. Five-place Common Logarithms of Numbers (*Continued*)

430–485

No.	L	0	1	2	3	4	5	6	7	8	9
430	63	347	357	367	377	387	397	407	417	428	438
431		448	458	468	478	488	498	508	518	528	538
432		548	558	568	579	589	599	609	619	629	639
433		649	659	669	679	689	699	709	719	729	739
434		749	759	769	779	789	799	809	819	829	839
435		849	859	869	879	889	899	909	919	929	939
436		949	959	969	979	988	998	*008	*018	*028	*038
437	64	048	058	068	078	088	098	108	118	128	137
438		147	157	167	177	187	197	207	217	227	237
439		246	256	266	276	286	296	306	316	326	335
440		345	355	365	375	385	395	404	414	424	434
441		444	454	464	473	483	493	503	513	523	532
442		542	552	562	572	582	591	601	611	621	631
443		640	650	660	670	680	689	699	709	719	729
444		738	748	758	768	777	787	797	807	816	826
445		836	846	856	865	875	885	895	904	914	924
446		933	943	953	963	972	982	992	*002	*011	*021
447	65	031	040	050	060	070	079	089	099	108	118
448		128	137	147	157	167	176	186	196	205	215
449		225	234	244	254	263	273	283	292	302	312
450		321	331	341	350	360	369	379	389	398	408
451		418	427	437	447	456	466	475	485	495	504
452		514	523	533	543	552	562	571	581	591	600
453		610	619	629	639	648	658	667	677	686	696
454		706	715	725	734	744	753	763	773	782	792
455		801	811	820	830	839	849	858	868	877	887
456		896	906	916	925	935	944	954	963	973	982
457		992	*001	*011	*020	*030	*039	*049	*058	*068	*077
458	66	087	096	106	115	124	134	143	153	162	172
459		181	191	200	210	219	229	238	247	257	266
460		276	285	295	304	314	323	332	342	351	361
461		370	380	389	398	408	417	427	436	445	455
462		464	474	483	492	502	511	521	530	539	549
463		558	567	577	586	596	605	614	624	633	642
464		652	661	671	680	689	699	708	717	727	736
465		745	755	764	773	783	792	801	811	820	829
466		839	848	857	867	876	885	894	904	913	922
467		932	941	950	960	969	978	987	997	*006	*015
468	67	025	034	043	052	062	071	080	090	099	108
469		117	127	136	145	154	164	173	182	191	201
470		210	219	228	238	247	256	265	274	284	293
471		302	311	321	330	339	348	357	367	376	385
472		394	403	413	422	431	440	449	459	468	477
473		486	495	504	514	523	532	541	550	560	569
474		578	587	596	605	614	624	633	642	651	660
475		669	679	688	697	706	715	724	733	742	752
476		761	770	779	788	797	806	815	825	834	843
477		852	861	870	879	888	897	906	916	925	934
478		943	952	961	970	979	988	997	*006	*015	*024
479	68	034	043	052	061	070	079	088	097	106	115
480		124	133	142	151	160	169	178	187	196	205
481		215	224	233	242	251	260	269	278	287	296
482		305	314	323	332	341	350	359	368	377	386
483		395	404	413	422	431	440	449	458	467	476
484		485	494	502	511	520	529	538	547	556	565
485		574	583	592	601	610	619	628	637	646	655
No.	L	0	1	2	3	4	5	6	7	8	9

Proportional parts

	10
1	1.0
2	2.0
3	3.0
4	4.0
5	5.0
6	6.0
7	7.0
8	8.0
9	9.0

	9
1	0.9
2	1.8
3	2.7
4	3.6
5	4.5
6	5.4
7	6.3
8	7.2
9	8.1

* Indicates change in the first two decimal places.

Table F-2. Five-place Common Logarithms of Numbers *(Continued)*

485–540

No.	L	0	1	2	3	4	5	6	7	8	9	Proportional parts
485	68	574	583	592	601	610	619	628	637	646	655	
486		664	673	682	690	699	708	717	726	735	744	
487		753	762	771	780	789	797	806	815	824	833	
488		842	851	860	869	878	886	895	904	913	922	
489		931	940	949	958	966	975	984	993	*002	*011	
490	69	020	028	037	046	055	064	073	082	090	099	
491		108	117	126	135	144	152	161	170	179	188	
492		197	205	214	223	232	241	249	258	267	276	
493		285	294	302	311	320	329	338	346	355	364	
494		373	381	390	399	408	417	425	434	443	452	
495		461	469	478	487	496	504	513	522	531	539	
496		548	557	566	574	583	592	601	609	618	627	
497		636	644	653	662	671	679	688	697	705	714	
498		723	732	740	749	758	767	775	784	793	801	
499		810	819	827	836	845	854	862	871	880	888	
500		897	906	914	923	932	940	949	958	966	975	
501		984	992	*001	*010	*018	*027	*036	*044	*053	*062	
502	70	070	079	088	096	105	114	122	131	140	148	
503		157	165	174	183	191	200	209	217	226	234	
504		243	252	260	269	278	286	295	303	312	321	
505		329	338	346	355	364	372	381	389	398	406	
506		415	424	432	441	449	458	467	475	484	492	
507		501	509	518	526	535	544	552	561	569	578	
508		586	595	603	612	621	629	638	646	655	663	
509		672	680	689	697	706	714	723	731	740	749	
510		757	766	774	783	791	800	808	817	825	834	
511		842	851	859	868	876	885	893	902	910	919	
512		927	935	944	952	961	969	978	986	995	*003	
513	71	012	020	029	037	046	054	063	071	079	088	
514		096	105	113	122	130	139	147	155	164	172	
515		181	189	198	206	214	223	231	240	248	257	
516		265	273	282	290	299	307	315	324	332	341	
517		349	357	366	374	383	391	399	408	416	425	
518		433	441	450	458	467	475	483	492	500	508	
519		517	525	533	542	550	559	567	575	584	592	
520		600	609	617	625	634	642	650	659	667	675	
521		684	692	700	709	717	725	734	742	750	759	
522		767	775	784	792	800	809	817	825	834	842	
523		850	858	867	875	883	892	900	908	917	925	
524		933	941	950	958	966	975	983	991	999	*008	
525	72	016	024	032	041	049	057	066	074	082	090	
526		099	107	115	123	132	140	148	156	165	173	
527		181	189	198	206	214	222	230	239	247	255	
528		263	272	280	288	296	305	313	321	329	337	
529		346	354	362	370	378	387	395	403	411	419	
530		428	436	444	452	460	469	477	485	493	501	
531		509	518	526	534	542	550	559	567	575	583	
532		591	599	607	616	624	632	640	648	656	665	
533		673	681	689	697	705	713	722	730	738	746	
534		754	762	770	779	787	795	803	811	819	827	
535		835	844	852	860	868	876	884	892	900	908	
536		916	925	933	941	949	957	965	973	981	989	
537		997	*006	*014	*022	*030	*038	*046	*054	*062	*070	
538	73	078	086	094	102	111	119	127	135	143	151	
539		159	167	175	183	191	199	207	215	223	231	
540		239	247	255	264	272	280	288	296	304	312	
No.	L	0	1	2	3	4	5	6	7	8	9	Proportional parts

Proportional parts:

	9
1	0.9
2	1.8
3	2.7
4	3.6
5	4.5
6	5.4
7	6.3
8	7.2
9	8.1

	8
1	0.8
2	1.6
3	2.4
4	3.2
5	4.0
6	4.8
7	5.6
8	6.4
9	7.2

* Indicates change in the first two decimal places.

Table F-2. Five-place Common Logarithms of Numbers (*Continued*)

540–595

No.	L	0	1	2	3	4	5	6	7	8	9
540	73	239	247	255	264	272	280	288	296	304	312
541		320	328	336	344	352	360	368	376	384	392
542		400	408	416	424	432	440	448	456	464	472
543		480	488	496	504	512	520	528	536	544	552
544		560	568	576	584	592	600	608	616	624	632
545		640	648	656	664	672	679	687	695	703	711
546		719	727	735	743	751	759	767	775	783	791
547		799	807	815	823	830	838	846	854	862	870
548		878	886	894	902	910	918	926	934	941	949
549		957	965	973	981	989	997	*005	*013	*020	*028
550	74	036	044	052	060	068	076	084	092	099	107
551		115	123	131	139	147	155	162	170	178	186
552		194	202	210	218	225	233	241	249	257	265
553		273	280	288	296	304	312	320	327	335	343
554		351	359	367	374	382	390	398	406	414	421
555		429	437	445	453	461	468	476	484	492	500
556		507	515	523	531	539	547	554	562	570	578
557		586	593	601	609	617	624	632	640	648	656
558		663	671	679	687	695	702	710	718	726	733
559		741	749	757	764	772	780	788	796	803	811
560		819	827	834	842	850	858	865	873	881	889
561		896	904	912	920	927	935	943	950	958	966
562		974	981	989	997	*005	*012	*020	*028	*035	*043
563	75	051	059	066	074	082	089	097	105	113	120
564		128	136	143	151	159	166	174	182	189	197
565		205	213	220	228	236	243	251	259	266	274
566		282	289	297	305	312	320	328	335	343	351
567		358	366	374	381	389	397	404	412	420	427
568		435	442	450	458	465	473	481	488	496	504
569		511	519	526	534	542	549	557	565	572	580
570		587	595	603	610	618	626	633	641	648	656
571		664	671	679	686	694	702	709	717	724	732
572		740	747	755	762	770	778	785	793	800	808
573		815	823	831	838	846	853	861	868	876	884
574		891	899	906	914	921	929	937	944	952	959
575		967	974	982	989	997	*005	*012	*020	*027	*035
576	76	042	050	057	065	072	080	087	095	103	110
577		118	125	133	140	148	155	163	170	178	185
578		193	200	208	215	223	230	238	245	253	260
579		268	275	283	290	298	305	313	320	328	335
580		343	350	358	365	373	380	388	395	403	410
581		418	425	433	440	448	455	462	470	477	485
582		492	500	507	515	522	530	537	545	552	559
583		567	574	582	589	597	604	612	619	626	634
584		641	649	656	664	671	678	686	693	701	708
585		716	723	730	738	745	753	760	768	775	782
586		790	797	805	812	819	827	834	842	849	856
587		864	871	879	886	893	901	908	916	923	930
588		938	945	953	960	967	975	982	989	997	*004
589	77	012	019	026	034	041	048	056	063	070	078
590		085	093	100	107	115	122	129	137	144	151
591		159	166	173	181	188	195	203	210	218	225
592		232	240	247	254	262	269	276	283	291	298
593		305	313	320	327	335	342	349	357	364	371
594		379	386	393	401	408	415	422	430	437	444
595		452	459	466	474	481	488	495	503	510	517
No.	L	0	1	2	3	4	5	6	7	8	9

Proportional parts

	9
1	0.9
2	1.8
3	2.7
4	3.6
5	4.5
6	5.4
7	6.3
8	7.2
9	8.1

	8
1	0.8
2	1.6
3	2.4
4	3.2
5	4.0
6	4.8
7	5.6
8	6.4
9	7.2

	7
1	0.7
2	1.4
3	2.1
4	2.8
5	3.5
6	4.2
7	4.9
8	5.6
9	6.3

Proportional parts

* Indicates change in the first two decimal places.

Table F-2. Five-place Common Logarithms of Numbers (*Continued*)

595–650

No.	L	0	1	2	3	4	5	6	7	8	9
595	77	452	459	466	474	481	488	495	503	510	517
596		525	532	539	546	554	561	568	576	583	590
597		597	605	612	619	627	634	641	648	656	663
598		670	677	685	692	699	706	714	721	728	735
599		743	750	757	764	772	779	786	793	801	808
600		815	822	830	837	844	851	859	866	873	880
601		887	895	902	909	916	924	931	938	945	952
602		960	967	974	981	989	996	*003	*010	*017	*025
603	78	032	039	046	053	061	068	075	082	089	097
604		104	111	118	125	132	140	147	154	161	168
605		176	183	190	197	204	211	219	226	233	240
606		247	254	262	269	276	283	290	297	305	312
607		319	326	333	340	347	355	362	369	376	383
608		390	398	405	412	419	426	433	440	447	455
609		462	469	476	483	490	497	505	512	519	526
610		533	540	547	554	561	569	576	583	590	597
611		604	611	618	625	633	640	647	654	661	668
612		675	682	689	696	704	711	718	725	732	739
613		746	753	760	767	774	781	789	796	803	810
614		817	824	831	838	845	852	859	866	873	880
615		888	895	902	909	916	923	930	937	944	951
616		958	965	972	979	986	993	*000	*007	*014	*021
617	79	029	036	043	050	057	064	071	078	085	092
618		099	106	113	120	127	134	141	148	155	162
619		169	176	183	190	197	204	211	218	225	232
620		239	246	253	260	267	274	281	288	295	302
621		309	316	323	330	337	344	351	358	365	372
622		379	386	393	400	407	414	421	428	435	442
623		449	456	463	470	477	484	491	498	505	512
624		518	525	532	539	546	553	560	567	574	581
625		588	595	602	609	616	623	630	637	644	651
626		657	664	671	678	685	692	699	706	713	720
627		727	734	741	748	754	761	768	775	782	789
628		796	803	810	817	824	831	837	844	851	858
629		865	872	879	886	893	900	906	913	920	927
630		934	941	948	955	962	969	975	982	989	996
631	80	003	010	017	024	030	037	044	051	058	065
632		072	079	085	092	099	106	113	120	127	134
633		140	147	154	161	168	175	182	188	195	202
634		209	216	223	229	236	243	250	257	264	271
635		277	284	291	298	305	312	318	325	332	339
636		346	353	359	366	373	380	387	393	400	407
637		414	421	428	434	441	448	455	462	468	475
638		482	489	496	502	509	516	523	530	536	543
639		550	557	564	570	577	584	591	598	604	611
640		618	625	632	638	645	652	659	665	672	679
641		686	693	699	706	713	720	726	733	740	747
642		754	760	767	774	781	787	794	801	808	814
643		821	828	835	841	848	855	862	868	875	882
644		889	895	902	909	916	922	929	936	943	949
645		956	963	969	976	983	990	996	*003	*010	*017
646	81	023	030	037	043	050	057	064	070	077	084
647		090	097	104	111	117	124	131	137	144	151
648		158	164	171	178	184	191	198	204	211	218
649		224	231	238	245	251	258	265	271	278	285
650		291	298	305	311	318	325	331	338	345	351
No.	L	0	1	2	3	4	5	6	7	8	9

Proportional parts

	8
1	0.8
2	1.6
3	2.4
4	3.2
5	4.0
6	4.8
7	5.6
8	6.4
9	7.2

	7
1	0.7
2	1.4
3	2.1
4	2.8
5	3.5
6	4.2
7	4.9
8	5.6
9	6.3

Proportional parts

* Indicates change in the first two decimal places.

Table F-2. Five-place Common Logarithms of Numbers *(Continued)*

650–705

No.	L	0	1	2	3	4	5	6	7	8	9
650	81	291	298	305	311	318	325	331	338	345	351
651		358	365	371	378	385	391	398	405	411	418
652		425	431	438	445	451	458	465	471	478	485
653		491	498	505	511	518	525	531	538	544	551
654		558	564	571	578	584	591	598	604	611	618
655		624	631	637	644	651	657	664	671	677	684
656		690	697	704	710	717	723	730	737	743	750
657		757	763	770	776	783	790	796	803	809	816
658		823	829	836	842	849	856	862	869	875	882
659		889	895	902	908	915	921	928	935	941	948
660		954	961	968	974	981	987	994	*000	*007	*014
661	82	020	027	033	040	046	053	060	066	073	079
662		086	092	099	105	112	119	125	132	138	145
663		151	158	164	171	178	184	191	197	204	210
664		217	223	230	236	243	250	256	263	269	276
665		282	289	295	302	308	315	321	328	334	341
666		347	354	360	367	374	380	387	393	400	406
667		413	419	426	432	439	445	452	458	465	471
668		478	484	491	497	504	510	517	523	530	536
669		543	549	556	562	569	575	582	588	595	601
670		607	614	620	627	633	640	646	653	659	666
671		672	679	685	692	698	705	711	718	724	730
672		737	743	750	756	763	769	776	782	789	795
673		802	808	814	821	827	834	840	847	853	860
674		866	872	879	885	892	898	905	911	918	924
675		930	937	943	950	956	963	969	975	982	988
676		995	*001	*008	*014	*020	*027	*033	*040	*046	*052
677	83	059	065	072	078	085	091	097	104	110	117
678		123	129	136	142	149	155	161	168	174	181
679		187	193	200	206	213	219	225	232	238	245
680		251	257	264	270	276	283	289	296	302	308
681		315	321	327	334	340	347	353	359	366	372
682		378	385	391	398	404	410	417	423	429	436
683		442	448	455	461	468	474	480	487	493	499
684		506	512	518	525	531	537	544	550	556	563
685		569	575	582	588	594	601	607	613	620	626
686		632	639	645	651	658	664	670	677	683	689
687		696	702	708	715	721	727	734	740	746	753
688		759	765	771	778	784	790	797	803	809	816
689		822	828	835	841	847	853	860	866	872	879
690		885	891	898	904	910	916	923	929	935	942
691		948	954	960	967	973	979	986	992	998	*004
692	84	011	017	023	029	036	042	048	055	061	067
693		073	080	086	092	098	105	111	117	123	130
694		136	142	148	155	161	167	173	180	186	192
695		198	205	211	217	223	230	236	242	248	255
696		261	267	273	280	286	292	298	305	311	317
697		323	330	336	342	348	354	361	367	373	379
698		386	392	398	404	410	417	423	429	435	442
699		448	454	460	466	473	479	485	491	497	504
700		510	516	522	528	535	541	547	553	559	566
701		572	578	584	590	597	603	609	615	621	628
702		634	640	646	652	658	665	671	677	683	689
703		696	702	708	714	720	726	733	739	745	751
704		757	763	770	776	782	788	794	800	807	813
705		819	825	831	837	844	850	856	862	868	874
No.	L	0	1	2	3	4	5	6	7	8	9

Proportional parts

	7
1	0.7
2	1.4
3	2.1
4	2.8
5	3.5
6	4.2
7	4.9
8	5.6
9	6.3

	6
1	0.6
2	1.2
3	1.8
4	2.4
5	3.0
6	3.6
7	4.2
8	4.8
9	5.4

Proportional parts

* Indicates change in the first two decimal places.

Table F-2. Five-place Common Logarithms of Numbers (*Continued*)

705–760

No.	L	0	1	2	3	4	5	6	7	8	9
705	84	819	825	831	837	844	850	856	862	868	874
706		880	887	893	899	905	911	917	924	930	936
707		942	948	954	960	967	973	979	985	991	997
708	85	003	009	016	022	028	034	040	046	052	059
709		065	071	077	083	089	095	101	107	114	120
710		126	132	138	144	150	156	163	169	175	181
711		187	193	199	205	211	217	224	230	236	242
712		248	254	260	266	272	278	285	291	297	303
713		309	315	321	327	333	339	345	352	358	364
714		370	376	382	388	394	400	406	412	418	425
715		431	437	443	449	455	461	467	473	479	485
716		491	497	503	510	516	522	528	534	540	546
717		552	558	564	570	576	582	588	594	600	606
718		612	618	625	631	637	643	649	655	661	667
719		673	679	685	691	697	703	709	715	721	727
720		733	739	745	751	757	763	769	775	781	788
721		794	800	806	812	818	824	830	836	842	848
722		854	860	866	872	878	884	890	896	902	908
723		914	920	926	932	938	944	950	956	962	968
724		974	980	986	992	998	*004	*010	*016	*022	*028
725	86	034	040	046	052	058	064	070	076	082	088
726		094	100	106	112	118	124	130	136	141	147
727		153	159	165	171	177	183	189	195	201	207
728		213	219	225	231	237	243	249	255	261	267
729		273	279	285	291	297	303	308	314	320	326
730		332	338	344	350	356	362	368	374	380	386
731		392	398	404	410	416	421	427	433	439	445
732		451	457	463	469	475	481	487	493	499	504
733		510	516	522	528	534	540	546	552	558	564
734		570	576	581	587	593	599	605	611	617	623
735		629	635	641	646	652	658	664	670	676	682
736		688	694	700	705	711	717	723	729	735	741
737		747	753	759	764	770	776	782	788	794	800
738		806	812	817	823	829	835	841	847	853	859
739		864	870	876	882	888	894	900	906	911	917
740		923	929	935	941	947	953	958	964	970	976
741		982	988	994	999	*005	*011	*017	*023	*029	*035
742	87	040	046	052	058	064	070	075	081	087	093
743		099	105	111	116	122	128	134	140	146	151
744		157	163	169	175	181	186	192	198	204	210
745		216	221	227	233	239	245	251	256	262	268
746		274	280	286	291	297	303	309	315	320	326
747		332	338	344	350	355	361	367	373	379	384
748		390	396	402	408	413	419	425	431	437	442
749		448	454	460	466	471	477	483	489	495	500
750		506	512	518	523	529	535	541	547	552	558
751		564	570	576	581	587	593	599	604	610	616
752		622	628	633	639	645	651	656	662	668	674
753		680	685	691	697	703	708	714	720	726	731
754		737	743	749	754	760	766	772	777	783	789
755		795	800	806	812	818	823	829	835	841	846
756		852	858	864	869	875	881	887	892	898	904
757		910	915	921	927	933	938	944	950	955	961
758		967	973	978	984	990	996	*001	*007	*013	*018
759	88	024	030	036	041	047	053	059	064	070	076
760		081	087	093	099	104	110	116	121	127	133
No.	L	0	1	2	3	4	5	6	7	8	9

Proportional parts

	6
1	0.6
2	1.2
3	1.8
4	2.4
5	3.0
6	3.6
7	4.2
8	4.8
9	5.4

* Indicates change in the first two decimal places.

Table F-2. Five-place Common Logarithms of Numbers (*Continued*)

760–815

No.	L	0	1	2	3	4	5	6	7	8	9
760	88	081	087	093	099	104	110	116	121	127	133
761		138	144	150	156	161	167	173	178	184	190
762		196	201	207	213	218	224	230	235	241	247
763		252	258	264	270	275	281	287	292	298	304
764		309	315	321	326	332	338	343	349	355	360
765		366	372	378	383	389	395	400	406	412	417
766		423	429	434	440	446	451	457	463	468	474
767		480	485	491	497	502	508	514	519	525	530
768		536	542	547	553	559	564	570	576	581	587
769		593	598	604	610	615	621	627	632	638	643
770		649	655	660	666	672	677	683	689	694	700
771		705	711	717	722	728	734	739	745	750	756
772		762	767	773	779	784	790	795	801	807	812
773		818	824	829	835	840	846	852	857	863	868
774		874	880	885	891	897	902	908	913	919	925
775		930	936	941	947	953	958	964	969	975	981
776		986	992	997	*003	*009	*014	*020	*025	*031	*037
777	89	042	048	053	059	064	070	076	081	087	092
778		098	104	109	115	120	126	131	137	143	148
779		154	159	165	170	176	182	187	193	198	204
780		209	215	221	226	232	237	243	248	254	260
781		265	271	276	282	287	293	298	304	310	315
782		321	326	332	337	343	348	354	360	365	371
783		376	382	387	393	398	404	409	415	421	426
784		432	437	443	448	454	459	465	470	476	481
785		487	493	498	504	509	515	520	526	531	537
786		542	548	553	559	564	570	575	581	586	592
787		597	603	609	614	620	625	631	636	642	647
788		653	658	664	669	675	680	686	691	697	702
789		708	713	719	724	730	735	741	746	752	757
790		763	768	774	779	785	790	796	801	807	812
791		818	823	829	834	840	845	851	856	862	867
792		873	878	883	889	894	900	905	911	916	922
793		927	933	938	944	949	955	960	966	971	977
794		982	988	993	998	*004	*009	*015	*020	*026	*031
795	90	037	042	048	053	059	064	069	075	080	086
796		091	097	102	108	113	119	124	129	135	140
797		146	151	157	162	168	173	179	184	189	195
798		200	206	211	217	222	227	233	238	244	249
799		255	260	266	271	276	282	287	293	298	304
800		309	314	320	325	331	336	342	347	352	358
801		363	369	374	380	385	390	396	401	407	412
802		417	423	428	434	439	445	450	455	461	466
803		472	477	482	488	493	499	504	509	515	520
804		526	531	536	542	547	553	558	563	569	574
805		580	585	590	596	601	607	612	617	623	628
806		634	639	644	650	655	660	666	671	677	682
807		687	693	698	704	709	714	720	725	730	736
808		741	747	752	757	763	768	773	779	784	789
809		795	800	806	811	816	822	827	832	838	843
810		849	854	859	865	870	875	881	886	891	897
811		902	907	913	918	924	929	934	940	945	950
812		956	961	966	972	977	982	988	993	998	*004
813	91	009	014	020	025	030	036	041	046	052	057
814		062	068	073	078	084	089	094	100	105	110
815		116	121	126	132	137	142	148	153	158	164
No.	L	0	1	2	3	4	5	6	7	8	9

Proportional parts

	6
1	0.6
2	1.2
3	1.8
4	2.4
5	3.0
6	3.6
7	4.2
8	4.8
9	5.4

	5
1	0.5
2	1.0
3	1.5
4	2.0
5	2.5
6	3.0
7	3.5
8	4.0
9	4.5

* Indicates change in the first two decimal places.

Table F-2. Five-place Common Logarithms of Numbers *(Continued)*

815–870

No.	L	0	1	2	3	4	5	6	7	8	9
815	91	116	121	126	132	137	142	148	153	158	164
816		169	174	180	185	190	196	201	206	212	217
817		222	228	233	238	243	249	254	259	265	270
818		275	281	286	291	297	302	307	312	318	323
819		328	334	339	344	350	355	360	365	371	376
820		381	387	392	397	403	408	413	418	424	429
821		434	440	445	450	455	461	466	471	477	482
822		487	492	498	503	508	514	519	524	529	535
823		540	545	551	556	561	566	572	577	582	587
824		593	598	603	609	614	619	624	630	635	640
825		645	651	656	661	666	672	677	682	687	693
826		698	703	709	714	719	724	730	735	740	745
827		751	756	761	766	772	777	782	787	793	798
828		803	808	814	819	824	829	834	840	845	850
829		855	861	866	871	876	882	887	892	897	903
830		908	913	918	924	929	934	939	944	950	955
831		960	965	971	976	981	986	991	997	*002	*007
832	92	012	018	023	028	033	038	044	049	054	059
833		065	070	075	080	085	091	096	101	106	111
834		117	122	127	132	137	143	148	153	158	163
835		169	174	179	184	189	195	200	205	210	215
836		221	226	231	236	241	247	252	257	262	267
837		273	278	283	288	293	298	304	309	314	319
838		324	330	335	340	345	350	355	361	366	371
839		376	381	387	392	397	402	407	412	418	423
840		428	433	438	443	449	454	459	464	469	474
841		480	485	490	495	500	505	511	516	521	526
842		531	536	542	547	552	557	562	567	572	578
843		583	588	593	598	603	609	614	619	624	629
844		634	639	645	650	655	660	665	670	675	681
845		686	691	696	701	706	711	717	722	727	732
846		737	742	747	752	758	763	768	773	778	783
847		788	793	799	804	809	814	819	824	829	834
848		840	845	850	855	860	865	870	875	881	886
849		891	896	901	906	911	916	921	927	932	937
850		942	947	952	957	962	967	973	978	983	988
851		993	998	*003	*008	*013	*018	*024	*029	*034	*039
852	93	044	049	054	059	064	069	075	080	085	090
853		095	100	105	110	115	120	125	131	136	141
854		146	151	156	161	166	171	176	181	186	192
855		197	202	207	212	217	222	227	232	237	242
856		247	252	258	263	268	273	278	283	288	293
857		298	303	308	313	318	323	328	334	339	344
858		349	354	359	364	369	374	379	384	389	394
859		399	404	409	414	420	425	430	435	440	445
860		450	455	460	465	470	475	480	485	490	495
861		500	505	510	515	520	526	531	536	541	546
862		551	556	561	566	571	576	581	586	591	596
863		601	606	611	616	621	626	631	636	641	646
864		651	656	661	666	671	677	682	687	692	697
865		702	707	712	717	722	727	732	737	742	747
866		752	757	762	767	772	777	782	787	792	797
867		802	807	812	817	822	827	832	837	842	847
868		852	857	862	867	872	877	882	887	892	897
869		902	907	912	917	922	927	932	937	942	947
870		952	957	962	967	972	977	982	987	992	997
No.	L	0	1	2	3	4	5	6	7	8	9

Proportional parts

	6
1	0.6
2	1.2
3	1.8
4	2.4
5	3.0
6	3.6
7	4.2
8	4.8
9	5.4

	5
1	0.5
2	1.0
3	1.5
4	2.0
5	2.5
6	3.0
7	3.5
8	4.0
9	4.5

Proportional parts

* Indicates change in the first two decimal places.

Table F-2. Five-place Common Logarithms of Numbers (*Continued*)

870–925

No.	L	0	1	2	3	4	5	6	7	8	9
870	93	952	957	962	967	972	977	982	987	992	997
871	94	002	007	012	017	022	027	032	037	042	047
872		052	057	062	067	072	077	082	087	091	096
873		101	106	111	116	121	126	131	136	141	146
874		151	156	161	166	171	176	181	186	191	196
875		201	206	211	216	221	226	231	236	240	245
876		250	255	260	265	270	275	280	285	290	295
877		300	305	310	315	320	325	330	335	340	345
878		349	354	359	364	369	374	379	384	389	394
879		399	404	409	414	419	424	429	433	438	443
880		448	453	458	463	468	473	478	483	488	493
881		498	503	507	512	517	522	527	532	537	542
882		547	552	557	562	567	571	576	581	586	591
883		596	601	606	611	616	621	626	630	635	640
884		645	650	655	660	665	670	675	680	685	689
885		694	699	704	709	714	719	724	729	734	738
886		743	748	753	758	763	768	773	778	783	787
887		792	797	802	807	812	817	822	827	832	836
888		841	846	851	856	861	866	871	876	880	885
889		890	895	900	905	910	915	919	924	929	934
890		939	944	949	954	959	963	968	973	978	983
891		988	993	998	*002	*007	*012	*017	*022	*027	*032
892	95	036	041	046	051	056	061	066	071	075	080
893		085	090	095	100	105	109	114	119	124	129
894		134	139	143	148	153	158	163	168	173	177
895		182	187	192	197	202	207	211	216	221	226
896		231	236	240	245	250	255	260	265	270	274
897		279	284	289	294	299	303	308	313	318	323
898		328	332	337	342	347	352	357	361	366	371
899		376	381	386	390	395	400	405	410	415	419
900		424	429	434	439	444	448	453	458	463	468
901		472	477	482	487	492	497	501	506	511	516
902		521	525	530	535	540	545	550	554	559	564
903		569	574	578	583	588	593	598	602	607	612
904		617	622	626	631	636	641	646	650	655	660
905		665	670	674	679	684	689	694	698	703	708
906		713	718	722	727	732	737	742	746	751	756
907		761	766	770	775	780	785	789	794	799	804
908		809	813	818	823	828	832	837	842	847	852
909		856	861	866	871	875	880	885	890	895	899
910		904	909	914	918	923	928	933	938	942	947
911		952	957	961	966	971	976	980	985	990	995
912		999	*004	*009	*014	*019	*023	*028	*033	*038	*042
913	96	047	052	057	061	066	071	076	080	085	090
914		095	099	104	109	114	118	123	128	133	137
915		142	147	152	156	161	166	171	175	180	185
916		190	194	199	204	209	213	218	223	227	232
917		237	242	246	251	256	261	265	270	275	280
918		284	289	294	298	303	308	313	317	322	327
919		332	336	341	346	350	355	360	365	369	374
920		379	384	388	393	398	402	407	412	417	421
921		426	431	435	440	445	450	454	459	464	468
922		473	478	483	487	492	497	501	506	511	515
923		520	525	530	534	539	544	548	553	558	563
924		567	572	577	581	586	591	595	600	605	609
925		614	619	624	628	633	638	642	647	652	656
No.	L	0	1	2	3	4	5	6	7	8	9

Proportional parts

	5
1	0.5
2	1.0
3	1.5
4	2.0
5	2.5
6	3.0
7	3.5
8	4.0
9	4.5

	4
1	0.4
2	0.8
3	1.2
4	1.6
5	2.0
6	2.4
7	2.8
8	3.2
9	3.6

Proportional parts

* Indicates change in the first two decimal places.

Table F-2. Five-place Common Logarithms of Numbers (*Continued*)

925–980

No.	L	0	1	2	3	4	5	6	7	8	9
925	96	614	619	624	628	633	638	642	647	652	656
926		661	666	670	675	680	685	689	694	699	703
927		708	713	717	722	727	731	736	741	745	750
928		755	759	764	769	774	778	783	788	792	797
929		802	806	811	816	820	825	830	834	839	844
930		848	853	858	862	867	872	876	881	886	890
931		895	900	904	909	914	918	923	928	932	937
932		942	946	951	956	960	965	970	974	979	984
933		988	993	997	*002	*007	*011	*016	*021	*025	*030
934	97	035	039	044	049	053	058	063	067	072	077
935		081	086	090	095	100	104	109	114	118	123
936		128	132	137	142	146	151	155	160	165	169
937		174	179	183	188	192	197	202	206	211	216
938		220	225	230	234	239	243	248	253	257	262
939		267	271	276	280	285	290	294	299	304	308
940		313	317	322	327	331	336	341	345	350	354
941		359	364	368	373	377	382	387	391	396	400
942		405	410	414	419	424	428	433	437	442	447
943		451	456	460	465	470	474	479	483	488	493
944		497	502	506	511	516	520	525	529	534	539
945		543	548	552	557	562	566	571	575	580	585
946		589	594	598	603	607	612	617	621	626	630
947		635	640	644	649	653	658	663	667	672	676
948		681	685	690	695	699	704	708	713	717	722
949		727	731	736	740	745	750	754	759	763	768
950		772	777	782	786	791	795	800	804	809	813
951		818	823	827	832	836	841	845	850	855	859
952		864	868	873	877	882	887	891	896	900	905
953		909	914	918	923	928	932	937	941	946	950
954		955	959	964	968	973	978	982	987	991	996
955	98	000	005	009	014	019	023	028	032	037	041
956		046	050	055	059	064	069	073	078	082	087
957		091	096	100	105	109	114	118	123	127	132
958		137	141	146	150	155	159	164	168	173	177
959		182	186	191	195	200	205	209	214	218	223
960		227	232	236	241	245	250	254	259	263	268
961		272	277	281	286	290	295	299	304	308	313
962		318	322	327	331	336	340	345	349	354	358
963		363	367	372	376	381	385	390	394	399	403
964		408	412	417	421	426	430	435	439	444	448
965		453	457	462	466	471	475	480	484	489	493
966		498	502	507	511	516	520	525	529	534	538
967		543	547	552	556	561	565	570	574	579	583
968		588	592	597	601	605	610	614	619	623	628
969		632	637	641	646	650	655	659	664	668	673
970		677	682	686	691	695	700	704	709	713	717
971		722	726	731	735	740	744	749	753	758	762
972		767	771	776	780	785	789	793	798	802	807
973		811	816	820	825	829	834	838	843	847	851
974		856	860	865	869	874	878	883	887	892	896
975		900	905	909	914	918	923	927	932	936	941
976		945	949	954	958	963	967	972	976	981	985
977		989	994	998	*003	*007	*012	*016	*021	*025	*029
978	99	034	038	043	047	052	056	061	065	069	074
979		078	083	087	092	096	100	105	109	114	118
980		123	127	131	136	140	145	149	154	158	162
No.	L	0	1	2	3	4	5	6	7	8	9

Proportional parts

	5
1	0.5
2	1.0
3	1.5
4	2.0
5	2.5
6	3.0
7	3.5
8	4.0
9	4.5

	4
1	0.4
2	0.8
3	1.2
4	1.6
5	2.0
6	2.4
7	2.8
8	3.2
9	3.6

Proportional parts

* Indicates change in the first two decimal places.

Table F-2. Five-place Common Logarithms of Numbers (*Continued*)

980–1000

No.	L	0	1	2	3	4	5	6	7	8	9
980	99	123	127	131	136	140	145	149	154	158	162
981		167	171	176	180	185	189	193	198	202	207
982		211	216	220	224	229	233	238	242	247	251
983		255	260	264	269	273	277	282	286	291	295
984		300	304	308	313	317	322	326	330	335	339
985		344	348	352	357	361	366	370	374	379	383
986		388	392	397	401	405	410	414	419	423	427
987		432	436	441	445	449	454	458	463	467	471
988		476	480	484	489	493	498	502	506	511	515
989		520	524	528	533	537	542	546	550	555	559
990		564	568	572	577	581	585	590	594	599	603
991		607	612	616	621	625	629	634	638	642	647
992		651	656	660	664	669	673	677	682	686	691
993		695	699	704	708	712	717	721	726	730	734
994		739	743	747	752	756	760	765	769	774	778
995		782	787	791	795	800	804	808	813	817	822
996		826	830	835	839	843	848	852	856	861	865
997		870	874	878	883	887	891	896	900	904	909
998		913	917	922	926	930	935	939	944	948	952
999		957	961	965	970	974	978	983	987	991	996
1000	00	000	004	009	013	017	022	026	030	035	039
No.	L	0	1	2	3	4	5	6	7	8	9

Proportional parts

	5
1	0.5
2	1.0
3	1.5
4	2.0
5	2.5
6	3.0
7	3.5
8	4.0
9	4.5

	4
1	0.4
2	0.8
3	1.2
4	1.6
5	2.0
6	2.4
7	2.8
8	3.2
9	3.6

Proportional parts

* Indicates change in the first two decimal places.

Table F-3. Natural Trigonometric Functions and Their Logarithms*

Deg	Radians	Nat sin	Log sin	Nat cos	Log cos	Nat tan	Log tan	Nat cot	Log cot	Radians	Deg
0° 00′	0.0000	0.0000	......	1.0000	0.0000	0.0000	......	....	......	1.5708	90° 00′
10	.0029	.0029	7.4637	1.0000	0.0000	.0029	7.4637	343.77	2.5363	1.5679	50
20	.0058	.0058	7.7648	1.0000	0.0000	.0058	7.7648	171.89	2.2352	1.5650	40
30	.0087	.0087	7.9408	1.0000	0.0000	.0087	7.9409	114.59	2.0591	1.5621	30
40	.0116	.0116	8.0658	0.9999	0.0000	.0116	8.0658	85.940	1.9342	1.5592	20
50	.0145	.0145	8.1627	.9999	0.0000	.0146	8.1627	68.750	1.8373	1.5563	10
1° 00′	.0175	.0175	8.2419	.9999	9.9999	.0175	8.2419	57.290	1.7581	1.5533	89° 00′
10	.0204	.0204	8.3088	.9998	9.9999	.0204	8.3089	49.104	1.6911	1.5504	50
20	.0233	.0233	8.3668	.9997	9.9999	.0233	8.3669	42.964	1.6331	1.5475	40
30	.0262	.0262	8.4179	.9997	9.9999	.0262	8.4181	38.188	1.5819	1.5446	30
40	.0291	.0291	8.4637	.9996	9.9998	.0291	8.4639	34.368	1.5362	1.5417	20
50	.0320	.0320	8.5050	.9995	9.9998	.0320	8.5053	31.242	1.4947	1.5388	10
2° 00′	.0349	.0349	8.5428	.9994	9.9997	.0349	8.5431	28.636	1.4569	1.5359	88° 00′
10	.0378	.0378	8.5776	.9993	9.9997	.0378	8.5779	26.432	1.4221	1.5330	50
20	.0407	.0407	8.6097	.9992	9.9996	.0408	8.6101	24.542	1.3899	1.5301	40
30	.0436	.0436	8.6397	.9991	9.9996	.0437	8.6401	22.904	1.3599	1.5272	30
40	.0465	.0465	8.6677	.9989	9.9995	.0466	8.6682	21.470	1.3318	1.5243	20
50	.0495	.0494	8.6940	.9988	9.9995	.0495	8.6945	20.206	1.3055	1.5213	10
3° 00′	.0524	.0523	8.7188	.9986	9.9994	.0524	8.7194	19.081	1.2806	1.5184	87° 00′
10	.0553	.0552	8.7423	.9985	9.9993	.0553	8.7429	18.075	1.2571	1.5155	50
20	.0582	.0581	8.7645	.9983	9.9993	.0582	8.7653	17.169	1.2348	1.5126	40
30	.0611	.0611	8.7857	.9981	9.9992	.0612	8.7865	16.350	1.2135	1.5097	30
40	.0640	.0640	8.8059	.9980	9.9991	.0641	8.8067	15.605	1.1933	1.5068	20
50	.0669	.0669	8.8251	.9978	9.9990	.0670	8.8261	14.924	1.1739	1.5039	10
4° 00′	.0698	.0698	8.8436	.9976	9.9989	.0699	8.8446	14.301	1.1554	1.5010	86° 00′
10	.0727	.0727	8.8613	.9974	9.9989	.0729	8.8624	13.727	1.1376	1.4981	50
20	.0756	.0756	8.8783	.9971	9.9988	.0758	8.8795	13.197	1.1205	1.4952	40
30	.0785	.0785	8.8946	.9969	9.9987	.0787	8.8960	12.706	1.1040	1.4923	30
40	.0814	.0814	8.9104	.9967	9.9986	.0816	8.9119	12.251	1.0882	1.4893	20
50	.0844	.0843	8.9256	.9964	9.9985	.0846	8.9272	11.826	1.0728	1.4864	10
5° 00′	.0873	.0872	8.9403	.9962	9.9983	.0875	8.9420	11.430	1.0581	1.4835	85° 00′
10	.0902	.0901	8.9545	.9959	9.9982	.0904	8.9563	11.059	1.0437	1.4806	50
20	.0931	.0930	8.9683	.9957	9.9981	.0934	8.9701	10.712	1.0299	1.4777	40
30	.0960	.0959	8.9816	.9954	9.9980	.0963	8.9836	10.385	1.0164	1.4748	30
40	.0989	.0987	8.9945	.9951	9.9979	.0992	8.9966	10.078	1.0034	1.4719	20
50	.1018	.1016	9.0070	.9948	9.9978	.1022	9.0093	9.7882	0.9907	1.4690	10

Deg	Radians	Nat cos	Log cos	Nat sin	Log sin	Nat cot	Log cot	Nat tan	Log tan	Radians	Deg
6° 00′	.1047	.1045	9.0192	.9945	9.9976	.1051	9.0216	9.5144	.9784	1.4661	84° 00′
10	.1076	.1074	9.0311	.9942	9.9975	.1081	9.0336	9.2553	.9664	1.4632	50
20	.1105	.1103	9.0426	.9939	9.9973	.1110	9.0453	9.0098	.9547	1.4603	40
30	.1134	.1132	9.0539	.9936	9.9972	.1139	9.0567	8.7769	.9433	1.4573	30
40	.1164	.1161	9.0648	.9932	9.9971	.1169	9.0678	8.5556	.9323	1.4544	20
50	.1193	.1190	9.0755	.9929	9.9969	.1198	9.0786	8.3450	.9214	1.4515	10
7° 00′	.1222	.1219	9.0859	.9926	9.9968	.1228	9.0891	8.1443	.9109	1.4486	83° 00′
10	.1251	.1248	9.0961	.9922	9.9966	.1257	9.0995	7.9530	.9005	1.4457	50
20	.1280	.1276	9.1060	.9918	9.9964	.1287	9.1096	7.7704	.8904	1.4428	40
30	.1309	.1305	9.1157	.9914	9.9963	.1317	9.1194	7.5958	.8806	1.4399	30
40	.1338	.1334	9.1252	.9911	9.9961	.1346	9.1291	7.4287	.8709	1.4370	20
50	.1367	.1363	9.1345	.9907	9.9959	.1376	9.1385	7.2687	.8615	1.4341	10
8° 00′	.1396	.1392	9.1436	.9903	9.9958	.1405	9.1478	7.1154	.8522	1.4312	82° 00′
10	.1425	.1421	9.1525	.9899	9.9956	.1435	9.1569	6.9682	.8431	1.4283	50
20	.1454	.1449	9.1612	.9894	9.9954	.1465	9.1658	6.8269	.8342	1.4254	40
30	.1484	.1478	9.1697	.9890	9.9952	.1495	9.1745	6.6912	.8255	1.4224	30
40	.1513	.1507	9.1781	.9886	9.9950	.1524	9.1831	6.5606	.8169	1.4195	20
50	.1542	.1536	9.1863	.9881	9.9948	.1554	9.1915	6.4348	.8085	1.4166	10
9° 00′	.1571	.1564	9.1943	.9877	9.9946	.1584	9.1997	6.3138	.8003	1.4137	81° 00′
10	.1600	.1593	9.2022	.9872	9.9944	.1614	9.2078	6.1970	.7922	1.4108	50
20	.1629	.1622	9.2100	.9868	9.9942	.1644	9.2158	6.0844	.7842	1.4079	40
30	.1658	.1651	9.2176	.9863	9.9940	.1673	9.2236	5.9758	.7764	1.4050	30
40	.1687	.1679	9.2251	.9858	9.9938	.1703	9.2313	5.8708	.7687	1.4021	20
50	.1716	.1708	9.2324	.9853	9.9936	.1733	9.2389	5.7694	.7611	1.3992	10
10° 00′	.1745	.1737	9.2397	.9848	9.9934	.1763	9.2463	5.6713	.7537	1.3963	80° 00′
10	.1774	.1765	9.2468	.9843	9.9931	.1793	9.2536	5.5764	.7464	1.3934	50
20	.1804	.1794	9.2538	.9838	9.9929	.1823	9.2609	5.4845	.7391	1.3904	40
30	.1833	.1822	9.2606	.9833	9.9927	.1853	9.2680	5.3955	.7320	1.3875	30
40	.1862	.1851	9.2674	.9827	9.9924	.1884	9.2750	5.3093	.7250	1.3846	20
50	.1891	.1880	9.2741	.9822	9.9922	.1914	9.2819	5.2257	.7181	1.3817	10
11° 00′	.1920	.1908	9.2806	.9816	9.9920	.1944	9.2887	5.1446	.7114	1.3788	79° 00′
10	.1949	.1937	9.2871	.9811	9.9917	.1974	9.2954	5.0658	.7047	1.3759	50
20	.1978	.1965	9.2934	.9805	9.9915	.2004	9.3020	4.9894	.6981	1.3730	40
30	.2007	.1994	9.2997	.9799	9.9912	.2035	9.3085	4.9152	.6915	1.3701	30
40	.2036	.2022	9.3058	.9793	9.9909	.2065	9.3149	4.8430	.6851	1.3672	20
50	.2065	.2051	9.3119	.9788	9.9907	.2095	9.3212	4.7729	.6788	1.3643	10
12° 00′	.2094	.2079	9.3179	.9782	9.9904	.2126	9.3275	4.7046	.6725	1.3614	78° 00′
Deg	Radians	Nat cos	Log cos	Nat sin	Log sin	Nat cot	Log cot	Nat tan	Log tan	Radians	Deg

* From R. H. Perry, *Engineering Manual*, McGraw-Hill, New York, 1959.

Table F-3. Natural Trigonometric Functions and Their Logarithms (*Continued*)

Deg	Radians	Nat sin	Log sin	Nat cos	Log cos	Nat tan	Log tan	Nat cot	Log cot	Radians	Deg
12° 00′	0.2094	0.2079	9.3179	0.9782	9.9904	0.2126	9.3275	4.7046	0.6725	1.3614	78° 00′
10	.2123	.2108	9.3238	.9775	9.9901	.2156	9.3337	4.6383	.6664	1.3584	50
20	.2153	.2136	9.3296	.9769	9.9899	.2186	9.3397	4.5736	.6603	1.3555	40
30	.2182	.2164	9.3353	.9763	9.9896	.2217	9.3458	4.5107	.6542	1.3526	30
40	.2211	.2193	9.3410	.9757	9.9893	.2248	9.3517	4.4494	.6483	1.3497	20
50	.2240	.2221	9.3466	.9750	9.9890	.2278	9.3576	4.3897	.6424	1.3468	10
13° 00′	.2269	.2250	9.3521	.9744	9.9887	.2309	9.3634	4.3315	.6366	1.3439	77° 00′
10	.2298	.2278	9.3575	.9737	9.9884	.2339	9.3691	4.2747	.6309	1.3410	50
20	.2327	.2306	9.3629	.9730	9.9881	.2370	9.3748	4.2193	.6252	1.3381	40
30	.2356	.2335	9.3682	.9724	9.9878	.2401	9.3804	4.1653	.6197	1.3352	30
40	.2385	.2363	9.3734	.9717	9.9875	.2432	9.3859	4.1126	.6141	1.3323	20
50	.2414	.2391	9.3786	.9710	9.9872	.2462	9.3914	4.0611	.6086	1.3294	10
14° 00′	.2443	.2419	9.3837	.9703	9.9869	.2493	9.3968	4.0108	.6032	1.3265	76° 00′
10	.2473	.2447	9.3887	.9696	9.9866	.2524	9.4021	3.9617	.5979	1.3235	50
20	.2502	.2476	9.3937	.9689	9.9863	.2555	9.4074	3.9136	.5926	1.3206	40
30	.2531	.2504	9.3986	.9682	9.9859	.2586	9.4127	3.8667	.5873	1.3177	30
40	.2560	.2532	9.4035	.9674	9.9856	.2617	9.4178	3.8208	.5822	1.3148	20
50	.2589	.2560	9.4083	.9667	9.9853	.2648	9.4230	3.7760	.5770	1.3119	10
15° 00′	.2618	.2588	9.4130	.9659	9.9849	.2680	9.4281	3.7321	.5720	1.3090	75° 00′
10	.2647	.2616	9.4177	.9652	9.9846	.2711	9.4331	3.6891	.5669	1.3061	50
20	.2676	.2644	9.4223	.9644	9.9843	.2742	9.4381	3.6471	.5619	1.3032	40
30	.2705	.2672	9.4269	.9636	9.9839	.2773	9.4430	3.6059	.5570	1.3003	30
40	.2734	.2700	9.4314	.9629	9.9836	.2805	9.4479	3.5656	.5521	1.2974	20
50	.2763	.2728	9.4359	.9621	9.9832	.2836	9.4527	3.5261	.5473	1.2945	10
16° 00′	.2793	.2756	9.4403	.9613	9.9828	.2868	9.4575	3.4874	.5425	1.2915	74° 00′
10	.2822	.2784	9.4447	.9605	9.9825	.2899	9.4622	3.4495	.5378	1.2886	50
20	.2851	.2812	9.4491	.9596	9.9821	.2931	9.4669	3.4124	.5331	1.2857	40
30	.2880	.2840	9.4533	.9588	9.9817	.2962	9.4716	3.3759	.5284	1.2828	30
40	.2909	.2868	9.4576	.9580	9.9814	.2994	9.4762	3.3402	.5238	1.2799	20
50	.2938	.2896	9.4618	.9572	9.9810	.3026	9.4808	3.3052	.5192	1.2770	10
17° 00′	.2967	.2924	9.4659	.9563	9.9806	.3057	9.4853	3.2709	.5147	1.2741	73° 00′
10	.2996	.2952	9.4701	.9555	9.9802	.3089	9.4898	3.2371	.5102	1.2712	50
20	.3025	.2979	9.4741	.9546	9.9798	.3121	9.4943	3.2041	.5057	1.2683	40
30	.3054	.3007	9.4781	.9537	9.9794	.3153	9.4987	3.1716	.5013	1.2654	30
40	.3083	.3035	9.4821	.9528	9.9790	.3185	9.5031	3.1397	.4969	1.2625	20
50	.3113	.3063	9.4861	.9520	9.9786	.3217	9.5075	3.1084	.4925	1.2595	10

Deg	Radians	Nat cos	Log cos	Nat sin	Log sin	Nat cot	Log cot	Nat tan	Log tan	Radians	Deg
18° 00′	.3142	.3090	9.4900	.9511	9.9782	.3249	9.5118	3.0777	.4882	1.2566	72° 00′
10	.3171	.3118	9.4939	.9502	9.9778	.3281	9.5161	3.0475	.4839	1.2537	50
20	.3200	.3145	9.4977	.9492	9.9774	.3314	9.5203	3.0178	.4797	1.2508	40
30	.3229	.3173	9.5015	.9483	9.9770	.3346	9.5245	2.9887	.4755	1.2479	30
40	.3258	.3201	9.5052	.9474	9.9765	.3378	9.5287	2.9600	.4713	1.2450	20
50	.3287	.3228	9.5090	.9465	9.9761	.3411	9.5329	2.9319	.4672	1.2421	10
19° 00′	.3316	.3256	9.5126	.9455	9.9757	.3443	9.5370	2.9042	.4630	1.2392	71° 00′
10	.3345	.3283	9.5163	.9446	9.9752	.3476	9.5411	2.8770	.4589	1.2363	50
20	.3374	.3311	9.5199	.9436	9.9748	.3509	9.5451	2.8502	.4549	1.2334	40
30	.3403	.3338	9.5235	.9426	9.9744	.3541	9.5492	2.8239	.4509	1.2305	30
40	.3432	.3366	9.5271	.9417	9.9739	.3574	9.5532	2.7980	.4469	1.2275	20
50	.3462	.3393	9.5306	.9407	9.9734	.3607	9.5571	2.7725	.4429	1.2246	10
20° 00′	.3491	.3420	9.5341	.9397	9.9730	.3640	9.5611	2.7475	.4389	1.2217	70° 00′
10	.3520	.3448	9.5375	.9387	9.9725	.3673	9.5650	2.7228	.4350	1.2188	50
20	.3549	.3475	9.5409	.9377	9.9721	.3706	9.5689	2.6985	.4311	1.2159	40
30	.3578	.3502	9.5443	.9367	9.9716	.3739	9.5727	2.6746	.4273	1.2130	30
40	.3607	.3529	9.5477	.9357	9.9711	.3772	9.5766	2.6511	.4234	1.2101	20
50	.3636	.3557	9.5510	.9346	9.9706	.3805	9.5804	2.6279	.4196	1.2072	10
21° 00′	.3665	.3584	9.5543	.9336	9.9702	.3839	9.5842	2.6051	.4158	1.2043	69° 00′
10	.3694	.3611	9.5576	.9325	9.9697	.3872	9.5879	2.5826	.4121	1.2014	50
20	.3723	.3638	9.5609	.9315	9.9692	.3906	9.5917	2.5605	.4083	1.1985	40
30	.3752	.3665	9.5641	.9304	9.9687	.3939	9.5954	2.5387	.4046	1.1956	30
40	.3782	.3692	9.5673	.9294	9.9682	.3973	9.5991	2.5172	.4009	1.1926	20
50	.3811	.3719	9.5704	.9283	9.9677	.4007	9.6028	2.4960	.3972	1.1897	10
22° 00′	.3840	.3746	9.5736	.9272	9.9672	.4040	9.6064	2.4751	.3936	1.1868	68° 00′
10	.3869	.3773	9.5767	.9261	9.9667	.4074	9.6100	2.4545	.3900	1.1839	50
20	.3898	.3800	9.5798	.9250	9.9661	.4108	9.6136	2.4342	.3864	1.1810	40
30	.3927	.3827	9.5828	.9239	9.9656	.4142	9.6172	2.4142	.3828	1.1781	30
40	.3956	.3854	9.5859	.9228	9.9651	.4176	9.6208	2.3945	.3792	1.1752	20
50	.3985	.3881	9.5889	.9216	9.9646	.4211	9.6243	2.3750	.3757	1.1723	10
23° 00′	.4014	.3907	9.5919	.9205	9.9640	.4245	9.6279	2.3559	.3722	1.1694	67° 00′
10	.4043	.3934	9.5948	.9194	9.9635	.4279	9.6314	2.3369	.3687	1.1665	50
20	.4072	.3961	9.5978	.9182	9.9629	.4314	9.6348	2.3183	.3652	1.1636	40
30	.4102	.3988	9.6007	.9171	9.9624	.4348	9.6383	2.2998	.3617	1.1606	30
40	.4131	.4014	9.6036	.9159	9.9619	.4383	9.6418	2.2817	.3583	1.1577	20
50	.4160	.4041	9.6065	.9147	9.9613	.4418	9.6452	2.2637	.3548	1.1548	10
24° 00′	.4189	.4067	9.6093	.9136	9.9607	.4452	9.6486	2.2460	.3514	1.1519	66° 00′
Deg	Radians	Nat cos	Log cos	Nat sin	Log sin	Nat cot	Log cot	Nat tan	Log tan	Radians	Deg

Table F-3. Natural Trigonometric Functions and Their Logarithms (*Continued*)

Deg	Radians	Nat sin	Log sin	Nat cos	Log cos	Nat tan	Log tan	Nat cot	Log cot	Radians	Deg
24° 00′	0.4189	0.4067	9.6093	0.9136	9.9607	0.4452	9.6486	2.2460	0.3514	1.1519	66° 00′
10	.4218	.4094	9.6121	.9124	9.9602	.4487	9.6520	2.2286	.3480	1.1490	50
20	.4247	.4120	9.6149	.9112	9.9596	.4522	9.6554	2.2113	.3447	1.1461	40
30	.4276	.4147	9.6177	.9100	9.9590	.4557	9.6587	2.1943	.3413	1.1432	30
40	.4305	.4173	9.6205	.9088	9.9584	.4592	9.6620	2.1775	.3380	1.1403	20
50	.4334	.4200	9.6232	.9075	9.9579	.4628	9.6654	2.1609	.3346	1.1374	10
25° 00′	.4363	.4226	9.6260	.9063	9.9573	.4663	9.6687	2.1445	.3313	1.1345	65° 00′
10	.4392	.4253	9.6287	.9051	9.9567	.4699	9.6720	2.1283	.3280	1.1316	50
20	.4422	.4279	9.6313	.9038	9.9561	.4734	9.6752	2.1123	.3248	1.1286	40
30	.4451	.4305	9.6340	.9026	9.9555	.4770	9.6785	2.0965	.3215	1.1257	30
40	.4480	.4331	9.6366	.9013	9.9549	.4806	9.6817	2.0809	.3183	1.1228	20
50	.4509	.4358	9.6392	.9001	9.9543	.4841	9.6850	2.0655	.3150	1.1199	10
26° 00′	.4538	.4384	9.6418	.8988	9.9537	.4877	9.6882	2.0503	.3118	1.1170	64° 00′
10	.4567	.4410	9.6444	.8975	9.9530	.4913	9.6914	2.0353	.3086	1.1141	50
20	.4596	.4436	9.6470	.8962	9.9524	.4950	9.6946	2.0204	.3054	1.1112	40
30	.4625	.4462	9.6495	.8949	9.9518	.4986	9.6977	2.0057	.3023	1.1083	30
40	.4654	.4488	9.6521	.8936	9.9512	.5022	9.7009	1.9912	.2991	1.1054	20
50	.4683	.4514	9.6546	.8923	9.9505	.5059	9.7040	1.9768	.2960	1.1025	10
27° 00′	.4712	.4540	9.6571	.8910	9.9499	.5095	9.7072	1.9626	.2928	1.0996	63° 00′
10	.4741	.4566	9.6595	.8897	9.9492	.5132	9.7103	1.9486	.2897	1.0966	50
20	.4771	.4592	9.6620	.8884	9.9486	.5169	9.7134	1.9347	.2866	1.0937	40
30	.4800	.4618	9.6644	.8870	9.9479	.5206	9.7165	1.9210	.2835	1.0908	30
40	.4829	.4643	9.6668	.8857	9.9473	.5243	9.7196	1.9074	.2805	1.0879	20
50	.4858	.4669	9.6692	.8843	9.9466	.5280	9.7226	1.8940	.2774	1.0850	10
28° 00′	.4887	.4695	9.6716	.8830	9.9459	.5317	9.7257	1.8807	.2743	1.0821	62° 00′
10	.4916	.4720	9.6740	.8816	9.9453	.5355	9.7287	1.8676	.2713	1.0792	50
20	.4945	.4746	9.6763	.8802	9.9446	.5392	9.7318	1.8546	.2683	1.0763	40
30	.4974	.4772	9.6787	.8788	9.9439	.5430	9.7348	1.8418	.2652	1.0734	30
40	.5003	.4797	9.6810	.8774	9.9432	.5467	9.7378	1.8291	.2622	1.0705	20
50	.5032	.4823	9.6833	.8760	9.9425	.5505	9.7408	1.8165	.2592	1.0676	10
29° 00′	.5061	.4848	9.6856	.8746	9.9418	.5543	9.7438	1.8041	.2563	1.0647	61° 00′
10	.5091	.4874	9.6878	.8732	9.9411	.5581	9.7467	1.7917	.2533	1.0617	50
20	.5120	.4899	9.6901	.8718	9.9404	.5619	9.7497	1.7796	.2503	1.0588	40
30	.5149	.4924	9.6923	.8704	9.9397	.5658	9.7526	1.7675	.2474	1.0559	30
40	.5178	.4950	9.6946	.8689	9.9390	.5696	9.7556	1.7556	.2444	1.0530	20
50	.5207	.4975	9.6968	.8675	9.9383	.5735	9.7585	1.7438	.2415	1.0501	10

Deg	Radians	Nat cos	Log cos	Nat sin	Log sin	Nat cot	Log cot	Nat tan	Log tan	Radians	Deg
30° 00′	.5236	.5000	9.6990	.8660	9.9375	.5774	9.7614	1.7321	.2386	1.0472	60° 00′
10	.5265	.5025	9.7012	.8646	9.9368	.5812	9.7644	1.7205	.2357	1.0443	50
20	.5294	.5050	9.7033	.8631	9.9361	.5851	9.7673	1.7090	.2328	1.0414	40
30	.5323	.5075	9.7055	.8616	9.9353	.5891	9.7702	1.6977	.2299	1.0385	30
40	.5352	.5100	9.7076	.8602	9.9346	.5930	9.7730	1.6864	.2270	1.0356	20
50	.5381	.5125	9.7097	.8587	9.9338	.5969	9.7759	1.6753	.2241	1.0327	10
31° 00′	.5411	.5150	9.7118	.8572	9.9331	.6009	9.7788	1.6643	.2212	1.0297	59° 00′
10	.5440	.5175	9.7139	.8557	9.9323	.6048	9.7816	1.6534	.2184	1.0268	50
20	.5469	.5200	9.7160	.8542	9.9315	.6088	9.7845	1.6426	.2155	1.0239	40
30	.5498	.5225	9.7181	.8526	9.9308	.6128	9.7873	1.6319	.2127	1.0210	30
40	.5527	.5250	9.7201	.8511	9.9300	.6168	9.7902	1.6213	.2099	1.0181	20
50	.5556	.5275	9.7222	.8496	9.9292	.6208	9.7930	1.6107	.2070	1.0152	10
32° 00′	.5585	.5299	9.7242	.8481	9.9284	.6249	9.7958	1.6003	.2042	1.0123	58° 00′
10	.5614	.5324	9.7262	.8465	9.9276	.6289	9.7986	1.5900	.2014	1.0094	50
20	.5643	.5348	9.7282	.8450	9.9268	.6330	9.8014	1.5798	.1986	1.0065	40
30	.5672	.5373	9.7302	.8434	9.9260	.6371	9.8042	1.5697	.1958	1.0036	30
40	.5701	.5398	9.7322	.8418	9.9252	.6412	9.8070	1.5597	.1930	1.0007	20
50	.5730	.5422	9.7342	.8403	9.9244	.6453	9.8098	1.5497	.1903	0.9977	10
33° 00′	.5760	.5446	9.7361	.8387	9.9236	.6494	9.8125	1.5399	.1875	.9948	57° 00′
10	.5789	.5471	9.7381	.8371	9.9228	.6536	9.8153	1.5301	.1847	.9919	50
20	.5818	.5495	9.7400	.8355	9.9219	.6577	9.8180	1.5204	.1820	.9890	40
30	.5847	.5519	9.7419	.8339	9.9211	.6619	9.8208	1.5108	.1792	.9861	30
40	.5876	.5544	9.7438	.8323	9.9203	.6661	9.8235	1.5013	.1765	.9832	20
50	.5905	.5568	9.7457	.8307	9.9194	.6703	9.8263	1.4919	.1737	.9803	10
34° 00′	.5934	.5592	9.7476	.8290	9.9186	.6745	9.8290	1.4826	.1710	.9774	56° 00′
10	.5963	.5616	9.7494	.8274	9.9177	.6788	9.8317	1.4733	.1683	.9745	50
20	.5992	.5640	9.7513	.8258	9.9169	.6830	9.8344	1.4641	.1656	.9716	40
30	.6021	.5664	9.7531	.8241	9.9160	.6873	9.8371	1.4550	.1629	.9687	30
40	.6050	.5688	9.7550	.8225	9.9151	.6916	9.8398	1.4460	.1602	.9657	20
50	.6080	.5712	9.7568	.8208	9.9143	.6959	9.8425	1.4370	.1575	.9628	10
35° 00′	.6109	.5736	9.7586	.8192	9.9134	.7002	9.8452	1.4282	.1548	.9599	55° 00′
10	.6138	.5760	9.7604	.8175	9.9125	.7046	9.8479	1.4193	.1521	.9570	50
20	.6167	.5783	9.7622	.8158	9.9116	.7089	9.8506	1.4106	.1494	.9541	40
30	.6196	.5807	9.7640	.8141	9.9107	.7133	9.8533	1.4020	.1467	.9512	30
40	.6225	.5831	9.7657	.8124	9.9098	.7177	9.8559	1.3934	.1441	.9483	20
50	.6254	.5854	9.7675	.8107	9.9089	.7221	9.8586	1.3848	.1414	.9454	10
36° 00′	.6283	.5878	9.7692	.8090	9.9080	.7265	9.8613	1.3764	.1387	.9425	54° 00′

Table F-3. Natural Trigonometric Functions and Their Logarithms (*Continued*)

Deg	Radians	Nat sin	Log sin	Nat cos	Log cos	Nat tan	Log tan	Nat cot	Log cot	Radians	Deg
36° 00′	0.6283	0.5878	9.7692	0.8090	9.9080	0.7265	9.8613	1.3764	0.1387	0.9425	54° 00′
10	.6312	.5901	9.7710	.8073	9.9070	.7310	9.8639	1.3680	.1361	.9396	50
20	.6341	.5925	9.7727	.8056	9.9061	.7355	9.8666	1.3597	.1334	.9367	40
30	.6370	.5948	9.7744	.8039	9.9052	.7400	9.8692	1.3514	.1308	.9338	30
40	.6400	.5972	9.7761	.8021	9.9042	.7445	9.8719	1.3432	.1282	.9308	20
50	.6429	.5995	9.7778	.8004	9.9033	.7490	9.8745	1.3351	.1255	.9279	10
37° 00′	.6458	.6018	9.7795	.7986	9.9024	.7536	9.8771	1.3270	.1229	.9250	53° 00′
10	.6487	.6041	9.7811	.7969	9.9014	.7581	9.8797	1.3190	.1203	.9221	50
20	.6516	.6065	9.7828	.7951	9.9004	.7627	9.8824	1.3111	.1176	.9192	40
30	.6545	.6088	9.7845	.7934	9.8995	.7673	9.8850	1.3032	.1150	.9163	30
40	.6574	.6111	9.7861	.7916	9.8985	.7720	9.8876	1.2954	.1124	.9134	20
50	.6603	.6134	9.7877	.7898	9.8975	.7766	9.8902	1.2876	.1098	.9105	10
38° 00′	.6632	.6157	9.7893	.7880	9.8965	.7813	9.8928	1.2799	.1072	.9076	52° 00′
10	.6661	.6180	9.7910	.7862	9.8955	.7860	9.8954	1.2723	.1046	.9047	50
20	.6690	.6202	9.7926	.7844	9.8946	.7907	9.8980	1.2647	.1020	.9018	40
30	.6720	.6225	9.7942	.7826	9.8935	.7954	9.9006	1.2572	.0994	.8988	30
40	.6749	.6248	9.7957	.7808	9.8925	.8002	9.9032	1.2497	.0968	.8959	20
50	.6778	.6271	9.7973	.7790	9.8915	.8050	9.9058	1.2423	.0942	.8930	10
39° 00′	.6807	.6293	9.7989	.7772	9.8905	.8098	9.9084	1.2349	.0916	.8901	51° 00′
10	.6836	.6316	9.8004	.7753	9.8895	.8146	9.9110	1.2276	.0891	.8872	50
20	.6865	.6338	9.8020	.7735	9.8884	.8195	9.9135	1.2203	.0865	.8843	40
30	.6894	.6361	9.8035	.7716	9.8874	.8243	9.9161	1.2131	.0839	.8814	30
40	.6923	.6383	9.8050	.7698	9.8864	.8292	9.9187	1.2059	.0813	.8785	20
50	.6952	.6406	9.8066	.7679	9.8853	.8342	9.9213	1.1988	.0788	.8756	10
40° 00′	.6981	.6428	9.8081	.7660	9.8843	.8391	9.9238	1.1918	.0762	.8727	50° 00′
10	.7010	.6450	9.8096	.7642	9.8832	.8441	9.9264	1.1847	.0736	.8698	50
20	.7039	.6472	9.8111	.7623	9.8821	.8491	9.9289	1.1778	.0711	.8668	40
30	.7069	.6495	9.8125	.7604	9.8811	.8541	9.9315	1.1709	.0685	.8639	30
40	.7098	.6517	9.8140	.7585	9.8800	.8591	9.9341	1.1640	.0659	.8610	20
50	.7127	.6539	9.8155	.7566	9.8789	.8642	9.9366	1.1572	.0634	.8581	10
41° 00′	.7156	.6561	9.8169	.7547	9.8778	.8693	9.9392	1.1504	.0608	.8552	49° 00′
10	.7185	.6583	9.8184	.7528	9.8767	.8744	9.9417	1.1436	.0583	.8523	50
20	.7214	.6604	9.8198	.7509	9.8756	.8796	9.9443	1.1369	.0557	.8494	40
30	.7243	.6626	9.8213	.7490	9.8745	.8847	9.9468	1.1303	.0532	.8465	30
40	.7272	.6648	9.8227	.7470	9.8733	.8899	9.9494	1.1237	.0507	.8436	20
50	.7301	.6670	9.8241	.7451	9.8722	.8952	9.9519	1.1171	.0481	.8407	10

Deg	Radians	Nat cos	Log cos	Nat sin	Log sin	Nat cot	Log cot	Nat tan	Log tan	Radians	Deg
42° 00′	.7330	.6691	9.8255	.7431	9.8711	.9004	9.9544	1.1106	.0456	.8378	48° 00′
10	.7359	.6713	9.8269	.7412	9.8699	.9057	9.9570	1.1041	.0430	.8348	50
20	.7389	.6734	9.8283	.7392	9.8688	.9110	9.9595	1.0977	.0405	.8319	40
30	.7418	.6756	9.8297	.7373	9.8676	.9163	9.9621	1.0913	.0380	.8290	30
40	.7447	.6777	9.8311	.7353	9.8665	.9217	9.9646	1.0850	.0354	.8261	20
50	.7476	.6799	9.8324	.7333	9.8653	.9271	9.9671	1.0786	.0329	.8232	10
43° 00′	.7505	.6820	9.8338	.7314	9.8641	.9325	9.9697	1.0724	.0303	.8203	47° 00′
10	.7534	.6841	9.8351	.7294	9.8630	.9380	9.9722	1.0661	.0278	.8174	50
20	.7563	.6862	9.8365	.7274	9.8618	.9435	9.9747	1.0599	.0253	.8145	40
30	.7592	.6884	9.8378	.7254	9.8606	.9490	9.9773	1.0538	.0228	.8116	30
40	.7621	.6905	9.8391	.7234	9.8594	.9545	9.9798	1.0477	.0202	.8087	20
50	.7650	.6926	9.8405	.7214	9.8582	.9601	9.9823	1.0416	.0177	.8058	10
44° 00′	.7679	.6947	9.8418	.7193	9.8569	.9657	9.9848	1.0355	.0152	.8029	46° 00′
10	.7709	.6968	9.8431	.7173	9.8557	.9713	9.9874	1.0295	.0126	.7999	50
20	.7738	.6988	9.8444	.7153	9.8545	.9770	9.9899	1.0236	.0101	.7970	40
30	.7767	.7009	9.8457	.7133	9.8532	.9827	9.9924	1.0176	.0076	.7941	30
40	.7796	.7030	9.8469	.7112	9.8520	.9884	9.9950	1.0117	.0051	.7912	20
50	.7825	.7051	9.8482	.7092	9.8507	.9942	9.9975	1.0058	.0025	.7883	10
45° 00′	.7854	.7071	9.8495	.7071	9.8495	1.0000	0.0000	1.0000	.0000	.7854	45° 00′

Table F-4. Values and Logarithms of Exponential and Hyperbolic Functions*

x	e^x		e^{-x}	sinh x		cosh x		tanh x
	Value	$\log_{10}$	(value)	Value	$\log_{10}$	Value	$\log_{10}$	(value)
0.00	1.0000	0.00000	1.00000	0.0000	$-\infty$	1.0000	0.00000	0.00000
0.01	1.0101	.00434	0.99005	.0100	$\bar{2}.00001$	1.0001	.00002	.01000
0.02	1.0202	.00869	.98020	.0200	$\bar{2}.30106$	1.0002	.00009	.02000
0.03	1.0305	.01303	.97045	.0300	$\bar{2}.47719$	1.0005	.00020	.02999
0.04	1.0408	.01737	.96079	.0400	$\bar{2}.60218$	1.0008	.00035	.03998
0.05	1.0513	.02171	.95123	.0500	$\bar{2}.69915$	1.0013	.00054	.04996
0.06	1.0618	.02606	.94176	.0600	$\bar{2}.77841$	1.0018	.00078	.05993
0.07	1.0725	.03040	.93239	.0701	$\bar{2}.84545$	1.0025	.00106	.06989
0.08	1.0833	.03474	.92312	.0801	$\bar{2}.90355$	1.0032	.00139	.07983
0.09	1.0942	.03909	.91393	.0901	$\bar{2}.95483$	1.0041	.00176	.08976
0.10	1.1052	.04343	.90484	.1002	$\bar{1}.00072$	1.0050	.00217	.09967
0.11	1.1163	.04777	.89583	.1102	$\bar{1}.04227$	1.0061	.00262	.10956
0.12	1.1275	.05212	.88692	.1203	$\bar{1}.08022$	1.0072	.00312	.11943
0.13	1.1388	.05646	.87809	.1304	$\bar{1}.11517$	1.0085	.00366	.12927
0.14	1.1503	.06080	.86936	.1405	$\bar{1}.14755$	1.0098	.00424	.13909
0.15	1.1618	.06514	.86071	.1506	$\bar{1}.17772$	1.0113	.00487	.14889
0.16	1.1735	.06949	.85214	.1607	$\bar{1}.20597$	1.0128	.00554	.15865
0.17	1.1853	.07383	.84366	.1708	$\bar{1}.23254$	1.0145	.00625	.16838
0.18	1.1972	.07817	.83527	.1810	$\bar{1}.25762$	1.0162	.00700	.17808
0.19	1.2092	.08252	.82696	.1911	$\bar{1}.28136$	1.0181	.00779	.18775
0.20	1.2214	.08686	.81873	.2013	$\bar{1}.30392$	1.0201	.00863	.19738
0.21	1.2337	.09120	.81058	.2115	$\bar{1}.32541$	1.0221	.00951	.20697
0.22	1.2461	.09554	.80252	.2218	$\bar{1}.34592$	1.0243	.01043	.21652
0.23	1.2586	.09989	.79453	.2320	$\bar{1}.36555$	1.0266	.01139	.22603
0.24	1.2712	.10423	.78663	.2423	$\bar{1}.38437$	1.0289	.01239	.23550
0.25	1.2840	.10857	.77880	.2526	$\bar{1}.40245$	1.0314	.01343	.24492
0.26	1.2969	.11292	.77105	.2629	$\bar{1}.41986$	1.0340	.01452	.25430
0.27	1.3100	.11726	.76338	.2733	$\bar{1}.43663$	1.0367	.01564	.26362
0.28	1.3231	.12160	.75578	.2837	$\bar{1}.45282$	1.0395	.01681	.27291
0.29	1.3364	.12595	.74826	.2941	$\bar{1}.46847$	1.0423	.01801	.28213
0.30	1.3499	.13029	.74082	.3045	$\bar{1}.48362$	1.0453	.01926	.29131
0.31	1.3634	.13463	.73345	.3150	$\bar{1}.49830$	1.0484	.02054	.30044
0.32	1.3771	.13897	.72615	.3255	$\bar{1}.51254$	1.0516	.02187	.30951
0.33	1.3910	.14332	.71892	.3360	$\bar{1}.52637$	1.0549	.02323	.31852
0.34	1.4049	.14766	.71177	.3466	$\bar{1}.53981$	1.0584	.02463	.32748
0.35	1.4191	.15200	.70469	.3572	$\bar{1}.55290$	1.0619	.02607	.33638
0.36	1.4333	.15635	.69768	.3678	$\bar{1}.56564$	1.0655	.02755	.34521
0.37	1.4477	.16069	.69073	.3785	$\bar{1}.57807$	1.0692	.02907	.35399
0.38	1.4623	.16503	.68386	.3892	$\bar{1}.59019$	1.0731	.03063	.36271
0.39	1.4770	.16937	.67706	.4000	$\bar{1}.60202$	1.0770	.03222	.37136
0.40	1.4918	.17372	.67032	.4108	$\bar{1}.61358$	1.0811	.03385	.37995
0.41	1.5068	.17806	.66365	.4216	$\bar{1}.62488$	1.0852	.03552	.38847
0.42	1.5220	.18240	.65705	.4325	$\bar{1}.63594$	1.0895	.03723	.39693
0.43	1.5373	.18675	.65051	.4434	$\bar{1}.64677$	1.0939	.03897	.40532
0.44	1.5527	.19109	.64404	.4543	$\bar{1}.65738$	1.0984	.04075	.41364
0.45	1.5683	.19543	.63763	.4653	$\bar{1}.66777$	1.1030	.04256	.42190
0.46	1.5841	.19978	.63128	.4764	$\bar{1}.67797$	1.1077	.04441	.43008
0.47	1.6000	.20412	.62500	.4875	$\bar{1}.68797$	1.1125	.04630	.43820
0.48	1.6161	.20846	.61878	.4986	$\bar{1}.69779$	1.1174	.04822	.44624
0.49	1.6323	.21280	.61263	.5098	$\bar{1}.70744$	1.1225	.05018	.45422
0.50	1.6487	.21715	.60653	.5211	$\bar{1}.71692$	1.1276	.05217	.46212

* From R. H. Perry, *Engineering Manual*, McGraw-Hill, New York, 1959.

Table F-4. Values and Logarithms of Exponential and Hyperbolic Functions (*Continued*)

x	e^x		e^{-x}	sinh x		cosh x		tanh x
	Value	$\log_{10}$	(value)	Value	$\log_{10}$	Value	$\log_{10}$	(value)
0.50	1.6487	0.21715	0.60653	0.5211	$\bar{1}.71692$	1.1276	0.05217	0.46212
0.51	1.6653	.22149	.60050	0.5324	$\bar{1}.72624$	1.1329	.05419	.46995
0.52	1.6820	.22583	.59452	0.5438	$\bar{1}.73540$	1.1383	.05625	.47770
0.53	1.6989	.23018	.58860	0.5552	$\bar{1}.74442$	1.1438	.05834	.48538
0.54	1.7160	.23452	.58275	0.5666	$\bar{1}.75330$	1.1494	.06046	.49299
0.55	1.7333	.23886	.57695	0.5782	$\bar{1}.76204$	1.1551	.06262	.50052
0.56	1.7507	.24320	.57121	0.5897	$\bar{1}.77065$	1.1609	.06481	.50798
0.57	1.7683	.24755	.56553	0.6014	$\bar{1}.77914$	1.1669	.06703	.51536
0.58	1.7860	.25189	.55990	0.6131	$\bar{1}.78751$	1.1730	.06929	.52267
0.59	1.8040	.25623	.55433	0.6248	$\bar{1}.79576$	1.1792	.07157	.52990
0.60	1.8221	.26058	.54881	0.6367	$\bar{1}.80390$	1.1855	.07389	.53705
0.61	1.8404	.26492	.54335	0.6485	$\bar{1}.81194$	1.1919	.07624	.54413
0.62	1.8589	.26926	.53794	0.6605	$\bar{1}.81987$	1.1984	.07861	.55113
0.63	1.8776	.27361	.53259	0.6725	$\bar{1}.82770$	1.2051	.08102	.55805
0.64	1.8965	.27795	.52729	0.6846	$\bar{1}.83543$	1.2119	.08346	.56490
0.65	1.9155	.28229	.52205	0.6967	$\bar{1}.84308$	1.2188	.08593	.57167
0.66	1.9348	.28664	.51685	0.7090	$\bar{1}.85063$	1.2258	.08843	.57836
0.67	1.9542	.29098	.51171	0.7213	$\bar{1}.85809$	1.2330	.09095	.58498
0.68	1.9739	.29532	.50662	0.7336	$\bar{1}.86548$	1.2402	.09351	.59152
0.69	1.9937	.29966	.50158	0.7461	$\bar{1}.87278$	1.2476	.09609	.59798
0.70	2.0138	.30401	.49659	0.7586	$\bar{1}.88000$	1.2552	.09870	.60437
0.71	2.0340	.30835	.49164	0.7712	$\bar{1}.88715$	1.2628	.10134	.61068
0.72	2.0544	.31269	.48675	0.7838	$\bar{1}.89423$	1.2706	.10401	.61691
0.73	2.0751	.31703	.48191	0.7966	$\bar{1}.90123$	1.2785	.10670	.62307
0.74	2.0959	.32138	.47711	0.8094	$\bar{1}.90817$	1.2865	.10942	.62915
0.75	2.1170	.32572	.47237	0.8223	$\bar{1}.91504$	1.2947	.11216	.63515
0.76	2.1383	.33006	.46767	0.8353	$\bar{1}.92185$	1.3030	.11493	.64108
0.77	2.1598	.33441	.46301	0.8484	$\bar{1}.92859$	1.3114	.11773	.64693
0.78	2.1815	.33875	.45841	0.8615	$\bar{1}.93527$	1.3199	.12055	.65271
0.79	2.2034	.34309	.45384	0.8748	$\bar{1}.94190$	1.3286	.12340	.65841
0.80	2.2255	.34744	.44933	0.8881	$\bar{1}.94846$	1.3374	.12627	.66404
0.81	2.2479	.35178	.44486	0.9015	$\bar{1}.95498$	1.3464	.12917	.66959
0.82	2.2705	.35612	.44043	0.9150	$\bar{1}.96144$	1.3555	.13209	.67507
0.83	2.2933	.36046	.43605	0.9286	$\bar{1}.96784$	1.3647	.13503	.68048
0.84	2.3164	.36481	.43171	0.9423	$\bar{1}.97420$	1.3740	.13800	.68581
0.85	2.3396	.36915	.42741	0.9561	$\bar{1}.98051$	1.3835	.14099	.69107
0.86	2.3632	.37349	.42316	0.9700	$\bar{1}.98677$	1.3932	.14400	.69626
0.87	2.3869	.37784	.41895	0.9840	$\bar{1}.99299$	1.4029	.14704	.70137
0.88	2.4109	.38218	.41478	0.9981	$\bar{1}.99916$	1.4128	.15009	.70642
0.89	2.4351	.38652	.41066	1.0122	0.00528	1.4229	.15317	.71139
0.90	2.4596	.39087	.40657	1.0265	.01137	1.4331	.15627	.71630
0.91	2.4843	.39521	.40252	1.0409	.01741	1.4434	.15939	.72113
0.92	2.5093	.39955	.39852	1.0554	.02341	1.4539	.16254	.72590
0.93	2.5345	.40389	.39455	1.0700	.02937	1.4645	.16570	.73059
0.94	2.5600	.40824	.39063	1.0847	.03530	1.4753	.16888	.73522
0.95	2.5857	.41258	.38674	1.0995	.04119	1.4862	.17208	.73978
0.96	2.6117	.41692	.38289	1.1144	.04704	1.4973	.17531	.74428
0.97	2.6379	.42127	.37908	1.1294	.05286	1.5085	.17855	.74870
0.98	2.6645	.42561	.37531	1.1446	.05864	1.5199	.18181	.75307
0.99	2.6912	.42995	.37158	1.1598	.06439	1.5314	.18509	.75736
1.00	2.7183	.43429	.36788	1.1752	.07011	1.5431	.18839	.76159

Table F-4. Values and Logarithms of Exponential and Hyperbolic Functions *(Continued)*

x	e^x		e^{-x} (value)	sinh x		cosh x		tanh x (value)
	Value	$\log_{10}$		Value	$\log_{10}$	Value	$\log_{10}$	
1.00	2.7183	0.43429	0.36788	1.1752	0.07011	1.5431	0.18839	0.76159
1.01	2.7456	.43864	.36422	1.1907	.07580	1.5549	.19171	.76576
1.02	2.7732	.44298	.36060	1.2063	.08146	1.5669	.19504	.76987
1.03	2.8011	.44732	.35701	1.2220	.08708	1.5790	.19839	.77391
1.04	2.8292	.45167	.35345	1.2379	.09268	1.5913	.20176	.77789
1.05	2.8577	.45601	.34994	1.2539	.09825	1.6038	.20515	.78181
1.06	2.8864	.46035	.34646	1.2700	.10379	1.6164	.20855	.78566
1.07	2.9154	.46470	.34301	1.2862	.10930	1.6292	.21197	.78946
1.08	2.9447	.46904	.33960	1.3025	.11479	1.6421	.21541	.79320
1.09	2.9743	.47338	.33622	1.3190	.12025	1.6552	.21886	.79688
1.10	3.0042	.47772	.33287	1.3356	.12569	1.6685	.22233	.80050
1.11	3.0344	.48207	.32956	1.3524	.13111	1.6820	.22582	.80406
1.12	3.0649	.48641	.32628	1.3693	.13649	1.6956	.22931	.80757
1.13	3.0957	.49075	.32303	1.3863	.14186	1.7093	.23283	.81102
1.14	3.1268	.49510	.31982	1.4035	.14720	1.7233	.23636	.81441
1.15	3.1582	.49944	.31664	1.4208	.15253	1.7374	.23990	.81775
1.16	3.1899	.50378	.31349	1.4382	.15783	1.7517	.24346	.82104
1.17	3.2220	.50812	.31037	1.4558	.16311	1.7662	.24703	.82427
1.18	3.2544	.51247	.30728	1.4735	.16836	1.7808	.25062	.82745
1.19	3.2871	.51681	.30422	1.4914	.17360	1.7957	.25422	.83058
1.20	3.3201	.52115	.30119	1.5095	.17882	1.8107	.25784	.83365
1.21	3.3535	.52550	.29820	1.5276	.18402	1.8258	.26146	.83668
1.22	3.3872	.52984	.29523	1.5460	.18920	1.8412	.26510	.83965
1.23	3.4212	.53418	.29229	1.5645	.19437	1.8568	.26876	.84258
1.24	3.4556	.53853	.28938	1.5831	.19951	1.8725	.27242	.84546
1.25	3.4903	.54287	.28650	1.6019	.20464	1.8884	.27610	.84828
1.26	3.5254	.54721	.28365	1.6209	.20975	1.9045	.27979	.85106
1.27	3.5609	.55155	.28083	1.6400	.21485	1.9208	.28349	.85380
1.28	3.5966	.55590	.27804	1.6593	.21993	1.9373	.28721	.85648
1.29	3.6328	.56024	.27527	1.6788	.22499	1.9540	.29093	.85913
1.30	3.6693	.56458	.27253	1.6984	.23004	1.9709	.29467	.86172
1.31	3.7062	.56893	.26982	1.7182	.23507	1.9880	.29842	.86428
1.32	3.7434	.57327	.26714	1.7381	.24009	2.0053	.30217	.86678
1.33	3.7810	.57761	.26448	1.7583	.24509	2.0228	.30594	.86925
1.34	3.8190	.58195	.26185	1.7786	.25008	2.0404	.30972	.87167
1.35	3.8574	.58630	.25924	1.7991	.25505	2.0583	.31352	.87405
1.36	3.8962	.59064	.25666	1.8198	.26002	2.0764	.31732	.87639
1.37	3.9354	.59498	.25411	1.8406	.26496	2.0947	.32113	.87869
1.38	3.9749	.59933	.25158	1.8617	.26990	2.1132	.32495	.88095
1.39	4.0149	.60367	.24908	1.8829	.27482	2.1320	.32878	.88317
1.40	4.0552	.60801	.24660	1.9043	.27974	2.1509	.33262	.88535
1.41	4.0960	.61236	.24414	1.9259	.28464	2.1700	.33647	.88749
1.42	4.1371	.61670	.24171	1.9477	.28952	2.1894	.34033	.88960
1.43	4.1787	.62104	.23931	1.9697	.29440	2.2090	.34420	.89167
1.44	4.2207	.62538	.23693	1.9919	.29926	2.2288	.34807	.89370
1.45	4.2631	.62973	.23457	2.0143	.30412	2.2488	.35196	.89569
1.46	4.3060	.63407	.23224	2.0369	.30896	2.2691	.35585	.89765
1.47	4.3492	.63841	.22993	2.0597	.31379	2.2896	.35976	.89958
1.48	4.3929	.64276	.22764	2.0827	.31862	2.3103	.36367	.90147
1.49	4.4371	.64710	.22537	2.1059	.32343	2.3312	.36759	.90332
1.50	4.4817	.65144	.22313	2.1293	.32823	2.3524	.37151	.90515

Table F-4. Values and Logarithms of Exponential and Hyperbolic Functions *(Continued)*

x	e^x Value	e^x $\log_{10}$	e^{-x} (value)	sinh x Value	sinh x $\log_{10}$	cosh x Value	cosh x $\log_{10}$	tanh x (value)
1.50	4.4817	0.65144	0.22313	2.1293	0.32823	2.3524	0.37151	0.90515
1.51	4.5267	.65578	.22091	2.1529	.33303	2.3738	.37545	.90694
1.52	4.5722	.66013	.21871	2.1768	.33781	2.3955	.37939	.90870
1.53	4.6182	.66447	.21654	2.2008	.34258	2.4174	.38334	.91042
1.54	4.6646	.66881	.21438	2.2251	.34735	2.4395	.38730	.91212
1.55	4.7115	.67316	.21225	2.2496	.35211	2.4619	.39126	.91379
1.56	4.7588	.67750	.21014	2.2743	.35686	2.4845	.39524	.91542
1.57	4.8066	.68184	.20805	2.2993	.36160	2.5073	.39921	.91703
1.58	4.8550	.68619	.20598	2.3245	.36633	2.5305	.40320	.91860
1.59	4.9037	.69053	.20393	2.3499	.37105	2.5538	.40719	.92015
1.60	4.9530	.69487	.20190	2.3756	.37577	2.5775	.41119	.92167
1.61	5.0028	.69921	.19989	2.4015	.38048	2.6013	.41520	.92316
1.62	5.0531	.70356	.19790	2.4276	.38518	2.6255	.41921	.92462
1.63	5.1039	.70790	.19593	2.4540	.38987	2.6499	.42323	.92606
1.64	5.1552	.71224	.19398	2.4806	.39456	2.6746	.42725	.92747
1.65	5.2070	.71659	.19205	2.5075	.39923	2.6995	.43129	.92886
1.66	5.2593	.72093	.19014	2.5346	.40391	2.7247	.43532	.93022
1.67	5.3122	.72527	.18825	2.5620	.40857	2.7502	.43937	.93155
1.68	5.3656	.72961	.18637	2.5896	.41323	2.7760	.44341	.93286
1.69	5.4195	.73396	.18452	2.6175	.41788	2.8020	.44747	.93415
1.70	5.4739	.73830	.18268	2.6456	.42253	2.8283	.45153	.93541
1.71	5.5290	.74264	.18087	2.6740	.42717	2.8549	.45559	.93665
1.72	5.5845	.74699	.17907	2.7027	.43180	2.8818	.45966	.93786
1.73	5.6407	.75133	.17728	2.7317	.43643	2.9090	.46374	.93906
1.74	5.6973	.75567	.17552	2.7609	.44105	2.9364	.46782	.94023
1.75	5.7546	.76002	.17377	2.7904	.44567	2.9642	.47191	.94138
1.76	5.8124	.76436	.17204	2.8202	.45028	2.9922	.47600	.94250
1.77	5.8709	.76870	.17033	2.8503	.45488	3.0206	.48009	.94361
1.78	5.9299	.77304	.16864	2.8806	.45948	3.0492	.48419	.94470
1.79	5.9895	.77739	.16696	2.9112	.46408	3.0782	.48830	.94576
1.80	6.0496	.78173	.16530	2.9422	.46867	3.1075	.49241	.94681
1.81	6.1104	.78607	.16365	2.9734	.47325	3.1371	.49652	.94783
1.82	6.1719	.79042	.16203	3.0049	.47783	3.1669	.50064	.94884
1.83	6.2339	.79476	.16041	3.0367	.48241	3.1972	.50476	.94983
1.84	6.2965	.79910	.15882	3.0689	.48698	3.2277	.50889	.95080
1.85	6.3598	.80344	.15724	3.1013	.49154	3.2585	.51302	.95175
1.86	6.4237	.80779	.15567	3.1340	.49610	3.2897	.51716	.95268
1.87	6.4883	.81213	.15412	3.1671	.50066	3.3212	.52130	.95359
1.88	6.5535	.81647	.15259	3.2005	.50521	3.3530	.52544	.95449
1.89	6.6194	.82082	.15107	3.2341	.50976	3.3852	.52959	.95537
1.90	6.6859	.82516	.14957	3.2682	.51430	3.4177	.53374	.95624
1.91	6.7531	.82950	.14808	3.3025	.51884	3.4506	.53789	.95709
1.92	6.8210	.83385	.14661	3.3372	.52338	3.4838	.54205	.95792
1.93	6.8895	.83819	.14515	3.3722	.52791	3.5173	.54621	.95873
1.94	6.9588	.84253	.14370	3.4075	.53244	3.5512	.55038	.95953
1.95	7.0287	.84687	.14227	3.4432	.53696	3.5855	.55455	.96032
1.96	7.0993	.85122	.14086	3.4792	.54148	3.6201	.55872	.96109
1.97	7.1707	.85556	.13946	3.5156	.54600	3.6551	.56290	.96185
1.98	7.2427	.85990	.13807	3.5523	.55051	3.6904	.56707	.96259
1.99	7.3155	.86425	.13670	3.5894	.55502	3.7261	.57126	.96331
2.00	7.3891	.86859	.13534	3.6269	.55953	3.7622	.57544	.96403

Table F-4. Values and Logarithms of Exponential and Hyperbolic Functions (*Continued*)

x	e^x		e^{-x} (value)	sinh x		cosh x		tanh x (value)
	Value	$\log_{10}$		Value	$\log_{10}$	Value	$\log_{10}$	
2.00	7.3891	0.86859	0.13534	3.6269	0.55953	3.7622	0.57544	0.96403
2.01	7.4633	0.87293	.13399	3.6647	.56403	3.7987	.57963	.96473
2.02	7.5383	0.87727	.13266	3.7028	.56853	3.8355	.58382	.96541
2.03	7.6141	0.88162	.13134	3.7414	.57303	3.8727	.58802	.96609
2.04	7.6906	0.88596	.13003	3.7803	.57753	3.9103	.59221	.96675
2.05	7.7679	0.89030	.12873	3.8196	.58202	3.9483	.59641	.96740
2.06	7.8460	0.89465	.12745	3.8593	.58650	3.9867	.60061	.96803
2.07	7.9248	0.89899	.12619	3.8993	.59099	4.0255	.60482	.96865
2.08	8.0045	0.90333	.12493	3.9398	.59547	4.0647	.60903	.96926
2.09	8.0849	0.90768	.12369	3.9806	.59995	4.1043	.61324	.96986
2.10	8.1662	0.91202	.12246	4.0219	.60443	4.1443	.61745	.97045
2.11	8.2482	0.91636	.12124	4.0635	.60890	4.1847	.62167	.97103
2.12	8.3311	0.92070	.12003	4.1056	.61337	4.2256	.62589	.97159
2.13	8.4149	0.92505	.11884	4.1480	.61784	4.2669	.63011	.97215
2.14	8.4994	0.92939	.11765	4.1909	.62231	4.3085	.63433	.97269
2.15	8.5849	0.93373	.11648	4.2342	.62677	4.3507	.63856	.97323
2.16	8.6711	0.93808	.11533	4.2779	.63123	4.3932	.64278	.97375
2.17	8.7583	0.94242	.11418	4.3221	.63569	4.4362	.64701	.97426
2.18	8.8463	0.94676	.11304	4.3666	.64015	4.4797	.65125	.97477
2.19	8.9352	0.95110	.11192	4.4116	.64460	4.5236	.65548	.97526
2.20	9.0250	0.95545	.11080	4.4571	.64905	4.5679	.65972	.97574
2.21	9.1157	0.95979	.10970	4.5030	.65350	4.6127	.66396	.97622
2.22	9.2073	0.96413	.10861	4.5494	.65795	4.6580	.66820	.97668
2.23	9.2999	0.96848	.10753	4.5962	.66240	4.7037	.67244	.97714
2.24	9.3933	0.97282	.10646	4.6434	.66684	4.7499	.67668	.97759
2.25	9.4877	0.97716	.10540	4.6912	.67128	4.7966	.68093	.97803
2.26	9.5831	0.98151	.10435	4.7394	.67572	4.8437	.68518	.97846
2.27	9.6794	0.98585	.10331	4.7880	.68016	4.8914	.68943	.97888
2.28	9.7767	0.99019	.10228	4.8372	.68459	4.9395	.69368	.97929
2.29	9.8749	0.99453	.10127	4.8868	.68903	4.9881	.69794	.97970
2.30	9.9742	0.99888	.10026	4.9370	.69346	5.0372	.70219	.98010
2.31	10.074	1.00322	.09926	4.9876	.69789	5.0868	.70645	.98049
2.32	10.176	1.00756	.09827	5.0387	.70232	5.1370	.71071	.98087
2.33	10.278	1.01191	.09730	5.0903	.70675	5.1876	.71497	.98124
2.34	10.381	1.01625	.09633	5.1425	.71117	5.2388	.71923	.98161
2.35	10.486	1.02059	.09537	5.1951	.71559	5.2905	.72349	.98197
2.36	10.591	1.02493	.09442	5.2483	.72002	5.3427	.72776	.98233
2.37	10.697	1.02928	.09348	5.3020	.72444	5.3954	.73203	.98267
2.38	10.805	1.03362	.09255	5.3562	.72885	5.4487	.73630	.98301
2.39	10.913	1.03796	.09163	5.4109	.73327	5.5026	.74056	.98335
2.40	11.023	1.04231	.09072	5.4662	.73769	5.5569	.74484	.98367
2.41	11.134	1.04665	.08982	5.5221	.74210	5.6119	.74911	.98400
2.42	11.246	1.05099	.08892	5.5785	.74652	5.6674	.75338	.98431
2.43	11.359	1.05534	.08804	5.6354	.75093	5.7235	.75766	.98462
2.44	11.473	1.05968	.08716	5.6929	.75534	5.7801	.76194	.98492
2.45	11.588	1.06402	.08629	5.7510	.75975	5.8373	.76621	.98522
2.46	11.705	1.06836	.08543	5.8097	.76415	5.8951	.77049	.98551
2.47	11.822	1.07271	.08458	5.8689	.76856	5.9535	.77477	.98579
2.48	11.941	1.07705	.08374	5.9288	.77296	6.0125	.77906	.98607
2.49	12.061	1.08139	.08291	5.9892	.77737	6.0721	.78334	.98635
2.50	12.182	1.08574	.08208	6.0502	.78177	6.1323	.78762	.98661

Table F-4. Values and Logarithms of Exponential and Hyperbolic Functions (*Continued*)

x	e^x		e^{-x} (value)	sinh x		cosh x		tanh x (value)
	Value	$\log_{10}$		Value	$\log_{10}$	Value	$\log_{10}$	
2.50	12.182	1.08574	0.08208	6.0502	0.78177	6.1323	0.78762	0.98661
2.51	12.305	1.09008	.08127	6.1118	.78617	6.1931	.79191	.98688
2.52	12.429	1.09442	.08046	6.1741	.79057	6.2545	.79619	.98714
2.53	12.554	1.09877	.07966	6.2369	.79497	6.3166	.80048	.98739
2.54	12.680	1.10311	.07887	6.3004	.79937	6.3793	.80477	.98764
2.55	12.807	1.10745	.07808	6.3645	.80377	6.4426	.80906	.98788
2.56	12.936	1.11179	.07730	6.4293	.80816	6.5066	.81335	.98812
2.57	13.066	1.11614	.07654	6.4946	.81256	6.5712	.81764	.98835
2.58	13.197	1.12048	.07577	6.5607	.81695	6.6365	.82194	.98858
2.59	13.330	1.12482	.07502	6.6274	.82134	6.7024	.82623	.98881
2.60	13.464	1.12917	.07427	6.6947	.82573	6.7690	.83052	.98903
2.61	13.599	1.13351	.07353	6.7628	.83012	6.8363	.83482	.98924
2.62	13.736	1.13785	.07280	6.8315	.83451	6.9043	.83912	.98946
2.63	13.874	1.14219	.07208	6.9008	.83890	6.9729	.84341	.98966
2.64	14.013	1.14654	.07136	6.9709	.84329	7.0423	.84771	.98987
2.65	14.154	1.15088	.07065	7.0417	.84768	7.1123	.85201	.99007
2.66	14.296	1.15522	.06995	7.1132	.85206	7.1831	.85631	.99026
2.67	14.440	1.15957	.06925	7.1854	.85645	7.2546	.86061	.99045
2.68	14.585	1.16391	.06856	7.2583	.86083	7.3268	.86492	.99064
2.69	14.732	1.16825	.06788	7.3319	.86522	7.3998	.86922	.99083
2.70	14.880	1.17260	.06721	7.4063	.86960	7.4735	.87352	.99101
2.71	15.029	1.17694	.06654	7.4814	.87398	7.5479	.87783	.99118
2.72	15.180	1.18128	.06587	7.5572	.87836	7.6231	.88213	.99136
2.73	15.333	1.18562	.06522	7.6338	.88274	7.6991	.88644	.99153
2.74	15.487	1.18997	.06457	7.7112	.88712	7.7758	.89074	.99170
2.75	15.643	1.19431	.06393	7.7894	.89150	7.8533	.89505	.99186
2.76	15.800	1.19865	.06329	7.8683	.89588	7.9316	.89936	.99202
2.77	15.959	1.20300	.06266	7.9480	.90026	8.0106	.90367	.99218
2.78	16.119	1.20734	.06204	8.0285	.90463	8.0905	.90798	.99233
2.79	16.281	1.21168	.06142	8.1098	.90901	8.1712	.91229	.99248
2.80	16.445	1.21602	.06081	8.1919	.91339	8.2527	.91660	.99263
2.81	16.610	1.22037	.06020	8.2749	.91776	8.3351	.92091	.99278
2.82	16.777	1.22471	.05961	8.3586	.92213	8.4182	.92522	.99292
2.83	16.945	1.22905	.05901	8.4432	.92651	8.5022	.92953	.99306
2.84	17.116	1.23340	.05843	8.5287	.93088	8.5871	.93385	.99320
2.85	17.288	1.23774	.05784	8.6150	.93525	8.6728	.93816	.99333
2.86	17.462	1.24208	.05727	8.7021	.93963	8.7594	.94247	.99346
2.87	17.637	1.24643	.05670	8.7902	.94400	8.8469	.94679	.99359
2.88	17.814	1.25077	.05613	8.8791	.94837	8.9352	.95110	.99372
2.89	17.993	1.25511	.05558	8.9689	.95274	9.0244	.95542	.99384
2.90	18.174	1.25945	.05502	9.0596	.95711	9.1146	.95974	.99396
2.91	18.357	1.26380	.05448	9.1512	.96148	9.2056	.96405	.99408
2.92	18.541	1.26814	.05393	9.2437	.96584	9.2976	.96837	.99420
2.93	18.728	1.27248	.05340	9.3371	.97021	9.3905	.97269	.99431
2.94	18.916	1.27683	.05287	9.4315	.97458	9.4844	.97701	.99443
2.95	19.106	1.28117	.05234	9.5268	.97895	9.5791	.98133	.99454
2.96	19.298	1.28551	.05182	9.6231	.98331	9.6749	.98565	.99464
2.97	19.492	1.28985	.05130	9.7203	.98768	9.7716	.98997	.99475
2.98	19.688	1.29420	.05079	9.8185	.99205	9.8693	.99429	.99485
2.99	19.886	1.29854	.05029	9.9177	.99641	9.9680	.99861	.99496
3.00	20.086	1.30288	.04979	10.018	1.00078	10.068	1.00293	.99505

Table F-4. Values and Logarithms of Exponential and Hyperbolic Functions (*Continued*)

x	e^x		e^{-x} (value)	sinh x		cosh x		tanh x (value)
	Value	$\log_{10}$		Value	$\log_{10}$	Value	$\log_{10}$	
3.00	20.086	1.30288	0.04979	10.018	1.00078	10.068	1.00293	0.99505
3.05	21.115	1.32460	.04736	10.534	1.02259	10.581	1.02454	0.99552
3.10	22.198	1.34631	.04505	11.076	1.04440	11.122	1.04616	0.99595
3.15	23.336	1.36803	.04285	11.647	1.06620	11.690	1.06779	0.99633
3.20	24.533	1.38974	.04076	12.246	1.08799	12.287	1.08943	0.99668
3.25	25.790	1.41146	.03877	12.876	1.10977	12.915	1.11108	0.99700
3.30	27.113	1.43317	.03688	13.538	1.13155	13.575	1.13273	0.99728
3.35	28.503	1.45489	.03508	14.234	1.15332	14.269	1.15439	0.99754
3.40	29.964	1.47660	.03337	14.965	1.17509	14.999	1.17605	0.99777
3.45	31.500	1.49832	.03175	15.734	1.19685	15.766	1.19772	0.99799
3.50	33.115	1.52003	.03020	16.543	1.21860	16.573	1.21940	0.99818
3.55	34.813	1.54175	.02872	17.392	1.24036	17.421	1.24107	0.99835
3.60	36.598	1.56346	.02732	18.286	1.26211	18.313	1.26275	0.99851
3.65	38.475	1.58517	.02599	19.224	1.28385	19.250	1.28444	0.99865
3.70	40.447	1.60689	.02472	20.211	1.30559	20.236	1.30612	0.99878
3.75	42.521	1.62860	.02352	21.249	1.32733	21.272	1.32781	0.99889
3.80	44.701	1.65032	.02237	22.339	1.34907	22.362	1.34951	0.99900
3.85	46.993	1.67203	.02128	23.486	1.37081	23.507	1.37120	0.99909
3.90	49.402	1.69375	.02024	24.691	1.39254	24.711	1.39290	0.99918
3.95	51.935	1.71546	.01925	25.958	1.41427	25.977	1.41459	0.99926
4.00	54.598	1.73718	.01832	27.290	1.43600	27.308	1.43629	0.99933
4.10	60.340	1.78061	.01657	30.162	1.47946	30.178	1.47970	0.99945
4.20	66.686	1.82404	.01500	33.336	1.52291	33.351	1.52310	0.99955
4.30	73.700	1.86747	.01357	36.843	1.56636	36.857	1.56652	0.99963
4.40	81.451	1.91090	.01227	40.719	1.60980	40.732	1.60993	0.99970
4.50	90.017	1.95433	.01111	45.003	1.65324	45.014	1.65335	0.99975
4.60	99.484	1.99775	.01005	49.737	1.69668	49.747	1.69677	0.99980
4.70	109.95	2.04118	.00910	54.969	1.74012	54.978	1.74019	0.99983
4.80	121.51	2.08461	.00823	60.751	1.78355	60.759	1.78361	0.99986
4.90	134.29	2.12804	.00745	67.141	1.82699	67.149	1.82704	0.99989
5.00	148.41	2.17147	.00674	74.203	1.87042	74.210	1.87046	0.99991
5.10	164.02	2.21490	.00610	82.008	1.91389	82.014	1.91389	0.99993
5.20	181.27	2.25833	.00552	90.633	1.95729	90.639	1.95731	0.99994
5.30	200.34	2.30176	.00499	100.17	2.00074	100.17	2.00074	0.99995
5.40	221.41	2.34519	.00452	110.70	2.04415	110.71	2.04417	0.99996
5.50	244.69	2.38862	.00409	122.34	2.08758	122.35	2.08760	0.99997
5.60	270.43	2.43205	.00370	135.21	2.13101	135.22	2.13103	0.99997
5.70	298.87	2.47548	.00335	149.43	2.17444	149.44	2.17445	0.99998
5.80	330.30	2.51891	.00303	165.15	2.21787	165.15	2.21788	0.99998
5.90	365.04	2.56234	.00274	182.52	2.26130	182.52	2.26131	0.99998
6.00	403.43	2.60577	.00248	201.71	2.30473	201.72	2.30474	0.99999
6.25	518.01	2.71434	.00193	259.01	2.41331	259.01	2.41331	0.99999
6.50	665.14	2.82291	.00150	332.57	2.52188	332.57	2.52189	1.00000
6.75	854.06	2.93149	.00117	427.03	2.63046	427.03	2.63046	1.00000
7.00	1096.6	3.04006	.00091	548.32	2.73903	548.32	2.73903	1.00000
7.50	1808.0	3.25721	.00055	904.02	2.95618	904.02	2.95618	1.00000
8.00	2981.0	3.47436	.00034	1490.5	3.17333	1490.5	3.17333	1.00000
8.50	4914.8	3.69150	.00020	2457.4	3.39047	2457.4	3.39047	1.00000
9.00	8103.1	3.90865	.00012	4051.5	3.60762	4051.5	3.60762	1.00000
9.50	13360.	4.12580	.00007	6679.9	3.82477	6679.9	3.82477	1.00000
10.00	22026.	4.34294	.00005	11013.	4.04191	11013.	4.04191	1.00000

Table F-5. Natural, Napierian, or Hyperbolic Logarithms*

N	0	1	2	3	4	5	6	7	8	9
0	— ∞	0.0000	0.6931	1.0986	1.3863	1.6094	1.7918	1.9459	2.0794	2.1972
10	2.3026	2.3979	2.4849	2.5649	2.6391	2.7081	2.7726	2.8332	2.8904	2.9444
20	2.9957	3.0445	3.0910	3.1355	3.1781	3.2189	3.2581	3.2958	3.3322	3.3673
30	3.4012	3.4340	3.4657	3.4965	3.5264	3.5553	3.5835	3.6109	3.6376	3.6636
40	3.6889	3.7136	3.7377	3.7612	3.7842	3.8067	3.8286	3.8501	3.8712	3.8918
50	3.9120	3.9318	3.9512	3.9703	3.9890	4.0073	4.0254	4.0431	4.0604	4.0775
60	4.0943	4.1109	4.1271	4.1431	4.1589	4.1744	4.1897	4.2047	4.2195	4.2341
70	4.2485	4.2627	4.2767	4.2905	4.3041	4.3175	4.3307	4.3438	4.3567	4.3694
80	4.3820	4.3944	4.4067	4.4188	4.4308	4.4427	4.4543	4.4659	4.4773	4.4886
90	4.4998	4.5109	4.5218	4.5326	4.5433	4.5539	4.5643	4.5747	4.5850	4.5951
100	4.6052	4.6151	4.6250	4.6347	4.6444	4.6540	4.6634	4.6728	4.6821	4.6913
110	4.7005	4.7095	4.7185	4.7274	4.7362	4.7449	4.7536	4.7622	4.7707	4.7791
120	4.7875	4.7958	4.8040	4.8122	4.8203	4.8283	4.8363	4.8442	4.8520	4.8598
130	4.8675	4.8752	4.8828	4.8903	4.8978	4.9053	4.9127	4.9200	4.9273	4.9345
140	4.9416	4.9488	4.9558	4.9628	4.9698	4.9767	4.9836	4.9904	4.9972	5.0039
150	5.0106	5.0173	5.0239	5.0304	5.0370	5.0434	5.0499	5.0562	5.0626	5.0689
160	5.0752	5.0814	5.0876	5.0938	5.0999	5.1059	5.1120	5.1180	5.1240	5.1299
170	5.1358	5.1417	5.1475	5.1533	5.1591	5.1648	5.1705	5.1761	5.1818	5.1874
180	5.1930	5.1985	5.2040	5.2095	5.2149	5.2204	5.2257	5.2311	5.2364	5.2417
190	5.2470	5.2523	5.2575	5.2627	5.2679	5.2730	5.2781	5.2832	5.2883	5.2933
200	5.2983	5.3033	5.3083	5.3132	5.3181	5.3230	5.3279	5.3327	5.3375	5.3423
210	5.3471	5.3519	5.3566	5.3613	5.3660	5.3706	5.3753	5.3799	5.3845	5.3891
220	5.3936	5.3982	5.4027	5.4072	5.4116	5.4161	5.4205	5.4250	5.4293	5.4337
230	5.4381	5.4424	5.4467	5.4510	5.4553	5.4596	5.4638	5.4681	5.4723	5.4765
240	5.4806	5.4848	5.4889	5.4931	5.4972	5.5013	5.5053	5.5094	5.5134	5.5175
250	5.5215	5.5255	5.5294	5.5334	5.5373	5.5413	5.5452	5.5491	5.5530	5.5568
260	5.5607	5.5645	5.5683	5.5722	5.5759	5.5797	5.5835	5.5872	5.5910	5.5947
270	5.5984	5.6021	5.6058	5.6095	5.6131	5.6168	5.6204	5.6240	5.6276	5.6312
280	5.6348	5.6384	5.6419	5.6454	5.6490	5.6525	5.6560	5.6595	5.6630	5.6664
290	5.6699	5.6733	5.6768	5.6802	5.6836	5.6870	5.6904	5.6937	5.6971	5.7004
300	5.7038	5.7071	5.7104	5.7137	5.7170	5.7203	5.7236	5.7268	5.7301	5.7333
310	5.7366	5.7398	5.7430	5.7462	5.7494	5.7526	5.7557	5.7589	5.7621	5.7652
320	5.7683	5.7714	5.7746	5.7777	5.7807	5.7838	5.7869	5.7900	5.7930	5.7961
330	5.7991	5.8021	5.8051	5.8081	5.8111	5.8141	5.8171	5.8201	5.8230	5.8260
340	5.8289	5.8319	5.8348	5.8377	5.8406	5.8435	5.8464	5.8493	5.8522	5.8551
350	5.8579	5.8608	5.8636	5.8665	5.8693	5.8721	5.8749	5.8777	5.8805	5.8833
360	5.8861	5.8889	5.8916	5.8944	5.8972	5.8999	5.9026	5.9054	5.9081	5.9108
370	5.9135	5.9162	5.9189	5.9216	5.9243	5.9269	5.9296	5.9322	5.9349	5.9375
380	5.9402	5.9428	5.9454	5.9480	5.9506	5.9532	5.9558	5.9584	5.9610	5.9636
390	5.9661	5.9687	5.9713	5.9738	5.9764	5.9789	5.9814	5.9839	5.9865	5.9890
400	5.9915	5.9940	5.9965	5.9989	6.0014	6.0039	6.0064	6.0088	6.0113	6.0137
410	6.0162	6.0186	6.0210	6.0234	6.0259	6.0283	6.0307	6.0331	6.0355	6.0379
420	6.0403	6.0426	6.0450	6.0474	6.0497	6.0521	6.0544	6.0568	6.0591	6.0615
430	6.0638	6.0661	6.0684	6.0707	6.0730	6.0753	6.0776	6.0799	6.0822	6.0845
440	6.0868	6.0890	6.0913	6.0936	6.0958	6.0981	6.1003	6.1026	6.1048	6.1070
450	6.1092	6.1115	6.1137	6.1159	6.1181	6.1203	6.1225	6.1247	6.1269	6.1291
460	6.1312	6.1334	6.1356	6.1377	6.1399	6.1420	6.1442	6.1463	6.1485	6.1506
470	6.1527	6.1549	6.1570	6.1591	6.1612	6.1633	6.1654	6.1675	6.1696	6.1717
480	6.1738	6.1759	6.1779	6.1800	6.1821	6.1841	6.1862	6.1883	6.1903	6.1924
490	6.1944	6.1964	6.1985	6.2005	6.2025	6.2046	6.2066	6.2086	6.2106	6.2126

NOTE 1: Moving the decimal point n places to the right (or left) in the number is equivalent to adding (or subtracting) n times 2.3026.

NOTE 2:

$$\log_e x = 2.3026 \log_{10} x$$
$$\log_{10} x = 0.4343 \log_e x$$
$$\log_e 10 = 2.3026$$
$$\log_{10} e = 0.4343$$

n	$n \times 2.3026$
1	2.3026 = 0.6974–3
2	4.6052 = 0.3948–5
3	6.9078 = 0.0922–7
4	9.2103 = 0.7897–10
5	11.5129 = 0.4871–12
6	13.8155 = 0.1845–14
7	16.1181 = 0.8819–17
8	18.4207 = 0.5793–19
9	20.7233 = 0.2767–21

* From A. E. Knowlton, *Standard Handbook for Electrical Engineers*, 9th ed., McGraw-Hill, New York, 1957.

Table F-5. Natural, Napierian, or Hyperbolic Logarithms (*Continued*)

N	0	1	2	3	4	5	6	7	8	9
500	6.2146	6.2166	6.2186	6.2206	6.2226	6.2246	6.2265	6.2285	6.2305	6.2324
510	6.2344	6.2364	6.2383	6.2403	6.2422	6.2442	6.2461	6.2480	6.2500	6.2519
520	6.2538	6.2558	6.2577	6.2596	6.2615	6.2634	6.2653	6.2672	6.2691	6.2710
530	6.2729	6.2748	6.2766	6.2785	6.2804	6.2823	6.2841	6.2860	6.2879	6.2897
540	6.2916	6.2934	6.2953	6.2971	6.2989	6.3008	6.3026	6.3044	6.3063	6.3081
550	6.3099	6.3117	6.3135	6.3154	6.3172	6.3190	6.3208	6.3226	6.3244	6.3261
560	6.3279	6.3297	6.3315	6.3333	6.3351	6.3368	6.3386	6.3404	6.3421	6.3439
570	6.3456	6.3474	6.3491	6.3509	6.3256	6.3544	6.3561	6.3578	6.3596	6.3613
580	6.3630	6.3648	6.3665	6.3682	6.3699	6.3716	6.3733	6.3750	6.3767	6.3784
590	6.3801	6.3818	6.3835	6.3852	6.3869	6.3886	6.3902	6.3919	6.3936	6.3953
600	6.3969	6.3986	6.4003	6.4019	6.4036	6.4052	6.4069	6.4085	6.4102	6.4118
610	6.4135	6.4151	6.4167	6.4184	6.4200	6.4216	6.4232	6.4249	6.4265	6.4281
620	6.4297	6.4313	6.4329	6.4345	6.4362	6.4378	6.4394	6.4409	6.4425	6.4441
630	6.4457	6.4473	6.4489	6.4505	6.4520	6.4536	6.4552	6.4568	6.4583	6.4599
640	6.4615	6.4630	6.4646	6.4661	6.4677	6.4693	6.4708	6.4723	6.4739	6.4754
650	6.4770	6.4785	6.4800	6.4816	6.4831	6.4846	6.4862	6.4877	6.4892	6.4907
660	6.4922	6.4938	6.4953	6.4968	6.4983	6.4998	6.5013	6.5028	6.5043	6.5058
670	6.5073	6.5088	6.5103	6.5117	6.5132	6.5147	6.5162	6.5177	6.5191	6.5206
680	6.5221	6.5236	6.5250	6.5265	6.5280	6.5294	6.5309	6.5323	6.5338	6.5352
690	6.5367	6.5381	6.5396	6.5410	6.5425	6.5439	6.5453	6.5468	6.5482	6.5497
700	6.5511	6.5525	6.5539	6.5554	6.5568	6.5582	6.5596	6.5610	6.5624	6.5639
710	6.5653	6.5667	6.5681	6.5695	6.5709	6.5723	6.5737	6.5751	6.5765	6.5779
720	6.5793	6.5806	6.5820	6.5834	6.5848	6.5862	6.5876	6.5889	6.5903	6.5917
730	6.5930	6.5944	6.5958	6.5971	6.5985	6.5999	6.6012	6.6026	6.6039	6.6053
740	6.6067	6.6080	6.6093	6.6107	6.6120	6.6134	6.6147	6.6161	6.6174	6.6187
750	6.6201	6.6214	6.6227	6.6241	6.6254	6.6267	6.6280	6.6294	6.6307	6.6320
760	6.6333	6.6346	6.6359	6.6373	6.6386	6.6399	6.6412	6.6425	6.6438	6.6451
770	6.6464	6.6477	6.6490	6.6503	6.6516	6.6529	6.6542	6.6554	6.6567	6.6580
780	6.6593	6.6606	6.6619	6.6631	6.6644	6.6657	6.6670	6.6682	6.6695	6.6708
790	6.6720	6.6733	6.6746	6.6758	6.6771	6.6783	6.6796	6.6809	6.6821	6.6834
800	6.6846	6.6859	6.6871	6.6884	6.6896	6.6908	6.6921	6.6933	6.6946	6.6958
810	6.6970	6.6983	6.6995	6.7007	6.7020	6.7032	6.7044	6.7056	6.7069	6.7081
820	6.7093	6.7105	6.7117	6.7130	6.7142	6.7154	6.7166	6.7178	6.7190	6.7202
830	6.7214	6.7226	6.7238	6.7250	6.7262	6.7274	6.7286	6.7298	6.7310	6.7322
840	6.7334	6.7346	6.7358	6.7370	6.7382	6.7393	6.7405	6.7417	6.6429	6.7441
850	6.7452	6.7464	6.7476	6.7488	6.7499	6.7511	6.7523	6.7534	6.7546	6.7558
860	6.7569	6.7581	6.7593	6.7604	6.7616	6.7627	6.7639	6.7650	6.7662	6.7673
870	6.7685	6.7696	6.7708	6.7719	6.7731	6.7742	6.7754	6.7765	6.7776	6.7788
880	6.7799	6.7811	6.7822	6.7833	6.7845	6.7856	6.7867	5.7878	6.7890	6.7901
890	6.7912	6.7923	6.7935	6.7946	6.7957	6.7968	6.7979	6.7991	6.8002	6.8013
900	6.8024	6.8035	6.8046	6.8057	6.8068	6.8079	6.8090	6.8101	6.8112	6.8123
910	6.8134	6.8145	6.8156	6.8167	6.8178	6.8189	6.8200	6.8211	6.8222	6.8233
920	6.8244	6.8255	6.8265	6.8276	6.8287	6.8298	6.8309	6.8320	6.8330	6.8341
930	6.8352	6.8363	6.8373	6.8384	6.8395	6.8405	6.8416	6.8427	6.8437	6.8448
940	6.8459	6.8469	6.8480	6.8491	6.8501	6.8512	6.8522	6.8533	6.8544	6.8554
950	6.8565	6.8575	6.8586	6.8596	6.8607	6.8617	6.8628	6.8638	6.8648	6.8659
960	6.8669	6.8680	6.8690	6.8701	6.8711	6.8721	6.8732	6.8742	6.8752	6.8763
970	6.8773	6.8783	6.8794	6.8804	6.8814	6.8824	6.8835	6.8845	6.8855	6.8865
980	6.8876	6.8886	6.8896	6.8906	6.8916	6.8926	6.8937	6.8947	6.8957	6.8967
990	6.8977	6.8987	6.8997	6.9007	6.9017	6.9027	6.9037	6.9047	6.9057	6.9068

Table F-6a. Sine Integral $Si(x)$*

$$Si(x) = \int_0^x \frac{\sin u}{u}\,du$$

x	$Si(x)$	(x)	$Si(x)$	x	$Si(x)$	x	$Si(x)$	x	$Si(x)$	x	$Si(x)$
0.0	0.00000	5.0	1.54993	10.0	1.65835	15.0	1.61819	20.0	1.54824	25.0	1.53148
0.1	0.09994	5.1	1.53125	10.1	1.65253	15.1	1.62226	20.1	1.55289	50.0	1.55162
0.2	0.19956	5.2	1.51367	10.2	1.64600	15.2	1.62575	20.2	1.55767		
0.3	0.29850	5.3	1.49732	10.3	1.63883	15.3	1.62865	20.3	1.56253		
0.4	0.39646	5.4	1.48230	10.4	1.63112	15.4	1.63093	20.4	1.56743		
0.5	0.49311	5.5	1.46872	10.5	1.62294	15.5	1.63258	20.5	1.57232		
0.6	0.58813	5.6	1.45667	10.6	1.61439	15.6	1.63359	20.6	1.57714		
0.7	0.68122	5.7	1.44620	10.7	1.60556	15.7	1.63396	20.7	1.58186		
0.8	0.77210	5.8	1.43736	10.8	1.59654	15.8	1.63370	20.8	1.58641		
0.9	0.86047	5.9	1.43018	10.9	1.58743	15.9	1.63280	20.9	1.59077		
1.0	0.94608	6.0	1.42469	11.0	1.57831	16.0	1.63130	21.0	1.59489		
1.1	1.02869	6.1	1.42087	11.1	1.56927	16.1	1.62921	21.1	1.59873		
1.2	1.10805	6.2	1.41871	11.2	1.56042	16.2	1.62657	21.2	1.60225		
1.3	1.18396	6.3	1.41817	11.3	1.55182	16.3	1.62339	21.3	1.60543		
1.4	1.25623	6.4	1.41922	11.4	1.54356	16.4	1.61973	21.4	1.60823		
1.5	1.32468	6.5	1.42179	11.5	1.53571	16.5	1.61563	21.5	1.61063		
1.6	1.38918	6.6	1.42582	11.6	1.52835	16.6	1.61112	21.6	1.61261		
1.7	1.44959	6.7	1.43121	11.7	1.52155	16.7	1.60627	21.7	1.61415		
1.8	1.50582	6.8	1.43787	11.8	1.51535	16.8	1.60111	21.8	1.61525		
1.9	1.55778	6.9	1.44570	11.9	1.50981	16.9	1.59572	21.9	1.61590		
2.0	1.60541	7.0	1.45460	12.0	1.50497	17.0	1.59014	22.0	1.61608		
2.1	1.64870	7.1	1.46443	12.1	1.50088	17.1	1.58443	22.1	1.61582		
2.2	1.68763	7.2	1.47509	12.2	1.49755	17.2	1.57865	22.2	1.61510		
2.3	1.72221	7.3	1.48644	12.3	1.49501	17.3	1.57285	22.3	1.61395		
2.4	1.75249	7.4	1.49834	12.4	1.49327	17.4	1.56711	22.4	1.61238		
2.5	1.77852	7.5	1.51068	12.5	1.49234	17.5	1.56146	22.5	1.61041		
2.6	1.80039	7.6	1.52331	12.6	1.49221	17.6	1.55598	22.6	1.60806		
2.7	1.81821	7.7	1.53611	12.7	1.49287	17.7	1.55070	22.7	1.60536		
2.8	1.83210	7.8	1.54894	12.8	1.49430	17.8	1.54568	22.8	1.60234		
2 9	1.84219	7.9	1.56167	12.9	1.49647	17.9	1.54097	22.9	1.59902		
3.0	1.84865	8.0	1.57419	13.0	1.49936	18.0	1.53661	23.0	1.59546		
3.1	1.85166	8.1	1.58637	13.1	1.50292	18.1	1.53264	23.1	1.59168		
3.2	1.85140	8.2	1.59810	13.2	1.50711	18.2	1.52909	23.2	1.58772		
3.3	1.84808	8.3	1.60928	13.3	1.51188	18.3	1.52600	23.3	1.58363		
3.4	1.84191	8.4	1.61981	13.4	1.51716	18.4	1.52339	23.4	1.57945		
3.5	1.83313	8.5	1.62960	13.5	1.52291	18.5	1.52128	23.5	1.57521		
3.6	1.82195	8.6	1.63857	13.6	1.52905	18.6	1.51969	23.6	1.57097		
3.7	1.80862	8.7	1.64665	13.7	1.53352	18.7	1.51863	23.7	1.56676		
3.8	1.79333	8.8	1.65379	13.8	1.54225	18.8	1.51810	23.8	1.56262		
3.9	1.77650	8.9	1.65993	13.9	1.54917	18.9	1.51810	23.9	1.55860		
4.0	1.75820	9.0	1.66504	14.0	1.55621	19.0	1.51863	24.0	1.55474		
4.1	1.73874	9.1	1.66908	14.1	1.56330	19.1	1.51967	24.1	1.55107		
4.2	1.71837	9.2	1.67205	14.2	1.57036	19.2	1.52122	24.2	1.54762		
4.3	1.69732	9.3	1.67393	14.3	1.57733	19.3	1.52324	24.3	1.54444		
4.4	1.67583	9.4	1.67473	14.4	1.58414	19.4	1.52572	24.4	1.54154		
4.5	1.65414	9.5	1.67446	14.5	1.59072	19.5	1.52863	24.5	1.53897		
4.6	1.63246	9.6	1.67316	14.6	1.59702	19.6	1.53192	24.6	1.53672		
4.7	1.61101	9.7	1.67084	14.7	1.60296	19.7	1.53357	24.7	1.53484		
4.8	1.58998	9.8	1.66757	14.8	1.60851	19.8	1.53954	24.8	1.53333		
4.9	1.56956	9.9	1.66338	14.9	1.61360	19.9	1.54378	24.9	1.53221		

* From P. O. Pedersen, *Radiation from a Vertical Antenna over Flat Perfectly Conducting Earth*, G. E. C. Gad, Copenhagen.

Table F-6b. $S_1(x)$ **and Cosine Integral** $Ci(x)$*

$$S_1(x) = \log_e x + 0.5772 - Ci(x) = \int_0^x \frac{1 - \cos x}{x}\,dx$$

$$Ci(x) = -\int_x^\infty \frac{\cos u}{u}\,du = \log_e x + 0.5772 - S_1(x)$$

x	$S_1(x)$	x	$S_1(x)$	x	$S_1(x)$	x	$S_1(x)$	x	$S_1(x)$	x	$S_1(x)$
0.0	0.00000	5.0	2.37669	10.0	2.92527	15.0	3.23899	20.0	3.52853	25.0	3.80295
0.1	0.00249	5.1	2.38994	10.1	2.94327	15.1	3.25090	20.1	3.53173	50.0	4.49486
0.2	0.00998	5.2	2.40113	10.2	2.96050	15.2	3.26308	20.2	3.53535		
0.3	0.02241	5.3	2.41044	10.3	2.97688	15.3	3.27552	20.3	3.53946		
0.4	0.03973	5.4	2.41801	10.4	2.99234	15.4	3.28814	20.4	3.54402		
0.5	0.06185	5.5	2.42402	10.5	3.00688	15.5	3.30087	20.5	3.54905		
0.6	0.08866	5.6	2.42866	10.6	3.02045	15.6	3.31363	20.6	3.55456		
0.7	0.12002	5.7	2.43210	10.7	3.03300	15.7	3.32641	20.7	3.56049		
0.8	0.15579	5.8	2.43452	10.8	3.04457	15.8	3.33911	20.8	3.56687		
0.9	0.19578	5.9	2.43610	10.9	3.05514	15.9	3.35167	20.9	3.57368		
1.0	0.23981	6.0	2.43704	11.0	3.06467	16.0	3.36401	21.0	3.58085		
1.1	0.28766	6.1	2.43749	11.1	3.07323	16.1	3.37612	21.1	3.58840		
1.2	0.33908	6.2	2.43764	11.2	3.08083	16.2	3.38790	21.2	3.59629		
1.3	0.39384	6.3	2.43766	11.3	3.08749	16.3	3.39932	21.3	3.60446		
1.4	0.45168	6.4	2.43770	11.4	3.09322	16.4	3.41032	21.4	3.61288		
1.5	0.51233	6.5	2.43792	11.5	3.09814	16.5	3.42088	21.5	3.62155		
1.6	0.57549	6.6	2.43847	11.6	3.10225	16.6	3.43096	21.6	3.63037		
1.7	0.64088	6.7	2.43947	11.7	3.10561	16.7	3.44050	21.7	3.63935		
1.8	0.70820	6.8	2.44106	11.8	3.10828	16.8	3.44947	21.8	3.64842		
1.9	0.77713	6.9	2.44335	11.9	3.11038	16.9	3.45788	21.9	3.65751		
2.0	0.84739	7.0	2.44643	12.0	3.11190	17.0	3.46568	22.0	3.66662		
2.1	0.91865	7.1	2.45040	12.1	3.11301	17.1	3.47288	22.1	3.67568		
2.2	0.99060	7.2	2.45534	12.2	3.11370	17.2	3.47945	22.2	3.68465		
2.3	1.06295	7.3	2.46130	12.3	3.11412	17.3	3.48543	22.3	3.69348		
2.4	1.13540	7.4	2.46834	12.4	3.11429	17.4	3.49077	22.4	3.70216		
2.5	1.20764	7.5	2.47649	12.5	3.11436	17.5	3.49553	22.5	3.71059		
2.6	1.27939	7.6	2.48577	12.6	3.11437	17.6	3.49969	22.6	3.71879		
2.7	1.35038	7.7	2.49619	12.7	3.11438	17.7	3.50330	22.7	3.72670		
2.8	1.42035	7.8	2.50775	12.8	3.11453	17.8	3.50639	22.8	3.73427		
2.9	1.48903	7.9	2.52044	12.9	3.11484	17.9	3.50895	22.9	3.74153		
3.0	1.55620	8.0	2.53423	13.0	3.11540	18.0	3.51107	23.0	3.74838		
3.1	1.62163	8.1	2.54906	13.1	3.11628	18.1	3.51276	23.1	3.75483		
3.2	1.68511	8.2	2.56491	13.2	3.11754	18.2	3.51404	23.2	3.76089		
3.3	1.74646	8.3	2.56171	13.3	3.11924	18.3	3.51500	23.3	3.76651		
3.4	1.80552	8.4	2.59938	13.4	3.12142	18.4	3.51568	23.4	3.77170		
3.5	1.86211	8.5	2.61786	13.5	3.12414	18.5	3.51610	23.5	3.77644		
3.6	1.91613	8.6	2.63704	13.6	3.12745	18.6	3.51633	23.6	3.78072		
3.7	1.96745	8.7	2.65686	13.7	3.13134	18.7	3.51645	23.7	3.78459		
3.8	2.01600	8.8	2.67721	13.8	3.13587	18.8	3.51648	23.8	3.78801		
3.9	2.06170	8.9	2.69799	13.9	3.14104	18.9	3.51648	23.9	3.79101		
4.0	2.10449	9.0	2.71909	14.0	3.14688	19.0	3.51660	24.0	3.79360		
4.1	2.14438	9.1	2.74042	14.1	3.15338	19.1	3.51661	24.1	3.79582		
4.2	2.18131	9.2	2.76186	14.2	3.16054	19.2	3.51685	24.2	3.79767		
4.3	2.21535	9.3	2.78332	14.3	3.16835	19.3	3.51727	24.3	3.79917		
4.4	2.24648	9.4	2.80468	14.4	3.17677	19.4	3.51790	24.4	3.80036		
4.5	2.27479	9.5	2.82583	14.5	3.18583	19.5	3.51879	24.5	3.80129		
4.6	2.30033	9.6	2.84669	14.6	3.19545	19.6	3.52002	24.6	3.80197		
4.7	2.32317	9.7	2.86713	14.7	3.20564	19.7	3.52156	24.7	3.80243		
4.8	2.34344	9.8	2.88712	14.8	3.21630	19.8	3.52348	24.8	3.80271		
4.9	2.36124	9.9	2.90651	14.9	3.22746	19.9	3.52578	24.9	3.80288		

* From P. O. Pedersen, *Radiation from a Vertical Antenna over Flat Perfectly Conducting Earth*, G. E. C. Gad, Copenhagen.

Table F-7. Exponential and Related Integrals†

The Exponential Integrals. Functions usually used in combination are (1) the exponential integrals $Ei(-x)$ and $\overline{Ei}(x)$ *or* (2) $E_0(x)$, $E_1(x)$, and $E_2(x)$.* $E_1(x) \equiv -Ei(-x)$. In all cases x is positive.

Small Values of the Argument. Where interpolation is unsatisfactory, use the formulas or series solutions given in the tables.

Large Values of the Argument for $E_1(x) = -Ei(-x)$.

a. For $5 < x < 16$ see special tables following main tables.

b. For $5 < x < 40$, $E_1(x) = -Ei(-x) = F_1e^{-x}/x$ and $\overline{Ei}(x) = F_2e^x/x$, where F_1 and F_2 are given below.

	5	10	15	20	25	30	35	40
F_1	0.8516	0.9156	0.9408	0.9549	0.9627	0.9687	0.9729	0.9753
F_2	1.354	1.1316	1.0781	1.0560	1.0440	1.0358	1.0305	1.0264

c. For large values of x, F_1 and F_2 can be computed from the semiconvergent series

$$F_1 = 1 - \frac{1!}{x} + \frac{2!}{x^2} - \frac{3!}{x^3} + \cdots \qquad F_2 = 1 + \frac{1!}{x} + \frac{2!}{x^2} + \frac{3!}{x^3} + \cdots$$

Large Values of the Argument for $E_0(x)$ and $E_2(x)$. Use the formulas given in the tables, i.e., $E_0(x) = e^{-x}/x$ and $E_2(x) = e^x - xE_1(x)$.

$$E_0(x) = e^{-x}/x$$

x	0	1	2	3	4	5	6	7	8	9
0.0	∞	99.00	49.01	32.35	24.02	19.02	15.70	13.32	11.54	10.15
1	9.048	8.144	7.391	6.755	6.210	5.738	5.326	4.963	4.640	4.352
2	4.094	3.860	3.648	3.454	3.278	3.115	2.966	2.827	2.699	2.580
3	2.469	2.366	2.269	2.179	2.093	2.013	1.938	1.867	1.800	1.736
4	1.676	1.619	1.564	1.513	1.464	1.417	1.372	1.330	1.289	1.250
5	1.213	1.177	1.143	1.111	1.079	1.049	1.020	*9921	*9653	*9395
6	0.9147	8907	8677	8454	8239	8031	7831	7637	7450	7269
7	7094	6925	6760	6601	6447	6298	6154	6013	5877	5745
8	5617	5492	5371	5254	5139	5028	4920	4816	4713	4614
9	4517	4423	4332	4243	4156	4071	3988	3908	3830	3753
1.0	3679	3606	3535	3466	3399	3333	3268	3206	3144	3085
1	3026	2969	2913	2859	2805	2753	2702	2653	2604	2556
2	2510	2464	2420	2376	2334	2292	2251	2211	2172	2134
3	2096	2060	2024	1989	1954	1920	1887	1855	1823	1792
4	1761	1732	1722	1673	1645	1618	1591	1564	1538	1513
5	1488	1463	1439	1415	1392	1369	1347	1325	1304	1283
6	1262	1242	1222	1202	1183	1164	1145	1127	1109	1092
7	1075	1058	1041	1025	1009	*9930	*9775	*9623	*9474	*9327
8	0.09183	9042	8903	8766	8631	8500	8370	8242	8116	7993
9	7872	7753	7636	7521	7407	7296	7187	7079	6973	6869
0.0	∞	9.048	4.094	2.469	1.676	1.213	*9147	*7094	*5617	*4517
1.0	0.3679	3026	2510	2096	1761	1488	1262	1075	*9183	*7872
2.0	$10^{-1} \times 0.6767$	5831	5037	4359	3780	3283	2857	2489	2172	1897
3.0	$10^{-2} \times 1.660$	1.453	1.274	1.118	0.9816	8628	7590	6682	5887	5190
4.0	$10^{-2} \times 0.4579$	4042	3570	3155	2790	2469	2185	1935	1714	1520
5.0	$10^{-3} \times 1.348$	1.195	1.061	0.9418	8364	7431	6603	5870	5220	4643
6.0	$10^{-3} \times 0.4131$	3677	3273	2915	2596	2313	2061	1837	1638	1461
7.0	$10^{-4} \times 1.303$	1.1621	1.037	0.9254	8260	7375	6585	5881	5253	4693
8.0	$10^{-4} \times 0.4193$	3747	3349	2994	2677	2394	2141	1915	1713	1533
9.0	$10^{-5} \times 1.371$	1.227	1.098	0.9831	8800	7879	7055	6318	5658	5068

* The functions are represented by the general formula $E_n(x) = \int_1^\infty e^{-xu}u^{-n}\,du$. These functions and the exponential integrals can be expressed in various alternative forms.

† From H. E. Etherington, *Nuclear Engineering Handbook*, McGraw-Hill, New York, 1958.

Table F-7. Exponential and Related Integrals (*Continued*)

$E_1(x) = -Ei(-x)$

x	0	1	2	3	4	5	6	7	8	9
0.0	∞	4.038	3.355	2.959	2.681	2.468	2.295	2.151	2.027	1.919
1	1.823	1.737	1.660	1.589	1.524	1.464	1.409	1.358	1.310	1.265
2	1.223	1.183	1.145	1.110	1.076	1.044	1.014	0.9849	9573	9309
3	0.9057	8815	8583	8361	8147	7942	7745	7554	7371	7194
4	7024	6859	6700	6546	6397	6253	6114	5979	5848	5721
5	5598	5478	5362	5250	5140	5034	4930	4830	4732	4636
6	4544	4454	4366	4280	4197	4115	4036	3959	3883	3810
7	3738	3668	3599	3532	3467	3403	3341	3280	3221	3163
8	3106	3050	2996	2943	2891	2840	2790	2742	2694	2647
9	2602	2557	2513	2470	2429	2387	2347	2308	2269	2231
1.0	2194	2157	2122	2087	2052	2019	1986	1953	1922	1890
1	1860	1830	1801	1772	1743	1716	1688	1662	1635	1609
2	1584	1559	1535	1511	1487	1464	1441	1419	1397	1376
3	1355	1334	1313	1293	1274	1254	1235	1216	1198	1180
4	1162	1145	1128	1111	1094	1078	1062	1046	1030	1015
5	1000	*9854	*9709	*9567	*9426	*9288	*9152	*9019	*8887	*8758
6	0.08631	8506	8383	8261	8142	8025	7909	7796	7684	7574
7	7465	7359	7254	7151	7049	6949	6850	6753	6658	6564
8	6471	6380	6290	6202	6115	6029	5945	5862	5780	5700
9	5620	5542	5465	5390	5315	5241	5169	5098	5027	4958
2.0	4890	4823	4757	4692	4627	4564	4502	4440	4380	4320
1	4261	4204	4147	4090	4035	3980	3927	3874	3821	3770
2	3719	3669	3620	3571	3523	3476	3430	3384	3339	3294
3	3250	3207	3164	3122	3081	3040	3000	2960	2921	2882
4	2844	2806	2769	2733	2697	2662	2627	2592	2558	2525
5	2491	2459	2427	2395	2364	2333	2303	2273	2243	2214
6	2185	2157	2129	2101	2074	2047	2021	1994	1969	1943
7	1918	1893	1869	1845	1821	1798	1775	1752	1730	1707
8	1686	1664	1643	1622	1601	1581	1560	1540	1521	1502
9	1482	1464	1445	1427	1409	1391	1373	1356	1338	1322
3.0	1305	1288	1272	1256	1240	1225	1209	1194	1179	1164
1	1149	1135	1121	1107	1093	1079	1066	1052	1039	1026
2	1013	1001	*9882	*9758	*9637	*9517	*9398	*9281	*9166	*9052
3	0.008939	8828	8718	8610	8503	8398	8294	8191	8090	7990
4	7891	7793	7697	7602	7508	7416	7324	7234	7145	7057
5	6970	6884	6800	6716	6634	6552	6472	6393	6314	6237
6	6160	6085	6011	5937	5864	5793	5722	5652	5583	5515
7	5448	5381	5316	5251	5187	5124	5062	5000	4939	4879
8	4820	4762	4704	4647	4591	4535	4480	4426	4372	4319
9	4267	4216	4165	4114	4065	4016	3967	3919	3872	3825
4.0	3779	3734	3689	3645	3601	3557	3515	3472	3431	3390
1	3349	3309	3269	3230	3191	3153	3115	3078	3041	3005
2	2969	2933	2898	2864	2829	2796	2762	2729	2697	2665
3	2633	2602	2571	2540	2510	2480	2450	2421	2393	2364
4	2336	2308	2281	2254	2227	2201	2175	2149	2123	2098
5	2073	2049	2025	2001	1977	1954	1931	1908	1885	1863
6	1841	1819	1798	1777	1756	1735	1715	1694	1674	1655
7	1635	1616	1597	1578	1560	1541	1523	1505	1488	1470
8	1453	1436	1419	1402	1386	1370	1354	1338	1322	1307
9	1291	1276	1261	1247	1232	1218	1204	1189	1176	1162
5.0	1148	1135	1122	1109	1096	1083	1070	1058	1045	1033

$$E_1(x) = \int_1^\infty e^{-xu}u^{-1}\,du = -Ei(-x) = \int_x^\infty e^{-u}u^{-1}\,du$$

$$= -\left[\ln \gamma x - \frac{x}{1 \cdot 1!} + \frac{x^2}{2 \cdot 2!} - \frac{x^3}{3 \cdot 3!} + \cdots\right]$$

where $\ln \gamma x = 0.5772 + \ln x$

Table F-7. Exponential and Related Integrals (*Continued*)

$\overline{Ei}(x) = Ei(x)$

x	0	1	2	3	4	5	6	7	8	9
0.0	– ∞	4.018	3.315	2.899	2.601	2.368	2.175	2.011	1.867	1.739
1	– 1.623	1.517	1.419	1.329	1.244	1.164	1.089	1.017	.9491	.8841
2	–0.8218	7619	7042	6485	5947	5425	4919	4427	3949	3482
3	– .3027	2582	2147	1721	1304	0894	0493	0098	*0290	*0672
4	+ .1048	1418	1783	2143	2498	2849	3195	3537	3876	4211
5	.4542	4870	5195	5517	5836	6153	6467	6778	7087	7394
6	.7699	8002	8302	8601	8898	9194	9488	9780	1.007	1.036
7	1.065	1.094	1.122	1.151	1.179	1.207	1.236	1.264	1.292	1.320
8	1.347	1.375	1.403	1.431	1.458	1.486	1.513	1.541	1.568	1.595
9	1.623	1.650	1.677	1.705	1.732	1.759	1.786	1.814	1.841	1.868
1.0	1.895	1.922	1.949	1.977	2.004	2.031	2.058	2.086	2.113	2.140
1	2.167	2.195	2.222	2.249	2.277	2.304	2.332	2.359	2.387	2.414
2	2.442	2.470	2.498	2.525	2.553	2.581	2.609	2.637	2.665	2.693
3	2.721	2.750	2.778	2.806	2.835	2.863	2.892	2.921	2.949	2.978
4	3.007	3.036	3.065	3.094	3.124	3.153	3.183	3.212	3.242	3.271
5	3.301	3.331	3.361	3.391	3.422	3.452	3.482	3.513	3.544	3.574
6	3.605	3.636	3.667	3.699	3.730	3.762	3.793	3.825	3.857	3.889
7	3.921	3.953	3.986	4.018	4.051	4.084	4.117	4.150	4.183	4.216
8	4.250	4.284	4.317	4.351	4.386	4.420	4.454	4.489	4.524	4.559
9	4.594	4.629	4.664	4.700	4.736	4.772	4.808	4.844	4.881	4.917
2.0	4.954	4.991	5.028	5.066	5.104	5.141	5.179	5.217	5.256	5.294
1	5.333	5.372	5.411	5.451	5.490	5.530	5.570	5.611	5.651	5.692
2	5.733	5.774	5.815	5.857	5.899	5.941	5.983	6.025	6.068	6.111
3	6.154	6.198	6.242	6.286	6.330	6.374	6.419	6.464	6.509	6.555
4	6.601	6.647	6.693	6.740	6.787	6.834	6.881	6.929	6.977	7.025
5	7.074	7.123	7.172	7.221	7.271	7.321	7.372	7.422	7.473	7.524
6	7.576	7.628	7.680	7.733	7.786	7.839	7.893	7.947	8.001	8.055
7	8.110	8.166	8.221	8.277	8.334	8.390	8.447	8.505	8.563	8.621
8	8.679	8.738	7.798	8.857	8.917	8.978	9.039	9.100	9.162	9.224
9	9.286	9.349	9.412	9.476	9.540	9.605	9.670	9.735	9.801	9.867
3.0	9.934	10.00	10.07	10.14	10.21	10.27	10.34	10.41	10.48	10.55
1	10.63	10.70	10.77	10.84	10.92	10.99	11.06	11.14	11.22	11.29
2	11.37	11.44	11.52	11.60	11.68	11.76	11.84	11.92	12.00	12.08
3	12.16	12.24	12.33	12.41	12.49	12.58	12.66	12.75	12.84	12.92
4	13.01	13.10	13.19	13.28	13.37	13.46	13.55	13.64	13.74	13.83
5	13.93	14.02	14.12	14.21	14.31	14.41	14.51	14.60	14.70	14.80
6	14.91	15.01	15.11	15.21	15.32	15.42	15.53	15.64	15.74	15.85
7	15.96	16.07	16.18	16.29	16.40	16.52	16.63	16.75	16.86	16.98
8	17.09	17.21	17.33	17.45	17.57	17.69	17.82	17.94	18.06	18.19
9	18.32	18.44	18.57	18.70	18.83	18.96	19.09	19.23	19.36	19.49
4.0	19.63	19.77	19.91	20.05	20.19	20.33	20.47	20.61	20.76	20.90
1	21.05	21.20	21.35	21.50	21.65	21.80	21.95	22.11	22.26	22.42
2	22.58	22.74	22.90	23.06	23.22	23.39	23.55	23.72	23.89	24.06
3	24.23	24.40	24.57	24.75	24.92	25.10	25.28	25.46	25.64	25.82
4	26.01	26.19	26.38	26.57	26.76	26.95	27.15	27.34	27.54	27.73
5	27.93	28.13	28.34	28.54	28.75	28.95	29.16	29.37	29.58	29.80
6	30.01	30.23	30.45	30.67	30.89	31.12	31.34	31.57	31.80	32.03
7	32.26	32.50	32.74	32.97	33.21	33.46	33.70	33.95	34.20	34.45
8	34.70	34.95	35.21	35.47	35.73	35.99	36.25	36.52	36.79	37.06
9	37.33	37.61	37.88	38.16	38.45	38.73	39.02	39.31	39.60	39.89
5.0	40.19	40.48	40.78	41.09	41.39	41.70	42.01	42.32	42.64	42.96

$$\overline{Ei}(x) = \ln \gamma x + \frac{x}{1 \cdot 1!} + \frac{x^2}{2 \cdot 2!} + \frac{x^3}{3 \cdot 3!} + \cdots$$

where $\ln \gamma x = 0.5772 + \ln x$

Table F-7. Exponential and Related Integrals (*Continued*)

$E_1(x) = -Ei(-x),\ 5 < x < 16$

x	0	1	2	3	4	5	6	7	8	9
5	$10^{-3} \times$ 1.148	1.021	9086	8086	7198	6409	5708	5085	4532	4039
6	$10^{-3} \times$ 0.3601	3211	2864	2555	2279	2034	1816	1621	1448	1293
7	$10^{-4} \times$ 1.155	1.032	9219	8239	7364	6583	5886	5263	4707	4210
8	$10^{-4} \times$ 0.3767	3370	3015	2699	2415	2162	1936	1733	1552	1390
9	$10^{-5} \times$ 1.245	1.115	9988	8948	8018	7185	6439	5771	5173	4637
10	$10^{-5} \times$ 0.4157	3727	3342	2997	2687	2410	2162	1939	1740	1561
11	$10^{-6} \times$ 1.400	1.256	1.127	1.012	0.9080	8149	7315	6566	5894	5291
12	$10^{-6} \times$ 0.4751	4266	3830	3440	3089	2774	2491	2238	2010	1805
13	$10^{-7} \times$ 1.622	1.457	1.309	1.176	1.057	0.9495	8532	7667	6890	6193
14	$10^{-7} \times$ 0.5566	5002	4500	4042	3633	3266	2936	2640	2373	2134

$\overline{Ei}(x) = Ei(x),\ 5 < x < 16$

x	0	1	2	3	4	5	6	7	8	9
5	$10^{2} \times$ 0.4019	4328	4662	5026	5419	5847	6310	6813	7360	7954
6	$10^{2} \times$ 0.8599	9300	1.006	1.089	1.179	1.277	1.384	1.501	1.627	1.765
7	$10^{3} \times$ 0.1915	2079	2257	2451	2663	2894	3146	3420	3720	4047
8	$10^{3} \times$ 0.4404	4793	5218	5682	6189	6743	7347	8007	8729	9517
9	$10^{4} \times$ 0.1038	1132	1235	1347	1471	1605	1752	1913	2089	2282
10	$10^{4} \times$ 0.2492	2723	2975	3251	3553	3884	4246	4642	5076	5551
11	$10^{4} \times$ 0.6071	6641	7265	7949	8698	9518	1.042	1.140	1.248	1.366
12	$10^{5} \times$ 0.1496	1638	1794	1964	2151	2357	2581	2828	3098	3395
13	$10^{5} \times$ 0.3720	4076	4467	4896	5367	5883	6449	7070	7751	8499
14	$10^{5} \times$ 0.9319	1.022	1.121	1.229	1.348	1.479	1.622	1.779	1.952	2.142

$$E_2(x) = \int_1^\infty e^{-xu}u^{-2}\,du = e^{x} - xE_1(x) = x[E_0(x) - E_1(x)]$$

x	0	1	2	3	4	5	6	7	8	9
0.0	1.000	0.9497	9131	8817	8535	8278	8040	7818	7610	7412
1	0.7225	7048	6878	6715	6560	6410	6267	6128	5995	5866
2	5742	5622	5505	5393	5283	5177	5074	4974	4877	4783
3	4691	4602	4515	4430	4348	4267	4189	4112	4038	3963
4	3894	3824	3756	3690	3626	3562	3500	3440	3381	3325
5	3266	3211	3157	3104	3052	3001	2951	2902	2855	2808
6	2762	2717	2673	2630	2587	2546	2505	2465	2426	2387
7	2349	2312	2276	2240	2205	2171	2137	2104	2072	2040
8	2009	1978	1948	1918	1889	1860	1832	1804	1777	1750
9	1724	1698	1673	1648	1623	1599	1576	1552	1530	1507
1.0	1485	1463	1442	1421	1400	1380	1360	1340	1321	1302
1	1283	1264	1246	1228	1211	1193	1176	1160	1143	1127
2	1111	1095	1080	1065	1050	1035	1020	1006	*9920	*9781
3	0.09645	9510	9378	9247	9119	8993	8868	8746	8625	8506
4	8389	8274	8160	8048	7938	7829	7722	7617	7513	7411
5	7310	7211	7113	7017	6922	6828	6736	6645	6555	6467
6	6380	6295	6210	6127	6045	5964	5884	5806	5729	5652
7	5577	5503	5430	5358	5287	5217	5148	5080	5013	4947
8	4882	4817	4754	4691	4630	4569	4509	4450	4392	4335
9	4278	4222	4167	4113	4059	4007	3955	3903	3852	3803
0	1.000	0.7225	5742	4691	3894	3266	2762	2349	2009	1724
1	$10^{-1} \times$ 1.485	1.283	1.111	0.9645	8389	7310	6380	5577	4882	4278
2	$10^{-1} \times$ 0.3753	3297	2898	2550	2246	1980	1746	1541	1362	1203
3	$10^{-2} \times$ 1.064	0.9417	8337	7384	6544	5802	5146	4567	4054	3600
4	$10^{-2} \times$ 0.3198	2842	2527	2247	1999	1779	1583	1410	1255	1118
5	$10^{-3} \times$ 0.9965	8881	7917	7060	6296	5617	5012	4473	3992	3564
6	$10^{-3} \times$ 0.3183	2842	2539	2268	2027	1812	1619	1448	1294	1157
7	$10^{-4} \times$ 1.035	0.9259	8283	7411	6632	5935	5313	4756	4258	3812
8	$10^{-4} \times$ 0.3414	3057	2738	2453	2198	1969	1764	1581	1417	1270
9	$10^{-5} \times$ 1.138	1.021	0.9149	8203	7356	6597	5916	5306	4760	4270

Table F-8. Complete Elliptic Integrals, K and E

$$K=\int_0^{\frac{\pi}{2}} \frac{dx}{\sqrt{1-k^2\sin^2 x}}; \quad E=\int_0^{\frac{\pi}{2}} \sqrt{1-k^2\sin^2 x}\, dx$$

$\sin^{-1}k$	K	E	$\sin^{-1}k$	K	E	$\sin^{-1}k$	K	E
0°	1.5708	1.5708	50°	1.9356	1.3055	81°.0	3.2553	1.0338
1	1.5709	1.5707	51	1.9539	1.2963	81.2	3.2771	1.0326
2	1.5713	1.5703	52	1.9729	1.2870	81.4	3.2995	1.0314
3	1.5719	1.5697	53	1.9927	1.2776	81.6	3.3223	1.0302
4	1.5727	1.5689	54	2.0133	1.2681	81.8	3.3458	1.0290
5	1.5738	1.5678	55	2.0347	1.2587	82.0	3.3699	1.0278
6	1.5751	1.5665	56	2.0571	1.2492	82.2	3.3946	1.0267
7	1.5767	1.5649	57	2.0804	1.2397	82.4	3.4199	1.0256
8	1.5785	1.5632	58	2.1047	1.2301	82.6	3.4460	1.0245
9	1.5805	1.5611	59	2.1300	1.2206	82.8	3.4728	1.0234
10	1.5828	1.5589	60	2.1565	1.2111	83.0	3.5004	1.0223
11	1.5854	1.5564	61	2.1842	1.2015	83.2	3.5288	1.0213
12	1.5882	1.5537	62	2.2132	1.1920	83.4	3.5581	1.0202
13	1.5913	1.5507	63	2.2435	1.1826	83.6	3.5884	1.0192
14	1.5946	1.5476	64	2.2754	1.1732	83.8	3.6196	1.0182
15	1.5981	1.5442	65	2.3088	1.1638	84.0	3.6519	1.0172
16	1.6020	1.5405	65.5	2.3261	1.1592	84.2	3.6852	1.0163
17	1.6061	1.5367	66.0	2.3439	1.1545	84.4	3.7198	1.0153
18	1.6105	1.5326	66.5	2.3622	1.1499	84.6	3.7557	1.0144
19	1.6151	1.5283	67.0	2.3809	1.1453	84.8	3.7930	1.0135
20	1.6200	1.5238	67.5	2.4001	1.1408	85.0	3.8317	1.0127
21	1.6252	1.5191	68.0	2.4198	1.1362	85.2	3.8721	1.0118
22	1.6307	1.5141	68.5	2.4401	1.1317	85.4	3.9142	1.0110
23	1.6365	1.5090	69.0	2.4610	1.1272	85.6	3.9583	1.0102
24	1.6426	1.5037	69.5	2.4825	1.1228	85.8	4.0044	1.0094
25	1.6490	1.4981	70.0	2.5046	1.1184	86.0	4.0528	1.0086
26	1.6557	1.4924	70.5	2.5273	1.1140	86.2	4.1037	1.0079
27	1.6627	1.4864	71.0	2.5507	1.1096	86.4	4.1574	1.0072
28	1.6701	1.4803	71.5	2.5749	1.1053	86.6	4.2142	1.0065
29	1.6777	1.4740	72.0	2.5998	1.1011	86.8	4.2744	1.0059
30	1.6858	1.4675	72.5	2.6256	1.0968	87.0	4.3387	1.0053
31	1.6941	1.4608	73.0	2.6521	1.0927	87.2	4.4073	1.0047
32	1.7028	1.4539	73.5	2.6796	1.0885	87.4	4.4811	1.0041
33	1.7119	1.4469	74.0	2.7081	1.0844	87.6	4.5609	1.0036
34	1.7214	1.4397	74.5	2.7375	1.0804	87.8	4.6477	1.0031
35	1.7312	1.4323	75.0	2.7681	1.0764	88.0	4.7427	1.0026
36	1.7415	1.4248	75.5	2.7998	1.0725	88.2	4.8478	1.0021
37	1.7522	1.4171	76.0	2.8327	1.0686	88.4	4.9654	1.0017
38	1.7633	1.4092	76.5	2.8669	1.0648	88.6	5.0988	1.0014
39	1.7748	1.4013	77.0	2.9026	1.0611	88.8	5.2527	1.0010
40	1.7868	1.3931	77.5	2.9397	1.0574	89.0	5.4349	1.0008
41	1.7992	1.3849	78.0	2.9786	1.0538	89.1	5.5402	1.0006
42	1.8122	1.3765	78.5	3.0192	1.0502	89.2	5.6579	1.0005
43	1.8256	1.3680	79.0	3.0617	1.0468	89.3	5.7914	1.0004
44	1.8396	1.3594	79.5	3.1064	1.0434	89.4	5.9455	1.0003
45	1.8541	1.3506	80.0	3.1534	1.0401	89.5	6.1278	1.0002
46	1.8691	1.3418	80.2	3.1729	1.0388	89.6	6.3509	1.0001
47	1.8848	1.3329	80.4	3.1928	1.0375	89.7	6.6385	1.0001
48	1.9011	1.3238	80.6	3.2132	1.0363	89.8	7.0440	1.0000
49	1.9180	1.3147	80.8	3.2340	1.0350	89.9	7.7371	1.0000

* From R. S. Burington, *Handbook of Mathematical Tables and Formulas*, 3d ed., McGraw-Hill, New York, 1948.

Table F-9a. Factorials and Their Reciprocals*

x	1	2	3	4	5	6	7	8	9	10
$x!$	1	2	6	24	120	720	5,040	40,320	362,880	3,628,800
$\frac{1}{x!}$	1	0.5	1.666667×10^{-1}	4.166667×10^{-2}	8.333333×10^{-3}	1.388889×10^{-3}	1.984127×10^{-4}	2.480159×10^{-5}	2.755732×10^{-6}	2.755732×10^{-7}

Table F-9b. Coefficients of the Binomial Expansion*

Example: $(1 + x)^6 = 1 + 6x + 15x^2 + 20x^3 + 15x^4 + 6x^5 + x^6$

n	0	1	2	3	4	5	6	7	8	9	10
1	1	1									
2	1	2	1								
3	1	3	3	1							
4	1	4	6	4	1						
5	1	5	10	10	5	1					
6	1	6	15	20	15	6	1				
7	1	7	21	35	35	21	7	1			
8	1	8	28	56	70	56	28	8	1		
9	1	9	36	84	126	126	84	36	9	1	
10	1	10	45	120	210	252	210	120	45	10	1

Each number in the table is the sum of the number above it and the number to the left of that number (see box: 36 + 84 = 120). The table can be extended indefinitely in this way.

*From H. E. Etherington, *Nuclear Engineering Handbook*, McGraw-Hill, New York, 1958.

Table F-10. Gamma and Factorial Functions: $\Gamma(x) = (y)!$*

x	y	0	1	2	3	4	5	6	7	8	9
1.00	0.00	1.0000	0.9994	9988	9983	9977	9971	9966	9960	9954	9949
1	1	0.9943	9938	9932	9927	9921	9916	9910	9905	9899	9894
2	2	9888	9883	9878	9872	9867	9862	9856	9851	9846	9841
3	3	9835	9830	9825	9820	9815	9810	9805	9800	9794	9789
4	4	9784	9779	9774	9769	9764	9759	9755	9750	9745	9740
5	5	9735	9730	9725	9721	9716	9711	9706	9702	9697	9692
6	6	9687	9683	9678	9673	9669	9664	9660	9655	9651	9646
7	7	9642	9637	9633	9628	9624	9619	9615	9610	9606	9602
8	8	9597	9593	9589	9584	9580	9576	9571	9567	9563	9559
9	9	9555	9550	9546	9542	9538	9534	9530	9526	9522	9518
1.10	0.10	9514	9509	9505	9501	9498	9494	9490	9486	9482	9478
1	1	9474	9470	9466	9462	9459	9455	9451	9447	9443	9440
2	2	9436	9432	9428	9425	9421	9417	9414	9410	9407	9403
3	3	9399	9396	9392	9389	9385	9382	9378	9375	9371	9368
4	4	9364	9361	9357	9354	9350	9347	9344	9340	9337	9334
5	5	9330	9327	9324	9321	9317	9314	9311	9308	9304	9301
6	6	9298	9295	9292	9289	9285	9282	9279	9276	9273	9270
7	7	9267	9264	9261	9258	9255	9252	9249	9246	9243	9240
8	8	9237	9234	9231	9229	9226	9223	9220	9217	9214	9212
9	9	9209	9206	9203	9201	9198	9195	9192	9190	9187	9184
1.20	0.20	9182	9179	9176	9174	9171	9169	9166	9163	9161	9158
1	1	9156	9153	9151	9148	9146	9143	9141	9138	9136	9133
2	2	9131	9129	9126	9124	9122	9119	9117	9114	9112	9110
3	3	9108	9105	9103	9101	9098	9096	9094	9092	9090	9087
4	4	9085	9083	9081	9079	9077	9074	9072	9070	9068	9066
5	5	9064	9062	9060	9058	9056	9054	9052	9050	9048	9046
6	6	9044	9042	9040	9038	9036	9034	9032	9031	9029	9027
7	7	9025	9023	9021	9020	9018	9016	9014	9012	9011	9009
8	8	9007	9005	9004	9002	9000	8999	8997	8995	8994	8992
9	9	8990	8989	8987	8986	8984	8982	8981	8979	8978	8976
1.30	0.30	8975	8973	8972	8970	8969	8967	8966	8964	8963	8961
1	1	8960	8959	8957	8956	8954	8953	8952	8950	8949	8948
2	2	8946	8945	8944	8943	8941	8940	8939	8937	8936	8935
3	3	8934	8933	8931	8930	8929	8928	8927	8926	8924	8923
4	4	8922	8921	8920	8919	8918	8917	8916	8915	8914	8913
5	5	8912	8911	8910	8909	8908	8907	8906	8905	8904	8903
6	6	8902	8901	8900	8899	8898	8897	8897	8896	8895	8894
7	7	8893	8892	8892	8891	8890	8889	8888	8888	8887	8886
8	8	8885	8885	8884	8883	8883	8882	8881	8880	8880	8879
9	9	8879	8878	8877	8877	8876	8875	8875	8874	8874	8873
1.40	0.40	8873	8872	8872	8871	8871	8870	8870	8869	8869	8868
1	1	8868	8867	8867	8866	8866	8865	8865	8865	8864	8864
2	2	8864	8863	8863	8863	8862	8862	8862	8861	8861	8861
3	3	8860	8860	8860	8860	8859	8859	8859	8859	8858	8858
4	4	8858	8858	8858	8858	8857	8857	8857	8857	8857	8857
5	5	8857	8857	8856	8856	8856	8856	8856	8856	8856	8856
6	6	8856	8856	8856	8856	8856	8856	8856	8856	8856	8856
7	7	8856	8856	8856	8857	8857	8857	8857	8857	8857	8857
8	8	8857	8858	8858	8858	8858	8858	8859	8859	8859	8859
9	9	8859	8860	8860	8860	8860	8861	8861	8861	8862	8862
1.50	0.50	8862	8863	8863	8863	8864	8864	8864	8865	8865	8866

$$\Gamma(x) = \int_0^\infty t^{x-1}e^{-t}\,dt = (x-1)! \qquad y! = \int_0^\infty t^y e^{-t}\,dt = \Gamma(1+y)$$

For higher values of argument

$$\Gamma(x) = (x-1)\Gamma(x-1) = (x-1)(x-2)\Gamma(x-2) = \cdots$$

$$x! = x(x-1)! = x(x-1)(x-2)! = \cdots$$

Example: $\Gamma(4.7) = (3.7)! = 3.7 \times 2.7 \times 1.7 \times 0.9085 = 15.43$.

* From H. E. Etherington, *Nuclear Engineering Handbook*, McGraw-Hill, New York, 1958.

Table F-10. Gamma and Factorial Functions: $\Gamma(x) = y!$ (*Continued*)

x	y	0	1	2	3	4	5	6	7	8	9
1.50	0.50	0.8862	8863	8863	8863	8864	8864	8864	8865	8865	8866
1	1	8866	8866	8867	8867	8868	8868	8869	8869	8869	8870
2	2	8870	8871	8871	8872	8872	8873	8873	8874	8875	8875
3	3	8876	8876	8877	8877	8878	8879	8879	8880	8880	8881
4	4	8882	8882	8883	8884	8884	8885	8886	8887	8887	8888
5	5	8889	8889	8890	8891	8892	8892	8893	8894	8895	8896
6	6	8896	8897	8898	8899	8900	8901	8901	8902	8903	8904
7	7	8905	8906	8907	8908	8909	8909	8910	8911	8912	8913
8	8	8914	8915	8916	8917	8918	8919	8920	8921	8922	8923
9	9	8924	8925	8926	8927	8929	8930	8931	8932	8933	8934
1.60	60	8935	8936	8937	8939	8940	8941	8942	8943	8944	8946
1	1	8947	8948	8949	8950	8952	8953	8954	8955	8957	8958
2	2	8959	8961	8962	8963	8964	8966	8967	8968	8970	8971
3	3	8972	8974	8975	8977	8978	8979	8981	8982	8984	8985
4	4	8986	8988	8989	8991	8992	8994	8995	8997	8998	9000
5	5	9001	9003	9004	9006	9007	9009	9010	9012	9014	9015
6	6	9017	9018	9020	9021	9023	9025	9026	9028	9030	9031
7	7	9033	9035	9036	9038	9040	9041	9043	9045	9047	9048
8	8	9050	9052	9054	9055	9057	9059	9061	9062	9064	9066
9	9	9068	9070	9071	9073	9075	9077	9079	9081	9083	9084
1.70	70	9086	9088	9090	9092	9094	9096	9098	9100	9102	9104
1	1	9106	9108	9110	9112	9114	9116	9118	9120	9122	9124
2	2	9126	9128	9130	9132	9134	9136	9138	9140	9142	9145
3	3	9147	9149	9151	9153	9155	9157	9160	9162	9164	9166
4	4	9168	9170	9173	9175	9177	9179	9182	9184	9186	9188
5	5	9191	9193	9195	9197	9200	9202	9204	9207	9209	9211
6	6	9214	9216	9218	9221	9223	9226	9228	9230	9233	9235
7	7	9238	9240	9242	9245	9247	9250	9252	9255	9257	9260
8	8	9262	9265	9267	9270	9272	9275	9277	9280	9283	9285
9	9	9288	9290	9293	9295	9298	9301	9303	9306	9309	9311
1.80	80	9314	9316	9319	9322	9325	9327	9330	9333	9335	9338
1	1	9341	9343	9346	9349	9352	9355	9357	9360	9363	9366
2	2	9368	9371	9374	9377	9380	9383	9385	9388	9291	9394
3	3	9397	9400	9403	9406	9408	9411	9414	9417	9420	9423
4	4	9426	9429	9432	9435	9438	9441	9444	9447	9450	9453
5	5	9456	9459	9462	9465	9468	9471	9474	9478	9481	9484
6	6	9487	9490	9493	9496	9499	9503	9506	9509	9512	9515
7	7	9518	9522	9525	9528	9531	9534	9538	9541	9544	9547
8	8	9551	9554	9557	9561	9564	9567	9570	9574	9577	9580
9	9	9584	9587	9591	9594	9597	9601	9604	9607	9611	9614
1.90	90	9618	9621	9625	9628	9631	9635	9638	9642	9645	9649
1	1	9652	9656	9659	9663	9666	9670	9673	9677	9681	9684
2	2	9688	9691	9695	9699	9702	9706	9709	9713	9717	9720
3	3	9724	9728	9731	9735	9739	9742	9746	9750	9754	9757
4	4	9761	9765	9768	9772	9776	9780	9784	9787	9791	9795
5	5	9799	9803	9806	9810	9814	9818	9822	9826	9830	9834
6	6	9837	9841	9845	9849	9853	9857	9861	9865	9869	9873
7	7	9877	9881	9885	9889	9893	9897	9901	9905	9909	9913
8	8	9917	9921	9925	9929	9933	9938	9942	9946	9950	9954
9	9	9958	9962	9966	9971	9975	9979	9983	9987	9992	9996
2.00	1.00	1.0000	0004	0008	0013	0017	0021	0026	0030	0034	0038

$\Gamma(3/2) = (1/2)! = 1/2\sqrt{\pi}$ $\quad$ $\Gamma(1/2) = (-1/2)! = \sqrt{\pi}$

Minimum value of the function: $\Gamma(1.461632) = (0.461632)! = 0.885603$.

Table F-11. Bessel Functions: $J_0(x)$ and $J_1(x)$*

$J_0(x)$

x	0	1	2	3	4	5	6	7	8	9
0.0	+1.0000	1.0000	0.9999	9998	9996	9994	9991	9988	9984	9980
1	+0.9975	9970	9964	9958	9951	9944	9936	9928	9919	9910
2	9900	9890	9879	9868	9857	9844	9832	9819	9805	9791
3	9776	9761	9746	9730	9713	9696	9679	9661	9642	9623
4	9604	9584	9564	9543	9522	9500	9478	9455	9432	9409
5	9385	9360	9335	9310	9284	9258	9231	9204	9177	9149
6	9120	9091	9062	9032	9002	8971	8940	8909	8877	8845
7	8812	8779	8745	8711	8677	8642	8607	8572	8536	8500
8	8463	8426	8388	8350	8312	8274	8235	8195	8156	8116
9	8075	8034	7993	7952	7910	7868	7825	7783	7739	7696
1.0	7652	7608	7563	7519	7473	7428	7382	7336	7290	7243
1	7196	7149	7101	7054	7006	6957	6909	6860	6810	6761
2	6711	6661	6611	6561	6510	6459	6408	6356	6305	6253
3	6201	6149	6096	6043	5990	5937	5884	5830	5777	5723
4	5669	5614	5560	5505	5450	5395	5340	5285	5230	5174
5	5118	5062	5006	4950	4894	4838	4781	4725	4668	4611
6	4554	4497	4440	4383	4325	4268	4210	4153	4095	4038
7	3980	3922	3864	3806	3748	3690	3632	3574	3516	3458
8	3400	3342	3284	3225	3167	3109	3051	2993	2934	2876
9	2818	2760	2702	2644	2586	2528	2470	2412	2354	2297
2.0	2239	2181	2124	2066	2009	1951	1894	1837	1780	1723
1	1666	1609	1553	1496	1440	1383	1327	1271	1215	1159
2	1104	1048	0993	0937	0882	0827	0773	0718	0664	0609
3	0555	0502	0448	0394	0341	0288	0235	0182	0130	0077
4	+0025	*0027	*0079	*0130	*0181	*0232	*0283	*0334	*0384	*0434
5	−0484	0533	0583	0632	0681	0729	0778	0826	0873	0921
6	−0968	1015	1062	1108	1154	1200	1245	1291	1336	1380
7	−1424	1469	1512	1556	1599	1641	1684	1726	1768	1809
8	−1850	1891	1932	1972	2012	2051	2090	2129	2167	2205
9	−2243	2280	2317	2354	2390	2426	2462	2497	2532	2566
3.0	−2601	2634	2668	2701	2733	2765	2797	2829	2860	2890
1	−2921	2951	2980	3009	3038	3066	3094	3122	3149	3176
2	−3202	3228	3253	3278	3303	3328	3351	3375	3398	3421
3	−3443	3465	3486	3507	3528	3548	3568	3587	3606	3625
4	−3643	3661	3678	3695	3711	3727	3743	3758	3773	3787
5	−3801	3815	3828	3841	3853	3865	3876	3887	3898	3908
6	−3918	3927	3936	3944	3953	3960	3967	3974	3981	3987
7	−3992	3997	4002	4007	4011	4014	4017	4020	4022	4024
8	−4026	4027	4027	4028	4027	4027	4026	4025	4023	4021
9	−4018	4015	4012	4008	4004	4000	3995	3990	3984	3978
4.0	−3971	3965	3958	3950	3942	3934	3925	3916	3907	3897
1	−3887	3876	3865	3854	3842	3831	3818	3806	3793	3779
2	−3766	3752	3737	3722	3707	3692	3676	3660	3644	3627
3	−3610	3593	3575	3557	3539	3520	3501	3482	3463	3443
4	−3423	3402	3381	3360	3339	3318	3296	3274	3251	3228
5	−3205	3182	3159	3135	3111	3087	3062	3037	3012	2987
6	−2961	2936	2910	2883	2857	2830	2803	2776	2749	2721
7	−2693	2665	2637	2609	2580	2551	2522	2493	2464	2434
8	−2404	2374	2344	2314	2283	2253	2222	2191	2160	2129
9	−2097	2066	2034	2002	1970	1938	1906	1874	1841	1809
5.0	−1776	1743	1710	1677	1644	1611	1578	1544	1511	1477

$$J_0(x) = 1 - \frac{x^2}{2^2(1!)^2} + \frac{x^4}{2^4(2!)^2} - \frac{x^6}{2^6(3!)^2} + \cdots$$

$$J'_0(x) = -J_1(x)$$

* From H. E. Etherington, *Nuclear Engineering Handbook*, McGraw-Hill, New York, 1958.

Table F-11. Bessel Functions: $J_0(x)$ and $J_1(x)$ (*Continued*)

$J_1(x)$

x	0	1	2	3	4	5	6	7	8	9
0.0	+0.00000	00500	01000	01500	02000	02499	02999	03498	03997	04495
1	0.04994	05492	05989	06486	06983	07479	07974	08469	08964	09457
2	0.09950	10442	1093	1142	1191	1240	1289	1338	1386	1435
3	0.1483	1531	1580	1628	1676	1723	1771	1819	1866	1913
4	1960	2007	2054	2101	2147	2194	2240	2286	2332	2377
5	2423	2468	2513	2558	2603	2647	2692	2736	2780	2823
6	2867	2910	2953	2996	3039	3081	3124	3166	3207	3249
7	3290	3331	3372	3412	3452	3492	3532	3572	3611	3650
8	3688	3727	3765	3803	3840	3878	3915	3951	3988	4024
9	4059	4095	4130	4165	4200	4234	4268	4302	4335	4368
1.0	4401	4433	4465	4497	4528	4559	4590	4620	4650	4680
1	4709	4738	4767	4795	4823	4850	4878	4904	4931	4957
2	4983	5008	5033	5058	5082	5106	5130	5153	5176	5198
3	5220	5242	5263	5284	5305	5325	5344	5364	5383	5401
4	5419	5437	5455	5472	5488	5504	5520	5536	5551	5565
5	5579	5593	5607	5620	5632	5644	5656	5667	5678	5689
6	5699	5709	5718	5727	5735	5743	5751	5758	5765	5772
7	5778	5783	5788	5793	5798	5802	5805	5808	5811	5813
8	5815	5817	5818	5818	5819	5818	5818	5817	5816	5814
9	5812	5809	5806	5803	5799	5794	5790	5785	5779	5773
2.0	5767	5761	5754	5746	5738	5730	5721	5712	5703	5693
1	5683	5672	5661	5650	5638	5626	5614	5601	5587	5574
2	5560	5545	5530	5515	5500	5484	5468	5451	5434	5416
3	5399	5381	5362	5343	5324	5305	5285	5265	5244	5223
4	5202	5180	5158	5136	5113	5091	5067	5044	5020	4996
5	4971	4946	4921	4895	4870	4843	4817	4790	4763	4736
6	4708	4680	4652	4624	4595	4566	4536	4507	4477	4446
7	4416	4385	4354	4323	4291	4260	4228	4195	4163	4130
8	4097	4064	4030	3997	3963	3928	3894	3859	3825	3790
9	3754	3719	3683	3647	3611	3575	3538	3502	3465	3428
3.0	3391	3353	3316	3278	3240	3202	3164	3125	3087	3048
1	3009	2970	2931	2892	2852	2813	2773	2733	2694	2654
2	2613	2573	2533	2492	2452	2411	2370	2330	2289	2248
3	2207	2165	2124	2083	2042	2000	1959	1917	1876	1834
4	1792	1751	1709	1667	1625	1583	1541	1500	1458	1416
5	1374	1332	1290	1248	1206	1164	1122	1080	1038	0996
6	0955	0913	0871	0829	07876	0746	0704	0663	0621	0580
7	0538	0497	0456	04145	0373	0332	0291	0250	0210	0169
8	+0128	0088	0047	00069	*0033	*0074	*0114	*0153	*0193	*0233
9	−0272	0312	0351	0390	0429	0468	0507	0546	0584	0622
4.0	−0660	0698	0736	0774	0811	0849	0886	0923	0960	0996
1	−1033	1069	1105	1141	1177	1212	1247	1282	1317	1352
2	−1386	1421	1455	1489	1522	1556	1589	1622	1654	1687
3	−1719	1751	1783	1814	1845	1876	1907	1938	1968	1998
4	−2028	2057	2086	2115	2144	2173	2201	2229	2256	2284
5	−2311	2337	2364	2390	2416	2442	2467	2492	2517	2541
6	−2566	2589	2613	2636	2659	2682	2704	2726	2748	2770
7	−2791	2812	2832	2852	2872	2892	2911	2930	2949	2967
8	−2985	3003	3020	3037	3054	3070	3086	3102	3117	3132
9	−3147	3161	3175	3189	3202	3216	3228	3241	3253	3264
5.0	−3276	3287	3298	3308	3318	3328	3337	3346	3355	3363

$$J_1(x) = \frac{x}{2 \cdot 0! \cdot 1!} - \frac{x^3}{2^3 \cdot 1! \cdot 2!} + \frac{x^5}{2^5 \cdot 2! \cdot 3!} - \cdots$$

$$J_1(x) = -J'_0(x)$$

Table F-11. Bessel Functions: $J_0(x)$ and $J_1(x)$ (*Continued*)

$J_0(x)$

x	0	1	2	3	4	5	6	7	8	9
5.0	−0.1776	1743	1710	1677	1644	1611	1578	1544	1511	1477
1	−1443	1410	1376	1342	1308	1274	1240	1206	1171	1137
2	−1103	1069	1034	1000	0965	0931	0896	0862	0827	0793
3	−0758	0723	0689	0654	0620	0585	0550	0516	0481	0447
4	−0412	0378	0343	0309	0274	0240	0205	0171	0137	0103
5	−0068	0034	0000	*0034	*0068	*0102	*0135	*0169	*0203	*0236
6	+0270	0303	0336	0370	0403	0436	0469	0501	0534	0567
7	0599	0632	0664	0696	0728	0760	0792	0823	0855	0886
8	0917	0948	0979	1010	1040	1071	1101	1131	1161	1191
9	1220	1250	1279	1308	1337	1366	1394	1423	1451	1479
6.0	1506	1534	1561	1589	1616	1642	1669	1695	1721	1747
1	1773	1798	1824	1849	1873	1898	1922	1947	1970	1994
2	2017	2041	2064	2086	2109	2131	2153	2175	2196	2217
3	2238	2259	2279	2299	2319	2339	2358	2377	2396	2415
4	2433	2451	2469	2486	2504	2521	2537	2554	2570	2585
5	2601	2616	2631	2646	2660	2674	2688	2702	2715	2728
6	2740	2753	2765	2777	2788	2799	2810	2821	2831	2841
7	2851	2860	2869	2878	2886	2895	2902	2910	2917	2924
8	2931	2937	2943	2949	2955	2960	2965	2969	2973	2977
9	2981	2984	2987	2990	2993	2995	2997	2998	2999	3000
7.0	3001	3001	3001	3001	3000	2999	2998	2997	2995	2993
1	2991	2988	2985	2982	2978	2974	2970	2966	2961	2956
2	2951	2945	2939	2933	2927	2920	2913	2906	2898	2890
3	2882	2874	2865	2856	2847	2837	2828	2818	2807	2797
4	2786	2775	2764	2752	2740	2728	2715	2703	2690	2677
5	2663	2650	2636	2622	2607	2593	2578	2563	2547	2532
6	2516	2500	2484	2467	2451	2434	2416	2399	2381	2364
7	2346	2327	2309	2290	2271	2252	2233	2214	2194	2174
8	2154	2134	2113	2093	2072	2051	2030	2009	1987	1965
9	1944	1922	1899	1877	1855	1832	1809	1786	1763	1740
8.0	1717	1693	1669	1645	1622	1597	1573	1549	1524	1500
1	1475	1450	1425	1400	1375	1350	1325	1299	1274	1248
2	1222	1196	1170	1144	1118	1092	1066	1039	1013	0987
3	0960	0934	0907	0880	0853	0826	0800	0773	0745	0719
4	0692	0665	0637	0610	0583	0556	0529	0501	0474	0447
5	0419	0392	0365	0337	0310	0283	0255	0228	0201	0174
6	+0146	0119	0092	0065	0037	0010	*0017	*0044	*0071	*0100
7	−0125	0152	0179	0206	0233	0260	0286	0313	0339	0366
8	−0392	0419	0445	0471	0497	0524	0549	0575	0601	0627
9	−0653	0678	0704	0729	0754	0779	0804	0829	0854	0879
9.0	−0903	0928	0952	0976	1000	1024	1048	1072	1096	1119
1	−1142	1166	1189	1211	1234	1257	1279	1302	1324	1346
2	−1367	1389	1411	1432	1453	1474	1495	1516	1536	1556
3	−1577	1597	1616	1636	1655	1674	1694	1712	1731	1749
4	−1768	1786	1804	1821	1839	1856	1873	1890	1907	1923
5	−1939	1955	1971	1987	2002	2017	2032	2047	2061	2076
6	−2090	2104	2117	2131	2144	2157	2169	2182	2194	2206
7	−2218	2230	2241	2252	2263	2273	2284	2294	2304	2313
8	−2323	2332	2341	2350	2358	2366	2374	2382	2389	2396
9	−2403	2410	2417	2423	2429	2434	2440	2445	2450	2455
10.0	−2459	2464	2468	2471	2475	2478	2481	2484	2486	2488

Zeros of $J_0(x)$	1	2	3	4	5	6	7	8
x =	2.4048	5.5201	8.6537	11.7915	14.9309	18.0711	21.2116	24.3525

Table F-11. Bessel Functions: $J_0(x)$ and $J_1(x)$ (*Continued*)

$J_1(x)$

x	0	1	2	3	4	5	6	7	8	9
5.0	−0.3276	3287	3298	3308	3318	3328	3337	3346	3355	3363
1	−3371	3379	3386	3393	3400	3406	3412	3417	3423	3428
2	−3432	3436	3440	3444	3447	3450	3453	3455	3457	3458
3	−3460	3460	3461	3461	3461	3461	3460	3459	3457	3456
4	−3453	3451	3448	3445	3442	3438	3434	3430	3425	3420
5	−3414	3409	3403	3396	3390	3383	3376	3368	3360	3352
6	−3343	3335	3325	3316	3306	3296	3286	3275	3264	3253
7	−3241	3230	3218	3205	3192	3179	3166	3153	3139	3125
8	−3110	3096	3081	3065	3050	3034	3018	3002	2985	2969
9	−2951	2934	2917	2899	2881	2862	2844	2825	2806	2786
6.0	−2767	2747	2727	2707	2686	2666	2645	2623	2602	2580
1	−2559	2537	2514	2492	2469	2446	2423	2400	2377	2353
2	−2329	2305	2281	2257	2232	2207	2182	2157	2132	2106
3	−2081	2055	2029	2003	1977	1950	1924	1897	1870	1843
4	−1816	1789	1762	1734	1707	1679	1651	1623	1595	1567
5	−1538	1510	1481	1453	1424	1395	1366	1337	1308	1279
6	−1250	1220	1191	1162	1132	1102	1073	1043	1013	0983
7	−0953	0923	0893	0863	0833	0803	0773	0743	0713	0682
8	−0652	0622	0592	0561	0531	0501	0470	0440	0410	0379
9	−0349	0319	0288	0258	0228	0198	0167	0137	0107	0077
7.0	−0047	0017	*0013	*0043	*0073	*0103	*0133	*0163	*0192	*0222
1	+0251	0282	0310	0340	0369	0398	0428	0457	0486	0514
2	0543	0572	0601	0629	0658	0686	0714	0742	0770	0798
3	0826	0853	0881	0908	0935	0963	0990	1016	1043	1070
4	1096	1123	1149	1175	1201	1226	1252	1277	1302	1328
5	1352	1377	1402	1426	1450	1475	1498	1522	1546	1569
6	1592	1615	1638	1660	1683	1705	1727	1749	1771	1792
7	1813	1834	1855	1875	1896	1916	1936	1956	1975	1994
8	2014	2032	2051	2069	2088	2106	2123	2141	2158	2175
9	2192	2208	2225	2241	2257	2272	2287	2303	2317	2332
8.0	2346	2360	2374	2388	2401	2414	2427	2440	2452	2464
1	2476	2488	2499	2510	2521	2531	2542	2552	2561	2571
2	2580	2589	2598	2606	2614	2622	2630	2637	2644	2651
3	2657	2664	2670	2675	2681	2686	2691	2696	2700	2704
4	2708	2711	2715	2718	2720	2723	2725	2727	2729	2730
5	2731	2732	2733	2733	2733	2733	2732	2731	2730	2729
6	2728	2726	2724	2721	2719	2716	2713	2709	2705	2701
7	2697	2693	2688	2683	2678	2672	2666	2660	2654	2648
8	2641	2634	2626	2619	2611	2603	2595	2586	2577	2568
9	2559	2550	2540	2530	2519	2509	2498	2487	2476	2465
9.0	2453	2441	2429	2417	2404	2391	2378	2365	2352	2338
1	2324	2310	2296	2281	2267	2252	2237	2221	2206	2190
2	2174	2158	2142	2125	2108	2091	2074	2057	2040	2022
3	2004	1986	1968	1950	1931	1912	1893	1874	1855	1836
4	1816	1797	1777	1757	1737	1716	1696	1675	1655	1634
5	1613	1591	1570	1549	1527	1506	1484	1462	1440	1418
6	1395	1373	1350	1328	1305	1282	1259	1236	1213	1190
7	1166	1143	1119	1096	1072	1048	1025	10006	0977	0953
8	0928	0904	0880	0856	0831	0807	0782	0758	0733	0708
9	0684	0659	0634	0609	0585	0560	0535	0510	0485	0460
10.0	+0435	0410	0385	0360	0334	0309	0284	0259	0234	0209

Zeros of $J_1(x)$	1	2	3	4	5	6	7	8
$x =$	3.8317	7.0156	10.1735	13.3237	16.4706	19.6159	22.7601	25.9037

Table F-11. Bessel Functions: $J_0(x)$ and $J_1(x)$ (*Continued*)

$J_0(x)$

x	0	1	2	3	4	5	6	7	8	9
10.0	−0.2459	2464	2468	2471	2475	2478	2481	2484	2486	2488
1	−2490	2492	2493	2495	2496	2496	2497	2497	2497	2497
2	−2496	2495	2494	2493	2492	2490	2488	2485	2483	2480
3	−2477	2474	2470	2467	2463	2458	2454	2449	2444	2439
4	−2434	2428	2422	2416	2410	2403	2396	2389	2382	2374
5	−2366	2358	2350	2342	2333	2324	2315	2306	2296	2286
6	−2276	2266	2256	2245	2234	2223	2212	2200	2188	2177
7	−2164	2152	2140	2127	2114	2101	2087	2074	2060	2046
8	−2032	2018	2003	1989	1974	1959	1943	1928	1912	1897
9	−1881	1865	1848	1832	1815	1798	1781	1764	1747	1730
11.0	−1712	1694	1676	1658	1640	1622	1603	1584	1566	1547
1	−1528	1508	1489	1470	1450	1430	1411	1391	1370	1350
2	−1330	1309	1289	1268	1247	1227	1206	1185	1163	1142
3	−1121	1099	1078	1056	1034	1012	0991	0969	0946	0924
4	−0902	0880	0858	0835	0813	0790	0767	0745	0722	0699
5	−0677	0654	0631	0608	0585	0562	0539	0516	0493	0469
6	−0446	0423	0400	0376	0353	0330	0307	0283	0260	0237
7	−0213	0190	0167	0143	0120	0097	0073	0050	0027	0004
8	+0020	0043	0066	0089	0112	0135	0159	0182	0205	0228
9	0250	0273	0296	0319	0342	0364	0387	0410	0432	0455
12.0	0477	0499	0521	0544	0566	0588	0610	0632	0653	0675
1	0697	0718	0740	0761	0782	0803	0824	0845	0866	0887
2	0908	0928	0949	0969	0989	1009	1029	1049	1069	1088
3	1108	1127	1147	1166	1185	1203	1222	1241	1259	1277
4	1296	1314	1331	1349	1367	1384	1401	1418	1435	1452
5	1469	1485	1502	1518	1534	1550	1565	1581	1596	1611
6	1626	1641	1655	1670	1684	1698	1712	1726	1739	1753
7	1766	1779	1792	1804	1817	1829	1841	1853	1864	1876
8	1887	1898	1909	1920	1930	1940	1950	1960	1970	1979
9	1988	1997	2006	2015	2023	2031	2039	2047	2055	2062
13.0	2069	2076	2083	2089	2096	2102	2108	2113	2119	2124
1	2129	2134	2138	2143	2147	2151	2154	2158	2161	2164
2	2167	2169	2172	2174	2176	2178	2179	2180	2182	2182
3	2183	2183	2184	2184	2183	2183	2182	2181	2180	2179
4	2177	2175	2173	2171	2169	2166	2163	2160	2157	2154
5	2150	2146	2142	2138	2133	2128	2123	2118	2113	2107
6	2101	2095	2089	2083	2076	2069	2062	2055	2048	2040
7	2032	2024	2016	2008	1999	1990	1981	1972	1963	1953
8	1943	1933	1923	1913	1903	1892	1881	1870	1859	1847
9	1836	1824	1812	1800	1788	1775	1763	1750	1737	1724
14.0	1711	1697	1684	1670	1656	1642	1628	1613	1599	1584
1	1570	1555	1539	1524	1509	1493	1478	1462	1446	1430
2	1414	1397	1381	1364	1348	1331	1314	1297	1280	1262
3	1245	1227	1210	1192	1174	1156	1138	1120	1102	1083
4	1065	1046	1028	1009	0990	0971	0952	0933	0914	0895
5	0875	0856	0837	0817	0798	0778	0758	0738	0719	0699
6	0679	0659	0639	0618	0598	0578	0558	0538	0517	0497
7	0476	0456	0436	0415	0394	0374	0353	0333	0312	0291
8	0271	0250	0229	0209	0188	0167	0147	0126	0105	0085
9	+0064	0043	0023	+0002	−0019	−0039	−0060	−0081	−0101	−0122

For $x > 15$

$$J_0(x) \approx \sqrt{\frac{2}{\pi x}}\left[\cos\left(x - \frac{\pi}{4}\right) + \frac{1}{8x}\sin\left(x - \frac{\pi}{4}\right)\right], \text{ error } < 0.0001$$

$$\sqrt{\frac{2}{\pi}} = 0.7979 \qquad \frac{\pi}{4} = 0.7854$$

Table F-11. Bessel Functions: $J_0(x)$ and $J_1(x)$ (*Continued*)

$J_1(x)$

x	0	1	2	3	4	5	6	7	8	9
10.0	+0.0435	0410	0385	0360	0334	0309	0284	0259	0234	0209
1	+0184	0159	0134	0109	0084	0059	0034	0009	*0016	*0041
2	−0066	0091	0116	0141	0165	0190	0215	0240	0264	0289
3	−0313	0338	0362	0386	0411	0435	0459	0483	0507	0531
4	−0555	0578	0602	0626	0649	0673	0696	0719	0742	0766
5	−0789	0811	0834	0857	0879	0902	0924	0946	0968	0990
6	−1012	1034	1056	1077	1099	1120	1141	1162	1183	1203
7	−1224	1244	1265	1285	1305	1325	1344	1364	1383	1403
8	−1422	1441	1459	1478	1496	1515	1533	1551	1568	1586
9	−1603	1621	1638	1655	1671	1688	1704	1720	1736	1752
11.0	−1768	1783	1798	1814	1828	1843	1857	1872	1886	1900
1	−1913	1927	1940	1953	1966	1979	1991	2003	2015	2027
2	−2039	2050	2061	2072	2083	2093	2104	2114	2123	2133
3	−2143	2152	2161	2169	2178	2186	2194	2202	2210	2217
4	−2225	2231	2238	2245	2251	2257	2263	2268	2274	2279
5	−2284	2288	2293	2297	2301	2305	2308	2312	2315	2317
6	−2320	2322	2324	2326	2328	2329	2331	2332	2332	2333
7	−2333	2333	2333	2332	2332	2331	2330	2328	2327	2325
8	−2323	2321	2318	2315	2312	2309	2306	2302	2298	2294
9	−2290	2285	2281	2276	2270	2265	2259	2253	2247	2241
12.0	−2234	2228	2221	2214	2206	2199	2191	2183	2175	2166
1	−2157	2149	2140	2130	2121	2111	2101	2091	2081	2070
2	−2060	2049	2038	2027	2015	2004	1992	1980	1968	1955
3	−1943	1930	1917	1904	1891	1877	1863	1850	1836	1821
4	−1807	1793	1778	1763	1748	1733	1718	1702	1687	1671
5	−1655	1639	1623	1606	1590	1573	1556	1539	1522	1505
6	−1487	1470	1452	1435	1417	1399	1380	1362	1344	1325
7	−1307	1288	1269	1250	1231	1212	1192	1173	1154	1134
8	−1114	1095	1075	1055	1035	1014	0994	0974	0954	0933
9	−0912	0892	0871	0850	0830	0809	0788	0767	0746	0724
13.0	−0703	0682	0661	0639	0618	0596	0575	0553	0532	0510
1	−0489	0467	0445	0423	0402	0380	0358	0336	0314	0293
2	−0271	0249	0227	0205	0183	0161	0139	0117	0096	0074
3	−0052	0030	0008	*0014	*0036	*0057	*0079	*0101	*0123	*0144
4	+0166	0188	0209	0231	0252	0274	0295	0317	0338	0359
5	0380	0402	0423	0444	0465	0486	0507	0528	0548	0569
6	0590	0610	0631	0651	0671	0692	0712	0732	0752	0772
7	0791	0811	0831	0850	0870	0889	0908	0927	0946	0965
8	0984	1003	1021	1040	1058	1076	1094	1112	1130	1148
9	1165	1183	1200	1217	1234	1251	1268	1285	1301	1318
14.0	1334	1350	1366	1382	1397	1413	1428	1443	1458	1473
1	1488	1502	1517	1531	1545	1559	1573	1586	1600	1613
2	1626	1639	1652	1664	1677	1689	1701	1713	1724	1736
3	1747	1758	1769	1780	1791	1801	1811	1821	1831	1841
4	1850	1860	1869	1878	1886	1895	1903	1911	1919	1927
5	1934	1942	1949	1956	1962	1969	1975	1981	1987	1993
6	1999	2004	2009	2014	2019	2023	2027	2031	2035	2039
7	2043	2046	2049	2052	2054	2057	2059	2061	2063	2065
8	2066	2067	2068	2069	2070	2070	2070	2070	2070	2069
9	+2069	2068	2067	2066	2064	2062	2061	2058	2056	2054

For $x > 15$

$$J_1(x) \approx \sqrt{\frac{2}{\pi x}}\left[\cos\left(x - \frac{3\pi}{4}\right) - \frac{3}{8x}\sin\left(x - \frac{3\pi}{4}\right)\right], \text{ error} < 0.0001$$

$$\sqrt{\frac{2}{\pi}} = 0.7979 \qquad \frac{3\pi}{4} = 2.3562$$

Table F-11. Bessel Functions: $N_0(x)$ and $N_1(x)$

$N_0(x)$

x	0	1	2	3	4	5	6	7	8	9
0.0	$-\infty$	3.005	2.564	2.305	2.122	1.979	1.863	1.764	1.678	1.602
1	−1.534	1.473	1.416	1.364	1.316	1.271	1.228	1.189	1.151	1.115
2	−1.081	1.049	1.0175	0.9877	0.9591	0.9316	0.9050	0.8794	0.8546	0.8306
3	−0.8073	7847	7627	7414	7206	7003	6806	6613	6424	6240
4	−6060	5884	5712	5542	5377	5214	5055	4898	4745	4594
5	−4445	4299	4156	4015	3876	3739	3604	3472	3341	3212
6	−3085	2960	2837	2715	2595	2476	2359	2244	2130	2018
7	−1907	1797	1689	1582	1476	1372	1269	1167	1066	0966
8	−0868	0771	0675	0580	0486	0393	0301	0210	0120	0032
9	+0056	0143	0229	0314	0398	0481	0563	0644	0725	0804
1.0	0883	0960	1037	1113	1188	1262	1336	1409	1480	1551
1	1622	1691	1760	1828	1895	1961	2026	2091	2155	2218
2	2281	2343	2404	2464	2523	2582	2640	2698	2754	2810
3	2865	2920	2974	3027	3079	3131	3182	3232	3282	3331
4	3379	3427	3473	3520	3565	3610	3654	3698	3741	3783
5	3824	3865	3906	3945	3984	4022	4060	4097	4133	4169
6	4204	4239	4273	4306	4338	4370	4401	4432	4462	4491
7	4520	4548	4576	4603	4629	4655	4680	4705	4728	4752
8	4774	4796	4818	4839	4859	4879	4898	4916	4934	4951
9	4968	4984	5000	5015	5029	5043	5056	5069	5081	5093
2.0	5104	5114	5124	5133	5142	5150	5158	5165	5172	5177
1	5183	5188	5192	5196	5199	5202	5204	5206	5207	5208
2	5208	5207	5207	5205	5203	5201	5198	5194	5190	5186
3	5181	5175	5169	5163	5156	5148	5141	5132	5123	5114
4	5104	5094	5083	5072	5060	5048	5036	5022	5009	4995
5	4981	4966	4951	4935	4919	4902	4885	4868	4850	4832
6	4813	4794	4775	4755	4735	4714	4693	4672	4650	4628
7	4605	4582	4559	4535	4511	4487	4462	4437	4411	4385
8	4359	4333	4306	4279	4251	4223	4195	4167	4138	4109
9	4079	4049	4019	3989	3958	3927	3896	3865	3833	3801
3.0	3769	3736	3703	3670	3637	3603	3569	3535	3500	3466
1	3431	3396	3361	3325	3289	3253	3217	3181	3144	3108
2	3071	3033	2996	2958	2921	2883	2845	2807	2768	2730
3	2691	2652	2613	2574	2535	2495	2456	2416	2376	2336
4	2296	2256	2216	2175	2135	2094	2054	2013	1972	1931
5	1890	1849	1808	1767	1726	1684	1643	1602	1560	1519
6	1477	1436	1394	1352	1311	1269	1227	1186	1144	1102
7	1061	1019	0977	0936	0894	0853	0811	0769	0728	0686
8	0645	0604	0562	0521	0480	0439	0397	0356	0315	0275
9	+0234	0193	0152	0112	0071	0031	*0009	*0050	*0090	*0130
4.0	−0169	0209	0249	0288	0328	0367	0406	0445	0484	0522
1	−0561	0599	0638	0676	0714	0751	0789	0826	0864	0901
2	−0938	0974	1011	1047	1083	1119	1155	1191	1226	1261
3	−1296	1331	1365	1400	1434	1467	1501	1535	1568	1601
4	−1633	1666	1698	1730	1762	1793	1825	1856	1886	1917
5	−1947	1977	2007	2036	2065	2094	2123	2151	2179	2207
6	−2235	2262	2289	2315	2342	2368	2394	2419	2444	2469
7	−2494	2518	2542	2566	2589	2612	2635	2658	2680	2702
8	−2723	2744	2765	2786	2806	2826	2845	2865	2884	2902
9	−2921	2939	2956	2973	2990	3007	3023	3039	3055	3070
5.0	−3085	3100	3114	3128	3142	3155	3168	3180	3193	3204

Linear interpolation is inaccurate for low values of the argument. For greater accuracy in this range, use the auxiliary functions, pages 1057 to 1059.

$$N'_0(x) = -N_1(x)$$

Table F-11. Bessel Functions: $N_0(x)$ and $N_1(x)$ (*Continued*)

$N_1(x)$

x	0	1	2	3	4	5	6	7	8	9
0.0	$-\infty$	63.68	31.86	21.26	15.96	12.79	10.68	9.167	8.038	7.160
1	−6.459	5.886	5.409	5.007	4.662	4.364	4.103	3.873	3.670	3.487
2	−3.324	3.176	3.042	2.919	2.807	2.704	2.609	2.521	2.440	2.364
3	−2.293	2.227	2.165	2.107	2.052	2.000	1.952	1.906	1.862	1.820
4	−1.781	1.743	1.708	1.673	1.641	1.610	1.580	1.551	1.523	1.497
5	−1.471	1.447	1.423	1.401	1.378	1.357	1.337	1.317	1.297	1.279
6	−1.260	1.243	1.226	1.209	1.193	1.177	1.161	1.146	1.132	1.117
7	−1.103	1.090	1.076	1.063	1.050	1.038	1.025	1.013	1.0013	*9896
8	−0.9781	9669	9558	9449	9342	9236	9132	9030	8929	8829
9	−8731	8634	8539	8444	8351	8258	8167	8077	7988	7900
1.0	−7812	7726	7640	7555	7471	7388	7305	7223	7142	7061
1	−6981	6902	6823	6745	6667	6590	6513	6437	6361	6286
2	−6211	6137	6063	5990	5916	5844	5771	5699	5628	5556
3	−5485	5415	5344	5274	5204	5135	5066	4997	4928	4860
4	−4791	4724	4656	4589	4521	4454	4388	4321	4255	4189
5	−4123	4057	3992	3927	3862	3797	3732	3668	3604	3540
6	−3476	3412	3349	3285	3222	3159	3096	3034	2972	2909
7	−2847	2785	2724	2662	2601	2540	2479	2418	2357	2297
8	−2237	2177	2117	2057	1997	1938	1879	1820	1761	1702
9	−1644	1586	1528	1470	1412	1355	1297	1240	1184	1127
2.0	−1070	1014	0958	0902	0846	0791	0736	0681	0626	0571
1	−0517	0463	0409	0355	0301	0248	0195	0142	0090	0037
2	+0015	0067	0118	0170	0221	0272	0323	0373	0423	0473
3	0523	0572	0621	0670	0719	0767	0815	0863	0911	0958
4	1005	1052	1098	1144	1190	1236	1281	1326	1371	1415
5	1459	1503	1547	1590	1633	1675	1718	1760	1801	1843
6	1884	1924	1965	2005	2045	2084	2123	2162	2200	2239
7	2276	2314	2351	2388	2424	2460	2496	2531	2566	2601
8	2635	2669	2703	2736	2769	2802	2834	2866	2897	2929
9	2959	2990	3020	3050	3079	3108	3136	3164	3192	3220
3.0	3247	3273	3300	3326	3351	3376	3401	3425	3449	3473
1	3496	3519	3542	3564	3585	3607	3627	3648	3668	3688
2	3707	3726	3745	3763	3780	3798	3815	3831	3847	3863
3	3879	3893	3908	3922	3936	3949	3962	3975	3987	3999
4	4010	4021	4032	4042	4052	4061	4070	4079	4087	4095
5	4102	4109	4115	4122	4127	4133	4138	4142	4147	4150
6	4154	4157	4160	4162	4164	4165	4166	4167	4167	4167
7	4167	4166	4165	4163	4161	4159	4156	4153	4149	4145
8	4141	4137	4132	4126	4120	4114	4108	4101	4094	4086
9	4078	4070	4061	4052	4043	4033	4023	4013	4002	3991
4.0	3979	3967	3955	3943	3930	3917	3903	3889	3875	3861
1	3846	3831	3815	3800	3783	3767	3750	3733	3716	3698
2	3680	3662	3643	3624	3605	3586	3566	3546	3525	3505
3	3484	3463	3441	3420	3397	3375	3353	3330	3307	3283
4	3260	3236	3212	3187	3163	3138	3113	3087	3062	3036
5	3010	2984	2957	2930	2904	2876	2849	2821	2794	2766
6	2737	2709	2680	2652	2623	2594	2564	2535	2505	2475
7	2445	2415	2384	2354	2323	2292	2261	2230	2199	2167
8	2136	2104	2072	2040	2008	1976	1943	1911	1878	1845
9	1812	1780	1746	1713	1680	1647	1613	1580	1546	1512
5.0	+1479	1445	1411	1377	1343	1309	1275	1240	1206	1172

Linear interpolation is inaccurate for low values of the argument. For greater accuracy in this range, use the auxiliary functions, pages 1057 to 1059.

$$N_1(x) = -N'_0(x)$$

Table F-11. Bessel Functions: $N_0(x)$ and $N_1(x)$ (*Continued*)

$N_0(x)$

x	0	1	2	3	4	5	6	7	8	9
5.0	−0.3085	3100	3114	3128	3142	3155	3168	3180	3193	3204
1	−3216	3227	3238	3249	3259	3269	3278	3287	3296	3304
2	−3313	3320	3328	3335	3341	3348	3354	3359	3365	3370
3	−3374	3379	3383	3386	3389	3392	3395	3397	3399	3400
4	−3402	3403	3403	3403	3403	3402	3402	3400	3399	3397
5	−3395	3392	3389	3386	3383	3379	3375	3370	3365	3360
6	−3354	3349	3342	3336	3329	3322	3315	3307	3299	3290
7	−3282	3273	3263	3254	3244	3233	3223	3212	3201	3189
8	−3177	3165	3153	3140	3127	3114	3101	3087	3073	3058
9	−3044	3029	3013	2998	2982	2966	2950	2933	2916	2899
6.0	−2882	2864	2846	2828	2810	2791	2772	2753	2734	2714
1	−2694	2674	2654	2633	2613	2592	2570	2549	2527	2505
2	−2483	2461	2438	2415	2393	2369	2346	2322	2299	2275
3	−2251	2226	2202	2177	2152	2127	2102	2077	2051	2025
4	−1999	1973	1947	1921	1894	1868	1841	1814	1787	1760
5	−1732	1705	1677	1650	1622	1594	1566	1538	1509	1481
6	−1452	1424	1395	1366	1337	1308	1279	1250	1221	1191
7	−1162	1132	1103	1073	1044	1014	0984	0954	0924	0894
8	−0864	0834	0804	0774	0744	0714	0684	0653	0623	0593
9	−0563	0532	0502	0472	0441	0411	0381	0350	0320	0290
7.0	−0259	0229	0199	0169	0139	0108	0078	0048	0018	*0012
1	+0042	0072	0102	0131	0161	0191	0221	0250	0280	0309
2	0339	0368	0397	0426	0455	0484	0513	0542	0571	0599
3	0628	0656	0684	0713	0741	0769	0797	0824	0852	0879
4	0907	0934	0961	0988	1015	1042	1068	1095	1121	1147
5	1173	1199	1225	1250	1276	1301	1326	1351	1375	1400
6	1424	1448	1472	1496	1520	1543	1567	1590	1613	1635
7	1658	1680	1702	1724	1746	1768	1789	1810	1831	1852
8	1872	1893	1913	1932	1952	1972	1991	2010	2028	2047
9	2065	2083	2101	2119	2136	2153	2170	2187	2203	2219
8.0	2235	2251	2266	2282	2296	2311	2326	2340	2354	2367
1	2381	2394	2407	2420	2432	2444	2456	2468	2479	2490
2	2501	2512	2522	2532	2542	2551	2561	2570	2578	2587
3	2595	2603	2611	2618	2625	2632	2639	2645	2651	2657
4	2662	2667	2672	2677	2681	2686	2689	2693	2696	2699
5	2702	2705	2707	2709	2710	2712	2713	2714	2714	2715
6	2715	2714	2714	2713	2712	2711	2709	2707	2705	2703
7	2700	2697	2694	2690	2687	2683	2678	2674	2669	2664
8	2659	2653	2647	2641	2635	2628	2621	2614	2607	2599
9	2592	2583	2575	2566	2558	2549	2539	2530	2520	2510
9.0	2499	2489	2478	2467	2456	2444	2433	2421	2408	2396
1	2383	2371	2357	2344	2331	2317	2303	2289	2274	2260
2	2245	2230	2215	2199	2184	2168	2152	2136	2119	2103
3	2086	2069	2052	2034	2017	1999	1981	1963	1945	1926
4	1907	1889	1870	1851	1831	1812	1792	1772	1752	1732
5	1712	1692	1671	1650	1630	1609	1588	1566	1545	1523
6	1502	1480	1458	1436	1414	1392	1369	1347	1324	1302
7	1279	1256	1233	1210	1186	1163	1140	1116	1093	1069
8	1045	1021	0998	0974	0949	0925	0901	0877	0853	0828
9	0804	0779	0755	0730	0705	0681	0656	0631	0606	0582
10.0	0557	0532	0507	0482	0457	0432	0407	0382	0357	0332

Zeros of $N_0(x)$	1	2	3	4	5	6	7	8
x =	0.8936	3.9577	7.0861	10.2223	13.3611	16.5009	19.6413	22.7820

Table F-11. Bessel Functions: $N_0(x)$ and $N_1(x)$ (*Continued*)

$N_1(x)$

x	0	1	2	3	4	5	6	7	8	9
5.0	+0.1479	1445	1411	1377	1343	1309	1275	1240	1206	1172
1	1137	1103	1069	1034	1000	0965	0930	0896	0861	0827
2	0792	0757	0723	0688	0653	0619	0584	0549	0515	0480
3	0445	0411	0376	0342	0307	0273	0238	0204	0170	0136
4	+0101	0067	0033	*0001	*0035	*0069	*0103	*0137	*0170	*0204
5	−0238	0271	0304	0338	0371	0404	0437	0470	0503	0535
6	−0568	0601	0633	0665	0697	0729	0761	0793	0824	0856
7	−0887	0918	0949	0980	1011	1042	1072	1102	1133	1163
8	−1192	1222	1251	1281	1310	1339	1368	1396	1425	1453
9	−1481	1509	1536	1564	1591	1618	1645	1671	1698	1724
6.0	−1750	1776	1801	1827	1852	1877	1902	1926	1950	1974
1	−1998	2022	2045	2068	2091	2114	2136	2158	2180	2201
2	−2223	2244	2265	2285	2306	2326	2346	2365	2385	2404
3	−2422	2441	2459	2477	2495	2512	2530	2547	2563	2580
4	−2596	2611	2627	2642	2657	2672	2686	2700	2714	2728
5	−2741	2754	2767	2779	2791	2803	2814	2826	2836	2847
6	−2857	2868	2877	2887	2896	2905	2913	2922	2930	2937
7	−2945	2952	2958	2965	2971	2977	2983	2988	2993	2997
8	−3002	3006	3010	3013	3016	3019	3022	3024	3026	3028
9	−3029	3030	3031	3032	3032	3032	3031	3031	3030	3028
7.0	−3027	3025	3023	3020	3017	3014	3011	3007	3003	2999
1	−2995	2990	2985	2980	2974	2968	2962	2955	2949	2942
2	−2934	2927	2919	2911	2902	2893	2885	2875	2866	2856
3	−2846	2836	2825	2814	2803	2792	2780	2768	2756	2744
4	−2731	2718	2705	2692	2678	2664	2650	2636	2621	2606
5	−2591	2576	2560	2545	2529	2512	2496	2479	2462	2445
6	−2428	2410	2393	2375	2357	2338	2320	2301	2282	2263
7	−2243	2224	2204	2184	2164	2143	2123	2102	2081	2060
8	−2039	2017	1996	1974	1952	1930	1908	1885	1863	1840
9	−1817	1794	1771	1748	1724	1701	1677	1653	1629	1605
8.0	−1581	1556	1532	1507	1482	1457	1432	1407	1382	1357
1	−1331	1306	1280	1255	1229	1203	1177	1151	1125	1099
2	−1072	1046	1020	0993	0967	0940	0913	0887	0860	0833
3	−0806	0779	0752	0725	0698	0671	0644	0617	0589	0562
4	−0535	0508	0480	0453	0426	0398	0371	0344	0316	0289
5	−0262	0234	0207	0180	0152	0125	0098	0071	0043	0016
6	+0011	0038	0065	0092	0119	0146	0173	0200	0227	0253
7	0280	0307	0333	0360	0386	0413	0439	0465	0491	0518
8	0544	0569	0595	0621	0647	0672	0698	0723	0748	0774
9	0799	0824	0849	0873	0898	0922	0947	0971	0995	1019
9.0	1043	1067	1091	1114	1137	1161	1184	1207	1229	1252
1	1275	1297	1319	1341	1363	1385	1406	1428	1449	1470
2	1491	1512	1532	1553	1573	1593	1613	1633	1652	1671
3	1691	1710	1728	1747	1765	1783	1801	1819	1837	1854
4	1871	1888	1905	1922	1938	1954	1970	1986	2001	2017
5	2032	2047	2061	2076	2090	2104	2118	2131	2145	2158
6	2171	2183	2196	2208	2220	2232	2243	2254	2265	2276
7	2287	2297	2307	2317	2326	2336	2345	2354	2362	2371
8	2379	2387	2394	2402	2409	2416	2423	2429	2435	2441
9	2447	2452	2458	2463	2467	2472	2476	2480	2484	2487
10.0	+2490	2493	2496	2498	2500	2502	2504	2506	2507	2508

Zeros of $N_1(x)$	1	2	3	4	5	6	7	8
$x =$	2.1971	5.4297	8.5960	11.7492	14.8974	18.0434	21.1881	24.3319

Table F-11. Bessel Functions: $N_0(x)$ and $N_1(x)$ (*Continued*)

$N_0(x)$

x	0	1	2	3	4	5	6	7	8	9
10.0	+0.0557	0532	0507	0482	0457	0432	0407	0382	0357	0332
1	0307	0281	0256	0231	0206	0181	0156	0131	0106	0081
2	+0056	0031	0006	*0019	*0044	*0069	*0094	*0119	*0143	*0168
3	−0193	0218	0242	0267	0291	0316	0340	0365	0389	0413
4	−0437	0462	0486	0510	0534	0557	0581	0605	0628	0652
5	−0675	0699	0722	0745	0768	0791	0814	0837	0859	0882
6	−0904	0926	0949	0971	0993	1015	1036	1058	1079	1101
7	−1122	1143	1164	1185	1205	1226	1246	1267	1287	1307
8	−1326	1346	1366	1385	1404	1423	1442	1461	1479	1498
9	−1516	1534	1552	1569	1587	1604	1622	1639	1655	1672
11.0	−1688	1705	1721	1737	1752	1768	1783	1798	1813	1828
1	−1843	1857	1871	1885	1899	1913	1926	1939	1952	1965
2	−1977	1990	2002	2014	2025	2037	2048	2059	2070	2081
3	−2091	2101	2111	2121	2130	2140	2149	2158	2166	2175
4	−2183	2191	2199	2206	2213	2220	2227	2234	2240	2246
5	−2252	2258	2263	2269	2274	2278	2283	2287	2291	2295
6	−2299	2302	2305	2308	2311	2313	2315	2317	2319	2321
7	−2322	2323	2324	2324	2325	2325	2324	2324	2324	2323
8	−2322	2320	2319	2317	2315	2313	2310	2308	2305	2302
9	−2298	2295	2291	2287	2283	2278	2273	2269	2263	2258
12.0	−2252	2247	2241	2234	2228	2221	2214	2207	2200	2192
1	−2184	2176	2168	2160	2151	2142	2133	2124	2115	2105
2	−2095	2085	2075	2064	2054	2043	2032	2021	2009	1998
3	−1986	1974	1962	1949	1937	1924	1911	1898	1885	1871
4	−1858	1844	1830	1816	1802	1787	1772	1758	1743	1727
5	−1712	1697	1681	1665	1649	1633	1617	1601	1584	1567
6	−1551	1534	1517	1499	1482	1464	1447	1429	1411	1393
7	−1375	1357	1338	1320	1301	1282	1264	1245	1226	1206
8	−1187	1168	1148	1129	1109	1089	1069	1049	1029	1009
9	−0989	0968	0948	0927	0907	0886	0866	0845	0824	0803
13.0	−0782	0761	0740	0719	0698	0676	0655	0634	0612	0591
1	−0569	0548	0526	0505	0483	0461	0439	0418	0396	0374
2	−0352	0331	0309	0287	0265	0243	0221	0199	0177	0156
3	−0134	0112	0090	0068	0046	0024	0002	*0019	*0041	*0063
4	+0085	0107	0128	0150	0172	0193	0215	0236	0258	0279
5	0301	0322	0343	0365	0386	0407	0428	0449	0470	0491
6	0512	0533	0554	0574	0595	0615	0636	0656	0677	0697
7	0717	0737	0757	0777	0796	0816	0836	0855	0875	0894
8	0913	0932	0951	0970	0989	1007	1026	1044	1062	1081
9	1099	1117	1134	1152	1169	1187	1204	1221	1238	1255
14.0	1272	1289	1305	1321	1337	1353	1369	1385	1401	1416
1	1431	1446	1461	1476	1491	1505	1520	1534	1548	1562
2	1575	1589	1602	1615	1628	1641	1654	1666	1679	1691
3	1703	1715	1726	1738	1749	1760	1771	1781	1792	1802
4	1812	1822	1832	1842	1851	1860	1869	1878	1886	1895
5	1903	1911	1919	1926	1934	1941	1948	1955	1962	1968
6	1974	1980	1986	1992	1997	2002	2007	2012	2017	2021
7	2025	2029	2033	2036	2040	2043	2046	2049	2051	2054
8	2056	2058	2059	2061	2062	2063	2064	2065	2065	2065
9	+2065	2065	2065	2064	2064	2063	2061	2060	2058	2057

For $x > 15$

$$N_0(x) = \sqrt{\frac{2}{\pi x}}\left[\sin\left(x - \frac{\pi}{4}\right) - \frac{1}{8x}\cos\left(x - \frac{\pi}{4}\right)\right], \text{ error } < 0.0001$$

$$\sqrt{\frac{2}{\pi}} = 0.7979 \qquad \frac{\pi}{4} = 0.7854$$

Table F-11. Bessel Functions: $N_0(x)$ and $N_1(x)$ (*Continued*)

$N_1(x)$

x	0	1	2	3	4	5	6	7	8	9
10.0	+0.2490	2493	2496	2498	2500	2502	2504	2506	2507	2508
1	2508	2509	2509	2509	2509	2508	2507	2506	2505	2504
2	2502	2500	2498	2495	2492	2489	2486	2483	2479	2475
3	2471	2466	2462	2457	2451	2446	2440	2435	2428	2422
4	2416	2409	2402	2394	2387	2379	2371	2363	2355	2346
5	2337	2328	2319	2309	2299	2289	2279	2269	2258	2247
6	2236	2225	2214	2202	2190	2178	2166	2153	2140	2128
7	2114	2101	2088	2074	2060	2046	2032	2017	2003	1988
8	1973	1958	1942	1927	1911	1895	1879	1863	1846	1830
9	1813	1796	1779	1762	1745	1727	1709	1692	1674	1655
11.0	1637	1619	1600	1581	1562	1543	1524	1505	1486	1466
1	1446	1427	1407	1387	1366	1346	1326	1305	1285	1264
2	1243	1222	1201	1180	1159	1137	1116	1095	1073	1051
3	1029	1008	0986	0964	0941	0919	0897	0875	0852	0830
4	0807	0785	0762	0740	0717	0694	0671	0648	0625	0602
5	0579	0556	0533	0510	0487	0464	0441	0417	0394	0371
6	0348	0324	0301	0278	0254	0231	0208	0184	0161	0138
7	0114	0091	0068	0045	0021	*0002	*0025	*0048	*0072	*0095
8	−0118	0141	0164	0187	0210	0233	0256	0279	0302	0324
9	−0347	0370	0392	0415	0437	0460	0482	0505	0527	0549
12.0	−0571	0593	0615	0637	0659	0681	0702	0723	0745	0766
1	−0787	0809	0830	0851	0871	0892	0913	0933	0954	0974
2	−0994	1014	1034	1054	1074	1093	1113	1132	1151	1171
3	−1189	1208	1227	1246	1264	1282	1300	1318	1336	1354
4	−1371	1389	1406	1423	1440	1457	1474	1490	1506	1522
5	−1538	1554	1570	1585	1601	1616	1631	1645	1660	1675
6	−1689	1703	1717	1730	1744	1757	1771	1783	1796	1809
7	−1821	1834	1846	1857	1869	1880	1892	1903	1914	1924
8	−1935	1945	1955	1965	1975	1984	1993	2002	2011	2020
9	−2028	2036	2044	2052	2060	2067	2074	2081	2088	2095
13.0	−2101	2107	2113	2118	2124	2129	2134	2139	2144	2148
1	−2152	2156	2160	2163	2167	2170	2172	2175	2178	2180
2	−2182	2183	2185	2186	2187	2188	2189	2189	2190	2190
3	−2190	2189	2188	2188	2187	2185	2184	2182	2180	2178
4	−2176	2173	2170	2167	2164	2161	2157	2153	2149	2145
5	−2140	2136	2131	2126	2120	2115	2109	2103	2097	2090
6	−2084	2077	2070	2063	2056	2048	2040	2032	2024	2016
7	−2007	1999	1990	1981	1971	1962	1952	1942	1932	1922
8	−1912	1901	1890	1879	1868	1857	1845	1834	1822	1810
9	−1798	1785	1773	1760	1747	1734	1721	1707	1694	1680
14.0	−1666	1652	1638	1624	1610	1595	1580	1565	1550	1535
1	−1520	1504	1489	1473	1457	1441	1425	1409	1392	1376
2	−1359	1342	1325	1308	1291	1274	1257	1239	1222	1204
3	−1186	1168	1150	1132	1114	1096	1077	1059	1040	1021
4	−1003	0984	0965	0946	0927	0907	0888	0869	0849	0830
5	−0810	0791	0771	0751	0732	0712	0692	0672	0652	0632
6	−0612	0591	0571	0551	0531	0510	0490	0469	0449	0428
7	−0408	0387	0367	0346	0326	0305	0284	0264	0243	0222
8	−0202	0181	0160	0140	0119	0098	0077	0057	0036	0015
9	+0005	0026	0047	0067	0088	0108	0129	0149	0170	0190

For $x > 15$

$$N_1(x) = \sqrt{\frac{2}{\pi x}}\left[\sin\left(x - \frac{3\pi}{4}\right) + \frac{3}{8x}\cos\left(x - \frac{3\pi}{4}\right)\right], \quad \text{error} < 0.0001$$

$$\sqrt{\frac{2}{\pi}} = 0.7979 \qquad \frac{3\pi}{4} = 2.3562$$

Table F-11. Bessel Functions: $I_0(x)$ and $I_1(x)$

$I_0(x)$

x	0	1	2	3	4	5	6	7	8	9
0.0	1.000	1.000	1.000	1.000	1.000	1.001	1.001	1.001	1.002	1.002
1	1.003	1.003	1.004	1.004	1.005	1.006	1.006	1.007	1.008	1.009
2	1.010	1.011	1.012	1.013	1.014	1.016	1.017	1.018	1.020	1.021
3	1.023	1.024	1.026	1.027	1.029	1.031	1.033	1.035	1.036	1.038
4	1.040	1.042	1.045	1.047	1.049	1.051	1.054	1.056	1.058	1.061
5	1.063	1.066	1.069	1.071	1.074	1.077	1.080	1.083	1.086	1.089
6	1.092	1.095	1.098	1.102	1.105	1.108	1.112	1.115	1.119	1.123
7	1.126	1.130	1.134	1.138	1.142	1.146	1.150	1.154	1.158	1.162
8	1.167	1.171	1.175	1.180	1.184	1.189	1.194	1.198	1.203	1.208
9	1.213	1.218	1.223	1.228	1.233	1.239	1.244	1.249	1.255	1.260
1.0	1.266	1.272	1.278	1.283	1.289	1.295	1.301	1.307	1.314	1.320
1	1.326	1.333	1.339	1.346	1.352	1.359	1.366	1.373	1.380	1.387
2	1.394	1.401	1.408	1.416	1.423	1.430	1.438	1.446	1.454	1.461
3	1.469	1.477	1.485	1.494	1.502	1.510	1.519	1.527	1.536	1.545
4	1.553	1.562	1.571	1.580	1.590	1.599	1.608	1.618	1.627	1.637
5	1.647	1.657	1.667	1.677	1.687	1.697	1.707	1.718	1.728	1.739
6	1.750	1.761	1.772	1.783	1.794	1.806	1.817	1.829	1.840	1.852
7	1.864	1.876	1.888	1.900	1.913	1.925	1.938	1.951	1.963	1.976
8	1.990	2.003	2.016	2.030	2.043	2.057	2.071	2.085	2.099	2.113
9	2.128	2.142	2.157	2.172	2.187	2.202	2.217	2.233	2.248	2.264
2.0	2.280	2.296	2.312	2.328	2.344	2.361	2.378	2.395	2.412	2.429
1	2.446	2.464	2.482	2.499	2.517	2.536	2.554	2.573	2.591	2.610
2	2.629	2.648	2.668	2.687	2.707	2.727	2.747	2.768	2.788	2.809
3	2.830	2.851	2.872	2.893	2.915	2.937	2.959	2.981	3.004	3.026
4	3.049	3.072	3.096	3.119	3.143	3.167	3.191	3.215	3.240	3.265
5	3.290	3.315	3.341	3.366	3.392	3.419	3.445	3.472	3.499	3.526
6	3.553	3.581	3.609	3.637	3.666	3.694	3.723	3.752	3.782	3.812
7	3.842	3.872	3.903	3.933	3.965	3.996	4.028	4.060	4.092	4.124
8	4.157	4.190	4.224	4.258	4.292	4.326	4.361	4.396	4.431	4.467
9	4.503	4.539	4.576	4.613	4.650	4.688	4.725	4.764	4.802	4.841
3.0	4.881	4.921	4.961	5.001	5.042	5.083	5.125	5.166	5.209	5.251
1	5.294	5.338	5.382	5.426	5.471	5.516	5.561	5.607	5.653	5.700
2	5.747	5.795	5.843	5.891	5.940	5.989	6.039	6.089	6.140	6.191
3	6.243	6.295	6.347	6.400	6.454	6.508	6.562	6.617	6.672	6.728
4	6.785	6.842	6.899	6.957	7.016	7.075	7.134	7.195	7.255	7.316
5	7.378	7.441	7.503	7.567	7.631	7.696	7.761	7.827	7.893	7.960
6	8.028	8.096	8.165	8.234	8.304	8.375	8.447	8.519	8.591	8.665
7	8.739	8.813	8.889	8.965	9.041	9.119	9.197	9.276	9.356	9.436
8	9.517	9.599	9.681	9.764	9.848	9.933	10.02	10.11	10.19	10.28
9	10.37	10.46	10.55	10.64	10.73	10.82	10.92	11.01	11.11	11.20
4.0	11.30	11.40	11.50	11.60	11.70	11.80	11.90	12.01	12.11	12.22
1	12.32	12.43	12.54	12.65	12.76	12.87	12.98	13.10	13.21	13.33
2	13.44	13.56	13.68	13.80	13.92	14.04	14.16	14.29	14.41	14.54
3	14.67	14.80	14.93	15.06	15.19	15.32	15.46	15.59	15.73	15.87
4	16.01	16.15	16.29	16.44	16.58	16.73	16.88	17.03	17.18	17.33
5	17.48	17.64	17.79	17.95	18.11	18.27	18.43	18.59	18.76	18.92
6	19.09	19.26	19.43	19.61	19.78	19.96	20.13	20.31	20.49	20.67
7	20.86	21.04	21.23	21.42	21.61	21.80	22.00	22.19	22.39	22.59
8	22.79	23.00	23.20	23.41	23.62	23.83	24.04	24.26	24.47	24.69
9	24.91	25.14	25.36	25.59	25.82	26.05	26.28	26.52	26.76	27.00
5.0	27.24	27.48	27.73	27.98	28.23	28.49	28.74	29.00	29.26	29.52

$$I_0(x) = 1 + \frac{x^2}{2^2(1!)^2} + \frac{x^4}{2^4(2!)^2} + \frac{x^6}{2^6(3!)^2} + \cdots$$

$$I'_0(x) = I_1(x)$$

Table F-11. Bessel Functions: $I_0(x)$ and $I_1(x)$ (*Continued*)

$I_1(x)$

x	0	1	2	3	4	5	6	7	8	9
0.0	0.0000	0050	0100	0150	0200	0250	0300	0350	0400	0450
1	0501	0551	0601	0651	0702	0752	0803	0853	0904	0954
2	1005	1056	1107	1158	1209	1260	1311	1362	1414	1465
3	1517	1569	1621	1673	1725	1777	1829	1882	1935	1987
4	2040	2093	2147	2200	2254	2307	2361	2415	2470	2524
5	2579	2634	2689	2744	2800	2855	2911	2967	3024	3080
6	3137	3194	3251	3309	3367	3425	3483	3542	3600	3659
7	3719	3778	3838	3899	3959	4020	4081	4142	4204	4266
8	4329	4391	4454	4518	4581	4646	4710	4775	4840	4905
9	4971	5038	5104	5171	5239	5306	5375	5443	5512	5582
1.0	5652	5722	5793	5864	5935	6008	6080	6153	6227	6300
1	6375	6450	6525	6601	6677	6754	6832	6910	6988	7067
2	7147	7227	7308	7389	7470	7553	7636	7719	7803	7888
3	7973	8059	8146	8233	8321	8409	8498	8588	8678	8769
4	8861	8953	9046	9140	9235	9330	9426	9522	9620	9718
5	9817	9916	1.002	1.012	1.022	1.032	1.043	1.053	1.064	1.074
6	1.085	1.096	1.106	1.117	1.128	1.139	1.151	1.162	1.173	1.185
7	1.196	1.208	1.220	1.232	1.244	1.256	1.268	1.280	1.292	1.305
8	1.317	1.330	1.343	1.355	1.368	1.381	1.395	1.408	1.421	1.435
9	1.448	1.462	1.476	1.490	1.504	1.518	1.532	1.547	1.561	1.576
2.0	1.591	1.606	1.621	1.636	1.651	1.666	1.682	1.698	1.713	1.729
1	1.745	1.762	1.778	1.795	1.811	1.828	1.845	1.862	1.879	1.897
2	1.914	1.932	1.950	1.968	1.986	2.004	2.022	2.041	2.060	2.079
3	2.098	2.117	2.136	2.156	2.176	2.196	2.216	2.236	2.257	2.277
4	2.298	2.319	2.340	2.362	2.383	2.405	2.427	2.449	2.471	2.494
5	2.517	2.540	2.563	2.586	2.610	2.633	2.657	2.682	2.706	2.731
6	2.755	2.780	2.806	2.831	2.857	2.883	2.909	2.935	2.962	2.989
7	3.016	3.043	3.071	3.099	3.127	3.155	3.184	3.213	3.242	3.271
8	3.301	3.331	3.361	3.392	3.422	3.453	3.485	3.516	3.548	3.580
9	3.613	3.645	3.678	3.712	3.745	3.779	3.813	3.848	3.883	3.918
3.0	3.953	3.989	4.025	4.062	4.098	4.136	4.173	4.211	4.249	4.287
1	4.326	4.365	4.405	4.445	4.485	4.526	4.567	4.608	4.650	4.692
2	4.734	4.777	4.820	4.864	4.908	4.953	4.997	5.043	5.088	5.134
3	5.181	5.228	5.275	5.323	5.371	5.420	5.469	5.519	5.569	5.619
4	5.670	5.722	5.773	5.826	5.879	5.932	5.986	6.040	6.095	6.150
5	6.206	6.262	6.319	6.376	6.434	6.493	6.552	6.611	6.671	6.732
6	6.793	6.854	6.917	6.979	7.043	7.107	7.171	7.237	7.302	7.369
7	7.436	7.503	7.572	7.640	7.710	7.780	7.851	7.922	7.994	8.067
8	8.140	8.215	8.289	8.365	8.441	8.518	8.595	8.674	8.753	8.832
9	8.913	8.994	9.076	9.159	9.242	9.326	9.411	9.497	9.584	9.671
4.0	9.759	9.848	9.938	10.03	10.12	10.21	10.31	10.40	10.50	10.59
1	10.69	10.79	10.88	10.98	11.08	11.18	11.29	11.39	11.49	11.60
2	11.71	11.81	11.92	12.03	12.14	12.25	12.36	12.48	12.59	12.71
3	12.82	12.94	13.06	13.18	13.30	13.42	13.54	13.67	13.79	13.92
4	14.05	14.17	14.30	14.44	14.57	14.70	14.84	14.97	15.11	15.25
5	15.39	15.53	15.67	15.82	15.96	16.11	16.26	16.41	16.56	16.71
6	16.86	17.02	17.17	17.33	17.49	17.65	17.81	17.98	18.14	18.31
7	18.48	18.65	18.82	18.99	19.17	19.35	19.52	19.70	19.88	20.07
8	20.25	20.44	20.63	20.82	21.01	21.20	21.40	21.60	21.80	22.00
9	22.20	22.40	22.61	22.82	23.03	23.24	23.46	23.67	23.89	24.11
5.0	24.34	24.56	24.79	25.02	25.25	25.48	25.72	25.95	26.19	26.44

$$I_1(x) = \frac{x}{2 \cdot 0!1!} + \frac{x^3}{2^3 \cdot 1!2!} + \frac{x^5}{2^5 \cdot 2!3!} + \cdots$$

$$I_1(x) = I'_0(x)$$

Table F-11. Bessel Functions: $I_0(x)$ and $I_1(x)$ (*Continued*)

$I_0(x)$

x	0	1	2	3	4	5	6	7	8	9
5.0	27.24	27.48	27.73	27.98	28.23	28.49	28.74	29.00	29.26	29.52
1	29.79	30.06	30.33	30.60	30.88	31.15	31.43	31.72	32.00	32.29
2	32.58	32.88	33.17	33.47	33.78	34.08	34.39	34.70	35.01	35.33
3	35.65	35.97	36.30	36.62	36.96	37.29	37.63	37.97	38.31	38.66
4	39.01	39.36	39.72	40.08	40.44	40.81	41.18	41.55	41.93	42.31
5	42.69	43.08	43.47	43.87	44.27	44.67	45.08	45.49	45.90	46.32
6	46.74	47.16	47.59	48.03	48.46	48.90	49.35	49.80	50.25	50.71
7	51.17	51.64	52.11	52.59	53.06	53.55	54.04	54.53	55.03	55.53
8	56.04	56.55	57.07	57.59	58.11	58.65	59.18	59.72	60.27	60.82
9	61.38	61.94	62.51	63.08	63.65	64.24	64.83	65.42	66.02	66.62
6.0	67.23	67.85	68.47	69.10	69.73	70.37	71.02	71.67	72.33	72.99
1	73.66	74.34	75.02	75.71	76.41	77.11	77.82	78.53	79.25	79.98
2	80.72	81.46	82.21	82.97	83.73	84.50	85.28	86.06	86.85	87.65
3	88.46	89.28	90.10	90.93	91.77	92.61	93.47	94.33	95.20	96.08
4	96.96	97.86	98.76	99.67	100.6	101.5	102.5	103.4	104.4	105.3
5	106.3	107.3	108.3	109.3	110.3	111.3	112.3	113.4	114.4	115.5
6	116.5	117.6	118.7	119.8	120.9	122.0	123.2	124.3	125.5	126.6
7	127.8	129.0	130.2	131.4	132.6	133.8	135.1	136.3	137.6	138.8
8	140.1	141.4	142.7	144.1	145.4	146.8	148.1	149.5	150.9	152.3
9	153.7	155.1	156.6	158.0	159.5	161.0	162.5	164.0	165.5	167.0
7.0	168.6	170.2	171.7	173.3	175.0	176.6	178.2	179.9	181.6	183.2
1	185.0	186.7	188.4	190.2	191.9	193.7	195.5	197.4	199.2	201.0
2	202.9	204.8	206.7	208.6	210.6	212.6	214.5	216.5	218.6	220.6
3	222.7	224.7	226.8	229.0	231.1	233.2	235.4	237.6	239.8	242.1
4	244.3	246.6	248.9	251.3	253.6	256.0	258.4	260.8	263.2	265.7
5	268.2	270.7	273.2	275.8	278.3	280.9	283.6	286.2	288.9	291.6
6	294.3	297.1	299.9	302.7	305.5	308.4	311.3	314.2	317.1	320.1
7	323.1	326.1	329.2	332.3	335.4	338.5	341.7	344.9	348.1	351.4
8	354.7	358.0	361.4	364.8	368.2	371.6	375.1	378.6	382.2	385.8
9	389.4	393.1	396.8	400.5	404.2	408.0	411.9	415.7	419.6	423.6
8.0	427.6	431.6	435.6	439.7	443.9	448.0	452.2	456.5	460.8	465.1
1	469.5	473.9	478.4	482.9	487.4	492.0	496.6	501.3	506.0	510.8
2	515.6	520.4	525.3	530.3	535.3	540.3	545.4	550.6	555.7	561.0
3	566.3	571.6	577.0	582.4	587.9	593.4	599.0	604.7	610.4	616.1
4	621.9	627.8	633.7	639.7	645.7	651.8	658.0	664.2	670.5	676.8
5	683.2	689.6	696.1	702.7	709.3	716.0	722.8	729.6	736.5	743.4
6	750.5	757.5	764.7	771.9	779.2	786.6	794.0	801.5	809.1	816.7
7	824.4	832.2	840.1	848.0	856.1	864.2	872.3	880.6	888.9	897.3
8	905.8	914.4	923.0	931.7	940.6	949.5	958.4	967.5	976.7	985.9
9	995.2	1005	1014	1024	1033	1043	1053	1063	1073	1083
9.0	1094	1104	1114	1125	1136	1146	1157	1168	1179	1190
1	1202	1213	1225	1236	1248	1260	1272	1284	1296	1308
2	1321	1333	1346	1359	1371	1384	1398	1411	1424	1438
3	1451	1465	1479	1493	1507	1522	1536	1551	1565	1580
4	1595	1610	1626	1641	1657	1673	1688	1704	1721	1737
5	1753	1770	1787	1804	1821	1838	1856	1874	1891	1909
6	1927	1946	1964	1983	2002	2021	2040	2060	2079	2099
7	2119	2139	2159	2180	2201	2222	2243	2264	2286	2307
8	2329	2352	2374	2397	2419	2442	2466	2489	2513	2537
9	2561	2585	2610	2635	2660	2685	2711	2737	2763	2789
10.0	2816	2843	2870	2897	2925	2952	2981	3009	3038	3067

For larger values of the argument, use the auxiliary functions, pages 1057 to 1059.

Table F-11. Bessel Functions: $I_0(x)$ and $I_1(x)$ (*Continued*)

$I_1(x)$

x	0	1	2	3	4	5	6	7	8	9
5.0	24.34	24.56	24.79	25.02	25.25	25.48	25.72	25.95	26.19	26.44
1	26.68	26.93	27.18	27.43	27.68	27.94	28.20	28.46	28.72	28.99
2	29.25	29.53	29.80	30.07	30.35	30.63	30.92	31.20	31.49	31.79
3	32.08	32.38	32.68	32.98	33.29	33.59	33.91	34.22	34.54	34.86
4	35.18	35.51	35.84	36.17	36.51	36.85	37.19	37.53	37.88	38.23
5	38.59	38.95	39.31	39.67	40.04	40.41	40.79	41.17	41.55	41.94
6	42.33	42.72	43.12	43.52	43.93	44.33	44.75	45.16	45.58	46.01
7	46.44	46.87	47.30	47.74	48.19	48.64	49.09	49.55	50.01	50.48
8	50.95	51.42	51.90	52.38	52.87	53.37	53.86	54.36	54.87	55.38
9	55.90	56.42	56.95	57.48	58.02	58.56	59.10	59.66	60.21	60.77
6.0	61.34	61.91	62.49	63.08	63.67	64.26	64.86	65.47	66.08	66.70
1	67.32	67.95	68.58	69.22	69.87	70.53	71.18	71.85	72.52	73.20
2	73.89	74.58	75.27	75.98	76.69	77.41	78.13	78.86	79.60	80.35
3	81.10	81.86	82.63	83.40	84.18	84.97	85.77	86.57	87.38	88.20
4	89.03	89.86	90.70	91.55	92.41	93.28	94.15	95.04	95.93	96.83
5	97.74	98.65	99.58	100.5	101.5	102.4	103.4	104.3	105.3	106.3
6	107.3	108.3	109.3	110.4	111.4	112.4	113.5	114.6	115.6	116.7
7	117.8	118.9	120.0	121.2	122.3	123.5	124.6	125.8	127.0	128.2
8	129.4	130.6	131.8	133.1	134.3	135.6	136.9	138.1	139.4	140.8
9	142.1	143.4	144.8	146.1	147.5	148.9	150.3	151.7	153.1	154.6
7.0	156.0	157.5	159.0	160.5	162.0	163.5	165.1	166.6	168.2	169.8
1	171.4	173.0	174.6	176.3	177.9	179.6	181.3	183.0	184.7	186.5
2	188.3	190.0	191.8	193.6	195.5	197.3	199.2	201.0	202.9	204.9
3	206.8	208.7	210.7	212.7	214.7	216.7	218.8	220.9	222.9	225.0
4	227.2	229.3	231.5	233.7	235.9	238.1	240.4	242.6	244.9	247.2
5	249.6	251.9	254.3	256.7	259.2	261.6	264.1	266.6	269.1	271.7
6	274.2	276.8	279.4	282.1	284.8	287.4	290.2	292.9	295.7	298.5
7	301.3	304.2	307.0	310.0	312.9	315.9	318.8	321.9	324.9	328.0
8	331.1	334.2	337.4	340.6	343.8	347.1	350.4	353.7	357.1	360.4
9	363.9	367.3	370.8	374.3	377.9	381.4	385.1	388.7	392.4	396.1
8.0	399.9	403.7	407.5	411.4	415.3	419.2	423.2	427.2	431.3	435.4
1	439.5	443.7	447.9	452.1	456.4	460.7	465.1	469.5	474.0	478.5
2	483.0	487.6	492.3	496.9	501.7	506.4	511.2	516.1	521.0	526.0
3	531.0	536.0	541.1	546.2	551.4	556.7	562.0	567.3	572.7	578.2
4	583.7	589.2	594.8	600.5	606.2	611.9	617.8	623.6	629.6	635.6
5	641.6	647.7	653.9	660.1	666.4	672.7	679.1	685.6	692.1	698.7
6	705.4	712.1	718.9	725.7	732.6	739.6	746.7	753.8	760.9	768.2
7	775.5	782.9	790.4	797.9	805.5	813.2	820.9	828.7	836.6	844.6
8	852.7	860.8	869.0	877.3	885.6	894.1	902.6	911.2	919.9	928.7
9	937.5	946.5	955.5	964.6	973.8	983.1	992.5	1002	1012	1021
9.0	1031	1041	1051	1061	1071	1081	1091	1102	1112	1123
1	1134	1144	1155	1166	1178	1189	1200	1212	1223	1235
2	1247	1259	1271	1283	1295	1307	1320	1332	1345	1358
3	1371	1384	1397	1411	1424	1438	1452	1465	1479	1494
4	1508	1522	1537	1552	1566	1581	1596	1612	1627	1643
5	1658	1674	1690	1707	1723	1739	1756	1773	1790	1807
6	1824	1842	1859	1877	1895	1913	1931	1950	1969	1987
7	2006	2026	2045	2065	2084	2104	2125	2145	2165	2186
8	2207	2228	2250	2271	2293	2315	2337	2359	2382	2405
9	2428	2451	2475	2498	2522	2547	2571	2596	2621	2646
10.0	2671	2697	2722	2749	2775	2802	2828	2856	2883	2911

For larger values of the argument, use the auxiliary functions, pages 1057 to 1059.

Table F-11. Bessel Functions: $K_0(x)$ and $K_1(x)$ (*Continued*)

$K_0(x)$

x	0	1	2	3	4	5	6	7	8	9
0.0	∞	4.721	4.028	3.624	3.337	3.114	2.933	2.780	2.647	2.531
1	2.427	2.333	2.248	2.170	2.097	2.030	1.967	1.909	1.854	1.802
2	1.753	1.706	1.662	1.620	1.580	1.542	1.505	1.470	1.436	1.404
3	1.372	1.342	1.314	1.286	1.259	1.233	1.208	1.183	1.160	1.137
4	1.115	1.093	1.072	1.052	1.032	1.013	0.9943	9761	9584	9412
5	0.9244	9081	8921	8766	8614	8466	8321	8180	8042	7907
6	7775	7646	7520	7397	7277	7159	7043	6930	6820	6711
7	6605	6501	6399	6300	6202	6106	6012	5920	5829	5740
8	5653	5568	5484	5402	5321	5242	5165	5088	5013	4940
9	4867	4796	4727	4658	4591	4524	4459	4396	4333	4271
1.0	4210	4151	4092	4034	3977	3922	3867	3813	3760	3707
1	3656	3605	3556	3507	3459	3411	3365	3319	3273	3229
2	3185	3142	3100	3058	3017	2976	2936	2897	2858	2820
3	2782	2746	2709	2673	2638	2603	2569	2535	2502	2469
4	2437	2405	2373	2342	2312	2282	2252	2223	2194	2166
5	2138	2111	2083	2057	2030	2004	1979	1953	1928	1904
6	1880	1856	1832	1809	1786	1763	1741	1719	1697	1676
7	1655	1634	1614	1593	1573	1554	1534	1515	1496	1478
8	1459	1441	1423	1406	1388	1371	1354	1337	1321	1305
9	1288	1273	1257	1242	1226	1211	1196	1182	1167	1153
2.0	1139	1125	1111	1098	1084	1071	1058	1045	1033	1020
1	1008	*9956	*9836	*9717	*9600	*9484	*9370	*9257	*9145	*9035
2	0.08927	8820	8714	8609	8506	8404	8304	8204	8106	8010
3	7914	7820	7726	7634	7544	7454	7365	7278	7191	7106
4	7022	6939	6856	6775	6695	6616	6538	6461	6384	6309
5	6235	6161	6089	6017	5946	5877	5808	5739	5672	5606
6	5540	5475	5411	5348	5285	5223	5162	5102	5042	4984
7	4926	4868	4811	4755	4700	4645	4592	4538	4485	4433
8	4382	4331	4281	4231	4182	4134	4086	4039	3992	3946
9	3901	3856	3811	3767	3724	3681	3638	3597	3555	3514
3.0	3474	3434	3395	3356	3317	3279	3241	3204	3168	3131
1	3095	3060	3025	2990	2956	2922	2889	2856	2824	2791
2	2759	2728	2697	2666	2636	2606	2576	2547	2518	2489
3	2461	2433	2405	2378	2351	2325	2298	2272	2246	2221
4	2196	2171	2146	2122	2098	2074	2051	2028	2005	1982
5	1960	1938	1916	1894	1873	1852	1831	1810	1790	1770
6	1750	1730	1711	1692	1673	1654	1635	1617	1599	1581
7	1563	1546	1528	1511	1494	1477	1461	1445	1428	1412
8	1397	1381	1366	1350	1335	1320	1306	1291	1277	1262
9	1248	1234	1221	1207	1194	1180	1167	1154	1141	1129
4.0	1116	1104	1091	1079	1067	1055	1044	1032	1021	1009
1	0.009980	9869	9760	9652	9545	9439	9334	9231	9128	9027
2	8927	8829	8731	8634	8539	8444	8351	8259	8167	8077
3	7988	7900	7813	7726	7641	7557	7473	7391	7309	7229
4	7149	7070	6992	6915	6839	6764	6689	6616	6543	6471
5	6400	6329	6260	6191	6123	6056	5989	5923	5858	5794
6	5730	5668	5605	5544	5483	5423	5363	5305	5246	5189
7	5132	5076	5020	4965	4911	4857	4804	4751	4699	4648
8	4597	4547	4497	4448	4399	4351	4304	4257	4210	4164
9	4119	4074	4030	3986	3942	3899	3857	3814	3773	3732
5.0	3691	3651	3611	3572	3533	3494	3456	3419	3382	3345

$K'_0(x) = -K_1(x)$

Table F-11. Bessel Functions: $K_0(x)$ and $K_1(x)$ (*Continued*)

$K_1(x)$

x	0	1	2	3	4	5	6	7	8	9
0.0	∞	99.97	49.95	33.27	24.92	19.91	16.56	14.17	12.37	10.97
1	9.854	8.935	8.169	7.519	6.962	6.477	6.053	5.678	5.345	5.046
2	4.776	4.532	4.309	4.106	3.919	3.747	3.588	3.440	3.303	3.175
3	3.056	2.944	2.839	2.740	2.647	2.559	2.476	2.397	2.323	2.252
4	2.184	2.120	2.059	2.001	1.945	1.892	1.840	1.792	1.745	1.700
5	1.656	1.615	1.575	1.536	1.499	1.464	1.429	1.396	1.364	1.333
6	1.303	1.274	1.246	1.219	1.192	1.167	1.142	1.118	1.095	1.072
7	1.050	1.029	1.008	9882	9686	9496	9311	9130	8955	8784
8	0.8618	8456	8298	8144	7993	7847	7704	7564	7428	7295
9	7165	7039	6915	6794	6675	6560	6447	6336	6228	6122
1.0	6019	5918	5819	5722	5627	5534	5443	5354	5267	5181
1	5098	5016	4935	4856	4779	4703	4629	4556	4485	4415
2	4346	4279	4212	4147	4084	4021	3960	3900	3841	3782
3	3725	3670	3615	3561	3508	3455	3404	3354	3305	3256
4	3208	3161	3115	3070	3026	2982	2939	2897	2855	2814
5	2774	2734	2695	2657	2620	2583	2546	2510	2475	2440
6	2406	2373	2340	2307	2275	2244	2213	2182	2152	2123
7	2094	2065	2037	2009	1982	1955	1928	1902	1876	1851
8	1826	1802	1777	1754	1730	1707	1684	1662	1640	1618
9	1597	1575	1555	1534	1514	1494	1474	1455	1436	1417
2.0	1399	1380	1362	1345	1327	1310	1293	1276	1260	1244
1	1227	1212	1196	1181	1166	1151	1136	1121	1107	1093
2	1079	1065	1052	1038	1025	1012	*9993	*9867	*9742	*9620
3	0.09498	9379	9261	9144	9029	8916	8804	8694	8586	8478
4	8372	8268	8165	8063	7963	7864	7767	7670	7575	7482
5	7389	7298	7208	7119	7031	6945	6859	6775	6692	6609
6	6528	6448	6369	6292	6215	6139	6064	5990	5917	5845
7	5774	5704	5634	5566	5498	5432	5366	5301	5237	5174
8	5111	5050	4989	4929	4869	4811	4753	4696	4639	4584
9	4529	4474	4421	4368	4316	4264	4213	4163	4113	4064
3.0	4016	3968	3921	3874	3828	3782	3738	3693	3649	3606
1	3563	3521	3480	3438	3398	3358	3318	3279	3240	3202
2	3164	3127	3090	3054	3018	2983	2948	2913	2879	2845
3	2812	2779	2746	2714	2682	2651	2620	2589	2559	2529
4	2500	2471	2442	2414	2385	2358	2330	2303	2276	2250
5	2224	2198	2173	2147	2123	2098	2074	2050	2026	2003
6	1979	1957	1934	1912	1890	1868	1846	1825	1804	1783
7	1763	1743	1722	1703	1683	1664	1645	1626	1607	1589
8	1571	1553	1535	1517	1500	1483	1466	1449	1432	1416
9	1400	1384	1368	1353	1337	1322	1307	1292	1277	1263
4.0	1248	1234	1220	1206	1193	1179	1166	1152	1139	1126
1	1114	1101	1089	1076	1064	1052	1040	1028	1017	10052
2	0.009938	9826	9715	9605	9497	9390	9284	9179	9076	8973
3	8872	8772	8674	8576	8479	8384	8290	8196	8104	8013
4	7923	7834	7746	7659	7573	7488	7404	7321	7239	7158
5	7078	6999	6920	6843	6766	6691	6616	6542	6469	6397
6	6325	6254	6185	6116	6047	5980	5913	5847	5782	5717
7	5654	5591	5529	5467	5406	5346	5286	5228	5169	5112
8	5055	4999	4943	4889	4834	4781	4727	4675	4623	4572
9	4521	4471	4421	4372	4324	4276	4229	4182	4136	4090
5.0	4045	4000	3956	3912	3869	3826	3784	3742	3700	3660

$K_1(x) = -K'_0(x)$

Table F-11. Bessel Functions: $K_0(x)$ and $K_1(x)$ (*Continued*)

$K_0(x)$

x	0	1	2	3	4	5	6	7	8	9
5.0	0.003691	3651	3611	3572	3533	3494	3456	3419	3382	3345
1	3308	3272	3237	3202	3167	3132	3098	3065	3031	2998
2	2966	2934	2902	2870	2839	2808	2778	2748	2718	2688
3	2659	2630	2602	2574	2546	2518	2491	2464	2437	2411
4	2385	2359	2333	2308	2283	2258	2234	2210	2186	2162
5	2139	2116	2093	2070	2048	2026	2004	1982	1961	1939
6	1918	1898	1877	1857	1837	1817	1798	1778	1759	1740
7	1721	1703	1684	1666	1648	1630	1613	1595	1578	1561
8	1544	1528	1511	1495	1479	1463	1447	1432	1416	1401
9	1386	1371	1356	1342	1327	1313	1299	1285	1271	1258
6.0	1244	1231	1217	1204	1191	1179	1166	1153	1141	1129
1	1117	1105	1093	1081	1070	1058	1047	1035	1024	1013
2	1002	*9918	*9811	*9706	*9602	*9499	*9398	*9297	*9197	*9099
3	0.0009001	8905	8810	8715	8622	8530	8438	8348	8259	8171
4	8083	7997	7911	7827	7743	7660	7578	7497	7417	7338
5	7259	7182	7105	7029	6954	6880	6806	6734	6662	6591
6	6520	6451	6382	6314	6246	6180	6114	6048	5984	5920
7	5857	5795	5733	5672	5611	5551	5492	5434	5376	5318
8	5262	5206	5150	5095	5041	4987	4934	4882	4830	4778
9	4728	4677	4627	4578	4529	4481	4434	4386	4340	4294
7.0	4248	4203	4158	4114	4070	4027	3984	3942	3900	3858
1	3817	3777	3737	3697	3658	3619	3580	3542	3505	3468
2	3431	3394	3358	3323	3287	3253	3218	3184	3150	3117
3	3084	3051	3019	2987	2955	2924	2893	2862	2832	2802
4	2772	2742	2713	2685	2656	2628	2600	2573	2545	2518
5	2492	2465	2439	2413	2388	2363	2338	2313	2288	2264
6	2240	2216	2193	2170	2147	2124	2102	2079	2057	2036
7	2014	1993	1972	1951	1930	1910	1890	1870	1850	1830
8	1811	1792	1773	1754	1736	1717	1699	1681	1664	1646
9	1629	1611	1594	1578	1561	1545	1528	1512	1496	1480
8.0	1465	1449	1434	1419	1404	1389	1374	1360	1346	1331
1	1317	1303	1290	1276	1263	1249	1236	1223	1210	1198
2	1185	1172	1160	1148	1136	1124	1112	1100	1089	1077
3	1066	1055	1043	1032	1022	1011	10002	*9897	*9793	*9690
4	.00009588	9487	9387	9288	9191	9094	8998	8904	8810	8717
5	8626	8535	8445	8356	8269	8182	8096	8011	7926	7843
6	7761	7679	7598	7519	7439	7361	7284	7208	7132	7057
7	6983	6909	6837	6765	6694	6624	6554	6485	6417	6350
8	6283	6217	6152	6088	6024	5961	5898	5836	5775	5714
9	5654	5595	5536	5478	5420	5364	5307	5252	5197	5142
9.0	5088	5035	4982	4930	4878	4827	4776	4726	4677	4628
1	4579	4531	4484	4437	4390	4344	4299	4254	4209	4165
2	4121	4078	4036	3993	3951	3910	3869	3829	3789	3749
3	3710	3671	3632	3594	3557	3519	3483	3446	3410	3374
4	3339	3304	3270	3235	3202	3168	3135	3102	3070	3038
5	3006	2974	2943	2912	2882	2852	2822	2793	2763	2734
6	2706	2678	2650	2622	2595	2567	2541	2514	2488	2462
7	2436	2411	2385	2360	2336	2311	2287	2263	2240	2216
8	2193	2170	2148	2125	2103	2081	2059	2038	2017	1995
9	1975	1954	1934	1913	1894	1874	1854	1835	1816	1797
10.0	1778	1759	1741	1723	1705	1687	1670	1652	1635	1618

For larger values of the argument, use the auxiliary functions, pages 1057 to 1059.

Table F-11. Bessel Functions: $K_0(x)$ and $K_1(x)$ (*Continued*)

$K_1(x)$

x	0	1	2	3	4	5	6	7	8	9
5.0	0.004045	4000	3956	3912	3869	3826	3784	3742	3700	3660
1	3619	3579	3540	3501	3462	3424	3386	3349	3312	3275
2	3239	3204	3168	3133	3099	3065	3031	2998	2965	2932
3	2900	2868	2836	2805	2774	2744	2714	2684	2655	2625
4	2597	2568	2540	2512	2485	2457	2430	2404	2377	2351
5	2326	2300	2275	2250	2225	2201	2177	2153	2130	2106
6	2083	2060	2038	2016	1994	1972	1950	1929	1908	1887
7	1866	1846	1826	1806	1786	1767	1748	1729	1710	1691
8	1673	1654	1636	1619	1601	1584	1566	1549	1532	1516
9	1499	1483	1467	1451	1435	1419	1404	1389	1374	1359
6.0	1344	1329	1315	1301	1286	1273	1259	1245	1232	1218
1	1205	1192	1179	1166	1154	1141	1129	1116	1104	1092
2	1081	1069	1057	1046	1034	1023	1012	1001	*9904	*9797
3	0.0009691	9586	9483	9380	9279	9178	9079	8981	8884	8788
4	8693	8599	8506	8414	8324	8234	8145	8057	7970	7884
5	7799	7715	7632	7549	7468	7387	7308	7229	7151	7074
6	6998	6922	6848	6774	6701	6629	6558	6487	6417	6348
7	6280	6212	6145	6079	6014	5949	5885	5822	5759	5697
8	5636	5576	5516	5456	5398	5340	5282	5226	5170	5114
9	5059	5005	4951	4898	4845	4793	4742	4691	4641	4591
7.0	4542	4493	4445	4397	4350	4304	4257	4212	4167	4122
1	4078	4034	3991	3948	3906	3864	3823	3782	3741	3701
2	3662	3623	3584	3545	3508	3470	3433	3396	3360	3324
3	3288	3253	3219	3184	3150	3116	3083	3050	3018	2985
4	2953	2922	2891	2860	2829	2799	2769	2740	2710	2682
5	2653	2625	2597	2569	2542	2514	2488	2461	2435	2409
6	2383	2358	2333	2308	2283	2259	2235	2211	2188	2164
7	2141	2118	2096	2074	2051	2030	2008	1987	1966	1945
8	1924	1903	1883	1863	1843	1824	1804	1785	1766	1747
9	1729	1710	1692	1674	1656	1639	1621	1604	1587	1570
8.0	1554	1537	1521	1505	1489	1473	1457	1442	1427	1411
1	1396	1382	1367	1352	1338	1324	1310	1296	1282	1269
2	1255	1242	1229	1216	1203	1190	1177	1165	1153	1140
3	1128	1116	1105	1093	1081	1070	1058	1047	1036	1025
4	1014	10036	*9930	*9825	*9721	*9618	*9516	*9415	*9316	*9217
5	00009120	9023	8928	8833	8740	8648	8556	8466	8376	8288
6	8200	8113	8028	7943	7859	7776	7694	7612	7532	7452
7	7374	7296	7219	7142	7067	6992	6918	6845	6773	6702
8	6631	6561	6492	6423	6355	6288	6222	6156	6091	6027
9	5964	5901	5838	5777	5716	5656	5596	5537	5479	5421
9.0	5364	5307	5251	5196	5141	5087	5033	4980	4928	4876
1	4825	4774	4723	4674	4624	4576	4528	4480	4433	4386
2	4340	4294	4249	4204	4160	4116	4073	4030	3988	3946
3	3904	3863	3822	3782	3742	3703	3664	3626	3587	3550
4	3512	3476	3439	3403	3367	3332	3297	3262	3228	3194
5	3160	3127	3094	3062	3029	2998	2966	2935	2904	2874
6	2843	2814	2784	2755	2726	2697	2669	2641	2613	2586
7	2559	2532	2505	2479	2453	2427	2402	2377	2352	2327
8	2302	2278	2254	2231	2207	2184	2161	2139	2116	2094
9	2072	2050	2029	2008	1987	1966	1945	1925	1905	1885
10.0	1865	1845	1826	1807	1788	1769	1751	1732	1714	1696

For larger values of the argument, use the auxiliary functions, pages 1057 to 1059.

Table F-11. Bessel Functions: Auxiliary Functions (*Continued*)

Auxiliary Functions $N_0(x)$ and $N_1(x)$ for Small Values of Argument

For small values of the argument, $N_0(x)$ and $N_1(x)$ are rapidly changing functions and linear interpolation is inaccurate. These tables of auxiliary functions can be used to give accurate interpolated values. For values of the argument above 0.1 the main tables are satisfactory if interpolation formulas are used.

$$N_0(x) = C_0 + D_0 \log x$$
$$N_1(x) = (C_1/x) + D_1 \log x$$

C_0

	0	1	2	3	4	5	6	7	8	9
0.0	−0.0738	0738	0737	0736	0735	0734	0732	0729	0727	0724
1	−0720	0717	0713	0708	0703	0698	0693	0687	0681	0674
2	−0667	0660	0652	0645	0636	0628	0619	0609	0600	0590
3	−0579	0569	0558	0547	0535	0523	0511	0498	0485	0472
4	−0458	0444	0430	0415	0400	0385	0369	0353	0337	0321

D_0

	0	1	2	3	4	5	6	7	8	9
0.0	1.4659	4658	4657	4655	4653	4650	4646	4641	4635	4629
1	1.4622	4614	4606	4597	4587	4576	4565	4553	4540	4527
2	1.4512	4498	4482	4465	4448	4431	4412	4393	4373	4352
3	1.4331	4309	4286	4262	4238	4213	4188	4161	4134	4107
4	1.4078	4049	4019	3989	3958	3926	3893	3860	3826	3792

C_1

	0	1	2	3	4	5	6	7	8	9
0.0	−0.6366	6366	6367	6368	6369	6371	6373	6376	6379	6382
1	−6386	6390	6394	6399	6404	6410	6416	6422	6429	6436
2	−6444	6452	6460	6468	6477	6487	6496	6506	6517	6527
3	−6538	6550	6561	6573	6586	6598	6611	6624	6638	6652
4	−6666	6681	6695	6710	6726	6741	6757	6773	6789	6806

D_1

	0	1	2	3	4	5	6	7	8	9
0.0	0.0000	0073	0146	0220	0293	0366	0440	0513	0586	0659
1	0732	0805	0878	0951	1024	1096	1169	1241	1314	1386
2	1459	1531	1603	1675	1746	1818	1890	1961	2032	2103
3	2174	2245	2316	2386	2456	2526	2596	2666	2735	2804
4	2873	2942	3011	3079	3148	3215	3283	3351	3418	3485

Auxiliary Functions $K_0(x)$ and $K_1(x)$ for Small Values of Argument

For small values of the argument, $K_0(x)$ and $K_1(x)$ are rapidly changing functions and linear interpolation is inaccurate. These tables of auxiliary functions can be used to give accurate interpolated values. For values of the argument above 0.1 the main tables are satisfactory if interpolation formulas are used.

$$K_0(x) = E_0 + F_0 \log x$$
$$K_1(x) = (E_1/x) + F_1 \log x$$

$E_0(x)$

	0	1	2	3	4	5	6	7	8	9
0.0	0.1159	1160	1160	1162	1164	1166	1169	1173	1177	1182
1	1187	1193	1200	1207	1214	1222	1231	1240	1250	1260
2	1271	1283	1295	1308	1321	1335	1349	1364	1380	1396
3	1412	1430	1448	1466	1485	1505	1525	1546	1567	1590
4	1612	1635	1659	1684	1709	1735	1761	1788	1816	1844

Table F-11. Bessel Functions: Auxiliary Functions (*Continued*)

$E_1(x)$

	0	1	2	3	4	5	6	7	8	9
0.0	1.0000	1.0000	9999	9997	9995	9992	9989	9985	9980	9975
1	0.9969	9963	9955	9948	9939	9930	9921	9910	9899	9888
2	9875	9863	9849	9835	9820	9804	9788	9771	9753	9735
3	9716	9696	9676	9654	9633	9610	9586	9562	9537	9512
4	9485	9458	9430	9401	9371	9341	9310	9278	9245	9211

$F_0(x)$

	0	1	2	3	4	5	6	7	8	9
0.0	−2.3026	3026	3028	3031	3035	3040	3047	3054	3063	3073
1	−2.3083	3096	3109	3123	3139	3156	3173	3193	3213	3234
2	−2.3257	3280	3305	3331	3359	3387	3417	3447	3479	3513
3	−2.3547	3582	3619	3657	3696	3736	3778	3821	3865	3910
4	−2.3956	4004	4053	4103	4154	4206	4260	4315	4371	4429

$F_1(x)$

	0	1	2	3	4	5	6	7	8	9
0.0	0.0000	0115	0230	0345	0461	0576	0691	0806	0922	1037
1	1153	1268	1384	1500	1616	1732	1848	1964	2081	2197
2	2314	2431	2548	2666	2783	2901	3019	3137	3255	3374
3	3493	3612	3731	3851	3971	4092	4212	4333	4454	4576
4	4698	4820	4943	5066	5189	5313	5437	5562	5687	5812

Examples of use of auxiliary functions for small values of argument:

Example 1. $N_0(0.115) = -0.0715 + 1.4610 \times \bar{1}.0607 = -0.0715 - 1.4610 + 0.0887 = -1.4438.$ Linear interpolation from the direct-reading table of N_0 would give the less accurate value

$$N_0(0.115) = -1.4444$$

Example 2.

$$N_1(0.115) = \frac{-0.6392}{0.115} + 0.08415 \times \bar{1}.0607 = -5.558 - 0.084 + 0.005 = -5.637$$

compared with the less accurate value of −5.648 obtained by linear interpolation of the table for $N_1(x)$.

Auxiliary functions $I_0(x)$, $I_1(x)$, $K_0(x)$, $K_1(x)$ for large values of argument

$e^{-x}I_0(x)$

	0	1	2	3	4	5	6	7	8	9
10.0	0.1278	1272	1265	1259	1253	1247	1241	1235	1229	1223
11.0	1217	1212	1206	1201	1195	1190	1185	1180	1174	1170
12.0	1164	1159	1154	1150	1145	1140	1136	1131	1126	1122
13.0	1118	1113	1109	1105	1100	1096	1092	1088	1084	1080
14.0	1076	1072	1068	1065	1061	1057	1053	1050	1046	1043
15.0	1039	1035	1032	1029	1025	1022	1018	1015	1012	1009
16.0	1005	1002	0999	0996	0993	0990	0987	0984	0981	0978
17.0	0975	0972	0969	0966	0963	0961	0958	0955	0952	0950
18.0	0950	0944	0942	0940	0937	0934	0931	0929	0926	0924
19.0	0921	0919	0917	0914	0912	0909	0907	0905	0902	0900
20	0898	0876	0856	0836	0819	0802	0786	0771	0757	0744
30	0731	0719	0708	0697	0687	0677	0667	0658	0649	0641

$I_0(x)$ = tabulated number $\times e^x$.

For greater values of x, $e^{-x}I_0(x) \approx \left(1 + \frac{1}{8x}\right)/\sqrt{2\pi x}$.

Table F-11. Bessel Functions: Auxiliary Functions (*Continued*)

$e^{-x}I_1(x)$

	0	1	2	3	4	5	6	7	8	9
10.0	0.1213	1207	1202	1196	1191	1186	1181	1175	1170	1165
11.0	1161	1156	1151	1146	1142	1137	1132	1128	1123	1119
12.0	1115	1110	1106	1102	1098	1094	1090	1086	1082	1078
13.0	1074	1070	1066	1062	1059	1055	1051	1048	1044	1040
14.0	1037	1034	1030	1027	1023	1020	1017	1013	1010	1007
15.0	1004	1001	0997	0994	0991	0988	0985	0982	0979	0976
16.0	0973	0971	0968	0965	0962	0959	0957	0954	0951	0948
17.0	0946	0943	0941	0938	0935	0933	0930	0928	0925	0923
18.0	0920	0918	0915	0913	0911	0908	0906	0904	0901	0899
19.0	0897	0895	0892	0890	0888	0886	0884	0881	0879	0877
20	0875	0855	0836	0818	0801	0786	0771	0757	0744	0731
30	0719	0708	0697	0687	0677	0667	0658	0649	0641	0633

$I_1(x)$ = tabulated number $\times\ e^x$.

For greater values of x, $e^{-x}I_1(x) \approx \left(1 - \frac{3}{8x}\right)\Big/\sqrt{2\pi x}$.

$e^xK_0(x)$

	0	1	2	3	4	5	6	7	8	9
10.0	0.3916	3897	3879	3860	3842	3824	3806	3789	3772	3755
11.0	3738	3721	3705	3689	3673	3657	3642	3627	3612	3597
12.0	3582	3567	3553	3539	3525	3511	3497	3484	3470	3457
13.0	3444	3431	3418	3406	3393	3381	3368	3356	3344	3333
14.0	3321	3309	3298	3286	3275	3264	3253	3242	3231	3221
15.0	3210	3200	3189	3179	3169	3159	3149	3139	3129	3119
16.0	3110	3100	3091	3081	3072	3063	3054	3045	3036	3027
17.0	3018	3009	3001	2992	2984	2975	2967	2959	2950	2942
18.0	2934	2926	2918	2910	2903	2895	2887	2879	2872	2864
19.0	2857	2850	2842	2835	2828	2821	2813	2806	2799	2792
20	2785	2719	2658	2599	2545	2494	2446	2401	2358	2318
30	2279	2242	2207	2174	2142	2111	2082	2054	2027	2001

$K_0(x)$ = tabulated number $\times\ e^{-x}$.

For greater values of x, $e^xK_0(x) \approx \sqrt{\frac{\pi}{2x}}\left(1 - \frac{1}{8x}\right)$.

$e^xK_1(x)$

	0	1	2	3	4	5	6	7	8	9
10.0	0.4108	4086	4064	4043	4023	4002	3982	3962	3943	3923
11.0	3904	3886	3867	3849	3831	3813	3796	3779	3762	3745
12.0	3728	3712	3696	3680	3664	3649	3633	3618	3603	3589
13.0	3574	3560	3545	3531	3518	3504	3490	3477	3464	3450
14.0	3437	3425	3412	3399	3387	3375	3363	3351	3339	3327
15.0	3315	3304	3292	3281	3270	3259	3248	3237	3226	3216
16.0	3205	3195	3185	3174	3164	3154	3144	3135	3125	3115
17.0	3106	3096	3087	3077	3068	3059	3050	3041	3032	3023
18.0	3015	3006	2997	2989	2980	2972	2964	2955	2947	2939
19.0	2931	2923	2915	2907	2900	2892	2884	2877	2869	2862
20	2854	2783	2717	2655	2598	2544	2493	2445	2400	2357
30	2317	2278	2241	2206	2173	2141	2110	2081	2053	2026

$K_1(x)$ = tabulated number $\times\ e^{-x}$.

For greater values of x, $e^xK_1(x) \approx \sqrt{\frac{\pi}{2x}}\left(1 + \frac{3}{8x}\right)$.

Example of use of auxiliary functions for large values of argument:

$$I_0(25) = 7.202 \times 10^{10} \times 0.0802 = 5.776 \times 10^9$$

Table F-12. Legendre Polynomials*

x	$P_2(x)$	$P_3(x)$	$P_4(x)$	$P_5(x)$	$P_6(x)$
0.00	−0.5000	0.0000	0.3750	0.0000	−0.3125
01	−4998	−0150	3746	0187	−3118
02	−4994	−0300	3735	0374	−3099
03	−4986	−0449	3716	0560	−3066
04	−4976	−0598	3690	0744	−3021
05	−4962	−0747	3657	0927	−2962
06	−4946	−0895	3616	1106	−2891
07	−4926	−1041	3567	1283	−2808
08	−4904	−1187	3512	1455	−2713
09	−4878	−1332	3449	1624	−2606
10	−4850	−1475	3379	1788	−2488
11	−4818	−1617	3303	1947	−2360
12	−4784	−1757	3219	2101	−2220
13	−4746	−1895	3129	2248	−2071
14	−4706	−2031	3032	2389	−1913
15	−4662	−2166	2928	2523	−1746
16	−4616	−2298	2819	2650	−1572
17	−4566	−2427	2703	2769	−1389
18	−4514	−2554	2581	2880	−1201
19	−4458	−2679	2453	2982	−1006
20	−4400	−2800	2320	3075	−0806
21	−4338	−2918	2181	3159	−0601
22	−4274	−3034	2037	3234	−0394
23	−4206	−3146	1889	3299	−0183
24	−4136	−3254	1735	3353	0029
25	−4062	−3359	1577	3397	0243
26	−3986	−3461	1415	3431	0456
27	−3906	−3558	1249	3453	0669
28	−3824	−3651	1079	3465	0879
29	−3738	−3740	0906	3465	1087
30	−3650	−3825	0729	3454	1292
31	−3558	−3905	0550	3431	1492
32	−3464	−3981	0369	3397	1686
33	−3366	−4052	0185	3351	1873
34	−3266	−4117	−0000	3294	2053
35	−3162	−4178	−0187	3225	2225
36	−3056	−4234	−0375	3144	2388
37	−2946	−4284	−0564	3051	2540
38	−2834	−4328	−0753	2948	2681
39	−2718	−4367	−0942	2833	2810
40	−2600	−4400	−1130	2706	2926
41	−2478	−4427	−1317	2569	3029
42	−2354	−4448	−1504	2421	3118
43	−2226	−4462	−1688	2263	3191
44	−2096	−4470	−1870	2095	3249
45	−1962	−4472	−2050	1917	3290
46	−1826	−4467	−2226	1730	3314
47	−1686	−4454	−2399	1534	3321
48	−1544	−4435	−2568	1330	3310
49	−1398	−4409	−2732	1118	3280

x	$P_2(x)$	$P_3(x)$	$P_4(x)$	$P_5(x)$	$P_6(x)$
0.50	−0.1250	−0.4375	−0.2891	+0.0898	+0.3232
51	−1098	−4334	−3044	0673	3166
52	−0944	−4285	−3191	0441	3080
53	−0786	−4228	−3332	+0204	2975
54	−0626	−4163	−3465	−0037	2851
55	−0462	−4091	−3590	−0282	2708
56	−0296	−4010	−3707	−0529	2546
57	−0126	−3920	−3815	−0779	2366
58	+0046	−3822	−3914	−1028	2168
59	0222	−3716	−4002	−1278	1953
60	0400	−3600	−4080	−1526	1721
61	0582	−3475	−4146	−1772	1473
62	0766	−3342	−4200	−2014	1211
63	0954	−3199	−4242	−2251	0935
64	1144	−3046	−4270	−2482	0646
65	1338	−2884	−4284	−2705	0347
66	1534	−2713	−4284	−2919	+0038
67	1734	−2531	−4268	−3122	−0278
68	1936	−2339	−4236	−3313	−0601
69	2142	−2137	−4187	−3490	−0926
70	2350	−1925	−4121	−3652	−1253
71	2562	−1702	−4036	−3796	−1578
72	2776	−1469	−3933	−3922	−1899
73	2994	−1225	−3810	−4026	−2214
74	3214	−0969	−3666	−4107	−2518
75	3438	−0703	−3501	−4164	−2808
76	3664	−0426	−3314	−4193	−3081
77	3894	−0137	−3104	−4193	−3333
78	4126	+0164	−2871	−4162	−3559
79	4362	0476	−2613	−4097	−3756
80	4600	0800	−2330	−3995	−3918
81	4842	1136	−2021	−3855	−4041
82	5086	1484	−1685	−3674	−4119
83	5334	1845	−1321	−3449	−4147
84	5584	2218	−0928	−3177	−4120
85	5838	2603	−0506	−2857	−4030
86	6094	3001	−0053	−2484	−3872
87	6354	3413	+0431	−2056	−3638
88	6616	3837	0947	−1570	−3322
89	6882	4274	1496	−1023	−2916
90	7150	4725	2079	−0411	−2412
91	7422	5189	2698	+0268	−1802
92	7696	5667	3352	1017	−1077
93	7974	6159	4044	1842	−0229
94	8254	6665	4773	2744	+0751
95	8538	7184	5541	3727	1875
96	8824	7718	6349	4796	3151
97	9114	8267	7198	5954	4590
98	9406	8830	8089	7204	6204
99	9702	9407	9022	8552	8003
1.00	1.0000	1.0000	1.0000	1.0000	1.0000

$P_1(x) = x$ for all values of x

* From H. E. Etherington, *Nuclear Engineering Handbook*, McGraw-Hill, New York, 1958.

Table F-13. Probability Function or Error Integral: erf x*

x	0	1	2	3	4	5	6	7	8	9
0.00	0.0000	0011	0023	0034	0045	0056	0068	0079	0090	0102
1	0113	0124	0135	0147	0158	0169	0181	0192	0203	0214
2	0226	0237	0248	0260	0271	0282	0293	0305	0316	0327
3	0338	0350	0361	0372	0384	0395	0406	0417	0429	0440
4	0451	0462	0474	0485	0496	0507	0519	0530	0541	0553
5	0564	0575	0586	0598	0609	0620	0631	0643	0654	0665
6	0676	0688	0699	0710	0721	0732	0744	0755	0766	0777
7	0789	0800	0811	0822	0834	0845	0856	0867	0878	0890
8	0901	0912	0923	0934	0946	0957	0968	0979	0990	1002
9	1013	1024	1035	1046	1058	1069	1080	1091	1102	1113
10	1125	1136	1147	1158	1169	1180	1192	1203	1214	1225
1	1236	1247	1259	1270	1281	1292	1303	1314	1325	1336
2	1348	1359	1370	1381	1392	1403	1414	1425	1436	1448
3	1459	1470	1481	1492	1503	1514	1525	1536	1547	1558
4	1569	1581	1592	1603	1614	1625	1636	1647	1658	1669
5	1680	1691	1702	1713	1724	1735	1746	1757	1768	1779
6	1790	1801	1812	1823	1834	1845	1856	1867	1878	1889
7	1900	1911	1922	1933	1944	1955	1966	1977	1988	1998
8	2009	2020	2031	2042	2053	2064	2075	2086	2097	2108
9	2118	2129	2140	2151	2162	2173	2184	2194	2205	2216
20	2227	2238	2249	2260	2270	2281	2292	2303	2314	2324
1	2335	2346	2357	2368	2378	2389	2400	2411	2421	2432
2	2443	2454	2464	2475	2486	2497	2507	2518	2529	2540
3	2550	2561	2572	2582	2593	2604	2614	2625	2636	2646
4	2657	2668	2678	2689	2700	2710	2721	2731	2742	2753
5	2763	2774	2784	2795	2806	2816	2827	2837	2848	2858
6	2869	2880	2890	2901	2911	2922	2932	2943	2953	2964
7	2974	2985	2995	3006	3016	3027	3037	3047	3058	3068
8	3079	3089	3100	3110	3120	3131	3141	3152	3162	3172
9	3183	3193	3204	3214	3224	3235	3245	3255	3266	3276
30	3286	3297	3307	3317	3327	3338	3348	3358	3369	3379
1	3389	3399	3410	3420	3430	3440	3450	3461	3471	3481
2	3491	3501	3512	3522	3532	3542	3552	3562	3573	3583
3	3593	3603	3613	3623	3633	3643	3653	3663	3674	3684
4	3694	3704	3714	3724	3734	3744	3754	3764	3774	3784
5	3794	3804	3814	3824	3834	3844	3854	3864	3873	3883
6	3893	3903	3913	3923	3933	3943	3953	3963	3972	3982
7	3992	4002	4012	4022	4031	4041	4051	4061	4071	4080
8	4090	4100	4110	4119	4129	4139	4149	4158	4168	4178
9	4187	4197	4207	4216	4226	4236	4245	4255	4265	4274
40	4284	4294	4303	4313	4322	4332	4341	4351	4361	4370
1	4380	4389	4399	4408	4418	4427	4437	4446	4456	4465
2	4475	4484	4494	4503	4512	4522	4531	4541	4550	4359
3	4569	4578	4588	4597	4606	4616	4625	4634	4644	4653
4	4662	4672	4681	4690	4699	4709	4718	4727	4736	4746
5	4755	4764	4773	4782	4792	4801	4810	4819	4828	4837
6	4847	4856	4865	4874	4883	4892	4901	4910	4919	4928
7	4937	4946	4956	4965	4974	4983	4992	5001	5010	5019
8	5027	5036	5045	5054	5063	5072	5081	5090	5099	5108
9	5117	5126	5134	5143	5152	5161	5170	5179	5187	5196
50	5205	5214	5223	5231	5240	5249	5258	5266	5275	5284

$$\text{erf } x = H(x) = \frac{2}{\sqrt{\pi}} \int_0^x e^{-t^2}\,dt$$

* From H. E. Etherington, *Nuclear Engineering Handbook*, McGraw-Hill, New York, 1958.

Table F-14. Normal-distribution Areas*

Fractional parts of the total area (1.000) under the normal curve between the mean and a perpendicular erected at various numbers of standard deviations (x/σ) from the mean. To illustrate the use of the table, 39.065 per cent of the total area under the curve will lie between the mean and a perpendicular erected at a distance of 1.23σ from the mean.

Each figure in the body of the table is preceded by a decimal point.

x/σ	0.00	0.01	0.02	0.03	0.04	0.05	0.06	0.07	0.08	0.09
0.0	00000	00399	00798	01197	01595	01994	02392	02790	03188	03586
0.1	03983	04380	04776	05172	05567	05962	06356	06749	07142	07535
0.2	07926	08317	08706	09095	09483	09871	10257	10642	11026	11409
0.3	11791	12172	12552	12930	13307	13683	14058	14431	14803	15173
0.4	15554	15910	16276	16640	17003	17364	17724	18082	18439	18793
0.5	19146	19497	19847	20194	20450	20884	21226	21566	21904	22240
0.6	22575	22907	23237	23565	23891	24215	24537	24857	25175	25490
0.7	25804	26115	26424	26730	27035	27337	27637	27935	28230	28524
0.8	28814	29103	29389	29673	29955	30234	30511	30785	31057	31327
0.9	31594	31859	32121	32381	32639	32894	33147	33398	33646	33891
1.0	34134	34375	34614	34850	35083	35313	35543	35769	35993	36214
1.1	36433	36650	36864	37076	37286	37493	37698	37900	38100	38298
1.2	38493	38686	38877	39065	39251	39435	39617	39796	39973	40147
1.3	40320	40490	40658	40824	40988	41149	41308	41466	41621	41774
1.4	41924	42073	42220	42364	42507	42647	42786	42922	43056	43189
1.5	43319	43448	43574	43699	43822	43943	44062	44179	44295	44408
1.6	44520	44630	44738	44845	44950	45053	45154	45254	45352	45449
1.7	45543	45637	45728	45818	45907	45994	46080	46164	46246	46327
1.8	46407	46485	46562	46638	46712	46784	46856	46926	46995	47062
1.9	47128	47193	47257	47320	47381	47441	47500	47558	47615	47670
2.0	47725	47778	47831	47882	47932	47982	48030	48077	48124	48169
2.1	48214	48257	48300	48341	48382	48422	48461	48500	48537	48574
2.2	48610	48645	48679	48713	48745	48778	48809	48840	48870	48899
2.3	48928	48956	48983	49010	49036	49061	49086	49111	49134	49158
2.4	49180	49202	49224	49245	49266	49286	49305	49324	49343	49361
2.5	49379	49396	49413	49430	49446	49461	49477	49492	49506	49520
2.6	49534	49547	49560	49573	49585	49598	49609	49621	49632	49643
2.7	49653	49664	49674	49683	49693	49702	49711	49720	49728	49736
2.8	49744	49752	49760	49767	49774	49781	49788	49795	49801	49807
2.9	49813	49819	49825	49831	49836	49841	49846	49851	49856	49861
3.0	49865									
3.5	4997674									
4.0	4999683									
4.5	4999966									
5.0	4999997133									

* This table was adapted, by permission, from F. C. Kent, *Elements of Statistics*, McGraw-Hill, New York, 1924.

Table F-15. Normal-curve Ordinates*

Ordinates (heights) of the unit normal curve. The height (y) at any number of standard deviations $\frac{x}{\sigma}$ from the mean is

$$y = 0.3989e^{-\frac{1}{2}\left(\frac{x}{\sigma}\right)^2}$$

To obtain answers in units of particular problems, multiply these ordinates by $\frac{Ni}{\sigma}$ where N is the number of cases, i the class interval, and σ the standard deviation.

Each figure in the body of the table is preceded by a decimal point.

x/σ	0.00	0.01	0.02	0.03	0.04	0.05	0.06	0.07	0.08	0.09
0.0	39894	39892	39886	39876	39862	39844	39822	39797	39767	39733
0.1	39695	39654	39608	39559	39505	39448	39387	39322	39253	39181
0.2	39104	39024	38940	38853	38762	38667	38568	38466	38361	38251
0.3	38139	38023	37903	37780	37654	37524	37391	37255	37115	36973
0.4	36827	36678	36526	36371	36213	36053	35889	35723	35553	35381
0.5	35207	35029	34849	34667	34482	34294	34105	33912	33718	33521
0.6	33322	33121	32918	32713	32506	32297	32086	31874	31659	31443
0.7	31225	31006	30785	30563	30339	30114	29887	29658	29430	29200
0.8	28969	28737	28504	28269	28034	27798	27562	27324	27086	26848
0.9	26609	26369	26129	25888	25647	25406	25164	24923	24681	24439
1.0	24197	23955	23713	23471	23230	22988	22747	22506	22265	22025
1.1	21785	21546	21307	21069	20831	20594	20357	20121	19886	19652
1.2	19419	19186	18954	18724	18494	18265	18037	17810	17585	17360
1.3	17137	16915	16694	16474	16256	16038	15822	15608	15395	15183
1.4	14973	14764	14556	14350	14146	13943	13742	13542	13344	13147
1.5	12952	12758	12566	12376	12188	12001	11816	11632	11450	11270
1.6	11092	10915	10741	10567	10396	10226	10059	09893	09728	09566
1.7	09405	09246	09089	08933	08780	08628	08478	08329	08183	08038
1.8	07895	07754	07614	07477	07341	07206	07074	06943	06814	06687
1.9	06562	06438	06316	06195	06077	05959	05844	05730	05618	05508
2.0	05399	05292	05186	05082	04980	04879	04780	04682	04586	04491
2.1	04398	04307	04217	04128	04041	03955	03871	03788	03706	03626
2.2	03547	03470	03394	03319	03246	03174	03103	03034	02965	02898
2.3	02833	02768	02705	02643	02582	02522	02463	02406	02349	02294
2.4	02239	02186	02134	02083	02033	01984	01936	01888	01842	01797
2.5	01753	01709	01667	01625	01585	01545	01506	01468	01431	01394
2.6	01358	01323	01289	01256	01223	01191	01160	01130	01100	01071
2.7	01042	01014	00987	00961	00935	00909	00885	00861	00837	00814
2.8	00792	00770	00748	00727	00707	00687	00668	00649	00631	00613
2.9	00595	00578	00562	00545	00530	00514	00499	00485	00470	00457
3.0	00443									
3.5	0008727									
4.0	0001338									
4.5	0000160									
5.0	000001487									

* This table was adapted, by permission, from F. C. Kent, *Elements of Statistics*, McGraw-Hill, New York, 1924.

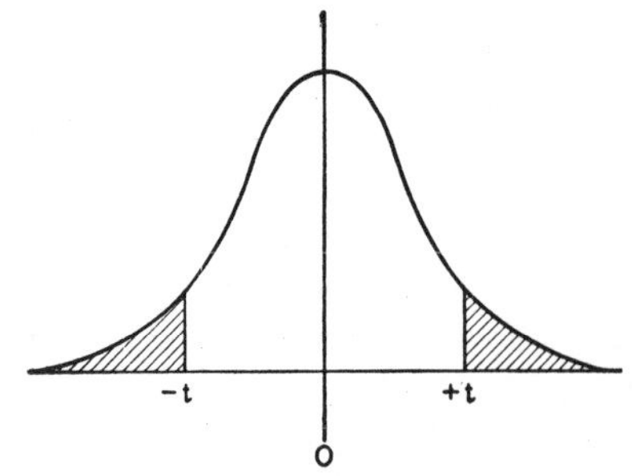

Table F-16. Distribution of t*

Values of t corresponding to certain selected probabilities (*i.e.*, tail areas under the curve). To illustrate: the probability is 0.05 that a sample with 20 degrees of freedom would have $t = 2.086$ or larger.

DF	Probability							
	0.80	0.40	0.20	0.10	0.05	0.02	0.01	0.001
1	0.325	1.376	3.078	6.314	12.706	31.821	63.657	636.619
2	0.289	1.061	1.886	2.920	4.303	6.965	9.925	31.598
3	0.277	0.978	1.638	2.353	3.182	4.541	5.841	12.941
4	0.271	0.941	1.533	2.132	2.776	3.747	4.604	8.610
5	0.267	0.920	1.476	2.015	2.571	3.365	4.032	6.859
6	0.265	0.906	1.440	1.943	2.447	3.143	3.707	5.959
7	0.263	0.896	1.415	1.895	2.365	2.998	3.499	5.405
8	0.262	0.889	1.397	1.860	2.306	2.896	3.355	5.041
9	0.261	0.883	1.383	1.833	2.262	2.821	3.250	4.781
10	0.260	0.879	1.372	1.812	2.228	2.764	3.169	4.587
11	0.260	0.876	1.363	1.796	2.201	2.718	3.106	4.437
12	0.259	0.873	1.356	1.782	2.179	2.681	3.055	4.318
13	0.259	0.870	1.350	1.771	2.160	2.650	3.012	4.221
14	0.258	0.868	1.345	1.761	2.145	2.624	2.977	4.140
15	0.258	0.866	1.341	1.753	2.131	2.602	2.947	4.073
16	0.258	0.865	1.337	1.746	2.120	2.583	2.921	4.015
17	0.257	0.863	1.333	1.740	2.110	2.567	2.898	3.965
18	0.257	0.862	1.330	1.734	2.101	2.552	2.878	3.922
19	0.257	0.861	1.328	1.729	2.093	2.539	2.861	3.883
20	0.257	0.860	1.325	1.725	2.086	2.528	2.845	3.850
21	0.257	0.859	1.323	1.721	2.080	2.518	2.831	3.819
22	0.256	0.858	1.321	1.717	2.074	2.508	2.819	3.792
23	0.256	0.858	1.319	1.714	2.069	2.500	2.807	3.767
24	0.256	0.857	1.318	1.711	2.064	2.492	2.797	3.745
25	0.256	0.856	1.316	1.708	2.060	2.485	2.787	3.725
26	0.256	0.856	1.315	1.706	2.056	2.479	2.779	3.707
27	0.256	0.855	1.314	1.703	2.052	2.473	2.771	3.690
28	0.256	0.855	1.313	1.701	2.048	2.467	2.763	3.674
29	0.256	0.854	1.311	1.699	2.045	2.462	2.756	3.659
30	0.256	0.854	1.310	1.697	2.042	2.457	2.750	3.646
40	0.255	0.851	1.303	1.684	2.021	2.423	2.704	3.551
60	0.254	0.848	1.296	1.671	2.000	2.390	2.660	3.460
120	0.254	0.845	1.289	1.658	1.980	2.358	2.617	3.373
∞	0.253	0.842	1.282	1.645	1.960	2.326	2.576	3.291

* This table is reproduced in abridged form from Table III of Fisher and Yates, *Statistical Tables for Biological, Agricultural, and Medical Research*, published by Oliver & Boyd, Ltd., Edinburgh, by permission of the authors and publishers.

Table F-17. Distribution of χ^2 *

Values of χ^2 corresponding to certain selected probabilities (*i.e.*, tail areas under the curve). To illustrate: the probability is 0.05 that a sample with 20 degrees of freedom, taken from a normal distribution, would have $\chi^2 = 31.410$ or larger.

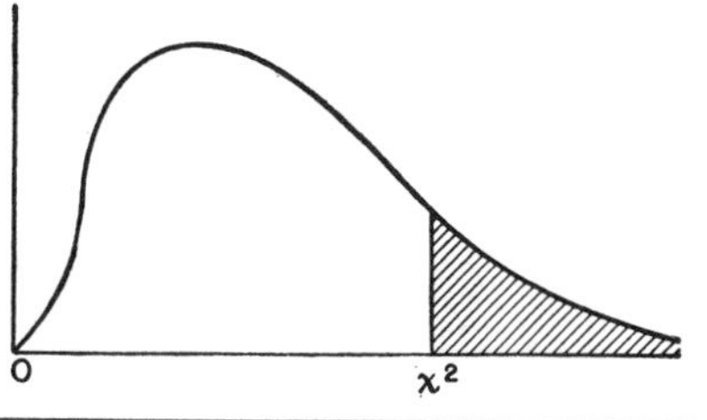

DF	Probability										
	0.99	0.98	0.95	0.90	0.80	0.20	0.10	0.05	0.02	0.01	0.001
1	0.0^3157	0.0^3628	0.00393	0.0158	0.0642	1.642	2.706	3.841	5.412	6.635	10.827
2	0.0201	0.0404	0.103	0.211	0.446	3.219	4.605	5.991	7.824	9.210	13.815
3	0.115	0.185	0.352	0.584	1.005	4.642	6.251	7.815	9.837	11.341	16.268
4	0.297	0.429	0.711	1.064	1.649	5.989	7.779	9.488	11.668	13.277	18.465
5	0.554	0.752	1.145	1.610	2.343	7.289	9.236	11.070	13.388	15.086	20.517
6	0.872	1.134	1.635	2.204	3.070	8.558	10.645	12.592	15.033	16.812	22.457
7	1.239	1.564	2.167	2.833	3.822	9.803	12.017	14.067	16.622	18.475	24.322
8	1.646	2.032	2.733	3.490	4.594	11.030	13.362	15.507	18.168	20.090	26.125
9	2.088	2.532	3.325	4.168	5.380	12.242	14.684	16.919	19.679	21.666	27.877
10	2.558	3.059	3.940	4.865	6.179	13.442	15.987	18.307	21.161	23.209	29.588
11	3.053	3.609	4.575	5.578	6.989	14.631	17.275	19.675	22.618	24.725	31.264
12	3.571	4.178	5.226	6.304	7.807	15.812	18.549	21.026	24.054	26.217	32.909
13	4.107	4.765	5.892	7.042	8.634	16.985	19.812	22.362	25.472	27.688	34.528
14	4.660	5.368	6.571	7.790	9.467	18.151	21.064	23.685	26.873	29.141	36.123
15	5.229	5.985	7.261	8.547	10.307	19.311	22.307	24.996	28.259	30.578	37.697
16	5.812	6.614	7.962	9.312	11.152	20.465	23.542	26.296	29.633	32.000	39.252
17	6.408	7.255	8.672	10.085	12.002	21.615	24.769	27.587	30.995	33.409	40.790
18	7.015	7.906	9.390	10.865	12.857	22.760	25.989	28.869	32.346	34.805	42.312
19	7.633	8.567	10.117	11.651	13.716	23.900	27.204	30.144	33.687	36.191	43.820
20	8.260	9.237	10.851	12.443	14.578	25.038	28.412	31.410	35.020	37.566	45.315
21	8.897	9.915	11.591	13.240	15.445	26.171	29.615	32.671	36.343	38.932	46.797
22	9.542	10.600	12.338	14.041	16.314	27.301	30.813	33.924	37.659	40.289	48.268
23	10.196	11.293	13.091	14.848	17.187	28.429	32.007	35.172	38.968	41.638	49.728
24	10.856	11.992	13.848	15.659	18.062	29.553	33.196	36.415	40.270	42.980	51.179
25	11.524	12.697	14.611	16.473	18.940	30.675	34.382	37.652	41.566	44.314	52.620
26	12.198	13.409	15.379	17.292	19.820	31.795	35.563	38.885	42.856	45.642	54.052
27	12.879	14.125	16.151	18.114	20.703	32.912	36.741	40.113	44.140	46.963	55.476
28	13.565	14.847	16.928	18.939	21.588	34.027	37.916	41.337	45.419	48.278	56.893
29	14.256	15.574	17.708	19.768	22.475	35.139	39.087	42.557	46.693	49.588	58.302
30	14.953	16.306	18.493	20.599	23.364	36.250	40.256	43.773	47.962	50.892	59.703

* This table is reproduced in abridged form from Table IV of Fisher and Yates, *Statistical Tables for Biological, Agricultural, and Medical Research,* published by Oliver & Boyd, Ltd., Edinburgh, by permission of the authors and publishers.

Table F-18. Distribution of F*

5 Per Cent (Roman Type) and 1 Per Cent (Boldface Type)

Values of F corresponding to two selected probabilities (*i.e.*, tail areas under the curve). To illustrate: the probability is 0.05 that the ratio of two mean squares obtained with 20 and 10 degrees of freedom in numerator and denominator, respectively, would yield $F = 2.77$ or larger.

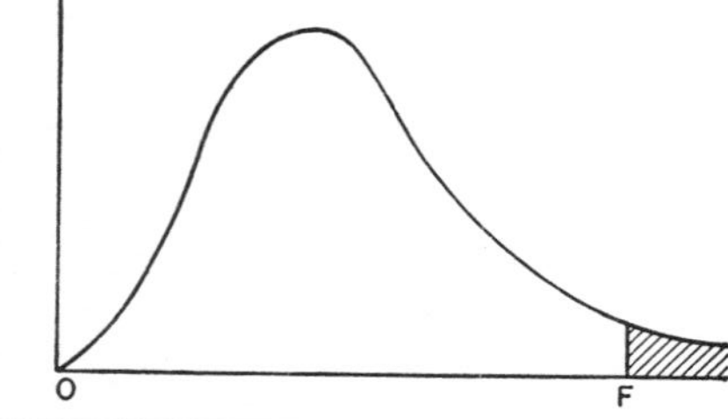

DF_2	DF_1 degrees of freedom for greater mean square (placed in the numerator)																							
	1	2	3	4	5	6	7	8	9	10	11	12	14	16	20	24	30	40	50	75	100	200	500	∞
1	161	200	216	225	230	234	237	239	241	242	243	244	245	246	248	249	250	251	252	253	253	254	254	254
	4,052	**4,999**	**5,403**	**5,625**	**5,764**	**5,859**	**5,928**	**5,981**	**6,022**	**6,056**	**6,082**	**6,106**	**6,142**	**6,169**	**6,208**	**6,234**	**6,258**	**6,286**	**6,302**	**6,323**	**6,334**	**6,352**	**6,361**	**6,366**
2	18.51	19.00	19.16	19.25	19.30	19.33	19.36	19.37	19.38	19.39	19.40	19.41	19.42	19.43	19.44	19.45	19.46	19.47	19.47	19.48	19.49	19.49	19.50	19.50
	98.49	**99.00**	**99.17**	**99.25**	**99.30**	**99.33**	**99.34**	**99.36**	**99.38**	**99.40**	**99.41**	**99.42**	**99.43**	**99.44**	**99.45**	**99.46**	**99.47**	**99.48**	**99.48**	**99.49**	**99.49**	**99.49**	**99.50**	**99.50**
3	10.13	9.55	9.28	9.12	9.01	8.94	8.88	8.84	8.81	8.78	8.76	8.74	8.71	8.69	8.66	8.64	8.62	8.60	8.58	8.57	8.56	8.54	8.54	8.53
	34.12	**30.82**	**29.46**	**28.71**	**28.24**	**27.91**	**27.67**	**27.49**	**27.34**	**27.23**	**27.13**	**27.05**	**26.92**	**26.83**	**26.69**	**26.60**	**26.50**	**26.41**	**26.35**	**26.27**	**26.23**	**26.18**	**26.14**	**26.12**
4	7.71	6.94	6.59	6.39	6.26	6.16	6.09	6.04	6.00	5.96	5.93	5.91	5.87	5.84	5.80	5.77	5.74	5.71	5.70	5.68	5.66	5.65	5.64	5.63
	21.20	**18.00**	**16.69**	**15.98**	**15.52**	**15.21**	**14.98**	**14.80**	**14.66**	**14.54**	**14.45**	**14.37**	**14.24**	**14.15**	**14.02**	**13.93**	**13.83**	**13.74**	**13.69**	**13.61**	**13.57**	**13.52**	**13.48**	**13.46**
5	6.61	5.79	5.41	5.19	5.05	4.95	4.88	4.82	4.78	4.74	4.70	4.68	4.64	4.60	4.56	4.53	4.50	4.46	4.44	4.42	4.40	4.38	4.37	4.36
	16.26	**13.27**	**12.06**	**11.39**	**10.97**	**10.67**	**10.45**	**10.27**	**10.15**	**10.05**	**9.96**	**9.89**	**9.77**	**9.68**	**9.55**	**9.47**	**9.38**	**9.29**	**9.24**	**9.17**	**9.13**	**9.07**	**9.04**	**9.02**
6	5.99	5.14	4.76	4.53	4.39	4.28	4.21	4.15	4.10	4.06	4.03	4.00	3.96	3.92	3.87	3.84	3.81	3.77	3.75	3.72	3.71	3.69	3.68	3.67
	13.74	**10.92**	**9.78**	**9.15**	**8.75**	**8.47**	**8.26**	**8.10**	**7.98**	**7.87**	**7.79**	**7.72**	**7.60**	**7.52**	**7.39**	**7.31**	**7.23**	**7.14**	**7.09**	**7.02**	**6.99**	**6.94**	**6.90**	**6.88**
7	5.59	4.74	4.35	4.12	3.97	3.87	3.79	3.73	3.68	3.63	3.60	3.57	3.52	3.49	3.44	3.41	3.38	3.34	3.32	3.29	3.28	3.25	3.24	3.23
	12.25	**9.55**	**8.45**	**7.85**	**7.46**	**7.19**	**7.00**	**6.84**	**6.71**	**6.62**	**6.54**	**6.47**	**6.35**	**6.27**	**6.15**	**6.07**	**5.98**	**5.90**	**5.85**	**5.78**	**5.75**	**5.70**	**5.67**	**5.65**
8	5.32	4.46	4.07	3.84	3.69	3.58	3.50	3.44	3.39	3.34	3.31	3.28	3.23	3.20	3.15	3.12	3.08	3.05	3.03	3.00	2.98	2.96	2.94	2.93
	11.26	**8.65**	**7.59**	**7.01**	**6.63**	**6.37**	**6.19**	**6.03**	**5.91**	**5.82**	**5.74**	**5.67**	**5.56**	**5.48**	**5.36**	**5.28**	**5.20**	**5.11**	**5.06**	**5.00**	**4.96**	**4.91**	**4.88**	**4.86**
9	5.12	4.26	3.86	3.63	3.48	3.37	3.29	3.23	3.18	3.13	3.10	3.07	3.02	2.98	2.93	2.90	2.86	2.82	2.80	2.77	2.76	2.73	2.72	2.71
	10.56	**8.02**	**6.99**	**6.42**	**6.06**	**5.80**	**5.62**	**5.47**	**5.35**	**5.26**	**5.18**	**5.11**	**5.00**	**4.92**	**4.80**	**4.73**	**4.64**	**4.56**	**4.51**	**4.45**	**4.41**	**4.36**	**4.33**	**4.31**
10	4.96	4.10	3.71	3.48	3.33	3.22	3.14	3.07	3.02	2.97	2.94	2.91	2.86	2.82	2.77	2.74	2.70	2.67	2.64	2.61	2.59	2.56	2.55	2.54
	10.04	**7.56**	**6.55**	**5.99**	**5.64**	**5.39**	**5.21**	**5.06**	**4.95**	**4.85**	**4.78**	**4.71**	**4.60**	**4.52**	**4.41**	**4.33**	**4.25**	**4.17**	**4.12**	**4.05**	**4.01**	**3.96**	**3.93**	**3.91**

* Reproduced by permission from George W. Snedecor, *Statistical Methods*, 5th ed., 1956, copyright by the Iowa State University Press, Ames, Iowa.

DF_2	DF_1 degrees of freedom for greater mean square (placed in the numerator)																							
	1	2	3	4	5	6	7	8	9	10	11	12	14	16	20	24	30	40	50	75	100	200	500	∞
11	4.84	3.98	3.59	3.36	3.20	3.09	3.01	2.95	2.90	2.86	2.82	2.79	2.74	2.70	2.65	2.61	2.57	2.53	2.50	2.47	2.45	2.42	2.41	2.40
	9.65	**7.20**	**6.22**	**5.67**	**5.32**	**5.07**	**4.88**	**4.74**	**4.63**	**4.54**	**4.46**	**4.40**	**4.29**	**4.21**	**4.10**	**4.02**	**3.94**	**3.86**	**3.80**	**3.74**	**3.70**	**3.66**	**3.62**	**3.60**
12	4.75	3.88	3.49	3.26	3.11	3.00	2.92	2.85	2.80	2.76	2.72	2.69	2.64	2.60	2.54	2.50	2.46	2.42	2.40	2.36	2.35	2.32	2.31	2.30
	9.33	**6.93**	**5.95**	**5.41**	**5.06**	**4.82**	**4.65**	**4.50**	**4.39**	**4.30**	**4.22**	**4.16**	**4.05**	**3.98**	**3.86**	**3.78**	**3.70**	**3.61**	**3.56**	**3.49**	**3.46**	**3.41**	**3.38**	**3.36**
13	4.67	3.80	3.41	3.18	3.02	2.92	2.84	2.77	2.72	2.67	2.63	2.60	2.55	2.51	2.46	2.42	2.38	2.34	2.32	2.28	2.26	2.24	2.22	2.21
	9.07	**6.70**	**5.74**	**5.20**	**4.86**	**4.62**	**4.44**	**4.30**	**4.19**	**4.10**	**4.02**	**3.96**	**3.85**	**3.78**	**3.67**	**3.59**	**3.51**	**3.42**	**3.37**	**3.30**	**3.27**	**3.21**	**3.18**	**3.16**
14	4.60	3.74	3.34	3.11	2.96	2.85	2.77	2.70	2.65	2.60	2.56	2.53	2.48	2.44	2.39	2.35	2.31	2.27	2.24	2.21	2.19	2.16	2.14	2.13
	8.86	**6.51**	**5.56**	**5.03**	**4.69**	**4.46**	**4.28**	**4.14**	**4.03**	**3.94**	**3.86**	**3.80**	**3.70**	**3.62**	**3.51**	**3.43**	**3.34**	**3.26**	**3.21**	**3.14**	**3.11**	**3.06**	**3.02**	**3.00**
15	4.54	3.68	3.29	3.06	2.90	2.79	2.70	2.64	2.59	2.55	2.51	2.48	2.43	2.39	2.33	2.29	2.25	2.21	2.18	2.15	2.12	2.10	2.08	2.07
	8.68	**6.36**	**5.42**	**4.89**	**4.56**	**4.32**	**4.14**	**4.00**	**3.89**	**3.80**	**3.73**	**3.67**	**3.56**	**3.48**	**3.36**	**3.29**	**3.20**	**3.12**	**3.07**	**3.00**	**2.97**	**2.92**	**2.89**	**2.87**
16	4.49	3.63	3.24	3.01	2.85	2.74	2.66	2.59	2.54	2.49	2.45	2.42	2.37	2.33	2.28	2.24	2.20	2.16	2.13	2.09	2.07	2.04	2.02	2.01
	8.53	**6.23**	**5.29**	**4.77**	**4.44**	**4.20**	**4.03**	**3.89**	**3.78**	**3.69**	**3.61**	**3.55**	**3.45**	**3.37**	**3.25**	**3.18**	**3.10**	**3.01**	**2.96**	**2.89**	**2.86**	**2.80**	**2.77**	**2.75**
17	4.45	3.59	3.20	2.96	2.81	2.70	2.62	2.55	2.50	2.45	2.41	2.38	2.33	2.29	2.23	2.19	2.15	2.11	2.08	2.04	2.02	1.99	1.97	1.96
	8.40	**6.11**	**5.18**	**4.67**	**4.34**	**4.10**	**3.93**	**3.79**	**3.68**	**3.59**	**3.52**	**3.45**	**3.35**	**3.27**	**3.16**	**3.08**	**3.00**	**2.92**	**2.86**	**2.79**	**2.76**	**2.70**	**2.67**	**2.65**
18	4.41	3.55	3.16	2.93	2.77	2.66	2.58	2.51	2.46	2.41	2.37	2.34	2.29	2.25	2.19	2.15	2.11	2.07	2.04	2.00	1.98	1.95	1.93	1.92
	8.28	**6.01**	**5.09**	**4.58**	**4.25**	**4.01**	**3.85**	**3.71**	**3.60**	**3.51**	**3.44**	**3.37**	**3.27**	**3.19**	**3.07**	**3.00**	**2.91**	**2.83**	**2.78**	**2.71**	**2.68**	**2.62**	**2.59**	**2.57**
19	4.38	3.52	3.13	2.90	2.74	2.63	2.55	2.48	2.43	2.38	2.34	2.31	2.26	2.21	2.15	2.11	2.07	2.02	2.00	1.96	1.94	1.91	1.90	1.88
	8.18	**5.93**	**5.01**	**4.50**	**4.17**	**3.94**	**3.77**	**3.63**	**3.52**	**3.43**	**3.36**	**3.30**	**3.19**	**3.12**	**3.00**	**2.92**	**2.84**	**2.76**	**2.70**	**2.63**	**2.60**	**2.54**	**2.51**	**2.49**
20	4.35	3.49	3.10	2.87	2.71	2.60	2.52	2.45	2.40	2.35	2.31	2.28	2.23	2.18	2.12	2.08	2.04	1.99	1.96	1.92	1.90	1.87	1.85	1.84
	8.10	**5.85**	**4.94**	**4.43**	**4.10**	**3.87**	**3.71**	**3.56**	**3.45**	**3.37**	**3.30**	**3.23**	**3.13**	**3.05**	**2.94**	**2.86**	**2.77**	**2.69**	**2.63**	**2.56**	**2.53**	**2.47**	**2.44**	**2.42**
21	4.32	3.47	3.07	2.84	2.68	2.57	2.49	2.42	2.37	2.32	2.28	2.25	2.20	2.15	2.09	2.05	2.00	1.96	1.93	1.89	1.87	1.84	1.82	1.81
	8.02	**5.78**	**4.87**	**4.37**	**4.04**	**3.81**	**3.65**	**3.51**	**3.40**	**3.31**	**3.24**	**3.17**	**3.07**	**2.99**	**2.88**	**2.80**	**2.72**	**2.63**	**2.58**	**2.51**	**2.47**	**2.42**	**2.38**	**2.36**
22	4.30	3.44	3.05	2.82	2.66	2.55	2.47	2.40	2.35	2.30	2.26	2.23	2.18	2.13	2.07	2.03	1.98	1.93	1.91	1.87	1.84	1.81	1.80	1.78
	7.94	**5.72**	**4.82**	**4.31**	**3.99**	**3.76**	**3.59**	**3.45**	**3.35**	**3.26**	**3.18**	**3.12**	**3.02**	**2.94**	**2.83**	**2.75**	**2.67**	**2.58**	**2.53**	**2.46**	**2.42**	**2.37**	**2.33**	**2.31**
23	4.28	3.42	3.03	2.80	2.64	2.53	2.45	2.38	2.32	2.28	2.24	2.20	2.14	2.10	2.04	2.00	1.96	1.91	1.88	1.84	1.82	1.79	1.77	1.76
	7.88	**5.66**	**4.76**	**4.26**	**3.94**	**3.71**	**3.54**	**3.41**	**3.30**	**3.21**	**3.14**	**3.07**	**2.97**	**2.89**	**2.78**	**2.70**	**2.62**	**2.53**	**2.48**	**2.41**	**2.37**	**2.32**	**2.28**	**2.26**
24	4.26	3.40	3.01	2.78	2.62	2.51	2.43	2.36	2.30	2.26	2.22	2.18	2.13	2.09	2.02	1.98	1.94	1.89	1.86	1.82	1.80	1.76	1.74	1.73
	7.82	**5.61**	**4.72**	**4.22**	**3.90**	**3.67**	**3.50**	**3.36**	**3.25**	**3.17**	**3.09**	**3.03**	**2.93**	**2.85**	**2.74**	**2.66**	**2.58**	**2.49**	**2.44**	**2.36**	**2.33**	**2.27**	**2.23**	**2.21**
25	4.24	3.38	2.99	2.76	2.60	2.49	2.41	2.34	2.28	2.24	2.20	2.16	2.11	2.06	2.00	1.96	1.92	1.87	1.84	1.80	1.77	1.74	1.72	1.71
	7.77	**5.57**	**4.68**	**4.18**	**3.86**	**3.63**	**3.46**	**3.32**	**3.21**	**3.13**	**3.05**	**2.99**	**2.89**	**2.81**	**2.70**	**2.62**	**2.54**	**2.45**	**2.40**	**2.32**	**2.29**	**2.23**	**2.19**	**2.17**

Table F-18. Distribution of F *(Continued)*

5 Per Cent (Roman Type) and 1 Per Cent (Boldface Type) Points for the Distribution of F

DF_2	DF_1 degrees of freedom for greater mean square (placed in the numerator)																							
	1	2	3	4	5	6	7	8	9	10	11	12	14	16	20	24	30	40	50	75	100	200	500	∞
26	4.22	3.37	2.98	2.74	2.59	2.47	2.39	2.32	2.27	2.22	2.18	2.15	2.10	2.05	1.99	1.95	1.90	1.85	1.82	1.78	1.76	1.72	1.70	1.69
	7.72	**5.53**	**4.64**	**4.14**	**3.82**	**3.59**	**3.42**	**3.29**	**3.17**	**3.09**	**3.02**	**2.96**	**2.86**	**2.77**	**2.66**	**2.58**	**2.50**	**2.41**	**2.36**	**2.28**	**2.25**	**2.19**	**2.15**	**2.13**
27	4.21	3.35	2.96	2.73	2.57	2.46	2.37	2.30	2.25	2.20	2.16	2.13	2.08	2.03	1.97	1.93	1.88	1.84	1.80	1.76	1.74	1.71	1.68	1.67
	7.68	**5.49**	**4.60**	**4.11**	**3.79**	**3.56**	**3.39**	**3.26**	**3.14**	**3.06**	**2.98**	**2.93**	**2.83**	**2.74**	**2.63**	**2.55**	**2.47**	**2.38**	**2.33**	**2.25**	**2.21**	**2.16**	**2.12**	**2.10**
28	4.20	3.34	2.95	2.71	2.56	2.44	2.36	2.29	2.24	2.19	2.15	2.12	2.06	2.02	1.96	1.91	1.87	1.81	1.78	1.75	1.72	1.69	1.67	1.65
	7.64	**5.45**	**4.57**	**4.07**	**3.76**	**3.53**	**3.36**	**3.23**	**3.11**	**3.03**	**2.95**	**2.90**	**2.80**	**2.71**	**2.60**	**2.52**	**2.44**	**2.35**	**2.30**	**2.22**	**2.18**	**2.13**	**2.09**	**2.06**
29	4.18	3.33	2.93	2.70	2.54	2.43	2.35	2.28	2.22	2.18	2.14	2.10	2.05	2.00	1.94	1.90	1.85	1.80	1.77	1.73	1.71	1.68	1.65	1.64
	7.60	**5.42**	**4.54**	**4.04**	**3.73**	**3.50**	**3.33**	**3.20**	**3.08**	**3.00**	**2.92**	**2.87**	**2.77**	**2.68**	**2.57**	**2.49**	**2.41**	**2.32**	**2.27**	**2.19**	**2.15**	**2.10**	**2.06**	**2.03**
30	4.17	3.32	2.92	2.69	2.53	2.42	2.34	2.27	2.21	2.16	2.12	2.09	2.04	1.99	1.93	1.89	1.84	1.79	1.76	1.72	1.69	1.66	1.64	1.62
	7.56	**5.39**	**4.51**	**4.02**	**3.70**	**3.47**	**3.30**	**3.17**	**3.06**	**2.98**	**2.90**	**2.84**	**2.74**	**2.66**	**2.55**	**2.47**	**2.38**	**2.29**	**2.24**	**2.16**	**2.13**	**2.07**	**2.03**	**2.01**
32	4.15	3.30	2.90	2.67	2.51	2.40	2.32	2.25	2.19	2.14	2.10	2.07	2.02	1.97	1.91	1.86	1.82	1.76	1.74	1.69	1.67	1.64	1.61	1.59
	7.50	**5.34**	**4.46**	**3.97**	**3.66**	**3.42**	**3.25**	**3.12**	**3.01**	**2.94**	**2.86**	**2.80**	**2.70**	**2.62**	**2.51**	**2.42**	**2.34**	**2.25**	**2.20**	**2.12**	**2.08**	**2.02**	**1.98**	**1.96**
34	4.13	3.28	2.88	2.65	2.49	2.38	2.30	2.23	2.17	2.12	2.08	2.05	2.00	1.95	1.89	1.84	1.80	1.74	1.71	1.67	1.64	1.61	1.59	1.57
	7.44	**5.29**	**4.42**	**3.93**	**3.61**	**3.38**	**3.21**	**3.08**	**2.97**	**2.89**	**2.82**	**2.76**	**2.66**	**2.58**	**2.47**	**2.38**	**2.30**	**2.21**	**2.15**	**2.08**	**2.04**	**1.98**	**1.94**	**1.91**
36	4.11	3.26	2.86	2.63	2.48	2.36	2.28	2.21	2.15	2.10	2.06	2.03	1.98	1.93	1.87	1.82	1.78	1.72	1.69	1.65	1.62	1.59	1.56	1.55
	7.39	**5.25**	**4.38**	**3.89**	**3.58**	**3.35**	**3.18**	**3.04**	**2.94**	**2.86**	**2.78**	**2.72**	**2.62**	**2.54**	**2.43**	**2.35**	**2.26**	**2.17**	**2.12**	**2.04**	**2.00**	**1.94**	**1.90**	**1.87**
38	4.10	3.25	2.85	2.62	2.46	2.35	2.26	2.19	2.14	2.09	2.05	2.02	1.96	1.92	1.85	1.80	1.76	1.71	1.67	1.63	1.60	1.57	1.54	1.53
	7.35	**5.21**	**4.34**	**3.86**	**3.54**	**3.32**	**3.15**	**3.02**	**2.91**	**2.82**	**2.75**	**2.69**	**2.59**	**2.51**	**2.40**	**2.32**	**2.22**	**2.14**	**2.08**	**2.00**	**1.97**	**1.90**	**1.86**	**1.84**
40	4.08	3.23	2.84	2.61	2.45	2.34	2.25	2.18	2.12	2.07	2.04	2.00	1.95	1.90	1.84	1.79	1.74	1.69	1.66	1.61	1.59	1.55	1.53	1.51
	7.31	**5.18**	**4.31**	**3.83**	**3.51**	**3.29**	**3.12**	**2.99**	**2.88**	**2.80**	**2.73**	**2.66**	**2.56**	**2.49**	**2.37**	**2.29**	**2.20**	**2.11**	**2.05**	**1.97**	**1.94**	**1.88**	**1.84**	**1.81**
42	4.07	3.22	2.83	2.59	2.44	2.32	2.24	2.17	2.11	2.06	2.02	1.99	1.94	1.89	1.82	1.78	1.73	1.68	1.64	1.60	1.57	1.54	1.51	1.49
	7.27	**5.15**	**4.29**	**3.80**	**3.49**	**3.26**	**3.10**	**2.96**	**2.86**	**2.77**	**2.70**	**2.64**	**2.54**	**2.46**	**2.35**	**2.26**	**2.17**	**2.08**	**2.02**	**1.94**	**1.91**	**1.85**	**1.80**	**1.78**
44	4.06	3.21	2.82	2.58	2.43	2.31	2.23	2.16	2.10	2.05	2.01	1.98	1.92	1.88	1.81	1.76	1.72	1.66	1.63	1.58	1.56	1.52	1.50	1.48
	7.24	**5.12**	**4.26**	**3.78**	**3.46**	**3.24**	**3.07**	**2.94**	**2.84**	**2.75**	**2.68**	**2.62**	**2.52**	**2.44**	**2.32**	**2.24**	**2.15**	**2.06**	**2.00**	**1.92**	**1.88**	**1.82**	**1.78**	**1.75**
46	4.05	3.20	2.81	2.57	2.42	2.30	2.22	2.14	2.09	2.04	2.00	1.97	1.91	1.87	1.80	1.75	1.71	1.65	1.62	1.57	1.54	1.51	1.48	1.46
	7.21	**5.10**	**4.24**	**3.76**	**3.44**	**3.22**	**3.05**	**2.92**	**2.82**	**2.73**	**2.66**	**2.60**	**2.50**	**2.42**	**2.30**	**2.22**	**2.13**	**2.04**	**1.98**	**1.90**	**1.86**	**1.80**	**1.76**	**1.72**
48	4.04	3.19	2.80	2.56	2.41	2.30	2.21	2.14	2.08	2.03	1.99	1.96	1.90	1.86	1.79	1.74	1.70	1.64	1.61	1.56	1.53	1.50	1.47	1.45
	7.19	**5.08**	**4.22**	**3.74**	**3.42**	**3.20**	**3.04**	**2.90**	**2.80**	**2.71**	**2.64**	**2.58**	**2.48**	**2.40**	**2.28**	**2.20**	**2.11**	**2.02**	**1.96**	**1.88**	**1.84**	**1.78**	**1.73**	**1.70**

DF_2	DF_1 degrees of freedom for greater mean square (placed in the numerator)																							
	1	2	3	4	5	6	7	8	9	10	11	12	14	16	20	24	30	40	50	75	100	200	500	∞
50	4.03	3.18	2.79	2.56	2.40	2.29	2.20	2.13	2.07	2.02	1.98	1.95	1.90	1.85	1.78	1.74	1.69	1.63	1.60	1.55	1.52	1.48	1.46	1.44
	7.17	**5.06**	**4.20**	**3.72**	**3.41**	**3.18**	**3.02**	**2.88**	**2.78**	**2.70**	**2.62**	**2.56**	**2.46**	**2.39**	**2.26**	**2.18**	**2.10**	**2.00**	**1.94**	**1.86**	**1.82**	**1.76**	**1.71**	**1.68**
55	4.02	3.17	2.78	2.54	2.38	2.27	2.18	2.11	2.05	2.00	1.97	1.93	1.88	1.83	1.76	1.72	1.67	1.61	1.58	1.52	1.50	1.46	1.43	1.41
	7.12	**5.01**	**4.16**	**3.68**	**3.37**	**3.15**	**2.98**	**2.85**	**2.75**	**2.66**	**2.59**	**2.53**	**2.43**	**2.35**	**2.23**	**2.15**	**2.06**	**1.96**	**1.90**	**1.82**	**1.78**	**1.71**	**1.66**	**1.64**
60	4.00	3.15	2.76	2.52	2.37	2.25	2.17	2.10	2.04	1.99	1.95	1.92	1.86	1.81	1.75	1.70	1.65	1.59	1.56	1.50	1.48	1.44	1.41	1.39
	7.08	**4.98**	**4.13**	**3.65**	**3.34**	**3.12**	**2.95**	**2.82**	**2.72**	**2.63**	**2.56**	**2.50**	**2.40**	**2.32**	**2.20**	**2.12**	**2.03**	**1.93**	**1.87**	**1.79**	**1.74**	**1.68**	**1.63**	**1.60**
65	3.99	3.14	2.75	2.51	2.36	2.24	2.15	2.08	2.02	1.98	1.94	1.90	1.85	1.80	1.73	1.68	1.63	1.57	1.54	1.49	1.46	1.42	1.39	1.37
	7.04	**4.95**	**4.10**	**3.62**	**3.31**	**3.09**	**2.93**	**2.79**	**2.70**	**2.61**	**2.54**	**2.47**	**2.37**	**2.30**	**2.18**	**2.09**	**2.00**	**1.90**	**1.84**	**1.76**	**1.71**	**1.64**	**1.60**	**1.56**
70	3.98	3.13	2.74	2.50	2.35	2.23	2.14	2.07	2.01	1.97	1.93	1.89	1.84	1.79	1.72	1.67	1.62	1.56	1.53	1.47	1.45	1.40	1.37	1.35
	7.01	**4.92**	**4.08**	**3.60**	**3.29**	**3.07**	**2.91**	**2.77**	**2.67**	**2.59**	**2.51**	**2.45**	**2.35**	**2.28**	**2.15**	**2.07**	**1.98**	**1.88**	**1.82**	**1.74**	**1.69**	**1.62**	**1.56**	**1.53**
80	3.96	3.11	2.72	2.48	2.33	2.21	2.12	2.05	1.99	1.95	1.91	1.88	1.82	1.77	1.70	1.65	1.60	1.54	1.51	1.45	1.42	1.38	1.35	1.32
	6.96	**4.88**	**4.04**	**3.56**	**3.25**	**3.04**	**2.87**	**2.74**	**2.64**	**2.55**	**2.48**	**2.41**	**2.32**	**2.24**	**2.11**	**2.03**	**1.94**	**1.84**	**1.78**	**1.70**	**1.65**	**1.57**	**1.52**	**1.49**
100	3.94	3.09	2.70	2.46	2.30	2.19	2.10	2.03	1.97	1.92	1.88	1.85	1.79	1.75	1.68	1.63	1.57	1.51	1.48	1.42	1.39	1.34	1.30	1.28
	6.90	**4.82**	**3.98**	**3.51**	**3.20**	**2.99**	**2.82**	**2.69**	**2.59**	**2.51**	**2.43**	**2.36**	**2.26**	**2.19**	**2.06**	**1.98**	**1.89**	**1.79**	**1.73**	**1.64**	**1.59**	**1.51**	**1.46**	**1.43**
125	3.92	3.07	2.68	2.44	2.29	2.17	2.08	2.01	1.95	1.90	1.86	1.83	1.77	1.72	1.65	1.60	1.55	1.49	1.45	1.39	1.36	1.31	1.27	1.25
	6.84	**4.78**	**3.94**	**3.47**	**3.17**	**2.95**	**2.79**	**2.65**	**2.56**	**2.47**	**2.40**	**2.33**	**2.23**	**2.15**	**2.03**	**1.94**	**1.85**	**1.75**	**1.68**	**1.59**	**1.54**	**1.46**	**1.40**	**1.37**
150	3.91	3.06	2.67	2.43	2.27	2.16	2.07	2.00	1.94	1.89	1.85	1.82	1.76	1.71	1.64	1.59	1.54	1.47	1.44	1.37	1.34	1.29	1.25	1.22
	6.81	**4.75**	**3.91**	**3.44**	**3.14**	**2.92**	**2.76**	**2.62**	**2.53**	**2.44**	**2.37**	**2.30**	**2.20**	**2.12**	**2.00**	**1.91**	**1.83**	**1.72**	**1.66**	**1.56**	**1.51**	**1.43**	**1.37**	**1.33**
200	3.89	3.04	2.65	2.41	2.26	2.14	2.05	1.98	1.92	1.87	1.83	1.80	1.74	1.69	1.62	1.57	1.52	1.45	1.42	1.35	1.32	1.26	1.22	1.19
	6.76	**4.71**	**3.88**	**3.41**	**3.11**	**2.90**	**2.73**	**2.60**	**2.50**	**2.41**	**2.34**	**2.28**	**2.17**	**2.09**	**1.97**	**1.88**	**1.79**	**1.69**	**1.62**	**1.53**	**1.48**	**1.39**	**1.33**	**1.28**
400	3.86	3.02	2.62	2.39	2.23	2.12	2.03	1.96	1.90	1.85	1.81	1.78	1.72	1.67	1.60	1.54	1.49	1.42	1.38	1.32	1.28	1.22	1.16	1.13
	6.70	**4.66**	**3.83**	**3.36**	**3.06**	**2.85**	**2.69**	**2.55**	**2.46**	**2.37**	**2.29**	**2.23**	**2.12**	**2.04**	**1.92**	**1.84**	**1.74**	**1.64**	**1.57**	**1.47**	**1.42**	**1.32**	**1.24**	**1.19**
1000	3.85	3.00	2.61	2.38	2.22	2.10	2.02	1.95	1.89	1.84	1.80	1.76	1.70	1.65	1.58	1.53	1.47	1.41	1.36	1.30	1.26	1.19	1.13	1.08
	6.66	**4.62**	**3.80**	**3.34**	**3.04**	**2.82**	**2.66**	**2.53**	**2.43**	**2.34**	**2.26**	**2.20**	**2.09**	**2.01**	**1.89**	**1.81**	**1.71**	**1.61**	**1.54**	**1.44**	**1.38**	**1.28**	**1.19**	**1.11**
∞	3.84	2.99	2.60	2.37	2.21	2.09	2.01	1.94	1.88	1.83	1.79	1.75	1.69	1.64	1.57	1.52	1.46	1.40	1.35	1.28	1.24	1.17	1.11	1.00
	6.64	**4.60**	**3.78**	**3.32**	**3.02**	**2.80**	**2.64**	**2.51**	**2.41**	**2.32**	**2.24**	**2.18**	**2.07**	**1.99**	**1.87**	**1.79**	**1.69**	**1.59**	**1.52**	**1.41**	**1.36**	**1.25**	**1.15**	**1.00**

Table F-19. Random Numbers*

10 09 73 25 33	76 52 01 35 86	34 67 35 48 76	80 95 90 91 17	39 29 27 49 45
37 54 20 48 05	64 89 47 42 96	24 80 52 40 37	20 63 61 04 02	00 82 29 16 65
08 42 26 89 53	19 64 50 93 03	23 20 90 25 60	15 95 33 47 64	35 08 03 36 06
99 01 90 25 29	09 37 67 07 15	38 31 13 11 65	88 67 67 43 97	04 43 62 76 59
12 80 79 99 70	80 15 73 61 47	64 03 23 66 53	98 95 11 68 77	12 17 17 68 33
66 06 57 47 17	34 07 27 68 50	36 69 73 61 70	65 81 33 98 85	11 19 92 91 70
31 06 01 08 05	45 57 18 24 06	35 30 34 26 14	86 79 90 74 39	23 40 30 97 32
85 26 97 76 02	02 05 16 56 92	68 66 57 48 18	73 05 38 52 47	18 62 38 85 79
63 57 33 21 35	05 32 54 70 48	90 55 35 75 48	28 46 82 87 09	83 49 12 56 24
73 79 64 57 53	03 52 96 47 78	35 80 83 42 82	60 93 52 03 44	35 27 38 84 35
98 52 01 77 67	14 90 56 86 07	22 10 94 05 58	60 97 09 34 33	50 50 07 39 98
11 80 50 54 31	39 80 82 77 32	50 72 56 82 48	29 40 52 42 01	52 77 56 78 51
83 45 29 96 34	06 28 89 80 83	13 74 67 00 78	18 47 54 06 10	68 71 17 78 17
88 68 54 02 00	86 50 75 84 01	36 76 66 79 51	90 36 47 64 93	29 60 91 10 62
99 59 46 73 48	87 51 76 49 69	91 82 60 89 28	93 78 56 13 68	23 47 83 41 13
65 48 11 76 74	17 46 85 09 50	58 04 77 69 74	73 03 95 71 86	40 21 81 65 44
80 12 43 56 35	17 72 70 80 15	45 31 82 23 74	21 11 57 82 53	14 38 55 37 63
74 35 09 98 17	77 40 27 72 14	43 23 60 02 10	45 52 16 42 37	96 28 60 26 55
69 91 62 68 03	66 25 22 91 48	36 93 68 72 03	76 62 11 39 90	94 40 05 64 18
09 89 32 05 05	14 22 56 85 14	46 42 75 67 88	96 29 77 88 22	54 38 21 45 98
91 49 91 45 23	68 47 92 76 86	46 16 28 35 54	94 75 08 99 23	37 08 92 00 48
80 33 69 45 98	26 94 03 68 58	70 29 73 41 35	53 14 03 33 40	42 05 08 23 41
44 10 48 19 49	85 15 74 79 54	32 97 92 65 75	57 60 04 08 81	22 22 20 64 13
12 55 07 37 42	11 10 00 20 40	12 86 07 46 97	96 64 48 94 39	28 70 72 58 15
63 60 64 93 29	16 50 53 44 84	40 21 95 25 63	43 65 17 70 82	07 20 73 17 90
61 19 69 04 46	26 45 74 77 74	51 92 43 37 29	65 39 45 95 93	42 58 26 05 27
15 47 44 52 66	95 27 07 99 53	59 36 78 38 48	82 39 61 01 18	33 21 15 94 66
94 55 72 85 73	67 89 75 43 87	54 62 24 44 31	91 19 04 25 92	92 92 74 59 73
42 48 11 62 13	97 34 40 87 21	16 86 84 87 67	03 07 11 20 59	25 70 14 66 70
23 52 37 83 17	73 20 88 98 37	68 93 59 14 16	26 25 22 96 63	05 52 28 25 62
04 49 35 24 94	75 24 63 38 24	45 86 25 10 25	61 96 27 93 35	65 33 71 24 72
00 54 99 76 54	64 05 18 81 59	96 11 96 38 96	54 69 28 23 91	23 28 72 95 29
35 96 31 53 07	26 89 80 93 54	33 35 13 54 62	77 97 45 00 24	90 10 33 93 33
59 80 80 83 91	45 42 72 68 42	83 60 94 97 00	13 02 12 48 92	78 56 52 01 06
46 05 88 52 36	01 39 09 22 86	77 28 14 40 77	93 91 08 36 47	70 61 74 29 41
32 17 90 05 97	87 37 92 52 41	05 56 70 70 07	86 74 31 71 57	85 39 41 18 38
69 23 46 14 06	20 11 74 52 04	15 95 66 00 00	18 74 39 24 23	97 11 89 63 38
19 56 54 14 30	01 75 87 53 79	40 41 92 15 85	66 67 43 68 06	84 96 28 52 07
45 15 51 49 38	19 47 60 72 46	43 66 79 45 43	59 04 79 00 33	20 82 66 95 41
94 86 43 19 94	36 16 81 08 51	34 88 88 15 53	01 54 03 54 56	05 01 45 11 76
98 08 62 48 26	45 24 02 84 04	44 99 90 88 96	39 09 47 34 07	35 44 13 18 80
33 18 51 62 32	41 94 15 09 49	89 43 54 85 81	88 69 54 19 94	37 54 87 30 43
80 95 10 04 06	96 38 27 07 74	20 15 12 33 87	25 01 62 52 98	94 62 46 11 71
79 75 24 91 40	71 96 12 82 96	69 86 10 25 91	74 85 22 05 39	00 38 75 95 79
18 63 33 25 37	98 14 50 65 71	31 01 02 46 74	05 45 56 14 27	77 93 89 19 36
74 02 94 39 02	77 55 73 22 70	97 79 01 71 19	52 52 75 80 21	80 81 45 17 48
54 17 84 56 11	80 99 33 71 43	05 33 51 29 69	56 12 71 92 55	36 04 09 03 24
11 66 44 98 83	52 07 98 48 27	59 38 17 15 39	09 97 33 34 40	88 46 12 33 56
48 32 47 79 28	31 24 96 47 10	02 29 53 68 70	32 30 75 75 46	15 02 00 99 94
69 07 49 41 38	87 63 79 19 76	35 58 40 44 01	10 51 82 16 15	01 84 87 69 38

* This table is reproduced with permission from tables of the RAND Corporation.

Table F-19. Random Numbers (*Continued*)

09 18 82 00 97	32 82 53 95 27	04 22 08 63 04	83 38 98 73 74	64 27 85 80 44
90 04 58 54 97	51 98 15 06 54	94 93 88 19 97	91 87 07 61 50	68 47 66 46 59
73 18 95 02 07	47 67 72 62 69	62 29 06 44 64	27 12 46 70 18	41 36 18 27 60
75 76 87 64 90	20 97 18 17 49	90 42 91 22 72	95 37 50 58 71	93 82 34 31 78
54 01 64 40 56	66 28 13 10 03	00 68 22 73 98	20 71 45 32 95	07 70 61 78 13
08 35 86 99 10	78 54 24 27 85	13 66 15 88 73	04 61 89 75 53	31 22 30 84 20
28 30 60 32 64	81 33 31 05 91	40 51 00 78 93	32 60 46 04 75	94 11 90 18 40
53 84 08 62 33	81 59 41 36 28	51 21 59 02 90	28 46 66 87 95	77 76 22 07 91
91 75 75 37 41	61 61 36 22 69	50 26 39 02 12	55 78 17 65 14	83 48 34 70 55
89 41 59 26 94	00 39 75 83 91	12 60 71 76 46	48 94 97 23 06	94 54 13 74 08
77 51 30 38 20	86 83 42 99 01	68 41 48 27 74	51 90 81 39 80	72 89 35 55 07
19 50 23 71 74	69 97 92 02 88	55 21 02 97 73	74 28 77 52 51	65 34 46 74 15
21 81 85 93 13	93 27 88 17 57	05 68 67 31 56	07 08 28 50 46	31 85 33 84 52
51 47 46 64 99	68 10 72 36 21	94 04 99 13 45	42 83 60 91 91	08 00 74 54 49
99 55 96 83 31	62 53 52 41 70	69 77 71 28 30	74 81 97 81 42	43 86 07 28 34
33 71 34 80 07	93 58 47 28 69	51 92 66 47 21	58 30 32 98 22	93 17 49 39 72
85 27 48 68 93	11 30 32 92 70	28 83 43 41 37	73 51 59 04 00	71 14 84 36 43
84 13 38 96 40	44 03 55 21 66	73 85 27 00 91	61 22 26 05 61	62 32 71 84 23
56 73 21 62 34	17 39 59 61 31	10 12 39 16 22	85 49 65 75 60	81 60 41 88 80
65 13 85 68 06	87 64 88 52 61	34 31 36 58 61	45 87 52 10 69	85 64 44 72 77
38 00 10 21 76	81 71 91 17 11	71 60 29 29 37	74 21 96 40 49	65 58 44 96 98
37 40 29 63 97	01 30 47 75 86	56 27 11 00 86	47 32 46 26 05	40 03 03 74 38
97 12 54 03 48	87 08 33 14 17	21 81 53 92 50	75 23 76 20 47	15 50 12 95 78
21 82 64 11 34	47 14 33 40 72	64 63 88 59 02	49 13 90 64 41	03 85 65 45 52
73 13 54 27 42	95 71 90 90 35	85 79 47 42 96	08 78 98 81 56	64 69 11 92 02
07 63 87 79 29	03 06 11 80 72	96 20 74 41 56	23 82 19 95 38	04 71 36 69 94
60 52 88 34 41	07 95 41 98 14	59 17 52 06 95	05 53 35 21 39	61 21 20 64 55
83 59 63 56 55	06 95 89 29 83	05 12 80 97 19	77 43 35 37 83	92 30 15 04 98
10 85 06 27 46	99 59 91 05 07	13 49 90 63 19	53 07 57 18 39	06 41 01 93 62
39 82 09 89 52	43 62 26 31 47	64 42 18 08 14	43 80 00 93 51	31 02 47 31 67
59 58 00 64 78	75 56 97 88 00	88 83 55 44 86	23 76 80 61 56	04 11 10 84 08
38 50 80 73 41	23 79 34 87 63	90 82 29 70 22	17 71 90 42 07	95 95 44 99 53
30 69 27 06 68	94 68 81 61 27	56 19 68 00 91	82 06 76 34 00	05 46 26 92 00
65 44 39 56 59	18 28 82 74 37	49 63 22 40 41	08 33 76 56 76	96 29 99 08 36
27 26 75 02 64	13 19 27 22 94	07 47 74 46 06	17 98 54 89 11	97 34 13 03 58
91 30 70 69 91	19 07 22 42 10	36 69 95 37 28	28 82 53 57 93	28 97 66 62 52
68 43 49 46 88	84 47 31 36 22	62 12 69 84 08	12 84 38 25 90	09 81 59 31 46
48 90 81 58 77	54 74 52 45 91	35 70 00 47 54	83 82 45 26 92	54 13 05 51 60
06 91 34 51 97	42 67 27 86 01	11 88 30 95 28	63 01 19 89 01	14 97 44 03 44
10 45 51 60 19	14 21 03 37 12	91 34 23 78 21	88 32 58 08 51	43 66 77 08 83
12 88 39 73 43	65 02 76 11 84	04 28 50 13 92	17 97 41 50 77	90 71 22 67 69
21 77 83 09 76	38 80 73 69 61	31 64 94 20 96	63 28 10 20 23	08 81 64 74 49
19 52 35 95 15	65 12 25 96 59	86 28 36 82 58	69 57 21 37 98	16 43 59 15 29
67 24 55 26 70	35 58 31 65 63	79 24 68 66 86	76 46 33 42 22	26 65 59 08 02
60 58 44 73 77	07 50 03 79 92	45 13 42 65 29	26 76 08 36 37	41 32 64 43 44
53 85 34 13 77	36 06 69 48 50	58 83 87 38 59	49 36 47 33 31	96 24 04 36 42
24 63 73 87 36	74 38 48 93 42	52 62 30 79 92	12 36 91 86 01	03 74 28 38 73
83 08 01 24 51	38 99 22 28 15	07 75 95 17 77	97 37 72 75 85	51 97 23 78 67
16 44 42 43 34	36 15 19 90 73	27 49 37 09 39	85 13 03 25 52	54 84 65 47 59
60 79 01 81 57	57 17 86 57 62	11 16 17 85 76	45 81 95 29 79	65 13 00 48 60

Table F-19. Random Numbers *(Continued)*

03 99 11 04 61	93 71 61 68 94	66 08 32 46 53	84 60 95 82 32	88 61 81 91 61
38 55 59 55 54	32 88 65 97 80	08 35 56 08 60	29 73 54 77 62	71 29 92 38 53
17 54 67 37 04	92 05 24 62 15	55 12 12 92 81	59 07 60 79 36	27 95 45 89 09
32 64 35 28 61	95 81 90 68 31	00 91 19 89 36	76 35 59 37 79	80 86 30 05 14
69 57 26 87 77	39 51 03 59 05	14 06 04 06 19	29 54 96 96 16	33 56 46 07 80
24 12 26 65 91	27 69 90 64 94	14 84 54 66 72	61 95 87 71 00	90 89 97 57 54
61 19 63 02 31	92 96 26 17 73	41 83 95 53 82	17 26 77 09 43	78 03 87 02 67
30 53 22 17 04	10 27 41 22 02	39 68 52 33 09	10 06 16 88 29	55 98 66 64 85
03 78 89 75 99	75 86 72 07 17	74 41 65 31 66	35 20 83 33 74	87 53 90 88 23
48 22 86 33 79	85 78 34 76 19	53 15 26 74 33	35 66 35 29 72	16 81 86 03 11
60 36 59 46 53	35 07 53 39 49	42 61 42 92 97	01 91 82 83 16	98 95 37 32 31
83 79 94 24 02	56 62 33 44 42	34 99 44 13 74	70 07 11 47 36	09 95 81 80 65
32 96 00 74 05	36 40 98 32 32	99 38 54 16 00	11 13 30 75 86	15 91 70 62 53
19 32 25 38 45	57 62 05 26 06	66 49 76 86 46	78 13 86 65 59	19 64 09 94 13
11 22 09 47 47	07 39 93 74 08	48 50 92 39 29	27 48 24 54 76	85 24 43 51 59
31 75 15 72 60	68 98 00 53 39	15 47 04 83 55	88 65 12 25 96	03 15 21 91 21
88 49 29 93 82	14 45 40 45 04	20 09 49 89 77	74 84 39 34 13	22 10 97 85 08
30 93 44 77 44	07 48 18 38 28	73 78 80 65 33	28 59 72 04 05	94 20 52 03 80
22 88 84 88 93	27 49 99 87 48	60 53 04 51 28	74 02 28 46 17	82 03 71 02 68
78 21 21 69 93	35 90 29 13 86	44 37 21 54 86	65 74 11 40 14	87 48 13 72 20
41 84 98 45 47	46 85 05 23 26	34 67 75 83 00	74 91 06 43 45	19 32 58 15 49
46 35 23 30 49	69 24 89 34 60	45 30 50 75 21	61 31 83 18 55	14 41 37 09 51
11 08 79 62 94	14 01 33 17 92	59 74 76 72 77	76 50 33 45 13	39 66 37 75 44
52 70 10 83 37	56 30 38 73 15	16 52 06 96 76	11 65 49 98 93	02 18 16 81 61
57 27 53 68 98	81 30 44 85 85	68 65 22 73 76	92 85 25 58 66	88 44 80 35 84
20 85 77 31 56	70 28 42 43 26	79 37 59 52 20	01 15 96 32 67	10 62 24 83 91
15 63 38 49 24	90 41 59 36 14	33 52 12 66 65	55 82 34 76 41	86 22 53 17 04
92 69 44 82 97	39 90 40 21 15	59 58 94 90 67	66 82 14 15 75	49 76 70 40 37
77 61 31 90 19	88 15 20 00 80	20 55 49 14 09	96 27 74 82 57	50 81 60 76 16
38 68 83 24 86	45 13 46 35 45	59 40 47 20 59	43 94 75 16 80	43 85 25 96 93
25 16 30 18 89	70 01 41 50 21	41 29 06 73 12	71 85 71 59 57	68 97 11 14 03
65 25 10 76 29	37 23 93 32 95	05 87 00 11 19	92 78 42 63 40	18 47 76 56 22
36 81 54 36 25	18 63 73 75 09	82 44 49 90 05	04 92 17 37 01	14 70 79 39 97
64 39 71 16 92	05 32 78 21 62	20 24 78 17 59	45 19 72 53 32	83 74 52 25 67
04 51 52 56 24	95 09 66 79 46	48 46 08 55 58	15 19 11 87 82	16 93 03 33 61
83 76 16 08 73	43 25 38 41 45	60 83 32 59 83	01 29 14 13 49	20 36 80 71 26
14 38 70 63 45	80 85 40 92 79	43 52 90 63 18	38 38 47 47 61	41 19 63 74 80
51 32 19 22 46	80 08 87 70 74	88 72 25 67 36	66 16 44 94 31	66 91 93 16 78
72 47 20 00 08	80 89 01 80 02	94 81 33 19 00	54 15 58 34 36	35 35 25 41 31
05 46 65 53 06	93 12 81 84 64	74 45 79 05 61	72 84 81 18 34	79 98 26 84 16
39 52 87 24 84	82 47 42 55 93	48 54 53 52 47	18 61 91 36 74	18 61 11 92 41
81 61 61 87 11	53 34 24 42 76	75 12 21 17 24	74 62 77 37 07	58 31 91 59 97
07 58 61 61 20	82 64 12 28 20	92 90 41 31 41	32 39 21 97 63	61 19 96 79 40
90 76 70 42 35	13 57 41 72 00	69 90 26 37 42	78 46 42 25 01	18 62 79 08 72
40 18 82 81 93	29 59 38 86 27	94 97 21 15 98	62 09 53 67 87	00 44 15 89 97
34 41 48 21 57	86 88 75 50 87	19 15 20 00 23	12 30 28 07 83	32 62 46 86 91
63 43 97 53 63	44 98 91 68 22	36 02 40 08 67	76 37 84 16 05	65 96 17 34 88
67 04 90 90 70	93 39 94 55 47	94 45 87 42 84	05 04 14 98 07	20 28 83 40 60
79 49 50 41 46	52 16 29 02 86	54 15 83 42 43	46 97 83 54 82	59 36 29 59 38
91 70 43 05 52	04 73 72 10 31	75 05 19 30 29	47 66 56 43 82	99 78 29 34 78

Table F-19. Random Numbers (*Continued*)

94 01 54 68 74	32 44 44 82 77	59 82 09 61 63	64 65 42 58 43	41 14 54 28 20
74 10 88 82 22	88 57 07 40 15	25 70 49 10 35	01 75 51 47 50	48 96 83 86 03
62 88 08 78 73	95 16 05 92 21	22 30 49 03 14	72 87 71 73 34	39 28 30 41 49
11 74 81 21 02	80 58 04 18 67	17 71 05 96 21	06 55 40 78 50	73 95 07 95 52
17 94 40 56 00	60 47 80 33 43	25 85 25 89 05	57 21 63 96 18	49 85 69 93 26
66 06 74 27 92	95 04 35 26 80	46 78 05 64 87	09 97 15 94 81	37 00 62 21 86
54 24 49 10 30	45 54 77 08 18	59 84 99 61 69	61 45 92 16 47	87 41 71 71 98
30 94 55 75 89	31 73 25 72 60	47 67 00 76 54	46 37 62 53 66	94 74 64 95 80
69 17 03 74 03	86 99 59 03 07	94 30 47 18 03	26 82 50 55 11	12 45 99 13 14
08 34 58 89 75	35 84 18 57 71	08 10 55 99 87	87 11 22 14 76	14 71 37 11 81
27 76 74 35 84	85 30 18 89 77	29 49 06 97 14	73 03 54 12 07	74 69 90 93 10
13 02 51 43 38	54 06 61 52 43	47 72 46 67 33	47 43 14 39 05	31 04 85 66 99
80 21 73 62 92	98 52 52 43 35	24 43 22 48 96	43 27 75 88 74	11 46 61 60 82
10 87 56 20 04	90 39 16 11 05	57 41 10 63 68	53 85 63 07 43	08 67 08 47 41
54 12 75 73 26	26 62 91 90 87	24 47 28 87 79	30 54 02 78 86	61 73 27 54 54
60 31 14 28 24	37 30 14 26 78	45 99 04 32 42	17 37 45 20 03	70 70 77 02 14
49 73 97 14 84	92 00 39 80 86	76 66 87 32 09	59 20 21 19 73	02 90 23 32 50
78 62 65 15 94	16 45 39 46 14	39 01 49 70 66	83 01 20 98 32	25 57 17 76 28
66 69 21 39 86	99 83 70 05 82	81 23 24 49 87	09 50 49 64 12	90 19 37 95 68
44 07 12 80 91	07 36 29 77 03	76 44 74 25 37	98 52 49 78 31	65 70 40 95 14
41 46 88 51 49	49 55 41 79 94	14 92 43 96 50	95 29 40 05 56	70 48 10 69 05
94 55 93 75 59	49 67 85 31 19	70 31 20 56 82	66 98 63 40 99	74 47 42 07 40
41 61 57 03 60	64 11 45 86 60	90 85 06 46 18	80 62 05 17 90	11 43 63 80 72
50 27 39 31 13	41 79 48 68 61	24 78 18 96 83	55 41 18 56 67	77 53 59 98 92
41 39 68 05 04	90 67 00 82 89	40 90 20 50 69	95 08 30 67 83	28 10 25 78 16
25 80 72 42 60	71 52 97 89 20	72 68 20 73 85	90 72 65 71 66	98 88 40 85 83
06 17 09 79 65	88 30 29 80 41	21 44 34 18 08	68 98 48 36 20	89 74 79 88 82
60 80 85 44 44	74 41 28 11 05	01 17 62 88 38	36 42 11 64 89	18 05 95 10 61
80 94 04 48 93	10 40 83 62 22	80 58 27 19 44	92 63 84 03 33	67 05 41 60 67
19 51 69 01 20	46 75 97 16 43	13 17 75 52 92	21 03 68 28 08	77 50 19 74 27
49 38 65 44 80	28 60 42 35 54	21 78 54 11 01	91 17 81 01 74	29 42 09 04 38
06 31 28 89 40	15 99 56 93 21	47 45 86 48 09	98 18 98 18 51	29 65 18 42 15
60 94 20 03 07	11 89 79 26 74	40 40 56 80 32	96 71 75 42 44	10 70 14 13 93
92 32 99 89 32	78 28 44 63 47	71 20 99 20 61	39 44 89 31 36	25 72 20 85 64
77 93 66 35 74	31 38 45 19 24	85 56 12 96 71	58 13 71 78 20	22 75 13 65 18
38 10 17 77 56	11 65 71 38 97	95 88 95 70 67	47 64 81 38 85	70 66 99 34 06
39 64 16 94 57	91 33 92 25 02	92 61 38 97 19	11 94 75 62 03	19 32 42 05 04
84 05 44 04 55	99 39 66 36 80	67 66 76 06 31	69 18 19 68 45	38 52 51 16 00
47 46 80 35 77	57 64 96 32 66	24 70 07 15 94	14 00 42 31 53	69 24 90 57 47
43 32 13 13 70	28 97 72 38 96	76 47 96 85 62	62 34 20 75 89	08 89 90 59 85
64 28 16 18 26	18 55 56 49 37	13 17 33 33 65	78 85 11 64 99	87 06 41 30 75
66 84 77 04 95	32 35 00 29 85	86 71 63 87 46	26 31 37 74 63	55 38 77 26 81
72 46 13 32 30	21 52 95 34 24	92 58 10 22 62	78 43 86 62 76	18 39 67 35 38
21 03 29 10 50	13 05 81 62 18	12 47 05 65 00	15 29 27 61 39	59 52 65 21 13
95 36 26 70 11	06 65 11 61 36	01 01 60 08 57	55 01 85 63 74	35 82 47 17 08
49 71 29 73 80	16 40 45 54 52	34 03 06 07 26	75 21 11 02 71	36 63 36 84 24
58 27 56 17 64	97 58 65 47 16	50 25 94 63 45	87 19 54 60 92	26 78 76 09 39
89 51 41 17 88	68 22 42 34 17	73 95 97 61 45	30 34 24 02 77	11 04 97 20 49
15 47 25 06 69	48 13 93 67 32	46 87 43 70 88	73 46 50 98 19	58 86 93 52 20
12 12 08 61 24	51 24 74 43 02	60 88 35 21 09	21 43 73 67 86	49 22 67 78 37

Table F-19. Random Numbers (*Continued*)

19 61 27 34 30	11 66 19 47 70	77 60 36 56 69	86 86 81 26 65	30 01 27 59 89
39 14 17 74 00	28 00 06 42 38	73 25 87 17 94	31 34 02 62 56	66 45 33 70 16
64 75 68 04 57	08 74 71 28 36	03 46 95 06 78	03 27 44 34 23	66 67 78 25 56
92 90 15 18 78	56 44 12 29 98	29 71 83 84 47	06 45 32 53 11	07 56 55 37 71
03 55 19 00 70	09 48 39 40 50	45 93 81 81 35	36 90 84 33 21	11 07 35 18 03
98 88 46 62 09	06 83 05 36 56	14 66 35 63 46	71 43 00 49 09	19 81 80 57 07
27 36 98 68 82	53 47 30 75 41	53 63 37 08 63	03 74 81 28 22	19 36 04 90 88
59 06 67 59 74	63 33 52 04 83	43 51 43 74 81	58 27 82 69 67	49 32 54 39 51
91 64 79 37 83	64 16 94 90 22	98 58 80 94 95	49 82 95 90 68	38 83 10 48 38
83 60 59 24 19	39 54 20 77 72	71 56 87 56 73	35 18 58 97 59	44 90 17 42 91
24 89 58 85 30	70 77 43 54 39	46 75 87 04 72	70 20 79 26 75	91 62 36 12 75
35 72 02 65 56	95 59 62 00 94	73 75 08 57 88	34 26 40 17 03	46 83 36 52 48
14 14 15 34 10	38 64 90 63 43	57 25 66 13 42	72 70 97 53 18	90 37 93 75 62
27 41 67 56 70	92 17 67 25 35	93 11 95 60 77	06 88 61 82 44	92 34 43 13 74
82 07 10 74 29	81 00 74 77 49	40 74 45 69 74	23 33 68 88 21	53 84 11 05 36
21 44 58 27 93	24 83 19 32 41	14 19 97 62 68	70 88 36 80 02	03 82 91 74 43
72 51 37 64 00	52 22 59 23 48	62 30 89 84 81	29 74 43 31 65	33 14 16 10 20
71 47 94 50 27	76 16 05 74 11	13 78 01 36 32	52 30 87 77 62	88 87 43 36 97
83 21 05 14 66	09 08 85 03 95	26 74 30 53 06	21 70 67 00 01	99 43 98 07 67
68 74 99 51 48	94 89 77 86 36	96 75 00 90 24	94 53 89 11 43	96 69 36 18 86
05 18 47 57 63	47 07 58 81 58	05 31 35 34 39	14 90 80 88 30	60 09 62 15 51
13 65 16 25 46	96 89 22 52 40	47 51 15 84 83	87 34 27 88 18	07 85 53 92 69
00 56 62 12 20	00 29 22 40 69	25 07 22 95 19	52 54 85 40 91	21 28 22 12 96
50 95 81 76 95	58 07 26 89 90	60 32 99 59 55	71 58 66 34 17	35 94 76 78 07
57 62 16 45 47	46 85 03 79 81	38 52 70 90 37	64 75 60 33 24	04 98 68 36 66
09 28 22 58 44	79 13 97 84 35	35 42 84 35 61	69 79 96 33 14	12 99 19 35 16
23 39 49 42 06	93 43 23 78 36	94 91 92 68 46	02 55 57 44 10	94 91 54 81 99
05 28 03 74 70	93 62 20 43 45	15 09 21 95 10	18 09 41 66 13	78 23 45 00 01
95 49 19 79 76	38 30 63 21 92	82 63 95 46 24	72 43 49 26 06	23 19 17 46 93
78 52 10 01 04	18 24 87 55 83	90 32 65 07 85	54 03 46 62 51	35 77 41 46 92
96 34 54 45 79	85 93 24 40 53	75 70 42 08 40	86 58 38 39 44	52 45 67 37 66
77 96 33 11 51	32 36 49 16 91	47 35 74 03 38	23 43 52 40 65	08 45 89 53 66
07 52 01 12 94	23 23 80 17 48	41 69 06 73 28	54 81 43 77 77	10 05 74 23 32
38 42 30 23 09	70 70 38 57 36	46 14 81 42 58	29 23 61 21 52	05 08 86 58 25
02 46 36 55 33	21 19 96 05 55	33 92 80 18 17	07 39 68 92 15	30 72 22 21 02
15 88 09 22 61	17 29 28 81 90	61 78 14 88 98	92 52 52 12 83	88 58 16 00 98
71 92 60 08 19	59 14 40 02 24	30 57 09 01 94	18 32 90 69 99	26 85 71 92 38
64 42 52 81 08	16 55 41 60 16	00 04 28 32 29	10 33 33 61 68	65 61 79 48 34
79 78 22 39 24	49 44 03 04 32	81 07 73 15 43	95 21 66 48 65	13 65 85 10 81
35 33 77 45 38	44 55 36 46 72	90 96 04 18 49	93 86 54 46 08	93 17 63 48 51
05 24 92 93 29	19 71 59 40 82	14 73 88 66 67	43 70 86 63 54	93 69 22 55 27
56 46 39 93 80	38 79 38 57 74	19 05 61 39 39	46 06 22 76 47	66 14 66 32 10
96 29 63 31 21	54 19 63 41 08	75 81 48 59 86	71 17 11 51 02	28 99 26 31 65
98 38 03 62 69	60 01 40 72 01	62 44 84 63 85	42 17 58 83 50	46 18 24 91 26
52 56 76 43 50	16 31 55 39 69	80 39 58 11 14	54 35 86 45 78	47 26 91 57 47
78 49 89 08 30	25 95 59 92 36	43 28 69 10 64	99 96 99 51 44	64 42 47 73 77
49 55 32 42 41	08 15 08 95 35	08 70 39 10 41	77 32 38 10 79	45 12 79 36 86
32 15 10 70 75	83 15 51 02 52	73 10 08 86 18	23 89 18 74 18	45 41 72 02 68
11 31 45 03 63	26 86 02 77 99	49 41 68 35 34	19 18 70 80 59	76 67 70 21 10
12 36 47 12 10	87 05 25 02 41	90 78 59 78 89	81 39 95 81 30	64 43 90 56 14

Table F-20. Normal Random Numbers

01	02	03	04	05	06	07	08	09	10
0.464	0.137	2.455	−0.323	−0.068	0.296	−0.288	1.298	0.241	−0.957
0.060	−2.526	−0.531	−0.194	0.543	−1.558	0.187	−1.190	0.022	0.525
1.486	−0.354	−0.634	0.697	0.926	1.375	0.785	−0.963	−0.853	−1.865
1.022	−0.472	1.279	3.521	0.571	−1.851	0.194	1.192	−0.501	−0.273
1.394	−0.555	0.046	0.321	2.945	1.974	−0.258	0.412	0.439	−0.035
0.906	−0.513	−0.525	0 595	0.881	−0.934	1.579	0.161	−1.885	0.371
1.179	−1.055	0.007	0.769	0.971	0.712	1.090	−0.631	−0.255	−0.702
−1.501	−0.488	−0.162	−0.136	1.033	0.203	0.448	0.748	−0.423	−0.432
−0.690	0.756	−1.618	−0.345	−0.511	−2.051	−0.457	−0.218	0.857	−0.465
1.372	0.225	0.378	0.761	0.181	−0.736	0.960	−1.530	−0.260	0.120
−0.482	1.678	−0.057	−1.229	−0.486	0.856	−0.491	−1.983	−2.830	−0.238
−1.376	−0.150	1.356	−0.561	−0.256	−0.212	0.219	0.779	0.953	−0.869
−1.010	0.598	−0.918	1.598	0.065	0.415	−0.169	0.313	−0.973	−1.016
−0.005	−0.899	0.012	−0.725	1.147	−0.121	1.096	0.481	−1.691	0.417
1.393	−1.163	−0.911	1.231	−0.199	−0.246	1.239	−2.574	−0.558	0.056
−1.787	−0.261	1.237	1.046	−0.508	−1.630	−0.146	−0.392	−0.627	0.561
−0.105	−0.357	−1.384	0.360	−0.992	−0.116	−1.698	−2.832	−1.108	−2.357
−1.339	1.827	−0.959	0.424	0.969	−1.141	−1.041	0.362	−1.726	1.956
1.041	0.535	0.731	1.377	0.983	−1.330	1.620	−1.040	0.524	−0.281
0.279	−2.056	0.717	−0.873	−1.096	−1.396	1.047	0.089	−0.573	0.932
−1.805	−2.008	−1.633	0.542	0.250	−0.166	0.032	0.079	0.471	−1.029
−1.186	1.180	1.114	0.882	1.265	−0.202	0.151	−0.376	−0.310	0.479
0.658	−1.141	1.151	−1.210	−0.927	0.425	0.290	−0.902	0.610	2.709
−0.439	0.358	−1.939	0.891	−0.227	0.602	0.873	−0.437	−0.220	−0.057
−1.399	−0.230	0.385	−0.649	−0.577	0.237	−0.289	0.513	0.738	−0.300
0.199	0.208	−1.083	−0.219	−0.291	1.221	1.119	0.004	−2.015	−0.594
0.159	0.272	−0.313	0.084	−2.828	−0.439	−0.792	−1.275	−0.623	−1.047
2.273	0.606	0.606	−0.747	0.247	1.291	0.063	−1.793	−0.699	−1.347
0.041	−0.307	0.121	0.790	−0.584	0.541	0.484	−0.986	0.481	0.996
−1.132	−2.098	0.921	0.145	0.446	−1.661	1.045	−1.363	−0.586	−1.023
0.768	0.079	−1.473	0.034	−2.127	0.665	0.084	−0.880	−0.579	0.551
0.375	−1.658	−0.851	0.234	−0.656	0.340	−0.086	−0.158	−0.120	0.418
−0.513	−0.344	0.210	−0.736	1.041	0.008	0.427	−0.831	0.191	0.074
0.292	−0.521	1.266	−1.206	−0.899	0.110	−0.528	−0.813	0.071	0.524
1.026	2.990	−0.574	−0.491	−1.114	1.297	−1.433	−1.345	−3.001	0.479
−1.334	1.278	−0.568	−0.109	−0.515	−0.566	2.923	0.500	0.359	0.326
−0.287	−0.144	−0.254	0.574	−0.451	−1.181	−1.190	−0.318	−0.094	1.114
0.161	−0.886	−0.921	−0.509	1.410	−0.518	0.192	−0.432	1.501	1.068
−1.346	0.193	−1.202	0.394	−1.045	0.843	0.942	1.045	0.031	0.772
1.250	−0.199	−0.288	1.810	1.378	0.584	1.216	0.733	0.402	0.226
0.630	−0.537	0.782	0.060	0.499	−0.431	1.705	1.164	0.884	−0.298
0.375	−1.941	0.247	−0.491	0.665	−0.135	−0.145	−0.498	0.457	1.064
−1.420	0.489	−1.711	−1.186	0.754	−0.732	−0.066	1.006	−0.798	0.162
−0.151	−0.243	−0.430	−0.762	0.298	1.049	1.810	2.885	−0.768	−0.129
−0.309	0.531	0.416	−1.541	1.456	2.040	−0.124	0.196	0.023	−1.204
0.424	−0.444	0.593	0.993	−0.106	0.116	0.484	−1.272	1.066	1.097
0.593	0.658	−1.127	−1.407	−1.579	−1.616	1.458	1.262	0.736	−0.916
0.862	−0.885	−0.142	−0.504	0.532	1.381	0.022	−0.281	−0.342	1.222
0.235	−0.628	−0.023	−0.463	−0.899	−0.394	−0.538	1.707	−0.188	−1.153
−0.853	0.402	0.777	0.833	0.410	−0.349	−1.094	0.580	1.395	1.298

Table F-20. Normal Random Numbers (*Continued*)

11	12	13	14	15	16	17	18	19	20
−1.329	−0.238	−0.838	−0.988	−0.445	0.964	−0.266	−0.322	−1.726	2.252
1.284	−0.229	1.058	0.090	0.050	0.523	0.016	0.277	1.639	0.554
0.619	0.628	0.005	0.973	−0.058	0.150	−0.635	−0.917	0.313	−1.203
0.699	−0.269	0.722	−0.994	−0.807	−1.203	1.163	1.244	1.306	−1.210
0.101	0.202	−0.150	0.731	0.420	0.116	−0.496	−0.037	−2.466	0.794
−1.381	0.301	0.522	0.233	0.791	−1.017	−0.182	0.926	−1.096	1.001
−0.574	1.366	−1.843	0.746	0.890	0.824	−1.249	−0.806	−0.240	0.217
0.096	0.210	1.091	0.990	0.900	−0.837	−1.097	−1.238	0.030	−0.311
1.389	−0.236	0.094	3.282	0.295	−0.416	0.313	0.720	0.007	0.354
1.249	0.706	1.453	0.366	−2.654	−1.400	0.212	0.307	−1.145	0.639
0.756	−0.397	−1.772	−0.257	1.120	1.188	−0.527	0.709	0.479	0.317
−0.860	0.412	−0.327	0.178	0.524	−0.672	−0.831	0.758	0.131	0.771
−0.778	−0.979	0.236	−1.033	1.497	−0.661	0.906	1.169	−1.582	1.303
0.037	0.062	0.426	1.220	0.471	0.784	−0.719	0.465	1.559	−1.326
2.619	−0.440	0.477	1.063	0.320	1.406	0.701	−0.128	0.518	−0.676
−0.420	−0.287	−0.050	−0.481	1.521	−1.367	0.609	0.292	0.048	0.592
1.048	0.220	1.121	−1.789	−1.211	−0.871	−0.740	0.513	−0.558	−0.395
1.000	−0.638	1.261	0.510	−0.150	0.034	0.054	−0.055	0.639	−0.825
0.170	−1.131	−0.985	0.102	−0.939	−1.457	1.766	1.087	−1.275	2.362
0.389	−0.435	0.171	0.891	1.158	1.041	1.048	−0.324	−0.404	1.060
−0.305	0.838	−2.019	−0.540	0.905	1.195	−1.190	0.106	0.571	0.298
−0.321	−0.039	1.799	−1.032	−2.225	−0.148	0.758	−0.862	0.158	−0.726
1.900	1.572	−0.244	−1.721	1.130	0.495	−0.484	0.014	−0.778	−1.483
−0.778	−0.288	−0.224	−1.324	−0.072	0.890	−0.410	0.752	0.376	−0.224
0.617	−1.718	−0.183	−0.100	1.719	0.696	−1.339	−0.614	1.071	−0.386
1.430	−0.953	0.770	−0.007	−1.872	1.075	−0.913	−1.168	1.775	0.238
0.267	−0.048	0.972	0.734	−1.408	−1.955	−0.848	2.002	0.232	−1.273
0.978	−0.520	−0.368	1.690	−1.479	0.985	1.475	−0.098	−1.633	2.399
−1.235	−1.168	0.325	1.421	2.652	−0.486	−1.253	0.270	−1.103	0.118
−0.258	0.638	2.309	0.741	−0.161	−0.679	0.336	1.973	0.370	−2.277
0.243	0.629	−1.516	−0.157	0.693	1.710	0.800	−0.265	1.218	0.655
−0.292	−1.455	−1.451	1.492	−0.713	0.821	−0.031	−0.780	1.330	0.977
−0.505	0.389	0.544	−0.042	1.615	−1.440	−0.989	−0.580	0.156	0.052
0.397	−0.287	1.712	0.289	−0.904	0.259	−0.600	−1.635	−0.009	−0.799
−0.605	−0.470	0.007	0.721	−1.117	0.635	0.592	−1.362	−1.441	0.672
1.360	0.182	−1.476	−0.599	−0.875	0.292	−0.700	0.058	−0.340	−0.639
0.480	−0.699	1.615	−0.225	1.014	−1.370	−1.097	0.294	0.309	−1.389
−0.027	−0.487	−1.000	−0.015	0.119	−1.990	−0.687	−1.964	−0.366	1.759
−1.482	−0.815	−0.121	1.884	−0.185	0.601	0.793	0.430	−1.181	0.426
−1.256	−0.567	−0.994	1.011	−1.071	−0.623	−0.420	−0.309	1.362	0.863
−1.132	2.039	1.934	−0.222	0.386	1.100	0.284	1.597	−1.718	−0.560
−0.780	−0.239	−0.497	−0.434	−0.284	−0.241	−0.333	1.348	−0.478	−0.169
−0.859	−0.215	0.241	1.471	0.389	−0.952	0.245	0.781	1.093	−0.240
0.447	1.479	0.067	0.426	−0.370	−0.675	−0.972	0.225	0.815	0.389
0.269	0.735	−0.066	−0.271	−1.439	1.036	−0.306	−1.439	−0.122	−0.336
0.097	−1.883	−0.218	0.202	−0.357	0.019	1.631	1.400	0.223	−0.793
−0.686	1.596	−0.286	0.722	0.655	−0.275	1.245	−1.504	0.066	−1.280
0.957	0.057	−1.153	0.701	−0.280	1.747	−0.745	1.338	−1.421	0.386
−0.976	−1.789	−0.696	−1.799	−0.354	0.071	2.355	0.135	−0.598	1.883
0.274	0.226	−0.909	−0.572	0.181	1.115	0.406	0.453	−1.218	−0.115

Table F-20. Normal Random Numbers (*Continued*)

21	22	23	24	25	26	27	28	29	30
-1.752	-0.329	-1.256	0.318	1.531	0.349	-0.958	-0.059	0.415	-1.084
-0.291	0.085	1.701	-1.087	-0.443	-0.292	0.248	-0.539	-1.382	0.318
-0.933	0.130	0.634	0.899	1.409	-0.883	-0.095	0.229	0.129	0.367
-0.450	-0.244	0.072	1.028	1.730	-0.056	-1.488	-0.078	-2.361	-0.992
0.512	-0.882	0.490	-1.304	-0.266	0.757	-0.361	0.194	-1.078	0.529
-0.702	0.472	0.429	-0.664	-0.592	1.443	-1.515	-1.209	-1.043	0.278
0.284	0.039	-0.518	1.351	1.473	0.889	0.300	0.339	-0.206	1.392
-0.509	1.420	-0.782	-0.429	-1.266	0.627	-1.165	0.819	-0.261	0.409
-1.776	-1.033	1.977	0.014	0.702	-0.435	-0.816	1.131	0.656	0.061
-0.044	1.807	0.342	-2.510	1.071	-1.220	-0.060	-0.764	0.079	-0.964
0.263	-0.578	1.612	-0.148	-0.383	-1.007	-0.414	0.638	-0.186	0.507
0.986	0.439	-0.192	-0.132	0.167	0.883	-0.400	-1.440	-0.385	-1.414
-0.441	-0.852	-1.446	-0.605	-0.348	1.018	0.963	-0.004	2.504	-0.847
-0.866	0.489	0.097	0.379	0.192	-0.842	0.065	1.420	0.426	-1.191
-1.215	0.675	1.621	0.394	-1.447	2.199	-0.321	-0.540	-0.037	0.185
-0.475	-1.210	0.183	0.526	0.495	1.297	-1.613	1.241	-1.016	-0.090
1.200	0.131	2.502	0.344	-1.060	-0.909	-1.695	-0.666	-0.838	-0.866
-0.498	-1.202	-0.057	-1.354	-1.441	-1.590	0.987	0.441	0.637	-1.116
-0.743	0.894	-0.028	1.119	-0.598	0.279	2.241	0.830	0.267	-0.156
0.779	-0.780	-0.954	0.705	-0.361	-0.734	1.365	1.297	-0.142	-1.387
-0.206	-0.195	1.017	-1.167	-0.079	-0.452	0.058	-1.068	-0.394	-0.406
-0.092	-0.927	-0.439	0.256	0.503	0.338	1.511	-0.465	-0.118	-0.454
-1.222	-1.582	1.786	-0.517	-1.080	-0.409	-0.474	-1.890	0.247	0.575
0.068	0.075	-1.383	-0.084	0.159	1.276	1.141	0.186	-0.973	-0.266
0.183	1.600	-0.335	1.553	0.889	0.896	-0.035	0.461	0.486	1.246
-0.811	-2.904	0.618	0.588	0.533	0.803	-0.696	0.690	0.820	0.557
-1.010	1.149	1.033	0.336	1.306	0.835	1.523	0.296	-0.426	0.004
1.453	1.210	-0.043	0.220	-0.256	-1.161	-2.030	-0.046	0.243	1.082
0.759	-0.838	-0.877	-0.177	1.183	-0.218	-3.154	-0.963	-0.822	-1.114
0.287	0.278	-0.454	0.897	-0.122	0.013	0.346	0.921	0.238	-0.586
-0.669	0.035	-2.077	1.077	0.525	-0.154	-1.036	0.015	-0.220	0.882
0.392	0.106	-1.430	-0.204	-0.326	0.825	-0.432	-0.094	-1.566	0.679
-0.337	0.199	-0.160	0.625	-0.891	-1.464	-0.318	1.297	0.932	-0.032
0.369	-1.990	-1.190	0.666	-1.614	0.082	0.922	-0.139	-0.833	0.091
-1.694	0.710	-0.655	-0.546	1.654	0.134	0.466	0.033	-0.039	0.838
0.985	0.340	0.276	0.911	-0.170	-0.551	1.000	-0.838	0.275	-0.304
-1.063	-0.594	-1.526	-0.787	0.873	-0.405	-1.324	0.162	-0.163	-2.716
0.033	-1.527	1.422	0.308	0.845	-0.151	0.741	0.064	1.212	0.823
0.597	0.362	-3.760	1.159	0.874	-0.794	-0.915	1.215	1.627	-1.248
-1.601	-0.570	0.133	-0.660	1.485	0.682	-0.898	0.686	0.658	0.346
-0.266	-1.309	0.597	0.989	0.934	1.079	-0.656	-0.999	-0.036	-0.537
0.901	1.531	-0.889	-1.019	0.084	1.531	-0.144	-1.920	0.678	-0.402
-1.433	-1.008	-0.990	0.090	0.940	0.207	-0.745	0.638	1.469	1.214
1.327	0.763	-1.724	-0.709	-1.100	-1.346	-0.946	-0.157	0.522	-1.264
-0.248	0.788	-0.577	0.122	-0.536	0.293	1.207	-2.243	1.642	1.353
-0.401	-0.679	0.921	0.476	1.121	-0.864	0.128	-0.551	-0.872	1.511
0.344	-0.324	0.686	-1.487	-0.126	0.803	-0.961	0.183	-0.358	-0.184
0.441	-0.372	-1.336	0.062	1.506	-0.315	-0.112	-0.452	1.594	-0.264
0.824	0.040	-1.734	0.251	0.054	-0.379	1.298	-0.126	0.104	-0.529
1.385	1.320	-0.509	-0.381	-1.671	-0.524	-0.805	1.348	0.676	0.799

Table F-20. Normal Random Numbers (*Continued*)

31	32	33	34	35	36	37	38	39	40
1.556	0.119	−0.078	0.164	−0.455	0.077	−0.043	−0.299	0.249	−0.182
0.647	1.029	1.186	0.887	1.204	−0.657	0.644	−0.410	−0.652	−0.165
0.329	0.407	1.169	−2.072	1.661	0.891	0.233	−1.628	−0.762	−0.717
−1.188	1.171	−1.170	−0.291	0.863	−0.045	−0.205	0.574	−0.926	1.407
−0.917	−0.616	−1.589	1.184	0.266	0.559	−1.833	−0.572	−0.648	−1.090
0.414	0.469	−0.182	0.397	1.649	1.198	0.067	−1.526	−0.081	−0.192
0.107	−0.187	1.343	0.472	−0.112	1.182	0.548	2.748	0.249	0.154
−0.497	1.907	0.191	0.136	−0.475	0.458	0.183	−1.640	−0.058	1.278
0.501	0.083	−0.321	1.133	1.126	−0.299	1.299	1.617	1.581	2.455
−1.382	−0.738	1.225	1.564	−0.363	−0.548	1.070	0.390	−1.398	0.524
−0.590	0.699	−0.162	−0.011	1.049	−0.689	1.225	0.339	−0.539	−0.445
−1.125	1.111	−1.065	0.534	0.102	0.425	−1.026	0.695	−0.057	0.795
0.849	0.169	−0.351	0.584	2.177	0.009	−0.696	−0.426	−0.692	−1.638
−1.233	−0.585	0.306	0.773	1.304	−1.304	0.282	−1.705	0.187	−0.880
0.104	−0.468	0.185	0.498	−0.624	−0.322	−0.875	1.478	−0.691	−0.281
0.261	−1.883	−0.181	1.675	−0.324	−1.029	−0.185	0.004	−0.101	−1.187
−0.007	1.280	0.568	−1.270	1.405	1.731	2.072	1.686	0.728	−0.417
0.794	−0.111	0.040	−0.536	−0.976	2.192	1.609	−0.190	−0.279	−1.611
0.431	−2.300	−1.081	−1.370	2.943	0.653	−2.523	0.756	0.886	−0.983
−0.149	1.294	−0.580	0.482	−1.449	−1.067	1.996	−0.274	0.721	0.490
−0.216	−1.647	1.043	0.481	−0.011	−0.587	−0.916	−1.016	−1.040	−1.117
1.604	−0.851	−0.317	−0.686	−0.008	1.939	0.078	−0.465	0.533	0.652
−0.212	0.005	0.535	0.837	0.362	1.103	0.219	0.488	1.332	−0.200
0.007	−0.076	1.484	0.455	−0.207	−0.554	1.120	0.913	−0.681	1.751
−0.217	0.937	0.860	0.323	1.321	−0.492	−1.386	−0.003	−0.230	0.539
−0.649	0.300	−0.698	0.900	0.569	0.842	0.804	1.025	0.603	−1.546
−1.541	0.193	2.047	−0.552	1.190	−0.087	2.062	−2.173	−0.791	−0.520
0.274	−0.530	0.112	0.385	0.656	0.436	0.882	0.312	−2.265	−0.218
0.876	−1.498	−0.128	−0.387	−1.259	−0.856	−0.353	0.714	0.863	1.169
−0.859	−1.083	1.288	−0.078	−0.081	0.210	0.572	1.194	−1.118	−1.543
−0.015	−0.567	0.113	2.127	−0.719	3.256	−0.721	−0.663	−0.779	−0.930
−1.529	−0.231	1.223	0.300	−0.995	−0.651	0.505	0.138	−0.064	1.341
0.278	−0.058	−2.740	−0.296	−1.180	0.574	1.452	0.846	−0.243	−1.208
1.428	0.322	2.302	−0.852	0.782	−1.322	−0.092	−0.546	0.560	−1.430
0.770	−1.874	0.347	0.994	−0.485	−1.179	0.048	−1.324	1.061	0.449
−0.303	−0.629	0.764	0.013	−1.192	−0.475	−1.085	−0.880	1.738	−1.225
−0.263	−2.105	0.509	−0.645	1.362	0.504	−0.755	1.274	1.448	0.604
0.997	−1.187	−0.242	0.121	2.510	−1.935	0.350	0.073	0.458	−0.446
−0.063	−0.475	−1.802	−0.476	0.193	−1.199	0.339	0.364	−0.684	1.353
−0.168	1.904	−0.485	−0.032	−0.554	0.056	−0.710	−0.778	0.722	−0.024
0.366	−0.491	0.301	−0.008	−0.894	−0.945	0.384	−1.748	−1.118	0.394
0.436	−0.464	0.539	0.942	−0.458	0.445	−1.883	1.228	1.113	−0.218
0.597	−1.471	−0.434	0.705	−0.788	0.575	0.086	0.504	1.445	−0.513
−0.805	−0.624	1.344	0.649	−1.124	0.680	−0.986	1.845	−1.152	−0.393
1.681	−1.910	0.440	0.067	−1.502	−0.755	−0.989	−0.054	−2.320	0.474
−0.007	−0.459	1.940	0.220	−1.259	−1.729	0.137	−0.520	−0.412	2.847
0.209	−0.633	0.299	0.174	1.975	−0.271	0.119	−0.199	0.007	2.315
1.254	1.672	−1.186	−1.310	0.474	0.878	−0.725	−0.191	0.642	−1.212
−1.016	−0.697	0.017	−0.263	−0.047	−1.294	−0.339	2.257	−0.078	−0.049
−1.169	−0.355	1.086	−0.199	0.031	0.396	−0.143	1.572	0.276	0.027

Table F-20. Normal Random Numbers *(Continued)*

41	42	43	44	45	46	47	48	49	50
-0.856	-0.063	0.787	-2.052	-1.192	-0.831	1.623	1.135	0.759	-0.189
-0.276	-1.110	0.752	-1.378	-0.583	0.360	0.365	1.587	0.621	1.344
0.379	-0.440	0.858	1.453	-1.356	0.503	-1.134	1.950	-1.816	-0.283
1.468	0.131	0.047	0.355	0.162	-1.491	-0.739	-1.182	-0.533	-0.497
-1.805	-0.772	1.286	-0.636	-1.312	-1.045	1.559	-0.871	-0.102	-0.123
2.285	0.554	0.418	-0.577	-1.489	-1.255	0.092	-0.597	-1.051	-0.980
-0.602	0.399	1.121	-1.026	0.087	1.018	-1.437	0.661	0.091	-0.637
0.229	-0.584	0.705	0.124	0.341	1.320	-0.824	-1.541	-0.163	2.329
1.382	-1.454	1.537	-1.299	0.363	-0.356	-0.025	0.294	2.194	-0.395
0.978	0.109	1.434	-1.094	-0.265	-0.857	-1.421	-1.773	0.570	-0.053
-0.678	-2.335	1.202	-1.697	0.547	-0.201	-0.373	-1.363	-0.081	0.958
-0.366	-1.084	-0.626	0.798	1.706	-1.160	-0.838	1.462	0.636	0.570
-1.074	-1.379	0.086	-0.331	-0.288	-0.309	-1.527	-0.408	0.183	0.856
-0.600	-0.096	0.696	0.446	1.417	-2.140	0.599	-0.157	1.485	1.387
0.918	1.163	-1.445	0.759	0.878	-1.781	-0.056	-2.141	-0.234	0.975
-0.791	-0.528	0.946	1.673	-0.680	-0.784	1.494	-0.086	-1.071	-1.196
0.598	-0.352	0.719	-0.341	0.056	-1.041	1.429	0.235	0.314	-1.693
0.567	-1.156	-0.125	-0.534	0.711	-0.511	0.187	-0.644	-1.090	-1,281
0.963	0.052	0.037	0.637	-1.335	0.055	0.010	-0.860	-0.621	0.713
0.489	-0.209	1.659	0.054	1.635	0.169	0.794	-1.550	1.845	-0.388
-1.627	-0.017	0.699	0.661	-0.073	0.188	1.183	-1.054	-1.615	-0.765
-1.096	1.215	0.320	0.738	1.865	-1.169	-0.667	-0.674	-0.062	1.378
-2.532	1.031	-0.799	1.665	-2.756	-0.151	-0.704	0.602	-0.672	1.264
0.024	-1.183	-0.927	-0.629	0.204	-0.825	0.496	2.543	0.262	-0.785
0.192	0.125	0.373	-0.931	-0.079	0.186	-0.306	0.621	-0.292	1.131
-1.324	-1.229	-0.648	-0.430	0.811	0.868	0.787	1.845	-0.374	-0.651
-0.726	-0.746	1.572	-1.420	1.509	-0.361	-0.310	-3.117	1.637	0.642
-1.618	1.082	-0.319	0.300	1.524	-0.418	-1.712	0.358	-1.032	0.537
1.695	0.843	2.049	0.388	-0.297	1.077	-0.462	0.655	0.940	-0.354
0.790	0.605	-3.077	1.009	-0.906	-1.004	0.693	-1.098	1.300	0.549
1.792	-0.895	-0.136	-1.765	1.077	0.418	-0.150	0.808	0.697	0.435
0.771	-0.741	-0.492	-0.770	-0.458	-0.021	1.385	-1.225	-0.066	-1.471
-1.438	0.423	-1.211	0.723	-0.731	0.883	-2.109	-2.455	-0.210	1.644
-0.294	1.266	-1.994	-0.730	0.545	0.397	1.069	-0.383	-0.097	-0.985
-1.966	0.909	0.400	0.685	-0.800	1.759	0.268	1.387	-0.414	1.615
-0.999	1.587	1.423	0.937	-0.943	0.090	1.185	-1.204	0.300	-1.354
0.581	0.481	-2.400	0.000	0.231	0.079	-2.842	-0.846	-0.508	-0.516
0.370	-1.452	-0.580	-1.462	-0.972	1.116	-0.994	0.374	-3.336	-0.058
0.834	-1.227	-0.709	-1.039	-0.014	-0.383	-0.512	-0.347	0.881	-0.638
-0.376	-0.813	0.660	-1.029	-0.137	0.371	0.376	0.968	1.338	-0.786
-1.621	0.815	-0.544	-0.376	-0.852	0.436	1.562	0.815	-1.048	0.188
0.163	-0.161	2.501	-0.265	-0.285	1.934	1.070	0.215	-0.876	0.073
1.786	-0.538	-0.437	0.324	0.105	-0.421	-0.410	-0.947	0.700	-1.006
2.140	1.218	-0.351	-0.068	0.254	0.448	-1.461	0.784	0.317	1.013
0.064	0.410	0.368	0.419	-0.982	1.371	0.100	-0.505	0.856	0.890
0.789	-0.131	1.330	0.506	-0.645	-1.414	2.426	1.389	-0.169	-0.194
-0.011	-0.372	-0.699	2.382	-1.395	-0.467	1.256	-0.585	-1.359	-1.804
-0.463	0.003	-1.470	1.493	0.960	0.364	-1.267	-0.007	0.616	0.624
-1.210	-0.669	0.009	1.284	-0.617	0.355	-0.589	-0.243	-0.015	-0.712
-1.157	0.481	0.560	1.287	1.129	-0.126	0.006	1.532	1.328	0.980

Table F-21. sin x/x*

x	0	1	2	3	4	5	6	7	8	9
0.0	+10000	10000	9999	9999	9997	9996	9994	9992	9989	9987
0.1	9983	9980	9976	9972	9967	9963	9957	9952	9946	9940
0.2	9933	9927	9919	9912	9904	9896	9889	9879	9870	9860
0.3	9851	9840	9830	9820	9808	9797	9785	9774	9761	9748
0.4	9735	9722	9709	9695	9680	9666	9651	9636	9620	9605
0.5	+9589	9572	9555	9538	9521	9503	9486	9467	9449	9430
0.6	9411	9391	9372	9351	9331	9311	9290	9269	9247	9225
0.7	9203	9181	9158	9135	9112	9089	9065	9041	9016	8992
0.8	8967	8942	8916	8891	8865	8839	8812	8785	8758	8731
0.9	8704	8676	8648	8620	8591	8562	8533	8504	8474	8445
1.0	+8415	8384	8354	8323	8292	8261	8230	8198	8166	8134
1.1	8102	8069	8037	8004	7970	7937	7903	7870	7836	7801
1.2	7767	7732	7698	7663	7627	7592	7556	7520	7484	7448
1.3	7412	7375	7339	7302	7265	7228	7190	7153	7115	7077
1.4	7039	7001	6962	6924	6885	6846	6807	6768	6729	6690
1.5	+6650	6610	6570	6530	6490	6450	6410	6369	6328	6288
1.6	6247	6206	6165	6124	6083	6042	6000	5959	5917	5875
1.7	5833	5791	5749	5707	5665	5623	5580	5538	5495	5453
1.8	5410	5368	5325	5282	5239	5196	5153	5110	5067	5024
1.9	4981	4937	4894	4851	4807	4764	4720	4677	4634	4590
2.0	+4546	4503	4459	4416	4372	4329	4285	4241	4198	4153
2.1	4111	4067	4023	3980	3936	3893	3849	3805	3762	3718
2.2	3675	3632	3588	3545	3501	3458	3415	3372	3328	3285
2.3	3242	3199	3156	3113	3070	3028	2984	2942	2899	2857
2.4	2814	2772	2730	2687	2645	2603	2561	2519	2477	2436
2.5	+2394	2352	2311	2269	2228	2187	2146	2105	2064	2023
2.6	1983	1942	1902	1861	1821	1781	1741	1702	1662	1622
2.7	1583	1544	1504	1465	1427	1388	1349	1311	1273	1234
2.8	1196	1159	1121	1083	1046	1009	972	935	898	861
2.9	825	789	753	717	681	646	610	575	540	505
3.0	+470	436	402	368	334	300	266	233	200	167
3.1	+134	+102	+69	+37	+5	−27	−58	−90	−121	−152
3.2	−182	213	243	273	303	333	362	392	421	449
3.3	478	506	535	562	590	618	645	672	699	725
3.4	752	778	804	829	855	880	905	930	954	978
x	0	1	2	3	4	5	6	7	8	9

* J. Sherman, *Z. Krist.*, **85**:404 (1933).

Table F-21. sin x/x (*Continued*)

x	0	1	2	3	4	5	6	7	8	9
3.5	−1002	1026	1050	1073	1096	1119	1141	1164	1186	1208
3.6	1229	1251	1272	1293	1313	1334	1354	1374	1393	1413
3.7	1432	1451	1470	1488	1506	1524	1542	1559	1576	1593
3.8	1610	1627	1643	1659	1675	1690	1705	1720	1735	1749
3.9	1764	1777	1791	1805	1818	1831	1844	1856	1868	1880
4.0	−1892	1903	1915	1926	1936	1947	1957	1967	1977	1987
4.1	1996	2005	2014	2022	2030	2039	2046	2054	2061	2068
4.2	2075	2082	2088	2094	2100	2106	2111	2116	2121	2126
4.3	2131	2135	2139	2143	2146	2150	2153	2156	2158	2161
4.4	2163	2165	2166	2168	2169	2170	2171	2172	2172	2172
4.5	−2172	2172	2172	2171	2170	2169	2168	2166	2164	2162
4.6	2160	2158	2155	2152	2150	2146	2143	2139	2136	2132
4.7	2127	2123	2119	2114	2109	2104	2098	2093	2087	2081
4.8	2075	2069	2063	2056	2049	2042	2035	2028	2020	2013
4.9	2005	1997	1989	1981	1972	1963	1955	1946	1937	1927
5.0	−1918	1908	1899	1889	1879	1868	1858	1848	1837	1826
5.1	1815	1804	1793	1782	1770	1759	1747	1735	1723	1711
5.2	1699	1687	1674	1662	1649	1636	1623	1610	1597	1584
5.3	1570	1557	1543	1530	1516	1502	1488	1474	1460	1445
5.4	1431	1417	1402	1387	1373	1358	1343	1328	1313	1298
5.5	−1283	1268	1252	1237	1221	1206	1190	1175	1159	1143
5.6	1127	1111	1095	1079	1063	1047	1031	1015	999	982
5.7	966	950	933	917	900	884	867	851	834	818
5.8	800	784	768	751	734	718	701	684	667	650
5.9	634	617	600	583	567	550	533	516	499	482
6.0	−466	449	432	416	399	382	365	348	332	315
6.1	299	282	265	249	232	216	200	183	167	150
6.2	−134	−118	−102	−85	−69	−53	−37	−21	−5	+11
6.3	+27	43	58	74	90	105	121	136	152	167
6.4	182	197	212	227	242	257	272	287	302	316
6.5	+331	346	360	374	388	403	417	431	445	458
6.6	472	486	499	513	526	539	552	566	579	591
6.7	604	617	630	642	654	667	679	691	703	715
6.8	727	738	750	761	773	784	795	806	817	828
6.9	838	849	859	870	880	890	900	910	919	929
x	0	1	2	3	4	5	6	7	8	9

Table F-21. sin x/x *(Continued)*

x	0	1	2	3	4	5	6	7	8	9
7.0	+939	948	957	966	975	984	993	1002	1010	1019
7.1	1027	1035	1043	1051	1058	1066	1074	1081	1088	1095
7.2	1102	1109	1116	1123	1129	1135	1142	1148	1153	1159
7.3	1165	1171	1176	1181	1186	1191	1196	1201	1206	1210
7.4	1214	1219	1223	1227	1231	1234	1238	1241	1244	1248
7.5	+1251	1254	1256	1259	1261	1264	1266	1268	1270	1272
7.6	1274	1275	1277	1278	1279	1280	1281	1282	1282	1283
7.7	1283	1284	1284	1284	1284	1283	1283	1282	1282	1281
7.8	1280	1279	1278	1277	1275	1274	1272	1270	1269	1267
7.9	1264	1262	1259	1257	1255	1252	1249	1246	1243	1240
8.0	+1237	1233	1230	1226	1222	1218	1214	1210	1206	1202
8.1	1197	1193	1188	1183	1179	1174	1169	1163	1158	1153
8.2	1147	1142	1136	1130	1124	1118	1112	1106	1100	1093
8.3	1087	1080	1074	1067	1060	1053	1046	1039	1032	1025
8.4	1017	1010	1002	995	987	979	972	964	956	948
8.5	+939	931	923	915	906	898	889	880	872	863
8.6	854	845	836	827	818	809	800	790	781	771
8.7	762	752	743	733	724	714	704	694	684	675
8.8	665	655	645	635	625	614	604	594	584	573
8.9	563	552	542	532	521	511	500	490	479	469
9.0	+458	447	437	426	415	404	394	383	372	361
9.1	351	340	329	318	307	296	286	275	264	253
9.2	242	231	220	210	199	188	177	166	156	145
9.3	134	123	112	101	91	80	69	58	48	37
9.4	+26	+16	+5	−6	−16	−27	−37	−48	−58	−69
9.5	−79	89	100	110	120	131	141	151	161	172
9.6	182	192	202	212	222	231	241	251	261	271
9.7	280	290	299	309	318	328	337	346	356	365
9.8	374	383	392	401	410	419	428	436	445	454
9.9	462	471	479	487	496	504	512	520	528	536
10.0	−544	552	560	567	575	582	590	597	604	612
10.1	619	626	633	640	647	653	660	667	673	680
10.2	686	692	699	705	711	717	723	728	734	740
10.3	745	751	756	761	767	772	777	782	787	791
10.4	796	801	805	809	814	818	822	826	830	834
x	0	1	2	3	4	5	6	7	8	9

Table F-21. sin x/x (*Continued*)

x	0	1	2	3	4	5	6	7	8	9
10.5	−838	842	845	849	852	855	859	862	865	868
10.6	871	873	876	879	881	883	886	888	890	892
10.7	894	896	898	899	901	902	904	905	906	907
10.8	908	909	910	911	911	912	912	913	913	913
10.9	913	913	913	913	913	912	912	911	911	910
11.0	−909	908	907	906	905	904	902	901	899	898
11.1	896	894	892	890	888	886	884	882	879	877
11.2	874	872	869	866	863	860	857	854	851	848
11.3	844	841	837	834	830	826	822	819	815	811
11.4	806	802	798	794	789	785	780	776	771	766
11.5	−761	756	751	746	741	736	731	726	720	715
11.6	709	704	698	693	687	681	675	669	663	657
11.7	651	645	639	633	626	620	614	607	601	594
11.8	588	581	574	568	561	554	547	540	533	526
11.9	519	512	505	498	491	484	476	469	462	454
12.0	−447	440	432	425	417	410	402	395	387	379
12.1	372	364	356	348	341	333	325	317	309	301
12.2	294	286	278	270	262	254	246	238	230	222
12.3	214	206	198	190	182	174	166	158	150	142
12.4	134	125	117	109	101	93	85	77	69	61
12.5	−53	−45	−37	−29	−21	−13	−5	+3	+11	+19
12.6	+27	35	42	50	58	66	74	82	89	97
12.7	105	113	120	128	136	143	151	158	166	173
12.8	181	188	196	203	210	218	225	232	240	247
12.9	254	261	268	275	282	289	296	303	310	316
13.0	+323	330	337	343	350	356	363	369	376	382
13.1	388	395	401	407	413	419	425	431	437	443
13.2	448	454	460	466	471	477	482	488	493	498
13.3	503	509	514	519	524	529	534	538	543	548
13.4	552	557	562	566	570	575	579	583	587	591
13.5	+595	599	603	607	611	614	618	622	625	628
13.6	632	635	638	641	644	647	650	653	656	659
13.7	661	664	666	669	671	673	676	678	680	682
13.8	684	686	688	689	691	692	694	695	697	698
13.9	699	700	702	703	703	704	705	706	706	707
x	0	1	2	3	4	5	6	7	8	9

Table F-21. sin x/x (*Continued*)

x	0	1	2	3	4	5	6	7	8	9
14.0	+708	708	708	709	709	709	709	709	709	709
14.1	709	708	708	708	707	707	706	705	705	704
14.2	703	702	701	700	699	697	696	695	693	692
14.3	690	688	687	685	683	681	679	677	675	673
14.4	671	668	666	663	661	658	656	653	650	648
14.5	+645	642	639	636	633	630	626	623	620	616
14.6	613	609	606	602	599	595	591	587	583	579
14.7	575	571	567	563	559	555	550	546	542	537
14.8	533	528	524	519	514	509	505	500	495	490
14.9	485	480	475	470	465	460	455	449	444	439
15.0	+434	428	423	417	412	406	401	395	390	384
15.1	378	373	367	361	355	349	344	338	332	326
15.2	320	314	308	302	296	290	284	278	272	265
15.3	259	253	247	241	234	228	222	216	209	203
15.4	197	190	184	178	171	165	159	152	146	140
15.5	+133	127	120	114	108	101	95	88	82	76
15.6	69	63	56	50	43	37	31	24	18	11
15.7	+5	−1	−8	−14	−20	−27	−33	−39	−46	−52
15.8	58	64	71	77	83	89	95	102	108	114
15.9	120	126	132	138	144	150	156	162	168	174
16.0	−180	186	192	197	203	209	215	220	226	232
16.1	237	243	248	254	259	265	270	276	281	286
16.2	292	297	302	307	312	318	323	328	333	337
16.3	342	347	352	357	362	366	371	376	380	385
16.4	389	393	398	402	407	411	415	419	423	427
16.5	−431	435	439	443	447	451	454	458	462	465
16.6	469	472	476	479	482	486	489	492	495	498
16.7	501	504	507	510	513	515	518	521	523	526
16.8	528	531	533	535	538	540	542	544	546	548
16.9	550	552	553	555	557	558	560	561	563	564
17.0	−566	567	568	569	570	571	572	573	574	575
17.1	575	576	577	577	578	578	579	579	579	579
17.2	580	580	580	580	580	579	579	579	579	578
17.3	578	577	577	576	576	575	574	573	572	571
17.4	570	569	568	567	566	565	563	562	561	559
x	0	1	2	3	4	5	6	7	8	9

Table F-21. sin x/x (*Continued*)

x	0	1	2	3	4	5	6	7	8	9
17.5	−557	556	554	553	551	549	547	545	543	541
17.6	539	537	535	533	530	528	526	523	521	518
17.7	516	513	510	508	505	502	499	496	493	490
17.8	487	484	481	478	475	471	468	465	461	458
17.9	454	451	447	444	440	436	433	429	425	421
18.0	−417	413	409	405	401	397	393	389	385	381
18.1	376	372	368	364	359	355	350	346	341	337
18.2	332	328	323	319	314	309	304	300	295	290
18.3	285	281	276	271	266	261	256	251	246	241
18.4	236	231	226	221	216	211	206	201	195	190
18.5	−185	180	175	170	164	159	154	149	143	138
18.6	133	128	122	117	112	106	101	96	90	85
18.7	80	74	69	64	58	53	48	42	37	32
18.8	−26	−21	−16	−10	−5	+0	+6	+11	+16	+21
18.9	+27	32	37	42	48	53	58	63	68	74
19.0	+79	84	89	94	99	104	110	115	120	125
19.1	130	135	140	145	150	155	159	164	169	174
19.2	179	184	188	193	198	202	207	212	216	221
19.3	226	230	235	239	244	248	252	257	261	265
19.4	270	274	278	282	286	290	295	299	303	307
19.5	+311	314	318	322	326	330	333	337	341	344
19.6	348	351	355	358	362	365	369	372	375	378
19.7	382	385	388	391	394	397	400	403	405	408
19.8	411	414	416	419	422	424	427	429	431	434
19.9	436	438	440	443	445	447	449	451	453	455
x	0	1	2	3	4	5	6	7	8	9

Table F-21. sin x/x (*Continued*)

x	0	2	4	6	8	x	0	2	4	6	8
20.0	+456	460	463	466	469	23.5	−425	425	425	424	424
20.1	472	475	477	479	481	23.6	423	423	422	421	419
20.2	483	485	486	487	488	23.7	418	416	415	413	411
20.6	489	490	490	490	490	23.8	408	406	403	401	398
20.4	490	490	489	488	487	23.9	395	392	388	385	381
20.5	+486	485	483	482	480	24.0	−377	373	369	365	361
20.6	478	475	473	470	467	24.1	356	352	347	342	337
20.7	464	461	458	454	450	24.2	332	327	321	316	310
20.8	447	442	438	434	429	24.3	304	299	293	287	280
20.9	424	420	415	409	404	24.4	274	268	261	255	248
21.0	+398	393	387	381	375	24.5	−241	235	228	221	214
21.1	369	362	356	349	342	24.6	206	199	192	185	177
21.2	335	328	321	314	307	24.7	170	162	155	147	139
21.3	299	292	284	276	268	24.8	132	124	116	108	100
21.4	260	252	244	236	228	24.9	93	85	77	69	61
21.5	+219	211	202	194	185	25.0	−53	45	37	29	21
21.6	176	168	159	150	141	25.1	−13	−5	+3	+11	+19
21.7	132	123	114	105	96	25.2	+27	35	42	50	58
21.8	87	78	69	60	51	25.3	66	74	81	89	96
21.9	42	32	23	14	5	25.4	104	111	119	126	134
22.0	−4	13	22	31	40	25.5	+141	148	155	162	169
22.1	49	58	67	76	85	25.6	176	183	189	196	203
22.2	93	102	111	119	128	25.7	209	215	222	228	234
22.3	136	145	153	161	169	25.8	240	246	251	257	263
22.4	178	185	193	201	209	25.9	268	273	278	284	288
22.5	−217	224	231	239	246	26.0	+293	298	303	307	311
22.6	253	260	267	274	280	26.1	315	320	323	327	331
22.7	287	293	299	305	311	26.2	334	338	341	344	347
22.8	317	323	329	334	339	26.3	350	352	355	357	359
22.9	344	349	354	359	364	26.4	361	363	365	367	368
23.0	−368	372	376	380	384	26.5	+370	371	372	373	373
23.1	388	391	394	397	400	26.6	374	374	375	375	375
23.2	403	406	408	410	413	26.7	375	374	374	373	372
23.3	415	416	418	419	421	26.8	371	370	369	368	366
23.4	422	423	423	424	424	26.9	365	363	361	359	357
x	0	2	4	6	8	x	0	2	4	6	8

Table F-21. sin x/x *(Continued)*

x	0	2	4	6	8	x	0	2	4	6	8
27.0	+354	352	349	346	343	30.5	−260	256	252	247	243
27.1	340	337	334	331	327	30.6	238	233	229	224	219
27.2	323	319	316	312	307	30.7	214	209	204	198	193
27.3	303	299	294	290	285	30.8	188	182	177	171	165
27.4	280	275	270	265	260	30.9	160	154	148	142	136
27.5	+254	249	243	238	232	31.0	−130	124	118	112	106
27.6	226	220	214	208	202	31.1	100	94	87	81	75
27.7	196	190	184	177	171	31.2	69	62	56	50	43
27.8	164	158	151	145	138	31.3	37	31	24	18	11
27.9	131	124	117	111	104	31.4	−5	+1	+8	+14	+20
28.0	+97	90	83	76	69	31.5	+27	33	39	45	52
28.1	62	55	48	41	34	31.6	58	64	70	76	82
28.2	+26	+19	+12	+5	−2	31.7	88	94	100	106	112
28.3	−9	16	23	30	37	31.8	118	124	129	135	140
28.4	44	51	58	65	72	31.9	146	151	157	162	167
28.5	−79	85	92	99	105	32.0	+172	177	182	187	192
28.6	112	118	125	131	138	32.1	197	202	206	211	215
28.7	144	150	156	162	168	32.2	219	224	228	232	236
28.8	174	180	186	192	197	32.3	239	243	247	250	254
28.9	203	208	213	219	224	32.4	257	260	263	266	269
29.0	−229	234	239	243	248	32.5	+272	275	277	280	282
29.1	253	257	261	266	270	32.6	284	286	288	290	292
29.2	274	278	281	285	288	32.7	293	295	296	297	299
29.3	292	295	298	301	304	32.8	300	300	301	302	302
29.4	307	310	312	315	317	32.9	303	303	303	303	303
29.5	−319	321	323	325	326	33.0	+303	303	302	302	301
29.6	328	329	330	331	332	33.1	300	299	298	297	296
29.7	333	334	334	335	335	33.2	294	293	291	290	288
29.8	335	335	335	335	334	33.3	286	284	281	279	277
29.9	334	333	332	332	331	33.4	274	272	269	266	263
30.0	−329	328	327	325	323	33.5	+260	257	254	250	247
30.1	321	320	317	315	313	33.6	243	240	236	232	228
30.2	311	308	305	302	300	33.7	224	220	216	212	208
30.3	296	293	290	287	283	33.8	203	199	194	190	185
30.4	280	276	272	268	264	33.9	180	175	171	166	161
x	0	2	4	6	8	x	0	2	4	6	8

Table F-21. sin x/x (*Continued*)

x	0	2	4	6	8	x	0	2	4	6	8
34.0	+156	151	145	140	135	37.0	−174	170	165	161	156
34.1	130	124	119	113	108	37.1	152	147	143	138	133
34.2	102	97	91	86	80	37.2	129	124	119	114	109
34.3	74	69	63	57	51	37.3	104	99	94	89	84
34.4	46	40	34	28	22	37.4	79	74	68	63	58
34.5	+17	+11	+5	−1	−7	37.5	−53	47	42	37	32
34.6	−12	18	24	30	35	37.6	26	21	16	10	5
34.7	41	47	52	58	63	37.7	+0	6	11	16	21
34.8	69	75	80	85	91	37.8	27	32	37	42	47
34.9	96	102	107	112	117	37.9	53	58	63	68	73
35.0	−122	127	132	137	142	38.0	+78	83	88	93	98
35.1	147	152	157	161	166	38.1	102	107	112	117	121
35.2	170	175	179	183	187	38.2	126	130	135	139	143
35.3	192	196	199	203	207	38.3	148	152	156	160	164
35.4	211	214	218	221	225	38.4	168	172	176	179	183
35.5	−228	231	234	237	240	38.5	+186	190	193	197	200
35.6	243	245	248	250	253	38.6	203	206	209	212	215
35.7	255	257	259	261	263	38.7	218	220	223	225	228
35.8	264	266	268	269	270	38.8	230	232	234	236	238
35.9	271	272	273	274	275	38.9	240	241	243	244	246
36.0	−275	276	276	277	277	39.0	+247	248	249	250	251
36.1	277	277	277	276	276	39.1	252	253	253	254	254
36.2	276	275	274	273	272	39.2	254	255	255	255	255
36.3	271	270	269	268	266	39.3	254	254	254	253	252
36.4	265	263	261	259	257	39.4	252	251	250	249	248
36.5	−255	253	251	248	246	39.5	+246	245	244	242	241
36.6	243	241	238	235	232	39.6	239	237	235	233	231
36.7	229	226	223	220	216	39.7	229	227	224	222	219
36.8	213	209	206	202	198	39.8	217	214	211	208	206
36.9	194	190	186	182	178	39.9	203	199	196	193	190
x	0	2	4	6	8	x	0	2	4	6	8

Table F-22.* Chebyshev Polynomials $T_n(x)$

n \ x	0.2	0.4	0.6	0.8	1.0
0	+1.00000 00000	+1.00000 00000	+1.00000 00000	+1.00000 00000	1
1	+0.20000 00000	+0.40000 00000	+0.60000 00000	+0.80000 00000	1
2	−0.92000 00000	−0.68000 00000	−0.28000 00000	+0.28000 00000	1
3	−0.56800 00000	−0.94400 00000	−0.93600 00000	−0.35200 00000	1
4	+0.69280 00000	−0.07520 00000	−0.84320 00000	−0.84320 00000	1
5	+0.84512 00000	+0.88384 00000	−0.07584 00000	−0.99712 00000	1
6	−0.35475 20000	+0.78227 20000	+0.75219 20000	−0.75219 20000	1
7	−0.98702 08000	−0.25802 24000	+0.97847 04000	−0.20638 72000	1
8	−0.04005 63200	−0.98868 99200	+0.42197 24800	+0.42197 24800	1
9	+0.97099 82720	−0.53292 95360	−0.47210 34240	+0.88154 31680	1
10	+0.42845 56288	+0.56234 62912	−0.98849 65888	+0.98849 65888	1
11	−0.79961 60205	+0.98280 65690	−0.71409 24826	+0.70005 13741	1
12	−0.74830 20370	+0.22389 89640	+0.13158 56097	+0.13158 56097	1

* From M. Abramowitz and I. A. Stegun (eds.), *Handbook of Mathematical Functions*, National Bureau of Standards, Washington, D.C., 1964.

BIBLIOGRAPHY

Short Tables of Transcendental Functions

Abramowitz, M., and I. A. Stegun: *Handbook of Mathematical Functions*, National Bureau of Standards, Washington, D.C., 1964.

Dwight, H. B.: *Mathematical Tables*, McGraw-Hill, New York, 1941.

Flügge, W.: *Four-place Tables of Transcendental Functions*, McGraw-Hill, New York, 1954.

Jahnke and F. Emde: *Tables of Functions with Formulae and Curves*. Dover, New York, 1954.

Statistical Tables

Beyer, W. H.: *CRC Handbook of Tables for Probability and Statistics*, Chemical Rubber Co., Cleveland, Ohio, 1966.

Burington, R. S., and D. C. May: *Handbook of Probability and Statistics*, 2d ed., McGraw-Hill, New York, 1967.

Hald, A.: *Statistical Tables and Formulas*, Wiley, New York, 1952.

Meredith, W.: *Mathematical and Statistical Tables*, McGraw-Hill, New York, 1967.

Owen, D. B.: *Handbook of Statistical Tables*, Addison-Wesley, Reading, Mass., 1962.

Pearson, E. S., and H. O. Hartley: *Biometrika Tables for Statisticians*, Cambridge, New York, 1956.

Indices to Numerical Tables

Etherington, Harold (ed.): *Nuclear Engineering Handbook*, McGraw-Hill, New York, 1958.

Fletcher, A.: Guide to Tables of Elliptic Functions, *Mathematical Tables and Other Aids to Computation*, vol. 3, no. 24, 1948.

———, J. C. P. Miller, and L. Rosenhead: *Index of Mathematical Tables*, Addison-Wesley, Reading, Mass., 1962.

Greenwood, J. A., and H. O Hartley: *Guide to Tables in Mathematical Statistics*, Princeton, Princeton, N.J., 1962.

GLOSSARY OF SYMBOLS AND NOTATIONS

The symbols and notations used in this handbook were chosen so as to permit reference to most standard textbooks while still maintaining consistency throughout the handbook. This glossary lists generally useful symbols whose definitions may not appear in their immediate context; each entry gives the handbook section or sections in which the symbol is defined.

Scalars and Matrices

α, β, . . . represent scalar (numerical) quantities, Chaps. 5, 6, and 12 to 16. α^* is the complex conjugate of α, and $|\alpha|$ is the absolute value of α, 1.3-2.

In Chaps. 13 and 14, A, B, . . . represent matrices, most frequently square matrices, with $A \equiv [a_{ik}]$, **13.2-1.**

$A^* \equiv [a_{ik}^*]$, complex conjugate of A, **13.3-1**
$\tilde{A} \equiv [a_{ki}]$, transpose of A, **13.3-1**
$A\dagger \equiv [a_{ki}^*]$, hermitian conjugate of A, **13.3-1**

$$x \equiv \{\xi_i\} \equiv \begin{bmatrix} \xi_1 \\ \xi_2 \\ \cdots \end{bmatrix}$$, column matrix, **13.2-1**

$\tilde{x} \equiv (\xi_i) \equiv (\xi_1, \xi_2, \ldots)$, row matrix, **13.2-1**

Vectors and Vector Components

a, **b**, . . . and **x**, **y**, . . . represent vectors, **5.1-1, 12.4-1, 14.2-1, 16.2-1, 16.7-3**
u, unit vector, **5.2-5, 14.2-5, 16.8-1**
i, j, k, right-handed rectangular cartesian base vectors, **5.2-3**
$\mathbf{e}_i$, $\mathbf{e}^i$, base vectors, **6.3-3, 14.2-4, 16.6-1**
$\mathbf{u}_i$, unit base vectors, **6.3-2, 16.8-3;**
orthogonal unit base vectors, **6.4-1, 14.7-4**

$\mathbf{r} \equiv x\mathbf{i} + y\mathbf{j} + z\mathbf{k}$ } position vectors in three-dimensional Euclidean
$\boldsymbol{\rho} \equiv \xi\mathbf{i} + \eta\mathbf{j} + \zeta\mathbf{k}$ } space, Chaps. 5, 15, and 17

$\mathbf{a} = \alpha_1\mathbf{e}_1 + \alpha_2\mathbf{e}_2 + \cdots$, vector, represented by the column matrix of the components: $a \equiv \{\alpha_1\} \equiv \begin{bmatrix} \alpha_1 \\ \alpha_2 \\ \cdots \end{bmatrix}$, **14.2-4, 14.5-2**

$\mathbf{a} = \hat{\alpha}_1\mathbf{u}_1 + \hat{\alpha}_2\mathbf{u}_2 + \cdots$, vector, expressed in terms of orthonormal base vectors and components, **14.7-4**

$\mathbf{a} = a^k\mathbf{e}_k = a^1\mathbf{e}_1 + a^2\mathbf{e}_2 + \cdots + a^n\mathbf{e}_n$, contravariant vector (function), represented by components a^k, **6.3-3, 16.6-1**

$\mathbf{a} = a_k\mathbf{e}^k = a_1\mathbf{e}^1 + a_2\mathbf{e}^2 + \cdots + a_n\mathbf{e}^n$, covariant vector (function), represented by components a_k, **6.3-3, 16.6-1**

$\mathbf{a} = \hat{a}_1\mathbf{u}_1 + \hat{a}_2\mathbf{u}_2 + \cdots + \hat{a}_n\mathbf{u}_n$, vector represented in terms of physical components $\hat{a}_k$, **6.3-2, 16.8-3**

$\mathbf{a} \cdot \mathbf{b}$, scalar product of vectors **a**, **b**, **5.2-6, 16.8-1**

$|\mathbf{a}| \equiv (\mathbf{a} \cdot \mathbf{a})^{1/2}$, absolute value (norm) of **a**, **5.2-5, 16.8-1**

$(\mathbf{a}, \mathbf{b})$, general inner product, **14.2-6**

$\|\mathbf{a}\| \equiv (\mathbf{a}, \mathbf{a})^{1/2}$, norm of **a**, **14.2-7**

(f, h), inner product of functions, **15.2-1**

$\|f\| \equiv (f, f)^{1/2}$, norm of a function, **15.2-1**

$\|A\|$, $\|A\|_{\mathrm{I}}$, $\|A\|_{\mathrm{II}}$, $\|A\|_p$, $\|A\|_1$, $\|A\|_2$, $\|x\|_\infty$, matrix norms, **13.2-1**

$\mathbf{a} \times \mathbf{b}$, vector product of three-dimensional vectors **a**, **b**, **5.2-7, 16.8-4**

$[\mathbf{abc}]$, scalar triple product, **5.2-8, 16.8-4**

See also Sec. 16.1-3 for dummy-index notation, and Sec. **14.7-7** for a comparison of notations.

Linear Operators and Tensors

A, **B**, . . . represent linear operators (Chaps. 14, 15, and 20) or tensors and dyadics (Chap. 16). Note the following:

L, **B**	linear differential operators operating on functions $y(x)$, $y(t)$, $\Phi(x)$, $\Phi(x^1, x^2, \ldots, x^n)$, **9.3-1, 10.4-2, 15.2-7, 15.4-1**
K	linear-integral-transformation operator, **15.3-1**
$\tilde{\mathbf{A}}$, $\tilde{\mathbf{B}}$, . . . ; $\tilde{\mathbf{L}}$, $\tilde{\mathbf{K}}$	transpose or adjoint of **A**, **B**, . . . ; **L**, **K**, **14.4-6, 15.3-1, 15.4-3**
$\mathbf{A}^\dagger$, $\mathbf{B}^\dagger$, . . . ; $\mathbf{L}^\dagger$, $\mathbf{K}^\dagger$	hermitian conjugate of **A**, **B**, . . . ; **L**, **K**, **14.4-3, 15.3-1, 15.4-3**
D	derivative operator, **20.4-2**

$\boldsymbol{\nabla}$ (del or nabla), vector differential operator, **5.5-2, 16.10-7**
$\boldsymbol{\nabla}^2 \equiv \boldsymbol{\nabla} \cdot \boldsymbol{\nabla}$ Laplacian operator, **5.5-5, 16.10-7**
E shift operator, **20.4-2**
∇ forward-difference operator, **20.4-2**
Δ backward-difference operator, **20.4-2**
δ central-difference operator, **20.4-2**
μ central-mean operator, **20.4-2**

Expected Values (Mean Values) and Averages

$E\{x\} = \xi$ expected value, ensemble average, **18.3-3, 18.4-4, 18.4-8, 18.9-2**
$<x>$ t average (a random variable), **18.9-4**
$\bar{x} = \frac{1}{n}(x_1 + x_2 + \cdots + x_n)$ statistical sample average (a random variable), **19.2-3**
$\text{Mean}\{x\}$ mean value over a group, **12.2-12**

$\arcsin z$, $\arccos z$, $\arctan z$ inverse trigonometric functions, **21.2-4**
$\arg z$ argument of z, **1.3-2**
B_k, $B_k^{(n)}$ Bernoulli numbers, **21.5-2**
$B_k^{(n)}(x)$ Bernoulli polynomial, **21.5-2**
$\text{ber}_m\, z$, $\text{bei}_m\, z$, **21.8-7**
$C(x)$, Fresnel integral, **21.3-2**
$\text{Ci}\, x$ cosine integral, **21.3-1**
$\text{cn}\, z$ (cosinus amplitudinis), elliptic function, **21.6-7**
$\cos z$ cosine function, **21.2-1**
$\cosh z$ hyperbolic cosine, **21.2-5**
$\cosh^{-1} z$ inverse hyperbolic cosine, **21.2-8**
dn (delta amplitudinis), elliptic function. **21.6-7**
$\det [a_{ik}]$ determinant, **1.5-1**
$\text{erf}\, x$ error function, **21.3-2**
$\text{erfc}\, x$ complementary error function, **21.3-2**
$E(k, \varphi)$ Legendre's normal elliptic integral of the second kind, **21.6-6**
$\text{Ei}\, x$, $\overline{\text{Ei}}\, x$ exponential integrals, **21.3-1**
$\mathbf{E}(k)$ Legendre's complete normal elliptic integral of the second kind, **21.6-6**
$F(a, b; c; z) \equiv {}_2F_1(a, b; c; z)$ hypergeometric function, **9.3-9, 9.3-11**
$F(a; c; z) \equiv {}_1F_1(a; c; z)$ confluent hypergeometric function, **9.3-10, 9.3-11**

$F(k, \varphi)$ Legendre's normal elliptic integral of the first kind, **21.6-6**
$g_{xy}(\omega), G_{xy}(\omega)$ spectral densities, **18.10-8**
$\mathrm{g}_{xy}(\omega), \mathrm{G}_{xy}(\omega)$ ensemble spectral densities, **18.10-6**
$h_j^{(1)}(z), h_j^{(2)}(z)$ spherical Bessel functions of the third kind, **21.8-8**
$H_n(z)$ Hermite polynomials, **21.7-1**
$H_m^{(1)}(z), H_m^{(2)}(z)$ Hankel functions, **21.8-1**
$\mathrm{her}_m\, z, \mathrm{hei}_m\, z$ **21.8-7**
$i = \sqrt{-1}$ unit imaginary number, **1.3-1**
$I_m(z)$ modified Bessel function, **21.8-6**
$I_z(p, q)$ incomplete beta-function ratio, **21.4-5**
$\mathrm{Im}\, z$ imaginary part of z, **1.3-1**
$\inf x$ greatest lower bound, **4.3-3**
$j_j(z)$ spherical Bessel function of the first kind, **21.8-8**
$J_m(z)$ Bessel function of the first kind, **21.8-1**
$\mathbf{K}(k)$ Legendre's complete elliptic integral of the first kind, **21.8-1**
$K_m(z)$ modified Hankel function, **21.8-6**
$\mathrm{ker}_m\, z, \mathrm{kei}_m\, z$ **21.8-7**
$L_n(z)$ Laguerre polynomial, **21.7-1**
L_n^m associated Laguerre polynomial or generalized Laguerre function, **21.7-5**
$L_2 \equiv L_2(V)$ class of quadratically integrable functions, **15.2-1**
$\mathrm{li}\,(z)$ logarithmic integral, **21.3-1**
$\lim z$ limit, **4.4-1**
$\mathrm{l.i.m.}\, x$ limit-in-mean, **15.2-2**
$\log_a z$ logarithm, **1.2-3**, **21.2-10**
$\max x, \min x$ maximum and minimum values, **4.3-3**
$n_j(z)$ spherical Bessel function of the second kind, **21.8-8**
$N_m(z)$ Neumann's Bessel function of the second kind, **21.8-1**
$o[g(x)], O[g(x)]$ asymptotic relations, **4.4-3**
$P_n(z)$ Legendre's polynomial of the first kind, **21.7-1**
$P_j^m(z)$ associated Legendre "polynomial" of the first kind, **21.8-10**
$Q_n(z)$ Legendre function of the second kind, **21.7-3**

$R_{xy}(\tau)$, $R_{xy}(t_1, t_2)$	t correlation functions, **18.10-7**
$\mathrm{R}_{xy}(\tau)$, $\mathrm{R}_{xy}(t_1, t_2)$	ensemble correlation functions, **18.9-3, 18.10-2**
$\mathrm{Re}\, z$	real part of z, **1.3-1**
$\mathrm{Res}_f\, a$	residue of $f(z)$ at $z = a$, **7.7-1**
$s(t)$	sampling function, **18.10-6**
$S(x)$	Fresnel integral, **21.3-2**
$S_k^{(n)}$	Stirling numbers, **21.5-1**
$\mathrm{Sgn}\, x$ or $\mathrm{sgn}\, x$	sign function, **21.9-1**
$\mathrm{Si}(x)$	sine integral, **21.3-1**
$\sin z$	sine function, **21.2-1**
$\sinh z$	hyperbolic sine, **21.2-5**
$\sinh^{-1} z$	inverse hyperbolic sine, **21.2-8**
$\sup x$	least upper bound, **4.3-3**
$T_n(z)$	Chebyshev polynomial of the first kind, **21.7-1**
$\tan z$	tangent function, **21.2-1**
$\tanh z$	hyperbolic tangent, **21.2-5**
$\tanh^{-1} z$	inverse hyperbolic tangent, **21.2-8**
$\mathrm{Tr}\, [a_{ik}]$	trace, **13.2-7**
$U_n(z)$	Chebyshev "polynomial" of the second kind, **21.7-4**
$Y_j(\vartheta, \varphi)$	spherical surface harmonic, **21.8-12** [$N_m(z)$ rather than $Y_m(z)$ is used for Neumann functions, 21.8-1]
$Z_m(z)$	cylinder function, **21.8-1**
$\mathrm{B}(p, q)$	beta function, **21.4-4**
$\mathrm{B}_z(p, q)$	incomplete beta function, **21.4-5**
$\Gamma(z)$	gamma function, **21.4-1**
$\Gamma_z(p)$	incomplete gamma function, **21.4-5**
$\Phi_{xy}(\omega)$, $\gamma_{xy}(\nu)$, $\Gamma_{xy}(\nu)$	spectral densities, **18.10-3, 18.10-6**
$\delta(x)$, $\delta_+(x)$, $\delta_-(x)$	impulse functions, 21.9-2
$\delta(x, \xi)$	multidimensional delta function, **15.3-1, 21.9-7**
Δy_k	forward difference, 20.4-1
∇y_k	backward difference, 20.4-1
δy_k	central difference, 20.4-1
$\zeta(z)$	Weierstrass zeta function, **21.6-3**
$\vartheta_i(z)$	Jacobi's theta functions, **21.6-8**
μy_k	central mean, **20.4-1**
$\sigma(z)$	Weierstrass sigma function, **21.6-3**
$\psi(z)$	psi function, **21.4-3**
$\wp(z)$	Weierstrass $\wp$ function, **21.6-2**

$\mathfrak{F}[f(t)] \equiv \mathfrak{F}[f(t); \nu]$	Fourier transform, **4.11-3**
$\mathfrak{F}_C[f(t)], \mathfrak{F}_S[f(t)]$	Fourier cosine and sine transforms **4.11-3**
$\mathcal{Z}[f_k; z]$	z transform, **8.7-3**
$\mathcal{L}[f(t)] \equiv \mathcal{L}[f(t); s] \equiv F(s)$	Laplace transform, **8.2-1**
$\mathcal{V}$	a vector space, **12.4-1, 14.2-1**
$n!$	factorial, **1.2-4**
$\binom{x}{n}$	binomial coefficient, **21.5-1**
$\int y(t)$	jump function, **20.4-5**
$\frac{\partial(y_1, y_2, \ldots, y_n)}{\partial(x_1, x_2, \ldots, x_n)}$	Jacobian, **4.5-6**
$\oplus$	direct sum, **12.7-5, 13.2-10, 14.8-2, 14.9-2**
$\otimes$	direct product, **12.7-2, 13.2-10, 14.10-7**
$\cap$	cap, **12.8-1,** 18.2-1
$\cup$	cup, **12.8-1**
$*$	convolution symbol, **4.6-18**
$\sum_{k=m}^{n}$	summation, **1.2-5**
$\prod_{k=m}^{n}$	product, **1.2-5**
$=$	equality symbol, **1.1-3, 12.1-3**
$\equiv$	identity symbol, **1.1-4**
$\triangleq$	identity by definition, **1.1-4**
$\approx$	approximate equality
$\simeq$	asymptotically equal, **4.4-3**
$\sim$	asymptotically proportional, **4.4-3**
$<, >, \leq, \geq$	inequality, inclusion, **1.1-5, 12.6-1**
$\subset, \supset$	inclusion, **4.3-2, 12.8-3**
$\in$	element of, **4.3-2**
$\ni$	such that
D, V	domain, region
S	surface, boundary surface or hypersurface
C	curve, boundary curve
$ds, d\mathbf{r}$	scalar and vector path elements (see index)
$dA, d\mathbf{A}$	scalar and vector surface elements (see index)

INDEX

References are to section numbers. References to essential definitions are printed in **boldface numbers** to permit the use of this index as a mathematical dictionary. Numbers preceded by letters (A-2) refer to the Appendixes.